# TABLES OF THE INCOMPLETE BETA-FUNCTION

P A

# TABLES OF THE INCOMPLETE BETA-FUNCTION

*Originally prepared under the direction of and edited by*

KARL PEARSON

SECOND EDITION

*with a new Introduction by*

E.S.PEARSON *and* N.L.JOHNSON



CAMBRIDGE

*Published for the Biometrika Trustees*

AT THE UNIVERSITY PRESS

1968

Published by the Syndics of the Cambridge University Press
Bentley House, P.O. Box 92, 200 Euston Road, London, N.W. 1
American Branch: 32 East 57th Street, New York, N.Y. 10022



Library of Congress Catalogue Card Number: 68-11285

Standard Book Number: 521 05922 4

First published 1934
Reprinted 1948 1956
Second edition 1968

Printed in Great Britain
at the University Printing House, Cambridge
(Brooke Crutchley, University Printer)

# PREFACE TO THE SECOND EDITION

Since Karl Pearson's original publication in 1934 there have been two reprints of these *Tables* in which no changes were made beyond the correction of a few identified errors. As another reprint is now overdue, the Biometrika Trustees agreed that the opportunity should be taken to revise the Introduction, leaving the main Tables I and II unaltered. The work of revision was entrusted to the undersigned.

The alterations which we have made consist in a complete re-casting of the Introduction and the addition of two short Tables, numbered III and IV. The changes to the Introduction may be summarized as follows:

(1) We have retained the first two and a half pages of Karl Pearson's Introduction in which he described the origin of the *Tables* and the methods used for computation.

(2) We have reduced very considerably the next 28 pages (originally pp. ix–xxxvi), which dealt primarily with methods of interpolation. In particular, more prominence has been given to the use of Lagrangian interpolation formulae. However, since the accuracy of a given formula is most readily assessed by a study of differences, we have also retained a number of the procedures originally recommended, based on Everett's central difference formulae. A considerable number of the earlier examples have also been used.

(3) The long section (pp. xxxvi–lix) containing examples of the application of Table I in answering a great variety of statistical questions, has been omitted because we think that the present-day statistician who turns to these *Tables* will know how he wants to use the incomplete beta-function integral. Further, the notation and methods of handling statistical techniques have changed very considerably in 34 years, so that a detailed section on applications would involve complete re-casting and re-illustration.

(4) In sections 4 and 5 we have, however, enlarged somewhat on some of the mathematical properties of the function which it may be useful to have available for reference. For example, since from time to time statisticians seem to forget that the sum of terms of the negative binomial expansion may be expressed as an incomplete beta-function, the relationship is set out in section 4·2. Again the incomplete beta-function is often used to provide an approximation to the probability integral of a distribution for which the moments and terminal points are alone readily available. Section 5 sets out the equations from which $p$ and $q$ can be estimated using such information.

(5) Finally, a section 6 has been given listing some recently published approximations to the incomplete beta function and giving a few comments on their accuracy and the work involved in using them. A list of references is also included.

The new Tables III and IV give some additional values of the integral computed a number of years ago but not hitherto published. Table III gives $I_x(p, 0{\cdot}5)$ for $p = 11{\cdot}5$, $12{\cdot}5$, $13{\cdot}5$ and $14{\cdot}5$, so making it possible to calculate values of the probability integral of the distribution of Student's $t$ up to 30 degrees of freedom, through the relation between $I_x(p, 0{\cdot}5)$ and the symmetrical beta distribution. Table IV gives additional values of $I_x(p, q)$ for small $q$ and $x$ near unity, a region where interpolation in Table I is difficult or impossible.

E. S. PEARSON
UNIVERSITY COLLEGE LONDON

N. L. JOHNSON
UNIVERSITY OF NORTH CAROLINA
CHAPEL HILL

*October 1967*

# PREFACE TO THE FIRST EDITION

Tables of the Incomplete B-Function have seemed to me essential to the modern theory of statistics ever since I personally learnt, about the year 1894, to appreciate two facts, namely how closely the sum of $n$ terms of a hypergeometrical series could be represented by the partial area of the curve

$$y = y_0 x^{p-1}(1-x)^{q-1},$$

and secondly how imperfect was Laplace's endeavour to represent such areas by a series based on the normal curve and its differential coefficients. Various methods were given in my lectures on statistics for evaluating the integral

$$B_x(p,q) = \int_0^x x^{p-1}(1-x)^{q-1}\,dx,$$

and were used in the Biometric Laboratory for many years. In 1921, I asked Mr Herbert E. Soper then a research assistant in that Laboratory to put together various possible methods for evaluating the Incomplete B-Function, and the results of his investigations were published in the Cambridge University Press *Tracts for Computers*, as No. VII. That *Tract* is an essential companion to the present volume, and will be of service to any one seeking values of the function outside the range of arguments in these tables. But the labour required to apply some of the methods of that *Tract*, and the relatively small degree of accuracy provided by others, only emphasized in my mind the already appreciated need for computing tables which would cover some of the field. Accordingly, when the *Tables of the Incomplete Γ-Function** had been finished and their publication rendered possible by a contribution from the Department of Industrial and Scientific Research, an application was made to the same Department for help in computing tables of the Incomplete B-Function. This was a still more serious undertaking, owing both to the extent of the computing work necessary—it being a table of triple, not double entry—and to the difficulty of eventually finding means for the publication of such a voluminous work as this promised to be. The Department of Scientific and Industrial Research again came to my aid, at first by granting payment for a definite research assistant for this work, and afterwards by a definite grant for the completion of the work of computing, which extended from 1923 to 1932. In supervision and proof-reading the aid of members of the Department of Applied Statistics at University College, London, has been frequently drawn upon and readily granted.

The present condition of our national finances did not justify the publication of this sister volume to the *Tables of the Incomplete Γ-Function* in the manner previously adopted, and it seemed for a time as if the printing of the manuscript must be indefinitely delayed. Arrangements have finally been made by which these tables appear as one of the *Biometrika* publications. As only a small edition can be issued the price must necessarily be heavy, but purchasers may be assured that the work is sold without profit, merely at cost price.

I have to thank most cordially Dr Ethel Elderton, Dr Brenda Stoessiger and Mr E. C. Fieller, for the heavy labour of proof-reading of the tables themselves; Dr Egon S. Pearson for much aid in the preparation of the Introduction, Mr E. C. Fieller for computing help therein, and Mr Walter Lewis and the Compositors and Readers of the University Press, Cambridge, for the rapidity and accuracy with which the work has been set up. Such errors as may be found must be due to false copying of figures by the computers on the original working sheets, as the latter have been compared throughout with the text of the tables.

I cannot hope that the work is wholly free from computing errors, and shall be very grateful for any such being pointed out to me, so that eventually a list of errata may be issued.

BIOMETRIC LABORATORY,
UNIVERSITY COLLEGE, LONDON
*1933*

KARL PEARSON

* Edited by Karl Pearson, London, H.M. Stationery Office, 1922; now, Cambridge University Press.

# CONTENTS


*Preface to the Second Edition* — *page* v

*Preface to the First Edition* — vi

*Introduction* — ix

1 Origin of the Tables and methods adopted for computing them — ix

2 Additional Tables included in this edition — xi
- 2·1 Table III. Values of $I_x(p, 0{\cdot}5)$ for $p = 11{\cdot}5,\ 12{\cdot}5,\ 13{\cdot}5,\ 14{\cdot}5$ — xi
- 2·2 Table IV, giving additional values of $I_x(p, q)$ for small $q$ and $x$ near unity — xii

3 Interpolation — xii
- 3·1 Trivariate interpolation — xii
  - 3·1·1 Description of the problem and notation — xii
  - 3·1·2 Scheme of computation and numerical examples — xiii
- 3·2 Bivariate interpolation — xvi
  - 3·2·1 A bivariate Lagrangian formula — xvi
  - 3·2·2 Numerical example — xvi
- 3·3 Special cases — xvi
  - 3·3·1 Applications in the analysis of variance — xvi
  - 3·3·2 Univariate interpolation formulae for $I_x(p+0{\cdot}5, q)$ — xvii
  - 3·3·3 Numerical example — xvii
  - 3·3·4 Univariate diagonal formulae for $I_x(p+0{\cdot}5,\ q+0{\cdot}5)$ — xviii
  - 3·3·5 Illustrations of diagonal interpolation — xix
- 3·4 Bivariate interpolation for $p$ and $q$ involving a horizontal or vertical 'slide' — xxi
- 3·5 Methods which may be used to overcome certain difficulties in interpolation — xxi
- 3·6 Inverse interpolation: existing tables of percentage points — xxii

4 Some special applications — xxiii
- 4·1 The binomial — xxiv
- 4·2 The negative binomial — xxiv
- 4·3 The relation between $I_{x'}(p, p)$ and $I_x(p, 0{\cdot}5)$ and some applications — xxiv

5 Relations between the moments of the beta distribution and the parameters $p, q$ — xxv
- 5·1 Moments of the distribution $f(x) = \{B(p,q)\}^{-1}x^{p-1}(1-x)^{q-1},\ 0 \leqslant x \leqslant 1$ — xxv
- 5·2 The general beta or Pearson Type I distribution — xxvi
- 5·3 Case ($a$). Estimation from the first four moments — xxvi
- 5·4 Case ($b$). Estimation using a known 'start' and the first three moments — xxvii
- 5·5 Case ($c$). Estimation using both known terminals and the first two moments — xxvii

6 Approximate formulae — *page* xxviii

6·1 Systematic computation and occasional approximation — xxviii

6·2 Cadwell's approximation for the symmetrical case $p = q$ — xxviii

6·3 Wise's approximations — xxix

6·4 Approximations based on the Cornish–Fisher expansion — xxix

6·5 Approximation of Peizer & Pratt — xxx

*Appendix* Some univariate interpolation formulae — xxxi

The Everett central difference formula — xxxi

Lagrangian interpolation formulae — xxxi

*List of References* — xxxii

TABLE I

Incomplete beta-function ratio $I_x(p,q)$ — *pages* 1–431

TABLE II

Frequency constants of the distribution $f(x) = \{B(p,q)\}^{-1}x^{p-1}(1-x)^{q-1}$ for various values of $p$ and $q$ — 433–494

TABLE III

Values of $I_x(p, 0{\cdot}5)$ for $p = 11{\cdot}5, 12{\cdot}5, 13{\cdot}5, 14{\cdot}5$ — 495–496

TABLE IV

Additional values of $I_x(p,q)$ for small $q$ and $x$ near unity — 497–505

## COMPUTERS AND COLLABORATORS

MARGARET T. BEER
SHEILA M. BURROUGH
A. E. R. CHURCH
E. C. FIELLER
MARIAN F. HOADLEY
MARY KINGSFORD
JOSEPH PEPPER
JOHANNES PRETORIUS
HAROLD W. P. RICHARDSON
BETTY O. SPIERS
BRENDA N. STOESSIGER
CATHERINE M. THOMPSON
JOHN WISHART

# INTRODUCTION*

## 1. Origin of the Tables and Methods Adopted for Computing Them

(from Karl Pearson's original Introduction of 1933)

The somewhat exaggerated use made by Laplace of the normal curve

$$y = y_0 e^{-\frac{1}{2}\frac{x^2}{\sigma^2}}$$

to represent almost any function $y = f(x)$

for very considerable distances from its mode, in particular the function

$$y = y_0 x^{p-1}(1-x)^{q-1},$$

led me many years ago to seek by Laplace's own methods for expansions of unimodal functions in the form

$$f(x) = y_0 e^{-\frac{1}{2}\frac{x^2}{\sigma^2}} \times \text{a polynomial in } x,$$

where $x$ is measured from the mode. Thus the partial area of $f(x)$, or what we may term the probability integral of the function, was expressed in what we should now call an incomplete normal moment series, or in another form a tetrachoric function series. But actual experience with the probability integrals of the curves

$$y = y_0 e^{-x} x^{p-1} \quad \text{and} \quad y = y_0 x^{p-1}(1-x)^{q-1}$$

obtained by such series was extremely unsatisfactory, and I was compelled to discard them, and to face the problem of the tabulation of the incomplete Γ- and B-functions. The work of computing the Incomplete Γ-function was first taken in hand, and the difficulties of the problem soon developed themselves; chief among these was the infinite range of $x$, which demanded as the power, $p-1$, of $x$ increased in the function either a change in argument intervals, or what amounts to the same thing the expression of $x$ in terms of the changing standard deviation. The latter course was chosen and after eight years of work the *Tables of the Incomplete Γ-Function* were published by H.M. Stationery Office in 1922.

In the case of the incomplete B-function the same problem arose, but in a less aggravated form, because the range of $x$ is finite. It could have been met in the same manner by expressing the variate $x$ in terms of the changing standard deviation of the curve instead of in terms of the range. But the variety of cases to which the tabled function can be applied—either directly or by transformation—raised new difficulties. In the case of either or both $p$ and $q$ being less than unity, the standard deviation, $\sigma$, of the curve

$$y = y_0 x^{p-1}(1-x)^{q-1}$$

given by

$$\sigma^2 = \frac{pq}{(p+q)^2(p+q+1)}$$

was not found to be wholly the best unit for the measurement of $x$, while in the case of transformed curves, the above expression is of course not their standard deviation. It was settled therefore to use the range, not the standard deviation, and, as increment of the argument $x$, to take $\frac{1}{100}$th part of the range.

To lessen the labour of computing the trivariate function, I avoided, except for testing purposes, quadrature and decided that a recurrence formula should be made the basis of the work. This required only the computing, which was easy, of the areas of the curve for the initial low values of $p$ and $q$. The function I proposed to have tabled was to be a *probability integral*; that is to say, if we represent by $B(p,q)$ the complete B-function, $= \int_0^1 x^{p-1}(1-x)^{q-1}dx$, and by $B_x(p,q)$ the incomplete B-function, or $\int_0^x x^{p-1}(1-x)^{q-1}dx$, we tabled the ratio

$$I_x(p,q) = B_x(p,q)/B(p,q) = \frac{\Gamma(p+q)}{\Gamma(p)\,\Gamma(q)}\int_0^x x^{p-1}(1-x)^{q-1}dx. \tag{1}$$

* [Section 1 of this Introduction has been taken unaltered from Karl Pearson's 1933 Introduction to these Tables. The remaining sections are ours, but we have drawn freely on illustrative examples and other material published in the original Introduction. E.S.P., N.L.J.]

The recurrence formula for $I_x(p,q)$ is the following:

$$I_x(p,q) = xI_x(p-1,q)+(1-x)\,I_x(p,q-1). \tag{2}$$

By aid of this formula $I_x(p,q)$ could be ultimately deduced from values of the function easy to integrate out.* In order to test the correctness of the results in any column of this $I_x(p, q)$ function for a given $p$ and $q$ an Euler–Maclaurin summation of the column was provided and was found very useful as a check. It runs†

Sum of column contents

$$= \frac{100q}{p+q} - 0{\cdot}5 + \frac{1}{B(p,q)}\left[\frac{1}{12}(0{\cdot}01)\,(x^{p-1}(1-x)^{q-1}) - \frac{(0{\cdot}01)^3}{720}\frac{d^2}{dx^2}(x^{p-1}(1-x)^{q-1}) + \frac{(0{\cdot}01)^5}{30240}\frac{d^4}{dx^4}(x^{p-1}(1-x)^{q-1}) - \frac{(0{\cdot}01)^7}{1209600}\frac{d^6}{dx^6}(x^{p-1}(1-x)^{q-1}) + \dots\right]_0^1. \tag{3}$$

At the head of each column of the table is given the value of the corresponding complete B-function, $B(p,q)$, so that it is possible to obtain rapidly, when it is required, the incomplete B-function itself, instead of the ratio.

In my original plan I proposed to take the argument intervals of $p$ and $q$ to be 0·5 from 10 to 50, and when either $p$ or $q$ were less than 10 to be 0·1, so that from 0 to 10 both $p$ and $q$ would proceed by 0·1. Here also $x$ was to advance by 0·005 instead of 0·01 and some portion of this was actually worked out. Further, to save labour in the use of the tables $p$ and $q$ were *both* to run from 0 to 50. But on reckoning out the space the printed tables would take, I found that it would extend to considerably over 2000 pages. The publication of such a table was wholly beyond any funds likely to be at my disposal, and accordingly the table had to be ruthlessly cut down. In the first place I discarded the idea of providing a table containing all the values of *both* $p$ and $q$ up to the limit of 50. I have had printed only the values of $p$ which are equal to or greater than the values of $q$. If the user of the tables requires $I_x(p,q)$ in which $p$ is less than $q$ then he must remember that

$$I_x(p,q) = 1-I_{1-x}(q,p) = 1-I_{1-x}(p',q'), \tag{4}$$

where $p' = q$ and $q' = p$, so that $p'$ is now greater than $q'$. This reduced the amount to be printed by almost a half.

In the next place the idea of publishing any differences whatever was dropped. It would have been needful to print three sets of differences, and any reasonable number of these would have been quite inadequate at certain parts of the table. When either $q$ or $p$ are low and fractional the differential coefficients of the curve at one or other terminal become infinite, and the differences may diverge. The only method of overcoming this difficulty is by the aid of auxiliary tables,‡ but that is not feasible when it is important to reduce the matter to be printed. Owing to the large number of differences required at some parts of the table, and to their total inadequacy at other parts I was not loath to omit them. As a matter of fact for many purposes we only need $p$ and $q$ to whole or half integers, and accordingly the interpolation requisite will often be with regard to $x$ alone.

In my opinion far more serious retrenchments were the following:

(*a*) The adoption throughout of 0·01 for the increment of $x$. When $p$ and $q$ approach 50, the standard deviation of the curve is about $\frac{1}{20}$th of the range and 99·9 % of the curve's area falls on less than a third of the range. It would accordingly have been more advantageous if this latter part of the table had proceeded by intervals of 0·005 in $x$, but this would have added upwards of 80 pages to the printed table. The adoption of a smaller interval in the case of $U$- and $J$-curves would also have been very advantageous.

(*b*) The adoption of 0·5 and, further on in the table, 1·0 for the increments of $p$ and $q$. This was again enforced by the limitation of space. The restriction affects peculiarly the table as applicable to $U$- and $J$-curves. In the case of $U$-curves, i.e. both $p$ and $q$ less than unity, interpolation becomes extremely difficult, and it is doubtful whether any table would be of much service which did not proceed by increments of 0·01

* Use was made of formulae of type $I_x(p+1,\ 0{\cdot}5) = I_x(p,\ 1{\cdot}5) - \dfrac{2\Gamma(p+\frac{3}{2})\,x^p\sqrt{(1-x)}}{\Gamma(p+1)\sqrt{\pi}}$, for the half-unit values of $p$ and $q$.

† It seems unnecessary here to enter into special variations of this formula, such as arise from altering the limits 0 and 1. When $p$ and $q$ are integers, the terms in the square brackets rapidly become negligible as $p$ and $q$ increase.

‡ [Table IV incorporated in the present Edition is an example of such a table. E.S.P., N.L.J.]

for $p$ and $q$. This would have involved an addition of some 5000 additional curves, or about 1666 pages of printed matter. Even with intervals of 0·02, we should have required upward of 200 additional pages. Again, an effective tabulation of $J$-curves with increment of $p$ as large even as 0·02 and 60 values of $q$ would have demanded space for 3000 additional curves or some 1000 additional pages. I was convinced at a very early stage of the work that the effective tabulation of $U$- and $J$-curves must be omitted from the present work, and left for others to undertake at a later date.

It may be asked why certain $J$- and $U$-curves have been included. The answer lies in the fact that $B_x(\frac{1}{2}, q)$ or $I_x(\frac{1}{2}, q)$ have special importance in practical statistics. For example, all symmetrical curves of the B-function type, i.e. $p = q$, can have their probability integrals determined by transformation to these types of $U$- or $J$-curves. Thus

$$\left.\begin{aligned} I_{x'}(p,p) &= \tfrac{1}{2}\{1+I_x(\tfrac{1}{2},p)\} \\ &= 1-\tfrac{1}{2}I_{1-x}(p,\tfrac{1}{2}) \end{aligned}\right\}, \qquad (5)$$

where $x = 4(x'-\frac{1}{2})^2$, or $x' = \frac{1}{2}(1+\sqrt{x})$ $(x' \geqslant 0{\cdot}5)$.*

This interchange may be of some service, as interpolating for $p$ in $I_{x'}(p,p)$ may involve extracting entries from several pages, while the interpolation for $I_{1-x}(p,\frac{1}{2})$ will probably need reference to one page only.

The function was computed to nine decimal places, but these were cut down to seven for publication. They might with but little recomputing of isolated values have been tabulated to eight decimals, but there seemed no particular advantage to be gained by incurring the additional cost of printing. The tables are intended in the first place for statisticians, and there are very few cases in statistical practice, wherein it is needful to ascertain a frequency or a probability to more than five figures. The additional two figures are given to provide greater accuracy for the purposes of interpolation. Should the reader feel that the tables fall short of the completeness desirable in dealing with such an important function, I may venture to remind him that the present is probably the first big attempt at tabling a trivariate function, that to provide a table which would effectively cover all regions of the B-function would not only have required another eight years of computing, but would have more than quadrupled the volume of the work, thus preventing or indefinitely delaying its publication; and finally that on studying the following account of the uses of the tables, he may convince himself that they are capable of giving at least a great deal of aid in a variety of inquiries.

## 2. Additional tables included in this edition

Two new tables have been incorporated in the present edition of the *Tables of the Incomplete Beta-Function*, using material computed some years ago. These tables help to cover certain regions of Table I where interpolation is difficult or impossible.

### 2·1 *Table III. Values of* $I_x(p, 0{\cdot}5)$ *for* $p = 11{\cdot}5, 12{\cdot}5, 13{\cdot}5, 14{\cdot}5$

This material was used by Karl Pearson and Brenda Stoessiger (1930) in their table of the function

$$P_x(n) = \tfrac{1}{2}\{1+I_x(0{\cdot}5, \tfrac{1}{2}(n-1))\}$$

which they prepared with the object of facilitating calculation of the probability integral of symmetrical beta distributions, through equation (5). The table later appeared as Table 25 of Pearson's *Tables for Statisticians and Biometricians* **2** (1931). We reproduce here the values of $I_x(p, 0{\cdot}5)$ for the four additional values of $p$ only, which are not included in Table I. This will make it possible with univariate interpolation to obtain the probability integral of Student's $t$-distribution to seven decimal places for all degrees of freedom up to 30.† The use of this transformation is discussed and illustrated in section 4·3 below.

* [Analogous results when $x' \leqslant 0{\cdot}5$ are given in equation (18) below. E.S.P., N.L.J.]

† There will of course be certain difficulties in interpolation when $x \to 1$ as indicated in section 4·3.

### 2·2 *Table IV, giving additional values of $I_x(p,q)$ for small q and x near unity*

When Catherine Thompson, under L. J. Comrie's direction, was preparing for *Biometrika* the table of percentage points of the beta distribution (Thompson, 1941; Pearson & Hartley, 1966, Table 16) it was found necessary to supplement Table I of the present volume by computing additional values of $I_x(p,q)$. The values of the integral, calculated to ten-decimal place accuracy, were for

$$p = 0{\cdot}5(0{\cdot}5)11{\cdot}0(1)16.$$

$$q = 0{\cdot}5; \quad x = 0{\cdot}9750,\ 0{\cdot}9850,\ 0{\cdot}9880(0{\cdot}0005)0{\cdot}9985,\ 0{\cdot}9988(0{\cdot}0001)0{\cdot}9999.$$

$$q = 1{\cdot}0(0{\cdot}5)3{\cdot}0; \quad x = 0{\cdot}975,\ 0{\cdot}985,\ 0{\cdot}988(0{\cdot}001)0{\cdot}999.$$

These results are now given in Table IV, with the following modifications:

(*a*) Values for $x = 0{\cdot}980$, taken from Table I, have been inserted.

(*b*) For $x = 0{\cdot}975, 0{\cdot}980, 0{\cdot}985$ seven decimal places are given.
For $x \geqslant 0{\cdot}988$ eight decimal places are given.

A copy of the original ten-figure table can be made available on request.
A description of the method of calculation used was given by Comrie & Hartley (1941).

## 3. Interpolation

If used to the full extent Table I is a table of triple entry, for $x$, for $p$ and for $q$. In a very considerable number of applications, however, the required values of $p$ or $q$, or both, correspond to those of tabled entries. In such cases only bivariate or univariate interpolation is needed; procedures which may then be followed are discussed in later sections. We shall start, however, by considering the case where interpolation is needed for all three variables.

### 3·1 *Trivariate interpolation*

#### 3·1·1 *Description of the problem and notation*

The problem is to find $I_x(p,q)$ for values of $x$, $p$ and $q$ none of which correspond to table entries. Following Karl Pearson's original notation we shall write:

$$z_{\theta\phi\chi} \text{ as the interpolate value for } I_x(p,q),$$

where $\theta_1, \theta_0; \phi_1, \phi_0; \chi_1, \chi_0$ are the argument interval ratios for $x$, $p$ and $q$ respectively. The position is then as illustrated in Fig. 1. Thus

$$\theta_0 = 1-\theta_1, \quad \phi_0 = 1-\phi_1, \quad \chi_0 = 1-\chi_1.$$

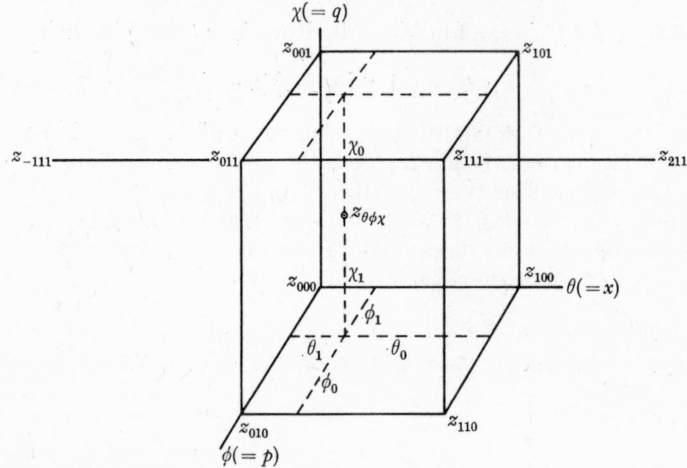


Fig. 1. Diagram of values of $z$ to assist the geometrical appreciation of trivariate interpolation. $z_{\theta\phi\chi}$ is the interpolate value and $\theta_1, \theta_0; \phi_1, \phi_0; \chi_1, \chi_0$ are the argument interval ratios of $z_{\theta\phi\chi}$.

Pearson measured $\theta_0$, $\phi_0$ and $\chi_0$ from the $z_{111}$ corner of the cell, in order to put the trivariate Everett interpolation formula into simplest shape, and we propose to retain this notation. Equation (viii) of his Introduction gave this formula in terms of the eight $z$-values at the corners of the cube in Fig. 1 and of 24 central second differences. An Everett expansion stopping at this point assumes that all fourth and higher central differences are negligible, to the number of decimal places retained.

Without a rather exhaustive exploration it would be difficult to say over what area of the table these differences are negligible, either to seven decimals or less. We shall do no more than quote a paragraph from p. ix of Karl Pearson's Introduction and refer the reader to later sections where the differences from various regions of Table 1 have been displayed. 'With the bivariate formula twelve tabular values must be used, while for the trivariate thirty-two are required. Hence, while it is relatively easy to use univariate interpolation formulae proceeding up to $\delta^4$, and to $\delta^6$, and possible though laborious to use bivariate formulae up to $\delta^4$ terms, it is for practical purposes of small use to provide trivariate formulae going as far as $\delta^4$. The number of terms to be dealt with becomes unmanageable. The only remedy is to ascertain what will be the extent of error we are introducing by neglecting the $\delta^4$ terms in the part of the table dealt with. For a very considerable proportion of the present table the fourth differences only affect the seventh decimal in the interpolate, and for most statistical purposes five-decimal accuracy is ample. A point may be borne in mind here, namely, that while in a bivariate formula the $\delta^2$ and $\delta^4$ terms are multiplied by the product of three proper fractions and the inverse factorial, in trivariate formulae they are multiplied by the product of four proper fractions as well as the inverse factorial.'

For computational purposes we believe it is more appropriate to convert this Everett formula into a trivariate Lagrangian. This expression makes use of four points on each of the 12 lines bounding the central cube of Fig. 1, e.g. for one such line (through (0, 1, 1) and parallel to the $x$-axis) we use values

$$z_{-111}, \quad z_{011}, \quad z_{111}, \quad z_{211}$$

as shown in the diagram. This means that 32 entries from the table are needed. Of course, if the interpolate is in a boundary panel of one or more of the variates, the formula will not be appropriate and some other method of interpolation will be necessary.

The most convenient scheme for calculation is probably one in which the table entries needed and the associated multiplying coefficients are classified according to the four values of $q$ required. For each $q$-value the table entries needed will then often be found on a single page (or on two facing pages) of the table. The Lagrangian may then be set out in the scheme shown in Table 1, the first quantity in each of the 32 cells being the $z$-entry, to be read from Table 1, and the second quantity the appropriate multiplying factor.

The quantity $\Psi$ appearing in the factors for the eight corners of the central cube is given by

$$\Psi = 1+\tfrac{1}{2}(\theta_1\theta_0+\phi_1\phi_0+\chi_1\chi_0). \tag{6}$$

The answer, $z_{\theta\phi\chi} = I_x(p,q)$, is the sum for the 32 cells of the product of the $z$'s and the factors (the 24 fringe factors being of course negative).

### 3·1·2 *Scheme of computation and numerical examples*

Anyone using this Lagrangian formula may choose his own method of calculating the multiplying factors. The procedure suggested to us by Miss Sheila Burrough of the Mathematics Department, University College London, who was using a hand calculating machine, is set out at the bottom of Table 1 and in Table 2. It is illustrated in detail on Example 2, in Table 3.

*Example* 1. Determine $I_{0\cdot6086}(34\cdot2228,\ 10\cdot2551)$.

This is the example taken as *Illustration* 1 on pp. x–xi of Karl Pearson's Introduction, using the same trivariate expression in Everett form, i.e. with second differences. Miss Burrough's result, using the Lagrangian scheme is 0·0099 4183. Certain small errors in computation occurred in Pearson's original calculations, so that he gave an answer of 0·0099 4274. When these errors are removed the Everett formula gives, as it should, the same result as that derived from the Lagrangian.

Table 1. *Trivariate Lagrangian. Schemes of* (*a*) *table entries and* (*b*) *multiplying factors*

$$\Upsilon = 1+\tfrac{1}{2}(\theta_1\theta_0+\phi_1\phi_0+\chi_1\chi_0)$$

| | 1st $p$-column | 2nd $p$-column | 3rd $p$-column | 4th $p$-column |
|---|---|---|---|---|
| | | *1st q-value* | | |
| 2nd $x$-line | — | $z_{00-1}$ $-\frac{1}{6}\chi_0(1-\chi_0^2)\,\theta_0\phi_0$ | $z_{01-1}$ $-\frac{1}{6}\chi_0(1-\chi_0^2)\,\theta_0\phi_1$ | — |
| 3rd $x$-line | — | $z_{10-1}$ $-\frac{1}{6}\chi_0(1-\chi_0^2)\,\theta_1\phi_0$ | $z_{11-1}$ $-\frac{1}{6}\chi_0(1-\chi_0^2)\,\theta_1\phi_1$ | — |
| | | *2nd q-value* | | |
| 1st $x$-line | — | $z_{-100}$ $-\frac{1}{6}\theta_0(1-\theta_0^2)\,\phi_0\chi_0$ | $z_{-110}$ $-\frac{1}{6}\theta_0(1-\theta_0^2)\,\phi_1\chi_0$ | — |
| 2nd $x$-line | $z_{0-10}$ $-\frac{1}{6}\phi_0(1-\phi_0^2)\,\theta_0\chi_0$ | $z_{000}$ $\theta_0\phi_0\chi_0\Upsilon$ | $z_{010}$ $\theta_0\phi_1\chi_0\Upsilon$ | $z_{020}$ $-\frac{1}{6}\phi_1(1-\phi_1^2)\,\theta_0\chi_0$ |
| 3rd $x$-line | $z_{1-10}$ $-\frac{1}{6}\phi_0(1-\phi_0^2)\,\theta_1\chi_0$ | $z_{100}$ $\theta_1\phi_0\chi_0\Upsilon$ | $z_{110}$ $\theta_1\phi_1\chi_0\Upsilon$ | $z_{120}$ $-\frac{1}{6}\phi_1(1-\phi_1^2)\,\theta_1\chi_0$ |
| 4th $x$-line | — | $z_{200}$ $-\frac{1}{6}\theta_1(1-\theta_1^2)\,\phi_0\chi_0$ | $z_{210}$ $-\frac{1}{6}\theta_1(1-\theta_1^2)\,\phi_1\chi_0$ | — |
| | | *3rd q-value* | | |
| 1st $x$-line | — | $z_{-101}$ $-\frac{1}{6}\theta_0(1-\theta_0^2)\,\phi_0\chi_1$ | $z_{-111}$ $-\frac{1}{6}\theta_0(1-\theta_0^2)\,\phi_1\chi_1$ | — |
| 2nd $x$-line | $z_{0-11}$ $-\frac{1}{6}\phi_0(1-\phi_0^2)\,\theta_0\chi_1$ | $z_{001}$ $\theta_0\phi_0\chi_1\Upsilon$ | $z_{011}$ $\theta_0\phi_1\chi_1\Upsilon$ | $z_{021}$ $-\frac{1}{6}\phi_1(1-\phi_1^2)\,\theta_0\chi_1$ |
| 3rd $x$-line | $z_{1-11}$ $-\frac{1}{6}\phi_0(1-\phi_0^2)\,\theta_1\chi_1$ | $z_{101}$ $\theta_1\phi_0\chi_1\Upsilon$ | $z_{111}$ $\theta_1\phi_1\chi_1\Upsilon$ | $z_{121}$ $-\frac{1}{6}\phi_1(1-\phi_1^2)\,\theta_1\chi_1$ |
| 4th $x$-line | — | $z_{201}$ $-\frac{1}{6}\theta_1(1-\theta_1^2)\,\phi_0\chi_1$ | $z_{211}$ $-\frac{1}{6}\theta_1(1-\theta_1^2)\,\phi_1\chi_1$ | — |
| | | *4th q-value* | | |
| 2nd $x$-line | — | $z_{002}$ $-\frac{1}{6}\chi_1(1-\chi_1^2)\,\theta_0\phi_0$ | $z_{012}$ $-\frac{1}{6}\chi_1(1-\chi_1^2)\,\theta_0\phi_1$ | — |
| 3rd $x$-line | — | $z_{102}$ $-\frac{1}{6}\chi_1(1-\chi_1^2)\,\theta_1\phi_0$ | $z_{112}$ $-\frac{1}{6}\chi_1(1-\chi_1^2)\,\theta_1\phi_1$ | — |

(*a*) Calculate $\Upsilon = 1+\frac{1}{2}(\theta_1\theta_0+\phi_1\phi_0+\chi_1\chi_0)$.

(*b*) Calculate $\theta_0\phi_0,\ \theta_0\phi_1,\ \theta_1\phi_0,\ \theta_1\phi_1$

and $\frac{1}{6}\phi_0(1-\phi_0^2),\ \frac{1}{6}\phi_1(1-\phi_1^2),\ \frac{1}{6}\theta_0(1-\theta_0^2),\ \frac{1}{6}\theta_1(1-\theta_1^2)$.

(*c*) Then proceed according to the scheme set out in Table 2.

*Example* 2. Determine $I_{0\cdot9082}$ (16·42, 2·69). Use $p = 15(1)18$, $q = 2{\cdot}0(0{\cdot}5)3{\cdot}5$.

$$\left.\begin{array}{lll}\theta_1 = 0{\cdot}82 & \phi_1 = 0{\cdot}42 & \chi_1 = 0{\cdot}38\\ \theta_0 = 0{\cdot}18 & \phi_0 = 0{\cdot}58 & \chi_0 = 0{\cdot}62\end{array}\right\}\ \Upsilon = 1{\cdot}3134$$

$$\theta_0\phi_0 = 0{\cdot}1044,\quad \theta_0\phi_1 = 0{\cdot}0756,\quad \theta_1\phi_0 = 0{\cdot}4756,\quad \theta_1\phi_1 = 0{\cdot}3444$$

$$\tfrac{1}{6}\phi_0(1-\phi_0^2) = 0{\cdot}064148,\quad \tfrac{1}{6}\phi_1(1-\phi_1^2) = 0{\cdot}057652,\quad \tfrac{1}{6}\theta_0(1-\theta_0^2) = 0{\cdot}029028,\quad \tfrac{1}{6}\theta_1(1-\theta_1^2) = 0{\cdot}044772$$

The computations follow the scheme of Tables 1 and 2. Miss Burrough found that the setting out and calculation took her in all one hour.

Table 2. *Operations to obtain the multiplying factors in the Lagrangian scheme of Table* 1

| Form and transfer to setting lever | Multiply successively by: | Record coefficient for: | |
|---|---|---|---|
| $\chi_0\Psi$ | $\theta_0\phi_0,\ \theta_0\phi_1,\ \theta_1\phi_0,\ \theta_1\phi_1$ | $z_{000},\ z_{010},\ z_{100},\ z_{110}$ | +ve |
| $\chi_1\Psi$ | Same | $z_{001},\ z_{011},\ z_{101},\ z_{111}$ | +ve |
| $\frac{1}{6}\chi_0(1-\chi_0^2)$ | Same | $z_{00-1},\ z_{01-1},\ z_{10-1},\ z_{11-1}$ | −ve |
| $\frac{1}{6}\chi_1(1-\chi_1^2)$ | Same | $z_{002},\ z_{012},\ z_{102},\ z_{112}$ | −ve |
| $\frac{1}{6}\phi_0(1-\phi_0^2)\ \chi_0$ | $\theta_0,\ \theta_1$ | $z_{0-10},\ z_{1-10}$ | −ve |
| $\frac{1}{6}\phi_0(1-\phi_0^2)\ \chi_1$ | Same | $z_{0-11},\ z_{1-11}$ | −ve |
| $\frac{1}{6}\phi_1(1-\phi_1^2)\ \chi_0$ | Same | $z_{020},\ z_{120}$ | −ve |
| $\frac{1}{6}\phi_1(1-\phi_1^2)\ \chi_1$ | Same | $z_{021},\ z_{121}$ | −ve |
| $\frac{1}{6}\theta_0(1-\theta_0^2)\ \chi_0$ | $\phi_0,\ \phi_1$ | $z_{-100},\ z_{-110}$ | −ve |
| $\frac{1}{6}\theta_0(1-\theta_0^2)\ \chi_1$ | Same | $z_{-101},\ z_{-111}$ | −ve |
| $\frac{1}{6}\theta_1(1-\theta_1^2)\ \chi_0$ | Same | $z_{200},\ z_{210}$ | −ve |
| $\frac{1}{6}\theta_1(1-\theta_1^2)\ \chi_1$ | Same | $z_{201},\ z_{211}$ | −ve |

Table 3. *The solution of Example* 2

| $q$ | $x$ | $p$ | | | |
|---|---|---|---|---|---|
| | | 15 | 16 | 17 | 18 |
| 2·0 | 0·90 | | 0·4817 852<br>−0·0066 4109 | 0·4502 839<br>−0·0048 0907 | |
| | 0·91 | | 0·5395 754<br>−0·0302 5387 | 0·5091 247<br>−0·0219 0797 | |
| 2·5 | 0·89 | | 0·5640 269<br>−0·0104 3847 | 0·5305 183<br>−0·0075 5889 | |
| | 0·90 | 0·6516 163<br>−0·0071 5892 | 0·6195 423<br>0·0850 1376 | 0·5879 695<br>0·0615 6168 | 0·5570 466<br>−0·0064 3396 |
| | 0·91 | 0·7050 654<br>−0·0326 1284 | 0·6759 755<br>0·3872 8488 | 0·6470 219<br>0·2804 4768 | 0·6183 494<br>−0·0293 1028 |
| | 0·92 | | 0·7322 864<br>−0·0161 0001 | 0·7065 964<br>−0·0116 5863 | |
| 3·0 | 0·89 | | 0·6827 240<br>−0·0063 9777 | 0·6511 660<br>−0·0046 3287 | |
| | 0·90 | 0·7617 972<br>−0·0043 8772 | 0·7337 960<br>0·0521 0520 | 0·7054 448<br>0·0377 3136 | 0·6769 268<br>−0·0039 4340 |
| | 0·91 | 0·8072 732<br>−0·0199 8852 | 0·7831 804<br>0·2373 6816 | 0·7585 154<br>0·1718 8728 | 0·7334 296<br>−0·0179 6436 |
| | 0·92 | | 0·8297 952<br>−0·0098 6775 | 0·8091 620<br>−0·0071 4561 | |
| 3·5 | 0·90 | | 0·8216 084<br>−0·0056 5723 | 0·7982 445<br>−0·0040 9661 | |
| | 0·91 | | 0·8612 783<br>−0·0257 7181 | 0·8419 952<br>−0·0186 6235 | |

Summing products gives $I_{0\cdot9082}(16{\cdot}42,\ 2{\cdot}69) = 0{\cdot}6986\ 0740$.

## 3·2 *Bivariate interpolation*

### 3·2·1 *A bivariate Lagrangian formula*

When it is only necessary to interpolate for two of the variables, the use of two series of univariate interpolations may be preferred. Some univariate Everett and Lagrangian formula have been listed in the Appendix. However, if fourth differences can be neglected a relatively straightforward bivariate Lagrangian procedure can be derived from the scheme of Table 1, putting $\theta_1 = 0$, $\theta_0 = 1$, or $\phi_1 = 0$, $\phi_0 = 1$ or $\chi_1 = 0$, $\chi_0 = 1$ as appropriate.

Suppose, for example, that it is the value of $q$ which corresponds to a table entry. Then $\chi_1 = 0$, $\chi_0 = 1$ and the blocks of Table 1 for 1st, 3rd and 4th $q$-values vanish. We are therefore left with the relatively simple scheme shown in Table 4, where

$$\Psi' = 1 + \tfrac{1}{2}(\theta_1\theta_0 + \phi_1\phi_0).$$

The first entry in each cell is to be taken from the Table of $I_x(p, q)$ and the second entry is its multiplying factor. $q$ corresponds to a table entry. The answer $z_{\theta\phi 0} = I_x(p, q)$ is the sum for the 12 cells of the product of the $z$'s and the factors.

Table 4. *Scheme for bivariate Lagrangian*

| | 1st $p$-column | 2nd $p$-column | 3rd $p$-column | 4th $p$-column |
|---|---|---|---|---|
| 1st $x$-line | | $z_{-100}$ $-\frac{1}{6}\theta_0\phi_0(1-\theta_0^2)$ | $z_{-110}$ $-\frac{1}{6}\theta_0\phi_1(1-\theta_0^2)$ | |
| 2nd $x$-line | $z_{0-10}$ $-\frac{1}{6}\theta_0\phi_0(1-\phi_0^2)$ | $z_{000}$ $\theta_0\phi_0\Psi'$ | $z_{010}$ $\theta_0\phi_1\Psi'$ | $z_{020}$ $-\frac{1}{6}\theta_0\phi_1(1-\phi_1^2)$ |
| 3rd $x$-like | $z_{1-10}$ $-\frac{1}{6}\theta_1\phi_0(1-\phi_0^2)$ | $z_{100}$ $\theta_1\phi_0\Psi'$ | $z_{110}$ $\theta_1\phi_1\Psi'$ | $z_{120}$ $-\frac{1}{6}\theta_1\phi_1(1-\phi_1^2)$ |
| 4th $x$-line | | $z_{200}$ $-\frac{1}{6}\theta_1\phi_0(1-\theta_1^2)$ | $z_{210}$ $-\frac{1}{6}\theta_1\phi_1(1-\theta_1^2)$ | |

*Note.* As a numerical check, the sum of the eight negative factors is equal to $1-\Psi'$.

### 3·2·2 *Numerical example*

*Example* 3. Determine $I_{0\cdot 8334}$ (12·5, 1·5).

Using the scheme of Table 4, it was found that the integral is 0·1997 755.

This example corresponds to *Illustration* 9 on pp. liii–liv of Karl Pearson's Introduction. He employs a second difference Everett formula, but has apparently again made some numerical slips.

We consider the example again below in § 3·3·3 where two univariate interpolations are performed. It appears that while the 4th differences for $x$ are negligible at this part of the Table, those for $p$ may just affect the last (7th) decimal place. However, the result obtained from the scheme of Table 4 in this case agrees with that found below, where 4th differences for $p$ are taken into account.

## 3·3 *Special cases*

### 3·3·1 *Applications in the analysis of variance*

Values of the incomplete beta-function are sometimes required by statisticians in connexion with the analysis of variance. This is essentially because, if $\chi_1^2$ and $\chi_2^2$ are two independent $\chi^2$-variables having $\nu_1$ and $\nu_2$ degrees of freedom, respectively, then the probability density distribution of

$$x = \frac{\chi_1^2}{\chi_1^2 + \chi_2^2}$$

has the form

$$f(x) = \{B(\tfrac{1}{2}\nu_1, \tfrac{1}{2}\nu_2)\}^{-1} x^{\frac{1}{2}\nu_1 - 1}(1-x)^{\frac{1}{2}\nu_2 - 1}, \quad 0 \leqslant x \leqslant 1. \tag{7}$$

The probability integral of this ratio is therefore an incomplete beta-function, for which $p = \frac{1}{2}\nu_1, q = \frac{1}{2}\nu_2$. It follows that the values of $p$ and $q$ with which the main Table must be entered will be integers or half integers. Since apart from the case where $q = 0{\cdot}5$, to which Table III refers, Table I gives half integer values only for $p, q \leqslant 10{\cdot}5$, interpolation for $p$ and/or $q$ may be needed at a half integer value. Probably the easiest procedure is then to use the very simple univariate formulae given below at an appropriate number of table entries for $x$; then to apply a further univariate formula to interpolate for $x$.

Two situations arise: when interpolation is necessary ($a$) for $p$ or $q$, ($b$) for $p$ and $q$. We deal with these in turn below. For convenient reference we have given in an Appendix the general univariate interpolation formulae both in terms of central differences, and as Lagrangians.

### 3·3·2 *Univariate interpolation formulae for $I_x(p+0{\cdot}5, q)$*

Here $x$, $p$ and $q$ correspond to table entries and $p \geqslant q$. If $p < q$ we must use the relation (4) of p. x above. In the notation of § 2·1 $\theta_1 = 0, \phi_1 = \phi_0 = 0{\cdot}5, \chi_1 = 0$. We may now use the Everett central difference equation (A 1) of the Appendix, substituting $\phi_1$, $\phi_0$ for $\theta_1$, $\theta_0$; as a result we obtain the simple formula

$$z_{0\cdot5} = \tfrac{1}{2}(z_0+z_1) - \tfrac{1}{16}(\delta^2 z_0+\delta^2 z_1) + \tfrac{3}{256}(\delta^4 z_0+\delta^4 z_1) - \tfrac{5}{2048}(\delta^6 z_0+\delta^6 z_1) + \tfrac{35}{65536}(\delta^8 z_0+\delta^8 z_1) - \dots \quad (8)$$

which is correct up to and including the ninth-order differences.

Alternatively, equations of Lagrangian form may be used. These may be derived from equations (A 4, 5 and 6) of the Appendix, putting $\theta = 0{\cdot}5$. Thus we have:

*Four-point Lagrangian*

$$z_{0\cdot5} = \tfrac{1}{16}\{9(z_0+z_1) - (z_{-1}+z_2)\}. \quad (9)$$

*Six-point Lagrangian*

$$z_{0\cdot5} = \tfrac{1}{256}\{150(z_0+z_1) - 25(z_{-1}+z_2) + 3(z_{-2}+z_3)\}. \quad (10)$$

*Eight-point Lagrangian*

$$z_{0\cdot5} = \tfrac{1}{2048}\{1225(z_0+z_1) - 245(z_{-1}+z_2) + 49(z_{-2}+z_3) - 5(z_{-3}+z_4)\}. \quad (11)$$

Expressions in Lagrangian form are easier to compute than those involving differences. But unless we have some knowledge of the magnitude of the differences in the part of Table I which we are entering, we can only judge whether a given order of Lagrangian gives the accuracy desired by comparing its result with that obtained from a higher-order formula.

### 3·3·3 *Numerical example*

*Example* 4. Consider again the case of Example 3 in which the bivariate scheme of Table 4 was used to determine $I_{0\cdot8334}(12{\cdot}5, 1{\cdot}5)$. Suppose now we use two univariate procedures, first determining $I_x(12{\cdot}5, 1{\cdot}5)$ (using one or other of the formulae of the preceding section), at an appropriate number of $x$-entries, and then interpolating for $x = 0{\cdot}8334$ between these values.

First consider the differences for $p$ when $x = 0{\cdot}83$, $q = 1{\cdot}5$. We have from pp. 37–8 of the main Table:

| $p$ | $I_{0\cdot83}(p, 1{\cdot}5)$ | $\delta^2$ | $\delta^4$ | $\delta^6$ |
|---|---|---|---|---|
| 12 | 0·2068 222 | 47 286 | 342 | −84 |
| 13 | 0·1766 323 | 41 685 | 439 | −59 |

The fourth differences are not negligible but the sixth differences are.

Using the Everett formula (8) we find

$$I_{0\cdot83}(12{\cdot}5,\ 1{\cdot}5) = 0{\cdot}1917\,272|5 - 5\,560|7 + 9|2 + 0|3 = 0{\cdot}1911\,721.$$

To obtain the same accuracy we must use the six-point Lagrangian of formula (10) and obtain, as we should, the same value, 0·1911 721.

We have now to interpolate for $x$ and the question arises as to the number of table entries at which we must follow the above procedure and find $I_x(12{\cdot}5, 1{\cdot}5)$. Will it suffice if we make calculations at $x = 0{\cdot}82$, 0·83, 0·84 and 0·85 only, enabling us to use an Everett central second-difference formula to find $I_{0\cdot8334}(12{\cdot}5, 1{\cdot}5)$ or must six points be used?

On examining the central differences for $x$, we find at $p = 12{\cdot}0$, $q = 1{\cdot}5$ and $p = 13{\cdot}0$, $q = 1{\cdot}5$

| $p$ | $x$ | $I_x(12{\cdot}0, 1{\cdot}5)$ | $\delta^2$ | $\delta^4$ |
|---|---|---|---|---|
| 12·0 | 0·83 | 0·2068 222 | 26 474 | −27 |
| | 0·84 | 0·2338 525 | 28 325 | −77 |
| 13·0 | 0·83 | 0·1766 323 | 27 613 | 40 |
| | 0·84 | 0·2020 264 | 29 970 | −15 |

It is not therefore quite clear that fourth differences can be neglected if seven-decimal place accuracy is required. Using the six point Lagrangian of equation (10), values of $I_x(12{\cdot}5, 1{\cdot}5)$ were therefore found at six values of $x$ as follows:

| $x$ | $I_x(12{\cdot}5, 1{\cdot}5)$ | $\delta^2$ | $\delta^4$ |
|---|---|---|---|
| 0·81 | 0·1466 356 | | |
| 0·82 | 0·1676 534 | 25 009 | |
| 0·83 | 0·1911 721 | 27 110 | 5 |
| 0·84 | 0·2174 018 | 29 216 | −48 |
| 0·85 | 0·2465 531 | 31 274 | |
| 0·86 | 0·2788 318 | | |

We now use the Everett formula (A 1) of the Appendix, obtaining the coefficients from A. J. Thompson's (1943) Table. We find

$$I_{0{\cdot}8334}(12{\cdot}5, 1{\cdot}5) = 0{\cdot}2000\,902|0 - 0{\cdot}0003\,147|3 - 0{\cdot}0000\,000|4 = 0{\cdot}1997\,754.$$

It is seen that the fourth-difference term, though less than a half unit in the seventh decimal place, just succeeds in reducing the seventh figure from 5 to 4. In Example 3, § 3·2, we attacked the same problem boldly using a bivariate Lagrangian which *assumed* that fourth differences of both $p$ and $x$ were negligible. The answer was 0·1997 755. The error is in this case probably smaller than might have been anticipated because at this point the $\delta_p^4$ change sign.

### 3·3·4 *Univariate diagonal formulae for* $I_x(p+0{\cdot}5, q+0{\cdot}5)$

Here $x$, $p$ and $q$ correspond to tabled entries and $p, q > 11{\cdot}0$. The problem is reduced to one of univariate interpolation by noting that the interpolate lies on a diagonal of table entries for $p$ and $q$. There is in fact a choice of two diagonals. Thus using the same $z$-notation and dropping the first subscript, since $x$ is a table entry and $\theta_1 = 0$, the position is as illustrated in Fig. 2.

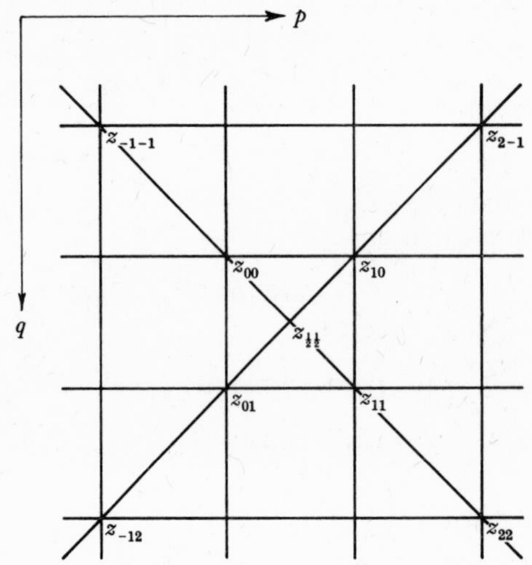


Fig. 2. Illustrating diagonal interpolation for $z_{\frac{1}{2}\frac{1}{2}}$.

Along what may be termed (*a*) the left-upper to right-lower diagonal $p-q$ remains constant while $p+q$ increases by 2 units at each step. On the other hand along (*b*) the right-upper to left-lower diagonal $p-q$ increases by 2 units at each step while $p+q$ remains constant. Since the asymmetry or skewness of the probability density distribution

$$f(x) = \frac{\Gamma(p+q)}{\Gamma(p)\,\Gamma(q)}\,x^{p-1}(1-x)^{q-1}, \quad 0 \leqslant x \leqslant 1, \tag{12}$$

is known to be sensitive to changes in $p-q$, particularly when $p$ and/or $q$ are small,* we might anticipate that $I_x(p,q)$ would change more rapidly along diagonal (*b*) than along diagonal (*a*). This suggests that the differences may converge more rapidly along (*a*) than along (*b*), so that diagonal interpolation will be most profitable along the former. The two examples given below, both taken from Karl Pearson's Introduction, seem to confirm this point although we have not established its universal correctness, since the shape of the distribution (12) does also change with $p+q$.

Using diagonal interpolation the formulae of § 3·3·2 will be applicable, with appropriate adjustment of the subscripts to the $z$'s.

### 3·3·5 *Illustrations of diagonal interpolation*

*Example* 5. Determine $I_{0\cdot19}(10\cdot5,\ 10\cdot5)$, using table entries for $p$ and $q$ at unit intervals.

(*a*) *Using left-upper to right-lower diagonal.* The tabled entries and their central differences are as follows:

| | | $z$ | $\delta^2$ | $\delta^4$ | $\delta^6$ | $\delta^8$ |
|---|---|---|---|---|---|---|
| $I_{0\cdot19}(6, 6)$ | $z_{-4,-4}$ | 90 095 | | | | |
| $I_{0\cdot19}(7, 7)$ | $z_{-3,-3}$ | 52 035 | 16 304 | | | |
| $I_{0\cdot19}(8, 8)$ | $z_{-2,-2}$ | 30 279 | 9 200 | 3 160 | | |
| $I_{0\cdot19}(9, 9)$ | $z_{-1,-1}$ | 17 723 | 5 256 | 1 719 | 677 | |
| $I_{0\cdot19}(10, 10)$ | $z_{0,0}$ | 10 423 | 3 031 | 955 | 346 | 174 |
| $I_{0\cdot19}(11, 11)$ | $z_{1,1}$ | 6 154 | 1 761 | 537 | 189 | 68 |
| $I_{0\cdot19}(12, 12)$ | $z_{2,2}$ | 3 646 | 1 028 | 308 | 100 | |
| $I_{0\cdot19}(13, 13)$ | $z_{3,3}$ | 2 166 | 603 | 179 | | |
| $I_{0\cdot19}(14, 14)$ | $z_{4,4}$ | 1 289 | 357 | | | |
| $I_{0\cdot19}(15, 15)$ | $z_{5,5}$ | 769 | | | | |

We may now use the Everett central difference formula (8), and find:

$$\begin{aligned} z_{\frac{1}{2},\frac{1}{2}} &= \tfrac{1}{2}(16577) - \tfrac{1}{16}(4792) + \tfrac{3}{256}(1492) - \tfrac{5}{2048}(535) + \tfrac{35}{65536}(242) \\ &= 8288|5 - 299|5 + 17|73 - 1|3 + 0|13, \\ &= 0\cdot0008\,288|5 \text{ by linear interpolation,} \\ &= 0\cdot0007\,989 \text{ up to third differences,} \\ &= 0\cdot0008\,006|73 \text{ up to fifth differences,} \\ &= 0\cdot0008\,005|43 \text{ up to seventh differences,} \\ &= 0\cdot0008\,005|56, \text{ up to ninth differences, or the correct value, i.e.} \\ &= 0\cdot0008\,006, \text{ up to seven decimal places using } \delta^8. \end{aligned}$$

In the present case $I_{0\cdot19}(10\cdot5,\ 10\cdot5)$ is given in the main Table, and has the value 0·0008 006. It is seen that if we went as far as $\delta^4$ we should only be in error by a unit in the seventh place of decimals. Equally, the six point Lagrangian of formula (10) could have been used, but unless the magnitude of the differences were known, the accuracy could only have been gauged by comparing the result with that obtained from the eight point formula (11).

* Note that the expression for the moment ratio $\sqrt{\beta_1} = \mu_3/\sigma^3$, quoted in equation (25) below, contains the factor $q-p$.

(*b*) *Using the right-upper to left-lower diagonal.* The tabled entries and their central differences are now as follows:

| | | $z$ | $\delta^2$ | $\delta^4$ | $\delta^6$ | $\delta^8$ | $\delta^{10}$ |
|---|---|---|---|---|---|---|---|
| $I_{0\cdot19}(16, 5)$ | $z_{6,-5}$ | 0 | | | | | |
| $I_{0\cdot19}(15, 6)$ | $z_{5,-4}$ | 1 | 8 | | | | |
| $I_{0\cdot19}(14, 7)$ | $z_{4,-3}$ | 10 | 65 | 321 | | | |
| $I_{0\cdot19}(13, 8)$ | $z_{3,-2}$ | 84 | 443 | 1 599 | 3 560 | | |
| $I_{0\cdot19}(12, 9)$ | $z_{2,-1}$ | 601 | 2 420 | 6 437 | 9 079 | + 42 | |
| $I_{0\cdot19}(11, 10)$ | $z_{1,0}$ | 3 538 | 10 834 | 20 354 | 14 640 | − 15 575 | − 8 909 |
| $I_{0\cdot19}(10, 11)$ | $z_{0,1}$ | 17 309 | 39 602 | 48 911 | 4 626 | − 40 101 | + 39 623 |
| $I_{0\cdot19}(9, 12)$ | $z_{-1,2}$ | 70 682 | 117 281 | 82 094 | − 45 489 | − 25 004 | |
| $I_{0\cdot19}(8, 13)$ | $z_{-2,3}$ | 241 336 | 277 054 | 69 788 | − 120 608 | | |
| $I_{0\cdot19}(7, 14)$ | $z_{-3,4}$ | 689 044 | 506 615 | − 63 126 | | | |
| $I_{0\cdot19}(6, 15)$ | $z_{-4,5}$ | 1 643 367 | 673 050 | | | | |
| $I_{0\cdot19}(5, 16)$ | $z_{-5,6}$ | 3 270 740 | | | | | |

It is seen that the differences are very large and change rapidly in magnitude, so that interpolation along this diagonal is bound to be less satisfactory than along the diagonal (*a*).

*Example* 6. Proceeding further into the Table, let us determine $I_{0\cdot58}(40\cdot5, 21\cdot5)$.

(*a*) *Using left-upper to right-lower diagonal.* The tabled entries, their central differences and the result of applying formula (8) up to the $\delta^4$ term are as follows:

| | $z$ | $\delta^2 z$ | $\delta^4 z$ |
|---|---|---|---|
| $I_{0\cdot58}(36, 17)$ | 0·0650 895 | | |
| $I_{0\cdot58}(37, 18)$ | 0·0751 282 | 5 924 | |
| $I_{0\cdot58}(38, 19)$ | 0·0857 593 | 5 437 | − 7 |
| $I_{0\cdot58}(39, 20)$ | 0·0969 341 | 4 943 | + 4 |
| $I_{0\cdot58}(40, 21)$ | 0·1086 032 | 4 453 | + 4 |
| $I_{0\cdot58}(41, 22)$ | 0·1207 176 | 3 967 | + 17 |
| $I_{0\cdot58}(42, 23)$ | 0·1332 287 | 3 498 | + 14 |
| $I_{0\cdot58}(43, 24)$ | 0·1460 896 | 3 043 | + 20 |
| $I_{0\cdot58}(44, 25)$ | 0·1592 548 | 2 608 | |
| $I_{0\cdot58}(45, 26)$ | 0·1726 808 | | |

$$z_{\frac{1}{2},\frac{1}{2}} = I_{0\cdot58}(40\cdot5, 21\cdot5) = \tfrac{1}{2}(0\cdot2293\,208) - \tfrac{1}{16}(8420) + \tfrac{3}{256}(21)$$
$$= 0\cdot1146\,604 - 0\cdot0000\,526|25 + 0\cdot0000\,000|25$$
$$= 0\cdot1146\,078.$$

This is a very satisfactory result since seven-figure accuracy is obtained using second differences only, or a four point Lagrangian.

(*b*) *Using right-upper to left-lower diagonal.* The tabled entries and central differences are as follows:

| | $z$ | $\delta^2$ | $\delta^4$ | $\delta^6$ | $\delta^8$ | $\delta^{10}$ |
|---|---|---|---|---|---|---|
| $I_{0\cdot58}(35, 27)$ | 0·5927 339 | | | | | |
| $I_{0\cdot58}(36, 26)$ | 0·4903 777 | 2 707 | | | | |
| $I_{0\cdot58}(37, 25)$ | 0·3882 922 | 68 321 | −12 184 | | | |
| $I_{0\cdot58}(38, 24)$ | 0·2930 388 | 121 751 | −20 993 | 5 881 | | |
| $I_{0\cdot58}(39, 23)$ | 0·2099 605 | 154 188 | −23 921 | 6 035 | −2 133 | |
| $I_{0\cdot58}(40, 22)$ | 0·1423 010 | 162 704 | −20 814 | 4 056 | − 921 | 25 |
| $I_{0\cdot58}(41, 21)$ | 0·0909 119 | 150 406 | −13 651 | 1 156 | + 316 | −313 |
| $I_{0\cdot58}(42, 20)$ | 0·0545 636 | 124 457 | − 5 332 | −1 428 | 1 240 | |
| $I_{0\cdot58}(43, 19)$ | 0·0306 610 | 93 176 | + 1 559 | −2 772 | | |
| $I_{0\cdot58}(44, 18)$ | 0·0160 760 | 63 454 | + 5 678 | | | |
| $I_{0\cdot58}(45, 17)$ | 0·0078 364 | 39 410 | | | | |
| $I_{0\cdot58}(46, 16)$ | 0·0035 378 | | | | | |

The correct answer would only be obtained here by using the $\delta^6$ term.

These two comparisons confirm the view that to calculate $I_x(p+0\cdot5, q+0\cdot5)$ univariate interpolation should be carried out along diagonal (*a*). If it is also necessary to interpolate for $x$, this computation will have to be carried out at an adequate number of neighbouring table entries for $x$, and a final univariate interpolation for $x$ performed.

### 3·4 *Bivariate interpolation for p and q involving a horizontal or vertical 'slide'*

Because of the rapid convergence of the differences along the left-upper to right-lower diagonal discussed in the preceding sections, it occurred to Karl Pearson that bivariate interpolation for $p$ and $q$ might be effectively carried out using oblique rather than rectangular axes. The procedure involved was discussed on pp. xxvi–xxxv of his 1933 Introduction. Formulae were given for both horizontal and vertical slides and the results compared, on some examples, with those obtained using the standard bivariate Everett formula based on rectangular axes, and taken to the $\delta^4$ term.* These slide formulae are undoubtedly of interest but as they do not seem to have given outstandingly better results in the examples chosen than those obtained from using rectangular axes, we have not felt it justifiable to include them in this shortened Introduction.

### 3·5 *Methods which may be used to overcome certain difficulties in interpolation*

At the limits of the Table considerable labour may occur in the work of interpolation and forward difference formulae may have to be employed. Two devices have proved useful in reducing the number of differences required in these boundary areas.

(1) Interpolation may be carried out to advantage in terms of

$$B_x(p, q) = \int_{0]}^{x} x^{p-1}(1-x)^{q-1}\,dx$$

instead of in terms of the ratio

$$I_x(p, q) = \frac{\Gamma(p+q)}{\Gamma(p)\,\Gamma(q)} B_x(p, q) = \frac{B_x(p, q)}{B(p, q)}.$$

Thus, if $x$ and $p$ are fixed and we are interpolating for $q$, all that is necessary is to multiply the appropriate table entries at $\ldots, q_{-1}, q_0, q_1, q_2, \ldots$ by the value of the complete B-function given at the head of each column of the table. When interpolation has given the answer required for $B_x(p, q)$, this must be divided by the value of $B(p, q)$ corresponding to the interpolated $q$. To find $B(p, q)$ use can be made of tables of the complete Γ-function.†

* The formula quoted by Pearson in his equation (xxii) is actually expressed in Lagrangian form.

† See, e.g. Brownlee (1923) and E. S. Pearson (1922) for extensive tables of log $\Gamma(x)$.

(2) A second device is to interpolate for $x^{-(p-1)} I_x(p,q)$ and after the result is obtained multiply up by the appropriate factor $x^{p-1}$. Two illustrations of this procedure are given below.

*Example* 7. To find $I_{0\cdot90}(0\cdot5, 3\cdot25) = 1 - I_{0\cdot10}(3\cdot25, 0\cdot5)$.

We have to interpolate for $p$ and proceed as follows, omitting a factor $10^{-7}$ in the later columns of our table:

| $p$ | $I_{0\cdot10}(p, 0\cdot5)$ | $I_{0\cdot10}(p, 0\cdot5) \times (0\cdot10)^{-(p-1)} \times 10^7$ | $\delta^2 \times 10^7$ |
|---|---|---|---|
| 2·5 | 0·0011 144 | $(0\cdot10)^{-2} \times (0\cdot10)^{0\cdot5} \times 11\,144 = (0\cdot10)^{-2} \times 3524$ | |
| 3·0 | 3 250 | $(0\cdot10)^{-2} \times 3\,250 = (0\cdot10)^{-2} \times 3250$ | $(0\cdot10)^{-2} \times 53$ |
| 3·5 | 958 | $(0\cdot10)^{-2} \times (0\cdot10)^{-0\cdot5} \times 958 = (0\cdot10)^{-2} \times 3029$ | $(0\cdot10)^{-2} \times 42$ |
| 4·0 | 285 | $(0\cdot10)^{-2} \times (0\cdot10)^{-1} \times 285 = (0\cdot10)^{-2} \times 2850$ (?) | |

We may now interpolate at $p = 3\cdot25$. Since the $\delta^4$ terms are negligible and $\delta^3 < 60$ in the units chosen, we may use the reduced second difference Everett formula (A 3) of the Appendix. Hence

$$\begin{aligned} I_{0\cdot10}(3\cdot25, 0\cdot5) &= 10^{-7} \times (0\cdot10)^{2\cdot25} \times (0\cdot10)^{-2}\{\tfrac{1}{2}(3250+3029) - \tfrac{1}{16}(53+42)\} \\ &= 10^{-7} \times (0\cdot10)^{0\cdot25} \times \{3139 - 6\} \\ &= 0\cdot562\,341 \times 0\cdot000\,3133 = 0\cdot0001\,762. \end{aligned}$$

Therefore $$I_{0\cdot90}(0\cdot5, 3\cdot25) = 0\cdot9998\,238.$$

Had we tried to interpolate directly for $I_x(p,q)$ we should have encountered great difficulties owing to the very slow convergence of differences. In multiplying by $x^{-(p-1)}$ we have, it is true, had to find two roots of 0·10, namely $(0\cdot10)^{0\cdot5} = 0\cdot3162\,278$ and $(0\cdot10)^{0\cdot25} = 0\cdot562\,341$, but have then required to undertake only the simplest of interpolation procedures.

*Example* 8. To find $I_{0\cdot8966}(0\cdot5, 3\cdot5) = 1 - I_{0\cdot1034}(3\cdot5, 0\cdot5)$.

We have here to interpolate for $x$ and proceed as follows:

| $x$ | $I_x(3\cdot5, 0\cdot5)$ | $x^{-2\cdot5}$ | $x^{-2\cdot5} I_x(3\cdot5, 0\cdot5)$ | $\delta^2$ |
|---|---|---|---|---|
| 0·09 | 0·0000 660 | 411·5226 | 0·02716\|0 | |
| 0·10 | 958 | 316·2278 | 3029\|5 | 6\|0 |
| 0·11 | 1 344 | 249·1829 | 3349\|0 | \|1 |
| 0·12 | 1 830 | 200·4688 | 3668\|6 | |

Taken together, the second differences make a negligible contribution when interpolating. Thus

$$\begin{aligned} I_{0\cdot1034}(3\cdot5, 0\cdot5) &= \{0\cdot66 \times 0\cdot0302\,9 + 0\cdot34 \times 0\cdot0334\,9\} \times (0\cdot1034)^{2\cdot5} \\ &= 0\cdot0313\,78 \times 0\cdot0034\,380 \\ &= 0\cdot0001\,079. \end{aligned}$$

Accordingly $$I_{0\cdot8966}(0\cdot5, 3\cdot5) = 0\cdot9998\,921.$$

In this case we have scarcely gained by the transformation. True, we have escaped with a linear interpolation, but it has been necessary to compute four power terms. On the other hand the fourth differences of $I_x(3\cdot5, 0\cdot5)$ are negligible and the two relevant second differences are so close together that the simple interpolation formula (A 3) of the Appendix can be used.

### 3·6 *Inverse interpolation: existing tables of percentage points*

By this we mean the determination of $x$, given $p$, $q$ and $I_x(p,q)$. Clearly situations could arise in which $p$ or $q$ has to be found, given values of the other three arguments, but the problem we have specified is that of most common occurrence. The general problem of inverse interpolation in a trivariate table presents obvious difficulties, but luckily the statistician is generally satisfied with an answer in a form which lends itself to standardized tabulation: he is satisfied with finding the so-called percentage points of the beta distribution.

If it is necessary to interpolate between these percentage points, this can be done following a method illustrated by Johnson *et al.* (1963, pp. 463–4, 466–8) using the normal function as an auxiliary and a divided difference formula.

There exist a number of detailed tables of such standard percentage points. We list here three of them.

(1) *Table* 16 *in Pearson & Hartley* (1954, 1966)

This gives for

$$\nu_1 = 2q = 1(1)10, 12, 15, 20, 24, 30, 40, 60, 120,$$

$$\nu_2 = 2p = 1(1)30, 40, 60, 120, \infty$$

the values of $x$, to five significant figures, for which

$$I_x(p, q) = 0{\cdot}50, 0{\cdot}25, 0{\cdot}10, 0{\cdot}05, 0{\cdot}025, 0{\cdot}01, 0{\cdot}005, 0{\cdot}0025, 0{\cdot}001.$$

The corresponding upper percentage points, $I_x(p, q) = 0{\cdot}75, 0{\cdot}90, \ldots, 0{\cdot}999$, are found by interchanging $\nu_1$ and $\nu_2$ and writing $1-x$ for $x$.

The introduction of $\nu_1$ and $\nu_2$ as arguments instead of $q$ and $p$ is due to the fact that the table has applications in the analysis of variance, where $\nu_1$ and $\nu_2$ are the degrees of freedom of two independent estimates of a variance. The argument intervals for the larger values of the $\nu$'s have been chosen to make possible harmonic interpolation.

The original table, going as far as $I_x(p, q) = 0{\cdot}005$ was computed by Catherine M. Thompson (1941), making use of the present *Tables of the Incomplete Beta-Function*, with certain necessary additional computation in regions where backward interpolation in this Table was impossible (see § 2·2 above). The final values for $I_x(p, q) = 0{\cdot}0025, 0{\cdot}001$ were obtained by D. E. Amos (1963).

(2) *Table* 3 *of H. L. Harter* (1964)

In the present notation, Harter's Table gives for

$$p, q = 1(1)40$$

the values of $x$, to seven significant figures, for which

$$I_x(p, q) = 0{\cdot}50,\ 0{\cdot}40,\ 0{\cdot}30,\ 0{\cdot}20,\ 0{\cdot}10,\ 0{\cdot}05,\ 0{\cdot}025,\ 0{\cdot}01,\ 0{\cdot}005,\ 0{\cdot}001,\ 0{\cdot}0005,\ 0{\cdot}0001.$$

It is of interest to note that these percentage points were derived from a freshly calculated table of the incomplete beta-function, for which $p, q = 1(1)40$, $x = 0{\cdot}00(0{\cdot}01)1{\cdot}00$. In view of the existence of Karl Pearson's Table, the author decided not to publish this master table of his.

(3) *Vogler's* (1964) *Table*

This gives for

$$\nu_1 = 2q = 1(1)10,\ 12,\ 15,\ 20,\ 24,\ 30,\ 40,\ 60,\ 120,$$
$$\nu_2 = 2p = 1(0{\cdot}1)2,\ 2{\cdot}2,\ 2{\cdot}5(0{\cdot}5)5,\ 6(1)10,\ 12,\ 15,\ 20,\ 24,\ 30,\ 40,\ 60,\ 120,\ \infty$$

the values of $x$, to six significant figures, for which

$$I_x(p, q) = 0{\cdot}50,\ 0{\cdot}25,\ 0{\cdot}10,\ 0{\cdot}05,\ 0{\cdot}025,\ 0{\cdot}01,\ 0{\cdot}005,\ 0{\cdot}001,\ 0{\cdot}0001.$$

The novel feature of this Table is that, for $\nu_2 = 2p < 5$ or $p < 2{\cdot}5$, the percentage points are given for certain fractional values of this argument.

## 4. Some special applications

No attempt will be made to set down the many applications of the present Tables but we here put on record two important classes of relationship: (*a*) that between the incomplete beta-function and the sums of terms in the binomial and negative binomial expansions, (*b*) that between certain symmetrical distributions (for which $p = q$) and the function $I_x(p, \frac{1}{2})$.

### 4·1 *The binomial*

The $(i+1)$th term in the expansion of the binomial $(Q+P)^N$, where $P+Q = 1$, may be written

$$b_1(i|N,P) = \binom{N}{i} P^i Q^{N-i}.$$

The sum of the first $k$ terms in this expansion is

$$\sum_{i=0}^{k-1} b_1(i|N,P) = 1 - I_P(k, N-k+1). \tag{13}$$

### 4·2 *The negative binomial*

The negative binomial expansion may be written as

$$(Q-P)^{-N} = Q^{-N}\left\{1 + NP/Q + \frac{N(N+1)}{2!}(P/Q)^2 + \dots\right\},$$

where now $Q-P = 1$.

The $(i+1)$th term in the expansion is then

$$b_2(i|N,P) = Q^{-N}\binom{N+i-1}{i}(P/Q)^i$$

and the sum of the first $k$ terms is

$$\sum_{i=0}^{k-1} b_2(i|N,P) = I_{Q^{-1}}(N,k). \tag{14}$$

### 4·3 *The relation between $I_{x'}(p,p)$ and $I_x(p, 0{\cdot}5)$ and some applications*

There are two important forms of distribution frequently occurring in statistical sampling theory whose probability integrals can be expressed in terms of incomplete beta-functions. In standard form these are the symmetrical distributions

$$f_1(u) = \{B(m_1+1, \tfrac{1}{2})\}^{-1}(1-u^2)^{m_1}, \quad -1 \leqslant u \leqslant 1, \tag{15}$$

$$f_2(v) = \{B(m_2-\tfrac{1}{2}, \tfrac{1}{2})\}^{-1}(1+v^2)^{-m_2}, \quad -\infty < v < \infty. \tag{16}$$

*Special case of* (15). Writing $u = r$, $m_1 = \frac{1}{2}(n-4)$, we have the distribution of the product moment correlation coefficient, $r$, in random samples of size $n$ from a bivariate normal population in which the correlation, $\rho$, is zero.

*Special case of* (16). Writing $v^2 = t^2/\nu$, $m_2 = \frac{1}{2}(\nu+1)$, we have the distribution of Student's $t$ based on $\nu$ degrees of freedom.

The distributions are also standard forms of the Type II and Type VII distributions in Karl Pearson's system of frequency curves.

Now make use of the transformation already referred to in equation (5), namely

(*a*) for $x' \geqslant 0{\cdot}5$
$$I_{x'}(p,p) = 1 - \tfrac{1}{2}I_{1-x}(p, \tfrac{1}{2}) = \tfrac{1}{2}\{1 + I_x(\tfrac{1}{2}, p)\}, \tag{17}$$
where
$$x = 4(x'-\tfrac{1}{2})^2, \quad x' = \tfrac{1}{2}(1+\sqrt{x}).$$

Note also (*b*) for $x' \leqslant 0{\cdot}5$
$$I_{x'}(p,p) = \tfrac{1}{2}I_x(p, \tfrac{1}{2}) = \tfrac{1}{2}\{1 - I_{1-x}(\tfrac{1}{2}, p)\}, \tag{18}$$
where
$$x = 1 - 4(\tfrac{1}{2}-x')^2, \quad x' = \tfrac{1}{2}\{1 - \sqrt{(1-x)}\}.$$

Then it can be shown that the probability integrals of $f_1(u)$ and $f_2(v)$ can each be expressed in terms of incomplete beta-functions, in two alternative forms:

$$\int_{-1}^{U} f_1(u)\,du = 1 - \tfrac{1}{2}I_{1-x}(m_1+1, \tfrac{1}{2}), \quad \text{where } x = U^2, \tag{19}$$

$$= I_{x'}(m_1+1, m_1+1), \quad \text{where } x' = \tfrac{1}{2}(1+U); \tag{20}$$

$$\int_{-\infty}^{V} f_2(v)\,dv = 1 - \tfrac{1}{2}I_{1-x}(m_2-\tfrac{1}{2}, \tfrac{1}{2}), \quad \text{where } x = \frac{V^2}{1+V^2}, \tag{21}$$

$$= I_{x'}(m_2-\tfrac{1}{2}, m_2-\tfrac{1}{2}), \quad \text{where } x' = \frac{1}{2}\left\{1 + \sqrt{\frac{V^2}{1+V^2}}\right\}. \tag{22}$$

Here, both $U$ and $V$ are positive so that we are using the transformation (*a*).

Within the limits of the main Table I, either form of integral may be used, i.e. that involving $I_x(p, \frac{1}{2})$ or $I_{x'}(p, p)$. However, according to the position of entry in the table interpolation for $x$ may be easier than for $x'$, or vice versa.

This is because $I_x(p, \frac{1}{2})$ is the probability integral of a $J$-shaped distribution with infinite ordinate at $x = 1$. Consequently interpolation in Table I becomes difficult or impossible when $x$ approaches unity. This means that interpolation near the terminals of the distribution (15) will be much easier using the relation (20) rather than (19). On the other hand, for distribution (16) the relation (21) can be used when $V^2$ is large. In certain extreme cases the new Table IV, giving additional values of $I_x(p, \frac{1}{2})$ for $x \geqslant 0{\cdot}975$, will be found helpful.

*Example* 9. To find the probability that Student's $t$, based on five degrees of freedom, is less than 0·4.

Here $\quad m_2 - \frac{1}{2} = \frac{1}{2}\nu = 2{\cdot}5; \quad V^2 = t^2/\nu = 0{\cdot}032; \quad 1-x = 0{\cdot}9689\,922; \quad x' = 0{\cdot}5880\,451.$

If we try to interpolate for $I_{1-x}(2{\cdot}5, 0{\cdot}5)$ we find that the differences are not converging. On the other hand we can interpolate for $I_{x'}(2{\cdot}5, 2{\cdot}5)$ using second differences only, i.e. using the first line of equation (A1) of the Appendix. We find

$$\Pr\{t \leqslant 0{\cdot}4 | \nu = 5\} = 0{\cdot}6471\,634.$$

The five-decimal place Table 9 of *Biometrika Tables for Statisticians*, Vol. 1, gives the probability as 0·64716.

On the other hand, if we need to find $\Pr\{t \leqslant 3{\cdot}0 | \nu = 15\}$ we have

$$m_2 - \tfrac{1}{2} = \tfrac{1}{2}\nu = 7{\cdot}5, \quad V^2 = 0{\cdot}6, \quad 1-x = 0{\cdot}625, \quad x' = 0{\cdot}8061\,862.$$

It is found that fourth differences are required to interpolate for $I_{x'}(7{\cdot}5, 7{\cdot}5)$ but only second differences for $I_{1-x}(7{\cdot}5, 0{\cdot}5)$. Using the latter function and equation (21) we find

$$\Pr\{t \leqslant 3{\cdot}0 | \nu = 15\} = 0{\cdot}9955\,137.$$

The value from the five-figure $t$-table is 0·99551.

## 5. Relations between the moments of the beta distribution and the parameters $p$, $q$

### 5·1 *Moments of the distribution* $f(x) = \{B(p,q)\}^{-1} x^{p-1}(1-x)^{q-1}$, $0 \leqslant x \leqslant 1$

For certain purposes it is necessary to relate the parameters of a beta distribution to its moments or moment ratios. Taking the distribution in its standard form

$$f(x) = \frac{\Gamma(p+q)}{\Gamma(p)\,\Gamma(q)} x^{p-1}(1-x)^{q-1}, \quad 0 \leqslant x \leqslant 1 \tag{23}$$

and denoting the $s$th moment about the origin by $\mu'_s$, the $s$th moment about the mean by $\mu_s$, we have

$$\mu'_s = \frac{\Gamma(p+s)\,\Gamma(p+q)}{\Gamma(p)\,\Gamma(p+q+s)}. \tag{24}$$

Proceeding from this result, it can be shown that:

$$\left.\begin{aligned} \text{mean } x &= \frac{p}{p+q}, \quad \operatorname{var} x = \mu_2 = \sigma^2 = \frac{pq}{(p+q)^2(p+q+1)}, \\ \sqrt{\beta_1} &= \frac{\mu_3}{\sigma^3} = \frac{2(q-p)\sqrt{(p+q+1)}}{(p+q+2)\sqrt{(pq)}}, \\ \beta_2 &= \frac{\mu_4}{\sigma^4} = \frac{3(p+q+1)\{2(p+q)^2+pq(p+q-6)\}}{pq(p+q+2)(p+q+3)}. \end{aligned}\right\} \tag{25}$$

If $p > q$ the distribution (23) is negatively skew with $\sqrt{\beta_1} < 0$ and the longer tail directed towards the start at $x = 0$.

Table II of the present volume gives a broad picture of the relation between $p$ and $q$ and the moments or moment ratios of the distributions whose probability integrals are given in the main Table I. The third column of Table II gives the positions of the mode or maximum ordinate* of the distribution, i.e.

$$\tilde{x} = (p-1)/(p+q-2).$$

### 5·2 *The general beta or Pearson Type I distribution*

In more general form the beta distribution may be written

$$f(X) = \frac{\Gamma(p+q)}{b^{p+q-1}\,\Gamma(p)\,\Gamma(q)}(a_1+X)^{p-1}(a_2-X)^{q-1}, \quad -a_1 \leqslant X \leqslant a_2, \tag{26}$$

where

$$b = a_1 + a_2 \tag{27}$$

is the range of the curve or distance between the two terminals. Equation (26) corresponds to the Type I curve of Karl Pearson's system. Its probability integral can be obtained from Table I by setting

$$x = \frac{a_1+X}{b}. \tag{28}$$

Two of the main uses of the curve (26), as well as of other members of the Pearson system, have lain in:

(1) The graduation and classification of observed distributions of univariate frequency data.

(2) The approximate representation of the sampling distributions of statistics for which only the moments or cumulants are known or readily expressible.

In both these problems it is customary to derive estimates of the parameters in (26) from the known moments and/or the terminal points of the distribution to be approximated.† Several alternative procedures have been used. It may be helpful to summarize three of these briefly.

### 5·3 *Case* (*a*). *Estimation from the first four moments*

Here we use the mean, the variance $\sigma^2$ and the two moment ratios $\beta_1$ and $\beta_2$ defined in equations (25).

Writing

$$\left.\begin{aligned} r &= \frac{6(\beta_2-\beta_1-1)}{3\beta_1-2\beta_2+6}, \\ \epsilon &= \frac{r^2}{4+\frac{1}{4}\beta_1(r+2)^2/(r+1)} \end{aligned}\right\} \tag{29}$$

(see Elderton, 1938, pp. 58–65), it may be shown by inverting the expression for $\beta_1$ and $\beta_2$, that $r = p+q$, $\epsilon = pq$, so that $p$ and $q$ are the roots of the quadratic

$$z^2 - rz + \epsilon = 0. \tag{30}$$

If $\sqrt{\beta_1} > 0$, then $p$ will be the smaller of the two roots, while if $\sqrt{\beta_1} < 0$, $p$ will be the larger root. If $\sqrt{\beta_1} = 0$ the distribution is symmetrical and $p = q$.

Further, it is found that

$$b = a_1 + a_2 = \sigma r\{(r+1)/\epsilon\}^{\frac{1}{2}}. \tag{31}$$

It is convenient to take the origin for $X$ at the mode of the curve. Then we find $a_1$ and $a_2$ from

$$\frac{a_1}{p-1} = \frac{a_2}{q-1} = \frac{b}{p+q-2} = \frac{b}{r-2}. \tag{32}$$

To 'place' the fitted distribution, it is necessary to relate the origin or mode to the mean of the data or distribution to which it is being fitted. Since the mean of the curve is at distance $bp/(p+q)$ from the 'start', the distance between mean and mode is

$$bp/(p+q) - a_1. \tag{33}$$

* For the $U$ curve with $p = q = 0{\cdot}5$ it is the minimum ordinate.

† In the case of problem (1) it may be possible to obtain improved estimates using the method of maximum likelihood, but this procedure would probably be based on the moment estimates as first approximations.

For the case of $\sqrt{\beta_1} > 0$ (positive skewness) the position is illustrated in Fig. 3. If the steps indicated above are gone through systematically in numerical terms, the solution is not difficult.

The integral under the fitted curve between the start at $X = -a_1$ and the ordinate at $X = X_0$ is then $I_{x_0}(p, q)$, where $x_0 = (a_1 + X_0)/b$.

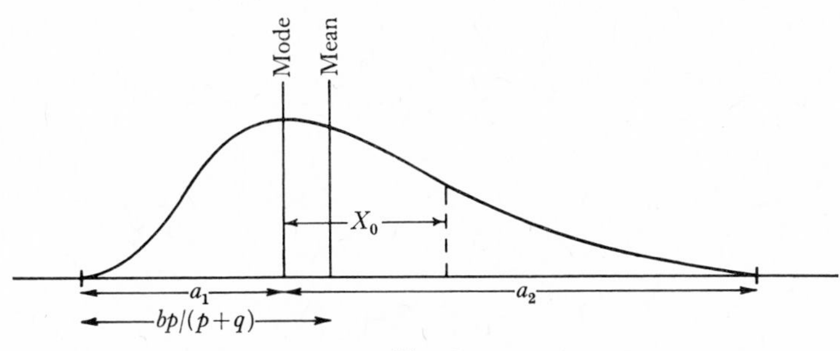


Fig. 3

### 5·4 *Case* (*b*). *Estimation using a known 'start' and the first three moments*

It will sometimes happen that one terminal of the distribution to be approximated to is definitely known and, if the curve comes down steeply at this end, it may be important to use this information. If the tail towards the other terminal is drawn out finely, it may be better to make use of the third moment (but not the fourth) in fitting, even if the position of this terminal is known.

The following solution of the estimation problem may then be adopted.

Take the known terminal as origin; the distribution may now be written

$$f(X) = \frac{\Gamma(p+q)}{b^{p+q-1}\,\Gamma(p)\,\Gamma(q)} X^{p-1}(b-X)^{q-1}, \quad 0 \leqslant X \leqslant b. \tag{34}$$

The solution is most easily obtained in terms of the moments about the origin, $\mu_1'$, $\mu_2'$ and $\mu_3'$. Write

$$\lambda_1 = \mu_1'^2/\mu_2', \quad \lambda_2 = \mu_2'\mu_1'/\mu_3'. \tag{35}$$

Then the estimates of $p$ and $q$ are determined from

$$p = \frac{2(\lambda_1 - \lambda_2)}{1 + \lambda_2 - 2\lambda_1}, \quad q = \frac{2(\lambda_1 - \lambda_2)}{2\lambda_2 - \lambda_1 - \lambda_1\lambda_2} - p. \tag{36}$$

The range, $b$, is given by

$$b = \frac{\mu_1'(1 + \lambda_2 - 2\lambda_1)}{2\lambda_2 - \lambda_1 - \lambda_1\lambda_2}. \tag{37}$$

This completes the solution.

### 5·5 *Case* (*c*). *Estimation using both known terminals and the first two moments*

The equation with origin at the 'start' or lower terminal will still be of the form of equation (34), but the range $b$ will now be known. Values for $p$ and $q$ may then be estimated as follows:

$$p = \frac{\mu_1'(\mu_1' - \mu_2'/b)}{\mu_2' - \mu_1'^2}, \quad q = \frac{(b - \mu_1')(\mu_1' - \mu_2'/b)}{\mu_2' - \mu_1'^2}. \tag{38}$$

We have not added any numerical illustration of these methods of fitting since this seems properly to belong to a text on frequency curves. It seemed however appropriate to collect together in these sections the relations between parameters of a general beta distribution (or Type I curve) and the moments.

## 6. Approximate formulae

### 6·1 *Systematic computation and occasional approximation*

The computational requirements called for in obtaining values of the incomplete beta-function range from those suitable for the preparation of large-scale tables to those more appropriate when an approximate value of $I_x(p,q)$ is required, for perhaps only a single combination of $x$, $p$ and $q$. We are concerned in this section with methods which have been suggested to meet the latter requirement. Before considering these, however, it should be mentioned that the systematic method of build-up used in preparing the present tables (see pp. ix–xi above) has inevitably been superseded now that hand-operated machines have been replaced by the digital computer. Recent methods used in large-scale computations have been described, for example by Amos (1963) and Harter (1964). In particular, Harter has pointed out the suitability of the formula

$$I_x(p,q) = 1-(1-x)^q \sum_{i=0}^{p-1} \binom{q+i+1}{i} x^i. \tag{39}$$

Since the first edition of these tables was published in 1934, a number of new approximations to the incomplete beta-function have been developed. Some of the more effective of these will be described in this Section. It should, however, be noted that none of these methods can be relied upon to provide more than three-figure accuracy over a wide range of values of $x$, $p$ and $q$. If an accuracy comparable with that of the tables is required, resort should be made to interpolation, as described in earlier sections, or to direct calculation, e.g. if $p$ and $q$ are integers, by using equation (39).

Some experimental calculations were carried out to see whether interpolation with respect to some function, such as (exact value)/(approximate value), could be used to improve values given by an approximate formula. The results of these calculations were not promising. Accuracy was generally not as good as that secured by interpolation, while the time consumed was often more than twice as great. However, approximations can be useful in special circumstances (e.g. for use in mathematical analysis), and for this reason we give the following summary.

### 6·2 *Cadwell's approximation for the symmetrical case $p = q$*

J. H. Cadwell (1952) has shown that the following approximation gives quite accurate results:

$$I_x(p,p) = \tfrac{1}{2} + L(p)\left\{\Phi(\zeta) - \frac{1}{2} + \frac{2(p-1)(2p+1)}{(4p-1)^4}\Psi(\zeta)\right\}, \tag{40}$$

where $\zeta$ is defined by the equation

$$x = \tfrac{1}{2} + (\tfrac{1}{3}\pi)^{\frac{1}{2}}\left[\Phi\left\{\left(\frac{3}{4p-1}\right)^{\frac{1}{2}}\zeta\right\} - \frac{1}{2}\right]$$

with

$$\Phi(\zeta) = (2\pi)^{-\frac{1}{2}}\int_{-\infty}^{\zeta} e^{-\frac{1}{2}u^2}du; \quad \Psi(\zeta) = (2\pi)^{-\frac{1}{2}}\int_0^{\zeta}\left(1-\frac{u^6}{15}\right)e^{-\frac{1}{2}u^2}du;$$

and where $L(p)$ is a function of $p$, which is very nearly equal to 1, as can be seen from the following table:

| $p$ | $L(p)$ | $p$ | $L(p)$ |
|---|---|---|---|
| 2·0 | 1·00090 | 3·0 | 1·00003 |
| 2·2 | 1·00054 | 3·2 | 1·00002 |
| 2·4 | 1·00023 | 3·4 | 1·00001 |
| 2·6 | 1·00012 | $\geqslant$ 3·6 | 1·00000 |
| 2·8 | 1·00006 | | |

Some values of $\Psi(\zeta)$ which will be found adequate for most purposes are:

| $\zeta$ | 0·0 | 0·1 | 0·2 | 0·3 | 0·4 | 0·5 | 0·6 | 0·7 | 0·8 | 0·9 |
|---|---|---|---|---|---|---|---|---|---|---|
| 0 | 0·000 | 0·040 | 0·079 | 0·118 | 0·156 | 0·191 | 0·226 | 0·258 | 0·288 | 0·315 |
| 1 | 0·339 | 0·360 | 0·377 | 0·391 | 0·400 | 0·406 | 0·406 | 0·403 | 0·395 | 0·383 |
| 2 | 0·367 | 0·348 | 0·326 | 0·302 | 0·272 | 0·249 | 0·222 | 0·196 | 0·171 | 0·147 |
| 3 | 0·125 | 0·105 | 0·087 | 0·071 | 0·057 | 0·046 | 0·036 | 0·028 | 0·022 | 0·017 |

This formula gives five-decimal place accuracy for $p \geqslant 5$. For $p \geqslant 4$, the simple formula

$$I_x(p,p) = \Phi(\zeta) \tag{41}$$

has an error less than 0·00045 (for $p \geqslant 6$, the error is less than 0·00022).

These formulae may also be used to obtain approximations to $I_x(p, \frac{1}{2})$ using the transformation (18), i.e.

$$I_x(p, \tfrac{1}{2}) = 2I_{x'}(p,p)$$

with $x' = \frac{1}{2}\{1-(1-x)^{\frac{1}{2}}\}$.

### 6·3 *Wise's approximations*

M. E. Wise (1950) obtained a formula which gives good results when $p$ or $q$ is large (and $p \geqslant q$, particularly if $p$ is large), by expanding $I_x(p,q)$ in powers of $(p+\frac{1}{2}q-\frac{1}{2})^{-1}$. The leading terms of this formula are

$$I_x(p,q) = \Pr\{\chi^2_{2q} > 2y\} + \frac{y^q e^{-y}}{\Gamma(q-1)}\left\{\frac{1}{24N^2}(q+1+y) - \frac{1}{5760N^4}\right.$$
$$\left.\times[(q-3)(q-2)(5q+7)(q+1+y)-(5q-7)(q+3+y)y^2]\right\}, \tag{42}$$

where $N = p+\frac{1}{2}q-\frac{1}{2}$, $y = -N\log_e x$ and $\chi^2_{2q}$ is a $\chi^2$-variable based on $2q$ degrees of freedom.

Later, the same author (Wise, 1960) presented an approximate normalizing transformation

$$I_x(p,q) = 1-\Phi(Y), \tag{43}$$

where
$$Y = -3\sqrt{q}(1-\tfrac{1}{9}q^{-1}) + 3(-\log_e x)^{\frac{1}{3}}\left\{N\sqrt{q}\left[1-\frac{(q-1)(q+\frac{1}{3})}{12N^2}\right]\right\}^{\frac{1}{3}},$$

and $\Phi(Y)$ is, as before, the probability integral of a standardized normal deviate.

In this formula, also, $p \geqslant q$.

### 6·4 *Approximations based on the Cornish–Fisher expansion*

A number of approximations have been put forward, using various forms of the Cornish–Fisher expansion, usually applied to the related distribution of $z = \frac{1}{2}\log_e F$ (see W. G. Cochran, 1940; A. H. Carter, 1947; D. H. Thomson, 1947; L. A. Aroian, 1950; J. Wishart, 1957).

Wishart (1957) obtained an expansion in terms of the normal integral

$$\Phi(u) = \frac{1}{\sqrt{(2\pi)}}\int_{-\infty}^{u} e^{-\frac{1}{2}t^2}dt$$

and its derivatives
$$\phi^{(r)} \equiv \phi^{(r)}(u) = \frac{d^{r+1}}{du^{r+1}}\Phi(u), \quad \text{with } \phi^{(0)} = \phi.$$

This is
$$I_x(p,q) \doteqdot \Phi(u) + \phi_1(u,v) + w^2\phi_2(u,v), \tag{44}$$

where
$$u = \left(\frac{pq}{p+q}\right)^{\frac{1}{2}}\log_e\left(\frac{qx}{p(1-x)}\right), \quad v = \left(\frac{pq}{p+q}\right)^{\frac{1}{2}}\left(\frac{1}{p}-\frac{1}{q}\right), \quad w^2 = \frac{1}{p}+\frac{1}{q},$$

$$\phi_1(u,v) = \tfrac{1}{36}(3\phi+\phi^{(2)})v + \tfrac{1}{144}(36\phi^{(1)}+21\phi^{(3)}+2\phi^{(5)})v^2 + \tfrac{1}{12960}(1620\phi^{(2)}+1269\phi^{(4)}+225\phi^{(6)}+10\phi^{(8)})v^3,$$
$$\phi_2(u,v) = \tfrac{1}{48}(6\phi^{(1)}+\phi^{(3)}) + \tfrac{1}{1440}(120\phi+270\phi^{(2)}+81\phi^{(4)}+5\phi^{(6)})v.$$

Tables of $\phi_1(u,v)$ and $\phi_2(u,v)$ are given by Bol'shev & Smirnov (1965).

In the same set of tables are given values of an auxiliary function

$$\gamma(y',q) = \frac{y'^q e^{-y'}}{\Gamma(q)}\{2y'^2-(q-1)y'-(q^2-1)\} \tag{45}$$

which is used in the following approximate formula (L. N. Bol'shev, 1963):

$$I_x(p,q) = \Pr\{\chi^2_{2q} \geqslant 2y'\} - \frac{1}{24N^2}\gamma(y',q), \tag{46}$$

where
$$y' = 2N(1-x)/(1+x),$$
and $N = (p+\frac{1}{2}q-\frac{1}{2})$ as before.

If $p$ is large compared with $q$, this approximation can give quite accurate results. The error is of order $p^{-4}$.

Noting that

$$\log_e x = \log_e\left\{1-\frac{1-x}{1+x}\right\} - \log_e\left\{1+\frac{1-x}{1+x}\right\}$$

$$\doteqdot -2(1-x)/(1+x)$$

if $(1-x)/(1+x)$ is small, i.e. $x$ near to 1 and that

$$\gamma(y', q) = -\frac{y'^q e^{-y'}}{\Gamma(q-1)}(y'+q+1) + \frac{2y'^{q+2}e^{-y'}}{\Gamma(q)},$$

it can be seen that there is some connexion between the formulae (42) of Wise (1950) and (46) of Bol'shev (1963).

### 6·5 *Approximation of Peizer & Pratt*

These authors have given (1966) a very widely applicable approximation by which the sums of terms of the binomial, negative binomial and Poisson series and the integral of the beta, inverted beta and gamma distributions can be obtained, after appropriate substitution, from the standard normal probability integral $\Phi(z)$ as defined above.

For the case of the incomplete beta-function, using the notation of the present Introduction, $I_x(p, q)$ is approximated by $\Phi(z_i)$ with

$$z_i = \frac{d_i}{|q-0{\cdot}5-n(1-x)|}\left\{\frac{2}{1+(6n)^{-1}}\left[(q-0{\cdot}5)\log_e\frac{q-0{\cdot}5}{n(1-x)} + (p-0{\cdot}5)\log_e\frac{p-0{\cdot}5}{nx}\right]\right\}^{\frac{1}{2}}, \tag{47}$$

where $n = p+q-1$.

$$i = 1: \quad d_1 = q-0{\cdot}5+\tfrac{1}{6}-(n+\tfrac{1}{3})(1-x) \quad \text{as a first approximation;}$$

$$i = 2: \quad d_2 = d_1 + \frac{1}{50}\left\{\frac{x}{q}-\frac{1-x}{p}+\frac{x-0{\cdot}5}{p+q}\right\} \quad \text{as a second approximation.}$$

For greater convenience and accuracy in computing the authors give the alternative form

$$z_i = d_i\left\{\frac{1+xg\left(\dfrac{q-0{\cdot}5}{n(1-x)}\right)+(1-x)\,g\left(\dfrac{p-0{\cdot}5}{nx}\right)}{(n+\frac{1}{6})\,x(1-x)}\right\}^{\frac{1}{2}}, \tag{48}$$

where the function $g(u)$ is defined by

$$g(u) = (1-u)^{-2}(1-u^2+2u\log_e u) \tag{49}$$

and is tabled. The expressions (47) and (48) can be shown to be algebraically identical.

The authors give the following overall statement of accuracy, besides some more detailed charts bearing on this:

$$|I_x(p,q)-\Phi(z_2)| < \begin{cases} 0{\cdot}001 & \text{if } p, q \geqslant 2{\cdot}0, \\ 0{\cdot}01 & \text{if } p, q \geqslant 1{\cdot}0, \end{cases}$$

and writing

$$R = (q-0{\cdot}5)\,x/\{(p-0{\cdot}5)(1-x)\},$$

then

$$\frac{|I_x(p,q)-\Phi(z_2)|}{I_x(p,q)} < \begin{cases} 0{\cdot}01 & \text{if } p, q \geqslant 3{\cdot}0 \text{ and } 0{\cdot}2 \leqslant R \leqslant 5, \\ 0{\cdot}02 & \text{if } p, q \geqslant 1{\cdot}75 \text{ and } 0{\cdot}125 \leqslant R \leqslant 8, \\ 0{\cdot}03 & \text{if } p, q \geqslant 1{\cdot}5 \text{ and } 0{\cdot}1 \leqslant R \leqslant 10. \end{cases}$$

## APPENDIX. SOME UNIVARIATE INTERPOLATION FORMULAE

As in the main text, we shall suppose that the interpolate value $z_{\theta_1}$, with $\theta_0 = 1-\theta_1$, lies between tabled entries $z_0$, $z_1$, where $\theta_1$ is the argument interval ratio measured from the $z_0$ point.

### *The Everett central difference formula*

This may be written as:

$$\begin{aligned} z_{\theta_1} = {} & \theta_0 z_0 + \theta_1 z_1 - \tfrac{1}{6}\theta_0\theta_1\{(1+\theta_0)\,\delta^2 z_0 + (1+\theta_1)\,\delta^2 z_1\} \\ & + \tfrac{1}{120}\theta_0\theta_1(1+\theta_0)\,(1+\theta_1)\,\{(2+\theta_0)\,\delta^4 z_0 + (2+\theta_1)\,\delta^4 z_1\} \\ & - \tfrac{1}{5040}\theta_0\theta_1(1+\theta_0)\,(1+\theta_1)\,(2+\theta_0)\,(2+\theta_1)\,\{(3+\theta_0)\,\delta^6 z_0 + (3+\theta_1)\,\delta^6 z_1\} + \ldots, \end{aligned} \tag{A1}$$

where

$$\delta^2 z_s = z_{s-1} + z_{s+1} - 2z_s, \quad \delta^4 z_s = \delta^2 z_{s-1} + \delta^2 z_{s+1} - 2\delta^2 z_s, \quad \text{etc.} \tag{A2}$$

It is easy to see how further terms of the series (A1) can be built up, the leading numerical factor for the term containing differences $\delta^v$ being $1/(v+1)\,!$.

We may note as a rough guide that it is unnecessary to retain in (A1) any second difference ($\delta^2$) under 4, any fourth difference ($\delta^4$) under 20 and any sixth difference ($\delta^6$) under 100, in terms of a unit in the last figure of the tabled entries. These limits involve a maximum error of about 0·5 in the final unit; clearly somewhat greater latitude may only involve a last place unit error.

Tables of the Everett coefficients up to those multiplying $\delta^6 z_0$ and $\delta^6 z_1$ have been given by A. J. Thompson (1943). Special end-panel formulae may be required when the interpolate has to be calculated near a boundary of the table (see, for example, K. Pearson, 1919, pp. 13–14).

It should be noted that if the $\delta^4$ and higher differences are negligible and if $\delta^3$ is less than 60 in last figure units, the equation (A1) can be reduced to the simple form

$$z_{\theta_1} = \theta_0 z_0 + \theta_1 z_1 - \tfrac{1}{4}\theta_0\theta_1(\delta^2 z_0 + \delta^2 z_1). \tag{A3}$$

We have given formula (A1) for the general situation, where $0 < \theta_1 < 1$, but in the main text we are primarily concerned with the special case where $\theta_1 = \theta_0 = \frac{1}{2}$.

### *Lagrangian interpolation formulae*

We shall give here formulae involving an even number of points only. For simplicity $\theta$ will be written for $\theta_1$ of the preceding section.

*Four-point Lagrangian*

$$z_\theta = \frac{(\theta+1)\,\theta(\theta-1)\,(\theta-2)}{6}\left\{-\frac{z_{-1}}{\theta+1} + \frac{3z_0}{\theta} - \frac{3z_1}{\theta-1} + \frac{z_2}{\theta-2}\right\}. \tag{A4}$$

*Six-point Lagrangian*

$$z_\theta = \frac{(\theta+2)\,(\theta+1)\,\theta(\theta-1)\,(\theta-2)\,(\theta-3)}{120}\left\{-\frac{z_{-2}}{\theta+2} + \frac{5z_{-1}}{\theta+1} - \frac{10z_0}{\theta} + \frac{10z_1}{\theta-1} - \frac{5z_2}{\theta-2} + \frac{z_3}{\theta-3}\right\}. \tag{A5}$$

*Eight-point Lagrangian*

$$\begin{aligned} z_\theta = {} & \frac{(\theta+3)\,(\theta+2)\,(\theta+1)\,\theta(\theta-1)\,(\theta-2)\,(\theta-3)\,(\theta-4)}{5040} \\ & \times\left\{-\frac{z_{-3}}{\theta+3} + \frac{7z_{-2}}{\theta+2} - \frac{21z_{-1}}{\theta+1} + \frac{35z_0}{\theta} - \frac{35z_1}{\theta-1} + \frac{21z_2}{\theta-2} - \frac{7z_3}{\theta-3} + \frac{z_4}{\theta-4}\right\}. \end{aligned} \tag{A6}$$

## List of References


Amos, D. E. (1963). Additional percentage points for the incomplete beta distribution. *Biometrika* **50**, 449–57.

Aroian, L. A. (1950). On the levels of significance of the incomplete beta function and the $F$-distribution. *Biometrika* **37**, 219–23.

Bol'shev, L. N. (1963). Asymptotically Pearson's transformations. *Teor. Veroyat. Primen.* **8**, 129–55 (in Russian).

Bol'shev, L. N. & Smirnov, N. V. (1965). *Tablitsi Matematicheskoi Statistiki.* Akad. Nauk, SSSR, Izdat. Nauka, Moscow (in Russian).

Brownlee, John (1923). *Log* $\Gamma(x)$ *from* $x = 1$ *to 50·9 by intervals of 0·01.* Tracts for Computers No. 9. Cambridge University Press.

Cadwell, J. H. (1952). An approximation to the symmetrical incomplete beta function. *Biometrika* **39**, 204–7.

Carter, A. H. (1947). Approximation to percentage points of the $z$ distribution. *Biometrika* **34**, 352–8.

Cochran, W. G. (1940). Note on an approximate formula for the significance levels of $z$. *Ann. Math. Statist.* **11**, 93–5.

Comrie, L. J. & Hartley, H. O. (1941). Tables of the percentage points of the incomplete beta-function. Description of the calculation and methods of interpolation. *Biometrika* **32**, 154–67.

Elderton, W. P. (1938). *Frequency Curves and Correlation.* Cambridge University Press. (3rd Edition.)

Harter, H. L. (1964). *New Tables of the Incomplete Gamma-Function Ratio and of Percentage Points of the Chi-square and Beta Distributions.* Washington: U.S. Government Printing Office.

Johnson, N. L., Nixon, Eric, Amos, D. E. & Pearson, E. S. (1963). Table of percentage points of Pearson curves, for given $\sqrt{\beta_1}$ and $\beta_2$, expressed in standard measure. *Biometrika* **50**, 459–98.

Pearson, E. S. (1922). *Table of Logarithms of the Complete Gamma-Function* (*for Arguments 2 to 1200, i.e. beyond Legendre's Range*). Tracts for Computers No. 8. Cambridge University Press.

Pearson, E. S. & Hartley, H. O. (1954, 1966). *Biometrika Tables for Statisticians* **1**. Cambridge University Press.

Pearson, Karl (1919). *On the Construction of Tables and on Interpolation. Part 1. Univariate Tables.* Tracts for Computers No. 2. Cambridge University Press. (Out of print.)

Pearson, Karl (1922). *Tables of the Incomplete Gamma-Function.* Cambridge University Press.

Pearson, Karl (1931). *Tables for Statisticians and Biometricians* **2** Cambridge University Press. (Out of print.)

Pearson, Karl & Stoessiger, Brenda (1930). Tables of the probability integrals of symmetrical frequency curves in the case of low powers such as arise in the theory of small samples. *Biometrika* **22**, 253–83.

Peizer, D. B. & Pratt, J. W. (1966). *Approximating the binomial, F and commonly used related distributions, I.* Harvard University Department of Statistics, Technical Report No. 12.

Thompson, A. J. (1921, 1943). *Table of Coefficients of Everett's Central-Difference Interpolation Formula.* Tracts for Computers No. 5. Cambridge University Press.

Thompson, Catherine M. (1941). Tables of percentage points of the incomplete beta-function. *Biometrika* **32**, 168–81.

Thomson, D. H. (1947). Approximate formulae for the percentage points of the incomplete beta-function and the $\chi^2$ distribution. *Biometrika* **34**, 368–72.

Vogler, L. E. (1964). *Percentage Points of the Beta Distribution.* National Bureau of Standards, Boulder, Colorado.

Wise, M. E. (1950). The incomplete beta-function, as a contour integral and a quickly converging series for its inverse. *Biometrika* **37**, 208–18.

Wise, M. E. (1960). On normalizing the incomplete beta-function for fitting to dosage response curves. *Biometrika* **47**, 173–5.

Wishart, John (1957). An approximate formula for the cumulative $z$-distribution. *Ann. Math. Statist.* **28**, 504–10.

# TABLES OF THE INCOMPLETE BETA-FUNCTION

## TABLE I

### THE $I_x(p, q)$ FUNCTION

The integral is given to seven decimal places; an *e* is affixed when the *exact* value is provided by those seven places. The corresponding value of the complete beta-function, $B(p, q)$ is given to eight significant figures at the top of each column.

$x$ = ·01 to ·60 $q = 0·5$ $p$ = 0·5 to 3

| $x$ | $p = 0·5$ | $p = 1$ | $p = 1·5$ | $p = 2$ | $p = 2·5$ | $p = 3$ |
|---|---|---|---|---|---|---|
| $B(p,q)$= | 3·1415 9265+ | 2·0000 0000 | 1·5707 9633 | 1·3333 3333 | 1·1780 9725− | 1·0666 6667 |
| ·01 | ·0637 686 | ·0050 126 | ·0004 257 | ·0000 376 | ·0000 034 | ·0000 003 |
| ·02 | ·0903 345− | ·0100 505+ | ·0012 077 | ·0001 510 | ·0000 193 | ·0000 025+ |
| ·03 | ·1108 247 | ·0151 142 | ·0022 255− | ·0003 409 | ·0000 535+ | ·0000 085+ |
| ·04 | ·1281 884 | ·0202 041 | ·0034 369 | ·0006 082 | ·0001 102 | ·0000 203 |
| ·05 | ·1435 663 | ·0253 206 | ·0048 182 | ·0009 536 | ·0001 933 | ·0000 398 |
| ·06 | ·1575 424 | ·0304 640 | ·0063 536 | ·0013 779 | ·0003 061 | ·0000 691 |
| ·07 | ·1704 634 | ·0356 349 | ·0080 318 | ·0018 821 | ·0004 517 | ·0001 101 |
| ·08 | ·1825 549 | ·0408 337 | ·0098 443 | ·0024 670 | ·0006 330 | ·0001 650+ |
| ·09 | ·1939 734 | ·0460 608 | ·0117 844 | ·0031 335+ | ·0008 531 | ·0002 359 |
| ·10 | ·2048 328 | ·0513 167 | ·0138 468 | ·0038 825+ | ·0011 144 | ·0003 250− |
| ·11 | ·2152 190 | ·0566 019 | ·0160 272 | ·0047 150− | ·0014 198 | ·0004 343 |
| ·12 | ·2251 989 | ·0619 168 | ·0183 220 | ·0056 319 | ·0017 718 | ·0005 662 |
| ·13 | ·2348 255− | ·0672 621 | ·0207 281 | ·0066 341 | ·0021 729 | ·0007 229 |
| ·14 | ·2441 418 | ·0726 382 | ·0232 430 | ·0077 228 | ·0026 257 | ·0009 067 |
| ·15 | ·2531 833 | ·0780 456 | ·0258 646 | ·0088 990 | ·0031 327 | ·0011 200 |
| ·16 | ·2619 798 | ·0834 849 | ·0285 911 | ·0101 636 | ·0036 963 | ·0013 651 |
| ·17 | ·2705 563 | ·0889 566 | ·0314 210 | ·0115 180 | ·0043 190 | ·0016 445+ |
| ·18 | ·2789 343 | ·0944 615− | ·0343 530 | ·0129 630 | ·0050 032 | ·0019 607 |
| ·19 | ·2871 326 | ·1000 000e | ·0373 861 | ·0145 000e | ·0057 515+ | ·0023 162 |
| ·20 | ·2951 672 | ·1055 728 | ·0405 193 | ·0161 301 | ·0065 663 | ·0027 137 |
| ·21 | ·3030 525+ | ·1111 806 | ·0437 521 | ·0178 545+ | ·0074 500− | ·0031 557 |
| ·22 | ·3108 011 | ·1168 239 | ·0470 837 | ·0196 745+ | ·0084 052 | ·0036 449 |
| ·23 | ·3184 242 | ·1225 036 | ·0505 139 | ·0215 915− | ·0094 344 | ·0041 841 |
| ·24 | ·3259 319 | ·1282 202 | ·0540 424 | ·0236 066 | ·0105 400 | ·0047 762 |
| ·25 | ·3333 333 | ·1339 746 | ·0576 689 | ·0257 214 | ·0117 248 | ·0054 240 |
| ·26 | ·3406 367 | ·1397 675− | ·0613 934 | ·0279 372 | ·0129 913 | ·0061 304 |
| ·27 | ·3478 494 | ·1455 996 | ·0652 160 | ·0302 556 | ·0143 420 | ·0068 984 |
| ·28 | ·3549 784 | ·1514 719 | ·0691 369 | ·0326 779 | ·0157 798 | ·0077 312 |
| ·29 | ·3620 301 | ·1573 850+ | ·0731 562 | ·0352 059 | ·0173 072 | ·0086 319 |
| ·30 | ·3690 101 | ·1633 400 | ·0772 743 | ·0378 410 | ·0189 271 | ·0096 037 |
| ·31 | ·3759 240 | ·1693 376 | ·0814 916 | ·0405 849 | ·0206 423 | ·0106 499 |
| ·32 | ·3827 767 | ·1753 789 | ·0858 087 | ·0434 395− | ·0224 556 | ·0117 740 |
| ·33 | ·3895 729 | ·1814 647 | ·0902 262 | ·0464 064 | ·0243 699 | ·0129 795− |
| ·34 | ·3963 171 | ·1875 962 | ·0947 447 | ·0494 875+ | ·0263 883 | ·0142 698 |
| ·35 | ·4030 133 | ·1937 742 | ·0993 650+ | ·0526 847 | ·0285 138 | ·0156 487 |
| ·36 | ·4096 655+ | ·2000 000e | ·1040 880 | ·0560 000e | ·0307 494 | ·0171 200e |
| ·37 | ·4162 774 | ·2062 746 | ·1089 147 | ·0594 354 | ·0330 985+ | ·0186 875+ |
| ·38 | ·4228 526 | ·2125 992 | ·1138 459 | ·0629 931 | ·0355 643 | ·0203 553 |
| ·39 | ·4293 943 | ·2189 750+ | ·1188 830 | ·0666 752 | ·0381 501 | ·0221 275− |
| ·40 | ·4359 058 | ·2254 033 | ·1240 271 | ·0704 840 | ·0408 594 | ·0240 082 |
| ·41 | ·4423 902 | ·2318 854 | ·1292 794 | ·0744 219 | ·0436 958 | ·0260 019 |
| ·42 | ·4488 506 | ·2384 227 | ·1346 415− | ·0784 915− | ·0466 629 | ·0281 131 |
| ·43 | ·4552 897 | ·2450 166 | ·1401 147 | ·0826 951 | ·0497 646 | ·0303 465− |
| ·44 | ·4617 105+ | ·2516 685+ | ·1457 008 | ·0870 356 | ·0530 046 | ·0327 067 |
| ·45 | ·4681 157 | ·2583 802 | ·1514 014 | ·0915 157 | ·0563 871 | ·0351 989 |
| ·46 | ·4745 080 | ·2651 531 | ·1572 183 | ·0961 383 | ·0599 161 | ·0378 282 |
| ·47 | ·4808 899 | ·2719 890 | ·1631 535+ | ·1009 064 | ·0635 961 | ·0405 998 |
| ·48 | ·4872 642 | ·2788 897 | ·1692 091 | ·1058 233 | ·0674 314 | ·0435 194 |
| ·49 | ·4936 334 | ·2858 572 | ·1753 872 | ·1108 922 | ·0714 267 | ·0465 925+ |
| ·50 | ·5000 000e | ·2928 932 | ·1816 901 | ·1161 165+ | ·0755 868 | ·0498 253 |
| ·51 | ·5063 666 | ·3000 000e | ·1881 204 | ·1215 000e | ·0799 167 | ·0532 237 |
| ·52 | ·5127 358 | ·3071 797 | ·1946 807 | ·1270 464 | ·0844 215+ | ·0567 944 |
| ·53 | ·5191 101 | ·3144 345+ | ·2013 737 | ·1327 597 | ·0891 068 | ·0605 439 |
| ·54 | ·5254 920 | ·3217 670 | ·2082 024 | ·1386 441 | ·0939 781 | ·0644 793 |
| ·55 | ·5318 843 | ·3291 796 | ·2151 699 | ·1447 040 | ·0990 414 | ·0686 078 |
| ·56 | ·5382 895− | ·3366 750+ | ·2222 797 | ·1509 441 | ·1043 027 | ·0729 370 |
| ·57 | ·5447 103 | ·3442 561 | ·2295 352 | ·1573 691 | ·1097 687 | ·0774 750− |
| ·58 | ·5511 494 | ·3519 259 | ·2369 403 | ·1639 844 | ·1154 461 | ·0822 299 |
| ·59 | ·5576 098 | ·3596 876 | ·2444 990 | ·1707 954 | ·1213 421 | ·0872 106 |
| ·60 | ·5640 942 | ·3675 445− | ·2522 155− | ·1778 078 | ·1274 640 | ·0924 263 |

$x = ·61$ to $1·00$ $q = 0·5$ $p = 0·5$ to $3$

| $x$ | $p = 0·5$ | $p = 1$ | $p = 1·5$ | $p = 2$ | $p = 2·5$ | $p = 3$ |
|---|---|---|---|---|---|---|
| $B(p,q)=$ | 3·1415 9265$^{+}$ | 2·0000 0000$^{e}$ | 1·5707 9633 | 1·3333 3333 | 1·1780 9725$^{-}$ | 1·0666 6667 |
| ·61 | ·5706 057 | ·3755 002 | ·2600 945$^{-}$ | ·1850 278 | ·1338 199 | ·0978 866 |
| ·62 | ·5771 474 | ·3835 586 | ·2681 408 | ·1924 618 | ·1404 181 | ·1036 017 |
| ·63 | ·5837 226 | ·3917 237 | ·2763 598 | ·2001 167 | ·1472 674 | ·1095 824 |
| ·64 | ·5903 345$^{-}$ | ·4000 000$^{e}$ | ·2847 570 | ·2080 000$^{e}$ | ·1543 773 | ·1158 400$^{e}$ |
| ·65 | ·5969 867 | ·4083 920 | ·2933 384 | ·2161 194 | ·1617 575$^{-}$ | ·1223 865$^{+}$ |
| ·66 | ·6036 829 | ·4169 048 | ·3021 105$^{+}$ | ·2244 834 | ·1694 187 | ·1292 348 |
| ·67 | ·6104 271 | ·4255 437 | ·3110 804 | ·2331 009 | ·1773 722 | ·1363 984 |
| ·68 | ·6172 233 | ·4343 146 | ·3202 554 | ·2419 815$^{+}$ | ·1856 299 | ·1438 917 |
| ·69 | ·6240 760 | ·4432 236 | ·3296 437 | ·2511 357 | ·1942 048 | ·1517 302 |
| ·70 | ·6309 899 | ·4522 774 | ·3392 541 | ·2605 745$^{+}$ | ·2031 107 | ·1599 305$^{+}$ |
| ·71 | ·6379 699 | ·4614 835$^{+}$ | ·3490 960 | ·2703 102 | ·2123 624 | ·1685 104 |
| ·72 | ·6450 216 | ·4708 497 | ·3591 800 | ·2803 556 | ·2219 760 | ·1774 888 |
| ·73 | ·6521 506 | ·4803 848 | ·3695 172 | ·2907 252 | ·2319 690 | ·1868 866 |
| ·74 | ·6593 633 | ·4900 980 | ·3801 201 | ·3014 343 | ·2423 601 | ·1967 260 |
| ·75 | ·6666 667 | ·5000 000$^{e}$ | ·3910 022 | ·3125 000$^{e}$ | ·2531 700 | ·2070 312 |
| ·76 | ·6740 681 | ·5101 021 | ·4021 785$^{-}$ | ·3239 408 | ·2644 211 | ·2178 289 |
| ·77 | ·6815 758 | ·5204 168 | ·4136 655$^{-}$ | ·3357 773 | ·2761 382 | ·2291 480 |
| ·78 | ·6891 989 | ·5309 584 | ·4254 815$^{-}$ | ·3480 322 | ·2883 484 | ·2410 204 |
| ·79 | ·6969 475$^{-}$ | ·5417 424 | ·4376 470 | ·3607 307 | ·3010 821 | ·2534 812 |
| ·80 | ·7048 328 | ·5527 864 | ·4501 849 | ·3739 010 | ·3143 726 | ·2665 697 |
| ·81 | ·7128 674 | ·5641 101 | ·4631 209 | ·3875 747 | ·3282 578 | ·2803 294 |
| ·82 | ·7210 657 | ·5757 359 | ·4764 843 | ·4017 877 | ·3427 799 | ·2948 095$^{-}$ |
| ·83 | ·7294 437 | ·5876 894 | ·4903 085$^{-}$ | ·4165 806 | ·3579 870 | ·3100 653 |
| ·84 | ·7380 202 | ·6000 000$^{e}$ | ·5046 316 | ·4320 000$^{e}$ | ·3739 339 | ·3261 600$^{e}$ |
| ·85 | ·7468 167 | ·6127 017 | ·5194 980 | ·4480 999 | ·3906 840 | ·3431 662 |
| ·86 | ·7558 582 | ·6258 343 | ·5349 594 | ·4649 430 | ·4083 108 | ·3611 681 |
| ·87 | ·7651 745$^{+}$ | ·6394 449 | ·5510 771 | ·4826 034 | ·4269 006 | ·3802 643 |
| ·88 | ·7748 011 | ·6535 898 | ·5679 242 | ·5011 694 | ·4465 564 | ·4005 719 |
| ·89 | ·7847 810 | ·6683 375$^{+}$ | ·5855 892 | ·5207 477 | ·4674 020 | ·4222 315$^{+}$ |
| ·90 | ·7951 672 | ·6837 722 | ·6041 813 | ·5414 697 | ·4895 897 | ·4454 156 |
| ·91 | ·8060 266 | ·7000 000$^{e}$ | ·6238 377 | ·5635 000$^{e}$ | ·5133 097 | ·4703 387 |
| ·92 | ·8174 451 | ·7171 573 | ·6447 345$^{-}$ | ·5870 496 | ·5388 053 | ·4972 754 |
| ·93 | ·8295 366 | ·7354 249 | ·6671 049 | ·6123 974 | ·5663 973 | ·5265 858 |
| ·94 | ·8424 576 | ·7550 510 | ·6912 688 | ·6399 250$^{+}$ | ·5965 238 | ·5587 612 |
| ·95 | ·8564 337 | ·7763 932 | ·7176 856 | ·6701 800 | ·6298 119 | ·5945 030 |
| ·96 | ·8718 116 | ·8000 000$^{e}$ | ·7470 601 | ·7040 000$^{e}$ | ·6672 191 | ·6348 800$^{e}$ |
| ·97 | ·8891 753 | ·8267 949 | ·7805 761 | ·7427 905$^{-}$ | ·7103 486 | ·6816 772 |
| ·98 | ·9096 655$^{+}$ | ·8585 786 | ·8205 388 | ·7892 822 | ·7623 093 | ·7383 493 |
| ·99 | ·9362 314 | ·9000 000$^{e}$ | ·8728 886 | ·8505 000$^{e}$ | ·8310 823 | ·8137 462 |
| 1·00 | 1·0000 000 | 1·0000 000 | 1·0000 000 | 1·0000 000 | 1·0000 000 | 1·0000 000 |

$x =$ ·02 to ·60 $q = 0·5$ $p =$ 3·5 to 6

| | $p = 3·5$ | $p = 4$ | $p = 4·5$ | $p = 5$ | $p = 5·5$ | $p = 6$ |
|---|---|---|---|---|---|---|
| $B(p, q) =$ | ·9817 4770 | ·9142 8571 | ·8590 2924 | ·8126 9841 | ·7731 2632 | ·7388 1674 |
| $x$ | | | | | | |
| ·02 | ·0000 003 | | | | | |
| ·03 | ·0000 014 | ·0000 002 | | | | |
| ·04 | ·0000 038 | ·0000 007 | ·0000 001 | | | |
| ·05 | ·0000 083 | ·0000 017 | ·0000 004 | ·0000 001 | | |
| ·06 | ·0000 158 | ·0000 036 | ·0000 008 | ·0000 002 | | |
| ·07 | ·0000 272 | ·0000 068 | ·0000 017 | ·0000 004 | ·0000 001 | |
| ·08 | ·0000 435$^{+}$ | ·0000 116 | ·0000 031 | ·0000 008 | ·0000 002 | ·0000 001 |
| ·09 | ·0000 660 | ·0000 186 | ·0000 053 | ·0000 015$^{+}$ | ·0000 004 | ·0000 001 |
| ·10 | ·0000 958 | ·0000 285$^{+}$ | ·0000 085$^{+}$ | ·0000 026 | ·0000 008 | ·0000 002 |
| ·11 | ·0001 344 | ·0000 419 | ·0000 132 | ·0000 042 | ·0000 013 | ·0000 004 |
| ·12 | ·0001 830 | ·0000 596 | ·0000 196 | ·0000 065$^{-}$ | ·0000 021 | ·0000 007 |
| ·13 | ·0002 432 | ·0000 825$^{+}$ | ·0000 282 | ·0000 097 | ·0000 033 | ·0000 012 |
| ·14 | ·0003 166 | ·0001 115$^{-}$ | ·0000 395$^{+}$ | ·0000 141 | ·0000 050$^{+}$ | ·0000 018 |
| ·15 | ·0004 049 | ·0001 476 | ·0000 542 | ·0000 200 | ·0000 074 | ·0000 028 |
| ·16 | ·0005 098 | ·0001 920 | ·0000 728 | ·0000 277 | ·0000 106 | ·0000 041 |
| ·17 | ·0006 331 | ·0002 458 | ·0000 960 | ·0000 377 | ·0000 149 | ·0000 059 |
| ·18 | ·0007 769 | ·0003 104 | ·0001 248 | ·0000 505$^{-}$ | ·0000 205$^{-}$ | ·0000 083 |
| ·19 | ·0009 431 | ·0003 872 | ·0001 600 | ·0000 664 | ·0000 277 | ·0000 116 |
| ·20 | ·0011 338 | ·0004 776 | ·0002 025$^{-}$ | ·0000 863 | ·0000 369 | ·0000 159 |
| ·21 | ·0013 512 | ·0005 834 | ·0002 535$^{-}$ | ·0001 107 | ·0000 486 | ·0000 214 |
| ·22 | ·0015 978 | ·0007 061 | ·0003 141 | ·0001 404 | ·0000 630 | ·0000 284 |
| ·23 | ·0018 757 | ·0008 477 | ·0003 856 | ·0001 763 | ·0000 809 | ·0000 373 |
| ·24 | ·0021 876 | ·0010 101 | ·0004 694 | ·0002 192 | ·0001 028 | ·0000 484 |
| ·25 | ·0025 360 | ·0011 953 | ·0005 670 | ·0002 703 | ·0001 294 | ·0000 622 |
| ·26 | ·0029 236 | ·0014 055$^{+}$ | ·0006 800 | ·0003 306 | ·0001 614 | ·0000 791 |
| ·27 | ·0033 532 | ·0016 430 | ·0008 101 | ·0004 015$^{-}$ | ·0001 998 | ·0000 998 |
| ·28 | ·0038 278 | ·0019 103 | ·0009 593 | ·0004 842 | ·0002 454 | ·0001 248 |
| ·29 | ·0043 503 | ·0022 098 | ·0011 295$^{+}$ | ·0005 802 | ·0002 993 | ·0001 549 |
| ·30 | ·0049 238 | ·0025 444 | ·0013 230 | ·0006 913 | ·0003 627 | ·0001 910 |
| ·31 | ·0055 517 | ·0029 167 | ·0015 419 | ·0008 191 | ·0004 369 | ·0002 339 |
| ·32 | ·0062 372 | ·0033 299 | ·0017 887 | ·0009 656 | ·0005 234 | ·0002 846 |
| ·33 | ·0069 839 | ·0037 871 | ·0020 661 | ·0011 328 | ·0006 236 | ·0003 444 |
| ·34 | ·0077 954 | ·0042 914 | ·0023 769 | ·0013 229 | ·0007 393 | ·0004 145$^{+}$ |
| ·35 | ·0086 754 | ·0048 466 | ·0027 239 | ·0015 384 | ·0008 723 | ·0004 963 |
| ·36 | ·0096 279 | ·0054 560$^{e}$ | ·0031 104 | ·0017 818 | ·0010 248 | ·0005 914 |
| ·37 | ·0106 569 | ·0061 236 | ·0035 397 | ·0020 560 | ·0011 990 | ·0007 015$^{+}$ |
| ·38 | ·0117 666 | ·0068 534 | ·0040 154 | ·0023 640 | ·0013 972 | ·0008 286 |
| ·39 | ·0129 614 | ·0076 494 | ·0045 412 | ·0027 088 | ·0016 222 | ·0009 747 |
| ·40 | ·0142 458 | ·0085 163 | ·0051 211 | ·0030 941 | ·0018 767 | ·0011 421 |
| ·41 | ·0156 244 | ·0094 584 | ·0057 593 | ·0035 234 | ·0021 640 | ·0013 334 |
| ·42 | ·0171 021 | ·0104 807 | ·0064 602 | ·0040 008 | ·0024 873 | ·0015 514 |
| ·43 | ·0186 841 | ·0115 882 | ·0072 287 | ·0045 304 | ·0028 502 | ·0017 990 |
| ·44 | ·0203 755$^{+}$ | ·0127 861 | ·0080 697 | ·0051 167 | ·0032 568 | ·0020 796 |
| ·45 | ·0221 819 | ·0140 801 | ·0089 885$^{+}$ | ·0057 646 | ·0037 112 | ·0023 968 |
| ·46 | ·0241 090 | ·0154 760 | ·0099 907 | ·0064 792 | ·0042 179 | ·0027 546 |
| ·47 | ·0261 625$^{+}$ | ·0169 797 | ·0110 821 | ·0072 660 | ·0047 819 | ·0031 570 |
| ·48 | ·0283 488 | ·0185 978 | ·0122 691 | ·0081 307 | ·0054 084 | ·0036 090 |
| ·49 | ·0306 742 | ·0203 368 | ·0135 582 | ·0090 797 | ·0061 032 | ·0041 153 |
| ·50 | ·0331 455$^{+}$ | ·0222 039 | ·0149 564 | ·0101 196 | ·0068 723 | ·0046 816 |
| ·51 | ·0357 696 | ·0242 063 | ·0164 710 | ·0112 573 | ·0077 223 | ·0053 137 |
| ·52 | ·0385 538 | ·0263 519 | ·0181 098 | ·0125 005$^{+}$ | ·0086 602 | ·0060 181 |
| ·53 | ·0415 056 | ·0286 487 | ·0198 811 | ·0138 572 | ·0096 936 | ·0068 017 |
| ·54 | ·0446 332 | ·0311 052 | ·0217 936 | ·0153 359 | ·0108 305$^{+}$ | ·0076 720 |
| ·55 | ·0479 448 | ·0337 304 | ·0238 564 | ·0169 456 | ·0120 798 | ·0086 372 |
| ·56 | ·0514 491 | ·0365 338 | ·0260 793 | ·0186 962 | ·0134 508 | ·0097 060 |
| ·57 | ·0551 552 | ·0395 252 | ·0284 726 | ·0205 978 | ·0149 534 | ·0108 880 |
| ·58 | ·0590 728 | ·0427 152 | ·0310 472 | ·0226 615$^{+}$ | ·0165 985$^{-}$ | ·0121 935$^{-}$ |
| ·59 | ·0632 120 | ·0461 148 | ·0338 148 | ·0248 990 | ·0183 976 | ·0136 335$^{-}$ |
| ·60 | ·0675 833 | ·0497 356 | ·0367 875$^{-}$ | ·0273 229 | ·0203 631 | ·0152 201 |

$x$ = ·61 to 1·00 $q$ = 0·5 $p$ = 3·5 to 6

| | $p$ = 3·5 | $p$ = 4 | $p$ = 4·5 | $p$ = 5 | $p$ = 5·5 | $p$ = 6 |
|---|---|---|---|---|---|---|
| $B(p, q)$ = | ·9817 4770 | ·9142 8571 | ·8590 2924 | ·8126 9841 | ·7731 2632 | ·7388 1674 |
| $x$ | | | | | | |
| ·61 | ·0721 979 | ·0535 899 | ·0399 784 | ·0299 465$^{-}$ | ·0225 083 | ·0169 663 |
| ·62 | ·0770 676 | ·0576 907 | ·0434 014 | ·0327 840 | ·0248 475$^{+}$ | ·0188 860 |
| ·63 | ·0822 049 | ·0620 519 | ·0470 711 | ·0358 507 | ·0273 962 | ·0209 946 |
| ·64 | ·0876 228 | ·0666 880$^{e}$ | ·0510 033 | ·0391 629 | ·0301 708 | ·0233 084 |
| ·65 | ·0933 354 | ·0716 146 | ·0552 145$^{+}$ | ·0427 380 | ·0331 891 | ·0258 452 |
| ·66 | ·0993 574 | ·0768 481 | ·0597 227 | ·0465 947 | ·0364 704 | ·0286 243 |
| ·67 | ·1057 046 | ·0824 061 | ·0645 469 | ·0507 532 | ·0400 352 | ·0316 664 |
| ·68 | ·1123 936 | ·0883 074 | ·0697 074 | ·0552 348 | ·0439 059 | ·0349 944 |
| ·69 | ·1194 425$^{+}$ | ·0945 721 | ·0752 260 | ·0600 628 | ·0481 065$^{+}$ | ·0386 326 |
| ·70 | ·1268 704 | ·1012 215$^{+}$ | ·0811 262 | ·0652 622 | ·0526 631 | ·0426 079 |
| ·71 | ·1346 977 | ·1082 788 | ·0874 332 | ·0708 600 | ·0576 040 | ·0469 493 |
| ·72 | ·1429 466 | ·1157 687 | ·0941 741 | ·0768 851 | ·0629 597 | ·0516 885$^{-}$ |
| ·73 | ·1516 408 | ·1237 181 | ·1013 783 | ·0833 692 | ·0687 636 | ·0568 600 |
| ·74 | ·1608 062 | ·1321 558 | ·1090 777 | ·0903 466 | ·0750 519 | ·0625 017 |
| ·75 | ·1704 707 | ·1411 133 | ·1173 068 | ·0978 546 | ·0818 642 | ·0686 550$^{+}$ |
| ·76 | ·1806 646 | ·1506 247 | ·1261 033 | ·1059 339 | ·0892 440 | ·0753 654 |
| ·77 | ·1914 214 | ·1607 275$^{+}$ | ·1355 083 | ·1146 292 | ·0972 389 | ·0826 831 |
| ·78 | ·2027 774 | ·1714 627 | ·1455 671 | ·1239 896 | ·1059 013 | ·0906 634 |
| ·79 | ·2147 730 | ·1828 753 | ·1563 295$^{-}$ | ·1340 690 | ·1152 891 | ·0993 677 |
| ·80 | ·2274 528 | ·1950 155$^{+}$ | ·1678 507 | ·1449 276 | ·1254 669 | ·1088 643 |
| ·81 | ·2408 665$^{-}$ | ·2079 389 | ·1801 920 | ·1566 321 | ·1365 063 | ·1192 294 |
| ·82 | ·2550 697 | ·2217 077 | ·1934 221 | ·1692 572 | ·1484 877 | ·1305 487 |
| ·83 | ·2701 255$^{-}$ | ·2363 922 | ·2076 183 | ·1828 871 | ·1615 019 | ·1429 189 |
| ·84 | ·2861 051 | ·2520 720$^{e}$ | ·2228 683 | ·1976 173 | ·1756 516 | ·1564 496 |
| ·85 | ·3030 905$^{+}$ | ·2688 382 | ·2392 724 | ·2135 568 | ·1910 543 | ·1712 665$^{-}$ |
| ·86 | ·3211 765$^{+}$ | ·2867 961 | ·2569 461 | ·2308 312 | ·2078 455$^{+}$ | ·1875 144 |
| ·87 | ·3404 738 | ·3060 685$^{+}$ | ·2760 240 | ·2495 869 | ·2261 829 | ·2053 619 |
| ·88 | ·3611 134 | ·3268 003 | ·2966 650$^{+}$ | ·2699 963 | ·2462 521 | ·2250 075$^{-}$ |
| ·89 | ·3832 528 | ·3491 653 | ·3190 589 | ·2922 651 | ·2682 745$^{-}$ | ·2466 879 |
| ·90 | ·4070 838 | ·3733 749 | ·3434 364 | ·3166 429 | ·2925 185$^{-}$ | ·2706 900 |
| ·91 | ·4328 453 | ·3996 915$^{-}$ | ·3700 831 | ·3434 386 | ·3193 155$^{-}$ | ·2973 674 |
| ·92 | ·4608 414 | ·4284 484 | ·3993 614 | ·3730 427 | ·3490 843 | ·3271 668 |
| ·93 | ·4914 709 | ·4600 818 | ·4317 438 | ·4059 641 | ·3823 694 | ·3606 677 |
| ·94 | ·5252 755$^{+}$ | ·4951 828 | ·4678 698 | ·4428 896 | ·4199 041 | ·3986 496 |
| ·95 | ·5630 278 | ·5345 921 | ·5086 465$^{-}$ | ·4847 912 | ·4627 245$^{-}$ | ·4422 114 |
| ·96 | ·6059 013 | ·5795 840$^{e}$ | ·5554 454 | ·5331 354 | ·5123 898 | ·4930 037 |
| ·97 | ·6558 521 | ·6322 773 | ·6105 421 | ·5903 492 | ·5714 749 | ·5537 459 |
| ·98 | ·7166 574 | ·6967 541 | ·6783 097 | ·6610 862 | ·6449 047 | ·6296 271 |
| ·99 | ·7979 717 | ·7834 244 | ·7698 750$^{-}$ | ·7571 581 | ·7451 499 | ·7337 548 |
| 1·00 | 1·0000 000 | 1·0000 000 | 1·0000 000 | 1·0000 000 | 1·0000 000 | 1·0000 000 |

$x =$ ·10 to ·70 $q =$ 0·5 $p =$ 6·5 to 9

| | $p = 6·5$ | $p = 7$ | $p = 7·5$ | $p = 8$ | $p = 8·5$ | $p = 9$ |
|---|---|---|---|---|---|---|
| $B(p, q) =$ | ·7086 9912 | ·6819 8468 | ·6580 7776 | ·6365 1904 | ·6169 4790 | ·5990 7674 |
| $x$ | | | | | | |
| ·10 | ·0000 001 | | | | | |
| ·11 | ·0000 001 | | | | | |
| ·12 | ·0000 002 | ·0000 001 | | | | |
| ·13 | ·0000 004 | ·0000 001 | | | | |
| ·14 | ·0000 007 | ·0000 002 | ·0000 001 | | | |
| ·15 | ·0000 010 | ·0000 004 | ·0000 001 | ·0000 001 | | |
| ·16 | ·0000 016 | ·0000 006 | ·0000 002 | ·0000 001 | | |
| ·17 | ·0000 023 | ·0000 009 | ·0000 004 | ·0000 001 | ·0000 001 | |
| ·18 | ·0000 034 | ·0000 014 | ·0000 006 | ·0000 002 | ·0000 001 | |
| ·19 | ·0000 049 | ·0000 021 | ·0000 009 | ·0000 004 | ·0000 002 | ·0000 001 |
| ·20 | ·0000 068 | ·0000 030 | ·0000 013 | ·0000 006 | ·0000 002 | ·0000 001 |
| ·21 | ·0000 094 | ·0000 042 | ·0000 019 | ·0000 008 | ·0000 004 | ·0000 002 |
| ·22 | ·0000 128 | ·0000 058 | ·0000 026 | ·0000 012 | ·0000 005+ | ·0000 003 |
| ·23 | ·0000 172 | ·0000 080 | ·0000 037 | ·0000 017 | ·0000 008 | ·0000 004 |
| ·24 | ·0000 228 | ·0000 108 | ·0000 051 | ·0000 024 | ·0000 012 | ·0000 006 |
| ·25 | ·0000 300 | ·0000 145− | ·0000 070 | ·0000 034 | ·0000 017 | ·0000 008 |
| ·26 | ·0000 389 | ·0000 192 | ·0000 095− | ·0000 047 | ·0000 023 | ·0000 012 |
| ·27 | ·0000 500− | ·0000 251 | ·0000 126 | ·0000 064 | ·0000 032 | ·0000 016 |
| ·28 | ·0000 637 | ·0000 325+ | ·0000 167 | ·0000 086 | ·0000 044 | ·0000 023 |
| ·29 | ·0000 804 | ·0000 419 | ·0000 218 | ·0000 114 | ·0000 060 | ·0000 031 |
| ·30 | ·0001 008 | ·0000 534 | ·0000 283 | ·0000 151 | ·0000 080 | ·0000 043 |
| ·31 | ·0001 255+ | ·0000 676 | ·0000 364 | ·0000 197 | ·0000 107 | ·0000 058 |
| ·32 | ·0001 553 | ·0000 849 | ·0000 465+ | ·0000 255+ | ·0000 140 | ·0000 077 |
| ·33 | ·0001 908 | ·0001 060 | ·0000 590 | ·0000 329 | ·0000 184 | ·0000 103 |
| ·34 | ·0002 331 | ·0001 314 | ·0000 742 | ·0000 420 | ·0000 238 | ·0000 135+ |
| ·35 | ·0002 832 | ·0001 620 | ·0000 929 | ·0000 533 | ·0000 307 | ·0000 177 |
| ·36 | ·0003 423 | ·0001 986 | ·0001 155− | ·0000 672 | ·0000 392 | ·0000 229 |
| ·37 | ·0004 116 | ·0002 421 | ·0001 427 | ·0000 843 | ·0000 499 | ·0000 295+ |
| ·38 | ·0004 928 | ·0002 938 | ·0001 755− | ·0001 050+ | ·0000 630 | ·0000 378 |
| ·39 | ·0005 873 | ·0003 547 | ·0002 147 | ·0001 302 | ·0000 791 | ·0000 481 |
| ·40 | ·0006 970 | ·0004 264 | ·0002 614 | ·0001 605+ | ·0000 988 | ·0000 608 |
| ·41 | ·0008 239 | ·0005 103 | ·0003 168 | ·0001 970 | ·0001 227 | ·0000 765+ |
| ·42 | ·0009 703 | ·0006 084 | ·0003 822 | ·0002 406 | ·0001 517 | ·0000 958 |
| ·43 | ·0011 386 | ·0007 224 | ·0004 593 | ·0002 925+ | ·0001 866 | ·0001 192 |
| ·44 | ·0013 316 | ·0008 547 | ·0005 497 | ·0003 542 | ·0002 286 | ·0001 477 |
| ·45 | ·0015 522 | ·0010 076 | ·0006 554 | ·0004 271 | ·0002 788 | ·0001 822 |
| ·46 | ·0018 038 | ·0011 840 | ·0007 787 | ·0005 131 | ·0003 387 | ·0002 238 |
| ·47 | ·0020 900 | ·0013 868 | ·0009 221 | ·0006 142 | ·0004 098 | ·0002 738 |
| ·48 | ·0024 147 | ·0016 194 | ·0010 882 | ·0007 326 | ·0004 940 | ·0003 335+ |
| ·49 | ·0027 823 | ·0018 855− | ·0012 803 | ·0008 709 | ·0005 934 | ·0004 048 |
| ·50 | ·0031 977 | ·0021 892 | ·0015 018 | ·0010 320 | ·0007 103 | ·0004 896 |
| ·51 | ·0036 661 | ·0025 351 | ·0017 565+ | ·0012 192 | ·0008 476 | ·0005 901 |
| ·52 | ·0041 931 | ·0029 281 | ·0020 489 | ·0014 361 | ·0010 082 | ·0007 088 |
| ·53 | ·0047 850+ | ·0033 739 | ·0023 836 | ·0016 869 | ·0011 957 | ·0008 487 |
| ·54 | ·0054 487 | ·0038 784 | ·0027 661 | ·0019 762 | ·0014 140 | ·0010 132 |
| ·55 | ·0061 916 | ·0044 483 | ·0032 021 | ·0023 090 | ·0016 676 | ·0012 059 |
| ·56 | ·0070 217 | ·0050 911 | ·0036 984 | ·0026 913 | ·0019 614 | ·0014 314 |
| ·57 | ·0079 480 | ·0058 147 | ·0042 621 | ·0031 294 | ·0023 012 | ·0016 945− |
| ·58 | ·0089 801 | ·0066 280 | ·0049 013 | ·0036 305+ | ·0026 933 | ·0020 007 |
| ·59 | ·0101 284 | ·0075 407 | ·0056 248 | ·0042 027 | ·0031 449 | ·0023 564 |
| ·60 | ·0114 043 | ·0085 635+ | ·0064 425+ | ·0048 549 | ·0036 639 | ·0027 688 |
| ·61 | ·0128 203 | ·0097 082 | ·0073 652 | ·0055 970 | ·0042 595− | ·0032 459 |
| ·62 | ·0143 899 | ·0109 874 | ·0084 049 | ·0064 400 | ·0049 416 | ·0037 968 |
| ·63 | ·0161 278 | ·0124 152 | ·0095 748 | ·0073 963 | ·0057 217 | ·0044 320 |
| ·64 | ·0180 501 | ·0140 071 | ·0108 895+ | ·0084 795− | ·0066 123 | ·0051 629 |
| ·65 | ·0201 741 | ·0157 799 | ·0123 651 | ·0097 048 | ·0076 277 | ·0060 028 |
| ·66 | ·0225 190 | ·0177 521 | ·0140 194 | ·0110 890 | ·0087 836 | ·0069 663 |
| ·67 | ·0251 054 | ·0199 440 | ·0158 718 | ·0126 510 | ·0100 978 | ·0080 700 |
| ·68 | ·0279 559 | ·0223 778 | ·0179 442 | ·0144 114 | ·0115 901 | ·0093 327 |
| ·69 | ·0310 952 | ·0250 780 | ·0202 603 | ·0163 933 | ·0132 826 | ·0107 755− |
| ·70 | ·0345 503 | ·0280 714 | ·0228 466 | ·0186 226 | ·0152 002 | ·0124 219 |

$x$ = ·71 to 1·00 $q = 0{\cdot}5$ $p = 6{\cdot}5$ to 9

| $x$ | $p = 6{\cdot}5$ | $p = 7$ | $p = 7{\cdot}5$ | $p = 8$ | $p = 8{\cdot}5$ | $p = 9$ |
|---|---|---|---|---|---|---|
| $B(p, q) =$ | ·7086 9912 | ·6819 8468 | ·6580 7776 | ·6365 1904 | ·6169 4790 | ·5990 7674 |
| ·71 | ·0383 506 | ·0313 875⁻ | ·0257 323 | ·0211 278 | ·0173 705⁻ | ·0142 986 |
| ·72 | ·0425 285⁻ | ·0350 587 | ·0289 496 | ·0239 405⁺ | ·0198 245⁺ | ·0164 358 |
| ·73 | ·0471 192 | ·0391 209 | ·0325 343 | ·0270 964 | ·0225 970 | ·0188 671 |
| ·74 | ·0521 618 | ·0436 136 | ·0365 260 | ·0306 347 | ·0257 270 | ·0216 307 |
| ·75 | ·0576 988 | ·0485 803 | ·0409 690 | ·0345 997 | ·0292 580 | ·0247 696 |
| ·76 | ·0637 776 | ·0540 694 | ·0459 120 | ·0390 404 | ·0332 393 | ·0283 323 |
| ·77 | ·0704 503 | ·0601 345⁻ | ·0514 098 | ·0440 122 | ·0377 260 | ·0323 739 |
| ·78 | ·0777 746 | ·0668 352 | ·0575 234 | ·0495 768 | ·0427 805⁻ | ·0369 566 |
| ·79 | ·0858 147 | ·0742 382 | ·0643 210 | ·0558 039 | ·0484 730 | ·0421 510 |
| ·80 | ·0946 423 | ·0824 179 | ·0718 796 | ·0627 720 | ·0548 834 | ·0480 375⁺ |
| ·81 | ·1043 378 | ·0914 580 | ·0802 856 | ·0705 698 | ·0621 022 | ·0547 079 |
| ·82 | ·1149 913 | ·1014 529 | ·0896 370 | ·0792 984 | ·0702 325⁺ | ·0622 672 |
| ·83 | ·1267 050⁻ | ·1125 097 | ·1000 452 | ·0890 728 | ·0793 927 | ·0708 361 |
| ·84 | ·1395 951 | ·1247 504 | ·1116 375⁺ | ·1000 251 | ·0897 187 | ·0805 539 |
| ·85 | ·1537 948 | ·1383 153 | ·1245 605⁻ | ·1123 074 | ·1013 679 | ·0915 823 |
| ·86 | ·1694 578 | ·1533 663 | ·1389 839 | ·1260 965⁺ | ·1145 234 | ·1041 103 |
| ·87 | ·1867 631 | ·1700 924 | ·1551 060 | ·1415 997 | ·1294 004 | ·1183 603 |
| ·88 | ·2059 217 | ·1887 165⁻ | ·1731 610 | ·1590 616 | ·1462 536 | ·1345 963 |
| ·89 | ·2271 852 | ·2095 046 | ·1934 288 | ·1787 752 | ·1653 885⁻ | ·1531 354 |
| ·90 | ·2508 583 | ·2327 788 | ·2162 484 | ·2010 959 | ·1871 760 | ·1743 635⁻ |
| ·91 | ·2773 168 | ·2589 365⁻ | ·2420 379 | ·2264 623 | ·2120 743 | ·1987 577 |
| ·92 | ·3070 344 | ·2884 781 | ·2713 244 | ·2554 269 | ·2406 613 | ·2269 203 |
| ·93 | ·3406 256 | ·3220 524 | ·3047 901 | ·2887 054 | ·2736 849 | ·2596 310 |
| ·94 | ·3789 153 | ·3605 294 | ·3433 496 | ·3272 560 | ·3121 466 | ·2979 337 |
| ·95 | ·4230 646 | ·4051 315⁺ | ·3882 859 | ·3724 217 | ·3574 488 | ·3432 896 |
| ·96 | ·4748 140 | ·4576 879 | ·4415 160 | ·4262 064 | ·4116 810 | ·3978 730 |
| ·97 | ·5370 247 | ·5211 995⁺ | ·5061 785⁻ | ·4918 845⁻ | ·4782 524 | ·4652 262 |
| ·98 | ·6151 439 | ·6013 664 | ·5882 217 | ·5756 491 | ·5635 970 | ·5520 214 |
| ·99 | ·7228 973 | ·7125 164 | ·7025 619 | ·6929 921 | ·6837 719 | ·6748 712 |
| 1·00 | 1·0000 000 | 1·0000 000 | 1·0000 000 | 1·0000 000 | 1·0000 000 | 1·0000 000 |

$x$ = ·21 to ·80 $q = 0·5$ $p$ = 9·5 to 13

| $x$ | $p = 9·5$ | $p = 10$ | $p = 10·5$ | $p = 11$ | $p = 12$ | $p = 13$ |
|---|---|---|---|---|---|---|
| $B(p, q)$ = | ·5826 7301 | ·5675 4639 | ·5535 3936 | ·5405 2037 | ·5170 1948 | ·4963 3870 |
| ·21 | ·0000 001 | | | | | |
| ·22 | ·0000 001 | ·0000 001 | | | | |
| ·23 | ·0000 002 | ·0000 001 | | | | |
| ·24 | ·0000 003 | ·0000 001 | ·0000 001 | | | |
| ·25 | ·0000 004 | ·0000 002 | ·0000 001 | | | |
| ·26 | ·0000 006 | ·0000 003 | ·0000 001 | ·0000 001 | | |
| ·27 | ·0000 008 | ·0000 004 | ·0000 002 | ·0000 001 | | |
| ·28 | ·0000 012 | ·0000 006 | ·0000 003 | ·0000 002 | | |
| ·29 | ·0000 016 | ·0000 009 | ·0000 005− | ·0000 002 | ·0000 001 | |
| ·30 | ·0000 023 | ·0000 012 | ·0000 007 | ·0000 004 | ·0000 001 | |
| ·31 | ·0000 031 | ·0000 017 | ·0000 009 | ·0000 005+ | ·0000 002 | |
| ·32 | ·0000 043 | ·0000 024 | ·0000 013 | ·0000 007 | ·0000 002 | ·0000 001 |
| ·33 | ·0000 058 | ·0000 032 | ·0000 018 | ·0000 010 | ·0000 003 | ·0000 001 |
| ·34 | ·0000 077 | ·0000 044 | ·0000 025− | ·0000 014 | ·0000 005− | ·0000 002 |
| ·35 | ·0000 102 | ·0000 059 | ·0000 034 | ·0000 020 | ·0000 007 | ·0000 002 |
| ·36 | ·0000 134 | ·0000 079 | ·0000 046 | ·0000 027 | ·0000 009 | ·0000 003 |
| ·37 | ·0000 175+ | ·0000 104 | ·0000 062 | ·0000 037 | ·0000 013 | ·0000 005− |
| ·38 | ·0000 227 | ·0000 137 | ·0000 082 | ·0000 050− | ·0000 018 | ·0000 007 |
| ·39 | ·0000 293 | ·0000 179 | ·0000 109 | ·0000 067 | ·0000 025− | ·0000 009 |
| ·40 | ·0000 375+ | ·0000 232 | ·0000 143 | ·0000 089 | ·0000 034 | ·0000 013 |
| ·41 | ·0000 478 | ·0000 299 | ·0000 187 | ·0000 117 | ·0000 046 | ·0000 018 |
| ·42 | ·0000 605+ | ·0000 383 | ·0000 243 | ·0000 154 | ·0000 062 | ·0000 025+ |
| ·43 | ·0000 763 | ·0000 489 | ·0000 313 | ·0000 201 | ·0000 083 | ·0000 034 |
| ·44 | ·0000 956 | ·0000 619 | ·0000 402 | ·0000 261 | ·0000 110 | ·0000 047 |
| ·45 | ·0001 193 | ·0000 782 | ·0000 513 | ·0000 337 | ·0000 146 | ·0000 063 |
| ·46 | ·0001 481 | ·0000 981 | ·0000 651 | ·0000 432 | ·0000 191 | ·0000 085− |
| ·47 | ·0001 831 | ·0001 227 | ·0000 822 | ·0000 552 | ·0000 249 | ·0000 113 |
| ·48 | ·0002 255+ | ·0001 526 | ·0001 034 | ·0000 701 | ·0000 324 | ·0000 150− |
| ·49 | ·0002 766 | ·0001 891 | ·0001 295− | ·0000 887 | ·0000 418 | ·0000 197 |
| ·50 | ·0003 379 | ·0002 334 | ·0001 615− | ·0001 118 | ·0000 537 | ·0000 259 |
| ·51 | ·0004 113 | ·0002 870 | ·0002 005− | ·0001 402 | ·0000 687 | ·0000 338 |
| ·52 | ·0004 989 | ·0003 516 | ·0002 480 | ·0001 751 | ·0000 875+ | ·0000 439 |
| ·53 | ·0006 032 | ·0004 291 | ·0003 056 | ·0002 179 | ·0001 110 | ·0000 567 |
| ·54 | ·0007 268 | ·0005 220 | ·0003 753 | ·0002 701 | ·0001 402 | ·0000 730 |
| ·55 | ·0008 732 | ·0006 330 | ·0004 593 | ·0003 336 | ·0001 764 | ·0000 936 |
| ·56 | ·0010 459 | ·0007 651 | ·0005 602 | ·0004 106 | ·0002 211 | ·0001 194 |
| ·57 | ·0012 492 | ·0009 220 | ·0006 812 | ·0005 037 | ·0002 761 | ·0001 518 |
| ·58 | ·0014 880 | ·0011 079 | ·0008 257 | ·0006 160 | ·0003 436 | ·0001 922 |
| ·59 | ·0017 678 | ·0013 276 | ·0009 980 | ·0007 509 | ·0004 262 | ·0002 426 |
| ·60 | ·0020 948 | ·0015 866 | ·0012 029 | ·0009 128 | ·0005 269 | ·0003 050+ |
| ·61 | ·0024 764 | ·0018 914 | ·0014 460 | ·0011 065− | ·0006 494 | ·0003 822 |
| ·62 | ·0029 207 | ·0022 491 | ·0017 336 | ·0013 375+ | ·0007 980 | ·0004 774 |
| ·63 | ·0034 370 | ·0026 682 | ·0020 734 | ·0016 126 | ·0009 778 | ·0005 945+ |
| ·64 | ·0040 359 | ·0031 582 | ·0024 738 | ·0019 394 | ·0011 948 | ·0007 381 |
| ·65 | ·0047 295− | ·0037 301 | ·0029 448 | ·0023 268 | ·0014 561 | ·0009 137 |
| ·66 | ·0055 313 | ·0043 964 | ·0034 977 | ·0027 851 | ·0017 700 | ·0011 279 |
| ·67 | ·0064 568 | ·0051 713 | ·0041 457 | ·0033 263 | ·0021 463 | ·0013 887 |
| ·68 | ·0075 235− | ·0060 711 | ·0049 037 | ·0039 642 | ·0025 965+ | ·0017 053 |
| ·69 | ·0087 513 | ·0071 145− | ·0057 892 | ·0047 147 | ·0031 341 | ·0020 890 |
| ·70 | ·0101 626 | ·0083 225+ | ·0068 218 | ·0055 964 | ·0037 749 | ·0025 530 |
| ·71 | ·0117 829 | ·0097 193 | ·0080 244 | ·0066 306 | ·0045 373 | ·0031 129 |
| ·72 | ·0136 410 | ·0113 325+ | ·0094 231 | ·0078 419 | ·0054 429 | ·0037 876 |
| ·73 | ·0157 696 | ·0131 934 | ·0110 479 | ·0092 587 | ·0065 170 | ·0045 989 |
| ·74 | ·0182 057 | ·0153 378 | ·0129 330 | ·0109 139 | ·0077 891 | ·0055 730 |
| ·75 | ·0209 915+ | ·0178 066 | ·0151 179 | ·0128 454 | ·0092 937 | ·0067 409 |
| ·76 | ·0241 746 | ·0206 463 | ·0176 480 | ·0150 970 | ·0110 712 | ·0081 391 |
| ·77 | ·0278 093 | ·0239 103 | ·0205 753 | ·0177 192 | ·0131 687 | ·0098 108 |
| ·78 | ·0319 575− | ·0276 597 | ·0239 598 | ·0207 707 | ·0156 415+ | ·0118 075− |
| ·79 | ·0366 896 | ·0319 644 | ·0278 706 | ·0243 194 | ·0185 543 | ·0141 897 |
| ·80 | ·0420 863 | ·0369 048 | ·0323 874 | ·0284 440 | ·0219 830 | ·0170 296 |

$x$ = ·81 to 1·00 $q = 0·5$ $p$ = 9·5 to 13

| $x$ | $p$ = 9·5 | $p$ = 10 | $p$ = 10·5 | $p$ = 11 | $p$ = 12 | $p$ = 13 |
|---|---|---|---|---|---|---|
| $B(p, q)$ = | ·5826 7301 | ·5675 4639 | ·5535 3936 | ·5405 2037 | ·5170 1948 | ·4963 3870 |
| ·81 | ·0482 400 | ·0425 736 | ·0376 025+ | ·0332 362 | ·0260 167 | ·0204 125+ |
| ·82 | ·0552 568 | ·0490 775− | ·0436 231 | ·0388 027 | ·0307 603 | ·0244 403 |
| ·83 | ·0632 595+ | ·0565 405− | ·0505 737 | ·0452 684 | ·0363 378 | ·0292 343 |
| ·84 | ·0723 900 | ·0651 067 | ·0586 000 | ·0527 799 | ·0428 960 | ·0349 395− |
| ·85 | ·0828 138 | ·0749 447 | ·0678 729 | ·0615 098 | ·0506 093 | ·0417 298 |
| ·86 | ·0947 249 | ·0862 527 | ·0785 943 | ·0716 629 | ·0596 861 | ·0498 152 |
| ·87 | ·1083 520 | ·0992 653 | ·0910 038 | ·0834 832 | ·0703 770 | ·0594 496 |
| ·88 | ·1239 679 | ·1142 629 | ·1053 887 | ·0972 642 | ·0829 853 | ·0709 434 |
| ·89 | ·1419 006 | ·1315 837 | ·1220 966 | ·1133 617 | ·0978 814 | ·0846 779 |
| ·90 | ·1625 500− | ·1516 409 | ·1415 531 | ·1322 131 | ·1155 229 | ·1011 275+ |
| ·91 | ·1864 114 | ·1749 472 | ·1642 873 | ·1543 630 | ·1364 828 | ·1208 898 |
| ·92 | ·2141 107 | ·2021 512 | ·1909 698 | ·1805 030 | ·1614 919 | ·1447 305− |
| ·93 | ·2464 587 | ·2340 939 | ·2224 710 | ·2115 320 | ·1915 031 | ·1736 523 |
| ·94 | ·2845 412 | ·2719 021 | ·2599 578 | ·2486 559 | ·2277 976 | ·2090 079 |
| ·95 | ·3298 768 | ·3171 516 | ·3050 620 | ·2935 620 | ·2721 706 | ·2526 955− |
| ·96 | ·3847 242 | ·3721 840 | ·3602 077 | ·3487 557 | ·3272 869 | ·3075 355+ |
| ·97 | ·4527 575− | ·4408 042 | ·4293 290 | ·4182 993 | ·3974 618 | ·3780 916 |
| ·98 | ·5408 843 | ·5301 526 | ·5197 973 | ·5097 927 | ·4907 470 | ·4728 599 |
| ·99 | ·6662 641 | ·6579 282 | ·6498 437 | ·6419 932 | ·6269 347 | ·6126 479 |
| 1·00 | 1·0000 000 | 1·0000 000 | 1·0000 000 | 1·0000 000 | 1·0000 000 | 1·0000 000 |

$x = {·34}$ to 1·00 $q = 0·5$ $p = 14$ to 19

| $x$ | $p = 14$ | $p = 15$ | $p = 16$ | $p = 17$ | $p = 18$ | $p = 19$ |
|---|---|---|---|---|---|---|
| $B(p, q) =$ | ·4779 5579 | ·4614 7455+ | ·4465 8828 | ·4330 5530 | ·4206 8229 | ·4093 1250− |
| ·34 | ·0000 001 | | | | | |
| ·35 | ·0000 001 | | | | | |
| ·36 | ·0000 001 | | | | | |
| ·37 | ·0000 002 | ·0000 001 | | | | |
| ·38 | ·0000 002 | ·0000 001 | | | | |
| ·39 | ·0000 004 | ·0000 001 | | | | |
| ·40 | ·0000 005+ | ·0000 002 | ·0000 001 | | | |
| ·41 | ·0000 007 | ·0000 003 | ·0000 001 | | | |
| ·42 | ·0000 010 | ·0000 004 | ·0000 002 | ·0000 001 | | |
| ·43 | ·0000 014 | ·0000 006 | ·0000 002 | ·0000 001 | | |
| ·44 | ·0000 020 | ·0000 008 | ·0000 004 | ·0000 002 | ·0000 001 | |
| ·45 | ·0000 027 | ·0000 012 | ·0000 005+ | ·0000 002 | ·0000 001 | |
| ·46 | ·0000 038 | ·0000 017 | ·0000 007 | ·0000 003 | ·0000 001 | ·0000 001 |
| ·47 | ·0000 051 | ·0000 023 | ·0000 011 | ·0000 005− | ·0000 002 | ·0000 001 |
| ·48 | ·0000 069 | ·0000 032 | ·0000 015+ | ·0000 007 | ·0000 003 | ·0000 002 |
| ·49 | ·0000 093 | ·0000 044 | ·0000 021 | ·0000 010 | ·0000 005− | ·0000 002 |
| ·50 | ·0000 125+ | ·0000 061 | ·0000 029 | ·0000 014 | ·0000 007 | ·0000 003 |
| ·51 | ·0000 166 | ·0000 082 | ·0000 041 | ·0000 020 | ·0000 010 | ·0000 005− |
| ·52 | ·0000 220 | ·0000 111 | ·0000 056 | ·0000 028 | ·0000 014 | ·0000 007 |
| ·53 | ·0000 291 | ·0000 149 | ·0000 077 | ·0000 040 | ·0000 020 | ·0000 011 |
| ·54 | ·0000 381 | ·0000 199 | ·0000 104 | ·0000 055− | ·0000 029 | ·0000 015+ |
| ·55 | ·0000 497 | ·0000 265+ | ·0000 141 | ·0000 076 | ·0000 041 | ·0000 022 |
| ·56 | ·0000 646 | ·0000 351 | ·0000 191 | ·0000 104 | ·0000 057 | ·0000 031 |
| ·57 | ·0000 837 | ·0000 462 | ·0000 256 | ·0000 142 | ·0000 079 | ·0000 044 |
| ·58 | ·0001 078 | ·0000 606 | ·0000 341 | ·0000 192 | ·0000 109 | ·0000 061 |
| ·59 | ·0001 384 | ·0000 791 | ·0000 453 | ·0000 260 | ·0000 149 | ·0000 086 |
| ·60 | ·0001 770 | ·0001 029 | ·0000 600 | ·0000 350− | ·0000 204 | ·0000 120 |
| ·61 | ·0002 255+ | ·0001 333 | ·0000 790 | ·0000 469 | ·0000 278 | ·0000 166 |
| ·62 | ·0002 863 | ·0001 721 | ·0001 036 | ·0000 625− | ·0000 377 | ·0000 228 |
| ·63 | ·0003 624 | ·0002 213 | ·0001 354 | ·0000 830 | ·0000 509 | ·0000 313 |
| ·64 | ·0004 571 | ·0002 836 | ·0001 763 | ·0001 098 | ·0000 685− | ·0000 427 |
| ·65 | ·0005 747 | ·0003 622 | ·0002 287 | ·0001 447 | ·0000 916 | ·0000 581 |
| ·66 | ·0007 204 | ·0004 611 | ·0002 957 | ·0001 899 | ·0001 221 | ·0000 787 |
| ·67 | ·0009 006 | ·0005 852 | ·0003 810 | ·0002 484 | ·0001 622 | ·0001 060 |
| ·68 | ·0011 226 | ·0007 405− | ·0004 893 | ·0003 238 | ·0002 146 | ·0001 424 |
| ·69 | ·0013 956 | ·0009 342 | ·0006 264 | ·0004 207 | ·0002 830 | ·0001 906 |
| ·70 | ·0017 305+ | ·0011 753 | ·0007 997 | ·0005 449 | ·0003 719 | ·0002 541 |
| ·71 | ·0021 406 | ·0014 748 | ·0010 179 | ·0007 036 | ·0004 871 | ·0003 376 |
| ·72 | ·0026 416 | ·0018 459 | ·0012 922 | ·0009 059 | ·0006 360 | ·0004 471 |
| ·73 | ·0032 525+ | ·0023 048 | ·0016 360 | ·0011 631 | ·0008 280 | ·0005 901 |
| ·74 | ·0039 962 | ·0028 711 | ·0020 662 | ·0014 892 | ·0010 748 | ·0007 767 |
| ·75 | ·0048 999 | ·0035 685+ | ·0026 033 | ·0019 019 | ·0013 914 | ·0010 191 |
| ·76 | ·0059 964 | ·0044 261 | ·0032 724 | ·0024 231 | ·0017 965+ | ·0013 336 |
| ·77 | ·0073 247 | ·0054 788 | ·0041 048 | ·0030 799 | ·0023 139 | ·0017 405− |
| ·78 | ·0089 319 | ·0067 691 | ·0051 384 | ·0039 061 | ·0029 732 | ·0022 658 |
| ·79 | ·0108 743 | ·0083 486 | ·0064 199 | ·0049 438 | ·0038 120 | ·0029 427 |
| ·80 | ·0132 192 | ·0102 798 | ·0080 067 | ·0062 450+ | ·0048 771 | ·0038 132 |
| ·81 | ·0160 478 | ·0126 386 | ·0099 692 | ·0078 745+ | ·0062 278 | ·0049 310 |
| ·82 | ·0194 573 | ·0155 171 | ·0123 939 | ·0099 129 | ·0079 383 | ·0063 641 |
| ·83 | ·0235 652 | ·0190 278 | ·0153 874 | ·0124 602 | ·0101 021 | ·0081 993 |
| ·84 | ·0285 131 | ·0233 077 | ·0190 809 | ·0156 413 | ·0128 371 | ·0105 470 |
| ·85 | ·0344 726 | ·0285 243 | ·0236 368 | ·0196 122 | ·0162 919 | ·0135 480 |
| ·86 | ·0416 527 | ·0348 837 | ·0292 563 | ·0245 681 | ·0206 548 | ·0173 828 |
| ·87 | ·0503 085+ | ·0426 397 | ·0361 903 | ·0307 546 | ·0261 647 | ·0222 824 |
| ·88 | ·0607 541 | ·0521 077 | ·0447 526 | ·0384 823 | ·0331 268 | ·0285 448 |
| ·89 | ·0733 788 | ·0636 817 | ·0553 390 | ·0481 460 | ·0419 325+ | ·0365 561 |
| ·90 | ·0886 700 | ·0778 587 | ·0684 528 | ·0602 521 | ·0530 885− | ·0468 204 |
| ·91 | ·1072 459 | ·0952 734 | ·0847 416 | ·0754 572 | ·0672 568 | ·0600 018 |
| ·92 | ·1299 031 | ·1167 490 | ·1050 507 | ·0946 246 | ·0853 147 | ·0769 875− |
| ·93 | ·1576 897 | ·1433 745+ | ·1305 053 | ·1189 109 | ·1084 452 | ·0989 825− |
| ·94 | ·1920 248 | ·1766 309 | ·1626 429 | ·1499 051 | ·1382 837 | ·1276 631 |
| ·95 | ·2349 057 | ·2186 090 | ·2036 432 | ·1898 700 | ·1771 703 | ·1654 407 |
| ·96 | ·2893 035+ | ·2724 259 | ·2567 635− | ·2421 974 | ·2286 252 | ·2159 579 |
| ·97 | ·3600 252 | ·3431 267 | ·3272 815+ | ·3123 920 | ·2983 739 | ·2851 541 |
| ·98 | ·4560 048 | ·4400 767 | ·4249 874 | ·4106 621 | ·3970 362 | ·3840 537 |
| ·99 | ·5990 480 | ·5860 650− | ·5736 402 | ·5617 240 | ·5502 740 | ·5392 534 |
| 1·00 | 1·0000 000 | 1·0000 000 | 1·0000 000 | 1·0000 000 | 1·0000 000 | 1·0000 000 |

$x = \cdot 48$ to $1 \cdot 00$ $q = 0 \cdot 5$ $p = 20$ to $25$

| | $p = 20$ | $p = 21$ | $p = 22$ | $p = 23$ | $p = 24$ | $p = 25$ |
|---|---|---|---|---|---|---|
| $B(p, q) =$ | ·3988 1731 | ·3890 9006 | ·3800 4145⁻ | ·3715 9608 | ·3636 8978 | ·3562 6754 |
| $x$ | | | | | | |
| ·48 | ·0000 001 | | | | | |
| ·49 | ·0000 001 | ·0000 001 | | | | |
| ·50 | ·0000 002 | ·0000 001 | | | | |
| ·51 | ·0000 002 | ·0000 001 | ·0000 001 | | | |
| ·52 | ·0000 004 | ·0000 002 | ·0000 001 | | | |
| ·53 | ·0000 005⁺ | ·0000 003 | ·0000 001 | ·0000 001 | | |
| ·54 | ·0000 008 | ·0000 004 | ·0000 002 | ·0000 001 | ·0000 001 | |
| ·55 | ·0000 012 | ·0000 006 | ·0000 003 | ·0000 002 | ·0000 001 | ·0000 001 |
| ·56 | ·0000 017 | ·0000 009 | ·0000 005⁺ | ·0000 003 | ·0000 002 | ·0000 001 |
| ·57 | ·0000 024 | ·0000 014 | ·0000 008 | ·0000 004 | ·0000 002 | ·0000 001 |
| ·58 | ·0000 035⁻ | ·0000 020 | ·0000 011 | ·0000 006 | ·0000 004 | ·0000 002 |
| ·59 | ·0000 050⁻ | ·0000 029 | ·0000 017 | ·0000 010 | ·0000 006 | ·0000 003 |
| ·60 | ·0000 070 | ·0000 041 | ·0000 024 | ·0000 014 | ·0000 008 | ·0000 005⁻ |
| ·61 | ·0000 099 | ·0000 059 | ·0000 035⁺ | ·0000 021 | ·0000 013 | ·0000 008 |
| ·62 | ·0000 138 | ·0000 084 | ·0000 051 | ·0000 031 | ·0000 019 | ·0000 011 |
| ·63 | ·0000 193 | ·0000 119 | ·0000 073 | ·0000 045⁺ | ·0000 028 | ·0000 017 |
| ·64 | ·0000 267 | ·0000 167 | ·0000 105⁻ | ·0000 066 | ·0000 041 | ·0000 026 |
| ·65 | ·0000 369 | ·0000 234 | ·0000 149 | ·0000 095⁻ | ·0000 061 | ·0000 039 |
| ·66 | ·0000 507 | ·0000 327 | ·0000 211 | ·0000 137 | ·0000 088 | ·0000 057 |
| ·67 | ·0000 694 | ·0000 455⁻ | ·0000 298 | ·0000 196 | ·0000 129 | ·0000 085⁻ |
| ·68 | ·0000 946 | ·0000 629 | ·0000 419 | ·0000 279 | ·0000 186 | ·0000 124 |
| ·69 | ·0001 285⁻ | ·0000 867 | ·0000 586 | ·0000 396 | ·0000 268 | ·0000 182 |
| ·70 | ·0001 738 | ·0001 190 | ·0000 816 | ·0000 560 | ·0000 384 | ·0000 264 |
| ·71 | ·0002 342 | ·0001 627 | ·0001 131 | ·0000 787 | ·0000 548 | ·0000 382 |
| ·72 | ·0003 146 | ·0002 216 | ·0001 562 | ·0001 103 | ·0000 779 | ·0000 550⁺ |
| ·73 | ·0004 211 | ·0003 008 | ·0002 150⁺ | ·0001 539 | ·0001 102 | ·0000 790 |
| ·74 | ·0005 618 | ·0004 068 | ·0002 949 | ·0002 139 | ·0001 553 | ·0001 128 |
| ·75 | ·0007 473 | ·0005 485⁻ | ·0004 030 | ·0002 963 | ·0002 180 | ·0001 605⁺ |
| ·76 | ·0009 910 | ·0007 372 | ·0005 488 | ·0004 090 | ·0003 050⁻ | ·0002 276 |
| ·77 | ·0013 106 | ·0009 878 | ·0007 452 | ·0005 627 | ·0004 251 | ·0003 215⁻ |
| ·78 | ·0017 285⁺ | ·0013 199 | ·0010 088 | ·0007 716 | ·0005 907 | ·0004 525⁻ |
| ·79 | ·0022 740 | ·0017 589 | ·0013 617 | ·0010 551 | ·0008 181 | ·0006 348 |
| ·80 | ·0029 845⁻ | ·0023 380 | ·0018 332 | ·0014 385⁺ | ·0011 297 | ·0008 877 |
| ·81 | ·0039 082 | ·0031 004 | ·0024 617 | ·0019 561 | ·0015 555⁺ | ·0012 378 |
| ·82 | ·0051 072 | ·0041 023 | ·0032 980 | ·0026 534 | ·0021 363 | ·0017 211 |
| ·83 | ·0066 615⁻ | ·0054 170 | ·0044 087 | ·0035 908 | ·0029 267 | ·0023 870 |
| ·84 | ·0086 739 | ·0071 398 | ·0058 819 | ·0048 492 | ·0040 007 | ·0033 027 |
| ·85 | ·0112 772 | ·0093 952 | ·0078 335⁺ | ·0065 363 | ·0054 577 | ·0045 599 |
| ·86 | ·0146 429 | ·0123 456 | ·0104 169 | ·0087 959 | ·0074 322 | ·0062 838 |
| ·87 | ·0189 937 | ·0162 040 | ·0138 348 | ·0118 204 | ·0101 060 | ·0086 455⁻ |
| ·88 | ·0246 188 | ·0212 503 | ·0183 565⁺ | ·0158 679 | ·0137 256 | ·0118 796 |
| ·89 | ·0318 971 | ·0278 542 | ·0243 417 | ·0212 866 | ·0186 267 | ·0163 087 |
| ·90 | ·0413 275⁻ | ·0365 075⁻ | ·0322 728 | ·0285 482 | ·0252 689 | ·0223 790 |
| ·91 | ·0535 734 | ·0478 699 | ·0428 032 | ·0382 974 | ·0342 862 | ·0307 120 |
| ·92 | ·0695 281 | ·0628 370 | ·0568 277 | ·0514 249 | ·0465 623 | ·0421 819 |
| ·93 | ·0904 137 | ·0826 441 | ·0755 903 | ·0691 794 | ·0633 468 | ·0580 356 |
| ·94 | ·1179 424 | ·1090 335⁻ | ·1008 584 | ·0933 485⁻ | ·0864 426 | ·0800 864 |
| ·95 | ·1545 908 | ·1445 411 | ·1352 212 | ·1265 685⁺ | ·1185 272 | ·1110 470 |
| ·96 | ·2041 173 | ·1930 344 | ·1826 482 | ·1729 041 | ·1637 531 | ·1551 511 |
| ·97 | ·2726 684 | ·2608 599 | ·2496 785⁺ | ·2390 790 | ·2290 210 | ·2194 680 |
| ·98 | ·3716 657 | ·3598 289 | ·3485 051 | ·3376 600 | ·3272 628 | ·3172 858 |
| ·99 | ·5286 301 | ·5183 759 | ·5084 660 | ·4988 782 | ·4895 926 | ·4805 913 |
| 1·00 | 1·0000 000 | 1·0000 000 | 1·0000 000 | 1·0000 000 | 1·0000 000 | 1·0000 000 |

$x$ = ·57 to 1·00 $q = 0·5$ $p$ = 26 to 31

| | $p = 26$ | $p = 27$ | $p = 28$ | $p = 29$ | $p = 30$ | $p = 31$ |
|---|---|---|---|---|---|---|
| $B(p, q)$ = | ·3492 8191 | ·3426 9168 | ·3364 6092 | ·3305 5810 | ·3249 5542 | ·3196 2828 |
| $x$ | | | | | | |
| ·57 | ·0000 001 | | | | | |
| ·58 | ·0000 001 | ·0000 001 | | | | |
| ·59 | ·0000 002 | ·0000 001 | ·0000 001 | | | |
| ·60 | ·0000 003 | ·0000 002 | ·0000 001 | ·0000 001 | | |
| ·61 | ·0000 004 | ·0000 003 | ·0000 002 | ·0000 001 | ·0000 001 | |
| ·62 | ·0000 007 | ·0000 004 | ·0000 003 | ·0000 002 | ·0000 001 | ·0000 001 |
| ·63 | ·0000 011 | ·0000 007 | ·0000 004 | ·0000 003 | ·0000 002 | ·0000 001 |
| ·64 | ·0000 016 | ·0000 010 | ·0000 006 | ·0000 004 | ·0000 003 | ·0000 002 |
| ·65 | ·0000 025$^-$ | ·0000 016 | ·0000 010 | ·0000 006 | ·0000 004 | ·0000 003 |
| ·66 | ·0000 037 | ·0000 024 | ·0000 016 | ·0000 010 | ·0000 007 | ·0000 004 |
| ·67 | ·0000 056 | ·0000 037 | ·0000 024 | ·0000 016 | ·0000 010 | ·0000 007 |
| ·68 | ·0000 083 | ·0000 055$^+$ | ·0000 037 | ·0000 025$^-$ | ·0000 017 | ·0000 011 |
| ·69 | ·0000 123 | ·0000 083 | ·0000 057 | ·0000 038 | ·0000 026 | ·0000 018 |
| ·70 | ·0000 181 | ·0000 125$^-$ | ·0000 086 | ·0000 059 | ·0000 041 | ·0000 028 |
| ·71 | ·0000 266 | ·0000 186 | ·0000 130 | ·0000 091 | ·0000 063 | ·0000 044 |
| ·72 | ·0000 389 | ·0000 275$^+$ | ·0000 195$^+$ | ·0000 138 | ·0000 098 | ·0000 069 |
| ·73 | ·0000 566 | ·0000 406 | ·0000 292 | ·0000 210 | ·0000 151 | ·0000 108 |
| ·74 | ·0000 820 | ·0000 597 | ·0000 434 | ·0000 316 | ·0000 230 | ·0000 168 |
| ·75 | ·0001 183 | ·0000 872 | ·0000 643 | ·0000 475$^-$ | ·0000 351 | ·0000 259 |
| ·76 | ·0001 699 | ·0001 270 | ·0000 949 | ·0000 710 | ·0000 531 | ·0000 398 |
| ·77 | ·0002 432 | ·0001 841 | ·0001 395$^-$ | ·0001 057 | ·0000 802 | ·0000 608 |
| ·78 | ·0003 468 | ·0002 660 | ·0002 041 | ·0001 567 | ·0001 204 | ·0000 925$^+$ |
| ·79 | ·0004 928 | ·0003 829 | ·0002 976 | ·0002 314 | ·0001 801 | ·0001 402 |
| ·80 | ·0006 980 | ·0005 492 | ·0004 323 | ·0003 405$^+$ | ·0002 683 | ·0002 115$^+$ |
| ·81 | ·0009 856 | ·0007 852 | ·0006 259 | ·0004 992 | ·0003 983 | ·0003 179 |
| ·82 | ·0013 875$^-$ | ·0011 192 | ·0009 032 | ·0007 293 | ·0005 892 | ·0004 762 |
| ·83 | ·0019 480 | ·0015 907 | ·0012 996 | ·0010 622 | ·0008 687 | ·0007 107 |
| ·84 | ·0027 282 | ·0022 548 | ·0018 646 | ·0015 426 | ·0012 769 | ·0010 573 |
| ·85 | ·0038 121 | ·0031 887 | ·0026 686 | ·0022 344 | ·0018 717 | ·0015 685$^+$ |
| ·86 | ·0053 160 | ·0044 996 | ·0038 106 | ·0032 285$^+$ | ·0027 366 | ·0023 207 |
| ·87 | ·0074 003 | ·0063 378 | ·0054 306 | ·0046 554 | ·0039 926 | ·0034 255$^+$ |
| ·88 | ·0102 876 | ·0089 136 | ·0077 268 | ·0067 011 | ·0058 141 | ·0050 465$^+$ |
| ·89 | ·0142 869 | ·0125 222 | ·0109 806 | ·0096 331 | ·0084 545$^+$ | ·0074 231 |
| ·90 | ·0198 302 | ·0175 803 | ·0155 929 | ·0138 362 | ·0122 824 | ·0109 074 |
| ·91 | ·0275 246 | ·0246 798 | ·0221 390 | ·0198 682 | ·0178 373 | ·0160 201 |
| ·92 | ·0382 326 | ·0346 691 | ·0314 513 | ·0285 439 | ·0259 152 | ·0235 370 |
| ·93 | ·0531 949 | ·0487 796 | ·0447 495$^-$ | ·0410 684 | ·0377 040 | ·0346 272 |
| ·94 | ·0742 310 | ·0688 328 | ·0638 524 | ·0592 545$^-$ | ·0550 069 | ·0510 808 |
| ·95 | ·1040 830 | ·0975 944 | ·0915 444 | ·0858 996 | ·0806 294 | ·0757 062 |
| ·96 | ·1470 584 | ·1394 388 | ·1322 594 | ·1254 903 | ·1191 040 | ·1130 753 |
| ·97 | ·2103 869 | ·2017 477 | ·1935 228 | ·1856 871 | ·1782 175$^+$ | ·1710 928 |
| ·98 | ·3077 039 | ·2984 943 | ·2896 360 | ·2811 098 | ·2728 983 | ·2649 851 |
| ·99 | ·4718 583 | ·4633 789 | ·4551 397 | ·4471 286 | ·4393 344 | ·4317 467 |
| 1·00 | 1·0000 000 | 1·0000 000 | 1·0000 000 | 1·0000 000 | 1·0000 000 | 1·0000 000 |

$x$ = ·63 to 1·00 $q$ = 0·5 $p$ = 32 to 37

| | $p$ = 32 | $p$ = 33 | $p$ = 34 | $p$ = 35 | $p$ = 36 | $p$ = 37 |
|---|---|---|---|---|---|---|
| $B(p, q)$ = | ·3145 5482 | ·3097 1551 | ·3050 9289 | ·3006 7126 | ·2964 3645+ | ·2923 7568 |
| $x$ | | | | | | |
| ·63 | ·0000 001 | | | | | |
| ·64 | ·0000 001 | ·0000 001 | | | | |
| ·65 | ·0000 002 | ·0000 001 | ·0000 001 | | | |
| ·66 | ·0000 003 | ·0000 002 | ·0000 001 | ·0000 001 | ·0000 001 | |
| ·67 | ·0000 005− | ·0000 003 | ·0000 002 | ·0000 001 | ·0000 001 | ·0000 001 |
| ·68 | ·0000 007 | ·0000 005− | ·0000 003 | ·0000 002 | ·0000 002 | ·0000 001 |
| ·69 | ·0000 012 | ·0000 008 | ·0000 006 | ·0000 004 | ·0000 003 | ·0000 002 |
| ·70 | ·0000 019 | ·0000 013 | ·0000 009 | ·0000 006 | ·0000 004 | ·0000 003 |
| ·71 | ·0000 031 | ·0000 022 | ·0000 015+ | ·0000 011 | ·0000 007 | ·0000 005+ |
| ·72 | ·0000 049 | ·0000 035− | ·0000 025− | ·0000 018 | ·0000 013 | ·0000 009 |
| ·73 | ·0000 078 | ·0000 056 | ·0000 040 | ·0000 029 | ·0000 021 | ·0000 015+ |
| ·74 | ·0000 122 | ·0000 089 | ·0000 065+ | ·0000 048 | ·0000 035− | ·0000 025+ |
| ·75 | ·0000 192 | ·0000 142 | ·0000 105− | ·0000 078 | ·0000 057 | ·0000 043 |
| ·76 | ·0000 298 | ·0000 223 | ·0000 167 | ·0000 126 | ·0000 094 | ·0000 071 |
| ·77 | ·0000 462 | ·0000 350+ | ·0000 266 | ·0000 202 | ·0000 154 | ·0000 117 |
| ·78 | ·0000 711 | ·0000 547 | ·0000 421 | ·0000 324 | ·0000 250− | ·0000 192 |
| ·79 | ·0001 092 | ·0000 851 | ·0000 663 | ·0000 517 | ·0000 403 | ·0000 314 |
| ·80 | ·0001 668 | ·0001 316 | ·0001 039 | ·0000 820 | ·0000 648 | ·0000 512 |
| ·81 | ·0002 539 | ·0002 029 | ·0001 621 | ·0001 296 | ·0001 037 | ·0000 829 |
| ·82 | ·0003 850− | ·0003 114 | ·0002 520 | ·0002 040 | ·0001 651 | ·0001 338 |
| ·83 | ·0005 817 | ·0004 763 | ·0003 901 | ·0003 196 | ·0002 620 | ·0002 148 |
| ·84 | ·0008 759 | ·0007 259 | ·0006 018 | ·0004 991 | ·0004 140 | ·0003 436 |
| ·85 | ·0013 150− | ·0011 029 | ·0009 253 | ·0007 766 | ·0006 519 | ·0005 475− |
| ·86 | ·0019 687 | ·0016 707 | ·0014 184 | ·0012 045+ | ·0010 232 | ·0008 695− |
| ·87 | ·0029 402 | ·0025 245+ | ·0021 684 | ·0018 631 | ·0016 013 | ·0013 767 |
| ·88 | ·0043 819 | ·0038 062 | ·0033 073 | ·0028 747 | ·0024 995− | ·0021 738 |
| ·89 | ·0065 199 | ·0057 286 | ·0050 350+ | ·0044 268 | ·0038 933 | ·0034 250− |
| ·90 | ·0096 897 | ·0086 110 | ·0076 549 | ·0068 070 | ·0060 548 | ·0053 872 |
| ·91 | ·0143 930 | ·0129 356 | ·0116 294 | ·0104 582 | ·0094 077 | ·0084 650− |
| ·92 | ·0213 844 | ·0194 350+ | ·0176 687 | ·0160 676 | ·0146 156 | ·0132 984 |
| ·93 | ·0318 120 | ·0292 347 | ·0268 742 | ·0247 112 | ·0227 283 | ·0209 099 |
| ·94 | ·0474 497 | ·0440 899 | ·0409 795− | ·0380 987 | ·0354 294 | ·0329 552 |
| ·95 | ·0711 046 | ·0668 014 | ·0627 753 | ·0590 067 | ·0554 777 | ·0521 717 |
| ·96 | ·1073 812 | ·1020 001 | ·0969 127 | ·0921 005− | ·0875 468 | ·0832 360 |
| ·97 | ·1642 933 | ·1578 009 | ·1515 986 | ·1456 708 | ·1400 031 | ·1345 817 |
| ·98 | ·2573 552 | ·2499 948 | ·2428 909 | ·2360 314 | ·2294 051 | ·2230 016 |
| ·99 | ·4243 560 | ·4171 535+ | ·4101 312 | ·4032 812 | ·3965 967 | ·3900 709 |
| 1·00 | 1·0000 000 | 1·0000 000 | 1·0000 000 | 1·0000 000 | 1·0000 000 | 1·0000 000 |

$x$ = ·68 to 1·00 $q = 0·5$ $p$ = 38 to 43

| | $p = 38$ | $p = 39$ | $p = 40$ | $p = 41$ | $p = 42$ | $p = 43$ |
|---|---|---|---|---|---|---|
| $B(p, q)$ = | ·2884 7734 | ·2847 3088 | ·2811 2669 | ·2776 5599 | ·2743 1074 | ·2710 8355⁻ |
| $x$ | | | | | | |
| ·68 | ·0000 001 | | | | | |
| ·69 | ·0000 001 | ·0000 001 | ·0000 001 | | | |
| ·70 | ·0000 002 | ·0000 001 | ·0000 001 | ·0000 001 | | |
| ·71 | ·0000 004 | ·0000 003 | ·0000 002 | ·0000 001 | ·0000 001 | ·0000 001 |
| ·72 | ·0000 006 | ·0000 005⁻ | ·0000 003 | ·0000 002 | ·0000 002 | ·0000 001 |
| ·73 | ·0000 011 | ·0000 008 | ·0000 006 | ·0000 004 | ·0000 003 | ·0000 002 |
| ·74 | ·0000 019 | ·0000 014 | ·0000 010 | ·0000 007 | ·0000 005⁺ | ·0000 004 |
| ·75 | ·0000 031 | ·0000 023 | ·0000 017 | ·0000 013 | ·0000 010 | ·0000 007 |
| ·76 | ·0000 053 | ·0000 040 | ·0000 030 | ·0000 022 | ·0000 017 | ·0000 013 |
| ·77 | ·0000 089 | ·0000 068 | ·0000 051 | ·0000 039 | ·0000 030 | ·0000 023 |
| ·78 | ·0000 148 | ·0000 114 | ·0000 088 | ·0000 068 | ·0000 052 | ·0000 040 |
| ·79 | ·0000 245⁺ | ·0000 192 | ·0000 150⁻ | ·0000 117 | ·0000 091 | ·0000 071 |
| ·80 | ·0000 405⁻ | ·0000 320 | ·0000 253 | ·0000 200 | ·0000 158 | ·0000 125⁺ |
| ·81 | ·0000 664 | ·0000 531 | ·0000 425⁺ | ·0000 341 | ·0000 273 | ·0000 219 |
| ·82 | ·0001 084 | ·0000 878 | ·0000 712 | ·0000 577 | ·0000 468 | ·0000 380 |
| ·83 | ·0001 762 | ·0001 445⁺ | ·0001 186 | ·0000 973 | ·0000 799 | ·0000 656 |
| ·84 | ·0002 852 | ·0002 368 | ·0001 967 | ·0001 634 | ·0001 358 | ·0001 128 |
| ·85 | ·0004 599 | ·0003 865⁻ | ·0003 248 | ·0002 731 | ·0002 296 | ·0001 931 |
| ·86 | ·0007 391 | ·0006 284 | ·0005 344 | ·0004 546 | ·0003 868 | ·0003 292 |
| ·87 | ·0011 839 | ·0010 184 | ·0008 762 | ·0007 541 | ·0006 492 | ·0005 590 |
| ·88 | ·0018 911 | ·0016 456 | ·0014 324 | ·0012 471 | ·0010 860 | ·0009 459 |
| ·89 | ·0030 139 | ·0026 528 | ·0023 355⁺ | ·0020 567 | ·0018 116 | ·0015 960 |
| ·90 | ·0047 945⁺ | ·0042 681 | ·0038 004 | ·0033 848 | ·0030 152 | ·0026 866 |
| ·91 | ·0076 187 | ·0068 587 | ·0061 760 | ·0055 625⁺ | ·0050 110 | ·0045 152 |
| ·92 | ·0121 029 | ·0110 175⁻ | ·0100 317 | ·0091 361 | ·0083 223 | ·0075 824 |
| ·93 | ·0192 416 | ·0177 105⁻ | ·0163 048 | ·0150 138 | ·0138 279 | ·0127 381 |
| ·94 | ·0306 608 | ·0285 325⁻ | ·0265 575⁻ | ·0247 242 | ·0230 219 | ·0214 409 |
| ·95 | ·0490 735⁻ | ·0461 689 | ·0434 449 | ·0408 894 | ·0384 914 | ·0362 403 |
| ·96 | ·0791 535⁺ | ·0752 859 | ·0716 206 | ·0681 459 | ·0648 509 | ·0617 253 |
| ·97 | ·1293 941 | ·1244 282 | ·1196 731 | ·1151 184 | ·1107 541 | ·1065 712 |
| ·98 | ·2168 109 | ·2108 239 | ·2050 319 | ·1994 266 | ·1940 004 | ·1887 461 |
| ·99 | ·3836 977 | ·3774 712 | ·3713 860 | ·3654 370 | ·3596 193 | ·3539 283 |
| 1·00 | 1·0000 000 | 1·0000 000 | 1·0000 000 | 1·0000 000 | 1·0000 000 | 1·0000 000 |

$x$ = ·72 to 1·00 $q$ = 0·5 $p$ = 44 to 50

| | $p$ = 44 | $p$ = 45 | $p$ = 46 | $p$ = 47 | $p$ = 48 | $p$ = 49 | $p$ = 50 |
|---|---|---|---|---|---|---|---|
| $B(p, q)$ = | ·2679 6765$^{-}$ | ·2649 5677 | ·2620 4516 | ·2592 2747 | ·2564 9876 | ·2538 5444 | ·2512 9026 |
| $x$ | | | | | | | |
| ·72 | ·0000 001 | ·0000 001 | | | | | |
| ·73 | ·0000 002 | ·0000 001 | ·0000 001 | ·0000 001 | | | |
| ·74 | ·0000 003 | ·0000 002 | ·0000 002 | ·0000 001 | ·0000 001 | ·0000 001 | |
| ·75 | ·0000 005$^{+}$ | ·0000 004 | ·0000 003 | ·0000 002 | ·0000 002 | ·0000 001 | ·0000 001 |
| ·76 | ·0000 010 | ·0000 007 | ·0000 005$^{+}$ | ·0000 004 | ·0000 003 | ·0000 002 | ·0000 002 |
| ·77 | ·0000 017 | ·0000 013 | ·0000 010 | ·0000 008 | ·0000 006 | ·0000 004 | ·0000 003 |
| ·78 | ·0000 031 | ·0000 024 | ·0000 019 | ·0000 014 | ·0000 011 | ·0000 009 | ·0000 007 |
| ·79 | ·0000 056 | ·0000 044 | ·0000 034 | ·0000 027 | ·0000 021 | ·0000 016 | ·0000 013 |
| ·80 | ·0000 099 | ·0000 079 | ·0000 062 | ·0000 049 | ·0000 039 | ·0000 031 | ·0000 025$^{-}$ |
| ·81 | ·0000 175$^{+}$ | ·0000 141 | ·0000 113 | ·0000 090 | ·0000 072 | ·0000 058 | ·0000 047 |
| ·82 | ·0000 308 | ·0000 250$^{+}$ | ·0000 203 | ·0000 165$^{-}$ | ·0000 134 | ·0000 109 | ·0000 088 |
| ·83 | ·0000 539 | ·0000 443 | ·0000 364 | ·0000 299 | ·0000 246 | ·0000 202 | ·0000 166 |
| ·84 | ·0000 938 | ·0000 780 | ·0000 649 | ·0000 540 | ·0000 449 | ·0000 374 | ·0000 311 |
| ·85 | ·0001 625$^{-}$ | ·0001 367 | ·0001 151 | ·0000 969 | ·0000 816 | ·0000 687 | ·0000 578 |
| ·86 | ·0002 802 | ·0002 386 | ·0002 032 | ·0001 731 | ·0001 474 | ·0001 256 | ·0001 071 |
| ·87 | ·0004 814 | ·0004 147 | ·0003 573 | ·0003 079 | ·0002 653 | ·0002 287 | ·0001 972 |
| ·88 | ·0008 240 | ·0007 180 | ·0006 258 | ·0005 455$^{+}$ | ·0004 756 | ·0004 147 | ·0003 617 |
| ·89 | ·0014 064 | ·0012 395$^{+}$ | ·0010 927 | ·0009 634 | ·0008 496 | ·0007 493 | ·0006 610 |
| ·90 | ·0023 943 | ·0021 342 | ·0019 027 | ·0016 966 | ·0015 132 | ·0013 497 | ·0012 042 |
| ·91 | ·0040 692 | ·0036 679 | ·0033 069 | ·0029 818 | ·0026 892 | ·0024 257 | ·0021 884 |
| ·92 | ·0069 097 | ·0062 978 | ·0057 411 | ·0052 345$^{+}$ | ·0047 734 | ·0043 536 | ·0039 714 |
| ·93 | ·0117 364 | ·0108 154 | ·0099 684 | ·0091 892 | ·0084 723 | ·0078 125$^{+}$ | ·0072 052 |
| ·94 | ·0199 719 | ·0186 068 | ·0173 379 | ·0161 581 | ·0150 608 | ·0140 401 | ·0130 905$^{-}$ |
| ·95 | ·0341 267 | ·0321 416 | ·0302 766 | ·0285 242 | ·0268 771 | ·0253 287 | ·0238 727 |
| ·96 | ·0587 597 | ·0559 450$^{-}$ | ·0532 729 | ·0507 356 | ·0483 257 | ·0460 363 | ·0438 609 |
| ·97 | ·1025 609 | ·0987 151 | ·0950 262 | ·0914 868 | ·0880 901 | ·0848 297 | ·0816 993 |
| ·98 | ·1836 567 | ·1787 258 | ·1734 947 | ·1693 151 | ·1648 239 | ·1604 684 | ·1562 435$^{+}$ |
| ·99 | ·3483 598 | ·3429 096 | ·3375 738 | ·3323 488 | ·3272 311 | ·3222 174 | ·3173 044 |
| 1·00 | 1·0000 000 | 1·0000 000 | 1·0000 000 | 1·0000 000 | 1·0000 000 | 1·0000 000 | 1·0000 000 |

$x = \cdot 01$ to $\cdot 60$ $q = 1$ $p = 1$ to $3\cdot 5$

| | $p = 1$ | $p = 1\cdot 5$ | $p = 2$ | $p = 2\cdot 5$ | $p = 3$ | $p = 3\cdot 5$ |
|---|---|---|---|---|---|---|
| $B(p,q) =$ | 1·0000 0000 | ·6666 6667 | ·5000 0000 | ·4000 0000 | ·3333 3333 | ·2857 1429 |
| $x$ | | | | | | |
| ·01 | ·0100 000$^e$ | ·0010 000$^e$ | ·0001 000$^e$ | ·0000 100$^e$ | ·0000 010$^e$ | ·0000 001 |
| ·02 | ·0200 000$^e$ | ·0028 284 | ·0004 000$^e$ | ·0000 566 | ·0000 080$^e$ | ·0000 011 |
| ·03 | ·0300 000$^e$ | ·0051 962 | ·0009 000$^e$ | ·0001 559 | ·0000 270$^e$ | ·0000 047 |
| ·04 | ·0400 000$^e$ | ·0080 000$^e$ | ·0016 000$^e$ | ·0003 200$^e$ | ·0000 640$^e$ | ·0000 128$^e$ |
| ·05 | ·0500 000$^e$ | ·0111 803 | ·0025 000$^e$ | ·0005 590 | ·0001 250$^e$ | ·0000 280 |
| ·06 | ·0600 000$^e$ | ·0146 969 | ·0036 000$^e$ | ·0008 818 | ·0002 160$^e$ | ·0000 529 |
| ·07 | ·0700 000$^e$ | ·0185 203 | ·0049 000$^e$ | ·0012 964 | ·0003 430$^e$ | ·0000 907 |
| ·08 | ·0800 000$^e$ | ·0226 274 | ·0064 000$^e$ | ·0018 102 | ·0005 120$^e$ | ·0001 448 |
| ·09 | ·0900 000$^e$ | ·0270 000$^e$ | ·0081 000$^e$ | ·0024 300$^e$ | ·0007 290$^e$ | ·0002 187$^e$ |
| ·10 | ·1000 000$^e$ | ·0316 228 | ·0100 000$^e$ | ·0031 623 | ·0010 000$^e$ | ·0003 162 |
| ·11 | ·1100 000$^e$ | ·0364 829 | ·0121 000$^e$ | ·0040 131 | ·0013 310$^e$ | ·0004 414 |
| ·12 | ·1200 000$^e$ | ·0415 692 | ·0144 000$^e$ | ·0049 883 | ·0017 280$^e$ | ·0005 986 |
| ·13 | ·1300 000$^e$ | ·0468 722 | ·0169 000$^e$ | ·0060 934 | ·0021 970$^e$ | ·0007 921 |
| ·14 | ·1400 000$^e$ | ·0523 832 | ·0196 000$^e$ | ·0073 336 | ·0027 440$^e$ | ·0010 267 |
| ·15 | ·1500 000$^e$ | ·0580 948 | ·0225 000$^e$ | ·0087 142 | ·0033 750$^e$ | ·0013 071 |
| ·16 | ·1600 000$^e$ | ·0640 000$^e$ | ·0256 000$^e$ | ·0102 400$^e$ | ·0040 960$^e$ | ·0016 384$^e$ |
| ·17 | ·1700 000$^e$ | ·0700 928 | ·0289 000$^e$ | ·0119 158 | ·0049 130$^e$ | ·0020 257 |
| ·18 | ·1800 000$^e$ | ·0763 675+ | ·0324 000$^e$ | ·0137 462 | ·0058 320$^e$ | ·0024 743 |
| ·19 | ·1900 000$^e$ | ·0828 191 | ·0361 000$^e$ | ·0157 356 | ·0068 590$^e$ | ·0029 898 |
| ·20 | ·2000 000$^e$ | ·0894 427 | ·0400 000$^e$ | ·0178 885+ | ·0080 000$^e$ | ·0035 777 |
| ·21 | ·2100 000$^e$ | ·0962 341 | ·0441 000$^e$ | ·0202 092 | ·0092 610$^e$ | ·0042 439 |
| ·22 | ·2200 000$^e$ | ·1031 891 | ·0484 000$^e$ | ·0227 016 | ·0106 480$^e$ | ·0049 944 |
| ·23 | ·2300 000$^e$ | ·1103 041 | ·0529 000$^e$ | ·0253 699 | ·0121 670$^e$ | ·0058 351 |
| ·24 | ·2400 000$^e$ | ·1175 755+ | ·0576 000$^e$ | ·0282 181 | ·0138 240$^e$ | ·0067 723 |
| ·25 | ·2500 000$^e$ | ·1250 000$^e$ | ·0625 000$^e$ | ·0312 500$^e$ | ·0156 250$^e$ | ·0078 125$^e$ |
| ·26 | ·2600 000$^e$ | ·1325 745+ | ·0676 000$^e$ | ·0344 694 | ·0175 760$^e$ | ·0089 620 |
| ·27 | ·2700 000$^e$ | ·1402 961 | ·0729 000$^e$ | ·0378 800 | ·0196 830$^e$ | ·0102 276 |
| ·28 | ·2800 000$^e$ | ·1481 621 | ·0784 000$^e$ | ·0414 854 | ·0219 520$^e$ | ·0116 159 |
| ·29 | ·2900 000$^e$ | ·1561 698 | ·0841 000$^e$ | ·0452 892 | ·0243 890$^e$ | ·0131 339 |
| ·30 | ·3000 000$^e$ | ·1643 168 | ·0900 000$^e$ | ·0492 950+ | ·0270 000$^e$ | ·0147 885+ |
| ·31 | ·3100 000$^e$ | ·1726 007 | ·0961 000$^e$ | ·0535 062 | ·0297 910$^e$ | ·0165 869 |
| ·32 | ·3200 000$^e$ | ·1810 193 | ·1024 000$^e$ | ·0579 262 | ·0327 680$^e$ | ·0185 364 |
| ·33 | ·3300 000$^e$ | ·1895 706 | ·1089 000$^e$ | ·0625 583 | ·0359 370$^e$ | ·0206 442 |
| ·34 | ·3400 000$^e$ | ·1982 524 | ·1156 000$^e$ | ·0674 058 | ·0393 040$^e$ | ·0229 180 |
| ·35 | ·3500 000$^e$ | ·2070 628 | ·1225 000$^e$ | ·0724 720 | ·0428 750$^e$ | ·0253 652 |
| ·36 | ·3600 000$^e$ | ·2160 000$^e$ | ·1296 000$^e$ | ·0777 600$^e$ | ·0466 560$^e$ | ·0279 936$^e$ |
| ·37 | ·3700 000$^e$ | ·2250 622 | ·1369 000$^e$ | ·0832 730 | ·0506 530$^e$ | ·0308 110 |
| ·38 | ·3800 000$^e$ | ·2342 477 | ·1444 000$^e$ | ·0890 141 | ·0548 720$^e$ | ·0338 254 |
| ·39 | ·3900 000$^e$ | ·2435 549 | ·1521 000$^e$ | ·0949 864 | ·0593 190$^e$ | ·0370 447 |
| ·40 | ·4000 000$^e$ | ·2529 822 | ·1600 000$^e$ | ·1011 929 | ·0640 000$^e$ | ·0404 772 |
| ·41 | ·4100 000$^e$ | ·2625 281 | ·1681 000$^e$ | ·1076 365+ | ·0689 210$^e$ | ·0441 310 |
| ·42 | ·4200 000$^e$ | ·2721 911 | ·1764 000$^e$ | ·1143 203 | ·0740 880$^e$ | ·0480 145+ |
| ·43 | ·4300 000$^e$ | ·2819 699 | ·1849 000$^e$ | ·1212 470 | ·0795 070$^e$ | ·0521 362 |
| ·44 | ·4400 000$^e$ | ·2918 630 | ·1936 000$^e$ | ·1284 197 | ·0851 840$^e$ | ·0565 047 |
| ·45 | ·4500 000$^e$ | ·3018 692 | ·2025 000$^e$ | ·1358 411 | ·0911 250$^e$ | ·0611 285+ |
| ·46 | ·4600 000$^e$ | ·3119 872 | ·2116 000$^e$ | ·1435 141 | ·0973 360$^e$ | ·0660 165− |
| ·47 | ·4700 000$^e$ | ·3222 158 | ·2209 000$^e$ | ·1514 414 | ·1038 230$^e$ | ·0711 775− |
| ·48 | ·4800 000$^e$ | ·3325 538 | ·2304 000$^e$ | ·1596 258 | ·1105 920$^e$ | ·0766 204 |
| ·49 | ·4900 000$^e$ | ·3430 000$^e$ | ·2401 000$^e$ | ·1680 700 | ·1176 490$^e$ | ·0823 543$^e$ |
| ·50 | ·5000 000$^e$ | ·3535 534 | ·2500 000$^e$ | ·1767 767 | ·1250 000$^e$ | ·0883 883 |
| ·51 | ·5100 000$^e$ | ·3642 128 | ·2601 000$^e$ | ·1857 486 | ·1326 510$^e$ | ·0947 318 |
| ·52 | ·5200 000$^e$ | ·3749 773 | ·2704 000$^e$ | ·1949 882 | ·1406 080$^e$ | ·1013 939 |
| ·53 | ·5300 000$^e$ | ·3858 458 | ·2809 000$^e$ | ·2044 983 | ·1488 770$^e$ | ·1083 841 |
| ·54 | ·5400 000$^e$ | ·3968 173 | ·2916 000$^e$ | ·2142 814 | ·1574 640$^e$ | ·1157 119 |
| ·55 | ·5500 000$^e$ | ·4078 909 | ·3025 000$^e$ | ·2243 400 | ·1663 750$^e$ | ·1233 870 |
| ·56 | ·5600 000$^e$ | ·4190 656 | ·3136 000$^e$ | ·2346 768 | ·1756 160$^e$ | ·1314 190 |
| ·57 | ·5700 000$^e$ | ·4303 406 | ·3249 000$^e$ | ·2452 941 | ·1851 930$^e$ | ·1398 176 |
| ·58 | ·5800 000$^e$ | ·4417 148 | ·3364 000$^e$ | ·2561 946 | ·1951 120$^e$ | ·1485 929 |
| ·59 | ·5900 000$^e$ | ·4531 876 | ·3481 000$^e$ | ·2673 807 | ·2053 790$^e$ | ·1577 546 |
| ·60 | ·6000 000$^e$ | ·4647 580 | ·3600 000$^e$ | ·2788 548 | ·2160 000$^e$ | ·1673 129 |

$x$ = ·61 to 1·00 | $q = 1$ | $p$ = 1 to 3·5

| | $p = 1$ | $p = 1·5$ | $p = 2$ | $p = 2·5$ | $p = 3$ | $p = 3·5$ |
|---|---|---|---|---|---|---|
| $B(p, q)$ = | 1·0000 0000 | ·6666 6667 | ·5000 0000 | ·4000 0000 | ·3333 3333 | ·2857 1429 |
| $x$ | | | | | | |
| ·61 | ·6100 000$^e$ | ·4764 252 | ·3721 000$^e$ | ·2906 194 | ·2269 810$^e$ | ·1772 778 |
| ·62 | ·6200 000$^e$ | ·4881 885− | ·3844 000$^e$ | ·3026 769 | ·2383 280$^e$ | ·1876 597 |
| ·63 | ·6300 000$^e$ | ·5000 470 | ·3969 000$^e$ | ·3150 296 | ·2500 470$^e$ | ·1984 687 |
| ·64 | ·6400 000$^e$ | ·5120 000$^e$ | ·4096 000$^e$ | ·3276 800$^e$ | ·2621 440$^e$ | ·2097 152$^e$ |
| ·65 | ·6500 000$^e$ | ·5240 468 | ·4225 000$^e$ | ·3406 304 | ·2746 250$^e$ | ·2214 098 |
| ·66 | ·6600 000$^e$ | ·5361 865+ | ·4356 000$^e$ | ·3538 831 | ·2874 960$^e$ | ·2335 629 |
| ·67 | ·6700 000$^e$ | ·5484 186 | ·4489 000$^e$ | ·3674 405− | ·3007 630$^e$ | ·2461 852 |
| ·68 | ·6800 000$^e$ | ·5607 424 | ·4624 000$^e$ | ·3813 048 | ·3144 320$^e$ | ·2592 873 |
| ·69 | ·6900 000$^e$ | ·5731 570 | ·4761 000$^e$ | ·3954 784 | ·3285 090$^e$ | ·2728 801 |
| ·70 | ·7000 000$^e$ | ·5856 620 | ·4900 000$^e$ | ·4099 634 | ·3430 000$^e$ | ·2869 744 |
| ·71 | ·7100 000$^e$ | ·5982 566 | ·5041 000$^e$ | ·4247 622 | ·3579 110$^e$ | ·3015 812 |
| ·72 | ·7200 000$^e$ | ·6109 403 | ·5184 000$^e$ | ·4398 770 | ·3732 480$^e$ | ·3167 114 |
| ·73 | ·7300 000$^e$ | ·6237 123 | ·5329 000$^e$ | ·4553 100 | ·3890 170$^e$ | ·3323 763 |
| ·74 | ·7400 000$^e$ | ·6365 721 | ·5476 000$^e$ | ·4710 633 | ·4052 240$^e$ | ·3485 869 |
| ·75 | ·7500 000$^e$ | ·6495 191 | ·5625 000$^e$ | ·4871 393 | ·4218 750$^e$ | ·3653 545− |
| ·76 | ·7600 000$^e$ | ·6625 526 | ·5776 000$^e$ | ·5035 400 | ·4389 760$^e$ | ·3826 904 |
| ·77 | ·7700 000$^e$ | ·6756 723 | ·5929 000$^e$ | ·5202 676 | ·4565 330$^e$ | ·4006 061 |
| ·78 | ·7800 000$^e$ | ·6888 773 | ·6084 000$^e$ | ·5373 243 | ·4745 520$^e$ | ·4191 130 |
| ·79 | ·7900 000$^e$ | ·7021 674 | ·6241 000$^e$ | ·5547 122 | ·4930 390$^e$ | ·4382 226 |
| ·80 | ·8000 000$^e$ | ·7155 418 | ·6400 000$^e$ | ·5724 334 | ·5120 000$^e$ | ·4579 467 |
| ·81 | ·8100 000$^e$ | ·7290 000$^e$ | ·6561 000$^e$ | ·5904 900$^e$ | ·5314 410$^e$ | ·4782 969$^e$ |
| ·82 | ·8200 000$^e$ | ·7425 416 | ·6724 000$^e$ | ·6088 841 | ·5513 680$^e$ | ·4992 850− |
| ·83 | ·8300 000$^e$ | ·7561 660 | ·6889 000$^e$ | ·6276 178 | ·5717 870$^e$ | ·5209 227 |
| ·84 | ·8400 000$^e$ | ·7698 727 | ·7056 000$^e$ | ·6466 931 | ·5927 040$^e$ | ·5432 222 |
| ·85 | ·8500 000$^e$ | ·7836 613 | ·7225 000$^e$ | ·6661 121 | ·6141 250$^e$ | ·5661 953 |
| ·86 | ·8600 000$^e$ | ·7975 312 | ·7396 000$^e$ | ·6858 768 | ·6360 560$^e$ | ·5898 541 |
| ·87 | ·8700 000$^e$ | ·8114 820 | ·7569 000$^e$ | ·7059 893 | ·6585 030$^e$ | ·6142 107 |
| ·88 | ·8800 000$^e$ | ·8255 132 | ·7744 000$^e$ | ·7264 516 | ·6814 720$^e$ | ·6392 774 |
| ·89 | ·8900 000$^e$ | ·8396 243 | ·7921 000$^e$ | ·7472 656 | ·7049 690$^e$ | ·6650 664 |
| ·90 | ·9000 000$^e$ | ·8538 150− | ·8100 000$^e$ | ·7684 335− | ·7290 000$^e$ | ·6915 901 |
| ·91 | ·9100 000$^e$ | ·8680 847 | ·8281 000$^e$ | ·7899 571 | ·7535 710$^e$ | ·7188 609 |
| ·92 | ·9200 000$^e$ | ·8824 330 | ·8464 000$^e$ | ·8118 384 | ·7786 880$^e$ | ·7468 913 |
| ·93 | ·9300 000$^e$ | ·8968 595+ | ·8649 000$^e$ | ·8340 794 | ·8043 570$^e$ | ·7756 938 |
| ·94 | ·9400 000$^e$ | ·9113 638 | ·8836 000$^e$ | ·8566 820 | ·8305 840$^e$ | ·8052 811 |
| ·95 | ·9500 000$^e$ | ·9259 455− | ·9025 000$^e$ | ·8796 482 | ·8573 750$^e$ | ·8356 658 |
| ·96 | ·9600 000$^e$ | ·9406 041 | ·9216 000$^e$ | ·9029 799 | ·8847 360$^e$ | ·8668 607 |
| ·97 | ·9700 000$^e$ | ·9553 392 | ·9409 000$^e$ | ·9266 790 | ·9126 730$^e$ | ·8988 787 |
| ·98 | ·9800 000$^e$ | ·9701 505+ | ·9604 000$^e$ | ·9507 475− | ·9411 920$^e$ | ·9317 325+ |
| ·99 | ·9900 000$^e$ | ·9850 376 | ·9801 000$^e$ | ·9751 872 | ·9702 990$^e$ | ·9654 353 |
| 1·00 | 1·0000 000 | 1·0000 000 | 1·0000 000 | 1·0000 000 | 1·0000 000 | 1·0000 000 |

$x$ = ·02 to ·60 $q = 1$ $p$ = 4 to 6·5

| | $p=4$ | $p=4{\cdot}5$ | $p=5$ | $p=5{\cdot}5$ | $p=6$ | $p=6{\cdot}5$ |
|---|---|---|---|---|---|---|
| $B(p, q)$ = | ·2500 0000 | ·2222 2222 | ·2000 0000 | ·1818 1818 | ·1666 6667 | ·1538 4615+ |
| $x$ | | | | | | |
| ·02 | ·0000 002 | | | | | |
| ·03 | ·0000 008 | ·0000 001 | | | | |
| ·04 | ·0000 026 | ·0000 005+ | ·0000 001 | | | |
| ·05 | ·0000 062 | ·0000 014 | ·0000 003 | ·0000 001 | | |
| ·06 | ·0000 130 | ·0000 032 | ·0000 008 | ·0000 002 | | |
| ·07 | ·0000 240 | ·0000 064 | ·0000 017 | ·0000 004 | ·0000 001 | |
| ·08 | ·0000 410 | ·0000 116 | ·0000 033 | ·0000 009 | ·0000 003 | ·0000 001 |
| ·09 | ·0000 656 | ·0000 197 | ·0000 059 | ·0000 018 | ·0000 005+ | ·0000 002 |
| ·10 | ·0001 000$^e$ | ·0000 316 | ·0000 100$^e$ | ·0000 032 | ·0000 010$^e$ | ·0000 003 |
| ·11 | ·0001 464 | ·0000 486 | ·0000 161 | ·0000 053 | ·0000 018 | ·0000 006 |
| ·12 | ·0002 074 | ·0000 718 | ·0000 249 | ·0000 086 | ·0000 030 | ·0000 010 |
| ·13 | ·0002 856 | ·0001 030 | ·0000 371 | ·0000 134 | ·0000 048 | ·0000 017 |
| ·14 | ·0003 842 | ·0001 437 | ·0000 538 | ·0000 201 | ·0000 075+ | ·0000 028 |
| ·15 | ·0005 062 | ·0001 961 | ·0000 759 | ·0000 294 | ·0000 114 | ·0000 044 |
| ·16 | ·0006 554 | ·0002 621 | ·0001 049 | ·0000 419 | ·0000 168 | ·0000 067 |
| ·17 | ·0008 352 | ·0003 444 | ·0001 420 | ·0000 585+ | ·0000 241 | ·0000 100 |
| ·18 | ·0010 498 | ·0004 454 | ·0001 890 | ·0000 802 | ·0000 340 | ·0000 144 |
| ·19 | ·0013 032 | ·0005 681 | ·0002 476 | ·0001 079 | ·0000 470 | ·0000 205+ |
| ·20 | ·0016 000$^e$ | ·0007 155+ | ·0003 200$^e$ | ·0001 431 | ·0000 640$^e$ | ·0000 286 |
| ·21 | ·0019 448 | ·0008 912 | ·0004 084 | ·0001 872 | ·0000 858 | ·0000 393 |
| ·22 | ·0023 426 | ·0010 988 | ·0005 154 | ·0002 417 | ·0001 134 | ·0000 532 |
| ·23 | ·0027 984 | ·0013 421 | ·0006 436 | ·0003 087 | ·0001 480 | ·0000 710 |
| ·24 | ·0033 178 | ·0016 254 | ·0007 963 | ·0003 901 | ·0001 911 | ·0000 936 |
| ·25 | ·0039 062 | ·0019 531 | ·0009 766 | ·0004 883 | ·0002 441 | ·0001 221 |
| ·26 | ·0045 698 | ·0023 301 | ·0011 881 | ·0006 058 | ·0003 089 | ·0001 575+ |
| ·27 | ·0053 144 | ·0027 614 | ·0014 349 | ·0007 456 | ·0003 874 | ·0002 013 |
| ·28 | ·0061 466 | ·0032 525− | ·0017 210 | ·0009 107 | ·0004 819 | ·0002 550− |
| ·29 | ·0070 728 | ·0038 088 | ·0020 511 | ·0011 046 | ·0005 948 | ·0003 203 |
| ·30 | ·0081 000$^e$ | ·0044 366 | ·0024 300$^e$ | ·0013 310 | ·0007 290$^e$ | ·0003 993 |
| ·31 | ·0092 352 | ·0051 419 | ·0028 629 | ·0015 940 | ·0008 875+ | ·0004 941 |
| ·32 | ·0104 858 | ·0059 316 | ·0033 554 | ·0018 981 | ·0010 737 | ·0006 074 |
| ·33 | ·0118 592 | ·0068 126 | ·0039 135+ | ·0022 482 | ·0012 915− | ·0007 419 |
| ·34 | ·0133 634 | ·0077 921 | ·0045 435+ | ·0026 493 | ·0015 448 | ·0009 008 |
| ·35 | ·0150 063 | ·0088 778 | ·0052 522 | ·0031 072 | ·0018 383 | ·0010 875+ |
| ·36 | ·0167 962 | ·0100 777 | ·0060 466 | ·0036 280 | ·0021 768 | ·0013 061 |
| ·37 | ·0187 416 | ·0114 001 | ·0069 344 | ·0042 180 | ·0025 657 | ·0015 607 |
| ·38 | ·0208 514 | ·0128 536 | ·0079 235+ | ·0048 844 | ·0030 109 | ·0018 561 |
| ·39 | ·0231 344 | ·0144 474 | ·0090 224 | ·0056 345− | ·0035 187 | ·0021 975− |
| ·40 | ·0256 000$^e$ | ·0161 909 | ·0102 400$^e$ | ·0064 763 | ·0040 960$^e$ | ·0025 905+ |
| ·41 | ·0282 576 | ·0180 937 | ·0115 856 | ·0074 184 | ·0047 501 | ·0030 416 |
| ·42 | ·0311 170 | ·0201 661 | ·0130 691 | ·0084 698 | ·0054 890 | ·0035 573 |
| ·43 | ·0341 880 | ·0224 186 | ·0147 008 | ·0096 400 | ·0063 214 | ·0041 452 |
| ·44 | ·0374 810 | ·0248 621 | ·0164 916 | ·0109 393 | ·0072 563 | ·0048 133 |
| ·45 | ·0410 062 | ·0275 078 | ·0184 528 | ·0123 785+ | ·0083 038 | ·0055 703 |
| ·46 | ·0447 746 | ·0303 676 | ·0205 963 | ·0139 691 | ·0094 743 | ·0064 258 |
| ·47 | ·0487 968 | ·0334 534 | ·0229 345+ | ·0157 231 | ·0107 792 | ·0073 899 |
| ·48 | ·0530 842 | ·0367 778 | ·0254 804 | ·0176 533 | ·0122 306 | ·0084 736 |
| ·49 | ·0576 480 | ·0403 536 | ·0282 475+ | ·0197 733 | ·0138 413 | ·0096 889 |
| ·50 | ·0625 000$^e$ | ·0441 942 | ·0312 500$^e$ | ·0220 971 | ·0156 250$^e$ | ·0110 485+ |
| ·51 | ·0676 520 | ·0483 132 | ·0345 025+ | ·0246 397 | ·0175 963 | ·0125 663 |
| ·52 | ·0731 162 | ·0527 248 | ·0380 204 | ·0274 169 | ·0197 706 | ·0142 568 |
| ·53 | ·0789 048 | ·0574 436 | ·0418 195+ | ·0304 451 | ·0221 644 | ·0161 359 |
| ·54 | ·0850 306 | ·0624 844 | ·0459 165+ | ·0337 416 | ·0247 949 | ·0182 205− |
| ·55 | ·0915 063 | ·0678 629 | ·0503 284 | ·0373 246 | ·0276 806 | ·0205 285+ |
| ·56 | ·0983 450− | ·0735 946 | ·0550 732 | ·0412 130 | ·0308 410 | ·0230 793 |
| ·57 | ·1055 600 | ·0796 961 | ·0601 692 | ·0454 268 | ·0342 964 | ·0258 932 |
| ·58 | ·1131 650− | ·0861 839 | ·0656 357 | ·0499 866 | ·0380 687 | ·0289 923 |
| ·59 | ·1211 736 | ·0930 752 | ·0714 924 | ·0549 144 | ·0421 805+ | ·0323 995− |
| ·60 | ·1296 000$^e$ | ·1003 877 | ·0777 600$^e$ | ·0602 326 | ·0466 560$^e$ | ·0361 396 |

$x$ = ·61 to 1·00 $q = 1$ $p$ = 4 to 6·5

| | $p = 4$ | $p = 4·5$ | $p = 5$ | $p = 5·5$ | $p = 6$ | $p = 6·5$ |
|---|---|---|---|---|---|---|
| $B(p, q)$ = | ·2500 0000 | ·2222 2222 | ·2000 0000 | ·1818 1818 | ·1666 6667 | ·1538 4615$^+$ |
| $x$ | | | | | | |
| ·61 | ·1384 584 | ·1081 395$^-$ | ·0844 596 | ·0659 651 | ·0515 204 | ·0402 387 |
| ·62 | ·1477 634 | ·1163 490 | ·0916 133 | ·0721 364 | ·0568 002 | ·0447 246 |
| ·63 | ·1575 296 | ·1250 353 | ·0992 437 | ·0787 722 | ·0625 235$^+$ | ·0496 265$^-$ |
| ·64 | ·1677 722 | ·1342 177 | ·1073 742 | ·0858 993 | ·0687 195$^-$ | ·0549 756 |
| ·65 | ·1785 062 | ·1439 163 | ·1160 291 | ·0935 456 | ·0754 189 | ·0608 047 |
| ·66 | ·1897 474 | ·1541 515$^-$ | ·1252 333 | ·1017 400 | ·0826 540 | ·0671 484 |
| ·67 | ·2015 112 | ·1649 440 | ·1350 125$^+$ | ·1105 125$^+$ | ·0904 584 | ·0740 434 |
| ·68 | ·2138 138 | ·1763 153 | ·1453 934 | ·1198 944 | ·0988 675$^-$ | ·0815 282 |
| ·69 | ·2266 712 | ·1882 872 | ·1564 031 | ·1299 182 | ·1079 182 | ·0896 436 |
| ·70 | ·2401 000$^e$ | ·2008 821 | ·1680 700$^e$ | ·1406 175$^-$ | ·1176 490$^e$ | ·0984 322 |
| ·71 | ·2541 168 | ·2141 226 | ·1804 229 | ·1520 271 | ·1281 003 | ·1079 392 |
| ·72 | ·2687 386 | ·2280 322 | ·1934 918 | ·1641 832 | ·1393 141 | ·1182 119 |
| ·73 | ·2839 824 | ·2426 347 | ·2073 072 | ·1771 233 | ·1513 342 | ·1293 000 |
| ·74 | ·2998 658 | ·2579 543 | ·2219 007 | ·1908 862 | ·1642 065$^-$ | ·1412 558 |
| ·75 | ·3164 062 | ·2740 159 | ·2373 047 | ·2055 119 | ·1779 785$^+$ | ·1541 339 |
| ·76 | ·3336 218 | ·2908 447 | ·2535 525$^+$ | ·2210 420 | ·1926 999 | ·1679 919 |
| ·77 | ·3515 304 | ·3084 667 | ·2706 784 | ·2375 193 | ·2084 224 | ·1828 899 |
| ·78 | ·3701 506 | ·3269 081 | ·2887 174 | ·2549 883 | ·2251 996 | ·1988 909 |
| ·79 | ·3895 008 | ·3461 959 | ·3077 056 | ·2734 948 | ·2430 875$^-$ | ·2160 609 |
| ·80 | ·4096 000$^e$ | ·3663 574 | ·3276 800$^e$ | ·2930 859 | ·2621 440$^e$ | ·2344 687 |
| ·81 | ·4304 672 | ·3874 205$^-$ | ·3486 784 | ·3138 106 | ·2824 295$^+$ | ·2541 866 |
| ·82 | ·4521 218 | ·4094 137 | ·3707 398 | ·3357 192 | ·3040 067 | ·2752 897 |
| ·83 | ·4745 832 | ·4323 659 | ·3939 041 | ·3588 637 | ·3269 404 | ·2978 569 |
| ·84 | ·4978 714 | ·4563 066 | ·4182 119 | ·3832 976 | ·3512 980 | ·3219 700 |
| ·85 | ·5220 062 | ·4812 660 | ·4437 053 | ·4090 761 | ·3771 495$^+$ | ·3477 147 |
| ·86 | ·5470 082 | ·5072 745$^-$ | ·4704 270 | ·4362 561 | ·4045 672 | ·3751 802 |
| ·87 | ·5728 976 | ·5343 633 | ·4984 209 | ·4648 961 | ·4336 262 | ·4044 596 |
| ·88 | ·5996 954 | ·5625 641 | ·5277 319 | ·4950 564 | ·4644 041 | ·4356 496 |
| ·89 | ·6274 224 | ·5919 091 | ·5584 059 | ·5267 991 | ·4969 813 | ·4688 512 |
| ·90 | ·6561 000$^e$ | ·6224 311 | ·5904 900$^e$ | ·5601 880 | ·5314 410$^e$ | ·5041 692 |
| ·91 | ·6857 496 | ·6541 634 | ·6240 321 | ·5952 887 | ·5678 693 | ·5417 127 |
| ·92 | ·7163 930 | ·6871 400 | ·6590 815$^+$ | ·6321 688 | ·6063 550$^+$ | ·5815 953 |
| ·93 | ·7480 520 | ·7213 952 | ·6956 884 | ·6708 976 | ·6469 902 | ·6239 347 |
| ·94 | ·7807 490 | ·7569 642 | ·7339 040 | ·7115 463 | ·6898 698 | ·6688 536 |
| ·95 | ·8145 062 | ·7938 825$^-$ | ·7737 809 | ·7541 884 | ·7350 919 | ·7164 789 |
| ·96 | ·8493 466 | ·8321 863 | ·8153 727 | ·7988 988 | ·7827 578 | ·7669 429 |
| ·97 | ·8852 928 | ·8719 123 | ·8587 340 | ·8457 549 | ·8329 720 | ·8203 823 |
| ·98 | ·9223 682 | ·9130 979 | ·9039 208 | ·8948 359 | ·8858 424 | ·8769 392 |
| ·99 | ·9605 960 | ·9557 810 | ·9509 900 | ·9462 232 | ·9414 801 | ·9367 609 |
| 1·00 | 1·0000 000 | 1·0000 000 | 1·0000 000 | 1·0000 000 | 1·0000 000 | 1·0000 000 |

$x$ = ·10 to ·70 $q = 1$ $p$ = 7 to 9·5

| $x$ | $p=7$ | $p=7{\cdot}5$ | $p=8$ | $p=8{\cdot}5$ | $p=9$ | $p=9{\cdot}5$ |
|---|---|---|---|---|---|---|
| $B(p,q)$ = | ·1428 5714 | ·1333 3333 | ·1250 0000 | ·1176 4706 | ·1111 1111 | ·1052 6316 |
| ·10 | ·0000 001[e] | | | | | |
| ·11 | ·0000 002 | ·0000 001 | | | | |
| ·12 | ·0000 004 | ·0000 001 | | | | |
| ·13 | ·0000 006 | ·0000 002 | ·0000 001 | | | |
| ·14 | ·0000 011 | ·0000 004 | ·0000 001 | ·0000 001 | | |
| ·15 | ·0000 017 | ·0000 007 | ·0000 003 | ·0000 001 | | |
| ·16 | ·0000 027 | ·0000 011 | ·0000 004 | ·0000 002 | ·0000 001 | |
| ·17 | ·0000 041 | ·0000 017 | ·0000 007 | ·0000 003 | ·0000 001 | |
| ·18 | ·0000 061 | ·0000 026 | ·0000 011 | ·0000 005− | ·0000 002 | ·0000 001 |
| ·19 | ·0000 089 | ·0000 039 | ·0000 017 | ·0000 007 | ·0000 003 | ·0000 001 |
| ·20 | ·0000 128[e] | ·0000 057 | ·0000 026 | ·0000 011 | ·0000 005+ | ·0000 002 |
| ·21 | ·0000 180 | ·0000 083 | ·0000 038 | ·0000 017 | ·0000 008 | ·0000 004 |
| ·22 | ·0000 249 | ·0000 117 | ·0000 055− | ·0000 026 | ·0000 012 | ·0000 006 |
| ·23 | ·0000 340 | ·0000 163 | ·0000 078 | ·0000 038 | ·0000 018 | ·0000 009 |
| ·24 | ·0000 459 | ·0000 225− | ·0000 110 | ·0000 054 | ·0000 026 | ·0000 013 |
| ·25 | ·0000 610 | ·0000 305+ | ·0000 153 | ·0000 076 | ·0000 038 | ·0000 019 |
| ·26 | ·0000 803 | ·0000 410 | ·0000 209 | ·0000 106 | ·0000 054 | ·0000 028 |
| ·27 | ·0001 046 | ·0000 544 | ·0000 282 | ·0000 147 | ·0000 076 | ·0000 040 |
| ·28 | ·0001 349 | ·0000 714 | ·0000 378 | ·0000 200 | ·0000 106 | ·0000 056 |
| ·29 | ·0001 725− | ·0000 929 | ·0000 500+ | ·0000 269 | ·0000 145+ | ·0000 078 |
| ·30 | ·0002 187[e] | ·0001 198 | ·0000 656 | ·0000 359 | ·0000 197 | ·0000 108 |
| ·31 | ·0002 751 | ·0001 532 | ·0000 853 | ·0000 475− | ·0000 264 | ·0000 147 |
| ·32 | ·0003 436 | ·0001 944 | ·0001 100 | ·0000 622 | ·0000 352 | ·0000 199 |
| ·33 | ·0004 262 | ·0002 448 | ·0001 406 | ·0000 808 | ·0000 464 | ·0000 267 |
| ·34 | ·0005 252 | ·0003 063 | ·0001 786 | ·0001 041 | ·0000 607 | ·0000 354 |
| ·35 | ·0006 434 | ·0003 806 | ·0002 252 | ·0001 332 | ·0000 788 | ·0000 466 |
| ·36 | ·0007 836 | ·0004 702 | ·0002 821 | ·0001 693 | ·0001 016 | ·0000 609 |
| ·37 | ·0009 493 | ·0005 774 | ·0003 512 | ·0002 137 | ·0001 300 | ·0000 791 |
| ·38 | ·0011 442 | ·0007 053 | ·0004 348 | ·0002 680 | ·0001 652 | ·0001 018 |
| ·39 | ·0013 723 | ·0008 570 | ·0005 352 | ·0003 342 | ·0002 087 | ·0001 304 |
| ·40 | ·0016 384[e] | ·0010 362 | ·0006 554 | ·0004 145− | ·0002 621 | ·0001 658 |
| ·41 | ·0019 475+ | ·0012 470 | ·0007 985− | ·0005 113 | ·0003 274 | ·0002 096 |
| ·42 | ·0023 054 | ·0014 941 | ·0009 683 | ·0006 275+ | ·0004 067 | ·0002 636 |
| ·43 | ·0027 182 | ·0017 824 | ·0011 688 | ·0007 664 | ·0005 026 | ·0003 296 |
| ·44 | ·0031 928 | ·0021 178 | ·0014 048 | ·0009 319 | ·0006 181 | ·0004 100 |
| ·45 | ·0037 367 | ·0025 067 | ·0016 815+ | ·0011 280 | ·0007 567 | ·0005 076 |
| ·46 | ·0043 582 | ·0029 559 | ·0020 048 | ·0013 597 | ·0009 222 | ·0006 255− |
| ·47 | ·0050 662 | ·0034 732 | ·0023 811 | ·0016 324 | ·0011 191 | ·0007 672 |
| ·48 | ·0058 707 | ·0040 673 | ·0028 179 | ·0019 523 | ·0013 526 | ·0009 371 |
| ·49 | ·0067 822 | ·0047 476 | ·0033 233 | ·0023 263 | ·0016 284 | ·0011 399 |
| ·50 | ·0078 125[e] | ·0055 243 | ·0039 062 | ·0027 621 | ·0019 531 | ·0013 811 |
| ·51 | ·0089 741 | ·0064 088 | ·0045 768 | ·0032 685− | ·0023 342 | ·0016 669 |
| ·52 | ·0102 807 | ·0074 135+ | ·0053 460 | ·0038 550+ | ·0027 799 | ·0020 046 |
| ·53 | ·0117 471 | ·0085 520 | ·0062 260 | ·0045 326 | ·0032 998 | ·0024 023 |
| ·54 | ·0133 893 | ·0098 391 | ·0072 302 | ·0053 131 | ·0039 043 | ·0028 691 |
| ·55 | ·0152 244 | ·0112 907 | ·0083 734 | ·0062 099 | ·0046 054 | ·0034 154 |
| ·56 | ·0172 709 | ·0129 244 | ·0096 717 | ·0072 377 | ·0054 162 | ·0040 531 |
| ·57 | ·0195 490 | ·0147 592 | ·0111 429 | ·0084 127 | ·0063 515− | ·0047 952 |
| ·58 | ·0220 798 | ·0168 155+ | ·0128 063 | ·0097 530 | ·0074 277 | ·0056 567 |
| ·59 | ·0248 865+ | ·0191 157 | ·0146 830 | ·0112 783 | ·0086 630 | ·0066 542 |
| ·60 | ·0279 936[e] | ·0216 837 | ·0167 962 | ·0130 102 | ·0100 777 | ·0078 061 |
| ·61 | ·0314 274 | ·0245 456 | ·0191 707 | ·0149 728 | ·0116 941 | ·0091 334 |
| ·62 | ·0352 161 | ·0277 292 | ·0218 340 | ·0171 921 | ·0135 371 | ·0106 591 |
| ·63 | ·0393 898 | ·0312 647 | ·0248 156 | ·0196 968 | ·0156 338 | ·0124 090 |
| ·64 | ·0439 805− | ·0351 844 | ·0281 475− | ·0225 180 | ·0180 144 | ·0144 115+ |
| ·65 | ·0490 223 | ·0395 230 | ·0318 645− | ·0256 900 | ·0207 119 | ·0166 985− |
| ·66 | ·0545 516 | ·0443 179 | ·0360 041 | ·0292 498 | ·0237 627 | ·0193 049 |
| ·67 | ·0606 071 | ·0496 091 | ·0406 068 | ·0332 381 | ·0272 065+ | ·0222 695+ |
| ·68 | ·0672 299 | ·0554 392 | ·0457 163 | ·0376 986 | ·0310 871 | ·0256 351 |
| ·69 | ·0744 635+ | ·0618 541 | ·0513 798 | ·0426 793 | ·0354 521 | ·0294 487 |
| ·70 | ·0823 543[e] | ·0689 026 | ·0576 480 | ·0482 318 | ·0403 536 | ·0337 622 |

$x$ = ·71 to 1·00 $q$ = 1 $p$ = 7 to 9·5

| | $p = 7$ | $p = 7·5$ | $p = 8$ | $p = 8·5$ | $p = 9$ | $p = 9·5$ |
|---|---|---|---|---|---|---|
| $B(p, q)$ = | ·1428 5714 | ·1333 3333 | ·1250 0000 | ·1176 4706 | ·1111 1111 | ·1052 6316 |
| $x$ | | | | | | |
| ·71 | ·0909 512 | ·0766 368 | ·0645 754 | ·0544 122 | ·0458 485$^{+}$ | ·0386 326 |
| ·72 | ·1003 061 | ·0851 126 | ·0722 204 | ·0612 811 | ·0519 987 | ·0441 224 |
| ·73 | ·1104 740 | ·0943 890 | ·0806 460 | ·0689 040 | ·0588 716 | ·0502 999 |
| ·74 | ·1215 128 | ·1045 293 | ·0899 195$^{-}$ | ·0773 517 | ·0665 404 | ·0572 402 |
| ·75 | ·1334 839 | ·1156 004 | ·1001 129 | ·0867 003 | ·0750 847 | ·0650 252 |
| ·76 | ·1464 519 | ·1276 738 | ·1113 035$^{-}$ | ·0970 321 | ·0845 906 | ·0737 444 |
| ·77 | ·1604 852 | ·1408 252 | ·1235 736 | ·1084 354 | ·0951 517 | ·0834 953 |
| ·78 | ·1756 557 | ·1551 349 | ·1370 114 | ·1210 052 | ·1068 689 | ·0943 841 |
| ·79 | ·1920 391 | ·1706 881 | ·1517 109 | ·1348 436 | ·1198 516 | ·1065 264 |
| ·80 | ·2097 152$^{e}$ | ·1875 750$^{-}$ | ·1677 722 | ·1500 600 | ·1342 177 | ·1200 480 |
| ·81 | ·2287 679 | ·2058 911 | ·1853 020 | ·1667 718 | ·1500 946 | ·1350 852 |
| ·82 | ·2492 855$^{-}$ | ·2257 376 | ·2044 141 | ·1851 048 | ·1676 196 | ·1517 860 |
| ·83 | ·2713 605$^{+}$ | ·2472 212 | ·2252 292 | ·2051 936 | ·1869 403 | ·1703 107 |
| ·84 | ·2950 903 | ·2704 548 | ·2478 759 | ·2271 820 | ·2082 157 | ·1908 329 |
| ·85 | ·3205 771 | ·2955 575$^{-}$ | ·2724 905$^{+}$ | ·2512 239 | ·2316 169 | ·2135 403 |
| ·86 | ·3479 278 | ·3226 550$^{-}$ | ·2992 179 | ·2774 833 | ·2573 274 | ·2386 356 |
| ·87 | ·3772 548 | ·3518 798 | ·3282 117 | ·3061 355$^{-}$ | ·2855 442 | ·2663 379 |
| ·88 | ·4086 756 | ·3833 717 | ·3596 345$^{+}$ | ·3373 671 | ·3164 784 | ·2968 830 |
| ·89 | ·4423 133 | ·4172 776 | ·3936 589 | ·3713 770 | ·3503 564 | ·3305 256 |
| ·90 | ·4782 969$^{e}$ | ·4537 523 | ·4304 672 | ·4083 771 | ·3874 205$^{-}$ | ·3675 393 |
| ·91 | ·5167 610 | ·4929 586 | ·4702 525$^{+}$ | ·4485 923 | ·4279 298 | ·4082 190 |
| ·92 | ·5578 466 | ·5350 677 | ·5132 189 | ·4922 623 | ·4721 614 | ·4528 813 |
| ·93 | ·6017 009 | ·5802 593 | ·5595 818 | ·5396 412 | ·5204 111 | ·5018 663 |
| ·94 | ·6484 776 | ·6287 224 | ·6095 689 | ·5909 990 | ·5729 948 | ·5555 391 |
| ·95 | ·6983 373 | ·6806 550$^{+}$ | ·6634 204 | ·6466 223 | ·6302 494 | ·6142 911 |
| ·96 | ·7514 475$^{-}$ | ·7362 652 | ·7213 896 | ·7068 145$^{+}$ | ·6925 340 | ·6785 420 |
| ·97 | ·8079 828 | ·7957 708 | ·7837 434 | ·7718 977 | ·7602 311 | ·7487 408 |
| ·98 | ·8681 255$^{+}$ | ·8594 004 | ·8507 630 | ·8422 124 | ·8337 478 | ·8253 682 |
| ·99 | ·9320 653 | ·9273 933 | ·9227 447 | ·9181 194 | ·9135 172 | ·9089 382 |
| 1·00 | 1·0000 000 | 1·0000 000 | 1·0000 000 | 1·0000 000 | 1·0000 000 | 1·0000 000 |

$x$ = ·19 to ·80 $q = 1$ $p$ = 10 to 14

| | $p = 10$ | $p = 10{\cdot}5$ | $p = 11$ | $p = 12$ | $p = 13$ | $p = 14$ |
|---|---|---|---|---|---|---|
| $B(p, q)$ = | ·1000 0000 | ·9523 8095 $\overset{+}{\times}\frac{1}{10}$ | ·9090 9091 $\times\frac{1}{10}$ | ·8363 6364 $\times\frac{1}{10}$ | ·7692 3077 $\times\frac{1}{10}$ | ·7142 8571 $\times\frac{1}{10}$ |
| $x$ | | | | | | |
| ·19 | ·0000 001 | | | | | |
| ·20 | ·0000 001 | | | | | |
| ·21 | ·0000 002 | ·0000 001 | | | | |
| ·22 | ·0000 003 | ·0000 001 | ·0000 001 | | | |
| ·23 | ·0000 004 | ·0000 002 | ·0000 001 | | | |
| ·24 | ·0000 006 | ·0000 003 | ·0000 002 | | | |
| ·25 | ·0000 010 | ·0000 005− | ·0000 002 | ·0000 001 | | |
| ·26 | ·0000 014 | ·0000 007 | ·0000 004 | ·0000 001 | | |
| ·27 | ·0000 021 | ·0000 011 | ·0000 006 | ·0000 002 | | |
| ·28 | ·0000 030 | ·0000 016 | ·0000 008 | ·0000 002 | ·0000 001 | |
| ·29 | ·0000 042 | ·0000 023 | ·0000 012 | ·0000 004 | ·0000 001 | |
| ·30 | ·0000 059 | ·0000 032 | ·0000 018 | ·0000 005+ | ·0000 002 | |
| ·31 | ·0000 082 | ·0000 046 | ·0000 025+ | ·0000 008 | ·0000 002 | ·0000 001 |
| ·32 | ·0000 113 | ·0000 064 | ·0000 036 | ·0000 012 | ·0000 004 | ·0000 001 |
| ·33 | ·0000 153 | ·0000 088 | ·0000 051 | ·0000 017 | ·0000 006 | ·0000 002 |
| ·34 | ·0000 206 | ·0000 120 | ·0000 070 | ·0000 024 | ·0000 008 | ·0000 003 |
| ·35 | ·0000 276 | ·0000 163 | ·0000 097 | ·0000 034 | ·0000 012 | ·0000 004 |
| ·36 | ·0000 366 | ·0000 219 | ·0000 132 | ·0000 047 | ·0000 017 | ·0000 006 |
| ·37 | ·0000 481 | ·0000 292 | ·0000 178 | ·0000 066 | ·0000 024 | ·0000 009 |
| ·38 | ·0000 628 | ·0000 387 | ·0000 239 | ·0000 091 | ·0000 034 | ·0000 013 |
| ·39 | ·0000 814 | ·0000 508 | ·0000 317 | ·0000 124 | ·0000 048 | ·0000 019 |
| ·40 | ·0001 049 | ·0000 663 | ·0000 419 | ·0000 168 | ·0000 067 | ·0000 027 |
| ·41 | ·0001 342 | ·0000 859 | ·0000 550+ | ·0000 226 | ·0000 093 | ·0000 038 |
| ·42 | ·0001 708 | ·0001 107 | ·0000 717 | ·0000 301 | ·0000 127 | ·0000 053 |
| ·43 | ·0002 161 | ·0001 417 | ·0000 929 | ·0000 400 | ·0000 172 | ·0000 074 |
| ·44 | ·0002 720 | ·0001 804 | ·0001 197 | ·0000 527 | ·0000 232 | ·0000 102 |
| ·45 | ·0003 405+ | ·0002 284 | ·0001 532 | ·0000 690 | ·0000 310 | ·0000 140 |
| ·46 | ·0004 242 | ·0002 877 | ·0001 951 | ·0000 898 | ·0000 413 | ·0000 190 |
| ·47 | ·0005 260 | ·0003 606 | ·0002 472 | ·0001 162 | ·0000 546 | ·0000 257 |
| ·48 | ·0006 493 | ·0004 498 | ·0003 116 | ·0001 496 | ·0000 718 | ·0000 345− |
| ·49 | ·0007 979 | ·0005 585+ | ·0003 910 | ·0001 916 | ·0000 939 | ·0000 460 |
| ·50 | ·0009 766 | ·0006 905+ | ·0004 883 | ·0002 441 | ·0001 221 | ·0000 610 |
| ·51 | ·0011 904 | ·0008 501 | ·0006 071 | ·0003 096 | ·0001 579 | ·0000 805+ |
| ·52 | ·0014 456 | ·0010 424 | ·0007 517 | ·0003 909 | ·0002 033 | ·0001 057 |
| ·53 | ·0017 489 | ·0012 732 | ·0009 269 | ·0004 913 | ·0002 604 | ·0001 380 |
| ·54 | ·0021 083 | ·0015 493 | ·0011 385− | ·0006 148 | ·0003 320 | ·0001 793 |
| ·55 | ·0025 330 | ·0018 785− | ·0013 931 | ·0007 662 | ·0004 214 | ·0002 318 |
| ·56 | ·0030 331 | ·0022 697 | ·0016 985+ | ·0009 512 | ·0005 327 | ·0002 983 |
| ·57 | ·0036 203 | ·0027 333 | ·0020 636 | ·0011 762 | ·0006 705− | ·0003 822 |
| ·58 | ·0043 080 | ·0032 809 | ·0024 987 | ·0014 492 | ·0008 406 | ·0004 875+ |
| ·59 | ·0051 112 | ·0039 260 | ·0030 156 | ·0017 792 | ·0010 497 | ·0006 193 |
| ·60 | ·0060 466 | ·0046 837 | ·0036 280 | ·0021 768 | ·0013 061 | ·0007 836 |
| ·61 | ·0071 334 | ·0055 714 | ·0043 514 | ·0026 543 | ·0016 192 | ·0009 877 |
| ·62 | ·0083 930 | ·0066 086 | ·0052 037 | ·0032 263 | ·0020 003 | ·0012 402 |
| ·63 | ·0098 493 | ·0078 176 | ·0062 051 | ·0039 092 | ·0024 628 | ·0015 516 |
| ·64 | ·0115 292 | ·0092 234 | ·0073 787 | ·0047 224 | ·0030 223 | ·0019 343 |
| ·65 | ·0134 627 | ·0108 540 | ·0087 508 | ·0056 880 | ·0036 972 | ·0024 032 |
| ·66 | ·0156 834 | ·0127 412 | ·0103 510 | ·0068 317 | ·0045 089 | ·0029 759 |
| ·67 | ·0182 284 | ·0149 206 | ·0122 130 | ·0081 827 | ·0054 824 | ·0036 732 |
| ·68 | ·0211 392 | ·0174 319 | ·0143 747 | ·0097 748 | ·0066 468 | ·0045 199 |
| ·69 | ·0244 619 | ·0203 196 | ·0168 787 | ·0116 463 | ·0080 360 | ·0055 448 |
| ·70 | ·0282 475+ | ·0236 336 | ·0197 733 | ·0138 413 | ·0096 889 | ·0067 822 |
| ·71 | ·0325 524 | ·0274 292 | ·0231 122 | ·0164 097 | ·0116 509 | ·0082 721 |
| ·72 | ·0374 391 | ·0317 681 | ·0269 561 | ·0194 084 | ·0139 741 | ·0100 613 |
| ·73 | ·0429 763 | ·0367 189 | ·0313 727 | ·0229 020 | ·0167 185− | ·0122 045+ |
| ·74 | ·0492 399 | ·0423 578 | ·0364 375+ | ·0269 638 | ·0199 532 | ·0147 654 |
| ·75 | ·0563 135+ | ·0487 689 | ·0422 351 | ·0316 764 | ·0237 573 | ·0178 179 |
| ·76 | ·0642 889 | ·0560 458 | ·0488 596 | ·0371 333 | ·0282 213 | ·0214 482 |
| ·77 | ·0732 668 | ·0642 914 | ·0564 154 | ·0434 399 | ·0334 487 | ·0257 555+ |
| ·78 | ·0833 578 | ·0736 196 | ·0650 191 | ·0507 149 | ·0395 576 | ·0308 549 |
| ·79 | ·0946 828 | ·0841 559 | ·0747 994 | ·0590 915+ | ·0466 823 | ·0368 790 |
| ·80 | ·1073 742 | ·0960 384 | ·0858 993 | ·0687 195− | ·0549 756 | ·0439 805− |

$x$ = ·81 to 1·00 $q = 1$ $p$ = 10 to 14

| | $p = 10$ | $p = 10·5$ | $p = 11$ | $p = 12$ | $p = 13$ | $p = 14$ |
|---|---|---|---|---|---|---|
| $B(p, q) =$ | ·1000 0000 | ·9523 8095 $\times \frac{1}{10}$ | ·9090 9091 $\times \frac{1}{10}$ | ·8363 6364 $\times \frac{1}{10}$ | ·7692 3077 $\times \frac{1}{10}$ | ·7142 8571 $\times \frac{1}{10}$ |
| $x$ | | | | | | |
| ·81 | ·1215 767 | ·1094 190 | ·0984 771 | ·0797 664 | ·0646 108 | ·0523 348 |
| ·82 | ·1374 480 | ·1244 645⁻ | ·1127 074 | ·0924 201 | ·0757 844 | ·0621 432 |
| ·83 | ·1551 604 | ·1413 579 | ·1287 831 | ·1068 900 | ·0887 187 | ·0736 365⁺ |
| ·84 | ·1749 012 | ·1602 996 | ·1469 170 | ·1234 103 | ·1036 647 | ·0870 783 |
| ·85 | ·1968 744 | ·1815 092 | ·1673 432 | ·1422 418 | ·1209 055⁻ | ·1027 697 |
| ·86 | ·2213 016 | ·2052 266 | ·1903 194 | ·1636 746 | ·1407 602 | ·1210 538 |
| ·87 | ·2484 234 | ·2317 139 | ·2161 284 | ·1880 317 | ·1635 876 | ·1423 212 |
| ·88 | ·2785 010 | ·2612 571 | ·2450 809 | ·2156 712 | ·1897 906 | ·1670 157 |
| ·89 | ·3118 172 | ·2941 678 | ·2775 173 | ·2469 904 | ·2198 215⁻ | ·1956 411 |
| ·90 | ·3486 784 | ·3307 854 | ·3138 106 | ·2824 295⁺ | ·2541 866 | ·2287 679 |
| ·91 | ·3894 161 | ·3714 793 | ·3543 687 | ·3224 755⁻ | ·2934 527 | ·2670 420 |
| ·92 | ·4343 885⁻ | ·4166 508 | ·3996 374 | ·3676 664 | ·3382 531 | ·3111 928 |
| ·93 | ·4839 823 | ·4667 356 | ·4501 035⁺ | ·4185 963 | ·3892 946 | ·3620 439 |
| ·94 | ·5386 151 | ·5222 067 | ·5062 982 | ·4759 203 | ·4473 651 | ·4205 232 |
| ·95 | ·5987 369 | ·5835 766 | ·5688 001 | ·5403 601 | ·5133 421 | ·4876 750⁻ |
| ·96 | ·6648 326 | ·6514 003 | ·6382 393 | ·6127 098 | ·5882 014 | ·5646 733 |
| ·97 | ·7374 241 | ·7262 785⁺ | ·7153 014 | ·6938 424 | ·6730 271 | ·6528 363 |
| ·98 | ·8170 728 | ·8088 608 | ·8007 314 | ·7847 167 | ·7690 224 | ·7536 419 |
| ·99 | ·9043 821 | ·8998 488 | ·8953 383 | ·8863 849 | ·8775 210 | ·8687 458 |
| 1·00 | 1·0000 000 | 1·0000 000 | 1·0000 000 | 1·0000 000 | 1·0000 000 | 1·0000 000 |

$x$ = ·33 to 1·00 $q = 1$ $p$ = 15 to 20

| $x$ | $p = 15$ | $p = 16$ | $p = 17$ | $p = 18$ | $p = 19$ | $p = 20$ |
|---|---|---|---|---|---|---|
| $B(p, q) =$ | ·6666 6667 × $\frac{1}{10}$ | ·6250 0000 × $\frac{1}{10}$ | ·5882 3529 × $\frac{1}{10}$ | ·5555 5556 × $\frac{1}{10}$ | ·5263 1579 × $\frac{1}{10}$ | ·5000 0000 × $\frac{1}{10}$ |
| ·33 | ·0000 001 | | | | | |
| ·34 | ·0000 001 | | | | | |
| ·35 | ·0000 001 | ·0000 001 | | | | |
| ·36 | ·0000 002 | ·0000 001 | | | | |
| ·37 | ·0000 003 | ·0000 001 | | | | |
| ·38 | ·0000 005$^-$ | ·0000 002 | ·0000 001 | | | |
| ·39 | ·0000 007 | ·0000 003 | ·0000 001 | | | |
| ·40 | ·0000 011 | ·0000 004 | ·0000 002 | ·0000 001 | | |
| ·41 | ·0000 016 | ·0000 006 | ·0000 003 | ·0000 001 | | |
| ·42 | ·0000 022 | ·0000 009 | ·0000 004 | ·0000 002 | ·0000 001 | |
| ·43 | ·0000 032 | ·0000 014 | ·0000 006 | ·0000 003 | ·0000 001 | |
| ·44 | ·0000 045$^-$ | ·0000 020 | ·0000 009 | ·0000 004 | ·0000 002 | ·0000 001 |
| ·45 | ·0000 063 | ·0000 028 | ·0000 013 | ·0000 006 | ·0000 003 | ·0000 001 |
| ·46 | ·0000 087 | ·0000 040 | ·0000 018 | ·0000 009 | ·0000 004 | ·0000 002 |
| ·47 | ·0000 121 | ·0000 057 | ·0000 027 | ·0000 013 | ·0000 006 | ·0000 003 |
| ·48 | ·0000 165$^+$ | ·0000 079 | ·0000 038 | ·0000 018 | ·0000 009 | ·0000 004 |
| ·49 | ·0000 225$^+$ | ·0000 110 | ·0000 054 | ·0000 027 | ·0000 013 | ·0000 006 |
| ·50 | ·0000 305$^+$ | ·0000 153 | ·0000 076 | ·0000 038 | ·0000 019 | ·0000 010 |
| ·51 | ·0000 411 | ·0000 209 | ·0000 107 | ·0000 054 | ·0000 028 | ·0000 014 |
| ·52 | ·0000 550$^-$ | ·0000 286 | ·0000 149 | ·0000 077 | ·0000 040 | ·0000 021 |
| ·53 | ·0000 731 | ·0000 388 | ·0000 205$^+$ | ·0000 109 | ·0000 058 | ·0000 031 |
| ·54 | ·0000 968 | ·0000 523 | ·0000 282 | ·0000 152 | ·0000 082 | ·0000 044 |
| ·55 | ·0001 275$^-$ | ·0000 701 | ·0000 386 | ·0000 212 | ·0000 117 | ·0000 064 |
| ·56 | ·0001 670 | ·0000 935$^+$ | ·0000 524 | ·0000 293 | ·0000 164 | ·0000 092 |
| ·57 | ·0002 178 | ·0001 242 | ·0000 708 | ·0000 403 | ·0000 230 | ·0000 131 |
| ·58 | ·0002 828 | ·0001 640 | ·0000 951 | ·0000 552 | ·0000 320 | ·0000 186 |
| ·59 | ·0003 654 | ·0002 156 | ·0001 272 | ·0000 750$^+$ | ·0000 443 | ·0000 261 |
| ·60 | ·0004 702 | ·0002 821 | ·0001 693 | ·0001 016 | ·0000 609 | ·0000 366 |
| ·61 | ·0006 025$^-$ | ·0003 675$^+$ | ·0002 242 | ·0001 368 | ·0000 834 | ·0000 509 |
| ·62 | ·0007 689 | ·0004 767 | ·0002 956 | ·0001 833 | ·0001 136 | ·0000 704 |
| ·63 | ·0009 775$^-$ | ·0006 158 | ·0003 880 | ·0002 444 | ·0001 540 | ·0000 970 |
| ·64 | ·0012 379 | ·0007 923 | ·0005 071 | ·0003 245$^+$ | ·0002 077 | ·0001 329 |
| ·65 | ·0015 621 | ·0010 153 | ·0006 600 | ·0004 290 | ·0002 788 | ·0001 812 |
| ·66 | ·0019 641 | ·0012 963 | ·0008 556 | ·0005 647 | ·0003 727 | ·0002 460 |
| ·67 | ·0024 611 | ·0016 489 | ·0011 048 | ·0007 402 | ·0004 959 | ·0003 323 |
| ·68 | ·0030 735$^+$ | ·0020 900 | ·0014 212 | ·0009 664 | ·0006 572 | ·0004 469 |
| ·69 | ·0038 259 | ·0026 399 | ·0018 215$^+$ | ·0012 569 | ·0008 672 | ·0005 984 |
| ·70 | ·0047 476 | ·0033 233 | ·0023 263 | ·0016 284 | ·0011 399 | ·0007 979 |
| ·71 | ·0058 732 | ·0041 700 | ·0029 607 | ·0021 021 | ·0014 925$^-$ | ·0010 597 |
| ·72 | ·0072 442 | ·0052 158 | ·0037 554 | ·0027 039 | ·0019 468 | ·0014 017 |
| ·73 | ·0089 093 | ·0065 038 | ·0047 478 | ·0034 659 | ·0025 301 | ·0018 470 |
| ·74 | ·0109 264 | ·0080 855$^+$ | ·0059 833 | ·0044 276 | ·0032 764 | ·0024 246 |
| ·75 | ·0133 635$^-$ | ·0100 226 | ·0075 169 | ·0056 377 | ·0042 283 | ·0031 712 |
| ·76 | ·0163 006 | ·0123 885$^-$ | ·0094 152 | ·0071 556 | ·0054 382 | ·0041 331 |
| ·77 | ·0198 317 | ·0152 704 | ·0117 582 | ·0090 538 | ·0069 715$^-$ | ·0053 680 |
| ·78 | ·0240 668 | ·0187 721 | ·0146 423 | ·0114 210 | ·0089 084 | ·0069 485$^+$ |
| ·79 | ·0291 344 | ·0230 162 | ·0181 828 | ·0143 644 | ·0113 479 | ·0089 648 |
| ·80 | ·0351 844 | ·0281 475$^-$ | ·0225 180 | ·0180 144 | ·0144 115$^+$ | ·0115 292 |
| ·81 | ·0423 912 | ·0343 368 | ·0278 128 | ·0225 284 | ·0182 480 | ·0147 809 |
| ·82 | ·0509 575$^-$ | ·0417 851 | ·0342 638 | ·0280 963 | ·0230 390 | ·0188 920 |
| ·83 | ·0611 183 | ·0507 282 | ·0421 044 | ·0349 467 | ·0290 057 | ·0240 748 |
| ·84 | ·0731 458 | ·0614 425$^-$ | ·0516 117 | ·0433 538 | ·0364 172 | ·0305 904 |
| ·85 | ·0873 542 | ·0742 511 | ·0631 134 | ·0536 464 | ·0455 994 | ·0387 595$^+$ |
| ·86 | ·1041 062 | ·0895 314 | ·0769 970 | ·0662 174 | ·0569 470 | ·0489 744 |
| ·87 | ·1238 194 | ·1077 229 | ·0937 189 | ·0815 355$^-$ | ·0709 359 | ·0617 142 |
| ·88 | ·1469 739 | ·1293 370 | ·1138 166 | ·1001 586 | ·0881 395$^+$ | ·0775 628 |
| ·89 | ·1741 206 | ·1549 673 | ·1379 209 | ·1227 496 | ·1092 472 | ·0972 300 |
| ·90 | ·2058 911 | ·1853 020 | ·1667 718 | ·1500 946 | ·1350 852 | ·1215 767 |
| ·91 | ·2430 082 | ·2211 374 | ·2012 351 | ·1831 239 | ·1666 428 | ·1516 449 |
| ·92 | ·2862 974 | ·2633 936 | ·2423 221 | ·2229 364 | ·2051 014 | ·1886 933 |
| ·93 | ·3367 009 | ·3131 318 | ·2912 126 | ·2708 277 | ·2518 698 | ·2342 389 |
| ·94 | ·3952 918 | ·3715 743 | ·3492 798 | ·3283 230 | ·3086 237 | ·2901 062 |
| ·95 | ·4632 912 | ·4401 267 | ·4181 203 | ·3972 143 | ·3773 536 | ·3584 859 |
| ·96 | ·5420 864 | ·5204 029 | ·4995 868 | ·4796 033 | ·4604 192 | ·4420 024 |
| ·97 | ·6332 512 | ·6142 537 | ·5958 260 | ·5779 513 | ·5606 127 | ·5437 943 |
| ·98 | ·7385 691 | ·7237 977 | ·7093 218 | ·6951 353 | ·6812 326 | ·6676 080 |
| ·99 | ·8600 584 | ·8514 578 | ·8429 432 | ·8345 138 | ·8261 686 | ·8179 069 |
| 1·00 | 1·0000 000 | 1·0000 000 | 1·0000 000 | 1·0000 000 | 1·0000 000 | 1·0000 000 |

$x$ = ·45 to 1·00 $q = 1$ $p$ = 21 to 26

| $x$ | $p = 21$ | $p = 22$ | $p = 23$ | $p = 24$ | $p = 25$ | $p = 26$ |
|---|---|---|---|---|---|---|
| $B(p, q)$ = | ·4761 9048 $\times \frac{1}{10}$ | ·4545 4545$^{+}$ $\times \frac{1}{10}$ | ·4347 8261 $\times \frac{1}{10}$ | ·4166 6667 $\times \frac{1}{10}$ | ·4000 0000 $\times \frac{1}{10}$ | ·3846 1538 $\times \frac{1}{10}$ |
| ·45 | ·0000 001 | | | | | |
| ·46 | ·0000 001 | | | | | |
| ·47 | ·0000 001 | ·0000 001 | | | | |
| ·48 | ·0000 002 | ·0000 001 | | | | |
| ·49 | ·0000 003 | ·0000 002 | ·0000 001 | | | |
| ·50 | ·0000 005$^{-}$ | ·0000 002 | ·0000 001 | ·0000 001 | | |
| ·51 | ·0000 007 | ·0000 004 | ·0000 002 | ·0000 001 | | |
| ·52 | ·0000 011 | ·0000 006 | ·0000 003 | ·0000 002 | ·0000 001 | |
| ·53 | ·0000 016 | ·0000 009 | ·0000 005$^{-}$ | ·0000 002 | ·0000 001 | ·0000 001 |
| ·54 | ·0000 024 | ·0000 013 | ·0000 007 | ·0000 004 | ·0000 002 | ·0000 001 |
| ·55 | ·0000 035$^{+}$ | ·0000 019 | ·0000 011 | ·0000 006 | ·0000 003 | ·0000 002 |
| ·56 | ·0000 052 | ·0000 029 | ·0000 016 | ·0000 009 | ·0000 005$^{+}$ | ·0000 003 |
| ·57 | ·0000 075$^{-}$ | ·0000 043 | ·0000 024 | ·0000 014 | ·0000 008 | ·0000 004 |
| ·58 | ·0000 108 | ·0000 062 | ·0000 036 | ·0000 021 | ·0000 012 | ·0000 007 |
| ·59 | ·0000 154 | ·0000 091 | ·0000 054 | ·0000 032 | ·0000 019 | ·0000 011 |
| ·60 | ·0000 219 | ·0000 132 | ·0000 079 | ·0000 047 | ·0000 028 | ·0000 017 |
| ·61 | ·0000 310 | ·0000 189 | ·0000 116 | ·0000 070 | ·0000 043 | ·0000 026 |
| ·62 | ·0000 437 | ·0000 271 | ·0000 168 | ·0000 104 | ·0000 065$^{-}$ | ·0000 040 |
| ·63 | ·0000 611 | ·0000 385$^{+}$ | ·0000 243 | ·0000 153 | ·0000 096 | ·0000 061 |
| ·64 | ·0000 851 | ·0000 544 | ·0000 348 | ·0000 223 | ·0000 143 | ·0000 091 |
| ·65 | ·0001 178 | ·0000 766 | ·0000 498 | ·0000 324 | ·0000 210 | ·0000 137 |
| ·66 | ·0001 623 | ·0001 071 | ·0000 707 | ·0000 467 | ·0000 308 | ·0000 203 |
| ·67 | ·0002 226 | ·0001 492 | ·0000 999 | ·0000 670 | ·0000 449 | ·0000 301 |
| ·68 | ·0003 039 | ·0002 066 | ·0001 405$^{+}$ | ·0000 955$^{+}$ | ·0000 650$^{-}$ | ·0000 442 |
| ·69 | ·0004 129 | ·0002 849 | ·0001 966 | ·0001 356 | ·0000 936 | ·0000 646 |
| ·70 | ·0005 585$^{+}$ | ·0003 910 | ·0002 737 | ·0001 916 | ·0001 341 | ·0000 939 |
| ·71 | ·0007 524 | ·0005 342 | ·0003 793 | ·0002 693 | ·0001 912 | ·0001 357 |
| ·72 | ·0010 092 | ·0007 266 | ·0005 232 | ·0003 767 | ·0002 712 | ·0001 953 |
| ·73 | ·0013 483 | ·0009 842 | ·0007 185$^{-}$ | ·0005 245$^{+}$ | ·0003 829 | ·0002 795$^{+}$ |
| ·74 | ·0017 942 | ·0013 277 | ·0009 825$^{-}$ | ·0007 270 | ·0005 380 | ·0003 981 |
| ·75 | ·0023 784 | ·0017 838 | ·0013 379 | ·0010 034 | ·0007 525$^{+}$ | ·0005 644 |
| ·76 | ·0031 411 | ·0023 873 | ·0018 143 | ·0013 789 | ·0010 479 | ·0007 964 |
| ·77 | ·0041 334 | ·0031 827 | ·0024 507 | ·0018 870 | ·0014 530 | ·0011 188 |
| ·78 | ·0054 198 | ·0042 275$^{-}$ | ·0032 974 | ·0025 720 | ·0020 062 | ·0015 648 |
| ·79 | ·0070 822 | ·0055 949 | ·0044 200 | ·0034 918 | ·0027 585$^{+}$ | ·0021 792 |
| ·80 | ·0092 234 | ·0073 787 | ·0059 030 | ·0047 224 | ·0037 779 | ·0030 223 |
| ·81 | ·0119 725$^{+}$ | ·0096 977 | ·0078 552 | ·0063 627 | ·0051 538 | ·0041 746 |
| ·82 | ·0154 914 | ·0127 030 | ·0104 164 | ·0085 415$^{-}$ | ·0070 040 | ·0057 433 |
| ·83 | ·0199 820 | ·0165 851 | ·0137 656 | ·0114 255$^{-}$ | ·0094 831 | ·0078 710 |
| ·84 | ·0256 960 | ·0215 846 | ·0181 311 | ·0152 301 | ·0127 933 | ·0107 464 |
| ·85 | ·0329 456 | ·0280 038 | ·0238 032 | ·0202 327 | ·0171 978 | ·0146 181 |
| ·86 | ·0421 180 | ·0362 215$^{-}$ | ·0311 505$^{-}$ | ·0267 894 | ·0230 389 | ·0198 134 |
| ·87 | ·0536 913 | ·0467 115$^{-}$ | ·0406 390 | ·0353 559 | ·0307 596 | ·0267 609 |
| ·88 | ·0682 553 | ·0600 646 | ·0528 569 | ·0465 140 | ·0409 324 | ·0360 205$^{-}$ |
| ·89 | ·0865 347 | ·0770 159 | ·0685 441 | ·0610 043 | ·0542 938 | ·0483 215$^{-}$ |
| ·90 | ·1094 190 | ·0984 771 | ·0886 294 | ·0797 664 | ·0717 898 | ·0646 108 |
| ·91 | ·1379 969 | ·1255 772 | ·1142 752 | ·1039 904 | ·0946 313 | ·0861 145$^{-}$ |
| ·92 | ·1735 979 | ·1597 100 | ·1469 332 | ·1351 786 | ·1243 643 | ·1144 151 |
| ·93 | ·2178 422 | ·2025 932 | ·1884 117 | ·1752 229 | ·1629 573 | ·1515 503 |
| ·94 | ·2726 999 | ·2563 379 | ·2409 576 | ·2265 001 | ·2129 101 | ·2001 355$^{+}$ |
| ·95 | ·3405 616 | ·3235 335$^{+}$ | ·3073 569 | ·2919 890 | ·2773 896 | ·2635 201 |
| ·96 | ·4243 223 | ·4073 494 | ·3910 555$^{-}$ | ·3754 132 | ·3603 967 | ·3459 808 |
| ·97 | ·5274 805$^{+}$ | ·5116 561 | ·4963 064 | ·4814 172 | ·4669 747 | ·4529 655$^{-}$ |
| ·98 | ·6542 558 | ·6411 707 | ·6283 473 | ·6157 803 | ·6034 647 | ·5913 954 |
| ·99 | ·8097 279 | ·8016 306 | ·7936 143 | ·7856 781 | ·7778 214 | ·7700 431 |
| 1·00 | 1·0000 000 | 1·0000 000 | 1·0000 000 | 1·0000 000 | 1·0000 000 | 1·0000 000 |

$x$ = ·54 to 1·00 $q = 1$ $p$ = 27 to 32

| $x$ | $p = 27$ | $p = 28$ | $p = 29$ | $p = 30$ | $p = 31$ | $p = 32$ |
|---|---|---|---|---|---|---|
| $B(p, q) =$ | ·3703 7037 × $\frac{1}{10}$ | ·3571 4286 × $\frac{1}{10}$ | ·3448 2759 × $\frac{1}{10}$ | ·3333 3333 × $\frac{1}{10}$ | ·3225 8065 × $\frac{1}{10}$ | ·3125 0000 × $\frac{1}{10}$ |
| ·54 | ·0000 001 | | | | | |
| ·55 | ·0000 001 | ·0000 001 | | | | |
| ·56 | ·0000 002 | ·0000 001 | | | | |
| ·57 | ·0000 003 | ·0000 001 | ·0000 001 | | | |
| ·58 | ·0000 004 | ·0000 002 | ·0000 001 | ·0000 001 | | |
| ·59 | ·0000 007 | ·0000 004 | ·0000 002 | ·0000 001 | ·0000 001 | |
| ·60 | ·0000 010 | ·0000 006 | ·0000 004 | ·0000 002 | ·0000 001 | ·0000 001 |
| ·61 | ·0000 016 | ·0000 010 | ·0000 006 | ·0000 004 | ·0000 002 | ·0000 001 |
| ·62 | ·0000 025$^{-}$ | ·0000 015$^{+}$ | ·0000 010 | ·0000 006 | ·0000 004 | ·0000 002 |
| ·63 | ·0000 038 | ·0000 024 | ·0000 015$^{+}$ | ·0000 010 | ·0000 006 | ·0000 004 |
| ·64 | ·0000 058 | ·0000 037 | ·0000 024 | ·0000 015$^{+}$ | ·0000 010 | ·0000 006 |
| ·65 | ·0000 089 | ·0000 058 | ·0000 038 | ·0000 024 | ·0000 016 | ·0000 010 |
| ·66 | ·0000 134 | ·0000 089 | ·0000 058 | ·0000 039 | ·0000 025$^{+}$ | ·0000 017 |
| ·67 | ·0000 201 | ·0000 135$^{-}$ | ·0000 090 | ·0000 061 | ·0000 041 | ·0000 027 |
| ·68 | ·0000 300 | ·0000 204 | ·0000 139 | ·0000 094 | ·0000 064 | ·0000 044 |
| ·69 | ·0000 446 | ·0000 307 | ·0000 212 | ·0000 146 | ·0000 101 | ·0000 070 |
| ·70 | ·0000 657 | ·0000 460 | ·0000 322 | ·0000 225$^{+}$ | ·0000 158 | ·0000 110 |
| ·71 | ·0000 964 | ·0000 684 | ·0000 486 | ·0000 345$^{-}$ | ·0000 245$^{-}$ | ·0000 174 |
| ·72 | ·0001 406 | ·0001 012 | ·0000 729 | ·0000 525$^{-}$ | ·0000 378 | ·0000 272 |
| ·73 | ·0002 040 | ·0001 489 | ·0001 087 | ·0000 794 | ·0000 579 | ·0000 423 |
| ·74 | ·0002 946 | ·0002 180 | ·0001 613 | ·0001 194 | ·0000 883 | ·0000 654 |
| ·75 | ·0004 233 | ·0003 175$^{-}$ | ·0002 381 | ·0001 786 | ·0001 339 | ·0001 005$^{-}$ |
| ·76 | ·0006 053 | ·0004 600 | ·0003 496 | ·0002 657 | ·0002 019 | ·0001 535$^{-}$ |
| ·77 | ·0008 615$^{-}$ | ·0006 633 | ·0005 108 | ·0003 933 | ·0003 028 | ·0002 332 |
| ·78 | ·0012 205$^{+}$ | ·0009 520 | ·0007 426 | ·0005 792 | ·0004 518 | ·0003 524 |
| ·79 | ·0017 216 | ·0013 601 | ·0010 744 | ·0008 488 | ·0006 706 | ·0005 297 |
| ·80 | ·0024 179 | ·0019 343 | ·0015 474 | ·0012 379 | ·0009 904 | ·0007 923 |
| ·81 | ·0033 814 | ·0027 389 | ·0022 185$^{+}$ | ·0017 970 | ·0014 556 | ·0011 790 |
| ·82 | ·0047 095$^{-}$ | ·0038 618 | ·0031 667 | ·0025 967 | ·0021 293 | ·0017 460 |
| ·83 | ·0065 329 | ·0054 223 | ·0045 005$^{+}$ | ·0037 354 | ·0031 004 | ·0025 734 |
| ·84 | ·0090 269 | ·0075 826 | ·0063 694 | ·0053 503 | ·0044 943 | ·0037 752 |
| ·85 | ·0124 254 | ·0105 616 | ·0089 774 | ·0076 308 | ·0064 861 | ·0055 132 |
| ·86 | ·0170 396 | ·0146 540 | ·0126 025$^{-}$ | ·0108 381 | ·0093 208 | ·0080 159 |
| ·87 | ·0232 820 | ·0202 553 | ·0176 221 | ·0153 313 | ·0133 382 | ·0116 042 |
| ·88 | ·0316 980 | ·0278 943 | ·0245 469 | ·0216 013 | ·0190 092 | ·0167 281 |
| ·89 | ·0430 061 | ·0382 754 | ·0340 651 | ·0303 180 | ·0269 830 | ·0240 149 |
| ·90 | ·0581 497 | ·0523 348 | ·0471 013 | ·0423 912 | ·0381 520 | ·0343 368 |
| ·91 | ·0783 642 | ·0713 114 | ·0648 934 | ·0590 530 | ·0537 382 | ·0489 018 |
| ·92 | ·1052 619 | ·0968 410 | ·0890 937 | ·0819 662 | ·0754 089 | ·0693 762 |
| ·93 | ·1409 417 | ·1310 758 | ·1219 005$^{+}$ | ·1133 675$^{-}$ | ·1054 317 | ·0980 515$^{+}$ |
| ·94 | ·1881 274 | ·1768 398 | ·1662 294 | ·1562 556 | ·1468 803 | ·1380 675$^{-}$ |
| ·95 | ·2503 441 | ·2378 269 | ·2259 355$^{+}$ | ·2146 388 | ·2039 068 | ·1937 115$^{-}$ |
| ·96 | ·3321 416 | ·3188 559 | ·3061 017 | ·2938 576 | ·2821 033 | ·2708 192 |
| ·97 | ·4393 765$^{+}$ | ·4261 952 | ·4134 093 | ·4010 071 | ·3889 769 | ·3773 076 |
| ·98 | ·5795 675$^{+}$ | ·5679 762 | ·5566 167 | ·5454 843 | ·5345 746 | ·5238 831 |
| ·99 | ·7623 427 | ·7547 193 | ·7471 721 | ·7397 004 | ·7323 034 | ·7249 803 |
| 1·00 | 1·0000 000 | 1·0000 000 | 1·0000 000 | 1·0000 000 | 1·0000 000 | 1·0000 000 |

$x$ = ·61 to 1·00 $q = 1$ $p$ = 33 to 38

| $x$ | $p = 33$ | $p = 34$ | $p = 35$ | $p = 36$ | $p = 37$ | $p = 38$ |
|---|---|---|---|---|---|---|
| $B(p, q)$ = | ·3030 3030 × $\frac{1}{10}$ | ·2941 1765 × $\frac{1}{10}$ | ·2857 1429 × $\frac{1}{10}$ | ·2777 7778 × $\frac{1}{10}$ | ·2702 7027 × $\frac{1}{10}$ | ·2631 5789 × $\frac{1}{10}$ |
| ·61 | ·0000 001 | ·0000 001 | | | | |
| ·62 | ·0000 001 | ·0000 001 | ·0000 001 | | | |
| ·63 | ·0000 002 | ·0000 002 | ·0000 001 | ·0000 001 | | |
| ·64 | ·0000 004 | ·0000 003 | ·0000 002 | ·0000 001 | ·0000 001 | |
| ·65 | ·0000 007 | ·0000 004 | ·0000 003 | ·0000 002 | ·0000 001 | ·0000 001 |
| ·66 | ·0000 011 | ·0000 007 | ·0000 005− | ·0000 003 | ·0000 002 | ·0000 001 |
| ·67 | ·0000 018 | ·0000 012 | ·0000 008 | ·0000 005+ | ·0000 004 | ·0000 002 |
| ·68 | ·0000 030 | ·0000 020 | ·0000 014 | ·0000 009 | ·0000 006 | ·0000 004 |
| ·69 | ·0000 048 | ·0000 033 | ·0000 023 | ·0000 016 | ·0000 011 | ·0000 008 |
| ·70 | ·0000 077 | ·0000 054 | ·0000 038 | ·0000 027 | ·0000 019 | ·0000 013 |
| ·71 | ·0000 123 | ·0000 088 | ·0000 062 | ·0000 044 | ·0000 031 | ·0000 022 |
| ·72 | ·0000 196 | ·0000 141 | ·0000 102 | ·0000 073 | ·0000 053 | ·0000 038 |
| ·73 | ·0000 309 | ·0000 225+ | ·0000 165− | ·0000 120 | ·0000 088 | ·0000 064 |
| ·74 | ·0000 484 | ·0000 358 | ·0000 265− | ·0000 196 | ·0000 145+ | ·0000 107 |
| ·75 | ·0000 753 | ·0000 565+ | ·0000 424 | ·0000 318 | ·0000 238 | ·0000 179 |
| ·76 | ·0001 166 | ·0000 886 | ·0000 674 | ·0000 512 | ·0000 389 | ·0000 296 |
| ·77 | ·0001 796 | ·0001 383 | ·0001 065− | ·0000 820 | ·0000 631 | ·0000 486 |
| ·78 | ·0002 749 | ·0002 144 | ·0001 672 | ·0001 304 | ·0001 017 | ·0000 794 |
| ·79 | ·0004 185− | ·0003 306 | ·0002 612 | ·0002 063 | ·0001 630 | ·0001 288 |
| ·80 | ·0006 338 | ·0005 071 | ·0004 056 | ·0003 245+ | ·0002 596 | ·0002 077 |
| ·81 | ·0009 550+ | ·0007 736 | ·0006 266 | ·0005 075+ | ·0004 111 | ·0003 330 |
| ·82 | ·0014 317 | ·0011 740 | ·0009 627 | ·0007 894 | ·0006 473 | ·0005 308 |
| ·83 | ·0021 359 | ·0017 728 | ·0014 714 | ·0012 213 | ·0010 137 | ·0008 413 |
| ·84 | ·0031 711 | ·0026 638 | ·0022 376 | ·0018 796 | ·0015 788 | ·0013 262 |
| ·85 | ·0046 862 | ·0039 833 | ·0033 858 | ·0028 779 | ·0024 462 | ·0020 793 |
| ·86 | ·0068 936 | ·0059 285+ | ·0050 985+ | ·0043 847 | ·0037 709 | ·0032 430 |
| ·87 | ·0100 957 | ·0087 832 | ·0076 414 | ·0066 480 | ·0057 838 | ·0050 319 |
| ·88 | ·0147 207 | ·0129 542 | ·0113 997 | ·0100 317 | ·0088 279 | ·0077 686 |
| ·89 | ·0213 732 | ·0190 222 | ·0169 297 | ·0150 675− | ·0134 100 | ·0119 349 |
| ·90 | ·0309 032 | ·0278 128 | ·0250 316 | ·0225 284 | ·0202 756 | ·0182 480 |
| ·91 | ·0445 006 | ·0404 956 | ·0368 510 | ·0335 344 | ·0305 163 | ·0277 698 |
| ·92 | ·0638 261 | ·0587 200 | ·0540 224 | ·0497 006 | ·0457 246 | ·0420 666 |
| ·93 | ·0911 879 | ·0848 048 | ·0788 684 | ·0733 476 | ·0682 133 | ·0634 384 |
| ·94 | ·1297 834 | ·1219 964 | ·1146 766 | ·1077 960 | ·1013 283 | ·0952 486 |
| ·95 | ·1840 259 | ·1748 246 | ·1660 834 | ·1577 792 | ·1498 903 | ·1423 957 |
| ·96 | ·2599 864 | ·2495 870 | ·2396 035− | ·2300 194 | ·2208 186 | ·2119 858 |
| ·97 | ·3659 883 | ·3550 087 | ·3443 584 | ·3340 277 | ·3240 068 | ·3142 866 |
| ·98 | ·5134 055− | ·5031 374 | ·4930 746 | ·4832 131 | ·4735 489 | ·4640 779 |
| ·99 | ·7177 305+ | ·7105 532 | ·7034 477 | ·6964 132 | ·6894 491 | ·6825 546 |
| 1·00 | 1·0000 000 | 1·0000 000 | 1·0000 000 | 1·0000 000 | 1·0000 000 | 1·0000 000 |

$x = \cdot 65$ to $1\cdot 00$ $q = 1$ $p = 39$ to $44$

| | $p = 39$ | $p = 40$ | $p = 41$ | $p = 42$ | $p = 43$ | $p = 44$ |
|---|---|---|---|---|---|---|
| $B(p, q) =$ | ·2564 1026 × $\frac{1}{10}$ | ·2500 0000 × $\frac{1}{10}$ | ·2439 0244 × $\frac{1}{10}$ | ·2380 9524 × $\frac{1}{10}$ | ·2325 5814 × $\frac{1}{10}$ | ·2272 7273 × $\frac{1}{10}$ |
| $x$ | | | | | | |
| ·65 | ·0000 001 | | | | | |
| ·66 | ·0000 001 | ·0000 001 | | | | |
| ·67 | ·0000 002 | ·0000 001 | ·0000 001 | | | |
| ·68 | ·0000 003 | ·0000 002 | ·0000 001 | ·0000 001 | ·0000 001 | |
| ·69 | ·0000 005+ | ·0000 004 | ·0000 002 | ·0000 002 | ·0000 001 | ·0000 001 |
| ·70 | ·0000 009 | ·0000 006 | ·0000 004 | ·0000 003 | ·0000 002 | ·0000 002 |
| ·71 | ·0000 016 | ·0000 011 | ·0000 008 | ·0000 006 | ·0000 004 | ·0000 003 |
| ·72 | ·0000 027 | ·0000 020 | ·0000 014 | ·0000 010 | ·0000 007 | ·0000 005+ |
| ·73 | ·0000 047 | ·0000 034 | ·0000 025− | ·0000 018 | ·0000 013 | ·0000 010 |
| ·74 | ·0000 079 | ·0000 059 | ·0000 044 | ·0000 032 | ·0000 024 | ·0000 018 |
| ·75 | ·0000 134 | ·0000 101 | ·0000 075+ | ·0000 057 | ·0000 042 | ·0000 032 |
| ·76 | ·0000 225− | ·0000 171 | ·0000 130 | ·0000 099 | ·0000 075− | ·0000 057 |
| ·77 | ·0000 374 | ·0000 288 | ·0000 222 | ·0000 171 | ·0000 132 | ·0000 101 |
| ·78 | ·0000 619 | ·0000 483 | ·0000 377 | ·0000 294 | ·0000 229 | ·0000 179 |
| ·79 | ·0001 017 | ·0000 804 | ·0000 635− | ·0000 502 | ·0000 396 | ·0000 313 |
| ·80 | ·0001 662 | ·0001 329 | ·0001 063 | ·0000 851 | ·0000 681 | ·0000 544 |
| ·81 | ·0002 697 | ·0002 185− | ·0001 770 | ·0001 433 | ·0001 161 | ·0000 940 |
| ·82 | ·0004 353 | ·0003 569 | ·0002 927 | ·0002 400 | ·0001 968 | ·0001 614 |
| ·83 | ·0006 983 | ·0005 796 | ·0004 811 | ·0003 993 | ·0003 314 | ·0002 751 |
| ·84 | ·0011 140 | ·0009 358 | ·0007 861 | ·0006 603 | ·0005 546 | ·0004 659 |
| ·85 | ·0017 674 | ·0015 023 | ·0012 770 | ·0010 854 | ·0009 226 | ·0007 842 |
| ·86 | ·0027 889 | ·0023 985− | ·0020 627 | ·0017 739 | ·0015 256 | ·0013 120 |
| ·87 | ·0043 777 | ·0038 086 | ·0033 135+ | ·0028 828 | ·0025 080 | ·0021 820 |
| ·88 | ·0068 363 | ·0060 160 | ·0052 941 | ·0046 588 | ·0040 997 | ·0036 078 |
| ·89 | ·0106 221 | ·0094 537 | ·0084 138 | ·0074 882 | ·0066 645+ | ·0059 314 |
| ·90 | ·0164 232 | ·0147 809 | ·0133 028 | ·0119 725+ | ·0107 753 | ·0096 977 |
| ·91 | ·0252 705+ | ·0229 962 | ·0209 265+ | ·0190 431 | ·0173 293 | ·0157 696 |
| ·92 | ·0387 013 | ·0356 052 | ·0327 568 | ·0301 362 | ·0277 253 | ·0255 073 |
| ·93 | ·0589 977 | ·0548 679 | ·0510 271 | ·0474 552 | ·0441 333 | ·0410 440 |
| ·94 | ·0895 337 | ·0841 616 | ·0791 119 | ·0743 652 | ·0699 033 | ·0657 091 |
| ·95 | ·1352 760 | ·1285 122 | ·1220 865+ | ·1159 822 | ·1101 831 | ·1046 740 |
| ·96 | ·2035 064 | ·1953 662 | ·1875 515+ | ·1800 494 | ·1728 475− | ·1659 336 |
| ·97 | ·3048 580 | ·2957 123 | ·2868 409 | ·2782 357 | ·2698 886 | ·2617 920 |
| ·98 | ·4547 963 | ·4457 004 | ·4367 864 | ·4280 507 | ·4194 897 | ·4110 999 |
| ·99 | ·6757 290 | ·6689 718 | ·6622 820 | ·6556 592 | ·6491 026 | ·6426 116 |
| 1·00 | 1·0000 000 | 1·0000 000 | 1·0000 000 | 1·0000 000 | 1·0000 000 | 1·0000 000 |

$x = \cdot 69$ to $1\cdot 00$ $q = 1$ $p = 45$ to $50$

| | $p = 45$ | $p = 46$ | $p = 47$ | $p = 48$ | $p = 49$ | $p = 50$ |
|---|---|---|---|---|---|---|
| $B(p, q) =$ | ·2222 2222 × $\frac{1}{10}$ | ·2173 9130 × $\frac{1}{10}$ | ·2127 6596 × $\frac{1}{10}$ | ·2083 3333 × $\frac{1}{10}$ | ·2040 8163 × $\frac{1}{10}$ | ·2000 0000 × $\frac{1}{10}$ |
| $x$ | | | | | | |
| ·69 | ·0000 001 | | | | | |
| ·70 | ·0000 001 | ·0000 001 | ·0000 001 | | | |
| ·71 | ·0000 002 | ·0000 001 | ·0000 001 | ·0000 001 | ·0000 001 | |
| ·72 | ·0000 004 | ·0000 003 | ·0000 002 | ·0000 001 | ·0000 001 | ·0000 001 |
| ·73 | ·0000 007 | ·0000 005+ | ·0000 004 | ·0000 003 | ·0000 002 | ·0000 001 |
| ·74 | ·0000 013 | ·0000 010 | ·0000 007 | ·0000 005+ | ·0000 004 | ·0000 003 |
| ·75 | ·0000 024 | ·0000 018 | ·0000 013 | ·0000 010 | ·0000 008 | ·0000 006 |
| ·76 | ·0000 043 | ·0000 033 | ·0000 025+ | ·0000 019 | ·0000 014 | ·0000 011 |
| ·77 | ·0000 078 | ·0000 060 | ·0000 046 | ·0000 036 | ·0000 027 | ·0000 021 |
| ·78 | ·0000 139 | ·0000 109 | ·0000 085− | ·0000 066 | ·0000 052 | ·0000 040 |
| ·79 | ·0000 247 | ·0000 195+ | ·0000 154 | ·0000 122 | ·0000 096 | ·0000 076 |
| ·80 | ·0000 436 | ·0000 348 | ·0000 279 | ·0000 223 | ·0000 178 | ·0000 143 |
| ·81 | ·0000 762 | ·0000 617 | ·0000 500− | ·0000 405− | ·0000 328 | ·0000 266 |
| ·82 | ·0001 323 | ·0001 085+ | ·0000 890 | ·0000 730 | ·0000 598 | ·0000 491 |
| ·83 | ·0002 283 | ·0001 895− | ·0001 573 | ·0001 305+ | ·0001 083 | ·0000 899 |
| ·84 | ·0003 914 | ·0003 287 | ·0002 761 | ·0002 320 | ·0001 948 | ·0001 637 |
| ·85 | ·0006 666 | ·0005 666 | ·0004 816 | ·0004 094 | ·0003 480 | ·0002 958 |
| ·86 | ·0011 283 | ·0009 704 | ·0008 345+ | ·0007 177 | ·0006 172 | ·0005 308 |
| ·87 | ·0018 983 | ·0016 515+ | ·0014 368 | ·0012 500+ | ·0010 875+ | ·0009 462 |
| ·88 | ·0031 748 | ·0027 938 | ·0024 586 | ·0021 636 | ·0019 039 | ·0016 755− |
| ·89 | ·0052 790 | ·0046 983 | ·0041 815− | ·0037 215+ | ·0033 122 | ·0029 478 |
| ·90 | ·0087 280 | ·0078 552 | ·0070 697 | ·0063 627 | ·0057 264 | ·0051 538 |
| ·91 | ·0143 504 | ·0130 588 | ·0118 835+ | ·0108 140 | ·0098 408 | ·0089 551 |
| ·92 | ·0234 667 | ·0215 894 | ·0198 622 | ·0182 732 | ·0168 114 | ·0154 665− |
| ·93 | ·0381 709 | ·0354 990 | ·0330 140 | ·0307 031 | ·0285 538 | ·0265 551 |
| ·94 | ·0617 666 | ·0580 606 | ·0545 769 | ·0513 023 | ·0482 242 | ·0453 307 |
| ·95 | ·0994 403 | ·0944 682 | ·0897 448 | ·0852 576 | ·0809 947 | ·0769 450− |
| ·96 | ·1592 962 | ·1529 244 | ·1468 074 | ·1409 351 | ·1352 977 | ·1298 858 |
| ·97 | ·2539 382 | ·2463 201 | ·2389 305− | ·2317 625+ | ·2248 097 | ·2180 654 |
| ·98 | ·4028 779 | ·3948 203 | ·3869 239 | ·3791 854 | ·3716 017 | ·3641 697 |
| ·99 | ·6361 855− | ·6298 236 | ·6235 254 | ·6172 901 | ·6111 172 | ·6050 061 |
| 1·00 | 1·0000 000 | 1·0000 000 | 1·0000 000 | 1·0000 000 | 1·0000 000 | 1·0000 000 |

$x=$ ·01 to ·60 $q=$ 1·5 $p=$ 1·5 to 4

| $x$ | $p=1·5$ | $p=2$ | $p=2·5$ | $p=3$ | $p=3·5$ | $p=4$ |
|---|---|---|---|---|---|---|
| $B(p, q)=$ | ·3926 9908 | ·2666 6667 | ·1963 4954 | ·1523 8095$^{+}$ | ·1227 1846 | ·1015 8730 |
| ·01 | ·0016 926 | ·0001 869 | ·0000 203 | ·0000 022 | ·0000 002 | |
| ·02 | ·0047 728 | ·0007 450$^{-}$ | ·0001 144 | ·0000 174 | ·0000 026 | ·0000 004 |
| ·03 | ·0087 414 | ·0016 705$^{+}$ | ·0003 141 | ·0000 584 | ·0000 108 | ·0000 020 |
| ·04 | ·0134 171 | ·0029 597 | ·0006 425$^{+}$ | ·0001 379 | ·0000 293 | ·0000 062 |
| ·05 | ·0186 930 | ·0046 086 | ·0011 183 | ·0002 683 | ·0000 638 | ·0000 151 |
| ·06 | ·0244 963 | ·0066 134 | ·0017 575$^{-}$ | ·0004 617 | ·0001 203 | ·0000 311 |
| ·07 | ·0307 722 | ·0089 702 | ·0025 741 | ·0007 303 | ·0002 054 | ·0000 574 |
| ·08 | ·0374 780 | ·0116 750$^{+}$ | ·0035 806 | ·0010 858 | ·0003 265$^{-}$ | ·0000 975$^{+}$ |
| ·09 | ·0445 784 | ·0147 239 | ·0047 883 | ·0015 399 | ·0004 910 | ·0001 555$^{+}$ |
| ·10 | ·0520 440 | ·0181 128 | ·0062 074 | ·0021 038 | ·0007 070 | ·0002 360 |
| ·11 | ·0598 494 | ·0218 377 | ·0078 471 | ·0027 887 | ·0009 828 | ·0003 441 |
| ·12 | ·0679 724 | ·0258 945$^{-}$ | ·0097 159 | ·0036 056 | ·0013 269 | ·0004 852 |
| ·13 | ·0763 934 | ·0302 790 | ·0118 216 | ·0045 652 | ·0017 484 | ·0006 653 |
| ·14 | ·0850 946 | ·0349 873 | ·0141 714 | ·0056 780 | ·0022 563 | ·0008 908 |
| ·15 | ·0940 602 | ·0400 149 | ·0167 718 | ·0069 542 | ·0028 599 | ·0011 686 |
| ·16 | ·1032 755$^{-}$ | ·0453 578 | ·0196 290 | ·0084 039 | ·0035 688 | ·0015 059 |
| ·17 | ·1127 270 | ·0510 117 | ·0227 484 | ·0100 369 | ·0043 927 | ·0019 103 |
| ·18 | ·1224 023 | ·0569 722 | ·0261 351 | ·0118 628 | ·0053 414 | ·0023 898 |
| ·19 | ·1322 897 | ·0632 350$^{e}$ | ·0297 938 | ·0138 908 | ·0064 247 | ·0029 529 |
| ·20 | ·1423 785$^{-}$ | ·0697 957 | ·0337 287 | ·0161 301 | ·0076 528 | ·0036 081 |
| ·21 | ·1526 583 | ·0766 499 | ·0379 437 | ·0185 895$^{-}$ | ·0090 357 | ·0043 646 |
| ·22 | ·1631 194 | ·0837 931 | ·0424 423 | ·0212 775$^{+}$ | ·0105 836 | ·0052 318 |
| ·23 | ·1737 527 | ·0912 208 | ·0472 276 | ·0242 026 | ·0123 066 | ·0062 193 |
| ·24 | ·1845 494 | ·0989 284 | ·0523 023 | ·0273 727 | ·0142 151 | ·0073 371 |
| ·25 | ·1955 011 | ·1069 113 | ·0576 689 | ·0307 958 | ·0163 192 | ·0085 954 |
| ·26 | ·2065 999 | ·1151 648 | ·0633 295$^{+}$ | ·0344 793 | ·0186 291 | ·0100 047 |
| ·27 | ·2178 381 | ·1236 843 | ·0692 860 | ·0384 306 | ·0211 551 | ·0115 757 |
| ·28 | ·2292 081 | ·1324 648 | ·0755 397 | ·0426 566 | ·0239 071 | ·0133 193 |
| ·29 | ·2407 030 | ·1415 017 | ·0820 920 | ·0471 641 | ·0268 954 | ·0152 466 |
| ·30 | ·2523 158 | ·1507 901 | ·0889 437 | ·0519 596 | ·0301 298 | ·0173 689 |
| ·31 | ·2640 397 | ·1603 249 | ·0960 955$^{-}$ | ·0570 492 | ·0336 202 | ·0196 978 |
| ·32 | ·2758 682 | ·1701 013 | ·1035 476 | ·0624 388 | ·0373 765$^{+}$ | ·0222 448 |
| ·33 | ·2877 950$^{+}$ | ·1801 141 | ·1113 002 | ·0681 339 | ·0414 083 | ·0250 215$^{+}$ |
| ·34 | ·2998 139 | ·1903 583 | ·1193 530 | ·0741 399 | ·0457 250$^{-}$ | ·0280 399 |
| ·35 | ·3119 188 | ·2008 287 | ·1277 055$^{+}$ | ·0804 617 | ·0503 360 | ·0313 119 |
| ·36 | ·3241 038 | ·2115 200$^{e}$ | ·1363 570 | ·0871 040$^{e}$ | ·0552 504 | ·0348 493 |
| ·37 | ·3363 631 | ·2224 269 | ·1453 064 | ·0940 711 | ·0604 772 | ·0386 642 |
| ·38 | ·3486 910 | ·2335 441 | ·1545 524 | ·1013 670 | ·0660 252 | ·0427 686 |
| ·39 | ·3610 818 | ·2448 660 | ·1640 934 | ·1089 955$^{-}$ | ·0719 029 | ·0471 744 |
| ·40 | ·3735 300 | ·2563 872 | ·1739 277 | ·1169 598 | ·0781 185$^{+}$ | ·0518 937 |
| ·41 | ·3860 303 | ·2681 020 | ·1840 529 | ·1252 630 | ·0846 801 | ·0569 383 |
| ·42 | ·3985 771 | ·2800 048 | ·1944 669 | ·1339 076 | ·0915 953 | ·0623 200 |
| ·43 | ·4111 652 | ·2920 898 | ·2051 669 | ·1428 961 | ·0988 717 | ·0680 506 |
| ·44 | ·4237 894 | ·3043 511 | ·2161 499 | ·1522 302 | ·1065 163 | ·0741 415$^{+}$ |
| ·45 | ·4364 443 | ·3167 827 | ·2274 128 | ·1619 116 | ·1145 358 | ·0806 043 |
| ·46 | ·4491 248 | ·3293 787 | ·2389 521 | ·1719 414 | ·1229 368 | ·0874 501 |
| ·47 | ·4618 257 | ·3421 329 | ·2507 640 | ·1823 204 | ·1317 252 | ·0946 898 |
| ·48 | ·4745 420 | ·3550 390 | ·2628 445$^{+}$ | ·1930 488 | ·1409 067 | ·1023 343 |
| ·49 | ·4872 685$^{-}$ | ·3680 907 | ·2751 892 | ·2041 266 | ·1504 866 | ·1103 938 |
| ·50 | ·5000 000$^{e}$ | ·3812 816 | ·2877 934 | ·2155 534 | ·1604 695$^{-}$ | ·1188 787 |
| ·51 | ·5127 315$^{+}$ | ·3946 050$^{e}$ | ·3006 523 | ·2273 282 | ·1708 597 | ·1277 985$^{-}$ |
| ·52 | ·5254 580 | ·4080 543 | ·3137 605$^{+}$ | ·2394 496 | ·1816 613 | ·1371 627 |
| ·53 | ·5381 743 | ·4216 227 | ·3271 125$^{+}$ | ·2519 157 | ·1928 773 | ·1469 802 |
| ·54 | ·5508 752 | ·4353 032 | ·3407 025$^{+}$ | ·2647 242 | ·2045 106 | ·1572 595$^{-}$ |
| ·55 | ·5635 557 | ·4490 888 | ·3545 243 | ·2778 723 | ·2165 635$^{-}$ | ·1680 085$^{-}$ |
| ·56 | ·5762 106 | ·4629 721 | ·3685 712 | ·2913 567 | ·2290 374 | ·1792 346 |
| ·57 | ·5888 348 | ·4769 459 | ·3828 364 | ·3051 734 | ·2419 335$^{-}$ | ·1909 447 |
| ·58 | ·6014 229 | ·4910 026 | ·3973 126 | ·3193 181 | ·2552 519 | ·2031 449 |
| ·59 | ·6139 697 | ·5051 345$^{+}$ | ·4119 924 | ·3337 857 | ·2689 924 | ·2158 406 |
| ·60 | ·6264 700 | ·5193 338 | ·4268 676 | ·3485 708 | ·2831 539 | ·2290 367 |

$x$ = ·61 to 1·00 $q = 1·5$ $p$ = 1·5 to 4

| | $p = 1·5$ | $p = 2$ | $p = 2·5$ | $p = 3$ | $p = 3·5$ | $p = 4$ |
|---|---|---|---|---|---|---|
| $B(p, q)$ = | ·3926 9908 | ·2666 6667 | ·1963 4954 | ·1523 8095+ | ·1227 1846 | ·1015 8730 |
| $x$ | | | | | | |
| ·61 | ·6389 182 | ·5335 923 | ·4419 299 | ·3636 671 | ·2977 344 | ·2427 370 |
| ·62 | ·6513 090 | ·5479 019 | ·4571 705− | ·3790 678 | ·3127 314 | ·2569 445+ |
| ·63 | ·6636 369 | ·5622 540 | ·4725 802 | ·3947 655+ | ·3281 413 | ·2716 615− |
| ·64 | ·6758 962 | ·5766 400$^e$ | ·4881 494 | ·4107 520$^e$ | ·3439 598 | ·2868 890 |
| ·65 | ·6880 812 | ·5910 510 | ·5038 679 | ·4270 184 | ·3601 815+ | ·3026 271 |
| ·66 | ·7001 861 | ·6054 778 | ·5197 252 | ·4435 552 | ·3768 001 | ·3188 748 |
| ·67 | ·7122 050− | ·6199 110 | ·5357 101 | ·4603 518 | ·3938 083 | ·3356 297 |
| ·68 | ·7241 318 | ·6343 409 | ·5518 112 | ·4773 972 | ·4111 976 | ·3528 885− |
| ·69 | ·7359 603 | ·6487 576 | ·5680 161 | ·4946 791 | ·4289 583 | ·3706 459 |
| ·70 | ·7476 842 | ·6631 506 | ·5843 121 | ·5121 846 | ·4470 796 | ·3888 957 |
| ·71 | ·7592 970 | ·6775 094 | ·6006 859 | ·5298 997 | ·4655 493 | ·4076 296 |
| ·72 | ·7707 919 | ·6918 229 | ·6171 234 | ·5478 094 | ·4843 539 | ·4268 380 |
| ·73 | ·7821 619 | ·7060 796 | ·6336 098 | ·5658 975+ | ·5034 782 | ·4465 091 |
| ·74 | ·7934 001 | ·7202 678 | ·6501 297 | ·5841 469 | ·5229 056 | ·4666 292 |
| ·75 | ·8044 989 | ·7343 750$^e$ | ·6666 667 | ·6025 391 | ·5426 177 | ·4871 826 |
| ·76 | ·8154 506 | ·7483 884 | ·6832 036 | ·6210 541 | ·5625 942 | ·5081 511 |
| ·77 | ·8262 473 | ·7622 946 | ·6997 222 | ·6396 709 | ·5828 130 | ·5295 139 |
| ·78 | ·8368 806 | ·7760 796 | ·7162 035+ | ·6583 665+ | ·6032 498 | ·5512 477 |
| ·79 | ·8473 417 | ·7897 285+ | ·7326 272 | ·6771 166 | ·6238 778 | ·5733 259 |
| ·80 | ·8576 215+ | ·8032 260 | ·7489 717 | ·6958 948 | ·6446 680 | ·5957 189 |
| ·81 | ·8677 103 | ·8165 557 | ·7652 143 | ·7146 727 | ·6655 882 | ·6183 933 |
| ·82 | ·8775 977 | ·8297 004 | ·7813 305+ | ·7334 200 | ·6866 036 | ·6413 118 |
| ·83 | ·8872 730 | ·8426 417 | ·7972 944 | ·7521 037 | ·7076 757 | ·6644 327 |
| ·84 | ·8967 245+ | ·8553 600$^e$ | ·8130 780 | ·7706 880$^e$ | ·7287 624 | ·6877 094 |
| ·85 | ·9059 398 | ·8678 344 | ·8286 514 | ·7891 342 | ·7498 173 | ·7110 898 |
| ·86 | ·9149 054 | ·8800 425− | ·8439 821 | ·8074 001 | ·7707 893 | ·7345 155+ |
| ·87 | ·9236 066 | ·8919 597 | ·8590 348 | ·8254 393 | ·7916 219 | ·7579 211 |
| ·88 | ·9320 276 | ·9035 594 | ·8737 710 | ·8432 009 | ·8122 521 | ·7812 328 |
| ·89 | ·9401 506 | ·9148 125− | ·8881 482 | ·8606 286 | ·8326 097 | ·8043 676 |
| ·90 | ·9479 560 | ·9256 865− | ·9021 194 | ·8776 594 | ·8526 158 | ·8272 309 |
| ·91 | ·9554 216 | ·9361 450$^e$ | ·9156 315+ | ·8942 224 | ·8721 808 | ·8497 147 |
| ·92 | ·9625 220 | ·9461 467 | ·9286 247 | ·9102 370 | ·8912 020 | ·8716 939 |
| ·93 | ·9692 278 | ·9556 440 | ·9410 296 | ·9256 099 | ·9095 605+ | ·8930 229 |
| ·94 | ·9755 037 | ·9645 804 | ·9527 649 | ·9402 312 | ·9271 156 | ·9135 283 |
| ·95 | ·9813 070 | ·9728 877 | ·9637 322 | ·9539 684 | ·9436 970 | ·9329 996 |
| ·96 | ·9865 829 | ·9804 800$^e$ | ·9738 084 | ·9666 560$^e$ | ·9590 921 | ·9511 731 |
| ·97 | ·9912 586 | ·9872 434 | ·9828 313 | ·9780 765− | ·9730 219 | ·9677 025− |
| ·98 | ·9952 272 | ·9930 138 | ·9905 689 | ·9879 205− | ·9850 906 | ·9820 972 |
| ·99 | ·9983 074 | ·9975 150$^e$ | ·9966 352 | ·9956 773 | ·9946 486 | ·9935 548 |
| 1·00 | 1·0000 000 | 1·0000 000 | 1·0000 000 | 1·0000 000 | 1·0000 000 | 1·0000 000 |

$x =$ ·02 to ·60 $q =$ 1·5 $p =$ 4·5 to 7

| | $p = 4·5$ | $p = 5$ | $p = 5·5$ | $p = 6$ | $p = 6·5$ | $p = 7$ |
|---|---|---|---|---|---|---|
| $B(p, q) =$ | ·8590 2924 × $\frac{1}{10}$ | ·7388 1674 × $\frac{1}{10}$ | ·6442 7193 × $\frac{1}{10}$ | ·5683 2057 × $\frac{1}{10}$ | ·5062 1366 × $\frac{1}{10}$ | ·4546 5645+ × $\frac{1}{10}$ |
| $x$ | | | | | | |
| ·02 | ·0000 001 | | | | | |
| ·03 | ·0000 004 | ·0000 001 | | | | |
| ·04 | ·0000 013 | ·0000 003 | ·0000 001 | | | |
| ·05 | ·0000 035+ | ·0000 008 | ·0000 002 | | | |
| ·06 | ·0000 080 | ·0000 021 | ·0000 005+ | ·0000 001 | | |
| ·07 | ·0000 160 | ·0000 044 | ·0000 012 | ·0000 003 | ·0000 001 | |
| ·08 | ·0000 290 | ·0000 086 | ·0000 025+ | ·0000 007 | ·0000 002 | ·0000 001 |
| ·09 | ·0000 490 | ·0000 154 | ·0000 048 | ·0000 015− | ·0000 005− | ·0000 001 |
| ·10 | ·0000 784 | ·0000 259 | ·0000 085+ | ·0000 028 | ·0000 009 | ·0000 003 |
| ·11 | ·0001 198 | ·0000 415+ | ·0000 144 | ·0000 049 | ·0000 017 | ·0000 006 |
| ·12 | ·0001 765− | ·0000 639 | ·0000 231 | ·0000 083 | ·0000 030 | ·0000 011 |
| ·13 | ·0002 518 | ·0000 949 | ·0000 356 | ·0000 133 | ·0000 050− | ·0000 019 |
| ·14 | ·0003 499 | ·0001 368 | ·0000 533 | ·0000 207 | ·0000 080 | ·0000 031 |
| ·15 | ·0004 750+ | ·0001 923 | ·0000 775+ | ·0000 312 | ·0000 125+ | ·0000 050+ |
| ·16 | ·0006 321 | ·0002 642 | ·0001 101 | ·0000 457 | ·0000 189 | ·0000 078 |
| ·17 | ·0008 265− | ·0003 561 | ·0001 529 | ·0000 654 | ·0000 279 | ·0000 119 |
| ·18 | ·0010 638 | ·0004 715+ | ·0002 083 | ·0000 917 | ·0000 403 | ·0000 177 |
| ·19 | ·0013 503 | ·0006 149 | ·0002 790 | ·0001 262 | ·0000 570 | ·0000 256 |
| ·20 | ·0016 926 | ·0007 907 | ·0003 681 | ·0001 708 | ·0000 791 | ·0000 365+ |
| ·21 | ·0020 977 | ·0010 040 | ·0004 789 | ·0002 277 | ·0001 080 | ·0000 511 |
| ·22 | ·0025 734 | ·0012 605+ | ·0006 153 | ·0002 995− | ·0001 454 | ·0000 704 |
| ·23 | ·0031 274 | ·0015 662 | ·0007 816 | ·0003 889 | ·0001 930 | ·0000 956 |
| ·24 | ·0037 684 | ·0019 275+ | ·0009 825+ | ·0004 994 | ·0002 532 | ·0001 281 |
| ·25 | ·0045 050+ | ·0023 516 | ·0012 233 | ·0006 345+ | ·0003 283 | ·0001 695− |
| ·26 | ·0053 468 | ·0028 459 | ·0015 096 | ·0007 985− | ·0004 213 | ·0002 218 |
| ·27 | ·0063 033 | ·0034 185+ | ·0018 477 | ·0009 958 | ·0005 354 | ·0002 872 |
| ·28 | ·0073 847 | ·0040 780 | ·0022 444 | ·0012 317 | ·0006 743 | ·0003 683 |
| ·29 | ·0086 016 | ·0048 335− | ·0027 070 | ·0015 117 | ·0008 421 | ·0004 681 |
| ·30 | ·0099 650− | ·0056 946 | ·0032 434 | ·0018 421 | ·0010 436 | ·0005 900 |
| ·31 | ·0114 862 | ·0066 715+ | ·0038 622 | ·0022 295+ | ·0012 839 | ·0007 378 |
| ·32 | ·0131 768 | ·0077 749 | ·0045 725− | ·0026 815+ | ·0015 688 | ·0009 158 |
| ·33 | ·0150 490 | ·0090 160 | ·0053 840 | ·0032 061 | ·0019 045+ | ·0011 290 |
| ·34 | ·0171 152 | ·0104 067 | ·0063 071 | ·0038 118 | ·0022 982 | ·0013 828 |
| ·35 | ·0193 881 | ·0119 591 | ·0073 529 | ·0045 083 | ·0027 576 | ·0016 832 |
| ·36 | ·0218 808 | ·0136 861 | ·0085 330 | ·0053 055+ | ·0032 909 | ·0020 371 |
| ·37 | ·0246 066 | ·0156 010 | ·0098 598 | ·0062 144 | ·0039 075− | ·0024 518 |
| ·38 | ·0275 791 | ·0177 177 | ·0113 464 | ·0072 464 | ·0046 171 | ·0029 358 |
| ·39 | ·0308 123 | ·0200 504 | ·0130 063 | ·0084 142 | ·0054 307 | ·0034 979 |
| ·40 | ·0343 200 | ·0226 139 | ·0148 541 | ·0097 308 | ·0063 598 | ·0041 482 |
| ·41 | ·0381 168 | ·0254 235+ | ·0169 046 | ·0112 104 | ·0074 170 | ·0048 973 |
| ·42 | ·0422 170 | ·0284 949 | ·0191 738 | ·0128 676 | ·0086 158 | ·0057 573 |
| ·43 | ·0466 352 | ·0318 441 | ·0216 778 | ·0147 184 | ·0099 705− | ·0067 407 |
| ·44 | ·0513 862 | ·0354 876 | ·0244 337 | ·0167 792 | ·0114 965+ | ·0078 614 |
| ·45 | ·0564 848 | ·0394 425− | ·0274 593 | ·0190 674 | ·0132 104 | ·0091 345+ |
| ·46 | ·0619 459 | ·0437 258 | ·0307 728 | ·0216 013 | ·0151 295+ | ·0105 760 |
| ·47 | ·0677 844 | ·0483 552 | ·0343 931 | ·0244 002 | ·0172 724 | ·0122 031 |
| ·48 | ·0740 152 | ·0533 484 | ·0383 397 | ·0274 839 | ·0196 587 | ·0140 344 |
| ·49 | ·0806 531 | ·0587 236 | ·0426 327 | ·0308 734 | ·0223 090 | ·0160 896 |
| ·50 | ·0877 129 | ·0644 991 | ·0472 926 | ·0345 904 | ·0252 452 | ·0183 898 |
| ·51 | ·0952 093 | ·0706 933 | ·0523 406 | ·0386 573 | ·0284 901 | ·0209 574 |
| ·52 | ·1031 566 | ·0773 249 | ·0577 983 | ·0430 976 | ·0320 678 | ·0238 163 |
| ·53 | ·1115 691 | ·0844 124 | ·0636 876 | ·0479 354 | ·0360 034 | ·0269 915− |
| ·54 | ·1204 608 | ·0919 746 | ·0700 309 | ·0531 954 | ·0403 231 | ·0305 096 |
| ·55 | ·1298 453 | ·1000 302 | ·0768 508 | ·0589 033 | ·0450 542 | ·0343 986 |
| ·56 | ·1397 359 | ·1085 977 | ·0841 704 | ·0650 853 | ·0502 250− | ·0386 879 |
| ·57 | ·1501 453 | ·1176 955+ | ·0920 128 | ·0717 683 | ·0558 649 | ·0434 082 |
| ·58 | ·1610 860 | ·1273 419 | ·1004 012 | ·0789 795+ | ·0620 043 | ·0485 919 |
| ·59 | ·1725 696 | ·1375 546 | ·1093 591 | ·0867 469 | ·0686 745− | ·0542 724 |
| ·60 | ·1846 073 | ·1483 512 | ·1189 096 | ·0950 988 | ·0759 075− | ·0604 847 |

$x$ = ·61 to 1·00 $q$ = 1·5 $p$ = 4·5 to 7

| $x$ | $p$ = 4·5 | $p$ = 5 | $p$ = 5·5 | $p$ = 6 | $p$ = 6·5 | $p$ = 7 |
|---|---|---|---|---|---|---|
| $B(p, q)$ = | ·8590 2924 × $\frac{1}{10}$ | ·7388 1674 × $\frac{1}{10}$ | ·6442 7193 × $\frac{1}{10}$ | ·5683 2057 × $\frac{1}{10}$ | ·5062 1366 × $\frac{1}{10}$ | ·4546 5645+ × $\frac{1}{10}$ |
| ·61 | ·1972 096 | ·1597 487 | ·1290 761 | ·1040 635+ | ·0837 363 | ·0672 649 |
| ·62 | ·2103 860 | ·1717 635+ | ·1398 814 | ·1136 701 | ·0921 946 | ·0746 507 |
| ·63 | ·2241 453 | ·1844 115− | ·1513 482 | ·1239 472 | ·1013 166 | ·0826 804 |
| ·64 | ·2384 955− | ·1977 076 | ·1634 986 | ·1349 239 | ·1111 371 | ·0913 938 |
| ·65 | ·2534 431 | ·2116 659 | ·1763 542 | ·1466 287 | ·1216 912 | ·1008 316 |
| ·66 | ·2689 938 | ·2262 996 | ·1899 359 | ·1590 899 | ·1330 141 | ·1110 351 |
| ·67 | ·2851 520 | ·2416 205− | ·2042 635− | ·1723 356 | ·1451 413 | ·1220 464 |
| ·68 | ·3019 207 | ·2576 393 | ·2193 560 | ·1863 929 | ·1581 079 | ·1339 081 |
| ·69 | ·3193 013 | ·2743 652 | ·2352 309 | ·2012 881 | ·1719 488 | ·1466 629 |
| ·70 | ·3372 936 | ·2918 056 | ·2519 045− | ·2170 463 | ·1866 982 | ·1603 538 |
| ·71 | ·3558 957 | ·3099 664 | ·2693 911 | ·2336 915− | ·2023 893 | ·1750 233 |
| ·72 | ·3751 036 | ·3288 512 | ·2877 033 | ·2512 456 | ·2190 543 | ·1907 133 |
| ·73 | ·3949 112 | ·3484 613 | ·3068 514 | ·2697 290 | ·2367 237 | ·2074 648 |
| ·74 | ·4153 103 | ·3687 957 | ·3268 431 | ·2891 593 | ·2554 260 | ·2253 174 |
| ·75 | ·4362 899 | ·3898 506 | ·3476 835+ | ·3095 517 | ·2751 874 | ·2443 089 |
| ·76 | ·4578 364 | ·4116 190 | ·3693 742 | ·3309 181 | ·2960 310 | ·2644 744 |
| ·77 | ·4799 329 | ·4340 904 | ·3919 133 | ·3532 668 | ·3179 768 | ·2858 463 |
| ·78 | ·5025 596 | ·4572 509 | ·4152 948 | ·3766 017 | ·3410 403 | ·3084 530 |
| ·79 | ·5256 927 | ·4810 820 | ·4395 079 | ·4009 220 | ·3652 324 | ·3323 184 |
| ·80 | ·5493 045− | ·5055 606 | ·4645 370 | ·4262 214 | ·3905 580 | ·3574 607 |
| ·81 | ·5733 629 | ·5306 587 | ·4903 602 | ·4524 871 | ·4170 159 | ·3838 916 |
| ·82 | ·5978 309 | ·5563 420 | ·5169 491 | ·4796 992 | ·4445 967 | ·4116 149 |
| ·83 | ·6226 659 | ·5825 700 | ·5442 680 | ·5078 293 | ·4732 823 | ·4406 250− |
| ·84 | ·6478 193 | ·6092 947 | ·5722 725− | ·5368 395− | ·5030 441 | ·4709 052 |
| ·85 | ·6732 356 | ·6364 599 | ·6009 084 | ·5666 809 | ·5338 413 | ·5024 260 |
| ·86 | ·6988 513 | ·6639 997 | ·6301 105− | ·5972 918 | ·5656 191 | ·5351 422 |
| ·87 | ·7245 942 | ·6918 376 | ·6598 007 | ·6285 958 | ·5983 058 | ·5689 903 |
| ·88 | ·7503 817 | ·7198 845− | ·6898 861 | ·6604 992 | ·6318 104 | ·6038 853 |
| ·89 | ·7761 192 | ·7480 363 | ·7202 562 | ·6928 880 | ·6660 184 | ·6397 158 |
| ·90 | ·8016 979 | ·7761 721 | ·7507 799 | ·7256 239 | ·7007 878 | ·6763 394 |
| ·91 | ·8269 920 | ·8041 498 | ·7813 011 | ·7585 394 | ·7359 425− | ·7135 751 |
| ·92 | ·8518 548 | ·8318 018 | ·8116 331 | ·7914 310 | ·7712 652 | ·7511 948 |
| ·93 | ·8761 134 | ·8589 288 | ·8415 513 | ·8240 505+ | ·8064 865− | ·7889 107 |
| ·94 | ·8995 608 | ·8852 900 | ·8707 814 | ·8560 916 | ·8412 695− | ·8263 578 |
| ·95 | ·9219 445− | ·9105 892 | ·8989 835− | ·8871 703 | ·8751 875+ | ·8630 684 |
| ·96 | ·9429 462 | ·9344 516 | ·9257 240 | ·9167 937 | ·9076 876 | ·8984 295− |
| ·97 | ·9621 475− | ·9563 819 | ·9504 273 | ·9443 028 | ·9380 252 | ·9316 097 |
| ·98 | ·9789 550+ | ·9756 769 | ·9722 740 | ·9687 560 | ·9651 314 | ·9614 082 |
| ·99 | ·9924 008 | ·9911 908 | ·9899 283 | ·9886 165− | ·9872 580 | ·9858 555− |
| 1·00 | 1·0000 000 | 1·0000 000 | 1·0000 000 | 1·0000 000 | 1·0000 000 | 1·0000 000 |

$x = \cdot 10$ to $\cdot 70$ $q = 1\cdot 5$ $p = 7\cdot 5$ to $10$

| | $p = 7\cdot 5$ | $p = 8$ | $p = 8\cdot 5$ | $p = 9$ | $p = 9\cdot 5$ | $p = 10$ |
|---|---|---|---|---|---|---|
| $B(p, q) =$ | $\cdot 4112\ 9860 \times \frac{1}{10}$ | $\cdot 3744\ 2296 \times \frac{1}{10}$ | $\cdot 3427\ 4883 \times \frac{1}{10}$ | $\cdot 3153\ 0355 \times \frac{1}{10}$ | $\cdot 2913\ 3651 \times \frac{1}{10}$ | $\cdot 2702\ 6018 \times \frac{1}{10}$ |
| $x$ | | | | | | |
| ·10 | ·0000 001 | | | | | |
| ·11 | ·0000 002 | ·0000 001 | | | | |
| ·12 | ·0000 004 | ·0000 001 | | | | |
| ·13 | ·0000 007 | ·0000 003 | ·0000 001 | | | |
| ·14 | ·0000 012 | ·0000 005− | ·0000 002 | ·0000 001 | | |
| ·15 | ·0000 020 | ·0000 008 | ·0000 003 | ·0000 001 | ·0000 001 | |
| ·16 | ·0000 032 | ·0000 013 | ·0000 005+ | ·0000 002 | ·0000 001 | |
| ·17 | ·0000 051 | ·0000 021 | ·0000 009 | ·0000 004 | ·0000 002 | ·0000 001 |
| ·18 | ·0000 077 | ·0000 034 | ·0000 015− | ·0000 006 | ·0000 003 | ·0000 001 |
| ·19 | ·0000 115+ | ·0000 052 | ·0000 023 | ·0000 010 | ·0000 005− | ·0000 002 |
| ·20 | ·0000 168 | ·0000 077 | ·0000 036 | ·0000 016 | ·0000 007 | ·0000 003 |
| ·21 | ·0000 241 | ·0000 114 | ·0000 054 | ·0000 025+ | ·0000 012 | ·0000 006 |
| ·22 | ·0000 340 | ·0000 164 | ·0000 079 | ·0000 038 | ·0000 018 | ·0000 009 |
| ·23 | ·0000 473 | ·0000 233 | ·0000 115− | ·0000 057 | ·0000 028 | ·0000 014 |
| ·24 | ·0000 647 | ·0000 326 | ·0000 164 | ·0000 082 | ·0000 041 | ·0000 021 |
| ·25 | ·0000 873 | ·0000 449 | ·0000 231 | ·0000 118 | ·0000 061 | ·0000 031 |
| ·26 | ·0001 165+ | ·0000 611 | ·0000 320 | ·0000 167 | ·0000 087 | ·0000 046 |
| ·27 | ·0001 538 | ·0000 822 | ·0000 439 | ·0000 234 | ·0000 124 | ·0000 066 |
| ·28 | ·0002 008 | ·0001 093 | ·0000 594 | ·0000 322 | ·0000 175− | ·0000 095− |
| ·29 | ·0002 597 | ·0001 439 | ·0000 796 | ·0000 439 | ·0000 242 | ·0000 134 |
| ·30 | ·0003 329 | ·0001 875+ | ·0001 055− | ·0000 593 | ·0000 332 | ·0000 186 |
| ·31 | ·0004 232 | ·0002 423 | ·0001 385+ | ·0000 791 | ·0000 451 | ·0000 257 |
| ·32 | ·0005 336 | ·0003 104 | ·0001 803 | ·0001 046 | ·0000 606 | ·0000 351 |
| ·33 | ·0006 680 | ·0003 946 | ·0002 327 | ·0001 371 | ·0000 807 | ·0000 474 |
| ·34 | ·0008 304 | ·0004 979 | ·0002 981 | ·0001 782 | ·0001 064 | ·0000 635− |
| ·35 | ·0010 255+ | ·0006 238 | ·0003 789 | ·0002 298 | ·0001 392 | ·0000 843 |
| ·36 | ·0012 586 | ·0007 764 | ·0004 782 | ·0002 942 | ·0001 807 | ·0001 109 |
| ·37 | ·0015 357 | ·0009 603 | ·0005 996 | ·0003 739 | ·0002 329 | ·0001 449 |
| ·38 | ·0018 633 | ·0011 807 | ·0007 471 | ·0004 721 | ·0002 980 | ·0001 879 |
| ·39 | ·0022 489 | ·0014 436 | ·0009 253 | ·0005 923 | ·0003 787 | ·0002 419 |
| ·40 | ·0027 008 | ·0017 556 | ·0011 396 | ·0007 387 | ·0004 783 | ·0003 094 |
| ·41 | ·0032 279 | ·0021 241 | ·0013 958 | ·0009 160 | ·0006 005− | ·0003 932 |
| ·42 | ·0038 403 | ·0025 576 | ·0017 009 | ·0011 297 | ·0007 495− | ·0004 967 |
| ·43 | ·0045 491 | ·0030 652 | ·0020 625− | ·0013 860 | ·0009 303 | ·0006 238 |
| ·44 | ·0053 663 | ·0036 574 | ·0024 892 | ·0016 920 | ·0011 488 | ·0007 792 |
| ·45 | ·0063 052 | ·0043 455− | ·0029 907 | ·0020 557 | ·0014 114 | ·0009 680 |
| ·46 | ·0073 801 | ·0051 420 | ·0035 777 | ·0024 862 | ·0017 257 | ·0011 966 |
| ·47 | ·0086 067 | ·0060 610 | ·0042 623 | ·0029 938 | ·0021 004 | ·0014 721 |
| ·48 | ·0100 020 | ·0071 174 | ·0050 578 | ·0035 898 | ·0025 450+ | ·0018 025− |
| ·49 | ·0115 844 | ·0083 280 | ·0059 789 | ·0042 872 | ·0030 707 | ·0021 972 |
| ·50 | ·0133 735− | ·0097 109 | ·0070 419 | ·0051 002 | ·0036 899 | ·0026 668 |
| ·51 | ·0153 906 | ·0112 857 | ·0082 645+ | ·0060 448 | ·0044 165− | ·0032 235+ |
| ·52 | ·0176 587 | ·0130 738 | ·0096 665− | ·0071 386 | ·0052 660 | ·0038 808 |
| ·53 | ·0202 021 | ·0150 983 | ·0112 691 | ·0084 010 | ·0062 561 | ·0046 542 |
| ·54 | ·0230 468 | ·0173 842 | ·0130 957 | ·0098 535+ | ·0074 060 | ·0055 610 |
| ·55 | ·0262 208 | ·0199 583 | ·0151 718 | ·0115 197 | ·0087 374 | ·0066 207 |
| ·56 | ·0297 533 | ·0228 494 | ·0175 249 | ·0134 255− | ·0102 741 | ·0078 549 |
| ·57 | ·0336 757 | ·0260 883 | ·0201 847 | ·0155 990 | ·0120 424 | ·0092 879 |
| ·58 | ·0380 211 | ·0297 081 | ·0231 834 | ·0180 710 | ·0140 713 | ·0109 465− |
| ·59 | ·0428 241 | ·0337 438 | ·0265 556 | ·0208 750− | ·0163 926 | ·0128 606 |
| ·60 | ·0481 215− | ·0382 328 | ·0303 385− | ·0240 472 | ·0190 410 | ·0150 630 |
| ·61 | ·0539 516 | ·0432 144 | ·0345 717 | ·0276 267 | ·0220 545+ | ·0175 899 |
| ·62 | ·0603 545+ | ·0487 306 | ·0392 976 | ·0316 558 | ·0254 744 | ·0204 812 |
| ·63 | ·0673 722 | ·0548 253 | ·0445 615− | ·0361 798 | ·0293 454 | ·0237 805− |
| ·64 | ·0750 480 | ·0615 447 | ·0504 112 | ·0412 472 | ·0337 161 | ·0275 352 |
| ·65 | ·0834 271 | ·0689 372 | ·0568 973 | ·0469 102 | ·0386 385+ | ·0317 972 |
| ·66 | ·0925 559 | ·0770 534 | ·0640 733 | ·0532 238 | ·0441 690 | ·0366 225− |
| ·67 | ·1024 824 | ·0859 459 | ·0719 955− | ·0602 469 | ·0503 677 | ·0420 719 |
| ·68 | ·1132 555+ | ·0956 691 | ·0807 226 | ·0680 415− | ·0572 989 | ·0482 110 |
| ·69 | ·1249 254 | ·1062 794 | ·0903 161 | ·0766 732 | ·0650 310 | ·0551 100 |
| ·70 | ·1375 427 | ·1178 345− | ·1008 400 | ·0862 107 | ·0736 368 | ·0628 442 |

$x = \cdot71$ to $1\cdot00$ $q = 1\cdot5$ $p = 7\cdot5$ to $10$

| | $p = 7\cdot5$ | $p = 8$ | $p = 8\cdot5$ | $p = 9$ | $p = 9\cdot5$ | $p = 10$ |
|---|---|---|---|---|---|---|
| $B(p, q) =$ | $\cdot4112\ 9860 \times \frac{1}{10}$ | $\cdot3744\ 2296 \times \frac{1}{10}$ | $\cdot3427\ 4883 \times \frac{1}{10}$ | $\cdot3153\ 0355 \bar{\times} \frac{1}{10}$ | $\cdot2913\ 3651 \times \frac{1}{10}$ | $\cdot2702\ 6018 \times \frac{1}{10}$ |
| $x$ | | | | | | |
| ·71 | ·1511 588 | ·1303 936 | ·1123 602 | ·0967 261 | ·0831 928 | ·0714 941 |
| ·72 | ·1658 250⁻ | ·1440 169 | ·1249 449 | ·1082 942 | ·0937 798 | ·0811 449 |
| ·73 | ·1815 925⁺ | ·1587 653 | ·1386 638 | ·1209 928 | ·1054 823 | ·0918 870 |
| ·74 | ·1985 120 | ·1746 999 | ·1535 879 | ·1349 019 | ·1183 885⁺ | ·1038 152 |
| ·75 | ·2166 328 | ·1918 816 | ·1697 891 | ·1501 036 | ·1325 897 | ·1170 293 |
| ·76 | ·2360 025⁻ | ·2103 703 | ·1873 393 | ·1666 812 | ·1481 798 | ·1316 328 |
| ·77 | ·2566 664 | ·2302 245⁻ | ·2063 101 | ·1847 189 | ·1652 549 | ·1477 329 |
| ·78 | ·2786 666 | ·2515 003 | ·2267 716 | ·2043 007 | ·1839 125⁺ | ·1654 397 |
| ·79 | ·3020 410 | ·2742 503 | ·2487 917 | ·2255 095⁻ | ·2042 503 | ·1848 650⁺ |
| ·80 | ·3268 224 | ·2985 229 | ·2724 346 | ·2484 259 | ·2263 649 | ·2061 217 |
| ·81 | ·3530 371 | ·3243 604 | ·2977 595⁻ | ·2731 265⁻ | ·2503 508 | ·2293 214 |
| ·82 | ·3807 040 | ·3517 979 | ·3248 191 | ·2996 824 | ·2762 979 | ·2545 735⁻ |
| ·83 | ·4098 320 | ·3808 611 | ·3536 573 | ·3281 568 | ·3042 897 | ·2819 821 |
| ·84 | ·4404 191 | ·4115 644 | ·3843 070 | ·3586 027 | ·3344 003 | ·3116 434 |
| ·85 | ·4724 492 | ·4439 082 | ·4167 870 | ·3910 593 | ·3666 910 | ·3436 421 |
| ·86 | ·5058 902 | ·4778 758 | ·4510 988 | ·4255 486 | ·4012 065⁻ | ·3780 472 |
| ·87 | ·5406 898 | ·5134 295⁺ | ·4872 222 | ·4620 705⁺ | ·4379 691 | ·4149 059 |
| ·88 | ·5767 725⁻ | ·5505 064 | ·5251 102 | ·5005 972 | ·4769 731 | ·4542 371 |
| ·89 | ·6140 336 | ·5890 124 | ·5646 826 | ·5410 659 | ·5181 766 | ·4960 229 |
| ·90 | ·6523 338 | ·6288 151 | ·6058 180 | ·5833 699 | ·5614 912 | ·5401 970 |
| ·91 | ·6914 911 | ·6697 350⁻ | ·6483 436 | ·6273 470 | ·6067 697 | ·5866 310 |
| ·92 | ·7312 700 | ·7115 334 | ·6920 213 | ·6727 643 | ·6537 884 | ·6351 153 |
| ·93 | ·7713 677 | ·7538 963 | ·7365 299 | ·7192 977 | ·7022 249 | ·6853 335⁻ |
| ·94 | ·8113 943 | ·7964 117 | ·7814 394 | ·7665 031 | ·7516 255⁺ | ·7368 270 |
| ·95 | ·8508 424 | ·8385 360 | ·8261 728 | ·8137 737 | ·8013 580 | ·7889 426 |
| ·96 | ·8890 407 | ·8795 405⁺ | ·8699 463 | ·8602 738 | ·8505 374 | ·8407 502 |
| ·97 | ·9250 698 | ·9184 180 | ·9116 653 | ·9048 222 | ·8978 981 | ·8909 017 |
| ·98 | ·9575 932 | ·9536 930 | ·9497 133 | ·9456 595⁺ | ·9415 367 | ·9373 494 |
| ·99 | ·9844 110 | ·9829 268 | ·9814 047 | ·9798 463 | ·9782 533 | ·9766 271 |
| 1·00 | 1·0000 000 | 1·0000 000 | 1·0000 000 | 1·0000 000 | 1·0000 000 | 1·0000 000 |

$x$ = ·18 to ·80 $q$ = 1·5 $p$ = 10·5 to 15

| | $p$ = 10·5 | $p$ = 11 | $p$ = 12 | $p$ = 13 | $p$ = 14 | $p$ = 15 |
|---|---|---|---|---|---|---|
| $B(p, q)$ = | ·2516 0880 × $\frac{1}{10}$ | ·2350 0886 × $\frac{1}{10}$ | ·2068 0779 × $\frac{1}{10}$ | ·1838 2915 × $\frac{1}{10}$ | ·1648 1234 × $\frac{1}{10}$ | ·1488 6276 × $\frac{1}{10}$ |
| $x$ | | | | | | |
| ·18 | ·0000 001 | | | | | |
| ·19 | ·0000 001 | | | | | |
| ·20 | ·0000 002 | ·0000 001 | | | | |
| ·21 | ·0000 003 | ·0000 001 | | | | |
| ·22 | ·0000 004 | ·0000 002 | | | | |
| ·23 | ·0000 007 | ·0000 003 | ·0000 001 | | | |
| ·24 | ·0000 010 | ·0000 005+ | ·0000 001 | | | |
| ·25 | ·0000 016 | ·0000 008 | ·0000 002 | ·0000 001 | | |
| ·26 | ·0000 024 | ·0000 012 | ·0000 003 | ·0000 001 | | |
| ·27 | ·0000 035+ | ·0000 019 | ·0000 005+ | ·0000 001 | | |
| ·28 | ·0000 051 | ·0000 028 | ·0000 008 | ·0000 002 | ·0000 001 | |
| ·29 | ·0000 074 | ·0000 040 | ·0000 012 | ·0000 004 | ·0000 001 | |
| ·30 | ·0000 104 | ·0000 058 | ·0000 018 | ·0000 006 | ·0000 002 | ·0000 001 |
| ·31 | ·0000 146 | ·0000 083 | ·0000 027 | ·0000 009 | ·0000 003 | ·0000 001 |
| ·32 | ·0000 203 | ·0000 117 | ·0000 039 | ·0000 013 | ·0000 004 | ·0000 001 |
| ·33 | ·0000 278 | ·0000 163 | ·0000 056 | ·0000 019 | ·0000 007 | ·0000 002 |
| ·34 | ·0000 378 | ·0000 225+ | ·0000 080 | ·0000 028 | ·0000 010 | ·0000 003 |
| ·35 | ·0000 509 | ·0000 308 | ·0000 112 | ·0000 041 | ·0000 015− | ·0000 005+ |
| ·36 | ·0000 680 | ·0000 417 | ·0000 156 | ·0000 058 | ·0000 022 | ·0000 008 |
| ·37 | ·0000 901 | ·0000 559 | ·0000 215+ | ·0000 083 | ·0000 032 | ·0000 012 |
| ·38 | ·0001 184 | ·0000 745− | ·0000 294 | ·0000 116 | ·0000 046 | ·0000 018 |
| ·39 | ·0001 544 | ·0000 984 | ·0000 399 | ·0000 161 | ·0000 065+ | ·0000 026 |
| ·40 | ·0001 999 | ·0001 291 | ·0000 537 | ·0000 223 | ·0000 092 | ·0000 038 |
| ·41 | ·0002 572 | ·0001 681 | ·0000 717 | ·0000 305− | ·0000 129 | ·0000 055− |
| ·42 | ·0003 289 | ·0002 176 | ·0000 950− | ·0000 413 | ·0000 180 | ·0000 078 |
| ·43 | ·0004 179 | ·0002 797 | ·0001 250+ | ·0000 557 | ·0000 248 | ·0000 110 |
| ·44 | ·0005 280 | ·0003 574 | ·0001 634 | ·0000 745+ | ·0000 339 | ·0000 154 |
| ·45 | ·0006 633 | ·0004 541 | ·0002 124 | ·0000 990 | ·0000 461 | ·0000 214 |
| ·46 | ·0008 290 | ·0005 738 | ·0002 743 | ·0001 307 | ·0000 622 | ·0000 295+ |
| ·47 | ·0010 308 | ·0007 211 | ·0003 521 | ·0001 715− | ·0000 833 | ·0000 404 |
| ·48 | ·0012 754 | ·0009 017 | ·0004 496 | ·0002 236 | ·0001 109 | ·0000 549 |
| ·49 | ·0015 707 | ·0011 219 | ·0005 710 | ·0002 899 | ·0001 468 | ·0000 742 |
| ·50 | ·0019 257 | ·0013 893 | ·0007 215+ | ·0003 737 | ·0001 931 | ·0000 996 |
| ·51 | ·0023 506 | ·0017 127 | ·0009 071 | ·0004 792 | ·0002 525+ | ·0001 328 |
| ·52 | ·0028 574 | ·0021 021 | ·0011 351 | ·0006 113 | ·0003 285− | ·0001 761 |
| ·53 | ·0034 594 | ·0025 691 | ·0014 138 | ·0007 760 | ·0004 249 | ·0002 322 |
| ·54 | ·0041 719 | ·0031 272 | ·0017 532 | ·0009 803 | ·0005 469 | ·0003 045− |
| ·55 | ·0050 123 | ·0037 915− | ·0021 647 | ·0012 327 | ·0007 004 | ·0003 971 |
| ·56 | ·0060 000 | ·0045 794 | ·0026 617 | ·0015 431 | ·0008 926 | ·0005 153 |
| ·57 | ·0071 571 | ·0055 107 | ·0032 598 | ·0019 234 | ·0011 323 | ·0006 653 |
| ·58 | ·0085 082 | ·0066 077 | ·0039 768 | ·0023 873 | ·0014 299 | ·0008 548 |
| ·59 | ·0100 808 | ·0078 956 | ·0048 332 | ·0029 510 | ·0017 978 | ·0010 932 |
| ·60 | ·0119 058 | ·0094 029 | ·0058 525+ | ·0036 335+ | ·0022 509 | ·0013 917 |
| ·61 | ·0140 172 | ·0111 614 | ·0070 617 | ·0044 567 | ·0028 065+ | ·0017 640 |
| ·62 | ·0164 529 | ·0132 066 | ·0084 913 | ·0054 461 | ·0034 854 | ·0022 263 |
| ·63 | ·0192 548 | ·0155 784 | ·0101 762 | ·0066 310 | ·0043 116 | ·0027 982 |
| ·64 | ·0224 689 | ·0183 207 | ·0121 554 | ·0080 452 | ·0053 134 | ·0035 027 |
| ·65 | ·0261 457 | ·0214 825+ | ·0144 733 | ·0097 274 | ·0065 240 | ·0043 674 |
| ·66 | ·0303 408 | ·0251 178 | ·0171 795+ | ·0117 220 | ·0079 815− | ·0054 245+ |
| ·67 | ·0351 144 | ·0292 859 | ·0203 298 | ·0140 792 | ·0097 303 | ·0067 124 |
| ·68 | ·0405 324 | ·0340 520 | ·0239 862 | ·0168 563 | ·0118 215+ | ·0082 756 |
| ·69 | ·0466 661 | ·0394 874 | ·0282 179 | ·0201 179 | ·0143 140 | ·0101 663 |
| ·70 | ·0535 923 | ·0456 699 | ·0331 014 | ·0239 369 | ·0172 750− | ·0124 451 |
| ·71 | ·0613 939 | ·0526 837 | ·0387 212 | ·0283 948 | ·0207 811 | ·0151 823 |
| ·72 | ·0701 599 | ·0606 201 | ·0451 705− | ·0335 833 | ·0249 196 | ·0184 590 |
| ·73 | ·0799 850+ | ·0695 773 | ·0525 510 | ·0396 040 | ·0297 891 | ·0223 683 |
| ·74 | ·0909 701 | ·0796 609 | ·0609 742 | ·0465 699 | ·0355 008 | ·0270 170 |
| ·75 | ·1032 217 | ·0909 833 | ·0705 609 | ·0546 059 | ·0421 794 | ·0325 267 |
| ·76 | ·1168 522 | ·1036 642 | ·0814 419 | ·0638 492 | ·0499 645+ | ·0390 353 |
| ·77 | ·1319 786 | ·1178 297 | ·0937 577 | ·0744 499 | ·0590 111 | ·0466 987 |
| ·78 | ·1487 229 | ·1336 125− | ·1076 589 | ·0865 716 | ·0694 909 | ·0556 921 |
| ·79 | ·1672 105+ | ·1511 504 | ·1233 053 | ·1003 910 | ·0815 925− | ·0662 113 |
| ·80 | ·1875 694 | ·1705 861 | ·1408 655− | ·1160 983 | ·0955 225− | ·0784 740 |

$x = \cdot 81$ to $1\cdot 00$ $q = 1\cdot 5$ $p = 10\cdot 5$ to $15$

| | $p = 10\cdot5$ | $p = 11$ | $p = 12$ | $p = 13$ | $p = 14$ | $p = 15$ |
|---|---|---|---|---|---|---|
| $B(p, q) =$ | ·2516 0880 × $\frac{1}{10}$ | ·2350 0886 × $\frac{1}{10}$ | ·2068 0779 × $\frac{1}{10}$ | ·1838 2915 × $\frac{1}{10}$ | ·1648 1234 × $\frac{1}{10}$ | ·1488 6276 × $\frac{1}{10}$ |
| $x$ | | | | | | |
| ·81 | ·2099 286 | ·1920 652 | ·1605 160 | ·1338 963 | ·1115 051 | ·0927 205− |
| ·82 | ·2344 164 | ·2157 347 | ·1824 393 | ·1539 995+ | ·1297 819 | ·1092 143 |
| ·83 | ·2611 580 | ·2417 407 | ·2068 222 | ·1766 323 | ·1506 109 | ·1282 418 |
| ·84 | ·2902 722 | ·2702 252 | ·2338 525+ | ·2020 264 | ·1742 643 | ·1501 112 |
| ·85 | ·3218 683 | ·3013 223 | ·2637 153 | ·2304 175+ | ·2010 258 | ·1751 506 |
| ·86 | ·3560 408 | ·3351 534 | ·2965 880 | ·2620 398 | ·2311 856 | ·2037 033 |
| ·87 | ·3928 636 | ·3718 209 | ·3326 332 | ·2971 193 | ·2650 339 | ·2361 227 |
| ·88 | ·4323 830 | ·4114 004 | ·3719 905+ | ·3358 649 | ·3028 516 | ·2727 623 |
| ·89 | ·4746 078 | ·4539 301 | ·4147 648 | ·3784 552 | ·3448 968 | ·3139 631 |
| ·90 | ·5194 974 | ·4993 986 | ·4610 110 | ·4250 227 | ·3913 874 | ·3600 345+ |
| ·91 | ·5669 463 | ·5477 269 | ·5107 149 | ·4756 307 | ·4424 761 | ·4112 278 |
| ·92 | ·6167 629 | ·5987 463 | ·5637 659 | ·5302 431 | ·4982 159 | ·4676 986 |
| ·93 | ·6686 422 | ·6521 674 | ·6199 209 | ·5886 821 | ·5585 126 | ·5294 529 |
| ·94 | ·7221 254 | ·7075 367 | ·6787 524 | ·6505 677 | ·6230 551 | ·5962 697 |
| ·95 | ·7765 432 | ·7641 736 | ·7395 734 | ·7152 295+ | ·6912 133 | ·6675 831 |
| ·96 | ·8309 243 | ·8210 705− | ·8013 191 | ·7815 678 | ·7618 772 | ·7422 992 |
| ·97 | ·8838 410 | ·8767 236 | ·8623 457 | ·8478 181 | ·8331 843 | ·8184 826 |
| ·98 | ·9331 019 | ·9287 983 | ·9200 372 | ·9110 937 | ·9019 919 | ·8927 536 |
| ·99 | ·9749 692 | ·9732 807 | ·9698 173 | ·9662 456 | ·9625 736 | ·9588 085+ |
| 1·00 | 1·0000 000 | 1·0000 000 | 1·0000 000 | 1·0000 000 | 1·0000 000 | 1·0000 000 |

$x = {\cdot}33$ to $1{\cdot}00$ $\qquad q = 1{\cdot}5$ $\qquad p = 16$ to $21$

| $x$ | $p = 16$ | $p = 17$ | $p = 18$ | $p = 19$ | $p = 20$ | $p = 21$ |
|---|---|---|---|---|---|---|
| $B(p, q) =$ | ·1353 2978 × $\frac{1}{10}$ | ·1237 3009 × $\frac{1}{10}$ | ·1136 9792 × $\frac{1}{10}$ | ·1049 5192 × $\frac{1}{10}$ | ·9727 2514 × $\frac{1}{10^2}$ | ·9048 6059 × $\frac{1}{10^2}$ |
| ·33 | ·0000 001 | | | | | |
| ·34 | ·0000 001 | | | | | |
| ·35 | ·0000 002 | ·0000 001 | | | | |
| ·36 | ·0000 003 | ·0000 001 | | | | |
| ·37 | ·0000 005− | ·0000 002 | ·0000 001 | | | |
| ·38 | ·0000 007 | ·0000 003 | ·0000 001 | | | |
| ·39 | ·0000 011 | ·0000 004 | ·0000 002 | ·0000 001 | | |
| ·40 | ·0000 016 | ·0000 006 | ·0000 003 | ·0000 001 | | |
| ·41 | ·0000 023 | ·0000 010 | ·0000 004 | ·0000 002 | ·0000 001 | |
| ·42 | ·0000 034 | ·0000 015− | ·0000 006 | ·0000 003 | ·0000 001 | |
| ·43 | ·0000 049 | ·0000 022 | ·0000 010 | ·0000 004 | ·0000 002 | ·0000 001 |
| ·44 | ·0000 070 | ·0000 032 | ·0000 014 | ·0000 006 | ·0000 003 | ·0000 001 |
| ·45 | ·0000 099 | ·0000 046 | ·0000 021 | ·0000 010 | ·0000 005− | ·0000 002 |
| ·46 | ·0000 140 | ·0000 066 | ·0000 031 | ·0000 015− | ·0000 007 | ·0000 003 |
| ·47 | ·0000 195+ | ·0000 094 | ·0000 046 | ·0000 022 | ·0000 011 | ·0000 005+ |
| ·48 | ·0000 271 | ·0000 134 | ·0000 066 | ·0000 032 | ·0000 016 | ·0000 008 |
| ·49 | ·0000 374 | ·0000 189 | ·0000 095− | ·0000 048 | ·0000 024 | ·0000 012 |
| ·50 | ·0000 513 | ·0000 263 | ·0000 135+ | ·0000 069 | ·0000 035+ | ·0000 018 |
| ·51 | ·0000 697 | ·0000 366 | ·0000 191 | ·0000 100 | ·0000 052 | ·0000 027 |
| ·52 | ·0000 943 | ·0000 504 | ·0000 269 | ·0000 143 | ·0000 076 | ·0000 041 |
| ·53 | ·0001 267 | ·0000 690 | ·0000 375+ | ·0000 204 | ·0000 111 | ·0000 060 |
| ·54 | ·0001 692 | ·0000 939 | ·0000 520 | ·0000 288 | ·0000 159 | ·0000 088 |
| ·55 | ·0002 248 | ·0001 270 | ·0000 717 | ·0000 404 | ·0000 228 | ·0000 128 |
| ·56 | ·0002 969 | ·0001 709 | ·0000 982 | ·0000 563 | ·0000 323 | ·0000 185− |
| ·57 | ·0003 902 | ·0002 285+ | ·0001 336 | ·0000 780 | ·0000 455+ | ·0000 265+ |
| ·58 | ·0005 101 | ·0003 039 | ·0001 809 | ·0001 075− | ·0000 638 | ·0000 378 |
| ·59 | ·0006 636 | ·0004 022 | ·0002 434 | ·0001 471 | ·0000 888 | ·0000 536 |
| ·60 | ·0008 590 | ·0005 294 | ·0003 258 | ·0002 003 | ·0001 230 | ·0000 754 |
| ·61 | ·0011 068 | ·0006 934 | ·0004 339 | ·0002 711 | ·0001 692 | ·0001 055+ |
| ·62 | ·0014 197 | ·0009 040 | ·0005 748 | ·0003 651 | ·0002 316 | ·0001 468 |
| ·63 | ·0018 130 | ·0011 729 | ·0007 578 | ·0004 890 | ·0003 152 | ·0002 030 |
| ·64 | ·0023 052 | ·0015 149 | ·0009 941 | ·0006 516 | ·0004 267 | ·0002 791 |
| ·65 | ·0029 188 | ·0019 479 | ·0012 982 | ·0008 642 | ·0005 746 | ·0003 817 |
| ·66 | ·0036 807 | ·0024 938 | ·0016 875− | ·0011 405− | ·0007 700 | ·0005 193 |
| ·67 | ·0046 230 | ·0031 794 | ·0021 837 | ·0014 981 | ·0010 266 | ·0007 028 |
| ·68 | ·0057 840 | ·0040 367 | ·0028 136 | ·0019 589 | ·0013 623 | ·0009 465+ |
| ·69 | ·0072 089 | ·0051 046 | ·0036 099 | ·0025 499 | ·0017 993 | ·0012 684 |
| ·70 | ·0089 515− | ·0064 295+ | ·0046 122 | ·0033 048 | ·0023 655− | ·0016 915+ |
| ·71 | ·0110 746 | ·0080 670 | ·0058 688 | ·0042 648 | ·0030 959 | ·0022 453 |
| ·72 | ·0136 523 | ·0100 833 | ·0074 380 | ·0054 806 | ·0040 341 | ·0029 666 |
| ·73 | ·0167 706 | ·0125 566 | ·0093 899 | ·0070 139 | ·0052 339 | ·0039 019 |
| ·74 | ·0205 298 | ·0155 793 | ·0118 081 | ·0089 399 | ·0067 616 | ·0051 094 |
| ·75 | ·0250 458 | ·0192 599 | ·0147 927 | ·0113 493 | ·0086 988 | ·0066 612 |
| ·76 | ·0304 522 | ·0237 252 | ·0184 623 | ·0143 514 | ·0111 449 | ·0086 471 |
| ·77 | ·0369 021 | ·0291 230 | ·0229 569 | ·0180 771 | ·0142 208 | ·0111 772 |
| ·78 | ·0445 703 | ·0356 241 | ·0284 409 | ·0226 824 | ·0180 726 | ·0143 870 |
| ·79 | ·0536 551 | ·0434 257 | ·0351 068 | ·0283 523 | ·0228 759 | ·0184 413 |
| ·80 | ·0643 805+ | ·0527 534 | ·0431 781 | ·0353 052 | ·0288 410 | ·0235 404 |
| ·81 | ·0769 977 | ·0638 643 | ·0529 134 | ·0437 967 | ·0362 179 | ·0299 256 |
| ·82 | ·0917 866 | ·0770 493 | ·0646 093 | ·0541 252 | ·0453 020 | ·0378 860 |
| ·83 | ·1090 565+ | ·0926 351 | ·0786 045+ | ·0666 356 | ·0564 400 | ·0477 661 |
| ·84 | ·1291 464 | ·1109 856 | ·0952 818 | ·0817 242 | ·0700 362 | ·0599 728 |
| ·85 | ·1524 235− | ·1325 018 | ·1150 703 | ·0998 420 | ·0865 572 | ·0749 829 |
| ·86 | ·1792 807 | ·1576 210 | ·1384 457 | ·1214 969 | ·1065 373 | ·0933 505− |
| ·87 | ·2101 315− | ·1868 125− | ·1659 283 | ·1472 543 | ·1305 804 | ·1157 115− |
| ·88 | ·2454 012 | ·2205 709 | ·1980 776 | ·1777 337 | ·1593 599 | ·1427 867 |
| ·89 | ·2855 145− | ·2594 039 | ·2354 821 | ·2136 002 | ·1936 129 | ·1753 794 |
| ·90 | ·3308 764 | ·3038 139 | ·2787 414 | ·2555 493 | ·2341 271 | ·2143 651 |
| ·91 | ·3818 441 | ·3542 692 | ·3284 381 | ·3042 789 | ·2817 154 | ·2606 693 |
| ·92 | ·4386 867 | ·4111 618 | ·3850 940 | ·3604 455− | ·3371 721 | ·3152 253 |
| ·93 | ·5015 266 | ·4747 435− | ·4491 026 | ·4245 942 | ·4012 016 | ·3789 025+ |
| ·94 | ·5702 521 | ·5450 313 | ·5206 264 | ·4970 486 | ·4743 022 | ·4523 861 |
| ·95 | ·6443 861 | ·6216 603 | ·5994 358 | ·5777 361 | ·5565 788 | ·5359 769 |
| ·96 | ·7228 777 | ·7036 505+ | ·6846 495+ | ·6659 018 | ·6474 305− | ·6292 546 |
| ·97 | ·8037 466 | ·7890 059 | ·7742 870 | ·7596 130 | ·7450 046 | ·7304 803 |
| ·98 | ·8833 983 | ·8739 436 | ·8644 054 | ·8547 984 | ·8451 357 | ·8354 296 |
| ·99 | ·9549 568 | ·9510 245+ | ·9470 170 | ·9429 394 | ·9387 963 | ·9345 921 |
| 1·00 | 1·0000 000 | 1·0000 000 | 1·0000 000 | 1·0000 000 | 1·0000 000 | 1·0000 000 |

$x$ = ·44 to 1·00 $q$ = 1·5 $p$ = 22 to 27

| $x$ | $p$ = 22 | $p$ = 23 | $p$ = 24 | $p$ = 25 | $p$ = 26 | $p$ = 27 |
|---|---|---|---|---|---|---|
| $B(p,q)$ = | ·8445 3655 × $\frac{1}{10^2}$ | ·7906 2997 × $\frac{1}{10^2}$ | ·7422 2405 × $\frac{1}{10^2}$ | ·6985 6381 × $\frac{1}{10^2}$ | ·6590 2246 × $\frac{1}{10^2}$ | ·6230 7578 × $\frac{1}{10^2}$ |
| ·44 | ·0000 001 | | | | | |
| ·45 | ·0000 001 | | | | | |
| ·46 | ·0000 002 | ·0000 001 | | | | |
| ·47 | ·0000 002 | ·0000 001 | ·0000 001 | | | |
| ·48 | ·0000 004 | ·0000 002 | ·0000 001 | | | |
| ·49 | ·0000 006 | ·0000 003 | ·0000 002 | ·0000 001 | | |
| ·50 | ·0000 009 | ·0000 005$^-$ | ·0000 002 | ·0000 001 | ·0000 001 | |
| ·51 | ·0000 014 | ·0000 007 | ·0000 004 | ·0000 002 | ·0000 001 | ·0000 001 |
| ·52 | ·0000 022 | ·0000 011 | ·0000 006 | ·0000 003 | ·0000 002 | ·0000 001 |
| ·53 | ·0000 032 | ·0000 018 | ·0000 009 | ·0000 005$^+$ | ·0000 003 | ·0000 001 |
| ·54 | ·0000 049 | ·0000 027 | ·0000 015$^-$ | ·0000 008 | ·0000 004 | ·0000 002 |
| ·55 | ·0000 072 | ·0000 040 | ·0000 023 | ·0000 013 | ·0000 007 | ·0000 004 |
| ·56 | ·0000 106 | ·0000 060 | ·0000 035$^-$ | ·0000 020 | ·0000 011 | ·0000 006 |
| ·57 | ·0000 155$^-$ | ·0000 090 | ·0000 052 | ·0000 030 | ·0000 018 | ·0000 010 |
| ·58 | ·0000 224 | ·0000 133 | ·0000 078 | ·0000 046 | ·0000 027 | ·0000 016 |
| ·59 | ·0000 323 | ·0000 194 | ·0000 117 | ·0000 070 | ·0000 042 | ·0000 025$^+$ |
| ·60 | ·0000 462 | ·0000 283 | ·0000 173 | ·0000 106 | ·0000 065$^-$ | ·0000 039 |
| ·61 | ·0000 657 | ·0000 409 | ·0000 255$^-$ | ·0000 158 | ·0000 098 | ·0000 061 |
| ·62 | ·0000 929 | ·0000 588 | ·0000 372 | ·0000 235$^-$ | ·0000 148 | ·0000 093 |
| ·63 | ·0001 306 | ·0000 839 | ·0000 539 | ·0000 346 | ·0000 222 | ·0000 142 |
| ·64 | ·0001 824 | ·0001 191 | ·0000 777 | ·0000 507 | ·0000 330 | ·0000 215$^-$ |
| ·65 | ·0002 533 | ·0001 680 | ·0001 113 | ·0000 737 | ·0000 488 | ·0000 323 |
| ·66 | ·0003 499 | ·0002 356 | ·0001 585$^+$ | ·0001 066 | ·0000 716 | ·0000 481 |
| ·67 | ·0004 808 | ·0003 286 | ·0002 244 | ·0001 531 | ·0001 044 | ·0000 712 |
| ·68 | ·0006 570 | ·0004 557 | ·0003 158 | ·0002 187 | ·0001 514 | ·0001 047 |
| ·69 | ·0008 933 | ·0006 287 | ·0004 421 | ·0003 107 | ·0002 182 | ·0001 531 |
| ·70 | ·0012 085$^+$ | ·0008 628 | ·0006 155$^-$ | ·0004 387 | ·0003 126 | ·0002 225$^+$ |
| ·71 | ·0016 270 | ·0011 780 | ·0008 522 | ·0006 162 | ·0004 452 | ·0003 215$^-$ |
| ·72 | ·0021 797 | ·0016 003 | ·0011 740 | ·0008 607 | ·0006 306 | ·0004 617 |
| ·73 | ·0029 065$^-$ | ·0021 633 | ·0016 089 | ·0011 958 | ·0008 882 | ·0006 594 |
| ·74 | ·0038 576 | ·0029 102 | ·0021 940 | ·0016 529 | ·0012 444 | ·0009 364 |
| ·75 | ·0050 967 | ·0038 966 | ·0029 769 | ·0022 728 | ·0017 342 | ·0013 225$^-$ |
| ·76 | ·0067 035$^+$ | ·0051 928 | ·0040 197 | ·0031 096 | ·0024 041 | ·0018 576 |
| ·77 | ·0087 779 | ·0068 884 | ·0054 018 | ·0042 333 | ·0033 156 | ·0025 954 |
| ·78 | ·0114 438 | ·0090 959 | ·0072 248 | ·0057 349 | ·0045 495$^-$ | ·0036 071 |
| ·79 | ·0148 546 | ·0119 567 | ·0096 176 | ·0077 312 | ·0062 111 | ·0049 872 |
| ·80 | ·0191 990 | ·0156 469 | ·0127 434 | ·0103 723 | ·0084 374 | ·0068 598 |
| ·81 | ·0247 074 | ·0203 847 | ·0168 072 | ·0138 490 | ·0114 049 | ·0093 872 |
| ·82 | ·0316 602 | ·0264 390 | ·0220 645$^-$ | ·0184 027 | ·0153 399 | ·0127 802 |
| ·83 | ·0403 954 | ·0341 386 | ·0288 326 | ·0243 368 | ·0205 307 | ·0173 109 |
| ·84 | ·0513 182 | ·0438 832 | ·0375 020 | ·0320 301 | ·0273 418 | ·0233 279 |
| ·85 | ·0649 105$^+$ | ·0561 544 | ·0485 499 | ·0419 514 | ·0362 305$^+$ | ·0312 742 |
| ·86 | ·0817 398 | ·0715 276 | ·0625 543 | ·0546 764 | ·0477 660 | ·0417 087 |
| ·87 | ·1024 675$^+$ | ·0906 834 | ·0802 083 | ·0709 052 | ·0626 495$^+$ | ·0553 290 |
| ·88 | ·1278 551 | ·1144 166 | ·1023 337 | ·0914 792 | ·0817 362 | ·0729 975$^+$ |
| ·89 | ·1587 653 | ·1436 426 | ·1298 909 | ·1173 968 | ·1060 547 | ·0957 662 |
| ·90 | ·1961 559 | ·1793 951 | ·1639 825$^+$ | ·1498 222 | ·1368 230 | ·1248 987 |
| ·91 | ·2410 613 | ·2228 126 | ·2058 452 | ·1900 832 | ·1754 529 | ·1618 834 |
| ·92 | ·2945 535$^-$ | ·2751 032 | ·2568 199 | ·2396 489 | ·2235 356 | ·2084 262 |
| ·93 | ·3576 707 | ·3374 763 | ·3182 872 | ·3000 696 | ·2827 884 | ·2664 078 |
| ·94 | ·4312 944 | ·4110 177 | ·3915 432 | ·3728 558 | ·3549 383 | ·3377 720 |
| ·95 | ·5159 391 | ·4964 706 | ·4775 734 | ·4592 471 | ·4414 889 | ·4242 942 |
| ·96 | ·6113 904 | ·5938 509 | ·5766 470 | ·5597 872 | ·5432 780 | ·5271 244 |
| ·97 | ·7160 562 | ·7017 469 | ·6875 651 | ·6735 222 | ·6596 282 | ·6458 918 |
| ·98 | ·8256 911 | ·8159 305$^-$ | ·8061 571 | ·7963 797 | ·7866 062 | ·7768 440 |
| ·99 | ·9303 308 | ·9260 163 | ·9216 521 | ·9172 414 | ·9127 876 | ·9082 935$^+$ |
| 1·00 | 1·0000 000 | 1·0000 000 | 1·0000 000 | 1·0000 000 | 1·0000 000 | 1·0000 000 |

$x = \cdot 53$ to $1 \cdot 00$ $\qquad q = 1 \cdot 5$ $\qquad p = 28$ to $33$

| | $p = 28$ | $p = 29$ | $p = 30$ | $p = 31$ | $p = 32$ | $p = 33$ |
|---|---|---|---|---|---|---|
| $B(p, q) =$ | $\cdot 5902\ 8232 \times \frac{1}{10^2}$ | $\cdot 5602\ 6797 \times \frac{1}{10^2}$ | $\cdot 5327\ 1380 \times \frac{1}{10^2}$ | $\cdot 5073\ 4648 \times \frac{1}{10^2}$ | $\cdot 4839\ 3049 \times \frac{1}{10^2}$ | $\cdot 4622\ 6196 \times \frac{1}{10^2}$ |
| $x$ | | | | | | |
| ·53 | ·0000 001 | | | | | |
| ·54 | ·0000 001 | ·0000 001 | | | | |
| ·55 | ·0000 002 | ·0000 001 | ·0000 001 | | | |
| ·56 | ·0000 004 | ·0000 002 | ·0000 001 | ·0000 001 | | |
| ·57 | ·0000 006 | ·0000 003 | ·0000 002 | ·0000 001 | ·0000 001 | |
| ·58 | ·0000 010 | ·0000 006 | ·0000 003 | ·0000 002 | ·0000 001 | ·0000 001 |
| ·59 | ·0000 015+ | ·0000 009 | ·0000 005+ | ·0000 003 | ·0000 002 | ·0000 001 |
| ·60 | ·0000 024 | ·0000 015− | ·0000 009 | ·0000 005+ | ·0000 003 | ·0000 002 |
| ·61 | ·0000 038 | ·0000 023 | ·0000 015− | ·0000 009 | ·0000 006 | ·0000 003 |
| ·62 | ·0000 059 | ·0000 037 | ·0000 023 | ·0000 015− | ·0000 009 | ·0000 006 |
| ·63 | ·0000 091 | ·0000 058 | ·0000 037 | ·0000 024 | ·0000 015+ | ·0000 010 |
| ·64 | ·0000 140 | ·0000 091 | ·0000 059 | ·0000 038 | ·0000 025− | ·0000 016 |
| ·65 | ·0000 213 | ·0000 141 | ·0000 093 | ·0000 061 | ·0000 040 | ·0000 027 |
| ·66 | ·0000 323 | ·0000 216 | ·0000 145+ | ·0000 097 | ·0000 065+ | ·0000 044 |
| ·67 | ·0000 485− | ·0000 330 | ·0000 225− | ·0000 153 | ·0000 104 | ·0000 071 |
| ·68 | ·0000 724 | ·0000 500+ | ·0000 345+ | ·0000 238 | ·0000 165− | ·0000 113 |
| ·69 | ·0001 074 | ·0000 753 | ·0000 528 | ·0000 370 | ·0000 259 | ·0000 181 |
| ·70 | ·0001 584 | ·0001 126 | ·0000 801 | ·0000 569 | ·0000 404 | ·0000 287 |
| ·71 | ·0002 320 | ·0001 674 | ·0001 207 | ·0000 870 | ·0000 626 | ·0000 451 |
| ·72 | ·0003 379 | ·0002 472 | ·0001 807 | ·0001 320 | ·0000 965− | ·0000 704 |
| ·73 | ·0004 892 | ·0003 628 | ·0002 689 | ·0001 992 | ·0001 475+ | ·0001 092 |
| ·74 | ·0007 042 | ·0005 293 | ·0003 977 | ·0002 987 | ·0002 242 | ·0001 682 |
| ·75 | ·0010 079 | ·0007 678 | ·0005 846 | ·0004 449 | ·0003 385+ | ·0002 574 |
| ·76 | ·0014 345+ | ·0011 073 | ·0008 543 | ·0006 588 | ·0005 079 | ·0003 913 |
| ·77 | ·0020 305+ | ·0015 878 | ·0012 410 | ·0009 696 | ·0007 572 | ·0005 911 |
| ·78 | ·0028 585− | ·0022 641 | ·0017 925− | ·0014 185− | ·0011 221 | ·0008 872 |
| ·79 | ·0040 024 | ·0032 105− | ·0025 741 | ·0020 630 | ·0016 527 | ·0013 235− |
| ·80 | ·0055 743 | ·0045 275+ | ·0036 757 | ·0029 829 | ·0024 197 | ·0019 620 |
| ·81 | ·0077 225+ | ·0063 501 | ·0052 193 | ·0042 880 | ·0035 215+ | ·0028 910 |
| ·82 | ·0106 423 | ·0088 580 | ·0073 696 | ·0061 288 | ·0050 949 | ·0042 339 |
| ·83 | ·0145 890 | ·0122 894 | ·0103 479 | ·0087 096 | ·0073 278 | ·0061 631 |
| ·84 | ·0198 938 | ·0169 576 | ·0144 487 | ·0123 061 | ·0104 772 | ·0089 170 |
| ·85 | ·0269 834 | ·0232 710 | ·0200 611 | ·0172 872 | ·0148 914 | ·0128 231 |
| ·86 | ·0364 029 | ·0317 585+ | ·0276 955− | ·0241 430 | ·0210 386 | ·0183 271 |
| ·87 | ·0488 422 | ·0430 979 | ·0380 142 | ·0335 177 | ·0295 426 | ·0260 303 |
| ·88 | ·0651 650+ | ·0581 494 | ·0518 691 | ·0462 504 | ·0412 262 | ·0367 358 |
| ·89 | ·0864 397 | ·0779 910 | ·0703 420 | ·0634 209 | ·0571 618 | ·0515 042 |
| ·90 | ·1139 681 | ·1039 549 | ·0947 877 | ·0863 996 | ·0787 287 | ·0717 169 |
| ·91 | ·1493 064 | ·1376 569 | ·1268 732 | ·1168 964 | ·1076 711 | ·0991 449 |
| ·92 | ·1942 683 | ·1810 103 | ·1686 027 | ·1569 974 | ·1461 484 | ·1360 113 |
| ·93 | ·2508 917 | ·2362 041 | ·2223 090 | ·2091 713 | ·1967 562 | ·1850 297 |
| ·94 | ·3213 368 | ·3056 118 | ·2905 756 | ·2762 059 | ·2624 805− | ·2493 771 |
| ·95 | ·4076 567 | ·3915 688 | ·3760 219 | ·3610 061 | ·3465 110 | ·3325 255+ |
| ·96 | ·5113 298 | ·4958 963 | ·4808 246 | ·4661 146 | ·4517 653 | ·4377 747 |
| ·97 | ·6323 207 | ·6189 217 | ·6057 006 | ·5926 623 | ·5798 113 | ·5671 509 |
| ·98 | ·7670 998 | ·7573 800 | ·7476 904 | ·7380 363 | ·7284 226 | ·7188 541 |
| ·99 | ·9037 620 | ·8991 957 | ·8945 970 | ·8899 685+ | ·8853 124 | ·8806 308 |
| 1·00 | 1·0000 000 | 1·0000 000 | 1·0000 000 | 1·0000 000 | 1·0000 000 | 1·0000 000 |

$x$ = ·59 to 1·00 $q$ = 1·5 $p$ = 34 to 39

| $x$ | $p = 34$ | $p = 35$ | $p = 36$ | $p = 37$ | $p = 38$ | $p = 39$ |
|---|---|---|---|---|---|---|
| $B(p, q) =$ | ·4421 6361 × $\frac{1}{10^2}$ | ·4234 8064 × $\frac{1}{10^2}$ | ·4060 7733 × $\frac{1}{10^2}$ | ·3898 3424 × $\frac{1}{10^2}$ | ·3746 4589 × $\frac{1}{10^2}$ | ·3604 1883 × $\frac{1}{10^2}$ |
| ·59 | ·0000 001 | | | | | |
| ·60 | ·0000 001 | ·0000 001 | | | | |
| ·61 | ·0000 002 | ·0000 001 | ·0000 001 | ·0000 001 | | |
| ·62 | ·0000 004 | ·0000 002 | ·0000 001 | ·0000 001 | ·0000 001 | |
| ·63 | ·0000 006 | ·0000 004 | ·0000 003 | ·0000 002 | ·0000 001 | ·0000 001 |
| ·64 | ·0000 011 | ·0000 007 | ·0000 004 | ·0000 003 | ·0000 002 | ·0000 001 |
| ·65 | ·0000 018 | ·0000 012 | ·0000 008 | ·0000 005+ | ·0000 003 | ·0000 002 |
| ·66 | ·0000 029 | ·0000 020 | ·0000 013 | ·0000 009 | ·0000 006 | ·0000 004 |
| ·67 | ·0000 048 | ·0000 033 | ·0000 022 | ·0000 015+ | ·0000 010 | ·0000 007 |
| ·68 | ·0000 078 | ·0000 054 | ·0000 037 | ·0000 026 | ·0000 018 | ·0000 012 |
| ·69 | ·0000 127 | ·0000 089 | ·0000 062 | ·0000 043 | ·0000 030 | ·0000 021 |
| ·70 | ·0000 204 | ·0000 144 | ·0000 102 | ·0000 073 | ·0000 051 | ·0000 036 |
| ·71 | ·0000 325− | ·0000 234 | ·0000 168 | ·0000 121 | ·0000 087 | ·0000 062 |
| ·72 | ·0000 514 | ·0000 375+ | ·0000 274 | ·0000 199 | ·0000 145+ | ·0000 106 |
| ·73 | ·0000 808 | ·0000 598 | ·0000 442 | ·0000 327 | ·0000 241 | ·0000 178 |
| ·74 | ·0001 262 | ·0000 946 | ·0000 709 | ·0000 531 | ·0000 398 | ·0000 298 |
| ·75 | ·0001 957 | ·0001 487 | ·0001 130 | ·0000 858 | ·0000 651 | ·0000 494 |
| ·76 | ·0003 014 | ·0002 321 | ·0001 787 | ·0001 375− | ·0001 058 | ·0000 813 |
| ·77 | ·0004 613 | ·0003 598 | ·0002 806 | ·0002 188 | ·0001 705− | ·0001 328 |
| ·78 | ·0007 013 | ·0005 542 | ·0004 377 | ·0003 457 | ·0002 729 | ·0002 154 |
| ·79 | ·0010 595− | ·0008 478 | ·0006 783 | ·0005 424 | ·0004 337 | ·0003 466 |
| ·80 | ·0015 904 | ·0012 887 | ·0010 439 | ·0008 454 | ·0006 844 | ·0005 539 |
| ·81 | ·0023 725− | ·0019 463 | ·0015 962 | ·0013 087 | ·0010 727 | ·0008 790 |
| ·82 | ·0035 171 | ·0029 208 | ·0024 247 | ·0020 124 | ·0016 696 | ·0013 849 |
| ·83 | ·0051 817 | ·0043 551 | ·0036 593 | ·0030 737 | ·0025 811 | ·0021 669 |
| ·84 | ·0075 866 | ·0064 526 | ·0054 864 | ·0046 636 | ·0039 630 | ·0033 668 |
| ·85 | ·0110 384 | ·0094 992 | ·0081 721 | ·0070 284 | ·0060 431 | ·0051 946 |
| ·86 | ·0159 599 | ·0138 941 | ·0120 922 | ·0105 210 | ·0091 515+ | ·0079 583 |
| ·87 | ·0229 282 | ·0201 898 | ·0177 732 | ·0156 417 | ·0137 622 | ·0121 055− |
| ·88 | ·0327 244 | ·0291 424 | ·0259 453 | ·0230 927 | ·0205 485+ | ·0182 802 |
| ·89 | ·0463 925+ | ·0417 763 | ·0376 092 | ·0338 489 | ·0304 571 | ·0273 986 |
| ·90 | ·0653 107 | ·0594 603 | ·0541 198 | ·0492 465+ | ·0448 013 | ·0407 480 |
| ·91 | ·0912 685− | ·0839 956 | ·0772 827 | ·0710 891 | ·0653 767 | ·0601 101 |
| ·92 | ·1265 439 | ·1177 058 | ·1094 586 | ·1017 658 | ·0945 927 | ·0879 067 |
| ·93 | ·1739 588 | ·1635 114 | ·1536 566 | ·1443 644 | ·1356 058 | ·1273 531 |
| ·94 | ·2368 732 | ·2249 467 | ·2135 757 | ·2027 385− | ·1924 138 | ·1825 809 |
| ·95 | ·3190 380 | ·3060 364 | ·2935 085+ | ·2814 417 | ·2698 233 | ·2586 405+ |
| ·96 | ·4241 402 | ·4108 586 | ·3979 261 | ·3853 385+ | ·3730 911 | ·3611 789 |
| ·97 | ·5546 844 | ·5424 140 | ·5303 416 | ·5184 688 | ·5067 966 | ·4953 256 |
| ·98 | ·7093 348 | ·6998 687 | ·6904 595− | ·6811 103 | ·6718 243 | ·6626 043 |
| ·99 | ·8759 258 | ·8711 994 | ·8664 534 | ·8616 895+ | ·8569 096 | ·8521 152 |
| 1·00 | 1·0000 000 | 1·0000 000 | 1·0000 000 | 1·0000 000 | 1·0000 000 | 1·0000 000 |

$x = \cdot 64$ to 1·00 $q = 1\cdot 5$ $p = 40$ to 45

| | $p = 40$ | $p = 41$ | $p = 42$ | $p = 43$ | $p = 44$ | $p = 45$ |
|---|---|---|---|---|---|---|
| $B(p, q) =$ | ·3470 6998 × $\frac{1}{10^2}$ | ·3345 2529 × $\frac{1}{10^2}$ | ·3227 1851 × $\frac{1}{10^2}$ | ·3115 9029 × $\frac{1}{10^2}$ | ·3010 8724 × $\frac{1}{10^2}$ | ·2911 6129 × $\frac{1}{10^2}$ |
| $x$ | | | | | | |
| ·64 | ·0000 001 | ·0000 001 | | | | |
| ·65 | ·0000 001 | ·0000 001 | ·0000 001 | | | |
| ·66 | ·0000 003 | ·0000 002 | ·0000 001 | ·0000 001 | ·0000 001 | |
| ·67 | ·0000 005− | ·0000 003 | ·0000 002 | ·0000 001 | ·0000 001 | ·0000 001 |
| ·68 | ·0000 008 | ·0000 006 | ·0000 004 | ·0000 003 | ·0000 002 | ·0000 001 |
| ·69 | ·0000 015− | ·0000 010 | ·0000 007 | ·0000 005+ | ·0000 004 | ·0000 002 |
| ·70 | ·0000 026 | ·0000 018 | ·0000 013 | ·0000 009 | ·0000 006 | ·0000 005− |
| ·71 | ·0000 045− | ·0000 032 | ·0000 023 | ·0000 017 | ·0000 012 | ·0000 009 |
| ·72 | ·0000 077 | ·0000 056 | ·0000 041 | ·0000 030 | ·0000 022 | ·0000 016 |
| ·73 | ·0000 132 | ·0000 097 | ·0000 072 | ·0000 053 | ·0000 039 | ·0000 029 |
| ·74 | ·0000 223 | ·0000 167 | ·0000 125− | ·0000 094 | ·0000 070 | ·0000 052 |
| ·75 | ·0000 375+ | ·0000 284 | ·0000 216 | ·0000 164 | ·0000 124 | ·0000 094 |
| ·76 | ·0000 625+ | ·0000 481 | ·0000 369 | ·0000 284 | ·0000 218 | ·0000 167 |
| ·77 | ·0001 035− | ·0000 806 | ·0000 627 | ·0000 488 | ·0000 380 | ·0000 296 |
| ·78 | ·0001 699 | ·0001 340 | ·0001 057 | ·0000 833 | ·0000 657 | ·0000 518 |
| ·79 | ·0002 770 | ·0002 213 | ·0001 767 | ·0001 411 | ·0001 126 | ·0000 899 |
| ·80 | ·0004 482 | ·0003 626 | ·0002 932 | ·0002 371 | ·0001 916 | ·0001 549 |
| ·81 | ·0007 200 | ·0005 897 | ·0004 828 | ·0003 953 | ·0003 235− | ·0002 647 |
| ·82 | ·0011 484 | ·0009 521 | ·0007 892 | ·0006 540 | ·0005 418 | ·0004 488 |
| ·83 | ·0018 187 | ·0015 261 | ·0012 802 | ·0010 737 | ·0009 004 | ·0007 548 |
| ·84 | ·0028 596 | ·0024 282 | ·0020 614 | ·0017 496 | ·0014 847 | ·0012 596 |
| ·85 | ·0044 641 | ·0038 355− | ·0032 946 | ·0028 294 | ·0024 293 | ·0020 855− |
| ·86 | ·0069 190 | ·0060 139 | ·0052 261 | ·0045 406 | ·0039 441 | ·0034 254 |
| ·87 | ·0106 457 | ·0093 598 | ·0082 274 | ·0072 305+ | ·0063 531 | ·0055 811 |
| ·88 | ·0162 584 | ·0144 571 | ·0128 525+ | ·0114 237 | ·0101 518 | ·0090 197 |
| ·89 | ·0246 417 | ·0221 573 | ·0199 193 | ·0179 037 | ·0160 890 | ·0144 556 |
| ·90 | ·0370 532 | ·0336 864 | ·0306 193 | ·0278 260 | ·0252 828 | ·0229 680 |
| ·91 | ·0552 560 | ·0507 836 | ·0466 641 | ·0428 707 | ·0393 786 | ·0361 646 |
| ·92 | ·0816 767 | ·0758 735− | ·0704 694 | ·0654 384 | ·0607 561 | ·0563 995− |
| ·93 | ·1195 797 | ·1122 601 | ·1053 698 | ·0988 856 | ·0927 852 | ·0870 473 |
| ·94 | ·1732 195+ | ·1643 098 | ·1558 325+ | ·1477 690 | ·1401 012 | ·1328 115+ |
| ·95 | ·2478 808 | ·2375 312 | ·2275 792 | ·2180 123 | ·2088 180 | ·1999 842 |
| ·96 | ·3495 966 | ·3383 385+ | ·3273 990 | ·3167 721 | ·3064 516 | ·2964 313 |
| ·97 | ·4840 560 | ·4729 879 | ·4621 208 | ·4514 543 | ·4409 875+ | ·4307 194 |
| ·98 | ·6534 529 | ·6443 723 | ·6353 649 | ·6264 325+ | ·6175 770 | ·6088 000 |
| ·99 | ·8473 079 | ·8424 892 | ·8376 605+ | ·8328 232 | ·8279 786 | ·8231 279 |
| 1·00 | 1·0000 000 | 1·0000 000 | 1·0000 000 | 1·0000 000 | 1·0000 000 | 1·0000 000 |

$x = \cdot 68$ to $1\cdot 00$ $q = 1\cdot 5$ $p = 46$ to $50$

| | $p = 46$ | $p = 47$ | $p = 48$ | $p = 49$ | $p = 50$ |
|---|---|---|---|---|---|
| $B(p, q) =$ | $\cdot 2817\ 6899 \times \frac{1}{10^2}$ | $\cdot 2728\ 7102 \times \frac{1}{10^2}$ | $\cdot 2644\ 3171 \times \frac{1}{10^2}$ | $\cdot 2564\ 1863 \times \frac{1}{10^2}$ | $\cdot 2488\ 0224 \times \frac{1}{10^2}$ |
| $x$ | | | | | |
| ·68 | ·0000 001 | ·0000 001 | | | |
| ·69 | ·0000 002 | ·0000 001 | ·0000 001 | ·0000 001 | |
| ·70 | ·0000 003 | ·0000 002 | ·0000 002 | ·0000 001 | ·0000 001 |
| ·71 | ·0000 006 | ·0000 004 | ·0000 003 | ·0000 002 | ·0000 002 |
| ·72 | ·0000 011 | ·0000 008 | ·0000 006 | ·0000 004 | ·0000 003 |
| ·73 | ·0000 021 | ·0000 016 | ·0000 012 | ·0000 009 | ·0000 006 |
| ·74 | ·0000 039 | ·0000 029 | ·0000 022 | ·0000 016 | ·0000 012 |
| ·75 | ·0000 071 | ·0000 054 | ·0000 041 | ·0000 031 | ·0000 023 |
| ·76 | ·0000 128 | ·0000 099 | ·0000 076 | ·0000 058 | ·0000 045⁻ |
| ·77 | ·0000 230 | ·0000 179 | ·0000 139 | ·0000 108 | ·0000 084 |
| ·78 | ·0000 408 | ·0000 321 | ·0000 253 | ·0000 199 | ·0000 157 |
| ·79 | ·0000 717 | ·0000 572 | ·0000 457 | ·0000 364 | ·0000 290 |
| ·80 | ·0001 252 | ·0001 011 | ·0000 817 | ·0000 660 | ·0000 533 |
| ·81 | ·0002 165⁺ | ·0001 771 | ·0001 448 | ·0001 184 | ·0000 968 |
| ·82 | ·0003 716 | ·0003 077 | ·0002 547 | ·0002 108 | ·0001 745⁻ |
| ·83 | ·0006 327 | ·0005 302 | ·0004 443 | ·0003 722 | ·0003 117 |
| ·84 | ·0010 685⁻ | ·0009 061 | ·0007 683 | ·0006 514 | ·0005 521 |
| ·85 | ·0017 899 | ·0015 359 | ·0013 178 | ·0011 304 | ·0009 695⁺ |
| ·86 | ·0029 742 | ·0025 821 | ·0022 412 | ·0019 450⁺ | ·0016 877 |
| ·87 | ·0049 020 | ·0043 048 | ·0037 796 | ·0033 180 | ·0029 123 |
| ·88 | ·0080 125⁻ | ·0071 164 | ·0063 195⁺ | ·0056 110 | ·0049 810 |
| ·89 | ·0129 856 | ·0116 632 | ·0104 737 | ·0094 040 | ·0084 423 |
| ·90 | ·0208 615⁻ | ·0189 450⁻ | ·0172 018 | ·0156 166 | ·0141 753 |
| ·91 | ·0332 074 | ·0304 871 | ·0279 853 | ·0256 849 | ·0235 702 |
| ·92 | ·0523 468 | ·0485 778 | ·0450 735⁻ | ·0418 159 | ·0387 883 |
| ·93 | ·0816 518 | ·0765 794 | ·0718 119 | ·0673 319 | ·0631 231 |
| ·94 | ·1258 831 | ·1192 996 | ·1130 453 | ·1071 050⁻ | ·1014 641 |
| ·95 | ·1914 988 | ·1833 501 | ·1755 264 | ·1680 165⁺ | ·1608 093 |
| ·96 | ·2867 050⁻ | ·2772 662 | ·2681 086 | ·2592 257 | ·2506 111 |
| ·97 | ·4206 486 | ·4107 737 | ·4010 932 | ·3916 053 | ·3823 081 |
| ·98 | ·6000 939 | ·5914 783 | ·5829 452 | ·5744 957 | ·5661 306 |
| ·99 | ·8182 723 | ·8134 131 | ·8085 513 | ·8036 879 | ·7988 241 |
| 1·00 | 1·0000 000 | 1·0000 000 | 1·0000 000 | 1·0000 000 | 1·0000 000 |

$x = \cdot 01$ to $\cdot 60$ $q = 2$ $p = 2$ to $4 \cdot 5$

| | $p = 2$ | $p = 2\cdot 5$ | $p = 3$ | $p = 3\cdot 5$ | $p = 4$ | $p = 4\cdot 5$ |
|---|---|---|---|---|---|---|
| $B(p, q) =$ | ·1666 6667 | ·1142 8571 | ·8333 3333 $\times \frac{1}{10}$ | ·6349 2063 $\times \frac{1}{10}$ | ·5000 0000 $\times \frac{1}{10}$ | ·4040 4040 $\times \frac{1}{10}$ |
| $x$ | | | | | | |
| ·01 | ·0002 980$^{e}$ | ·0000 347 | ·0000 040 | ·0000 004 | | |
| ·02 | ·0011 840$^{e}$ | ·0001 952 | ·0000 315$^{+}$ | ·0000 050$^{+}$ | ·0000 008 | ·0000 001 |
| ·03 | ·0026 460$^{e}$ | ·0005 339 | ·0001 056 | ·0000 206 | ·0000 040 | ·0000 008 |
| ·04 | ·0046 720$^{e}$ | ·0010 880 | ·0002 483 | ·0000 558 | ·0000 124 | ·0000 027 |
| ·05 | ·0072 500$^{e}$ | ·0018 867 | ·0004 812 | ·0001 209 | ·0000 300$^{e}$ | ·0000 074 |
| ·06 | ·0103 680$^{e}$ | ·0029 541 | ·0008 251 | ·0002 270 | ·0000 617 | ·0000 166 |
| ·07 | ·0140 140$^{e}$ | ·0043 106 | ·0013 000 | ·0003 861 | ·0001 133 | ·0000 329 |
| ·08 | ·0181 760$^{e}$ | ·0059 736 | ·0019 251 | ·0006 111 | ·0001 917 | ·0000 595$^{+}$ |
| ·09 | ·0228 420$^{e}$ | ·0079 582 | ·0027 192 | ·0009 153 | ·0003 044 | ·0001 003 |
| ·10 | ·0280 000$^{e}$ | ·0102 774 | ·0037 000$^{e}$ | ·0013 123 | ·0004 600$^{e}$ | ·0001 597 |
| ·11 | ·0336 380$^{e}$ | ·0129 423 | ·0048 848 | ·0018 165$^{+}$ | ·0006 676 | ·0002 430 |
| ·12 | ·0397 440$^{e}$ | ·0159 626 | ·0062 899 | ·0024 423 | ·0009 373 | ·0003 563 |
| ·13 | ·0463 060$^{e}$ | ·0193 465$^{-}$ | ·0079 312 | ·0032 042 | ·0012 795$^{+}$ | ·0005 061 |
| ·14 | ·0533 120$^{e}$ | ·0231 010 | ·0098 235$^{+}$ | ·0041 171 | ·0017 057 | ·0007 000 |
| ·15 | ·0607 500$^{e}$ | ·0272 319 | ·0119 812 | ·0051 958 | ·0022 275$^{e}$ | ·0009 460 |
| ·16 | ·0686 080$^{e}$ | ·0317 440$^{e}$ | ·0144 179 | ·0064 553 | ·0028 574 | ·0012 530 |
| ·17 | ·0768 740$^{e}$ | ·0366 410 | ·0171 464 | ·0079 103 | ·0036 081 | ·0016 306 |
| ·18 | ·0855 360$^{e}$ | ·0419 258 | ·0201 787 | ·0095 756 | ·0044 930 | ·0020 888 |
| ·19 | ·0945 820$^{e}$ | ·0476 003 | ·0235 264 | ·0114 658 | ·0055 256 | ·0026 386 |
| ·20 | ·1040 000$^{e}$ | ·0536 656 | ·0272 000$^{e}$ | ·0135 953 | ·0067 200$^{e}$ | ·0032 915$^{-}$ |
| ·21 | ·1137 780$^{e}$ | ·0601 222 | ·0312 096 | ·0159 784 | ·0080 904 | ·0040 595$^{+}$ |
| ·22 | ·1239 040$^{e}$ | ·0669 698 | ·0355 643 | ·0186 289 | ·0096 513 | ·0049 554 |
| ·23 | ·1343 660$^{e}$ | ·0742 071 | ·0402 728 | ·0215 607 | ·0114 175$^{+}$ | ·0059 923 |
| ·24 | ·1451 520$^{e}$ | ·0818 326 | ·0453 427 | ·0247 868 | ·0134 038 | ·0071 841 |
| ·25 | ·1562 500$^{e}$ | ·0898 437 | ·0507 812 | ·0283 203 | ·0156 250$^{e}$ | ·0085 449 |
| ·26 | ·1676 480$^{e}$ | ·0982 377 | ·0565 947 | ·0321 737 | ·0180 962 | ·0100 895$^{-}$ |
| ·27 | ·1793 340$^{e}$ | ·1070 109 | ·0627 888 | ·0363 591 | ·0208 325$^{-}$ | ·0118 328 |
| ·28 | ·1912 960$^{e}$ | ·1161 591 | ·0693 683 | ·0408 880 | ·0238 487 | ·0137 904 |
| ·29 | ·2035 220$^{e}$ | ·1256 776 | ·0763 376 | ·0457 716 | ·0271 596 | ·0159 780 |
| ·30 | ·2160 000$^{e}$ | ·1355 613 | ·0837 000$^{e}$ | ·0510 204 | ·0307 800$^{e}$ | ·0184 117 |
| ·31 | ·2287 180$^{e}$ | ·1458 044 | ·0914 584 | ·0566 444 | ·0347 244 | ·0211 077 |
| ·32 | ·2416 640$^{e}$ | ·1564 007 | ·0996 147 | ·0626 530 | ·0390 070 | ·0240 825$^{-}$ |
| ·33 | ·2548 260$^{e}$ | ·1673 434 | ·1081 704 | ·0690 550$^{-}$ | ·0436 419 | ·0273 526 |
| ·34 | ·2681 920$^{e}$ | ·1786 254 | ·1171 259 | ·0758 585$^{-}$ | ·0486 426 | ·0309 347 |
| ·35 | ·2817 500$^{e}$ | ·1902 389 | ·1264 812 | ·0830 710 | ·0540 225$^{e}$ | ·0348 454 |
| ·36 | ·2954 880$^{e}$ | ·2021 760$^{e}$ | ·1362 355$^{+}$ | ·0906 993 | ·0597 943 | ·0391 015$^{-}$ |
| ·37 | ·3093 940$^{e}$ | ·2144 280 | ·1463 872 | ·0987 493 | ·0659 705$^{-}$ | ·0437 193 |
| ·38 | ·3234 560$^{e}$ | ·2269 861 | ·1569 339 | ·1072 264 | ·0725 627 | ·0487 153 |
| ·39 | ·3376 620$^{e}$ | ·2398 407 | ·1678 728 | ·1161 351 | ·0795 824 | ·0541 056 |
| ·40 | ·3520 000$^{e}$ | ·2529 822 | ·1792 000$^{e}$ | ·1254 792 | ·0870 400$^{e}$ | ·0599 062 |
| ·41 | ·3664 580$^{e}$ | ·2664 004 | ·1909 112 | ·1352 614 | ·0949 456 | ·0661 325$^{-}$ |
| ·42 | ·3810 240$^{e}$ | ·2800 847 | ·2030 011 | ·1454 840 | ·1033 083 | ·0727 996 |
| ·43 | ·3956 860$^{e}$ | ·2940 241 | ·2154 640 | ·1561 480 | ·1121 367 | ·0799 222 |
| ·44 | ·4104 320$^{e}$ | ·3082 073 | ·2282 931 | ·1672 538 | ·1214 383 | ·0875 144 |
| ·45 | ·4252 500$^{e}$ | ·3226 227 | ·2414 812 | ·1788 009 | ·1312 200$^{e}$ | ·0955 897 |
| ·46 | ·4401 280$^{e}$ | ·3372 581 | ·2550 203 | ·1907 876 | ·1414 876 | ·1041 608 |
| ·47 | ·4550 540$^{e}$ | ·3521 013 | ·2689 016 | ·2032 117 | ·1522 460 | ·1132 398 |
| ·48 | ·4700 160$^{e}$ | ·3671 393 | ·2831 155$^{+}$ | ·2160 695$^{-}$ | ·1634 992 | ·1228 378 |
| ·49 | ·4850 020$^{e}$ | ·3823 592 | ·2976 520 | ·2293 567 | ·1752 500$^{-}$ | ·1329 651 |
| ·50 | ·5000 000$^{e}$ | ·3977 476 | ·3125 000$^{e}$ | ·2430 680 | ·1875 000$^{e}$ | ·1436 311 |
| ·51 | ·5149 980$^{e}$ | ·4132 905$^{+}$ | ·3276 480 | ·2571 967 | ·2002 499 | ·1548 438 |
| ·52 | ·5299 840$^{e}$ | ·4289 741 | ·3430 835$^{+}$ | ·2717 356 | ·2134 992 | ·1666 104 |
| ·53 | ·5449 460$^{e}$ | ·4447 838 | ·3587 936 | ·2866 759 | ·2272 459 | ·1789 367 |
| ·54 | ·5598 720$^{e}$ | ·4607 049 | ·3747 643 | ·3020 082 | ·2414 868 | ·1918 272 |
| ·55 | ·5747 500$^{e}$ | ·4767 225$^{+}$ | ·3909 812 | ·3177 215$^{+}$ | ·2562 175$^{e}$ | ·2052 851 |
| ·56 | ·5895 680$^{e}$ | ·4928 212 | ·4074 291 | ·3338 042 | ·2714 321 | ·2193 120 |
| ·57 | ·6043 140$^{e}$ | ·5089 853 | ·4240 920 | ·3502 432 | ·2871 232 | ·2339 079 |
| ·58 | ·6189 760$^{e}$ | ·5251 989 | ·4409 531 | ·3670 244 | ·3032 821 | ·2490 714 |
| ·59 | ·6335 420$^{e}$ | ·5414 459 | ·4579 952 | ·3841 325$^{-}$ | ·3198 983 | ·2647 990 |
| ·60 | ·6480 000$^{e}$ | ·5577 096 | ·4752 000$^{e}$ | ·4015 509 | ·3369 600$^{e}$ | ·2810 856 |

$x$ = ·61 to 1·00 $q = 2$ $p$ = 2 to 4·5

| | $p = 2$ | $p = 2·5$ | $p = 3$ | $p = 3·5$ | $p = 4$ | $p = 4·5$ |
|---|---|---|---|---|---|---|
| $B(p, q)$ = | ·1666 6667 | ·1142 8571 | ·8333 3333 × $\frac{1}{10}$ | ·6349 2063 × $\frac{1}{10}$ | ·5000 0000 × $\frac{1}{10}$ | ·4040 4040 × $\frac{1}{10}$ |
| $x$ | | | | | | |
| ·61 | ·6623 380$^e$ | ·5739 733 | ·4925 488 | ·4192 621 | ·3544 535$^+$ | ·2979 243 |
| ·62 | ·6765 440$^e$ | ·5902 199 | ·5100 219 | ·4372 470 | ·3723 637 | ·3153 058 |
| ·63 | ·6906 060$^e$ | ·6064 320 | ·5275 992 | ·4554 856 | ·3906 734 | ·3332 189 |
| ·64 | ·7045 120$^e$ | ·6225 920$^e$ | ·5452 595$^+$ | ·4739 564 | ·4093 641 | ·3516 504 |
| ·65 | ·7182 500$^e$ | ·6386 820 | ·5629 812 | ·4926 367 | ·4284 150$^e$ | ·3705 846 |
| ·66 | ·7318 080$^e$ | ·6546 838 | ·5807 419 | ·5115 027 | ·4478 038 | ·3900 033 |
| ·67 | ·7451 740$^e$ | ·6705 789 | ·5985 184 | ·5305 289 | ·4675 060 | ·4098 859 |
| ·68 | ·7583 360$^e$ | ·6863 487 | ·6162 867 | ·5496 890 | ·4874 954 | ·4302 094 |
| ·69 | ·7712 820$^e$ | ·7019 741 | ·6340 224 | ·5689 549 | ·5077 435$^+$ | ·4509 480 |
| ·70 | ·7840 000$^e$ | ·7174 360 | ·6517 000$^e$ | ·5882 975$^-$ | ·5282 200$^e$ | ·4720 729 |
| ·71 | ·7964 780$^e$ | ·7327 148 | ·6692 936 | ·6076 861 | ·5488 923 | ·4935 527 |
| ·72 | ·8087 040$^e$ | ·7477 909 | ·6867 763 | ·6270 886 | ·5697 257 | ·5153 528 |
| ·73 | ·8206 660$^e$ | ·7626 442 | ·7041 208 | ·6464 718 | ·5906 834 | ·5374 358 |
| ·74 | ·8323 520$^e$ | ·7772 545$^-$ | ·7212 987 | ·6658 009 | ·6117 262 | ·5597 608 |
| ·75 | ·8437 500$^e$ | ·7916 013 | ·7382 812 | ·6850 396 | ·6328 125$^e$ | ·5822 837 |
| ·76 | ·8548 480$^e$ | ·8056 640 | ·7550 387 | ·7041 503 | ·6538 986 | ·6049 570 |
| ·77 | ·8656 340$^e$ | ·8194 215$^+$ | ·7715 408 | ·7230 940 | ·6749 384 | ·6277 297 |
| ·78 | ·8760 960$^e$ | ·8328 527 | ·7877 563 | ·7418 300 | ·6958 831 | ·6505 472 |
| ·79 | ·8862 220$^e$ | ·8459 361 | ·8036 536 | ·7603 163 | ·7166 815$^-$ | ·6733 510 |
| ·80 | ·8960 000$^e$ | ·8586 501 | ·8192 000$^e$ | ·7785 094 | ·7372 800$^e$ | ·6960 790 |
| ·81 | ·9054 180$^e$ | ·8709 727 | ·8343 624 | ·7963 643 | ·7576 223 | ·7186 650$^+$ |
| ·82 | ·9144 640$^e$ | ·8828 819 | ·8491 067 | ·8138 345$^-$ | ·7776 494 | ·7410 387 |
| ·83 | ·9231 260$^e$ | ·8943 553 | ·8633 984 | ·8308 718 | ·7972 998 | ·7631 258 |
| ·84 | ·9313 920$^e$ | ·9053 703 | ·8772 019 | ·8474 266 | ·8165 090 | ·7848 474 |
| ·85 | ·9392 500$^e$ | ·9159 041 | ·8904 812 | ·8634 478 | ·8352 100$^e$ | ·8061 205$^+$ |
| ·86 | ·9466 880$^e$ | ·9259 337 | ·9031 995$^+$ | ·8788 826 | ·8533 327 | ·8268 574 |
| ·87 | ·9536 940$^e$ | ·9354 358 | ·9153 192 | ·8936 766 | ·8708 044 | ·8469 659 |
| ·88 | ·9602 560$^e$ | ·9443 871 | ·9268 019 | ·9077 739 | ·8875 491 | ·8663 487 |
| ·89 | ·9663 620$^e$ | ·9527 637 | ·9376 088 | ·9211 170 | ·9034 883 | ·8849 041 |
| ·90 | ·9720 000$^e$ | ·9605 418 | ·9477 000$^e$ | ·9336 467 | ·9185 400$^e$ | ·9025 251 |
| ·91 | ·9771 580$^e$ | ·9676 974 | ·9570 352 | ·9453 021 | ·9326 195$^-$ | ·9190 996 |
| ·92 | ·9818 240$^e$ | ·9742 060 | ·9655 731 | ·9560 209 | ·9456 387 | ·9345 104 |
| ·93 | ·9859 860$^e$ | ·9800 432 | ·9732 720 | ·9657 388 | ·9575 066 | ·9486 347 |
| ·94 | ·9896 320$^e$ | ·9851 843 | ·9800 891 | ·9743 901 | ·9681 287 | ·9613 445$^+$ |
| ·95 | ·9927 500$^e$ | ·9896 042 | ·9859 812 | ·9819 073 | ·9774 075$^e$ | ·9725 061 |
| ·96 | ·9953 280$^e$ | ·9932 779 | ·9909 043 | ·9882 212 | ·9852 420 | ·9819 798 |
| ·97 | ·9973 540$^e$ | ·9961 800 | ·9948 136 | ·9932 609 | ·9915 279 | ·9896 205$^-$ |
| ·98 | ·9988 160$^e$ | ·9982 849 | ·9976 635$^+$ | ·9969 538 | ·9961 576 | ·9952 767 |
| ·99 | ·9997 020$^e$ | ·9995 669 | ·9994 080 | ·9992 256 | ·9990 199 | ·9987 911 |
| 1·00 | 1·0000 000 | 1·0000 000 | 1·0000 000 | 1·0000 000 | 1·0000 000 | 1·0000 000 |

$x$ = ·03 to ·60 $q = 2$ $p$ = 5 to 7·5

| | $p = 5$ | $p = 5·5$ | $p = 6$ | $p = 6·5$ | $p = 7$ | $p = 7·5$ |
|---|---|---|---|---|---|---|
| $B(p, q) =$ | ·3333 3333 × $\frac{1}{10}$ | ·2797 2028 × $\frac{1}{10}$ | ·2380 9524 × $\frac{1}{10}$ | ·2051 2821 × $\frac{1}{10}$ | ·1785 7143 × $\frac{1}{10}$ | ·1568 6275 × $\frac{1}{10}$ |
| $x$ | | | | | | |
| ·03 | ·0000 001 | | | | | |
| ·04 | ·0000 006 | ·0000 001 | | | | |
| ·05 | ·0000 018 | ·0000 004 | ·0000 001 | | | |
| ·06 | ·0000 044 | ·0000 012 | ·0000 003 | ·0000 001 | | |
| ·07 | ·0000 095− | ·0000 027 | ·0000 008 | ·0000 002 | ·0000 001 | |
| ·08 | ·0000 184 | ·0000 056 | ·0000 017 | ·0000 005+ | ·0000 002 | |
| ·09 | ·0000 328 | ·0000 106 | ·0000 034 | ·0000 011 | ·0000 004 | ·0000 001 |
| ·10 | ·0000 550$^{e}$ | ·0000 188 | ·0000 064$^{e}$ | ·0000 022 | ·0000 007 | ·0000 002 |
| ·11 | ·0000 878 | ·0000 315− | ·0000 112 | ·0000 040 | ·0000 014 | ·0000 005− |
| ·12 | ·0001 344 | ·0000 503 | ·0000 188 | ·0000 070 | ·0000 026 | ·0000 009 |
| ·13 | ·0001 986 | ·0000 774 | ·0000 300 | ·0000 116 | ·0000 044 | ·0000 017 |
| ·14 | ·0002 850+ | ·0001 153 | ·0000 464 | ·0000 186 | ·0000 074 | ·0000 029 |
| ·15 | ·0003 987 | ·0001 669 | ·0000 695− | ·0000 288 | ·0000 119 | ·0000 049 |
| ·16 | ·0005 453 | ·0002 357 | ·0001 013 | ·0000 434 | ·0000 185− | ·0000 078 |
| ·17 | ·0007 312 | ·0003 258 | ·0001 443 | ·0000 636 | ·0000 279 | ·0000 122 |
| ·18 | ·0009 637 | ·0004 417 | ·0002 014 | ·0000 913 | ·0000 413 | ·0000 186 |
| ·19 | ·0012 504 | ·0005 888 | ·0002 757 | ·0001 285− | ·0000 596 | ·0000 276 |
| ·20 | ·0016 000$^{e}$ | ·0007 728 | ·0003 712$^{e}$ | ·0001 775− | ·0000 845− | ·0000 401 |
| ·21 | ·0020 216 | ·0010 004 | ·0004 923 | ·0002 411 | ·0001 176 | ·0000 572 |
| ·22 | ·0025 253 | ·0012 787 | ·0006 440 | ·0003 228 | ·0001 611 | ·0000 801 |
| ·23 | ·0031 216 | ·0016 159 | ·0008 320 | ·0004 263 | ·0002 176 | ·0001 106 |
| ·24 | ·0038 221 | ·0020 207 | ·0010 625+ | ·0005 561 | ·0002 899 | ·0001 505+ |
| ·25 | ·0046 387 | ·0025 024 | ·0013 428 | ·0007 172 | ·0003 815− | ·0002 022 |
| ·26 | ·0055 842 | ·0030 716 | ·0016 805+ | ·0009 152 | ·0004 964 | ·0002 683 |
| ·27 | ·0066 722 | ·0037 391 | ·0020 843 | ·0011 565+ | ·0006 391 | ·0003 519 |
| ·28 | ·0079 168 | ·0045 170 | ·0025 637 | ·0014 484 | ·0008 150− | ·0004 569 |
| ·29 | ·0093 326 | ·0054 179 | ·0031 288 | ·0017 986 | ·0010 298 | ·0005 876 |
| ·30 | ·0109 350$^{e}$ | ·0064 552 | ·0037 908$^{e}$ | ·0022 161 | ·0012 903 | ·0007 487 |
| ·31 | ·0127 400 | ·0076 432 | ·0045 618 | ·0027 104 | ·0016 040 | ·0009 459 |
| ·32 | ·0147 640 | ·0089 971 | ·0054 546 | ·0032 921 | ·0019 791 | ·0011 856 |
| ·33 | ·0170 239 | ·0105 326 | ·0064 832 | ·0039 728 | ·0024 250− | ·0014 751 |
| ·34 | ·0195 372 | ·0122 663 | ·0076 622 | ·0047 651 | ·0029 518 | ·0018 223 |
| ·35 | ·0223 218 | ·0142 156 | ·0090 075+ | ·0056 824 | ·0035 708 | ·0022 362 |
| ·36 | ·0253 958 | ·0163 984 | ·0105 356 | ·0067 393 | ·0042 944 | ·0027 271 |
| ·37 | ·0287 777 | ·0188 335− | ·0122 642 | ·0079 516 | ·0051 358 | ·0033 059 |
| ·38 | ·0324 864 | ·0215 401 | ·0142 116 | ·0093 360 | ·0061 098 | ·0039 850− |
| ·39 | ·0365 408 | ·0245 382 | ·0163 973 | ·0109 104 | ·0072 321 | ·0047 778 |
| ·40 | ·0409 600$^{e}$ | ·0278 483 | ·0188 416$^{e}$ | ·0126 936 | ·0085 197 | ·0056 992 |
| ·41 | ·0457 632 | ·0314 912 | ·0215 655− | ·0147 059 | ·0099 909 | ·0067 652 |
| ·42 | ·0509 696 | ·0354 883 | ·0245 909 | ·0169 683 | ·0116 653 | ·0079 933 |
| ·43 | ·0565 983 | ·0398 614 | ·0279 404 | ·0195 031 | ·0135 637 | ·0094 023 |
| ·44 | ·0626 682 | ·0446 324 | ·0316 375+ | ·0223 337 | ·0157 085− | ·0110 128 |
| ·45 | ·0691 980 | ·0498 236 | ·0357 062 | ·0254 843 | ·0181 230 | ·0128 466 |
| ·46 | ·0762 063 | ·0554 573 | ·0401 710 | ·0289 803 | ·0208 321 | ·0149 271 |
| ·47 | ·0837 109 | ·0615 559 | ·0450 571 | ·0328 479 | ·0238 619 | ·0172 793 |
| ·48 | ·0917 294 | ·0681 419 | ·0503 900 | ·0371 144 | ·0272 400 | ·0199 299 |
| ·49 | ·1002 787 | ·0752 373 | ·0561 956 | ·0418 076 | ·0309 948 | ·0229 070 |
| ·50 | ·1093 750$^{e}$ | ·0828 641 | ·0625 000$^{e}$ | ·0469 563 | ·0351 562 | ·0262 403 |
| ·51 | ·1190 337 | ·0910 438 | ·0693 294 | ·0525 898 | ·0397 553 | ·0299 611 |
| ·52 | ·1292 694 | ·0997 975+ | ·0767 100 | ·0587 380 | ·0448 239 | ·0341 022 |
| ·53 | ·1400 955− | ·1091 457 | ·0846 679 | ·0654 311 | ·0503 951 | ·0386 979 |
| ·54 | ·1515 245− | ·1191 078 | ·0932 289 | ·0726 997 | ·0565 026 | ·0437 838 |
| ·55 | ·1635 674 | ·1297 029 | ·1024 184 | ·0805 744 | ·0631 811 | ·0493 967 |
| ·56 | ·1762 342 | ·1409 484 | ·1122 612 | ·0890 860 | ·0704 655− | ·0555 749 |
| ·57 | ·1895 330 | ·1528 610 | ·1227 813 | ·0982 649 | ·0783 914 | ·0623 574 |
| ·58 | ·2034 706 | ·1654 558 | ·1340 018 | ·1081 411 | ·0869 946 | ·0697 844 |
| ·59 | ·2180 519 | ·1787 463 | ·1459 446 | ·1187 441 | ·0963 108 | ·0778 965− |
| ·60 | ·2332 800$^{e}$ | ·1927 444 | ·1586 304$^{e}$ | ·1301 025− | ·1063 757 | ·0867 350− |

$x = \cdot 61$ to $1\cdot 00$ $q = 2$ $p = 5$ to $7\cdot 5$

| $x$ | $p = 5$ | $p = 5\cdot 5$ | $p = 6$ | $p = 6\cdot 5$ | $p = 7$ | $p = 7\cdot 5$ |
|---|---|---|---|---|---|---|
| $B(p, q) =$ | ·3333 3333 × 1/10 | ·2797 2028 × 1/10 | ·2380 9524 × 1/10 | ·2051 2821 × 1/10 | ·1785 7143 × 1/10 | ·1568 6275 × 1/10 |
| ·61 | ·2491 559 | ·2074 602 | ·1720 781 | ·1422 438 | ·1172 243 | ·0963 415+ |
| ·62 | ·2656 785+ | ·2229 014 | ·1863 048 | ·1551 942 | ·1288 911 | ·1067 575+ |
| ·63 | ·2828 444 | ·2390 737 | ·2013 257 | ·1689 782 | ·1414 094 | ·1180 242 |
| ·64 | ·3006 477 | ·2559 801 | ·2171 535+ | ·1836 184 | ·1548 112 | ·1301 822 |
| ·65 | ·3190 799 | ·2736 209 | ·2337 986 | ·1991 352 | ·1691 269 | ·1432 710 |
| ·66 | ·3381 298 | ·2919 937 | ·2512 680 | ·2155 463 | ·1843 844 | ·1573 287 |
| ·67 | ·3577 832 | ·3110 927 | ·2695 660 | ·2328 664 | ·2006 096 | ·1723 915− |
| ·68 | ·3780 227 | ·3309 086 | ·2886 930 | ·2511 069 | ·2178 248 | ·1884 932 |
| ·69 | ·3988 280 | ·3514 287 | ·3086 459 | ·2702 753 | ·2360 494 | ·2056 647 |
| ·70 | ·4201 750e | ·3726 362 | ·3294 172e | ·2903 750+ | ·2552 983 | ·2239 333 |
| ·71 | ·4420 362 | ·3945 102 | ·3509 948 | ·3114 046 | ·2755 821 | ·2433 220 |
| ·72 | ·4643 802 | ·4170 253 | ·3733 617 | ·3333 576 | ·2969 061 | ·2638 490 |
| ·73 | ·4871 718 | ·4401 514 | ·3964 957 | ·3562 216 | ·3192 698 | ·2855 268 |
| ·74 | ·5103 715+ | ·4638 534 | ·4203 686 | ·3799 780 | ·3426 661 | ·3083 613 |
| ·75 | ·5339 355+ | ·4880 907 | ·4449 463 | ·4046 015+ | ·3670 807 | ·3323 513 |
| ·76 | ·5578 156 | ·5128 174 | ·4701 878 | ·4300 593 | ·3924 912 | ·3574 868 |
| ·77 | ·5819 586 | ·5379 813 | ·4960 453 | ·4563 103 | ·4188 665− | ·3837 487 |
| ·78 | ·6063 066 | ·5635 242 | ·5224 631 | ·4833 049 | ·4461 654 | ·4111 075− |
| ·79 | ·6307 966 | ·5893 812 | ·5493 776 | ·5109 839 | ·4743 366 | ·4395 218 |
| ·80 | ·6553 600e | ·6154 804 | ·5767 168e | ·5392 781 | ·5033 165− | ·4689 374 |
| ·81 | ·6799 230 | ·6417 427 | ·6043 992 | ·5681 070 | ·5330 293 | ·4992 860 |
| ·82 | ·7044 057 | ·6680 812 | ·6323 339 | ·5973 788 | ·5633 852 | ·5304 833 |
| ·83 | ·7287 225+ | ·6944 012 | ·6604 196 | ·6269 887 | ·5942 795+ | ·5624 282 |
| ·84 | ·7527 815− | ·7205 994 | ·6885 441 | ·6568 187 | ·6255 915+ | ·5950 005− |
| ·85 | ·7764 843 | ·7465 639 | ·7165 841 | ·6867 365− | ·6571 830 | ·6280 596 |
| ·86 | ·7997 259 | ·7721 732 | ·7444 037 | ·7165 942 | ·6888 971 | ·6614 427 |
| ·87 | ·8223 945+ | ·7972 968 | ·7718 546 | ·7462 280 | ·7205 567 | ·6949 627 |
| ·88 | ·8443 711 | ·8217 937 | ·7987 750+ | ·7754 564 | ·7519 631 | ·7284 062 |
| ·89 | ·8655 292 | ·8455 126 | ·8249 889 | ·8040 798 | ·7828 946 | ·7615 316 |
| ·90 | ·8857 350e | ·8682 914 | ·8503 056e | ·8318 792 | ·8131 047 | ·7940 665− |
| ·91 | ·9048 466 | ·8899 566 | ·8745 186 | ·8586 147 | ·8423 205− | ·8257 056 |
| ·92 | ·9227 141 | ·9103 231 | ·8974 054 | ·8840 248 | ·8702 407 | ·8561 083 |
| ·93 | ·9391 793 | ·9291 931 | ·9187 261 | ·9078 250+ | ·8965 343 | ·8848 954 |
| ·94 | ·9540 752 | ·9463 566 | ·9382 229 | ·9297 065− | ·9208 382 | ·9116 474 |
| ·95 | ·9672 262 | ·9615 902 | ·9556 195− | ·9493 346 | ·9427 553 | ·9359 006 |
| ·96 | ·9784 472 | ·9746 566 | ·9706 197 | ·9663 480 | ·9618 528 | ·9571 447 |
| ·97 | ·9875 441 | ·9853 045− | ·9829 070 | ·9803 568 | ·9776 592 | ·9748 192 |
| ·98 | ·9943 129 | ·9932 679 | ·9921 435− | ·9909 413 | ·9896 631 | ·9883 105− |
| ·99 | ·9985 396 | ·9982 654 | ·9979 690 | ·9976 504 | ·9973 099 | ·9969 478 |
| 1·00 | 1·0000 000 | 1·0000 000 | 1·0000 000 | 1·0000 000 | 1·0000 000 | 1·0000 000 |

$x = \cdot 10$ to $\cdot 70$ $\quad q = 2 \quad$ $p = 8$ to $10\cdot 5$

| | $p = 8$ | $p = 8\cdot 5$ | $p = 9$ | $p = 9\cdot 5$ | $p = 10$ | $p = 10\cdot 5$ |
|---|---|---|---|---|---|---|
| $B(p, q) =$ | ·1388 8889 × $\frac{1}{10}$ | ·1238 3901 × $\frac{1}{10}$ | ·1111 1111 × $\frac{1}{10}$ | ·1002 5063 × $\frac{1}{10}$ | ·9090 9091 × $\frac{1}{10^2}$ | ·8281 5735 × $\frac{1}{10^2}$ |
| $x$ | | | | | | |
| ·10 | ·0000 001 | | | | | |
| ·11 | ·0000 002 | ·0000 001 | | | | |
| ·12 | ·0000 003 | ·0000 001 | | | | |
| ·13 | ·0000 006 | ·0000 002 | ·0000 001 | | | |
| ·14 | ·0000 012 | ·0000 005− | ·0000 002 | ·0000 001 | | |
| ·15 | ·0000 020 | ·0000 008 | ·0000 003 | ·0000 001 | ·0000 001 | |
| ·16 | ·0000 033 | ·0000 014 | ·0000 006 | ·0000 002 | ·0000 001 | |
| ·17 | ·0000 053 | ·0000 023 | ·0000 010 | ·0000 004 | ·0000 002 | ·0000 001 |
| ·18 | ·0000 083 | ·0000 037 | ·0000 017 | ·0000 007 | ·0000 003 | ·0000 001 |
| ·19 | ·0000 127 | ·0000 058 | ·0000 027 | ·0000 012 | ·0000 006 | ·0000 003 |
| ·20 | ·0000 189 | ·0000 089 | ·0000 042 | ·0000 020 | ·0000 009 | ·0000 004 |
| ·21 | ·0000 277 | ·0000 134 | ·0000 064 | ·0000 031 | ·0000 015− | ·0000 007 |
| ·22 | ·0000 397 | ·0000 196 | ·0000 097 | ·0000 048 | ·0000 023 | ·0000 011 |
| ·23 | ·0000 561 | ·0000 283 | ·0000 143 | ·0000 072 | ·0000 036 | ·0000 018 |
| ·24 | ·0000 779 | ·0000 402 | ·0000 207 | ·0000 106 | ·0000 055− | ·0000 028 |
| ·25 | ·0001 068 | ·0000 563 | ·0000 296 | ·0000 155− | ·0000 081 | ·0000 042 |
| ·26 | ·0001 445+ | ·0000 776 | ·0000 416 | ·0000 222 | ·0000 119 | ·0000 063 |
| ·27 | ·0001 932 | ·0001 057 | ·0000 577 | ·0000 314 | ·0000 171 | ·0000 093 |
| ·28 | ·0002 554 | ·0001 423 | ·0000 791 | ·0000 439 | ·0000 243 | ·0000 134 |
| ·29 | ·0003 342 | ·0001 895+ | ·0001 072 | ·0000 605+ | ·0000 341 | ·0000 192 |
| ·30 | ·0004 330 | ·0002 498 | ·0001 437 | ·0000 825− | ·0000 472 | ·0000 270 |
| ·31 | ·0005 561 | ·0003 260 | ·0001 906 | ·0001 112 | ·0000 648 | ·0000 376 |
| ·32 | ·0007 081 | ·0004 217 | ·0002 505+ | ·0001 485− | ·0000 878 | ·0000 518 |
| ·33 | ·0008 945− | ·0005 409 | ·0003 263 | ·0001 964 | ·0001 179 | ·0000 707 |
| ·34 | ·0011 215− | ·0006 883 | ·0004 214 | ·0002 574 | ·0001 569 | ·0000 955− |
| ·35 | ·0013 962 | ·0008 693 | ·0005 399 | ·0003 346 | ·0002 069 | ·0001 277 |
| ·36 | ·0017 265+ | ·0010 901 | ·0006 865+ | ·0004 314 | ·0002 706 | ·0001 694 |
| ·37 | ·0021 215+ | ·0013 578 | ·0008 668 | ·0005 522 | ·0003 510 | ·0002 227 |
| ·38 | ·0025 913 | ·0016 805− | ·0010 871 | ·0007 017 | ·0004 520 | ·0002 906 |
| ·39 | ·0031 470 | ·0020 672 | ·0013 546 | ·0008 857 | ·0005 780 | ·0003 764 |
| ·40 | ·0038 011 | ·0025 284 | ·0016 777 | ·0011 108 | ·0007 340 | ·0004 841 |
| ·41 | ·0045 674 | ·0030 754 | ·0020 658 | ·0013 846 | ·0009 262 | ·0006 184 |
| ·42 | ·0054 610 | ·0037 211 | ·0025 295− | ·0017 157 | ·0011 615− | ·0007 848 |
| ·43 | ·0064 986 | ·0044 799 | ·0030 809 | ·0021 142 | ·0014 480 | ·0009 899 |
| ·44 | ·0076 984 | ·0053 675− | ·0037 335− | ·0025 913 | ·0017 950+ | ·0012 412 |
| ·45 | ·0090 802 | ·0064 014 | ·0045 022 | ·0031 598 | ·0022 133 | ·0015 475+ |
| ·46 | ·0106 653 | ·0076 007 | ·0054 040 | ·0038 341 | ·0027 149 | ·0019 190 |
| ·47 | ·0124 771 | ·0089 865− | ·0064 574 | ·0046 303 | ·0033 137 | ·0023 673 |
| ·48 | ·0145 405+ | ·0105 816 | ·0076 828 | ·0055 664 | ·0040 254 | ·0029 058 |
| ·49 | ·0168 823 | ·0124 108 | ·0091 028 | ·0066 627 | ·0048 673 | ·0035 496 |
| ·50 | ·0195 312 | ·0145 012 | ·0107 422 | ·0079 411 | ·0058 594 | ·0043 158 |
| ·51 | ·0225 178 | ·0168 817 | ·0126 278 | ·0094 265− | ·0070 235+ | ·0052 241 |
| ·52 | ·0258 745+ | ·0195 836 | ·0147 891 | ·0111 457 | ·0083 842 | ·0062 961 |
| ·53 | ·0296 356 | ·0226 402 | ·0172 578 | ·0131 284 | ·0099 686 | ·0075 564 |
| ·54 | ·0338 373 | ·0260 873 | ·0200 681 | ·0154 069 | ·0118 066 | ·0090 324 |
| ·55 | ·0385 176 | ·0299 626 | ·0232 571 | ·0180 164 | ·0139 312 | ·0107 543 |
| ·56 | ·0437 162 | ·0343 065+ | ·0268 642 | ·0209 950+ | ·0163 785− | ·0127 559 |
| ·57 | ·0494 745+ | ·0391 612 | ·0309 316 | ·0243 838 | ·0191 878 | ·0150 741 |
| ·58 | ·0558 355+ | ·0445 712 | ·0355 042 | ·0282 271 | ·0224 018 | ·0177 497 |
| ·59 | ·0628 434 | ·0505 830 | ·0406 295− | ·0325 722 | ·0260 670 | ·0208 272 |
| ·60 | ·0705 439 | ·0572 451 | ·0463 574 | ·0374 695+ | ·0302 331 | ·0243 552 |
| ·61 | ·0789 834 | ·0646 077 | ·0527 406 | ·0429 727 | ·0349 538 | ·0283 862 |
| ·62 | ·0882 094 | ·0727 227 | ·0598 339 | ·0491 385+ | ·0402 864 | ·0329 772 |
| ·63 | ·0982 697 | ·0816 430 | ·0676 944 | ·0560 264 | ·0462 917 | ·0381 892 |
| ·64 | ·1092 123 | ·0914 231 | ·0763 810 | ·0636 989 | ·0530 344 | ·0440 877 |
| ·65 | ·1210 850+ | ·1021 176 | ·0859 544 | ·0722 209 | ·0605 823 | ·0507 425− |
| ·66 | ·1339 351 | ·1137 819 | ·0964 765− | ·0816 597 | ·0690 068 | ·0582 274 |
| ·67 | ·1478 086 | ·1264 709 | ·1080 099 | ·0920 844 | ·0783 820 | ·0666 203 |
| ·68 | ·1627 501 | ·1402 390 | ·1206 179 | ·1035 657 | ·0887 848 | ·0760 029 |
| ·69 | ·1788 018 | ·1551 392 | ·1343 634 | ·1161 752 | ·1002 940 | ·0864 600 |
| ·70 | ·1960 032 | ·1712 228 | ·1493 083 | ·1299 847 | ·1129 901 | ·0980 793 |

$x$ = ·71 to 1·00 $q = 2$ $p$ = 8 to 10·5

| $x$ | $p = 8$ | $p = 8{\cdot}5$ | $p = 9$ | $p = 9{\cdot}5$ | $p = 10$ | $p = 10{\cdot}5$ |
|---|---|---|---|---|---|---|
| $B(p, q) =$ | ·1388 8889 × $\frac{1}{10}$ | ·1238 3901 × $\frac{1}{10}$ | ·1111 1111 × $\frac{1}{10}$ | ·1002 5063 × $\frac{1}{10}$ | ·9090 9091 × $\frac{1}{10^2}$ | ·8281 5735 × $\frac{1}{10^2}$ |
| ·71 | ·2143 902 | ·1885 381 | ·1655 131 | ·1450 655+ | ·1269 545− | ·1109 510 |
| ·72 | ·2339 941 | ·2071 300 | ·1830 354 | ·1614 878 | ·1422 684 | ·1251 663 |
| ·73 | ·2548 414 | ·2270 386 | ·2019 295+ | ·1793 192 | ·1590 122 | ·1408 171 |
| ·74 | ·2769 520 | ·2482 988 | ·2222 450− | ·1986 236 | ·1772 637 | ·1579 945− |
| ·75 | ·3003 387 | ·2709 385+ | ·2440 252 | ·2194 602 | ·1970 973 | ·1767 874 |
| ·76 | ·3250 062 | ·2949 777 | ·2673 064 | ·2418 817 | ·2185 822 | ·1972 811 |
| ·77 | ·3509 491 | ·3204 267 | ·2921 157 | ·2659 324 | ·2417 805− | ·2195 550− |
| ·78 | ·3781 516 | ·3472 850− | ·3184 694 | ·2916 468 | ·2667 448 | ·2436 808 |
| ·79 | ·4065 852 | ·3755 394 | ·3463 711 | ·3190 467 | ·2935 166 | ·2697 196 |
| ·80 | ·4362 076 | ·4051 620 | ·3758 096 | ·3481 392 | ·3221 225+ | ·2977 190 |
| ·81 | ·4669 611 | ·4361 083 | ·4067 565− | ·3789 139 | ·3525 723 | ·3277 099 |
| ·82 | ·4987 704 | ·4683 152 | ·4391 632 | ·4113 399 | ·3848 545− | ·3597 024 |
| ·83 | ·5315 410 | ·5016 983 | ·4729 588 | ·4453 624 | ·4189 331 | ·3936 816 |
| ·84 | ·5651 570 | ·5361 495+ | ·5080 464 | ·4808 989 | ·4547 432 | ·4296 030 |
| ·85 | ·5994 792 | ·5715 343 | ·5442 998 | ·5178 352 | ·4921 860 | ·4673 863 |
| ·86 | ·6343 420 | ·6076 884 | ·5815 600 | ·5560 210 | ·5311 238 | ·5069 098 |
| ·87 | ·6695 518 | ·6444 152 | ·6196 308 | ·5952 651 | ·5713 739 | ·5480 035− |
| ·88 | ·7048 837 | ·6814 815+ | ·6582 750+ | ·6353 297 | ·6127 021 | ·5904 410 |
| ·89 | ·7400 787 | ·7186 146 | ·6972 092 | ·6759 248 | ·6548 161 | ·6339 315+ |
| ·90 | ·7748 410 | ·7554 975+ | ·7360 989 | ·7167 017 | ·6973 569 | ·6781 101 |
| ·91 | ·8088 343 | ·7917 654 | ·7745 529 | ·7572 463 | ·7398 906 | ·7225 272 |
| ·92 | ·8416 790 | ·8270 006 | ·8121 175+ | ·7970 710 | ·7818 992 | ·7666 374 |
| ·93 | ·8729 476 | ·8607 276 | ·8482 701 | ·8356 073 | ·8227 699 | ·8097 863 |
| ·94 | ·9021 620 | ·8924 085+ | ·8824 120 | ·8721 963 | ·8617 842 | ·8511 970 |
| ·95 | ·9287 886 | ·9214 367 | ·9138 616 | ·9060 794 | ·8981 054 | ·8899 543 |
| ·96 | ·9522 342 | ·9471 315− | ·9418 462 | ·9363 879 | ·9307 657 | ·9249 884 |
| ·97 | ·9718 418 | ·9687 316 | ·9654 934 | ·9621 319 | ·9586 514 | ·9550 563 |
| ·98 | ·9868 851 | ·9853 885+ | ·9838 224 | ·9821 881 | ·9804 874 | ·9787 216 |
| ·99 | ·9965 643 | ·9961 595+ | ·9957 338 | ·9952 873 | ·9948 203 | ·9943 329 |
| 1·00 | 1·0000 000 | 1·0000 000 | 1·0000 000 | 1·0000 000 | 1·0000 000 | 1·0000 000 |

$x$ = ·18 to ·80 $q = 2$ $p$ = 11 to 16

| $x$ | $p = 11$ | $p = 12$ | $p = 13$ | $p = 14$ | $p = 15$ | $p = 16$ |
|---|---|---|---|---|---|---|
| $B(p,q)$ = | ·7575 7576 $\times\frac{1}{10^2}$ | ·6410 2564 $\times\frac{1}{10^2}$ | ·5494 5055 $\times\frac{1}{10^2}$ | ·4761 9048 $\times\frac{1}{10^2}$ | ·4166 6667 $\times\frac{1}{10^2}$ | ·3676 4706 $\times\frac{1}{10^2}$ |
| ·18 | ·0000 001 | | | | | |
| ·19 | ·0000 001 | | | | | |
| ·20 | ·0000 002 | | | | | |
| ·21 | ·0000 003 | ·0000 001 | | | | |
| ·22 | ·0000 006 | ·0000 001 | | | | |
| ·23 | ·0000 009 | ·0000 002 | ·0000 001 | | | |
| ·24 | ·0000 014 | ·0000 004 | ·0000 001 | | | |
| ·25 | ·0000 022 | ·0000 006 | ·0000 002 | | | |
| ·26 | ·0000 034 | ·0000 009 | ·0000 003 | ·0000 001 | | |
| ·27 | ·0000 050+ | ·0000 015− | ·0000 004 | ·0000 001 | | |
| ·28 | ·0000 074 | ·0000 022 | ·0000 007 | ·0000 002 | ·0000 001 | |
| ·29 | ·0000 107 | ·0000 034 | ·0000 010 | ·0000 003 | ·0000 001 | |
| ·30 | ·0000 154 | ·0000 050− | ·0000 016 | ·0000 005+ | ·0000 002 | ·0000 001 |
| ·31 | ·0000 218 | ·0000 073 | ·0000 024 | ·0000 008 | ·0000 003 | ·0000 001 |
| ·32 | ·0000 306 | ·0000 106 | ·0000 036 | ·0000 012 | ·0000 004 | ·0000 001 |
| ·33 | ·0000 423 | ·0000 151 | ·0000 053 | ·0000 019 | ·0000 007 | ·0000 002 |
| ·34 | ·0000 580 | ·0000 213 | ·0000 078 | ·0000 028 | ·0000 010 | ·0000 004 |
| ·35 | ·0000 787 | ·0000 297 | ·0000 112 | ·0000 042 | ·0000 016 | ·0000 006 |
| ·36 | ·0001 058 | ·0000 411 | ·0000 159 | ·0000 061 | ·0000 023 | ·0000 009 |
| ·37 | ·0001 411 | ·0000 564 | ·0000 224 | ·0000 088 | ·0000 035− | ·0000 014 |
| ·38 | ·0001 866 | ·0000 765+ | ·0000 312 | ·0000 127 | ·0000 051 | ·0000 021 |
| ·39 | ·0002 448 | ·0001 030 | ·0000 431 | ·0000 180 | ·0000 075− | ·0000 031 |
| ·40 | ·0003 188 | ·0001 376 | ·0000 591 | ·0000 252 | ·0000 107 | ·0000 046 |
| ·41 | ·0004 122 | ·0001 823 | ·0000 802 | ·0000 351 | ·0000 153 | ·0000 067 |
| ·42 | ·0005 294 | ·0002 398 | ·0001 081 | ·0000 485− | ·0000 217 | ·0000 096 |
| ·43 | ·0006 756 | ·0003 133 | ·0001 445+ | ·0000 663 | ·0000 303 | ·0000 138 |
| ·44 | ·0008 568 | ·0004 065− | ·0001 918 | ·0000 901 | ·0000 422 | ·0000 197 |
| ·45 | ·0010 803 | ·0005 240 | ·0002 529 | ·0001 215− | ·0000 581 | ·0000 277 |
| ·46 | ·0013 542 | ·0006 714 | ·0003 312 | ·0001 626 | ·0000 795+ | ·0000 387 |
| ·47 | ·0016 885− | ·0008 552 | ·0004 309 | ·0002 161 | ·0001 080 | ·0000 537 |
| ·48 | ·0020 942 | ·0010 830 | ·0005 572 | ·0002 854 | ·0001 456 | ·0000 740 |
| ·49 | ·0025 844 | ·0013 641 | ·0007 163 | ·0003 744 | ·0001 950− | ·0001 012 |
| ·50 | ·0031 738 | ·0017 090 | ·0009 155+ | ·0004 883 | ·0002 594 | ·0001 373 |
| ·51 | ·0038 795− | ·0021 302 | ·0011 638 | ·0006 330 | ·0003 430 | ·0001 852 |
| ·52 | ·0047 206 | ·0026 423 | ·0014 716 | ·0008 160 | ·0004 507 | ·0002 481 |
| ·53 | ·0057 190 | ·0032 620 | ·0018 512 | ·0010 460 | ·0005 888 | ·0003 303 |
| ·54 | ·0068 993 | ·0040 084 | ·0023 173 | ·0013 338 | ·0007 648 | ·0004 370 |
| ·55 | ·0082 891 | ·0049 038 | ·0028 867 | ·0016 920 | ·0009 880 | ·0005 749 |
| ·56 | ·0099 193 | ·0059 733 | ·0035 794 | ·0021 357 | ·0012 695+ | ·0007 521 |
| ·57 | ·0118 244 | ·0072 457 | ·0044 183 | ·0026 828 | ·0016 229 | ·0009 784 |
| ·58 | ·0140 425− | ·0087 533 | ·0054 300 | ·0033 541 | ·0020 642 | ·0012 661 |
| ·59 | ·0166 159 | ·0105 328 | ·0066 448 | ·0041 743 | ·0026 127 | ·0016 299 |
| ·60 | ·0195 910 | ·0126 253 | ·0080 976 | ·0051 720 | ·0032 913 | ·0020 876 |
| ·61 | ·0230 189 | ·0150 767 | ·0098 283 | ·0063 804 | ·0041 270 | ·0026 608 |
| ·62 | ·0269 549 | ·0179 380 | ·0118 817 | ·0078 379 | ·0051 517 | ·0033 752 |
| ·63 | ·0314 597 | ·0212 660 | ·0143 088 | ·0095 886 | ·0064 025− | ·0042 614 |
| ·64 | ·0365 983 | ·0251 230 | ·0171 667 | ·0116 831 | ·0079 228 | ·0053 558 |
| ·65 | ·0424 413 | ·0295 776 | ·0205 195− | ·0141 788 | ·0097 629 | ·0067 013 |
| ·66 | ·0490 639 | ·0347 049 | ·0244 383 | ·0171 411 | ·0119 809 | ·0083 481 |
| ·67 | ·0565 463 | ·0405 863 | ·0290 020 | ·0206 435+ | ·0146 433 | ·0103 552 |
| ·68 | ·0649 735+ | ·0473 099 | ·0342 977 | ·0247 688 | ·0178 263 | ·0127 907 |
| ·69 | ·0744 352 | ·0549 707 | ·0404 209 | ·0296 093 | ·0216 165− | ·0157 337 |
| ·70 | ·0850 251 | ·0636 699 | ·0474 756 | ·0352 676 | ·0261 116 | ·0192 751 |
| ·71 | ·0968 402 | ·0735 154 | ·0555 747 | ·0418 569 | ·0314 217 | ·0235 187 |
| ·72 | ·1099 810 | ·0846 207 | ·0648 396 | ·0495 017 | ·0376 696 | ·0285 825+ |
| ·73 | ·1245 495− | ·0971 047 | ·0754 004 | ·0583 375+ | ·0449 919 | ·0346 001 |
| ·74 | ·1406 489 | ·1110 907 | ·0873 950− | ·0685 113 | ·0535 392 | ·0417 212 |
| ·75 | ·1583 818 | ·1267 054 | ·1009 684 | ·0801 808 | ·0634 764 | ·0501 130 |
| ·76 | ·1778 488 | ·1440 771 | ·1162 717 | ·0935 140 | ·0749 828 | ·0599 602 |
| ·77 | ·1991 465+ | ·1633 340 | ·1334 604 | ·1086 883 | ·0882 513 | ·0714 657 |
| ·78 | ·2223 652 | ·1846 021 | ·1526 923 | ·1258 881 | ·1034 874 | ·0848 500+ |
| ·79 | ·2475 860 | ·2080 021 | ·1741 250− | ·1453 033 | ·1209 078 | ·1003 506 |
| ·80 | ·2748 779 | ·2336 462 | ·1979 121 | ·1671 258 | ·1407 375− | ·1182 195− |

$x = \cdot 81$ to $1\cdot 00$ | $q = 2$ | $p = 11$ to $16$

| | $p = 11$ | $p = 12$ | $p = 13$ | $p = 14$ | $p = 15$ | $p = 16$ |
|---|---|---|---|---|---|---|
| $B(p, q) =$ | ·7575 7576 × $\frac{1}{10^2}$ | ·6410 2564 × $\frac{1}{10^2}$ | ·5494 5055 × $\frac{1}{10^2}$ | ·4761 9048 × $\frac{1}{10^2}$ | ·4166 6667 × $\frac{1}{10^2}$ | ·3676 4706 × $\frac{1}{10^2}$ |
| $x$ | | | | | | |
| ·81 | ·3042 942 | ·2616 339 | ·2241 995+ | ·1915 452 | ·1632 060 | ·1387 208 |
| ·82 | ·3358 680 | ·2920 474 | ·2531 200 | ·2187 442 | ·1885 426 | ·1621 263 |
| ·83 | ·3696 076 | ·3249 456 | ·2847 870 | ·2488 915− | ·2169 700 | ·1887 089 |
| ·84 | ·4054 910 | ·3603 581 | ·3192 871 | ·2821 337 | ·2486 957 | ·2187 351 |
| ·85 | ·4434 596 | ·3982 769 | ·3566 712 | ·3185 860 | ·2839 012 | ·2524 537 |
| ·86 | ·4834 112 | ·4386 481 | ·3969 438 | ·3583 192 | ·3227 293 | ·2900 816 |
| ·87 | ·5251 919 | ·4813 611 | ·4400 505+ | ·4013 457 | ·3652 673 | ·3317 865+ |
| ·88 | ·5685 876 | ·5262 376 | ·4858 640 | ·4476 022 | ·4115 268 | ·3776 640 |
| ·89 | ·6133 132 | ·5730 177 | ·5341 661 | ·4969 284 | ·4614 195+ | ·4277 098 |
| ·90 | ·6590 023 | ·6213 450− | ·5846 291 | ·5490 430 | ·5147 278 | ·4817 852 |
| ·91 | ·7051 936 | ·6707 490 | ·6367 923 | ·6035 148 | ·5710 692 | ·5395 754 |
| ·92 | ·7513 183 | ·7206 261 | ·6900 363 | ·6597 288 | ·6298 543 | ·6005 374 |
| ·93 | ·7966 833 | ·7702 172 | ·7435 526 | ·7168 470 | ·6902 368 | ·6638 394 |
| ·94 | ·8404 550+ | ·8185 829 | ·7963 099 | ·7737 627 | ·7510 544 | ·7282 856 |
| ·95 | ·8816 401 | ·8645 761 | ·8470 144 | ·8290 475− | ·8107 597 | ·7922 280 |
| ·96 | ·9190 646 | ·9068 104 | ·8940 661 | ·8808 904 | ·8673 382 | ·8534 608 |
| ·97 | ·9513 509 | ·9436 256 | ·9355 077 | ·9270 275+ | ·9182 142 | ·9090 954 |
| ·98 | ·9768 922 | ·9730 487 | ·9689 682 | ·9646 617 | ·9601 398 | ·9554 130 |
| ·99 | ·9938 255− | ·9927 511 | ·9915 988 | ·9903 702 | ·9890 671 | ·9876 910 |
| 1·00 | 1·0000 000 | 1·0000 000 | 1·0000 000 | 1·0000 000 | 1·0000 000 | 1·0000 000 |

$x$ = ·33 to 1·00 $q = 2$ $p$ = 17 to 22

| | $p = 17$ | $p = 18$ | $p = 19$ | $p = 20$ | $p = 21$ | $p = 22$ |
|---|---|---|---|---|---|---|
| $B(p, q) =$ | ·3267 9739 $\times \frac{1}{10^2}$ | ·2923 9766 $\times \frac{1}{10^2}$ | ·2631 5789 $\times \frac{1}{10^2}$ | ·2380 9524 $\times \frac{1}{10^2}$ | ·2164 5022 $\times \frac{1}{10^2}$ | ·1976 2846 $\times \frac{1}{10^2}$ |
| $x$ | | | | | | |
| ·33 | ·0000 001 | | | | | |
| ·34 | ·0000 001 | | | | | |
| ·35 | ·0000 002 | ·0000 001 | | | | |
| ·36 | ·0000 003 | ·0000 001 | | | | |
| ·37 | ·0000 005+ | ·0000 002 | ·0000 001 | | | |
| ·38 | ·0000 008 | ·0000 003 | ·0000 001 | ·0000 001 | | |
| ·39 | ·0000 013 | ·0000 005+ | ·0000 002 | ·0000 001 | | |
| ·40 | ·0000 019 | ·0000 008 | ·0000 003 | ·0000 001 | ·0000 001 | |
| ·41 | ·0000 029 | ·0000 012 | ·0000 005+ | ·0000 002 | ·0000 001 | |
| ·42 | ·0000 043 | ·0000 019 | ·0000 008 | ·0000 004 | ·0000 002 | ·0000 001 |
| ·43 | ·0000 063 | ·0000 028 | ·0000 013 | ·0000 006 | ·0000 003 | ·0000 001 |
| ·44 | ·0000 091 | ·0000 042 | ·0000 020 | ·0000 009 | ·0000 004 | ·0000 002 |
| ·45 | ·0000 132 | ·0000 062 | ·0000 030 | ·0000 014 | ·0000 007 | ·0000 003 |
| ·46 | ·0000 188 | ·0000 091 | ·0000 044 | ·0000 021 | ·0000 010 | ·0000 005− |
| ·47 | ·0000 267 | ·0000 132 | ·0000 065+ | ·0000 032 | ·0000 016 | ·0000 008 |
| ·48 | ·0000 375+ | ·0000 190 | ·0000 096 | ·0000 048 | ·0000 024 | ·0000 012 |
| ·49 | ·0000 523 | ·0000 270 | ·0000 139 | ·0000 071 | ·0000 037 | ·0000 019 |
| ·50 | ·0000 725− | ·0000 381 | ·0000 200 | ·0000 105− | ·0000 055− | ·0000 029 |
| ·51 | ·0000 997 | ·0000 535+ | ·0000 286 | ·0000 153 | ·0000 082 | ·0000 043 |
| ·52 | ·0001 361 | ·0000 745− | ·0000 407 | ·0000 222 | ·0000 120 | ·0000 065+ |
| ·53 | ·0001 847 | ·0001 030 | ·0000 573 | ·0000 318 | ·0000 176 | ·0000 097 |
| ·54 | ·0002 490 | ·0001 415− | ·0000 802 | ·0000 453 | ·0000 256 | ·0000 144 |
| ·55 | ·0003 336 | ·0001 930 | ·0001 114 | ·0000 642 | ·0000 369 | ·0000 212 |
| ·56 | ·0004 442 | ·0002 617 | ·0001 538 | ·0000 902 | ·0000 528 | ·0000 308 |
| ·57 | ·0005 881 | ·0003 526 | ·0002 109 | ·0001 258 | ·0000 749 | ·0000 445+ |
| ·58 | ·0007 743 | ·0004 723 | ·0002 873 | ·0001 745− | ·0001 057 | ·0000 639 |
| ·59 | ·0010 138 | ·0006 289 | ·0003 892 | ·0002 403 | ·0001 481 | ·0000 911 |
| ·60 | ·0013 203 | ·0008 328 | ·0005 240 | ·0003 291 | ·0002 062 | ·0001 290 |
| ·61 | ·0017 105+ | ·0010 968 | ·0007 016 | ·0004 478 | ·0002 853 | ·0001 814 |
| ·62 | ·0022 049 | ·0014 367 | ·0009 339 | ·0006 058 | ·0003 922 | ·0002 535− |
| ·63 | ·0028 282 | ·0018 722 | ·0012 365− | ·0008 149 | ·0005 360 | ·0003 519 |
| ·64 | ·0036 103 | ·0024 274 | ·0016 283 | ·0010 900 | ·0007 282 | ·0004 857 |
| ·65 | ·0045 868 | ·0031 316 | ·0021 331 | ·0014 500− | ·0009 837 | ·0006 662 |
| ·66 | ·0058 006 | ·0040 204 | ·0027 802 | ·0019 186 | ·0013 214 | ·0009 086 |
| ·67 | ·0073 025+ | ·0051 370 | ·0036 054 | ·0025 253 | ·0017 654 | ·0012 320 |
| ·68 | ·0091 525− | ·0065 329 | ·0046 527 | ·0033 068 | ·0023 459 | ·0016 613 |
| ·69 | ·0114 209 | ·0082 701 | ·0059 752 | ·0043 084 | ·0031 008 | ·0022 279 |
| ·70 | ·0141 905− | ·0104 218 | ·0076 373 | ·0055 855− | ·0040 774 | ·0029 715− |
| ·71 | ·0175 569 | ·0130 750− | ·0097 160 | ·0072 057 | ·0053 342 | ·0039 422 |
| ·72 | ·0216 309 | ·0163 313 | ·0123 037 | ·0092 511 | ·0069 434 | ·0052 027 |
| ·73 | ·0265 400 | ·0203 100 | ·0155 094 | ·0118 205+ | ·0089 930 | ·0068 307 |
| ·74 | ·0324 294 | ·0251 489 | ·0194 621 | ·0150 323 | ·0115 904 | ·0089 221 |
| ·75 | ·0394 640 | ·0310 074 | ·0243 126 | ·0190 273 | ·0148 651 | ·0115 947 |
| ·76 | ·0478 294 | ·0380 677 | ·0302 366 | ·0239 718 | ·0189 724 | ·0149 920 |
| ·77 | ·0577 330 | ·0465 368 | ·0374 367 | ·0300 609 | ·0240 976 | ·0192 872 |
| ·78 | ·0694 043 | ·0566 480 | ·0461 453 | ·0375 220 | ·0304 595+ | ·0246 885− |
| ·79 | ·0830 954 | ·0686 619 | ·0566 259 | ·0466 171 | ·0383 148 | ·0314 436 |
| ·80 | ·0990 792 | ·0828 662 | ·0691 753 | ·0576 461 | ·0479 615+ | ·0398 450− |
| ·81 | ·1176 483 | ·0995 755+ | ·0841 233 | ·0709 482 | ·0597 429 | ·0502 343 |
| ·82 | ·1391 110 | ·1191 284 | ·1018 323 | ·0869 030 | ·0740 489 | ·0630 067 |
| ·83 | ·1637 861 | ·1418 834 | ·1226 942 | ·1059 289 | ·0913 179 | ·0786 134 |
| ·84 | ·1919 954 | ·1682 127 | ·1471 254 | ·1284 798 | ·1120 344 | ·0975 625− |
| ·85 | ·2240 527 | ·1984 917 | ·1755 579 | ·1550 381 | ·1367 242 | ·1204 162 |
| ·86 | ·2602 498 | ·2330 852 | ·2084 259 | ·1861 027 | ·1659 448 | ·1477 835+ |
| ·87 | ·3008 377 | ·2723 284 | ·2461 474 | ·2221 711 | ·2002 687 | ·1803 063 |
| ·88 | ·3460 023 | ·3165 011 | ·2890 977 | ·2637 135− | ·2402 585+ | ·2186 352 |
| ·89 | ·3958 330 | ·3657 938 | ·3375 737 | ·3111 359 | ·2864 298 | ·2633 942 |
| ·90 | ·4502 839 | ·4202 650− | ·3917 470 | ·3647 300 | ·3391 989 | ·3151 267 |
| ·91 | ·5091 247 | ·4797 847 | ·4516 019 | ·4246 058 | ·3988 110 | ·3742 199 |
| ·92 | ·5718 802 | ·5439 647 | ·5168 556 | ·4906 027 | ·4652 423 | ·4407 997 |
| ·93 | ·6377 555+ | ·6120 706 | ·5868 565+ | ·5621 733 | ·5380 701 | ·5145 867 |
| ·94 | ·7055 453 | ·6829 119 | ·6604 546 | ·6382 337 | ·6163 017 | ·5947 039 |
| ·95 | ·7735 226 | ·7547 072 | ·7358 395+ | ·7169 718 | ·6981 513 | ·6794 204 |
| ·96 | ·8393 058 | ·8249 177 | ·8103 378 | ·7956 044 | ·7807 531 | ·7658 170 |
| ·97 | ·8996 973 | ·8900 449 | ·8801 620 | ·8700 709 | ·8597 932 | ·8493 491 |
| ·98 | ·9504 912 | ·9453 841 | ·9401 010 | ·9346 512 | ·9290 433 | ·9232 858 |
| ·99 | ·9862 435+ | ·9847 262 | ·9831 407 | ·9814 883 | ·9797 707 | ·9779 893 |
| 1·00 | 1·0000 000 | 1·0000 000 | 1·0000 000 | 1·0000 000 | 1·0000 000 | 1·0000 000 |

$x$ = ·43 to 1·00 $q = 2$ $p$ = 23 to 28

| $x$ | $p = 23$ | $p = 24$ | $p = 25$ | $p = 26$ | $p = 27$ | $p = 28$ |
|---|---|---|---|---|---|---|
| $B(p, q)$ = | ·1811 5942 × $\frac{1}{10^2}$ | ·1666 6667 × $\frac{1}{10^2}$ | ·1538 4615+ × $\frac{1}{10^2}$ | ·1424 5014 × $\frac{1}{10^2}$ | ·1322 7513 × $\frac{1}{10^2}$ | ·1231 5271 × $\frac{1}{10^2}$ |
| ·43 | ·0000 001 | | | | | |
| ·44 | ·0000 001 | | | | | |
| ·45 | ·0000 001 | ·0000 001 | | | | |
| ·46 | ·0000 002 | ·0000 001 | ·0000 001 | | | |
| ·47 | ·0000 004 | ·0000 002 | ·0000 001 | | | |
| ·48 | ·0000 006 | ·0000 003 | ·0000 002 | ·0000 001 | | |
| ·49 | ·0000 010 | ·0000 005− | ·0000 002 | ·0000 001 | ·0000 001 | |
| ·50 | ·0000 015− | ·0000 008 | ·0000 004 | ·0000 002 | ·0000 001 | ·0000 001 |
| ·51 | ·0000 023 | ·0000 012 | ·0000 006 | ·0000 003 | ·0000 002 | ·0000 001 |
| ·52 | ·0000 035+ | ·0000 019 | ·0000 010 | ·0000 006 | ·0000 003 | ·0000 002 |
| ·53 | ·0000 054 | ·0000 030 | ·0000 016 | ·0000 009 | ·0000 005− | ·0000 003 |
| ·54 | ·0000 081 | ·0000 046 | ·0000 026 | ·0000 014 | ·0000 008 | ·0000 004 |
| ·55 | ·0000 121 | ·0000 069 | ·0000 040 | ·0000 023 | ·0000 013 | ·0000 007 |
| ·56 | ·0000 180 | ·0000 105− | ·0000 061 | ·0000 035+ | ·0000 020 | ·0000 012 |
| ·57 | ·0000 264 | ·0000 157 | ·0000 093 | ·0000 055− | ·0000 032 | ·0000 019 |
| ·58 | ·0000 386 | ·0000 233 | ·0000 140 | ·0000 084 | ·0000 051 | ·0000 030 |
| ·59 | ·0000 560 | ·0000 343 | ·0000 210 | ·0000 128 | ·0000 078 | ·0000 048 |
| ·60 | ·0000 806 | ·0000 502 | ·0000 313 | ·0000 194 | ·0000 121 | ·0000 075− |
| ·61 | ·0001 152 | ·0000 730 | ·0000 462 | ·0000 292 | ·0000 184 | ·0000 116 |
| ·62 | ·0001 635+ | ·0001 053 | ·0000 678 | ·0000 435+ | ·0000 279 | ·0000 179 |
| ·63 | ·0002 307 | ·0001 510 | ·0000 987 | ·0000 644 | ·0000 420 | ·0000 273 |
| ·64 | ·0003 234 | ·0002 150− | ·0001 427 | ·0000 946 | ·0000 627 | ·0000 415− |
| ·65 | ·0004 505− | ·0003 041 | ·0002 050+ | ·0001 381 | ·0000 928 | ·0000 624 |
| ·66 | ·0006 237 | ·0004 275+ | ·0002 926 | ·0002 000 | ·0001 366 | ·0000 932 |
| ·67 | ·0008 584 | ·0005 973 | ·0004 150− | ·0002 879 | ·0001 996 | ·0001 382 |
| ·68 | ·0011 747 | ·0008 293 | ·0005 847 | ·0004 118 | ·0002 896 | ·0002 035− |
| ·69 | ·0015 982 | ·0011 448 | ·0008 189 | ·0005 851 | ·0004 175+ | ·0002 976 |
| ·70 | ·0021 621 | ·0015 710 | ·0011 399 | ·0008 261 | ·0005 980 | ·0004 324 |
| ·71 | ·0029 090 | ·0021 435− | ·0015 773 | ·0011 592 | ·0008 510 | ·0006 241 |
| ·72 | ·0038 924 | ·0029 080 | ·0021 697 | ·0016 169 | ·0012 035+ | ·0008 949 |
| ·73 | ·0051 804 | ·0039 233 | ·0029 674 | ·0022 417 | ·0016 915− | ·0012 750+ |
| ·74 | ·0068 578 | ·0052 638 | ·0040 351 | ·0030 895− | ·0023 628 | ·0018 052 |
| ·75 | ·0090 305+ | ·0070 237 | ·0054 559 | ·0042 331 | ·0032 806 | ·0025 398 |
| ·76 | ·0118 293 | ·0093 212 | ·0073 356 | ·0057 662 | ·0045 276 | ·0035 514 |
| ·77 | ·0154 148 | ·0123 034 | ·0098 078 | ·0078 093 | ·0062 113 | ·0049 353 |
| ·78 | ·0199 824 | ·0161 521 | ·0130 400 | ·0105 155− | ·0084 706 | ·0068 165+ |
| ·79 | ·0257 686 | ·0210 905+ | ·0172 408 | ·0140 779 | ·0114 831 | ·0093 572 |
| ·80 | ·0330 566 | ·0273 897 | ·0226 674 | ·0187 384 | ·0154 743 | ·0127 663 |
| ·81 | ·0421 822 | ·0353 765+ | ·0296 342 | ·0247 969 | ·0207 279 | ·0173 100 |
| ·82 | ·0535 404 | ·0454 406 | ·0385 220 | ·0326 218 | ·0275 976 | ·0233 252 |
| ·83 | ·0675 892 | ·0580 414 | ·0497 865+ | ·0426 609 | ·0365 191 | ·0312 327 |
| ·84 | ·0848 534 | ·0737 137 | ·0639 664 | ·0554 512 | ·0480 233 | ·0415 528 |
| ·85 | ·1059 242 | ·0930 705+ | ·0816 896 | ·0716 289 | ·0627 484 | ·0549 203 |
| ·86 | ·1314 549 | ·1168 017 | ·1036 749 | ·0919 343 | ·0814 491 | ·0720 978 |
| ·87 | ·1621 495+ | ·1456 664 | ·1307 285− | ·1172 127 | ·1050 017 | ·0939 847 |
| ·88 | ·1987 418 | ·1804 745+ | ·1637 294 | ·1484 044 | ·1343 996 | ·1216 190 |
| ·89 | ·2419 607 | ·2220 555+ | ·2036 017 | ·1865 209 | ·1707 343 | ·1561 638 |
| ·90 | ·2924 770 | ·2712 059 | ·2512 643 | ·2325 989 | ·2151 540 | ·1988 721 |
| ·91 | ·3508 249 | ·3286 098 | ·3075 517 | ·2876 224 | ·2687 891 | ·2510 161 |
| ·92 | ·4172 904 | ·3947 214 | ·3730 929 | ·3523 986 | ·3326 277 | ·3137 648 |
| ·93 | ·4917 545− | ·4695 973 | ·4481 325− | ·4273 717 | ·4073 216 | ·3879 844 |
| ·94 | ·5734 791 | ·5526 604 | ·5322 753 | ·5123 470 | ·4928 938 | ·4739 305+ |
| ·95 | ·6608 173 | ·6423 759 | ·6241 265+ | ·6060 962 | ·5883 086 | ·5707 845+ |
| ·96 | ·7508 265− | ·7358 100 | ·7207 934 | ·7058 009 | ·6908 546 | ·6759 746 |
| ·97 | ·8387 578 | ·8280 376 | ·8172 057 | ·8062 785+ | ·7952 715− | ·7841 992 |
| ·98 | ·9173 870 | ·9113 549 | ·9051 971 | ·8989 211 | ·8925 340 | ·8860 428 |
| ·99 | ·9761 456 | ·9742 409 | ·9722 767 | ·9702 544 | ·9681 752 | ·9660 407 |
| 1·00 | 1·0000 000 | 1·0000 000 | 1·0000 000 | 1·0000 000 | 1·0000 000 | 1·0000 000 |

$x = \cdot 52$ to $1\cdot 00$ $q = 2$ $p = 29$ to $34$

| | $p = 29$ | $p = 30$ | $p = 31$ | $p = 32$ | $p = 33$ | $p = 34$ |
|---|---|---|---|---|---|---|
| $B(p, q) =$ | ·1149 4253 × $\frac{1}{10^2}$ | ·1075 2688 × $\frac{1}{10^2}$ | ·1008 0645$^{+}$ × $\frac{1}{10^2}$ | ·9469 6970 × $\frac{1}{10^3}$ | ·8912 6560 × $\frac{1}{10^3}$ | ·8403 3613 × $\frac{1}{10^3}$ |
| $x$ | | | | | | |
| ·52 | ·0000 001 | | | | | |
| ·53 | ·0000 001 | ·0000 001 | | | | |
| ·54 | ·0000 002 | ·0000 001 | ·0000 001 | | | |
| ·55 | ·0000 004 | ·0000 002 | ·0000 001 | ·0000 001 | | |
| ·56 | ·0000 007 | ·0000 004 | ·0000 002 | ·0000 001 | ·0000 001 | |
| ·57 | ·0000 011 | ·0000 007 | ·0000 004 | ·0000 002 | ·0000 001 | ·0000 001 |
| ·58 | ·0000 018 | ·0000 011 | ·0000 006 | ·0000 004 | ·0000 002 | ·0000 001 |
| ·59 | ·0000 029 | ·0000 018 | ·0000 011 | ·0000 007 | ·0000 004 | ·0000 002 |
| ·60 | ·0000 046 | ·0000 029 | ·0000 018 | ·0000 011 | ·0000 007 | ·0000 004 |
| ·61 | ·0000 073 | ·0000 046 | ·0000 029 | ·0000 018 | ·0000 011 | ·0000 007 |
| ·62 | ·0000 115$^{-}$ | ·0000 073 | ·0000 047 | ·0000 030 | ·0000 019 | ·0000 012 |
| ·63 | ·0000 178 | ·0000 116 | ·0000 075$^{+}$ | ·0000 049 | ·0000 032 | ·0000 020 |
| ·64 | ·0000 274 | ·0000 181 | ·0000 119 | ·0000 079 | ·0000 052 | ·0000 034 |
| ·65 | ·0000 419 | ·0000 281 | ·0000 188 | ·0000 126 | ·0000 084 | ·0000 056 |
| ·66 | ·0000 635$^{-}$ | ·0000 432 | ·0000 294 | ·0000 200 | ·0000 136 | ·0000 092 |
| ·67 | ·0000 956 | ·0000 660 | ·0000 456 | ·0000 314 | ·0000 217 | ·0000 149 |
| ·68 | ·0001 428 | ·0001 001 | ·0000 701 | ·0000 491 | ·0000 343 | ·0000 240 |
| ·69 | ·0002 119 | ·0001 508 | ·0001 072 | ·0000 761 | ·0000 540 | ·0000 383 |
| ·70 | ·0003 123 | ·0002 254 | ·0001 625$^{+}$ | ·0001 171 | ·0000 843 | ·0000 606 |
| ·71 | ·0004 572 | ·0003 346 | ·0002 447 | ·0001 788 | ·0001 305$^{-}$ | ·0000 952 |
| ·72 | ·0006 647 | ·0004 933 | ·0003 657 | ·0002 710 | ·0002 006 | ·0001 484 |
| ·73 | ·0009 601 | ·0007 223 | ·0005 429 | ·0004 078 | ·0003 060 | ·0002 295$^{-}$ |
| ·74 | ·0013 778 | ·0010 506 | ·0008 004 | ·0006 093 | ·0004 635$^{-}$ | ·0003 523 |
| ·75 | ·0019 644 | ·0015 179 | ·0011 719 | ·0009 041 | ·0006 969 | ·0005 368 |
| ·76 | ·0027 830 | ·0021 788 | ·0017 044 | ·0013 322 | ·0010 404 | ·0008 120 |
| ·77 | ·0039 177 | ·0031 071 | ·0024 621 | ·0019 494 | ·0015 424 | ·0012 194 |
| ·78 | ·0054 802 | ·0044 020 | ·0035 330 | ·0028 332 | ·0022 704 | ·0018 181 |
| ·79 | ·0076 178 | ·0061 963 | ·0050 359 | ·0040 896 | ·0033 187 | ·0026 912 |
| ·80 | ·0105 225$^{-}$ | ·0086 656 | ·0071 305$^{+}$ | ·0058 629 | ·0048 171 | ·0039 551 |
| ·81 | ·0144 426 | ·0120 400 | ·0100 289 | ·0083 475$^{-}$ | ·0069 429 | ·0057 707 |
| ·82 | ·0196 966 | ·0166 186 | ·0140 106 | ·0118 029 | ·0099 361 | ·0083 589 |
| ·83 | ·0266 882 | ·0227 862 | ·0194 396 | ·0165 724 | ·0141 182 | ·0120 195$^{-}$ |
| ·84 | ·0359 235$^{-}$ | ·0310 318 | ·0267 858 | ·0231 041 | ·0199 148 | ·0171 546 |
| ·85 | ·0480 289 | ·0419 692 | ·0366 467 | ·0319 767 | ·0278 831 | ·0242 982 |
| ·86 | ·0637 684 | ·0563 582 | ·0497 729 | ·0439 269 | ·0387 423 | ·0341 484 |
| ·87 | ·0840 575$^{+}$ | ·0751 231 | ·0670 911 | ·0598 778 | ·0534 061 | ·0476 051 |
| ·88 | ·1099 703 | ·0993 660 | ·0897 232 | ·0809 638 | ·0730 146 | ·0658 074 |
| ·89 | ·1427 329 | ·1303 673 | ·1189 950$^{+}$ | ·1085 472 | ·0989 581 | ·0901 651 |
| ·90 | ·1836 950$^{+}$ | ·1695 646 | ·1564 234 | ·1442 147 | ·1328 836 | ·1223 765$^{-}$ |
| ·91 | ·2342 651 | ·2184 960 | ·2036 678 | ·1897 389 | ·1766 674 | ·1644 119 |
| ·92 | ·2957 911 | ·2786 851 | ·2624 230 | ·2469 793 | ·2323 270 | ·2184 384 |
| ·93 | ·3693 585$^{+}$ | ·3514 392 | ·3342 186 | ·3176 869 | ·3018 320 | ·2866 401 |
| ·94 | ·4554 685$^{-}$ | ·4375 157 | ·4200 776 | ·4031 570 | ·3867 545$^{+}$ | ·3708 691 |
| ·95 | ·5535 421 | ·5365 969 | ·5199 624 | ·5036 499 | ·4876 687 | ·4720 265$^{-}$ |
| ·96 | ·6611 797 | ·6464 868 | ·6319 115$^{-}$ | ·6174 678 | ·6031 685$^{+}$ | ·5890 253 |
| ·97 | ·7730 755$^{-}$ | ·7619 134 | ·7507 253 | ·7395 228 | ·7283 168 | ·7171 175$^{+}$ |
| ·98 | ·8794 543 | ·8727 749 | ·8660 109 | ·8591 683 | ·8522 531 | ·8452 708 |
| ·99 | ·9638 520 | ·9616 105$^{-}$ | ·9593 174 | ·9569 740 | ·9545 816 | ·9521 413 |
| 1·00 | 1·0000 000 | 1·0000 000 | 1·0000 000 | 1·0000 000 | 1·0000 000 | 1·0000 000 |

$x$ = ·58 to 1·00 $q = 2$ $p$ = 35 to 40

| $x$ | $p = 35$ | $p = 36$ | $p = 37$ | $p = 38$ | $p = 39$ | $p = 40$ |
|---|---|---|---|---|---|---|
| $B(p, q)$ = | ·7936 5079 × $\frac{1}{10^3}$ | ·7507 5075+ × $\frac{1}{10^3}$ | ·7112 3755+ × $\frac{1}{10^3}$ | ·6747 6383 × $\frac{1}{10^3}$ | ·6410 2564 × $\frac{1}{10^3}$ | ·6097 5610 × $\frac{1}{10^3}$ |
| ·58 | ·0000 001 | | | | | |
| ·59 | ·0000 001 | ·0000 001 | ·0000 001 | | | |
| ·60 | ·0000 003 | ·0000 002 | ·0000 001 | ·0000 001 | | |
| ·61 | ·0000 004 | ·0000 003 | ·0000 002 | ·0000 001 | ·0000 001 | |
| ·62 | ·0000 008 | ·0000 005− | ·0000 003 | ·0000 002 | ·0000 001 | ·0000 001 |
| ·63 | ·0000 013 | ·0000 009 | ·0000 006 | ·0000 004 | ·0000 002 | ·0000 001 |
| ·64 | ·0000 022 | ·0000 015− | ·0000 010 | ·0000 006 | ·0000 004 | ·0000 003 |
| ·65 | ·0000 038 | ·0000 025+ | ·0000 017 | ·0000 011 | ·0000 007 | ·0000 005− |
| ·66 | ·0000 062 | ·0000 042 | ·0000 029 | ·0000 019 | ·0000 013 | ·0000 009 |
| ·67 | ·0000 103 | ·0000 071 | ·0000 048 | ·0000 033 | ·0000 023 | ·0000 016 |
| ·68 | ·0000 168 | ·0000 117 | ·0000 082 | ·0000 057 | ·0000 040 | ·0000 028 |
| ·69 | ·0000 271 | ·0000 192 | ·0000 136 | ·0000 096 | ·0000 068 | ·0000 048 |
| ·70 | ·0000 436 | ·0000 313 | ·0000 225− | ·0000 161 | ·0000 116 | ·0000 083 |
| ·71 | ·0000 694 | ·0000 506 | ·0000 368 | ·0000 268 | ·0000 195− | ·0000 141 |
| ·72 | ·0001 097 | ·0000 810 | ·0000 598 | ·0000 441 | ·0000 325+ | ·0000 240 |
| ·73 | ·0001 720 | ·0001 288 | ·0000 964 | ·0000 721 | ·0000 539 | ·0000 403 |
| ·74 | ·0002 676 | ·0002 031 | ·0001 541 | ·0001 168 | ·0000 885− | ·0000 670 |
| ·75 | ·0004 132 | ·0003 178 | ·0002 443 | ·0001 877 | ·0001 441 | ·0001 106 |
| ·76 | ·0006 333 | ·0004 936 | ·0003 845− | ·0002 993 | ·0002 329 | ·0001 811 |
| ·77 | ·0009 634 | ·0007 607 | ·0006 003 | ·0004 734 | ·0003 731 | ·0002 939 |
| ·78 | ·0014 549 | ·0011 635+ | ·0009 299 | ·0007 428 | ·0005 930 | ·0004 732 |
| ·79 | ·0021 809 | ·0017 662 | ·0014 296 | ·0011 564 | ·0009 349 | ·0007 555− |
| ·80 | ·0032 452 | ·0026 611 | ·0021 808 | ·0017 862 | ·0014 622 | ·0011 963 |
| ·81 | ·0047 933 | ·0039 790 | ·0033 011 | ·0027 372 | ·0022 684 | ·0018 789 |
| ·82 | ·0070 276 | ·0059 047 | ·0049 584 | ·0041 614 | ·0034 907 | ·0029 266 |
| ·83 | ·0102 263 | ·0086 954 | ·0073 895+ | ·0062 763 | ·0053 281 | ·0045 208 |
| ·84 | ·0147 679 | ·0127 058 | ·0109 255− | ·0093 896 | ·0080 655− | ·0069 247 |
| ·85 | ·0211 613 | ·0184 188 | ·0160 229 | ·0139 314 | ·0121 068 | ·0105 161 |
| ·86 | ·0300 814 | ·0264 839 | ·0233 040 | ·0204 955− | ·0180 166 | ·0158 300 |
| ·87 | ·0424 099 | ·0377 608 | ·0336 038 | ·0298 895− | ·0265 729 | ·0236 136 |
| ·88 | ·0592 785− | ·0533 688 | ·0480 239 | ·0431 933 | ·0388 305− | ·0348 927 |
| ·89 | ·0821 092 | ·0747 346 | ·0679 889 | ·0618 230 | ·0561 909 | ·0510 498 |
| ·90 | ·1126 420 | ·1036 306 | ·0952 951 | ·0875 904 | ·0804 737 | ·0739 044 |
| ·91 | ·1529 315− | ·1421 857 | ·1321 355− | ·1227 426 | ·1139 701 | ·1057 824 |
| ·92 | ·2052 852 | ·1928 384 | ·1810 693 | ·1699 491 | ·1594 493 | ·1495 417 |
| ·93 | ·2720 961 | ·2581 837 | ·2448 858 | ·2321 845− | ·2200 614 | ·2084 978 |
| ·94 | ·3554 975+ | ·3406 354 | ·3262 770 | ·3124 153 | ·2990 424 | ·2861 495+ |
| ·95 | ·4567 293 | ·4417 818 | ·4271 872 | ·4129 477 | ·3990 641 | ·3855 365− |
| ·96 | ·5750 484 | ·5612 472 | ·5476 301 | ·5342 043 | ·5209 764 | ·5079 520 |
| ·97 | ·7059 347 | ·6947 775+ | ·6836 544 | ·6725 734 | ·6615 419 | ·6505 670 |
| ·98 | ·8382 269 | ·8311 266 | ·8239 750+ | ·8167 771 | ·8095 375− | ·8022 607 |
| ·99 | ·9496 544 | ·9471 220 | ·9445 452 | ·9419 253 | ·9392 634 | ·9365 605− |
| 1·00 | 1·0000 000 | 1·0000 000 | 1·0000 000 | 1·0000 000 | 1·0000 000 | 1·0000 000 |

$x$ = ·62 to 1·00 $q = 2$ $p$ = 41 to 45

| $x$ | $p = 41$ | $p = 42$ | $p = 43$ | $p = 44$ | $p = 45$ |
|---|---|---|---|---|---|
| $B(p, q) =$ | ·5807 2009 × $\frac{1}{10^3}$ | ·5537 0986 × $\frac{1}{10^3}$ | ·5285 4123 × $\frac{1}{10^3}$ | ·5050 5051 × $\frac{1}{10^3}$ | ·4830 9179 × $\frac{1}{10^3}$ |
| ·62 | ·0000 001 | | | | |
| ·63 | ·0000 001 | ·0000 001 | | | |
| ·64 | ·0000 002 | ·0000 001 | ·0000 001 | | |
| ·65 | ·0000 003 | ·0000 002 | ·0000 001 | ·0000 001 | ·0000 001 |
| ·66 | ·0000 006 | ·0000 004 | ·0000 003 | ·0000 002 | ·0000 001 |
| ·67 | ·0000 011 | ·0000 007 | ·0000 005+ | ·0000 003 | ·0000 002 |
| ·68 | ·0000 019 | ·0000 013 | ·0000 009 | ·0000 006 | ·0000 004 |
| ·69 | ·0000 034 | ·0000 024 | ·0000 017 | ·0000 012 | ·0000 008 |
| ·70 | ·0000 059 | ·0000 042 | ·0000 030 | ·0000 022 | ·0000 016 |
| ·71 | ·0000 103 | ·0000 075− | ·0000 054 | ·0000 039 | ·0000 028 |
| ·72 | ·0000 177 | ·0000 130 | ·0000 096 | ·0000 070 | ·0000 052 |
| ·73 | ·0000 301 | ·0000 224 | ·0000 167 | ·0000 125− | ·0000 093 |
| ·74 | ·0000 507 | ·0000 384 | ·0000 290 | ·0000 219 | ·0000 166 |
| ·75 | ·0000 849 | ·0000 651 | ·0000 499 | ·0000 382 | ·0000 292 |
| ·76 | ·0001 407 | ·0001 093 | ·0000 849 | ·0000 659 | ·0000 511 |
| ·77 | ·0002 314 | ·0001 821 | ·0001 433 | ·0001 126 | ·0000 885+ |
| ·78 | ·0003 774 | ·0003 008 | ·0002 397 | ·0001 909 | ·0001 519 |
| ·79 | ·0006 101 | ·0004 925+ | ·0003 974 | ·0003 205+ | ·0002 584 |
| ·80 | ·0009 783 | ·0007 997 | ·0006 533 | ·0005 336 | ·0004 356 |
| ·81 | ·0015 555+ | ·0012 872 | ·0010 647 | ·0008 803 | ·0007 275− |
| ·82 | ·0024 525+ | ·0020 543 | ·0017 199 | ·0014 394 | ·0012 041 |
| ·83 | ·0038 341 | ·0032 502 | ·0027 540 | ·0023 326 | ·0019 748 |
| ·84 | ·0059 425+ | ·0050 974 | ·0043 705+ | ·0037 458 | ·0032 091 |
| ·85 | ·0091 302 | ·0079 235+ | ·0068 734 | ·0059 600 | ·0051 660 |
| ·86 | ·0139 026 | ·0122 046 | ·0107 095+ | ·0093 939 | ·0082 367 |
| ·87 | ·0209 746 | ·0186 226 | ·0165 277 | ·0146 628 | ·0130 034 |
| ·88 | ·0313 409 | ·0281 390 | ·0252 543 | ·0226 567 | ·0203 189 |
| ·89 | ·0463 598 | ·0420 840 | ·0381 878 | ·0346 396 | ·0314 100 |
| ·90 | ·0678 443 | ·0622 571 | ·0571 089 | ·0523 678 | ·0480 038 |
| ·91 | ·0981 454 | ·0910 262 | ·0843 935− | ·0782 173 | ·0724 693 |
| ·92 | ·1401 989 | ·1313 939 | ·1231 004 | ·1152 930 | ·1079 469 |
| ·93 | ·1974 749 | ·1869 735+ | ·1769 747 | ·1674 595+ | ·1584 093 |
| ·94 | ·2737 273 | ·2617 656 | ·2502 538 | ·2391 811 | ·2285 363 |
| ·95 | ·3723 640 | ·3595 449 | ·3470 768 | ·3349 567 | ·3231 808 |
| ·96 | ·4951 360 | ·4825 325+ | ·4701 451 | ·4579 766 | ·4460 294 |
| ·97 | ·6396 552 | ·6288 127 | ·6180 449 | ·6073 574 | ·5967 548 |
| ·98 | ·7949 512 | ·7876 132 | ·7802 508 | ·7728 677 | ·7654 679 |
| ·99 | ·9338 177 | ·9310 361 | ·9282 168 | ·9253 607 | ·9224 690 |
| 1·00 | 1·0000 000 | 1·0000 000 | 1·0000 000 | 1·0000 000 | 1·0000 000 |

$x$ = ·66 to 1·00 $q = 2$ $p$ = 46 to 50

| $x$ | $p = 46$ | $p = 47$ | $p = 48$ | $p = 49$ | $p = 50$ |
|---|---|---|---|---|---|
| $B(p, q) =$ | ·4625 3469 × $\frac{1}{10^3}$ | ·4432 6241 × $\frac{1}{10^3}$ | ·4251 7007 × $\frac{1}{10^3}$ | ·4081 6327 × $\frac{1}{10^3}$ | ·3921 5686 × $\frac{1}{10^3}$ |
| ·66 | ·0000 001 | ·0000 001 | | | |
| ·67 | ·0000 002 | ·0000 001 | ·0000 001 | ·0000 001 | |
| ·68 | ·0000 003 | ·0000 002 | ·0000 001 | ·0000 001 | ·0000 001 |
| ·69 | ·0000 006 | ·0000 004 | ·0000 003 | ·0000 002 | ·0000 001 |
| ·70 | ·0000 011 | ·0000 008 | ·0000 006 | ·0000 004 | ·0000 003 |
| ·71 | ·0000 021 | ·0000 015− | ·0000 011 | ·0000 008 | ·0000 006 |
| ·72 | ·0000 038 | ·0000 028 | ·0000 020 | ·0000 015+ | ·0000 011 |
| ·73 | ·0000 069 | ·0000 052 | ·0000 038 | ·0000 029 | ·0000 021 |
| ·74 | ·0000 125+ | ·0000 094 | ·0000 071 | ·0000 054 | ·0000 041 |
| ·75 | ·0000 224 | ·0000 171 | ·0000 131 | ·0000 100 | ·0000 076 |
| ·76 | ·0000 396 | ·0000 307 | ·0000 238 | ·0000 184 | ·0000 143 |
| ·77 | ·0000 695+ | ·0000 546 | ·0000 429 | ·0000 336 | ·0000 264 |
| ·78 | ·0001 209 | ·0000 962 | ·0000 765− | ·0000 608 | ·0000 483 |
| ·79 | ·0002 083 | ·0001 678 | ·0001 351 | ·0001 087 | ·0000 875+ |
| ·80 | ·0003 554 | ·0002 899 | ·0002 364 | ·0001 927 | ·0001 570 |
| ·81 | ·0006 010 | ·0004 963 | ·0004 097 | ·0003 381 | ·0002 789 |
| ·82 | ·0010 069 | ·0008 417 | ·0007 033 | ·0005 875− | ·0004 906 |
| ·83 | ·0016 713 | ·0014 139 | ·0011 958 | ·0010 109 | ·0008 543 |
| ·84 | ·0027 482 | ·0023 527 | ·0020 134 | ·0017 224 | ·0014 730 |
| ·85 | ·0044 761 | ·0038 769 | ·0033 568 | ·0029 055− | ·0025 140 |
| ·86 | ·0072 194 | ·0063 255+ | ·0055 404 | ·0048 512 | ·0042 463 |
| ·87 | ·0115 277 | ·0102 159 | ·0090 503 | ·0080 151 | ·0070 962 |
| ·88 | ·0182 159 | ·0163 250+ | ·0146 256 | ·0130 990 | ·0117 282 |
| ·89 | ·0284 717 | ·0257 997 | ·0233 711 | ·0211 647 | ·0191 608 |
| ·90 | ·0439 889 | ·0402 970 | ·0369 036 | ·0337 859 | ·0309 227 |
| ·91 | ·0671 224 | ·0621 509 | ·0575 305+ | ·0532 385− | ·0492 530 |
| ·92 | ·1010 383 | ·0945 442 | ·0884 425+ | ·0827 120 | ·0773 324 |
| ·93 | ·1498 056 | ·1416 302 | ·1338 653 | ·1264 935− | ·1194 978 |
| ·94 | ·2183 077 | ·2084 839 | ·1990 530 | ·1900 033 | ·1813 229 |
| ·95 | ·3117 452 | ·3006 452 | ·2898 758 | ·2794 318 | ·2693 074 |
| ·96 | ·4343 052 | ·4228 053 | ·4115 305+ | ·4004 812 | ·3896 574 |
| ·97 | ·5862 417 | ·5758 224 | ·5655 006 | ·5552 799 | ·5451 634 |
| ·98 | ·7580 550− | ·7506 324 | ·7432 034 | ·7357 714 | ·7283 394 |
| ·99 | ·9195 425+ | ·9165 823 | ·9135 894 | ·9105 647 | ·9075 091 |
| 1·00 | 1·0000 000 | 1·0000 000 | 1·0000 000 | 1·0000 000 | 1·0000 000 |

$x$ = ·01 to ·60 $q = 2·5$ $p$ = 2·5 to 5

| $x$ | $p = 2·5$ | $p = 3$ | $p = 3·5$ | $p = 4$ | $p = 4·5$ | $p = 5$ |
|---|---|---|---|---|---|---|
| $B(p, q) =$ | ·7363 1078 $\times \frac{1}{10}$ | ·5079 3651 $\times \frac{1}{10}$ | ·3681 5539 $\times \frac{1}{10}$ | ·2770 5628 $\times \frac{1}{10}$ | ·2147 5731 $\times \frac{1}{10}$ | ·1704 9617 $\times \frac{1}{10}$ |
| ·01 | ·0000 537 | ·0000 065− | ·0000 008 | ·0000 001 | | |
| ·02 | ·0003 007 | ·0000 513 | ·0000 086 | ·0000 014 | ·0000 002 | |
| ·03 | ·0008 198 | ·0001 712 | ·0000 350+ | ·0000 070 | ·0000 014 | ·0000 003 |
| ·04 | ·0016 645− | ·0004 013 | ·0000 947 | ·0000 220 | ·0000 050+ | ·0000 011 |
| ·05 | ·0028 758 | ·0007 746 | ·0002 044 | ·0000 530 | ·0000 136 | ·0000 034 |
| ·06 | ·0044 861 | ·0013 230 | ·0003 822 | ·0001 086 | ·0000 305− | ·0000 084 |
| ·07 | ·0065 218 | ·0020 762 | ·0006 476 | ·0001 987 | ·0000 602 | ·0000 180 |
| ·08 | ·0090 042 | ·0030 625− | ·0010 207 | ·0003 347 | ·0001 083 | ·0000 347 |
| ·09 | ·0119 506 | ·0043 085+ | ·0015 224 | ·0005 293 | ·0001 816 | ·0000 616 |
| ·10 | ·0153 747 | ·0058 392 | ·0021 738 | ·0007 964 | ·0002 879 | ·0001 030 |
| ·11 | ·0192 876 | ·0076 779 | ·0029 963 | ·0011 508 | ·0004 362 | ·0001 636 |
| ·12 | ·0236 975− | ·0098 465− | ·0040 114 | ·0016 085+ | ·0006 366 | ·0002 493 |
| ·13 | ·0286 103 | ·0123 651 | ·0052 404 | ·0021 862 | ·0009 003 | ·0003 668 |
| ·14 | ·0340 299 | ·0152 523 | ·0067 046 | ·0029 014 | ·0012 395+ | ·0005 239 |
| ·15 | ·0399 583 | ·0185 255− | ·0084 247 | ·0037 721 | ·0016 675− | ·0007 293 |
| ·16 | ·0463 959 | ·0222 001 | ·0104 212 | ·0048 169 | ·0021 984 | ·0009 927 |
| ·17 | ·0533 411 | ·0262 903 | ·0127 139 | ·0060 549 | ·0028 473 | ·0013 249 |
| ·18 | ·0607 913 | ·0308 087 | ·0153 223 | ·0075 052 | ·0036 303 | ·0017 376 |
| ·19 | ·0687 422 | ·0357 667 | ·0182 650+ | ·0091 875− | ·0045 641 | ·0022 437 |
| ·20 | ·0771 886 | ·0411 741 | ·0215 599 | ·0111 213 | ·0056 660 | ·0028 568 |
| ·21 | ·0861 238 | ·0470 391 | ·0252 242 | ·0133 263 | ·0069 543 | ·0035 917 |
| ·22 | ·0955 402 | ·0533 689 | ·0292 740 | ·0158 220 | ·0084 475+ | ·0044 640 |
| ·23 | ·1054 291 | ·0601 690 | ·0337 248 | ·0186 278 | ·0101 648 | ·0054 903 |
| ·24 | ·1157 809 | ·0674 439 | ·0385 909 | ·0217 628 | ·0121 258 | ·0066 880 |
| ·25 | ·1265 850− | ·0751 965+ | ·0438 857 | ·0252 457 | ·0143 502 | ·0080 751 |
| ·26 | ·1378 301 | ·0834 285+ | ·0496 214 | ·0290 949 | ·0168 582 | ·0096 706 |
| ·27 | ·1495 041 | ·0921 404 | ·0558 093 | ·0333 282 | ·0196 699 | ·0114 941 |
| ·28 | ·1615 940 | ·1013 313 | ·0624 595− | ·0379 626 | ·0228 056 | ·0135 657 |
| ·29 | ·1740 864 | ·1109 992 | ·0695 808 | ·0430 149 | ·0262 856 | ·0159 061 |
| ·30 | ·1869 670 | ·1211 409 | ·0771 809 | ·0485 005+ | ·0301 298 | ·0185 364 |
| ·31 | ·2002 209 | ·1317 520 | ·0852 664 | ·0544 346 | ·0343 580 | ·0214 781 |
| ·32 | ·2138 328 | ·1428 268 | ·0938 425+ | ·0608 310 | ·0389 898 | ·0247 529 |
| ·33 | ·2277 868 | ·1543 587 | ·1029 132 | ·0677 028 | ·0440 442 | ·0283 827 |
| ·34 | ·2420 664 | ·1663 399 | ·1124 811 | ·0750 619 | ·0495 396 | ·0323 895− |
| ·35 | ·2566 548 | ·1787 614 | ·1225 476 | ·0829 192 | ·0554 939 | ·0367 951 |
| ·36 | ·2715 347 | ·1916 134 | ·1331 128 | ·0912 844 | ·0619 243 | ·0416 215− |
| ·37 | ·2866 884 | ·2048 850− | ·1441 754 | ·1001 659 | ·0688 471 | ·0468 900 |
| ·38 | ·3020 977 | ·2185 640 | ·1557 328 | ·1095 708 | ·0762 775+ | ·0526 219 |
| ·39 | ·3177 444 | ·2326 377 | ·1677 811 | ·1195 051 | ·0842 301 | ·0588 377 |
| ·40 | ·3336 096 | ·2470 920 | ·1803 149 | ·1299 730 | ·0927 180 | ·0655 576 |
| ·41 | ·3496 744 | ·2619 123 | ·1933 277 | ·1409 776 | ·1017 533 | ·0728 007 |
| ·42 | ·3659 195− | ·2770 829 | ·2068 115− | ·1525 204 | ·1113 467 | ·0805 856 |
| ·43 | ·3823 255− | ·2925 871 | ·2207 568 | ·1646 013 | ·1215 075− | ·0889 297 |
| ·44 | ·3988 726 | ·3084 076 | ·2351 531 | ·1772 186 | ·1322 436 | ·0978 493 |
| ·45 | ·4155 411 | ·3245 263 | ·2499 882 | ·1903 692 | ·1435 613 | ·1073 595+ |
| ·46 | ·4323 110 | ·3409 241 | ·2652 489 | ·2040 481 | ·1554 653 | ·1174 741 |
| ·47 | ·4491 620 | ·3575 814 | ·2809 205+ | ·2182 489 | ·1679 584 | ·1282 052 |
| ·48 | ·4660 741 | ·3744 778 | ·2969 871 | ·2329 632 | ·1810 417 | ·1395 635+ |
| ·49 | ·4830 269 | ·3915 922 | ·3134 313 | ·2481 810 | ·1947 144 | ·1515 578 |
| ·50 | ·5000 000$^e$ | ·4089 029 | ·3302 347 | ·2638 908 | ·2089 738 | ·1641 950− |
| ·51 | ·5169 731 | ·4263 876 | ·3473 776 | ·2800 789 | ·2238 151 | ·1774 800 |
| ·52 | ·5339 259 | ·4440 233 | ·3648 389 | ·2967 302 | ·2392 314 | ·1914 156 |
| ·53 | ·5508 380 | ·4617 867 | ·3825 964 | ·3138 276 | ·2552 136 | ·2060 025− |
| ·54 | ·5676 890 | ·4796 537 | ·4006 270 | ·3313 524 | ·2717 505+ | ·2212 386 |
| ·55 | ·5844 589 | ·4976 001 | ·4189 059 | ·3492 838 | ·2888 286 | ·2371 197 |
| ·56 | ·6011 274 | ·5156 008 | ·4374 078 | ·3675 997 | ·3064 321 | ·2536 388 |
| ·57 | ·6176 745+ | ·5336 308 | ·4561 059 | ·3862 758 | ·3245 428 | ·2707 863 |
| ·58 | ·6340 805+ | ·5516 645− | ·4749 725+ | ·4052 863 | ·3431 402 | ·2885 496 |
| ·59 | ·6503 256 | ·5696 759 | ·4939 790 | ·4246 035− | ·3622 012 | ·3069 134 |
| ·60 | ·6663 904 | ·5876 390 | ·5130 958 | ·4441 981 | ·3817 004 | ·3258 593 |

$x$ = ·61 to 1·00 $q = 2·5$ $p$ = 2·5 to 5

| $x$ | $p = 2·5$ | $p = 3$ | $p = 3·5$ | $p = 4$ | $p = 4·5$ | $p = 5$ |
|---|---|---|---|---|---|---|
| $B(p, q)$ = | ·7363 1078 × $\frac{1}{10}$ | ·5079 3651 × $\frac{1}{10}$ | ·3681 5539 × $\frac{1}{10}$ | ·2770 5628 × $\frac{1}{10}$ | ·2147 5731 × $\frac{1}{10}$ | ·1704 9617 × $\frac{1}{10}$ |
| ·61 | ·6822 556 | ·6055 273 | ·5322 924 | ·4640 391 | ·4016 101 | ·3453 658 |
| ·62 | ·6979 023 | ·6233 144 | ·5515 373 | ·4840 939 | ·4218 998 | ·3654 083 |
| ·63 | ·7133 116 | ·6409 736 | ·5707 986 | ·5043 281 | ·4425 369 | ·3859 590 |
| ·64 | ·7284 653 | ·6584 781 | ·5900 433 | ·5247 060 | ·4634 861 | ·4069 866 |
| ·65 | ·7433 452 | ·6758 011 | ·6092 379 | ·5451 902 | ·4847 097 | ·4284 567 |
| ·66 | ·7579 336 | ·6929 160 | ·6283 482 | ·5657 420 | ·5061 677 | ·4503 316 |
| ·67 | ·7722 132 | ·7097 960 | ·6473 396 | ·5863 211 | ·5278 177 | ·4725 699 |
| ·68 | ·7861 672 | ·7264 146 | ·6661 769 | ·6068 862 | ·5496 149 | ·4951 272 |
| ·69 | ·7997 791 | ·7427 455$^{-}$ | ·6848 247 | ·6273 946 | ·5715 124 | ·5179 555$^{-}$ |
| ·70 | ·8130 330 | ·7587 624 | ·7032 470 | ·6478 024 | ·5934 610 | ·5410 034 |
| ·71 | ·8259 136 | ·7744 398 | ·7214 080 | ·6680 648 | ·6154 094 | ·5642 163 |
| ·72 | ·8384 060 | ·7897 521 | ·7392 714 | ·6881 361 | ·6373 044 | ·5875 363 |
| ·73 | ·8504 959 | ·8046 744 | ·7568 011 | ·7079 698 | ·6590 909 | ·6109 025$^{-}$ |
| ·74 | ·8621 699 | ·8191 823 | ·7739 612 | ·7275 185$^{-}$ | ·6807 120 | ·6342 506 |
| ·75 | ·8734 150$^{+}$ | ·8332 520 | ·7907 157 | ·7467 346 | ·7021 092 | ·6575 136 |
| ·76 | ·8842 191 | ·8468 603 | ·8070 291 | ·7655 701 | ·7232 229 | ·6806 218 |
| ·77 | ·8945 709 | ·8599 848 | ·8228 666 | ·7839 765$^{+}$ | ·7439 918 | ·7035 027 |
| ·78 | ·9044 598 | ·8726 042 | ·8381 936 | ·8019 058 | ·7643 541 | ·7260 817 |
| ·79 | ·9138 762 | ·8846 979 | ·8529 765$^{+}$ | ·8193 098 | ·7842 469 | ·7482 819 |
| ·80 | ·9228 114 | ·8962 464 | ·8671 827 | ·8361 409 | ·8036 071 | ·7700 249 |
| ·81 | ·9312 578 | ·9072 316 | ·8807 805$^{+}$ | ·8523 523 | ·8223 712 | ·7912 305$^{+}$ |
| ·82 | ·9392 087 | ·9176 365$^{-}$ | ·8937 398 | ·8678 980 | ·8404 762 | ·8118 179 |
| ·83 | ·9466 589 | ·9274 456 | ·9060 318 | ·8827 334 | ·8578 596 | ·8317 056 |
| ·84 | ·9536 041 | ·9366 451 | ·9176 295$^{-}$ | ·8968 154 | ·8744 598 | ·8508 121 |
| ·85 | ·9600 417 | ·9452 230 | ·9285 080 | ·9101 030 | ·8902 171 | ·8690 566 |
| ·86 | ·9659 701 | ·9531 692 | ·9386 448 | ·9225 576 | ·9050 737 | ·8863 595$^{+}$ |
| ·87 | ·9713 897 | ·9604 757 | ·9480 199 | ·9341 436 | ·9189 746 | ·9026 438 |
| ·88 | ·9763 025$^{+}$ | ·9671 370 | ·9566 165$^{-}$ | ·9448 285$^{+}$ | ·9318 683 | ·9178 352 |
| ·89 | ·9807 124 | ·9731 505$^{-}$ | ·9644 211 | ·9545 844 | ·9437 079 | ·9318 641 |
| ·90 | ·9846 253 | ·9785 163 | ·9714 243 | ·9633 877 | ·9544 517 | ·9446 662 |
| ·91 | ·9880 494 | ·9832 380 | ·9776 213 | ·9712 209 | ·9640 646 | ·9561 845$^{+}$ |
| ·92 | ·9909 958 | ·9873 232 | ·9830 123 | ·9780 729 | ·9725 197 | ·9663 712 |
| ·93 | ·9934 782 | ·9907 838 | ·9876 039 | ·9839 406 | ·9797 996 | ·9751 898 |
| ·94 | ·9955 139 | ·9936 370 | ·9914 100 | ·9888 305$^{+}$ | ·9858 990 | ·9826 181 |
| ·95 | ·9971 242 | ·9959 061 | ·9944 529 | ·9927 608 | ·9908 275$^{-}$ | ·9886 522 |
| ·96 | ·9983 355$^{+}$ | ·9976 218 | ·9967 658 | ·9957 638 | ·9946 130 | ·9933 113 |
| ·97 | ·9991 802 | ·9988 244 | ·9983 955$^{-}$ | ·9978 907 | ·9973 080 | ·9966 455$^{-}$ |
| ·98 | ·9996 993 | ·9995 672 | ·9994 071 | ·9992 178 | ·9989 980 | ·9987 469 |
| ·99 | ·9999 463 | ·9999 224 | ·9998 933 | ·9998 587 | ·9998 184 | ·9997 720 |
| 1·00 | 1·0000 000 | 1·0000 000 | 1·0000 000 | 1·0000 000 | 1·0000 000 | 1·0000 000 |

$x = \cdot03$ to $\cdot60$ $\qquad q = 2\cdot5$ $\qquad p = 5\cdot5$ to 8

| $x$ | $p = 5\cdot5$ | $p = 6$ | $p = 6\cdot5$ | $p = 7$ | $p = 7\cdot5$ | $p = 8$ |
|---|---|---|---|---|---|---|
| $B(p, q) =$ | $\cdot1380\ 5827 \times \frac{1}{10}$ | $\cdot1136\ 6411 \times \frac{1}{10}$ | $\cdot9491\ 5061 \times \frac{1}{10^2}$ | $\cdot8023\ 3492 \times \frac{1}{10^2}$ | $\cdot6854\ 9767 \times \frac{1}{10^2}$ | $\cdot5911\ 9415 \times \frac{1}{10^2}$ |
| ·03 | ·0000 001 | | | | | |
| ·04 | ·0000 003 | ·0000 001 | | | | |
| ·05 | ·0000 009 | ·0000 002 | ·0000 001 | | | |
| ·06 | ·0000 023 | ·0000 006 | ·0000 002 | | | |
| ·07 | ·0000 053 | ·0000 016 | ·0000 005⁻ | ·0000 001 | | |
| ·08 | ·0000 110 | ·0000 035⁻ | ·0000 011 | ·0000 003 | ·0000 001 | |
| ·09 | ·0000 207 | ·0000 069 | ·0000 023 | ·0000 008 | ·0000 002 | ·0000 001 |
| ·10 | ·0000 365⁻ | ·0000 128 | ·0000 045⁻ | ·0000 016 | ·0000 005⁺ | ·0000 002 |
| ·11 | ·0000 608 | ·0000 224 | ·0000 082 | ·0000 030 | ·0000 011 | ·0000 004 |
| ·12 | ·0000 967 | ·0000 372 | ·0000 142 | ·0000 054 | ·0000 020 | ·0000 008 |
| ·13 | ·0001 481 | ·0000 593 | ·0000 236 | ·0000 093 | ·0000 037 | ·0000 014 |
| ·14 | ·0002 194 | ·0000 912 | ·0000 376 | ·0000 154 | ·0000 063 | ·0000 026 |
| ·15 | ·0003 160 | ·0001 359 | ·0000 580 | ·0000 246 | ·0000 104 | ·0000 044 |
| ·16 | ·0004 442 | ·0001 972 | ·0000 870 | ·0000 381 | ·0000 166 | ·0000 072 |
| ·17 | ·0006 109 | ·0002 795⁺ | ·0001 270 | ·0000 574 | ·0000 258 | ·0000 115⁺ |
| ·18 | ·0008 242 | ·0003 880 | ·0001 814 | ·0000 843 | ·0000 390 | ·0000 179 |
| ·19 | ·0010 932 | ·0005 285⁺ | ·0002 538 | ·0001 212 | ·0000 576 | ·0000 272 |
| ·20 | ·0014 277 | ·0007 080 | ·0003 488 | ·0001 708 | ·0000 832 | ·0000 404 |
| ·21 | ·0018 387 | ·0009 342 | ·0004 715⁻ | ·0002 366 | ·0001 181 | ·0000 587 |
| ·22 | ·0023 384 | ·0012 157 | ·0006 278 | ·0003 224 | ·0001 647 | ·0000 837 |
| ·23 | ·0029 398 | ·0015 623 | ·0008 248 | ·0004 329 | ·0002 261 | ·0001 175⁺ |
| ·24 | ·0036 569 | ·0019 846 | ·0010 701 | ·0005 737 | ·0003 060 | ·0001 624 |
| ·25 | ·0045 050⁺ | ·0024 947 | ·0013 725⁻ | ·0007 508 | ·0004 086 | ·0002 214 |
| ·26 | ·0055 002 | ·0031 052 | ·0017 418 | ·0009 715⁻ | ·0005 391 | ·0002 978 |
| ·27 | ·0066 597 | ·0038 304 | ·0021 889 | ·0012 438 | ·0007 033 | ·0003 958 |
| ·28 | ·0080 015⁺ | ·0046 852 | ·0027 259 | ·0015 770 | ·0009 078 | ·0005 203 |
| ·29 | ·0095 448 | ·0056 861 | ·0033 659 | ·0019 813 | ·0011 605⁺ | ·0006 767 |
| ·30 | ·0113 093 | ·0068 504 | ·0041 233 | ·0024 681 | ·0014 700 | ·0008 717 |
| ·31 | ·0133 159 | ·0081 966 | ·0050 138 | ·0030 500⁺ | ·0018 463 | ·0011 127 |
| ·32 | ·0155 860 | ·0097 444 | ·0060 543 | ·0037 409 | ·0023 002 | ·0014 082 |
| ·33 | ·0181 418 | ·0115 143 | ·0072 628 | ·0045 562 | ·0028 443 | ·0017 679 |
| ·34 | ·0210 061 | ·0135 282 | ·0086 589 | ·0055 122 | ·0034 921 | ·0022 027 |
| ·35 | ·0242 022 | ·0158 087 | ·0102 632 | ·0066 271 | ·0042 587 | ·0027 249 |
| ·36 | ·0277 539 | ·0183 793 | ·0120 976 | ·0079 203 | ·0051 606 | ·0033 482 |
| ·37 | ·0316 851 | ·0212 644 | ·0141 852 | ·0094 125⁻ | ·0062 160 | ·0040 876 |
| ·38 | ·0360 202 | ·0244 891 | ·0165 503 | ·0111 260 | ·0074 444 | ·0049 599 |
| ·39 | ·0407 836 | ·0280 794 | ·0192 183 | ·0130 847 | ·0088 670 | ·0059 836 |
| ·40 | ·0459 996 | ·0320 615⁺ | ·0222 157 | ·0153 135⁺ | ·0105 068 | ·0071 788 |
| ·41 | ·0516 926 | ·0364 624 | ·0255 700 | ·0178 390 | ·0123 881 | ·0085 672 |
| ·42 | ·0578 864 | ·0413 092 | ·0293 094 | ·0206 891 | ·0145 373 | ·0101 728 |
| ·43 | ·0646 045⁺ | ·0466 292 | ·0334 631 | ·0238 928 | ·0169 821 | ·0120 211 |
| ·44 | ·0718 701 | ·0524 500⁺ | ·0380 609 | ·0274 804 | ·0197 519 | ·0141 395⁺ |
| ·45 | ·0797 052 | ·0587 988 | ·0431 331 | ·0314 835⁻ | ·0228 777 | ·0165 576 |
| ·46 | ·0881 313 | ·0657 028 | ·0487 104 | ·0359 343 | ·0263 920 | ·0193 065⁻ |
| ·47 | ·0971 688 | ·0731 885⁺ | ·0548 237 | ·0408 662 | ·0303 287 | ·0224 194 |
| ·48 | ·1068 366 | ·0812 821 | ·0615 041 | ·0463 133 | ·0347 230 | ·0259 314 |
| ·49 | ·1171 527 | ·0900 087 | ·0687 824 | ·0523 100 | ·0396 114 | ·0298 792 |
| ·50 | ·1281 332 | ·0993 927 | ·0766 892 | ·0588 912 | ·0450 313 | ·0343 011 |
| ·51 | ·1397 926 | ·1094 569 | ·0852 544 | ·0660 921 | ·0510 212 | ·0392 370 |
| ·52 | ·1521 435⁻ | ·1202 230 | ·0945 072 | ·0739 478 | ·0576 199 | ·0447 283 |
| ·53 | ·1651 964 | ·1317 109 | ·1044 757 | ·0824 928 | ·0648 671 | ·0508 174 |
| ·54 | ·1789 595⁻ | ·1439 387 | ·1151 867 | ·0917 613 | ·0728 024 | ·0575 478 |
| ·55 | ·1934 386 | ·1569 223 | ·1266 656 | ·1017 866 | ·0814 654 | ·0649 639 |
| ·56 | ·2086 370 | ·1706 753 | ·1389 357 | ·1126 008 | ·0908 954 | ·0731 102 |
| ·57 | ·2245 549 | ·1852 085⁺ | ·1520 182 | ·1242 344 | ·1011 309 | ·0820 316 |
| ·58 | ·2411 898 | ·2005 302 | ·1659 319 | ·1367 161 | ·1122 094 | ·0917 727 |
| ·59 | ·2585 359 | ·2166 452 | ·1806 927 | ·1500 723 | ·1241 666 | ·1023 776 |
| ·60 | ·2765 841 | ·2335 551 | ·1963 135⁻ | ·1643 269 | ·1370 367 | ·1138 893 |

$x = \cdot 61$ to $1\cdot 00$ $q = 2\cdot 5$ $p = 5\cdot 5$ to 8

| $x$ | $p = 5\cdot 5$ | $p = 6$ | $p = 6\cdot 5$ | $p = 7$ | $p = 7\cdot 5$ | $p = 8$ |
|---|---|---|---|---|---|---|
| $B(p, q) =$ | $\cdot 1380\ 5827 \times \frac{1}{10}$ | $\cdot 1136\ 6411 \times \frac{1}{10}$ | $\cdot 9491\ 5061 \times \frac{1}{10^2}$ | $\cdot 8023\ 3492 \times \frac{1}{10^2}$ | $\cdot 6854\ 9767 \times \frac{1}{10^2}$ | $\cdot 5911\ 9415^{+} \times \frac{1}{10^2}$ |
| ·61 | ·2953 218 | ·2512 579 | ·2128 035$^-$ | ·1795 007 | ·1508 512 | ·1263 490 |
| ·62 | ·3147 328 | ·2697 478 | ·2301 683 | ·1956 109 | ·1656 391 | ·1397 964 |
| ·63 | ·3347 971 | ·2890 146 | ·2484 093 | ·2126 710 | ·1814 256 | ·1542 681 |
| ·64 | ·3554 906 | ·3090 440 | ·2675 233 | ·2306 899 | ·1982 322 | ·1697 976 |
| ·65 | ·3767 853 | ·3298 169 | ·2875 023 | ·2496 720 | ·2160 760 | ·1864 148 |
| ·66 | ·3986 489 | ·3513 094 | ·3083 331 | ·2696 161 | ·2349 688 | ·2041 448 |
| ·67 | ·4210 448 | ·3734 926 | ·3299 967 | ·2905 154 | ·2549 169 | ·2230 074 |
| ·68 | ·4439 321 | ·3963 322 | ·3524 683 | ·3123 565$^+$ | ·2759 202 | ·2430 165$^+$ |
| ·69 | ·4672 651 | ·4197 886 | ·3757 171 | ·3351 196 | ·2979 717 | ·2641 792 |
| ·70 | ·4909 940 | ·4438 163 | ·3997 053 | ·3587 775$^+$ | ·3210 565$^+$ | ·2864 946 |
| ·71 | ·5150 641 | ·4683 641 | ·4243 884 | ·3832 953 | ·3451 518 | ·3099 538 |
| ·72 | ·5394 161 | ·4933 749 | ·4497 148 | ·4086 297 | ·3702 256 | ·3345 381 |
| ·73 | ·5639 862 | ·5187 856 | ·4756 253 | ·4347 290 | ·3962 365$^-$ | ·3602 188 |
| ·74 | ·5887 061 | ·5445 268 | ·5020 532 | ·4615 324 | ·4231 325$^+$ | ·3869 560 |
| ·75 | ·6135 028 | ·5705 231 | ·5289 239 | ·4889 696 | ·4508 511 | ·4146 976 |
| ·76 | ·6382 992 | ·5966 929 | ·5561 548 | ·5169 605$^-$ | ·4793 183 | ·4433 788 |
| ·77 | ·6630 138 | ·6229 485$^-$ | ·5836 553 | ·5454 150$^-$ | ·5084 478 | ·4729 212 |
| ·78 | ·6875 610 | ·6491 961 | ·6113 265$^-$ | ·5742 326 | ·5381 413 | ·5032 315$^+$ |
| ·79 | ·7118 517 | ·6753 364 | ·6390 617 | ·6033 026 | ·5682 873 | ·5342 016 |
| ·80 | ·7357 930 | ·7012 642 | ·6667 460 | ·6325 035$^-$ | ·5987 613 | ·5657 074 |
| ·81 | ·7592 891 | ·7268 693 | ·6942 572 | ·6617 035$^+$ | ·6294 254 | ·5976 083 |
| ·82 | ·7822 413 | ·7520 366 | ·7214 653 | ·6907 607 | ·6601 283 | ·6297 474 |
| ·83 | ·8045 490 | ·7766 466 | ·7482 337 | ·7195 230 | ·6907 054 | ·6619 504 |
| ·84 | ·8261 099 | ·8005 765$^-$ | ·7744 193 | ·7478 291 | ·7209 793 | ·6940 267 |
| ·85 | ·8468 208 | ·8237 002 | ·7998 739 | ·7755 091 | ·7507 602 | ·7257 689 |
| ·86 | ·8665 788 | ·8458 900 | ·8244 445$^-$ | ·8023 853 | ·7798 469 | ·7569 540 |
| ·87 | ·8852 820 | ·8670 175$^+$ | ·8479 751 | ·8282 740 | ·8080 280 | ·7873 442 |
| ·88 | ·9028 305$^-$ | ·8869 549 | ·8703 080 | ·8529 866 | ·8350 838 | ·8166 889 |
| ·89 | ·9191 282 | ·9055 767 | ·8912 861 | ·8763 320 | ·8607 884 | ·8447 269 |
| ·90 | ·9340 845$^+$ | ·9227 620 | ·9107 548 | ·8981 197 | ·8849 127 | ·8711 892 |
| ·91 | ·9476 159 | ·9383 964 | ·9285 653 | ·9181 625$^+$ | ·9072 286 | ·8958 040 |
| ·92 | ·9596 488 | ·9523 760 | ·9445 781 | ·9362 815$^-$ | ·9275 134 | ·9183 016 |
| ·93 | ·9701 222 | ·9646 100 | ·9586 677 | ·9523 111 | ·9455 567 | ·9384 220 |
| ·94 | ·9789 920 | ·9750 265$^-$ | ·9707 286 | ·9661 064 | ·9611 686 | ·9559 247 |
| ·95 | ·9862 353 | ·9835 781 | ·9806 829 | ·9775 526 | ·9741 908 | ·9706 018 |
| ·96 | ·9918 574 | ·9902 506 | ·9884 906 | ·9865 778 | ·9845 127 | ·9822 963 |
| ·97 | ·9959 016 | ·9950 752 | ·9941 653 | ·9931 712 | ·9920 925$^-$ | ·9909 286 |
| ·98 | ·9984 636 | ·9981 471 | ·9977 969 | ·9974 123 | ·9969 928 | ·9965 379 |
| ·99 | ·9997 195$^-$ | ·9996 605$^-$ | ·9995 948 | ·9995 224 | ·9994 430 | ·9993 565$^-$ |
| 1·00 | 1·0000 000 | 1·0000 000 | 1·0000 000 | 1·0000 000 | 1·0000 000 | 1·0000 000 |

$x = \cdot 10$ to $\cdot 70$ $q = 2\cdot 5$ $p = 8\cdot 5$ to $11$

| | $p = 8\cdot 5$ | $p = 9$ | $p = 9\cdot 5$ | $p = 10$ | $p = 10\cdot 5$ | $p = 11$ |
|---|---|---|---|---|---|---|
| $B(p, q) =$ | $\cdot 5141\ 2325 \times \frac{1}{10^3}$ | $\cdot 4504\ 3364 \times \frac{1}{10^3}$ | $\cdot 3972\ 7706 \times \frac{1}{10^3}$ | $\cdot 3525\ 1328 \times \frac{1}{10^3}$ | $\cdot 3145\ 1100 \times \frac{1}{10^3}$ | $\cdot 2820\ 1063 \times \frac{1}{10^3}$ |
| $x$ | | | | | | |
| ·10 | ·0000 001 | | | | | |
| ·11 | ·0000 001 | | | | | |
| ·12 | ·0000 003 | ·0000 001 | | | | |
| ·13 | ·0000 006 | ·0000 002 | ·0000 001 | | | |
| ·14 | ·0000 010 | ·0000 004 | ·0000 002 | ·0000 001 | | |
| ·15 | ·0000 018 | ·0000 008 | ·0000 003 | ·0000 001 | ·0000 001 | |
| ·16 | ·0000 031 | ·0000 013 | ·0000 006 | ·0000 002 | ·0000 001 | |
| ·17 | ·0000 051 | ·0000 023 | ·0000 010 | ·0000 004 | ·0000 002 | ·0000 001 |
| ·18 | ·0000 082 | ·0000 038 | ·0000 017 | ·0000 008 | ·0000 004 | ·0000 002 |
| ·19 | ·0000 128 | ·0000 060 | ·0000 028 | ·0000 013 | ·0000 006 | ·0000 003 |
| ·20 | ·0000 195⁻ | ·0000 094 | ·0000 045⁻ | ·0000 022 | ·0000 010 | ·0000 005⁻ |
| ·21 | ·0000 290 | ·0000 143 | ·0000 070 | ·0000 034 | ·0000 017 | ·0000 008 |
| ·22 | ·0000 424 | ·0000 214 | ·0000 108 | ·0000 054 | ·0000 027 | ·0000 013 |
| ·23 | ·0000 608 | ·0000 314 | ·0000 161 | ·0000 083 | ·0000 042 | ·0000 022 |
| ·24 | ·0000 859 | ·0000 453 | ·0000 238 | ·0000 124 | ·0000 065⁻ | ·0000 034 |
| ·25 | ·0001 195⁻ | ·0000 642 | ·0000 344 | ·0000 184 | ·0000 098 | ·0000 052 |
| ·26 | ·0001 639 | ·0000 898 | ·0000 491 | ·0000 267 | ·0000 145⁺ | ·0000 079 |
| ·27 | ·0002 219 | ·0001 239 | ·0000 690 | ·0000 383 | ·0000 212 | ·0000 117 |
| ·28 | ·0002 970 | ·0001 689 | ·0000 957 | ·0000 541 | ·0000 305⁻ | ·0000 171 |
| ·29 | ·0003 930 | ·0002 274 | ·0001 312 | ·0000 754 | ·0000 433 | ·0000 247 |
| ·30 | ·0005 148 | ·0003 030 | ·0001 777 | ·0001 039 | ·0000 606 | ·0000 353 |
| ·31 | ·0006 679 | ·0003 995⁺ | ·0002 382 | ·0001 416 | ·0000 839 | ·0000 496 |
| ·32 | ·0008 587 | ·0005 218 | ·0003 160 | ·0001 908 | ·0001 149 | ·0000 690 |
| ·33 | ·0010 946 | ·0006 753 | ·0004 152 | ·0002 546 | ·0001 557 | ·0000 950⁻ |
| ·34 | ·0013 840 | ·0008 665⁺ | ·0005 408 | ·0003 365⁺ | ·0002 088 | ·0001 293 |
| ·35 | ·0017 368 | ·0011 031 | ·0006 984 | ·0004 409 | ·0002 775⁺ | ·0001 743 |
| ·36 | ·0021 639 | ·0013 936 | ·0008 947 | ·0005 727 | ·0003 656 | ·0002 328 |
| ·37 | ·0026 777 | ·0017 480 | ·0011 375⁻ | ·0007 380 | ·0004 776 | ·0003 083 |
| ·38 | ·0032 921 | ·0021 775⁻ | ·0014 357 | ·0009 439 | ·0006 190 | ·0004 049 |
| ·39 | ·0040 226 | ·0026 949 | ·0017 998 | ·0011 986 | ·0007 961 | ·0005 275⁻ |
| ·40 | ·0048 864 | ·0033 147 | ·0022 416 | ·0015 115⁺ | ·0010 166 | ·0006 821 |
| ·41 | ·0059 027 | ·0040 530 | ·0027 744 | ·0018 937 | ·0012 893 | ·0008 756 |
| ·42 | ·0070 922 | ·0049 278 | ·0034 134 | ·0023 578 | ·0016 244 | ·0011 164 |
| ·43 | ·0084 779 | ·0059 591 | ·0041 758 | ·0029 180 | ·0020 338 | ·0014 142 |
| ·44 | ·0100 848 | ·0071 689 | ·0050 806 | ·0035 906 | ·0025 311 | ·0017 800 |
| ·45 | ·0119 398 | ·0085 815⁺ | ·0061 492 | ·0043 941 | ·0031 320 | ·0022 271 |
| ·46 | ·0140 723 | ·0102 235⁺ | ·0074 052 | ·0053 490 | ·0038 540 | ·0027 704 |
| ·47 | ·0165 135⁺ | ·0121 238 | ·0088 746 | ·0064 784 | ·0047 173 | ·0034 270 |
| ·48 | ·0192 971 | ·0143 138 | ·0105 860 | ·0078 079 | ·0057 445⁺ | ·0042 167 |
| ·49 | ·0224 589 | ·0168 273 | ·0125 709 | ·0093 659 | ·0069 608 | ·0051 615⁻ |
| ·50 | ·0260 366 | ·0197 007 | ·0148 633 | ·0111 838 | ·0083 945⁻ | ·0062 865⁺ |
| ·51 | ·0300 704 | ·0229 728 | ·0175 000⁻ | ·0132 957 | ·0100 768 | ·0076 200 |
| ·52 | ·0346 023 | ·0266 852 | ·0205 209 | ·0157 391 | ·0120 424 | ·0091 933 |
| ·53 | ·0396 760 | ·0308 817 | ·0239 687 | ·0185 548 | ·0143 293 | ·0110 415⁺ |
| ·54 | ·0453 373 | ·0356 085⁻ | ·0278 889 | ·0217 866 | ·0169 791 | ·0132 033 |
| ·55 | ·0516 333 | ·0409 140 | ·0323 302 | ·0254 820 | ·0200 371 | ·0157 213 |
| ·56 | ·0586 124 | ·0468 489 | ·0373 436 | ·0296 915⁺ | ·0235 524 | ·0186 422 |
| ·57 | ·0663 240 | ·0534 656 | ·0429 830 | ·0344 692 | ·0275 778 | ·0220 170 |
| ·58 | ·0748 184 | ·0608 180 | ·0493 047 | ·0398 720 | ·0321 701 | ·0259 010 |
| ·59 | ·0841 461 | ·0689 615⁺ | ·0563 672 | ·0459 601 | ·0373 898 | ·0303 537 |
| ·60 | ·0943 574 | ·0779 524 | ·0642 308 | ·0527 966 | ·0433 008 | ·0354 391 |
| ·61 | ·1055 022 | ·0878 473 | ·0729 576 | ·0604 469 | ·0499 708 | ·0412 256 |
| ·62 | ·1176 293 | ·0987 029 | ·0826 105⁻ | ·0689 787 | ·0574 706 | ·0477 853 |
| ·63 | ·1307 859 | ·1105 754 | ·0932 529 | ·0784 613 | ·0658 736 | ·0551 946 |
| ·64 | ·1450 166 | ·1235 195⁺ | ·1049 484 | ·0889 652 | ·0752 558 | ·0635 332 |
| ·65 | ·1603 634 | ·1375 882 | ·1177 597 | ·1005 613 | ·0856 948 | ·0728 838 |
| ·66 | ·1768 644 | ·1528 317 | ·1317 479 | ·1133 205⁺ | ·0972 695⁺ | ·0833 316 |
| ·67 | ·1945 529 | ·1692 964 | ·1469 718 | ·1273 124 | ·1100 588 | ·0949 636 |
| ·68 | ·2134 570 | ·1870 245⁺ | ·1634 864 | ·1426 042 | ·1241 411 | ·1078 675⁻ |
| ·69 | ·2335 985⁻ | ·2060 523 | ·1813 426 | ·1592 602 | ·1395 928 | ·1221 306 |
| ·70 | ·2549 916 | ·2264 094 | ·2005 851 | ·1773 399 | ·1564 873 | ·1378 389 |

$x = \cdot71$ to $1\cdot00$ $q = 2\cdot5$ $p = 8\cdot5$ to $11$

| | $p = 8\cdot5$ | $p = 9$ | $p = 9\cdot5$ | $p = 10$ | $p = 10\cdot5$ | $p = 11$ |
|---|---|---|---|---|---|---|
| $B(p, q) =$ | $\cdot5141\ 2325 \times \frac{1}{10^2}$ | $\cdot4504\ 3364 \times \frac{1}{10^2}$ | $\cdot3972\ 7706 \times \frac{1}{10^2}$ | $\cdot3525\ 1328 \times \frac{1}{10^2}$ | $\cdot3145\ 1100 \times \frac{1}{10^2}$ | $\cdot2820\ 1063 \times \frac{1}{10^2}$ |
| $x$ | | | | | | |
| ·71 | ·2776 422 | ·2481 177 | ·2212 519 | ·1968 969 | ·1748 931 | ·1550 751 |
| ·72 | ·3015 470 | ·2711 898 | ·2433 722 | ·2179 772 | ·1948 728 | ·1739 172 |
| ·73 | ·3266 918 | ·2956 278 | ·2669 653 | ·2406 178 | ·2164 806 | ·1944 368 |
| ·74 | ·3530 509 | ·3214 219 | ·2920 387 | ·2648 442 | ·2397 609 | ·2166 965+ |
| ·75 | ·3805 856 | ·3485 491 | ·3185 866 | ·2906 691 | ·2647 454 | ·2407 477 |
| ·76 | ·4092 433 | ·3769 714 | ·3465 881 | ·3180 901 | ·2914 515− | ·2666 279 |
| ·77 | ·4389 562 | ·4066 346 | ·3760 049 | ·3470 872 | ·3198 788 | ·2943 580 |
| ·78 | ·4696 400 | ·4374 667 | ·4067 799 | ·3776 208 | ·3500 074 | ·3239 389 |
| ·79 | ·5011 932 | ·4693 763 | ·4388 352 | ·4096 289 | ·3817 940 | ·3553 484 |
| ·80 | ·5334 960 | ·5022 511 | ·4720 697 | ·4430 252 | ·4151 697 | ·3885 374 |
| ·81 | ·5664 089 | ·5359 568 | ·5063 578 | ·4776 961 | ·4500 363 | ·4234 262 |
| ·82 | ·5997 726 | ·5703 357 | ·5415 472 | ·5134 985− | ·4862 636 | ·4599 010 |
| ·83 | ·6334 072 | ·6052 055+ | ·5774 572 | ·5502 575+ | ·5236 864 | ·4978 097 |
| ·84 | ·6671 117 | ·6403 589 | ·6138 779 | ·5877 644 | ·5621 010 | ·5369 581 |
| ·85 | ·7006 642 | ·6755 625+ | ·6505 682 | ·6257 744 | ·6012 633 | ·5771 066 |
| ·86 | ·7338 222 | ·7105 573 | ·6872 560 | ·6640 058 | ·6408 858 | ·6179 665+ |
| ·87 | ·7663 232 | ·7450 587 | ·7236 372 | ·7021 388 | ·6806 366 | ·6591 975− |
| ·88 | ·7978 869 | ·7787 579 | ·7593 773 | ·7398 154 | ·7201 380 | ·7004 056 |
| ·89 | ·8282 167 | ·8113 242 | ·7941 123 | ·7766 410 | ·7589 668 | ·7411 428 |
| ·90 | ·8570 033 | ·8424 073 | ·8274 521 | ·8121 863 | ·7966 566 | ·7809 075+ |
| ·91 | ·8839 290 | ·8716 429 | ·8589 846 | ·8459 918 | ·8327 012 | ·8191 480 |
| ·92 | ·9086 741 | ·8986 587 | ·8882 832 | ·8775 752 | ·8665 616 | ·8552 689 |
| ·93 | ·9309 248 | ·9230 833 | ·9149 158 | ·9064 408 | ·8976 767 | ·8886 417 |
| ·94 | ·9503 848 | ·9445 594 | ·9384 593 | ·9320 955− | ·9254 792 | ·9186 219 |
| ·95 | ·9667 899 | ·9627 604 | ·9585 183 | ·9540 695− | ·9494 196 | ·9445 747 |
| ·96 | ·9799 300 | ·9774 154 | ·9747 543 | ·9719 488 | ·9690 011 | ·9659 136 |
| ·97 | ·9896 796 | ·9883 454 | ·9869 262 | ·9854 221 | ·9838 336 | ·9821 612 |
| ·98 | ·9960 473 | ·9955 204 | ·9949 570 | ·9943 570 | ·9937 199 | ·9930 458 |
| ·99 | ·9992 626 | ·9991 614 | ·9990 525+ | ·9989 360 | ·9988 117 | ·9986 795− |
| 1·00 | 1·0000 000 | 1·0000 000 | 1·0000 000 | 1·0000 000 | 1·0000 000 | 1·0000 000 |

$x = \cdot 19$ to $\cdot 80$ $\qquad q = 2\cdot 5$ $\qquad p = 12$ to $17$

| $x$ | $p = 12$ | $p = 13$ | $p = 14$ | $p = 15$ | $p = 16$ | $p = 17$ |
|---|---|---|---|---|---|---|
| $B(p, q) =$ | $\cdot 2297\ 8644 \times \frac{1}{10^2}$ | $\cdot 1901\ 6809 \times \frac{1}{10^2}$ | $\cdot 1594\ 9581 \times \frac{1}{10^2}$ | $\cdot 1353\ 2978 \times \frac{1}{10^2}$ | $\cdot 1159\ 9696 \times \frac{1}{10^2}$ | $\cdot 1003\ 2169 \times \frac{1}{10^2}$ |
| ·19 | ·0000 001 | | | | | |
| ·20 | ·0000 001 | | | | | |
| ·21 | ·0000 002 | | | | | |
| ·22 | ·0000 003 | ·0000 001 | | | | |
| ·23 | ·0000 006 | ·0000 001 | | | | |
| ·24 | ·0000 009 | ·0000 002 | ·0000 001 | | | |
| ·25 | ·0000 015− | ·0000 004 | ·0000 001 | | | |
| ·26 | ·0000 023 | ·0000 007 | ·0000 002 | ·0000 001 | | |
| ·27 | ·0000 035+ | ·0000 011 | ·0000 003 | ·0000 001 | | |
| ·28 | ·0000 054 | ·0000 017 | ·0000 005+ | ·0000 002 | | |
| ·29 | ·0000 080 | ·0000 026 | ·0000 008 | ·0000 003 | ·0000 001 | |
| ·30 | ·0000 119 | ·0000 040 | ·0000 013 | ·0000 004 | ·0000 001 | |
| ·31 | ·0000 172 | ·0000 059 | ·0000 020 | ·0000 007 | ·0000 002 | ·0000 001 |
| ·32 | ·0000 247 | ·0000 088 | ·0000 031 | ·0000 011 | ·0000 004 | ·0000 001 |
| ·33 | ·0000 351 | ·0000 129 | ·0000 047 | ·0000 017 | ·0000 006 | ·0000 002 |
| ·34 | ·0000 492 | ·0000 186 | ·0000 070 | ·0000 026 | ·0000 010 | ·0000 004 |
| ·35 | ·0000 683 | ·0000 265+ | ·0000 102 | ·0000 039 | ·0000 015+ | ·0000 006 |
| ·36 | ·0000 938 | ·0000 375− | ·0000 149 | ·0000 059 | ·0000 023 | ·0000 009 |
| ·37 | ·0001 276 | ·0000 524 | ·0000 214 | ·0000 087 | ·0000 035− | ·0000 014 |
| ·38 | ·0001 721 | ·0000 726 | ·0000 304 | ·0000 127 | ·0000 052 | ·0000 022 |
| ·39 | ·0002 301 | ·0000 996 | ·0000 428 | ·0000 183 | ·0000 078 | ·0000 033 |
| ·40 | ·0003 050+ | ·0001 354 | ·0000 597 | ·0000 261 | ·0000 114 | ·0000 049 |
| ·41 | ·0004 013 | ·0001 825− | ·0000 824 | ·0000 370 | ·0000 165+ | ·0000 074 |
| ·42 | ·0005 240 | ·0002 441 | ·0001 129 | ·0000 519 | ·0000 238 | ·0000 108 |
| ·43 | ·0006 793 | ·0003 239 | ·0001 534 | ·0000 722 | ·0000 338 | ·0000 158 |
| ·44 | ·0008 748 | ·0004 266 | ·0002 067 | ·0000 996 | ·0000 477 | ·0000 228 |
| ·45 | ·0011 190 | ·0005 580 | ·0002 764 | ·0001 362 | ·0000 667 | ·0000 325+ |
| ·46 | ·0014 225− | ·0007 249 | ·0003 670 | ·0001 848 | ·0000 925+ | ·0000 461 |
| ·47 | ·0017 973 | ·0009 356 | ·0004 839 | ·0002 488 | ·0001 273 | ·0000 648 |
| ·48 | ·0022 578 | ·0012 000 | ·0006 337 | ·0003 327 | ·0001 738 | ·0000 904 |
| ·49 | ·0028 203 | ·0015 298 | ·0008 245− | ·0004 418 | ·0002 356 | ·0001 251 |
| ·50 | ·0035 040 | ·0019 389 | ·0010 660 | ·0005 828 | ·0003 170 | ·0001 717 |
| ·51 | ·0043 307 | ·0024 435− | ·0013 699 | ·0007 637 | ·0004 237 | ·0002 340 |
| ·52 | ·0053 254 | ·0030 626 | ·0017 502 | ·0009 947 | ·0005 625− | ·0003 167 |
| ·53 | ·0065 165+ | ·0038 185− | ·0022 235− | ·0012 876 | ·0007 420 | ·0004 257 |
| ·54 | ·0079 362 | ·0047 365+ | ·0028 093 | ·0016 571 | ·0009 727 | ·0005 684 |
| ·55 | ·0096 208 | ·0058 462 | ·0035 305+ | ·0021 205+ | ·0012 674 | ·0007 543 |
| ·56 | ·0116 108 | ·0071 810 | ·0044 141 | ·0026 986 | ·0016 419 | ·0009 946 |
| ·57 | ·0139 514 | ·0087 794 | ·0054 911 | ·0034 160 | ·0021 149 | ·0013 038 |
| ·58 | ·0166 928 | ·0106 845− | ·0067 976 | ·0043 016 | ·0027 092 | ·0016 990 |
| ·59 | ·0198 903 | ·0129 452 | ·0083 748 | ·0053 893 | ·0034 518 | ·0022 014 |
| ·60 | ·0236 045− | ·0156 161 | ·0102 700 | ·0067 187 | ·0043 748 | ·0028 367 |
| ·61 | ·0279 017 | ·0187 581 | ·0125 370 | ·0083 355+ | ·0055 163 | ·0036 354 |
| ·62 | ·0328 536 | ·0224 387 | ·0152 365− | ·0102 926 | ·0069 209 | ·0046 345− |
| ·63 | ·0385 378 | ·0267 323 | ·0184 366 | ·0126 504 | ·0086 405+ | ·0058 775− |
| ·64 | ·0450 372 | ·0317 200 | ·0222 137 | ·0154 777 | ·0107 356 | ·0074 161 |
| ·65 | ·0524 401 | ·0374 906 | ·0266 523 | ·0188 526 | ·0132 758 | ·0093 110 |
| ·66 | ·0608 399 | ·0441 398 | ·0318 460 | ·0228 627 | ·0163 408 | ·0116 328 |
| ·67 | ·0703 345− | ·0517 702 | ·0378 970 | ·0276 061 | ·0200 217 | ·0144 637 |
| ·68 | ·0810 255− | ·0604 914 | ·0449 170 | ·0331 918 | ·0244 213 | ·0178 982 |
| ·69 | ·0930 177 | ·0704 188 | ·0530 263 | ·0397 397 | ·0296 551 | ·0220 445− |
| ·70 | ·1064 176 | ·0816 734 | ·0623 539 | ·0473 812 | ·0358 523 | ·0270 255− |
| ·71 | ·1213 325− | ·0943 805+ | ·0730 367 | ·0562 589 | ·0431 555− | ·0329 798 |
| ·72 | ·1378 681 | ·1086 684 | ·0852 187 | ·0665 260 | ·0517 213 | ·0400 627 |
| ·73 | ·1561 277 | ·1246 663 | ·0990 494 | ·0783 455+ | ·0617 203 | ·0484 461 |
| ·74 | ·1762 087 | ·1425 026 | ·1146 821 | ·0918 892 | ·0733 358 | ·0583 191 |
| ·75 | ·1982 010 | ·1623 022 | ·1322 715+ | ·1073 353 | ·0867 629 | ·0698 872 |
| ·76 | ·2221 833 | ·1841 831 | ·1519 706 | ·1248 661 | ·1022 068 | ·0833 712 |
| ·77 | ·2482 199 | ·2082 528 | ·1739 272 | ·1446 647 | ·1198 793 | ·0990 053 |
| ·78 | ·2763 573 | ·2346 045− | ·1982 795− | ·1669 102 | ·1399 954 | ·1170 338 |
| ·79 | ·3066 194 | ·2633 114 | ·2251 504 | ·1917 732 | ·1627 684 | ·1377 064 |
| ·80 | ·3390 030 | ·2944 221 | ·2546 421 | ·2194 085+ | ·1884 029 | ·1612 730 |

$x = \cdot 81$ to $1\cdot 00$ $q = 2\cdot 5$ $p = 12$ to $17$

| | $p = 12$ | $p = 13$ | $p = 14$ | $p = 15$ | $p = 16$ | $p = 17$ |
|---|---|---|---|---|---|---|
| $B(p, q) =$ | $\cdot 2297\ 8644 \times \frac{1}{10^2}$ | $\cdot 1901\ 6809 \times \frac{1}{10^2}$ | $\cdot 1594\ 9581 \times \frac{1}{10^2}$ | $\cdot 1353\ 2978 \times \frac{1}{10^2}$ | $\cdot 1159\ 9696 \times \frac{1}{10^2}$ | $\cdot 1003\ 2169 \times \frac{1}{10^2}$ |
| $x$ | | | | | | |
| ·81 | ·3734 733 | ·3279 536 | ·2868 284 | ·2499 479 | ·2170 874 | ·1879 750⁻ |
| ·82 | ·4099 579 | ·3638 854 | ·3217 468 | ·2834 909 | ·2489 841 | ·2180 359 |
| ·83 | ·4483 418 | ·4021 512 | ·3593 893 | ·3200 943 | ·2842 178 | ·2516 488 |
| ·84 | ·4884 612 | ·4426 317 | ·3996 929 | ·3597 598 | ·3228 617 | ·2889 615⁻ |
| ·85 | ·5300 979 | ·4851 459 | ·4425 279 | ·4024 213 | ·3649 216 | ·3300 586 |
| ·86 | ·5729 735⁺ | ·5294 428 | ·4876 868 | ·4479 291 | ·4103 183 | ·3749 407 |
| ·87 | ·6167 441 | ·5751 929 | ·5348 722 | ·4960 348 | ·4588 674 | ·4235 002 |
| ·88 | ·6609 958 | ·6219 801 | ·5836 847 | ·5463 740 | ·5102 573 | ·4754 949 |
| ·89 | ·7052 412 | ·6692 948 | ·6336 110 | ·5984 497 | ·5640 269 | ·5305 183 |
| ·90 | ·7489 179 | ·7165 283 | ·6840 143 | ·6516 163 | ·6195 423 | ·5879 695⁻ |
| ·91 | ·7913 890 | ·7629 708 | ·7341 263 | ·7050 654 | ·6759 755⁻ | ·6470 219 |
| ·92 | ·8319 486 | ·8078 122 | ·7830 445⁻ | ·7578 168 | ·7322 864 | ·7065 964 |
| ·93 | ·8698 312 | ·8501 508 | ·8297 361 | ·8087 163 | ·7872 130 | ·7653 401 |
| ·94 | ·9042 298 | ·8890 100 | ·8730 527 | ·8564 458 | ·8392 741 | ·8216 196 |
| ·95 | ·9343 246 | ·9233 699 | ·9117 620 | ·8995 531 | ·8867 947 | ·8735 380 |
| ·96 | ·9593 299 | ·9522 194 | ·9446 057 | ·9365 134 | ·9279 680 | ·9189 953 |
| ·97 | ·9785 667 | ·9746 443 | ·9704 005⁻ | ·9658 429 | ·9609 800 | ·9558 208 |
| ·98 | ·9915 856 | ·9899 758 | ·9882 161 | ·9863 069 | ·9842 487 | ·9820 426 |
| ·99 | ·9983 908 | ·9980 694 | ·9977 144 | ·9973 254 | ·9969 017 | ·9964 429 |
| 1·00 | 1·0000 000 | 1·0000 000 | 1·0000 000 | 1·0000 000 | 1·0000 000 | 1·0000 000 |

$x = \cdot 33$ to $\cdot 90$ $\qquad q = 2\cdot 5$ $\qquad p = 18$ to $23$

| $x$ | $p = 18$ | $p = 19$ | $p = 20$ | $p = 21$ | $p = 22$ | $p = 23$ |
|---|---|---|---|---|---|---|
| $B(p, q) =$ | ·8745 9936 × $\frac{1}{10^3}$ | ·7679 4090 × $\frac{1}{10^3}$ | ·6786 4545 × $\frac{1}{10^3}$ | ·6032 4040 × $\frac{1}{10^3}$ | ·5390 6589 × $\frac{1}{10^3}$ | ·4840 5916 × $\frac{1}{10^3}$ |
| ·33 | ·0000 001 | | | | | |
| ·34 | ·0000 001 | | | | | |
| ·35 | ·0000 002 | ·0000 001 | | | | |
| ·36 | ·0000 004 | ·0000 001 | ·0000 001 | | | |
| ·37 | ·0000 006 | ·0000 002 | ·0000 001 | | | |
| ·38 | ·0000 009 | ·0000 004 | ·0000 001 | ·0000 001 | | |
| ·39 | ·0000 014 | ·0000 006 | ·0000 002 | ·0000 001 | | |
| ·40 | ·0000 021 | ·0000 009 | ·0000 004 | ·0000 002 | ·0000 001 | |
| ·41 | ·0000 033 | ·0000 014 | ·0000 006 | ·0000 003 | ·0000 001 | ·0000 001 |
| ·42 | ·0000 049 | ·0000 022 | ·0000 010 | ·0000 004 | ·0000 002 | ·0000 001 |
| ·43 | ·0000 073 | ·0000 034 | ·0000 016 | ·0000 007 | ·0000 003 | ·0000 002 |
| ·44 | ·0000 108 | ·0000 051 | ·0000 024 | ·0000 011 | ·0000 005+ | ·0000 002 |
| ·45 | ·0000 158 | ·0000 077 | ·0000 037 | ·0000 018 | ·0000 009 | ·0000 004 |
| ·46 | ·0000 229 | ·0000 113 | ·0000 056 | ·0000 027 | ·0000 013 | ·0000 007 |
| ·47 | ·0000 329 | ·0000 166 | ·0000 084 | ·0000 042 | ·0000 021 | ·0000 011 |
| ·48 | ·0000 468 | ·0000 242 | ·0000 124 | ·0000 064 | ·0000 033 | ·0000 017 |
| ·49 | ·0000 661 | ·0000 348 | ·0000 183 | ·0000 096 | ·0000 050− | ·0000 026 |
| ·50 | ·0000 926 | ·0000 498 | ·0000 267 | ·0000 142 | ·0000 076 | ·0000 040 |
| ·51 | ·0001 287 | ·0000 705+ | ·0000 385+ | ·0000 210 | ·0000 114 | ·0000 062 |
| ·52 | ·0001 776 | ·0000 992 | ·0000 553 | ·0000 307 | ·0000 170 | ·0000 094 |
| ·53 | ·0002 432 | ·0001 385+ | ·0000 786 | ·0000 445− | ·0000 251 | ·0000 141 |
| ·54 | ·0003 309 | ·0001 919 | ·0001 110 | ·0000 640 | ·0000 368 | ·0000 211 |
| ·55 | ·0004 471 | ·0002 641 | ·0001 555− | ·0000 913 | ·0000 534 | ·0000 312 |
| ·56 | ·0006 002 | ·0003 609 | ·0002 163 | ·0001 293 | ·0000 770 | ·0000 458 |
| ·57 | ·0008 006 | ·0004 899 | ·0002 988 | ·0001 817 | ·0001 102 | ·0000 667 |
| ·58 | ·0010 614 | ·0006 607 | ·0004 100 | ·0002 537 | ·0001 566 | ·0000 964 |
| ·59 | ·0013 986 | ·0008 855+ | ·0005 589 | ·0003 517 | ·0002 207 | ·0001 382 |
| ·60 | ·0018 323 | ·0011 795+ | ·0007 569 | ·0004 843 | ·0003 091 | ·0001 968 |
| ·61 | ·0023 868 | ·0015 617 | ·0010 186 | ·0006 625+ | ·0004 298 | ·0002 781 |
| ·62 | ·0030 918 | ·0020 556 | ·0013 625− | ·0009 005+ | ·0005 936 | ·0003 904 |
| ·63 | ·0039 832 | ·0026 903 | ·0018 115+ | ·0012 164 | ·0008 146 | ·0005 443 |
| ·64 | ·0051 042 | ·0035 013 | ·0023 944 | ·0016 329 | ·0011 107 | ·0007 537 |
| ·65 | ·0065 065+ | ·0045 317 | ·0031 467 | ·0021 790 | ·0015 050− | ·0010 370 |
| ·66 | ·0082 514 | ·0058 337 | ·0041 120 | ·0028 905− | ·0020 267 | ·0014 177 |
| ·67 | ·0104 113 | ·0074 700 | ·0053 437 | ·0038 122 | ·0027 128 | ·0019 260 |
| ·68 | ·0130 712 | ·0095 152 | ·0069 063 | ·0049 992 | ·0036 097 | ·0026 004 |
| ·69 | ·0163 298 | ·0120 580 | ·0088 778 | ·0065 189 | ·0047 750− | ·0034 896 |
| ·70 | ·0203 015− | ·0152 025− | ·0113 514 | ·0084 534 | ·0062 800 | ·0046 548 |
| ·71 | ·0251 176 | ·0190 703 | ·0144 377 | ·0109 019 | ·0082 122 | ·0061 723 |
| ·72 | ·0309 278 | ·0238 026 | ·0182 674 | ·0139 832 | ·0106 782 | ·0081 364 |
| ·73 | ·0379 009 | ·0295 614 | ·0229 930 | ·0178 384 | ·0138 068 | ·0106 630 |
| ·74 | ·0462 262 | ·0365 318 | ·0287 915+ | ·0226 342 | ·0177 523 | ·0138 933 |
| ·75 | ·0561 136 | ·0449 225+ | ·0358 666 | ·0285 653 | ·0226 981 | ·0179 977 |
| ·76 | ·0677 931 | ·0549 671 | ·0444 498 | ·0358 571 | ·0288 603 | ·0231 801 |
| ·77 | ·0815 142 | ·0669 237 | ·0548 020 | ·0447 683 | ·0364 905+ | ·0296 820 |
| ·78 | ·0975 433 | ·0810 739 | ·0672 136 | ·0555 918 | ·0458 792 | ·0377 869 |
| ·79 | ·1161 605+ | ·0977 208 | ·0820 034 | ·0686 553 | ·0573 572 | ·0478 231 |
| ·80 | ·1376 540 | ·1171 843 | ·0995 156 | ·0843 206 | ·0712 963 | ·0601 664 |
| ·81 | ·1623 133 | ·1397 951 | ·1201 155− | ·1029 794 | ·0881 077 | ·0752 403 |
| ·82 | ·1904 191 | ·1658 862 | ·1441 810 | ·1250 479 | ·1082 381 | ·0935 143 |
| ·83 | ·2222 313 | ·1957 800 | ·1720 922 | ·1509 568 | ·1321 613 | ·1154 975− |
| ·84 | ·2579 727 | ·2297 730 | ·2042 151 | ·1811 363 | ·1603 654 | ·1417 283 |
| ·85 | ·2978 104 | ·2681 151 | ·2408 814 | ·2159 967 | ·1933 337 | ·1727 568 |
| ·86 | ·3418 314 | ·3109 846 | ·2823 620 | ·2559 003 | ·2315 179 | ·2091 192 |
| ·87 | ·3900 159 | ·3584 569 | ·3288 329 | ·3011 271 | ·2753 014 | ·2513 011 |
| ·88 | ·4422 048 | ·4104 683 | ·3803 353 | ·3518 294 | ·3249 525+ | ·2996 882 |
| ·89 | ·4980 644 | ·4667 733 | ·4367 257 | ·4079 776 | ·3805 642 | ·3545 028 |
| ·90 | ·5570 466 | ·5268 969 | ·4976 199 | ·4692 945− | ·4419 806 | ·4157 221 |

$x = \cdot 91$ to $1 \cdot 00$ $q = 2 \cdot 5$ $p = 18$ to $23$

| | $p = 18$ | $p = 19$ | $p = 20$ | $p = 21$ | $p = 22$ | $p = 23$ |
|---|---|---|---|---|---|---|
| $B(p, q) =$ | $\cdot 8745\ 9936 \times \frac{1}{10^3}$ | $\cdot 7679\ 4090 \times \frac{1}{10^3}$ | $\cdot 6786\ 4545 \times \frac{1}{10^3}$ | $\cdot 6032\ 4040 \times \frac{1}{10^3}$ | $\cdot 5390\ 6589 \times \frac{1}{10^3}$ | $\cdot 4840\ 5916 \times \frac{1}{10^3}$ |
| $x$ | | | | | | |
| ·91 | ·6183 494 | ·5900 830 | ·5623 299 | ·5351 805− | ·5087 098 | ·4829 790 |
| ·92 | ·6808 762 | ·6552 418 | ·6297 962 | ·6046 305+ | ·5798 244 | ·5554 467 |
| ·93 | ·7432 035+ | ·7209 009 | ·6985 219 | ·6761 486 | ·6538 551 | ·6317 086 |
| ·94 | ·8035 600 | ·7851 693 | ·7665 173 | ·7476 694 | ·7286 869 | ·7096 268 |
| ·95 | ·8598 329 | ·8457 281 | ·8312 706 | ·8165 059 | ·8014 776 | ·7862 272 |
| ·96 | ·9096 215− | ·8998 727 | ·8897 750+ | ·8793 542 | ·8686 356 | ·8576 442 |
| ·97 | ·9503 748 | ·9446 519 | ·9386 625+ | ·9324 171 | ·9259 262 | ·9192 008 |
| ·98 | ·9796 898 | ·9771 920 | ·9745 509 | ·9717 685− | ·9688 469 | ·9657 886 |
| ·99 | ·9959 486 | ·9954 186 | ·9948 523 | ·9942 497 | ·9936 105+ | ·9929 346 |
| 1·00 | 1·0000 000 | 1·0000 000 | 1·0000 000 | 1·0000 000 | 1·0000 000 | 1·0000 000 |

$x =$ ·43 to 1·00 $q = 2{\cdot}5$ $p =$ 24 to 29

| | $p = 24$ | $p = 25$ | $p = 26$ | $p = 27$ | $p = 28$ | $p = 29$ |
|---|---|---|---|---|---|---|
| $B(p, q) =$ | ·4366 0238 $\times \frac{1}{10^3}$ | ·3954 1348 $\times \frac{1}{10^3}$ | ·3594 6680 $\times \frac{1}{10^3}$ | ·3279 3462 $\times \frac{1}{10^3}$ | ·3001 4355+ $\times \frac{1}{10^3}$ | ·2755 4162 $\times \frac{1}{10^3}$ |
| $x$ | | | | | | |
| ·43 | ·0000 001 | | | | | |
| ·44 | ·0000 001 | ·0000 001 | | | | |
| ·45 | ·0000 002 | ·0000 001 | | | | |
| ·46 | ·0000 003 | ·0000 002 | ·0000 001 | | | |
| ·47 | ·0000 005+ | ·0000 003 | ·0000 001 | ·0000 001 | | |
| ·48 | ·0000 008 | ·0000 004 | ·0000 002 | ·0000 001 | ·0000 001 | |
| ·49 | ·0000 014 | ·0000 007 | ·0000 004 | ·0000 002 | ·0000 001 | |
| ·50 | ·0000 021 | ·0000 011 | ·0000 006 | ·0000 003 | ·0000 002 | ·0000 001 |
| ·51 | ·0000 033 | ·0000 018 | ·0000 010 | ·0000 005+ | ·0000 003 | ·0000 001 |
| ·52 | ·0000 052 | ·0000 028 | ·0000 016 | ·0000 009 | ·0000 005− | ·0000 003 |
| ·53 | ·0000 079 | ·0000 044 | ·0000 025− | ·0000 014 | ·0000 008 | ·0000 004 |
| ·54 | ·0000 121 | ·0000 069 | ·0000 039 | ·0000 022 | ·0000 013 | ·0000 007 |
| ·55 | ·0000 182 | ·0000 106 | ·0000 061 | ·0000 036 | ·0000 021 | ·0000 012 |
| ·56 | ·0000 272 | ·0000 161 | ·0000 095− | ·0000 056 | ·0000 033 | ·0000 019 |
| ·57 | ·0000 403 | ·0000 243 | ·0000 146 | ·0000 088 | ·0000 052 | ·0000 031 |
| ·58 | ·0000 592 | ·0000 363 | ·0000 222 | ·0000 136 | ·0000 083 | ·0000 050+ |
| ·59 | ·0000 863 | ·0000 538 | ·0000 335− | ·0000 208 | ·0000 129 | ·0000 080 |
| ·60 | ·0001 250− | ·0000 792 | ·0000 501 | ·0000 317 | ·0000 200 | ·0000 126 |
| ·61 | ·0001 796 | ·0001 157 | ·0000 744 | ·0000 478 | ·0000 306 | ·0000 196 |
| ·62 | ·0002 562 | ·0001 677 | ·0001 096 | ·0000 715+ | ·0000 466 | ·0000 303 |
| ·63 | ·0003 628 | ·0002 414 | ·0001 603 | ·0001 062 | ·0000 703 | ·0000 465− |
| ·64 | ·0005 104 | ·0003 449 | ·0002 326 | ·0001 566 | ·0001 053 | ·0000 706 |
| ·65 | ·0007 130 | ·0004 893 | ·0003 351 | ·0002 291 | ·0001 564 | ·0001 066 |
| ·66 | ·0009 896 | ·0006 894 | ·0004 793 | ·0003 327 | ·0002 305+ | ·0001 595+ |
| ·67 | ·0013 645− | ·0009 647 | ·0006 808 | ·0004 796 | ·0003 374 | ·0002 369 |
| ·68 | ·0018 693 | ·0013 412 | ·0009 604 | ·0006 866 | ·0004 901 | ·0003 492 |
| ·69 | ·0025 449 | ·0018 523 | ·0013 457 | ·0009 760 | ·0007 067 | ·0005 110 |
| ·70 | ·0034 430 | ·0025 417 | ·0018 730 | ·0013 778 | ·0010 120 | ·0007 422 |
| ·71 | ·0046 295− | ·0034 656 | ·0025 897 | ·0019 319 | ·0014 389 | ·0010 702 |
| ·72 | ·0061 869 | ·0046 956 | ·0035 574 | ·0026 906 | ·0020 318 | ·0015 321 |
| ·73 | ·0082 184 | ·0063 223 | ·0048 551 | ·0037 223 | ·0028 494 | ·0021 780 |
| ·74 | ·0108 515+ | ·0084 599 | ·0065 838 | ·0051 155+ | ·0039 686 | ·0030 744 |
| ·75 | ·0142 425+ | ·0112 501 | ·0088 711 | ·0069 840 | ·0054 899 | ·0043 094 |
| ·76 | ·0185 816 | ·0148 683 | ·0118 769 | ·0094 723 | ·0075 432 | ·0059 986 |
| ·77 | ·0240 976 | ·0195 288 | ·0157 998 | ·0127 628 | ·0102 943 | ·0082 918 |
| ·78 | ·0310 632 | ·0254 910 | ·0208 838 | ·0170 830 | ·0139 536 | ·0113 819 |
| ·79 | ·0397 999 | ·0330 655+ | ·0274 261 | ·0227 139 | ·0187 845+ | ·0155 140 |
| ·80 | ·0506 818 | ·0426 199 | ·0357 834 | ·0299 987 | ·0251 138 | ·0209 966 |
| ·81 | ·0641 380 | ·0545 831 | ·0463 793 | ·0393 508 | ·0333 414 | ·0282 131 |
| ·82 | ·0806 533 | ·0694 482 | ·0597 087 | ·0512 616 | ·0439 501 | ·0376 335+ |
| ·83 | ·1007 644 | ·0877 717 | ·0763 408 | ·0663 057 | ·0575 139 | ·0498 257 |
| ·84 | ·1250 521 | ·1101 685+ | ·0969 163 | ·0851 421 | ·0747 024 | ·0654 632 |
| ·85 | ·1541 258 | ·1372 996 | ·1221 393 | ·1085 095+ | ·0962 806 | ·0853 292 |
| ·86 | ·1886 001 | ·1698 508 | ·1527 589 | ·1372 119 | ·1230 986 | ·1103 110 |
| ·87 | ·2290 590 | ·2084 990 | ·1895 386 | ·1720 913 | ·1560 689 | ·1413 827 |
| ·88 | ·2760 057 | ·2538 625+ | ·2332 073 | ·2139 822 | ·1961 241 | ·1795 671 |
| ·89 | ·3297 955+ | ·3064 317 | ·2843 902 | ·2636 416 | ·2441 494 | ·2258 719 |
| ·90 | ·3905 481 | ·3664 755+ | ·3435 103 | ·3216 491 | ·3008 810 | ·2811 884 |
| ·91 | ·4580 370 | ·4339 211 | ·4106 590 | ·3882 692 | ·3667 625+ | ·3461 430 |
| ·92 | ·5315 565+ | ·5082 039 | ·4854 304 | ·4632 701 | ·4417 500− | ·4208 908 |
| ·93 | ·6097 691 | ·5880 901 | ·5667 190 | ·5456 972 | ·5250 608 | ·5048 409 |
| ·94 | ·6905 417 | ·6714 806 | ·6524 880 | ·6336 051 | ·6148 690 | ·5963 136 |
| ·95 | ·7707 945+ | ·7552 172 | ·7395 308 | ·7237 689 | ·7079 633 | ·6921 436 |
| ·96 | ·8464 044 | ·8349 397 | ·8232 732 | ·8114 272 | ·7994 234 | ·7872 823 |
| ·97 | ·9122 518 | ·9050 899 | ·8977 260 | ·8901 710 | ·8824 355+ | ·8745 301 |
| ·98 | ·9625 960 | ·9592 716 | ·9558 183 | ·9522 388 | ·9485 361 | ·9447 129 |
| ·99 | ·9922 218 | ·9914 720 | ·9906 851 | ·9898 612 | ·9890 002 | ·9881 022 |
| 1·00 | 1·0000 000 | 1·0000 000 | 1·0000 000 | 1·0000 000 | 1·0000 000 | 1·0000 000 |

$x = \cdot 51$ to $1\cdot 00$ $q = 2\cdot 5$ $p = 30$ to $36$

| | $p = 30$ | $p = 31$ | $p = 32$ | $p = 33$ | $p = 34$ | $p = 35$ | $p = 36$ |
|---|---|---|---|---|---|---|---|
| $B(p, q) =$ | $\cdot 2536\ 7324 \times \frac{1}{10^3}$ | $\cdot 2341\ 5991 \times \frac{1}{10^3}$ | $\cdot 2166\ 8529 \times \frac{1}{10^3}$ | $\cdot 2009\ 8346 \times \frac{1}{10^3}$ | $\cdot 1868\ 2970 \times \frac{1}{10^3}$ | $\cdot 1740\ 3314 \times \frac{1}{10^3}$ | $\cdot 1624\ 3093 \times \frac{1}{10^3}$ |
| $x$ | | | | | | | |
| ·51 | ·0000 001 | | | | | | |
| ·52 | ·0000 001 | ·0000 001 | | | | | |
| ·53 | ·0000 002 | ·0000 001 | ·0000 001 | | | | |
| ·54 | ·0000 004 | ·0000 002 | ·0000 001 | ·0000 001 | | | |
| ·55 | ·0000 007 | ·0000 004 | ·0000 002 | ·0000 001 | ·0000 001 | | |
| ·56 | ·0000 011 | ·0000 007 | ·0000 004 | ·0000 002 | ·0000 001 | ·0000 001 | |
| ·57 | ·0000 019 | ·0000 011 | ·0000 007 | ·0000 004 | ·0000 002 | ·0000 001 | ·0000 001 |
| ·58 | ·0000 031 | ·0000 019 | ·0000 011 | ·0000 007 | ·0000 004 | ·0000 002 | ·0000 001 |
| ·59 | ·0000 049 | ·0000 030 | ·0000 019 | ·0000 012 | ·0000 007 | ·0000 004 | ·0000 003 |
| ·60 | ·0000 079 | ·0000 050⁻ | ·0000 031 | ·0000 019 | ·0000 012 | ·0000 008 | ·0000 005⁻ |
| ·61 | ·0000 125⁺ | ·0000 080 | ·0000 051 | ·0000 032 | ·0000 021 | ·0000 013 | ·0000 008 |
| ·62 | ·0000 197 | ·0000 128 | ·0000 083 | ·0000 053 | ·0000 035⁻ | ·0000 022 | ·0000 014 |
| ·63 | ·0000 306 | ·0000 202 | ·0000 133 | ·0000 087 | ·0000 057 | ·0000 038 | ·0000 025⁻ |
| ·64 | ·0000 473 | ·0000 317 | ·0000 212 | ·0000 141 | ·0000 094 | ·0000 063 | ·0000 042 |
| ·65 | ·0000 725⁺ | ·0000 493 | ·0000 335⁻ | ·0000 227 | ·0000 154 | ·0000 104 | ·0000 070 |
| ·66 | ·0001 102 | ·0000 760 | ·0000 524 | ·0000 361 | ·0000 248 | ·0000 170 | ·0000 117 |
| ·67 | ·0001 662 | ·0001 164 | ·0000 814 | ·0000 569 | ·0000 397 | ·0000 277 | ·0000 193 |
| ·68 | ·0002 485⁺ | ·0001 766 | ·0001 254 | ·0000 889 | ·0000 629 | ·0000 445⁺ | ·0000 315⁻ |
| ·69 | ·0003 689 | ·0002 660 | ·0001 916 | ·0001 378 | ·0000 990 | ·0000 711 | ·0000 510 |
| ·70 | ·0005 435⁺ | ·0003 975⁺ | ·0002 904 | ·0002 119 | ·0001 544 | ·0001 124 | ·0000 818 |
| ·71 | ·0007 948 | ·0005 895⁺ | ·0004 367 | ·0003 232 | ·0002 389 | ·0001 764 | ·0001 301 |
| ·72 | ·0011 537 | ·0008 677 | ·0006 517 | ·0004 890 | ·0003 664 | ·0002 743 | ·0002 052 |
| ·73 | ·0016 625⁺ | ·0012 674 | ·0009 651 | ·0007 340 | ·0005 576 | ·0004 232 | ·0003 209 |
| ·74 | ·0023 784 | ·0018 377 | ·0014 182 | ·0010 932 | ·0008 418 | ·0006 475⁺ | ·0004 976 |
| ·75 | ·0033 782 | ·0026 449 | ·0020 683 | ·0016 156 | ·0012 606 | ·0009 826 | ·0007 652 |
| ·76 | ·0047 640 | ·0037 787 | ·0029 937 | ·0023 691 | ·0018 729 | ·0014 791 | ·0011 670 |
| ·77 | ·0066 702 | ·0053 590 | ·0043 006 | ·0034 474 | ·0027 606 | ·0022 084 | ·0017 650⁺ |
| ·78 | ·0092 722 | ·0075 444 | ·0061 315⁻ | ·0049 777 | ·0040 369 | ·0032 707 | ·0026 475⁻ |
| ·79 | ·0127 966 | ·0105 425⁺ | ·0086 757 | ·0071 317 | ·0058 565⁺ | ·0048 047 | ·0039 382 |
| ·80 | ·0175 324 | ·0146 225⁻ | ·0121 819 | ·0101 379 | ·0084 284 | ·0070 005⁻ | ·0058 092 |
| ·81 | ·0238 442 | ·0201 285⁺ | ·0169 732 | ·0142 976 | ·0120 318 | ·0101 156 | ·0084 969 |
| ·82 | ·0321 860 | ·0274 957 | ·0234 636 | ·0200 022 | ·0170 349 | ·0144 944 | ·0123 218 |
| ·83 | ·0431 145⁻ | ·0372 656 | ·0321 762 | ·0277 540 | ·0239 167 | ·0205 912 | ·0177 128 |
| ·84 | ·0573 009 | ·0501 017 | ·0437 618 | ·0381 866 | ·0332 906 | ·0289 965⁻ | ·0252 349 |
| ·85 | ·0755 390 | ·0668 012 | ·0590 147 | ·0520 860 | ·0459 289 | ·0404 644 | ·0356 205⁺ |
| ·86 | ·0987 448 | ·0883 006 | ·0788 839 | ·0704 059 | ·0627 835⁻ | ·0559 390 | ·0498 004 |
| ·87 | ·1279 448 | ·1156 693 | ·1044 728 | ·0942 753 | ·0850 002 | ·0765 748 | ·0689 306 |
| ·88 | ·1642 434 | ·1500 842 | ·1370 213 | ·1249 870 | ·1139 155⁻ | ·1037 427 | ·0944 070 |
| ·89 | ·2087 636 | ·1927 759 | ·1778 584 | ·1639 594 | ·1510 271 | ·1390 095⁻ | ·1278 555⁻ |
| ·90 | ·2625 483 | ·2449 335⁻ | ·2283 130 | ·2126 534 | ·1979 191 | ·1840 732 | ·1710 779 |
| ·91 | ·3264 087 | ·3075 526 | ·2895 633 | ·2724 256 | ·2561 215⁻ | ·2406 302 | ·2259 289 |
| ·92 | ·4007 077 | ·3812 109 | ·3624 059 | ·3442 943 | ·3268 743 | ·3101 408 | ·2940 863 |
| ·93 | ·4850 636 | ·4657 512 | ·4469 215⁺ | ·4285 891 | ·4107 650⁻ | ·3934 572 | ·3766 712 |
| ·94 | ·5779 693 | ·5598 635⁻ | ·5420 205⁻ | ·5244 619 | ·5072 066 | ·4902 710 | ·4736 693 |
| ·95 | ·6763 375⁺ | ·6605 709 | ·6448 679 | ·6292 508 | ·6137 402 | ·5983 550⁻ | ·5831 127 |
| ·96 | ·7750 240 | ·7626 676 | ·7502 315⁻ | ·7377 332 | ·7251 895⁺ | ·7126 163 | ·7000 287 |
| ·97 | ·8664 652 | ·8582 511 | ·8498 979 | ·8414 155⁺ | ·8328 136 | ·8241 016 | ·8152 888 |
| ·98 | ·9407 725⁻ | ·9367 178 | ·9325 519 | ·9282 779 | ·9238 990 | ·9194 184 | ·9148 392 |
| ·99 | ·9871 671 | ·9861 951 | ·9851 863 | ·9841 408 | ·9830 586 | ·9819 400 | ·9807 851 |
| 1·00 | 1·0000 000 | 1·0000 000 | 1·0000 000 | 1·0000 000 | 1·0000 000 | 1·0000 000 | 1·0000 000 |

$x$ = ·58 to 1·00 $q = 2·5$ $p$ = 37 to 43

| $x$ | $p = 37$ | $p = 38$ | $p = 39$ | $p = 40$ | $p = 41$ | $p = 42$ | $p = 43$ |
|---|---|---|---|---|---|---|---|
| $B(p, q)$ = | ·1518 8347 × $\frac{1}{10^3}$ | ·1422 7059 × $\frac{1}{10^3}$ | ·1334 8846 × $\frac{1}{10^3}$ | ·1254 4698 × $\frac{1}{10^3}$ | ·1180 6775 × $\frac{1}{10^3}$ | ·1112 8225 × $\frac{1}{10^3}$ | ·1050 3043 × $\frac{1}{10^3}$ |
| ·58 | ·0000 001 | ·0000 001 | | | | | |
| ·59 | ·0000 002 | ·0000 001 | ·0000 001 | | | | |
| ·60 | ·0000 003 | ·0000 002 | ·0000 001 | ·0000 001 | | | |
| ·61 | ·0000 005$^{+}$ | ·0000 003 | ·0000 002 | ·0000 001 | ·0000 001 | ·0000 001 | |
| ·62 | ·0000 009 | ·0000 006 | ·0000 004 | ·0000 002 | ·0000 002 | ·0000 001 | ·0000 001 |
| ·63 | ·0000 016 | ·0000 011 | ·0000 007 | ·0000 004 | ·0000 003 | ·0000 002 | ·0000 001 |
| ·64 | ·0000 028 | ·0000 018 | ·0000 012 | ·0000 008 | ·0000 005$^{+}$ | ·0000 004 | ·0000 002 |
| ·65 | ·0000 047 | ·0000 032 | ·0000 022 | ·0000 014 | ·0000 010 | ·0000 007 | ·0000 004 |
| ·66 | ·0000 080 | ·0000 055$^{-}$ | ·0000 038 | ·0000 026 | ·0000 018 | ·0000 012 | ·0000 008 |
| ·67 | ·0000 134 | ·0000 093 | ·0000 065$^{-}$ | ·0000 045$^{-}$ | ·0000 031 | ·0000 022 | ·0000 015$^{-}$ |
| ·68 | ·0000 222 | ·0000 157 | ·0000 110 | ·0000 078 | ·0000 055$^{-}$ | ·0000 038 | ·0000 027 |
| ·69 | ·0000 365$^{+}$ | ·0000 261 | ·0000 187 | ·0000 133 | ·0000 095$^{+}$ | ·0000 068 | ·0000 048 |
| ·70 | ·0000 594 | ·0000 431 | ·0000 313 | ·0000 227 | ·0000 164 | ·0000 119 | ·0000 086 |
| ·71 | ·0000 959 | ·0000 706 | ·0000 519 | ·0000 382 | ·0000 280 | ·0000 206 | ·0000 151 |
| ·72 | ·0001 533 | ·0001 145$^{-}$ | ·0000 854 | ·0000 636 | ·0000 474 | ·0000 353 | ·0000 262 |
| ·73 | ·0002 431 | ·0001 840 | ·0001 391 | ·0001 051 | ·0000 794 | ·0000 599 | ·0000 451 |
| ·74 | ·0003 820 | ·0002 931 | ·0002 246 | ·0001 720 | ·0001 316 | ·0001 007 | ·0000 769 |
| ·75 | ·0005 954 | ·0004 628 | ·0003 595$^{-}$ | ·0002 790 | ·0002 163 | ·0001 676 | ·0001 298 |
| ·76 | ·0009 199 | ·0007 245$^{+}$ | ·0005 701 | ·0004 483 | ·0003 523 | ·0002 766 | ·0002 170 |
| ·77 | ·0014 094 | ·0011 244 | ·0008 964 | ·0007 140 | ·0005 683 | ·0004 520 | ·0003 593 |
| ·78 | ·0021 411 | ·0017 301 | ·0013 968 | ·0011 269 | ·0009 085$^{-}$ | ·0007 319 | ·0005 892 |
| ·79 | ·0032 251 | ·0026 389 | ·0021 575$^{-}$ | ·0017 626 | ·0014 389 | ·0011 738 | ·0009 570 |
| ·80 | ·0048 164 | ·0039 900 | ·0033 028 | ·0027 319 | ·0022 580 | ·0018 651 | ·0015 395$^{-}$ |
| ·81 | ·0071 311 | ·0059 800 | ·0050 108 | ·0041 956 | ·0035 105$^{-}$ | ·0029 352 | ·0024 526 |
| ·82 | ·0104 661 | ·0088 828 | ·0075 332 | ·0063 839 | ·0054 062 | ·0045 751 | ·0038 693 |
| ·83 | ·0152 242 | ·0130 748 | ·0112 205$^{-}$ | ·0096 222 | ·0082 458 | ·0070 617 | ·0060 437 |
| ·84 | ·0219 435$^{-}$ | ·0190 666 | ·0165 547 | ·0143 634 | ·0124 538 | ·0107 910 | ·0093 444 |
| ·85 | ·0313 317 | ·0275 384 | ·0241 869 | ·0212 285$^{-}$ | ·0186 195$^{+}$ | ·0163 208 | ·0142 971 |
| ·86 | ·0443 013 | ·0393 803 | ·0349 813 | ·0310 525$^{+}$ | ·0275 471 | ·0244 222 | ·0216 388 |
| ·87 | ·0620 030 | ·0557 317 | ·0500 603 | ·0449 364 | ·0403 115$^{-}$ | ·0361 405$^{+}$ | ·0323 822 |
| ·88 | ·0858 493 | ·0780 132 | ·0708 452 | ·0642 948 | ·0583 143 | ·0528 589 | ·0478 867 |
| ·89 | ·1175 147 | ·1079 384 | ·0990 790 | ·0908 909 | ·0833 302 | ·0763 550$^{+}$ | ·0699 254 |
| ·90 | ·1588 947 | ·1474 854 | ·1368 117 | ·1268 358 | ·1175 209 | ·1088 307 | ·1007 302 |
| ·91 | ·2119 933 | ·1987 978 | ·1863 159 | ·1745 205$^{+}$ | ·1633 842 | ·1528 794 | ·1429 786 |
| ·92 | ·2787 006 | ·2639 720 | ·2498 868 | ·2364 300 | ·2235 854 | ·2113 362 | ·1996 643 |
| ·93 | ·3604 097 | ·3446 734 | ·3294 610 | ·3147 693 | ·3005 937 | ·2869 280 | ·2737 650$^{+}$ |
| ·94 | ·4574 134 | ·4415 134 | ·4259 775$^{-}$ | ·4108 120 | ·3960 219 | ·3816 105$^{+}$ | ·3675 800 |
| ·95 | ·5680 291 | ·5531 188 | ·5383 949 | ·5238 692 | ·5095 523 | ·4954 536 | ·4815 816 |
| ·96 | ·6874 411 | ·6748 671 | ·6623 195$^{+}$ | ·6498 106 | ·6373 517 | ·6249 536 | ·6126 264 |
| ·97 | ·8063 842 | ·7973 966 | ·7883 344 | ·7792 061 | ·7700 195$^{+}$ | ·7607 826 | ·7515 027 |
| ·98 | ·9101 647 | ·9053 979 | ·9005 420 | ·8956 002 | ·8905 757 | ·8854 714 | ·8802 907 |
| ·99 | ·9795 942 | ·9783 673 | ·9771 048 | ·9758 069 | ·9744 737 | ·9731 055$^{+}$ | ·9717 027 |
| 1·00 | 1·0000 000 | 1·0000 000 | 1·0000 000 | 1·0000 000 | 1·0000 000 | 1·0000 000 | 1·0000 000 |

$x$ = ·63 to 1·00 $q$ = 2·5 $p$ = 44 to 50

| $x$ | $p$ = 44 | $p$ = 45 | $p$ = 46 | $p$ = 47 | $p$ = 48 | $p$ = 49 | $p$ = 50 |
|---|---|---|---|---|---|---|---|
| $B(p, q)$ = | ·9925 9531 × $\frac{1}{10^{4}}$ | ·9392 2997 × $\frac{1}{10^{4}}$ | ·8897 9681 × $\frac{1}{10^{4}}$ | ·8439 3100 × $\frac{1}{10^{4}}$ | ·8013 0822 × $\frac{1}{10^{4}}$ | ·7616 3950 × $\frac{1}{10^{4}}$ | ·7246 6671 × $\frac{1}{10^{4}}$ |
| ·63 | ·0000 001 | ·0000 001 | | | | | |
| ·64 | ·0000 002 | ·0000 001 | ·0000 001 | | | | |
| ·65 | ·0000 003 | ·0000 002 | ·0000 001 | ·0000 001 | ·0000 001 | | |
| ·66 | ·0000 006 | ·0000 004 | ·0000 003 | ·0000 002 | ·0000 001 | ·0000 001 | ·0000 001 |
| ·67 | ·0000 010 | ·0000 007 | ·0000 005− | ·0000 003 | ·0000 002 | ·0000 002 | ·0000 001 |
| ·68 | ·0000 019 | ·0000 013 | ·0000 009 | ·0000 007 | ·0000 005− | ·0000 003 | ·0000 002 |
| ·69 | ·0000 035− | ·0000 025− | ·0000 018 | ·0000 012 | ·0000 009 | ·0000 006 | ·0000 004 |
| ·70 | ·0000 062 | ·0000 045− | ·0000 032 | ·0000 023 | ·0000 017 | ·0000 012 | ·0000 009 |
| ·71 | ·0000 111 | ·0000 081 | ·0000 059 | ·0000 043 | ·0000 032 | ·0000 023 | ·0000 017 |
| ·72 | ·0000 195− | ·0000 145− | ·0000 107 | ·0000 080 | ·0000 059 | ·0000 044 | ·0000 032 |
| ·73 | ·0000 340 | ·0000 256 | ·0000 193 | ·0000 145− | ·0000 109 | ·0000 082 | ·0000 061 |
| ·74 | ·0000 587 | ·0000 448 | ·0000 342 | ·0000 261 | ·0000 199 | ·0000 151 | ·0000 115+ |
| ·75 | ·0001 005− | ·0000 777 | ·0000 601 | ·0000 464 | ·0000 358 | ·0000 276 | ·0000 213 |
| ·76 | ·0001 702 | ·0001 333 | ·0001 044 | ·0000 817 | ·0000 639 | ·0000 500− | ·0000 391 |
| ·77 | ·0002 854 | ·0002 265+ | ·0001 797 | ·0001 425+ | ·0001 129 | ·0000 894 | ·0000 708 |
| ·78 | ·0004 740 | ·0003 811 | ·0003 062 | ·0002 459 | ·0001 974 | ·0001 584 | ·0001 270 |
| ·79 | ·0007 797 | ·0006 348 | ·0005 166 | ·0004 201 | ·0003 415− | ·0002 774 | ·0002 253 |
| ·80 | ·0012 699 | ·0010 469 | ·0008 625+ | ·0007 103 | ·0005 845+ | ·0004 808 | ·0003 953 |
| ·81 | ·0020 481 | ·0017 092 | ·0014 256 | ·0011 884 | ·0009 901 | ·0008 245+ | ·0006 862 |
| ·82 | ·0032 704 | ·0027 625− | ·0023 321 | ·0019 677 | ·0016 594 | ·0013 987 | ·0011 783 |
| ·83 | ·0051 694 | ·0044 189 | ·0037 752 | ·0032 236 | ·0027 511 | ·0023 467 | ·0020 007 |
| ·84 | ·0080 868 | ·0069 945− | ·0060 463 | ·0052 239 | ·0045 110 | ·0038 935− | ·0033 589 |
| ·85 | ·0125 169 | ·0109 522 | ·0095 778 | ·0083 716 | ·0073 135− | ·0063 860 | ·0055 736 |
| ·86 | ·0191 615+ | ·0169 585− | ·0150 007 | ·0132 621 | ·0117 191 | ·0103 508 | ·0091 379 |
| ·87 | ·0289 984 | ·0259 542 | ·0232 174 | ·0207 588 | ·0185 515− | ·0165 711 | ·0147 955− |
| ·88 | ·0433 585− | ·0392 378 | ·0354 908 | ·0320 859 | ·0289 939 | ·0261 879 | ·0236 431 |
| ·89 | ·0640 034 | ·0585 531 | ·0535 407 | ·0489 342 | ·0447 035+ | ·0408 206 | ·0372 590 |
| ·90 | ·0931 855+ | ·0861 638 | ·0796 335+ | ·0735 647 | ·0679 284 | ·0626 972 | ·0578 450+ |
| ·91 | ·1336 546 | ·1248 805+ | ·1166 299 | ·1088 771 | ·1015 968 | ·0947 647 | ·0883 572 |
| ·92 | ·1885 517 | ·1779 795+ | ·1679 289 | ·1583 808 | ·1493 162 | ·1407 162 | ·1325 620 |
| ·93 | ·2610 964 | ·2489 130 | ·2372 047 | ·2259 609 | ·2151 705+ | ·2048 218 | ·1949 029 |
| ·94 | ·3539 313 | ·3406 641 | ·3277 772 | ·3152 686 | ·3031 352 | ·2913 734 | ·2799 788 |
| ·95 | ·4679 434 | ·4545 454 | ·4413 931 | ·4284 910 | ·4158 427 | ·4034 514 | ·3913 193 |
| ·96 | ·6003 794 | ·5882 214 | ·5761 608 | ·5642 050+ | ·5523 611 | ·5406 357 | ·5290 347 |
| ·97 | ·7421 873 | ·7328 432 | ·7234 774 | ·7140 963 | ·7047 062 | ·6953 132 | ·6859 230 |
| ·98 | ·8750 364 | ·8697 117 | ·8643 193 | ·8588 625− | ·8533 441 | ·8477 672 | ·8421 344 |
| ·99 | ·9702 655− | ·9687 941 | ·9672 889 | ·9657 501 | ·9641 781 | ·9625 732 | ·9609 357 |
| 1·00 | 1·0000 000 | 1·0000 000 | 1·0000 000 | 1·0000 000 | 1·0000 000 | 1·0000 000 | 1·0000 000 |

$x$ = ·01 to ·60 $q = 3$ $p$ = 3 to 5·5

| | $p=3$ | $p=3·5$ | $p=4$ | $p=4·5$ | $p=5$ | $p=5·5$ |
|---|---|---|---|---|---|---|
| $B(p,q)$ = | ·3333 3333 × $\frac{1}{10}$ | ·2308 8023 × $\frac{1}{10}$ | ·1666 6667 × $\frac{1}{10}$ | ·1243 2012 × $\frac{1}{10}$ | ·9523 8095+ × $\frac{1}{10^2}$ | ·7459 2075− × $\frac{1}{10^2}$ |
| $x$ | | | | | | |
| ·01 | ·0000 099 | ·0000 012 | ·0000 001 | | | |
| ·02 | ·0000 776 | ·0000 136 | ·0000 023 | ·0000 004 | ·0000 001 | |
| ·03 | ·0002 580 | ·0000 552 | ·0000 116 | ·0000 024 | ·0000 005− | ·0000 001 |
| ·04 | ·0006 022 | ·0001 487 | ·0000 360 | ·0000 086 | ·0000 020 | ·0000 005− |
| ·05 | ·0011 581 | ·0003 195+ | ·0000 864 | ·0000 230 | ·0000 060 | ·0000 016 |
| ·06 | ·0019 703 | ·0005 951 | ·0001 762 | ·0000 513 | ·0000 147 | ·0000 042 |
| ·07 | ·0030 799 | ·0010 042 | ·0003 210 | ·0001 009 | ·0000 313 | ·0000 096 |
| ·08 | ·0045 253 | ·0015 764 | ·0005 384 | ·0001 809 | ·0000 600 | ·0000 196 |
| ·09 | ·0063 413 | ·0023 415− | ·0008 477 | ·0003 020 | ·0001 061 | ·0000 369 |
| ·10 | ·0085 600$^e$ | ·0033 295− | ·0012 700$^e$ | ·0004 767 | ·0001 765$^e$ | ·0000 646 |
| ·11 | ·0112 105− | ·0045 702 | ·0018 273 | ·0007 190 | ·0002 791 | ·0001 071 |
| ·12 | ·0143 189 | ·0060 928 | ·0025 431 | ·0010 447 | ·0004 234 | ·0001 697 |
| ·13 | ·0179 086 | ·0079 258 | ·0034 413 | ·0014 707 | ·0006 202 | ·0002 586 |
| ·14 | ·0220 003 | ·0100 970 | ·0045 469 | ·0020 156 | ·0008 817 | ·0003 813 |
| ·15 | ·0266 119 | ·0126 330 | ·0058 852 | ·0026 991 | ·0012 216 | ·0005 467 |
| ·16 | ·0317 587 | ·0155 592 | ·0074 816 | ·0035 420 | ·0016 551 | ·0007 647 |
| ·17 | ·0374 538 | ·0188 998 | ·0093 619 | ·0045 663 | ·0021 984 | ·0010 467 |
| ·18 | ·0437 073 | ·0226 774 | ·0115 516 | ·0057 948 | ·0028 695− | ·0014 053 |
| ·19 | ·0505 275+ | ·0269 133 | ·0140 760 | ·0072 508 | ·0036 873 | ·0018 545+ |
| ·20 | ·0579 200$^e$ | ·0316 269 | ·0169 600$^e$ | ·0089 586 | ·0046 720$^e$ | ·0024 099 |
| ·21 | ·0658 883 | ·0368 364 | ·0202 280 | ·0109 427 | ·0058 450− | ·0030 882 |
| ·22 | ·0744 338 | ·0425 576 | ·0239 035− | ·0132 279 | ·0072 285− | ·0039 075+ |
| ·23 | ·0835 557 | ·0488 052 | ·0280 093 | ·0158 393 | ·0088 458 | ·0048 873 |
| ·24 | ·0932 512 | ·0555 915+ | ·0325 671 | ·0188 019 | ·0107 209 | ·0060 481 |
| ·25 | ·1035 156 | ·0629 272 | ·0375 977 | ·0221 405+ | ·0128 784 | ·0074 120 |
| ·26 | ·1143 424 | ·0708 212 | ·0431 203 | ·0258 797 | ·0153 436 | ·0090 017 |
| ·27 | ·1257 232 | ·0792 800 | ·0491 530 | ·0300 436 | ·0181 420 | ·0108 413 |
| ·28 | ·1376 478 | ·0883 088 | ·0557 124 | ·0346 555+ | ·0212 996 | ·0129 558 |
| ·29 | ·1501 045+ | ·0979 103 | ·0628 136 | ·0397 384 | ·0248 421 | ·0153 708 |
| ·30 | ·1630 800$^e$ | ·1080 855+ | ·0704 700$^e$ | ·0453 138 | ·0287 955$^e$ | ·0181 128 |
| ·31 | ·1765 593 | ·1188 335+ | ·0786 932 | ·0514 027 | ·0331 855− | ·0212 087 |
| ·32 | ·1905 263 | ·1301 513 | ·0874 932 | ·0580 245+ | ·0380 373 | ·0246 859 |
| ·33 | ·2049 631 | ·1420 341 | ·0968 779 | ·0651 975− | ·0433 757 | ·0285 720 |
| ·34 | ·2198 509 | ·1544 752 | ·1068 534 | ·0729 384 | ·0492 247 | ·0328 949 |
| ·35 | ·2351 694 | ·1674 658 | ·1174 239 | ·0812 625+ | ·0556 075+ | ·0376 820 |
| ·36 | ·2508 973 | ·1809 954 | ·1285 914 | ·0901 833 | ·0625 462 | ·0429 610 |
| ·37 | ·2670 122 | ·1950 518 | ·1403 559 | ·0997 123 | ·0700 617 | ·0487 587 |
| ·38 | ·2834 907 | ·2096 209 | ·1527 154 | ·1098 594 | ·0781 734 | ·0551 015− |
| ·39 | ·3003 084 | ·2246 868 | ·1656 655+ | ·1206 323 | ·0868 994 | ·0620 149 |
| ·40 | ·3174 400$^e$ | ·2402 319 | ·1792 000$^e$ | ·1320 365− | ·0962 560$^e$ | ·0695 236 |
| ·41 | ·3348 596 | ·2562 371 | ·1933 103 | ·1440 754 | ·1062 575+ | ·0776 507 |
| ·42 | ·3525 403 | ·2726 816 | ·2079 858 | ·1567 500+ | ·1169 164 | ·0864 182 |
| ·43 | ·3704 549 | ·2895 431 | ·2232 135+ | ·1700 592 | ·1282 428 | ·0958 464 |
| ·44 | ·3885 753 | ·3067 978 | ·2389 786 | ·1839 991 | ·1402 448 | ·1059 537 |
| ·45 | ·4068 731 | ·3244 205− | ·2552 639 | ·1985 635+ | ·1529 277 | ·1167 565+ |
| ·46 | ·4253 194 | ·3423 846 | ·2720 502 | ·2137 438 | ·1662 945+ | ·1282 691 |
| ·47 | ·4438 849 | ·3606 624 | ·2893 163 | ·2295 284 | ·1803 454 | ·1405 030 |
| ·48 | ·4625 400 | ·3792 249 | ·3070 388 | ·2459 036 | ·1950 779 | ·1534 675+ |
| ·49 | ·4812 550− | ·3980 420 | ·3251 924 | ·2628 528 | ·2104 864 | ·1671 689 |
| ·50 | ·5000 000$^e$ | ·4170 825+ | ·3437 500$^e$ | ·2803 568 | ·2265 625$^e$ | ·1816 104 |
| ·51 | ·5187 450+ | ·4363 144 | ·3626 824 | ·2983 938 | ·2432 946 | ·1967 923 |
| ·52 | ·5374 600 | ·4557 046 | ·3819 588 | ·3169 394 | ·2606 679 | ·2127 113 |
| ·53 | ·5561 151 | ·4752 195+ | ·4015 466 | ·3359 666 | ·2786 646 | ·2293 608 |
| ·54 | ·5746 806 | ·4948 247 | ·4214 115− | ·3554 459 | ·2972 634 | ·2467 304 |
| ·55 | ·5931 269 | ·5144 852 | ·4415 177 | ·3753 452 | ·3164 401 | ·2648 061 |
| ·56 | ·6114 247 | ·5341 656 | ·4618 279 | ·3956 300 | ·3361 667 | ·2835 701 |
| ·57 | ·6295 451 | ·5538 299 | ·4823 037 | ·4162 635− | ·3564 123 | ·3030 004 |
| ·58 | ·6474 597 | ·5734 422 | ·5029 051 | ·4372 064 | ·3771 426 | ·3230 712 |
| ·59 | ·6651 404 | ·5929 660 | ·5235 912 | ·4584 175+ | ·3983 201 | ·3437 523 |
| ·60 | ·6825 600$^e$ | ·6123 651 | ·5443 200$^e$ | ·4798 533 | ·4199 040$^e$ | ·3650 098 |

$x$ = ·61 to 1·00 $q = 3$ $p$ = 3 to 5·5

| $x$ | $p = 3$ | $p = 3·5$ | $p = 4$ | $p = 4·5$ | $p = 5$ | $p = 5·5$ |
|---|---|---|---|---|---|---|
| $B(p, q)$ = | ·3333 3333 $\times\frac{1}{10}$ | ·2308 8023 $\times\frac{1}{10}$ | ·1666 6667 $\times\frac{1}{10}$ | ·1243 2012 $\times\frac{1}{10}$ | ·9523 8095 $\overset{+}{\times}\frac{1}{10^2}$ | ·7459 2075 $\overset{-}{\times}\frac{1}{10^2}$ |
| ·61 | ·6996 916 | ·6316 032 | ·5650 488 | ·5014 684 | ·4418 506 | ·3868 052 |
| ·62 | ·7165 093 | ·6506 442 | ·5857 340 | ·5232 156 | ·4641 129 | ·4090 962 |
| ·63 | ·7329 878 | ·6694 521 | ·6063 315$^-$ | ·5450 459 | ·4866 413 | ·4318 361 |
| ·64 | ·7491 027 | ·6879 917 | ·6267 968 | ·5669 088 | ·5093 831 | ·4549 745$^-$ |
| ·65 | ·7648 306 | ·7062 279 | ·6470 852 | ·5887 528 | ·5322 833 | ·4784 566 |
| ·66 | ·7801 491 | ·7241 266 | ·6671 517 | ·6105 247 | ·5552 843 | ·5022 241 |
| ·67 | ·7950 369 | ·7416 542 | ·6869 517 | ·6321 707 | ·5783 261 | ·5262 150$^-$ |
| ·68 | ·8094 737 | ·7587 783 | ·7064 407 | ·6536 362 | ·6013 469 | ·5503 634 |
| ·69 | ·8234 407 | ·7754 672 | ·7255 745$^+$ | ·6748 662 | ·6242 831 | ·5746 006 |
| ·70 | ·8369 200$^e$ | ·7916 906 | ·7443 100$^e$ | ·6958 053 | ·6470 695$^e$ | ·5988 546 |
| ·71 | ·8498 955$^-$ | ·8074 195$^-$ | ·7626 045$^+$ | ·7163 981 | ·6696 397 | ·6230 506 |
| ·72 | ·8623 522 | ·8226 263 | ·7804 168 | ·7365 897 | ·6919 265$^+$ | ·6471 117 |
| ·73 | ·8742 768 | ·8372 849 | ·7977 066 | ·7563 257 | ·7138 622 | ·6709 586 |
| ·74 | ·8856 576 | ·8513 711 | ·8144 354 | ·7755 524 | ·7353 788 | ·6945 107 |
| ·75 | ·8964 844 | ·8648 625$^+$ | ·8305 664 | ·7942 178 | ·7564 087 | ·7176 860 |
| ·76 | ·9067 488 | ·8777 387 | ·8460 648 | ·8122 711 | ·7768 850$^-$ | ·7404 022 |
| ·77 | ·9164 443 | ·8899 815$^-$ | ·8608 980 | ·8296 636 | ·7967 419 | ·7625 766 |
| ·78 | ·9255 662 | ·9015 749 | ·8750 359 | ·8463 488 | ·8159 155$^-$ | ·7841 274 |
| ·79 | ·9341 117 | ·9125 055$^+$ | ·8884 513 | ·8622 831 | ·8343 438 | ·8049 737 |
| ·80 | ·9420 800$^e$ | ·9227 626 | ·9011 200$^e$ | ·8774 259 | ·8519 680$^e$ | ·8250 368 |
| ·81 | ·9494 725$^-$ | ·9323 382 | ·9130 210 | ·8917 403 | ·8687 323 | ·8442 407 |
| ·82 | ·9562 927 | ·9412 270 | ·9241 369 | ·9051 931 | ·8845 853 | ·8625 130 |
| ·83 | ·9625 462 | ·9494 273 | ·9344 543 | ·9177 560 | ·8994 799 | ·8797 857 |
| ·84 | ·9682 413 | ·9569 402 | ·9439 641 | ·9294 054 | ·9133 749 | ·8959 964 |
| ·85 | ·9733 881 | ·9637 705$^+$ | ·9526 614 | ·9401 230 | ·9262 348 | ·9110 891 |
| ·86 | ·9779 997 | ·9699 265$^+$ | ·9605 463 | ·9498 969 | ·9380 315$^-$ | ·9250 156 |
| ·87 | ·9820 914 | ·9754 203 | ·9676 241 | ·9587 213 | ·9487 442 | ·9377 361 |
| ·88 | ·9856 811 | ·9802 680 | ·9739 053 | ·9665 977 | ·9583 612 | ·9492 212 |
| ·89 | ·9887 895$^+$ | ·9844 895$^+$ | ·9794 064 | ·9735 351 | ·9668 799 | ·9594 526 |
| ·90 | ·9914 400$^e$ | ·9881 094 | ·9841 500$^e$ | ·9795 510 | ·9743 085$^e$ | ·9684 250$^+$ |
| ·91 | ·9936 587 | ·9911 564 | ·9881 652 | ·9846 713 | ·9806 665$^+$ | ·9761 470 |
| ·92 | ·9954 747 | ·9936 642 | ·9914 879 | ·9889 319 | ·9859 860 | ·9826 432 |
| ·93 | ·9969 201 | ·9956 709 | ·9941 611 | ·9923 783 | ·9903 124 | ·9879 554 |
| ·94 | ·9980 297 | ·9972 198 | ·9962 357 | ·9950 673 | ·9937 060 | ·9921 447 |
| ·95 | ·9988 419 | ·9983 595$^-$ | ·9977 702 | ·9970 668 | ·9962 430 | ·9952 930 |
| ·96 | ·9993 978 | ·9991 436 | ·9988 316 | ·9984 571 | ·9980 162 | ·9975 051 |
| ·97 | ·9997 420 | ·9996 317 | ·9994 956 | ·9993 314 | ·9991 370 | ·9989 106 |
| ·98 | ·9999 224 | ·9998 888 | ·9998 471 | ·9997 965$^+$ | ·9997 364 | ·9996 660 |
| ·99 | ·9999 901 | ·9999 858 | ·9999 804 | ·9999 739 | ·9999 660 | ·9999 568 |
| 1·00 | 1·0000 000 | 1·0000 000 | 1·0000 000 | 1·0000 000 | 1·0000 000 | 1·0000 000 |

$x$ = ·04 to ·60 $q = 3$ $p$ = 6 to 8·5

| $x$ | $p = 6$ | $p = 6·5$ | $p = 7$ | $p = 7·5$ | $p = 8$ | $p = 8·5$ |
|---|---|---|---|---|---|---|
| $B(p, q)$ = | ·5952 3810 × $\frac{1}{10^2}$ | ·4826 5460 × $\frac{1}{10^2}$ | ·3968 2540 × $\frac{1}{10^2}$ | ·3302 3736 × $\frac{1}{10^2}$ | ·2777 7778 × $\frac{1}{10^2}$ | ·2358 8383 × $\frac{1}{10^2}$ |
| ·04 | ·0000 001 | | | | | |
| ·05 | ·0000 004 | ·0000 001 | | | | |
| ·06 | ·0000 012 | ·0000 003 | ·0000 001 | | | |
| ·07 | ·0000 029 | ·0000 009 | ·0000 003 | ·0000 001 | | |
| ·08 | ·0000 064 | ·0000 020 | ·0000 007 | ·0000 002 | ·0000 001 | |
| ·09 | ·0000 127 | ·0000 043 | ·0000 015− | ·0000 005− | ·0000 002 | ·0000 001 |
| ·10 | ·0000 234 | ·0000 084 | ·0000 030 | ·0000 011 | ·0000 004 | ·0000 001 |
| ·11 | ·0000 407 | ·0000 153 | ·0000 057 | ·0000 021 | ·0000 008 | ·0000 003 |
| ·12 | ·0000 673 | ·0000 265− | ·0000 103 | ·0000 040 | ·0000 015+ | ·0000 006 |
| ·13 | ·0001 067 | ·0000 437 | ·0000 177 | ·0000 072 | ·0000 029 | ·0000 011 |
| ·14 | ·0001 633 | ·0000 694 | ·0000 292 | ·0000 122 | ·0000 051 | ·0000 021 |
| ·15 | ·0002 423 | ·0001 065− | ·0000 464 | ·0000 201 | ·0000 087 | ·0000 037 |
| ·16 | ·0003 499 | ·0001 588 | ·0000 715+ | ·0000 320 | ·0000 142 | ·0000 063 |
| ·17 | ·0004 935+ | ·0002 308 | ·0001 071 | ·0000 494 | ·0000 226 | ·0000 103 |
| ·18 | ·0006 816 | ·0003 278 | ·0001 565+ | ·0000 742 | ·0000 350+ | ·0000 164 |
| ·19 | ·0009 239 | ·0004 564 | ·0002 238 | ·0001 090 | ·0000 528 | ·0000 254 |
| ·20 | ·0012 314 | ·0006 240 | ·0003 139 | ·0001 568 | ·0000 779 | ·0000 385+ |
| ·21 | ·0016 164 | ·0008 390 | ·0004 323 | ·0002 213 | ·0001 127 | ·0000 570 |
| ·22 | ·0020 926 | ·0011 114 | ·0005 861 | ·0003 070 | ·0001 599 | ·0000 829 |
| ·23 | ·0026 751 | ·0014 524 | ·0007 828 | ·0004 192 | ·0002 232 | ·0001 182 |
| ·24 | ·0033 805+ | ·0018 742 | ·0010 316 | ·0005 642 | ·0003 068 | ·0001 660 |
| ·25 | ·0042 267 | ·0023 909 | ·0013 428 | ·0007 493 | ·0004 158 | ·0002 295+ |
| ·26 | ·0052 329 | ·0030 177 | ·0017 279 | ·0009 831 | ·0005 562 | ·0003 130 |
| ·27 | ·0064 199 | ·0037 714 | ·0021 999 | ·0012 752 | ·0007 350+ | ·0004 215− |
| ·28 | ·0078 097 | ·0046 704 | ·0027 735− | ·0016 367 | ·0009 605− | ·0005 608 |
| ·29 | ·0094 256 | ·0057 345+ | ·0034 646 | ·0020 802 | ·0012 420 | ·0007 378 |
| ·30 | ·0112 922 | ·0069 851 | ·0042 909 | ·0026 196 | ·0015 904 | ·0009 607 |
| ·31 | ·0134 351 | ·0084 448 | ·0052 716 | ·0032 706 | ·0020 179 | ·0012 388 |
| ·32 | ·0158 811 | ·0101 381 | ·0064 277 | ·0040 504 | ·0025 384 | ·0015 829 |
| ·33 | ·0186 577 | ·0120 906 | ·0077 818 | ·0049 782 | ·0031 673 | ·0020 052 |
| ·34 | ·0217 935− | ·0143 292 | ·0093 580 | ·0060 746 | ·0039 219 | ·0025 196 |
| ·35 | ·0253 175+ | ·0168 822 | ·0111 822 | ·0073 623 | ·0048 213 | ·0031 419 |
| ·36 | ·0292 594 | ·0197 791 | ·0132 818 | ·0088 658 | ·0058 864 | ·0038 893 |
| ·37 | ·0336 492 | ·0230 502 | ·0156 858 | ·0106 113 | ·0071 403 | ·0047 816 |
| ·38 | ·0385 171 | ·0267 269 | ·0184 246 | ·0126 269 | ·0086 079 | ·0058 401 |
| ·39 | ·0438 932 | ·0308 411 | ·0215 299 | ·0149 425+ | ·0103 163 | ·0070 886 |
| ·40 | ·0498 074 | ·0354 256 | ·0250 348 | ·0175 898 | ·0122 946 | ·0085 529 |
| ·41 | ·0562 892 | ·0405 133 | ·0289 732 | ·0206 019 | ·0145 738 | ·0102 612 |
| ·42 | ·0633 676 | ·0461 373 | ·0333 803 | ·0240 137 | ·0171 871 | ·0122 440 |
| ·43 | ·0710 705− | ·0523 308 | ·0382 916 | ·0278 616 | ·0201 696 | ·0145 340 |
| ·44 | ·0794 247 | ·0591 265+ | ·0437 436 | ·0321 828 | ·0235 583 | ·0171 662 |
| ·45 | ·0884 559 | ·0665 568 | ·0497 728 | ·0370 162 | ·0273 918 | ·0201 780 |
| ·46 | ·0981 878 | ·0746 531 | ·0564 157 | ·0424 011 | ·0317 105+ | ·0236 089 |
| ·47 | ·1086 426 | ·0834 458 | ·0637 089 | ·0483 776 | ·0365 560 | ·0275 003 |
| ·48 | ·1198 402 | ·0929 639 | ·0716 881 | ·0549 862 | ·0419 713 | ·0318 958 |
| ·49 | ·1317 981 | ·1032 346 | ·0803 884 | ·0622 675+ | ·0480 003 | ·0368 406 |
| ·50 | ·1445 312 | ·1142 834 | ·0898 437 | ·0702 618 | ·0546 875$^e$ | ·0423 815+ |
| ·51 | ·1580 516 | ·1261 331 | ·1000 864 | ·0790 088 | ·0620 778 | ·0485 665+ |
| ·52 | ·1723 681 | ·1388 041 | ·1111 469 | ·0885 472 | ·0702 161 | ·0554 447 |
| ·53 | ·1874 861 | ·1523 138 | ·1230 533 | ·0989 143 | ·0791 470 | ·0630 655− |
| ·54 | ·2034 075+ | ·1666 763 | ·1358 313 | ·1101 457 | ·0889 141 | ·0714 788 |
| ·55 | ·2201 303 | ·1819 019 | ·1495 031 | ·1222 746 | ·0995 597 | ·0807 342 |
| ·56 | ·2376 483 | ·1979 971 | ·1640 878 | ·1353 313 | ·1111 243 | ·0908 804 |
| ·57 | ·2559 510 | ·2149 641 | ·1796 003 | ·1493 433 | ·1236 462 | ·1019 650− |
| ·58 | ·2750 235− | ·2328 005+ | ·1960 513 | ·1643 337 | ·1371 607 | ·1140 335− |
| ·59 | ·2948 461 | ·2514 990 | ·2134 467 | ·1803 219 | ·1516 993 | ·1271 290 |
| ·60 | ·3153 946 | ·2710 469 | ·2317 870 | ·1973 221 | ·1672 898 | ·1412 913 |

$x$ = ·61 to 1·00 $q = 3$ $p$ = 6 to 8·5

| $x$ | $p = 6$ | $p = 6·5$ | $p = 7$ | $p = 7·5$ | $p = 8$ | $p = 8·5$ |
|---|---|---|---|---|---|---|
| $B(p, q)$ = | ·5952 3810 × $\frac{1}{10^2}$ | ·4826 5460 × $\frac{1}{10^2}$ | ·3968 2540 × $\frac{1}{10^2}$ | ·3302 3736 × $\frac{1}{10^2}$ | ·2777 7778 × $\frac{1}{10^2}$ | ·2358 8383 × $\frac{1}{10^2}$ |
| ·61 | ·3366 393 | ·2914 263 | ·2510 674 | ·2153 432 | ·1839 547 | ·1565 564 |
| ·62 | ·3585 458 | ·3126 134 | ·2712 770 | ·2343 882 | ·2017 113 | ·1729 553 |
| ·63 | ·3810 745⁻ | ·3345 787 | ·2923 984 | ·2544 535⁺ | ·2205 708 | ·1905 137 |
| ·64 | ·4041 805⁻ | ·3572 863 | ·3144 075⁺ | ·2755 288 | ·2405 373 | ·2092 507 |
| ·65 | ·4278 137 | ·3806 941 | ·3372 733 | ·2975 960 | ·2616 074 | ·2291 786 |
| ·66 | ·4519 187 | ·4047 537 | ·3609 571 | ·3206 292 | ·2837 696 | ·2503 011 |
| ·67 | ·4764 353 | ·4294 099 | ·3854 128 | ·3445 939 | ·3070 034 | ·2726 133 |
| ·68 | ·5012 977 | ·4546 013 | ·4105 864 | ·3694 467 | ·3312 788 | ·2961 002 |
| ·69 | ·5264 356 | ·4802 598 | ·4364 159 | ·3951 353 | ·3565 555⁺ | ·3207 365⁺ |
| ·70 | ·5517 738 | ·5063 107 | ·4628 312 | ·4215 975⁻ | ·3827 828 | ·3464 851 |
| ·71 | ·5772 327 | ·5326 733 | ·4897 540 | ·4487 614 | ·4098 985⁺ | ·3732 967 |
| ·72 | ·6027 284 | ·5592 605⁺ | ·5170 982 | ·4765 453 | ·4378 290 | ·4011 090 |
| ·73 | ·6281 732 | ·5859 796 | ·5447 693 | ·5048 573 | ·4664 888 | ·4298 463 |
| ·74 | ·6534 761 | ·6127 322 | ·5726 655⁺ | ·5335 958 | ·4957 800 | ·4594 186 |
| ·75 | ·6785 431 | ·6394 149 | ·6006 775⁻ | ·5626 490 | ·5255 928 | ·4897 214 |
| ·76 | ·7032 777 | ·6659 199 | ·6286 889 | ·5918 960 | ·5558 051 | ·5206 356 |
| ·77 | ·7275 817 | ·6921 354 | ·6565 772 | ·6212 065⁻ | ·5862 827 | ·5520 271 |
| ·78 | ·7513 559 | ·7179 464 | ·6842 140 | ·6504 419 | ·6168 803 | ·5837 474 |
| ·79 | ·7745 009 | ·7432 358 | ·7114 664 | ·6794 559 | ·6474 414 | ·6156 334 |
| ·80 | ·7969 178 | ·7678 851 | ·7381 975⁺ | ·7080 955⁺ | ·6777 995⁺ | ·6475 088 |
| ·81 | ·8185 090 | ·7917 753 | ·7642 679 | ·7362 023 | ·7077 796 | ·6791 845⁻ |
| ·82 | ·8391 800 | ·8147 888 | ·7895 369 | ·7636 138 | ·7371 990 | ·7104 601 |
| ·83 | ·8588 397 | ·8368 102 | ·8138 644 | ·7901 653 | ·7658 695⁻ | ·7411 259 |
| ·84 | ·8774 020 | ·8577 280 | ·8371 123 | ·8156 916 | ·7935 995⁻ | ·7709 649 |
| ·85 | ·8947 872 | ·8774 362 | ·8591 466 | ·8400 298 | ·8201 965⁻ | ·7997 554 |
| ·86 | ·9109 236 | ·8958 366 | ·8798 399 | ·8630 214 | ·8454 702 | ·8272 748 |
| ·87 | ·9257 486 | ·9128 400 | ·8990 736 | ·8845 160 | ·8692 358 | ·8533 029 |
| ·88 | ·9392 108 | ·9283 694 | ·9167 411 | ·9043 738 | ·8913 182 | ·8776 267 |
| ·89 | ·9512 719 | ·9423 616 | ·9327 504 | ·9224 703 | ·9115 565⁺ | ·9000 462 |
| ·90 | ·9619 082 | ·9547 704 | ·9470 279 | ·9387 000 | ·9298 092 | ·9203 798 |
| ·91 | ·9711 132 | ·9655 691 | ·9595 219 | ·9529 814 | ·9459 600 | ·9384 720 |
| ·92 | ·9788 995⁺ | ·9747 537 | ·9702 068 | ·9652 621 | ·9599 246 | ·9542 011 |
| ·93 | ·9853 013 | ·9823 462 | ·9790 877 | ·9755 247 | ·9716 579 | ·9674 889 |
| ·94 | ·9903 771 | ·9883 984 | ·9862 047 | ·9837 933 | ·9811 622 | ·9783 102 |
| ·95 | ·9942 118 | ·9929 950⁺ | ·9916 390 | ·9901 403 | ·9884 964 | ·9867 051 |
| ·96 | ·9969 203 | ·9962 588 | ·9955 176 | ·9946 942 | ·9937 863 | ·9927 917 |
| ·97 | ·9986 501 | ·9983 540 | ·9980 204 | ·9976 479 | ·9972 350⁺ | ·9967 804 |
| ·98 | ·9995 845⁺ | ·9994 915⁻ | ·9993 861 | ·9992 679 | ·9991 361 | ·9989 903 |
| ·99 | ·9999 461 | ·9999 337 | ·9999 197 | ·9999 039 | ·9998 862 | ·9998 664 |
| 1·00 | 1·0000 000 | 1·0000 000 | 1·0000 000 | 1·0000 000 | 1·0000 000 | 1·0000 000 |

$x$ = ·11 to ·70 $q = 3$ $p$ = 9 to 12

| | $p = 9$ | $p = 9{\cdot}5$ | $p = 10$ | $p = 10{\cdot}5$ | $p = 11$ | $p = 12$ |
|---|---|---|---|---|---|---|
| $B(p, q) =$ | ·2020 2020 $\times \frac{1}{10^2}$ | ·1743 4892 $\times \frac{1}{10^2}$ | ·1515 1515$^{+}$ $\times \frac{1}{10^2}$ | ·1325 0518 $\times \frac{1}{10^2}$ | ·1165 5012 $\times \frac{1}{10^2}$ | ·9157 5092 $\times \frac{1}{10^3}$ |
| $x$ | | | | | | |
| ·11 | ·0000 001 | | | | | |
| ·12 | ·0000 002 | ·0000 001 | | | | |
| ·13 | ·0000 005$^{-}$ | ·0000 002 | ·0000 001 | | | |
| ·14 | ·0000 009 | ·0000 004 | ·0000 001 | ·0000 001 | | |
| ·15 | ·0000 016 | ·0000 007 | ·0000 003 | ·0000 001 | ·0000 001 | |
| ·16 | ·0000 028 | ·0000 012 | ·0000 005$^{+}$ | ·0000 002 | ·0000 001 | |
| ·17 | ·0000 047 | ·0000 021 | ·0000 010 | ·0000 004 | ·0000 002 | |
| ·18 | ·0000 077 | ·0000 036 | ·0000 016 | ·0000 008 | ·0000 003 | ·0000 001 |
| ·19 | ·0000 122 | ·0000 058 | ·0000 028 | ·0000 013 | ·0000 006 | ·0000 001 |
| ·20 | ·0000 189 | ·0000 093 | ·0000 045$^{+}$ | ·0000 022 | ·0000 011 | ·0000 002 |
| ·21 | ·0000 287 | ·0000 144 | ·0000 072 | ·0000 036 | ·0000 018 | ·0000 004 |
| ·22 | ·0000 427 | ·0000 219 | ·0000 112 | ·0000 057 | ·0000 029 | ·0000 007 |
| ·23 | ·0000 623 | ·0000 327 | ·0000 171 | ·0000 089 | ·0000 046 | ·0000 012 |
| ·24 | ·0000 894 | ·0000 479 | ·0000 256 | ·0000 136 | ·0000 072 | ·0000 020 |
| ·25 | ·0001 261 | ·0000 690 | ·0000 376 | ·0000 204 | ·0000 111 | ·0000 032 |
| ·26 | ·0001 754 | ·0000 978 | ·0000 544 | ·0000 301 | ·0000 166 | ·0000 050$^{+}$ |
| ·27 | ·0002 406 | ·0001 368 | ·0000 774 | ·0000 437 | ·0000 246 | ·0000 077 |
| ·28 | ·0003 259 | ·0001 886 | ·0001 087 | ·0000 625$^{-}$ | ·0000 358 | ·0000 116 |
| ·29 | ·0004 363 | ·0002 569 | ·0001 507 | ·0000 881 | ·0000 513 | ·0000 173 |
| ·30 | ·0005 777 | ·0003 459 | ·0002 064 | ·0001 227 | ·0000 727 | ·0000 253 |
| ·31 | ·0007 571 | ·0004 608 | ·0002 794 | ·0001 688 | ·0001 017 | ·0000 366 |
| ·32 | ·0009 826 | ·0006 075$^{-}$ | ·0003 742 | ·0002 297 | ·0001 405$^{+}$ | ·0000 521 |
| ·33 | ·0012 638 | ·0007 933 | ·0004 961 | ·0003 091 | ·0001 920 | ·0000 735$^{-}$ |
| ·34 | ·0016 116 | ·0010 266 | ·0006 515$^{-}$ | ·0004 120 | ·0002 598 | ·0001 024 |
| ·35 | ·0020 384 | ·0013 171 | ·0008 479 | ·0005 440 | ·0003 479 | ·0001 411 |
| ·36 | ·0025 585$^{-}$ | ·0016 763 | ·0010 942 | ·0007 118 | ·0004 616 | ·0001 925$^{+}$ |
| ·37 | ·0031 880 | ·0021 171 | ·0014 007 | ·0009 236 | ·0006 072 | ·0002 601 |
| ·38 | ·0039 450$^{+}$ | ·0026 543 | ·0017 794 | ·0011 888 | ·0007 918 | ·0003 483 |
| ·39 | ·0048 497 | ·0033 048 | ·0022 439 | ·0015 185$^{+}$ | ·0010 245$^{-}$ | ·0004 624 |
| ·40 | ·0059 245$^{-}$ | ·0040 877 | ·0028 102 | ·0019 255$^{+}$ | ·0013 153 | ·0006 087 |
| ·41 | ·0071 941 | ·0050 240 | ·0034 960 | ·0024 247 | ·0016 766 | ·0007 950$^{-}$ |
| ·42 | ·0086 857 | ·0061 376 | ·0043 216 | ·0030 330 | ·0021 221 | ·0010 304 |
| ·43 | ·0104 290 | ·0074 547 | ·0053 098 | ·0037 698 | ·0026 683 | ·0013 259 |
| ·44 | ·0124 564 | ·0090 043 | ·0064 860 | ·0046 570 | ·0033 337 | ·0016 945$^{-}$ |
| ·45 | ·0148 026 | ·0108 180 | ·0078 785$^{-}$ | ·0057 192 | ·0041 394 | ·0021 510 |
| ·46 | ·0175 050$^{+}$ | ·0129 305$^{-}$ | ·0095 184 | ·0069 843 | ·0051 097 | ·0027 130 |
| ·47 | ·0206 038 | ·0153 792 | ·0114 400 | ·0084 829 | ·0062 717 | ·0034 009 |
| ·48 | ·0241 413 | ·0182 045$^{+}$ | ·0136 810 | ·0102 492 | ·0076 559 | ·0042 380 |
| ·49 | ·0281 626 | ·0214 499 | ·0162 820 | ·0123 207 | ·0092 962 | ·0052 508 |
| ·50 | ·0327 148 | ·0251 613 | ·0192 871 | ·0147 386 | ·0112 305$^{-}$ | ·0064 697 |
| ·51 | ·0378 473 | ·0293 879 | ·0227 437 | ·0175 476 | ·0135 002 | ·0079 289 |
| ·52 | ·0436 112 | ·0341 812 | ·0267 022 | ·0207 963 | ·0161 510 | ·0096 669 |
| ·53 | ·0500 591 | ·0395 950$^{+}$ | ·0312 165$^{+}$ | ·0245 369 | ·0192 327 | ·0117 264 |
| ·54 | ·0572 449 | ·0456 857 | ·0363 433 | ·0288 252 | ·0227 991 | ·0141 554 |
| ·55 | ·0652 235$^{+}$ | ·0525 112 | ·0421 420 | ·0337 206 | ·0269 082 | ·0170 062 |
| ·56 | ·0740 499 | ·0601 308 | ·0486 745$^{-}$ | ·0392 859 | ·0316 222 | ·0203 367 |
| ·57 | ·0837 790 | ·0686 051 | ·0560 047 | ·0455 868 | ·0370 072 | ·0242 097 |
| ·58 | ·0944 650$^{-}$ | ·0779 948 | ·0641 984 | ·0526 919 | ·0431 329 | ·0286 935$^{+}$ |
| ·59 | ·1061 607 | ·0883 607 | ·0733 223 | ·0606 720 | ·0500 726 | ·0338 613 |
| ·60 | ·1189 168 | ·0997 626 | ·0834 433 | ·0695 996 | ·0579 024 | ·0397 916 |
| ·61 | ·1327 812 | ·1122 588 | ·0946 285$^{+}$ | ·0795 485$^{-}$ | ·0667 007 | ·0465 674 |
| ·62 | ·1477 979 | ·1259 049 | ·1069 435$^{+}$ | ·0905 924 | ·0765 479 | ·0542 761 |
| ·63 | ·1640 065$^{+}$ | ·1407 534 | ·1204 520 | ·1028 046 | ·0875 249 | ·0630 091 |
| ·64 | ·1814 410 | ·1568 521 | ·1352 146 | ·1162 569 | ·0997 128 | ·0728 604 |
| ·65 | ·2001 289 | ·1742 434 | ·1512 876 | ·1310 181 | ·1131 914 | ·0839 266 |
| ·66 | ·2200 899 | ·1929 630 | ·1687 217 | ·1471 529 | ·1280 380 | ·0963 048 |
| ·67 | ·2413 356 | ·2130 387 | ·1875 609 | ·1647 207 | ·1443 261 | ·1100 919 |
| ·68 | ·2638 673 | ·2344 892 | ·2078 409 | ·1837 736 | ·1621 233 | ·1253 830 |
| ·69 | ·2876 760 | ·2573 225$^{+}$ | ·2295 875$^{+}$ | ·2043 551 | ·1814 903 | ·1422 692 |
| ·70 | ·3127 405$^{-}$ | ·2815 350$^{-}$ | ·2528 153 | ·2264 983 | ·2024 783 | ·1608 358 |

$x$ = ·71 to 1·00 $q = 3$ $p$ = 9 to 12

| $x$ | $p = 9$ | $p = 9·5$ | $p = 10$ | $p = 10·5$ | $p = 11$ | $p = 12$ |
|---|---|---|---|---|---|---|
| $B(p, q) =$ | ·2020 2020 × $\frac{1}{10^2}$ | ·1743 4892 × $\frac{1}{10^2}$ | ·1515 1515$^+$ × $\frac{1}{10^2}$ | ·1325 0518 × $\frac{1}{10^2}$ | ·1165 5012 × $\frac{1}{10^2}$ | ·9157 5092 × $\frac{1}{10^3}$ |
| ·71 | ·3390 267 | ·3071 096 | ·2775 258 | ·2502 236 | ·2251 270 | ·1811 596 |
| ·72 | ·3664 868 | ·3340 151 | ·3037 057 | ·2755 374 | ·2494 628 | ·2033 070 |
| ·73 | ·3950 578 | ·3622 040 | ·3313 255⁻ | ·3024 295⁺ | ·2754 960 | ·2273 303 |
| ·74 | ·4246 609 | ·3916 119 | ·3603 376 | ·3308 713 | ·3032 185⁺ | ·2532 653 |
| ·75 | ·4552 009 | ·4221 561 | ·3906 750⁺ | ·3608 139 | ·3326 017 | ·2811 276 |
| ·76 | ·4865 654 | ·4537 346 | ·4222 494 | ·3921 858 | ·3635 933 | ·3109 094 |
| ·77 | ·5186 243 | ·4862 253 | ·4549 502 | ·4248 912 | ·3961 154 | ·3425 756 |
| ·78 | ·5512 299 | ·5194 852 | ·4886 432 | ·4588 083 | ·4300 620 | ·3760 608 |
| ·79 | ·5842 166 | ·5533 502 | ·5231 696 | ·4937 878 | ·4652 970 | ·4112 651 |
| ·80 | ·6174 015⁺ | ·5876 349 | ·5583 457 | ·5296 517 | ·5016 522 | ·4480 510 |
| ·81 | ·6505 852 | ·6221 331 | ·5939 627 | ·5661 927 | ·5389 257 | ·4862 403 |
| ·82 | ·6835 525⁺ | ·6566 185⁻ | ·6297 869 | ·6031 736 | ·5768 815⁻ | ·5256 113 |
| ·83 | ·7160 746 | ·6908 461 | ·6655 606 | ·6403 281 | ·6152 486 | ·5658 971 |
| ·84 | ·7479 110 | ·7245 543 | ·7010 041 | ·6773 621 | ·6537 220 | ·6067 838 |
| ·85 | ·7788 120 | ·7574 674 | ·7358 181 | ·7139 552 | ·6919 643 | ·6479 112 |
| ·86 | ·8085 227 | ·7892 993 | ·7696 869 | ·7497 648 | ·7296 083 | ·6888 739 |
| ·87 | ·8367 871 | ·8197 580 | ·8022 834 | ·7844 299 | ·7662 615⁺ | ·7292 245⁻ |
| ·88 | ·8633 530 | ·8485 511 | ·8332 749 | ·8175 779 | ·8015 124 | ·7684 795⁻ |
| ·89 | ·8879 783 | ·8753 928 | ·8623 305⁻ | ·8488 321 | ·8349 386 | ·8061 273 |
| ·90 | ·9104 381 | ·9000 120 | ·8891 300 | ·8778 218 | ·8661 172 | ·8416 400 |
| ·91 | ·9305 334 | ·9221 616 | ·9133 755⁺ | ·9041 945⁺ | ·8946 391 | ·8744 890 |
| ·92 | ·9481 000 | ·9416 307 | ·9348 040 | ·9276 313 | ·9201 251 | ·9041 652 |
| ·93 | ·9630 207 | ·9582 572 | ·9532 032 | ·9478 642 | ·9422 468 | ·9302 047 |
| ·94 | ·9752 372 | ·9719 434 | ·9684 300 | ·9646 986 | ·9607 515⁻ | ·9522 214 |
| ·95 | ·9847 647 | ·9826 739 | ·9804 317 | ·9780 379 | ·9754 922 | ·9699 464 |
| ·96 | ·9917 087 | ·9905 356 | ·9892 710 | ·9879 137 | ·9864 627 | ·9832 766 |
| ·97 | ·9962 828 | ·9957 410 | ·9951 539 | ·9945 204 | ·9938 398 | ·9923 333 |
| ·98 | ·9988 298 | ·9986 542 | ·9984 630 | ·9982 556 | ·9980 316 | ·9975 319 |
| ·99 | ·9998 446 | ·9998 206 | ·9997 944 | ·9997 658 | ·9997 347 | ·9996 649 |
| 1·00 | 1·0000 000 | 1·0000 000 | 1·0000 000 | 1·0000 000 | 1·0000 000 | 1·0000 000 |

$x$ = ·20 to ·80 $q = 3$ $p$ = 13 to 18

| $x$ | $p = 13$ | $p = 14$ | $p = 15$ | $p = 16$ | $p = 17$ | $p = 18$ |
|---|---|---|---|---|---|---|
| $B(p, q) =$ | ·7326 0073 × $\frac{1}{10^3}$ | ·5952 3810 × $\frac{1}{10^3}$ | ·4901 9608 × $\frac{1}{10^3}$ | ·4084 9673 × $\frac{1}{10^3}$ | ·3439 9725 × $\frac{1}{10^3}$ | ·2923 9766 × $\frac{1}{10^3}$ |
| ·20 | ·0000 001 | | | | | |
| ·21 | ·0000 001 | | | | | |
| ·22 | ·0000 002 | | | | | |
| ·23 | ·0000 003 | ·0000 001 | | | | |
| ·24 | ·0000 006 | ·0000 002 | | | | |
| ·25 | ·0000 009 | ·0000 003 | ·0000 001 | | | |
| ·26 | ·0000 015⁻ | ·0000 004 | ·0000 001 | | | |
| ·27 | ·0000 024 | ·0000 007 | ·0000 002 | ·0000 001 | | |
| ·28 | ·0000 037 | ·0000 012 | ·0000 004 | ·0000 001 | | |
| ·29 | ·0000 058 | ·0000 019 | ·0000 006 | ·0000 002 | ·0000 001 | |
| ·30 | ·0000 087 | ·0000 030 | ·0000 010 | ·0000 003 | ·0000 001 | |
| ·31 | ·0000 130 | ·0000 046 | ·0000 016 | ·0000 006 | ·0000 002 | ·0000 001 |
| ·32 | ·0000 192 | ·0000 070 | ·0000 025⁺ | ·0000 009 | ·0000 003 | ·0000 001 |
| ·33 | ·0000 278 | ·0000 104 | ·0000 039 | ·0000 014 | ·0000 005⁺ | ·0000 002 |
| ·34 | ·0000 399 | ·0000 154 | ·0000 059 | ·0000 023 | ·0000 009 | ·0000 003 |
| ·35 | ·0000 566 | ·0000 225⁺ | ·0000 089 | ·0000 035⁻ | ·0000 014 | ·0000 005⁺ |
| ·36 | ·0000 795⁻ | ·0000 325⁺ | ·0000 132 | ·0000 053 | ·0000 021 | ·0000 009 |
| ·37 | ·0001 104 | ·0000 464 | ·0000 194 | ·0000 080 | ·0000 033 | ·0000 014 |
| ·38 | ·0001 517 | ·0000 655⁺ | ·0000 281 | ·0000 119 | ·0000 051 | ·0000 021 |
| ·39 | ·0002 066 | ·0000 915⁺ | ·0000 403 | ·0000 176 | ·0000 076 | ·0000 033 |
| ·40 | ·0002 789 | ·0001 267 | ·0000 571 | ·0000 256 | ·0000 114 | ·0000 050⁺ |
| ·41 | ·0003 733 | ·0001 738 | ·0000 803 | ·0000 368 | ·0000 168 | ·0000 076 |
| ·42 | ·0004 954 | ·0002 362 | ·0001 118 | ·0000 525⁺ | ·0000 245⁺ | ·0000 114 |
| ·43 | ·0006 525⁺ | ·0003 184 | ·0001 542 | ·0000 742 | ·0000 355⁻ | ·0000 169 |
| ·44 | ·0008 530 | ·0004 258 | ·0002 110 | ·0001 038 | ·0000 508 | ·0000 247 |
| ·45 | ·0011 070 | ·0005 650⁻ | ·0002 862 | ·0001 440 | ·0000 721 | ·0000 359 |
| ·46 | ·0014 268 | ·0007 441 | ·0003 852 | ·0001 981 | ·0001 013 | ·0000 515⁺ |
| ·47 | ·0018 268 | ·0009 731 | ·0005 146 | ·0002 703 | ·0001 412 | ·0000 734 |
| ·48 | ·0023 240 | ·0012 639 | ·0006 824 | ·0003 660 | ·0001 952 | ·0001 035⁺ |
| ·49 | ·0029 382 | ·0016 307 | ·0008 985⁻ | ·0004 918 | ·0002 677 | ·0001 449 |
| ·50 | ·0036 926 | ·0020 905⁻ | ·0011 749 | ·0006 561 | ·0003 643 | ·0002 012 |
| ·51 | ·0046 140 | ·0026 633 | ·0015 263 | ·0008 692 | ·0004 921 | ·0002 772 |
| ·52 | ·0057 331 | ·0033 729 | ·0019 702 | ·0011 436 | ·0006 600 | ·0003 790 |
| ·53 | ·0070 851 | ·0042 467 | ·0025 275⁻ | ·0014 948 | ·0008 790 | ·0005 143 |
| ·54 | ·0087 098 | ·0053 169 | ·0032 229 | ·0019 414 | ·0011 629 | ·0006 930 |
| ·55 | ·0106 524 | ·0066 202 | ·0040 857 | ·0025 059 | ·0015 283 | ·0009 274 |
| ·56 | ·0129 635⁻ | ·0081 993 | ·0051 502 | ·0032 150⁺ | ·0019 959 | ·0012 328 |
| ·57 | ·0156 994 | ·0101 023 | ·0064 561 | ·0041 007 | ·0025 903 | ·0016 281 |
| ·58 | ·0189 228 | ·0123 840 | ·0080 496 | ·0052 006 | ·0033 415⁺ | ·0021 364 |
| ·59 | ·0227 025⁺ | ·0151 060 | ·0099 837 | ·0065 586 | ·0042 853 | ·0027 861 |
| ·60 | ·0271 140 | ·0183 372 | ·0123 188 | ·0082 264 | ·0054 639 | ·0036 115⁻ |
| ·61 | ·0322 391 | ·0221 542 | ·0151 236 | ·0102 631 | ·0069 276 | ·0046 536 |
| ·62 | ·0381 662 | ·0266 415⁻ | ·0184 754 | ·0127 373 | ·0087 350⁺ | ·0059 617 |
| ·63 | ·0449 900 | ·0318 915⁻ | ·0224 606 | ·0157 269 | ·0109 544 | ·0075 940 |
| ·64 | ·0528 107 | ·0380 048 | ·0271 753 | ·0193 203 | ·0136 647 | ·0096 192 |
| ·65 | ·0617 341 | ·0450 897 | ·0327 254 | ·0236 169 | ·0169 564 | ·0121 177 |
| ·66 | ·0718 702 | ·0532 623 | ·0392 266 | ·0287 279 | ·0209 326 | ·0151 825⁻ |
| ·67 | ·0833 323 | ·0626 450⁻ | ·0468 044 | ·0347 762 | ·0257 099 | ·0189 208 |
| ·68 | ·0962 358 | ·0733 663 | ·0555 935⁺ | ·0418 966 | ·0314 185⁻ | ·0234 551 |
| ·69 | ·1106 963 | ·0855 593 | ·0657 370 | ·0502 360 | ·0382 033 | ·0289 240 |
| ·70 | ·1268 277 | ·0993 597 | ·0773 853 | ·0599 522 | ·0462 237 | ·0354 831 |
| ·71 | ·1447 400 | ·1149 039 | ·0906 940 | ·0712 132 | ·0556 528 | ·0433 053 |
| ·72 | ·1645 361 | ·1323 265⁻ | ·1058 225⁺ | ·0841 953 | ·0666 773 | ·0525 804 |
| ·73 | ·1863 092 | ·1517 569 | ·1229 303 | ·0990 812 | ·0794 950⁺ | ·0635 151 |
| ·74 | ·2101 390 | ·1733 158 | ·1421 739 | ·1160 562 | ·0943 132 | ·0763 305⁺ |
| ·75 | ·2360 878 | ·1971 111 | ·1637 024 | ·1353 050⁺ | ·1113 448 | ·0912 604 |
| ·76 | ·2641 963 | ·2232 326 | ·1876 526 | ·1570 064 | ·1308 039 | ·1085 472 |
| ·77 | ·2944 791 | ·2517 472 | ·2141 432 | ·1813 273 | ·1529 006 | ·1284 369 |
| ·78 | ·3269 198 | ·2826 928 | ·2432 676 | ·2084 157 | ·1778 332 | ·1511 725⁻ |
| ·79 | ·3614 657 | ·3160 716 | ·2750 872 | ·2383 925⁺ | ·2057 801 | ·1769 853 |
| ·80 | ·3980 232 | ·3518 437 | ·3096 225⁻ | ·2713 419 | ·2368 893 | ·2060 847 |

$x$ = ·81 to 1·00 $q = 3$ $p$ = 13 to 18

| $x$ | $p = 13$ | $p = 14$ | $p = 15$ | $p = 16$ | $p = 17$ | $p = 18$ |
|---|---|---|---|---|---|---|
| $B(p, q)$ = | ·7326 0073 × $\frac{1}{10^3}$ | ·5952 3810 × $\frac{1}{10^3}$ | ·4901 9608 × $\frac{1}{10^3}$ | ·4084 9673 × $\frac{1}{10^3}$ | ·3439 9725$^{-}$ × $\frac{1}{10^3}$ | ·2923 9766 × $\frac{1}{10^3}$ |
| ·81 | ·4364 525$^{+}$ | ·3899 202 | ·3468 445$^{-}$ | ·3073 010 | ·2712 670 | ·2386 456 |
| ·82 | ·4765 629 | ·4301 555$^{+}$ | ·3866 652 | ·3462 482 | ·3089 635$^{+}$ | ·2747 932 |
| ·83 | ·5181 084 | ·4723 415$^{-}$ | ·4289 283 | ·3880 910 | ·3499 592 | ·3145 863 |
| ·84 | ·5607 843 | ·5162 002 | ·4733 995$^{+}$ | ·4326 532 | ·3941 480 | ·3579 983 |
| ·85 | ·6042 252 | ·5613 793 | ·5197 576 | ·4796 620 | ·4413 206 | ·4048 963 |
| ·86 | ·6480 036 | ·6074 478 | ·5675 872 | ·5287 364 | ·4911 483 | ·4550 195$^{-}$ |
| ·87 | ·6916 319 | ·6538 947 | ·6163 731 | ·5793 769 | ·5431 668 | ·5079 578 |
| ·88 | ·7345 656 | ·7001 300 | ·6654 976 | ·6309 576 | ·5967 629 | ·5631 315$^{+}$ |
| ·89 | ·7762 116 | ·7454 904 | ·7142 426 | ·6827 240 | ·6511 660 | ·6197 751 |
| ·90 | ·8159 389 | ·7892 493 | ·7617 972 | ·7337 960 | ·7054 448 | ·6769 268 |
| ·91 | ·8530 963 | ·8306 340 | ·8072 732 | ·7831 804 | ·7585 154 | ·7334 296 |
| ·92 | ·8870 349 | ·8688 504 | ·8497 307 | ·8297 952 | ·8091 620 | ·7879 462 |
| ·93 | ·9171 390 | ·9031 186 | ·8882 169 | ·8725 105$^{-}$ | ·8560 776 | ·8389 971 |
| ·94 | ·9428 667 | ·9327 204 | ·9218 205$^{-}$ | ·9102 084 | ·8979 286 | ·8850 276 |
| ·95 | ·9637 998 | ·9570 621 | ·9497 470 | ·9418 711 | ·9334 536 | ·9245 163 |
| ·96 | ·9797 082 | ·9757 555$^{-}$ | ·9714 188 | ·9667 005$^{-}$ | ·9616 047 | ·9561 372 |
| ·97 | ·9906 286 | ·9887 205$^{+}$ | ·9866 054 | ·9842 801 | ·9817 426 | ·9789 916 |
| ·98 | ·9969 606 | ·9963 146 | ·9955 912 | ·9947 876 | ·9939 017 | ·9929 313 |
| ·99 | ·9995 842 | ·9994 921 | ·9993 878 | ·9992 708 | ·9991 406 | ·9989 964 |
| 1·00 | 1·0000 000 | 1·0000 000 | 1·0000 000 | 1·0000 000 | 1·0000 000 | 1·0000 000 |

$x$ = ·33 to ·90 $q = 3$ $p$ = 19 to 24

| $x$ | $p = 19$ | $p = 20$ | $p = 21$ | $p = 22$ | $p = 23$ | $p = 24$ |
|---|---|---|---|---|---|---|
| $B(p, q) =$ | ·2506 2657 $\times \frac{1}{10^3}$ | ·2164 5022 $\times \frac{1}{10^3}$ | ·1882 1758 $\times \frac{1}{10^3}$ | ·1646 9038 $\times \frac{1}{10^3}$ | ·1449 2754 $\times \frac{1}{10^3}$ | ·1282 0513 $\times \frac{1}{10^3}$ |
| ·33 | ·0000 001 | | | | | |
| ·34 | ·0000 001 | | | | | |
| ·35 | ·0000 002 | ·0000 001 | | | | |
| ·36 | ·0000 003 | ·0000 001 | ·0000 001 | | | |
| ·37 | ·0000 006 | ·0000 002 | ·0000 001 | | | |
| ·38 | ·0000 009 | ·0000 004 | ·0000 002 | ·0000 001 | | |
| ·39 | ·0000 014 | ·0000 006 | ·0000 003 | ·0000 001 | | |
| ·40 | ·0000 022 | ·0000 010 | ·0000 004 | ·0000 002 | ·0000 001 | |
| ·41 | ·0000 034 | ·0000 015+ | ·0000 007 | ·0000 003 | ·0000 001 | ·0000 001 |
| ·42 | ·0000 053 | ·0000 024 | ·0000 011 | ·0000 005+ | ·0000 002 | ·0000 001 |
| ·43 | ·0000 080 | ·0000 038 | ·0000 018 | ·0000 008 | ·0000 004 | ·0000 002 |
| ·44 | ·0000 120 | ·0000 058 | ·0000 028 | ·0000 013 | ·0000 006 | ·0000 003 |
| ·45 | ·0000 178 | ·0000 088 | ·0000 043 | ·0000 021 | ·0000 010 | ·0000 005− |
| ·46 | ·0000 261 | ·0000 131 | ·0000 066 | ·0000 033 | ·0000 016 | ·0000 008 |
| ·47 | ·0000 379 | ·0000 195+ | ·0000 100 | ·0000 051 | ·0000 026 | ·0000 013 |
| ·48 | ·0000 547 | ·0000 287 | ·0000 150+ | ·0000 079 | ·0000 041 | ·0000 021 |
| ·49 | ·0000 781 | ·0000 419 | ·0000 224 | ·0000 119 | ·0000 063 | ·0000 034 |
| ·50 | ·0001 106 | ·0000 606 | ·0000 330 | ·0000 179 | ·0000 097 | ·0000 052 |
| ·51 | ·0001 554 | ·0000 868 | ·0000 482 | ·0000 267 | ·0000 148 | ·0000 081 |
| ·52 | ·0002 166 | ·0001 233 | ·0000 699 | ·0000 395− | ·0000 222 | ·0000 125− |
| ·53 | ·0002 995+ | ·0001 737 | ·0001 003 | ·0000 578 | ·0000 331 | ·0000 190 |
| ·54 | ·0004 111 | ·0002 429 | ·0001 429 | ·0000 838 | ·0000 490 | ·0000 285+ |
| ·55 | ·0005 602 | ·0003 370 | ·0002 019 | ·0001 206 | ·0000 718 | ·0000 426 |
| ·56 | ·0007 580 | ·0004 642 | ·0002 831 | ·0001 721 | ·0001 043 | ·0000 630 |
| ·57 | ·0010 187 | ·0006 348 | ·0003 940 | ·0002 438 | ·0001 503 | ·0000 924 |
| ·58 | ·0013 598 | ·0008 620 | ·0005 443 | ·0003 426 | ·0002 149 | ·0001 344 |
| ·59 | ·0018 034 | ·0011 625+ | ·0007 466 | ·0004 779 | ·0003 049 | ·0001 940 |
| ·60 | ·0023 765+ | ·0015 575+ | ·0010 170 | ·0006 618 | ·0004 293 | ·0002 777 |
| ·61 | ·0031 123 | ·0020 731 | ·0013 759 | ·0009 100 | ·0006 000 | ·0003 945− |
| ·62 | ·0040 511 | ·0027 419 | ·0018 490 | ·0012 427 | ·0008 326 | ·0005 562 |
| ·63 | ·0052 417 | ·0036 038 | ·0024 687 | ·0016 855− | ·0011 472 | ·0007 786 |
| ·64 | ·0067 425+ | ·0047 076 | ·0032 750+ | ·0022 708 | ·0015 697 | ·0010 820 |
| ·65 | ·0086 231 | ·0061 125+ | ·0043 174 | ·0030 395+ | ·0021 333 | ·0014 931 |
| ·66 | ·0109 657 | ·0078 897 | ·0056 565− | ·0040 422 | ·0028 799 | ·0020 461 |
| ·67 | ·0138 667 | ·0101 240 | ·0073 657 | ·0053 416 | ·0038 622 | ·0027 847 |
| ·68 | ·0174 383 | ·0129 162 | ·0095 337 | ·0070 146 | ·0051 458 | ·0037 645+ |
| ·69 | ·0218 099 | ·0163 844 | ·0122 665− | ·0091 545+ | ·0068 120 | ·0050 552 |
| ·70 | ·0271 294 | ·0206 662 | ·0156 896 | ·0118 741 | ·0089 605+ | ·0067 437 |
| ·71 | ·0335 644 | ·0259 204 | ·0199 504 | ·0153 080 | ·0117 123 | ·0089 373 |
| ·72 | ·0413 029 | ·0323 284 | ·0252 206 | ·0196 156 | ·0152 131 | ·0117 677 |
| ·73 | ·0505 535+ | ·0400 956 | ·0316 979 | ·0249 838 | ·0196 368 | ·0153 942 |
| ·74 | ·0615 447 | ·0494 515− | ·0396 076 | ·0316 294 | ·0251 888 | ·0200 083 |
| ·75 | ·0745 235− | ·0606 494 | ·0492 033 | ·0398 012 | ·0321 085+ | ·0258 373 |
| ·76 | ·0897 527 | ·0739 653 | ·0607 670 | ·0497 810 | ·0406 726 | ·0331 483 |
| ·77 | ·1075 069 | ·0896 943 | ·0746 071 | ·0618 835− | ·0511 957 | ·0422 505− |
| ·78 | ·1280 665− | ·1081 467 | ·0910 555+ | ·0764 548 | ·0640 309 | ·0534 975+ |
| ·79 | ·1517 098 | ·1296 403 | ·1104 620 | ·0938 681 | ·0795 672 | ·0672 871 |
| ·80 | ·1787 028 | ·1544 915− | ·1331 855− | ·1145 174 | ·0982 252 | ·0840 581 |
| ·81 | ·2092 864 | ·1830 021 | ·1595 829 | ·1388 066 | ·1204 480 | ·1042 844 |
| ·82 | ·2436 602 | ·2154 439 | ·1899 928 | ·1671 353 | ·1466 882 | ·1284 637 |
| ·83 | ·2819 647 | ·2520 386 | ·2247 161 | ·1998 786 | ·1773 894 | ·1571 003 |
| ·84 | ·3242 587 | ·2929 341 | ·2639 901 | ·2373 617 | ·2129 604 | ·1906 809 |
| ·85 | ·3704 955+ | ·3381 769 | ·3079 590 | ·2798 276 | ·2537 421 | ·2296 413 |
| ·86 | ·4204 964 | ·3876 813 | ·3566 382 | ·3273 985+ | ·2999 664 | ·2743 234 |
| ·87 | ·4739 224 | ·4411 948 | ·4098 744 | ·3800 305+ | ·3517 060 | ·3249 208 |
| ·88 | ·5302 475− | ·4982 634 | ·4673 028 | ·4374 627 | ·4088 162 | ·3814 152 |
| ·89 | ·5887 329 | ·5581 972 | ·5283 028 | ·4991 629 | ·4708 706 | ·4435 010 |
| ·90 | ·6484 088 | ·6200 409 | ·5919 567 | ·5642 737 | ·5370 940 | ·5105 052 |

$x = \cdot 91$ to $1\cdot 00$ $q = 3$ $p = 19$ to $24$

| | $p = 19$ | $p = 20$ | $p = 21$ | $p = 22$ | $p = 23$ | $p = 24$ |
|---|---|---|---|---|---|---|
| $B(p, q) =$ | ·2506 2657 $\times \frac{1}{10^3}$ | ·2164 5022 $\times \frac{1}{10^3}$ | ·1822 1758 $\times \frac{1}{10^3}$ | ·1646 9038 $\times \frac{1}{10^3}$ | ·1449 2754 $\times \frac{1}{10^3}$ | ·1282 0513 $\times \frac{1}{10^3}$ |
| $x$ | | | | | | |
| ·91 | ·7080 651 | ·6825 538 | ·6570 169 | ·6315 652 | ·6062 985+ | ·5813 066 |
| ·92 | ·7662 590 | ·7442 065− | ·7218 894 | ·6994 022 | ·6768 332 | ·6542 643 |
| ·93 | ·8213 473 | ·8032 051 | ·7846 456 | ·7657 415+ | ·7465 624 | ·7271 749 |
| ·94 | ·8715 532 | ·8575 540 | ·8430 789 | ·8281 764 | ·8128 946 | ·7972 805+ |
| ·95 | ·9150 825− | ·9051 770 | ·8948 257 | ·8840 554 | ·8728 935+ | ·8613 676 |
| ·96 | ·9503 052 | ·9441 172 | ·9375 826 | ·9307 120 | ·9235 166 | ·9160 083 |
| ·97 | ·9760 268 | ·9728 481 | ·9694 564 | ·9658 532 | ·9620 404 | ·9580 203 |
| ·98 | ·9918 747 | ·9907 302 | ·9894 965− | ·9881 723 | ·9867 566 | ·9852 485+ |
| ·99 | ·9988 379 | ·9986 644 | ·9984 754 | ·9982 706 | ·9980 493 | ·9978 112 |
| 1·00 | 1·0000 000 | 1·0000 000 | 1·0000 000 | 1·0000 000 | 1·0000 000 | 1·0000 000 |

$x$ = ·43 to 1·00 $q = 3$ $p$ = 25 to 30

| $x$ | $p = 25$ | $p = 26$ | $p = 27$ | $p = 28$ | $p = 29$ | $p = 30$ |
|---|---|---|---|---|---|---|
| $B(p, q) =$ | ·1139 6011 × $\frac{1}{10^3}$ | ·1017 5010 × $\frac{1}{10^3}$ | ·9122 4229 × $\frac{1}{10^4}$ | ·8210 1806 × $\frac{1}{10^4}$ | ·7415 6470 × $\frac{1}{10^4}$ | ·6720 4301 × $\frac{1}{10^4}$ |
| ·43 | ·0000 001 | | | | | |
| ·44 | ·0000 001 | ·0000 001 | | | | |
| ·45 | ·0000 002 | ·0000 001 | ·0000 001 | | | |
| ·46 | ·0000 004 | ·0000 002 | ·0000 001 | | | |
| ·47 | ·0000 007 | ·0000 003 | ·0000 002 | ·0000 001 | | |
| ·48 | ·0000 011 | ·0000 006 | ·0000 003 | ·0000 001 | ·0000 001 | |
| ·49 | ·0000 018 | ·0000 009 | ·0000 005$^{-}$ | ·0000 003 | ·0000 001 | ·0000 001 |
| ·50 | ·0000 028 | ·0000 015$^{+}$ | ·0000 008 | ·0000 004 | ·0000 002 | ·0000 001 |
| ·51 | ·0000 045$^{-}$ | ·0000 024 | ·0000 013 | ·0000 007 | ·0000 004 | ·0000 002 |
| ·52 | ·0000 070 | ·0000 039 | ·0000 022 | ·0000 012 | ·0000 007 | ·0000 004 |
| ·53 | ·0000 108 | ·0000 062 | ·0000 035$^{-}$ | ·0000 020 | ·0000 011 | ·0000 006 |
| ·54 | ·0000 166 | ·0000 096 | ·0000 056 | ·0000 032 | ·0000 018 | ·0000 011 |
| ·55 | ·0000 252 | ·0000 149 | ·0000 088 | ·0000 051 | ·0000 030 | ·0000 018 |
| ·56 | ·0000 380 | ·0000 228 | ·0000 137 | ·0000 082 | ·0000 049 | ·0000 029 |
| ·57 | ·0000 567 | ·0000 346 | ·0000 211 | ·0000 129 | ·0000 078 | ·0000 047 |
| ·58 | ·0000 838 | ·0000 522 | ·0000 324 | ·0000 201 | ·0000 124 | ·0000 076 |
| ·59 | ·0001 230 | ·0000 779 | ·0000 492 | ·0000 310 | ·0000 195$^{-}$ | ·0000 122 |
| ·60 | ·0001 791 | ·0001 152 | ·0000 740 | ·0000 474 | ·0000 303 | ·0000 193 |
| ·61 | ·0002 587 | ·0001 692 | ·0001 104 | ·0000 719 | ·0000 467 | ·0000 303 |
| ·62 | ·0003 706 | ·0002 463 | ·0001 633 | ·0001 081 | ·0000 714 | ·0000 470 |
| ·63 | ·0005 270 | ·0003 559 | ·0002 397 | ·0001 612 | ·0001 081 | ·0000 724 |
| ·64 | ·0007 439 | ·0005 102 | ·0003 491 | ·0002 383 | ·0001 624 | ·0001 104 |
| ·65 | ·0010 423 | ·0007 258 | ·0005 043 | ·0003 496 | ·0002 419 | ·0001 671 |
| ·66 | ·0014 499 | ·0010 250$^{-}$ | ·0007 229 | ·0005 088 | ·0003 574 | ·0002 506 |
| ·67 | ·0020 027 | ·0014 368 | ·0010 285$^{+}$ | ·0007 347 | ·0005 238 | ·0003 727 |
| ·68 | ·0027 470 | ·0019 997 | ·0014 525$^{-}$ | ·0010 528 | ·0007 616 | ·0005 499 |
| ·69 | ·0037 419 | ·0027 633 | ·0020 361 | ·0014 972 | ·0010 987 | ·0008 049 |
| ·70 | ·0050 625$^{+}$ | ·0037 916 | ·0028 335$^{+}$ | ·0021 132 | ·0015 729 | ·0011 687 |
| ·71 | ·0068 029 | ·0051 663 | ·0039 148 | ·0029 605$^{+}$ | ·0022 345$^{+}$ | ·0016 836 |
| ·72 | ·0090 803 | ·0069 905$^{+}$ | ·0053 701 | ·0041 171 | ·0031 504 | ·0024 064 |
| ·73 | ·0120 389 | ·0093 937 | ·0073 141 | ·0056 835$^{+}$ | ·0044 082 | ·0034 130 |
| ·74 | ·0158 553 | ·0125 362 | ·0098 911 | ·0077 887 | ·0061 219 | ·0048 034 |
| ·75 | ·0207 420 | ·0166 147 | ·0132 812 | ·0105 959 | ·0084 380 | ·0067 080 |
| ·76 | ·0269 532 | ·0218 683 | ·0177 066 | ·0143 093 | ·0115 430 | ·0092 956 |
| ·77 | ·0347 887 | ·0285 834 | ·0234 378 | ·0191 823 | ·0156 714 | ·0127 816 |
| ·78 | ·0445 969 | ·0370 990 | ·0308 007 | ·0255 242 | ·0211 145$^{+}$ | ·0174 378 |
| ·79 | ·0567 774 | ·0478 105$^{-}$ | ·0401 817 | ·0337 086 | ·0282 295$^{+}$ | ·0236 026 |
| ·80 | ·0717 800 | ·0611 716 | ·0520 322 | ·0441 790 | ·0374 477 | ·0316 913 |
| ·81 | ·0901 009 | ·0776 931 | ·0668 697 | ·0574 534 | ·0492 813 | ·0422 055$^{-}$ |
| ·82 | ·1122 742 | ·0979 367 | ·0852 757 | ·0741 246 | ·0643 276 | ·0557 400 |
| ·83 | ·1388 569 | ·1225 036 | ·1078 862 | ·0948 551 | ·0832 668 | ·0729 851 |
| ·84 | ·1704 066 | ·1520 137 | ·1353 753 | ·1203 637 | ·1068 532 | ·0947 218 |
| ·85 | ·2074 486 | ·1870 756 | ·1684 265$^{+}$ | ·1514 006 | ·1358 949 | ·1218 060 |
| ·86 | ·2504 326 | ·2282 428 | ·2076 917 | ·1887 085$^{+}$ | ·1712 169 | ·1551 367 |
| ·87 | ·2996 758 | ·2759 556 | ·2537 316 | ·2329 645$^{+}$ | ·2136 066 | ·1956 038 |
| ·88 | ·3552 929 | ·3304 663 | ·3069 383 | ·2847 000 | ·2637 324 | ·2440 084 |
| ·89 | ·4171 120 | ·3917 470 | ·3674 356 | ·3441 957 | ·3220 348 | ·3009 514 |
| ·90 | ·4845 811 | ·4593 829 | ·4349 600 | ·4113 512 | ·3885 856 | ·3666 835$^{+}$ |
| ·91 | ·5566 686 | ·5324 545$^{-}$ | ·5087 246 | ·4855 308 | ·4629 169 | ·4409 190 |
| ·92 | ·6317 706 | ·6094 208 | ·5872 774 | ·5653 964 | ·5438 279 | ·5226 165$^{+}$ |
| ·93 | ·7076 419 | ·6880 230 | ·6683 739 | ·6487 466 | ·6291 895$^{-}$ | ·6097 469 |
| ·94 | ·7813 802 | ·7652 382 | ·7488 975$^{+}$ | ·7323 995$^{+}$ | ·7157 837 | ·6990 876 |
| ·95 | ·8495 056 | ·8373 351 | ·8248 838 | ·8121 788 | ·7992 470 | ·7861 145$^{-}$ |
| ·96 | ·9081 997 | ·9001 038 | ·8917 338 | ·8831 034 | ·8742 265$^{-}$ | ·8651 169 |
| ·97 | ·9537 958 | ·9493 703 | ·9447 474 | ·9399 309 | ·9349 252 | ·9297 349 |
| ·98 | ·9836 475$^{+}$ | ·9819 530 | ·9801 646 | ·9782 822 | ·9763 056 | ·9742 350$^{-}$ |
| ·99 | ·9975 559 | ·9972 829 | ·9969 918 | ·9966 823 | ·9963 540 | ·9960 066 |
| 1·00 | 1·0000 000 | 1·0000 000 | 1·0000 000 | 1·0000 000 | 1·0000 000 | 1·0000 000 |

$x$ = ·50 to 1·00 $q = 3$ $p$ = 31 to 36

| $x$ | $p = 31$ | $p = 32$ | $p = 33$ | $p = 34$ | $p = 35$ | $p = 36$ |
|---|---|---|---|---|---|---|
| $B(p, q) =$ | ·6109 4819 × $\frac{1}{10^4}$ | ·5570 4100 × $\frac{1}{10^4}$ | ·5092 9463 × $\frac{1}{10^4}$ | ·4668 5341 × $\frac{1}{10^4}$ | ·4290 0043 × $\frac{1}{10^4}$ | ·3951 3197 × $\frac{1}{10^4}$ |
| ·50 | ·0000 001 | | | | | |
| ·51 | ·0000 001 | ·0000 001 | | | | |
| ·52 | ·0000 002 | ·0000 001 | ·0000 001 | | | |
| ·53 | ·0000 004 | ·0000 002 | ·0000 001 | ·0000 001 | | |
| ·54 | ·0000 006 | ·0000 003 | ·0000 002 | ·0000 001 | ·0000 001 | |
| ·55 | ·0000 010 | ·0000 006 | ·0000 004 | ·0000 002 | ·0000 001 | ·0000 001 |
| ·56 | ·0000 017 | ·0000 010 | ·0000 006 | ·0000 004 | ·0000 002 | ·0000 001 |
| ·57 | ·0000 029 | ·0000 017 | ·0000 010 | ·0000 006 | ·0000 004 | ·0000 002 |
| ·58 | ·0000 047 | ·0000 029 | ·0000 018 | ·0000 011 | ·0000 007 | ·0000 004 |
| ·59 | ·0000 076 | ·0000 048 | ·0000 030 | ·0000 019 | ·0000 012 | ·0000 007 |
| ·60 | ·0000 123 | ·0000 078 | ·0000 050$^-$ | ·0000 031 | ·0000 020 | ·0000 013 |
| ·61 | ·0000 196 | ·0000 127 | ·0000 082 | ·0000 053 | ·0000 034 | ·0000 022 |
| ·62 | ·0000 309 | ·0000 203 | ·0000 133 | ·0000 087 | ·0000 057 | ·0000 037 |
| ·63 | ·0000 484 | ·0000 323 | ·0000 215$^+$ | ·0000 143 | ·0000 095$^+$ | ·0000 063 |
| ·64 | ·0000 750$^-$ | ·0000 508 | ·0000 344 | ·0000 232 | ·0000 157 | ·0000 106 |
| ·65 | ·0001 152 | ·0000 793 | ·0000 545$^-$ | ·0000 374 | ·0000 256 | ·0000 175$^+$ |
| ·66 | ·0001 754 | ·0001 225$^+$ | ·0000 855$^-$ | ·0000 595$^+$ | ·0000 414 | ·0000 288 |
| ·67 | ·0002 648 | ·0001 878 | ·0001 330 | ·0000 940 | ·0000 664 | ·0000 468 |
| ·68 | ·0003 964 | ·0002 853 | ·0002 050$^-$ | ·0001 471 | ·0001 054 | ·0000 754 |
| ·69 | ·0005 886 | ·0004 297 | ·0003 132 | ·0002 280 | ·0001 657 | ·0001 203 |
| ·70 | ·0008 668 | ·0006 419 | ·0004 746 | ·0003 504 | ·0002 584 | ·0001 902 |
| ·71 | ·0012 663 | ·0009 509 | ·0007 130 | ·0005 338 | ·0003 991 | ·0002 980 |
| ·72 | ·0018 350$^+$ | ·0013 971 | ·0010 621 | ·0008 062 | ·0006 112 | ·0004 627 |
| ·73 | ·0026 381 | ·0020 359 | ·0015 688 | ·0012 072 | ·0009 277 | ·0007 120 |
| ·74 | ·0037 626 | ·0029 427 | ·0022 981 | ·0017 922 | ·0013 958 | ·0010 857 |
| ·75 | ·0053 240 | ·0042 190 | ·0033 385$^-$ | ·0026 381 | ·0020 818 | ·0016 408 |
| ·76 | ·0074 737 | ·0059 997 | ·0048 095$^-$ | ·0038 501 | ·0030 781 | ·0024 578 |
| ·77 | ·0104 081 | ·0084 626 | ·0068 710 | ·0055 711 | ·0045 113 | ·0036 487 |
| ·78 | ·0143 787 | ·0118 387 | ·0097 337 | ·0079 922 | ·0065 540 | ·0053 681 |
| ·79 | ·0197 036 | ·0164 246 | ·0136 724 | ·0113 663 | ·0094 374 | ·0078 265$^-$ |
| ·80 | ·0267 791 | ·0225 959 | ·0190 401 | ·0160 231 | ·0134 675$^+$ | ·0113 062 |
| ·81 | ·0360 919 | ·0308 205$^-$ | ·0262 837 | ·0223 863 | ·0190 436 | ·0161 813 |
| ·82 | ·0482 287 | ·0416 720 | ·0359 596 | ·0309 915$^-$ | ·0266 780 | ·0229 388 |
| ·83 | ·0638 823 | ·0558 396 | ·0487 470 | ·0425 033 | ·0370 162 | ·0322 017 |
| ·84 | ·0838 520 | ·0741 324 | ·0654 576 | ·0577 291 | ·0508 553 | ·0447 514 |
| ·85 | ·1090 321 | ·0974 738 | ·0870 352 | ·0776 246 | ·0691 551 | ·0615 447 |
| ·86 | ·1403 858 | ·1268 815$^+$ | ·1145 420 | ·1032 869 | ·0930 381 | ·0837 205$^+$ |
| ·87 | ·1788 971 | ·1634 246 | ·1491 222 | ·1359 250$^-$ | ·1237 680 | ·1125 871 |
| ·88 | ·2254 942 | ·2081 506 | ·1919 343 | ·1767 990 | ·1626 966 | ·1495 772 |
| ·89 | ·2809 362 | ·2619 734 | ·2440 417 | ·2271 153 | ·2111 646 | ·1961 573 |
| ·90 | ·3456 575$^+$ | ·3255 132 | ·3062 503 | ·2878 629 | ·2703 408 | ·2536 698 |
| ·91 | ·4195 664 | ·3988 819 | ·3788 826 | ·3595 803 | ·3409 819 | ·3230 902 |
| ·92 | ·5018 010 | ·4814 153 | ·4614 882 | ·4420 442 | ·4231 035$^+$ | ·4046 823 |
| ·93 | ·5904 600 | ·5713 658 | ·5524 985$^-$ | ·5338 884 | ·5155 629 | ·4975 464 |
| ·94 | ·6823 470 | ·6655 956 | ·6488 651 | ·6321 854 | ·6155 841 | ·5990 872 |
| ·95 | ·7728 069 | ·7593 490 | ·7457 650$^+$ | ·7320 781 | ·7183 106 | ·7044 842 |
| ·96 | ·8557 887 | ·8462 558 | ·8365 324 | ·8266 321 | ·8165 687 | ·8063 559 |
| ·97 | ·9243 646 | ·9188 193 | ·9131 043 | ·9072 247 | ·9011 860 | ·8949 937 |
| ·98 | ·9720 705$^+$ | ·9698 125$^-$ | ·9674 613 | ·9650 175$^-$ | ·9624 817 | ·9598 546 |
| ·99 | ·9956 397 | ·9952 530 | ·9948 463 | ·9944 192 | ·9939 716 | ·9935 031 |
| 1·00 | 1·0000 000 | 1·0000 000 | 1·0000 000 | 1·0000 000 | 1·0000 000 | 1·0000 000 |

$x$ = ·56 to 1·00 $q = 3$ $p$ = 37 to 43

| | $p = 37$ | $p = 38$ | $p = 39$ | $p = 40$ | $p = 41$ | $p = 42$ | $p = 43$ |
|---|---|---|---|---|---|---|---|
| $B(p, q)$ = | ·3647 3721 × $\frac{1}{10^{4}}$ | ·3373 8192 × $\frac{1}{10^{4}}$ | ·3126 9543 × $\frac{1}{10^{4}}$ | ·2903 6005 × $\frac{1}{10^{4}}$ | ·2701 0237 × $\frac{1}{10^{4}}$ | ·2516 8630 × $\frac{1}{10^{4}}$ | ·2349 0721 × $\frac{1}{10^{4}}$ |
| $x$ | | | | | | | |
| ·56 | ·0000 001 | | | | | | |
| ·57 | ·0000 001 | ·0000 001 | | | | | |
| ·58 | ·0000 002 | ·0000 002 | ·0000 001 | ·0000 001 | | | |
| ·59 | .0000 004 | ·0000 003 | ·0000 002 | ·0000 001 | ·0000 001 | | |
| ·60 | ·0000 008 | ·0000 005+ | ·0000 003 | ·0000 002 | ·0000 001 | ·0000 001 | |
| ·61 | ·0000 014 | ·0000 009 | ·0000 006 | ·0000 004 | ·0000 002 | ·0000 001 | ·0000 001 |
| ·62 | ·0000 024 | ·0000 016 | ·0000 010 | ·0000 007 | ·0000 004 | ·0000 003 | ·0000 002 |
| ·63 | ·0000 042 | ·0000 028 | ·0000 018 | ·0000 012 | ·0000 008 | ·0000 005+ | ·0000 003 |
| ·64 | ·0000 071 | ·0000 048 | ·0000 032 | ·0000 021 | ·0000 014 | ·0000 010 | ·0000 006 |
| ·65 | ·0000 120 | ·0000 082 | ·0000 056 | ·0000 038 | ·0000 026 | ·0000 018 | ·0000 012 |
| ·66 | ·0000 200 | ·0000 138 | ·0000 096 | ·0000 066 | ·0000 046 | ·0000 032 | ·0000 022 |
| ·67 | ·0000 330 | ·0000 232 | ·0000 163 | ·0000 114 | ·0000 080 | ·0000 056 | ·0000 039 |
| ·68 | ·0000 539 | ·0000 385− | ·0000 274 | ·0000 195+ | ·0000 139 | ·0000 099 | ·0000 070 |
| ·69 | ·0000 872 | ·0000 632 | ·0000 457 | ·0000 330 | ·0000 238 | ·0000 172 | ·0000 124 |
| ·70 | ·0001 399 | ·0001 028 | ·0000 754 | ·0000 553 | ·0000 405− | ·0000 296 | ·0000 216 |
| ·71 | ·0002 223 | ·0001 656 | ·0001 232 | ·0000 916 | ·0000 680 | ·0000 504 | ·0000 374 |
| ·72 | ·0003 499 | ·0002 643 | ·0001 994 | ·0001 503 | ·0001 131 | ·0000 851 | ·0000 640 |
| ·73 | ·0005 458 | ·0004 179 | ·0003 196 | ·0002 442 | ·0001 864 | ·0001 421 | ·0001 083 |
| ·74 | ·0008 435− | ·0006 545+ | ·0005 074 | ·0003 929 | ·0003 039 | ·0002 349 | ·0001 814 |
| ·75 | ·0012 917 | ·0010 157 | ·0007 978 | ·0006 260 | ·0004 907 | ·0003 843 | ·0003 007 |
| ·76 | ·0019 602 | ·0015 616 | ·0012 427 | ·0009 879 | ·0007 846 | ·0006 225+ | ·0004 935− |
| ·77 | ·0029 476 | ·0023 785− | ·0019 173 | ·0015 439 | ·0012 420 | ·0009 982 | ·0008 016 |
| ·78 | ·0043 917 | ·0035 890 | ·0029 298 | ·0023 894 | ·0019 467 | ·0015 846 | ·0012 887 |
| ·79 | ·0064 831 | ·0053 645− | ·0044 343 | ·0036 617 | ·0030 209 | ·0024 899 | ·0020 505+ |
| ·80 | ·0094 811 | ·0079 421 | ·0066 461 | ·0055 562 | ·0046 406 | ·0038 724 | ·0032 286 |
| ·81 | ·0137 341 | ·0116 447 | ·0098 632 | ·0083 462 | ·0070 559 | ·0059 599 | ·0050 298 |
| ·82 | ·0197 023 | ·0169 050− | ·0144 904 | ·0124 089 | ·0106 168 | ·0090 755+ | ·0077 515+ |
| ·83 | ·0279 836 | ·0242 934 | ·0210 693 | ·0182 560 | ·0158 043 | ·0136 701 | ·0118 144 |
| ·84 | ·0393 392 | ·0345 473 | ·0303 102 | ·0265 685+ | ·0232 684 | ·0203 610 | ·0178 025+ |
| ·85 | ·0547 164 | ·0485 987 | ·0431 249 | ·0382 336 | ·0338 681 | ·0299 764 | ·0265 109 |
| ·86 | ·0752 622 | ·0675 949 | ·0606 539 | ·0543 786 | ·0487 119 | ·0436 009 | ·0389 961 |
| ·87 | ·1023 193 | ·0929 034 | ·0842 804 | ·0763 937 | ·0691 892 | ·0626 156 | ·0566 242 |
| ·88 | ·1373 908 | ·1260 871 | ·1156 163 | ·1059 295− | ·0969 789 | ·0887 181 | ·0811 024 |
| ·89 | ·1820 588 | ·1688 329 | ·1564 422 | ·1448 491 | ·1340 153 | ·1239 028 | ·1144 742 |
| ·90 | ·2378 323 | ·2228 081 | ·2085 747 | ·1951 077 | ·1823 813 | ·1703 689 | ·1590 429 |
| ·91 | ·3059 043 | ·2894 197 | ·2736 293 | ·2585 231 | ·2440 891 | ·2303 134 | ·2171 806 |
| ·92 | ·3867 933 | ·3694 457 | ·3526 460 | ·3363 977 | ·3207 018 | ·3055 571 | ·2909 606 |
| ·93 | ·4798 601 | ·4625 228 | ·4455 505+ | ·4289 569 | ·4127 531 | ·3969 485+ | ·3815 504 |
| ·94 | ·5827 186 | ·5665 004 | ·5504 529 | ·5345 947 | ·5189 426 | ·5035 120 | ·4883 165+ |
| ·95 | ·6906 193 | ·6767 358 | ·6628 522 | ·6489 864 | ·6351 553 | ·6213 748 | ·6076 599 |
| ·96 | ·7960 068 | ·7855 347 | ·7749 524 | ·7642 724 | ·7535 069 | ·7426 680 | ·7317 670 |
| ·97 | ·8886 535+ | ·8821 711 | ·8755 523 | ·8688 027 | ·8619 283 | ·8549 348 | ·8478 281 |
| ·98 | ·9571 370 | ·9543 298 | ·9514 339 | ·9484 505− | ·9453 805− | ·9422 251 | ·9389 856 |
| ·99 | ·9930 135+ | ·9925 026 | ·9919 702 | ·9914 161 | ·9908 402 | ·9902 421 | ·9896 219 |
| 1·00 | 1·0000 000 | 1·0000 000 | 1·0000 000 | 1·0000 000 | 1·0000 000 | 1·0000 000 | 1·0000 000 |

$x$ = ·61 to 1·00 $q = 3$ $p$ = 44 to 50

| | $p = 44$ | $p = 45$ | $p = 46$ | $p = 47$ | $p = 48$ | $p = 49$ | $p = 50$ |
|---|---|---|---|---|---|---|---|
| $B(p, q)$ = | ·2195 8718 × $\frac{1}{10^{4}}$ | ·2055 7097 × $\frac{1}{10^{4}}$ | ·1927 2279 × $\frac{1}{10^{4}}$ | ·1809 2343 × $\frac{1}{10^{4}}$ | ·1700 6803 × $\frac{1}{10^{4}}$ | ·1600 6403 × $\frac{1}{10^{4}}$ | ·1508 2956 × $\frac{1}{10^{4}}$ |
| $x$ | | | | | | | |
| ·61 | ·0000 001 | | | | | | |
| ·62 | ·0000 001 | ·0000 001 | | | | | |
| ·63 | ·0000 002 | ·0000 001 | ·0000 001 | ·0000 001 | | | |
| ·64 | ·0000 004 | ·0000 003 | ·0000 002 | ·0000 001 | ·0000 001 | ·0000 001 | |
| ·65 | ·0000 008 | ·0000 005+ | ·0000 004 | ·0000 003 | ·0000 002 | ·0000 001 | ·0000 001 |
| ·66 | ·0000 015− | ·0000 010 | ·0000 007 | ·0000 005− | ·0000 003 | ·0000 002 | ·0000 002 |
| ·67 | ·0000 027 | ·0000 019 | ·0000 013 | ·0000 009 | ·0000 007 | ·0000 005− | ·0000 003 |
| ·68 | ·0000 050− | ·0000 035+ | ·0000 025− | ·0000 018 | ·0000 012 | ·0000 009 | ·0000 006 |
| ·69 | ·0000 089 | ·0000 064 | ·0000 046 | ·0000 033 | ·0000 024 | ·0000 017 | ·0000 012 |
| ·70 | ·0000 158 | ·0000 115+ | ·0000 084 | ·0000 061 | ·0000 045− | ·0000 032 | ·0000 024 |
| ·71 | ·0000 277 | ·0000 205− | ·0000 151 | ·0000 112 | ·0000 083 | ·0000 061 | ·0000 045− |
| ·72 | ·0000 480 | ·0000 360 | ·0000 270 | ·0000 202 | ·0000 151 | ·0000 113 | ·0000 085− |
| ·73 | ·0000 824 | ·0000 627 | ·0000 476 | ·0000 361 | ·0000 274 | ·0000 208 | ·0000 158 |
| ·74 | ·0001 399 | ·0001 078 | ·0000 830 | ·0000 639 | ·0000 491 | ·0000 378 | ·0000 290 |
| ·75 | ·0002 351 | ·0001 836 | ·0001 433 | ·0001 118 | ·0000 871 | ·0000 678 | ·0000 528 |
| ·76 | ·0003 909 | ·0003 093 | ·0002 446 | ·0001 933 | ·0001 526 | ·0001 204 | ·0000 949 |
| ·77 | ·0006 431 | ·0005 156 | ·0004 130 | ·0003 306 | ·0002 644 | ·0002 113 | ·0001 688 |
| ·78 | ·0010 472 | ·0008 502 | ·0006 898 | ·0005 592 | ·0004 530 | ·0003 667 | ·0002 967 |
| ·79 | ·0016 872 | ·0013 872 | ·0011 396 | ·0009 355+ | ·0007 674 | ·0006 291 | ·0005 154 |
| ·80 | ·0026 896 | ·0022 388 | ·0018 621 | ·0015 477 | ·0012 854 | ·0010 669 | ·0008 849 |
| ·81 | ·0042 414 | ·0035 737 | ·0030 089 | ·0025 315+ | ·0021 284 | ·0017 882 | ·0015 014 |
| ·82 | ·0066 153 | ·0056 413 | ·0048 071 | ·0040 933 | ·0034 831 | ·0029 619 | ·0025 171 |
| ·83 | ·0102 025− | ·0088 038 | ·0075 912 | ·0065 411 | ·0056 324 | ·0048 467 | ·0041 680 |
| ·84 | ·0155 535− | ·0135 784 | ·0118 455+ | ·0103 267 | ·0089 966 | ·0078 327 | ·0068 151 |
| ·85 | ·0234 283 | ·0206 889 | ·0182 570 | ·0161 000 | ·0141 885+ | ·0124 961 | ·0109 987 |
| ·86 | ·0348 518 | ·0311 257 | ·0277 788 | ·0247 754 | ·0220 825− | ·0196 701 | ·0175 108 |
| ·87 | ·0511 692 | ·0462 076 | ·0416 992 | ·0376 064 | ·0338 941 | ·0305 298 | ·0274 835− |
| ·88 | ·0740 889 | ·0676 365+ | ·0617 061 | ·0562 603 | ·0512 642 | ·0466 844 | ·0424 896 |
| ·89 | ·1056 924 | ·0975 213 | ·0899 258 | ·0828 720 | ·0763 269 | ·0702 590 | ·0646 382 |
| ·90 | ·1483 754 | ·1383 382 | ·1289 033 | ·1200 427 | ·1117 288 | ·1039 345− | ·0966 333 |
| ·91 | ·2046 739 | ·1927 755− | ·1814 667 | ·1707 283 | ·1605 405− | ·1508 833 | ·1417 366 |
| ·92 | ·2769 072 | ·2633 904 | ·2504 022 | ·2379 336 | ·2259 743 | ·2145 133 | ·2035 388 |
| ·93 | ·3665 640 | ·3519 932 | ·3378 401 | ·3241 054 | ·3107 886 | ·2978 879 | ·2854 006 |
| ·94 | ·4733 684 | ·4586 785− | ·4442 562 | ·4301 099 | ·4162 465− | ·4026 719 | ·3893 909 |
| ·95 | ·5940 247 | ·5804 825+ | ·5670 456 | ·5537 256 | ·5405 331 | ·5274 781 | ·5145 695+ |
| ·96 | ·7208 154 | ·7098 240 | ·6988 032 | ·6877 633 | ·6767 140 | ·6656 647 | ·6546 244 |
| ·97 | ·8406 140 | ·8332 982 | ·8258 865+ | ·8183 846 | ·8107 981 | ·8031 325+ | ·7953 935− |
| ·98 | ·9356 633 | ·9322 594 | ·9287 753 | ·9252 124 | ·9215 723 | ·9178 562 | ·9140 659 |
| ·99 | ·9889 793 | ·9883 142 | ·9876 264 | ·9869 160 | ·9861 827 | ·9854 265+ | ·9846 474 |
| 1·00 | 1·0000 000 | 1·0000 000 | 1·0000 000 | 1·0000 000 | 1·0000 000 | 1·0000 000 | 1·0000 000 |

$x$ = ·01 to ·60 $q$ = 3·5 $p$ = 3·5 to 6

| $x$ | $p$ = 3·5 | $p$ = 4 | $p$ = 4·5 | $p$ = 5 | $p$ = 5·5 | $p$ = 6 |
|---|---|---|---|---|---|---|
| $B(p, q)$ = | ·1533 9808 × $\frac{1}{10}$ | ·1065 6011 × $\frac{1}{10}$ | ·7669 9039 × $\frac{1}{10^2}$ | ·5683 2057 × $\frac{1}{10^2}$ | ·4314 3210 × $\frac{1}{10^2}$ | ·3343 0622 × $\frac{1}{10^2}$ |
| ·01 | ·0000 018 | ·0000 002 | | | | |
| ·02 | ·0000 203 | ·0000 036 | ·0000 006 | ·0000 001 | | |
| ·03 | ·0000 821 | ·0000 179 | ·0000 038 | ·0000 008 | ·0000 002 | |
| ·04 | ·0002 203 | ·0000 554 | ·0000 137 | ·0000 033 | ·0000 008 | ·0000 002 |
| ·05 | ·0004 715+ | ·0001 324 | ·0000 365− | ·0000 099 | ·0000 026 | ·0000 007 |
| ·06 | ·0008 747 | ·0002 689 | ·0000 811 | ·0000 241 | ·0000 070 | ·0000 020 |
| ·07 | ·0014 700 | ·0004 879 | ·0001 589 | ·0000 509 | ·0000 161 | ·0000 050+ |
| ·08 | ·0022 981 | ·0008 148 | ·0002 835− | ·0000 971 | ·0000 328 | ·0000 109 |
| ·09 | ·0033 994 | ·0012 776 | ·0004 712 | ·0001 711 | ·0000 613 | ·0000 217 |
| ·10 | ·0048 140 | ·0019 058 | ·0007 405+ | ·0002 832 | ·0001 069 | ·0000 399 |
| ·11 | ·0065 804 | ·0027 304 | ·0011 121 | ·0004 459 | ·0001 764 | ·0000 690 |
| ·12 | ·0087 361 | ·0037 833 | ·0016 086 | ·0006 733 | ·0002 781 | ·0001 135+ |
| ·13 | ·0113 166 | ·0050 974 | ·0022 544 | ·0009 818 | ·0004 219 | ·0001 792 |
| ·14 | ·0143 557 | ·0067 055+ | ·0030 758 | ·0013 893 | ·0006 193 | ·0002 729 |
| ·15 | ·0178 848 | ·0086 407 | ·0041 001 | ·0019 160 | ·0008 836 | ·0004 029 |
| ·16 | ·0219 331 | ·0109 358 | ·0053 559 | ·0025 836 | ·0012 301 | ·0005 790 |
| ·17 | ·0265 272 | ·0136 229 | ·0068 729 | ·0034 155+ | ·0016 755− | ·0008 126 |
| ·18 | ·0316 912 | ·0167 334 | ·0086 813 | ·0044 368 | ·0022 385+ | ·0011 168 |
| ·19 | ·0374 464 | ·0202 976 | ·0108 117 | ·0056 739 | ·0029 397 | ·0015 062 |
| ·20 | ·0438 114 | ·0243 445+ | ·0132 951 | ·0071 543 | ·0038 011 | ·0019 973 |
| ·21 | ·0508 020 | ·0289 016 | ·0161 623 | ·0089 068 | ·0048 467 | ·0026 084 |
| ·22 | ·0584 312 | ·0339 947 | ·0194 439 | ·0109 608 | ·0061 016 | ·0033 596 |
| ·23 | ·0667 088 | ·0396 476 | ·0231 699 | ·0133 465+ | ·0075 927 | ·0042 726 |
| ·24 | ·0756 421 | ·0458 824 | ·0273 697 | ·0160 946 | ·0093 480 | ·0053 710 |
| ·25 | ·0852 353 | ·0527 186 | ·0320 715− | ·0192 360 | ·0113 966 | ·0066 800 |
| ·26 | ·0954 899 | ·0601 737 | ·0373 024 | ·0228 014 | ·0137 688 | ·0082 262 |
| ·27 | ·1064 045− | ·0682 626 | ·0430 882 | ·0268 216 | ·0164 954 | ·0100 380 |
| ·28 | ·1179 748 | ·0769 977 | ·0494 530 | ·0313 267 | ·0196 079 | ·0121 448 |
| ·29 | ·1301 940 | ·0863 889 | ·0564 190 | ·0363 461 | ·0231 383 | ·0145 775− |
| ·30 | ·1430 526 | ·0964 432 | ·0640 066 | ·0419 084 | ·0271 185+ | ·0173 678 |
| ·31 | ·1565 382 | ·1071 650+ | ·0722 339 | ·0480 410 | ·0315 805− | ·0205 484 |
| ·32 | ·1706 363 | ·1185 560 | ·0811 167 | ·0547 699 | ·0365 558 | ·0241 525+ |
| ·33 | ·1853 298 | ·1306 150− | ·0906 684 | ·0621 193 | ·0420 756 | ·0282 140 |
| ·34 | ·2005 991 | ·1433 378 | ·1008 998 | ·0701 119 | ·0481 700 | ·0327 667 |
| ·35 | ·2164 227 | ·1567 177 | ·1118 190 | ·0787 680 | ·0548 681 | ·0378 445− |
| ·36 | ·2327 766 | ·1707 450− | ·1234 311 | ·0881 060 | ·0621 977 | ·0434 809 |
| ·37 | ·2496 350− | ·1854 072 | ·1357 386 | ·0981 414 | ·0701 849 | ·0497 089 |
| ·38 | ·2669 701 | ·2006 891 | ·1487 407 | ·1088 874 | ·0788 540 | ·0565 605− |
| ·39 | ·2847 524 | ·2165 728 | ·1624 338 | ·1203 544 | ·0882 272 | ·0640 666 |
| ·40 | ·3029 506 | ·2330 378 | ·1768 111 | ·1325 496 | ·0983 242 | ·0722 568 |
| ·41 | ·3215 320 | ·2500 607 | ·1918 625+ | ·1454 773 | ·1091 623 | ·0811 585+ |
| ·42 | ·3404 622 | ·2676 161 | ·2075 752 | ·1591 384 | ·1207 557 | ·0907 975− |
| ·43 | ·3597 059 | ·2856 759 | ·2239 328 | ·1735 305+ | ·1331 157 | ·1011 968 |
| ·44 | ·3792 263 | ·3042 096 | ·2409 160 | ·1886 478 | ·1462 503 | ·1123 770 |
| ·45 | ·3989 858 | ·3231 847 | ·2585 024 | ·2044 808 | ·1601 639 | ·1243 557 |
| ·46 | ·4189 460 | ·3425 666 | ·2766 664 | ·2210 166 | ·1748 575− | ·1371 472 |
| ·47 | ·4390 676 | ·3623 187 | ·2953 797 | ·2382 386 | ·1903 279 | ·1507 621 |
| ·48 | ·4593 106 | ·3824 026 | ·3146 108 | ·2561 263 | ·2065 682 | ·1652 073 |
| ·49 | ·4796 350− | ·4027 783 | ·3343 255− | ·2746 558 | ·2235 674 | ·1804 858 |
| ·50 | ·5000 000e | ·4234 041 | ·3544 869 | ·2937 995+ | ·2413 101 | ·1965 961 |
| ·51 | ·5203 650+ | ·4442 373 | ·3750 556 | ·3135 262 | ·2597 767 | ·2135 322 |
| ·52 | ·5406 894 | ·4652 335+ | ·3959 895+ | ·3338 009 | ·2789 434 | ·2312 835+ |
| ·53 | ·5609 324 | ·4863 479 | ·4172 446 | ·3545 855+ | ·2987 819 | ·2498 345− |
| ·54 | ·5810 540 | ·5075 344 | ·4387 744 | ·3758 383 | ·3192 595+ | ·2691 645+ |
| ·55 | ·6010 142 | ·5287 465− | ·4605 307 | ·3975 144 | ·3403 393 | ·2892 480 |
| ·56 | ·6207 737 | ·5499 371 | ·4824 634 | ·4195 659 | ·3619 798 | ·3100 540 |
| ·57 | ·6402 941 | ·5710 590 | ·5045 211 | ·4419 417 | ·3841 356 | ·3315 465− |
| ·58 | ·6595 378 | ·5920 649 | ·5266 508 | ·4645 885− | ·4067 572 | ·3536 840 |
| ·59 | ·6784 680 | ·6129 075+ | ·5487 986 | ·4874 499 | ·4297 909 | ·3764 200 |
| ·60 | ·6970 494 | ·6335 401 | ·5709 098 | ·5104 678 | ·4531 795+ | ·3997 027 |

$x$ = ·61 to 1·00 $q$ = 3·5 $p$ = 3·5 to 6

| | $p$ = 3·5 | $p$ = 4 | $p$ = 4·5 | $p$ = 5 | $p$ = 5·5 | $p$ = 6 |
|---|---|---|---|---|---|---|
| $B(p, q)$ = | ·1533 9808 × $\frac{1}{10}$ | ·1065 6011 × $\frac{1}{10}$ | ·7669 9039 × $\frac{1}{10^2}$ | ·5683 2057 × $\frac{1}{10^2}$ | ·4314 3210 × $\frac{1}{10^2}$ | ·3343 0622 × $\frac{1}{10^2}$ |
| $x$ | | | | | | |
| ·61 | ·7152 476 | ·6539 163 | ·5929 289 | ·5335 816 | ·4768 622 | ·4234 754 |
| ·62 | ·7330 299 | ·6739 907 | ·6148 005− | ·5567 294 | ·5007 748 | ·4476 764 |
| ·63 | ·7503 650+ | ·6937 187 | ·6364 686 | ·5798 476 | ·5248 501 | ·4722 394 |
| ·64 | ·7672 234 | ·7130 571 | ·6578 780 | ·6028 717 | ·5490 185+ | ·4970 937 |
| ·65 | ·7835 773 | ·7319 638 | ·6789 737 | ·6257 363 | ·5732 077 | ·5221 645+ |
| ·66 | ·7994 009 | ·7503 987 | ·6997 016 | ·6483 759 | ·5973 437 | ·5473 733 |
| ·67 | ·8146 702 | ·7683 231 | ·7200 089 | ·6707 246 | ·6213 507 | ·5726 380 |
| ·68 | ·8293 637 | ·7857 007 | ·7398 441 | ·6927 172 | ·6451 522 | ·5978 740 |
| ·69 | ·8434 618 | ·8024 972 | ·7591 575− | ·7142 893 | ·6686 709 | ·6229 941 |
| ·70 | ·8569 474 | ·8186 809 | ·7779 015+ | ·7353 776 | ·6918 293 | ·6479 092 |
| ·71 | ·8698 060 | ·8342 225+ | ·7960 310 | ·7559 207 | ·7145 506 | ·6725 293 |
| ·72 | ·8820 252 | ·8490 958 | ·8135 034 | ·7758 592 | ·7367 589 | ·6967 636 |
| ·73 | ·8935 955+ | ·8632 774 | ·8302 793 | ·7951 362 | ·7583 801 | ·7205 215+ |
| ·74 | ·9045 101 | ·8767 472 | ·8463 226 | ·8136 980 | ·7793 423 | ·7437 135+ |
| ·75 | ·9147 647 | ·8894 882 | ·8616 008 | ·8314 946 | ·7995 763 | ·7662 517 |
| ·76 | ·9243 579 | ·9014 873 | ·8760 855+ | ·8484 796 | ·8190 168 | ·7880 508 |
| ·77 | ·9332 912 | ·9127 346 | ·8897 524 | ·8646 113 | ·8376 025− | ·8090 288 |
| ·78 | ·9415 688 | ·9232 243 | ·9025 816 | ·8798 529 | ·8552 771 | ·8291 084 |
| ·79 | ·9491 980 | ·9329 543 | ·9145 582 | ·8941 731 | ·8719 899 | ·8482 174 |
| ·80 | ·9561 886 | ·9419 266 | ·9256 723 | ·9075 463 | ·8876 964 | ·8662 899 |
| ·81 | ·9625 536 | ·9501 472 | ·9359 190 | ·9199 530 | ·9023 593 | ·8832 671 |
| ·82 | ·9683 088 | ·9576 262 | ·9452 990 | ·9313 807 | ·9159 486 | ·8990 988 |
| ·83 | ·9734 728 | ·9643 779 | ·9538 186 | ·9418 236 | ·9284 427 | ·9137 435− |
| ·84 | ·9780 669 | ·9704 207 | ·9614 898 | ·9512 833 | ·9398 290 | ·9271 702 |
| ·85 | ·9821 152 | ·9757 774 | ·9683 305+ | ·9597 693 | ·9501 040 | ·9393 589 |
| ·86 | ·9856 443 | ·9804 746 | ·9743 644 | ·9672 985+ | ·9592 745− | ·9503 013 |
| ·87 | ·9886 834 | ·9845 432 | ·9796 213 | ·9738 963 | ·9673 571 | ·9600 020 |
| ·88 | ·9912 639 | ·9880 178 | ·9841 365− | ·9795 959 | ·9743 797 | ·9684 790 |
| ·89 | ·9934 196 | ·9909 368 | ·9879 513 | ·9844 388 | ·9803 808 | ·9757 640 |
| ·90 | ·9951 860 | ·9933 422 | ·9911 126 | ·9884 746 | ·9854 098 | ·9819 034 |
| ·91 | ·9966 006 | ·9952 791 | ·9936 723 | ·9917 606 | ·9895 272 | ·9869 579 |
| ·92 | ·9977 019 | ·9967 956 | ·9956 873 | ·9943 616 | ·9928 043 | ·9910 028 |
| ·93 | ·9985 300 | ·9979 419 | ·9972 189 | ·9963 492 | ·9953 221 | ·9941 275+ |
| ·94 | ·9991 253 | ·9987 704 | ·9983 317 | ·9978 013 | ·9971 713 | ·9964 348 |
| ·95 | ·9995 285− | ·9993 345− | ·9990 934 | ·9988 004 | ·9984 505+ | ·9980 392 |
| ·96 | ·9997 797 | ·9996 878 | ·9995 730 | ·9994 327 | ·9992 644 | ·9990 655− |
| ·97 | ·9999 179 | ·9998 832 | ·9998 396 | ·9997 860 | ·9997 214 | ·9996 447 |
| ·98 | ·9999 797 | ·9999 711 | ·9999 601 | ·9999 466 | ·9999 302 | ·9999 106 |
| ·99 | ·9999 982 | ·9999 974 | ·9999 964 | ·9999 951 | ·9999 936 | ·9999 918 |
| 1·00 | 1·0000 000 | 1·0000 000 | 1·0000 000 | 1·0000 000 | 1·0000 000 | 1·0000 000 |

$x = \cdot05$ to $\cdot60$ $q = 3\cdot5$ $p = 6\cdot5$ to $9$

| $x$ | $p = 6\cdot5$ | $p = 7$ | $p = 7\cdot5$ | $p = 8$ | $p = 8\cdot5$ | $p = 9$ |
|---|---|---|---|---|---|---|
| $B(p, q) =$ | ·2636 5295 $\times \frac{1}{10^2}$ | ·2111 4077 $\times \frac{1}{10^2}$ | ·1713 7442 $\times \frac{1}{10^2}$ | ·1407 6051 $\times \frac{1}{10^2}$ | ·1168 4619 $\times \frac{1}{10^2}$ | ·9792 0356 $\times \frac{1}{10^3}$ |
| ·05 | ·0000 002 | | | | | |
| ·06 | ·0000 006 | ·0000 002 | | | | |
| ·07 | ·0000 016 | ·0000 005$^-$ | ·0000 001 | | | |
| ·08 | ·0000 036 | ·0000 012 | ·0000 004 | ·0000 001 | | |
| ·09 | ·0000 076 | ·0000 026 | ·0000 009 | ·0000 003 | ·0000 001 | |
| ·10 | ·0000 147 | ·0000 054 | ·0000 020 | ·0000 007 | ·0000 003 | ·0000 001 |
| ·11 | ·0000 267 | ·0000 102 | ·0000 039 | ·0000 015$^-$ | ·0000 006 | ·0000 002 |
| ·12 | ·0000 459 | ·0000 184 | ·0000 073 | ·0000 029 | ·0000 011 | ·0000 004 |
| ·13 | ·0000 754 | ·0000 314 | ·0000 130 | ·0000 053 | ·0000 022 | ·0000 009 |
| ·14 | ·0001 191 | ·0000 515$^-$ | ·0000 221 | ·0000 094 | ·0000 040 | ·0000 017 |
| ·15 | ·0001 819 | ·0000 814 | ·0000 361 | ·0000 159 | ·0000 070 | ·0000 030 |
| ·16 | ·0002 699 | ·0001 247 | ·0000 571 | ·0000 260 | ·0000 118 | ·0000 053 |
| ·17 | ·0003 903 | ·0001 858 | ·0000 878 | ·0000 412 | ·0000 192 | ·0000 089 |
| ·18 | ·0005 517 | ·0002 702 | ·0001 313 | ·0000 633 | ·0000 304 | ·0000 145$^-$ |
| ·19 | ·0007 641 | ·0003 843 | ·0001 918 | ·0000 951 | ·0000 468 | ·0000 229 |
| ·20 | ·0010 393 | ·0005 361 | ·0002 744 | ·0001 395$^+$ | ·0000 705$^-$ | ·0000 354 |
| ·21 | ·0013 903 | ·0007 347 | ·0003 852 | ·0002 006 | ·0001 038 | ·0000 534 |
| ·22 | ·0018 321 | ·0009 906 | ·0005 315$^+$ | ·0002 832 | ·0001 500$^+$ | ·0000 790 |
| ·23 | ·0023 814 | ·0013 161 | ·0007 218 | ·0003 932 | ·0002 129 | ·0001 146 |
| ·24 | ·0030 568 | ·0017 250$^+$ | ·0009 662 | ·0005 375$^-$ | ·0002 972 | ·0001 634 |
| ·25 | ·0038 785$^+$ | ·0022 331 | ·0012 761 | ·0007 243 | ·0004 086 | ·0002 292 |
| ·26 | ·0048 688 | ·0028 577 | ·0016 648 | ·0009 634 | ·0005 541 | ·0003 169 |
| ·27 | ·0060 517 | ·0036 183 | ·0021 473 | ·0012 659 | ·0007 418 | ·0004 323 |
| ·28 | ·0074 529 | ·0045 360 | ·0027 404 | ·0016 447 | ·0009 811 | ·0005 821 |
| ·29 | ·0090 999 | ·0056 342 | ·0034 629 | ·0021 144 | ·0012 833 | ·0007 747 |
| ·30 | ·0110 219 | ·0069 380 | ·0043 356 | ·0026 916 | ·0016 611 | ·0010 196 |
| ·31 | ·0132 495$^-$ | ·0084 745$^-$ | ·0053 813 | ·0033 948 | ·0021 291 | ·0013 281 |
| ·32 | ·0158 148 | ·0102 727 | ·0066 249 | ·0042 448 | ·0027 039 | ·0017 131 |
| ·33 | ·0187 511 | ·0123 632 | ·0080 935$^+$ | ·0052 644 | ·0034 042 | ·0021 897 |
| ·34 | ·0220 927 | ·0147 787 | ·0098 163 | ·0064 786 | ·0042 510 | ·0027 746 |
| ·35 | ·0258 749 | ·0175 532 | ·0118 244 | ·0079 148 | ·0052 675$^-$ | ·0034 872 |
| ·36 | ·0301 336 | ·0207 221 | ·0141 509 | ·0096 028 | ·0064 792 | ·0043 489 |
| ·37 | ·0349 051 | ·0243 221 | ·0168 310 | ·0115 744 | ·0079 144 | ·0053 837 |
| ·38 | ·0402 257 | ·0283 911 | ·0199 013 | ·0138 638 | ·0096 036 | ·0066 183 |
| ·39 | ·0461 318 | ·0329 676 | ·0234 003 | ·0165 074 | ·0115 799 | ·0080 818 |
| ·40 | ·0526 591 | ·0380 908 | ·0273 677 | ·0195 436 | ·0138 789 | ·0098 063 |
| ·41 | ·0598 428 | ·0438 000 | ·0318 446 | ·0230 127 | ·0165 388 | ·0118 265$^-$ |
| ·42 | ·0677 168 | ·0501 346 | ·0368 727 | ·0269 568 | ·0196 000 | ·0141 800 |
| ·43 | ·0763 137 | ·0571 335$^-$ | ·0424 947 | ·0314 194 | ·0231 051 | ·0169 070 |
| ·44 | ·0856 642 | ·0648 349 | ·0487 533 | ·0364 455$^-$ | ·0270 990 | ·0200 506 |
| ·45 | ·0957 970 | ·0732 760 | ·0556 914 | ·0420 808 | ·0316 280 | ·0236 562 |
| ·46 | ·1067 380 | ·0824 922 | ·0633 512 | ·0483 719 | ·0367 406 | ·0277 718 |
| ·47 | ·1185 107 | ·0925 173 | ·0717 742 | ·0553 654 | ·0424 861 | ·0324 474 |
| ·48 | ·1311 349 | ·1033 824 | ·0810 007 | ·0631 079 | ·0489 149 | ·0377 350$^-$ |
| ·49 | ·1446 270 | ·1151 161 | ·0910 691 | ·0716 453 | ·0560 779 | ·0436 881 |
| ·50 | ·1589 996 | ·1277 437 | ·1020 155$^-$ | ·0810 224 | ·0640 260 | ·0503 615$^+$ |
| ·51 | ·1742 608 | ·1412 866 | ·1138 734 | ·0912 823 | ·0728 099 | ·0578 107 |
| ·52 | ·1904 140 | ·1557 624 | ·1266 728 | ·1024 660 | ·0824 790 | ·0660 912 |
| ·53 | ·2074 580 | ·1711 839 | ·1404 403 | ·1146 116 | ·0930 811 | ·0752 586 |
| ·54 | ·2253 861 | ·1875 590 | ·1551 976 | ·1277 539 | ·1046 619 | ·0853 670 |
| ·55 | ·2441 861 | ·2048 904 | ·1709 618 | ·1419 235$^-$ | ·1172 640 | ·0964 692 |
| ·56 | ·2638 404 | ·2231 746 | ·1877 446 | ·1571 463 | ·1309 264 | ·1086 154 |
| ·57 | ·2843 251 | ·2424 023 | ·2055 516 | ·1734 429 | ·1456 838 | ·1218 526 |
| ·58 | ·3056 106 | ·2625 575$^-$ | ·2243 821 | ·1908 279 | ·1615 653 | ·1362 237 |
| ·59 | ·3276 606 | ·2836 174 | ·2442 281 | ·2093 091 | ·1785 945$^-$ | ·1517 666 |
| ·60 | ·3504 331 | ·3055 524 | ·2650 745$^+$ | ·2288 871 | ·1967 877 | ·1685 132 |

$x$ = ·61 to 1·00 $q = 3·5$ $p$ = 6·5 to 9

| $x$ | $p = 6·5$ | $p = 7$ | $p = 7·5$ | $p = 8$ | $p = 8·5$ | $p = 9$ |
|---|---|---|---|---|---|---|
| $B(p, q) =$ | ·2636 5295 $\times \frac{1}{10^2}$ | ·2111 4077 $\times \frac{1}{10^2}$ | ·1713 7442 $\times \frac{1}{10^2}$ | ·1407 6051 $\times \frac{1}{10^2}$ | ·1168 4619 $\times \frac{1}{10^2}$ | ·9792 0356 $\times \frac{1}{10^3}$ |
| ·61 | ·3738 793 | ·3283 253 | ·2868 983 | ·2495 545+ | ·2161 539 | ·1864 887 |
| ·62 | ·3979 443 | ·3518 915− | ·3096 683 | ·2712 954 | ·2366 935+ | ·2057 102 |
| ·63 | ·4225 670 | ·3761 991 | ·3333 447 | ·2940 846 | ·2583 979 | ·2261 862 |
| ·64 | ·4476 803 | ·4011 884 | ·3578 790 | ·3178 877 | ·2812 485+ | ·2479 152 |
| ·65 | ·4732 108 | ·4267 922 | ·3832 136 | ·3426 601 | ·3052 161 | ·2708 849 |
| ·66 | ·4990 801 | ·4529 358 | ·4092 822 | ·3683 469 | ·3302 602 | ·2950 717 |
| ·67 | ·5252 039 | ·4795 375+ | ·4360 092 | ·3948 826 | ·3563 286 | ·3204 392 |
| ·68 | ·5514 934 | ·5065 084 | ·4633 100 | ·4221 910 | ·3833 570 | ·3469 377 |
| ·69 | ·5778 552 | ·5337 530 | ·4910 913 | ·4501 851 | ·4112 685+ | ·3745 039 |
| ·70 | ·6041 921 | ·5611 697 | ·5192 514 | ·4787 672 | ·4399 734 | ·4030 599 |
| ·71 | ·6304 035+ | ·5886 514 | ·5476 805+ | ·5078 291 | ·4693 694 | ·4325 128 |
| ·72 | ·6563 866 | ·6160 861 | ·5762 615+ | ·5372 527 | ·4993 415− | ·4627 551 |
| ·73 | ·6820 363 | ·6433 575+ | ·6048 704 | ·5669 101 | ·5297 622 | ·4936 639 |
| ·74 | ·7072 471 | ·6703 464 | ·6333 773 | ·5966 649 | ·5604 925− | ·5251 017 |
| ·75 | ·7319 132 | ·6969 312 | ·6616 477 | ·6263 728 | ·5913 822 | ·5569 168 |
| ·76 | ·7559 299 | ·7229 891 | ·6895 431 | ·6558 826 | ·6222 712 | ·5889 439 |
| ·77 | ·7791 946 | ·7483 976 | ·7169 229 | ·6850 380 | ·6529 905+ | ·6210 053 |
| ·78 | ·8016 080 | ·7730 358 | ·7436 453 | ·7136 788 | ·6833 641 | ·6529 122 |
| ·79 | ·8230 749 | ·7967 853 | ·7695 695+ | ·7416 427 | ·7132 105+ | ·6844 668 |
| ·80 | ·8435 064 | ·8195 326 | ·7945 573 | ·7687 675+ | ·7423 451 | ·7154 642 |
| ·81 | ·8628 199 | ·8411 700 | ·8184 749 | ·7948 933 | ·7705 824 | ·7456 954 |
| ·82 | ·8809 416 | ·8615 979 | ·8411 952 | ·8198 648 | ·7977 391 | ·7749 496 |
| ·83 | ·8978 072 | ·8807 260 | ·8625 999 | ·8435 342 | ·8236 371 | ·8030 183 |
| ·84 | ·9133 635− | ·8984 757 | ·8825 820 | ·8657 638 | ·8481 068 | ·8296 990 |
| ·85 | ·9275 695+ | ·9147 814 | ·9010 481 | ·8864 296 | ·8709 905+ | ·8547 995− |
| ·86 | ·9403 983 | ·9295 931 | ·9179 211 | ·9054 236 | ·8921 472 | ·8781 423 |
| ·87 | ·9518 375− | ·9428 774 | ·9331 422 | ·9226 581 | ·9114 558 | ·8995 702 |
| ·88 | ·9618 911 | ·9546 199 | ·9466 743 | ·9380 682 | ·9288 198 | ·9189 509 |
| ·89 | ·9705 804 | ·9648 265− | ·9585 033 | ·9516 155+ | ·9441 717 | ·9361 835− |
| ·90 | ·9779 443 | ·9735 250− | ·9686 411 | ·9632 914 | ·9574 774 | ·9512 030 |
| ·91 | ·9840 407 | ·9807 663 | ·9771 276 | ·9731 197 | ·9687 397 | ·9639 868 |
| ·92 | ·9889 462 | ·9866 251 | ·9840 315+ | ·9811 592 | ·9780 029 | ·9745 591 |
| ·93 | ·9927 563 | ·9912 004 | ·9894 523 | ·9875 059 | ·9853 554 | ·9829 963 |
| ·94 | ·9955 848 | ·9946 151 | ·9935 198 | ·9922 936 | ·9909 317 | ·9894 296 |
| ·95 | ·9975 621 | ·9970 149 | ·9963 936 | ·9956 943 | ·9949 134 | ·9940 476 |
| ·96 | ·9988 334 | ·9985 659 | ·9982 606 | ·9979 152 | ·9975 274 | ·9970 952 |
| ·97 | ·9995 548 | ·9994 505+ | ·9993 309 | ·9991 949 | ·9990 414 | ·9988 694 |
| ·98 | ·9998 875+ | ·9998 606 | ·9998 296 | ·9997 942 | ·9997 540 | ·9997 087 |
| ·99 | ·9999 896 | ·9999 871 | ·9999 842 | ·9999 808 | ·9999 769 | ·9999 726 |
| 1·00 | 1·0000 000 | 1·0000 000 | 1·0000 000 | 1·0000 000 | 1·0000 000 | 1·0000 000 |

$x$ = ·11 to ·70 $q = 3·5$ $p$ = 9·5 to 13

| | $p = 9·5$ | $p = 10$ | $p = 10·5$ | $p = 11$ | $p = 12$ | $p = 13$ |
|---|---|---|---|---|---|---|
| $B(p, q) =$ | ·8276 6053 × $\frac{1}{10^3}$ | ·7050 2657 × $\frac{1}{10^3}$ | ·6048 2885+ × $\frac{1}{10^3}$ | ·5222 4190 × $\frac{1}{10^3}$ | ·3961 8351 × $\frac{1}{10^3}$ | ·3067 2272 × $\frac{1}{10^3}$ |
| $x$ | | | | | | |
| ·11 | ·0000 001 | | | | | |
| ·12 | ·0000 002 | ·0000 001 | | | | |
| ·13 | ·0000 004 | ·0000 001 | ·0000 001 | | | |
| ·14 | ·0000 007 | ·0000 003 | ·0000 001 | | | |
| ·15 | ·0000 013 | ·0000 006 | ·0000 002 | ·0000 001 | | |
| ·16 | ·0000 024 | ·0000 011 | ·0000 005− | ·0000 002 | | |
| ·17 | ·0000 041 | ·0000 019 | ·0000 009 | ·0000 004 | ·0000 001 | |
| ·18 | ·0000 069 | ·0000 032 | ·0000 015+ | ·0000 007 | ·0000 002 | |
| ·19 | ·0000 112 | ·0000 054 | ·0000 026 | ·0000 013 | ·0000 003 | ·0000 001 |
| ·20 | ·0000 177 | ·0000 088 | ·0000 044 | ·0000 022 | ·0000 005+ | ·0000 001 |
| ·21 | ·0000 274 | ·0000 139 | ·0000 071 | ·0000 036 | ·0000 009 | ·0000 002 |
| ·22 | ·0000 414 | ·0000 216 | ·0000 112 | ·0000 058 | ·0000 015+ | ·0000 004 |
| ·23 | ·0000 614 | ·0000 327 | ·0000 174 | ·0000 092 | ·0000 025+ | ·0000 007 |
| ·24 | ·0000 894 | ·0000 487 | ·0000 264 | ·0000 142 | ·0000 041 | ·0000 012 |
| ·25 | ·0001 280 | ·0000 711 | ·0000 393 | ·0000 217 | ·0000 065+ | ·0000 019 |
| ·26 | ·0001 804 | ·0001 022 | ·0000 576 | ·0000 324 | ·0000 101 | ·0000 031 |
| ·27 | ·0002 506 | ·0001 447 | ·0000 831 | ·0000 476 | ·0000 154 | ·0000 049 |
| ·28 | ·0003 436 | ·0002 019 | ·0001 182 | ·0000 689 | ·0000 232 | ·0000 077 |
| ·29 | ·0004 653 | ·0002 782 | ·0001 657 | ·0000 983 | ·0000 342 | ·0000 118 |
| ·30 | ·0006 227 | ·0003 786 | ·0002 293 | ·0001 383 | ·0000 498 | ·0000 177 |
| ·31 | ·0008 244 | ·0005 094 | ·0003 135− | ·0001 922 | ·0000 715− | ·0000 262 |
| ·32 | ·0010 801 | ·0006 780 | ·0004 238 | ·0002 639 | ·0001 013 | ·0000 384 |
| ·33 | ·0014 016 | ·0008 932 | ·0005 668 | ·0003 584 | ·0001 418 | ·0000 554 |
| ·34 | ·0018 023 | ·0011 655− | ·0007 506 | ·0004 816 | ·0001 962 | ·0000 790 |
| ·35 | ·0022 976 | ·0015 071 | ·0009 846 | ·0006 408 | ·0002 687 | ·0001 113 |
| ·36 | ·0029 051 | ·0019 321 | ·0012 798 | ·0008 446 | ·0003 641 | ·0001 551 |
| ·37 | ·0036 449 | ·0024 570 | ·0016 495+ | ·0011 033 | ·0004 886 | ·0002 138 |
| ·38 | ·0045 395+ | ·0031 002 | ·0021 088 | ·0014 291 | ·0006 498 | ·0002 919 |
| ·39 | ·0056 141 | ·0038 830 | ·0026 751 | ·0018 362 | ·0008 564 | ·0003 947 |
| ·40 | ·0068 965+ | ·0048 294 | ·0033 686 | ·0023 410 | ·0011 194 | ·0005 290 |
| ·41 | ·0084 178 | ·0059 662 | ·0042 120 | ·0029 628 | ·0014 515− | ·0007 028 |
| ·42 | ·0102 118 | ·0073 231 | ·0052 311 | ·0037 232 | ·0018 677 | ·0009 260 |
| ·43 | ·0123 154 | ·0089 333 | ·0064 549 | ·0046 474 | ·0023 856 | ·0012 104 |
| ·44 | ·0147 687 | ·0108 330 | ·0079 157 | ·0057 634 | ·0030 257 | ·0015 702 |
| ·45 | ·0176 147 | ·0130 621 | ·0096 492 | ·0071 028 | ·0038 117 | ·0020 222 |
| ·46 | ·0208 995− | ·0156 635− | ·0116 949 | ·0087 012 | ·0047 707 | ·0025 860 |
| ·47 | ·0246 720 | ·0186 838 | ·0140 960 | ·0105 977 | ·0059 335+ | ·0032 846 |
| ·48 | ·0289 839 | ·0221 729 | ·0168 994 | ·0128 357 | ·0073 352 | ·0041 449 |
| ·49 | ·0338 893 | ·0261 838 | ·0201 558 | ·0154 624 | ·0090 150− | ·0051 975+ |
| ·50 | ·0394 446 | ·0307 726 | ·0239 196 | ·0185 296 | ·0110 168 | ·0064 778 |
| ·51 | ·0457 080 | ·0359 983 | ·0282 487 | ·0220 929 | ·0133 894 | ·0080 259 |
| ·52 | ·0527 391 | ·0419 222 | ·0332 047 | ·0262 124 | ·0161 866 | ·0098 871 |
| ·53 | ·0605 982 | ·0486 078 | ·0388 518 | ·0309 516 | ·0194 671 | ·0121 123 |
| ·54 | ·0693 463 | ·0561 200 | ·0452 574 | ·0363 783 | ·0232 950− | ·0147 581 |
| ·55 | ·0790 438 | ·0645 250− | ·0524 908 | ·0425 633 | ·0277 392 | ·0178 873 |
| ·56 | ·0897 500− | ·0738 889 | ·0606 230 | ·0495 804 | ·0328 738 | ·0215 689 |
| ·57 | ·1015 224 | ·0842 777 | ·0697 262 | ·0575 056 | ·0387 773 | ·0258 782 |
| ·58 | ·1144 159 | ·0957 560 | ·0798 727 | ·0664 169 | ·0455 328 | ·0308 965− |
| ·59 | ·1284 813 | ·1083 860 | ·0911 337 | ·0763 927 | ·0532 267 | ·0367 113 |
| ·60 | ·1437 649 | ·1222 266 | ·1035 793 | ·0875 116 | ·0619 488 | ·0434 157 |
| ·61 | ·1603 073 | ·1373 324 | ·1172 761 | ·0998 507 | ·0717 906 | ·0511 079 |
| ·62 | ·1781 419 | ·1537 523 | ·1322 868 | ·1134 848 | ·0828 450− | ·0598 906 |
| ·63 | ·1972 943 | ·1715 280 | ·1486 686 | ·1284 846 | ·0952 043 | ·0698 696 |
| ·64 | ·2177 805− | ·1906 932 | ·1664 716 | ·1449 156 | ·1089 593 | ·0811 532 |
| ·65 | ·2396 063 | ·2112 717 | ·1857 373 | ·1628 359 | ·1241 974 | ·0938 500+ |
| ·66 | ·2627 660 | ·2332 763 | ·2064 972 | ·1822 951 | ·1410 003 | ·1080 678 |
| ·67 | ·2872 408 | ·2567 073 | ·2287 708 | ·2033 319 | ·1594 427 | ·1239 108 |
| ·68 | ·3129 984 | ·2815 510 | ·2525 641 | ·2259 723 | ·1795 893 | ·1414 780 |
| ·69 | ·3399 915− | ·3077 784 | ·2778 679 | ·2502 276 | ·2014 925+ | ·1608 596 |
| ·70 | ·3681 569 | ·3353 439 | ·3046 560 | ·2760 924 | ·2251 900 | ·1821 350− |

$x = \cdot 71$ to $1\cdot 00$ $q = 3\cdot 5$ $p = 9\cdot 5$ to $13$

| $x$ | $p = 9\cdot 5$ | $p = 10$ | $p = 10\cdot 5$ | $p = 11$ | $p = 12$ | $p = 13$ |
|---|---|---|---|---|---|---|
| $B(p, q) =$ | ·8276 6053 $\times \frac{1}{10^3}$ | ·7050 2657 $\times \frac{1}{10^3}$ | ·6048 2885 $^{+}\times \frac{1}{10^3}$ | ·5222 4190 $\times \frac{1}{10^3}$ | ·3961 8351 $\times \frac{1}{10^3}$ | ·3067 2272 $\times \frac{1}{10^3}$ |
| ·71 | ·3974 153 | ·3641 842 | ·3328 839 | ·3035 425+ | ·2507 016 | ·2053 685+ |
| ·72 | ·4276 701 | ·3942 173 | ·3624 868 | ·3325 333 | ·2780 270 | ·2306 066 |
| ·73 | ·4588 070 | ·4253 414 | ·3933 789 | ·3629 972 | ·3071 424 | ·2578 739 |
| ·74 | ·4906 945− | ·4574 348 | ·4254 517 | ·3948 428 | ·3379 980 | ·2871 692 |
| ·75 | ·5231 833 | ·4903 549 | ·4585 738 | ·4279 531 | ·3705 151 | ·3184 619 |
| ·76 | ·5561 072 | ·5239 390 | ·4925 898 | ·4621 843 | ·4045 841 | ·3516 878 |
| ·77 | ·5892 838 | ·5580 041 | ·5273 207 | ·4973 655+ | ·4400 620 | ·3867 459 |
| ·78 | ·6225 156 | ·5923 481 | ·5625 638 | ·5332 980 | ·4767 711 | ·4234 944 |
| ·79 | ·6555 917 | ·6267 508 | ·5980 942 | ·5697 563 | ·5144 976 | ·4617 485− |
| ·80 | ·6882 900 | ·6609 764 | ·6336 659 | ·6064 886 | ·5529 915− | ·5012 776 |
| ·81 | ·7203 797 | ·6947 755+ | ·6690 145− | ·6432 191 | ·5919 674 | ·5418 048 |
| ·82 | ·7516 246 | ·7278 884 | ·7038 596 | ·6796 506 | ·6311 059 | ·5830 062 |
| ·83 | ·7817 866 | ·7600 490 | ·7379 095+ | ·7154 683 | ·6700 568 | ·6245 128 |
| ·84 | ·8106 301 | ·7909 895− | ·7708 655− | ·7503 445− | ·7084 432 | ·6659 133 |
| ·85 | ·8379 272 | ·8204 457 | ·8024 276 | ·7839 449 | ·7458 678 | ·7067 595+ |
| ·86 | ·8634 624 | ·8481 632 | ·8323 017 | ·8159 357 | ·7819 210 | ·7465 740 |
| ·87 | ·8870 394 | ·8739 041 | ·8602 070 | ·8459 922 | ·8161 900 | ·7848 603 |
| ·88 | ·9084 867 | ·8974 547 | ·8858 848 | ·8738 088 | ·8482 712 | ·8211 163 |
| ·89 | ·9276 652 | ·9186 338 | ·9091 084 | ·8991 098 | ·8777 843 | ·8548 504 |
| ·90 | ·9444 748 | ·9373 013 | ·9296 930 | ·9216 620 | ·9043 875+ | ·8856 016 |
| ·91 | ·9588 618 | ·9533 672 | ·9475 073 | ·9412 875− | ·9277 966 | ·9129 623 |
| ·92 | ·9708 254 | ·9668 004 | ·9624 843 | ·9578 779 | ·9478 036 | ·9366 043 |
| ·93 | ·9804 246 | ·9776 374 | ·9746 323 | ·9714 077 | ·9642 974 | ·9563 071 |
| ·94 | ·9877 834 | ·9859 895+ | ·9840 451 | ·9819 475− | ·9772 844 | ·9719 880 |
| ·95 | ·9930 936 | ·9920 487 | ·9909 099 | ·9896 750− | ·9869 074 | ·9837 306 |
| ·96 | ·9966 165− | ·9960 893 | ·9955 118 | ·9948 823 | ·9934 602 | ·9918 106 |
| ·97 | ·9986 779 | ·9984 660 | ·9982 326 | ·9979 768 | ·9973 945+ | ·9967 120 |
| ·98 | ·9996 580 | ·9996 017 | ·9995 393 | ·9994 705+ | ·9993 128 | ·9991 261 |
| ·99 | ·9999 677 | ·9999 622 | ·9999 561 | ·9999 494 | ·9999 338 | ·9999 152 |
| 1·00 | 1·0000 000 | 1·0000 000 | 1·0000 000 | 1·0000 000 | 1·0000 000 | 1·0000 000 |

$x$ = ·21 to ·80 $q$ = 3·5 $p$ = 14 to 19

| $x$ | $p$ = 14 | $p$ = 15 | $p$ = 16 | $p$ = 17 | $p$ = 18 | $p$ = 19 |
|---|---|---|---|---|---|---|
| $B(p, q)$ = | ·2416 6032 × $\frac{1}{10^3}$ | ·1933 2826 × $\frac{1}{10^3}$ | ·1567 5264 × $\frac{1}{10^3}$ | ·1286 1755+ × $\frac{1}{10^3}$ | ·1066 5846 × $\frac{1}{10^3}$ | ·8929 5453 × $\frac{1}{10^4}$ |
| ·21 | ·0000 001 | | | | | |
| ·22 | ·0000 001 | | | | | |
| ·23 | ·0000 002 | | | | | |
| ·24 | ·0000 003 | ·0000 001 | | | | |
| ·25 | ·0000 006 | ·0000 002 | | | | |
| ·26 | ·0000 010 | ·0000 003 | ·0000 001 | | | |
| ·27 | ·0000 016 | ·0000 005− | ·0000 002 | | | |
| ·28 | ·0000 025+ | ·0000 008 | ·0000 003 | ·0000 001 | | |
| ·29 | ·0000 040 | ·0000 013 | ·0000 005− | ·0000 001 | | |
| ·30 | ·0000 062 | ·0000 022 | ·0000 007 | ·0000 003 | ·0000 001 | |
| ·31 | ·0000 095+ | ·0000 034 | ·0000 012 | ·0000 004 | ·0000 002 | ·0000 001 |
| ·32 | ·0000 144 | ·0000 053 | ·0000 020 | ·0000 007 | ·0000 003 | ·0000 001 |
| ·33 | ·0000 214 | ·0000 082 | ·0000 031 | ·0000 012 | ·0000 004 | ·0000 002 |
| ·34 | ·0000 315− | ·0000 124 | ·0000 049 | ·0000 019 | ·0000 007 | ·0000 003 |
| ·35 | ·0000 456 | ·0000 185+ | ·0000 075− | ·0000 030 | ·0000 012 | ·0000 005− |
| ·36 | ·0000 654 | ·0000 273 | ·0000 113 | ·0000 046 | ·0000 019 | ·0000 008 |
| ·37 | ·0000 926 | ·0000 397 | ·0000 169 | ·0000 071 | ·0000 030 | ·0000 012 |
| ·38 | ·0001 298 | ·0000 572 | ·0000 250− | ·0000 108 | ·0000 047 | ·0000 020 |
| ·39 | ·0001 801 | ·0000 814 | ·0000 365− | ·0000 162 | ·0000 072 | ·0000 032 |
| ·40 | ·0002 474 | ·0001 146 | ·0000 527 | ·0000 240 | ·0000 109 | ·0000 049 |
| ·41 | ·0003 368 | ·0001 599 | ·0000 753 | ·0000 352 | ·0000 164 | ·0000 076 |
| ·42 | ·0004 544 | ·0002 210 | ·0001 066 | ·0000 510 | ·0000 243 | ·0000 115− |
| ·43 | ·0006 079 | ·0003 026 | ·0001 494 | ·0000 732 | ·0000 357 | ·0000 173 |
| ·44 | ·0008 067 | ·0004 107 | ·0002 074 | ·0001 040 | ·0000 518 | ·0000 257 |
| ·45 | ·0010 620 | ·0005 528 | ·0002 855− | ·0001 464 | ·0000 746 | ·0000 378 |
| ·46 | ·0013 878 | ·0007 381 | ·0003 895+ | ·0002 041 | ·0001 063 | ·0000 550− |
| ·47 | ·0018 003 | ·0009 780 | ·0005 271 | ·0002 821 | ·0001 500+ | ·0000 793 |
| ·48 | ·0023 191 | ·0012 862 | ·0007 078 | ·0003 867 | ·0002 100 | ·0001 134 |
| ·49 | ·0029 673 | ·0016 793 | ·0009 430 | ·0005 258 | ·0002 914 | ·0001 605+ |
| ·50 | ·0037 719 | ·0021 773 | ·0012 472 | ·0007 094 | ·0004 010 | ·0002 254 |
| ·51 | ·0047 645− | ·0028 041 | ·0016 377 | ·0009 499 | ·0005 475+ | ·0003 138 |
| ·52 | ·0059 814 | ·0035 878 | ·0021 356 | ·0012 625+ | ·0007 418 | ·0004 333 |
| ·53 | ·0074 645+ | ·0045 614 | ·0027 663 | ·0016 662 | ·0009 974 | ·0005 937 |
| ·54 | ·0092 616 | ·0057 635+ | ·0035 597 | ·0021 837 | ·0013 314 | ·0008 073 |
| ·55 | ·0114 268 | ·0072 390 | ·0045 518 | ·0028 429 | ·0017 648 | ·0010 895− |
| ·56 | ·0140 208 | ·0090 391 | ·0057 843 | ·0036 768 | ·0023 231 | ·0014 597 |
| ·57 | ·0171 117 | ·0112 226 | ·0073 063 | ·0047 252 | ·0030 376 | ·0019 421 |
| ·58 | ·0207 749 | ·0138 561 | ·0091 744 | ·0060 347 | ·0039 459 | ·0025 661 |
| ·59 | ·0250 933 | ·0170 147 | ·0114 539 | ·0076 604 | ·0050 931 | ·0033 680 |
| ·60 | ·0301 574 | ·0207 819 | ·0142 191 | ·0096 661 | ·0065 326 | ·0043 914 |
| ·61 | ·0360 653 | ·0252 507 | ·0175 543 | ·0121 259 | ·0083 277 | ·0056 889 |
| ·62 | ·0429 220 | ·0305 228 | ·0215 541 | ·0151 246 | ·0105 522 | ·0073 235− |
| ·63 | ·0508 394 | ·0367 095− | ·0263 240 | ·0187 588 | ·0132 918 | ·0093 693 |
| ·64 | ·0599 350− | ·0439 304 | ·0319 802 | ·0231 372 | ·0166 453 | ·0119 135− |
| ·65 | ·0703 308 | ·0523 134 | ·0386 502 | ·0283 815+ | ·0207 253 | ·0150 575+ |
| ·66 | ·0821 523 | ·0619 939 | ·0464 718 | ·0346 266 | ·0256 590 | ·0189 184 |
| ·67 | ·0955 263 | ·0731 126 | ·0555 926 | ·0420 201 | ·0315 892 | ·0236 298 |
| ·68 | ·1105 785− | ·0858 147 | ·0661 688 | ·0507 222 | ·0386 739 | ·0293 431 |
| ·69 | ·1274 313 | ·1002 469 | ·0783 635− | ·0609 046 | ·0470 864 | ·0362 276 |
| ·70 | ·1462 007 | ·1165 548 | ·0923 441 | ·0727 485− | ·0570 144 | ·0444 708 |
| ·71 | ·1669 923 | ·1348 796 | ·1082 796 | ·0864 427 | ·0686 584 | ·0542 779 |
| ·72 | ·1898 980 | ·1553 538 | ·1263 367 | ·1021 800 | ·0822 294 | ·0658 699 |
| ·73 | ·2149 913 | ·1780 969 | ·1466 752 | ·1201 534 | ·0979 452 | ·0794 816 |
| ·74 | ·2423 225+ | ·2032 099 | ·1694 426 | ·1405 505− | ·1160 262 | ·0953 576 |
| ·75 | ·2719 143 | ·2307 695+ | ·1947 679 | ·1635 477 | ·1366 892 | ·1137 475+ |
| ·76 | ·3037 557 | ·2608 222 | ·2227 545+ | ·1893 025+ | ·1601 403 | ·1348 987 |
| ·77 | ·3377 976 | ·2933 770 | ·2534 726 | ·2179 451 | ·1865 660 | ·1590 483 |
| ·78 | ·3739 471 | ·3283 990 | ·2869 502 | ·2495 686 | ·2161 231 | ·1864 122 |
| ·79 | ·4120 629 | ·3658 020 | ·3231 650− | ·2842 187 | ·2489 265− | ·2171 733 |
| ·80 | ·4519 505+ | ·4054 421 | ·3620 343 | ·3218 820 | ·2850 364 | ·2514 660 |

$x$ = ·81 to 1·00 $q$ = 3·5 $p$ = 14 to 19

| | $p$ = 14 | $p$ = 15 | $p$ = 16 | $p$ = 17 | $p$ = 18 | $p$ = 19 |
|---|---|---|---|---|---|---|
| $B(p, q)$ = | ·2416 6032 × $\frac{1}{10^3}$ | ·1933 2826 × $\frac{1}{10^3}$ | ·1567 5264 × $\frac{1}{10^3}$ | ·1286 1755$^+$ × $\frac{1}{10^3}$ | ·1066 5846 × $\frac{1}{10^3}$ | ·8929 5453 × $\frac{1}{10^4}$ |
| $x$ | | | | | | |
| ·81 | ·4933 593 | ·4471 111 | ·4034 066 | ·3624 746 | ·3244 440 | ·2893 607 |
| ·82 | ·5359 795+ | ·4905 316 | ·4470 531 | ·4058 300 | ·3670 560 | ·3308 454 |
| ·83 | ·5794 418 | ·5353 528 | ·4926 598 | ·4516 879 | ·4126 803 | ·3758 073 |
| ·84 | ·6233 181 | ·5811 487 | ·5398 228 | ·4996 850− | ·4610 110 | ·4240 129 |
| ·85 | ·6671 248 | ·6274 193 | ·5880 446 | ·5493 467 | ·5116 163 | ·4750 911 |
| ·86 | ·7103 298 | ·6735 937 | ·6367 352 | ·6000 839 | ·5639 286 | ·5285 164 |
| ·87 | ·7523 619 | ·7190 394 | ·6852 170 | ·6511 938 | ·6172 407 | ·5835 988 |
| ·88 | ·7926 245+ | ·7630 744 | ·7327 364 | ·7018 674 | ·6707 079 | ·6394 791 |
| ·89 | ·8305 141 | ·8049 870 | ·7784 814 | ·7512 055− | ·7233 599 | ·6951 354 |
| ·90 | ·8654 429 | ·8440 602 | ·8216 084 | ·7982 445+ | ·7741 247 | ·7494 020 |
| ·91 | ·8968 671 | ·8796 049 | ·8612 783 | ·8419 952 | ·8218 671 | ·8010 065+ |
| ·92 | ·9243 195− | ·9109 993 | ·8967 022 | ·8814 938 | ·8654 444 | ·8486 282 |
| ·93 | ·9474 471 | ·9377 360 | ·9271 994 | ·9158 692 | ·9037 826 | ·8909 809 |
| ·94 | ·9660 518 | ·9594 755− | ·9522 634 | ·9444 248 | ·9359 729 | ·9269 247 |
| ·95 | ·9801 321 | ·9761 032 | ·9716 378 | ·9667 328 | ·9613 878 | ·9556 048 |
| ·96 | ·9899 224 | ·9877 860 | ·9853 933 | ·9827 374 | ·9798 127 | ·9766 151 |
| ·97 | ·9959 227 | ·9950 203 | ·9939 991 | ·9928 537 | ·9915 793 | ·9901 715+ |
| ·98 | ·9989 079 | ·9986 559 | ·9983 677 | ·9980 412 | ·9976 742 | ·9972 646 |
| ·99 | ·9998 932 | ·9998 675− | ·9998 378 | ·9998 039 | ·9997 653 | ·9997 219 |
| 1·00 | 1·0000 000 | 1·0000 000 | 1·0000 000 | 1·0000 000 | 1·0000 000 | 1·0000 000 |

$x$ = ·33 to ·90 $q = 3·5$ $p$ = 20 to 25

| | $p = 20$ | $p = 21$ | $p = 22$ | $p = 23$ | $p = 24$ | $p = 25$ |
|---|---|---|---|---|---|---|
| $B(p, q)$ = | ·7540 5050 × $\frac{1}{10^4}$ | ·6417 4510 × $\frac{1}{10^4}$ | ·5500 6723 × $\frac{1}{10^4}$ | ·4745 6781 × $\frac{1}{10^4}$ | ·4118 8904 × $\frac{1}{10^4}$ | ·3594 6680 × $\frac{1}{10^4}$ |
| $x$ | | | | | | |
| ·33 | ·0000 001 | | | | | |
| ·34 | ·0000 001 | | | | | |
| ·35 | ·0000 002 | ·0000 001 | | | | |
| ·36 | ·0000 003 | ·0000 001 | | | | |
| ·37 | ·0000 005+ | ·0000 002 | ·0000 001 | | | |
| ·38 | ·0000 009 | ·0000 004 | ·0000 002 | ·0000 001 | | |
| ·39 | ·0000 014 | ·0000 006 | ·0000 003 | ·0000 001 | | |
| ·40 | ·0000 022 | ·0000 010 | ·0000 004 | ·0000 002 | ·0000 001 | |
| ·41 | ·0000 035− | ·0000 016 | ·0000 007 | ·0000 003 | ·0000 001 | ·0000 001 |
| ·42 | ·0000 054 | ·0000 025+ | ·0000 012 | ·0000 005+ | ·0000 003 | ·0000 001 |
| ·43 | ·0000 083 | ·0000 040 | ·0000 019 | ·0000 009 | ·0000 004 | ·0000 002 |
| ·44 | ·0000 126 | ·0000 062 | ·0000 030 | ·0000 015− | ·0000 007 | ·0000 003 |
| ·45 | ·0000 190 | ·0000 095+ | ·0000 048 | ·0000 024 | ·0000 012 | ·0000 006 |
| ·46 | ·0000 283 | ·0000 145+ | ·0000 074 | ·0000 038 | ·0000 019 | ·0000 010 |
| ·47 | ·0000 417 | ·0000 218 | ·0000 114 | ·0000 059 | ·0000 031 | ·0000 016 |
| ·48 | ·0000 609 | ·0000 325+ | ·0000 173 | ·0000 092 | ·0000 048 | ·0000 025+ |
| ·49 | ·0000 880 | ·0000 480 | ·0000 261 | ·0000 141 | ·0000 076 | ·0000 041 |
| ·50 | ·0001 260 | ·0000 701 | ·0000 389 | ·0000 214 | ·0000 118 | ·0000 065− |
| ·51 | ·0001 789 | ·0001 015+ | ·0000 574 | ·0000 323 | ·0000 181 | ·0000 101 |
| ·52 | ·0002 519 | ·0001 457 | ·0000 839 | ·0000 481 | ·0000 275+ | ·0000 157 |
| ·53 | ·0003 516 | ·0002 073 | ·0001 216 | ·0000 711 | ·0000 414 | ·0000 240 |
| ·54 | ·0004 870 | ·0002 924 | ·0001 748 | ·0001 041 | ·0000 618 | ·0000 365+ |
| ·55 | ·0006 692 | ·0004 091 | ·0002 491 | ·0001 510 | ·0000 912 | ·0000 549 |
| ·56 | ·0009 126 | ·0005 680 | ·0003 520 | ·0002 172 | ·0001 336 | ·0000 819 |
| ·57 | ·0012 355− | ·0007 824 | ·0004 934 | ·0003 099 | ·0001 940 | ·0001 210 |
| ·58 | ·0016 606 | ·0010 697 | ·0006 862 | ·0004 385− | ·0002 792 | ·0001 772 |
| ·59 | ·0022 162 | ·0014 518 | ·0009 471 | ·0006 154 | ·0003 985+ | ·0002 572 |
| ·60 | ·0029 376 | ·0019 563 | ·0012 974 | ·0008 571 | ·0005 643 | ·0003 703 |
| ·61 | ·0038 675+ | ·0026 176 | ·0017 643 | ·0011 847 | ·0007 927 | ·0005 287 |
| ·62 | ·0050 583 | ·0034 783 | ·0023 822 | ·0016 253 | ·0011 050+ | ·0007 489 |
| ·63 | ·0065 729 | ·0045 910 | ·0031 937 | ·0022 134 | ·0015 287 | ·0010 524 |
| ·64 | ·0084 866 | ·0060 193 | ·0042 522 | ·0029 927 | ·0020 991 | ·0014 676 |
| ·65 | ·0108 887 | ·0078 403 | ·0056 229 | ·0040 179 | ·0028 612 | ·0020 310 |
| ·66 | ·0138 842 | ·0101 464 | ·0073 857 | ·0053 566 | ·0038 718 | ·0027 898 |
| ·67 | ·0175 954 | ·0130 469 | ·0096 367 | ·0070 922 | ·0052 020 | ·0038 037 |
| ·68 | ·0221 633 | ·0166 708 | ·0124 912 | ·0093 262 | ·0069 400 | ·0051 484 |
| ·69 | ·0277 491 | ·0211 678 | ·0160 860 | ·0121 811 | ·0091 939 | ·0069 180 |
| ·70 | ·0345 350− | ·0267 105+ | ·0205 813 | ·0158 034 | ·0120 953 | ·0092 292 |
| ·71 | ·0427 242 | ·0334 958 | ·0261 635+ | ·0203 661 | ·0158 024 | ·0122 248 |
| ·72 | ·0525 412 | ·0417 449 | ·0330 463 | ·0260 715− | ·0205 038 | ·0160 775− |
| ·73 | ·0642 297 | ·0517 040 | ·0414 718 | ·0331 534 | ·0264 210 | ·0209 943 |
| ·74 | ·0780 505− | ·0636 422 | ·0517 108 | ·0418 783 | ·0338 113 | ·0272 199 |
| ·75 | ·0942 773 | ·0778 493 | ·0640 615− | ·0525 455+ | ·0429 698 | ·0350 399 |
| ·76 | ·1131 910 | ·0946 308 | ·0788 459 | ·0654 861 | ·0542 290 | ·0447 824 |
| ·77 | ·1350 716 | ·1143 019 | ·0964 052 | ·0810 589 | ·0679 578 | ·0568 191 |
| ·78 | ·1601 886 | ·1371 773 | ·1170 917 | ·0996 446 | ·0845 567 | ·0715 623 |
| ·79 | ·1887 876 | ·1635 598 | ·1412 573 | ·1216 361 | ·1044 505+ | ·0894 597 |
| ·80 | ·2210 759 | ·1937 248 | ·1692 391 | ·1474 246 | ·1280 760 | ·1109 848 |
| ·81 | ·2572 041 | ·2279 014 | ·2013 406 | ·1773 815+ | ·1558 653 | ·1366 217 |
| ·82 | ·2972 458 | ·2662 502 | ·2378 080 | ·2118 352 | ·1882 224 | ·1668 431 |
| ·83 | ·3411 757 | ·3088 385− | ·2788 034 | ·2510 414 | ·2254 943 | ·2020 815− |
| ·84 | ·3888 453 | ·3556 119 | ·3243 724 | ·2951 494 | ·2679 338 | ·2426 914 |
| ·85 | ·4399 596 | ·4063 652 | ·3744 105− | ·3441 624 | ·3156 569 | ·2889 033 |
| ·86 | ·4940 548 | ·4607 132 | ·4286 258 | ·3978 949 | ·3685 936 | ·3407 696 |
| ·87 | ·5504 792 | ·5180 635− | ·4865 044 | ·4559 280 | ·4264 350− | ·3981 033 |
| ·88 | ·6083 819 | ·5775 956 | ·5472 784 | ·5175 676 | ·4885 802 | ·4604 140 |
| ·89 | ·6667 103 | ·6382 497 | ·6099 043 | ·5818 102 | ·5540 885+ | ·5268 463 |
| ·90 | ·7242 238 | ·6987 308 | ·6730 558 | ·6473 224 | ·6216 450− | ·5961 280 |

$x = \cdot 91$ to $1\cdot 00$ $q = 3\cdot 5$ $p = 20$ to $25$

| | $p = 20$ | $p = 21$ | $p = 22$ | $p = 23$ | $p = 24$ | $p = 25$ |
|---|---|---|---|---|---|---|
| $B(p, q) =$ | $\cdot 7540\ 5050 \times \frac{1}{10^4}$ | $\cdot 6417\ 4510 \times \frac{1}{10^4}$ | $\cdot 5500\ 6723 \times \frac{1}{10^4}$ | $\cdot 4745\ 6781 \times \frac{1}{10^4}$ | $\cdot 4118\ 8904 \times \frac{1}{10^4}$ | $\cdot 3594\ 6680 \times \frac{1}{10^4}$ |
| $x$ | | | | | | |
| ·91 | ·7795 256 | ·7575 345+ | ·7351 403 | ·7124 458 | ·6895 490 | ·6665 425− |
| ·92 | ·8311 216 | ·8130 023 | ·7943 481 | ·7752 360 | ·7557 416 | ·7359 386 |
| ·93 | ·8775 088 | ·8634 136 | ·8487 445− | ·8335 520 | ·8178 872 | ·8018 014 |
| ·94 | ·9173 002 | ·9071 224 | ·8964 162 | ·8852 089 | ·8735 288 | ·8614 060 |
| ·95 | ·9493 881 | ·9427 440 | ·9356 807 | ·9282 080 | ·9203 373 | ·9120 813 |
| ·96 | ·9731 415+ | ·9693 900 | ·9653 599 | ·9610 512 | ·9564 654 | ·9516 043 |
| ·97 | ·9886 263 | ·9869 400 | ·9851 096 | ·9831 323 | ·9810 059 | ·9787 284 |
| ·98 | ·9968 103 | ·9963 095− | ·9957 602 | ·9951 608 | ·9945 095− | ·9938 047 |
| ·99 | ·9996 732 | ·9996 189 | ·9995 588 | ·9994 926 | ·9994 199 | ·9993 404 |
| 1·00 | 1·0000 000 | 1·0000 000 | 1·0000 000 | 1·0000 000 | 1·0000 000 | 1·0000 000 |

$x$ = ·42 to 1·00 $q = 3{\cdot}5$ $p$ = 26 to 31

| $x$ | $p = 26$ | $p = 27$ | $p = 28$ | $p = 29$ | $p = 30$ | $p = 31$ |
|---|---|---|---|---|---|---|
| $B(p, q) =$ | ·3153 2175 $\times \frac{1}{10^4}$ | ·2779 1070 $\times \frac{1}{10^4}$ | ·2460 1931 $\times \frac{1}{10^4}$ | ·2186 8383 $\times \frac{1}{10^4}$ | ·1951 3326 $\times \frac{1}{10^4}$ | ·1747 4620 $\times \frac{1}{10^4}$ |
| ·42 | ·0000 001 | | | | | |
| ·43 | ·0000 001 | | | | | |
| ·44 | ·0000 002 | ·0000 001 | | | | |
| ·45 | ·0000 003 | ·0000 001 | ·0000 001 | | | |
| ·46 | ·0000 005⁻ | ·0000 002 | ·0000 001 | ·0000 001 | | |
| ·47 | ·0000 008 | ·0000 004 | ·0000 002 | ·0000 001 | ·0000 001 | |
| ·48 | ·0000 013 | ·0000 007 | ·0000 004 | ·0000 002 | ·0000 001 | ·0000 001 |
| ·49 | ·0000 022 | ·0000 012 | ·0000 006 | ·0000 003 | ·0000 002 | ·0000 001 |
| ·50 | ·0000 035⁺ | ·0000 019 | ·0000 010 | ·0000 006 | ·0000 003 | ·0000 002 |
| ·51 | ·0000 056 | ·0000 031 | ·0000 017 | ·0000 010 | ·0000 005⁺ | ·0000 003 |
| ·52 | ·0000 089 | ·0000 050⁺ | ·0000 028 | ·0000 016 | ·0000 009 | ·0000 005⁺ |
| ·53 | ·0000 139 | ·0000 080 | ·0000 046 | ·0000 026 | ·0000 015⁺ | ·0000 009 |
| ·54 | ·0000 215⁺ | ·0000 127 | ·0000 074 | ·0000 043 | ·0000 025⁺ | ·0000 015⁻ |
| ·55 | ·0000 330 | ·0000 197 | ·0000 118 | ·0000 070 | ·0000 042 | ·0000 025⁻ |
| ·56 | ·0000 500⁺ | ·0000 305⁻ | ·0000 185⁺ | ·0000 112 | ·0000 068 | ·0000 041 |
| ·57 | ·0000 752 | ·0000 466 | ·0000 288 | ·0000 178 | ·0000 109 | ·0000 067 |
| ·58 | ·0001 121 | ·0000 707 | ·0000 445⁻ | ·0000 279 | ·0000 175⁻ | ·0000 109 |
| ·59 | ·0001 655⁻ | ·0001 062 | ·0000 679 | ·0000 433 | ·0000 276 | ·0000 175⁺ |
| ·60 | ·0002 422 | ·0001 580 | ·0001 028 | ·0000 667 | ·0000 432 | ·0000 279 |
| ·61 | ·0003 515⁺ | ·0002 331 | ·0001 541 | ·0001 016 | ·0000 669 | ·0000 439 |
| ·62 | ·0005 059 | ·0003 409 | ·0002 290 | ·0001 535⁺ | ·0001 027 | ·0000 685⁻ |
| ·63 | ·0007 223 | ·0004 944 | ·0003 375⁻ | ·0002 298 | ·0001 561 | ·0001 058 |
| ·64 | ·0010 230 | ·0007 111 | ·0004 930 | ·0003 409 | ·0002 352 | ·0001 620 |
| ·65 | ·0014 374 | ·0010 145⁺ | ·0007 142 | ·0005 015⁺ | ·0003 514 | ·0002 456 |
| ·66 | ·0020 042 | ·0014 359 | ·0010 261 | ·0007 314 | ·0005 202 | ·0003 692 |
| ·67 | ·0027 732 | ·0020 163 | ·0014 623 | ·0010 579 | ·0007 636 | ·0005 500⁺ |
| ·68 | ·0038 082 | ·0028 093 | ·0020 671 | ·0015 174 | ·0011 114 | ·0008 123 |
| ·69 | ·0051 906 | ·0038 841 | ·0028 991 | ·0021 588 | ·0016 039 | ·0011 892 |
| ·70 | ·0070 223 | ·0053 290 | ·0040 339 | ·0030 464 | ·0022 955⁺ | ·0017 261 |
| ·71 | ·0094 306 | ·0072 560 | ·0055 690 | ·0042 644 | ·0032 582 | ·0024 843 |
| ·72 | ·0125 719 | ·0098 051 | ·0076 286 | ·0059 216 | ·0045 866 | ·0035 453 |
| ·73 | ·0166 367 | ·0131 498 | ·0103 687 | ·0081 572 | ·0064 036 | ·0050 169 |
| ·74 | ·0218 546 | ·0175 024 | ·0139 836 | ·0111 472 | ·0088 673 | ·0070 396 |
| ·75 | ·0284 977 | ·0231 193 | ·0187 119 | ·0151 113 | ·0121 780 | ·0097 947 |
| ·76 | ·0368 851 | ·0303 060 | ·0248 430 | ·0203 203 | ·0165 868 | ·0135 129 |
| ·77 | ·0473 847 | ·0394 216 | ·0327 224 | ·0271 033 | ·0224 037 | ·0184 834 |
| ·78 | ·0604 130 | ·0508 804 | ·0427 565⁻ | ·0358 541 | ·0300 061 | ·0250 645⁺ |
| ·79 | ·0764 326 | ·0651 517 | ·0554 146 | ·0470 354 | ·0398 453 | ·0336 917 |
| ·80 | ·0959 445⁺ | ·0827 554 | ·0712 270 | ·0611 809 | ·0524 512 | ·0448 855⁻ |
| ·81 | ·1194 756 | ·1042 519 | ·0907 789 | ·0788 914 | ·0684 324 | ·0592 547 |
| ·82 | ·1475 589 | ·1302 254 | ·1146 958 | ·1008 246 | ·0884 697 | ·0774 944 |
| ·83 | ·1807 055⁺ | ·1612 576 | ·1436 211 | ·1276 759 | ·1133 005⁻ | ·1003 745⁺ |
| ·84 | ·2193 673 | ·1978 913 | ·1781 811 | ·1601 462 | ·1436 910 | ·1287 167 |
| ·85 | ·2638 887 | ·2405 818 | ·2189 367 | ·1988 955⁺ | ·1803 920 | ·1633 534 |
| ·86 | ·3144 481 | ·2896 351 | ·2663 200 | ·2444 787 | ·2240 760 | ·2050 674 |
| ·87 | ·3709 899 | ·3451 331 | ·3205 547 | ·2972 624 | ·2752 511 | ·2545 055⁻ |
| ·88 | ·4331 492 | ·4068 492 | ·3815 622 | ·3573 228 | ·3341 533 | ·3120 650⁻ |
| ·89 | ·5001 761 | ·4741 573 | ·4488 564 | ·4243 281 | ·4006 161 | ·3777 536 |
| ·90 | ·5708 663 | ·5459 446 | ·5214 382 | ·4974 132 | ·4739 267 | ·4510 274 |
| ·91 | ·6435 130 | ·6205 410 | ·5977 010 | ·5750 608 | ·5526 821 | ·5306 204 |
| ·92 | ·7158 979 | ·6956 877 | ·6753 727 | ·6550 141 | ·6346 696 | ·6143 929 |
| ·93 | ·7853 456 | ·7685 702 | ·7515 246 | ·7342 567 | ·7168 132 | ·6992 388 |
| ·94 | ·8488 709 | ·8359 549 | ·8226 898 | ·8091 072 | ·7952 389 | ·7811 164 |
| ·95 | ·9034 538 | ·8944 695⁺ | ·8851 442 | ·8754 942 | ·8655 364 | ·8552 881 |
| ·96 | ·9464 711 | ·9410 693 | ·9354 035⁻ | ·9294 786 | ·9233 005⁻ | ·9168 751 |
| ·97 | ·9762 983 | ·9737 145⁺ | ·9709 761 | ·9680 828 | ·9650 342 | ·9618 307 |
| ·98 | ·9930 450⁻ | ·9922 289 | ·9913 550⁺ | ·9904 222 | ·9894 292 | ·9883 749 |
| ·99 | ·9992 539 | ·9991 599 | ·9990 583 | ·9989 488 | ·9988 310 | ·9987 046 |
| 1·00 | 1·0000 000 | 1·0000 000 | 1·0000 000 | 1·0000 000 | 1·0000 000 | 1·0000 000 |

$x$ = ·50 to 1·00 $q = 3·5$ $p$ = 32 to 37

| | $p = 32$ | $p = 33$ | $p = 34$ | $p = 35$ | $p = 36$ | $p = 37$ |
|---|---|---|---|---|---|---|
| $B(p, q) =$ | ·1570 1833 $\times \frac{1}{10^4}$ | ·1415 3765 $\times \frac{1}{10^4}$ | ·1279 6554 $\times \frac{1}{10^4}$ | ·1160 2209 $\times \frac{1}{10^4}$ | ·1054 7463 $\times \frac{1}{10^4}$ | ·9612 8778 $\times \frac{1}{10^5}$ |
| $x$ | | | | | | |
| ·50 | ·0000 001 | | | | | |
| ·51 | ·0000 002 | ·0000 001 | | | | |
| ·52 | ·0000 003 | ·0000 002 | ·0000 001 | | | |
| ·53 | ·0000 005$^{-}$ | ·0000 003 | ·0000 002 | ·0000 001 | ·0000 001 | |
| ·54 | ·0000 009 | ·0000 005$^{-}$ | ·0000 003 | ·0000 002 | ·0000 001 | ·0000 001 |
| ·55 | ·0000 015$^{-}$ | ·0000 009 | ·0000 005$^{+}$ | ·0000 003 | ·0000 002 | ·0000 001 |
| ·56 | ·0000 025$^{-}$ | ·0000 015$^{-}$ | ·0000 009 | ·0000 005$^{+}$ | ·0000 003 | ·0000 002 |
| ·57 | ·0000 041 | ·0000 025$^{+}$ | ·0000 015$^{+}$ | ·0000 009 | ·0000 006 | ·0000 003 |
| ·58 | ·0000 068 | ·0000 042 | ·0000 026 | ·0000 016 | ·0000 010 | ·0000 006 |
| ·59 | ·0000 111 | ·0000 070 | ·0000 044 | ·0000 028 | ·0000 018 | ·0000 011 |
| ·60 | ·0000 180 | ·0000 116 | ·0000 074 | ·0000 048 | ·0000 030 | ·0000 019 |
| ·61 | ·0000 288 | ·0000 188 | ·0000 123 | ·0000 080 | ·0000 052 | ·0000 034 |
| ·62 | ·0000 456 | ·0000 303 | ·0000 201 | ·0000 133 | ·0000 088 | ·0000 058 |
| ·63 | ·0000 716 | ·0000 483 | ·0000 326 | ·0000 219 | ·0000 147 | ·0000 099 |
| ·64 | ·0001 113 | ·0000 763 | ·0000 522 | ·0000 357 | ·0000 243 | ·0000 166 |
| ·65 | ·0001 714 | ·0001 193 | ·0000 829 | ·0000 575$^{+}$ | ·0000 399 | ·0000 276 |
| ·66 | ·0002 615$^{-}$ | ·0001 848 | ·0001 304 | ·0000 919 | ·0000 646 | ·0000 454 |
| ·67 | ·0003 954 | ·0002 837 | ·0002 032 | ·0001 452 | ·0001 037 | ·0000 739 |
| ·68 | ·0005 925$^{-}$ | ·0004 313 | ·0003 134 | ·0002 274 | ·0001 647 | ·0001 191 |
| ·69 | ·0008 799 | ·0006 499 | ·0004 791 | ·0003 526 | ·0002 591 | ·0001 901 |
| ·70 | ·0012 954 | ·0009 704 | ·0007 256 | ·0005 416 | ·0004 037 | ·0003 004 |
| ·71 | ·0018 905$^{+}$ | ·0014 360 | ·0010 888 | ·0008 242 | ·0006 229 | ·0004 701 |
| ·72 | ·0027 351 | ·0021 062 | ·0016 190 | ·0012 425$^{+}$ | ·0009 521 | ·0007 284 |
| ·73 | ·0039 229 | ·0030 619 | ·0023 857 | ·0018 559 | ·0014 414 | ·0011 179 |
| ·74 | ·0055 780 | ·0044 120 | ·0034 837 | ·0027 463 | ·0021 616 | ·0016 990 |
| ·75 | ·0078 631 | ·0063 012 | ·0050 411 | ·0040 265$^{-}$ | ·0032 112 | ·0025 572 |
| ·76 | ·0109 883 | ·0089 197 | ·0072 284 | ·0058 486 | ·0047 250$^{+}$ | ·0038 118 |
| ·77 | ·0152 214 | ·0125 134 | ·0102 702 | ·0084 160 | ·0068 863 | ·0056 266 |
| ·78 | ·0208 992 | ·0173 965$^{+}$ | ·0144 574 | ·0119 963 | ·0099 396 | ·0082 239 |
| ·79 | ·0284 383 | ·0239 639 | ·0201 614 | ·0169 365$^{-}$ | ·0142 068 | ·0119 007 |
| ·80 | ·0383 448 | ·0327 034 | ·0278 484 | ·0236 788 | ·0201 049 | ·0170 472 |
| ·81 | ·0512 212 | ·0442 057 | ·0380 927 | ·0327 770 | ·0281 638 | ·0241 676 |
| ·82 | ·0677 688 | ·0591 708 | ·0515 864 | ·0449 098 | ·0390 440 | ·0339 000 |
| ·83 | ·0887 808 | ·0784 063 | ·0691 430 | ·0608 892 | ·0535 492 | ·0470 340 |
| ·84 | ·1151 239 | ·1028 139 | ·0916 902 | ·0816 592 | ·0726 313 | ·0645 213 |
| ·85 | ·1477 026 | ·1333 601 | ·1202 454 | ·1082 783 | ·0973 796 | ·0874 724 |
| ·86 | ·1874 017 | ·1710 223 | ·1558 689 | ·1418 787 | ·1289 877 | ·1171 316 |
| ·87 | ·2350 012 | ·2167 068 | ·1995 850$^{-}$ | ·1835 937 | ·1686 875$^{-}$ | ·1548 185$^{-}$ |
| ·88 | ·2910 597 | ·2711 310 | ·2522 651 | ·2344 424 | ·2176 382 | ·2018 235$^{+}$ |
| ·89 | ·3557 652 | ·3346 665$^{+}$ | ·3144 662 | ·2951 660 | ·2767 618 | ·2592 446 |
| ·90 | ·4287 560 | ·4071 457 | ·3862 230 | ·3660 081 | ·3465 150$^{+}$ | ·3277 530 |
| ·91 | ·5089 253 | ·4876 403 | ·4668 036 | ·4464 480 | ·4266 013 | ·4072 866 |
| ·92 | ·5942 340 | ·5742 388 | ·5544 496 | ·5349 049 | ·5156 394 | ·4966 843 |
| ·93 | ·6815 766 | ·6638 675$^{+}$ | ·6461 503 | ·6284 618 | ·6108 365$^{-}$ | ·5933 066 |
| ·94 | ·7667 706 | ·7522 321 | ·7375 306 | ·7226 950$^{+}$ | ·7077 535$^{-}$ | ·6927 331 |
| ·95 | ·8447 671 | ·8339 913 | ·8229 787 | ·8117 475$^{+}$ | ·8003 158 | ·7887 015$^{-}$ |
| ·96 | ·9102 094 | ·9033 103 | ·8961 855$^{+}$ | ·8888 427 | ·8812 902 | ·8735 362 |
| ·97 | ·9584 728 | ·9549 610 | ·9512 966 | ·9474 808 | ·9435 150$^{+}$ | ·9394 011 |
| ·98 | ·9872 585$^{-}$ | ·9860 789 | ·9848 353 | ·9835 269 | ·9821 532 | ·9807 134 |
| ·99 | ·9985 694 | ·9984 251 | ·9982 715$^{-}$ | ·9981 081 | ·9979 349 | ·9977 515$^{+}$ |
| 1·00 | 1·0000 000 | 1·0000 000 | 1·0000 000 | 1·0000 000 | 1·0000 000 | 1·0000 000 |

$x$ = ·55 to 1·00 $q$ = 3·5 $p$ = 38 to 43

| $x$ | $p$ = 38 | $p$ = 39 | $p$ = 40 | $p$ = 41 | $p$ = 42 | $p$ = 43 |
|---|---|---|---|---|---|---|
| $B(p, q)$ = | ·8782 1352 × $\frac{1}{10^5}$ | ·8041 4732 × $\frac{1}{10^5}$ | ·7379 2343 × $\frac{1}{10^5}$ | ·6785 5028 × $\frac{1}{10^5}$ | ·6251 8115 $\stackrel{+}{\times}$ $\frac{1}{10^5}$ | ·5770 9030 × $\frac{1}{10^5}$ |
| ·55 | ·0000 001 | | | | | |
| ·56 | ·0000 001 | ·0000 001 | | | | |
| ·57 | ·0000 002 | ·0000 001 | ·0000 001 | | | |
| ·58 | ·0000 004 | ·0000 002 | ·0000 001 | ·0000 001 | ·0000 001 | |
| ·59 | ·0000 007 | ·0000 004 | ·0000 003 | ·0000 002 | ·0000 001 | ·0000 001 |
| ·60 | ·0000 012 | ·0000 008 | ·0000 005− | ·0000 003 | ·0000 002 | ·0000 001 |
| ·61 | ·0000 022 | ·0000 014 | ·0000 009 | ·0000 006 | ·0000 004 | ·0000 002 |
| ·62 | ·0000 038 | ·0000 025+ | ·0000 017 | ·0000 011 | ·0000 007 | ·0000 005− |
| ·63 | ·0000 066 | ·0000 044 | ·0000 029 | ·0000 020 | ·0000 013 | ·0000 009 |
| ·64 | ·0000 113 | ·0000 077 | ·0000 052 | ·0000 035+ | ·0000 024 | ·0000 016 |
| ·65 | ·0000 190 | ·0000 131 | ·0000 090 | ·0000 062 | ·0000 043 | ·0000 029 |
| ·66 | ·0000 318 | ·0000 223 | ·0000 156 | ·0000 109 | ·0000 076 | ·0000 053 |
| ·67 | ·0000 526 | ·0000 374 | ·0000 265+ | ·0000 188 | ·0000 133 | ·0000 094 |
| ·68 | ·0000 860 | ·0000 620 | ·0000 447 | ·0000 321 | ·0000 231 | ·0000 166 |
| ·69 | ·0001 393 | ·0001 019 | ·0000 744 | ·0000 543 | ·0000 396 | ·0000 288 |
| ·70 | ·0002 232 | ·0001 656 | ·0001 228 | ·0000 909 | ·0000 672 | ·0000 496 |
| ·71 | ·0003 542 | ·0002 666 | ·0002 003 | ·0001 504 | ·0001 127 | ·0000 844 |
| ·72 | ·0005 565+ | ·0004 246 | ·0003 235+ | ·0002 462 | ·0001 871 | ·0001 421 |
| ·73 | ·0008 657 | ·0006 695+ | ·0005 171 | ·0003 989 | ·0003 074 | ·0002 366 |
| ·74 | ·0013 334 | ·0010 451 | ·0008 181 | ·0006 396 | ·0004 995+ | ·0003 896 |
| ·75 | ·0020 336 | ·0016 151 | ·0012 810 | ·0010 149 | ·0008 031 | ·0006 348 |
| ·76 | ·0030 708 | ·0024 707 | ·0019 853 | ·0015 934 | ·0012 773 | ·0010 229 |
| ·77 | ·0045 911 | ·0037 413 | ·0030 450+ | ·0024 754 | ·0020 100 | ·0016 303 |
| ·78 | ·0067 953 | ·0056 076 | ·0046 219 | ·0038 049 | ·0031 288 | ·0025 701 |
| ·79 | ·0099 557 | ·0083 181 | ·0069 414 | ·0057 859 | ·0048 174 | ·0040 067 |
| ·80 | ·0144 358 | ·0122 092 | ·0103 137 | ·0087 026 | ·0073 351 | ·0061 759 |
| ·81 | ·0207 120 | ·0177 287 | ·0151 574 | ·0129 445+ | ·0110 428 | ·0094 106 |
| ·82 | ·0293 969 | ·0254 614 | ·0220 274 | ·0190 356 | ·0164 327 | ·0141 713 |
| ·83 | ·0412 609 | ·0361 540 | ·0316 436 | ·0276 660 | ·0241 633 | ·0210 830 |
| ·84 | ·0572 485+ | ·0507 375+ | ·0449 177 | ·0397 234 | ·0350 943 | ·0309 743 |
| ·85 | ·0784 823 | ·0703 380 | ·0629 716 | ·0563 188 | ·0503 191 | ·0449 158 |
| ·86 | ·1062 465− | ·0962 693 | ·0871 390 | ·0787 961 | ·0711 838 | ·0642 475− |
| ·87 | ·1419 372 | ·1299 932 | ·1189 358 | ·1087 147 | ·0992 800 | ·0905 833 |
| ·88 | ·1869 663 | ·1730 318 | ·1599 833 | ·1477 830 | ·1363 921 | ·1257 715− |
| ·89 | ·2426 009 | ·2268 135+ | ·2118 620 | ·1977 235+ | ·1843 730 | ·1717 838 |
| ·90 | ·3097 262 | ·2924 348 | ·2758 749 | ·2600 395− | ·2449 186 | ·2304 998 |
| ·91 | ·3885 226 | ·3703 240 | ·3527 017 | ·3356 631 | ·3192 126 | ·3033 515+ |
| ·92 | ·4780 674 | ·4598 129 | ·4419 423 | ·4244 737 | ·4074 227 | ·3908 020 |
| ·93 | ·5759 023 | ·5586 514 | ·5415 796 | ·5247 106 | ·5080 658 | ·4916 648 |
| ·94 | ·6776 599 | ·6625 589 | ·6474 541 | ·6323 682 | ·6173 227 | ·6023 382 |
| ·95 | ·7769 223 | ·7649 960 | ·7529 396 | ·7407 702 | ·7285 044 | ·7161 583 |
| ·96 | ·8655 894 | ·8574 587 | ·8491 527 | ·8406 807 | ·8320 516 | ·8232 746 |
| ·97 | ·9351 409 | ·9307 368 | ·9261 908 | ·9215 057 | ·9166 840 | ·9117 286 |
| ·98 | ·9792 071 | ·9776 338 | ·9759 931 | ·9742 848 | ·9725 085+ | ·9706 642 |
| ·99 | ·9975 577 | ·9973 531 | ·9971 377 | ·9969 110 | ·9966 730 | ·9964 233 |
| 1·00 | 1·0000 000 | 1·0000 000 | 1·0000 000 | 1·0000 000 | 1·0000 000 | 1·0000 000 |

$x$ = ·60 to 1·00 $q = 3·5$ $p$ = 44 to 50

| $x$ | $p = 44$ | $p = 45$ | $p = 46$ | $p = 47$ | $p = 48$ | $p = 49$ | $p = 50$ |
|---|---|---|---|---|---|---|---|
| $B(p, q) =$ | ·5336 5339 $\times \frac{1}{10^5}$ | ·4943 3156 $\times \frac{1}{10^5}$ | ·4586 5815+ $\times \frac{1}{10^5}$ | ·4262 2778 $\times \frac{1}{10^5}$ | ·3966 8724 $\times \frac{1}{10^5}$ | ·3697 2791 $\times \frac{1}{10^5}$ | ·3450 7938 $\times \frac{1}{10^5}$ |
| ·60 | ·0000 001 | ·0000 001 | | | | | |
| ·61 | ·0000 002 | ·0000 001 | ·0000 001 | | | | |
| ·62 | ·0000 003 | ·0000 002 | ·0000 001 | ·0000 001 | ·0000 001 | | |
| ·63 | ·0000 006 | ·0000 004 | ·0000 003 | ·0000 002 | ·0000 001 | ·0000 001 | |
| ·64 | ·0000 011 | ·0000 007 | ·0000 005− | ·0000 003 | ·0000 002 | ·0000 001 | ·0000 001 |
| ·65 | ·0000 020 | ·0000 014 | ·0000 009 | ·0000 006 | ·0000 004 | ·0000 003 | ·0000 002 |
| ·66 | ·0000 037 | ·0000 026 | ·0000 018 | ·0000 012 | ·0000 009 | ·0000 006 | ·0000 004 |
| ·67 | ·0000 066 | ·0000 047 | ·0000 033 | ·0000 023 | ·0000 016 | ·0000 011 | ·0000 008 |
| ·68 | ·0000 119 | ·0000 085− | ·0000 061 | ·0000 043 | ·0000 031 | ·0000 022 | ·0000 016 |
| ·69 | ·0000 210 | ·0000 152 | ·0000 110 | ·0000 080 | ·0000 058 | ·0000 042 | ·0000 030 |
| ·70 | ·0000 366 | ·0000 270 | ·0000 198 | ·0000 146 | ·0000 107 | ·0000 079 | ·0000 058 |
| ·71 | ·0000 631 | ·0000 472 | ·0000 352 | ·0000 263 | ·0000 196 | ·0000 146 | ·0000 108 |
| ·72 | ·0001 078 | ·0000 816 | ·0000 618 | ·0000 467 | ·0000 353 | ·0000 266 | ·0000 201 |
| ·73 | ·0001 819 | ·0001 397 | ·0001 072 | ·0000 821 | ·0000 629 | ·0000 481 | ·0000 368 |
| ·74 | ·0003 036 | ·0002 363 | ·0001 838 | ·0001 428 | ·0001 108 | ·0000 859 | ·0000 666 |
| ·75 | ·0005 012 | ·0003 953 | ·0003 115− | ·0002 452 | ·0001 929 | ·0001 516 | ·0001 190 |
| ·76 | ·0008 182 | ·0006 538 | ·0005 220 | ·0004 163 | ·0003 317 | ·0002 641 | ·0002 101 |
| ·77 | ·0013 210 | ·0010 693 | ·0008 647 | ·0006 986 | ·0005 639 | ·0004 548 | ·0003 664 |
| ·78 | ·0021 090 | ·0017 288 | ·0014 159 | ·0011 585− | ·0009 470 | ·0007 735+ | ·0006 313 |
| ·79 | ·0033 290 | ·0027 632 | ·0022 914 | ·0018 984 | ·0015 715− | ·0012 997 | ·0010 741 |
| ·80 | ·0051 947 | ·0043 652 | ·0036 646 | ·0030 738 | ·0025 759 | ·0021 569 | ·0018 046 |
| ·81 | ·0080 117 | ·0068 143 | ·0057 904 | ·0049 160 | ·0041 701 | ·0035 345− | ·0029 933 |
| ·82 | ·0122 091 | ·0105 087 | ·0090 370 | ·0077 645− | ·0066 656 | ·0057 175+ | ·0049 005− |
| ·83 | ·0183 776 | ·0160 047 | ·0139 257 | ·0121 063 | ·0105 159 | ·0091 272 | ·0079 157 |
| ·84 | ·0273 123 | ·0240 614 | ·0211 790 | ·0186 262 | ·0163 678 | ·0143 719 | ·0126 098 |
| ·85 | ·0400 559 | ·0356 904 | ·0317 735− | ·0282 632 | ·0251 207 | ·0223 105+ | ·0198 000 |
| ·86 | ·0579 354 | ·0521 987 | ·0469 909 | ·0422 689 | ·0379 919 | ·0341 222 | ·0306 244 |
| ·87 | ·0825 773 | ·0752 163 | ·0684 564 | ·0622 557 | ·0565 742 | ·0513 738 | ·0466 186 |
| ·88 | ·1158 819 | ·1066 846 | ·0981 414 | ·0902 147 | ·0828 682 | ·0760 666 | ·0697 758 |
| ·89 | ·1599 279 | ·1487 767 | ·1383 007 | ·1284 704 | ·1192 560 | ·1106 281 | ·1025 575+ |
| ·90 | ·2167 683 | ·2037 079 | ·1913 005− | ·1795 269 | ·1683 670 | ·1578 000 | ·1478 045+ |
| ·91 | ·2880 788 | ·2733 909 | ·2592 825− | ·2457 460 | ·2327 725+ | ·2203 518 | ·2084 723 |
| ·92 | ·3746 220 | ·3588 906 | ·3436 137 | ·3287 951 | ·3144 367 | ·3005 391 | ·2871 009 |
| ·93 | ·4755 250− | ·4596 622 | ·4440 901 | ·4288 211 | ·4138 656 | ·3992 325− | ·3849 294 |
| ·94 | ·5874 338 | ·5726 276 | ·5579 366 | ·5433 765− | ·5289 620 | ·5147 067 | ·5006 230 |
| ·95 | ·7037 475+ | ·6912 874 | ·6787 927 | ·6662 776 | ·6537 559 | ·6412 407 | ·6287 446 |
| ·96 | ·8143 588 | ·8053 133 | ·7961 472 | ·7868 695+ | ·7774 892 | ·7680 150+ | ·7584 558 |
| ·97 | ·9066 423 | ·9014 283 | ·8960 898 | ·8906 300 | ·8850 523 | ·8793 601 | ·8735 570 |
| ·98 | ·9687 516 | ·9667 708 | ·9647 218 | ·9626 046 | ·9604 194 | ·9581 663 | ·9558 457 |
| ·99 | ·9961 617 | ·9958 880 | ·9956 020 | ·9953 035+ | ·9949 923 | ·9946 681 | ·9943 307 |
| 1·00 | 1·0000 000 | 1·0000 000 | 1·0000 000 | 1·0000 000 | 1·0000 000 | 1·0000 000 | 1·0000 000 |

$x$ = ·01 to ·60 $q = 4$ $p$ = 4 to 6·5

| | $p=4$ | $p=4{\cdot}5$ | $p=5$ | $p=5{\cdot}5$ | $p=6$ | $p=6{\cdot}5$ |
|---|---|---|---|---|---|---|
| $B(p, q)$ = | ·7142 8571 $\times \frac{1}{10^2}$ | ·4972 8050 $\times \frac{1}{10^2}$ | ·3571 4286 $\times \frac{1}{10^2}$ | ·2632 6615 $\times \frac{1}{10^2}$ | ·1984 1270 $\times \frac{1}{10^2}$ | ·1524 1724 $\times \frac{1}{10^2}$ |
| $x$ | | | | | | |
| ·01 | ·0000 003 | | | | | |
| ·02 | ·0000 053 | ·0000 010 | ·0000 002 | | | |
| ·03 | ·0000 264 | ·0000 058 | ·0000 013 | ·0000 003 | ·0000 001 | |
| ·04 | ·0000 813 | ·0000 207 | ·0000 052 | ·0000 013 | ·0000 003 | ·0000 001 |
| ·05 | ·0001 936 | ·0000 551 | ·0000 154 | ·0000 042 | ·0000 012 | ·0000 003 |
| ·06 | ·0003 915− | ·0001 220 | ·0000 373 | ·0000 113 | ·0000 033 | ·0000 010 |
| ·07 | ·0007 072 | ·0002 379 | ·0000 786 | ·0000 256 | ·0000 082 | ·0000 026 |
| ·08 | ·0011 763 | ·0004 228 | ·0001 493 | ·0000 519 | ·0000 178 | ·0000 060 |
| ·09 | ·0018 366 | ·0006 997 | ·0002 619 | ·0000 965+ | ·0000 351 | ·0000 126 |
| ·10 | ·0027 280$^e$ | ·0010 948 | ·0004 316 | ·0001 676 | ·0000 642 | ·0000 243 |
| ·11 | ·0038 916 | ·0016 369 | ·0006 765+ | ·0002 754 | ·0001 106 | ·0000 439 |
| ·12 | ·0053 693 | ·0023 572 | ·0010 169 | ·0004 322 | ·0001 813 | ·0000 752 |
| ·13 | ·0072 028 | ·0032 889 | ·0014 759 | ·0006 525+ | ·0002 847 | ·0001 228 |
| ·14 | ·0094 339 | ·0044 670 | ·0020 790 | ·0009 533 | ·0004 315+ | ·0001 931 |
| ·15 | ·0121 032 | ·0059 276 | ·0028 539 | ·0013 539 | ·0006 340 | ·0002 936 |
| ·16 | ·0152 503 | ·0077 080 | ·0038 303 | ·0018 757 | ·0009 068 | ·0004 335− |
| ·17 | ·0189 131 | ·0098 458 | ·0050 399 | ·0025 425+ | ·0012 664 | ·0006 238 |
| ·18 | ·0231 276 | ·0123 790 | ·0065 160 | ·0033 805+ | ·0017 318 | ·0008 773 |
| ·19 | ·0279 276 | ·0153 452 | ·0082 929 | ·0044 178 | ·0023 240 | ·0012 091 |
| ·20 | ·0333 440$^e$ | ·0187 815+ | ·0104 064$^e$ | ·0056 843 | ·0030 664 | ·0016 360 |
| ·21 | ·0394 053 | ·0227 243 | ·0128 926 | ·0072 118 | ·0039 844 | ·0021 773 |
| ·22 | ·0461 368 | ·0272 084 | ·0157 883 | ·0090 337 | ·0051 056 | ·0028 544 |
| ·23 | ·0535 606 | ·0322 673 | ·0191 302 | ·0111 847 | ·0064 598 | ·0036 908 |
| ·24 | ·0616 955− | ·0379 325− | ·0229 548 | ·0137 004 | ·0080 784 | ·0047 125− |
| ·25 | ·0705 566 | ·0442 333 | ·0272 980 | ·0166 173 | ·0099 945+ | ·0059 475− |
| ·26 | ·0801 558 | ·0511 967 | ·0321 948 | ·0199 724 | ·0122 430 | ·0074 259 |
| ·27 | ·0905 009 | ·0588 469 | ·0376 789 | ·0238 028 | ·0148 598 | ·0091 799 |
| ·28 | ·1015 962 | ·0672 052 | ·0437 826 | ·0281 456 | ·0178 821 | ·0112 435− |
| ·29 | ·1134 424 | ·0762 897 | ·0505 362 | ·0330 373 | ·0213 477 | ·0136 523 |
| ·30 | ·1260 360$^e$ | ·0861 154 | ·0579 676 | ·0385 136 | ·0252 948 | ·0164 436 |
| ·31 | ·1393 702 | ·0966 937 | ·0661 027 | ·0446 090 | ·0297 621 | ·0196 557 |
| ·32 | ·1534 344 | ·1080 324 | ·0749 644 | ·0513 568 | ·0347 877 | ·0233 281 |
| ·33 | ·1682 141 | ·1201 357 | ·0845 724 | ·0587 880 | ·0404 096 | ·0275 007 |
| ·34 | ·1836 917 | ·1330 038 | ·0949 435− | ·0669 319 | ·0466 645− | ·0322 141 |
| ·35 | ·1998 457 | ·1466 333 | ·1060 909 | ·0758 150− | ·0535 882 | ·0375 087 |
| ·36 | ·2166 517 | ·1610 168 | ·1180 242 | ·0854 611 | ·0612 147 | ·0434 246 |
| ·37 | ·2340 816 | ·1761 428 | ·1307 490 | ·0958 908 | ·0695 762 | ·0500 012 |
| ·38 | ·2521 046 | ·1919 964 | ·1442 673 | ·1071 215+ | ·0787 022 | ·0572 769 |
| ·39 | ·2706 869 | ·2085 583 | ·1585 766 | ·1191 669 | ·0886 197 | ·0652 882 |
| ·40 | ·2897 920$^e$ | ·2258 059 | ·1736 704$^e$ | ·1320 365− | ·0993 526 | ·0740 700 |
| ·41 | ·3093 807 | ·2437 124 | ·1895 380 | ·1457 360 | ·1109 212 | ·0836 546 |
| ·42 | ·3294 116 | ·2622 478 | ·2061 644 | ·1602 666 | ·1233 422 | ·0940 716 |
| ·43 | ·3498 411 | ·2813 784 | ·2235 301 | ·1756 252 | ·1366 281 | ·1053 473 |
| ·44 | ·3706 237 | ·3010 672 | ·2416 115+ | ·1918 036 | ·1507 869 | ·1175 044 |
| ·45 | ·3917 122 | ·3212 740 | ·2603 807 | ·2087 894 | ·1658 220 | ·1305 615− |
| ·46 | ·4130 579 | ·3419 558 | ·2798 056 | ·2265 650− | ·1817 320 | ·1445 326 |
| ·47 | ·4346 107 | ·3630 666 | ·2998 501 | ·2451 079 | ·1985 102 | ·1594 270 |
| ·48 | ·4563 199 | ·3845 578 | ·3204 741 | ·2643 908 | ·2161 445− | ·1752 488 |
| ·49 | ·4781 337 | ·4063 787 | ·3416 336 | ·2843 817 | ·2346 175+ | ·1919 967 |
| ·50 | ·5000 000$^e$ | ·4284 763 | ·3632 812 | ·3050 434 | ·2539 062 | ·2096 634 |
| ·51 | ·5218 663 | ·4507 960 | ·3853 661 | ·3263 342 | ·2739 820 | ·2282 357 |
| ·52 | ·5436 801 | ·4732 815+ | ·4078 342 | ·3482 078 | ·2948 105− | ·2476 940 |
| ·53 | ·5653 893 | ·4958 754 | ·4306 287 | ·3706 135+ | ·3163 517 | ·2680 126 |
| ·54 | ·5869 421 | ·5185 191 | ·4536 899 | ·3934 963 | ·3385 600 | ·2891 591 |
| ·55 | ·6082 878 | ·5411 538 | ·4769 563 | ·4167 973 | ·3613 846 | ·3110 944 |
| ·56 | ·6293 763 | ·5637 198 | ·5003 641 | ·4404 540 | ·3847 691 | ·3337 729 |
| ·57 | ·6501 589 | ·5861 581 | ·5238 478 | ·4644 003 | ·4086 522 | ·3571 427 |
| ·58 | ·6705 884 | ·6084 094 | ·5473 412 | ·4885 673 | ·4329 677 | ·3811 453 |
| ·59 | ·6906 193 | ·6304 154 | ·5707 766 | ·5128 835+ | ·4576 451 | ·4057 158 |
| ·60 | ·7102 080$^e$ | ·6521 187 | ·5940 864$^e$ | ·5372 751 | ·4826 097 | ·4307 838 |

$x = \cdot 61$ to $1\cdot 00$ $q = 4$ $p = 4$ to $6\cdot 5$

| $x$ | $p = 4$ | $p = 4\cdot 5$ | $p = 5$ | $p = 5\cdot 5$ | $p = 6$ | $p = 6\cdot 5$ |
|---|---|---|---|---|---|---|
| $B(p, q) =$ | $\cdot 7142\ 8571 \times \frac{1}{10^2}$ | $\cdot 4972\ 8050 \times \frac{1}{10^2}$ | $\cdot 3571\ 4286 \times \frac{1}{10^2}$ | $\cdot 2632\ 6615 \times \frac{1}{10^2}$ | $\cdot 1984\ 1270 \times \frac{1}{10^2}$ | $\cdot 1524\ 1724 \times \frac{1}{10^2}$ |
| ·61 | ·7293 131 | ·6734 633 | ·6172 027 | ·5616 666 | ·5077 830 | ·4562 729 |
| ·62 | ·7478 954 | ·6943 947 | ·6400 580 | ·5859 812 | ·5330 834 | ·4821 015⁻ |
| ·63 | ·7659 184 | ·7148 604 | ·6625 859 | ·6101 414 | ·5584 267 | ·5081 832 |
| ·64 | ·7833 483 | ·7348 104 | ·6847 209 | ·6340 695⁻ | ·5837 263 | ·5344 275⁺ |
| ·65 | ·8001 543 | ·7541 970 | ·7063 994 | ·6576 878 | ·6088 944 | ·5607 400 |
| ·66 | ·8163 083 | ·7729 754 | ·7275 601 | ·6809 200 | ·6338 421 | ·5870 234 |
| ·67 | ·8317 859 | ·7911 043 | ·7481 442 | ·7036 908 | ·6584 802 | ·6131 781 |
| ·68 | ·8465 656 | ·8085 455⁻ | ·7680 957 | ·7259 272 | ·6827 203 | ·6391 029 |
| ·69 | ·8606 298 | ·8252 646 | ·7873 623 | ·7475 588 | ·7064 750⁺ | ·6646 961 |
| ·70 | ·8739 640$^e$ | ·8412 313 | ·8058 956 | ·7685 183 | ·7296 591 | ·6898 560 |
| ·71 | ·8865 576 | ·8564 193 | ·8236 514 | ·7887 424 | ·7521 900 | ·7144 824 |
| ·72 | ·8984 038 | ·8708 067 | ·8405 901 | ·8081 721 | ·7739 888 | ·7384 769 |
| ·73 | ·9094 991 | ·8843 762 | ·8566 771 | ·8267 535⁻ | ·7949 811 | ·7617 445⁺ |
| ·74 | ·9198 442 | ·8971 151 | ·8718 832 | ·8444 379 | ·8150 974 | ·7841 944 |
| ·75 | ·9294 434 | ·9090 155⁻ | ·8861 847 | ·8611 831 | ·8342 743 | ·8057 411 |
| ·76 | ·9383 045⁺ | ·9200 744 | ·8995 638 | ·8769 531 | ·8524 552 | ·8263 051 |
| ·77 | ·9464 394 | ·9302 939 | ·9120 090 | ·8917 190 | ·8695 907 | ·8458 147 |
| ·78 | ·9538 632 | ·9396 809 | ·9235 147 | ·9054 591 | ·8856 398 | ·8642 063 |
| ·79 | ·9605 947 | ·9482 472 | ·9340 820 | ·9181 598 | ·9005 700 | ·8814 257 |
| ·80 | ·9666 560$^e$ | ·9560 096 | ·9437 184$^e$ | ·9298 150⁺ | ·9143 583 | ·8974 290 |
| ·81 | ·9720 724 | ·9629 896 | ·9524 378 | ·9404 273 | ·9269 914 | ·9121 834 |
| ·82 | ·9768 724 | ·9692 134 | ·9602 607 | ·9500 073 | ·9384 662 | ·9256 680 |
| ·83 | ·9810 869 | ·9747 115⁺ | ·9672 137 | ·9585 741 | ·9487 901 | ·9378 743 |
| ·84 | ·9847 497 | ·9795 188 | ·9733 297 | ·9661 552 | ·9579 813 | ·9488 068 |
| ·85 | ·9878 968 | ·9836 738 | ·9786 475⁺ | ·9727 861 | ·9660 685⁻ | ·9584 836 |
| ·86 | ·9905 661 | ·9872 188 | ·9832 113 | ·9785 104 | ·9730 910 | ·9669 360 |
| ·87 | ·9927 972 | ·9901 990 | ·9870 703 | ·9833 789 | ·9790 985⁻ | ·9742 088 |
| ·88 | ·9946 307 | ·9926 624 | ·9902 784 | ·9874 494 | ·9841 503 | ·9803 598 |
| ·89 | ·9961 084 | ·9946 588 | ·9928 932 | ·9907 862 | ·9883 149 | ·9854 595⁻ |
| ·90 | ·9972 720$^e$ | ·9962 399 | ·9949 756 | ·9934 584 | ·9916 689 | ·9895 896 |
| ·91 | ·9981 634 | ·9974 578 | ·9965 887 | ·9955 398 | ·9942 959 | ·9928 425⁻ |
| ·92 | ·9988 237 | ·9983 649 | ·9977 967 | ·9971 072 | ·9962 849 | ·9953 189 |
| ·93 | ·9992 928 | ·9990 128 | ·9986 641 | ·9982 388 | ·9977 287 | ·9971 263 |
| ·94 | ·9996 085⁺ | ·9994 512 | ·9992 544 | ·9990 129 | ·9987 217 | ·9983 760 |
| ·95 | ·9998 064 | ·9997 275⁺ | ·9996 282 | ·9995 058 | ·9993 574 | ·9991 803 |
| ·96 | ·9999 187 | ·9998 851 | ·9998 426 | ·9997 899 | ·9997 257 | ·9996 487 |
| ·97 | ·9999 736 | ·9999 626 | ·9999 485⁺ | ·9999 310 | ·9999 096 | ·9998 837 |
| ·98 | ·9999 947 | ·9999 924 | ·9999 895⁻ | ·9999 859 | ·9999 814 | ·9999 760 |
| ·99 | ·9999 997 | ·9999 995⁺ | ·9999 993 | ·9999 991 | ·9999 988 | ·9999 984 |
| 1·00 | 1·0000 000 | 1·0000 000 | 1·0000 000 | 1·0000 000 | 1·0000 000 | 1·0000 000 |

$x$ = ·05 to ·70 $q = 4$ $p$ = 7 to 9·5

| $x$ | $p = 7$ | $p = 7{\cdot}5$ | $p = 8$ | $p = 8{\cdot}5$ | $p = 9$ | $p = 9{\cdot}5$ |
|---|---|---|---|---|---|---|
| $B(p, q) =$ | ·1190 4762 $\times \frac{1}{10^2}$ | ·9435 3531 $\times \frac{1}{10^3}$ | ·7575 7576 $\times \frac{1}{10^3}$ | ·6153 4911 $\times \frac{1}{10^3}$ | ·5050 5051 $\times \frac{1}{10^3}$ | ·4184 3740 $\times \frac{1}{10^3}$ |
| ·05 | ·0000 001 | | | | | |
| ·06 | ·0000 003 | ·0000 001 | | | | |
| ·07 | ·0000 008 | ·0000 003 | ·0000 001 | | | |
| ·08 | ·0000 020 | ·0000 007 | ·0000 002 | ·0000 001 | | |
| ·09 | ·0000 045− | ·0000 016 | ·0000 006 | ·0000 002 | ·0000 001 | |
| ·10 | ·0000 091 | ·0000 034 | ·0000 012 | ·0000 005− | ·0000 002 | ·0000 001 |
| ·11 | ·0000 173 | ·0000 067 | ·0000 026 | ·0000 010 | ·0000 004 | ·0000 001 |
| ·12 | ·0000 308 | ·0000 125+ | ·0000 051 | ·0000 020 | ·0000 008 | ·0000 003 |
| ·13 | ·0000 525− | ·0000 222 | ·0000 093 | ·0000 039 | ·0000 016 | ·0000 007 |
| ·14 | ·0000 856 | ·0000 376 | ·0000 164 | ·0000 071 | ·0000 030 | ·0000 013 |
| ·15 | ·0001 346 | ·0000 611 | ·0000 276 | ·0000 123 | ·0000 055− | ·0000 024 |
| ·16 | ·0002 051 | ·0000 962 | ·0000 448 | ·0000 207 | ·0000 095− | ·0000 043 |
| ·17 | ·0003 042 | ·0001 470 | ·0000 705− | ·0000 336 | ·0000 159 | ·0000 075− |
| ·18 | ·0004 401 | ·0002 188 | ·0001 079 | ·0000 528 | ·0000 257 | ·0000 124 |
| ·19 | ·0006 229 | ·0003 181 | ·0001 611 | ·0000 810 | ·0000 405− | ·0000 201 |
| ·20 | ·0008 644 | ·0004 527 | ·0002 352 | ·0001 213 | ·0000 622 | ·0000 317 |
| ·21 | ·0011 783 | ·0006 321 | ·0003 364 | ·0001 778 | ·0000 934 | ·0000 487 |
| ·22 | ·0015 804 | ·0008 674 | ·0004 724 | ·0002 555− | ·0001 373 | ·0000 733 |
| ·23 | ·0020 885+ | ·0011 717 | ·0006 522 | ·0003 605+ | ·0001 980 | ·0001 081 |
| ·24 | ·0027 228 | ·0015 598 | ·0008 867 | ·0005 005+ | ·0002 807 | ·0001 565+ |
| ·25 | ·0035 057 | ·0020 489 | ·0011 883 | ·0006 844 | ·0003 917 | ·0002 228 |
| ·26 | ·0044 618 | ·0026 582 | ·0015 716 | ·0009 228 | ·0005 384 | ·0003 123 |
| ·27 | ·0056 181 | ·0034 095− | ·0020 534 | ·0012 282 | ·0007 301 | ·0004 315− |
| ·28 | ·0070 039 | ·0043 266 | ·0026 526 | ·0016 152 | ·0009 774 | ·0005 881 |
| ·29 | ·0086 507 | ·0054 361 | ·0033 905+ | ·0021 003 | ·0012 930 | ·0007 915+ |
| ·30 | ·0105 921 | ·0067 668 | ·0042 909 | ·0027 025+ | ·0016 917 | ·0010 529 |
| ·31 | ·0128 637 | ·0083 500− | ·0053 801 | ·0034 433 | ·0021 902 | ·0013 853 |
| ·32 | ·0155 029 | ·0102 193 | ·0066 870 | ·0043 465+ | ·0028 080 | ·0018 040 |
| ·33 | ·0185 489 | ·0124 106 | ·0082 432 | ·0054 390 | ·0035 670 | ·0023 264 |
| ·34 | ·0220 422 | ·0149 620 | ·0100 828 | ·0067 501 | ·0044 918 | ·0029 725+ |
| ·35 | ·0260 243 | ·0179 136 | ·0122 423 | ·0083 120 | ·0056 098 | ·0037 653 |
| ·36 | ·0305 376 | ·0213 070 | ·0147 609 | ·0101 597 | ·0069 513 | ·0047 303 |
| ·37 | ·0356 252 | ·0251 856 | ·0176 797 | ·0123 311 | ·0085 500− | ·0058 962 |
| ·38 | ·0413 301 | ·0295 939 | ·0210 423 | ·0148 665+ | ·0104 420 | ·0072 950− |
| ·39 | ·0476 949 | ·0345 773 | ·0248 940 | ·0178 092 | ·0126 670 | ·0089 615+ |
| ·40 | ·0547 619 | ·0401 818 | ·0292 815− | ·0212 045− | ·0152 673 | ·0109 344 |
| ·41 | ·0625 719 | ·0464 535− | ·0342 530 | ·0251 001 | ·0182 882 | ·0132 552 |
| ·42 | ·0711 643 | ·0534 380 | ·0398 575+ | ·0295 455+ | ·0217 779 | ·0159 689 |
| ·43 | ·0805 763 | ·0611 804 | ·0461 445+ | ·0345 920 | ·0257 867 | ·0191 237 |
| ·44 | ·0908 427 | ·0697 243 | ·0531 634 | ·0402 918 | ·0303 675− | ·0227 708 |
| ·45 | ·1019 949 | ·0791 116 | ·0609 632 | ·0466 981 | ·0355 749 | ·0269 641 |
| ·46 | ·1140 612 | ·0893 816 | ·0695 918 | ·0538 643 | ·0414 650− | ·0317 600 |
| ·47 | ·1270 655− | ·1005 708 | ·0790 955− | ·0618 434 | ·0480 949 | ·0372 174 |
| ·48 | ·1410 272 | ·1127 123 | ·0895 181 | ·0706 877 | ·0555 222 | ·0433 965− |
| ·49 | ·1559 607 | ·1258 348 | ·1009 009 | ·0804 478 | ·0638 044 | ·0503 588 |
| ·50 | ·1718 750$^{e}$ | ·1399 626 | ·1132 812 | ·0911 721 | ·0729 980 | ·0581 667 |
| ·51 | ·1887 732 | ·1551 145+ | ·1266 924 | ·1029 060 | ·0831 583 | ·0668 821 |
| ·52 | ·2066 520 | ·1713 036 | ·1411 628 | ·1156 913 | ·0943 380 | ·0765 664 |
| ·53 | ·2255 015− | ·1885 364 | ·1567 149 | ·1295 651 | ·1065 866 | ·0872 792 |
| ·54 | ·2453 048 | ·2068 129 | ·1733 651 | ·1445 592 | ·1199 498 | ·0990 774 |
| ·55 | ·2660 379 | ·2261 255− | ·1911 227 | ·1606 994 | ·1344 681 | ·1120 147 |
| ·56 | ·2876 693 | ·2464 586 | ·2099 895+ | ·1780 042 | ·1501 761 | ·1261 399 |
| ·57 | ·3101 599 | ·2677 890 | ·2299 590 | ·1964 846 | ·1671 016 | ·1414 964 |
| ·58 | ·3334 628 | ·2900 844 | ·2510 159 | ·2161 430 | ·1852 645+ | ·1581 208 |
| ·59 | ·3575 238 | ·3133 043 | ·2731 357 | ·2369 724 | ·2046 760 | ·1760 416 |
| ·60 | ·3822 806 | ·3373 991 | ·2962 843 | ·2589 560 | ·2253 373 | ·1952 786 |
| ·61 | ·4076 639 | ·3623 103 | ·3204 173 | ·2820 663 | ·2472 392 | ·2158 413 |
| ·62 | ·4335 970 | ·3879 704 | ·3454 804 | ·3062 647 | ·2703 611 | ·2377 280 |
| ·63 | ·4599 962 | ·4143 032 | ·3714 088 | ·3315 011 | ·2946 700 | ·2609 244 |
| ·64 | ·4867 716 | ·4412 240 | ·3981 272 | ·3577 136 | ·3201 202 | ·2854 035− |
| ·65 | ·5138 270 | ·4686 396 | ·4255 501 | ·3848 283 | ·3466 527 | ·3111 236 |
| ·66 | ·5410 612 | ·4964 494 | ·4535 820 | ·4127 590 | ·3741 947 | ·3380 283 |
| ·67 | ·5683 680 | ·5245 453 | ·4821 177 | ·4414 077 | ·4026 596 | ·3660 460 |
| ·68 | ·5956 374 | ·5528 130 | ·5110 427 | ·4706 649 | ·4319 466 | ·3950 887 |
| ·69 | ·6227 567 | ·5811 322 | ·5402 343 | ·5004 096 | ·4619 412 | ·4250 526 |
| ·70 | ·6496 107 | ·6093 785− | ·5695 623 | ·5305 105− | ·4925 158 | ·4558 178 |

$x =$ ·71 to 1·00 $q = 4$ $p =$ 7 to 9·5

| | $p = 7$ | $p = 7{\cdot}5$ | $p = 8$ | $p = 8{\cdot}5$ | $p = 9$ | $p = 9{\cdot}5$ |
|---|---|---|---|---|---|---|
| $B(p, q) =$ | ·1190 4762 $\times \frac{1}{10^2}$ | ·9435 3531 $\times \frac{1}{10^3}$ | ·7575 7576 $\times \frac{1}{10^3}$ | ·6153 4911 $\times \frac{1}{10^3}$ | ·5050 5051 $\times \frac{1}{10^3}$ | ·4184 3740 $\times \frac{1}{10^3}$ |
| $x$ | | | | | | |
| ·71 | ·6760 836 | ·6374 233 | ·5988 899 | ·5608 266 | ·5235 296 | ·4872 487 |
| ·72 | ·7020 595$^{-}$ | ·6651 360 | ·6280 749 | ·5912 085$^{-}$ | ·5548 303 | ·5191 943 |
| ·73 | ·7274 239 | ·6923 850$^{-}$ | ·6569 714 | ·6214 995$^{+}$ | ·5862 547 | ·5514 897 |
| ·74 | ·7520 651 | ·7190 388 | ·6854 310 | ·6515 375$^{+}$ | ·6176 308 | ·5839 569 |
| ·75 | ·7758 751 | ·7449 681 | ·7133 045$^{+}$ | ·6811 564 | ·6487 786 | ·6164 063 |
| ·76 | ·7987 513 | ·7700 469 | ·7404 442 | ·7101 882 | ·6795 133 | ·6486 393 |
| ·77 | ·8205 976 | ·7941 548 | ·7667 052 | ·7384 654 | ·7096 466 | ·6804 502 |
| ·78 | ·8413 261 | ·8171 781 | ·7919 480 | ·7658 234 | ·7389 900 | ·7116 290 |
| ·79 | ·8608 582 | ·8390 121 | ·8160 407 | ·7921 026 | ·7673 576 | ·7419 646 |
| ·80 | ·8791 261 | ·8595 623 | ·8388 608 | ·8171 516 | ·7945 689 | ·7712 483 |
| ·81 | ·8960 739 | ·8787 470 | ·8602 980 | ·8408 301 | ·8204 525$^{+}$ | ·7992 777 |
| ·82 | ·9116 589 | ·8964 982 | ·8802 561 | ·8630 114 | ·8448 495$^{-}$ | ·8258 606 |
| ·83 | ·9258 528 | ·9127 637 | ·8986 556 | ·8835 853 | ·8676 168 | ·8508 196 |
| ·84 | ·9386 423 | ·9275 084 | ·9154 354 | ·9024 614 | ·8886 315$^{+}$ | ·8739 963 |
| ·85 | ·9500 302 | ·9407 156 | ·9305 551 | ·9195 715$^{+}$ | ·9077 937 | ·8952 559 |
| ·86 | ·9600 358 | ·9523 880 | ·9439 967 | ·9348 721 | ·9250 303 | ·9144 919 |
| ·87 | ·9686 952 | ·9625 487 | ·9557 655$^{+}$ | ·9483 468 | ·9402 983 | ·9316 302 |
| ·88 | ·9760 612 | ·9712 415$^{+}$ | ·9658 920 | ·9600 077 | ·9535 873 | ·9466 329 |
| ·89 | ·9822 028 | ·9785 307 | ·9744 317 | ·9698 974 | ·9649 218 | ·9595 019 |
| ·90 | ·9872 048 | ·9845 007 | ·9814 652 | ·9780 886 | ·9743 625$^{+}$ | ·9702 809 |
| ·91 | ·9911 662 | ·9892 550$^{-}$ | ·9870 977 | ·9846 845$^{+}$ | ·9820 069 | ·9790 574 |
| ·92 | ·9941 987 | ·9929 144 | ·9914 567 | ·9898 173 | ·9879 882 | ·9859 624 |
| ·93 | ·9964 239 | ·9956 142 | ·9946 902 | ·9936 454 | ·9924 734 | ·9911 682 |
| ·94 | ·9979 707 | ·9975 010 | ·9969 622 | ·9963 496 | ·9956 587 | ·9948 852 |
| ·95 | ·9989 715$^{+}$ | ·9987 283 | ·9984 477 | ·9981 271 | ·9977 636 | ·9973 545$^{-}$ |
| ·96 | ·9995 574 | ·9994 505$^{-}$ | ·9993 266 | ·9991 841 | ·9990 218 | ·9988 382 |
| ·97 | ·9998 529 | ·9998 166 | ·9997 744 | ·9997 256 | ·9996 696 | ·9996 060 |
| ·98 | ·9999 695$^{-}$ | ·9999 618 | ·9999 528 | ·9999 424 | ·9999 304 | ·9999 166 |
| ·99 | ·9999 980 | ·9999 975$^{-}$ | ·9999 969 | ·9999 962 | ·9999 954 | ·9999 944 |
| 1·00 | 1·0000 000 | 1·0000 000 | 1·0000 000 | 1·0000 000 | 1·0000 000 | 1·0000 000 |

$x$ = ·11 to ·70 $q = 4$ $p$ = 10 to 14

| | $p$ = 10 | $p$ = 10·5 | $p$ = 11 | $p$ = 12 | $p$ = 13 | $p$ = 14 |
|---|---|---|---|---|---|---|
| $B(p, q)$ = | ·3496 5035 $\times \frac{1}{10^3}$ | ·2944 5595 $\times \frac{1}{10^3}$ | ·2497 5025 $\times \frac{1}{10^3}$ | ·1831 5018 $\times \frac{1}{10^3}$ | ·1373 6264 $\times \frac{1}{10^3}$ | ·1050 4202 $\times \frac{1}{10^3}$ |
| $x$ | | | | | | |
| ·11 | ·0000 001 | | | | | |
| ·12 | ·0000 001 | | | | | |
| ·13 | ·0000 003 | ·0000 001 | | | | |
| ·14 | ·0000 006 | ·0000 002 | ·0000 001 | | | |
| ·15 | ·0000 011 | ·0000 005− | ·0000 002 | | | |
| ·16 | ·0000 020 | ·0000 009 | ·0000 004 | ·0000 001 | | |
| ·17 | ·0000 035− | ·0000 016 | ·0000 008 | ·0000 002 | | |
| ·18 | ·0000 060 | ·0000 029 | ·0000 014 | ·0000 003 | ·0000 001 | |
| ·19 | ·0000 099 | ·0000 049 | ·0000 024 | ·0000 006 | ·0000 001 | |
| ·20 | ·0000 161 | ·0000 081 | ·0000 041 | ·0000 010 | ·0000 002 | ·0000 001 |
| ·21 | ·0000 253 | ·0000 131 | ·0000 067 | ·0000 018 | ·0000 005− | ·0000 001 |
| ·22 | ·0000 390 | ·0000 206 | ·0000 108 | ·0000 030 | ·0000 008 | ·0000 002 |
| ·23 | ·0000 587 | ·0000 317 | ·0000 171 | ·0000 049 | ·0000 014 | ·0000 004 |
| ·24 | ·0000 868 | ·0000 479 | ·0000 263 | ·0000 079 | ·0000 023 | ·0000 007 |
| ·25 | ·0001 261 | ·0000 710 | ·0000 398 | ·0000 124 | ·0000 038 | ·0000 011 |
| ·26 | ·0001 802 | ·0001 035− | ·0000 592 | ·0000 191 | ·0000 061 | ·0000 019 |
| ·27 | ·0002 536 | ·0001 484 | ·0000 864 | ·0000 290 | ·0000 096 | ·0000 031 |
| ·28 | ·0003 520 | ·0002 096 | ·0001 243 | ·0000 432 | ·0000 148 | ·0000 050− |
| ·29 | ·0004 820 | ·0002 921 | ·0001 762 | ·0000 634 | ·0000 225− | ·0000 079 |
| ·30 | ·0006 520 | ·0004 018 | ·0002 465− | ·0000 917 | ·0000 336 | ·0000 122 |
| ·31 | ·0008 717 | ·0005 459 | ·0003 404 | ·0001 307 | ·0000 495+ | ·0000 185+ |
| ·32 | ·0011 530 | ·0007 334 | ·0004 645+ | ·0001 841 | ·0000 719 | ·0000 278 |
| ·33 | ·0015 095− | ·0009 748 | ·0006 268 | ·0002 561 | ·0001 031 | ·0000 410 |
| ·34 | ·0019 572 | ·0012 826 | ·0008 369 | ·0003 521 | ·0001 461 | ·0000 599 |
| ·35 | ·0025 146 | ·0016 715− | ·0011 062 | ·0004 789 | ·0002 044 | ·0000 862 |
| ·36 | ·0032 028 | ·0021 585− | ·0014 485− | ·0006 447 | ·0002 829 | ·0001 227 |
| ·37 | ·0040 459 | ·0027 635− | ·0018 795+ | ·0008 593 | ·0003 875− | ·0001 726 |
| ·38 | ·0050 712 | ·0035 092 | ·0024 180 | ·0011 348 | ·0005 253 | ·0002 402 |
| ·39 | ·0063 089 | ·0044 213 | ·0030 854 | ·0014 854 | ·0007 053 | ·0003 309 |
| ·40 | ·0077 930 | ·0055 291 | ·0039 064 | ·0019 278 | ·0009 385− | ·0004 514 |
| ·41 | ·0095 608 | ·0068 652 | ·0049 091 | ·0024 818 | ·0012 377 | ·0006 100 |
| ·42 | ·0116 532 | ·0084 661 | ·0061 252 | ·0031 702 | ·0016 189 | ·0008 169 |
| ·43 | ·0141 149 | ·0103 720 | ·0075 903 | ·0040 196 | ·0021 004 | ·0010 847 |
| ·44 | ·0169 939 | ·0126 270 | ·0093 442 | ·0050 603 | ·0027 042 | ·0014 283 |
| ·45 | ·0203 418 | ·0152 794 | ·0114 305+ | ·0063 268 | ·0034 559 | ·0018 659 |
| ·46 | ·0242 138 | ·0183 811 | ·0138 976 | ·0078 579 | ·0043 851 | ·0024 190 |
| ·47 | ·0286 678 | ·0219 881 | ·0167 979 | ·0096 975+ | ·0055 260 | ·0031 130 |
| ·48 | ·0337 648 | ·0261 599 | ·0201 881 | ·0118 941 | ·0069 176 | ·0039 777 |
| ·49 | ·0395 680 | ·0309 594 | ·0241 294 | ·0145 013 | ·0086 041 | ·0050 477 |
| ·50 | ·0461 426 | ·0364 526 | ·0286 865+ | ·0175 781 | ·0106 354 | ·0063 629 |
| ·51 | ·0535 551 | ·0427 082 | ·0339 282 | ·0211 886 | ·0130 670 | ·0079 692 |
| ·52 | ·0618 728 | ·0497 968 | ·0399 264 | ·0254 018 | ·0159 608 | ·0099 186 |
| ·53 | ·0711 627 | ·0577 903 | ·0467 556 | ·0302 919 | ·0193 847 | ·0122 698 |
| ·54 | ·0814 908 | ·0667 614 | ·0544 926 | ·0359 375− | ·0234 128 | ·0150 886 |
| ·55 | ·0929 213 | ·0767 824 | ·0632 154 | ·0424 213 | ·0281 253 | ·0184 480 |
| ·56 | ·1055 154 | ·0879 241 | ·0730 024 | ·0498 295− | ·0336 084 | ·0224 284 |
| ·57 | ·1193 300 | ·1002 553 | ·0839 312 | ·0582 509 | ·0399 538 | ·0271 176 |
| ·58 | ·1344 168 | ·1138 406 | ·0960 776 | ·0677 763 | ·0472 578 | ·0326 108 |
| ·59 | ·1508 209 | ·1287 401 | ·1095 141 | ·0784 965− | ·0556 210 | ·0390 098 |
| ·60 | ·1685 797 | ·1450 070 | ·1243 088 | ·0905 019 | ·0651 467 | ·0464 229 |
| ·61 | ·1877 210 | ·1626 871 | ·1405 231 | ·1038 804 | ·0759 403 | ·0549 637 |
| ·62 | ·2082 624 | ·1818 164 | ·1582 109 | ·1187 157 | ·0881 069 | ·0647 500+ |
| ·63 | ·2302 093 | ·2024 201 | ·1774 161 | ·1350 855− | ·1017 502 | ·0759 024 |
| ·64 | ·2535 542 | ·2245 107 | ·1981 713 | ·1530 594 | ·1169 699 | ·0885 424 |
| ·65 | ·2782 749 | ·2480 866 | ·2204 957 | ·1726 965− | ·1338 596 | ·1027 902 |
| ·66 | ·3043 339 | ·2731 307 | ·2443 933 | ·1940 432 | ·1525 044 | ·1187 620 |
| ·67 | ·3316 770 | ·2996 086 | ·2698 512 | ·2171 306 | ·1729 772 | ·1365 675+ |
| ·68 | ·3602 327 | ·3274 678 | ·2968 377 | ·2419 722 | ·1953 366 | ·1563 061 |
| ·69 | ·3899 116 | ·3566 364 | ·3253 010 | ·2685 612 | ·2196 230 | ·1780 633 |
| ·70 | ·4206 056 | ·3870 219 | ·3551 674 | ·2968 679 | ·2458 559 | ·2019 070 |

| | $p = 10$ | $p = 10{\cdot}5$ | $p = 11$ | $p = 12$ | $p = 13$ | $p = 14$ |
|---|---|---|---|---|---|---|
| $B(p, q)$ = | ·3496 5035 $\times\frac{1}{10^3}$ | ·2944 5595 $\times\frac{1}{10^3}$ | ·2497 5025 $\times\frac{1}{10^3}$ | ·1831 5018 $\times\frac{1}{10^3}$ | ·1373 6264 $\times\frac{1}{10^3}$ | ·1050 4202 $\times\frac{1}{10^3}$ |
| $x$ | | | | | | |
| ·71 | ·4521 885⁻ | ·4185 114 | ·3863 406 | ·3268 381 | ·2740 297 | ·2278 832 |
| ·72 | ·4845 154 | ·4509 704 | ·4187 006 | ·3583 904 | ·3041 112 | ·2560 115⁻ |
| ·73 | ·5174 238 | ·4842 435⁻ | ·4521 033 | ·3914 146 | ·3360 362 | ·2862 807 |
| ·74 | ·5507 345⁺ | ·5181 546 | ·4863 804 | ·4257 705⁻ | ·3697 063 | ·3186 448 |
| ·75 | ·5842 527 | ·5525 082 | ·5213 400 | ·4612 869 | ·4049 871 | ·3530 181 |
| ·76 | ·6177 699 | ·5870 905⁻ | ·5567 675⁺ | ·4977 616 | ·4417 059 | ·3892 723 |
| ·77 | ·6510 664 | ·6216 716 | ·5924 277 | ·5349 617 | ·4796 507 | ·4272 329 |
| ·78 | ·6839 137 | ·6560 084 | ·6280 664 | ·5726 251 | ·5185 700 | ·4666 770 |
| ·79 | ·7160 781 | ·6898 474 | ·6634 141 | ·6104 628 | ·5581 734 | ·5073 320 |
| ·80 | ·7473 243 | ·7229 290 | ·6981 899 | ·6481 621 | ·5981 343 | ·5488 762 |
| ·81 | ·7774 195⁻ | ·7549 915⁺ | ·7321 057 | ·6853 912 | ·6380 929 | ·5909 401 |
| ·82 | ·8061 382 | ·7857 770 | ·7648 720 | ·7218 051 | ·6776 615⁻ | ·6331 104 |
| ·83 | ·8332 673 | ·8150 361 | ·7962 041 | ·7570 519 | ·7164 315⁺ | ·6749 362 |
| ·84 | ·8586 111 | ·8425 348 | ·8258 289 | ·7907 817 | ·7539 821 | ·7159 370 |
| ·85 | ·8819 973 | ·8680 608 | ·8534 924 | ·8226 552 | ·7898 907 | ·7556 140 |
| ·86 | ·9032 822 | ·8914 301 | ·8789 679 | ·8523 547 | ·8237 456 | ·7934 639 |
| ·87 | ·9223 564 | ·9124 942 | ·9020 641 | ·8795 949 | ·8551 597 | ·8289 953 |
| ·88 | ·9391 499 | ·9311 463 | ·9226 334 | ·9041 349 | ·8837 866 | ·8617 478 |
| ·89 | ·9536 368 | ·9473 282 | ·9405 800 | ·9257 902 | ·9093 365⁺ | ·8913 135⁻ |
| ·90 | ·9658 393 | ·9610 350⁻ | ·9558 671 | ·9444 444 | ·9315 938 | ·9173 594 |
| ·91 | ·9758 301 | ·9723 198 | ·9685 229 | ·9600 598 | ·9504 331 | ·9396 512 |
| ·92 | ·9837 335⁻ | ·9812 959 | ·9786 448 | ·9726 864 | ·9658 343 | ·9580 756 |
| ·93 | ·9897 245⁻ | ·9881 370 | ·9864 010 | ·9824 673 | ·9778 943 | ·9726 600 |
| ·94 | ·9940 250⁻ | ·9930 740 | ·9920 286 | ·9896 401 | ·9868 337 | ·9835 869 |
| ·95 | ·9968 970 | ·9963 886 | ·9958 268 | ·9945 327 | ·9929 961 | ·9911 994 |
| ·96 | ·9986 318 | ·9984 012 | ·9981 450⁺ | ·9975 503 | ·9968 366 | ·9959 934 |
| ·97 | ·9995 342 | ·9994 535⁻ | ·9993 633 | ·9991 524 | ·9988 967 | ·9985 914 |
| ·98 | ·9999 010 | ·9998 834 | ·9998 636 | ·9998 170 | ·9997 599 | ·9996 910 |
| ·99 | ·9999 933 | ·9999 921 | ·9999 908 | ·9999 875⁺ | ·9999 835⁻ | ·9999 786 |
| 1·00 | 1·0000 000 | 1·0000 000 | 1·0000 000 | 1·0000 000 | 1·0000 000 | 1·0000 000 |

$x$ = ·22 to ·80 $q = 4$ $p$ = 15 to 20

| $x$ | $p = 15$ | $p = 16$ | $p = 17$ | $p = 18$ | $p = 19$ | $p = 20$ |
|---|---|---|---|---|---|---|
| $B(p, q)$ = | ·8169 9346 × $\frac{1}{10^4}$ | ·6449 9484 × $\frac{1}{10^4}$ | ·5159 9587 × $\frac{1}{10^4}$ | ·4177 1094 × $\frac{1}{10^4}$ | ·3417 6350 × $\frac{1}{10^4}$ | ·2823 2637 × $\frac{1}{10^4}$ |
| ·22 | ·0000 001 | | | | | |
| ·23 | ·0000 001 | | | | | |
| ·24 | ·0000 002 | ·0000 001 | | | | |
| ·25 | ·0000 003 | ·0000 001 | | | | |
| ·26 | ·0000 006 | ·0000 002 | ·0000 001 | | | |
| ·27 | ·0000 010 | ·0000 003 | ·0000 001 | | | |
| ·28 | ·0000 017 | ·0000 006 | ·0000 002 | ·0000 001 | | |
| ·29 | ·0000 027 | ·0000 009 | ·0000 003 | ·0000 001 | | |
| ·30 | ·0000 044 | ·0000 015+ | ·0000 005+ | ·0000 002 | ·0000 001 | |
| ·31 | ·0000 068 | ·0000 025+ | ·0000 009 | ·0000 003 | ·0000 001 | |
| ·32 | ·0000 106 | ·0000 040 | ·0000 015+ | ·0000 006 | ·0000 002 | ·0000 001 |
| ·33 | ·0000 161 | ·0000 063 | ·0000 024 | ·0000 009 | ·0000 004 | ·0000 001 |
| ·34 | ·0000 243 | ·0000 097 | ·0000 039 | ·0000 015+ | ·0000 006 | ·0000 002 |
| ·35 | ·0000 360 | ·0000 149 | ·0000 061 | ·0000 025− | ·0000 010 | ·0000 004 |
| ·36 | ·0000 526 | ·0000 224 | ·0000 094 | ·0000 039 | ·0000 016 | ·0000 007 |
| ·37 | ·0000 761 | ·0000 332 | ·0000 144 | ·0000 062 | ·0000 026 | ·0000 011 |
| ·38 | ·0001 087 | ·0000 487 | ·0000 216 | ·0000 095+ | ·0000 042 | ·0000 018 |
| ·39 | ·0001 536 | ·0000 706 | ·0000 322 | ·0000 146 | ·0000 065+ | ·0000 029 |
| ·40 | ·0002 148 | ·0001 013 | ·0000 473 | ·0000 220 | ·0000 101 | ·0000 046 |
| ·41 | ·0002 975− | ·0001 437 | ·0000 688 | ·0000 327 | ·0000 154 | ·0000 072 |
| ·42 | ·0004 079 | ·0002 018 | ·0000 990 | ·0000 482 | ·0000 233 | ·0000 112 |
| ·43 | ·0005 543 | ·0002 806 | ·0001 409 | ·0000 702 | ·0000 347 | ·0000 171 |
| ·44 | ·0007 466 | ·0003 866 | ·0001 986 | ·0001 012 | ·0000 512 | ·0000 258 |
| ·45 | ·0009 971 | ·0005 279 | ·0002 772 | ·0001 445− | ·0000 748 | ·0000 385− |
| ·46 | ·0013 208 | ·0007 145+ | ·0003 834 | ·0002 042 | ·0001 080 | ·0000 568 |
| ·47 | ·0017 358 | ·0009 591 | ·0005 256 | ·0002 859 | ·0001 545− | ·0000 830 |
| ·48 | ·0022 641 | ·0012 771 | ·0007 145+ | ·0003 968 | ·0002 189 | ·0001 200 |
| ·49 | ·0029 316 | ·0016 873 | ·0009 633 | ·0005 459 | ·0003 073 | ·0001 720 |
| ·50 | ·0037 689 | ·0022 125+ | ·0012 884 | ·0007 448 | ·0004 277 | ·0002 441 |
| ·51 | ·0048 122 | ·0028 801 | ·0017 100 | ·0010 079 | ·0005 902 | ·0003 435+ |
| ·52 | ·0061 034 | ·0037 227 | ·0022 526 | ·0013 533 | ·0008 077 | ·0004 791 |
| ·53 | ·0076 909 | ·0047 787 | ·0029 459 | ·0018 030 | ·0010 964 | ·0006 627 |
| ·54 | ·0096 304 | ·0060 935− | ·0038 254 | ·0023 845+ | ·0014 767 | ·0009 092 |
| ·55 | ·0119 850− | ·0077 194 | ·0049 334 | ·0031 307 | ·0019 740 | ·0012 373 |
| ·56 | ·0148 260 | ·0097 172 | ·0063 198 | ·0040 815+ | ·0026 192 | ·0016 710 |
| ·57 | ·0182 332 | ·0121 562 | ·0080 429 | ·0052 845+ | ·0034 502 | ·0022 396 |
| ·58 | ·0222 951 | ·0151 154 | ·0101 704 | ·0067 961 | ·0045 129 | ·0029 795− |
| ·59 | ·0271 091 | ·0186 834 | ·0127 802 | ·0086 826 | ·0058 621 | ·0039 353 |
| ·60 | ·0327 813 | ·0229 593 | ·0159 612 | ·0110 213 | ·0075 634 | ·0051 610 |
| ·61 | ·0394 261 | ·0280 525+ | ·0198 138 | ·0139 013 | ·0096 936 | ·0067 216 |
| ·62 | ·0471 657 | ·0340 829 | ·0244 507 | ·0174 249 | ·0123 428 | ·0086 945− |
| ·63 | ·0561 289 | ·0411 802 | ·0299 966 | ·0217 077 | ·0156 153 | ·0111 710 |
| ·64 | ·0664 502 | ·0494 835− | ·0365 887 | ·0268 797 | ·0196 303 | ·0142 581 |
| ·65 | ·0782 675− | ·0591 398 | ·0443 756 | ·0330 853 | ·0245 236 | ·0180 797 |
| ·66 | ·0917 200 | ·0703 027 | ·0535 169 | ·0404 832 | ·0304 472 | ·0227 777 |
| ·67 | ·1069 457 | ·0831 298 | ·0641 812 | ·0492 453 | ·0375 703 | ·0285 131 |
| ·68 | ·1240 781 | ·0977 800 | ·0765 443 | ·0595 558 | ·0460 782 | ·0354 664 |
| ·69 | ·1432 421 | ·1144 102 | ·0907 861 | ·0716 089 | ·0561 712 | ·0438 373 |
| ·70 | ·1645 505− | ·1331 710 | ·1070 868 | ·0856 057 | ·0680 628 | ·0538 438 |
| ·71 | ·1880 983 | ·1542 017 | ·1256 225− | ·1017 505+ | ·0819 765+ | ·0657 202 |
| ·72 | ·2139 586 | ·1776 249 | ·1465 596 | ·1202 454 | ·0981 415+ | ·0797 138 |
| ·73 | ·2421 761 | ·2035 405− | ·1700 482 | ·1412 843 | ·1167 870 | ·0960 803 |
| ·74 | ·2727 623 | ·2320 187 | ·1962 153 | ·1650 453 | ·1381 351 | ·1150 774 |
| ·75 | ·3056 892 | ·2630 931 | ·2251 560 | ·1916 821 | ·1623 925− | ·1369 567 |
| ·76 | ·3408 836 | ·2967 531 | ·2569 253 | ·2213 146 | ·1897 397 | ·1619 538 |
| ·77 | ·3782 223 | ·3329 364 | ·2915 282 | ·2540 172 | ·2203 198 | ·1902 760 |
| ·78 | ·4175 269 | ·3715 225− | ·3289 108 | ·2898 084 | ·2542 252 | ·2220 879 |
| ·79 | ·4585 606 | ·4123 253 | ·3689 508 | ·3286 381 | ·2914 831 | ·2574 961 |
| ·80 | ·5010 255− | ·4550 887 | ·4114 489 | ·3703 760 | ·3320 414 | ·2965 314 |

$x$ = ·81 to 1·00 $q = 4$ $p$ = 15 to 20

| | $p = 15$ | $p = 16$ | $p = 17$ | $p = 18$ | $p = 19$ | $p = 20$ |
|---|---|---|---|---|---|---|
| $B(p, q)$ = | ·8169 9346 × $\frac{1}{10^4}$ | ·6449 9484 × $\frac{1}{10^4}$ | ·5159 9587 × $\frac{1}{10^4}$ | ·4177 1094 × $\frac{1}{10^4}$ | ·3417 6350 × $\frac{1}{10^4}$ | ·2823 2637 × $\frac{1}{10^4}$ |
| $x$ | | | | | | |
| ·81 | ·5445 619 | ·4994 823 | ·4561 214 | ·4148 010 | ·3757 532 | ·3391 305+ |
| ·82 | ·5887 503 | ·5450 999 | ·5025 954 | ·4615 910 | ·4223 634 | ·3851 179 |
| ·83 | ·6331 149 | ·5914 608 | ·5504 055+ | ·5103 163 | ·4714 965+ | ·4341 887 |
| ·84 | ·6771 310 | ·6380 145+ | ·5989 959 | ·5604 363 | ·5226 479 | ·4858 937 |
| ·85 | ·7202 355+ | ·6841 495+ | ·6477 252 | ·6113 008 | ·5751 800 | ·5396 296 |
| ·86 | ·7618 411 | ·7292 065− | ·6958 783 | ·6621 581 | ·6283 255− | ·5946 353 |
| ·87 | ·8013 544 | ·7724 973 | ·7426 843 | ·7121 699 | ·6811 977 | ·6499 973 |
| ·88 | ·8381 978 | ·8133 289 | ·7873 410 | ·7604 359 | ·7328 133 | ·7046 673 |
| ·89 | ·8718 357 | ·8510 334 | ·8290 480 | ·8060 279 | ·7821 255− | ·7574 934 |
| ·90 | ·9018 032 | ·8850 024 | ·8670 467 | ·8480 347 | ·8280 721 | ·8072 690 |
| ·91 | ·9277 372 | ·9147 271 | ·9006 680 | ·8856 165+ | ·8696 369 | ·8527 994 |
| ·92 | ·9494 080 | ·9398 390 | ·9293 848 | ·9180 697 | ·9059 249 | ·8929 874 |
| ·93 | ·9667 490 | ·9601 523 | ·9528 671 | ·9448 962 | ·9362 477 | ·9269 348 |
| ·94 | ·9798 809 | ·9757 006 | ·9710 343 | ·9658 739 | ·9602 146 | ·9540 550− |
| ·95 | ·9891 268 | ·9867 640 | ·9840 985− | ·9811 194 | ·9778 175+ | ·9741 855− |
| ·96 | ·9950 104 | ·9938 780 | ·9925 871 | ·9911 291 | ·9894 961 | ·9876 810 |
| ·97 | ·9982 318 | ·9978 133 | ·9973 312 | ·9967 810 | ·9961 584 | ·9954 590 |
| ·98 | ·9996 090 | ·9995 125+ | ·9994 003 | ·9992 709 | ·9991 230 | ·9989 552 |
| ·99 | ·9999 726 | ·9999 656 | ·9999 574 | ·9999 478 | ·9999 367 | ·9999 239 |
| 1·00 | 1·0000 000 | 1·0000 000 | 1·0000 000 | 1·0000 000 | 1·0000 000 | 1·0000 000 |

$x$ = ·33 to ·90 $q = 4$ $p$ = 21 to 26

| $x$ | $p = 21$ | $p = 22$ | $p = 23$ | $p = 24$ | $p = 25$ | $p = 26$ |
|---|---|---|---|---|---|---|
| $B(p, q) =$ | ·2352 7197 × $\frac{1}{10^4}$ | ·1976 2846 × $\frac{1}{10^4}$ | ·1672 2408 × $\frac{1}{10^4}$ | ·1424 5014 × $\frac{1}{10^4}$ | ·1221 0012 × $\frac{1}{10^4}$ | ·1052 5873 × $\frac{1}{10^4}$ |
| ·33 | ·0000 001 | | | | | |
| ·34 | ·0000 001 | | | | | |
| ·35 | ·0000 002 | ·0000 001 | | | | |
| ·36 | ·0000 003 | ·0000 001 | | | | |
| ·37 | ·0000 005− | ·0000 002 | ·0000 001 | | | |
| ·38 | ·0000 008 | ·0000 003 | ·0000 001 | ·0000 001 | | |
| ·39 | ·0000 013 | ·0000 006 | ·0000 003 | ·0000 001 | | |
| ·40 | ·0000 021 | ·0000 010 | ·0000 004 | ·0000 002 | ·0000 001 | |
| ·41 | ·0000 034 | ·0000 016 | ·0000 007 | ·0000 003 | ·0000 002 | ·0000 001 |
| ·42 | ·0000 053 | ·0000 025+ | ·0000 012 | ·0000 006 | ·0000 003 | ·0000 001 |
| ·43 | ·0000 084 | ·0000 041 | ·0000 020 | ·0000 009 | ·0000 005− | ·0000 002 |
| ·44 | ·0000 129 | ·0000 064 | ·0000 032 | ·0000 016 | ·0000 008 | ·0000 004 |
| ·45 | ·0000 197 | ·0000 100 | ·0000 051 | ·0000 026 | ·0000 013 | ·0000 006 |
| ·46 | ·0000 297 | ·0000 154 | ·0000 080 | ·0000 041 | ·0000 021 | ·0000 011 |
| ·47 | ·0000 443 | ·0000 235+ | ·0000 124 | ·0000 065+ | ·0000 034 | ·0000 018 |
| ·48 | ·0000 654 | ·0000 355− | ·0000 192 | ·0000 103 | ·0000 055+ | ·0000 029 |
| ·49 | ·0000 957 | ·0000 530 | ·0000 292 | ·0000 160 | ·0000 087 | ·0000 048 |
| ·50 | ·0001 386 | ·0000 783 | ·0000 440 | ·0000 246 | ·0000 137 | ·0000 076 |
| ·51 | ·0001 988 | ·0001 145+ | ·0000 656 | ·0000 375− | ·0000 213 | ·0000 121 |
| ·52 | ·0002 827 | ·0001 659 | ·0000 970 | ·0000 564 | ·0000 327 | ·0000 189 |
| ·53 | ·0003 984 | ·0002 383 | ·0001 419 | ·0000 841 | ·0000 497 | ·0000 292 |
| ·54 | ·0005 567 | ·0003 392 | ·0002 057 | ·0001 242 | ·0000 747 | ·0000 448 |
| ·55 | ·0007 714 | ·0004 785+ | ·0002 955− | ·0001 817 | ·0001 113 | ·0000 679 |
| ·56 | ·0010 603 | ·0006 695+ | ·0004 208 | ·0002 634 | ·0001 642 | ·0001 020 |
| ·57 | ·0014 460 | ·0009 290 | ·0005 942 | ·0003 784 | ·0002 401 | ·0001 517 |
| ·58 | ·0019 567 | ·0012 788 | ·0008 319 | ·0005 390 | ·0003 478 | ·0002 236 |
| ·59 | ·0026 280 | ·0017 464 | ·0011 554 | ·0007 612 | ·0004 996 | ·0003 267 |
| ·60 | ·0035 034 | ·0023 668 | ·0015 918 | ·0010 661 | ·0007 113 | ·0004 729 |
| ·61 | ·0046 368 | ·0031 833 | ·0021 758 | ·0014 811 | ·0010 044 | ·0006 786 |
| ·62 | ·0060 932 | ·0042 500+ | ·0029 514 | ·0020 412 | ·0014 064 | ·0009 656 |
| ·63 | ·0079 512 | ·0056 329 | ·0039 732 | ·0027 912 | ·0019 534 | ·0013 623 |
| ·64 | ·0103 042 | ·0074 122 | ·0053 089 | ·0037 872 | ·0026 916 | ·0019 063 |
| ·65 | ·0132 629 | ·0096 847 | ·0070 417 | ·0050 997 | ·0036 796 | ·0026 458 |
| ·66 | ·0169 565− | ·0125 656 | ·0092 725− | ·0068 155− | ·0049 912 | ·0036 427 |
| ·67 | ·0215 344 | ·0161 908 | ·0121 223 | ·0090 409 | ·0067 183 | ·0049 754 |
| ·68 | ·0271 679 | ·0207 188 | ·0157 355− | ·0119 048 | ·0089 743 | ·0067 424 |
| ·69 | ·0340 503 | ·0263 326 | ·0202 813 | ·0155 612 | ·0118 972 | ·0090 657 |
| ·70 | ·0423 975+ | ·0332 405+ | ·0259 565+ | ·0201 927 | ·0156 536 | ·0120 950+ |
| ·71 | ·0524 470 | ·0416 767 | ·0329 870 | ·0260 126 | ·0204 418 | ·0160 119 |
| ·72 | ·0644 557 | ·0519 005− | ·0416 280 | ·0332 671 | ·0264 948 | ·0210 336 |
| ·73 | ·0786 971 | ·0641 945− | ·0521 639 | ·0422 361 | ·0340 829 | ·0274 168 |
| ·74 | ·0954 552 | ·0788 605+ | ·0649 059 | ·0532 325− | ·0435 144 | ·0354 601 |
| ·75 | ·1150 184 | ·0962 141 | ·0801 877 | ·0666 001 | ·0551 356 | ·0455 054 |
| ·76 | ·1376 690 | ·1165 759 | ·0983 591 | ·0827 085− | ·0693 272 | ·0579 371 |
| ·77 | ·1636 721 | ·1402 607 | ·1197 758 | ·1019 450− | ·0864 990 | ·0731 784 |
| ·78 | ·1932 608 | ·1675 635− | ·1447 863 | ·1247 028 | ·1070 795− | ·0916 838 |
| ·79 | ·2266 190 | ·1987 413 | ·1737 147 | ·1513 649 | ·1315 015+ | ·1139 264 |
| ·80 | ·2638 622 | ·2339 933 | ·2068 397 | ·1822 833 | ·1601 827 | ·1403 805− |
| ·81 | ·3050 165− | ·2734 366 | ·2443 688 | ·2177 527 | ·1934 989 | ·1714 958 |
| ·82 | ·3499 954 | ·3170 806 | ·2864 100 | ·2579 796 | ·2317 526 | ·2076 658 |
| ·83 | ·3985 783 | ·3647 994 | ·3329 397 | ·3030 470 | ·2751 347 | ·2491 874 |
| ·84 | ·4503 891 | ·4163 047 | ·3837 696 | ·3528 754 | ·3236 804 | ·2962 137 |
| ·85 | ·5048 790 | ·4711 213 | ·4385 144 | ·4071 834 | ·3772 232 | ·3487 011 |
| ·86 | ·5613 157 | ·5285 673 | ·4965 632 | ·4654 496 | ·4353 472 | ·4063 526 |
| ·87 | ·6187 813 | ·5877 437 | ·5570 588 | ·5268 809 | ·4973 442 | ·4685 637 |
| ·88 | ·6761 835+ | ·6475 370 | ·6188 905+ | ·5903 935+ | ·5621 814 | ·5343 756 |
| ·89 | ·7322 824 | ·7066 393 | ·6807 047 | ·6546 123 | ·6284 873 | ·6024 458 |
| ·90 | ·7857 378 | ·7635 914 | ·7409 416 | ·7178 980 | ·6945 663 | ·6710 480 |

$x = \cdot 91$ to $1\cdot 00$ $q = 4$ $p = 21$ to $26$

| | $p = 21$ | $p = 22$ | $p = 23$ | $p = 24$ | $p = 25$ | $p = 26$ |
|---|---|---|---|---|---|---|
| $B(p, q) =$ | $\cdot 2352\ 7197 \times \frac{1}{10^4}$ | $\cdot 1976\ 2846 \times \frac{1}{10^4}$ | $\cdot 1672\ 2408 \times \frac{1}{10^4}$ | $\cdot 1424\ 5014 \times \frac{1}{10^4}$ | $\cdot 1221\ 0012 \times \frac{1}{10^4}$ | $\cdot 1052\ 5873 \times \frac{1}{10^4}$ |
| $x$ | | | | | | |
| ·91 | ·8351 790 | ·8168 538 | ·7979 038 | ·7784 100 | ·7584 533 | ·7381 134 |
| ·92 | ·8792 996 | ·8649 078 | ·8498 618 | ·8342 140 | ·8180 185+ | ·8013 307 |
| ·93 | ·9169 745+ | ·9063 882 | ·8952 004 | ·8834 386 | ·8711 328 | ·8583 152 |
| ·94 | ·9473 964 | ·9402 432 | ·9326 023 | ·9244 830 | ·9158 968 | ·9068 573 |
| ·95 | ·9702 175+ | ·9659 094 | ·9612 586 | ·9562 641 | ·9509 261 | ·9452 466 |
| ·96 | ·9856 770 | ·9834 784 | ·9810 800 | ·9784 771 | ·9756 660 | ·9726 435+ |
| ·97 | ·9946 790 | ·9938 142 | ·9928 610 | ·9918 158 | ·9906 752 | ·9894 360 |
| ·98 | ·9987 660 | ·9985 541 | ·9983 182 | ·9980 568 | ·9977 686 | ·9974 523 |
| ·99 | ·9999 095− | ·9998 931 | ·9998 746 | ·9998 540 | ·9998 310 | ·9998 055+ |
| 1·00 | 1·0000 000 | 1·0000 000 | 1·0000 000 | 1·0000 000 | 1·0000 000 | 1·0000 000 |

$x = \cdot 42$ to $1\cdot 00$ | $q = 4$ | $p = 27$ to $32$

| $x$ | $p = 27$ | $p = 28$ | $p = 29$ | $p = 30$ | $p = 31$ | $p = 32$ |
|---|---|---|---|---|---|---|
| $B(p, q) =$ | $\cdot 9122\ 4229 \times \frac{1}{10^5}$ | $\cdot 7945\ 3361 \times \frac{1}{10^5}$ | $\cdot 6952\ 1691 \times \frac{1}{10^5}$ | $\cdot 6109\ 4819 \times \frac{1}{10^5}$ | $\cdot 5390\ 7193 \times \frac{1}{10^5}$ | $\cdot 4774\ 6371 \times \frac{1}{10^5}$ |
| ·42 | ·0000 001 | | | | | |
| ·43 | ·0000 001 | | | | | |
| ·44 | ·0000 002 | ·0000 001 | | | | |
| ·45 | ·0000 003 | ·0000 002 | ·0000 001 | | | |
| ·46 | ·0000 006 | ·0000 003 | ·0000 001 | ·0000 001 | | |
| ·47 | ·0000 009 | ·0000 005− | ·0000 003 | ·0000 001 | ·0000 001 | |
| ·48 | ·0000 016 | ·0000 008 | ·0000 004 | ·0000 002 | ·0000 001 | ·0000 001 |
| ·49 | ·0000 026 | ·0000 014 | ·0000 008 | ·0000 004 | ·0000 002 | ·0000 001 |
| ·50 | ·0000 042 | ·0000 023 | ·0000 013 | ·0000 007 | ·0000 004 | ·0000 002 |
| ·51 | ·0000 068 | ·0000 038 | ·0000 021 | ·0000 012 | ·0000 007 | ·0000 004 |
| ·52 | ·0000 109 | ·0000 062 | ·0000 036 | ·0000 020 | ·0000 012 | ·0000 007 |
| ·53 | ·0000 171 | ·0000 100 | ·0000 058 | ·0000 034 | ·0000 020 | ·0000 011 |
| ·54 | ·0000 267 | ·0000 159 | ·0000 094 | ·0000 056 | ·0000 033 | ·0000 019 |
| ·55 | ·0000 413 | ·0000 250+ | ·0000 151 | ·0000 091 | ·0000 055− | ·0000 033 |
| ·56 | ·0000 631 | ·0000 390 | ·0000 240 | ·0000 147 | ·0000 090 | ·0000 055− |
| ·57 | ·0000 956 | ·0000 600 | ·0000 376 | ·0000 235− | ·0000 146 | ·0000 091 |
| ·58 | ·0001 433 | ·0000 915+ | ·0000 583 | ·0000 370 | ·0000 235− | ·0000 148 |
| ·59 | ·0002 129 | ·0001 383 | ·0000 896 | ·0000 579 | ·0000 373 | ·0000 240 |
| ·60 | ·0003 133 | ·0002 070 | ·0001 363 | ·0000 895− | ·0000 586 | ·0000 383 |
| ·61 | ·0004 570 | ·0003 068 | ·0002 054 | ·0001 371 | ·0000 913 | ·0000 606 |
| ·62 | ·0006 607 | ·0004 507 | ·0003 066 | ·0002 079 | ·0001 407 | ·0000 949 |
| ·63 | ·0009 470 | ·0006 562 | ·0004 534 | ·0003 124 | ·0002 147 | ·0001 472 |
| ·64 | ·0013 457 | ·0009 470 | ·0006 646 | ·0004 651 | ·0003 246 | ·0002 261 |
| ·65 | ·0018 963 | ·0013 549 | ·0009 654 | ·0006 860 | ·0004 862 | ·0003 438 |
| ·66 | ·0026 500− | ·0019 220 | ·0013 900 | ·0010 026 | ·0007 213 | ·0005 177 |
| ·67 | ·0036 730 | ·0027 033 | ·0019 841 | ·0014 523 | ·0010 604 | ·0007 725− |
| ·68 | ·0050 496 | ·0037 707 | ·0028 078 | ·0020 853 | ·0015 448 | ·0011 418 |
| ·69 | ·0068 865+ | ·0052 158 | ·0039 395+ | ·0029 678 | ·0022 302 | ·0016 721 |
| ·70 | ·0093 166 | ·0071 556 | ·0054 808 | ·0041 871 | ·0031 910 | ·0024 263 |
| ·71 | ·0125 037 | ·0097 362 | ·0075 607 | ·0058 563 | ·0045 252 | ·0034 887 |
| ·72 | ·0166 478 | ·0131 392 | ·0103 424 | ·0081 203 | ·0063 604 | ·0049 707 |
| ·73 | ·0219 891 | ·0175 866 | ·0140 284 | ·0111 623 | ·0088 607 | ·0070 180 |
| ·74 | ·0288 121 | ·0233 460 | ·0188 678 | ·0152 110 | ·0122 344 | ·0098 186 |
| ·75 | ·0374 493 | ·0307 360 | ·0251 615− | ·0205 481 | ·0167 421 | ·0136 113 |
| ·76 | ·0482 818 | ·0401 284 | ·0332 679 | ·0275 145+ | ·0227 047 | ·0186 955+ |
| ·77 | ·0617 381 | ·0519 502 | ·0436 061 | ·0365 165− | ·0305 115+ | ·0254 403 |
| ·78 | ·0782 895− | ·0666 811 | ·0566 565− | ·0480 284 | ·0406 254 | ·0342 924 |
| ·79 | ·0984 400 | ·0848 464 | ·0729 569 | ·0625 925− | ·0535 858 | ·0457 820 |
| ·80 | ·1227 108 | ·1070 044 | ·0930 931 | ·0808 127 | ·0700 060 | ·0605 240 |
| ·81 | ·1516 168 | ·1337 258 | ·1176 813 | ·1033 409 | ·0905 636 | ·0792 124 |
| ·82 | ·1856 356 | ·1655 636 | ·1473 411 | ·1308 529 | ·1159 805+ | ·1026 050+ |
| ·83 | ·2251 662 | ·2030 133 | ·1826 564 | ·1640 123 | ·1469 902 | ·1314 946 |
| ·84 | ·2704 796 | ·2464 610 | ·2241 238 | ·2034 195− | ·1842 887 | ·1666 637 |
| ·85 | ·3216 599 | ·2961 210 | ·2720 871 | ·2495 449 | ·2284 680 | ·2088 189 |
| ·86 | ·3785 401 | ·3519 637 | ·3266 591 | ·3026 460 | ·2799 295+ | ·2585 028 |
| ·87 | ·4406 355+ | ·4136 383 | ·3876 342 | ·3626 702 | ·3387 797 | ·3159 836 |
| ·88 | ·5070 831 | ·4803 971 | ·4543 974 | ·4291 507 | ·4047 119 | ·3811 246 |
| ·89 | ·5765 947 | ·5510 308 | ·5258 413 | ·5011 034 | ·4768 850− | ·4532 447 |
| ·90 | ·6474 392 | ·6238 304 | ·6003 059 | ·5769 437 | ·5538 150+ | ·5309 849 |
| ·91 | ·7174 684 | ·6965 940 | ·6755 631 | ·6544 451 | ·6333 060 | ·6122 079 |
| ·92 | ·7842 064 | ·7667 016 | ·7488 717 | ·7307 713 | ·7124 537 | ·6939 706 |
| ·93 | ·8450 193 | ·8312 802 | ·8171 338 | ·8026 168 | ·7877 658 | ·7726 178 |
| ·94 | ·8973 797 | ·8874 809 | ·8771 791 | ·8664 936 | ·8554 448 | ·8440 538 |
| ·95 | ·9392 284 | ·9328 760 | ·9261 945+ | ·9191 905+ | ·9118 713 | ·9042 452 |
| ·96 | ·9694 071 | ·9659 550− | ·9622 858 | ·9583 991 | ·9542 947 | ·9499 731 |
| ·97 | ·9880 954 | ·9866 504 | ·9850 987 | ·9834 378 | ·9816 656 | ·9797 802 |
| ·98 | ·9971 065+ | ·9967 300 | ·9963 215+ | ·9958 798 | ·9954 036 | ·9948 918 |
| ·99 | ·9997 774 | ·9997 465− | ·9997 125+ | ·9996 755− | ·9996 351 | ·9995 913 |
| 1·00 | 1·0000 000 | 1·0000 000 | 1·0000 000 | 1·0000 000 | 1·0000 000 | 1·0000 000 |

$x$ = ·49 to 1·00 $q = 4$ $p$ = 33 to 38

| $x$ | $p = 33$ | $p = 34$ | $p = 35$ | $p = 36$ | $p = 37$ | $p = 38$ |
|---|---|---|---|---|---|---|
| $B(p, q)$ = | ·4244 1219 × $\frac{1}{10^5}$ | ·3785 2979 × $\frac{1}{10^5}$ | ·3386 8455 × $\frac{1}{10^5}$ | ·3039 4767 × $\frac{1}{10^5}$ | ·2735 5291 × $\frac{1}{10^5}$ | ·2468 6482 × $\frac{1}{10^5}$ |
| ·49 | ·0000 001 | | | | | |
| ·50 | ·0000 001 | ·0000 001 | | | | |
| ·51 | ·0000 002 | ·0000 001 | ·0000 001 | | | |
| ·52 | ·0000 004 | ·0000 002 | ·0000 001 | ·0000 001 | | |
| ·53 | ·0000 007 | ·0000 004 | ·0000 002 | ·0000 001 | ·0000 001 | |
| ·54 | ·0000 011 | ·0000 007 | ·0000 004 | ·0000 002 | ·0000 001 | ·0000 001 |
| ·55 | ·0000 020 | ·0000 012 | ·0000 007 | ·0000 004 | ·0000 002 | ·0000 001 |
| ·56 | ·0000 033 | ·0000 020 | ·0000 012 | ·0000 007 | ·0000 004 | ·0000 003 |
| ·57 | ·0000 056 | ·0000 035− | ·0000 021 | ·0000 013 | ·0000 008 | ·0000 005− |
| ·58 | ·0000 093 | ·0000 059 | ·0000 037 | ·0000 023 | ·0000 014 | ·0000 009 |
| ·59 | ·0000 154 | ·0000 098 | ·0000 063 | ·0000 040 | ·0000 025+ | ·0000 016 |
| ·60 | ·0000 250− | ·0000 162 | ·0000 105+ | ·0000 068 | ·0000 044 | ·0000 028 |
| ·61 | ·0000 402 | ·0000 266 | ·0000 175+ | ·0000 115+ | ·0000 076 | ·0000 050− |
| ·62 | ·0000 639 | ·0000 429 | ·0000 288 | ·0000 193 | ·0000 129 | ·0000 086 |
| ·63 | ·0001 007 | ·0000 687 | ·0000 468 | ·0000 318 | ·0000 216 | ·0000 146 |
| ·64 | ·0001 571 | ·0001 089 | ·0000 753 | ·0000 520 | ·0000 358 | ·0000 247 |
| ·65 | ·0002 425+ | ·0001 707 | ·0001 199 | ·0000 841 | ·0000 588 | ·0000 411 |
| ·66 | ·0003 708 | ·0002 650− | ·0001 890 | ·0001 345− | ·0000 955+ | ·0000 678 |
| ·67 | ·0005 614 | ·0004 072 | ·0002 947 | ·0002 129 | ·0001 535+ | ·0001 105+ |
| ·68 | ·0008 420 | ·0006 196 | ·0004 550+ | ·0003 336 | ·0002 441 | ·0001 783 |
| ·69 | ·0012 508 | ·0009 338 | ·0006 957 | ·0005 173 | ·0003 840 | ·0002 845+ |
| ·70 | ·0018 408 | ·0013 937 | ·0010 531 | ·0007 942 | ·0005 979 | ·0004 494 |
| ·71 | ·0026 837 | ·0020 603 | ·0015 785+ | ·0012 072 | ·0009 216 | ·0007 023 |
| ·72 | ·0038 763 | ·0030 167 | ·0023 431 | ·0018 166 | ·0014 059 | ·0010 863 |
| ·73 | ·0055 467 | ·0043 751 | ·0034 443 | ·0027 066 | ·0021 231 | ·0016 627 |
| ·74 | ·0078 633 | ·0062 848 | ·0050 137 | ·0039 924 | ·0031 737 | ·0025 187 |
| ·75 | ·0110 431 | ·0089 418 | ·0072 268 | ·0058 303 | ·0046 957 | ·0037 757 |
| ·76 | ·0153 629 | ·0125 998 | ·0103 146 | ·0084 290 | ·0068 765− | ·0056 009 |
| ·77 | ·0211 693 | ·0175 817 | ·0145 756 | ·0120 624 | ·0099 660 | ·0082 208 |
| ·78 | ·0288 895− | ·0242 921 | ·0203 897 | ·0170 850− | ·0142 924 | ·0119 377 |
| ·79 | ·0390 389 | ·0332 277 | ·0282 317 | ·0239 466 | ·0202 793 | ·0171 472 |
| ·80 | ·0522 272 | ·0449 864 | ·0386 826 | ·0332 073 | ·0284 621 | ·0243 581 |
| ·81 | ·0691 560 | ·0602 697 | ·0524 368 | ·0455 482 | ·0395 035+ | ·0342 104 |
| ·82 | ·0906 088 | ·0798 777 | ·0703 018 | ·0617 764 | ·0542 031 | ·0474 894 |
| ·83 | ·1174 275+ | ·1046 904 | ·0931 858 | ·0828 185− | ·0734 966 | ·0651 320 |
| ·84 | ·1504 707 | ·1356 320 | ·1220 678 | ·1096 971 | ·0984 399 | ·0882 171 |
| ·85 | ·1905 513 | ·1736 123 | ·1579 437 | ·1434 839 | ·1301 688 | ·1179 332 |
| ·86 | ·2383 483 | ·2194 397 | ·2017 435+ | ·1852 203 | ·1698 262 | ·1555 138 |
| ·87 | ·2942 916 | ·2737 039 | ·2542 123 | ·2358 010 | ·2184 484 | ·2021 275+ |
| ·88 | ·3584 217 | ·3366 270 | ·3157 553 | ·2958 140 | ·2768 032 | ·2587 173 |
| ·89 | ·4302 324 | ·4078 895+ | ·3862 498 | ·3653 396 | ·3451 787 | ·3257 807 |
| ·90 | ·5085 114 | ·4864 466 | ·4648 360 | ·4437 194 | ·4231 307 | ·4030 984 |
| ·91 | ·5912 086 | ·5703 621 | ·5497 178 | ·5293 214 | ·5092 138 | ·4894 323 |
| ·92 | ·6753 720 | ·6567 058 | ·6380 176 | ·6193 508 | ·6007 462 | ·5822 422 |
| ·93 | ·7572 094 | ·7415 770 | ·7257 560 | ·7097 813 | ·6936 868 | ·6775 053 |
| ·94 | ·8323 425+ | ·8203 331 | ·8080 481 | ·7955 105− | ·7827 430 | ·7697 684 |
| ·95 | ·8963 212 | ·8881 090 | ·8796 191 | ·8708 624 | ·8618 502 | ·8525 945+ |
| ·96 | ·9454 355− | ·9406 833 | ·9357 188 | ·9305 442 | ·9251 627 | ·9195 776 |
| ·97 | ·9777 799 | ·9756 632 | ·9734 289 | ·9710 759 | ·9686 032 | ·9660 102 |
| ·98 | ·9943 432 | ·9937 567 | ·9931 312 | ·9924 656 | ·9917 591 | ·9910 105− |
| ·99 | ·9995 438 | ·9994 926 | ·9994 374 | ·9993 780 | ·9993 144 | ·9992 463 |
| 1·00 | 1·0000 000 | 1·0000 000 | 1·0000 000 | 1·0000 000 | 1·0000 000 | 1·0000 000 |

$x$ = ·55 to 1·00 $q = 4$ $p$ = 39 to 44

| | $p = 39$ | $p = 40$ | $p = 41$ | $p = 42$ | $p = 43$ | $p = 44$ |
|---|---|---|---|---|---|---|
| $B(p, q)$ = | ·2233 5388 × $\frac{1}{10^5}$ | ·2025 7678 × $\frac{1}{10^5}$ | ·1841 6071 × $\frac{1}{10^5}$ | ·1677 9087 × $\frac{1}{10^5}$ | ·1532 0036 × $\frac{1}{10^5}$ | ·1401 6203 × $\frac{1}{10^5}$ |
| $x$ | | | | | | |
| ·55 | ·0000 001 | ·0000 001 | | | | |
| ·56 | ·0000 002 | ·0000 001 | ·0000 001 | | | |
| ·57 | ·0000 003 | ·0000 002 | ·0000 001 | ·0000 001 | | |
| ·58 | ·0000 006 | ·0000 003 | ·0000 002 | ·0000 001 | ·0000 001 | ·0000 001 |
| ·59 | ·0000 010 | ·0000 006 | ·0000 004 | ·0000 003 | ·0000 002 | ·0000 001 |
| ·60 | ·0000 018 | ·0000 012 | ·0000 008 | ·0000 005− | ·0000 003 | ·0000 002 |
| ·61 | ·0000 033 | ·0000 021 | ·0000 014 | ·0000 009 | ·0000 006 | ·0000 004 |
| ·62 | ·0000 057 | ·0000 038 | ·0000 025+ | ·0000 017 | ·0000 011 | ·0000 007 |
| ·63 | ·0000 099 | ·0000 067 | ·0000 045+ | ·0000 030 | ·0000 020 | ·0000 014 |
| ·64 | ·0000 169 | ·0000 116 | ·0000 080 | ·0000 054 | ·0000 037 | ·0000 025+ |
| ·65 | ·0000 287 | ·0000 200 | ·0000 139 | ·0000 096 | ·0000 067 | ·0000 046 |
| ·66 | ·0000 480 | ·0000 339 | ·0000 239 | ·0000 169 | ·0000 119 | ·0000 083 |
| ·67 | ·0000 794 | ·0000 570 | ·0000 408 | ·0000 292 | ·0000 209 | ·0000 149 |
| ·68 | ·0001 300 | ·0000 946 | ·0000 688 | ·0000 499 | ·0000 362 | ·0000 262 |
| ·69 | ·0002 105− | ·0001 555− | ·0001 147 | ·0000 844 | ·0000 621 | ·0000 456 |
| ·70 | ·0003 372 | ·0002 526 | ·0001 890 | ·0001 412 | ·0001 053 | ·0000 784 |
| ·71 | ·0005 344 | ·0004 060 | ·0003 080 | ·0002 333 | ·0001 765− | ·0001 333 |
| ·72 | ·0008 380 | ·0006 454 | ·0004 964 | ·0003 812 | ·0002 924 | ·0002 240 |
| ·73 | ·0013 001 | ·0010 150− | ·0007 913 | ·0006 160 | ·0004 789 | ·0003 718 |
| ·74 | ·0019 957 | ·0015 790 | ·0012 475− | ·0009 842 | ·0007 755− | ·0006 102 |
| ·75 | ·0030 312 | ·0024 299 | ·0019 451 | ·0015 549 | ·0012 414 | ·0009 898 |
| ·76 | ·0045 549 | ·0036 988 | ·0029 994 | ·0024 290 | ·0019 644 | ·0015 868 |
| ·77 | ·0067 710 | ·0055 688 | ·0045 736 | ·0037 513 | ·0030 729 | ·0025 140 |
| ·78 | ·0099 560 | ·0082 913 | ·0068 955− | ·0057 271 | ·0047 507 | ·0039 359 |
| ·79 | ·0144 775− | ·0122 062 | ·0102 773 | ·0086 419 | ·0072 577 | ·0060 879 |
| ·80 | ·0208 157 | ·0177 638 | ·0151 392 | ·0128 858 | ·0109 544 | ·0093 014 |
| ·81 | ·0295 844 | ·0255 491 | ·0220 354 | ·0189 811 | ·0163 303 | ·0140 334 |
| ·82 | ·0415 496 | ·0363 043 | ·0316 805+ | ·0276 116 | ·0240 368 | ·0209 009 |
| ·83 | ·0576 414 | ·0509 459 | ·0449 718 | ·0396 505+ | ·0349 184 | ·0307 167 |
| ·84 | ·0789 520 | ·0705 706 | ·0630 023 | ·0561 797 | ·0500 393 | ·0445 216 |
| ·85 | ·1067 120 | ·0964 402 | ·0870 544 | ·0784 927 | ·0706 954 | ·0636 054 |
| ·86 | ·1422 334 | ·1299 337 | ·1185 627 | ·1080 680 | ·0983 980 | ·0895 015+ |
| ·87 | ·1868 074 | ·1724 536 | ·1590 292 | ·1464 955− | ·1348 122 | ·1239 386 |
| ·88 | ·2415 452 | ·2252 713 | ·2098 762 | ·1953 372 | ·1816 290 | ·1687 242 |
| ·89 | ·3071 534 | ·2893 000 | ·2722 186 | ·2559 039 | ·2403 466 | ·2255 347 |
| ·90 | ·3836 460 | ·3647 922 | ·3465 511 | ·3289 329 | ·3119 439 | ·2955 870 |
| ·91 | ·4700 101 | ·4509 762 | ·4323 564 | ·4141 725+ | ·3964 433 | ·3791 840 |
| ·92 | ·5638 745− | ·5456 763 | ·5276 784 | ·5099 087 | ·4923 928 | ·4751 540 |
| ·93 | ·6612 685+ | ·6450 067 | ·6287 489 | ·6125 229 | ·5963 548 | ·5802 695− |
| ·94 | ·7566 095− | ·7432 886 | ·7298 278 | ·7162 489 | ·7025 729 | ·6888 207 |
| ·95 | ·8431 074 | ·8334 013 | ·8234 890 | ·8133 833 | ·8030 971 | ·7926 435+ |
| ·96 | ·9137 926 | ·9078 118 | ·9016 396 | ·8952 807 | ·8887 402 | ·8820 232 |
| ·97 | ·9632 965− | ·9604 617 | ·9575 057 | ·9544 286 | ·9512 305+ | ·9479 120 |
| ·98 | ·9902 190 | ·9893 836 | ·9885 035+ | ·9875 780 | ·9866 061 | ·9855 873 |
| ·99 | ·9991 735+ | ·9990 959 | ·9990 134 | ·9989 257 | ·9988 326 | ·9987 341 |
| 1·00 | 1·0000 000 | 1·0000 000 | 1·0000 000 | 1·0000 000 | 1·0000 000 | 1·0000 000 |

$x$ = ·59 to 1·00 $q = 4$ $p$ = 45 to 50

| | $p = 45$ | $p = 46$ | $p = 47$ | $p = 48$ | $p = 49$ | $p = 50$ |
|---|---|---|---|---|---|---|
| $B(p, q)$ = | ·1284 8186 × $\frac{1}{10^5}$ | ·1179 9354 × $\frac{1}{10^5}$ | ·1085 5406 × $\frac{1}{10^5}$ | ·1000 4002 × $\frac{1}{10^5}$ | ·9234 4630 × $\frac{1}{10^6}$ | ·8537 5224 × $\frac{1}{10^6}$ |
| $x$ | | | | | | |
| ·59 | ·0000 001 | | | | | |
| ·60 | ·0000 001 | ·0000 001 | ·0000 001 | | | |
| ·61 | ·0000 002 | ·0000 002 | ·0000 001 | ·0000 001 | | |
| ·62 | ·0000 005⁻ | ·0000 003 | ·0000 002 | ·0000 001 | ·0000 001 | ·0000 001 |
| ·63 | ·0000 009 | ·0000 006 | ·0000 004 | ·0000 003 | ·0000 002 | ·0000 001 |
| ·64 | ·0000 017 | ·0000 012 | ·0000 008 | ·0000 005⁺ | ·0000 004 | ·0000 002 |
| ·65 | ·0000 032 | ·0000 022 | ·0000 015⁺ | ·0000 010 | ·0000 007 | ·0000 005⁻ |
| ·66 | ·0000 059 | ·0000 041 | ·0000 029 | ·0000 020 | ·0000 014 | ·0000 010 |
| ·67 | ·0000 106 | ·0000 075⁺ | ·0000 054 | ·0000 038 | ·0000 027 | ·0000 019 |
| ·68 | ·0000 190 | ·0000 137 | ·0000 099 | ·0000 071 | ·0000 051 | ·0000 037 |
| ·69 | ·0000 335⁻ | ·0000 245⁺ | ·0000 179 | ·0000 131 | ·0000 096 | ·0000 070 |
| ·70 | ·0000 584 | ·0000 434 | ·0000 322 | ·0000 239 | ·0000 177 | ·0000 131 |
| ·71 | ·0001 006 | ·0000 758 | ·0000 571 | ·0000 429 | ·0000 322 | ·0000 242 |
| ·72 | ·0001 713 | ·0001 309 | ·0000 999 | ·0000 762 | ·0000 580 | ·0000 441 |
| ·73 | ·0002 884 | ·0002 234 | ·0001 728 | ·0001 336 | ·0001 031 | ·0000 795⁺ |
| ·74 | ·0004 796 | ·0003 765⁻ | ·0002 952 | ·0002 312 | ·0001 809 | ·0001 414 |
| ·75 | ·0007 882 | ·0006 270 | ·0004 982 | ·0003 954 | ·0003 135⁺ | ·0002 483 |
| ·76 | ·0012 802 | ·0010 316 | ·0008 304 | ·0006 678 | ·0005 364 | ·0004 304 |
| ·77 | ·0020 544 | ·0016 769 | ·0013 672 | ·0011 136 | ·0009 060 | ·0007 365⁻ |
| ·78 | ·0032 571 | ·0026 923 | ·0022 230 | ·0018 336 | ·0015 109 | ·0012 437 |
| ·79 | ·0051 008 | ·0042 689 | ·0035 689 | ·0029 806 | ·0024 868 | ·0020 728 |
| ·80 | ·0078 889 | ·0066 835⁺ | ·0056 564 | ·0047 822 | ·0040 391 | ·0034 083 |
| ·81 | ·0120 461 | ·0103 290 | ·0088 475⁺ | ·0075 709 | ·0064 722 | ·0055 277 |
| ·82 | ·0181 542 | ·0157 517 | ·0136 532 | ·0118 226 | ·0102 277 | ·0088 398 |
| ·83 | ·0269 915⁻ | ·0236 934 | ·0207 775⁺ | ·0182 029 | ·0159 323 | ·0139 324 |
| ·84 | ·0395 707 | ·0351 346 | ·0311 654 | ·0276 184 | ·0244 527 | ·0216 307 |
| ·85 | ·0571 679 | ·0513 313 | ·0460 466 | ·0412 679 | ·0369 521 | ·0330 591 |
| ·86 | ·0813 289 | ·0738 319 | ·0669 640 | ·0606 806 | ·0549 391 | ·0496 991 |
| ·87 | ·1138 336 | ·1044 561 | ·0957 657 | ·0877 224 | ·0802 873 | ·0734 228 |
| ·88 | ·1565 937 | ·1452 072 | ·1345 336 | ·1245 412 | ·1151 984 | ·1064 734 |
| ·89 | ·2114 532 | ·1980 852 | ·1854 117 | ·1734 124 | ·1620 655⁺ | ·1513 485⁺ |
| ·90 | ·2798 622 | ·2647 663 | ·2502 939 | ·2364 374 | ·2231 871 | ·2105 317 |
| ·91 | ·3624 072 | ·3461 226 | ·3303 371 | ·3150 554 | ·3002 799 | ·2860 110 |
| ·92 | ·4582 129 | ·4415 880 | ·4252 957 | ·4093 500⁻ | ·3937 630 | ·3785 451 |
| ·93 | ·5642 901 | ·5484 386 | ·5327 353 | ·5171 990 | ·5018 473 | ·4866 960 |
| ·94 | ·6750 121 | ·6611 668 | ·6473 034 | ·6334 400 | ·6195 939 | ·6057 817 |
| ·95 | ·7820 355⁻ | ·7712 860 | ·7604 080 | ·7494 142 | ·7383 174 | ·7271 300 |
| ·96 | ·8751 352 | ·8680 820 | ·8608 692 | ·8535 030 | ·8459 895⁻ | ·8383 349 |
| ·97 | ·9444 736 | ·9409 160 | ·9372 401 | ·9334 468 | ·9295 374 | ·9255 131 |
| ·98 | ·9845 207 | ·9834 058 | ·9822 419 | ·9810 285⁺ | ·9797 651 | ·9784 511 |
| ·99 | ·9986 299 | ·9985 199 | ·9984 038 | ·9982 816 | ·9981 531 | ·9980 180 |
| 1·00 | 1·0000 000 | 1·0000 000 | 1·0000 000 | 1·0000 000 | 1·0000 000 | 1·0000 000 |

$x$ = ·01 to ·60 $q = 4·5$ $p$ = 4·5 to 7

| | $p = 4·5$ | $p = 5$ | $p = 5·5$ | $p = 6$ | $p = 6·5$ | $p = 7$ |
|---|---|---|---|---|---|---|
| $B(p, q) =$ | ·3355 5830 $\times \frac{1}{10^2}$ | ·2340 1435 $\overset{+}{\times} \frac{1}{10^2}$ | ·1677 7915 $\bar{\times} \frac{1}{10^2}$ | ·1231 6545 $\bar{\times} \frac{1}{10^2}$ | ·9227 8532 $\times \frac{1}{10^3}$ | ·7038 0256 $\times \frac{1}{10^3}$ |
| $x$ | | | | | | |
| ·01 | ·0000 001 | | | | | |
| ·02 | ·0000 014 | ·0000 003 | | | | |
| ·03 | ·0000 085+ | ·0000 019 | ·0000 004 | ·0000 001 | | |
| ·04 | ·0000 302 | ·0000 078 | ·0000 020 | ·0000 005− | ·0000 001 | |
| ·05 | ·0000 800 | ·0000 230 | ·0000 065+ | ·0000 018 | ·0000 005− | ·0000 001 |
| ·06 | ·0001 763 | ·0000 556 | ·0000 172 | ·0000 053 | ·0000 016 | ·0000 005− |
| ·07 | ·0003 424 | ·0001 164 | ·0000 389 | ·0000 128 | ·0000 042 | ·0000 013 |
| ·08 | ·0006 058 | ·0002 201 | ·0000 786 | ·0000 277 | ·0000 096 | ·0000 033 |
| ·09 | ·0009 983 | ·0003 845− | ·0001 456 | ·0000 543 | ·0000 200 | ·0000 073 |
| ·10 | ·0015 552 | ·0006 309 | ·0002 517 | ·0000 990 | ·0000 384 | ·0000 147 |
| ·11 | ·0023 151 | ·0009 844 | ·0004 117 | ·0001 697 | ·0000 690 | ·0000 278 |
| ·12 | ·0033 192 | ·0014 731 | ·0006 430 | ·0002 767 | ·0001 175+ | ·0000 494 |
| ·13 | ·0046 106 | ·0021 283 | ·0009 664 | ·0004 326 | ·0001 912 | ·0000 836 |
| ·14 | ·0062 342 | ·0029 842 | ·0014 054 | ·0006 525− | ·0002 991 | ·0001 356 |
| ·15 | ·0082 355− | ·0040 775+ | ·0019 864 | ·0009 541 | ·0004 526 | ·0002 123 |
| ·16 | ·0106 606 | ·0054 472 | ·0027 390 | ·0013 579 | ·0006 649 | ·0003 220 |
| ·17 | ·0135 554 | ·0071 339 | ·0036 950+ | ·0018 873 | ·0009 521 | ·0004 750+ |
| ·18 | ·0169 648 | ·0091 798 | ·0048 893 | ·0025 681 | ·0013 324 | ·0006 838 |
| ·19 | ·0209 329 | ·0116 278 | ·0063 584 | ·0034 293 | ·0018 271 | ·0009 629 |
| ·20 | ·0255 016 | ·0145 216 | ·0081 412 | ·0045 021 | ·0024 597 | ·0013 293 |
| ·21 | ·0307 110 | ·0179 044 | ·0102 782 | ·0058 206 | ·0032 567 | ·0018 027 |
| ·22 | ·0365 983 | ·0218 196 | ·0128 109 | ·0074 208 | ·0042 474 | ·0024 052 |
| ·23 | ·0431 978 | ·0263 092 | ·0157 819 | ·0093 411 | ·0054 635+ | ·0031 618 |
| ·24 | ·0505 404 | ·0314 141 | ·0192 342 | ·0116 214 | ·0069 393 | ·0041 001 |
| ·25 | ·0586 534 | ·0371 731 | ·0232 108 | ·0143 033 | ·0087 116 | ·0052 506 |
| ·26 | ·0675 599 | ·0436 231 | ·0277 545− | ·0174 294 | ·0108 191 | ·0066 464 |
| ·27 | ·0772 790 | ·0507 982 | ·0329 070 | ·0210 432 | ·0133 026 | ·0083 230 |
| ·28 | ·0878 252 | ·0587 293 | ·0387 088 | ·0251 885− | ·0162 045+ | ·0103 187 |
| ·29 | ·0992 085+ | ·0674 441 | ·0451 987 | ·0299 088 | ·0195 685+ | ·0126 738 |
| ·30 | ·1114 342 | ·0769 665− | ·0524 132 | ·0352 474 | ·0234 393 | ·0154 308 |
| ·31 | ·1245 026 | ·0873 163 | ·0603 863 | ·0412 464 | ·0278 619 | ·0186 338 |
| ·32 | ·1384 093 | ·0985 090 | ·0691 489 | ·0479 466 | ·0328 817 | ·0223 283 |
| ·33 | ·1531 449 | ·1105 555+ | ·0787 285− | ·0553 867 | ·0385 436 | ·0265 610 |
| ·34 | ·1686 953 | ·1234 622 | ·0891 486 | ·0636 032 | ·0448 917 | ·0313 790 |
| ·35 | ·1850 416 | ·1372 303 | ·1004 288 | ·0726 295+ | ·0519 688 | ·0368 299 |
| ·36 | ·2021 598 | ·1518 561 | ·1125 841 | ·0824 959 | ·0598 158 | ·0429 607 |
| ·37 | ·2200 219 | ·1673 307 | ·1256 246 | ·0932 289 | ·0684 713 | ·0498 177 |
| ·38 | ·2385 951 | ·1836 402 | ·1395 556 | ·1048 508 | ·0779 711 | ·0574 458 |
| ·39 | ·2578 423 | ·2007 655+ | ·1543 771 | ·1173 792 | ·0883 474 | ·0658 882 |
| ·40 | ·2777 227 | ·2186 823 | ·1700 836 | ·1308 270 | ·0996 289 | ·0751 853 |
| ·41 | ·2981 915− | ·2373 613 | ·1866 642 | ·1452 017 | ·1118 396 | ·0853 747 |
| ·42 | ·3192 003 | ·2567 683 | ·2041 024 | ·1605 052 | ·1249 988 | ·0964 903 |
| ·43 | ·3406 976 | ·2768 644 | ·2223 759 | ·1767 339 | ·1391 205− | ·1085 617 |
| ·44 | ·3626 291 | ·2976 060 | ·2414 569 | ·1938 778 | ·1542 130 | ·1216 138 |
| ·45 | ·3849 375− | ·3189 453 | ·2613 120 | ·2119 210 | ·1702 787 | ·1356 663 |
| ·46 | ·4075 636 | ·3408 302 | ·2819 023 | ·2308 414 | ·1873 136 | ·1507 328 |
| ·47 | ·4304 463 | ·3632 050+ | ·3031 835+ | ·2506 102 | ·2053 069 | ·1668 210 |
| ·48 | ·4535 226 | ·3860 105+ | ·3251 063 | ·2711 929 | ·2242 412 | ·1839 314 |
| ·49 | ·4767 288 | ·4091 844 | ·3476 165− | ·2925 481 | ·2440 919 | ·2020 578 |
| ·50 | ·5000 000 | ·4326 616 | ·3706 550+ | ·3146 289 | ·2648 273 | ·2211 863 |
| ·51 | ·5232 712 | ·4563 747 | ·3941 589 | ·3373 819 | ·2864 088 | ·2412 952 |
| ·52 | ·5464 774 | ·4802 544 | ·4180 611 | ·3607 484 | ·3087 905− | ·2623 551 |
| ·53 | ·5695 537 | ·5042 299 | ·4422 910 | ·3846 641 | ·3319 195− | ·2843 284 |
| ·54 | ·5924 364 | ·5282 295− | ·4667 750+ | ·4090 596 | ·3557 361 | ·3071 693 |
| ·55 | ·6150 625+ | ·5521 808 | ·4914 370 | ·4338 610 | ·3801 741 | ·3308 242 |
| ·56 | ·6373 709 | ·5760 114 | ·5161 988 | ·4589 901 | ·4051 611 | ·3552 313 |
| ·57 | ·6593 024 | ·5996 493 | ·5409 807 | ·4843 651 | ·4306 188 | ·3803 211 |
| ·58 | ·6807 997 | ·6230 235− | ·5657 019 | ·5099 009 | ·4564 635+ | ·4060 166 |
| ·59 | ·7018 085+ | ·6460 642 | ·5902 813 | ·5355 101 | ·4826 068 | ·4322 341 |
| ·60 | ·7222 773 | ·6687 036 | ·6146 382 | ·5611 032 | ·5089 561 | ·4588 829 |

$x$ = ·61 to 1·00 $q = 4\cdot5$ $p$ = 4·5 to 7

| $x$ | $p = 4\cdot5$ | $p = 5$ | $p = 5\cdot5$ | $p = 6$ | $p = 6\cdot5$ | $p = 7$ |
|---|---|---|---|---|---|---|
| $B(p, q)$ = | ·3355 5830 $\times \frac{1}{10^2}$ | ·2340 1435 $\overset{+}{\times} \frac{1}{10^2}$ | ·1677 7915 $\overset{-}{\times} \frac{1}{10^2}$ | ·1231 6545 $\overset{-}{\times} \frac{1}{10^2}$ | ·9227 8532 $\times \frac{1}{10^3}$ | ·7038 0256 $\times \frac{1}{10^3}$ |
| ·61 | ·7421 577 | ·6908 763 | ·6386 924 | ·5865 899 | ·5354 153 | ·4858 667 |
| ·62 | ·7614 049 | ·7125 194 | ·6623 655− | ·6118 791 | ·5618 854 | ·5130 838 |
| ·63 | ·7799 781 | ·7335 736 | ·6855 807 | ·6368 800 | ·5882 657 | ·5404 280 |
| ·64 | ·7978 402 | ·7539 831 | ·7082 644 | ·6615 029 | ·6144 541 | ·5677 897 |
| ·65 | ·8149 584 | ·7736 961 | ·7303 457 | ·6856 600 | ·6403 485− | ·5950 563 |
| ·66 | ·8313 047 | ·7926 653 | ·7517 579 | ·7092 660 | ·6658 474 | ·6221 137 |
| ·67 | ·8468 551 | ·8108 482 | ·7724 387 | ·7322 388 | ·6908 512 | ·6488 474 |
| ·68 | ·8615 907 | ·8282 074 | ·7923 304 | ·7545 007 | ·7152 626 | ·6751 432 |
| ·69 | ·8754 974 | ·8447 110 | ·8113 812 | ·7759 787 | ·7389 881 | ·7008 888 |
| ·70 | ·8885 658 | ·8603 325− | ·8295 449 | ·7966 055+ | ·7619 390 | ·7259 748 |
| ·71 | ·9007 915− | ·8750 513 | ·8467 816 | ·8163 199 | ·7840 320 | ·7502 961 |
| ·72 | ·9121 748 | ·8888 528 | ·8630 583 | ·8350 678 | ·8051 902 | ·7737 530 |
| ·73 | ·9227 210 | ·9017 285+ | ·8783 490 | ·8528 026 | ·8253 445+ | ·7962 525− |
| ·74 | ·9324 401 | ·9136 759 | ·8926 346 | ·8694 857 | ·8444 339 | ·8177 095− |
| ·75 | ·9413 466 | ·9246 987 | ·9059 040 | ·8850 869 | ·8624 063 | ·8380 480 |
| ·76 | ·9494 596 | ·9348 064 | ·9181 533 | ·8995 850+ | ·8792 197 | ·8572 020 |
| ·77 | ·9568 022 | ·9440 148 | ·9293 863 | ·9129 680 | ·8948 422 | ·8751 168 |
| ·78 | ·9634 017 | ·9523 451 | ·9396 143 | ·9252 330 | ·9092 529 | ·8917 496 |
| ·79 | ·9692 890 | ·9598 242 | ·9488 562 | ·9363 868 | ·9224 421 | ·9070 704 |
| ·80 | ·9744 984 | ·9664 840 | ·9571 380 | ·9464 452 | ·9344 117 | ·9210 627 |
| ·81 | ·9790 671 | ·9723 615− | ·9644 926 | ·9554 336 | ·9451 748 | ·9337 235− |
| ·82 | ·9830 352 | ·9774 978 | ·9709 596 | ·9633 859 | ·9547 563 | ·9450 641 |
| ·83 | ·9864 446 | ·9819 380 | ·9765 843 | ·9703 449 | ·9631 922 | ·9551 097 |
| ·84 | ·9893 394 | ·9857 306 | ·9814 177 | ·9763 610 | ·9705 290 | ·9638 993 |
| ·85 | ·9917 645+ | ·9889 269 | ·9855 154 | ·9814 917 | ·9768 236 | ·9714 852 |
| ·86 | ·9937 658 | ·9915 802 | ·9889 370 | ·9858 012 | ·9821 416 | ·9779 320 |
| ·87 | ·9953 894 | ·9937 452 | ·9917 452 | ·9893 586 | ·9865 572 | ·9833 160 |
| ·88 | ·9966 808 | ·9954 772 | ·9940 047 | ·9922 374 | ·9901 511 | ·9877 233 |
| ·89 | ·9976 849 | ·9968 315− | ·9957 814 | ·9945 140 | ·9930 093 | ·9912 484 |
| ·90 | ·9984 448 | ·9978 622 | ·9971 413 | ·9962 663 | ·9952 216 | ·9939 922 |
| ·91 | ·9990 017 | ·9986 217 | ·9981 490 | ·9975 720 | ·9968 793 | ·9960 595− |
| ·92 | ·9993 942 | ·9991 600 | ·9988 670 | ·9985 074 | ·9980 733 | ·9975 568 |
| ·93 | ·9996 576 | ·9995 232 | ·9993 541 | ·9991 455− | ·9988 923 | ·9985 893 |
| ·94 | ·9998 237 | ·9997 534 | ·9996 645+ | ·9995 543 | ·9994 197 | ·9992 579 |
| ·95 | ·9999 200 | ·9998 877 | ·9998 465+ | ·9997 952 | ·9997 323 | ·9996 562 |
| ·96 | ·9999 698 | ·9999 574 | ·9999 416 | ·9999 217 | ·9998 973 | ·9998 675+ |
| ·97 | ·9999 915− | ·9999 879 | ·9999 834 | ·9999 776 | ·9999 705+ | ·9999 618 |
| ·98 | ·9999 986 | ·9999 980 | ·9999 972 | ·9999 962 | ·9999 950+ | ·9999 935+ |
| ·99 | ·9999 999 | ·9999 999 | ·9999 999 | ·9999 998 | ·9999 998 | ·9999 997 |
| 1·00 | 1·0000 000 | 1·0000 000 | 1·0000 000 | 1·0000 000 | 1·0000 000 | 1·0000 000 |

$x$ = ·06 to ·70 $q = 4·5$ $p$ = 7·5 to 10

| $x$ | $p = 7·5$ | $p = 8$ | $p = 8·5$ | $p = 9$ | $p = 9·5$ | $p = 10$ |
|---|---|---|---|---|---|---|
| $B(p, q) =$ | ·5452 8223 $\times\frac{1}{10^3}$ | ·4284 0156 $\times\frac{1}{10^3}$ | ·3408 0140 $\times\frac{1}{10^3}$ | ·2741 7700 $\times\frac{1}{10^3}$ | ·2228 3168 $\times\frac{1}{10^3}$ | ·1827 8467 $\times\frac{1}{10^3}$ |
| ·06 | ·0000 001 | | | | | |
| ·07 | ·0000 004 | ·0000 001 | | | | |
| ·08 | ·0000 011 | ·0000 004 | ·0000 001 | | | |
| ·09 | ·0000 026 | ·0000 009 | ·0000 003 | ·0000 001 | | |
| ·10 | ·0000 056 | ·0000 021 | ·0000 008 | ·0000 003 | ·0000 001 | |
| ·11 | ·0000 111 | ·0000 044 | ·0000 017 | ·0000 007 | ·0000 003 | ·0000 001 |
| ·12 | ·0000 205+ | ·0000 085− | ·0000 035− | ·0000 014 | ·0000 006 | ·0000 002 |
| ·13 | ·0000 362 | ·0000 155+ | ·0000 066 | ·0000 028 | ·0000 012 | ·0000 005− |
| ·14 | ·0000 609 | ·0000 271 | ·0000 119 | ·0000 052 | ·0000 023 | ·0000 010 |
| ·15 | ·0000 986 | ·0000 454 | ·0000 207 | ·0000 094 | ·0000 042 | ·0000 019 |
| ·16 | ·0001 544 | ·0000 734 | ·0000 346 | ·0000 162 | ·0000 075+ | ·0000 035− |
| ·17 | ·0002 347 | ·0001 149 | ·0000 558 | ·0000 269 | ·0000 129 | ·0000 061 |
| ·18 | ·0003 475− | ·0001 750+ | ·0000 875− | ·0000 434 | ·0000 214 | ·0000 105− |
| ·19 | ·0005 025+ | ·0002 599 | ·0001 334 | ·0000 680 | ·0000 344 | ·0000 173 |
| ·20 | ·0007 115− | ·0003 775− | ·0001 987 | ·0001 038 | ·0000 539 | ·0000 278 |
| ·21 | ·0009 883 | ·0005 371 | ·0002 896 | ·0001 550+ | ·0000 824 | ·0000 436 |
| ·22 | ·0013 490 | ·0007 501 | ·0004 138 | ·0002 266 | ·0001 233 | ·0000 667 |
| ·23 | ·0018 124 | ·0010 300 | ·0005 808 | ·0003 251 | ·0001 808 | ·0001 000 |
| ·24 | ·0023 997 | ·0013 925+ | ·0008 018 | ·0004 584 | ·0002 604 | ·0001 470 |
| ·25 | ·0031 350− | ·0018 559 | ·0010 902 | ·0006 359 | ·0003 685+ | ·0002 123 |
| ·26 | ·0040 449 | ·0024 410 | ·0014 617 | ·0008 692 | ·0005 135+ | ·0003 016 |
| ·27 | ·0051 593 | ·0031 713 | ·0019 345− | ·0011 718 | ·0007 053 | ·0004 220 |
| ·28 | ·0065 104 | ·0040 734 | ·0025 293 | ·0015 597 | ·0009 556 | ·0005 821 |
| ·29 | ·0081 336 | ·0051 766 | ·0032 699 | ·0020 512 | ·0012 786 | ·0007 924 |
| ·30 | ·0100 667 | ·0065 134 | ·0041 828 | ·0026 677 | ·0016 907 | ·0010 654 |
| ·31 | ·0123 503 | ·0081 189 | ·0052 976 | ·0034 332 | ·0022 111 | ·0014 158 |
| ·32 | ·0150 271 | ·0100 315+ | ·0066 473 | ·0043 750+ | ·0028 616 | ·0018 610 |
| ·33 | ·0181 421 | ·0122 923 | ·0082 677 | ·0055 235+ | ·0036 674 | ·0024 212 |
| ·34 | ·0217 419 | ·0149 447 | ·0101 979 | ·0069 125− | ·0046 568 | ·0031 195− |
| ·35 | ·0258 749 | ·0180 351 | ·0124 801 | ·0085 790 | ·0058 614 | ·0039 822 |
| ·36 | ·0305 903 | ·0216 116 | ·0151 592 | ·0105 635− | ·0073 166 | ·0050 394 |
| ·37 | ·0359 379 | ·0257 244 | ·0182 831 | ·0129 098 | ·0090 610 | ·0063 245+ |
| ·38 | ·0419 678 | ·0304 250− | ·0219 020 | ·0156 648 | ·0111 372 | ·0078 748 |
| ·39 | ·0487 297 | ·0357 659 | ·0260 683 | ·0188 786 | ·0135 912 | ·0097 313 |
| ·40 | ·0562 722 | ·0418 003 | ·0308 362 | ·0226 039 | ·0164 724 | ·0119 392 |
| ·41 | ·0646 425+ | ·0485 811 | ·0362 614 | ·0268 959 | ·0198 337 | ·0145 473 |
| ·42 | ·0738 857 | ·0561 608 | ·0424 000 | ·0318 119 | ·0237 308 | ·0176 084 |
| ·43 | ·0840 438 | ·0645 906 | ·0493 088 | ·0374 109 | ·0282 226 | ·0211 787 |
| ·44 | ·0951 556 | ·0739 195+ | ·0570 439 | ·0437 529 | ·0333 698 | ·0253 178 |
| ·45 | ·1072 557 | ·0841 943 | ·0656 605− | ·0508 983 | ·0392 353 | ·0300 884 |
| ·46 | ·1203 739 | ·0954 579 | ·0752 119 | ·0589 074 | ·0458 832 | ·0355 557 |
| ·47 | ·1345 346 | ·1077 495+ | ·0857 489 | ·0678 394 | ·0533 781 | ·0417 869 |
| ·48 | ·1497 561 | ·1211 032 | ·0973 187 | ·0777 517 | ·0617 846 | ·0488 507 |
| ·49 | ·1660 502 | ·1355 474 | ·1099 643 | ·0886 992 | ·0711 661 | ·0568 163 |
| ·50 | ·1834 214 | ·1511 043 | ·1237 237 | ·1007 329 | ·0815 842 | ·0657 528 |
| ·51 | ·2018 664 | ·1677 889 | ·1386 287 | ·1138 995+ | ·0930 976 | ·0757 279 |
| ·52 | ·2213 740 | ·1856 083 | ·1547 044 | ·1282 401 | ·1057 610 | ·0868 075+ |
| ·53 | ·2419 242 | ·2045 615+ | ·1719 680 | ·1437 891 | ·1196 242 | ·0990 539 |
| ·54 | ·2634 884 | ·2246 382 | ·1904 282 | ·1605 735− | ·1347 305+ | ·1125 249 |
| ·55 | ·2860 286 | ·2458 189 | ·2100 845+ | ·1786 115+ | ·1511 162 | ·1272 726 |
| ·56 | ·3094 979 | ·2680 739 | ·2309 264 | ·1979 122 | ·1688 088 | ·1433 419 |
| ·57 | ·3338 399 | ·2913 635− | ·2529 328 | ·2184 738 | ·1878 263 | ·1607 695− |
| ·58 | ·3589 893 | ·3156 374 | ·2760 712 | ·2402 836 | ·2081 760 | ·1795 820 |
| ·59 | ·3848 715+ | ·3408 348 | ·3002 979 | ·2633 169 | ·2298 531 | ·1997 952 |
| ·60 | ·4114 035− | ·3668 846 | ·3255 572 | ·2875 361 | ·2528 403 | ·2214 123 |
| ·61 | ·4384 937 | ·3937 050− | ·3517 811 | ·3128 906 | ·2771 064 | ·2444 229 |
| ·62 | ·4660 429 | ·4212 042 | ·3788 901 | ·3393 165− | ·3026 058 | ·2688 021 |
| ·63 | ·4939 449 | ·4492 810 | ·4067 925+ | ·3667 359 | ·3292 782 | ·2945 090 |
| ·64 | ·5220 870 | ·4778 250− | ·4353 852 | ·3950 574 | ·3570 475− | ·3214 863 |
| ·65 | ·5503 513 | ·5067 176 | ·4645 540 | ·4241 762 | ·3858 223 | ·3496 596 |
| ·66 | ·5786 153 | ·5358 330 | ·4941 745+ | ·4539 742 | ·4154 956 | ·3789 369 |
| ·67 | ·6067 533 | ·5650 390 | ·5241 132 | ·4843 211 | ·4459 453 | ·4092 085+ |
| ·68 | ·6346 377 | ·5941 985− | ·5542 279 | ·5150 750+ | ·4770 345− | ·4403 474 |
| ·69 | ·6621 401 | ·6231 706 | ·5843 699 | ·5460 840 | ·5086 126 | ·4722 092 |
| ·70 | ·6891 327 | ·6518 125− | ·6143 849 | ·5771 867 | ·5405 165+ | ·5046 339 |

$x = \cdot71$ to $1\cdot00$ $q = 4\cdot5$ $p = 7\cdot5$ to $10$

| | $p = 7\cdot5$ | $p = 8$ | $p = 8\cdot5$ | $p = 9$ | $p = 9\cdot5$ | $p = 10$ |
|---|---|---|---|---|---|---|
| $B(p, q) =$ | $\cdot5452\ 8223 \times \frac{1}{10^3}$ | $\cdot4284\ 0156 \times \frac{1}{10^3}$ | $\cdot3408\ 0140 \times \frac{1}{10^3}$ | $\cdot2741\ 7700 \times \frac{1}{10^3}$ | $\cdot2228\ 3168 \times \frac{1}{10^3}$ | $\cdot1827\ 8467 \times \frac{1}{10^3}$ |
| $x$ | | | | | | |
| ·71 | ·7154 901 | ·6799 806 | ·6441 151 | ·6082 150− | ·5725 722 | ·5374 460 |
| ·72 | ·7410 902 | ·7075 329 | ·6734 005+ | ·6389 951 | ·6045 960 | ·5704 573 |
| ·73 | ·7658 165+ | ·7343 300 | ·7020 818 | ·6693 502 | ·6363 976 | ·6034 678 |
| ·74 | ·7895 592 | ·7602 379 | ·7300 018 | ·6991 025− | ·6677 819 | ·6362 689 |
| ·75 | ·8122 167 | ·7851 292 | ·7570 080 | ·7280 761 | ·6985 519 | ·6686 458 |
| ·76 | ·8336 973 | ·8088 854 | ·7829 550+ | ·7560 994 | ·7285 116 | ·7003 809 |
| ·77 | ·8539 207 | ·8313 987 | ·8077 068 | ·7830 082 | ·7574 695+ | ·7312 573 |
| ·78 | ·8728 192 | ·8525 740 | ·8311 391 | ·8086 484 | ·7852 419 | ·7610 623 |
| ·79 | ·8903 389 | ·8723 306 | ·8531 419 | ·8328 792 | ·8116 564 | ·7895 923 |
| ·80 | ·9064 408 | ·8906 036 | ·8736 216 | ·8555 758 | ·8365 553 | ·8166 559 |
| ·81 | ·9211 018 | ·9073 458 | ·8925 031 | ·8766 322 | ·8597 997 | ·8420 794 |
| ·82 | ·9343 153 | ·9225 282 | ·9097 316 | ·8959 641 | ·8812 724 | ·8657 104 |
| ·83 | ·9460 915+ | ·9361 419 | ·9252 743 | ·9135 108 | ·9008 814 | ·8874 223 |
| ·84 | ·9564 575+ | ·9481 976 | ·9391 214 | ·9292 379 | ·9185 628 | ·9071 181 |
| ·85 | ·9654 572 | ·9587 268 | ·9512 872 | ·9431 377 | ·9342 832 | ·9247 339 |
| ·86 | ·9731 507 | ·9677 809 | ·9618 102 | ·9552 315− | ·9480 416 | ·9402 419 |
| ·87 | ·9796 133 | ·9754 305− | ·9707 528 | ·9655 686 | ·9598 700 | ·9536 523 |
| ·88 | ·9849 339 | ·9817 647 | ·9782 002 | ·9742 270 | ·9698 345+ | ·9650 144 |
| ·89 | ·9892 136 | ·9868 888 | ·9842 590 | ·9813 112 | ·9780 337 | ·9744 167 |
| ·90 | ·9925 635+ | ·9909 221 | ·9890 549 | ·9869 502 | ·9845 969 | ·9819 853 |
| ·91 | ·9951 016 | ·9939 949 | ·9927 290 | ·9912 942 | ·9896 810 | ·9878 807 |
| ·92 | ·9969 500− | ·9962 450+ | ·9954 342 | ·9945 101 | ·9934 655+ | ·9922 934 |
| ·93 | ·9982 315− | ·9978 135− | ·9973 301 | ·9967 763 | ·9961 468 | ·9954 366 |
| ·94 | ·9990 657 | ·9988 401 | ·9985 777 | ·9982 754 | ·9979 300 | ·9975 383 |
| ·95 | ·9995 654 | ·9994 581 | ·9993 328 | ·9991 876 | ·9990 208 | ·9988 307 |
| ·96 | ·9998 318 | ·9997 894 | ·9997 396 | ·9996 817 | ·9996 147 | ·9995 380 |
| ·97 | ·9999 513 | ·9999 388 | ·9999 240 | ·9999 067 | ·9998 867 | ·9998 635+ |
| ·98 | ·9999 917 | ·9999 895+ | ·9999 870 | ·9999 839 | ·9999 804 | ·9999 763 |
| ·99 | ·9999 996 | ·9999 995+ | ·9999 994 | ·9999 992 | ·9999 991 | ·9999 989 |
| 1·00 | 1·0000 000 | 1·0000 000 | 1·0000 000 | 1·0000 000 | 1·0000 000 | 1·0000 000 |

$x = \cdot 12$ to $\cdot 70$ $\qquad q = 4\cdot 5$ $\qquad p = 10\cdot 5$ to $15$

| $x$ | $p = 10\cdot 5$ | $p = 11$ | $p = 12$ | $p = 13$ | $p = 14$ | $p = 15$ |
|---|---|---|---|---|---|---|
| $B(p, q) =$ | $\cdot 1512\ 0721 \times \frac{1}{10^3}$ | $\cdot 1260\ 5839 \times \frac{1}{10^3}$ | $\cdot 8946\ 0793 \times \frac{1}{10^4}$ | $\cdot 6506\ 2395 \times \frac{1}{10^4}$ | $\cdot 4833\ 2065 \times \frac{1}{10^4}$ | $\cdot 3657\ 5617 \times \frac{1}{10^4}$ |
| ·12 | ·0000 001 | | | | | |
| ·13 | ·0000 002 | ·0000 001 | | | | |
| ·14 | ·0000 004 | ·0000 002 | | | | |
| ·15 | ·0000 008 | ·0000 004 | ·0000 001 | | | |
| ·16 | ·0000 016 | ·0000 007 | ·0000 002 | | | |
| ·17 | ·0000 029 | ·0000 014 | ·0000 003 | ·0000 001 | | |
| ·18 | ·0000 051 | ·0000 025$^-$ | ·0000 006 | ·0000 001 | | |
| ·19 | ·0000 087 | ·0000 043 | ·0000 011 | ·0000 003 | | |
| ·20 | ·0000 143 | ·0000 073 | ·0000 019 | ·0000 005$^-$ | ·0000 001 | |
| ·21 | ·0000 229 | ·0000 120 | ·0000 032 | ·0000 009 | ·0000 002 | ·0000 001 |
| ·22 | ·0000 359 | ·0000 192 | ·0000 054 | ·0000 015$^+$ | ·0000 004 | ·0000 001 |
| ·23 | ·0000 550$^-$ | ·0000 301 | ·0000 089 | ·0000 026 | ·0000 007 | ·0000 002 |
| ·24 | ·0000 825$^+$ | ·0000 461 | ·0000 142 | ·0000 043 | ·0000 013 | ·0000 004 |
| ·25 | ·0001 216 | ·0000 693 | ·0000 222 | ·0000 070 | ·0000 022 | ·0000 007 |
| ·26 | ·0001 762 | ·0001 024 | ·0000 341 | ·0000 112 | ·0000 036 | ·0000 012 |
| ·27 | ·0002 511 | ·0001 487 | ·0000 514 | ·0000 175$^-$ | ·0000 059 | ·0000 019 |
| ·28 | ·0003 527 | ·0002 126 | ·0000 762 | ·0000 269 | ·0000 093 | ·0000 032 |
| ·29 | ·0004 884 | ·0002 996 | ·0001 112 | ·0000 406 | ·0000 146 | ·0000 052 |
| ·30 | ·0006 677 | ·0004 164 | ·0001 598 | ·0000 603 | ·0000 225$^-$ | ·0000 083 |
| ·31 | ·0009 017 | ·0005 715$^-$ | ·0002 265$^-$ | ·0000 883 | ·0000 340 | ·0000 129 |
| ·32 | ·0012 039 | ·0007 750$^-$ | ·0003 168 | ·0001 275$^-$ | ·0000 506 | ·0000 198 |
| ·33 | ·0015 900 | ·0010 391 | ·0004 379 | ·0001 816 | ·0000 743 | ·0000 300 |
| ·34 | ·0020 787 | ·0013 785$^-$ | ·0005 982 | ·0002 555$^+$ | ·0001 076 | ·0000 448 |
| ·35 | ·0026 915$^-$ | ·0018 103 | ·0008 082 | ·0003 552 | ·0001 540 | ·0000 659 |
| ·36 | ·0034 531 | ·0023 547 | ·0010 807 | ·0004 883 | ·0002 176 | ·0000 958 |
| ·37 | ·0043 918 | ·0030 352 | ·0014 308 | ·0006 641 | ·0003 041 | ·0001 375$^+$ |
| ·38 | ·0055 396 | ·0038 784 | ·0018 767 | ·0008 941 | ·0004 202 | ·0001 951 |
| ·39 | ·0069 324 | ·0049 153 | ·0024 394 | ·0011 922 | ·0005 748 | ·0002 738 |
| ·40 | ·0086 101 | ·0061 803 | ·0031 438 | ·0015 749 | ·0007 784 | ·0003 802 |
| ·41 | ·0106 169 | ·0077 124 | ·0040 185$^-$ | ·0020 622 | ·0010 442 | ·0005 225$^-$ |
| ·42 | ·0130 010 | ·0095 550$^+$ | ·0050 964 | ·0026 775$^+$ | ·0013 881 | ·0007 112 |
| ·43 | ·0158 150$^-$ | ·0117 558 | ·0064 148 | ·0034 483 | ·0018 293 | ·0009 591 |
| ·44 | ·0191 155$^-$ | ·0143 673 | ·0080 160 | ·0044 064 | ·0023 905$^+$ | ·0012 818 |
| ·45 | ·0229 629 | ·0174 463 | ·0099 473 | ·0055 885$^-$ | ·0030 989 | ·0016 986 |
| ·46 | ·0274 215$^+$ | ·0210 543 | ·0122 611 | ·0070 366 | ·0039 862 | ·0022 323 |
| ·47 | ·0325 586 | ·0252 567 | ·0150 154 | ·0087 981 | ·0050 892 | ·0029 103 |
| ·48 | ·0384 443 | ·0301 229 | ·0182 733 | ·0109 265$^+$ | ·0064 506 | ·0037 651 |
| ·49 | ·0451 508 | ·0357 258 | ·0221 033 | ·0134 813 | ·0081 192 | ·0048 348 |
| ·50 | ·0527 519 | ·0421 412 | ·0265 790 | ·0165 284 | ·0101 502 | ·0061 637 |
| ·51 | ·0613 217 | ·0494 468 | ·0317 787 | ·0201 398 | ·0126 059 | ·0078 030 |
| ·52 | ·0709 340 | ·0577 218 | ·0377 849 | ·0243 940 | ·0155 559 | ·0098 112 |
| ·53 | ·0816 612 | ·0670 458 | ·0446 838 | ·0293 752 | ·0190 772 | ·0122 548 |
| ·54 | ·0935 729 | ·0774 975$^-$ | ·0525 643 | ·0351 735$^-$ | ·0232 540 | ·0152 084 |
| ·55 | ·1067 347 | ·0891 534 | ·0615 170 | ·0418 837 | ·0281 781 | ·0187 555$^-$ |
| ·56 | ·1212 071 | ·1020 868 | ·0716 331 | ·0496 049 | ·0339 479 | ·0229 880 |
| ·57 | ·1370 433 | ·1163 660 | ·0830 029 | ·0584 393 | ·0406 684 | ·0280 067 |
| ·58 | ·1542 886 | ·1320 527 | ·0957 143 | ·0684 908 | ·0484 501 | ·0339 207 |
| ·59 | ·1729 782 | ·1492 002 | ·1098 511 | ·0798 638 | ·0574 079 | ·0408 467 |
| ·60 | ·1931 359 | ·1678 520 | ·1254 907 | ·0926 607 | ·0676 594 | ·0489 084 |
| ·61 | ·2147 726 | ·1880 398 | ·1427 026 | ·1069 807 | ·0793 237 | ·0582 352 |
| ·62 | ·2378 846 | ·2097 815$^+$ | ·1615 456 | ·1229 167 | ·0925 187 | ·0689 603 |
| ·63 | ·2624 526 | ·2330 800 | ·1820 660 | ·1405 533 | ·1073 592 | ·0812 188 |
| ·64 | ·2884 402 | ·2579 208 | ·2042 947 | ·1599 637 | ·1239 534 | ·0951 451 |
| ·65 | ·3157 926 | ·2842 713 | ·2282 454 | ·1812 070 | ·1424 004 | ·1108 699 |
| ·66 | ·3444 362 | ·3120 787 | ·2539 121 | ·2043 250$^-$ | ·1627 863 | ·1285 169 |
| ·67 | ·3742 777 | ·3412 692 | ·2812 665$^-$ | ·2293 391 | ·1851 809 | ·1481 984 |
| ·68 | ·4052 040 | ·3717 473 | ·3102 568 | ·2562 475$^+$ | ·2096 334 | ·1700 114 |
| ·69 | ·4370 817 | ·4033 949 | ·3408 052 | ·2850 221 | ·2361 689 | ·1940 331 |
| ·70 | ·4697 584 | ·4360 714 | ·3728 070 | ·3156 054 | ·2647 840 | ·2203 152 |

$x = \cdot 71$ to $1\cdot 00$ $q = 4\cdot 5$ $p = 10\cdot 5$ to $15$

| $x$ | $p = 10\cdot 5$ | $p = 11$ | $p = 12$ | $p = 13$ | $p = 14$ | $p = 15$ |
|---|---|---|---|---|---|---|
| $B(p, q) =$ | $\cdot 1512\ 0721 \times \frac{1}{10^3}$ | $\cdot 1260\ 5839 \times \frac{1}{10^3}$ | $\cdot 8946\ 0793 \times \frac{1}{10^4}$ | $\cdot 6506\ 2395 \times \frac{1}{10^4}$ | $\cdot 4833\ 2065 \times \frac{1}{10^4}$ | $\cdot 3657\ 5617 \times \frac{1}{10^4}$ |
| ·71 | ·5030 626 | ·4696 140 | ·4061 294 | ·3479 088 | ·2954 430 | ·2488 796 |
| ·72 | ·5368 054 | ·5038 386 | ·4406 113 | ·3818 100 | ·3280 746 | ·2797 128 |
| ·73 | ·5707 826 | ·5385 407 | ·4760 632 | ·4171 521 | ·3625 687 | ·3127 613 |
| ·74 | ·6047 761 | ·5734 981 | ·5122 681 | ·4537 424 | ·3987 732 | ·3479 267 |
| ·75 | ·6385 574 | ·6084 726 | ·5489 832 | ·4913 529 | ·4364 932 | ·3850 623 |
| ·76 | ·6718 904 | ·6432 137 | ·5859 426 | ·5297 215− | ·4754 897 | ·4239 695− |
| ·77 | ·7045 353 | ·6774 622 | ·6228 601 | ·5685 539 | ·5154 799 | ·4643 963 |
| ·78 | ·7362 527 | ·7109 542 | ·6594 339 | ·6075 272 | ·5561 396 | ·5060 367 |
| ·79 | ·7668 083 | ·7434 267 | ·6953 516 | ·6462 949 | ·5971 062 | ·5485 323 |
| ·80 | ·7959 774 | ·7746 224 | ·7302 963 | ·6844 925+ | ·6379 841 | ·5914 757 |
| ·81 | ·8235 505− | ·8042 960 | ·7639 535+ | ·7217 453 | ·6783 519 | ·6344 162 |
| ·82 | ·8493 381 | ·8322 197 | ·7960 192 | ·7576 769 | ·7177 713 | ·6768 682 |
| ·83 | ·8731 761 | ·8581 901 | ·8262 075− | ·7919 194 | ·7557 982 | ·7183 225− |
| ·84 | ·8949 312 | ·8820 343 | ·8542 597 | ·8241 243 | ·7919 953 | ·7582 599 |
| ·85 | ·9145 049 | ·9036 156 | ·8799 534 | ·8539 743 | ·8259 469 | ·7961 678 |
| ·86 | ·9318 380 | ·9228 390 | ·9031 105+ | ·8811 954 | ·8572 742 | ·8315 589 |
| ·87 | ·9469 138 | ·9396 564 | ·9236 058 | ·9055 689 | ·8856 520 | ·8639 923 |
| ·88 | ·9597 606 | ·9540 697 | ·9413 739 | ·9269 430 | ·9108 248 | ·8930 947 |
| ·89 | ·9704 519 | ·9661 329 | ·9564 146 | ·9452 425+ | ·9326 224 | ·9185 825− |
| ·90 | ·9791 065+ | ·9759 530 | ·9687 964 | ·9604 769 | ·9509 735+ | ·9402 822 |
| ·91 | ·9858 853 | ·9836 873 | ·9786 572 | ·9727 446 | ·9659 157 | ·9581 477 |
| ·92 | ·9909 870 | ·9895 401 | ·9862 012 | ·9822 334 | ·9776 003 | ·9722 722 |
| ·93 | ·9946 407 | ·9937 545+ | ·9916 925+ | ·9892 156 | ·9862 918 | ·9828 929 |
| ·94 | ·9970 969 | ·9966 028 | ·9954 437 | ·9940 364 | ·9923 573 | ·9903 844 |
| ·95 | ·9986 153 | ·9983 729 | ·9977 996 | ·9970 961 | ·9962 479 | ·9952 407 |
| ·96 | ·9994 506 | ·9993 517 | ·9991 161 | ·9988 239 | ·9984 678 | ·9980 405+ |
| ·97 | ·9998 370 | ·9998 069 | ·9997 345+ | ·9996 439 | ·9995 322 | ·9993 969 |
| ·98 | ·9999 716 | ·9999 662 | ·9999 531 | ·9999 366 | ·9999 160 | ·9998 908 |
| ·99 | ·9999 986 | ·9999 984 | ·9999 977 | ·9999 969 | ·9999 959 | ·9999 946 |
| 1·00 | 1·0000 000 | 1·0000 000 | 1·0000 000 | 1·0000 000 | 1·0000 000 | 1·0000 000 |

$x$ = ·23 to ·80 $q = 4{\cdot}5$ $p$ = 16 to 21

| $x$ | $p = 16$ | $p = 17$ | $p = 18$ | $p = 19$ | $p = 20$ | $p = 21$ |
|---|---|---|---|---|---|---|
| $B(p, q) =$ | ·2813 5090 $\times \frac{1}{10^4}$ | ·2195 9094 $\times \frac{1}{10^4}$ | ·1736 3005 $\times \frac{1}{10^4}$ | ·1389 0404 $\times \frac{1}{10^4}$ | ·1123 0539 $\times \frac{1}{10^4}$ | ·9167 7872 $\times \frac{1}{10^5}$ |
| ·23 | ·0000 001 | | | | | |
| ·24 | ·0000 001 | | | | | |
| ·25 | ·0000 002 | ·0000 001 | | | | |
| ·26 | ·0000 004 | ·0000 001 | | | | |
| ·27 | ·0000 006 | ·0000 002 | ·0000 001 | | | |
| ·28 | ·0000 011 | ·0000 004 | ·0000 001 | | | |
| ·29 | ·0000 018 | ·0000 006 | ·0000 002 | ·0000 001 | | |
| ·30 | ·0000 030 | ·0000 011 | ·0000 004 | ·0000 001 | | |
| ·31 | ·0000 048 | ·0000 018 | ·0000 007 | ·0000 002 | ·0000 001 | |
| ·32 | ·0000 077 | ·0000 029 | ·0000 011 | ·0000 004 | ·0000 002 | ·0000 001 |
| ·33 | ·0000 120 | ·0000 047 | ·0000 019 | ·0000 007 | ·0000 003 | ·0000 001 |
| ·34 | ·0000 184 | ·0000 075+ | ·0000 030 | ·0000 012 | ·0000 005− | ·0000 002 |
| ·35 | ·0000 279 | ·0000 117 | ·0000 049 | ·0000 020 | ·0000 008 | ·0000 003 |
| ·36 | ·0000 417 | ·0000 180 | ·0000 077 | ·0000 033 | ·0000 014 | ·0000 006 |
| ·37 | ·0000 615+ | ·0000 273 | ·0000 120 | ·0000 052 | ·0000 023 | ·0000 010 |
| ·38 | ·0000 896 | ·0000 408 | ·0000 184 | ·0000 082 | ·0000 037 | ·0000 016 |
| ·39 | ·0001 290 | ·0000 602 | ·0000 279 | ·0000 128 | ·0000 058 | ·0000 026 |
| ·40 | ·0001 837 | ·0000 879 | ·0000 417 | ·0000 196 | ·0000 092 | ·0000 043 |
| ·41 | ·0002 587 | ·0001 268 | ·0000 617 | ·0000 297 | ·0000 142 | ·0000 068 |
| ·42 | ·0003 605+ | ·0001 810 | ·0000 901 | ·0000 445+ | ·0000 218 | ·0000 106 |
| ·43 | ·0004 975+ | ·0002 557 | ·0001 303 | ·0000 659 | ·0000 331 | ·0000 165− |
| ·44 | ·0006 802 | ·0003 575+ | ·0001 863 | ·0000 964 | ·0000 495− | ·0000 252 |
| ·45 | ·0009 214 | ·0004 951 | ·0002 638 | ·0001 395− | ·0000 732 | ·0000 382 |
| ·46 | ·0012 372 | ·0006 793 | ·0003 699 | ·0001 998 | ·0001 072 | ·0000 572 |
| ·47 | ·0016 472 | ·0009 237 | ·0005 137 | ·0002 835− | ·0001 553 | ·0000 846 |
| ·48 | ·0021 753 | ·0012 452 | ·0007 069 | ·0003 983 | ·0002 228 | ·0001 239 |
| ·49 | ·0028 500+ | ·0016 647 | ·0009 643 | ·0005 544 | ·0003 165+ | ·0001 796 |
| ·50 | ·0037 055− | ·0022 074 | ·0013 042 | ·0007 648 | ·0004 454 | ·0002 578 |
| ·51 | ·0047 820 | ·0029 043 | ·0017 495− | ·0010 460 | ·0006 211 | ·0003 665+ |
| ·52 | ·0061 269 | ·0037 920 | ·0023 279 | ·0014 185+ | ·0008 585+ | ·0005 164 |
| ·53 | ·0077 952 | ·0049 145+ | ·0030 735− | ·0019 080 | ·0011 765− | ·0007 210 |
| ·54 | ·0098 500+ | ·0063 235+ | ·0040 272 | ·0025 460 | ·0015 988 | ·0009 979 |
| ·55 | ·0123 638 | ·0080 794 | ·0052 378 | ·0033 711 | ·0021 552 | ·0013 695− |
| ·56 | ·0154 184 | ·0102 521 | ·0067 633 | ·0044 298 | ·0028 822 | ·0018 639 |
| ·57 | ·0191 055+ | ·0129 220 | ·0086 717 | ·0057 780 | ·0038 247 | ·0025 165+ |
| ·58 | ·0235 272 | ·0161 804 | ·0110 419 | ·0074 821 | ·0050 370 | ·0033 708 |
| ·59 | ·0287 956 | ·0201 302 | ·0139 650− | ·0096 202 | ·0065 846 | ·0044 801 |
| ·60 | ·0350 327 | ·0248 861 | ·0175 447 | ·0122 833 | ·0085 450+ | ·0059 095+ |
| ·61 | ·0423 696 | ·0305 746 | ·0218 983 | ·0155 766 | ·0110 101 | ·0077 370 |
| ·62 | ·0509 459 | ·0373 339 | ·0271 568 | ·0196 201 | ·0140 866 | ·0100 555− |
| ·63 | ·0609 077 | ·0453 126 | ·0334 649 | ·0245 495+ | ·0178 982 | ·0129 745+ |
| ·64 | ·0724 057 | ·0546 691 | ·0409 805+ | ·0305 164 | ·0225 857 | ·0166 218 |
| ·65 | ·0855 930 | ·0655 690 | ·0498 737 | ·0376 880 | ·0283 083 | ·0211 445− |
| ·66 | ·1006 216 | ·0781 833 | ·0603 250+ | ·0462 468 | ·0352 435+ | ·0267 105− |
| ·67 | ·1176 385− | ·0926 844 | ·0725 230 | ·0563 882 | ·0435 866 | ·0335 085+ |
| ·68 | ·1367 818 | ·1092 427 | ·0866 607 | ·0683 191 | ·0535 492 | ·0417 481 |
| ·69 | ·1581 755+ | ·1280 215+ | ·1029 316 | ·0822 534 | ·0653 571 | ·0516 584 |
| ·70 | ·1819 239 | ·1491 713 | ·1215 242 | ·0984 082 | ·0792 462 | ·0634 855+ |
| ·71 | ·2081 056 | ·1728 234 | ·1426 155+ | ·1169 976 | ·0954 583 | ·0774 892 |
| ·72 | ·2367 675+ | ·1990 830 | ·1663 640 | ·1382 256 | ·1142 340 | ·0939 371 |
| ·73 | ·2679 181 | ·2280 216 | ·1929 010 | ·1622 777 | ·1358 048 | ·1130 976 |
| ·74 | ·3015 209 | ·2596 686 | ·2223 216 | ·1893 109 | ·1603 832 | ·1352 306 |
| ·75 | ·3374 887 | ·2940 035− | ·2546 749 | ·2194 430 | ·1881 516 | ·1605 760 |
| ·76 | ·3756 779 | ·3309 478 | ·2899 540 | ·2527 407 | ·2192 488 | ·1893 405− |
| ·77 | ·4158 838 | ·3703 579 | ·3280 858 | ·2892 071 | ·2537 560 | ·2216 815+ |
| ·78 | ·4578 377 | ·4120 185− | ·3689 215− | ·3287 694 | ·2916 817 | ·2576 907 |
| ·79 | ·5012 052 | ·4556 380 | ·4122 286 | ·3712 670 | ·3329 463 | ·2973 751 |
| ·80 | ·5455 874 | ·5008 463 | ·4576 844 | ·4164 407 | ·3773 677 | ·3406 392 |

$x = ·81$ to $1·00$ $q = 4·5$ $p = 16$ to $21$

| | $p = 16$ | $p = 17$ | $p = 18$ | $p = 19$ | $p = 20$ | $p = 21$ |
|---|---|---|---|---|---|---|
| $B(p, q) =$ | $·2813\ 5090 \times \frac{1}{10^4}$ | $·2195\ 9094 \times \frac{1}{10^4}$ | $·1736\ 3005 \times \frac{1}{10^4}$ | $·1389\ 0404 \times \frac{1}{10^4}$ | $·1123\ 0539 \times \frac{1}{10^4}$ | $·9167\ 7872 \times \frac{1}{10^5}$ |
| $x$ | | | | | | |
| ·81 | ·5905 244 | ·5471 949 | ·5048 722 | ·4639 250+ | ·4246 481 | ·3872 662 |
| ·82 | ·6355 015− | ·5941 606 | ·5532 818 | ·5132 432 | ·4743 637 | ·4369 033 |
| ·83 | ·6799 598 | ·6411 536 | ·6023 131 | ·5638 071 | ·5259 598 | ·4890 492 |
| ·84 | ·7233 099 | ·6875 299 | ·6512 869 | ·6149 231 | ·5787 506 | ·5430 484 |
| ·85 | ·7649 493 | ·7326 089 | ·6994 600 | ·6658 047 | ·6319 279 | ·5980 935+ |
| ·86 | ·8042 836 | ·7756 957 | ·7460 483 | ·7155 938 | ·6845 783 | ·6532 372 |
| ·87 | ·8407 516 | ·8161 090 | ·7902 562 | ·7633 907 | ·7357 122 | ·7074 179 |
| ·88 | ·8738 517 | ·8532 136 | ·8313 129 | ·8082 929 | ·7843 035+ | ·7594 986 |
| ·89 | ·9031 714 | ·8864 551 | ·8685 147 | ·8494 429 | ·8293 423 | ·8083 222 |
| ·90 | ·9284 148 | ·9153 978 | ·9012 705− | ·8860 836 | ·8698 976 | ·8527 810 |
| ·91 | ·9494 294 | ·9397 604 | ·9291 500− | ·9176 171 | ·9051 888 | ·8918 999 |
| ·92 | ·9662 266 | ·9594 480 | ·9519 277 | ·9436 638 | ·9346 604 | ·9249 277 |
| ·93 | ·9789 943 | ·9745 756 | ·9696 200 | ·9641 153 | ·9580 529 | ·9514 281 |
| ·94 | ·9880 971 | ·9854 768 | ·9825 066 | ·9791 716 | ·9754 594 | ·9713 591 |
| ·95 | ·9940 606 | ·9926 942 | ·9911 289 | ·9893 527 | ·9873 544 | ·9851 239 |
| ·96 | ·9975 346 | ·9969 427 | ·9962 575+ | ·9954 718 | ·9945 786 | ·9935 711 |
| ·97 | ·9992 349 | ·9990 435− | ·9988 196 | ·9985 601 | ·9982 621 | ·9979 224 |
| ·98 | ·9998 603 | ·9998 239 | ·9997 809 | ·9997 306 | ·9996 722 | ·9996 050− |
| ·99 | ·9999 930 | ·9999 911 | ·9999 889 | ·9999 862 | ·9999 831 | ·9999 794 |
| 1·00 | 1·0000 000 | 1·0000 000 | 1·0000 000 | 1·0000 000 | 1·0000 000 | 1·0000 000 |

$x$ = ·34 to ·90 $q = 4{\cdot}5$ $p$ = 22 to 27

| $x$ | $p = 22$ | $p = 23$ | $p = 24$ | $p = 25$ | $p = 26$ | $p = 27$ |
|---|---|---|---|---|---|---|
| $B(p, q) =$ | ·7549 9424 $\times \frac{1}{10^5}$ | ·6267 8767 $\times \frac{1}{10^5}$ | ·5242 2241 $\times \frac{1}{10^5}$ | ·4414 5045$^{\pm}$ $\times \frac{1}{10^5}$ | ·3741 1055$^{\pm}$ $\times \frac{1}{10^5}$ | ·3189 1392 $\times \frac{1}{10^5}$ |
| ·34 | ·0000 001 | | | | | |
| ·35 | ·0000 001 | ·0000 001 | | | | |
| ·36 | ·0000 002 | ·0000 001 | | | | |
| ·37 | ·0000 004 | ·0000 002 | ·0000 001 | | | |
| ·38 | ·0000 007 | ·0000 003 | ·0000 001 | ·0000 001 | | |
| ·39 | ·0000 012 | ·0000 005+ | ·0000 002 | ·0000 001 | | |
| ·40 | ·0000 020 | ·0000 009 | ·0000 004 | ·0000 002 | ·0000 001 | |
| ·41 | ·0000 032 | ·0000 015+ | ·0000 007 | ·0000 003 | ·0000 002 | ·0000 001 |
| ·42 | ·0000 052 | ·0000 025− | ·0000 012 | ·0000 006 | ·0000 003 | ·0000 001 |
| ·43 | ·0000 082 | ·0000 040 | ·0000 020 | ·0000 010 | ·0000 005− | ·0000 002 |
| ·44 | ·0000 128 | ·0000 065− | ·0000 032 | ·0000 016 | ·0000 008 | ·0000 004 |
| ·45 | ·0000 198 | ·0000 102 | ·0000 052 | ·0000 027 | ·0000 014 | ·0000 007 |
| ·46 | ·0000 303 | ·0000 160 | ·0000 084 | ·0000 044 | ·0000 023 | ·0000 012 |
| ·47 | ·0000 458 | ·0000 246 | ·0000 132 | ·0000 070 | ·0000 037 | ·0000 020 |
| ·48 | ·0000 685− | ·0000 376 | ·0000 206 | ·0000 112 | ·0000 061 | ·0000 033 |
| ·49 | ·0001 013 | ·0000 568 | ·0000 317 | ·0000 176 | ·0000 097 | ·0000 054 |
| ·50 | ·0001 483 | ·0000 849 | ·0000 483 | ·0000 274 | ·0000 155− | ·0000 087 |
| ·51 | ·0002 150+ | ·0001 255− | ·0000 729 | ·0000 421 | ·0000 242 | ·0000 139 |
| ·52 | ·0003 088 | ·0001 837 | ·0001 087 | ·0000 641 | ·0000 376 | ·0000 220 |
| ·53 | ·0004 393 | ·0002 662 | ·0001 606 | ·0000 964 | ·0000 576 | ·0000 343 |
| ·54 | ·0006 193 | ·0003 823 | ·0002 348 | ·0001 436 | ·0000 875− | ·0000 530 |
| ·55 | ·0008 653 | ·0005 439 | ·0003 402 | ·0002 118 | ·0001 313 | ·0000 811 |
| ·56 | ·0011 987 | ·0007 668 | ·0004 882 | ·0003 094 | ·0001 953 | ·0001 228 |
| ·57 | ·0016 466 | ·0010 718 | ·0006 943 | ·0004 478 | ·0002 876 | ·0001 840 |
| ·58 | ·0022 432 | ·0014 852 | ·0009 787 | ·0006 420 | ·0004 195− | ·0002 730 |
| ·59 | ·0030 316 | ·0020 410 | ·0013 676 | ·0009 123 | ·0006 061 | ·0004 011 |
| ·60 | ·0040 647 | ·0027 817 | ·0018 947 | ·0012 849 | ·0008 678 | ·0005 839 |
| ·61 | ·0054 077 | ·0037 607 | ·0026 032 | ·0017 941 | ·0012 315+ | ·0008 421 |
| ·62 | ·0071 396 | ·0050 442 | ·0035 473 | ·0024 839 | ·0017 323 | ·0012 035+ |
| ·63 | ·0093 556 | ·0067 130 | ·0047 948 | ·0034 101 | ·0024 156 | ·0017 048 |
| ·64 | ·0121 687 | ·0088 654 | ·0064 295+ | ·0046 432 | ·0033 399 | ·0023 935+ |
| ·65 | ·0157 119 | ·0116 190 | ·0085 538 | ·0062 708 | ·0045 791 | ·0033 315+ |
| ·66 | ·0201 400 | ·0151 137 | ·0112 914 | ·0084 009 | ·0062 260 | ·0045 974 |
| ·67 | ·0256 308 | ·0195 131 | ·0147 904 | ·0111 648 | ·0083 956 | ·0062 904 |
| ·68 | ·0323 859 | ·0250 068 | ·0192 254 | ·0147 208 | ·0112 287 | ·0085 345+ |
| ·69 | ·0406 309 | ·0318 115− | ·0248 000 | ·0192 566 | ·0148 961 | ·0114 824 |
| ·70 | ·0506 143 | ·0401 710 | ·0317 483 | ·0249 926 | ·0196 015− | ·0153 197 |
| ·71 | ·0626 047 | ·0503 555+ | ·0403 351 | ·0321 831 | ·0255 849 | ·0202 695+ |
| ·72 | ·0768 876 | ·0626 591 | ·0508 556 | ·0411 178 | ·0331 249 | ·0265 954 |
| ·73 | ·0937 586 | ·0773 952 | ·0636 321 | ·0521 199 | ·0425 395− | ·0346 043 |
| ·74 | ·1135 154 | ·0948 898 | ·0790 094 | ·0655 441 | ·0541 848 | ·0446 474 |
| ·75 | ·1364 474 | ·1154 719 | ·0973 464 | ·0817 698 | ·0684 517 | ·0571 186 |
| ·76 | ·1628 218 | ·1394 612 | ·1190 055− | ·1011 920 | ·0857 583 | ·0724 498 |
| ·77 | ·1928 680 | ·1671 519 | ·1443 373 | ·1242 081 | ·1065 387 | ·0911 018 |
| ·78 | ·2267 589 | ·1987 938 | ·1736 616 | ·1511 998 | ·1312 267 | ·1135 505− |
| ·79 | ·2645 904 | ·2345 700 | ·2072 449 | ·1825 100 | ·1602 337 | ·1402 665+ |
| ·80 | ·3063 592 | ·2745 722 | ·2452 730 | ·2184 154 | ·1939 212 | ·1716 880 |
| ·81 | ·3519 403 | ·3187 742 | ·2878 215− | ·2590 935+ | ·2325 661 | ·2081 864 |
| ·82 | ·4010 661 | ·3670 046 | ·3348 238 | ·3045 873 | ·2763 221 | ·2500 247 |
| ·83 | ·4533 074 | ·4189 222 | ·3860 394 | ·3547 666 | ·3251 762 | ·2973 100 |
| ·84 | ·5080 603 | ·4739 945+ | ·4410 248 | ·4092 915− | ·3789 036 | ·3499 416 |
| ·85 | ·5645 410 | ·5314 843 | ·4991 102 | ·4675 791 | ·4370 256 | ·4075 590 |
| ·86 | ·6217 916 | ·5904 461 | ·5593 867 | ·5287 804 | ·4987 738 | ·4694 944 |
| ·87 | ·6786 991 | ·6497 389 | ·6207 094 | ·5917 706 | ·5630 691 | ·5347 374 |
| ·88 | ·7340 322 | ·7080 564 | ·6817 193 | ·6551 626 | ·6285 210 | ·6019 204 |
| ·89 | ·7864 962 | ·7639 807 | ·7408 926 | ·7173 475− | ·6934 586 | ·6693 355+ |
| ·90 | ·8348 084 | ·8160 598 | ·7966 184 | ·7765 693 | ·7559 990 | ·7349 936 |

$x = \cdot 91$ to $1\cdot 00$ $q = 4\cdot 5$ $p = 22$ to $27$

| $x$ | $p = 22$ | $p = 23$ | $p = 24$ | $p = 25$ | $p = 26$ | $p = 27$ |
|---|---|---|---|---|---|---|
| $B(p, q) =$ | $\cdot 7549\,9424 \times \frac{1}{10^5}$ | $\cdot 6267\,8767 \times \frac{1}{10^5}$ | $\cdot 5242\,2241 \times \frac{1}{10^5}$ | $\cdot 4414\,5045 \overset{+}{\times} \frac{1}{10^5}$ | $\cdot 3741\,1055 \overset{+}{\times} \frac{1}{10^5}$ | $\cdot 3189\,1392 \times \frac{1}{10^5}$ |
| ·91 | ·8777 916 | ·8629 105⁻ | ·8473 079 | ·8310 390 | ·8141 617 | ·7967 358 |
| ·92 | ·9144 814 | ·9033 417 | ·8915 337 | ·8790 861 | ·8660 311 | ·8524 036 |
| ·93 | ·9442 402 | ·9364 921 | ·9281 897 | ·9193 425⁺ | ·9099 627 | ·9000 653 |
| ·94 | ·9668 626 | ·9619 633 | ·9566 573 | ·9509 422 | ·9448 179 | ·9382 861 |
| ·95 | ·9826 517 | ·9799 296 | ·9769 499 | ·9737 065⁺ | ·9701 939 | ·9664 077 |
| ·96 | ·9924 426 | ·9911 870 | ·9897 981 | ·9882 704 | ·9865 984 | ·9847 772 |
| ·97 | ·9975 381 | ·9971 059 | ·9966 229 | ·9960 861 | ·9954 924 | ·9948 391 |
| ·98 | ·9995 281 | ·9994 407 | ·9993 421 | ·9992 313 | ·9991 076 | ·9989 700 |
| ·99 | ·9999 752 | ·9999 704 | ·9999 649 | ·9999 586 | ·9999 516 | ·9999 437 |
| 1·00 | 1·0000 000 | 1·0000 000 | 1·0000 000 | 1·0000 000 | 1·0000 000 | 1·0000 000 |

$x$ = ·42 to 1·00 $q$ = 4·5 $p$ = 28 to 33

| $x$ | $p$ = 28 | $p$ = 29 | $p$ = 30 | $p$ = 31 | $p$ = 32 | $p$ = 33 |
|---|---|---|---|---|---|---|
| $B(p, q)$ = | ·2733 5478 × $\frac{1}{10^5}$ | ·2355 0566 × $\frac{1}{10^5}$ | ·2038 7057 × $\frac{1}{10^5}$ | ·1772 7876 × $\frac{1}{10^5}$ | ·1548 0680 × $\frac{1}{10^5}$ | ·1357 2103 × $\frac{1}{10^5}$ |
| ·42 | ·0000 001 | | | | | |
| ·43 | ·0000 001 | ·0000 001 | | | | |
| ·44 | ·0000 002 | ·0000 001 | | | | |
| ·45 | ·0000 003 | ·0000 002 | ·0000 001 | | | |
| ·46 | ·0000 006 | ·0000 003 | ·0000 002 | ·0000 001 | | |
| ·47 | ·0000 010 | ·0000 005+ | ·0000 003 | ·0000 001 | ·0000 001 | |
| ·48 | ·0000 018 | ·0000 009 | ·0000 005+ | ·0000 003 | ·0000 001 | ·0000 001 |
| ·49 | ·0000 029 | ·0000 016 | ·0000 009 | ·0000 005− | ·0000 003 | ·0000 001 |
| ·50 | ·0000 049 | ·0000 027 | ·0000 015+ | ·0000 008 | ·0000 005− | ·0000 003 |
| ·51 | ·0000 079 | ·0000 045+ | ·0000 026 | ·0000 014 | ·0000 008 | ·0000 005− |
| ·52 | ·0000 128 | ·0000 074 | ·0000 043 | ·0000 025− | ·0000 014 | ·0000 008 |
| ·53 | ·0000 204 | ·0000 120 | ·0000 071 | ·0000 042 | ·0000 024 | ·0000 014 |
| ·54 | ·0000 321 | ·0000 193 | ·0000 116 | ·0000 069 | ·0000 041 | ·0000 025− |
| ·55 | ·0000 499 | ·0000 306 | ·0000 187 | ·0000 114 | ·0000 069 | ·0000 042 |
| ·56 | ·0000 769 | ·0000 480 | ·0000 299 | ·0000 185+ | ·0000 115− | ·0000 071 |
| ·57 | ·0001 173 | ·0000 745− | ·0000 472 | ·0000 298 | ·0000 187 | ·0000 118 |
| ·58 | ·0001 770 | ·0001 144 | ·0000 737 | ·0000 473 | ·0000 303 | ·0000 194 |
| ·59 | ·0002 645+ | ·0001 738 | ·0001 139 | ·0000 744 | ·0000 484 | ·0000 315− |
| ·60 | ·0003 914 | ·0002 615+ | ·0001 742 | ·0001 157 | ·0000 766 | ·0000 506 |
| ·61 | ·0005 738 | ·0003 897 | ·0002 638 | ·0001 780 | ·0001 198 | ·0000 804 |
| ·62 | ·0008 332 | ·0005 749 | ·0003 955− | ·0002 712 | ·0001 855− | ·0001 265+ |
| ·63 | ·0011 989 | ·0008 403 | ·0005 872 | ·0004 091 | ·0002 842 | ·0001 969 |
| ·64 | ·0017 093 | ·0012 167 | ·0008 634 | ·0006 109 | ·0004 310 | ·0003 033 |
| ·65 | ·0024 154 | ·0017 456 | ·0012 576 | ·0009 034 | ·0006 472 | ·0004 624 |
| ·66 | ·0033 831 | ·0024 816 | ·0018 147 | ·0013 232 | ·0009 622 | ·0006 979 |
| ·67 | ·0046 971 | ·0034 962 | ·0025 944 | ·0019 198 | ·0014 167 | ·0010 428 |
| ·68 | ·0064 650− | ·0048 818 | ·0036 752 | ·0027 591 | ·0020 658 | ·0015 427 |
| ·69 | ·0088 216 | ·0067 561 | ·0051 589 | ·0039 283 | ·0029 833 | ·0022 599 |
| ·70 | ·0119 340 | ·0092 677 | ·0071 760 | ·0055 411 | ·0042 674 | ·0032 783 |
| ·71 | ·0160 064 | ·0126 012 | ·0098 917 | ·0077 436 | ·0060 462 | ·0047 092 |
| ·72 | ·0212 847 | ·0169 830 | ·0135 120 | ·0107 213 | ·0084 852 | ·0066 991 |
| ·73 | ·0280 607 | ·0226 867 | ·0182 903 | ·0147 065− | ·0117 949 | ·0094 370 |
| ·74 | ·0366 748 | ·0300 376 | ·0245 334 | ·0199 850− | ·0162 392 | ·0131 641 |
| ·75 | ·0475 169 | ·0394 155+ | ·0326 062 | ·0269 033 | ·0221 433 | ·0181 828 |
| ·76 | ·0610 241 | ·0512 552 | ·0429 348 | ·0358 735+ | ·0299 011 | ·0248 655+ |
| ·77 | ·0776 745+ | ·0660 431 | ·0560 061 | ·0473 759 | ·0399 803 | ·0336 629 |
| ·78 | ·0979 758 | ·0843 090 | ·0723 624 | ·0619 568 | ·0529 242 | ·0451 081 |
| ·79 | ·1224 476 | ·1066 111 | ·0925 902 | ·0802 215+ | ·0693 471 | ·0598 166 |
| ·80 | ·1515 958 | ·1335 128 | ·1173 005+ | ·1028 175+ | ·0899 230 | ·0784 791 |
| ·81 | ·1858 790 | ·1655 513 | ·1470 987 | ·1304 084 | ·1153 628 | ·1018 430 |
| ·82 | ·2256 655+ | ·2031 942 | ·1825 438 | ·1636 349 | ·1463 790 | ·1306 815+ |
| ·83 | ·2711 829 | ·2467 867 | ·2240 941 | ·2030 617 | ·1836 340 | ·1657 453 |
| ·84 | ·3224 599 | ·2964 897 | ·2720 419 | ·2491 099 | ·2276 721 | ·2076 948 |
| ·85 | ·3792 657 | ·3522 101 | ·3264 374 | ·3019 748 | ·2788 340 | ·2570 129 |
| ·86 | ·4410 500− | ·4135 300 | ·3870 064 | ·3615 350− | ·3371 563 | ·3138 976 |
| ·87 | ·5068 937 | ·4796 416 | ·4530 708 | ·4272 573 | ·4022 640 | ·3781 416 |
| ·88 | ·5754 774 | ·5492 989 | ·5234 814 | ·4981 114 | ·4732 652 | ·4490 091 |
| ·89 | ·6450 828 | ·6207 998 | ·5965 796 | ·5725 087 | ·5486 669 | ·5251 269 |
| ·90 | ·7136 380 | ·6920 156 | ·6702 067 | ·6482 887 | ·6263 355− | ·6044 165− |
| ·91 | ·7788 227 | ·7604 841 | ·7417 819 | ·7227 774 | ·7035 307 | ·6841 006 |
| ·92 | ·8382 411 | ·8235 830 | ·8084 699 | ·7929 437 | ·7770 470 | ·7608 223 |
| ·93 | ·8896 674 | ·8787 887 | ·8674 504 | ·8556 756 | ·8434 887 | ·8309 152 |
| ·94 | ·9313 504 | ·9240 158 | ·9162 892 | ·9081 788 | ·8996 943 | ·8908 466 |
| ·95 | ·9623 445− | ·9580 020 | ·9533 787 | ·9484 742 | ·9432 888 | ·9378 239 |
| ·96 | ·9828 023 | ·9806 693 | ·9783 746 | ·9759 146 | ·9732 864 | ·9704 874 |
| ·97 | ·9941 232 | ·9933 420 | ·9924 928 | ·9915 729 | ·9905 799 | ·9895 113 |
| ·98 | ·9988 177 | ·9986 498 | ·9984 654 | ·9982 636 | ·9980 435+ | ·9978 042 |
| ·99 | ·9999 348 | ·9999 250− | ·9999 140 | ·9999 019 | ·9998 886 | ·9998 740 |
| 1·00 | 1·0000 000 | 1·0000 000 | 1·0000 000 | 1·0000 000 | 1·0000 000 | 1·0000 000 |

$x$ = ·49 to 1·00 $q$ = 4·5 $p$ = 34 to 39

| | $p$ = 34 | $p$ = 35 | $p$ = 36 | $p$ = 37 | $p$ = 38 | $p$ = 39 |
|---|---|---|---|---|---|---|
| $B(p, q)$ = | ·1194 3451 × $\frac{1}{10^8}$ | ·1054 7463 × $\frac{1}{10^8}$ | ·9345 8534 × $\frac{1}{10^9}$ | ·8307 4252 × $\frac{1}{10^9}$ | ·7406 6201 × $\frac{1}{10^9}$ | ·6622 3897 × $\frac{1}{10^9}$ |
| $x$ | | | | | | |
| ·49 | ·0000 001 | | | | | |
| ·50 | ·0000 001 | ·0000 001 | | | | |
| ·51 | ·0000 003 | ·0000 001 | ·0000 001 | | | |
| ·52 | ·0000 005⁻ | ·0000 003 | ·0000 002 | ·0000 001 | | |
| ·53 | ·0000 008 | ·0000 005⁻ | ·0000 003 | ·0000 002 | ·0000 001 | ·0000 001 |
| ·54 | ·0000 015⁻ | ·0000 009 | ·0000 005⁺ | ·0000 003 | ·0000 002 | ·0000 001 |
| ·55 | ·0000 025⁺ | ·0000 015⁺ | ·0000 009 | ·0000 006 | ·0000 003 | ·0000 002 |
| ·56 | ·0000 043 | ·0000 027 | ·0000 016 | ·0000 010 | ·0000 006 | ·0000 004 |
| ·57 | ·0000 074 | ·0000 046 | ·0000 029 | ·0000 018 | ·0000 011 | ·0000 007 |
| ·58 | ·0000 123 | ·0000 078 | ·0000 050⁻ | ·0000 031 | ·0000 020 | ·0000 012 |
| ·59 | ·0000 204 | ·0000 132 | ·0000 085⁻ | ·0000 055⁻ | ·0000 035⁺ | ·0000 022 |
| ·60 | ·0000 333 | ·0000 219 | ·0000 144 | ·0000 094 | ·0000 061 | ·0000 040 |
| ·61 | ·0000 538 | ·0000 360 | ·0000 240 | ·0000 159 | ·0000 106 | ·0000 070 |
| ·62 | ·0000 861 | ·0000 584 | ·0000 396 | ·0000 267 | ·0000 180 | ·0000 121 |
| ·63 | ·0001 361 | ·0000 939 | ·0000 646 | ·0000 443 | ·0000 304 | ·0000 208 |
| ·64 | ·0002 129 | ·0001 491 | ·0001 042 | ·0000 727 | ·0000 506 | ·0000 351 |
| ·65 | ·0003 296 | ·0002 344 | ·0001 663 | ·0001 177 | ·0000 832 | ·0000 587 |
| ·66 | ·0005 050⁻ | ·0003 645⁺ | ·0002 626 | ·0001 887 | ·0001 354 | ·0000 969 |
| ·67 | ·0007 657 | ·0005 610 | ·0004 101 | ·0002 991 | ·0002 178 | ·0001 582 |
| ·68 | ·0011 494 | ·0008 543 | ·0006 336 | ·0004 690 | ·0003 464 | ·0002 554 |
| ·69 | ·0017 079 | ·0012 877 | ·0009 689 | ·0007 274 | ·0005 451 | ·0004 077 |
| ·70 | ·0025 125⁻ | ·0019 212 | ·0014 660 | ·0011 163 | ·0008 484 | ·0006 435⁺ |
| ·71 | ·0036 593 | ·0028 371 | ·0021 950⁺ | ·0016 948 | ·0013 060 | ·0010 046 |
| ·72 | ·0052 767 | ·0041 471 | ·0032 525⁻ | ·0025 458 | ·0019 888 | ·0015 508 |
| ·73 | ·0075 332 | ·0060 003 | ·0047 694 | ·0037 835⁻ | ·0029 957 | ·0023 676 |
| ·74 | ·0106 472 | ·0085 930 | ·0069 208 | ·0055 631 | ·0044 634 | ·0035 747 |
| ·75 | ·0148 973 | ·0121 796 | ·0099 375⁺ | ·0080 924 | ·0065 777 | ·0053 371 |
| ·76 | ·0206 326 | ·0170 845⁻ | ·0141 182 | ·0116 447 | ·0095 869 | ·0078 790 |
| ·77 | ·0282 826 | ·0237 133 | ·0198 431 | ·0165 733 | ·0138 174 | ·0114 999 |
| ·78 | ·0383 649 | ·0325 638 | ·0275 865⁺ | ·0233 267 | ·0196 898 | ·0165 917 |
| ·79 | ·0514 890 | ·0442 330 | ·0379 275⁻ | ·0324 619 | ·0277 356 | ·0236 579 |
| ·80 | ·0683 529 | ·0594 181 | ·0515 555⁻ | ·0446 538 | ·0386 102 | ·0333 300 |
| ·81 | ·0897 304 | ·0789 093 | ·0692 676 | ·0606 986 | ·0531 012 | ·0463 804 |
| ·82 | ·1164 444 | ·1035 682 | ·0919 538 | ·0815 041 | ·0721 248 | ·0637 254 |
| ·83 | ·1493 229 | ·1342 892 | ·1205 634 | ·1080 634 | ·0967 070 | ·0864 130 |
| ·84 | ·1891 341 | ·1719 381 | ·1560 490 | ·1414 046 | ·1279 396 | ·1155 873 |
| ·85 | ·2364 978 | ·2172 649 | ·1992 821 | ·1825 106 | ·1669 064 | ·1524 211 |
| ·86 | ·2917 736 | ·2707 883 | ·2509 362 | ·2322 036 | ·2145 696 | ·1980 075⁺ |
| ·87 | ·3549 292 | ·3326 556 | ·3113 398 | ·2909 920 | ·2716 149 | ·2532 041 |
| ·88 | ·4253 998 | ·4024 849 | ·3803 033 | ·3588 858 | ·3382 554 | ·3184 286 |
| ·89 | ·5019 542 | ·4792 075⁺ | ·4569 385⁻ | ·4351 922 | ·4140 071 | ·3934 158 |
| ·90 | ·5825 971 | ·5609 382 | ·5394 959 | ·5183 216 | ·4974 621 | ·4769 594 |
| ·91 | ·6645 438 | ·6449 152 | ·6252 670 | ·6056 487 | ·5861 074 | ·5666 869 |
| ·92 | ·7443 125⁻ | ·7275 599 | ·7106 063 | ·6934 925⁺ | ·6762 585⁻ | ·6589 428 |
| ·93 | ·8179 816 | ·8047 152 | ·7911 437 | ·7772 951 | ·7631 976 | ·7488 794 |
| ·94 | ·8816 476 | ·8721 105⁻ | ·8622 490 | ·8520 781 | ·8416 130 | ·8308 697 |
| ·95 | ·9320 817 | ·9260 650⁻ | ·9197 775⁺ | ·9132 237 | ·9064 086 | ·8993 380 |
| ·96 | ·9675 153 | ·9643 684 | ·9610 452 | ·9575 449 | ·9538 667 | ·9500 103 |
| ·97 | ·9883 649 | ·9871 384 | ·9858 297 | ·9844 368 | ·9829 579 | ·9813 913 |
| ·98 | ·9975 448 | ·9972 645⁻ | ·9969 622 | ·9966 373 | ·9962 887 | ·9959 156 |
| ·99 | ·9998 579 | ·9998 404 | ·9998 214 | ·9998 007 | ·9997 783 | ·9997 540 |
| 1·00 | 1·0000 000 | 1·0000 000 | 1·0000 000 | 1·0000 000 | 1·0000 000 | 1·0000 000 |

$x$ = ·54 to 1·00 $q = 4{\cdot}5$ $p$ = 40 to 45

| $x$ | $p = 40$ | $p = 41$ | $p = 42$ | $p = 43$ | $p = 44$ | $p = 45$ |
|---|---|---|---|---|---|---|
| $B(p, q) =$ | ·5937 3149 $\times \frac{1}{10^{8}}$ | ·5336 9123 $\times \frac{1}{10^{8}}$ | ·4809 0858 $\times \frac{1}{10^{8}}$ | ·4343 6904 $\times \frac{1}{10^{8}}$ | ·3932 1829 $\times \frac{1}{10^{8}}$ | ·3567 3412 $\times \frac{1}{10^{8}}$ |
| ·54 | ·0000 001 | | | | | |
| ·55 | ·0000 001 | ·0000 001 | | | | |
| ·56 | ·0000 002 | ·0000 001 | ·0000 001 | | | |
| ·57 | ·0000 004 | ·0000 003 | ·0000 002 | ·0000 001 | ·0000 001 | |
| ·58 | ·0000 008 | ·0000 005− | ·0000 003 | ·0000 002 | ·0000 001 | ·0000 001 |
| ·59 | ·0000 014 | ·0000 009 | ·0000 006 | ·0000 004 | ·0000 002 | ·0000 001 |
| ·60 | ·0000 026 | ·0000 017 | ·0000 011 | ·0000 007 | ·0000 005− | ·0000 003 |
| ·61 | ·0000 046 | ·0000 031 | ·0000 020 | ·0000 013 | ·0000 009 | ·0000 006 |
| ·62 | ·0000 082 | ·0000 055− | ·0000 037 | ·0000 024 | ·0000 016 | ·0000 011 |
| ·63 | ·0000 142 | ·0000 097 | ·0000 066 | ·0000 045− | ·0000 030 | ·0000 020 |
| ·64 | ·0000 243 | ·0000 168 | ·0000 116 | ·0000 080 | ·0000 055+ | ·0000 038 |
| ·65 | ·0000 413 | ·0000 290 | ·0000 204 | ·0000 143 | ·0000 100 | ·0000 070 |
| ·66 | ·0000 693 | ·0000 494 | ·0000 352 | ·0000 250+ | ·0000 178 | ·0000 126 |
| ·67 | ·0001 148 | ·0000 831 | ·0000 601 | ·0000 433 | ·0000 312 | ·0000 225− |
| ·68 | ·0001 880 | ·0001 381 | ·0001 013 | ·0000 742 | ·0000 542 | ·0000 396 |
| ·69 | ·0003 044 | ·0002 269 | ·0001 688 | ·0001 254 | ·0000 930 | ·0000 689 |
| ·70 | ·0004 873 | ·0003 684 | ·0002 780 | ·0002 095− | ·0001 576 | ·0001 184 |
| ·71 | ·0007 713 | ·0005 913 | ·0004 525− | ·0003 457 | ·0002 638 | ·0002 010 |
| ·72 | ·0012 072 | ·0009 381 | ·0007 278 | ·0005 638 | ·0004 361 | ·0003 369 |
| ·73 | ·0018 680 | ·0014 713 | ·0011 571 | ·0009 085+ | ·0007 123 | ·0005 577 |
| ·74 | ·0028 580 | ·0022 812 | ·0018 180 | ·0014 466 | ·0011 494 | ·0009 120 |
| ·75 | ·0043 231 | ·0034 960 | ·0028 228 | ·0022 758 | ·0018 321 | ·0014 729 |
| ·76 | ·0064 645+ | ·0052 955− | ·0043 311 | ·0035 371 | ·0028 846 | ·0023 492 |
| ·77 | ·0095 553 | ·0079 269 | ·0065 660 | ·0054 308 | ·0044 856 | ·0036 998 |
| ·78 | ·0139 584 | ·0117 246 | ·0098 335+ | ·0082 356 | ·0068 877 | ·0057 528 |
| ·79 | ·0201 474 | ·0171 315+ | ·0145 455+ | ·0123 324 | ·0104 417 | ·0088 292 |
| ·80 | ·0287 267 | ·0247 219 | ·0212 445+ | ·0182 308 | ·0156 236 | ·0133 719 |
| ·81 | ·0404 480 | ·0352 224 | ·0306 282 | ·0265 969 | ·0230 657 | ·0199 779 |
| ·82 | ·0562 198 | ·0495 266 | ·0435 697 | ·0382 780 | ·0335 856 | ·0294 318 |
| ·83 | ·0771 022 | ·0686 980 | ·0611 271 | ·0543 196 | ·0482 095− | ·0427 347 |
| ·84 | ·1042 801 | ·0939 511 | ·0845 340 | ·0759 644 | ·0681 801 | ·0611 211 |
| ·85 | ·1390 037 | ·1266 010 | ·1151 587 | ·1046 222 | ·0949 373 | ·0860 503 |
| ·86 | ·1824 859 | ·1679 694 | ·1544 194 | ·1417 953 | ·1300 549 | ·1191 551 |
| ·87 | ·2357 492 | ·2192 347 | ·2036 406 | ·1889 431 | ·1751 156 | ·1621 287 |
| ·88 | ·2994 152 | ·2812 193 | ·2638 400 | ·2472 718 | ·2315 050+ | ·2165 266 |
| ·89 | ·3734 449 | ·3541 156 | ·3354 439 | ·3174 413 | ·3001 148 | ·2834 676 |
| ·90 | ·4568 509 | ·4371 698 | ·4179 447 | ·3992 002 | ·3809 570 | ·3632 321 |
| ·91 | ·5474 282 | ·5283 693 | ·5095 452 | ·4909 878 | ·4727 260 | ·4547 858 |
| ·92 | ·6415 828 | ·6242 141 | ·6068 708 | ·5895 853 | ·5723 882 | ·5553 084 |
| ·93 | ·7343 684 | ·7196 924 | ·7048 785+ | ·6899 535+ | ·6749 435+ | ·6598 738 |
| ·94 | ·8198 648 | ·8086 150+ | ·7971 375− | ·7854 495+ | ·7735 686 | ·7615 121 |
| ·95 | ·8920 181 | ·8844 557 | ·8766 581 | ·8686 331 | ·8603 889 | ·8519 338 |
| ·96 | ·9459 760 | ·9417 642 | ·9373 757 | ·9328 117 | ·9280 736 | ·9231 632 |
| ·97 | ·9797 353 | ·9779 884 | ·9761 493 | ·9742 166 | ·9721 894 | ·9700 666 |
| ·98 | ·9955 171 | ·9950 925− | ·9946 408 | ·9941 613 | ·9936 531 | ·9931 154 |
| ·99 | ·9997 278 | ·9996 997 | ·9996 694 | ·9996 370 | ·9996 022 | ·9995 651 |
| 1·00 | 1·0000 000 | 1·0000 000 | 1·0000 000 | 1·0000 000 | 1·0000 000 | 1·0000 000 |

$x$ = ·59 to 1·00 $q$ = 4·5 $p$ = 46 to 50

| $x$ | $p$ = 46 | $p$ = 47 | $p$ = 48 | $p$ = 49 | $p$ = 50 |
|---|---|---|---|---|---|
| $B(p, q)$ = | ·3243 0374 × $\frac{1}{10^8}$ | ·2954 0539 × $\frac{1}{10^8}$ | ·2695 9327 × $\frac{1}{10^8}$ | ·2464 8527 × $\frac{1}{10^8}$ | ·2257 5287 × $\frac{1}{10^8}$ |
| ·59 | ·0000 001 | ·0000 001 | | | |
| ·60 | ·0000 002 | ·0000 001 | ·0000 001 | | |
| ·61 | ·0000 004 | ·0000 002 | ·0000 002 | ·0000 001 | ·0000 001 |
| ·62 | ·0000 007 | ·0000 005⁻ | ·0000 003 | ·0000 002 | ·0000 001 |
| ·63 | ·0000 014 | ·0000 009 | ·0000 006 | ·0000 004 | ·0000 003 |
| ·64 | ·0000 026 | ·0000 018 | ·0000 012 | ·0000 008 | ·0000 006 |
| ·65 | ·0000 049 | ·0000 034 | ·0000 024 | ·0000 016 | ·0000 011 |
| ·66 | ·0000 089 | ·0000 063 | ·0000 044 | ·0000 031 | ·0000 022 |
| ·67 | ·0000 161 | ·0000 116 | ·0000 083 | ·0000 059 | ·0000 042 |
| ·68 | ·0000 289 | ·0000 210 | ·0000 153 | ·0000 111 | ·0000 081 |
| ·69 | ·0000 510 | ·0000 377 | ·0000 278 | ·0000 205⁻ | ·0000 151 |
| ·70 | ·0000 888 | ·0000 666 | ·0000 498 | ·0000 372 | ·0000 278 |
| ·71 | ·0001 529 | ·0001 162 | ·0000 882 | ·0000 668 | ·0000 506 |
| ·72 | ·0002 598 | ·0002 002 | ·0001 540 | ·0001 183 | ·0000 908 |
| ·73 | ·0004 361 | ·0003 405⁺ | ·0002 656 | ·0002 069 | ·0001 609 |
| ·74 | ·0007 227 | ·0005 719 | ·0004 520 | ·0003 568 | ·0002 814 |
| ·75 | ·0011 826 | ·0009 482 | ·0007 594 | ·0006 074 | ·0004 853 |
| ·76 | ·0019 107 | ·0015 520 | ·0012 592 | ·0010 204 | ·0008 259 |
| ·77 | ·0030 477 | ·0025 074 | ·0020 604 | ·0016 911 | ·0013 864 |
| ·78 | ·0047 987 | ·0039 978 | ·0033 266 | ·0027 650⁻ | ·0022 956 |
| ·79 | ·0074 563 | ·0062 891 | ·0052 984 | ·0044 587 | ·0037 479 |
| ·80 | ·0114 305⁻ | ·0097 591 | ·0083 225⁻ | ·0070 894 | ·0060 324 |
| ·81 | ·0172 823 | ·0149 327 | ·0128 878 | ·0111 107 | ·0095 684 |
| ·82 | ·0257 607 | ·0225 214 | ·0196 673 | ·0171 564 | ·0149 503 |
| ·83 | ·0378 371 | ·0334 629 | ·0295 619 | ·0260 880 | ·0229 987 |
| ·84 | ·0547 304 | ·0489 537 | ·0437 400 | ·0390 411 | ·0348 121 |
| ·85 | ·0779 087 | ·0704 619 | ·0636 607 | ·0574 582 | ·0518 095⁻ |
| ·86 | ·1090 521 | ·0997 024 | ·0910 630 | ·0830 913 | ·0757 459 |
| ·87 | ·1499 513 | ·1385 509 | ·1278 939 | ·1179 463 | ·1086 737 |
| ·88 | ·2023 204 | ·1888 677 | ·1761 477 | ·1641 380 | ·1528 145⁺ |
| ·89 | ·2674 992 | ·2522 061 | ·2375 816 | ·2236 167 | ·2103 002 |
| ·90 | ·3460 389 | ·3293 877 | ·3132 856 | ·2977 371 | ·2827 438 |
| ·91 | ·4371 905⁺ | ·4199 605⁺ | ·4031 136 | ·3866 650⁺ | ·3706 277 |
| ·92 | ·5383 728 | ·5216 066 | ·5050 330 | ·4886 735⁻ | ·4725 477 |
| ·93 | ·6447 690 | ·6296 526 | ·6145 475⁺ | ·5994 755⁻ | ·5844 573 |
| ·94 | ·7492 976 | ·7369 423 | ·7244 635⁻ | ·7118 781 | ·6992 028 |
| ·95 | ·8432 767 | ·8344 268 | ·8253 932 | ·8161 856 | ·8068 135⁺ |
| ·96 | ·9180 825⁺ | ·9128 340 | ·9074 202 | ·9018 440 | ·8961 085⁻ |
| ·97 | ·9678 473 | ·9655 308 | ·9631 164 | ·9606 037 | ·9579 923 |
| ·98 | ·9925 476 | ·9919 487 | ·9913 181 | ·9906 551 | ·9899 589 |
| ·99 | ·9995 254 | ·9994 832 | ·9994 383 | ·9993 906 | ·9993 400 |
| 1·00 | 1·0000 000 | 1·0000 000 | 1·0000 000 | 1·0000 000 | 1·0000 000 |

$x = \cdot 02$ to $\cdot 60$ $q = 5$ $p = 5$ to $7\cdot 5$

| | $p = 5$ | $p = 5\cdot 5$ | $p = 6$ | $p = 6\cdot 5$ | $p = 7$ | $p = 7\cdot 5$ |
|---|---|---|---|---|---|---|
| $B(p, q) =$ | $\cdot 1587\ 3016 \times \frac{1}{10^2}$ | $\cdot 1108\ 4890 \times \frac{1}{10^2}$ | $\cdot 7936\ 5079 \times \frac{1}{10^3}$ | $\cdot 5806\ 3711 \times \frac{1}{10^3}$ | $\cdot 4329\ 0043 \times \frac{1}{10^3}$ | $\cdot 3281\ 8619 \times \frac{1}{10^3}$ |
| $x$ | | | | | | |
| ·02 | ·0000 004 | ·0000 001 | | | | |
| ·03 | ·0000 028 | ·0000 006 | ·0000 001 | | | |
| ·04 | ·0000 113 | ·0000 029 | ·0000 007 | ·0000 002 | | |
| ·05 | ·0000 332 | ·0000 096 | ·0000 028 | ·0000 008 | ·0000 002 | ·0000 001 |
| ·06 | ·0000 798 | ·0000 254 | ·0000 079 | ·0000 024 | ·0000 007 | ·0000 002 |
| ·07 | ·0001 666 | ·0000 572 | ·0000 193 | ·0000 064 | ·0000 021 | ·0000 007 |
| ·08 | ·0003 136 | ·0001 149 | ·0000 415− | ·0000 147 | ·0000 052 | ·0000 018 |
| ·09 | ·0005 453 | ·0002 119 | ·0000 810 | ·0000 305+ | ·0000 114 | ·0000 042 |
| ·10 | ·0008 909 | ·0003 646 | ·0001 469 | ·0000 584 | ·0000 229 | ·0000 089 |
| ·11 | ·0013 838 | ·0005 936 | ·0002 507 | ·0001 044 | ·0000 429 | ·0000 175− |
| ·12 | ·0020 615− | ·0009 230 | ·0004 069 | ·0001 769 | ·0000 760 | ·0000 323 |
| ·13 | ·0029 649 | ·0013 808 | ·0006 332 | ·0002 864 | ·0001 279 | ·0000 565+ |
| ·14 | ·0041 384 | ·0019 986 | ·0009 505− | ·0004 459 | ·0002 066 | ·0000 947 |
| ·15 | ·0056 287 | ·0028 117 | ·0013 832 | ·0006 713 | ·0003 219 | ·0001 527 |
| ·16 | ·0074 847 | ·0038 587 | ·0019 593 | ·0009 815+ | ·0004 858 | ·0002 379 |
| ·17 | ·0097 568 | ·0051 808 | ·0027 098 | ·0013 985− | ·0007 131 | ·0003 598 |
| ·18 | ·0124 962 | ·0068 224 | ·0036 694 | ·0019 475− | ·0010 214 | ·0005 300 |
| ·19 | ·0157 541 | ·0088 297 | ·0048 757 | ·0026 570 | ·0014 309 | ·0007 625− |
| ·20 | ·0195 814 | ·0112 506 | ·0063 694 | ·0035 589 | ·0019 654 | ·0010 739 |
| ·21 | ·0240 280 | ·0141 343 | ·0081 935+ | ·0046 883 | ·0026 515− | ·0014 839 |
| ·22 | ·0291 417 | ·0175 304 | ·0103 936 | ·0060 831 | ·0035 193 | ·0020 149 |
| ·23 | ·0349 682 | ·0214 888 | ·0130 167 | ·0077 843 | ·0046 020 | ·0026 926 |
| ·24 | ·0415 503 | ·0260 588 | ·0161 116 | ·0098 356 | ·0059 361 | ·0035 460 |
| ·25 | ·0489 273 | ·0312 883 | ·0197 277 | ·0122 827 | ·0075 612 | ·0046 073 |
| ·26 | ·0571 345− | ·0372 238 | ·0239 148 | ·0151 734 | ·0095 196 | ·0059 122 |
| ·27 | ·0662 028 | ·0439 094 | ·0287 224 | ·0185 569 | ·0118 563 | ·0074 993 |
| ·28 | ·0761 583 | ·0513 861 | ·0341 994 | ·0224 834 | ·0146 187 | ·0094 105+ |
| ·29 | ·0870 218 | ·0596 916 | ·0403 932 | ·0270 037 | ·0178 560 | ·0116 907 |
| ·30 | ·0988 087 | ·0688 598 | ·0473 490 | ·0321 685− | ·0216 192 | ·0143 873 |
| ·31 | ·1115 286 | ·0789 198 | ·0551 097 | ·0380 276 | ·0259 599 | ·0175 500+ |
| ·32 | ·1251 852 | ·0898 962 | ·0637 149 | ·0446 299 | ·0309 308 | ·0212 307 |
| ·33 | ·1397 759 | ·1018 081 | ·0732 005− | ·0520 222 | ·0365 839 | ·0254 824 |
| ·34 | ·1552 923 | ·1146 689 | ·0835 979 | ·0602 487 | ·0429 711 | ·0303 595+ |
| ·35 | ·1717 193 | ·1284 861 | ·0949 341 | ·0693 508 | ·0501 427 | ·0359 166 |
| ·36 | ·1890 360 | ·1432 612 | ·1072 304 | ·0793 658 | ·0581 470 | ·0422 081 |
| ·37 | ·2072 151 | ·1589 890 | ·1205 026 | ·0903 267 | ·0670 298 | ·0492 878 |
| ·38 | ·2262 237 | ·1756 580 | ·1347 603 | ·1022 617 | ·0768 336 | ·0572 076 |
| ·39 | ·2460 227 | ·1932 500+ | ·1500 068 | ·1151 933 | ·0875 966 | ·0660 176 |
| ·40 | ·2665 677 | ·2117 404 | ·1662 386 | ·1291 382 | ·0993 526 | ·0757 644 |
| ·41 | ·2878 090 | ·2310 979 | ·1834 452 | ·1441 064 | ·1121 300 | ·0864 912 |
| ·42 | ·3096 920 | ·2512 848 | ·2016 092 | ·1601 012 | ·1259 511 | ·0982 366 |
| ·43 | ·3321 576 | ·2722 571 | ·2207 058 | ·1771 186 | ·1408 320 | ·1110 338 |
| ·44 | ·3551 423 | ·2939 650+ | ·2407 033 | ·1951 471 | ·1567 813 | ·1249 104 |
| ·45 | ·3785 791 | ·3163 528 | ·2615 627 | ·2141 676 | ·1738 004 | ·1398 868 |
| ·46 | ·4023 977 | ·3393 595+ | ·2832 382 | ·2341 530 | ·1918 826 | ·1559 764 |
| ·47 | ·4265 251 | ·3629 191 | ·3056 772 | ·2550 683 | ·2110 130 | ·1731 846 |
| ·48 | ·4508 861 | ·3869 611 | ·3288 205− | ·2768 707 | ·2311 679 | ·1915 083 |
| ·49 | ·4754 037 | ·4114 111 | ·3526 028 | ·2995 097 | ·2523 153 | ·2109 355+ |
| ·50 | ·5000 000 | ·4361 909 | ·3769 531 | ·3229 271 | ·2744 141 | ·2314 449 |
| ·51 | ·5245 963 | ·4612 197 | ·4017 953 | ·3470 575+ | ·2974 144 | ·2530 054 |
| ·52 | ·5491 139 | ·4864 143 | ·4270 483 | ·3718 286 | ·3212 580 | ·2755 766 |
| ·53 | ·5734 749 | ·5116 897 | ·4526 270 | ·3971 615− | ·3458 780 | ·2991 077 |
| ·54 | ·5976 023 | ·5369 600 | ·4784 429 | ·4229 716 | ·3711 994 | ·3235 386 |
| ·55 | ·6214 209 | ·5621 389 | ·5044 046 | ·4491 689 | ·3971 396 | ·3487 993 |
| ·56 | ·6448 577 | ·5871 404 | ·5304 187 | ·4756 587 | ·4236 090 | ·3748 107 |
| ·57 | ·6678 424 | ·6118 794 | ·5563 906 | ·5023 426 | ·4505 114 | ·4014 846 |
| ·58 | ·6903 080 | ·6362 726 | ·5822 251 | ·5291 191 | ·4777 449 | ·4287 246 |
| ·59 | ·7121 910 | ·6602 391 | ·6078 272 | ·5558 845+ | ·5052 028 | ·4564 267 |
| ·60 | ·7334 323 | ·6837 007 | ·6331 033 | ·5825 339 | ·5327 742 | ·4844 800 |

$x$ = ·61 to 1·00 $q = 5$ $p$ = 5 to 7·5

| $x$ | $p = 5$ | $p = 5·5$ | $p = 6$ | $p = 6·5$ | $p = 7$ | $p = 7·5$ |
|---|---|---|---|---|---|---|
| $B(p, q) =$ | $·1587\ 3016 \times \frac{1}{10^2}$ | $·1108\ 4890 \times \frac{1}{10^2}$ | $·7936\ 5079 \times \frac{1}{10^3}$ | $·5806\ 3711 \times \frac{1}{10^3}$ | $·4329\ 0043 \times \frac{1}{10^3}$ | $·3281\ 8619 \times \frac{1}{10^3}$ |
| ·61 | ·7539 773 | ·7065 830 | ·6579 615+ | ·6089 621 | ·5603 455− | ·5127 679 |
| ·62 | ·7737 763 | ·7288 159 | ·6823 130 | ·6350 644 | ·5878 009 | ·5411 687 |
| ·63 | ·7927 849 | ·7503 340 | ·7060 723 | ·6607 382 | ·6150 242 | ·5695 573 |
| ·64 | ·8109 640 | ·7710 771 | ·7291 585− | ·6858 833 | ·6418 992 | ·5978 059 |
| ·65 | ·8282 807 | ·7909 911 | ·7514 955+ | ·7104 032 | ·6683 115+ | ·6257 860 |
| ·66 | ·8447 077 | ·8100 277 | ·7730 134 | ·7342 063 | ·6941 496 | ·6533 689 |
| ·67 | ·8602 241 | ·8281 458 | ·7936 486 | ·7572 064 | ·7193 060 | ·6804 283 |
| ·68 | ·8748 148 | ·8453 106 | ·8133 446 | ·7793 242 | ·7436 783 | ·7068 406 |
| ·69 | ·8884 714 | ·8614 950− | ·8320 525+ | ·8004 873 | ·7671 708 | ·7324 873 |
| ·70 | ·9011 913 | ·8766 789 | ·8497 317 | ·8206 321 | ·7896 954 | ·7572 560 |
| ·71 | ·9129 782 | ·8908 500− | ·8663 497 | ·8397 034 | ·8111 725− | ·7810 421 |
| ·72 | ·9238 417 | ·9040 031 | ·8818 829 | ·8576 558 | ·8315 324 | ·8037 502 |
| ·73 | ·9337 972 | ·9161 408 | ·8963 169 | ·8744 538 | ·8507 158 | ·8252 952 |
| ·74 | ·9428 655+ | ·9272 727 | ·9096 458 | ·8900 724 | ·8686 748 | ·8456 036 |
| ·75 | ·9510 727 | ·9374 159 | ·9218 731 | ·9044 972 | ·8853 736 | ·8646 149 |
| ·76 | ·9584 497 | ·9465 941 | ·9330 110 | ·9177 247 | ·9007 887 | ·8822 820 |
| ·77 | ·9650 318 | ·9548 373 | ·9430 804 | ·9297 621 | ·9149 093 | ·8985 724 |
| ·78 | ·9708 583 | ·9621 817 | ·9521 103 | ·9406 271 | ·9277 377 | ·9134 683 |
| ·79 | ·9759 720 | ·9686 690 | ·9601 376 | ·9503 479 | ·9392 889 | ·9269 674 |
| ·80 | ·9804 186 | ·9743 458 | ·9672 065+ | ·9589 624 | ·9495 904 | ·9390 824 |
| ·81 | ·9842 459 | ·9792 626 | ·9733 675+ | ·9665 176 | ·9586 818 | ·9498 412 |
| ·82 | ·9875 038 | ·9834 739 | ·9786 771 | ·9730 688 | ·9666 138 | ·9592 861 |
| ·83 | ·9902 432 | ·9870 364 | ·9831 962 | ·9786 789 | ·9734 478 | ·9674 733 |
| ·84 | ·9925 153 | ·9900 092 | ·9869 899 | ·9834 168 | ·9792 542 | ·9744 715− |
| ·85 | ·9943 713 | ·9924 520 | ·9901 259 | ·9873 568 | ·9841 116 | ·9803 606 |
| ·86 | ·9958 616 | ·9944 251 | ·9926 737 | ·9905 766 | ·9881 044 | ·9852 302 |
| ·87 | ·9970 351 | ·9959 877 | ·9947 033 | ·9931 564 | ·9913 223 | ·9891 774 |
| ·88 | ·9979 385+ | ·9971 976 | ·9962 839 | ·9951 771 | ·9938 572 | ·9923 048 |
| ·89 | ·9986 162 | ·9981 104 | ·9974 830 | ·9967 188 | ·9958 022 | ·9947 181 |
| ·90 | ·9991 091 | ·9987 780 | ·9983 651 | ·9978 592 | ·9972 490 | ·9965 233 |
| ·91 | ·9994 547 | ·9992 487 | ·9989 904 | ·9986 722 | ·9982 862 | ·9978 246 |
| ·92 | ·9996 864 | ·9995 661 | ·9994 143 | ·9992 263 | ·9989 970 | ·9987 213 |
| ·93 | ·9998 334 | ·9997 684 | ·9996 861 | ·9995 835− | ·9994 577 | ·9993 056 |
| ·94 | ·9999 202 | ·9998 885+ | ·9998 483 | ·9997 978 | ·9997 356 | ·9996 600 |
| ·95 | ·9999 668 | ·9999 534 | ·9999 363 | ·9999 148 | ·9998 881 | ·9998 554 |
| ·96 | ·9999 887 | ·9999 841 | ·9999 782 | ·9999 707 | ·9999 614 | ·9999 499 |
| ·97 | ·9999 972 | ·9999 961 | ·9999 946 | ·9999 927 | ·9999 904 | ·9999 874 |
| ·98 | ·9999 996 | ·9999 995− | ·9999 993 | ·9999 990 | ·9999 987 | ·9999 983 |
| ·99 | 1·0000 000 | 1·0000 000 | 1·0000 000 | 1·0000 000 | 1·0000 000 | ·9999 999 |
| 1·00 | | | | | | 1·0000 000 |

$x =$ ·06 to ·60 $q = 5$ $p = 8$ to 10·5

| | $p = 8$ | $p = 8{\cdot}5$ | $p = 9$ | $p = 9{\cdot}5$ | $p = 10$ | $p = 10{\cdot}5$ |
|---|---|---|---|---|---|---|
| $B(p, q) =$ | ·2525 2525+ $\times \frac{1}{10^3}$ | ·1969 1172 $\times \frac{1}{10^3}$ | ·1554 0016 $\times \frac{1}{10^3}$ | ·1239 8145+ $\times \frac{1}{10^3}$ | ·9990 0100 $\times \frac{1}{10^4}$ | ·8122 9227 $\times \frac{1}{10^4}$ |
| $x$ | | | | | | |
| ·06 | ·0000 001 | | | | | |
| ·07 | ·0000 002 | ·0000 001 | | | | |
| ·08 | ·0000 006 | ·0000 002 | ·0000 001 | | | |
| ·09 | ·0000 015+ | ·0000 006 | ·0000 002 | ·0000 001 | | |
| ·10 | ·0000 034 | ·0000 013 | ·0000 005− | ·0000 002 | ·0000 001 | |
| ·11 | ·0000 070 | ·0000 028 | ·0000 011 | ·0000 004 | ·0000 002 | ·0000 001 |
| ·12 | ·0000 136 | ·0000 057 | ·0000 023 | ·0000 010 | ·0000 004 | ·0000 002 |
| ·13 | ·0000 247 | ·0000 107 | ·0000 046 | ·0000 020 | ·0000 008 | ·0000 004 |
| ·14 | ·0000 430 | ·0000 193 | ·0000 086 | ·0000 038 | ·0000 017 | ·0000 007 |
| ·15 | ·0000 717 | ·0000 334 | ·0000 154 | ·0000 071 | ·0000 032 | ·0000 015− |
| ·16 | ·0001 153 | ·0000 554 | ·0000 264 | ·0000 125+ | ·0000 059 | ·0000 027 |
| ·17 | ·0001 797 | ·0000 890 | ·0000 437 | ·0000 213 | ·0000 103 | ·0000 050− |
| ·18 | ·0002 723 | ·0001 387 | ·0000 701 | ·0000 352 | ·0000 175+ | ·0000 087 |
| ·19 | ·0004 024 | ·0002 105+ | ·0001 093 | ·0000 563 | ·0000 288 | ·0000 147 |
| ·20 | ·0005 812 | ·0003 119 | ·0001 660 | ·0000 877 | ·0000 460 | ·0000 240 |
| ·21 | ·0008 226 | ·0004 521 | ·0002 465+ | ·0001 334 | ·0000 718 | ·0000 383 |
| ·22 | ·0011 427 | ·0006 425+ | ·0003 585− | ·0001 985+ | ·0001 092 | ·0000 597 |
| ·23 | ·0015 607 | ·0008 969 | ·0005 114 | ·0002 895+ | ·0001 628 | ·0000 910 |
| ·24 | ·0020 985+ | ·0012 314 | ·0007 170 | ·0004 145+ | ·0002 381 | ·0001 359 |
| ·25 | ·0027 815+ | ·0016 651 | ·0009 891 | ·0005 834 | ·0003 419 | ·0001 991 |
| ·26 | ·0036 381 | ·0022 200 | ·0013 443 | ·0008 083 | ·0004 829 | ·0002 867 |
| ·27 | ·0047 002 | ·0029 214 | ·0018 020 | ·0011 037 | ·0006 717 | ·0004 063 |
| ·28 | ·0060 031 | ·0037 979 | ·0023 846 | ·0014 868 | ·0009 211 | ·0005 672 |
| ·29 | ·0075 855+ | ·0048 815+ | ·0031 178 | ·0019 776 | ·0012 464 | ·0007 809 |
| ·30 | ·0094 894 | ·0062 080 | ·0040 310 | ·0025 994 | ·0016 657 | ·0010 611 |
| ·31 | ·0117 598 | ·0078 164 | ·0051 568 | ·0033 790 | ·0022 001 | ·0014 242 |
| ·32 | ·0144 450+ | ·0097 495− | ·0065 319 | ·0043 465+ | ·0028 742 | ·0018 896 |
| ·33 | ·0175 957 | ·0120 533 | ·0081 965− | ·0055 363 | ·0037 162 | ·0024 801 |
| ·34 | ·0212 648 | ·0147 773 | ·0101 946 | ·0069 862 | ·0047 579 | ·0032 218 |
| ·35 | ·0255 075− | ·0179 736 | ·0125 739 | ·0087 382 | ·0060 353 | ·0041 448 |
| ·36 | ·0303 799 | ·0216 971 | ·0153 856 | ·0108 384 | ·0075 886 | ·0052 832 |
| ·37 | ·0359 393 | ·0260 050+ | ·0186 840 | ·0133 365− | ·0094 620 | ·0066 755+ |
| ·38 | ·0422 430 | ·0309 562 | ·0225 264 | ·0162 862 | ·0117 042 | ·0083 644 |
| ·39 | ·0493 480 | ·0366 105− | ·0269 726 | ·0197 446 | ·0143 677 | ·0103 974 |
| ·40 | ·0573 099 | ·0430 284 | ·0320 843 | ·0237 720 | ·0175 095+ | ·0128 263 |
| ·41 | ·0661 826 | ·0502 704 | ·0379 249 | ·0284 314 | ·0211 901 | ·0157 074 |
| ·42 | ·0760 168 | ·0583 957 | ·0445 582 | ·0337 882 | ·0254 733 | ·0191 014 |
| ·43 | ·0868 601 | ·0674 620 | ·0520 483 | ·0399 092 | ·0304 262 | ·0230 730 |
| ·44 | ·0987 553 | ·0775 240 | ·0604 581 | ·0468 622 | ·0361 181 | ·0276 905+ |
| ·45 | ·1117 400 | ·0886 330 | ·0698 492 | ·0547 151 | ·0426 201 | ·0330 255− |
| ·46 | ·1258 456 | ·1008 359 | ·0802 801 | ·0635 349 | ·0500 043 | ·0391 519 |
| ·47 | ·1410 967 | ·1141 738 | ·0918 057 | ·0733 869 | ·0583 426 | ·0461 455+ |
| ·48 | ·1575 100 | ·1286 816 | ·1044 764 | ·0843 333 | ·0677 063 | ·0540 831 |
| ·49 | ·1750 940 | ·1443 868 | ·1183 363 | ·0964 325+ | ·0781 644 | ·0630 412 |
| ·50 | ·1938 477 | ·1613 085− | ·1334 229 | ·1097 376 | ·0897 827 | ·0730 951 |
| ·51 | ·2137 607 | ·1794 567 | ·1497 655+ | ·1242 952 | ·1026 224 | ·0843 176 |
| ·52 | ·2348 123 | ·1988 316 | ·1673 846 | ·1401 443 | ·1167 390 | ·0967 775+ |
| ·53 | ·2569 713 | ·2194 227 | ·1862 905+ | ·1573 152 | ·1321 804 | ·1105 385+ |
| ·54 | ·2801 956 | ·2412 081 | ·2064 825+ | ·1758 280 | ·1489 863 | ·1256 574 |
| ·55 | ·3044 320 | ·2641 544 | ·2279 482 | ·1956 915+ | ·1671 861 | ·1421 824 |
| ·56 | ·3296 164 | ·2882 158 | ·2506 627 | ·2169 024 | ·1867 979 | ·1601 520 |
| ·57 | ·3556 739 | ·3133 346 | ·2745 878 | ·2394 442 | ·2078 269 | ·1795 930 |
| ·58 | ·3825 187 | ·3394 403 | ·2996 720 | ·2632 861 | ·2302 648 | ·2005 190 |
| ·59 | ·4100 553 | ·3664 504 | ·3258 498 | ·2883 828 | ·2540 880 | ·2229 293 |
| ·60 | ·4381 782 | ·3942 704 | ·3530 418 | ·3146 737 | ·2792 570 | ·2468 070 |

$x = \cdot 61$ to $1\cdot 00$ $q = 5$ $p = 8$ to $10\cdot 5$

| $x$ | $p = 8$ | $p = 8\cdot 5$ | $p = 9$ | $p = 9\cdot 5$ | $p = 10$ | $p = 10\cdot 5$ |
|---|---|---|---|---|---|---|
| $B(p, q) =$ | $\cdot 2525\ 2525 \overset{+}{\times} \frac{1}{10^3}$ | $\cdot 1969\ 1172 \times \frac{1}{10^3}$ | $\cdot 1554\ 0016 \times \frac{1}{10^3}$ | $\cdot 1239\ 8145 \overset{+}{\times} \frac{1}{10^3}$ | $\cdot 9990\ 0100 \times \frac{1}{10^4}$ | $\cdot 8122\ 9227 \times \frac{1}{10^4}$ |
| ·61 | ·4667 735⁻ | ·4227 942 | ·3811 551 | ·3420 826 | ·3057 158 | ·2721 184 |
| ·62 | ·4957 191 | ·4519 052 | ·4100 831 | ·3705 178 | ·3333 912 | ·2988 113 |
| ·63 | ·5248 865⁻ | ·4814 765⁻ | ·4397 064 | ·3998 722 | ·3621 925⁻ | ·3268 149 |
| ·64 | ·5541 413 | ·5113 727 | ·4698 937 | ·4300 238 | ·3920 115⁻ | ·3560 391 |
| ·65 | ·5833 450⁺ | ·5414 508 | ·5005 027 | ·4608 362 | ·4227 230 | ·3863 739 |
| ·66 | ·6123 567 | ·5715 616 | ·5313 816 | ·4921 603 | ·4541 854 | ·4176 902 |
| ·67 | ·6410 338 | ·6015 515⁻ | ·5623 703 | ·5238 347 | ·4862 415⁺ | ·4498 401 |
| ·68 | ·6692 347 | ·6312 644 | ·5933 026 | ·5556 881 | ·5187 203 | ·4826 576 |
| ·69 | ·6968 205⁺ | ·6605 432 | ·6240 079 | ·5875 411 | ·5514 381 | ·5159 606 |
| ·70 | ·7236 555⁻ | ·6892 323 | ·6543 136 | ·6192 080 | ·5842 012 | ·5495 522 |
| ·71 | ·7496 105⁺ | ·7171 796 | ·6840 471 | ·6504 996 | ·6168 081 | ·5832 231 |
| ·72 | ·7745 643 | ·7442 385⁺ | ·7130 388 | ·6812 262 | ·6490 522 | ·6167 545⁺ |
| ·73 | ·7984 048 | ·7702 704 | ·7411 243 | ·7111 996 | ·6807 252 | ·6499 214 |
| ·74 | ·8210 314 | ·7951 465⁻ | ·7681 472 | ·7402 372 | ·7116 199 | ·6824 957 |
| ·75 | ·8423 563 | ·8187 503 | ·7939 619 | ·7681 643 | ·7415 346 | ·7142 503 |
| ·76 | ·8623 060 | ·8409 795⁺ | ·8184 357 | ·7948 179 | ·7702 759 | ·7449 633 |
| ·77 | ·8808 224 | ·8617 478 | ·8414 519 | ·8200 494 | ·7976 633 | ·7744 225⁻ |
| ·78 | ·8978 640 | ·8809 865⁻ | ·8629 117 | ·8437 278 | ·8235 322 | ·8024 295⁺ |
| ·79 | ·9134 068 | ·8986 458 | ·8827 365⁻ | ·8657 427 | ·8477 382 | ·8288 047 |
| ·80 | ·9274 445ᵉ | ·9146 962 | ·9008 694 | ·8860 067 | ·8701 604 | ·8533 911 |
| ·81 | ·9399 888 | ·9291 291 | ·9172 769 | ·9044 573 | ·8907 040 | ·8760 588 |
| ·82 | ·9510 694 | ·9419 567 | ·9319 498 | ·9210 594 | ·9093 037 | ·8967 085⁺ |
| ·83 | ·9607 331 | ·9532 123 | ·9449 033 | ·9358 056 | ·9259 252 | ·9152 748 |
| ·84 | ·9690 432 | ·9629 499 | ·9561 774 | ·9487 173 | ·9405 668 | ·9317 281 |
| ·85 | ·9760 781 | ·9712 422 | ·9658 354 | ·9598 443 | ·9532 597 | ·9460 768 |
| ·86 | ·9819 293 | ·9781 801 | ·9739 635⁻ | ·9692 637 | ·9640 681 | ·9583 670 |
| ·87 | ·9866 999 | ·9838 694 | ·9806 677 | ·9770 783 | ·9730 872 | ·9686 824 |
| ·88 | ·9905 014 | ·9884 292 | ·9860 717 | ·9834 136 | ·9804 411 | ·9771 416 |
| ·89 | ·9934 514 | ·9919 878 | ·9903 132 | ·9884 143 | ·9862 788 | ·9838 949 |
| ·90 | ·9956 707 | ·9946 798 | ·9935 398 | ·9922 399 | ·9907 698 | ·9891 195⁻ |
| ·91 | ·9972 793 | ·9966 420 | ·9959 047 | ·9950 594 | ·9940 980 | ·9930 128 |
| ·92 | ·9983 938 | ·9980 090 | ·9975 614 | ·9970 453 | ·9964 551 | ·9957 853 |
| ·93 | ·9991 240 | ·9989 094 | ·9986 584 | ·9983 675⁺ | ·9980 331 | ·9976 514 |
| ·94 | ·9995 692 | ·9994 614 | ·9993 346 | ·9991 868 | ·9990 160 | ·9988 200 |
| ·95 | ·9998 161 | ·9997 690 | ·9997 134 | ·9996 483 | ·9995 726 | ·9994 853 |
| ·96 | ·9999 360 | ·9999 193 | ·9998 994 | ·9998 760 | ·9998 487 | ·9998 170 |
| ·97 | ·9999 839 | ·9999 796 | ·9999 744 | ·9999 684 | ·9999 612 | ·9999 529 |
| ·98 | ·9999 977 | ·9999 971 | ·9999 964 | ·9999 955⁺ | ·9999 945⁻ | ·9999 933 |
| ·99 | ·9999 999 | ·9999 999 | ·9999 999 | ·9999 998 | ·9999 998 | ·9999 998 |
| 1·00 | 1·0000 000 | 1·0000 000 | 1·0000 000 | 1·0000 000 | 1·0000 000 | 1·0000 000 |

$x$ = ·12 to ·70 $q = 5$ $p$ = 11 to 16

| | $p = 11$ | $p = 12$ | $p = 13$ | $p = 14$ | $p = 15$ | $p = 16$ |
|---|---|---|---|---|---|---|
| $B(p, q)$ = | ·6660 0067 × $\frac{1}{10^4}$ | ·4578 7546 × $\frac{1}{10^4}$ | ·3232 0621 × $\frac{1}{10^4}$ | ·2334 2670 × $\frac{1}{10^4}$ | ·1719 9862 × $\frac{1}{10^4}$ | ·1289 9897 × $\frac{1}{10^4}$ |
| $x$ | | | | | | |
| ·12 | ·0000 001 | | | | | |
| ·13 | ·0000 001 | | | | | |
| ·14 | ·0000 003 | ·0000 001 | | | | |
| ·15 | ·0000 007 | ·0000 001 | | | | |
| ·16 | ·0000 013 | ·0000 003 | ·0000 001 | | | |
| ·17 | ·0000 024 | ·0000 005+ | ·0000 001 | | | |
| ·18 | ·0000 043 | ·0000 010 | ·0000 002 | ·0000 001 | | |
| ·19 | ·0000 074 | ·0000 019 | ·0000 005− | ·0000 001 | | |
| ·20 | ·0000 125− | ·0000 033 | ·0000 009 | ·0000 002 | ·0000 001 | |
| ·21 | ·0000 204 | ·0000 057 | ·0000 015+ | ·0000 004 | ·0000 001 | |
| ·22 | ·0000 325− | ·0000 095− | ·0000 027 | ·0000 008 | ·0000 002 | ·0000 001 |
| ·23 | ·0000 506 | ·0000 154 | ·0000 046 | ·0000 014 | ·0000 004 | ·0000 001 |
| ·24 | ·0000 771 | ·0000 245− | ·0000 076 | ·0000 023 | ·0000 007 | ·0000 002 |
| ·25 | ·0001 153 | ·0000 381 | ·0000 124 | ·0000 039 | ·0000 012 | ·0000 004 |
| ·26 | ·0001 693 | ·0000 582 | ·0000 196 | ·0000 065+ | ·0000 021 | ·0000 007 |
| ·27 | ·0002 444 | ·0000 871 | ·0000 305+ | ·0000 105+ | ·0000 036 | ·0000 012 |
| ·28 | ·0003 474 | ·0001 284 | ·0000 466 | ·0000 166 | ·0000 059 | ·0000 020 |
| ·29 | ·0004 866 | ·0001 861 | ·0000 699 | ·0000 259 | ·0000 094 | ·0000 034 |
| ·30 | ·0006 722 | ·0002 658 | ·0001 033 | ·0000 395− | ·0000 149 | ·0000 056 |
| ·31 | ·0009 169 | ·0003 745− | ·0001 502 | ·0000 594 | ·0000 231 | ·0000 089 |
| ·32 | ·0012 356 | ·0005 206 | ·0002 155+ | ·0000 878 | ·0000 353 | ·0000 140 |
| ·33 | ·0016 463 | ·0007 148 | ·0003 050+ | ·0001 281 | ·0000 531 | ·0000 217 |
| ·34 | ·0021 700 | ·0009 702 | ·0004 263 | ·0001 844 | ·0000 787 | ·0000 332 |
| ·35 | ·0028 314 | ·0013 023 | ·0005 887 | ·0002 621 | ·0001 151 | ·0000 499 |
| ·36 | ·0036 589 | ·0017 298 | ·0008 038 | ·0003 679 | ·0001 661 | ·0000 741 |
| ·37 | ·0046 850+ | ·0022 748 | ·0010 858 | ·0005 105− | ·0002 368 | ·0001 085+ |
| ·38 | ·0059 467 | ·0029 633 | ·0014 517 | ·0007 006 | ·0003 336 | ·0001 570 |
| ·39 | ·0074 855+ | ·0038 254 | ·0019 222 | ·0009 515+ | ·0004 648 | ·0002 244 |
| ·40 | ·0093 477 | ·0048 957 | ·0025 214 | ·0012 794 | ·0006 407 | ·0003 170 |
| ·41 | ·0115 843 | ·0062 138 | ·0032 779 | ·0017 038 | ·0008 741 | ·0004 432 |
| ·42 | ·0142 514 | ·0078 243 | ·0042 252 | ·0022 484 | ·0011 809 | ·0006 130 |
| ·43 | ·0174 098 | ·0097 774 | ·0054 015− | ·0029 409 | ·0015 805+ | ·0008 396 |
| ·44 | ·0211 247 | ·0121 287 | ·0068 510 | ·0038 143 | ·0020 964 | ·0011 389 |
| ·45 | ·0254 659 | ·0149 394 | ·0086 235− | ·0049 068 | ·0027 564 | ·0015 307 |
| ·46 | ·0305 067 | ·0182 764 | ·0107 751 | ·0062 628 | ·0035 941 | ·0020 391 |
| ·47 | ·0363 239 | ·0222 119 | ·0133 684 | ·0079 330 | ·0046 485+ | ·0026 931 |
| ·48 | ·0429 969 | ·0268 234 | ·0164 724 | ·0099 751 | ·0059 654 | ·0035 275− |
| ·49 | ·0506 066 | ·0321 929 | ·0201 626 | ·0124 540 | ·0075 976 | ·0045 833 |
| ·50 | ·0592 346 | ·0384 064 | ·0245 209 | ·0154 419 | ·0096 054 | ·0059 090 |
| ·51 | ·0689 623 | ·0455 532 | ·0296 350− | ·0190 187 | ·0120 575+ | ·0075 606 |
| ·52 | ·0798 689 | ·0537 247 | ·0355 980 | ·0232 719 | ·0150 310 | ·0096 030 |
| ·53 | ·0920 308 | ·0630 135− | ·0425 080 | ·0282 960 | ·0186 116 | ·0121 102 |
| ·54 | ·1055 192 | ·0735 116 | ·0504 661 | ·0341 925− | ·0228 939 | ·0151 657 |
| ·55 | ·1203 993 | ·0853 092 | ·0595 764 | ·0410 686 | ·0279 810 | ·0188 633 |
| ·56 | ·1367 278 | ·0984 926 | ·0699 436 | ·0490 369 | ·0339 841 | ·0233 066 |
| ·57 | ·1545 517 | ·1131 424 | ·0816 713 | ·0582 132 | ·0410 218 | ·0286 096 |
| ·58 | ·1739 062 | ·1293 316 | ·0948 606 | ·0687 157 | ·0492 190 | ·0348 955+ |
| ·59 | ·1948 127 | ·1471 230 | ·1096 072 | ·0806 623 | ·0587 055− | ·0422 964 |
| ·60 | ·2172 777 | ·1665 674 | ·1259 991 | ·0941 686 | ·0696 137 | ·0509 520 |
| ·61 | ·2412 907 | ·1877 007 | ·1441 141 | ·1093 455− | ·0820 769 | ·0610 074 |
| ·62 | ·2668 227 | ·2105 420 | ·1640 167 | ·1262 954 | ·0962 261 | ·0726 117 |
| ·63 | ·2938 252 | ·2350 915+ | ·1857 552 | ·1451 097 | ·1121 868 | ·0859 144 |
| ·64 | ·3222 290 | ·2613 279 | ·2093 590 | ·1658 650+ | ·1300 757 | ·1010 625+ |
| ·65 | ·3519 434 | ·2892 070 | ·2348 354 | ·1886 196 | ·1499 964 | ·1181 966 |
| ·66 | ·3828 561 | ·3186 597 | ·2621 669 | ·2134 092 | ·1720 349 | ·1374 459 |
| ·67 | ·4148 327 | ·3495 910 | ·2913 085− | ·2402 440 | ·1962 555+ | ·1589 240 |
| ·68 | ·4477 179 | ·3818 793 | ·3221 856 | ·2691 042 | ·2226 958 | ·1827 227 |
| ·69 | ·4813 356 | ·4153 755+ | ·3546 922 | ·2999 373 | ·2513 618 | ·2089 068 |
| ·70 | ·5154 911 | ·4499 041 | ·3886 896 | ·3326 549 | ·2822 235+ | ·2375 078 |

$x = \cdot 71$ to $1\cdot 00$ $q = 5$ $p = 11$ to $16$

| | $p = 11$ | $p = 12$ | $p = 13$ | $p = 14$ | $p = 15$ | $p = 16$ |
|---|---|---|---|---|---|---|
| $B(p, q) =$ | ·6660 0067 × $\frac{1}{10^5}$ | ·4578 7546 × $\frac{1}{10^5}$ | ·3232 0621 × $\frac{1}{10^5}$ | ·2334 2670 × $\frac{1}{10^5}$ | ·1719 9862 × $\frac{1}{10^5}$ | ·1289 9897 × $\frac{1}{10^5}$ |
| $x$ | | | | | | |
| ·71 | ·5499 725+ | ·4852 636 | ·4240 057 | ·3671 302 | ·3152 110 | ·2685 183 |
| ·72 | ·5845 538 | ·5212 280 | ·4604 353 | ·4031 966 | ·3502 100 | ·3018 862 |
| ·73 | ·6189 973 | ·5575 499 | ·4977 412 | ·4406 469 | ·3870 598 | ·3375 096 |
| ·74 | ·6530 577 | ·5939 630 | ·5356 562 | ·4792 333 | ·4255 508 | ·3752 325− |
| ·75 | ·6864 859 | ·6301 862 | ·5738 864 | ·5186 693 | ·4654 243 | ·4148 415+ |
| ·76 | ·7190 339 | ·6659 286 | ·6121 151 | ·5586 329 | ·5063 730 | ·4560 642 |
| ·77 | ·7504 591 | ·7008 947 | ·6500 086 | ·5987 702 | ·5480 442 | ·4985 694 |
| ·78 | ·7805 297 | ·7347 907 | ·6872 221 | ·6387 022 | ·5900 436 | ·5419 690 |
| ·79 | ·8090 302 | ·7673 310 | ·7234 079 | ·6780 320 | ·6319 430 | ·5858 233 |
| ·80 | ·8357 663 | ·7982 454 | ·7582 232 | ·7163 538 | ·6732 881 | ·6296 483 |
| ·81 | ·8605 703 | ·8272 863 | ·7913 396 | ·7532 637 | ·7136 103 | ·6729 260 |
| ·82 | ·8833 060 | ·8542 358 | ·8224 525− | ·7883 709 | ·7524 392 | ·7151 181 |
| ·83 | ·9038 726 | ·8789 131 | ·8512 912 | ·8213 109 | ·7893 176 | ·7556 819 |
| ·84 | ·9222 087 | ·9011 804 | ·8776 286 | ·8517 580 | ·8238 177 | ·7940 892 |
| ·85 | ·9382 946 | ·9209 487 | ·9012 900 | ·8794 386 | ·8555 581 | ·8298 468 |
| ·86 | ·9521 541 | ·9381 822 | ·9221 610 | ·9041 434 | ·8842 211 | ·8625 191 |
| ·87 | ·9638 542 | ·9529 005+ | ·9401 942 | ·9257 383 | ·9095 684 | ·8917 492 |
| ·88 | ·9735 041 | ·9651 798 | ·9554 126 | ·9441 729 | ·9314 559 | ·9172 806 |
| ·89 | ·9812 519 | ·9751 511 | ·9679 115+ | ·9594 857 | ·9498 442 | ·9389 750+ |
| ·90 | ·9872 795+ | ·9829 960 | ·9778 558 | ·9718 061 | ·9648 058 | ·9568 255+ |
| ·91 | ·9917 963 | ·9889 400 | ·9854 744 | ·9813 503 | ·9765 251 | ·9709 633 |
| ·92 | ·9950 303 | ·9932 428 | ·9910 501 | ·9884 122 | ·9852 918 | ·9816 556 |
| ·93 | ·9972 188 | ·9961 862 | ·9949 058 | ·9933 486 | ·9914 866 | ·9892 932 |
| ·94 | ·9985 967 | ·9980 593 | ·9973 858 | ·9965 579 | ·9955 573 | ·9943 659 |
| ·95 | ·9993 853 | ·9991 427 | ·9988 354 | ·9984 536 | ·9979 872 | ·9974 261 |
| ·96 | ·9997 806 | ·9996 914 | ·9995 772 | ·9994 338 | ·9992 569 | ·9990 417 |
| ·97 | ·9999 433 | ·9999 196 | ·9998 889 | ·9998 500− | ·9998 014 | ·9997 418 |
| ·98 | ·9999 919 | ·9999 884 | ·9999 838 | ·9999 779 | ·9999 706 | ·9999 614 |
| ·99 | ·9999 997 | ·9999 996 | ·9999 994 | ·9999 992 | ·9999 990 | ·9999 986 |
| 1·00 | 1·0000 000 | 1·0000 000 | 1·0000 000 | 1·0000 000 | 1·0000 000 | 1·0000 000 |

$x = \cdot 24$ to $\cdot 80$ $q = 5$ $p = 17$ to $22$

| | $p = 17$ | $p = 18$ | $p = 19$ | $p = 20$ | $p = 21$ | $p = 22$ |
|---|---|---|---|---|---|---|
| $B(p, q) =$ | $\cdot 9828\ 4928 \times \frac{1}{10^5}$ | $\cdot 7594\ 7444 \times \frac{1}{10^5}$ | $\cdot 5943\ 7130 \times \frac{1}{10^5}$ | $\cdot 4705\ 4395 \times \frac{1}{10^5}$ | $\cdot 3764\ 3516 \times \frac{1}{10^5}$ | $\cdot 3040\ 4378 \times \frac{1}{10^5}$ |
| $x$ | | | | | | |
| ·24 | ·0000 001 | | | | | |
| ·25 | ·0000 001 | | | | | |
| ·26 | ·0000 002 | ·0000 001 | | | | |
| ·27 | ·0000 004 | ·0000 001 | | | | |
| ·28 | ·0000 007 | ·0000 002 | ·0000 001 | | | |
| ·29 | ·0000 012 | ·0000 004 | ·0000 001 | ·0000 001 | | |
| ·30 | ·0000 020 | ·0000 007 | ·0000 003 | ·0000 001 | | |
| ·31 | ·0000 034 | ·0000 013 | ·0000 005− | ·0000 002 | ·0000 001 | |
| ·32 | ·0000 055+ | ·0000 021 | ·0000 008 | ·0000 003 | ·0000 001 | |
| ·33 | ·0000 088 | ·0000 035+ | ·0000 014 | ·0000 006 | ·0000 002 | ·0000 001 |
| ·34 | ·0000 138 | ·0000 057 | ·0000 023 | ·0000 010 | ·0000 004 | ·0000 002 |
| ·35 | ·0000 214 | ·0000 091 | ·0000 038 | ·0000 016 | ·0000 007 | ·0000 003 |
| ·36 | ·0000 327 | ·0000 143 | ·0000 062 | ·0000 027 | ·0000 011 | ·0000 005− |
| ·37 | ·0000 492 | ·0000 221 | ·0000 098 | ·0000 043 | ·0000 019 | ·0000 008 |
| ·38 | ·0000 731 | ·0000 337 | ·0000 154 | ·0000 070 | ·0000 031 | ·0000 014 |
| ·39 | ·0001 071 | ·0000 507 | ·0000 238 | ·0000 110 | ·0000 051 | ·0000 023 |
| ·40 | ·0001 552 | ·0000 753 | ·0000 362 | ·0000 172 | ·0000 082 | ·0000 038 |
| ·41 | ·0002 223 | ·0001 104 | ·0000 544 | ·0000 266 | ·0000 129 | ·0000 062 |
| ·42 | ·0003 149 | ·0001 602 | ·0000 808 | ·0000 404 | ·0000 201 | ·0000 099 |
| ·43 | ·0004 413 | ·0002 298 | ·0001 186 | ·0000 607 | ·0000 309 | ·0000 156 |
| ·44 | ·0006 123 | ·0003 261 | ·0001 722 | ·0000 902 | ·0000 469 | ·0000 242 |
| ·45 | ·0008 413 | ·0004 580 | ·0002 472 | ·0001 324 | ·0000 704 | ·0000 372 |
| ·46 | ·0011 450+ | ·0006 370 | ·0003 513 | ·0001 923 | ·0001 045− | ·0000 564 |
| ·47 | ·0015 444 | ·0008 774 | ·0004 943 | ·0002 763 | ·0001 533 | ·0000 845+ |
| ·48 | ·0020 647 | ·0011 974 | ·0006 886 | ·0003 929 | ·0002 226 | ·0001 253 |
| ·49 | ·0027 371 | ·0016 196 | ·0009 504 | ·0005 534 | ·0003 200 | ·0001 838 |
| ·50 | ·0035 987 | ·0021 718 | ·0012 997 | ·0007 719 | ·0004 553 | ·0002 668 |
| ·51 | ·0046 938 | ·0028 877 | ·0017 619 | ·0010 669 | ·0006 415+ | ·0003 833 |
| ·52 | ·0060 748 | ·0038 085− | ·0023 681 | ·0014 614 | ·0008 956 | ·0005 454 |
| ·53 | ·0078 030 | ·0049 830 | ·0031 563 | ·0019 843 | ·0012 389 | ·0007 686 |
| ·54 | ·0099 492 | ·0064 694 | ·0041 728 | ·0026 715+ | ·0016 987 | ·0010 733 |
| ·55 | ·0125 948 | ·0083 360 | ·0054 731 | ·0035 670 | ·0023 090 | ·0014 853 |
| ·56 | ·0158 324 | ·0106 620 | ·0071 232 | ·0047 242 | ·0031 121 | ·0020 374 |
| ·57 | ·0197 659 | ·0135 389 | ·0092 008 | ·0062 074 | ·0041 600 | ·0027 707 |
| ·58 | ·0245 110 | ·0170 707 | ·0117 964 | ·0080 933 | ·0055 159 | ·0037 363 |
| ·59 | ·0301 948 | ·0213 748 | ·0150 146 | ·0104 721 | ·0072 560 | ·0049 971 |
| ·60 | ·0369 556 | ·0265 819 | ·0189 745− | ·0134 491 | ·0094 708 | ·0066 292 |
| ·61 | ·0449 419 | ·0328 361 | ·0238 105+ | ·0171 458 | ·0122 673 | ·0087 246 |
| ·62 | ·0543 105− | ·0402 940 | ·0296 725+ | ·0217 009 | ·0157 700 | ·0113 924 |
| ·63 | ·0652 248 | ·0491 235− | ·0367 254 | ·0272 703 | ·0201 222 | ·0147 611 |
| ·64 | ·0778 519 | ·0595 019 | ·0451 481 | ·0340 277 | ·0254 873 | ·0189 802 |
| ·65 | ·0923 592 | ·0716 134 | ·0551 319 | ·0421 636 | ·0320 484 | ·0242 211 |
| ·66 | ·1089 101 | ·0856 449 | ·0668 777 | ·0518 837 | ·0400 084 | ·0306 779 |
| ·67 | ·1276 589 | ·1017 824 | ·0805 924 | ·0634 062 | ·0495 885+ | ·0385 673 |
| ·68 | ·1487 457 | ·1202 049 | ·0964 843 | ·0769 586 | ·0610 256 | ·0481 274 |
| ·69 | ·1722 894 | ·1410 784 | ·1147 572 | ·0927 720 | ·0745 683 | ·0596 152 |
| ·70 | ·1983 815− | ·1645 488 | ·1356 030 | ·1110 752 | ·0904 719 | ·0733 025− |
| ·71 | ·2270 785− | ·1907 334 | ·1591 939 | ·1320 865+ | ·1089 911 | ·0894 699 |
| ·72 | ·2583 947 | ·2197 129 | ·1856 729 | ·1560 044 | ·1303 708 | ·1083 991 |
| ·73 | ·2922 950+ | ·2515 221 | ·2151 436 | ·1829 965+ | ·1548 357 | ·1303 626 |
| ·74 | ·3286 880 | ·2861 409 | ·2476 594 | ·2131 881 | ·1825 775+ | ·1556 111 |
| ·75 | ·3674 201 | ·3234 856 | ·2832 123 | ·2466 484 | ·2137 409 | ·1843 592 |
| ·76 | ·4082 709 | ·3634 014 | ·3217 226 | ·2833 781 | ·2484 079 | ·2167 682 |
| ·77 | ·4509 499 | ·4056 554 | ·3630 282 | ·3232 952 | ·2865 819 | ·2529 280 |
| ·78 | ·4950 962 | ·4499 329 | ·4068 772 | ·3662 235+ | ·3281 717 | ·2928 379 |
| ·79 | ·5402 801 | ·4958 352 | ·4529 213 | ·4118 820 | ·3729 768 | ·3363 873 |
| ·80 | ·5860 084 | ·5428 819 | ·5007 138 | ·4598 773 | ·4206 743 | ·3833 381 |

$x$ = ·81 to 1·00 $q = 5$ $p$ = 17 to 22

| $x$ | $p = 17$ | $p = 18$ | $p = 19$ | $p = 20$ | $p = 21$ | $p = 22$ |
|---|---|---|---|---|---|---|
| $B(p, q)$ = | ·9828 4928 × $\frac{1}{10^5}$ | ·7594 7444 × $\frac{1}{10^5}$ | ·5943 7130 × $\frac{1}{10^5}$ | ·4705 4395 × $\frac{1}{10^5}$ | ·3764 3516 × $\frac{1}{10^5}$ | ·3040 4378 × $\frac{1}{10^5}$ |
| ·81 | ·6317 331 | ·5905 160 | ·5497 111 | ·5097 008 | ·4708 108 | ·4333 097 |
| ·82 | ·6768 640 | ·6381 149 | ·5992 796 | ·5607 305+ | ·5227 982 | ·4857 690 |
| ·83 | ·7207 849 | ·6850 053 | ·6487 088 | ·6122 403 | ·5759 178 | ·5400 277 |
| ·84 | ·7628 742 | ·7304 842 | ·6972 304 | ·6634 165− | ·6293 321 | ·5952 477 |
| ·85 | ·8025 286 | ·7738 444 | ·7440 448 | ·7133 825− | ·6821 070 | ·6504 591 |
| ·86 | ·8391 894 | ·8144 050− | ·7883 539 | ·7612 333 | ·7332 448 | ·7045 899 |
| ·87 | ·8723 707 | ·8515 446 | ·8293 995+ | ·8060 772 | ·7817 288 | ·7565 107 |
| ·88 | ·9016 879 | ·8847 376 | ·8665 067 | ·8470 860 | ·8265 777 | ·8050 928 |
| ·89 | ·9268 831 | ·9135 890 | ·8991 280 | ·8835 482 | ·8669 090 | ·8492 793 |
| ·90 | ·9478 476 | ·9378 663 | ·9268 869 | ·9149 251 | ·9020 064 | ·8881 649 |
| ·91 | ·9646 367 | ·9575 249 | ·9496 150− | ·9409 016 | ·9313 865+ | ·9210 786 |
| ·92 | ·9774 739 | ·9727 216 | ·9673 779 | ·9614 266 | ·9548 565− | ·9476 606 |
| ·93 | ·9867 434 | ·9838 141 | ·9804 844 | ·9767 359 | ·9725 526 | ·9679 211 |
| ·94 | ·9929 660 | ·9913 404 | ·9894 729 | ·9873 478 | ·9849 507 | ·9822 683 |
| ·95 | ·9967 597 | ·9959 777 | ·9950 697 | ·9940 255− | ·9928 351 | ·9914 888 |
| ·96 | ·9987 835+ | ·9984 774 | ·9981 181 | ·9977 006 | ·9972 197 | ·9966 700 |
| ·97 | ·9996 695− | ·9995 828 | ·9994 801 | ·9993 594 | ·9992 190 | ·9990 569 |
| ·98 | ·9999 502 | ·9999 366 | ·9999 203 | ·9999 010 | ·9998 783 | ·9998 518 |
| ·99 | ·9999 982 | ·9999 977 | ·9999 971 | ·9999 964 | ·9999 955+ | ·9999 945− |
| 1·00 | 1·0000 000 | 1·0000 000 | 1·0000 000 | 1·0000 000 | 1·0000 000 | 1·0000 000 |

$x$ = ·34 to ·90 $q = 5$ $p$ = 23 to 28

| $x$ | $p = 23$ | $p = 24$ | $p = 25$ | $p = 26$ | $p = 27$ | $p = 28$ |
|---|---|---|---|---|---|---|
| $B(p, q)$ = | ·2477 3938 × $\frac{1}{10^5}$ | ·2035 0020 × $\frac{1}{10^5}$ | ·1684 1396 × $\frac{1}{10^5}$ | ·1403 4497 × $\frac{1}{10^5}$ | ·1177 0868 × $\frac{1}{10^5}$ | ·9931 6701 × $\frac{1}{10^6}$ |
| ·34 | ·0000 001 | | | | | |
| ·35 | ·0000 001 | | | | | |
| ·36 | ·0000 002 | ·0000 001 | | | | |
| ·37 | ·0000 004 | ·0000 002 | ·0000 001 | | | |
| ·38 | ·0000 006 | ·0000 003 | ·0000 001 | ·0000 001 | | |
| ·39 | ·0000 011 | ·0000 005− | ·0000 002 | ·0000 001 | | |
| ·40 | ·0000 018 | ·0000 008 | ·0000 004 | ·0000 002 | ·0000 001 | |
| ·41 | ·0000 030 | ·0000 014 | ·0000 007 | ·0000 003 | ·0000 001 | ·0000 001 |
| ·42 | ·0000 049 | ·0000 024 | ·0000 011 | ·0000 006 | ·0000 003 | ·0000 001 |
| ·43 | ·0000 078 | ·0000 039 | ·0000 019 | ·0000 010 | ·0000 005− | ·0000 002 |
| ·44 | ·0000 124 | ·0000 064 | ·0000 032 | ·0000 016 | ·0000 008 | ·0000 004 |
| ·45 | ·0000 195+ | ·0000 102 | ·0000 053 | ·0000 027 | ·0000 014 | ·0000 007 |
| ·46 | ·0000 303 | ·0000 161 | ·0000 086 | ·0000 045+ | ·0000 024 | ·0000 012 |
| ·47 | ·0000 463 | ·0000 252 | ·0000 137 | ·0000 074 | ·0000 040 | ·0000 021 |
| ·48 | ·0000 701 | ·0000 390 | ·0000 216 | ·0000 119 | ·0000 065+ | ·0000 036 |
| ·49 | ·0001 049 | ·0000 596 | ·0000 337 | ·0000 189 | ·0000 106 | ·0000 059 |
| ·50 | ·0001 554 | ·0000 900 | ·0000 519 | ·0000 297 | ·0000 170 | ·0000 097 |
| ·51 | ·0002 276 | ·0001 344 | ·0000 790 | ·0000 462 | ·0000 269 | ·0000 156 |
| ·52 | ·0003 301 | ·0001 987 | ·0001 190 | ·0000 710 | ·0000 421 | ·0000 249 |
| ·53 | ·0004 741 | ·0002 908 | ·0001 775− | ·0001 078 | ·0000 652 | ·0000 392 |
| ·54 | ·0006 742 | ·0004 212 | ·0002 618 | ·0001 620 | ·0000 998 | ·0000 612 |
| ·55 | ·0009 499 | ·0006 042 | ·0003 824 | ·0002 409 | ·0001 511 | ·0000 943 |
| ·56 | ·0013 261 | ·0008 585− | ·0005 530 | ·0003 546 | ·0002 263 | ·0001 439 |
| ·57 | ·0018 348 | ·0012 086 | ·0007 921 | ·0005 167 | ·0003 356 | ·0002 171 |
| ·58 | ·0025 165− | ·0016 859 | ·0011 239 | ·0007 458 | ·0004 928 | ·0003 243 |
| ·59 | ·0034 220 | ·0023 311 | ·0015 801 | ·0010 662 | ·0007 164 | ·0004 794 |
| ·60 | ·0046 142 | ·0031 950− | ·0022 015+ | ·0015 101 | ·0010 314 | ·0007 016 |
| ·61 | ·0061 706 | ·0043 417 | ·0030 401 | ·0021 191 | ·0014 709 | ·0010 169 |
| ·62 | ·0081 848 | ·0058 503 | ·0041 616 | ·0029 471 | ·0020 783 | ·0014 598 |
| ·63 | ·0107 696 | ·0078 176 | ·0056 479 | ·0040 622 | ·0029 096 | ·0020 758 |
| ·64 | ·0140 586 | ·0103 609 | ·0076 000 | ·0055 502 | ·0040 366 | ·0029 244 |
| ·65 | ·0182 083 | ·0136 203 | ·0101 411 | ·0075 177 | ·0055 502 | ·0040 819 |
| ·66 | ·0234 000 | ·0177 613 | ·0134 195− | ·0100 954 | ·0075 639 | ·0056 457 |
| ·67 | ·0298 405− | ·0229 766 | ·0176 114 | ·0134 415+ | ·0102 179 | ·0077 381 |
| ·68 | ·0377 620 | ·0294 877 | ·0229 234 | ·0177 455− | ·0136 828 | ·0105 109 |
| ·69 | ·0474 217 | ·0375 449 | ·0295 941 | ·0232 303 | ·0181 637 | ·0141 499 |
| ·70 | ·0590 987 | ·0474 269 | ·0378 949 | ·0301 549 | ·0239 034 | ·0188 791 |
| ·71 | ·0730 899 | ·0594 375− | ·0481 287 | ·0388 148 | ·0311 846 | ·0249 646 |
| ·72 | ·0897 032 | ·0739 011 | ·0606 273 | ·0495 411 | ·0403 310 | ·0327 173 |
| ·73 | ·1092 489 | ·0911 555− | ·0757 459 | ·0626 970 | ·0517 059 | ·0424 936 |
| ·74 | ·1320 277 | ·1115 410 | ·0938 541 | ·0786 716 | ·0657 082 | ·0546 940 |
| ·75 | ·1583 163 | ·1353 873 | ·1153 243 | ·0978 696 | ·0827 645+ | ·0697 574 |
| ·76 | ·1883 500+ | ·1629 961 | ·1405 155+ | ·1206 967 | ·1033 171 | ·0881 518 |
| ·77 | ·2223 030 | ·1946 207 | ·1697 527 | ·1475 406 | ·1278 060 | ·1103 592 |
| ·78 | ·2602 666 | ·2304 425+ | ·2033 026 | ·1787 465− | ·1566 460 | ·1368 537 |
| ·79 | ·3022 261 | ·2705 452 | ·2413 461 | ·2145 879 | ·1901 969 | ·1680 733 |
| ·80 | ·3480 384 | ·3148 874 | ·2839 465− | ·2552 333 | ·2287 288 | ·2043 839 |
| ·81 | ·3974 109 | ·3632 758 | ·3310 182 | ·3007 090 | ·2723 815− | ·2460 369 |
| ·82 | ·4498 844 | ·4153 415+ | ·3822 955+ | ·3508 622 | ·3211 214 | ·2931 210 |
| ·83 | ·5048 227 | ·4705 208 | ·4373 052 | ·4053 252 | ·3746 981 | ·3455 117 |
| ·84 | ·5614 112 | ·5280 455− | ·4953 471 | ·4634 857 | ·4326 048 | ·4028 218 |
| ·85 | ·6186 674 | ·5869 448 | ·5554 866 | ·5244 687 | ·4940 474 | ·4643 585− |
| ·86 | ·6754 662 | ·6460 639 | ·6165 635+ | ·5871 340 | ·5579 309 | ·5290 954 |
| ·87 | ·7305 820 | ·7041 008 | ·6772 225− | ·6500 968 | ·6228 669 | ·5956 672 |
| ·88 | ·7827 485+ | ·7596 659 | ·7359 678 | ·7117 767 | ·6872 135− | ·6623 955+ |
| ·89 | ·8307 361 | ·8113 625− | ·7912 462 | ·7704 782 | ·7491 510 | ·7273 578 |
| ·90 | ·8734 426 | ·8578 881 | ·8415 559 | ·8245 051 | ·8067 985+ | ·7885 017 |

$x = \cdot 91$ to $1 \cdot 00$ $q = 5$ $p = 23$ to $28$

| | $p = 23$ | $p = 24$ | $p = 25$ | $p = 26$ | $p = 27$ | $p = 28$ |
|---|---|---|---|---|---|---|
| $B(p, q) =$ | $\cdot 2477\ 3938 \times \frac{1}{10^5}$ | $\cdot 2035\ 0020 \times \frac{1}{10^5}$ | $\cdot 1684\ 1396 \times \frac{1}{10^5}$ | $\cdot 1403\ 4497 \times \frac{1}{10^5}$ | $\cdot 1177\ 0868 \times \frac{1}{10^5}$ | $\cdot 9931\ 6701 \times \frac{1}{10^6}$ |
| $x$ | | | | | | |
| ·91 | ·9099 929 | ·8981 504 | ·8855 777 | ·8723 059 | ·8583 705+ | ·8438 106 |
| ·92 | ·9398 367 | ·9313 869 | ·9223 174 | ·9126 385− | ·9023 639 | ·8915 109 |
| ·93 | ·9628 307 | ·9572 732 | ·9512 434 | ·9447 384 | ·9377 581 | ·9303 046 |
| ·94 | ·9792 883 | ·9760 000 | ·9723 938 | ·9684 616 | ·9641 967 | ·9595 938 |
| ·95 | ·9899 773 | ·9882 916 | ·9864 233 | ·9843 645− | ·9821 077 | ·9796 461 |
| ·96 | ·9960 464 | ·9953 437 | ·9945 565+ | ·9936 800 | ·9927 091 | ·9916 389 |
| ·97 | ·9988 710 | ·9986 593 | ·9984 198 | ·9981 503 | ·9978 487 | ·9975 127 |
| ·98 | ·9998 212 | ·9997 859 | ·9997 455+ | ·9996 997 | ·9996 478 | ·9995 894 |
| ·99 | ·9999 933 | ·9999 919 | ·9999 903 | ·9999 884 | ·9999 863 | ·9999 839 |
| 1·00 | 1·0000 000 | 1·0000 000 | 1·0000 000 | 1·0000 000 | 1·0000 000 | 1·0000 000 |

$x$ = ·42 to 1·00 $q = 5$ $p$ = 29 to 34

| $x$ | $p = 29$ | $p = 30$ | $p = 31$ | $p = 32$ | $p = 33$ | $p = 34$ |
|---|---|---|---|---|---|---|
| $B(p, q)$ = | ·8426 8716 × $\frac{1}{10^8}$ | ·7187 6258 × $\frac{1}{10^8}$ | ·6160 8221 × $\frac{1}{10^8}$ | ·5305 1524 × $\frac{1}{10^8}$ | ·4588 2399 × $\frac{1}{10^8}$ | ·3984 5241 × $\frac{1}{10^8}$ |
| ·42 | ·0000 001 | | | | | |
| ·43 | ·0000 001 | ·0000 001 | | | | |
| ·44 | ·0000 002 | ·0000 001 | ·0000 001 | | | |
| ·45 | ·0000 004 | ·0000 002 | ·0000 001 | | | |
| ·46 | ·0000 006 | ·0000 003 | ·0000 002 | ·0000 001 | | |
| ·47 | ·0000 011 | ·0000 006 | ·0000 003 | ·0000 002 | ·0000 001 | |
| ·48 | ·0000 019 | ·0000 010 | ·0000 006 | ·0000 003 | ·0000 002 | ·0000 001 |
| ·49 | ·0000 033 | ·0000 018 | ·0000 010 | ·0000 005+ | ·0000 003 | ·0000 002 |
| ·50 | ·0000 055− | ·0000 031 | ·0000 017 | ·0000 010 | ·0000 005+ | ·0000 003 |
| ·51 | ·0000 090 | ·0000 052 | ·0000 030 | ·0000 017 | ·0000 010 | ·0000 005+ |
| ·52 | ·0000 146 | ·0000 086 | ·0000 050+ | ·0000 029 | ·0000 017 | ·0000 010 |
| ·53 | ·0000 235+ | ·0000 141 | ·0000 084 | ·0000 050− | ·0000 029 | ·0000 017 |
| ·54 | ·0000 374 | ·0000 228 | ·0000 138 | ·0000 083 | ·0000 050+ | ·0000 030 |
| ·55 | ·0000 587 | ·0000 364 | ·0000 225− | ·0000 138 | ·0000 085− | ·0000 052 |
| ·56 | ·0000 911 | ·0000 575− | ·0000 362 | ·0000 227 | ·0000 142 | ·0000 088 |
| ·57 | ·0001 399 | ·0000 898 | ·0000 575− | ·0000 367 | ·0000 233 | ·0000 148 |
| ·58 | ·0002 126 | ·0001 388 | ·0000 904 | ·0000 586 | ·0000 379 | ·0000 245− |
| ·59 | ·0003 195+ | ·0002 123 | ·0001 405+ | ·0000 927 | ·0000 610 | ·0000 400 |
| ·60 | ·0004 755− | ·0003 211 | ·0002 161 | ·0001 450− | ·0000 970 | ·0000 647 |
| ·61 | ·0007 004 | ·0004 807 | ·0003 288 | ·0002 242 | ·0001 524 | ·0001 033 |
| ·62 | ·0010 216 | ·0007 124 | ·0004 951 | ·0003 431 | ·0002 370 | ·0001 633 |
| ·63 | ·0014 755+ | ·0010 452 | ·0007 379 | ·0005 194 | ·0003 645− | ·0002 550+ |
| ·64 | ·0021 108 | ·0015 184 | ·0010 886 | ·0007 781 | ·0005 545+ | ·0003 941 |
| ·65 | ·0029 911 | ·0021 843 | ·0015 899 | ·0011 538 | ·0008 348 | ·0006 024 |
| ·66 | ·0041 987 | ·0031 120 | ·0022 992 | ·0016 935+ | ·0012 438 | ·0009 110 |
| ·67 | ·0058 393 | ·0043 916 | ·0032 923 | ·0024 608 | ·0018 340 | ·0013 631 |
| ·68 | ·0080 459 | ·0061 385− | ·0046 685+ | ·0035 400 | ·0026 766 | ·0020 184 |
| ·69 | ·0109 847 | ·0084 994 | ·0065 560 | ·0050 420 | ·0038 667 | ·0029 575+ |
| ·70 | ·0148 596 | ·0116 578 | ·0091 178 | ·0071 104 | ·0055 295− | ·0042 887 |
| ·71 | ·0199 175− | ·0158 397 | ·0125 585+ | ·0099 283 | ·0078 274 | ·0061 549 |
| ·72 | ·0264 523 | ·0213 193 | ·0171 308 | ·0137 260 | ·0109 681 | ·0087 417 |
| ·73 | ·0348 080 | ·0284 237 | ·0231 417 | ·0187 883 | ·0152 131 | ·0122 868 |
| ·74 | ·0453 792 | ·0375 355− | ·0309 572 | ·0254 612 | ·0208 857 | ·0170 895− |
| ·75 | ·0586 084 | ·0490 933 | ·0410 055+ | ·0341 570 | ·0283 785− | ·0235 193 |
| ·76 | ·0749 797 | ·0635 880 | ·0537 761 | ·0453 567 | ·0381 582 | ·0320 242 |
| ·77 | ·0950 060 | ·0815 534 | ·0698 138 | ·0596 079 | ·0507 670 | ·0431 344 |
| ·78 | ·1192 103 | ·1035 503 | ·0897 068 | ·0775 156 | ·0668 179 | ·0574 622 |
| ·79 | ·1480 988 | ·1301 425+ | ·1140 656 | ·0997 260 | ·0869 817 | ·0756 934 |
| ·80 | ·1821 257 | ·1618 631 | ·1434 917 | ·1268 982 | ·1119 640 | ·0985 685− |
| ·81 | ·2216 493 | ·1991 707 | ·1785 354 | ·1596 640 | ·1424 675− | ·1268 499 |
| ·82 | ·2668 806 | ·2423 956 | ·2196 409 | ·1985 744 | ·1791 406 | ·1612 733 |
| ·83 | ·3178 263 | ·2916 779 | ·2670 810 | ·2440 313 | ·2225 087 | ·2024 796 |
| ·84 | ·3742 301 | ·3469 004 | ·3208 825+ | ·2962 075− | ·2728 896 | ·2509 284 |
| ·85 | ·4355 177 | ·4076 218 | ·3807 487 | ·3549 593 | ·3302 981 | ·3067 952 |
| ·86 | ·5007 544 | ·4730 192 | ·4459 866 | ·4197 389 | ·3943 442 | ·3698 576 |
| ·87 | ·5686 229 | ·5418 490 | ·5154 500+ | ·4895 194 | ·4641 398 | ·4393 831 |
| ·88 | ·6374 358 | ·6124 416 | ·5875 140 | ·5627 473 | ·5382 282 | ·5140 361 |
| ·89 | ·7051 910 | ·6827 413 | ·6600 971 | ·6373 434 | ·6145 612 | ·5918 273 |
| ·90 | ·7696 821 | ·7504 083 | ·7307 490 | ·7107 726 | ·6905 464 | ·6701 364 |
| ·91 | ·8286 683 | ·8129 883 | ·7968 169 | ·7802 021 | ·7631 926 | ·7458 379 |
| ·92 | ·8800 998 | ·8681 535+ | ·8556 975+ | ·8427 594 | ·8293 684 | ·8155 554 |
| ·93 | ·9223 827 | ·9139 991 | ·9051 627 | ·8958 846 | ·8861 773 | ·8760 553 |
| ·94 | ·9546 489 | ·9493 596 | ·9437 247 | ·9377 444 | ·9314 203 | ·9247 551 |
| ·95 | ·9769 735+ | ·9740 844 | ·9709 737 | ·9676 373 | ·9640 715− | ·9602 734 |
| ·96 | ·9904 648 | ·9891 822 | ·9877 867 | ·9862 741 | ·9846 406 | ·9828 823 |
| ·97 | ·9971 403 | ·9967 292 | ·9962 773 | ·9957 824 | ·9952 423 | ·9946 549 |
| ·98 | ·9995 241 | ·9994 512 | ·9993 703 | ·9992 807 | ·9991 819 | ·9990 734 |
| ·99 | ·9999 812 | ·9999 782 | ·9999 747 | ·9999 709 | ·9999 666 | ·9999 619 |
| 1·00 | 1·0000 000 | 1·0000 000 | 1·0000 000 | 1·0000 000 | 1·0000 000 | 1·0000 000 |

$x$ = ·49 to 1·00 $q = 5$ $p$ = 35 to 40

| $x$ | $p = 35$ | $p = 36$ | $p = 37$ | $p = 38$ | $p = 39$ | $p = 40$ |
|---|---|---|---|---|---|---|
| $B(p, q) =$ | ·3473 6877 $\times \frac{1}{10^8}$ | ·3039 4767 $\times \frac{1}{10^8}$ | ·2668 8088 $\times \frac{1}{10^8}$ | ·2351 0935 $\bar{\times} \frac{1}{10^8}$ | ·2077 7105 $\overset{+}{\times} \frac{1}{10^8}$ | ·1841 6071 $\times \frac{1}{10^8}$ |
| ·49 | ·0000 001 | | | | | |
| ·50 | ·0000 002 | ·0000 001 | ·0000 001 | | | |
| ·51 | ·0000 003 | ·0000 002 | ·0000 001 | ·0000 001 | | |
| ·52 | ·0000 006 | ·0000 003 | ·0000 002 | ·0000 001 | ·0000 001 | |
| ·53 | ·0000 010 | ·0000 006 | ·0000 004 | ·0000 002 | ·0000 001 | ·0000 001 |
| ·54 | ·0000 018 | ·0000 011 | ·0000 006 | ·0000 004 | ·0000 002 | ·0000 001 |
| ·55 | ·0000 032 | ·0000 019 | ·0000 012 | ·0000 007 | ·0000 004 | ·0000 003 |
| ·56 | ·0000 055− | ·0000 034 | ·0000 021 | ·0000 013 | ·0000 008 | ·0000 005− |
| ·57 | ·0000 093 | ·0000 059 | ·0000 037 | ·0000 023 | ·0000 015− | ·0000 009 |
| ·58 | ·0000 157 | ·0000 101 | ·0000 065− | ·0000 041 | ·0000 026 | ·0000 017 |
| ·59 | ·0000 262 | ·0000 171 | ·0000 111 | ·0000 072 | ·0000 047 | ·0000 030 |
| ·60 | ·0000 430 | ·0000 285+ | ·0000 189 | ·0000 125− | ·0000 082 | ·0000 054 |
| ·61 | ·0000 699 | ·0000 471 | ·0000 317 | ·0000 213 | ·0000 142 | ·0000 095+ |
| ·62 | ·0001 122 | ·0000 769 | ·0000 525+ | ·0000 358 | ·0000 244 | ·0000 166 |
| ·63 | ·0001 780 | ·0001 239 | ·0000 861 | ·0000 596 | ·0000 412 | ·0000 284 |
| ·64 | ·0002 793 | ·0001 975− | ·0001 393 | ·0000 980 | ·0000 688 | ·0000 482 |
| ·65 | ·0004 335+ | ·0003 112 | ·0002 229 | ·0001 593 | ·0001 136 | ·0000 808 |
| ·66 | ·0006 655− | ·0004 849 | ·0003 526 | ·0002 557 | ·0001 851 | ·0001 337 |
| ·67 | ·0010 105+ | ·0007 473 | ·0005 514 | ·0004 059 | ·0002 981 | ·0002 186 |
| ·68 | ·0015 181 | ·0011 391 | ·0008 527 | ·0006 368 | ·0004 747 | ·0003 531 |
| ·69 | ·0022 563 | ·0017 172 | ·0013 039 | ·0009 879 | ·0007 469 | ·0005 636 |
| ·70 | ·0033 180 | ·0025 609 | ·0019 720 | ·0015 152 | ·0011 618 | ·0008 890 |
| ·71 | ·0048 277 | ·0037 778 | ·0029 495− | ·0022 978 | ·0017 864 | ·0013 861 |
| ·72 | ·0069 501 | ·0055 127 | ·0043 628 | ·0034 454 | ·0027 153 | ·0021 357 |
| ·73 | ·0098 993 | ·0079 573 | ·0063 821 | ·0051 078 | ·0040 797 | ·0032 523 |
| ·74 | ·0139 498 | ·0113 608 | ·0092 322 | ·0074 867 | ·0060 590 | ·0048 942 |
| ·75 | ·0194 462 | ·0160 422 | ·0132 056 | ·0108 481 | ·0088 939 | ·0072 779 |
| ·76 | ·0268 139 | ·0224 015+ | ·0186 755− | ·0155 376 | ·0129 017 | ·0106 930 |
| ·77 | ·0365 659 | ·0309 301 | ·0261 083 | ·0219 942 | ·0184 929 | ·0155 203 |
| ·78 | ·0493 062 | ·0422 176 | ·0360 740 | ·0307 640 | ·0261 863 | ·0222 494 |
| ·79 | ·0657 264 | ·0569 527 | ·0492 513 | ·0425 094 | ·0366 227 | ·0314 952 |
| ·80 | ·0865 913 | ·0759 145− | ·0664 240 | ·0580 108 | ·0505 718 | ·0440 102 |
| ·81 | ·1127 114 | ·0999 504 | ·0884 655+ | ·0781 570 | ·0689 282 | ·0606 862 |
| ·82 | ·1448 984 | ·1299 365− | ·1163 045− | ·1039 177 | ·0926 915− | ·0825 418 |
| ·83 | ·1838 996 | ·1667 158 | ·1508 685+ | ·1362 933 | ·1229 225+ | ·1106 865− |
| ·84 | ·2303 107 | ·2110 125+ | ·1930 009 | ·1762 355− | ·1606 701 | ·1462 542 |
| ·85 | ·2844 675− | ·2633 199 | ·2433 473 | ·2245 352 | ·2068 617 | ·1902 985− |
| ·86 | ·3463 216 | ·3237 674 | ·3022 157 | ·2816 774 | ·2621 552 | ·2436 442 |
| ·87 | ·4153 109 | ·3919 746 | ·3694 162 | ·3476 687 | ·3267 567 | ·3066 973 |
| ·88 | ·4902 424 | ·4669 110 | ·4440 980 | ·4218 523 | ·4002 155− | ·3792 222 |
| ·89 | ·5692 137 | ·5467 876 | ·5246 106 | ·5027 393 | ·4812 249 | ·4601 131 |
| ·90 | ·6496 064 | ·6290 177 | ·6084 290 | ·5878 959 | ·5674 709 | ·5472 031 |
| ·91 | ·7281 871 | ·7102 892 | ·6921 924 | ·6739 440 | ·6555 899 | ·6371 747 |
| ·92 | ·8013 524 | ·7867 922 | ·7719 086 | ·7567 352 | ·7413 064 | ·7256 560 |
| ·93 | ·8655 344 | ·8546 316 | ·8433 655+ | ·8317 553 | ·8198 212 | ·8075 842 |
| ·94 | ·9177 527 | ·9104 181 | ·9027 576 | ·8947 783 | ·8864 881 | ·8778 962 |
| ·95 | ·9562 407 | ·9519 717 | ·9474 657 | ·9427 221 | ·9377 414 | ·9325 244 |
| ·96 | ·9809 958 | ·9789 777 | ·9768 251 | ·9745 352 | ·9721 055+ | ·9695 338 |
| ·97 | ·9940 182 | ·9933 299 | ·9925 881 | ·9917 908 | ·9909 359 | ·9900 217 |
| ·98 | ·9989 546 | ·9988 248 | ·9986 835− | ·9985 300 | ·9983 638 | ·9981 842 |
| ·99 | ·9999 566 | ·9999 508 | ·9999 445− | ·9999 375− | ·9999 299 | ·9999 215+ |
| 1·00 | 1·0000 000 | 1·0000 000 | 1·0000 000 | 1·0000 000 | 1·0000 000 | 1·0000 000 |

$x$ = ·54 to 1·00 $q = 5$ $p$ = 41 to 45

| | $p = 41$ | $p = 42$ | $p = 43$ | $p = 44$ | $p = 45$ |
|---|---|---|---|---|---|
| $B(p, q)$ = | ·1636 9841 × $\frac{1}{10^{8}}$ | ·1459 0510 × $\frac{1}{10^{8}}$ | ·1303 8328 × $\frac{1}{10^{8}}$ | ·1168 0169 × $\frac{1}{10^{8}}$ | ·1048 8315 × $\frac{1}{10^{8}}$ |
| $x$ | | | | | |
| ·54 | ·0000 001 | | | | |
| ·55 | ·0000 002 | ·0000 001 | ·0000 001 | | |
| ·56 | ·0000 003 | ·0000 002 | ·0000 001 | ·0000 001 | |
| ·57 | ·0000 006 | ·0000 004 | ·0000 002 | ·0000 001 | ·0000 001 |
| ·58 | ·0000 011 | ·0000 007 | ·0000 004 | ·0000 003 | ·0000 002 |
| ·59 | ·0000 020 | ·0000 013 | ·0000 008 | ·0000 005+ | ·0000 003 |
| ·60 | ·0000 035+ | ·0000 023 | ·0000 015+ | ·0000 010 | ·0000 006 |
| ·61 | ·0000 064 | ·0000 042 | ·0000 028 | ·0000 019 | ·0000 012 |
| ·62 | ·0000 112 | ·0000 076 | ·0000 051 | ·0000 035− | ·0000 023 |
| ·63 | ·0000 196 | ·0000 135− | ·0000 092 | ·0000 063 | ·0000 043 |
| ·64 | ·0000 337 | ·0000 235+ | ·0000 164 | ·0000 114 | ·0000 079 |
| ·65 | ·0000 574 | ·0000 407 | ·0000 288 | ·0000 203 | ·0000 143 |
| ·66 | ·0000 964 | ·0000 693 | ·0000 498 | ·0000 357 | ·0000 256 |
| ·67 | ·0001 599 | ·0001 168 | ·0000 851 | ·0000 619 | ·0000 450− |
| ·68 | ·0002 621 | ·0001 942 | ·0001 436 | ·0001 061 | ·0000 782 |
| ·69 | ·0004 244 | ·0003 190 | ·0002 394 | ·0001 793 | ·0001 341 |
| ·70 | ·0006 790 | ·0005 177 | ·0003 940 | ·0002 993 | ·0002 270 |
| ·71 | ·0010 734 | ·0008 298 | ·0006 403 | ·0004 933 | ·0003 794 |
| ·72 | ·0016 767 | ·0013 140 | ·0010 279 | ·0008 028 | ·0006 260 |
| ·73 | ·0025 878 | ·0020 554 | ·0016 297 | ·0012 901 | ·0010 196 |
| ·74 | ·0039 461 | ·0031 760 | ·0025 518 | ·0020 470 | ·0016 395− |
| ·75 | ·0059 447 | ·0048 473 | ·0039 458 | ·0032 068 | ·0026 022 |
| ·76 | ·0088 466 | ·0073 063 | ·0060 243 | ·0049 593 | ·0040 763 |
| ·77 | ·0130 026 | ·0108 748 | ·0090 803 | ·0075 701 | ·0063 015− |
| ·78 | ·0188 715+ | ·0159 797 | ·0135 093 | ·0114 032 | ·0096 110 |
| ·79 | ·0270 395− | ·0231 760 | ·0198 331 | ·0169 466 | ·0144 590 |
| ·80 | ·0382 360 | ·0331 660 | ·0287 236 | ·0248 392 | ·0214 491 |
| ·81 | ·0533 426 | ·0468 139 | ·0410 220 | ·0358 942 | ·0313 630 |
| ·82 | ·0733 868 | ·0651 472 | ·0577 474 | ·0511 150− | ·0451 821 |
| ·83 | ·0995 150− | ·0893 380 | ·0800 867 | ·0716 938 | ·0640 944 |
| ·84 | ·1329 339 | ·1206 532 | ·1093 550− | ·0989 816 | ·0894 759 |
| ·85 | ·1748 119 | ·1603 640 | ·1469 137 | ·1344 175− | ·1228 300 |
| ·86 | ·2261 328 | ·2096 037 | ·1940 349 | ·1794 003 | ·1656 703 |
| ·87 | ·2875 005− | ·2691 698 | ·2517 033 | ·2350 939 | ·2193 301 |
| ·88 | ·3589 007 | ·3392 730 | ·3203 558 | ·3021 600 | ·2846 920 |
| ·89 | ·4394 447 | ·4192 552 | ·3995 753 | ·3804 308 | ·3618 433 |
| ·90 | ·5271 379 | ·5073 174 | ·4877 800 | ·4685 607 | ·4496 909 |
| ·91 | ·6187 411 | ·6003 299 | ·5819 801 | ·5637 284 | ·5456 095+ |
| ·92 | ·7098 178 | ·6938 250+ | ·6777 105− | ·6615 059 | ·6452 425− |
| ·93 | ·7950 657 | ·7822 877 | ·7692 724 | ·7560 422 | ·7426 196 |
| ·94 | ·8690 121 | ·8598 463 | ·8504 099 | ·8407 145+ | ·8307 724 |
| ·95 | ·9270 726 | ·9213 881 | ·9154 736 | ·9093 321 | ·9029 673 |
| ·96 | ·9668 180 | ·9639 565− | ·9609 478 | ·9577 909 | ·9544 846 |
| ·97 | ·9890 462 | ·9880 077 | ·9869 044 | ·9857 346 | ·9844 968 |
| ·98 | ·9979 906 | ·9977 823 | ·9975 588 | ·9973 194 | ·9970 634 |
| ·99 | ·9999 124 | ·9999 026 | ·9998 919 | ·9998 803 | ·9998 678 |
| 1·00 | 1·0000 000 | 1·0000 000 | 1·0000 000 | 1·0000 000 | 1·0000 000 |

$x$ = ·57 to 1·00 $q = 5$ $p$ = 46 to 50

| $x$ | $p = 46$ | $p = 47$ | $p = 48$ | $p = 49$ | $p = 50$ |
|---|---|---|---|---|---|
| $B(p, q) =$ | ·9439 4835 $\times \frac{1}{10^7}$ | ·8514 0439 $\times \frac{1}{10^7}$ | ·7695 3858 $\times \frac{1}{10^7}$ | ·6969 4060 $\times \frac{1}{10^7}$ | ·6324 0907 $\times \frac{1}{10^7}$ |
| ·57 | ·0000 001 | | | | |
| ·58 | ·0000 001 | ·0000 001 | | | |
| ·59 | ·0000 002 | ·0000 001 | ·0000 001 | ·0000 001 | |
| ·60 | ·0000 004 | ·0000 003 | ·0000 002 | ·0000 001 | ·0000 001 |
| ·61 | ·0000 008 | ·0000 005+ | ·0000 004 | ·0000 002 | ·0000 002 |
| ·62 | ·0000 016 | ·0000 010 | ·0000 007 | ·0000 005− | ·0000 003 |
| ·63 | ·0000 029 | ·0000 020 | ·0000 014 | ·0000 009 | ·0000 006 |
| ·64 | ·0000 055− | ·0000 038 | ·0000 026 | ·0000 018 | ·0000 012 |
| ·65 | ·0000 101 | ·0000 071 | ·0000 050− | ·0000 035− | ·0000 024 |
| ·66 | ·0000 183 | ·0000 130 | ·0000 093 | ·0000 066 | ·0000 047 |
| ·67 | ·0000 326 | ·0000 236 | ·0000 171 | ·0000 123 | ·0000 089 |
| ·68 | ·0000 575+ | ·0000 423 | ·0000 310 | ·0000 227 | ·0000 166 |
| ·69 | ·0001 001 | ·0000 746 | ·0000 556 | ·0000 413 | ·0000 307 |
| ·70 | ·0001 719 | ·0001 300 | ·0000 982 | ·0000 740 | ·0000 557 |
| ·71 | ·0002 914 | ·0002 234 | ·0001 711 | ·0001 308 | ·0000 999 |
| ·72 | ·0004 874 | ·0003 789 | ·0002 941 | ·0002 280 | ·0001 765+ |
| ·73 | ·0008 046 | ·0006 340 | ·0004 989 | ·0003 920 | ·0003 077 |
| ·74 | ·0013 111 | ·0010 470 | ·0008 349 | ·0006 649 | ·0005 288 |
| ·75 | ·0021 084 | ·0017 058 | ·0013 782 | ·0011 120 | ·0008 961 |
| ·76 | ·0033 456 | ·0027 420 | ·0022 441 | ·0018 343 | ·0014 974 |
| ·77 | ·0052 378 | ·0043 476 | ·0036 038 | ·0029 833 | ·0024 665+ |
| ·78 | ·0080 889 | ·0067 984 | ·0057 061 | ·0047 832 | ·0040 045+ |
| ·79 | ·0123 191 | ·0104 816 | ·0089 064 | ·0075 582 | ·0064 063 |
| ·80 | ·0184 960 | ·0159 281 | ·0136 989 | ·0117 669 | ·0100 952 |
| ·81 | ·0273 666 | ·0238 480 | ·0207 553 | ·0180 415+ | ·0156 639 |
| ·82 | ·0398 846 | ·0351 629 | ·0309 617 | ·0272 296 | ·0239 194 |
| ·83 | ·0572 262 | ·0510 299 | ·0454 493 | ·0404 314 | ·0359 266 |
| ·84 | ·0807 813 | ·0728 427 | ·0656 068 | ·0590 222 | ·0530 395+ |
| ·85 | ·1121 052 | ·1021 964 | ·0930 571 | ·0846 414 | ·0769 040 |
| ·86 | ·1528 129 | ·1407 940 | ·1295 782 | ·1191 287 | ·1094 085+ |
| ·87 | ·2043 965− | ·1902 744 | ·1769 427 | ·1643 775− | ·1525 534 |
| ·88 | ·2679 538 | ·2519 434 | ·2366 552 | ·2220 803 | ·2082 075+ |
| ·89 | ·3438 299 | ·3264 039 | ·3095 748 | ·2933 488 | ·2777 288 |
| ·90 | ·4311 984 | ·4131 080 | ·3954 409 | ·3782 155+ | ·3614 471 |
| ·91 | ·5276 557 | ·5098 970 | ·4923 613 | ·4750 740 | ·4580 583 |
| ·92 | ·6289 501 | ·6126 578 | ·5963 932 | ·5801 827 | ·5640 517 |
| ·93 | ·7290 269 | ·7152 865− | ·7014 204 | ·6874 503 | ·6733 975− |
| ·94 | ·8205 960 | ·8701 985− | ·7995 930 | ·7887 930 | ·7778 123 |
| ·95 | ·8963 832 | ·8895 844 | ·8825 759 | ·8753 630 | ·8679 513 |
| ·96 | ·9510 285+ | ·9474 222 | ·9436 654 | ·9397 584 | ·9357 014 |
| ·97 | ·9831 894 | ·9818 109 | ·9803 600 | ·9788 353 | ·9772 356 |
| ·98 | ·9967 903 | ·9964 993 | ·9961 899 | ·9958 614 | ·9955 132 |
| ·99 | ·9998 543 | ·9998 398 | ·9998 242 | ·9998 075+ | ·9997 896 |
| 1·00 | 1·0000 000 | 1·0000 000 | 1·0000 000 | 1·0000 000 | 1·0000 000 |

$x$ = ·02 to ·60 $q = 5{\cdot}5$ $p = 5{\cdot}5$ to 8

| $x$ | $p = 5{\cdot}5$ | $p = 6$ | $p = 6{\cdot}5$ | $p = 7$ | $p = 7{\cdot}5$ | $p = 8$ |
|---|---|---|---|---|---|---|
| $B(p, q) =$ | ·7550 0617 $\times \frac{1}{10^3}$ | ·5278 5192 $\times \frac{1}{10^3}$ | ·3775 0308 $\times \frac{1}{10^3}$ | ·2754 0100 $\times \frac{1}{10^3}$ | ·2044 8084 $\times \frac{1}{10^3}$ | ·1542 2456 $\times \frac{1}{10^3}$ |
| ·02 | ·0000 001 | | | | | |
| ·03 | ·0000 009 | ·0000 002 | | | | |
| ·04 | ·0000 042 | ·0000 011 | ·0000 003 | ·0000 001 | | |
| ·05 | ·0000 139 | ·0000 041 | ·0000 012 | ·0000 003 | ·0000 001 | |
| ·06 | ·0000 363 | ·0000 116 | ·0000 037 | ·0000 011 | ·0000 003 | ·0000 001 |
| ·07 | ·0000 814 | ·0000 281 | ·0000 096 | ·0000 032 | ·0000 011 | ·0000 003 |
| ·08 | ·0001 630 | ·0000 602 | ·0000 219 | ·0000 079 | ·0000 028 | ·0000 010 |
| ·09 | ·0002 991 | ·0001 171 | ·0000 451 | ·0000 172 | ·0000 065$^{-}$ | ·0000 024 |
| ·10 | ·0005 124 | ·0002 113 | ·0000 858 | ·0000 344 | ·0000 136 | ·0000 053 |
| ·11 | ·0008 304 | ·0003 589 | ·0001 528 | ·0000 642 | ·0000 267 | ·0000 109 |
| ·12 | ·0012 853 | ·0005 798 | ·0002 577 | ·0001 130 | ·0000 490 | ·0000 210 |
| ·13 | ·0019 139 | ·0008 980 | ·0004 151 | ·0001 894 | ·0000 854 | ·0000 381 |
| ·14 | ·0027 574 | ·0013 416 | ·0006 433 | ·0003 045$^{-}$ | ·0001 424 | ·0000 659 |
| ·15 | ·0038 611 | ·0019 432 | ·0009 638 | ·0004 719 | ·0002 284 | ·0001 094 |
| ·16 | ·0052 739 | ·0027 392 | ·0014 024 | ·0007 087 | ·0003 541 | ·0001 750$^{+}$ |
| ·17 | ·0070 476 | ·0037 702 | ·0019 883 | ·0010 352 | ·0005 328 | ·0002 714 |
| ·18 | ·0092 365$^{-}$ | ·0050 806 | ·0027 552 | ·0014 752 | ·0007 809 | ·0004 091 |
| ·19 | ·0118 967 | ·0067 178 | ·0037 403 | ·0020 563 | ·0011 177 | ·0006 013 |
| ·20 | ·0150 854 | ·0087 326 | ·0049 848 | ·0028 100 | ·0015 661 | ·0008 640 |
| ·21 | ·0188 600 | ·0111 778 | ·0065 334 | ·0037 715$^{-}$ | ·0021 527 | ·0012 163 |
| ·22 | ·0232 773 | ·0141 083 | ·0084 340 | ·0049 799 | ·0029 077 | ·0016 806 |
| ·23 | ·0283 932 | ·0175 800 | ·0107 373 | ·0064 780 | ·0038 651 | ·0022 830 |
| ·24 | ·0342 612 | ·0216 497 | ·0134 966 | ·0083 120 | ·0050 630 | ·0030 532 |
| ·25 | ·0409 321 | ·0263 735$^{-}$ | ·0167 667 | ·0105 313 | ·0065 429 | ·0040 248 |
| ·26 | ·0484 533 | ·0318 069 | ·0206 040 | ·0131 881 | ·0083 503 | ·0052 352 |
| ·27 | ·0568 679 | ·0380 036 | ·0250 652 | ·0163 368 | ·0105 339 | ·0067 260 |
| ·28 | ·0662 140 | ·0450 148 | ·0302 072 | ·0200 336 | ·0131 455$^{-}$ | ·0085 423 |
| ·29 | ·0765 244 | ·0528 888 | ·0360 857 | ·0243 362 | ·0162 397 | ·0107 329 |
| ·30 | ·0878 258 | ·0616 695$^{-}$ | ·0427 552 | ·0293 024 | ·0198 733 | ·0133 501 |
| ·31 | ·1001 384 | ·0713 966 | ·0502 676 | ·0349 903 | ·0241 046 | ·0164 490 |
| ·32 | ·1134 755$^{+}$ | ·0821 043 | ·0586 717 | ·0414 566 | ·0289 934 | ·0200 876 |
| ·33 | ·1278 433 | ·0938 210 | ·0680 125$^{+}$ | ·0487 568 | ·0345 993 | ·0243 255$^{+}$ |
| ·34 | ·1432 404 | ·1065 687 | ·0783 303 | ·0569 435$^{+}$ | ·0409 820 | ·0292 243 |
| ·35 | ·1596 577 | ·1203 623 | ·0896 599 | ·0660 662 | ·0481 997 | ·0348 460 |
| ·36 | ·1770 786 | ·1352 096 | ·1020 304 | ·0761 703 | ·0563 087 | ·0412 527 |
| ·37 | ·1954 786 | ·1511 107 | ·1154 640 | ·0872 961 | ·0653 625$^{+}$ | ·0485 059 |
| ·38 | ·2148 256 | ·1680 574 | ·1299 758 | ·0994 782 | ·0754 108 | ·0566 652 |
| ·39 | ·2350 800 | ·1860 339 | ·1455 731 | ·1127 450$^{+}$ | ·0864 986 | ·0657 877 |
| ·40 | ·2561 949 | ·2050 159 | ·1622 553 | ·1271 175$^{+}$ | ·0986 654 | ·0759 272 |
| ·41 | ·2781 166 | ·2249 710 | ·1800 132 | ·1426 092 | ·1119 445$^{-}$ | ·0871 326 |
| ·42 | ·3007 846 | ·2458 585$^{+}$ | ·1988 288 | ·1592 249 | ·1263 618 | ·0994 477 |
| ·43 | ·3241 326 | ·2676 302 | ·2186 757 | ·1769 611 | ·1419 355$^{+}$ | ·1129 099 |
| ·44 | ·3480 884 | ·2902 298 | ·2395 182 | ·1958 048 | ·1586 751 | ·1275 491 |
| ·45 | ·3725 750$^{-}$ | ·3135 941 | ·2613 120 | ·2157 338 | ·1765 811 | ·1433 871 |
| ·46 | ·3975 107 | ·3376 527 | ·2840 043 | ·2367 160 | ·1956 439 | ·1604 366 |
| ·47 | ·4228 105$^{+}$ | ·3623 293 | ·3075 336 | ·2587 099 | ·2158 441 | ·1787 009 |
| ·48 | ·4483 860 | ·3875 414 | ·3318 307 | ·2816 642 | ·2371 519 | ·1981 725$^{-}$ |
| ·49 | ·4741 465$^{+}$ | ·4132 019 | ·3568 186 | ·3055 184 | ·2595 268 | ·2188 332 |
| ·50 | ·5000 000$^{e}$ | ·4392 190 | ·3824 137 | ·3302 026 | ·2829 175$^{+}$ | ·2406 535$^{-}$ |
| ·51 | ·5258 535$^{-}$ | ·4654 975$^{-}$ | ·4085 256 | ·3556 384 | ·3072 626 | ·2635 921 |
| ·52 | ·5516 140 | ·4919 394 | ·4350 587 | ·3817 390 | ·3324 901 | ·2875 963 |
| ·53 | ·5771 895$^{-}$ | ·5184 450$^{-}$ | ·4619 126 | ·4084 102 | ·3585 181 | ·3126 013 |
| ·54 | ·6024 893 | ·5449 133 | ·4889 828 | ·4355 511 | ·3852 554 | ·3385 312 |
| ·55 | ·6274 250$^{+}$ | ·5712 434 | ·5161 621 | ·4630 548 | ·4126 020 | ·3652 986 |
| ·56 | ·6519 116 | ·5973 352 | ·5433 414 | ·4908 095$^{+}$ | ·4404 502 | ·3928 058 |
| ·57 | ·6758 674 | ·6230 904 | ·5704 105$^{+}$ | ·5186 996 | ·4686 852 | ·4209 451 |
| ·58 | ·6992 154 | ·6484 132 | ·5972 596 | ·5466 066 | ·4971 861 | ·4495 995$^{+}$ |
| ·59 | ·7218 834 | ·6732 113 | ·6237 800 | ·5744 107 | ·5258 275$^{+}$ | ·4786 446 |
| ·60 | ·7438 051 | ·6973 970 | ·6498 655$^{+}$ | ·6019 914 | ·5544 807 | ·5079 487 |

$x$ = ·61 to ·99 $q = 5·5$ $p$ = 5·5 to 8

| $x$ | $p = 5·5$ | $p = 6$ | $p = 6·5$ | $p = 7$ | $p = 7·5$ | $p = 8$ |
|---|---|---|---|---|---|---|
| $B(p, q)$ = | ·7550 0617 × $\frac{1}{10^3}$ | ·5278 5192 × $\frac{1}{10^3}$ | ·3775 0308 × $\frac{1}{10^3}$ | ·2754 0100 × $\frac{1}{10^3}$ | ·2044 8084 × $\frac{1}{10^3}$ | ·1542 2456 × $\frac{1}{10^3}$ |
| ·61 | ·7649 200 | ·7208 875+ | ·6754 132 | ·6292 294 | ·5830 146 | ·5373 749 |
| ·62 | ·7851 744 | ·7436 061 | ·7003 246 | ·6560 076 | ·6112 976 | ·5667 823 |
| ·63 | ·8045 214 | ·7654 825+ | ·7245 068 | ·6822 124 | ·6391 989 | ·5960 278 |
| ·64 | ·8229 214 | ·7864 539 | ·7478 731 | ·7077 348 | ·6665 901 | ·6249 673 |
| ·65 | ·8403 423 | ·8064 650+ | ·7703 444 | ·7324 720 | ·6933 468 | ·6534 579 |
| ·66 | ·8567 596 | ·8254 690 | ·7918 495− | ·7563 282 | ·7193 498 | ·6813 598 |
| ·67 | ·8721 567 | ·8434 274 | ·8123 259 | ·7792 160 | ·7444 869 | ·7085 376 |
| ·68 | ·8865 245− | ·8603 108 | ·8317 207 | ·8010 572 | ·7686 541 | ·7348 624 |
| ·69 | ·8998 616 | ·8760 987 | ·8499 908 | ·8217 836 | ·7917 571 | ·7602 136 |
| ·70 | ·9121 742 | ·8907 798 | ·8671 037 | ·8413 383 | ·8137 124 | ·7844 806 |
| ·71 | ·9234 756 | ·9043 517 | ·8830 370 | ·8596 756 | ·8344 484 | ·8075 641 |
| ·72 | ·9337 860 | ·9168 210 | ·8977 792 | ·8767 620 | ·8539 063 | ·8293 778 |
| ·73 | ·9431 321 | ·9282 029 | ·9113 295− | ·8925 763 | ·8720 410 | ·8498 498 |
| ·74 | ·9515 467 | ·9385 206 | ·9236 974 | ·9071 097 | ·8888 214 | ·8689 231 |
| ·75 | ·9590 679 | ·9478 055− | ·9349 025− | ·9203 661 | ·9042 310 | ·8865 569 |
| ·76 | ·9657 388 | ·9560 958 | ·9449 742 | ·9323 613 | ·9182 678 | ·9027 270 |
| ·77 | ·9716 068 | ·9634 362 | ·9539 509 | ·9431 228 | ·9309 440 | ·9174 262 |
| ·78 | ·9767 227 | ·9698 776 | ·9618 793 | ·9526 894 | ·9422 861 | ·9306 640 |
| ·79 | ·9811 400 | ·9754 752 | ·9688 135− | ·9611 102 | ·9523 338 | ·9424 665− |
| ·80 | ·9849 146 | ·9802 886 | ·9748 140 | ·9684 434 | ·9611 394 | ·9528 754 |
| ·81 | ·9881 033 | ·9843 803 | ·9799 469 | ·9747 555+ | ·9687 663 | ·9619 477 |
| ·82 | ·9907 635+ | ·9878 151 | ·9842 822 | ·9801 199 | ·9752 882 | ·9697 534 |
| ·83 | ·9929 524 | ·9906 585+ | ·9878 932 | ·9846 152 | ·9807 869 | ·9763 748 |
| ·84 | ·9947 261 | ·9929 765− | ·9908 546 | ·9883 241 | ·9853 511 | ·9819 039 |
| ·85 | ·9961 389 | ·9948 338 | ·9932 416 | ·9913 315− | ·9890 739 | ·9864 408 |
| ·86 | ·9972 426 | ·9962 933 | ·9951 284 | ·9937 228 | ·9920 516 | ·9900 909 |
| ·87 | ·9980 861 | ·9974 153 | ·9965 873 | ·9955 824 | ·9943 807 | ·9929 627 |
| ·88 | ·9987 147 | ·9982 562 | ·9976 871 | ·9969 923 | ·9961 567 | ·9951 650− |
| ·89 | ·9991 696 | ·9988 682 | ·9984 919 | ·9980 300 | ·9974 713 | ·9968 045− |
| ·90 | ·9994 876 | ·9992 985− | ·9990 610 | ·9987 678 | ·9984 112 | ·9979 832 |
| ·91 | ·9997 009 | ·9995 887 | ·9994 470 | ·9992 710 | ·9990 559 | ·9987 962 |
| ·92 | ·9998 370 | ·9997 748 | ·9996 959 | ·9995 974 | ·9994 762 | ·9993 292 |
| ·93 | ·9999 186 | ·9998 870 | ·9998 467 | ·9997 962 | ·9997 337 | ·9996 574 |
| ·94 | ·9999 637 | ·9999 494 | ·9999 311 | ·9999 079 | ·9998 791 | ·9998 438 |
| ·95 | ·9999 861 | ·9999 806 | ·9999 735− | ·9999 644 | ·9999 530 | ·9999 391 |
| ·96 | ·9999 958 | ·9999 941 | ·9999 918 | ·9999 890 | ·9999 854 | ·9999 810 |
| ·97 | ·9999 991 | ·9999 987 | ·9999 982 | ·9999 976 | ·9999 968 | ·9999 959 |
| ·98 | ·9999 999 | ·9999 999 | ·9999 998 | ·9999 997 | ·9999 996 | ·9999 995+ |
| ·99 | 1·0000 000 | 1·0000 000 | 1·0000 000 | 1·0000 000 | 1·0000 000 | 1·0000 000 |

$x$ = ·07 to ·70 $q = 5{\cdot}5$ $p$ = 8·5 to 11

| | $p = 8{\cdot}5$ | $p = 9$ | $p = 9{\cdot}5$ | $p = 10$ | $p = 10{\cdot}5$ | $p = 11$ |
|---|---|---|---|---|---|---|
| $B(p, q) =$ | ·1179 6971 $\times \frac{1}{10^3}$ | ·9139 2333 $\times \frac{1}{10^4}$ | ·7162 4469 $\times \frac{1}{10^4}$ | ·5672 6275$\overset{+}{\times} \frac{1}{10^4}$ | ·4536 2164 $\times \frac{1}{10^4}$ | ·3659 7597 $\times \frac{1}{10^4}$ |
| $x$ | | | | | | |
| ·07 | ·0000 001 | | | | | |
| ·08 | ·0000 003 | ·0000 001 | | | | |
| ·09 | ·0000 009 | ·0000 003 | ·0000 001 | | | |
| ·10 | ·0000 021 | ·0000 008 | ·0000 003 | ·0000 001 | | |
| ·11 | ·0000 045− | ·0000 018 | ·0000 007 | ·0000 003 | ·0000 001 | |
| ·12 | ·0000 089 | ·0000 038 | ·0000 016 | ·0000 006 | ·0000 003 | ·0000 001 |
| ·13 | ·0000 168 | ·0000 074 | ·0000 032 | ·0000 014 | ·0000 006 | ·0000 003 |
| ·14 | ·0000 302 | ·0000 137 | ·0000 062 | ·0000 028 | ·0000 012 | ·0000 005+ |
| ·15 | ·0000 519 | ·0000 244 | ·0000 114 | ·0000 053 | ·0000 024 | ·0000 011 |
| ·16 | ·0000 857 | ·0000 416 | ·0000 200 | ·0000 096 | ·0000 045+ | ·0000 021 |
| ·17 | ·0001 369 | ·0000 685− | ·0000 340 | ·0000 167 | ·0000 082 | ·0000 040 |
| ·18 | ·0002 123 | ·0001 092 | ·0000 557 | ·0000 282 | ·0000 142 | ·0000 071 |
| ·19 | ·0003 204 | ·0001 693 | ·0000 887 | ·0000 462 | ·0000 239 | ·0000 123 |
| ·20 | ·0004 722 | ·0002 559 | ·0001 375+ | ·0000 734 | ·0000 389 | ·0000 205+ |
| ·21 | ·0006 808 | ·0003 779 | ·0002 081 | ·0001 138 | ·0000 618 | ·0000 334 |
| ·22 | ·0009 624 | ·0005 465+ | ·0003 079 | ·0001 723 | ·0000 957 | ·0000 529 |
| ·23 | ·0013 362 | ·0007 755− | ·0004 466 | ·0002 553 | ·0001 450+ | ·0000 819 |
| ·24 | ·0018 245− | ·0010 811 | ·0006 357 | ·0003 712 | ·0002 153 | ·0001 241 |
| ·25 | ·0024 534 | ·0014 831 | ·0008 897 | ·0005 300 | ·0003 137 | ·0001 845− |
| ·26 | ·0032 528 | ·0020 044 | ·0012 257 | ·0007 443 | ·0004 491 | ·0002 693 |
| ·27 | ·0042 563 | ·0026 714 | ·0016 641 | ·0010 293 | ·0006 326 | ·0003 865− |
| ·28 | ·0055 018 | ·0035 148 | ·0022 286 | ·0014 033 | ·0008 779 | ·0005 460 |
| ·29 | ·0070 311 | ·0045 689 | ·0029 469 | ·0018 876 | ·0012 014 | ·0007 601 |
| ·30 | ·0088 899 | ·0058 724 | ·0038 505− | ·0025 075− | ·0016 225+ | ·0010 437 |
| ·31 | ·0111 278 | ·0074 681 | ·0049 753 | ·0032 920 | ·0021 645+ | ·0014 148 |
| ·32 | ·0137 980 | ·0094 030 | ·0063 613 | ·0042 745− | ·0028 542 | ·0018 948 |
| ·33 | ·0169 571 | ·0117 282 | ·0080 530 | ·0054 925+ | ·0037 228 | ·0025 087 |
| ·34 | ·0206 645− | ·0144 985+ | ·0100 994 | ·0069 883 | ·0048 057 | ·0032 858 |
| ·35 | ·0249 819 | ·0177 724 | ·0125 536 | ·0088 088 | ·0061 432 | ·0042 598 |
| ·36 | ·0299 730 | ·0216 116 | ·0154 729 | ·0110 054 | ·0077 802 | ·0054 690 |
| ·37 | ·0357 025− | ·0260 803 | ·0189 184 | ·0136 342 | ·0097 666 | ·0069 568 |
| ·38 | ·0422 353 | ·0312 450− | ·0229 545+ | ·0167 554 | ·0121 573 | ·0087 717 |
| ·39 | ·0496 361 | ·0371 732 | ·0276 487 | ·0204 336 | ·0150 118 | ·0109 674 |
| ·40 | ·0579 679 | ·0439 332 | ·0330 706 | ·0247 368 | ·0183 943 | ·0136 029 |
| ·41 | ·0672 914 | ·0515 929 | ·0392 914 | ·0297 360 | ·0223 734 | ·0167 421 |
| ·42 | ·0776 640 | ·0602 190 | ·0463 827 | ·0355 048 | ·0270 213 | ·0204 539 |
| ·43 | ·0891 383 | ·0698 755+ | ·0544 163 | ·0421 183 | ·0324 136 | ·0248 117 |
| ·44 | ·1017 616 | ·0806 232 | ·0634 622 | ·0496 522 | ·0386 280 | ·0298 927 |
| ·45 | ·1155 747 | ·0925 183 | ·0735 880 | ·0581 818 | ·0457 442 | ·0357 773 |
| ·46 | ·1306 106 | ·1056 109 | ·0848 578 | ·0677 811 | ·0538 422 | ·0425 486 |
| ·47 | ·1468 936 | ·1199 443 | ·0973 304 | ·0785 209 | ·0630 014 | ·0502 908 |
| ·48 | ·1644 386 | ·1355 537 | ·1110 585+ | ·0904 682 | ·0732 991 | ·0590 886 |
| ·49 | ·1832 499 | ·1524 648 | ·1260 872 | ·1036 841 | ·0848 096 | ·0690 254 |
| ·50 | ·2033 206 | ·1706 932 | ·1424 524 | ·1182 230 | ·0976 021 | ·0801 821 |
| ·51 | ·2246 320 | ·1902 428 | ·1601 802 | ·1341 305− | ·1117 395− | ·0926 355− |
| ·52 | ·2471 529 | ·2111 053 | ·1792 848 | ·1514 424 | ·1272 764 | ·1064 565+ |
| ·53 | ·2708 395+ | ·2332 596 | ·1997 683 | ·1701 829 | ·1442 580 | ·1217 085− |
| ·54 | ·2956 349 | ·2566 706 | ·2216 189 | ·1903 636 | ·1627 177 | ·1384 452 |
| ·55 | ·3214 692 | ·2812 894 | ·2448 103 | ·2119 818 | ·1826 763 | ·1567 090 |
| ·56 | ·3482 598 | ·3070 526 | ·2693 013 | ·2350 199 | ·2041 399 | ·1765 294 |
| ·57 | ·3759 116 | ·3338 824 | ·2950 349 | ·2594 439 | ·2270 985+ | ·1979 204 |
| ·58 | ·4043 178 | ·3616 869 | ·3219 383 | ·2852 028 | ·2515 254 | ·2208 798 |
| ·59 | ·4333 604 | ·3903 602 | ·3499 224 | ·3122 286 | ·2773 753 | ·2453 869 |
| ·60 | ·4629 113 | ·4197 836 | ·3788 829 | ·3404 351 | ·3045 841 | ·2714 019 |
| ·61 | ·4928 335+ | ·4498 260 | ·4086 999 | ·3697 188 | ·3330 683 | ·2988 640 |
| ·62 | ·5229 827 | ·4803 453 | ·4392 395+ | ·3999 589 | ·3627 246 | ·3276 915− |
| ·63 | ·5532 085+ | ·5111 898 | ·4703 543 | ·4310 179 | ·3934 307 | ·3577 808 |
| ·64 | ·5833 564 | ·5421 997 | ·5018 852 | ·4627 429 | ·4250 450− | ·3890 069 |
| ·65 | ·6132 693 | ·5732 093 | ·5336 629 | ·4949 669 | ·4574 083 | ·4212 235− |
| ·66 | ·6427 902 | ·6040 487 | ·5655 101 | ·5275 107 | ·4903 449 | ·4542 638 |
| ·67 | ·6717 636 | ·6345 461 | ·5972 436 | ·5601 847 | ·5236 648 | ·4879 426 |
| ·68 | ·7000 377 | ·6645 304 | ·6286 767 | ·5927 919 | ·5571 654 | ·5220 576 |
| ·69 | ·7274 671 | ·6938 334 | ·6596 222 | ·6251 299 | ·5906 346 | ·5563 921 |
| ·70 | ·7539 142 | ·7222 924 | ·6898 949 | ·6569 948 | ·6238 539 | ·5907 178 |

$x = \cdot 71$ to $\cdot 99$ $q = 5\cdot 5$ $p = 8\cdot 5$ to $11$

| $x$ | $p = 8\cdot 5$ | $p = 9$ | $p = 9\cdot 5$ | $p = 10$ | $p = 10\cdot 5$ | $p = 11$ |
|---|---|---|---|---|---|---|
| $B(p, q) =$ | $\cdot 1179\ 6971 \times \frac{1}{10^3}$ | $\cdot 9139\ 2333 \times \frac{1}{10^4}$ | $\cdot 7162\ 4469 \times \frac{1}{10^4}$ | $\cdot 5672\ 6275^{+} \times \frac{1}{10^4}$ | $\cdot 4536\ 2164 \times \frac{1}{10^4}$ | $\cdot 3659\ 7597 \times \frac{1}{10^4}$ |
| ·71 | ·7792 517 | ·7497 528 | ·7193 146 | ·6881 839 | ·6566 015+ | ·6247 986 |
| ·72 | ·8033 647 | ·7760 707 | ·7477 094 | ·7184 989 | ·6886 563 | ·6583 940 |
| ·73 | ·8261 520 | ·8011 149 | ·7749 183 | ·7477 502 | ·7198 017 | ·6912 636 |
| ·74 | ·8475 283 | ·8247 697 | ·8007 943 | ·7757 595+ | ·7498 295+ | ·7231 715+ |
| ·75 | ·8674 253 | ·8469 367 | ·8252 069 | ·8023 640 | ·7785 445+ | ·7538 911 |
| ·76 | ·8857 927 | ·8675 364 | ·8480 452 | ·8274 191 | ·8057 681 | ·7832 098 |
| ·77 | ·9025 994 | ·8865 101 | ·8692 196 | ·8508 019 | ·8313 422 | ·8109 338 |
| ·78 | ·9178 338 | ·9038 206 | ·8886 636 | ·8724 138 | ·8551 332 | ·8368 927 |
| ·79 | ·9315 035+ | ·9194 531 | ·9063 356 | ·8921 824 | ·8770 349 | ·8609 437 |
| ·80 | ·9436 358 | ·9334 155− | ·9222 197 | ·9100 636 | ·8969 713 | ·8829 754 |
| ·81 | ·9542 763 | ·9457 377 | ·9363 258 | ·9260 426 | ·9148 985− | ·9029 108 |
| ·82 | ·9634 880 | ·9564 713 | ·9486 892 | ·9401 343 | ·9308 060 | ·9207 097 |
| ·83 | ·9713 498 | ·9656 879 | ·9593 701 | ·9523 827 | ·9447 172 | ·9363 700 |
| ·84 | ·9779 543 | ·9734 773 | ·9684 517 | ·9628 599 | ·9566 884 | ·9499 278 |
| ·85 | ·9834 059 | ·9799 453 | ·9760 375+ | ·9716 636 | ·9668 076 | ·9614 564 |
| ·86 | ·9878 178 | ·9852 106 | ·9822 491 | ·9789 150− | ·9751 915+ | ·9710 643 |
| ·87 | ·9913 091 | ·9894 014 | ·9872 220 | ·9847 540 | ·9819 819 | ·9788 914 |
| ·88 | ·9940 019 | ·9926 524 | ·9911 018 | ·9893 358 | ·9873 409 | ·9851 039 |
| ·89 | ·9960 180 | ·9951 002 | ·9940 397 | ·9928 250+ | ·9914 451 | ·9898 889 |
| ·90 | ·9974 756 | ·9968 799 | ·9961 877 | ·9953 905− | ·9944 796 | ·9934 467 |
| ·91 | ·9984 865− | ·9981 210 | ·9976 940 | ·9971 994 | ·9966 312 | ·9959 833 |
| ·92 | ·9991 529 | ·9989 437 | ·9986 979 | ·9984 117 | ·9980 810 | ·9977 019 |
| ·93 | ·9995 654 | ·9994 557 | ·9993 261 | ·9991 744 | ·9989 981 | ·9987 950− |
| ·94 | ·9998 011 | ·9997 497 | ·9996 888 | ·9996 171 | ·9995 333 | ·9994 362 |
| ·95 | ·9999 220 | ·9999 015− | ·9998 770 | ·9998 480 | ·9998 139 | ·9997 742 |
| ·96 | ·9999 756 | ·9999 690 | ·9999 612 | ·9999 518 | ·9999 407 | ·9999 278 |
| ·97 | ·9999 946 | ·9999 932 | ·9999 914 | ·9999 893 | ·9999 868 | ·9999 838 |
| ·98 | ·9999 994 | ·9999 992 | ·9999 990 | ·9999 988 | ·9999 985− | ·9999 981 |
| ·99 | 1·0000 000 | 1·0000 000 | 1·0000 000 | 1·0000 000 | 1·0000 000 | 1·0000 000 |

$x$ = ·14 to ·70 $q = 5{\cdot}5$ $p$ = 12 to 17

| $x$ | $p = 12$ | $p = 13$ | $p = 14$ | $p = 15$ | $p = 16$ | $p = 17$ |
|---|---|---|---|---|---|---|
| $B(p, q) =$ | ·2439 8398 × $\frac{1}{10^4}$ | ·1673 0330 × $\frac{1}{10^4}$ | ·1175 6448 × $\frac{1}{10^4}$ | ·8440 5269 × $\frac{1}{10^5}$ | ·6175 9953 × $\frac{1}{10^5}$ | ·4596 0895 × $\frac{1}{10^5}$ |
| ·14 | ·0000 001 | | | | | |
| ·15 | ·0000 002 | | | | | |
| ·16 | ·0000 005⁻ | ·0000 001 | | | | |
| ·17 | ·0000 009 | ·0000 002 | | | | |
| ·18 | ·0000 017 | ·0000 004 | ·0000 001 | | | |
| ·19 | ·0000 032 | ·0000 008 | ·0000 002 | | | |
| ·20 | ·0000 056 | ·0000 015⁻ | ·0000 004 | ·0000 001 | | |
| ·21 | ·0000 096 | ·0000 027 | ·0000 007 | ·0000 002 | ·0000 001 | |
| ·22 | ·0000 159 | ·0000 047 | ·0000 013 | ·0000 004 | ·0000 001 | |
| ·23 | ·0000 257 | ·0000 079 | ·0000 024 | ·0000 007 | ·0000 002 | ·0000 001 |
| ·24 | ·0000 406 | ·0000 130 | ·0000 041 | ·0000 013 | ·0000 004 | ·0000 001 |
| ·25 | ·0000 628 | ·0000 210 | ·0000 069 | ·0000 022 | ·0000 007 | ·0000 002 |
| ·26 | ·0000 953 | ·0000 330 | ·0000 113 | ·0000 038 | ·0000 013 | ·0000 004 |
| ·27 | ·0001 419 | ·0000 511 | ·0000 181 | ·0000 063 | ·0000 022 | ·0000 007 |
| ·28 | ·0002 077 | ·0000 775⁺ | ·0000 284 | ·0000 103 | ·0000 037 | ·0000 013 |
| ·29 | ·0002 993 | ·0001 156 | ·0000 439 | ·0000 164 | ·0000 061 | ·0000 022 |
| ·30 | ·0004 250⁻ | ·0001 697 | ·0000 666 | ·0000 258 | ·0000 098 | ·0000 037 |
| ·31 | ·0005 949 | ·0002 453 | ·0000 995⁻ | ·0000 397 | ·0000 157 | ·0000 061 |
| ·32 | ·0008 218 | ·0003 497 | ·0001 463 | ·0000 603 | ·0000 245⁺ | ·0000 099 |
| ·33 | ·0011 213 | ·0004 917 | ·0002 120 | ·0000 901 | ·0000 378 | ·0000 156 |
| ·34 | ·0015 120 | ·0006 827 | ·0003 032 | ·0001 326 | ·0000 573 | ·0000 244 |
| ·35 | ·0020 163 | ·0009 366 | ·0004 279 | ·0001 926 | ·0000 856 | ·0000 376 |
| ·36 | ·0026 605⁻ | ·0012 703 | ·0005 966 | ·0002 761 | ·0001 261 | ·0000 569 |
| ·37 | ·0034 754 | ·0017 043 | ·0008 222 | ·0003 908 | ·0001 834 | ·0000 850⁺ |
| ·38 | ·0044 968 | ·0022 631 | 0011 205⁺ | ·0005 468 | ·0002 634 | ·0001 254 |
| ·39 | ·0057 653 | ·0029 757 | ·0015 111 | ·0007 564 | ·0003 737 | ·0001 825⁻ |
| ·40 | ·0073 274 | ·0038 759 | ·0020 174 | ·0010 351 | ·0005 242 | ·0002 624 |
| ·41 | ·0092 352 | ·0050 031 | ·0026 674 | ·0014 019 | ·0007 274 | ·0003 731 |
| ·42 | ·0115 465⁺ | ·0064 025⁺ | ·0034 942 | ·0018 800 | ·0009 987 | ·0005 245⁻ |
| ·43 | ·0143 255⁻ | ·0081 255⁻ | ·0045 366 | ·0024 974 | ·0013 575⁻ | ·0007 295⁻ |
| ·44 | ·0176 417 | ·0102 299 | ·0058 399 | ·0032 874 | ·0018 273 | ·0010 042 |
| ·45 | ·0215 708 | ·0127 805⁺ | ·0074 557 | ·0042 893 | ·0024 369 | ·0013 689 |
| ·46 | ·0261 934 | ·0158 487 | ·0094 430 | ·0055 492 | ·0032 207 | ·0018 483 |
| ·47 | ·0315 949 | ·0195 126 | ·0118 682 | ·0071 205⁺ | ·0042 197 | ·0024 728 |
| ·48 | ·0378 646 | ·0238 568 | ·0148 056 | ·0090 646 | ·0054 821 | ·0032 790 |
| ·49 | ·0450 951 | ·0289 721 | ·0183 371 | ·0114 509 | ·0070 645⁻ | ·0043 106 |
| ·50 | ·0533 805⁺ | ·0349 545⁻ | ·0225 523 | ·0143 580 | ·0090 317 | ·0056 196 |
| ·51 | ·0628 157 | ·0419 045⁻ | ·0275 482 | ·0178 731 | ·0114 584 | ·0072 669 |
| ·52 | ·0734 942 | ·0499 261 | ·0334 284 | ·0220 922 | ·0144 288 | ·0093 232 |
| ·53 | ·0855 069 | ·0591 250⁻ | ·0403 025⁺ | ·0271 201 | ·0180 374 | ·0118 696 |
| ·54 | ·0989 400 | ·0696 074 | ·0482 848 | ·0330 697 | ·0223 886 | ·0149 987 |
| ·55 | ·1138 726 | ·0814 776 | ·0574 928 | ·0400 610 | ·0275 973 | ·0188 142 |
| ·56 | ·1303 750⁻ | ·0948 361 | ·0680 453 | ·0482 201 | ·0337 873 | ·0234 318 |
| ·57 | ·1485 059 | ·1097 772 | ·0800 604 | ·0576 773 | ·0410 915⁻ | ·0289 786 |
| ·58 | ·1683 103 | ·1263 861 | ·0936 530 | ·0685 654 | ·0496 494 | ·0355 924 |
| ·59 | ·1898 172 | ·1447 363 | ·1089 316 | ·0810 168 | ·0596 061 | ·0434 210 |
| ·60 | ·2130 374 | ·1648 867 | ·1259 958 | ·0951 608 | ·0711 096 | ·0526 202 |
| ·61 | ·2379 610 | ·1868 787 | ·1449 322 | ·1111 204 | ·0843 076 | ·0633 517 |
| ·62 | ·2645 561 | ·2107 331 | ·1658 116 | ·1290 081 | ·0993 445⁺ | ·0757 805⁻ |
| ·63 | ·2927 663 | ·2364 475⁺ | ·1886 848 | ·1489 224 | ·1163 570 | ·0900 705⁺ |
| ·64 | ·3225 105⁺ | ·2639 937 | ·2135 792 | ·1709 429 | ·1354 695⁺ | ·1063 814 |
| ·65 | ·3536 811 | ·2933 152 | ·2404 950⁺ | ·1951 262 | ·1567 896 | ·1248 624 |
| ·66 | ·3861 442 | ·3243 257 | ·2694 023 | ·2215 012 | ·1804 021 | ·1456 477 |
| ·67 | ·4197 395⁻ | ·3569 074 | ·3002 376 | ·2500 647 | ·2063 640 | ·1688 497 |
| ·68 | ·4542 813 | ·3909 105⁺ | ·3329 019 | ·2807 769 | ·2346 985⁻ | ·1945 526 |
| ·69 | ·4895 601 | ·4261 533 | ·3672 582 | ·3135 584 | ·2653 897 | ·2228 056 |
| ·70 | ·5253 446 | ·4624 228 | ·4031 311 | ·3482 864 | ·2983 776 | ·2536 157 |

$x = \cdot71$ to $1\cdot00$ $q = 5\cdot5$ $p = 12$ to $17$

| | $p = 12$ | $p = 13$ | $p = 14$ | $p = 15$ | $p = 16$ | $p = 17$ |
|---|---|---|---|---|---|---|
| $B(p, q) =$ | $\cdot2439\ 8398 \times \frac{1}{10^4}$ | $\cdot1673\ 0330 \times \frac{1}{10^4}$ | $\cdot1175\ 6448 \times \frac{1}{10^4}$ | $\cdot8440\ 5269 \times \frac{1}{10^5}$ | $\cdot6175\ 9953 \times \frac{1}{10^5}$ | $\cdot4596\ 0895^{+} \times \frac{1}{10^5}$ |
| $x$ | | | | | | |
| ·71 | ·5613 845+ | ·4994 766 | ·4403 068 | ·3847 929 | ·3335 536 | ·2869 418 |
| ·72 | ·5974 149 | ·5370 455+ | ·4785 337 | ·4228 638 | ·3707 569 | ·3226 882 |
| ·73 | ·6331 595+ | ·5748 375− | ·5175 249 | ·4622 387 | ·4097 721 | ·3606 995− |
| ·74 | ·6683 366 | ·6125 421 | ·5569 622 | ·5026 130 | ·4503 290 | ·4007 573 |
| ·75 | ·7026 642 | ·6498 363 | ·5965 006 | ·5436 410 | ·4921 029 | ·4425 781 |
| ·76 | ·7358 657 | ·6863 911 | ·6357 747 | ·5849 415− | ·5347 182 | ·4858 133 |
| ·77 | ·7676 769 | ·7218 786 | ·6744 069 | ·6261 044 | ·5777 537 | ·5300 527 |
| ·78 | ·7978 517 | ·7559 804 | ·7120 154 | ·6667 001 | ·6207 503 | ·5748 293 |
| ·79 | ·8261 693 | ·7883 957 | ·7482 249 | ·7062 895− | ·6632 218 | ·6196 292 |
| ·80 | ·8524 395+ | ·8188 501 | ·7826 769 | ·7444 367 | ·7046 668 | ·6639 027 |
| ·81 | ·8765 089 | ·8471 038 | ·8150 410 | ·7807 222 | ·7445 846 | ·7070 806 |
| ·82 | ·8982 654 | ·8729 595− | ·8450 256 | ·8147 573 | ·7824 912 | ·7485 917 |
| ·83 | ·9176 424 | ·8962 695− | ·8723 893 | ·8461 980 | ·8179 375− | ·7878 842 |
| ·84 | ·9346 209 | ·9169 414 | ·8969 501 | ·8747 596 | ·8505 277 | ·8244 480 |
| ·85 | ·9492 310 | ·9349 425− | ·9185 931 | ·9002 293 | ·8799 373 | ·8578 381 |
| ·86 | ·9615 508 | ·9503 010 | ·9372 773 | ·9224 767 | ·9059 297 | ·8876 969 |
| ·87 | ·9717 042 | ·9631 066 | ·9530 375+ | ·9414 617 | ·9283 693 | ·9137 755+ |
| ·88 | ·9798 563 | ·9735 067 | ·9659 849 | ·9572 380 | ·9472 317 | ·9359 495+ |
| ·89 | ·9862 067 | ·9817 007 | ·9763 021 | ·9699 529 | ·9626 069 | ·9542 302 |
| ·90 | ·9909 817 | ·9879 312 | ·9842 354 | ·9798 401 | ·9746 976 | ·9687 676 |
| ·91 | ·9944 240 | ·9924 728 | ·9900 827 | ·9872 085+ | ·9838 084 | ·9798 441 |
| ·92 | ·9967 819 | ·9956 180 | ·9941 766 | ·9924 242 | ·9903 284 | ·9878 580 |
| ·93 | ·9982 978 | ·9976 621 | ·9968 661 | ·9958 880 | ·9947 054 | ·9932 964 |
| ·94 | ·9991 966 | ·9988 870 | ·9984 953 | ·9980 086 | ·9974 139 | ·9966 977 |
| ·95 | ·9996 755− | ·9995 465+ | ·9993 816 | ·9991 745+ | ·9989 188 | ·9986 076 |
| ·96 | ·9998 953 | ·9998 525− | ·9997 971 | ·9997 268 | ·9996 391 | ·9995 313 |
| ·97 | ·9999 763 | ·9999 664 | ·9999 533 | ·9999 366 | ·9999 156 | ·9998 894 |
| ·98 | ·9999 972 | ·9999 960 | ·9999 944 | ·9999 923 | ·9999 897 | ·9999 864 |
| ·99 | ·9999 999 | ·9999 999 | ·9999 999 | ·9999 998 | ·9999 997 | ·9999 997 |
| 1·00 | 1·0000 000 | 1·0000 000 | 1·0000 000 | 1·0000 000 | 1·0000 000 | 1·0000 000 |

$x$ = ·25 to ·80 $q$ = 5·5 $p$ = 18 to 23

| | $p$ = 18 | $p$ = 19 | $p$ = 20 | $p$ = 21 | $p$ = 22 | $p$ = 23 |
|---|---|---|---|---|---|---|
| $B(p, q)$ = | ·3472 6010 × $\frac{1}{10^5}$ | ·2659 8646 × $\frac{1}{10^5}$ | ·2062 7521 × $\frac{1}{10^5}$ | ·1617 8448 × $\frac{1}{10^5}$ | ·1282 0657 × $\frac{1}{10^5}$ | ·1025 6525$^+$ × $\frac{1}{10^5}$ |
| $x$ | | | | | | |
| ·25 | ·0000 001 | | | | | |
| ·26 | ·0000 001 | | | | | |
| ·27 | ·0000 002 | ·0000 001 | | | | |
| ·28 | ·0000 004 | ·0000 002 | ·0000 001 | | | |
| ·29 | ·0000 008 | ·0000 003 | ·0000 001 | | | |
| ·30 | ·0000 014 | ·0000 005+ | ·0000 002 | ·0000 001 | | |
| ·31 | ·0000 023 | ·0000 009 | ·0000 003 | ·0000 001 | | |
| ·32 | ·0000 039 | ·0000 015+ | ·0000 006 | ·0000 002 | ·0000 001 | |
| ·33 | ·0000 064 | ·0000 026 | ·0000 010 | ·0000 004 | ·0000 002 | ·0000 001 |
| ·34 | ·0000 103 | ·0000 043 | ·0000 018 | ·0000 007 | ·0000 003 | ·0000 001 |
| ·35 | ·0000 163 | ·0000 070 | ·0000 030 | ·0000 013 | ·0000 005+ | ·0000 002 |
| ·36 | ·0000 254 | ·0000 112 | ·0000 049 | ·0000 021 | ·0000 009 | ·0000 004 |
| ·37 | ·0000 390 | ·0000 177 | ·0000 080 | ·0000 036 | ·0000 016 | ·0000 007 |
| ·38 | ·0000 590 | ·0000 275+ | ·0000 127 | ·0000 058 | ·0000 027 | ·0000 012 |
| ·39 | ·0000 882 | ·0000 422 | ·0000 200 | ·0000 094 | ·0000 044 | ·0000 020 |
| ·40 | ·0001 300 | ·0000 638 | ·0000 310 | ·0000 150− | ·0000 072 | ·0000 034 |
| ·41 | ·0001 893 | ·0000 952 | ·0000 474 | ·0000 234 | ·0000 115+ | ·0000 056 |
| ·42 | ·0002 725+ | ·0001 403 | ·0000 716 | ·0000 362 | ·0000 182 | ·0000 091 |
| ·43 | ·0003 879 | ·0002 043 | ·0001 067 | ·0000 553 | ·0000 284 | ·0000 145+ |
| ·44 | ·0005 462 | ·0002 943 | ·0001 572 | ·0000 833 | ·0000 438 | ·0000 229 |
| ·45 | ·0007 611 | ·0004 192 | ·0002 289 | ·0001 240 | ·0000 667 | ·0000 356 |
| ·46 | ·0010 500− | ·0005 909 | ·0003 297 | ·0001 825+ | ·0001 003 | ·0000 548 |
| ·47 | ·0014 345− | ·0008 244 | ·0004 698 | ·0002 656 | ·0001 491 | ·0000 832 |
| ·48 | ·0019 415− | ·0011 390 | ·0006 626 | ·0003 825− | ·0002 192 | ·0001 248 |
| ·49 | ·0026 040 | ·0015 587 | ·0009 252 | ·0005 449 | ·0003 187 | ·0001 851 |
| ·50 | ·0034 619 | ·0021 134 | ·0012 794 | ·0007 686 | ·0004 584 | ·0002 717 |
| ·51 | ·0045 634 | ·0028 398 | ·0017 527 | ·0010 735− | ·0006 528 | ·0003 944 |
| ·52 | ·0059 654 | ·0037 829 | ·0023 792 | ·0014 850+ | ·0009 204 | ·0005 668 |
| ·53 | ·0077 354 | ·0049 965+ | ·0032 011 | ·0020 354 | ·0012 852 | ·0008 063 |
| ·54 | ·0099 518 | ·0065 451 | ·0042 698 | ·0027 647 | ·0017 778 | ·0011 359 |
| ·55 | ·0127 048 | ·0085 046 | ·0056 474 | ·0037 223 | ·0024 367 | ·0015 849 |
| ·56 | ·0160 977 | ·0109 638 | ·0074 079 | ·0049 686 | ·0033 098 | ·0021 909 |
| ·57 | ·0202 466 | ·0140 251 | ·0096 389 | ·0065 763 | ·0044 565+ | ·0030 011 |
| ·58 | ·0252 812 | ·0178 056 | ·0124 428 | ·0086 325+ | ·0059 490 | ·0040 742 |
| ·59 | ·0313 440 | ·0224 372 | ·0159 376 | ·0112 401 | ·0078 746 | ·0054 828 |
| ·60 | ·0385 900 | ·0280 673 | ·0202 584 | ·0145 189 | ·0103 372 | ·0073 150− |
| ·61 | ·0471 849 | ·0348 577 | ·0255 571 | ·0186 073 | ·0134 594 | ·0096 769 |
| ·62 | ·0573 035− | ·0429 838 | ·0320 029 | ·0236 629 | ·0173 840 | ·0126 949 |
| ·63 | ·0691 265− | ·0526 330 | ·0397 811 | ·0298 627 | ·0222 751 | ·0165 171 |
| ·64 | ·0828 371 | ·0640 016 | ·0490 919 | ·0374 026 | ·0283 184 | ·0213 153 |
| ·65 | ·0986 164 | ·0772 914 | ·0601 473 | ·0464 963 | ·0357 218 | ·0272 858 |
| ·66 | ·1166 380 | ·0927 050− | ·0731 681 | ·0573 725− | ·0447 135− | ·0346 495+ |
| ·67 | ·1370 619 | ·1104 396 | ·0883 781 | ·0702 711 | ·0555 398 | ·0436 510 |
| ·68 | ·1600 272 | ·1306 806 | ·1059 986 | ·0854 384 | ·0684 616 | ·0545 561 |
| ·69 | ·1856 446 | ·1535 933 | ·1262 401 | ·1031 198 | ·0837 482 | ·0676 478 |
| ·70 | ·2139 883 | ·1793 142 | ·1492 938 | ·1235 513 | ·1016 702 | ·0832 204 |
| ·71 | ·2450 872 | ·2079 412 | ·1753 212 | ·1469 499 | ·1224 898 | ·1015 709 |
| ·72 | ·2789 174 | ·2395 237 | ·2044 426 | ·1735 010 | ·1464 493 | ·1229 880 |
| ·73 | ·3153 939 | ·2740 525+ | ·2367 256 | ·2033 460 | ·1737 574 | ·1477 396 |
| ·74 | ·3543 640 | ·3114 502 | ·2721 728 | ·2365 678 | ·2045 742 | ·1760 562 |
| ·75 | ·3956 023 | ·3515 625− | ·3107 097 | ·2731 763 | ·2389 941 | ·2081 135+ |
| ·76 | ·4388 071 | ·3941 512 | ·3521 746 | ·3130 944 | ·2770 290 | ·2440 127 |
| ·77 | ·4836 003 | ·4388 899 | ·3963 091 | ·3561 447 | ·3185 911 | ·2837 601 |
| ·78 | ·5295 296 | ·4853 624 | ·4427 526 | ·4020 390 | ·3634 774 | ·3272 470 |
| ·79 | ·5760 751 | ·5330 654 | ·4910 404 | ·4503 707 | ·4113 568 | ·3742 316 |
| ·80 | ·6226 591 | ·5814 154 | ·5406 059 | ·5006 125+ | ·4617 618 | ·4243 239 |

$x = \cdot 81$ to $1\cdot 00$ $q = 5\cdot 5$ $p = 18$ to $23$

| | $p = 18$ | $p = 19$ | $p = 20$ | $p = 21$ | $p = 22$ | $p = 23$ |
|---|---|---|---|---|---|---|
| $B(p, q) =$ | ·3472 6010 × $\frac{1}{10^5}$ | ·2659 8646 × $\frac{1}{10^5}$ | ·2062 7521 × $\frac{1}{10^5}$ | ·1617 8448 × $\frac{1}{10^5}$ | ·1282 0657 × $\frac{1}{10^5}$ | ·1025 6525$^{+}$ × $\frac{1}{10^5}$ |
| $x$ | | | | | | |
| ·81 | ·6686 610 | ·6297 612 | ·5907 897 | ·5521 202 | ·5140 860 | ·4769 768 |
| ·82 | ·7134 359 | ·6774 012 | ·6408 545$^{-}$ | ·6041 433 | ·5675 894 | ·5314 841 |
| ·83 | ·7563 371 | ·7236 070 | ·6900 070 | ·6558 442 | ·6214 129 | ·5869 895$^{-}$ |
| ·84 | ·7967 423 | ·7676 512 | ·7374 271 | ·7063 265$^{+}$ | ·6746 039 | ·6425 064 |
| ·85 | ·8340 813 | ·8088 398 | ·7823 031 | ·7546 716 | ·7261 520 | ·6969 519 |
| ·86 | ·8678 661 | ·8465 480 | ·8238 722 | ·7999 833 | ·7750 365$^{-}$ | ·7491 938 |
| ·87 | ·8977 180 | ·8802 554 | ·8614 648 | ·8414 387 | ·8202 826 | ·7981 119 |
| ·88 | ·9233 931 | ·9095 811 | ·8945 478 | ·8783 419 | ·8610 247 | ·8426 685$^{+}$ |
| ·89 | ·9448 015$^{+}$ | ·9343 121 | ·9227 654 | ·9101 767 | ·8965 718 | ·8819 868 |
| ·90 | ·9620 179 | ·9544 245$^{-}$ | ·9459 718 | ·9366 527 | ·9264 683 | ·9154 274 |
| ·91 | ·9752 816 | ·9700 918 | ·9642 505$^{+}$ | ·9577 390 | ·9505 437 | ·9426 567 |
| ·92 | ·9849 836 | ·9816 780 | ·9779 166 | ·9736 775$^{-}$ | ·9689 418 | ·9636 938 |
| ·93 | ·9916 390 | ·9897 124 | ·9874 962 | ·9849 714 | ·9821 202 | ·9789 263 |
| ·94 | ·9958 462 | ·9948 457 | ·9936 826 | ·9923 432 | ·9908 143 | ·9890 833 |
| ·95 | ·9982 337 | ·9977 896 | ·9972 679 | ·9966 607 | ·9959 602 | ·9951 587 |
| ·96 | ·9994 003 | ·9992 432 | ·9990 566 | ·9988 372 | ·9985 814 | ·9982 856 |
| ·97 | ·9998 573 | ·9998 184 | ·9997 717 | ·9997 163 | ·9996 509 | ·9995 746 |
| ·98 | ·9999 823 | ·9999 772 | ·9999 711 | ·9999 638 | ·9999 551 | ·9999 448 |
| ·99 | ·9999 995$^{+}$ | ·9999 994 | ·9999 993 | ·9999 991 | ·9999 988 | ·9999 985$^{+}$ |
| 1·00 | 1·0000 000 | 1·0000 000 | 1·0000 000 | 1·0000 000 | 1·0000 000 | 1·0000 000 |

$x$ = ·35 to ·90 $q = 5{\cdot}5$ $p$ = 24 to 29

| $x$ | $p = 24$ | $p = 25$ | $p = 26$ | $p = 27$ | $p = 28$ | $p = 29$ |
|---|---|---|---|---|---|---|
| $B(p, q)$ = | ·8277 1960 × $\frac{1}{10^8}$ | ·6733 9900 × $\frac{1}{10^8}$ | ·5519 6639 × $\frac{1}{10^8}$ | ·4555 9131 × $\frac{1}{10^8}$ | ·3784 9124 × $\frac{1}{10^8}$ | ·3163 5089 × $\frac{1}{10^8}$ |
| ·35 | ·0000 001 | | | | | |
| ·36 | ·0000 002 | ·0000 001 | | | | |
| ·37 | ·0000 003 | ·0000 001 | ·0000 001 | | | |
| ·38 | ·0000 005+ | ·0000 002 | ·0000 001 | | | |
| ·39 | ·0000 009 | ·0000 004 | ·0000 002 | ·0000 001 | | |
| ·40 | ·0000 016 | ·0000 008 | ·0000 004 | ·0000 002 | ·0000 001 | |
| ·41 | ·0000 027 | ·0000 013 | ·0000 006 | ·0000 003 | ·0000 001 | ·0000 001 |
| ·42 | ·0000 045+ | ·0000 022 | ·0000 011 | ·0000 005+ | ·0000 003 | ·0000 001 |
| ·43 | ·0000 074 | ·0000 037 | ·0000 019 | ·0000 009 | ·0000 005− | ·0000 002 |
| ·44 | ·0000 119 | ·0000 061 | ·0000 032 | ·0000 016 | ·0000 008 | ·0000 004 |
| ·45 | ·0000 189 | ·0000 100 | ·0000 052 | ·0000 027 | ·0000 014 | ·0000 007 |
| ·46 | ·0000 297 | ·0000 160 | ·0000 086 | ·0000 046 | ·0000 024 | ·0000 013 |
| ·47 | ·0000 461 | ·0000 254 | ·0000 139 | ·0000 076 | ·0000 041 | ·0000 022 |
| ·48 | ·0000 706 | ·0000 397 | ·0000 222 | ·0000 124 | ·0000 069 | ·0000 038 |
| ·49 | ·0001 069 | ·0000 614 | ·0000 350+ | ·0000 199 | ·0000 113 | ·0000 063 |
| ·50 | ·0001 600 | ·0000 937 | ·0000 546 | ·0000 316 | ·0000 183 | ·0000 105− |
| ·51 | ·0002 369 | ·0001 414 | ·0000 840 | ·0000 497 | ·0000 292 | ·0000 171 |
| ·52 | ·0003 469 | ·0002 111 | ·0001 278 | ·0000 770 | ·0000 462 | ·0000 276 |
| ·53 | ·0005 028 | ·0003 118 | ·0001 923 | ·0001 181 | ·0000 721 | ·0000 439 |
| ·54 | ·0007 214 | ·0004 556 | ·0002 863 | ·0001 790 | ·0001 114 | ·0000 690 |
| ·55 | ·0010 248 | ·0006 590 | ·0004 215+ | ·0002 683 | ·0001 701 | ·0001 073 |
| ·56 | ·0014 417 | ·0009 435+ | ·0006 143 | ·0003 980 | ·0002 567 | ·0001 649 |
| ·57 | ·0020 092 | ·0013 378 | ·0008 862 | ·0005 842 | ·0003 834 | ·0002 506 |
| ·58 | ·0027 741 | ·0018 786 | ·0012 658 | ·0008 488 | ·0005 666 | ·0003 767 |
| ·59 | ·0037 955+ | ·0026 134 | ·0017 904 | ·0012 208 | ·0008 287 | ·0005 602 |
| ·60 | ·0051 469 | ·0036 021 | ·0025 084 | ·0017 386 | ·0011 997 | ·0008 245− |
| ·61 | ·0069 182 | ·0049 198 | ·0034 814 | ·0024 521 | ·0017 195+ | ·0012 009 |
| ·62 | ·0092 188 | ·0066 595+ | ·0047 872 | ·0034 254 | ·0024 404 | ·0017 315+ |
| ·63 | ·0121 799 | ·0089 351 | ·0065 229 | ·0047 402 | ·0034 299 | ·0024 717 |
| ·64 | ·0159 564 | ·0118 837 | ·0088 079 | ·0064 987 | ·0047 746 | ·0034 937 |
| ·65 | ·0207 296 | ·0156 690 | ·0117 876 | ·0088 280 | ·0065 836 | ·0048 903 |
| ·66 | ·0267 078 | ·0204 834 | ·0156 359 | ·0118 828 | ·0089 929 | ·0067 791 |
| ·67 | ·0341 270 | ·0265 495− | ·0205 587 | ·0158 502 | ·0121 697 | ·0093 074 |
| ·68 | ·0432 503 | ·0341 208 | ·0267 954 | ·0209 519 | ·0163 161 | ·0126 571 |
| ·69 | ·0543 650+ | ·0434 814 | ·0346 200 | ·0274 473 | ·0216 733 | ·0170 490 |
| ·70 | ·0677 788 | ·0549 429 | ·0443 405− | ·0356 343 | ·0285 242 | ·0227 472 |
| ·71 | ·0838 125+ | ·0688 400 | ·0562 960 | ·0458 483 | ·0371 942 | ·0300 622 |
| ·72 | ·1027 910 | ·0855 225− | ·0708 511 | ·0584 595+ | ·0480 506 | ·0393 516 |
| ·73 | ·1250 306 | ·1053 447 | ·0883 873 | ·0738 659 | ·0614 985− | ·0510 193 |
| ·74 | ·1508 241 | ·1286 513 | ·1092 900 | ·0924 829 | ·0779 728 | ·0655 097 |
| ·75 | ·1804 218 | ·1557 588 | ·1339 320 | ·1147 287 | ·0979 257 | ·0832 982 |
| ·76 | ·2140 110 | ·1869 344 | ·1626 521 | ·1410 036 | ·1218 085+ | ·1048 757 |
| ·77 | ·2516 928 | ·2223 713 | ·1957 298 | ·1716 654 | ·1500 475− | ·1307 265− |
| ·78 | ·2934 582 | ·2621 613 | ·2333 557 | ·2069 986 | ·1830 136 | ·1612 986 |
| ·79 | ·3391 644 | ·3062 670 | ·2756 000 | ·2471 800 | ·2209 862 | ·1969 674 |
| ·80 | ·3885 137 | ·3544 941 | ·3223 795− | ·2922 412 | ·2641 121 | ·2379 923 |
| ·81 | ·4410 373 | ·4064 680 | ·3734 266 | ·3420 310 | ·3123 621 | ·2844 680 |
| ·82 | ·4960 853 | ·4616 156 | ·4282 628 | ·3961 799 | ·3654 873 | ·3362 746 |
| ·83 | ·5528 280 | ·5191 575+ | ·4861 807 | ·4540 727 | ·4229 814 | ·3930 283 |
| ·84 | ·6102 694 | ·5781 129 | ·5462 394 | ·5148 318 | ·4840 523 | ·4540 423 |
| ·85 | ·6672 756 | ·6373 211 | ·6072 768 | ·5773 191 | ·5476 111 | ·5183 010 |
| ·86 | ·7226 208 | ·6954 832 | ·6679 439 | ·6401 609 | ·6122 854 | ·5844 597 |
| ·87 | ·7750 496 | ·7512 233 | ·7267 632 | ·7017 999 | ·6764 621 | ·6508 754 |
| ·88 | ·8233 546 | ·8031 716 | ·7822 135+ | ·7605 783 | ·7383 662 | ·7156 781 |
| ·89 | ·8664 664 | ·8500 633 | ·8328 368 | ·8148 517 | ·7961 771 | ·7768 856 |
| ·90 | ·9035 465+ | ·8908 488 | ·8773 638 | ·8631 268 | ·8481 779 | ·8325 617 |

$x = \cdot 91$ to $1\cdot 00$ $q = 5\cdot 5$ $p = 24$ to $29$

| | $p = 24$ | $p = 25$ | $p = 26$ | $p = 27$ | $p = 28$ | $p = 29$ |
|---|---|---|---|---|---|---|
| $B(p, q) =$ | $\cdot 8277\ 1960 \times \frac{1}{10^8}$ | $\cdot 6733\ 9900 \times \frac{1}{10^8}$ | $\cdot 5519\ 6639 \times \frac{1}{10^8}$ | $\cdot 4555\ 9131 \times \frac{1}{10^8}$ | $\cdot 3784\ 9124 \times \frac{1}{10^8}$ | $\cdot 3163\ 5089 \times \frac{1}{10^8}$ |
| $x$ | | | | | | |
| ·91 | ·9340 753 | ·9248 021 | ·9148 444 | ·9042 147 | ·8929 294 | ·8810 093 |
| ·92 | ·9579 210 | ·9516 142 | ·9447 675+ | ·9373 784 | ·9294 474 | ·9209 783 |
| ·93 | ·9753 747 | ·9714 525− | ·9671 482 | ·9624 524 | ·9573 574 | ·9518 576 |
| ·94 | ·9871 377 | ·9849 660 | ·9825 571 | ·9799 008 | ·9769 878 | ·9738 095− |
| ·95 | ·9942 482 | ·9932 212 | ·9920 698 | ·9907 867 | ·9893 646 | ·9877 964 |
| ·96 | ·9979 461 | ·9975 591 | ·9971 207 | ·9966 269 | ·9960 739 | ·9954 578 |
| ·97 | ·9994 860 | ·9993 840 | ·9992 673 | ·9991 344 | ·9989 841 | ·9988 148 |
| ·98 | ·9999 327 | ·9999 187 | ·9999 025− | ·9998 838 | ·9998 625+ | ·9998 383 |
| ·99 | ·9999 982 | ·9999 978 | ·9999 973 | ·9999 968 | ·9999 962 | ·9999 955− |
| 1·00 | 1·0000 000 | 1·0000 000 | 1·0000 000 | 1·0000 000 | 1·0000 000 | 1·0000 000 |

$x$ = ·42 to 1·00 $q = 5·5$ $p$ = 30 to 35

| | $p = 30$ | $p = 31$ | $p = 32$ | $p = 33$ | $p = 34$ | $p = 35$ |
|---|---|---|---|---|---|---|
| $B(p, q) =$ | ·2659 1814 $\times \frac{1}{10^8}$ | ·2247 1955 $\times \frac{1}{10^8}$ | ·1908 5770 $\times \frac{1}{10^8}$ | ·1628 6524 $\times \frac{1}{10^8}$ | ·1395 9878 $\times \frac{1}{10^8}$ | ·1201 6097 $\times \frac{1}{10^8}$ |
| $x$ | | | | | | |
| ·42 | ·0000 001 | | | | | |
| ·43 | ·0000 001 | ·0000 001 | | | | |
| ·44 | ·0000 002 | ·0000 001 | ·0000 001 | | | |
| ·45 | ·0000 004 | ·0000 002 | ·0000 001 | ·0000 001 | | |
| ·46 | ·0000 007 | ·0000 004 | ·0000 002 | ·0000 001 | ·0000 001 | |
| ·47 | ·0000 012 | ·0000 006 | ·0000 003 | ·0000 002 | ·0000 001 | ·0000 001 |
| ·48 | ·0000 021 | ·0000 011 | ·0000 006 | ·0000 003 | ·0000 002 | ·0000 001 |
| ·49 | ·0000 036 | ·0000 020 | ·0000 011 | ·0000 006 | ·0000 003 | ·0000 002 |
| ·50 | ·0000 060 | ·0000 034 | ·0000 019 | ·0000 011 | ·0000 006 | ·0000 003 |
| ·51 | ·0000 100 | ·0000 058 | ·0000 034 | ·0000 019 | ·0000 011 | ·0000 006 |
| ·52 | ·0000 164 | ·0000 097 | ·0000 057 | ·0000 034 | ·0000 020 | ·0000 012 |
| ·53 | ·0000 266 | ·0000 161 | ·0000 097 | ·0000 058 | ·0000 035− | ·0000 021 |
| ·54 | ·0000 426 | ·0000 262 | ·0000 161 | ·0000 098 | ·0000 060 | ·0000 036 |
| ·55 | ·0000 674 | ·0000 422 | ·0000 263 | ·0000 164 | ·0000 101 | ·0000 063 |
| ·56 | ·0001 055− | ·0000 672 | ·0000 427 | ·0000 270 | ·0000 170 | ·0000 107 |
| ·57 | ·0001 631 | ·0001 058 | ·0000 684 | ·0000 440 | ·0000 283 | ·0000 181 |
| ·58 | ·0002 494 | ·0001 645+ | ·0001 082 | ·0000 709 | ·0000 463 | ·0000 301 |
| ·59 | ·0003 772 | ·0002 531 | ·0001 692 | ·0001 127 | ·0000 749 | ·0000 496 |
| ·60 | ·0005 643 | ·0003 849 | ·0002 616 | ·0001 772 | ·0001 196 | ·0000 805+ |
| ·61 | ·0008 354 | ·0005 790 | ·0003 999 | ·0002 753 | ·0001 890 | ·0001 293 |
| ·62 | ·0012 238 | ·0008 618 | ·0006 048 | ·0004 231 | ·0002 950+ | ·0002 051 |
| ·63 | ·0017 744 | ·0012 692 | ·0009 048 | ·0006 429 | ·0004 554 | ·0003 216 |
| ·64 | ·0025 468 | ·0018 499 | ·0013 391 | ·0009 662 | ·0006 950+ | ·0004 985+ |
| ·65 | ·0036 188 | ·0026 684 | ·0019 610 | ·0014 365+ | ·0010 491 | ·0007 640 |
| ·66 | ·0050 912 | ·0038 101 | ·0028 418 | ·0021 129 | ·0015 662 | ·0011 576 |
| ·67 | ·0070 921 | ·0053 853 | ·0040 756 | ·0030 748 | ·0023 128 | ·0017 347 |
| ·68 | ·0097 829 | ·0075 353 | ·0057 850+ | ·0044 275− | ·0033 785− | ·0025 708 |
| ·69 | ·0133 631 | ·0104 383 | ·0081 273 | ·0063 084 | ·0048 822 | ·0037 679 |
| ·70 | ·0180 759 | ·0143 154 | ·0113 010 | ·0088 942 | ·0069 797 | ·0054 621 |
| ·71 | ·0242 128 | ·0194 367 | ·0155 534 | ·0124 086 | ·0098 713 | ·0078 314 |
| ·72 | ·0321 165+ | ·0261 259 | ·0211 865− | ·0171 300 | ·0138 111 | ·0111 052 |
| ·73 | ·0421 825− | ·0347 640 | ·0285 623 | ·0233 985− | ·0191 148 | ·0155 739 |
| ·74 | ·0548 558 | ·0457 894 | ·0381 064 | ·0316 214 | ·0261 681 | ·0215 986 |
| ·75 | ·0706 252 | ·0596 947 | ·0503 068 | ·0422 758 | ·0354 312 | ·0296 183 |
| ·76 | ·0900 099 | ·0770 172 | ·0657 093 | ·0559 068 | ·0474 410 | ·0401 554 |
| ·77 | ·1135 408 | ·0983 228 | ·0849 041 | ·0731 186 | ·0628 063 | ·0538 149 |
| ·78 | ·1417 326 | ·1241 819 | ·1085 052 | ·0945 579 | ·0821 954 | ·0712 765− |
| ·79 | ·1750 482 | ·1551 346 | ·1371 192 | ·1208 857 | ·1063 124 | ·0932 757 |
| ·80 | ·2138 539 | ·1916 466 | ·1713 019 | ·1527 373 | ·1358 605− | ·1205 720 |
| ·81 | ·2583 679 | ·2340 556 | ·2115 039 | ·1906 684 | ·1714 902 | ·1538 998 |
| ·82 | ·3086 030 | ·2825 088 | ·2580 054 | ·2350 871 | ·2137 314 | ·1939 020 |
| ·83 | ·3643 095+ | ·3368 974 | ·3108 426 | ·2861 761 | ·2629 110 | ·2410 453 |
| ·84 | ·4249 222 | ·3967 922 | ·3697 330 | ·3438 069 | ·3190 593 | ·2955 199 |
| ·85 | ·4895 214 | ·4613 894 | ·4340 061 | ·4074 571 | ·3818 132 | ·3571 310 |
| ·86 | ·5568 162 | ·5294 768 | ·5025 520 | ·4761 403 | ·4503 290 | ·4251 933 |
| ·87 | ·6251 608 | ·5994 334 | ·5738 014 | ·5483 656 | ·5232 189 | ·4984 456 |
| ·88 | ·6926 145+ | ·6692 742 | ·6457 531 | ·6221 438 | ·5985 345+ | ·5750 086 |
| ·89 | ·7570 519 | ·7367 522 | ·7160 628 | ·6950 599 | ·6738 182 | ·6524 111 |
| ·90 | ·8163 262 | ·7995 224 | ·7822 038 | ·7644 250+ | ·7462 422 | ·7277 118 |
| ·91 | ·8684 788 | ·8553 657 | ·8417 006 | ·8275 166 | ·8128 490 | ·7977 350− |
| ·92 | ·9119 776 | ·9024 549 | ·8924 223 | ·8818 943 | ·8708 877 | ·8594 215− |
| ·93 | ·9459 491 | ·9396 300 | ·9329 001 | ·9257 611 | ·9182 166 | ·9102 715− |
| ·94 | ·9703 583 | ·9666 275− | ·9626 115+ | ·9583 056 | ·9537 061 | ·9488 104 |
| ·95 | ·9860 756 | ·9841 955− | ·9821 501 | ·9799 338 | ·9775 412 | ·9749 674 |
| ·96 | ·9947 744 | ·9940 200 | ·9931 907 | ·9922 826 | ·9912 919 | ·9902 149 |
| ·97 | ·9986 252 | ·9984 136 | ·9981 786 | ·9979 186 | ·9976 319 | ·9973 171 |
| ·98 | ·9998 108 | ·9997 799 | ·9997 451 | ·9997 063 | ·9996 631 | ·9996 151 |
| ·99 | ·9999 947 | ·9999 937 | ·9999 927 | ·9999 915− | ·9999 902 | ·9999 887 |
| 1·00 | 1·0000 000 | 1·0000 000 | 1·0000 000 | 1·0000 000 | 1·0000 000 | 1·0000 000 |

$x$ = ·48 to 1·00 $q$ = 5·5 $p$ = 36 to 40

| | $p$ = 36 | $p$ = 37 | $p$ = 38 | $p$ = 39 | $p$ = 40 |
|---|---|---|---|---|---|
| $B(p, q)$ = | ·1038 4282 × $\frac{1}{10^6}$ | ·9008 0514 × $\frac{1}{10^7}$ | ·7842 3036 × $\frac{1}{10^7}$ | ·6850 7480 × $\frac{1}{10^7}$ | ·6004 0263 × $\frac{1}{10^7}$ |
| $x$ | | | | | |
| ·48 | ·0000 001 | | | | |
| ·49 | ·0000 001 | ·0000 001 | | | |
| ·50 | ·0000 002 | ·0000 001 | ·0000 001 | | |
| ·51 | ·0000 004 | ·0000 002 | ·0000 001 | ·0000 001 | |
| ·52 | ·0000 007 | ·0000 004 | ·0000 002 | ·0000 001 | ·0000 001 |
| ·53 | ·0000 012 | ·0000 007 | ·0000 004 | ·0000 003 | ·0000 001 |
| ·54 | ·0000 022 | ·0000 013 | ·0000 008 | ·0000 005⁻ | ·0000 003 |
| ·55 | ·0000 039 | ·0000 024 | ·0000 015⁻ | ·0000 009 | ·0000 005⁺ |
| ·56 | ·0000 067 | ·0000 042 | ·0000 026 | ·0000 016 | ·0000 010 |
| ·57 | ·0000 115⁺ | ·0000 073 | ·0000 047 | ·0000 030 | ·0000 019 |
| ·58 | ·0000 196 | ·0000 127 | ·0000 082 | ·0000 053 | ·0000 034 |
| ·59 | ·0000 327 | ·0000 215⁺ | ·0000 142 | ·0000 093 | ·0000 061 |
| ·60 | ·0000 541 | ·0000 362 | ·0000 242 | ·0000 161 | ·0000 107 |
| ·61 | ·0000 882 | ·0000 600 | ·0000 407 | ·0000 276 | ·0000 186 |
| ·62 | ·0001 422 | ·0000 983 | ·0000 678 | ·0000 467 | ·0000 320 |
| ·63 | ·0002 265⁺ | ·0001 591 | ·0001 115⁻ | ·0000 779 | ·0000 543 |
| ·64 | ·0003 566 | ·0002 544 | ·0001 810 | ·0001 285⁻ | ·0000 910 |
| ·65 | ·0005 548 | ·0004 018 | ·0002 903 | ·0002 092 | ·0001 505⁻ |
| ·66 | ·0008 533 | ·0006 273 | ·0004 601 | ·0003 366 | ·0002 457 |
| ·67 | ·0012 976 | ·0009 681 | ·0007 205⁻ | ·0005 349 | ·0003 963 |
| ·68 | ·0019 509 | ·0014 767 | ·0011 150⁺ | ·0008 399 | ·0006 313 |
| ·69 | ·0029 002 | ·0022 267 | ·0017 054 | ·0013 031 | ·0009 935⁺ |
| ·70 | ·0042 633 | ·0033 192 | ·0025 779 | ·0019 976 | ·0015 445⁺ |
| ·71 | ·0061 968 | ·0048 912 | ·0038 515⁺ | ·0030 259 | ·0023 721 |
| ·72 | ·0089 064 | ·0071 254 | ·0056 872 | ·0045 290 | ·0035 989 |
| ·73 | ·0126 567 | ·0102 609 | ·0082 993 | ·0066 977 | ·0053 937 |
| ·74 | ·0177 824 | ·0146 054 | ·0119 685⁻ | ·0097 861 | ·0079 848 |
| ·75 | ·0246 981 | ·0205 467 | ·0170 544 | ·0141 251 | ·0116 746 |
| ·76 | ·0339 065⁻ | ·0285 637 | ·0240 092 | ·0201 380 | ·0168 564 |
| ·77 | ·0460 014 | ·0392 329 | ·0333 874 | ·0283 532 | ·0240 297 |
| ·78 | ·0616 647 | ·0532 303 | ·0458 514 | ·0394 143 | ·0338 140 |
| ·79 | ·0816 526 | ·0713 225⁺ | ·0621 693 | ·0540 819 | ·0469 556 |
| ·80 | ·1067 687 | ·0943 457 | ·0831 986 | ·0732 249 | ·0643 253 |
| ·81 | ·1378 197 | ·1231 667 | ·1098 542 | ·0977 942 | ·0868 984 |
| ·82 | ·1755 513 | ·1586 228 | ·1430 532 | ·1287 742 | ·1157 144 |
| ·83 | ·2205 634 | ·2014 384 | ·1836 340 | ·1671 065⁻ | ·1518 057 |
| ·84 | ·2732 045⁺ | ·2521 166 | ·2322 482 | ·2135 825⁻ | ·1960 941 |
| ·85 | ·3334 536 | ·3108 122 | ·2892 263 | ·2687 055⁺ | ·2492 503 |
| ·86 | ·4007 973 | ·3771 942 | ·3544 267 | ·3325 280 | ·3115 222 |
| ·87 | ·4741 219 | ·4503 150⁻ | ·4270 840 | ·4044 796 | ·3825 446 |
| ·88 | ·5516 440 | ·5285 130 | ·5056 821 | ·4832 116 | ·4611 561 |
| ·89 | ·6309 091 | ·6093 802 | ·5878 892 | ·5664 971 | ·5452 614 |
| ·90 | ·7088 902 | ·6898 334 | ·6705 963 | ·6512 326 | ·6317 944 |
| ·91 | ·7822 129 | ·7663 221 | ·7501 028 | ·7335 953 | ·7168 403 |
| ·92 | ·8475 163 | ·8351 944 | ·8224 795⁺ | ·8093 966 | ·7959 715⁻ |
| ·93 | ·9019 325⁺ | ·8932 079 | ·8841 072 | ·8746 412 | ·8648 221 |
| ·94 | ·9436 167 | ·9381 244 | ·9323 337 | ·9262 459 | ·9198 630 |
| ·95 | ·9722 079 | ·9692 587 | ·9661 162 | ·9627 773 | ·9592 393 |
| ·96 | ·9890 481 | ·9877 880 | ·9864 312 | ·9849 743 | ·9834 144 |
| ·97 | ·9969 725⁺ | ·9965 964 | ·9961 873 | ·9957 434 | ·9952 632 |
| ·98 | ·9995 621 | ·9995 036 | ·9994 393 | ·9993 688 | ·9992 918 |
| ·99 | ·9999 870 | ·9999 851 | ·9999 831 | ·9999 808 | ·9999 782 |
| 1·00 | 1·0000 000 | 1·0000 000 | 1·0000 000 | 1·0000 000 | 1·0000 000 |

$x = \cdot 53$ to $1\cdot 00$ $q = 5\cdot 5$ $p = 41$ to $45$

| | $p = 41$ | $p = 42$ | $p = 43$ | $p = 44$ | $p = 45$ |
|---|---|---|---|---|---|
| $B(p, q) =$ | $\cdot 5278\ 2649 \times \frac{1}{10^7}$ | $\cdot 4653\ 9540 \times \frac{1}{10^7}$ | $\cdot 4115\ 0751 \times \frac{1}{10^7}$ | $\cdot 3648\ 4171 \times \frac{1}{10^7}$ | $\cdot 3243\ 0374 \times \frac{1}{10^7}$ |
| $x$ | | | | | |
| ·53 | ·0000 001 | ·0000 001 | | | |
| ·54 | ·0000 002 | ·0000 001 | ·0000 001 | | |
| ·55 | ·0000 003 | ·0000 002 | ·0000 001 | ·0000 001 | |
| ·56 | ·0000 006 | ·0000 004 | ·0000 002 | ·0000 001 | ·0000 001 |
| ·57 | ·0000 012 | ·0000 007 | ·0000 005⁻ | ·0000 003 | ·0000 002 |
| ·58 | ·0000 022 | ·0000 014 | ·0000 009 | ·0000 006 | ·0000 004 |
| ·59 | ·0000 040 | ·0000 026 | ·0000 017 | ·0000 011 | ·0000 007 |
| ·60 | ·0000 071 | ·0000 047 | ·0000 031 | ·0000 020 | ·0000 013 |
| ·61 | ·0000 126 | ·0000 084 | ·0000 057 | ·0000 038 | ·0000 025⁺ |
| ·62 | ·0000 219 | ·0000 150⁻ | ·0000 102 | ·0000 070 | ·0000 047 |
| ·63 | ·0000 378 | ·0000 262 | ·0000 182 | ·0000 126 | ·0000 087 |
| ·64 | ·0000 643 | ·0000 453 | ·0000 319 | ·0000 224 | ·0000 157 |
| ·65 | ·0001 080 | ·0000 773 | ·0000 552 | ·0000 394 | ·0000 280 |
| ·66 | ·0001 790 | ·0001 301 | ·0000 944 | ·0000 683 | ·0000 494 |
| ·67 | ·0002 929 | ·0002 161 | ·0001 591 | ·0001 169 | ·0000 857 |
| ·68 | ·0004 735⁻ | ·0003 544 | ·0002 647 | ·0001 974 | ·0001 469 |
| ·69 | ·0007 558 | ·0005 739 | ·0004 348 | ·0003 289 | ·0002 483 |
| ·70 | ·0011 917 | ·0009 176 | ·0007 052 | ·0005 409 | ·0004 141 |
| ·71 | ·0018 556 | ·0014 487 | ·0011 289 | ·0008 780 | ·0006 817 |
| ·72 | ·0028 539 | ·0022 586 | ·0017 840 | ·0014 066 | ·0011 071 |
| ·73 | ·0043 347 | ·0034 767 | ·0027 833 | ·0022 241 | ·0017 742 |
| ·74 | ·0065 018 | ·0052 840 | ·0042 863 | ·0034 707 | ·0028 054 |
| ·75 | ·0096 299 | ·0079 282 | ·0065 151 | ·0053 443 | ·0043 765⁻ |
| ·76 | ·0140 817 | ·0117 416 | ·0097 725⁺ | ·0081 194 | ·0067 346 |
| ·77 | ·0203 261 | ·0171 613 | ·0144 633 | ·0121 684 | ·0102 206 |
| ·78 | ·0289 543 | ·0247 477 | ·0211 151 | ·0179 851 | ·0152 940 |
| ·79 | ·0406 926 | ·0352 017 | ·0303 991 | ·0262 081 | ·0225 585⁺ |
| ·80 | ·0564 046 | ·0493 726 | ·0431 442 | ·0376 401 | ·0327 865⁻ |
| ·81 | ·0770 800 | ·0682 541 | ·0603 393 | ·0532 573 | ·0469 342 |
| ·82 | ·1038 006 | ·0929 590 | ·0831 165⁻ | ·0742 009 | ·0661 425⁻ |
| ·83 | ·1376 774 | ·1246 639 | ·1127 053 | ·1017 410 | ·0917 100 |
| ·84 | ·1797 512 | ·1645 165⁻ | ·1503 481 | ·1372 013 | ·1250 284 |
| ·85 | ·2308 529 | ·2134 987 | ·1971 673 | ·1818 328 | ·1674 654 |
| ·86 | ·2914 248 | ·2722 440 | ·2539 812 | ·2366 315⁺ | ·2201 848 |
| ·87 | ·3613 143 | ·3408 168 | ·3210 732 | ·3020 987 | ·2839 026 |
| ·88 | ·4395 637 | ·4184 768 | ·3979 322 | ·3779 610 | ·3585 888 |
| ·89 | ·5242 353 | ·5034 683 | ·4830 053 | ·4628 873 | ·4431 512 |
| ·90 | ·6123 319 | ·5928 932 | ·5735 239 | ·5542 672 | ·5351 637 |
| ·91 | ·6998 779 | ·6827 480 | ·6654 896 | ·6481 408 | ·6307 389 |
| ·92 | ·7822 309 | ·7682 021 | ·7539 127 | ·7393 908 | ·7246 642 |
| ·93 | ·8546 631 | ·8441 781 | ·8333 824 | ·8222 917 | ·8109 224 |
| ·94 | ·9131 881 | ·9062 251 | ·8989 785⁺ | ·8914 540 | ·8836 574 |
| ·95 | ·9555 002 | ·9515 581 | ·9474 118 | ·9430 607 | ·9385 043 |
| ·96 | ·9817 484 | ·9799 735⁻ | ·9780 870 | ·9760 865⁻ | ·9739 695⁺ |
| ·97 | ·9947 449 | ·9941 871 | ·9935 879 | ·9929 460 | ·9922 596 |
| ·98 | ·9992 078 | ·9991 164 | ·9990 173 | ·9989 100 | ·9987 942 |
| ·99 | ·9999 754 | ·9999 724 | ·9999 690 | ·9999 654 | ·9999 614 |
| 1·00 | 1·0000 000 | 1·0000 000 | 1·0000 000 | 1·0000 000 | 1·0000 000 |

$x$ = ·56 to 1·00 $q$ = 5·5 $p$ = 46 to 50

| $x$ | $p = 46$ | $p = 47$ | $p = 48$ | $p = 49$ | $p = 50$ |
|---|---|---|---|---|---|
| $B(p, q) =$ | $\cdot 2889\ 8353 \times \frac{1}{10^7}$ | $\cdot 2581\ 2121 \times \frac{1}{10^7}$ | $\cdot 2310\ 7994 \times \frac{1}{10^7}$ | $\cdot 2073\ 2406 \times \frac{1}{10^7}$ | $\cdot 1864\ 0145^{+} \times \frac{1}{10^7}$ |
| ·56 | ·0000 001 | | | | |
| ·57 | ·0000 001 | ·0000 001 | | | |
| ·58 | ·0000 002 | ·0000 001 | ·0000 001 | ·0000 001 | |
| ·59 | ·0000 005− | ·0000 003 | ·0000 002 | ·0000 001 | ·0000 001 |
| ·60 | ·0000 009 | ·0000 006 | ·0000 004 | ·0000 002 | ·0000 002 |
| ·61 | ·0000 017 | ·0000 011 | ·0000 007 | ·0000 005− | ·0000 003 |
| ·62 | ·0000 032 | ·0000 022 | ·0000 015− | ·0000 010 | ·0000 007 |
| ·63 | ·0000 060 | ·0000 041 | ·0000 028 | ·0000 019 | ·0000 013 |
| ·64 | ·0000 110 | ·0000 077 | ·0000 054 | ·0000 037 | ·0000 026 |
| ·65 | ·0000 199 | ·0000 141 | ·0000 100 | ·0000 071 | ·0000 050− |
| ·66 | ·0000 356 | ·0000 256 | ·0000 184 | ·0000 132 | ·0000 095− |
| ·67 | ·0000 628 | ·0000 459 | ·0000 335− | ·0000 244 | ·0000 177 |
| ·68 | ·0001 091 | ·0000 809 | ·0000 599 | ·0000 443 | ·0000 327 |
| ·69 | ·0001 871 | ·0001 408 | ·0001 058 | ·0000 793 | ·0000 594 |
| ·70 | ·0003 166 | ·0002 416 | ·0001 840 | ·0001 400 | ·0001 063 |
| ·71 | ·0005 283 | ·0004 088 | ·0003 158 | ·0002 436 | ·0001 876 |
| ·72 | ·0008 699 | ·0006 823 | ·0005 344 | ·0004 179 | ·0003 263 |
| ·73 | ·0014 129 | ·0011 234 | ·0008 918 | ·0007 068 | ·0005 594 |
| ·74 | ·0022 639 | ·0018 240 | ·0014 673 | ·0011 786 | ·0009 453 |
| ·75 | ·0035 780 | ·0029 206 | ·0023 803 | ·0019 371 | ·0015 741 |
| ·76 | ·0055 768 | ·0046 109 | ·0038 065− | ·0031 378 | ·0025 829 |
| ·77 | ·0085 709 | ·0071 763 | ·0059 996 | ·0050 087 | ·0041 756 |
| ·78 | ·0129 850− | ·0110 078 | ·0093 180 | ·0078 763 | ·0066 485+ |
| ·79 | ·0193 870 | ·0166 365− | ·0142 555− | ·0121 982 | ·0104 236 |
| ·80 | ·0285 153 | ·0247 640 | ·0214 757 | ·0185 985− | ·0160 852 |
| ·81 | ·0413 004 | ·0362 905+ | ·0318 440 | ·0279 047 | ·0244 208 |
| ·82 | ·0588 737 | ·0523 303 | ·0464 510 | ·0411 780 | ·0364 570 |
| ·83 | ·0825 516 | ·0742 065+ | ·0666 169 | ·0597 270 | ·0534 832 |
| ·84 | ·1137 807 | ·1034 084 | ·0938 615− | ·0850 902 | ·0770 457 |
| ·85 | ·1540 319 | ·1414 964 | ·1298 210 | ·1189 666 | ·1088 931 |
| ·86 | ·2046 262 | ·1899 369 | ·1760 945+ | ·1630 741 | ·1508 481 |
| ·87 | ·2664 889 | ·2498 570 | ·2340 018 | ·2189 146 | ·2045 832 |
| ·88 | ·3398 366 | ·3217 203 | ·3042 516 | ·2874 380 | ·2712 832 |
| ·89 | ·4238 295− | ·4049 509 | ·3865 403 | ·3686 187 | ·3512 036 |
| ·90 | ·5162 512 | ·4975 649 | ·4791 369 | ·4609 970 | ·4431 716 |
| ·91 | ·6133 195+ | ·5959 172 | ·5785 649 | ·5612 939 | ·5441 339 |
| ·92 | ·7097 609 | ·6947 085+ | ·6795 345− | ·6642 656 | ·6489 282 |
| ·93 | ·7992 917 | ·7874 170 | ·7753 161 | ·7630 073 | ·7505 088 |
| ·94 | ·8755 958 | ·8672 766 | ·8587 078 | ·8498 981 | ·8408 563 |
| ·95 | ·9337 429 | ·9287 771 | ·9236 079 | ·9182 368 | ·9126 657 |
| ·96 | ·9717 341 | ·9693 781 | ·9668 997 | ·9642 975+ | ·9615 699 |
| ·97 | ·9915 272 | ·9907 473 | ·9899 184 | ·9890 390 | ·9881 076 |
| ·98 | ·9986 692 | ·9985 348 | ·9983 905− | ·9982 358 | ·9980 702 |
| ·99 | ·9999 570 | ·9999 523 | ·9999 471 | ·9999 416 | ·9999 355+ |
| 1·00 | 1·0000 000 | 1·0000 000 | 1·0000 000 | 1·0000 000 | 1·0000 000 |

$x =$ ·03 to ·60 $q = 6$ $p =$ 6 to 8·5

| | $p = 6$ | $p = 6{\cdot}5$ | $p = 7$ | $p = 7{\cdot}5$ | $p = 8$ | $p = 8{\cdot}5$ |
|---|---|---|---|---|---|---|
| $B(p, q) =$ | ·3607 5036 $\times \frac{1}{10^3}$ | ·2524 5092 $\times \frac{1}{10^3}$ | ·1803 7518 $\times \frac{1}{10^3}$ | ·1312 7448 $\times \frac{1}{10^3}$ | ·9712 5097 $\times \frac{1}{10^4}$ | ·7293 0265 $\overset{+}{\times} \frac{1}{10^4}$ |
| $x$ | | | | | | |
| ·03 | ·0000 003 | ·0000 001 | | | | |
| ·04 | ·0000 016 | ·0000 004 | ·0000 001 | | | |
| ·05 | ·0000 058 | ·0000 017 | ·0000 005$^{-}$ | ·0000 001 | | |
| ·06 | ·0000 166 | ·0000 053 | ·0000 017 | ·0000 005$^{+}$ | ·0000 002 | ·0000 001 |
| ·07 | ·0000 399 | ·0000 139 | ·0000 048 | ·0000 016 | ·0000 005$^{+}$ | ·0000 002 |
| ·08 | ·0000 850$^{+}$ | ·0000 316 | ·0000 116 | ·0000 042 | ·0000 015$^{-}$ | ·0000 005$^{+}$ |
| ·09 | ·0001 646 | ·0000 648 | ·0000 252 | ·0000 096 | ·0000 037 | ·0000 014 |
| ·10 | ·0002 957 | ·0001 227 | ·0000 502 | ·0000 203 | ·0000 081 | ·0000 032 |
| ·11 | ·0005 000$^{-}$ | ·0002 174 | ·0000 932 | ·0000 395$^{-}$ | ·0000 165$^{+}$ | ·0000 068 |
| ·12 | ·0008 040 | ·0003 649 | ·0001 633 | ·0000 722 | ·0000 315$^{+}$ | ·0000 136 |
| ·13 | ·0012 394 | ·0005 852 | ·0002 724 | ·0001 253 | ·0000 569 | ·0000 256 |
| ·14 | ·0018 431 | ·0009 024 | ·0004 357 | ·0002 078 | ·0000 980 | ·0000 457 |
| ·15 | ·0026 569 | ·0013 455$^{+}$ | ·0006 721 | ·0003 316 | ·0001 618 | ·0000 781 |
| ·16 | ·0037 274 | ·0019 482 | ·0010 045$^{-}$ | ·0005 115$^{+}$ | ·0002 576 | ·0001 284 |
| ·17 | ·0051 058 | ·0027 488 | ·0014 599 | ·0007 659 | ·0003 974 | ·0002 041 |
| ·18 | ·0068 470 | ·0037 902 | ·0020 700 | ·0011 168 | ·0005 959 | ·0003 148 |
| ·19 | ·0090 095$^{+}$ | ·0051 199 | ·0028 708 | ·0015 904 | ·0008 714 | ·0004 727 |
| ·20 | ·0116 542 | ·0067 894 | ·0039 031 | ·0022 170 | ·0012 456 | ·0006 929 |
| ·21 | ·0148 440 | ·0088 539 | ·0052 119 | ·0030 316 | ·0017 444 | ·0009 938 |
| ·22 | ·0186 427 | ·0113 717 | ·0068 464 | ·0040 734 | ·0023 975$^{+}$ | ·0013 973 |
| ·23 | ·0231 144 | ·0144 036 | ·0088 599 | ·0053 861 | ·0032 395$^{-}$ | ·0019 294 |
| ·24 | ·0283 222 | ·0180 121 | ·0113 088 | ·0070 178 | ·0043 090 | ·0026 202 |
| ·25 | ·0343 275$^{+}$ | ·0222 606 | ·0142 528 | ·0090 207 | ·0056 493 | ·0035 040 |
| ·26 | ·0411 890 | ·0272 129 | ·0177 536 | ·0114 504 | ·0073 081 | ·0046 199 |
| ·27 | ·0489 618 | ·0329 317 | ·0218 748 | ·0143 660 | ·0093 373 | ·0060 115$^{-}$ |
| ·28 | ·0576 964 | ·0394 782 | ·0266 804 | ·0178 295$^{-}$ | ·0117 928 | ·0077 267 |
| ·29 | ·0674 377 | ·0469 106 | ·0322 347 | ·0219 045$^{-}$ | ·0147 338 | ·0098 182 |
| ·30 | ·0782 248 | ·0552 840 | ·0386 008 | ·0266 563 | ·0182 228 | ·0123 425$^{-}$ |
| ·31 | ·0900 894 | ·0646 484 | ·0458 401 | ·0321 505$^{+}$ | ·0223 247 | ·0153 600 |
| ·32 | ·1030 559 | ·0750 489 | ·0540 108 | ·0384 525$^{+}$ | ·0271 061 | ·0189 344 |
| ·33 | ·1171 403 | ·0865 238 | ·0631 675$^{+}$ | ·0456 261 | ·0326 344 | ·0231 323 |
| ·34 | ·1323 501 | ·0991 047 | ·0733 600 | ·0537 329 | ·0389 772 | ·0280 222 |
| ·35 | ·1486 837 | ·1128 153 | ·0846 321 | ·0628 311 | ·0462 011 | ·0336 737 |
| ·36 | ·1661 304 | ·1276 707 | ·0970 210 | ·0729 747 | ·0543 707 | ·0401 570 |
| ·37 | ·1846 699 | ·1436 773 | ·1105 567 | ·0842 119 | ·0635 477 | ·0475 416 |
| ·38 | ·2042 725$^{-}$ | ·1608 319 | ·1252 603 | ·0965 849 | ·0737 896 | ·0558 951 |
| ·39 | ·2248 992 | ·1791 218 | ·1411 446 | ·1101 282 | ·0851 487 | ·0652 824 |
| ·40 | ·2465 019 | ·1985 241 | ·1582 123 | ·1248 682 | ·0976 709 | ·0757 644 |
| ·41 | ·2690 235$^{+}$ | ·2190 061 | ·1764 563 | ·1408 223 | ·1113 948 | ·0873 967 |
| ·42 | ·2923 988 | ·2405 251 | ·1958 591 | ·1579 978 | ·1263 506 | ·1002 286 |
| ·43 | ·3165 543 | ·2630 287 | ·2163 926 | ·1763 916 | ·1425 591 | ·1143 017 |
| ·44 | ·3414 096 | ·2864 549 | ·2380 178 | ·1959 899 | ·1600 308 | ·1296 490 |
| ·45 | ·3668 774 | ·3107 326 | ·2606 851 | ·2167 674 | ·1787 653 | ·1462 935$^{-}$ |
| ·46 | ·3928 649 | ·3357 825$^{-}$ | ·2843 345$^{-}$ | ·2386 872 | ·1987 505$^{-}$ | ·1642 475$^{-}$ |
| ·47 | ·4192 743 | ·3615 170 | ·3088 958 | ·2617 009 | ·2199 623 | ·1835 115$^{+}$ |
| ·48 | ·4460 035$^{-}$ | ·3878 418 | ·3342 890 | ·2857 484 | ·2423 639 | ·2040 737 |
| ·49 | ·4729 477 | ·4146 562 | ·3604 252 | ·3107 587 | ·2659 063 | ·2259 090 |
| ·50 | ·5000 000[e] | ·4418 541 | ·3872 070 | ·3366 495$^{-}$ | ·2905 273 | ·2489 790 |
| ·51 | ·5270 523 | ·4693 252 | ·4145 297 | ·3633 285$^{+}$ | ·3161 529 | ·2732 313 |
| ·52 | ·5539 965$^{+}$ | ·4969 562 | ·4422 820 | ·3906 940 | ·3426 966 | ·2986 000 |
| ·53 | ·5807 258 | ·5246 314 | ·4703 473 | ·4186 353 | ·3700 606 | ·3250 053 |
| ·54 | ·6071 351 | ·5522 345$^{-}$ | ·4986 046 | ·4470 344 | ·3981 365$^{-}$ | ·3523 543 |
| ·55 | ·6331 226 | ·5796 493 | ·5269 302 | ·4757 668 | ·4268 060 | ·3805 412 |
| ·56 | ·6585 904 | ·6067 611 | ·5551 986 | ·5047 029 | ·4559 424 | ·4094 486 |
| ·57 | ·6834 457 | ·6334 581 | ·5832 839 | ·5337 095$^{-}$ | ·4854 116 | ·4389 483 |
| ·58 | ·7076 012 | ·6596 321 | ·6110 616 | ·5626 509 | ·5150 736 | ·4689 025$^{-}$ |
| ·59 | ·7309 765$^{-}$ | ·6851 798 | ·6384 093 | ·5913 910 | ·5447 841 | ·4991 654 |
| ·60 | ·7534 981 | ·7100 040 | ·6652 086 | ·6197 944 | ·5743 964 | ·5295 848 |

$x$ = ·61 to ·99 $q = 6$ $p$ = 6 to 8·5

| | $p = 6$ | $p = 6·5$ | $p = 7$ | $p = 7·5$ | $p = 8$ | $p = 8·5$ |
|---|---|---|---|---|---|---|
| $B(p, q) =$ | ·3607 5036 $\times \frac{1}{10^3}$ | ·2524 5092 $\times \frac{1}{10^3}$ | ·1803 7518 $\times \frac{1}{10^3}$ | ·1312 7448 $\times \frac{1}{10^3}$ | ·9712 5097 $\times \frac{1}{10^4}$ | ·7293 0265$^+$ $\times \frac{1}{10^4}$ |
| $x$ | | | | | | |
| ·61 | ·7751 008 | ·7340 145$^-$ | ·6913 462 | ·6477 283 | ·6037 629 | ·5600 040 |
| ·62 | ·7957 275$^+$ | ·7571 289 | ·7167 154 | ·6750 640 | ·6327 368 | ·5902 636 |
| ·63 | ·8153 301 | ·7792 734 | ·7412 169 | ·7016 784 | ·6611 747 | ·6202 037 |
| ·64 | ·8338 696 | ·8003 838 | ·7647 603 | ·7274 558 | ·6889 374 | ·6496 659 |
| ·65 | ·8513 163 | ·8204 056 | ·7872 646 | ·7522 887 | ·7158 928 | ·6784 954 |
| ·66 | ·8676 499 | ·8392 948 | ·8086 598 | ·7760 800 | ·7419 168 | ·7065 437 |
| ·67 | ·8828 597 | ·8570 181 | ·8288 870 | ·7987 434 | ·7668 955$^-$ | ·7336 701 |
| ·68 | ·8969 441 | ·8735 528 | ·8478 991 | ·8202 049 | ·7907 265$^+$ | ·7597 439 |
| ·69 | ·9099 106 | ·8888 874 | ·8656 613 | ·8404 034 | ·8133 206 | ·7846 467 |
| ·70 | ·9217 752 | ·9030 210 | ·8821 513 | ·8592 915$^-$ | ·8346 025$^+$ | ·8082 737 |
| ·71 | ·9325 623 | ·9159 630 | ·8973 592 | ·8768 359 | ·8545 121 | ·8305 356 |
| ·72 | ·9423 036 | ·9277 329 | ·9112 877 | ·8930 178 | ·8730 051 | ·8513 596 |
| ·73 | ·9510 382 | ·9383 599 | ·9239 511 | ·9078 324 | ·8900 536 | ·8706 907 |
| ·74 | ·9588 110 | ·9478 817 | ·9353 756 | ·9212 894 | ·9056 461 | ·8884 923 |
| ·75 | ·9656 725$^-$ | ·9563 442 | ·9455 978 | ·9334 119 | ·9197 874 | ·9047 465$^-$ |
| ·76 | ·9716 778 | ·9638 002 | ·9546 644 | ·9442 358 | ·9324 984 | ·9194 543 |
| ·77 | ·9768 856 | ·9703 086 | ·9626 311 | ·9538 093 | ·9438 151 | ·9326 352 |
| ·78 | ·9813 573 | ·9759 335$^+$ | ·9695 610 | ·9621 912 | ·9537 876 | ·9443 261 |
| ·79 | ·9851 560 | ·9807 426 | ·9755 239 | ·9694 498 | ·9624 793 | ·9545 810 |
| ·80 | ·9883 458 | ·9848 064 | ·9805 947 | ·9756 616 | ·9699 647 | ·9634 686 |
| ·81 | ·9909 905$^-$ | ·9881 969 | ·9848 518 | ·9809 093 | ·9763 279 | ·9710 711 |
| ·82 | ·9931 530 | ·9909 863 | ·9883 759 | ·9852 803 | ·9816 607 | ·9774 820 |
| ·83 | ·9948 942 | ·9932 461 | ·9912 483 | ·9888 647 | ·9860 607 | ·9828 038 |
| ·84 | ·9962 726 | ·9950 458 | ·9935 497 | ·9917 539 | ·9896 286 | ·9871 452 |
| ·85 | ·9973 431 | ·9964 518 | ·9953 584 | ·9940 381 | ·9924 664 | ·9906 188 |
| ·86 | ·9981 569 | ·9975 269 | ·9967 496 | ·9958 054 | ·9946 747 | ·9933 378 |
| ·87 | ·9987 606 | ·9983 291 | ·9977 936 | ·9971 394 | ·9963 514 | ·9954 143 |
| ·88 | ·9991 960 | ·9989 111 | ·9985 554 | ·9981 183 | ·9975 889 | ·9969 556 |
| ·89 | ·9995 000$^+$ | ·9993 197 | ·9990 933 | ·9988 135$^-$ | ·9984 727 | ·9980 627 |
| ·90 | ·9997 043 | ·9995 958 | ·9994 588 | ·9992 885$^+$ | ·9990 800 | ·9988 277 |
| ·91 | ·9998 354 | ·9997 740 | ·9996 960 | ·9995 985$^+$ | ·9994 785$^-$ | ·9993 324 |
| ·92 | ·9999 150$^-$ | ·9998 827 | ·9998 416 | ·9997 898 | ·9997 257 | ·9996 474 |
| ·93 | ·9999 601 | ·9999 447 | ·9999 249 | ·9998 999 | ·9998 688 | ·9998 306 |
| ·94 | ·9999 834 | ·9999 769 | ·9999 686 | ·9999 579 | ·9999 446 | ·9999 281 |
| ·95 | ·9999 942 | ·9999 919 | ·9999 889 | ·9999 851 | ·9999 803 | ·9999 743 |
| ·96 | ·9999 984 | ·9999 978 | ·9999 969 | ·9999 959 | ·9999 945$^-$ | ·9999 928 |
| ·97 | ·9999 997 | ·9999 996 | ·9999 994 | ·9999 992 | ·9999 990 | ·9999 986 |
| ·98 | 1·0000 000 | 1·0000 000 | ·9999 999 | ·9999 999 | ·9999 999 | ·9999 999 |
| ·99 | | | 1·0000 000 | 1·0000 000 | 1·0000 000 | 1·0000 000 |

$x$ = ·07 to ·70 $q = 6$ $p$ = 9 to 12

| $x$ | $p = 9$ | $p = 9{\cdot}5$ | $p = 10$ | $p = 10{\cdot}5$ | $p = 11$ | $p = 12$ |
|---|---|---|---|---|---|---|
| $B(p, q) =$ | ·5550 0056 × $\frac{1}{10^4}$ | ·4275 2225− × $\frac{1}{10^4}$ | ·3330 0033 × $\frac{1}{10^4}$ | ·2620 2976 × $\frac{1}{10^4}$ | ·2081 2521 × $\frac{1}{10^4}$ | ·1346 6925+ × $\frac{1}{10^4}$ |
| ·07 | ·0000 001 | | | | | |
| ·08 | ·0000 002 | ·0000 001 | | | | |
| ·09 | ·0000 005+ | ·0000 002 | ·0000 001 | | | |
| ·10 | ·0000 013 | ·0000 005− | ·0000 002 | ·0000 001 | | |
| ·11 | ·0000 028 | ·0000 011 | ·0000 005− | ·0000 002 | ·0000 001 | |
| ·12 | ·0000 058 | ·0000 025− | ·0000 010 | ·0000 004 | ·0000 002 | |
| ·13 | ·0000 114 | ·0000 050+ | ·0000 022 | ·0000 010 | ·0000 004 | ·0000 001 |
| ·14 | ·0000 211 | ·0000 097 | ·0000 044 | ·0000 020 | ·0000 009 | ·0000 002 |
| ·15 | ·0000 374 | ·0000 177 | ·0000 083 | ·0000 039 | ·0000 018 | ·0000 004 |
| ·16 | ·0000 634 | ·0000 310 | ·0000 151 | ·0000 073 | ·0000 035− | ·0000 008 |
| ·17 | ·0001 038 | ·0000 524 | ·0000 262 | ·0000 130 | ·0000 064 | ·0000 015+ |
| ·18 | ·0001 647 | ·0000 855− | ·0000 440 | ·0000 225+ | ·0000 114 | ·0000 029 |
| ·19 | ·0002 541 | ·0001 354 | ·0000 716 | ·0000 376 | ·0000 196 | ·0000 052 |
| ·20 | ·0003 819 | ·0002 088 | ·0001 132 | ·0000 610 | ·0000 326 | ·0000 092 |
| ·21 | ·0005 611 | ·0003 141 | ·0001 745+ | ·0000 963 | ·0000 527 | ·0000 156 |
| ·22 | ·0008 071 | ·0004 623 | ·0002 628 | ·0001 483 | ·0000 831 | ·0000 257 |
| ·23 | ·0011 389 | ·0006 667 | ·0003 873 | ·0002 234 | ·0001 280 | ·0000 413 |
| ·24 | ·0015 791 | ·0009 439 | ·0005 599 | ·0003 298 | ·0001 930 | ·0000 649 |
| ·25 | ·0021 542 | ·0013 136 | ·0007 949 | ·0004 777 | ·0002 852 | ·0000 999 |
| ·26 | ·0028 949 | ·0017 993 | ·0011 100 | ·0006 800 | ·0004 139 | ·0001 507 |
| ·27 | ·0038 365+ | ·0024 288 | ·0015 262 | ·0009 524 | ·0005 905+ | ·0002 231 |
| ·28 | ·0050 189 | ·0032 340 | ·0020 685− | ·0013 139 | ·0008 293 | ·0003 246 |
| ·29 | ·0064 865− | ·0042 514 | ·0027 660 | ·0017 873 | ·0011 476 | ·0004 649 |
| ·30 | ·0082 885+ | ·0055 223 | ·0036 525+ | ·0023 994 | ·0015 663 | ·0006 560 |
| ·31 | ·0104 789 | ·0070 931 | ·0047 665+ | ·0031 815+ | ·0021 103 | ·0009 126 |
| ·32 | ·0131 156 | ·0090 147 | ·0061 515− | ·0041 696 | ·0028 087 | ·0012 528 |
| ·33 | ·0162 610 | ·0113 430 | ·0078 560 | ·0054 048 | ·0036 955− | ·0016 985− |
| ·34 | ·0199 807 | ·0141 384 | ·0099 337 | ·0069 335− | ·0048 097 | ·0022 756 |
| ·35 | ·0243 434 | ·0174 656 | ·0124 432 | ·0088 071 | ·0061 955+ | ·0030 149 |
| ·36 | ·0294 202 | ·0213 931 | ·0154 480 | ·0110 828 | ·0079 030 | ·0039 521 |
| ·37 | ·0352 836 | ·0259 924 | ·0190 160 | ·0138 228 | ·0099 875− | ·0051 285+ |
| ·38 | ·0420 064 | ·0313 376 | ·0232 190 | ·0170 942 | ·0125 102 | ·0065 911 |
| ·39 | ·0496 612 | ·0375 043 | ·0281 322 | ·0209 691 | ·0155 377 | ·0083 932 |
| ·40 | ·0583 189 | ·0445 689 | ·0338 333 | ·0255 233 | ·0191 419 | ·0105 942 |
| ·41 | ·0680 476 | ·0526 072 | ·0404 016 | ·0308 363 | ·0233 994 | ·0132 599 |
| ·42 | ·0789 110 | ·0616 932 | ·0479 172 | ·0369 899 | ·0283 910 | ·0164 623 |
| ·43 | ·0909 679 | ·0718 980 | ·0564 592 | ·0440 677 | ·0342 010 | ·0202 796 |
| ·44 | ·1042 701 | ·0832 884 | ·0661 050− | ·0521 536 | ·0409 160 | ·0247 951 |
| ·45 | ·1188 614 | ·0959 254 | ·0769 287 | ·0613 304 | ·0486 241 | ·0300 975+ |
| ·46 | ·1347 765− | ·1098 627 | ·0889 995− | ·0716 789 | ·0574 134 | ·0362 794 |
| ·47 | ·1520 393 | ·1251 455− | ·1023 801 | ·0832 755+ | ·0673 703 | ·0434 364 |
| ·48 | ·1706 624 | ·1418 087 | ·1171 252 | ·0961 914 | ·0785 785− | ·0516 659 |
| ·49 | ·1906 456 | ·1598 760 | ·1332 802 | ·1104 903 | ·0911 166 | ·0610 655+ |
| ·50 | ·2119 751 | ·1793 583 | ·1508 789 | ·1262 267 | ·1050 568 | ·0717 316 |
| ·51 | ·2346 231 | ·2002 526 | ·1699 428 | ·1434 444 | ·1204 623 | ·0837 568 |
| ·52 | ·2585 468 | ·2225 413 | ·1904 791 | ·1621 747 | ·1373 862 | ·0972 287 |
| ·53 | ·2836 887 | ·2461 910 | ·2124 798 | ·1824 343 | ·1558 687 | ·1122 268 |
| ·54 | ·3099 757 | ·2711 522 | ·2359 206 | ·2042 246 | ·1759 360 | ·1288 208 |
| ·55 | ·3373 200 | ·2973 588 | ·2607 598 | ·2275 294 | ·1975 976 | ·1470 678 |
| ·56 | ·3656 193 | ·3247 283 | ·2869 379 | ·2523 147 | ·2208 455− | ·1670 102 |
| ·57 | ·3947 574 | ·3531 615+ | ·3143 773 | ·2785 270 | ·2456 523 | ·1886 730 |
| ·58 | ·4246 049 | ·3825 436 | ·3429 821 | ·3060 933 | ·2719 702 | ·2120 620 |
| ·59 | ·4550 210 | ·4127 445+ | ·3726 385− | ·3349 203 | ·2997 299 | ·2371 611 |
| ·60 | ·4858 546 | ·4436 204 | ·4032 156 | ·3648 950+ | ·3288 404 | ·2639 312 |
| ·61 | ·5169 458 | ·4750 147 | ·4345 661 | ·3958 851 | ·3591 887 | ·2923 084 |
| ·62 | ·5481 284 | ·5067 602 | ·4665 283 | ·4277 396 | ·3906 401 | ·3222 029 |
| ·63 | ·5792 314 | ·5386 811 | ·4989 270 | ·4602 906 | ·4230 393 | ·3534 986 |
| ·64 | ·6100 817 | ·5705 947 | ·5315 764 | ·4933 547 | ·4562 113 | ·3860 533 |
| ·65 | ·6405 063 | ·6023 147 | ·5642 821 | ·5267 354 | ·4899 636 | ·4196 988 |
| ·66 | ·6703 348 | ·6336 534 | ·5968 440 | ·5602 259 | ·5240 881 | ·4542 424 |
| ·67 | ·6994 022 | ·6644 244 | ·6290 592 | ·5936 116 | ·5583 644 | ·4894 692 |
| ·68 | ·7275 509 | ·6944 461 | ·6607 251 | ·6266 738 | ·5925 628 | ·5251 440 |
| ·69 | ·7546 337 | ·7235 440 | ·6916 431 | ·6591 931 | ·6264 477 | ·5610 153 |
| ·70 | ·7805 158 | ·7515 540 | ·7216 214 | ·6909 535− | ·6597 823 | ·5968 189 |

$x = \cdot 71$ to $\cdot 99$ $q = 6$ $p = 9$ to $12$

| $x$ | $p = 9$ | $p = 9\cdot 5$ | $p = 10$ | $p = 10\cdot 5$ | $p = 11$ | $p = 12$ |
|---|---|---|---|---|---|---|
| $B(p, q) =$ | $\cdot 5550\ 0056 \times \frac{1}{10^4}$ | $\cdot 4275\ 2225 \bar{\times} \frac{1}{10^4}$ | $\cdot 3330\ 0033 \times \frac{1}{10^4}$ | $\cdot 2620\ 2976 \times \frac{1}{10^4}$ | $\cdot 2081\ 2521 \times \frac{1}{10^4}$ | $\cdot 1346\ 6925 \overset{+}{\times} \frac{1}{10^4}$ |
| ·71 | ·8050 772 | ·7783 252 | ·7504 792 | ·7217 456 | ·6923 323 | ·6322 823 |
| ·72 | ·8282 145⁺ | ·8037 222 | ·7780 491 | ·7513 713 | ·7238 704 | ·6671 305⁺ |
| ·73 | ·8498 427 | ·8276 281 | ·8041 810 | ·7796 473 | ·7541 814 | ·7010 909 |
| ·74 | ·8698 964 | ·8499 459 | ·8287 445⁺ | ·8064 089 | ·7830 659 | ·7338 992 |
| ·75 | ·8883 310 | ·8706 009 | ·8516 319 | ·8315 133 | ·8103 454 | ·7653 056 |
| ·76 | ·9051 233 | ·8895 416 | ·8727 600 | ·8548 428 | ·8358 657 | ·7950 808 |
| ·77 | ·9202 715⁺ | ·9067 404 | ·8920 716 | ·8763 073 | ·8595 007 | ·8230 214 |
| ·78 | ·9337 949 | ·9221 945⁺ | ·9095 371 | ·8958 462 | ·8811 555⁻ | ·8489 552 |
| ·79 | ·9457 333 | ·9359 250⁻ | ·9251 544 | ·9134 297 | ·9007 683 | ·8727 465⁻ |
| ·80 | ·9561 456 | ·9479 762 | ·9389 486 | ·9290 592 | ·9183 121 | ·8942 988 |
| ·81 | ·9651 082 | ·9584 145⁻ | ·9509 714 | ·9427 669 | ·9337 952 | ·9135 585⁺ |
| ·82 | ·9727 128 | ·9673 260 | ·9612 991 | ·9546 148 | ·9472 604 | ·9305 160 |
| ·83 | ·9790 640 | ·9748 141 | ·9700 304 | ·9646 924 | ·9587 836 | ·9452 056 |
| ·84 | ·9842 764 | ·9809 968 | ·9772 829 | ·9731 138 | ·9684 710 | ·9577 045⁺ |
| ·85 | ·9884 717 | ·9860 026 | ·9831 899 | ·9800 137 | ·9764 556 | ·9681 296 |
| ·86 | ·9917 752 | ·9899 675⁻ | ·9878 962 | ·9855 434 | ·9828 923 | ·9766 329 |
| ·87 | ·9943 125⁺ | ·9930 306 | ·9915 532 | ·9898 654 | ·9879 524 | ·9833 956 |
| ·88 | ·9962 068 | ·9953 306 | ·9943 149 | ·9931 479 | ·9918 176 | ·9886 211 |
| ·89 | ·9975 751 | ·9970 014 | ·9963 325⁺ | ·9955 596 | ·9946 737 | ·9925 262 |
| ·90 | ·9985 259 | ·9981 689 | ·9977 503 | ·9972 639 | ·9967 032 | ·9953 325⁺ |
| ·91 | ·9991 568 | ·9989 479 | ·9987 015⁺ | ·9984 137 | ·9980 801 | ·9972 575⁻ |
| ·92 | ·9995 526 | ·9994 392 | ·9993 048 | ·9991 469 | ·9989 628 | ·9985 052 |
| ·93 | ·9997 841 | ·9997 282 | ·9996 615⁺ | ·9995 828 | ·9994 906 | ·9992 592 |
| ·94 | ·9999 080 | ·9998 837 | ·9998 545⁻ | ·9998 198 | ·9997 790 | ·9996 758 |
| ·95 | ·9999 669 | ·9999 580 | ·9999 472 | ·9999 343 | ·9999 191 | ·9998 803 |
| ·96 | ·9999 907 | ·9999 881 | ·9999 850⁺ | ·9999 813 | ·9999 768 | ·9999 654 |
| ·97 | ·9999 982 | ·9999 977 | ·9999 971 | ·9999 964 | ·9999 955⁻ | ·9999 932 |
| ·98 | ·9999 998 | ·9999 998 | ·9999 997 | ·9999 997 | ·9999 996 | ·9999 993 |
| ·99 | 1·0000 000 | 1·0000 000 | 1·0000 000 | 1·0000 000 | 1·0000 000 | 1·0000 000 |

$x$ = ·15 to ·70 $q = 6$ $p$ = 13 to 18

| | $p = 13$ | $p = 14$ | $p = 15$ | $p = 16$ | $p = 17$ | $p = 18$ |
|---|---|---|---|---|---|---|
| $B(p, q) =$ | ·8977 9502 × $\frac{1}{10^5}$ | ·6142 8080 × $\frac{1}{10^5}$ | ·4299 9656 × $\frac{1}{10^5}$ | ·3071 4040 × $\frac{1}{10^5}$ | ·2233 7484 × $\frac{1}{10^5}$ | ·1651 0314 × $\frac{1}{10^5}$ |
| $x$ | | | | | | |
| ·15 | ·0000 001 | | | | | |
| ·16 | ·0000 002 | | | | | |
| ·17 | ·0000 004 | ·0000 001 | | | | |
| ·18 | ·0000 007 | ·0000 002 | | | | |
| ·19 | ·0000 014 | ·0000 004 | ·0000 001 | | | |
| ·20 | ·0000 025$^+$ | ·0000 007 | ·0000 002 | | | |
| ·21 | ·0000 045$^-$ | ·0000 013 | ·0000 004 | ·0000 001 | | |
| ·22 | ·0000 078 | ·0000 023 | ·0000 007 | ·0000 002 | ·0000 001 | |
| ·23 | ·0000 130 | ·0000 040 | ·0000 012 | ·0000 004 | ·0000 001 | |
| ·24 | ·0000 214 | ·0000 069 | ·0000 022 | ·0000 007 | ·0000 002 | ·0000 001 |
| ·25 | ·0000 342 | ·0000 115$^+$ | ·0000 038 | ·0000 012 | ·0000 004 | ·0000 001 |
| ·26 | ·0000 537 | ·0000 188 | ·0000 065$^-$ | ·0000 022 | ·0000 007 | ·0000 002 |
| ·27 | ·0000 825$^-$ | ·0000 299 | ·0000 107 | ·0000 038 | ·0000 013 | ·0000 004 |
| ·28 | ·0001 244 | ·0000 468 | ·0000 173 | ·0000 063 | ·0000 023 | ·0000 008 |
| ·29 | ·0001 845$^-$ | ·0000 719 | ·0000 275$^+$ | ·0000 104 | ·0000 039 | ·0000 014 |
| ·30 | ·0002 691 | ·0001 084 | ·0000 429 | ·0000 168 | ·0000 065$^-$ | ·0000 025$^-$ |
| ·31 | ·0003 866 | ·0001 608 | ·0000 658 | ·0000 265$^+$ | ·0000 106 | ·0000 042 |
| ·32 | ·0005 474 | ·0002 349 | ·0000 992 | ·0000 413 | ·0000 170 | ·0000 069 |
| ·33 | ·0007 648 | ·0003 383 | ·0001 472 | ·0000 631 | ·0000 267 | ·0000 112 |
| ·34 | ·0010 551 | ·0004 804 | ·0002 153 | ·0000 951 | ·0000 415$^-$ | ·0000 179 |
| ·35 | ·0014 379 | ·0006 736 | ·0003 106 | ·0001 412 | ·0000 633 | ·0000 281 |
| ·36 | ·0019 372 | ·0009 328 | ·0004 421 | ·0002 066 | ·0000 953 | ·0000 435$^-$ |
| ·37 | ·0025 816 | ·0012 768 | ·0006 216 | ·0002 984 | ·0001 414 | ·0000 662 |
| ·38 | ·0034 047 | ·0017 282 | ·0008 635$^+$ | ·0004 255$^-$ | ·0002 070 | ·0000 995$^+$ |
| ·39 | ·0044 459 | ·0023 143 | ·0011 861 | ·0005 994 | ·0002 991 | ·0001 476 |
| ·40 | ·0057 505$^-$ | ·0030 678 | ·0016 115$^+$ | ·0008 348 | ·0004 271 | ·0002 160 |
| ·41 | ·0073 705$^+$ | ·0040 272 | ·0021 669 | ·0011 499 | ·0006 026 | ·0003 122 |
| ·42 | ·0093 648 | ·0052 373 | ·0028 846 | ·0015 671 | ·0008 408 | ·0004 461 |
| ·43 | ·0117 991 | ·0067 499 | ·0038 034 | ·0021 140 | ·0011 606 | ·0006 300 |
| ·44 | ·0147 464 | ·0086 244 | ·0049 687 | ·0028 240 | ·0015 855$^-$ | ·0008 802 |
| ·45 | ·0182 868 | ·0109 278 | ·0064 336 | ·0037 370 | ·0021 444 | ·0012 169 |
| ·46 | ·0225 071 | ·0137 352 | ·0082 590 | ·0049 003 | ·0028 724 | ·0016 653 |
| ·47 | ·0275 003 | ·0171 297 | ·0105 147 | ·0063 693 | ·0038 121 | ·0022 567 |
| ·48 | ·0333 653 | ·0212 024 | ·0132 792 | ·0082 083 | ·0050 137 | ·0030 292 |
| ·49 | ·0402 050$^+$ | ·0260 520 | ·0166 402 | ·0104 912 | ·0065 366 | ·0040 289 |
| ·50 | ·0481 262 | ·0317 841 | ·0206 947 | ·0133 018 | ·0084 503 | ·0053 110 |
| ·51 | ·0572 371 | ·0385 101 | ·0255 484 | ·0167 344 | ·0108 345$^-$ | ·0069 406 |
| ·52 | ·0676 460 | ·0463 464 | ·0313 150$^+$ | ·0208 933 | ·0137 804 | ·0089 939 |
| ·53 | ·0794 589 | ·0554 124 | ·0381 160 | ·0258 933 | ·0173 908 | ·0115 591 |
| ·54 | ·0927 776 | ·0658 285$^-$ | ·0460 786 | ·0318 587 | ·0217 803 | ·0147 373 |
| ·55 | ·1076 967 | ·0777 141 | ·0553 342 | ·0389 223 | ·0270 749 | ·0186 424 |
| ·56 | ·1243 009 | ·0911 847 | ·0660 164 | ·0472 241 | ·0334 118 | ·0234 019 |
| ·57 | ·1426 623 | ·1063 492 | ·0782 584 | ·0569 094 | ·0409 377 | ·0291 562 |
| ·58 | ·1628 374 | ·1233 063 | ·0921 896 | ·0681 261 | ·0498 077 | ·0360 582 |
| ·59 | ·1848 640 | ·1421 413 | ·1079 326 | ·0810 218 | ·0601 827 | ·0442 715$^-$ |
| ·60 | ·2087 584 | ·1629 225$^-$ | ·1255 990 | ·0957 402 | ·0722 264 | ·0539 686 |
| ·61 | ·2345 126 | ·1856 974 | ·1452 854 | ·1124 170 | ·0861 017 | ·0653 281 |
| ·62 | ·2620 921 | ·2104 893 | ·1670 693 | ·1311 754 | ·1019 667 | ·0785 311 |
| ·63 | ·2914 336 | ·2372 937 | ·1910 042 | ·1521 209 | ·1199 694 | ·0937 564 |
| ·64 | ·3224 434 | ·2660 752 | ·2171 154 | ·1753 363 | ·1402 420 | ·1111 755$^+$ |
| ·65 | ·3549 966 | ·2967 646 | ·2453 957 | ·2008 760 | ·1628 951 | ·1309 465$^+$ |
| ·66 | ·3889 368 | ·3292 574 | ·2758 017 | ·2287 608 | ·1880 115$^+$ | ·1532 069 |
| ·67 | ·4240 762 | ·3634 115$^+$ | ·3082 501 | ·2589 725$^-$ | ·2156 390 | ·1780 663 |
| ·68 | ·4601 973 | ·3990 475$^+$ | ·3426 150$^-$ | ·2914 495$^-$ | ·2457 842 | ·2055 988 |
| ·69 | ·4970 552 | ·4359 486 | ·3787 267 | ·3260 825$^+$ | ·2784 067 | ·2358 349 |
| ·70 | ·5343 801 | ·4738 625$^+$ | ·4163 708 | ·3627 119 | ·3134 128 | ·2687 536 |

$x = ·71$ to $1·00$ $q = 6$ $p = 13$ to $18$

| | $p = 13$ | $p = 14$ | $p = 15$ | $p = 16$ | $p = 17$ | $p = 18$ |
|---|---|---|---|---|---|---|
| $B(p, q) =$ | ·8977 9502 × $\frac{1}{10^5}$ | ·6142 8080 × $\frac{1}{10^5}$ | ·4299 9656 × $\frac{1}{10^5}$ | ·3071 4040 × $\frac{1}{10^5}$ | ·2233 7484 × $\frac{1}{10^5}$ | ·1651 0314 × $\frac{1}{10^5}$ |
| $x$ | | | | | | |
| ·71 | ·5718 821 | ·5125 041 | ·4552 891 | ·4011 255+ | ·3506 519 | ·3042 755+ |
| ·72 | ·6092 559 | ·5515 593 | ·4951 815− | ·4410 588 | ·3899 129 | ·3422 569 |
| ·73 | ·6461 865− | ·5906 908 | ·5357 104 | ·4821 962 | ·4309 229 | ·3824 847 |
| ·74 | ·6823 560 | ·6295 441 | ·5765 058 | ·5241 748 | ·4733 482 | ·4246 743 |
| ·75 | ·7174 508 | ·6677 554 | ·6171 727 | ·5665 899 | ·5167 974 | ·4684 695− |
| ·76 | ·7511 690 | ·7049 604 | ·6572 994 | ·6090 030 | ·5608 273 | ·5134 450+ |
| ·77 | ·7832 284 | ·7408 030 | ·6964 685− | ·6509 517 | ·6049 513 | ·5591 132 |
| ·78 | ·8133 740 | ·7749 462 | ·7342 676 | ·6919 619 | ·6486 514 | ·6049 334 |
| ·79 | ·8413 854 | ·8070 812 | ·7703 021 | ·7315 616 | ·6913 925− | ·6503 254 |
| ·80 | ·8670 837 | ·8369 377 | ·8042 078 | ·7692 959 | ·7326 384 | ·6946 871 |
| ·81 | ·8903 369 | ·8642 930 | ·8356 633 | ·8047 432 | ·7718 713 | ·7374 138 |
| ·82 | ·9110 645+ | ·8889 797 | ·8644 024 | ·8375 312 | ·8086 111 | ·7779 218 |
| ·83 | ·9292 401 | ·9108 922 | ·8902 245− | ·8673 522 | ·8424 358 | ·8156 726 |
| ·84 | ·9448 924 | ·9299 909 | ·9130 032 | ·8939 769 | ·8730 005− | ·8501 979 |
| ·85 | ·9581 036 | ·9463 039 | ·9326 920 | ·9172 652 | ·9000 548 | ·8811 232 |
| ·86 | ·9690 068 | ·9599 259 | ·9493 273 | ·9371 741 | ·9234 562 | ·9081 891 |
| ·87 | ·9777 794 | ·9710 141 | ·9630 262 | ·9537 602 | ·9431 795+ | ·9312 670 |
| ·88 | ·9846 361 | ·9797 805− | ·9739 815+ | ·9671 774 | ·9593 187 | ·9503 690 |
| ·89 | ·9898 186 | ·9864 820 | ·9824 518 | ·9776 694 | ·9720 829 | ·9656 485+ |
| ·90 | ·9935 849 | ·9914 070 | ·9887 469 | ·9855 547 | ·9817 840 | ·9773 923 |
| ·91 | ·9961 970 | ·9948 608 | ·9932 106 | ·9912 083 | ·9888 169 | ·9860 006 |
| ·92 | ·9979 088 | ·9971 491 | ·9962 005+ | ·9950 369 | ·9936 319 | ·9919 591 |
| ·93 | ·9989 545+ | ·9985 621 | ·9980 668 | ·9974 527 | ·9967 030 | ·9958 008 |
| ·94 | ·9995 384 | ·9993 596 | ·9991 315− | ·9988 455+ | ·9984 928 | ·9980 636 |
| ·95 | ·9998 280 | ·9997 593 | ·9996 707 | ·9995 585− | ·9994 185+ | ·9992 465− |
| ·96 | ·9999 499 | ·9999 292 | ·9999 023 | ·9998 679 | ·9998 245+ | ·9997 707 |
| ·97 | ·9999 901 | ·9999 859 | ·9999 804 | ·9999 732 | ·9999 641 | ·9999 526 |
| ·98 | ·9999 990 | ·9999 986 | ·9999 981 | ·9999 973 | ·9999 964 | ·9999 952 |
| ·99 | 1·0000 000 | 1·0000 000 | 1·0000 000 | 1·0000 000 | ·9999 999 | ·9999 999 |
| 1·00 | | | | | 1·0000 000 | 1·0000 000 |

$x = \cdot 26$ to $\cdot 80$ $q = 6$ $p = 19$ to $24$

| $x$ | $p = 19$ | $p = 20$ | $p = 21$ | $p = 22$ | $p = 23$ | $p = 24$ |
|---|---|---|---|---|---|---|
| $B(p, q) =$ | ·1238 2735 $\times \frac{1}{10^5}$ | ·9410 8790 $\times \frac{1}{10^6}$ | ·7239 1377 $\times \frac{1}{10^6}$ | ·5630 4404 $\times \frac{1}{10^6}$ | ·4423 9175 $\times \frac{1}{10^6}$ | ·3508 6242 $\times \frac{1}{10^6}$ |
| ·26 | ·0000 001 | | | | | |
| ·27 | ·0000 002 | ·0000 001 | | | | |
| ·28 | ·0000 003 | ·0000 001 | | | | |
| ·29 | ·0000 005+ | ·0000 002 | ·0000 001 | | | |
| ·30 | ·0000 009 | ·0000 003 | ·0000 001 | | | |
| ·31 | ·0000 016 | ·0000 006 | ·0000 002 | ·0000 001 | | |
| ·32 | ·0000 028 | ·0000 011 | ·0000 004 | ·0000 002 | ·0000 001 | |
| ·33 | ·0000 046 | ·0000 019 | ·0000 008 | ·0000 003 | ·0000 001 | |
| ·34 | ·0000 076 | ·0000 032 | ·0000 013 | ·0000 006 | ·0000 002 | ·0000 001 |
| ·35 | ·0000 123 | ·0000 054 | ·0000 023 | ·0000 010 | ·0000 004 | ·0000 002 |
| ·36 | ·0000 196 | ·0000 088 | ·0000 039 | ·0000 017 | ·0000 007 | ·0000 003 |
| ·37 | ·0000 307 | ·0000 141 | ·0000 064 | ·0000 029 | ·0000 013 | ·0000 006 |
| ·38 | ·0000 474 | ·0000 223 | ·0000 104 | ·0000 048 | ·0000 022 | ·0000 010 |
| ·39 | ·0000 720 | ·0000 348 | ·0000 167 | ·0000 079 | ·0000 037 | ·0000 018 |
| ·40 | ·0001 081 | ·0000 536 | ·0000 263 | ·0000 128 | ·0000 062 | ·0000 030 |
| ·41 | ·0001 601 | ·0000 813 | ·0000 409 | ·0000 205− | ·0000 101 | ·0000 050− |
| ·42 | ·0002 342 | ·0001 218 | ·0000 628 | ·0000 321 | ·0000 163 | ·0000 082 |
| ·43 | ·0003 385+ | ·0001 802 | ·0000 951 | ·0000 498 | ·0000 259 | ·0000 133 |
| ·44 | ·0004 837 | ·0002 633 | ·0001 421 | ·0000 761 | ·0000 405− | ·0000 214 |
| ·45 | ·0006 836 | ·0003 804 | ·0002 099 | ·0001 149 | ·0000 625− | ·0000 337 |
| ·46 | ·0009 558 | ·0005 435− | ·0003 064 | ·0001 714 | ·0000 952 | ·0000 525− |
| ·47 | ·0013 226 | ·0007 680 | ·0004 422 | ·0002 527 | ·0001 433 | ·0000 807 |
| ·48 | ·0018 121 | ·0010 741 | ·0006 313 | ·0003 682 | ·0002 132 | ·0001 226 |
| ·49 | ·0024 589 | ·0014 871 | ·0008 918 | ·0005 307 | ·0003 136 | ·0001 840 |
| ·50 | ·0033 054 | ·0020 387 | ·0012 470 | ·0007 569 | ·0004 561 | ·0002 731 |
| ·51 | ·0044 030 | ·0027 683 | ·0017 262 | ·0010 682 | ·0006 563 | ·0004 006 |
| ·52 | ·0058 135− | ·0037 245− | ·0023 666 | ·0014 924 | ·0009 345+ | ·0005 813 |
| ·53 | ·0076 098 | ·0049 658 | ·0032 142 | ·0020 648 | ·0013 171 | ·0008 347 |
| ·54 | ·0098 776 | ·0065 628 | ·0043 253 | ·0028 294 | ·0018 380 | ·0011 863 |
| ·55 | ·0127 162 | ·0085 991 | ·0057 685+ | ·0038 411 | ·0025 400 | ·0016 689 |
| ·56 | ·0162 393 | ·0111 726 | ·0076 260 | ·0051 670 | ·0034 770 | ·0023 249 |
| ·57 | ·0205 754 | ·0143 972 | ·0099 952 | ·0068 887 | ·0047 155− | ·0032 075+ |
| ·58 | ·0258 683 | ·0184 028 | ·0129 903 | ·0091 036 | ·0063 370 | ·0043 836 |
| ·59 | ·0322 761 | ·0233 365− | ·0167 435− | ·0119 275− | ·0084 402 | ·0059 355− |
| ·60 | ·0399 709 | ·0293 622 | ·0214 057 | ·0154 951 | ·0111 427 | ·0079 636 |
| ·61 | ·0491 362 | ·0366 600 | ·0271 468 | ·0199 622 | ·0145 834 | ·0105 891 |
| ·62 | ·0599 648 | ·0454 245+ | ·0341 558 | ·0255 057 | ·0189 238 | ·0139 558 |
| ·63 | ·0726 549 | ·0558 626 | ·0426 387 | ·0323 240 | ·0243 489 | ·0182 323 |
| ·64 | ·0874 057 | ·0681 896 | ·0528 168 | ·0406 356 | ·0310 679 | ·0236 134 |
| ·65 | ·1044 114 | ·0826 247 | ·0649 230 | ·0506 773 | ·0393 132 | ·0303 207 |
| ·66 | ·1238 550− | ·0993 847 | ·0791 968 | ·0627 004 | ·0493 382 | ·0386 021 |
| ·67 | ·1458 999 | ·1186 770 | ·0958 778 | ·0769 653 | ·0614 141 | ·0487 297 |
| ·68 | ·1706 822 | ·1406 907 | ·1151 978 | ·0937 353 | ·0758 238 | ·0609 963 |
| ·69 | ·1983 008 | ·1655 869 | ·1373 711 | ·1132 668 | ·0928 548 | ·0757 088 |
| ·70 | ·2288 084 | ·1934 884 | ·1625 835− | ·1357 992 | ·1127 890 | ·0931 804 |
| ·71 | ·2622 018 | ·2244 684 | ·1909 800 | ·1615 421 | ·1358 909 | ·1137 194 |
| ·72 | ·2984 134 | ·2585 388 | ·2226 518 | ·1906 610 | ·1623 928 | ·1376 151 |
| ·73 | ·3373 026 | ·2956 399 | ·2576 228 | ·2232 625+ | ·1924 789 | ·1651 215+ |
| ·74 | ·3786 504 | ·3356 302 | ·2958 365+ | ·2593 779 | ·2262 669 | ·1964 381 |
| ·75 | ·4221 552 | ·3782 785+ | ·3371 441 | ·2989 479 | ·2637 900 | ·2316 893 |
| ·76 | ·4674 317 | ·4232 588 | ·3812 946 | ·3418 083 | ·3049 783 | ·2709 025+ |
| ·77 | ·5140 137 | ·4701 484 | ·4279 281 | ·3876 781 | ·3496 418 | ·3139 870 |
| ·78 | ·5613 610 | ·5184 308 | ·4765 738 | ·4361 519 | ·3974 571 | ·3607 139 |
| ·79 | ·6088 706 | ·5675 030 | ·5266 525− | ·4866 968 | ·4479 579 | ·4107 013 |
| ·80 | ·6558 924 | ·6166 894 | ·5774 864 | ·5386 567 | ·5005 331 | ·4634 039 |

$x = \cdot 81$ to $1\cdot 00$ $q = 6$ $p = 19$ to $24$

| | $p = 19$ | $p = 20$ | $p = 21$ | $p = 22$ | $p = 23$ | $p = 24$ |
|---|---|---|---|---|---|---|
| $B(p, q) =$ | $\cdot 1238\ 2735^{+} \times \frac{1}{10^5}$ | $\cdot 9410\ 8790 \times \frac{1}{10^6}$ | $\cdot 7239\ 1377 \times \frac{1}{10^6}$ | $\cdot 5630\ 4404 \times \frac{1}{10^6}$ | $\cdot 4423\ 9175^{-} \times \frac{1}{10^6}$ | $\cdot 3508\ 6242 \times \frac{1}{10^6}$ |
| $x$ | | | | | | |
| ·81 | ·7017 503 | ·6652 609 | ·6283 154 | ·5912 643 | ·5544 321 | ·5181 124 |
| ·82 | ·7457 662 | ·7124 598 | ·6783 207 | ·6436 614 | ·6087 815+ | ·5739 623 |
| ·83 | ·7872 888 | ·7575 305+ | ·7266 564 | ·6949 295− | ·6626 113 | ·6299 559 |
| ·84 | ·8257 231 | ·7997 540 | ·7724 865+ | ·7441 283 | ·7148 936 | ·6849 979 |
| ·85 | ·8605 614 | ·8384 846 | ·8150 280 | ·7903 426 | ·7645 913 | ·7379 444 |
| ·86 | ·8914 121 | ·8731 871 | ·8535 952 | ·8327 344 | ·8107 169 | ·7876 655− |
| ·87 | ·9180 242 | ·9034 711 | ·8876 446 | ·8705 972 | ·8523 952 | ·8331 170 |
| ·88 | ·9403 055− | ·9291 191 | ·9168 142 | ·9034 076 | ·8889 285+ | ·8734 170 |
| ·89 | ·9583 313 | ·9501 051 | ·9409 536 | ·9308 694 | ·9198 547 | ·9079 206 |
| ·90 | ·9723 417 | ·9666 001 | ·9601 407 | ·9529 431 | ·9449 931 | ·9362 826 |
| ·91 | ·9827 259 | ·9789 617 | ·9746 799 | ·9698 558 | ·9644 681 | ·9584 995+ |
| ·92 | ·9899 926 | ·9877 073 | ·9850 792 | ·9820 857 | ·9787 058 | ·9749 203 |
| ·93 | ·9947 286 | ·9934 691 | ·9920 050− | ·9903 191 | ·9883 949 | ·9862 164 |
| ·94 | ·9975 482 | ·9969 361 | ·9962 170 | ·9953 801 | ·9944 146 | ·9933 097 |
| ·95 | ·9990 376 | ·9987 870 | ·9984 894 | ·9981 394 | ·9977 313 | ·9972 593 |
| ·96 | ·9997 046 | ·9996 244 | ·9995 282 | ·9994 139 | ·9992 792 | ·9991 218 |
| ·97 | ·9999 385− | ·9999 211 | ·9999 000 | ·9998 747 | ·9998 446 | ·9998 091 |
| ·98 | ·9999 937 | ·9999 918 | ·9999 896 | ·9999 868 | ·9999 835− | ·9999 795+ |
| ·99 | ·9999 999 | ·9999 999 | ·9999 998 | ·9999 998 | ·9999 997 | ·9999 996 |
| 1·00 | 1·0000 000 | 1·0000 000 | 1·0000 000 | 1·0000 000 | 1·0000 000 | 1·0000 000 |

$x = \cdot 35$ to 1·00 $q = 6$ $p = 25$ to 30

| | $p = 25$ | $p = 26$ | $p = 27$ | $p = 28$ | $p = 29$ | $p = 30$ |
|---|---|---|---|---|---|---|
| $B(p, q) =$ | ·2806 8994 × $\frac{1}{10^8}$ | ·2263 6285 $\stackrel{+}{\times}$ $\frac{1}{10^8}$ | ·1839 1982 × $\frac{1}{10^8}$ | ·1504 7985 $\stackrel{+}{\times}$ $\frac{1}{10^8}$ | ·1239 2458 × $\frac{1}{10^8}$ | ·1026 8037 × $\frac{1}{10^8}$ |
| $x$ | | | | | | |
| ·35 | ·0000 001 | | | | | |
| ·36 | ·0000 001 | ·0000 001 | | | | |
| ·37 | ·0000 003 | ·0000 001 | | | | |
| ·38 | ·0000 005⁻ | ·0000 002 | ·0000 001 | | | |
| ·39 | ·0000 008 | ·0000 004 | ·0000 002 | ·0000 001 | | |
| ·40 | ·0000 014 | ·0000 007 | ·0000 003 | ·0000 001 | ·0000 001 | |
| ·41 | ·0000 024 | ·0000 012 | ·0000 006 | ·0000 003 | ·0000 001 | ·0000 001 |
| ·42 | ·0000 041 | ·0000 021 | ·0000 010 | ·0000 005⁺ | ·0000 002 | ·0000 001 |
| ·43 | ·0000 068 | ·0000 035⁻ | ·0000 018 | ·0000 009 | ·0000 004 | ·0000 002 |
| ·44 | ·0000 112 | ·0000 058 | ·0000 030 | ·0000 016 | ·0000 008 | ·0000 004 |
| ·45 | ·0000 181 | ·0000 096 | ·0000 051 | ·0000 027 | ·0000 014 | ·0000 007 |
| ·46 | ·0000 288 | ·0000 157 | ·0000 085⁻ | ·0000 046 | ·0000 025⁻ | ·0000 013 |
| ·47 | ·0000 452 | ·0000 252 | ·0000 139 | ·0000 077 | ·0000 042 | ·0000 023 |
| ·48 | ·0000 701 | ·0000 398 | ·0000 225⁺ | ·0000 127 | ·0000 071 | ·0000 039 |
| ·49 | ·0001 073 | ·0000 622 | ·0000 359 | ·0000 206 | ·0000 118 | ·0000 067 |
| ·50 | ·0001 625⁻ | ·0000 961 | ·0000 565⁺ | ·0000 331 | ·0000 193 | ·0000 112 |
| ·51 | ·0002 430 | ·0001 466 | ·0000 879 | ·0000 525⁻ | ·0000 312 | ·0000 184 |
| ·52 | ·0003 594 | ·0002 210 | ·0001 351 | ·0000 822 | ·0000 498 | ·0000 300 |
| ·53 | ·0005 258 | ·0003 293 | ·0002 052 | ·0001 272 | ·0000 785⁻ | ·0000 482 |
| ·54 | ·0007 610 | ·0004 855⁻ | ·0003 080 | ·0001 945⁻ | ·0001 222 | ·0000 765⁻ |
| ·55 | ·0010 900 | ·0007 079 | ·0004 573 | ·0002 940 | ·0001 881 | ·0001 198 |
| ·56 | ·0015 452 | ·0010 213 | ·0006 715⁺ | ·0004 394 | ·0002 861 | ·0001 855⁺ |
| ·57 | ·0021 689 | ·0014 585⁻ | ·0009 757 | ·0006 495⁻ | ·0004 304 | ·0002 839 |
| ·58 | ·0030 145⁺ | ·0020 617 | ·0014 027 | ·0009 498 | ·0006 401 | ·0004 296 |
| ·59 | ·0041 498 | ·0028 855⁺ | ·0019 962 | ·0013 743 | ·0009 418 | ·0006 427 |
| ·60 | ·0056 588 | ·0039 993 | ·0028 121 | ·0019 679 | ·0013 709 | ·0009 510 |
| ·61 | ·0076 450⁺ | ·0054 899 | ·0039 225⁺ | ·0027 893 | ·0019 746 | ·0013 920 |
| ·62 | ·0102 340 | ·0074 650⁻ | ·0054 180 | ·0039 139 | ·0028 148 | ·0020 159 |
| ·63 | ·0135 760 | ·0100 559 | ·0074 118 | ·0054 375⁻ | ·0039 716 | ·0028 888 |
| ·64 | ·0178 485⁺ | ·0134 211 | ·0100 427 | ·0074 801 | ·0055 472 | ·0040 968 |
| ·65 | ·0232 578 | ·0177 488 | ·0134 793 | ·0101 902 | ·0076 705⁻ | ·0057 503 |
| ·66 | ·0300 400 | ·0232 588 | ·0179 226 | ·0137 484 | ·0105 015⁺ | ·0079 891 |
| ·67 | ·0384 607 | ·0302 044 | ·0236 088 | ·0183 715⁻ | ·0142 359 | ·0109 872 |
| ·68 | ·0488 129 | ·0388 714 | ·0308 110 | ·0243 150⁻ | ·0191 089 | ·0149 584 |
| ·69 | ·0614 132 | ·0495 765⁺ | ·0398 386 | ·0318 751 | ·0253 991 | ·0201 602 |
| ·70 | ·0765 948 | ·0626 628 | ·0510 350⁻ | ·0413 882 | ·0334 296 | ·0268 981 |
| ·71 | ·0946 981 | ·0784 920 | ·0647 728 | ·0532 284 | ·0435 683 | ·0355 270 |
| ·72 | ·1160 586 | ·0974 337 | ·0814 449 | ·0678 012 | ·0562 235⁻ | ·0464 503 |
| ·73 | ·1409 901 | ·1198 510 | ·1014 518 | ·0855 331 | ·0718 373 | ·0601 156 |
| ·74 | ·1697 663 | ·1460 817 | ·1251 846 | ·1068 570 | ·0908 728 | ·0770 051 |
| ·75 | ·2025 981 | ·1764 160 | ·1530 031 | ·1321 917 | ·1137 959 | ·0976 202 |
| ·76 | ·2396 097 | ·2110 706 | ·1852 097 | ·1619 158 | ·1410 512 | ·1224 600 |
| ·77 | ·2808 131 | ·2501 604 | ·2220 189 | ·1963 372 | ·1730 310 | ·1519 911 |
| ·78 | ·3260 834 | ·2936 693 | ·2635 242 | ·2356 567 | ·2100 385⁻ | ·1866 111 |
| ·79 | ·3751 367 | ·3414 214 | ·3096 643 | ·2799 302 | ·2522 456 | ·2266 039 |
| ·80 | ·4275 124 | ·3930 566 | ·3601 910 | ·3290 296 | ·2996 488 | ·2720 917 |
| ·81 | ·4825 645⁺ | ·4480 120 | ·4146 422 | ·3826 072 | ·3520 252 | ·3229 828 |
| ·82 | ·5394 623 | ·5055 143 | ·4723 236 | ·4400 671 | ·4088 935⁺ | ·3789 239 |
| ·83 | ·5972 053 | ·5645 857 | ·5323 048 | ·5005 500⁻ | ·4694 870 | ·4392 594 |
| ·84 | ·6546 537 | ·6240 669 | ·5934 329 | ·5629 351 | ·5327 423 | ·5030 076 |
| ·85 | ·7105 757 | ·6826 597 | ·6543 678 | ·6258 664 | ·5973 141 | ·5688 603 |
| ·86 | ·7637 112 | ·7389 904 | ·7136 421 | ·6878 055⁺ | ·6616 184 | ·6352 145⁻ |
| ·87 | ·8128 507 | ·7916 927 | ·7697 453 | ·7471 152 | ·7239 112 | ·7002 431 |
| ·88 | ·8569 231 | ·8395 055⁺ | ·8212 305⁻ | ·8021 703 | ·7824 022 | ·7620 069 |
| ·89 | ·8950 864 | ·8813 795⁻ | ·8668 344 | ·8514 919 | ·8353 988 | ·8186 065⁻ |
| ·90 | ·9268 099 | ·9165 794 | ·9056 013 | ·8938 914 | ·8814 704 | ·8683 642 |
| ·91 | ·9519 366 | ·9447 698 | ·9369 939 | ·9286 074 | ·9196 129 | ·9100 167 |
| ·92 | ·9707 121 | ·9660 662 | ·9609 700 | ·9554 133 | ·9493 882 | ·9428 894 |
| ·93 | ·9837 683 | ·9810 362 | ·9780 067 | ·9746 676 | ·9710 076 | ·9670 170 |
| ·94 | ·9920 548 | ·9906 392 | ·9890 526 | ·9872 851 | ·9853 269 | ·9831 689 |
| ·95 | ·9967 175⁺ | ·9960 999 | ·9954 003 | ·9946 125⁺ | ·9937 306 | ·9927 483 |
| ·96 | ·9989 392 | ·9987 288 | ·9984 880 | ·9982 140 | ·9979 041 | ·9975 552 |
| ·97 | ·9997 674 | ·9997 189 | ·9996 628 | ·9995 983 | ·9995 245⁺ | ·9994 407 |
| ·98 | ·9999 749 | ·9999 694 | ·9999 629 | ·9999 555⁻ | ·9999 468 | ·9999 369 |
| ·99 | ·9999 995⁺ | ·9999 994 | ·9999 993 | ·9999 991 | ·9999 989 | ·9999 987 |
| 1·00 | 1·0000 000 | 1·0000 000 | 1·0000 000 | 1·0000 000 | 1·0000 000 | 1·0000 000 |

$x = \cdot 42$ to $1\cdot 00$ | $q = 6$ | $p = 31$ to $36$

| | $p = 31$ | $p = 32$ | $p = 33$ | $p = 34$ | $p = 35$ | $p = 36$ |
|---|---|---|---|---|---|---|
| $B(p, q) =$ | ·8556 6974 × $\frac{1}{10^7}$ | ·7169 1248 × $\frac{1}{10^7}$ | ·6037 1577 × $\frac{1}{10^7}$ | ·5108 3642 × $\frac{1}{10^7}$ | ·4342 1096 × $\frac{1}{10^7}$ | ·3706 6789 × $\frac{1}{10^7}$ |
| $x$ | | | | | | |
| ·42 | ·0000 001 | | | | | |
| ·43 | ·0000 001 | ·0000 001 | | | | |
| ·44 | ·0000 002 | ·0000 001 | ·0000 001 | | | |
| ·45 | ·0000 004 | ·0000 002 | ·0000 001 | ·0000 001 | | |
| ·46 | ·0000 007 | ·0000 004 | ·0000 002 | ·0000 001 | ·0000 001 | |
| ·47 | ·0000 012 | ·0000 007 | ·0000 004 | ·0000 002 | ·0000 001 | ·0000 001 |
| ·48 | ·0000 022 | ·0000 012 | ·0000 007 | ·0000 004 | ·0000 002 | ·0000 001 |
| ·49 | ·0000 038 | ·0000 021 | ·0000 012 | ·0000 007 | ·0000 004 | ·0000 002 |
| ·50 | ·0000 065− | ·0000 037 | ·0000 021 | ·0000 012 | ·0000 007 | ·0000 004 |
| ·51 | ·0000 109 | ·0000 064 | ·0000 037 | ·0000 022 | ·0000 013 | ·0000 007 |
| ·52 | ·0000 180 | ·0000 108 | ·0000 064 | ·0000 038 | ·0000 023 | ·0000 013 |
| ·53 | ·0000 295− | ·0000 180 | ·0000 109 | ·0000 066 | ·0000 040 | ·0000 024 |
| ·54 | ·0000 476 | ·0000 296 | ·0000 183 | ·0000 113 | ·0000 069 | ·0000 042 |
| ·55 | ·0000 760 | ·0000 480 | ·0000 302 | ·0000 190 | ·0000 119 | ·0000 074 |
| ·56 | ·0001 198 | ·0000 771 | ·0000 494 | ·0000 315+ | ·0000 201 | ·0000 127 |
| ·57 | ·0001 866 | ·0001 221 | ·0000 796 | ·0000 517 | ·0000 335+ | ·0000 216 |
| ·58 | ·0002 871 | ·0001 912 | ·0001 268 | ·0000 838 | ·0000 552 | ·0000 363 |
| ·59 | ·0004 368 | ·0002 957 | ·0001 995− | ·0001 341 | ·0000 899 | ·0000 600 |
| ·60 | ·0006 570 | ·0004 522 | ·0003 101 | ·0002 119 | ·0001 444 | ·0000 980 |
| ·61 | ·0009 774 | ·0006 836 | ·0004 765− | ·0003 310 | ·0002 291 | ·0001 581 |
| ·62 | ·0014 380 | ·0010 219 | ·0007 237 | ·0005 107 | ·0003 593 | ·0002 519 |
| ·63 | ·0020 930 | ·0015 107 | ·0010 866 | ·0007 789 | ·0005 566 | ·0003 965+ |
| ·64 | ·0030 139 | ·0022 090 | ·0016 134 | ·0011 744 | ·0008 522 | ·0006 165+ |
| ·65 | ·0042 942 | ·0031 950+ | ·0023 690 | ·0017 507 | ·0012 897 | ·0009 472 |
| ·66 | ·0060 545+ | ·0045 718 | ·0034 403 | ·0025 803 | ·0019 293 | ·0014 382 |
| ·67 | ·0084 479 | ·0064 722 | ·0049 416 | ·0037 607 | ·0028 531 | ·0021 582 |
| ·68 | ·0116 656 | ·0090 654 | ·0070 210 | ·0054 202 | ·0041 715− | ·0032 011 |
| ·69 | ·0159 429 | ·0125 636 | ·0098 676 | ·0077 254 | ·0060 300 | ·0046 931 |
| ·70 | ·0215 640 | ·0172 279 | ·0137 184 | ·0108 895− | ·0086 181 | ·0068 009 |
| ·71 | ·0288 661 | ·0233 742 | ·0188 656 | ·0151 795− | ·0121 775− | ·0097 416 |
| ·72 | ·0382 409 | ·0313 767 | ·0256 623 | ·0209 245+ | ·0170 117 | ·0137 920 |
| ·73 | ·0501 327 | ·0416 697 | ·0345 264 | ·0285 217 | ·0234 937 | ·0192 988 |
| ·74 | ·0650 326 | ·0547 440 | ·0459 409 | ·0384 395+ | ·0320 722 | ·0266 872 |
| ·75 | ·0834 665+ | ·0711 392 | ·0604 490 | ·0512 166 | ·0432 740 | ·0364 660 |
| ·76 | ·1059 759 | ·0914 273 | ·0786 427 | ·0674 543 | ·0577 006 | ·0492 288 |
| ·77 | ·1330 903 | ·1161 894 | ·1011 422 | ·0878 004 | ·0760 165− | ·0656 466 |
| ·78 | ·1652 921 | ·1459 813 | ·1285 653 | ·1129 227 | ·0989 270 | ·0864 510 |
| ·79 | ·2029 709 | ·1812 895− | ·1614 848 | ·1434 686 | ·1271 428 | ·1124 029 |
| ·80 | ·2463 717 | ·2224 770 | ·2003 744 | ·1800 132 | ·1613 288 | ·1442 460 |
| ·81 | ·2955 378 | ·2697 218 | ·2455 435− | ·2229 917 | ·2020 384 | ·1826 417 |
| ·82 | ·3502 530 | ·3229 508 | ·2970 650− | ·2726 225− | ·2496 322 | ·2280 869 |
| ·83 | ·4099 891 | ·3817 763 | ·3547 008 | ·3288 232 | ·3041 862 | ·2808 162 |
| ·84 | ·4738 676 | ·4454 420 | ·4178 336 | ·3911 288 | ·3653 979 | ·3406 962 |
| ·85 | ·5406 435+ | ·5127 909 | ·4854 170 | ·4586 237 | ·4325 003 | ·4071 232 |
| ·86 | ·6087 226 | ·5822 649 | ·5559 560 | ·5299 022 | ·5042 009 | ·4789 402 |
| ·87 | ·6762 200 | ·6519 489 | ·6275 337 | ·6030 741 | ·5786 649 | ·5543 952 |
| ·88 | ·7410 677 | ·7196 693 | ·6978 963 | ·6758 331 | ·6535 622 | ·6311 641 |
| ·89 | ·8011 705− | ·7831 495− | ·7646 048 | ·7455 992 | ·7261 968 | ·7064 618 |
| ·90 | ·8546 027 | ·8402 197 | ·8252 524 | ·8097 408 | ·7937 273 | ·7772 564 |
| ·91 | ·8998 287 | ·8890 623 | ·8777 340 | ·8658 634 | ·8534 725− | ·8405 860 |
| ·92 | ·9359 141 | ·9284 617 | ·9205 342 | ·9121 359 | ·9032 732 | ·8939 547 |
| ·93 | ·9626 872 | ·9580 111 | ·9529 827 | ·9475 978 | ·9418 533 | ·9357 478 |
| ·94 | ·9808 022 | ·9782 188 | ·9754 109 | ·9723 715+ | ·9690 944 | ·9655 738 |
| ·95 | ·9916 596 | ·9904 584 | ·9891 391 | ·9876 958 | ·9861 231 | ·9844 155− |
| ·96 | ·9971 645− | ·9967 288 | ·9962 453 | ·9957 108 | ·9951 222 | ·9944 764 |
| ·97 | ·9993 458 | ·9992 389 | ·9991 190 | ·9989 851 | ·9988 360 | ·9986 709 |
| ·98 | ·9999 256 | ·9999 127 | ·9998 981 | ·9998 816 | ·9998 630 | ·9998 423 |
| ·99 | ·9999 985− | ·9999 982 | ·9999 979 | ·9999 975+ | ·9999 971 | ·9999 967 |
| 1·00 | 1·0000 000 | 1·0000 000 | 1·0000 000 | 1·0000 000 | 1·0000 000 | 1·0000 000 |

$x$ = ·48 to 1·00 $q = 6$ $p$ = 37 to 43

| | $p = 37$ | $p = 38$ | $p = 39$ | $p = 40$ | $p = 41$ | $p = 42$ | $p = 43$ |
|---|---|---|---|---|---|---|---|
| $B(p, q)$ = | ·3177 1534 × $\frac{1}{10^7}$ | ·2733 8296 × $\frac{1}{10^7}$ | ·2361 0346 × $\frac{1}{10^7}$ | ·2046 2301 × $\frac{1}{10^7}$ | ·1779 3305 × $\frac{1}{10^7}$ | ·1552 1819 × $\frac{1}{10^7}$ | ·1358 1592 × $\frac{1}{10^7}$ |
| $x$ | | | | | | | |
| ·48 | ·0000 001 | | | | | | |
| ·49 | ·0000 001 | ·0000 001 | | | | | |
| ·50 | ·0000 002 | ·0000 001 | ·0000 001 | | | | |
| ·51 | ·0000 004 | ·0000 002 | ·0000 001 | ·0000 001 | | | |
| ·52 | ·0000 008 | ·0000 005$^-$ | ·0000 003 | ·0000 002 | ·0000 001 | ·0000 001 | |
| ·53 | ·0000 014 | ·0000 009 | ·0000 005$^+$ | ·0000 003 | ·0000 002 | ·0000 001 | ·0000 001 |
| ·54 | ·0000 026 | ·0000 016 | ·0000 010 | ·0000 006 | ·0000 003 | ·0000 002 | ·0000 001 |
| ·55 | ·0000 046 | ·0000 028 | ·0000 018 | ·0000 011 | ·0000 007 | ·0000 004 | ·0000 002 |
| ·56 | ·0000 081 | ·0000 051 | ·0000 032 | ·0000 020 | ·0000 013 | ·0000 008 | ·0000 005$^-$ |
| ·57 | ·0000 139 | ·0000 089 | ·0000 057 | ·0000 037 | ·0000 023 | ·0000 015$^-$ | ·0000 009 |
| ·58 | ·0000 238 | ·0000 155$^+$ | ·0000 101 | ·0000 066 | ·0000 042 | ·0000 027 | ·0000 018 |
| ·59 | ·0000 400 | ·0000 265$^+$ | ·0000 176 | ·0000 116 | ·0000 077 | ·0000 050$^+$ | ·0000 033 |
| ·60 | ·0000 664 | ·0000 448 | ·0000 302 | ·0000 203 | ·0000 136 | ·0000 091 | ·0000 061 |
| ·61 | ·0001 088 | ·0000 747 | ·0000 511 | ·0000 349 | ·0000 238 | ·0000 161 | ·0000 109 |
| ·62 | ·0001 762 | ·0001 228 | ·0000 854 | ·0000 593 | ·0000 410 | ·0000 283 | ·0000 195$^+$ |
| ·63 | ·0002 816 | ·0001 995$^-$ | ·0001 409 | ·0000 993 | ·0000 698 | ·0000 490 | ·0000 343 |
| ·64 | ·0004 447 | ·0003 199 | ·0002 295$^+$ | ·0001 643 | ·0001 173 | ·0000 835$^+$ | ·0000 594 |
| ·65 | ·0006 937 | ·0005 066 | ·0003 691 | ·0002 682 | ·0001 944 | ·0001 406 | ·0001 015$^-$ |
| ·66 | ·0010 691 | ·0007 925$^+$ | ·0005 860 | ·0004 322 | ·0003 180 | ·0002 335$^-$ | ·0001 710 |
| ·67 | ·0016 280 | ·0012 247 | ·0009 189 | ·0006 878 | ·0005 136 | ·0003 826 | ·0002 845$^-$ |
| ·68 | ·0024 496 | ·0018 695$^+$ | ·0014 232 | ·0010 807 | ·0008 188 | ·0006 189 | ·0004 668 |
| ·69 | ·0036 424 | ·0028 195$^+$ | ·0021 770 | ·0016 768 | ·0012 886 | ·0009 880 | ·0007 559 |
| ·70 | ·0053 522 | ·0042 011 | ·0032 893 | ·0025 692 | ·0020 022 | ·0015 568 | ·0012 080 |
| ·71 | ·0077 719 | ·0061 844 | ·0049 090 | ·0038 873 | ·0030 713 | ·0024 213 | ·0019 048 |
| ·72 | ·0111 518 | ·0089 940 | ·0072 360 | ·0058 079 | ·0046 512 | ·0037 168 | ·0029 639 |
| ·73 | ·0158 113 | ·0129 214 | ·0105 341 | ·0085 680 | ·0069 534 | ·0056 309 | ·0045 506 |
| ·74 | ·0221 489 | ·0183 367 | ·0151 445$^+$ | ·0124 794 | ·0102 608 | ·0084 187 | ·0068 933 |
| ·75 | ·0306 509 | ·0257 002 | ·0214 986 | ·0179 435$^-$ | ·0149 438 | ·0124 196 | ·0103 012 |
| ·76 | ·0418 960 | ·0355 700 | ·0301 296 | ·0254 648 | ·0214 765$^-$ | ·0180 756 | ·0151 833 |
| ·77 | ·0565 528 | ·0486 043 | ·0416 787 | ·0356 623 | ·0304 505$^+$ | ·0259 481 | ·0220 685$^+$ |
| ·78 | ·0753 680 | ·0655 552 | ·0568 940 | ·0492 722 | ·0425 840 | ·0367 311 | ·0316 223 |
| ·79 | ·0991 410 | ·0872 484 | ·0766 170 | ·0671 414 | ·0587 200 | ·0512 558 | ·0446 570 |
| ·80 | ·1286 816 | ·1145 474 | ·1017 523 | ·0902 039 | ·0798 103 | ·0704 814 | ·0621 299 |
| ·81 | ·1647 482 | ·1482 959 | ·1332 160 | ·1194 354 | ·1068 777 | ·0954 656 | ·0851 213 |
| ·82 | ·2079 661 | ·1892 374 | ·1718 591 | ·1557 820 | ·1409 509 | ·1273 062 | ·1147 856 |
| ·83 | ·2587 251 | ·2379 117 | ·2183 635$^+$ | ·2000 584 | ·1829 661 | ·1670 493 | ·1522 656 |
| ·84 | ·3170 650$^-$ | ·2945 323 | ·2731 143 | ·2528 167 | ·2336 354 | ·2155 583 | ·1985 658 |
| ·85 | ·3825 568 | ·3588 536 | ·3360 548 | ·3141 914 | ·2932 844 | ·2733 464 | ·2543 815$^-$ |
| ·86 | ·4541 988 | ·4300 458 | ·4065 411 | ·3837 356 | ·3616 712 | ·3403 817 | ·3198 932 |
| ·87 | ·5303 479 | ·5065 996 | ·4832 200 | ·4602 721 | ·4378 118 | ·4158 883 | ·3945 443 |
| ·88 | ·6087 161 | ·5862 925$^-$ | ·5639 632 | ·5417 943 | ·5198 471 | ·4981 782 | ·4768 395$^+$ |
| ·89 | ·6864 582 | ·6662 491 | ·6458 964 | ·6254 603 | ·6049 986 | ·5845 668 | ·5642 177 |
| ·90 | ·7603 736 | ·7431 259 | ·7255 604 | ·7077 246 | ·6896 660 | ·6714 311 | ·6530 660 |
| ·91 | ·8272 306 | ·8134 348 | ·7992 287 | ·7846 439 | ·7697 126 | ·7544 682 | ·7389 442 |
| ·92 | ·8841 911 | ·8739 946 | ·8633 795$^+$ | ·8523 616 | ·8409 581 | ·8291 875$^-$ | ·8170 693 |
| ·93 | ·9292 811 | ·9224 543 | ·9152 699 | ·9077 319 | ·8998 453 | ·8916 163 | ·8830 522 |
| ·94 | ·9618 048 | ·9577 832 | ·9535 055$^+$ | ·9489 690 | ·9441 716 | ·9391 120 | ·9337 899 |
| ·95 | ·9825 680 | ·9805 757 | ·9784 340 | ·9761 385$^+$ | ·9736 852 | ·9710 704 | ·9682 905$^+$ |
| ·96 | ·9937 704 | ·9930 010 | ·9921 651 | ·9912 599 | ·9902 822 | ·9892 292 | ·9880 979 |
| ·97 | ·9984 884 | ·9982 874 | ·9980 669 | ·9978 255$^+$ | ·9975 622 | ·9972 755$^+$ | ·9969 644 |
| ·98 | ·9998 191 | ·9997 933 | ·9997 647 | ·9997 331 | ·9996 983 | ·9996 599 | ·9996 179 |
| ·99 | ·9999 962 | ·9999 956 | ·9999 949 | ·9999 942 | ·9999 934 | ·9999 924 | ·9999 914 |
| 1·00 | 1·0000 000 | 1·0000 000 | 1·0000 000 | 1·0000 000 | 1·0000 000 | 1·0000 000 | 1·0000 000 |

$x$ = ·54 to 1·00 $q = 6$ $p$ = 44 to 50

| $x$ | $p$ = 44 | $p$ = 45 | $p$ = 46 | $p$ = 47 | $p$ = 48 | $p$ = 49 | $p$ = 50 |
|---|---|---|---|---|---|---|---|
| $B(p, q)$ = | ·1191 8540 × $\frac{1}{10^7}$ | ·1048 8315 × $\frac{1}{10^7}$ | ·9254 3956 × $\frac{1}{10^8}$ | ·8186 5807 × $\frac{1}{10^8}$ | ·7259 7980 × $\frac{1}{10^8}$ | ·6453 1537 × $\frac{1}{10^8}$ | ·5749 1733 × $\frac{1}{10^8}$ |
| ·54 | ·0000 001 | | | | | | |
| ·55 | ·0000 002 | ·0000 001 | ·0000 001 | | | | |
| ·56 | ·0000 003 | ·0000 002 | ·0000 001 | ·0000 001 | | | |
| ·57 | ·0000 006 | ·0000 004 | ·0000 002 | ·0000 001 | ·0000 001 | ·0000 001 | |
| ·58 | ·0000 011 | ·0000 007 | ·0000 005− | ·0000 003 | ·0000 002 | ·0000 001 | ·0000 001 |
| ·59 | ·0000 022 | ·0000 014 | ·0000 009 | ·0000 006 | ·0000 004 | ·0000 003 | ·0000 002 |
| ·60 | ·0000 040 | ·0000 027 | ·0000 018 | ·0000 012 | ·0000 008 | ·0000 005+ | ·0000 003 |
| ·61 | ·0000 074 | ·0000 050− | ·0000 034 | ·0000 023 | ·0000 015+ | ·0000 010 | ·0000 007 |
| ·62 | ·0000 134 | ·0000 092 | ·0000 063 | ·0000 043 | ·0000 029 | ·0000 020 | ·0000 014 |
| ·63 | ·0000 239 | ·0000 167 | ·0000 116 | ·0000 080 | ·0000 056 | ·0000 039 | ·0000 027 |
| ·64 | ·0000 421 | ·0000 298 | ·0000 210 | ·0000 148 | ·0000 104 | ·0000 073 | ·0000 051 |
| ·65 | ·0000 731 | ·0000 525+ | ·0000 377 | ·0000 270 | ·0000 193 | ·0000 137 | ·0000 098 |
| ·66 | ·0001 250+ | ·0000 912 | ·0000 664 | ·0000 483 | ·0000 350+ | ·0000 254 | ·0000 183 |
| ·67 | ·0002 110 | ·0001 562 | ·0001 155− | ·0000 852 | ·0000 627 | ·0000 461 | ·0000 338 |
| ·68 | ·0003 514 | ·0002 640 | ·0001 979 | ·0001 481 | ·0001 106 | ·0000 825+ | ·0000 614 |
| ·69 | ·0005 772 | ·0004 398 | ·0003 345+ | ·0002 540 | ·0001 925− | ·0001 456 | ·0001 100 |
| ·70 | ·0009 354 | ·0007 229 | ·0005 576 | ·0004 293 | ·0003 300 | ·0002 532 | ·0001 940 |
| ·71 | ·0014 955− | ·0011 718 | ·0009 165− | ·0007 155− | ·0005 576 | ·0004 338 | ·0003 370 |
| ·72 | ·0023 588 | ·0018 736 | ·0014 855− | ·0011 756 | ·0009 288 | ·0007 326 | ·0005 769 |
| ·73 | ·0036 703 | ·0029 546 | ·0023 741 | ·0019 043 | ·0015 248 | ·0012 190 | ·0009 729 |
| ·74 | ·0056 333 | ·0045 949 | ·0037 411 | ·0030 406 | ·0024 671 | ·0019 986 | ·0016 164 |
| ·75 | ·0085 276 | ·0070 462 | ·0058 118 | ·0047 853 | ·0039 335+ | ·0032 281 | ·0026 451 |
| ·76 | ·0127 295+ | ·0106 528 | ·0088 990 | ·0074 213 | ·0061 788 | ·0051 361 | ·0042 628 |
| ·77 | ·0187 339 | ·0158 744 | ·0134 280 | ·0113 395+ | ·0095 603 | ·0080 476 | ·0067 639 |
| ·78 | ·0271 741 | ·0233 102 | ·0199 615+ | ·0170 656 | ·0145 666 | ·0124 142 | ·0105 641 |
| ·79 | ·0388 378 | ·0337 183 | ·0292 245− | ·0252 884 | ·0218 482 | ·0188 473 | ·0162 347 |
| ·80 | ·0546 717 | ·0480 272 | ·0421 210 | ·0368 824 | ·0322 457 | ·0281 499 | ·0245 390 |
| ·81 | ·0757 682 | ·0673 312 | ·0597 379 | ·0529 188 | ·0468 078 | ·0413 422 | ·0364 633 |
| ·82 | ·1033 249 | ·0928 592 | ·0833 238 | ·0746 548 | ·0667 901 | ·0596 692 | ·0532 342 |
| ·83 | ·1385 684 | ·1259 078 | ·1142 320 | ·1034 876 | ·0936 211 | ·0845 789 | ·0763 080 |
| ·84 | ·1826 323 | ·1677 273 | ·1538 159 | ·1408 602 | ·1288 197 | ·1176 521 | ·1073 141 |
| ·85 | ·2363 869 | ·2193 533 | ·2032 661 | ·1881 057 | ·1738 484 | ·1604 673 | ·1479 328 |
| ·86 | ·3002 242 | ·2813 866 | ·2633 863 | ·2462 234 | ·2298 930 | ·2143 860 | ·1996 892 |
| ·87 | ·3738 157 | ·3537 326 | ·3343 189 | ·3155 931 | ·2975 685+ | ·2802 537 | ·2636 527 |
| ·88 | ·4558 780 | ·4353 356 | ·4152 498 | ·3956 531 | ·3765 733 | ·3580 342 | ·3400 550− |
| ·89 | ·5440 012 | ·5239 638 | ·5041 491 | ·4845 971 | ·4653 447 | ·4464 251 | ·4278 685+ |
| ·90 | ·6346 155− | ·6161 230 | ·5976 305+ | ·5791 783 | ·5608 045+ | ·5425 456 | ·5244 358 |
| ·91 | ·7231 748 | ·7071 939 | ·6910 355+ | ·6747 330 | ·6583 196 | ·6418 275− | ·6252 883 |
| ·92 | ·8046 243 | ·7918 737 | ·7788 398 | ·7655 453 | ·7520 131 | ·7382 667 | ·7243 295− |
| ·93 | ·8741 615+ | ·8649 536 | ·8554 387 | ·8456 281 | ·8355 335+ | ·8251 677 | ·8145 438 |
| ·94 | ·9282 054 | ·9223 594 | ·9162 536 | ·9098 903 | ·9032 725− | ·8964 037 | ·8892 882 |
| ·95 | ·9653 426 | ·9622 238 | ·9589 318 | ·9554 644 | ·9518 200 | ·9479 972 | ·9439 949 |
| ·96 | ·9868 856 | ·9855 896 | ·9842 072 | ·9827 358 | ·9811 729 | ·9795 164 | ·9777 638 |
| ·97 | ·9966 275+ | ·9962 636 | ·9958 714 | ·9954 495+ | ·9949 969 | ·9945 120 | ·9939 937 |
| ·98 | ·9995 720 | ·9995 218 | ·9994 672 | ·9994 078 | ·9993 434 | ·9992 738 | ·9991 986 |
| ·99 | ·9999 903 | ·9999 891 | ·9999 878 | ·9999 863 | ·9999 847 | ·9999 829 | ·9999 810 |
| 1·00 | 1·0000 000 | 1·0000 000 | 1·0000 000 | 1·0000 000 | 1·0000 000 | 1·0000 000 | 1·0000 000 |

$x$ = ·03 to ·60 $q$ = 6·5 $p$ = 6·5 to 9

| $x$ | $p$ = 6·5 | $p$ = 7 | $p$ = 7·5 | $p$ = 8 | $p$ = 8·5 | $p$ = 9 |
|---|---|---|---|---|---|---|
| $B(p, q)$ = | ·1730 2225 × $\frac{1}{10^3}$ | ·1211 7644 × $\frac{1}{10^3}$ | ·8651 1124 × $\frac{1}{10^4}$ | ·6283 2229 × $\frac{1}{10^4}$ | ·4634 5245 × $\frac{1}{10^4}$ | ·3466 6057 × $\frac{1}{10^4}$ |
| ·03 | ·0000 001 | | | | | |
| ·04 | ·0000 006 | ·0000 002 | | | | |
| ·05 | ·0000 024 | ·0000 007 | ·0000 002 | ·0000 001 | | |
| ·06 | ·0000 076 | ·0000 025− | ·0000 008 | ·0000 002 | ·0000 001 | |
| ·07 | ·0000 196 | ·0000 069 | ·0000 024 | ·0000 008 | ·0000 003 | ·0000 001 |
| ·08 | ·0000 445− | ·0000 166 | ·0000 061 | ·0000 022 | ·0000 008 | ·0000 003 |
| ·09 | ·0000 908 | ·0000 360 | ·0000 140 | ·0000 054 | ·0000 021 | ·0000 008 |
| ·10 | ·0001 711 | ·0000 714 | ·0000 294 | ·0000 119 | ·0000 048 | ·0000 019 |
| ·11 | ·0003 019 | ·0001 320 | ·0000 569 | ·0000 243 | ·0000 102 | ·0000 043 |
| ·12 | ·0005 043 | ·0002 301 | ·0001 036 | ·0000 461 | ·0000 203 | ·0000 088 |
| ·13 | ·0008 048 | ·0003 820 | ·0001 789 | ·0000 828 | ·0000 379 | ·0000 172 |
| ·14 | ·0012 353 | ·0006 081 | ·0002 954 | ·0001 418 | ·0000 673 | ·0000 317 |
| ·15 | ·0018 330 | ·0009 334 | ·0004 691 | ·0002 330 | ·0001 144 | ·0000 557 |
| ·16 | ·0026 412 | ·0013 880 | ·0007 200 | ·0003 691 | ·0001 872 | ·0000 940 |
| ·17 | ·0037 085− | ·0020 074 | ·0010 727 | ·0005 665− | ·0002 960 | ·0001 531 |
| ·18 | ·0050 884 | ·0028 321 | ·0015 562 | ·0008 452 | ·0004 542 | ·0002 417 |
| ·19 | ·0068 397 | ·0039 082 | ·0022 049 | ·0012 296 | ·0006 785− | ·0003 707 |
| ·20 | ·0090 250+ | ·0052 867 | ·0030 579 | ·0017 485+ | ·0009 893 | ·0005 544 |
| ·21 | ·0117 106 | ·0070 235+ | ·0041 599 | ·0024 358 | ·0014 114 | ·0008 100 |
| ·22 | ·0149 651 | ·0091 789 | ·0055 603 | ·0033 303 | ·0019 740 | ·0011 589 |
| ·23 | ·0188 590 | ·0118 171 | ·0073 137 | ·0044 759 | ·0027 110 | ·0016 265+ |
| ·24 | ·0234 636 | ·0150 051 | ·0094 791 | ·0059 217 | ·0036 616 | ·0022 429 |
| ·25 | ·0288 494 | ·0188 125− | ·0121 195+ | ·0077 217 | ·0048 699 | ·0030 428 |
| ·26 | ·0350 856 | ·0233 099 | ·0153 015− | ·0099 346 | ·0063 854 | ·0040 662 |
| ·27 | ·0422 387 | ·0285 687 | ·0190 942 | ·0126 235+ | ·0082 625+ | ·0053 585− |
| ·28 | ·0503 710 | ·0346 590 | ·0235 686 | ·0158 549 | ·0105 605+ | ·0069 700 |
| ·29 | ·0595 402 | ·0416 494 | ·0287 968 | ·0196 987 | ·0133 432 | ·0089 565+ |
| ·30 | ·0697 976 | ·0496 054 | ·0348 505+ | ·0242 267 | ·0166 781 | ·0113 787 |
| ·31 | ·0811 875+ | ·0585 882 | ·0418 003 | ·0295 122 | ·0206 363 | ·0143 018 |
| ·32 | ·0937 462 | ·0686 536 | ·0497 143 | ·0356 287 | ·0252 912 | ·0177 952 |
| ·33 | ·1075 009 | ·0798 511 | ·0586 568 | ·0426 490 | ·0307 180 | ·0219 321 |
| ·34 | ·1224 692 | ·0922 225− | ·0686 876 | ·0506 437 | ·0369 924 | ·0267 879 |
| ·35 | ·1386 584 | ·1058 011 | ·0798 602 | ·0596 803 | ·0441 893 | ·0324 402 |
| ·36 | ·1560 651 | ·1206 108 | ·0922 210 | ·0698 217 | ·0523 823 | ·0389 672 |
| ·37 | ·1746 748 | ·1366 654 | ·1058 081 | ·0811 249 | ·0616 416 | ·0464 468 |
| ·38 | ·1944 616 | ·1539 675+ | ·1206 501 | ·0936 401 | ·0720 330 | ·0549 551 |
| ·39 | ·2153 885− | ·1725 088 | ·1367 657 | ·1074 090 | ·0836 166 | ·0645 651 |
| ·40 | ·2374 070 | ·1922 689 | ·1541 621 | ·1224 639 | ·0964 456 | ·0753 455− |
| ·41 | ·2604 580 | ·2132 157 | ·1728 350+ | ·1388 267 | ·1105 643 | ·0873 588 |
| ·42 | ·2844 717 | ·2353 050− | ·1927 680 | ·1565 078 | ·1260 076 | ·1006 603 |
| ·43 | ·3093 686 | ·2584 808 | ·2139 317 | ·1755 054 | ·1427 995− | ·1152 964 |
| ·44 | ·3350 600 | ·2826 758 | ·2362 845− | ·1958 048 | ·1609 517 | ·1313 031 |
| ·45 | ·3614 487 | ·3078 114 | ·2597 715− | ·2173 780 | ·1804 633 | ·1487 052 |
| ·46 | ·3884 302 | ·3337 988 | ·2843 256 | ·2401 832 | ·2013 195+ | ·1675 141 |
| ·47 | ·4158 939 | ·3605 395− | ·3098 675+ | ·2641 650+ | ·2234 914 | ·1877 280 |
| ·48 | ·4437 238 | ·3879 264 | ·3363 064 | ·2892 544 | ·2469 352 | ·2093 300 |
| ·49 | ·4718 000 | ·4158 450+ | ·3635 406 | ·3153 690 | ·2715 924 | ·2322 879 |
| ·50 | ·5000 000[e] | ·4441 743 | ·3914 588 | ·3424 139 | ·2973 897 | ·2565 535+ |
| ·51 | ·5282 000 | ·4727 881 | ·4199 407 | ·3702 821 | ·3242 394 | ·2820 628 |
| ·52 | ·5562 762 | ·5015 569 | ·4488 589 | ·3988 558 | ·3520 400 | ·3087 356 |
| ·53 | ·5841 061 | ·5303 487 | ·4780 797 | ·4280 074 | ·3806 768 | ·3364 759 |
| ·54 | ·6115 698 | ·5590 310 | ·5074 651 | ·4576 011 | ·4100 232 | ·3651 731 |
| ·55 | ·6385 513 | ·5874 717 | ·5368 742 | ·4874 938 | ·4399 419 | ·3947 018 |
| ·56 | ·6649 400 | ·6155 414 | ·5661 645+ | ·5175 378 | ·4702 864 | ·4249 243 |
| ·57 | ·6906 314 | ·6431 145− | ·5951 945+ | ·5475 816 | ·5009 029 | ·4556 910 |
| ·58 | ·7155 283 | ·6700 702 | ·6238 246 | ·5774 725+ | ·5316 317 | ·4868 426 |
| ·59 | ·7395 420 | ·6962 948 | ·6519 191 | ·6070 582 | ·5623 100 | ·5182 120 |
| ·60 | ·7625 930 | ·7216 821 | ·6793 481 | ·6361 887 | ·5927 734 | ·5496 267 |

$x$ = ·61 to ·98 $q = 6·5$ $p$ = 6·5 to 9

| $x$ | $p = 6·5$ | $p = 7$ | $p = 7·5$ | $p = 8$ | $p = 8·5$ | $p = 9$ |
|---|---|---|---|---|---|---|
| $B(p, q)$ = | ·1730 2225 $\times \frac{1}{10^3}$ | ·1211 7644 $\times \frac{1}{10^3}$ | ·8651 1124 $\times \frac{1}{10^4}$ | ·6283 2229 $\times \frac{1}{10^4}$ | ·4634 5245 $\times \frac{1}{10^4}$ | ·3466 6057 $\times \frac{1}{10^4}$ |
| ·61 | ·7846 115+ | ·7461 352 | ·7059 887 | ·6647 187 | ·6228 582 | ·5809 105+ |
| ·62 | ·8055 384 | ·7695 671 | ·7317 269 | ·6925 089 | ·6524 041 | ·6118 867 |
| ·63 | ·8253 252 | ·7919 019 | ·7564 585− | ·7194 285− | ·6812 560 | ·6423 801 |
| ·64 | ·8439 349 | ·8130 753 | ·7800 908 | ·7453 564 | ·7092 664 | ·6722 200 |
| ·65 | ·8613 416 | ·8330 353 | ·8025 434 | ·7701 832 | ·7362 975+ | ·7012 423 |
| ·66 | ·8775 308 | ·8517 425− | ·8237 493 | ·7938 124 | ·7622 232 | ·7292 927 |
| ·67 | ·8924 991 | ·8691 703 | ·8436 551 | ·8161 615+ | ·7869 309 | ·7562 285− |
| ·68 | ·9062 538 | ·8853 051 | ·8622 219 | ·8371 634 | ·8103 230 | ·7819 209 |
| ·69 | ·9188 125− | ·9001 455+ | ·8794 253 | ·8567 666 | ·8323 183 | ·8062 573 |
| ·70 | ·9302 024 | ·9137 027 | ·8952 554 | ·8749 361 | ·8528 530 | ·8291 430 |
| ·71 | ·9404 598 | ·9259 994 | ·9097 165+ | ·8916 531 | ·8718 817 | ·8505 020 |
| ·72 | ·9496 290 | ·9370 691 | ·9228 266 | ·9069 155+ | ·8893 773 | ·8702 790 |
| ·73 | ·9577 613 | ·9469 554 | ·9346 168 | ·9207 369 | ·9053 313 | ·8884 390 |
| ·74 | ·9649 144 | ·9557 110 | ·9451 302 | ·9331 461 | ·9197 537 | ·9049 683 |
| ·75 | ·9711 506 | ·9633 961 | ·9544 207 | ·9441 863 | ·9326 718 | ·9198 739 |
| ·76 | ·9765 364 | ·9700 775+ | ·9625 519 | ·9539 134 | ·9441 297 | ·9331 829 |
| ·77 | ·9811 410 | ·9758 275− | ·9695 957 | ·9623 952 | ·9541 865+ | ·9449 416 |
| ·78 | ·9850 349 | ·9807 217 | ·9756 302 | ·9697 090 | ·9629 150− | ·9552 136 |
| ·79 | ·9882 894 | ·9848 385+ | ·9807 388 | ·9759 404 | ·9703 994 | ·9640 781 |
| ·80 | ·9909 750− | ·9882 571 | ·9850 078 | ·9811 808 | ·9767 334 | ·9716 277 |
| ·81 | ·9931 603 | ·9910 564 | ·9885 254 | ·9855 258 | ·9820 181 | ·9779 660 |
| ·82 | ·9949 116 | ·9933 136 | ·9913 794 | ·9890 728 | ·9863 589 | ·9832 045+ |
| ·83 | ·9962 915+ | ·9951 031 | ·9936 558 | ·9919 193 | ·9898 637 | ·9874 599 |
| ·84 | ·9973 588 | ·9964 954 | ·9954 375+ | ·9941 607 | ·9926 402 | ·9908 514 |
| ·85 | ·9981 670 | ·9975 560 | ·9968 030 | ·9958 887 | ·9947 934 | ·9934 972 |
| ·86 | ·9987 647 | ·9983 451 | ·9978 249 | ·9971 895+ | ·9964 239 | ·9955 125− |
| ·87 | ·9991 952 | ·9989 166 | ·9985 693 | ·9981 426 | ·9976 255− | ·9970 063 |
| ·88 | ·9994 957 | ·9993 179 | ·9990 950+ | ·9988 196 | ·9984 838 | ·9980 795+ |
| ·89 | ·9996 981 | ·9995 898 | ·9994 532 | ·9992 834 | ·9990 753 | ·9988 233 |
| ·90 | ·9998 289 | ·9997 664 | ·9996 871 | ·9995 881 | ·9994 660 | ·9993 172 |
| ·91 | ·9999 092 | ·9998 754 | ·9998 324 | ·9997 783 | ·9997 112 | ·9996 291 |
| ·92 | ·9999 555+ | ·9999 387 | ·9999 172 | ·9998 900 | ·9998 561 | ·9998 143 |
| ·93 | ·9999 804 | ·9999 728 | ·9999 631 | ·9999 507 | ·9999 353 | ·9999 161 |
| ·94 | ·9999 924 | ·9999 895− | ·9999 856 | ·9999 807 | ·9999 745+ | ·9999 669 |
| ·95 | ·9999 976 | ·9999 966 | ·9999 953 | ·9999 937 | ·9999 917 | ·9999 891 |
| ·96 | ·9999 994 | ·9999 992 | ·9999 988 | ·9999 984 | ·9999 979 | ·9999 973 |
| ·97 | ·9999 999 | ·9999 999 | ·9999 998 | ·9999 997 | ·9999 997 | ·9999 995+ |
| ·98 | 1·0000 000 | 1·0000 000 | 1·0000 000 | 1·0000 000 | 1·0000 000 | 1·0000 000 |

$x$ = ·08 to ·70 $q = 6{\cdot}5$ $p$ = 9·5 to 13

| $x$ | $p = 9{\cdot}5$ | $p = 10$ | $p = 10{\cdot}5$ | $p = 11$ | $p = 12$ | $p = 13$ |
|---|---|---|---|---|---|---|
| $B(p, q) =$ | ·2626 2305+ × $\frac{1}{10^4}$ | ·2012 8678 × $\frac{1}{10^4}$ | ·1559 3244 × $\frac{1}{10^4}$ | ·1219 9199 × $\frac{1}{10^4}$ | ·7668 0679 × $\frac{1}{10^5}$ | ·4973 8819 × $\frac{1}{10^5}$ |
| ·08 | ·0000 001 | | | | | |
| ·09 | ·0000 003 | ·0000 001 | | | | |
| ·10 | ·0000 008 | ·0000 003 | ·0000 001 | | | |
| ·11 | ·0000 018 | ·0000 007 | ·0000 003 | ·0000 001 | | |
| ·12 | ·0000 038 | ·0000 016 | ·0000 007 | ·0000 003 | ·0000 001 | |
| ·13 | ·0000 077 | ·0000 034 | ·0000 015+ | ·0000 007 | ·0000 001 | |
| ·14 | ·0000 147 | ·0000 068 | ·0000 031 | ·0000 014 | ·0000 003 | ·0000 001 |
| ·15 | ·0000 268 | ·0000 128 | ·0000 061 | ·0000 029 | ·0000 006 | ·0000 001 |
| ·16 | ·0000 468 | ·0000 231 | ·0000 113 | ·0000 055− | ·0000 013 | ·0000 003 |
| ·17 | ·0000 785+ | ·0000 399 | ·0000 201 | ·0000 101 | ·0000 025− | ·0000 006 |
| ·18 | ·0001 275− | ·0000 667 | ·0000 346 | ·0000 178 | ·0000 046 | ·0000 012 |
| ·19 | ·0002 008 | ·0001 078 | ·0000 575− | ·0000 304 | ·0000 084 | ·0000 022 |
| ·20 | ·0003 079 | ·0001 696 | ·0000 927 | ·0000 503 | ·0000 145+ | ·0000 041 |
| ·21 | ·0004 608 | ·0002 600 | ·0001 456 | ·0000 809 | ·0000 245+ | ·0000 073 |
| ·22 | ·0006 745− | ·0003 893 | ·0002 230 | ·0001 269 | ·0000 403 | ·0000 125− |
| ·23 | ·0009 674 | ·0005 707 | ·0003 342 | ·0001 943 | ·0000 645− | ·0000 209 |
| ·24 | ·0013 619 | ·0008 204 | ·0004 905− | ·0002 912 | ·0001 007 | ·0000 341 |
| ·25 | ·0018 848 | ·0011 582 | ·0007 064 | ·0004 279 | ·0001 541 | ·0000 542 |
| ·26 | ·0025 672 | ·0016 080 | ·0009 998 | ·0006 174 | ·0002 310 | ·0000 845+ |
| ·27 | ·0034 457 | ·0021 982 | ·0013 921 | ·0008 756 | ·0003 400 | ·0001 291 |
| ·28 | ·0045 615+ | ·0029 620 | ·0019 093 | ·0012 224 | ·0004 919 | ·0001 935+ |
| ·29 | ·0059 618 | ·0039 376 | ·0025 819 | ·0016 816 | ·0007 002 | ·0002 851 |
| ·30 | ·0076 988 | ·0051 688 | ·0034 454 | ·0022 813 | ·0009 818 | ·0004 133 |
| ·31 | ·0098 302 | ·0067 050+ | ·0045 409 | ·0030 548 | ·0013 574 | ·0005 901 |
| ·32 | ·0124 189 | ·0086 011 | ·0059 149 | ·0040 408 | ·0018 519 | ·0008 304 |
| ·33 | ·0155 325− | ·0109 176 | ·0076 200 | ·0052 836 | ·0024 948 | ·0011 527 |
| ·34 | ·0192 430 | ·0137 202 | ·0097 144 | ·0068 335+ | ·0033 213 | ·0015 798 |
| ·35 | ·0236 261 | ·0170 798 | ·0122 622 | ·0087 468 | ·0043 719 | ·0021 390 |
| ·36 | ·0287 603 | ·0210 717 | ·0153 330 | ·0110 859 | ·0056 937 | ·0028 627 |
| ·37 | ·0347 260 | ·0257 749 | ·0190 016 | ·0139 195− | ·0073 397 | ·0037 894 |
| ·38 | ·0416 043 | ·0312 713 | ·0233 471 | ·0173 216 | ·0093 702 | ·0049 638 |
| ·39 | ·0494 762 | ·0376 449 | ·0284 529 | ·0213 716 | ·0118 518 | ·0064 374 |
| ·40 | ·0584 206 | ·0449 803 | ·0344 048 | ·0261 538 | ·0148 580 | ·0082 687 |
| ·41 | ·0685 133 | ·0533 614 | ·0412 907 | ·0317 560 | ·0184 687 | ·0105 240 |
| ·42 | ·0798 252 | ·0628 701 | ·0491 990 | ·0382 687 | ·0227 699 | ·0132 768 |
| ·43 | ·0924 211 | ·0735 849 | ·0582 168 | ·0457 842 | ·0278 527 | ·0166 082 |
| ·44 | ·1063 576 | ·0855 786 | ·0684 290 | ·0543 945− | ·0338 129 | ·0206 065− |
| ·45 | ·1216 819 | ·0989 173 | ·0799 162 | ·0641 903 | ·0407 496 | ·0253 666 |
| ·46 | ·1384 302 | ·1136 583 | ·0927 527 | ·0752 591 | ·0487 636 | ·0309 895+ |
| ·47 | ·1566 261 | ·1298 483 | ·1070 050− | ·0876 828 | ·0579 562 | ·0375 811 |
| ·48 | ·1762 793 | ·1475 219 | ·1227 296 | ·1015 366 | ·0684 272 | ·0452 506 |
| ·49 | ·1973 847 | ·1667 000 | ·1399 714 | ·1168 859 | ·0802 726 | ·0541 093 |
| ·50 | ·2199 211 | ·1873 883 | ·1587 616 | ·1337 852 | ·0935 828 | ·0642 687 |
| ·51 | ·2438 504 | ·2095 760 | ·1791 161 | ·1522 751 | ·1084 400 | ·0758 376 |
| ·52 | ·2691 175+ | ·2332 348 | ·2010 338 | ·1723 812 | ·1249 154 | ·0889 205+ |
| ·53 | ·2956 498 | ·2583 182 | ·2244 956 | ·1941 116 | ·1430 674 | ·1036 145− |
| ·54 | ·3233 572 | ·2847 607 | ·2494 630 | ·2174 556 | ·1629 384 | ·1200 061 |
| ·55 | ·3521 327 | ·3124 778 | ·2758 773 | ·2423 819 | ·1845 527 | ·1381 689 |
| ·56 | ·3818 530 | ·3413 664 | ·3036 592 | ·2688 381 | ·2079 143 | ·1581 599 |
| ·57 | ·4123 797 | ·3713 047 | ·3327 088 | ·2967 495− | ·2330 047 | ·1800 169 |
| ·58 | ·4435 605− | ·4021 539 | ·3629 057 | ·3260 187 | ·2597 812 | ·2037 553 |
| ·59 | ·4752 311 | ·4337 588 | ·3941 102 | ·3565 263 | ·2881 756 | ·2293 655− |
| ·60 | ·5072 172 | ·4659 500+ | ·4261 639 | ·3881 308 | ·3180 934 | ·2568 107 |
| ·61 | ·5393 365− | ·4985 458 | ·4588 919 | ·4206 699 | ·3494 134 | ·2860 249 |
| ·62 | ·5714 015+ | ·5313 541 | ·4921 043 | ·4539 623 | ·3819 879 | ·3169 111 |
| ·63 | ·6032 224 | ·5641 761 | ·5255 994 | ·4878 099 | ·4156 438 | ·3493 412 |
| ·64 | ·6346 091 | ·5968 082 | ·5591 660 | ·5219 998 | ·4501 836 | ·3831 553 |
| ·65 | ·6653 754 | ·6290 459 | ·5925 869 | ·5563 081 | ·4853 887 | ·4181 629 |
| ·66 | ·6953 408 | ·6606 868 | ·6256 422 | ·5905 030 | ·5210 210 | ·4541 446 |
| ·67 | ·7243 341 | ·6915 340 | ·6581 132 | ·6243 489 | ·5568 278 | ·4908 540 |
| ·68 | ·7521 962 | ·7213 996 | ·6897 863 | ·6576 101 | ·5925 449 | ·5280 219 |
| ·69 | ·7787 825− | ·7501 078 | ·7204 567 | ·6900 559 | ·6279 022 | ·5653 601 |
| ·70 | ·8039 656 | ·7774 985+ | ·7499 321 | ·7214 643 | ·6626 284 | ·6025 667 |

$x = \cdot 71$ to $\cdot 99$ $q = 6\cdot 5$ $p = 9\cdot 5$ to $13$

| | $p = 9\cdot 5$ | $p = 10$ | $p = 10\cdot 5$ | $p = 11$ | $p = 12$ | $p = 13$ |
|---|---|---|---|---|---|---|
| $B(p, q) =$ | $\cdot 2626\ 2305 \times \frac{1}{10^4}$ | $\cdot 2012\ 8678 \times \frac{1}{10^4}$ | $\cdot 1559\ 3244 \times \frac{1}{10^4}$ | $\cdot 1219\ 9199 \times \frac{1}{10^4}$ | $\cdot 7668\ 0679 \times \frac{1}{10^5}$ | $\cdot 4973\ 8819 \times \frac{1}{10^5}$ |
| $x$ | | | | | | |
| ·71 | ·8276 373 | ·8034 298 | ·7780 369 | ·7516 267 | ·6964 565⁻ | ·6393 323 |
| ·72 | ·8497 103 | ·8277 806 | ·8046 152 | ·7803 523 | ·7291 298 | ·6753 462 |
| ·73 | ·8701 198 | ·8504 530 | ·8295 339 | ·8074 719 | ·7604 075⁺ | ·7103 036 |
| ·74 | ·8888 243 | ·8713 740 | ·8526 856 | ·8328 413 | ·7900 701 | ·7439 128 |
| ·75 | ·9058 056 | ·8904 964 | ·8739 903 | ·8563 451 | ·8179 248 | ·7759 027 |
| ·76 | ·9210 694 | ·9077 996 | ·8933 971 | ·8778 981 | ·8438 103 | ·8060 297 |
| ·77 | ·9346 441 | ·9232 895⁻ | ·9108 847 | ·8974 477 | ·8676 004 | ·8340 844 |
| ·78 | ·9465 797 | ·9369 976 | ·9264 614 | ·9149 745⁺ | ·8892 075⁺ | ·8598 975⁺ |
| ·79 | ·9569 460 | ·9489 800 | ·9401 646 | ·9304 923 | ·9085 845⁺ | ·8833 449 |
| ·80 | ·9658 307 | ·9593 149 | ·9520 588 | ·9440 470 | ·9257 255⁻ | ·9043 504 |
| ·81 | ·9733 366 | ·9681 006 | ·9622 333 | ·9557 145⁺ | ·9406 654 | ·9228 887 |
| ·82 | ·9795 784 | ·9754 519 | ·9707 993 | ·9655 983 | ·9534 784 | ·9389 850⁻ |
| ·83 | ·9846 798 | ·9814 968 | ·9778 862 | ·9738 253 | ·9642 742 | ·9527 134 |
| ·84 | ·9887 700 | ·9863 727 | ·9836 370 | ·9805 415⁺ | ·9731 942 | ·9641 938 |
| ·85 | ·9919 801 | ·9902 222 | ·9882 042 | ·9859 073 | ·9804 059 | ·9735 863 |
| ·86 | ·9944 394 | ·9931 888 | ·9917 447 | ·9900 914 | ·9860 957 | ·9810 845⁻ |
| ·87 | ·9962 730 | ·9954 135⁻ | ·9944 152 | ·9932 656 | ·9904 626 | ·9869 063 |
| ·88 | ·9975 980 | ·9970 303 | ·9963 671 | ·9955 991 | ·9937 100 | ·9912 856 |
| ·89 | ·9985 214 | ·9981 635⁻ | ·9977 430 | ·9972 533 | ·9960 381 | ·9944 610 |
| ·90 | ·9991 381 | ·9989 246 | ·9986 723 | ·9983 768 | ·9976 373 | ·9966 667 |
| ·91 | ·9995 297 | ·9994 104 | ·9992 688 | ·9991 020 | ·9986 810 | ·9981 222 |
| ·92 | ·9997 634 | ·9997 021 | ·9996 288 | ·9995 421 | ·9993 212 | ·9990 250⁻ |
| ·93 | ·9998 926 | ·9998 642 | ·9998 300 | ·9997 893 | ·9996 849 | ·9995 433 |
| ·94 | ·9999 574 | ·9999 459 | ·9999 320 | ·9999 153 | ·9998 722 | ·9998 131 |
| ·95 | ·9999 859 | ·9999 821 | ·9999 773 | ·9999 717 | ·9999 568 | ·9999 363 |
| ·96 | ·9999 964 | ·9999 954 | ·9999 942 | ·9999 927 | ·9999 888 | ·9999 834 |
| ·97 | ·9999 994 | ·9999 992 | ·9999 990 | ·9999 988 | ·9999 981 | ·9999 972 |
| ·98 | 1·0000 000 | ·9999 999 | ·9999 999 | ·9999 999 | ·9999 999 | ·9999 998 |
| ·99 | | 1·0000 000 | 1·0000 000 | 1·0000 000 | 1·0000 000 | 1·0000 000 |

$x$ = ·16 to ·80 $q = 6·5$ $p$ = 14 to 19

| $x$ | $p = 14$ | $p = 15$ | $p = 16$ | $p = 17$ | $p = 18$ | $p = 19$ |
|---|---|---|---|---|---|---|
| $B(p, q) =$ | ·3315 9213 $\times \frac{1}{10^5}$ | ·2264 5316 $\times \frac{1}{10^5}$ | ·1579 9058 $\times \frac{1}{10^5}$ | ·1123 4885 $\overset{+}{\times} \frac{1}{10^5}$ | ·8127 3640 $\times \frac{1}{10^6}$ | ·5971 1245 $\overset{+}{\times} \frac{1}{10^6}$ |
| ·16 | ·0000 001 | | | | | |
| ·17 | ·0000 001 | | | | | |
| ·18 | ·0000 003 | ·0000 001 | | | | |
| ·19 | ·0000 006 | ·0000 001 | | | | |
| ·20 | ·0000 011 | ·0000 003 | ·0000 001 | | | |
| ·21 | ·0000 021 | ·0000 006 | ·0000 002 | | | |
| ·22 | ·0000 038 | ·0000 011 | ·0000 003 | ·0000 001 | | |
| ·23 | ·0000 066 | ·0000 021 | ·0000 006 | ·0000 002 | ·0000 001 | |
| ·24 | ·0000 113 | ·0000 037 | ·0000 012 | ·0000 004 | ·0000 001 | |
| ·25 | ·0000 187 | ·0000 063 | ·0000 021 | ·0000 007 | ·0000 002 | ·0000 001 |
| ·26 | ·0000 303 | ·0000 107 | ·0000 037 | ·0000 013 | ·0000 004 | ·0000 001 |
| ·27 | ·0000 480 | ·0000 176 | ·0000 063 | ·0000 022 | ·0000 008 | ·0000 003 |
| ·28 | ·0000 747 | ·0000 283 | ·0000 106 | ·0000 039 | ·0000 014 | ·0000 005$^{+}$ |
| ·29 | ·0001 139 | ·0000 447 | ·0000 173 | ·0000 066 | ·0000 025$^{-}$ | ·0000 009 |
| ·30 | ·0001 706 | ·0000 692 | ·0000 277 | ·0000 109 | ·0000 042 | ·0000 016 |
| ·31 | ·0002 516 | ·0001 054 | ·0000 435$^{-}$ | ·0000 177 | ·0000 071 | ·0000 028 |
| ·32 | ·0003 652 | ·0001 579 | ·0000 672 | ·0000 282 | ·0000 117 | ·0000 048 |
| ·33 | ·0005 225$^{-}$ | ·0002 328 | ·0001 021 | ·0000 442 | ·0000 189 | ·0000 080 |
| ·34 | ·0007 372 | ·0003 382 | ·0001 528 | ·0000 681 | ·0000 299 | ·0000 130 |
| ·35 | ·0010 268 | ·0004 846 | ·0002 252 | ·0001 032 | ·0000 467 | ·0000 209 |
| ·36 | ·0014 124 | ·0006 852 | ·0003 274 | ·0001 543 | ·0000 718 | ·0000 330 |
| ·37 | ·0019 200 | ·0009 566 | ·0004 695$^{-}$ | ·0002 273 | ·0001 087 | ·0000 514 |
| ·38 | ·0025 810 | ·0013 198 | ·0006 648 | ·0003 303 | ·0001 621 | ·0000 787 |
| ·39 | ·0034 324 | ·0018 000 | ·0009 300 | ·0004 740 | ·0002 386 | ·0001 188 |
| ·40 | ·0045 179 | ·0024 282 | ·0012 858 | ·0006 718 | ·0003 467 | ·0001 769 |
| ·41 | ·0058 886 | ·0032 414 | ·0017 581 | ·0009 409 | ·0004 975$^{-}$ | ·0002 601 |
| ·42 | ·0076 029 | ·0042 836 | ·0023 784 | ·0013 031 | ·0007 054 | ·0003 776 |
| ·43 | ·0097 274 | ·0056 063 | ·0031 845$^{-}$ | ·0017 851 | ·0009 887 | ·0005 416 |
| ·44 | ·0123 372 | ·0072 693 | ·0042 218 | ·0024 200 | ·0013 707 | ·0007 679 |
| ·45 | ·0155 156 | ·0093 411 | ·0055 438 | ·0032 476 | ·0018 800 | ·0010 766 |
| ·46 | ·0193 544 | ·0118 996 | ·0072 130 | ·0043 161 | ·0025 524 | ·0014 932 |
| ·47 | ·0239 533 | ·0150 319 | ·0093 014 | ·0056 823 | ·0034 309 | ·0020 495$^{-}$ |
| ·48 | ·0294 192 | ·0188 348 | ·0118 914 | ·0074 129 | ·0045 678 | ·0027 848 |
| ·49 | ·0358 655$^{+}$ | ·0234 141 | ·0150 758 | ·0095 855$^{+}$ | ·0060 249 | ·0037 471 |
| ·50 | ·0434 105$^{-}$ | ·0288 843 | ·0189 580 | ·0122 888 | ·0078 754 | ·0049 944 |
| ·51 | ·0521 758 | ·0353 674 | ·0236 520 | ·0156 233 | ·0102 039 | ·0065 955$^{+}$ |
| ·52 | ·0622 843 | ·0429 921 | ·0292 817 | ·0197 016 | ·0131 082 | ·0086 321 |
| ·53 | ·0738 579 | ·0518 911 | ·0359 798 | ·0246 480 | ·0166 991 | ·0111 989 |
| ·54 | ·0870 143 | ·0621 998 | ·0438 867 | ·0305 982 | ·0211 008 | ·0144 052 |
| ·55 | ·1018 647 | ·0740 530 | ·0531 479 | ·0376 978 | ·0264 509 | ·0183 751 |
| ·56 | ·1185 095$^{-}$ | ·0875 822 | ·0639 124 | ·0461 010 | ·0328 995$^{+}$ | ·0232 478 |
| ·57 | ·1370 356 | ·1029 116 | ·0763 289 | ·0559 683 | ·0406 080 | ·0291 773 |
| ·58 | ·1575 123 | ·1201 546 | ·0905 424 | ·0674 634 | ·0497 469 | ·0363 315$^{+}$ |
| ·59 | ·1799 876 | ·1394 096 | ·1066 902 | ·0807 498 | ·0604 934 | ·0448 904 |
| ·60 | ·2044 848 | ·1607 552 | ·1248 969 | ·0959 862 | ·0730 277 | ·0550 436 |
| ·61 | ·2309 988 | ·1842 462 | ·1452 701 | ·1133 220 | ·0875 285$^{+}$ | ·0669 869 |
| ·62 | ·2594 933 | ·2099 089 | ·1678 945$^{-}$ | ·1328 911 | ·1041 678 | ·0809 179 |
| ·63 | ·2898 983 | ·2377 372 | ·1928 265$^{+}$ | ·1548 068 | ·1231 051 | ·0970 304 |
| ·64 | ·3221 079 | ·2676 885$^{-}$ | ·2200 897 | ·1791 547 | ·1444 803 | ·1155 080 |
| ·65 | ·3559 792 | ·2996 806 | ·2496 688 | ·2059 866 | ·1684 070 | ·1365 165$^{+}$ |
| ·66 | ·3913 322 | ·3335 897 | ·2815 059 | ·2353 141 | ·1949 643 | ·1601 961 |
| ·67 | ·4279 506 | ·3692 483 | ·3154 965$^{-}$ | ·2671 030 | ·2241 895$^{-}$ | ·1866 520 |
| ·68 | ·4655 835$^{-}$ | ·4064 454 | ·3514 864 | ·3012 676 | ·2560 707 | ·2159 459 |
| ·69 | ·5039 485$^{-}$ | ·4449 276 | ·3892 708 | ·3376 666 | ·2905 398 | ·2480 864 |
| ·70 | ·5427 360 | ·4844 011 | ·4285 941 | ·3761 006 | ·3274 669 | ·2830 211 |
| ·71 | ·5816 149 | ·5245 365$^{+}$ | ·4691 515$^{-}$ | ·4163 107 | ·3666 559 | ·3206 286 |
| ·72 | ·6202 387 | ·5649 737 | ·5105 930 | ·4579 797 | ·4078 422 | ·3607 130 |
| ·73 | ·6582 534 | ·6053 294 | ·5525 290 | ·5007 350$^{-}$ | ·4506 929 | ·4030 000 |
| ·74 | ·6953 057 | ·6452 056 | ·5945 377 | ·5441 548 | ·4948 092 | ·4471 359 |
| ·75 | ·7310 522 | ·6841 994 | ·6361 753 | ·5877 760 | ·5397 325$^{+}$ | ·4926 900 |
| ·76 | ·7651 685$^{-}$ | ·7219 140 | ·6769 870 | ·6311 053 | ·5849 538 | ·5391 611 |
| ·77 | ·7973 585$^{+}$ | ·7579 701 | ·7165 203 | ·6736 328 | ·6299 253 | ·5859 871 |
| ·78 | ·8273 635$^{-}$ | ·7920 175$^{+}$ | ·7543 387 | ·7148 467 | ·6740 769 | ·6325 597 |
| ·79 | ·8549 697 | ·8237 468 | ·7900 366 | ·7542 510 | ·7168 341 | ·6782 426 |
| ·80 | ·8800 157 | ·8528 999 | ·8232 533 | ·7913 832 | ·7576 384 | ·7223 938 |

$x$ = ·81 to ·99 $q = 6{\cdot}5$ $p$ = 14 to 19

| | $p = 14$ | $p = 15$ | $p = 16$ | $p = 17$ | $p = 18$ | $p = 19$ |
|---|---|---|---|---|---|---|
| $B(p, q) =$ | ·3315 9213 $\times \frac{1}{10^5}$ | ·2264 5316 $\times \frac{1}{10^5}$ | ·1579 9058 $\times \frac{1}{10^5}$ | ·1123 4885 $\overset{+}{\times} \frac{1}{10^5}$ | ·8127 3640 $\times \frac{1}{10^6}$ | ·5971 1245 $\overset{+}{\times} \frac{1}{10^6}$ |
| $x$ | | | | | | |
| ·81 | ·9023 977 | ·8792 793 | ·8536 873 | ·8258 321 | ·7959 696 | ·7643 900 |
| ·82 | ·9220 723 | ·9027 556 | ·8811 080 | ·8572 551 | ·8313 676 | ·8036 537 |
| ·83 | ·9390 583 | ·9232 720 | ·9053 652 | ·8853 934 | ·8634 538 | ·8396 799 |
| ·84 | ·9534 348 | ·9408 468 | ·9263 957 | ·9100 841 | ·8919 494 | ·8720 617 |
| ·85 | ·9653 374 | ·9555 712 | ·9442 261 | ·9312 679 | ·9166 899 | ·9005 124 |
| ·86 | ·9749 514 | ·9676 050− | ·9589 704 | ·9489 921 | ·9376 345+ | ·9248 824 |
| ·87 | ·9825 034 | ·9771 680 | ·9708 241 | ·9634 078 | ·9548 681 | ·9451 685− |
| ·88 | ·9882 495− | ·9845 281 | ·9800 526 | ·9747 602 | ·9685 961 | ·9615 143 |
| ·89 | ·9924 635+ | ·9899 874 | ·9869 755+ | ·9833 735+ | ·9791 306 | ·9742 006 |
| ·90 | ·9954 235+ | ·9938 652 | ·9919 484 | ·9896 304 | ·9868 691 | ·9836 246 |
| ·91 | ·9973 987 | ·9964 816 | ·9953 410 | ·9939 463 | ·9922 664 | ·9902 707 |
| ·92 | ·9986 371 | ·9981 401 | ·9975 151 | ·9967 426 | ·9958 019 | ·9946 719 |
| ·93 | ·9993 559 | ·9991 132 | ·9988 046 | ·9984 190 | ·9979 444 | ·9973 682 |
| ·94 | ·9997 340 | ·9996 305− | ·9994 975− | ·9993 295− | ·9991 205− | ·9988 640 |
| ·95 | ·9999 086 | ·9998 719 | ·9998 242 | ·9997 634 | ·9996 869 | ·9995 921 |
| ·96 | ·9999 759 | ·9999 660 | ·9999 529 | ·9999 360 | ·9999 146 | ·9998 877 |
| ·97 | ·9999 958 | ·9999 941 | ·9999 917 | ·9999 886 | ·9999 847 | ·9999 797 |
| ·98 | ·9999 997 | ·9999 995+ | ·9999 993 | ·9999 991 | ·9999 987 | ·9999 983 |
| ·99 | 1·0000 000 | 1·0000 000 | 1·0000 000 | 1·0000 000 | 1·0000 000 | 1·0000 000 |

$x = \cdot 27$ to $\cdot 80$ $\qquad q = 6\cdot 5 \qquad$ $p = 20$ to $25$

| $x$ | $p = 20$ | $p = 21$ | $p = 22$ | $p = 23$ | $p = 24$ | $p = 25$ |
|---|---|---|---|---|---|---|
| $B(p, q) =$ | ·4449 0732 × $\frac{1}{10^8}$ | ·3357 7911 × $\frac{1}{10^8}$ | ·2564 1314 × $\frac{1}{10^8}$ | ·1979 3295 × $\frac{1}{10^8}$ | ·1543 2060 × $\frac{1}{10^8}$ | ·1214 3261 × $\frac{1}{10^8}$ |
| ·27 | ·0000 001 | | | | | |
| ·28 | ·0000 002 | ·0000 001 | | | | |
| ·29 | ·0000 003 | ·0000 001 | | | | |
| ·30 | ·0000 006 | ·0000 002 | ·0000 001 | | | |
| ·31 | ·0000 011 | ·0000 004 | ·0000 002 | ·0000 001 | | |
| ·32 | ·0000 019 | ·0000 008 | ·0000 003 | ·0000 001 | | |
| ·33 | ·0000 033 | ·0000 014 | ·0000 006 | ·0000 002 | ·0000 001 | |
| ·34 | ·0000 056 | ·0000 024 | ·0000 010 | ·0000 004 | ·0000 002 | ·0000 001 |
| ·35 | ·0000 093 | ·0000 041 | ·0000 018 | ·0000 008 | ·0000 003 | ·0000 001 |
| ·36 | ·0000 150+ | ·0000 068 | ·0000 030 | ·0000 013 | ·0000 006 | ·0000 003 |
| ·37 | ·0000 240 | ·0000 111 | ·0000 051 | ·0000 023 | ·0000 011 | ·0000 005− |
| ·38 | ·0000 378 | ·0000 180 | ·0000 085− | ·0000 040 | ·0000 018 | ·0000 008 |
| ·39 | ·0000 585+ | ·0000 286 | ·0000 138 | ·0000 066 | ·0000 032 | ·0000 015− |
| ·40 | ·0000 894 | ·0000 447 | ·0000 222 | ·0000 109 | ·0000 053 | ·0000 026 |
| ·41 | ·0001 346 | ·0000 690 | ·0000 351 | ·0000 177 | ·0000 089 | ·0000 044 |
| ·42 | ·0002 001 | ·0001 051 | ·0000 547 | ·0000 282 | ·0000 145− | ·0000 074 |
| ·43 | ·0002 937 | ·0001 578 | ·0000 841 | ·0000 444 | ·0000 233 | ·0000 121 |
| ·44 | ·0004 259 | ·0002 340 | ·0001 275+ | ·0000 689 | ·0000 370 | ·0000 197 |
| ·45 | ·0006 104 | ·0003 429 | ·0001 910 | ·0001 055+ | ·0000 579 | ·0000 315+ |
| ·46 | ·0008 649 | ·0004 964 | ·0002 825+ | ·0001 595+ | ·0000 894 | ·0000 498 |
| ·47 | ·0012 123 | ·0007 106 | ·0004 130 | ·0002 382 | ·0001 364 | ·0000 775+ |
| ·48 | ·0016 813 | ·0010 059 | ·0005 968 | ·0003 513 | ·0002 054 | ·0001 192 |
| ·49 | ·0023 079 | ·0014 088 | ·0008 528 | ·0005 123 | ·0003 055+ | ·0001 810 |
| ·50 | ·0031 369 | ·0019 527 | ·0012 056 | ·0007 386 | ·0004 493 | ·0002 715+ |
| ·51 | ·0042 225+ | ·0026 795− | ·0016 864 | ·0010 533 | ·0006 533 | ·0004 025− |
| ·52 | ·0056 307 | ·0036 408 | ·0023 350+ | ·0014 863 | ·0009 394 | ·0005 898 |
| ·53 | ·0074 399 | ·0048 998 | ·0032 010 | ·0020 755− | ·0013 363 | ·0008 548 |
| ·54 | ·0097 429 | ·0065 330 | ·0043 456 | ·0028 691 | ·0018 812 | ·0012 254 |
| ·55 | ·0126 476 | ·0086 312 | ·0058 437 | ·0039 272 | ·0026 211 | ·0017 382 |
| ·56 | ·0162 783 | ·0113 020 | ·0077 854 | ·0053 238 | ·0036 157 | ·0024 399 |
| ·57 | ·0207 758 | ·0146 700 | ·0102 782 | ·0071 490 | ·0049 389 | ·0033 904 |
| ·58 | ·0262 983 | ·0188 787 | ·0134 482 | ·0095 111 | ·0066 816 | ·0046 643 |
| ·59 | ·0330 198 | ·0240 901 | ·0174 417 | ·0125 386 | ·0089 539 | ·0063 543 |
| ·60 | ·0411 295− | ·0304 852 | ·0224 260 | ·0163 816 | ·0118 877 | ·0085 735− |
| ·61 | ·0508 293 | ·0382 627 | ·0285 894 | ·0212 135+ | ·0156 383 | ·0114 581 |
| ·62 | ·0623 302 | ·0476 366 | ·0361 406 | ·0272 313 | ·0203 865+ | ·0151 703 |
| ·63 | ·0758 482 | ·0588 335+ | ·0453 069 | ·0346 547 | ·0263 390 | ·0198 995+ |
| ·64 | ·0915 982 | ·0720 878 | ·0563 308 | ·0437 252 | ·0337 285− | ·0258 643 |
| ·65 | ·1097 873 | ·0876 355− | ·0694 657 | ·0547 027 | ·0428 121 | ·0333 121 |
| ·66 | ·1306 066 | ·1057 070 | ·0849 692 | ·0678 605+ | ·0538 686 | ·0425 176 |
| ·67 | ·1542 216 | ·1265 180 | ·1030 952 | ·0834 786 | ·0671 926 | ·0537 804 |
| ·68 | ·1807 627 | ·1502 589 | ·1240 838 | ·1018 349 | ·0830 878 | ·0674 184 |
| ·69 | ·2103 140 | ·1770 838 | ·1481 498 | ·1231 942 | ·1018 571 | ·0837 607 |
| ·70 | ·2429 029 | ·2070 974 | ·1754 693 | ·1477 946 | ·1237 899 | ·1031 358 |
| ·71 | ·2784 895− | ·2403 430 | ·2061 656 | ·1758 331 | ·1491 471 | ·1258 581 |
| ·72 | ·3169 573 | ·2767 896 | ·2402 943 | ·2074 485+ | ·1781 444 | ·1522 103 |
| ·73 | ·3581 059 | ·3163 208 | ·2778 287 | ·2427 046 | ·2109 326 | ·1824 239 |
| ·74 | ·4016 455− | ·3587 253 | ·3186 460 | ·2815 727 | ·2475 780 | ·2166 571 |
| ·75 | ·4471 950− | ·4036 903 | ·3625 162 | ·3239 156 | ·2880 421 | ·2549 713 |
| ·76 | ·4942 844 | ·4507 988 | ·4090 940 | ·3694 745+ | ·3321 633 | ·2973 083 |
| ·77 | ·5423 612 | ·4995 314 | ·4579 151 | ·4178 595− | ·3796 411 | ·3434 691 |
| ·78 | ·5908 022 | ·5492 743 | ·5083 989 | ·4685 455+ | ·4300 263 | ·3930 960 |
| ·79 | ·6389 302 | ·5993 327 | ·5598 577 | ·5208 762 | ·4827 167 | ·4456 623 |
| ·80 | ·6860 362 | ·6489 515− | ·6115 135+ | ·5740 756 | ·5369 632 | ·5004 694 |

$x$ = ·81 to 1·00 $q = 6{\cdot}5$ $p$ = 20 to 25

| | $p = 20$ | $p = 21$ | $p = 22$ | $p = 23$ | $p = 24$ | $p = 25$ |
|---|---|---|---|---|---|---|
| $B(p, q) =$ | ·4449 0732 $\times \frac{1}{10^8}$ | ·3357 7911 $\times \frac{1}{10^8}$ | ·2564 1314 $\times \frac{1}{10^8}$ | ·1979 3295 $\times \frac{1}{10^8}$ | ·1543 2060 $\times \frac{1}{10^8}$ | ·1214 3261 $\times \frac{1}{10^8}$ |
| $x$ | | | | | | |
| ·81 | ·7314 059 | ·6973 416 | ·6625 231 | ·6272 693 | ·5918 852 | ·5566 559 |
| ·82 | ·7743 498 | ·7437 126 | ·7120 105⁻ | ·6795 157 | ·6464 982 | ·6132 194 |
| ·83 | ·8142 355⁻ | ·7873 090 | ·7591 066 | ·7298 467 | ·6997 535⁺ | ·6690 522 |
| ·84 | ·8505 201 | ·8274 492 | ·8029 939 | ·7773 159 | ·7505 885⁻ | ·7229 924 |
| ·85 | ·8827 810 | ·8635 646 | ·8429 527 | ·8210 526 | ·7979 860 | ·7738 863 |
| ·86 | ·9107 410 | ·8952 349 | ·8784 071 | ·8603 173 | ·8410 398 | ·8206 618 |
| ·87 | ·9342 870 | ·9222 167 | ·9089 653 | ·8945 544 | ·8790 187 | ·8624 053 |
| ·88 | ·9534 783 | ·9444 620 | ·9344 495⁺ | ·9234 358 | ·9114 260 | ·8984 355⁺ |
| ·89 | ·9685 427 | ·9621 224 | ·9549 119 | ·9468 901 | ·9380 435⁺ | ·9283 657 |
| ·90 | ·9798 594 | ·9755 387 | ·9706 317 | ·9651 112 | ·9589 548 | ·9521 442 |
| ·91 | ·9879 289 | ·9852 118 | ·9820 917 | ·9785 425⁺ | ·9745 405⁻ | ·9700 640 |
| ·92 | ·9933 315⁺ | ·9917 592 | ·9899 338 | ·9878 346 | ·9854 415⁺ | ·9827 353 |
| ·93 | ·9966 772 | ·9958 578 | ·9948 961 | ·9937 782 | ·9924 900 | ·9910 174 |
| ·94 | ·9985 531 | ·9981 805⁺ | ·9977 386 | ·9972 192 | ·9966 143 | ·9959 154 |
| ·95 | ·9994 758 | ·9993 351 | ·9991 663 | ·9989 660 | ·9987 301 | ·9984 546 |
| ·96 | ·9998 545⁻ | ·9998 138 | ·9997 645⁺ | ·9997 054 | ·9996 350⁻ | ·9995 519 |
| ·97 | ·9999 735⁻ | ·9999 658 | ·9999 563 | ·9999 449 | ·9999 311 | ·9999 147 |
| ·98 | ·9999 978 | ·9999 971 | ·9999 962 | ·9999 952 | ·9999 940 | ·9999 925⁻ |
| ·99 | 1·0000 000 | 1·0000 000 | 1·0000 000 | ·9999 999 | ·9999 999 | ·9999 999 |
| 1·00 | | | | 1·0000 000 | 1·0000 000 | 1·0000 000 |

$x$ = ·35 to 1·00 $q = 6{\cdot}5$ $p$ = 26 to 31

| $x$ | $p = 26$ | $p = 27$ | $p = 28$ | $p = 29$ | $p = 30$ | $p = 31$ |
|---|---|---|---|---|---|---|
| $B(p, q)$ = | ·9637 5084 $\times \frac{1}{10^7}$ | ·7710 0067 $\times \frac{1}{10^7}$ | ·6214 0353 $\times \frac{1}{10^7}$ | ·5043 2750 $\stackrel{+}{\times} \frac{1}{10^7}$ | ·4119 8585 $\stackrel{-}{\times} \frac{1}{10^7}$ | ·3386 1850 $\stackrel{+}{\times} \frac{1}{10^7}$ |
| ·35 | ·0000 001 | | | | | |
| ·36 | ·0000 001 | | | | | |
| ·37 | ·0000 002 | ·0000 001 | | | | |
| ·38 | ·0000 004 | ·0000 002 | ·0000 001 | | | |
| ·39 | ·0000 007 | ·0000 003 | ·0000 002 | ·0000 001 | | |
| ·40 | ·0000 012 | ·0000 006 | ·0000 003 | ·0000 001 | ·0000 001 | |
| ·41 | ·0000 022 | ·0000 011 | ·0000 005+ | ·0000 003 | ·0000 001 | ·0000 001 |
| ·42 | ·0000 037 | ·0000 019 | ·0000 009 | ·0000 005− | ·0000 002 | ·0000 001 |
| ·43 | ·0000 063 | ·0000 032 | ·0000 017 | ·0000 008 | ·0000 004 | ·0000 002 |
| ·44 | ·0000 104 | ·0000 055− | ·0000 029 | ·0000 015− | ·0000 008 | ·0000 004 |
| ·45 | ·0000 171 | ·0000 092 | ·0000 049 | ·0000 026 | ·0000 014 | ·0000 007 |
| ·46 | ·0000 275+ | ·0000 152 | ·0000 083 | ·0000 045+ | ·0000 024 | ·0000 013 |
| ·47 | ·0000 438 | ·0000 246 | ·0000 138 | ·0000 076 | ·0000 042 | ·0000 023 |
| ·48 | ·0000 688 | ·0000 394 | ·0000 225− | ·0000 128 | ·0000 072 | ·0000 041 |
| ·49 | ·0001 066 | ·0000 624 | ·0000 363 | ·0000 210 | ·0000 121 | ·0000 069 |
| ·50 | ·0001 630 | ·0000 973 | ·0000 578 | ·0000 341 | ·0000 201 | ·0000 117 |
| ·51 | ·0002 464 | ·0001 500+ | ·0000 908 | ·0000 547 | ·0000 328 | ·0000 196 |
| ·52 | ·0003 681 | ·0002 284 | ·0001 409 | ·0000 865+ | ·0000 529 | ·0000 321 |
| ·53 | ·0005 434 | ·0003 435+ | ·0002 160 | ·0001 351 | ·0000 841 | ·0000 521 |
| ·54 | ·0007 934 | ·0005 108 | ·0003 271 | ·0002 084 | ·0001 321 | ·0000 834 |
| ·55 | ·0011 457 | ·0007 509 | ·0004 895+ | ·0003 175+ | ·0002 050− | ·0001 317 |
| ·56 | ·0016 367 | ·0010 917 | ·0007 243 | ·0004 782 | ·0003 142 | ·0002 055+ |
| ·57 | ·0023 136 | ·0015 700 | ·0010 598 | ·0007 118 | ·0004 759 | ·0003 167 |
| ·58 | ·0032 369 | ·0022 339 | ·0015 337 | ·0010 477 | ·0007 124 | ·0004 823 |
| ·59 | ·0044 831 | ·0031 456 | ·0021 957 | ·0015 251 | ·0010 545− | ·0007 259 |
| ·60 | ·0061 474 | ·0043 839 | ·0031 102 | ·0021 959 | ·0015 433 | ·0010 799 |
| ·61 | ·0083 472 | ·0060 481 | ·0043 599 | ·0031 279 | ·0022 338 | ·0015 885− |
| ·62 | ·0112 247 | ·0082 610 | ·0060 491 | ·0044 084 | ·0031 983 | ·0023 104 |
| ·63 | ·0149 502 | ·0111 725− | ·0083 077 | ·0061 484 | ·0045 300 | ·0033 235+ |
| ·64 | ·0197 240 | ·0149 629 | ·0112 951 | ·0084 866 | ·0063 483 | ·0047 289 |
| ·65 | ·0257 785− | ·0198 458 | ·0152 040 | ·0115 942 | ·0088 028 | ·0066 558 |
| ·66 | ·0333 778 | ·0260 695+ | ·0202 635− | ·0156 788 | ·0120 790 | ·0092 676 |
| ·67 | ·0428 172 | ·0339 181 | ·0267 411 | ·0209 880 | ·0164 023 | ·0127 667 |
| ·68 | ·0544 190 | ·0437 095+ | ·0349 436 | ·0278 119 | ·0220 426 | ·0174 003 |
| ·69 | ·0685 270 | ·0557 923 | ·0452 154 | ·0364 838 | ·0293 164 | ·0234 642 |
| ·70 | ·0854 972 | ·0705 383 | ·0579 341 | ·0473 780 | ·0385 874 | ·0313 058 |
| ·71 | ·1056 851 | ·0883 324 | ·0735 023 | ·0609 047 | ·0502 640 | ·0413 241 |
| ·72 | ·1294 297 | ·1095 581 | ·0923 360 | ·0775 004 | ·0647 929 | ·0539 661 |
| ·73 | ·1570 340 | ·1345 786 | ·1148 470 | ·0976 135+ | ·0826 471 | ·0697 187 |
| ·74 | ·1887 416 | ·1637 144 | ·1414 216 | ·1216 845− | ·1043 090 | ·0890 939 |
| ·75 | ·2247 115− | ·1972 158 | ·1723 933 | ·1501 195− | ·1302 459 | ·1126 081 |
| ·76 | ·2649 909 | ·2352 339 | ·2080 118 | ·1832 591 | ·1608 793 | ·1407 524 |
| ·77 | ·3094 890 | ·2777 896 | ·2484 089 | ·2213 420 | ·1965 477 | ·1739 560 |
| ·78 | ·3579 531 | ·3247 431 | ·2935 626 | ·2644 645+ | ·2374 635+ | ·2125 416 |
| ·79 | ·4099 492 | ·3757 677 | ·3432 635+ | ·3125 414 | ·2836 678 | ·2566 758 |
| ·80 | ·4648 514 | ·4303 294 | ·3970 859 | ·3652 672 | ·3349 845+ | ·3063 170 |
| ·81 | ·5218 424 | ·4876 782 | ·4543 681 | ·4220 871 | ·3909 805− | ·3611 647 |
| ·82 | ·5799 272 | ·5468 527 | ·5142 069 | ·4821 791 | ·4509 354 | ·4206 186 |
| ·83 | ·6379 641 | ·6067 025+ | ·5754 699 | ·5444 549 | ·5138 302 | ·4837 516 |
| ·84 | ·6947 119 | ·6659 311 | ·6368 305− | ·6075 844 | ·5783 584 | ·5493 078 |
| ·85 | ·7488 949 | ·7231 585+ | ·6968 264 | ·6700 476 | ·6429 687 | ·6157 318 |
| ·86 | ·7992 813 | ·7770 045− | ·7539 438 | ·7302 160 | ·7059 400 | ·6812 352 |
| ·87 | ·8447 719 | ·8261 855− | ·8067 215− | ·7864 615− | ·7654 924 | ·7439 047 |
| ·88 | ·8844 889 | ·8696 196 | ·8538 692 | ·8372 863 | ·8199 257 | ·8018 475− |
| ·89 | ·9178 575+ | ·9065 269 | ·8943 884 | ·8814 631 | ·8677 779 | ·8533 650+ |
| ·90 | ·9446 661 | ·9365 122 | ·9276 788 | ·9181 671 | ·9079 830 | ·8971 369 |
| ·91 | ·9650 943 | ·9596 151 | ·9536 134 | ·9470 790 | ·9400 050+ | ·9323 875− |
| ·92 | ·9796 979 | ·9763 123 | ·9725 631 | ·9684 364 | ·9639 197 | ·9590 025− |
| ·93 | ·9893 465+ | ·9874 639 | ·9853 565− | ·9830 116 | ·9804 172 | ·9775 621 |
| ·94 | ·9951 139 | ·9942 012 | ·9931 684 | ·9920 068 | ·9907 079 | ·9892 631 |
| ·95 | ·9981 354 | ·9977 679 | ·9973 478 | ·9968 702 | ·9963 305− | ·9957 237 |
| ·96 | ·9994 547 | ·9993 416 | ·9992 109 | ·9990 608 | ·9988 893 | ·9986 945+ |
| ·97 | ·9998 953 | ·9998 724 | ·9998 458 | ·9998 149 | ·9997 792 | ·9997 382 |
| ·98 | ·9999 907 | ·9999 885+ | ·9999 860 | ·9999 830 | ·9999 796 | ·9999 756 |
| ·99 | ·9999 999 | ·9999 998 | ·9999 998 | ·9999 998 | ·9999 997 | ·9999 997 |
| 1·00 | 1·0000 000 | 1·0000 000 | 1·0000 000 | 1·0000 000 | 1·0000 000 | 1·0000 000 |

$x$ = ·42 to 1·00 $q$ = 6·5 $p$ = 32 to 37

| $x$ | $p$ = 32 | $p$ = 33 | $p$ = 34 | $p$ = 35 | $p$ = 36 | $p$ = 37 |
|---|---|---|---|---|---|---|
| $B(p, q)$ = | ·2799 2463 × $\frac{1}{10^7}$ | ·2326 6463 × $\frac{1}{10^7}$ | ·1943 7804 × $\frac{1}{10^7}$ | ·1631 8157 × $\frac{1}{10^7}$ | ·1376 2301 × $\frac{1}{10^7}$ | ·1165 7478 × $\frac{1}{10^7}$ |
| ·42 | ·0000 001 | | | | | |
| ·43 | ·0000 001 | ·0000 001 | | | | |
| ·44 | ·0000 002 | ·0000 001 | ·0000 001 | | | |
| ·45 | ·0000 004 | ·0000 002 | ·0000 001 | ·0000 001 | | |
| ·46 | ·0000 007 | ·0000 004 | ·0000 002 | ·0000 001 | ·0000 001 | |
| ·47 | ·0000 013 | ·0000 007 | ·0000 004 | ·0000 002 | ·0000 001 | ·0000 001 |
| ·48 | ·0000 023 | ·0000 013 | ·0000 007 | ·0000 004 | ·0000 002 | ·0000 001 |
| ·49 | ·0000 040 | ·0000 023 | ·0000 013 | ·0000 007 | ·0000 004 | ·0000 002 |
| ·50 | ·0000 068 | ·0000 040 | ·0000 023 | ·0000 013 | ·0000 008 | ·0000 004 |
| ·51 | ·0000 116 | ·0000 069 | ·0000 041 | ·0000 024 | ·0000 014 | ·0000 008 |
| ·52 | ·0000 195− | ·0000 117 | ·0000 071 | ·0000 042 | ·0000 025+ | ·0000 015− |
| ·53 | ·0000 322 | ·0000 198 | ·0000 121 | ·0000 074 | ·0000 045− | ·0000 027 |
| ·54 | ·0000 524 | ·0000 328 | ·0000 205− | ·0000 127 | ·0000 079 | ·0000 049 |
| ·55 | ·0000 843 | ·0000 537 | ·0000 341 | ·0000 216 | ·0000 136 | ·0000 086 |
| ·56 | ·0001 339 | ·0000 869 | ·0000 561 | ·0000 362 | ·0000 232 | ·0000 148 |
| ·57 | ·0002 099 | ·0001 386 | ·0000 911 | ·0000 597 | ·0000 390 | ·0000 254 |
| ·58 | ·0003 252 | ·0002 184 | ·0001 461 | ·0000 974 | ·0000 647 | ·0000 428 |
| ·59 | ·0004 976 | ·0003 398 | ·0002 312 | ·0001 567 | ·0001 059 | ·0000 713 |
| ·60 | ·0007 526 | ·0005 224 | ·0003 613 | ·0002 490 | ·0001 710 | ·0001 171 |
| ·61 | ·0011 249 | ·0007 936 | ·0005 578 | ·0003 907 | ·0002 727 | ·0001 898 |
| ·62 | ·0016 623 | ·0011 914 | ·0008 508 | ·0006 054 | ·0004 294 | ·0003 036 |
| ·63 | ·0024 286 | ·0017 679 | ·0012 823 | ·0009 268 | ·0006 677 | ·0004 795+ |
| ·64 | ·0035 085+ | ·0025 933 | ·0019 099 | ·0014 018 | ·0010 255+ | ·0007 479 |
| ·65 | ·0050 126 | ·0037 610 | ·0028 118 | ·0020 951 | ·0015 560 | ·0011 520 |
| ·66 | ·0070 828 | ·0053 930 | ·0040 919 | ·0030 943 | ·0023 323 | ·0017 526 |
| ·67 | ·0098 987 | ·0076 468 | ·0058 866 | ·0045 165− | ·0034 542 | ·0026 338 |
| ·68 | ·0136 834 | ·0107 215+ | ·0083 717 | ·0065 154 | ·0050 548 | ·0039 098 |
| ·69 | ·0187 097 | ·0148 653 | ·0117 706 | ·0092 898 | ·0073 090 | ·0057 335− |
| ·70 | ·0253 044 | ·0203 813 | ·0163 608 | ·0130 912 | ·0104 428 | ·0083 057 |
| ·71 | ·0338 506 | ·0276 324 | ·0224 817 | ·0182 331 | ·0147 426 | ·0118 857 |
| ·72 | ·0447 878 | ·0370 436 | ·0305 385+ | ·0250 972 | ·0205 638 | ·0168 010 |
| ·73 | ·0586 065− | ·0491 003 | ·0410 042 | ·0341 380 | ·0283 381 | ·0234 572 |
| ·74 | ·0758 372 | ·0643 411 | ·0544 161 | ·0458 835+ | ·0385 772 | ·0323 445+ |
| ·75 | ·0970 328 | ·0833 435+ | ·0713 655− | ·0609 287 | ·0518 710 | ·0440 399 |
| ·76 | ·1227 421 | ·1067 016 | ·0924 791 | ·0799 214 | ·0688 778 | ·0592 024 |
| ·77 | ·1534 740 | ·1349 923 | ·1183 895+ | ·1035 374 | ·0903 041 | ·0785 577 |
| ·78 | ·1896 536 | ·1687 325+ | ·1496 944 | ·1324 424 | ·1168 713 | ·1028 703 |
| ·79 | ·2315 689 | ·2083 254 | ·1869 027 | ·1672 410 | ·1492 675− | ·1328 990 |
| ·80 | ·2793 139 | ·2539 986 | ·2303 710 | ·2084 112 | ·1880 827 | ·1693 353 |
| ·81 | ·3327 292 | ·3057 376 | ·2802 306 | ·2562 278 | ·2337 302 | ·2127 231 |
| ·82 | ·3913 482 | ·3632 212 | ·3363 131 | ·3106 791 | ·2863 561 | ·2633 641 |
| ·83 | ·4543 571 | ·4257 663 | ·3980 809 | ·3713 848 | ·3457 452 | ·3212 130 |
| ·84 | ·5205 759 | ·4922 928 | ·4645 755− | ·4375 266 | ·4112 350+ | ·3857 761 |
| ·85 | ·5884 729 | ·5613 206 | ·5343 945− | ·5078 049 | ·4816 523 | ·4560 262 |
| ·86 | ·6562 195+ | ·6310 085− | ·6057 133 | ·5804 405+ | ·5552 905− | ·5303 570 |
| ·87 | ·7217 913 | ·6992 459 | ·6763 624 | ·6532 332 | ·6299 488 | ·6065 964 |
| ·88 | ·7831 162 | ·7637 995− | ·7439 677 | ·7236 926 | ·7030 468 | ·6821 027 |
| ·89 | ·8382 618 | ·8225 096 | ·8061 535+ | ·7892 419 | ·7718 253 | ·7539 563 |
| ·90 | ·8856 436 | ·8735 218 | ·8607 938 | ·8474 856 | ·8336 261 | ·8192 468 |
| ·91 | ·9242 256 | ·9155 218 | ·9062 813 | ·8965 121 | ·8862 252 | ·8754 339 |
| ·92 | ·9536 761 | ·9479 335+ | ·9417 699 | ·9351 820 | ·9281 687 | ·9207 308 |
| ·93 | ·9744 357 | ·9710 285+ | ·9673 317 | ·9633 375− | ·9590 391 | ·9544 309 |
| ·94 | ·9876 640 | ·9859 025− | ·9839 707 | ·9818 611 | ·9795 664 | ·9770 799 |
| ·95 | ·9950 451 | ·9942 895− | ·9934 521 | ·9925 278 | ·9915 118 | ·9903 992 |
| ·96 | ·9984 744 | ·9982 267 | ·9979 493 | ·9976 399 | ·9972 963 | ·9969 159 |
| ·97 | ·9996 914 | ·9996 382 | ·9995 780 | ·9995 102 | ·9994 341 | ·9993 489 |
| ·98 | ·9999 710 | ·9999 657 | ·9999 596 | ·9999 528 | ·9999 449 | ·9999 361 |
| ·99 | ·9999 996 | ·9999 995+ | ·9999 994 | ·9999 993 | ·9999 992 | ·9999 990 |
| 1·00 | 1·0000 000 | 1·0000 000 | 1·0000 000 | 1·0000 000 | 1·0000 000 | 1·0000 000 |

$x = \cdot 48$ to $1\cdot 00$ $\qquad q = 6\cdot 5 \qquad$ $p = 38$ to $43$

| | $p = 38$ | $p = 39$ | $p = 40$ | $p = 41$ | $p = 42$ | $p = 43$ |
|---|---|---|---|---|---|---|
| $B(p, q) =$ | ·9915 5563 × $\frac{1}{10^8}$ | ·8467 2166 × $\frac{1}{10^8}$ | ·7257 6142 × $\frac{1}{10^8}$ | ·6243 1090 × $\frac{1}{10^8}$ | ·5388 7888 × $\frac{1}{10^8}$ | ·4666 5800 × $\frac{1}{10^8}$ |
| $x$ | | | | | | |
| ·48 | ·0000 001 | | | | | |
| ·49 | ·0000 001 | ·0000 001 | | | | |
| ·50 | ·0000 002 | ·0000 001 | ·0000 001 | | | |
| ·51 | ·0000 005− | ·0000 003 | ·0000 002 | ·0000 001 | ·0000 001 | |
| ·52 | ·0000 009 | ·0000 005+ | ·0000 003 | ·0000 002 | ·0000 001 | ·0000 001 |
| ·53 | ·0000 016 | ·0000 010 | ·0000 006 | ·0000 004 | ·0000 002 | ·0000 001 |
| ·54 | ·0000 030 | ·0000 018 | ·0000 011 | ·0000 007 | ·0000 004 | ·0000 003 |
| ·55 | ·0000 054 | ·0000 033 | ·0000 021 | ·0000 013 | ·0000 008 | ·0000 005− |
| ·56 | ·0000 095− | ·0000 060 | ·0000 038 | ·0000 024 | ·0000 015+ | ·0000 010 |
| ·57 | ·0000 165− | ·0000 107 | ·0000 069 | ·0000 044 | ·0000 028 | ·0000 018 |
| ·58 | ·0000 283 | ·0000 186 | ·0000 122 | ·0000 080 | ·0000 052 | ·0000 034 |
| ·59 | ·0000 479 | ·0000 320 | ·0000 214 | ·0000 142 | ·0000 095− | ·0000 063 |
| ·60 | ·0000 799 | ·0000 544 | ·0000 369 | ·0000 250− | ·0000 169 | ·0000 114 |
| ·61 | ·0001 316 | ·0000 911 | ·0000 628 | ·0000 432 | ·0000 297 | ·0000 203 |
| ·62 | ·0002 140 | ·0001 504 | ·0001 054 | ·0000 737 | ·0000 514 | ·0000 357 |
| ·63 | ·0003 433 | ·0002 451 | ·0001 745+ | ·0001 239 | ·0000 878 | ·0000 620 |
| ·64 | ·0005 438 | ·0003 943 | ·0002 851 | ·0002 056 | ·0001 479 | ·0001 062 |
| ·65 | ·0008 504 | ·0006 260 | ·0004 596 | ·0003 365− | ·0002 458 | ·0001 791 |
| ·66 | ·0013 132 | ·0009 811 | ·0007 311 | ·0005 434 | ·0004 028 | ·0002 980 |
| ·67 | ·0020 024 | ·0015 181 | ·0011 479 | ·0008 658 | ·0006 514 | ·0004 889 |
| ·68 | ·0030 155− | ·0023 193 | ·0017 791 | ·0013 613 | ·0010 391 | ·0007 913 |
| ·69 | ·0044 848 | ·0034 985− | ·0027 219 | ·0021 124 | ·0016 355− | ·0012 633 |
| ·70 | ·0065 874 | ·0052 105− | ·0041 107 | ·0032 350− | ·0025 398 | ·0019 894 |
| ·71 | ·0095 558 | ·0076 621 | ·0061 280 | ·0048 890 | ·0038 913 | ·0030 902 |
| ·72 | ·0136 891 | ·0111 243 | ·0090 172 | ·0072 914 | ·0058 822 | ·0047 347 |
| ·73 | ·0193 646 | ·0159 445+ | ·0130 958 | ·0107 303 | ·0087 718 | ·0071 549 |
| ·74 | ·0270 468 | ·0225 590 | ·0187 697 | ·0155 800 | ·0129 031 | ·0106 627 |
| ·75 | ·0372 936 | ·0315 015− | ·0265 447 | ·0223 160 | ·0187 191 | ·0156 681 |
| ·76 | ·0507 561 | ·0434 077 | ·0370 354 | ·0315 265+ | ·0267 781 | ·0226 968 |
| ·77 | ·0681 685+ | ·0590 110 | ·0509 653 | ·0439 183 | ·0377 642 | ·0324 050− |
| ·78 | ·0903 261 | ·0791 255+ | ·0691 570 | ·0603 124 | ·0524 882 | ·0455 861 |
| ·79 | ·1180 458 | ·1046 134 | ·0925 052 | ·0816 246 | ·0718 758 | ·0631 657 |
| ·80 | ·1521 080 | ·1363 313 | ·1219 301 | ·1088 250+ | ·0969 345+ | ·0861 765− |
| ·81 | ·1931 780 | ·1750 551 | ·1583 053 | ·1428 725+ | ·1286 950+ | ·1157 074 |
| ·82 | ·2417 081 | ·2213 800 | ·2023 602 | ·1846 195− | ·1681 206 | ·1528 199 |
| ·83 | ·2978 246 | ·2756 025+ | ·2545 571 | ·2346 875+ | ·2159 835+ | ·1984 262 |
| ·84 | ·3612 116 | ·3375 910 | ·3149 515− | ·2933 194 | ·2727 110 | ·2531 329 |
| ·85 | ·4310 063 | ·4066 612 | ·3830 495+ | ·3602 200 | ·3382 118 | ·3170 551 |
| ·86 | ·5057 268 | ·4814 789 | ·4576 850− | ·4344 086 | ·4117 055+ | ·3896 241 |
| ·87 | ·5832 598 | ·5600 183 | ·5369 468 | ·5141 145+ | ·4915 858 | ·4694 192 |
| ·88 | ·6609 322 | ·6396 058 | ·6181 918 | ·5967 564 | ·5753 629 | ·5540 712 |
| ·89 | ·7356 889 | ·7170 778 | ·6981 780 | ·6790 443 | ·6597 309 | ·6402 911 |
| ·90 | ·8043 817 | ·7890 668 | ·7733 396 | ·7572 388 | ·7408 043 | ·7240 762 |
| ·91 | ·8641 541 | ·8524 038 | ·8402 031 | ·8275 738 | ·8145 395− | ·8011 250+ |
| ·92 | ·9128 707 | ·9045 928 | ·8959 030 | ·8868 093 | ·8773 207 | ·8674 481 |
| ·93 | ·9495 083 | ·9442 676 | ·9387 064 | ·9328 234 | ·9266 182 | ·9200 917 |
| ·94 | ·9743 951 | ·9715 062 | ·9684 076 | ·9650 944 | ·9615 623 | ·9578 072 |
| ·95 | ·9891 850+ | ·9878 647 | ·9864 334 | ·9848 867 | ·9832 203 | ·9814 299 |
| ·96 | ·9964 965+ | ·9960 357 | ·9955 308 | ·9949 795+ | ·9943 793 | ·9937 276 |
| ·97 | ·9992 541 | ·9991 488 | ·9990 322 | ·9989 036 | ·9987 621 | ·9986 069 |
| ·98 | ·9999 262 | ·9999 150+ | ·9999 026 | ·9998 887 | ·9998 732 | ·9998 561 |
| ·99 | ·9999 989 | ·9999 987 | ·9999 985− | ·9999 983 | ·9999 980 | ·9999 977 |
| 1·00 | 1·0000 000 | 1·0000 000 | 1·0000 000 | 1·0000 000 | 1·0000 000 | 1·0000 000 |

$x = \cdot 53$ to $1\cdot 00$ $q = 6\cdot 5$ $p = 44$ to $50$

| | $p = 44$ | $p = 45$ | $p = 46$ | $p = 47$ | $p = 48$ | $p = 49$ | $p = 50$ |
|---|---|---|---|---|---|---|---|
| $B(p, q) =$ | ·4053 7968 × $\frac{1}{10^8}$ | ·3532 0210 × $\frac{1}{10^8}$ | ·3086 2319 × $\frac{1}{10^8}$ | ·2704 1270 × $\frac{1}{10^8}$ | ·2375 5882 × $\frac{1}{10^8}$ | ·2092 2612 × $\frac{1}{10^8}$ | ·1847 2216 × $\frac{1}{10^8}$ |
| $x$ | | | | | | | |
| ·53 | ·0000 001 | | | | | | |
| ·54 | ·0000 002 | ·0000 001 | ·0000 001 | | | | |
| ·55 | ·0000 003 | ·0000 002 | ·0000 001 | ·0000 001 | | | |
| ·56 | ·0000 006 | ·0000 004 | ·0000 002 | ·0000 001 | ·0000 001 | ·0000 001 | |
| ·57 | ·0000 012 | ·0000 007 | ·0000 005− | ·0000 003 | ·0000 002 | ·0000 001 | ·0000 001 |
| ·58 | ·0000 022 | ·0000 014 | ·0000 009 | ·0000 006 | ·0000 004 | ·0000 002 | ·0000 002 |
| ·59 | ·0000 041 | ·0000 027 | ·0000 018 | ·0000 012 | ·0000 008 | ·0000 005+ | ·0000 003 |
| ·60 | ·0000 076 | ·0000 051 | ·0000 034 | ·0000 023 | ·0000 015+ | ·0000 010 | ·0000 007 |
| ·61 | ·0000 139 | ·0000 094 | ·0000 064 | ·0000 044 | ·0000 030 | ·0000 020 | ·0000 013 |
| ·62 | ·0000 248 | ·0000 172 | ·0000 119 | ·0000 082 | ·0000 056 | ·0000 039 | ·0000 027 |
| ·63 | ·0000 437 | ·0000 308 | ·0000 216 | ·0000 151 | ·0000 106 | ·0000 074 | ·0000 051 |
| ·64 | ·0000 760 | ·0000 543 | ·0000 387 | ·0000 275+ | ·0000 196 | ·0000 139 | ·0000 098 |
| ·65 | ·0001 302 | ·0000 944 | ·0000 684 | ·0000 494 | ·0000 356 | ·0000 256 | ·0000 184 |
| ·66 | ·0002 199 | ·0001 619 | ·0001 190 | ·0000 872 | ·0000 638 | ·0000 466 | ·0000 340 |
| ·67 | ·0003 661 | ·0002 736 | ·0002 040 | ·0001 518 | ·0001 128 | ·0000 836 | ·0000 619 |
| ·68 | ·0006 012 | ·0004 558 | ·0003 449 | ·0002 604 | ·0001 963 | ·0001 476 | ·0001 109 |
| ·69 | ·0009 736 | ·0007 488 | ·0005 747 | ·0004 402 | ·0003 365− | ·0002 568 | ·0001 956 |
| ·70 | ·0015 548 | ·0012 126 | ·0009 438 | ·0007 331 | ·0005 684 | ·0004 399 | ·0003 398 |
| ·71 | ·0024 487 | ·0019 362 | ·0015 279 | ·0012 034 | ·0009 460 | ·0007 423 | ·0005 814 |
| ·72 | ·0038 029 | ·0030 481 | ·0024 382 | ·0019 465+ | ·0015 511 | ·0012 338 | ·0009 797 |
| ·73 | ·0058 236 | ·0047 303 | ·0038 346 | ·0031 026 | ·0025 056 | ·0020 200 | ·0016 256 |
| ·74 | ·0087 928 | ·0072 361 | ·0059 433 | ·0048 723 | ·0039 870 | ·0032 568 | ·0026 558 |
| ·75 | ·0130 871 | ·0109 095− | ·0090 766 | ·0075 376 | ·0062 483 | ·0051 705− | ·0042 714 |
| ·76 | ·0191 982 | ·0162 069 | ·0136 557 | ·0114 850− | ·0096 421 | ·0080 811 | ·0067 615+ |
| ·77 | ·0277 505+ | ·0237 187 | ·0202 347 | ·0172 312 | ·0146 480 | ·0124 309 | ·0105 322 |
| ·78 | ·0395 139 | ·0341 855− | ·0295 214 | ·0254 484 | ·0218 997 | ·0188 145+ | ·0161 380 |
| ·79 | ·0554 046 | ·0485 069 | ·0423 917 | ·0369 831 | ·0322 103 | ·0280 078 | ·0243 151 |
| ·80 | ·0764 692 | ·0677 326 | ·0598 892 | ·0528 641 | ·0465 865− | ·0409 889 | ·0360 081 |
| ·81 | ·1038 419 | ·0930 295− | ·0832 009 | ·0742 879 | ·0662 236 | ·0589 430 | ·0523 838 |
| ·82 | ·1386 684 | ·1256 138 | ·1136 006 | ·1025 719 | ·0924 702 | ·0832 376 | ·0748 171 |
| ·83 | ·1819 897 | ·1666 422 | ·1523 468 | ·1390 629 | ·1267 471 | ·1153 537 | ·1048 357 |
| ·84 | ·2345 838 | ·2170 550− | ·2005 311 | ·1849 915− | ·1704 107 | ·1567 594 | ·1440 052 |
| ·85 | ·2967 718 | ·2773 758 | ·2588 742 | ·2412 676 | ·2245 506 | ·2087 130 | ·1937 400 |
| ·86 | ·3682 051 | ·3474 823 | ·3274 825− | ·3082 261 | ·2897 277 | ·2719 962 | ·2550 354 |
| ·87 | ·4476 675+ | ·4263 781 | ·4055 925− | ·3853 469 | ·3656 720 | ·3465 935+ | ·3281 322 |
| ·88 | ·5329 380 | ·5120 161 | ·4913 545+ | ·4709 984 | ·4509 888 | ·4313 627 | ·4121 532 |
| ·89 | ·6207 767 | ·6012 379 | ·5817 230 | ·5622 780 | ·5429 469 | ·5237 708 | ·5047 884 |
| ·90 | ·7070 953 | ·6899 022 | ·6725 371 | ·6550 398 | ·6374 496 | ·6198 043 | ·6021 410 |
| ·91 | ·7873 564 | ·7732 608 | ·7588 661 | ·7442 007 | ·7292 935+ | ·7141 735+ | ·6988 700 |
| ·92 | ·8572 035− | ·8466 003 | ·8356 532 | ·8243 776 | ·8127 902 | ·8009 082 | ·7887 498 |
| ·93 | ·9132 457 | ·9060 831 | ·8986 077 | ·8908 243 | ·8827 387 | ·8743 575+ | ·8656 881 |
| ·94 | ·9538 260 | ·9496 159 | ·9451 747 | ·9405 008 | ·9355 933 | ·9304 515+ | ·9250 758 |
| ·95 | ·9795 114 | ·9774 611 | ·9752 751 | ·9729 502 | ·9704 831 | ·9678 708 | ·9651 106 |
| ·96 | ·9930 219 | ·9922 598 | ·9914 388 | ·9905 564 | ·9896 101 | ·9885 976 | ·9875 165− |
| ·97 | ·9984 370 | ·9982 517 | ·9980 500− | ·9978 309 | ·9975 935+ | ·9973 369 | ·9970 600 |
| ·98 | ·9998 372 | ·9998 163 | ·9997 934 | ·9997 682 | ·9997 407 | ·9997 106 | ·9996 778 |
| ·99 | ·9999 974 | ·9999 970 | ·9999 966 | ·9999 962 | ·9999 957 | ·9999 952 | ·9999 946 |
| 1·00 | 1·0000 000 | 1·0000 000 | 1·0000 000 | 1·0000 000 | 1·0000 000 | 1·0000 000 | 1·0000 000 |

$x$ = ·04 to ·60 $q = 7$ $p$ = 7 to 9·5

| $x$ | $p = 7$ | $p = 7·5$ | $p = 8$ | $p = 8·5$ | $p = 9$ | $p = 9·5$ |
|---|---|---|---|---|---|---|
| $B(p, q)$ = | ·8325 0083 × $\frac{1}{10^4}$ | ·5834 4212 × $\frac{1}{10^4}$ | ·4162 5042 × $\frac{1}{10^4}$ | ·3017 8041 × $\frac{1}{10^4}$ | ·2220 0022 × $\frac{1}{10^4}$ | ·1654 9248 × $\frac{1}{10^4}$ |
| ·04 | ·0000 002 | ·0000 001 | | | | |
| ·05 | ·0000 010 | ·0000 003 | ·0000 001 | | | |
| ·06 | ·0000 035− | ·0000 011 | ·0000 004 | ·0000 001 | | |
| ·07 | ·0000 097 | ·0000 034 | ·0000 012 | ·0000 004 | ·0000 001 | |
| ·08 | ·0000 233 | ·0000 087 | ·0000 032 | ·0000 012 | ·0000 004 | ·0000 002 |
| ·09 | ·0000 503 | ·0000 200 | ·0000 078 | ·0000 030 | ·0000 012 | ·0000 004 |
| ·10 | ·0000 993 | ·0000 416 | ·0000 172 | ·0000 070 | ·0000 028 | ·0000 011 |
| ·11 | ·0001 827 | ·0000 802 | ·0000 348 | ·0000 149 | ·0000 063 | ·0000 027 |
| ·12 | ·0003 171 | ·0001 454 | ·0000 658 | ·0000 294 | ·0000 130 | ·0000 057 |
| ·13 | ·0005 239 | ·0002 498 | ·0001 176 | ·0000 548 | ·0000 252 | ·0000 115+ |
| ·14 | ·0008 298 | ·0004 104 | ·0002 004 | ·0000 968 | ·0000 462 | ·0000 219 |
| ·15 | ·0012 675+ | ·0006 485− | ·0003 276 | ·0001 637 | ·0000 809 | ·0000 396 |
| ·16 | ·0018 758 | ·0009 904 | ·0005 165+ | ·0002 663 | ·0001 359 | ·0000 687 |
| ·17 | ·0026 995− | ·0014 682 | ·0007 887 | ·0004 190 | ·0002 203 | ·0001 147 |
| ·18 | ·0037 897 | ·0021 193 | ·0011 708 | ·0006 396 | ·0003 458 | ·0001 852 |
| ·19 | ·0052 035+ | ·0029 875− | ·0016 945− | ·0009 505− | ·0005 277 | ·0002 903 |
| ·20 | ·0070 036 | ·0041 222 | ·0023 972 | ·0013 788 | ·0007 850− | ·0004 428 |
| ·21 | ·0092 574 | ·0055 789 | ·0033 221 | ·0019 567 | ·0011 409 | ·0006 590 |
| ·22 | ·0120 368 | ·0074 185− | ·0045 182 | ·0027 220 | ·0016 235+ | ·0009 594 |
| ·23 | ·0154 169 | ·0097 070 | ·0060 403 | ·0037 183 | ·0022 662 | ·0013 686 |
| ·24 | ·0194 752 | ·0125 150− | ·0079 489 | ·0049 949 | ·0031 078 | ·0019 161 |
| ·25 | ·0242 901 | ·0159 165− | ·0103 095+ | ·0066 071 | ·0041 930 | ·0026 369 |
| ·26 | ·0299 400 | ·0199 884 | ·0131 924 | ·0086 157 | ·0055 723 | ·0035 716 |
| ·27 | ·0365 018 | ·0248 092 | ·0166 717 | ·0110 869 | ·0073 020 | ·0047 665+ |
| ·28 | ·0440 494 | ·0304 579 | ·0208 246 | ·0140 914 | ·0094 445− | ·0062 741 |
| ·29 | ·0526 525− | ·0370 124 | ·0257 302 | ·0177 045+ | ·0120 671 | ·0081 528 |
| ·30 | ·0623 752 | ·0445 486 | ·0314 685+ | ·0220 043 | ·0152 425+ | ·0104 669 |
| ·31 | ·0732 747 | ·0531 388 | ·0381 192 | ·0270 714 | ·0190 474 | ·0132 864 |
| ·32 | ·0853 996 | ·0628 501 | ·0457 600 | ·0329 874 | ·0235 618 | ·0166 860 |
| ·33 | ·0987 895+ | ·0737 433 | ·0544 656 | ·0398 339 | ·0288 685+ | ·0207 450− |
| ·34 | ·1134 732 | ·0858 713 | ·0643 058 | ·0476 909 | ·0350 512 | ·0255 462 |
| ·35 | ·1294 682 | ·0992 779 | ·0753 446 | ·0566 352 | ·0421 938 | ·0311 750− |
| ·36 | ·1467 798 | ·1139 967 | ·0876 380 | ·0667 393 | ·0503 786 | ·0377 177 |
| ·37 | ·1654 004 | ·1300 498 | ·1012 332 | ·0780 696 | ·0596 849 | ·0452 610 |
| ·38 | ·1853 096 | ·1474 471 | ·1161 672 | ·0906 848 | ·0701 875− | ·0538 895+ |
| ·39 | ·2064 732 | ·1661 855− | ·1324 652 | ·1046 346 | ·0819 548 | ·0636 851 |
| ·40 | ·2288 440 | ·1862 481 | ·1501 401 | ·1199 579 | ·0950 474 | ·0747 245+ |
| ·41 | ·2523 614 | ·2076 043 | ·1691 911 | ·1366 818 | ·1095 164 | ·0870 778 |
| ·42 | ·2769 524 | ·2302 092 | ·1896 034 | ·1548 205− | ·1254 018 | ·1008 066 |
| ·43 | ·3025 317 | ·2540 040 | ·2113 473 | ·1743 737 | ·1427 310 | ·1159 625+ |
| ·44 | ·3290 026 | ·2789 161 | ·2343 784 | ·1953 265+ | ·1615 177 | ·1325 852 |
| ·45 | ·3562 582 | ·3048 598 | ·2586 371 | ·2176 483 | ·1817 605− | ·1507 007 |
| ·46 | ·3841 825− | ·3317 369 | ·2840 492 | ·2412 926 | ·2034 419 | ·1703 204 |
| ·47 | ·4126 515+ | ·3594 375+ | ·3105 262 | ·2661 967 | ·2265 281 | ·1914 396 |
| ·48 | ·4415 349 | ·3878 418 | ·3379 660 | ·2922 824 | ·2509 681 | ·2140 361 |
| ·49 | ·4706 973 | ·4168 207 | ·3662 539 | ·3194 557 | ·2766 936 | ·2380 701 |
| ·50 | ·5000 000$^e$ | ·4462 376 | ·3952 637 | ·3476 083 | ·3036 194 | ·2634 833 |
| ·51 | ·5293 027 | ·4759 500+ | ·4248 593 | ·3766 179 | ·3316 436 | ·2901 989 |
| ·52 | ·5584 651 | ·5058 114 | ·4548 962 | ·4063 499 | ·3606 485+ | ·3181 218 |
| ·53 | ·5873 485− | ·5356 726 | ·4852 232 | ·4366 590 | ·3905 019 | ·3471 390 |
| ·54 | ·6158 175+ | ·5653 845− | ·5156 842 | ·4673 906 | ·4210 583 | ·3771 209 |
| ·55 | ·6437 418 | ·5947 988 | ·5461 207 | ·4983 829 | ·4521 604 | ·4079 221 |
| ·56 | ·6709 974 | ·6237 708 | ·5763 732 | ·5294 691 | ·4836 415+ | ·4393 831 |
| ·57 | ·6974 683 | ·6521 610 | ·6062 839 | ·5604 795+ | ·5153 275+ | ·4713 328 |
| ·58 | ·7230 476 | ·6798 365+ | ·6356 985− | ·5912 442 | ·5470 392 | ·5035 900 |
| ·59 | ·7476 386 | ·7066 731 | ·6644 682 | ·6215 949 | ·5785 949 | ·5359 663 |
| ·60 | ·7711 560 | ·7325 564 | ·6924 522 | ·6513 678 | ·6098 132 | ·5682 688 |

$x$ = ·61 to ·98 | $q = 7$ | $p$ = 7 to 9·5

| | $p = 7$ | $p = 7·5$ | $p = 8$ | $p = 8·5$ | $p = 9$ | $p = 9·5$ |
|---|---|---|---|---|---|---|
| $B(p, q)$ = | ·8325 0083 × $\frac{1}{10^4}$ | ·5834 4212 × $\frac{1}{10^4}$ | ·4162 5042 × $\frac{1}{10^4}$ | ·3017 8041 × $\frac{1}{10^4}$ | ·2220 0022 × $\frac{1}{10^4}$ | ·1654 9248 × $\frac{1}{10^4}$ |
| $x$ | | | | | | |
| ·61 | ·7935 268 | ·7573 837 | ·7195 189 | ·6804 056 | ·6405 154 | ·6003 031 |
| ·62 | ·8146 904 | ·7810 645− | ·7455 481 | ·7085 601 | ·6705 286 | ·6318 762 |
| ·63 | ·8345 996 | ·8035 218 | ·7704 324 | ·7356 941 | ·6996 880 | ·6627 993 |
| ·64 | ·8532 202 | ·8246 932 | ·7940 784 | ·7616 833 | ·7278 396 | ·6928 914 |
| ·65 | ·8705 318 | ·8445 303 | ·8164 081 | ·7864 181 | ·7548 425− | ·7219 819 |
| ·66 | ·8865 268 | ·8630 004 | ·8373 594 | ·8098 051 | ·7805 710 | ·7499 135+ |
| ·67 | ·9012 105− | ·8800 851 | ·8568 865+ | ·8317 682 | ·8049 167 | ·7765 447 |
| ·68 | ·9146 004 | ·8957 811 | ·8749 607 | ·8522 492 | ·8277 896 | ·8017 522 |
| ·69 | ·9267 253 | ·9100 992 | ·8915 699 | ·8712 089 | ·8491 197 | ·8254 328 |
| ·70 | ·9376 248 | ·9230 637 | ·9067 181 | ·8886 267 | ·8688 574 | ·8475 049 |
| ·71 | ·9473 475+ | ·9347 113 | ·9204 253 | ·9045 004 | ·8869 743 | ·8679 096 |
| ·72 | ·9559 506 | ·9450 905+ | ·9327 259 | ·9188 458 | ·9034 627 | ·8866 112 |
| ·73 | ·9634 982 | ·9542 596 | ·9436 682 | ·9316 960 | ·9183 353 | ·9035 977 |
| ·74 | ·9700 600 | ·9622 859 | ·9533 123 | ·9430 996 | ·9316 242 | ·9188 796 |
| ·75 | ·9757 099 | ·9692 436 | ·9617 292 | ·9531 193 | ·9433 797 | ·9324 897 |
| ·76 | ·9805 248 | ·9752 127 | ·9689 984 | ·9618 307 | ·9536 684 | ·9444 813 |
| ·77 | ·9845 831 | ·9802 770 | ·9752 064 | ·9693 194 | ·9625 714 | ·9549 262 |
| ·78 | ·9879 632 | ·9845 225− | ·9804 446 | ·9756 793 | ·9701 817 | ·9639 126 |
| ·79 | ·9907 426 | ·9880 359 | ·9848 073 | ·9810 104 | ·9766 018 | ·9715 424 |
| ·80 | ·9929 964 | ·9909 030 | ·9883 901 | ·9854 161 | ·9819 412 | ·9779 281 |
| ·81 | ·9947 965− | ·9932 071 | ·9912 874 | ·9890 013 | ·9863 134 | ·9831 898 |
| ·82 | ·9962 103 | ·9950 281 | ·9935 914 | ·9918 698 | ·9898 332 | ·9874 519 |
| ·83 | ·9973 005+ | ·9964 409 | ·9953 898 | ·9941 226 | ·9926 144 | ·9908 401 |
| ·84 | ·9981 242 | ·9975 147 | ·9967 649 | ·9958 556 | ·9947 668 | ·9934 782 |
| ·85 | ·9987 325− | ·9983 124 | ·9977 925+ | ·9971 583 | ·9963 944 | ·9954 850− |
| ·86 | ·9991 702 | ·9988 898 | ·9985 408 | ·9981 125+ | ·9975 936 | ·9969 722 |
| ·87 | ·9994 761 | ·9992 958 | ·9990 699 | ·9987 912 | ·9984 515− | ·9980 423 |
| ·88 | ·9996 829 | ·9995 717 | ·9994 316 | ·9992 578 | ·9990 446 | ·9987 865− |
| ·89 | ·9998 173 | ·9997 520 | ·9996 694 | ·9995 662 | ·9994 390 | ·9992 841 |
| ·90 | ·9999 007 | ·9998 646 | ·9998 186 | ·9997 609 | ·9996 894 | ·9996 017 |
| ·91 | ·9999 497 | ·9999 311 | ·9999 073 | ·9998 773 | ·9998 398 | ·9997 936 |
| ·92 | ·9999 767 | ·9999 679 | ·9999 566 | ·9999 423 | ·9999 243 | ·9999 020 |
| ·93 | ·9999 903 | ·9999 866 | ·9999 818 | ·9999 757 | ·9999 680 | ·9999 584 |
| ·94 | ·9999 965+ | ·9999 952 | ·9999 934 | ·9999 911 | ·9999 883 | ·9999 847 |
| ·95 | ·9999 990 | ·9999 986 | ·9999 980 | ·9999 974 | ·9999 965− | ·9999 954 |
| ·96 | ·9999 998 | ·9999 997 | ·9999 996 | ·9999 994 | ·9999 992 | ·9999 990 |
| ·97 | 1·0000 000 | 1·0000 000 | ·9999 999 | ·9999 999 | ·9999 999 | ·9999 998 |
| ·98 | | | 1·0000 000 | 1·0000 000 | 1·0000 000 | 1·0000 000 |

$x$ = ·08 to ·70 $q = 7$ $p$ = 10 to 14

| $x$ | $p = 10$ | $p = 10{\cdot}5$ | $p = 11$ | $p = 12$ | $p = 13$ | $p = 14$ |
|---|---|---|---|---|---|---|
| $B(p, q) =$ | ·1248 7512 $\times \frac{1}{10^4}$ | ·9528 3550+ $\times \frac{1}{10^5}$ | ·7345 5956 $\times \frac{1}{10^5}$ | ·4488 9751 $\times \frac{1}{10^5}$ | ·2835 1422 $\times \frac{1}{10^5}$ | ·1842 8424 $\times \frac{1}{10^5}$ |
| ·08 | ·0000 001 | | | | | |
| ·09 | ·0000 002 | ·0000 001 | | | | |
| ·10 | ·0000 005− | ·0000 002 | ·0000 001 | | | |
| ·11 | ·0000 011 | ·0000 005− | ·0000 002 | | | |
| ·12 | ·0000 025− | ·0000 011 | ·0000 005− | ·0000 001 | | |
| ·13 | ·0000 052 | ·0000 023 | ·0000 010 | ·0000 002 | | |
| ·14 | ·0000 103 | ·0000 048 | ·0000 022 | ·0000 005− | ·0000 001 | |
| ·15 | ·0000 192 | ·0000 093 | ·0000 044 | ·0000 010 | ·0000 002 | |
| ·16 | ·0000 344 | ·0000 171 | ·0000 084 | ·0000 020 | ·0000 005− | ·0000 001 |
| ·17 | ·0000 592 | ·0000 303 | ·0000 154 | ·0000 039 | ·0000 010 | ·0000 002 |
| ·18 | ·0000 983 | ·0000 518 | ·0000 271 | ·0000 072 | ·0000 019 | ·0000 005− |
| ·19 | ·0001 583 | ·0000 856 | ·0000 460 | ·0000 130 | ·0000 036 | ·0000 010 |
| ·20 | ·0002 476 | ·0001 373 | ·0000 756 | ·0000 225− | ·0000 065+ | ·0000 018 |
| ·21 | ·0003 774 | ·0002 144 | ·0001 209 | ·0000 377 | ·0000 115− | ·0000 034 |
| ·22 | ·0005 621 | ·0003 267 | ·0001 885+ | ·0000 615− | ·0000 196 | ·0000 061 |
| ·23 | ·0008 195− | ·0004 868 | ·0002 871 | ·0000 978 | ·0000 325+ | ·0000 106 |
| ·24 | ·0011 714 | ·0007 105+ | ·0004 278 | ·0001 520 | ·0000 527 | ·0000 179 |
| ·25 | ·0016 445− | ·0010 175+ | ·0006 250+ | ·0002 312 | ·0000 835− | ·0000 295+ |
| ·26 | ·0022 702 | ·0014 318 | ·0008 965+ | ·0003 446 | ·0001 293 | ·0000 475+ |
| ·27 | ·0030 857 | ·0019 822 | ·0012 642 | ·0005 042 | ·0001 963 | ·0000 749 |
| ·28 | ·0041 338 | ·0027 028 | ·0017 546 | ·0007 250+ | ·0002 926 | ·0001 156 |
| ·29 | ·0054 633 | ·0036 333 | ·0023 992 | ·0010 259 | ·0004 285− | ·0001 753 |
| ·30 | ·0071 295+ | ·0048 197 | ·0032 353 | ·0014 298 | ·0006 173 | ·0002 610 |
| ·31 | ·0091 936 | ·0063 140 | ·0043 061 | ·0019 646 | ·0008 757 | ·0003 824 |
| ·32 | ·0117 228 | ·0081 749 | ·0056 612 | ·0026 635− | ·0012 246 | ·0005 516 |
| ·33 | ·0147 901 | ·0104 671 | ·0073 567 | ·0035 657 | ·0016 891 | ·0007 840 |
| ·34 | ·0184 736 | ·0132 618 | ·0094 554 | ·0047 167 | ·0023 000 | ·0010 991 |
| ·35 | ·0228 559 | ·0166 359 | ·0120 267 | ·0061 690 | ·0030 938 | ·0015 207 |
| ·36 | ·0280 230 | ·0206 714 | ·0151 462 | ·0079 820 | ·0041 133 | ·0020 778 |
| ·37 | ·0340 635+ | ·0254 549 | ·0188 956 | ·0102 223 | ·0054 087 | ·0028 056 |
| ·38 | ·0410 670 | ·0310 765− | ·0233 618 | ·0129 640 | ·0070 372 | ·0037 456 |
| ·39 | ·0491 230 | ·0376 284 | ·0286 360 | ·0162 879 | ·0090 643 | ·0049 468 |
| ·40 | ·0583 189 | ·0452 038 | ·0348 127 | ·0202 816 | ·0115 629 | ·0064 659 |
| ·41 | ·0687 387 | ·0538 953 | ·0419 885+ | ·0250 386 | ·0146 145− | ·0083 680 |
| ·42 | ·0804 607 | ·0637 929 | ·0502 603 | ·0306 575− | ·0183 077 | ·0107 269 |
| ·43 | ·0935 561 | ·0749 825− | ·0597 237 | ·0372 405+ | ·0227 389 | ·0136 252 |
| ·44 | ·1080 866 | ·0875 435− | ·0704 711 | ·0448 925+ | ·0280 107 | ·0171 544 |
| ·45 | ·1241 030 | ·1015 471 | ·0825 896 | ·0537 190 | ·0342 313 | ·0214 144 |
| ·46 | ·1416 430 | ·1170 540 | ·0961 590 | ·0638 240 | ·0415 129 | ·0265 129 |
| ·47 | ·1607 297 | ·1341 126 | ·1112 492 | ·0753 084 | ·0499 701 | ·0325 647 |
| ·48 | ·1813 698 | ·1527 568 | ·1279 183 | ·0882 670 | ·0597 181 | ·0396 899 |
| ·49 | ·2035 528 | ·1730 044 | ·1462 103 | ·1027 865− | ·0708 699 | ·0480 128 |
| ·50 | ·2272 491 | ·1948 550− | ·1661 530 | ·1189 423 | ·0835 342 | ·0576 591 |
| ·51 | ·2524 102 | ·2182 892 | ·1877 557 | ·1367 963 | ·0978 123 | ·0687 542 |
| ·52 | ·2789 672 | ·2432 672 | ·2110 083 | ·1563 941 | ·1137 950− | ·0814 197 |
| ·53 | ·3068 315+ | ·2697 278 | ·2358 790 | ·1777 625− | ·1315 598 | ·0957 705+ |
| ·54 | ·3358 949 | ·2975 886 | ·2623 138 | ·2009 070 | ·1511 675− | ·1119 115+ |
| ·55 | ·3660 301 | ·3267 454 | ·2902 355− | ·2258 100 | ·1726 590 | ·1299 338 |
| ·56 | ·3970 919 | ·3570 730 | ·3195 435− | ·2524 288 | ·1960 525+ | ·1499 107 |
| ·57 | ·4289 189 | ·3884 263 | ·3501 143 | ·2806 945+ | ·2213 407 | ·1718 943 |
| ·58 | ·4613 352 | ·4206 413 | ·3818 019 | ·3105 111 | ·2484 882 | ·1959 118 |
| ·59 | ·4941 528 | ·4535 374 | ·4144 394 | ·3417 553 | ·2774 299 | ·2219 615+ |
| ·60 | ·5271 741 | ·4869 193 | ·4478 406 | ·3742 769 | ·3080 695− | ·2500 107 |
| ·61 | ·5601 952 | ·5205 801 | ·4818 027 | ·4078 999 | ·3402 788 | ·2799 921 |
| ·62 | ·5930 085− | ·5543 043 | ·5161 085+ | ·4424 244 | ·3738 981 | ·3118 028 |
| ·63 | ·6254 064 | ·5878 711 | ·5505 306 | ·4776 288 | ·4087 365+ | ·3453 027 |
| ·64 | ·6571 848 | ·6210 582 | ·5848 344 | ·5132 732 | ·4445 745− | ·3803 147 |
| ·65 | ·6881 463 | ·6536 457 | ·6187 824 | ·5491 031 | ·4811 658 | ·4166 254 |
| ·66 | ·7181 038 | ·6854 197 | ·6521 385− | ·5848 538 | ·5182 420 | ·4539 872 |
| ·67 | ·7468 837 | ·7161 768 | ·6846 723 | ·6202 553 | ·5555 162 | ·4921 217 |
| ·68 | ·7743 289 | ·7457 271 | ·7161 638 | ·6550 375− | ·5926 886 | ·5307 235− |
| ·69 | ·8003 019 | ·7738 985+ | ·7464 071 | ·6889 357 | ·6294 527 | ·5694 665− |
| ·70 | ·8246 866 | ·8005 395− | ·7752 153 | ·7216 964 | ·6655 015+ | ·6080 098 |

$x = \cdot 71$ to $\cdot 99$ $q = 7$ $p = 10$ to $14$

| $x$ | $p = 10$ | $p = 10 \cdot 5$ | $p = 11$ | $p = 12$ | $p = 13$ | $p = 14$ |
|---|---|---|---|---|---|---|
| $B(p, q) =$ | $\cdot 1248\ 7512 \times \frac{1}{10^4}$ | $\cdot 9528\ 3550 \times \frac{1}{10^5}$ | $\cdot 7345\ 5956 \times \frac{1}{10^5}$ | $\cdot 4488\ 9751 \times \frac{1}{10^5}$ | $\cdot 2835\ 1422 \times \frac{1}{10^5}$ | $\cdot 1842\ 8424 \times \frac{1}{10^5}$ |
| ·71 | ·8473 907 | ·8255 220 | ·8024 238 | ·7530 828 | ·7005 346 | ·6460 057 |
| ·72 | ·8683 469 | ·8487 440 | ·8278 935− | ·7828 799 | ·7342 651 | ·6831 075+ |
| ·73 | ·8875 136 | ·8701 311 | ·8515 139 | ·8108 997 | ·7664 271 | ·7189 783 |
| ·74 | ·9048 755− | ·8896 372 | ·8732 050− | ·8369 855− | ·7967 818 | ·7533 000 |
| ·75 | ·9204 427 | ·9072 456 | ·8929 184 | ·8610 152 | ·8251 241 | ·7857 819 |
| ·76 | ·9342 504 | ·9229 681 | ·9106 381 | ·8829 043 | ·8512 879 | ·8161 693 |
| ·77 | ·9463 564 | ·9368 439 | ·9263 796 | ·9026 072 | ·8751 501 | ·8442 503 |
| ·78 | ·9568 399 | ·9489 380 | ·9401 893 | ·9201 178 | ·8966 342 | ·8698 628 |
| ·79 | ·9657 978 | ·9593 388 | ·9521 416 | ·9354 686 | ·9157 112 | ·8928 989 |
| ·80 | ·9733 427 | ·9681 543 | ·9623 366 | ·9487 290 | ·9323 999 | ·9133 075− |
| ·81 | ·9795 984 | ·9755 094 | ·9708 958 | ·9600 017 | ·9467 654 | ·9310 956 |
| ·82 | ·9846 971 | ·9815 412 | ·9779 585− | ·9694 188 | ·9589 151 | ·9463 267 |
| ·83 | ·9887 751 | ·9863 950+ | ·9836 765+ | ·9771 365− | ·9689 941 | ·9591 168 |
| ·84 | ·9919 693 | ·9902 199 | ·9882 096 | ·9833 288 | ·9771 790 | ·9696 289 |
| ·85 | ·9944 137 | ·9931 643 | ·9917 200 | ·9881 815− | ·9836 698 | ·9780 649 |
| ·86 | ·9962 360 | ·9953 722 | ·9943 679 | ·9918 850− | ·9886 820 | ·9846 562 |
| ·87 | ·9975 547 | ·9969 793 | ·9963 064 | ·9946 280 | ·9924 377 | ·9896 526 |
| ·88 | ·9984 771 | ·9981 099 | ·9976 779 | ·9965 911 | ·9951 565+ | ·9933 114 |
| ·89 | ·9990 973 | ·9988 744 | ·9986 107 | ·9979 414 | ·9970 479 | ·9958 856 |
| ·90 | ·9994 955− | ·9993 679 | ·9992 162 | ·9988 279 | ·9983 036 | ·9976 139 |
| ·91 | ·9997 373 | ·9996 694 | ·9995 882 | ·9993 784 | ·9990 921 | ·9987 113 |
| ·92 | ·9998 747 | ·9998 416 | ·9998 018 | ·9996 981 | ·9995 549 | ·9993 625− |
| ·93 | ·9999 465+ | ·9999 321 | ·9999 146 | ·9998 687 | ·9998 047 | ·9997 178 |
| ·94 | ·9999 803 | ·9999 748 | ·9999 682 | ·9999 506 | ·9999 259 | ·9998 919 |
| ·95 | ·9999 940 | ·9999 923 | ·9999 903 | ·9999 848 | ·9999 769 | ·9999 661 |
| ·96 | ·9999 986 | ·9999 982 | ·9999 978 | ·9999 965− | ·9999 946 | ·9999 920 |
| ·97 | ·9999 998 | ·9999 997 | ·9999 997 | ·9999 995− | ·9999 992 | ·9999 988 |
| ·98 | 1·0000 000 | 1·0000 000 | 1·0000 000 | 1·0000 000 | ·9999 999 | ·9999 999 |
| ·99 | | | | | 1·0000 000 | 1·0000 000 |

$x = \cdot 17$ to $\cdot 80$ $\qquad$ $q = 7$ $\qquad$ $p = 15$ to $20$

| | $p = 15$ | $p = 16$ | $p = 17$ | $p = 18$ | $p = 19$ | $p = 20$ |
|---|---|---|---|---|---|---|
| $B(p, q) =$ | ·1228 5616 $\times \frac{1}{10^5}$ | ·8376 5564 $\times \frac{1}{10^6}$ | ·5827 1696 $\times \frac{1}{10^6}$ | ·4127 5785 $\bar{\times} \frac{1}{10^6}$ | ·2971 8565 $\stackrel{+}{\times} \frac{1}{10^6}$ | ·2171 7413 $\times \frac{1}{10^6}$ |
| $x$ | | | | | | |
| ·17 | ·0000 001 | | | | | |
| ·18 | ·0000 001 | | | | | |
| ·19 | ·0000 003 | ·0000 001 | | | | |
| ·20 | ·0000 005+ | ·0000 001 | | | | |
| ·21 | ·0000 010 | ·0000 003 | ·0000 001 | | | |
| ·22 | ·0000 019 | ·0000 006 | ·0000 002 | | | |
| ·23 | ·0000 034 | ·0000 011 | ·0000 003 | ·0000 001 | | |
| ·24 | ·0000 060 | ·0000 020 | ·0000 006 | ·0000 002 | ·0000 001 | |
| ·25 | ·0000 102 | ·0000 035− | ·0000 012 | ·0000 004 | ·0000 001 | |
| ·26 | ·0000 171 | ·0000 061 | ·0000 021 | ·0000 007 | ·0000 002 | ·0000 001 |
| ·27 | ·0000 280 | ·0000 103 | ·0000 037 | ·0000 013 | ·0000 005− | ·0000 002 |
| ·28 | ·0000 449 | ·0000 171 | ·0000 064 | ·0000 024 | ·0000 009 | ·0000 003 |
| ·29 | ·0000 704 | ·0000 278 | ·0000 108 | ·0000 041 | ·0000 016 | ·0000 006 |
| ·30 | ·0001 084 | ·0000 442 | ·0000 178 | ·0000 071 | ·0000 028 | ·0000 011 |
| ·31 | ·0001 640 | ·0000 691 | ·0000 287 | ·0000 118 | ·0000 048 | ·0000 019 |
| ·32 | ·0002 440 | ·0001 061 | ·0000 455− | ·0000 192 | ·0000 080 | ·0000 033 |
| ·33 | ·0003 574 | ·0001 602 | ·0000 708 | ·0000 309 | ·0000 133 | ·0000 057 |
| ·34 | ·0005 158 | ·0002 381 | ·0001 083 | ·0000 486 | ·0000 216 | ·0000 095− |
| ·35 | ·0007 341 | ·0003 487 | ·0001 632 | ·0000 754 | ·0000 344 | ·0000 155+ |
| ·36 | ·0010 310 | ·0005 034 | ·0002 422 | ·0001 150+ | ·0000 540 | ·0000 250+ |
| ·37 | ·0014 297 | ·0007 169 | ·0003 544 | ·0001 728 | ·0000 833 | ·0000 397 |
| ·38 | ·0019 587 | ·0010 081 | ·0005 114 | ·0002 560 | ·0001 267 | ·0000 620 |
| ·39 | ·0026 528 | ·0014 002 | ·0007 286 | ·0003 742 | ·0001 899 | ·0000 953 |
| ·40 | ·0035 533 | ·0019 222 | ·0010 251 | ·0005 396 | ·0002 807 | ·0001 444 |
| ·41 | ·0047 093 | ·0026 092 | ·0014 253 | ·0007 686 | ·0004 096 | ·0002 159 |
| ·42 | ·0061 783 | ·0035 038 | ·0019 593 | ·0010 816 | ·0005 901 | ·0003 185+ |
| ·43 | ·0080 267 | ·0046 565− | ·0026 638 | ·0015 046 | ·0008 399 | ·0004 639 |
| ·44 | ·0103 304 | ·0061 268 | ·0035 837 | ·0020 697 | ·0011 816 | ·0006 674 |
| ·45 | ·0131 749 | ·0079 841 | ·0047 722 | ·0028 168 | ·0016 435+ | ·0009 488 |
| ·46 | ·0166 558 | ·0103 078 | ·0062 927 | ·0037 939 | ·0022 613 | ·0013 337 |
| ·47 | ·0208 782 | ·0131 885− | ·0082 190 | ·0050 590 | ·0030 787 | ·0018 540 |
| ·48 | ·0259 563 | ·0167 274 | ·0106 362 | ·0066 806 | ·0041 490 | ·0025 500+ |
| ·49 | ·0320 128 | ·0210 368 | ·0136 417 | ·0087 392 | ·0055 362 | ·0034 712 |
| ·50 | ·0391 769 | ·0262 394 | ·0173 448 | ·0113 279 | ·0073 166 | ·0046 777 |
| ·51 | ·0475 833 | ·0324 673 | ·0218 672 | ·0145 532 | ·0095 796 | ·0062 421 |
| ·52 | ·0573 695− | ·0398 609 | ·0273 423 | ·0185 350+ | ·0124 287 | ·0082 507 |
| ·53 | ·0686 729 | ·0485 665− | ·0339 139 | ·0234 072 | ·0159 824 | ·0108 046 |
| ·54 | ·0816 284 | ·0587 343 | ·0417 355− | ·0293 163 | ·0203 745+ | ·0140 211 |
| ·55 | ·0963 640 | ·0705 152 | ·0509 671 | ·0364 210 | ·0257 538 | ·0180 342 |
| ·56 | ·1129 972 | ·0840 571 | ·0617 731 | ·0448 898 | ·0322 836 | ·0229 948 |
| ·57 | ·1316 309 | ·0995 007 | ·0743 186 | ·0548 988 | ·0401 397 | ·0290 704 |
| ·58 | ·1523 485− | ·1169 751 | ·0887 648 | ·0666 280 | ·0495 089 | ·0364 443 |
| ·59 | ·1752 097 | ·1365 926 | ·1052 646 | ·0802 574 | ·0605 851 | ·0453 132 |
| ·60 | ·2002 460 | ·1584 437 | ·1239 567 | ·0959 615− | ·0735 653 | ·0558 840 |
| ·61 | ·2274 565− | ·1825 911 | ·1449 602 | ·1139 037 | ·0886 444 | ·0683 705− |
| ·62 | ·2568 041 | ·2090 652 | ·1683 678 | ·1342 298 | ·1060 091 | ·0829 870 |
| ·63 | ·2882 122 | ·2378 585− | ·1942 395− | ·1570 607 | ·1258 306 | ·0999 424 |
| ·64 | ·3215 630 | ·2689 214 | ·2225 968 | ·1824 851 | ·1482 565+ | ·1194 324 |
| ·65 | ·3566 950+ | ·3021 584 | ·2534 162 | ·2105 518 | ·1734 027 | ·1416 304 |
| ·66 | ·3934 042 | ·3374 254 | ·2866 247 | ·2412 626 | ·2013 440 | ·1666 779 |
| ·67 | ·4314 440 | ·3745 284 | ·3220 949 | ·2745 655− | ·2321 058 | ·1946 743 |
| ·68 | ·4705 288 | ·4132 234 | ·3596 429 | ·3103 488 | ·2656 555− | ·2256 667 |
| ·69 | ·5103 371 | ·4532 182 | ·3990 266 | ·3484 372 | ·3018 949 | ·2596 394 |
| ·70 | ·5505 181 | ·4941 763 | ·4399 472 | ·3885 891 | ·3406 549 | ·2965 050− |
| ·71 | ·5906 979 | ·5357 219 | ·4820 516 | ·4304 965+ | ·3816 911 | ·3360 965+ |
| ·72 | ·6304 882 | ·5774 480 | ·5249 381 | ·4737 874 | ·4246 827 | ·3781 624 |
| ·73 | ·6694 960 | ·6189 250+ | ·5681 644 | ·5180 309 | ·4692 343 | ·4223 638 |
| ·74 | ·7073 335+ | ·6597 122 | ·6112 576 | ·5627 459 | ·5148 811 | ·4682 759 |
| ·75 | ·7436 296 | ·6993 697 | ·6537 266 | ·6074 123 | ·5610 981 | ·5153 932 |
| ·76 | ·7780 405− | ·7374 715− | ·6950 769 | ·6514 852 | ·6073 124 | ·5631 395+ |
| ·77 | ·8102 605− | ·7736 194 | ·7348 258 | ·6944 119 | ·6529 203 | ·6108 828 |
| ·78 | ·8400 319 | ·8074 565− | ·7725 194 | ·7356 504 | ·6973 068 | ·6579 540 |
| ·79 | ·8671 535+ | ·8386 792 | ·8077 490 | ·7746 901 | ·7398 680 | ·7036 713 |
| ·80 | ·8914 875+ | ·8670 492 | ·8401 670 | ·8110 711 | ·7800 353 | ·7473 661 |

$x$ = ·81 to ·99 $q = 7$ $p$ = 15 to 20

| $x$ | $p = 15$ | $p = 16$ | $p = 17$ | $p = 18$ | $p = 19$ | $p = 20$ |
|---|---|---|---|---|---|---|
| $B(p, q)$ = | ·1228 5616 × $\frac{1}{10^5}$ | ·8376 5564 × $\frac{1}{10^6}$ | ·5827 1696 × $\frac{1}{10^6}$ | ·4127 5785 $\bar{\times}$ $\frac{1}{10^6}$ | ·2971 8565 $\overset{+}{\times}$ $\frac{1}{10^6}$ | ·2171 7413 × $\frac{1}{10^6}$ |
| ·81 | ·9129 635⁻ | ·8924 016 | ·8695 009 | ·8444 043 | ·8173 001 | ·7884 126 |
| ·82 | ·9315 803 | ·9146 515⁻ | ·8955 642 | ·8743 886 | ·8512 366 | ·8262 567 |
| ·83 | ·9474 051 | ·9337 961 | ·9182 649 | ·9008 242 | ·8815 231 | ·8604 444 |
| ·84 | ·9605 688 | ·9499 141 | ·9376 079 | ·9236 223 | ·9079 584 | ·8906 457 |
| ·85 | ·9712 590 | ·9631 599 | ·9536 941 | ·9428 085⁻ | ·9304 714 | ·9166 734 |
| ·86 | ·9797 101 | ·9737 551 | ·9667 132 | ·9585 199 | ·9491 248 | ·9384 935⁺ |
| ·87 | ·9861 912 | ·9819 751 | ·9769 317 | ·9709 953 | ·9641 091 | ·9562 261 |
| ·88 | ·9909 918 | ·9881 341 | ·9846 762 | ·9805 594 | ·9757 289 | ·9701 357 |
| ·89 | ·9944 079 | ·9925 667 | ·9903 135⁻ | ·9876 003 | ·9843 807 | ·9806 104 |
| ·90 | ·9967 272 | ·9956 100 | ·9942 274 | ·9925 439 | ·9905 236 | ·9881 313 |
| ·91 | ·9982 162 | ·9975 855⁻ | ·9967 963 | ·9958 247 | ·9946 458 | ·9932 342 |
| ·92 | ·9991 095⁻ | ·9987 837 | ·9983 715⁺ | ·9978 585⁺ | ·9972 293 | ·9964 675⁺ |
| ·93 | ·9996 022 | ·9994 517 | ·9992 593 | ·9990 172 | ·9987 170 | ·9983 497 |
| ·94 | ·9998 463 | ·9997 863 | ·9997 086 | ·9996 099 | ·9994 862 | ·9993 332 |
| ·95 | ·9999 513 | ·9999 316 | ·9999 060 | ·9998 730 | ·9998 312 | ·9997 790 |
| ·96 | ·9999 884 | ·9999 836 | ·9999 772 | ·9999 690 | ·9999 584 | ·9999 450⁺ |
| ·97 | ·9999 982 | ·9999 975⁻ | ·9999 965⁻ | ·9999 952 | ·9999 935⁻ | ·9999 913 |
| ·98 | ·9999 999 | ·9999 998 | ·9999 998 | ·9999 997 | ·9999 996 | ·9999 994 |
| ·99 | 1·0000 000 | 1·0000 000 | 1·0000 000 | 1·0000 000 | 1·0000 000 | 1·0000 000 |

$x$ = ·27 to ·90 $q = 7$ $p$ = 21 to 26

| $x$ | $p = 21$ | $p = 22$ | $p = 23$ | $p = 24$ | $p = 25$ | $p = 26$ |
|---|---|---|---|---|---|---|
| $B(p, q) =$ | $\cdot 1608\ 6973 \times \frac{1}{10^8}$ | $\cdot 1206\ 5229 \times \frac{1}{10^8}$ | $\cdot 9152\ 9327 \times \frac{1}{10^7}$ | $\cdot 7017\ 2484 \times \frac{1}{10^7}$ | $\cdot 5432\ 7084 \times \frac{1}{10^7}$ | $\cdot 4244\ 3035 \times \frac{1}{10^7}$ |
| ·27 | ·0000 001 | | | | | |
| ·28 | ·0000 001 | | | | | |
| ·29 | ·0000 002 | ·0000 001 | | | | |
| ·30 | ·0000 004 | ·0000 002 | ·0000 001 | | | |
| ·31 | ·0000 008 | ·0000 003 | ·0000 001 | | | |
| ·32 | ·0000 014 | ·0000 005+ | ·0000 002 | ·0000 001 | | |
| ·33 | ·0000 024 | ·0000 010 | ·0000 004 | ·0000 002 | ·0000 001 | |
| ·34 | ·0000 041 | ·0000 018 | ·0000 008 | ·0000 003 | ·0000 001 | ·0000 001 |
| ·35 | ·0000 069 | ·0000 031 | ·0000 013 | ·0000 006 | ·0000 003 | ·0000 001 |
| ·36 | ·0000 115− | ·0000 052 | ·0000 024 | ·0000 011 | ·0000 005− | ·0000 002 |
| ·37 | ·0000 187 | ·0000 088 | ·0000 041 | ·0000 019 | ·0000 009 | ·0000 004 |
| ·38 | ·0000 300 | ·0000 144 | ·0000 069 | ·0000 032 | ·0000 015+ | ·0000 007 |
| ·39 | ·0000 474 | ·0000 233 | ·0000 114 | ·0000 055+ | ·0000 026 | ·0000 013 |
| ·40 | ·0000 736 | ·0000 371 | ·0000 186 | ·0000 092 | ·0000 045+ | ·0000 022 |
| ·41 | ·0001 127 | ·0000 583 | ·0000 299 | ·0000 152 | ·0000 077 | ·0000 038 |
| ·42 | ·0001 702 | ·0000 901 | ·0000 473 | ·0000 246 | ·0000 127 | ·0000 065+ |
| ·43 | ·0002 537 | ·0001 374 | ·0000 738 | ·0000 394 | ·0000 208 | ·0000 109 |
| ·44 | ·0003 732 | ·0002 068 | ·0001 137 | ·0000 620 | ·0000 335+ | ·0000 180 |
| ·45 | ·0005 424 | ·0003 073 | ·0001 726 | ·0000 962 | ·0000 532 | ·0000 293 |
| ·46 | ·0007 790 | ·0004 509 | ·0002 588 | ·0001 474 | ·0000 833 | ·0000 468 |
| ·47 | ·0011 058 | ·0006 536 | ·0003 832 | ·0002 229 | ·0001 287 | ·0000 738 |
| ·48 | ·0015 523 | ·0009 366 | ·0005 604 | ·0003 328 | ·0001 962 | ·0001 149 |
| ·49 | ·0021 557 | ·0013 270 | ·0008 101 | ·0004 908 | ·0002 953 | ·0001 764 |
| ·50 | ·0029 623 | ·0018 596 | ·0011 578 | ·0007 155− | ·0004 390 | ·0002 675+ |
| ·51 | ·0040 293 | ·0025 784 | ·0016 366 | ·0010 309 | ·0006 449 | ·0004 007 |
| ·52 | ·0054 263 | ·0035 381 | ·0022 884 | ·0014 690 | ·0009 364 | ·0005 930 |
| ·53 | ·0072 371 | ·0048 061 | ·0031 663 | ·0020 705− | ·0013 445− | ·0008 674 |
| ·54 | ·0095 611 | ·0064 645− | ·0043 363 | ·0028 873 | ·0019 092 | ·0012 543 |
| ·55 | ·0125 146 | ·0086 115+ | ·0058 794 | ·0039 847 | ·0026 820 | ·0017 937 |
| ·56 | ·0162 325+ | ·0113 637 | ·0078 935+ | ·0054 433 | ·0037 282 | ·0025 372 |
| ·57 | ·0208 681 | ·0148 569 | ·0104 961 | ·0073 620 | ·0051 290 | ·0035 507 |
| ·58 | ·0265 936 | ·0192 478 | ·0138 253 | ·0098 598 | ·0069 848 | ·0049 171 |
| ·59 | ·0335 996 | ·0247 140 | ·0180 418 | ·0130 782 | ·0094 175+ | ·0067 394 |
| ·60 | ·0420 927 | ·0314 536 | ·0233 293 | ·0171 830 | ·0125 733 | ·0091 437 |
| ·61 | ·0522 933 | ·0396 841 | ·0298 949 | ·0223 656 | ·0166 246 | ·0122 821 |
| ·62 | ·0644 311 | ·0496 395− | ·0379 675− | ·0288 431 | ·0217 716 | ·0163 351 |
| ·63 | ·0787 400 | ·0615 661 | ·0477 957 | ·0368 572 | ·0282 432 | ·0215 139 |
| ·64 | ·0954 508 | ·0757 173 | ·0596 435+ | ·0466 727 | ·0362 960 | ·0280 610 |
| ·65 | ·1147 828 | ·0923 459 | ·0737 844 | ·0585 721 | ·0462 121 | ·0362 499 |
| ·66 | ·1369 343 | ·1116 948 | ·0904 935+ | ·0728 504 | ·0582 949 | ·0463 826 |
| ·67 | ·1620 715− | ·1339 865− | ·1100 376 | ·0898 060 | ·0728 620 | ·0587 850+ |
| ·68 | ·1903 167 | ·1594 106 | ·1326 629 | ·1097 296 | ·0902 362 | ·0737 995− |
| ·69 | ·2217 362 | ·1881 107 | ·1585 814 | ·1328 909 | ·1107 328 | ·0917 744 |
| ·70 | ·2563 285+ | ·2201 697 | ·1879 555+ | ·1595 230 | ·1346 445+ | ·1130 500+ |
| ·71 | ·2940 127 | ·2555 962 | ·2208 817 | ·1898 046 | ·1622 237 | ·1379 415+ |
| ·72 | ·3346 194 | ·2943 111 | ·2573 740 | ·2238 415− | ·1936 623 | ·1667 183 |
| ·73 | ·3778 837 | ·3361 360 | ·2973 486 | ·2616 473 | ·2290 698 | ·1995 807 |
| ·74 | ·4234 416 | ·3807 851 | ·3406 103 | ·3031 256 | ·2684 522 | ·2366 358 |
| ·75 | ·4708 309 | ·4278 602 | ·3868 426 | ·3480 543 | ·3116 902 | ·2778 717 |
| ·76 | ·5194 967 | ·4768 515− | ·4356 019 | ·3960 741 | ·3585 226 | ·3231 341 |
| ·77 | ·5688 032 | ·5271 444 | ·4863 188 | ·4466 825− | ·4085 325+ | ·3721 069 |
| ·78 | ·6180 504 | ·5780 327 | ·5383 061 | ·4992 358 | ·4611 423 | ·4242 982 |
| ·79 | ·6664 974 | ·6287 392 | ·5907 752 | ·5529 597 | ·5156 168 | ·4790 358 |
| ·80 | ·7133 902 | ·6784 435+ | ·6428 614 | ·6069 699 | ·5710 784 | ·5354 741 |
| ·81 | ·7579 941 | ·7263 155− | ·6936 576 | ·6603 040 | ·6265 335+ | ·5926 144 |
| ·82 | ·7996 282 | ·7715 542 | ·7422 551 | ·7119 624 | ·6809 124 | ·6493 407 |
| ·83 | ·8377 004 | ·8134 294 | ·7877 903 | ·7609 585− | ·7331 204 | ·7044 695+ |
| ·84 | ·8717 402 | ·8513 223 | ·8294 937 | ·8063 744 | ·7820 991 | ·7568 139 |
| ·85 | ·9014 266 | ·8847 640 | ·8667 381 | ·8474 190 | ·8268 925+ | ·8052 576 |
| ·86 | ·9266 077 | ·9134 655− | ·8990 807 | ·8834 825+ | ·8667 146 | ·8488 332 |
| ·87 | ·9473 105+ | ·9373 378 | ·9262 953 | ·9141 821 | ·9010 090 | ·8867 979 |
| ·88 | ·9637 371 | ·9564 976 | ·9483 893 | ·9393 926 | ·9294 963 | ·9186 974 |
| ·89 | ·9762 482 | ·9712 565− | ·9656 023 | ·9592 573 | ·9521 985+ | ·9444 084 |
| ·90 | ·9853 322 | ·9820 933 | ·9783 833 | ·9741 732 | ·9694 369 | ·9641 511 |

$x = \cdot 91$ to $\cdot 99$ $q = 7$ $p = 21$ to $26$

| | $p = 21$ | $p = 22$ | $p = 23$ | $p = 24$ | $p = 25$ | $p = 26$ |
|---|---|---|---|---|---|---|
| $B(p, q) =$ | $\cdot 1608\ 6973 \times \frac{1}{10^6}$ | $\cdot 1206\ 5229 \times \frac{1}{10^6}$ | $\cdot 9152\ 9327 \times \frac{1}{10^7}$ | $\cdot 7017\ 2484 \times \frac{1}{10^7}$ | $\cdot 5432\ 7084 \times \frac{1}{10^7}$ | $\cdot 4244\ 3035 \times \frac{1}{10^7}$ |
| $x$ | | | | | | |
| ·91 | ·9915 644 | ·9896 106 | ·9873 478 | ·9847 514 | ·9817 981 | ·9784 655+ |
| ·92 | ·9955 564 | ·9944 788 | ·9932 170 | ·9917 532 | ·9900 699 | ·9881 496 |
| ·93 | ·9979 055+ | ·9973 745− | ·9967 459 | ·9960 088 | ·9951 520 | ·9941 639 |
| ·94 | ·9991 463 | ·9989 203 | ·9986 499 | ·9983 295+ | ·9979 530 | ·9975 142 |
| ·95 | ·9997 146 | ·9996 358 | ·9995 406 | ·9994 265+ | ·9992 911 | ·9991 315+ |
| ·96 | ·9999 284 | ·9999 078 | ·9998 826 | ·9998 522 | ·9998 157 | ·9997 722 |
| ·97 | ·9999 886 | ·9999 852 | ·9999 809 | ·9999 758 | ·9999 695+ | ·9999 620 |
| ·98 | ·9999 992 | ·9999 990 | ·9999 986 | ·9999 983 | ·9999 978 | ·9999 972 |
| ·99 | 1·0000 000 | 1·0000 000 | 1·0000 000 | 1·0000 000 | 1·0000 000 | 1·0000 000 |

$x$ = ·36 to 1·00 $q = 7$ $p$ = 27 to 32

| | $p = 27$ | $p = 28$ | $p = 29$ | $p = 30$ | $p = 31$ | $p = 32$ |
|---|---|---|---|---|---|---|
| $B(p, q)$ = | ·3343 9967 × $\frac{1}{10^7}$ | ·2655 5268 × $\frac{1}{10^7}$ | ·2124 4214 × $\frac{1}{10^7}$ | ·1711 3395 $\bar{\times}$ $\frac{1}{10^7}$ | ·1387 5725 $\overset{+}{\times}$ $\frac{1}{10^7}$ | ·1131 9671 × $\frac{1}{10^7}$ |
| $x$ | | | | | | |
| ·36 | ·0000 001 | | | | | |
| ·37 | ·0000 002 | ·0000 001 | | | | |
| ·38 | ·0000 003 | ·0000 001 | ·0000 001 | | | |
| ·39 | ·0000 006 | ·0000 003 | ·0000 001 | ·0000 001 | | |
| ·40 | ·0000 011 | ·0000 005+ | ·0000 003 | ·0000 001 | ·0000 001 | |
| ·41 | ·0000 019 | ·0000 009 | ·0000 005− | ·0000 002 | ·0000 001 | ·0000 001 |
| ·42 | ·0000 033 | ·0000 017 | ·0000 009 | ·0000 004 | ·0000 002 | ·0000 001 |
| ·43 | ·0000 057 | ·0000 030 | ·0000 015+ | ·0000 008 | ·0000 004 | ·0000 002 |
| ·44 | ·0000 096 | ·0000 051 | ·0000 027 | ·0000 014 | ·0000 007 | ·0000 004 |
| ·45 | ·0000 160 | ·0000 087 | ·0000 047 | ·0000 025+ | ·0000 013 | ·0000 007 |
| ·46 | ·0000 261 | ·0000 145− | ·0000 080 | ·0000 044 | ·0000 024 | ·0000 013 |
| ·47 | ·0000 421 | ·0000 238 | ·0000 134 | ·0000 075+ | ·0000 042 | ·0000 023 |
| ·48 | ·0000 668 | ·0000 387 | ·0000 222 | ·0000 127 | ·0000 072 | ·0000 041 |
| ·49 | ·0001 048 | ·0000 618 | ·0000 363 | ·0000 212 | ·0000 123 | ·0000 071 |
| ·50 | ·0001 620 | ·0000 976 | ·0000 584 | ·0000 348 | ·0000 206 | ·0000 122 |
| ·51 | ·0002 474 | ·0001 519 | ·0000 928 | ·0000 563 | ·0000 341 | ·0000 205− |
| ·52 | ·0003 732 | ·0002 335+ | ·0001 453 | ·0000 900 | ·0000 554 | ·0000 340 |
| ·53 | ·0005 561 | ·0003 545+ | ·0002 248 | ·0001 418 | ·0000 890 | ·0000 556 |
| ·54 | ·0008 190 | ·0005 317 | ·0003 434 | ·0002 206 | ·0001 410 | ·0000 898 |
| ·55 | ·0011 923 | ·0007 881 | ·0005 181 | ·0003 389 | ·0002 206 | ·0001 429 |
| ·56 | ·0017 163 | ·0011 544 | ·0007 724 | ·0005 142 | ·0003 406 | ·0002 247 |
| ·57 | ·0024 434 | ·0016 720 | ·0011 381 | ·0007 708 | ·0005 196 | ·0003 487 |
| ·58 | ·0034 410 | ·0023 947 | ·0016 578 | ·0011 419 | ·0007 829 | ·0005 344 |
| ·59 | ·0047 947 | ·0033 923 | ·0023 876 | ·0016 722 | ·0011 657 | ·0008 090 |
| ·60 | ·0066 111 | ·0047 538 | ·0034 007 | ·0024 208 | ·0017 153 | ·0012 101 |
| ·61 | ·0090 218 | ·0065 912 | ·0047 907 | ·0034 652 | ·0024 950− | ·0017 885+ |
| ·62 | ·0121 866 | ·0090 430 | ·0066 763 | ·0049 053 | ·0035 878 | ·0026 127 |
| ·63 | ·0162 961 | ·0122 784 | ·0092 049 | ·0068 679 | ·0051 012 | ·0037 727 |
| ·64 | ·0215 744 | ·0165 005− | ·0125 573 | ·0095 115+ | ·0071 724 | ·0053 855+ |
| ·65 | ·0282 802 | ·0219 487 | ·0169 513 | ·0130 310 | ·0099 731 | ·0076 008 |
| ·66 | ·0367 062 | ·0289 006 | ·0226 449 | ·0176 619 | ·0137 154 | ·0106 066 |
| ·67 | ·0471 769 | ·0376 711 | ·0299 375− | ·0236 839 | ·0186 560 | ·0146 353 |
| ·68 | ·0600 432 | ·0486 102 | ·0391 697 | ·0314 221 | ·0251 000 | ·0199 689 |
| ·69 | ·0756 743 | ·0620 965+ | ·0507 203 | ·0412 467 | ·0334 025− | ·0269 424 |
| ·70 | ·0944 455+ | ·0785 283 | ·0649 987 | ·0535 685+ | ·0439 672 | ·0359 454 |
| ·71 | ·1167 226 | ·0983 093 | ·0824 344 | ·0688 312 | ·0572 414 | ·0474 199 |
| ·72 | ·1428 417 | ·1218 304 | ·1034 604 | ·0874 976 | ·0737 057 | ·0618 536 |
| ·73 | ·1730 859 | ·1494 467 | ·1284 921 | ·1100 305− | ·0938 581 | ·0797 672 |
| ·74 | ·2076 585− | ·1814 501 | ·1579 000 | ·1368 673 | ·1181 903 | ·1016 943 |
| ·75 | ·2466 545+ | ·2180 388 | ·1919 781 | ·1683 886 | ·1471 581 | ·1281 534 |
| ·76 | ·2900 323 | ·2592 843 | ·2309 084 | ·2048 808 | ·1811 436 | ·1596 117 |
| ·77 | ·3375 867 | ·3050 993 | ·2747 236 | ·2464 951 | ·2204 120 | ·1964 408 |
| ·78 | ·3889 279 | ·3552 083 | ·3232 709 | ·2932 057 | ·2650 647 | ·2388 664 |
| ·79 | ·4434 678 | ·4091 249 | ·3761 802 | ·3447 692 | ·3149 916 | ·2869 141 |
| ·80 | ·5004 175− | ·4661 399 | ·4328 417 | ·4006 917 | ·3698 277 | ·3403 575+ |
| ·81 | ·5587 997 | ·5253 231 | ·4923 965+ | ·4602 079 | ·4289 206 | ·3986 728 |
| ·82 | ·6174 776 | ·5855 437 | ·5537 467 | ·5222 786 | ·4913 140 | ·4610 086 |
| ·83 | ·6752 015+ | ·6455 108 | ·6155 867 | ·5856 111 | ·5557 553 | ·5261 789 |
| ·84 | ·7306 730 | ·7038 349 | ·6764 601 | ·6487 077 | ·6207 333 | ·5926 867 |
| ·85 | ·7826 241 | ·7591 105− | ·7348 410 | ·7099 439 | ·6845 489 | ·6587 852 |
| ·86 | ·8299 064 | ·8100 123 | ·7892 371 | ·7676 740 | ·7454 208 | ·7225 789 |
| ·87 | ·8715 810 | ·8554 005− | ·8383 069 | ·8203 586 | ·8016 206 | ·7821 632 |
| ·88 | ·9070 014 | ·8944 216 | ·8809 793 | ·8667 026 | ·8516 264 | ·8357 916 |
| ·89 | ·9358 753 | ·9265 931 | ·9165 617 | ·9057 867 | ·8942 789 | ·8820 546 |
| ·90 | ·9582 962 | ·9518 557 | ·9448 171 | ·9371 719 | ·9289 149 | ·9200 454 |
| ·91 | ·9747 331 | ·9705 818 | ·9659 946 | ·9609 566 | ·9554 551 | ·9494 797 |
| ·92 | ·9859 753 | ·9835 303 | ·9807 989 | ·9777 662 | ·9744 180 | ·9707 415− |
| ·93 | ·9930 329 | ·9917 473 | ·9902 956 | ·9886 661 | ·9868 475+ | ·9848 290 |
| ·94 | ·9970 065+ | ·9964 232 | ·9957 575− | ·9950 021 | ·9941 501 | ·9931 943 |
| ·95 | ·9989 449 | ·9987 283 | ·9984 784 | ·9981 919 | ·9978 653 | ·9974 950− |
| ·96 | ·9997 208 | ·9996 606 | ·9995 903 | ·9995 089 | ·9994 151 | ·9993 077 |
| ·97 | ·9999 530 | ·9999 424 | ·9999 299 | ·9999 152 | ·9998 981 | ·9998 783 |
| ·98 | ·9999 965+ | ·9999 957 | ·9999 947 | ·9999 936 | ·9999 922 | ·9999 906 |
| ·99 | 1·0000 000 | 1·0000 000 | ·9999 999 | ·9999 999 | ·9999 999 | ·9999 999 |
| 1·00 | | | 1·0000 000 | 1·0000 000 | 1·0000 000 | 1·0000 000 |

$x = \cdot 42$ to $1\cdot 00$ $q = 7$ $p = 33$ to $38$

| | $p = 33$ | $p = 34$ | $p = 35$ | $p = 36$ | $p = 37$ | $p = 38$ |
|---|---|---|---|---|---|---|
| $B(p, q) =$ | $\cdot 9287\,9350^{-} \times \frac{1}{10^8}$ | $\cdot 7662\,5464 \times \frac{1}{10^8}$ | $\cdot 6354\,3067 \times \frac{1}{10^8}$ | $\cdot 5295\,2556 \times \frac{1}{10^8}$ | $\cdot 4433\,2373 \times \frac{1}{10^8}$ | $\cdot 3727\,9495^{+} \times \frac{1}{10^8}$ |
| $x$ | | | | | | |
| ·42 | ·0000 001 | | | | | |
| ·43 | ·0000 001 | ·0000 001 | | | | |
| ·44 | ·0000 002 | ·0000 001 | ·0000 001 | | | |
| ·45 | ·0000 004 | ·0000 002 | ·0000 001 | ·0000 001 | | |
| ·46 | ·0000 007 | ·0000 004 | ·0000 002 | ·0000 001 | ·0000 001 | |
| ·47 | ·0000 013 | ·0000 007 | ·0000 004 | ·0000 002 | ·0000 001 | ·0000 001 |
| ·48 | ·0000 023 | ·0000 013 | ·0000 007 | ·0000 004 | ·0000 002 | ·0000 001 |
| ·49 | ·0000 041 | ·0000 024 | ·0000 013 | ·0000 008 | ·0000 004 | ·0000 002 |
| ·50 | ·0000 071 | ·0000 042 | ·0000 024 | ·0000 014 | ·0000 008 | ·0000 005− |
| ·51 | ·0000 123 | ·0000 073 | ·0000 044 | ·0000 026 | ·0000 015+ | ·0000 009 |
| ·52 | ·0000 208 | ·0000 126 | ·0000 076 | ·0000 046 | ·0000 028 | ·0000 017 |
| ·53 | ·0000 346 | ·0000 214 | ·0000 132 | ·0000 081 | ·0000 050− | ·0000 030 |
| ·54 | ·0000 569 | ·0000 359 | ·0000 226 | ·0000 141 | ·0000 088 | ·0000 055− |
| ·55 | ·0000 922 | ·0000 593 | ·0000 379 | ·0000 242 | ·0000 154 | ·0000 097 |
| ·56 | ·0001 475+ | ·0000 965+ | ·0000 629 | ·0000 408 | ·0000 264 | ·0000 170 |
| ·57 | ·0002 330 | ·0001 550+ | ·0001 028 | ·0000 679 | ·0000 447 | ·0000 293 |
| ·58 | ·0003 632 | ·0002 459 | ·0001 658 | ·0001 114 | ·0000 746 | ·0000 498 |
| ·59 | ·0005 591 | ·0003 849 | ·0002 639 | ·0001 803 | ·0001 228 | ·0000 833 |
| ·60 | ·0008 501 | ·0005 948 | ·0004 146 | ·0002 880 | ·0001 994 | ·0001 375+ |
| ·61 | ·0012 768 | ·0009 079 | ·0006 432 | ·0004 540 | ·0003 194 | ·0002 240 |
| ·62 | ·0018 949 | ·0013 689 | ·0009 852 | ·0007 066 | ·0005 050+ | ·0003 598 |
| ·63 | ·0027 789 | ·0020 389 | ·0014 904 | ·0010 857 | ·0007 882 | ·0005 704 |
| ·64 | ·0040 276 | ·0030 004 | ·0022 271 | ·0016 473 | ·0012 143 | ·0008 923 |
| ·65 | ·0057 696 | ·0043 630 | ·0032 873 | ·0024 683 | ·0018 472 | ·0013 780 |
| ·66 | ·0081 700 | ·0062 695+ | ·0047 938 | ·0036 529 | ·0027 744 | ·0021 006 |
| ·67 | ·0114 364 | ·0089 034 | ·0069 068 | ·0053 398 | ·0041 149 | ·0031 611 |
| ·68 | ·0158 256 | ·0124 959 | ·0098 321 | ·0077 102 | ·0060 268 | ·0046 965− |
| ·69 | ·0216 492 | ·0173 329 | ·0138 290 | ·0109 968 | ·0087 170 | ·0068 888 |
| ·70 | ·0292 773 | ·0237 609 | ·0192 181 | ·0154 929 | ·0124 507 | ·0099 758 |
| ·71 | ·0391 391 | ·0321 908 | ·0263 870 | ·0215 598 | ·0175 613 | ·0142 620 |
| ·72 | ·0517 200 | ·0430 973 | ·0357 933 | ·0296 329 | ·0244 582 | ·0201 282 |
| ·73 | ·0675 522 | ·0570 140 | ·0479 635− | ·0402 240 | ·0336 326 | ·0280 406 |
| ·74 | ·0871 984 | ·0745 211 | ·0634 844 | ·0539 171 | ·0456 574 | ·0385 540 |
| ·75 | ·1112 273 | ·0962 246 | ·0829 869 | ·0713 567 | ·0611 803 | ·0523 103 |
| ·76 | ·1401 791 | ·1227 251 | ·1071 192 | ·0932 255+ | ·0809 064 | ·0700 257 |
| ·77 | ·1745 221 | ·1545 762 | ·1365 074 | ·1202 094 | ·1055 684 | ·0924 667 |
| ·78 | ·2146 002 | ·1922 311 | ·1717 042 | ·1529 485− | ·1358 808 | ·1204 092 |
| ·79 | ·2605 740 | ·2359 819 | ·2131 257 | ·1919 739 | ·1724 790 | ·1545 805+ |
| ·80 | ·3123 609 | ·2858 914 | ·2609 789 | ·2376 323 | ·2158 421 | ·1955 832 |
| ·81 | ·3695 782 | ·3417 268 | ·3151 860 | ·2900 026 | ·2662 043 | ·2438 017 |
| ·82 | ·4314 988 | ·4029 010 | ·3753 126 | ·3488 120 | ·3234 597 | ·2992 997 |
| ·83 | ·4970 276 | ·4684 329 | ·4405 109 | ·4133 628 | ·3870 744 | ·3617 168 |
| ·84 | ·5647 102 | ·5369 372 | ·5094 909 | ·4824 837 | ·4560 167 | ·4301 792 |
| ·85 | ·6327 799 | ·6066 565+ | ·5805 331 | ·5545 216 | ·5287 269 | ·5032 459 |
| ·86 | ·6992 517 | ·6755 428 | ·6515 549 | ·6273 889 | ·6031 423 | ·5789 088 |
| ·87 | ·7620 614 | ·7413 931 | ·7202 384 | ·6986 788 | ·6767 958 | ·6546 703 |
| ·88 | ·8192 441 | ·8020 348 | ·7842 181 | ·7658 516 | ·7469 954 | ·7277 110 |
| ·89 | ·8691 352 | ·8555 462 | ·8413 178 | ·8264 836 | ·8110 808 | ·7951 493 |
| ·90 | ·9105 661 | ·9004 836 | ·8898 080 | ·8785 528 | ·8667 349 | ·8543 740 |
| ·91 | ·9430 226 | ·9360 783 | ·9286 437 | ·9207 185+ | ·9123 046 | ·9034 063 |
| ·92 | ·9667 249 | ·9623 578 | ·9576 310 | ·9525 369 | ·9470 693 | ·9412 233 |
| ·93 | ·9825 997 | ·9801 496 | ·9774 689 | ·9745 484 | ·9713 797 | ·9679 549 |
| ·94 | ·9921 273 | ·9909 419 | ·9896 311 | ·9881 876 | ·9866 047 | ·9848 754 |
| ·95 | ·9970 772 | ·9966 081 | ·9960 839 | ·9955 004 | ·9948 538 | ·9941 399 |
| ·96 | ·9991 852 | ·9990 462 | ·9988 892 | ·9987 127 | ·9985 150+ | ·9982 945− |
| ·97 | ·9998 555+ | ·9998 294 | ·9997 996 | ·9997 658 | ·9997 274 | ·9996 842 |
| ·98 | ·9999 888 | ·9999 866 | ·9999 842 | ·9999 813 | ·9999 781 | ·9999 744 |
| ·99 | ·9999 999 | ·9999 999 | ·9999 998 | ·9999 998 | ·9999 998 | ·9999 997 |
| 1·00 | 1·0000 000 | 1·0000 000 | 1·0000 000 | 1·0000 000 | 1·0000 000 | 1·0000 000 |

$x = \cdot 48$ to $1\cdot 00$ $q = 7$ $p = 39$ to $44$

| $x$ | $p = 39$ | $p = 40$ | $p = 41$ | $p = 42$ | $p = 43$ | $p = 44$ |
|---|---|---|---|---|---|---|
| $B(p, q) =$ | $\cdot 3148\ 0463 \times \frac{1}{10^8}$ | $\cdot 2668\ 9957 \times \frac{1}{10^8}$ | $\cdot 2271\ 4857 \times \frac{1}{10^8}$ | $\cdot 1940\ 2274 \times \frac{1}{10^8}$ | $\cdot 1663\ 0521 \times \frac{1}{10^8}$ | $\cdot 1430\ 2248 \times \frac{1}{10^8}$ |
| ·48 | ·0000 001 | | | | | |
| ·49 | ·0000 001 | ·0000 001 | | | | |
| ·50 | ·0000 003 | ·0000 002 | ·0000 001 | ·0000 001 | | |
| ·51 | ·0000 005+ | ·0000 003 | ·0000 002 | ·0000 001 | ·0000 001 | |
| ·52 | ·0000 010 | ·0000 006 | ·0000 004 | ·0000 002 | ·0000 001 | ·0000 001 |
| ·53 | ·0000 019 | ·0000 011 | ·0000 007 | ·0000 004 | ·0000 002 | ·0000 001 |
| ·54 | ·0000 034 | ·0000 021 | ·0000 013 | ·0000 008 | ·0000 005− | ·0000 003 |
| ·55 | ·0000 061 | ·0000 039 | ·0000 024 | ·0000 015+ | ·0000 009 | ·0000 006 |
| ·56 | ·0000 109 | ·0000 070 | ·0000 045− | ·0000 029 | ·0000 018 | ·0000 011 |
| ·57 | ·0000 192 | ·0000 125− | ·0000 081 | ·0000 053 | ·0000 034 | ·0000 022 |
| ·58 | ·0000 331 | ·0000 220 | ·0000 145+ | ·0000 096 | ·0000 063 | ·0000 041 |
| ·59 | ·0000 564 | ·0000 380 | ·0000 256 | ·0000 171 | ·0000 115− | ·0000 077 |
| ·60 | ·0000 946 | ·0000 649 | ·0000 444 | ·0000 302 | ·0000 206 | ·0000 139 |
| ·61 | ·0001 566 | ·0001 091 | ·0000 758 | ·0000 525+ | ·0000 363 | ·0000 250+ |
| ·62 | ·0002 555+ | ·0001 810 | ·0001 278 | ·0000 900 | ·0000 632 | ·0000 443 |
| ·63 | ·0004 115− | ·0002 960 | ·0002 123 | ·0001 519 | ·0001 084 | ·0000 771 |
| ·64 | ·0006 537 | ·0004 775+ | ·0003 478 | ·0002 527 | ·0001 831 | ·0001 323 |
| ·65 | ·0010 249 | ·0007 600 | ·0005 621 | ·0004 145+ | ·0003 050− | ·0002 238 |
| ·66 | ·0015 856 | ·0011 935− | ·0008 958 | ·0006 706 | ·0005 008 | ·0003 730 |
| ·67 | ·0024 212 | ·0018 492 | ·0014 084 | ·0010 699 | ·0008 107 | ·0006 128 |
| ·68 | ·0036 490 | ·0028 272 | ·0021 845− | ·0016 835− | ·0012 942 | ·0009 925− |
| ·69 | ·0054 281 | ·0042 652 | ·0033 425− | ·0026 126 | ·0020 370 | ·0015 845− |
| ·70 | ·0079 699 | ·0063 497 | ·0050 454 | ·0039 989 | ·0031 616 | ·0024 937 |
| ·71 | ·0115 496 | ·0093 276 | ·0075 132 | ·0060 366 | ·0048 384 | ·0038 689 |
| ·72 | ·0165 184 | ·0135 195− | ·0110 363 | ·0089 869 | ·0073 004 | ·0059 168 |
| ·73 | ·0233 138 | ·0193 325− | ·0159 901 | ·0131 931 | ·0108 596 | ·0089 185+ |
| ·74 | ·0324 675+ | ·0272 706 | ·0228 481 | ·0190 964 | ·0159 236 | ·0132 481 |
| ·75 | ·0446 074 | ·0379 414 | ·0321 920 | ·0272 489 | ·0230 120 | ·0193 909 |
| ·76 | ·0604 506 | ·0520 540 | ·0447 154 | ·0383 219 | ·0327 686 | ·0279 592 |
| ·77 | ·0807 854 | ·0704 071 | ·0612 171 | ·0531 052 | ·0459 668 | ·0397 032 |
| ·78 | ·1064 358 | ·0938 598 | ·0825 791 | ·0724 926 | ·0635 011 | ·0555 092 |
| ·79 | ·1382 082 | ·1232 842 | ·1097 257 | ·0974 470 | ·0863 611 | ·0763 812 |
| ·80 | ·1768 170 | ·1594 944 | ·1435 576 | ·1289 423 | ·1155 799 | ·1033 982 |
| ·81 | ·2227 904 | ·2031 530 | ·1848 607 | ·1678 756 | ·1521 523 | ·1376 393 |
| ·82 | ·2763 604 | ·2546 563 | ·2341 893 | ·2149 504 | ·1969 207 | ·1800 735− |
| ·83 | ·3373 467 | ·3140 077 | ·2917 306 | ·2705 348 | ·2504 290 | ·2314 127 |
| ·84 | ·4050 488 | ·3806 917 | ·3571 627 | ·3345 060 | ·3127 556 | ·2919 358 |
| ·85 | ·4781 672 | ·4535 708 | ·4295 279 | ·4061 007 | ·3833 428 | ·3612 994 |
| ·86 | ·5547 773 | ·5308 314 | ·5071 490 | ·4838 016 | ·4608 544 | ·4383 662 |
| ·87 | ·6323 817 | ·6100 075− | ·5876 220 | ·5652 967 | ·5430 988 | ·5210 920 |
| ·88 | ·7080 613 | ·6881 093 | ·6679 178 | ·6475 490 | ·6270 639 | ·6065 216 |
| ·89 | ·7787 315+ | ·7618 717 | ·7446 156 | ·7270 103 | ·7091 031 | ·6909 419 |
| ·90 | ·8414 926 | ·8281 158 | ·8142 708 | ·7999 869 | ·7852 948 | ·7702 268 |
| ·91 | ·8940 304 | ·8841 856 | ·8738 830 | ·8631 357 | ·8519 584 | ·8403 679 |
| ·92 | ·9349 958 | ·9283 850+ | ·9213 909 | ·9140 146 | ·9062 590 | ·8981 282 |
| ·93 | ·9642 670 | ·9603 095+ | ·9560 770 | ·9515 648 | ·9467 689 | ·9416 864 |
| ·94 | ·9829 932 | ·9809 517 | ·9787 449 | ·9763 669 | ·9738 123 | ·9710 759 |
| ·95 | ·9933 546 | ·9924 938 | ·9915 534 | ·9905 292 | ·9894 173 | ·9882 136 |
| ·96 | ·9980 493 | ·9977 777 | ·9974 779 | ·9971 479 | ·9967 859 | ·9963 899 |
| ·97 | ·9996 357 | ·9995 814 | ·9995 208 | ·9994 535− | ·9993 788 | ·9992 963 |
| ·98 | ·9999 702 | ·9999 655− | ·9999 601 | ·9999 541 | ·9999 474 | ·9999 399 |
| ·99 | ·9999 997 | ·9999 996 | ·9999 996 | ·9999 995− | ·9999 994 | ·9999 993 |
| 1·00 | 1·0000 000 | 1·0000 000 | 1·0000 000 | 1·0000 000 | 1·0000 000 | 1·0000 000 |

$x = \cdot 53$ to $1 \cdot 00$ $\qquad q = 7 \qquad p = 45$ to $50$

| | $p = 45$ | $p = 46$ | $p = 47$ | $p = 48$ | $p = 49$ | $p = 50$ |
|---|---|---|---|---|---|---|
| $B(p, q) =$ | $\cdot 1233\ 9194 \times \frac{1}{10^8}$ | $\cdot 1067\ 8149 \times \frac{1}{10^8}$ | $\cdot 9267\ 8272 \times \frac{1}{10^9}$ | $\cdot 8066\ 4422 \times \frac{1}{10^9}$ | $\cdot 7039\ 8041 \times \frac{1}{10^9}$ | $\cdot 6159\ 8286 \times \frac{1}{10^9}$ |
| $x$ | | | | | | |
| ·53 | ·0000 001 | ·0000 001 | | | | |
| ·54 | ·0000 002 | ·0000 001 | ·0000 001 | | | |
| ·55 | ·0000 004 | ·0000 002 | ·0000 001 | ·0000 001 | ·0000 001 | |
| ·56 | ·0000 007 | ·0000 005− | ·0000 003 | ·0000 002 | ·0000 001 | ·0000 001 |
| ·57 | ·0000 014 | ·0000 009 | ·0000 006 | ·0000 004 | ·0000 002 | ·0000 001 |
| ·58 | ·0000 027 | ·0000 018 | ·0000 011 | ·0000 007 | ·0000 005− | ·0000 003 |
| ·59 | ·0000 051 | ·0000 034 | ·0000 022 | ·0000 015− | ·0000 010 | ·0000 006 |
| ·60 | ·0000 094 | ·0000 064 | ·0000 043 | ·0000 029 | ·0000 019 | ·0000 013 |
| ·61 | ·0000 172 | ·0000 118 | ·0000 081 | ·0000 055+ | ·0000 038 | ·0000 026 |
| ·62 | ·0000 309 | ·0000 216 | ·0000 150+ | ·0000 104 | ·0000 072 | ·0000 050− |
| ·63 | ·0000 548 | ·0000 388 | ·0000 274 | ·0000 193 | ·0000 136 | ·0000 096 |
| ·64 | ·0000 954 | ·0000 686 | ·0000 493 | ·0000 353 | ·0000 252 | ·0000 180 |
| ·65 | ·0001 638 | ·0001 197 | ·0000 872 | ·0000 634 | ·0000 460 | ·0000 334 |
| ·66 | ·0002 772 | ·0002 055+ | ·0001 521 | ·0001 123 | ·0000 827 | ·0000 608 |
| ·67 | ·0004 621 | ·0003 477 | ·0002 611 | ·0001 956 | ·0001 463 | ·0001 092 |
| ·68 | ·0007 593 | ·0005 797 | ·0004 416 | ·0003 357 | ·0002 547 | ·0001 928 |
| ·69 | ·0012 296 | ·0009 522 | ·0007 357 | ·0005 673 | ·0004 366 | ·0003 353 |
| ·70 | ·0019 625− | ·0015 410 | ·0012 075− | ·0009 442 | ·0007 369 | ·0005 740 |
| ·71 | ·0030 868 | ·0024 574 | ·0019 522 | ·0015 478 | ·0012 247 | ·0009 673 |
| ·72 | ·0047 847 | ·0038 609 | ·0031 090 | ·0024 986 | ·0020 041 | ·0016 045− |
| ·73 | ·0073 083 | ·0059 760 | ·0048 767 | ·0039 717 | ·0032 284 | ·0026 195− |
| ·74 | ·0109 983 | ·0091 114 | ·0075 330 | ·0062 159 | ·0051 194 | ·0042 086 |
| ·75 | ·0163 047 | ·0136 815− | ·0114 574 | ·0095 764 | ·0079 894 | ·0066 533 |
| ·76 | ·0238 057 | ·0202 281 | ·0171 545− | ·0145 203 | ·0122 681 | ·0103 468 |
| ·77 | ·0342 226 | ·0294 398 | ·0252 768 | ·0216 620 | ·0185 307 | ·0158 243 |
| ·78 | ·0484 254 | ·0421 634 | ·0366 419 | ·0317 853 | ·0275 237 | ·0237 925+ |
| ·79 | ·0674 220 | ·0594 005+ | ·0522 370 | ·0458 553 | ·0401 837 | ·0351 544 |
| ·80 | ·0923 240 | ·0822 834 | ·0732 032 | ·0650 117 | ·0576 394 | ·0510 193 |
| ·81 | ·1242 808 | ·1120 176 | ·1007 888 | ·0905 324 | ·0811 863 | ·0726 889 |
| ·82 | ·1643 749 | ·1497 857 | ·1362 621 | ·1237 572 | ·1122 213 | ·1016 036 |
| ·83 | ·2134 769 | ·1966 053 | ·1807 753 | ·1659 591 | ·1521 244 | ·1392 356 |
| ·84 | ·2720 625− | ·2531 430 | ·2351 778 | ·2181 605− | ·2020 791 | ·1869 167 |
| ·85 | ·3400 075− | ·3194 963 | ·2997 877 | ·2808 968 | ·2628 324 | ·2455 974 |
| ·86 | ·4163 890 | ·3949 687 | ·3741 443 | ·3539 491 | ·3344 103 | ·3155 493 |
| ·87 | ·4993 353 | ·4778 832 | ·4567 855− | ·4360 873 | ·4158 289 | ·3960 460 |
| ·88 | ·5859 793 | ·5654 917 | ·5451 111 | ·5248 866 | ·5048 643 | ·4850 872 |
| ·89 | ·6725 743 | ·6540 475+ | ·6354 080 | ·6167 010 | ·5979 707 | ·5792 594 |
| ·90 | ·7548 165− | ·7390 979 | ·7231 059 | ·7068 758 | ·6904 428 | ·6738 421 |
| ·91 | ·8283 823 | ·8160 211 | ·8033 051 | ·7902 564 | ·7768 978 | ·7632 530 |
| ·92 | ·8896 279 | ·8807 648 | ·8715 473 | ·8619 845+ | ·8520 871 | ·8418 665− |
| ·93 | ·9363 151 | ·9306 537 | ·9247 019 | ·9184 601 | ·9119 297 | ·9051 127 |
| ·94 | ·9681 529 | ·9650 390 | ·9617 300 | ·9582 226 | ·9545 134 | ·9505 999 |
| ·95 | ·9869 141 | ·9855 150− | ·9840 124 | ·9824 028 | ·9806 825+ | ·9788 481 |
| ·96 | ·9959 579 | ·9954 879 | ·9949 778 | ·9944 256 | ·9938 292 | ·9931 866 |
| ·97 | ·9992 053 | ·9991 053 | ·9989 956 | ·9988 756 | ·9987 447 | ·9986 022 |
| ·98 | ·9999 315+ | ·9999 222 | ·9999 119 | ·9999 006 | ·9998 880 | ·9998 742 |
| ·99 | ·9999 992 | ·9999 991 | ·9999 990 | ·9999 988 | ·9999 987 | ·9999 985− |
| 1·00 | 1·0000 000 | 1·0000 000 | 1·0000 000 | 1·0000 000 | 1·0000 000 | 1·0000 000 |

$x$ = ·04 to ·70 $q = 7·5$ $p$ = 7·5 to 10

| | $p = 7·5$ | $p = 8$ | $p = 8·5$ | $p = 9$ | $p = 9·5$ | $p = 10$ |
|---|---|---|---|---|---|---|
| $B(p, q) =$ | ·4016 5879 × $\frac{1}{10^4}$ | ·2816 6171 × $\frac{1}{10^4}$ | ·2008 2939 × $\frac{1}{10^4}$ | ·1453 7379 × $\frac{1}{10^4}$ | ·1066 9062 × $\frac{1}{10^4}$ | ·7929 4794 × $\frac{1}{10^5}$ |
| $x$ | | | | | | |
| ·04 | ·0000 001 | | | | | |
| ·05 | ·0000 004 | ·0000 001 | | | | |
| ·06 | ·0000 016 | ·0000 005+ | ·0000 002 | ·0000 001 | | |
| ·07 | ·0000 048 | ·0000 017 | ·0000 006 | ·0000 002 | ·0000 001 | |
| ·08 | ·0000 123 | ·0000 046 | ·0000 017 | ·0000 006 | ·0000 002 | ·0000 001 |
| ·09 | ·0000 279 | ·0000 111 | ·0000 044 | ·0000 017 | ·0000 007 | ·0000 003 |
| ·10 | ·0000 577 | ·0000 243 | ·0000 101 | ·0000 041 | ·0000 017 | ·0000 007 |
| ·11 | ·0001 108 | ·0000 489 | ·0000 213 | ·0000 092 | ·0000 039 | ·0000 017 |
| ·12 | ·0001 998 | ·0000 920 | ·0000 418 | ·0000 188 | ·0000 084 | ·0000 037 |
| ·13 | ·0003 417 | ·0001 636 | ·0000 774 | ·0000 362 | ·0000 168 | ·0000 077 |
| ·14 | ·0005 586 | ·0002 774 | ·0001 361 | ·0000 661 | ·0000 317 | ·0000 151 |
| ·15 | ·0008 783 | ·0004 512 | ·0002 290 | ·0001 150− | ·0000 572 | ·0000 282 |
| ·16 | ·0013 348 | ·0007 077 | ·0003 708 | ·0001 922 | ·0000 986 | ·0000 501 |
| ·17 | ·0019 688 | ·0010 752 | ·0005 804 | ·0003 099 | ·0001 638 | ·0000 858 |
| ·18 | ·0028 278 | ·0015 880 | ·0008 814 | ·0004 840 | ·0002 632 | ·0001 418 |
| ·19 | ·0039 661 | ·0022 866 | ·0013 031 | ·0007 348 | ·0004 102 | ·0002 270 |
| ·20 | ·0054 448 | ·0032 182 | ·0018 804 | ·0010 872 | ·0006 224 | ·0003 531 |
| ·21 | ·0073 312 | ·0044 367 | ·0026 546 | ·0015 716 | ·0009 215− | ·0005 354 |
| ·22 | ·0096 984 | ·0060 027 | ·0036 734 | ·0022 246 | ·0013 342 | ·0007 931 |
| ·23 | ·0126 246 | ·0079 827 | ·0049 911 | ·0030 885− | ·0018 928 | ·0011 498 |
| ·24 | ·0161 917 | ·0104 494 | ·0066 688 | ·0042 124 | ·0026 356 | ·0016 345− |
| ·25 | ·0204 845− | ·0134 804 | ·0087 736 | ·0056 522 | ·0036 070 | ·0022 817 |
| ·26 | ·0255 892 | ·0171 573 | ·0113 784 | ·0074 699 | ·0048 582 | ·0031 321 |
| ·27 | ·0315 922 | ·0215 651 | ·0145 615+ | ·0097 343 | ·0064 469 | ·0042 329 |
| ·28 | ·0385 780 | ·0267 902 | ·0184 054 | ·0125 197 | ·0084 378 | ·0056 381 |
| ·29 | ·0466 280 | ·0329 198 | ·0229 958 | ·0159 059 | ·0109 016 | ·0074 084 |
| ·30 | ·0558 188 | ·0400 396 | ·0284 203 | ·0199 770 | ·0139 152 | ·0096 113 |
| ·31 | ·0662 204 | ·0482 326 | ·0347 674 | ·0248 203 | ·0175 607 | ·0123 208 |
| ·32 | ·0778 947 | ·0575 776 | ·0421 243 | ·0305 256 | ·0219 246 | ·0156 169 |
| ·33 | ·0908 939 | ·0681 467 | ·0505 761 | ·0371 829 | ·0270 969 | ·0195 851 |
| ·34 | ·1052 591 | ·0800 047 | ·0602 030 | ·0448 816 | ·0331 694 | ·0243 151 |
| ·35 | ·1210 189 | ·0932 066 | ·0710 797 | ·0537 084 | ·0402 349 | ·0298 998 |
| ·36 | ·1381 888 | ·1077 963 | ·0832 726 | ·0637 457 | ·0483 847 | ·0364 343 |
| ·37 | ·1567 695− | ·1238 056 | ·0968 389 | ·0750 696 | ·0577 077 | ·0440 139 |
| ·38 | ·1767 469 | ·1412 524 | ·1118 243 | ·0877 481 | ·0682 879 | ·0527 325− |
| ·39 | ·1980 915− | ·1601 398 | ·1282 618 | ·1018 393 | ·0802 026 | ·0626 807 |
| ·40 | ·2207 580 | ·1804 558 | ·1461 705+ | ·1173 896 | ·0935 206 | ·0739 440 |
| ·41 | ·2446 858 | ·2021 718 | ·1655 541 | ·1344 321 | ·1083 000 | ·0866 004 |
| ·42 | ·2697 991 | ·2252 432 | ·1864 000 | ·1529 851 | ·1245 866 | ·1007 184 |
| ·43 | ·2960 075− | ·2496 089 | ·2086 789 | ·1730 507 | ·1424 119 | ·1163 552 |
| ·44 | ·3232 069 | ·2751 915+ | ·2323 440 | ·1946 140 | ·1617 916 | ·1335 542 |
| ·45 | ·3512 809 | ·3018 985− | ·2573 312 | ·2176 421 | ·1827 241 | ·1523 435− |
| ·46 | ·3801 019 | ·3296 221 | ·2835 594 | ·2420 838 | ·2051 896 | ·1727 340 |
| ·47 | ·4095 323 | ·3582 413 | ·3109 306 | ·2678 693 | ·2291 492 | ·1947 181 |
| ·48 | ·4394 271 | ·3876 228 | ·3393 313 | ·2949 106 | ·2545 443 | ·2182 684 |
| ·49 | ·4696 348 | ·4176 227 | ·3686 331 | ·3231 019 | ·2812 964 | ·2433 369 |
| ·50 | ·5000 000[e] | ·4480 881 | ·3986 948 | ·3523 208 | ·3093 080 | ·2698 545+ |
| ·51 | ·5303 652 | ·4788 597 | ·4293 636 | ·3824 292 | ·3384 621 | ·2977 311 |
| ·52 | ·5605 729 | ·5097 730 | ·4604 771 | ·4132 750+ | ·3686 245+ | ·3268 557 |
| ·53 | ·5904 677 | ·5406 616 | ·4918 660 | ·4446 943 | ·3996 444 | ·3570 976 |
| ·54 | ·6198 981 | ·5713 586 | ·5233 557 | ·4765 132 | ·4313 564 | ·3883 071 |
| ·55 | ·6487 191 | ·6016 993 | ·5547 693 | ·5085 505− | ·4635 828 | ·4203 178 |
| ·56 | ·6767 931 | ·6315 236 | ·5859 302 | ·5406 199 | ·4961 362 | ·4529 484 |
| ·57 | ·7039 925+ | ·6606 778 | ·6166 640 | ·5725 335− | ·5288 217 | ·4860 051 |
| ·58 | ·7302 009 | ·6890 171 | ·6468 019 | ·6041 038 | ·5614 405− | ·5192 848 |
| ·59 | ·7553 142 | ·7164 073 | ·6761 825− | ·6351 473 | ·5937 924 | ·5525 780 |
| ·60 | ·7792 420 | ·7427 266 | ·7046 546 | ·6654 867 | ·6256 796 | ·5856 720 |
| ·61 | ·8019 085+ | ·7678 671 | ·7320 789 | ·6949 541 | ·6569 094 | ·6183 548 |
| ·62 | ·8232 531 | ·7917 362 | ·7583 305− | ·7233 934 | ·6872 975− | ·6504 185− |
| ·63 | ·8432 305+ | ·8142 572 | ·7832 999 | ·7506 627 | ·7166 712 | ·6816 626 |
| ·64 | ·8618 112 | ·8353 704 | ·8068 951 | ·7766 363 | ·7448 721 | ·7118 982 |
| ·65 | ·8789 811 | ·8550 335− | ·8290 418 | ·8012 066 | ·7717 586 | ·7409 504 |
| ·66 | ·8947 409 | ·8732 212 | ·8496 849 | ·8242 855+ | ·7972 079 | ·7686 620 |
| ·67 | ·9091 061 | ·8899 253 | ·8687 883 | ·8458 054 | ·8211 184 | ·7948 958 |
| ·68 | ·9221 053 | ·9051 542 | ·8863 350− | ·8657 196 | ·8434 106 | ·8195 372 |
| ·69 | ·9337 796 | ·9189 319 | ·9023 266 | ·8840 028 | ·8640 279 | ·8424 953 |
| ·70 | ·9441 812 | ·9312 968 | ·9167 828 | ·9006 506 | ·8829 376 | ·8637 050+ |

$x$ = ·71 to ·97 $q$ = 7·5 $p$ = 7·5 to 10

| $x$ | $p$ = 7·5 | $p$ = 8 | $p$ = 8·5 | $p$ = 9 | $p$ = 9·5 | $p$ = 10 |
|---|---|---|---|---|---|---|
| $B(p, q)$ = | ·4016 5879 × $\frac{1}{10^4}$ | ·2816 6171 × $\frac{1}{10^4}$ | ·2008 2939 × $\frac{1}{10^4}$ | ·1453 7379 × $\frac{1}{10^4}$ | ·1066 9062 × $\frac{1}{10^4}$ | ·7929 4794 × $\frac{1}{10^5}$ |
| ·71 | ·9533 720 | ·9423 006 | ·9297 398 | ·9156 790 | ·9001 301 | ·8831 267 |
| ·72 | ·9614 220 | ·9520 067 | ·9412 495+ | ·9291 229 | ·9156 185+ | ·9007 471 |
| ·73 | ·9684 078 | ·9604 882 | ·9513 772 | ·9410 349 | ·9294 377 | ·9165 778 |
| ·74 | ·9744 108 | ·9678 266 | ·9601 999 | ·9514 834 | ·9416 422 | ·9306 550− |
| ·75 | ·9795 155+ | ·9741 092 | ·9678 046 | ·9605 504 | ·9523 049 | ·9430 369 |
| ·76 | ·9838 083 | ·9794 279 | ·9742 855− | ·9683 291 | ·9615 136 | ·9538 020 |
| ·77 | ·9873 754 | ·9838 765+ | ·9797 420 | ·9749 215− | ·9693 695− | ·9630 461 |
| ·78 | ·9903 016 | ·9875 496 | ·9842 765+ | ·9804 357 | ·9759 832 | ·9708 793 |
| ·79 | ·9926 688 | ·9905 402 | ·9879 922 | ·9849 831 | ·9814 725+ | ·9774 225− |
| ·80 | ·9945 552 | ·9929 384 | ·9909 909 | ·9886 763 | ·9859 588 | ·9828 040 |
| ·81 | ·9960 339 | ·9948 300 | ·9933 709 | ·9916 259 | ·9895 644 | ·9871 561 |
| ·82 | ·9971 722 | ·9962 953 | ·9952 258 | ·9939 389 | ·9924 093 | ·9906 113 |
| ·83 | ·9980 312 | ·9974 077 | ·9966 427 | ·9957 166 | ·9946 090 | ·9932 992 |
| ·84 | ·9986 652 | ·9982 338 | ·9977 012 | ·9970 526 | ·9962 722 | ·9953 438 |
| ·85 | ·9991 217 | ·9988 321 | ·9984 725− | ·9980 319 | ·9974 986 | ·9968 604 |
| ·86 | ·9994 414 | ·9992 536 | ·9990 190 | ·9987 298 | ·9983 778 | ·9979 541 |
| ·87 | ·9996 583 | ·9995 412 | ·9993 941 | ·9992 117 | ·9989 883 | ·9987 179 |
| ·88 | ·9998 002 | ·9997 304 | ·9996 422 | ·9995 323 | ·9993 969 | ·9992 321 |
| ·89 | ·9998 892 | ·9998 498 | ·9997 997 | ·9997 368 | ·9996 590 | ·9995 638 |
| ·90 | ·9999 423 | ·9999 214 | ·9998 946 | ·9998 610 | ·9998 190 | ·9997 673 |
| ·91 | ·9999 721 | ·9999 619 | ·9999 486 | ·9999 319 | ·9999 109 | ·9998 850− |
| ·92 | ·9999 877 | ·9999 832 | ·9999 772 | ·9999 696 | ·9999 601 | ·9999 482 |
| ·93 | ·9999 952 | ·9999 934 | ·9999 910 | ·9999 880 | ·9999 841 | ·9999 793 |
| ·94 | ·9999 984 | ·9999 978 | ·9999 970 | ·9999 959 | ·9999 946 | ·9999 929 |
| ·95 | ·9999 996 | ·9999 994 | ·9999 992 | ·9999 989 | ·9999 985+ | ·9999 980 |
| ·96 | ·9999 999 | ·9999 999 | ·9999 998 | ·9999 998 | ·9999 997 | ·9999 996 |
| ·97 | 1·0000 000 | 1·0000 000 | 1·0000 000 | 1·0000 000 | 1·0000 000 | 1·0000 000 |

$x = \cdot 09$ to $\cdot 70$ $q = 7\cdot 5$ $p = 10\cdot 5$ to $15$

| $x$ | $p = 10\cdot 5$ | $p = 11$ | $p = 12$ | $p = 13$ | $p = 14$ | $p = 15$ |
|---|---|---|---|---|---|---|
| $B(p, q) =$ | ·5962 1226 × $\frac{1}{10^5}$ | ·4531 1311 × $\frac{1}{10^5}$ | ·2694 1860 × $\frac{1}{10^5}$ | ·1657 9606 × $\frac{1}{10^5}$ | ·1051 3897 × $\frac{1}{10^5}$ | ·6846 2583 × $\frac{1}{10^6}$ |
| ·09 | ·0000 001 | | | | | |
| ·10 | ·0000 003 | ·0000 001 | | | | |
| ·11 | ·0000 007 | ·0000 003 | | | | |
| ·12 | ·0000 016 | ·0000 007 | ·0000 001 | | | |
| ·13 | ·0000 035+ | ·0000 016 | ·0000 003 | ·0000 001 | | |
| ·14 | ·0000 071 | ·0000 033 | ·0000 007 | ·0000 001 | | |
| ·15 | ·0000 137 | ·0000 067 | ·0000 015+ | ·0000 003 | ·0000 001 | |
| ·16 | ·0000 253 | ·0000 126 | ·0000 031 | ·0000 007 | ·0000 002 | |
| ·17 | ·0000 446 | ·0000 230 | ·0000 060 | ·0000 015+ | ·0000 004 | ·0000 001 |
| ·18 | ·0000 757 | ·0000 401 | ·0000 110 | ·0000 030 | ·0000 008 | ·0000 002 |
| ·19 | ·0001 245+ | ·0000 678 | ·0000 196 | ·0000 055+ | ·0000 015+ | ·0000 004 |
| ·20 | ·0001 987 | ·0001 109 | ·0000 338 | ·0000 100 | ·0000 029 | ·0000 008 |
| ·21 | ·0003 085+ | ·0001 764 | ·0000 564 | ·0000 176 | ·0000 054 | ·0000 016 |
| ·22 | ·0004 675+ | ·0002 735− | ·0000 916 | ·0000 299 | ·0000 095+ | ·0000 030 |
| ·23 | ·0006 927 | ·0004 141 | ·0001 449 | ·0000 494 | ·0000 165− | ·0000 054 |
| ·24 | ·0010 053 | ·0006 136 | ·0002 238 | ·0000 796 | ·0000 277 | ·0000 094 |
| ·25 | ·0014 316 | ·0008 914 | ·0003 384 | ·0001 253 | ·0000 454 | ·0000 161 |
| ·26 | ·0020 030 | ·0012 712 | ·0005 015− | ·0001 929 | ·0000 726 | ·0000 268 |
| ·27 | ·0027 569 | ·0017 821 | ·0007 294 | ·0002 912 | ·0001 137 | ·0000 435+ |
| ·28 | ·0037 373 | ·0024 588 | ·0010 426 | ·0004 313 | ·0001 745+ | ·0000 692 |
| ·29 | ·0049 946 | ·0033 423 | ·0014 664 | ·0006 277 | ·0002 629 | ·0001 080 |
| ·30 | ·0065 864 | ·0044 803 | ·0020 314 | ·0008 988 | ·0003 891 | ·0001 652 |
| ·31 | ·0085 770 | ·0059 273 | ·0027 741 | ·0012 671 | ·0005 664 | ·0002 483 |
| ·32 | ·0110 380 | ·0077 452 | ·0037 377 | ·0017 607 | ·0008 118 | ·0003 671 |
| ·33 | ·0140 474 | ·0100 031 | ·0049 726 | ·0024 133 | ·0011 464 | ·0005 343 |
| ·34 | ·0176 891 | ·0127 772 | ·0065 363 | ·0032 650+ | ·0015 967 | ·0007 661 |
| ·35 | ·0220 527 | ·0161 503 | ·0084 944 | ·0043 634 | ·0021 946 | ·0010 831 |
| ·36 | ·0272 316 | ·0202 114 | ·0109 200 | ·0057 633 | ·0029 787 | ·0015 108 |
| ·37 | ·0333 229 | ·0250 544 | ·0138 942 | ·0075 282 | ·0039 950+ | ·0020 809 |
| ·38 | ·0404 246 | ·0307 777 | ·0175 050+ | ·0097 295− | ·0052 974 | ·0028 313 |
| ·39 | ·0486 353 | ·0374 822 | ·0218 476 | ·0124 474 | ·0069 482 | ·0038 078 |
| ·40 | ·0580 511 | ·0452 699 | ·0270 228 | ·0157 703 | ·0090 189 | ·0050 645− |
| ·41 | ·0687 645+ | ·0542 422 | ·0331 358 | ·0197 949 | ·0115 902 | ·0066 644 |
| ·42 | ·0808 618 | ·0644 976 | ·0402 955+ | ·0246 247 | ·0147 520 | ·0086 804 |
| ·43 | ·0944 207 | ·0761 297 | ·0486 118 | ·0303 697 | ·0186 036 | ·0111 952 |
| ·44 | ·1095 086 | ·0892 247 | ·0581 941 | ·0371 450+ | ·0232 526 | ·0143 020 |
| ·45 | ·1261 797 | ·1038 592 | ·0691 489 | ·0450 687 | ·0288 145− | ·0181 041 |
| ·46 | ·1444 737 | ·1200 975+ | ·0815 772 | ·0542 599 | ·0354 109 | ·0227 148 |
| ·47 | ·1644 128 | ·1379 894 | ·0955 718 | ·0648 367 | ·0431 685− | ·0282 561 |
| ·48 | ·1860 006 | ·1575 679 | ·1112 147 | ·0769 134 | ·0522 164 | ·0348 580 |
| ·49 | ·2092 207 | ·1788 469 | ·1285 740 | ·0905 970 | ·0626 840 | ·0426 563 |
| ·50 | ·2340 348 | ·2018 199 | ·1477 014 | ·1059 850+ | ·0746 977 | ·0517 910 |
| ·51 | ·2603 825+ | ·2264 577 | ·1686 290 | ·1231 612 | ·0883 784 | ·0624 030 |
| ·52 | ·2881 810 | ·2527 080 | ·1913 676 | ·1421 930 | ·1038 368 | ·0746 313 |
| ·53 | ·3173 245− | ·2804 942 | ·2159 036 | ·1631 277 | ·1211 709 | ·0886 094 |
| ·54 | ·3476 854 | ·3097 154 | ·2421 980 | ·1859 897 | ·1404 610 | ·1044 609 |
| ·55 | ·3791 154 | ·3402 466 | ·2701 844 | ·2107 774 | ·1617 667 | ·1222 955+ |
| ·56 | ·4114 463 | ·3719 398 | ·2997 686 | ·2374 608 | ·1851 222 | ·1422 046 |
| ·57 | ·4444 932 | ·4046 252 | ·3308 284 | ·2659 794 | ·2105 336 | ·1642 561 |
| ·58 | ·4780 559 | ·4381 131 | ·3632 137 | ·2962 411 | ·2379 750+ | ·1884 905− |
| ·59 | ·5119 227 | ·4721 968 | ·3967 481 | ·3281 212 | ·2673 865− | ·2149 159 |
| ·60 | ·5458 733 | ·5066 555+ | ·4312 307 | ·3614 627 | ·2986 715+ | ·2435 050− |
| ·61 | ·5796 825+ | ·5412 577 | ·4664 384 | ·3960 771 | ·3316 966 | ·2741 909 |
| ·62 | ·6131 241 | ·5757 651 | ·5021 298 | ·4317 467 | ·3662 904 | ·3068 655− |
| ·63 | ·6459 747 | ·6099 371 | ·5380 486 | ·4682 268 | ·4022 453 | ·3413 773 |
| ·64 | ·6780 179 | ·6435 347 | ·5739 284 | ·5052 500+ | ·4393 189 | ·3775 319 |
| ·65 | ·7090 485− | ·6763 256 | ·6094 976 | ·5425 305+ | ·4772 375+ | ·4150 926 |
| ·66 | ·7388 756 | ·7080 879 | ·6444 852 | ·5797 694 | ·5157 007 | ·4537 830 |
| ·67 | ·7673 267 | ·7386 153 | ·6786 254 | ·6166 609 | ·5543 865− | ·4932 909 |
| ·68 | ·7942 508 | ·7677 205+ | ·7116 643 | ·6528 988 | ·5929 579 | ·5332 739 |
| ·69 | ·8195 208 | ·7952 391 | ·7433 647 | ·6881 833 | ·6310 705− | ·5733 662 |
| ·70 | ·8430 360 | ·8210 328 | ·7735 115− | ·7222 280 | ·6683 804 | ·6131 867 |

$x$ = ·71 to ·98 $q = 7·5$ $p$ = 10·5 to 15

| $x$ | $p$ = 10·5 | $p$ = 11 | $p$ = 12 | $p$ = 13 | $p$ = 14 | $p$ = 15 |
|---|---|---|---|---|---|---|
| $B(p, q)$ = | ·5962 1226 × $\frac{1}{10^5}$ | ·4531 1311 × $\frac{1}{10^5}$ | ·2694 1860 × $\frac{1}{10^5}$ | ·1657 9606 × $\frac{1}{10^5}$ | ·1051 3897 × $\frac{1}{10^5}$ | ·6846 2583 × $\frac{1}{10^6}$ |
| ·71 | ·8647 231 | ·8449 917 | ·8019 165+ | ·7547 671 | ·7045 530 | ·6523 482 |
| ·72 | ·8845 376 | ·8670 365+ | ·8284 227 | ·7855 613 | ·7392 709 | ·6904 677 |
| ·73 | ·9024 637 | ·8871 192 | ·8529 070 | ·8144 041 | ·7722 434 | ·7271 766 |
| ·74 | ·9185 135+ | ·9052 234 | ·8752 836 | ·8411 272 | ·8032 136 | ·7621 315+ |
| ·75 | ·9327 262 | ·9213 639 | ·8955 042 | ·8656 038 | ·8319 659 | ·7950 243 |
| ·76 | ·9451 657 | ·9355 851 | ·9135 591 | ·8877 520 | ·8583 320 | ·8255 917 |
| ·77 | ·9559 180 | ·9479 585− | ·9294 761 | ·9075 360 | ·8821 952 | ·8536 234 |
| ·78 | ·9650 884 | ·9585 802 | ·9433 182 | ·9249 657 | ·9034 932 | ·8789 686 |
| ·79 | ·9727 979 | ·9675 671 | ·9551 808 | ·9400 952 | ·9222 189 | ·9015 397 |
| ·80 | ·9791 788 | ·9750 526 | ·9651 872 | ·9530 198 | ·9384 190 | ·9213 152 |
| ·81 | ·9843 715− | ·9811 822 | ·9734 840 | ·9638 709 | ·9521 910 | ·9383 378 |
| ·82 | ·9885 195− | ·9861 089 | ·9802 354 | ·9728 103 | ·9636 775− | ·9527 115+ |
| ·83 | ·9917 661 | ·9899 886 | ·9856 172 | ·9800 235+ | ·9730 594 | ·9645 956 |
| ·84 | ·9942 506 | ·9929 754 | ·9898 104 | ·9857 118 | ·9805 475− | ·9741 954 |
| ·85 | ·9961 045− | ·9952 175− | ·9929 957 | ·9900 843 | ·9863 723 | ·9817 521 |
| ·86 | ·9974 492 | ·9968 533 | ·9953 473 | ·9933 505− | ·9907 746 | ·9875 309 |
| ·87 | ·9983 938 | ·9980 091 | ·9970 281 | ·9957 122 | ·9939 951 | ·9918 076 |
| ·88 | ·9990 334 | ·9987 961 | ·9981 858 | ·9973 578 | ·9962 648 | ·9948 564 |
| ·89 | ·9994 483 | ·9993 096 | ·9989 498 | ·9984 560 | ·9977 968 | ·9969 378 |
| ·90 | ·9997 043 | ·9996 283 | ·9994 292 | ·9991 529 | ·9987 800 | ·9982 885− |
| ·91 | ·9998 531 | ·9998 145+ | ·9997 125− | ·9995 694 | ·9993 740 | ·9991 137 |
| ·92 | ·9999 336 | ·9999 157 | ·9998 682 | ·9998 007 | ·9997 076 | ·9995 822 |
| ·93 | ·9999 733 | ·9999 660 | ·9999 463 | ·9999 181 | ·9998 788 | ·9998 252 |
| ·94 | ·9999 908 | ·9999 883 | ·9999 813 | ·9999 712 | ·9999 570 | ·9999 374 |
| ·95 | ·9999 975− | ·9999 967 | ·9999 947 | ·9999 918 | ·9999 876 | ·9999 819 |
| ·96 | ·9999 995− | ·9999 993 | ·9999 989 | ·9999 983 | ·9999 974 | ·9999 961 |
| ·97 | ·9999 999 | ·9999 999 | ·9999 999 | ·9999 998 | ·9999 997 | ·9999 995− |
| ·98 | 1·0000 000 | 1·0000 000 | 1·0000 000 | 1·0000 000 | 1·0000 000 | 1·0000 000 |

$x = \cdot 19$ to $\cdot 80$ $q = 7\cdot 5$ $p = 16$ to $21$

| | $p = 16$ | $p = 17$ | $p = 18$ | $p = 19$ | $p = 20$ | $p = 21$ |
|---|---|---|---|---|---|---|
| $B(p, q) =$ | ·4564 1722 × $\frac{1}{10^8}$ | ·3107 5215+ × $\frac{1}{10^8}$ | ·2156 2394 × $\frac{1}{10^8}$ | ·1522 0514 × $\frac{1}{10^8}$ | ·1091 2821 × $\frac{1}{10^8}$ | ·7936 5971 × $\frac{1}{10^7}$ |
| $x$ | | | | | | |
| ·19 | ·0000 001 | | | | | |
| ·20 | ·0000 002 | ·0000 001 | | | | |
| ·21 | ·0000 005− | ·0000 001 | | | | |
| ·22 | ·0000 009 | ·0000 003 | ·0000 001 | | | |
| ·23 | ·0000 017 | ·0000 005+ | ·0000 002 | ·0000 001 | | |
| ·24 | ·0000 032 | ·0000 010 | ·0000 003 | ·0000 001 | | |
| ·25 | ·0000 056 | ·0000 019 | ·0000 006 | ·0000 002 | ·0000 001 | |
| ·26 | ·0000 097 | ·0000 035− | ·0000 012 | ·0000 004 | ·0000 001 | |
| ·27 | ·0000 164 | ·0000 061 | ·0000 022 | ·0000 008 | ·0000 003 | ·0000 001 |
| ·28 | ·0000 270 | ·0000 104 | ·0000 039 | ·0000 015− | ·0000 005+ | ·0000 002 |
| ·29 | ·0000 436 | ·0000 173 | ·0000 068 | ·0000 026 | ·0000 010 | ·0000 004 |
| ·30 | ·0000 689 | ·0000 283 | ·0000 115− | ·0000 046 | ·0000 018 | ·0000 007 |
| ·31 | ·0001 070 | ·0000 454 | ·0000 190 | ·0000 078 | ·0000 032 | ·0000 013 |
| ·32 | ·0001 632 | ·0000 714 | ·0000 308 | ·0000 131 | ·0000 055+ | ·0000 023 |
| ·33 | ·0002 447 | ·0001 104 | ·0000 491 | ·0000 215+ | ·0000 093 | ·0000 040 |
| ·34 | ·0003 613 | ·0001 678 | ·0000 768 | ·0000 347 | ·0000 155+ | ·0000 068 |
| ·35 | ·0005 255− | ·0002 510 | ·0001 182 | ·0000 550− | ·0000 253 | ·0000 115− |
| ·36 | ·0007 534 | ·0003 700 | ·0001 791 | ·0000 856 | ·0000 405− | ·0000 189 |
| ·37 | ·0010 657 | ·0005 375− | ·0002 673 | ·0001 313 | ·0000 637 | ·0000 306 |
| ·38 | ·0014 881 | ·0007 703 | ·0003 932 | ·0001 982 | ·0000 987 | ·0000 487 |
| ·39 | ·0020 523 | ·0010 895+ | ·0005 705− | ·0002 950− | ·0001 507 | ·0000 762 |
| ·40 | ·0027 973 | ·0015 220 | ·0008 168 | ·0004 329 | ·0002 268 | ·0001 176 |
| ·41 | ·0037 697 | ·0021 007 | ·0011 548 | ·0006 270 | ·0003 365− | ·0001 787 |
| ·42 | ·0050 252 | ·0028 664 | ·0016 130 | ·0008 965− | ·0004 926 | ·0002 678 |
| ·43 | ·0066 291 | ·0038 680 | ·0022 268 | ·0012 663 | ·0007 119 | ·0003 961 |
| ·44 | ·0086 571 | ·0051 643 | ·0030 399 | ·0017 676 | ·0010 162 | ·0005 782 |
| ·45 | ·0111 959 | ·0068 244 | ·0041 050− | ·0024 394 | ·0014 334 | ·0008 336 |
| ·46 | ·0143 438 | ·0089 288 | ·0054 855+ | ·0033 297 | ·0019 987 | ·0011 875− |
| ·47 | ·0182 101 | ·0115 704 | ·0072 565− | ·0044 968 | ·0027 560 | ·0016 719 |
| ·48 | ·0229 154 | ·0148 541 | ·0095 052 | ·0060 106 | ·0037 593 | ·0023 275+ |
| ·49 | ·0285 902 | ·0188 978 | ·0123 327 | ·0079 540 | ·0050 745+ | ·0032 050+ |
| ·50 | ·0353 745+ | ·0238 317 | ·0158 535+ | ·0104 239 | ·0067 804 | ·0043 666 |
| ·51 | ·0434 150+ | ·0297 971 | ·0201 964 | ·0135 320 | ·0089 704 | ·0058 878 |
| ·52 | ·0528 635+ | ·0369 458 | ·0255 038 | ·0174 054 | ·0117 535+ | ·0078 594 |
| ·53 | ·0638 735+ | ·0454 375+ | ·0319 305− | ·0221 866 | ·0152 557 | ·0103 884 |
| ·54 | ·0765 967 | ·0554 374 | ·0396 426 | ·0280 334 | ·0196 198 | ·0135 998 |
| ·55 | ·0911 791 | ·0671 125− | ·0488 148 | ·0351 169 | ·0250 058 | ·0176 372 |
| ·56 | ·1077 560 | ·0806 278 | ·0596 274 | ·0436 204 | ·0315 898 | ·0226 632 |
| ·57 | ·1264 474 | ·0961 414 | ·0722 620 | ·0537 356 | ·0395 629 | ·0288 590 |
| ·58 | ·1473 523 | ·1137 990 | ·0868 971 | ·0656 595+ | ·0491 278 | ·0364 232 |
| ·59 | ·1705 434 | ·1337 280 | ·1037 018 | ·0795 891 | ·0604 957 | ·0455 694 |
| ·60 | ·1960 618 | ·1560 316 | ·1228 300 | ·0957 154 | ·0738 811 | ·0565 227 |
| ·61 | ·2239 118 | ·1807 818 | ·1444 130 | ·1142 168 | ·0894 957 | ·0695 148 |
| ·62 | ·2540 565− | ·2080 136 | ·1685 522 | ·1352 512 | ·1075 412 | ·0847 775− |
| ·63 | ·2864 135+ | ·2377 190 | ·1953 119 | ·1589 477 | ·1282 009 | ·1025 350− |
| ·64 | ·3208 527 | ·2698 414 | ·2247 114 | ·1853 982 | ·1516 302 | ·1229 949 |
| ·65 | ·3571 943 | ·3042 716 | ·2567 190 | ·2146 481 | ·1779 468 | ·1463 379 |
| ·66 | ·3952 088 | ·3408 446 | ·2912 453 | ·2466 886 | ·2072 207 | ·1727 060 |
| ·67 | ·4346 187 | ·3793 385+ | ·3281 394 | ·2814 485+ | ·2394 637 | ·2021 916 |
| ·68 | ·4751 019 | ·4194 749 | ·3671 855+ | ·3187 888 | ·2746 205− | ·2348 248 |
| ·69 | ·5162 966 | ·4609 213 | ·4081 030 | ·3584 979 | ·3125 609 | ·2705 630 |
| ·70 | ·5578 089 | ·5032 964 | ·4505 475+ | ·4002 896 | ·3530 736 | ·3092 807 |
| ·71 | ·5992 212 | ·5461 771 | ·4941 160 | ·4438 046 | ·3958 632 | ·3507 624 |
| ·72 | ·6401 028 | ·5891 083 | ·5383 538 | ·4886 144 | ·4405 504 | ·3946 974 |
| ·73 | ·6800 218 | ·6316 143 | ·5827 655+ | ·5342 289 | ·4866 757 | ·4406 798 |
| ·74 | ·7185 571 | ·6732 125+ | ·6268 276 | ·5801 078 | ·5337 076 | ·4882 122 |
| ·75 | ·7553 120 | ·7134 280 | ·6700 041 | ·6256 756 | ·5810 554 | ·5367 142 |
| ·76 | ·7899 266 | ·7518 095− | ·7117 641 | ·6703 394 | ·6280 862 | ·5855 372 |
| ·77 | ·8220 897 | ·7879 446 | ·7516 002 | ·7135 092 | ·6741 451 | ·6339 840 |
| ·78 | ·8515 500− | ·8214 753 | ·7890 476 | ·7546 203 | ·7185 803 | ·6813 330 |
| ·79 | ·8781 241 | ·8521 107 | ·8237 026 | ·7931 560 | ·7607 686 | ·7268 671 |
| ·80 | ·9017 028 | ·8796 389 | ·8552 388 | ·8286 698 | ·8001 431 | ·7699 047 |

$x = \cdot 81$ to $\cdot 99$ $q = 7\cdot 5$ $p = 16$ to $21$

| | $p = 16$ | $p = 17$ | $p = 18$ | $p = 19$ | $p = 20$ | $p = 21$ |
|---|---|---|---|---|---|---|
| $B(p, q) =$ | $\cdot 4564\ 1722 \times \frac{1}{10^8}$ | $\cdot 3107\ 5215^+ \times \frac{1}{10^8}$ | $\cdot 2156\ 2394 \times \frac{1}{10^8}$ | $\cdot 1522\ 0514 \times \frac{1}{10^8}$ | $\cdot 1091\ 2821 \times \frac{1}{10^8}$ | $\cdot 7936\ 5971 \times \frac{1}{10^7}$ |
| $x$ | | | | | | |
| ·81 | ·9222 542 | ·9039 340 | ·8834 207 | ·8608 049 | ·8362 191 | ·8098 324 |
| ·82 | ·9398 229 | ·9249 607 | ·9081 139 | ·8893 111 | ·8686 181 | ·8461 351 |
| ·83 | ·9545 264 | ·9427 738 | ·9292 894 | ·9140 558 | ·8970 863 | ·8784 242 |
| ·84 | ·9665 474 | ·9575 133 | ·9470 231 | ·9350 292 | ·9215 078 | ·9064 584 |
| ·85 | ·9761 232 | ·9693 949 | ·9614 892 | ·9523 426 | ·9419 084 | ·9301 568 |
| ·86 | ·9835 324 | ·9786 968 | ·9729 480 | ·9662 189 | ·9584 519 | ·9496 016 |
| ·87 | ·9890 797 | ·9857 424 | ·9817 287 | ·9769 759 | ·9714 263 | ·9650 291 |
| ·88 | ·9930 799 | ·9908 815+ | ·9882 073 | ·9850 041 | ·9812 210 | ·9768 100 |
| ·89 | ·9958 419 | ·9944 704 | ·9927 830 | ·9907 390 | ·9882 974 | ·9854 181 |
| ·90 | ·9976 545− | ·9968 521 | ·9958 538 | ·9946 309 | ·9931 537 | ·9913 922 |
| ·91 | ·9987 741 | ·9983 396 | ·9977 930 | ·9971 160 | ·9962 892 | ·9952 922 |
| ·92 | ·9994 169 | ·9992 029 | ·9989 308 | ·9985 901 | ·9981 694 | ·9976 566 |
| ·93 | ·9997 537 | ·9996 603 | ·9995 402 | ·9993 882 | ·9991 984 | ·9989 645+ |
| ·94 | ·9999 110 | ·9998 761 | ·9998 308 | ·9997 728 | ·9996 996 | ·9996 084 |
| ·95 | ·9999 740 | ·9999 634 | ·9999 496 | ·9999 317 | ·9999 089 | ·9998 803 |
| ·96 | ·9999 944 | ·9999 921 | ·9999 890 | ·9999 849 | ·9999 797 | ·9999 731 |
| ·97 | ·9999 993 | ·9999 989 | ·9999 985+ | ·9999 980 | ·9999 972 | ·9999 963 |
| ·98 | 1·0000 000 | ·9999 999 | ·9999 999 | ·9999 999 | ·9999 998 | ·9999 998 |
| ·99 | | 1·0000 000 | 1·0000 000 | 1·0000 000 | 1·0000 000 | 1·0000 000 |

$x$ = ·28 to ·90 $q = 7{\cdot}5$ $p$ = 22 to 27

| | $p = 22$ | $p = 23$ | $p = 24$ | $p = 25$ | $p = 26$ | $p = 27$ |
|---|---|---|---|---|---|---|
| $B(p, q) =$ | ·5848 0189 × $\frac{1}{10^7}$ | ·4361 2345 × $\frac{1}{10^7}$ | ·3288 7998 × $\frac{1}{10^7}$ | ·2505 7522 × $\frac{1}{10^7}$ | ·1927 5017 × $\frac{1}{10^7}$ | ·1495 9715 × $\frac{1}{10^7}$ |
| $x$ | | | | | | |
| ·28 | ·0000 001 | | | | | |
| ·29 | ·0000 001 | ·0000 001 | | | | |
| ·30 | ·0000 003 | ·0000 001 | | | | |
| ·31 | ·0000 005+ | ·0000 002 | ·0000 001 | | | |
| ·32 | ·0000 009 | ·0000 004 | ·0000 002 | ·0000 001 | | |
| ·33 | ·0000 017 | ·0000 007 | ·0000 003 | ·0000 001 | ·0000 001 | |
| ·34 | ·0000 030 | ·0000 013 | ·0000 006 | ·0000 002 | ·0000 001 | |
| ·35 | ·0000 052 | ·0000 023 | ·0000 010 | ·0000 004 | ·0000 002 | ·0000 001 |
| ·36 | ·0000 087 | ·0000 040 | ·0000 018 | ·0000 008 | ·0000 004 | ·0000 002 |
| ·37 | ·0000 145+ | ·0000 068 | ·0000 032 | ·0000 015− | ·0000 007 | ·0000 003 |
| ·38 | ·0000 238 | ·0000 115− | ·0000 055+ | ·0000 026 | ·0000 012 | ·0000 006 |
| ·39 | ·0000 382 | ·0000 189 | ·0000 093 | ·0000 045+ | ·0000 022 | ·0000 011 |
| ·40 | ·0000 603 | ·0000 307 | ·0000 155− | ·0000 077 | ·0000 038 | ·0000 019 |
| ·41 | ·0000 940 | ·0000 490 | ·0000 253 | ·0000 130 | ·0000 066 | ·0000 033 |
| ·42 | ·0001 442 | ·0000 769 | ·0000 407 | ·0000 214 | ·0000 111 | ·0000 058 |
| ·43 | ·0002 182 | ·0001 192 | ·0000 645+ | ·0000 347 | ·0000 185− | ·0000 098 |
| ·44 | ·0003 258 | ·0001 820 | ·0001 008 | ·0000 554 | ·0000 302 | ·0000 164 |
| ·45 | ·0004 802 | ·0002 741 | ·0001 552 | ·0000 872 | ·0000 486 | ·0000 269 |
| ·46 | ·0006 988 | ·0004 076 | ·0002 358 | ·0001 353 | ·0000 771 | ·0000 437 |
| ·47 | ·0010 047 | ·0005 984 | ·0003 535+ | ·0002 073 | ·0001 206 | ·0000 697 |
| ·48 | ·0014 276 | ·0008 679 | ·0005 234 | ·0003 132 | ·0001 861 | ·0001 098 |
| ·49 | ·0020 054 | ·0012 439 | ·0007 653 | ·0004 673 | ·0002 833 | ·0001 706 |
| ·50 | ·0027 861 | ·0017 624 | ·0011 058 | ·0006 887 | ·0004 259 | ·0002 616 |
| ·51 | ·0038 291 | ·0024 690 | ·0015 793 | ·0010 027 | ·0006 321 | ·0003 959 |
| ·52 | ·0052 077 | ·0034 214 | ·0022 300 | ·0014 427 | ·0009 269 | ·0005 916 |
| ·53 | ·0070 103 | ·0046 910 | ·0031 143 | ·0020 523 | ·0013 432 | ·0008 733 |
| ·54 | ·0093 429 | ·0063 650− | ·0043 024 | ·0028 870 | ·0019 239 | ·0012 739 |
| ·55 | ·0123 301 | ·0085 488 | ·0058 814 | ·0040 169 | ·0027 249 | ·0018 366 |
| ·56 | ·0161 170 | ·0113 680 | ·0079 570 | ·0055 295− | ·0038 166 | ·0026 177 |
| ·57 | ·0208 692 | ·0149 696 | ·0106 564 | ·0075 320 | ·0052 881 | ·0036 893 |
| ·58 | ·0267 737 | ·0195 234 | ·0141 298 | ·0101 543 | ·0072 490 | ·0051 427 |
| ·59 | ·0340 371 | ·0252 227 | ·0185 525− | ·0135 512 | ·0098 333 | ·0070 913 |
| ·60 | ·0428 840 | ·0322 831 | ·0241 249 | ·0179 043 | ·0132 016 | ·0096 745+ |
| ·61 | ·0535 539 | ·0409 412 | ·0310 731 | ·0234 232 | ·0175 436 | ·0130 603 |
| ·62 | ·0662 955− | ·0514 511 | ·0396 465+ | ·0303 456 | ·0230 796 | ·0174 485+ |
| ·63 | ·0813 606 | ·0640 794 | ·0501 154 | ·0389 356 | ·0300 610 | ·0230 722 |
| ·64 | ·0989 958 | ·0790 984 | ·0627 652 | ·0494 809 | ·0387 684 | ·0301 985− |
| ·65 | ·1194 326 | ·0967 772 | ·0778 894 | ·0622 873 | ·0495 092 | ·0391 270 |
| ·66 | ·1428 755+ | ·1173 704 | ·0957 798 | ·0776 707 | ·0626 111 | ·0501 870 |
| ·67 | ·1694 898 | ·1411 061 | ·1167 146 | ·0959 463 | ·0784 137 | ·0637 302 |
| ·68 | ·1993 877 | ·1681 708 | ·1409 443 | ·1174 160 | ·0972 570 | ·0801 218 |
| ·69 | ·2326 149 | ·1986 945− | ·1686 749 | ·1423 515− | ·1194 659 | ·0997 271 |
| ·70 | ·2691 373 | ·2327 345− | ·2000 511 | ·1709 765+ | ·1453 327 | ·1228 944 |
| ·71 | ·3088 293 | ·2702 604 | ·2351 376 | ·2034 465− | ·1750 957 | ·1499 343 |
| ·72 | ·3514 645+ | ·3111 400 | ·2739 013 | ·2398 278 | ·2089 163 | ·1810 960 |
| ·73 | ·3967 100 | ·3551 286 | ·3161 957 | ·2800 773 | ·2468 556 | ·2165 408 |
| ·74 | ·4441 250− | ·4018 614 | ·3617 477 | ·3240 241 | ·2888 507 | ·2563 152 |
| ·75 | ·4931 647 | ·4508 524 | ·4101 498 | ·3713 552 | ·3346 943 | ·3003 246 |
| ·76 | ·5431 908 | ·5014 989 | ·4608 584 | ·4216 064 | ·3840 186 | ·3483 103 |
| ·77 | ·5934 881 | ·5530 935+ | ·5131 995− | ·4741 615− | ·4362 868 | ·3998 325− |
| ·78 | ·6432 875− | ·6048 442 | ·5663 843 | ·5282 609 | ·4907 932 | ·4542 622 |
| ·79 | ·6917 951 | ·6559 021 | ·6195 332 | ·5830 203 | ·5466 754 | ·5107 848 |
| ·80 | ·7382 265− | ·7053 963 | ·6717 097 | ·6374 616 | ·6029 396 | ·5684 176 |
| ·81 | ·7818 436 | ·7524 745− | ·7219 625+ | ·6905 543 | ·6584 990 | ·6260 430 |
| ·82 | ·8219 927 | ·7963 468 | ·7693 741 | ·7412 662 | ·7122 252 | ·6824 581 |
| ·83 | ·8581 402 | ·8363 303 | ·8131 123 | ·7886 221 | ·7630 102 | ·7364 379 |
| ·84 | ·8899 041 | ·8718 900 | ·8524 817 | ·8317 634 | ·8098 352 | ·7868 105+ |
| ·85 | ·9170 762 | ·9026 727 | ·8869 697 | ·8700 072 | ·8518 403 | ·8325 380 |
| ·86 | ·9396 343 | ·9285 299 | ·9162 813 | ·9028 946 | ·8883 887 | ·8727 949 |
| ·87 | ·9577 408 | ·9495 266 | ·9403 605+ | ·9302 264 | ·9191 173 | ·9070 361 |
| ·88 | ·9717 267 | ·9659 318 | ·9593 911 | ·9520 764 | ·9439 659 | ·9350 444 |
| ·89 | ·9820 624 | ·9781 935− | ·9737 770 | ·9687 817 | ·9631 801 | ·9569 482 |
| ·90 | ·9893 162 | ·9868 957 | ·9841 016 | ·9809 058 | ·9772 819 | ·9732 049 |

$x = \cdot91$ to $\cdot99$ $q = 7\cdot5$ $p = 22$ to $27$

| | $p = 22$ | $p = 23$ | $p = 24$ | $p = 25$ | $p = 26$ | $p = 27$ |
|---|---|---|---|---|---|---|
| $B(p, q) =$ | $\cdot5848\ 0189 \times \frac{1}{10^7}$ | $\cdot4361\ 2345 \times \frac{1}{10^7}$ | $\cdot3288\ 7998 \times \frac{1}{10^7}$ | $\cdot2505\ 7522 \times \frac{1}{10^7}$ | $\cdot1927\ 5017 \times \frac{1}{10^7}$ | $\cdot1495\ 9715 \times \frac{1}{10^7}$ |
| $x$ | | | | | | |
| ·91 | ·9941 042 | ·9927 036 | ·9910 690 | ·9891 785+ | ·9870 109 | ·9845 453 |
| ·92 | ·9970 388 | ·9963 025− | ·9954 336 | ·9944 177 | ·9932 401 | ·9918 859 |
| ·93 | ·9986 798 | ·9983 366 | ·9979 274 | ·9974 437 | ·9968 769 | ·9962 180 |
| ·94 | ·9994 962 | ·9993 596 | ·9991 949 | ·9989 981 | ·9987 651 | ·9984 912 |
| ·95 | ·9998 446 | ·9998 006 | ·9997 471 | ·9996 825− | ·9996 051 | ·9995 133 |
| ·96 | ·9999 647 | ·9999 544 | ·9999 416 | ·9999 260 | ·9999 071 | ·9998 845+ |
| ·97 | ·9999 951 | ·9999 936 | ·9999 917 | ·9999 894 | ·9999 866 | ·9999 831 |
| ·98 | ·9999 997 | ·9999 996 | ·9999 995+ | ·9999 994 | ·9999 992 | ·9999 990 |
| ·99 | 1·0000 000 | 1·0000 000 | 1·0000 000 | 1·0000 000 | 1·0000 000 | 1·0000 000 |

$x$ = ·36 to ·99 $q = 7{\cdot}5$ $p$ = 28 to 33

| $x$ | $p = 28$ | $p = 29$ | $p = 30$ | $p = 31$ | $p = 32$ | $p = 33$ |
|---|---|---|---|---|---|---|
| $B(p, q) =$ | ·1170 7603 × $\frac{1}{10^7}$ | ·9234 1655 × $\frac{1}{10^8}$ | ·7336 7342 × $\frac{1}{10^8}$ | ·5869 3874 × $\frac{1}{10^8}$ | ·4726 0002 × $\frac{1}{10^8}$ | ·3828 6584 × $\frac{1}{10^8}$ |
| ·36 | ·0000 001 | | | | | |
| ·37 | ·0000 001 | ·0000 001 | | | | |
| ·38 | ·0000 003 | ·0000 001 | ·0000 001 | | | |
| ·39 | ·0000 005+ | ·0000 002 | ·0000 001 | ·0000 001 | | |
| ·40 | ·0000 009 | ·0000 005− | ·0000 002 | ·0000 001 | ·0000 001 | |
| ·41 | ·0000 017 | ·0000 008 | ·0000 004 | ·0000 002 | ·0000 001 | |
| ·42 | ·0000 030 | ·0000 015+ | ·0000 008 | ·0000 004 | ·0000 002 | ·0000 001 |
| ·43 | ·0000 052 | ·0000 027 | ·0000 014 | ·0000 007 | ·0000 004 | ·0000 002 |
| ·44 | ·0000 088 | ·0000 047 | ·0000 025+ | ·0000 013 | ·0000 007 | ·0000 004 |
| ·45 | ·0000 148 | ·0000 081 | ·0000 044 | ·0000 024 | ·0000 013 | ·0000 007 |
| ·46 | ·0000 246 | ·0000 137 | ·0000 076 | ·0000 042 | ·0000 023 | ·0000 013 |
| ·47 | ·0000 401 | ·0000 229 | ·0000 130 | ·0000 073 | ·0000 041 | ·0000 023 |
| ·48 | ·0000 644 | ·0000 376 | ·0000 218 | ·0000 126 | ·0000 072 | ·0000 041 |
| ·49 | ·0001 021 | ·0000 608 | ·0000 360 | ·0000 212 | ·0000 124 | ·0000 072 |
| ·50 | ·0001 597 | ·0000 969 | ·0000 585− | ·0000 351 | ·0000 210 | ·0000 125− |
| ·51 | ·0002 464 | ·0001 525− | ·0000 938 | ·0000 574 | ·0000 350− | ·0000 212 |
| ·52 | ·0003 753 | ·0002 367 | ·0001 484 | ·0000 926 | ·0000 575+ | ·0000 355+ |
| ·53 | ·0005 644 | ·0003 626 | ·0002 317 | ·0001 473 | ·0000 932 | ·0000 587 |
| ·54 | ·0008 383 | ·0005 486 | ·0003 570 | ·0002 311 | ·0001 489 | ·0000 955+ |
| ·55 | ·0012 304 | ·0008 196 | ·0005 430 | ·0003 579 | ·0002 348 | ·0001 533 |
| ·56 | ·0017 846 | ·0012 098 | ·0008 157 | ·0005 472 | ·0003 654 | ·0002 428 |
| ·57 | ·0025 586 | ·0017 645− | ·0012 104 | ·0008 261 | ·0005 612 | ·0003 795− |
| ·58 | ·0036 269 | ·0025 437 | ·0017 745+ | ·0012 318 | ·0008 510 | ·0005 853 |
| ·59 | ·0050 841 | ·0036 249 | ·0025 710 | ·0018 145+ | ·0012 746 | ·0008 913 |
| ·60 | ·0070 488 | ·0051 076 | ·0036 819 | ·0026 411 | ·0018 857 | ·0013 404 |
| ·61 | ·0096 672 | ·0071 169 | ·0052 125− | ·0037 991 | ·0027 562 | ·0019 908 |
| ·62 | ·0131 168 | ·0098 076 | ·0072 961 | ·0054 015+ | ·0039 806 | ·0029 207 |
| ·63 | ·0176 094 | ·0133 688 | ·0100 985− | ·0075 917 | ·0056 814 | ·0042 334 |
| ·64 | ·0233 933 | ·0180 269 | ·0138 226 | ·0105 488 | ·0080 143 | ·0060 628 |
| ·65 | ·0307 540 | ·0240 481 | ·0187 122 | ·0144 925− | ·0111 745+ | ·0085 798 |
| ·66 | ·0400 130 | ·0317 393 | ·0250 548 | ·0196 872 | ·0154 017 | ·0119 987 |
| ·67 | ·0515 238 | ·0414 470 | ·0331 822 | ·0264 451 | ·0209 848 | ·0165 832 |
| ·68 | ·0656 648 | ·0535 519 | ·0434 689 | ·0351 270 | ·0282 650+ | ·0226 511 |
| ·69 | ·0828 285− | ·0684 616 | ·0563 266 | ·0461 393 | ·0376 361 | ·0305 772 |
| ·70 | ·1034 063 | ·0865 978 | ·0721 947 | ·0599 280 | ·0495 409 | ·0407 930 |
| ·71 | ·1277 690 | ·1083 784 | ·0915 252 | ·0769 669 | ·0644 632 | ·0537 823 |
| ·72 | ·1562 432 | ·1341 952 | ·1147 626 | ·0977 396 | ·0829 131 | ·0700 696 |
| ·73 | ·1890 835− | ·1643 866 | ·1423 169 | ·1227 154 | ·1054 060 | ·0902 035− |
| ·74 | ·2264 429 | ·1992 057 | ·1745 326 | ·1523 185+ | ·1324 334 | ·1147 294 |
| ·75 | ·2683 418 | ·2387 862 | ·2116 511 | ·1868 904 | ·1644 260 | ·1441 554 |
| ·76 | ·3146 387 | ·2831 076 | ·2537 728 | ·2266 479 | ·2017 105+ | ·1789 084 |
| ·77 | ·3650 051 | ·3319 625+ | ·3008 171 | ·2716 391 | ·2444 611 | ·2192 833 |
| ·78 | ·4189 083 | ·3849 306 | ·3524 879 | ·3216 997 | ·2926 495+ | ·2653 878 |
| ·79 | ·4756 053 | ·4413 619 | ·4082 461 | ·3764 164 | ·3459 984 | ·3170 871 |
| ·80 | ·5341 512 | ·5003 744 | ·4672 964 | ·4351 005+ | ·4039 432 | ·3739 543 |
| ·81 | ·5934 248 | ·5608 707 | ·5285 915+ | ·4967 804 | ·4656 107 | ·4352 348 |
| ·82 | ·6521 729 | ·6215 740 | ·5908 591 | ·5602 158 | ·5298 196 | ·4998 319 |
| ·83 | ·7090 733 | ·6810 882 | ·6526 543 | ·6239 409 | ·5951 116 | ·5663 229 |
| ·84 | ·7628 137 | ·7379 770 | ·7124 381 | ·6863 372 | ·6598 154 | ·6330 118 |
| ·85 | ·8121 813 | ·7908 612 | ·7686 774 | ·7457 355+ | ·7221 461 | ·6980 223 |
| ·86 | ·8561 558 | ·8385 242 | ·8199 624 | ·8005 406 | ·7803 357 | ·7594 299 |
| ·87 | ·8939 952 | ·8800 158 | ·8651 278 | ·8493 688 | ·8327 837 | ·8154 238 |
| ·88 | ·9253 033 | ·9147 413 | ·9033 634 | ·8911 815+ | ·8782 137 | ·8644 840 |
| ·89 | ·9500 666 | ·9425 203 | ·9342 986 | ·9253 959 | ·9158 111 | ·9055 480 |
| ·90 | ·9686 523 | ·9636 038 | ·9580 417 | ·9519 512 | ·9453 205− | ·9381 406 |
| ·91 | ·9817 614 | ·9786 400 | ·9751 629 | ·9713 131 | ·9670 752 | ·9624 354 |
| ·92 | ·9903 401 | ·9885 878 | ·9866 143 | ·9844 054 | ·9819 470 | ·9792 260 |
| ·93 | ·9954 577 | ·9945 864 | ·9935 946 | ·9924 723 | ·9912 098 | ·9897 971 |
| ·94 | ·9981 719 | ·9978 020 | ·9973 763 | ·9968 895+ | ·9963 360 | ·9957 100 |
| ·95 | ·9994 050− | ·9992 783 | ·9991 309 | ·9989 605+ | ·9987 647 | ·9985 410 |
| ·96 | ·9998 576 | ·9998 257 | ·9997 882 | ·9997 445− | ·9996 937 | ·9996 350+ |
| ·97 | ·9999 790 | ·9999 741 | ·9999 682 | ·9999 613 | ·9999 532 | ·9999 438 |
| ·98 | ·9999 987 | ·9999 984 | ·9999 980 | ·9999 976 | ·9999 971 | ·9999 964 |
| ·99 | 1·0000 000 | 1·0000 000 | 1·0000 000 | 1·0000 000 | 1·0000 000 | 1·0000 000 |

$x$ = ·43 to 1·00 $q$ = 7·5 $p$ = 34 to 39

| $x$ | $p$ = 34 | $p$ = 35 | $p$ = 36 | $p$ = 37 | $p$ = 38 | $p$ = 39 |
|---|---|---|---|---|---|---|
| $B(p, q)$ = | ·3119 6476 × $\frac{1}{10^8}$ | ·2555 8559 × $\frac{1}{10^8}$ | ·2104 8225 × $\frac{1}{10^8}$ | ·1741 9221 × $\frac{1}{10^8}$ | ·1448 3397 × $\frac{1}{10^8}$ | ·1209 6024 × $\frac{1}{10^8}$ |
| ·43 | ·0000 001 | | | | | |
| ·44 | ·0000 002 | ·0000 001 | ·0000 001 | | | |
| ·45 | ·0000 004 | ·0000 002 | ·0000 001 | ·0000 001 | | |
| ·46 | ·0000 007 | ·0000 004 | ·0000 002 | ·0000 001 | ·0000 001 | |
| ·47 | ·0000 013 | ·0000 007 | ·0000 004 | ·0000 002 | ·0000 001 | ·0000 001 |
| ·48 | ·0000 023 | ·0000 013 | ·0000 007 | ·0000 004 | ·0000 002 | ·0000 001 |
| ·49 | ·0000 042 | ·0000 024 | ·0000 014 | ·0000 008 | ·0000 005− | ·0000 003 |
| ·50 | ·0000 074 | ·0000 044 | ·0000 026 | ·0000 015− | ·0000 009 | ·0000 005+ |
| ·51 | ·0000 128 | ·0000 077 | ·0000 046 | ·0000 027 | ·0000 016 | ·0000 010 |
| ·52 | ·0000 219 | ·0000 134 | ·0000 082 | ·0000 050− | ·0000 030 | ·0000 018 |
| ·53 | ·0000 368 | ·0000 230 | ·0000 143 | ·0000 088 | ·0000 055− | ·0000 034 |
| ·54 | ·0000 610 | ·0000 388 | ·0000 246 | ·0000 155+ | ·0000 097 | ·0000 061 |
| ·55 | ·0000 997 | ·0000 645+ | ·0000 416 | ·0000 267 | ·0000 171 | ·0000 109 |
| ·56 | ·0001 607 | ·0001 059 | ·0000 695+ | ·0000 455− | ·0000 296 | ·0000 192 |
| ·57 | ·0002 555− | ·0001 713 | ·0001 144 | ·0000 761 | ·0000 505− | ·0000 334 |
| ·58 | ·0004 008 | ·0002 734 | ·0001 857 | ·0001 257 | ·0000 848 | ·0000 570 |
| ·59 | ·0006 207 | ·0004 305− | ·0002 974 | ·0002 047 | ·0001 404 | ·0000 960 |
| ·60 | ·0009 488 | ·0006 688 | ·0004 697 | ·0003 287 | ·0002 292 | ·0001 593 |
| ·61 | ·0014 319 | ·0010 258 | ·0007 321 | ·0005 206 | ·0003 689 | ·0002 605+ |
| ·62 | ·0021 341 | ·0015 532 | ·0011 262 | ·0008 136 | ·0005 857 | ·0004 203 |
| ·63 | ·0031 415− | ·0023 220 | ·0017 099 | ·0012 547 | ·0009 175− | ·0006 687 |
| ·64 | ·0045 677 | ·0034 280 | ·0025 631 | ·0019 096 | ·0014 179 | ·0010 494 |
| ·65 | ·0065 610 | ·0049 979 | ·0037 932 | ·0028 688 | ·0021 624 | ·0016 246 |
| ·66 | ·0093 104 | ·0071 969 | ·0055 430 | ·0042 543 | ·0032 543 | ·0024 814 |
| ·67 | ·0130 533 | ·0102 362 | ·0079 981 | ·0062 279 | ·0048 335− | ·0037 394 |
| ·68 | ·0180 817 | ·0143 805− | ·0113 963 | ·0090 006 | ·0070 854 | ·0055 602 |
| ·69 | ·0247 471 | ·0199 553 | ·0160 350− | ·0128 415+ | ·0102 509 | ·0081 577 |
| ·70 | ·0334 634 | ·0273 517 | ·0222 791 | ·0180 871 | ·0146 372 | ·0118 092 |
| ·71 | ·0447 051 | ·0370 282 | ·0305 654 | ·0251 483 | ·0206 265− | ·0168 668 |
| ·72 | ·0590 009 | ·0495 079 | ·0414 035+ | ·0345 148 | ·0286 836 | ·0237 670 |
| ·73 | ·0769 197 | ·0653 686 | ·0553 704 | ·0467 538 | ·0393 587 | ·0330 369 |
| ·74 | ·0990 479 | ·0852 252 | ·0730 967 | ·0625 011 | ·0532 830 | ·0452 948 |
| ·75 | ·1259 579 | ·1097 006 | ·0952 432 | ·0824 424 | ·0711 552 | ·0612 418 |
| ·76 | ·1581 653 | ·1393 868 | ·1224 646 | ·1072 817 | ·0937 155+ | ·0816 417 |
| ·77 | ·1960 777 | ·1747 934 | ·1553 609 | ·1376 962 | ·1217 048 | ·1072 852 |
| ·78 | ·2399 352 | ·2162 868 | ·1944 154 | ·1742 755− | ·1558 066 | ·1389 368 |
| ·79 | ·2897 484 | ·2640 218 | ·2399 234 | ·2174 483 | ·1965 738 | ·1772 621 |
| ·80 | ·3452 376 | ·3178 724 | ·2919 144 | ·2673 986 | ·2443 405− | ·2227 386 |
| ·81 | ·4057 840 | ·3773 683 | ·3500 771 | ·3239 798 | ·2991 275− | ·2755 537 |
| ·82 | ·4703 985+ | ·4416 490 | ·4136 963 | ·3866 365− | ·3605 494 | ·3354 989 |
| ·83 | ·5377 218 | ·5094 445− | ·4816 156 | ·4543 472 | ·4277 383 | ·4018 752 |
| ·84 | ·6060 620 | ·5790 963 | ·5522 385+ | ·5256 045+ | ·4993 017 | ·4734 279 |
| ·85 | ·6734 781 | ·6486 271 | ·6235 809 | ·5984 477 | ·5733 315− | ·5483 309 |
| ·86 | ·7379 095+ | ·7158 639 | ·6933 836 | ·6705 599 | ·6474 832 | ·6242 426 |
| ·87 | ·7973 458 | ·7786 112 | ·7592 851 | ·7394 355+ | ·7191 327 | ·6984 478 |
| ·88 | ·8500 220 | ·8348 625− | ·8190 446 | ·8026 116 | ·7856 100 | ·7680 895+ |
| ·89 | ·8946 146 | ·8830 236 | ·8707 918 | ·8579 399 | ·8444 923 | ·8304 767 |
| ·90 | ·9304 059 | ·9221 139 | ·9132 651 | ·9038 633 | ·8939 151 | ·8834 303 |
| ·91 | ·9573 815+ | ·9519 033 | ·9459 923 | ·9396 420 | ·9328 481 | ·9256 081 |
| ·92 | ·9762 295− | ·9729 457 | ·9693 635+ | ·9654 729 | ·9612 647 | ·9567 310 |
| ·93 | ·9882 245− | ·9864 824 | ·9845 614 | ·9824 522 | ·9801 462 | ·9776 347 |
| ·94 | ·9950 056 | ·9942 170 | ·9933 379 | ·9923 624 | ·9912 844 | ·9900 977 |
| ·95 | ·9982 865+ | ·9979 986 | ·9976 743 | ·9973 105+ | ·9969 042 | ·9964 522 |
| ·96 | ·9995 676 | ·9994 905− | ·9994 027 | ·9993 032 | ·9991 910 | ·9990 648 |
| ·97 | ·9999 328 | ·9999 201 | ·9999 056 | ·9998 889 | ·9998 698 | ·9998 482 |
| ·98 | ·9999 957 | ·9999 948 | ·9999 938 | ·9999 927 | ·9999 914 | ·9999 898 |
| ·99 | 1·0000 000 | 1·0000 000 | 1·0000 000 | ·9999 999 | ·9999 999 | ·9999 999 |
| 1·00 | | | | 1·0000 000 | 1·0000 000 | 1·0000 000 |

$x$ = ·48 to 1·00 $q = 7{\cdot}5$ $p$ = 40 to 45

| | $p = 40$ | $p = 41$ | $p = 42$ | $p = 43$ | $p = 44$ | $p = 45$ |
|---|---|---|---|---|---|---|
| $B(p, q) =$ | ·1014 5052 × $\frac{1}{10^8}$ | ·8543 2018 × $\frac{1}{10^9}$ | ·7222 0881 × $\frac{1}{10^9}$ | ·6127 8324 × $\frac{1}{10^9}$ | ·5217 7582 × $\frac{1}{10^9}$ | ·4457 8905 + × $\frac{1}{10^9}$ |
| $x$ | | | | | | |
| ·48 | ·0000 001 | | | | | |
| ·49 | ·0000 001 | ·0000 001 | | | | |
| ·50 | ·0000 003 | ·0000 002 | ·0000 001 | ·0000 001 | | |
| ·51 | ·0000 006 | ·0000 003 | ·0000 002 | ·0000 001 | ·0000 001 | |
| ·52 | ·0000 011 | ·0000 007 | ·0000 004 | ·0000 002 | ·0000 001 | ·0000 001 |
| ·53 | ·0000 021 | ·0000 013 | ·0000 008 | ·0000 005− | ·0000 003 | ·0000 002 |
| ·54 | ·0000 038 | ·0000 024 | ·0000 015− | ·0000 009 | ·0000 006 | ·0000 003 |
| ·55 | ·0000 069 | ·0000 044 | ·0000 028 | ·0000 018 | ·0000 011 | ·0000 007 |
| ·56 | ·0000 125− | ·0000 080 | ·0000 052 | ·0000 033 | ·0000 021 | ·0000 014 |
| ·57 | ·0000 220 | ·0000 144 | ·0000 094 | ·0000 062 | ·0000 040 | ·0000 026 |
| ·58 | ·0000 382 | ·0000 255+ | ·0000 170 | ·0000 113 | ·0000 075− | ·0000 049 |
| ·59 | ·0000 654 | ·0000 444 | ·0000 301 | ·0000 203 | ·0000 137 | ·0000 092 |
| ·60 | ·0001 103 | ·0000 762 | ·0000 525− | ·0000 360 | ·0000 247 | ·0000 168 |
| ·61 | ·0001 834 | ·0001 287 | ·0000 901 | ·0000 629 | ·0000 438 | ·0000 304 |
| ·62 | ·0003 007 | ·0002 144 | ·0001 525− | ·0001 081 | ·0000 765− | ·0000 539 |
| ·63 | ·0004 859 | ·0003 520 | ·0002 542 | ·0001 831 | ·0001 315+ | ·0000 943 |
| ·64 | ·0007 743 | ·0005 696 | ·0004 178 | ·0003 056 | ·0002 229 | ·0001 622 |
| ·65 | ·0012 169 | ·0009 087 | ·0006 767 | ·0005 025+ | ·0003 722 | ·0002 750− |
| ·66 | ·0018 863 | ·0014 297 | ·0010 806 | ·0008 145− | ·0006 123 | ·0004 592 |
| ·67 | ·0028 842 | ·0022 181 | ·0017 011 | ·0013 011 | ·0009 926 | ·0007 553 |
| ·68 | ·0043 503 | ·0033 938 | ·0026 403 | ·0020 486 | ·0015 855− | ·0012 240 |
| ·69 | ·0064 726 | ·0051 209 | ·0040 404 | ·0031 795+ | ·0024 957 | ·0019 541 |
| ·70 | ·0094 996 | ·0076 202 | ·0060 961 | ·0048 641 | ·0038 713 | ·0030 737 |
| ·71 | ·0137 526 | ·0111 821 | ·0090 678 | ·0073 343 | ·0059 175− | ·0047 629 |
| ·72 | ·0196 371 | ·0161 803 | ·0132 968 | ·0108 995− | ·0089 124 | ·0072 704 |
| ·73 | ·0276 528 | ·0230 837 | ·0192 195+ | ·0159 621 | ·0132 247 | ·0109 312 |
| ·74 | ·0383 982 | ·0324 655+ | ·0273 793 | ·0230 330 | ·0193 305+ | ·0161 860 |
| ·75 | ·0525 675− | ·0450 046 | ·0384 332 | ·0327 419 | ·0278 282 | ·0235 985+ |
| ·76 | ·0709 362 | ·0614 778 | ·0531 499 | ·0458 412 | ·0394 469 | ·0338 693 |
| ·77 | ·0943 317 | ·0827 366 | ·0723 929 | ·0631 957 | ·0550 433 | ·0478 386 |
| ·78 | ·1235 852 | ·1096 652 | ·0970 863 | ·0857 562 | ·0755 829 | ·0664 755− |
| ·79 | ·1594 631 | ·1431 170 | ·1281 564 | ·1145 083 | ·1020 965+ | ·0908 427 |
| ·80 | ·2025 769 | ·1838 266 | ·1664 481 | ·1503 938 | ·1356 089 | ·1220 336 |
| ·81 | ·2532 765+ | ·2322 998 | ·2126 149 | ·1942 025− | ·1770 340 | ·1610 731 |
| ·82 | ·3115 339 | ·2886 893 | ·2669 870 | ·2464 369 | ·2270 386 | ·2087 821 |
| ·83 | ·3768 312 | ·3526 667 | ·3294 306 | ·3071 598 | ·2858 809 | ·2656 103 |
| ·84 | ·4480 717 | ·4233 113 | ·3992 153 | ·3758 421 | ·3532 408 | ·3314 511 |
| ·85 | ·5235 387 | ·4990 409 | ·4749 166 | ·4512 373 | ·4280 675+ | ·4054 638 |
| ·86 | ·6009 246 | ·5776 123 | ·5543 854 | ·5313 188 | ·5084 829 | ·4859 428 |
| ·87 | ·6774 527 | ·6562 187 | ·6348 164 | ·6133 148 | ·5917 807 | ·5702 783 |
| ·88 | ·7501 018 | ·7317 004 | ·7129 399 | ·6938 756 | ·6745 631 | ·6550 575− |
| ·89 | ·8159 238 | ·8008 671 | ·7853 421 | ·7693 865− | ·7530 394 | ·7363 412 |
| ·90 | ·8724 212 | ·8609 030 | ·8488 931 | ·8364 114 | ·8234 798 | ·8101 220 |
| ·91 | ·9179 217 | ·9097 903 | ·9012 178 | ·8922 094 | ·8827 727 | ·8729 166 |
| ·92 | ·9518 647 | ·9466 603 | ·9411 131 | ·9352 199 | ·9289 786 | ·9223 883 |
| ·93 | ·9749 097 | ·9719 636 | ·9687 895− | ·9653 806 | ·9617 312 | ·9578 358 |
| ·94 | ·9887 963 | ·9873 742 | ·9858 255− | ·9841 444 | ·9823 253 | ·9803 627 |
| ·95 | ·9959 513 | ·9953 981 | ·9947 892 | ·9941 212 | ·9933 907 | ·9925 942 |
| ·96 | ·9989 234 | ·9987 656 | ·9985 902 | ·9983 957 | ·9981 807 | ·9979 439 |
| ·97 | ·9998 237 | ·9997 961 | ·9997 651 | ·9997 303 | ·9996 915+ | ·9996 483 |
| ·98 | ·9999 881 | ·9999 861 | ·9999 838 | ·9999 813 | ·9999 784 | ·9999 752 |
| ·99 | ·9999 999 | ·9999 999 | ·9999 999 | ·9999 999 | ·9999 998 | ·9999 998 |
| 1·00 | 1·0000 000 | 1·0000 000 | 1·0000 000 | 1·0000 000 | 1·0000 000 | 1·0000 000 |

$x$ = ·53 to 1·00 $q$ = 7·5 $p$ = 46 to 50

| $x$ | $p$ = 46 | $p$ = 47 | $p$ = 48 | $p$ = 49 | $p$ = 50 |
|---|---|---|---|---|---|
| $B(p, q)$ = | ·3821 0490 × $\frac{1}{10^9}$ | ·3285 3880 × $\frac{1}{10^9}$ | ·2833 2703 × $\frac{1}{10^9}$ | ·2450 3960 × $\frac{1}{10^9}$ | ·2125 1222 × $\frac{1}{10^9}$ |
| ·53 | ·0000 001 | ·0000 001 | | | |
| ·54 | ·0000 002 | ·0000 001 | ·0000 001 | | |
| ·55 | ·0000 004 | ·0000 003 | ·0000 002 | ·0000 001 | ·0000 001 |
| ·56 | ·0000 009 | ·0000 005+ | ·0000 003 | ·0000 002 | ·0000 001 |
| ·57 | ·0000 017 | ·0000 011 | ·0000 007 | ·0000 005− | ·0000 003 |
| ·58 | ·0000 033 | ·0000 021 | ·0000 014 | ·0000 009 | ·0000 006 |
| ·59 | ·0000 062 | ·0000 041 | ·0000 027 | ·0000 018 | ·0000 012 |
| ·60 | ·0000 115− | ·0000 078 | ·0000 053 | ·0000 036 | ·0000 024 |
| ·61 | ·0000 210 | ·0000 145+ | ·0000 100 | ·0000 069 | ·0000 047 |
| ·62 | ·0000 379 | ·0000 266 | ·0000 187 | ·0000 130 | ·0000 091 |
| ·63 | ·0000 674 | ·0000 480 | ·0000 342 | ·0000 243 | ·0000 172 |
| ·64 | ·0001 178 | ·0000 853 | ·0000 616 | ·0000 444 | ·0000 320 |
| ·65 | ·0002 027 | ·0001 490 | ·0001 093 | ·0000 800 | ·0000 585− |
| ·66 | ·0003 435+ | ·0002 564 | ·0001 909 | ·0001 419 | ·0001 052 |
| ·67 | ·0005 734 | ·0004 343 | ·0003 282 | ·0002 475− | ·0001 862 |
| ·68 | ·0009 427 | ·0007 244 | ·0005 554 | ·0004 249 | ·0003 244 |
| ·69 | ·0015 265+ | ·0011 897 | ·0009 252 | ·0007 180 | ·0005 561 |
| ·70 | ·0024 347 | ·0019 243 | ·0015 175− | ·0011 942 | ·0009 379 |
| ·71 | ·0038 248 | ·0030 646 | ·0024 502 | ·0019 549 | ·0015 566 |
| ·72 | ·0059 174 | ·0048 055+ | ·0038 943 | ·0031 494 | ·0025 419 |
| ·73 | ·0090 151 | ·0074 187 | ·0060 922 | ·0049 927 | ·0040 836 |
| ·74 | ·0135 229 | ·0112 737 | ·0093 792 | ·0077 874 | ·0064 532 |
| ·75 | ·0199 681 | ·0168 604 | ·0142 074 | ·0119 482 | ·0100 290 |
| ·76 | ·0290 180 | ·0248 101 | ·0211 698 | ·0180 285− | ·0153 244 |
| ·77 | ·0414 897 | ·0359 103 | ·0310 199 | ·0267 445− | ·0230 156 |
| ·78 | ·0583 456 | ·0511 082 | ·0446 823 | ·0389 914 | ·0339 637 |
| ·79 | ·0806 680 | ·0714 942 | ·0632 446 | ·0558 448 | ·0492 236 |
| ·80 | ·1096 047 | ·0982 566 | ·0879 226 | ·0785 358 | ·0700 303 |
| ·81 | ·1462 774 | ·1325 994 | ·1199 880 | ·1083 894 | ·0977 484 |
| ·82 | ·1916 494 | ·1756 155− | ·1606 493 | ·1467 152 | ·1337 735+ |
| ·83 | ·2463 555+ | ·2281 158 | ·2108 831 | ·1946 431 | ·1793 759 |
| ·84 | ·3105 039 | ·2904 219 | ·2712 201 | ·2529 064 | ·2354 822 |
| ·85 | ·3834 753 | ·3621 442 | ·3415 051 | ·3215 863 | ·3024 094 |
| ·86 | ·4637 584 | ·4419 838 | ·4206 680 | ·3998 539 | ·3795 793 |
| ·87 | ·5488 692 | ·5276 113 | ·5065 592 | ·4857 636 | ·4652 715+ |
| ·88 | ·6354 131 | ·6156 833 | ·5959 200 | ·5761 731 | ·5564 907 |
| ·89 | ·7193 332 | ·7020 572 | ·6845 550+ | ·6668 688 | ·6490 399 |
| ·90 | ·7963 635+ | ·7822 312 | ·7677 530 | ·7529 581 | ·7378 764 |
| ·91 | ·8626 520 | ·8519 914 | ·8409 486 | ·8295 389 | ·8177 787 |
| ·92 | ·9154 495+ | ·9081 638 | ·9005 339 | ·8925 638 | ·8842 587 |
| ·93 | ·9536 898 | ·9492 892 | ·9446 307 | ·9397 116 | ·9345 299 |
| ·94 | ·9782 514 | ·9759 864 | ·9735 628 | ·9709 761 | ·9682 221 |
| ·95 | ·9917 283 | ·9907 894 | ·9897 741 | ·9886 789 | ·9875 005− |
| ·96 | ·9976 837 | ·9973 986 | ·9970 871 | ·9967 475− | ·9963 782 |
| ·97 | ·9996 004 | ·9995 473 | ·9994 887 | ·9994 241 | ·9993 532 |
| ·98 | ·9999 715+ | ·9999 675− | ·9999 629 | ·9999 579 | ·9999 523 |
| ·99 | ·9999 998 | ·9999 997 | ·9999 997 | ·9999 996 | ·9999 996 |
| 1·00 | 1·0000 000 | 1·0000 000 | 1·0000 000 | 1·0000 000 | 1·0000 000 |

$x$ = ·05 to ·70 $q = 8$ $p$ = 8 to 10·5

| $x$ | $p = 8$ | $p = 8·5$ | $p = 9$ | $p = 9·5$ | $p = 10$ | $p = 10·5$ |
|---|---|---|---|---|---|---|
| $B(p, q) =$ | ·1942 5019 × $\frac{1}{10^4}$ | ·1362 8793 × $\frac{1}{10^4}$ | ·9712 5097 × $\frac{1}{10^5}$ | ·7020 8932 × $\frac{1}{10^5}$ | ·5141 9169 × $\frac{1}{10^5}$ | ·3811 3420 × $\frac{1}{10^5}$ |
| ·05 | ·0000 002 | ·0000 001 | | | | |
| ·06 | ·0000 007 | ·0000 002 | ·0000 001 | | | |
| ·07 | ·0000 024 | ·0000 008 | ·0000 003 | ·0000 001 | | |
| ·08 | ·0000 065⁻ | ·0000 024 | ·0000 009 | ·0000 003 | ·0000 001 | |
| ·09 | ·0000 155⁻ | ·0000 062 | ·0000 025⁻ | ·0000 010 | ·0000 004 | ·0000 001 |
| ·10 | ·0000 336 | ·0000 142 | ·0000 059 | ·0000 024 | ·0000 010 | ·0000 004 |
| ·11 | ·0000 673 | ·0000 298 | ·0000 130 | ·0000 056 | ·0000 024 | ·0000 010 |
| ·12 | ·0001 261 | ·0000 583 | ·0000 266 | ·0000 120 | ·0000 054 | ·0000 024 |
| ·13 | ·0002 233 | ·0001 073 | ·0000 510 | ·0000 240 | ·0000 112 | ·0000 051 |
| ·14 | ·0003 767 | ·0001 877 | ·0000 925⁻ | ·0000 451 | ·0000 218 | ·0000 104 |
| ·15 | ·0006 096 | ·0003 143 | ·0001 602 | ·0000 808 | ·0000 404 | ·0000 200 |
| ·16 | ·0009 515⁻ | ·0005 063 | ·0002 664 | ·0001 387 | ·0000 715⁺ | ·0000 366 |
| ·17 | ·0014 384 | ·0007 884 | ·0004 274 | ·0002 292 | ·0001 218 | ·0000 641 |
| ·18 | ·0021 136 | ·0011 913 | ·0006 640 | ·0003 663 | ·0002 002 | ·0001 084 |
| ·19 | ·0030 279 | ·0017 522 | ·0010 028 | ·0005 680 | ·0003 187 | ·0001 773 |
| ·20 | ·0042 397 | ·0025 153 | ·0014 759 | ·0008 573 | ·0004 932 | ·0002 813 |
| ·21 | ·0058 149 | ·0035 323 | ·0021 224 | ·0012 624 | ·0007 439 | ·0004 345⁺ |
| ·22 | ·0078 264 | ·0048 622 | ·0029 881 | ·0018 180 | ·0010 959 | ·0006 548 |
| ·23 | ·0103 536 | ·0065 715⁻ | ·0041 263 | ·0025 652 | ·0015 800 | ·0009 649 |
| ·24 | ·0134 815⁺ | ·0087 335⁻ | ·0055 975⁺ | ·0035 523 | ·0022 337 | ·0013 925⁺ |
| ·25 | ·0172 998 | ·0114 280 | ·0074 697 | ·0048 347 | ·0031 008 | ·0019 718 |
| ·26 | ·0219 012 | ·0147 407 | ·0098 178 | ·0064 756 | ·0042 326 | ·0027 432 |
| ·27 | ·0273 800 | ·0187 616 | ·0127 231 | ·0085 452 | ·0056 878 | ·0037 542 |
| ·28 | ·0338 305⁻ | ·0235 840 | ·0162 726 | ·0111 208 | ·0075 326 | ·0050 598 |
| ·29 | ·0413 451 | ·0293 030 | ·0205 578 | ·0142 864 | ·0098 407 | ·0067 227 |
| ·30 | ·0500 125⁺ | ·0360 140 | ·0256 735⁺ | ·0181 310 | ·0126 927 | ·0088 131 |
| ·31 | ·0599 156 | ·0438 104 | ·0317 165⁺ | ·0227 488 | ·0161 757 | ·0114 088 |
| ·32 | ·0711 294 | ·0527 821 | ·0387 834 | ·0282 367 | ·0203 822 | ·0145 947 |
| ·33 | ·0837 194 | ·0630 134 | ·0469 693 | ·0346 936 | ·0254 092 | ·0184 618 |
| ·34 | ·0977 397 | ·0745 808 | ·0563 653 | ·0422 180 | ·0313 568 | ·0231 069 |
| ·35 | ·1132 311 | ·0875 511 | ·0670 569 | ·0509 066 | ·0383 263 | ·0286 306 |
| ·36 | ·1302 201 | ·1019 798 | ·0791 215⁺ | ·0608 521 | ·0464 185⁻ | ·0351 364 |
| ·37 | ·1487 170 | ·1179 087 | ·0926 268 | ·0721 406 | ·0557 319 | ·0427 286 |
| ·38 | ·1687 154 | ·1353 649 | ·1076 281 | ·0848 502 | ·0663 602 | ·0515 105⁻ |
| ·39 | ·1901 914 | ·1543 589 | ·1241 671 | ·0990 479 | ·0783 902 | ·0615 820 |
| ·40 | ·2131 032 | ·1748 841 | ·1422 697 | ·1147 883 | ·0918 993 | ·0730 376 |
| ·41 | ·2373 908 | ·1969 153 | ·1619 449 | ·1321 111 | ·1069 532 | ·0859 638 |
| ·42 | ·2629 766 | ·2204 087 | ·1831 832 | ·1510 395⁻ | ·1236 042 | ·1004 365⁻ |
| ·43 | ·2897 659 | ·2453 016 | ·2059 560 | ·1715 783 | ·1418 880 | ·1165 187 |
| ·44 | ·3176 477 | ·2715 125⁻ | ·2302 149 | ·1937 132 | ·1618 231 | ·1342 582 |
| ·45 | ·3464 961 | ·2989 419 | ·2558 915⁻ | ·2174 092 | ·1834 078 | ·1536 850⁺ |
| ·46 | ·3761 718 | ·3274 731 | ·2828 977 | ·2426 106 | ·2066 202 | ·1748 100 |
| ·47 | ·4065 240 | ·3569 733 | ·3111 262 | ·2692 404 | ·2314 160 | ·1976 227 |
| ·48 | ·4373 921 | ·3872 958 | ·3404 517 | ·2972 007 | ·2577 291 | ·2220 899 |
| ·49 | ·4686 084 | ·4182 812 | ·3707 319 | ·3263 735⁺ | ·2854 705⁺ | ·2481 553 |
| ·50 | ·5000 000[e] | ·4497 601 | ·4018 097 | ·3566 217 | ·3145 294 | ·2757 383 |
| ·51 | ·5313 916 | ·4815 552 | ·4335 151 | ·3877 906 | ·3447 737 | ·3047 349 |
| ·52 | ·5626 079 | ·5134 841 | ·4656 674 | ·4197 102 | ·3760 513 | ·3350 175⁺ |
| ·53 | ·5934 760 | ·5453 618 | ·4980 782 | ·4521 971 | ·4081 923 | ·3664 366 |
| ·54 | ·6238 282 | ·5770 037 | ·5305 540 | ·4850 576 | ·4410 109 | ·3988 219 |
| ·55 | ·6535 039 | ·6082 281 | ·5628 993 | ·5180 904 | ·4743 082 | ·4319 851 |
| ·56 | ·6823 523 | ·6388 591 | ·5949 196 | ·5510 897 | ·5078 754 | ·4657 224 |
| ·57 | ·7102 341 | ·6687 292 | ·6264 243 | ·5838 487 | ·5414 970 | ·4998 171 |
| ·58 | ·7370 234 | ·6976 815⁺ | ·6572 300 | ·6161 631 | ·5749 542 | ·5340 439 |
| ·59 | ·7626 092 | ·7255 726 | ·6871 633 | ·6478 340 | ·6080 290 | ·5681 724 |
| ·60 | ·7868 968 | ·7522 737 | ·7160 634 | ·6786 717 | ·6405 077 | ·6019 708 |
| ·61 | ·8098 086 | ·7776 729 | ·7437 842 | ·7084 987 | ·6721 845⁺ | ·6352 104 |
| ·62 | ·8312 846 | ·8016 764 | ·7701 973 | ·7371 523 | ·7028 656 | ·6676 701 |
| ·63 | ·8512 830 | ·8242 093 | ·7951 929 | ·7644 876 | ·7323 719 | ·6991 395⁻ |
| ·64 | ·8697 799 | ·8452 163 | ·8186 814 | ·7903 794 | ·7605 426 | ·7294 238 |
| ·65 | ·8867 689 | ·8646 621 | ·8405 946 | ·8147 240 | ·7872 377 | ·7583 466 |
| ·66 | ·9022 603 | ·8825 306 | ·8608 860 | ·8374 408 | ·8123 400 | ·7857 536 |
| ·67 | ·9162 806 | ·8988 252 | ·8795 305⁺ | ·8584 726 | ·8357 571 | ·8115 150⁻ |
| ·68 | ·9288 706 | ·9135 670 | ·8965 247 | ·8777 863 | ·8574 221 | ·8355 273 |
| ·69 | ·9400 844 | ·9267 943 | ·9118 853 | ·8953 722 | ·8772 945⁻ | ·8577 154 |
| ·70 | ·9499 875⁻ | ·9385 603 | ·9256 484 | ·9112 437 | ·8953 599 | ·8780 324 |

$x = \cdot 71$ to $\cdot 97$ | $q = 8$ | $p = 8$ to $10\cdot 5$

| | $p = 8$ | $p = 8\cdot 5$ | $p = 9$ | $p = 9\cdot 5$ | $p = 10$ | $p = 10\cdot 5$ |
|---|---|---|---|---|---|---|
| $B(p, q) =$ | $\cdot 1942\ 5019 \times \frac{1}{10^4}$ | $\cdot 1362\ 8793 \times \frac{1}{10^4}$ | $\cdot 9712\ 5097 \times \frac{1}{10^5}$ | $\cdot 7020\ 8932 \times \frac{1}{10^5}$ | $\cdot 5141\ 9169 \times \frac{1}{10^5}$ | $\cdot 3811\ 3420 \times \frac{1}{10^5}$ |
| $x$ | | | | | | |
| ·71 | ·9586 549 | ·9489 321 | ·9378 675⁺ | ·9254 355⁺ | ·9116 293 | ·8964 606 |
| ·72 | ·9661 695⁺ | ·9579 879 | ·9486 116 | ·9380 024 | ·9261 375⁺ | ·9130 101 |
| ·73 | ·9726 200 | ·9658 154 | ·9579 632 | ·9490 166 | ·9389 418 | ·9277 175⁺ |
| ·74 | ·9780 988 | ·9725 095⁻ | ·9660 154 | ·9585 657 | ·9501 190 | ·9406 443 |
| ·75 | ·9827 002 | ·9781 696 | ·9728 700 | ·9667 496 | ·9597 632 | ·9518 736 |
| ·76 | ·9865 185⁻ | ·9828 978 | ·9786 345⁻ | ·9736 779 | ·9679 823 | ·9615 075⁺ |
| ·77 | ·9896 464 | ·9867 968 | ·9834 192 | ·9794 665⁺ | ·9748 947 | ·9696 633 |
| ·78 | ·9921 736 | ·9899 674 | ·9873 354 | ·9842 353 | ·9806 264 | ·9764 699 |
| ·79 | ·9941 851 | ·9925 071 | ·9904 926 | ·9881 046 | ·9853 067 | ·9820 637 |
| ·80 | ·9957 603 | ·9945 086 | ·9929 964 | ·9911 925⁺ | ·9890 657 | ·9865 849 |
| ·81 | ·9969 721 | ·9960 581 | ·9949 469 | ·9936 131 | ·9920 307 | ·9901 734 |
| ·82 | ·9978 864 | ·9972 344 | ·9964 368 | ·9954 735⁺ | ·9943 237 | ·9929 657 |
| ·83 | ·9985 616 | ·9981 084 | ·9975 506 | ·9968 728 | ·9960 588 | ·9950 916 |
| ·84 | ·9990 485⁺ | ·9987 424 | ·9983 634 | ·9979 001 | ·9973 404 | ·9966 713 |
| ·85 | ·9993 904 | ·9991 903 | ·9989 410 | ·9986 345⁻ | ·9982 619 | ·9978 139 |
| ·86 | ·9996 233 | ·9994 972 | ·9993 392 | ·9991 437 | ·9989 047 | ·9986 157 |
| ·87 | ·9997 767 | ·9997 005⁺ | ·9996 045⁻ | ·9994 849 | ·9993 380 | ·9991 592 |
| ·88 | ·9998 739 | ·9998 300 | ·9997 744 | ·9997 048 | ·9996 187 | ·9995 134 |
| ·89 | ·9999 327 | ·9999 088 | ·9998 784 | ·9998 401 | ·9997 924 | ·9997 338 |
| ·90 | ·9999 664 | ·9999 542 | ·9999 387 | ·9999 190 | ·9998 944 | ·9998 639 |
| ·91 | ·9999 845⁺ | ·9999 788 | ·9999 715⁻ | ·9999 622 | ·9999 504 | ·9999 358 |
| ·92 | ·9999 935⁺ | ·9999 911 | ·9999 880 | ·9999 840 | ·9999 789 | ·9999 726 |
| ·93 | ·9999 976 | ·9999 967 | ·9999 956 | ·9999 940 | ·9999 921 | ·9999 897 |
| ·94 | ·9999 993 | ·9999 990 | ·9999 986 | ·9999 981 | ·9999 975⁺ | ·9999 967 |
| ·95 | ·9999 998 | ·9999 997 | ·9999 997 | ·9999 995⁺ | ·9999 994 | ·9999 992 |
| ·96 | 1·0000 000 | 1·0000 000 | ·9999 999 | ·9999 999 | ·9999 999 | ·9999 998 |
| ·97 | | | 1·0000 000 | 1·0000 000 | 1·0000 000 | 1·0000 000 |

$x$ = ·09 to ·70 $q = 8$ $p$ = 11 to 16

| $x$ | $p = 11$ | $p = 12$ | $p = 13$ | $p = 14$ | $p = 15$ | $p = 16$ |
|---|---|---|---|---|---|---|
| $B(p, q)$ = | ·2856 6205+ × $\frac{1}{10^5}$ | ·1653 8329 × $\frac{1}{10^5}$ | ·9922 9975+ × $\frac{1}{10^6}$ | ·6142 8080 × $\frac{1}{10^6}$ | ·3909 0596 × $\frac{1}{10^6}$ | ·2549 3867 × $\frac{1}{10^6}$ |
| ·09 | ·0000 001 | | | | | |
| ·10 | ·0000 002 | | | | | |
| ·11 | ·0000 004 | ·0000 001 | | | | |
| ·12 | ·0000 010 | ·0000 002 | | | | |
| ·13 | ·0000 024 | ·0000 005− | ·0000 001 | | | |
| ·14 | ·0000 049 | ·0000 011 | ·0000 002 | | | |
| ·15 | ·0000 098 | ·0000 023 | ·0000 005+ | ·0000 001 | | |
| ·16 | ·0000 185+ | ·0000 047 | ·0000 011 | ·0000 003 | ·0000 001 | |
| ·17 | ·0000 335− | ·0000 089 | ·0000 023 | ·0000 006 | ·0000 001 | |
| ·18 | ·0000 582 | ·0000 164 | ·0000 045+ | ·0000 012 | ·0000 003 | ·0000 001 |
| ·19 | ·0000 978 | ·0000 291 | ·0000 084 | ·0000 024 | ·0000 007 | ·0000 002 |
| ·20 | ·0001 591 | ·0000 498 | ·0000 152 | ·0000 045+ | ·0000 013 | ·0000 004 |
| ·21 | ·0002 518 | ·0000 826 | ·0000 264 | ·0000 082 | ·0000 025+ | ·0000 008 |
| ·22 | ·0003 881 | ·0001 334 | ·0000 446 | ·0000 146 | ·0000 047 | ·0000 015− |
| ·23 | ·0005 845− | ·0002 098 | ·0000 733 | ·0000 250+ | ·0000 084 | ·0000 027 |
| ·24 | ·0008 612 | ·0003 222 | ·0001 174 | ·0000 418 | ·0000 146 | ·0000 050− |
| ·25 | ·0012 440 | ·0004 844 | ·0001 837 | ·0000 681 | ·0000 247 | ·0000 088 |
| ·26 | ·0017 639 | ·0007 136 | ·0002 812 | ·0001 083 | ·0000 408 | ·0000 151 |
| ·27 | ·0024 586 | ·0010 319 | ·0004 219 | ·0001 686 | ·0000 660 | ·0000 253 |
| ·28 | ·0033 724 | ·0014 663 | ·0006 212 | ·0002 572 | ·0001 043 | ·0000 415+ |
| ·29 | ·0045 572 | ·0020 500− | ·0008 987 | ·0003 851 | ·0001 616 | ·0000 666 |
| ·30 | ·0060 725+ | ·0028 226 | ·0012 789 | ·0005 664 | ·0002 458 | ·0001 047 |
| ·31 | ·0079 857 | ·0038 311 | ·0017 919 | ·0008 194 | ·0003 671 | ·0001 615+ |
| ·32 | ·0103 719 | ·0051 302 | ·0024 744 | ·0011 669 | ·0005 393 | ·0002 447 |
| ·33 | ·0133 140 | ·0067 826 | ·0033 700 | ·0016 374 | ·0007 798 | ·0003 647 |
| ·34 | ·0169 019 | ·0088 597 | ·0045 303 | ·0022 657 | ·0011 108 | ·0005 348 |
| ·35 | ·0212 315+ | ·0114 409 | ·0060 153 | ·0030 938 | ·0015 600 | ·0007 726 |
| ·36 | ·0264 042 | ·0146 140 | ·0078 936 | ·0041 715− | ·0021 616 | ·0011 003 |
| ·37 | ·0325 250+ | ·0184 743 | ·0102 430 | ·0055 574 | ·0029 569 | ·0015 457 |
| ·38 | ·0397 012 | ·0231 241 | ·0131 503 | ·0073 194 | ·0039 958 | ·0021 434 |
| ·39 | ·0480 401 | ·0286 713 | ·0167 110 | ·0095 348 | ·0053 368 | ·0029 355− |
| ·40 | ·0576 473 | ·0352 279 | ·0210 289 | ·0122 911 | ·0070 484 | ·0039 727 |
| ·41 | ·0686 241 | ·0429 087 | ·0262 151 | ·0156 853 | ·0092 095− | ·0053 153 |
| ·42 | ·0810 647 | ·0518 285+ | ·0323 865− | ·0198 239 | ·0119 095− | ·0070 342 |
| ·43 | ·0950 544 | ·0621 005− | ·0396 644 | ·0248 220 | ·0152 487 | ·0092 111 |
| ·44 | ·1106 660 | ·0738 328 | ·0481 724 | ·0308 023 | ·0193 380 | ·0119 398 |
| ·45 | ·1279 578 | ·0871 264 | ·0580 341 | ·0378 932 | ·0242 982 | ·0153 254 |
| ·46 | ·1469 711 | ·1020 717 | ·0693 699 | ·0462 271 | ·0302 586 | ·0194 852 |
| ·47 | ·1677 276 | ·1187 454 | ·0822 945+ | ·0559 377 | ·0373 562 | ·0245 473 |
| ·48 | ·1902 275+ | ·1372 081 | ·0969 133 | ·0671 571 | ·0457 327 | ·0306 499 |
| ·49 | ·2144 478 | ·1575 005+ | ·1133 189 | ·0800 128 | ·0555 328 | ·0379 398 |
| ·50 | ·2403 412 | ·1796 417 | ·1315 880 | ·0946 236 | ·0669 003 | ·0465 698 |
| ·51 | ·2678 349 | ·2036 260 | ·1517 773 | ·1110 960 | ·0799 748 | ·0566 961 |
| ·52 | ·2968 307 | ·2294 211 | ·1739 206 | ·1295 201 | ·0948 878 | ·0684 749 |
| ·53 | ·3272 050+ | ·2569 670 | ·1980 256 | ·1499 657 | ·1117 581 | ·0820 580 |
| ·54 | ·3588 102 | ·2861 747 | ·2240 714 | ·1724 779 | ·1306 871 | ·0975 888 |
| ·55 | ·3914 755− | ·3169 260 | ·2520 059 | ·1970 734 | ·1517 542 | ·1151 966 |
| ·56 | ·4250 094 | ·3490 739 | ·2817 445+ | ·2237 376 | ·1750 119 | ·1349 917 |
| ·57 | ·4592 024 | ·3824 440 | ·3131 696 | ·2524 212 | ·2004 814 | ·1570 597 |
| ·58 | ·4938 302 | ·4168 362 | ·3461 300 | ·2830 384 | ·2281 486 | ·1814 557 |
| ·59 | ·5286 573 | ·4520 275− | ·3804 424 | ·3154 653 | ·2579 605− | ·2081 997 |
| ·60 | ·5634 408 | ·4877 753 | ·4158 929 | ·3495 400 | ·2898 224 | ·2372 709 |
| ·61 | ·5979 356 | ·5238 217 | ·4522 400 | ·3850 633 | ·3235 966 | ·2686 045− |
| ·62 | ·6318 979 | ·5598 979 | ·4892 180 | ·4218 002 | ·3591 017 | ·3020 878 |
| ·63 | ·6650 906 | ·5957 297 | ·5265 423 | ·4594 836 | ·3961 132 | ·3375 590 |
| ·64 | ·6972 877 | ·6310 425− | ·5639 140 | ·4978 182 | ·4343 663 | ·3748 061 |
| ·65 | ·7282 784 | ·6655 670 | ·6010 266 | ·5364 862 | ·4735 593 | ·4135 690 |
| ·66 | ·7578 715+ | ·6990 455− | ·6375 723 | ·5751 534 | ·5133 587 | ·4535 414 |
| ·67 | ·7858 991 | ·7312 366 | ·6732 489 | ·6134 769 | ·5534 061 | ·4943 764 |
| ·68 | ·8122 194 | ·7619 212 | ·7077 668 | ·6511 129 | ·5933 260 | ·5356 931 |
| ·69 | ·8367 194 | ·7909 064 | ·7408 558 | ·6877 251 | ·6327 348 | ·5770 847 |
| ·70 | ·8593 165+ | ·8180 305− | ·7722 718 | ·7229 932 | ·6712 507 | ·6181 284 |

$x = \cdot 71$ to $\cdot 98$ $q = 8$ $p = 11$ to $16$

| $x$ | $p = 11$ | $p = 12$ | $p = 13$ | $p = 14$ | $p = 15$ | $p = 16$ |
|---|---|---|---|---|---|---|
| $B(p, q) =$ | ·2856 6205+ × $\frac{1}{10^5}$ | ·1653 8329 × $\frac{1}{10^5}$ | ·9922 9975+ × $\frac{1}{10^6}$ | ·6142 8080 × $\frac{1}{10^6}$ | ·3909 0596 × $\frac{1}{10^6}$ | ·2549 3867 × $\frac{1}{10^6}$ |
| ·71 | ·8799 597 | ·8431 654 | ·8018 024 | ·7566 214 | ·7085 036 | ·6583 969 |
| ·72 | ·8986 292 | ·8662 194 | ·8292 722 | ·7883 461 | ·7441 459 | ·6974 705− |
| ·73 | ·9153 363 | ·8871 384 | ·8545 463 | ·8179 430 | ·7778 623 | ·7349 492 |
| ·74 | ·9301 214 | ·9059 060 | ·8775 337 | ·8452 330 | ·8093 791 | ·7704 657 |
| ·75 | ·9430 520 | ·9225 428 | ·8981 881 | ·8700 866 | ·8384 724 | ·8036 967 |
| ·76 | ·9542 197 | ·9371 040 | ·9165 081 | ·8924 268 | ·8649 741 | ·8343 735− |
| ·77 | ·9637 363 | ·9496 766 | ·9325 355− | ·9122 299 | ·8887 769 | ·8622 907 |
| ·78 | ·9717 302 | ·9603 755− | ·9463 524 | ·9295 247 | ·9098 363 | ·8873 127 |
| ·79 | ·9783 420 | ·9693 386 | ·9580 769 | ·9443 895− | ·9281 699 | ·9093 769 |
| ·80 | ·9837 199 | ·9767 217 | ·9678 573 | ·9569 474 | ·9438 554 | ·9284 942 |
| ·81 | ·9880 151 | ·9826 925+ | ·9758 664 | ·9673 599 | ·9570 246 | ·9447 462 |
| ·82 | ·9913 779 | ·9874 253 | ·9822 935− | ·9758 194 | ·9678 564 | ·9582 795+ |
| ·83 | ·9939 538 | ·9910 948 | ·9873 377 | ·9825 401 | ·9765 672 | ·9692 961 |
| ·84 | ·9958 795− | ·9938 713 | ·9912 006 | ·9877 491 | ·9834 002 | ·9780 425− |
| ·85 | ·9972 806 | ·9959 157 | ·9940 789 | ·9916 768 | ·9886 141 | ·9847 960 |
| ·86 | ·9982 696 | ·9973 757 | ·9961 586 | ·9945 483 | ·9924 709 | ·9898 507 |
| ·87 | ·9989 439 | ·9983 828 | ·9976 099 | ·9965 755− | ·9952 255+ | ·9935 030 |
| ·88 | ·9993 858 | ·9990 504 | ·9985 832 | ·9979 506 | ·9971 155+ | ·9960 377 |
| ·89 | ·9996 625− | ·9994 731 | ·9992 064 | ·9988 411 | ·9983 534 | ·9977 169 |
| ·90 | ·9998 265+ | ·9997 267 | ·9995 844 | ·9993 873 | ·9991 213 | ·9987 702 |
| ·91 | ·9999 178 | ·9998 693 | ·9997 993 | ·9997 014 | ·9995 677 | ·9993 893 |
| ·92 | ·9999 648 | ·9999 434 | ·9999 124 | ·9998 684 | ·9998 077 | ·9997 257 |
| ·93 | ·9999 867 | ·9999 784 | ·9999 663 | ·9999 489 | ·9999 246 | ·9998 915+ |
| ·94 | ·9999 957 | ·9999 930 | ·9999 890 | ·9999 832 | ·9999 750− | ·9999 636 |
| ·95 | ·9999 989 | ·9999 982 | ·9999 971 | ·9999 956 | ·9999 934 | ·9999 903 |
| ·96 | ·9999 998 | ·9999 997 | ·9999 995− | ·9999 992 | ·9999 987 | ·9999 981 |
| ·97 | 1·0000 000 | 1·0000 000 | ·9999 999 | ·9999 999 | ·9999 999 | ·9999 998 |
| ·98 | | | 1·0000 000 | 1·0000 000 | 1·0000 000 | 1·0000 000 |

$x$ = ·20 to ·80 $q = 8$ $p$ = 17 to 22

| $x$ | $p = 17$ | $p = 18$ | $p = 19$ | $p = 20$ | $p = 21$ | $p = 22$ |
|---|---|---|---|---|---|---|
| $B(p, q)$ = | ·1699 5911 × $\frac{1}{10^6}$ | ·1155 7220 × $\frac{1}{10^6}$ | ·8001 1522 × $\frac{1}{10^7}$ | ·5630 4404 × $\frac{1}{10^7}$ | ·4021 7432 × $\frac{1}{10^7}$ | ·2912 2968 × $\frac{1}{10^7}$ |
| ·20 | ·0000 001 | | | | | |
| ·21 | ·0000 002 | ·0000 001 | | | | |
| ·22 | ·0000 005$^-$ | ·0000 001 | | | | |
| ·23 | ·0000 009 | ·0000 003 | ·0000 001 | | | |
| ·24 | ·0000 017 | ·0000 006 | ·0000 002 | ·0000 001 | | |
| ·25 | ·0000 031 | ·0000 011 | ·0000 004 | ·0000 001 | | |
| ·26 | ·0000 055$^-$ | ·0000 020 | ·0000 007 | ·0000 002 | ·0000 001 | |
| ·27 | ·0000 096 | ·0000 036 | ·0000 013 | ·0000 005$^-$ | ·0000 002 | ·0000 001 |
| ·28 | ·0000 163 | ·0000 063 | ·0000 024 | ·0000 009 | ·0000 003 | ·0000 001 |
| ·29 | ·0000 270 | ·0000 108 | ·0000 042 | ·0000 016 | ·0000 006 | ·0000 002 |
| ·30 | ·0000 439 | ·0000 181 | ·0000 074 | ·0000 030 | ·0000 012 | ·0000 005$^-$ |
| ·31 | ·0000 699 | ·0000 298 | ·0000 125$^+$ | ·0000 052 | ·0000 021 | ·0000 009 |
| ·32 | ·0001 093 | ·0000 480 | ·0000 208 | ·0000 089 | ·0000 038 | ·0000 016 |
| ·33 | ·0001 678 | ·0000 760 | ·0000 340 | ·0000 150$^+$ | ·0000 065$^+$ | ·0000 028 |
| ·34 | ·0002 534 | ·0001 182 | ·0000 544 | ·0000 247 | ·0000 111 | ·0000 049 |
| ·35 | ·0003 765$^+$ | ·0001 808 | ·0000 856 | ·0000 401 | ·0000 185$^+$ | ·0000 085$^-$ |
| ·36 | ·0005 511 | ·0002 720 | ·0001 325$^-$ | ·0000 637 | ·0000 303 | ·0000 142 |
| ·37 | ·0007 952 | ·0004 031 | ·0002 016 | ·0000 996 | ·0000 487 | ·0000 235$^+$ |
| ·38 | ·0011 316 | ·0005 887 | ·0003 023 | ·0001 533 | ·0000 769 | ·0000 381 |
| ·39 | ·0015 893 | ·0008 480 | ·0004 466 | ·0002 323 | ·0001 195$^-$ | ·0000 608 |
| ·40 | ·0022 041 | ·0012 054 | ·0006 506 | ·0003 469 | ·0001 829 | ·0000 954 |
| ·41 | ·0030 202 | ·0016 918 | ·0009 353 | ·0005 109 | ·0002 759 | ·0001 475$^+$ |
| ·42 | ·0040 907 | ·0023 454 | ·0013 274 | ·0007 422 | ·0004 104 | ·0002 247 |
| ·43 | ·0054 792 | ·0032 136 | ·0018 606 | ·0010 645$^-$ | ·0006 023 | ·0003 373 |
| ·44 | ·0072 603 | ·0043 536 | ·0025 773 | ·0015 077 | ·0008 724 | ·0004 997 |
| ·45 | ·0095 212 | ·0058 338 | ·0035 291 | ·0021 100 | ·0012 478 | ·0007 305$^+$ |
| ·46 | ·0123 612 | ·0077 349 | ·0047 792 | ·0029 186 | ·0017 632 | ·0010 545$^+$ |
| ·47 | ·0158 933 | ·0101 511 | ·0064 027 | ·0039 919 | ·0024 623 | ·0015 037 |
| ·48 | ·0202 428 | ·0131 904 | ·0084 889 | ·0054 007 | ·0033 995$^+$ | ·0021 188 |
| ·49 | ·0255 478 | ·0169 754 | ·0111 414 | ·0072 296 | ·0046 419 | ·0029 513 |
| ·50 | ·0319 573 | ·0216 426 | ·0144 796 | ·0095 786 | ·0062 705$^-$ | ·0040 650$^+$ |
| ·51 | ·0396 300 | ·0273 423 | ·0186 386 | ·0125 643 | ·0083 822 | ·0055 383 |
| ·52 | ·0487 312 | ·0342 371 | ·0237 690 | ·0163 202 | ·0110 912 | ·0074 657 |
| ·53 | ·0594 303 | ·0424 994 | ·0300 364 | ·0209 975$^-$ | ·0145 301 | ·0099 598 |
| ·54 | ·0718 963 | ·0523 095$^-$ | ·0376 194 | ·0267 642 | ·0188 508 | ·0131 531 |
| ·55 | ·0862 933 | ·0638 508 | ·0467 071 | ·0338 043 | ·0242 240 | ·0171 984 |
| ·56 | ·1027 756 | ·0773 058 | ·0574 960 | ·0423 155$^-$ | ·0308 390 | ·0222 698 |
| ·57 | ·1214 810 | ·0928 507 | ·0701 850$^-$ | ·0525 057 | ·0389 015$^+$ | ·0285 623 |
| ·58 | ·1425 255$^+$ | ·1106 486 | ·0849 699 | ·0645 892 | ·0486 311 | ·0362 901 |
| ·59 | ·1659 963 | ·1308 433 | ·1020 375$^-$ | ·0787 805$^-$ | ·0602 563 | ·0456 840 |
| ·60 | ·1919 452 | ·1535 517 | ·1215 571 | ·0952 879 | ·0740 098 | ·0569 873 |
| ·61 | ·2203 832 | ·1788 562 | ·1436 736 | ·1143 054 | ·0901 207 | ·0704 504 |
| ·62 | ·2512 742 | ·2067 973 | ·1684 978 | ·1360 037 | ·1088 061 | ·0863 228 |
| ·63 | ·2845 308 | ·2373 669 | ·1960 984 | ·1605 207 | ·1302 619 | ·1048 444 |
| ·64 | ·3200 108 | ·2705 015$^+$ | ·2264 933 | ·1879 514 | ·1546 512 | ·1262 350$^+$ |
| ·65 | ·3575 155$^+$ | ·3060 782 | ·2596 418 | ·2183 378 | ·1820 936 | ·1506 819 |
| ·66 | ·3967 897 | ·3439 105$^-$ | ·2954 379 | ·2516 595$^-$ | ·2126 529 | ·1783 271 |
| ·67 | ·4375 235$^+$ | ·3837 474 | ·3337 057 | ·2878 253 | ·2463 266 | ·2092 543 |
| ·68 | ·4793 571 | ·4252 744 | ·3741 963 | ·3266 669 | ·2830 348 | ·2434 751 |
| ·69 | ·5218 867 | ·4681 173 | ·4165 884 | ·3679 342 | ·3226 128 | ·2809 172 |
| ·70 | ·5646 740 | ·5118 485$^+$ | ·4604 905$^-$ | ·4112 948 | ·3648 049 | ·3214 144 |
| ·71 | ·6072 568 | ·5559 963 | ·5054 478 | ·4563 359 | ·4092 622 | ·3646 991 |
| ·72 | ·6491 614 | ·6000 567 | ·5509 520 | ·5025 709 | ·4555 445$^-$ | ·4103 991 |
| ·73 | ·6899 173 | ·6435 080 | ·5964 541 | ·5494 497 | ·5031 269 | ·4580 394 |
| ·74 | ·7290 716 | ·6858 269 | ·6413 810 | ·5963 737 | ·5514 114 | ·5070 485$^+$ |
| ·75 | ·7662 042 | ·7265 062 | ·6851 542 | ·6427 139 | ·5997 432 | ·5567 724 |
| ·76 | ·8009 423 | ·7650 726 | ·7272 101 | ·6878 332 | ·6474 324 | ·6064 930 |
| ·77 | ·8329 738 | ·8011 045$^+$ | ·7670 222 | ·7311 101 | ·6937 795$^+$ | ·6554 534 |
| ·78 | ·8620 582 | ·8342 485$^-$ | ·8041 213 | ·7719 645$^+$ | ·7381 034 | ·7028 878 |
| ·79 | ·8880 350$^+$ | ·8642 326 | ·8381 160 | ·8098 826 | ·7797 717 | ·7480 549 |
| ·80 | ·9108 287 | ·8908 772 | ·8687 088 | ·8444 403 | ·8182 303 | ·7902 729 |

$x = \cdot 81$ to $\cdot 99$ $q = 8$ $p = 17$ to $22$

| $x$ | $p = 17$ | $p = 18$ | $p = 19$ | $p = 20$ | $p = 21$ | $p = 22$ |
|---|---|---|---|---|---|---|
| $B(p, q) =$ | ·1699 5911 × $\frac{1}{10^6}$ | ·1155 7220 × $\frac{1}{10^6}$ | ·8001 1522 × $\frac{1}{10^7}$ | ·5630 4404 × $\frac{1}{10^7}$ | ·4021 7432 × $\frac{1}{10^7}$ | ·2912 2968 × $\frac{1}{10^7}$ |
| ·81 | ·9304 496 | ·9141 010 | ·8957 088 | ·8753 226 | ·8530 302 | ·8289 544 |
| ·82 | ·9469 908 | ·9339 224 | ·9190 389 | ·9023 381 | ·8838 503 | ·8636 370 |
| ·83 | ·9606 208 | ·9504 554 | ·9387 369 | ·9254 272 | ·9105 136 | ·8940 093 |
| ·84 | ·9715 729 | ·9639 008 | ·9549 500+ | ·9446 613 | ·9329 940 | ·9199 265+ |
| ·85 | ·9801 307 | ·9745 324 | ·9679 232 | ·9602 358 | ·9514 144 | ·9414 168 |
| ·86 | ·9866 115− | ·9826 786 | ·9779 811 | ·9724 528 | ·9660 345+ | ·9586 749 |
| ·87 | ·9913 487 | ·9887 028 | ·9855 056 | ·9816 993 | ·9772 287 | ·9720 429 |
| ·88 | ·9946 744 | ·9929 806 | ·9909 104 | ·9884 174 | ·9854 558 | ·9819 808 |
| ·89 | ·9969 025+ | ·9958 793 | ·9946 144 | ·9930 740 | ·9912 231 | ·9890 268 |
| ·90 | ·9983 159 | ·9977 387 | ·9970 172 | ·9961 286 | ·9950 490 | ·9937 534 |
| ·91 | ·9991 560 | ·9988 561 | ·9984 772 | ·9980 053 | ·9974 257 | ·9967 223 |
| ·92 | ·9996 174 | ·9994 767 | ·9992 969 | ·9990 705+ | ·9987 894 | ·9984 446 |
| ·93 | ·9998 473 | ·9997 892 | ·9997 141 | ·9996 186 | ·9994 987 | ·9993 500− |
| ·94 | ·9999 483 | ·9999 280 | ·9999 015+ | ·9998 674 | ·9998 242 | ·9997 699 |
| ·95 | ·9999 861 | ·9999 804 | ·9999 730 | ·9999 633 | ·9999 508 | ·9999 351 |
| ·96 | ·9999 973 | ·9999 962 | ·9999 947 | ·9999 927 | ·9999 901 | ·9999 868 |
| ·97 | ·9999 997 | ·9999 996 | ·9999 994 | ·9999 991 | ·9999 988 | ·9999 984 |
| ·98 | 1·0000 000 | 1·0000 000 | 1·0000 000 | 1·0000 000 | ·9999 999 | ·9999 999 |
| ·99 | | | | | 1·0000 000 | 1·0000 000 |

$x = \cdot 29$ to $\cdot 90$ $q = 8$ $p = 23$ to $28$

| $x$ | $p = 23$ | $p = 24$ | $p = 25$ | $p = 26$ | $p = 27$ | $p = 28$ |
|---|---|---|---|---|---|---|
| $B(p, q) =$ | $\cdot 2135\ 6843 \times \frac{1}{10^7}$ | $\cdot 1584\ 5400 \times \frac{1}{10^7}$ | $\cdot 1188\ 4050^{-} \times \frac{1}{10^7}$ | $\cdot 9003\ 0680 \times \frac{1}{10^8}$ | $\cdot 6884\ 6990 \times \frac{1}{10^8}$ | $\cdot 5311\ 0535^{+} \times \frac{1}{10^8}$ |
| ·29 | ·0000 001 | | | | | |
| ·30 | ·0000 002 | ·0000 001 | | | | |
| ·31 | ·0000 003 | ·0000 001 | ·0000 001 | | | |
| ·32 | ·0000 007 | ·0000 003 | ·0000 001 | | | |
| ·33 | ·0000 012 | ·0000 005+ | ·0000 002 | ·0000 001 | | |
| ·34 | ·0000 022 | ·0000 010 | ·0000 004 | ·0000 002 | ·0000 001 | |
| ·35 | ·0000 038 | ·0000 017 | ·0000 008 | ·0000 003 | ·0000 001 | ·0000 001 |
| ·36 | ·0000 066 | ·0000 031 | ·0000 014 | ·0000 006 | ·0000 003 | ·0000 001 |
| ·37 | ·0000 113 | ·0000 053 | ·0000 025+ | ·0000 012 | ·0000 005+ | ·0000 002 |
| ·38 | ·0000 187 | ·0000 091 | ·0000 044 | ·0000 021 | ·0000 010 | ·0000 005− |
| ·39 | ·0000 307 | ·0000 153 | ·0000 076 | ·0000 037 | ·0000 018 | ·0000 009 |
| ·40 | ·0000 493 | ·0000 253 | ·0000 128 | ·0000 065− | ·0000 032 | ·0000 016 |
| ·41 | ·0000 781 | ·0000 410 | ·0000 213 | ·0000 110 | ·0000 056 | ·0000 029 |
| ·42 | ·0001 218 | ·0000 654 | ·0000 349 | ·0000 184 | ·0000 097 | ·0000 050+ |
| ·43 | ·0001 871 | ·0001 029 | ·0000 561 | ·0000 304 | ·0000 163 | ·0000 087 |
| ·44 | ·0002 835+ | ·0001 595− | ·0000 889 | ·0000 492 | ·0000 271 | ·0000 148 |
| ·45 | ·0004 237 | ·0002 436 | ·0001 389 | ·0000 786 | ·0000 442 | ·0000 246 |
| ·46 | ·0006 248 | ·0003 670 | ·0002 138 | ·0001 236 | ·0000 710 | ·0000 405− |
| ·47 | ·0009 098 | ·0005 457 | ·0003 247 | ·0001 917 | ·0001 124 | ·0000 655− |
| ·48 | ·0013 084 | ·0008 011 | ·0004 865+ | ·0002 933 | ·0001 755+ | ·0001 044 |
| ·49 | ·0018 593 | ·0011 614 | ·0007 197 | ·0004 426 | ·0002 703 | ·0001 640 |
| ·50 | ·0026 114 | ·0016 634 | ·0010 512 | ·0006 594 | ·0004 107 | ·0002 541 |
| ·51 | ·0036 264 | ·0023 546 | ·0015 168 | ·0009 699 | ·0006 159 | ·0003 886 |
| ·52 | ·0049 806 | ·0032 950+ | ·0021 629 | ·0014 093 | ·0009 120 | ·0005 863 |
| ·53 | ·0067 669 | ·0045 596 | ·0030 485− | ·0020 234 | ·0013 338 | ·0008 735+ |
| ·54 | ·0090 974 | ·0062 407 | ·0042 482 | ·0028 710 | ·0019 271 | ·0012 852 |
| ·55 | ·0121 048 | ·0084 507 | ·0058 548 | ·0040 273 | ·0027 516 | ·0018 680 |
| ·56 | ·0159 443 | ·0113 238 | ·0079 817 | ·0055 861 | ·0038 834 | ·0026 827 |
| ·57 | ·0207 939 | ·0150 182 | ·0107 658 | ·0076 633 | ·0054 187 | ·0038 076 |
| ·58 | ·0268 549 | ·0197 169 | ·0143 694 | ·0103 994 | ·0074 769 | ·0053 424 |
| ·59 | ·0343 507 | ·0256 289 | ·0189 823 | ·0139 627 | ·0102 038 | ·0074 111 |
| ·60 | ·0435 241 | ·0329 877 | ·0248 219 | ·0185 507 | ·0137 748 | ·0101 664 |
| ·61 | ·0546 337 | ·0420 492 | ·0321 336 | ·0243 915+ | ·0183 973 | ·0137 929 |
| ·62 | ·0679 478 | ·0530 880 | ·0411 878 | ·0317 438 | ·0243 120 | ·0185 098 |
| ·63 | ·0837 364 | ·0663 911 | ·0522 764 | ·0408 943 | ·0317 930 | ·0245 726 |
| ·64 | ·1022 621 | ·0822 499 | ·0657 065− | ·0521 541 | ·0411 454 | ·0322 733 |
| ·65 | ·1237 678 | ·1009 493 | ·0817 913 | ·0658 518 | ·0527 017 | ·0419 382 |
| ·66 | ·1484 637 | ·1227 552 | ·1008 387 | ·0823 236 | ·0668 137 | ·0539 232 |
| ·67 | ·1765 128 | ·1478 996 | ·1231 372 | ·1019 010 | ·0838 420 | ·0686 056 |
| ·68 | ·2080 152 | ·1765 638 | ·1489 390 | ·1248 943 | ·1041 420 | ·0863 718 |
| ·69 | ·2429 931 | ·2088 614 | ·1784 415+ | ·1515 747 | ·1280 456 | ·1076 014 |
| ·70 | ·2813 767 | ·2448 206 | ·2117 678 | ·1821 524 | ·1558 404 | ·1326 467 |
| ·71 | ·3229 920 | ·2843 677 | ·2489 459 | ·2167 546 | ·1877 454 | ·1618 089 |
| ·72 | ·3675 521 | ·3273 131 | ·2898 909 | ·2554 025+ | ·2238 855+ | ·1953 101 |
| ·73 | ·4146 528 | ·3733 413 | ·3343 880 | ·2979 901 | ·2642 659 | ·2332 647 |
| ·74 | ·4637 746 | ·4220 058 | ·3820 819 | ·3442 659 | ·3087 480 | ·2756 505+ |
| ·75 | ·5142 900 | ·4727 310 | ·4324 708 | ·3938 210 | ·3570 294 | ·3222 818 |
| ·76 | ·5654 792 | ·5248 219 | ·4849 101 | ·4460 839 | ·4086 315− | ·3727 882 |
| ·77 | ·6165 525− | ·5774 824 | ·5386 239 | ·5003 250+ | ·4628 952 | ·4266 021 |
| ·78 | ·6666 799 | ·6298 422 | ·5927 282 | ·5556 736 | ·5189 896 | ·4829 577 |
| ·79 | ·7150 262 | ·6809 922 | ·6462 634 | ·6111 456 | ·5759 332 | ·5409 035− |
| ·80 | ·7607 906 | ·7300 265− | ·6982 369 | ·6656 843 | ·6326 309 | ·5993 327 |
| ·81 | ·8032 480 | ·7760 886 | ·7476 732 | ·7182 120 | ·6879 237 | ·6570 296 |
| ·82 | ·8417 883 | ·8184 196 | ·7936 683 | ·7676 894 | ·7406 513 | ·7127 319 |
| ·83 | ·8759 521 | ·8564 032 | ·8354 451 | ·8131 792 | ·7897 230 | ·7652 069 |
| ·84 | ·9054 573 | ·8896 040 | ·8724 032 | ·8539 089 | ·8341 912 | ·8133 342 |
| ·85 | ·9302 150+ | ·9177 956 | ·9041 602 | ·8893 248 | ·8733 197 | ·8561 883 |
| ·86 | ·9503 317 | ·9409 728 | ·9305 766 | ·9191 326 | ·9066 409 | ·8931 129 |
| ·87 | ·9660 957 | ·9593 469 | ·9517 630 | ·9433 175+ | ·9339 918 | ·9237 749 |
| ·88 | ·9779 498 | ·9733 230 | ·9680 638 | ·9621 398 | ·9555 232 | ·9481 910 |
| ·89 | ·9864 501 | ·9834 589 | ·9800 203 | ·9761 030 | ·9716 779 | ·9667 186 |
| ·90 | ·9922 164 | ·9904 121 | ·9883 145+ | ·9858 982 | ·9831 380 | ·9800 098 |

$x = \cdot 91$ to $\cdot 99$ $q = 8$ $p = 23$ to $28$

| | $p = 23$ | $p = 24$ | $p = 25$ | $p = 26$ | $p = 27$ | $p = 28$ |
|---|---|---|---|---|---|---|
| $B(p, q) =$ | ·2135 6843 × $\frac{1}{10^7}$ | ·1584 5400 × $\frac{1}{10^7}$ | ·1188 4050$^-$ × $\frac{1}{10^7}$ | ·9003 0680 × $\frac{1}{10^8}$ | ·6884 6990 × $\frac{1}{10^8}$ | ·5311 0535$^+$ × $\frac{1}{10^8}$ |
| $x$ | | | | | | |
| ·91 | ·9958 786 | ·9948 771 | ·9937 000 | ·9923 289 | ·9907 453 | ·9889 306 |
| ·92 | ·9980 264 | ·9975 245$^+$ | ·9969 281 | ·9962 259 | ·9954 058 | ·9944 558 |
| ·93 | ·9991 677 | ·9989 466 | ·9986 810 | ·9983 648 | ·9979 915$^+$ | ·9975 544 |
| ·94 | ·9997 027 | ·9996 203 | ·9995 203 | ·9993 999 | ·9992 563 | ·9990 863 |
| ·95 | ·9999 154 | ·9998 909 | ·9998 609 | ·9998 244 | ·9997 805$^-$ | ·9997 279 |
| ·96 | ·9999 826 | ·9999 774 | ·9999 709 | ·9999 630 | ·9999 533 | ·9999 416 |
| ·97 | ·9999 979 | ·9999 972 | ·9999 964 | ·9999 954 | ·9999 941 | ·9999 925$^+$ |
| ·98 | ·9999 999 | ·9999 999 | ·9999 998 | ·9999 998 | ·9999 997 | ·9999 996 |
| ·99 | 1·0000 000 | 1·0000 000 | 1·0000 000 | 1·0000 000 | 1·0000 000 | 1·0000 000 |

$x$ = ·36 to ·99 $q = 8$ $p$ = 29 to 34

| $x$ | $p = 29$ | $p = 30$ | $p = 31$ | $p = 32$ | $p = 33$ | $p = 34$ |
|---|---|---|---|---|---|---|
| $B(p, q)$ = | ·4130 8194 × $\frac{1}{10^8}$ | ·3237 6693 × $\frac{1}{10^8}$ | ·2556 0547 × $\frac{1}{10^8}$ | ·2031 7358 × $\frac{1}{10^8}$ | ·1625 3886 × $\frac{1}{10^8}$ | ·1308 2396 × $\frac{1}{10^8}$ |
| ·36 | ·0000 001 | | | | | |
| ·37 | ·0000 001 | ·0000 001 | | | | |
| ·38 | ·0000 002 | ·0000 001 | | | | |
| ·39 | ·0000 004 | ·0000 002 | ·0000 001 | | | |
| ·40 | ·0000 008 | ·0000 004 | ·0000 002 | ·0000 001 | | |
| ·41 | ·0000 015− | ·0000 007 | ·0000 004 | ·0000 002 | ·0000 001 | |
| ·42 | ·0000 026 | ·0000 013 | ·0000 007 | ·0000 004 | ·0000 002 | ·0000 001 |
| ·43 | ·0000 046 | ·0000 024 | ·0000 013 | ·0000 007 | ·0000 003 | ·0000 002 |
| ·44 | ·0000 080 | ·0000 043 | ·0000 023 | ·0000 012 | ·0000 007 | ·0000 003 |
| ·45 | ·0000 137 | ·0000 075+ | ·0000 041 | ·0000 022 | ·0000 012 | ·0000 007 |
| ·46 | ·0000 229 | ·0000 129 | ·0000 072 | ·0000 040 | ·0000 022 | ·0000 012 |
| ·47 | ·0000 379 | ·0000 218 | ·0000 125− | ·0000 071 | ·0000 040 | ·0000 023 |
| ·48 | ·0000 617 | ·0000 362 | ·0000 212 | ·0000 123 | ·0000 071 | ·0000 041 |
| ·49 | ·0000 989 | ·0000 593 | ·0000 353 | ·0000 209 | ·0000 124 | ·0000 073 |
| ·50 | ·0001 563 | ·0000 955+ | ·0000 581 | ·0000 351 | ·0000 211 | ·0000 127 |
| ·51 | ·0002 436 | ·0001 518 | ·0000 941 | ·0000 580 | ·0000 356 | ·0000 218 |
| ·52 | ·0003 746 | ·0002 380 | ·0001 504 | ·0000 945+ | ·0000 591 | ·0000 368 |
| ·53 | ·0005 686 | ·0003 680 | ·0002 369 | ·0001 517 | ·0000 967 | ·0000 613 |
| ·54 | ·0008 520 | ·0005 615+ | ·0003 681 | ·0002 401 | ·0001 558 | ·0001 006 |
| ·55 | ·0012 605+ | ·0008 458 | ·0005 644 | ·0003 748 | ·0002 476 | ·0001 629 |
| ·56 | ·0018 421 | ·0012 578 | ·0008 543 | ·0005 772 | ·0003 882 | ·0002 598 |
| ·57 | ·0026 597 | ·0018 475+ | ·0012 765+ | ·0008 775+ | ·0006 004 | ·0004 089 |
| ·58 | ·0037 949 | ·0026 806 | ·0018 836 | ·0013 169 | ·0009 164 | ·0006 347 |
| ·59 | ·0053 515− | ·0038 430 | ·0027 453 | ·0019 514 | ·0013 806 | ·0009 723 |
| ·60 | ·0074 601 | ·0054 444 | ·0039 528 | ·0028 557 | ·0020 534 | ·0014 700 |
| ·61 | ·0102 821 | ·0076 235+ | ·0056 234 | ·0041 278 | ·0030 159 | ·0021 938 |
| ·62 | ·0140 131 | ·0105 521 | ·0079 057 | ·0058 944 | ·0043 746 | ·0032 324 |
| ·63 | ·0188 865+ | ·0144 397 | ·0109 844 | ·0083 161 | ·0062 673 | ·0047 028 |
| ·64 | ·0251 755+ | ·0195 365− | ·0150 854 | ·0115 934 | ·0088 697 | ·0067 568 |
| ·65 | ·0331 928 | ·0261 361 | ·0204 791 | ·0159 717 | ·0124 010 | ·0095 877 |
| ·66 | ·0432 886 | ·0345 755+ | ·0274 831 | ·0217 451 | ·0171 296 | ·0134 371 |
| ·67 | ·0558 451 | ·0452 319 | ·0364 619 | ·0292 591 | ·0233 776 | ·0186 011 |
| ·68 | ·0712 671 | ·0585 167 | ·0478 234 | ·0389 100 | ·0315 230 | ·0254 343 |
| ·69 | ·0899 682 | ·0748 645+ | ·0620 113 | ·0511 400 | ·0419 978 | ·0343 517 |
| ·70 | ·1123 523 | ·0947 172 | ·0794 922 | ·0664 281 | ·0552 829 | ·0458 263 |
| ·71 | ·1387 903 | ·1185 022 | ·1007 365+ | ·0852 747 | ·0718 954 | ·0603 811 |
| ·72 | ·1695 922 | ·1466 057 | ·1261 937 | ·1081 785− | ·0923 701 | ·0785 737 |
| ·73 | ·2049 761 | ·1793 408 | ·1562 605− | ·1356 073 | ·1172 324 | ·1009 734 |
| ·74 | ·2450 354 | ·2169 117 | ·1912 441 | ·1679 612 | ·1469 628 | ·1281 280 |
| ·75 | ·2897 058 | ·2593 765+ | ·2313 219 | ·2055 298 | ·1819 542 | ·1605 218 |
| ·76 | ·3387 370 | ·3066 115+ | ·2764 992 | ·2484 462 | ·2224 621 | ·1985 252 |
| ·77 | ·3916 701 | ·3582 798 | ·3265 702 | ·2966 405− | ·2685 533 | ·2423 385+ |
| ·78 | ·4478 266 | ·4138 100 | ·3810 860 | ·3497 977 | ·3200 543 | ·2919 332 |
| ·79 | ·5063 116 | ·4723 877 | ·4393 345+ | ·4073 262 | ·3765 083 | ·3469 977 |
| ·80 | ·5660 345+ | ·5329 659 | ·5003 383 | ·4683 421 | ·4371 459 | ·4068 950− |
| ·81 | ·6257 493 | ·5942 964 | ·5628 750+ | ·5316 766 | ·5008 779 | ·4706 392 |
| ·82 | ·6841 146 | ·6549 841 | ·6255 235− | ·5959 108 | ·5663 166 | ·5369 018 |
| ·83 | ·7397 715+ | ·7135 642 | ·6867 367 | ·6594 419 | ·6318 315− | ·6040 537 |
| ·84 | ·7914 343 | ·7685 981 | ·7449 397 | ·7205 792 | ·6956 402 | ·6702 477 |
| ·85 | ·8379 862 | ·8187 799 | ·7986 452 | ·7776 662 | ·7559 333 | ·7335 417 |
| ·86 | ·8785 703 | ·8630 448 | ·8465 774 | ·8292 177 | ·8110 224 | ·7920 553 |
| ·87 | ·9126 641 | ·9006 644 | ·8877 887 | ·8740 574 | ·8594 979 | ·8441 443 |
| ·88 | ·9401 256 | ·9313 148 | ·9217 522 | ·9114 369 | ·9003 738 | ·8885 731 |
| ·89 | ·9612 013 | ·9551 057 | ·9484 148 | ·9411 151 | ·9331 974 | ·9246 557 |
| ·90 | ·9764 905+ | ·9725 586 | ·9681 943 | ·9633 794 | ·9580 981 | ·9523 366 |
| ·91 | ·9868 663 | ·9845 345− | ·9819 173 | ·9789 979 | ·9757 602 | ·9721 888 |
| ·92 | ·9933 632 | ·9921 155− | ·9906 997 | ·9891 030 | ·9873 128 | ·9853 164 |
| ·93 | ·9970 463 | ·9964 597 | ·9957 869 | ·9950 198 | ·9941 504 | ·9931 703 |
| ·94 | ·9988 866 | ·9986 535+ | ·9983 833 | ·9980 720 | ·9977 153 | ·9973 089 |
| ·95 | ·9996 654 | ·9995 917 | ·9995 054 | ·9994 049 | ·9992 885− | ·9991 545− |
| ·96 | ·9999 275+ | ·9999 108 | ·9998 910 | ·9998 676 | ·9998 403 | ·9998 086 |
| ·97 | ·9999 906 | ·9999 884 | ·9999 857 | ·9999 825− | ·9999 786 | ·9999 742 |
| ·98 | ·9999 995+ | ·9999 994 | ·9999 993 | ·9999 991 | ·9999 989 | ·9999 986 |
| ·99 | 1·0000 000 | 1·0000 000 | 1·0000 000 | 1·0000 000 | 1·0000 000 | 1·0000 000 |

$x$ = ·43 to ·99 $q = 8$ $p$ = 35 to 40

| | $p = 35$ | $p = 36$ | $p = 37$ | $p = 38$ | $p = 39$ | $p = 40$ |
|---|---|---|---|---|---|---|
| $B(p, q)$ = | ·1059 0511 × $\frac{1}{10^8}$ | ·8620 1836 × $\frac{1}{10^9}$ | ·7052 8775 × $\frac{1}{10^9}$ | ·5799 0326 × $\frac{1}{10^9}$ | ·4790 5052 × $\frac{1}{10^9}$ | ·3975 1000 × $\frac{1}{10^9}$ |
| $x$ | | | | | | |
| ·43 | ·0000 001 | | | | | |
| ·44 | ·0000 002 | ·0000 001 | | | | |
| ·45 | ·0000 004 | ·0000 002 | ·0000 001 | ·0000 001 | | |
| ·46 | ·0000 007 | ·0000 004 | ·0000 002 | ·0000 001 | ·0000 001 | |
| ·47 | ·0000 013 | ·0000 007 | ·0000 004 | ·0000 002 | ·0000 001 | ·0000 001 |
| ·48 | ·0000 023 | ·0000 013 | ·0000 008 | ·0000 004 | ·0000 002 | ·0000 001 |
| ·49 | ·0000 042 | ·0000 025− | ·0000 014 | ·0000 008 | ·0000 005− | ·0000 003 |
| ·50 | ·0000 075+ | ·0000 045− | ·0000 026 | ·0000 016 | ·0000 009 | ·0000 005+ |
| ·51 | ·0000 132 | ·0000 080 | ·0000 048 | ·0000 029 | ·0000 017 | ·0000 010 |
| ·52 | ·0000 228 | ·0000 141 | ·0000 086 | ·0000 053 | ·0000 032 | ·0000 020 |
| ·53 | ·0000 387 | ·0000 243 | ·0000 152 | ·0000 095+ | ·0000 059 | ·0000 037 |
| ·54 | ·0000 647 | ·0000 415− | ·0000 264 | ·0000 168 | ·0000 106 | ·0000 067 |
| ·55 | ·0001 066 | ·0000 695+ | ·0000 452 | ·0000 292 | ·0000 188 | ·0000 121 |
| ·56 | ·0001 732 | ·0001 149 | ·0000 760 | ·0000 500+ | ·0000 328 | ·0000 215− |
| ·57 | ·0002 773 | ·0001 872 | ·0001 259 | ·0000 844 | ·0000 563 | ·0000 375− |
| ·58 | ·0004 378 | ·0003 007 | ·0002 057 | ·0001 402 | ·0000 952 | ·0000 645− |
| ·59 | ·0006 819 | ·0004 762 | ·0003 313 | ·0002 296 | ·0001 586 | ·0001 092 |
| ·60 | ·0010 479 | ·0007 439 | ·0005 261 | ·0003 707 | ·0002 602 | ·0001 821 |
| ·61 | ·0015 891 | ·0011 464 | ·0008 239 | ·0005 899 | ·0004 209 | ·0002 993 |
| ·62 | ·0023 785− | ·0017 432 | ·0012 727 | ·0009 258 | ·0006 711 | ·0004 848 |
| ·63 | ·0035 142 | ·0026 157 | ·0019 395 | ·0014 329 | ·0010 550− | ·0007 742 |
| ·64 | ·0051 261 | ·0038 737 | ·0029 163 | ·0021 877 | ·0016 355− | ·0012 186 |
| ·65 | ·0073 826 | ·0056 626 | ·0043 272 | ·0032 950− | ·0025 004 | ·0018 913 |
| ·66 | ·0104 984 | ·0081 710 | ·0063 361 | ·0048 960 | ·0037 705− | ·0028 943 |
| ·67 | ·0147 420 | ·0116 393 | ·0091 562 | ·0071 778 | ·0056 081 | ·0043 677 |
| ·68 | ·0204 416 | ·0163 675+ | ·0130 585− | ·0103 826 | ·0082 279 | ·0064 996 |
| ·69 | ·0279 896 | ·0227 219 | ·0183 804 | ·0148 180 | ·0119 071 | ·0095 381 |
| ·70 | ·0378 438 | ·0311 386 | ·0255 322 | ·0208 653 | ·0169 967 | ·0138 026 |
| ·71 | ·0505 228 | ·0421 235+ | ·0350 005− | ·0289 863 | ·0239 297 | ·0196 951 |
| ·72 | ·0665 952 | ·0562 458 | ·0473 453 | ·0397 245− | ·0332 268 | ·0277 087 |
| ·73 | ·0866 607 | ·0741 228 | ·0631 905− | ·0537 000 | ·0454 957 | ·0384 316 |
| ·74 | ·1113 206 | ·0963 957 | ·0832 038 | ·0715 948 | ·0614 217 | ·0525 424 |
| ·75 | ·1411 381 | ·1236 927 | ·1080 646 | ·0941 260 | ·0817 464 | ·0707 951 |
| ·76 | ·1765 878 | ·1565 809 | ·1384 190 | ·1220 046 | ·1072 317 | ·0939 890 |
| ·77 | ·2179 974 | ·1955 061 | ·1748 205− | ·1558 791 | ·1386 076 | ·1229 214 |
| ·78 | ·2654 828 | ·2407 252 | ·2176 595− | ·1962 644 | ·1765 021 | ·1583 208 |
| ·79 | ·3188 846 | ·2922 333 | ·2670 849 | ·2434 590 | ·2213 563 | ·2007 612 |
| ·80 | ·3777 118 | ·3496 959 | ·3229 251 | ·2974 567 | ·2733 288 | ·2505 619 |
| ·81 | ·4411 031 | ·4123 940 | ·3846 180 | ·3578 629 | ·3321 991 | ·3076 803 |
| ·82 | ·5078 158 | ·4791 951 | ·4511 627 | ·4238 274 | ·3972 833 | ·3716 105− |
| ·83 | ·5762 514 | ·5485 604 | ·5211 078 | ·4940 113 | ·4673 783 | ·4413 053 |
| ·84 | ·6445 266 | ·6185 997 | ·5925 865− | ·5666 013 | ·5407 529 | ·5151 431 |
| ·85 | ·7105 904 | ·6871 801 | ·6634 121 | ·6393 872 | ·6152 042 | ·5909 592 |
| ·86 | ·7723 852 | ·7520 857 | ·7312 337 | ·7099 082 | ·6881 898 | ·6661 597 |
| ·87 | ·8280 365+ | ·8112 200 | ·7937 448 | ·7756 652 | ·7570 383 | ·7379 243 |
| ·88 | ·8760 505+ | ·8628 267 | ·8489 269 | ·8343 810 | ·8192 226 | ·8034 890 |
| ·89 | ·9154 886 | ·9056 980 | ·8952 901 | ·8842 746 | ·8726 649 | ·8604 776 |
| ·90 | ·9460 837 | ·9393 306 | ·9320 711 | ·9243 014 | ·9160 205− | ·9072 300 |
| ·91 | ·9682 697 | ·9639 901 | ·9593 384 | ·9543 045+ | ·9488 799 | ·9430 574 |
| ·92 | ·9831 015+ | ·9806 564 | ·9779 694 | ·9750 297 | ·9718 270 | ·9683 516 |
| ·93 | ·9920 712 | ·9908 446 | ·9894 821 | ·9879 752 | ·9863 156 | ·9844 952 |
| ·94 | ·9968 482 | ·9963 286 | ·9957 452 | ·9950 930 | ·9943 670 | ·9935 621 |
| ·95 | ·9990 009 | ·9988 259 | ·9986 273 | ·9984 029 | ·9981 505+ | ·9978 677 |
| ·96 | ·9997 718 | ·9997 294 | ·9996 809 | ·9996 254 | ·9995 624 | ·9994 910 |
| ·97 | ·9999 689 | ·9999 628 | ·9999 558 | ·9999 476 | ·9999 383 | ·9999 276 |
| ·98 | ·9999 984 | ·9999 980 | ·9999 976 | ·9999 971 | ·9999 966 | ·9999 960 |
| ·99 | 1·0000 000 | 1·0000 000 | 1·0000 000 | 1·0000 000 | 1·0000 000 | 1·0000 000 |

$x$ = ·48 to ·99 $q = 8$ $p$ = 41 to 45

| | $p = 41$ | $p = 42$ | $p = 43$ | $p = 44$ | $p = 45$ |
|---|---|---|---|---|---|
| $B(p, q)$ = | ·3312 5834 × $\frac{1}{10^9}$ | ·2771 7534 × $\frac{1}{10^9}$ | ·2328 2729 × $\frac{1}{10^9}$ | ·1963 0536 × $\frac{1}{10^9}$ | ·1661 0454 × $\frac{1}{10^9}$ |
| $x$ | | | | | |
| ·48 | ·0000 001 | | | | |
| ·49 | ·0000 002 | ·0000 001 | | | |
| ·50 | ·0000 003 | ·0000 002 | ·0000 001 | ·0000 001 | |
| ·51 | ·0000 006 | ·0000 004 | ·0000 002 | ·0000 001 | ·0000 001 |
| ·52 | ·0000 012 | ·0000 007 | ·0000 004 | ·0000 003 | ·0000 002 |
| ·53 | ·0000 023 | ·0000 014 | ·0000 009 | ·0000 005+ | ·0000 003 |
| ·54 | ·0000 042 | ·0000 026 | ·0000 017 | ·0000 010 | ·0000 006 |
| ·55 | ·0000 077 | ·0000 049 | ·0000 031 | ·0000 020 | ·0000 013 |
| ·56 | ·0000 140 | ·0000 091 | ·0000 059 | ·0000 038 | ·0000 024 |
| ·57 | ·0000 249 | ·0000 164 | ·0000 108 | ·0000 071 | ·0000 047 |
| ·58 | ·0000 435− | ·0000 292 | ·0000 196 | ·0000 131 | ·0000 087 |
| ·59 | ·0000 749 | ·0000 512 | ·0000 349 | ·0000 237 | ·0000 161 |
| ·60 | ·0001 270 | ·0000 883 | ·0000 612 | ·0000 423 | ·0000 292 |
| ·61 | ·0002 121 | ·0001 499 | ·0001 056 | ·0000 742 | ·0000 520 |
| ·62 | ·0003 492 | ·0002 507 | ·0001 794 | ·0001 281 | ·0000 912 |
| ·63 | ·0005 663 | ·0004 129 | ·0003 002 | ·0002 177 | ·0001 574 |
| ·64 | ·0009 051 | ·0006 702 | ·0004 949 | ·0003 644 | ·0002 675+ |
| ·65 | ·0014 261 | ·0010 720 | ·0008 036 | ·0006 006 | ·0004 478 |
| ·66 | ·0022 148 | ·0016 898 | ·0012 855+ | ·0009 753 | ·0007 379 |
| ·67 | ·0033 911 | ·0026 251 | ·0020 264 | ·0015 599 | ·0011 976 |
| ·68 | ·0051 188 | ·0040 195− | ·0031 474 | ·0024 578 | ·0019 143 |
| ·69 | ·0076 175− | ·0060 660 | ·0048 170 | ·0038 149 | ·0030 135− |
| ·70 | ·0111 754 | ·0090 225− | ·0072 642 | ·0058 331 | ·0046 719 |
| ·71 | ·0161 623 | ·0132 259 | ·0107 935− | ·0087 854 | ·0071 328 |
| ·72 | ·0230 405− | ·0191 055− | ·0158 000 | ·0130 327 | ·0107 233 |
| ·73 | ·0323 724 | ·0271 940 | ·0227 837 | ·0190 401 | ·0158 725+ |
| ·74 | ·0448 219 | ·0381 333 | ·0323 588 | ·0273 900 | ·0231 282 |
| ·75 | ·0611 443 | ·0526 705− | ·0452 558 | ·0387 896 | ·0331 684 |
| ·76 | ·0821 634 | ·0716 414 | ·0623 119 | ·0540 673 | ·0468 045+ |
| ·77 | ·1087 294 | ·0959 359 | ·0844 430 | ·0741 528 | ·0649 689 |
| ·78 | ·1416 576 | ·1264 413 | ·1125 945− | ·1000 357 | ·0886 814 |
| ·79 | ·1816 437 | ·1639 624 | ·1476 661 | ·1326 963 | ·1189 887 |
| ·80 | ·2291 610 | ·2091 173 | ·1904 098 | ·1730 075− | ·1568 708 |
| ·81 | ·2843 446 | ·2622 155− | ·2413 035− | ·2216 073 | ·2031 152 |
| ·82 | ·3468 747 | ·3231 283 | ·3004 109 | ·2787 502 | ·2581 626 |
| ·83 | ·4158 776 | ·3911 693 | ·3672 435− | ·3441 523 | ·3219 374 |
| ·84 | ·4898 662 | ·4650 086 | ·4406 481 | ·4168 542 | ·3936 875− |
| ·85 | ·5667 445− | ·5426 479 | ·5187 521 | ·4951 342 | ·4718 652 |
| ·86 | ·6438 982 | ·6214 847 | ·5989 964 | ·5765 082 | ·5540 915+ |
| ·87 | ·7183 850+ | ·6984 835+ | ·6782 835+ | ·6578 486 | ·6372 419 |
| ·88 | ·7872 205− | ·7704 599 | ·7532 524 | ·7356 447 | ·7176 848 |
| ·89 | ·8477 328 | ·8344 533 | ·8206 648 | ·8063 953 | ·7916 750− |
| ·90 | ·8979 341 | ·8881 394 | ·8778 549 | ·8670 921 | ·8558 645+ |
| ·91 | ·9368 317 | ·9301 990 | ·9231 574 | ·9157 063 | ·9078 472 |
| ·92 | ·9645 948 | ·9605 484 | ·9562 052 | ·9515 591 | ·9466 046 |
| ·93 | ·9825 059 | ·9803 400 | ·9779 901 | ·9754 488 | ·9727 094 |
| ·94 | ·9926 730 | ·9916 947 | ·9906 217 | ·9894 490 | ·9881 712 |
| ·95 | ·9975 520 | ·9972 008 | ·9968 117 | ·9963 818 | ·9959 084 |
| ·96 | ·9994 105− | ·9993 200 | ·9992 186 | ·9991 054 | ·9989 795+ |
| ·97 | ·9999 154 | ·9999 015+ | ·9998 858 | ·9998 681 | ·9998 483 |
| ·98 | ·9999 953 | ·9999 944 | ·9999 935+ | ·9999 924 | ·9999 912 |
| ·99 | 1·0000 000 | 1·0000 000 | 1·0000 000 | 1·0000 000 | 1·0000 000 |

$x = \cdot 52$ to $1 \cdot 00$ $q = 8$ $p = 46$ to $50$

| | $p = 46$ | $p = 47$ | $p = 48$ | $p = 49$ | $p = 50$ |
|---|---|---|---|---|---|
| $B(p, q) =$ | $\cdot 1410\ 3215 \times \frac{1}{10^9}$ | $\cdot 1201\ 3850 \times \frac{1}{10^9}$ | $\cdot 1026\ 6381 \times \frac{1}{10^9}$ | $\cdot 8799\ 7551 \times \frac{1}{10^{10}}$ | $\cdot 7564\ 7018 \times \frac{1}{10^{10}}$ |
| $x$ | | | | | |
| ·52 | ·0000 001 | ·0000 001 | | | |
| ·53 | ·0000 002 | ·0000 001 | ·0000 001 | | |
| ·54 | ·0000 004 | ·0000 002 | ·0000 002 | ·0000 001 | ·0000 001 |
| ·55 | ·0000 008 | ·0000 005+ | ·0000 003 | ·0000 002 | ·0000 001 |
| ·56 | ·0000 016 | ·0000 010 | ·0000 006 | ·0000 004 | ·0000 003 |
| ·57 | ·0000 030 | ·0000 020 | ·0000 013 | ·0000 008 | ·0000 005+ |
| ·58 | ·0000 058 | ·0000 039 | ·0000 025+ | ·0000 017 | ·0000 011 |
| ·59 | ·0000 109 | ·0000 073 | ·0000 049 | ·0000 033 | ·0000 022 |
| ·60 | ·0000 200 | ·0000 137 | ·0000 094 | ·0000 064 | ·0000 044 |
| ·61 | ·0000 363 | ·0000 253 | ·0000 176 | ·0000 122 | ·0000 084 |
| ·62 | ·0000 647 | ·0000 458 | ·0000 324 | ·0000 228 | ·0000 160 |
| ·63 | ·0001 135+ | ·0000 817 | ·0000 586 | ·0000 419 | ·0000 300 |
| ·64 | ·0001 959 | ·0001 431 | ·0001 043 | ·0000 758 | ·0000 550+ |
| ·65 | ·0003 329 | ·0002 469 | ·0001 827 | ·0001 349 | ·0000 993 |
| ·66 | ·0005 569 | ·0004 193 | ·0003 149 | ·0002 359 | ·0001 764 |
| ·67 | ·0009 172 | ·0007 007 | ·0005 340 | ·0004 060 | ·0003 081 |
| ·68 | ·0014 872 | ·0011 526 | ·0008 912 | ·0006 875+ | ·0005 292 |
| ·69 | ·0023 745− | ·0018 665− | ·0014 637 | ·0011 453 | ·0008 942 |
| ·70 | ·0037 326 | ·0029 751 | ·0023 658 | ·0018 772 | ·0014 862 |
| ·71 | ·0057 769 | ·0046 677 | ·0037 630 | ·0030 269 | ·0024 296 |
| ·72 | ·0088 018 | ·0072 078 | ·0058 892 | ·0048 014 | ·0039 063 |
| ·73 | ·0132 005− | ·0109 530 | ·0090 681 | ·0074 914 | ·0061 760 |
| ·74 | ·0194 838 | ·0163 766 | ·0137 348 | ·0114 948 | ·0096 004 |
| ·75 | ·0282 967 | ·0240 868 | ·0204 592 | ·0173 418 | ·0146 697 |
| ·76 | ·0404 262 | ·0348 410 | ·0299 640 | ·0257 170 | ·0220 282 |
| ·77 | ·0567 972 | ·0495 475+ | ·0431 338 | ·0374 751 | ·0324 954 |
| ·78 | ·0784 475− | ·0692 502 | ·0610 079 | ·0536 414 | ·0470 747 |
| ·79 | ·1064 752 | ·0950 852 | ·0847 469 | ·0753 886 | ·0669 394 |
| ·80 | ·1419 533 | ·1282 033 | ·1155 650− | ·1039 799 | ·0933 877 |
| ·81 | ·1858 067 | ·1696 533 | ·1546 203 | ·1406 679 | ·1277 519 |
| ·82 | ·2386 548 | ·2202 241 | ·2028 601 | ·1865 451 | ·1712 556 |
| ·83 | ·3006 310 | ·2802 555+ | ·2608 251 | ·2423 460 | ·2248 172 |
| ·84 | ·3712 004 | ·3494 368 | ·3284 325+ | ·3082 160 | ·2888 081 |
| ·85 | ·4490 099 | ·4266 265+ | ·4047 671 | ·3834 769 | ·3627 950− |
| ·86 | ·5318 143 | ·5097 405+ | ·4879 297 | ·4664 370 | ·4453 127 |
| ·87 | ·6165 253 | ·5957 591 | ·5750 017 | ·5543 093 | ·5337 351 |
| ·88 | ·6994 217 | ·6809 044 | ·6621 823 | ·6433 041 | ·6243 181 |
| ·89 | ·7765 360 | ·7610 119 | ·7451 377 | ·7289 493 | ·7124 834 |
| ·90 | ·8441 879 | ·8320 797 | ·8195 593 | ·8066 476 | ·7933 671 |
| ·91 | ·8995 828 | ·8909 178 | ·8818 583 | ·8724 119 | ·8625 876 |
| ·92 | ·9413 374 | ·9357 542 | ·9298 526 | ·9236 314 | ·9170 902 |
| ·93 | ·9697 655+ | ·9666 111 | ·9632 405+ | ·9596 488 | ·9558 312 |
| ·94 | ·9867 833 | ·9852 801 | ·9836 566 | ·9819 080 | ·9800 296 |
| ·95 | ·9953 887 | ·9948 199 | ·9941 990 | ·9935 232 | ·9927 895− |
| ·96 | ·9988 399 | ·9986 854 | ·9985 150+ | ·9983 276 | ·9981 219 |
| ·97 | ·9998 260 | ·9998 011 | ·9997 733 | ·9997 424 | ·9997 082 |
| ·98 | ·9999 898 | ·9999 883 | ·9999 865+ | ·9999 846 | ·9999 823 |
| ·99 | ·9999 999 | ·9999 999 | ·9999 999 | ·9999 999 | ·9999 999 |
| 1·00 | 1·0000 000 | 1·0000 000 | 1·0000 000 | 1·0000 000 | 1·0000 000 |

$x$ = ·05 to ·70 $q = 8{\cdot}5$ $p$ = 8·5 to 11

| $x$ | $p = 8{\cdot}5$ | $p = 9$ | $p = 9{\cdot}5$ | $p = 10$ | $p = 10{\cdot}5$ | $p = 11$ |
|---|---|---|---|---|---|---|
| $B(p, q) =$ | ·9413 8778 $\times \frac{1}{10^5}$ | ·6607 8995 $\bar{\times} \frac{1}{10^5}$ | ·4706 9389 $\times \frac{1}{10^5}$ | ·3398 3483 $\times \frac{1}{10^5}$ | ·2484 2178 $\times \frac{1}{10^5}$ | ·1836 9450 $\overset{+}{\times} \frac{1}{10^5}$ |
| ·05 | ·0000 001 | | | | | |
| ·06 | ·0000 003 | ·0000 001 | | | | |
| ·07 | ·0000 012 | ·0000 004 | ·0000 001 | ·0000 001 | | |
| ·08 | ·0000 034 | ·0000 013 | ·0000 005$^{-}$ | ·0000 002 | ·0000 001 | |
| ·09 | ·0000 086 | ·0000 035$^{-}$ | ·0000 014 | ·0000 005$^{+}$ | ·0000 002 | ·0000 001 |
| ·10 | ·0000 196 | ·0000 083 | ·0000 035$^{-}$ | ·0000 014 | ·0000 006 | ·0000 002 |
| ·11 | ·0000 410 | ·0000 182 | ·0000 080 | ·0000 035$^{-}$ | ·0000 015$^{-}$ | ·0000 006 |
| ·12 | ·0000 797 | ·0000 370 | ·0000 169 | ·0000 077 | ·0000 035$^{-}$ | ·0000 015$^{+}$ |
| ·13 | ·0001 461 | ·0000 704 | ·0000 336 | ·0000 159 | ·0000 074 | ·0000 034 |
| ·14 | ·0002 544 | ·0001 272 | ·0000 629 | ·0000 308 | ·0000 149 | ·0000 072 |
| ·15 | ·0004 238 | ·0002 192 | ·0001 122 | ·0000 568 | ·0000 285$^{+}$ | ·0000 142 |
| ·16 | ·0006 793 | ·0003 627 | ·0001 915$^{+}$ | ·0001 001 | ·0000 519 | ·0000 266 |
| ·17 | ·0010 524 | ·0005 788 | ·0003 149 | ·0001 696 | ·0000 905$^{+}$ | ·0000 479 |
| ·18 | ·0015 821 | ·0008 947 | ·0005 006 | ·0002 773 | ·0001 522 | ·0000 828 |
| ·19 | ·0023 151 | ·0013 441 | ·0007 721 | ·0004 392 | ·0002 476 | ·0001 383 |
| ·20 | ·0033 062 | ·0019 680 | ·0011 592 | ·0006 761 | ·0003 908 | ·0002 239 |
| ·21 | ·0046 187 | ·0028 151 | ·0016 979 | ·0010 142 | ·0006 003 | ·0003 523 |
| ·22 | ·0063 244 | ·0039 423 | ·0024 321 | ·0014 859 | ·0008 997 | ·0005 402 |
| ·23 | ·0085 025$^{+}$ | ·0054 149 | ·0034 131 | ·0021 308 | ·0013 184 | ·0008 089 |
| ·24 | ·0112 398 | ·0073 060 | ·0047 006 | ·0029 956 | ·0018 922 | ·0011 853 |
| ·25 | ·0146 290 | ·0096 967 | ·0063 625$^{-}$ | ·0041 354 | ·0026 643 | ·0017 024 |
| ·26 | ·0187 680 | ·0126 753 | ·0084 747 | ·0056 133 | ·0036 856 | ·0024 002 |
| ·27 | ·0237 581 | ·0163 358 | ·0111 210 | ·0075 007 | ·0050 152 | ·0033 261 |
| ·28 | ·0297 020 | ·0207 776 | ·0143 918 | ·0098 772 | ·0067 206 | ·0045 360 |
| ·29 | ·0367 024 | ·0261 029 | ·0183 839 | ·0128 298 | ·0088 775$^{-}$ | ·0060 937 |
| ·30 | ·0448 594 | ·0324 158 | ·0231 985$^{-}$ | ·0164 526 | ·0115 700 | ·0080 720 |
| ·31 | ·0542 682 | ·0398 198 | ·0289 400 | ·0208 455$^{-}$ | ·0148 896 | ·0105 519 |
| ·32 | ·0650 174 | ·0484 160 | ·0357 143 | ·0261 126 | ·0189 344 | ·0136 228 |
| ·33 | ·0771 858 | ·0583 002 | ·0436 262 | ·0323 611 | ·0238 084 | ·0173 813 |
| ·34 | ·0908 411 | ·0695 614 | ·0527 778 | ·0396 988 | ·0296 193 | ·0219 306 |
| ·35 | ·1060 372 | ·0822 787 | ·0632 657 | ·0482 324 | ·0364 772 | ·0273 791 |
| ·36 | ·1228 123 | ·0965 194 | ·0751 786 | ·0580 649 | ·0444 926 | ·0338 386 |
| ·37 | ·1411 875$^{+}$ | ·1123 364 | ·0885 953 | ·0692 932 | ·0537 736 | ·0414 228 |
| ·38 | ·1611 655$^{-}$ | ·1297 668 | ·1035 814 | ·0820 055$^{+}$ | ·0644 242 | ·0502 443 |
| ·39 | ·1827 289 | ·1488 295$^{+}$ | ·1201 879 | ·0962 787 | ·0765 408 | ·0604 128 |
| ·40 | ·2058 405$^{+}$ | ·1695 243 | ·1384 485$^{+}$ | ·1121 761 | ·0902 101 | ·0720 324 |
| ·41 | ·2304 421 | ·1918 302 | ·1583 783 | ·1297 446 | ·1055 062 | ·0851 982 |
| ·42 | ·2564 552 | ·2157 051 | ·1799 715$^{-}$ | ·1490 128 | ·1224 878 | ·0999 940 |
| ·43 | ·2837 815$^{-}$ | ·2410 854 | ·2032 008 | ·1699 892 | ·1411 962 | ·1164 893 |
| ·44 | ·3123 034 | ·2678 858 | ·2280 168 | ·1926 601 | ·1616 522 | ·1347 363 |
| ·45 | ·3418 860 | ·2960 005$^{+}$ | ·2543 469 | ·2169 892 | ·1838 550$^{-}$ | ·1547 677 |
| ·46 | ·3723 785$^{-}$ | ·3253 035$^{+}$ | ·2820 965$^{-}$ | ·2429 160 | ·2077 802 | ·1765 940 |
| ·47 | ·4036 162 | ·3556 507 | ·3111 487 | ·2703 564 | ·2333 787 | ·2002 019 |
| ·48 | ·4354 232 | ·3868 811 | ·3413 662 | ·2992 025$^{+}$ | ·2605 761 | ·2255 525$^{+}$ |
| ·49 | ·4676 147 | ·4188 199 | ·3725 924 | ·3293 236 | ·2892 728 | ·2525 805$^{-}$ |
| ·50 | ·5000 000$^{e}$ | ·4512 804 | ·4046 540 | ·3605 675$^{-}$ | ·3193 444 | ·2811 937 |
| ·51 | ·5323 853 | ·4840 669 | ·4373 629 | ·3927 624 | ·3506 425$^{+}$ | ·3112 731 |
| ·52 | ·5645 768 | ·5169 782 | ·4705 197 | ·4257 194 | ·3829 971 | ·3426 739 |
| ·53 | ·5963 838 | ·5498 105$^{-}$ | ·5039 163 | ·4592 354 | ·4162 181 | ·3752 270 |
| ·54 | ·6276 215$^{+}$ | ·5823 606 | ·5373 396 | ·4930 960 | ·4500 987 | ·4087 409 |
| ·55 | ·6581 140 | ·6144 297 | ·5705 750$^{-}$ | ·5270 793 | ·4844 182 | ·4430 046 |
| ·56 | ·6876 966 | ·6458 258 | ·6034 100 | ·5609 597 | ·5189 460 | ·4777 910 |
| ·57 | ·7162 185$^{+}$ | ·6763 675$^{+}$ | ·6356 379 | ·5945 117 | ·5534 457 | ·5128 605$^{-}$ |
| ·58 | ·7435 448 | ·7058 866 | ·6670 610 | ·6275 138 | ·5876 788 | ·5479 655$^{+}$ |
| ·59 | ·7695 579 | ·7342 304 | ·6974 940 | ·6597 529 | ·6214 098 | ·5828 549 |
| ·60 | ·7941 595$^{-}$ | ·7612 642 | ·7267 675$^{+}$ | ·6910 273 | ·6544 099 | ·6172 786 |
| ·61 | ·8172 711 | ·7868 731 | ·7547 300 | ·7211 510 | ·6864 615$^{-}$ | ·6509 926 |
| ·62 | ·8388 345$^{+}$ | ·8109 633 | ·7812 505$^{-}$ | ·7499 562 | ·7173 624 | ·6837 636 |
| ·63 | ·8588 125$^{-}$ | ·8334 627 | ·8062 202 | ·7772 967 | ·7469 294 | ·7153 736 |
| ·64 | ·8771 877 | ·8543 220 | ·8295 541 | ·8030 494 | ·7750 011 | ·7456 241 |
| ·65 | ·8939 628 | ·8735 141 | ·8511 913 | ·8271 168 | ·8014 413 | ·7743 398 |
| ·66 | ·9091 589 | ·8910 338 | ·8710 955$^{+}$ | ·8494 273 | ·8261 407 | ·8013 719 |
| ·67 | ·9228 142 | ·9068 968 | ·8892 546 | ·8699 365$^{-}$ | ·8490 184 | ·8266 005$^{-}$ |
| ·68 | ·9349 826 | ·9211 384 | ·9056 796 | ·8886 260 | ·8700 224 | ·8499 363 |
| ·69 | ·9457 318 | ·9338 117 | ·9204 036 | ·9055 036 | ·8891 299 | ·8713 216 |
| ·70 | ·9551 406 | ·9449 854 | ·9334 797 | ·9206 013 | ·9063 466 | ·8907 307 |

$x = \cdot 71$ to $\cdot 97$ $q = 8\cdot 5$ $p = 8\cdot 5$ to $11$

| | $p = 8\cdot 5$ | $p = 9$ | $p = 9\cdot 5$ | $p = 10$ | $p = 10\cdot 5$ | $p = 11$ |
|---|---|---|---|---|---|---|
| $B(p, q) =$ | ·9413 8778 × $\frac{1}{10^5}$ | ·6607 8995$^{-}$ × $\frac{1}{10^5}$ | ·4706 9389 × $\frac{1}{10^5}$ | ·3398 3483 × $\frac{1}{10^5}$ | ·2484 2178 × $\frac{1}{10^5}$ | ·1836 9450$^{+}$ × $\frac{1}{10^5}$ |
| $x$ | | | | | | |
| ·71 | ·9632 976 | ·9547 418 | ·9449 790 | ·9339 734 | ·9217 048 | ·9081 687 |
| ·72 | ·9702 980 | ·9631 740 | ·9549 877 | ·9456 944 | ·9352 617 | ·9236 702 |
| ·73 | ·9762 419 | ·9703 835⁻ | ·9636 048 | ·9558 559 | ·9470 967 | ·9372 970 |
| ·74 | ·9812 320 | ·9764 775⁺ | ·9709 386 | ·9645 637 | ·9573 081 | ·9491 352 |
| ·75 | ·9853 710 | ·9815 666 | ·9771 044 | ·9719 341 | ·9660 099 | ·9592 916 |
| ·76 | ·9887 602 | ·9857 616 | ·9822 210 | ·9780 913 | ·9733 277 | ·9678 898 |
| ·77 | ·9914 975⁻ | ·9891 719 | ·9864 080 | ·9831 630 | ·9793 953 | ·9750 659 |
| ·78 | ·9936 756 | ·9919 033 | ·9897 833 | ·9872 780 | ·9843 504 | ·9809 645⁺ |
| ·79 | ·9953 813 | ·9940 559 | ·9924 604 | ·9905 629 | ·9883 313 | ·9857 338 |
| ·80 | ·9966 938 | ·9957 230 | ·9945 468 | ·9931 392 | ·9914 732 | ·9895 219 |
| ·81 | ·9976 849 | ·9969 896 | ·9961 420 | ·9951 213 | ·9939 056 | ·9924 728 |
| ·82 | ·9984 179 | ·9979 321 | ·9973 363 | ·9966 144 | ·9957 493 | ·9947 234 |
| ·83 | ·9989 476 | ·9986 174 | ·9982 100 | ·9977 133 | ·9971 145⁺ | ·9964 001 |
| ·84 | ·9993 207 | ·9991 031 | ·9988 330 | ·9985 016 | ·9980 998 | ·9976 174 |
| ·85 | ·9995 762 | ·9994 376 | ·9992 646 | ·9990 511 | ·9987 906 | ·9984 760 |
| ·86 | ·9997 456 | ·9996 607 | ·9995 541 | ·9994 218 | ·9992 594 | ·9990 622 |
| ·87 | ·9998 539 | ·9998 042 | ·9997 414 | ·9996 630 | ·9995 662 | ·9994 480 |
| ·88 | ·9999 203 | ·9998 926 | ·9998 575⁻ | ·9998 133 | ·9997 586 | ·9996 913 |
| ·89 | ·9999 590 | ·9999 445⁺ | ·9999 260 | ·9999 026 | ·9998 735⁻ | ·9998 374 |
| ·90 | ·9999 804 | ·9999 733 | ·9999 642 | ·9999 527 | ·9999 382 | ·9999 203 |
| ·91 | ·9999 914 | ·9999 882 | ·9999 841 | ·9999 789 | ·9999 724 | ·9999 641 |
| ·92 | ·9999 966 | ·9999 953 | ·9999 937 | ·9999 916 | ·9999 889 | ·9999 855⁻ |
| ·93 | ·9999 988 | ·9999 984 | ·9999 978 | ·9999 970 | ·9999 961 | ·9999 949 |
| ·94 | ·9999 997 | ·9999 995⁺ | ·9999 994 | ·9999 991 | ·9999 988 | ·9999 985⁻ |
| ·95 | ·9999 999 | ·9999 999 | ·9999 999 | ·9999 998 | ·9999 997 | ·9999 996 |
| ·96 | 1·0000 000 | 1·0000 000 | 1·0000 000 | 1·0000 000 | 1·0000 000 | ·9999 999 |
| ·97 | | | | | | 1·0000 000 |

$x$ = ·11 to ·70 $q = 8·5$ $p$ = 12 to 17

| $x$ | $p$ = 12 | $p$ = 13 | $p$ = 14 | $p$ = 15 | $p$ = 16 | $p$ = 17 |
|---|---|---|---|---|---|---|
| $B(p, q)$ = | ·1036 2254 × $\frac{1}{10^6}$ | ·6065 7097 × $\frac{1}{10^6}$ | ·3667 6384 × $\frac{1}{10^6}$ | ·2282 0861 × $\frac{1}{10^6}$ | ·1456 6507 × $\frac{1}{10^6}$ | ·9512 8210 × $\frac{1}{10^7}$ |
| ·11 | ·0000 001 | | | | | |
| ·12 | ·0000 003 | ·0000 001 | | | | |
| ·13 | ·0000 007 | ·0000 001 | | | | |
| ·14 | ·0000 016 | ·0000 004 | ·0000 001 | | | |
| ·15 | ·0000 034 | ·0000 008 | ·0000 002 | | | |
| ·16 | ·0000 069 | ·0000 017 | ·0000 004 | ·0000 001 | | |
| ·17 | ·0000 131 | ·0000 035− | ·0000 009 | ·0000 002 | ·0000 001 | |
| ·18 | ·0000 240 | ·0000 067 | ·0000 018 | ·0000 005− | ·0000 001 | |
| ·19 | ·0000 422 | ·0000 125+ | ·0000 036 | ·0000 010 | ·0000 003 | ·0000 001 |
| ·20 | ·0000 718 | ·0000 224 | ·0000 068 | ·0000 020 | ·0000 006 | ·0000 002 |
| ·21 | ·0001 186 | ·0000 388 | ·0000 124 | ·0000 039 | ·0000 012 | ·0000 004 |
| ·22 | ·0001 903 | ·0000 652 | ·0000 218 | ·0000 071 | ·0000 023 | ·0000 007 |
| ·23 | ·0002 976 | ·0001 065− | ·0000 372 | ·0000 127 | ·0000 043 | ·0000 014 |
| ·24 | ·0004 546 | ·0001 696 | ·0000 617 | ·0000 220 | ·0000 077 | ·0000 026 |
| ·25 | ·0006 794 | ·0002 638 | ·0001 000 | ·0000 371 | ·0000 135− | ·0000 048 |
| ·26 | ·0009 951 | ·0004 015− | ·0001 581 | ·0000 609 | ·0000 230 | ·0000 085+ |
| ·27 | ·0014 305− | ·0005 988 | ·0002 447 | ·0000 978 | ·0000 384 | ·0000 148 |
| ·28 | ·0020 208 | ·0008 763 | ·0003 710 | ·0001 537 | ·0000 625− | ·0000 249 |
| ·29 | ·0028 083 | ·0012 601 | ·0005 521 | ·0002 368 | ·0000 996 | ·0000 412 |
| ·30 | ·0038 435+ | ·0017 822 | ·0008 070 | ·0003 577 | ·0001 556 | ·0000 665− |
| ·31 | ·0051 852 | ·0024 817 | ·0011 602 | ·0005 310 | ·0002 384 | ·0001 052 |
| ·32 | ·0069 009 | ·0034 056 | ·0016 418 | ·0007 750+ | ·0003 590 | ·0001 634 |
| ·33 | ·0090 674 | ·0046 092 | ·0022 891 | ·0011 134 | ·0005 314 | ·0002 493 |
| ·34 | ·0117 704 | ·0061 568 | ·0031 471 | ·0015 756 | ·0007 742 | ·0003 739 |
| ·35 | ·0151 040 | ·0081 226 | ·0042 694 | ·0021 983 | ·0011 109 | ·0005 520 |
| ·36 | ·0191 707 | ·0105 900 | ·0057 188 | ·0030 257 | ·0015 714 | ·0008 025− |
| ·37 | ·0240 797 | ·0136 523 | ·0075 682 | ·0041 112 | ·0021 925+ | ·0011 499 |
| ·38 | ·0299 460 | ·0174 117 | ·0099 009 | ·0055 177 | ·0030 193 | ·0016 249 |
| ·39 | ·0368 881 | ·0219 792 | ·0128 103 | ·0073 188 | ·0041 062 | ·0022 661 |
| ·40 | ·0450 266 | ·0274 729 | ·0164 005− | ·0095 989 | ·0055 179 | ·0031 204 |
| ·41 | ·0544 814 | ·0340 163 | ·0207 849 | ·0124 538 | ·0073 302 | ·0042 448 |
| ·42 | ·0653 689 | ·0417 372 | ·0260 858 | ·0159 906 | ·0096 307 | ·0057 074 |
| ·43 | ·0777 991 | ·0507 644 | ·0324 327 | ·0203 273 | ·0125 193 | ·0075 881 |
| ·44 | ·0918 727 | ·0612 252 | ·0399 606 | ·0255 917 | ·0161 083 | ·0099 797 |
| ·45 | ·1076 774 | ·0732 426 | ·0488 071 | ·0319 205− | ·0205 220 | ·0129 883 |
| ·46 | ·1252 849 | ·0869 314 | ·0591 103 | ·0394 567 | ·0258 958 | ·0167 336 |
| ·47 | ·1447 480 | ·1023 950+ | ·0710 050− | ·0483 481 | ·0323 750− | ·0213 485+ |
| ·48 | ·1660 968 | ·1197 214 | ·0846 188 | ·0587 432 | ·0401 127 | ·0269 782 |
| ·49 | ·1893 372 | ·1389 797 | ·1000 689 | ·0707 885− | ·0492 674 | ·0337 789 |
| ·50 | ·2144 475+ | ·1602 163 | ·1174 570 | ·0846 240 | ·0599 993 | ·0419 155− |
| ·51 | ·2413 775− | ·1834 515+ | ·1368 657 | ·1003 790 | ·0724 666 | ·0515 586 |
| ·52 | ·2700 469 | ·2086 770 | ·1583 537 | ·1181 670 | ·0868 213 | ·0628 811 |
| ·53 | ·3003 450+ | ·2358 529 | ·1819 523 | ·1380 812 | ·1032 036 | ·0760 535+ |
| ·54 | ·3321 312 | ·2649 061 | ·2076 614 | ·1601 891 | ·1217 366 | ·0912 390 |
| ·55 | ·3652 355− | ·2957 294 | ·2354 462 | ·1845 284 | ·1425 212 | ·1085 873 |
| ·56 | ·3994 611 | ·3281 810 | ·2652 351 | ·2111 017 | ·1656 296 | ·1282 288 |
| ·57 | ·4345 867 | ·3620 856 | ·2969 182 | ·2398 735+ | ·1911 003 | ·1502 680 |
| ·58 | ·4703 697 | ·3972 357 | ·3303 462 | ·2707 668 | ·2189 327 | ·1747 765+ |
| ·59 | ·5065 511 | ·4333 949 | ·3653 314 | ·3036 611 | ·2490 828 | ·2017 873 |
| ·60 | ·5428 594 | ·4703 007 | ·4016 491 | ·3383 914 | ·2814 596 | ·2312 884 |
| ·61 | ·5790 165− | ·5076 701 | ·4390 404 | ·3747 491 | ·3159 226 | ·2632 177 |
| ·62 | ·6147 428 | ·5452 043 | ·4772 170 | ·4124 834 | ·3522 812 | ·2974 595+ |
| ·63 | ·6497 634 | ·5825 948 | ·5158 655+ | ·4513 049 | ·3902 951 | ·3338 419 |
| ·64 | ·6838 137 | ·6195 308 | ·5546 545− | ·4908 904 | ·4296 768 | ·3721 361 |
| ·65 | ·7166 451 | ·6557 050− | ·5932 414 | ·5308 893 | ·4700 960 | ·4120 575− |
| ·66 | ·7480 304 | ·6908 217 | ·6312 806 | ·5709 314 | ·5111 857 | ·4532 697 |
| ·67 | ·7777 687 | ·7246 031 | ·6684 316 | ·6106 352 | ·5525 498 | ·4953 901 |
| ·68 | ·8056 892 | ·7567 963 | ·7043 680 | ·6496 179 | ·5937 728 | ·5379 974 |
| ·69 | ·8316 550− | ·7871 787 | ·7387 852 | ·6875 053 | ·6344 306 | ·5806 427 |
| ·70 | ·8555 650− | ·8155 639 | ·7714 089 | ·7239 422 | ·6741 022 | ·6228 605− |

$x$ = ·71 to ·98 $q = 8·5$ $p$ = 12 to 17

| $x$ | $p = 12$ | $p = 13$ | $p = 14$ | $p = 15$ | $p = 16$ | $p = 17$ |
|---|---|---|---|---|---|---|
| $B(p, q) =$ | ·1036 2254 × $\frac{1}{10^6}$ | ·6065 7097 × $\frac{1}{10^8}$ | ·3667 6384 × $\frac{1}{10^8}$ | ·2282 0861 × $\frac{1}{10^8}$ | ·1456 6507 × $\frac{1}{10^8}$ | ·9512 8210 × $\frac{1}{10^7}$ |
| ·71 | ·8773 556 | ·8418 049 | ·8020 019 | ·7586 023 | ·7123 818 | ·6641 824 |
| ·72 | ·8970 009 | ·8657 978 | ·8303 703 | ·7911 976 | ·7488 910 | ·7041 519 |
| ·73 | ·9145 117 | ·8874 827 | ·8563 681 | ·8214 864 | ·7832 909 | ·7423 383 |
| ·74 | ·9299 338 | ·9068 441 | ·8799 001 | ·8492 803 | ·8152 923 | ·7783 515+ |
| ·75 | ·9433 447 | ·9239 095− | ·9009 236 | ·8744 488 | ·8446 646 | ·8118 554 |
| ·76 | ·9548 504 | ·9387 468 | ·9194 473 | ·8969 219 | ·8712 430 | ·8425 790 |
| ·77 | ·9645 803 | ·9514 601 | ·9355 292 | ·9166 909 | ·8949 326 | ·8703 254 |
| ·78 | ·9726 823 | ·9621 847 | ·9492 725+ | ·9338 057 | ·9157 094 | ·8949 779 |
| ·79 | ·9793 177 | ·9710 809 | ·9608 199 | ·9483 711 | ·9336 192 | ·9165 024 |
| ·80 | ·9846 549 | ·9783 279 | ·9703 461 | ·9605 399 | ·9487 725+ | ·9349 458 |
| ·81 | ·9888 650− | ·9841 161 | ·9780 503 | ·9705 049 | ·9613 373 | ·9504 307 |
| ·82 | ·9921 155+ | ·9886 406 | ·9841 473 | ·9784 888 | ·9715 290 | ·9631 467 |
| ·83 | ·9945 670 | ·9920 946 | ·9888 587 | ·9847 339 | ·9795 987 | ·9733 384 |
| ·84 | ·9963 683 | ·9946 633 | ·9924 047 | ·9894 912 | ·9858 202 | ·9812 911 |
| ·85 | ·9976 540 | ·9965 185+ | ·9949 966 | ·9930 099 | ·9904 769 | ·9873 146 |
| ·86 | ·9985 421 | ·9978 153 | ·9968 296 | ·9955 278 | ·9938 484 | ·9917 272 |
| ·87 | ·9991 334 | ·9986 886 | ·9980 785− | ·9972 632 | ·9961 994 | ·9948 400 |
| ·88 | ·9995 106 | ·9992 523 | ·9988 938 | ·9984 093 | ·9977 698 | ·9969 432 |
| ·89 | ·9997 398 | ·9995 986 | ·9994 004 | ·9991 295− | ·9987 678 | ·9982 951 |
| ·90 | ·9998 712 | ·9997 993 | ·9996 974 | ·9995 565+ | ·9993 663 | ·9991 149 |
| ·91 | ·9999 415− | ·9999 080 | ·9998 599 | ·9997 928 | ·9997 011 | ·9995 786 |
| ·92 | ·9999 761 | ·9999 621 | ·9999 417 | ·9999 130 | ·9998 733 | ·9998 196 |
| ·93 | ·9999 915− | ·9999 863 | ·9999 788 | ·9999 681 | ·9999 531 | ·9999 326 |
| ·94 | ·9999 974 | ·9999 959 | ·9999 935+ | ·9999 902 | ·9999 854 | ·9999 789 |
| ·95 | ·9999 994 | ·9999 990 | ·9999 985− | ·9999 976 | ·9999 964 | ·9999 948 |
| ·96 | ·9999 999 | ·9999 998 | ·9999 997 | ·9999 996 | ·9999 994 | ·9999 991 |
| ·97 | 1·0000 000 | 1·0000 000 | 1·0000 000 | 1·0000 000 | ·9999 999 | ·9999 999 |
| ·98 | | | | | 1·0000 000 | 1·0000 000 |

$x$ = ·21 to ·80 $q = 8{\cdot}5$ $p$ = 18 to 23

| | $p = 18$ | $p = 19$ | $p = 20$ | $p = 21$ | $p = 22$ | $p = 23$ |
|---|---|---|---|---|---|---|
| $B(p, q)$ = | ·6341 8806 × $\frac{1}{10^7}$ | ·4307 6925+ × $\frac{1}{10^7}$ | ·2976 2239 × $\frac{1}{10^7}$ | ·2088 5782 × $\frac{1}{10^7}$ | ·1486 7845− × $\frac{1}{10^7}$ | ·1072 4347 × $\frac{1}{10^7}$ |
| $x$ | | | | | | |
| ·21 | ·0000 001 | | | | | |
| ·22 | ·0000 002 | ·0000 001 | | | | |
| ·23 | ·0000 005− | ·0000 001 | | | | |
| ·24 | ·0000 009 | ·0000 003 | ·0000 001 | | | |
| ·25 | ·0000 017 | ·0000 006 | ·0000 002 | ·0000 001 | | |
| ·26 | ·0000 031 | ·0000 011 | ·0000 004 | ·0000 001 | | |
| ·27 | ·0000 056 | ·0000 021 | ·0000 008 | ·0000 003 | ·0000 001 | |
| ·28 | ·0000 098 | ·0000 038 | ·0000 015− | ·0000 005+ | ·0000 002 | ·0000 001 |
| ·29 | ·0000 167 | ·0000 067 | ·0000 027 | ·0000 010 | ·0000 004 | ·0000 002 |
| ·30 | ·0000 280 | ·0000 116 | ·0000 047 | ·0000 019 | ·0000 008 | ·0000 003 |
| ·31 | ·0000 457 | ·0000 196 | ·0000 083 | ·0000 034 | ·0000 014 | ·0000 006 |
| ·32 | ·0000 732 | ·0000 323 | ·0000 141 | ·0000 061 | ·0000 026 | ·0000 011 |
| ·33 | ·0001 151 | ·0000 524 | ·0000 236 | ·0000 105− | ·0000 046 | ·0000 020 |
| ·34 | ·0001 778 | ·0000 834 | ·0000 386 | ·0000 176 | ·0000 080 | ·0000 036 |
| ·35 | ·0002 700 | ·0001 303 | ·0000 620 | ·0000 292 | ·0000 136 | ·0000 062 |
| ·36 | ·0004 035+ | ·0002 001 | ·0000 979 | ·0000 474 | ·0000 226 | ·0000 107 |
| ·37 | ·0005 939 | ·0003 024 | ·0001 520 | ·0000 755+ | ·0000 371 | ·0000 180 |
| ·38 | ·0008 613 | ·0004 502 | ·0002 323 | ·0001 184 | ·0000 597 | ·0000 298 |
| ·39 | ·0012 318 | ·0006 603 | ·0003 495− | ·0001 828 | ·0000 946 | ·0000 484 |
| ·40 | ·0017 382 | ·0009 550+ | ·0005 181 | ·0002 778 | ·0001 473 | ·0000 773 |
| ·41 | ·0024 217 | ·0013 628 | ·0007 573 | ·0004 159 | ·0002 260 | ·0001 215+ |
| ·42 | ·0033 326 | ·0019 197 | ·0010 920 | ·0006 140 | ·0003 415+ | ·0001 881 |
| ·43 | ·0045 322 | ·0026 706 | ·0015 541 | ·0008 940 | ·0005 088 | ·0002 867 |
| ·44 | ·0060 934 | ·0036 709 | ·0021 843 | ·0012 849 | ·0007 478 | ·0004 309 |
| ·45 | ·0081 025− | ·0049 878 | ·0030 329 | ·0018 233 | ·0010 846 | ·0006 388 |
| ·46 | ·0106 597 | ·0067 015− | ·0041 620 | ·0025 557 | ·0015 530 | ·0009 345− |
| ·47 | ·0138 797 | ·0089 068 | ·0056 468 | ·0035 401 | ·0021 963 | ·0013 494 |
| ·48 | ·0178 923 | ·0117 138 | ·0075 775− | ·0048 475+ | ·0030 691 | ·0019 245+ |
| ·49 | ·0228 413 | ·0152 488 | ·0100 599 | ·0065 639 | ·0042 391 | ·0027 115+ |
| ·50 | ·0288 845− | ·0196 542 | ·0132 173 | ·0087 919 | ·0057 890 | ·0037 757 |
| ·51 | ·0361 911 | ·0250 882 | ·0171 904 | ·0116 522 | ·0078 189 | ·0051 974 |
| ·52 | ·0449 400 | ·0317 234 | ·0221 378 | ·0152 842 | ·0104 475− | ·0070 750− |
| ·53 | ·0553 157 | ·0397 450+ | ·0282 350+ | ·0198 471 | ·0138 138 | ·0095 261 |
| ·54 | ·0675 046 | ·0493 479 | ·0356 729 | ·0255 193 | ·0180 782 | ·0126 901 |
| ·55 | ·0816 897 | ·0607 319 | ·0446 552 | ·0324 971 | ·0234 220 | ·0167 290 |
| ·56 | ·0980 442 | ·0740 977 | ·0553 942 | ·0409 926 | ·0300 473 | ·0218 284 |
| ·57 | ·1167 254 | ·0896 398 | ·0681 067 | ·0512 302 | ·0381 750− | ·0281 966 |
| ·58 | ·1378 672 | ·1075 400 | ·0830 069 | ·0634 417 | ·0480 411 | ·0360 637 |
| ·59 | ·1615 723 | ·1279 592 | ·1002 992 | ·0778 600 | ·0598 926 | ·0456 779 |
| ·60 | ·1879 050+ | ·1510 292 | ·1201 699 | ·0947 111 | ·0739 803 | ·0573 014 |
| ·61 | ·2168 838 | ·1768 437 | ·1427 780 | ·1142 053 | ·0905 513 | ·0712 033 |
| ·62 | ·2484 747 | ·2054 498 | ·1682 445+ | ·1365 270 | ·1098 390 | ·0876 516 |
| ·63 | ·2825 858 | ·2368 397 | ·1966 434 | ·1618 233 | ·1320 521 | ·1069 022 |
| ·64 | ·3190 632 | ·2709 438 | ·2279 909 | ·1901 923 | ·1573 616 | ·1291 869 |
| ·65 | ·3576 890 | ·3076 247 | ·2622 374 | ·2216 726 | ·1858 886 | ·1546 996 |
| ·66 | ·3981 814 | ·3466 738 | ·2992 598 | ·2562 315− | ·2176 905− | ·1835 816 |
| ·67 | ·4401 973 | ·3878 102 | ·3388 559 | ·2937 566 | ·2527 486 | ·2159 066 |
| ·68 | ·4833 376 | ·4306 820 | ·3807 423 | ·3340 487 | ·2909 572 | ·2516 655+ |
| ·69 | ·5271 554 | ·4748 716 | ·4245 553 | ·3768 177 | ·3321 148 | ·2907 545+ |
| ·70 | ·5711 666 | ·5199 035− | ·4698 545+ | ·4216 824 | ·3759 189 | ·3329 636 |
| ·71 | ·6148 632 | ·5652 562 | ·5161 322 | ·4681 750− | ·4219 647 | ·3779 705− |
| ·72 | ·6577 284 | ·6103 765− | ·5628 252 | ·5157 494 | ·4697 496 | ·4253 389 |
| ·73 | ·6992 536 | ·6546 969 | ·6093 312 | ·5637 953 | ·5186 823 | ·4745 228 |
| ·74 | ·7389 553 | ·6976 550− | ·6550 286 | ·6116 564 | ·5680 982 | ·5248 766 |
| ·75 | ·7763 926 | ·7387 134 | ·6992 989 | ·6586 527 | ·6172 807 | ·5756 736 |
| ·76 | ·8111 834 | ·7773 808 | ·7415 501 | ·7041 070 | ·6654 871 | ·6261 300 |
| ·77 | ·8430 186 | ·8132 314 | ·7812 416 | ·7473 723 | ·7119 790 | ·6754 353 |
| ·78 | ·8716 732 | ·8459 216 | ·8179 065+ | ·7878 603 | ·7560 543 | ·7227 881 |
| ·79 | ·8970 145− | ·8752 042 | ·8511 727 | ·8250 685+ | ·7970 811 | ·7674 335+ |
| ·80 | ·9190 044 | ·9009 375− | ·8807 786 | ·8586 038 | ·8345 284 | ·8087 019 |

$x = \cdot 81$ to $\cdot 98$ $q = 8\cdot 5$ $p = 18$ to $23$

| $x$ | $p = 18$ | $p = 19$ | $p = 20$ | $p = 21$ | $p = 22$ | $p = 23$ |
|---|---|---|---|---|---|---|
| $B(p, q) =$ | ·6341 8806 × $\frac{1}{10^7}$ | ·4307 6925$\overset{+}{\times}\frac{1}{10^7}$ | ·2976 2239 × $\frac{1}{10^7}$ | ·2088 5782 × $\frac{1}{10^7}$ | ·1486 7845$\overset{-}{\times}\frac{1}{10^7}$ | ·1072 4347 × $\frac{1}{10^7}$ |
| ·81 | ·9376 988 | ·9230 889 | ·9065 837 | ·8882 009 | ·8679 930 | ·8460 445+ |
| ·82 | ·9532 408 | ·9417 334 | ·9285 727 | ·9137 339 | ·8972 205− | ·8790 632 |
| ·83 | ·9658 501 | ·9570 451 | ·9468 521 | ·9352 193 | ·9221 159 | ·9075 323 |
| ·84 | ·9758 082 | ·9692 836 | ·9616 395− | ·9528 105− | ·9427 455− | ·9314 086 |
| ·85 | ·9834 408 | ·9787 761 | ·9732 459 | ·9667 825+ | ·9593 266 | ·9508 285+ |
| ·86 | ·9890 981 | ·9858 950+ | ·9820 530 | ·9775 098 | ·9722 072 | ·9660 924 |
| ·87 | ·9931 355+ | ·9910 348 | ·9884 857 | ·9854 363 | ·9818 359 | ·9776 357 |
| ·88 | ·9958 949 | ·9945 880 | ·9929 839 | ·9910 431 | ·9887 251 | ·9859 899 |
| ·89 | ·9976 888 | ·9969 243 | ·9959 754 | ·9948 141 | ·9934 114 | ·9917 374 |
| ·90 | ·9987 888 | ·9983 730 | ·9978 511 | ·9972 052 | ·9964 163 | ·9954 642 |
| ·91 | ·9994 179 | ·9992 107 | ·9989 478 | ·9986 188 | ·9982 125− | ·9977 167 |
| ·92 | ·9997 485+ | ·9996 559 | ·9995 370 | ·9993 865+ | ·9991 987 | ·9989 670 |
| ·93 | ·9999 051 | ·9998 689 | ·9998 220 | ·9997 620 | ·9996 862 | ·9995 917 |
| ·94 | ·9999 700 | ·9999 581 | ·9999 426 | ·9999 226 | ·9998 970 | ·9998 648 |
| ·95 | ·9999 925+ | ·9999 895− | ·9999 855− | ·9999 802 | ·9999 734 | ·9999 648 |
| ·96 | ·9999 987 | ·9999 981 | ·9999 974 | ·9999 964 | ·9999 952 | ·9999 935+ |
| ·97 | ·9999 999 | ·9999 998 | ·9999 997 | ·9999 996 | ·9999 995− | ·9999 993 |
| ·98 | 1·0000 000 | 1·0000 000 | 1·0000 000 | 1·0000 000 | 1·0000 000 | 1·0000 000 |

$x$ = ·29 to ·90 $q = 8{\cdot}5$ $p$ = 24 to 29

| | $p = 24$ | $p = 25$ | $p = 26$ | $p = 27$ | $p = 28$ | $p = 29$ |
|---|---|---|---|---|---|---|
| $B(p, q)$ = | ·7830 4756 × $\frac{1}{10^8}$ | ·5782 5051 × $\frac{1}{10^8}$ | ·4315 3023 × $\frac{1}{10^8}$ | ·3252 1119 × $\frac{1}{10^8}$ | ·2473 4372 × $\frac{1}{10^8}$ | ·1897 4313 × $\frac{1}{10^8}$ |
| $x$ | | | | | | |
| ·29 | ·0000 001 | | | | | |
| ·30 | ·0000 001 | | | | | |
| ·31 | ·0000 002 | ·0000 001 | | | | |
| ·32 | ·0000 005− | ·0000 002 | ·0000 001 | | | |
| ·33 | ·0000 009 | ·0000 004 | ·0000 002 | ·0000 001 | | |
| ·34 | ·0000 016 | ·0000 007 | ·0000 003 | ·0000 001 | ·0000 001 | |
| ·35 | ·0000 028 | ·0000 013 | ·0000 006 | ·0000 003 | ·0000 001 | |
| ·36 | ·0000 050+ | ·0000 023 | ·0000 011 | ·0000 005− | ·0000 002 | ·0000 001 |
| ·37 | ·0000 087 | ·0000 041 | ·0000 020 | ·0000 009 | ·0000 004 | ·0000 002 |
| ·38 | ·0000 147 | ·0000 072 | ·0000 035+ | ·0000 017 | ·0000 008 | ·0000 004 |
| ·39 | ·0000 246 | ·0000 124 | ·0000 062 | ·0000 030 | ·0000 015− | ·0000 007 |
| ·40 | ·0000 402 | ·0000 207 | ·0000 106 | ·0000 054 | ·0000 027 | ·0000 014 |
| ·41 | ·0000 648 | ·0000 342 | ·0000 179 | ·0000 093 | ·0000 048 | ·0000 025− |
| ·42 | ·0001 026 | ·0000 555− | ·0000 298 | ·0000 158 | ·0000 084 | ·0000 044 |
| ·43 | ·0001 601 | ·0000 886 | ·0000 486 | ·0000 265− | ·0000 143 | ·0000 077 |
| ·44 | ·0002 460 | ·0001 393 | ·0000 782 | ·0000 436 | ·0000 241 | ·0000 132 |
| ·45 | ·0003 728 | ·0002 157 | ·0001 238 | ·0000 705+ | ·0000 399 | ·0000 224 |
| ·46 | ·0005 572 | ·0003 294 | ·0001 932 | ·0001 124 | ·0000 650− | ·0000 373 |
| ·47 | ·0008 216 | ·0004 960 | ·0002 971 | ·0001 766 | ·0001 042 | ·0000 611 |
| ·48 | ·0011 959 | ·0007 369 | ·0004 505− | ·0002 734 | ·0001 647 | ·0000 986 |
| ·49 | ·0017 190 | ·0010 806 | ·0006 740 | ·0004 173 | ·0002 566 | ·0001 567 |
| ·50 | ·0024 408 | ·0015 647 | ·0009 953 | ·0006 284 | ·0003 941 | ·0002 455− |
| ·51 | ·0034 245+ | ·0022 378 | ·0014 510 | ·0009 340 | ·0005 971 | ·0003 792 |
| ·52 | ·0047 494 | ·0031 622 | ·0020 893 | ·0013 704 | ·0008 927 | ·0005 778 |
| ·53 | ·0065 125+ | ·0044 162 | ·0029 719 | ·0019 856 | ·0013 176 | ·0008 688 |
| ·54 | ·0088 318 | ·0060 972 | ·0041 775− | ·0028 418 | ·0019 202 | ·0012 893 |
| ·55 | ·0118 476 | ·0083 238 | ·0058 043 | ·0040 188 | ·0027 640 | ·0018 890 |
| ·56 | ·0157 250− | ·0112 390 | ·0079 731 | ·0056 167 | ·0039 306 | ·0027 334 |
| ·57 | ·0206 543 | ·0150 117 | ·0108 306 | ·0077 598 | ·0055 233 | ·0039 070 |
| ·58 | ·0268 515− | ·0198 387 | ·0145 510 | ·0105 995+ | ·0076 710 | ·0055 175+ |
| ·59 | ·0345 565− | ·0259 443 | ·0193 388 | ·0143 173 | ·0105 317 | ·0076 999 |
| ·60 | ·0440 308 | ·0335 802 | ·0254 288 | ·0191 271 | ·0142 957 | ·0106 205+ |
| ·61 | ·0555 525+ | ·0430 221 | ·0330 855− | ·0252 757 | ·0191 884 | ·0144 805− |
| ·62 | ·0694 097 | ·0545 653 | ·0426 008 | ·0330 429 | ·0254 710 | ·0195 189 |
| ·63 | ·0858 911 | ·0685 175+ | ·0542 886 | ·0427 385+ | ·0334 407 | ·0260 141 |
| ·64 | ·1052 751 | ·0851 892 | ·0684 777 | ·0546 972 | ·0434 278 | ·0342 834 |
| ·65 | ·1278 160 | ·1048 810 | ·0855 009 | ·0692 700 | ·0557 894 | ·0446 799 |
| ·66 | ·1537 290 | ·1278 692 | ·1056 814 | ·0868 133 | ·0709 012 | ·0575 862 |
| ·67 | ·1831 732 | ·1543 883 | ·1293 167 | ·1076 731 | ·0891 439 | ·0734 039 |
| ·68 | ·2162 347 | ·1846 127 | ·1566 589 | ·1321 670 | ·1108 863 | ·0925 393 |
| ·69 | ·2529 098 | ·2186 367 | ·1878 938 | ·1605 621 | ·1364 647 | ·1153 837 |
| ·70 | ·2930 898 | ·2564 558 | ·2231 189 | ·1930 515+ | ·1661 580 | ·1422 899 |
| ·71 | ·3365 489 | ·2979 492 | ·2623 217 | ·2297 294 | ·2001 609 | ·1735 439 |
| ·72 | ·3829 364 | ·3428 660 | ·3053 601 | ·2705 661 | ·2385 557 | ·2093 348 |
| ·73 | ·4317 745− | ·3908 162 | ·3519 469 | ·3153 872 | ·2812 852 | ·2497 226 |
| ·74 | ·4824 631 | ·4412 690 | ·4016 402 | ·3638 557 | ·3281 284 | ·2946 085− |
| ·75 | ·5342 927 | ·4935 583 | ·4538 423 | ·4154 629 | ·3786 826 | ·3437 085+ |
| ·76 | ·5864 648 | ·5468 988 | ·5078 075+ | ·4695 282 | ·4323 547 | ·3965 354 |
| ·77 | ·6381 211 | ·6004 104 | ·5626 620 | ·5252 112 | ·4883 638 | ·4523 915− |
| ·78 | ·6883 793 | ·6531 532 | ·6174 340 | ·5815 362 | ·5457 581 | ·5103 760 |
| ·79 | ·7363 745− | ·7041 701 | ·6710 962 | ·6374 308 | ·6034 474 | ·5694 095− |
| ·80 | ·7813 035− | ·7525 351 | ·7226 160 | ·6917 763 | ·6602 513 | ·6282 759 |
| ·81 | ·8224 689 | ·7974 051 | ·7710 130 | ·7434 687 | ·7149 604 | ·6856 833 |
| ·82 | ·8593 192 | ·8380 696 | ·8154 176 | ·7914 849 | ·7664 088 | ·7403 385+ |
| ·83 | ·8914 809 | ·8739 949 | ·8551 275+ | ·8349 503 | ·8135 512 | ·7910 325− |
| ·84 | ·9187 803 | ·9048 576 | ·8896 540 | ·8731 991 | ·8555 374 | ·8367 277 |
| ·85 | ·9412 497 | ·9305 633 | ·9187 549 | ·9058 223 | ·8917 762 | ·8766 389 |
| ·86 | ·9591 189 | ·9512 475− | ·9424 472 | ·9326 959 | ·9219 803 | ·9102 964 |
| ·87 | ·9727 899 | ·9672 566 | ·9609 985+ | ·9539 834 | ·9461 850− | ·9375 830 |
| ·88 | ·9827 980 | ·9791 115− | ·9748 940 | ·9701 120 | ·9647 350− | ·9587 357 |
| ·89 | ·9897 618 | ·9874 540 | ·9847 838 | ·9817 219 | ·9782 398 | ·9743 107 |
| ·90 | ·9943 279 | ·9929 857 | ·9914 154 | ·9895 943 | ·9875 001 | ·9851 105− |

$x = \cdot 91$ to $\cdot 99$ — $q = 8\cdot 5$ — $p = 24$ to $29$

| | $p = 24$ | $p = 25$ | $p = 26$ | $p = 27$ | $p = 28$ | $p = 29$ |
|---|---|---|---|---|---|---|
| $B(p, q) =$ | $\cdot 7830\ 4756 \times \frac{1}{10^8}$ | $\cdot 5782\ 5051 \times \frac{1}{10^8}$ | $\cdot 4315\ 3023 \times \frac{1}{10^8}$ | $\cdot 3252\ 1119 \times \frac{1}{10^8}$ | $\cdot 2473\ 4372 \times \frac{1}{10^8}$ | $\cdot 1897\ 4313 \times \frac{1}{10^8}$ |
| $x$ | | | | | | |
| ·91 | ·9971 184 | ·9964 038 | ·9955 584 | ·9945 672 | ·9934 147 | ·9920 850− |
| ·92 | ·9986 843 | ·9983 430 | ·9979 348 | ·9974 509 | ·9968 820 | ·9962 185− |
| ·93 | ·9994 752 | ·9993 330 | ·9991 611 | ·9989 551 | ·9987 103 | ·9984 216 |
| ·94 | ·9998 246 | ·9997 750− | ·9997 144 | ·9996 410 | ·9995 528 | ·9994 478 |
| ·95 | ·9999 539 | ·9999 403 | ·9999 236 | ·9999 031 | ·9998 781 | ·9998 482 |
| ·96 | ·9999 914 | ·9999 888 | ·9999 856 | ·9999 815+ | ·9999 766 | ·9999 705+ |
| ·97 | ·9999 991 | ·9999 988 | ·9999 984 | ·9999 980 | ·9999 974 | ·9999 967 |
| ·98 | 1·0000 000 | 1·0000 000 | ·9999 999 | ·9999 999 | ·9999 999 | ·9999 999 |
| ·99 | | | 1·0000 000 | 1·0000 000 | 1·0000 000 | 1·0000 000 |

$x = \cdot 37$ to $\cdot 99$ $q = 8\cdot 5$ $p = 30$ to $35$

| x | p = 30 | p = 31 | p = 32 | p = 33 | p = 34 | p = 35 |
|---|---|---|---|---|---|---|
| B (p, q) = | ·1467 3468 × 1/10⁸ | ·1143 3872 × 1/10⁸ | ·8973 4182 × 1/10⁹ | ·7090 1082 × 1/10⁹ | ·5637 9174 × 1/10⁹ | ·4510 3339 × 1/10⁹ |
| ·37 | ·0000 001 | | | | | |
| ·38 | ·0000 002 | ·0000 001 | | | | |
| ·39 | ·0000 004 | ·0000 002 | ·0000 001 | | | |
| ·40 | ·0000 007 | ·0000 003 | ·0000 002 | ·0000 001 | | |
| ·41 | ·0000 013 | ·0000 006 | ·0000 003 | ·0000 002 | ·0000 001 | |
| ·42 | ·0000 023 | ·0000 012 | ·0000 006 | ·0000 003 | ·0000 002 | ·0000 001 |
| ·43 | ·0000 041 | ·0000 022 | ·0000 012 | ·0000 006 | ·0000 003 | ·0000 002 |
| ·44 | ·0000 072 | ·0000 039 | ·0000 021 | ·0000 011 | ·0000 006 | ·0000 003 |
| ·45 | ·0000 125+ | ·0000 069 | ·0000 038 | ·0000 021 | ·0000 011 | ·0000 006 |
| ·46 | ·0000 213 | ·0000 121 | ·0000 068 | ·0000 038 | ·0000 021 | ·0000 012 |
| ·47 | ·0000 356 | ·0000 206 | ·0000 119 | ·0000 068 | ·0000 039 | ·0000 022 |
| ·48 | ·0000 587 | ·0000 347 | ·0000 204 | ·0000 119 | ·0000 069 | ·0000 040 |
| ·49 | ·0000 951 | ·0000 574 | ·0000 344 | ·0000 206 | ·0000 122 | ·0000 072 |
| ·50 | ·0001 520 | ·0000 936 | ·0000 573 | ·0000 349 | ·0000 211 | ·0000 127 |
| ·51 | ·0002 394 | ·0001 502 | ·0000 938 | ·0000 582 | ·0000 360 | ·0000 221 |
| ·52 | ·0003 717 | ·0002 377 | ·0001 512 | ·0000 957 | ·0000 603 | ·0000 378 |
| ·53 | ·0005 693 | ·0003 710 | ·0002 404 | ·0001 550+ | ·0000 994 | ·0000 635− |
| ·54 | ·0008 604 | ·0005 709 | ·0003 768 | ·0002 474 | ·0001 617 | ·0001 051 |
| ·55 | ·0012 833 | ·0008 669 | ·0005 825− | ·0003 893 | ·0002 590 | ·0001 715− |
| ·56 | ·0018 896 | ·0012 990 | ·0008 882 | ·0006 042 | ·0004 091 | ·0002 757 |
| ·57 | ·0027 475− | ·0019 213 | ·0013 364 | ·0009 249 | ·0006 371 | ·0004 368 |
| ·58 | ·0039 455− | ·0028 057 | ·0019 848 | ·0013 970 | ·0009 786 | ·0006 824 |
| ·59 | ·0055 971 | ·0040 462 | ·0029 099 | ·0020 823 | ·0014 830 | ·0010 515− |
| ·60 | ·0078 451 | ·0057 635− | ·0042 124 | ·0030 636 | ·0022 176 | ·0015 981 |
| ·61 | ·0108 660 | ·0081 099 | ·0060 219 | ·0044 498 | ·0032 728 | ·0023 965− |
| ·62 | ·0148 742 | ·0112 746 | ·0085 029 | ·0063 817 | ·0047 676 | ·0035 461 |
| ·63 | ·0201 253 | ·0154 879 | ·0118 595− | ·0090 378 | ·0068 562 | ·0051 785+ |
| ·64 | ·0269 175+ | ·0210 248 | ·0163 410 | ·0126 409 | ·0097 345+ | ·0074 642 |
| ·65 | ·0355 912 | ·0282 067 | ·0222 454 | ·0174 624 | ·0136 469 | ·0106 198 |
| ·66 | ·0465 255+ | ·0374 005− | ·0299 209 | ·0238 274 | ·0188 916 | ·0149 154 |
| ·67 | ·0601 307 | ·0490 145− | ·0397 647 | ·0321 148 | ·0258 245+ | ·0206 804 |
| ·68 | ·0768 368 | ·0634 896 | ·0522 178 | ·0427 564 | ·0348 605+ | ·0283 069 |
| ·69 | ·0970 760 | ·0812 856 | ·0677 543 | ·0562 294 | ·0464 699 | ·0382 504 |
| ·70 | ·1212 614 | ·1028 614 | ·0868 652 | ·0730 436 | ·0611 695+ | ·0510 242 |
| ·71 | ·1497 585+ | ·1286 489 | ·1100 351 | ·0937 218 | ·0795 069 | ·0671 881 |
| ·72 | ·1828 546 | ·1590 224 | ·1377 118 | ·1187 720 | ·1020 361 | ·0873 282 |
| ·73 | ·2207 231 | ·1942 610 | ·1702 701 | ·1486 521 | ·1292 844 | ·1120 271 |
| ·74 | ·2633 887 | ·2345 105− | ·2079 704 | ·1837 278 | ·1617 110 | ·1418 247 |
| ·75 | ·3106 942 | ·2797 432 | ·2509 139 | ·2242 243 | ·1996 577 | ·1771 684 |
| ·76 | ·3622 724 | ·3297 225+ | ·2989 996 | ·2701 777 | ·2432 947 | ·2183 568 |
| ·77 | ·4175 294 | ·3839 746 | ·3518 865− | ·3213 878 | ·2925 664 | ·2654 787 |
| ·78 | ·4756 406 | ·4417 736 | ·4089 663 | ·3773 791 | ·3471 414 | ·3183 534 |
| ·79 | ·5355 652 | ·5021 439 | ·4693 534 | ·4373 774 | ·4063 753 | ·3764 811 |
| ·80 | ·5960 800 | ·5638 841 | ·5318 960 | ·5003 076 | ·4692 936 | ·4390 094 |
| ·81 | ·6558 359 | ·6256 153 | ·5952 145− | ·5648 183 | ·5346 018 | ·5047 274 |
| ·82 | ·7134 322 | ·6858 533 | ·6577 672 | ·6293 389 | ·6007 296 | ·5720 951 |
| ·83 | ·7675 082 | ·7431 018 | ·7179 434 | ·6921 679 | ·6659 121 | ·6393 126 |
| ·84 | ·8168 414 | ·7959 607 | ·7741 775− | ·7515 910 | ·7283 063 | ·7044 327 |
| ·85 | ·8604 447 | ·8432 383 | ·8250 745− | ·8060 167 | ·7861 359 | ·7655 096 |
| ·86 | ·8976 497 | ·8840 544 | ·8695 338 | ·8541 192 | ·8378 499 | ·8207 718 |
| ·87 | ·9281 638 | ·9179 204 | ·9068 527 | ·8949 669 | ·8822 762 | ·8687 997 |
| ·88 | ·9520 911 | ·9447 819 | ·9367 937 | ·9281 166 | ·9187 452 | ·9086 793 |
| ·89 | ·9699 094 | ·9650 129 | ·9596 007 | ·9536 549 | ·9471 605− | ·9401 054 |
| ·90 | ·9824 036 | ·9793 584 | ·9759 546 | ·9721 732 | ·9679 964 | ·9634 082 |
| ·91 | ·9905 620 | ·9888 296 | ·9868 717 | ·9846 724 | ·9822 163 | ·9794 881 |
| ·92 | ·9954 501 | ·9945 666 | ·9935 570 | ·9924 105+ | ·9911 160 | ·9896 624 |
| ·93 | ·9980 837 | ·9976 909 | ·9972 372 | ·9967 164 | ·9961 220 | ·9954 472 |
| ·94 | ·9993 235+ | ·9991 775− | ·9990 070 | ·9988 092 | ·9985 809 | ·9983 191 |
| ·95 | ·9998 123 | ·9997 697 | ·9997 195− | ·9996 605+ | ·9995 918 | ·9995 122 |
| ·96 | ·9999 632 | ·9999 545− | ·9999 441 | ·9999 317 | ·9999 171 | ·9999 001 |
| ·97 | ·9999 958 | ·9999 948 | ·9999 936 | ·9999 921 | ·9999 903 | ·9999 882 |
| ·98 | ·9999 998 | ·9999 998 | ·9999 997 | ·9999 997 | ·9999 996 | ·9999 995− |
| ·99 | 1·0000 000 | 1·0000 000 | 1·0000 000 | 1·0000 000 | 1·0000 000 | 1·0000 000 |

$x$ = ·43 to ·99 $q$ = 8·5 $p$ = 36 to 40

| $x$ | $p$ = 36 | $p$ = 37 | $p$ = 38 | $p$ = 39 | $p$ = 40 |
|---|---|---|---|---|---|
| $B(p, q)$ = | ·3629 0043 × $\frac{1}{10^9}$ | ·2935 8237 × $\frac{1}{10^9}$ | ·2387 3731 × $\frac{1}{10^9}$ | ·1950 9716 × $\frac{1}{10^9}$ | ·1601 8503 × $\frac{1}{10^9}$ |
| ·43 | ·0000 001 | | | | |
| ·44 | ·0000 002 | ·0000 001 | | | |
| ·45 | ·0000 003 | ·0000 002 | ·0000 001 | ·0000 001 | |
| ·46 | ·0000 007 | ·0000 004 | ·0000 002 | ·0000 001 | ·0000 001 |
| ·47 | ·0000 012 | ·0000 007 | ·0000 004 | ·0000 002 | ·0000 001 |
| ·48 | ·0000 023 | ·0000 013 | ·0000 008 | ·0000 004 | ·0000 002 |
| ·49 | ·0000 042 | ·0000 025⁻ | ·0000 015⁻ | ·0000 008 | ·0000 005⁻ |
| ·50 | ·0000 076 | ·0000 046 | ·0000 027 | ·0000 016 | ·0000 010 |
| ·51 | ·0000 135⁺ | ·0000 082 | ·0000 050⁺ | ·0000 030 | ·0000 018 |
| ·52 | ·0000 236 | ·0000 146 | ·0000 091 | ·0000 056 | ·0000 034 |
| ·53 | ·0000 404 | ·0000 256 | ·0000 161 | ·0000 101 | ·0000 063 |
| ·54 | ·0000 681 | ·0000 439 | ·0000 282 | ·0000 180 | ·0000 115⁻ |
| ·55 | ·0001 131 | ·0000 742 | ·0000 485⁺ | ·0000 316 | ·0000 205⁺ |
| ·56 | ·0001 850⁻ | ·0001 236 | ·0000 822 | ·0000 545⁺ | ·0000 360 |
| ·57 | ·0002 982 | ·0002 027 | ·0001 372 | ·0000 926 | ·0000 622 |
| ·58 | ·0004 738 | ·0003 276 | ·0002 256 | ·0001 548 | ·0001 058 |
| ·59 | ·0007 423 | ·0005 219 | ·0003 655⁻ | ·0002 550⁻ | ·0001 772 |
| ·60 | ·0011 468 | ·0008 195⁺ | ·0005 834 | ·0004 137 | ·0002 924 |
| ·61 | ·0017 474 | ·0012 689 | ·0009 179 | ·0006 615⁺ | ·0004 751 |
| ·62 | ·0026 265⁺ | ·0019 376 | ·0014 239 | ·0010 425⁺ | ·0007 606 |
| ·63 | ·0038 952 | ·0029 182 | ·0021 779 | ·0016 195⁺ | ·0012 001 |
| ·64 | ·0056 998 | ·0043 353 | ·0032 851 | ·0024 802 | ·0018 661 |
| ·65 | ·0082 305⁻ | ·0063 539 | ·0048 869 | ·0037 451 | ·0028 602 |
| ·66 | ·0117 288 | ·0091 874 | ·0071 702 | ·0055 760 | ·0043 215⁻ |
| ·67 | ·0164 952 | ·0131 070 | ·0103 767 | ·0081 864 | ·0064 367 |
| ·68 | ·0228 955⁺ | ·0184 491 | ·0148 127 | ·0118 519 | ·0094 514 |
| ·69 | ·0313 636 | ·0256 218 | ·0208 568 | ·0169 201 | ·0136 813 |
| ·70 | ·0424 006 | ·0351 066 | ·0289 657 | ·0238 188 | ·0195 230 |
| ·71 | ·0565 675⁺ | ·0474 559 | ·0396 754 | ·0330 609 | ·0274 615⁻ |
| ·72 | ·0744 693 | ·0632 820 | ·0535 945⁻ | ·0452 428 | ·0380 732 |
| ·73 | ·0967 298 | ·0832 363 | ·0713 894 | ·0610 342 | ·0520 212 |
| ·74 | ·1239 554 | ·1079 773 | ·0937 568 | ·0811 567 | ·0700 395⁻ |
| ·75 | ·1566 871 | ·1381 259 | ·1213 832 | ·1063 479 | ·0929 028 |
| ·76 | ·1953 427 | ·1742 081 | ·1548 899 | ·1373 103 | ·1213 805⁺ |
| ·77 | ·2401 516 | ·2165 868 | ·1947 640 | ·1746 438 | ·1561 720 |
| ·78 | ·2910 870 | ·2653 885⁺ | ·2412 805⁻ | ·2187 649 | ·1978 254 |
| ·79 | ·3478 040 | ·3204 293 | ·2944 196 | ·2698 165⁺ | ·2466 423 |
| ·80 | ·4095 904 | ·3811 520 | ·3537 897 | ·3275 795⁻ | ·3025 790 |
| ·81 | ·4753 439 | ·4465 847 | ·4185 678 | ·3913 952 | ·3651 526 |
| ·82 | ·5435 833 | ·5153 329 | ·4874 719 | ·4601 167 | ·4333 718 |
| ·83 | ·6125 041 | ·5856 174 | ·5587 780 | ·5321 045⁺ | ·5057 081 |
| ·84 | ·6800 816 | ·6553 653 | ·6303 951 | ·6052 804 | ·5801 270 |
| ·85 | ·7442 203 | ·7223 544 | ·7000 010 | ·6772 505⁻ | ·6541 937 |
| ·86 | ·8029 375⁻ | ·7844 046 | ·7652 356 | ·7454 966 | ·7252 565⁺ |
| ·87 | ·8545 628 | ·8395 963 | ·8239 360 | ·8076 225⁺ | ·7907 005⁻ |
| ·88 | ·8979 231 | ·8864 857 | ·8743 807 | ·8616 257 | ·8482 428 |
| ·89 | ·9324 809 | ·9242 814 | ·9155 046 | ·9061 515⁺ | ·8962 265⁻ |
| ·90 | ·9583 939 | ·9529 408 | ·9470 382 | ·9406 774 | ·9338 518 |
| ·91 | ·9764 735⁻ | ·9731 586 | ·9695 307 | ·9655 777 | ·9612 886 |
| ·92 | ·9880 385⁻ | ·9862 332 | ·9842 358 | ·9820 354 | ·9796 217 |
| ·93 | ·9946 852 | ·9938 289 | ·9928 711 | ·9918 045⁺ | ·9906 219 |
| ·94 | ·9980 202 | ·9976 808 | ·9972 970 | ·9968 650⁺ | ·9963 809 |
| ·95 | ·9994 203 | ·9993 148 | ·9991 943 | ·9990 572 | ·9989 019 |
| ·96 | ·9998 802 | ·9998 571 | ·9998 304 | ·9997 998 | ·9997 648 |
| ·97 | ·9999 857 | ·9999 828 | ·9999 794 | ·9999 755⁻ | ·9999 709 |
| ·98 | ·9999 994 | ·9999 992 | ·9999 991 | ·9999 989 | ·9999 987 |
| ·99 | 1·0000 000 | 1·0000 000 | 1·0000 000 | 1·0000 000 | 1·0000 000 |

$x$ = ·47 to ·99 $q = 8{\cdot}5$ $p$ = 41 to 45

| $x$ | $p = 41$ | $p = 42$ | $p = 43$ | $p = 44$ | $p = 45$ |
|---|---|---|---|---|---|
| $B(p, q) =$ | ·1321 1137 $\times \frac{1}{10^9}$ | ·1094 2558 $\times \frac{1}{10^9}$ | ·9100 7411 $\times \frac{1}{10^{10}}$ | ·7598 6771 $\times \frac{1}{10^{10}}$ | ·6368 4151 $\times \frac{1}{10^{10}}$ |
| ·47 | ·0000 001 | | | | |
| ·48 | ·0000 001 | ·0000 001 | | | |
| ·49 | ·0000 003 | ·0000 002 | ·0000 001 | ·0000 001 | |
| ·50 | ·0000 006 | ·0000 003 | ·0000 002 | ·0000 001 | ·0000 001 |
| ·51 | ·0000 011 | ·0000 007 | ·0000 004 | ·0000 002 | ·0000 001 |
| ·52 | ·0000 021 | ·0000 013 | ·0000 008 | ·0000 005⁻ | ·0000 003 |
| ·53 | ·0000 039 | ·0000 025⁻ | ·0000 015⁺ | ·0000 009 | ·0000 006 |
| ·54 | ·0000 073 | ·0000 046 | ·0000 029 | ·0000 018 | ·0000 011 |
| ·55 | ·0000 133 | ·0000 085⁺ | ·0000 055⁻ | ·0000 035⁺ | ·0000 022 |
| ·56 | ·0000 237 | ·0000 155⁺ | ·0000 102 | ·0000 066 | ·0000 043 |
| ·57 | ·0000 417 | ·0000 278 | ·0000 185⁺ | ·0000 123 | ·0000 081 |
| ·58 | ·0000 721 | ·0000 490 | ·0000 331 | ·0000 224 | ·0000 150⁺ |
| ·59 | ·0001 228 | ·0000 848 | ·0000 583 | ·0000 400 | ·0000 274 |
| ·60 | ·0002 059 | ·0001 445⁺ | ·0001 011 | ·0000 705⁺ | ·0000 491 |
| ·61 | ·0003 400 | ·0002 425⁺ | ·0001 725⁻ | ·0001 223 | ·0000 864 |
| ·62 | ·0005 531 | ·0004 008 | ·0002 896 | ·0002 086 | ·0001 498 |
| ·63 | ·0008 863 | ·0006 524 | ·0004 788 | ·0003 503 | ·0002 556 |
| ·64 | ·0013 993 | ·0010 460 | ·0007 794 | ·0005 791 | ·0004 290 |
| ·65 | ·0021 772 | ·0016 520 | ·0012 497 | ·0009 426 | ·0007 089 |
| ·66 | ·0033 383 | ·0025 707 | ·0019 736 | ·0015 107 | ·0011 532 |
| ·67 | ·0050 446 | ·0039 412 | ·0030 700 | ·0023 844 | ·0018 468 |
| ·68 | ·0075 130 | ·0059 537 | ·0047 041 | ·0037 061 | ·0029 118 |
| ·69 | ·0110 276 | ·0088 616 | ·0071 001 | ·0056 728 | ·0045 200 |
| ·70 | ·0159 522 | ·0129 954 | ·0105 560 | ·0085 506 | ·0069 075⁺ |
| ·71 | ·0227 405⁻ | ·0187 754 | ·0154 575⁻ | ·0126 909 | ·0103 918 |
| ·72 | ·0319 432 | ·0267 222 | ·0222 918 | ·0185 456 | ·0153 885⁺ |
| ·73 | ·0442 081 | ·0374 612 | ·0316 564 | ·0266 799 | ·0224 277 |
| ·74 | ·0602 702 | ·0517 186 | ·0442 603 | ·0377 786 | ·0321 645⁻ |
| ·75 | ·0809 282 | ·0703 045⁻ | ·0609 139 | ·0526 425⁻ | ·0453 815⁻ |
| ·76 | ·1070 039 | ·0940 789 | ·0825 019 | ·0721 687 | ·0629 768 |
| ·77 | ·1392 819 | ·1238 974 | ·1099 360 | ·0973 107 | ·0859 321 |
| ·78 | ·1784 301 | ·1605 345⁻ | ·1440 833 | ·1290 132 | ·1152 549 |
| ·79 | ·2249 020 | ·2045 854 | ·1856 692 | ·1681 190 | ·1518 910 |
| ·80 | ·2788 285⁻ | ·2563 524 | ·2351 607 | ·2152 503 | ·1966 070 |
| ·81 | ·3399 106 | ·3157 244 | ·2926 352 | ·2706 710 | ·2498 474 |
| ·82 | ·4073 290 | ·3820 674 | ·3576 539 | ·3341 432 | ·3115 782 |
| ·83 | ·4796 910 | ·4541 468 | ·4291 590 | ·4048 017 | ·3811 392 |
| ·84 | ·5550 365⁻ | ·5301 051 | ·5054 230 | ·4810 739 | ·4571 342 |
| ·85 | ·6309 208 | ·6075 201 | ·5840 777 | ·5606 762 | ·5373 943 |
| ·86 | ·7045 863 | ·6835 582 | ·6622 447 | ·6407 180 | ·6190 495⁺ |
| ·87 | ·7732 178 | ·7552 257 | ·7367 772 | ·7179 277 | ·6987 333 |
| ·88 | ·8342 577 | ·8196 996 | ·8046 007 | ·7889 962 | ·7729 236 |
| ·89 | ·8857 369 | ·8746 935⁺ | ·8631 097 | ·8510 020 | ·8383 893 |
| ·90 | ·9265 569 | ·9187 906 | ·9105 526 | ·9018 454 | ·8926 730 |
| ·91 | ·9566 538 | ·9516 645⁺ | ·9463 136 | ·9405 949 | ·9345 038 |
| ·92 | ·9769 848 | ·9741 151 | ·9710 035⁻ | ·9676 415⁻ | ·9640 212 |
| ·93 | ·9893 158 | ·9878 790 | ·9863 041 | ·9845 840 | ·9827 116 |
| ·94 | ·9958 405⁺ | ·9952 396 | ·9945 739 | ·9938 390 | ·9930 304 |
| ·95 | ·9987 267 | ·9985 298 | ·9983 094 | ·9980 634 | ·9977 900 |
| ·96 | ·9997 248 | ·9996 794 | ·9996 281 | ·9995 702 | ·9995 051 |
| ·97 | ·9999 657 | ·9999 597 | ·9999 528 | ·9999 449 | ·9999 360 |
| ·98 | ·9999 984 | ·9999 981 | ·9999 978 | ·9999 974 | ·9999 970 |
| ·99 | 1·0000 000 | 1·0000 000 | 1·0000 000 | 1·0000 000 | 1·0000 000 |

$x = \cdot 51$ to $\cdot 99$ $q = 8\cdot 5$ $p = 46$ to $50$

| $x$ | $p = 46$ | $p = 47$ | $p = 48$ | $p = 49$ | $p = 50$ |
|---|---|---|---|---|---|
| $B(p, q) =$ | $\cdot 5356\ 6108 \times \frac{1}{10^{10}}$ | $\cdot 4521\ 1761 \times \frac{1}{10^{10}}$ | $\cdot 3828\ 7437 \times \frac{1}{10^{10}}$ | $\cdot 3252\ 7380 \times \frac{1}{10^{10}}$ | $\cdot 2771\ 8985 \times \frac{1}{10^{10}}$ |
| ·51 | ·0000 001 | | | | |
| ·52 | ·0000 002 | ·0000 001 | ·0000 001 | | |
| ·53 | ·0000 004 | ·0000 002 | ·0000 001 | ·0000 001 | |
| ·54 | ·0000 007 | ·0000 004 | ·0000 003 | ·0000 002 | ·0000 001 |
| ·55 | ·0000 014 | ·0000 009 | ·0000 006 | ·0000 004 | ·0000 002 |
| ·56 | ·0000 028 | ·0000 018 | ·0000 012 | ·0000 007 | ·0000 005⁻ |
| ·57 | ·0000 054 | ·0000 035⁺ | ·0000 023 | ·0000 015⁺ | ·0000 010 |
| ·58 | ·0000 101 | ·0000 067 | ·0000 045⁺ | ·0000 030 | ·0000 020 |
| ·59 | ·0000 187 | ·0000 127 | ·0000 086 | ·0000 058 | ·0000 039 |
| ·60 | ·0000 340 | ·0000 235⁺ | ·0000 162 | ·0000 112 | ·0000 077 |
| ·61 | ·0000 609 | ·0000 428 | ·0000 300 | ·0000 210 | ·0000 147 |
| ·62 | ·0001 073 | ·0000 767 | ·0000 546 | ·0000 388 | ·0000 275⁺ |
| ·63 | ·0001 859 | ·0001 349 | ·0000 976 | ·0000 705⁻ | ·0000 508 |
| ·64 | ·0003 170 | ·0002 336 | ·0001 717 | ·0001 259 | ·0000 921 |
| ·65 | ·0005 317 | ·0003 978 | ·0002 968 | ·0002 209 | ·0001 641 |
| ·66 | ·0008 779 | ·0006 666 | ·0005 049 | ·0003 814 | ·0002 875⁺ |
| ·67 | ·0014 266 | ·0010 991 | ·0008 447 | ·0006 476 | ·0004 954 |
| ·68 | ·0022 817 | ·0017 834 | ·0013 904 | ·0010 814 | ·0008 392 |
| ·69 | ·0035 920 | ·0028 473 | ·0022 515⁻ | ·0017 761 | ·0013 979 |
| ·70 | ·0055 657 | ·0044 732 | ·0035 865⁺ | ·0028 688 | ·0022 895⁺ |
| ·71 | ·0084 873 | ·0069 147 | ·0056 200 | ·0045 571 | ·0036 870 |
| ·72 | ·0127 366 | ·0105 159 | ·0086 619 | ·0071 184 | ·0058 369 |
| ·73 | ·0188 063 | ·0157 317 | ·0131 290 | ·0109 322 | ·0090 831 |
| ·74 | ·0273 018 | ·0231 462 | ·0195 668 | ·0165 042 | ·0138 909 |
| ·75 | ·0390 281 | ·0334 862 | ·0286 665⁻ | ·0244 869 | ·0208 724 |
| ·76 | ·0548 267 | ·0476 227 | ·0412 740 | ·0356 951 | ·0308 061 |
| ·77 | ·0757 104 | ·0665 564 | ·0583 830 | ·0511 061 | ·0446 453 |
| ·78 | ·1027 348 | ·0913 770 | ·0811 042 | ·0718 393 | ·0635 067 |
| ·79 | ·1369 341 | ·1231 917 | ·1106 028 | ·0991 037 | ·0886 288 |
| ·80 | ·1792 066 | ·1630 166 | ·1479 978 | ·1341 054 | ·1212 904 |
| ·81 | ·2301 691 | ·2116 308 | ·1942 187 | ·1779 111 | ·1626 802 |
| ·82 | ·2899 910 | ·2694 034 | ·2498 277 | ·2312 674 | ·2137 185⁺ |
| ·83 | ·3582 260 | ·3361 072 | ·3148 191 | ·2943 892 | ·2748 369 |
| ·84 | ·4336 734 | ·4107 531 | ·3884 278 | ·3667 444 | ·3457 424 |
| ·85 | ·5143 065⁻ | ·4914 821 | ·4689 856 | ·4468 757 | ·4252 057 |
| ·86 | ·5973 087 | ·5755 633 | ·5538 779 | ·5323 146 | ·5109 316 |
| ·87 | ·6792 509 | ·6595 378 | ·6396 506 | ·6196 453 | ·5995 767 |
| ·88 | ·7564 223 | ·7395 336 | ·7223 000 | ·7047 648 | ·6869 719 |
| ·89 | ·8252 931 | ·8117 372 | ·7977 472 | ·7833 505⁺ | ·7685 764 |
| ·90 | ·8830 421 | ·8729 610 | ·8624 402 | ·8514 920 | ·8401 304 |
| ·91 | ·9280 372 | ·9211 931 | ·9139 711 | ·9063 722 | ·8983 987 |
| ·92 | ·9601 355⁻ | ·9559 778 | ·9515 422 | ·9468 240 | ·9418 187 |
| ·93 | ·9806 801 | ·9784 827 | ·9761 131 | ·9735 650⁻ | ·9708 325⁺ |
| ·94 | ·9921 437 | ·9911 742 | ·9901 175⁺ | ·9889 691 | ·9877 242 |
| ·95 | ·9974 869 | ·9971 520 | ·9967 831 | ·9963 779 | ·9959 340 |
| ·96 | ·9994 323 | ·9993 509 | ·9992 604 | ·9991 598 | ·9990 486 |
| ·97 | ·9999 260 | ·9999 146 | ·9999 018 | ·9998 875⁺ | ·9998 715⁻ |
| ·98 | ·9999 965⁻ | ·9999 959 | ·9999 952 | ·9999 945⁻ | ·9999 936 |
| ·99 | 1·0000 000 | 1·0000 000 | 1·0000 000 | 1·0000 000 | 1·0000 000 |

$x = \cdot 06$ to $\cdot 70$ $q = 9$ $p = 9$ to $12$

| $x$ | $p = 9$ | $p = 9\cdot5$ | $p = 10$ | $p = 10\cdot5$ | $p = 11$ | $p = 12$ |
|---|---|---|---|---|---|---|
| $B(p, q) =$ | $\cdot 4570\ 5928 \times \frac{1}{10^5}$ | $\cdot 3209\ 5512 \times \frac{1}{10^5}$ | $\cdot 2285\ 2964 \times \frac{1}{10^5}$ | $\cdot 1648\ 1479 \times \frac{1}{10^5}$ | $\cdot 1202\ 7876 \times \frac{1}{10^5}$ | $\cdot 6615\ 3317 \times \frac{1}{10^6}$ |
| ·06 | ·0000 002 | ·0000 001 | | | | |
| ·07 | ·0000 006 | ·0000 002 | ·0000 001 | | | |
| ·08 | ·0000 018 | ·0000 007 | ·0000 003 | ·0000 001 | | |
| ·09 | ·0000 048 | ·0000 019 | ·0000 008 | ·0000 003 | ·0000 001 | |
| ·10 | ·0000 115− | ·0000 049 | ·0000 020 | ·0000 009 | ·0000 004 | ·0000 001 |
| ·11 | ·0000 250− | ·0000 111 | ·0000 049 | ·0000 021 | ·0000 009 | ·0000 002 |
| ·12 | ·0000 505− | ·0000 235− | ·0000 108 | ·0000 049 | ·0000 022 | ·0000 004 |
| ·13 | ·0000 958 | ·0000 463 | ·0000 222 | ·0000 105− | ·0000 049 | ·0000 011 |
| ·14 | ·0001 721 | ·0000 863 | ·0000 428 | ·0000 210 | ·0000 102 | ·0000 024 |
| ·15 | ·0002 950+ | ·0001 530 | ·0000 786 | ·0000 399 | ·0000 201 | ·0000 050− |
| ·16 | ·0004 856 | ·0002 600 | ·0001 378 | ·0000 723 | ·0000 376 | ·0000 099 |
| ·17 | ·0007 711 | ·0004 253 | ·0002 322 | ·0001 255+ | ·0000 673 | ·0000 188 |
| ·18 | ·0011 859 | ·0006 726 | ·0003 776 | ·0002 100 | ·0001 157 | ·0000 343 |
| ·19 | ·0017 723 | ·0010 321 | ·0005 949 | ·0003 397 | ·0001 922 | ·0000 601 |
| ·20 | ·0025 815− | ·0015 412 | ·0009 109 | ·0005 333 | ·0003 095− | ·0001 017 |
| ·21 | ·0036 733 | ·0022 456 | ·0013 591 | ·0008 148 | ·0004 843 | ·0001 670 |
| ·22 | ·0051 170 | ·0031 993 | ·0019 805− | ·0012 146 | ·0007 385− | ·0002 665− |
| ·23 | ·0069 908 | ·0044 657 | ·0028 245+ | ·0017 700 | ·0010 997 | ·0004 144 |
| ·24 | ·0093 819 | ·0061 170 | ·0039 492 | ·0025 264 | ·0016 024 | ·0006 295− |
| ·25 | ·0123 848 | ·0082 344 | ·0054 218 | ·0035 375− | ·0022 884 | ·0009 354 |
| ·26 | ·0161 012 | ·0109 078 | ·0073 184 | ·0048 660 | ·0032 081 | ·0013 622 |
| ·27 | ·0206 378 | ·0142 344 | ·0097 243 | ·0065 839 | ·0044 203 | ·0019 467 |
| ·28 | ·0261 050− | ·0183 183 | ·0127 329 | ·0087 722 | ·0059 933 | ·0027 339 |
| ·29 | ·0326 144 | ·0232 682 | ·0164 451 | ·0115 209 | ·0080 047 | ·0037 768 |
| ·30 | ·0402 769 | ·0291 961 | ·0209 680 | ·0149 280 | ·0105 412 | ·0051 382 |
| ·31 | ·0491 999 | ·0362 150+ | ·0264 132 | ·0190 987 | ·0136 982 | ·0068 899 |
| ·32 | ·0594 848 | ·0444 367 | ·0328 950+ | ·0241 441 | ·0175 793 | ·0091 139 |
| ·33 | ·0712 244 | ·0539 687 | ·0405 282 | ·0301 791 | ·0222 947 | ·0119 016 |
| ·34 | ·0844 999 | ·0649 124 | ·0494 254 | ·0373 208 | ·0279 599 | ·0153 538 |
| ·35 | ·0993 789 | ·0773 594 | ·0596 947 | ·0456 857 | ·0346 936 | ·0195 794 |
| ·36 | ·1159 125− | ·0913 894 | ·0714 363 | ·0553 875− | ·0426 158 | ·0246 946 |
| ·37 | ·1341 335+ | ·1070 674 | ·0847 405+ | ·0665 340 | ·0518 448 | ·0308 214 |
| ·38 | ·1540 544 | ·1244 411 | ·0996 840 | ·0792 241 | ·0624 947 | ·0380 849 |
| ·39 | ·1756 661 | ·1435 387 | ·1163 278 | ·0935 451 | ·0746 723 | ·0466 117 |
| ·40 | ·1989 365− | ·1643 673 | ·1347 141 | ·1095 695− | ·0884 741 | ·0565 264 |
| ·41 | ·2238 105+ | ·1869 111 | ·1548 647 | ·1273 522 | ·1039 827 | ·0679 490 |
| ·42 | ·2502 097 | ·2111 305+ | ·1767 785− | ·1469 280 | ·1212 645− | ·0809 916 |
| ·43 | ·2780 325− | ·2369 616 | ·2004 302 | ·1683 092 | ·1403 660 | ·0957 546 |
| ·44 | ·3071 557 | ·2643 160 | ·2257 694 | ·1914 836 | ·1613 115− | ·1123 234 |
| ·45 | ·3374 356 | ·2930 817 | ·2527 203 | ·2164 136 | ·1841 009 | ·1307 650− |
| ·46 | ·3687 099 | ·3231 239 | ·2811 814 | ·2430 344 | ·2087 079 | ·1511 243 |
| ·47 | ·4008 001 | ·3542 865+ | ·3110 266 | ·2712 547 | ·2350 781 | ·1734 218 |
| ·48 | ·4335 145+ | ·3863 949 | ·3421 061 | ·3009 563 | ·2631 292 | ·1976 502 |
| ·49 | ·4666 509 | ·4192 577 | ·3742 489 | ·3319 955− | ·2927 504 | ·2237 730 |
| ·50 | ·5000 000$^e$ | ·4526 706 | ·4072 647 | ·3642 045− | ·3238 029 | ·2517 223 |
| ·51 | ·5333 491 | ·4864 192 | ·4409 471 | ·3973 939 | ·3561 221 | ·2813 990 |
| ·52 | ·5664 855− | ·5202 827 | ·4750 771 | ·4313 554 | ·3895 188 | ·3126 719 |
| ·53 | ·5991 999 | ·5540 378 | ·5094 263 | ·4658 652 | ·4237 823 | ·3453 791 |
| ·54 | ·6312 901 | ·5874 626 | ·5437 616 | ·5006 879 | ·4586 840 | ·3793 297 |
| ·55 | ·6625 644 | ·6203 404 | ·5778 491 | ·5355 805+ | ·4939 810 | ·4143 062 |
| ·56 | ·6928 443 | ·6524 634 | ·6114 580 | ·5702 973 | ·5294 206 | ·4500 681 |
| ·57 | ·7219 675+ | ·6836 363 | ·6443 652 | ·6045 940 | ·5647 452 | ·4863 557 |
| ·58 | ·7497 903 | ·7136 795+ | ·6763 592 | ·6382 326 | ·5996 970 | ·5228 955− |
| ·59 | ·7761 895− | ·7424 321 | ·7072 437 | ·6709 856 | ·6340 232 | ·5594 050− |
| ·60 | ·8010 635+ | ·7697 541 | ·7368 412 | ·7026 408 | ·6674 810 | ·5955 987 |
| ·61 | ·8243 339 | ·7955 285− | ·7649 957 | ·7330 044 | ·6998 422 | ·6311 942 |
| ·62 | ·8459 456 | ·8196 625− | ·7915 752 | ·7619 054 | ·7308 978 | ·6659 179 |
| ·63 | ·8658 665− | ·8420 885− | ·8164 735− | ·7891 974 | ·7604 618 | ·6995 110 |
| ·64 | ·8840 875+ | ·8627 642 | ·8396 114 | ·8147 616 | ·7883 748 | ·7317 352 |
| ·65 | ·9006 211 | ·8816 722 | ·8609 369 | ·8385 083 | ·8145 064 | ·7623 776 |
| ·66 | ·9155 001 | ·8988 193 | ·8804 257 | ·8603 770 | ·8387 573 | ·7912 553 |
| ·67 | ·9287 756 | ·9142 348 | ·8980 795+ | ·8803 373 | ·8610 600 | ·8182 183 |
| ·68 | ·9405 152 | ·9279 688 | ·9139 254 | ·8983 875+ | ·8813 795− | ·8431 528 |
| ·69 | ·9508 001 | ·9400 897 | ·9280 133 | ·9145 537 | ·8997 122 | ·8659 824 |
| ·70 | ·9597 231 | ·9506 821 | ·9404 141 | ·9288 872 | ·9160 848 | ·8866 685+ |

$x$ = ·71 to ·96 $q = 9$ $p$ = 9 to 12

| | $p = 9$ | $p = 9·5$ | $p = 10$ | $p = 10·5$ | $p = 11$ | $p = 12$ |
|---|---|---|---|---|---|---|
| $B(p, q)$ = | ·4570 5928 × $\frac{1}{10^5}$ | ·3209 5512 × $\frac{1}{10^5}$ | ·2285 2964 × $\frac{1}{10^5}$ | ·1648 1479 × $\frac{1}{10^5}$ | ·1202 7876 × $\frac{1}{10^5}$ | ·6615 3317 × $\frac{1}{10^6}$ |
| $x$ | | | | | | |
| ·71 | ·9673 856 | ·9598 433 | ·9512 162 | ·9414 623 | ·9305 518 | ·9052 098 |
| ·72 | ·9738 950+ | ·9676 808 | ·9605 229 | ·9523 730 | ·9431 927 | ·9216 401 |
| ·73 | ·9793 622 | ·9743 093 | ·9684 487 | ·9617 295+ | ·9541 083 | ·9360 264 |
| ·74 | ·9838 988 | ·9798 474 | ·9751 161 | ·9696 546 | ·9634 175− | ·9484 645− |
| ·75 | ·9876 152 | ·9844 149 | ·9806 522 | ·9762 795+ | ·9712 522 | ·9590 748 |
| ·76 | ·9906 181 | ·9881 302 | ·9851 855+ | ·9817 407 | ·9777 537 | ·9679 978 |
| ·77 | ·9930 092 | ·9911 079 | ·9888 428 | ·9861 756 | ·9830 683 | ·9753 882 |
| ·78 | ·9948 830 | ·9934 568 | ·9917 466 | ·9897 197 | ·9873 430 | ·9814 101 |
| ·79 | ·9963 267 | ·9952 780 | ·9940 125+ | ·9925 030 | ·9907 217 | ·9862 313 |
| ·80 | ·9974 185+ | ·9966 641 | ·9957 480 | ·9946 483 | ·9933 423 | ·9900 182 |
| ·81 | ·9982 277 | ·9976 978 | ·9970 502 | ·9962 681 | ·9953 336 | ·9929 318 |
| ·82 | ·9988 141 | ·9984 516 | ·9980 058 | ·9974 641 | ·9968 128 | ·9951 231 |
| ·83 | ·9992 289 | ·9989 880 | ·9986 900 | ·9983 256 | ·9978 848 | ·9967 305+ |
| ·84 | ·9995 144 | ·9993 594 | ·9991 665+ | ·9989 293 | ·9986 406 | ·9978 775+ |
| ·85 | ·9997 050− | ·9996 088 | ·9994 885+ | ·9993 396 | ·9991 573 | ·9986 711 |
| ·86 | ·9998 279 | ·9997 707 | ·9996 987 | ·9996 090 | ·9994 986 | ·9992 014 |
| ·87 | ·9999 042 | ·9998 718 | ·9998 306 | ·9997 791 | ·9997 153 | ·9995 421 |
| ·88 | ·9999 495+ | ·9999 321 | ·9999 098 | ·9998 818 | ·9998 469 | ·9997 514 |
| ·89 | ·9999 750+ | ·9999 662 | ·9999 549 | ·9999 407 | ·9999 228 | ·9998 733 |
| ·90 | ·9999 885+ | ·9999 844 | ·9999 791 | ·9999 724 | ·9999 639 | ·9999 401 |
| ·91 | ·9999 952 | ·9999 934 | ·9999 912 | ·9999 883 | ·9999 846 | ·9999 742 |
| ·92 | ·9999 982 | ·9999 975+ | ·9999 967 | ·9999 955+ | ·9999 941 | ·9999 901 |
| ·93 | ·9999 994 | ·9999 992 | ·9999 989 | ·9999 985+ | ·9999 981 | ·9999 967 |
| ·94 | ·9999 998 | ·9999 998 | ·9999 997 | ·9999 996 | ·9999 995− | ·9999 991 |
| ·95 | 1·0000 000 | 1·0000 000 | ·9999 999 | ·9999 999 | ·9999 999 | ·9999 998 |
| ·96 | | | 1·0000 000 | 1·0000 000 | 1·0000 000 | 1·0000 000 |

$x$ = ·12 to ·70 $q = 9$ $p$ = 13 to 18

| $x$ | $p = 13$ | $p = 14$ | $p = 15$ | $p = 16$ | $p = 17$ | $p = 18$ |
|---|---|---|---|---|---|---|
| $B(p, q)$ = | ·3780 1895 $\times \frac{1}{10^6}$ | ·2233 7484 $\times \frac{1}{10^6}$ | ·1359 6729 $\times \frac{1}{10^6}$ | ·8497 9557 $\times \frac{1}{10^7}$ | ·5438 6917 $\times \frac{1}{10^7}$ | ·3556 0676 $\times \frac{1}{10^7}$ |
| ·12 | ·0000 001 | | | | | |
| ·13 | ·0000 002 | | | | | |
| ·14 | ·0000 005+ | ·0000 001 | | | | |
| ·15 | ·0000 012 | ·0000 003 | ·0000 001 | | | |
| ·16 | ·0000 025+ | ·0000 006 | ·0000 002 | | | |
| ·17 | ·0000 051 | ·0000 014 | ·0000 004 | ·0000 001 | | |
| ·18 | ·0000 099 | ·0000 028 | ·0000 008 | ·0000 002 | ·0000 001 | |
| ·19 | ·0000 182 | ·0000 054 | ·0000 016 | ·0000 004 | ·0000 001 | |
| ·20 | ·0000 325− | ·0000 101 | ·0000 031 | ·0000 009 | ·0000 003 | ·0000 001 |
| ·21 | ·0000 559 | ·0000 183 | ·0000 058 | ·0000 018 | ·0000 006 | ·0000 002 |
| ·22 | ·0000 934 | ·0000 319 | ·0000 107 | ·0000 035− | ·0000 011 | ·0000 004 |
| ·23 | ·0001 518 | ·0000 542 | ·0000 189 | ·0000 065− | ·0000 022 | ·0000 007 |
| ·24 | ·0002 403 | ·0000 894 | ·0000 325+ | ·0000 116 | ·0000 041 | ·0000 014 |
| ·25 | ·0003 716 | ·0001 440 | ·0000 545+ | ·0000 202 | ·0000 074 | ·0000 026 |
| ·26 | ·0005 623 | ·0002 263 | ·0000 891 | ·0000 343 | ·0000 130 | ·0000 048 |
| ·27 | ·0008 336 | ·0003 481 | ·0001 422 | ·0000 569 | ·0000 223 | ·0000 086 |
| ·28 | ·0012 128 | ·0005 248 | ·0002 220 | ·0000 921 | ·0000 375− | ·0000 150+ |
| ·29 | ·0017 334 | ·0007 761 | ·0003 398 | ·0001 458 | ·0000 615− | ·0000 255− |
| ·30 | ·0024 367 | ·0011 275− | ·0005 103 | ·0002 264 | ·0000 986 | ·0000 423 |
| ·31 | ·0033 723 | ·0016 108 | ·0007 527 | ·0003 448 | ·0001 551 | ·0000 686 |
| ·32 | ·0045 990 | ·0022 652 | ·0010 916 | ·0005 157 | ·0002 393 | ·0001 093 |
| ·33 | ·0061 854 | ·0031 383 | ·0015 581 | ·0007 585+ | ·0003 627 | ·0001 706 |
| ·34 | ·0082 103 | ·0042 869 | ·0021 906 | ·0010 978 | ·0005 405− | ·0002 618 |
| ·35 | ·0107 627 | ·0057 779 | ·0030 363 | ·0015 649 | ·0007 925− | ·0003 949 |
| ·36 | ·0139 420 | ·0076 889 | ·0041 514 | ·0021 987 | ·0011 443 | ·0005 860 |
| ·37 | ·0178 570 | ·0101 083 | ·0056 029 | ·0030 469 | ·0016 283 | ·0008 564 |
| ·38 | ·0226 254 | ·0131 357 | ·0074 689 | ·0041 671 | ·0022 851 | ·0012 334 |
| ·39 | ·0283 723 | ·0168 814 | ·0098 392 | ·0056 279 | ·0031 643 | ·0017 514 |
| ·40 | ·0352 279 | ·0214 658 | ·0128 154 | ·0075 098 | ·0043 264 | ·0024 538 |
| ·41 | ·0433 260 | ·0270 180 | ·0165 109 | ·0099 055+ | ·0058 432 | ·0033 939 |
| ·42 | ·0528 006 | ·0336 741 | ·0210 506 | ·0129 211 | ·0077 995− | ·0046 361 |
| ·43 | ·0637 832 | ·0415 753 | ·0265 692 | ·0166 751 | ·0102 934 | ·0062 579 |
| ·44 | ·0763 989 | ·0508 648 | ·0332 098 | ·0212 986 | ·0134 372 | ·0083 504 |
| ·45 | ·0907 630 | ·0616 846 | ·0411 221 | ·0269 339 | ·0173 569 | ·0110 192 |
| ·46 | ·1069 769 | ·0741 720 | ·0504 588 | ·0337 330 | ·0221 923 | ·0143 853 |
| ·47 | ·1251 243 | ·0884 554 | ·0613 728 | ·0418 553 | ·0280 954 | ·0185 849 |
| ·48 | ·1452 670 | ·1046 499 | ·0740 130 | ·0514 642 | ·0352 291 | ·0237 690 |
| ·49 | ·1674 414 | ·1228 528 | ·0885 196 | ·0627 239 | ·0437 641 | ·0301 019 |
| ·50 | ·1916 552 | ·1431 394 | ·1050 198 | ·0757 948 | ·0538 761 | ·0377 593 |
| ·51 | ·2178 843 | ·1655 580 | ·1236 222 | ·0908 285− | ·0657 412 | ·0469 258 |
| ·52 | ·2460 713 | ·1901 267 | ·1444 120 | ·1079 622 | ·0795 313 | ·0577 901 |
| ·53 | ·2761 230 | ·2168 291 | ·1674 457 | ·1273 135+ | ·0954 084 | ·0705 412 |
| ·54 | ·3079 109 | ·2456 117 | ·1927 464 | ·1489 739 | ·1135 182 | ·0853 622 |
| ·55 | ·3412 711 | ·2763 821 | ·2202 995+ | ·1730 032 | ·1339 838 | ·1024 239 |
| ·56 | ·3760 057 | ·3090 078 | ·2500 496 | ·1994 241 | ·1568 988 | ·1218 779 |
| ·57 | ·4118 857 | ·3433 160 | ·2818 971 | ·2282 170 | ·1823 205+ | ·1438 485− |
| ·58 | ·4486 540 | ·3790 954 | ·3156 978 | ·2593 161 | ·2102 641 | ·1684 256 |
| ·59 | ·4860 303 | ·4160 987 | ·3512 620 | ·2926 064 | ·2406 963 | ·1956 566 |
| ·60 | ·5237 164 | ·4540 459 | ·3883 565− | ·3279 223 | ·2735 315− | ·2255 396 |
| ·61 | ·5614 021 | ·4926 299 | ·4267 069 | ·3650 470 | ·3086 281 | ·2580 171 |
| ·62 | ·5987 719 | ·5315 227 | ·4660 027 | ·4037 150+ | ·3457 875+ | ·2929 712 |
| ·63 | ·6355 125+ | ·5703 818 | ·5059 024 | ·4436 154 | ·3847 541 | ·3302 208 |
| ·64 | ·6713 195+ | ·6088 591 | ·5460 417 | ·4843 969 | ·4252 179 | ·3695 200 |
| ·65 | ·7059 048 | ·6466 083 | ·5860 411 | ·5256 759 | ·4668 197 | ·4105 602 |
| ·66 | ·7390 031 | ·6832 942 | ·6255 161 | ·5670 447 | ·5091 580 | ·4529 738 |
| ·67 | ·7703 784 | ·7186 009 | ·6640 866 | ·6080 822 | ·5517 979 | ·4963 412 |
| ·68 | ·7998 293 | ·7522 400 | ·7013 875+ | ·6483 653 | ·5942 827 | ·5402 000 |
| ·69 | ·8271 932 | ·7839 581 | ·7370 789 | ·6874 807 | ·6361 465+ | ·5840 575− |
| ·70 | ·8523 495+ | ·8135 426 | ·7708 550+ | ·7250 370 | ·6769 281 | ·6274 043 |

$x = \cdot 71$ to $\cdot 97$ $q = 9$ $p = 13$ to $18$

| | $p = 13$ | $p = 14$ | $p = 15$ | $p = 16$ | $p = 17$ | $p = 18$ |
|---|---|---|---|---|---|---|
| $B(p, q) =$ | $\cdot 3780\ 1895^{+} \times \frac{1}{10^8}$ | $\cdot 2233\ 7484 \times \frac{1}{10^8}$ | $\cdot 1359\ 6729 \times \frac{1}{10^8}$ | $\cdot 8497\ 9557 \times \frac{1}{10^7}$ | $\cdot 5438\ 6917 \times \frac{1}{10^7}$ | $\cdot 3556\ 0676 \times \frac{1}{10^7}$ |
| $x$ | | | | | | |
| ·71 | ·8752 216 | ·8408 276 | ·8024 536 | ·7606 772 | ·7161 852 | ·6697 304 |
| ·72 | ·8957 771 | ·8656 964 | ·8316 623 | ·7940 886 | ·7535 090 | ·7105 423 |
| ·73 | ·9140 268 | ·8880 842 | ·8583 243 | ·8250 130 | ·7885 372 | ·7493 793 |
| ·74 | ·9300 225− | ·9079 772 | ·8823 417 | ·8532 540 | ·8209 665+ | ·7858 303 |
| ·75 | ·9438 532 | ·9254 115+ | ·9036 767 | ·8786 817 | ·8505 623 | ·8195 483 |
| ·76 | ·9556 403 | ·9404 690 | ·9223 502 | ·9012 358 | ·8771 654 | ·8502 631 |
| ·77 | ·9655 321 | ·9532 726 | ·9384 386 | ·9209 246 | ·9006 959 | ·8777 899 |
| ·78 | ·9736 974 | ·9639 794 | ·9520 679 | ·9378 218 | ·9211 538 | ·9020 346 |
| ·79 | ·9803 188 | ·9727 737 | ·9634 069 | ·9520 606 | ·9386 152 | ·9229 949 |
| ·80 | ·9855 860 | ·9798 583 | ·9726 577 | ·9638 250+ | ·9532 258 | ·9407 560 |
| ·81 | ·9896 893 | ·9854 467 | ·9800 465+ | ·9733 395− | ·9651 904 | ·9554 834 |
| ·82 | ·9928 137 | ·9897 548 | ·9858 131 | ·9808 570 | ·9747 611 | ·9674 101 |
| ·83 | ·9951 337 | ·9929 928 | ·9902 005− | ·9866 467 | ·9822 223 | ·9768 219 |
| ·84 | ·9968 092 | ·9953 596 | ·9934 461 | ·9909 815+ | ·9878 761 | ·9840 401 |
| ·85 | ·9979 823 | ·9970 364 | ·9957 731 | ·9941 265+ | ·9920 271 | ·9894 029 |
| ·86 | ·9987 754 | ·9981 836 | ·9973 838 | ·9963 292 | ·9949 687 | ·9932 481 |
| ·87 | ·9992 909 | ·9989 379 | ·9984 553 | ·9978 115+ | ·9969 714 | ·9958 964 |
| ·88 | ·9996 112 | ·9994 119 | ·9991 363 | ·9987 645+ | ·9982 737 | ·9976 385+ |
| ·89 | ·9997 999 | ·9996 945− | ·9995 470 | ·9993 456 | ·9990 769 | ·9987 252 |
| ·90 | ·9999 046 | ·9998 528 | ·9997 797 | ·9996 787 | ·9995 425− | ·9993 621 |
| ·91 | ·9999 585− | ·9999 353 | ·9999 022 | ·9998 561 | ·9997 931 | ·9997 087 |
| ·92 | ·9999 838 | ·9999 746 | ·9999 612 | ·9999 424 | ·9999 164 | ·9998 812 |
| ·93 | ·9999 946 | ·9999 914 | ·9999 867 | ·9999 800 | ·9999 707 | ·9999 580 |
| ·94 | ·9999 985− | ·9999 976 | ·9999 962 | ·9999 942 | ·9999 915− | ·9999 877 |
| ·95 | ·9999 997 | ·9999 995− | ·9999 992 | ·9999 987 | ·9999 981 | ·9999 972 |
| ·96 | 1·0000 000 | ·9999 999 | ·9999 999 | ·9999 998 | ·9999 997 | ·9999 996 |
| ·97 | | 1·0000 000 | 1·0000 000 | 1·0000 000 | 1·0000 000 | 1·0000 000 |

$x$ = ·22 to ·80 $q = 9$ $p$ = 19 to 24

| $x$ | $p = 19$ | $p = 20$ | $p = 21$ | $p = 22$ | $p = 23$ | $p = 24$ |
|---|---|---|---|---|---|---|
| $B(p, q)$ = | ·2370 7118 × $\frac{1}{10^7}$ | ·1608 6973 × $\frac{1}{10^7}$ | ·1109 4464 × $\frac{1}{10^7}$ | ·7766 1247 × $\frac{1}{10^8}$ | ·5511 4433 × $\frac{1}{10^8}$ | ·3961 3499 × $\frac{1}{10^8}$ |
| ·22 | ·0000 001 | | | | | |
| ·23 | ·0000 002 | ·0000 001 | | | | |
| ·24 | ·0000 005$^{-}$ | ·0000 002 | ·0000 001 | | | |
| ·25 | ·0000 009 | ·0000 003 | ·0000 001 | | | |
| ·26 | ·0000 018 | ·0000 006 | ·0000 002 | ·0000 001 | | |
| ·27 | ·0000 033 | ·0000 012 | ·0000 005$^{-}$ | ·0000 002 | ·0000 001 | |
| ·28 | ·0000 059 | ·0000 023 | ·0000 009 | ·0000 003 | ·0000 001 | |
| ·29 | ·0000 104 | ·0000 042 | ·0000 017 | ·0000 007 | ·0000 003 | ·0000 001 |
| ·30 | ·0000 178 | ·0000 074 | ·0000 031 | ·0000 012 | ·0000 005$^{-}$ | ·0000 002 |
| ·31 | ·0000 299 | ·0000 129 | ·0000 055$^{-}$ | ·0000 023 | ·0000 009 | ·0000 004 |
| ·32 | ·0000 491 | ·0000 218 | ·0000 095$^{+}$ | ·0000 041 | ·0000 018 | ·0000 007 |
| ·33 | ·0000 791 | ·0000 362 | ·0000 163 | ·0000 073 | ·0000 032 | ·0000 014 |
| ·34 | ·0001 249 | ·0000 588 | ·0000 273 | ·0000 126 | ·0000 057 | ·0000 026 |
| ·35 | ·0001 939 | ·0000 939 | ·0000 449 | ·0000 212 | ·0000 099 | ·0000 046 |
| ·36 | ·0002 957 | ·0001 472 | ·0000 724 | ·0000 352 | ·0000 169 | ·0000 081 |
| ·37 | ·0004 439 | ·0002 270 | ·0001 146 | ·0000 572 | ·0000 283 | ·0000 138 |
| ·38 | ·0006 561 | ·0003 443 | ·0001 785$^{+}$ | ·0000 915$^{-}$ | ·0000 464 | ·0000 233 |
| ·39 | ·0009 554 | ·0005 143 | ·0002 735$^{-}$ | ·0001 437 | ·0000 748 | ·0000 385$^{-}$ |
| ·40 | ·0013 719 | ·0007 569 | ·0004 125$^{+}$ | ·0002 223 | ·0001 185$^{+}$ | ·0000 626 |
| ·41 | ·0019 433 | ·0010 982 | ·0006 130 | ·0003 384 | ·0001 848 | ·0001 000 |
| ·42 | ·0027 170 | ·0015 716 | ·0008 981 | ·0005 075$^{+}$ | ·0002 838 | ·0001 572 |
| ·43 | ·0037 515$^{-}$ | ·0022 199 | ·0012 979 | ·0007 504 | ·0004 293 | ·0002 433 |
| ·44 | ·0051 174 | ·0030 960 | ·0018 508 | ·0010 942 | ·0006 402 | ·0003 710 |
| ·45 | ·0068 996 | ·0042 653 | ·0026 057 | ·0015 744 | ·0009 415$^{-}$ | ·0005 576 |
| ·46 | ·0091 980 | ·0058 071 | ·0036 234 | ·0022 362 | ·0013 661 | ·0008 266 |
| ·47 | ·0121 284 | ·0078 160 | ·0049 785$^{+}$ | ·0031 369 | ·0019 565$^{+}$ | ·0012 088 |
| ·48 | ·0158 233 | ·0104 036 | ·0067 615$^{-}$ | ·0043 473 | ·0027 671 | ·0017 448 |
| ·49 | ·0204 320 | ·0136 988 | ·0090 798 | ·0059 543 | ·0038 658 | ·0024 866 |
| ·50 | ·0261 195$^{-}$ | ·0178 491 | ·0120 598 | ·0080 624 | ·0053 369 | ·0035 002 |
| ·51 | ·0330 651 | ·0230 197 | ·0158 473 | ·0107 959 | ·0072 829 | ·0048 680 |
| ·52 | ·0414 600 | ·0293 929 | ·0206 081 | ·0142 997 | ·0098 265$^{+}$ | ·0066 914 |
| ·53 | ·0515 040 | ·0371 659 | ·0265 271 | ·0187 405$^{-}$ | ·0131 129 | ·0090 928 |
| ·54 | ·0634 005$^{+}$ | ·0465 478 | ·0338 072 | ·0243 063 | ·0173 102 | ·0122 182 |
| ·55 | ·0773 514 | ·0577 552 | ·0426 661 | ·0312 056 | ·0226 103 | ·0162 385$^{-}$ |
| ·56 | ·0935 499 | ·0710 067 | ·0533 329 | ·0396 652 | ·0292 280 | ·0213 502 |
| ·57 | ·1121 732 | ·0865 162 | ·0660 419 | ·0499 257 | ·0373 990 | ·0277 752 |
| ·58 | ·1333 742 | ·1044 845$^{-}$ | ·0810 260 | ·0622 370 | ·0473 765$^{-}$ | ·0357 595$^{-}$ |
| ·59 | ·1572 727 | ·1250 909 | ·0985 087 | ·0768 506 | ·0594 256 | ·0455 690 |
| ·60 | ·1839 466 | ·1484 831 | ·1186 938 | ·0940 112 | ·0738 164 | ·0574 849 |
| ·61 | ·2134 231 | ·1747 672 | ·1417 550$^{+}$ | ·1139 462 | ·0908 144 | ·0717 959 |
| ·62 | ·2456 713 | ·2039 976 | ·1678 249 | ·1368 541 | ·1106 697 | ·0887 886 |
| ·63 | ·2805 955$^{+}$ | ·2361 678 | ·1969 826 | ·1628 915$^{-}$ | ·1336 041 | ·1087 353 |
| ·64 | ·3180 304 | ·2712 020 | ·2292 437 | ·1921 606 | ·1597 971 | ·1318 801 |
| ·65 | ·3577 388 | ·3089 484 | ·2645 492 | ·2246 957 | ·1893 709 | ·1584 233 |
| ·66 | ·3994 116 | ·3491 759 | ·3027 581 | ·2604 516 | ·2223 757 | ·1885 047 |
| ·67 | ·4426 715$^{-}$ | ·3915 722 | ·3436 412 | ·2992 935$^{+}$ | ·2587 759 | ·2221 867 |
| ·68 | ·4870 789 | ·4357 470 | ·3868 791 | ·3409 898 | ·2984 379 | ·2594 382 |
| ·69 | ·5321 421 | ·4812 376 | ·4320 639 | ·3852 084 | ·3411 217 | ·3001 210 |
| ·70 | ·5773 301 | ·5275 195$^{+}$ | ·4787 051 | ·4315 179 | ·3864 755$^{+}$ | ·3439 791 |
| ·71 | ·6220 885$^{-}$ | ·5740 202 | ·5262 404 | ·4793 934 | ·4340 370 | ·3906 329 |
| ·72 | ·6658 570 | ·6201 369 | ·5740 510 | ·5282 285$^{-}$ | ·4832 391 | ·4395 798 |
| ·73 | ·7080 895$^{-}$ | ·6652 568 | ·6214 817 | ·5773 523 | ·5334 234 | ·4902 013 |
| ·74 | ·7482 735$^{-}$ | ·7087 795$^{+}$ | ·6678 638 | ·6260 518 | ·5838 597 | ·5417 777 |
| ·75 | ·7859 498 | ·7501 408 | ·7125 414 | ·6735 992 | ·6337 719 | ·5935 117 |
| ·76 | ·8207 304 | ·7888 351 | ·7548 984 | ·7192 811 | ·6823 687 | ·6445 574 |
| ·77 | ·8523 133 | ·8244 366 | ·7943 854 | ·7624 311 | ·7288 790 | ·6940 578 |
| ·78 | ·8804 937 | ·8566 173 | ·8305 442 | ·8024 598 | ·7725 882 | ·7411 841 |
| ·79 | ·9051 703 | ·8851 599 | ·8630 284 | ·8388 839 | ·8128 738 | ·7851 787 |
| ·80 | ·9263 466 | ·9099 653 | ·8916 183 | ·8713 492 | ·8492 375$^{+}$ | ·8253 953 |

$x$ = ·81 to ·98 $q = 9$ $p$ = 19 to 24

| $x$ | $p = 19$ | $p = 20$ | $p = 21$ | $p = 22$ | $p = 23$ | $p = 24$ |
|---|---|---|---|---|---|---|
| $B(p, q) =$ | ·2370 7118 × $\frac{1}{10^7}$ | ·1608 6973 × $\frac{1}{10^7}$ | ·1109 4464 × $\frac{1}{10^7}$ | ·7766 1247 × $\frac{1}{10^8}$ | ·5511 4433 × $\frac{1}{10^8}$ | ·3961 3499 × $\frac{1}{10^8}$ |
| ·81 | ·9441 263 | ·9310 535$^{+}$ | ·9162 291 | ·8996 469 | ·8813 311 | ·8613 350$^{+}$ |
| ·82 | ·9587 033 | ·9485 576 | ·9369 103 | ·9237 211 | ·9089 732 | ·8926 735$^{+}$ |
| ·83 | ·9703 475$^{-}$ | ·9627 110 | ·9538 375$^{-}$ | ·9436 667 | ·9321 552 | ·9192 773 |
| ·84 | ·9793 857 | ·9738 298 | ·9672 961 | ·9597 169 | ·9510 354 | ·9412 064 |
| ·85 | ·9861 810 | ·9822 892 | ·9776 580 | ·9722 218 | ·9659 208 | ·9587 020 |
| ·86 | ·9911 107 | ·9884 986 | ·9853 537 | ·9816 186 | ·9772 384 | ·9721 613 |
| ·87 | ·9945 456 | ·9928 756 | ·9908 415$^{+}$ | ·9883 977 | ·9854 984 | ·9820 987 |
| ·88 | ·9968 311 | ·9958 215$^{-}$ | ·9945 776 | ·9930 660 | ·9912 520 | ·9891 006 |
| ·89 | ·9982 730 | ·9977 011 | ·9969 885$^{+}$ | ·9961 127 | ·9950 498 | ·9937 748 |
| ·90 | ·9991 276 | ·9988 277 | ·9984 498 | ·9979 802 | ·9974 038 | ·9967 046 |
| ·91 | ·9995 979 | ·9994 546 | ·9992 720 | ·9990 425$^{+}$ | ·9987 577 | ·9984 085$^{-}$ |
| ·92 | ·9998 345$^{-}$ | ·9997 734 | ·9996 947 | ·9995 946 | ·9994 692 | ·9993 136 |
| ·93 | ·9999 409 | ·9999 184 | ·9998 890 | ·9998 513 | ·9998 034 | ·9997 434 |
| ·94 | ·9999 825$^{+}$ | ·9999 756 | ·9999 665$^{+}$ | ·9999 547 | ·9999 396 | ·9999 205$^{-}$ |
| ·95 | ·9999 960 | ·9999 944 | ·9999 922 | ·9999 893 | ·9999 856 | ·9999 809 |
| ·96 | ·9999 994 | ·9999 991 | ·9999 987 | ·9999 983 | ·9999 976 | ·9999 968 |
| ·97 | ·9999 999 | ·9999 999 | ·9999 999 | ·9999 998 | ·9999 998 | ·9999 997 |
| ·98 | 1·0000 000 | 1·0000 000 | 1·0000 000 | 1·0000 000 | 1·0000 000 | 1·0000 000 |

$x = {\cdot}30$ to ${\cdot}98$ $\qquad q = 9 \qquad p = 25$ to $30$

| | $p = 25$ | $p = 26$ | $p = 27$ | $p = 28$ | $p = 29$ | $p = 30$ |
|---|---|---|---|---|---|---|
| $B(p, q) =$ | $\cdot 2880\ 9817 \times \frac{1}{10^8}$ | $\cdot 2118\ 3689 \times \frac{1}{10^8}$ | $\cdot 1573\ 6455 \times \frac{1}{10^8}$ | $\cdot 1180\ 2341 \times \frac{1}{10^8}$ | $\cdot 8931\ 5014 \times \frac{1}{10^9}$ | $\cdot 6816\ 1458 \times \frac{1}{10^9}$ |
| $x$ | | | | | | |
| ·30 | ·0000 001 | | | | | |
| ·31 | ·0000 002 | ·0000 001 | | | | |
| ·32 | ·0000 003 | ·0000 001 | ·0000 001 | | | |
| ·33 | ·0000 006 | ·0000 003 | ·0000 001 | | | |
| ·34 | ·0000 011 | ·0000 005+ | ·0000 002 | ·0000 001 | | |
| ·35 | ·0000 021 | ·0000 010 | ·0000 004 | ·0000 002 | ·0000 001 | |
| ·36 | ·0000 038 | ·0000 018 | ·0000 008 | ·0000 004 | ·0000 002 | ·0000 001 |
| ·37 | ·0000 067 | ·0000 032 | ·0000 015+ | ·0000 007 | ·0000 003 | ·0000 002 |
| ·38 | ·0000 116 | ·0000 057 | ·0000 028 | ·0000 014 | ·0000 007 | ·0000 003 |
| ·39 | ·0000 196 | ·0000 099 | ·0000 050− | ·0000 025− | ·0000 012 | ·0000 006 |
| ·40 | ·0000 327 | ·0000 170 | ·0000 087 | ·0000 045− | ·0000 023 | ·0000 011 |
| ·41 | ·0000 536 | ·0000 285− | ·0000 150+ | ·0000 078 | ·0000 041 | ·0000 021 |
| ·42 | ·0000 862 | ·0000 469 | ·0000 253 | ·0000 136 | ·0000 072 | ·0000 038 |
| ·43 | ·0001 366 | ·0000 760 | ·0000 420 | ·0000 230 | ·0000 125+ | ·0000 068 |
| ·44 | ·0002 130 | ·0001 213 | ·0000 685+ | ·0000 384 | ·0000 214 | ·0000 118 |
| ·45 | ·0003 273 | ·0001 905+ | ·0001 100 | ·0000 631 | ·0000 359 | ·0000 203 |
| ·46 | ·0004 957 | ·0002 948 | ·0001 739 | ·0001 019 | ·0000 592 | ·0000 342 |
| ·47 | ·0007 402 | ·0004 495+ | ·0002 709 | ·0001 620 | ·0000 962 | ·0000 568 |
| ·48 | ·0010 905− | ·0006 759 | ·0004 157 | ·0002 538 | ·0001 539 | ·0000 927 |
| ·49 | ·0015 854 | ·0010 026 | ·0006 291 | ·0003 919 | ·0002 425− | ·0001 490 |
| ·50 | ·0022 757 | ·0014 675+ | ·0009 391 | ·0005 966 | ·0003 764 | ·0002 360 |
| ·51 | ·0032 260 | ·0021 205+ | ·0013 833 | ·0008 959 | ·0005 763 | ·0003 683 |
| ·52 | ·0045 177 | ·0030 257 | ·0020 111 | ·0013 272 | ·0008 700 | ·0005 666 |
| ·53 | ·0062 520 | ·0042 645+ | ·0028 871 | ·0019 407 | ·0012 958 | ·0008 597 |
| ·54 | ·0085 520 | ·0059 388 | ·0040 934 | ·0028 016 | ·0019 048 | ·0012 869 |
| ·55 | ·0115 658 | ·0081 735+ | ·0057 336 | ·0039 941 | ·0027 640 | ·0019 008 |
| ·56 | ·0154 681 | ·0111 200 | ·0079 359 | ·0056 245− | ·0039 602 | ·0027 712 |
| ·57 | ·0204 612 | ·0149 581 | ·0108 562 | ·0078 253 | ·0056 041 | ·0039 888 |
| ·58 | ·0267 757 | ·0198 976 | ·0146 809 | ·0107 587 | ·0078 339 | ·0056 695+ |
| ·59 | ·0346 684 | ·0261 791 | ·0196 292 | ·0146 198 | ·0108 198 | ·0079 593 |
| ·60 | ·0444 197 | ·0340 721 | ·0259 532 | ·0196 385− | ·0147 671 | ·0110 380 |
| ·61 | ·0563 276 | ·0438 725+ | ·0339 372 | ·0260 809 | ·0199 194 | ·0151 240 |
| ·62 | ·0707 003 | ·0558 968 | ·0438 946 | ·0342 484 | ·0265 590 | ·0204 764 |
| ·63 | ·0878 455+ | ·0704 735+ | ·0561 617 | ·0444 737 | ·0350 065− | ·0273 968 |
| ·64 | ·1080 576 | ·0879 324 | ·0710 891 | ·0571 154 | ·0456 170 | ·0362 280 |
| ·65 | ·1316 021 | ·1085 895+ | ·0890 288 | ·0725 471 | ·0587 731 | ·0473 502 |
| ·66 | ·1586 983 | ·1327 309 | ·1103 191 | ·0911 445− | ·0748 735− | ·0611 722 |
| ·67 | ·1895 004 | ·1605 926 | ·1352 649 | ·1132 673 | ·0943 180 | ·0781 196 |
| ·68 | ·2240 784 | ·1923 395+ | ·1641 163 | ·1392 381 | ·1174 874 | ·0986 168 |
| ·69 | ·2624 004 | ·2280 444 | ·1970 448 | ·1693 173 | ·1447 191 | ·1230 642 |
| ·70 | ·3043 157 | ·2676 667 | ·2341 188 | ·2036 772 | ·1762 797 | ·1518 110 |
| ·71 | ·3495 437 | ·3110 349 | ·2752 809 | ·2423 740 | ·2123 347 | ·1851 233 |
| ·72 | ·3976 669 | ·3578 329 | ·3203 276 | ·2853 227 | ·2529 182 | ·2231 507 |
| ·73 | ·4481 317 | ·4075 934 | ·3688 950+ | ·3322 748 | ·2979 042 | ·2658 921 |
| ·74 | ·5002 568 | ·4596 992 | ·4204 519 | ·3828 035+ | ·3469 838 | ·3131 651 |
| ·75 | ·5532 514 | ·5133 938 | ·4743 027 | ·4362 975− | ·3996 496 | ·3645 813 |
| ·76 | ·6062 421 | ·5678 041 | ·5296 027 | ·4919 672 | ·4551 919 | ·4195 326 |
| ·77 | ·6583 080 | ·6219 719 | ·5853 843 | ·5488 644 | ·5127 097 | ·4771 908 |
| ·78 | ·7085 238 | ·6748 968 | ·6405 972 | ·6059 165− | ·5711 367 | ·5365 248 |
| ·79 | ·7560 065− | ·7255 857 | ·6941 587 | ·6619 751 | ·6292 857 | ·5963 372 |
| ·80 | ·7999 636 | ·7731 078 | ·7450 124 | ·7158 765− | ·6859 081 | ·6553 196 |
| ·81 | ·8397 393 | ·8166 491 | ·7921 913 | ·7665 105+ | ·7397 659 | ·7121 267 |
| ·82 | ·8748 526 | ·8555 632 | ·8348 791 | ·8128 926 | ·7897 125+ | ·7654 614 |
| ·83 | ·9050 259 | ·8894 119 | ·8724 648 | ·8542 310 | ·8347 729 | ·8141 674 |
| ·84 | ·9301 979 | ·9179 916 | ·9045 836 | ·8899 837 | ·8742 158 | ·8573 169 |
| ·85 | ·9505 207 | ·9413 413 | ·9311 381 | ·9198 956 | ·9076 092 | ·8942 848 |
| ·86 | ·9663 394 | ·9597 305− | ·9522 979 | ·9440 120 | ·9348 502 | ·9247 974 |
| ·87 | ·9781 551 | ·9736 262 | ·9684 737 | ·9626 629 | ·9561 630 | ·9489 482 |
| ·88 | ·9865 761 | ·9836 438 | ·9802 693 | ·9764 199 | ·9720 646 | ·9671 746 |
| ·89 | ·9922 618 | ·9904 844 | ·9884 156 | ·9860 290 | ·9832 979 | ·9801 968 |
| ·90 | ·9958 656 | ·9948 689 | ·9936 958 | ·9923 272 | ·9907 435+ | ·9889 250+ |
| ·91 | ·9979 847 | ·9974 757 | ·9968 700 | ·9961 554 | ·9953 194 | ·9943 488 |
| ·92 | ·9991 228 | ·9988 910 | ·9986 122 | ·9982 797 | ·9978 864 | ·9974 247 |
| ·93 | ·9996 691 | ·9995 778 | ·9994 667 | ·9993 329 | ·9991 728 | ·9989 829 |
| ·94 | ·9998 964 | ·9998 667 | ·9998 300 | ·9997 854 | ·9997 315− | ·9996 668 |
| ·95 | ·9999 749 | ·9999 674 | ·9999 580 | ·9999 465+ | ·9999 325− | ·9999 154 |
| ·96 | ·9999 958 | ·9999 945− | ·9999 928 | ·9999 908 | ·9999 883 | ·9999 852 |
| ·97 | ·9999 996 | ·9999 995− | ·9999 993 | ·9999 991 | ·9999 989 | ·9999 985+ |
| ·98 | 1·0000 000 | 1·0000 000 | 1·0000 000 | 1·0000 000 | 1·0000 000 | 1·0000 000 |

$x$ = ·37 to ·99 $q = 9$ $p$ = 31 to 36

| | $p = 31$ | $p = 32$ | $p = 33$ | $p = 34$ | $p = 35$ | $p = 36$ |
|---|---|---|---|---|---|---|
| $B(p, q)$ = | ·5243 1891 × $\frac{1}{10^9}$ | ·4063 4716 × $\frac{1}{10^9}$ | ·3171 4900 × $\frac{1}{10^9}$ | ·2491 8850 × $\frac{1}{10^9}$ | ·1970 3277 × $\frac{1}{10^9}$ | ·1567 3061 × $\frac{1}{10^9}$ |
| $x$ | | | | | | |
| ·37 | ·0000 001 | | | | | |
| ·38 | ·0000 001 | ·0000 001 | | | | |
| ·39 | ·0000 003 | ·0000 001 | ·0000 001 | | | |
| ·40 | ·0000 006 | ·0000 003 | ·0000 001 | ·0000 001 | | |
| ·41 | ·0000 011 | ·0000 005+ | ·0000 003 | ·0000 001 | ·0000 001 | |
| ·42 | ·0000 020 | ·0000 010 | ·0000 005+ | ·0000 003 | ·0000 001 | ·0000 001 |
| ·43 | ·0000 036 | ·0000 019 | ·0000 010 | ·0000 005+ | ·0000 003 | ·0000 001 |
| ·44 | ·0000 065+ | ·0000 036 | ·0000 019 | ·0000 010 | ·0000 006 | ·0000 003 |
| ·45 | ·0000 114 | ·0000 064 | ·0000 035+ | ·0000 020 | ·0000 011 | ·0000 006 |
| ·46 | ·0000 196 | ·0000 112 | ·0000 064 | ·0000 036 | ·0000 020 | ·0000 011 |
| ·47 | ·0000 333 | ·0000 194 | ·0000 113 | ·0000 065− | ·0000 037 | ·0000 021 |
| ·48 | ·0000 555− | ·0000 330 | ·0000 195+ | ·0000 115+ | ·0000 067 | ·0000 039 |
| ·49 | ·0000 910 | ·0000 553 | ·0000 334 | ·0000 201 | ·0000 120 | ·0000 071 |
| ·50 | ·0001 470 | ·0000 911 | ·0000 561 | ·0000 344 | ·0000 210 | ·0000 127 |
| ·51 | ·0002 340 | ·0001 478 | ·0000 928 | ·0000 580 | ·0000 361 | ·0000 223 |
| ·52 | ·0003 668 | ·0002 361 | ·0001 512 | ·0000 963 | ·0000 610 | ·0000 385− |
| ·53 | ·0005 670 | ·0003 718 | ·0002 425− | ·0001 573 | ·0001 016 | ·0000 653 |
| ·54 | ·0008 642 | ·0005 771 | ·0003 833 | ·0002 533 | ·0001 665+ | ·0001 090 |
| ·55 | ·0012 994 | ·0008 833 | ·0005 973 | ·0004 018 | ·0002 690 | ·0001 792 |
| ·56 | ·0019 277 | ·0013 335+ | ·0009 176 | ·0006 282 | ·0004 280 | ·0002 902 |
| ·57 | ·0028 225− | ·0019 862 | ·0013 903 | ·0009 683 | ·0006 711 | ·0004 631 |
| ·58 | ·0040 794 | ·0029 192 | ·0020 780 | ·0014 718 | ·0010 375+ | ·0007 281 |
| ·59 | ·0058 215+ | ·0042 348 | ·0030 646 | ·0022 067 | ·0015 815+ | ·0011 284 |
| ·60 | ·0082 039 | ·0060 646 | ·0044 602 | ·0032 641 | ·0023 776 | ·0017 241 |
| ·61 | ·0114 187 | ·0085 753 | ·0064 071 | ·0047 639 | ·0035 257 | ·0025 978 |
| ·62 | ·0156 995− | ·0119 735+ | ·0090 859 | ·0068 616 | ·0051 580 | ·0038 604 |
| ·63 | ·0213 242 | ·0165 112 | ·0127 210 | ·0097 542 | ·0074 454 | ·0056 584 |
| ·64 | ·0286 167 | ·0224 883 | ·0175 856 | ·0136 872 | ·0106 052 | ·0081 819 |
| ·65 | ·0379 453 | ·0302 545+ | ·0240 058 | ·0189 594 | ·0149 075+ | ·0116 718 |
| ·66 | ·0497 179 | ·0402 071 | ·0323 608 | ·0259 267 | ·0206 811 | ·0164 277 |
| ·67 | ·0643 725+ | ·0527 851 | ·0430 806 | ·0350 024 | ·0283 165− | ·0228 130 |
| ·68 | ·0823 629 | ·0684 579 | ·0566 388 | ·0466 533 | ·0382 656 | ·0312 582 |
| ·69 | ·1041 378 | ·0877 085− | ·0735 382 | ·0613 904 | ·0510 361 | ·0422 587 |
| ·70 | ·1301 153 | ·1110 092 | ·0942 913 | ·0797 518 | ·0671 794 | ·0563 672 |
| ·71 | ·1606 511 | ·1387 920 | ·1193 920 | ·1022 788 | ·0872 696 | ·0741 772 |
| ·72 | ·1960 027 | ·1714 119 | ·1492 802 | ·1294 824 | ·1118 740 | ·0962 981 |
| ·73 | ·2362 915+ | ·2091 068 | ·1843 007 | ·1618 023 | ·1415 141 | ·1233 185− |
| ·74 | ·2814 656 | ·2519 545− | ·2246 566 | ·1995 592 | ·1766 172 | ·1557 596 |
| ·75 | ·3312 665− | ·2998 323 | ·2703 628 | ·2429 025+ | ·2174 614 | ·1940 192 |
| ·76 | ·3852 046 | ·3523 826 | ·3212 017 | ·2917 593 | ·2641 182 | ·2383 092 |
| ·77 | ·4425 481 | ·4089 893 | ·3766 890 | ·3457 884 | ·3163 965− | ·2885 917 |
| ·78 | ·5023 283 | ·4687 716 | ·4360 538 | ·4043 472 | ·3737 970 | ·3445 213 |
| ·79 | ·5633 666 | ·5305 981 | ·4982 393 | ·4664 785+ | ·4354 838 | ·4054 012 |
| ·80 | ·6243 234 | ·5931 271 | ·5619 309 | ·5309 237 | ·5002 813 | ·4701 642 |
| ·81 | ·6837 689 | ·6548 713 | ·6256 126 | ·5961 677 | ·5667 054 | ·5373 862 |
| ·82 | ·7402 726 | ·7142 875− | ·6876 527 | ·6605 176 | ·6330 312 | ·6053 407 |
| ·83 | ·7925 042 | ·7698 836 | ·7464 147 | ·7222 134 | ·6973 998 | ·6720 971 |
| ·84 | ·8393 366 | ·8203 354 | ·8003 842 | ·7795 623 | ·7579 566 | ·7356 595+ |
| ·85 | ·8799 389 | ·8645 980 | ·8482 983 | ·8310 848 | ·8130 106 | ·7941 361 |
| ·86 | ·9138 466 | ·9019 986 | ·8892 619 | ·8756 530 | ·8611 955− | ·8459 201 |
| ·87 | ·9409 975− | ·9322 953 | ·9228 316 | ·9126 022 | ·9016 087 | ·8898 582 |
| ·88 | ·9617 239 | ·9556 895− | ·9490 516 | ·9417 942 | ·9339 050− | ·9253 756 |
| ·89 | ·9767 008 | ·9727 863 | ·9684 316 | ·9636 162 | ·9583 222 | ·9525 335+ |
| ·90 | ·9868 520 | ·9845 047 | ·9818 640 | ·9789 113 | ·9756 285+ | ·9719 987 |
| ·91 | ·9932 299 | ·9919 491 | ·9904 921 | ·9888 448 | ·9869 930 | ·9849 227 |
| ·92 | ·9968 867 | ·9962 640 | ·9955 479 | ·9947 294 | ·9937 991 | ·9927 477 |
| ·93 | ·9987 592 | ·9984 974 | ·9981 931 | ·9978 415+ | ·9974 376 | ·9969 761 |
| ·94 | ·9995 898 | ·9994 987 | ·9993 917 | ·9992 668 | ·9991 216 | ·9989 541 |
| ·95 | ·9998 949 | ·9998 704 | ·9998 413 | ·9998 070 | ·9997 667 | ·9997 196 |
| ·96 | ·9999 814 | ·9999 768 | ·9999 714 | ·9999 649 | ·9999 571 | ·9999 480 |
| ·97 | ·9999 982 | ·9999 977 | ·9999 971 | ·9999 964 | ·9999 956 | ·9999 946 |
| ·98 | ·9999 999 | ·9999 999 | ·9999 999 | ·9999 999 | ·9999 998 | ·9999 998 |
| ·99 | 1·0000 000 | 1·0000 000 | 1·0000 000 | 1·0000 000 | 1·0000 000 | 1·0000 000 |

$x$ = ·43 to ·99 $q = 9$ $p$ = 37 to 43

| $x$ | $p = 37$ | $p = 38$ | $p = 39$ | $p = 40$ | $p = 41$ | $p = 42$ | $p = 43$ |
|---|---|---|---|---|---|---|---|
| $B(p, q) =$ | ·1253 8449 × $\frac{1}{10^9}$ | ·1008 5274 × $\frac{1}{10^9}$ | ·8154 0514 × $\frac{1}{10^{10}}$ | ·6625 1667 × $\frac{1}{10^{10}}$ | ·5408 2994 × $\frac{1}{10^{10}}$ | ·4434 8055 × $\frac{1}{10^{10}}$ | ·3652 1928 × $\frac{1}{10^{10}}$ |
| ·43 | ·0000 001 | | | | | | |
| ·44 | ·0000 002 | ·0000 001 | | | | | |
| ·45 | ·0000 003 | ·0000 002 | ·0000 001 | ·0000 001 | | | |
| ·46 | ·0000 006 | ·0000 003 | ·0000 002 | ·0000 001 | ·0000 001 | | |
| ·47 | ·0000 012 | ·0000 007 | ·0000 004 | ·0000 002 | ·0000 001 | ·0000 001 | |
| ·48 | ·0000 023 | ·0000 013 | ·0000 008 | ·0000 004 | ·0000 003 | ·0000 001 | ·0000 001 |
| ·49 | ·0000 042 | ·0000 025− | ·0000 015− | ·0000 009 | ·0000 005− | ·0000 003 | ·0000 002 |
| ·50 | ·0000 077 | ·0000 046 | ·0000 028 | ·0000 017 | ·0000 010 | ·0000 006 | ·0000 003 |
| ·51 | ·0000 137 | ·0000 084 | ·0000 051 | ·0000 031 | ·0000 019 | ·0000 011 | ·0000 007 |
| ·52 | ·0000 242 | ·0000 151 | ·0000 094 | ·0000 058 | ·0000 036 | ·0000 022 | ·0000 014 |
| ·53 | ·0000 418 | ·0000 266 | ·0000 169 | ·0000 107 | ·0000 067 | ·0000 042 | ·0000 026 |
| ·54 | ·0000 710 | ·0000 461 | ·0000 298 | ·0000 192 | ·0000 123 | ·0000 079 | ·0000 050+ |
| ·55 | ·0001 189 | ·0000 785+ | ·0000 517 | ·0000 339 | ·0000 221 | ·0000 144 | ·0000 093 |
| ·56 | ·0001 960 | ·0001 318 | ·0000 882 | ·0000 589 | ·0000 391 | ·0000 259 | ·0000 171 |
| ·57 | ·0003 181 | ·0002 176 | ·0001 483 | ·0001 006 | ·0000 681 | ·0000 459 | ·0000 308 |
| ·58 | ·0005 087 | ·0003 539 | ·0002 453 | ·0001 693 | ·0001 165− | ·0000 798 | ·0000 545+ |
| ·59 | ·0008 016 | ·0005 671 | ·0003 996 | ·0002 805+ | ·0001 962 | ·0001 368 | ·0000 950+ |
| ·60 | ·0012 449 | ·0008 952 | ·0006 412 | ·0004 576 | ·0003 253 | ·0002 305+ | ·0001 628 |
| ·61 | ·0019 060 | ·0013 927 | ·0010 137 | ·0007 351 | ·0005 311 | ·0003 825− | ·0002 745− |
| ·62 | ·0028 770 | ·0021 356 | ·0015 791 | ·0011 633 | ·0008 539 | ·0006 247 | ·0004 555− |
| ·63 | ·0042 824 | ·0032 281 | ·0024 241 | ·0018 136 | ·0013 521 | ·0010 046 | ·0007 440 |
| ·64 | ·0062 863 | ·0048 108 | ·0036 677 | ·0027 860 | ·0021 089 | ·0015 910 | ·0011 964 |
| ·65 | ·0091 012 | ·0070 690 | ·0054 700 | ·0042 175− | ·0032 405− | ·0024 815+ | ·0018 942 |
| ·66 | ·0129 965+ | ·0102 424 | ·0080 419 | ·0062 917 | ·0049 056 | ·0038 122 | ·0029 531 |
| ·67 | ·0183 063 | ·0146 339 | ·0116 554 | ·0092 504 | ·0073 169 | ·0057 686 | ·0045 337 |
| ·68 | ·0254 343 | ·0206 178 | ·0166 530 | ·0134 039 | ·0107 527 | ·0085 981 | ·0068 538 |
| ·69 | ·0348 564 | ·0286 445+ | ·0234 559 | ·0191 414 | ·0155 690 | ·0126 230 | ·0102 032 |
| ·70 | ·0471 167 | ·0392 413 | ·0325 679 | ·0269 383 | ·0222 094 | ·0182 533 | ·0149 566 |
| ·71 | ·0628 159 | ·0530 053 | ·0445 734 | ·0373 587 | ·0312 117 | ·0259 958 | ·0215 872 |
| ·72 | ·0825 913 | ·0705 886 | ·0601 273 | ·0510 501 | ·0432 074 | ·0364 589 | ·0306 744 |
| ·73 | ·1070 839 | ·0926 703 | ·0799 331 | ·0687 277 | ·0589 118 | ·0503 480 | ·0429 057 |
| ·74 | ·1368 951 | ·1199 170 | ·1047 082 | ·0911 451 | ·0791 011 | ·0684 495− | ·0590 659 |
| ·75 | ·1725 306 | ·1529 294 | ·1351 337 | ·1190 490 | ·1045 729 | ·0915 973 | ·0800 119 |
| ·76 | ·2143 356 | ·1921 761 | ·1717 895− | ·1531 174 | ·1360 884 | ·1206 211 | ·1066 269 |
| ·77 | ·2624 243 | ·2379 189 | ·2150 773 | ·1938 815− | ·1742 965− | ·1562 736 | ·1397 525+ |
| ·78 | ·3166 117 | ·2901 353 | ·2651 360 | ·2416 366 | ·2196 413 | ·1991 373 | ·1800 979 |
| ·79 | ·3763 548 | ·3484 467 | ·3217 577 | ·2963 484 | ·2722 604 | ·2495 179 | ·2281 290 |
| ·80 | ·4407 164 | ·4120 645− | ·3843 173 | ·3575 663 | ·3318 852 | ·3073 316 | ·2839 473 |
| ·81 | ·5083 603 | ·4797 657 | ·4517 281 | ·4243 590 | ·3977 563 | ·3720 035+ | ·3471 705+ |
| ·82 | ·5775 887 | ·5499 117 | ·5224 386 | ·4952 895+ | ·4685 748 | ·4423 945− | ·4168 374 |
| ·83 | ·6464 289 | ·6205 179 | ·5944 842 | ·5684 438 | ·5425 075+ | ·5167 800 | ·4913 588 |
| ·84 | ·7127 678 | ·6893 812 | ·6656 007 | ·6415 274 | ·6172 617 | ·5929 012 | ·5685 407 |
| ·85 | ·7745 275− | ·7542 564 | ·7333 986 | ·7120 327 | ·6902 395− | ·6681 007 | ·6456 984 |
| ·86 | ·8298 640 | ·8130 702 | ·7955 870 | ·7774 671 | ·7587 675− | ·7395 479 | ·7198 707 |
| ·87 | ·8773 634 | ·8641 427 | ·8502 191 | ·8356 208 | ·8203 801 | ·8045 336 | ·7881 211 |
| ·88 | ·9162 017 | ·9063 832 | ·8959 240 | ·8848 318 | ·8731 184 | ·8607 994 | ·8478 938 |
| ·89 | ·9462 367 | ·9394 209 | ·9320 777 | ·9242 017 | ·9157 902 | ·9068 431 | ·8973 635− |
| ·90 | ·9680 060 | ·9636 355+ | ·9588 740 | ·9537 096 | ·9481 321 | ·9421 328 | ·9357 050+ |
| ·91 | ·9826 202 | ·9800 718 | ·9772 645− | ·9741 858 | ·9708 240 | ·9671 677 | ·9632 068 |
| ·92 | ·9915 655− | ·9902 426 | ·9887 694 | ·9871 359 | ·9853 326 | ·9833 499 | ·9811 783 |
| ·93 | ·9964 515+ | ·9958 582 | ·9951 902 | ·9944 415+ | ·9936 061 | ·9926 774 | ·9916 493 |
| ·94 | ·9987 615+ | ·9985 414 | ·9982 910 | ·9980 072 | ·9976 872 | ·9973 276 | ·9969 253 |
| ·95 | ·9996 650+ | ·9996 019 | ·9995 293 | ·9994 463 | ·9993 516 | ·9992 440 | ·9991 224 |
| ·96 | ·9999 373 | ·9999 249 | ·9999 104 | ·9998 936 | ·9998 743 | ·9998 521 | ·9998 268 |
| ·97 | ·9999 935− | ·9999 921 | ·9999 905− | ·9999 886 | ·9999 864 | ·9999 838 | ·9999 809 |
| ·98 | ·9999 998 | ·9999 997 | ·9999 996 | ·9999 996 | ·9999 995− | ·9999 994 | ·9999 993 |
| ·99 | 1·0000 000 | 1·0000 000 | 1·0000 000 | 1·0000 000 | 1·0000 000 | 1·0000 000 | 1·0000 000 |

| $x$ | $p = 44$ | $p = 45$ | $p = 46$ | $p = 47$ | $p = 48$ | $p = 49$ | $p = 50$ |
|---|---|---|---|---|---|---|---|
| $B(p, q) =$ | ·3020 0825 $\times\frac{1}{10^{10}}$ | ·2507 2383 $\times\frac{1}{10^{10}}$ | ·2089 3652 $\times\frac{1}{10^{10}}$ | ·1747 4691 $\times\frac{1}{10^{10}}$ | ·1466 6259 $\times\frac{1}{10^{10}}$ | ·1235 0533 $\times\frac{1}{10^{10}}$ | ·1043 4071 $\times\frac{1}{10^{10}}$ |
| ·49 | ·0000 001 | ·0000 001 | | | | | |
| ·50 | ·0000 002 | ·0000 001 | ·0000 001 | | | | |
| ·51 | ·0000 004 | ·0000 002 | ·0000 001 | ·0000 001 | ·0000 001 | | |
| ·52 | ·0000 008 | ·0000 005+ | ·0000 003 | ·0000 002 | ·0000 001 | ·0000 001 | |
| ·53 | ·0000 016 | ·0000 010 | ·0000 006 | ·0000 004 | ·0000 002 | ·0000 001 | ·0000 001 |
| ·54 | ·0000 032 | ·0000 020 | ·0000 013 | ·0000 008 | ·0000 005− | ·0000 003 | ·0000 002 |
| ·55 | ·0000 060 | ·0000 039 | ·0000 025− | ·0000 016 | ·0000 010 | ·0000 007 | ·0000 004 |
| ·56 | ·0000 112 | ·0000 074 | ·0000 048 | ·0000 031 | ·0000 020 | ·0000 013 | ·0000 009 |
| ·57 | ·0000 206 | ·0000 138 | ·0000 092 | ·0000 061 | ·0000 040 | ·0000 026 | ·0000 017 |
| ·58 | ·0000 371 | ·0000 252 | ·0000 171 | ·0000 115+ | ·0000 077 | ·0000 052 | ·0000 035− |
| ·59 | ·0000 658 | ·0000 454 | ·0000 313 | ·0000 214 | ·0000 147 | ·0000 100 | ·0000 068 |
| ·60 | ·0001 146 | ·0000 804 | ·0000 563 | ·0000 393 | ·0000 273 | ·0000 190 | ·0000 131 |
| ·61 | ·0001 964 | ·0001 401 | ·0000 996 | ·0000 706 | ·0000 499 | ·0000 352 | ·0000 248 |
| ·62 | ·0003 311 | ·0002 399 | ·0001 733 | ·0001 249 | ·0000 897 | ·0000 643 | ·0000 460 |
| ·63 | ·0005 493 | ·0004 043 | ·0002 967 | ·0002 171 | ·0001 585− | ·0001 154 | ·0000 838 |
| ·64 | ·0008 969 | ·0006 703 | ·0004 995+ | ·0003 712 | ·0002 751 | ·0002 034 | ·0001 500− |
| ·65 | ·0014 415− | ·0010 937 | ·0008 274 | ·0006 242 | ·0004 697 | ·0003 525+ | ·0002 639 |
| ·66 | ·0022 807 | ·0017 561 | ·0013 484 | ·0010 325− | ·0007 885+ | ·0006 006 | ·0004 564 |
| ·67 | ·0035 523 | ·0027 753 | ·0021 621 | ·0016 798 | ·0013 017 | ·0010 061 | ·0007 758 |
| ·68 | ·0054 471 | ·0043 166 | ·0034 112 | ·0026 885− | ·0021 133 | ·0016 571 | ·0012 962 |
| ·69 | ·0082 228 | ·0066 079 | ·0052 955+ | ·0042 325+ | ·0033 742 | ·0026 832 | ·0021 286 |
| ·70 | ·0122 195+ | ·0099 552 | ·0080 885− | ·0065 544 | ·0052 979 | ·0042 716 | ·0034 360 |
| ·71 | ·0178 746 | ·0147 595− | ·0121 545+ | ·0099 834 | ·0081 795− | ·0066 852 | ·0054 511 |
| ·72 | ·0257 347 | ·0215 315+ | ·0179 672 | ·0149 546 | ·0124 163 | ·0102 841 | ·0084 983 |
| ·73 | ·0364 620 | ·0309 028 | ·0261 232 | ·0220 272 | ·0185 283 | ·0155 483 | ·0130 178 |
| ·74 | ·0508 302 | ·0436 276 | ·0373 502 | ·0318 971 | ·0271 749 | ·0230 981 | ·0195 887 |
| ·75 | ·0697 063 | ·0605 718 | ·0525 030 | ·0453 990 | ·0391 641 | ·0337 085− | ·0289 488 |
| ·76 | ·0940 126 | ·0826 827 | ·0725 411 | ·0634 931 | ·0554 461 | ·0483 111 | ·0420 032 |
| ·77 | ·1246 646 | ·1109 346 | ·0984 830 | ·0872 278 | ·0770 862 | ·0679 757 | ·0598 152 |
| ·78 | ·1624 842 | ·1462 476 | ·1313 316 | ·1176 737 | ·1052 072 | ·0938 627 | ·0835 694 |
| ·79 | ·2080 881 | ·1893 773 | ·1719 678 | ·1558 225− | ·1408 966 | ·1271 399 | ·1144 978 |
| ·80 | ·2617 593 | ·2407 816 | ·2210 160 | ·2024 534 | ·1850 757 | ·1688 566 | ·1537 628 |
| ·81 | ·3233 135− | ·3004 758 | ·2786 887 | ·2579 720 | ·2383 352 | ·2197 784 | ·2022 933 |
| ·82 | ·3919 817 | ·3678 943 | ·3446 312 | ·3222 379 | ·3007 499 | ·2801 930 | ·2605 843 |
| ·83 | ·4663 337 | ·4417 863 | ·4177 899 | ·3944 091 | ·3716 998 | ·3497 097 | ·3284 779 |
| ·84 | ·5442 708 | ·5201 775− | ·4963 412 | ·4728 365− | ·4497 318 | ·4270 893 | ·4049 643 |
| ·85 | ·6231 138 | ·6004 265+ | ·5777 140 | ·5550 509 | ·5325 083 | ·5101 536 | ·4880 498 |
| ·86 | ·6997 999 | ·6794 007 | ·6587 386 | ·6378 789 | ·6168 860 | ·5958 232 | ·5747 517 |
| ·87 | ·7711 856 | ·7537 730 | ·7359 308 | ·7177 084 | ·6991 566 | ·6803 264 | ·6612 695+ |
| ·88 | ·8344 239 | ·8204 152 | ·8058 960 | ·7908 970 | ·7754 512 | ·7595 936 | ·7433 605− |
| ·89 | ·8873 570 | ·8768 320 | ·8657 994 | ·8542 728 | ·8422 679 | ·8298 029 | ·8168 977 |
| ·90 | ·9288 437 | ·9215 458 | ·9138 100 | ·9056 370 | ·8970 292 | ·8879 911 | ·8785 287 |
| ·91 | ·9589 318 | ·9543 341 | ·9494 065+ | ·9441 425+ | ·9385 370 | ·9325 857 | ·9262 859 |
| ·92 | ·9788 088 | ·9762 324 | ·9734 408 | ·9704 259 | ·9671 800 | ·9636 961 | ·9599 677 |
| ·93 | ·9905 153 | ·9892 689 | ·9879 036 | ·9864 132 | ·9847 911 | ·9830 311 | ·9811 271 |
| ·94 | ·9964 767 | ·9959 784 | ·9954 267 | ·9948 179 | ·9941 482 | ·9934 138 | ·9926 107 |
| ·95 | ·9989 854 | ·9988 315+ | ·9986 594 | ·9984 674 | ·9982 540 | ·9980 174 | ·9977 560 |
| ·96 | ·9997 979 | ·9997 652 | ·9997 282 | ·9996 864 | ·9996 396 | ·9995 871 | ·9995 285+ |
| ·97 | ·9999 775+ | ·9999 736 | ·9999 692 | ·9999 642 | ·9999 584 | ·9999 520 | ·9999 446 |
| ·98 | ·9999 991 | ·9999 990 | ·9999 988 | ·9999 986 | ·9999 983 | ·9999 981 | ·9999 978 |
| ·99 | 1·0000 000 | 1·0000 000 | 1·0000 000 | 1·0000 000 | 1·0000 000 | 1·0000 000 | 1·0000 000 |

$x$ = ·06 to ·70 $q = 9{\cdot}5$ $p$ = 9·5 to 13

| $x$ | $p = 9{\cdot}5$ | $p = 10$ | $p = 10{\cdot}5$ | $p = 11$ | $p = 12$ | $p = 13$ |
|---|---|---|---|---|---|---|
| $B(p, q) =$ | $\cdot 2222\ 7212 \times \frac{1}{10^5}$ | $\cdot 1561\ 4033 \times \frac{1}{10^5}$ | $\cdot 1111\ 3606 \times \frac{1}{10^5}$ | $\cdot 8007\ 1963 \times \frac{1}{10^6}$ | $\cdot 4296\ 5443 \times \frac{1}{10^6}$ | $\cdot 2398\ 0713 \times \frac{1}{10^6}$ |
| ·06 | ·0000 001 | | | | | |
| ·07 | ·0000 003 | ·0000 001 | | | | |
| ·08 | ·0000 010 | ·0000 004 | ·0000 001 | ·0000 001 | | |
| ·09 | ·0000 027 | ·0000 011 | ·0000 004 | ·0000 002 | | |
| ·10 | ·0000 067 | ·0000 029 | ·0000 012 | ·0000 005+ | ·0000 001 | |
| ·11 | ·0000 152 | ·0000 068 | ·0000 030 | ·0000 013 | ·0000 002 | |
| ·12 | ·0000 320 | ·0000 149 | ·0000 069 | ·0000 031 | ·0000 006 | ·0000 001 |
| ·13 | ·0000 628 | ·0000 305− | ·0000 146 | ·0000 069 | ·0000 015+ | ·0000 003 |
| ·14 | ·0001 165+ | ·0000 586 | ·0000 292 | ·0000 144 | ·0000 034 | ·0000 008 |
| ·15 | ·0002 056 | ·0001 070 | ·0000 551 | ·0000 281 | ·0000 071 | ·0000 018 |
| ·16 | ·0003 476 | ·0001 866 | ·0000 992 | ·0000 522 | ·0000 141 | ·0000 037 |
| ·17 | ·0005 657 | ·0003 128 | ·0001 713 | ·0000 929 | ·0000 267 | ·0000 074 |
| ·18 | ·0008 899 | ·0005 061 | ·0002 850+ | ·0001 590 | ·0000 483 | ·0000 142 |
| ·19 | ·0013 585− | ·0007 932 | ·0004 586 | ·0002 628 | ·0000 841 | ·0000 261 |
| ·20 | ·0020 180 | ·0012 081 | ·0007 162 | ·0004 208 | ·0001 416 | ·0000 462 |
| ·21 | ·0029 247 | ·0017 928 | ·0010 884 | ·0006 548 | ·0002 312 | ·0000 792 |
| ·22 | ·0041 447 | ·0025 985+ | ·0016 136 | ·0009 930 | ·0003 669 | ·0001 316 |
| ·23 | ·0057 542 | ·0036 859 | ·0023 386 | ·0014 706 | ·0005 674 | ·0002 125− |
| ·24 | ·0078 394 | ·0051 254 | ·0033 195+ | ·0021 309 | ·0008 569 | ·0003 345+ |
| ·25 | ·0104 958 | ·0069 979 | ·0046 222 | ·0030 263 | ·0012 661 | ·0005 144 |
| ·26 | ·0138 273 | ·0093 935+ | ·0063 224 | ·0042 184 | ·0018 332 | ·0007 737 |
| ·27 | ·0179 450+ | ·0124 120 | ·0085 063 | ·0057 793 | ·0026 047 | ·0011 404 |
| ·28 | ·0229 655+ | ·0161 609 | ·0112 692 | ·0077 910 | ·0036 364 | ·0016 492 |
| ·29 | ·0290 086 | ·0207 546 | ·0147 155+ | ·0103 454 | ·0049 941 | ·0023 429 |
| ·30 | ·0361 950+ | ·0263 122 | ·0189 575− | ·0135 440 | ·0067 537 | ·0032 736 |
| ·31 | ·0446 435+ | ·0329 556 | ·0241 133 | ·0174 970 | ·0090 019 | ·0045 030 |
| ·32 | ·0544 683 | ·0408 066 | ·0303 053 | ·0223 216 | ·0118 356 | ·0061 032 |
| ·33 | ·0657 756 | ·0499 845− | ·0376 576 | ·0281 403 | ·0153 615− | ·0081 574 |
| ·34 | ·0786 609 | ·0606 027 | ·0462 934 | ·0350 791 | ·0196 953 | ·0107 599 |
| ·35 | ·0932 057 | ·0727 658 | ·0563 322 | ·0432 644 | ·0249 602 | ·0140 157 |
| ·36 | ·1094 748 | ·0865 665− | ·0678 862 | ·0528 207 | ·0312 847 | ·0180 401 |
| ·37 | ·1275 135+ | ·1020 820 | ·0810 574 | ·0638 667 | ·0388 009 | ·0229 573 |
| ·38 | ·1473 453 | ·1193 717 | ·0959 342 | ·0765 127 | ·0476 413 | ·0288 990 |
| ·39 | ·1689 699 | ·1384 739 | ·1125 881 | ·0908 567 | ·0579 358 | ·0360 023 |
| ·40 | ·1923 621 | ·1594 040 | ·1310 709 | ·1069 810 | ·0698 084 | ·0444 071 |
| ·41 | ·2174 706 | ·1821 521 | ·1514 116 | ·1249 493 | ·0833 732 | ·0542 527 |
| ·42 | ·2442 178 | ·2066 820 | ·1736 144 | ·1448 030 | ·0987 312 | ·0656 747 |
| ·43 | ·2725 002 | ·2329 302 | ·1976 569 | ·1665 589 | ·1159 658 | ·0788 010 |
| ·44 | ·3021 890 | ·2608 058 | ·2234 884 | ·1902 069 | ·1351 397 | ·0937 476 |
| ·45 | ·3331 321 | ·2901 909 | ·2510 297 | ·2157 081 | ·1562 912 | ·1106 145− |
| ·46 | ·3651 559 | ·3209 418 | ·2801 730 | ·2429 940 | ·1794 311 | ·1294 813 |
| ·47 | ·3980 682 | ·3528 911 | ·3107 827 | ·2719 658 | ·2045 404 | ·1504 033 |
| ·48 | ·4316 610 | ·3858 496 | ·3426 968 | ·3024 951 | ·2315 680 | ·1734 078 |
| ·49 | ·4657 143 | ·4196 096 | ·3757 291 | ·3344 248 | ·2604 301 | ·1984 904 |
| ·50 | ·5000 000[e] | ·4539 484 | ·4096 722 | ·3675 710 | ·2910 093 | ·2256 128 |
| ·51 | ·5342 857 | ·4886 321 | ·4443 006 | ·4017 262 | ·3231 553 | ·2547 005− |
| ·52 | ·5683 390 | ·5234 200 | ·4793 749 | ·4366 619 | ·3566 867 | ·2856 420 |
| ·53 | ·6019 318 | ·5580 688 | ·5146 464 | ·4721 332 | ·3913 927 | ·3182 890 |
| ·54 | ·6348 441 | ·5923 373 | ·5498 612 | ·5078 830 | ·4270 371 | ·3524 569 |
| ·55 | ·6668 679 | ·6259 907 | ·5847 655+ | ·5436 470 | ·4633 618 | ·3879 272 |
| ·56 | ·6978 110 | ·6588 053 | ·6191 104 | ·5791 590 | ·5000 919 | ·4244 511 |
| ·57 | ·7274 998 | ·6905 719 | ·6526 565+ | ·6141 560 | ·5369 412 | ·4617 533 |
| ·58 | ·7557 822 | ·7211 001 | ·6851 788 | ·6483 836 | ·5736 178 | ·4995 373 |
| ·59 | ·7825 294 | ·7502 211 | ·7164 703 | ·6816 010 | ·6098 305+ | ·5374 919 |
| ·60 | ·8076 379 | ·7777 905+ | ·7463 467 | ·7135 858 | ·6452 952 | ·5752 974 |
| ·61 | ·8310 301 | ·8036 903 | ·7746 483 | ·7441 382 | ·6797 407 | ·6126 332 |
| ·62 | ·8526 547 | ·8278 299 | ·8012 437 | ·7730 847 | ·7129 148 | ·6491 848 |
| ·63 | ·8724 865− | ·8501 473 | ·8260 303 | ·8002 810 | ·7445 895+ | ·6846 515− |
| ·64 | ·8905 252 | ·8706 086 | ·8489 365− | ·8256 142 | ·7745 660 | ·7187 533 |
| ·65 | ·9067 943 | ·8892 073 | ·8699 208 | ·8490 037 | ·8026 782 | ·7512 375+ |
| ·66 | ·9213 391 | ·9059 631 | ·8889 717 | ·8704 021 | ·8287 957 | ·7818 846 |
| ·67 | ·9342 244 | ·9209 200 | ·9061 064 | ·8897 945+ | ·8528 260 | ·8105 125− |
| ·68 | ·9455 317 | ·9341 434 | ·9213 687 | ·9071 971 | ·8747 146 | ·8369 807 |
| ·69 | ·9553 565− | ·9457 177 | ·9348 263 | ·9226 549 | ·8944 450− | ·8611 924 |
| ·70 | ·9638 050− | ·9557 431 | ·9465 675− | ·9362 394 | ·9120 371 | ·8830 951 |

$x$ = ·71 to ·96 $q = 9·5$ $p$ = 9·5 to 13

| | $p = 9·5$ | $p = 10$ | $p = 10·5$ | $p = 11$ | $p = 12$ | $p = 13$ |
|---|---|---|---|---|---|---|
| $B(p, q)$ = | ·2222 7212 × $\frac{1}{10^5}$ | ·1561 4033 × $\frac{1}{10^5}$ | ·1111 3606 × $\frac{1}{10^5}$ | ·8007 1963 × $\frac{1}{10^6}$ | ·4296 5443 × $\frac{1}{10^6}$ | ·2398 0713 × $\frac{1}{10^6}$ |
| $x$ | | | | | | |
| ·71 | ·9709 914 | ·9643 319 | ·9566 983 | ·9480 446 | ·9275 448 | ·9026 802 |
| ·72 | ·9770 345⁻ | ·9716 053 | ·9653 381 | ·9581 834 | ·9410 523 | ·9199 811 |
| ·73 | ·9820 550⁻ | ·9776 900 | ·9726 162 | ·9667 839 | ·9526 704 | ·9350 697 |
| ·74 | ·9861 727 | ·9827 148 | ·9786 679 | ·9739 841 | ·9625 310 | ·9480 524 |
| ·75 | ·9895 042 | ·9868 077 | ·9836 307 | ·9799 287 | ·9707 827 | ·9590 644 |
| ·76 | ·9921 606 | ·9900 930 | ·9876 407 | ·9847 642 | ·9775 849 | ·9682 638 |
| ·77 | ·9942 458 | ·9926 889 | ·9908 302 | ·9886 356 | ·9831 029 | ·9758 250⁺ |
| ·78 | ·9958 553 | ·9947 057 | ·9933 242 | ·9916 826 | ·9875 026 | ·9819 326 |
| ·79 | ·9970 753 | ·9962 442 | ·9952 391 | ·9940 370 | ·9909 459 | ·9867 743 |
| ·80 | ·9979 820 | ·9973 949 | ·9966 803 | ·9958 203 | ·9935 872 | ·9905 353 |
| ·81 | ·9986 415⁺ | ·9982 371 | ·9977 417 | ·9971 419 | ·9955 692 | ·9933 931 |
| ·82 | ·9991 101 | ·9988 390 | ·9985 051 | ·9980 982 | ·9970 213 | ·9955 128 |
| ·83 | ·9994 343 | ·9992 583 | ·9990 400 | ·9987 724 | ·9980 575⁻ | ·9970 438 |
| ·84 | ·9996 524 | ·9995 418 | ·9994 040 | ·9992 339 | ·9987 754 | ·9981 175⁻ |
| ·85 | ·9997 944 | ·9997 276 | ·9996 438 | ·9995 398 | ·9992 570 | ·9988 462 |
| ·86 | ·9998 835⁻ | ·9998 448 | ·9997 961 | ·9997 353 | ·9995 682 | ·9993 228 |
| ·87 | ·9999 372 | ·9999 159 | ·9998 889 | ·9998 551 | ·9997 612 | ·9996 218 |
| ·88 | ·9999 680 | ·9999 569 | ·9999 429 | ·9999 251 | ·9998 753 | ·9998 006 |
| ·89 | ·9999 848 | ·9999 794 | ·9999 725⁺ | ·9999 638 | ·9999 391 | ·9999 017 |
| ·90 | ·9999 933 | ·9999 909 | ·9999 878 | ·9999 838 | ·9999 726 | ·9999 552 |
| ·91 | ·9999 973 | ·9999 963 | ·9999 951 | ·9999 934 | ·9999 888 | ·9999 815⁻ |
| ·92 | ·9999 990 | ·9999 987 | ·9999 982 | ·9999 976 | ·9999 959 | ·9999 932 |
| ·93 | ·9999 997 | ·9999 996 | ·9999 995⁻ | ·9999 993 | ·9999 987 | ·9999 979 |
| ·94 | ·9999 999 | ·9999 999 | ·9999 999 | ·9999 998 | ·9999 997 | ·9999 994 |
| ·95 | 1·0000 000 | 1·0000 000 | 1·0000 000 | 1·0000 000 | ·9999 999 | ·9999 999 |
| ·96 | | | | | 1·0000 000 | 1·0000 000 |

$x$ = ·13 to ·70 $q$ = 9·5 $p$ = 14 to 19

| $x$ | $p$ = 14 | $p$ = 15 | $p$ = 16 | $p$ = 17 | $p$ = 18 | $p$ = 19 |
|---|---|---|---|---|---|---|
| $B(p, q)$ = | ·1385 5523 $\times\frac{1}{10^6}$ | ·8254 3540 $\times\frac{1}{10^7}$ | ·5053 6861 $\times\frac{1}{10^7}$ | ·3170 9403 $\times\frac{1}{10^7}$ | ·2034 1881 $\times\frac{1}{10^7}$ | ·1331 4686 $\times\frac{1}{10^7}$ |
| ·13 | ·0000 001 | | | | | |
| ·14 | ·0000 002 | | | | | |
| ·15 | ·0000 004 | ·0000 001 | | | | |
| ·16 | ·0000 009 | ·0000 002 | ·0000 001 | | | |
| ·17 | ·0000 020 | ·0000 005+ | ·0000 001 | | | |
| ·18 | ·0000 041 | ·0000 011 | ·0000 003 | ·0000 001 | | |
| ·19 | ·0000 079 | ·0000 023 | ·0000 007 | ·0000 002 | ·0000 001 | |
| ·20 | ·0000 147 | ·0000 046 | ·0000 014 | ·0000 004 | ·0000 001 | |
| ·21 | ·0000 264 | ·0000 086 | ·0000 027 | ·0000 009 | ·0000 003 | ·0000 001 |
| ·22 | ·0000 459 | ·0000 157 | ·0000 052 | ·0000 017 | ·0000 005+ | ·0000 002 |
| ·23 | ·0000 775+ | ·0000 276 | ·0000 096 | ·0000 033 | ·0000 011 | ·0000 004 |
| ·24 | ·0001 272 | ·0000 472 | ·0000 172 | ·0000 061 | ·0000 021 | ·0000 007 |
| ·25 | ·0002 036 | ·0000 787 | ·0000 298 | ·0000 111 | ·0000 040 | ·0000 014 |
| ·26 | ·0003 182 | ·0001 278 | ·0000 503 | ·0000 194 | ·0000 074 | ·0000 027 |
| ·27 | ·0004 865+ | ·0002 028 | ·0000 828 | ·0000 331 | ·0000 130 | ·0000 050+ |
| ·28 | ·0007 289 | ·0003 148 | ·0001 331 | ·0000 552 | ·0000 225+ | ·0000 090 |
| ·29 | ·0010 714 | ·0004 788 | ·0002 096 | ·0000 900 | ·0000 380 | ·0000 158 |
| ·30 | ·0015 470 | ·0007 145+ | ·0003 232 | ·0001 435+ | ·0000 626 | ·0000 269 |
| ·31 | ·0021 964 | ·0010 473 | ·0004 892 | ·0002 242 | ·0001 010 | ·0000 448 |
| ·32 | ·0030 694 | ·0015 092 | ·0007 270 | ·0003 438 | ·0001 598 | ·0000 731 |
| ·33 | ·0042 257 | ·0021 404 | ·0010 624 | ·0005 176 | ·0002 480 | ·0001 169 |
| ·34 | ·0057 355− | ·0029 900 | ·0015 276 | ·0007 662 | ·0003 779 | ·0001 835+ |
| ·35 | ·0076 806 | ·0041 171 | ·0021 631 | ·0011 159 | ·0005 661 | ·0002 828 |
| ·36 | ·0101 545− | ·0055 921 | ·0030 189 | ·0016 004 | ·0008 344 | ·0004 284 |
| ·37 | ·0132 622 | ·0074 970 | ·0041 552 | ·0022 618 | ·0012 110 | ·0006 386 |
| ·38 | ·0171 201 | ·0099 266 | ·0056 441 | ·0031 522 | ·0017 318 | ·0009 372 |
| ·39 | ·0218 552 | ·0129 880 | ·0075 701 | ·0043 346 | ·0024 419 | ·0013 551 |
| ·40 | ·0276 031 | ·0168 006 | ·0100 310 | ·0058 846 | ·0033 968 | ·0019 317 |
| ·41 | ·0345 067 | ·0214 955− | ·0131 380 | ·0078 910 | ·0046 641 | ·0027 163 |
| ·42 | ·0427 131 | ·0272 141 | ·0170 157 | ·0104 569 | ·0063 248 | ·0037 698 |
| ·43 | ·0523 711 | ·0341 061 | ·0218 017 | ·0136 999 | ·0084 743 | ·0051 662 |
| ·44 | ·0636 269 | ·0423 272 | ·0276 446 | ·0177 522 | ·0112 233 | ·0069 940 |
| ·45 | ·0766 204 | ·0520 354 | ·0347 030 | ·0227 599 | ·0146 983 | ·0093 575+ |
| ·46 | ·0914 810 | ·0633 879 | ·0431 421 | ·0288 815+ | ·0190 417 | ·0123 780 |
| ·47 | ·1083 222 | ·0765 359 | ·0531 306 | ·0362 861 | ·0244 107 | ·0161 936 |
| ·48 | ·1272 375+ | ·0916 205− | ·0648 364 | ·0451 502 | ·0309 761 | ·0209 597 |
| ·49 | ·1482 954 | ·1087 669 | ·0784 221 | ·0556 541 | ·0389 196 | ·0268 475− |
| ·50 | ·1715 349 | ·1280 794 | ·0940 393 | ·0679 774 | ·0484 309 | ·0340 426 |
| ·51 | ·1969 614 | ·1496 360 | ·1118 230 | ·0822 934 | ·0597 033 | ·0427 419 |
| ·52 | ·2245 436 | ·1734 828 | ·1318 853 | ·0987 633 | ·0729 281 | ·0531 498 |
| ·53 | ·2542 108 | ·1996 299 | ·1543 095− | ·1175 292 | ·0882 888 | ·0654 733 |
| ·54 | ·2858 509 | ·2280 465+ | ·1791 440 | ·1387 077 | ·1059 543 | ·0799 153 |
| ·55 | ·3193 107 | ·2586 587 | ·2063 968 | ·1623 825+ | ·1260 707 | ·0966 683 |
| ·56 | ·3543 961 | ·2913 465+ | ·2360 311 | ·1885 981 | ·1487 544 | ·1159 054 |
| ·57 | ·3908 742 | ·3259 439 | ·2679 611 | ·2173 531 | ·1740 832 | ·1377 725+ |
| ·58 | ·4284 771 | ·3622 388 | ·3020 502 | ·2485 953 | ·2020 895− | ·1623 787 |
| ·59 | ·4669 061 | ·3999 756 | ·3381 096 | ·2822 175− | ·2327 529 | ·1897 875+ |
| ·60 | ·5058 381 | ·4388 594 | ·3758 995− | ·3180 550+ | ·2659 950+ | ·2200 087 |
| ·61 | ·5449 320 | ·4785 607 | ·4151 318 | ·3558 853 | ·3016 747 | ·2529 906 |
| ·62 | ·5838 370 | ·5187 227 | ·4554 749 | ·3954 290 | ·3395 864 | ·2886 145− |
| ·63 | ·6222 007 | ·5589 692 | ·4965 598 | ·4363 542 | ·3794 599 | ·3266 904 |
| ·64 | ·6596 777 | ·5989 143 | ·5379 888 | ·4782 818 | ·4209 631 | ·3669 561 |
| ·65 | ·6959 389 | ·6381 715+ | ·5793 451 | ·5207 944 | ·4637 075+ | ·4090 785+ |
| ·66 | ·7306 792 | ·6763 649 | ·6202 040 | ·5634 463 | ·5072 563 | ·4526 582 |
| ·67 | ·7636 258 | ·7131 389 | ·6601 445− | ·6057 755+ | ·5511 347 | ·4972 376 |
| ·68 | ·7945 447 | ·7481 681 | ·6987 616 | ·6473 171 | ·5948 436 | ·5423 119 |
| ·69 | ·8232 462 | ·7811 665+ | ·7356 784 | ·6876 173 | ·6378 741 | ·5873 433 |
| ·70 | ·8495 892 | ·8118 951 | ·7705 572 | ·7262 482 | ·6797 237 | ·6317 777 |

$x = \cdot 71$ to $\cdot 97$ $q = 9\cdot 5$ $p = 14$ to $19$

| | $p = 14$ | $p = 15$ | $p = 16$ | $p = 17$ | $p = 18$ | $p = 19$ |
|---|---|---|---|---|---|---|
| $B(p, q) =$ | $\cdot 1385\ 5523 \times \frac{1}{10^6}$ | $\cdot 8254\ 3540 \times \frac{1}{10^7}$ | $\cdot 5053\ 6861 \times \frac{1}{10^7}$ | $\cdot 3170\ 9403 \times \frac{1}{10^7}$ | $\cdot 2034\ 1881 \times \frac{1}{10^7}$ | $\cdot 1331\ 4686 \times \frac{1}{10^7}$ |
| $x$ | | | | | | |
| ·71 | ·8734 835− | ·8401 679 | ·8031 100 | ·7628 210 | ·7199 132 | ·6750 627 |
| ·72 | ·8948 900 | ·8658 561 | ·8331 059 | ·7969 988 | ·7580 031 | ·7166 676 |
| ·73 | ·9138 203 | ·8888 901 | ·8603 783 | ·8285 075+ | ·7936 090 | ·7561 027 |
| ·74 | ·9303 328 | ·9092 592 | ·8848 278 | ·8571 439 | ·8264 149 | ·7929 373 |
| ·75 | ·9445 292 | ·9270 091 | ·9064 230 | ·8827 811 | ·8561 840 | ·8268 163 |
| ·76 | ·9565 478 | ·9422 376 | ·9251 989 | ·9053 701 | ·8827 653 | ·8574 730 |
| ·77 | ·9665 570 | ·9550 878 | ·9412 521 | ·9249 389 | ·9060 973 | ·8847 381 |
| ·78 | ·9747 474 | ·9657 402 | ·9547 334 | ·9415 872 | ·9262 061 | ·9085 435+ |
| ·79 | ·9813 239 | ·9744 038 | ·9658 390 | ·9554 783 | ·9432 009 | ·9289 216 |
| ·80 | ·9864 975− | ·9813 060 | ·9747 993 | ·9668 286 | ·9572 638 | ·9459 985− |
| ·81 | ·9904 780 | ·9866 831 | ·9818 674 | ·9758 944 | ·9686 373 | ·9599 831 |
| ·82 | ·9934 670 | ·9907 709 | ·9873 074 | ·9829 585− | ·9776 093 | ·9711 516 |
| ·83 | ·9956 523 | ·9937 962 | ·9913 826 | ·9883 151 | ·9844 960 | ·9798 294 |
| ·84 | ·9972 034 | ·9959 695− | ·9943 456 | ·9922 569 | ·9896 251 | ·9863 705− |
| ·85 | ·9982 687 | ·9974 799 | ·9964 295− | ·9950 622 | ·9933 190 | ·9911 376 |
| ·86 | ·9989 738 | ·9984 913 | ·9978 413 | ·9969 853 | ·9958 811 | ·9944 831 |
| ·87 | ·9994 212 | ·9991 406 | ·9987 583 | ·9982 489 | ·9975 842 | ·9967 327 |
| ·88 | ·9996 918 | ·9995 379 | ·9993 257 | ·9990 398 | ·9986 624 | ·9981 735− |
| ·89 | ·9998 465+ | ·9997 676 | ·9996 577 | ·9995 078 | ·9993 077 | ·9990 455+ |
| ·90 | ·9999 295− | ·9998 922 | ·9998 396 | ·9997 671 | ·9996 693 | ·9995 396 |
| ·91 | ·9999 706 | ·9999 546 | ·9999 317 | ·9999 000 | ·9998 566 | ·9997 984 |
| ·92 | ·9999 891 | ·9999 830 | ·9999 742 | ·9999 619 | ·9999 448 | ·9999 217 |
| ·93 | ·9999 965+ | ·9999 945+ | ·9999 916 | ·9999 875− | ·9999 817 | ·9999 738 |
| ·94 | ·9999 991 | ·9999 986 | ·9999 978 | ·9999 966 | ·9999 950+ | ·9999 928 |
| ·95 | ·9999 998 | ·9999 997 | ·9999 995+ | ·9999 993 | ·9999 990 | ·9999 985− |
| ·96 | 1·0000 000 | 1·0000 000 | ·9999 999 | ·9999 999 | ·9999 999 | ·9999 998 |
| ·97 | | | 1·0000 000 | 1·0000 000 | 1·0000 000 | 1·0000 000 |

$x$ = ·22 to ·80 $q$ = 9·5 $p$ = 20 to 25

| | $p = 20$ | $p = 21$ | $p = 22$ | $p = 23$ | $p = 24$ | $p = 25$ |
|---|---|---|---|---|---|---|
| $B(p, q)$ = | ·8876 4573 $\times \frac{1}{10^8}$ | ·6017 9371 $\times \frac{1}{10^8}$ | ·4143 4977 $\times \frac{1}{10^8}$ | ·2893 8714 $\times \frac{1}{10^8}$ | ·2047 9705+ $\times \frac{1}{10^8}$ | ·1467 2028 $\times \frac{1}{10^8}$ |
| $x$ | | | | | | |
| ·22 | ·0000 001 | | | | | |
| ·23 | ·0000 001 | | | | | |
| ·24 | ·0000 003 | ·0000 001 | | | | |
| ·25 | ·0000 005+ | ·0000 002 | ·0000 001 | | | |
| ·26 | ·0000 010 | ·0000 004 | ·0000 001 | | | |
| ·27 | ·0000 019 | ·0000 007 | ·0000 003 | ·0000 001 | | |
| ·28 | ·0000 036 | ·0000 014 | ·0000 005+ | ·0000 002 | ·0000 001 | |
| ·29 | ·0000 065− | ·0000 026 | ·0000 010 | ·0000 004 | ·0000 002 | ·0000 001 |
| ·30 | ·0000 114 | ·0000 048 | ·0000 020 | ·0000 008 | ·0000 003 | ·0000 001 |
| ·31 | ·0000 196 | ·0000 085− | ·0000 036 | ·0000 015+ | ·0000 006 | ·0000 003 |
| ·32 | ·0000 330 | ·0000 147 | ·0000 065− | ·0000 028 | ·0000 012 | ·0000 005+ |
| ·33 | ·0000 544 | ·0000 249 | ·0000 113 | ·0000 051 | ·0000 022 | ·0000 010 |
| ·34 | ·0000 879 | ·0000 415+ | ·0000 194 | ·0000 089 | ·0000 041 | ·0000 018 |
| ·35 | ·0001 393 | ·0000 677 | ·0000 325+ | ·0000 154 | ·0000 073 | ·0000 034 |
| ·36 | ·0002 169 | ·0001 084 | ·0000 535+ | ·0000 261 | ·0000 126 | ·0000 060 |
| ·37 | ·0003 321 | ·0001 704 | ·0000 864 | ·0000 433 | ·0000 215+ | ·0000 106 |
| ·38 | ·0005 002 | ·0002 635− | ·0001 372 | ·0000 706 | ·0000 360 | ·0000 182 |
| ·39 | ·0007 417 | ·0004 008 | ·0002 140 | ·0001 130 | ·0000 591 | ·0000 306 |
| ·40 | ·0010 835+ | ·0006 001 | ·0003 284 | ·0001 778 | ·0000 952 | ·0000 505+ |
| ·41 | ·0015 605− | ·0008 852 | ·0004 962 | ·0002 752 | ·0001 510 | ·0000 821 |
| ·42 | ·0022 167 | ·0012 871 | ·0007 387 | ·0004 193 | ·0002 356 | ·0001 311 |
| ·43 | ·0031 073 | ·0018 458 | ·0010 837 | ·0006 294 | ·0003 619 | ·0002 061 |
| ·44 | ·0043 005+ | ·0026 118 | ·0015 680 | ·0009 312 | ·0005 475+ | ·0003 189 |
| ·45 | ·0058 790 | ·0036 483 | ·0022 383 | ·0013 586 | ·0008 164 | ·0004 860 |
| ·46 | ·0079 413 | ·0050 331 | ·0031 538 | ·0019 554 | ·0012 004 | ·0007 300 |
| ·47 | ·0106 038 | ·0068 601 | ·0043 883 | ·0027 777 | ·0017 410 | ·0010 811 |
| ·48 | ·0140 009 | ·0092 412 | ·0060 317 | ·0038 960 | ·0024 919 | ·0015 793 |
| ·49 | ·0182 858 | ·0123 077 | ·0081 927 | ·0053 973 | ·0035 214 | ·0022 766 |
| ·50 | ·0236 299 | ·0162 109 | ·0110 000 | ·0073 878 | ·0049 143 | ·0032 395[e] |
| ·51 | ·0302 217 | ·0211 226 | ·0146 038 | ·0099 947 | ·0067 753 | ·0045 519 |
| ·52 | ·0382 641 | ·0272 337 | ·0191 763 | ·0133 677 | ·0092 309 | ·0063 179 |
| ·53 | ·0479 713 | ·0347 529 | ·0249 116 | ·0176 804 | ·0124 315− | ·0086 643 |
| ·54 | ·0595 638 | ·0439 034 | ·0320 238 | ·0231 303 | ·0165 530 | ·0117 433 |
| ·55 | ·0732 624 | ·0549 180 | ·0407 448 | ·0299 377 | ·0217 971 | ·0157 341 |
| ·56 | ·0892 805+ | ·0680 338 | ·0513 198 | ·0383 436 | ·0283 914 | ·0208 443 |
| ·57 | ·1078 162 | ·0834 842 | ·0640 013 | ·0486 053 | ·0365 864 | ·0273 093 |
| ·58 | ·1290 425+ | ·1014 902 | ·0790 416 | ·0609 909 | ·0466 523 | ·0353 906 |
| ·59 | ·1530 973 | ·1222 500− | ·0966 834 | ·0757 712 | ·0588 731 | ·0453 723 |
| ·60 | ·1800 732 | ·1459 283 | ·1171 491 | ·0932 100 | ·0735 383 | ·0575 551 |
| ·61 | ·2100 077 | ·1726 448 | ·1406 283 | ·1135 526 | ·0909 326 | ·0722 475− |
| ·62 | ·2428 739 | ·2024 621 | ·1672 653 | ·1370 121 | ·1113 232 | ·0897 552 |
| ·63 | ·2785 730 | ·2353 756 | ·1971 459 | ·1637 557 | ·1349 458 | ·1103 673 |
| ·64 | ·3169 287 | ·2713 036 | ·2302 845− | ·1938 893 | ·1619 882 | ·1343 406 |
| ·65 | ·3576 842 | ·3100 801 | ·2666 131 | ·2274 434 | ·1925 738 | ·1618 813 |
| ·66 | ·4005 028 | ·3514 505+ | ·3059 721 | ·2643 594 | ·2267 450+ | ·1931 272 |
| ·67 | ·4449 716 | ·3950 707 | ·3481 044 | ·3044 791 | ·2644 482 | ·2281 284 |
| ·68 | ·4906 097 | ·4405 102 | ·3926 532 | ·3475 372 | ·3055 204 | ·2668 299 |
| ·69 | ·5368 790 | ·4872 600 | ·4391 650− | ·3931 577 | ·3496 809 | ·3090 572 |
| ·70 | ·5832 007 | ·5347 452 | ·4870 973 | ·4408 572 | ·3965 270 | ·3545 056 |
| ·71 | ·6289 728 | ·5823 415− | ·5358 322 | ·4900 523 | ·4455 363 | ·4027 361 |
| ·72 | ·6735 918 | ·6293 959 | ·5846 949 | ·5400 753 | ·4960 764 | ·4531 775− |
| ·73 | ·7164 744 | ·6752 511 | ·6329 775− | ·5901 947 | ·5474 212 | ·5051 379 |
| ·74 | ·7570 811 | ·7192 706 | ·6799 658 | ·6396 426 | ·5987 759 | ·5578 241 |
| ·75 | ·7949 370 | ·7608 659 | ·7249 696 | ·6876 456 | ·6493 074 | ·6103 701 |
| ·76 | ·8296 515+ | ·7995 209 | ·7673 528 | ·7334 593 | ·6981 806 | ·6618 730 |
| ·77 | ·8609 339 | ·8348 147 | ·8065 625+ | ·7764 033 | ·7445 984 | ·7114 351 |
| ·78 | ·8886 034 | ·8664 399 | ·8421 551 | ·8158 943 | ·7878 410 | ·7582 097 |
| ·79 | ·9125 943 | ·8942 139 | ·8738 160 | ·8514 757 | ·8273 044 | ·8014 462 |
| ·80 | ·9329 545+ | ·9180 844 | ·9013 732 | ·8828 389 | ·8625 318 | ·8405 325− |

$x = \cdot 81$ to $\cdot 98$ $q = 9\cdot 5$ $p = 20$ to $25$

| $x$ | $p = 20$ | $p = 21$ | $p = 22$ | $p = 23$ | $p = 24$ | $p = 25$ |
|---|---|---|---|---|---|---|
| $B(p, q) =$ | $\cdot 8876\ 4573 \times \frac{1}{10^8}$ | $\cdot 6017\ 9371 \times \frac{1}{10^8}$ | $\cdot 4143\ 4977 \times \frac{1}{10^8}$ | $\cdot 2893\ 8714 \times \frac{1}{10^8}$ | $\cdot 2047\ 9705^{+} \times \frac{1}{10^8}$ | $\cdot 1467\ 2028 \times \frac{1}{10^8}$ |
| ·81 | ·9498 372 | ·9381 263 | ·9248 010 | ·9098 372 | ·8932 373 | ·8750 292 |
| ·82 | ·9634 874 | ·9545 318 | ·9442 157 | ·9324 883 | ·9193 178 | ·9046 932 |
| ·83 | ·9742 232 | ·9675 926 | ·9598 615+ | ·9509 656 | ·9408 532 | ·9294 873 |
| ·84 | ·9824 135+ | ·9776 770 | ·9720 880 | ·9655 793 | ·9580 914 | ·9495 740 |
| ·85 | ·9884 538 | ·9852 031 | ·9813 217 | ·9767 477 | ·9714 230 | ·9652 940 |
| ·86 | ·9927 429 | ·9906 102 | ·9880 338 | ·9849 620 | ·9813 440 | ·9771 305− |
| ·87 | ·9956 606 | ·9943 315− | ·9927 070 | ·9907 478 | ·9884 132 | ·9856 629 |
| ·88 | ·9975 507 | ·9967 698 | ·9958 044 | ·9946 267 | ·9932 073 | ·9915 158 |
| ·89 | ·9987 078 | ·9982 795− | ·9977 440 | ·9970 833 | ·9962 779 | ·9953 073 |
| ·90 | ·9993 708 | ·9991 542 | ·9988 804 | ·9985 388 | ·9981 177 | ·9976 045+ |
| ·91 | ·9997 219 | ·9996 226 | ·9994 957 | ·9993 356 | ·9991 360 | ·9988 901 |
| ·92 | ·9998 909 | ·9998 506 | ·9997 984 | ·9997 319 | ·9996 481 | ·9995 437 |
| ·93 | ·9999 632 | ·9999 491 | ·9999 307 | ·9999 070 | ·9998 768 | ·9998 387 |
| ·94 | ·9999 898 | ·9999 858 | ·9999 804 | ·9999 735+ | ·9999 646 | ·9999 532 |
| ·95 | ·9999 978 | ·9999 970 | ·9999 958 | ·9999 942 | ·9999 922 | ·9999 896 |
| ·96 | ·9999 997 | ·9999 996 | ·9999 994 | ·9999 991 | ·9999 988 | ·9999 984 |
| ·97 | 1·0000 000 | 1·0000 000 | 1·0000 000 | ·9999 999 | ·9999 999 | ·9999 999 |
| ·98 | | | | 1·0000 000 | 1·0000 000 | 1·0000 000 |

$x = \cdot 30$ to $\cdot 98$ $q = 9{\cdot}5$ $p = 26$ to $31$

| | $p = 26$ | $p = 27$ | $p = 28$ | $p = 29$ | $p = 30$ | $p = 31$ |
|---|---|---|---|---|---|---|
| $B(p, q) =$ | ·1063 1904 × $\frac{1}{10^8}$ | ·7786 7467 × $\frac{1}{10^9}$ | ·5760 0592 × $\frac{1}{10^9}$ | ·4300 8442 × $\frac{1}{10^9}$ | ·3239 5969 × $\frac{1}{10^9}$ | ·2460 4534 × $\frac{1}{10^9}$ |
| $x$ | | | | | | |
| ·30 | ·0000 001 | | | | | |
| ·31 | ·0000 001 | | | | | |
| ·32 | ·0000 002 | ·0000 001 | | | | |
| ·33 | ·0000 004 | ·0000 002 | ·0000 001 | | | |
| ·34 | ·0000 008 | ·0000 004 | ·0000 002 | ·0000 001 | | |
| ·35 | ·0000 016 | ·0000 007 | ·0000 003 | ·0000 001 | ·0000 001 | |
| ·36 | ·0000 029 | ·0000 013 | ·0000 006 | ·0000 003 | ·0000 001 | ·0000 001 |
| ·37 | ·0000 051 | ·0000 025− | ·0000 012 | ·0000 006 | ·0000 003 | ·0000 001 |
| ·38 | ·0000 091 | ·0000 045− | ·0000 022 | ·0000 011 | ·0000 005+ | ·0000 003 |
| ·39 | ·0000 157 | ·0000 080 | ·0000 040 | ·0000 020 | ·0000 010 | ·0000 005− |
| ·40 | ·0000 266 | ·0000 139 | ·0000 072 | ·0000 037 | ·0000 019 | ·0000 010 |
| ·41 | ·0000 442 | ·0000 236 | ·0000 125+ | ·0000 066 | ·0000 034 | ·0000 018 |
| ·42 | ·0000 723 | ·0000 396 | ·0000 215− | ·0000 116 | ·0000 062 | ·0000 033 |
| ·43 | ·0001 163 | ·0000 651 | ·0000 362 | ·0000 199 | ·0000 109 | ·0000 059 |
| ·44 | ·0001 841 | ·0001 054 | ·0000 599 | ·0000 338 | ·0000 189 | ·0000 105+ |
| ·45 | ·0002 868 | ·0001 679 | ·0000 975− | ·0000 562 | ·0000 322 | ·0000 183 |
| ·46 | ·0004 401 | ·0002 632 | ·0001 562 | ·0000 920 | ·0000 538 | ·0000 313 |
| ·47 | ·0006 656 | ·0004 064 | ·0002 463 | ·0001 481 | ·0000 885− | ·0000 525+ |
| ·48 | ·0009 923 | ·0006 185− | ·0003 825+ | ·0002 349 | ·0001 432 | ·0000 868 |
| ·49 | ·0014 593 | ·0009 279 | ·0005 855+ | ·0003 668 | ·0002 283 | ·0001 411 |
| ·50 | ·0021 174 | ·0013 729 | ·0008 835− | ·0005 645− | ·0003 582 | ·0002 259 |
| ·51 | ·0030 325− | ·0020 042 | ·0013 147 | ·0008 563 | ·0005 540 | ·0003 562 |
| ·52 | ·0042 882 | ·0028 876 | ·0019 301 | ·0012 810 | ·0008 445+ | ·0005 533 |
| ·53 | ·0059 889 | ·0041 073 | ·0027 962 | ·0018 903 | ·0012 694 | ·0008 472 |
| ·54 | ·0082 630 | ·0057 693 | ·0039 987 | ·0027 524 | ·0018 821 | ·0012 790 |
| ·55 | ·0112 657 | ·0080 046 | ·0056 463 | ·0039 555+ | ·0027 530 | ·0019 043 |
| ·56 | ·0151 810 | ·0109 727 | ·0078 742 | ·0056 122 | ·0039 743 | ·0027 971 |
| ·57 | ·0202 234 | ·0148 641 | ·0108 475+ | ·0078 631 | ·0056 634 | ·0040 543 |
| ·58 | ·0266 380 | ·0199 018 | ·0147 649 | ·0108 810 | ·0079 681 | ·0057 999 |
| ·59 | ·0346 986 | ·0263 423 | ·0198 600 | ·0148 743 | ·0110 707 | ·0081 907 |
| ·60 | ·0447 046 | ·0344 736 | ·0264 024 | ·0200 897 | ·0151 918 | ·0114 205− |
| ·61 | ·0569 743 | ·0446 118 | ·0346 967 | ·0268 124 | ·0205 933 | ·0157 247 |
| ·62 | ·0718 365+ | ·0570 949 | ·0450 778 | ·0353 654 | ·0275 788 | ·0213 832 |
| ·63 | ·0896 182 | ·0722 727 | ·0579 049 | ·0461 053 | ·0364 927 | ·0287 209 |
| ·64 | ·1106 299 | ·0904 941 | ·0735 502 | ·0594 142 | ·0477 154 | ·0381 068 |
| ·65 | ·1351 482 | ·1120 908 | ·0923 853 | ·0756 884 | ·0616 544 | ·0499 477 |
| ·66 | ·1633 957 | ·1373 577 | ·1147 625− | ·0953 225+ | ·0787 315+ | ·0646 790 |
| ·67 | ·1955 206 | ·1665 309 | ·1409 932 | ·1186 887 | ·0993 646 | ·0827 490 |
| ·68 | ·2315 752 | ·1997 646 | ·1713 235+ | ·1461 126 | ·1239 443 | ·1045 988 |
| ·69 | ·2714 965+ | ·2371 069 | ·2059 078 | ·1778 453 | ·1528 069 | ·1306 353 |
| ·70 | ·3150 896 | ·2784 782 | ·2447 821 | ·2140 345− | ·1862 025+ | ·1612 002 |
| ·71 | ·3620 159 | ·3236 528 | ·2878 401 | ·2546 942 | ·2242 629 | ·1965 348 |
| ·72 | ·4117 886 | ·3722 463 | ·3348 129 | ·2996 791 | ·2669 682 | ·2367 434 |
| ·73 | ·4637 763 | ·4237 113 | ·3852 562 | ·3486 621 | ·3141 186 | ·2817 570 |
| ·74 | ·5172 163 | ·4773 426 | ·4385 469 | ·4011 229 | ·3653 120 | ·3313 036 |
| ·75 | ·5712 382 | ·5322 943 | ·4938 914 | ·4563 457 | ·4199 328 | ·3848 854 |
| ·76 | ·6248 973 | ·5876 087 | ·5503 477 | ·5134 328 | ·4771 543 | ·4417 706 |
| ·77 | ·6772 173 | ·6422 559 | ·6068 607 | ·5713 328 | ·5359 580 | ·5010 018 |
| ·78 | ·7272 391 | ·6951 844 | ·6623 106 | ·6288 850+ | ·5951 713 | ·5614 238 |
| ·79 | ·7740 727 | ·7453 779 | ·7155 725+ | ·6848 783 | ·6535 225+ | ·6217 330 |
| ·80 | ·8169 492 | ·7919 146 | ·7655 820 | ·7381 208 | ·7097 126 | ·6805 469 |
| ·81 | ·8552 661 | ·8340 246 | ·8114 024 | ·7875 158 | ·7624 966 | ·7364 891 |
| ·82 | ·8886 236 | ·8711 386 | ·8522 872 | ·8321 365− | ·8107 697 | ·7882 847 |
| ·83 | ·9168 461 | ·9029 238 | ·8877 305− | ·8712 918 | ·8536 486 | ·8348 556 |
| ·84 | ·9399 868 | ·9293 008 | ·9174 986 | ·9045 753 | ·8905 379 | ·8754 055+ |
| ·85 | ·9583 132 | ·9504 395+ | ·9416 400 | ·9318 899 | ·9211 731 | ·9094 829 |
| ·86 | ·9722 748 | ·9667 338 | ·9604 683 | ·9534 442 | ·9456 330 | ·9370 120 |
| ·87 | ·9824 565+ | ·9787 550+ | ·9745 209 | ·9697 190 | ·9643 168 | ·9582 853 |
| ·88 | ·9895 211 | ·9871 921 | ·9844 972 | ·9814 058 | ·9778 881 | ·9739 153 |
| ·89 | ·9941 497 | ·9927 826 | ·9911 829 | ·9893 270 | ·9871 910 | ·9847 514 |
| ·90 | ·9969 856 | ·9962 465− | ·9953 718 | ·9943 457 | ·9931 515− | ·9917 722 |
| ·91 | ·9985 903 | ·9982 282 | ·9977 950− | ·9972 811 | ·9966 764 | ·9959 702 |
| ·92 | ·9994 150− | ·9992 578 | ·9990 678 | ·9988 398 | ·9985 687 | ·9982 485− |
| ·93 | ·9997 913 | ·9997 327 | ·9996 612 | ·9995 744 | ·9994 700 | ·9993 455− |
| ·94 | ·9999 389 | ·9999 210 | ·9998 989 | ·9998 718 | ·9998 389 | ·9997 992 |
| ·95 | ·9999 863 | ·9999 822 | ·9999 770 | ·9999 705+ | ·9999 626 | ·9999 530 |
| ·96 | ·9999 979 | ·9999 973 | ·9999 964 | ·9999 954 | ·9999 941 | ·9999 925+ |
| ·97 | ·9999 998 | ·9999 998 | ·9999 997 | ·9999 996 | ·9999 995+ | ·9999 994 |
| ·98 | 1·0000 000 | 1·0000 000 | 1·0000 000 | 1·0000 000 | 1·0000 000 | 1·0000 000 |

$x$ = ·37 to ·99 $q$ = 9·5 $p$ = 32 to 37

| $x$ | $p$ = 32 | $p$ = 33 | $p$ = 34 | $p$ = 35 | $p$ = 36 | $p$ = 37 |
|---|---|---|---|---|---|---|
| $B(p, q)$ = | ·1883 3100 $\times\frac{1}{10^9}$ | ·1452 1908 $\times\frac{1}{10^9}$ | ·1127 5835 $\times\frac{1}{10^9}$ | ·8813 2961 $\times\frac{1}{10^{10}}$ | ·6931 8059 $\times\frac{1}{10^{10}}$ | ·5484 5058 $\times\frac{1}{10^{10}}$ |
| ·37 | ·0000 001 | | | | | |
| ·38 | ·0000 001 | ·0000 001 | | | | |
| ·39 | ·0000 002 | ·0000 001 | ·0000 001 | | | |
| ·40 | ·0000 005− | ·0000 002 | ·0000 001 | ·0000 001 | | |
| ·41 | ·0000 009 | ·0000 005− | ·0000 002 | ·0000 001 | ·0000 001 | |
| ·42 | ·0000 017 | ·0000 009 | ·0000 005− | ·0000 002 | ·0000 001 | ·0000 001 |
| ·43 | ·0000 032 | ·0000 017 | ·0000 009 | ·0000 005− | ·0000 003 | ·0000 001 |
| ·44 | ·0000 058 | ·0000 032 | ·0000 017 | ·0000 009 | ·0000 005+ | ·0000 003 |
| ·45 | ·0000 103 | ·0000 058 | ·0000 032 | ·0000 018 | ·0000 010 | ·0000 005+ |
| ·46 | ·0000 181 | ·0000 104 | ·0000 059 | ·0000 034 | ·0000 019 | ·0000 011 |
| ·47 | ·0000 310 | ·0000 182 | ·0000 106 | ·0000 061 | ·0000 035+ | ·0000 020 |
| ·48 | ·0000 523 | ·0000 313 | ·0000 186 | ·0000 110 | ·0000 065+ | ·0000 038 |
| ·49 | ·0000 867 | ·0000 530 | ·0000 322 | ·0000 195− | ·0000 117 | ·0000 070 |
| ·50 | ·0001 416 | ·0000 882 | ·0000 547 | ·0000 337 | ·0000 207 | ·0000 126 |
| ·51 | ·0002 276 | ·0001 446 | ·0000 914 | ·0000 574 | ·0000 359 | ·0000 224 |
| ·52 | ·0003 603 | ·0002 333 | ·0001 502 | ·0000 963 | ·0000 614 | ·0000 389 |
| ·53 | ·0005 620 | ·0003 707 | ·0002 432 | ·0001 587 | ·0001 031 | ·0000 667 |
| ·54 | ·0008 640 | ·0005 804 | ·0003 878 | ·0002 578 | ·0001 705+ | ·0001 123 |
| ·55 | ·0013 095− | ·0008 954 | ·0006 090 | ·0004 121 | ·0002 775+ | ·0001 860 |
| ·56 | ·0019 572 | ·0013 619 | ·0009 426 | ·0006 492 | ·0004 449 | ·0003 035+ |
| ·57 | ·0028 856 | ·0020 425+ | ·0014 382 | ·0010 076 | ·0007 025+ | ·0004 876 |
| ·58 | ·0041 975+ | ·0030 213 | ·0021 634 | ·0015 414 | ·0010 930 | ·0007 715+ |
| ·59 | ·0060 255+ | ·0044 088 | ·0032 092 | ·0023 245+ | ·0016 758 | ·0012 027 |
| ·60 | ·0085 372 | ·0063 478 | ·0046 957 | ·0034 567 | ·0025 327 | ·0018 474 |
| ·61 | ·0119 407 | ·0090 192 | ·0067 781 | ·0050 693 | ·0037 737 | ·0027 969 |
| ·62 | ·0164 887 | ·0126 480 | ·0096 534 | ·0073 327 | ·0055 443 | ·0041 738 |
| ·63 | ·0224 822 | ·0175 078 | ·0135 667 | ·0104 631 | ·0080 329 | ·0061 405− |
| ·64 | ·0302 711 | ·0239 242 | ·0188 159 | ·0147 293 | ·0114 787 | ·0089 071 |
| ·65 | ·0402 519 | ·0322 756 | ·0257 556 | ·0204 580 | ·0161 784 | ·0127 398 |
| ·66 | ·0528 612 | ·0429 897 | ·0347 963 | ·0280 368 | ·0224 921 | ·0179 685+ |
| ·67 | ·0685 642 | ·0565 359 | ·0464 011 | ·0379 133 | ·0308 453 | ·0249 917 |
| ·68 | ·0878 369 | ·0734 111 | ·0610 749 | ·0505 892 | ·0417 272 | ·0342 782 |
| ·69 | ·1111 422 | ·0941 192 | ·0793 479 | ·0666 077 | ·0556 820 | ·0463 633 |
| ·70 | ·1388 997 | ·1191 429 | ·1017 509 | ·0865 329 | ·0732 932 | ·0618 372 |
| ·71 | ·1714 499 | ·1489 087 | ·1287 822 | ·1109 199 | ·0951 577 | ·0813 242 |
| ·72 | ·2090 145+ | ·1837 466 | ·1608 676 | ·1402 766 | ·1218 505+ | ·1054 514 |
| ·73 | ·2516 556 | ·2238 447 | ·1983 134 | ·1750 161 | ·1538 788 | ·1348 053 |
| ·74 | ·2992 370 | ·2692 046 | ·2412 563 | ·2154 040 | ·1916 274 | ·1698 784 |
| ·75 | ·3513 925+ | ·3196 005− | ·2896 148 | ·2615 032 | ·2352 992 | ·2110 059 |
| ·76 | ·4075 056 | ·3745 469 | ·3430 464 | ·3131 209 | ·2848 541 | ·2582 991 |
| ·77 | ·4667 053 | ·4332 823 | ·4009 176 | ·3697 667 | ·3399 552 | ·3115 805− |
| ·78 | ·5278 831 | ·4947 722 | ·4622 935− | ·4306 266 | ·3999 279 | ·3703 293 |
| ·79 | ·5897 333 | ·5577 386 | ·5259 523 | ·4945 633 | ·4637 439 | ·4336 478 |
| ·80 | ·6508 167 | ·6207 149 | ·5904 306 | ·5601 464 | ·5300 352 | ·5002 586 |
| ·81 | ·7096 469 | ·6821 295+ | ·6540 992 | ·6257 186 | ·5971 474 | ·5685 405− |
| ·82 | ·7647 916 | ·7404 101 | ·7152 676 | ·6894 966 | ·6632 322 | ·6366 103 |
| ·83 | ·8149 806 | ·7941 024 | ·7723 101 | ·7497 005− | ·7263 771 | ·7024 480 |
| ·84 | ·8592 090 | ·8419 901 | ·8238 007 | ·8047 019 | ·7847 626 | ·7640 590 |
| ·85 | ·8968 216 | ·8832 009 | ·8686 411 | ·8531 714 | ·8368 287 | ·8196 576 |
| ·86 | ·9275 650+ | ·9172 826 | ·9061 620 | ·8942 074 | ·8814 296 | ·8678 461 |
| ·87 | ·9515 990 | ·9442 369 | ·9361 820 | ·9274 223 | ·9179 506 | ·9077 645− |
| ·88 | ·9694 607 | ·9644 994 | ·9590 089 | ·9529 694 | ·9463 638 | ·9391 784 |
| ·89 | ·9819 849 | ·9788 686 | ·9753 807 | ·9715 004 | ·9672 083 | ·9624 863 |
| ·90 | ·9901 904 | ·9883 887 | ·9863 495− | ·9840 553 | ·9814 892 | ·9786 344 |
| ·91 | ·9951 513 | ·9942 082 | ·9931 289 | ·9919 013 | ·9905 128 | ·9889 509 |
| ·92 | ·9978 732 | ·9974 362 | ·9969 305+ | ·9963 491 | ·9956 842 | ·9949 282 |
| ·93 | ·9991 979 | ·9990 242 | ·9988 211 | ·9985 849 | ·9983 119 | ·9979 981 |
| ·94 | ·9997 517 | ·9996 952 | ·9996 283 | ·9995 498 | ·9994 580 | ·9993 514 |
| ·95 | ·9999 413 | ·9999 272 | ·9999 105− | ·9998 906 | ·9998 670 | ·9998 394 |
| ·96 | ·9999 906 | ·9999 882 | ·9999 854 | ·9999 820 | ·9999 779 | ·9999 731 |
| ·97 | ·9999 992 | ·9999 990 | ·9999 987 | ·9999 984 | ·9999 980 | ·9999 976 |
| ·98 | 1·0000 000 | 1·0000 000 | 1·0000 000 | 1·0000 000 | ·9999 999 | ·9999 999 |
| ·99 | | | | | 1·0000 000 | 1·0000 000 |

$x$ = ·43 to ·99 $q = 9·5$ $p$ = 38 to 43

| $x$ | $p = 38$ | $p = 39$ | $p = 40$ | $p = 41$ | $p = 42$ | $p = 43$ |
|---|---|---|---|---|---|---|
| $B(p, q)$ = | ·4364 0154 $\times\frac{1}{10^{10}}$ | ·3491 2123 $\times\frac{1}{10^{10}}$ | ·2807 3666 $\times\frac{1}{10^{10}}$ | ·2268 5791 $\times\frac{1}{10^{10}}$ | ·1841 8167 $\times\frac{1}{10^{10}}$ | ·1502 0641 $\times\frac{1}{10^{10}}$ |
| ·43 | ·0000 001 | | | | | |
| ·44 | ·0000 001 | ·0000 001 | | | | |
| ·45 | ·0000 003 | ·0000 002 | ·0000 001 | | | |
| ·46 | ·0000 006 | ·0000 003 | ·0000 002 | ·0000 001 | ·0000 001 | |
| ·47 | ·0000 012 | ·0000 007 | ·0000 004 | ·0000 002 | ·0000 001 | ·0000 001 |
| ·48 | ·0000 022 | ·0000 013 | ·0000 007 | ·0000 004 | ·0000 002 | ·0000 001 |
| ·49 | ·0000 042 | ·0000 025− | ·0000 015− | ·0000 009 | ·0000 005+ | ·0000 003 |
| ·50 | ·0000 077 | ·0000 046 | ·0000 028 | ·0000 017 | ·0000 010 | ·0000 006 |
| ·51 | ·0000 139 | ·0000 086 | ·0000 053 | ·0000 032 | ·0000 020 | ·0000 012 |
| ·52 | ·0000 246 | ·0000 155− | ·0000 097 | ·0000 060 | ·0000 038 | ·0000 023 |
| ·53 | ·0000 429 | ·0000 275− | ·0000 175+ | ·0000 112 | ·0000 071 | ·0000 045− |
| ·54 | ·0000 736 | ·0000 480 | ·0000 312 | ·0000 202 | ·0000 130 | ·0000 084 |
| ·55 | ·0001 242 | ·0000 825+ | ·0000 546 | ·0000 360 | ·0000 236 | ·0000 155− |
| ·56 | ·0002 062 | ·0001 394 | ·0000 939 | ·0000 630 | ·0000 421 | ·0000 281 |
| ·57 | ·0003 370 | ·0002 319 | ·0001 589 | ·0001 085+ | ·0000 738 | ·0000 500+ |
| ·58 | ·0005 423 | ·0003 795+ | ·0002 646 | ·0001 837 | ·0001 271 | ·0000 876 |
| ·59 | ·0008 594 | ·0006 116 | ·0004 335+ | ·0003 061 | ·0002 154 | ·0001 510 |
| ·60 | ·0013 418 | ·0009 706 | ·0006 993 | ·0005 019 | ·0003 590 | ·0002 558 |
| ·61 | ·0020 641 | ·0015 171 | ·0011 107 | ·0008 101 | ·0005 888 | ·0004 264 |
| ·62 | ·0031 288 | ·0023 360 | ·0017 374 | ·0012 873 | ·0009 505− | ·0006 993 |
| ·63 | ·0046 743 | ·0035 441 | ·0026 768 | ·0020 143 | ·0015 104 | ·0011 287 |
| ·64 | ·0068 832 | ·0052 981 | ·0040 626 | ·0031 038 | ·0023 630 | ·0017 929 |
| ·65 | ·0099 913 | ·0078 051 | ·0060 744 | ·0047 104 | ·0036 400 | ·0028 034 |
| ·66 | ·0142 971 | ·0113 319 | ·0089 484 | ·0070 409 | ·0055 210 | ·0043 149 |
| ·67 | ·0201 688 | ·0162 146 | ·0129 879 | ·0103 666 | ·0082 462 | ·0065 381 |
| ·68 | ·0280 493 | ·0228 661 | ·0185 734 | ·0150 341 | ·0121 283 | ·0097 526 |
| ·69 | ·0384 563 | ·0317 801 | ·0261 695− | ·0214 755− | ·0175 652 | ·0143 210 |
| ·70 | ·0519 758 | ·0435 287 | ·0363 270 | ·0302 145+ | ·0250 488 | ·0207 009 |
| ·71 | ·0692 460 | ·0587 524 | ·0496 780 | ·0418 661 | ·0351 698 | ·0294 532 |
| ·72 | ·0909 314 | ·0781 386 | ·0669 203 | ·0571 267 | ·0486 134 | ·0412 434 |
| ·73 | ·1176 830 | ·1023 878 | ·0887 888 | ·0767 520 | ·0661 435+ | ·0568 320 |
| ·74 | ·1500 868 | ·1321 649 | ·1160 123 | ·1015 194 | ·0885 712 | ·0770 503 |
| ·75 | ·1886 002 | ·1680 371 | ·1492 535+ | ·1321 722 | ·1167 053 | ·1027 574 |
| ·76 | ·2334 809 | ·2103 999 | ·1890 353 | ·1693 477 | ·1512 832 | ·1347 757 |
| ·77 | ·2847 127 | ·2593 968 | ·2356 551 | ·2134 893 | ·1928 832 | ·1738 053 |
| ·78 | ·3419 385+ | ·3148 403 | ·2890 970 | ·2647 503 | ·2418 228 | ·2203 201 |
| ·79 | ·4044 099 | ·3761 453 | ·3489 497 | ·3228 997 | ·2980 537 | ·2744 529 |
| ·80 | ·4709 648 | ·4422 877 | ·4143 460 | ·3872 425− | ·3610 645− | ·3358 837 |
| ·81 | ·5400 457 | ·5118 021 | ·4839 387 | ·4565 733 | ·4298 120 | ·4037 485− |
| ·82 | ·6097 654 | ·5828 286 | ·5559 264 | ·5291 789 | ·5026 988 | ·4765 907 |
| ·83 | ·6780 241 | ·6532 177 | ·6281 411 | ·6029 046 | ·5776 158 | ·5523 781 |
| ·84 | ·7426 728 | ·7206 900 | ·6981 999 | ·6752 938 | ·6520 636 | ·6286 011 |
| ·85 | ·8017 091 | ·7830 403 | ·7637 133 | ·7437 944 | ·7233 533 | ·7024 619 |
| ·86 | ·8534 807 | ·8383 629 | ·8225 280 | ·8060 162 | ·7888 721 | ·7711 442 |
| ·87 | 8968 668 | ·8852 650+ | ·8729 716 | ·8600 037 | ·8463 825+ | ·8321 338 |
| ·88 | ·9314 027 | ·9230 295− | ·9140 551 | ·9044 794 | ·8943 058 | ·8835 412 |
| ·89 | ·9573 183 | ·9516 900 | ·9455 890 | ·9390 053 | ·9319 310 | ·9243 606 |
| ·90 | ·9754 747 | ·9719 950+ | ·9681 807 | ·9640 183 | ·9594 955+ | ·9546 013 |
| ·91 | ·9872 031 | ·9852 568 | ·9830 996 | ·9807 195+ | ·9781 046 | ·9752 434 |
| ·92 | ·9940 728 | ·9931 098 | ·9920 307 | ·9908 271 | ·9894 901 | ·9880 112 |
| ·93 | ·9976 392 | ·9972 308 | ·9967 682 | ·9962 465− | ·9956 608 | ·9950 058 |
| ·94 | ·9992 281 | ·9990 863 | ·9989 240 | ·9987 390 | ·9985 290 | ·9982 917 |
| ·95 | ·9998 072 | ·9997 697 | ·9997 263 | ·9996 763 | ·9996 190 | ·9995 535− |
| ·96 | ·9999 674 | ·9999 607 | ·9999 528 | ·9999 437 | ·9999 331 | ·9999 209 |
| ·97 | ·9999 970 | ·9999 964 | ·9999 956 | ·9999 947 | ·9999 937 | ·9999 924 |
| ·98 | ·9999 999 | ·9999 999 | ·9999 999 | ·9999 998 | ·9999 998 | ·9999 998 |
| ·99 | 1·0000 000 | 1·0000 000 | 1·0000 000 | 1·0000 000 | 1·0000 000 | 1·0000 000 |

$x = \cdot48$ to $\cdot99$ $q = 9\cdot5$ $p = 44$ to $50$

| $x$ | $p = 44$ | $p = 45$ | $p = 46$ | $p = 47$ | $p = 48$ | $p = 49$ | $p = 50$ |
|---|---|---|---|---|---|---|---|
| $B(p, q) =$ | $\cdot1230\ 2620 \times \frac{1}{10^{10}}$ | $\cdot1011\ 8043 \times \frac{1}{10^{10}}$ | $\cdot8354\ 3471 \times \frac{1}{10^{11}}$ | $\cdot6924\ 3237 \times \frac{1}{10^{11}}$ | $\cdot5760\ 0569 \times \frac{1}{10^{11}}$ | $\cdot4808\ 3953 \times \frac{1}{10^{11}}$ | $\cdot4027\ 5448 \times \frac{1}{10^{11}}$ |
| ·48 | ·0000 001 | | | | | | |
| ·49 | ·0000 002 | ·0000 001 | ·0000 001 | | | | |
| ·50 | ·0000 004 | ·0000 002 | ·0000 001 | ·0000 001 | | | |
| ·51 | ·0000 007 | ·0000 004 | ·0000 003 | ·0000 002 | ·0000 001 | ·0000 001 | |
| ·52 | ·0000 014 | ·0000 009 | ·0000 005+ | ·0000 003 | ·0000 002 | ·0000 001 | ·0000 001 |
| ·53 | ·0000 028 | ·0000 018 | ·0000 011 | ·0000 007 | ·0000 004 | ·0000 003 | ·0000 002 |
| ·54 | ·0000 054 | ·0000 034 | ·0000 022 | ·0000 014 | ·0000 009 | ·0000 006 | ·0000 003 |
| ·55 | ·0000 101 | ·0000 066 | ·0000 043 | ·0000 027 | ·0000 018 | ·0000 011 | ·0000 007 |
| ·56 | ·0000 186 | ·0000 123 | ·0000 081 | ·0000 053 | ·0000 035+ | ·0000 023 | ·0000 015− |
| ·57 | ·0000 338 | ·0000 228 | ·0000 153 | ·0000 102 | ·0000 068 | ·0000 045+ | ·0000 030 |
| ·58 | ·0000 602 | ·0000 412 | ·0000 282 | ·0000 192 | ·0000 130 | ·0000 088 | ·0000 059 |
| ·59 | ·0001 055− | ·0000 735− | ·0000 510 | ·0000 353 | ·0000 244 | ·0000 168 | ·0000 115+ |
| ·60 | ·0001 817 | ·0001 287 | ·0000 908 | ·0000 639 | ·0000 448 | ·0000 314 | ·0000 219 |
| ·61 | ·0003 078 | ·0002 215− | ·0001 589 | ·0001 136 | ·0000 810 | ·0000 576 | ·0000 409 |
| ·62 | ·0005 129 | ·0003 749 | ·0002 732 | ·0001 985+ | ·0001 438 | ·0001 039 | ·0000 749 |
| ·63 | ·0008 407 | ·0006 242 | ·0004 620 | ·0003 410 | ·0002 510 | ·0001 842 | ·0001 348 |
| ·64 | ·0013 559 | ·0010 222 | ·0007 683 | ·0005 758 | ·0004 303 | ·0003 207 | ·0002 384 |
| ·65 | ·0021 521 | ·0016 470 | ·0012 566 | ·0009 560 | ·0007 253 | ·0005 488 | ·0004 141 |
| ·66 | ·0033 615− | ·0026 107 | ·0020 215+ | ·0015 608 | ·0012 018 | ·0009 229 | ·0007 069 |
| ·67 | ·0051 674 | ·0040 716 | ·0031 987 | ·0025 059 | ·0019 577 | ·0015 254 | ·0011 855− |
| ·68 | ·0078 177 | ·0062 478 | ·0049 787 | ·0039 562 | ·0031 351 | ·0024 779 | ·0019 535+ |
| ·69 | ·0116 401 | ·0094 328 | ·0076 222 | ·0061 420 | ·0049 359 | ·0039 564 | ·0031 632 |
| ·70 | ·0170 558 | ·0140 113 | ·0114 776 | ·0093 763 | ·0076 394 | ·0062 082 | ·0050 326 |
| ·71 | ·0245 921 | ·0204 740 | ·0169 979 | ·0140 738 | ·0116 222 | ·0095 733 | ·0078 663 |
| ·72 | ·0348 880 | ·0294 282 | ·0247 545+ | ·0207 677 | ·0173 781 | ·0145 054 | ·0120 782 |
| ·73 | ·0486 909 | ·0415 999 | ·0354 456 | ·0301 228 | ·0255 345+ | ·0215 919 | ·0182 145+ |
| ·74 | ·0668 397 | ·0578 241 | ·0498 925− | ·0429 384 | ·0368 618 | ·0315 688 | ·0269 726 |
| ·75 | ·0902 287 | ·0790 169 | ·0690 197 | ·0601 363 | ·0522 689 | ·0453 234 | ·0392 106 |
| ·76 | ·1197 500+ | ·1061 244 | ·0938 130 | ·0827 273 | ·0727 785+ | ·0638 785− | ·0559 411 |
| ·77 | ·1562 116 | ·1400 473 | ·1252 498 | ·1117 503 | ·0994 758 | ·0883 508 | ·0782 985+ |
| ·78 | ·2002 326 | ·1815 375− | ·1642 009 | ·1481 796 | ·1334 230 | ·1198 746 | ·1074 737 |
| ·79 | ·2521 228 | ·2310 741 | ·2113 047 | ·1928 010 | ·1755 394 | ·1594 879 | ·1446 075− |
| ·80 | ·3117 570 | ·2887 270 | ·2668 229 | ·2460 617 | ·2264 489 | ·2079 802 | ·1906 422 |
| ·81 | ·3784 637 | ·3540 266 | ·3304 937 | ·3079 098 | ·2863 085− | ·2657 130 | ·2461 367 |
| ·82 | ·4509 502 | ·4258 632 | ·4014 062 | ·3776 457 | ·3546 385− | ·3324 317 | ·3110 633 |
| ·83 | ·5272 901 | ·5024 445− | ·4779 273 | ·4538 179 | ·4301 881 | ·4071 023 | ·3846 172 |
| ·84 | ·6049 967 | ·5813 387 | ·5577 123 | ·5341 988 | ·5108 755− | ·4878 145− | ·4650 830 |
| ·85 | ·6811 941 | ·6596 241 | ·6378 265− | ·6158 748 | ·5938 414 | ·5717 966 | ·5498 080 |
| ·86 | ·7528 846 | ·7341 476 | ·7149 902 | ·6954 704 | ·6756 475− | ·6555 809 | ·6353 300 |
| ·87 | ·8172 870 | ·8018 750+ | ·7859 339 | ·7695 024 | ·7526 217 | ·7353 347 | ·7176 862 |
| ·88 | ·8721 958 | ·8602 831 | ·8478 198 | ·8348 255− | ·8213 224 | ·8073 355+ | ·7928 919 |
| ·89 | ·9162 912 | ·9077 220 | ·8986 548 | ·8890 939 | ·8790 457 | ·8685 193 | ·8575 255+ |
| ·90 | ·9493 257 | ·9436 604 | ·9375 986 | ·9311 348 | ·9242 653 | ·9169 880 | ·9093 023 |
| ·91 | ·9721 250+ | ·9687 391 | ·9650 759 | ·9611 265− | ·9568 825− | ·9523 366 | ·9474 822 |
| ·92 | ·9863 816 | ·9845 928 | ·9826 362 | ·9805 035+ | ·9781 866 | ·9756 776 | ·9729 689 |
| ·93 | ·9942 763 | ·9934 668 | ·9925 717 | ·9915 855− | ·9905 024 | ·9893 168 | ·9880 229 |
| ·94 | ·9980 245+ | ·9977 249 | ·9973 900 | ·9970 171 | ·9966 031 | ·9961 451 | ·9956 398 |
| ·95 | ·9994 790 | ·9993 945+ | ·9992 992 | ·9991 918 | ·9990 714 | ·9989 367 | ·9987 866 |
| ·96 | ·9999 069 | ·9998 908 | ·9998 725− | ·9998 516 | ·9998 280 | ·9998 013 | ·9997 711 |
| ·97 | ·9999 910 | ·9999 894 | ·9999 875− | ·9999 853 | ·9999 828 | ·9999 799 | ·9999 767 |
| ·98 | ·9999 997 | ·9999 997 | ·9999 996 | ·9999 995+ | ·9999 994 | ·9999 993 | ·9999 992 |
| ·99 | 1·0000 000 | 1·0000 000 | 1·0000 000 | 1·0000 000 | 1·0000 000 | 1·0000 000 | 1·0000 000 |

$x$ = ·07 to ·70 $q = 10$ $p$ = 10 to 14

| $x$ | $p = 10$ | $p = 10{\cdot}5$ | $p = 11$ | $p = 12$ | $p = 13$ | $p = 14$ |
|---|---|---|---|---|---|---|
| $B(p, q) =$ | ·1082 5088$\times\frac{1}{10^5}$ | ·7606 8365$\bar{\times}\frac{1}{10^6}$ | ·5412 5441$\times\frac{1}{10^6}$ | ·2835 1422$\times\frac{1}{10^6}$ | ·1546 4412$\times\frac{1}{10^6}$ | ·8740 7545$\bar{\times}\frac{1}{10^7}$ |
| ·07 | ·0000 001 | ·0000 001 | | | | |
| ·08 | ·0000 005+ | ·0000 002 | ·0000 001 | | | |
| ·09 | ·0000 015− | ·0000 006 | ·0000 002 | | | |
| ·10 | ·0000 039 | ·0000 017 | ·0000 007 | ·0000 001 | | |
| ·11 | ·0000 093 | ·0000 042 | ·0000 018 | ·0000 004 | ·0000 001 | |
| ·12 | ·0000 203 | ·0000 095− | ·0000 044 | ·0000 009 | ·0000 002 | |
| ·13 | ·0000 413 | ·0000 201 | ·0000 097 | ·0000 022 | ·0000 005− | ·0000 001 |
| ·14 | ·0000 790 | ·0000 398 | ·0000 199 | ·0000 048 | ·0000 011 | ·0000 003 |
| ·15 | ·0001 435+ | ·0000 748 | ·0000 386 | ·0000 100 | ·0000 025+ | ·0000 006 |
| ·16 | ·0002 491 | ·0001 340 | ·0000 714 | ·0000 198 | ·0000 053 | ·0000 014 |
| ·17 | ·0004 154 | ·0002 303 | ·0001 265− | ·0000 371 | ·0000 106 | ·0000 029 |
| ·18 | ·0006 686 | ·0003 811 | ·0002 152 | ·0000 669 | ·0000 201 | ·0000 059 |
| ·19 | ·0010 423 | ·0006 101 | ·0003 538 | ·0001 159 | ·0000 368 | ·0000 114 |
| ·20 | ·0015 791 | ·0009 477 | ·0005 634 | ·0001 941 | ·0000 648 | ·0000 210 |
| ·21 | ·0023 310 | ·0014 324 | ·0008 721 | ·0003 151 | ·0001 103 | ·0000 376 |
| ·22 | ·0033 605+ | ·0021 122 | ·0013 153 | ·0004 972 | ·0001 823 | ·0000 650− |
| ·23 | ·0047 410 | ·0030 445− | ·0019 372 | ·0007 647 | ·0002 927 | ·0001 090 |
| ·24 | ·0065 569 | ·0042 977 | ·0027 914 | ·0011 483 | ·0004 582 | ·0001 779 |
| ·25 | ·0089 033 | ·0059 512 | ·0039 421 | ·0016 871 | ·0007 005− | ·0002 831 |
| ·26 | ·0118 854 | ·0080 950− | ·0054 642 | ·0024 287 | ·0010 475+ | ·0004 398 |
| ·27 | ·0156 176 | ·0108 299 | ·0074 436 | ·0034 309 | ·0015 349 | ·0006 686 |
| ·28 | ·0202 213 | ·0142 665+ | ·0099 772 | ·0047 620 | ·0022 065+ | ·0009 957 |
| ·29 | ·0258 233 | ·0185 236 | ·0131 721 | ·0065 015− | ·0031 161 | ·0014 547 |
| ·30 | ·0325 534 | ·0237 267 | ·0171 448 | ·0087 402 | ·0043 277 | ·0020 875+ |
| ·31 | ·0405 410 | ·0300 056 | ·0220 195− | ·0115 801 | ·0059 167 | ·0029 456 |
| ·32 | ·0499 125+ | ·0374 921 | ·0279 259 | ·0151 338 | ·0079 701 | ·0040 908 |
| ·33 | ·0607 877 | ·0463 164 | ·0349 974 | ·0195 232 | ·0105 869 | ·0055 963 |
| ·34 | ·0732 761 | ·0566 042 | ·0433 674 | ·0248 784 | ·0138 774 | ·0075 477 |
| ·35 | ·0874 736 | ·0684 732 | ·0531 666 | ·0313 349 | ·0179 630 | ·0100 427 |
| ·36 | ·1034 592 | ·0820 289 | ·0645 194 | ·0390 316 | ·0229 742 | ·0131 916 |
| ·37 | ·1212 913 | ·0973 619 | ·0775 400 | ·0481 073 | ·0290 496 | ·0171 166 |
| ·38 | ·1410 055+ | ·1145 436 | ·0923 288 | ·0586 976 | ·0363 329 | ·0219 506 |
| ·39 | ·1626 116 | ·1336 233 | ·1089 687 | ·0709 309 | ·0449 701 | ·0278 360 |
| ·40 | ·1860 920 | ·1546 255− | ·1275 212 | ·0849 243 | ·0551 065− | ·0349 221 |
| ·41 | ·2114 003 | ·1775 471 | ·1480 239 | ·1007 797 | ·0668 820 | ·0433 622 |
| ·42 | ·2384 607 | ·2023 562 | ·1704 869 | ·1185 796 | ·0804 278 | ·0533 107 |
| ·43 | ·2671 682 | ·2289 903 | ·1948 909 | ·1383 832 | ·0958 612 | ·0649 183 |
| ·44 | ·2973 894 | ·2573 565+ | ·2211 858 | ·1602 229 | ·1132 814 | ·0783 281 |
| ·45 | ·3289 641 | ·2873 316 | ·2492 894 | ·1841 009 | ·1327 651 | ·0936 708 |
| ·46 | ·3617 076 | ·3187 634 | ·2790 878 | ·2099 875+ | ·1543 618 | ·1110 593 |
| ·47 | ·3954 137 | ·3514 726 | ·3104 359 | ·2378 184 | ·1780 905+ | ·1305 839 |
| ·48 | ·4298 582 | ·3852 557 | ·3431 591 | ·2674 945− | ·2039 362 | ·1523 073 |
| ·49 | ·4648 028 | ·4198 879 | ·3770 561 | ·2988 817 | ·2318 471 | ·1762 600 |
| ·50 | ·5000 000[e] | ·4551 280 | ·4119 015− | ·3318 119 | ·2617 336 | ·2024 364 |
| ·51 | ·5351 972 | ·4907 221 | ·4474 504 | ·3660 852 | ·2934 668 | ·2307 915+ |
| ·52 | ·5701 418 | ·5264 088 | ·4834 428 | ·4014 727 | ·3268 800 | ·2612 384 |
| ·53 | ·6045 863 | ·5619 244 | ·5196 084 | ·4377 206 | ·3617 697 | ·2936 476 |
| ·54 | ·6382 924 | ·5970 078 | ·5556 725+ | ·4745 548 | ·3978 986 | ·3278 466 |
| ·55 | ·6710 359 | ·6314 063 | ·5913 612 | ·5116 865− | ·4349 995+ | ·3636 217 |
| ·56 | ·7026 106 | ·6648 796 | ·6264 070 | ·5488 179 | ·4727 805+ | ·4007 205+ |
| ·57 | ·7328 318 | ·6972 052 | ·6605 546 | ·5856 491 | ·5109 308 | ·4388 564 |
| ·58 | ·7615 393 | ·7281 821 | ·6935 656 | ·6218 841 | ·5491 275− | ·4777 140 |
| ·59 | ·7885 997 | ·7576 343 | ·7252 234 | ·6572 378 | ·5870 428 | ·5169 557 |
| ·60 | ·8139 080 | ·7854 139 | ·7553 372 | ·6914 418 | ·6243 517 | ·5562 293 |
| ·61 | ·8373 884 | ·8114 026 | ·7837 454 | ·7242 504 | ·6607 396 | ·5951 768 |
| ·62 | ·8589 945− | ·8355 136 | ·8103 177 | ·7554 458 | ·6959 097 | ·6334 426 |
| ·63 | ·8787 087 | ·8576 914 | ·8349 573 | ·7848 422 | ·7295 902 | ·6706 831 |
| ·64 | ·8965 408 | ·8779 117 | ·8576 011 | ·8122 894 | ·7615 402 | ·7065 750+ |
| ·65 | ·9125 264 | ·8961 801 | ·8782 194 | ·8376 748 | ·7915 553 | ·7408 238 |
| ·66 | ·9267 239 | ·9125 304 | ·8968 153 | ·8609 249 | ·8194 715− | ·7731 712 |
| ·67 | ·9392 123 | ·9270 217 | ·9134 220 | ·8820 048 | ·8451 681 | ·8034 009 |
| ·68 | ·9500 875− | ·9397 355+ | ·9281 009 | ·9009 175+ | ·8685 693 | ·8313 439 |
| ·69 | ·9594 590 | ·9507 723 | ·9409 375+ | ·9177 014 | ·8896 439 | ·8568 813 |
| ·70 | ·9674 466 | ·9602 476 | ·9520 381 | ·9324 272 | ·9084 039 | ·8799 455+ |

$x = \cdot 71$ to $\cdot 96$ $q = 10$ $p = 10$ to $14$

| | $p = 10$ | $p = 10{\cdot}5$ | $p = 11$ | $p = 12$ | $p = 13$ | $p = 14$ |
|---|---|---|---|---|---|---|
| $B(p, q) =$ | ·1082 5088$\times\frac{1}{10^5}$ | ·7606 8365$\overline{\times}\frac{1}{10^6}$ | ·5412 5441$\times\frac{1}{10^6}$ | ·2835 1422$\times\frac{1}{10^6}$ | ·1546 4412$\times\frac{1}{10^6}$ | ·8740 7545$\overline{\times}\frac{1}{10^7}$ |
| $x$ | | | | | | |
| ·71 | ·9741 767 | ·9682 883 | ·9615 255$^-$ | ·9451 939 | ·9249 020 | ·9005 204 |
| ·72 | ·9797 787 | ·9750 286 | ·9695 346 | ·9561 242 | ·9392 270 | ·9186 384 |
| ·73 | ·9843 824 | ·9806 062 | ·9762 084 | ·9653 593 | ·9514 995$^+$ | ·9343 774 |
| ·74 | ·9881 146 | ·9851 590 | ·9816 933 | ·9730 538 | ·9618 657 | ·9478 547 |
| ·75 | ·9910 967 | ·9888 215$^-$ | ·9861 356 | ·9793 704 | ·9704 911 | ·9592 212 |
| ·76 | ·9934 431 | ·9917 224 | ·9896 777 | ·9844 745$^-$ | ·9775 543 | ·9686 538 |
| ·77 | ·9952 590 | ·9939 823 | ·9924 551 | ·9885 297 | ·9832 403 | ·9763 477 |
| ·78 | ·9966 395$^-$ | ·9957 115$^-$ | ·9945 942 | ·9916 937 | ·9877 345$^+$ | ·9825 084 |
| ·79 | ·9976 690 | ·9970 093 | ·9962 101 | ·9941 145$^+$ | ·9912 174 | ·9873 442 |
| ·80 | ·9984 209 | ·9979 632 | ·9974 052 | ·9959 278 | ·9938 594 | ·9910 592 |
| ·81 | ·9989 577 | ·9986 485$^-$ | ·9982 691 | ·9972 550$^-$ | ·9958 175$^+$ | ·9938 471 |
| ·82 | ·9993 314 | ·9991 285$^+$ | ·9988 781 | ·9982 022 | ·9972 323 | ·9958 863 |
| ·83 | ·9995 846 | ·9994 557 | ·9992 956 | ·9988 596 | ·9982 262 | ·9973 365$^+$ |
| ·84 | ·9997 509 | ·9996 719 | ·9995 733 | ·9993 019 | ·9989 031 | ·9983 361 |
| ·85 | ·9998 565$^-$ | ·9998 100 | ·9997 516 | ·9995 895$^+$ | ·9993 484 | ·9990 016 |
| ·86 | ·9999 210 | ·9998 949 | ·9998 619 | ·9997 694 | ·9996 302 | ·9994 277 |
| ·87 | ·9999 587 | ·9999 448 | ·9999 271 | ·9998 770 | ·9998 008 | ·9996 887 |
| ·88 | ·9999 797 | ·9999 727 | ·9999 638 | ·9999 383 | ·9998 990 | ·9998 406 |
| ·89 | ·9999 907 | ·9999 874 | ·9999 832 | ·9999 711 | ·9999 523 | ·9999 239 |
| ·90 | ·9999 961 | ·9999 947 | ·9999 928 | ·9999 876 | ·9999 793 | ·9999 666 |
| ·91 | ·9999 985$^+$ | ·9999 980 | ·9999 972 | ·9999 952 | ·9999 919 | ·9999 868 |
| ·92 | ·9999 995$^-$ | ·9999 993 | ·9999 991 | ·9999 983 | ·9999 972 | ·9999 954 |
| ·93 | ·9999 999 | ·9999 998 | ·9999 997 | ·9999 995$^+$ | ·9999 992 | ·9999 986 |
| ·94 | 1·0000 000 | 1·0000 000 | ·9999 999 | ·9999 999 | ·9999 998 | ·9999 997 |
| ·95 | | | 1·0000 000 | 1·0000 000 | 1·0000 000 | ·9999 999 |
| ·96 | | | | | | 1·0000 000 |

$x$ = ·14 to ·70 $q = 10$ $p$ = 15 to 20

| $x$ | $p = 15$ | $p = 16$ | $p = 17$ | $p = 18$ | $p = 19$ | $p = 20$ |
|---|---|---|---|---|---|---|
| $B(p, q) =$ | ·5098 7734 $\times \frac{1}{10^7}$ | ·3059 2641 $\times \frac{1}{10^7}$ | ·1882 6240 $\times \frac{1}{10^7}$ | ·1185 3559 $\times \frac{1}{10^7}$ | ·7620 1449 $\times \frac{1}{10^8}$ | ·4992 5087 $\times \frac{1}{10^8}$ |
| ·14 | ·0000 001 | | | | | |
| ·15 | ·0000 001 | | | | | |
| ·16 | ·0000 004 | ·0000 001 | | | | |
| ·17 | ·0000 008 | ·0000 002 | ·0000 001 | | | |
| ·18 | ·0000 017 | ·0000 005− | ·0000 001 | | | |
| ·19 | ·0000 034 | ·0000 010 | ·0000 003 | ·0000 001 | | |
| ·20 | ·0000 067 | ·0000 021 | ·0000 006 | ·0000 002 | ·0000 001 | |
| ·21 | ·0000 125− | ·0000 041 | ·0000 013 | ·0000 004 | ·0000 001 | |
| ·22 | ·0000 226 | ·0000 077 | ·0000 026 | ·0000 008 | ·0000 003 | ·0000 001 |
| ·23 | ·0000 396 | ·0000 141 | ·0000 049 | ·0000 017 | ·0000 006 | ·0000 002 |
| ·24 | ·0000 674 | ·0000 250− | ·0000 091 | ·0000 032 | ·0000 011 | ·0000 004 |
| ·25 | ·0001 117 | ·0000 431 | ·0000 163 | ·0000 061 | ·0000 022 | ·0000 008 |
| ·26 | ·0001 803 | ·0000 723 | ·0000 284 | ·0000 110 | ·0000 042 | ·0000 016 |
| ·27 | ·0002 843 | ·0001 183 | ·0000 482 | ·0000 193 | ·0000 076 | ·0000 030 |
| ·28 | ·0004 387 | ·0001 891 | ·0000 799 | ·0000 332 | ·0000 136 | ·0000 055− |
| ·29 | ·0006 631 | ·0002 959 | ·0001 294 | ·0000 556 | ·0000 235+ | ·0000 098 |
| ·30 | ·0009 835− | ·0004 535+ | ·0002 051 | ·0000 911 | ·0000 398 | ·0000 171 |
| ·31 | ·0014 325− | ·0006 820 | ·0003 184 | ·0001 461 | ·0000 659 | ·0000 293 |
| ·32 | ·0020 513 | ·0010 071 | ·0004 850+ | ·0002 295− | ·0001 068 | ·0000 490 |
| ·33 | ·0028 907 | ·0014 621 | ·0007 255+ | ·0003 538 | ·0001 697 | ·0000 802 |
| ·34 | ·0040 120 | ·0020 886 | ·0010 668 | ·0005 355+ | ·0002 645+ | ·0001 288 |
| ·35 | ·0054 885+ | ·0029 382 | ·0015 435− | ·0007 969 | ·0004 049 | ·0002 027 |
| ·36 | ·0074 059 | ·0040 733 | ·0021 987 | ·0011 666 | ·0006 092 | ·0003 136 |
| ·37 | ·0098 630 | ·0055 688 | ·0030 863 | ·0016 815− | ·0009 018 | ·0004 767 |
| ·38 | ·0129 720 | ·0075 130 | ·0042 717 | ·0023 879 | ·0013 142 | ·0007 129 |
| ·39 | ·0168 580 | ·0100 076 | ·0058 332 | ·0033 433 | ·0018 867 | ·0010 496 |
| ·40 | ·0216 581 | ·0131 691 | ·0078 635− | ·0046 177 | ·0026 702 | ·0015 222 |
| ·41 | ·0275 200 | ·0171 275− | ·0104 697 | ·0062 950− | ·0037 275− | ·0021 762 |
| ·42 | ·0345 998 | ·0220 262 | ·0137 747 | ·0084 743 | ·0051 351 | ·0030 683 |
| ·43 | ·0430 593 | ·0280 203 | ·0179 160 | ·0112 709 | ·0069 848 | ·0042 688 |
| ·44 | ·0530 619 | ·0352 744 | ·0230 456 | ·0148 163 | ·0093 849 | ·0058 631 |
| ·45 | ·0647 690 | ·0439 597 | ·0293 282 | ·0192 582 | ·0124 610 | ·0079 534 |
| ·46 | ·0783 350+ | ·0542 500− | ·0369 388 | ·0247 599 | ·0163 565− | ·0106 598 |
| ·47 | ·0939 020 | ·0663 173 | ·0460 597 | ·0314 981 | ·0212 321 | ·0141 216 |
| ·48 | ·1115 943 | ·0803 266 | ·0568 759 | ·0396 603 | ·0272 651 | ·0184 971 |
| ·49 | ·1315 124 | ·0964 303 | ·0695 705+ | ·0494 415+ | ·0346 467 | ·0239 633 |
| ·50 | ·1537 281 | ·1147 615− | ·0843 188 | ·0610 391 | ·0435 793 | ·0307 142 |
| ·51 | ·1782 786 | ·1354 280 | ·1012 815− | ·0746 472 | ·0542 719 | ·0389 583 |
| ·52 | ·2051 618 | ·1585 060 | ·1205 982 | ·0904 503 | ·0669 349 | ·0489 148 |
| ·53 | ·2343 327 | ·1840 337 | ·1423 798 | ·1086 157 | ·0817 732 | ·0608 078 |
| ·54 | ·2657 005+ | ·2120 063 | ·1667 018 | ·1292 856 | ·0989 784 | ·0748 603 |
| ·55 | ·2991 267 | ·2423 712 | ·1935 968 | ·1525 690 | ·1187 211 | ·0912 864 |
| ·56 | ·3344 253 | ·2750 248 | ·2230 493 | ·1785 339 | ·1411 409 | ·1102 819 |
| ·57 | ·3713 639 | ·3098 107 | ·2549 899 | ·2071 991 | ·1663 380 | ·1320 146 |
| ·58 | ·4096 672 | ·3465 197 | ·2892 924 | ·2385 283 | ·1943 636 | ·1566 144 |
| ·59 | ·4490 213 | ·3848 912 | ·3257 713 | ·2724 242 | ·2252 121 | ·1841 624 |
| ·60 | ·4890 802 | ·4246 170 | ·3641 828 | ·3087 255+ | ·2588 139 | ·2146 816 |
| ·61 | ·5294 736 | ·4653 472 | ·4042 268 | ·3472 050− | ·2950 301 | ·2481 275+ |
| ·62 | ·5698 155− | ·5066 973 | ·4455 516 | ·3875 711 | ·3336 492 | ·2843 816 |
| ·63 | ·6097 143 | ·5482 577 | ·4877 613 | ·4294 713 | ·3743 873 | ·3232 461 |
| ·64 | ·6487 830 | ·5896 040 | ·5304 250+ | ·4724 992 | ·4168 904 | ·3644 426 |
| ·65 | ·6866 499 | ·6303 090 | ·5730 877 | ·5162 031 | ·4607 406 | ·4076 133 |
| ·66 | ·7229 685− | ·6699 544 | ·6152 836 | ·5600 983 | ·5054 648 | ·4523 266 |
| ·67 | ·7574 272 | ·7081 434 | ·6565 493 | ·6036 807 | ·5505 476 | ·4980 858 |
| ·68 | ·7897 579 | ·7445 123 | ·6964 388 | ·6464 424 | ·5954 461 | ·5443 424 |
| ·69 | ·8197 425+ | ·7787 413 | ·7345 370 | ·6878 883 | ·6396 070 | ·5905 125− |
| ·70 | ·8472 184 | ·8105 640 | ·7704 732 | ·7275 525+ | ·6824 858 | ·6359 959 |

$x$ = ·71 to ·97 $q$ = 10 $p$ = 15 to 20

| | $p$ = 15 | $p$ = 16 | $p$ = 17 | $p$ = 18 | $p$ = 19 | $p$ = 20 |
|---|---|---|---|---|---|---|
| $B(p, q)$ = | ·5098 7734 × $\frac{1}{10^7}$ | ·3059 2641 × $\frac{1}{10^7}$ | ·1882 6240 × $\frac{1}{10^7}$ | ·1185 3559 × $\frac{1}{10^7}$ | ·7620 1449 × $\frac{1}{10^8}$ | ·4992 5087 × $\frac{1}{10^8}$ |
| $x$ | | | | | | |
| ·71 | ·8720 810 | ·8397 739 | ·8039 332 | ·7650 144 | ·7235 659 | ·6801 976 |
| ·72 | ·8942 851 | ·8662 301 | ·8346 682 | ·7999 129 | ·7623 773 | ·7225 500− |
| ·73 | ·9138 430 | ·8898 589 | ·8625 021 | ·8319 589 | ·7985 142 | ·7625 347 |
| ·74 | ·9308 213 | ·9106 538 | ·8873 351 | ·8609 438 | ·8316 495+ | ·7997 033 |
| ·75 | ·9453 351 | ·9286 717 | ·9091 444 | ·8867 454 | ·8615 465− | ·8336 951 |
| ·76 | ·9575 410 | ·9440 277 | ·9279 808 | ·9093 285+ | ·8880 450− | ·8642 498 |
| ·77 | ·9676 286 | ·9568 867 | ·9439 628 | ·9287 430 | ·9111 642 | ·8912 168 |
| ·78 | ·9758 115+ | ·9674 538 | ·9572 678 | ·9451 165− | ·9308 995− | ·9145 574 |
| ·79 | ·9823 174 | ·9759 635− | ·9681 203 | ·9586 440 | ·9474 145+ | ·9343 410 |
| ·80 | ·9873 789 | ·9826 681 | ·9767 797 | ·9695 749 | ·9609 293 | ·9507 365− |
| ·81 | ·9912 250− | ·9878 267 | ·9835 258 | ·9781 978 | ·9717 242 | ·9639 968 |
| ·82 | ·9940 731 | ·9916 942 | ·9886 463 | ·9848 238 | ·9801 221 | ·9744 405− |
| ·83 | ·9961 234 | ·9945 123 | ·9924 230 | ·9897 709 | ·9864 689 | ·9824 300 |
| ·84 | ·9975 537 | ·9965 022 | ·9951 220 | ·9933 489 | ·9911 148 | ·9883 492 |
| ·85 | ·9985 174 | ·9978 587 | ·9969 840 | ·9958 468 | ·9943 970 | ·9925 808 |
| ·86 | ·9991 416 | ·9987 478 | ·9982 188 | ·9975 229 | ·9966 252 | ·9954 875− |
| ·87 | ·9995 283 | ·9993 051 | ·9990 017 | ·9985 981 | ·9980 712 | ·9973 958 |
| ·88 | ·9997 561 | ·9996 371 | ·9994 735− | ·9992 533 | ·9989 626 | ·9985 857 |
| ·89 | ·9998 825− | ·9998 234 | ·9997 413 | ·9996 295+ | ·9994 803 | ·9992 846 |
| ·90 | ·9999 479 | ·9999 210 | ·9998 832 | ·9998 311 | ·9997 607 | ·9996 674 |
| ·91 | ·9999 792 | ·9999 681 | ·9999 523 | ·9999 304 | ·9999 005− | ·9998 604 |
| ·92 | ·9999 926 | ·9999 886 | ·9999 829 | ·9999 747 | ·9999 635+ | ·9999 483 |
| ·93 | ·9999 978 | ·9999 965+ | ·9999 947 | ·9999 922 | ·9999 886 | ·9999 837 |
| ·94 | ·9999 995− | ·9999 991 | ·9999 987 | ·9999 980 | ·9999 971 | ·9999 958 |
| ·95 | ·9999 999 | ·9999 998 | ·9999 998 | ·9999 996 | ·9999 994 | ·9999 992 |
| ·96 | 1·0000 000 | 1·0000 000 | 1·0000 000 | 1·0000 000 | ·9999 999 | ·9999 999 |
| ·97 | | | | | 1·0000 000 | 1·0000 000 |

$x$ = ·23 to ·80 $q$ = 10 $p$ = 21 to 26

| | $p$ = 21 | $p$ = 22 | $p$ = 23 | $p$ = 24 | $p$ = 25 | $p$ = 26 |
|---|---|---|---|---|---|---|
| $B(p, q)$ = | ·3328 3392 × $\frac{1}{10^8}$ | ·2254 6814 × $\frac{1}{10^8}$ | ·1550 0934 × $\frac{1}{10^8}$ | ·1080 3682 × $\frac{1}{10^8}$ | ·7626 1281 × $\frac{1}{10^9}$ | ·5447 2344 × $\frac{1}{10^9}$ |
| $x$ | | | | | | |
| ·23 | ·0000 001 | | | | | |
| ·24 | ·0000 001 | | | | | |
| ·25 | ·0000 003 | ·0000 001 | | | | |
| ·26 | ·0000 006 | ·0000 002 | ·0000 001 | | | |
| ·27 | ·0000 011 | ·0000 004 | ·0000 002 | ·0000 001 | | |
| ·28 | ·0000 022 | ·0000 008 | ·0000 003 | ·0000 001 | | |
| ·29 | ·0000 040 | ·0000 016 | ·0000 007 | ·0000 003 | ·0000 001 | |
| ·30 | ·0000 073 | ·0000 031 | ·0000 013 | ·0000 005+ | ·0000 002 | ·0000 001 |
| ·31 | ·0000 129 | ·0000 056 | ·0000 024 | ·0000 010 | ·0000 004 | ·0000 002 |
| ·32 | ·0000 222 | ·0000 099 | ·0000 044 | ·0000 019 | ·0000 008 | ·0000 004 |
| ·33 | ·0000 374 | ·0000 172 | ·0000 078 | ·0000 035+ | ·0000 016 | ·0000 007 |
| ·34 | ·0000 618 | ·0000 293 | ·0000 137 | ·0000 064 | ·0000 029 | ·0000 013 |
| ·35 | ·0001 001 | ·0000 488 | ·0000 236 | ·0000 112 | ·0000 053 | ·0000 025− |
| ·36 | ·0001 592 | ·0000 798 | ·0000 396 | ·0000 194 | ·0000 094 | ·0000 045+ |
| ·37 | ·0002 486 | ·0001 280 | ·0000 652 | ·0000 328 | ·0000 164 | ·0000 081 |
| ·38 | ·0003 816 | ·0002 017 | ·0001 054 | ·0000 545− | ·0000 279 | ·0000 141 |
| ·39 | ·0005 761 | ·0003 124 | ·0001 674 | ·0000 888 | ·0000 466 | ·0000 242 |
| ·40 | ·0008 564 | ·0004 759 | ·0002 615− | ·0001 421 | ·0000 765− | ·0000 408 |
| ·41 | ·0012 539 | ·0007 138 | ·0004 017 | ·0002 237 | ·0001 233 | ·0000 673 |
| ·42 | ·0018 096 | ·0010 544 | ·0006 075− | ·0003 463 | ·0001 955− | ·0001 093 |
| ·43 | ·0025 754 | ·0015 351 | ·0009 048 | ·0005 277 | ·0003 048 | ·0001 744 |
| ·44 | ·0036 162 | ·0022 039 | ·0013 282 | ·0007 922 | ·0004 679 | ·0002 738 |
| ·45 | ·0050 121 | ·0031 214 | ·0019 224 | ·0011 718 | ·0007 073 | ·0004 231 |
| ·46 | ·0068 601 | ·0043 632 | ·0027 448 | ·0017 089 | ·0010 538 | ·0006 439 |
| ·47 | ·0092 758 | ·0060 222 | ·0038 674 | ·0024 583 | ·0015 477 | ·0009 657 |
| ·48 | ·0123 946 | ·0082 100 | ·0053 797 | ·0034 895+ | ·0022 420 | ·0014 277 |
| ·49 | ·0163 727 | ·0110 593 | ·0073 906 | ·0048 896 | ·0032 045− | ·0020 815+ |
| ·50 | ·0213 870 | ·0147 247 | ·0100 308 | ·0067 655− | ·0045 206 | ·0029 941 |
| ·51 | ·0276 339 | ·0193 833 | ·0134 541 | ·0092 469 | ·0062 966 | ·0042 503 |
| ·52 | ·0353 275+ | ·0252 342 | ·0178 385+ | ·0124 879 | ·0086 622 | ·0059 567 |
| ·53 | ·0446 958 | ·0324 968 | ·0233 864 | ·0166 684 | ·0117 727 | ·0082 438 |
| ·54 | ·0559 759 | ·0414 079 | ·0303 229 | ·0219 948 | ·0158 111 | ·0112 698 |
| ·55 | ·0694 073 | ·0522 166 | ·0388 937 | ·0286 989 | ·0209 890 | ·0152 220 |
| ·56 | ·0852 243 | ·0651 783 | ·0493 601 | ·0370 357 | ·0275 460 | ·0203 185+ |
| ·57 | ·1036 463 | ·0805 464 | ·0619 930 | ·0472 794 | ·0357 476 | ·0268 081 |
| ·58 | ·1248 673 | ·0985 625+ | ·0770 644 | ·0597 163 | ·0458 812 | ·0349 681 |
| ·59 | ·1490 444 | ·1194 449 | ·0948 370 | ·0746 371 | ·0582 500− | ·0451 009 |
| ·60 | ·1762 865− | ·1433 764 | ·1155 524 | ·0923 254 | ·0731 631 | ·0575 267 |
| ·61 | ·2066 423 | ·1704 908 | ·1394 170 | ·1130 448 | ·0909 251 | ·0725 746 |
| ·62 | ·2400 900 | ·2008 604 | ·1665 879 | ·1370 242 | ·1118 211 | ·0905 699 |
| ·63 | ·2765 286 | ·2344 829 | ·1971 577 | ·1644 414 | ·1361 009 | ·1118 188 |
| ·64 | ·3157 710 | ·2712 712 | ·2311 405+ | ·1954 068 | ·1639 611 | ·1365 907 |
| ·65 | ·3575 409 | ·3110 451 | ·2684 591 | ·2299 466 | ·1955 260 | ·1650 982 |
| ·66 | ·4014 733 | ·3535 259 | ·3089 348 | ·2679 886 | ·2308 299 | ·1974 762 |
| ·67 | ·4471 190 | ·3983 366 | ·3522 816 | ·3093 503 | ·2697 998 | ·2337 614 |
| ·68 | ·4939 541 | ·4450 055+ | ·3981 039 | ·3537 309 | ·3122 421 | ·2738 733 |
| ·69 | ·5413 934 | ·4929 761 | ·4459 012 | ·4007 094 | ·3578 336 | ·3175 989 |
| ·70 | ·5888 087 | ·5416 215− | ·4950 777 | ·4497 481 | ·4061 184 | ·3645 829 |
| ·71 | ·6355 500+ | ·5902 646 | ·5449 586 | ·5002 042 | ·4565 126 | ·4143 241 |
| ·72 | ·6809 703 | ·6382 026 | ·5948 128 | ·5513 476 | ·5083 170 | ·4661 814 |
| ·73 | ·7244 504 | ·6847 339 | ·6438 801 | ·6023 868 | ·5607 379 | ·5193 889 |
| ·74 | ·7654 251 | ·7291 880 | ·6914 027 | ·6525 002 | ·6129 169 | ·5730 803 |
| ·75 | ·8034 066 | ·7709 548 | ·7366 590 | ·7008 722 | ·6639 670 | ·6263 237 |
| ·76 | ·8380 055− | ·8095 116 | ·7789 973 | ·7467 317 | ·7130 142 | ·6781 638 |
| ·77 | ·8689 456 | ·8444 473 | ·8178 666 | ·7893 905+ | ·7592 416 | ·7276 695+ |
| ·78 | ·8960 745− | ·8754 793 | ·8528 432 | ·8282 782 | ·8019 322 | ·7739 844 |
| ·79 | ·9193 654 | ·9024 643 | ·8836 503 | ·8629 712 | ·8405 086 | ·8163 748 |
| ·80 | ·9389 129 | ·9254 001 | ·9101 676 | ·8932 132 | ·8745 632 | ·8542 721 |

$x = \cdot 81$ to $\cdot 98$ $q = 10$ $p = 21$ to $26$

| $x$ | $p = 21$ | $p = 22$ | $p = 23$ | $p = 24$ | $p = 25$ | $p = 26$ |
|---|---|---|---|---|---|---|
| $B(p, q) =$ | $\cdot 3328\ 3392 \times \frac{1}{10^8}$ | $\cdot 2254\ 6814 \times \frac{1}{10^8}$ | $\cdot 1550\ 0934 \times \frac{1}{10^8}$ | $\cdot 1080\ 3682 \times \frac{1}{10^8}$ | $\cdot 7626\ 1281 \times \frac{1}{10^9}$ | $\cdot 5447\ 2344 \times \frac{1}{10^9}$ |
| ·81 | ·9549 209 | ·9444 188 | ·9324 322 | ·9189 237 | ·9038 787 | ·8873 051 |
| ·82 | ·9676 850+ | ·9597 715+ | ·9506 278 | ·9401 960 | ·9284 342 | ·9153 175− |
| ·83 | ·9775 693 | ·9718 059 | ·9650 652 | ·9572 813 | ·9483 979 | ·9383 703 |
| ·84 | ·9849 807 | ·9809 385− | ·9761 540 | ·9705 624 | ·9641 040 | ·9567 261 |
| ·85 | ·9903 424 | ·9876 243 | ·9843 688 | ·9805 187 | ·9760 190 | ·9708 174 |
| ·86 | ·9940 687 | ·9923 257 | ·9902 135− | ·9876 862 | ·9846 976 | ·9812 022 |
| ·87 | ·9965 437 | ·9954 848 | ·9941 865+ | ·9926 151 | ·9907 353 | ·9885 111 |
| ·88 | ·9981 047 | ·9975 001 | ·9967 503 | ·9958 323 | ·9947 216 | ·9933 922 |
| ·89 | ·9990 320 | ·9987 109 | ·9983 082 | ·9978 095+ | ·9971 993 | ·9964 606 |
| ·90 | ·9995 456 | ·9993 891 | ·9991 906 | ·9989 420 | ·9986 343 | ·9982 578 |
| ·91 | ·9998 074 | ·9997 386 | ·9996 503 | ·9995 385+ | ·9993 987 | ·9992 256 |
| ·92 | ·9999 280 | ·9999 013 | ·9998 668 | ·9998 225+ | ·9997 665+ | ·9996 965− |
| ·93 | ·9999 770 | ·9999 682 | ·9999 567 | ·9999 418 | ·9999 227 | ·9998 985+ |
| ·94 | ·9999 941 | ·9999 917 | ·9999 886 | ·9999 845− | ·9999 792 | ·9999 724 |
| ·95 | ·9999 988 | ·9999 984 | ·9999 977 | ·9999 969 | ·9999 958 | ·9999 944 |
| ·96 | ·9999 999 | ·9999 998 | ·9999 997 | ·9999 996 | ·9999 994 | ·9999 992 |
| ·97 | 1·0000 000 | 1·0000 000 | 1·0000 000 | 1·0000 000 | 1·0000 000 | ·9999 999 |
| ·98 | | | | | | 1·0000 000 |

$x$ = ·31 to ·98 $q = 10$ $p$ = 27 to 32

| | $p = 27$ | $p = 28$ | $p = 29$ | $p = 30$ | $p = 31$ | $p = 32$ |
|---|---|---|---|---|---|---|
| $B(p, q)$ = | ·3934 1137 × $\frac{1}{10^9}$ | ·2870 8397 × $\frac{1}{10^9}$ | ·2115 3556 × $\frac{1}{10^9}$ | ·1572 9567 × $\frac{1}{10^9}$ | ·1179 7175 × $\frac{1}{10^9}$ | ·8919 8156 × $\frac{1}{10^{10}}$ |
| $x$ | | | | | | |
| ·31 | ·0000 001 | | | | | |
| ·32 | ·0000 001 | ·0000 001 | | | | |
| ·33 | ·0000 003 | ·0000 001 | ·0000 001 | | | |
| ·34 | ·0000 006 | ·0000 003 | ·0000 001 | ·0000 001 | | |
| ·35 | ·0000 011 | ·0000 005+ | ·0000 002 | ·0000 001 | | |
| ·36 | ·0000 022 | ·0000 010 | ·0000 005− | ·0000 002 | ·0000 001 | |
| ·37 | ·0000 040 | ·0000 019 | ·0000 009 | ·0000 004 | ·0000 002 | ·0000 001 |
| ·38 | ·0000 071 | ·0000 035+ | ·0000 017 | ·0000 009 | ·0000 004 | ·0000 002 |
| ·39 | ·0000 125− | ·0000 064 | ·0000 032 | ·0000 016 | ·0000 008 | ·0000 004 |
| ·40 | ·0000 215+ | ·0000 113 | ·0000 059 | ·0000 030 | ·0000 016 | ·0000 008 |
| ·41 | ·0000 365− | ·0000 196 | ·0000 104 | ·0000 055+ | ·0000 029 | ·0000 015+ |
| ·42 | ·0000 606 | ·0000 333 | ·0000 182 | ·0000 098 | ·0000 053 | ·0000 028 |
| ·43 | ·0000 989 | ·0000 557 | ·0000 311 | ·0000 172 | ·0000 095− | ·0000 052 |
| ·44 | ·0001 588 | ·0000 914 | ·0000 522 | ·0000 296 | ·0000 167 | ·0000 093 |
| ·45 | ·0002 509 | ·0001 476 | ·0000 862 | ·0000 499 | ·0000 287 | ·0000 164 |
| ·46 | ·0003 901 | ·0002 345− | ·0001 398 | ·0000 828 | ·0000 487 | ·0000 285− |
| ·47 | ·0005 974 | ·0003 667 | ·0002 233 | ·0001 351 | ·0000 811 | ·0000 484 |
| ·48 | ·0009 015− | ·0005 647 | ·0003 511 | ·0002 167 | ·0001 329 | ·0000 810 |
| ·49 | ·0013 408 | ·0008 569 | ·0005 435+ | ·0003 423 | ·0002 142 | ·0001 331 |
| ·50 | ·0019 666 | ·0012 816 | ·0008 290 | ·0005 325+ | ·0003 398 | ·0002 154 |
| ·51 | ·0028 455− | ·0018 902 | ·0012 463 | ·0008 161 | ·0005 308 | ·0003 431 |
| ·52 | ·0040 628 | ·0027 497 | ·0018 474 | ·0012 327 | ·0008 171 | ·0005 382 |
| ·53 | ·0057 262 | ·0039 470 | ·0027 009 | ·0018 356 | ·0012 393 | ·0008 316 |
| ·54 | ·0079 687 | ·0055 918 | ·0038 958 | ·0026 957 | ·0018 532 | ·0012 662 |
| ·55 | ·0109 522 | ·0078 211 | ·0055 454 | ·0039 053 | ·0027 327 | ·0019 005− |
| ·56 | ·0148 702 | ·0108 021 | ·0077 917 | ·0055 827 | ·0039 745− | ·0028 125− |
| ·57 | ·0199 488 | ·0147 357 | ·0108 091 | ·0078 764 | ·0057 032 | ·0041 049 |
| ·58 | ·0264 475+ | ·0198 582 | ·0148 080 | ·0109 699 | ·0080 759 | ·0059 101 |
| ·59 | ·0346 575+ | ·0264 420 | ·0200 369 | ·0150 851 | ·0112 870 | ·0083 956 |
| ·60 | ·0448 973 | ·0347 938 | ·0267 831 | ·0204 851 | ·0155 726 | ·0117 694 |
| ·61 | ·0575 060 | ·0452 502 | ·0353 712 | ·0274 748 | ·0212 129 | ·0162 842 |
| ·62 | ·0728 333 | ·0581 710 | ·0461 584 | ·0363 992 | ·0285 333 | ·0222 406 |
| ·63 | ·0912 257 | ·0739 275− | ·0595 267 | ·0476 386 | ·0379 023 | ·0299 876 |
| ·64 | ·1130 101 | ·0928 880 | ·0758 705− | ·0615 992 | ·0497 255− | ·0399 201 |
| ·65 | ·1384 739 | ·1153 995+ | ·0955 803 | ·0786 997 | ·0644 357 | ·0524 723 |
| ·66 | ·1678 428 | ·1417 654 | ·1190 221 | ·0993 531 | ·0824 772 | ·0681 053 |
| ·67 | ·2012 575+ | ·1722 208 | ·1465 129 | ·1239 431 | ·1042 848 | ·0872 899 |
| ·68 | ·2387 510 | ·2069 069 | ·1782 926 | ·1527 964 | ·1302 576 | ·1104 817 |
| ·69 | ·2802 271 | ·2458 451 | ·2144 960 | ·1861 522 | ·1607 277 | ·1380 917 |
| ·70 | ·3254 436 | ·2889 137 | ·2551 235+ | ·2241 297 | ·1959 254 | ·1704 505+ |
| ·71 | ·3740 015+ | ·3358 296 | ·3000 161 | ·2666 972 | ·2359 438 | ·2077 698 |
| ·72 | ·4253 424 | ·3861 369 | ·3488 356 | ·3136 438 | ·2807 043 | ·2501 025− |
| ·73 | ·4787 556 | ·4392 058 | ·4010 543 | ·3645 605+ | ·3299 279 | ·2973 062 |
| ·74 | ·5333 969 | ·4942 426 | ·4559 553 | ·4188 299 | ·3831 152 | ·3490 134 |
| ·75 | ·5883 185− | ·5503 132 | ·5126 473 | ·4756 308 | ·4395 397 | ·4046 129 |
| ·76 | ·6425 091 | ·6063 791 | ·5700 942 | ·5339 594 | ·4982 582 | ·4632 481 |
| ·77 | ·6949 439 | ·6613 456 | ·6271 594 | ·5926 666 | ·5581 393 | ·5238 348 |
| ·78 | ·7446 392 | ·7141 202 | ·6826 639 | ·6505 133 | ·6179 126 | ·5851 015+ |
| ·79 | ·7907 094 | ·7636 752 | ·7354 534 | ·7062 390 | ·6762 358 | ·6456 519 |
| ·80 | ·8324 202 | ·8091 115− | ·7844 708 | ·7586 405+ | ·7317 771 | ·7040 471 |
| ·81 | ·8692 334 | ·8497 161 | ·8288 256 | ·8066 528 | ·7833 048 | ·7589 025− |
| ·82 | ·9008 385+ | ·8850 083 | ·8678 550+ | ·8494 242 | ·8297 769 | ·8089 888 |
| ·83 | ·9271 663 | ·9147 673 | ·9011 683 | ·8863 781 | ·8704 196 | ·8533 284 |
| ·84 | ·9483 833 | ·9390 393 | ·9286 676 | ·9172 515− | ·9047 851 | ·8912 731 |
| ·85 | ·9648 655− | ·9581 200 | ·9505 434 | ·9421 046 | ·9327 797 | ·9225 525− |
| ·86 | ·9771 556 | ·9725 155+ | ·9672 424 | ·9613 001 | ·9546 566 | ·9472 845− |
| ·87 | ·9859 063 | ·9828 846 | ·9794 108 | ·9754 507 | ·9709 718 | ·9659 438 |
| ·88 | ·9918 175− | ·9899 698 | ·9878 212 | ·9853 436 | ·9825 092 | ·9792 909 |
| ·89 | ·9955 757 | ·9945 255+ | ·9932 905+ | ·9918 502 | ·9901 838 | ·9882 700 |
| ·90 | ·9978 016 | ·9972 541 | ·9966 031 | ·9958 353 | ·9949 369 | ·9938 937 |
| ·91 | ·9990 136 | ·9987 564 | ·9984 470 | ·9980 782 | ·9976 419 | ·9971 295+ |
| ·92 | ·9996 097 | ·9995 033 | ·9993 740 | ·9992 180 | ·9990 315+ | ·9988 101 |
| ·93 | ·9998 683 | ·9998 308 | ·9997 848 | ·9997 286 | ·9996 608 | ·9995 793 |
| ·94 | ·9999 639 | ·9999 532 | ·9999 399 | ·9999 235+ | ·9999 035− | ·9998 792 |
| ·95 | ·9999 925+ | ·9999 902 | ·9999 874 | ·9999 838 | ·9999 793 | ·9999 739 |
| ·96 | ·9999 990 | ·9999 987 | ·9999 982 | ·9999 977 | ·9999 971 | ·9999 963 |
| ·97 | ·9999 999 | ·9999 999 | ·9999 999 | ·9999 998 | ·9999 998 | ·9999 997 |
| ·98 | 1·0000 000 | 1·0000 000 | 1·0000 000 | 1·0000 000 | 1·0000 000 | 1·0000 000 |

$x = {}$·38 to ·98 $q = 10$ $p = 33$ to 38

| | $p = 33$ | $p = 34$ | $p = 35$ | $p = 36$ | $p = 37$ | $p = 38$ |
|---|---|---|---|---|---|---|
| $B(p, q) =$ | ·6796 0500 × $\frac{1}{10^{10}}$ | ·5215 5732 × $\frac{1}{10^{10}}$ | ·4030 2157 × $\frac{1}{10^{10}}$ | ·3134 6122 × $\frac{1}{10^{10}}$ | ·2453 1748 × $\frac{1}{10^{10}}$ | ·1931 2227 × $\frac{1}{10^{10}}$ |
| $x$ | | | | | | |
| ·38 | ·0000 001 | | | | | |
| ·39 | ·0000 002 | ·0000 001 | | | | |
| ·40 | ·0000 004 | ·0000 002 | ·0000 001 | ·0000 001 | | |
| ·41 | ·0000 008 | ·0000 004 | ·0000 002 | ·0000 001 | ·0000 001 | |
| ·42 | ·0000 015+ | ·0000 008 | ·0000 004 | ·0000 002 | ·0000 001 | ·0000 001 |
| ·43 | ·0000 028 | ·0000 015+ | ·0000 008 | ·0000 004 | ·0000 002 | ·0000 001 |
| ·44 | ·0000 052 | ·0000 029 | ·0000 016 | ·0000 009 | ·0000 005− | ·0000 003 |
| ·45 | ·0000 093 | ·0000 053 | ·0000 030 | ·0000 017 | ·0000 009 | ·0000 005+ |
| ·46 | ·0000 165+ | ·0000 095+ | ·0000 055− | ·0000 031 | ·0000 018 | ·0000 010 |
| ·47 | ·0000 287 | ·0000 169 | ·0000 099 | ·0000 058 | ·0000 034 | ·0000 019 |
| ·48 | ·0000 490 | ·0000 295+ | ·0000 177 | ·0000 105+ | ·0000 062 | ·0000 037 |
| ·49 | ·0000 823 | ·0000 505+ | ·0000 309 | ·0000 188 | ·0000 114 | ·0000 068 |
| ·50 | ·0001 358 | ·0000 851 | ·0000 530 | ·0000 329 | ·0000 203 | ·0000 125− |
| ·51 | ·0002 205− | ·0001 409 | ·0000 895+ | ·0000 566 | ·0000 356 | ·0000 223 |
| ·52 | ·0003 524 | ·0002 295− | ·0001 486 | ·0000 957 | ·0000 614 | ·0000 392 |
| ·53 | ·0005 547 | ·0003 679 | ·0002 428 | ·0001 593 | ·0001 041 | ·0000 677 |
| ·54 | ·0008 601 | ·0005 810 | ·0003 903 | ·0002 609 | ·0001 736 | ·0001 149 |
| ·55 | ·0013 140 | ·0009 035+ | ·0006 180 | ·0004 205+ | ·0002 848 | ·0001 920 |
| ·56 | ·0019 787 | ·0013 845− | ·0009 636 | ·0006 673 | ·0004 599 | ·0003 155+ |
| ·57 | ·0029 376 | ·0020 908 | ·0014 803 | ·0010 429 | ·0007 312 | ·0005 104 |
| ·58 | ·0043 006 | ·0031 125+ | ·0022 410 | ·0016 056 | ·0011 449 | ·0008 127 |
| ·59 | ·0062 099 | ·0045 686 | ·0033 439 | ·0024 355+ | ·0017 656 | ·0012 742 |
| ·60 | ·0088 457 | ·0066 131 | ·0049 189 | ·0036 410 | ·0026 826 | ·0019 676 |
| ·61 | ·0124 322 | ·0094 416 | ·0071 344 | ·0053 651 | ·0040 160 | ·0029 929 |
| ·62 | ·0172 418 | ·0132 973 | ·0102 044 | ·0077 937 | ·0059 253 | ·0044 852 |
| ·63 | ·0235 989 | ·0184 764 | ·0143 949 | ·0111 624 | ·0086 168 | ·0066 230 |
| ·64 | ·0318 797 | ·0253 304 | ·0200 293 | ·0157 643 | ·0123 522 | ·0096 373 |
| ·65 | ·0425 090 | ·0342 667 | ·0274 910 | ·0219 543 | ·0174 557 | ·0138 203 |
| ·66 | ·0559 522 | ·0457 435+ | ·0372 223 | ·0301 521 | ·0243 192 | ·0195 331 |
| ·67 | ·0727 008 | ·0602 604 | ·0497 189 | ·0408 399 | ·0334 038 | ·0272 097 |
| ·68 | ·0932 520 | ·0783 404 | ·0655 165− | ·0545 538 | ·0452 356 | ·0373 579 |
| ·69 | ·1180 801 | ·1005 063 | ·0851 706 | ·0718 679 | ·0603 943 | ·0505 519 |
| ·70 | ·1476 028 | ·1272 475− | ·1092 271 | ·0933 691 | ·0794 934 | ·0674 177 |
| ·71 | ·1821 402 | ·1589 804 | ·1381 843 | ·1196 222 | ·1031 484 | ·0886 069 |
| ·72 | ·2218 722 | ·1960 031 | ·1724 469 | ·1511 253 | ·1319 357 | ·1147 585+ |
| ·73 | ·2667 947 | ·2384 468 | ·2122 750− | ·1882 567 | ·1663 401 | ·1464 492 |
| ·74 | ·3166 806 | ·2862 291 | ·2577 300 | ·2312 177 | ·2066 938 | ·1841 318 |
| ·75 | ·3710 503 | ·3390 134 | ·3086 254 | ·2799 738 | ·2531 130 | ·2280 671 |
| ·76 | ·4291 570 | ·3961 815+ | ·3644 863 | ·3342 038 | ·3054 354 | ·2782 532 |
| ·77 | ·4899 913 | ·4568 246 | ·4245 262 | ·3932 612 | ·3631 687 | ·3343 613 |
| ·78 | ·5523 110 | ·5197 590 | ·4876 474 | ·4561 596 | ·4254 591 | ·3956 878 |
| ·79 | ·6146 952 | ·5835 697 | ·5524 717 | ·5215 869 | ·4910 881 | ·4611 334 |
| ·80 | ·6756 239 | ·6466 838 | ·6174 033 | ·5879 555+ | ·5585 077 | ·5292 190 |
| ·81 | ·7335 774 | ·7074 695+ | ·6807 244 | ·6534 901 | ·6259 154 | ·5981 470 |
| ·82 | ·7871 483 | ·7643 548 | ·7407 165+ | ·7163 489 | ·6913 721 | ·6659 092 |
| ·83 | ·8351 531 | ·8159 534 | ·7957 993 | ·7747 699 | ·7529 519 | ·7304 382 |
| ·84 | ·8767 309 | ·8611 839 | ·8446 675+ | ·8272 263 | ·8089 129 | ·7897 878 |
| ·85 | ·9114 143 | ·8993 649 | ·8864 118 | ·8725 704 | ·8578 640 | ·8423 228 |
| ·86 | ·9391 613 | ·9302 701 | ·9205 997 | ·9101 446 | ·8989 053 | ·8868 884 |
| ·87 | ·9603 392 | ·9541 334 | ·9473 052 | ·9398 371 | ·9317 155+ | ·9229 311 |
| ·88 | ·9756 621 | ·9715 980 | ·9670 748 | ·9620 709 | ·9565 666 | ·9505 446 |
| ·89 | ·9860 878 | ·9836 159 | ·9808 336 | ·9777 206 | ·9742 574 | ·9704 254 |
| ·90 | ·9926 908 | ·9913 128 | ·9897 444 | ·9879 698 | ·9859 734 | ·9837 396 |
| ·91 | ·9965 321 | ·9958 403 | ·9950 440 | ·9941 331 | ·9930 969 | ·9919 247 |
| ·92 | ·9985 491 | ·9982 436 | ·9978 880 | ·9974 768 | ·9970 039 | ·9964 630 |
| ·93 | ·9994 823 | ·9993 674 | ·9992 324 | ·9990 744 | ·9988 908 | ·9986 785+ |
| ·94 | ·9998 499 | ·9998 150− | ·9997 734 | ·9997 242 | ·9996 664 | ·9995 989 |
| ·95 | ·9999 672 | ·9999 592 | ·9999 496 | ·9999 381 | ·9999 244 | ·9999 083 |
| ·96 | ·9999 953 | ·9999 940 | ·9999 926 | ·9999 908 | ·9999 886 | ·9999 861 |
| ·97 | ·9999 996 | ·9999 995+ | ·9999 994 | ·9999 993 | ·9999 991 | ·9999 989 |
| ·98 | 1·0000 000 | 1·0000 000 | 1·0000 000 | 1·0000 000 | 1·0000 000 | 1·0000 000 |

$x$ = ·43 to ·99 $q$ = 10 $p$ = 39 to 44

| | $p$ = 39 | $p$ = 40 | $p$ = 41 | $p$ = 42 | $p$ = 43 | $p$ = 44 |
|---|---|---|---|---|---|---|
| $B(p, q)$ = | ·1528 8846 $\times\frac{1}{10^{10}}$ | ·1216 8674 $\times\frac{1}{10^{10}}$ | ·9734 9389 $\times\frac{1}{10^{11}}$ | ·7826 1273 $\times\frac{1}{10^{11}}$ | ·6321 1028 $\times\frac{1}{10^{11}}$ | ·5128 4419 $\times\frac{1}{10^{11}}$ |
| $x$ | | | | | | |
| ·43 | ·0000 001 | | | | | |
| ·44 | ·0000 001 | ·0000 001 | | | | |
| ·45 | ·0000 003 | ·0000 002 | ·0000 001 | | | |
| ·46 | ·0000 006 | ·0000 003 | ·0000 002 | ·0000 001 | ·0000 001 | |
| ·47 | ·0000 011 | ·0000 006 | ·0000 004 | ·0000 002 | ·0000 001 | ·0000 001 |
| ·48 | ·0000 022 | ·0000 013 | ·0000 007 | ·0000 004 | ·0000 002 | ·0000 001 |
| ·49 | ·0000 041 | ·0000 024 | ·0000 015− | ·0000 009 | ·0000 005+ | ·0000 003 |
| ·50 | ·0000 076 | ·0000 046 | ·0000 028 | ·0000 017 | ·0000 010 | ·0000 006 |
| ·51 | ·0000 139 | ·0000 086 | ·0000 053 | ·0000 033 | ·0000 020 | ·0000 012 |
| ·52 | ·0000 249 | ·0000 157 | ·0000 099 | ·0000 062 | ·0000 039 | ·0000 024 |
| ·53 | ·0000 438 | ·0000 282 | ·0000 181 | ·0000 116 | ·0000 074 | ·0000 047 |
| ·54 | ·0000 758 | ·0000 497 | ·0000 325+ | ·0000 212 | ·0000 137 | ·0000 089 |
| ·55 | ·0001 288 | ·0000 861 | ·0000 573 | ·0000 380 | ·0000 251 | ·0000 165+ |
| ·56 | ·0002 155+ | ·0001 466 | ·0000 993 | ·0000 670 | ·0000 450+ | ·0000 302 |
| ·57 | ·0003 547 | ·0002 454 | ·0001 692 | ·0001 161 | ·0000 794 | ·0000 541 |
| ·58 | ·0005 744 | ·0004 043 | ·0002 834 | ·0001 979 | ·0001 377 | ·0000 955− |
| ·59 | ·0009 156 | ·0006 552 | ·0004 670 | ·0003 316 | ·0002 346 | ·0001 654 |
| ·60 | ·0014 371 | ·0010 453 | ·0007 573 | ·0005 466 | ·0003 931 | ·0002 817 |
| ·61 | ·0022 210 | ·0016 415+ | ·0012 085− | ·0008 863 | ·0006 477 | ·0004 717 |
| ·62 | ·0033 809 | ·0025 382 | ·0018 982 | ·0014 142 | ·0010 499 | ·0007 767 |
| ·63 | ·0050 694 | ·0038 647 | ·0029 351 | ·0022 208 | ·0016 744 | ·0012 581 |
| ·64 | ·0074 882 | ·0057 954 | ·0044 683 | ·0034 325− | ·0026 275− | ·0020 044 |
| ·65 | ·0108 977 | ·0085 596 | ·0066 979 | ·0052 222 | ·0040 574 | ·0031 418 |
| ·66 | ·0156 261 | ·0124 524 | ·0098 865− | ·0078 212 | ·0061 661 | ·0048 450+ |
| ·67 | ·0220 768 | ·0178 441 | ·0143 701 | ·0115 316 | ·0092 223 | ·0073 512 |
| ·68 | ·0307 323 | ·0251 872 | ·0205 682 | ·0167 377 | ·0135 749 | ·0109 740 |
| ·69 | ·0421 521 | ·0350 188 | ·0289 894 | ·0239 158 | ·0196 649 | ·0161 178 |
| ·70 | ·0569 628 | ·0479 554 | ·0402 316 | ·0336 381 | ·0280 337 | ·0232 894 |
| ·71 | ·0758 372 | ·0646 784 | ·0549 731 | ·0465 697 | ·0393 247 | ·0331 042 |
| ·72 | ·0994 618 | ·0859 065+ | ·0739 508 | ·0634 530 | ·0542 750+ | ·0462 837 |
| ·73 | ·1284 899 | ·1123 541 | ·0979 247 | ·0850 790 | ·0736 922 | ·0636 400 |
| ·74 | ·1634 817 | ·1446 742 | ·1276 252 | ·1122 395− | ·0984 144 | ·0860 425− |
| ·75 | ·2048 338 | ·1833 876 | ·1636 839 | ·1456 622 | ·1292 497 | ·1143 638 |
| ·76 | ·2527 019 | ·2288 016 | ·2065 504 | ·1859 274 | ·1668 953 | ·1494 034 |
| ·77 | ·3069 260 | ·2809 257 | ·2564 010 | ·2333 717 | ·2118 393 | ·1917 891 |
| ·78 | ·3669 664 | ·3393 939 | ·3130 483 | ·2879 879 | ·2642 521 | ·2418 631 |
| ·79 | ·4318 645+ | ·4034 061 | ·3758 655+ | ·3493 325+ | ·3238 798 | ·2995 635+ |
| ·80 | ·5002 387 | ·4717 042 | ·4437 404 | ·4164 587 | ·3899 564 | ·3643 170 |
| ·81 | ·5703 274 | ·5425 934 | ·5150 744 | ·4878 909 | ·4611 540 | ·4349 643 |
| ·82 | ·6400 845− | ·6140 214 | ·5878 410 | ·5616 606 | ·5355 924 | ·5097 425+ |
| ·83 | ·7073 260 | ·6837 160 | ·6597 106 | ·6354 124 | ·6109 233 | ·5863 430 |
| ·84 | ·7699 179 | ·7493 754 | ·7282 372 | ·7065 834 | ·6844 966 | ·6620 605− |
| ·85 | ·8259 842 | ·8088 915− | ·7910 937 | ·7726 447 | ·7536 028 | ·7340 294 |
| ·86 | ·8741 062 | ·8605 767 | ·8463 234 | ·8313 748 | ·8157 643 | ·7995 293 |
| ·87 | ·9134 785− | ·9033 570 | ·8925 700 | ·8811 253 | ·8690 347 | ·8563 143 |
| ·88 | ·9439 901 | ·9368 911 | ·9292 384 | ·9210 257 | ·9122 499 | ·9029 108 |
| ·89 | ·9662 071 | ·9615 865+ | ·9565 489 | ·9510 813 | ·9451 723 | ·9388 126 |
| ·90 | ·9812 531 | ·9784 987 | ·9754 621 | ·9721 291 | ·9684 867 | ·9645 224 |
| ·91 | ·9906 053 | ·9891 275+ | ·9874 802 | ·9856 521 | ·9836 320 | ·9814 090 |
| ·92 | ·9958 475− | ·9951 506 | ·9943 651 | ·9934 839 | ·9924 995− | ·9914 042 |
| ·93 | ·9984 344 | ·9981 549 | ·9978 364 | ·9974 753 | ·9970 675− | ·9966 088 |
| ·94 | ·9995 205− | ·9994 297 | ·9993 251 | ·9992 053 | ·9990 685− | ·9989 130 |
| ·95 | ·9998 894 | ·9998 672 | ·9998 414 | ·9998 116 | ·9997 771 | ·9997 375+ |
| ·96 | ·9999 831 | ·9999 795− | ·9999 753 | ·9999 704 | ·9999 646 | ·9999 579 |
| ·97 | ·9999 986 | ·9999 983 | ·9999 980 | ·9999 976 | ·9999 971 | ·9999 965− |
| ·98 | 1·0000 000 | 1·0000 000 | 1·0000 000 | ·9999 999 | ·9999 999 | ·9999 999 |
| ·99 | | | | 1·0000 000 | 1·0000 000 | 1·0000 000 |

$x$ = ·48 to ·99 $q$ = 10 $p$ = 45 to 50

| $x$ | $p$ = 45 | $p$ = 46 | $p$ = 47 | $p$ = 48 | $p$ = 49 | $p$ = 50 |
|---|---|---|---|---|---|---|
| $B(p, q)$ = | ·4178 7305$\bar{\times}\frac{1}{10^{11}}$ | ·3418 9613$\times\frac{1}{10^{11}}$ | ·2808 4325$\bar{\times}\frac{1}{10^{11}}$ | ·2315 7250$\overset{+}{\times}\frac{1}{10^{11}}$ | ·1916 4621$\times\frac{1}{10^{11}}$ | ·1591 6380$\times\frac{1}{10^{11}}$ |
| ·48 | ·0000 001 | | | | | |
| ·49 | ·0000 002 | ·0000 001 | ·0000 001 | | | |
| ·50 | ·0000 004 | ·0000 002 | ·0000 001 | ·0000 001 | | |
| ·51 | ·0000 007 | ·0000 005$^{-}$ | ·0000 003 | ·0000 002 | ·0000 001 | ·0000 001 |
| ·52 | ·0000 015$^{+}$ | ·0000 009 | ·0000 006 | ·0000 004 | ·0000 002 | ·0000 001 |
| ·53 | ·0000 030 | ·0000 019 | ·0000 012 | ·0000 007 | ·0000 005$^{-}$ | ·0000 003 |
| ·54 | ·0000 057 | ·0000 037 | ·0000 023 | ·0000 015$^{-}$ | ·0000 010 | ·0000 006 |
| ·55 | ·0000 108 | ·0000 071 | ·0000 046 | ·0000 030 | ·0000 019 | ·0000 013 |
| ·56 | ·0000 201 | ·0000 134 | ·0000 089 | ·0000 059 | ·0000 039 | ·0000 025$^{+}$ |
| ·57 | ·0000 368 | ·0000 249 | ·0000 168 | ·0000 113 | ·0000 076 | ·0000 051 |
| ·58 | ·0000 660 | ·0000 454 | ·0000 312 | ·0000 213 | ·0000 146 | ·0000 099 |
| ·59 | ·0001 162 | ·0000 814 | ·0000 568 | ·0000 395$^{+}$ | ·0000 274 | ·0000 190 |
| ·60 | ·0002 012 | ·0001 432 | ·0001 016 | ·0000 719 | ·0000 507 | ·0000 357 |
| ·61 | ·0003 423 | ·0002 477 | ·0001 786 | ·0001 284 | ·0000 921 | ·0000 658 |
| ·62 | ·0005 727 | ·0004 210 | ·0003 085$^{-}$ | ·0002 253 | ·0001 641 | ·0001 192 |
| ·63 | ·0009 422 | ·0007 033 | ·0005 234 | ·0003 884 | ·0002 874 | ·0002 120 |
| ·64 | ·0015 242 | ·0011 553 | ·0008 730 | ·0006 578 | ·0004 942 | ·0003 703 |
| ·65 | ·0024 250$^{-}$ | ·0018 658 | ·0014 313 | ·0010 947 | ·0008 350$^{-}$ | ·0006 351 |
| ·66 | ·0037 948 | ·0029 630 | ·0023 066 | ·0017 905$^{-}$ | ·0013 859 | ·0010 699 |
| ·67 | ·0058 411 | ·0046 271 | ·0036 545$^{-}$ | ·0028 781 | ·0022 603 | ·0017 704 |
| ·68 | ·0088 436 | ·0071 053 | ·0056 919 | ·0045 467 | ·0036 221 | ·0028 778 |
| ·69 | ·0131 698 | ·0107 288 | ·0087 149 | ·0070 593 | ·0057 027 | ·0045 948 |
| ·70 | ·0192 892 | ·0159 290 | ·0131 166 | ·0107 710 | ·0088 212 | ·0072 056 |
| ·71 | ·0277 842 | ·0232 516 | ·0194 038 | ·0161 488 | ·0134 043 | ·0110 979 |
| ·72 | ·0393 531 | ·0333 651 | ·0282 101 | ·0237 878 | ·0200 068 | ·0167 844 |
| ·73 | ·0548 010 | ·0470 580 | ·0402 997 | ·0344 214 | ·0293 257 | ·0249 225$^{+}$ |
| ·74 | ·0750 146 | ·0652 219 | ·0565 574 | ·0489 180 | ·0422 048 | ·0363 246 |
| ·75 | ·1009 158 | ·0888 126 | ·0779 592 | ·0682 604 | ·0596 224 | ·0519 540 |
| ·76 | ·1333 904 | ·1187 866 | ·1055 162 | ·0934 993 | ·0826 542 | ·0728 979 |
| ·77 | ·1731 926 | ·1560 094 | ·1401 896 | ·1256 758 | ·1124 048 | ·1003 092 |
| ·78 | ·2208 277 | ·2011 386 | ·1827 763 | ·1657 111 | ·1499 044 | ·1353 107 |
| ·79 | ·2764 244 | ·2544 885$^{+}$ | ·2337 687 | ·2142 655$^{+}$ | ·1959 691 | ·1788 602 |
| ·80 | ·3396 099 | ·3158 911 | ·2932 036 | ·2715 780 | ·2510 337 | ·2315 795$^{+}$ |
| ·81 | ·4094 115$^{+}$ | ·3845 742 | ·3605 198 | ·3373 047 | ·3149 747 | ·2935 652 |
| ·82 | ·4842 098 | ·4590 857 | ·4344 531 | ·4103 865$^{+}$ | ·3869 517 | ·3642 055$^{+}$ |
| ·83 | ·5617 684 | ·5372 921 | ·5130 020 | ·4889 806 | ·4653 045$^{+}$ | ·4420 440 |
| ·84 | ·6393 592 | ·6164 763 | ·5934 939 | ·5704 920 | ·5475 476 | ·5247 342 |
| ·85 | ·7139 890 | ·6935 478 | ·6727 732 | ·6517 335$^{-}$ | ·6304 965$^{+}$ | ·6091 295$^{+}$ |
| ·86 | ·7827 113 | ·7653 551 | ·7475 084 | ·7292 213 | ·7105 456 | ·6915 344 |
| ·87 | ·8429 840 | ·8290 670 | ·8145 904 | ·7995 840 | ·7840 805$^{+}$ | ·7681 151 |
| ·88 | ·8930 113 | ·8825 575$^{-}$ | ·8715 582 | ·8600 254 | ·8479 735$^{+}$ | ·8354 200 |
| ·89 | ·9319 948 | ·9247 133 | ·9169 648 | ·9087 482 | ·9000 642 | ·8909 159 |
| ·90 | ·9602 248 | ·9555 833 | ·9505 887 | ·9452 327 | ·9395 085$^{+}$ | ·9334 106 |
| ·91 | ·9789 722 | ·9763 113 | ·9734 161 | ·9702 770 | ·9668 848 | ·9632 309 |
| ·92 | ·9901 905$^{-}$ | ·9888 505$^{+}$ | ·9873 765$^{+}$ | ·9857 608 | ·9839 956 | ·9820 734 |
| ·93 | ·9960 950$^{+}$ | ·9955 216 | ·9948 840 | ·9941 775$^{+}$ | ·9933 973 | ·9925 384 |
| ·94 | ·9987 369 | ·9985 383 | ·9983 150$^{+}$ | ·9980 650$^{+}$ | ·9977 860 | ·9974 754 |
| ·95 | ·9996 922 | ·9996 406 | ·9995 819 | ·9995 155$^{+}$ | ·9994 406 | ·9993 564 |
| ·96 | ·9999 502 | ·9999 413 | ·9999 311 | ·9999 195$^{-}$ | ·9999 062 | ·9998 911 |
| ·97 | ·9999 958 | ·9999 950$^{-}$ | ·9999 941 | ·9999 930 | ·9999 918 | ·9999 904 |
| ·98 | ·9999 999 | ·9999 999 | ·9999 998 | ·9999 998 | ·9999 998 | ·9999 997 |
| ·99 | 1·0000 000 | 1·0000 000 | 1·0000 000 | 1·0000 000 | 1·0000 000 | 1·0000 000 |

$x$ = ·07 to ·70 $q$ = 10·5 $p$ = 10·5 to 15

| $x$ | $p$ = 10·5 | $p$ = 11 | $p$ = 12 | $p$ = 13 | $p$ = 14 | $p$ = 15 |
|---|---|---|---|---|---|---|
| $B(p, q)$ = | ·5278 9627$\times\frac{1}{10^6}$ | ·3710 6519$\times\frac{1}{10^6}$ | ·1898 4731$\times\frac{1}{10^6}$ | ·1012 5190$\times\frac{1}{10^6}$ | ·5601 1688$\times\frac{1}{10^7}$ | ·3200 6679$\times\frac{1}{10^7}$ |
| ·07 | ·0000 001 | | | | | |
| ·08 | ·0000 003 | ·0000 001 | | | | |
| ·09 | ·0000 008 | ·0000 003 | ·0000 001 | | | |
| ·10 | ·0000 023 | ·0000 010 | ·0000 002 | | | |
| ·11 | ·0000 057 | ·0000 026 | ·0000 005+ | ·0000 001 | | |
| ·12 | ·0000 129 | ·0000 060 | ·0000 013 | ·0000 003 | ·0000 001 | |
| ·13 | ·0000 272 | ·0000 132 | ·0000 031 | ·0000 007 | ·0000 001 | |
| ·14 | ·0000 536 | ·0000 271 | ·0000 067 | ·0000 016 | ·0000 004 | ·0000 001 |
| ·15 | ·0001 002 | ·0000 524 | ·0000 139 | ·0000 036 | ·0000 009 | ·0000 002 |
| ·16 | ·0001 787 | ·0000 964 | ·0000 273 | ·0000 075− | ·0000 020 | ·0000 005+ |
| ·17 | ·0003 054 | ·0001 697 | ·0000 510 | ·0000 148 | ·0000 042 | ·0000 012 |
| ·18 | ·0005 028 | ·0002 873 | ·0000 913 | ·0000 281 | ·0000 084 | ·0000 024 |
| ·19 | ·0008 006 | ·0004 696 | ·0001 574 | ·0000 511 | ·0000 161 | ·0000 049 |
| ·20 | ·0012 369 | ·0007 440 | ·0002 621 | ·0000 894 | ·0000 296 | ·0000 096 |
| ·21 | ·0018 596 | ·0011 454 | ·0004 232 | ·0001 514 | ·0000 527 | ·0000 179 |
| ·22 | ·0027 273 | ·0017 180 | ·0006 641 | ·0002 487 | ·0000 905+ | ·0000 321 |
| ·23 | ·0039 098 | ·0025 164 | ·0010 157 | ·0003 972 | ·0001 510 | ·0000 560 |
| ·24 | ·0054 891 | ·0036 061 | ·0015 167 | ·0006 183 | ·0002 451 | ·0000 947 |
| ·25 | ·0075 590 | ·0050 643 | ·0022 157 | ·0009 397 | ·0003 876 | ·0001 559 |
| ·26 | ·0102 249 | ·0069 803 | ·0031 714 | ·0013 971 | ·0005 987 | ·0002 502 |
| ·27 | ·0136 032 | ·0094 552 | ·0044 543 | ·0020 351 | ·0009 046 | ·0003 923 |
| ·28 | ·0178 191 | ·0126 015− | ·0061 466 | ·0029 084 | ·0013 392 | ·0006 016 |
| ·29 | ·0230 055+ | ·0165 416 | ·0083 429 | ·0040 829 | ·0019 448 | ·0009 039 |
| ·30 | ·0293 000 | ·0214 066 | ·0111 496 | ·0056 364 | ·0027 738 | ·0013 323 |
| ·31 | ·0368 420 | ·0273 336 | ·0146 847 | ·0076 593 | ·0038 899 | ·0019 285− |
| ·32 | ·0457 696 | ·0344 633 | ·0190 764 | ·0102 546 | ·0053 687 | ·0027 443 |
| ·33 | ·0562 154 | ·0429 368 | ·0244 614 | ·0135 377 | ·0072 987 | ·0038 426 |
| ·34 | ·0683 033 | ·0528 918 | ·0309 821 | ·0176 355− | ·0097 815− | ·0052 991 |
| ·35 | ·0821 437 | ·0644 588 | ·0387 847 | ·0226 849 | ·0129 321 | ·0072 023 |
| ·36 | ·0978 300 | ·0777 570 | ·0480 147 | ·0288 310 | ·0168 780 | ·0096 550− |
| ·37 | ·1154 349 | ·0928 902 | ·0588 140 | ·0362 242 | ·0217 581 | ·0127 736 |
| ·38 | ·1350 066 | ·1099 427 | ·0713 158 | ·0450 174 | ·0277 211 | ·0166 885+ |
| ·39 | ·1565 659 | ·1289 755+ | ·0856 413 | ·0553 615+ | ·0349 227 | ·0215 425+ |
| ·40 | ·1801 039 | ·1500 231 | ·1018 943 | ·0674 019 | ·0435 226 | ·0274 894 |
| ·41 | ·2055 800 | ·1730 900 | ·1201 571 | ·0812 735− | ·0536 811 | ·0346 916 |
| ·42 | ·2329 213 | ·1981 492 | ·1404 868 | ·0970 958 | ·0655 538 | ·0433 168 |
| ·43 | ·2620 221 | ·2251 403 | ·1629 109 | ·1149 682 | ·0792 879 | ·0535 343 |
| ·44 | ·2927 449 | ·2539 688 | ·1874 245+ | ·1349 655− | ·0950 158 | ·0655 102 |
| ·45 | ·3249 218 | ·2845 066 | ·2139 882 | ·1571 326 | ·1128 509 | ·0794 024 |
| ·46 | ·3583 573 | ·3165 932 | ·2425 257 | ·1814 817 | ·1328 813 | ·0953 549 |
| ·47 | ·3928 310 | ·3500 375− | ·2729 240 | ·2079 880 | ·1551 651 | ·1134 916 |
| ·48 | ·4281 024 | ·3846 212 | ·3050 336 | ·2365 882 | ·1797 258 | ·1339 110 |
| ·49 | ·4639 146 | ·4201 028 | ·3386 697 | ·2671 783 | ·2065 480 | ·1566 796 |
| ·50 | ·5000 000$^e$ | ·4562 215− | ·3736 154 | ·2996 141 | ·2355 745− | ·1818 270 |
| ·51 | ·5360 854 | ·4927 030 | ·4096 246 | ·3337 118 | ·2667 041 | ·2093 407 |
| ·52 | ·5718 976 | ·5292 648 | ·4464 273 | ·3692 504 | ·2997 911 | ·2391 632 |
| ·53 | ·6071 690 | ·5656 221 | ·4837 343 | ·4059 750+ | ·3346 458 | ·2711 883 |
| ·54 | ·6416 427 | ·6014 939 | ·5212 438 | ·4436 018 | ·3710 364 | ·3052 610 |
| ·55 | ·6750 782 | ·6366 087 | ·5586 476 | ·4818 234 | ·4086 927 | ·3411 774 |
| ·56 | ·7072 551 | ·6707 103 | ·5956 382 | ·5203 159 | ·4473 112 | ·3786 867 |
| ·57 | ·7379 779 | ·7035 626 | ·6319 154 | ·5587 457 | ·4865 609 | ·4174 956 |
| ·58 | ·7670 787 | ·7349 545+ | ·6671 931 | ·5967 777 | ·5260 914 | ·4572 733 |
| ·59 | ·7944 200 | ·7647 036 | ·7012 056 | ·6340 830 | ·5655 405− | ·4976 589 |
| ·60 | ·8198 961 | ·7926 590 | ·7337 135+ | ·6703 471 | ·6045 435− | ·5382 699 |
| ·61 | ·8434 341 | ·8187 037 | ·7645 081 | ·7052 769 | ·6427 424 | ·5787 115+ |
| ·62 | ·8649 934 | ·8427 552 | ·7934 158 | ·7386 080 | ·6797 951 | ·6185 875+ |
| ·63 | ·8845 651 | ·8647 660 | ·8203 007 | ·7701 105− | ·7153 839 | ·6575 104 |
| ·64 | ·9021 700 | ·8847 226 | ·8450 662 | ·7995 936 | ·7492 239 | ·6951 124 |
| ·65 | ·9178 563 | ·9026 437 | ·8676 558 | ·8269 094 | ·7810 697 | ·7310 554 |
| ·66 | ·9316 967 | ·9185 779 | ·8880 520 | ·8519 551 | ·8107 213 | ·7650 401 |
| ·67 | ·9437 846 | ·9326 002 | ·9062 747 | ·8746 732 | ·8380 275+ | ·7968 143 |
| ·68 | ·9542 304 | ·9448 084 | ·9223 784 | ·8950 512 | ·8628 891 | ·8261 784 |
| ·69 | ·9631 580 | ·9553 191 | ·9364 481 | ·9131 189 | ·8852 583 | ·8529 899 |
| ·70 | ·9707 000 | ·9642 632 | ·9485 953 | ·9289 453 | ·9051 385− | ·8771 655− |

$x$ = ·71 to ·95 $q$ = 10·5 $p$ = 10·5 to 15

| | $p$ = 10·5 | $p$ = 11 | $p$ = 12 | $p$ = 13 | $p$ = 14 | $p$ = 15 |
|---|---|---|---|---|---|---|
| $B(p, q)$ = | ·5278 9627 $\times \frac{1}{10^6}$ | ·3710 6519 $\times \frac{1}{10^6}$ | ·1898 4731 $\times \frac{1}{10^6}$ | ·1012 5190 $\times \frac{1}{10^6}$ | ·5601 1688 $\times \frac{1}{10^7}$ | ·3200 6679 $\times \frac{1}{10^7}$ |
| $x$ | | | | | | |
| ·71 | ·9769 945− | ·9717 812 | ·9589 526 | ·9426 336 | ·9225 801 | ·8986 806 |
| ·72 | ·9821 809 | ·9780 195− | ·9676 687 | ·9543 161 | ·9376 768 | ·9175 670 |
| ·73 | ·9863 968 | ·9831 258 | ·9749 028 | ·9641 479 | ·9505 594 | ·9339 087 |
| ·74 | ·9897 751 | ·9872 456 | ·9808 198 | ·9723 003 | ·9613 887 | ·9478 350+ |
| ·75 | ·9924 410 | ·9905 188 | ·9855 848 | ·9789 547 | ·9703 483 | ·9595 135+ |
| ·76 | ·9945 109 | ·9930 771 | ·9893 590 | ·9842 961 | ·9776 365+ | ·9691 408 |
| ·77 | ·9960 902 | ·9950 419 | ·9922 960 | ·9885 076 | ·9834 590 | ·9769 336 |
| ·78 | ·9972 727 | ·9965 227 | ·9945 383 | ·9917 650+ | ·9880 211 | ·9831 193 |
| ·79 | ·9981 404 | ·9976 161 | ·9962 154 | ·9942 328 | ·9915 219 | ·9879 271 |
| ·80 | ·9987 631 | ·9984 059 | ·9974 422 | ·9960 608 | ·9941 481 | ·9915 797 |
| ·81 | ·9991 994 | ·9989 628 | ·9983 180 | ·9973 823 | ·9960 705− | ·9942 869 |
| ·82 | ·9994 972 | ·9993 451 | ·9989 269 | ·9983 123 | ·9974 402 | ·9962 397 |
| ·83 | ·9996 946 | ·9996 002 | ·9993 379 | ·9989 479 | ·9983 877 | ·9976 071 |
| ·84 | ·9998 213 | ·9997 648 | ·9996 065+ | ·9993 683 | ·9990 219 | ·9985 335+ |
| ·85 | ·9998 998 | ·9998 674 | ·9997 758 | ·9996 364 | ·9994 312 | ·9991 385+ |
| ·86 | ·9999 464 | ·9999 287 | ·9998 782 | ·9998 005− | ·9996 847 | ·9995 177 |
| ·87 | ·9999 728 | ·9999 637 | ·9999 374 | ·9998 964 | ·9998 346 | ·9997 444 |
| ·88 | ·9999 871 | ·9999 827 | ·9999 698 | ·9999 495− | ·9999 185+ | ·9998 729 |
| ·89 | ·9999 943 | ·9999 923 | ·9999 865− | ·9999 771 | ·9999 628 | ·9999 413 |
| ·90 | ·9999 977 | ·9999 969 | ·9999 944 | ·9999 905+ | ·9999 844 | ·9999 752 |
| ·91 | ·9999 992 | ·9999 989 | ·9999 979 | ·9999 965− | ·9999 941 | ·9999 906 |
| ·92 | ·9999 997 | ·9999 996 | ·9999 993 | ·9999 988 | ·9999 981 | ·9999 969 |
| ·93 | ·9999 999 | ·9999 999 | ·9999 998 | ·9999 997 | ·9999 995− | ·9999 991 |
| ·94 | 1·0000 000 | 1·0000 000 | 1·0000 000 | ·9999 999 | ·9999 999 | ·9999 998 |
| ·95 | | | | 1·0000 000 | 1·0000 000 | 1·0000 000 |

$x = \cdot 15$ to $\cdot 70$ $q = 10\cdot 5$ $p = 16$ to $21$

| | $p = 16$ | $p = 17$ | $p = 18$ | $p = 19$ | $p = 20$ | $p = 21$ |
|---|---|---|---|---|---|---|
| $B(p, q) =$ | ·1882 7458$\times\frac{1}{10^7}$ | ·1136 7522$\times\frac{1}{10^7}$ | ·7027 1954$\times\frac{1}{10^8}$ | ·4438 2286$\times\frac{1}{10^8}$ | ·2858 5201$\times\frac{1}{10^8}$ | ·1874 4394$\times\frac{1}{10^8}$ |
| $x$ | | | | | | |
| ·15 | ·0000 001 | | | | | |
| ·16 | ·0000 001 | | | | | |
| ·17 | ·0000 003 | ·0000 001 | | | | |
| ·18 | ·0000 007 | ·0000 002 | ·0000 001 | | | |
| ·19 | ·0000 015⁻ | ·0000 004 | ·0000 001 | | | |
| ·20 | ·0000 030 | ·0000 009 | ·0000 003 | ·0000 001 | | |
| ·21 | ·0000 059 | ·0000 019 | ·0000 006 | ·0000 002 | ·0000 001 | |
| ·22 | ·0000 111 | ·0000 038 | ·0000 013 | ·0000 004 | ·0000 001 | |
| ·23 | ·0000 203 | ·0000 072 | ·0000 025⁺ | ·0000 009 | ·0000 003 | ·0000 001 |
| ·24 | ·0000 358 | ·0000 132 | ·0000 048 | ·0000 017 | ·0000 006 | ·0000 002 |
| ·25 | ·0000 613 | ·0000 236 | ·0000 089 | ·0000 033 | ·0000 012 | ·0000 004 |
| ·26 | ·0001 023 | ·0000 409 | ·0000 161 | ·0000 062 | ·0000 024 | ·0000 009 |
| ·27 | ·0001 663 | ·0000 691 | ·0000 282 | ·0000 113 | ·0000 045⁻ | ·0000 017 |
| ·28 | ·0002 643 | ·0001 138 | ·0000 481 | ·0000 200 | ·0000 082 | ·0000 033 |
| ·29 | ·0004 109 | ·0001 831 | ·0000 801 | ·0000 344 | ·0000 146 | ·0000 061 |
| ·30 | ·0006 260 | ·0002 882 | ·0001 303 | ·0000 579 | ·0000 253 | ·0000 109 |
| ·31 | ·0009 354 | ·0004 447 | ·0002 076 | ·0000 953 | ·0000 431 | ·0000 192 |
| ·32 | ·0013 726 | ·0006 730 | ·0003 240 | ·0001 534 | ·0000 715⁺ | ·0000 329 |
| ·33 | ·0019 799 | ·0010 002 | ·0004 962 | ·0002 421 | ·0001 163 | ·0000 551 |
| ·34 | ·0028 099 | ·0014 610 | ·0007 461 | ·0003 748 | ·0001 854 | ·0000 904 |
| ·35 | ·0039 268 | ·0020 997 | ·0011 029 | ·0005 698 | ·0002 900 | ·0001 455⁺ |
| ·36 | ·0054 079 | ·0029 711 | ·0016 036 | ·0008 515⁻ | ·0004 454 | ·0002 297 |
| ·37 | ·0073 440 | ·0041 422 | ·0022 956 | ·0012 517 | ·0006 723 | ·0003 561 |
| ·38 | ·0098 410 | ·0056 939 | ·0032 374 | ·0018 113 | ·0009 984 | ·0005 428 |
| ·39 | ·0130 194 | ·0077 217 | ·0045 010 | ·0025 820 | ·0014 594 | ·0008 136 |
| ·40 | ·0170 144 | ·0103 365⁺ | ·0061 727 | ·0036 281 | ·0021 014 | ·0012 006 |
| ·41 | ·0219 749 | ·0136 654 | ·0083 547 | ·0050 281 | ·0029 822 | ·0017 450⁻ |
| ·42 | ·0280 622 | ·0178 511 | ·0111 659 | ·0068 762 | ·0041 737 | ·0024 995⁻ |
| ·43 | ·0354 467 | ·0230 510 | ·0147 423 | ·0092 839 | ·0057 633 | ·0035 303 |
| ·44 | ·0443 055⁻ | ·0294 357 | ·0192 367 | ·0123 808 | ·0078 558 | ·0049 192 |
| ·45 | ·0548 178 | ·0371 860 | ·0248 178 | ·0163 146 | ·0105 750⁺ | ·0067 653 |
| ·46 | ·0671 600 | ·0464 896 | ·0316 678 | ·0212 513 | ·0140 639 | ·0091 873 |
| ·47 | ·0815 003 | ·0575 368 | ·0399 800 | ·0273 732 | ·0184 854 | ·0123 240 |
| ·48 | ·0979 922 | ·0705 144 | ·0499 544 | ·0348 772 | ·0240 215⁺ | ·0163 357 |
| ·49 | ·1167 683 | ·0856 001 | ·0617 930 | ·0439 708 | ·0308 715⁻ | ·0214 039 |
| ·50 | ·1379 332 | ·1029 553 | ·0756 931 | ·0548 678 | ·0392 489 | ·0277 299 |
| ·51 | ·1615 571 | ·1227 179 | ·0918 407 | ·0677 823 | ·0493 776 | ·0355 326 |
| ·52 | ·1876 698 | ·1449 947 | ·1104 027 | ·0829 213 | ·0614 858 | ·0450 448 |
| ·53 | ·2162 553 | ·1698 540 | ·1315 184 | ·1004 772 | ·0757 994 | ·0565 076 |
| ·54 | ·2472 472 | ·1973 190 | ·1552 912 | ·1206 183 | ·0925 333 | ·0701 635⁺ |
| ·55 | ·2805 261 | ·2273 615⁺ | ·1817 807 | ·1434 801 | ·1118 821 | ·0862 483 |
| ·56 | ·3159 183 | ·2598 974 | ·2109 945⁻ | ·1691 553 | ·1340 104 | ·1049 807 |
| ·57 | ·3531 958 | ·2947 834 | ·2428 823 | ·1976 851 | ·1590 415⁻ | ·1265 519 |
| ·58 | ·3920 796 | ·3318 162 | ·2773 310 | ·2290 510 | ·1870 474 | ·1511 134 |
| ·59 | ·4322 437 | ·3707 329 | ·3141 611 | ·2631 679 | ·2180 390 | ·1787 655⁻ |
| ·60 | ·4733 217 | ·4112 150⁺ | ·3531 270 | ·2998 797 | ·2519 571 | ·2095 456 |
| ·61 | ·5149 155⁻ | ·4528 937 | ·3939 183 | ·3389 565⁺ | ·2886 665⁻ | ·2434 180 |
| ·62 | ·5566 047 | ·4953 580 | ·4361 648 | ·3800 957 | ·3279 514 | ·2802 655⁻ |
| ·63 | ·5979 587 | ·5381 650⁺ | ·4794 441 | ·4229 253 | ·3695 149 | ·3198 834 |
| ·64 | ·6385 479 | ·5808 521 | ·5232 921 | ·4670 111 | ·4129 814 | ·3619 774 |
| ·65 | ·6779 568 | ·6229 500⁻ | ·5672 151 | ·5118 673 | ·4579 032 | ·4061 651 |
| ·66 | ·7157 958 | ·6639 970 | ·6107 052 | ·5569 692 | ·5037 706 | ·4519 818 |
| ·67 | ·7517 133 | ·7035 538 | ·6532 555⁺ | ·6017 696 | ·5500 263 | ·4988 909 |
| ·68 | ·7854 050⁻ | ·7412 169 | ·6943 774 | ·6457 165⁻ | ·5960 823 | ·5462 992 |
| ·69 | ·8166 233 | ·7766 315⁻ | ·7336 167 | ·6882 719 | ·6413 401 | ·5935 753 |
| ·70 | ·8451 830 | ·8095 026 | ·7705 689 | ·7289 315⁺ | ·6852 123 | ·6400 722 |

$x = ·71$ to $·96$ $q = 10·5$ $p = 16$ to $21$

| | $p = 16$ | $p = 17$ | $p = 18$ | $p = 19$ | $p = 20$ | $p = 21$ |
|---|---|---|---|---|---|---|
| $B(p, q) =$ | $·1882\ 7458 \times \frac{1}{10^7}$ | $·1136\ 7522 \times \frac{1}{10^7}$ | $·7027\ 1954 \times \frac{1}{10^8}$ | $·4438\ 2286 \times \frac{1}{10^8}$ | $·2858\ 5201 \times \frac{1}{10^8}$ | $·1874\ 4394 \times \frac{1}{10^8}$ |
| $x$ | | | | | | |
| ·71 | ·8709 651 | ·8396 033 | ·8048 932 | ·7672 423 | ·7271 442 | ·6851 514 |
| ·72 | ·8939 179 | ·8667 806 | ·8363 229 | ·8028 194 | ·7666 357 | ·7282 085+ |
| ·73 | ·9140 555+ | ·8909 576 | ·8646 734 | ·8353 593 | ·8032 604 | ·7686 979 |
| ·74 | ·9314 532 | ·9121 328 | ·8898 461 | ·8646 498 | ·8366 819 | ·8061 550+ |
| ·75 | ·9462 409 | ·9303 759 | ·9118 279 | ·8905 750+ | ·8666 655+ | ·8402 156 |
| ·76 | ·9585 947 | ·9458 208 | ·9306 875+ | ·9131 160 | ·8930 846 | ·8706 293 |
| ·77 | ·9687 269 | ·9586 556 | ·9465 672 | ·9323 465+ | ·9159 216 | ·8972 670 |
| ·78 | ·9768 744 | ·9691 113 | ·9596 721 | ·9484 238 | ·9352 633 | ·9201 222 |
| ·79 | ·9832 886 | ·9774 484 | ·9702 565− | ·9615 762 | ·9512 900 | ·9393 040 |
| ·80 | ·9882 236 | ·9839 446 | ·9786 084 | ·9720 865− | ·9642 601 | ·9550 249 |
| ·81 | ·9919 272 | ·9888 810 | ·9850 346 | ·9802 748 | ·9744 917 | ·9675 823 |
| ·82 | ·9946 319 | ·9925 307 | ·9898 448 | ·9864 800 | ·9823 414 | ·9773 356 |
| ·83 | ·9965 489 | ·9951 492 | ·9933 382 | ·9910 417 | ·9881 825+ | ·9846 822 |
| ·84 | ·9978 635− | ·9969 664 | ·9957 918 | ·9942 844 | ·9923 850+ | ·9900 318 |
| ·85 | ·9987 322 | ·9981 817 | ·9974 523 | ·9965 051 | ·9952 974 | ·9937 833 |
| ·86 | ·9992 830 | ·9989 613 | ·9985 301 | ·9979 635+ | ·9972 326 | ·9963 055− |
| ·87 | ·9996 162 | ·9994 384 | ·9991 974 | ·9988 770 | ·9984 588 | ·9979 223 |
| ·88 | ·9998 072 | ·9997 151 | ·9995 888 | ·9994 189 | ·9991 948 | ·9989 038 |
| ·89 | ·9999 101 | ·9998 658 | ·9998 045− | ·9997 210 | ·9996 095+ | ·9994 632 |
| ·90 | ·9999 616 | ·9999 422 | ·9999 149 | ·9998 774 | ·9998 267 | ·9997 595− |
| ·91 | ·9999 853 | ·9999 776 | ·9999 667 | ·9999 516 | ·9999 309 | ·9999 031 |
| ·92 | ·9999 951 | ·9999 924 | ·9999 886 | ·9999 832 | ·9999 758 | ·9999 658 |
| ·93 | ·9999 986 | ·9999 978 | ·9999 967 | ·9999 951 | ·9999 929 | ·9999 898 |
| ·94 | ·9999 997 | ·9999 995− | ·9999 992 | ·9999 988 | ·9999 983 | ·9999 975+ |
| ·95 | ·9999 999 | ·9999 999 | ·9999 999 | ·9999 998 | ·9999 997 | ·9999 996 |
| ·96 | 1·0000 000 | 1·0000 000 | 1·0000 000 | 1·0000 000 | 1·0000 000 | 1·0000 000 |

$x$ = ·24 to ·80 $q$ = 10·5 $p$ = 22 to 27

| | $p = 22$ | $p = 23$ | $p = 24$ | $p = 25$ | $p = 26$ | $p = 27$ |
|---|---|---|---|---|---|---|
| $B(p, q)$ = | ·1249 6263 × $\frac{1}{10^8}$ | ·8459 0088 × $\frac{1}{10^9}$ | ·5807 6777 × $\frac{1}{10^9}$ | ·4040 1236 × $\frac{1}{10^9}$ | ·2845 1575− × $\frac{1}{10^9}$ | ·2026 6875+ × $\frac{1}{10^9}$ |
| $x$ | | | | | | |
| ·24 | ·0000 001 | | | | | |
| ·25 | ·0000 002 | ·0000 001 | | | | |
| ·26 | ·0000 003 | ·0000 001 | | | | |
| ·27 | ·0000 007 | ·0000 003 | ·0000 001 | | | |
| ·28 | ·0000 013 | ·0000 005+ | ·0000 002 | ·0000 001 | | |
| ·29 | ·0000 025+ | ·0000 010 | ·0000 004 | ·0000 002 | ·0000 001 | |
| ·30 | ·0000 047 | ·0000 020 | ·0000 008 | ·0000 003 | ·0000 001 | ·0000 001 |
| ·31 | ·0000 084 | ·0000 037 | ·0000 016 | ·0000 007 | ·0000 003 | ·0000 001 |
| ·32 | ·0000 149 | ·0000 067 | ·0000 030 | ·0000 013 | ·0000 006 | ·0000 002 |
| ·33 | ·0000 258 | ·0000 119 | ·0000 054 | ·0000 025− | ·0000 011 | ·0000 005− |
| ·34 | ·0000 435+ | ·0000 207 | ·0000 097 | ·0000 045+ | ·0000 021 | ·0000 010 |
| ·35 | ·0000 721 | ·0000 353 | ·0000 171 | ·0000 082 | ·0000 039 | ·0000 018 |
| ·36 | ·0001 169 | ·0000 588 | ·0000 293 | ·0000 144 | ·0000 070 | ·0000 034 |
| ·37 | ·0001 862 | ·0000 962 | ·0000 492 | ·0000 248 | ·0000 124 | ·0000 062 |
| ·38 | ·0002 913 | ·0001 545− | ·0000 810 | ·0000 420 | ·0000 216 | ·0000 110 |
| ·39 | ·0004 478 | ·0002 436 | ·0001 310 | ·0000 697 | ·0000 368 | ·0000 192 |
| ·40 | ·0006 773 | ·0003 776 | ·0002 082 | ·0001 136 | ·0000 614 | ·0000 329 |
| ·41 | ·0010 082 | ·0005 757 | ·0003 251 | ·0001 817 | ·0001 006 | ·0000 552 |
| ·42 | ·0014 782 | ·0008 640 | ·0004 996 | ·0002 859 | ·0001 620 | ·0000 910 |
| ·43 | ·0021 357 | ·0012 771 | ·0007 554 | ·0004 423 | ·0002 565+ | ·0001 474 |
| ·44 | ·0030 425− | ·0018 602 | ·0011 251 | ·0006 736 | ·0003 995− | ·0002 348 |
| ·45 | ·0042 755− | ·0026 712 | ·0016 511 | ·0010 103 | ·0006 124 | ·0003 679 |
| ·46 | ·0059 292 | ·0037 834 | ·0023 885+ | ·0014 929 | ·0009 244 | ·0005 674 |
| ·47 | ·0081 181 | ·0052 877 | ·0034 079 | ·0021 747 | ·0013 749 | ·0008 616 |
| ·48 | ·0109 776 | ·0072 952 | ·0047 975− | ·0031 240 | ·0020 156 | ·0012 891 |
| ·49 | ·0146 662 | ·0099 391 | ·0066 660 | ·0044 274 | ·0029 137 | ·0019 009 |
| ·50 | ·0193 649 | ·0133 764 | ·0091 453 | ·0061 924 | ·0041 549 | ·0027 639 |
| ·51 | ·0252 775+ | ·0177 889 | ·0123 923 | ·0085 505− | ·0058 467 | ·0039 639 |
| ·52 | ·0326 279 | ·0233 830 | ·0165 900 | ·0116 594 | ·0081 212 | ·0056 091 |
| ·53 | ·0416 575− | ·0303 882 | ·0219 486 | ·0157 050− | ·0111 384 | ·0078 338 |
| ·54 | ·0526 192 | ·0390 543 | ·0287 037 | ·0209 019 | ·0150 880 | ·0108 014 |
| ·55 | ·0657 717 | ·0496 464 | ·0371 142 | ·0274 932 | ·0201 908 | ·0147 070 |
| ·56 | ·0813 699 | ·0624 383 | ·0474 577 | ·0357 478 | ·0266 984 | ·0197 791 |
| ·57 | ·0996 551 | ·0777 037 | ·0600 232 | ·0459 562 | ·0348 911 | ·0262 795− |
| ·58 | ·1208 432 | ·0957 052 | ·0751 030 | ·0584 238 | ·0450 737 | ·0345 015+ |
| ·59 | ·1451 118 | ·1166 822 | ·0929 805− | ·0734 611 | ·0575 685− | ·0447 657 |
| ·60 | ·1725 870 | ·1408 362 | ·1139 171 | ·0913 723 | ·0727 052 | ·0574 125+ |
| ·61 | ·2033 300 | ·1683 168 | ·1381 370 | ·1124 401 | ·0908 084 | ·0727 917 |
| ·62 | ·2373 254 | ·1992 064 | ·1658 108 | ·1369 096 | ·1121 819 | ·0912 488 |
| ·63 | ·2744 705+ | ·2335 060 | ·1970 387 | ·1649 703 | ·1370 900 | ·1131 076 |
| ·64 | ·3145 680 | ·2711 236 | ·2318 349 | ·1967 369 | ·1657 384 | ·1386 505− |
| ·65 | ·3573 219 | ·3118 644 | ·2701 127 | ·2322 317 | ·1982 525− | ·1680 959 |
| ·66 | ·4023 385− | ·3554 256 | ·3116 742 | ·2713 682 | ·2346 576 | ·2015 756 |
| ·67 | ·4491 314 | ·4013 961 | ·3562 033 | ·3139 386 | ·2748 606 | ·2391 118 |
| ·68 | ·4971 325− | ·4492 620 | ·4032 647 | ·3596 056 | ·3186 358 | ·2805 970 |
| ·69 | ·5457 081 | ·4984 175− | ·4523 091 | ·4079 010 | ·3656 156 | ·3257 779 |
| ·70 | ·5941 797 | ·5481 829 | ·5026 862 | ·4582 320 | ·4152 893 | ·3742 460 |
| ·71 | ·6418 488 | ·5978 278 | ·5536 633 | ·5098 944 | ·4670 096 | ·4254 362 |
| ·72 | ·6880 247 | ·6465 989 | ·6044 526 | ·5620 955+ | ·5200 096 | ·4786 359 |
| ·73 | ·7320 534 | ·6937 515+ | ·6542 424 | ·6139 842 | ·5734 280 | ·5330 045+ |
| ·74 | ·7733 458 | ·7385 830 | ·7022 332 | ·6646 868 | ·6263 445− | ·5876 040 |
| ·75 | ·8114 041 | ·7804 645− | ·7476 752 | ·7133 489 | ·6778 212 | ·6414 395+ |
| ·76 | ·8458 429 | ·8188 708 | ·7899 052 | ·7591 774 | ·7269 502 | ·6935 082 |
| ·77 | ·8764 050− | ·8534 046 | ·8283 792 | ·8014 820 | ·7729 011 | ·7428 527 |
| ·78 | ·9029 694 | ·8838 129 | ·8626 991 | ·8397 114 | ·8149 675+ | ·7886 152 |
| ·79 | ·9255 515+ | ·9099 956 | ·8926 305− | ·8734 818 | ·8526 059 | ·8300 880 |
| ·80 | ·9442 946 | ·9320 034 | ·9181 091 | ·9025 938 | ·8854 649 | ·8667 548 |

$x = \cdot 81$ to $\cdot 97$ $q = 10\cdot 5$ $p = 22$ to $27$

| $x$ | $p = 22$ | $p = 23$ | $p = 24$ | $p = 25$ | $p = 26$ | $p = 27$ |
|---|---|---|---|---|---|---|
| $B(p, q) =$ | ·1249 6263 $\times \frac{1}{10^8}$ | ·8459 0088 $\times \frac{1}{10^9}$ | ·5807 6777 $\times \frac{1}{10^9}$ | ·4040 1236 $\times \frac{1}{10^9}$ | ·2845 1575 $\bar{\times} \frac{1}{10^9}$ | ·2026 6875 $\overset{+}{\times} \frac{1}{10^9}$ |
| ·81 | ·9594 538 | ·9500 267 | ·9392 367 | ·9270 373 | ·9134 007 | ·8983 193 |
| ·82 | ·9713 741 | ·9643 746 | ·9562 644 | ·9469 816 | ·9364 771 | ·9247 162 |
| ·83 | ·9804 627 | ·9754 482 | ·9695 670 | ·9627 535− | ·9549 492 | ·9461 049 |
| ·84 | ·9871 608 | ·9837 077 | ·9796 091 | ·9748 035− | ·9692 328 | ·9628 437 |
| ·85 | ·9919 140 | ·9896 391 | ·9869 067 | ·9836 648 | ·9798 620 | ·9754 486 |
| ·86 | ·9951 474 | ·9937 215− | ·9919 886 | ·9899 085− | ·9874 398 | ·9845 409 |
| ·87 | ·9972 443 | ·9963 998 | ·9953 615+ | ·9941 007 | ·9925 869 | ·9907 888 |
| ·88 | ·9985 318 | ·9980 632 | ·9974 805+ | ·9967 647 | ·9958 955+ | ·9948 511 |
| ·89 | ·9992 741 | ·9990 331 | ·9987 300 | ·9983 535+ | ·9978 911 | ·9973 292 |
| ·90 | ·9996 716 | ·9995 583 | ·9994 142 | ·9992 333 | ·9990 085− | ·9987 323 |
| ·91 | ·9998 665− | ·9998 187 | ·9997 573 | ·9996 792 | ·9995 812 | ·9994 594 |
| ·92 | ·9999 524 | ·9999 348 | ·9999 119 | ·9998 824 | ·9998 450+ | ·9997 980 |
| ·93 | ·9999 857 | ·9999 802 | ·9999 729 | ·9999 635+ | ·9999 515− | ·9999 361 |
| ·94 | ·9999 965+ | ·9999 951 | ·9999 933 | ·9999 909 | ·9999 878 | ·9999 838 |
| ·95 | ·9999 994 | ·9999 991 | ·9999 988 | ·9999 983 | ·9999 977 | ·9999 969 |
| ·96 | ·9999 999 | ·9999 999 | ·9999 999 | ·9999 998 | ·9999 997 | ·9999 996 |
| ·97 | 1·0000 000 | 1·0000 000 | 1·0000 000 | 1·0000 000 | 1·0000 000 | 1·0000 000 |

$x$ = ·32 to ·98 $q$ = 10·5 $p$ = 28 to 33

| | $p = 28$ | $p = 29$ | $p = 30$ | $p = 31$ | $p = 32$ | $p = 33$ |
|---|---|---|---|---|---|---|
| $B(p, q)$ = | ·1459 2150$\overset{+}{\times}\frac{1}{10^9}$ | ·1061 2473$\times\frac{1}{10^9}$ | ·7791 4357$\times\frac{1}{10^{10}}$ | ·5771 4338$\times\frac{1}{10^{10}}$ | ·4311 1915$\overset{+}{\times}\frac{1}{10^{10}}$ | ·3246 0736$\times\frac{1}{10^{10}}$ |
| $x$ | | | | | | |
| ·32 | ·0000 001 | | | | | |
| ·33 | ·0000 002 | ·0000 001 | | | | |
| ·34 | ·0000 004 | ·0000 002 | ·0000 001 | | | |
| ·35 | ·0000 008 | ·0000 004 | ·0000 002 | ·0000 001 | | |
| ·36 | ·0000 016 | ·0000 008 | ·0000 004 | ·0000 002 | ·0000 001 | |
| ·37 | ·0000 030 | ·0000 015− | ·0000 007 | ·0000 003 | ·0000 002 | ·0000 001 |
| ·38 | ·0000 056 | ·0000 028 | ·0000 014 | ·0000 007 | ·0000 003 | ·0000 002 |
| ·39 | ·0000 099 | ·0000 051 | ·0000 026 | ·0000 013 | ·0000 007 | ·0000 003 |
| ·40 | ·0000 174 | ·0000 092 | ·0000 048 | ·0000 025− | ·0000 013 | ·0000 007 |
| ·41 | ·0000 300 | ·0000 162 | ·0000 087 | ·0000 046 | ·0000 024 | ·0000 013 |
| ·42 | ·0000 507 | ·0000 280 | ·0000 153 | ·0000 084 | ·0000 045+ | ·0000 024 |
| ·43 | ·0000 840 | ·0000 475− | ·0000 266 | ·0000 148 | ·0000 082 | ·0000 045+ |
| ·44 | ·0001 368 | ·0000 791 | ·0000 454 | ·0000 259 | ·0000 146 | ·0000 082 |
| ·45 | ·0002 192 | ·0001 295+ | ·0000 760 | ·0000 443 | ·0000 256 | ·0000 147 |
| ·46 | ·0003 453 | ·0002 085+ | ·0001 250− | ·0000 744 | ·0000 440 | ·0000 258 |
| ·47 | ·0005 355− | ·0003 302 | ·0002 021 | ·0001 228 | ·0000 741 | ·0000 445− |
| ·48 | ·0008 177 | ·0005 146 | ·0003 215− | ·0001 995− | ·0001 229 | ·0000 753 |
| ·49 | ·0012 301 | ·0007 898 | ·0005 034 | ·0003 186 | ·0002 004 | ·0001 252 |
| ·50 | ·0018 237 | ·0011 941 | ·0007 762 | ·0005 010 | ·0003 213 | ·0002 048 |
| ·51 | ·0026 658 | ·0017 792 | ·0011 788 | ·0007 757 | ·0005 071 | ·0003 295− |
| ·52 | ·0038 432 | ·0026 133 | ·0017 643 | ·0011 830 | ·0007 881 | ·0005 218 |
| ·53 | ·0054 661 | ·0037 855− | ·0026 029 | ·0017 777 | ·0012 063 | ·0008 136 |
| ·54 | ·0076 722 | ·0054 091 | ·0037 866 | ·0026 331 | ·0018 193 | ·0012 494 |
| ·55 | ·0106 297 | ·0076 263 | ·0054 334 | ·0038 453 | ·0027 042 | ·0018 902 |
| ·56 | ·0145 409 | ·0106 123 | ·0076 916 | ·0055 380 | ·0039 625− | ·0028 182 |
| ·57 | ·0196 438 | ·0145 781 | ·0107 448 | ·0078 679 | ·0057 255− | ·0041 418 |
| ·58 | ·0262 121 | ·0197 731 | ·0148 150− | ·0110 286 | ·0081 596 | ·0060 015+ |
| ·59 | ·0345 544 | ·0264 856 | ·0201 655− | ·0152 558 | ·0114 714 | ·0085 757 |
| ·60 | ·0450 085− | ·0350 410 | ·0271 013 | ·0208 290 | ·0159 123 | ·0120 865− |
| ·61 | ·0579 347 | ·0457 970 | ·0359 675+ | ·0280 728 | ·0217 813 | ·0168 041 |
| ·62 | ·0737 039 | ·0591 353 | ·0471 438 | ·0373 548 | ·0294 256 | ·0230 501 |
| ·63 | ·0926 826 | ·0754 490 | ·0610 352 | ·0490 789 | ·0392 381 | ·0311 979 |
| ·64 | ·1152 144 | ·0951 263 | ·0780 584 | ·0636 758 | ·0516 501 | ·0416 688 |
| ·65 | ·1415 972 | ·1185 291 | ·0986 230 | ·0815 866 | ·0671 195− | ·0549 241 |
| ·66 | ·1720 591 | ·1459 687 | ·1231 081 | ·1032 422 | ·0861 126 | ·0714 508 |
| ·67 | ·2067 327 | ·1776 782 | ·1518 347 | ·1290 364 | ·1090 806 | ·0917 408 |
| ·68 | ·2456 295+ | ·2137 841 | ·1850 354 | ·1592 957 | ·1364 289 | ·1162 632 |
| ·69 | ·2886 182 | ·2542 786 | ·2228 224 | ·1942 444 | ·1684 827 | ·1454 300 |
| ·70 | ·3354 068 | ·2989 951 | ·2651 573 | ·2339 702 | ·2054 490 | ·1795 572 |
| ·71 | ·3855 333 | ·3475 900 | ·3118 251 | ·2783 909 | ·2473 780 | ·2188 219 |
| ·72 | ·4383 655− | ·3995 333 | ·3624 150+ | ·3272 270 | ·2941 275− | ·2632 208 |
| ·73 | ·4931 125− | ·4541 109 | ·4163 130 | ·3799 829 | ·3453 345− | ·3125 322 |
| ·74 | ·5488 491 | ·5104 403 | ·4727 069 | ·4359 421 | ·4003 988 | ·3662 883 |
| ·75 | ·6045 525− | ·5675 008 | ·5306 088 | ·4941 779 | ·4584 816 | ·4237 613 |
| ·76 | ·6591 497 | ·6241 776 | ·5888 920 | ·5535 829 | ·5185 244 | ·4839 698 |
| ·77 | ·7115 746 | ·6793 190 | ·6463 459 | ·6129 168 | ·5792 881 | ·5457 068 |
| ·78 | ·7608 282 | ·7318 007 | ·7017 422 | ·6708 722 | ·6394 146 | ·6075 933 |
| ·79 | ·8060 398 | ·7805 958 | ·7539 104 | ·7261 532 | ·6975 050+ | ·6681 541 |
| ·80 | ·8465 203 | ·8248 404 | ·8018 148 | ·7775 612 | ·7522 123 | ·7259 128 |
| ·81 | ·8818 051 | ·8638 901 | ·8446 253 | ·8240 794 | ·8023 373 | ·7794 978 |
| ·82 | ·9116 790 | ·8973 613 | ·8817 748 | ·8649 466 | ·8469 187 | ·8277 472 |
| ·83 | ·9361 813 | ·9251 501 | ·9129 948 | ·8997 112 | ·8853 070 | ·8698 022 |
| ·84 | ·9555 885− | ·9474 264 | ·9383 242 | ·9282 572 | ·9172 095+ | ·9051 744 |
| ·85 | ·9703 774 | ·9646 042 | ·9580 896 | ·9507 986 | ·9427 020 | ·9337 768 |
| ·86 | ·9811 708 | ·9772 890 | ·9728 572 | ·9678 389 | ·9622 005+ | ·9559 120 |
| ·87 | ·9886 740 | ·9862 098 | ·9833 637 | ·9801 035+ | ·9763 979 | ·9722 170 |
| ·88 | ·9936 086 | ·9921 443 | ·9904 335+ | ·9884 514 | ·9861 725− | ·9835 717 |
| ·89 | ·9966 531 | ·9958 472 | ·9948 950+ | ·9937 793 | ·9924 819 | ·9909 844 |
| ·90 | ·9983 962 | ·9979 912 | ·9975 072 | ·9969 337 | ·9962 594 | ·9954 723 |
| ·91 | ·9993 096 | ·9991 271 | ·9989 065+ | ·9986 422 | ·9983 281 | ·9979 573 |
| ·92 | ·9997 396 | ·9996 676 | ·9995 797 | ·9994 732 | ·9993 452 | ·9991 925− |
| ·93 | ·9999 169 | ·9998 929 | ·9998 633 | ·9998 271 | ·9997 830 | ·9997 299 |
| ·94 | ·9999 787 | ·9999 723 | ·9999 643 | ·9999 544 | ·9999 422 | ·9999 274 |
| ·95 | ·9999 959 | ·9999 947 | ·9999 931 | ·9999 911 | ·9999 886 | ·9999 855+ |
| ·96 | ·9999 995− | ·9999 993 | ·9999 991 | ·9999 989 | ·9999 985+ | ·9999 981 |
| ·97 | 1·0000 000 | 1·0000 000 | ·9999 999 | ·9999 999 | ·9999 999 | ·9999 999 |
| ·98 | | | 1·0000 000 | 1·0000 000 | 1·0000 000 | 1·0000 000 |

$x$ = ·38 to ·98 $q$ = 10·5 $p$ = 34 to 39

| $x$ | $p$ = 34 | $p$ = 35 | $p$ = 36 | $p$ = 37 | $p$ = 38 | $p$ = 39 |
|---|---|---|---|---|---|---|
| $B(p, q)$ = | ·2462 5386 × $\frac{1}{10^{10}}$ | ·1881 4902 × $\frac{1}{10^{10}}$ | ·1447 3001 × $\frac{1}{10^{10}}$ | ·1120 4904 × $\frac{1}{10^{10}}$ | ·8728 0307 × $\frac{1}{10^{11}}$ | ·6838 4571 × $\frac{1}{10^{11}}$ |
| ·38 | ·0000 001 | | | | | |
| ·39 | ·0000 002 | ·0000 001 | | | | |
| ·40 | ·0000 003 | ·0000 002 | ·0000 001 | | | |
| ·41 | ·0000 007 | ·0000 003 | ·0000 002 | ·0000 001 | | |
| ·42 | ·0000 013 | ·0000 007 | ·0000 004 | ·0000 002 | ·0000 001 | ·0000 001 |
| ·43 | ·0000 025− | ·0000 013 | ·0000 007 | ·0000 004 | ·0000 002 | ·0000 001 |
| ·44 | ·0000 046 | ·0000 026 | ·0000 014 | ·0000 008 | ·0000 004 | ·0000 002 |
| ·45 | ·0000 084 | ·0000 048 | ·0000 027 | ·0000 015+ | ·0000 008 | ·0000 005− |
| ·46 | ·0000 151 | ·0000 088 | ·0000 051 | ·0000 029 | ·0000 017 | ·0000 009 |
| ·47 | ·0000 265+ | ·0000 157 | ·0000 093 | ·0000 054 | ·0000 032 | ·0000 018 |
| ·48 | ·0000 458 | ·0000 277 | ·0000 167 | ·0000 100 | ·0000 060 | ·0000 035+ |
| ·49 | ·0000 778 | ·0000 480 | ·0000 295− | ·0000 180 | ·0000 110 | ·0000 066 |
| ·50 | ·0001 297 | ·0000 817 | ·0000 512 | ·0000 319 | ·0000 198 | ·0000 122 |
| ·51 | ·0002 128 | ·0001 367 | ·0000 873 | ·0000 555− | ·0000 351 | ·0000 221 |
| ·52 | ·0003 434 | ·0002 248 | ·0001 463 | ·0000 948 | ·0000 611 | ·0000 392 |
| ·53 | ·0005 455+ | ·0003 637 | ·0002 412 | ·0001 592 | ·0001 045+ | ·0000 683 |
| ·54 | ·0008 530 | ·0005 792 | ·0003 912 | ·0002 629 | ·0001 758 | ·0001 170 |
| ·55 | ·0013 137 | ·0009 080 | ·0006 243 | ·0004 271 | ·0002 908 | ·0001 971 |
| ·56 | ·0019 930 | ·0014 017 | ·0009 807 | ·0006 828 | ·0004 730 | ·0003 263 |
| ·57 | ·0029 792 | ·0021 314 | ·0015 170 | ·0010 744 | ·0007 573 | ·0005 314 |
| ·58 | ·0043 895− | ·0031 933 | ·0023 112 | ·0016 645+ | ·0011 932 | ·0008 514 |
| ·59 | ·0063 755− | ·0047 146 | ·0034 687 | ·0025 396 | ·0018 508 | ·0013 427 |
| ·60 | ·0091 302 | ·0068 608 | ·0051 296 | ·0038 167 | ·0028 268 | ·0020 843 |
| ·61 | ·0128 940 | ·0098 423 | ·0074 756 | ·0056 509 | ·0042 520 | ·0031 854 |
| ·62 | ·0179 594 | ·0139 212 | ·0107 380 | ·0082 436 | ·0063 000 | ·0047 937 |
| ·63 | ·0246 744 | ·0194 162 | ·0152 044 | ·0118 507 | ·0091 955− | ·0071 044 |
| ·64 | ·0334 418 | ·0267 053 | ·0212 237 | ·0167 897 | ·0132 234 | ·0103 703 |
| ·65 | ·0447 151 | ·0362 251 | ·0292 088 | ·0234 446 | ·0187 360 | ·0149 102 |
| ·66 | ·0589 883 | ·0484 648 | ·0396 341 | ·0322 678 | ·0261 577 | ·0211 170 |
| ·67 | ·0767 787 | ·0639 531 | ·0530 276 | ·0437 757 | ·0359 854 | ·0294 610 |
| ·68 | ·0986 029 | ·0832 385+ | ·0699 549 | ·0585 384 | ·0487 819 | ·0404 888 |
| ·69 | ·1249 446 | ·1068 601 | ·0909 949 | ·0771 591 | ·0651 612 | ·0548 131 |
| ·70 | ·1562 153 | ·1353 106 | ·1167 053 | ·1002 449 | ·0857 642 | ·0730 935+ |
| ·71 | ·1927 104 | ·1689 912 | ·1475 795− | ·1283 654 | ·1112 208 | ·0960 050− |
| ·72 | ·2345 619 | ·2081 620 | ·1839 948 | ·1620 027 | ·1421 027 | ·1241 928 |
| ·73 | ·2816 931 | ·2528 903 | ·2261 572 | ·2014 922 | ·1788 637 | ·1582 152 |
| ·74 | ·3337 799 | ·3030 022 | ·2740 448 | ·2469 615+ | ·2217 741 | ·1984 757 |
| ·75 | ·3902 247 | ·3580 443 | ·3273 580 | ·2982 700 | ·2708 525+ | ·2451 487 |
| ·76 | ·4501 482 | ·4172 616 | ·3854 838 | ·3549 595− | ·3258 046 | ·2981 075− |
| ·77 | ·5124 053 | ·4795 984 | ·4474 805− | ·4162 235− | ·3859 760 | ·3568 628 |
| ·78 | ·5756 273 | ·5437 272 | ·5120 913 | ·4809 037 | ·4503 313 | ·4205 233 |
| ·79 | ·6382 917 | ·6081 087 | ·5777 921 | ·5475 218 | ·5174 683 | ·4877 905− |
| ·80 | ·6988 164 | ·6710 824 | ·6428 730 | ·6143 501 | ·5856 730 | ·5569 960 |
| ·81 | ·7556 721 | ·7309 809 | ·7055 525+ | ·6795 203 | ·6530 201 | ·6261 887 |
| ·82 | ·8075 008 | ·7862 601 | ·7641 151 | ·7411 642 | ·7175 124 | ·6932 693 |
| ·83 | ·8532 285+ | ·8356 288 | ·8170 560 | ·7975 726 | ·7772 494 | ·7561 640 |
| ·84 | ·8921 546 | ·8781 622 | ·8632 183 | ·8473 528 | ·8306 040 | ·8130 178 |
| ·85 | ·9240 065− | ·9133 812 | ·9018 983 | ·8895 622 | ·8763 843 | ·8623 827 |
| ·86 | ·9489 470 | ·9412 835− | ·9329 039 | ·9237 958 | ·9139 517 | ·9033 693 |
| ·87 | ·9675 324 | ·9623 181 | ·9565 503 | ·9502 082 | ·9432 738 | ·9357 327 |
| ·88 | ·9806 242 | ·9773 056 | ·9735 926 | ·9694 629 | ·9648 957 | ·9598 717 |
| ·89 | ·9892 680 | ·9873 136 | ·9851 020 | ·9826 143 | ·9798 317 | ·9767 361 |
| ·90 | ·9945 600 | ·9935 096 | ·9923 075+ | ·9909 402 | ·9893 937 | ·9876 538 |
| ·91 | ·9975 227 | ·9970 168 | ·9964 314 | ·9957 582 | ·9949 882 | ·9941 124 |
| ·92 | ·9990 115+ | ·9987 985+ | ·9985 494 | ·9982 597 | ·9979 247 | ·9975 395+ |
| ·93 | ·9996 663 | ·9995 906 | ·9995 011 | ·9993 959 | ·9992 729 | ·9991 300 |
| ·94 | ·9999 094 | ·9998 879 | ·9998 621 | ·9998 314 | ·9997 952 | ·9997 527 |
| ·95 | ·9999 818 | ·9999 772 | ·9999 717 | ·9999 651 | ·9999 572 | ·9999 478 |
| ·96 | ·9999 976 | ·9999 970 | ·9999 962 | ·9999 953 | ·9999 942 | ·9999 928 |
| ·97 | ·9999 998 | ·9999 998 | ·9999 997 | ·9999 997 | ·9999 996 | ·9999 995+ |
| ·98 | 1·0000 000 | 1·0000 000 | 1·0000 000 | 1·0000 000 | 1·0000 000 | 1·0000 000 |

$x$ = ·43 to ·98 $q$ = 10·5 $p$ = 40 to 45

| | $p$ = 40 | $p$ = 41 | $p$ = 42 | $p$ = 43 | $p$ = 44 | $p$ = 45 |
|---|---|---|---|---|---|---|
| $B(p, q)$ = | ·5387 8753 × $\frac{1}{10^{11}}$ | ·4267 6240 × $\frac{1}{10^{11}}$ | ·3397 5259 × $\frac{1}{10^{11}}$ | ·2718 0207 × $\frac{1}{10^{11}}$ | ·2184 5774 × $\frac{1}{10^{11}}$ | ·1763 6955 × $\frac{1}{10^{11}}$ |
| $x$ | | | | | | |
| ·43 | ·0000 001 | | | | | |
| ·44 | ·0000 001 | ·0000 001 | | | | |
| ·45 | ·0000 003 | ·0000 001 | ·0000 001 | | | |
| ·46 | ·0000 005+ | ·0000 003 | ·0000 002 | ·0000 001 | ·0000 001 | |
| ·47 | ·0000 011 | ·0000 006 | ·0000 004 | ·0000 002 | ·0000 001 | ·0000 001 |
| ·48 | ·0000 021 | ·0000 012 | ·0000 007 | ·0000 004 | ·0000 002 | ·0000 001 |
| ·49 | ·0000 040 | ·0000 024 | ·0000 014 | ·0000 009 | ·0000 005+ | ·0000 003 |
| ·50 | ·0000 075+ | ·0000 046 | ·0000 028 | ·0000 017 | ·0000 010 | ·0000 006 |
| ·51 | ·0000 138 | ·0000 086 | ·0000 054 | ·0000 033 | ·0000 020 | ·0000 013 |
| ·52 | ·0000 250+ | ·0000 159 | ·0000 101 | ·0000 064 | ·0000 040 | ·0000 025+ |
| ·53 | ·0000 445− | ·0000 288 | ·0000 186 | ·0000 119 | ·0000 076 | ·0000 049 |
| ·54 | ·0000 776 | ·0000 512 | ·0000 336 | ·0000 220 | ·0000 144 | ·0000 093 |
| ·55 | ·0001 330 | ·0000 893 | ·0000 598 | ·0000 398 | ·0000 265− | ·0000 175+ |
| ·56 | ·0002 240 | ·0001 532 | ·0001 043 | ·0000 708 | ·0000 478 | ·0000 322 |
| ·57 | ·0003 712 | ·0002 582 | ·0001 789 | ·0001 235+ | ·0000 849 | ·0000 582 |
| ·58 | ·0006 050− | ·0004 280 | ·0003 017 | ·0002 118 | ·0001 481 | ·0001 032 |
| ·59 | ·0009 699 | ·0006 978 | ·0005 000− | ·0003 569 | ·0002 538 | ·0001 799 |
| ·60 | ·0015 303 | ·0011 189 | ·0008 150− | ·0005 913 | ·0004 275− | ·0003 079 |
| ·61 | ·0023 763 | ·0017 655− | ·0013 066 | ·0009 633 | ·0007 077 | ·0005 180 |
| ·62 | ·0036 323 | ·0027 412 | ·0020 607 | ·0015 434 | ·0011 518 | ·0008 566 |
| ·63 | ·0054 662 | ·0041 890 | ·0031 979 | ·0024 323 | ·0018 434 | ·0013 923 |
| ·64 | ·0080 995+ | ·0063 011 | ·0048 834 | ·0037 708 | ·0029 014 | ·0022 249 |
| ·65 | ·0118 177 | ·0093 301 | ·0073 386 | ·0057 512 | ·0044 915+ | ·0034 959 |
| ·66 | ·0169 796 | ·0136 005− | ·0108 535− | ·0086 304 | ·0068 389 | ·0054 013 |
| ·67 | ·0240 249 | ·0195 177 | ·0157 981 | ·0127 423 | ·0102 426 | ·0082 061 |
| ·68 | ·0334 759 | ·0275 745− | ·0226 317 | ·0185 104 | ·0150 887 | ·0122 597 |
| ·69 | ·0459 336 | ·0383 516 | ·0319 078 | ·0264 559 | ·0218 630 | ·0180 096 |
| ·70 | ·0620 636 | ·0525 088 | ·0442 708 | ·0371 999 | ·0311 567 | ·0260 131 |
| ·71 | ·0825 701 | ·0707 660 | ·0604 431 | ·0514 560 | ·0436 655+ | ·0369 400 |
| ·72 | ·1081 565− | ·0938 681 | ·0811 968 | ·0700 099 | ·0601 757 | ·0515 664 |
| ·73 | ·1394 701 | ·1225 362 | ·1073 102 | ·0936 811 | ·0815 337 | ·0707 516 |
| ·74 | ·1770 352 | ·1574 011 | ·1395 053 | ·1232 670 | ·1085 959 | ·0953 953 |
| ·75 | ·2211 749 | ·1989 242 | ·1783 695− | ·1594 665− | ·1421 570 | ·1263 720 |
| ·76 | ·2719 302 | ·2473 104 | ·2242 639 | ·2027 867 | ·1828 579 | ·1644 419 |
| ·77 | ·3289 850+ | ·3024 210 | ·2772 273 | ·2534 402 | ·2310 776 | ·2101 407 |
| ·78 | ·3916 095+ | ·3637 005+ | ·3368 874 | ·3112 426 | ·2868 204 | ·2636 582 |
| ·79 | ·4586 339 | ·4301 297 | ·4023 937 | ·3755 262 | ·3496 115− | ·3247 186 |
| ·80 | ·5284 660 | ·5002 213 | ·4723 899 | ·4450 887 | ·4184 223 | ·3924 833 |
| ·81 | ·5991 612 | ·5720 695− | ·5450 406 | ·5181 951 | ·4916 461 | ·4654 984 |
| ·82 | ·6685 476 | ·6434 612 | ·6181 240 | ·5926 480 | ·5671 424 | ·5417 121 |
| ·83 | ·7344 001 | ·7120 459 | ·6891 927 | ·6659 343 | ·6423 648 | ·6185 783 |
| ·84 | ·7946 469 | ·7755 504 | ·7557 925+ | ·7354 419 | ·7145 707 | ·6932 536 |
| ·85 | ·8475 823 | ·8320 141 | ·8157 150− | ·7987 270 | ·7810 971 | ·7628 761 |
| ·86 | ·8920 515+ | ·8800 066 | ·8672 477 | ·8537 932 | ·8396 660 | ·8248 934 |
| ·87 | ·9275 737 | ·9187 896 | ·9093 767 | ·8993 351 | ·8886 689 | ·8773 857 |
| ·88 | ·9543 737 | ·9483 864 | ·9418 967 | ·9348 941 | ·9273 703 | ·9193 198 |
| ·89 | ·9733 099 | ·9695 364 | ·9653 998 | ·9608 855− | ·9559 801 | ·9506 717 |
| ·90 | ·9857 065− | ·9835 377 | ·9811 335− | ·9784 802 | ·9755 648 | ·9723 743 |
| ·91 | ·9931 212 | ·9920 051 | ·9907 540 | ·9893 581 | ·9878 071 | ·9860 910 |
| ·92 | ·9970 988 | ·9965 971 | ·9960 285+ | ·9953 871 | ·9946 667 | ·9938 608 |
| ·93 | ·9989 646 | ·9987 744 | ·9985 564 | ·9983 079 | ·9980 257 | ·9977 065+ |
| ·94 | ·9997 030 | ·9996 451 | ·9995 782 | ·9995 010 | ·9994 124 | ·9993 111 |
| ·95 | ·9999 367 | ·9999 237 | ·9999 085− | ·9998 907 | ·9998 701 | ·9998 464 |
| ·96 | ·9999 912 | ·9999 893 | ·9999 871 | ·9999 844 | ·9999 813 | ·9999 777 |
| ·97 | ·9999 994 | ·9999 992 | ·9999 991 | ·9999 989 | ·9999 986 | ·9999 984 |
| ·98 | 1·0000 000 | 1·0000 000 | 1·0000 000 | 1·0000 000 | 1·0000 000 | 1·0000 000 |

$x = \cdot48$ to $\cdot99$ $q = 10\cdot5$ $p = 46$ to $50$

| $x$ | $p = 46$ | $p = 47$ | $p = 48$ | $p = 49$ | $p = 50$ |
|---|---|---|---|---|---|
| $B(p, q) =$ | $\cdot1430\ 0234 \times \frac{1}{10^{11}}$ | $\cdot1164\ 2668 \times \frac{1}{10^{11}}$ | $\cdot9516\ 6158 \times \frac{1}{10^{12}}$ | $\cdot7808\ 5052 \times \frac{1}{10^{12}}$ | $\cdot6430\ 5337 \times \frac{1}{10^{12}}$ |
| ·48 | ·0000 001 | | | | |
| ·49 | ·0000 002 | ·0000 001 | ·0000 001 | | |
| ·50 | ·0000 004 | ·0000 002 | ·0000 001 | ·0000 001 | |
| ·51 | ·0000 008 | ·0000 005⁻ | ·0000 003 | ·0000 002 | ·0000 001 |
| ·52 | ·0000 016 | ·0000 010 | ·0000 006 | ·0000 004 | ·0000 002 |
| ·53 | ·0000 031 | ·0000 020 | ·0000 012 | ·0000 008 | ·0000 005⁻ |
| ·54 | ·0000 060 | ·0000 039 | ·0000 025⁺ | ·0000 016 | ·0000 010 |
| ·55 | ·0000 115⁺ | ·0000 076 | ·0000 050⁻ | ·0000 032 | ·0000 021 |
| ·56 | ·0000 216 | ·0000 145⁻ | ·0000 096 | ·0000 064 | ·0000 042 |
| ·57 | ·0000 397 | ·0000 270 | ·0000 183 | ·0000 124 | ·0000 084 |
| ·58 | ·0000 717 | ·0000 496 | ·0000 343 | ·0000 236 | ·0000 162 |
| ·59 | ·0001 270 | ·0000 894 | ·0000 628 | ·0000 439 | ·0000 306 |
| ·60 | ·0002 211 | ·0001 582 | ·0001 129 | ·0000 803 | ·0000 569 |
| ·61 | ·0003 780 | ·0002 749 | ·0001 993 | ·0001 440 | ·0001 038 |
| ·62 | ·0006 349 | ·0004 691 | ·0003 455⁻ | ·0002 537 | ·0001 858 |
| ·63 | ·0010 481 | ·0007 865⁻ | ·0005 883 | ·0004 388 | ·0003 263 |
| ·64 | ·0017 006 | ·0012 957 | ·0009 841 | ·0007 453 | ·0005 628 |
| ·65 | ·0027 122 | ·0020 975⁺ | ·0016 173 | ·0012 433 | ·0009 531 |
| ·66 | ·0042 522 | ·0033 371 | ·0026 111 | ·0020 371 | ·0015 848 |
| ·67 | ·0065 537 | ·0052 179 | ·0041 420 | ·0032 785⁺ | ·0025 878 |
| ·68 | ·0099 297 | ·0080 182 | ·0064 556 | ·0051 828 | ·0041 494 |
| ·69 | ·0147 895⁺ | ·0121 088 | ·0098 852 | ·0080 473 | ·0065 332 |
| ·70 | ·0216 524 | ·0179 696 | ·0148 705⁺ | ·0122 718 | ·0101 001 |
| ·71 | ·0311 568 | ·0262 027 | ·0219 744 | ·0183 781 | ·0153 296 |
| ·72 | ·0440 591 | ·0375 375⁺ | ·0318 929 | ·0270 244 | ·0228 394 |
| ·73 | ·0612 190 | ·0528 230 | ·0454 551 | ·0390 120 | ·0333 967 |
| ·74 | ·0835 645⁺ | ·0730 017 | ·0636 054 | ·0552 759 | ·0479 170 |
| ·75 | ·1120 339 | ·0990 595⁺ | ·0873 619 | ·0768 522 | ·0674 418 |
| ·76 | ·1474 909 | ·1319 477 | ·1177 471 | ·1048 186 | ·0930 880 |
| ·77 | ·1906 158 | ·1724 767 | ·1556 865⁺ | ·1401 993 | ·1259 621 |
| ·78 | ·2417 776 | ·2211 860 | ·2018 782 | ·1838 374 | ·1670 374 |
| ·79 | ·3009 017 | ·2782 006 | ·2566 417 | ·2362 394 | ·2169 967 |
| ·80 | ·3673 512 | ·3430 933 | ·3197 644 | ·2974 076 | ·2760 545⁺ |
| ·81 | ·4398 475⁺ | ·4147 793 | ·3903 699 | ·3666 851 | ·3437 809 |
| ·82 | ·5164 571 | ·4914 710 | ·4668 412 | ·4426 475⁻ | ·4189 623 |
| ·83 | ·5946 676 | ·5707 232 | ·5468 322 | ·5230 781 | ·4995 398 |
| ·84 | ·6715 670 | ·6495 881 | ·6273 940 | ·6050 613 | ·5826 648 |
| ·85 | ·7441 187 | ·7248 821 | ·7052 260 | ·6852 116 | ·6649 010 |
| ·86 | ·8095 070 | ·7935 419 | ·7770 367 | ·7600 328 | ·7425 744 |
| ·87 | ·8654 969 | ·8530 177 | ·8399 662 | ·8263 641 | ·8122 360 |
| ·88 | ·9107 398 | ·9016 301 | ·8919 932 | ·8818 343 | ·8711 612 |
| ·89 | ·9449 499 | ·9388 057 | ·9322 321 | ·9252 237 | ·9177 769 |
| ·90 | ·9688 968 | ·9651 206 | ·9610 350⁺ | ·9566 303 | ·9518 975⁺ |
| ·91 | ·9841 996 | ·9821 230 | ·9798 514 | ·9773 751 | ·9746 847 |
| ·92 | ·9929 628 | ·9919 661 | ·9908 637 | ·9896 488 | ·9883 144 |
| ·93 | ·9973 471 | ·9969 438 | ·9964 929 | ·9959 906 | ·9954 328 |
| ·94 | ·9991 959 | ·9990 651 | ·9989 174 | ·9987 511 | ·9985 644 |
| ·95 | ·9998 190 | ·9997 876 | ·9997 518 | ·9997 111 | ·9996 648 |
| ·96 | ·9999 735⁺ | ·9999 686 | ·9999 630 | ·9999 565⁺ | ·9999 491 |
| ·97 | ·9999 980 | ·9999 977 | ·9999 972 | ·9999 967 | ·9999 961 |
| ·98 | 1·0000 000 | 1·0000 000 | ·9999 999 | ·9999 999 | ·9999 999 |
| ·99 | | | 1·0000 000 | 1·0000 000 | 1·0000 000 |

$x$ = ·08 to ·70 $q = 11$ $p$ = 11 to 16

| $x$ | $p = 11$ | $p = 12$ | $p = 13$ | $p = 14$ | $p = 15$ | $p = 16$ |
|---|---|---|---|---|---|---|
| $B(p, q) =$ | ·2577 4020 $\times \frac{1}{10^8}$ | ·1288 7010 $\times \frac{1}{10^8}$ | ·6723 6573 $\times \frac{1}{10^7}$ | ·3641 9810 $\times \frac{1}{10^7}$ | ·2039 5094 $\times \frac{1}{10^7}$ | ·1176 6400 $\times \frac{1}{10^7}$ |
| ·08 | ·0000 001 | | | | | |
| ·09 | ·0000 005− | ·0000 001 | | | | |
| ·10 | ·0000 014 | ·0000 002 | | | | |
| ·11 | ·0000 035− | ·0000 007 | ·0000 001 | | | |
| ·12 | ·0000 082 | ·0000 018 | ·0000 004 | ·0000 001 | | |
| ·13 | ·0000 179 | ·0000 042 | ·0000 010 | ·0000 002 | | |
| ·14 | ·0000 364 | ·0000 092 | ·0000 023 | ·0000 005+ | ·0000 001 | |
| ·15 | ·0000 701 | ·0000 190 | ·0000 050− | ·0000 013 | ·0000 003 | ·0000 001 |
| ·16 | ·0001 283 | ·0000 371 | ·0000 104 | ·0000 028 | ·0000 007 | ·0000 002 |
| ·17 | ·0002 247 | ·0000 690 | ·0000 205+ | ·0000 059 | ·0000 017 | ·0000 005− |
| ·18 | ·0003 784 | ·0001 229 | ·0000 386 | ·0000 118 | ·0000 035− | ·0000 010 |
| ·19 | ·0006 154 | ·0002 108 | ·0000 699 | ·0000 225− | ·0000 070 | ·0000 022 |
| ·20 | ·0009 697 | ·0003 492 | ·0001 217 | ·0000 412 | ·0000 136 | ·0000 044 |
| ·21 | ·0014 848 | ·0005 607 | ·0002 049 | ·0000 727 | ·0000 251 | ·0000 085− |
| ·22 | ·0022 152 | ·0008 752 | ·0003 347 | ·0001 243 | ·0000 450− | ·0000 159 |
| ·23 | ·0032 270 | ·0013 310 | ·0005 315+ | ·0002 062 | ·0000 779 | ·0000 288 |
| ·24 | ·0045 989 | ·0019 765− | ·0008 226 | ·0003 327 | ·0001 311 | ·0000 505− |
| ·25 | ·0064 227 | ·0028 710 | ·0012 431 | ·0005 231 | ·0002 145+ | ·0000 859 |
| ·26 | ·0088 032 | ·0040 861 | ·0018 376 | ·0008 032 | ·0003 422 | ·0001 425− |
| ·27 | ·0118 575+ | ·0057 061 | ·0026 611 | ·0012 066 | ·0005 333 | ·0002 303 |
| ·28 | ·0157 139 | ·0078 285+ | ·0037 807 | ·0017 755− | ·0008 130 | ·0003 638 |
| ·29 | ·0205 098 | ·0105 639 | ·0052 760 | ·0025 629 | ·0012 141 | ·0005 621 |
| ·30 | ·0263 899 | ·0140 351 | ·0072 399 | ·0036 333 | ·0017 784 | ·0008 510 |
| ·31 | ·0335 028 | ·0183 761 | ·0097 791 | ·0050 640 | ·0025 582 | ·0012 636 |
| ·32 | ·0419 973 | ·0237 301 | ·0130 133 | ·0069 460 | ·0036 176 | ·0018 425− |
| ·33 | ·0520 190 | ·0302 468 | ·0170 747 | ·0093 842 | ·0050 335+ | ·0026 407 |
| ·34 | ·0637 053 | ·0380 795+ | ·0221 062 | ·0124 976 | ·0068 971 | ·0037 235+ |
| ·35 | ·0771 815+ | ·0473 812 | ·0282 594 | ·0164 185+ | ·0093 140 | ·0051 697 |
| ·36 | ·0925 560 | ·0583 004 | ·0356 916 | ·0212 916 | ·0124 047 | ·0070 726 |
| ·37 | ·1099 160 | ·0709 765− | ·0445 625+ | ·0272 716 | ·0163 042 | ·0095 409 |
| ·38 | ·1293 231 | ·0855 353 | ·0550 298 | ·0345 207 | ·0211 605− | ·0126 990 |
| ·39 | ·1508 102 | ·1020 838 | ·0672 445− | ·0432 053 | ·0271 334 | ·0166 867 |
| ·40 | ·1743 779 | ·1207 057 | ·0813 462 | ·0534 917 | ·0343 915+ | ·0216 581 |
| ·41 | ·1999 925+ | ·1414 570 | ·0974 578 | ·0655 414 | ·0431 088 | ·0277 798 |
| ·42 | ·2275 849 | ·1643 618 | ·1156 801 | ·0795 058 | ·0534 604 | ·0352 285+ |
| ·43 | ·2570 493 | ·1894 097 | ·1360 870 | ·0955 208 | ·0656 177 | ·0441 872 |
| ·44 | ·2882 449 | ·2165 526 | ·1587 207 | ·1137 009 | ·0797 430 | ·0548 406 |
| ·45 | ·3209 966 | ·2457 040 | ·1835 876 | ·1341 334 | ·0959 830 | ·0673 702 |
| ·46 | ·3550 980 | ·2767 383 | ·2106 550+ | ·1568 733 | ·1144 627 | ·0819 478 |
| ·47 | ·3903 150+ | ·3094 918 | ·2398 491 | ·1819 386 | ·1352 792 | ·0987 294 |
| ·48 | ·4263 902 | ·3437 644 | ·2710 538 | ·2093 056 | ·1584 957 | ·1178 478 |
| ·49 | ·4630 479 | ·3793 231 | ·3041 104 | ·2389 067 | ·1841 356 | ·1394 059 |
| ·50 | ·5000 000[e] | ·4159 060 | ·3388 197 | ·2706 281 | ·2121 781 | ·1634 698 |
| ·51 | ·5369 521 | ·4532 273 | ·3749 447 | ·3043 096 | ·2425 544 | ·1900 625− |
| ·52 | ·5736 098 | ·4909 840 | ·4122 141 | ·3397 458 | ·2751 455− | ·2191 585+ |
| ·53 | ·6096 850− | ·5288 617 | ·4503 285− | ·3766 885− | ·3097 813 | ·2506 799 |
| ·54 | ·6449 020 | ·5665 423 | ·4889 662 | ·4148 512 | ·3462 419 | ·2844 935+ |
| ·55 | ·6790 034 | ·6037 108 | ·5277 907 | ·4539 147 | ·3842 601 | ·3204 101 |
| ·56 | ·7117 551 | ·6400 627 | ·5664 585+ | ·4935 338 | ·4235 261 | ·3581 855− |
| ·57 | ·7429 507 | ·6753 110 | ·6046 275− | ·5333 459 | ·4636 937 | ·3975 240 |
| ·58 | ·7724 151 | ·7091 921 | ·6419 650− | ·5729 796 | ·5043 884 | ·4380 835+ |
| ·59 | ·8000 075− | ·7414 719 | ·6781 560 | ·6120 638 | ·5452 164 | ·4794 831 |
| ·60 | ·8256 221 | ·7719 500+ | ·7129 107 | ·6502 381 | ·5857 750− | ·5213 118 |
| ·61 | ·8491 898 | ·8004 634 | ·7459 711 | ·6871 613 | ·6256 631 | ·5631 399 |
| ·62 | ·8706 769 | ·8268 891 | ·7771 169 | ·7225 207 | ·6644 927 | ·6045 304 |
| ·63 | ·8900 840 | ·8511 445+ | ·8061 694 | ·7560 395− | ·7018 992 | ·6450 518 |
| ·64 | ·9074 440 | ·8731 883 | ·8329 950− | ·7874 838 | ·7375 515+ | ·6842 904 |
| ·65 | ·9228 185− | ·8930 182 | ·8575 062 | ·8166 674 | ·7711 612 | ·7218 630 |
| ·66 | ·9362 947 | ·9106 690 | ·8796 618 | ·8434 550− | ·8024 896 | ·7574 276 |
| ·67 | ·9479 810 | ·9262 089 | ·8994 654 | ·8677 641 | ·8313 529 | ·7906 938 |
| ·68 | ·9580 027 | ·9397 354 | ·9169 623 | ·8895 644 | ·8576 263 | ·8214 298 |
| ·69 | ·9664 972 | ·9513 705+ | ·9322 352 | ·9088 755+ | ·8812 443 | ·8494 684 |
| ·70 | ·9736 101 | ·9612 552 | ·9453 998 | ·9257 635+ | ·9022 000 | ·8747 092 |

$x = \cdot 71$ to $\cdot 95$ $q = 11$ $p = 11$ to $16$

| | $p = 11$ | $p = 12$ | $p = 13$ | $p = 14$ | $p = 15$ | $p = 16$ |
|---|---|---|---|---|---|---|
| $B(p, q) =$ | ·2577 4020 $\times \frac{1}{10^6}$ | ·1288 7010 $\times \frac{1}{10^6}$ | ·6723 6573 $\times \frac{1}{10^7}$ | ·3641 9810 $\times \frac{1}{10^7}$ | ·2039 5094 $\times \frac{1}{10^7}$ | ·1176 6400 $\times \frac{1}{10^7}$ |
| $x$ | | | | | | |
| ·71 | ·9794 902 | ·9695 443 | ·9565 980 | ·9403 355$^-$ | ·9205 417 | ·8971 190 |
| ·72 | ·9842 861 | ·9764 008 | ·9659 921 | ·9527 331 | ·9363 677 | ·9167 291 |
| ·73 | ·9881 425$^-$ | ·9819 910 | ·9737 583 | ·9631 254 | ·9498 192 | ·9336 299 |
| ·74 | ·9911 968 | ·9864 796 | ·9800 800 | ·9717 014 | ·9610 726 | ·9479 637 |
| ·75 | ·9935 773 | ·9900 256 | ·9851 419 | ·9786 618 | ·9703 301 | ·9599 155$^+$ |
| ·76 | ·9954 011 | ·9927 787 | ·9891 249 | ·9842 118 | ·9778 108 | ·9697 029 |
| ·77 | ·9967 730 | ·9948 771 | ·9922 006 | ·9885 544 | ·9837 415$^+$ | ·9775 649 |
| ·78 | ·9977 848 | ·9964 448 | ·9945 285$^+$ | ·9918 841 | ·9883 481 | ·9837 514 |
| ·79 | ·9985 152 | ·9975 910 | ·9962 526 | ·9943 818 | ·9918 483 | ·9885 125$^-$ |
| ·80 | ·9990 303 | ·9984 098 | ·9974 997 | ·9962 116 | ·9944 451 | ·9920 897 |
| ·81 | ·9993 846 | ·9989 800 | ·9983 791 | ·9975 180 | ·9963 223 | ·9947 082 |
| ·82 | ·9996 216 | ·9993 661 | ·9989 820 | ·9984 248 | ·9976 415$^-$ | ·9965 710 |
| ·83 | ·9997 753 | ·9996 196 | ·9993 827 | ·9990 349 | ·9985 399 | ·9978 552 |
| ·84 | ·9998 717 | ·9997 805$^+$ | ·9996 402 | ·9994 315$^+$ | ·9991 311 | ·9987 104 |
| ·85 | ·9999 299 | ·9998 789 | ·9997 993 | ·9996 796 | ·9995 053 | ·9992 583 |
| ·86 | ·9999 636 | ·9999 364 | ·9998 935$^+$ | ·9998 283 | ·9997 322 | ·9995 944 |
| ·87 | ·9999 821 | ·9999 685$^-$ | ·9999 467 | ·9999 131 | ·9998 631 | ·9997 906 |
| ·88 | ·9999 918 | ·9999 854 | ·9999 750$^+$ | ·9999 589 | ·9999 345$^+$ | ·9998 988 |
| ·89 | ·9999 965$^+$ | ·9999 937 | ·9999 892 | ·9999 820 | ·9999 710 | ·9999 548 |
| ·90 | ·9999 986 | ·9999 975$^+$ | ·9999 957 | ·9999 928 | ·9999 883 | ·9999 816 |
| ·91 | ·9999 995$^+$ | ·9999 991 | ·9999 985$^-$ | ·9999 974 | ·9999 958 | ·9999 933 |
| ·92 | ·9999 999 | ·9999 997 | ·9999 995$^+$ | ·9999 992 | ·9999 987 | ·9999 979 |
| ·93 | 1·0000 000 | ·9999 999 | ·9999 999 | ·9999 998 | ·9999 997 | ·9999 994 |
| ·94 | | 1·0000 000 | 1·0000 000 | 1·0000 000 | ·9999 999 | ·9999 999 |
| ·95 | | | | | 1·0000 000 | 1·0000 000 |

$x$ = ·17 to ·80 $q$ = 11 $p$ = 17 to 22

| | $p$ = 17 | $p$ = 18 | $p$ = 19 | $p$ = 20 | $p$ = 21 | $p$ = 22 |
|---|---|---|---|---|---|---|
| $B(p, q)$ = | ·6972 6816 $\times\frac{1}{10^8}$ | ·4233 4138 $\times\frac{1}{10^8}$ | ·2627 6362 $\times\frac{1}{10^8}$ | ·1664 1696 $\times\frac{1}{10^8}$ | ·1073 6578 $\times\frac{1}{10^8}$ | ·7045 8793 $\times\frac{1}{10^9}$ |
| $x$ | | | | | | |
| ·17 | ·0000 001 | | | | | |
| ·18 | ·0000 003 | ·0000 001 | | | | |
| ·19 | ·0000 006 | ·0000 002 | ·0000 001 | | | |
| ·20 | ·0000 014 | ·0000 004 | ·0000 001 | | | |
| ·21 | ·0000 028 | ·0000 009 | ·0000 003 | ·0000 001 | | |
| ·22 | ·0000 055− | ·0000 019 | ·0000 006 | ·0000 002 | ·0000 001 | |
| ·23 | ·0000 104 | ·0000 037 | ·0000 013 | ·0000 004 | ·0000 001 | |
| ·24 | ·0000 190 | ·0000 070 | ·0000 025+ | ·0000 009 | ·0000 003 | ·0000 001 |
| ·25 | ·0000 337 | ·0000 130 | ·0000 049 | ·0000 018 | ·0000 007 | ·0000 002 |
| ·26 | ·0000 581 | ·0000 232 | ·0000 091 | ·0000 035+ | ·0000 013 | ·0000 005+ |
| ·27 | ·0000 974 | ·0000 404 | ·0000 165− | ·0000 066 | ·0000 026 | ·0000 010 |
| ·28 | ·0001 594 | ·0000 685+ | ·0000 289 | ·0000 120 | ·0000 049 | ·0000 020 |
| ·29 | ·0002 549 | ·0001 134 | ·0000 496 | ·0000 213 | ·0000 090 | ·0000 038 |
| ·30 | ·0003 989 | ·0001 834 | ·0000 829 | ·0000 369 | ·0000 162 | ·0000 070 |
| ·31 | ·0006 114 | ·0002 903 | ·0001 355− | ·0000 622 | ·0000 282 | ·0000 126 |
| ·32 | ·0009 194 | ·0004 503 | ·0002 167 | ·0001 027 | ·0000 479 | ·0000 221 |
| ·33 | ·0013 575+ | ·0006 850+ | ·0003 398 | ·0001 659 | ·0000 798 | ·0000 379 |
| ·34 | ·0019 701 | ·0010 233 | ·0005 225+ | ·0002 626 | ·0001 301 | ·0000 636 |
| ·35 | ·0028 126 | ·0015 024 | ·0007 890 | ·0004 079 | ·0002 079 | ·0001 045+ |
| ·36 | ·0039 533 | ·0021 698 | ·0011 710 | ·0006 223 | ·0003 259 | ·0001 684 |
| ·37 | ·0054 745+ | ·0030 849 | ·0017 096 | ·0009 328 | ·0005 018 | ·0002 663 |
| ·38 | ·0074 741 | ·0043 207 | ·0024 566 | ·0013 755+ | ·0007 593 | ·0004 136 |
| ·39 | ·0100 661 | ·0059 652 | ·0034 773 | ·0019 964 | ·0011 300 | ·0006 313 |
| ·40 | ·0133 813 | ·0081 231 | ·0048 514 | ·0028 539 | ·0016 554 | ·0009 477 |
| ·41 | ·0175 669 | ·0109 164 | ·0066 750− | ·0040 207 | ·0023 883 | ·0014 003 |
| ·42 | ·0227 853 | ·0144 849 | ·0090 620 | ·0055 857 | ·0033 956 | ·0020 377 |
| ·43 | ·0292 126 | ·0189 858 | ·0121 453 | ·0076 557 | ·0047 599 | ·0029 218 |
| ·44 | ·0370 354 | ·0245 927 | ·0160 763 | ·0103 569 | ·0065 821 | ·0041 303 |
| ·45 | ·0464 471 | ·0314 932 | ·0210 255− | ·0138 358 | ·0089 828 | ·0057 590 |
| ·46 | ·0576 429 | ·0398 861 | ·0271 801 | ·0182 591 | ·0121 037 | ·0079 238 |
| ·47 | ·0708 144 | ·0499 768 | ·0347 421 | ·0238 132 | ·0161 084 | ·0107 627 |
| ·48 | ·0861 424 | ·0619 717 | ·0439 243 | ·0307 021 | ·0211 822 | ·0144 366 |
| ·49 | ·1037 899 | ·0760 722 | ·0549 452 | ·0391 444 | ·0275 308 | ·0191 303 |
| ·50 | ·1238 943 | ·0924 667 | ·0680 230 | ·0493 686 | ·0353 778 | ·0250 512 |
| ·51 | ·1465 598 | ·1113 226 | ·0833 678 | ·0616 071 | ·0449 603 | ·0324 275+ |
| ·52 | ·1718 495+ | ·1327 779 | ·1011 733 | ·0760 892 | ·0565 236 | ·0415 047 |
| ·53 | ·1997 789 | ·1569 322 | ·1216 074 | ·0930 316 | ·0703 138 | ·0525 398 |
| ·54 | ·2303 093 | ·1838 384 | ·1448 028 | ·1126 293 | ·0865 687 | ·0657 947 |
| ·55 | ·2633 441 | ·2134 953 | ·1708 469 | ·1350 447 | ·1055 079 | ·0815 268 |
| ·56 | ·2987 256 | ·2458 412 | ·1997 731 | ·1603 970 | ·1273 210 | ·0999 782 |
| ·57 | ·3362 344 | ·2807 492 | ·2315 524 | ·1887 511 | ·1521 561 | ·1213 639 |
| ·58 | ·3755 912 | ·3180 248 | ·2660 871 | ·2201 085+ | ·1801 072 | ·1458 584 |
| ·59 | ·4164 612 | ·3574 061 | ·3032 065+ | ·2543 985− | ·2112 033 | ·1735 824 |
| ·60 | ·4584 602 | ·3985 663 | ·3426 654 | ·2914 719 | ·2453 977 | ·2045 892 |
| ·61 | ·5011 638 | ·4411 198 | ·3841 448 | ·3310 981 | ·2825 603 | ·2388 532 |
| ·62 | ·5441 185− | ·4846 305− | ·4272 576 | ·3729 647 | ·3224 723 | ·2762 598 |
| ·63 | ·5868 543 | ·5286 226 | ·4715 555+ | ·4166 810 | ·3648 246 | ·3165 982 |
| ·64 | ·6288 989 | ·5725 950− | ·5165 414 | ·4617 858 | ·4092 205− | ·3595 587 |
| ·65 | ·6697 916 | ·6160 356 | ·5616 824 | ·5077 582 | ·4551 822 | ·4047 342 |
| ·66 | ·7090 986 | ·6584 385+ | ·6064 275− | ·5540 332 | ·5021 628 | ·4516 263 |
| ·67 | ·7464 261 | ·6993 201 | ·6502 252 | ·6000 192 | ·5495 621 | ·4996 577 |
| ·68 | ·7814 327 | ·7382 358 | ·6925 431 | ·6451 189 | ·5967 461 | ·5481 891 |
| ·69 | ·8138 396 | ·7747 947 | ·7328 865+ | ·6887 506 | ·6430 699 | ·5965 408 |
| ·70 | ·8434 384 | ·8086 726 | ·7708 166 | ·7303 704 | ·6879 019 | ·6440 177 |
| ·71 | ·8700 951 | ·8396 217 | ·8059 655+ | ·7694 928 | ·7306 494 | ·6899 378 |
| ·72 | ·8937 521 | ·8674 771 | ·8380 492 | ·8057 094 | ·7707 824 | ·7336 601 |
| ·73 | ·9144 254 | ·8921 594 | ·8668 752 | ·8387 033 | ·8078 550− | ·7746 123 |
| ·74 | ·9322 003 | ·9136 736 | ·8923 473 | ·8682 599 | ·8415 228 | ·8123 158 |
| ·75 | ·9472 227 | ·9321 034 | ·9144 642 | ·8942 719 | ·8715 556 | ·8464 054 |
| ·76 | ·9596 896 | ·9476 029 | ·9333 138 | ·9167 384 | ·8978 425+ | ·8766 431 |
| ·77 | ·9698 364 | ·9603 849 | ·9490 642 | ·9357 593 | ·9203 921 | ·9029 248 |
| ·78 | ·9779 250− | ·9707 071 | ·9619 494 | ·9515 232 | ·9393 245− | ·9252 785+ |
| ·79 | ·9842 301 | ·9788 570 | ·9722 541 | ·9642 924 | ·9548 577 | ·9438 551 |
| ·80 | ·9890 277 | ·9851 371 | ·9802 956 | ·9743 837 | ·9672 896 | ·9589 117 |

$x$ = ·81 to ·96 $q$ = 11 $p$ = 17 to 22

| | $p$ = 17 | $p$ = 18 | $p$ = 19 | $p$ = 20 | $p$ = 21 | $p$ = 22 |
|---|---|---|---|---|---|---|
| $B(p, q)$ = | ·6972 6816 × $\frac{1}{10^8}$ | ·4233 4138 × $\frac{1}{10^8}$ | ·2627 6362 × $\frac{1}{10^8}$ | ·1664 1696 × $\frac{1}{10^8}$ | ·1073 6578 × $\frac{1}{10^8}$ | ·7045 8793 × $\frac{1}{10^9}$ |
| $x$ | | | | | | |
| ·81 | ·9925 835+ | ·9898 502 | ·9864 063 | ·9821 485− | ·9769 752 | ·9707 895+ |
| ·82 | ·9951 445+ | ·9932 868 | ·9909 171 | ·9879 513 | ·9843 034 | ·9798 877 |
| ·83 | ·9969 318 | ·9957 144 | ·9941 427 | ·9921 515+ | ·9896 725+ | ·9866 352 |
| ·84 | ·9981 363 | ·9973 703 | ·9963 694 | ·9950 862 | ·9934 693 | ·9914 644 |
| ·85 | ·9989 172 | ·9984 566 | ·9978 477 | ·9970 576 | ·9960 503 | ·9947 864 |
| ·86 | ·9994 018 | ·9991 387 | ·9987 868 | ·9983 249 | ·9977 291 | ·9969 726 |
| ·87 | ·9996 880 | ·9995 463 | ·9993 546 | ·9990 999 | ·9987 676 | ·9983 409 |
| ·88 | ·9998 478 | ·9997 764 | ·9996 788 | ·9995 476 | ·9993 745− | ·9991 495+ |
| ·89 | ·9999 313 | ·9998 981 | ·9998 522 | ·9997 897 | ·9997 064 | ·9995 969 |
| ·90 | ·9999 717 | ·9999 577 | ·9999 380 | ·9999 109 | ·9998 744 | ·9998 259 |
| ·91 | ·9999 896 | ·9999 843 | ·9999 767 | ·9999 663 | ·9999 520 | ·9999 328 |
| ·92 | ·9999 967 | ·9999 949 | ·9999 924 | ·9999 889 | ·9999 840 | ·9999 774 |
| ·93 | ·9999 991 | ·9999 986 | ·9999 979 | ·9999 969 | ·9999 955+ | ·9999 936 |
| ·94 | ·9999 998 | ·9999 997 | ·9999 995+ | ·9999 993 | ·9999 990 | ·9999 986 |
| ·95 | 1·0000 000 | 1·0000 000 | ·9999 999 | ·9999 999 | ·9999 998 | ·9999 998 |
| ·96 | | | 1·0000 000 | 1·0000 000 | 1·0000 000 | 1·0000 000 |

$x$ = ·25 to ·80 $q$ = 11 $p$ = 23 to 28

| $x$ | $p$ = 23 | $p$ = 24 | $p$ = 25 | $p$ = 26 | $p$ = 27 | $p$ = 28 |
|---|---|---|---|---|---|---|
| $B(p, q)$ = | ·4697 2528$\times\frac{1}{10^9}$ | ·3177 5534$\times\frac{1}{10^9}$ | ·2178 8938$\times\frac{1}{10^9}$ | ·1513 1207$\times\frac{1}{10^9}$ | ·1063 2740$\times\frac{1}{10^9}$ | ·7554 8414$\times\frac{1}{10^{10}}$ |
| ·25 | ·0000 001 | | | | | |
| ·26 | ·0000 002 | ·0000 001 | | | | |
| ·27 | ·0000 004 | ·0000 001 | ·0000 001 | | | |
| ·28 | ·0000 008 | ·0000 003 | ·0000 001 | | | |
| ·29 | ·0000 016 | ·0000 006 | ·0000 003 | ·0000 001 | | |
| ·30 | ·0000 030 | ·0000 013 | ·0000 005+ | ·0000 002 | ·0000 001 | |
| ·31 | ·0000 055+ | ·0000 024 | ·0000 010 | ·0000 004 | ·0000 002 | ·0000 001 |
| ·32 | ·0000 100 | ·0000 045+ | ·0000 020 | ·0000 009 | ·0000 004 | ·0000 002 |
| ·33 | ·0000 177 | ·0000 082 | ·0000 038 | ·0000 017 | ·0000 008 | ·0000 003 |
| ·34 | ·0000 307 | ·0000 146 | ·0000 069 | ·0000 032 | ·0000 015− | ·0000 007 |
| ·35 | ·0000 519 | ·0000 255− | ·0000 124 | ·0000 059 | ·0000 028 | ·0000 013 |
| ·36 | ·0000 860 | ·0000 434 | ·0000 216 | ·0000 107 | ·0000 052 | ·0000 025+ |
| ·37 | ·0001 396 | ·0000 723 | ·0000 371 | ·0000 188 | ·0000 095− | ·0000 047 |
| ·38 | ·0002 225+ | ·0001 183 | ·0000 623 | ·0000 324 | ·0000 167 | ·0000 085+ |
| ·39 | ·0003 483 | ·0001 900 | ·0001 025+ | ·0000 548 | ·0000 290 | ·0000 152 |
| ·40 | ·0005 360 | ·0002 997 | ·0001 658 | ·0000 908 | ·0000 492 | ·0000 265− |
| ·41 | ·0008 111 | ·0004 645+ | ·0002 632 | ·0001 476 | ·0000 820 | ·0000 452 |
| ·42 | ·0012 082 | ·0007 083 | ·0004 108 | ·0002 359 | ·0001 342 | ·0000 757 |
| ·43 | ·0017 721 | ·0010 628 | ·0006 307 | ·0003 706 | ·0002 158 | ·0001 245+ |
| ·44 | ·0025 611 | ·0015 705+ | ·0009 530 | ·0005 727 | ·0003 409 | ·0002 012 |
| ·45 | ·0036 489 | ·0022 865− | ·0014 180 | ·0008 708 | ·0005 298 | ·0003 196 |
| ·46 | ·0051 271 | ·0032 813 | ·0020 785− | ·0013 038 | ·0008 104 | ·0004 994 |
| ·47 | ·0071 082 | ·0046 438 | ·0030 029 | ·0019 232 | ·0012 205+ | ·0007 680 |
| ·48 | ·0097 270 | ·0064 835+ | ·0042 779 | ·0027 958 | ·0018 107 | ·0011 628 |
| ·49 | ·0131 431 | ·0089 338 | ·0060 118 | ·0040 074 | ·0026 474 | ·0017 342 |
| ·50 | ·0175 410 | ·0121 533 | ·0083 369 | ·0056 655− | ·0038 160 | ·0025 488 |
| ·51 | ·0231 305+ | ·0163 276 | ·0114 124 | ·0079 030 | ·0054 248 | ·0036 928 |
| ·52 | ·0301 449 | ·0216 695+ | ·0154 260 | ·0108 807 | ·0076 081 | ·0052 761 |
| ·53 | ·0388 377 | ·0284 181 | ·0205 948 | ·0147 898 | ·0105 299 | ·0074 359 |
| ·54 | ·0494 777 | ·0368 355+ | ·0271 643 | ·0198 528 | ·0143 861 | ·0103 408 |
| ·55 | ·0623 419 | ·0472 025+ | ·0354 065− | ·0263 235− | ·0194 064 | ·0141 930 |
| ·56 | ·0777 063 | ·0598 112 | ·0456 145+ | ·0344 843 | ·0258 541 | ·0192 312 |
| ·57 | ·0958 344 | ·0749 558 | ·0580 962 | ·0446 423 | ·0340 241 | ·0257 301 |
| ·58 | ·1169 649 | ·0929 205+ | ·0731 640 | ·0571 218 | ·0442 386 | ·0339 988 |
| ·59 | ·1412 968 | ·1139 663 | ·0911 226 | ·0722 537 | ·0568 393 | ·0443 764 |
| ·60 | ·1689 745− | ·1383 148 | ·1122 541 | ·0903 632 | ·0721 768 | ·0572 236 |
| ·61 | ·2000 731 | ·1661 321 | ·1368 013 | ·1117 529 | ·0905 966 | ·0729 115+ |
| ·62 | ·2345 845− | ·1975 116 | ·1649 492 | ·1366 850+ | ·1124 214 | ·0918 062 |
| ·63 | ·2724 052 | ·2324 586 | ·1968 063 | ·1653 609 | ·1379 309 | ·1142 496 |
| ·64 | ·3133 282 | ·2708 765− | ·2323 869 | ·1979 003 | ·1673 398 | ·1405 372 |
| ·65 | ·3570 379 | ·3125 559 | ·2715 955− | ·2343 214 | ·2007 748 | ·1708 935− |
| ·66 | ·4031 112 | ·3571 695+ | ·3142 140 | ·2745 232 | ·2382 519 | ·2054 464 |
| ·67 | ·4510 236 | ·4042 714 | ·3598 958 | ·3182 714 | ·2796 568 | ·2442 029 |
| ·68 | ·5001 619 | ·4533 040 | ·4081 642 | ·3651 911 | ·3247 303 | ·2870 268 |
| ·69 | ·5498 425+ | ·5036 112 | ·4584 202 | ·4147 656 | ·3730 587 | ·3336 225− |
| ·70 | ·5993 357 | ·5544 594 | ·5099 571 | ·4663 448 | ·4240 745− | ·3835 262 |
| ·71 | ·6478 939 | ·6050 638 | ·5619 840 | ·5191 626 | ·4770 659 | ·4361 074 |
| ·72 | ·6947 828 | ·6546 210 | ·6136 558 | ·5723 630 | ·5311 972 | ·4905 803 |
| ·73 | ·7393 146 | ·7023 441 | ·6641 104 | ·6250 356 | ·5855 400 | ·5460 297 |
| ·74 | ·7808 784 | ·7475 000+ | ·7125 084 | ·6762 571 | ·6391 135− | ·6014 470 |
| ·75 | ·8189 688 | ·7894 446 | ·7580 752 | ·7251 373 | ·6909 326 | ·6557 778 |
| ·76 | ·8532 081 | ·8276 538 | ·8001 403 | ·7708 659 | ·7400 603 | ·7079 768 |
| ·77 | ·8833 614 | ·8617 481 | ·8381 716 | ·8127 561 | ·7856 593 | ·7570 672 |
| ·78 | ·9093 428 | ·8915 086 | ·8718 018 | ·8502 820 | ·8270 406 | ·8021 981 |
| ·79 | ·9312 121 | ·9168 815− | ·9008 432 | ·8831 048 | ·8637 018 | ·8426 962 |
| ·80 | ·9491 629 | ·9379 729 | ·9252 910 | ·9110 872 | ·8953 538 | ·8781 053 |

$x$ = ·81 to ·97 $q$ = 11 $p$ = 23 to 28

| | $p$ = 23 | $p$ = 24 | $p$ = 25 | $p$ = 26 | $p$ = 27 | $p$ = 28 |
|---|---|---|---|---|---|---|
| $B(p, q)$ = | ·4697 2528$\times\frac{1}{10^9}$ | ·3177 5534$\times\frac{1}{10^9}$ | ·2178 8938$\times\frac{1}{10^9}$ | ·1513 1207$\times\frac{1}{10^9}$ | ·1063 2740$\times\frac{1}{10^9}$ | ·7554 8414$\times\frac{1}{10^{10}}$ |
| $x$ | | | | | | |
| ·81 | ·9635 016 | ·9550 318 | ·9453 127 | ·9342 913 | ·9219 303 | ·9082 096 |
| ·82 | ·9746 209 | ·9684 244 | ·9612 262 | ·9529 626 | ·9435 803 | ·9330 373 |
| ·83 | ·9829 683 | ·9786 015+ | ·9734 669 | ·9675 005− | ·9606 437 | ·9528 447 |
| ·84 | ·9890 147 | ·9860 623 | ·9825 490 | ·9784 173 | ·9736 119 | ·9680 803 |
| ·85 | ·9932 238 | ·9913 180 | ·9890 232 | ·9862 923 | ·9830 783 | ·9793 345+ |
| ·86 | ·9960 263 | ·9948 587 | ·9934 361 | ·9917 234 | ·9896 839 | ·9872 803 |
| ·87 | ·9978 008 | ·9971 267 | ·9962 958 | ·9952 838 | ·9940 647 | ·9926 113 |
| ·88 | ·9988 616 | ·9984 981 | ·9980 449 | ·9974 866 | ·9968 063 | ·9959 859 |
| ·89 | ·9994 551 | ·9992 741 | ·9990 459 | ·9987 615− | ·9984 111 | ·9979 836 |
| ·90 | ·9997 623 | ·9996 803 | ·9995 757 | ·9994 439 | ·9992 797 | ·9990 771 |
| ·91 | ·9999 073 | ·9998 741 | ·9998 314 | ·9997 768 | ·9997 081 | ·9996 225− |
| ·92 | ·9999 685+ | ·9999 569 | ·9999 416 | ·9999 220 | ·9998 970 | ·9998 655+ |
| ·93 | ·9999 910 | ·9999 876 | ·9999 830 | ·9999 771 | ·9999 695+ | ·9999 598 |
| ·94 | ·9999 980 | ·9999 972 | ·9999 961 | ·9999 947 | ·9999 928 | ·9999 904 |
| ·95 | ·9999 997 | ·9999 995+ | ·9999 993 | ·9999 991 | ·9999 988 | ·9999 983 |
| ·96 | 1·0000 000 | ·9999 999 | ·9999 999 | ·9999 999 | ·9999 999 | ·9999 998 |
| ·97 | | 1·0000 000 | 1·0000 000 | 1·0000 000 | 1·0000 000 | 1·0000 000 |

$x$ = ·32 to ·98 $q = 11$ $p$ = 29 to 34

| $x$ | $p = 29$ | $p = 30$ | $p = 31$ | $p = 32$ | $p = 33$ | $p = 34$ |
|---|---|---|---|---|---|---|
| $B(p, q) =$ | ·5423 9887 $\times \frac{1}{10^{10}}$ | ·3932 3918 $\times \frac{1}{10^{10}}$ | ·2877 3599 $\times \frac{1}{10^{10}}$ | ·2123 7656 $\times \frac{1}{10^{10}}$ | ·1580 4767 $\times \frac{1}{10^{10}}$ | ·1185 3576 $\times \frac{1}{10^{10}}$ |
| ·32 | ·0000 001 | | | | | |
| ·33 | ·0000 002 | ·0000 001 | | | | |
| ·34 | ·0000 003 | ·0000 001 | ·0000 001 | | | |
| ·35 | ·0000 006 | ·0000 003 | ·0000 001 | ·0000 001 | | |
| ·36 | ·0000 012 | ·0000 006 | ·0000 003 | ·0000 001 | ·0000 001 | |
| ·37 | ·0000 023 | ·0000 011 | ·0000 006 | ·0000 003 | ·0000 001 | ·0000 001 |
| ·38 | ·0000 043 | ·0000 022 | ·0000 011 | ·0000 005+ | ·0000 003 | ·0000 001 |
| ·39 | ·0000 079 | ·0000 041 | ·0000 021 | ·0000 011 | ·0000 005+ | ·0000 003 |
| ·40 | ·0000 141 | ·0000 075− | ·0000 039 | ·0000 020 | ·0000 011 | ·0000 005+ |
| ·41 | ·0000 247 | ·0000 134 | ·0000 072 | ·0000 038 | ·0000 020 | ·0000 011 |
| ·42 | ·0000 423 | ·0000 235− | ·0000 129 | ·0000 071 | ·0000 038 | ·0000 021 |
| ·43 | ·0000 713 | ·0000 404 | ·0000 228 | ·0000 128 | ·0000 071 | ·0000 039 |
| ·44 | ·0001 178 | ·0000 684 | ·0000 394 | ·0000 226 | ·0000 128 | ·0000 072 |
| ·45 | ·0001 912 | ·0001 135+ | ·0000 669 | ·0000 391 | ·0000 228 | ·0000 131 |
| ·46 | ·0003 052 | ·0001 851 | ·0001 115− | ·0000 666 | ·0000 396 | ·0000 234 |
| ·47 | ·0004 793 | ·0002 969 | ·0001 825+ | ·0001 114 | ·0000 676 | ·0000 407 |
| ·48 | ·0007 407 | ·0004 682 | ·0002 938 | ·0001 831 | ·0001 134 | ·0000 698 |
| ·49 | ·0011 270 | ·0007 268 | ·0004 654 | ·0002 959 | ·0001 870 | ·0001 174 |
| ·50 | ·0016 889 | ·0011 107 | ·0007 252 | ·0004 703 | ·0003 031 | ·0001 941 |
| ·51 | ·0024 940 | ·0016 719 | ·0011 128 | ·0007 356 | ·0004 832 | ·0003 155− |
| ·52 | ·0036 303 | ·0024 795− | ·0016 815+ | ·0011 327 | ·0007 582 | ·0005 044 |
| ·53 | ·0052 105− | ·0036 243 | ·0025 034 | ·0017 176 | ·0011 711 | ·0007 936 |
| ·54 | ·0073 761 | ·0052 231 | ·0036 730 | ·0025 659 | ·0017 812 | ·0012 291 |
| ·55 | ·0103 016 | ·0074 233 | ·0053 125− | ·0037 771 | ·0026 687 | ·0018 744 |
| ·56 | ·0141 978 | ·0104 071 | ·0075 768 | ·0054 805− | ·0039 397 | ·0028 154 |
| ·57 | ·0193 141 | ·0143 959 | ·0106 580 | ·0078 402 | ·0057 321 | ·0041 663 |
| ·58 | ·0259 387 | ·0196 518 | ·0147 899 | ·0110 604 | ·0082 213 | ·0060 756 |
| ·59 | ·0343 972 | ·0264 792 | ·0202 504 | ·0153 900 | ·0116 261 | ·0087 325+ |
| ·60 | ·0450 474 | ·0352 225− | ·0273 625+ | ·0211 253 | ·0162 135− | ·0123 733 |
| ·61 | ·0582 708 | ·0462 604 | ·0364 919 | ·0286 109 | ·0223 012 | ·0172 859 |
| ·62 | ·0744 601 | ·0599 970 | ·0480 408 | ·0382 367 | ·0302 587 | ·0238 134 |
| ·63 | ·0940 021 | ·0768 476 | ·0624 379 | ·0504 313 | ·0405 033 | ·0323 533 |
| ·64 | ·1172 572 | ·0972 203 | ·0801 222 | ·0656 494 | ·0534 923 | ·0433 540 |
| ·65 | ·1445 338 | ·1214 919 | ·1015 222 | ·0843 547 | ·0697 087 | ·0573 040 |
| ·66 | ·1760 622 | ·1499 811 | ·1270 298 | ·1069 955− | ·0896 407 | ·0747 157 |
| ·67 | ·2119 652 | ·1829 179 | ·1569 690 | ·1339 749 | ·1137 545− | ·0961 014 |
| ·68 | ·2522 319 | ·2204 125− | ·1915 629 | ·1656 170 | ·1424 602 | ·1219 418 |
| ·69 | ·2966 933 | ·2624 255+ | ·2308 992 | ·2021 289 | ·1760 738 | ·1526 479 |
| ·70 | ·3450 054 | ·3087 427 | ·2748 975+ | ·2435 634 | ·2147 752 | ·1885 169 |
| ·71 | ·3966 409 | ·3589 572 | ·3232 833 | ·2897 844 | ·2585 676 | ·2296 873 |
| ·72 | ·4508 918 | ·4124 624 | ·3755 701 | ·3404 392 | ·3072 404 | ·2760 940 |
| ·73 | ·5068 864 | ·4684 584 | ·4310 552 | ·3949 429 | ·3603 429 | ·3274 310 |
| ·74 | ·5636 192 | ·5259 740 | ·4888 307 | ·4524 782 | ·4171 708 | ·3831 260 |
| ·75 | ·6199 952 | ·5839 041 | ·5478 130 | ·5120 130 | ·4767 723 | ·4423 326 |
| ·76 | ·6748 850− | ·6410 628 | ·6067 897 | ·5723 397 | ·5379 759 | ·5039 452 |
| ·77 | ·7271 884 | ·6962 484 | ·6644 833 | ·6321 341 | ·5994 413 | ·5666 395− |
| ·78 | ·7759 006 | ·7483 154 | ·7196 267 | ·6900 312 | ·6597 328 | ·6289 385+ |
| ·79 | ·8201 752 | ·7962 486 | ·7710 459 | ·7447 132 | ·7174 094 | ·6893 031 |
| ·80 | ·8593 784 | ·8392 309 | ·8177 401 | ·7950 015+ | ·7711 260 | ·7462 376 |
| ·81 | ·8931 266 | ·8766 966 | ·8589 522 | ·8399 427 | ·8197 333 | ·7984 032 |
| ·82 | ·9213 045+ | ·9083 660 | ·8942 200 | ·8788 784 | ·8623 670 | ·8447 248 |
| ·83 | ·9440 597 | ·9342 538 | ·9234 020 | ·9114 895+ | ·8985 123 | ·8844 773 |
| ·84 | ·9617 742 | ·9546 506 | ·9466 721 | ·9378 083 | ·9280 359 | ·9173 396 |
| ·85 | ·9750 159 | ·9700 792 | ·9644 843 | ·9581 945− | ·9511 775− | ·9434 056 |
| ·86 | ·9844 750+ | ·9812 305+ | ·9775 102 | ·9732 786 | ·9685 022 | ·9631 497 |
| ·87 | ·9908 952 | ·9888 874 | ·9865 584 | ·9838 785+ | ·9808 184 | ·9773 494 |
| ·88 | ·9950 062 | ·9938 466 | ·9924 862 | ·9909 027 | ·9890 739 | ·9869 767 |
| ·89 | ·9974 674 | ·9968 495+ | ·9961 163 | ·9952 532 | ·9942 450− | ·9930 758 |
| ·90 | ·9988 297 | ·9985 303 | ·9981 709 | ·9977 432 | ·9972 380 | ·9966 455− |
| ·91 | ·9995 167 | ·9993 872 | ·9992 301 | ·9990 411 | ·9988 153 | ·9985 475+ |
| ·92 | ·9998 262 | ·9997 776 | ·9997 179 | ·9996 453 | ·9995 576 | ·9994 524 |
| ·93 | ·9999 475+ | ·9999 322 | ·9999 132 | ·9998 899 | ·9998 613 | ·9998 268 |
| ·94 | ·9999 874 | ·9999 836 | ·9999 788 | ·9999 728 | ·9999 654 | ·9999 564 |
| ·95 | ·9999 978 | ·9999 971 | ·9999 962 | ·9999 951 | ·9999 937 | ·9999 920 |
| ·96 | ·9999 998 | ·9999 997 | ·9999 996 | ·9999 994 | ·9999 993 | ·9999 991 |
| ·97 | 1·0000 000 | 1·0000 000 | 1·0000 000 | 1·0000 000 | 1·0000 000 | ·9999 999 |
| ·98 | | | | | | 1·0000 000 |

$x$ = ·38 to ·98 $q = 11$ $p$ = 35 to 40

| | $p = 35$ | $p = 36$ | $p = 37$ | $p = 38$ | $p = 39$ | $p = 40$ |
|---|---|---|---|---|---|---|
| $B(p, q) =$ | ·8956 0349 $\times \frac{1}{10^{11}}$ | ·6814 3744 $\times \frac{1}{10^{11}}$ | ·5219 5208 $\times \frac{1}{10^{11}}$ | ·4023 3806 $\times \frac{1}{10^{11}}$ | ·3120 1727 $\times \frac{1}{10^{11}}$ | ·2433 7347 $\times \frac{1}{10^{11}}$ |
| $x$ | | | | | | |
| ·38 | ·0000 001 | | | | | |
| ·39 | ·0000 001 | ·0000 001 | | | | |
| ·40 | ·0000 003 | ·0000 001 | ·0000 001 | | | |
| ·41 | ·0000 006 | ·0000 003 | ·0000 002 | ·0000 001 | | |
| ·42 | ·0000 011 | ·0000 006 | ·0000 003 | ·0000 002 | ·0000 001 | |
| ·43 | ·0000 021 | ·0000 012 | ·0000 006 | ·0000 003 | ·0000 002 | ·0000 001 |
| ·44 | ·0000 041 | ·0000 023 | ·0000 013 | ·0000 007 | ·0000 004 | ·0000 002 |
| ·45 | ·0000 075+ | ·0000 043 | ·0000 024 | ·0000 014 | ·0000 008 | ·0000 004 |
| ·46 | ·0000 137 | ·0000 080 | ·0000 046 | ·0000 027 | ·0000 015+ | ·0000 009 |
| ·47 | ·0000 244 | ·0000 145+ | ·0000 086 | ·0000 051 | ·0000 030 | ·0000 017 |
| ·48 | ·0000 427 | ·0000 260 | ·0000 157 | ·0000 095− | ·0000 057 | ·0000 034 |
| ·49 | ·0000 733 | ·0000 455− | ·0000 281 | ·0000 172 | ·0000 105+ | ·0000 064 |
| ·50 | ·0001 235+ | ·0000 782 | ·0000 492 | ·0000 308 | ·0000 192 | ·0000 119 |
| ·51 | ·0002 047 | ·0001 321 | ·0000 848 | ·0000 542 | ·0000 344 | ·0000 218 |
| ·52 | ·0003 336 | ·0002 194 | ·0001 436 | ·0000 935− | ·0000 605+ | ·0000 390 |
| ·53 | ·0005 347 | ·0003 583 | ·0002 388 | ·0001 584 | ·0001 045+ | ·0000 687 |
| ·54 | ·0008 433 | ·0005 754 | ·0003 905+ | ·0002 638 | ·0001 773 | ·0001 186 |
| ·55 | ·0013 090 | ·0009 092 | ·0006 282 | ·0004 319 | ·0002 955+ | ·0002 013 |
| ·56 | ·0020 006 | ·0014 140 | ·0009 942 | ·0006 956 | ·0004 844 | ·0003 357 |
| ·57 | ·0030 113 | ·0021 649 | ·0015 484 | ·0011 021 | ·0007 807 | ·0005 505+ |
| ·58 | ·0044 651 | ·0032 641 | ·0023 740 | ·0017 183 | ·0012 378 | ·0008 877 |
| ·59 | ·0065 232 | ·0048 473 | ·0035 838 | ·0026 369 | ·0019 312 | ·0014 080 |
| ·60 | ·0093 915+ | ·0070 913 | ·0053 278 | ·0039 837 | ·0029 651 | ·0021 971 |
| ·61 | ·0133 268 | ·0102 218 | ·0078 015+ | ·0059 262 | ·0044 812 | ·0033 737 |
| ·62 | ·0186 420 | ·0145 196 | ·0112 538 | ·0086 817 | ·0066 674 | ·0050 983 |
| ·63 | ·0257 087 | ·0203 266 | ·0159 940 | ·0125 267 | ·0097 675+ | ·0075 835− |
| ·64 | ·0349 571 | ·0280 477 | ·0223 973 | ·0178 037 | ·0140 901 | ·0111 040 |
| ·65 | ·0468 694 | ·0381 491 | ·0309 064 | ·0249 263 | ·0200 163 | ·0160 065− |
| ·66 | ·0619 679 | ·0511 506 | ·0420 279 | ·0343 797 | ·0280 035− | ·0227 161 |
| ·67 | ·0807 952 | ·0676 092 | ·0563 219 | ·0467 149 | ·0385 843 | ·0317 401 |
| ·68 | ·1038 857 | ·0880 995+ | ·0743 830 | ·0625 350− | ·0523 581 | ·0436 634 |
| ·69 | ·1317 299 | ·1131 727 | ·0968 114 | ·0824 709 | ·0699 721 | ·0591 366 |
| ·70 | ·1647 300 | ·1433 217 | ·1241 732 | ·1071 466 | ·0920 914 | ·0788 506 |
| ·71 | ·2031 514 | ·1789 279 | ·1569 519 | ·1371 318 | ·1193 564 | ·1034 998 |
| ·72 | ·2470 728 | ·2202 075− | ·1954 914 | ·1728 862 | ·1523 274 | ·1337 295+ |
| ·73 | ·2963 388 | ·2671 567 | ·2399 362 | ·2146 947 | ·1914 194 | ·1700 718 |
| ·74 | ·3505 230 | ·3195 036 | ·2901 731 | ·2626 023 | ·2368 310 | ·2128 702 |
| ·75 | ·4089 058 | ·3766 728 | ·3457 829 | ·3163 539 | ·2884 739 | ·2622 023 |
| ·76 | ·4704 751 | ·4377 700 | ·4060 097 | ·3753 481 | ·3459 130 | ·3178 063 |
| ·77 | ·5339 534 | ·5015 942 | ·4697 563 | ·4386 155− | ·4083 269 | ·3790 246 |
| ·78 | ·5978 545− | ·5666 816 | ·5356 126 | ·5048 292 | ·4744 994 | ·4447 762 |
| ·79 | ·6605 685− | ·6313 823 | ·6019 206 | ·5723 553 | ·5428 522 | ·5135 685+ |
| ·80 | ·7204 707 | ·6939 677 | ·6668 757 | ·6393 443 | ·6115 232 | ·5835 594 |
| ·81 | ·7760 442 | ·7527 589 | ·7286 587 | ·7038 614 | ·6784 900 | ·6526 696 |
| ·82 | ·8260 033 | ·8062 655+ | ·7855 847 | ·7640 431 | ·7417 305+ | ·7187 429 |
| ·83 | ·8694 020 | ·8533 146 | ·8362 529 | ·8182 644 | ·7994 049 | ·7797 378 |
| ·84 | ·9057 121 | ·8931 543 | ·8796 757 | ·8652 936 | ·8500 335+ | ·8339 282 |
| ·85 | ·9348 565+ | ·9255 136 | ·9153 662 | ·9044 097 | ·8926 458 | ·8800 827 |
| ·86 | ·9571 927 | ·9506 059 | ·9433 678 | ·9354 607 | ·9268 711 | ·9175 899 |
| ·87 | ·9734 436 | ·9690 748 | ·9642 181 | ·9588 508 | ·9529 524 | ·9465 050− |
| ·88 | ·9845 885+ | ·9818 864 | ·9788 480 | ·9754 516 | ·9716 762 | ·9675 020 |
| ·89 | ·9917 292 | ·9901 882 | ·9884 358 | ·9864 547 | ·9842 274 | ·9817 369 |
| ·90 | ·9959 554 | ·9951 568 | ·9942 385− | ·9931 886 | ·9919 950+ | ·9906 454 |
| ·91 | ·9982 322 | ·9978 633 | ·9974 343 | ·9969 385− | ·9963 685− | ·9957 168 |
| ·92 | ·9993 273 | ·9991 793 | ·9990 052 | ·9988 018 | ·9985 655− | ·9982 923 |
| ·93 | ·9997 851 | ·9997 354 | ·9996 763 | ·9996 064 | ·9995 244 | ·9994 285+ |
| ·94 | ·9999 454 | ·9999 321 | ·9999 162 | ·9998 972 | ·9998 746 | ·9998 479 |
| ·95 | ·9999 898 | ·9999 873 | ·9999 841 | ·9999 803 | ·9999 758 | ·9999 704 |
| ·96 | ·9999 988 | ·9999 985− | ·9999 981 | ·9999 976 | ·9999 970 | ·9999 963 |
| ·97 | ·9999 999 | ·9999 999 | ·9999 999 | ·9999 999 | ·9999 998 | ·9999 998 |
| ·98 | 1·0000 000 | 1·0000 000 | 1·0000 000 | 1·0000 000 | 1·0000 000 | 1·0000 000 |

$x$ = ·43 to ·98 $q = 11$ $p$ = 41 to 45

| $x$ | $p = 41$ | $p = 42$ | $p = 43$ | $p = 44$ | $p = 45$ |
|---|---|---|---|---|---|
| $B(p, q) =$ | $\cdot 1908\ 8115 \overset{+}{\times} \frac{1}{10^{11}}$ | $\cdot 1505\ 0245 \overset{-}{\times} \frac{1}{10^{11}}$ | $\cdot 1192\ 6609 \times \frac{1}{10^{11}}$ | $\cdot 9497\ 1147 \times \frac{1}{10^{12}}$ | $\cdot 7597\ 6917 \times \frac{1}{10^{12}}$ |
| ·43 | ·0000 001 | | | | |
| ·44 | ·0000 001 | ·0000 001 | | | |
| ·45 | ·0000 002 | ·0000 001 | ·0000 001 | | |
| ·46 | ·0000 005⁻ | ·0000 003 | ·0000 002 | ·0000 001 | ·0000 001 |
| ·47 | ·0000 010 | ·0000 006 | ·0000 003 | ·0000 002 | ·0000 001 |
| ·48 | ·0000 020 | ·0000 012 | ·0000 007 | ·0000 004 | ·0000 002 |
| ·49 | ·0000 039 | ·0000 023 | ·0000 014 | ·0000 008 | ·0000 005⁺ |
| ·50 | ·0000 074 | ·0000 045⁺ | ·0000 028 | ·0000 017 | ·0000 010 |
| ·51 | ·0000 137 | ·0000 086 | ·0000 054 | ·0000 033 | ·0000 021 |
| ·52 | ·0000 251 | ·0000 160 | ·0000 102 | ·0000 065⁻ | ·0000 041 |
| ·53 | ·0000 449 | ·0000 292 | ·0000 190 | ·0000 123 | ·0000 079 |
| ·54 | ·0000 790 | ·0000 524 | ·0000 346 | ·0000 228 | ·0000 149 |
| ·55 | ·0001 365⁻ | ·0000 922 | ·0000 620 | ·0000 415⁺ | ·0000 277 |
| ·56 | ·0002 317 | ·0001 592 | ·0001 090 | ·0000 743 | ·0000 505⁻ |
| ·57 | ·0003 865⁺ | ·0002 703 | ·0001 882 | ·0001 306 | ·0000 902 |
| ·58 | ·0006 339 | ·0004 508 | ·0003 193 | ·0002 253 | ·0001 584 |
| ·59 | ·0010 222 | ·0007 391 | ·0005 322 | ·0003 818 | ·0002 729 |
| ·60 | ·0016 212 | ·0011 914 | ·0008 720 | ·0006 359 | ·0004 620 |
| ·61 | ·0025 293 | ·0018 885⁺ | ·0014 046 | ·0010 408 | ·0007 684 |
| ·62 | ·0038 823 | ·0029 444 | ·0022 245⁻ | ·0016 744 | ·0012 557 |
| ·63 | ·0058 636 | ·0045 157 | ·0034 644 | ·0026 481 | ·0020 169 |
| ·64 | ·0087 152 | ·0068 134 | ·0053 065⁻ | ·0041 177 | ·0031 840 |
| ·65 | ·0127 485⁻ | ·0101 143 | ·0079 944 | ·0062 960 | ·0049 411 |
| ·66 | ·0183 540 | ·0147 729 | ·0118 466 | ·0094 661 | ·0075 378 |
| ·67 | ·0260 080 | ·0212 308 | ·0172 680 | ·0139 954 | ·0113 045⁺ |
| ·68 | ·0362 730 | ·0300 217 | ·0247 587 | ·0203 476 | ·0166 663 |
| ·69 | ·0497 909 | ·0417 697 | ·0349 172 | ·0290 894 | ·0241 543 |
| ·70 | ·0672 649 | ·0571 769 | ·0484 339 | ·0408 906 | ·0344 102 |
| ·71 | ·0894 270 | ·0769 984 | ·0660 730 | ·0565 121 | ·0481 810 |
| ·72 | ·1169 915⁻ | ·1020 007 | ·0886 375⁺ | ·0767 785⁻ | ·0662 994 |
| ·73 | ·1505 920 | ·1329 035⁺ | ·1169 165⁻ | ·1025 318 | ·0896 445⁻ |
| ·74 | ·1907 065⁺ | ·1703 051 | ·1516 135⁻ | ·1345 650⁺ | ·1190 819 |
| ·75 | ·2375 727 | ·2145 951 | ·1932 587 | ·1735 350⁺ | ·1553 802 |
| ·76 | ·2911 049 | ·2658 623 | ·2421 102 | ·2198 606 | ·1991 078 |
| ·77 | ·3508 212 | ·3238 078 | ·2980 550⁺ | ·2736 139 | ·2505 170 |
| ·78 | ·4157 960 | ·3876 782 | ·3605 245⁻ | ·3344 190 | ·3094 289 |
| ·79 | ·4846 509 | ·4562 340 | ·4284 397 | ·4013 757 | ·3751 359 |
| ·80 | ·5555 956 | ·5277 682 | ·5002 059 | ·4730 281 | ·4463 444 |
| ·81 | ·6265 265⁺ | ·6001 858 | ·5737 697 | ·5473 967 | ·5211 795⁺ |
| ·82 | ·6951 806 | ·6711 470 | ·6467 472 | ·6220 863 | ·5972 686 |
| ·83 | ·7593 331 | ·7382 666 | ·7166 182 | ·6944 715⁻ | ·6719 119 |
| ·84 | ·8170 177 | ·7993 482 | ·7809 719 | ·7619 461 | ·7423 322 |
| ·85 | ·8667 343 | ·8526 209 | ·8377 682 | ·8222 074 | ·8059 746 |
| ·86 | ·9076 126 | ·8969 393 | ·8855 748 | ·8735 284 | ·8608 140 |
| ·87 | ·9394 934 | ·9319 056 | ·9237 323 | ·9149 680 | ·9056 101 |
| ·88 | ·9629 104 | ·9578 842 | ·9524 081 | ·9464 684 | ·9400 536 |
| ·89 | ·9789 663 | ·9758 989 | ·9725 190 | ·9688 113 | ·9647 615⁻ |
| ·90 | ·9891 271 | ·9874 273 | ·9855 332 | ·9834 321 | ·9811 114 |
| ·91 | ·9949 755⁻ | ·9941 364 | ·9931 910 | ·9921 306 | ·9909 464 |
| ·92 | ·9979 781 | ·9976 186 | ·9972 091 | ·9967 447 | ·9962 203 |
| ·93 | ·9993 171 | ·9991 881 | ·9990 397 | ·9988 695⁺ | ·9986 753 |
| ·94 | ·9998 165⁺ | ·9997 798 | ·9997 371 | ·9996 877 | ·9996 306 |
| ·95 | ·9999 639 | ·9999 563 | ·9999 473 | ·9999 368 | ·9999 246 |
| ·96 | ·9999 955⁻ | ·9999 945⁻ | ·9999 933 | ·9999 919 | ·9999 902 |
| ·97 | ·9999 997 | ·9999 997 | ·9999 996 | ·9999 995⁻ | ·9999 994 |
| ·98 | 1·0000 000 | 1·0000 000 | 1·0000 000 | 1·0000 000 | 1·0000 000 |

$x$ = ·47 to ·98 $q$ = 11 $p$ = 46 to 50

| $x$ | $p$ = 46 | $p$ = 47 | $p$ = 48 | $p$ = 49 | $p$ = 50 |
|---|---|---|---|---|---|
| $B(p, q)$ = | ·6105 2880 $\times \frac{1}{10^{12}}$ | ·4927 0745 $\times \frac{1}{10^{12}}$ | ·3992 6294 $\times \frac{1}{10^{12}}$ | ·3248 2408 $\times \frac{1}{10^{12}}$ | ·2652 7300 $\times \frac{1}{10^{12}}$ |
| ·47 | ·0000 001 | | | | |
| ·48 | ·0000 001 | ·0000 001 | | | |
| ·49 | ·0000 003 | ·0000 002 | ·0000 001 | ·0000 001 | |
| ·50 | ·0000 006 | ·0000 004 | ·0000 002 | ·0000 001 | ·0000 001 |
| ·51 | ·0000 013 | ·0000 008 | ·0000 005⁻ | ·0000 003 | ·0000 002 |
| ·52 | ·0000 026 | ·0000 016 | ·0000 010 | ·0000 006 | ·0000 004 |
| ·53 | ·0000 051 | ·0000 032 | ·0000 021 | ·0000 013 | ·0000 008 |
| ·54 | ·0000 097 | ·0000 063 | ·0000 041 | ·0000 027 | ·0000 017 |
| ·55 | ·0000 184 | ·0000 122 | ·0000 081 | ·0000 053 | ·0000 035⁻ |
| ·56 | ·0000 342 | ·0000 230 | ·0000 155⁻ | ·0000 104 | ·0000 069 |
| ·57 | ·0000 621 | ·0000 426 | ·0000 292 | ·0000 199 | ·0000 135⁺ |
| ·58 | ·0001 109 | ·0000 774 | ·0000 539 | ·0000 374 | ·0000 258 |
| ·59 | ·0001 944 | ·0001 380 | ·0000 976 | ·0000 688 | ·0000 484 |
| ·60 | ·0003 345⁻ | ·0002 413 | ·0001 736 | ·0001 244 | ·0000 889 |
| ·61 | ·0005 653 | ·0004 145⁺ | ·0003 029 | ·0002 207 | ·0001 603 |
| ·62 | ·0009 385⁺ | ·0006 991 | ·0005 191 | ·0003 842 | ·0002 835⁺ |
| ·63 | ·0015 309 | ·0011 581 | ·0008 733 | ·0006 565⁺ | ·0004 921 |
| ·64 | ·0024 537 | ·0018 847 | ·0014 430 | ·0011 014 | ·0008 382 |
| ·65 | ·0038 648 | ·0030 130 | ·0023 416 | ·0018 143 | ·0014 016 |
| ·66 | ·0059 824 | ·0047 326 | ·0037 323 | ·0029 345⁺ | ·0023 006 |
| ·67 | ·0091 010 | ·0073 036 | ·0058 432 | ·0046 608 | ·0037 070 |
| ·68 | ·0136 068 | ·0110 740 | ·0089 853 | ·0072 691 | ·0058 638 |
| ·69 | ·0199 924 | ·0164 964 | ·0135 709 | ·0111 318 | ·0091 053 |
| ·70 | ·0288 658 | ·0241 410 | ·0201 300 | ·0167 374 | ·0138 779 |
| ·71 | ·0409 515⁻ | ·0347 027 | ·0293 220 | ·0247 059 | ·0207 596 |
| ·72 | ·0570 778 | ·0489 948 | ·0419 369 | ·0357 964 | ·0304 731 |
| ·73 | ·0781 461 | ·0679 276 | ·0588 809 | ·0509 010 | ·0438 868 |
| ·74 | ·1050 783 | ·0924 629 | ·0811 412 | ·0710 177 | ·0619 975⁺ |
| ·75 | ·1387 383 | ·1235 435⁺ | ·1097 228 | ·0971 977 | ·0858 868 |
| ·76 | ·1798 307 | ·1619 952 | ·1455 562 | ·1304 597 | ·1166 449 |
| ·77 | ·2287 802 | ·2084 044 | ·1893 768 | ·1716 732 | ·1552 595⁺ |
| ·78 | ·2856 050⁺ | ·2629 827 | ·2415 829 | ·2214 137 | ·2024 710 |
| ·79 | ·3498 000 | ·3254 334 | ·3020 881 | ·2798 031 | ·2586 051 |
| ·80 | ·4202 538 | ·3948 437 | ·3701 906 | ·3463 592 | ·3234 033 |
| ·81 | ·4952 245⁻ | ·4696 306 | ·4444 887 | ·4198 810 | ·3958 810 |
| ·82 | ·5723 956 | ·5475 660 | ·5228 737 | ·4984 077 | ·4742 513 |
| ·83 | ·6490 266 | ·6259 024 | ·6026 257 | ·5792 811 | ·5559 508 |
| ·84 | ·7221 953 | ·7016 030 | ·6806 253 | ·6593 328 | ·6377 971 |
| ·85 | ·7891 106 | ·7716 600 | ·7536 710 | ·7351 948 | ·7162 850⁺ |
| ·86 | ·8474 498 | ·8334 580 | ·8188 648 | ·8037 001 | ·7879 969 |
| ·87 | ·8956 595⁻ | ·8851 205⁺ | ·8740 008 | ·8623 111 | ·8500 657 |
| ·88 | ·9331 540 | ·9257 625⁺ | ·9178 741 | ·9094 860 | ·9005 981 |
| ·89 | ·9603 562 | ·9555 831 | ·9504 313 | ·9448 909 | ·9389 536 |
| ·90 | ·9785 586 | ·9757 616 | ·9727 087 | ·9693 887 | ·9657 909 |
| ·91 | ·9896 292 | ·9881 700 | ·9865 597 | ·9847 889 | ·9828 487 |
| ·92 | ·9956 308 | ·9949 704 | ·9942 336 | ·9934 146 | ·9925 073 |
| ·93 | ·9984 546 | ·9982 046 | ·9979 227 | ·9976 060 | ·9972 512 |
| ·94 | ·9995 651 | ·9994 901 | ·9994 046 | ·9993 075⁻ | ·9991 976 |
| ·95 | ·9999 104 | ·9998 940 | ·9998 751 | ·9998 533 | ·9998 285⁻ |
| ·96 | ·9999 882 | ·9999 860 | ·9999 833 | ·9999 802 | ·9999 767 |
| ·97 | ·9999 992 | ·9999 991 | ·9999 989 | ·9999 987 | ·9999 984 |
| ·98 | 1·0000 000 | 1·0000 000 | 1·0000 000 | 1·0000 000 | 1·0000 000 |

$x$ = ·09 to ·70 $q = 12$ $p$ = 12 to 17

| $x$ | $p = 12$ | $p = 13$ | $p = 14$ | $p = 15$ | $p = 16$ | $p = 17$ |
|---|---|---|---|---|---|---|
| $B(p, q) =$ | ·6163 3525$^+\times\frac{1}{10^7}$ | ·3081 6763$\times\frac{1}{10^7}$ | ·1602 4717$\times\frac{1}{10^7}$ | ·8628 6935$^+\times\frac{1}{10^8}$ | ·4793 7186$\times\frac{1}{10^8}$ | ·2739 2678$\times\frac{1}{10^8}$ |
| ·09 | ·0000 001 | | | | | |
| ·10 | ·0000 005$^-$ | ·0000 001 | | | | |
| ·11 | ·0000 013 | ·0000 003 | ·0000 001 | | | |
| ·12 | ·0000 033 | ·0000 007 | ·0000 002 | | | |
| ·13 | ·0000 078 | ·0000 018 | ·0000 004 | ·0000 001 | | |
| ·14 | ·0000 169 | ·0000 043 | ·0000 011 | ·0000 003 | ·0000 001 | |
| ·15 | ·0000 344 | ·0000 094 | ·0000 025$^-$ | ·0000 006 | ·0000 002 | |
| ·16 | ·0000 663 | ·0000 193 | ·0000 055$^-$ | ·0000 015$^+$ | ·0000 004 | ·0000 001 |
| ·17 | ·0001 220 | ·0000 378 | ·0000 113 | ·0000 033 | ·0000 009 | ·0000 003 |
| ·18 | ·0002 149 | ·0000 704 | ·0000 223 | ·0000 069 | ·0000 021 | ·0000 006 |
| ·19 | ·0003 646 | ·0001 258 | ·0000 421 | ·0000 137 | ·0000 043 | ·0000 013 |
| ·20 | ·0005 974 | ·0002 168 | ·0000 763 | ·0000 261 | ·0000 087 | ·0000 028 |
| ·21 | ·0009 489 | ·0003 611 | ·0001 333 | ·0000 479 | ·0000 168 | ·0000 057 |
| ·22 | ·0014 648 | ·0005 833 | ·0002 253 | ·0000 847 | ·0000 310 | ·0000 111 |
| ·23 | ·0022 031 | ·0009 160 | ·0003 695$^-$ | ·0001 450$^-$ | ·0000 555$^+$ | ·0000 208 |
| ·24 | ·0032 352 | ·0014 016 | ·0005 892 | ·0002 410 | ·0000 962 | ·0000 375$^+$ |
| ·25 | ·0046 468 | ·0020 940 | ·0009 158 | ·0003 898 | ·0001 619 | ·0000 658 |
| ·26 | ·0065 390 | ·0030 599 | ·0013 900 | ·0006 147 | ·0002 652 | ·0001 119 |
| ·27 | ·0090 279 | ·0043 801 | ·0020 634 | ·0009 464 | ·0004 237 | ·0001 855$^+$ |
| ·28 | ·0122 443 | ·0061 505$^+$ | ·0030 005$^-$ | ·0014 255$^-$ | ·0006 611 | ·0002 999 |
| ·29 | ·0163 325$^+$ | ·0084 824 | ·0042 795$^+$ | ·0021 030 | ·0010 090 | ·0004 736 |
| ·30 | ·0214 480 | ·0115 023 | ·0059 940 | ·0030 431 | ·0015 086 | ·0007 318 |
| ·31 | ·0277 547 | ·0153 515$^+$ | ·0082 531 | ·0043 237 | ·0022 122 | ·0011 077 |
| ·32 | ·0354 211 | ·0201 838 | ·0111 821 | ·0060 382 | ·0031 851 | ·0016 444 |
| ·33 | ·0446 165$^-$ | ·0261 635$^-$ | ·0149 213 | ·0082 965$^+$ | ·0045 071 | ·0023 969 |
| ·34 | ·0555 050$^+$ | ·0334 618 | ·0196 254 | ·0112 247 | ·0062 739 | ·0034 334 |
| ·35 | ·0682 414 | ·0422 531 | ·0254 606 | ·0149 653 | ·0085 982 | ·0048 376 |
| ·36 | ·0829 644 | ·0527 098 | ·0326 022 | ·0196 758 | ·0116 098 | ·0067 096 |
| ·37 | ·0997 917 | ·0649 973 | ·0412 301 | ·0255 268 | ·0154 557 | ·0091 675$^+$ |
| ·38 | ·1188 140 | ·0792 678 | ·0515 246 | ·0326 988 | ·0202 990 | ·0123 475$^+$ |
| ·39 | ·1400 904 | ·0956 544 | ·0636 604 | ·0413 790 | ·0263 167 | ·0164 038 |
| ·40 | ·1636 434 | ·1142 651 | ·0778 011 | ·0517 553 | ·0336 970 | ·0215 076 |
| ·41 | ·1894 561 | ·1351 771 | ·0940 920 | ·0640 119 | ·0426 350$^-$ | ·0278 448 |
| ·42 | ·2174 692 | ·1584 315$^+$ | ·1126 546 | ·0783 219 | ·0533 278 | ·0356 131 |
| ·43 | ·2475 798 | ·1840 289 | ·1335 793 | ·0948 412 | ·0659 684 | ·0450 176 |
| ·44 | ·2796 419 | ·2119 260 | ·1569 199 | ·1137 009 | ·0807 391 | ·0562 650$^+$ |
| ·45 | ·3134 674 | ·2420 335$^-$ | ·1826 884 | ·1350 004 | ·0978 038 | ·0695 576 |
| ·46 | ·3488 293 | ·2742 152 | ·2108 506 | ·1588 011 | ·1173 003 | ·0850 853 |
| ·47 | ·3854 657 | ·3082 889 | ·2413 232 | ·1851 199 | ·1393 329 | ·1030 181 |
| ·48 | ·4230 852 | ·3440 288 | ·2739 728 | ·2139 247 | ·1639 647 | ·1234 971 |
| ·49 | ·4613 734 | ·3811 693 | ·3086 154 | ·2451 307 | ·1912 110 | ·1466 262 |
| ·50 | ·5000 000$^e$ | ·4194 099 | ·3450 190 | ·2785 985$^+$ | ·2210 342 | ·1724 642 |
| ·51 | ·5386 266 | ·4584 224 | ·3829 072 | ·3141 343 | ·2533 391 | ·2010 172 |
| ·52 | ·5769 148 | ·4978 585$^-$ | ·4219 644 | ·3514 913 | ·2879 716 | ·2322 330 |
| ·53 | ·6145 343 | ·5373 576 | ·4618 431 | ·3903 741 | ·3247 178 | ·2659 965$^+$ |
| ·54 | ·6511 707 | ·5765 567 | ·5021 722 | ·4304 442 | ·3633 069 | ·3021 280 |
| ·55 | ·6865 326 | ·6150 988 | ·5425 659 | ·4713 283 | ·4034 151 | ·3403 832 |
| ·56 | ·7203 581 | ·6526 423 | ·5826 346 | ·5126 268 | ·4446 726 | ·3804 559 |
| ·57 | ·7524 202 | ·6888 693 | ·6219 943 | ·5539 250$^+$ | ·4866 726 | ·4219 841 |
| ·58 | ·7825 308 | ·7234 932 | ·6602 774 | ·5948 040 | ·5289 814 | ·4645 575$^+$ |
| ·59 | ·8105 439 | ·7562 648 | ·6971 424 | ·6348 527 | ·5711 512 | ·5077 283 |
| ·60 | ·8363 566 | ·7869 782 | ·7322 822 | ·6736 793 | ·6127 323 | ·5510 234 |
| ·61 | ·8599 096 | ·8154 736 | ·7654 318 | ·7109 220 | ·6532 870 | ·5939 589 |
| ·62 | ·8811 860 | ·8416 397 | ·7963 745$^-$ | ·7462 594 | ·6924 024 | ·6360 545$^+$ |
| ·63 | ·9002 083 | ·8654 139 | ·8249 454 | ·7794 183 | ·7297 027 | ·6768 488 |
| ·64 | ·9170 356 | ·8867 810 | ·8510 340 | ·8101 803 | ·7648 599 | ·7159 140 |
| ·65 | ·9317 586 | ·9057 702 | ·8745 842 | ·8383 862 | ·7976 031 | ·7528 691 |
| ·66 | ·9444 950$^-$ | ·9224 517 | ·8955 928 | ·8639 377 | ·8277 243 | ·7873 916 |
| ·67 | ·9553 835$^+$ | ·9369 306 | ·9141 056 | ·8867 972 | ·8550 831 | ·8192 263 |
| ·68 | ·9645 789 | ·9493 415$^+$ | ·9302 129 | ·9069 852 | ·8796 074 | ·8481 915$^+$ |
| ·69 | ·9722 453 | ·9598 422 | ·9440 425$^+$ | ·9245 751 | ·9012 920 | ·8741 818 |
| ·70 | ·9785 520 | ·9686 063 | ·9557 535$^+$ | ·9396 874 | ·9201 940 | ·8971 673 |

$x$ = ·71 to ·94 $q$ = 12 $p$ = 12 to 17

| $x$ | $p$ = 12 | $p$ = 13 | $p$ = 14 | $p$ = 15 | $p$ = 16 | $p$ = 17 |
|---|---|---|---|---|---|---|
| $B(p, q)$ = | ·6163 3525$\overset{+}{\times}\frac{1}{10^7}$ | ·3081 6763$\times\frac{1}{10^7}$ | ·1602 4717$\times\frac{1}{10^7}$ | ·8628 6935$\overset{+}{\times}\frac{1}{10^8}$ | ·4793 7186$\times\frac{1}{10^8}$ | ·2739 2678$\times\frac{1}{10^8}$ |
| ·71 | ·9836 675$^{-}$ | ·9758 173 | ·9655 276 | ·9524 817 | ·9364 265$^{+}$ | ·9171 904 |
| ·72 | ·9877 557 | ·9816 619 | ·9735 618 | ·9631 475$^{-}$ | ·9501 503 | ·9343 588 |
| ·73 | ·9909 721 | ·9863 244 | ·9800 607 | ·9718 955$^{-}$ | ·9615 638 | ·9488 364 |
| ·74 | ·9934 610 | ·9899 819 | ·9852 290 | ·9789 483 | ·9708 923 | ·9608 324 |
| ·75 | ·9953 532 | ·9928 003 | ·9892 657 | ·9845 318 | ·9783 777 | ·9705 890 |
| ·76 | ·9967 648 | ·9949 312 | ·9923 586 | ·9888 671 | ·9842 677 | ·9783 689 |
| ·77 | ·9977 969 | ·9965 097 | ·9946 800 | ·9921 642 | ·9888 063 | ·9844 432 |
| ·78 | ·9985 352 | ·9976 537 | ·9963 844 | ·9946 164 | ·9922 261 | ·9890 799 |
| ·79 | ·9990 511 | ·9984 634 | ·9976 063 | ·9963 971 | ·9947 413 | ·9925 340 |
| ·80 | ·9994 026 | ·9990 220 | ·9984 599 | ·9976 570 | ·9965 435$^{+}$ | ·9950 404 |
| ·81 | ·9996 354 | ·9993 967 | ·9990 398 | ·9985 235$^{-}$ | ·9977 986 | ·9968 077 |
| ·82 | ·9997 851 | ·9996 405$^{+}$ | ·9994 217 | ·9991 012 | ·9986 458 | ·9980 156 |
| ·83 | ·9998 780 | ·9997 938 | ·9996 648 | ·9994 736 | ·9991 985$^{-}$ | ·9988 131 |
| ·84 | ·9999 337 | ·9998 867 | ·9998 139 | ·9997 046 | ·9995 456 | ·9993 201 |
| ·85 | ·9999 656 | ·9999 407 | ·9999 015$^{+}$ | ·9998 421 | ·9997 545$^{+}$ | ·9996 289 |
| ·86 | ·9999 831 | ·9999 706 | ·9999 507 | ·9999 201 | ·9998 745$^{-}$ | ·9998 083 |
| ·87 | ·9999 922 | ·9999 863 | ·9999 768 | ·9999 620 | ·9999 397 | ·9999 070 |
| ·88 | ·9999 967 | ·9999 941 | ·9999 898 | ·9999 832 | ·9999 731 | ·9999 580 |
| ·89 | ·9999 987 | ·9999 976 | ·9999 959 | ·9999 932 | ·9999 890 | ·9999 826 |
| ·90 | ·9999 995$^{+}$ | ·9999 992 | ·9999 985$^{+}$ | ·9999 975$^{-}$ | ·9999 959 | ·9999 935$^{-}$ |
| ·91 | ·9999 999 | ·9999 997 | ·9999 995$^{+}$ | ·9999 992 | ·9999 987 | ·9999 978 |
| ·92 | 1·0000 000 | ·9999 999 | ·9999 999 | ·9999 998 | ·9999 996 | ·9999 994 |
| ·93 | | 1·0000 000 | 1·0000 000 | ·9999 999 | ·9999 999 | ·9999 999 |
| ·94 | | | | 1·0000 000 | 1·0000 000 | 1·0000 000 |

$x$ = ·17 to ·80 $q$ = 12 $p$ = 18 to 23

| $x$ | $p$ = 18 | $p$ = 19 | $p$ = 20 | $p$ = 21 | $p$ = 22 | $p$ = 23 |
|---|---|---|---|---|---|---|
| $B(p, q)$ = | ·1605 7777 × $\frac{1}{10^8}$ | ·9634 6660 × $\frac{1}{10^9}$ | ·5905 1179 × $\frac{1}{10^9}$ | ·3690 6987 × $\frac{1}{10^9}$ | ·2348 6264 × $\frac{1}{10^9}$ | ·1519 6994 × $\frac{1}{10^9}$ |
| ·17 | ·0000 001 | | | | | |
| ·18 | ·0000 002 | | | | | |
| ·19 | ·0000 004 | ·0000 001 | | | | |
| ·20 | ·0000 009 | ·0000 003 | ·0000 001 | | | |
| ·21 | ·0000 019 | ·0000 006 | ·0000 002 | ·0000 001 | | |
| ·22 | ·0000 039 | ·0000 013 | ·0000 005− | ·0000 002 | | |
| ·23 | ·0000 076 | ·0000 027 | ·0000 010 | ·0000 003 | ·0000 001 | |
| ·24 | ·0000 143 | ·0000 054 | ·0000 020 | ·0000 007 | ·0000 003 | ·0000 001 |
| ·25 | ·0000 262 | ·0000 102 | ·0000 039 | ·0000 015− | ·0000 006 | ·0000 002 |
| ·26 | ·0000 463 | ·0000 188 | ·0000 075− | ·0000 029 | ·0000 011 | ·0000 004 |
| ·27 | ·0000 796 | ·0000 335+ | ·0000 139 | ·0000 056 | ·0000 023 | ·0000 009 |
| ·28 | ·0001 333 | ·0000 582 | ·0000 249 | ·0000 105+ | ·0000 044 | ·0000 018 |
| ·29 | ·0002 179 | ·0000 984 | ·0000 437 | ·0000 191 | ·0000 082 | ·0000 035− |
| ·30 | ·0003 479 | ·0001 624 | ·0000 745+ | ·0000 337 | ·0000 150− | ·0000 066 |
| ·31 | ·0005 437 | ·0002 620 | ·0001 242 | ·0000 579 | ·0000 266 | ·0000 121 |
| ·32 | ·0008 324 | ·0004 138 | ·0002 022 | ·0000 973 | ·0000 461 | ·0000 216 |
| ·33 | ·0012 499 | ·0006 401 | ·0003 224 | ·0001 599 | ·0000 781 | ·0000 377 |
| ·34 | ·0018 427 | ·0009 714 | ·0005 036 | ·0002 571 | ·0001 294 | ·0000 642 |
| ·35 | ·0026 697 | ·0014 473 | ·0007 717 | ·0004 052 | ·0002 098 | ·0001 071 |
| ·36 | ·0038 041 | ·0021 190 | ·0011 611 | ·0006 266 | ·0003 333 | ·0001 750+ |
| ·37 | ·0053 355− | ·0030 511 | ·0017 166 | ·0009 513 | ·0005 197 | ·0002 803 |
| ·38 | ·0073 709 | ·0043 240 | ·0024 960 | ·0014 192 | ·0007 957 | ·0004 403 |
| ·39 | ·0100 363 | ·0060 353 | ·0035 716 | ·0020 822 | ·0011 971 | ·0006 794 |
| ·40 | ·0134 769 | ·0083 016 | ·0050 330 | ·0030 064 | ·0017 712 | ·0010 301 |
| ·41 | ·0178 571 | ·0112 596 | ·0069 886 | ·0042 744 | ·0025 787 | ·0015 358 |
| ·42 | ·0233 588 | ·0150 667 | ·0095 677 | ·0059 878 | ·0036 968 | ·0022 534 |
| ·43 | ·0301 795− | ·0199 000 | ·0129 207 | ·0082 691 | ·0052 211 | ·0032 552 |
| ·44 | ·0385 285+ | ·0259 553 | ·0172 202 | ·0112 629 | ·0072 686 | ·0046 324 |
| ·45 | ·0486 222 | ·0334 440 | ·0226 595+ | ·0151 373 | ·0099 792 | ·0064 976 |
| ·46 | ·0606 778 | ·0425 890 | ·0294 509 | ·0200 834 | ·0135 172 | ·0089 866 |
| ·47 | ·0749 062 | ·0536 192 | ·0378 220 | ·0263 138 | ·0180 717 | ·0122 610 |
| ·48 | ·0915 039 | ·0667 625− | ·0480 111 | ·0340 601 | ·0238 559 | ·0165 089 |
| ·49 | ·1106 437 | ·0822 374 | ·0602 600 | ·0435 681 | ·0311 049 | ·0219 443 |
| ·50 | ·1324 654 | ·1002 442 | ·0748 064 | ·0550 921 | ·0400 717 | ·0288 063 |
| ·51 | ·1570 669 | ·1209 543 | ·0918 742 | ·0688 864 | ·0510 215+ | ·0373 550− |
| ·52 | ·1844 945+ | ·1445 003 | ·1116 630 | ·0851 961 | ·0642 242 | ·0478 661 |
| ·53 | ·2147 363 | ·1709 657 | ·1343 367 | ·1042 459 | ·0799 440 | ·0606 241 |
| ·54 | ·2477 148 | ·2003 753 | ·1600 121 | ·1262 281 | ·0984 288 | ·0759 113 |
| ·55 | ·2832 836 | ·2326 871 | ·1887 480 | ·1512 900 | ·1198 965+ | ·0939 970 |
| ·56 | ·3212 255 | ·2677 864 | ·2205 351 | ·1795 209 | ·1445 221 | ·1151 231 |
| ·57 | ·3612 531 | ·3054 818 | ·2552 876 | ·2109 410 | ·1724 229 | ·1394 899 |
| ·58 | ·4030 138 | ·3455 046 | ·2928 382 | ·2454 912 | ·2036 454 | ·1672 396 |
| ·59 | ·4460 962 | ·3875 114 | ·3329 351 | ·2830 251 | ·2381 536 | ·1984 423 |
| ·60 | ·4900 406 | ·4310 905− | ·3752 430 | ·3233 049 | ·2758 186 | ·2330 810 |
| ·61 | ·5343 517 | ·4757 710 | ·4193 486 | ·3660 012 | ·3164 135− | ·2710 407 |
| ·62 | ·5785 134 | ·5210 362 | ·4647 690 | ·4106 963 | ·3596 104 | ·3121 005+ |
| ·63 | ·6220 051 | ·5663 388 | ·5109 654 | ·4568 933 | ·4049 841 | ·3559 299 |
| ·64 | ·6643 191 | ·6111 191 | ·5573 591 | ·5040 292 | ·4520 198 | ·4020 908 |
| ·65 | ·7049 774 | ·6548 241 | ·6033 511 | ·5514 919 | ·5001 267 | ·4500 456 |
| ·66 | ·7435 475+ | ·6969 267 | ·6483 429 | ·5986 417 | ·5486 564 | ·4991 710 |
| ·67 | ·7796 573 | ·7369 447 | ·6917 593 | ·6448 342 | ·5969 260 | ·5487 782 |
| ·68 | ·8130 057 | ·7744 577 | ·7330 692 | ·6894 458 | ·6442 437 | ·5981 375+ |
| ·69 | ·8433 718 | ·8091 214 | ·7718 064 | ·7318 981 | ·6899 373 | ·6465 079 |
| ·70 | ·8706 189 | ·8406 782 | ·8075 859 | ·7716 807 | ·7333 818 | ·6931 680 |
| ·71 | ·8946 955− | ·8689 638 | ·8401 172 | ·8083 716 | ·7740 258 | ·7374 475+ |
| ·72 | ·9156 319 | ·8939 088 | ·8692 129 | ·8416 524 | ·8114 145+ | ·7787 577 |
| ·73 | ·9335 336 | ·9155 359 | ·8947 911 | ·8713 183 | ·8452 077 | ·8166 166 |
| ·74 | ·9485 711 | ·9339 529 | ·9168 727 | ·8972 818 | ·8751 906 | ·8506 694 |
| ·75 | ·9609 676 | ·9493 417 | ·9355 743 | ·9195 696 | ·9012 785+ | ·8807 011 |
| ·76 | ·9709 851 | ·9619 440 | ·9510 947 | ·9383 141 | ·9235 131 | ·9066 399 |
| ·77 | ·9789 098 | ·9720 453 | ·9636 995+ | ·9537 388 | ·9420 516 | ·9285 529 |
| ·78 | ·9850 379 | ·9799 584 | ·9737 027 | ·9661 395− | ·9571 500+ | ·9466 324 |
| ·79 | ·9896 618 | ·9860 062 | ·9814 463 | ·9758 627 | ·9691 411 | ·9611 760 |
| ·80 | ·9930 597 | ·9905 069 | ·9872 823 | ·9832 837 | ·9784 093 | ·9725 600 |

$x = \cdot 81$ to $\cdot 95$ $q = 12$ $p = 18$ to $23$

| | $p = 18$ | $p = 19$ | $p = 20$ | $p = 21$ | $p = 22$ | $p = 23$ |
|---|---|---|---|---|---|---|
| $B(p, q) =$ | $\cdot 1605\ 7777 \times \frac{1}{10^8}$ | $\cdot 9634\ 6660 \times \frac{1}{10^9}$ | $\cdot 5905\ 1179 \times \frac{1}{10^9}$ | $\cdot 3690\ 6987 \times \frac{1}{10^9}$ | $\cdot 2348\ 6264 \times \frac{1}{10^9}$ | $\cdot 1519\ 6994 \times \frac{1}{10^9}$ |
| $x$ | | | | | | |
| ·81 | ·9954 858 | ·9937 607 | ·9915 544 | ·9887 843 | ·9853 653 | ·9812 112 |
| ·82 | ·9971 644 | ·9960 399 | ·9945 839 | ·9927 334 | ·9904 212 | ·9875 771 |
| ·83 | ·9982 863 | ·9975 819 | ·9966 588 | ·9954 711 | ·9939 690 | ·9920 989 |
| ·84 | ·9990 081 | ·9985 859 | ·9980 260 | ·9972 969 | ·9963 637 | ·9951 879 |
| ·85 | ·9994 531 | ·9992 123 | ·9988 891 | ·9984 633 | ·9979 117 | ·9972 085+ |
| ·86 | ·9997 146 | ·9995 847 | ·9994 083 | ·9991 732 | ·9988 651 | ·9984 677 |
| ·87 | ·9998 601 | ·9997 944 | ·9997 041 | ·9995 824 | ·9994 210 | ·9992 103 |
| ·88 | ·9999 363 | ·9999 054 | ·9998 624 | ·9998 039 | ·9997 254 | ·9996 217 |
| ·89 | ·9999 733 | ·9999 600 | ·9999 413 | ·9999 154 | ·9998 804 | ·9998 336 |
| ·90 | ·9999 899 | ·9999 847 | ·9999 773 | ·9999 670 | ·9999 529 | ·9999 339 |
| ·91 | ·9999 966 | ·9999 948 | ·9999 923 | ·9999 886 | ·9999 836 | ·9999 767 |
| ·92 | ·9999 990 | ·9999 985− | ·9999 977 | ·9999 966 | ·9999 951 | ·9999 930 |
| ·93 | ·9999 998 | ·9999 996 | ·9999 994 | ·9999 992 | ·9999 988 | ·9999 982 |
| ·94 | 1·0000 000 | ·9999 999 | ·9999 999 | ·9999 998 | ·9999 998 | ·9999 997 |
| ·95 | | 1·0000 000 | 1·0000 000 | 1·0000 000 | 1·0000 000 | 1·0000 000 |

$x$ = ·25 to ·80 $q$ = 12 $p$ = 24 to 29

| | $p$ = 24 | $p$ = 25 | $p$ = 26 | $p$ = 27 | $p$ = 28 | $p$ = 29 |
|---|---|---|---|---|---|---|
| $B(p, q)$ = | ·9986 5964 × $\frac{1}{10^{10}}$ | ·6657 7309 × $\frac{1}{10^{10}}$ | ·4498 4668 × $\frac{1}{10^{10}}$ | ·3077 8984 × $\frac{1}{10^{10}}$ | ·2130 8527 × $\frac{1}{10^{10}}$ | ·1491 5969 × $\frac{1}{10^{10}}$ |
| $x$ | | | | | | |
| ·25 | ·0000 001 | | | | | |
| ·26 | ·0000 002 | ·0000 001 | | | | |
| ·27 | ·0000 004 | ·0000 001 | ·0000 001 | | | |
| ·28 | ·0000 007 | ·0000 003 | ·0000 001 | | | |
| ·29 | ·0000 015− | ·0000 006 | ·0000 002 | ·0000 001 | | |
| ·30 | ·0000 029 | ·0000 012 | ·0000 005+ | ·0000 002 | ·0000 001 | |
| ·31 | ·0000 054 | ·0000 024 | ·0000 010 | ·0000 005− | ·0000 002 | ·0000 001 |
| ·32 | ·0000 100 | ·0000 046 | ·0000 021 | ·0000 009 | ·0000 004 | ·0000 002 |
| ·33 | ·0000 179 | ·0000 084 | ·0000 039 | ·0000 018 | ·0000 008 | ·0000 004 |
| ·34 | ·0000 315− | ·0000 153 | ·0000 073 | ·0000 035− | ·0000 016 | ·0000 008 |
| ·35 | ·0000 540 | ·0000 269 | ·0000 133 | ·0000 065− | ·0000 031 | ·0000 015+ |
| ·36 | ·0000 908 | ·0000 465+ | ·0000 236 | ·0000 118 | ·0000 059 | ·0000 029 |
| ·37 | ·0001 493 | ·0000 786 | ·0000 409 | ·0000 211 | ·0000 108 | ·0000 054 |
| ·38 | ·0002 407 | ·0001 301 | ·0000 695+ | ·0000 368 | ·0000 193 | ·0000 100 |
| ·39 | ·0003 809 | ·0002 111 | ·0001 157 | ·0000 628 | ·0000 338 | ·0000 180 |
| ·40 | ·0005 918 | ·0003 362 | ·0001 889 | ·0001 051 | ·0000 579 | ·0000 316 |
| ·41 | ·0009 038 | ·0005 258 | ·0003 027 | ·0001 725+ | ·0000 974 | ·0000 545− |
| ·42 | ·0013 572 | ·0008 083 | ·0004 763 | ·0002 779 | ·0001 606 | ·0000 920 |
| ·43 | ·0020 055+ | ·0012 219 | ·0007 367 | ·0004 398 | ·0002 601 | ·0001 524 |
| ·44 | ·0029 178 | ·0018 175+ | ·0011 204 | ·0006 839 | ·0004 136 | ·0002 479 |
| ·45 | ·0041 815− | ·0026 615+ | ·0016 766 | ·0010 459 | ·0006 464 | ·0003 961 |
| ·46 | ·0059 057 | ·0038 390 | ·0024 700 | ·0015 738 | ·0009 936 | ·0006 219 |
| ·47 | ·0082 239 | ·0054 567 | ·0035 839 | ·0023 313 | ·0015 028 | ·0009 603 |
| ·48 | ·0112 957 | ·0076 465− | ·0051 241 | ·0034 012 | ·0022 372 | ·0014 590 |
| ·49 | ·0153 090 | ·0105 674 | ·0072 218 | ·0048 889 | ·0032 800 | ·0021 820 |
| ·50 | ·0204 798 | ·0144 084 | ·0100 369 | ·0069 265− | ·0047 377 | ·0032 133 |
| ·51 | ·0270 515+ | ·0193 884 | ·0137 605+ | ·0096 760 | ·0067 443 | ·0046 617 |
| ·52 | ·0352 918 | ·0257 562 | ·0186 160 | ·0133 322 | ·0094 653 | ·0066 645+ |
| ·53 | ·0454 873 | ·0337 878 | ·0248 588 | ·0181 242 | ·0131 007 | ·0093 923 |
| ·54 | ·0579 364 | ·0437 813 | ·0327 742 | ·0243 157 | ·0178 872 | ·0130 521 |
| ·55 | ·0729 395− | ·0560 496 | ·0426 728 | ·0322 029 | ·0240 985− | ·0178 899 |
| ·56 | ·0907 859 | ·0709 105− | ·0548 830 | ·0421 103 | ·0320 435− | ·0241 914 |
| ·57 | ·1117 402 | ·0886 733 | ·0697 400 | ·0543 822 | ·0420 618 | ·0322 802 |
| ·58 | ·1360 256 | ·1096 237 | ·0875 729 | ·0693 725− | ·0545 156 | ·0425 133 |
| ·59 | ·1638 071 | ·1340 065− | ·1086 878 | ·0874 299 | ·0697 780 | ·0552 719 |
| ·60 | ·1951 745+ | ·1620 064 | ·1333 491 | ·1088 802 | ·0882 175+ | ·0709 495− |
| ·61 | ·2301 263 | ·1937 296 | ·1617 587 | ·1340 055− | ·1101 788 | ·0899 347 |
| ·62 | ·2685 567 | ·2291 859 | ·1940 356 | ·1630 222 | ·1359 601 | ·1125 901 |
| ·63 | ·3102 455+ | ·2682 730 | ·2301 955+ | ·1960 576 | ·1657 887 | ·1392 276 |
| ·64 | ·3548 537 | ·3107 657 | ·2701 341 | ·2331 282 | ·1997 954 | ·1700 817 |
| ·65 | ·4019 242 | ·3563 092 | ·3136 135− | ·2741 199 | ·2379 907 | ·2052 808 |
| ·66 | ·4508 905+ | ·4044 205+ | ·3602 554 | ·3187 742 | ·2802 428 | ·2448 214 |
| ·67 | ·5010 909 | ·4544 965+ | ·4095 422 | ·3666 801 | ·3262 626 | ·2885 445− |
| ·68 | ·5517 908 | ·5058 303 | ·4608 257 | ·4172 752 | ·3755 957 | ·3361 193 |
| ·69 | ·6022 100 | ·5576 351 | ·5133 456 | ·4698 566 | ·4276 240 | ·3870 355− |
| ·70 | ·6515 554 | ·6090 759 | ·5662 566 | ·5236 020 | ·4815 793 | ·4406 071 |
| ·71 | ·6990 563 | ·6593 053 | ·6186 639 | ·5776 005− | ·5365 675− | ·4959 888 |
| ·72 | ·7439 994 | ·7075 032 | ·6696 639 | ·6308 933 | ·5916 056 | ·5522 058 |
| ·73 | ·7857 630 | ·7529 168 | ·7183 889 | ·6825 197 | ·6456 674 | ·6081 965+ |
| ·74 | ·8238 454 | ·7948 978 | ·7640 512 | ·7315 674 | ·6977 361 | ·6628 657 |
| ·75 | ·8578 870 | ·8329 340 | ·8059 849 | ·7772 218 | ·7468 608 | ·7151 444 |
| ·76 | ·8876 832 | ·8666 729 | ·8436 792 | ·8188 107 | ·7922 106 | ·7640 524 |
| ·77 | ·9131 878 | ·8959 341 | ·8768 031 | ·8558 401 | ·8331 223 | ·8087 575− |
| ·78 | ·9345 052 | ·9207 104 | ·9052 162 | ·8880 175+ | ·8691 373 | ·8486 252 |
| ·79 | ·9518 742 | ·9411 577 | ·9289 666 | ·9152 610 | ·9000 224 | ·8832 545− |
| ·80 | ·9656 426 | ·9575 723 | ·9482 753 | ·9376 910 | ·9257 738 | ·9124 948 |

$x = \cdot 81$ to $\cdot 96$ $q = 12$ $p = 24$ to $29$

| | $p = 24$ | $p = 25$ | $p = 26$ | $p = 27$ | $p = 28$ | $p = 29$ |
|---|---|---|---|---|---|---|
| $B(p, q) =$ | $\cdot 9986\ 5964 \times \frac{1}{10^{10}}$ | $\cdot 6657\ 7309 \times \frac{1}{10^{10}}$ | $\cdot 4498\ 4668 \times \frac{1}{10^{10}}$ | $\cdot 3077\ 8984 \times \frac{1}{10^{10}}$ | $\cdot 2130\ 8527 \times \frac{1}{10^{10}}$ | $\cdot 1491\ 5969 \times \frac{1}{10^{10}}$ |
| $x$ | | | | | | |
| ·81 | ·9762 371 | ·9703 615⁻ | ·9635 081 | ·9556 084 | ·9466 026 | ·9364 422 |
| ·82 | ·9841 297 | ·9800 070 | ·9751 390 | ·9694 585⁻ | ·9629 027 | ·9554 150⁻ |
| ·83 | ·9898 043 | ·9870 270 | ·9837 075⁻ | ·9797 866 | ·9752 065⁻ | ·9699 115⁺ |
| ·84 | ·9937 278 | ·9919 392 | ·9897 757 | ·9871 895⁻ | ·9841 320 | ·9805 548 |
| ·85 | ·9963 250⁻ | ·9952 297 | ·9938 891 | ·9922 675⁻ | ·9903 275⁺ | ·9880 308 |
| ·86 | ·9979 624 | ·9973 288 | ·9965 440 | ·9955 836 | ·9944 211 | ·9930 287 |
| ·87 | ·9989 395⁻ | ·9985 958 | ·9981 652 | ·9976 322 | ·9969 794 | ·9961 885⁻ |
| ·88 | ·9994 869 | ·9993 138 | ·9990 946 | ·9988 200 | ·9984 799 | ·9980 630 |
| ·89 | ·9997 721 | ·9996 922 | ·9995 898 | ·9994 601 | ·9992 977 | ·9990 964 |
| ·90 | ·9999 085⁺ | ·9998 752 | ·9998 321 | ·9997 769 | ·9997 069 | ·9996 192 |
| ·91 | ·9999 675⁺ | ·9999 553 | ·9999 392 | ·9999 184 | ·9998 918 | ·9998 580 |
| ·92 | ·9999 901 | ·9999 862 | ·9999 811 | ·9999 743 | ·9999 656 | ·9999 545⁻ |
| ·93 | ·9999 975⁻ | ·9999 965⁻ | ·9999 951 | ·9999 933 | ·9999 910 | ·9999 879 |
| ·94 | ·9999 995⁺ | ·9999 993 | ·9999 990 | ·9999 987 | ·9999 982 | ·9999 975⁺ |
| ·95 | ·9999 999 | ·9999 999 | ·9999 999 | ·9999 998 | ·9999 997 | ·9999 996 |
| ·96 | 1·0000 000 | 1·0000 000 | 1·0000 000 | 1·0000 000 | 1·0000 000 | 1·0000 000 |

$x$ = ·32 to ·97 $q$ = 12 $p$ = 30 to 35

| | $p$ = 30 | $p$ = 31 | $p$ = 32 | $p$ = 33 | $p$ = 34 | $p$ = 35 |
|---|---|---|---|---|---|---|
| $B(p, q)$ = | ·1055 0320 $\times\frac{1}{10^{10}}$ | ·7535 9425 $\stackrel{+}{\times}\frac{1}{10^{11}}$ | ·5432 8888 $\times\frac{1}{10^{11}}$ | ·3951 1919 $\times\frac{1}{10^{11}}$ | ·2897 5407 $\times\frac{1}{10^{11}}$ | ·2141 6605 $\stackrel{+}{\times}\frac{1}{10^{11}}$ |
| $x$ | | | | | | |
| ·32 | ·0000 001 | | | | | |
| ·33 | ·0000 002 | ·0000 001 | | | | |
| ·34 | ·0000 004 | ·0000 002 | ·0000 001 | | | |
| ·35 | ·0000 007 | ·0000 003 | ·0000 002 | ·0000 001 | | |
| ·36 | ·0000 014 | ·0000 007 | ·0000 003 | ·0000 002 | ·0000 001 | |
| ·37 | ·0000 027 | ·0000 014 | ·0000 007 | ·0000 003 | ·0000 002 | ·0000 001 |
| ·38 | ·0000 052 | ·0000 026 | ·0000 013 | ·0000 007 | ·0000 003 | ·0000 002 |
| ·39 | ·0000 095+ | ·0000 050− | ·0000 026 | ·0000 013 | ·0000 007 | ·0000 003 |
| ·40 | ·0000 171 | ·0000 092 | ·0000 049 | ·0000 026 | ·0000 014 | ·0000 007 |
| ·41 | ·0000 302 | ·0000 166 | ·0000 091 | ·0000 049 | ·0000 027 | ·0000 014 |
| ·42 | ·0000 523 | ·0000 295− | ·0000 165− | ·0000 091 | ·0000 050+ | ·0000 028 |
| ·43 | ·0000 886 | ·0000 511 | ·0000 292 | ·0000 166 | ·0000 094 | ·0000 053 |
| ·44 | ·0001 474 | ·0000 869 | ·0000 509 | ·0000 296 | ·0000 171 | ·0000 098 |
| ·45 | ·0002 407 | ·0001 451 | ·0000 868 | ·0000 516 | ·0000 304 | ·0000 178 |
| ·46 | ·0003 860 | ·0002 378 | ·0001 454 | ·0000 882 | ·0000 532 | ·0000 319 |
| ·47 | ·0006 087 | ·0003 828 | ·0002 390 | ·0001 482 | ·0000 912 | ·0000 558 |
| ·48 | ·0009 438 | ·0006 058 | ·0003 860 | ·0002 443 | ·0001 535+ | ·0000 959 |
| ·49 | ·0014 398 | ·0009 428 | ·0006 129 | ·0003 957 | ·0002 537 | ·0001 617 |
| ·50 | ·0021 620 | ·0014 436 | ·0009 570 | ·0006 300 | ·0004 120 | ·0002 678 |
| ·51 | ·0031 967 | ·0021 755+ | ·0014 700 | ·0009 865− | ·0006 577 | ·0004 357 |
| ·52 | ·0046 557 | ·0032 281 | ·0022 223 | ·0015 195+ | ·0010 323 | ·0006 969 |
| ·53 | ·0066 813 | ·0047 177 | ·0033 077 | ·0023 035− | ·0015 938 | ·0010 960 |
| ·54 | ·0094 508 | ·0067 930 | ·0048 485− | ·0034 375+ | ·0024 217 | ·0016 956 |
| ·55 | ·0131 799 | ·0096 396 | ·0070 014 | ·0050 517 | ·0036 219 | ·0025 811 |
| ·56 | ·0181 263 | ·0134 845+ | ·0099 627 | ·0073 126 | ·0053 338 | ·0038 672 |
| ·57 | ·0245 900 | ·0185 992 | ·0139 728 | ·0104 293 | ·0077 362 | ·0057 045+ |
| ·58 | ·0329 114 | ·0253 004 | ·0193 196 | ·0146 583 | ·0110 536 | ·0082 864 |
| ·59 | ·0434 669 | ·0339 481 | ·0263 393 | ·0203 069 | ·0155 614 | ·0118 557 |
| ·60 | ·0566 587 | ·0449 402 | ·0354 143 | ·0277 339 | ·0215 897 | ·0167 104 |
| ·61 | ·0729 017 | ·0587 019 | ·0469 664 | ·0373 470 | ·0295 232 | ·0232 066 |
| ·62 | ·0926 047 | ·0756 704 | ·0614 456 | ·0495 946 | ·0397 977 | ·0317 585+ |
| ·63 | ·1161 470 | ·0962 746 | ·0793 126 | ·0649 531 | ·0528 912 | ·0428 337 |
| ·64 | ·1438 516 | ·1209 090 | ·1010 155+ | ·0839 072 | ·0693 080 | ·0569 417 |
| ·65 | ·1759 547 | ·1499 033 | ·1269 613 | ·1069 229 | ·0895 563 | ·0746 159 |
| ·66 | ·2125 757 | ·1834 901 | ·1574 819 | ·1344 159 | ·1141 178 | ·0963 869 |
| ·67 | ·2536 877 | ·2217 705+ | ·1927 980 | ·1667 136 | ·1434 116 | ·1227 482 |
| ·68 | ·2990 931 | ·2646 835− | ·2329 822 | ·2040 151 | ·1777 517 | ·1541 146 |
| ·69 | ·3484 064 | ·3119 792 | ·2779 256 | ·2463 515+ | ·2173 034 | ·1907 756 |
| ·70 | ·4010 478 | ·3632 027 | ·3273 109 | ·2935 502 | ·2620 402 | ·2328 471 |
| ·71 | ·4562 496 | ·4176 894 | ·3805 969 | ·3452 084 | ·3117 073 | ·2802 261 |
| ·72 | ·5130 776 | ·4745 755+ | ·4370 173 | ·4006 798 | ·3657 958 | ·3325 533 |
| ·73 | ·5704 672 | ·5328 260 | ·4955 976 | ·4590 788 | ·4235 339 | ·3891 912 |
| ·74 | ·6272 739 | ·5912 786 | ·5551 905+ | ·5193 054 | ·4838 987 | ·4492 210 |
| ·75 | ·6823 343 | ·6487 040 | ·6145 312 | ·5800 915− | ·5456 518 | ·5114 653 |
| ·76 | ·7345 349 | ·7038 761 | ·6723 074 | ·6400 678 | ·6073 984 | ·5745 368 |
| ·77 | ·7828 804 | ·7556 491 | ·7272 406 | ·6978 468 | ·6676 691 | ·6369 145− |
| ·78 | ·8265 570 | ·8030 324 | ·7781 721 | ·7521 154 | ·7250 165+ | ·6970 409 |
| ·79 | ·8649 832 | ·8452 564 | ·8241 423 | ·8017 284 | ·7781 191 | ·7534 335− |
| ·80 | ·8978 420 | ·8818 216 | ·8644 576 | ·8457 913 | ·8258 805+ | ·8047 986 |
| ·81 | ·9250 905− | ·9125 242 | ·8987 337 | ·8837 236 | ·8675 128 | ·8501 337 |
| ·82 | ·9469 462 | ·9374 555− | ·9269 116 | ·9152 936 | ·9025 912 | ·8888 054 |
| ·83 | ·9638 497 | ·9569 736 | ·9492 413 | ·9406 174 | ·9310 736 | ·9205 894 |
| ·84 | ·9764 101 | ·9716 520 | ·9662 370 | ·9601 248 | ·9532 792 | ·9456 685− |
| ·85 | ·9853 380 | ·9822 100 | ·9786 077 | ·9744 931 | ·9698 300 | ·9645 840 |
| ·86 | ·9913 769 | ·9894 356 | ·9871 736 | ·9845 596 | ·9815 622 | ·9781 505− |
| ·87 | ·9952 394 | ·9941 108 | ·9927 806 | ·9912 255+ | ·9894 216 | ·9873 445− |
| ·88 | ·9975 571 | ·9969 486 | ·9962 231 | ·9953 652 | ·9943 586 | ·9931 861 |
| ·89 | ·9988 492 | ·9985 486 | ·9981 861 | ·9977 526 | ·9972 381 | ·9966 322 |
| ·90 | ·9995 103 | ·9993 763 | ·9992 130 | ·9990 155+ | ·9987 785+ | ·9984 962 |
| ·91 | ·9998 156 | ·9997 629 | ·9996 980 | ·9996 185+ | ·9995 221 | ·9994 061 |
| ·92 | ·9999 403 | ·9999 225+ | ·9999 004 | ·9998 729 | ·9998 393 | ·9997 983 |
| ·93 | ·9999 840 | ·9999 791 | ·9999 728 | ·9999 650+ | ·9999 554 | ·9999 434 |
| ·94 | ·9999 967 | ·9999 956 | ·9999 942 | ·9999 925+ | ·9999 903 | ·9999 876 |
| ·95 | ·9999 995+ | ·9999 993 | ·9999 991 | ·9999 989 | ·9999 985+ | ·9999 981 |
| ·96 | 1·0000 000 | ·9999 999 | ·9999 999 | ·9999 999 | ·9999 999 | ·9999 998 |
| ·97 | | 1·0000 000 | 1·0000 000 | 1·0000 000 | 1·0000 000 | 1·0000 000 |

$x$ = ·38 to ·97 $q$ = 12 $p$ = 36 to 40

| $x$ | $p$ = 36 | $p$ = 37 | $p$ = 38 | $p$ = 39 | $p$ = 40 |
|---|---|---|---|---|---|
| $B(p, q)$ = | ·1594 8536 $\times \frac{1}{10^{11}}$ | ·1196 1402 $\times \frac{1}{10^{11}}$ | ·9032 0789 $\times \frac{1}{10^{12}}$ | ·6864 3800 $\times \frac{1}{10^{12}}$ | ·5249 2317 $\times \frac{1}{10^{12}}$ |
| ·38 | ·0000 001 | | | | |
| ·39 | ·0000 002 | ·0000 001 | | | |
| ·40 | ·0000 004 | ·0000 002 | ·0000 001 | | |
| ·41 | ·0000 008 | ·0000 004 | ·0000 002 | ·0000 001 | ·0000 001 |
| ·42 | ·0000 015+ | ·0000 008 | ·0000 004 | ·0000 002 | ·0000 001 |
| ·43 | ·0000 029 | ·0000 016 | ·0000 009 | ·0000 005− | ·0000 003 |
| ·44 | ·0000 056 | ·0000 032 | ·0000 018 | ·0000 010 | ·0000 006 |
| ·45 | ·0000 104 | ·0000 060 | ·0000 035− | ·0000 020 | ·0000 011 |
| ·46 | ·0000 190 | ·0000 112 | ·0000 066 | ·0000 039 | ·0000 023 |
| ·47 | ·0000 339 | ·0000 205+ | ·0000 123 | ·0000 074 | ·0000 044 |
| ·48 | ·0000 595+ | ·0000 367 | ·0000 226 | ·0000 138 | ·0000 084 |
| ·49 | ·0001 024 | ·0000 645+ | ·0000 404 | ·0000 252 | ·0000 156 |
| ·50 | ·0001 730 | ·0001 111 | ·0000 710 | ·0000 451 | ·0000 285+ |
| ·51 | ·0002 870 | ·0001 879 | ·0001 224 | ·0000 793 | ·0000 511 |
| ·52 | ·0004 677 | ·0003 121 | ·0002 072 | ·0001 368 | ·0000 899 |
| ·53 | ·0007 493 | ·0005 094 | ·0003 444 | ·0002 317 | ·0001 550+ |
| ·54 | ·0011 803 | ·0008 170 | ·0005 625+ | ·0003 853 | ·0002 626 |
| ·55 | ·0018 287 | ·0012 885− | ·0009 030 | ·0006 297 | ·0004 369 |
| ·56 | ·0027 878 | ·0019 986 | ·0014 253 | ·0010 113 | ·0007 140 |
| ·57 | ·0041 825− | ·0030 498 | ·0022 123 | ·0015 967 | ·0011 469 |
| ·58 | ·0061 770 | ·0045 798 | ·0033 779 | ·0024 791 | ·0018 107 |
| ·59 | ·0089 823 | ·0067 689 | ·0050 748 | ·0037 859 | ·0028 110 |
| ·60 | ·0128 628 | ·0098 488 | ·0075 028 | ·0056 877 | ·0042 915− |
| ·61 | ·0181 425+ | ·0141 095+ | ·0109 180 | ·0084 077 | ·0064 444 |
| ·62 | ·0252 077 | ·0199 052 | ·0156 403 | ·0122 306 | ·0095 203 |
| ·63 | ·0345 061 | ·0276 566 | ·0220 585+ | ·0175 109 | ·0138 377 |
| ·64 | ·0465 399 | ·0378 486 | ·0306 324 | ·0246 772 | ·0197 909 |
| ·65 | ·0618 525+ | ·0510 214 | ·0418 881 | ·0342 330 | ·0278 537 |
| ·66 | ·0810 065+ | ·0677 538 | ·0564 066 | ·0467 495+ | ·0385 782 |
| ·67 | ·1045 525+ | ·0886 364 | ·0748 023 | ·0628 504 | ·0525 840 |
| ·68 | ·1329 898 | ·1142 356 | ·0976 914 | ·0831 848 | ·0705 379 |
| ·69 | ·1667 187 | ·1450 474 | ·1256 487 | ·1083 890 | ·0931 207 |
| ·70 | ·2059 895+ | ·1814 446 | ·1591 552 | ·1390 361 | ·1209 804 |
| ·71 | ·2508 496 | ·2236 193 | ·1985 379 | ·1755 753 | ·1546 734 |
| ·72 | ·3010 965+ | ·2715 271 | ·2439 076 | ·2182 652 | ·1945 952 |
| ·73 | ·3562 419 | ·3248 394 | ·2951 003 | ·2671 065− | ·2409 071 |
| ·74 | ·4154 945+ | ·3829 109 | ·3516 307 | ·3217 828 | ·2934 655+ |
| ·75 | ·4777 672 | ·4447 711 | ·4126 668 | ·3816 186 | ·3517 645+ |
| ·76 | ·5417 128 | ·5091 440 | ·4770 330 | ·4455 642 | ·4149 023 |
| ·77 | ·6057 908 | ·5745 029 | ·5432 488 | ·5122 168 | ·4815 826 |
| ·78 | ·6683 618 | ·6391 570 | ·6096 049 | ·5798 817 | ·5501 585− |
| ·79 | ·7278 027 | ·7013 675− | ·6742 749 | ·6466 761 | ·6187 235+ |
| ·80 | ·7826 324 | ·7594 810 | ·7354 537 | ·7106 676 | ·6852 460 |
| ·81 | ·8316 325+ | ·8120 675− | ·7915 083 | ·7700 349 | ·7477 355− |
| ·82 | ·8739 482 | ·8580 428 | ·8411 228 | ·8232 322 | ·8044 241 |
| ·83 | ·9091 527 | ·8967 597 | ·8834 155+ | ·8691 337 | ·8539 364 |
| ·84 | ·9372 662 | ·9280 517 | ·9180 104 | ·9071 341 | ·8954 212 |
| ·85 | ·9587 234 | ·9522 198 | ·9450 483 | ·9371 879 | ·9286 221 |
| ·86 | ·9742 942 | ·9699 645+ | ·9651 340 | ·9597 772 | ·9538 710 |
| ·87 | ·9849 694 | ·9822 718 | ·9792 270 | ·9758 113 | ·9720 015− |
| ·88 | ·9918 302 | ·9902 723 | ·9884 938 | ·9864 757 | ·9841 989 |
| ·89 | ·9959 233 | ·9950 997 | ·9941 487 | ·9930 574 | ·9918 122 |
| ·90 | ·9981 623 | ·9977 699 | ·9973 118 | ·9967 801 | ·9961 666 |
| ·91 | ·9992 672 | ·9991 022 | ·9989 075+ | ·9986 790 | ·9984 124 |
| ·92 | ·9997 488 | ·9996 893 | ·9996 183 | ·9995 341 | ·9994 348 |
| ·93 | ·9999 289 | ·9999 112 | ·9998 899 | ·9998 643 | ·9998 338 |
| ·94 | ·9999 843 | ·9999 802 | ·9999 752 | ·9999 692 | ·9999 619 |
| ·95 | ·9999 975+ | ·9999 969 | ·9999 960 | ·9999 950+ | ·9999 938 |
| ·96 | ·9999 998 | ·9999 997 | ·9999 996 | ·9999 995+ | ·9999 994 |
| ·97 | 1·0000 000 | 1·0000 000 | 1·0000 000 | 1·0000 000 | 1·0000 000 |

$x = \cdot 42$ to $\cdot 98$ | $q = 12$ | $p = 41$ to $45$

| | $p = 41$ | $p = 42$ | $p = 43$ | $p = 44$ | $p = 45$ |
|---|---|---|---|---|---|
| $B(p, q) =$ | ·4037 8706 $\times \frac{1}{10^{12}}$ | ·3123 6357 $\times \frac{1}{10^{12}}$ | ·2429 4945 $\times \frac{1}{10^{12}}$ | ·1899 4229 $\times \frac{1}{10^{12}}$ | ·1492 4037 $\times \frac{1}{10^{12}}$ |
| $x$ | | | | | |
| ·42 | ·0000 001 | | | | |
| ·43 | ·0000 001 | ·0000 001 | | | |
| ·44 | ·0000 003 | ·0000 002 | ·0000 001 | ·0000 001 | |
| ·45 | ·0000 006 | ·0000 004 | ·0000 002 | ·0000 001 | ·0000 001 |
| ·46 | ·0000 013 | ·0000 008 | ·0000 004 | ·0000 002 | ·0000 001 |
| ·47 | ·0000 026 | ·0000 015+ | ·0000 009 | ·0000 005+ | ·0000 003 |
| ·48 | ·0000 051 | ·0000 030 | ·0000 018 | ·0000 011 | ·0000 006 |
| ·49 | ·0000 096 | ·0000 059 | ·0000 036 | ·0000 022 | ·0000 013 |
| ·50 | ·0000 179 | ·0000 112 | ·0000 070 | ·0000 044 | ·0000 027 |
| ·51 | ·0000 328 | ·0000 209 | ·0000 133 | ·0000 084 | ·0000 053 |
| ·52 | ·0000 588 | ·0000 382 | ·0000 248 | ·0000 160 | ·0000 103 |
| ·53 | ·0001 033 | ·0000 685− | ·0000 452 | ·0000 297 | ·0000 195− |
| ·54 | ·0001 782 | ·0001 203 | ·0000 809 | ·0000 542 | ·0000 361 |
| ·55 | ·0003 017 | ·0002 074 | ·0001 420 | ·0000 968 | ·0000 657 |
| ·56 | ·0005 018 | ·0003 511 | ·0002 446 | ·0001 697 | ·0001 172 |
| ·57 | ·0008 199 | ·0005 836 | ·0004 136 | ·0002 919 | ·0002 052 |
| ·58 | ·0013 165− | ·0009 529 | ·0006 868 | ·0004 929 | ·0003 524 |
| ·59 | ·0020 776 | ·0015 288 | ·0011 202 | ·0008 175− | ·0005 942 |
| ·60 | ·0032 234 | ·0024 106 | ·0017 952 | ·0013 315− | ·0009 837 |
| ·61 | ·0049 175+ | ·0037 362 | ·0028 269 | ·0021 303 | ·0015 991 |
| ·62 | ·0073 779 | ·0056 931 | ·0043 751 | ·0033 488 | ·0025 534 |
| ·63 | ·0108 873 | ·0085 298 | ·0066 556 | ·0051 728 | ·0040 051 |
| ·64 | ·0158 036 | ·0125 671 | ·0099 533 | ·0078 525− | ·0061 718 |
| ·65 | ·0225 669 | ·0182 085− | ·0146 335+ | ·0117 154 | ·0093 444 |
| ·66 | ·0317 020 | ·0259 461 | ·0211 522 | ·0171 789 | ·0139 010 |
| ·67 | ·0438 139 | ·0363 615− | ·0300 606 | ·0247 591 | ·0203 191 |
| ·68 | ·0595 731 | ·0501 167 | ·0420 021 | ·0350 727 | ·0291 827 |
| ·69 | ·0796 885− | ·0679 337 | ·0576 985+ | ·0488 297 | ·0411 803 |
| ·70 | ·1048 658 | ·0905 591 | ·0779 216 | ·0668 123 | ·0570 916 |
| ·71 | ·1357 519 | ·1187 134 | ·1034 477 | ·0898 364 | ·0777 563 |
| ·72 | ·1728 661 | ·1530 238 | ·1349 957 | ·1186 948 | ·1040 241 |
| ·73 | ·2165 220 | ·1939 450+ | ·1731 473 | ·1540 811 | ·1366 832 |
| ·74 | ·2667 482 | ·2416 730 | ·2182 575+ | ·1964 975− | ·1763 694 |
| ·75 | ·3232 166 | ·2960 612 | ·2703 606 | ·2461 542 | ·2234 607 |
| ·76 | ·3851 909 | ·3565 521 | ·3290 860 | ·3028 719 | ·2779 685+ |
| ·77 | ·4515 074 | ·4221 365+ | ·3935 978 | ·3660 015− | ·3394 401 |
| ·78 | ·5205 987 | ·4913 562 | ·4625 732 | ·4343 793 | ·4068 902 |
| ·79 | ·5905 683 | ·5623 581 | ·5342 352 | ·5063 347 | ·4787 830 |
| ·80 | ·6593 159 | ·6330 064 | ·6064 463 | ·5797 626 | ·5530 790 |
| ·81 | ·7247 058 | ·7010 470 | ·6768 643 | ·6522 654 | ·6273 591 |
| ·82 | ·7847 603 | ·7643 099 | ·7431 486 | ·7213 574 | ·6990 214 |
| ·83 | ·8378 538 | ·8209 240 | ·8031 920 | ·7847 095+ | ·7655 339 |
| ·84 | ·8828 766 | ·8695 121 | ·8553 456 | ·8404 017 | ·8247 106 |
| ·85 | ·9193 390 | ·9093 313 | ·8985 968 | ·8871 384 | ·8749 638 |
| ·86 | ·9473 948 | ·9403 310 | ·9326 652 | ·9243 860 | ·9154 859 |
| ·87 | ·9677 754 | ·9631 124 | ·9579 930 | ·9523 997 | ·9463 171 |
| ·88 | ·9816 443 | ·9787 931 | ·9756 269 | ·9721 279 | ·9682 789 |
| ·89 | ·9903 991 | ·9888 041 | ·9870 127 | ·9850 106 | ·9827 832 |
| ·90 | ·9954 627 | ·9946 591 | ·9937 465+ | ·9927 151 | ·9915 547 |
| ·91 | ·9981 031 | ·9977 461 | ·9973 361 | ·9968 676 | ·9963 347 |
| ·92 | ·9993 182 | ·9991 823 | ·9990 244 | ·9988 420 | ·9986 323 |
| ·93 | ·9997 976 | ·9997 549 | ·9997 049 | ·9996 464 | ·9995 784 |
| ·94 | ·9999 532 | ·9999 428 | ·9999 305− | ·9999 159 | ·9998 988 |
| ·95 | ·9999 923 | ·9999 905− | ·9999 883 | ·9999 858 | ·9999 827 |
| ·96 | ·9999 992 | ·9999 990 | ·9999 988 | ·9999 985+ | ·9999 982 |
| ·97 | 1·0000 000 | 1·0000 000 | ·9999 999 | ·9999 999 | ·9999 999 |
| ·98 | | | 1·0000 000 | 1·0000 000 | 1·0000 000 |

$x$ = ·46 to ·98 $q$ = 12 $p$ = 46 to 50

| | $p$ = 46 | $p$ = 47 | $p$ = 48 | $p$ = 49 | $p$ = 50 |
|---|---|---|---|---|---|
| $B(p, q)$ = | ·1178 2135 × $\frac{1}{10^{12}}$ | ·9344 4517 × $\frac{1}{10^{13}}$ | ·7443 8852 × $\frac{1}{10^{13}}$ | ·5955 1082 × $\frac{1}{10^{13}}$ | ·4783 6115+ × $\frac{1}{10^{13}}$ |
| $x$ | | | | | |
| ·46 | ·0000 001 | | | | |
| ·47 | ·0000 002 | ·0000 001 | ·0000 001 | | |
| ·48 | ·0000 004 | ·0000 002 | ·0000 001 | ·0000 001 | |
| ·49 | ·0000 008 | ·0000 005− | ·0000 003 | ·0000 002 | ·0000 001 |
| ·50 | ·0000 017 | ·0000 010 | ·0000 006 | ·0000 004 | ·0000 002 |
| ·51 | ·0000 033 | ·0000 021 | ·0000 013 | ·0000 008 | ·0000 005− |
| ·52 | ·0000 066 | ·0000 042 | ·0000 027 | ·0000 017 | ·0000 011 |
| ·53 | ·0000 127 | ·0000 082 | ·0000 053 | ·0000 034 | ·0000 022 |
| ·54 | ·0000 240 | ·0000 159 | ·0000 105− | ·0000 069 | ·0000 045− |
| ·55 | ·0000 444 | ·0000 299 | ·0000 201 | ·0000 134 | ·0000 090 |
| ·56 | ·0000 807 | ·0000 553 | ·0000 378 | ·0000 257 | ·0000 175− |
| ·57 | ·0001 437 | ·0001 002 | ·0000 697 | ·0000 483 | ·0000 333 |
| ·58 | ·0002 510 | ·0001 781 | ·0001 259 | ·0000 887 | ·0000 623 |
| ·59 | ·0004 303 | ·0003 104 | ·0002 232 | ·0001 599 | ·0001 142 |
| ·60 | ·0007 240 | ·0005 309 | ·0003 880 | ·0002 826 | ·0002 051 |
| ·61 | ·0011 960 | ·0008 912 | ·0006 618 | ·0004 898 | ·0003 613 |
| ·62 | ·0019 398 | ·0014 683 | ·0011 076 | ·0008 327 | ·0006 240 |
| ·63 | ·0030 897 | ·0023 750− | ·0018 194 | ·0013 891 | ·0010 572 |
| ·64 | ·0048 333 | ·0037 718 | ·0029 334 | ·0022 739 | ·0017 571 |
| ·65 | ·0074 265+ | ·0058 818 | ·0046 427 | ·0036 528 | ·0028 649 |
| ·66 | ·0112 086 | ·0090 068 | ·0072 135− | ·0057 586 | ·0045 829 |
| ·67 | ·0166 171 | ·0135 437 | ·0110 025+ | ·0089 098 | ·0071 928 |
| ·68 | ·0241 984 | ·0199 986 | ·0164 743 | ·0135 286 | ·0110 759 |
| ·69 | ·0346 121 | ·0289 962 | ·0242 144 | ·0201 587 | ·0167 322 |
| ·70 | ·0486 239 | ·0412 790 | ·0349 343 | ·0294 752 | ·0247 960 |
| ·71 | ·0670 829 | ·0576 926 | ·0494 652 | ·0422 850− | ·0360 426 |
| ·72 | ·0908 791 | ·0791 515+ | ·0687 314 | ·0595 096 | ·0513 794 |
| ·73 | ·1208 782 | ·1065 815+ | ·0937 024 | ·0821 460 | ·0718 160 |
| ·74 | ·1578 337 | ·1408 373 | ·1253 163 | ·1111 987 | ·0984 064 |
| ·75 | ·2022 801 | ·1825 960 | ·1643 777 | ·1475 827 | ·1321 587 |
| ·76 | ·2544 154 | ·2322 346 | ·2114 318 | ·1919 985− | ·1739 136 |
| ·77 | ·3139 883 | ·2897 040 | ·2666 287 | ·2447 890 | ·2241 972 |
| ·78 | ·3802 075− | ·3544 180 | ·3295 943 | ·3057 946 | ·2830 634 |
| ·79 | ·4516 965+ | ·4251 813 | ·3993 317 | ·3742 307 | ·3499 493 |
| ·80 | ·5265 139 | ·5001 799 | ·4741 820 | ·4486 175− | ·4235 746 |
| ·81 | ·6022 535+ | ·5770 552 | ·5518 675+ | ·5267 901 | ·5019 174 |
| ·82 | ·6762 288 | ·6530 695− | ·6296 342 | ·6060 134 | ·5822 963 |
| ·83 | ·7457 277 | ·7253 574 | ·7044 930 | ·6832 070 | ·6615 734 |
| ·84 | ·8083 081 | ·7912 353 | ·7735 377 | ·7552 649 | ·7364 701 |
| ·85 | ·8620 858 | ·8485 220 | ·8342 943 | ·8194 294 | ·8039 577 |
| ·86 | ·9059 609 | ·8958 105− | ·8850 381 | ·8736 508 | ·8616 592 |
| ·87 | ·9397 316 | ·9326 321 | ·9250 101 | ·9168 592 | ·9081 760 |
| ·88 | ·9640 639 | ·9594 678 | ·9544 765+ | ·9490 777 | ·9432 601 |
| ·89 | ·9803 162 | ·9775 956 | ·9746 075− | ·9713 387 | ·9677 763 |
| ·90 | ·9902 551 | ·9888 058 | ·9871 961 | ·9854 153 | ·9834 529 |
| ·91 | ·9957 312 | ·9950 507 | ·9942 865+ | ·9934 317 | ·9924 793 |
| ·92 | ·9983 922 | ·9981 184 | ·9978 076 | ·9974 562 | ·9970 603 |
| ·93 | ·9994 998 | ·9994 091 | ·9993 051 | ·9991 861 | ·9990 507 |
| ·94 | ·9998 788 | ·9998 554 | ·9998 284 | ·9997 971 | ·9997 612 |
| ·95 | ·9999 791 | ·9999 748 | ·9999 698 | ·9999 640 | ·9999 572 |
| ·96 | ·9999 978 | ·9999 973 | ·9999 968 | ·9999 961 | ·9999 953 |
| ·97 | ·9999 999 | ·9999 999 | ·9999 998 | ·9999 998 | ·9999 998 |
| ·98 | 1·0000 000 | 1·0000 000 | 1·0000 000 | 1·0000 000 | 1·0000 000 |

$x$ = ·10 to ·70 $q$ = 13 $p$ = 13 to 18

| $x$ | $p$ = 13 | $p$ = 14 | $p$ = 15 | $p$ = 16 | $p$ = 17 | $p$ = 18 |
|---|---|---|---|---|---|---|
| $B(p, q)$ = | ·1479 2046 × $\frac{1}{10^7}$ | ·7396 0230 × $\frac{1}{10^8}$ | ·3834 9749 × $\frac{1}{10^8}$ | ·2054 4508 × $\frac{1}{10^8}$ | ·1133 4901 × $\frac{1}{10^8}$ | ·6423 1107 × $\frac{1}{10^9}$ |
| ·10 | ·0000 002 | | | | | |
| ·11 | ·0000 005− | ·0000 001 | | | | |
| ·12 | ·0000 014 | ·0000 003 | ·0000 001 | | | |
| ·13 | ·0000 034 | ·0000 008 | ·0000 002 | | | |
| ·14 | ·0000 078 | ·0000 020 | ·0000 005+ | ·0000 001 | | |
| ·15 | ·0000 169 | ·0000 047 | ·0000 012 | ·0000 003 | ·0000 001 | |
| ·16 | ·0000 344 | ·0000 101 | ·0000 029 | ·0000 008 | ·0000 002 | ·0000 001 |
| ·17 | ·0000 664 | ·0000 207 | ·0000 063 | ·0000 018 | ·0000 005+ | ·0000 001 |
| ·18 | ·0001 224 | ·0000 403 | ·0000 129 | ·0000 040 | ·0000 012 | ·0000 004 |
| ·19 | ·0002 166 | ·0000 753 | ·0000 254 | ·0000 083 | ·0000 027 | ·0000 008 |
| ·20 | ·0003 690 | ·0001 348 | ·0000 479 | ·0000 165+ | ·0000 056 | ·0000 018 |
| ·21 | ·0006 080 | ·0002 330 | ·0000 867 | ·0000 315− | ·0000 111 | ·0000 039 |
| ·22 | ·0009 712 | ·0003 894 | ·0001 517 | ·0000 576 | ·0000 213 | ·0000 077 |
| ·23 | ·0015 081 | ·0006 313 | ·0002 569 | ·0001 018 | ·0000 394 | ·0000 149 |
| ·24 | ·0022 818 | ·0009 954 | ·0004 221 | ·0001 744 | ·0000 704 | ·0000 278 |
| ·25 | ·0033 704 | ·0015 295− | ·0006 748 | ·0002 901 | ·0001 218 | ·0000 501 |
| ·26 | ·0048 691 | ·0022 945+ | ·0010 514 | ·0004 696 | ·0002 049 | ·0000 875+ |
| ·27 | ·0068 899 | ·0033 666 | ·0015 999 | ·0007 413 | ·0003 356 | ·0001 487 |
| ·28 | ·0095 630 | ·0048 380 | ·0023 810 | ·0011 426 | ·0005 359 | ·0002 460 |
| ·29 | ·0130 355− | ·0068 187 | ·0034 706 | ·0017 229 | ·0008 359 | ·0003 971 |
| ·30 | ·0174 697 | ·0094 367 | ·0049 612 | ·0025 444 | ·0012 756 | ·0006 262 |
| ·31 | ·0230 415− | ·0128 375+ | ·0069 630 | ·0036 850− | ·0019 066 | ·0009 662 |
| ·32 | ·0299 357 | ·0171 832 | ·0096 046 | ·0052 394 | ·0027 948 | ·0014 604 |
| ·33 | ·0383 424 | ·0226 503 | ·0130 333 | ·0073 207 | ·0040 218 | ·0021 646 |
| ·34 | ·0484 512 | ·0294 262 | ·0174 132 | ·0100 613 | ·0056 869 | ·0031 497 |
| ·35 | ·0604 449 | ·0377 051 | ·0229 242 | ·0136 123 | ·0079 087 | ·0045 034 |
| ·36 | ·0744 932 | ·0476 829 | ·0297 584 | ·0181 433 | ·0108 257 | ·0063 319 |
| ·37 | ·0907 452 | ·0595 507 | ·0381 156 | ·0238 398 | ·0145 963 | ·0087 620 |
| ·38 | ·1093 229 | ·0734 880 | ·0481 987 | ·0309 009 | ·0193 978 | ·0119 411 |
| ·39 | ·1303 145− | ·0896 555+ | ·0602 068 | ·0395 338 | ·0254 245+ | ·0160 377 |
| ·40 | ·1537 678 | ·1081 877 | ·0743 283 | ·0499 495− | ·0328 843 | ·0212 399 |
| ·41 | ·1796 859 | ·1291 855+ | ·0907 331 | ·0623 552 | ·0419 941 | ·0277 532 |
| ·42 | ·2080 232 | ·1527 094 | ·1095 647 | ·0769 473 | ·0529 735− | ·0357 969 |
| ·43 | ·2386 827 | ·1787 738 | ·1309 322 | ·0939 029 | ·0660 383 | ·0455 988 |
| ·44 | ·2715 160 | ·2073 422 | ·1549 030 | ·1133 713 | ·0813 918 | ·0573 883 |
| ·45 | ·3063 240 | ·2383 244 | ·1814 962 | ·1354 654 | ·0992 161 | ·0713 894 |
| ·46 | ·3428 601 | ·2715 750− | ·2106 771 | ·1602 536 | ·1196 628 | ·0878 109 |
| ·47 | ·3808 351 | ·3068 938 | ·2423 536 | ·1877 527 | ·1428 434 | ·1068 367 |
| ·48 | ·4199 229 | ·3440 288 | ·2763 747 | ·2179 215− | ·1688 208 | ·1286 160 |
| ·49 | ·4597 693 | ·3826 808 | ·3125 302 | ·2506 575− | ·1976 015+ | ·1532 530 |
| ·50 | ·5000 000$^e$ | ·4225 095− | ·3505 540 | ·2857 941 | ·2291 292 | ·1807 973 |
| ·51 | ·5402 307 | ·4631 422 | ·3901 283 | ·3231 016 | ·2632 803 | ·2112 357 |
| ·52 | ·5800 771 | ·5041 830 | ·4308 910 | ·3622 897 | ·2998 625− | ·2444 859 |
| ·53 | ·6191 649 | ·5452 237 | ·4724 444 | ·4030 129 | ·3386 152 | ·2803 921 |
| ·54 | ·6571 399 | ·5858 547 | ·5143 659 | ·4448 788 | ·3792 134 | ·3187 240 |
| ·55 | ·6936 760 | ·6256 765− | ·5562 198 | ·4874 577 | ·4212 741 | ·3591 784 |
| ·56 | ·7284 840 | ·6643 103 | ·5975 696 | ·5302 949 | ·4643 658 | ·4013 840 |
| ·57 | ·7613 173 | ·7014 084 | ·6379 906 | ·5729 238 | ·5080 198 | ·4449 101 |
| ·58 | ·7919 768 | ·7366 631 | ·6770 823 | ·6148 799 | ·5517 445+ | ·4892 776 |
| ·59 | ·8203 141 | ·7698 137 | ·7144 797 | ·6557 150+ | ·5950 405− | ·5339 733 |
| ·60 | ·8462 322 | ·8006 522 | ·7498 630 | ·6950 107 | ·6374 158 | ·5784 657 |
| ·61 | ·8696 855+ | ·8290 266 | ·7829 658 | ·7323 911 | ·6784 025+ | ·6222 227 |
| ·62 | ·8906 771 | ·8548 421 | ·8135 807 | ·7675 329 | ·7175 711 | ·6647 292 |
| ·63 | ·9092 548 | ·8780 603 | ·8415 628 | ·8001 745+ | ·7545 440 | ·7055 046 |
| ·64 | ·9255 068 | ·8986 966 | ·8668 307 | ·8301 213 | ·7890 066 | ·7441 191 |
| ·65 | ·9395 551 | ·9168 153 | ·8893 651 | ·8572 484 | ·8207 156 | ·7802 072 |
| ·66 | ·9515 488 | ·9325 238 | ·9092 045+ | ·8815 012 | ·8495 039 | ·8134 788 |
| ·67 | ·9616 576 | ·9459 654 | ·9264 399 | ·9028 922 | ·8752 824 | ·8437 261 |
| ·68 | ·9700 643 | ·9573 118 | ·9412 073 | ·9214 953 | ·8980 381 | ·8708 277 |
| ·69 | ·9769 585+ | ·9667 546 | ·9536 789 | ·9374 390 | ·9178 292 | ·8947 474 |
| ·70 | ·9825 303 | ·9744 972 | ·9640 543 | ·9508 962 | ·9347 775+ | ·9155 299 |

$x$ = ·71 to ·93 $q = 13$ $p$ = 13 to 18

| | $p = 13$ | $p = 14$ | $p = 15$ | $p = 16$ | $p = 17$ | $p = 18$ |
|---|---|---|---|---|---|---|
| $B(p, q)$ = | ·1479 2046 × $\frac{1}{10^7}$ | ·7396 0230 × $\frac{1}{10^8}$ | ·3834 9749 × $\frac{1}{10^8}$ | ·2054 4508 × $\frac{1}{10^8}$ | ·1133 4901 × $\frac{1}{10^8}$ | ·6423 1107 × $\frac{1}{10^9}$ |
| $x$ | | | | | | |
| ·71 | ·9869 645+ | ·9807 478 | ·9725 507 | ·9620 747 | ·9490 582 | ·9332 930 |
| ·72 | ·9904 370 | ·9857 119 | ·9793 939 | ·9712 057 | ·9608 886 | ·9482 167 |
| ·73 | ·9931 101 | ·9895 867 | ·9848 101 | ·9785 336 | ·9705 154 | ·9605 303 |
| ·74 | ·9951 309 | ·9925 564 | ·9890 183 | ·9843 056 | ·9782 025+ | ·9704 984 |
| ·75 | ·9966 296 | ·9947 886 | ·9922 244 | ·9887 627 | ·9842 193 | ·9784 064 |
| ·76 | ·9977 182 | ·9964 319 | ·9946 164 | ·9921 327 | ·9888 294 | ·9845 468 |
| ·77 | ·9984 919 | ·9976 152 | ·9963 614 | ·9946 238 | ·9922 822 | ·9892 066 |
| ·78 | ·9990 288 | ·9984 471 | ·9976 043 | ·9964 211 | ·9948 060 | ·9926 570 |
| ·79 | ·9993 920 | ·9990 170 | ·9984 668 | ·9976 845− | ·9966 029 | ·9951 453 |
| ·80 | ·9996 310 | ·9993 968 | ·9990 488 | ·9985 477 | ·9978 463 | ·9968 890 |
| ·81 | ·9997 834 | ·9996 422 | ·9994 296 | ·9991 197 | ·9986 804 | ·9980 734 |
| ·82 | ·9998 776 | ·9997 955+ | ·9996 706 | ·9994 861 | ·9992 214 | ·9988 511 |
| ·83 | ·9999 336 | ·9998 879 | ·9998 175− | ·9997 123 | ·9995 594 | ·9993 430 |
| ·84 | ·9999 656 | ·9999 414 | ·9999 035− | ·9998 462 | ·9997 620 | ·9996 414 |
| ·85 | ·9999 831 | ·9999 709 | ·9999 516 | ·9999 220 | ·9998 780 | ·9998 143 |
| ·86 | ·9999 922 | ·9999 864 | ·9999 771 | ·9999 627 | ·9999 411 | ·9999 094 |
| ·87 | ·9999 966 | ·9999 940 | ·9999 899 | ·9999 834 | ·9999 734 | ·9999 587 |
| ·88 | ·9999 986 | ·9999 976 | ·9999 959 | ·9999 931 | ·9999 889 | ·9999 826 |
| ·89 | ·9999 995+ | ·9999 991 | ·9999 985− | ·9999 974 | ·9999 958 | ·9999 933 |
| ·90 | ·9999 998 | ·9999 997 | ·9999 995− | ·9999 991 | ·9999 986 | ·9999 977 |
| ·91 | 1·0000 000 | ·9999 999 | ·9999 998 | ·9999 997 | ·9999 996 | ·9999 993 |
| ·92 | | 1·0000 000 | 1·0000 000 | ·9999 999 | ·9999 999 | ·9999 998 |
| ·93 | | | | 1·0000 000 | 1·0000 000 | 1·0000 000 |

$x$ = ·18 to ·70 $q = 13$ $p$ = 19 to 24

| | $p = 19$ | $p = 20$ | $p = 21$ | $p = 22$ | $p = 23$ | $p = 24$ |
|---|---|---|---|---|---|---|
| $B(p, q) =$ | ·3729 5481 × $\frac{1}{10^9}$ | ·2214 4192 × $\frac{1}{10^9}$ | ·1342 0722 × $\frac{1}{10^9}$ | ·8289 2697 × $\frac{1}{10^{10}}$ | ·5210 3981 × $\frac{1}{10^{10}}$ | ·3328 8655 × $\frac{1}{10^{10}}$ |
| $x$ | | | | | | |
| ·18 | ·0000 001 | | | | | |
| ·19 | ·0000 003 | ·0000 001 | | | | |
| ·20 | ·0000 006 | ·0000 002 | ·0000 001 | | | |
| ·21 | ·0000 013 | ·0000 004 | ·0000 001 | | | |
| ·22 | ·0000 027 | ·0000 010 | ·0000 003 | ·0000 001 | | |
| ·23 | ·0000 055+ | ·0000 020 | ·0000 007 | ·0000 003 | ·0000 001 | |
| ·24 | ·0000 108 | ·0000 041 | ·0000 015+ | ·0000 006 | ·0000 002 | ·0000 001 |
| ·25 | ·0000 202 | ·0000 080 | ·0000 031 | ·0000 012 | ·0000 004 | ·0000 002 |
| ·26 | ·0000 367 | ·0000 151 | ·0000 061 | ·0000 024 | ·0000 010 | ·0000 004 |
| ·27 | ·0000 646 | ·0000 276 | ·0000 116 | ·0000 048 | ·0000 019 | ·0000 008 |
| ·28 | ·0001 108 | ·0000 490 | ·0000 213 | ·0000 091 | ·0000 038 | ·0000 016 |
| ·29 | ·0001 850+ | ·0000 847 | ·0000 381 | ·0000 169 | ·0000 074 | ·0000 032 |
| ·30 | ·0003 016 | ·0001 426 | ·0000 664 | ·0000 304 | ·0000 137 | ·0000 061 |
| ·31 | ·0004 803 | ·0002 346 | ·0001 127 | ·0000 533 | ·0000 249 | ·0000 114 |
| ·32 | ·0007 487 | ·0003 771 | ·0001 868 | ·0000 912 | ·0000 439 | ·0000 208 |
| ·33 | ·0011 432 | ·0005 933 | ·0003 029 | ·0001 523 | ·0000 755− | ·0000 369 |
| ·34 | ·0017 120 | ·0009 145− | ·0004 806 | ·0002 488 | ·0001 270 | ·0000 640 |
| ·35 | ·0025 169 | ·0013 825+ | ·0007 473 | ·0003 979 | ·0002 089 | ·0001 082 |
| ·36 | ·0036 356 | ·0020 519 | ·0011 397 | ·0006 236 | ·0003 365+ | ·0001 792 |
| ·37 | ·0051 642 | ·0029 922 | ·0017 064 | ·0009 588 | ·0005 313 | ·0002 906 |
| ·38 | ·0072 185+ | ·0042 905+ | ·0025 103 | ·0014 473 | ·0008 230 | ·0004 620 |
| ·39 | ·0099 362 | ·0060 538 | ·0036 311 | ·0021 464 | ·0012 515+ | ·0007 204 |
| ·40 | ·0134 769 | ·0084 105+ | ·0051 681 | ·0031 299 | ·0018 700 | ·0011 031 |
| ·41 | ·0180 220 | ·0115 123 | ·0072 420 | ·0044 906 | ·0027 473 | ·0016 596 |
| ·42 | ·0237 734 | ·0155 341 | ·0099 973 | ·0063 430 | ·0039 710 | ·0024 550− |
| ·43 | ·0309 505− | ·0206 735+ | ·0136 030 | ·0088 253 | ·0056 503 | ·0035 728 |
| ·44 | ·0397 858 | ·0271 491 | ·0182 528 | ·0121 017 | ·0079 189 | ·0051 183 |
| ·45 | ·0505 194 | ·0351 965− | ·0241 639 | ·0163 624 | ·0109 367 | ·0072 213 |
| ·46 | ·0633 911 | ·0450 634 | ·0315 742 | ·0218 234 | ·0148 915+ | ·0100 392 |
| ·47 | ·0786 314 | ·0570 025− | ·0407 375− | ·0287 246 | ·0199 989 | ·0137 581 |
| ·48 | ·0964 522 | ·0712 628 | ·0519 174 | ·0373 254 | ·0265 008 | ·0185 941 |
| ·49 | ·1170 351 | ·0880 798 | ·0653 788 | ·0478 991 | ·0346 622 | ·0247 920 |
| ·50 | ·1405 208 | ·1076 636 | ·0813 778 | ·0607 247 | ·0447 655+ | ·0326 227 |
| ·51 | ·1669 978 | ·1301 872 | ·1001 498 | ·0760 770 | ·0571 032 | ·0423 779 |
| ·52 | ·1964 928 | ·1557 745− | ·1218 969 | ·0942 140 | ·0719 670 | ·0543 629 |
| ·53 | ·2289 617 | ·1844 879 | ·1467 742 | ·1153 640 | ·0896 362 | ·0688 862 |
| ·54 | ·2642 836 | ·2163 187 | ·1748 771 | ·1397 108 | ·1103 630 | ·0862 468 |
| ·55 | ·3022 573 | ·2511 781 | ·2062 285− | ·1673 791 | ·1343 571 | ·1067 192 |
| ·56 | ·3426 011 | ·2888 920 | ·2407 687 | ·1984 202 | ·1617 695− | ·1305 367 |
| ·57 | ·3849 559 | ·3291 986 | ·2783 478 | ·2328 001 | ·1926 767 | ·1578 740 |
| ·58 | ·4288 929 | ·3717 500− | ·3187 213 | ·2703 894 | ·2270 665+ | ·1888 293 |
| ·59 | ·4739 239 | ·4161 185+ | ·3615 502 | ·3109 576 | ·2648 263 | ·2234 084 |
| ·60 | ·5195 156 | ·4618 066 | ·4064 059 | ·3541 710 | ·3057 350− | ·2615 108 |
| ·61 | ·5651 065+ | ·5082 609 | ·4527 796 | ·3995 968 | ·3494 599 | ·3029 198 |
| ·62 | ·6101 258 | ·5548 902 | ·5000 965+ | ·4467 118 | ·3955 595+ | ·3472 985− |
| ·63 | ·6540 133 | ·6010 856 | ·5477 344 | ·4949 168 | ·4434 917 | ·3941 906 |
| ·64 | ·6962 391 | ·6462 423 | ·5950 456 | ·5435 563 | ·4926 288 | ·4430 297 |
| ·65 | ·7363 231 | ·6897 829 | ·6413 811 | ·5919 420 | ·5422 783 | ·4931 544 |
| ·66 | ·7738 511 | ·7311 783 | ·6861 158 | ·6393 796 | ·5917 087 | ·5438 305+ |
| ·67 | ·8084 882 | ·7699 677 | ·7286 736 | ·6851 969 | ·6401 787 | ·5942 798 |
| ·68 | ·8399 893 | ·8057 749 | ·7685 496 | ·7287 717 | ·6869 688 | ·6437 118 |
| ·69 | ·8682 033 | ·8383 203 | ·8053 294 | ·7695 579 | ·7314 124 | ·6913 596 |
| ·70 | ·8930 744 | ·8674 279 | ·8387 037 | ·8071 071 | ·7729 254 | ·7365 144 |

$x = \cdot 71$ to $\cdot 95$ $q = 13$ $p = 19$ to $24$

| | $p = 19$ | $p = 20$ | $p = 21$ | $p = 22$ | $p = 23$ | $p = 24$ |
|---|---|---|---|---|---|---|
| $B(p, q) =$ | $\cdot 3729\ 5481 \times \frac{1}{10^9}$ | $\cdot 2214\ 4192 \times \frac{1}{10^9}$ | $\cdot 1342\ 0722 \times \frac{1}{10^9}$ | $\cdot 8289\ 2697 \times \frac{1}{10^{10}}$ | $\cdot 5210\ 3981 \times \frac{1}{10^{10}}$ | $\cdot 3328\ 8655 \times \frac{1}{10^{10}}$ |
| $x$ | | | | | | |
| ·71 | ·9146 376 | ·8930 267 | ·8684 767 | ·8410 859 | ·8110 308 | ·7785 582 |
| ·72 | ·9330 105− | ·9151 472 | ·8945 686 | ·8712 855− | ·8453 777 | ·8169 918 |
| ·73 | ·9483 818 | ·9339 123 | ·9170 119 | ·8976 248 | ·8757 526 | ·8514 554 |
| ·74 | ·9609 965+ | ·9495 244 | ·9359 413 | ·9201 461 | ·9020 822 | ·8817 406 |
| ·75 | ·9711 402 | ·9622 487 | ·9515 789 | ·9390 038 | ·9244 281 | ·9077 929 |
| ·76 | ·9791 221 | ·9723 955+ | ·9642 160 | ·9544 473 | ·9429 735+ | ·9297 038 |
| ·77 | ·9852 595+ | ·9803 007 | ·9741 915− | ·9667 993 | ·9580 026 | ·9476 952 |
| ·78 | ·9898 633 | ·9863 080 | ·9818 709 | ·9764 323 | ·9698 763 | ·9620 947 |
| ·79 | ·9932 261 | ·9907 523 | ·9876 255− | ·9837 438 | ·9790 045+ | ·9733 072 |
| ·80 | ·9956 125+ | ·9939 465− | ·9918 139 | ·9891 330 | ·9858 184 | ·9817 832 |
| ·81 | ·9972 540 | ·9961 711 | ·9947 676 | ·9929 812 | ·9907 449 | ·9879 884 |
| ·82 | ·9983 451 | ·9976 681 | ·9967 799 | ·9956 353 | ·9941 848 | ·9923 749 |
| ·83 | ·9990 436 | ·9986 382 | ·9980 998 | ·9973 975+ | ·9964 968 | ·9953 591 |
| ·84 | ·9994 725+ | ·9992 411 | ·9989 300 | ·9985 194 | ·9979 864 | ·9973 050− |
| ·85 | ·9997 240 | ·9995 988 | ·9994 284 | ·9992 009 | ·9989 021 | ·9985 155+ |
| ·86 | ·9998 639 | ·9998 001 | ·9997 124 | ·9995 938 | ·9994 361 | ·9992 298 |
| ·87 | ·9999 373 | ·9999 070 | ·9998 648 | ·9998 071 | ·9997 295+ | ·9996 268 |
| ·88 | ·9999 733 | ·9999 600 | ·9999 413 | ·9999 154 | ·9998 801 | ·9998 329 |
| ·89 | ·9999 896 | ·9999 843 | ·9999 767 | ·9999 661 | ·9999 516 | ·9999 318 |
| ·90 | ·9999 964 | ·9999 945− | ·9999 917 | ·9999 879 | ·9999 825− | ·9999 751 |
| ·91 | ·9999 989 | ·9999 983 | ·9999 974 | ·9999 962 | ·9999 944 | ·9999 920 |
| ·92 | ·9999 997 | ·9999 996 | ·9999 993 | ·9999 990 | ·9999 985+ | ·9999 978 |
| ·93 | ·9999 999 | ·9999 999 | ·9999 999 | ·9999 998 | ·9999 997 | ·9999 995+ |
| ·94 | 1·0000 000 | 1·0000 000 | 1·0000 000 | 1·0000 000 | ·9999 999 | ·9999 999 |
| ·95 | | | | | 1·0000 000 | 1·0000 000 |

$x = \cdot 25$ to $\cdot 96$ $q = 13$ $p = 25$ to $30$

| | $p = 25$ | $p = 26$ | $p = 27$ | $p = 28$ | $p = 29$ | $p = 30$ |
|---|---|---|---|---|---|---|
| $B(p, q) =$ | $\cdot 2159\ 2641 \times \frac{1}{10^{10}}$ | $\cdot 1420\ 5685 \times \frac{1}{10^{10}}$ | $\cdot 9470\ 4565^{e} \times \frac{1}{10^{11}}$ | $\cdot 6392\ 5581 \times \frac{1}{10^{11}}$ | $\cdot 4365\ 6495 \times \frac{1}{10^{11}}$ | $\cdot 3014\ 3770 \times \frac{1}{10^{11}}$ |
| $x$ | | | | | | |
| ·25 | ·0000 001 | | | | | |
| ·26 | ·0000 001 | ·0000 001 | | | | |
| ·27 | ·0000 003 | ·0000 001 | | | | |
| ·28 | ·0000 007 | ·0000 003 | ·0000 001 | | | |
| ·29 | ·0000 014 | ·0000 006 | ·0000 002 | ·0000 001 | | |
| ·30 | ·0000 027 | ·0000 012 | ·0000 005+ | ·0000 002 | ·0000 001 | |
| ·31 | ·0000 052 | ·0000 023 | ·0000 010 | ·0000 005− | ·0000 002 | ·0000 001 |
| ·32 | ·0000 098 | ·0000 045+ | ·0000 021 | ·0000 009 | ·0000 004 | ·0000 002 |
| ·33 | ·0000 178 | ·0000 085+ | ·0000 040 | ·0000 019 | ·0000 009 | ·0000 004 |
| ·34 | ·0000 318 | ·0000 156 | ·0000 076 | ·0000 037 | ·0000 017 | ·0000 008 |
| ·35 | ·0000 554 | ·0000 280 | ·0000 140 | ·0000 069 | ·0000 034 | ·0000 017 |
| ·36 | ·0000 943 | ·0000 490 | ·0000 252 | ·0000 128 | ·0000 065− | ·0000 032 |
| ·37 | ·0001 570 | ·0000 839 | ·0000 443 | ·0000 232 | ·0000 120 | ·0000 062 |
| ·38 | ·0002 562 | ·0001 405− | ·0000 762 | ·0000 409 | ·0000 218 | ·0000 115− |
| ·39 | ·0004 097 | ·0002 304 | ·0001 282 | ·0000 706 | ·0000 385+ | ·0000 208 |
| ·40 | ·0006 429 | ·0003 705+ | ·0002 113 | ·0001 193 | ·0000 667 | ·0000 370 |
| ·41 | ·0009 907 | ·0005 848 | ·0003 415+ | ·0001 975− | ·0001 131 | ·0000 642 |
| ·42 | ·0014 999 | ·0009 062 | ·0005 418 | ·0003 207 | ·0001 881 | ·0001 093 |
| ·43 | ·0022 328 | ·0013 800 | ·0008 441 | ·0005 112 | ·0003 067 | ·0001 824 |
| ·44 | ·0032 698 | ·0020 661 | ·0012 921 | ·0008 001 | ·0004 909 | ·0002 985+ |
| ·45 | ·0047 134 | ·0030 432 | ·0019 447 | ·0012 306 | ·0007 716 | ·0004 796 |
| ·46 | ·0066 911 | ·0044 117 | ·0028 793 | ·0018 610 | ·0011 919 | ·0007 567 |
| ·47 | ·0093 584 | ·0062 979 | ·0041 956 | ·0027 684 | ·0018 101 | ·0011 734 |
| ·48 | ·0129 014 | ·0088 572 | ·0060 201 | ·0040 530 | ·0027 041 | ·0017 888 |
| ·49 | ·0175 375− | ·0122 765− | ·0085 088 | ·0058 421 | ·0039 754 | ·0026 823 |
| ·50 | ·0235 155+ | ·0167 762 | ·0118 514 | ·0082 945+ | ·0057 539 | ·0039 579 |
| ·51 | ·0311 130 | ·0226 103 | ·0162 725+ | ·0116 037 | ·0082 021 | ·0057 494 |
| ·52 | ·0406 317 | ·0300 642 | ·0220 328 | ·0160 004 | ·0115 192 | ·0082 247 |
| ·53 | ·0523 900 | ·0394 503 | ·0294 270 | ·0217 537 | ·0159 438 | ·0115 905− |
| ·54 | ·0667 127 | ·0511 010 | ·0387 797 | ·0291 692 | ·0217 553 | ·0160 952 |
| ·55 | ·0839 179 | ·0653 576 | ·0504 380 | ·0385 852 | ·0292 723 | ·0220 307 |
| ·56 | ·1043 012 | ·0825 572 | ·0647 605+ | ·0503 650+ | ·0388 486 | ·0297 308 |
| ·57 | ·1281 177 | ·1030 153 | ·0821 030 | ·0648 853 | ·0508 651 | ·0395 668 |
| ·58 | ·1555 630 | ·1270 072 | ·1028 006 | ·0825 209 | ·0657 177 | ·0519 391 |
| ·59 | ·1867 536 | ·1547 467 | ·1271 468 | ·1036 256 | ·0838 006 | ·0672 638 |
| ·60 | ·2217 090 | ·1863 650+ | ·1553 711 | ·1285 097 | ·1054 856 | ·0859 548 |
| ·61 | ·2603 356 | ·2218 906 | ·1876 154 | ·1574 152 | ·1310 978 | ·1084 013 |
| ·62 | ·3024 157 | ·2612 312 | ·2239 118 | ·1904 901 | ·1608 881 | ·1349 404 |
| ·63 | ·3476 011 | ·3041 610 | ·2641 628 | ·2277 643 | ·1950 058 | ·1658 280 |
| ·64 | ·3954 147 | ·3503 137 | ·3081 269 | ·2691 276 | ·2334 710 | ·2012 080 |
| ·65 | ·4452 586 | ·3991 828 | ·3554 108 | ·3143 137 | ·2761 522 | ·2410 831 |
| ·66 | ·4964 311 | ·4501 314 | ·4054 699 | ·3628 927 | ·3227 484 | ·2852 897 |
| ·67 | ·5481 513 | ·5024 103 | ·4576 193 | ·4142 716 | ·3727 817 | ·3334 806 |
| ·68 | ·5995 897 | ·5551 852 | ·5110 540 | ·4677 074 | ·4255 992 | ·3851 172 |
| ·69 | ·6499 050+ | ·6075 716 | ·5648 800 | ·5223 306 | ·4803 891 | ·4394 745− |
| ·70 | ·6982 828 | ·6586 750− | ·6181 531 | ·5771 809 | ·5362 088 | ·4956 605− |
| ·71 | ·7439 748 | ·7076 347 | ·6699 248 | ·6312 512 | ·5920 251 | ·5526 502 |
| ·72 | ·7863 350− | ·7536 671 | ·7192 904 | ·6835 387 | ·6467 655− | ·6093 329 |
| ·73 | ·8248 500− | ·7961 055− | ·7654 373 | ·7330 994 | ·6993 756 | ·6645 704 |
| ·74 | ·8591 615− | ·8344 328 | ·8076 878 | ·7791 004 | ·7488 793 | ·7172 619 |
| ·75 | ·8890 782 | ·8683 048 | ·8455 341 | ·8208 658 | ·7944 354 | ·7664 101 |
| ·76 | ·9145 764 | ·8975 611 | ·8786 610 | ·8579 129 | ·8353 864 | ·8111 820 |
| ·77 | ·9357 901 | ·9222 231 | ·9069 550+ | ·8899 735+ | ·8712 938 | ·8509 587 |
| ·78 | ·9529 902 | ·9424 799 | ·9304 982 | ·9169 988 | ·9019 566 | ·8853 687 |
| ·79 | ·9665 558 | ·9586 620 | ·9495 478 | ·9391 475− | ·9274 099 | ·9143 003 |
| ·80 | ·9769 411 | ·9712 079 | ·9645 045+ | ·9567 584 | ·9479 057 | ·9378 929 |
| ·81 | ·9846 393 | ·9806 244 | ·9758 713 | ·9703 103 | ·9638 753 | ·9565 062 |
| ·82 | ·9901 487 | ·9874 469 | ·9842 090 | ·9803 739 | ·9758 813 | ·9706 730 |
| ·83 | ·9939 426 | ·9922 026 | ·9900 919 | ·9875 614 | ·9845 609 | ·9810 400 |
| ·84 | ·9964 465− | ·9953 791 | ·9940 688 | ·9924 789 | ·9905 710 | ·9883 053 |
| ·85 | ·9980 226 | ·9974 026 | ·9966 323 | ·9956 866 | ·9945 382 | ·9931 582 |
| ·86 | ·9989 637 | ·9986 249 | ·9981 991 | ·9976 702 | ·9970 204 | ·9962 303 |
| ·87 | ·9994 928 | ·9993 202 | ·9991 008 | ·9988 250− | ·9984 822 | ·9980 607 |
| ·88 | ·9997 706 | ·9996 895+ | ·9995 852 | ·9994 525+ | ·9992 858 | ·9990 784 |
| ·89 | ·9999 055− | ·9998 707 | ·9998 256 | ·9997 675+ | ·9996 937 | ·9996 008 |
| ·90 | ·9999 651 | ·9999 518 | ·9999 343 | ·9999 116 | ·9998 823 | ·9998 451 |
| ·91 | ·9999 887 | ·9999 843 | ·9999 783 | ·9999 705+ | ·9999 604 | ·9999 474 |
| ·92 | ·9999 969 | ·9999 956 | ·9999 939 | ·9999 917 | ·9999 887 | ·9999 848 |
| ·93 | ·9999 993 | ·9999 990 | ·9999 986 | ·9999 981 | ·9999 974 | ·9999 964 |
| ·94 | ·9999 999 | ·9999 998 | ·9999 998 | ·9999 997 | ·9999 995+ | ·9999 994 |
| ·95 | 1·0000 000 | 1·0000 000 | 1·0000 000 | 1·0000 000 | ·9999 999 | ·9999 999 |
| ·96 | | | | | 1·0000 000 | 1·0000 000 |

$x = \cdot 32$ to $\cdot 96$ $q = 13$ $p = 31$ to $36$

| $x$ | $p = 31$ | $p = 32$ | $p = 33$ | $p = 34$ | $p = 35$ | $p = 36$ |
|---|---|---|---|---|---|---|
| $B(p, q) =$ | $\cdot 2103\ 0537 \times \frac{1}{10^{11}}$ | $\cdot 1481\ 6969 \times \frac{1}{10^{11}}$ | $\cdot 1053\ 6512 \times \frac{1}{10^{11}}$ | $\cdot 7558\ 8018 \times \frac{1}{10^{12}}$ | $\cdot 5468\ 0694 \times \frac{1}{10^{12}}$ | $\cdot 3987\ 1339 \times \frac{1}{10^{12}}$ |
| ·32 | ·0000 001 | | | | | |
| ·33 | ·0000 002 | ·0000 001 | | | | |
| ·34 | ·0000 004 | ·0000 002 | ·0000 001 | | | |
| ·35 | ·0000 008 | ·0000 004 | ·0000 002 | ·0000 001 | | |
| ·36 | ·0000 016 | ·0000 008 | ·0000 004 | ·0000 002 | ·0000 001 | |
| ·37 | ·0000 031 | ·0000 016 | ·0000 008 | ·0000 004 | ·0000 002 | ·0000 001 |
| ·38 | ·0000 060 | ·0000 031 | ·0000 016 | ·0000 008 | ·0000 004 | ·0000 002 |
| ·39 | ·0000 112 | ·0000 059 | ·0000 031 | ·0000 016 | ·0000 009 | ·0000 004 |
| ·40 | ·0000 203 | ·0000 111 | ·0000 060 | ·0000 032 | ·0000 017 | ·0000 009 |
| ·41 | ·0000 361 | ·0000 202 | ·0000 112 | ·0000 062 | ·0000 034 | ·0000 018 |
| ·42 | ·0000 630 | ·0000 360 | ·0000 204 | ·0000 115+ | ·0000 064 | ·0000 036 |
| ·43 | ·0001 075+ | ·0000 629 | ·0000 365+ | ·0000 211 | ·0000 120 | ·0000 069 |
| ·44 | ·0001 800 | ·0001 077 | ·0000 640 | ·0000 377 | ·0000 221 | ·0000 128 |
| ·45 | ·0002 956 | ·0001 808 | ·0001 097 | ·0000 661 | ·0000 396 | ·0000 235+ |
| ·46 | ·0004 765− | ·0002 977 | ·0001 846 | ·0001 136 | ·0000 695− | ·0000 422 |
| ·47 | ·0007 544 | ·0004 812 | ·0003 047 | ·0001 916 | ·0001 196 | ·0000 742 |
| ·48 | ·0011 736 | ·0007 641 | ·0004 938 | ·0003 169 | ·0002 020 | ·0001 279 |
| ·49 | ·0017 952 | ·0011 922 | ·0007 860 | ·0005 145+ | ·0003 346 | ·0002 162 |
| ·50 | ·0027 008 | ·0018 289 | ·0012 294 | ·0008 207 | ·0005 443 | ·0003 586 |
| ·51 | ·0039 982 | ·0027 594 | ·0018 907 | ·0012 865− | ·0008 696 | ·0005 841 |
| ·52 | ·0058 263 | ·0040 964 | ·0028 595− | ·0019 824 | ·0013 654 | ·0009 345+ |
| ·53 | ·0083 603 | ·0059 855+ | ·0042 550− | ·0030 042 | ·0021 074 | ·0014 691 |
| ·54 | ·0118 162 | ·0086 110 | ·0062 312 | ·0044 788 | ·0031 985+ | ·0022 701 |
| ·55 | ·0164 547 | ·0122 007 | ·0089 837 | ·0065 709 | ·0047 755− | ·0034 494 |
| ·56 | ·0225 824 | ·0170 298 | ·0127 542 | ·0094 892 | ·0070 156 | ·0051 553 |
| ·57 | ·0305 507 | ·0234 222 | ·0178 353 | ·0134 927 | ·0101 438 | ·0075 804 |
| ·58 | ·0407 508 | ·0317 497 | ·0245 713 | ·0188 939 | ·0144 387 | ·0109 688 |
| ·59 | ·0536 044 | ·0424 257 | ·0333 570 | ·0260 608 | ·0202 367 | ·0156 224 |
| ·60 | ·0695 490 | ·0558 951 | ·0446 306 | ·0354 143 | ·0279 327 | ·0219 047 |
| ·61 | ·0890 185+ | ·0726 182 | ·0588 624 | ·0474 201 | ·0379 768 | ·0302 414 |
| ·62 | ·1124 178 | ·0930 484 | ·0765 359 | ·0625 754 | ·0508 650− | ·0411 152 |
| ·63 | ·1400 933 | ·1176 044 | ·0981 235− | ·0813 875+ | ·0671 226 | ·0550 545− |
| ·64 | ·1723 004 | ·1466 378 | ·1240 548 | ·1043 460 | ·0872 804 | ·0726 138 |
| ·65 | ·2091 702 | ·1803 971 | ·1546 811 | ·1318 874 | ·1118 424 | ·0943 459 |
| ·66 | ·2506 778 | ·2189 912 | ·1902 356 | ·1643 556 | ·1412 462 | ·1207 647 |
| ·67 | ·2966 163 | ·2623 563 | ·2307 942 | ·2019 579 | ·1758 187 | ·1523 009 |
| ·68 | ·3465 784 | ·3102 276 | ·2762 396 | ·2447 235− | ·2157 286 | ·1892 522 |
| ·69 | ·3999 509 | ·3621 231 | ·3262 339 | ·2924 654 | ·2609 416 | ·2317 325− |
| ·70 | ·4559 232 | ·4173 395− | ·3802 027 | ·3447 540 | ·3111 819 | ·2796 242 |
| ·71 | ·5135 116 | ·4749 663 | ·4373 365+ | ·4009 041 | ·3659 074 | ·3325 407 |
| ·72 | ·5716 008 | ·5339 174 | ·4966 109 | ·4599 827 | ·4243 025− | ·3898 048 |
| ·73 | ·6289 994 | ·5929 809 | ·5568 273 | ·5208 381 | ·4852 934 | ·4504 495+ |
| ·74 | ·6845 063 | ·6508 842 | ·6166 737 | ·5821 522 | ·5475 901 | ·5132 452 |
| ·75 | ·7369 836 | ·7063 705+ | ·6748 008 | ·6425 135+ | ·6097 514 | ·5767 554 |
| ·76 | ·7854 286 | ·7582 795+ | ·7299 087 | ·7005 062 | ·6702 736 | ·6394 190 |
| ·77 | ·8290 375+ | ·8056 242 | ·7808 354 | ·7548 072 | ·7276 918 | ·6996 546 |
| ·78 | ·8672 547 | ·8476 565+ | ·8266 375− | ·8042 809 | ·7806 881 | ·7559 763 |
| ·79 | ·8998 011 | ·8839 128 | ·8666 540 | ·8480 617 | ·8281 898 | ·8071 085− |
| ·80 | ·9266 787 | ·9142 344 | ·9005 458 | ·8856 128 | ·8694 499 | ·8520 864 |
| ·81 | ·9481 496 | ·9387 606 | ·9283 036 | ·9167 533 | ·9040 956 | ·8903 276 |
| ·82 | ·9646 938 | ·9578 930 | ·9502 251 | ·9416 510 | ·9321 388 | ·9216 645− |
| ·83 | ·9769 487 | ·9722 385− | ·9668 629 | ·9607 787 | ·9539 465+ | ·9463 316 |
| ·84 | ·9856 408 | ·9825 362 | ·9789 504 | ·9748 430 | ·9701 750+ | ·9649 096 |
| ·85 | ·9915 160 | ·9895 797 | ·9873 167 | ·9846 937 | ·9816 773 | ·9782 342 |
| ·86 | ·9952 790 | ·9941 443 | ·9928 024 | ·9912 288 | ·9893 978 | ·9872 833 |
| ·87 | ·9975 472 | ·9969 275+ | ·9961 863 | ·9953 069 | ·9942 718 | ·9930 625− |
| ·88 | ·9988 228 | ·9985 108 | ·9981 333 | ·9976 804 | ·9971 411 | ·9965 038 |
| ·89 | ·9994 851 | ·9993 422 | ·9991 673 | ·9989 551 | ·9986 996 | ·9983 942 |
| ·90 | ·9997 982 | ·9997 397 | ·9996 673 | ·9995 784 | ·9994 702 | ·9993 394 |
| ·91 | ·9999 308 | ·9999 098 | ·9998 836 | ·9998 511 | ·9998 110 | ·9997 621 |
| ·92 | ·9999 798 | ·9999 735− | ·9999 654 | ·9999 553 | ·9999 428 | ·9999 273 |
| ·93 | ·9999 952 | ·9999 937 | ·9999 917 | ·9999 891 | ·9999 859 | ·9999 819 |
| ·94 | ·9999 991 | ·9999 988 | ·9999 985− | ·9999 980 | ·9999 974 | ·9999 966 |
| ·95 | ·9999 999 | ·9999 999 | ·9999 998 | ·9999 997 | ·9999 997 | ·9999 996 |
| ·96 | 1·0000 000 | 1·0000 000 | 1·0000 000 | 1·0000 000 | 1·0000 000 | 1·0000 000 |

$x$ = ·38 to ·97 $q = 13$ $p$ = 37 to 43

| $x$ | $p = 37$ | $p = 38$ | $p = 39$ | $p = 40$ | $p = 41$ | $p = 42$ | $p = 43$ |
|---|---|---|---|---|---|---|---|
| $B(p, q)$ = | ·2929 3229 $\times\frac{1}{10^{12}}$ | ·2167 6989 $\times\frac{1}{10^{12}}$ | ·1615 1482 $\times\frac{1}{10^{12}}$ | ·1211 3612 $\times\frac{1}{10^{12}}$ | ·9142 3484 $\times\frac{1}{10^{13}}$ | ·6941 4127 $\times\frac{1}{10^{13}}$ | ·5300 7152 $\times\frac{1}{10^{13}}$ |
| ·38 | ·0000 001 | ·0000 001 | | | | | |
| ·39 | ·0000 002 | ·0000 001 | ·0000 001 | | | | |
| ·40 | ·0000 005− | ·0000 002 | ·0000 001 | ·0000 001 | | | |
| ·41 | ·0000 010 | ·0000 005+ | ·0000 003 | ·0000 001 | ·0000 001 | | |
| ·42 | ·0000 020 | ·0000 011 | ·0000 006 | ·0000 003 | ·0000 002 | ·0000 001 | |
| ·43 | ·0000 039 | ·0000 022 | ·0000 012 | ·0000 007 | ·0000 004 | ·0000 002 | ·0000 001 |
| ·44 | ·0000 074 | ·0000 043 | ·0000 024 | ·0000 014 | ·0000 008 | ·0000 004 | ·0000 002 |
| ·45 | ·0000 139 | ·0000 082 | ·0000 048 | ·0000 028 | ·0000 016 | ·0000 009 | ·0000 005+ |
| ·46 | ·0000 255− | ·0000 153 | ·0000 091 | ·0000 054 | ·0000 032 | ·0000 019 | ·0000 011 |
| ·47 | ·0000 458 | ·0000 280 | ·0000 171 | ·0000 104 | ·0000 062 | ·0000 037 | ·0000 022 |
| ·48 | ·0000 805− | ·0000 504 | ·0000 313 | ·0000 194 | ·0000 119 | ·0000 073 | ·0000 045− |
| ·49 | ·0001 388 | ·0000 886 | ·0000 563 | ·0000 355+ | ·0000 223 | ·0000 139 | ·0000 087 |
| ·50 | ·0002 349 | ·0001 529 | ·0000 990 | ·0000 638 | ·0000 409 | ·0000 260 | ·0000 165+ |
| ·51 | ·0003 900 | ·0002 589 | ·0001 709 | ·0001 122 | ·0000 733 | ·0000 476 | ·0000 308 |
| ·52 | ·0006 358 | ·0004 300 | ·0002 893 | ·0001 936 | ·0001 289 | ·0000 854 | ·0000 563 |
| ·53 | ·0010 180 | ·0007 014 | ·0004 806 | ·0003 276 | ·0002 222 | ·0001 499 | ·0001 007 |
| ·54 | ·0016 017 | ·0011 237 | ·0007 840 | ·0005 442 | ·0003 758 | ·0002 583 | ·0001 767 |
| ·55 | ·0024 770 | ·0017 687 | ·0012 561 | ·0008 875− | ·0006 239 | ·0004 365− | ·0003 039 |
| ·56 | ·0037 664 | ·0027 363 | ·0019 773 | ·0014 215− | ·0010 168 | ·0007 239 | ·0005 130 |
| ·57 | ·0056 323 | ·0041 617 | ·0030 587 | ·0022 366 | ·0016 274 | ·0011 786 | ·0008 496 |
| ·58 | ·0082 854 | ·0062 243 | ·0046 513 | ·0034 583 | ·0025 587 | ·0018 843 | ·0013 813 |
| ·59 | ·0119 925− | ·0091 562 | ·0069 544 | ·0052 556 | ·0039 526 | ·0029 588 | ·0022 050− |
| ·60 | ·0170 824 | ·0132 505+ | ·0102 254 | ·0078 518 | ·0060 004 | ·0045 645− | ·0034 568 |
| ·61 | ·0239 500− | ·0188 675+ | ·0147 882 | ·0115 341 | ·0089 536 | ·0069 188 | ·0053 230 |
| ·62 | ·0330 554 | ·0264 377 | ·0210 390 | ·0166 619 | ·0131 340 | ·0103 065− | ·0080 525+ |
| ·63 | ·0449 173 | ·0364 595+ | ·0294 485+ | ·0236 725+ | ·0189 420 | ·0150 895− | ·0119 690 |
| ·64 | ·0600 983 | ·0494 906 | ·0405 578 | ·0330 817 | ·0268 616 | ·0217 156 | ·0174 811 |
| ·65 | ·0791 823 | ·0661 294 | ·0549 656 | ·0454 764 | ·0374 581 | ·0307 207 | ·0250 902 |
| ·66 | ·1027 410 | ·0869 873 | ·0733 065− | ·0614 988 | ·0513 679 | ·0427 245− | ·0353 899 |
| ·67 | ·1312 916 | ·1126 501 | ·0962 162 | ·0818 176 | ·0692 764 | ·0584 145− | ·0490 577 |
| ·68 | ·1652 469 | ·1436 291 | ·1242 869 | ·1070 873 | ·0918 827 | ·0785 176 | ·0668 326 |
| ·69 | ·2048 601 | ·1803 046 | ·1580 107 | ·1378 948 | ·1198 509 | ·1037 565+ | ·0894 786 |
| ·70 | ·2501 703 | ·2228 658 | ·1977 169 | ·1746 959 | ·1537 469 | ·1347 906 | ·1177 299 |
| ·71 | ·3009 535− | ·2712 530 | ·2435 064 | ·2177 448 | ·1939 669 | ·1721 434 | ·1522 216 |
| ·72 | ·3566 870 | ·3251 088 | ·2951 926 | ·2670 253 | ·2406 607 | ·2161 224 | ·1934 069 |
| ·73 | ·4165 348 | ·3837 475− | ·3522 544 | ·3221 906 | ·2936 601 | ·2667 370 | ·2414 678 |
| ·74 | ·4793 583 | ·4461 491 | ·4138 139 | ·3825 233 | ·3524 218 | ·3236 271 | ·2962 310 |
| ·75 | ·5437 593 | ·5109 862 | ·4786 443 | ·4469 243 | ·4159 974 | ·3860 133 | ·3571 001 |
| ·76 | ·6081 530 | ·5766 842 | ·5452 154 | ·5139 403 | ·4830 404 | ·4526 832 | ·4230 199 |
| ·77 | ·6708 697 | ·6415 169 | ·6117 779 | ·5818 329 | ·5518 581 | ·5220 221 | ·4924 845+ |
| ·78 | ·7302 761 | ·7037 284 | ·6764 821 | ·6486 909 | ·6205 106 | ·5920 967 | ·5636 015+ |
| ·79 | ·7849 029 | ·7616 710 | ·7375 221 | ·7125 744 | ·6869 531 | ·6607 882 | ·6342 120 |
| ·80 | ·8335 653 | ·8139 430 | ·7932 879 | ·7716 795+ | ·7492 068 | ·7259 667 | ·7020 626 |
| ·81 | ·8754 582 | ·8595 077 | ·8425 079 | ·8245 011 | ·8055 400 | ·7856 863 | ·7650 101 |
| ·82 | ·9102 126 | ·8977 764 | ·8843 585− | ·8699 703 | ·8546 325− | ·8383 744 | ·8212 338 |
| ·83 | ·9379 044 | ·9286 413 | ·9185 250− | ·9075 449 | ·8956 974 | ·8829 860 | ·8694 210 |
| ·84 | ·9590 124 | ·9524 521 | ·9452 012 | ·9372 364 | ·9285 388 | ·9190 945+ | ·9088 947 |
| ·85 | ·9743 320 | ·9699 395− | ·9650 267 | ·9595 661 | ·9535 320 | ·9469 019 | ·9396 561 |
| ·86 | ·9848 587 | ·9820 972 | ·9789 724 | ·9754 582 | ·9715 294 | ·9671 616 | ·9623 321 |
| ·87 | ·9916 597 | ·9900 434 | ·9881 933 | ·9860 883 | ·9837 076 | ·9810 303 | ·9780 354 |
| ·88 | ·9957 560 | ·9948 845+ | ·9938 755− | ·9927 143 | ·9913 859 | ·9898 747 | ·9881 650− |
| ·89 | ·9980 318 | ·9976 047 | ·9971 045− | ·9965 223 | ·9958 488 | ·9950 738 | ·9941 871 |
| ·90 | ·9991 824 | ·9989 954 | ·9987 738 | ·9985 131 | ·9982 081 | ·9978 532 | ·9974 425+ |
| ·91 | ·9997 027 | ·9996 311 | ·9995 454 | ·9994 435− | ·9993 228 | ·9991 809 | ·9990 149 |
| ·92 | ·9999 082 | ·9998 850+ | ·9998 570 | ·9998 232 | ·9997 828 | ·9997 347 | ·9996 779 |
| ·93 | ·9999 770 | ·9999 709 | ·9999 634 | ·9999 543 | ·9999 434 | ·9999 302 | ·9999 144 |
| ·94 | ·9999 956 | ·9999 944 | ·9999 929 | ·9999 910 | ·9999 887 | ·9999 860 | ·9999 826 |
| ·95 | ·9999 994 | ·9999 992 | ·9999 990 | ·9999 988 | ·9999 985− | ·9999 981 | ·9999 976 |
| ·96 | 1·0000 000 | ·9999 999 | ·9999 999 | ·9999 999 | ·9999 999 | ·9999 998 | ·9999 998 |
| ·97 | | 1·0000 000 | 1·0000 000 | 1·0000 000 | 1·0000 000 | 1·0000 000 | 1·0000 000 |

$x$ = ·43 to ·97 $q$ = 13 $p$ = 44 to 50

| $x$ | $p$ = 44 | $p$ = 45 | $p$ = 46 | $p$ = 47 | $p$ = 48 | $p$ = 49 | $p$ = 50 |
|---|---|---|---|---|---|---|---|
| $B(p, q)$ = | ·4070 1920 × $\frac{1}{10^{13}}$ | ·3141 9026 × $\frac{1}{10^{13}}$ | ·2437 6830 × $\frac{1}{10^{13}}$ | ·1900 5664 × $\frac{1}{10^{13}}$ | ·1488 7770 × $\frac{1}{10^{13}}$ | ·1171 4967 × $\frac{1}{10^{13}}$ | ·9258 6029 × $\frac{1}{10^{14}}$ |
| ·43 | ·0000 001 | | | | | | |
| ·44 | ·0000 001 | ·0000 001 | | | | | |
| ·45 | ·0000 003 | ·0000 002 | ·0000 001 | ·0000 001 | | | |
| ·46 | ·0000 006 | ·0000 004 | ·0000 002 | ·0000 001 | ·0000 001 | | |
| ·47 | ·0000 013 | ·0000 008 | ·0000 005− | ·0000 003 | ·0000 002 | ·0000 001 | ·0000 001 |
| ·48 | ·0000 027 | ·0000 016 | ·0000 010 | ·0000 006 | ·0000 004 | ·0000 002 | ·0000 001 |
| ·49 | ·0000 054 | ·0000 033 | ·0000 020 | ·0000 012 | ·0000 008 | ·0000 005− | ·0000 003 |
| ·50 | ·0000 104 | ·0000 066 | ·0000 041 | ·0000 026 | ·0000 016 | ·0000 010 | ·0000 006 |
| ·51 | ·0000 198 | ·0000 127 | ·0000 081 | ·0000 052 | ·0000 033 | ·0000 021 | ·0000 013 |
| ·52 | ·0000 369 | ·0000 241 | ·0000 157 | ·0000 102 | ·0000 066 | ·0000 042 | ·0000 027 |
| ·53 | ·0000 673 | ·0000 448 | ·0000 297 | ·0000 196 | ·0000 129 | ·0000 085− | ·0000 055+ |
| ·54 | ·0001 203 | ·0000 816 | ·0000 551 | ·0000 370 | ·0000 248 | ·0000 166 | ·0000 110 |
| ·55 | ·0002 107 | ·0001 455− | ·0001 000 | ·0000 685− | ·0000 467 | ·0000 317 | ·0000 215− |
| ·56 | ·0003 619 | ·0002 542 | ·0001 779 | ·0001 239 | ·0000 860 | ·0000 595+ | ·0000 410 |
| ·57 | ·0006 098 | ·0004 358 | ·0003 102 | ·0002 199 | ·0001 553 | ·0001 093 | ·0000 766 |
| ·58 | ·0010 082 | ·0007 328 | ·0005 304 | ·0003 824 | ·0002 747 | ·0001 966 | ·0001 402 |
| ·59 | ·0016 361 | ·0012 089 | ·0008 897 | ·0006 522 | ·0004 763 | ·0003 466 | ·0002 513 |
| ·60 | ·0026 066 | ·0019 574 | ·0014 641 | ·0010 908 | ·0008 097 | ·0005 988 | ·0004 413 |
| ·61 | ·0040 778 | ·0031 111 | ·0023 642 | ·0017 897 | ·0013 498 | ·0010 144 | ·0007 597 |
| ·62 | ·0062 651 | ·0048 547 | ·0037 470 | ·0028 811 | ·0022 072 | ·0016 849 | ·0012 817 |
| ·63 | ·0094 544 | ·0074 382 | ·0058 292 | ·0045 512 | ·0035 404 | ·0027 444 | ·0021 202 |
| ·64 | ·0140 148 | ·0111 914 | ·0089 025− | ·0070 554 | ·0055 715+ | ·0043 844 | ·0034 385+ |
| ·65 | ·0204 090 | ·0165 364 | ·0133 479 | ·0107 348 | ·0086 026 | ·0068 701 | ·0054 683 |
| ·66 | ·0291 982 | ·0239 971 | ·0196 490 | ·0160 307 | ·0130 328 | ·0105 596 | ·0085 275+ |
| ·67 | ·0410 392 | ·0342 015+ | ·0283 987 | ·0234 965+ | ·0193 735− | ·0159 205− | ·0130 403 |
| ·68 | ·0566 695− | ·0478 737 | ·0402 976 | ·0338 019 | ·0282 571 | ·0235 440 | ·0195 542 |
| ·69 | ·0768 774 | ·0658 113 | ·0561 396 | ·0477 251 | ·0404 368 | ·0341 506 | ·0287 509 |
| ·70 | ·1024 546 | ·0888 457 | ·0767 792 | ·0661 291 | ·0567 707 | ·0485 821 | ·0414 462 |
| ·71 | ·1341 299 | ·1177 816 | ·1030 790 | ·0899 169 | ·0781 859 | ·0677 746 | ·0585 724 |
| ·72 | ·1724 875+ | ·1533 178 | ·1358 350− | ·1199 636 | ·1056 186 | ·0927 081 | ·0811 360 |
| ·73 | ·2178 734 | ·1959 521 | ·1756 821 | ·1570 250− | ·1399 279 | ·1243 268 | ·1101 489 |
| ·74 | ·2703 003 | ·2458 783 | ·2229 867 | ·2016 278 | ·1817 868 | ·1634 339 | ·1465 268 |
| ·75 | ·3293 637 | ·3028 879 | ·2777 360 | ·2539 510 | ·2315 576 | ·2105 639 | ·1909 626 |
| ·76 | ·3941 844 | ·3662 926 | ·3394 421 | ·3137 123 | ·2891 649 | ·2658 450− | ·2437 815− |
| ·77 | ·4633 934 | ·4348 842 | ·4070 781 | ·3800 821 | ·3539 878 | ·3288 721 | ·3047 968 |
| ·78 | ·5351 726 | ·5069 505− | ·4790 670 | ·4516 442 | ·4247 933 | ·3986 135+ | ·3731 925+ |
| ·79 | ·6073 578 | ·5803 571 | ·5533 384 | ·5264 254 | ·4997 357 | ·4733 797 | ·4474 593 |
| ·80 | ·6776 026 | ·6526 979 | ·6274 611 | ·6020 049 | ·5764 403 | ·5508 757 | ·5254 155+ |
| ·81 | ·7435 886 | ·7215 050+ | ·6988 473 | ·6757 068 | ·6521 773 | ·6283 537 | ·6043 308 |
| ·82 | ·8032 560 | ·7844 938 | ·7650 061 | ·7448 575− | ·7241 173 | ·7028 586 | ·6811 574 |
| ·83 | ·8550 200 | ·8398 074 | ·8238 139 | ·8070 763 | ·7896 371 | ·7715 440 | ·7528 490 |
| ·84 | ·8979 358 | ·8862 198 | ·8737 539 | ·8605 510 | ·8466 288 | ·8320 106 | ·8167 241 |
| ·85 | ·9317 785− | ·9232 563 | ·9140 807 | ·9042 469 | ·8937 540 | ·8826 053 | ·8708 082 |
| ·86 | ·9570 196 | ·9512 049 | ·9448 707 | ·9380 023 | ·9305 873 | ·9226 162 | ·9140 822 |
| ·87 | ·9747 028 | ·9710 126 | ·9669 461 | ·9624 853 | ·9576 135− | ·9523 154 | ·9465 773 |
| ·88 | ·9862 405+ | ·9840 851 | ·9816 826 | ·9790 168 | ·9760 720 | ·9728 327 | ·9692 840 |
| ·89 | ·9931 777 | ·9920 343 | ·9907 453 | ·9892 988 | ·9876 828 | ·9858 849 | ·9838 930 |
| ·90 | ·9969 698 | ·9964 283 | ·9958 110 | ·9951 104 | ·9943 190 | ·9934 286 | ·9924 311 |
| ·91 | ·9988 216 | ·9985 978 | ·9983 398 | ·9980 438 | ·9977 056 | ·9973 210 | ·9968 852 |
| ·92 | ·9996 110 | ·9995 327 | ·9994 415− | ·9993 357 | ·9992 134 | ·9990 728 | ·9989 118 |
| ·93 | ·9998 956 | ·9998 734 | ·9998 473 | ·9998 166 | ·9997 808 | ·9997 392 | ·9996 910 |
| ·94 | ·9999 786 | ·9999 738 | ·9999 681 | ·9999 614 | ·9999 534 | ·9999 440 | ·9999 331 |
| ·95 | ·9999 970 | ·9999 963 | ·9999 954 | ·9999 944 | ·9999 931 | ·9999 917 | ·9999 900 |
| ·96 | ·9999 997 | ·9999 997 | ·9999 996 | ·9999 995+ | ·9999 994 | ·9999 993 | ·9999 991 |
| ·97 | 1·0000 000 | 1·0000 000 | 1·0000 000 | 1·0000 000 | 1·0000 000 | 1·0000 000 | 1·0000 000 |

$x$ = ·10 to ·70 $q = 14$ $p$ = 14 to 19

| $x$ | $p = 14$ | $p = 15$ | $p = 16$ | $p = 17$ | $p = 18$ | $p = 19$ |
|---|---|---|---|---|---|---|
| $B(p, q) =$ | ·3561 0481 × $\frac{1}{10^8}$ | ·1780 5241 × $\frac{1}{10^8}$ | ·9209 6072 × $\frac{1}{10^9}$ | ·4911 7905+ × $\frac{1}{10^9}$ | ·2693 5625+ × $\frac{1}{10^9}$ | ·1515 1289 × $\frac{1}{10^9}$ |
| ·10 | ·0000 001 | | | | | |
| ·11 | ·0000 002 | | | | | |
| ·12 | ·0000 006 | ·0000 001 | | | | |
| ·13 | ·0000 015− | ·0000 004 | ·0000 001 | | | |
| ·14 | ·0000 036 | ·0000 009 | ·0000 002 | ·0000 001 | | |
| ·15 | ·0000 083 | ·0000 023 | ·0000 006 | ·0000 002 | | |
| ·16 | ·0000 179 | ·0000 053 | ·0000 015+ | ·0000 004 | ·0000 001 | |
| ·17 | ·0000 362 | ·0000 114 | ·0000 035− | ·0000 010 | ·0000 003 | ·0000 001 |
| ·18 | ·0000 699 | ·0000 232 | ·0000 075− | ·0000 023 | ·0000 007 | ·0000 002 |
| ·19 | ·0001 289 | ·0000 451 | ·0000 153 | ·0000 051 | ·0000 016 | ·0000 005+ |
| ·20 | ·0002 285+ | ·0000 840 | ·0000 300 | ·0000 105− | ·0000 036 | ·0000 012 |
| ·21 | ·0003 905− | ·0001 505+ | ·0000 565− | ·0000 207 | ·0000 074 | ·0000 026 |
| ·22 | ·0006 454 | ·0002 603 | ·0001 022 | ·0000 391 | ·0000 146 | ·0000 054 |
| ·23 | ·0010 346 | ·0004 357 | ·0001 786 | ·0000 714 | ·0000 279 | ·0000 107 |
| ·24 | ·0016 129 | ·0007 079 | ·0003 024 | ·0001 261 | ·0000 514 | ·0000 205+ |
| ·25 | ·0024 500− | ·0011 186 | ·0004 972 | ·0002 157 | ·0000 915− | ·0000 380 |
| ·26 | ·0036 333 | ·0017 227 | ·0007 954 | ·0003 585− | ·0001 580 | ·0000 682 |
| ·27 | ·0052 692 | ·0025 906 | ·0012 406 | ·0005 799 | ·0002 651 | ·0001 188 |
| ·28 | ·0074 840 | ·0038 098 | ·0018 895− | ·0009 149 | ·0004 333 | ·0002 011 |
| ·29 | ·0104 244 | ·0054 872 | ·0028 145+ | ·0014 097 | ·0006 907 | ·0003 317 |
| ·30 | ·0142 565+ | ·0077 498 | ·0041 060 | ·0021 247 | ·0010 758 | ·0005 338 |
| ·31 | ·0191 640 | ·0107 453 | ·0058 737 | ·0031 364 | ·0016 390 | ·0008 395+ |
| ·32 | ·0253 448 | ·0146 415− | ·0082 480 | ·0045 398 | ·0024 458 | ·0012 918 |
| ·33 | ·0330 071 | ·0196 246 | ·0113 810 | ·0064 503 | ·0035 789 | ·0019 470 |
| ·34 | ·0423 632 | ·0258 962 | ·0154 452 | ·0090 047 | ·0051 404 | ·0028 777 |
| ·35 | ·0536 230 | ·0336 688 | ·0206 321 | ·0123 619 | ·0072 539 | ·0041 748 |
| ·36 | ·0669 863 | ·0431 604 | ·0271 494 | ·0167 023 | ·0100 652 | ·0059 503 |
| ·37 | ·0826 346 | ·0545 876 | ·0352 165+ | ·0222 258 | ·0137 436 | ·0083 385+ |
| ·38 | ·1007 226 | ·0681 578 | ·0450 585− | ·0291 489 | ·0184 800 | ·0114 979 |
| ·39 | ·1213 695− | ·0840 603 | ·0568 991 | ·0376 996 | ·0244 858 | ·0156 106 |
| ·40 | ·1446 518 | ·1024 577 | ·0709 528 | ·0481 117 | ·0319 886 | ·0208 816 |
| ·41 | ·1705 958 | ·1234 768 | ·0874 151 | ·0606 167 | ·0412 272 | ·0275 361 |
| ·42 | ·1991 729 | ·1472 002 | ·1064 535− | ·0754 351 | ·0524 450− | ·0358 154 |
| ·43 | ·2302 954 | ·1736 584 | ·1281 977 | ·0927 668 | ·0658 810 | ·0459 706 |
| ·44 | ·2638 151 | ·2028 244 | ·1527 306 | ·1127 809 | ·0817 611 | ·0582 549 |
| ·45 | ·2995 240 | ·2346 087 | ·1800 799 | ·1356 048 | ·1002 864 | ·0729 146 |
| ·46 | ·3371 573 | ·2688 580 | ·2102 116 | ·1613 152 | ·1216 229 | ·0901 777 |
| ·47 | ·3763 986 | ·3053 548 | ·2430 256 | ·1899 290 | ·1458 901 | ·1102 430 |
| ·48 | ·4168 872 | ·3438 207 | ·2783 531 | ·2213 963 | ·1731 506 | ·1332 674 |
| ·49 | ·4582 276 | ·3839 219 | ·3159 570 | ·2555 957 | ·2034 010 | ·1593 544 |
| ·50 | ·5000 000$^e$ | ·4252 770 | ·3555 356 | ·2923 324 | ·2365 648 | ·1885 428 |
| ·51 | ·5417 724 | ·4674 668 | ·3967 279 | ·3313 385+ | ·2724 881 | ·2207 979 |
| ·52 | ·5831 128 | ·5100 463 | ·4391 231 | ·3722 780 | ·3109 378 | ·2560 042 |
| ·53 | ·6236 014 | ·5525 576 | ·4822 716 | ·4147 531 | ·3516 034 | ·2939 618 |
| ·54 | ·6628 427 | ·5945 434 | ·5256 976 | ·4583 149 | ·3941 031 | ·3343 861 |
| ·55 | ·7004 760 | ·6355 607 | ·5689 143 | ·5024 763 | ·4379 922 | ·3769 115+ |
| ·56 | ·7361 849 | ·6751 941 | ·6114 385− | ·5467 265− | ·4827 758 | ·4210 989 |
| ·57 | ·7697 046 | ·7130 675+ | ·6528 057 | ·5905 478 | ·5279 236 | ·4664 475− |
| ·58 | ·8008 271 | ·7488 543 | ·6925 850+ | ·6334 320 | ·5728 872 | ·5124 096 |
| ·59 | ·8294 042 | ·7822 851 | ·7303 914 | ·6748 975+ | ·6171 186 | ·5584 088 |
| ·60 | ·8553 482 | ·8131 542 | ·7658 968 | ·7145 044 | ·6600 889 | ·6038 596 |
| ·61 | ·8786 305+ | ·8413 213 | ·7988 385− | ·7518 685− | ·7013 066 | ·6481 886 |
| ·62 | ·8992 774 | ·8667 127 | ·8290 244 | ·7866 721 | ·7403 338 | ·6908 548 |
| ·63 | ·9173 654 | ·8893 184 | ·8563 352 | ·8186 725− | ·7768 004 | ·7313 691 |
| ·64 | ·9330 137 | ·9091 878 | ·8807 239 | ·8477 057 | ·8104 145+ | ·7693 114 |
| ·65 | ·9463 770 | ·9264 229 | ·9022 118 | ·8736 881 | ·8409 698 | ·8043 435− |
| ·66 | ·9576 368 | ·9411 698 | ·9208 825+ | ·8966 138 | ·8683 479 | ·8362 190 |
| ·67 | ·9669 929 | ·9536 104 | ·9368 734 | ·9165 484 | ·8925 170 | ·8647 875+ |
| ·68 | ·9746 552 | ·9639 519 | ·9503 658 | ·9336 209 | ·9135 271 | ·8899 950+ |
| ·69 | ·9808 360 | ·9724 173 | ·9615 740 | ·9480 132 | ·9315 008 | ·9118 786 |
| ·70 | ·9857 435− | ·9792 367 | ·9707 346 | ·9599 475− | ·9466 222 | ·9305 579 |

$x$ = ·71 to ·93 $q$ = 14 $p$ = 14 to 19

| | $p$ = 14 | $p$ = 15 | $p$ = 16 | $p$ = 17 | $p$ = 18 | $p$ = 19 |
|---|---|---|---|---|---|---|
| $B(p, q)$ = | ·3561 0481 × $\frac{1}{10^8}$ | ·1780 5241 × $\frac{1}{10^8}$ | ·9209 6072 × $\frac{1}{10^9}$ | ·4911 7905$^+$ × $\frac{1}{10^9}$ | ·2693 5625$^+$ × $\frac{1}{10^9}$ | ·1515 1289 × $\frac{1}{10^9}$ |
| $x$ | | | | | | |
| ·71 | ·9895 756 | ·9846 383 | ·9780 949 | ·9696 742 | ·9591 237 | ·9462 227 |
| ·72 | ·9925 160 | ·9888 418 | ·9839 037 | ·9774 595$^-$ | ·9692 715$^-$ | ·9591 184 |
| ·73 | ·9947 308 | ·9920 522 | ·9884 022 | ·9835 727 | ·9773 513 | ·9695 295$^+$ |
| ·74 | ·9963 667 | ·9944 561 | ·9918 170 | ·9882 772 | ·9836 547 | ·9777 636 |
| ·75 | ·9975 500$^+$ | ·9962 186 | ·9943 546 | ·9918 208 | ·9884 672 | ·9841 354 |
| ·76 | ·9983 871 | ·9974 822 | ·9961 983 | ·9944 297 | ·9920 578 | ·9889 532 |
| ·77 | ·9989 654 | ·9983 665$^-$ | ·9975 056 | ·9963 043 | ·9946 718 | ·9925 070 |
| ·78 | ·9993 546 | ·9989 696 | ·9984 089 | ·9976 163 | ·9965 252 | ·9950 596 |
| ·79 | ·9996 095$^+$ | ·9993 696 | ·9990 157 | ·9985 090 | ·9978 026 | ·9968 415$^+$ |
| ·80 | ·9997 715$^-$ | ·9996 269 | ·9994 111 | ·9990 981 | ·9986 563 | ·9980 475$^+$ |
| ·81 | ·9998 711 | ·9997 872 | ·9996 604 | ·9994 742 | ·9992 080 | ·9988 368 |
| ·82 | ·9999 301 | ·9998 834 | ·9998 119 | ·9997 056 | ·9995 518 | ·9993 346 |
| ·83 | ·9999 638 | ·9999 389 | ·9999 004 | ·9998 424 | ·9997 575$^+$ | ·9996 361 |
| ·84 | ·9999 821 | ·9999 696 | ·9999 498 | ·9999 198 | ·9998 752 | ·9998 108 |
| ·85 | ·9999 917 | ·9999 857 | ·9999 761 | ·9999 614 | ·9999 393 | ·9999 070 |
| ·86 | ·9999 964 | ·9999 937 | ·9999 893 | ·9999 826 | ·9999 723 | ·9999 572 |
| ·87 | ·9999 985$^+$ | ·9999 974 | ·9999 956 | ·9999 927 | ·9999 883 | ·9999 817 |
| ·88 | ·9999 994 | ·9999 990 | ·9999 983 | ·9999 972 | ·9999 954 | ·9999 928 |
| ·89 | ·9999 998 | ·9999 997 | ·9999 994 | ·9999 990 | ·9999 984 | ·9999 974 |
| ·90 | ·9999 999 | ·9999 999 | ·9999 998 | ·9999 997 | ·9999 995$^-$ | ·9999 992 |
| ·91 | 1·0000 000 | 1·0000 000 | 1·0000 000 | ·9999 999 | ·9999 999 | ·9999 998 |
| ·92 | | | | 1·0000 000 | 1·0000 000 | ·9999 999 |
| ·93 | | | | | | 1·0000 000 |

$x$ = ·18 to ·80 $q = 14$ $p$ = 20 to 25

| $x$ | $p = 20$ | $p = 21$ | $p = 22$ | $p = 23$ | $p = 24$ | $p = 25$ |
|---|---|---|---|---|---|---|
| $B(p, q) =$ | ·8723 4696 $\times \frac{1}{10^{10}}$ | ·5131 4527 $\times \frac{1}{10^{10}}$ | ·3078 8716 $\times \frac{1}{10^{10}}$ | ·1881 5327 $\times \frac{1}{10^{10}}$ | ·1169 6014 $\times \frac{1}{10^{10}}$ | ·7386 9561 $\times \frac{1}{10^{11}}$ |
| ·18 | ·0000 001 | | | | | |
| ·19 | ·0000 002 | | | | | |
| ·20 | ·0000 004 | ·0000 001 | | | | |
| ·21 | ·0000 009 | ·0000 003 | ·0000 001 | | | |
| ·22 | ·0000 019 | ·0000 007 | ·0000 002 | ·0000 001 | | |
| ·23 | ·0000 040 | ·0000 015− | ·0000 005+ | ·0000 002 | ·0000 001 | |
| ·24 | ·0000 080 | ·0000 031 | ·0000 012 | ·0000 004 | ·0000 002 | ·0000 001 |
| ·25 | ·0000 155− | ·0000 062 | ·0000 024 | ·0000 009 | ·0000 004 | ·0000 001 |
| ·26 | ·0000 289 | ·0000 120 | ·0000 049 | ·0000 020 | ·0000 008 | ·0000 003 |
| ·27 | ·0000 522 | ·0000 225+ | ·0000 096 | ·0000 040 | ·0000 017 | ·0000 007 |
| ·28 | ·0000 916 | ·0000 410 | ·0000 180 | ·0000 078 | ·0000 033 | ·0000 014 |
| ·29 | ·0001 563 | ·0000 724 | ·0000 330 | ·0000 148 | ·0000 065+ | ·0000 029 |
| ·30 | ·0002 600 | ·0001 244 | ·0000 586 | ·0000 272 | ·0000 124 | ·0000 056 |
| ·31 | ·0004 221 | ·0002 086 | ·0001 014 | ·0000 486 | ·0000 230 | ·0000 107 |
| ·32 | ·0006 698 | ·0003 414 | ·0001 712 | ·0000 846 | ·0000 412 | ·0000 198 |
| ·33 | ·0010 400 | ·0005 461 | ·0002 823 | ·0001 437 | ·0000 722 | ·0000 358 |
| ·34 | ·0015 820 | ·0008 551 | ·0004 549 | ·0002 385− | ·0001 233 | ·0000 629 |
| ·35 | ·0023 598 | ·0013 117 | ·0007 177 | ·0003 870 | ·0002 058 | ·0001 080 |
| ·36 | ·0034 553 | ·0019 733 | ·0011 095+ | ·0006 148 | ·0003 360 | ·0001 813 |
| ·37 | ·0049 703 | ·0029 141 | ·0016 823 | ·0009 572 | ·0005 372 | ·0002 977 |
| ·38 | ·0070 293 | ·0042 275+ | ·0025 038 | ·0014 617 | ·0008 418 | ·0004 787 |
| ·39 | ·0097 809 | ·0060 296 | ·0036 608 | ·0021 911 | ·0012 940 | ·0007 546 |
| ·40 | ·0133 990 | ·0084 604 | ·0052 621 | ·0032 269 | ·0019 526 | ·0011 668 |
| ·41 | ·0180 821 | ·0116 864 | ·0074 409 | ·0046 717 | ·0028 946 | ·0017 713 |
| ·42 | ·0240 523 | ·0159 004 | ·0103 571 | ·0066 531 | ·0042 182 | ·0026 416 |
| ·43 | ·0315 513 | ·0213 207 | ·0141 983 | ·0093 260 | ·0060 467 | ·0038 728 |
| ·44 | ·0408 357 | ·0281 893 | ·0191 802 | ·0128 739 | ·0085 307 | ·0055 846 |
| ·45 | ·0521 696 | ·0367 665− | ·0255 442 | ·0175 101 | ·0118 513 | ·0079 255− |
| ·46 | ·0658 160 | ·0473 254 | ·0335 543 | ·0234 764 | ·0162 203 | ·0110 745+ |
| ·47 | ·0820 255+ | ·0601 428 | ·0434 912 | ·0310 403 | ·0218 807 | ·0152 439 |
| ·48 | ·1010 250+ | ·0754 891 | ·0556 440 | ·0404 895+ | ·0291 039 | ·0206 786 |
| ·49 | ·1230 043 | ·0936 153 | ·0703 001 | ·0521 247 | ·0381 851 | ·0276 548 |
| ·50 | ·1481 032 | ·1147 405+ | ·0877 326 | ·0662 491 | ·0494 359 | ·0364 757 |
| ·51 | ·1763 987 | ·1390 367 | ·1081 864 | ·0831 556 | ·0631 745+ | ·0474 644 |
| ·52 | ·2078 939 | ·1666 153 | ·1318 627 | ·1031 128 | ·0797 128 | ·0609 539 |
| ·53 | ·2425 091 | ·1975 137 | ·1589 033 | ·1263 478 | ·0993 409 | ·0772 739 |
| ·54 | ·2800 751 | ·2316 840 | ·1893 764 | ·1530 302 | ·1223 099 | ·0967 351 |
| ·55 | ·3203 315+ | ·2689 851 | ·2232 624 | ·1832 550+ | ·1488 139 | ·1196 107 |
| ·56 | ·3629 279 | ·3091 779 | ·2604 445− | ·2170 275− | ·1789 715+ | ·1461 166 |
| ·57 | ·4074 304 | ·3519 249 | ·3007 012 | ·2542 507 | ·2128 087 | ·1763 916 |
| ·58 | ·4533 326 | ·3967 958 | ·3437 051 | ·2947 169 | ·2502 441 | ·2104 781 |
| ·59 | ·5000 698 | ·4432 767 | ·3890 259 | ·3381 041 | ·2910 789 | ·2483 055+ |
| ·60 | ·5470 384 | ·4907 854 | ·4361 396 | ·3839 778 | ·3349 910 | ·2896 782 |
| ·61 | ·5936 168 | ·5386 903 | ·4844 438 | ·4318 001 | ·3815 368 | ·3342 683 |
| ·62 | ·6391 883 | ·5863 334 | ·5332 772 | ·4809 445− | ·4301 590 | ·3816 165+ |
| ·63 | ·6831 642 | ·6330 552 | ·5819 440 | ·5307 166 | ·4802 020 | ·4311 397 |
| ·64 | ·7250 065+ | ·6782 206 | ·6297 415− | ·5803 809 | ·5309 345− | ·4821 473 |
| ·65 | ·7642 473 | ·7212 441 | ·6759 884 | ·6291 899 | ·5815 774 | ·5338 658 |
| ·66 | ·8005 051 | ·7616 128 | ·7200 535+ | ·6764 163 | ·6313 371 | ·5854 691 |
| ·67 | ·8334 970 | ·7989 053 | ·7613 815+ | ·7213 846 | ·6794 400 | ·6361 147 |
| ·68 | ·8630 446 | ·8328 062 | ·7995 152 | ·7635 003 | ·7251 680 | ·6849 829 |
| ·69 | ·8890 755+ | ·8631 142 | ·8341 117 | ·8022 749 | ·7678 912 | ·7313 155− |
| ·70 | ·9116 189 | ·8897 443 | ·8649 532 | ·8373 448 | ·8070 957 | ·7744 518 |
| ·71 | ·9307 959 | ·9127 233 | ·8919 485− | ·8684 823 | ·8424 043 | ·8138 598 |
| ·72 | ·9468 065− | ·9321 799 | ·9151 294 | ·8955 990 | ·8735 889 | ·8491 578 |
| ·73 | ·9599 129 | ·9483 296 | ·9346 393 | ·9187 399 | ·9005 731 | ·8801 278 |
| ·74 | ·9704 214 | ·9614 566 | ·9507 158 | ·9380 711 | ·9234 252 | ·9067 166 |
| ·75 | ·9786 638 | ·9718 926 | ·9636 704 | ·9538 598 | ·9423 431 | ·9290 268 |
| ·76 | ·9849 794 | ·9799 962 | ·9738 644 | ·9664 506 | ·9576 314 | ·9472 982 |
| ·77 | ·9896 995+ | ·9861 327 | ·9816 860 | ·9762 388 | ·9696 738 | ·9618 806 |
| ·78 | ·9931 343 | ·9906 563 | ·9875 270 | ·9836 439 | ·9789 031 | ·9732 022 |
| ·79 | ·9955 628 | ·9938 960 | ·9917 640 | ·9890 845+ | ·9857 713 | ·9817 360 |
| ·80 | ·9972 273 | ·9961 447 | ·9947 423 | ·9929 575+ | ·9907 227 | ·9879 664 |

$x$ = ·81 to ·94 $q$ = 14 $p$ = 20 to 25

| | $p$ = 20 | $p$ = 21 | $p$ = 22 | $p$ = 23 | $p$ = 24 | $p$ = 25 |
|---|---|---|---|---|---|---|
| $B(p, q)$ = | ·8723 4696 × $\frac{1}{10^{10}}$ | ·5131 4527 × $\frac{1}{10^{10}}$ | ·3078 8716 × $\frac{1}{10^{10}}$ | ·1881 5327 × $\frac{1}{10^{10}}$ | ·1169 6014 × $\frac{1}{10^{10}}$ | ·7386 9561 × $\frac{1}{10^{11}}$ |
| $x$ | | | | | | |
| ·81 | ·9983 303 | ·9976 534 | ·9967 657 | ·9956 217 | ·9941 714 | ·9923 603 |
| ·82 | ·9990 346 | ·9986 288 | ·9980 899 | ·9973 870 | ·9964 848 | ·9953 443 |
| ·83 | ·9994 665⁻ | ·9992 341 | ·9989 219 | ·9985 096 | ·9979 740 | ·9972 887 |
| ·84 | ·9997 196 | ·9995 933 | ·9994 215⁻ | ·9991 919 | ·9988 900 | ·9984 990 |
| ·85 | ·9998 608 | ·9997 959 | ·9997 067 | ·9995 860 | ·9994 254 | ·9992 150⁺ |
| ·86 | ·9999 352 | ·9999 040 | ·9998 606 | ·9998 011 | ·9997 211 | ·9996 151 |
| ·87 | ·9999 719 | ·9999 580 | ·9999 384 | ·9999 113 | ·9998 743 | ·9998 247 |
| ·88 | ·9999 888 | ·9999 831 | ·9999 750⁺ | ·9999 636 | ·9999 479 | ·9999 267 |
| ·89 | ·9999 960 | ·9999 939 | ·9999 908 | ·9999 865⁺ | ·9999 805⁻ | ·9999 722 |
| ·90 | ·9999 987 | ·9999 980 | ·9999 970 | ·9999 956 | ·9999 935⁺ | ·9999 907 |
| ·91 | ·9999 996 | ·9999 994 | ·9999 992 | ·9999 987 | ·9999 981 | ·9999 973 |
| ·92 | ·9999 999 | ·9999 999 | ·9999 998 | ·9999 997 | ·9999 995⁺ | ·9999 993 |
| ·93 | 1·0000 000 | 1·0000 000 | 1·0000 000 | ·9999 999 | ·9999 999 | ·9999 999 |
| ·94 | | | | 1·0000000 | 1·0000 000 | 1·0000 000 |

$x = \cdot25$ to $\cdot95$ $q = 14$ $p = 26$ to $31$

| | $p = 26$ | $p = 27$ | $p = 28$ | $p = 29$ | $p = 30$ | $p = 31$ |
|---|---|---|---|---|---|---|
| $B(p, q) =$ | ·4735 2283 $\times \frac{1}{10^{11}}$ | ·3077 8984 $\times \frac{1}{10^{11}}$ | ·2026 9087 $\times \frac{1}{10^{11}}$ | ·1351 2725 $\times \frac{1}{10^{11}}$ | ·9113 2328 $\times \frac{1}{10^{12}}$ | ·6213 5678 $\times \frac{1}{10^{12}}$ |
| $x$ | | | | | | |
| ·25 | ·0000 001 | | | | | |
| ·26 | ·0000 001 | | | | | |
| ·27 | ·0000 003 | ·0000 001 | | | | |
| ·28 | ·0000 006 | ·0000 002 | ·0000 001 | | | |
| ·29 | ·0000 012 | ·0000 005+ | ·0000 002 | ·0000 001 | | |
| ·30 | ·0000 025+ | ·0000 011 | ·0000 005− | ·0000 002 | ·0000 001 | |
| ·31 | ·0000 049 | ·0000 022 | ·0000 010 | ·0000 004 | ·0000 002 | ·0000 001 |
| ·32 | ·0000 094 | ·0000 044 | ·0000 021 | ·0000 009 | ·0000 004 | ·0000 002 |
| ·33 | ·0000 175+ | ·0000 085− | ·0000 041 | ·0000 019 | ·0000 009 | ·0000 004 |
| ·34 | ·0000 317 | ·0000 158 | ·0000 078 | ·0000 038 | ·0000 018 | ·0000 009 |
| ·35 | ·0000 560 | ·0000 287 | ·0000 146 | ·0000 073 | ·0000 036 | ·0000 018 |
| ·36 | ·0000 967 | ·0000 509 | ·0000 266 | ·0000 137 | ·0000 070 | ·0000 035+ |
| ·37 | ·0001 630 | ·0000 882 | ·0000 473 | ·0000 251 | ·0000 132 | ·0000 068 |
| ·38 | ·0002 690 | ·0001 495− | ·0000 822 | ·0000 447 | ·0000 241 | ·0000 129 |
| ·39 | ·0004 348 | ·0002 478 | ·0001 397 | ·0000 780 | ·0000 431 | ·0000 236 |
| ·40 | ·0006 890 | ·0004 024 | ·0002 325+ | ·0001 330 | ·0000 754 | ·0000 423 |
| ·41 | ·0010 712 | ·0006 407 | ·0003 792 | ·0002 222 | ·0001 290 | ·0000 742 |
| ·42 | ·0016 351 | ·0010 010 | ·0006 064 | ·0003 638 | ·0002 162 | ·0001 273 |
| ·43 | ·0024 519 | ·0015 354 | ·0009 516 | ·0005 840 | ·0003 551 | ·0002 140 |
| ·44 | ·0036 143 | ·0023 138 | ·0014 662 | ·0009 200 | ·0005 720 | ·0003 525− |
| ·45 | ·0052 402 | ·0034 277 | ·0022 193 | ·0014 231 | ·0009 042 | ·0005 695− |
| ·46 | ·0074 766 | ·0049 940 | ·0033 022 | ·0021 626 | ·0014 035− | ·0009 029 |
| ·47 | ·0105 025+ | ·0071 599 | ·0048 324 | ·0032 306 | ·0021 403 | ·0014 057 |
| ·48 | ·0145 315− | ·0101 055+ | ·0069 582 | ·0047 461 | ·0032 083 | ·0021 503 |
| ·49 | ·0198 119 | ·0140 473 | ·0098 626 | ·0068 602 | ·0047 294 | ·0032 330 |
| ·50 | ·0266 260 | ·0192 387 | ·0137 666 | ·0097 602 | ·0068 591 | ·0047 799 |
| ·51 | ·0352 859 | ·0259 693 | ·0189 302 | ·0136 734 | ·0097 906 | ·0069 524 |
| ·52 | ·0461 268 | ·0345 617 | ·0256 523 | ·0188 684 | ·0137 594 | ·0099 515+ |
| ·53 | ·0594 968 | ·0453 640 | ·0342 672 | ·0256 552 | ·0190 448 | ·0140 230 |
| ·54 | ·0757 434 | ·0587 401 | ·0451 375− | ·0343 817 | ·0259 699 | ·0194 592 |
| ·55 | ·0951 968 | ·0750 553 | ·0586 438 | ·0454 266 | ·0348 985− | ·0265 988 |
| ·56 | ·1181 504 | ·0946 589 | ·0751 696 | ·0591 884 | ·0462 270 | ·0358 234 |
| ·57 | ·1448 398 | ·1178 630 | ·0950 826 | ·0760 691 | ·0603 731 | ·0475 495− |
| ·58 | ·1754 203 | ·1449 200 | ·1187 124 | ·0964 546 | ·0777 581 | ·0622 150+ |
| ·59 | ·2099 464 | ·1759 986 | ·1463 256 | ·1206 904 | ·0987 855− | ·0802 612 |
| ·60 | ·2483 529 | ·2111 602 | ·1781 000 | ·1490 542 | ·1238 145− | ·1021 083 |
| ·61 | ·2904 410 | ·2503 390 | ·2140 987 | ·1817 284 | ·1531 308 | ·1281 270 |
| ·62 | ·3358 701 | ·2933 259 | ·2542 483 | ·2187 715− | ·1869 157 | ·1586 065− |
| ·63 | ·3841 576 | ·3397 595− | ·2983 213 | ·2600 945+ | ·2252 159 | ·1937 206 |
| ·64 | ·4346 872 | ·3891 255+ | ·3459 262 | ·3054 424 | ·2679 180 | ·2334 957 |
| ·65 | ·4867 268 | ·4407 662 | ·3965 078 | ·3543 834 | ·3147 283 | ·2777 829 |
| ·66 | ·5394 543 | ·4938 996 | ·4493 573 | ·4063 103 | ·3651 633 | ·3262 382 |
| ·67 | ·5919 923 | ·5476 492 | ·5036 346 | ·4604 531 | ·4185 522 | ·3783 134 |
| ·68 | ·6434 477 | ·6010 817 | ·5584 019 | ·5159 050+ | ·4740 529 | ·4332 611 |
| ·69 | ·6929 549 | ·6532 517 | ·6126 661 | ·5716 603 | ·5306 827 | ·4901 558 |
| ·70 | ·7397 188 | ·7032 491 | ·6654 286 | ·6266 627 | ·5873 620 | ·5470 304 |
| ·71 | ·7830 545− | ·7502 469 | ·7157 381 | ·6798 613 | ·6429 701 | ·6054 271 |
| ·72 | ·8224 204 | ·7935 440 | ·7627 425+ | ·7302 689 | ·6964 068 | ·6614 612 |
| ·73 | ·8574 418 | ·8326 006 | ·8057 353 | ·7770 182 | ·7466 573 | ·7148 896 |
| ·74 | ·8879 228 | ·8670 617 | ·8441 918 | ·8194 105+ | ·7928 519 | ·7646 820 |
| ·75 | ·9138 463 | ·8967 683 | ·8777 926 | ·8569 533 | ·8343 175+ | ·8099 841 |
| ·76 | ·9353 613 | ·9217 532 | ·9064 315+ | ·8893 807 | ·8706 130 | ·8501 688 |
| ·77 | ·9527 594 | ·9422 244 | ·9302 067 | ·9166 567 | ·9015 462 | ·8848 692 |
| ·78 | ·9664 433 | ·9585 354 | ·9493 973 | ·9389 604 | ·9271 702 | ·9139 888 |
| ·79 | ·9768 905− | ·9711 485+ | ·9644 283 | ·9566 544 | ·9477 601 | ·9376 887 |
| ·80 | ·9846 147 | ·9805 926 | ·9758 258 | ·9702 418 | ·9637 720 | ·9563 533 |
| ·81 | ·9901 305− | ·9874 212 | ·9841 701 | ·9803 141 | ·9757 906 | ·9705 388 |
| ·82 | ·9939 228 | ·9921 743 | ·9900 502 | ·9874 998 | ·9844 710 | ·9809 111 |
| ·83 | ·9964 241 | ·9953 476 | ·9940 239 | ·9924 152 | ·9904 814 | ·9881 809 |
| ·84 | ·9979 998 | ·9973 709 | ·9965 881 | ·9956 254 | ·9944 542 | ·9930 440 |
| ·85 | ·9989 431 | ·9985 965+ | ·9981 600 | ·9976 168 | ·9969 480 | ·9961 332 |
| ·86 | ·9994 765− | ·9992 976 | ·9990 698 | ·9987 829 | ·9984 255+ | ·9979 850+ |
| ·87 | ·9997 591 | ·9996 735+ | ·9995 632 | ·9994 227 | ·9992 456 | ·9990 248 |
| ·88 | ·9998 982 | ·9998 606 | ·9998 117 | ·9997 486 | ·9996 681 | ·9995 667 |
| ·89 | ·9999 611 | ·9999 462 | ·9999 265+ | ·9999 009 | ·9998 679 | ·9998 258 |
| ·90 | ·9999 868 | ·9999 815+ | ·9999 745+ | ·9999 653 | ·9999 533 | ·9999 378 |
| ·91 | ·9999 961 | ·9999 945+ | ·9999 923 | ·9999 895− | ·9999 857 | ·9999 807 |
| ·92 | ·9999 990 | ·9999 986 | ·9999 981 | ·9999 973 | ·9999 963 | ·9999 950+ |
| ·93 | ·9999 998 | ·9999 997 | ·9999 996 | ·9999 995− | ·9999 992 | ·9999 990 |
| ·94 | 1·0000 000 | 1·0000 000 | ·9999 999 | ·9999 999 | ·9999 999 | ·9999 998 |
| ·95 | | | 1·0000 000 | 1·0000 000 | 1·0000 000 | 1·0000 000 |

$x = \cdot 32$ to $\cdot 96$ $q = 14$ $p = 32$ to $37$

| $x$ | $p = 32$ | $p = 33$ | $p = 34$ | $p = 35$ | $p = 36$ | $p = 37$ |
|---|---|---|---|---|---|---|
| $B(p, q) =$ | $\cdot 4280\ 4578 \times \frac{1}{10^{12}}$ | $\cdot 2977\ 7098 \times \frac{1}{10^{12}}$ | $\cdot 2090\ 7324 \times \frac{1}{10^{12}}$ | $\cdot 1480\ 9355^{-} \times \frac{1}{10^{12}}$ | $\cdot 1057\ 8110 \times \frac{1}{10^{12}}$ | $\cdot 7616\ 2395^{+} \times \frac{1}{10^{13}}$ |
| ·32 | ·0000 001 | | | | | |
| ·33 | ·0000 002 | ·0000 001 | | | | |
| ·34 | ·0000 004 | ·0000 002 | ·0000 001 | | | |
| ·35 | ·0000 009 | ·0000 004 | ·0000 002 | ·0000 001 | | |
| ·36 | ·0000 018 | ·0000 009 | ·0000 004 | ·0000 002 | ·0000 001 | |
| ·37 | ·0000 035⁺ | ·0000 018 | ·0000 009 | ·0000 005⁻ | ·0000 002 | ·0000 001 |
| ·38 | ·0000 068 | ·0000 036 | ·0000 019 | ·0000 010 | ·0000 005⁻ | ·0000 003 |
| ·39 | ·0000 128 | ·0000 069 | ·0000 037 | ·0000 020 | ·0000 010 | ·0000 005⁺ |
| ·40 | ·0000 236 | ·0000 130 | ·0000 071 | ·0000 039 | ·0000 021 | ·0000 011 |
| ·41 | ·0000 423 | ·0000 240 | ·0000 135⁻ | ·0000 075⁻ | ·0000 042 | ·0000 023 |
| ·42 | ·0000 744 | ·0000 431 | ·0000 248 | ·0000 141 | ·0000 080 | ·0000 045⁺ |
| ·43 | ·0001 279 | ·0000 758 | ·0000 446 | ·0000 260 | ·0000 151 | ·0000 087 |
| ·44 | ·0002 154 | ·0001 306 | ·0000 786 | ·0000 469 | ·0000 278 | ·0000 164 |
| ·45 | ·0003 557 | ·0002 204 | ·0001 355⁺ | ·0000 828 | ·0000 502 | ·0000 302 |
| ·46 | ·0005 761 | ·0003 647 | ·0002 291 | ·0001 429 | ·0000 885⁺ | ·0000 545⁻ |
| ·47 | ·0009 157 | ·0005 919 | ·0003 797 | ·0002 419 | ·0001 530 | ·0000 962 |
| ·48 | ·0014 294 | ·0009 429 | ·0006 174 | ·0004 013 | ·0002 591 | ·0001 662 |
| ·49 | ·0021 922 | ·0014 750⁺ | ·0009 852 | ·0006 534 | ·0004 304 | ·0002 817 |
| ·50 | ·0033 044 | ·0022 669 | ·0015 438 | ·0010 441 | ·0007 013 | ·0004 681 |
| ·51 | ·0048 978 | ·0034 243 | ·0023 768 | ·0016 383 | ·0011 217 | ·0007 632 |
| ·52 | ·0071 411 | ·0050 859 | ·0035 962 | ·0025 254 | ·0017 618 | ·0012 213 |
| ·53 | ·0102 454 | ·0074 299 | ·0053 498 | ·0038 259 | ·0027 182 | ·0019 191 |
| ·54 | ·0144 690 | ·0106 797 | ·0078 273 | ·0056 981 | ·0041 212 | ·0029 622 |
| ·55 | ·0201 197 | ·0151 085⁻ | ·0112 666 | ·0083 456 | ·0061 423 | ·0044 929 |
| ·56 | ·0275 542 | ·0210 422 | ·0159 589 | ·0120 238 | ·0090 017 | ·0066 982 |
| ·57 | ·0371 748 | ·0288 588 | ·0222 514 | ·0170 451 | ·0129 753 | ·0098 178 |
| ·58 | ·0494 196 | ·0389 833 | ·0305 457 | ·0237 808 | ·0183 998 | ·0141 517 |
| ·59 | ·0647 486 | ·0518 781 | ·0412 930 | ·0326 599 | ·0256 745⁺ | ·0200 649 |
| ·60 | ·0836 230 | ·0680 261 | ·0549 813 | ·0441 619 | ·0352 590 | ·0279 884 |
| ·61 | ·1064 786 | ·0879 083 | ·0721 179 | ·0588 029 | ·0476 639 | ·0384 155⁻ |
| ·62 | ·1336 944 | ·1119 742 | ·0932 026 | ·0771 143 | ·0634 347 | ·0518 906 |
| ·63 | ·1655 576 | ·1406 070 | ·1186 958 | ·0996 137 | ·0831 268 | ·0689 893 |
| ·64 | ·2022 268 | ·1740 849 | ·1489 789 | ·1267 674 | ·1072 721 | ·0902 896 |
| ·65 | ·2436 979 | ·2125 420 | ·1843 129 | ·1589 482 | ·1363 374 | ·1163 331 |
| ·66 | ·2897 742 | ·2559 311 | ·2247 954 | ·1963 887 | ·1706 765⁺ | ·1475 785⁻ |
| ·67 | ·3400 475⁺ | ·3039 939 | ·2703 220 | ·2391 359 | ·2104 804 | ·1843 481 |
| ·68 | ·3938 904 | ·3562 421 | ·3205 562 | ·2870 114 | ·2557 284 | ·2267 743 |
| ·69 | ·4504 657 | ·4119 538 | ·3749 124 | ·3395 815⁻ | ·3061 483 | ·2747 490 |
| ·70 | ·5087 531 | ·4701 880 | ·4325 578 | ·3961 450⁺ | ·3611 888 | ·3278 832 |
| ·71 | ·5675 935⁻ | ·5298 190 | ·4924 336 | ·4557 410 | ·4200 129 | ·3854 857 |
| ·72 | ·6257 489 | ·5895 903 | ·5533 001 | ·5171 808 | ·4815 155⁺ | ·4465 635⁺ |
| ·73 | ·6819 743 | ·6481 846 | ·6138 010 | ·5791 040 | ·5443 673 | ·5098 525⁺ |
| ·74 | ·7350 946 | ·7043 052 | ·6725 454 | ·6400 570 | ·6070 860 | ·5738 768 |
| ·75 | ·7840 807 | ·7567 607 | ·7281 989 | ·6985 870 | ·6681 291 | ·6370 367 |
| ·76 | ·8281 153 | ·8045 457 | ·7795 763 | ·7533 436 | ·7260 017 | ·6977 180 |
| ·77 | ·8666 428 | ·8469 071 | ·8257 241 | ·8031 767 | ·7793 666 | ·7544 123 |
| ·78 | ·8993 957 | ·8833 889 | ·8659 851 | ·8472 198 | ·8271 462 | ·8058 348 |
| ·79 | ·9263 957 | ·9138 500⁻ | ·9000 344 | ·8849 471 | ·8686 010 | ·8510 244 |
| ·80 | ·9479 295⁺ | ·9384 528 | ·9278 848 | ·9161 978 | ·9033 755⁺ | ·8894 135⁻ |
| ·81 | ·9645 010 | ·9576 235⁻ | ·9498 581 | ·9411 633 | ·9315 045⁻ | ·9208 557 |
| ·82 | ·9767 678 | ·9719 901 | ·9665 291 | ·9603 388 | ·9533 775⁻ | ·9456 078 |
| ·83 | ·9854 707 | ·9823 073 | ·9786 475⁻ | ·9744 483 | ·9696 685⁻ | ·9642 686 |
| ·84 | ·9913 628 | ·9893 768 | ·9870 514 | ·9843 512 | ·9812 405⁺ | ·9776 840 |
| ·85 | ·9951 502 | ·9939 751 | ·9925 829 | ·9909 471 | ·9890 401 | ·9868 339 |
| ·86 | ·9974 473 | ·9967 970 | ·9960 175⁻ | ·9950 907 | ·9939 977 | ·9927 182 |
| ·87 | ·9987 522 | ·9984 186 | ·9980 141 | ·9975 276 | ·9969 471 | ·9962 598 |
| ·88 | ·9994 400 | ·9992 832 | ·9990 909 | ·9988 569 | ·9985 745⁺ | ·9982 363 |
| ·89 | ·9997 726 | ·9997 060 | ·9996 234 | ·9995 218 | ·9993 977 | ·9992 475⁻ |
| ·90 | ·9999 180 | ·9998 929 | ·9998 615⁻ | ·9998 223 | ·9997 740 | ·9997 149 |
| ·91 | ·9999 744 | ·9999 662 | ·9999 558 | ·9999 428 | ·9999 265⁺ | ·9999 064 |
| ·92 | ·9999 933 | ·9999 911 | ·9999 882 | ·9999 846 | ·9999 800 | ·9999 742 |
| ·93 | ·9999 986 | ·9999 981 | ·9999 975⁻ | ·9999 967 | ·9999 956 | ·9999 943 |
| ·94 | ·9999 998 | ·9999 997 | ·9999 996 | ·9999 995⁻ | ·9999 993 | ·9999 991 |
| ·95 | 1·0000 000 | 1·0000 000 | 1·0000 000 | ·9999 999 | ·9999 999 | ·9999 999 |
| ·96 | | | | 1·0000 000 | 1·0000 000 | 1·0000 000 |

$x$ = ·37 to ·96 $q = 14$ $p$ = 38 to 43

| | $p = 38$ | $p = 39$ | $p = 40$ | $p = 41$ | $p = 42$ | $p = 43$ |
|---|---|---|---|---|---|---|
| $B(p, q) =$ | ·5525 5071 $\times\frac{1}{10^{13}}$ | ·4037 8706 $\times\frac{1}{10^{13}}$ | ·2971 2632 $\times\frac{1}{10^{13}}$ | ·2200 9357 $\times\frac{1}{10^{13}}$ | ·1640 6976 $\times\frac{1}{10^{13}}$ | ·1230 5232 $\times\frac{1}{10^{13}}$ |
| $x$ | | | | | | |
| ·37 | ·0000 001 | | | | | |
| ·38 | ·0000 001 | ·0000 001 | | | | |
| ·39 | ·0000 003 | ·0000 001 | ·0000 001 | | | |
| ·40 | ·0000 006 | ·0000 003 | ·0000 002 | ·0000 001 | | |
| ·41 | ·0000 012 | ·0000 007 | ·0000 004 | ·0000 002 | ·0000 001 | ·0000 001 |
| ·42 | ·0000 025+ | ·0000 014 | ·0000 008 | ·0000 004 | ·0000 002 | ·0000 001 |
| ·43 | ·0000 050− | ·0000 028 | ·0000 016 | ·0000 009 | ·0000 005+ | ·0000 003 |
| ·44 | ·0000 096 | ·0000 056 | ·0000 032 | ·0000 019 | ·0000 011 | ·0000 006 |
| ·45 | ·0000 181 | ·0000 108 | ·0000 064 | ·0000 037 | ·0000 022 | ·0000 013 |
| ·46 | ·0000 333 | ·0000 203 | ·0000 122 | ·0000 074 | ·0000 044 | ·0000 026 |
| ·47 | ·0000 601 | ·0000 373 | ·0000 230 | ·0000 141 | ·0000 086 | ·0000 052 |
| ·48 | ·0001 060 | ·0000 672 | ·0000 423 | ·0000 265+ | ·0000 165+ | ·0000 103 |
| ·49 | ·0001 832 | ·0001 185− | ·0000 762 | ·0000 487 | ·0000 310 | ·0000 196 |
| ·50 | ·0003 105+ | ·0002 048 | ·0001 343 | ·0000 876 | ·0000 568 | ·0000 367 |
| ·51 | ·0005 161 | ·0003 469 | ·0002 319 | ·0001 542 | ·0001 020 | ·0000 671 |
| ·52 | ·0008 415− | ·0005 764 | ·0003 926 | ·0002 660 | ·0001 793 | ·0001 203 |
| ·53 | ·0013 468 | ·0009 397 | ·0006 520 | ·0004 500− | ·0003 090 | ·0002 111 |
| ·54 | ·0021 165+ | ·0015 036 | ·0010 622 | ·0007 465− | ·0005 219 | ·0003 631 |
| ·55 | ·0032 670 | ·0023 621 | ·0016 985+ | ·0012 149 | ·0008 646 | ·0006 123 |
| ·56 | ·0049 549 | ·0036 448 | ·0026 665+ | ·0019 406 | ·0014 053 | ·0010 127 |
| ·57 | ·0073 857 | ·0055 251 | ·0041 111 | ·0030 431 | ·0022 414 | ·0016 429 |
| ·58 | ·0108 222 | ·0082 304 | ·0062 261 | ·0046 858 | ·0035 091 | ·0026 155− |
| ·59 | ·0155 923 | ·0120 508 | ·0092 647 | ·0070 868 | ·0053 943 | ·0040 867 |
| ·60 | ·0220 932 | ·0173 461 | ·0135 484 | ·0105 292 | ·0081 433 | ·0062 687 |
| ·61 | ·0307 918 | ·0245 504 | ·0194 740 | ·0153 711 | ·0120 747 | ·0094 415+ |
| ·62 | ·0422 185− | ·0341 703 | ·0275 171 | ·0220 515+ | ·0175 884 | ·0139 648 |
| ·63 | ·0569 533 | ·0467 765+ | ·0382 280 | ·0310 922 | ·0251 712 | ·0202 864 |
| ·64 | ·0756 019 | ·0629 860 | ·0522 205− | ·0430 913 | ·0353 960 | ·0289 467 |
| ·65 | ·0987 618 | ·0834 331 | ·0701 483 | ·0587 067 | ·0489 116 | ·0405 741 |
| ·66 | ·1269 775− | ·1087 293 | ·0926 710 | ·0786 279 | ·0664 207 | ·0558 703 |
| ·67 | ·1606 878 | ·1394 122 | ·1204 059 | ·1035 332 | ·0886 440 | ·0755 805+ |
| ·68 | ·2001 679 | ·1758 860 | ·1538 704 | ·1340 343 | ·1162 690 | ·1004 494 |
| ·69 | ·2454 712 | ·2183 585− | ·1934 147 | ·1706 099 | ·1498 854 | ·1311 593 |
| ·70 | ·2963 780 | ·2667 797 | ·2391 545+ | ·2135 322 | ·1899 097 | ·1682 558 |
| ·71 | ·3523 582 | ·3207 912 | ·2909 078 | ·2627 949 | ·2365 060 | ·2120 635+ |
| ·72 | ·4125 562 | ·3796 944 | ·3481 471 | ·3180 509 | ·2895 109 | ·2626 018 |
| ·73 | ·4758 041 | ·4424 457 | ·4099 768 | ·3785 713 | ·3483 761 | ·3195 108 |
| ·74 | ·5406 676 | ·5076 856 | ·4751 434 | ·4432 358 | ·4121 375+ | ·3820 018 |
| ·75 | ·6055 240 | ·5738 041 | ·5420 842 | ·5105 625− | ·4794 252 | ·4488 439 |
| ·76 | ·6686 699 | ·6390 408 | ·6090 167 | ·5787 824 | ·5485 186 | ·5183 989 |
| ·77 | ·7284 464 | ·7016 126 | ·6740 633 | ·6459 561 | ·6174 513 | ·5887 089 |
| ·78 | ·7833 714 | ·7598 557 | ·7353 995− | ·7101 239 | ·6841 579 | ·6576 355+ |
| ·79 | ·8322 602 | ·8123 652 | ·7914 091 | ·7694 733 | ·7466 495− | ·7230 376 |
| ·80 | ·8743 194 | ·8581 131 | ·8408 264 | ·8225 025− | ·8031 953 | ·7829 688 |
| ·81 | ·9091 996 | ·8965 282 | ·8828 430 | ·8681 554 | ·8524 863 | ·8358 658 |
| ·82 | ·9369 981 | ·9275 230 | ·9171 635+ | ·9059 079 | ·8937 519 | ·8806 986 |
| ·83 | ·9582 119 | ·9514 651 | ·9439 987 | ·9357 875− | ·9268 112 | ·9170 549 |
| ·84 | ·9736 469 | ·9690 956 | ·9639 981 | ·9583 246 | ·9520 478 | ·9451 433 |
| ·85 | ·9842 998 | ·9814 088 | ·9781 324 | ·9744 423 | ·9703 113 | ·9657 130 |
| ·86 | ·9912 313 | ·9895 151 | ·9875 471 | ·9853 046 | ·9827 646 | ·9799 040 |
| ·87 | ·9954 516 | ·9945 080 | ·9934 135− | ·9921 517 | ·9907 059 | ·9890 588 |
| ·88 | ·9978 341 | ·9973 590 | ·9968 017 | ·9961 518 | ·9953 985+ | ·9945 305+ |
| ·89 | ·9990 668 | ·9988 509 | ·9985 948 | ·9982 927 | ·9979 386 | ·9975 260 |
| ·90 | ·9996 429 | ·9995 560 | ·9994 517 | ·9993 274 | ·9991 799 | ·9990 062 |
| ·91 | ·9998 816 | ·9998 514 | ·9998 147 | ·9997 704 | ·9997 173 | ·9996 541 |
| ·92 | ·9999 671 | ·9999 583 | ·9999 475− | ·9999 343 | ·9999 183 | ·9998 991 |
| ·93 | ·9999 927 | ·9999 906 | ·9999 881 | ·9999 850− | ·9999 811 | ·9999 765− |
| ·94 | ·9999 988 | ·9999 984 | ·9999 980 | ·9999 974 | ·9999 967 | ·9999 959 |
| ·95 | ·9999 999 | ·9999 998 | ·9999 998 | ·9999 997 | ·9999 996 | ·9999 995+ |
| ·96 | 1·0000 000 | 1·0000 000 | 1·0000 000 | 1·0000 000 | 1·0000 000 | 1·0000 000 |

$x$ = ·42 to ·97 $q$ = 14 $p$ = 44 to 50

| $x$ | $p$ = 44 | $p$ = 45 | $p$ = 46 | $p$ = 47 | $p$ = 48 | $p$ = 49 | $p$ = 50 |
|---|---|---|---|---|---|---|---|
| $B(p, q)$ = | ·9282 8940 × $\frac{1}{10^{14}}$ | ·7042 1955 × $\frac{1}{10^{14}}$ | ·5371 1660 × $\frac{1}{10^{14}}$ | ·4117 8940 × $\frac{1}{10^{14}}$ | ·3172 8035 × $\frac{1}{10^{14}}$ | ·2456 3640 × $\frac{1}{10^{14}}$ | ·1910 5054 × $\frac{1}{10^{14}}$ |
| ·42 | ·0000 001 | | | | | | |
| ·43 | ·0000 002 | ·0000 001 | | | | | |
| ·44 | ·0000 003 | ·0000 002 | ·0000 001 | ·0000 001 | | | |
| ·45 | ·0000 007 | ·0000 004 | ·0000 002 | ·0000 001 | ·0000 001 | | |
| ·46 | ·0000 015+ | ·0000 009 | ·0000 005+ | ·0000 003 | ·0000 002 | ·0000 001 | ·0000 001 |
| ·47 | ·0000 032 | ·0000 019 | ·0000 011 | ·0000 007 | ·0000 004 | ·0000 002 | ·0000 001 |
| ·48 | ·0000 063 | ·0000 039 | ·0000 024 | ·0000 014 | ·0000 009 | ·0000 005+ | ·0000 003 |
| ·49 | ·0000 123 | ·0000 077 | ·0000 048 | ·0000 030 | ·0000 019 | ·0000 011 | ·0000 007 |
| ·50 | ·0000 236 | ·0000 151 | ·0000 096 | ·0000 061 | ·0000 038 | ·0000 024 | ·0000 015+ |
| ·51 | ·0000 439 | ·0000 286 | ·0000 186 | ·0000 120 | ·0000 077 | ·0000 050− | ·0000 032 |
| ·52 | ·0000 803 | ·0000 533 | ·0000 353 | ·0000 232 | ·0000 152 | ·0000 099 | ·0000 065− |
| ·53 | ·0001 435+ | ·0000 971 | ·0000 655− | ·0000 439 | ·0000 293 | ·0000 195+ | ·0000 129 |
| ·54 | ·0002 514 | ·0001 733 | ·0001 189 | ·0000 813 | ·0000 553 | ·0000 375− | ·0000 253 |
| ·55 | ·0004 316 | ·0003 028 | ·0002 116 | ·0001 472 | ·0001 020 | ·0000 704 | ·0000 484 |
| ·56 | ·0007 263 | ·0005 186 | ·0003 687 | ·0002 610 | ·0001 840 | ·0001 292 | ·0000 904 |
| ·57 | ·0011 987 | ·0008 706 | ·0006 296 | ·0004 535− | ·0003 253 | ·0002 324 | ·0001 654 |
| ·58 | ·0019 404 | ·0014 332 | ·0010 540 | ·0007 720 | ·0005 631 | ·0004 092 | ·0002 962 |
| ·59 | ·0030 820 | ·0023 140 | ·0017 300 | ·0012 881 | ·0009 553 | ·0007 057 | ·0005 194 |
| ·60 | ·0048 039 | ·0036 653 | ·0027 848 | ·0021 072 | ·0015 882 | ·0011 925− | ·0008 920 |
| ·61 | ·0073 497 | ·0056 967 | ·0043 970 | ·0033 802 | ·0025 883 | ·0019 745+ | ·0015 007 |
| ·62 | ·0110 389 | ·0086 889 | ·0068 110 | ·0053 176 | ·0041 357 | ·0032 044 | ·0024 738 |
| ·63 | ·0162 785+ | ·0130 076 | ·0103 516 | ·0082 054 | ·0064 794 | ·0050 974 | ·0039 958 |
| ·64 | ·0235 712 | ·0191 145− | ·0154 381 | ·0124 204 | ·0099 548 | ·0079 494 | ·0063 255+ |
| ·65 | ·0335 163 | ·0275 734 | ·0225 945− | ·0184 436 | ·0149 992 | ·0121 541 | ·0098 140 |
| ·66 | ·0468 018 | ·0390 482 | ·0324 525− | ·0268 691 | ·0221 647 | ·0182 190 | ·0149 239 |
| ·67 | ·0641 819 | ·0542 884 | ·0457 448 | ·0384 028 | ·0321 232 | ·0267 763 | ·0222 434 |
| ·68 | ·0864 398 | ·0740 986 | ·0632 823 | ·0538 486 | ·0456 593 | ·0385 824 | ·0324 934 |
| ·69 | ·1143 319 | ·0992 905+ | ·0859 137 | ·0740 753 | ·0636 473 | ·0545 033 | ·0465 201 |
| ·70 | ·1485 154 | ·1306 145+ | ·1144 639 | ·0999 635− | ·0870 056 | ·0754 786 | ·0652 689 |
| ·71 | ·1894 628 | ·1686 752 | ·1496 523 | ·1323 291 | ·1166 275+ | ·1024 602 | ·0897 327 |
| ·72 | ·2373 698 | ·2138 352 | ·1919 952 | ·1718 263 | ·1532 882 | ·1363 257 | ·1208 726 |
| ·73 | ·2920 687 | ·2661 172 | ·2416 997 | ·2188 376 | ·1975 319 | ·1777 665+ | ·1595 098 |
| ·74 | ·3529 594 | ·3251 183 | ·2985 641 | ·2733 607 | ·2495 515− | ·2271 609 | ·2061 960 |
| ·75 | ·4189 739 | ·3899 524 | ·3618 983 | ·3349 114 | ·3090 730 | ·2844 457 | ·2610 749 |
| ·76 | ·4885 874 | ·4592 366 | ·4304 859 | ·4024 603 | ·3752 694 | ·3490 075+ | ·3237 533 |
| ·77 | ·5598 864 | ·5311 359 | ·5026 026 | ·4744 229 | ·4467 228 | ·4196 171 | ·3932 085− |
| ·78 | ·6306 937 | ·6034 702 | ·5761 015− | ·5487 209 | ·5214 568 | ·4944 313 | ·4677 588 |
| ·79 | ·6987 448 | ·6738 834 | ·6485 690 | ·6229 188 | ·5970 504 | ·5710 795+ | ·5451 193 |
| ·80 | ·7618 956 | ·7400 560 | ·7175 370 | ·6944 306 | ·6708 325+ | ·6468 412 | ·6225 560 |
| ·81 | ·8183 332 | ·7999 358 | ·7807 290 | ·7607 748 | ·7401 413 | ·7189 016 | ·6971 332 |
| ·82 | ·8667 590 | ·8519 512 | ·8363 011 | ·8198 413 | ·8026 109 | ·7846 555+ | ·7660 259 |
| ·83 | ·9065 090 | ·8951 697 | ·8830 392 | ·8701 255+ | ·8564 425− | ·8420 097 | ·8268 524 |
| ·84 | ·9375 901 | ·9293 709 | ·9204 722 | ·9108 848 | ·9006 038 | ·8896 289 | ·8779 641 |
| ·85 | ·9606 228 | ·9550 178 | ·9488 773 | ·9421 827 | ·9349 184 | ·9270 714 | ·9186 319 |
| ·86 | ·9767 002 | ·9731 309 | ·9691 745− | ·9648 104 | ·9600 191 | ·9547 827 | ·9490 847 |
| ·87 | ·9871 925− | ·9850 891 | ·9827 305+ | ·9800 986 | ·9771 756 | ·9739 437 | ·9703 861 |
| ·88 | ·9935 357 | ·9924 016 | ·9911 154 | ·9896 635+ | ·9880 325+ | ·9862 086 | ·9841 776 |
| ·89 | ·9970 477 | ·9964 962 | ·9958 636 | ·9951 415− | ·9943 210 | ·9933 931 | ·9923 480 |
| ·90 | ·9988 026 | ·9985 651 | ·9982 897 | ·9979 718 | ·9976 065+ | ·9971 887 | ·9967 130 |
| ·91 | ·9995 792 | ·9994 909 | ·9993 873 | ·9992 664 | ·9991 259 | ·9989 635− | ·9987 764 |
| ·92 | ·9998 761 | ·9998 486 | ·9998 160 | ·9997 776 | ·9997 325− | ·9996 797 | ·9996 183 |
| ·93 | ·9999 708 | ·9999 640 | ·9999 558 | ·9999 461 | ·9999 345+ | ·9999 208 | ·9999 047 |
| ·94 | ·9999 949 | ·9999 936 | ·9999 921 | ·9999 902 | ·9999 880 | ·9999 854 | ·9999 822 |
| ·95 | ·9999 994 | ·9999 992 | ·9999 990 | ·9999 988 | ·9999 985+ | ·9999 982 | ·9999 978 |
| ·96 | 1·0000 000 | ·9999 999 | ·9999 999 | ·9999 999 | ·9999 999 | ·9999 999 | ·9999 998 |
| ·97 | | 1·0000 000 | 1·0000 000 | 1·0000 000 | 1·0000 000 | 1·0000 000 | 1·0000 000 |

$x$ = ·11 to ·70 $q = 15$ $p$ = 15 to 20

| | $p = 15$ | $p = 16$ | $p = 17$ | $p = 18$ | $p = 19$ | $p = 20$ |
|---|---|---|---|---|---|---|
| $B(p, q)$ = | ·8595 6334 $\times \frac{1}{10^9}$ | ·4297 8167 $\times \frac{1}{10^9}$ | ·2218 2280 $\times \frac{1}{10^9}$ | ·1178 4336 $\times \frac{1}{10^9}$ | ·6427 8197 $\times \frac{1}{10^{10}}$ | ·3592 0169 $\times \frac{1}{10^{10}}$ |
| $x$ | | | | | | |
| ·11 | ·0000 001 | | | | | |
| ·12 | ·0000 002 | ·0000 001 | | | | |
| ·13 | ·0000 006 | ·0000 002 | | | | |
| ·14 | ·0000 017 | ·0000 004 | ·0000 001 | | | |
| ·15 | ·0000 041 | ·0000 011 | ·0000 003 | ·0000 001 | | |
| ·16 | ·0000 093 | ·0000 028 | ·0000 008 | ·0000 002 | ·0000 001 | |
| ·17 | ·0000 198 | ·0000 062 | ·0000 019 | ·0000 006 | ·0000 002 | |
| ·18 | ·0000 400 | ·0000 133 | ·0000 043 | ·0000 014 | ·0000 004 | ·0000 001 |
| ·19 | ·0000 769 | ·0000 270 | ·0000 093 | ·0000 031 | ·0000 010 | ·0000 003 |
| ·20 | ·0001 418 | ·0000 524 | ·0000 189 | ·0000 066 | ·0000 023 | ·0000 008 |
| ·21 | ·0002 513 | ·0000 974 | ·0000 368 | ·0000 136 | ·0000 049 | ·0000 017 |
| ·22 | ·0004 297 | ·0001 742 | ·0000 688 | ·0000 266 | ·0000 100 | ·0000 037 |
| ·23 | ·0007 112 | ·0003 011 | ·0001 243 | ·0000 501 | ·0000 197 | ·0000 076 |
| ·24 | ·0011 423 | ·0005 040 | ·0002 168 | ·0000 911 | ·0000 374 | ·0000 151 |
| ·25 | ·0017 843 | ·0008 190 | ·0003 665+ | ·0001 602 | ·0000 686 | ·0000 288 |
| ·26 | ·0027 162 | ·0012 948 | ·0006 019 | ·0002 734 | ·0001 216 | ·0000 530 |
| ·27 | ·0040 370 | ·0019 956 | ·0009 622 | ·0004 533 | ·0002 091 | ·0000 946 |
| ·28 | ·0058 674 | ·0030 033 | ·0014 996 | ·0007 319 | ·0003 497 | ·0001 638 |
| ·29 | ·0083 508 | ·0044 200 | ·0022 827 | ·0011 524 | ·0005 697 | ·0002 762 |
| ·30 | ·0116 538 | ·0063 703 | ·0033 984 | ·0017 726 | ·0009 054 | ·0004 536 |
| ·31 | ·0159 649 | ·0090 019 | ·0049 547 | ·0026 669 | ·0014 060 | ·0007 271 |
| ·32 | ·0214 916 | ·0124 860 | ·0070 826 | ·0039 296 | ·0021 359 | ·0011 389 |
| ·33 | ·0284 571 | ·0170 161 | ·0099 370 | ·0056 771 | ·0031 779 | ·0017 455+ |
| ·34 | ·0370 938 | ·0228 057 | ·0136 970 | ·0080 497 | ·0046 362 | ·0026 204 |
| ·35 | ·0476 367 | ·0300 837 | ·0185 645+ | ·0112 126 | ·0066 380 | ·0038 572 |
| ·36 | ·0603 150+ | ·0390 891 | ·0247 615+ | ·0153 559 | ·0093 363 | ·0055 725− |
| ·37 | ·0753 424 | ·0500 631 | ·0325 256 | ·0206 929 | ·0129 097 | ·0079 079 |
| ·38 | ·0929 070 | ·0632 409 | ·0421 038 | ·0274 571 | ·0175 624 | ·0110 319 |
| ·39 | ·1131 615− | ·0788 414 | ·0537 449 | ·0358 969 | ·0235 222 | ·0151 400 |
| ·40 | ·1362 129 | ·0970 568 | ·0676 898 | ·0462 691 | ·0310 366 | ·0204 540 |
| ·41 | ·1621 144 | ·1180 418 | ·0841 610 | ·0588 301 | ·0403 667 | ·0272 188 |
| ·42 | ·1908 573 | ·1419 031 | ·1033 516 | ·0738 258 | ·0517 798 | ·0356 978 |
| ·43 | ·2223 662 | ·1686 902 | ·1254 139 | ·0914 802 | ·0655 397 | ·0461 663 |
| ·44 | ·2564 962 | ·1983 875− | ·1504 478 | ·1119 832 | ·0818 954 | ·0589 019 |
| ·45 | ·2930 325− | ·2309 086 | ·1784 915− | ·1354 787 | ·1010 684 | ·0741 741 |
| ·46 | ·3316 934 | ·2660 932 | ·2095 131 | ·1620 524 | ·1232 401 | ·0922 310 |
| ·47 | ·3721 359 | ·3037 075− | ·2434 049 | ·1917 220 | ·1485 381 | ·1132 864 |
| ·48 | ·4139 645+ | ·3434 466 | ·2799 804 | ·2244 289 | ·1770 249 | ·1375 050− |
| ·49 | ·4567 414 | ·3849 414 | ·3189 751 | ·2600 323 | ·2086 865+ | ·1649 886 |
| ·50 | ·5000 000$^{e}$ | ·4277 678 | ·3600 501 | ·2983 074 | ·2434 251 | ·1957 642 |
| ·51 | ·5432 586 | ·4714 585+ | ·4027 997 | ·3389 471 | ·2810 540 | ·2297 729 |
| ·52 | ·5860 355− | ·5155 176 | ·4467 626 | ·3815 667 | ·3212 967 | ·2668 634 |
| ·53 | ·6278 641 | ·5594 356 | ·4914 348 | ·4257 141 | ·3637 905− | ·3067 882 |
| ·54 | ·6683 066 | ·6027 065− | ·5362 864 | ·4708 821 | ·4080 939 | ·3492 053 |
| ·55 | ·7069 675+ | ·6448 436 | ·5807 783 | ·5165 246 | ·4536 987 | ·3936 835− |
| ·56 | ·7435 038 | ·6853 951 | ·6243 809 | ·5620 747 | ·5000 453 | ·4397 137 |
| ·57 | ·7776 338 | ·7239 577 | ·6665 914 | ·6069 643 | ·5465 420 | ·4867 241 |
| ·58 | ·8091 427 | ·7601 885− | ·7069 508 | ·6506 441 | ·5925 856 | ·5340 993 |
| ·59 | ·8378 856 | ·7938 130 | ·7450 576 | ·6926 026 | ·6375 831 | ·5812 027 |
| ·60 | ·8637 871 | ·8246 309 | ·7805 803 | ·7323 838 | ·6809 741 | ·6273 998 |
| ·61 | ·8868 385+ | ·8525 185+ | ·8132 650+ | ·7696 012 | ·7222 503 | ·6720 832 |
| ·62 | ·9070 930 | ·8774 269 | ·8429 401 | ·8039 497 | ·7609 736 | ·7146 952 |
| ·63 | ·9246 576 | ·8993 783 | ·8695 171 | ·8352 119 | ·7967 901 | ·7547 485+ |
| ·64 | ·9396 850− | ·9184 590 | ·8929 878 | ·8632 614 | ·8294 394 | ·7918 436 |
| ·65 | ·9523 633 | ·9348 103 | ·9134 175+ | ·8880 608 | ·8587 597 | ·8256 804 |
| ·66 | ·9629 062 | ·9486 182 | ·9309 367 | ·9096 565− | ·8846 877 | ·8560 656 |
| ·67 | ·9715 429 | ·9601 020 | ·9457 293 | ·9281 692 | ·9072 533 | ·8829 137 |
| ·68 | ·9785 084 | ·9695 027 | ·9580 206 | ·9437 827 | ·9265 706 | ·9062 423 |
| ·69 | ·9840 351 | ·9770 722 | ·9680 639 | ·9567 293 | ·9428 256 | ·9261 631 |
| ·70 | ·9883 462 | ·9830 627 | ·9761 281 | ·9672 763 | ·9562 608 | ·9428 682 |

$x = \cdot 71$ to $\cdot 92$     $q = 15$     $p = 15$ to $20$

| | $p = 15$ | $p = 16$ | $p = 17$ | $p = 18$ | $p = 19$ | $p = 20$ |
|---|---|---|---|---|---|---|
| $B(p, q) =$ | $\cdot 8595\ 6334 \times \frac{1}{10^9}$ | $\cdot 4297\ 8167 \times \frac{1}{10^9}$ | $\cdot 2218\ 2280 \times \frac{1}{10^9}$ | $\cdot 1178\ 4336 \times \frac{1}{10^9}$ | $\cdot 6427\ 8197 \times \frac{1}{10^{10}}$ | $\cdot 3592\ 0169 \times \frac{1}{10^{10}}$ |
| $x$ | | | | | | |
| ·71 | ·9916 492 | ·9877 184 | ·9824 856 | ·9757 107 | ·9671 592 | ·9566 138 |
| ·72 | ·9941 326 | ·9912 685+ | ·9874 020 | ·9823 254 | ·9758 275− | ·9677 016 |
| ·73 | ·9959 630 | ·9939 216 | ·9911 274 | ·9874 078 | ·9825 807 | ·9764 604 |
| ·74 | ·9972 838 | ·9958 624 | ·9938 903 | ·9912 290 | ·9877 280 | ·9832 283 |
| ·75 | ·9982 157 | ·9972 505− | ·9958 931 | ·9940 366 | ·9915 613 | ·9883 369 |
| ·76 | ·9988 577 | ·9982 195− | ·9973 099 | ·9960 494 | ·9943 463 | ·9920 983 |
| ·77 | ·9992 888 | ·9988 786 | ·9982 865+ | ·9974 551 | ·9963 171 | ·9947 950+ |
| ·78 | ·9995 703 | ·9993 148 | ·9989 411 | ·9984 096 | ·9976 726 | ·9966 742 |
| ·79 | ·9997 487 | ·9995 948 | ·9993 668 | ·9990 383 | ·9985 770 | ·9979 440 |
| ·80 | ·9998 582 | ·9997 688 | ·9996 346 | ·9994 390 | ·9991 607 | ·9987 740 |
| ·81 | ·9999 231 | ·9998 731 | ·9997 973 | ·9996 854 | ·9995 241 | ·9992 973 |
| ·82 | ·9999 600 | ·9999 333 | ·9998 924 | ·9998 311 | ·9997 417 | ·9996 144 |
| ·83 | ·9999 802 | ·9999 666 | ·9999 455+ | ·9999 135+ | ·9998 664 | ·9997 984 |
| ·84 | ·9999 907 | ·9999 842 | ·9999 739 | ·9999 581 | ·9999 345+ | ·9999 001 |
| ·85 | ·9999 959 | ·9999 929 | ·9999 882 | ·9999 809 | ·9999 698 | ·9999 534 |
| ·86 | ·9999 983 | ·9999 970 | ·9999 950+ | ·9999 918 | ·9999 870 | ·9999 797 |
| ·87 | ·9999 994 | ·9999 989 | ·9999 981 | ·9999 968 | ·9999 948 | ·9999 918 |
| ·88 | ·9999 998 | ·9999 996 | ·9999 993 | ·9999 988 | ·9999 981 | ·9999 970 |
| ·89 | ·9999 999 | ·9999 999 | ·9999 998 | ·9999 996 | ·9999 994 | ·9999 990 |
| ·90 | 1·0000 000 | 1·0000 000 | ·9999 999 | ·9999 999 | ·9999 998 | ·9999 997 |
| ·91 | | | 1·0000 000 | 1·0000 000 | 1·0000 000 | ·9999 999 |
| ·92 | | | | | | 1·0000 000 |

$x = \cdot 19$ to $\cdot 80$ $q = 15$ $p = 21$ to $26$

| $x$ | $p = 21$ | $p = 22$ | $p = 23$ | $p = 24$ | $p = 25$ | $p = 26$ |
|---|---|---|---|---|---|---|
| $B(p, q) =$ | ·2052 5811 × $\frac{1}{10^{10}}$ | ·1197 3390 × $\frac{1}{10^{10}}$ | ·7119 3127 × $\frac{1}{10^{11}}$ | ·4309 0577 × $\frac{1}{10^{11}}$ | ·2651 7278 × $\frac{1}{10^{11}}$ | ·1657 3299 × $\frac{1}{10^{11}}$ |
| ·19 | ·0000 001 | | | | | |
| ·20 | ·0000 003 | ·0000 001 | | | | |
| ·21 | ·0000 006 | ·0000 002 | ·0000 001 | | | |
| ·22 | ·0000 013 | ·0000 005⁻ | ·0000 002 | ·0000 001 | | |
| ·23 | ·0000 029 | ·0000 011 | ·0000 004 | ·0000 001 | ·0000 001 | |
| ·24 | ·0000 060 | ·0000 023 | ·0000 009 | ·0000 003 | ·0000 001 | |
| ·25 | ·0000 118 | ·0000 048 | ·0000 019 | ·0000 007 | ·0000 003 | ·0000 001 |
| ·26 | ·0000 227 | ·0000 095⁺ | ·0000 039 | ·0000 016 | ·0000 006 | ·0000 003 |
| ·27 | ·0000 420 | ·0000 183 | ·0000 079 | ·0000 033 | ·0000 014 | ·0000 006 |
| ·28 | ·0000 754 | ·0000 341 | ·0000 152 | ·0000 067 | ·0000 029 | ·0000 012 |
| ·29 | ·0001 315⁻ | ·0000 615⁺ | ·0000 284 | ·0000 129 | ·0000 058 | ·0000 025⁺ |
| ·30 | ·0002 232 | ·0001 080 | ·0000 514 | ·0000 241 | ·0000 112 | ·0000 051 |
| ·31 | ·0003 693 | ·0001 845⁻ | ·0000 907 | ·0000 440 | ·0000 210 | ·0000 099 |
| ·32 | ·0005 966 | ·0003 074 | ·0001 559 | ·0000 779 | ·0000 384 | ·0000 187 |
| ·33 | ·0009 419 | ·0004 999 | ·0002 613 | ·0001 346 | ·0000 684 | ·0000 343 |
| ·34 | ·0014 553 | ·0007 950⁺ | ·0004 277 | ·0002 268 | ·0001 186 | ·0000 613 |
| ·35 | ·0022 026 | ·0012 374 | ·0006 846 | ·0003 734 | ·0002 009 | ·0001 067 |
| ·36 | ·0032 690 | ·0018 869 | ·0010 728 | ·0006 013 | ·0003 325⁻ | ·0001 816 |
| ·37 | ·0047 618 | ·0028 217 | ·0016 470 | ·0009 479 | ·0005 383 | ·0003 019 |
| ·38 | ·0068 132 | ·0041 413 | ·0024 799 | ·0014 643 | ·0008 533 | ·0004 910 |
| ·39 | ·0095 826 | ·0059 703 | ·0036 650⁺ | ·0022 187 | ·0013 256 | ·0007 822 |
| ·40 | ·0132 579 | ·0084 604 | ·0053 203 | ·0032 997 | ·0020 200 | ·0012 214 |
| ·41 | ·0180 547 | ·0117 926 | ·0075 912 | ·0048 202 | ·0030 213 | ·0018 708 |
| ·42 | ·0242 153 | ·0161 775⁺ | ·0106 534 | ·0069 210 | ·0044 389 | ·0028 127 |
| ·43 | ·0320 043 | ·0218 549 | ·0147 134 | ·0097 734 | ·0064 100 | ·0041 539 |
| ·44 | ·0417 028 | ·0290 902 | ·0200 090 | ·0135 812 | ·0091 031 | ·0060 294 |
| ·45 | ·0535 999 | ·0381 693 | ·0268 067 | ·0185 812 | ·0127 206 | ·0086 064 |
| ·46 | ·0679 820 | ·0493 911 | ·0353 972 | ·0250 417 | ·0174 994 | ·0120 871 |
| ·47 | ·0851 203 | ·0630 569 | ·0460 881 | ·0332 582 | ·0237 106 | ·0167 103 |
| ·48 | ·1052 567 | ·0794 581 | ·0591 944 | ·0435 474 | ·0316 556 | ·0227 510 |
| ·49 | ·1285 882 | ·0988 613 | ·0750 256 | ·0562 369 | ·0416 600 | ·0305 175⁻ |
| ·50 | ·1552 523 | ·1214 925⁻ | ·0938 708 | ·0716 533 | ·0540 645⁺ | ·0403 452 |
| ·51 | ·1853 122 | ·1475 206 | ·1159 818 | ·0901 062 | ·0692 117 | ·0525 881 |
| ·52 | ·2187 443 | ·1770 411 | ·1415 555⁺ | ·1118 710 | ·0874 308 | ·0676 049 |
| ·53 | ·2554 292 | ·2100 620 | ·1707 163 | ·1371 699 | ·1090 188 | ·0857 435⁻ |
| ·54 | ·2951 455⁺ | ·2464 917 | ·2034 994 | ·1661 522 | ·1342 204 | ·1073 210 |
| ·55 | ·3375 692 | ·2861 312 | ·2398 369 | ·1988 766 | ·1632 069 | ·1326 024 |
| ·56 | ·3822 779 | ·3286 712 | ·2795 480 | ·2352 943 | ·1960 561 | ·1617 776 |
| ·57 | ·4287 604 | ·3736 950⁻ | ·3223 339 | ·2752 381 | ·2327 341 | ·1949 395⁺ |
| ·58 | ·4764 318 | ·4206 866 | ·3677 794 | ·3184 146 | ·2730 812 | ·2320 636 |
| ·59 | ·5246 530 | ·4690 459 | ·4153 597 | ·3644 046 | ·3168 040 | ·2729 924 |
| ·60 | ·5727 541 | ·5181 083 | ·4644 561 | ·4126 700 | ·3634 733 | ·3174 252 |
| ·61 | ·6200 600 | ·5671 697 | ·5143 756 | ·4625 684 | ·4125 314 | ·3649 162 |
| ·62 | ·6659 177 | ·6155 143 | ·5643 778 | ·5133 746 | ·4633 066 | ·4148 807 |
| ·63 | ·7097 220 | ·6624 441 | ·6137 050⁻ | ·5643 089 | ·5150 363 | ·4666 111 |
| ·64 | ·7509 393 | ·7073 081 | ·6616 143 | ·6145 695⁺ | ·5668 976 | ·5193 018 |
| ·65 | ·7891 277 | ·7495 289 | ·7074 103 | ·6633 688 | ·6180 427 | ·5720 821 |
| ·66 | ·8239 517 | ·7886 263 | ·7504 749 | ·7099 680 | ·6676 384 | ·6240 558 |
| ·67 | ·8551 909 | ·8242 338 | ·7902 936 | ·7537 119 | ·7149 048 | ·6743 437 |
| ·68 | ·8827 427 | ·8561 099 | ·8264 748 | ·7940 566 | ·7591 531 | ·7221 273 |
| ·69 | ·9066 179 | ·8841 410 | ·8587 625⁺ | ·8305 924 | ·7998 166 | ·7666 894 |
| ·70 | ·9269 310 | ·9083 377 | ·8870 398 | ·8630 566 | ·8364 752 | ·8074 482 |
| ·71 | ·9438 856 | ·9288 238 | ·9113 248 | ·8913 378 | ·8688 692 | ·8439 829 |
| ·72 | ·9577 555⁺ | ·9458 202 | ·9317 583 | ·9154 708 | ·8969 032 | ·8760 480 |
| ·73 | ·9688 651 | ·9596 241 | ·9485 854 | ·9356 221 | ·9206 386 | ·9035 755⁻ |
| ·74 | ·9775 676 | ·9705 862 | ·9621 323 | ·9520 684 | ·9402 769 | ·9266 649 |
| ·75 | ·9842 258 | ·9790 870 | ·9727 802 | ·9651 709 | ·9561 349 | ·9455 628 |
| ·76 | ·9891 938 | ·9855 147 | ·9809 393 | ·9753 454 | ·9686 141 | ·9606 334 |
| ·77 | ·9928 027 | ·9902 459 | ·9870 242 | ·9830 336 | ·9781 684 | ·9723 243 |
| ·78 | ·9953 502 | ·9936 291 | ·9914 324 | ·9886 759 | ·9852 717 | ·9811 295⁻ |
| ·79 | ·9970 939 | ·9959 746 | ·9945 277 | ·9926 889 | ·9903 888 | ·9875 541 |
| ·80 | ·9982 481 | ·9975 470 | ·9966 291 | ·9954 478 | ·9939 515⁺ | ·9920 841 |

$x = \cdot 81$ to $\cdot 93$ $q = 15$ $p = 21$ to $26$

| | $p = 21$ | $p = 22$ | $p = 23$ | $p = 24$ | $p = 25$ | $p = 26$ |
|---|---|---|---|---|---|---|
| $B(p, q) =$ | ·2052 5811 × $\frac{1}{10^{10}}$ | ·1197 3390 × $\frac{1}{10^{10}}$ | ·7119 3127 × $\frac{1}{10^{11}}$ | ·4309 0577 × $\frac{1}{10^{11}}$ | ·2651 7278 × $\frac{1}{10^{11}}$ | ·1657 3299 × $\frac{1}{10^{11}}$ |
| $x$ | | | | | | |
| ·81 | ·9989 850− | ·9985 633 | ·9980 044 | ·9972 761 | ·9963 421 | ·9951 619 |
| ·82 | ·9994 370 | ·9991 945+ | ·9988 692 | ·9984 400 | ·9978 828 | ·9971 700 |
| ·83 | ·9997 025− | ·9995 698 | ·9993 896 | ·9991 489 | ·9988 327 | ·9984 232 |
| ·84 | ·9998 510 | ·9997 823 | ·9996 878 | ·9995 602 | ·9993 904 | ·9991 679 |
| ·85 | ·9999 298 | ·9998 963 | ·9998 498 | ·9997 861 | ·9997 005− | ·9995 869 |
| ·86 | ·9999 691 | ·9999 539 | ·9999 325+ | ·9999 029 | ·9998 626 | ·9998 086 |
| ·87 | ·9999 874 | ·9999 811 | ·9999 720 | ·9999 593 | ·9999 418 | ·9999 180 |
| ·88 | ·9999 953 | ·9999 929 | ·9999 894 | ·9999 844 | ·9999 775− | ·9999 680 |
| ·89 | ·9999 984 | ·9999 976 | ·9999 964 | ·9999 946 | ·9999 922 | ·9999 888 |
| ·90 | ·9999 995+ | ·9999 993 | ·9999 989 | ·9999 984 | ·9999 976 | ·9999 965+ |
| ·91 | ·9999 999 | ·9999 998 | ·9999 997 | ·9999 996 | ·9999 994 | ·9999 991 |
| ·92 | 1·0000 000 | 1·0000 000 | ·9999 999 | ·9999 999 | ·9999 999 | ·9999 998 |
| ·93 | | | 1·0000 000 | 1·0000 000 | 1·0000 000 | 1·0000 000 |

$x$ = ·26 to ·94 $q = 15$ $p$ = 27 to 32

| | $p = 27$ | $p = 28$ | $p = 29$ | $p = 30$ | $p = 31$ | $p = 32$ |
|---|---|---|---|---|---|---|
| $B(p, q)$ = | ·1050 9897 $\times \frac{1}{10^{11}}$ | ·6756 3623 $\times \frac{1}{10^{12}}$ | ·4399 4917 $\times \frac{1}{10^{12}}$ | ·2899 6650 $\times \frac{1}{10^{12}}$ | ·1933 1100 $\times \frac{1}{10^{12}}$ | ·1302 7480 $\times \frac{1}{10^{12}}$ |
| $x$ | | | | | | |
| ·26 | ·0000 001 | | | | | |
| ·27 | ·0000 002 | ·0000 001 | | | | |
| ·28 | ·0000 005$^+$ | ·0000 002 | ·0000 001 | | | |
| ·29 | ·0000 011 | ·0000 005$^-$ | ·0000 002 | ·0000 001 | | |
| ·30 | ·0000 023 | ·0000 010 | ·0000 005$^-$ | ·0000 002 | ·0000 001 | |
| ·31 | ·0000 046 | ·0000 021 | ·0000 010 | ·0000 004 | ·0000 002 | ·0000 001 |
| ·32 | ·0000 090 | ·0000 043 | ·0000 020 | ·0000 009 | ·0000 004 | ·0000 002 |
| ·33 | ·0000 170 | ·0000 083 | ·0000 040 | ·0000 019 | ·0000 009 | ·0000 004 |
| ·34 | ·0000 313 | ·0000 158 | ·0000 079 | ·0000 039 | ·0000 019 | ·0000 009 |
| ·35 | ·0000 560 | ·0000 291 | ·0000 149 | ·0000 076 | ·0000 038 | ·0000 019 |
| ·36 | ·0000 980 | ·0000 523 | ·0000 276 | ·0000 144 | ·0000 075$^-$ | ·0000 038 |
| ·37 | ·0001 673 | ·0000 917 | ·0000 497 | ·0000 267 | ·0000 142 | ·0000 075$^-$ |
| ·38 | ·0002 792 | ·0001 570 | ·0000 874 | ·0000 481 | ·0000 263 | ·0000 142 |
| ·39 | ·0004 562 | ·0002 631 | ·0001 502 | ·0000 849 | ·0000 475$^-$ | ·0000 263 |
| ·40 | ·0007 300 | ·0004 315$^+$ | ·0002 524 | ·0001 462 | ·0000 839 | ·0000 477 |
| ·41 | ·0011 450$^+$ | ·0006 932 | ·0004 153 | ·0002 464 | ·0001 448 | ·0000 844 |
| ·42 | ·0017 619 | ·0010 917 | ·0006 695$^+$ | ·0004 066 | ·0002 446 | ·0001 459 |
| ·43 | ·0026 614 | ·0016 868 | ·0010 582 | ·0006 574 | ·0004 047 | ·0002 469 |
| ·44 | ·0039 487 | ·0025 585$^-$ | ·0016 409 | ·0010 423 | ·0006 560 | ·0004 093 |
| ·45 | ·0057 581 | ·0038 118 | ·0024 980 | ·0016 214 | ·0010 428 | ·0006 649 |
| ·46 | ·0082 568 | ·0055 813 | ·0037 352 | ·0024 761 | ·0016 266 | ·0010 593 |
| ·47 | ·0116 486 | ·0080 360 | ·0054 891 | ·0037 142 | ·0024 907 | ·0016 560 |
| ·48 | ·0161 754 | ·0113 824 | ·0079 315$^+$ | ·0054 754 | ·0037 463 | ·0025 416 |
| ·49 | ·0221 177 | ·0158 676 | ·0112 738 | ·0079 362 | ·0055 375$^+$ | ·0038 314 |
| ·50 | ·0297 919 | ·0217 793 | ·0157 697 | ·0113 144 | ·0080 472 | ·0056 758 |
| ·51 | ·0395 449 | ·0294 437 | ·0217 162 | ·0158 727 | ·0115 017 | ·0082 658 |
| ·52 | ·0517 442 | ·0392 201 | ·0294 513 | ·0219 192 | ·0161 747 | ·0118 386 |
| ·53 | ·0667 651 | ·0514 911 | ·0393 482 | ·0298 056 | ·0223 878 | ·0166 809 |
| ·54 | ·0849 738 | ·0666 491 | ·0518 061 | ·0399 214 | ·0305 088 | ·0231 305$^+$ |
| ·55 | ·1067 062 | ·0850 781 | ·0672 349 | ·0526 835$^+$ | ·0409 454 | ·0315 738 |
| ·56 | ·1322 454 | ·1071 320 | ·0860 368 | ·0685 205$^+$ | ·0541 338 | ·0424 388 |
| ·57 | ·1617 966 | ·1331 096 | ·1085 822 | ·0878 523 | ·0705 221 | ·0561 827 |
| ·58 | ·1954 633 | ·1632 279 | ·1351 831 | ·1110 646 | ·0905 478 | ·0732 739 |
| ·59 | ·2332 249 | ·1975 962 | ·1660 648 | ·1384 803 | ·1146 105$^-$ | ·0941 671 |
| ·60 | ·2749 192 | ·2361 915$^+$ | ·2013 366 | ·1703 277 | ·1430 400 | ·1192 732 |
| ·61 | ·3202 311 | ·2788 395$^-$ | ·2409 661 | ·2067 104 | ·1760 629 | ·1489 250$^-$ |
| ·62 | ·3686 899 | ·3252 021 | ·2847 585$^-$ | ·2475 782 | ·2137 689 | ·1833 406 |
| ·63 | ·4196 760 | ·3747 748 | ·3323 431 | ·2927 060 | ·2560 814 | ·2225 876 |
| ·64 | ·4724 384 | ·4268 940 | ·3831 714 | ·3416 802 | ·3027 338 | ·2665 513 |
| ·65 | ·5261 216 | ·4807 567 | ·4365 261 | ·3938 968 | ·3532 570 | ·3149 113 |
| ·66 | ·5798 027 | ·5354 512 | ·4915 433 | ·4485 741 | ·4069 799 | ·3671 300 |
| ·67 | ·6325 345$^+$ | ·5899 975$^+$ | ·5472 479 | ·5047 783 | ·4630 449 | ·4224 557 |
| ·68 | ·6833 927 | ·6433 957 | ·6025 987 | ·5614 640 | ·5204 391 | ·4799 435$^+$ |
| ·69 | ·7315 237 | ·6946 779 | ·6565 424 | ·6175 259 | ·5780 412 | ·5384 928 |
| ·70 | ·7761 885$^-$ | ·7429 605$^+$ | ·7080 712 | ·6718 584 | ·6346 800 | ·5969 019 |
| ·71 | ·8167 995$^-$ | ·7874 917 | ·7562 789 | ·7234 193 | ·6892 016 | ·6539 352 |
| ·72 | ·8529 469 | ·8276 897 | ·8004 119 | ·7712 905$^-$ | ·7405 383 | ·7083 972 |
| ·73 | ·8844 122 | ·8631 695$^-$ | ·8399 086 | ·8147 307 | ·7877 736 | ·7592 078 |
| ·74 | ·9111 680 | ·8937 542 | ·8744 249 | ·8532 159 | ·8301 971 | ·8054 704 |
| ·75 | ·9333 641 | ·9194 713 | ·9038 418 | ·8864 607 | ·8673 416 | ·8465 263 |
| ·76 | ·9513 022 | ·9405 332 | ·9282 566 | ·9144 222 | ·8990 013 | ·8819 887 |
| ·77 | ·9654 013 | ·9573 066 | ·9479 571 | ·9372 826 | ·9252 275$^+$ | ·9117 530 |
| ·78 | ·9761 588 | ·9702 713 | ·9633 829 | ·9554 161 | ·9463 021 | ·9359 827 |
| ·79 | ·9841 089 | ·9799 760 | ·9750 785$^-$ | ·9693 416 | ·9626 945$^-$ | ·9550 718 |
| ·80 | ·9897 858 | ·9869 938 | ·9836 434 | ·9796 691 | ·9750 060 | ·9695 907 |
| ·81 | ·9936 912 | ·9918 822 | ·9896 842 | ·9870 445$^-$ | ·9839 084 | ·9802 210 |
| ·82 | ·9962 708 | ·9951 511 | ·9937 738 | ·9920 993 | ·9900 854 | ·9876 883 |
| ·83 | ·9979 004 | ·9972 414 | ·9964 209 | ·9954 112 | ·9941 821 | ·9927 011 |
| ·84 | ·9988 808 | ·9985 136 | ·9980 515$^+$ | ·9974 759 | ·9967 668 | ·9959 022 |
| ·85 | ·9994 383 | ·9992 466 | ·9990 021 | ·9986 940 | ·9983 099 | ·9978 359 |
| ·86 | ·9997 370 | ·9996 436 | ·9995 231 | ·9993 695$^-$ | ·9991 756 | ·9989 337 |
| ·87 | ·9998 863 | ·9998 443 | ·9997 895$^-$ | ·9997 188 | ·9996 285$^+$ | ·9995 146 |
| ·88 | ·9999 551 | ·9999 379 | ·9999 152 | ·9998 855$^+$ | ·9998 473 | ·9997 984 |
| ·89 | ·9999 841 | ·9999 777 | ·9999 693 | ·9999 581 | ·9999 436 | ·9999 248 |
| ·90 | ·9999 950$^+$ | ·9999 930 | ·9999 902 | ·9999 865$^+$ | ·9999 816 | ·9999 753 |
| ·91 | ·9999 987 | ·9999 981 | ·9999 973 | ·9999 963 | ·9999 949 | ·9999 930 |
| ·92 | ·9999 997 | ·9999 996 | ·9999 994 | ·9999 991 | ·9999 988 | ·9999 984 |
| ·93 | ·9999 999 | ·9999 999 | ·9999 999 | ·9999 998 | ·9999 998 | ·9999 997 |
| ·94 | 1·0000 000 | 1·0000 000 | 1·0000 000 | 1·0000 000 | 1·0000 000 | 1·0000 000 |

$x$ = ·32 to ·95 $q = 15$ $p$ = 33 to 38

| $x$ | $p = 33$ | $p = 34$ | $p = 35$ | $p = 36$ | $p = 37$ | $p = 38$ |
|---|---|---|---|---|---|---|
| $B(p, q) =$ | ·8869 7739 × $\frac{1}{10^{13}}$ | ·6097 9695+ × $\frac{1}{10^{13}}$ | ·4231 2442 × $\frac{1}{10^{13}}$ | ·2961 8709 × $\frac{1}{10^{13}}$ | ·2090 7324 × $\frac{1}{10^{13}}$ | ·1487 6365+ × $\frac{1}{10^{13}}$ |
| ·32 | ·0000 001 | | | | | |
| ·33 | ·0000 002 | ·0000 001 | | | | |
| ·34 | ·0000 004 | ·0000 002 | ·0000 001 | | | |
| ·35 | ·0000 009 | ·0000 005− | ·0000 002 | ·0000 001 | ·0000 001 | |
| ·36 | ·0000 019 | ·0000 010 | ·0000 005− | ·0000 002 | ·0000 001 | ·0000 001 |
| ·37 | ·0000 039 | ·0000 020 | ·0000 010 | ·0000 005+ | ·0000 003 | ·0000 001 |
| ·38 | ·0000 076 | ·0000 041 | ·0000 021 | ·0000 011 | ·0000 006 | ·0000 003 |
| ·39 | ·0000 145− | ·0000 079 | ·0000 043 | ·0000 023 | ·0000 012 | ·0000 006 |
| ·40 | ·0000 269 | ·0000 150+ | ·0000 083 | ·0000 046 | ·0000 025+ | ·0000 014 |
| ·41 | ·0000 487 | ·0000 279 | ·0000 159 | ·0000 090 | ·0000 050+ | ·0000 028 |
| ·42 | ·0000 863 | ·0000 506 | ·0000 294 | ·0000 170 | ·0000 098 | ·0000 056 |
| ·43 | ·0001 494 | ·0000 897 | ·0000 534 | ·0000 316 | ·0000 185+ | ·0000 108 |
| ·44 | ·0002 532 | ·0001 554 | ·0000 947 | ·0000 572 | ·0000 344 | ·0000 205+ |
| ·45 | ·0004 204 | ·0002 637 | ·0001 642 | ·0001 015− | ·0000 623 | ·0000 380 |
| ·46 | ·0006 842 | ·0004 385− | ·0002 789 | ·0001 761 | ·0001 104 | ·0000 688 |
| ·47 | ·0010 920 | ·0007 145− | ·0004 640 | ·0002 992 | ·0001 916 | ·0001 219 |
| ·48 | ·0017 103 | ·0011 419 | ·0007 568 | ·0004 980 | ·0003 255+ | ·0002 114 |
| ·49 | ·0026 296 | ·0017 910 | ·0012 108 | ·0008 128 | ·0005 419 | ·0003 590 |
| ·50 | ·0039 714 | ·0027 576 | ·0019 008 | ·0013 011 | ·0008 846 | ·0005 976 |
| ·51 | ·0058 935− | ·0041 703 | ·0029 296 | ·0020 437 | ·0014 163 | ·0009 752 |
| ·52 | ·0085 973 | ·0061 968 | ·0044 345+ | ·0031 516 | ·0022 251 | ·0015 609 |
| ·53 | ·0123 329 | ·0090 509 | ·0065 951 | ·0047 730 | ·0034 316 | ·0024 518 |
| ·54 | ·0174 031 | ·0129 982 | ·0096 401 | ·0071 014 | ·0051 974 | ·0037 802 |
| ·55 | ·0241 644 | ·0183 604 | ·0138 537 | ·0103 836 | ·0077 328 | ·0057 232 |
| ·56 | ·0330 243 | ·0255 155+ | ·0195 792 | ·0149 251 | ·0113 052 | ·0085 111 |
| ·57 | ·0444 334 | ·0348 951 | ·0272 196 | ·0210 946 | ·0162 455+ | ·0124 358 |
| ·58 | ·0588 719 | ·0469 749 | ·0372 334 | ·0293 233 | ·0229 512 | ·0178 570 |
| ·59 | ·0768 286 | ·0622 590 | ·0501 234 | ·0400 993 | ·0318 852 | ·0252 051 |
| ·60 | ·0987 743 | ·0812 571 | ·0664 190 | ·0539 550+ | ·0435 684 | ·0349 783 |
| ·61 | ·1251 285− | ·1044 543 | ·0866 503 | ·0714 456 | ·0585 639 | ·0477 327 |
| ·62 | ·1562 214 | ·1322 743 | ·1113 135− | ·0931 195+ | ·0774 525+ | ·0640 636 |
| ·63 | ·1922 548 | ·1650 379 | ·1408 310 | ·1194 804 | ·1007 987 | ·0845 759 |
| ·64 | ·2332 634 | ·2029 210 | ·1755 057 | ·1509 416 | ·1291 069 | ·1098 451 |
| ·65 | ·2790 820 | ·2459 128 | ·2154 752 | ·1877 770 | ·1627 716 | ·1403 682 |
| ·66 | ·3293 224 | ·2937 832 | ·2606 691 | ·2300 716 | ·2020 239 | ·1765 081 |
| ·67 | ·3833 633 | ·3460 597 | ·3107 749 | ·2776 777 | ·2468 789 | ·2184 358 |
| ·68 | ·4403 591 | ·4020 221 | ·3652 187 | ·3301 818 | ·2970 914 | ·2660 759 |
| ·69 | ·4992 657 | ·4607 162 | ·4231 644 | ·3868 894 | ·3521 259 | ·3190 629 |
| ·70 | ·5588 877 | ·5209 888 | ·4835 356 | ·4468 316 | ·4111 471 | ·3767 164 |
| ·71 | ·6179 415+ | ·5815 442 | ·5450 613 | ·5087 973 | ·4730 369 | ·4380 401 |
| ·72 | ·6751 313 | ·6410 186 | ·6063 440 | ·5713 920 | ·5364 400 | ·5017 526 |
| ·73 | ·7292 315+ | ·6980 653 | ·6659 458 | ·6331 196 | ·5998 375− | ·5663 485− |
| ·74 | ·7791 675− | ·7514 457 | ·7224 847 | ·6924 810 | ·6616 439 | ·6301 901 |
| ·75 | ·8240 849 | ·8001 134 | ·7747 318 | ·7480 811 | ·7203 200 | ·6916 210 |
| ·76 | ·8634 024 | ·8432 841 | ·8216 984 | ·7987 312 | ·7744 880 | ·7490 917 |
| ·77 | ·8968 385− | ·8804 822 | ·8627 019 | ·8435 348 | ·8230 366 | ·8012 809 |
| ·78 | ·9244 120 | ·9115 581 | ·8974 037 | ·8819 470 | ·8652 023 | ·8471 995+ |
| ·79 | ·9464 152 | ·9366 752 | ·9258 123 | ·9137 979 | ·9006 155− | ·8862 609 |
| ·80 | ·9633 631 | ·9562 674 | ·9482 535+ | ·9392 779 | ·9293 050+ | ·9183 079 |
| ·81 | ·9759 275− | ·9709 743 | ·9653 102 | ·9588 871 | ·9516 611 | ·9435 934 |
| ·82 | ·9848 626 | ·9815 626 | ·9777 423 | ·9733 566 | ·9683 618 | ·9627 164 |
| ·83 | ·9909 342 | ·9888 454 | ·9863 979 | ·9835 539 | ·9802 754 | ·9765 246 |
| ·84 | ·9948 581 | ·9936 090 | ·9921 278 | ·9903 858 | ·9883 535+ | ·9860 005− |
| ·85 | ·9972 568 | ·9965 557 | ·9957 144 | ·9947 133 | ·9935 314 | ·9921 466 |
| ·86 | ·9986 345+ | ·9982 682 | ·9978 233 | ·9972 877 | ·9966 480 | ·9958 897 |
| ·87 | ·9993 721 | ·9991 956 | ·9989 787 | ·9987 146 | ·9983 955+ | ·9980 128 |
| ·88 | ·9997 366 | ·9996 591 | ·9995 628 | ·9994 442 | ·9992 993 | ·9991 234 |
| ·89 | ·9999 007 | ·9998 702 | ·9998 319 | ·9997 841 | ·9997 251 | ·9996 527 |
| ·90 | ·9999 670 | ·9999 565− | ·9999 431 | ·9999 262 | ·9999 050+ | ·9998 788 |
| ·91 | ·9999 906 | ·9999 875− | ·9999 835− | ·9999 783 | ·9999 719 | ·9999 637 |
| ·92 | ·9999 978 | ·9999 970 | ·9999 960 | ·9999 947 | ·9999 931 | ·9999 910 |
| ·93 | ·9999 996 | ·9999 994 | ·9999 992 | ·9999 990 | ·9999 987 | ·9999 982 |
| ·94 | ·9999 999 | ·9999 999 | ·9999 999 | ·9999 999 | ·9999 998 | ·9999 998 |
| ·95 | 1·0000 000 | 1·0000 000 | 1·0000 000 | 1·0000 000 | 1·0000 000 | 1·0000 000 |

$x = \cdot37$ to $\cdot96$ $q = 15$ $p = 39$ to $44$

| | $p = 39$ | $p = 40$ | $p = 41$ | $p = 42$ | $p = 43$ | $p = 44$ |
|---|---|---|---|---|---|---|
| $B(p, q) =$ | $\cdot1066\ 6073 \times \frac{1}{10^{13}}$ | $\cdot7703\ 2751 \times \frac{1}{10^{14}}$ | $\cdot5602\ 3819 \times \frac{1}{10^{14}}$ | $\cdot4101\ 7439 \times \frac{1}{10^{14}}$ | $\cdot3022\ 3376 \times \frac{1}{10^{14}}$ | $\cdot2240\ 6986 \times \frac{1}{10^{14}}$ |
| $x$ | | | | | | |
| ·37 | ·0000 001 | | | | | |
| ·38 | ·0000 002 | ·0000 001 | | | | |
| ·39 | ·0000 003 | ·0000 002 | ·0000 001 | | | |
| ·40 | ·0000 007 | ·0000 004 | ·0000 002 | ·0000 001 | ·0000 001 | |
| ·41 | ·0000 015+ | ·0000 008 | ·0000 005− | ·0000 003 | ·0000 001 | ·0000 001 |
| ·42 | ·0000 032 | ·0000 018 | ·0000 010 | ·0000 006 | ·0000 003 | ·0000 002 |
| ·43 | ·0000 063 | ·0000 036 | ·0000 021 | ·0000 012 | ·0000 007 | ·0000 004 |
| ·44 | ·0000 121 | ·0000 072 | ·0000 042 | ·0000 024 | ·0000 014 | ·0000 008 |
| ·45 | ·0000 230 | ·0000 139 | ·0000 083 | ·0000 049 | ·0000 029 | ·0000 017 |
| ·46 | ·0000 426 | ·0000 262 | ·0000 160 | ·0000 097 | ·0000 059 | ·0000 035+ |
| ·47 | ·0000 770 | ·0000 484 | ·0000 302 | ·0000 188 | ·0000 116 | ·0000 071 |
| ·48 | ·0001 364 | ·0000 875− | ·0000 558 | ·0000 354 | ·0000 223 | ·0000 140 |
| ·49 | ·0002 363 | ·0001 547 | ·0001 006 | ·0000 651 | ·0000 419 | ·0000 268 |
| ·50 | ·0004 012 | ·0002 677 | ·0001 776 | ·0001 172 | ·0000 769 | ·0000 502 |
| ·51 | ·0006 673 | ·0004 540 | ·0003 071 | ·0002 066 | ·0001 382 | ·0000 920 |
| ·52 | ·0010 884 | ·0007 544 | ·0005 200 | ·0003 565− | ·0002 431 | ·0001 649 |
| ·53 | ·0017 411 | ·0012 292 | ·0008 630 | ·0006 026 | ·0004 186 | ·0002 893 |
| ·54 | ·0027 329 | ·0019 644 | ·0014 042 | ·0009 983 | ·0007 061 | ·0004 970 |
| ·55 | ·0042 107 | ·0030 802 | ·0022 409 | ·0016 216 | ·0011 674 | ·0008 363 |
| ·56 | ·0063 699 | ·0047 404 | ·0035 085+ | ·0025 831 | ·0018 921 | ·0013 792 |
| ·57 | ·0094 642 | ·0071 623 | ·0053 911 | ·0040 367 | ·0030 074 | ·0022 296 |
| ·58 | ·0138 139 | ·0106 270 | ·0081 317 | ·0061 902 | ·0046 888 | ·0035 345− |
| ·59 | ·0198 118 | ·0154 875+ | ·0120 432 | ·0093 172 | ·0071 727 | ·0054 955− |
| ·60 | ·0279 254 | ·0221 746 | ·0175 164 | ·0137 672 | ·0107 678 | ·0083 822 |
| ·61 | ·0386 916 | ·0311 968 | ·0250 247 | ·0199 742 | ·0158 665− | ·0125 449 |
| ·62 | ·0527 041 | ·0431 331 | ·0351 221 | ·0284 593 | ·0229 514 | ·0184 246 |
| ·63 | ·0705 901 | ·0586 162 | ·0484 323 | ·0398 257 | ·0325 961 | ·0265 586 |
| ·64 | ·0929 758 | ·0783 039 | ·0656 274 | ·0547 441 | ·0454 570 | ·0375 781 |
| ·65 | ·1204 409 | ·1028 385+ | ·0873 924 | ·0739 241 | ·0622 516 | ·0521 943 |
| ·66 | ·1534 633 | ·1327 939 | ·1143 775− | ·0980 722 | ·0837 235+ | ·0711 701 |
| ·67 | ·1923 580 | ·1686 138 | ·1471 372 | ·1278 345− | ·1105 907 | ·0952 758 |
| ·68 | ·2372 151 | ·2105 448 | ·1860 614 | ·1637 279 | ·1434 787 | ·1252 263 |
| ·69 | ·2878 445+ | ·2585 713 | ·2313 033 | ·2060 637 | ·1828 433 | ·1616 048 |
| ·70 | ·3437 353 | ·3123 611 | ·2827 124 | ·2548 716 | ·2288 869 | ·2047 754 |
| ·71 | ·4040 379 | ·3712 302 | ·3397 839 | ·3098 333 | ·2814 801 | ·2547 951 |
| ·72 | ·4675 763 | ·4341 361 | ·4016 322 | ·3702 383 | ·3401 001 | ·3113 356 |
| ·73 | ·5328 947 | ·4997 069 | ·4670 003 | ·4349 717 | ·4037 973 | ·3736 306 |
| ·74 | ·5983 389 | ·5663 081 | ·5343 093 | ·5025 446 | ·4712 035− | ·4404 600 |
| ·75 | ·6621 668 | ·6321 461 | ·6017 502 | ·5711 690 | ·5405 877 | ·5101 842 |
| ·76 | ·7226 795− | ·6954 004 | ·6674 121 | ·6388 776 | ·6099 627 | ·5808 327 |
| ·77 | ·7783 572 | ·7543 696 | ·7294 345− | ·7036 784 | ·6772 354 | ·6502 451 |
| ·78 | ·8279 839 | ·8076 153 | ·7861 672 | ·7637 252 | ·7403 854 | ·7162 533 |
| ·79 | ·8707 428 | ·8540 827 | ·8363 147 | ·8174 850+ | ·7976 511 | ·7768 808 |
| ·80 | ·9062 689 | ·8931 804 | ·8790 448 | ·8638 749 | ·8476 937 | ·8305 341 |
| ·81 | ·9346 510 | ·9248 075+ | ·9140 436 | ·9023 477 | ·8897 162 | ·8761 534 |
| ·82 | ·9563 816 | ·9493 223 | ·9415 077 | ·9329 117 | ·9235 133 | ·9132 975+ |
| ·83 | ·9722 645+ | ·9674 593 | ·9620 751 | ·9560 802 | ·9494 459 | ·9421 466 |
| ·84 | ·9832 957 | ·9802 081 | ·9767 067 | ·9727 613 | ·9683 424 | ·9634 221 |
| ·85 | ·9905 360 | ·9886 754 | ·9865 405− | ·9841 061 | ·9813 471 | ·9782 385− |
| ·86 | ·9949 972 | ·9939 542 | ·9927 433 | ·9913 462 | ·9897 443 | ·9879 182 |
| ·87 | ·9975 572 | ·9970 185− | ·9963 858 | ·9956 474 | ·9947 909 | ·9938 031 |
| ·88 | ·9989 117 | ·9986 585+ | ·9983 577 | ·9980 026 | ·9975 859 | ·9970 999 |
| ·89 | ·9995 645− | ·9994 578 | ·9993 297 | ·9991 766 | ·9989 951 | ·9987 809 |
| ·90 | ·9998 465+ | ·9998 071 | ·9997 591 | ·9997 012 | ·9996 317 | ·9995 488 |
| ·91 | ·9999 536 | ·9999 411 | ·9999 258 | ·9999 070 | ·9998 842 | ·9998 568 |
| ·92 | ·9999 884 | ·9999 851 | ·9999 811 | ·9999 760 | ·9999 699 | ·9999 624 |
| ·93 | ·9999 977 | ·9999 970 | ·9999 962 | ·9999 951 | ·9999 938 | ·9999 922 |
| ·94 | ·9999 997 | ·9999 996 | ·9999 994 | ·9999 993 | ·9999 991 | ·9999 988 |
| ·95 | 1·0000 000 | 1·0000 000 | ·9999 999 | ·9999 999 | ·9999 999 | ·9999 999 |
| ·96 | | | 1·0000 000 | 1·0000 000 | 1·0000 000 | 1·0000 000 |

$x$ = ·42 to ·96 $q = 15$ $p$ = 45 to 50

| $x$ | $p = 45$ | $p = 46$ | $p = 47$ | $p = 48$ | $p = 49$ | $p = 50$ |
|---|---|---|---|---|---|---|
| $B(p, q)$ = | ·1671 0294 $\times \frac{1}{10^{14}}$ | ·1253 2721 $\times \frac{1}{10^{14}}$ | ·9450 9042 $\times \frac{1}{10^{15}}$ | ·7164 3951 $\times \frac{1}{10^{15}}$ | ·5458 5868 $\times \frac{1}{10^{15}}$ | ·4179 2305 $\times \frac{1}{10^{15}}$ |
| ·42 | ·0000 001 | | | | | |
| ·43 | ·0000 002 | ·0000 001 | ·0000 001 | | | |
| ·44 | ·0000 005⁻ | ·0000 003 | ·0000 002 | ·0000 001 | | |
| ·45 | ·0000 010 | ·0000 006 | ·0000 003 | ·0000 002 | ·0000 001 | ·0000 001 |
| ·46 | ·0000 021 | ·0000 013 | ·0000 008 | ·0000 004 | ·0000 003 | ·0000 002 |
| ·47 | ·0000 044 | ·0000 027 | ·0000 016 | ·0000 010 | ·0000 006 | ·0000 003 |
| ·48 | ·0000 087 | ·0000 054 | ·0000 034 | ·0000 021 | ·0000 013 | ·0000 008 |
| ·49 | ·0000 171 | ·0000 108 | ·0000 068 | ·0000 043 | ·0000 027 | ·0000 017 |
| ·50 | ·0000 327 | ·0000 211 | ·0000 136 | ·0000 087 | ·0000 056 | ·0000 035⁺ |
| ·51 | ·0000 610 | ·0000 402 | ·0000 264 | ·0000 172 | ·0000 112 | ·0000 073 |
| ·52 | ·0001 114 | ·0000 748 | ·0000 501 | ·0000 333 | ·0000 221 | ·0000 146 |
| ·53 | ·0001 990 | ·0001 362 | ·0000 928 | ·0000 630 | ·0000 426 | ·0000 286 |
| ·54 | ·0003 481 | ·0002 427 | ·0001 684 | ·0001 164 | ·0000 801 | ·0000 549 |
| ·55 | ·0005 962 | ·0004 231 | ·0002 989 | ·0002 103 | ·0001 473 | ·0001 028 |
| ·56 | ·0010 005⁺ | ·0007 225⁺ | ·0005 195⁻ | ·0003 719 | ·0002 651 | ·0001 882 |
| ·57 | ·0016 453 | ·0012 085⁺ | ·0008 839 | ·0006 437 | ·0004 668 | ·0003 372 |
| ·58 | ·0026 519 | ·0019 808 | ·0014 731 | ·0010 909 | ·0008 046 | ·0005 911 |
| ·59 | ·0041 911 | ·0031 820 | ·0024 055⁺ | ·0018 109 | ·0013 578 | ·0010 140 |
| ·60 | ·0064 955⁻ | ·0050 112 | ·0038 496 | ·0029 450⁺ | ·0022 440 | ·0017 032 |
| ·61 | ·0098 741 | ·0077 380 | ·0060 385⁻ | ·0046 929 | ·0036 327 | ·0028 012 |
| ·62 | ·0147 250⁺ | ·0117 177 | ·0092 857 | ·0073 287 | ·0057 614 | ·0045 121 |
| ·63 | ·0215 447 | ·0174 033 | ·0140 001 | ·0112 174 | ·0089 530 | ·0071 189 |
| ·64 | ·0309 312 | ·0253 537 | ·0206 977 | ·0168 302 | ·0136 331 | ·0110 024 |
| ·65 | ·0435 770 | ·0362 331 | ·0300 068 | ·0247 541 | ·0203 441 | ·0166 586 |
| ·66 | ·0602 487 | ·0507 980 | ·0426 621 | ·0356 930 | ·0297 519 | ·0247 104 |
| ·67 | ·0817 499 | ·0698 682 | ·0594 846 | ·0504 554 | ·0426 413 | ·0359 100 |
| ·68 | ·1088 654 | ·0942 788 | ·0813 411 | ·0699 230 | ·0598 940 | ·0511 258 |
| ·69 | ·1422 874 | ·1248 115⁺ | ·1090 833 | ·0949 981 | ·0824 448 | ·0713 081 |
| ·70 | ·1825 272 | ·1621 082 | ·1434 648 | ·1265 270 | ·1112 125⁻ | ·0974 294 |
| ·71 | ·2298 203 | ·2065 716 | ·1850 413 | ·1652 013 | ·1470 064 | ·1303 970 |
| ·72 | ·2840 355⁻ | ·2582 642 | ·2340 616 | ·2114 450⁺ | ·1904 116 | ·1709 407 |
| ·73 | ·3446 020 | ·3168 184 | ·2903 636 | ·2652 990 | ·2416 653 | ·2194 833 |
| ·74 | ·4104 712 | ·3813 753 | ·3532 915⁺ | ·3263 191 | ·3005 380 | ·2760 091 |
| ·75 | ·4801 263 | ·4505 693 | ·4216 548 | ·3935 094 | ·3662 434 | ·3399 513 |
| ·76 | ·5516 496 | ·5225 703 | ·4937 439 | ·4653 100 | ·4373 974 | ·4101 228 |
| ·77 | ·6228 500⁻ | ·5951 931 | ·5674 159 | ·5396 565⁺ | ·5120 474 | ·4847 145⁻ |
| ·78 | ·6914 410 | ·6660 663 | ·6402 503 | ·6141 157 | ·5877 852 | ·5613 794 |
| ·79 | ·7552 513 | ·7328 480 | ·7097 629 | ·6860 933 | ·6619 404 | ·6374 079 |
| ·80 | ·8124 385⁻ | ·7934 582 | ·7736 527 | ·7530 886 | ·7318 391 | ·7099 825⁺ |
| ·81 | ·8616 721 | ·8462 929 | ·8300 444 | ·8129 628 | ·7950 912 | ·7764 792 |
| ·82 | ·9022 552 | ·8903 835⁻ | ·8776 859 | ·8641 724 | ·8498 593 | ·8347 693 |
| ·83 | ·9341 606 | ·9254 699 | ·9160 614 | ·9059 262 | ·8950 604 | ·8834 650⁺ |
| ·84 | ·9579 739 | ·9519 736 | ·9453 994 | ·9382 321 | ·9304 556 | ·9220 569 |
| ·85 | ·9747 554 | ·9708 737 | ·9665 700 | ·9618 223 | ·9566 096 | ·9509 130 |
| ·86 | ·9858 479 | ·9835 137 | ·9808 952 | ·9779 725⁺ | ·9747 260 | ·9711 362 |
| ·87 | ·9926 703 | ·9913 781 | ·9899 118 | ·9882 561 | ·9863 955⁻ | ·9843 143 |
| ·88 | ·9965 361 | ·9958 856 | ·9951 390 | ·9942 862 | ·9933 169 | ·9922 202 |
| ·89 | ·9985 295⁺ | ·9982 363 | ·9978 959 | ·9975 026 | ·9970 506 | ·9965 333 |
| ·90 | ·9994 504 | ·9993 343 | ·9991 981 | ·9990 389 | ·9988 539 | ·9986 398 |
| ·91 | ·9998 239 | ·9997 846 | ·9997 379 | ·9996 828 | ·9996 181 | ·9995 423 |
| ·92 | ·9999 533 | ·9999 423 | ·9999 291 | ·9999 134 | ·9998 947 | ·9998 726 |
| ·93 | ·9999 902 | ·9999 878 | ·9999 849 | ·9999 814 | ·9999 771 | ·9999 721 |
| ·94 | ·9999 985⁺ | ·9999 981 | ·9999 977 | ·9999 971 | ·9999 964 | ·9999 955⁺ |
| ·95 | ·9999 999 | ·9999 998 | ·9999 998 | ·9999 997 | ·9999 996 | ·9999 995⁺ |
| ·96 | 1·0000 000 | 1·0000 000 | 1·0000 000 | 1·0000 000 | 1·0000 000 | 1·0000 000 |

$x = \cdot 12$ to $\cdot 70$ $q = 16$ $p = 16$ to $21$

| | $p = 16$ | $p = 17$ | $p = 18$ | $p = 19$ | $p = 20$ | $p = 21$ |
|---|---|---|---|---|---|---|
| $B(p, q) =$ | $\cdot 2079\ 5887 \times \frac{1}{10^9}$ | $\cdot 1039\ 7944 \times \frac{1}{10^9}$ | $\cdot 5356\ 5164 \times \frac{1}{10^{10}}$ | $\cdot 2835\ 8028 \times \frac{1}{10^{10}}$ | $\cdot 1539\ 4358 \times \frac{1}{10^{10}}$ | $\cdot 8552\ 4212 \times \frac{1}{10^{11}}$ |
| $x$ | | | | | | |
| ·12 | ·0000 001 | | | | | |
| ·13 | ·0000 003 | ·0000 001 | | | | |
| ·14 | ·0000 008 | ·0000 002 | ·0000 001 | | | |
| ·15 | ·0000 020 | ·0000 006 | ·0000 002 | | | |
| ·16 | ·0000 049 | ·0000 014 | ·0000 004 | ·0000 001 | | |
| ·17 | ·0000 109 | ·0000 034 | ·0000 011 | ·0000 003 | ·0000 001 | |
| ·18 | ·0000 229 | ·0000 077 | ·0000 025+ | ·0000 008 | ·0000 003 | ·0000 001 |
| ·19 | ·0000 460 | ·0000 162 | ·0000 056 | ·0000 019 | ·0000 006 | ·0000 002 |
| ·20 | ·0000 882 | ·0000 327 | ·0000 118 | ·0000 042 | ·0000 015− | ·0000 005− |
| ·21 | ·0001 620 | ·0000 631 | ·0000 240 | ·0000 089 | ·0000 032 | ·0000 012 |
| ·22 | ·0002 867 | ·0001 168 | ·0000 464 | ·0000 180 | ·0000 069 | ·0000 026 |
| ·23 | ·0004 898 | ·0002 083 | ·0000 865− | ·0000 351 | ·0000 139 | ·0000 054 |
| ·24 | ·0008 104 | ·0003 592 | ·0001 554 | ·0000 658 | ·0000 273 | ·0000 111 |
| ·25 | ·0013 016 | ·0006 003 | ·0002 703 | ·0001 190 | ·0000 513 | ·0000 217 |
| ·26 | ·0020 340 | ·0009 743 | ·0004 556 | ·0002 084 | ·0000 934 | ·0000 411 |
| ·27 | ·0030 980 | ·0015 388 | ·0007 464 | ·0003 542 | ·0001 646 | ·0000 751 |
| ·28 | ·0046 072 | ·0023 697 | ·0011 905− | ·0005 851 | ·0002 818 | ·0001 332 |
| ·29 | ·0066 999 | ·0035 637 | ·0018 517 | ·0009 415− | ·0004 691 | ·0002 294 |
| ·30 | ·0095 404 | ·0052 410 | ·0028 131 | ·0014 777 | ·0007 609 | ·0003 845− |
| ·31 | ·0133 190 | ·0075 476 | ·0041 799 | ·0022 659 | ·0012 041 | ·0006 281 |
| ·32 | ·0182 496 | ·0106 560 | ·0060 821 | ·0033 986 | ·0018 620 | ·0010 015+ |
| ·33 | ·0245 671 | ·0147 650− | ·0086 761 | ·0049 923 | ·0028 170 | ·0015 607 |
| ·34 | ·0325 216 | ·0200 974 | ·0121 459 | ·0071 895− | ·0041 739 | ·0023 796 |
| ·35 | ·0423 708 | ·0268 967 | ·0167 020 | ·0101 604 | ·0060 633 | ·0035 539 |
| ·36 | ·0543 718 | ·0354 212 | ·0225 794 | ·0141 038 | ·0086 438 | ·0052 039 |
| ·37 | ·0687 698 | ·0459 359 | ·0300 328 | ·0192 452 | ·0121 027 | ·0074 779 |
| ·38 | ·0857 871 | ·0587 035− | ·0393 307 | ·0258 344 | ·0166 568 | ·0105 538 |
| ·39 | ·1056 111 | ·0739 727 | ·0507 465− | ·0341 397 | ·0225 499 | ·0146 399 |
| ·40 | ·1283 817 | ·0919 666 | ·0645 481 | ·0444 412 | ·0300 489 | ·0199 743 |
| ·41 | ·1541 813 | ·1128 693 | ·0809 862 | ·0570 207 | ·0394 375+ | ·0268 217 |
| ·42 | ·1830 246 | ·1368 143 | ·1002 809 | ·0721 503 | ·0510 078 | ·0354 682 |
| ·43 | ·2148 516 | ·1638 721 | ·1226 087 | ·0900 794 | ·0650 489 | ·0462 135+ |
| ·44 | ·2495 232 | ·1940 409 | ·1480 886 | ·1110 204 | ·0818 341 | ·0593 606 |
| ·45 | ·2868 201 | ·2272 394 | ·1767 710 | ·1351 346 | ·1016 063 | ·0752 028 |
| ·46 | ·3264 454 | ·2633 019 | ·2086 272 | ·1625 181 | ·1245 631 | ·0940 093 |
| ·47 | ·3680 302 | ·3019 788 | ·2435 427 | ·1931 903 | ·1508 413 | ·1160 092 |
| ·48 | ·4111 438 | ·3429 388 | ·2813 137 | ·2270 835+ | ·1805 027 | ·1413 748 |
| ·49 | ·4553 054 | ·3857 770 | ·3216 472 | ·2640 373 | ·2135 225− | ·1702 060 |
| ·50 | ·5000 000[e] | ·4300 250+ | ·3641 662 | ·3037 957 | ·2497 799 | ·2025 161 |
| ·51 | ·5446 946 | ·4751 661 | ·4084 188 | ·3460 100 | ·2890 538 | ·2382 204 |
| ·52 | ·5888 562 | ·5206 513 | ·4538 907 | ·3902 455+ | ·3310 221 | ·2771 288 |
| ·53 | ·6319 698 | ·5659 183 | ·5000 223 | ·4359 934 | ·3752 670 | ·3189 432 |
| ·54 | ·6735 546 | ·6104 112 | ·5462 278 | ·4826 862 | ·4212 850− | ·3632 608 |
| ·55 | ·7131 799 | ·6535 992 | ·5919 156 | ·5297 180 | ·4685 025− | ·4095 825− |
| ·56 | ·7504 768 | ·6949 946 | ·6365 098 | ·5764 655− | ·5162 947 | ·4573 273 |
| ·57 | ·7851 484 | ·7341 689 | ·6794 709 | ·6223 115+ | ·5640 089 | ·5058 521 |
| ·58 | ·8169 754 | ·7707 650+ | ·7203 142 | ·6666 682 | ·6109 893 | ·5544 751 |
| ·59 | ·8458 187 | ·8045 066 | ·7586 260 | ·7089 984 | ·6566 022 | ·6025 030 |
| ·60 | ·8716 183 | ·8352 031 | ·7940 754 | ·7488 349 | ·7002 608 | ·6492 581 |
| ·61 | ·8943 889 | ·8627 506 | ·8264 224 | ·7857 953 | ·7414 476 | ·6941 064 |
| ·62 | ·9142 129 | ·8871 292 | ·8555 210 | ·8195 930 | ·7797 318 | ·7364 825− |
| ·63 | ·9312 302 | ·9083 964 | ·8813 181 | ·8500 428 | ·8147 839 | ·7759 110 |
| ·64 | ·9456 282 | ·9266 777 | ·9038 478 | ·8770 608 | ·8463 826 | ·8120 230 |
| ·65 | ·9576 292 | ·9421 551 | ·9232 221 | ·9006 603 | ·8744 173 | ·8445 659 |
| ·66 | ·9674 784 | ·9550 542 | ·9396 190 | ·9209 424 | ·8988 843 | ·8734 072 |
| ·67 | ·9754 329 | ·9656 307 | ·9532 684 | ·9380 834 | ·9198 774 | ·8985 309 |
| ·68 | ·9817 504 | ·9741 569 | ·9644 371 | ·9523 198 | ·9375 750+ | ·9200 287 |
| ·69 | ·9866 810 | ·9809 097 | ·9734 138 | ·9639 315− | ·9522 233 | ·9380 856 |
| ·70 | ·9904 596 | ·9861 601 | ·9804 950− | ·9732 247 | ·9641 178 | ·9529 618 |

$x = \cdot 71$ to $\cdot 91$ $q = 16$ $p = 16$ to $21$

| | $p = 16$ | $p = 17$ | $p = 18$ | $p = 19$ | $p = 20$ | $p = 21$ |
|---|---|---|---|---|---|---|
| $B(p, q) =$ | $\cdot 2079\ 5887 \times \frac{1}{10^9}$ | $\cdot 1039\ 7944 \times \frac{1}{10^9}$ | $\cdot 5356\ 5164 \times \frac{1}{10^{10}}$ | $\cdot 2835\ 8028 \times \frac{1}{10^{10}}$ | $\cdot 1539\ 4358 \times \frac{1}{10^{10}}$ | $\cdot 8552\ 4212 \times \frac{1}{10^{11}}$ |
| $x$ | | | | | | |
| ·71 | ·9933 001 | ·9901 639 | ·9859 725⁻ | ·9805 166 | ·9735 848 | ·9649 720 |
| ·72 | ·9953 928 | ·9931 554 | ·9901 230 | ·9861 203 | ·9809 630 | ·9744 649 |
| ·73 | ·9969 020 | ·9953 429 | ·9932 004 | ·9903 331 | ·9865 875⁻ | ·9818 024 |
| ·74 | ·9979 660 | ·9969 063 | ·9954 302 | ·9934 277 | ·9907 758 | ·9873 417 |
| ·75 | ·9986 984 | ·9979 970 | ·9970 069 | ·9956 455⁺ | ·9938 184 | ·9914 202 |
| ·76 | ·9991 896 | ·9987 385⁺ | ·9980 931 | ·9971 939 | ·9959 709 | ·9943 444 |
| ·77 | ·9995 102 | ·9992 288 | ·9988 208 | ·9982 450⁻ | ·9974 515⁻ | ·9963 823 |
| ·78 | ·9997 133 | ·9995 434 | ·9992 940 | ·9989 373 | ·9984 394 | ·9977 598 |
| ·79 | ·9998 380 | ·9997 390 | ·9995 919 | ·9993 787 | ·9990 774 | ·9986 609 |
| ·80 | ·9999 118 | ·9998 564 | ·9997 729 | ·9996 505⁻ | ·9994 752 | ·9992 298 |
| ·81 | ·9999 540 | ·9999 242 | ·9998 788 | ·9998 115⁻ | ·9997 138 | ·9995 753 |
| ·82 | ·9999 771 | ·9999 618 | ·9999 383 | ·9999 029 | ·9998 510 | ·9997 765⁻ |
| ·83 | ·9999 891 | ·9999 817 | ·9999 701 | ·9999 525⁺ | ·9999 263 | ·9998 883 |
| ·84 | ·9999 951 | ·9999 917 | ·9999 864 | ·9999 781 | ·9999 656 | ·9999 473 |
| ·85 | ·9999 980 | ·9999 965⁺ | ·9999 942 | ·9999 905⁺ | ·9999 849 | ·9999 767 |
| ·86 | ·9999 992 | ·9999 986 | ·9999 977 | ·9999 962 | ·9999 939 | ·9999 904 |
| ·87 | ·9999 997 | ·9999 995⁺ | ·9999 991 | ·9999 986 | ·9999 977 | ·9999 964 |
| ·88 | ·9999 999 | ·9999 998 | ·9999 997 | ·9999 995⁺ | ·9999 992 | ·9999 988 |
| ·89 | 1·0000 000 | 1·0000 000 | ·9999 999 | ·9999 999 | ·9999 998 | ·9999 996 |
| ·90 | | | 1·0000 000 | 1·0000 000 | ·9999 999 | ·9999 999 |
| ·91 | | | | | 1·0000 000 | 1·0000 000 |

$x = \cdot 19$ to $\cdot 80$ | $q = 16$ | $p = 22$ to $27$

| $x$ | $p = 22$ | $p = 23$ | $p = 24$ | $p = 25$ | $p = 26$ | $p = 27$ |
|---|---|---|---|---|---|---|
| $B(p, q) =$ | ·4854 0769 × $\frac{1}{10^{11}}$ | ·2810 2550 $\stackrel{+}{\times}$ $\frac{1}{10^{11}}$ | ·1657 3299 × $\frac{1}{10^{11}}$ | ·9943 9793 × $\frac{1}{10^{12}}$ | ·6063 4020 × $\frac{1}{10^{12}}$ | ·3753 5346 × $\frac{1}{10^{12}}$ |
| ·19 | ·0000 001 | | | | | |
| ·20 | ·0000 002 | ·0000 001 | | | | |
| ·21 | ·0000 004 | ·0000 001 | | | | |
| ·22 | ·0000 009 | ·0000 003 | ·0000 001 | | | |
| ·23 | ·0000 021 | ·0000 008 | ·0000 003 | ·0000 001 | | |
| ·24 | ·0000 044 | ·0000 017 | ·0000 007 | ·0000 003 | ·0000 001 | |
| ·25 | ·0000 090 | ·0000 037 | ·0000 015⁻ | ·0000 006 | ·0000 002 | ·0000 001 |
| ·26 | ·0000 177 | ·0000 075⁺ | ·0000 031 | ·0000 013 | ·0000 005⁺ | ·0000 002 |
| ·27 | ·0000 337 | ·0000 148 | ·0000 064 | ·0000 028 | ·0000 012 | ·0000 005⁻ |
| ·28 | ·0000 618 | ·0000 282 | ·0000 127 | ·0000 056 | ·0000 025⁻ | ·0000 011 |
| ·29 | ·0001 102 | ·0000 521 | ·0000 242 | ·0000 111 | ·0000 050⁺ | ·0000 022 |
| ·30 | ·0001 909 | ·0000 933 | ·0000 449 | ·0000 213 | ·0000 100 | ·0000 046 |
| ·31 | ·0003 220 | ·0001 624 | ·0000 807 | ·0000 395⁺ | ·0000 191 | ·0000 091 |
| ·32 | ·0005 295⁻ | ·0002 754 | ·0001 411 | ·0000 713 | ·0000 355⁺ | ·0000 175⁻ |
| ·33 | ·0008 500⁻ | ·0004 556 | ·0002 405⁺ | ·0001 252 | ·0000 643 | ·0000 326 |
| ·34 | ·0013 338 | ·0007 358 | ·0003 999 | ·0002 143 | ·0001 133 | ·0000 592 |
| ·35 | ·0020 482 | ·0011 619 | ·0006 494 | ·0003 579 | ·0001 946 | ·0001 045⁺ |
| ·36 | ·0030 811 | ·0017 958 | ·0010 313 | ·0005 841 | ·0003 265⁻ | ·0001 802 |
| ·37 | ·0045 445⁻ | ·0027 191 | ·0016 032 | ·0009 323 | ·0005 351 | ·0003 034 |
| ·38 | ·0065 781 | ·0040 372 | ·0024 420 | ·0014 570 | ·0008 581 | ·0004 992 |
| ·39 | ·0093 515⁻ | ·0058 827 | ·0036 477 | ·0022 312 | ·0013 473 | ·0008 037 |
| ·40 | ·0130 660 | ·0084 186 | ·0053 472 | ·0033 509 | ·0020 732 | ·0012 673 |
| ·41 | ·0179 545⁻ | ·0118 402 | ·0076 984 | ·0049 389 | ·0031 287 | ·0019 584 |
| ·42 | ·0242 796 | ·0163 764 | ·0108 923 | ·0071 493 | ·0046 341 | ·0029 682 |
| ·43 | ·0323 291 | ·0222 882 | ·0151 547 | ·0101 702 | ·0067 409 | ·0044 156 |
| ·44 | ·0424 091 | ·0298 651 | ·0207 461 | ·0142 260 | ·0096 359 | ·0064 511 |
| ·45 | ·0548 344 | ·0394 192 | ·0279 583 | ·0195 775⁺ | ·0135 434 | ·0092 615⁻ |
| ·46 | ·0699 155⁻ | ·0512 756 | ·0371 093 | ·0265 199 | ·0187 262 | ·0130 728 |
| ·47 | ·0879 445⁻ | ·0657 606 | ·0485 343 | ·0353 778 | ·0254 840 | ·0181 512 |
| ·48 | ·1091 781 | ·0831 866 | ·0625 742 | ·0464 965⁺ | ·0341 489 | ·0248 027 |
| ·49 | ·1338 202 | ·1038 350⁻ | ·0795 600 | ·0602 310 | ·0450 771 | ·0333 678 |
| ·50 | ·1620 043 | ·1279 375⁺ | ·0997 954 | ·0769 300 | ·0586 376 | ·0442 148 |
| ·51 | ·1937 775⁻ | ·1556 576 | ·1235 374 | ·0969 178 | ·0751 962 | ·0577 271 |
| ·52 | ·2290 867 | ·1870 717 | ·1509 754 | ·1204 740 | ·0950 968 | ·0742 875⁺ |
| ·53 | ·2677 691 | ·2221 543 | ·1822 116 | ·1478 110 | ·1186 392 | ·0942 584 |
| ·54 | ·3095 470 | ·2607 651 | ·2172 432 | ·1790 527 | ·1460 561 | ·1179 582 |
| ·55 | ·3540 294 | ·3026 428 | ·2559 480 | ·2142 145⁻ | ·1774 890 | ·1456 368 |
| ·56 | ·4007 186 | ·3474 035⁺ | ·2980 755⁻ | ·2531 870 | ·2129 669 | ·1774 494 |
| ·57 | ·4490 245⁺ | ·3945 476 | ·3432 445⁻ | ·2957 250⁺ | ·2523 873 | ·2134 333 |
| ·58 | ·4982 840 | ·4434 720 | ·3909 479 | ·3414 439 | ·2955 042 | ·2534 870 |
| ·59 | ·5477 856 | ·4934 910 | ·4405 656 | ·3898 233 | ·3419 226 | ·2973 566 |
| ·60 | ·5967 982 | ·5438 614 | ·4913 848 | ·4402 202 | ·3911 022 | ·3446 290 |
| ·61 | ·6446 011 | ·5938 131 | ·5426 277 | ·4918 902 | ·4423 703 | ·3947 360 |
| ·62 | ·6905 146 | ·6425 826 | ·5934 836 | ·5440 163 | ·4949 448 | ·4469 679 |
| ·63 | ·7339 282 | ·6894 456 | ·6431 450⁺ | ·5957 448 | ·5479 653 | ·5004 983 |
| ·64 | ·7743 256 | ·7337 495⁺ | ·6908 447 | ·6462 238 | ·6005 319 | ·5544 182 |
| ·65 | ·8113 030 | ·7749 405⁺ | ·7358 904 | ·6946 437 | ·6517 472 | ·6077 782 |
| ·66 | ·8445 817 | ·8125 854 | ·7776 955⁻ | ·7402 761 | ·7007 612 | ·6596 353 |
| ·67 | ·8740 128 | ·8463 855⁻ | ·8158 032 | ·7825 067 | ·7468 129 | ·7091 010 |
| ·68 | ·8995 747 | ·8761 827 | ·8499 024 | ·8208 626 | ·7892 673 | ·7553 874 |
| ·69 | ·9213 628 | ·9019 567 | ·8798 338 | ·8550 284 | ·8276 433 | ·7978 463 |
| ·70 | ·9395 745⁺ | ·9238 141 | ·9055 869 | ·8848 534 | ·8616 318 | ·8359 988 |
| ·71 | ·9544 890 | ·9419 714 | ·9272 877 | ·9103 463 | ·8911 009 | ·8695 535⁺ |
| ·72 | ·9664 444 | ·9567 323 | ·9451 791 | ·9316 618 | ·9160 900 | ·8984 099 |
| ·73 | ·9758 143 | ·9684 625⁻ | ·9595 956 | ·9490 772 | ·9367 917 | ·9226 493 |
| ·74 | ·9829 853 | ·9775 635⁻ | ·9709 348 | ·9629 637 | ·9535 260 | ·9425 130 |
| ·75 | ·9883 369 | ·9844 477 | ·9796 285⁺ | ·9737 551 | ·9667 070 | ·9583 713 |
| ·76 | ·9922 253 | ·9895 167 | ·9861 156 | ·9819 152 | ·9768 076 | ·9706 863 |
| ·77 | ·9949 709 | ·9931 432 | ·9908 180 | ·9879 086 | ·9843 242 | ·9799 719 |
| ·78 | ·9968 510 | ·9956 589 | ·9941 227 | ·9921 755⁻ | ·9897 453 | ·9867 563 |
| ·79 | ·9980 968 | ·9973 473 | ·9963 690 | ·9951 132 | ·9935 258 | ·9915 482 |
| ·80 | ·9988 932 | ·9984 404 | ·9978 419 | ·9970 638 | ·9960 679 | ·9948 115⁻ |

$x$ = ·81 to ·93 $q = 16$ $p$ = 22 to 27

| | $p = 22$ | $p = 23$ | $p = 24$ | $p = 25$ | $p = 26$ | $p = 27$ |
|---|---|---|---|---|---|---|
| $B(p, q)$ = | ·4854 0769 $\times \frac{1}{10^{11}}$ | ·2810 2550$^{+}$ $\times \frac{1}{10^{11}}$ | ·1657 3299 $\times \frac{1}{10^{11}}$ | ·9943 9793 $\times \frac{1}{10^{12}}$ | ·6063 4020 $\times \frac{1}{10^{12}}$ | ·3753 5346 $\times \frac{1}{10^{12}}$ |
| $x$ | | | | | | |
| ·81 | ·9993 830 | ·9991 211 | ·9987 705$^{+}$ | ·9983 091 | ·9977 112 | ·9969 474 |
| ·82 | ·9996 717 | ·9995 273 | ·9993 315$^{+}$ | ·9990 708 | ·9987 286 | ·9982 862 |
| ·83 | ·9998 341 | ·9997 585$^{+}$ | ·9996 549 | ·9995 151 | ·9993 295$^{+}$ | ·9990 866 |
| ·84 | ·9999 209 | ·9998 836 | ·9998 318 | ·9997 612 | ·9996 663 | ·9995 405$^{+}$ |
| ·85 | ·9999 646 | ·9999 474 | ·9999 232 | ·9998 898 | ·9998 444 | ·9997 835$^{-}$ |
| ·86 | ·9999 853 | ·9999 779 | ·9999 674 | ·9999 528 | ·9999 326 | ·9999 052 |
| ·87 | ·9999 944 | ·9999 915$^{-}$ | ·9999 873 | ·9999 814 | ·9999 731 | ·9999 618 |
| ·88 | ·9999 981 | ·9999 970 | ·9999 955$^{+}$ | ·9999 933 | ·9999 903 | ·9999 861 |
| ·89 | ·9999 994 | ·9999 991 | ·9999 986 | ·9999 979 | ·9999 969 | ·9999 955$^{-}$ |
| ·90 | ·9999 998 | ·9999 997 | ·9999 996 | ·9999 994 | ·9999 991 | ·9999 987 |
| ·91 | 1·0000 000 | ·9999 999 | ·9999 999 | ·9999 999 | ·9999 998 | ·9999 997 |
| ·92 | | 1·0000 000 | 1·0000 000 | 1·0000 000 | 1·0000 000 | ·9999 999 |
| ·93 | | | | | | 1·0000 000 |

$x$ = ·26 to ·94 | $q = 16$ | $p$ = 28 to 33

| $x$ | $p = 28$ | $p = 29$ | $p = 30$ | $p = 31$ | $p = 32$ | $p = 33$ |
|---|---|---|---|---|---|---|
| $B(p, q)$ = | ·2356 8706 $\times \frac{1}{10^{12}}$ | ·1499 8267 $\times \frac{1}{10^{12}}$ | ·9665 5500− $\times \frac{1}{10^{13}}$ | ·6303 6195+ $\times \frac{1}{10^{13}}$ | ·4157 7065+ $\times \frac{1}{10^{13}}$ | ·2771 8043 $\times \frac{1}{10^{13}}$ |
| ·26 | ·0000 001 | | | | | |
| ·27 | ·0000 002 | ·0000 001 | | | | |
| ·28 | ·0000 005− | ·0000 002 | ·0000 001 | | | |
| ·29 | ·0000 010 | ·0000 004 | ·0000 002 | ·0000 001 | | |
| ·30 | ·0000 021 | ·0000 009 | ·0000 004 | ·0000 002 | ·0000 001 | |
| ·31 | ·0000 043 | ·0000 020 | ·0000 009 | ·0000 004 | ·0000 002 | ·0000 001 |
| ·32 | ·0000 085+ | ·0000 041 | ·0000 019 | ·0000 009 | ·0000 004 | ·0000 002 |
| ·33 | ·0000 163 | ·0000 081 | ·0000 040 | ·0000 019 | ·0000 009 | ·0000 004 |
| ·34 | ·0000 305+ | ·0000 156 | ·0000 079 | ·0000 039 | ·0000 019 | ·0000 010 |
| ·35 | ·0000 555− | ·0000 291 | ·0000 151 | ·0000 078 | ·0000 040 | ·0000 020 |
| ·36 | ·0000 983 | ·0000 531 | ·0000 283 | ·0000 150− | ·0000 078 | ·0000 041 |
| ·37 | ·0001 700 | ·0000 942 | ·0000 517 | ·0000 280 | ·0000 151 | ·0000 080 |
| ·38 | ·0002 871 | ·0001 633 | ·0000 919 | ·0000 512 | ·0000 283 | ·0000 155− |
| ·39 | ·0004 740 | ·0002 765− | ·0001 596 | ·0000 912 | ·0000 516 | ·0000 290 |
| ·40 | ·0007 658 | ·0004 578 | ·0002 708 | ·0001 587 | ·0000 921 | ·0000 530 |
| ·41 | ·0012 119 | ·0007 419 | ·0004 496 | ·0002 698 | ·0001 604 | ·0000 945− |
| ·42 | ·0018 799 | ·0011 779 | ·0007 305+ | ·0004 487 | ·0002 731 | ·0001 647 |
| ·43 | ·0028 602 | ·0018 331 | ·0011 630 | ·0007 307 | ·0004 549 | ·0002 808 |
| ·44 | ·0042 712 | ·0027 983 | ·0018 149 | ·0011 659 | ·0007 422 | ·0004 684 |
| ·45 | ·0062 641 | ·0041 928 | ·0027 785− | ·0018 239 | ·0011 864 | ·0007 651 |
| ·46 | ·0090 274 | ·0061 696 | ·0041 751 | ·0027 989 | ·0018 595+ | ·0012 248 |
| ·47 | ·0127 902 | ·0089 206 | ·0061 612 | ·0042 159 | ·0028 591 | ·0019 226 |
| ·48 | ·0178 241 | ·0126 800 | ·0089 336 | ·0062 362 | ·0043 150+ | ·0029 605+ |
| ·49 | ·0244 427 | ·0177 266 | ·0127 335− | ·0090 635+ | ·0063 952 | ·0044 747 |
| ·50 | ·0329 970 | ·0243 834 | ·0178 489 | ·0129 480 | ·0093 119 | ·0066 416 |
| ·51 | ·0438 682 | ·0330 137 | ·0246 146 | ·0181 893 | ·0133 268 | ·0096 845− |
| ·52 | ·0574 551 | ·0440 133 | ·0334 081 | ·0251 361 | ·0187 533 | ·0138 784 |
| ·53 | ·0741 578 | ·0577 973 | ·0446 412 | ·0341 821 | ·0259 565+ | ·0195 534 |
| ·54 | ·0943 560 | ·0747 831 | ·0587 467 | ·0457 573 | ·0353 490 | ·0270 939 |
| ·55 | ·1183 854 | ·0953 677 | ·0761 598 | ·0603 133 | ·0473 805+ | ·0369 333 |
| ·56 | ·1465 098 | ·1199 017 | ·0972 940 | ·0783 035− | ·0625 230 | ·0495 436 |
| ·57 | ·1788 941 | ·1486 600 | ·1225 126 | ·1001 567 | ·0812 479 | ·0654 177 |
| ·58 | ·2155 782 | ·1818 123 | ·1520 982 | ·1262 471 | ·1039 983 | ·0850 452 |
| ·59 | ·2564 548 | ·2193 949 | ·1862 199 | ·1568 600 | ·1311 559 | ·1088 817 |
| ·60 | ·3012 540 | ·2612 870 | ·2249 033 | ·1921 580 | ·1630 041 | ·1373 122 |
| ·61 | ·3495 364 | ·3071 940 | ·2680 054 | ·2321 478 | ·1996 909 | ·1706 115+ |
| ·62 | ·4006 969 | ·3566 403 | ·3151 967 | ·2766 542 | ·2411 950+ | ·2089 050+ |
| ·63 | ·4539 806 | ·4089 747 | ·3659 553 | ·3253 020 | ·2872 977 | ·2521 318 |
| ·64 | ·5085 095− | ·4633 878 | ·4195 730 | ·3775 109 | ·3375 654 | ·3000 167 |
| ·65 | ·5633 207 | ·5189 426 | ·4751 766 | ·4325 047 | ·3913 470 | ·3520 543 |
| ·66 | ·6174 127 | ·5746 171 | ·5317 625− | ·4893 364 | ·4477 862 | ·4075 085+ |
| ·67 | ·6697 969 | ·6293 557 | ·5882 452 | ·5469 291 | ·5058 529 | ·4654 313 |
| ·68 | ·7195 501 | ·6821 256 | ·6435 139 | ·6041 300 | ·5643 903 | ·5247 003 |
| ·69 | ·7658 641 | ·7319 744 | ·6964 953 | ·6597 745+ | ·6221 772 | ·5840 746 |
| ·70 | ·8080 873 | ·7780 825− | ·7462 153 | ·7127 547 | ·6779 989 | ·6422 655+ |
| ·71 | ·8457 556 | ·8198 073 | ·7918 548 | ·7620 854 | ·7307 218 | ·6980 155+ |
| ·72 | ·8786 082 | ·8567 133 | ·8327 949 | ·8069 630 | ·7793 646 | ·7501 793 |
| ·73 | ·9065 897 | ·8885 858 | ·8686 449 | ·8468 097 | ·8231 572 | ·7977 973 |
| ·74 | ·9298 357 | ·9154 289 | ·8992 535− | ·8812 988 | ·8615 834 | ·8401 553 |
| ·75 | ·9486 463 | ·9374 452 | ·9246 991 | ·9103 597 | ·8944 013 | ·8768 222 |
| ·76 | ·9634 496 | ·9550 033 | ·9452 638 | ·9341 608 | ·9216 395− | ·9076 626 |
| ·77 | ·9747 589 | ·9685 945− | ·9613 928 | ·9530 747 | ·9435 707 | ·9328 223 |
| ·78 | ·9831 296 | ·9787 853 | ·9736 441 | ·9676 288 | ·9606 667 | ·9526 907 |
| ·79 | ·9891 181 | ·9861 697 | ·9826 358 | ·9784 482 | ·9735 391 | ·9678 431 |
| ·80 | ·9932 479 | ·9913 270 | ·9889 955− | ·9861 976 | ·9828 762 | ·9789 736 |
| ·81 | ·9959 850− | ·9947 878 | ·9933 166 | ·9915 290 | ·9893 805+ | ·9868 244 |
| ·82 | ·9977 219 | ·9970 112 | ·9961 271 | ·9950 396 | ·9937 164 | ·9921 227 |
| ·83 | ·9987 729 | ·9983 730 | ·9978 695+ | ·9972 427 | ·9964 706 | ·9955 294 |
| ·84 | ·9993 762 | ·9991 643 | ·9988 941 | ·9985 538 | ·9981 295+ | ·9976 061 |
| ·85 | ·9997 029 | ·9995 978 | ·9994 622 | ·9992 894 | ·9990 714 | ·9987 992 |
| ·86 | ·9998 686 | ·9998 202 | ·9997 571 | ·9996 757 | ·9995 718 | ·9994 406 |
| ·87 | ·9999 466 | ·9999 261 | ·9998 992 | ·9998 640 | ·9998 186 | ·9997 605+ |
| ·88 | ·9999 803 | ·9999 725− | ·9999 620 | ·9999 483 | ·9999 303 | ·9999 070 |
| ·89 | ·9999 935+ | ·9999 908 | ·9999 872 | ·9999 824 | ·9999 761 | ·9999 678 |
| ·90 | ·9999 981 | ·9999 973 | ·9999 963 | ·9999 948 | ·9999 928 | ·9999 903 |
| ·91 | ·9999 995+ | ·9999 993 | ·9999 991 | ·9999 987 | ·9999 982 | ·9999 975− |
| ·92 | ·9999 999 | ·9999 999 | ·9999 998 | ·9999 997 | ·9999 996 | ·9999 995− |
| ·93 | 1·0000 000 | 1·0000 000 | 1·0000 000 | 1·0000 000 | ·9999 999 | ·9999 999 |
| ·94 | | | | | 1·0000 000 | 1·0000 000 |

$x$ = ·32 to ·95 $q = 16$ $p$ = 34 to 39

| | $p = 34$ | $p = 35$ | $p = 36$ | $p = 37$ | $p = 38$ | $p = 39$ |
|---|---|---|---|---|---|---|
| $B(p, q) =$ | ·1866 7254 $\times\frac{1}{10^{13}}$ | ·1269 3733 $\times\frac{1}{10^{13}}$ | ·8711 3851 $\times\frac{1}{10^{14}}$ | ·6030 9589 $\times\frac{1}{10^{14}}$ | ·4210 2921 $\times\frac{1}{10^{14}}$ | ·2962 7981 $\times\frac{1}{10^{14}}$ |
| $x$ | | | | | | |
| ·32 | ·0000 001 | | | | | |
| ·33 | ·0000 002 | ·0000 001 | | | | |
| ·34 | ·0000 005⁻ | ·0000 002 | ·0000 001 | ·0000 001 | | |
| ·35 | ·0000 010 | ·0000 005⁻ | ·0000 002 | ·0000 001 | ·0000 001 | |
| ·36 | ·0000 021 | ·0000 011 | ·0000 005⁺ | ·0000 003 | ·0000 001 | ·0000 001 |
| ·37 | ·0000 042 | ·0000 022 | ·0000 012 | ·0000 006 | ·0000 003 | ·0000 002 |
| ·38 | ·0000 084 | ·0000 045⁺ | ·0000 024 | ·0000 013 | ·0000 007 | ·0000 004 |
| ·39 | ·0000 161 | ·0000 089 | ·0000 049 | ·0000 026 | ·0000 014 | ·0000 008 |
| ·40 | ·0000 302 | ·0000 171 | ·0000 096 | ·0000 053 | ·0000 030 | ·0000 016 |
| ·41 | ·0000 552 | ·0000 320 | ·0000 184 | ·0000 105⁺ | ·0000 060 | ·0000 034 |
| ·42 | ·0000 985⁺ | ·0000 585⁻ | ·0000 344 | ·0000 201 | ·0000 117 | ·0000 067 |
| ·43 | ·0001 718 | ·0001 043 | ·0000 629 | ·0000 376 | ·0000 223 | ·0000 132 |
| ·44 | ·0002 931 | ·0001 820 | ·0001 121 | ·0000 686 | ·0000 417 | ·0000 251 |
| ·45 | ·0004 894 | ·0003 105⁺ | ·0001 956 | ·0001 223 | ·0000 759 | ·0000 468 |
| ·46 | ·0008 002 | ·0005 187 | ·0003 337 | ·0002 131 | ·0001 352 | ·0000 852 |
| ·47 | ·0012 823 | ·0008 486 | ·0005 574 | ·0003 635⁺ | ·0002 354 | ·0001 515⁻ |
| ·48 | ·0020 149 | ·0013 607 | ·0009 121 | ·0006 071 | ·0004 013 | ·0002 635⁺ |
| ·49 | ·0031 060 | ·0021 395⁻ | ·0014 629 | ·0009 932 | ·0006 698 | ·0004 487 |
| ·50 | ·0046 996 | ·0033 002 | ·0023 007 | ·0015 926 | ·0010 951 | ·0007 481 |
| ·51 | ·0069 825⁺ | ·0049 966 | ·0035 497 | ·0025 043 | ·0017 550⁺ | ·0012 221 |
| ·52 | ·0101 912 | ·0074 280 | ·0053 753 | ·0038 632 | ·0027 581 | ·0019 566 |
| ·53 | ·0146 172 | ·0108 468 | ·0079 921 | ·0058 487 | ·0042 521 | ·0030 719 |
| ·54 | ·0206 099 | ·0155 638 | ·0116 711 | ·0086 932 | ·0064 332 | ·0047 311 |
| ·55 | ·0285 755⁻ | ·0219 507 | ·0167 455⁻ | ·0126 898 | ·0095 548 | ·0071 500⁻ |
| ·56 | ·0389 712 | ·0304 387 | ·0236 127 | ·0181 974 | ·0139 354 | ·0106 066 |
| ·57 | ·0522 930 | ·0415 114 | ·0327 322 | ·0256 429 | ·0199 639 | ·0154 490 |
| ·58 | ·0690 557 | ·0556 903 | ·0446 162 | ·0355 169 | ·0280 997 | ·0220 997 |
| ·59 | ·0897 664 | ·0735 128 | ·0598 133 | ·0483 628 | ·0388 681 | ·0310 551 |
| ·60 | ·1148 902 | ·0955 017 | ·0788 830 | ·0647 572 | ·0528 456 | ·0428 775⁺ |
| ·61 | ·1448 102 | ·1221 278 | ·1023 618 | ·0852 806 | ·0706 369 | ·0581 783 |
| ·62 | ·1797 853 | ·1537 660 | ·1307 204 | ·1104 786 | ·0928 409 | ·0775 889 |
| ·63 | ·2199 071 | ·1906 489 | ·1643 166 | ·1408 150⁻ | ·1200 065⁺ | ·1017 224 |
| ·64 | ·2650 622 | ·2328 219 | ·2033 450⁻ | ·1766 193 | ·1525 806 | ·1311 229 |
| ·65 | ·3149 048 | ·2801 044 | ·2477 898 | ·2180 335⁻ | ·1908 506 | ·1662 072 |
| ·66 | ·3688 419 | ·3320 631 | ·2973 860 | ·2649 629 | ·2348 883 | ·2072 038 |
| ·67 | ·4260 387 | ·3880 016 | ·3515 947 | ·3170 385⁺ | ·2844 996 | ·2540 929 |
| ·68 | ·4854 433 | ·4469 714 | ·4095 987 | ·3735 964 | ·3391 898 | ·3065 579 |
| ·69 | ·5458 335⁺ | ·5078 061 | ·4703 219 | ·4336 811 | ·3981 495⁻ | ·3639 550⁻ |
| ·70 | ·6058 825⁻ | ·5691 784 | ·5324 744 | ·4960 762 | ·4602 682 | ·4253 084 |
| ·71 | ·6642 389 | ·6296 774 | ·5946 222 | ·5593 624 | ·5241 790 | ·4893 381 |
| ·72 | ·7196 143 | ·6878 986 | ·6552 768 | ·6220 025⁻ | ·5883 325⁺ | ·5545 208 |
| ·73 | ·7708 696 | ·7425 402 | ·7129 966 | ·6824 436 | ·6510 979 | ·6191 831 |
| ·74 | ·8170 908 | ·7924 932 | ·7664 900 | ·7392 300 | ·7108 796 | ·6816 190 |
| ·75 | ·8576 450⁺ | ·8369 167 | ·8147 078 | ·7911 109 | ·7662 384 | ·7402 205⁺ |
| ·76 | ·8922 118 | ·8752 886 | ·8569 148 | ·8371 324 | ·8160 026 | ·7936 050⁺ |
| ·77 | ·9207 841 | ·9074 252 | ·8927 304 | ·8767 008 | ·8593 542 | ·8407 249 |
| ·78 | ·9436 415⁻ | ·9334 692 | ·9221 343 | ·9096 093 | ·8958 791 | ·8809 422 |
| ·79 | ·9612 978 | ·9538 459 | ·9454 358 | ·9360 235⁺ | ·9255 734 | ·9140 590 |
| ·80 | ·9744 323 | ·9691 966 | ·9632 128 | ·9564 313 | ·9488 066 | ·9402 991 |
| ·81 | ·9838 129 | ·9802 974 | ·9762 294 | ·9715 615⁻ | ·9662 475⁺ | ·9602 442 |
| ·82 | ·9902 219 | ·9879 755⁺ | ·9853 441 | ·9822 873 | ·9787 646 | ·9747 356 |
| ·83 | ·9943 931 | ·9930 339 | ·9914 223 | ·9895 274 | ·9873 169 | ·9847 580 |
| ·84 | ·9969 666 | ·9961 924 | ·9952 633 | ·9941 577 | ·9928 526 | ·9913 235⁻ |
| ·85 | ·9984 627 | ·9980 504 | ·9975 498 | ·9969 471 | ·9962 270 | ·9953 734 |
| ·86 | ·9992 765⁻ | ·9990 730 | ·9988 231 | ·9985 186 | ·9981 505⁺ | ·9977 091 |
| ·87 | ·9996 871 | ·9995 950⁺ | ·9994 806 | ·9993 395⁺ | ·9991 670 | ·9989 578 |
| ·88 | ·9998 773 | ·9998 395⁺ | ·9997 921 | ·9997 330 | ·9996 598 | ·9995 700 |
| ·89 | ·9999 571 | ·9999 433 | ·9999 258 | ·9999 037 | ·9998 761 | ·9998 418 |
| ·90 | ·9999 869 | ·9999 825⁺ | ·9999 769 | ·9999 697 | ·9999 606 | ·9999 492 |
| ·91 | ·9999 966 | ·9999 954 | ·9999 939 | ·9999 919 | ·9999 894 | ·9999 861 |
| ·92 | ·9999 993 | ·9999 990 | ·9999 987 | ·9999 982 | ·9999 977 | ·9999 969 |
| ·93 | ·9999 999 | ·9999 998 | ·9999 998 | ·9999 997 | ·9999 996 | ·9999 995⁻ |
| ·94 | 1·0000 000 | 1·0000 000 | 1·0000 000 | 1·0000 000 | 1·0000 000 | ·9999 999 |
| ·95 | | | | | | 1·0000 000 |

$x = \cdot 37$ to $\cdot 95$ $q = 16$ $p = 40$ to $45$

| $x$ | $p = 40$ | $p = 41$ | $p = 42$ | $p = 43$ | $p = 44$ | $p = 45$ |
|---|---|---|---|---|---|---|
| $B(p, q) =$ | $\cdot 2100\ 8932 \times \frac{1}{10^{14}}$ | $\cdot 1500\ 6380 \times \frac{1}{10^{14}}$ | $\cdot 1079\ 4063 \times \frac{1}{10^{14}}$ | $\cdot 7816\ 3903 \times \frac{1}{10^{15}}$ | $\cdot 5696\ 6913 \times \frac{1}{10^{15}}$ | $\cdot 4177\ 5736 \times \frac{1}{10^{15}}$ |
| ·37 | ·0000 001 | | | | | |
| ·38 | ·0000 002 | ·0000 001 | | | | |
| ·39 | ·0000 004 | ·0000 002 | ·0000 001 | ·0000 001 | | |
| ·40 | ·0000 009 | ·0000 005− | ·0000 003 | ·0000 001 | ·0000 001 | |
| ·41 | ·0000 019 | ·0000 010 | ·0000 006 | ·0000 003 | ·0000 002 | ·0000 001 |
| ·42 | ·0000 039 | ·0000 022 | ·0000 012 | ·0000 007 | ·0000 004 | ·0000 002 |
| ·43 | ·0000 077 | ·0000 045− | ·0000 026 | ·0000 015− | ·0000 009 | ·0000 005− |
| ·44 | ·0000 151 | ·0000 090 | ·0000 053 | ·0000 031 | ·0000 018 | ·0000 011 |
| ·45 | ·0000 287 | ·0000 175− | ·0000 106 | ·0000 064 | ·0000 038 | ·0000 023 |
| ·46 | ·0000 533 | ·0000 332 | ·0000 205+ | ·0000 126 | ·0000 077 | ·0000 047 |
| ·47 | ·0000 969 | ·0000 615+ | ·0000 389 | ·0000 244 | ·0000 153 | ·0000 095− |
| ·48 | ·0001 720 | ·0001 116 | ·0000 719 | ·0000 461 | ·0000 294 | ·0000 187 |
| ·49 | ·0002 987 | ·0001 977 | ·0001 301 | ·0000 851 | ·0000 554 | ·0000 359 |
| ·50 | ·0005 079 | ·0003 428 | ·0002 300 | ·0001 535− | ·0001 019 | ·0000 673 |
| ·51 | ·0008 457 | ·0005 818 | ·0003 979 | ·0002 707 | ·0001 831 | ·0001 233 |
| ·52 | ·0013 796 | ·0009 670 | ·0006 739 | ·0004 671 | ·0003 221 | ·0002 209 |
| ·53 | ·0022 059 | ·0015 747 | ·0011 178 | ·0007 892 | ·0005 542 | ·0003 873 |
| ·54 | ·0034 584 | ·0025 135− | ·0018 165+ | ·0013 057 | ·0009 337 | ·0006 643 |
| ·55 | ·0053 186 | ·0039 336 | ·0028 932 | ·0021 166 | ·0015 405− | ·0011 156 |
| ·56 | ·0080 255− | ·0060 380 | ·0045 178 | ·0033 625+ | ·0024 898 | ·0018 345+ |
| ·57 | ·0118 857 | ·0090 930 | ·0069 188 | ·0052 369 | ·0039 438 | ·0029 554 |
| ·58 | ·0172 811 | ·0134 384 | ·0103 942 | ·0079 979 | ·0061 233 | ·0046 653 |
| ·59 | ·0246 724 | ·0194 944 | ·0153 217 | ·0119 806 | ·0093 217 | ·0072 181 |
| ·60 | ·0345 964 | ·0277 644 | ·0221 655+ | ·0176 064 | ·0139 167 | ·0109 482 |
| ·61 | ·0476 555− | ·0388 295− | ·0314 759 | ·0253 882 | ·0203 794 | ·0162 823 |
| ·62 | ·0644 957 | ·0533 337 | ·0438 814 | ·0359 280 | ·0292 767 | ·0237 471 |
| ·63 | ·0857 731 | ·0719 570 | ·0600 684 | ·0499 037 | ·0412 660 | ·0339 691 |
| ·64 | ·1121 080 | ·0953 750− | ·0807 479 | ·0680 432 | ·0570 757 | ·0476 637 |
| ·65 | ·1440 282 | ·1242 057 | ·1066 071 | ·0910 827 | ·0774 717 | ·0656 086 |
| ·66 | ·1819 044 | ·1589 453 | ·1382 484 | ·1197 100 | ·1032 064 | ·0886 008 |
| ·67 | ·2258 848 | ·1998 981 | ·1761 171 | ·1544 934 | ·1349 516 | ·1173 950+ |
| ·68 | ·2758 337 | ·2471 066 | ·2204 254 | ·1958 025− | ·1732 181 | ·1526 252 |
| ·69 | ·3312 860 | ·3002 914 | ·2710 808 | ·2437 272 | ·2182 693 | ·1947 149 |
| ·70 | ·3914 242 | ·3588 107 | ·3276 290 | ·2980 063 | ·2700 371 | ·2437 841 |
| ·71 | ·4550 868 | ·4216 489 | ·3892 224 | ·3579 771 | ·3280 543 | ·2995 665− |
| ·72 | ·5208 131 | ·4874 424 | ·4546 253 | ·4225 582 | ·3914 159 | ·3613 494 |
| ·73 | ·5869 245+ | ·5545 450− | ·5222 602 | ·4902 752 | ·4587 812 | ·4279 528 |
| ·74 | ·6516 382 | ·6211 327 | ·5902 998 | ·5593 348 | ·5284 273 | ·4977 587 |
| ·75 | ·7132 019 | ·6853 390 | ·6567 965− | ·6277 443 | ·5983 543 | ·5687 973 |
| ·76 | ·7700 359 | ·7454 062 | ·7198 393 | ·6934 690 | ·6664 362 | ·6388 875− |
| ·77 | ·8208 632 | ·7998 346 | ·7777 187 | ·7546 075+ | ·7306 042 | ·7058 207 |
| ·78 | ·8648 103 | ·8475 088 | ·8290 764 | ·8095 644 | ·7890 359 | ·7675 650+ |
| ·79 | ·9014 639 | ·8877 826 | ·8730 201 | ·8571 926 | ·8403 271 | ·8224 612 |
| ·80 | ·9308 753 | ·9205 092 | ·9091 824 | ·8968 846 | ·8836 145+ | ·8693 793 |
| ·81 | ·9535 112 | ·9460 124 | ·9377 161 | ·9285 961 | ·9186 320 | ·9078 096 |
| ·82 | ·9701 612 | ·9650 036 | ·9592 270 | ·9527 986 | ·9456 884 | ·9378 704 |
| ·83 | ·9818 172 | ·9784 611 | ·9746 563 | ·9703 706 | ·9655 725− | ·9602 325− |
| ·84 | ·9895 450+ | ·9874 909 | ·9851 342 | ·9824 475− | ·9794 034 | ·9759 747 |
| ·85 | ·9943 687 | ·9931 944 | ·9918 312 | ·9902 586 | ·9884 556 | ·9864 005+ |
| ·86 | ·9971 834 | ·9965 618 | ·9958 316 | ·9949 794 | ·9939 908 | ·9928 508 |
| ·87 | ·9987 056 | ·9984 041 | ·9980 457 | ·9976 226 | ·9971 261 | ·9965 468 |
| ·88 | ·9994 607 | ·9993 283 | ·9991 692 | ·9989 792 | ·9987 537 | ·9984 876 |
| ·89 | ·9997 996 | ·9997 479 | ·9996 851 | ·9996 092 | ·9995 180 | ·9994 093 |
| ·90 | ·9999 350− | ·9999 174 | ·9998 958 | ·9998 694 | ·9998 373 | ·9997 986 |
| ·91 | ·9999 821 | ·9999 770 | ·9999 707 | ·9999 629 | ·9999 534 | ·9999 417 |
| ·92 | ·9999 960 | ·9999 948 | ·9999 933 | ·9999 914 | ·9999 891 | ·9999 862 |
| ·93 | ·9999 993 | ·9999 991 | ·9999 988 | ·9999 985− | ·9999 980 | ·9999 975− |
| ·94 | ·9999 999 | ·9999 999 | ·9999 998 | ·9999 998 | ·9999 997 | ·9999 997 |
| ·95 | 1·0000 000 | 1·0000 000 | 1·0000 000 | 1·0000 000 | 1·0000 000 | 1·0000 000 |

$x = \cdot 42$ to $\cdot 96$ $q = 16$ $p = 46$ to $50$

| | $p = 46$ | $p = 47$ | $p = 48$ | $p = 49$ | $p = 50$ |
|---|---|---|---|---|---|
| $B(p, q) =$ | $\cdot 3081\ 8166 \times \frac{1}{10^{15}}$ | $\cdot 2286\ 5091 \times \frac{1}{10^{15}}$ | $\cdot 1705\ 8084 \times \frac{1}{10^{15}}$ | $\cdot 1279\ 3563 \times \frac{1}{10^{15}}$ | $\cdot 9644\ 3780 \times \frac{1}{10^{16}}$ |
| $x$ | | | | | |
| ·42 | ·0000 001 | ·0000 001 | | | |
| ·43 | ·0000 003 | ·0000 002 | ·0000 001 | | |
| ·44 | ·0000 006 | ·0000 004 | ·0000 002 | ·0000 001 | ·0000 001 |
| ·45 | ·0000 013 | ·0000 008 | ·0000 005⁻ | ·0000 003 | ·0000 002 |
| ·46 | ·0000 028 | ·0000 017 | ·0000 010 | ·0000 006 | ·0000 004 |
| ·47 | ·0000 059 | ·0000 036 | ·0000 022 | ·0000 013 | ·0000 008 |
| ·48 | ·0000 118 | ·0000 074 | ·0000 046 | ·0000 029 | ·0000 018 |
| ·49 | ·0000 231 | ·0000 148 | ·0000 094 | ·0000 060 | ·0000 038 |
| ·50 | ·0000 442 | ·0000 289 | ·0000 188 | ·0000 122 | ·0000 079 |
| ·51 | ·0000 826 | ·0000 550⁺ | ·0000 365⁺ | ·0000 241 | ·0000 159 |
| ·52 | ·0001 508 | ·0001 024 | ·0000 693 | ·0000 466 | ·0000 313 |
| ·53 | ·0002 693 | ·0001 864 | ·0001 284 | ·0000 880 | ·0000 601 |
| ·54 | ·0004 704 | ·0003 315⁻ | ·0002 325⁺ | ·0001 624 | ·0001 129 |
| ·55 | ·0008 040 | ·0005 767 | ·0004 118 | ·0002 928 | ·0002 073 |
| ·56 | ·0013 453 | ·0009 819 | ·0007 135⁻ | ·0005 162 | ·0003 719 |
| ·57 | ·0022 043 | ·0016 365⁻ | ·0012 096 | ·0008 902 | ·0006 524 |
| ·58 | ·0035 378 | ·0026 706 | ·0020 072 | ·0015 021 | ·0011 195⁻ |
| ·59 | ·0055 633 | ·0042 686 | ·0032 610 | ·0024 807 | ·0018 794 |
| ·60 | ·0085 734 | ·0066 839 | ·0051 884 | ·0040 106 | ·0030 877 |
| ·61 | ·0129 500⁺ | ·0102 545⁺ | ·0080 855⁻ | ·0063 489 | ·0049 653 |
| ·62 | ·0191 759 | ·0154 176 | ·0123 438 | ·0098 425⁺ | ·0078 170 |
| ·63 | ·0278 398 | ·0227 191 | ·0184 635⁻ | ·0149 446 | ·0120 491 |
| ·64 | ·0396 321 | ·0328 157 | ·0270 609 | ·0222 269 | ·0181 861 |
| ·65 | ·0553 271 | ·0464 650⁺ | ·0388 662 | ·0323 835⁻ | ·0268 798 |
| ·66 | ·0757 478 | ·0644 987 | ·0547 048 | ·0462 208 | ·0389 072 |
| ·67 | ·1017 112 | ·0877 764 | ·0754 605⁻ | ·0646 301 | ·0551 525⁻ |
| ·68 | ·1339 544 | ·1171 181 | ·1020 157 | ·0885 367 | ·0765 652 |
| ·69 | ·1730 448 | ·1532 168 | ·1351 690 | ·1188 245⁻ | ·1040 944 |
| ·70 | ·2192 813 | ·1965 364 | ·1755 336 | ·1562 372 | ·1385 949 |
| ·71 | ·2725 980 | ·2472 065⁺ | ·2234 250⁻ | ·2012 636 | ·1807 123 |
| ·72 | ·3324 855⁺ | ·3049 268 | ·2787 519 | ·2540 166 | ·2307 554 |
| ·73 | ·3979 465⁻ | ·3688 991 | ·3409 271 | ·3141 264 | ·2885 727 |
| ·74 | ·4674 991 | ·4378 051 | ·4088 187 | ·3806 657 | ·3534 550⁺ |
| ·75 | ·5392 403 | ·5098 439 | ·4807 603 | ·4521 311 | ·4240 861 |
| ·76 | ·6109 713 | ·5828 368 | ·5546 303 | ·5264 944 | ·4985 653 |
| ·77 | ·6803 763 | ·6543 954 | ·6280 055⁻ | ·6013 351 | ·5745 124 |
| ·78 | ·7452 353 | ·7221 386 | ·6983 736 | ·6740 441 | ·6492 579 |
| ·79 | ·8036 424 | ·7839 277 | ·7633 825⁻ | ·7420 796 | ·7200 986 |
| ·80 | ·8541 951 | ·8380 866 | ·8210 870 | ·8032 374 | ·7845 865⁻ |
| ·81 | ·8961 214 | ·8835 668 | ·8701 520 | ·8558 905⁻ | ·8408 023 |
| ·82 | ·9293 228 | ·9200 281 | ·9099 741 | ·8991 534 | ·8875 643 |
| ·83 | ·9543 228 | ·9478 184 | ·9406 967 | ·9329 385⁺ | ·9245 280 |
| ·84 | ·9721 345⁺ | ·9678 569 | ·9631 169 | ·9578 911 | ·9521 576 |
| ·85 | ·9840 715⁺ | ·9814 463 | ·9785 027 | ·9752 187 | ·9715 729 |
| ·86 | ·9915 436 | ·9900 528 | ·9883 616 | ·9864 526 | ·9843 083 |
| ·87 | ·9958 749 | ·9950 997 | ·9942 100 | ·9931 941 | ·9920 397 |
| ·88 | ·9981 754 | ·9978 110 | ·9973 880 | ·9968 995⁻ | ·9963 380 |
| ·89 | ·9992 803 | ·9991 280 | ·9989 492 | ·9987 403 | ·9984 976 |
| ·90 | ·9997 522 | ·9996 968 | ·9996 310 | ·9995 533 | ·9994 619 |
| ·91 | ·9999 276 | ·9999 105⁺ | ·9998 900 | ·9998 655⁺ | ·9998 365⁻ |
| ·92 | ·9999 827 | ·9999 784 | ·9999 732 | ·9999 669 | ·9999 594 |
| ·93 | ·9999 968 | ·9999 960 | ·9999 949 | ·9999 937 | ·9999 922 |
| ·94 | ·9999 996 | ·9999 995⁻ | ·9999 993 | ·9999 991 | ·9999 989 |
| ·95 | 1·0000 000 | 1·0000 000 | ·9999 999 | ·9999 999 | ·9999 999 |
| ·96 | | | 1·0000 000 | 1·0000 000 | 1·0000 000 |

$x$ = ·13 to ·70 $q = 17$ $p$ = 17 to 22

| $x$ | $p = 17$ | $p = 18$ | $p = 19$ | $p = 20$ | $p = 21$ | $p = 22$ |
|---|---|---|---|---|---|---|
| $B(p, q)$ = | ·5041 4272$\times\frac{1}{10^{10}}$ | ·2520 7136$\times\frac{1}{10^{10}}$ | ·1296 3670$\times\frac{1}{10^{10}}$ | ·6841 9369$\times\frac{1}{10^{11}}$ | ·3698 3443$\times\frac{1}{10^{11}}$ | ·2043 8218$\times\frac{1}{10^{11}}$ |
| ·13 | ·0000 001 | | | | | |
| ·14 | ·0000 004 | ·0000 001 | | | | |
| ·15 | ·0000 010 | ·0000 003 | ·0000 001 | | | |
| ·16 | ·0000 025$^{+}$ | ·0000 008 | ·0000 002 | ·0000 001 | | |
| ·17 | ·0000 060 | ·0000 019 | ·0000 006 | ·0000 002 | ·0000 001 | |
| ·18 | ·0000 132 | ·0000 044 | ·0000 015$^{-}$ | ·0000 005$^{-}$ | ·0000 001 | |
| ·19 | ·0000 275$^{+}$ | ·0000 098 | ·0000 034 | ·0000 011 | ·0000 004 | ·0000 001 |
| ·20 | ·0000 549 | ·0000 205$^{-}$ | ·0000 074 | ·0000 026 | ·0000 009 | ·0000 003 |
| ·21 | ·0001 046 | ·0000 409 | ·0000 156 | ·0000 058 | ·0000 021 | ·0000 008 |
| ·22 | ·0001 915$^{+}$ | ·0000 783 | ·0000 313 | ·0000 122 | ·0000 047 | ·0000 018 |
| ·23 | ·0003 378 | ·0001 443 | ·0000 602 | ·0000 246 | ·0000 098 | ·0000 039 |
| ·24 | ·0005 758 | ·0002 563 | ·0001 115$^{-}$ | ·0000 475$^{-}$ | ·0000 198 | ·0000 081 |
| ·25 | ·0009 510 | ·0004 404 | ·0001 993 | ·0000 883 | ·0000 384 | ·0000 164 |
| ·26 | ·0015 253 | ·0007 337 | ·0003 450$^{-}$ | ·0001 588 | ·0000 717 | ·0000 318 |
| ·27 | ·0023 808 | ·0011 877 | ·0005 792 | ·0002 766 | ·0001 295$^{-}$ | ·0000 595$^{+}$ |
| ·28 | ·0036 227 | ·0018 715$^{-}$ | ·0009 453 | ·0004 676 | ·0002 268 | ·0001 080 |
| ·29 | ·0053 827 | ·0028 757 | ·0015 024 | ·0007 688 | ·0003 858 | ·0001 901 |
| ·30 | ·0078 207 | ·0043 154 | ·0023 290 | ·0012 313 | ·0006 385$^{+}$ | ·0003 252 |
| ·31 | ·0111 258 | ·0063 331 | ·0035 267 | ·0019 241 | ·0010 299 | ·0005 415$^{-}$ |
| ·32 | ·0155 159 | ·0091 009 | ·0052 234 | ·0029 377 | ·0016 211 | ·0008 788 |
| ·33 | ·0212 344 | ·0128 203 | ·0075 756 | ·0043 873 | ·0024 935$^{-}$ | ·0013 923 |
| ·34 | ·0285 458 | ·0177 219 | ·0107 705$^{-}$ | ·0064 167 | ·0037 522 | ·0021 561 |
| ·35 | ·0377 286 | ·0240 613 | ·0150 258 | ·0092 002 | ·0055 301 | ·0032 668 |
| ·36 | ·0490 656 | ·0321 144 | ·0205 876 | ·0129 436 | ·0079 902 | ·0048 483 |
| ·37 | ·0628 330 | ·0421 689 | ·0277 270 | ·0178 837 | ·0113 281 | ·0070 544 |
| ·38 | ·0792 871 | ·0545 141 | ·0367 327 | ·0242 857 | ·0157 719 | ·0100 717 |
| ·39 | ·0986 506 | ·0694 291 | ·0479 026 | ·0324 374 | ·0215 809 | ·0141 210 |
| ·40 | ·1210 987 | ·0871 683 | ·0615 320 | ·0426 421 | ·0290 414 | ·0194 561 |
| ·41 | ·1467 452 | ·1079 474 | ·0779 006 | ·0552 074 | ·0384 598 | ·0263 617 |
| ·42 | ·1756 310 | ·1319 280 | ·0972 569 | ·0704 324 | ·0501 532 | ·0351 465$^{-}$ |
| ·43 | ·2077 145$^{-}$ | ·1592 042 | ·1198 030 | ·0885 932 | ·0644 368 | ·0461 354 |
| ·44 | ·2428 653 | ·1897 904 | ·1456 792 | ·1099 259 | ·0816 093 | ·0596 572 |
| ·45 | ·2808 620 | ·2236 119 | ·1749 494 | ·1346 107 | ·1019 363 | ·0760 302 |
| ·46 | ·3213 939 | ·2604 999 | ·2075 897 | ·1627 553 | ·1256 325$^{-}$ | ·0955 453 |
| ·47 | ·3640 672 | ·3001 892 | ·2434 798 | ·1943 814 | ·1528 441 | ·1184 473 |
| ·48 | ·4084 156 | ·3423 226 | ·2823 983 | ·2294 126 | ·1836 329 | ·1449 164 |
| ·49 | ·4539 149 | ·3864 584 | ·3240 236 | ·2676 680 | ·2179 624 | ·1750 499 |
| ·50 | ·5000 000$^{e}$ | ·4320 831 | ·3679 394 | ·3088 597 | ·2556 879 | ·2088 461 |
| ·51 | ·5460 851 | ·4786 286 | ·4136 455$^{-}$ | ·3525 956 | ·2965 517 | ·2461 924 |
| ·52 | ·5915 844 | ·5254 914 | ·4605 734 | ·3983 888 | ·3401 840 | ·2868 573 |
| ·53 | ·6359 328 | ·5720 549 | ·5081 060 | ·4456 716 | ·3861 093 | ·3304 894 |
| ·54 | ·6786 061 | ·6177 121 | ·5556 002 | ·4938 152 | ·4337 602 | ·3766 221 |
| ·55 | ·7191 380 | ·6618 879 | ·6024 114 | ·5421 524 | ·4824 959 | ·4246 860 |
| ·56 | ·7571 347 | ·7040 598 | ·6479 183 | ·5900 039 | ·5316 262 | ·4740 269 |
| ·57 | ·7922 855$^{+}$ | ·7437 752 | ·6915 458 | ·6367 050$^{-}$ | ·5804 382 | ·5239 303 |
| ·58 | ·8243 690 | ·7806 660 | ·7327 869 | ·6816 319 | ·6282 261 | ·5736 504 |
| ·59 | ·8532 548 | ·8144 570 | ·7712 190 | ·7242 261 | ·6743 196 | ·6224 407 |
| ·60 | ·8789 013 | ·8449 709 | ·8065 165$^{+}$ | ·7640 142 | ·7181 118 | ·6695 864 |
| ·61 | ·9013 494 | ·8721 278 | ·8384 581 | ·8006 240 | ·7590 821 | ·7144 345$^{+}$ |
| ·62 | ·9207 129 | ·8959 400 | ·8669 281 | ·8337 935$^{+}$ | ·7968 153 | ·7564 210 |
| ·63 | ·9371 670 | ·9165 029 | ·8919 127 | ·8633 750$^{+}$ | ·8310 133 | ·7950 918 |
| ·64 | ·9509 344 | ·9339 832 | ·9134 911 | ·8893 321 | ·8615 008 | ·8301 177 |
| ·65 | ·9622 714 | ·9486 042 | ·9318 238 | ·9117 315$^{+}$ | ·8882 236 | ·8613 014 |
| ·66 | ·9714 542 | ·9606 302 | ·9471 363 | ·9307 306 | ·9112 407 | ·8885 766 |
| ·67 | ·9787 656 | ·9703 515$^{+}$ | ·9597 030 | ·9465 606 | ·9307 108 | ·9120 005$^{-}$ |
| ·68 | ·9844 841 | ·9780 690 | ·9698 293 | ·9595 079 | ·9468 746 | ·9317 386 |
| ·69 | ·9888 742 | ·9840 814 | ·9778 349 | ·9698 953 | ·9600 343 | ·9480 461 |
| ·70 | ·9921 793 | ·9886 740 | ·9840 392 | ·9780 628 | ·9705 325$^{-}$ | ·9612 451 |

$x = \cdot 71$ to $\cdot 90$ $q = 17$ $p = 17$ to 22

| | $p = 17$ | $p = 18$ | $p = 19$ | $p = 20$ | $p = 21$ | $p = 22$ |
|---|---|---|---|---|---|---|
| $B(p, q) =$ | $\cdot 5041\ 4272 \times \frac{1}{10^{10}}$ | $\cdot 2520\ 7136 \times \frac{1}{10^{10}}$ | $\cdot 1296\ 3670 \times \frac{1}{10^{10}}$ | $\cdot 6841\ 9369 \times \frac{1}{10^{11}}$ | $\cdot 3698\ 3443 \times \frac{1}{10^{11}}$ | $\cdot 2043\ 8218 \times \frac{1}{10^{11}}$ |
| $x$ | | | | | | |
| ·71 | ·9946 173 | ·9921 103 | ·9887 481 | ·9843 508 | ·9787 309 | ·9717 008 |
| ·72 | ·9963 773 | ·9946 261 | ·9922 445⁻ | ·9890 857 | ·9849 919 | ·9797 986 |
| ·73 | ·9976 192 | ·9964 261 | ·9947 810 | ·9925 688 | ·9896 618 | ·9859 230 |
| ·74 | ·9984 747 | ·9976 831 | ·9965 767 | ·9950 685⁻ | ·9930 595⁺ | ·9904 402 |
| ·75 | ·9990 490 | ·9985 385⁺ | ·9978 153 | ·9968 160 | ·9954 671 | ·9936 845⁺ |
| ·76 | ·9994 242 | ·9991 048 | ·9986 461 | ·9980 041 | ·9971 258 | ·9959 497 |
| ·77 | ·9996 622 | ·9994 687 | ·9991 872 | ·9987 880 | ·9982 347 | ·9974 840 |
| ·78 | ·9998 085⁻ | ·9996 953 | ·9995 285⁺ | ·9992 889 | ·9989 525⁺ | ·9984 902 |
| ·79 | ·9998 954 | ·9998 316 | ·9997 365⁺ | ·9995 981 | ·9994 013 | ·9991 274 |
| ·80 | ·9999 451 | ·9999 107 | ·9998 586 | ·9997 819 | ·9996 715⁺ | ·9995 158 |
| ·81 | ·9999 725⁻ | ·9999 547 | ·9999 275⁻ | ·9998 869 | ·9998 277 | ·9997 432 |
| ·82 | ·9999 868 | ·9999 781 | ·9999 646 | ·9999 441 | ·9999 139 | ·9998 703 |
| ·83 | ·9999 940 | ·9999 900 | ·9999 836 | ·9999 739 | ·9999 593 | ·9999 380 |
| ·84 | ·9999 975⁻ | ·9999 957 | ·9999 929 | ·9999 885⁺ | ·9999 819 | ·9999 721 |
| ·85 | ·9999 990 | ·9999 983 | ·9999 971 | ·9999 953 | ·9999 925⁻ | ·9999 883 |
| ·86 | ·9999 996 | ·9999 994 | ·9999 989 | ·9999 982 | ·9999 971 | ·9999 955⁻ |
| ·87 | ·9999 999 | ·9999 998 | ·9999 996 | ·9999 994 | ·9999 990 | ·9999 984 |
| ·88 | 1·0000 000 | ·9999 999 | ·9999 999 | ·9999 998 | ·9999 997 | ·9999 995⁻ |
| ·89 | | 1·0000 000 | 1·0000 000 | ·9999 999 | ·9999 999 | ·9999 999 |
| ·90 | | | | 1·0000 000 | 1·0000 000 | 1·0000 000 |

$x$ = ·20 to ·80 $q = 17$ $p$ = 23 to 28

| | $p = 23$ | $p = 24$ | $p = 25$ | $p = 26$ | $p = 27$ | $p = 28$ |
|---|---|---|---|---|---|---|
| $B(p, q) =$ | ·1152 9256 $\times \frac{1}{10^{11}}$ | ·6629 3196 $\times \frac{1}{10^{12}}$ | ·3880 5773 $\times \frac{1}{10^{12}}$ | ·2309 8674 $\times \frac{1}{10^{12}}$ | ·1396 6640 $\times \frac{1}{10^{12}}$ | ·8570 4384 $\times \frac{1}{10^{13}}$ |
| $x$ | | | | | | |
| ·20 | ·0000 001 | | | | | |
| ·21 | ·0000 003 | ·0000 001 | | | | |
| ·22 | ·0000 007 | ·0000 002 | ·0000 001 | | | |
| ·23 | ·0000 015− | ·0000 006 | ·0000 002 | ·0000 001 | | |
| ·24 | ·0000 033 | ·0000 013 | ·0000 005+ | ·0000 002 | ·0000 001 | |
| ·25 | ·0000 069 | ·0000 028 | ·0000 011 | ·0000 005− | ·0000 002 | ·0000 001 |
| ·26 | ·0000 138 | ·0000 059 | ·0000 025+ | ·0000 010 | ·0000 004 | ·0000 002 |
| ·27 | ·0000 269 | ·0000 120 | ·0000 052 | ·0000 023 | ·0000 010 | ·0000 004 |
| ·28 | ·0000 506 | ·0000 233 | ·0000 106 | ·0000 047 | ·0000 021 | ·0000 009 |
| ·29 | ·0000 921 | ·0000 439 | ·0000 206 | ·0000 096 | ·0000 044 | ·0000 020 |
| ·30 | ·0001 629 | ·0000 803 | ·0000 390 | ·0000 187 | ·0000 088 | ·0000 041 |
| ·31 | ·0002 799 | ·0001 425− | ·0000 714 | ·0000 353 | ·0000 172 | ·0000 083 |
| ·32 | ·0004 685+ | ·0002 459 | ·0001 272 | ·0000 649 | ·0000 326 | ·0000 162 |
| ·33 | ·0007 647 | ·0004 135− | ·0002 203 | ·0001 158 | ·0000 601 | ·0000 308 |
| ·34 | ·0012 187 | ·0006 783 | ·0003 720 | ·0002 013 | ·0001 075− | ·0000 567 |
| ·35 | ·0018 986 | ·0010 866 | ·0006 129 | ·0003 410 | ·0001 873 | ·0001 016 |
| ·36 | ·0028 947 | ·0017 021 | ·0009 866 | ·0005 641 | ·0003 184 | ·0001 776 |
| ·37 | ·0043 232 | ·0026 096 | ·0015 529 | ·0009 117 | ·0005 285− | ·0003 026 |
| ·38 | ·0063 303 | ·0039 196 | ·0023 928 | ·0014 413 | ·0008 572 | ·0005 037 |
| ·39 | ·0090 956 | ·0057 724 | ·0036 123 | ·0022 307 | ·0013 602 | ·0008 196 |
| ·40 | ·0128 336 | ·0083 418 | ·0053 472 | ·0033 828 | ·0021 135− | ·0013 049 |
| ·41 | ·0177 940 | ·0118 376 | ·0077 674 | ·0050 306 | ·0032 180 | ·0020 344 |
| ·42 | ·0242 598 | ·0165 066 | ·0110 794 | ·0073 411 | ·0048 048 | ·0031 084 |
| ·43 | ·0325 425− | ·0226 315− | ·0155 286 | ·0105 196 | ·0070 403 | ·0046 576 |
| ·44 | ·0429 736 | ·0305 262 | ·0213 981 | ·0148 113 | ·0101 296 | ·0068 489 |
| ·45 | ·0558 941 | ·0405 294 | ·0290 059 | ·0205 015+ | ·0143 195− | ·0098 890 |
| ·46 | ·0716 396 | ·0529 932 | ·0386 977 | ·0279 131 | ·0198 993 | ·0140 285− |
| ·47 | ·0905 233 | ·0682 692 | ·0508 367 | ·0373 998 | ·0271 981 | ·0195 619 |
| ·48 | ·1128 169 | ·0866 907 | ·0657 897 | ·0493 365− | ·0365 789 | ·0268 264 |
| ·49 | ·1387 303 | ·1085 534 | ·0839 090 | ·0641 047 | ·0484 289 | ·0361 959 |
| ·50 | ·1683 918 | ·1340 936 | ·1055 118 | ·0820 747 | ·0631 447 | ·0480 709 |
| ·51 | ·2018 303 | ·1634 668 | ·1308 578 | ·1035 836 | ·0811 139 | ·0628 635+ |
| ·52 | ·2389 602 | ·1967 275− | ·1601 258 | ·1289 119 | ·1026 922 | ·0809 784 |
| ·53 | ·2795 719 | ·2338 126 | ·1933 918 | ·1582 581 | ·1281 782 | ·1027 886 |
| ·54 | ·3233 279 | ·2745 289 | ·2306 099 | ·1917 151 | ·1577 870 | ·1286 087 |
| ·55 | ·3697 665+ | ·3185 482 | ·2715 980 | ·2292 490 | ·1916 235− | ·1586 663 |
| ·56 | ·4183 126 | ·3654 083 | ·3160 309 | ·2706 827 | ·2296 601 | ·1930 739 |
| ·57 | ·4682 957 | ·4145 237 | ·3634 403 | ·3156 875− | ·2717 182 | ·2318 038 |
| ·58 | ·5189 755− | ·4652 039 | ·4132 247 | ·3637 821 | ·3174 582 | ·2746 686 |
| ·59 | ·5695 713 | ·5166 790 | ·4646 681 | ·4143 425− | ·3663 783 | ·3213 096 |
| ·60 | ·6192 964 | ·5681 317 | ·5169 671 | ·4666 212 | ·4178 243 | ·3711 962 |
| ·61 | ·6673 922 | ·6187 340 | ·5692 649 | ·5197 760 | ·4710 104 | ·4236 355+ |
| ·62 | ·7131 624 | ·6676 845− | ·6206 906 | ·5729 072 | ·5250 502 | ·4777 960 |
| ·63 | ·7560 027 | ·7142 454 | ·6704 002 | ·6250 993 | ·5789 969 | ·5327 409 |
| ·64 | ·7954 252 | ·7577 762 | ·7176 173 | ·6754 666 | ·6318 892 | ·5874 725− |
| ·65 | ·8310 751 | ·7977 604 | ·7616 696 | ·7231 967 | ·6828 003 | ·6409 824 |
| ·66 | ·8627 396 | ·8338 246 | ·8020 181 | ·7675 907 | ·7308 859 | ·6923 050+ |
| ·67 | ·8903 475+ | ·8657 479 | ·8382 783 | ·8080 947 | ·7754 268 | ·7405 689 |
| ·68 | ·9139 607 | ·8934 621 | ·8702 302 | ·8443 221 | ·8158 630 | ·7850 429 |
| ·69 | ·9337 584 | ·9170 418 | ·8978 176 | ·8760 636 | ·8518 162 | ·8251 711 |
| ·70 | ·9500 158 | ·9366 871 | ·9211 370 | ·9032 854 | ·8830 995− | ·8605 958 |
| ·71 | ·9630 793 | ·9526 997 | ·9404 172 | ·9261 155− | ·9097 125+ | ·8911 650+ |
| ·72 | ·9733 400 | ·9654 549 | ·9559 929 | ·9448 201 | ·9318 252 | ·9169 245− |
| ·73 | ·9812 087 | ·9753 731 | ·9682 732 | ·9597 732 | ·9497 497 | ·9380 965+ |
| ·74 | ·9870 923 | ·9828 913 | ·9777 101 | ·9714 223 | ·9639 058 | ·9550 476 |
| ·75 | ·9913 753 | ·9884 386 | ·9847 678 | ·9802 526 | ·9747 823 | ·9682 483 |
| ·76 | ·9944 057 | ·9924 161 | ·9898 959 | ·9867 547 | ·9828 983 | ·9782 306 |
| ·77 | ·9964 856 | ·9951 821 | ·9935 092 | ·9913 966 | ·9887 689 | ·9855 466 |
| ·78 | ·9978 673 | ·9970 435− | ·9959 725+ | ·9946 026 | ·9928 764 | ·9907 321 |
| ·79 | ·9987 535+ | ·9982 528 | ·9975 935− | ·9967 392 | ·9956 491 | ·9942 776 |
| ·80 | ·9993 008 | ·9990 090 | ·9986 199 | ·9981 095+ | ·9974 499 | ·9966 095+ |

$x$ = ·81 to ·92 $q = 17$ $p$ = 23 to 28

| | $p = 23$ | $p = 24$ | $p = 25$ | $p = 26$ | $p = 27$ | $p = 28$ |
|---|---|---|---|---|---|---|
| $B(p, q)$ = | ·1152 9256 $\times \frac{1}{10^{11}}$ | ·6629 3196 $\times \frac{1}{10^{12}}$ | ·3880 5773 $\times \frac{1}{10^{12}}$ | ·2309 8674 $\times \frac{1}{10^{12}}$ | ·1396 6640 $\times \frac{1}{10^{12}}$ | ·8570 4384 $\times \frac{1}{10^{13}}$ |
| $x$ | | | | | | |
| ·81 | ·9996 250− | ·9994 626 | ·9992 435− | ·9989 523 | ·9985 714 | ·9980 800 |
| ·82 | ·9998 086 | ·9997 227 | ·9996 054 | ·9994 476 | ·9992 385+ | ·9989 655+ |
| ·83 | ·9999 075+ | ·9998 646 | ·9998 052 | ·9997 243 | ·9996 159 | ·9994 726 |
| ·84 | ·9999 580 | ·9999 378 | ·9999 095+ | ·9998 706 | ·9998 178 | ·9997 472 |
| ·85 | ·9999 822 | ·9999 733 | ·9999 608 | ·9999 433 | ·9999 193 | ·9998 869 |
| ·86 | ·9999 930 | ·9999 894 | ·9999 843 | ·9999 770 | ·9999 670 | ·9999 532 |
| ·87 | ·9999 975− | ·9999 962 | ·9999 942 | ·9999 915− | ·9999 876 | ·9999 823 |
| ·88 | ·9999 992 | ·9999 987 | ·9999 981 | ·9999 972 | ·9999 958 | ·9999 940 |
| ·89 | ·9999 998 | ·9999 996 | ·9999 994 | ·9999 992 | ·9999 988 | ·9999 982 |
| ·90 | ·9999 999 | ·9999 999 | ·9999 999 | ·9999 998 | ·9999 997 | ·9999 995+ |
| ·91 | 1·0000 000 | 1·0000 000 | 1·0000 000 | 1·0000 000 | ·9999 999 | ·9999 999 |
| ·92 | | | | | 1·0000 000 | 1·0000 000 |

$x$ = ·26 to ·93 $q = 17$ $p$ = 29 to 34

| | $p = 29$ | $p = 30$ | $p = 31$ | $p = 32$ | $p = 33$ | $p = 34$ |
|---|---|---|---|---|---|---|
| $B(p, q)$ = | ·5332 7172 $\times \frac{1}{10^{13}}$ | ·3361 9304 $\times \frac{1}{10^{13}}$ | ·2145 9130 $\times \frac{1}{10^{13}}$ | ·1385 9022 $\times \frac{1}{10^{13}}$ | ·9050 7897 $\times \frac{1}{10^{14}}$ | ·5973 5212 $\times \frac{1}{10^{14}}$ |
| $x$ | | | | | | |
| ·26 | ·0000 001 | | | | | |
| ·27 | ·0000 002 | ·0000 001 | | | | |
| ·28 | ·0000 004 | ·0000 002 | ·0000 001 | | | |
| ·29 | ·0000 009 | ·0000 004 | ·0000 002 | ·0000 001 | | |
| ·30 | ·0000 019 | ·0000 009 | ·0000 004 | ·0000 002 | ·0000 001 | |
| ·31 | ·0000 040 | ·0000 019 | ·0000 009 | ·0000 004 | ·0000 002 | ·0000 001 |
| ·32 | ·0000 080 | ·0000 039 | ·0000 019 | ·0000 009 | ·0000 004 | ·0000 002 |
| ·33 | ·0000 156 | ·0000 078 | ·0000 039 | ·0000 019 | ·0000 009 | ·0000 004 |
| ·34 | ·0000 296 | ·0000 152 | ·0000 078 | ·0000 039 | ·0000 020 | ·0000 010 |
| ·35 | ·0000 545+ | ·0000 289 | ·0000 152 | ·0000 079 | ·0000 041 | ·0000 021 |
| ·36 | ·0000 979 | ·0000 534 | ·0000 288 | ·0000 154 | ·0000 081 | ·0000 043 |
| ·37 | ·0001 713 | ·0000 959 | ·0000 532 | ·0000 292 | ·0000 159 | ·0000 085+ |
| ·38 | ·0002 926 | ·0001 682 | ·0000 957 | ·0000 539 | ·0000 301 | ·0000 166 |
| ·39 | ·0004 883 | ·0002 878 | ·0001 679 | ·0000 970 | ·0000 555− | ·0000 315− |
| ·40 | ·0007 966 | ·0004 811 | ·0002 877 | ·0001 703 | ·0000 999 | ·0000 581 |
| ·41 | ·0012 718 | ·0007 867 | ·0004 817 | ·0002 921 | ·0001 755+ | ·0001 045+ |
| ·42 | ·0019 887 | ·0012 590 | ·0007 890 | ·0004 898 | ·0003 012 | ·0001 837 |
| ·43 | ·0030 476 | ·0019 734 | ·0012 651 | ·0008 033 | ·0005 055− | ·0003 153 |
| ·44 | ·0045 805+ | ·0030 318 | ·0019 869 | ·0012 899 | ·0008 298 | ·0005 293 |
| ·45 | ·0067 561 | ·0045 684 | ·0030 589 | ·0020 291 | ·0013 339 | ·0008 694 |
| ·46 | ·0097 847 | ·0067 555+ | ·0046 189 | ·0031 288 | ·0021 007 | ·0013 984 |
| ·47 | ·0139 220 | ·0098 088 | ·0068 446 | ·0047 323 | ·0032 431 | ·0022 039 |
| ·48 | ·0194 703 | ·0139 912 | ·0099 586 | ·0070 239 | ·0049 110 | ·0034 050− |
| ·49 | ·0267 766 | ·0196 146 | ·0142 336 | ·0102 360 | ·0072 977 | ·0051 600 |
| ·50 | ·0362 271 | ·0270 380 | ·0199 930 | ·0146 525− | ·0106 471 | ·0076 733 |
| ·51 | ·0482 371 | ·0366 621 | ·0276 104 | ·0206 114 | ·0152 572 | ·0112 026 |
| ·52 | ·0632 352 | ·0489 182 | ·0375 028 | ·0285 030 | ·0214 832 | ·0160 630 |
| ·53 | ·0816 427 | ·0642 520 | ·0501 191 | ·0387 627 | ·0297 343 | ·0226 293 |
| ·54 | ·1038 489 | ·0831 019 | ·0659 234 | ·0518 591 | ·0404 671 | ·0313 328 |
| ·55 | ·1301 819 | ·1058 720 | ·0853 706 | ·0682 751 | ·0541 713 | ·0426 532 |
| ·56 | ·1608 781 | ·1329 011 | ·1088 781 | ·0884 819 | ·0713 490 | ·0571 028 |
| ·57 | ·1960 520 | ·1644 301 | ·1367 925+ | ·1129 083 | ·0924 873 | ·0752 038 |
| ·58 | ·2356 689 | ·2005 692 | ·1693 539 | ·1419 046 | ·1180 237 | ·0974 571 |
| ·59 | ·2795 246 | ·2412 697 | ·2066 617 | ·1757 043 | ·1483 071 | ·1243 054 |
| ·60 | ·3272 325+ | ·2863 008 | ·2486 437 | ·2143 878 | ·1835 576 | ·1560 906 |
| ·61 | ·3782 233 | ·3352 383 | ·2950 330 | ·2578 496 | ·2238 267 | ·1930 103 |
| ·62 | ·4317 568 | ·3874 640 | ·3453 562 | ·3057 750− | ·2689 644 | ·2350 764 |
| ·63 | ·4869 474 | ·4421 803 | ·3989 353 | ·3576 294 | ·3185 953 | ·2820 806 |
| ·64 | ·5428 020 | ·4984 396 | ·4549 052 | ·4126 629 | ·3721 103 | ·3335 730 |
| ·65 | ·5982 685− | ·5551 863 | ·5122 477 | ·4699 325− | ·4286 751 | ·3888 555− |
| ·66 | ·6522 911 | ·6113 114 | ·5698 399 | ·5283 416 | ·4872 584 | ·4469 968 |
| ·67 | ·7038 686 | ·6657 129 | ·6265 142 | ·5866 960 | ·5466 786 | ·5068 675− |
| ·68 | ·7521 094 | ·7173 588 | ·6811 256 | ·6437 703 | ·6056 679 | ·5671 960 |
| ·69 | ·7962 801 | ·7653 468 | ·7326 194 | ·6983 823 | ·6629 469 | ·6266 418 |
| ·70 | ·8358 418 | ·8089 539 | ·7800 941 | ·7494 655+ | ·7173 055+ | ·6838 786 |
| ·71 | ·8704 713 | ·8476 725+ | ·8228 522 | ·7961 344 | ·7676 800 | ·7376 820 |
| ·72 | ·9000 653 | ·8812 296 | ·8604 350− | ·8377 353 | ·8132 196 | ·7870 101 |
| ·73 | ·9247 286 | ·9095 860 | ·8926 364 | ·8738 770 | ·8533 355− | ·8310 697 |
| ·74 | ·9447 467 | ·9329 185− | ·9194 974 | ·9044 398 | ·8877 258 | ·8693 607 |
| ·75 | ·9605 475− | ·9515 854 | ·9412 790 | ·9295 596 | ·9163 752 | ·9016 927 |
| ·76 | ·9726 560 | ·9660 819 | ·9584 208 | ·9495 933 | ·9395 299 | ·9281 736 |
| ·77 | ·9816 476 | ·9769 890 | ·9714 887 | ·9650 676 | ·9576 512 | ·9491 718 |
| ·78 | ·9881 038 | ·9849 227 | ·9811 180 | ·9766 187 | ·9713 545+ | ·9652 577 |
| ·79 | ·9925 750− | ·9904 877 | ·9879 594 | ·9849 312 | ·9813 427 | ·9771 333 |
| ·80 | ·9955 530 | ·9942 415+ | ·9926 327 | ·9906 814 | ·9883 398 | ·9855 583 |
| ·81 | ·9974 545− | ·9966 683 | ·9956 918 | ·9944 927 | ·9930 357 | ·9912 834 |
| ·82 | ·9986 137 | ·9981 662 | ·9976 034 | ·9969 037 | ·9960 431 | ·9949 953 |
| ·83 | ·9992 857 | ·9990 449 | ·9987 385+ | ·9983 530 | ·9978 730 | ·9972 814 |
| ·84 | ·9996 539 | ·9995 323 | ·9993 758 | ·9991 764 | ·9989 251 | ·9986 117 |
| ·85 | ·9998 435+ | ·9997 863 | ·9997 118 | ·9996 157 | ·9994 932 | ·9993 386 |
| ·86 | ·9999 346 | ·9999 097 | ·9998 770 | ·9998 343 | ·9997 791 | ·9997 088 |
| ·87 | ·9999 750+ | ·9999 651 | ·9999 520 | ·9999 347 | ·9999 120 | ·9998 828 |
| ·88 | ·9999 914 | ·9999 879 | ·9999 831 | ·9999 768 | ·9999 684 | ·9999 575− |
| ·89 | ·9999 974 | ·9999 963 | ·9999 947 | ·9999 927 | ·9999 900 | ·9999 863 |
| ·90 | ·9999 993 | ·9999 990 | ·9999 986 | ·9999 980 | ·9999 972 | ·9999 962 |
| ·91 | ·9999 998 | ·9999 998 | ·9999 997 | ·9999 995+ | ·9999 994 | ·9999 991 |
| ·92 | 1·0000 000 | 1·0000 000 | ·9999 999 | ·9999 999 | ·9999 999 | ·9999 998 |
| ·93 | | | 1·0000 000 | 1·0000 000 | 1·0000 000 | 1·0000 000 |

$x$ = ·32 to ·94 $q$ = 17 $p$ = 35 to 40

| | $p$ = 35 | $p$ = 36 | $p$ = 37 | $p$ = 38 | $p$ = 39 | $p$ = 40 |
|---|---|---|---|---|---|---|
| $B(p, q)$ = | ·3982 3475 $\times \frac{1}{10^{14}}$ | ·2680 4262 $\times \frac{1}{10^{14}}$ | ·1820 6668 $\times \frac{1}{10^{14}}$ | ·1247 4939 $\times \frac{1}{10^{14}}$ | ·8619 0491 $\times \frac{1}{10^{15}}$ | ·6002 5520 $\times \frac{1}{10^{15}}$ |
| $x$ | | | | | | |
| ·32 | ·0000 001 | | | | | |
| ·33 | ·0000 002 | ·0000 001 | | | | |
| ·34 | ·0000 005⁻ | ·0000 002 | ·0000 001 | ·0000 001 | | |
| ·35 | ·0000 010 | ·0000 005⁺ | ·0000 003 | ·0000 001 | ·0000 001 | |
| ·36 | ·0000 022 | ·0000 011 | ·0000 006 | ·0000 003 | ·0000 001 | ·0000 001 |
| ·37 | ·0000 046 | ·0000 024 | ·0000 013 | ·0000 007 | ·0000 003 | ·0000 002 |
| ·38 | ·0000 091 | ·0000 050⁻ | ·0000 027 | ·0000 014 | ·0000 008 | ·0000 004 |
| ·39 | ·0000 177 | ·0000 099 | ·0000 055⁻ | ·0000 030 | ·0000 016 | ·0000 009 |
| ·40 | ·0000 335⁻ | ·0000 192 | ·0000 109 | ·0000 061 | ·0000 034 | ·0000 019 |
| ·41 | ·0000 617 | ·0000 362 | ·0000 210 | ·0000 121 | ·0000 070 | ·0000 040 |
| ·42 | ·0001 110 | ·0000 666 | ·0000 396 | ·0000 234 | ·0000 137 | ·0000 080 |
| ·43 | ·0001 950⁺ | ·0001 197 | ·0000 729 | ·0000 441 | ·0000 265⁻ | ·0000 158 |
| ·44 | ·0003 348 | ·0002 101 | ·0001 308 | ·0000 809 | ·0000 497 | ·0000 303 |
| ·45 | ·0005 620 | ·0003 605⁻ | ·0002 295⁻ | ·0001 450⁺ | ·0000 910 | ·0000 567 |
| ·46 | ·0009 234 | ·0006 049 | ·0003 934 | ·0002 539 | ·0001 628 | ·0001 037 |
| ·47 | ·0014 856 | ·0009 936 | ·0006 597 | ·0004 348 | ·0002 847 | ·0001 851 |
| ·48 | ·0023 420 | ·0015 984 | ·0010 829 | ·0007 285⁻ | ·0004 867 | ·0003 231 |
| ·49 | ·0036 195⁺ | ·0025 196 | ·0017 411 | ·0011 947 | ·0008 143 | ·0005 513 |
| ·50 | ·0054 868 | ·0038 937 | ·0027 432 | ·0019 191 | ·0013 336 | ·0009 208 |
| ·51 | ·0081 617 | ·0059 018 | ·0042 370 | ·0030 209 | ·0021 394 | ·0015 055⁺ |
| ·52 | ·0119 182 | ·0087 776 | ·0064 187 | ·0046 616 | ·0033 632 | ·0024 111 |
| ·53 | ·0170 915⁺ | ·0128 148 | ·0095 407 | ·0070 551 | ·0051 830 | ·0037 838 |
| ·54 | ·0240 791 | ·0183 714 | ·0139 194 | ·0104 758 | ·0078 332 | ·0058 208 |
| ·55 | ·0333 370 | ·0258 708 | ·0199 394 | ·0152 663 | ·0116 140 | ·0087 810 |
| ·56 | ·0453 706 | ·0357 971 | ·0280 533 | ·0218 414 | ·0168 981 | ·0129 942 |
| ·57 | ·0607 161 | ·0486 830 | ·0387 758 | ·0306 866 | ·0241 345⁻ | ·0188 675⁺ |
| ·58 | ·0799 151 | ·0650 895⁺ | ·0526 690 | ·0423 499 | ·0338 448 | ·0268 881 |
| ·59 | ·1034 804 | ·0855 769 | ·0703 191 | ·0574 242 | ·0466 128 | ·0376 172 |
| ·60 | ·1318 550⁺ | ·1106 662 | ·0923 026 | ·0765 198 | ·0630 629 | ·0516 763 |
| ·61 | ·1653 661 | ·1407 944 | ·1191 440 | ·1002 263 | ·0838 275⁺ | ·0697 204 |
| ·62 | ·2041 784 | ·1762 644 | ·1512 658 | ·1290 643 | ·1095 037 | ·0924 006 |
| ·63 | ·2482 509 | ·2171 952 | ·1889 345⁺ | ·1634 311 | ·1405 989 | ·1203 134 |
| ·64 | ·2973 026 | ·2634 778 | ·2322 088 | ·2035 426 | ·1774 715⁺ | ·1539 407 |
| ·65 | ·3507 926 | ·3147 416 | ·2808 938 | ·2493 787 | ·2202 687 | ·1935 845⁻ |
| ·66 | ·4079 193 | ·3703 380 | ·3345 105⁻ | ·3006 389 | ·2688 710 | ·2393 024 |
| ·67 | ·4676 417 | ·4293 462 | ·3922 847 | ·3567 156 | ·3228 501 | ·2908 516 |
| ·68 | ·5287 242 | ·4906 040 | ·4531 616 | ·4166 906 | ·3814 482 | ·3476 515⁺ |
| ·69 | ·5898 027 | ·5527 637 | ·5158 481 | ·4793 615⁺ | ·4435 855⁻ | ·4087 727 |
| ·70 | ·6494 686 | ·6143 703 | ·5788 821 | ·5432 979 | ·5079 011 | ·4729 580 |
| ·71 | ·7063 607 | ·6739 565⁺ | ·6407 242 | ·6069 261 | ·5728 256 | ·5386 813 |
| ·72 | ·7592 589 | ·7301 439 | ·6998 643 | ·6686 354 | ·6366 833 | ·6042 396 |
| ·73 | ·8071 667 | ·7817 408 | ·7549 306 | ·7268 958 | ·6978 133 | ·6678 734 |
| ·74 | ·8493 751 | ·8278 250⁺ | ·8047 903 | ·7803 735⁺ | ·7546 974 | ·7279 020 |
| ·75 | ·8854 987 | ·8678 010 | ·8486 284 | ·8280 309 | ·8060 783 | ·7828 592 |
| ·76 | ·9154 812 | ·9014 252 | ·8859 949 | ·8691 968 | ·8510 548 | ·8316 102 |
| ·77 | ·9395 700 | ·9287 969 | ·9168 148 | ·9035 989 | ·8891 379 | ·8734 347 |
| ·78 | ·9582 642 | ·9503 156 | ·9413 602 | ·9313 544 | ·9202 637 | ·9080 639 |
| ·79 | ·9722 429 | ·9666 134 | ·9601 895⁺ | ·9529 201 | ·9447 593 | ·9356 673 |
| ·80 | ·9822 860 | ·9784 714 | ·9740 633 | ·9690 120 | ·9632 694 | ·9567 906 |
| ·81 | ·9891 960 | ·9867 324 | ·9838 499 | ·9805 055⁻ | ·9766 558 | ·9722 583 |
| ·82 | ·9937 317 | ·9922 220 | ·9904 337 | ·9883 333 | ·9858 857 | ·9830 553 |
| ·83 | ·9965 593 | ·9956 860 | ·9946 391 | ·9933 943 | ·9919 261 | ·9902 076 |
| ·84 | ·9982 246 | ·9977 508 | ·9971 759 | ·9964 842 | ·9956 585⁻ | ·9946 803 |
| ·85 | ·9991 454 | ·9989 061 | ·9986 122 | ·9982 544 | ·9978 223 | ·9973 042 |
| ·86 | ·9996 198 | ·9995 082 | ·9993 697 | ·9991 990 | ·9989 904 | ·9987 374 |
| ·87 | ·9998 454 | ·9997 979 | ·9997 383 | ·9996 641 | ·9995 723 | ·9994 596 |
| ·88 | ·9999 433 | ·9999 252 | ·9999 021 | ·9998 730 | ·9998 367 | ·9997 916 |
| ·89 | ·9999 816 | ·9999 755⁻ | ·9999 676 | ·9999 575⁺ | ·9999 448 | ·9999 288 |
| ·90 | ·9999 948 | ·9999 930 | ·9999 907 | ·9999 877 | ·9999 839 | ·9999 790 |
| ·91 | ·9999 988 | ·9999 983 | ·9999 978 | ·9999 970 | ·9999 960 | ·9999 948 |
| ·92 | ·9999 998 | ·9999 997 | ·9999 996 | ·9999 994 | ·9999 992 | ·9999 990 |
| ·93 | 1·0000 000 | 1·0000 000 | ·9999 999 | ·9999 999 | ·9999 999 | ·9999 998 |
| ·94 | | | 1·0000 000 | 1·0000 000 | 1·0000 000 | 1·0000 000 |

$x = \cdot 37$ to $\cdot 95$ $\qquad q = 17 \qquad p = 41$ to $45$

| | $p = 41$ | $p = 42$ | $p = 43$ | $p = 44$ | $p = 45$ |
|---|---|---|---|---|---|
| $B(p, q) =$ | $\cdot 4212\ 3172 \times \frac{1}{10^{15}}$ | $\cdot 2977\ 6725^{+} \times \frac{1}{10^{15}}$ | $\cdot 2119\ 6991 \times \frac{1}{10^{15}}$ | $\cdot 1519\ 1177 \times \frac{1}{10^{15}}$ | $\cdot 1095\ 7570 \times \frac{1}{10^{15}}$ |
| $x$ | | | | | |
| ·37 | ·0000 001 | | | | |
| ·38 | ·0000 002 | ·0000 001 | ·0000 001 | | |
| ·39 | ·0000 005− | ·0000 003 | ·0000 001 | ·0000 001 | |
| ·40 | ·0000 010 | ·0000 006 | ·0000 003 | ·0000 002 | ·0000 001 |
| ·41 | ·0000 022 | ·0000 013 | ·0000 007 | ·0000 004 | ·0000 002 |
| ·42 | ·0000 046 | ·0000 027 | ·0000 015+ | ·0000 009 | ·0000 005− |
| ·43 | ·0000 093 | ·0000 055+ | ·0000 032 | ·0000 019 | ·0000 011 |
| ·44 | ·0000 184 | ·0000 111 | ·0000 066 | ·0000 039 | ·0000 023 |
| ·45 | ·0000 351 | ·0000 216 | ·0000 132 | ·0000 081 | ·0000 049 |
| ·46 | ·0000 656 | ·0000 413 | ·0000 258 | ·0000 160 | ·0000 099 |
| ·47 | ·0001 196 | ·0000 768 | ·0000 491 | ·0000 311 | ·0000 197 |
| ·48 | ·0002 131 | ·0001 397 | ·0000 910 | ·0000 590 | ·0000 380 |
| ·49 | ·0003 710 | ·0002 481 | ·0001 650− | ·0001 091 | ·0000 717 |
| ·50 | ·0006 318 | ·0004 309 | ·0002 922 | ·0001 970 | ·0001 321 |
| ·51 | ·0010 529 | ·0007 320 | ·0005 059 | ·0003 478 | ·0002 378 |
| ·52 | ·0017 179 | ·0012 168 | ·0008 570 | ·0006 002 | ·0004 182 |
| ·53 | ·0027 455+ | ·0019 805− | ·0014 206 | ·0010 134 | ·0007 191 |
| ·54 | ·0042 994 | ·0031 573 | ·0023 056 | ·0016 745+ | ·0012 098 |
| ·55 | ·0065 997 | ·0049 318 | ·0036 649 | ·0027 089 | ·0019 919 |
| ·56 | ·0099 335− | ·0075 506 | ·0057 078 | ·0042 919 | ·0032 107 |
| ·57 | ·0146 645− | ·0113 338 | ·0087 122 | ·0066 617 | ·0050 680 |
| ·58 | ·0212 392 | ·0166 843 | ·0130 360 | ·0101 327 | ·0078 364 |
| ·59 | ·0301 869 | ·0240 922 | ·0191 264 | ·0151 065− | ·0118 723 |
| ·60 | ·0421 115+ | ·0341 331 | ·0275 224 | ·0220 802 | ·0176 274 |
| ·61 | ·0576 730 | ·0474 561 | ·0388 497 | ·0316 462 | ·0256 543 |
| ·62 | ·0775 552 | ·0647 592 | ·0538 033 | ·0444 832 | ·0366 035− |
| ·63 | ·1024 215+ | ·0867 509 | ·0731 174 | ·0613 324 | ·0512 080 |
| ·64 | ·1328 570 | ·1140 977 | ·0975 181 | ·0829 588 | ·0702 526 |
| ·65 | ·1693 019 | ·1473 587 | ·1276 621 | ·1100 955− | ·0945 251 |
| ·66 | ·2119 810 | ·1869 119 | ·1640 632 | ·1433 719 | ·1247 497 |
| ·67 | ·2608 369 | ·2328 794 | ·2070 120 | ·1832 321 | ·1615 058 |
| ·68 | ·3154 772 | ·2850 606 | ·2564 980 | ·2298 484 | ·2051 370 |
| ·69 | ·3751 435− | ·3428 840 | ·3121 454 | ·2830 438 | ·2556 618 |
| ·70 | ·4387 138 | ·4053 883 | ·3731 737 | ·3422 327 | ·3126 981 |
| ·71 | ·5047 419 | ·4712 413 | ·4383 947 | ·4063 960 | ·3754 154 |
| ·72 | ·5715 364 | ·5388 013 | ·5062 532 | ·4740 988 | ·4425 259 |
| ·73 | ·6372 747 | ·6062 208 | ·5749 155− | ·5435 592 | ·5123 455− |
| ·74 | ·7001 420 | ·6715 830 | ·6423 985− | ·6127 660 | ·5828 641 |
| ·75 | ·7584 792 | ·7330 585− | ·7067 299 | ·6796 360 | ·6519 263 |
| ·76 | ·8109 213 | ·7890 616 | ·7661 194 | ·7421 954 | ·7174 015+ |
| ·77 | ·8565 067 | ·8383 854 | ·8191 165+ | ·7987 587 | ·7773 829 |
| ·78 | ·8947 418 | ·8802 954 | ·8647 346 | ·8480 809 | ·8303 674 |
| ·79 | ·9256 115− | ·9145 673 | ·9025 186 | ·8894 584 | ·8753 890 |
| ·80 | ·9495 343 | ·9414 639 | ·9325 481 | ·9227 614 | ·9120 850− |
| ·81 | ·9672 716 | ·9616 561 | ·9553 747 | ·9483 936 | ·9406 826 |
| ·82 | ·9798 060 | ·9761 018 | ·9719 072 | ·9671 878 | ·9619 107 |
| ·83 | ·9882 107 | ·9859 065− | ·9832 654 | ·9802 576 | ·9768 533 |
| ·84 | ·9935 300 | ·9921 867 | ·9906 284 | ·9888 324 | ·9867 752 |
| ·85 | ·9966 878 | ·9959 593 | ·9951 042 | ·9941 069 | ·9929 509 |
| ·86 | ·9984 328 | ·9980 687 | ·9976 362 | ·9971 258 | ·9965 273 |
| ·87 | ·9993 224 | ·9991 564 | ·9989 570 | ·9987 190 | ·9984 366 |
| ·88 | ·9997 360 | ·9996 680 | ·9995 853 | ·9994 855+ | ·9993 658 |
| ·89 | ·9999 089 | ·9998 843 | ·9998 540 | ·9998 171 | ·9997 722 |
| ·90 | ·9999 728 | ·9999 651 | ·9999 555+ | ·9999 437 | ·9999 292 |
| ·91 | ·9999 932 | ·9999 912 | ·9999 886 | ·9999 854 | ·9999 815+ |
| ·92 | ·9999 986 | ·9999 982 | ·9999 976 | ·9999 970 | ·9999 961 |
| ·93 | ·9999 998 | ·9999 997 | ·9999 996 | ·9999 995+ | ·9999 994 |
| ·94 | 1·0000 000 | 1·0000 000 | 1·0000 000 | ·9999 999 | ·9999 999 |
| ·95 | | | | 1·0000 000 | 1·0000 000 |

$x$ = ·41 to ·95 $q = 17$ $p$ = 46 to 50

| | $p = 46$ | $p = 47$ | $p = 48$ | $p = 49$ | $p = 50$ |
|---|---|---|---|---|---|
| $B(p, q)$ = | ·7953 0751 $\times\frac{1}{10^{16}}$ | ·5807 0072 $\times\frac{1}{10^{16}}$ | ·4264 5209 $\times\frac{1}{10^{16}}$ | ·3149 1847 $\times\frac{1}{10^{16}}$ | ·2338 0310 $\times\frac{1}{10^{16}}$ |
| $x$ | | | | | |
| ·41 | ·0000 001 | ·0000 001 | | | |
| ·42 | ·0000 003 | ·0000 002 | ·0000 001 | | |
| ·43 | ·0000 006 | ·0000 004 | ·0000 002 | ·0000 001 | ·0000 001 |
| ·44 | ·0000 014 | ·0000 008 | ·0000 005− | ·0000 003 | ·0000 002 |
| ·45 | ·0000 029 | ·0000 018 | ·0000 010 | ·0000 006 | ·0000 004 |
| ·46 | ·0000 061 | ·0000 037 | ·0000 023 | ·0000 014 | ·0000 008 |
| ·47 | ·0000 124 | ·0000 077 | ·0000 048 | ·0000 030 | ·0000 018 |
| ·48 | ·0000 244 | ·0000 156 | ·0000 099 | ·0000 062 | ·0000 039 |
| ·49 | ·0000 469 | ·0000 305+ | ·0000 198 | ·0000 128 | ·0000 082 |
| ·50 | ·0000 882 | ·0000 585+ | ·0000 387 | ·0000 254 | ·0000 166 |
| ·51 | ·0001 617 | ·0001 095− | ·0000 737 | ·0000 494 | ·0000 330 |
| ·52 | ·0002 898 | ·0001 999 | ·0001 372 | ·0000 937 | ·0000 637 |
| ·53 | ·0005 077 | ·0003 567 | ·0002 494 | ·0001 735+ | ·0001 202 |
| ·54 | ·0008 697 | ·0006 221 | ·0004 429 | ·0003 139 | ·0002 214 |
| ·55 | ·0014 573 | ·0010 611 | ·0007 689 | ·0005 547 | ·0003 983 |
| ·56 | ·0023 899 | ·0017 704 | ·0013 053 | ·0009 581 | ·0007 002 |
| ·57 | ·0038 366 | ·0028 906 | ·0021 677 | ·0016 184 | ·0012 030 |
| ·58 | ·0060 310 | ·0046 196 | ·0035 224 | ·0026 739 | ·0020 210 |
| ·59 | ·0092 856 | ·0072 287 | ·0056 019 | ·0043 222 | ·0033 206 |
| ·60 | ·0140 058 | ·0110 770 | ·0087 216 | ·0068 372 | ·0053 374 |
| ·61 | ·0206 996 | ·0166 260 | ·0132 952 | ·0105 862 | ·0083 940 |
| ·62 | ·0299 810 | ·0244 469 | ·0198 477 | ·0160 457 | ·0129 188 |
| ·63 | ·0425 618 | ·0352 200 | ·0290 201 | ·0238 121 | ·0194 598 |
| ·64 | ·0592 292 | ·0497 204 | ·0415 630 | ·0346 020 | ·0286 923 |
| ·65 | ·0808 058 | ·0687 865+ | ·0583 144 | ·0492 386 | ·0414 130 |
| ·66 | ·1080 891 | ·0932 684 | ·0801 567 | ·0686 185+ | ·0585 167 |
| ·67 | ·1417 736 | ·1239 545+ | ·1079 515− | ·0936 554 | ·0809 495− |
| ·68 | ·1823 586 | ·1614 816 | ·1424 525+ | ·1251 995− | ·1096 365+ |
| ·69 | ·2300 506 | ·2062 321 | ·1842 025+ | ·1639 353 | ·1453 846 |
| ·70 | ·2846 731 | ·2582 321 | ·2334 225+ | ·2102 669 | ·1887 653 |
| ·71 | ·3455 984 | ·3170 647 | ·2899 092 | ·2642 020 | ·2399 900 |
| ·72 | ·4117 168 | ·3818 156 | ·3529 578 | ·3252 542 | ·2987 946 |
| ·73 | ·4814 578 | ·4510 669 | ·4213 292 | ·3923 844 | ·3643 553 |
| ·74 | ·5528 692 | ·5229 525+ | ·4932 777 | ·4639 986 | ·4352 573 |
| ·75 | ·6237 548 | ·5952 771 | ·5666 479 | ·5380 187 | ·5095 355+ |
| ·76 | ·6918 583 | ·6656 931 | ·6390 380 | ·6120 276 | ·5847 966 |
| ·77 | ·7550 714 | ·7319 159 | ·7080 165+ | ·6834 798 | ·6584 173 |
| ·78 | ·8116 383 | ·7919 484 | ·7713 619 | ·7499 520 | ·7277 993 |
| ·79 | ·8603 222 | ·8442 794 | ·8272 910 | ·8093 966 | ·7906 441 |
| ·80 | ·9005 070 | ·8880 229 | ·8746 357 | ·8603 561 | ·8452 022 |
| ·81 | ·9322 160 | ·9229 726 | ·9129 367 | ·9020 979 | ·8904 518 |
| ·82 | ·9560 449 | ·9495 618 | ·9424 361 | ·9346 452 | ·9261 706 |
| ·83 | ·9730 231 | ·9687 383 | ·9639 712 | ·9586 957 | ·9528 872 |
| ·84 | ·9844 327 | ·9817 805+ | ·9787 944 | ·9754 498 | ·9717 231 |
| ·85 | ·9916 190 | ·9900 931 | ·9883 545+ | ·9863 842 | ·9841 625− |
| ·86 | ·9958 296 | ·9950 208 | ·9940 885+ | ·9930 195+ | ·9917 999 |
| ·87 | ·9981 036 | ·9977 131 | ·9972 577 | ·9967 294 | ·9961 198 |
| ·88 | ·9992 229 | ·9990 535− | ·9988 536 | ·9986 191 | ·9983 454 |
| ·89 | ·9997 181 | ·9996 532 | ·9995 757 | ·9994 839 | ·9993 754 |
| ·90 | ·9999 115− | ·9998 900 | ·9998 641 | ·9998 330 | ·9997 959 |
| ·91 | ·9999 767 | ·9999 707 | ·9999 634 | ·9999 546 | ·9999 440 |
| ·92 | ·9999 950+ | ·9999 937 | ·9999 921 | ·9999 901 | ·9999 876 |
| ·93 | ·9999 992 | ·9999 990 | ·9999 987 | ·9999 983 | ·9999 979 |
| ·94 | ·9999 999 | ·9999 999 | ·9999 998 | ·9999 998 | ·9999 998 |
| ·95 | 1·0000 000 | 1·0000 000 | 1·0000 000 | 1·0000 000 | 1·0000 000 |

$x$ = ·13 to ·70 $q$ = 18 $p$ = 18 to 23

| $x$ | $p$ = 18 | $p$ = 19 | $p$ = 20 | $p$ = 21 | $p$ = 22 | $p$ = 23 |
|---|---|---|---|---|---|---|
| $B(p, q)$ = | ·1224 3466×$\frac{1}{10^{10}}$ | ·6121 7330×$\frac{1}{10^{11}}$ | ·3143 5926×$\frac{1}{10^{11}}$ | ·1654 5224×$\frac{1}{10^{11}}$ | ·8908 9670×$\frac{1}{10^{12}}$ | ·4899 9319×$\frac{1}{10^{12}}$ |
| ·13 | ·0000 001 | | | | | |
| ·14 | ·0000 002 | | | | | |
| ·15 | ·0000 005$^-$ | ·0000 001 | | | | |
| ·16 | ·0000 013 | ·0000 004 | ·0000 001 | | | |
| ·17 | ·0000 033 | ·0000 010 | ·0000 003 | ·0000 001 | | |
| ·18 | ·0000 076 | ·0000 026 | ·0000 008 | ·0000 003 | ·0000 001 | |
| ·19 | ·0000 165$^+$ | ·0000 059 | ·0000 020 | ·0000 007 | ·0000 002 | ·0000 001 |
| ·20 | ·0000 342 | ·0000 128 | ·0000 047 | ·0000 017 | ·0000 006 | ·0000 002 |
| ·21 | ·0000 677 | ·0000 265$^+$ | ·0000 102 | ·0000 038 | ·0000 014 | ·0000 005$^+$ |
| ·22 | ·0001 281 | ·0000 526 | ·0000 211 | ·0000 083 | ·0000 032 | ·0000 012 |
| ·23 | ·0002 333 | ·0001 000 | ·0000 419 | ·0000 172 | ·0000 069 | ·0000 027 |
| ·24 | ·0004 097 | ·0001 831 | ·0000 800 | ·0000 343 | ·0000 144 | ·0000 059 |
| ·25 | ·0006 957 | ·0003 234 | ·0001 471 | ·0000 655$^+$ | ·0000 287 | ·0000 123 |
| ·26 | ·0011 453 | ·0005 531 | ·0002 613 | ·0001 210 | ·0000 550$^-$ | ·0000 245$^+$ |
| ·27 | ·0018 320 | ·0009 175$^-$ | ·0004 496 | ·0002 159 | ·0001 018 | ·0000 471 |
| ·28 | ·0028 522 | ·0014 792 | ·0007 508 | ·0003 735$^+$ | ·0001 824 | ·0000 875$^-$ |
| ·29 | ·0043 297 | ·0023 223 | ·0012 193 | ·0006 275$^+$ | ·0003 170 | ·0001 573 |
| ·30 | ·0064 185$^+$ | ·0035 559 | ·0019 287 | ·0010 256 | ·0005 353 | ·0002 746 |
| ·31 | ·0093 046 | ·0053 179 | ·0029 762 | ·0016 333 | ·0008 799 | ·0004 659 |
| ·32 | ·0132 065$^+$ | ·0077 780 | ·0044 866 | ·0025 380 | ·0014 098 | ·0007 697 |
| ·33 | ·0183 736 | ·0111 389 | ·0066 153 | ·0038 537 | ·0022 046 | ·0012 399 |
| ·34 | ·0250 822 | ·0156 365$^-$ | ·0095 514 | ·0057 240 | ·0033 691 | ·0019 498 |
| ·35 | ·0336 284 | ·0215 367 | ·0135 180 | ·0083 258 | ·0050 375$^-$ | ·0029 972 |
| ·36 | ·0443 193 | ·0291 310 | ·0187 711 | ·0118 713 | ·0073 766 | ·0045 082 |
| ·37 | ·0574 603 | ·0387 283 | ·0255 962 | ·0166 073 | ·0105 890 | ·0066 415$^+$ |
| ·38 | ·0733 416 | ·0506 441 | ·0343 018 | ·0228 133 | ·0149 135$^+$ | ·0095 919 |
| ·39 | ·0922 219 | ·0651 871 | ·0452 098 | ·0307 962 | ·0206 243 | ·0135 918 |
| ·40 | ·1143 126 | ·0826 443 | ·0586 430 | ·0408 820 | ·0280 265$^-$ | ·0189 108 |
| ·41 | ·1397 616 | ·1032 636 | ·0749 104 | ·0534 046 | ·0374 493 | ·0258 526 |
| ·42 | ·1686 385$^-$ | ·1272 372 | ·0942 904 | ·0686 908 | ·0492 351 | ·0347 494 |
| ·43 | ·2009 230 | ·1546 846 | ·1170 125$^+$ | ·0870 443 | ·0637 262 | ·0459 515$^-$ |
| ·44 | ·2364 963 | ·1856 387 | ·1432 395$^+$ | ·1087 266 | ·0812 478 | ·0598 142 |
| ·45 | ·2751 370 | ·2200 338 | ·1730 511 | ·1339 380 | ·1020 887 | ·0766 817 |
| ·46 | ·3165 224 | ·2576 987 | ·2064 293 | ·1627 990 | ·1264 820 | ·0968 671 |
| ·47 | ·3602 345$^-$ | ·2983 545$^-$ | ·2432 487 | ·1953 343 | ·1545 842 | ·1206 319 |
| ·48 | ·4057 719 | ·3416 176 | ·2832 710 | ·2314 592 | ·1864 569 | ·1481 641 |
| ·49 | ·4525 657 | ·3870 092 | ·3261 452 | ·2709 720 | ·2220 517 | ·1795 578 |
| ·50 | ·5000 000$^e$ | ·4339 697 | ·3714 147 | ·3135 513 | ·2611 987 | ·2147 953 |
| ·51 | ·5474 343 | ·4818 778 | ·4185 295$^-$ | ·3587 604 | ·3036 021 | ·2537 339 |
| ·52 | ·5942 281 | ·5300 739 | ·4668 650$^+$ | ·4060 581 | ·3488 417 | ·2960 986 |
| ·53 | ·6397 655$^+$ | ·5778 855$^+$ | ·5157 450$^+$ | ·4548 162 | ·3963 826 | ·3414 816 |
| ·54 | ·6834 776 | ·6246 540 | ·5644 682 | ·5043 425$^-$ | ·4455 911 | ·3893 501 |
| ·55 | ·7248 630 | ·6697 598 | ·6123 365$^-$ | ·5539 082 | ·4957 582 | ·4390 620 |
| ·56 | ·7635 037 | ·7126 461 | ·6586 835$^+$ | ·6027 783 | ·5461 277 | ·4898 890 |
| ·57 | ·7990 770 | ·7528 386 | ·7029 011 | ·6502 421 | ·5959 280 | ·5410 461 |
| ·58 | ·8313 615$^+$ | ·7899 602 | ·7444 623 | ·6956 431 | ·6444 062 | ·5917 253 |
| ·59 | ·8602 384 | ·8237 405$^-$ | ·7829 396 | ·7384 054 | ·6908 599 | ·6411 316 |
| ·60 | ·8856 874 | ·8540 190 | ·8180 171 | ·7780 550$^-$ | ·7346 675$^+$ | ·6885 191 |
| ·61 | ·9077 781 | ·8807 433 | ·8494 968 | ·8142 351 | ·7753 129 | ·7332 238 |
| ·62 | ·9266 584 | ·9039 609 | ·8772 973 | ·8467 142 | ·8124 028 | ·7746 914 |
| ·63 | ·9425 397 | ·9238 077 | ·9014 476 | ·8753 869 | ·8456 777 | ·8124 980 |
| ·64 | ·9556 807 | ·9404 925$^-$ | ·9220 747 | ·9002 681 | ·8750 140 | ·8463 620 |
| ·65 | ·9663 716 | ·9542 799 | ·9393 879 | ·9214 804 | ·9004 177 | ·8761 478 |
| ·66 | ·9749 178 | ·9654 721 | ·9536 600 | ·9392 374 | ·9220 128 | ·9018 599 |
| ·67 | ·9816 264 | ·9743 917 | ·9652 074 | ·9538 235$^+$ | ·9400 219 | ·9236 294 |
| ·68 | ·9867 935$^-$ | ·9813 649 | ·9743 707 | ·9655 719 | ·9547 453 | ·9416 942 |
| ·69 | ·9906 954 | ·9867 086 | ·9814 965$^+$ | ·9748 432 | ·9665 361 | ·9563 750$^+$ |
| ·70 | ·9935 815$^-$ | ·9907 188 | ·9869 220 | ·9820 051 | ·9757 771 | ·9680 487 |

$x$ = ·71 to ·90 $q = 18$ $p$ = 18 to 23

| | $p = 18$ | $p = 19$ | $p = 20$ | $p = 21$ | $p = 22$ | $p = 23$ |
|---|---|---|---|---|---|---|
| $B(p, q)$ = | ·1224 3466 × $\frac{1}{10^{10}}$ | ·6121 7330 × $\frac{1}{10^{11}}$ | ·3143 5926 × $\frac{1}{10^{11}}$ | ·1654 5224 × $\frac{1}{10^{11}}$ | ·8908 9670 × $\frac{1}{10^{12}}$ | ·4899 9319 × $\frac{1}{10^{12}}$ |
| $x$ | | | | | | |
| ·71 | ·9956 703 | ·9936 629 | ·9909 624 | ·9874 152 | ·9828 580 | ·9771 222 |
| ·72 | ·9971 478 | ·9957 749 | ·9939 019 | ·9914 071 | ·9881 567 | ·9840 080 |
| ·73 | ·9981 680 | ·9972 536 | ·9959 887 | ·9942 804 | ·9920 239 | ·9891 038 |
| ·74 | ·9988 547 | ·9982 624 | ·9974 320 | ·9962 951 | ·9947 729 | ·9927 759 |
| ·75 | ·9993 043 | ·9989 320 | ·9984 030 | ·9976 691 | ·9966 729 | ·9953 485+ |
| ·76 | ·9995 903 | ·9993 637 | ·9990 374 | ·9985 786 | ·9979 477 | ·9970 976 |
| ·77 | ·9997 667 | ·9996 334 | ·9994 390 | ·9991 620 | ·9987 761 | ·9982 493 |
| ·78 | ·9998 719 | ·9997 963 | ·9996 847 | ·9995 236 | ·9992 963 | ·9989 819 |
| ·79 | ·9999 323 | ·9998 912 | ·9998 297 | ·9997 397 | ·9996 111 | ·9994 310 |
| ·80 | ·9999 658 | ·9999 443 | ·9999 119 | ·9998 638 | ·9997 942 | ·9996 955+ |
| ·81 | ·9999 835− | ·9999 728 | ·9999 565+ | ·9999 320 | ·9998 961 | ·9998 446 |
| ·82 | ·9999 924 | ·9999 874 | ·9999 796 | ·9999 678 | ·9999 503 | ·9999 248 |
| ·83 | ·9999 967 | ·9999 945− | ·9999 910 | ·9999 856 | ·9999 775+ | ·9999 656 |
| ·84 | ·9999 987 | ·9999 977 | ·9999 963 | ·9999 940 | ·9999 905− | ·9999 853 |
| ·85 | ·9999 995+ | ·9999 991 | ·9999 986 | ·9999 976 | ·9999 962 | ·9999 941 |
| ·86 | ·9999 998 | ·9999 997 | ·9999 995− | ·9999 992 | ·9999 986 | ·9999 979 |
| ·87 | ·9999 999 | ·9999 999 | ·9999 998 | ·9999 997 | ·9999 996 | ·9999 993 |
| ·88 | 1·0000 000 | 1·0000 000 | 1·0000 000 | ·9999 999 | ·9999 999 | ·9999 998 |
| ·89 | | | | 1·0000 000 | 1·0000 000 | ·9999 999 |
| ·90 | | | | | | 1·0000 000 |

$x$ = ·20 to ·80 $q = 18$ $p$ = 24 to 29

| $x$ | $p = 24$ | $p = 25$ | $p = 26$ | $p = 27$ | $p = 28$ | $p = 29$ |
|---|---|---|---|---|---|---|
| $B(p, q)$ = | ·2748 7423 $\times \frac{1}{10^{12}}$ | ·1570 7099 $\times \frac{1}{10^{12}}$ | ·9132 0341 $\times \frac{1}{10^{13}}$ | ·5396 2020 $\times \frac{1}{10^{13}}$ | ·3237 7212 $\times \frac{1}{10^{13}}$ | ·1970 7868 $\times \frac{1}{10^{13}}$ |
| ·20 | ·0000 001 | | | | | |
| ·21 | ·0000 002 | ·0000 001 | | | | |
| ·22 | ·0000 005$^{-}$ | ·0000 002 | ·0000 001 | | | |
| ·23 | ·0000 011 | ·0000 004 | ·0000 002 | ·0000 001 | | |
| ·24 | ·0000 024 | ·0000 010 | ·0000 004 | ·0000 001 | ·0000 001 | |
| ·25 | ·0000 052 | ·0000 022 | ·0000 009 | ·0000 004 | ·0000 001 | ·0000 001 |
| ·26 | ·0000 108 | ·0000 046 | ·0000 020 | ·0000 008 | ·0000 003 | ·0000 001 |
| ·27 | ·0000 215$^{-}$ | ·0000 096 | ·0000 042 | ·0000 018 | ·0000 008 | ·0000 003 |
| ·28 | ·0000 413 | ·0000 192 | ·0000 088 | ·0000 040 | ·0000 018 | ·0000 008 |
| ·29 | ·0000 768 | ·0000 369 | ·0000 175$^{-}$ | ·0000 082 | ·0000 038 | ·0000 017 |
| ·30 | ·0001 386 | ·0000 689 | ·0000 337 | ·0000 163 | ·0000 078 | ·0000 037 |
| ·31 | ·0002 427 | ·0001 245$^{+}$ | ·0000 630 | ·0000 314 | ·0000 155$^{-}$ | ·0000 075$^{+}$ |
| ·32 | ·0004 135$^{+}$ | ·0002 188 | ·0001 141 | ·0000 587 | ·0000 298 | ·0000 150$^{-}$ |
| ·33 | ·0006 862 | ·0003 741 | ·0002 010 | ·0001 066 | ·0000 558 | ·0000 288 |
| ·34 | ·0011 106 | ·0006 231 | ·0003 447 | ·0001 881 | ·0001 014 | ·0000 540 |
| ·35 | ·0017 553 | ·0010 128 | ·0005 761 | ·0003 234 | ·0001 793 | ·0000 982 |
| ·36 | ·0027 123 | ·0016 078 | ·0009 398 | ·0005 421 | ·0003 088 | ·0001 738 |
| ·37 | ·0041 014 | ·0024 958 | ·0014 978 | ·0008 871 | ·0005 189 | ·0002 999 |
| ·38 | ·0060 751 | ·0037 921 | ·0023 346 | ·0014 186 | ·0008 514 | ·0005 050$^{-}$ |
| ·39 | ·0088 220 | ·0056 440 | ·0035 619 | ·0022 189 | ·0013 653 | ·0008 303 |
| ·40 | ·0125 694 | ·0082 361 | ·0053 241 | ·0033 977 | ·0021 420 | ·0013 348 |
| ·41 | ·0175 838 | ·0117 921 | ·0078 028 | ·0050 977 | ·0032 904 | ·0020 994 |
| ·42 | ·0241 686 | ·0165 769 | ·0112 201 | ·0074 993 | ·0049 525$^{+}$ | ·0032 335$^{+}$ |
| ·43 | ·0326 591 | ·0228 947 | ·0158 409 | ·0108 246 | ·0073 094 | ·0048 802 |
| ·44 | ·0434 130 | ·0310 846 | ·0219 716 | ·0153 400 | ·0105 850$^{-}$ | ·0072 225$^{-}$ |
| ·45 | ·0567 980 | ·0415 123 | ·0299 564 | ·0213 561 | ·0150 492 | ·0104 880 |
| ·46 | ·0731 752 | ·0545 573 | ·0401 694 | ·0292 236 | ·0210 182 | ·0149 521 |
| ·47 | ·0928 797 | ·0705 969 | ·0530 024 | ·0393 261 | ·0288 511 | ·0209 387 |
| ·48 | ·1161 979 | ·0899 857 | ·0688 481 | ·0520 681 | ·0389 424 | ·0288 169 |
| ·49 | ·1433 455$^{+}$ | ·1130 329 | ·0880 795$^{+}$ | ·0678 577 | ·0517 102 | ·0389 940 |
| ·50 | ·1744 444 | ·1399 781 | ·1110 264 | ·0870 856 | ·0675 782 | ·0519 027 |
| ·51 | ·2095 030 | ·1709 669 | ·1379 491 | ·1100 999 | ·0869 540 | ·0679 828 |
| ·52 | ·2484 005$^{-}$ | ·2060 286 | ·1690 126 | ·1371 788 | ·1102 026 | ·0876 582 |
| ·53 | ·2908 771 | ·2450 590 | ·2042 626 | ·1685 030 | ·1376 172 | ·1113 092 |
| ·54 | ·3365 323 | ·2878 080 | ·2436 053 | ·2041 289 | ·1693 896 | ·1392 409 |
| ·55 | ·3848 308 | ·3338 760 | ·2867 939 | ·2439 672 | ·2055 818 | ·1716 519 |
| ·56 | ·4351 175$^{-}$ | ·3827 194 | ·3334 233 | ·2877 674 | ·2461 023 | ·2086 037 |
| ·57 | ·4866 415$^{-}$ | ·4336 650$^{-}$ | ·3829 346 | ·3351 116 | ·2906 892 | ·2499 952 |
| ·58 | ·5385 863 | ·4859 344 | ·4346 304 | ·3854 181 | ·3389 033 | ·2955 449 |
| ·59 | ·5901 060 | ·5386 765$^{-}$ | ·4876 995$^{+}$ | ·4379 578 | ·3901 321 | ·3447 830 |
| ·60 | ·6403 641 | ·5910 053 | ·5412 517 | ·4918 807 | ·4436 069 | ·3970 571 |
| ·61 | ·6885 728 | ·6420 427 | ·5943 587 | ·5462 529 | ·4984 321 | ·4515 507 |
| ·62 | ·7340 288 | ·6909 603 | ·6461 001 | ·6001 011 | ·5536 252 | ·5073 152 |
| ·63 | ·7761 445$^{+}$ | ·7370 191 | ·6956 088 | ·6524 624 | ·6081 654 | ·5633 148 |
| ·64 | ·8144 711 | ·7796 038 | ·7421 144 | ·7024 333 | ·6610 474 | ·6184 790 |
| ·65 | ·8487 122 | ·8182 473 | ·7849 796 | ·7492 168 | ·7113 348 | ·6717 616 |
| ·66 | ·8787 279 | ·8526 466 | ·8237 276 | ·7921 614 | ·7582 102 | ·7221 977 |
| ·67 | ·9045 285$^{-}$ | ·8826 659 | ·8580 574 | ·8307 893 | ·8010 166 | ·7689 578 |
| ·68 | ·9262 599 | ·9083 304 | ·8878 478 | ·8648 126 | ·8392 863 | ·8113 897 |
| ·69 | ·9441 817 | ·9298 089 | ·9131 478 | ·8941 350$^{+}$ | ·8727 562 | ·8490 486 |
| ·70 | ·9586 403 | ·9473 893 | ·9341 581 | ·9188 405$^{+}$ | ·9013 671 | ·8817 095$^{+}$ |
| ·71 | ·9700 397 | ·9614 492 | ·9512 024 | ·9391 703 | ·9252 488 | ·9093 633 |
| ·72 | ·9788 132 | ·9724 235$^{-}$ | ·9646 945$^{+}$ | ·9554 911 | ·9446 925$^{-}$ | ·9321 969 |
| ·73 | ·9853 965$^{+}$ | ·9807 732 | ·9751 032 | ·9682 578 | ·9601 142 | ·9505 601 |
| ·74 | ·9902 059 | ·9869 570 | ·9829 180 | ·9779 748 | ·9720 137 | ·9649 243 |
| ·75 | ·9936 211 | ·9914 077 | ·9886 189 | ·9851 598 | ·9809 319 | ·9758 358 |
| ·76 | ·9959 740 | ·9945 153 | ·9926 527 | ·9903 117 | ·9874 122 | ·9838 707 |
| ·77 | ·9975 438 | ·9966 158 | ·9954 154 | ·9938 867 | ·9919 685$^{+}$ | ·9895 947 |
| ·78 | ·9985 555$^{-}$ | ·9979 872 | ·9972 426 | ·9962 820 | ·9950 610 | ·9935 304 |
| ·79 | ·9991 836 | ·9988 497 | ·9984 065$^{-}$ | ·9978 274 | ·9970 820 | ·9961 355$^{-}$ |
| ·80 | ·9995 582 | ·9993 706 | ·9991 183 | ·9987 847 | ·9983 496 | ·9977 903 |

$x = \cdot81$ to $\cdot91$ $q = 18$ $p = 24$ to $29$

| | $p = 24$ | $p = 25$ | $p = 26$ | $p = 27$ | $p = 28$ | $p = 29$ |
|---|---|---|---|---|---|---|
| $B(p, q) =$ | $\cdot2748\ 7423 \times \frac{1}{10^{12}}$ | $\cdot1570\ 7099 \times \frac{1}{10^{12}}$ | $\cdot9132\ 0341 \times \frac{1}{10^{13}}$ | $\cdot5396\ 2020 \times \frac{1}{10^{13}}$ | $\cdot3237\ 7212 \times \frac{1}{10^{13}}$ | $\cdot1970\ 7868 \times \frac{1}{10^{13}}$ |
| $x$ | | | | | | |
| ·81 | ·9997 720 | ·9996 716 | ·9995 349 | ·9993 519 | ·9991 102 | ·9987 956 |
| ·82 | ·9998 884 | ·9998 374 | ·9997 673 | ·9996 721 | ·9995 449 | ·9993 773 |
| ·83 | ·9999 484 | ·9999 241 | ·9998 901 | ·9998 435+ | ·9997 804 | ·9996 963 |
| ·84 | ·9999 777 | ·9999 668 | ·9999 514 | ·9999 300 | ·9999 008 | ·9998 613 |
| ·85 | ·9999 910 | ·9999 865− | ·9999 800 | ·9999 709 | ·9999 583 | ·9999 411 |
| ·86 | ·9999 967 | ·9999 949 | ·9999 924 | ·9999 889 | ·9999 839 | ·9999 770 |
| ·87 | ·9999 989 | ·9999 983 | ·9999 974 | ·9999 961 | ·9999 943 | ·9999 918 |
| ·88 | ·9999 997 | ·9999 995− | ·9999 992 | ·9999 988 | ·9999 982 | ·9999 974 |
| ·89 | ·9999 999 | ·9999 999 | ·9999 998 | ·9999 997 | ·9999 995+ | ·9999 993 |
| ·90 | 1·0000 000 | 1·0000 000 | ·9999 999 | ·9999 999 | ·9999 999 | ·9999 998 |
| ·91 | | | 1·0000 000 | 1·0000 000 | 1·0000 000 | 1·0000 000 |

$x$ = ·26 to ·93 $q$ = 18 $p$ = 30 to 35

| | $p$ = 30 | $p$ = 31 | $p$ = 32 | $p$ = 33 | $p$ = 34 | $p$ = 35 |
|---|---|---|---|---|---|---|
| $B(p, q)$ = | ·1216 0174 $\times \frac{1}{10^{13}}$ | ·7600 1087 $\times \frac{1}{10^{14}}$ | ·4808 2320 $\times \frac{1}{10^{14}}$ | ·3077 2685 $\times \frac{1}{10^{14}}$ | ·1991 1737 $\times \frac{1}{10^{14}}$ | ·1301 9213 $\times \frac{1}{10^{14}}$ |
| $x$ | | | | | | |
| ·26 | ·0000 001 | | | | | |
| ·27 | ·0000 001 | ·0000 001 | | | | |
| ·28 | ·0000 003 | ·0000 001 | ·0000 001 | | | |
| ·29 | ·0000 008 | ·0000 003 | ·0000 002 | ·0000 001 | | |
| ·30 | ·0000 017 | ·0000 008 | ·0000 004 | ·0000 002 | ·0000 001 | |
| ·31 | ·0000 036 | ·0000 017 | ·0000 008 | ·0000 004 | ·0000 002 | ·0000 001 |
| ·32 | ·0000 074 | ·0000 036 | ·0000 018 | ·0000 008 | ·0000 004 | ·0000 002 |
| ·33 | ·0000 147 | ·0000 075− | ·0000 037 | ·0000 018 | ·0000 009 | ·0000 004 |
| ·34 | ·0000 284 | ·0000 148 | ·0000 076 | ·0000 039 | ·0000 020 | ·0000 010 |
| ·35 | ·0000 532 | ·0000 285− | ·0000 151 | ·0000 079 | ·0000 041 | ·0000 021 |
| ·36 | ·0000 967 | ·0000 532 | ·0000 290 | ·0000 157 | ·0000 084 | ·0000 044 |
| ·37 | ·0001 714 | ·0000 969 | ·0000 542 | ·0000 301 | ·0000 165+ | ·0000 090 |
| ·38 | ·0002 962 | ·0001 718 | ·0000 987 | ·0000 561 | ·0000 316 | ·0000 177 |
| ·39 | ·0004 994 | ·0002 972 | ·0001 750+ | ·0001 021 | ·0000 590 | ·0000 338 |
| ·40 | ·0008 226 | ·0005 016 | ·0003 028 | ·0001 811 | ·0001 073 | ·0000 630 |
| ·41 | ·0013 249 | ·0008 274 | ·0005 116 | ·0003 133 | ·0001 901 | ·0001 144 |
| ·42 | ·0020 883 | ·0013 347 | ·0008 446 | ·0005 295− | ·0003 289 | ·0002 025+ |
| ·43 | ·0032 233 | ·0021 071 | ·0013 639 | ·0008 746 | ·0005 558 | ·0003 502 |
| ·44 | ·0048 757 | ·0032 580 | ·0021 558 | ·0014 133 | ·0009 182 | ·0005 915+ |
| ·45 | ·0072 322 | ·0049 369 | ·0033 376 | ·0022 356 | ·0014 842 | ·0009 770 |
| ·46 | ·0105 260 | ·0073 362 | ·0050 642 | ·0034 639 | ·0023 485+ | ·0015 789 |
| ·47 | ·0150 398 | ·0106 963 | ·0075 354 | ·0052 605− | ·0036 405− | ·0024 984 |
| ·48 | ·0211 076 | ·0153 101 | ·0110 013 | ·0078 343 | ·0055 311 | ·0038 727 |
| ·49 | ·0291 105+ | ·0215 233 | ·0157 667 | ·0114 476 | ·0082 409 | ·0058 840 |
| ·50 | ·0394 703 | ·0297 317 | ·0221 921 | ·0164 196 | ·0120 465− | ·0087 666 |
| ·51 | ·0526 356 | ·0403 733 | ·0306 900 | ·0231 279 | ·0172 845+ | ·0128 143 |
| ·52 | ·0690 630 | ·0539 141 | ·0417 168 | ·0320 047 | ·0243 527 | ·0183 841 |
| ·53 | ·0891 923 | ·0708 279 | ·0557 573 | ·0435 265− | ·0337 048 | ·0258 966 |
| ·54 | ·1134 170 | ·0915 699 | ·0733 030 | ·0581 985− | ·0458 403 | ·0358 301 |
| ·55 | ·1420 509 | ·1165 448 | ·0948 234 | ·0765 299 | ·0612 854 | ·0487 086 |
| ·56 | ·1752 945+ | ·1460 713 | ·1207 320 | ·0990 035− | ·0805 672 | ·0650 807 |
| ·57 | ·2132 022 | ·1803 460 | ·1513 478 | ·1260 378 | ·1041 792 | ·0854 900 |
| ·58 | ·2556 551 | ·2194 086 | ·1868 569 | ·1579 469 | ·1325 412 | ·1104 382 |
| ·59 | ·3023 425+ | ·2631 134 | ·2272 757 | ·1948 986 | ·1659 554 | ·1403 406 |
| ·60 | ·3527 546 | ·3111 103 | ·2724 213 | ·2368 758 | ·2045 617 | ·1754 791 |
| ·61 | ·4061 889 | ·3628 381 | ·3218 926 | ·2836 469 | ·2482 986 | ·2159 550− |
| ·62 | ·4617 717 | ·4175 339 | ·3750 655− | ·3347 471 | ·2968 722 | ·2616 486 |
| ·63 | ·5184 950+ | ·4742 579 | ·4311 054 | ·3894 766 | ·3497 401 | ·3121 891 |
| ·64 | ·5752 648 | ·5319 354 | ·4889 973 | ·4469 180 | ·4061 138 | ·3669 417 |
| ·65 | ·6309 602 | ·5894 109 | ·5475 934 | ·5059 720 | ·4649 812 | ·4250 152 |
| ·66 | ·6844 964 | ·6455 132 | ·6056 749 | ·5654 133 | ·5251 517 | ·4852 927 |
| ·67 | ·7348 869 | ·6991 239 | ·6620 227 | ·6239 592 | ·5853 189 | ·5464 854 |
| ·68 | ·7812 998 | ·7492 441 | ·7154 925− | ·6803 486 | ·6441 398 | ·6072 068 |
| ·69 | ·8231 010 | ·7950 517 | ·7650 842 | ·7334 217 | ·7003 199 | ·6660 596 |
| ·70 | ·8598 828 | ·8359 462 | ·8100 020 | ·7821 931 | ·7526 987 | ·7217 297 |
| ·71 | ·8914 730 | ·8715 730 | ·8496 958 | ·8259 112 | ·8003 247 | ·7730 752 |
| ·72 | ·9179 260 | ·9018 285+ | ·8838 824 | ·8640 968 | ·8425 125+ | ·8192 015+ |
| ·73 | ·9394 971 | ·9268 447 | ·9125 435− | ·8965 573 | ·8788 757 | ·8595 143 |
| ·74 | ·9566 028 | ·9469 554 | ·9359 013 | ·9233 757 | ·9093 318 | ·8937 431 |
| ·75 | ·9697 732 | ·9626 496 | ·9543 771 | ·9448 766 | ·9340 806 | ·9219 352 |
| ·76 | ·9796 014 | ·9745 181 | ·9685 361 | ·9615 746 | ·9535 584 | ·9444 199 |
| ·77 | ·9866 954 | ·9831 979 | ·9790 279 | ·9741 113 | ·9683 752 | ·9617 500− |
| ·78 | ·9916 367 | ·9893 226 | ·9865 278 | ·9831 896 | ·9792 446 | ·9746 289 |
| ·79 | ·9949 495− | ·9934 816 | ·9916 860 | ·9895 139 | ·9869 139 | ·9838 330 |
| ·80 | ·9970 806 | ·9961 910 | ·9950 891 | ·9937 392 | ·9921 030 | ·9901 396 |
| ·81 | ·9983 914 | ·9978 785− | ·9972 352 | ·9964 373 | ·9954 580 | ·9942 683 |
| ·82 | ·9991 593 | ·9988 792 | ·9985 236 | ·9980 771 | ·9975 224 | ·9968 401 |
| ·83 | ·9995 856 | ·9994 416 | ·9992 565+ | ·9990 213 | ·9987 255+ | ·9983 573 |
| ·84 | ·9998 086 | ·9997 394 | ·9996 493 | ·9995 334 | ·9993 860 | ·9992 001 |
| ·85 | ·9999 179 | ·9998 870 | ·9998 463 | ·9997 933 | ·9997 251 | ·9996 382 |
| ·86 | ·9999 676 | ·9999 549 | ·9999 380 | ·9999 158 | ·9998 868 | ·9998 494 |
| ·87 | ·9999 883 | ·9999 836 | ·9999 773 | ·9999 688 | ·9999 576 | ·9999 430 |
| ·88 | ·9999 962 | ·9999 947 | ·9999 925+ | ·9999 896 | ·9999 858 | ·9999 807 |
| ·89 | ·9999 989 | ·9999 985− | ·9999 978 | ·9999 970 | ·9999 958 | ·9999 942 |
| ·90 | ·9999 997 | ·9999 996 | ·9999 995− | ·9999 992 | ·9999 989 | ·9999 985+ |
| ·91 | ·9999 999 | ·9999 999 | ·9999 999 | ·9999 998 | ·9999 998 | ·9999 997 |
| ·92 | 1·0000 000 | 1·0000 000 | 1·0000 000 | 1·0000 000 | 1·0000 000 | ·9999 999 |
| ·93 | | | | | | 1·0000 000 |

$x$ = ·32 to ·93 $q$ = 18 $p$ = 36 to 40

| | $p$ = 36 | $p$ = 37 | $p$ = 38 | $p$ = 39 | $p$ = 40 |
|---|---|---|---|---|---|
| $B(p, q)$ = | ·8597 5934 × $\frac{1}{10^{15}}$ | ·5731 7289 × $\frac{1}{10^{15}}$ | ·3855 8904 × $\frac{1}{10^{15}}$ | ·2616 4970 × $\frac{1}{10^{15}}$ | ·1790 2348 × $\frac{1}{10^{15}}$ |
| $x$ | | | | | |
| ·32 | ·0000 001 | | | | |
| ·33 | ·0000 002 | ·0000 001 | | | |
| ·34 | ·0000 005⁻ | ·0000 002 | ·0000 001 | ·0000 001 | |
| ·35 | ·0000 011 | ·0000 006 | ·0000 003 | ·0000 001 | ·0000 001 |
| ·36 | ·0000 023 | ·0000 012 | ·0000 006 | ·0000 003 | ·0000 002 |
| ·37 | ·0000 048 | ·0000 026 | ·0000 014 | ·0000 007 | ·0000 004 |
| ·38 | ·0000 098 | ·0000 054 | ·0000 029 | ·0000 016 | ·0000 009 |
| ·39 | ·0000 192 | ·0000 108 | ·0000 061 | ·0000 034 | ·0000 019 |
| ·40 | ·0000 367 | ·0000 212 | ·0000 122 | ·0000 069 | ·0000 039 |
| ·41 | ·0000 682 | ·0000 404 | ·0000 237 | ·0000 138 | ·0000 080 |
| ·42 | ·0001 237 | ·0000 749 | ·0000 451 | ·0000 269 | ·0000 159 |
| ·43 | ·0002 188 | ·0001 356 | ·0000 834 | ·0000 510 | ·0000 309 |
| ·44 | ·0003 779 | ·0002 396 | ·0001 507 | ·0000 941 | ·0000 584 |
| ·45 | ·0006 379 | ·0004 133 | ·0002 657 | ·0001 696 | ·0001 075⁺ |
| ·46 | ·0010 530 | ·0006 968 | ·0004 576 | ·0002 984 | ·0001 933 |
| ·47 | ·0017 009 | ·0011 490 | ·0007 705⁺ | ·0005 130 | ·0003 392 |
| ·48 | ·0026 901 | ·0018 544 | ·0012 689 | ·0008 622 | ·0005 818 |
| ·49 | ·0041 682 | ·0029 304 | ·0020 452 | ·0014 174 | ·0009 757 |
| ·50 | ·0063 302 | ·0045 367 | ·0032 279 | ·0022 808 | ·0016 008 |
| ·51 | ·0094 272 | ·0068 840 | ·0049 911 | ·0035 938 | ·0025 705⁺ |
| ·52 | ·0137 730 | ·0102 430 | ·0075 639 | ·0055 476 | ·0040 421 |
| ·53 | ·0197 481 | ·0149 507 | ·0112 397 | ·0083 931 | ·0062 267 |
| ·54 | ·0277 991 | ·0214 145⁻ | ·0163 827 | ·0124 499 | ·0094 005⁺ |
| ·55 | ·0384 316 | ·0301 101 | ·0234 304 | ·0181 130 | ·0139 136 |
| ·56 | ·0521 959 | ·0415 732 | ·0328 912 | ·0258 542 | ·0201 958 |
| ·57 | ·0696 630 | ·0563 815⁻ | ·0453 327 | ·0362 175⁻ | ·0287 570 |
| ·58 | ·0913 918 | ·0751 282 | ·0613 613 | ·0498 044 | ·0401 795⁺ |
| ·59 | ·1178 875⁻ | ·0983 845⁻ | ·0815 907 | ·0672 498 | ·0551 005⁻ |
| ·60 | ·1495 539 | ·1266 534 | ·1066 000 | ·0891 851 | ·0741 816 |
| ·61 | ·1866 424 | ·1603 180 | ·1368 822 | ·1161 909 | ·0980 674 |
| ·62 | ·2292 026 | ·1995 866 | ·1727 881 | ·1487 400 | ·1273 311 |
| ·63 | ·2770 414 | ·2444 418 | ·2144 679 | ·1871 364 | ·1624 119 |
| ·64 | ·3296 947 | ·2945 998 | ·2618 192 | ·2314 540 | ·2035 492 |
| ·65 | ·3864 195⁻ | ·3494 855⁻ | ·3144 481 | ·2814 853 | ·2507 200 |
| ·66 | ·4462 081 | ·4082 309 | ·3716 496 | ·3367 049 | ·3035 880 |
| ·67 | ·5078 295⁻ | ·4696 997 | ·4324 150⁻ | ·3962 586 | ·3614 742 |
| ·68 | ·5698 939 | ·5325 396 | ·4954 679 | ·4589 816 | ·4233 560 |
| ·69 | ·6309 378 | ·5952 600 | ·5593 315⁻ | ·5234 502 | ·4879 002 |
| ·70 | ·6895 219 | ·6563 299 | ·6224 203 | ·5880 645⁺ | ·5535 326 |
| ·71 | ·7443 308 | ·7142 849 | ·6831 508 | ·6511 565⁻ | ·6185 387 |
| ·72 | ·7942 654 | ·7678 331 | ·7400 577 | ·7111 129 | ·6811 884 |
| ·73 | ·8385 154 | ·8159 475⁺ | ·7919 035⁺ | ·7664 992 | ·7398 702 |
| ·74 | ·8766 044 | ·8579 327 | ·8377 673 | ·8161 691 | ·7932 197 |
| ·75 | ·9084 016 | ·8934 583 | ·8771 015⁻ | ·8593 457 | ·8402 241 |
| ·76 | ·9341 011 | ·9225 557 | ·9097 495⁺ | ·8956 628 | ·8802 902 |
| ·77 | ·9541 708 | ·9455 789 | ·9359 235⁺ | ·9251 628 | ·9132 653 |
| ·78 | ·9692 800 | ·9631 376 | ·9561 453 | ·9482 514 | ·9394 101 |
| ·79 | ·9802 169 | ·9760 112 | ·9711 620 | ·9656 175⁻ | ·9593 279 |
| ·80 | ·9878 060 | ·9850 575⁻ | ·9818 484 | ·9781 326 | ·9738 642 |
| ·81 | ·9928 364 | ·9911 290 | ·9891 105⁺ | ·9867 441 | ·9839 918 |
| ·82 | ·9960 088 | ·9950 053 | ·9938 044 | ·9923 790 | ·9907 007 |
| ·83 | ·9979 032 | ·9973 483 | ·9966 761 | ·9958 686 | ·9949 062 |
| ·84 | ·9989 683 | ·9986 815⁻ | ·9983 299 | ·9979 025⁻ | ·9973 869 |
| ·85 | ·9995 283 | ·9993 909 | ·9992 205⁻ | ·9990 107 | ·9987 548 |
| ·86 | ·9998 016 | ·9997 412 | ·9996 653 | ·9995 708 | ·9994 541 |
| ·87 | ·9999 241 | ·9999 000 | ·9998 693 | ·9998 307 | ·9997 825⁻ |
| ·88 | ·9999 740 | ·9999 654 | ·9999 543 | ·9999 402 | ·9999 224 |
| ·89 | ·9999 922 | ·9999 895⁻ | ·9999 860 | ·9999 814 | ·9999 756 |
| ·90 | ·9999 980 | ·9999 972 | ·9999 963 | ·9999 950⁺ | ·9999 934 |
| ·91 | ·9999 996 | ·9999 994 | ·9999 992 | ·9999 989 | ·9999 985⁺ |
| ·92 | ·9999 999 | ·9999 999 | ·9999 999 | ·9999 998 | ·9999 997 |
| ·93 | 1·0000 000 | 1·0000 000 | 1·0000 000 | 1·0000 000 | 1·0000 000 |

$x$ = ·36 to ·94 $q$ = 18 $p$ = 41 to 45

| | $p$ = 41 | $p$ = 42 | $p$ = 43 | $p$ = 44 | $p$ = 45 |
|---|---|---|---|---|---|
| $B(p,q)$ = | ·1234 6447 × $\frac{1}{10^{15}}$ | ·8579 7344 × $\frac{1}{10^{16}}$ | ·6005 8140 × $\frac{1}{10^{16}}$ | ·4233 6066 × $\frac{1}{10^{16}}$ | ·3004 4950+ × $\frac{1}{10^{16}}$ |
| $x$ | | | | | |
| ·36 | ·0000 001 | | | | |
| ·37 | ·0000 002 | ·0000 001 | ·0000 001 | | |
| ·38 | ·0000 005− | ·0000 002 | ·0000 001 | ·0000 001 | |
| ·39 | ·0000 010 | ·0000 006 | ·0000 003 | ·0000 002 | ·0000 001 |
| ·40 | ·0000 022 | ·0000 012 | ·0000 007 | ·0000 004 | ·0000 002 |
| ·41 | ·0000 046 | ·0000 026 | ·0000 015− | ·0000 008 | ·0000 005− |
| ·42 | ·0000 094 | ·0000 055− | ·0000 032 | ·0000 018 | ·0000 011 |
| ·43 | ·0000 186 | ·0000 111 | ·0000 066 | ·0000 039 | ·0000 023 |
| ·44 | ·0000 360 | ·0000 220 | ·0000 134 | ·0000 081 | ·0000 049 |
| ·45 | ·0000 677 | ·0000 424 | ·0000 263 | ·0000 163 | ·0000 100 |
| ·46 | ·0001 243 | ·0000 795− | ·0000 505− | ·0000 319 | ·0000 200 |
| ·47 | ·0002 228 | ·0001 455− | ·0000 944 | ·0000 609 | ·0000 390 |
| ·48 | ·0003 901 | ·0002 599 | ·0001 721 | ·0001 133 | ·0000 741 |
| ·49 | ·0006 673 | ·0004 535+ | ·0003 064 | ·0002 058 | ·0001 374 |
| ·50 | ·0011 163 | ·0007 736 | ·0005 329 | ·0003 649 | ·0002 485+ |
| ·51 | ·0018 269 | ·0012 904 | ·0009 060 | ·0006 325− | ·0004 391 |
| ·52 | ·0029 265− | ·0021 058 | ·0015 064 | ·0010 714 | ·0007 578 |
| ·53 | ·0045 905+ | ·0033 638 | ·0024 505− | ·0017 751 | ·0012 788 |
| ·54 | ·0070 540 | ·0052 615+ | ·0039 018 | ·0028 772 | ·0021 102 |
| ·55 | ·0106 224 | ·0080 616 | ·0060 831 | ·0045 647 | ·0034 070 |
| ·56 | ·0156 804 | ·0121 033 | ·0092 893 | ·0070 904 | ·0053 833 |
| ·57 | ·0226 972 | ·0178 110 | ·0138 985− | ·0107 867 | ·0083 277 |
| ·58 | ·0322 246 | ·0256 977 | ·0203 798 | ·0160 760 | ·0126 153 |
| ·59 | ·0448 859 | ·0363 605− | ·0292 945+ | ·0234 774 | ·0187 193 |
| ·60 | ·0613 536 | ·0504 654 | ·0412 882 | ·0336 050− | ·0272 140 |
| ·61 | ·0823 136 | ·0687 192 | ·0570 701 | ·0471 548 | ·0387 696 |
| ·62 | ·1084 162 | ·0918 266 | ·0773 777 | ·0648 778 | ·0541 336 |
| ·63 | ·1402 154 | ·1204 335+ | ·1029 266 | ·0875 367 | ·0740 951 |
| ·64 | ·1781 000 | ·1550 592 | ·1343 444 | ·1158 456 | ·0994 321 |
| ·65 | ·2222 237 | ·1960 209 | ·1720 954 | ·1503 954 | ·1308 408 |
| ·66 | ·2724 416 | ·2433 615+ | ·2164 001 | ·1915 705+ | ·1688 515− |
| ·67 | ·3282 639 | ·2967 870 | ·2671 613 | ·2394 646 | ·2137 382 |
| ·68 | ·3888 347 | ·3556 270 | ·3239 057 | ·2938 074 | ·2654 329 |
| ·69 | ·4529 456 | ·4188 265+ | ·3857 554 | ·3539 148 | ·3234 564 |
| ·70 | ·5190 869 | ·4849 774 | ·4514 363 | ·4186 752 | ·3868 821 |
| ·71 | ·5855 376 | ·5523 917 | ·5193 326 | ·4865 809 | ·4543 429 |
| ·72 | ·6504 858 | ·6192 142 | ·5875 851 | ·5558 089 | ·5240 905+ |
| ·73 | ·7121 694 | ·6835 633 | ·6542 284 | ·6243 477 | ·5941 071 |
| ·74 | ·7690 195− | ·7436 860 | ·7173 512 | ·6901 591 | ·6622 624 |
| ·75 | ·8197 878 | ·7981 055+ | ·7752 616 | ·7513 552 | ·7264 980 |
| ·76 | ·8636 416 | ·8457 424 | ·8266 329 | ·8063 679 | ·7850 160 |
| ·77 | ·9002 109 | ·8859 910 | ·8706 099 | ·8540 841 | ·8364 428 |
| ·78 | ·9295 831 | ·9187 398 | ·9068 587 | ·8939 276 | ·8799 443 |
| ·79 | ·9522 475− | ·9443 346 | ·9355 533 | ·9258 734 | ·9152 716 |
| ·80 | ·9689 982 | ·9634 914 | ·9573 027 | ·9503 944 | ·9427 325+ |
| ·81 | ·9808 150− | ·9771 748 | ·9730 328 | ·9683 513 | ·9630 943 |
| ·82 | ·9887 397 | ·9864 649 | ·9838 445− | ·9808 463 | ·9774 379 |
| ·83 | ·9937 680 | ·9924 315+ | ·9908 733 | ·9890 686 | ·9869 920 |
| ·84 | ·9967 698 | ·9960 365+ | ·9951 712 | ·9941 570 | ·9929 759 |
| ·85 | ·9984 447 | ·9980 719 | ·9976 267 | ·9970 988 | ·9964 766 |
| ·86 | ·9993 111 | ·9991 372 | ·9989 270 | ·9986 749 | ·9983 742 |
| ·87 | ·9997 226 | ·9996 490 | ·9995 591 | ·9994 499 | ·9993 181 |
| ·88 | ·9999 000 | ·9998 721 | ·9998 377 | ·9997 955− | ·9997 439 |
| ·89 | ·9999 683 | ·9999 591 | ·9999 475− | ·9999 332 | ·9999 154 |
| ·90 | ·9999 914 | ·9999 887 | ·9999 854 | ·9999 812 | ·9999 760 |
| ·91 | ·9999 980 | ·9999 974 | ·9999 966 | ·9999 956 | ·9999 944 |
| ·92 | ·9999 997 | ·9999 995+ | ·9999 994 | ·9999 992 | ·9999 989 |
| ·93 | 1·0000 000 | ·9999 999 | ·9999 999 | ·9999 999 | ·9999 999 |
| ·94 | | 1·0000 000 | 1·0000 000 | 1·0000 000 | 1·0000 000 |

$x$ = ·40 to ·95 $q$ = 18 $p$ = 46 to 50

| $x$ | $p$ = 46 | $p$ = 47 | $p$ = 48 | $p$ = 49 | $p$ = 50 |
|---|---|---|---|---|---|
| $B(p, q)$ = | ·2146 0679 $\times \frac{1}{10^{18}}$ | ·1542 4863 $\times \frac{1}{10^{18}}$ | ·1115 3362 $\times \frac{1}{10^{18}}$ | ·8111 5363 $\times \frac{1}{10^{17}}$ | ·5932 3176 $\times \frac{1}{10^{17}}$ |
| ·40 | ·0000 001 | ·0000 001 | | | |
| ·41 | ·0000 003 | ·0000 001 | ·0000 001 | | |
| ·42 | ·0000 006 | ·0000 003 | ·0000 002 | ·0000 001 | ·0000 001 |
| ·43 | ·0000 013 | ·0000 008 | ·0000 005$^{-}$ | ·0000 003 | ·0000 001 |
| ·44 | ·0000 029 | ·0000 017 | ·0000 010 | ·0000 006 | ·0000 004 |
| ·45 | ·0000 061 | ·0000 037 | ·0000 022 | ·0000 014 | ·0000 008 |
| ·46 | ·0000 125$^{+}$ | ·0000 078 | ·0000 048 | ·0000 030 | ·0000 018 |
| ·47 | ·0000 249 | ·0000 158 | ·0000 100 | ·0000 063 | ·0000 039 |
| ·48 | ·0000 483 | ·0000 313 | ·0000 201 | ·0000 129 | ·0000 082 |
| ·49 | ·0000 913 | ·0000 603 | ·0000 396 | ·0000 259 | ·0000 169 |
| ·50 | ·0001 684 | ·0001 134 | ·0000 761 | ·0000 507 | ·0000 337 |
| ·51 | ·0003 032 | ·0002 082 | ·0001 423 | ·0000 968 | ·0000 655$^{+}$ |
| ·52 | ·0005 332 | ·0003 732 | ·0002 599 | ·0001 801 | ·0001 243 |
| ·53 | ·0009 164 | ·0006 533 | ·0004 635$^{-}$ | ·0003 272 | ·0002 299 |
| ·54 | ·0015 396 | ·0011 175$^{+}$ | ·0008 072 | ·0005 803 | ·0004 152 |
| ·55 | ·0025 296 | ·0018 688 | ·0013 738 | ·0010 052 | ·0007 321 |
| ·56 | ·0040 662 | ·0030 560 | ·0022 857 | ·0017 016 | ·0012 610 |
| ·57 | ·0063 965$^{+}$ | ·0048 889 | ·0037 188 | ·0028 156 | ·0021 222 |
| ·58 | ·0098 499 | ·0076 532 | ·0059 183 | ·0045 556 | ·0034 911 |
| ·59 | ·0148 515$^{-}$ | ·0117 261 | ·0092 152 | ·0072 091 | ·0056 148 |
| ·60 | ·0219 307 | ·0175 892 | ·0140 422 | ·0111 602 | ·0088 311 |
| ·61 | ·0317 223 | ·0258 348 | ·0209 443 | ·0169 047 | ·0135 855$^{+}$ |
| ·62 | ·0449 556 | ·0371 623 | ·0305 828 | ·0250 587 | ·0204 455$^{+}$ |
| ·63 | ·0624 278 | ·0523 609 | ·0437 248 | ·0363 571 | ·0301 051 |
| ·64 | ·0849 591 | ·0722 731 | ·0612 175$^{-}$ | ·0516 359 | ·0433 762 |
| ·65 | ·1133 285$^{+}$ | ·0977 388 | ·0839 403 | ·0717 947 | ·0611 611 |
| ·66 | ·1481 923 | ·1295 181 | ·1127 353 | ·0977 356 | ·0844 011 |
| ·67 | ·1899 899 | ·1681 982 | ·1483 168 | ·1302 785$^{+}$ | ·1139 999 |
| ·68 | ·2388 491 | ·2140 915$^{+}$ | ·1911 670 | ·1700 574 | ·1507 227 |
| ·69 | ·2945 006 | ·2671 373 | ·2414 275$^{+}$ | ·2174 050$^{-}$ | ·1950 787 |
| ·70 | ·3562 194 | ·3268 232 | ·2988 030 | ·2722 422 | ·2471 991 |
| ·71 | ·4228 070 | ·3921 417 | ·3624 943 | ·3339 895$^{+}$ | ·3067 297 |
| ·72 | ·4926 259 | ·4615 990 | ·4311 794 | ·4015 204 | ·3727 572 |
| ·73 | ·5636 918 | ·5332 831 | ·5030 555$^{+}$ | ·4731 743 | ·4437 932 |
| ·74 | ·6338 201 | ·6049 946 | ·5759 482 | ·5468 413 | ·5178 294 |
| ·75 | ·7008 122 | ·6744 284 | ·6474 833 | ·6201 171 | ·5924 717 |
| ·76 | ·7626 581 | ·7393 865$^{+}$ | ·7153 029 | ·6905 168 | ·6651 440 |
| ·77 | ·8177 274 | ·7979 908 | ·7772 967 | ·7557 188 | ·7333 395$^{-}$ |
| ·78 | ·8649 170 | ·8488 639 | ·8318 135$^{-}$ | ·8138 040 | ·7948 829 |
| ·79 | ·9037 323 | ·8912 472 | ·8778 164 | ·8634 482 | ·8481 593 |
| ·80 | ·9342 874 | ·9250 345$^{+}$ | ·9149 548 | ·9040 350$^{+}$ | ·8922 685$^{-}$ |
| ·81 | ·9572 274 | ·9507 190 | ·9435 404 | ·9356 663 | ·9270 755$^{+}$ |
| ·82 | ·9735 871 | ·9692 626 | ·9644 338 | ·9590 719 | ·9531 496 |
| ·83 | ·9846 173 | ·9819 179 | ·9788 669 | ·9754 378 | ·9716 042 |
| ·84 | ·9916 090 | ·9900 364 | ·9882 377 | ·9861 917 | ·9838 767 |
| ·85 | ·9957 479 | ·9948 997 | ·9939 179 | ·9927 879 | ·9914 941 |
| ·86 | ·9980 180 | ·9975 984 | ·9971 070 | ·9965 347 | ·9958 719 |
| ·87 | ·9991 602 | ·9989 721 | ·9987 492 | ·9984 867 | ·9981 790 |
| ·88 | ·9996 814 | ·9996 060 | ·9995 157 | ·9994 081 | ·9992 806 |
| ·89 | ·9998 937 | ·9998 673 | ·9998 352 | ·9997 966 | ·9997 502 |
| ·90 | ·9999 696 | ·9999 616 | ·9999 519 | ·9999 400 | ·9999 256 |
| ·91 | ·9999 928 | ·9999 908 | ·9999 883 | ·9999 853 | ·9999 816 |
| ·92 | ·9999 986 | ·9999 982 | ·9999 977 | ·9999 971 | ·9999 964 |
| ·93 | ·9999 998 | ·9999 997 | ·9999 997 | ·9999 996 | ·9999 995$^{-}$ |
| ·94 | 1·0000 000 | 1·0000 000 | 1·0000 000 | 1·0000 000 | ·9999 999 |
| ·95 | | | | | 1·0000 000 |

$x =$ ·14 to ·70 $q = 19$ $p =$ 19 to 24

| $x$ | $p = 19$ | $p = 20$ | $p = 21$ | $p = 22$ | $p = 23$ | $p = 24$ |
|---|---|---|---|---|---|---|
| $B(p, q) =$ | ·2978 1404 $\times \frac{1}{10^{11}}$ | ·1489 0702 $\times \frac{1}{10^{11}}$ | ·7636 2574 $\times \frac{1}{10^{12}}$ | ·4009 0351 $\times \frac{1}{10^{12}}$ | ·2151 1896 $\times \frac{1}{10^{12}}$ | ·1178 0324 $\times \frac{1}{10^{12}}$ |
| ·14 | ·0000 001 | | | | | |
| ·15 | ·0000 002 | ·0000 001 | | | | |
| ·16 | ·0000 007 | ·0000 002 | ·0000 001 | | | |
| ·17 | ·0000 018 | ·0000 006 | ·0000 002 | ·0000 001 | | |
| ·18 | ·0000 044 | ·0000 015− | ·0000 005− | ·0000 002 | ·0000 001 | |
| ·19 | ·0000 099 | ·0000 035+ | ·0000 012 | ·0000 004 | ·0000 001 | |
| ·20 | ·0000 214 | ·0000 080 | ·0000 029 | ·0000 011 | ·0000 004 | ·0000 001 |
| ·21 | ·0000 438 | ·0000 172 | ·0000 066 | ·0000 025+ | ·0000 009 | ·0000 003 |
| ·22 | ·0000 858 | ·0000 354 | ·0000 143 | ·0000 056 | ·0000 022 | ·0000 008 |
| ·23 | ·0001 613 | ·0000 694 | ·0000 292 | ·0000 121 | ·0000 049 | ·0000 019 |
| ·24 | ·0002 918 | ·0001 308 | ·0000 574 | ·0000 247 | ·0000 104 | ·0000 043 |
| ·25 | ·0005 096 | ·0002 377 | ·0001 086 | ·0000 486 | ·0000 214 | ·0000 092 |
| ·26 | ·0008 611 | ·0004 173 | ·0001 980 | ·0000 921 | ·0000 421 | ·0000 189 |
| ·27 | ·0014 113 | ·0007 093 | ·0003 491 | ·0001 685+ | ·0000 799 | ·0000 372 |
| ·28 | ·0022 481 | ·0011 701 | ·0005 966 | ·0002 983 | ·0001 465+ | ·0000 707 |
| ·29 | ·0034 866 | ·0018 768 | ·0009 898 | ·0005 121 | ·0002 602 | ·0001 300 |
| ·30 | ·0052 735− | ·0029 321 | ·0015 975+ | ·0008 540 | ·0004 484 | ·0002 315+ |
| ·31 | ·0077 897 | ·0044 684 | ·0025 121 | ·0013 859 | ·0007 511 | ·0004 003 |
| ·32 | ·0112 522 | ·0066 516 | ·0038 544 | ·0021 920 | ·0012 249 | ·0006 731 |
| ·33 | ·0159 138 | ·0096 838 | ·0057 776 | ·0033 837 | ·0019 473 | ·0011 024 |
| ·34 | ·0220 595+ | ·0138 042 | ·0084 712 | ·0051 039 | ·0030 222 | ·0017 605+ |
| ·35 | ·0300 009 | ·0192 870 | ·0121 622 | ·0075 312 | ·0045 841 | ·0027 454 |
| ·36 | ·0400 666 | ·0264 374 | ·0171 151 | ·0108 825− | ·0068 029 | ·0041 849 |
| ·37 | ·0525 900 | ·0355 839 | ·0236 286 | ·0154 136 | ·0098 872 | ·0062 422 |
| ·38 | ·0678 942 | ·0470 669 | ·0320 297 | ·0214 176 | ·0140 857 | ·0091 191 |
| ·39 | ·0862 742 | ·0612 249 | ·0426 634 | ·0292 195+ | ·0196 866 | ·0130 592 |
| ·40 | ·1079 789 | ·0783 774 | ·0558 802 | ·0391 680 | ·0270 136 | ·0183 471 |
| ·41 | ·1331 919 | ·0988 059 | ·0720 191 | ·0516 229 | ·0364 184 | ·0253 060 |
| ·42 | ·1620 143 | ·1227 344 | ·0913 891 | ·0669 398 | ·0482 694 | ·0342 909 |
| ·43 | ·1944 496 | ·1503 105− | ·1142 488 | ·0854 509 | ·0629 363 | ·0456 783 |
| ·44 | ·2303 934 | ·1815 872 | ·1407 853 | ·1074 443 | ·0807 715− | ·0598 507 |
| ·45 | ·2696 267 | ·2165 101 | ·1710 954 | ·1331 417 | ·1020 887 | ·0771 788 |
| ·46 | ·3118 165− | ·2549 074 | ·2051 689 | ·1626 780 | ·1271 401 | ·0979 991 |
| ·47 | ·3565 217 | ·2964 870 | ·2428 761 | ·1960 814 | ·1560 932 | ·1225 900 |
| ·48 | ·4032 057 | ·3408 397 | ·2839 618 | ·2332 593 | ·1890 098 | ·1511 476 |
| ·49 | ·4512 546 | ·3874 488 | ·3280 456 | ·2739 887 | ·2258 289 | ·1837 624 |
| ·50 | ·5000 000[e] | ·4357 073 | ·3746 293 | ·3179 140 | ·2663 546 | ·2203 995+ |
| ·51 | ·5487 454 | ·4849 396 | ·4231 118 | ·3645 520 | ·3102 511 | ·2608 846 |
| ·52 | ·5967 943 | ·5344 282 | ·4728 106 | ·4133 055+ | ·3570 462 | ·3048 962 |
| ·53 | ·6434 783 | ·5834 437 | ·5229 888 | ·4634 839 | ·4061 428 | ·3519 679 |
| ·54 | ·6881 835+ | ·6312 745− | ·5728 858 | ·5143 302 | ·4568 393 | ·4014 981 |
| ·55 | ·7303 733 | ·6772 567 | ·6217 499 | ·5650 537 | ·5083 574 | ·4527 704 |
| ·56 | ·7696 066 | ·7208 005− | ·6688 707 | ·6148 638 | ·5598 749 | ·5049 816 |
| ·57 | ·8055 504 | ·7614 112 | ·7136 085− | ·6630 059 | ·6105 632 | ·5572 769 |
| ·58 | ·8379 857 | ·7987 059 | ·7554 195+ | ·7087 939 | ·6596 251 | ·6087 888 |
| ·59 | ·8668 081 | ·8324 220 | ·7938 752 | ·7516 389 | ·7063 309 | ·6586 787 |
| ·60 | ·8920 211 | ·8624 195− | ·8286 737 | ·7910 712 | ·7500 504 | ·7061 759 |
| ·61 | ·9137 258 | ·8886 765− | ·8596 443 | ·8267 551 | ·7902 779 | ·7506 129 |
| ·62 | ·9321 058 | ·9112 786 | ·8867 441 | ·8584 944 | ·8266 493 | ·7914 535− |
| ·63 | ·9474 100 | ·9304 039 | ·9100 476 | ·8862 308 | ·8589 496 | ·8283 117 |
| ·64 | ·9599 334 | ·9463 043 | ·9297 313 | ·9100 330 | ·8871 115− | ·8609 609 |
| ·65 | ·9699 991 | ·9592 852 | ·9460 535+ | ·9300 810 | ·9112 044 | ·8893 321 |
| ·66 | ·9779 405− | ·9696 851 | ·9593 329 | ·9466 441 | ·9314 174 | ·9135 030 |
| ·67 | ·9840 862 | ·9778 562 | ·9699 254 | ·9600 573 | ·9480 361 | ·9336 786 |
| ·68 | ·9887 478 | ·9841 471 | ·9782 031 | ·9706 966 | ·9614 158 | ·9501 659 |
| ·69 | ·9922 103 | ·9888 891 | ·9845 348 | ·9789 552 | ·9719 554 | ·9633 455+ |
| ·70 | ·9947 265+ | ·9923 852 | ·9892 712 | ·9852 230 | ·9800 707 | ·9736 416 |

$x = \cdot 71$ to $\cdot 89$ $q = 19$ $p = 19$ to $24$

| | $p = 19$ | $p = 20$ | $p = 21$ | $p = 22$ | $p = 23$ | $p = 24$ |
|---|---|---|---|---|---|---|
| $B(p, q) =$ | ·2978 1404 $\times \frac{1}{10^{11}}$ | ·1489 0702 $\times \frac{1}{10^{11}}$ | ·7636 2574 $\times \frac{1}{10^{12}}$ | ·4009 0351 $\times \frac{1}{10^{12}}$ | ·2151 1896 $\times \frac{1}{10^{12}}$ | ·1178 0324 $\times \frac{1}{10^{12}}$ |
| $x$ | | | | | | |
| ·71 | ·9965 134 | ·9949 036 | ·9927 320 | ·9898 685+ | ·9861 721 | ·9814 937 |
| ·72 | ·9977 519 | ·9966 739 | ·9951 992 | ·9932 273 | ·9906 459 | ·9873 327 |
| ·73 | ·9985 887 | ·9978 867 | ·9969 130 | ·9955 930 | ·9938 409 | ·9915 609 |
| ·74 | ·9991 389 | ·9986 951 | ·9980 711 | ·9972 136 | ·9960 598 | ·9945 378 |
| ·75 | ·9994 904 | ·9992 186 | ·9988 312 | ·9982 916 | ·9975 559 | ·9965 722 |
| ·76 | ·9997 082 | ·9995 472 | ·9993 147 | ·9989 866 | ·9985 333 | ·9979 191 |
| ·77 | ·9998 387 | ·9997 467 | ·9996 123 | ·9994 199 | ·9991 507 | ·9987 811 |
| ·78 | ·9999 142 | ·9998 637 | ·9997 889 | ·9996 805− | ·9995 268 | ·9993 131 |
| ·79 | ·9999 562 | ·9999 296 | ·9998 897 | ·9998 312 | ·9997 472 | ·9996 288 |
| ·80 | ·9999 786 | ·9999 653 | ·9999 450− | ·9999 148 | ·9998 710 | ·9998 084 |
| ·81 | ·9999 901 | ·9999 837 | ·9999 739 | ·9999 591 | ·9999 374 | ·9999 059 |
| ·82 | ·9999 956 | ·9999 928 | ·9999 883 | ·9999 814 | ·9999 712 | ·9999 563 |
| ·83 | ·9999 982 | ·9999 970 | ·9999 950+ | ·9999 921 | ·9999 876 | ·9999 809 |
| ·84 | ·9999 993 | ·9999 988 | ·9999 980 | ·9999 968 | ·9999 950− | ·9999 922 |
| ·85 | ·9999 998 | ·9999 996 | ·9999 993 | ·9999 988 | ·9999 981 | ·9999 971 |
| ·86 | ·9999 999 | ·9999 999 | ·9999 998 | ·9999 996 | ·9999 994 | ·9999 990 |
| ·87 | 1·0000 000 | 1·0000 000 | ·9999 999 | ·9999 999 | ·9999 998 | ·9999 997 |
| ·88 | | | 1·0000 000 | 1·0000 000 | ·9999 999 | ·9999 999 |
| ·89 | | | | | 1·0000 000 | 1·0000 000 |

$x$ = ·21 to ·91 $q = 19$ $p$ = 25 to 30

| | $p = 25$ | $p = 26$ | $p = 27$ | $p = 28$ | $p = 29$ | $p = 30$ |
|---|---|---|---|---|---|---|
| $B(p, q)$ = | ·6575 0645$^{+}$ × $\frac{1}{10^{13}}$ | ·3735 8321 × $\frac{1}{10^{13}}$ | ·2158 4808 × $\frac{1}{10^{13}}$ | ·1266 9344 × $\frac{1}{10^{13}}$ | ·7547 6941 × $\frac{1}{10^{14}}$ | ·4560 0652 × $\frac{1}{10^{14}}$ |
| $x$ | | | | | | |
| ·21 | ·0000 001 | | | | | |
| ·22 | ·0000 003 | ·0000 001 | | | | |
| ·23 | ·0000 008 | ·0000 003 | ·0000 001 | | | |
| ·24 | ·0000 018 | ·0000 007 | ·0000 003 | ·0000 001 | | |
| ·25 | ·0000 039 | ·0000 016 | ·0000 007 | ·0000 003 | ·0000 001 | |
| ·26 | ·0000 084 | ·0000 036 | ·0000 016 | ·0000 007 | ·0000 003 | ·0000 001 |
| ·27 | ·0000 171 | ·0000 077 | ·0000 034 | ·0000 015$^{+}$ | ·0000 007 | ·0000 003 |
| ·28 | ·0000 336 | ·0000 157 | ·0000 073 | ·0000 033 | ·0000 015$^{-}$ | ·0000 007 |
| ·29 | ·0000 639 | ·0000 310 | ·0000 148 | ·0000 070 | ·0000 032 | ·0000 015$^{-}$ |
| ·30 | ·0001 177 | ·0000 589 | ·0000 291 | ·0000 142 | ·0000 068 | ·0000 032 |
| ·31 | ·0002 100 | ·0001 086 | ·0000 553 | ·0000 278 | ·0000 138 | ·0000 068 |
| ·32 | ·0003 642 | ·0001 941 | ·0001 020 | ·0000 529 | ·0000 271 | ·0000 137 |
| ·33 | ·0006 144 | ·0003 374 | ·0001 828 | ·0000 977 | ·0000 516 | ·0000 269 |
| ·34 | ·0010 098 | ·0005 708 | ·0003 182 | ·0001 751 | ·0000 952 | ·0000 511 |
| ·35 | ·0016 192 | ·0009 412 | ·0005 396 | ·0003 054 | ·0001 707 | ·0000 943 |
| ·36 | ·0025 356 | ·0015 143 | ·0008 921 | ·0005 188 | ·0002 980 | ·0001 692 |
| ·37 | ·0038 820 | ·0023 800 | ·0014 395$^{-}$ | ·0008 595$^{+}$ | ·0005 070 | ·0002 956 |
| ·38 | ·0058 163 | ·0036 576 | ·0022 694 | ·0013 902 | ·0008 414 | ·0005 033 |
| ·39 | ·0085 359 | ·0055 018 | ·0034 992 | ·0021 975$^{+}$ | ·0013 635$^{+}$ | ·0008 364 |
| ·40 | ·0122 805$^{-}$ | ·0081 067 | ·0052 813 | ·0033 977 | ·0021 600 | ·0013 575$^{+}$ |
| ·41 | ·0173 328 | ·0117 101 | ·0078 088 | ·0051 429 | ·0033 473 | ·0021 541 |
| ·42 | ·0240 168 | ·0165 947 | ·0113 194 | ·0076 266 | ·0050 786 | ·0033 442 |
| ·43 | ·0326 916 | ·0230 867 | ·0160 973 | ·0110 882 | ·0075 496 | ·0050 836 |
| ·44 | ·0437 417 | ·0315 504 | ·0224 726 | ·0158 155$^{+}$ | ·0110 034 | ·0075 719 |
| ·45 | ·0575 622 | ·0423 790 | ·0308 164 | ·0221 445$^{-}$ | ·0157 334 | ·0110 578 |
| ·46 | ·0745 405$^{+}$ | ·0559 801 | ·0415 316 | ·0304 544 | ·0220 831 | ·0158 423 |
| ·47 | ·0950 337 | ·0727 571 | ·0550 387 | ·0411 592 | ·0304 423 | ·0222 790 |
| ·48 | ·1193 434 | ·0930 858 | ·0717 566 | ·0546 932 | ·0412 376 | ·0307 700 |
| ·49 | ·1476 904 | ·1172 888 | ·0920 790 | ·0714 909 | ·0549 175$^{+}$ | ·0417 559 |
| ·50 | ·1801 888 | ·1456 076 | ·1163 466 | ·0919 624 | ·0719 325$^{+}$ | ·0557 014 |
| ·51 | ·2168 249 | ·1781 757 | ·1448 186 | ·1164 649 | ·0927 087 | ·0730 729 |
| ·52 | ·2574 398 | ·2149 947 | ·1776 431 | ·1452 717 | ·1176 172 | ·0943 112 |
| ·53 | ·3017 207 | ·2559 154 | ·2148 316 | ·1785 408 | ·1469 420 | ·1197 996 |
| ·54 | ·3492 007 | ·3006 268 | ·2562 377 | ·2162 876 | ·1808 461 | ·1498 287 |
| ·55 | ·3992 679 | ·3486 546 | ·3015 453 | ·2583 617 | ·2193 423 | ·1845 612 |
| ·56 | ·4511 862 | ·3993 705$^{+}$ | ·3502 652 | ·3044 335$^{+}$ | ·2622 684 | ·2239 999 |
| ·57 | ·5041 237 | ·4520 124 | ·4017 451 | ·3539 911 | ·3092 728 | ·2679 625$^{-}$ |
| ·58 | ·5571 900 | ·5057 150$^{-}$ | ·4551 903 | ·4063 497 | ·3598 117 | ·3160 659 |
| ·59 | ·6094 778 | ·5595 487 | ·5096 964 | ·4606 750$^{+}$ | ·4131 593 | ·3677 244 |
| ·60 | ·6601 077 | ·6125 653 | ·5642 914 | ·5160 176 | ·4684 334 | ·4221 619 |
| ·61 | ·7082 705$^{+}$ | ·6638 449 | ·6179 840 | ·5713 588 | ·5246 336 | ·4784 402 |
| ·62 | ·7532 661 | ·7125 430 | ·6698 151 | ·6256 629 | ·5806 908 | ·5355 016 |
| ·63 | ·7945 335$^{-}$ | ·7579 313 | ·7189 078 | ·6779 331 | ·6355 243 | ·5922 235$^{-}$ |
| ·64 | ·8316 724 | ·7994 315$^{-}$ | ·7645 121 | ·7272 648 | ·6881 019 | ·6474 806 |
| ·65 | ·8644 524 | ·8366 370 | ·8060 399 | ·7728 931 | ·7374 971 | ·7002 092 |
| ·66 | ·8928 118 | ·8693 232 | ·8430 882 | ·8142 297 | ·7829 388 | ·7494 684 |
| ·67 | ·9168 444 | ·8974 447 | ·8754 484 | ·8508 859 | ·8238 496 | ·7944 919 |
| ·68 | ·9367 786 | ·9211 207 | ·9031 021 | ·8826 811 | ·8598 678 | ·8347 261 |
| ·69 | ·9529 492 | ·9406 108 | ·9262 033 | ·9096 347 | ·8908 530 | ·8698 499 |
| ·70 | ·9657 659 | ·9562 835$^{+}$ | ·9450 506 | ·9319 456 | ·9168 748 | ·8997 772 |
| ·71 | ·9756 808 | ·9685 821 | ·9600 527 | ·9499 595$^{+}$ | ·9381 866 | ·9246 397 |
| ·72 | ·9831 582 | ·9779 883 | ·9716 891 | ·9641 301 | ·9551 888 | ·9447 552 |
| ·73 | ·9886 482 | ·9849 911 | ·9804 731 | ·9749 762 | ·9683 839 | ·9605 844 |
| ·74 | ·9925 668 | ·9900 581 | ·9869 164 | ·9830 417 | ·9783 312 | ·9726 818 |
| ·75 | ·9952 811 | ·9936 155$^{+}$ | ·9915 016 | ·9888 592 | ·9856 033 | ·9816 458 |
| ·76 | ·9971 022 | ·9960 343 | ·9946 609 | ·9929 212 | ·9907 491 | ·9880 736 |
| ·77 | ·9982 831 | ·9976 235$^{+}$ | ·9967 641 | ·9956 611 | ·9942 658 | ·9925 246 |
| ·78 | ·9990 214 | ·9986 301 | ·9981 135$^{-}$ | ·9974 420 | ·9965 814 | ·9954 936 |
| ·79 | ·9994 652 | ·9992 429 | ·9989 456 | ·9985 543 | ·9980 463 | ·9973 960 |
| ·80 | ·9997 208 | ·9996 003 | ·9994 372 | ·9992 197 | ·9989 338 | ·9985 632 |
| ·81 | ·9998 614 | ·9997 994 | ·9997 144 | ·9995 996 | ·9994 468 | ·9992 463 |
| ·82 | ·9999 349 | ·9999 047 | ·9998 629 | ·9998 056 | ·9997 285$^{+}$ | ·9996 261 |
| ·83 | ·9999 713 | ·9999 575$^{-}$ | ·9999 381 | ·9999 113 | ·9998 747 | ·9998 256 |
| ·84 | ·9999 881 | ·9999 823 | ·9999 739 | ·9999 622 | ·9999 460 | ·9999 241 |
| ·85 | ·9999 955$^{-}$ | ·9999 932 | ·9999 898 | ·9999 851 | ·9999 785$^{-}$ | ·9999 694 |
| ·86 | ·9999 984 | ·9999 976 | ·9999 964 | ·9999 946 | ·9999 921 | ·9999 887 |
| ·87 | ·9999 995$^{-}$ | ·9999 992 | ·9999 988 | ·9999 982 | ·9999 974 | ·9999 962 |
| ·88 | ·9999 999 | ·9999 998 | ·9999 997 | ·9999 995$^{-}$ | ·9999 992 | ·9999 989 |
| ·89 | 1·0000 000 | ·9999 999 | ·9999 999 | ·9999 999 | ·9999 998 | ·9999 997 |
| ·90 | | 1·0000 000 | 1·0000 000 | 1·0000 000 | 1·0000 000 | ·9999 999 |
| ·91 | | | | | | 1·0000 000 |

$x = \cdot 27$ to $\cdot 92$ $q = 19$ $p = 31$ to $36$

| | $p = 31$ | $p = 32$ | $p = 33$ | $p = 34$ | $p = 35$ | $p = 36$ |
|---|---|---|---|---|---|---|
| $B(p, q) =$ | $\cdot 2791\ 8767 \times \frac{1}{10^{14}}$ | $\cdot 1730\ 9635 \overset{+}{\times} \frac{1}{10^{14}}$ | $\cdot 1086\ 0948 \times \frac{1}{10^{14}}$ | $\cdot 6892\ 5244 \times \frac{1}{10^{15}}$ | $\cdot 4421\ 6195 \bar{\times} \frac{1}{10^{15}}$ | $\cdot 2865\ 8645 \bar{\times} \frac{1}{10^{15}}$ |
| $x$ | | | | | | |
| ·27 | ·0000 001 | | | | | |
| ·28 | ·0000 003 | ·0000 001 | ·0000 001 | | | |
| ·29 | ·0000 007 | ·0000 003 | ·0000 001 | ·0000 001 | | |
| ·30 | ·0000 015+ | ·0000 007 | ·0000 003 | ·0000 001 | ·0000 001 | |
| ·31 | ·0000 033 | ·0000 016 | ·0000 007 | ·0000 004 | ·0000 002 | ·0000 001 |
| ·32 | ·0000 069 | ·0000 034 | ·0000 017 | ·0000 008 | ·0000 004 | ·0000 002 |
| ·33 | ·0000 139 | ·0000 071 | ·0000 036 | ·0000 018 | ·0000 009 | ·0000 004 |
| ·34 | ·0000 271 | ·0000 143 | ·0000 074 | ·0000 038 | ·0000 019 | ·0000 010 |
| ·35 | ·0000 515+ | ·0000 278 | ·0000 149 | ·0000 079 | ·0000 041 | ·0000 022 |
| ·36 | ·0000 950− | ·0000 528 | ·0000 290 | ·0000 158 | ·0000 085+ | ·0000 046 |
| ·37 | ·0001 704 | ·0000 972 | ·0000 549 | ·0000 307 | ·0000 170 | ·0000 094 |
| ·38 | ·0002 978 | ·0001 744 | ·0001 011 | ·0000 580 | ·0000 330 | ·0000 186 |
| ·39 | ·0005 075− | ·0003 047 | ·0001 811 | ·0001 066 | ·0000 622 | ·0000 360 |
| ·40 | ·0008 440 | ·0005 193 | ·0003 164 | ·0001 909 | ·0001 142 | ·0000 677 |
| ·41 | ·0013 714 | ·0008 641 | ·0005 391 | ·0003 332 | ·0002 041 | ·0001 239 |
| ·42 | ·0021 787 | ·0014 049 | ·0008 972 | ·0005 676 | ·0003 559 | ·0002 212 |
| ·43 | ·0033 870 | ·0022 339 | ·0014 591 | ·0009 442 | ·0006 056 | ·0003 851 |
| ·44 | ·0051 561 | ·0034 760 | ·0023 209 | ·0015 354 | ·0010 068 | ·0006 546 |
| ·45 | ·0076 913 | ·0052 968 | ·0036 131 | ·0024 422 | ·0016 363 | ·0010 872 |
| ·46 | ·0112 490 | ·0079 092 | ·0055 087 | ·0038 022 | ·0026 017 | ·0017 654 |
| ·47 | ·0161 402 | ·0115 796 | ·0082 305− | ·0057 978 | ·0040 491 | ·0028 045+ |
| ·48 | ·0227 308 | ·0166 315− | ·0120 570 | ·0086 635+ | ·0061 723 | ·0043 616 |
| ·49 | ·0314 373 | ·0234 453 | ·0173 265− | ·0126 928 | ·0092 203 | ·0066 437 |
| ·50 | ·0427 166 | ·0324 543 | ·0244 369 | ·0182 417 | ·0135 042 | ·0099 172 |
| ·51 | ·0570 501 | ·0441 336 | ·0338 408 | ·0257 282 | ·0194 004 | ·0145 135+ |
| ·52 | ·0749 206 | ·0589 828 | ·0460 333 | ·0356 266 | ·0273 502 | ·0208 332 |
| ·53 | ·0967 829 | ·0775 009 | ·0615 329 | ·0484 537 | ·0378 519 | ·0293 431 |
| ·54 | ·1230 297 | ·1001 554 | ·0808 552 | ·0647 483 | ·0514 459 | ·0405 684 |
| ·55 | ·1539 538 | ·1273 451 | ·1044 783 | ·0850 415− | ·0686 917 | ·0550 747 |
| ·56 | ·1897 113 | ·1593 604 | ·1328 034 | ·1098 194 | ·0901 344 | ·0734 415− |
| ·57 | ·2302 874 | ·1963 434 | ·1661 120 | ·1394 809 | ·1162 648 | ·0962 260 |
| ·58 | ·2754 698 | ·2382 524 | ·2045 241 | ·1742 913 | ·1474 730 | ·1239 189 |
| ·59 | ·3248 339 | ·2848 350+ | ·2479 611 | ·2143 387 | ·1839 995+ | ·1568 936 |
| ·60 | ·3777 412 | ·3356 133 | ·2961 183 | ·2594 957 | ·2258 890 | ·1953 550− |
| ·61 | ·4333 554 | ·3898 849 | ·3484 521 | ·3093 922 | ·2729 517 | ·2392 911 |
| ·62 | ·4906 738 | ·4467 427 | ·4041 843 | ·3634 057 | ·3247 380 | ·2884 345+ |
| ·63 | ·5485 762 | ·5051 120 | ·4623 269 | ·4206 698 | ·3805 319 | ·3422 404 |
| ·64 | ·6058 843 | ·5638 050− | ·5217 257 | ·4801 054 | ·4393 665− | ·3998 846 |
| ·65 | ·6614 298 | ·6215 871 | ·5811 218 | ·5404 726 | ·5000 625+ | ·4602 874 |
| ·66 | ·7141 236 | ·6772 510 | ·6392 262 | ·6004 408 | ·5612 905− | ·5221 625− |
| ·67 | ·7630 205− | ·7296 912 | ·6947 996 | ·6586 710 | ·6216 498 | ·5840 891 |
| ·68 | ·8073 718 | ·7779 704 | ·7467 314 | ·7139 021 | ·6797 596 | ·6446 026 |
| ·69 | ·8466 625− | ·8213 732 | ·7941 082 | ·7650 338 | ·7343 518 | ·7022 935− |
| ·70 | ·8806 279 | ·8594 401 | ·8362 660 | ·8111 958 | ·7843 560 | ·7559 057 |
| ·71 | ·9092 503 | ·8919 795+ | ·8728 197 | ·8517 962 | ·8289 671 | ·8044 225+ |
| ·72 | ·9327 357 | ·9190 568 | ·9036 680 | ·8865 445− | ·8676 884 | ·8471 300 |
| ·73 | ·9514 747 | ·9409 633 | ·9289 737 | ·9154 472 | ·9003 453 | ·8836 512 |
| ·74 | ·9659 930 | ·9581 691 | ·9491 228 | ·9387 772 | ·9270 683 | ·9139 477 |
| ·75 | ·9768 967 | ·9712 668 | ·9646 693 | ·9570 221 | ·9482 504 | ·9382 882 |
| ·76 | ·9848 203 | ·9809 121 | ·9762 711 | ·9708 201 | ·9644 840 | ·9571 921 |
| ·77 | ·9903 795− | ·9877 686 | ·9846 274 | ·9808 894 | ·9764 873 | ·9713 545+ |
| ·78 | ·9941 360 | ·9924 622 | ·9904 222 | ·9879 631 | ·9850 296 | ·9815 647 |
| ·79 | ·9965 740 | ·9955 475− | ·9942 804 | ·9927 335− | ·9908 644 | ·9886 284 |
| ·80 | ·9980 887 | ·9974 888 | ·9967 389 | ·9958 117 | ·9946 773 | ·9933 030 |
| ·81 | ·9989 864 | ·9986 537 | ·9982 326 | ·9977 054 | ·9970 523 | ·9962 513 |
| ·82 | ·9994 916 | ·9993 174 | ·9990 942 | ·9988 112 | ·9984 564 | ·9980 159 |
| ·83 | ·9997 603 | ·9996 747 | ·9995 636 | ·9994 211 | ·9992 403 | ·9990 130 |
| ·84 | ·9998 945+ | ·9998 553 | ·9998 038 | ·9997 369 | ·9996 510 | ·9995 418 |
| ·85 | ·9999 570 | ·9999 404 | ·9999 184 | ·9998 894 | ·9998 517 | ·9998 032 |
| ·86 | ·9999 840 | ·9999 775+ | ·9999 689 | ·9999 574 | ·9999 423 | ·9999 226 |
| ·87 | ·9999 946 | ·9999 923 | ·9999 893 | ·9999 852 | ·9999 797 | ·9999 725− |
| ·88 | ·9999 984 | ·9999 977 | ·9999 967 | ·9999 954 | ·9999 936 | ·9999 913 |
| ·89 | ·9999 996 | ·9999 994 | ·9999 991 | ·9999 988 | ·9999 983 | ·9999 976 |
| ·90 | ·9999 999 | ·9999 999 | ·9999 998 | ·9999 997 | ·9999 996 | ·9999 994 |
| ·91 | 1·0000 000 | 1·0000 000 | 1·0000 000 | ·9999 999 | ·9999 999 | ·9999 999 |
| ·92 | | | | 1·0000 000 | 1·0000 000 | 1·0000 000 |

$x = \cdot 32$ to $\cdot 93$ $q = 19$ $p = 37$ to $43$

| $x$ | $p = 37$ | $p = 38$ | $p = 39$ | $p = 40$ | $p = 41$ | $p = 42$ | $p = 43$ |
|---|---|---|---|---|---|---|---|
| $B(p, q) =$ | $\cdot 1875\ 8386 \times \frac{1}{10^{15}}$ | $\cdot 1239\ 3933 \times \frac{1}{10^{15}}$ | $\cdot 8262\ 6222 \times \frac{1}{10^{16}}$ | $\cdot 5555\ 9011 \times \frac{1}{10^{16}}$ | $\cdot 3766\ 7126 \times \frac{1}{10^{16}}$ | $\cdot 2573\ 9203 \times \frac{1}{10^{16}}$ | $\cdot 1772\ 2074 \times \frac{1}{10^{16}}$ |
| ·32 | ·0000 001 | | | | | | |
| ·33 | ·0000 002 | ·0000 001 | | | | | |
| ·34 | ·0000 005− | ·0000 002 | ·0000 001 | ·0000 001 | | | |
| ·35 | ·0000 011 | ·0000 006 | ·0000 003 | ·0000 001 | ·0000 001 | | |
| ·36 | ·0000 024 | ·0000 013 | ·0000 007 | ·0000 003 | ·0000 002 | ·0000 001 | |
| ·37 | ·0000 051 | ·0000 028 | ·0000 015− | ·0000 008 | ·0000 004 | ·0000 002 | ·0000 001 |
| ·38 | ·0000 104 | ·0000 058 | ·0000 032 | ·0000 017 | ·0000 009 | ·0000 005+ | ·0000 003 |
| ·39 | ·0000 206 | ·0000 117 | ·0000 066 | ·0000 037 | ·0000 021 | ·0000 011 | ·0000 006 |
| ·40 | ·0000 398 | ·0000 232 | ·0000 134 | ·0000 077 | ·0000 044 | ·0000 025− | ·0000 014 |
| ·41 | ·0000 746 | ·0000 446 | ·0000 264 | ·0000 156 | ·0000 091 | ·0000 053 | ·0000 030 |
| ·42 | ·0001 364 | ·0000 834 | ·0000 506 | ·0000 305+ | ·0000 183 | ·0000 109 | ·0000 064 |
| ·43 | ·0002 429 | ·0001 520 | ·0000 944 | ·0000 582 | ·0000 356 | ·0000 217 | ·0000 131 |
| ·44 | ·0004 222 | ·0002 702 | ·0001 716 | ·0001 082 | ·0000 677 | ·0000 421 | ·0000 260 |
| ·45 | ·0007 165+ | ·0004 686 | ·0003 042 | ·0001 960 | ·0001 254 | ·0000 798 | ·0000 504 |
| ·46 | ·0011 883 | ·0007 938 | ·0005 263 | ·0003 465− | ·0002 265+ | ·0001 471 | ·0000 949 |
| ·47 | ·0019 271 | ·0013 141 | ·0008 895+ | ·0005 979 | ·0003 991 | ·0002 647 | ·0001 744 |
| ·48 | ·0030 578 | ·0021 276 | ·0014 696 | ·0010 079 | ·0006 867 | ·0004 647 | ·0003 125+ |
| ·49 | ·0047 499 | ·0033 705+ | ·0023 744 | ·0016 611 | ·0011 543 | ·0007 969 | ·0005 467 |
| ·50 | ·0072 269 | ·0052 274 | ·0037 541 | ·0026 774 | ·0018 968 | ·0013 352 | ·0009 340 |
| ·51 | ·0107 751 | ·0079 409 | ·0058 108 | ·0042 231 | ·0030 489 | ·0021 872 | ·0015 594 |
| ·52 | ·0157 499 | ·0118 206 | ·0088 096 | ·0065 212 | ·0047 957 | ·0035 046 | ·0025 454 |
| ·53 | ·0225 787 | ·0172 494 | ·0130 869 | ·0098 626 | ·0073 847 | ·0054 949 | ·0040 640 |
| ·54 | ·0317 576 | ·0246 851 | ·0190 569 | ·0146 150− | ·0111 369 | ·0084 343 | ·0063 493 |
| ·55 | ·0438 406 | ·0346 560 | ·0272 117 | ·0212 275+ | ·0164 552 | ·0126 781 | ·0097 103 |
| ·56 | ·0594 194 | ·0477 470 | ·0381 142 | ·0302 301 | ·0238 282 | ·0186 692 | ·0145 421 |
| ·57 | ·0790 929 | ·0645 760 | ·0523 818 | ·0422 231 | ·0338 270 | ·0269 401 | ·0213 322 |
| ·58 | ·1034 268 | ·0857 593 | ·0706 582 | ·0578 572 | ·0470 915− | ·0381 061 | ·0306 610 |
| ·59 | ·1329 048 | ·1118 661 | ·0935 734 | ·0777 995− | ·0643 049 | ·0528 477 | ·0431 909 |
| ·60 | ·1678 743 | ·1433 646 | ·1216 928 | ·1026 883 | ·0861 544 | ·0718 788 | ·0596 426 |
| ·61 | ·2084 916 | ·1805 639 | ·1554 584 | ·1330 759 | ·1132 786 | ·0959 004 | ·0807 566 |
| ·62 | ·2546 723 | ·2235 563 | ·1951 261 | ·1693 640 | ·1462 039 | ·1255 405− | ·1072 386 |
| ·63 | ·3060 549 | ·2721 677 | ·2407 061 | ·2117 372 | ·1852 742 | ·1612 831 | ·1396 912 |
| ·64 | ·3619 821 | ·3259 235− | ·2919 145− | ·2601 030 | ·2305 819 | ·2033 937 | ·1785 360 |
| ·65 | ·4215 068 | ·3840 362 | ·3481 434 | ·3140 452 | ·2819 077 | ·2518 473 | ·2239 341 |
| ·66 | ·4834 257 | ·4454 219 | ·4084 581 | ·3728 023 | ·3386 797 | ·3062 715− | ·2757 152 |
| ·67 | ·5463 406 | ·5087 451 | ·4716 246 | ·4352 750− | ·3999 613 | ·3659 138 | ·3333 255− |
| ·68 | ·6087 424 | ·5724 946 | ·5361 704 | ·5000 698 | ·4644 746 | ·4296 434 | ·3958 073 |
| ·69 | ·6691 131 | ·6350 808 | ·6004 753 | ·5655 770 | ·5306 613 | ·4959 925+ | ·4618 190 |
| ·70 | ·7260 330 | ·6949 492 | ·6628 838 | ·6300 784 | ·5967 810 | ·5632 399 | ·5296 988 |
| ·71 | ·7782 826 | ·7506 944 | ·7218 284 | ·6918 744 | ·6610 367 | ·6295 297 | ·5975 725+ |
| ·72 | ·8249 269 | ·8011 635− | ·7759 493 | ·7494 163 | ·7217 157 | ·6930 153 | ·6634 948 |
| ·73 | ·8653 712 | ·8455 350− | ·8241 953 | ·8014 275+ | ·7773 278 | ·7520 114 | ·7256 100 |
| ·74 | ·8993 838 | ·8833 635+ | ·8658 930 | ·8469 979 | ·8267 235+ | ·8051 338 | ·7823 103 |
| ·75 | ·9270 807 | ·9145 859 | ·9007 759 | ·8856 379 | ·8691 754 | ·8514 079 | ·8323 713 |
| ·76 | ·9488 794 | ·9394 882 | ·9289 701 | ·9172 869 | ·9044 121 | ·8903 313 | ·8750 437 |
| ·77 | ·9654 261 | ·9586 405+ | ·9509 407 | ·9422 753 | ·9326 005+ | ·9218 803 | ·9100 881 |
| ·78 | ·9775 107 | ·9728 104 | ·9674 074 | ·9612 480 | ·9542 817 | ·9464 625− | ·9377 496 |
| ·79 | ·9859 788 | ·9828 673 | ·9792 448 | ·9750 623 | ·9702 712 | ·9648 245− | ·9586 775+ |
| ·80 | ·9916 539 | ·9896 928 | ·9873 808 | ·9846 774 | ·9815 416 | ·9779 315+ | ·9738 058 |
| ·81 | ·9952 781 | ·9941 062 | ·9927 074 | ·9910 515− | ·9891 065+ | ·9868 395+ | ·9842 162 |
| ·82 | ·9974 740 | ·9968 134 | ·9960 152 | ·9950 586 | ·9939 212 | ·9925 791 | ·9910 068 |
| ·83 | ·9987 300 | ·9983 808 | ·9979 537 | ·9974 357 | ·9968 122 | ·9960 675− | ·9951 844 |
| ·84 | ·9994 041 | ·9992 323 | ·9990 195+ | ·9987 583 | ·9984 401 | ·9980 556 | ·9975 941 |
| ·85 | ·9997 413 | ·9996 632 | ·9995 653 | ·9994 438 | ·9992 939 | ·9991 106 | ·9988 880 |
| ·86 | ·9998 972 | ·9998 647 | ·9998 236 | ·9997 718 | ·9997 073 | ·9996 275+ | ·9995 294 |
| ·87 | ·9999 630 | ·9999 509 | ·9999 352 | ·9999 154 | ·9998 903 | ·9998 590 | ·9998 200 |
| ·88 | ·9999 882 | ·9999 841 | ·9999 788 | ·9999 721 | ·9999 634 | ·9999 525− | ·9999 387 |
| ·89 | ·9999 967 | ·9999 955+ | ·9999 940 | ·9999 919 | ·9999 893 | ·9999 860 | ·9999 818 |
| ·90 | ·9999 992 | ·9999 989 | ·9999 985+ | ·9999 980 | ·9999 974 | ·9999 965− | ·9999 954 |
| ·91 | ·9999 998 | ·9999 998 | ·9999 997 | ·9999 996 | ·9999 995− | ·9999 993 | ·9999 990 |
| ·92 | 1·0000 000 | 1·0000 000 | 1·0000 000 | ·9999 999 | ·9999 999 | ·9999 999 | ·9999 998 |
| ·93 | | | | 1·0000 000 | 1·0000 000 | 1·0000 000 | 1·0000 000 |

$x$ = ·37 to ·94 $q = 19$ $p$ = 44 to 50

| $x$ | $p = 44$ | $p = 45$ | $p = 46$ | $p = 47$ | $p = 48$ | $p = 49$ | $p = 50$ |
|---|---|---|---|---|---|---|---|
| $B(p, q) =$ | $\cdot 1229\ 1116 \times \frac{1}{10^{16}}$ | $\cdot 8584\ 2715 \times \frac{1}{10^{17}}$ | $\cdot 6035\ 8159 \times \frac{1}{10^{17}}$ | $\cdot 4271\ 5005 \times \frac{1}{10^{17}}$ | $\cdot 3041\ 8261 \times \frac{1}{10^{17}}$ | $\cdot 2179\ 2187 \times \frac{1}{10^{17}}$ | $\cdot 1570\ 3194 \times \frac{1}{10^{17}}$ |
| ·37 | ·0000 001 | | | | | | |
| ·38 | ·0000 001 | ·0000 001 | | | | | |
| ·39 | ·0000 003 | ·0000 002 | ·0000 001 | ·0000 001 | | | |
| ·40 | ·0000 008 | ·0000 004 | ·0000 002 | ·0000 001 | ·0000 001 | | |
| ·41 | ·0000 017 | ·0000 010 | ·0000 006 | ·0000 003 | ·0000 002 | ·0000 001 | ·0000 001 |
| ·42 | ·0000 038 | ·0000 022 | ·0000 013 | ·0000 007 | ·0000 004 | ·0000 002 | ·0000 001 |
| ·43 | ·0000 079 | ·0000 047 | ·0000 028 | ·0000 016 | ·0000 010 | ·0000 006 | ·0000 003 |
| ·44 | ·0000 160 | ·0000 098 | ·0000 059 | ·0000 036 | ·0000 021 | ·0000 013 | ·0000 008 |
| ·45 | ·0000 316 | ·0000 197 | ·0000 122 | ·0000 076 | ·0000 046 | ·0000 028 | ·0000 017 |
| ·46 | ·0000 609 | ·0000 388 | ·0000 246 | ·0000 155+ | ·0000 097 | ·0000 061 | ·0000 038 |
| ·47 | ·0001 142 | ·0000 744 | ·0000 481 | ·0000 310 | ·0000 199 | ·0000 126 | ·0000 080 |
| ·48 | ·0002 089 | ·0001 388 | ·0000 917 | ·0000 603 | ·0000 394 | ·0000 256 | ·0000 166 |
| ·49 | ·0003 728 | ·0002 528 | ·0001 704 | ·0001 142 | ·0000 762 | ·0000 506 | ·0000 334 |
| ·50 | ·0006 495− | ·0004 490 | ·0003 087 | ·0002 111 | ·0001 436 | ·0000 971 | ·0000 654 |
| ·51 | ·0011 052 | ·0007 788 | ·0005 457 | ·0003 804 | ·0002 637 | ·0001 819 | ·0001 249 |
| ·52 | ·0018 379 | ·0013 195− | ·0009 421 | ·0006 690 | ·0004 726 | ·0003 322 | ·0002 324 |
| ·53 | ·0029 882 | ·0021 848 | ·0015 886 | ·0011 490 | ·0008 268 | ·0005 920 | ·0004 218 |
| ·54 | ·0047 522 | ·0035 369 | ·0026 181 | ·0019 278 | ·0014 123 | ·0010 296 | ·0007 470 |
| ·55 | ·0073 948 | ·0056 003 | ·0042 185− | ·0031 611 | ·0023 568 | ·0017 486 | ·0012 912 |
| ·56 | ·0112 633 | ·0086 761 | ·0066 478 | ·0050 674 | ·0038 435− | ·0029 010 | ·0021 794 |
| ·57 | ·0167 976 | ·0131 555+ | ·0102 492 | ·0079 443 | ·0061 273 | ·0047 033 | ·0035 934 |
| ·58 | ·0245 353 | ·0195 289 | ·0154 637 | ·0121 833 | ·0095 520 | ·0074 535+ | ·0057 893 |
| ·59 | ·0351 084 | ·0283 889 | ·0228 385+ | ·0182 825− | ·0145 649 | ·0115 490 | ·0091 160 |
| ·60 | ·0492 275+ | ·0404 221 | ·0330 255+ | ·0268 510 | ·0217 275− | ·0175 006 | ·0140 328 |
| ·61 | ·0676 519 | ·0563 878 | ·0467 682 | ·0386 042 | ·0317 169 | ·0259 401 | ·0211 218 |
| ·62 | ·0911 415+ | ·0770 785− | ·0648 718 | ·0543 422 | ·0453 136 | ·0376 167 | ·0310 917 |
| ·63 | ·1203 941 | ·1032 634 | ·0881 542 | ·0749 107 | ·0633 719 | ·0533 764 | ·0447 660 |
| ·64 | ·1559 674 | ·1356 147 | ·1173 787 | ·1011 407 | ·0867 683 | ·0741 207 | ·0630 526 |
| ·65 | ·1981 956 | ·1746 214 | ·1531 689 | ·1337 684 | ·1163 285+ | ·1007 417 | ·0868 885− |
| ·66 | ·2471 060 | ·2204 995− | ·1959 150+ | ·1733 401 | ·1527 344 | ·1340 348 | ·1171 594 |
| ·67 | ·3023 514 | ·2731 090 | ·2456 797 | ·2201 108 | ·1964 188 | ·1745 925+ | ·1545 970 |
| ·68 | ·3631 673 | ·3318 923 | ·3021 185− | ·2739 498 | ·2474 593 | ·2226 907 | ·1996 610 |
| ·69 | ·4283 687 | ·3958 459 | ·3644 288 | ·3342 685− | ·3054 878 | ·2781 821 | ·2524 200 |
| ·70 | ·4963 917 | ·4635 388 | ·4313 430 | ·3999 871 | ·3696 318 | ·3404 149 | ·3124 502 |
| ·71 | ·5653 850− | ·5331 828 | ·5011 738 | ·4695 545+ | ·4385 070 | ·4081 970 | ·3787 714 |
| ·72 | ·6333 428 | ·6027 521 | ·5719 168 | ·5410 278 | ·5102 703 | ·4798 203 | ·4498 426 |
| ·73 | ·6982 692 | ·6701 454 | ·6414 029 | ·6122 106 | ·5827 387 | ·5531 563 | ·5236 283 |
| ·74 | ·7583 510 | ·7333 679 | ·7074 855+ | ·6808 379 | ·6535 665+ | ·6258 180 | ·5977 410 |
| ·75 | ·8121 173 | ·7907 125− | ·7682 374 | ·7447 852 | ·7204 597 | ·6953 741 | ·6696 485− |
| ·76 | ·8585 615+ | ·8409 106 | ·8221 300 | ·8022 716 | ·7813 991 | ·7595 873 | ·7369 209 |
| ·77 | ·8972 072 | ·8832 314 | ·8681 655− | ·8520 253 | ·8348 377 | ·8166 404 | ·7974 812 |
| ·78 | ·9281 088 | ·9175 126 | ·9059 416 | ·8933 845− | ·8798 389 | ·8653 112 | ·8498 170 |
| ·79 | ·9517 887 | ·9441 201 | ·9356 386 | ·9263 164 | ·9161 314 | ·9050 679 | ·8931 171 |
| ·80 | ·9691 235+ | ·9638 453 | ·9579 337 | ·9513 539 | ·9440 741 | ·9360 663 | ·9273 067 |
| ·81 | ·9812 019 | ·9777 614 | ·9738 600 | ·9694 632 | ·9645 379 | ·9590 523 | ·9529 767 |
| ·82 | ·9891 779 | ·9870 647 | ·9846 388 | ·9818 710 | ·9787 323 | ·9751 935− | ·9712 256 |
| ·83 | ·9941 448 | ·9929 288 | ·9915 158 | ·9898 842 | ·9880 112 | ·9858 738 | ·9834 479 |
| ·84 | ·9970 441 | ·9963 932 | ·9956 278 | ·9947 331 | ·9936 939 | ·9924 935+ | ·9911 148 |
| ·85 | ·9986 196 | ·9982 982 | ·9979 156 | ·9974 633 | ·9969 315− | ·9963 099 | ·9955 875+ |
| ·86 | ·9994 098 | ·9992 648 | ·9990 903 | ·9988 814 | ·9986 330 | ·9983 392 | ·9979 938 |
| ·87 | ·9997 719 | ·9997 129 | ·9996 410 | ·9995 541 | ·9994 494 | ·9993 243 | ·9991 754 |
| ·88 | ·9999 215+ | ·9999 002 | ·9998 739 | ·9998 418 | ·9998 027 | ·9997 553 | ·9996 984 |
| ·89 | ·9999 764 | ·9999 677 | ·9999 614 | ·9999 510 | ·9999 383 | ·9999 227 | ·9999 037 |
| ·90 | ·9999 940 | ·9999 922 | ·9999 899 | ·9999 871 | ·9999 836 | ·9999 792 | ·9999 739 |
| ·91 | ·9999 987 | ·9999 983 | ·9999 978 | ·9999 972 | ·9999 964 | ·9999 954 | ·9999 942 |
| ·92 | ·9999 998 | ·9999 997 | ·9999 996 | ·9999 995+ | ·9999 994 | ·9999 992 | ·9999 990 |
| ·93 | 1·0000 000 | 1·0000 000 | 1·0000 000 | ·9999 999 | ·9999 999 | ·9999 999 | ·9999 999 |
| ·94 | | | | 1·0000 000 | 1·0000 000 | 1·0000 000 | 1·0000 000 |

$x$ = ·15 to ·70 $q = 20$ $p$ = 20 to 25

| | $p = 20$ | $p = 21$ | $p = 22$ | $p = 23$ | $p = 24$ | $p = 25$ |
|---|---|---|---|---|---|---|
| $B(p, q)$ = | ·7254 4446 $\times \frac{1}{10^{12}}$ | ·3627 2223 $\times \frac{1}{10^{12}}$ | ·1857 8456 $\times \frac{1}{10^{12}}$ | ·9731 5720 $\times \frac{1}{10^{13}}$ | ·5205 2594 $\times \frac{1}{10^{13}}$ | ·2839 2324 $\times \frac{1}{10^{13}}$ |
| $x$ | | | | | | |
| ·15 | ·0000 001 | | | | | |
| ·16 | ·0000 004 | ·0000 001 | | | | |
| ·17 | ·0000 010 | ·0000 003 | ·0000 001 | | | |
| ·18 | ·0000 025+ | ·0000 009 | ·0000 003 | ·0000 001 | | |
| ·19 | ·0000 060 | ·0000 021 | ·0000 007 | ·0000 003 | ·0000 001 | |
| ·20 | ·0000 134 | ·0000 050+ | ·0000 019 | ·0000 007 | ·0000 002 | ·0000 001 |
| ·21 | ·0000 284 | ·0000 112 | ·0000 043 | ·0000 016 | ·0000 006 | ·0000 002 |
| ·22 | ·0000 576 | ·0000 238 | ·0000 096 | ·0000 038 | ·0000 015− | ·0000 006 |
| ·23 | ·0001 117 | ·0000 482 | ·0000 204 | ·0000 085− | ·0000 034 | ·0000 014 |
| ·24 | ·0002 081 | ·0000 936 | ·0000 413 | ·0000 178 | ·0000 076 | ·0000 032 |
| ·25 | ·0003 736 | ·0001 749 | ·0000 802 | ·0000 361 | ·0000 160 | ·0000 069 |
| ·26 | ·0006 480 | ·0003 150+ | ·0001 501 | ·0000 702 | ·0000 322 | ·0000 146 |
| ·27 | ·0010 884 | ·0005 487 | ·0002 712 | ·0001 315+ | ·0000 627 | ·0000 294 |
| ·28 | ·0017 737 | ·0009 262 | ·0004 741 | ·0002 383 | ·0001 176 | ·0000 571 |
| ·29 | ·0028 105− | ·0015 178 | ·0008 038 | ·0004 178 | ·0002 135− | ·0001 073 |
| ·30 | ·0043 369 | ·0024 194 | ·0013 236 | ·0007 110 | ·0003 754 | ·0001 950− |
| ·31 | ·0065 276 | ·0037 569 | ·0021 209 | ·0011 758 | ·0006 407 | ·0003 435+ |
| ·32 | ·0095 960 | ·0056 917 | ·0033 119 | ·0018 927 | ·0010 634 | ·0005 879 |
| ·33 | ·0137 956 | ·0084 236 | ·0050 468 | ·0029 702 | ·0017 187 | ·0009 788 |
| ·34 | ·0194 178 | ·0121 931 | ·0075 142 | ·0045 495− | ·0027 088 | ·0015 875− |
| ·35 | ·0267 867 | ·0172 808 | ·0109 435+ | ·0068 099 | ·0041 680 | ·0025 113 |
| ·36 | ·0362 504 | ·0240 038 | ·0156 062 | ·0099 721 | ·0062 683 | ·0038 794 |
| ·37 | ·0481 684 | ·0327 084 | ·0218 127 | ·0142 996 | ·0092 234 | ·0058 583 |
| ·38 | ·0628 956 | ·0437 587 | ·0299 073 | ·0200 979 | ·0132 911 | ·0086 567 |
| ·39 | ·0807 634 | ·0575 224 | ·0402 577 | ·0277 093 | ·0187 727 | ·0125 283 |
| ·40 | ·1020 586 | ·0743 516 | ·0532 414 | ·0375 047 | ·0260 101 | ·0177 723 |
| ·41 | ·1270 024 | ·0945 623 | ·0692 280 | ·0498 704 | ·0353 774 | ·0247 311 |
| ·42 | ·1557 295+ | ·1184 121 | ·0885 582 | ·0651 907 | ·0472 688 | ·0337 826 |
| ·43 | ·1882 702 | ·1460 780 | ·1115 206 | ·0838 275+ | ·0620 824 | ·0453 297 |
| ·44 | ·2245 367 | ·1776 359 | ·1383 286 | ·1060 966 | ·0801 989 | ·0597 829 |
| ·45 | ·2643 150+ | ·2130 443 | ·1690 979 | ·1322 428 | ·1019 576 | ·0775 402 |
| ·46 | ·3072 637 | ·2521 325+ | ·2038 271 | ·1624 161 | ·1276 309 | ·0989 621 |
| ·47 | ·3529 196 | ·2945 965+ | ·2423 835− | ·1966 496 | ·1573 980 | ·1243 449 |
| ·48 | ·4007 111 | ·3400 015− | ·2844 955+ | ·2348 429 | ·1913 214 | ·1538 928 |
| ·49 | ·4499 785− | ·3877 927 | ·3297 527 | ·2767 516 | ·2293 271 | ·1876 924 |
| ·50 | ·5000 000$^{e}$ | ·4373 147 | ·3776 143 | ·3219 845− | ·2711 920 | ·2256 904 |
| ·51 | ·5500 215+ | ·4878 357 | ·4274 267 | ·3700 107 | ·3165 389 | ·2676 790 |
| ·52 | ·5992 889 | ·5385 793 | ·4784 479 | ·4201 751 | ·3648 412 | ·3132 885+ |
| ·53 | ·6470 804 | ·5887 573 | ·5298 788 | ·4717 229 | ·4154 380 | ·3619 909 |
| ·54 | ·6927 363 | ·6376 050+ | ·5808 986 | ·5238 314 | ·4675 581 | ·4131 137 |
| ·55 | ·7356 850− | ·6844 142 | ·6307 019 | ·5756 469 | ·5203 525− | ·4658 644 |
| ·56 | ·7754 633 | ·7285 626 | ·6785 351 | ·6263 246 | ·5729 337 | ·5193 648 |
| ·57 | ·8117 298 | ·7695 376 | ·7237 290 | ·6750 677 | ·6244 176 | ·5726 913 |
| ·58 | ·8442 705− | ·8069 531 | ·7657 262 | ·7211 637 | ·6739 663 | ·6249 202 |
| ·59 | ·8729 976 | ·8405 574 | ·8041 008 | ·7640 151 | ·7208 272 | ·6751 739 |
| ·60 | ·8979 414 | ·8702 343 | ·8385 691 | ·8031 616 | ·7643 673 | ·7226 634 |
| ·61 | ·9192 366 | ·8959 956 | ·8689 918 | ·8382 934 | ·8040 980 | ·7667 253 |
| ·62 | ·9371 044 | ·9179 675− | ·8953 677 | ·8692 547 | ·8396 902 | ·8068 491 |
| ·63 | ·9518 316 | ·9363 715+ | ·9178 194 | ·8960 376 | ·8709 790 | ·8426 942 |
| ·64 | ·9637 496 | ·9515 030 | ·9365 738 | ·9187 674 | ·8979 571 | ·8740 946 |
| ·65 | ·9732 133 | ·9637 074 | ·9519 381 | ·9376 813 | ·9207 591 | ·9010 518 |
| ·66 | ·9805 822 | ·9733 574 | ·9642 749 | ·9531 034 | ·9396 392 | ·9237 179 |
| ·67 | ·9862 044 | ·9808 323 | ·9739 765+ | ·9654 162 | ·9549 428 | ·9423 703 |
| ·68 | ·9904 040 | ·9864 997 | ·9814 427 | ·9750 341 | ·9670 763 | ·9573 810 |
| ·69 | ·9934 724 | ·9907 018 | ·9870 604 | ·9823 778 | ·9764 778 | ·9691 839 |
| ·70 | ·9956 631 | ·9937 455− | ·9911 887 | ·9878 533 | ·9835 898 | ·9782 426 |

$x = \cdot 71$ to $\cdot 88$ $\quad q = 20 \quad p = 20$ to $25$

| | $p = 20$ | $p = 21$ | $p = 22$ | $p = 23$ | $p = 24$ | $p = 25$ |
|---|---|---|---|---|---|---|
| $B(p, q) =$ | $\cdot 7254\ 4446 \times \frac{1}{10^{12}}$ | $\cdot 3627\ 2223 \times \frac{1}{10^{12}}$ | $\cdot 1857\ 8456 \times \frac{1}{10^{12}}$ | $\cdot 9731\ 5720 \times \frac{1}{10^{13}}$ | $\cdot 5205\ 2594 \times \frac{1}{10^{13}}$ | $\cdot 2839\ 2324 \times \frac{1}{10^{13}}$ |
| $x$ | | | | | | |
| ·71 | ·9971 895[e] | ·9958 968 | ·9941 486 | ·9918 354 | ·9888 363 | ·9850 212 |
| ·72 | ·9982 263 | ·9973 787 | ·9962 163 | ·9946 566 | ·9926 059 | ·9899 605+ |
| ·73 | ·9989 116 | ·9983 720 | ·9976 217 | ·9966 009 | ·9952 401 | ·9934 603 |
| ·74 | ·9993 520 | ·9990 189 | ·9985 496 | ·9979 022 | ·9970 275− | ·9958 677 |
| ·75 | ·9996 264 | ·9994 276 | ·9991 436 | ·9987 467 | ·9982 030 | ·9974 725+ |
| ·76 | ·9997 919 | ·9996 774 | ·9995 116 | ·9992 768 | ·9989 509 | ·9985 072 |
| ·77 | ·9998 883 | ·9998 248 | ·9997 317 | ·9995 981 | ·9994 102 | ·9991 509 |
| ·78 | ·9999 424 | ·9999 086 | ·9998 585− | ·9997 855− | ·9996 816 | ·9995 363 |
| ·79 | ·9999 716 | ·9999 544 | ·9999 285+ | ·9998 905− | ·9998 355+ | ·9997 577 |
| ·80 | ·9999 866 | ·9999 783 | ·9999 656 | ·9999 467 | ·9999 190 | ·9998 794 |
| ·81 | ·9999 940 | ·9999 902 | ·9999 843 | ·9999 754 | ·9999 622 | ·9999 430 |
| ·82 | ·9999 975− | ·9999 958 | ·9999 932 | ·9999 893 | ·9999 833 | ·9999 746 |
| ·83 | ·9999 990 | ·9999 983 | ·9999 973 | ·9999 956 | ·9999 931 | ·9999 894 |
| ·84 | ·9999 996 | ·9999 994 | ·9999 990 | ·9999 983 | ·9999 974 | ·9999 959 |
| ·85 | ·9999 999 | ·9999 998 | ·9999 996 | ·9999 994 | ·9999 991 | ·9999 985+ |
| ·86 | 1·0000 000 | ·9999 999 | ·9999 999 | ·9999 998 | ·9999 997 | ·9999 995+ |
| ·87 | | 1·0000 000 | 1·0000 000 | ·9999 999 | ·9999 999 | ·9999 999 |
| ·88 | | | | 1·0000 000 | 1·0000 000 | 1·0000 000 |

$x$ = ·21 to ·90 $q = 20$ $p$ = 26 to 31

| | $p = 26$ | $p = 27$ | $p = 28$ | $p = 29$ | $p = 30$ | $p = 31$ |
|---|---|---|---|---|---|---|
| $B(p, q)$ = | ·1577 3513 $\times \frac{1}{10^{13}}$ | ·8915 4641 $\times \frac{1}{10^{14}}$ | ·5121 6496 $\times \frac{1}{10^{14}}$ | ·2987 6289 $\times \frac{1}{10^{14}}$ | ·1768 1885 $\overset{+}{\times} \frac{1}{10^{14}}$ | ·1060 9131 $\times \frac{1}{10^{14}}$ |
| $x$ | | | | | | |
| ·21 | ·0000 001 | | | | | |
| ·22 | ·0000 002 | ·0000 001 | | | | |
| ·23 | ·0000 005+ | ·0000 002 | ·0000 001 | | | |
| ·24 | ·0000 013 | ·0000 005+ | ·0000 002 | ·0000 001 | | |
| ·25 | ·0000 030 | ·0000 013 | ·0000 005+ | ·0000 002 | ·0000 001 | |
| ·26 | ·0000 065− | ·0000 028 | ·0000 012 | ·0000 005+ | ·0000 002 | ·0000 001 |
| ·27 | ·0000 136 | ·0000 062 | ·0000 028 | ·0000 012 | ·0000 005+ | ·0000 002 |
| ·28 | ·0000 273 | ·0000 129 | ·0000 060 | ·0000 027 | ·0000 012 | ·0000 006 |
| ·29 | ·0000 531 | ·0000 259 | ·0000 125− | ·0000 059 | ·0000 028 | ·0000 013 |
| ·30 | ·0000 997 | ·0000 503 | ·0000 250− | ·0000 123 | ·0000 059 | ·0000 028 |
| ·31 | ·0001 814 | ·0000 944 | ·0000 485− | ·0000 246 | ·0000 123 | ·0000 061 |
| ·32 | ·0003 202 | ·0001 718 | ·0000 910 | ·0000 476 | ·0000 245+ | ·0000 125+ |
| ·33 | ·0005 491 | ·0003 036 | ·0001 656 | ·0000 892 | ·0000 475− | ·0000 250− |
| ·34 | ·0009 165+ | ·0005 217 | ·0002 929 | ·0001 624 | ·0000 889 | ·0000 482 |
| ·35 | ·0014 907 | ·0008 725+ | ·0005 039 | ·0002 873 | ·0001 618 | ·0000 901 |
| ·36 | ·0023 657 | ·0014 226 | ·0008 442 | ·0004 946 | ·0002 863 | ·0001 639 |
| ·37 | ·0036 670 | ·0022 636 | ·0013 790 | ·0008 296 | ·0004 932 | ·0002 898 |
| ·38 | ·0055 573 | ·0035 188 | ·0021 991 | ·0013 573 | ·0008 278 | ·0004 992 |
| ·39 | ·0082 421 | ·0053 489 | ·0034 266 | ·0021 681 | ·0013 558 | ·0008 383 |
| ·40 | ·0119 729 | ·0079 580 | ·0052 218 | ·0033 847 | ·0021 684 | ·0013 738 |
| ·41 | ·0170 487 | ·0115 972 | ·0077 892 | ·0051 684 | ·0033 900 | ·0021 990 |
| ·42 | ·0238 137 | ·0165 670 | ·0113 816 | ·0077 258 | ·0051 845− | ·0034 411 |
| ·43 | ·0326 512 | ·0232 155− | ·0163 029 | ·0113 135+ | ·0077 625− | ·0052 685− |
| ·44 | ·0439 727 | ·0319 326 | ·0229 071 | ·0162 410 | ·0113 863 | ·0078 974 |
| ·45 | ·0582 015+ | ·0431 397 | ·0315 923 | ·0228 699 | ·0163 732 | ·0115 982 |
| ·46 | ·0757 518 | ·0572 729 | ·0427 909 | ·0316 087 | ·0230 948 | ·0166 981 |
| ·47 | ·0970 034 | ·0747 621 | ·0569 526 | ·0429 022 | ·0319 719 | ·0235 811 |
| ·48 | ·1222 732 | ·0960 046 | ·0745 227 | ·0572 144 | ·0434 633 | ·0326 824 |
| ·49 | ·1517 866 | ·1213 357 | ·0959 148 | ·0750 062 | ·0580 486 | ·0444 768 |
| ·50 | ·1856 490 | ·1509 978 | ·1214 801 | ·0967 063 | ·0762 039 | ·0594 602 |
| ·51 | ·2238 224 | ·1851 105+ | ·1514 742 | ·1226 791 | ·0983 720 | ·0781 243 |
| ·52 | ·2661 075+ | ·2236 446 | ·1860 256 | ·1531 896 | ·1249 280 | ·1009 244 |
| ·53 | ·3121 354 | ·2664 026 | ·2251 076 | ·1883 697 | ·1561 418 | ·1282 431 |
| ·54 | ·3613 697 | ·3130 090 | ·2685 172 | ·2281 885− | ·1921 430 | ·1603 508 |
| ·55 | ·4131 200 | ·3629 114 | ·3158 640 | ·2724 292 | ·2328 886 | ·1973 679 |
| ·56 | ·4665 673 | ·4153 944 | ·3665 716 | ·3206 782 | ·2781 397 | ·2392 312 |
| ·57 | ·5207 994 | ·4696 060 | ·4198 916 | ·3723 255+ | ·3274 494 | ·2856 697 |
| ·58 | ·5748 540 | ·5245 952 | ·4749 321 | ·4265 815+ | ·3801 650− | ·3361 930 |
| ·59 | ·6277 676 | ·5793 584 | ·5306 982 | ·4825 073 | ·4354 463 | ·3900 952 |
| ·60 | ·6786 242 | ·6328 911 | ·5861 417 | ·5390 584 | ·4922 998 | ·4464 764 |
| ·61 | ·7266 019 | ·6842 409 | ·6402 169 | ·5951 394 | ·5496 267 | ·5042 809 |
| ·62 | ·7710 128 | ·7325 576 | ·6919 377 | ·6496 638 | ·6062 822 | ·5623 510 |
| ·63 | ·8113 319 | ·7771 350+ | ·7404 303 | ·7016 151 | ·6611 402 | ·6194 915+ |
| ·64 | ·8472 159 | ·8174 425+ | ·7849 785+ | ·7501 030 | ·7131 589 | ·6745 401 |
| ·65 | ·8785 066 | ·8531 433 | ·8250 557 | ·7944 102 | ·7614 398 | ·7264 363 |
| ·66 | ·9052 237 | ·8840 976 | ·8603 425+ | ·8340 253 | ·8052 759 | ·7742 841 |
| ·67 | ·9275 448 | ·9103 530 | ·8907 289 | ·8686 587 | ·8441 837 | ·8173 998 |
| ·68 | ·9457 777 | ·9321 215+ | ·9163 006 | ·8982 421 | ·8779 170 | ·8553 425+ |
| ·69 | ·9603 262 | ·9497 481 | ·9373 130 | ·9229 104 | ·9064 616 | ·8879 239 |
| ·70 | ·9716 549 | ·9636 736 | ·9541 552 | ·9429 711 | ·9300 129 | ·9151 974 |
| ·71 | ·9802 539 | ·9743 955+ | ·9673 091 | ·9588 636 | ·9489 386 | ·9374 290 |
| ·72 | ·9866 083 | ·9824 309 | ·9773 067 | ·9711 137 | ·9637 333 | ·9550 540 |
| ·73 | ·9911 736 | ·9882 845− | ·9846 912 | ·9802 882 | ·9749 682 | ·9686 250− |
| ·74 | ·9943 572 | ·9924 226 | ·9899 836 | ·9869 540 | ·9832 432 | ·9787 581 |
| ·75 | ·9965 083 | ·9952 566 | ·9936 572 | ·9916 438 | ·9891 443 | ·9860 824 |
| ·76 | ·9979 137 | ·9971 330 | ·9961 222 | ·9948 326 | ·9932 105− | ·9911 968 |
| ·77 | ·9987 996 | ·9983 314 | ·9977 173 | ·9969 234 | ·9959 117 | ·9946 393 |
| ·78 | ·9993 370 | ·9990 678 | ·9987 101 | ·9982 418 | ·9976 372 | ·9968 669 |
| ·79 | ·9996 496 | ·9995 018 | ·9993 028 | ·9990 389 | ·9986 939 | ·9982 487 |
| ·80 | ·9998 236 | ·9997 463 | ·9996 410 | ·9994 995+ | ·9993 123 | ·9990 676 |
| ·81 | ·9999 158 | ·9998 775− | ·9998 247 | ·9997 529 | ·9996 566 | ·9995 293 |
| ·82 | ·9999 620 | ·9999 442 | ·9999 193 | ·9998 849 | ·9998 383 | ·9997 759 |
| ·83 | ·9999 840 | ·9999 762 | ·9999 651 | ·9999 498 | ·9999 287 | ·9999 000 |
| ·84 | ·9999 937 | ·9999 905+ | ·9999 860 | ·9999 796 | ·9999 707 | ·9999 585+ |
| ·85 | ·9999 977 | ·9999 965+ | ·9999 948 | ·9999 924 | ·9999 889 | ·9999 841 |
| ·86 | ·9999 992 | ·9999 988 | ·9999 982 | ·9999 974 | ·9999 962 | ·9999 945− |
| ·87 | ·9999 998 | ·9999 997 | ·9999 995− | ·9999 992 | ·9999 988 | ·9999 983 |
| ·88 | ·9999 999 | ·9999 999 | ·9999 999 | ·9999 998 | ·9999 997 | ·9999 995+ |
| ·89 | 1·0000 000 | 1·0000 000 | 1·0000 000 | ·9999 999 | ·9999 999 | ·9999 999 |
| ·90 | | | | 1·0000 000 | 1·0000 000 | 1·0000 000 |

$x$ = ·27 to ·91 $q$ = 20 $p$ = 32 to 37

| | $p$ = 32 | $p$ = 33 | $p$ = 34 | $p$ = 35 | $p$ = 36 | $p$ = 37 |
|---|---|---|---|---|---|---|
| $B(p, q)$ = | ·6448 6876 $\times\frac{1}{10^{15}}$ | ·3968 4232 $\times\frac{1}{10^{15}}$ | ·2470 9050 $\times\frac{1}{10^{15}}$ | ·1555 7550 $\times\frac{1}{10^{15}}$ | ·9900 2590 $\times\frac{1}{10^{16}}$ | ·6364 4522 $\times\frac{1}{10^{16}}$ |
| $x$ | | | | | | |
| ·27 | ·0000 001 | | | | | |
| ·28 | ·0000 002 | ·0000 001 | | | | |
| ·29 | ·0000 006 | ·0000 003 | ·0000 001 | ·0000 001 | | |
| ·30 | ·0000 013 | ·0000 006 | ·0000 003 | ·0000 001 | ·0000 001 | |
| ·31 | ·0000 030 | ·0000 014 | ·0000 007 | ·0000 003 | ·0000 002 | ·0000 001 |
| ·32 | ·0000 063 | ·0000 032 | ·0000 016 | ·0000 008 | ·0000 004 | ·0000 002 |
| ·33 | ·0000 130 | ·0000 067 | ·0000 034 | ·0000 017 | ·0000 009 | ·0000 004 |
| ·34 | ·0000 258 | ·0000 137 | ·0000 072 | ·0000 037 | ·0000 019 | ·0000 010 |
| ·35 | ·0000 496 | ·0000 271 | ·0000 146 | ·0000 078 | ·0000 041 | ·0000 022 |
| ·36 | ·0000 928 | ·0000 520 | ·0000 288 | ·0000 158 | ·0000 086 | ·0000 046 |
| ·37 | ·0001 685− | ·0000 969 | ·0000 552 | ·0000 312 | ·0000 174 | ·0000 097 |
| ·38 | ·0002 978 | ·0001 758 | ·0001 028 | ·0000 595+ | ·0000 342 | ·0000 194 |
| ·39 | ·0005 128 | ·0003 105− | ·0001 861 | ·0001 106 | ·0000 651 | ·0000 380 |
| ·40 | ·0008 611 | ·0005 343 | ·0003 283 | ·0001 998 | ·0001 205+ | ·0000 721 |
| ·41 | ·0014 114 | ·0008 968 | ·0005 643 | ·0003 518 | ·0002 174 | ·0001 332 |
| ·42 | ·0022 601 | ·0014 696 | ·0009 464 | ·0006 039 | ·0003 819 | ·0002 395+ |
| ·43 | ·0035 387 | ·0023 533 | ·0015 501 | ·0010 117 | ·0006 546 | ·0004 199 |
| ·44 | ·0054 214 | ·0036 851 | ·0024 813 | ·0016 556 | ·0010 950+ | ·0007 182 |
| ·45 | ·0081 324 | ·0056 468 | ·0038 843 | ·0026 479 | ·0017 895+ | ·0011 994 |
| ·46 | ·0119 521 | ·0084 727 | ·0059 506 | ·0041 422 | ·0028 587 | ·0019 567 |
| ·47 | ·0172 203 | ·0124 557 | ·0089 270 | ·0063 417 | ·0044 670 | ·0031 209 |
| ·48 | ·0243 359 | ·0179 509 | ·0131 214 | ·0095 079 | ·0068 318 | ·0048 693 |
| ·49 | ·0337 507 | ·0253 744 | ·0189 068 | ·0139 667 | ·0102 320 | ·0074 361 |
| ·50 | ·0459 573 | ·0351 971 | ·0267 194 | ·0201 118 | ·0150 145− | ·0111 207 |
| ·51 | ·0614 689 | ·0479 311 | ·0370 517 | ·0284 026 | ·0215 970 | ·0162 942 |
| ·52 | ·0807 924 | ·0641 080 | ·0504 369 | ·0393 553 | ·0304 647 | ·0234 016 |
| ·53 | ·1043 943 | ·0842 494 | ·0674 254 | ·0535 258 | ·0421 600 | ·0329 568 |
| ·54 | ·1326 609 | ·1088 303 | ·0885 526 | ·0714 835+ | ·0572 626 | ·0455 303 |
| ·55 | ·1658 577 | ·1382 369 | ·1142 990 | ·0937 757 | ·0763 602 | ·0617 264 |
| ·56 | ·2040 881 | ·1727 228 | ·1450 453 | ·1208 845+ | ·1000 096 | ·0821 499 |
| ·57 | ·2472 594 | ·2123 660 | ·1810 254 | ·1531 783 | ·1286 889 | ·1073 626 |
| ·58 | ·2950 580 | ·2570 338 | ·2222 819 | ·1908 622 | ·1627 460 | ·1378 319 |
| ·59 | ·3469 385+ | ·3063 578 | ·2686 300 | ·2339 315− | ·2023 459 | ·1738 751 |
| ·60 | ·4021 311 | ·3597 260 | ·3196 339 | ·2821 359 | ·2474 235+ | ·2156 039 |
| ·61 | ·4596 664 | ·4162 928 | ·3746 016 | ·3349 581 | ·2976 480 | ·2628 770 |
| ·62 | ·5184 198 | ·4750 103 | ·4326 006 | ·3916 128 | ·3524 051 | ·3152 666 |
| ·63 | ·5771 711 | ·5346 788 | ·4924 954 | ·4510 690 | ·4108 024 | ·3720 458 |
| ·64 | ·6346 754 | ·5940 135+ | ·5530 066 | ·5120 961 | ·4717 000 | ·4322 016 |
| ·65 | ·6897 391 | ·6517 230 | ·6127 854 | ·5733 324 | ·5337 666 | ·4944 757 |
| ·66 | ·7412 929 | ·7065 902 | ·6704 994 | ·6333 684 | ·5955 584 | ·5574 333 |
| ·67 | ·7884 560 | ·7575 494 | ·7249 195+ | ·6908 405+ | ·6556 125+ | ·6195 528 |
| ·68 | ·8305 834 | ·8037 508 | ·7749 992 | ·7445 225+ | ·7125 481 | ·6793 303 |
| ·69 | ·8672 932 | ·8446 058 | ·8199 385+ | ·7934 066 | ·7651 616 | ·7353 865+ |
| ·70 | ·8984 702 | ·8798 090 | ·8592 250+ | ·8367 643 | ·8125 067 | ·7865 646 |
| ·71 | ·9242 487 | ·9093 343 | ·8926 482 | ·8741 807 | ·8539 508 | ·8320 070 |
| ·72 | ·9449 748 | ·9334 089 | ·9202 868 | ·9055 593 | ·8891 991 | ·8712 029 |
| ·73 | ·9611 563 | ·9524 670 | ·9424 717 | ·9310 975+ | ·9182 870 | ·9039 998 |
| ·74 | ·9734 050− | ·9670 916 | ·9597 299 | ·9512 379 | ·9415 424 | ·9305 812 |
| ·75 | ·9823 785+ | ·9779 512 | ·9727 189 | ·9666 018 | ·9595 234 | ·9514 127 |
| ·76 | ·9887 285+ | ·9857 387 | ·9821 582 | ·9779 164 | ·9729 426 | ·9671 674 |
| ·77 | ·9930 590 | ·9911 198 | ·9887 668 | ·9859 425+ | ·9825 873 | ·9786 402 |
| ·78 | ·9958 979 | ·9946 932 | ·9932 126 | ·9914 123 | ·9892 459 | ·9866 641 |
| ·79 | ·9976 815− | ·9969 672 | ·9960 782 | ·9949 833 | ·9936 487 | ·9920 380 |
| ·80 | ·9987 518 | ·9983 492 | ·9978 417 | ·9972 088 | ·9964 277 | ·9954 729 |
| ·81 | ·9993 629 | ·9991 482 | ·9988 740 | ·9985 279 | ·9980 954 | ·9975 601 |
| ·82 | ·9996 934 | ·9995 855+ | ·9994 462 | ·9992 680 | ·9990 426 | ·9987 603 |
| ·83 | ·9998 617 | ·9998 110 | ·9997 448 | ·9996 590 | ·9995 492 | ·9994 099 |
| ·84 | ·9999 420 | ·9999 199 | ·9998 906 | ·9998 523 | ·9998 026 | ·9997 389 |
| ·85 | ·9999 776 | ·9999 687 | ·9999 568 | ·9999 410 | ·9999 204 | ·9998 935+ |
| ·86 | ·9999 921 | ·9999 888 | ·9999 844 | ·9999 785+ | ·9999 707 | ·9999 604 |
| ·87 | ·9999 975− | ·9999 964 | ·9999 950− | ·9999 930 | ·9999 903 | ·9999 868 |
| ·88 | ·9999 993 | ·9999 990 | ·9999 986 | ·9999 980 | ·9999 972 | ·9999 961 |
| ·89 | ·9999 998 | ·9999 998 | ·9999 996 | ·9999 995− | ·9999 993 | ·9999 990 |
| ·90 | 1·0000 000 | ·9999 999 | ·9999 999 | ·9999 999 | ·9999 998 | ·9999 998 |
| ·91 | | 1·0000 000 | 1·0000 000 | 1·0000 000 | 1·0000 000 | 1·0000 000 |

$x = \cdot 32$ to $\cdot 92$   $q = 20$   $p = 38$ to $43$

| $x$ | $p = 38$ | $p = 39$ | $p = 40$ | $p = 41$ | $p = 42$ | $p = 43$ |
|---|---|---|---|---|---|---|
| $B(p, q) =$ | ·4131 3111 × $\frac{1}{10^{16}}$ | ·2706 7211 × $\frac{1}{10^{16}}$ | ·1789 1885+ × $\frac{1}{10^{16}}$ | ·1192 7923 × $\frac{1}{10^{16}}$ | ·8017 1288 × $\frac{1}{10^{17}}$ | ·5430 9582 × $\frac{1}{10^{17}}$ |
| ·32 | ·0000 001 | | | | | |
| ·33 | ·0000 002 | ·0000 001 | | | | |
| ·34 | ·0000 005− | ·0000 002 | ·0000 001 | ·0000 001 | | |
| ·35 | ·0000 011 | ·0000 006 | ·0000 003 | ·0000 002 | ·0000 001 | |
| ·36 | ·0000 025− | ·0000 013 | ·0000 007 | ·0000 004 | ·0000 002 | ·0000 001 |
| ·37 | ·0000 053 | ·0000 029 | ·0000 016 | ·0000 008 | ·0000 004 | ·0000 002 |
| ·38 | ·0000 110 | ·0000 061 | ·0000 034 | ·0000 019 | ·0000 010 | ·0000 006 |
| ·39 | ·0000 220 | ·0000 126 | ·0000 072 | ·0000 041 | ·0000 023 | ·0000 013 |
| ·40 | ·0000 428 | ·0000 252 | ·0000 147 | ·0000 085+ | ·0000 049 | ·0000 028 |
| ·41 | ·0000 809 | ·0000 488 | ·0000 292 | ·0000 173 | ·0000 102 | ·0000 060 |
| ·42 | ·0001 490 | ·0000 919 | ·0000 563 | ·0000 342 | ·0000 207 | ·0000 124 |
| ·43 | ·0002 672 | ·0001 687 | ·0001 057 | ·0000 658 | ·0000 406 | ·0000 249 |
| ·44 | ·0004 673 | ·0003 017 | ·0001 933 | ·0001 230 | ·0000 777 | ·0000 488 |
| ·45 | ·0007 974 | ·0005 261 | ·0003 446 | ·0002 241 | ·0001 447 | ·0000 928 |
| ·46 | ·0013 287 | ·0008 954 | ·0005 990 | ·0003 978 | ·0002 625− | ·0001 720 |
| ·47 | ·0021 633 | ·0014 882 | ·0010 163 | ·0006 892 | ·0004 642 | ·0003 106 |
| ·48 | ·0034 436 | ·0024 171 | ·0016 843 | ·0011 655+ | ·0008 011 | ·0005 471 |
| ·49 | ·0053 627 | ·0038 387 | ·0027 281 | ·0019 254 | ·0013 499 | ·0009 403 |
| ·50 | ·0081 740 | ·0059 641 | ·0043 207 | ·0031 088 | ·0022 220 | ·0015 780 |
| ·51 | ·0122 011 | ·0090 699 | ·0066 949 | ·0049 084 | ·0035 750+ | ·0025 874 |
| ·52 | ·0178 427 | ·0135 068 | ·0101 537 | ·0075 819 | ·0056 248 | ·0041 467 |
| ·53 | ·0255 743 | ·0197 052 | ·0150 792 | ·0114 628 | ·0086 579 | ·0064 988 |
| ·54 | ·0359 415+ | ·0281 746 | ·0219 372 | ·0169 691 | ·0130 431 | ·0099 639 |
| ·55 | ·0495 447 | ·0394 949 | ·0312 746 | ·0246 059 | ·0192 384 | ·0149 507 |
| ·56 | ·0670 126 | ·0542 973 | ·0437 077 | ·0349 607 | ·0277 925− | ·0219 623 |
| ·57 | ·0889 644 | ·0732 339 | ·0598 993 | ·0486 882 | ·0393 365+ | ·0315 947 |
| ·58 | ·1159 614 | ·0969 341 | ·0805 218 | ·0664 811 | ·0545 636 | ·0445 245+ |
| ·59 | ·1484 514 | ·1259 514 | ·1062 091 | ·0890 284 | ·0741 943 | ·0614 829 |
| ·60 | ·1867 082 | ·1607 020 | ·1374 965+ | ·1169 597 | ·0989 273 | ·0832 134 |
| ·61 | ·2307 749 | ·2014 015− | ·1747 545+ | ·1507 789 | ·1293 763 | ·1104 146 |
| ·62 | ·2804 167 | ·2480 063 | ·2181 222 | ·1907 932 | ·1659 972 | ·1436 689 |
| ·63 | ·3350 909 | ·3001 686 | ·2674 490 | ·2370 443 | ·2090 127 | ·1833 637 |
| ·64 | ·3939 414 | ·3572 117 | ·3222 526 | ·2892 511 | ·2583 425− | ·2296 121 |
| ·65 | ·4558 219 | ·4181 344 | ·3817 032 | ·3467 748 | ·3135 502 | ·2821 845+ |
| ·66 | ·5193 494 | ·4816 463 | ·4446 394 | ·4086 131 | ·3738 169 | ·3404 623 |
| ·67 | ·5829 863 | ·5462 369 | ·5096 195− | ·4734 323 | ·4379 512 | ·4034 247 |
| ·68 | ·6451 429 | ·6102 717 | ·5750 071 | ·5396 367 | ·5044 388 | ·4696 767 |
| ·69 | ·7042 918 | ·6721 087 | ·6390 839 | ·6054 729 | ·5715 340 | ·5375 223 |
| ·70 | ·7590 800 | ·7302 211 | ·7001 783 | ·6691 591 | ·6373 834 | ·6050 780 |
| ·71 | ·8084 264 | ·7833 130 | ·7567 958 | ·7290 257 | ·7001 718 | ·6704 180 |
| ·72 | ·8515 918 | ·8304 119 | ·8077 331 | ·7836 483 | ·7582 710 | ·7317 337 |
| ·73 | ·8882 143 | ·8709 292 | ·8521 637 | ·8319 580 | ·8103 724 | ·7874 866 |
| ·74 | ·9183 046 | ·9046 776 | ·8896 809 | ·8733 119 | ·8555 856 | ·8365 340 |
| ·75 | ·9422 060 | ·9318 485− | ·9202 958 | ·9075 157 | ·8934 888 | ·8782 094 |
| ·76 | ·9605 244 | ·9529 514 | ·9443 919 | ·9347 967 | ·9241 250+ | ·9123 455+ |
| ·77 | ·9740 403 | ·9687 274 | ·9626 434 | ·9557 335+ | ·9479 473 | ·9392 397 |
| ·78 | ·9836 163 | ·9800 503 | ·9759 138 | ·9711 548 | ·9657 225− | ·9595 684 |
| ·79 | ·9901 122 | ·9878 300 | ·9851 488 | ·9820 245+ | ·9784 125− | ·9742 682 |
| ·80 | ·9943 169 | ·9929 297 | ·9912 792 | ·9893 317 | ·9870 517 | ·9844 025− |
| ·81 | ·9969 039 | ·9961 065+ | ·9951 461 | ·9939 986 | ·9926 383 | ·9910 381 |
| ·82 | ·9984 098 | ·9979 788 | ·9974 532 | ·9968 174 | ·9960 545+ | ·9951 459 |
| ·83 | ·9992 350− | ·9990 172 | ·9987 483 | ·9984 192 | ·9980 194 | ·9975 374 |
| ·84 | ·9996 578 | ·9995 557 | ·9994 281 | ·9992 700 | ·9990 757 | ·9988 386 |
| ·85 | ·9998 590 | ·9998 149 | ·9997 592 | ·9996 894 | ·9996 026 | ·9994 954 |
| ·86 | ·9999 470 | ·9999 297 | ·9999 076 | ·9998 796 | ·9998 443 | ·9998 002 |
| ·87 | ·9999 821 | ·9999 760 | ·9999 681 | ·9999 580 | ·9999 451 | ·9999 289 |
| ·88 | ·9999 946 | ·9999 927 | ·9999 903 | ·9999 870 | ·9999 829 | ·9999 776 |
| ·89 | ·9999 986 | ·9999 981 | ·9999 974 | ·9999 965+ | ·9999 954 | ·9999 939 |
| ·90 | ·9999 997 | ·9999 996 | ·9999 994 | ·9999 992 | ·9999 989 | ·9999 986 |
| ·91 | ·9999 999 | ·9999 999 | ·9999 999 | ·9999 999 | ·9999 998 | ·9999 997 |
| ·92 | 1·0000 000 | 1·0000 000 | 1·0000 000 | 1·0000 000 | 1·0000 000 | 1·0000 000 |

$x$ = ·37 to ·93 $q$ = 20 $p$ = 44 to 50

| | $p$ = 44 | $p$ = 45 | $p$ = 46 | $p$ = 47 | $p$ = 48 | $p$ = 49 | $p$ = 50 |
|---|---|---|---|---|---|---|---|
| $B(p, q)$ = | ·3706 8445⁺ × $\frac{1}{10^{17}}$ | ·2548 4556 × $\frac{1}{10^{17}}$ | ·1764 3154 × $\frac{1}{10^{17}}$ | ·1229 6744 × $\frac{1}{10^{17}}$ | ·8626 0740 × $\frac{1}{10^{18}}$ | ·6088 9934 × $\frac{1}{10^{18}}$ | ·4324 0678 × $\frac{1}{10^{18}}$ |
| $x$ | | | | | | | |
| ·37 | ·0000 001 | ·0000 001 | | | | | |
| ·38 | ·0000 003 | ·0000 002 | ·0000 001 | | | | |
| ·39 | ·0000 007 | ·0000 004 | ·0000 002 | ·0000 001 | ·0000 001 | | |
| ·40 | ·0000 016 | ·0000 009 | ·0000 005+ | ·0000 003 | ·0000 002 | ·0000 001 | |
| ·41 | ·0000 035− | ·0000 020 | ·0000 012 | ·0000 007 | ·0000 004 | ·0000 002 | ·0000 001 |
| ·42 | ·0000 074 | ·0000 044 | ·0000 026 | ·0000 015+ | ·0000 009 | ·0000 005+ | ·0000 003 |
| ·43 | ·0000 152 | ·0000 092 | ·0000 056 | ·0000 033 | ·0000 020 | ·0000 012 | ·0000 007 |
| ·44 | ·0000 304 | ·0000 189 | ·0000 116 | ·0000 071 | ·0000 043 | ·0000 026 | ·0000 016 |
| ·45 | ·0000 592 | ·0000 375− | ·0000 236 | ·0000 148 | ·0000 092 | ·0000 057 | ·0000 035+ |
| ·46 | ·0001 120 | ·0000 725− | ·0000 466 | ·0000 298 | ·0000 190 | ·0000 120 | ·0000 076 |
| ·47 | ·0002 065+ | ·0001 365− | ·0000 897 | ·0000 586 | ·0000 380 | ·0000 246 | ·0000 158 |
| ·48 | ·0003 712 | ·0002 504 | ·0001 679 | ·0001 119 | ·0000 742 | ·0000 490 | ·0000 321 |
| ·49 | ·0006 509 | ·0004 478 | ·0003 063 | ·0002 084 | ·0001 410 | ·0000 949 | ·0000 635+ |
| ·50 | ·0011 138 | ·0007 814 | ·0005 450+ | ·0003 781 | ·0002 608 | ·0001 790 | ·0001 222 |
| ·51 | ·0018 611 | ·0013 308 | ·0009 461 | ·0006 689 | ·0004 704 | ·0003 290 | ·0002 290 |
| ·52 | ·0030 385− | ·0022 134 | ·0016 031 | ·0011 548 | ·0008 273 | ·0005 897 | ·0004 182 |
| ·53 | ·0048 488 | ·0035 967 | ·0026 529 | ·0019 461 | ·0014 200 | ·0010 308 | ·0007 446 |
| ·54 | ·0075 665+ | ·0057 129 | ·0042 893 | ·0032 030 | ·0023 793 | ·0017 584 | ·0012 932 |
| ·55 | ·0115 506 | ·0088 729 | ·0067 784 | ·0051 506 | ·0038 934 | ·0029 283 | ·0021 916 |
| ·56 | ·0172 548 | ·0134 802 | ·0104 739 | ·0080 951 | ·0062 244 | ·0047 621 | ·0036 257 |
| ·57 | ·0252 319 | ·0200 391 | ·0158 294 | ·0124 388 | ·0097 249 | ·0075 656 | ·0058 576 |
| ·58 | ·0361 290 | ·0291 570 | ·0234 058 | ·0186 924 | ·0148 534 | ·0117 455− | ·0092 439 |
| ·59 | ·0506 693 | ·0415 343 | ·0338 691 | ·0274 786 | ·0221 839 | ·0178 236 | ·0142 535− |
| ·60 | ·0696 191 | ·0579 403 | ·0479 744 | ·0395 250+ | ·0324 060 | ·0264 438 | ·0214 794 |
| ·61 | ·0937 372 | ·0791 709 | ·0665 339 | ·0556 413 | ·0463 108 | ·0383 662 | ·0316 409 |
| ·62 | ·1237 085+ | ·1059 891 | ·0903 645+ | ·0766 760 | ·0647 583 | ·0544 445+ | ·0455 704 |
| ·63 | ·1600 650− | ·1390 484 | ·1202 176 | ·1034 540 | ·0886 236 | ·0755 822 | ·0641 802 |
| ·64 | ·2031 000 | ·1788 053 | ·1566 917 | ·1366 934 | ·1187 203 | ·1026 645− | ·0884 042 |
| ·65 | ·2527 884 | ·2254 300 | ·2001 386 | ·1769 090 | ·1557 058 | ·1364 684 | ·1191 154 |
| ·66 | ·3087 212 | ·2787 258 | ·2505 701 | ·2243 119 | ·1999 756 | ·1775 557 | ·1570 210 |
| ·67 | ·3700 705+ | ·3380 732 | ·3075 834 | ·2787 174 | ·2515 589 | ·2261 600 | ·2025 442 |
| ·68 | ·4355 937 | ·4024 093 | ·3703 162 | ·3394 790 | ·3100 327 | ·2820 833 | ·2557 081 |
| ·69 | ·5036 847 | ·4702 547 | ·4374 487 | ·4054 628 | ·3744 705+ | ·3446 211 | ·3160 388 |
| ·70 | ·5724 721 | ·5397 921 | ·5072 574 | ·4750 763 | ·4434 430 | ·4125 346 | ·3825 092 |
| ·71 | ·6399 584 | ·6089 935− | ·5777 258 | ·5463 561 | ·5150 799 | ·4840 838 | ·4535 432 |
| ·72 | ·7041 842 | ·6757 833 | ·6467 006 | ·6171 122 | ·5871 965− | ·5571 312 | ·5270 904 |
| ·73 | ·7633 979 | ·7382 197 | ·7120 792 | ·6851 147 | ·6574 731 | ·6293 076 | ·6007 742 |
| ·74 | ·8162 064 | ·7946 684 | ·7720 009 | ·7482 985− | ·7236 682 | ·6982 271 | ·6721 007 |
| ·75 | ·8616 864 | ·8439 429 | ·8250 165+ | ·8049 587 | ·7838 340 | ·7617 190 | ·7387 014 |
| ·76 | ·8994 374 | ·8853 909 | ·8702 083 | ·8539 035− | ·8365 024 | ·8180 428 | ·7985 736 |
| ·77 | ·9295 722 | ·9189 138 | ·9072 417 | ·8945 419 | ·8808 100 | ·8660 510 | ·8502 799 |
| ·78 | ·9526 473 | ·9449 177 | ·9363 429 | ·9268 921 | ·9165 404 | ·9052 700 | ·8930 703 |
| ·79 | ·9695 475− | ·9642 077 | ·9582 082 | ·9515 109 | ·9440 812 | ·9358 884 | ·9269 065− |
| ·80 | ·9813 467 | ·9778 464 | ·9738 639 | ·9693 619 | ·9643 043 | ·9586 567 | ·9523 867 |
| ·81 | ·9891 693 | ·9870 018 | ·9845 048 | ·9816 469 | ·9783 962 | ·9747 208 | ·9705 895− |
| ·82 | ·9940 717 | ·9928 104 | ·9913 395+ | ·9896 352 | ·9876 727 | ·9854 264 | ·9828 703 |
| ·83 | ·9969 607 | ·9962 753 | ·9954 662 | ·9945 172 | ·9934 112 | ·9921 298 | ·9906 539 |
| ·84 | ·9985 515+ | ·9982 062 | ·9977 936 | ·9973 040 | ·9967 264 | ·9960 491 | ·9952 596 |
| ·85 | ·9993 641 | ·9992 042 | ·9990 109 | ·9987 787 | ·9985 017 | ·9981 729 | ·9977 851 |
| ·86 | ·9997 456 | ·9996 783 | ·9995 959 | ·9994 959 | ·9993 751 | ·9992 301 | ·9990 570 |
| ·87 | ·9999 084 | ·9998 830 | ·9998 516 | ·9998 129 | ·9997 656 | ·9997 083 | ·9996 390 |
| ·88 | ·9999 709 | ·9999 624 | ·9999 518 | ·9999 386 | ·9999 223 | ·9999 022 | ·9998 778 |
| ·89 | ·9999 920 | ·9999 895+ | ·9999 864 | ·9999 825+ | ·9999 777 | ·9999 716 | ·9999 641 |
| ·90 | ·9999 981 | ·9999 975+ | ·9999 968 | ·9999 958 | ·9999 946 | ·9999 930 | ·9999 911 |
| ·91 | ·9999 996 | ·9999 995+ | ·9999 994 | ·9999 992 | ·9999 989 | ·9999 986 | ·9999 982 |
| ·92 | ·9999 999 | ·9999 999 | ·9999 999 | ·9999 999 | ·9999 998 | ·9999 998 | ·9999 997 |
| ·93 | 1·0000 000 | 1·0000 000 | 1·0000 000 | 1·0000 000 | 1·0000 000 | 1·0000 000 | 1·0000 000 |

$x = \cdot 15$ to $\cdot 70$ | $q = 21$ | $p = 21$ to $26$

| | $p = 21$ | $p = 22$ | $p = 23$ | $p = 24$ | $p = 25$ | $p = 26$ |
|---|---|---|---|---|---|---|
| $B(p, q) =$ | $\cdot 1769\ 3767 \times \frac{1}{10^{12}}$ | $\cdot 8846\ 8836 \times \frac{1}{10^{13}}$ | $\cdot 4526\ 3125 \overset{+}{\times} \frac{1}{10^{13}}$ | $\cdot 2366\ 0270 \times \frac{1}{10^{13}}$ | $\cdot 1261\ 8811 \times \frac{1}{10^{13}}$ | $\cdot 6858\ 0493 \times \frac{1}{10^{14}}$ |
| $x$ | | | | | | |
| ·15 | ·0000 001 | | | | | |
| ·16 | ·0000 002 | ·0000 001 | | | | |
| ·17 | ·0000 005+ | ·0000 002 | ·0000 001 | | | |
| ·18 | ·0000 014 | ·0000 005− | ·0000 002 | ·0000 001 | | |
| ·19 | ·0000 036 | ·0000 013 | ·0000 005− | ·0000 002 | ·0000 001 | |
| ·20 | ·0000 084 | ·0000 032 | ·0000 012 | ·0000 004 | ·0000 002 | ·0000 001 |
| ·21 | ·0000 184 | ·0000 073 | ·0000 028 | ·0000 011 | ·0000 004 | ·0000 001 |
| ·22 | ·0000 386 | ·0000 160 | ·0000 065+ | ·0000 026 | ·0000 010 | ·0000 004 |
| ·23 | ·0000 774 | ·0000 335− | ·0000 142 | ·0000 059 | ·0000 024 | ·0000 010 |
| ·24 | ·0001 486 | ·0000 670 | ·0000 296 | ·0000 129 | ·0000 055− | ·0000 023 |
| ·25 | ·0002 742 | ·0001 287 | ·0000 592 | ·0000 268 | ·0000 119 | ·0000 052 |
| ·26 | ·0004 882 | ·0002 380 | ·0001 138 | ·0000 534 | ·0000 247 | ·0000 112 |
| ·27 | ·0008 401 | ·0004 248 | ·0002 107 | ·0001 027 | ·0000 492 | ·0000 232 |
| ·28 | ·0014 008 | ·0007 336 | ·0003 770 | ·0001 903 | ·0000 944 | ·0000 461 |
| ·29 | ·0022 676 | ·0012 283 | ·0006 529 | ·0003 409 | ·0001 750+ | ·0000 885− |
| ·30 | ·0035 699 | ·0019 975− | ·0010 969 | ·0005 918 | ·0003 140 | ·0001 640 |
| ·31 | ·0054 747 | ·0031 606 | ·0017 911 | ·0009 973 | ·0005 462 | ·0002 945− |
| ·32 | ·0081 904 | ·0048 731 | ·0028 464 | ·0016 340 | ·0009 227 | ·0005 130 |
| ·33 | ·0119 691 | ·0073 312 | ·0044 093 | ·0026 066 | ·0015 160 | ·0008 682 |
| ·34 | ·0171 059 | ·0107 754 | ·0066 663 | ·0040 543 | ·0024 262 | ·0014 298 |
| ·35 | ·0239 349 | ·0154 905+ | ·0098 481 | ·0061 560 | ·0037 869 | ·0022 944 |
| ·36 | ·0328 214 | ·0218 036 | ·0142 314 | ·0091 350+ | ·0057 714 | ·0035 918 |
| ·37 | ·0441 488 | ·0300 770 | ·0201 373 | ·0132 615+ | ·0085 975+ | ·0054 913 |
| ·38 | ·0583 028 | ·0406 976 | ·0279 258 | ·0188 522 | ·0125 310 | ·0082 073 |
| ·39 | ·0756 504 | ·0540 608 | ·0379 864 | ·0262 661 | ·0178 860 | ·0120 032 |
| ·40 | ·0965 172 | ·0705 517 | ·0507 235+ | ·0358 955+ | ·0250 216 | ·0171 924 |
| ·41 | ·1211 632 | ·0905 215− | ·0665 373 | ·0481 530 | ·0343 340 | ·0241 357 |
| ·42 | ·1497 587 | ·1142 624 | ·0858 008 | ·0634 523 | ·0462 439 | ·0332 344 |
| ·43 | ·1823 633 | ·1419 829 | ·1088 343 | ·0821 858 | ·0611 778 | ·0449 176 |
| ·44 | ·2189 086 | ·1737 838 | ·1358 790 | ·1046 981 | ·0795 456 | ·0596 248 |
| ·45 | ·2591 880 | ·2096 384 | ·1670 708 | ·1312 586 | ·1017 134 | ·0777 819 |
| ·46 | ·3028 532 | ·2493 791 | ·2024 191 | ·1620 335− | ·1279 749 | ·0997 745− |
| ·47 | ·3494 202 | ·2926 907 | ·2417 889 | ·1970 618 | ·1585 218 | ·1259 170 |
| ·48 | ·3982 827 | ·3391 134 | ·2848 927 | ·2362 356 | ·1934 174 | ·1564 224 |
| ·49 | ·4487 348 | ·3880 539 | ·3312 897 | ·2792 888 | ·2325 746 | ·1913 727 |
| ·50 | ·5000 000[e] | ·4388 072 | ·3803 958 | ·3257 939 | ·2757 422 | ·2306 956 |
| ·51 | ·5512 652 | ·4905 844 | ·4315 033 | ·3751 707 | ·3224 998 | ·2741 479 |
| ·52 | ·6017 173 | ·5425 480 | ·4838 090 | ·4267 045− | ·3722 648 | ·3213 093 |
| ·53 | ·6505 798 | ·5938 503 | ·5364 504 | ·4795 746 | ·4243 103 | ·3715 881 |
| ·54 | ·6971 468 | ·6436 726 | ·5885 456 | ·5328 914 | ·4777 936 | ·4242 386 |
| ·55 | ·7408 120 | ·6912 625+ | ·6392 355− | ·5857 381 | ·5317 950− | ·4783 912 |
| ·56 | ·7810 914 | ·7359 667 | ·6877 242 | ·6372 164 | ·5853 617 | ·5330 922 |
| ·57 | ·8176 367 | ·7772 564 | ·7333 153 | ·6864 893 | ·6375 561 | ·5873 507 |
| ·58 | ·8502 413 | ·8147 450− | ·7754 408 | ·7328 215+ | ·6875 030 | ·6401 904 |
| ·59 | ·8788 368 | ·8481 950+ | ·8136 813 | ·7756 111 | ·7344 319 | ·6906 995+ |
| ·60 | ·9034 828 | ·8775 173 | ·8477 750+ | ·8144 119 | ·7777 125+ | ·7380 772 |
| ·61 | ·9243 496 | ·9027 601 | ·8776 181 | ·8489 452 | ·8168 794 | ·7816 712 |
| ·62 | ·9416 972 | ·9240 920 | ·9032 538 | ·8790 997 | ·8516 444 | ·8210 044 |
| ·63 | ·9558 512 | ·9417 794 | ·9248 550− | ·9049 209 | ·8818 970 | ·8557 879 |
| ·64 | ·9671 786 | ·9561 609 | ·9426 992 | ·9265 920 | ·9076 930 | ·8859 212 |
| ·65 | ·9760 651 | ·9676 206 | ·9571 419 | ·9444 079 | ·9292 333 | ·9114 789 |
| ·66 | ·9828 941 | ·9765 636 | ·9685 871 | ·9587 448 | ·9468 357 | ·9326 876 |
| ·67 | ·9880 309 | ·9833 929 | ·9774 606 | ·9700 297 | ·9609 021 | ·9498 942 |
| ·68 | ·9918 096 | ·9884 922 | ·9841 856 | ·9787 106 | ·9718 851 | ·9635 308 |
| ·69 | ·9945 253 | ·9922 111 | ·9891 628 | ·9852 305− | ·9802 560 | ·9740 778 |
| ·70 | ·9964 301 | ·9948 577 | ·9927 564 | ·9900 064 | ·9864 773 | ·9820 306 |

$x = \cdot71$ to $\cdot88$ $q = 21$ $p = 21$ to $26$

| | $p = 21$ | $p = 22$ | $p = 23$ | $p = 24$ | $p = 25$ | $p = 26$ |
|---|---|---|---|---|---|---|
| $B(p, q) =$ | $\cdot1769\ 3767 \times \frac{1}{10^{12}}$ | $\cdot8846\ 8836 \times \frac{1}{10^{13}}$ | $\cdot4526\ 3125^{+} \times \frac{1}{10^{13}}$ | $\cdot2366\ 0270 \times \frac{1}{10^{13}}$ | $\cdot1261\ 8811 \times \frac{1}{10^{13}}$ | $\cdot6858\ 0493 \times \frac{1}{10^{14}}$ |
| $x$ | | | | | | |
| ·71 | ·9977 324 | ·9966 931 | ·9952 844 | ·9934 144 | ·9909 804 | ·9878 697 |
| ·72 | ·9985 992 | ·9979 320 | ·9970 149 | ·9957 804 | ·9941 508 | ·9920 389 |
| ·73 | ·9991 599 | ·9987 446 | ·9981 658 | ·9973 758 | ·9963 186 | ·9949 295⁻ |
| ·74 | ·9995 118 | ·9992 616 | ·9989 082 | ·9984 192 | ·9977 558 | ·9968 722 |
| ·75 | ·9997 258 | ·9995 802 | ·9993 718 | ·9990 796 | ·9986 779 | ·9981 355⁻ |
| ·76 | ·9998 514 | ·9997 699 | ·9996 515⁺ | ·9994 834 | ·9992 491 | ·9989 286 |
| ·77 | ·9999 226 | ·9998 787 | ·9998 141 | ·9997 212 | ·9995 901 | ·9994 083 |
| ·78 | ·9999 614 | ·9999 387 | ·9999 050⁺ | ·9998 558 | ·9997 856 | ·9996 869 |
| ·79 | ·9999 816 | ·9999 704 | ·9999 536 | ·9999 288 | ·9998 929 | ·9998 418 |
| ·80 | ·9999 916 | ·9999 864 | ·9999 785⁻ | ·9999 666 | ·9999 492 | ·9999 240 |
| ·81 | ·9999 964 | ·9999 941 | ·9999 906 | ·9999 852 | ·9999 772 | ·9999 655⁻ |
| ·82 | ·9999 986 | ·9999 976 | ·9999 961 | ·9999 938 | ·9999 903 | ·9999 853 |
| ·83 | ·9999 995⁻ | ·9999 991 | ·9999 985⁻ | ·9999 976 | ·9999 962 | ·9999 941 |
| ·84 | ·9999 998 | ·9999 997 | ·9999 995⁻ | ·9999 991 | ·9999 986 | ·9999 978 |
| ·85 | ·9999 999 | ·9999 999 | ·9999 998 | ·9999 997 | ·9999 995⁺ | ·9999 993 |
| ·86 | 1·0000 000 | 1·0000 000 | ·9999 999 | ·9999 999 | ·9999 999 | ·9999 998 |
| ·87 | | | 1·0000 000 | 1·0000 000 | 1·0000 000 | ·9999 999 |
| ·88 | | | | | | 1·0000 000 |

$x = \cdot 21$ to $\cdot 89$ | $q = 21$ | $p = 27$ to $32$

| | $p = 27$ | $p = 28$ | $p = 29$ | $p = 30$ | $p = 31$ | $p = 32$ |
|---|---|---|---|---|---|---|
| $B(p, q) =$ | ·3793 8145 $\overset{+}{\times} \frac{1}{10^{14}}$ | ·2134 0207 $\times \frac{1}{10^{14}}$ | ·1219 4404 $\times \frac{1}{10^{14}}$ | ·7072 7542 $\times \frac{1}{10^{15}}$ | ·4160 4436 $\times \frac{1}{10^{15}}$ | ·2480 2645 $\overset{-}{\times} \frac{1}{10^{15}}$ |
| $x$ | | | | | | |
| ·21 | ·0000 001 | | | | | |
| ·22 | ·0000 001 | ·0000 001 | | | | |
| ·23 | ·0000 004 | ·0000 002 | ·0000 001 | | | |
| ·24 | ·0000 010 | ·0000 004 | ·0000 002 | ·0000 001 | | |
| ·25 | ·0000 022 | ·0000 009 | ·0000 004 | ·0000 002 | ·0000 001 | |
| ·26 | ·0000 050+ | ·0000 022 | ·0000 010 | ·0000 004 | ·0000 002 | ·0000 001 |
| ·27 | ·0000 108 | ·0000 049 | ·0000 022 | ·0000 010 | ·0000 004 | ·0000 002 |
| ·28 | ·0000 222 | ·0000 105+ | ·0000 049 | ·0000 023 | ·0000 010 | ·0000 005− |
| ·29 | ·0000 440 | ·0000 216 | ·0000 105− | ·0000 050+ | ·0000 024 | ·0000 011 |
| ·30 | ·0000 844 | ·0000 428 | ·0000 214 | ·0000 106 | ·0000 052 | ·0000 025− |
| ·31 | ·0001 564 | ·0000 819 | ·0000 423 | ·0000 216 | ·0000 109 | ·0000 054 |
| ·32 | ·0002 810 | ·0001 518 | ·0000 809 | ·0000 426 | ·0000 221 | ·0000 114 |
| ·33 | ·0004 899 | ·0002 727 | ·0001 497 | ·0000 812 | ·0000 435+ | ·0000 231 |
| ·34 | ·0008 304 | ·0004 757 | ·0002 689 | ·0001 501 | ·0000 828 | ·0000 452 |
| ·35 | ·0013 702 | ·0008 071 | ·0004 692 | ·0002 694 | ·0001 529 | ·0000 858 |
| ·36 | ·0022 035+ | ·0013 335+ | ·0007 966 | ·0004 700 | ·0002 741 | ·0001 580 |
| ·37 | ·0034 579 | ·0021 482 | ·0013 175+ | ·0007 982 | ·0004 779 | ·0002 830 |
| ·38 | ·0053 004 | ·0033 776 | ·0021 250+ | ·0013 208 | ·0008 114 | ·0004 930 |
| ·39 | ·0079 441 | ·0051 884 | ·0033 460 | ·0021 320 | ·0013 428 | ·0008 365+ |
| ·40 | ·0116 517 | ·0077 938 | ·0051 483 | ·0033 604 | ·0021 684 | ·0013 840 |
| ·41 | ·0167 380 | ·0114 582 | ·0077 472 | ·0051 764 | ·0034 197 | ·0022 348 |
| ·42 | ·0235 673 | ·0164 996 | ·0114 108 | ·0077 995+ | ·0052 717 | ·0035 250− |
| ·43 | ·0325 474 | ·0232 880 | ·0164 626 | ·0115 035+ | ·0079 495+ | ·0054 354 |
| ·44 | ·0441 172 | ·0322 395+ | ·0232 804 | ·0166 197 | ·0117 352 | ·0081 995− |
| ·45 | ·0587 287 | ·0438 037 | ·0322 901 | ·0235 358 | ·0169 701 | ·0121 094 |
| ·46 | ·0768 236 | ·0584 460 | ·0439 538 | ·0326 900 | ·0240 543 | ·0175 191 |
| ·47 | ·0988 049 | ·0766 232 | ·0587 510 | ·0445 581 | ·0334 403 | ·0248 437 |
| ·48 | ·1250 051 | ·0987 543 | ·0771 535+ | ·0596 346 | ·0456 195− | ·0345 520 |
| ·49 | ·1556 538 | ·1251 869 | ·0995 948 | ·0784 062 | ·0611 022 | ·0471 530 |
| ·50 | ·1908 467 | ·1561 634 | ·1264 349 | ·1013 194 | ·0803 898 | ·0631 735+ |
| ·51 | ·2305 196 | ·1917 873 | ·1579 243 | ·1287 437 | ·1039 402 | ·0831 292 |
| ·52 | ·2744 303 | ·2319 960 | ·1941 689 | ·1609 333 | ·1321 290 | ·1074 874 |
| ·53 | ·3221 509 | ·2765 405+ | ·2351 003 | ·1979 898 | ·1652 088 | ·1366 260 |
| ·54 | ·3730 730 | ·3249 773 | ·2804 544 | ·2398 312 | ·2032 702 | ·1707 899 |
| ·55 | ·4264 253 | ·3766 727 | ·3297 632 | ·2861 696 | ·2462 089 | ·2100 508 |
| ·56 | ·4813 051 | ·4308 224 | ·3823 589 | ·3365 025− | ·2937 031 | ·2542 725+ |
| ·57 | ·5367 205− | ·4864 841 | ·4373 959 | ·3901 189 | ·3452 058 | ·3030 888 |
| ·58 | ·5916 404 | ·5426 229 | ·4938 856 | ·4461 229 | ·3999 524 | ·3558 967 |
| ·59 | ·6450 497 | ·5981 656 | ·5507 457 | ·5034 729 | ·4569 881 | ·4118 678 |
| ·60 | ·6960 027 | ·6520 583 | ·6068 584 | ·5610 349 | ·5152 115+ | ·4699 794 |
| ·61 | ·7436 734 | ·7033 254 | ·6611 329 | ·6176 455− | ·5734 333 | ·5290 642 |
| ·62 | ·7873 946 | ·7511 210 | ·7125 673 | ·6721 789 | ·6304 443 | ·5878 750+ |
| ·63 | ·8266 863 | ·7947 716 | ·7603 037 | ·7236 132 | ·6850 882 | ·6451 589 |
| ·64 | ·8612 689 | ·8338 044 | ·8036 719 | ·7710 872 | ·7363 302 | ·6997 345+ |
| ·65 | ·8910 614 | ·8679 594 | ·8422 172 | ·8139 451 | ·7833 170 | ·7505 647 |
| ·66 | ·9161 670 | ·8971 867 | ·8757 118 | ·8517 636 | ·8254 206 | ·7968 172 |
| ·67 | ·9368 456 | ·9216 271 | ·9041 475+ | ·8843 595− | ·8622 628 | ·8379 065+ |
| ·68 | ·9534 798 | ·9415 825− | ·9277 135+ | ·9117 786 | ·8937 191 | ·8735 157 |
| ·69 | ·9665 356 | ·9574 766 | ·9467 611 | ·9342 682 | ·9199 015− | ·9035 929 |
| ·70 | ·9765 235− | ·9698 130 | ·9617 604 | ·9522 362 | ·9411 245+ | ·9283 282 |
| ·71 | ·9839 622 | ·9791 328 | ·9732 547 | ·9662 031 | ·9578 586 | ·9481 117 |
| ·72 | ·9893 487 | ·9859 769 | ·9818 152 | ·9767 523 | ·9706 768 | ·9634 802 |
| ·73 | ·9931 353 | ·9908 554 | ·9880 023 | ·9844 831 | ·9802 014 | ·9750 592 |
| ·74 | ·9957 153 | ·9942 250+ | ·9923 346 | ·9899 708 | ·9870 555+ | ·9835 064 |
| ·75 | ·9974 157 | ·9964 761 | ·9952 680 | ·9937 371 | ·9918 234 | ·9894 622 |
| ·76 | ·9984 977 | ·9979 276 | ·9971 848 | ·9962 310 | ·9950 228 | ·9935 121 |
| ·77 | ·9991 606 | ·9988 286 | ·9983 904 | ·9978 203 | ·9970 887 | ·9961 619 |
| ·78 | ·9995 507 | ·9993 657 | ·9991 185− | ·9987 926 | ·9983 689 | ·9978 253 |
| ·79 | ·9997 704 | ·9996 722 | ·9995 392 | ·9993 617 | ·9991 280 | ·9988 242 |
| ·80 | ·9998 885− | ·9998 390 | ·9997 711 | ·9996 793 | ·9995 570 | ·9993 959 |
| ·81 | ·9999 488 | ·9999 252 | ·9998 925− | ·9998 477 | ·9997 872 | ·9997 066 |
| ·82 | ·9999 779 | ·9999 673 | ·9999 525− | ·9999 319 | ·9999 039 | ·9998 660 |
| ·83 | ·9999 911 | ·9999 867 | ·9999 804 | ·9999 716 | ·9999 594 | ·9999 428 |
| ·84 | ·9999 967 | ·9999 949 | ·9999 925− | ·9999 890 | ·9999 841 | ·9999 774 |
| ·85 | ·9999 988 | ·9999 982 | ·9999 974 | ·9999 961 | ·9999 943 | ·9999 918 |
| ·86 | ·9999 996 | ·9999 994 | ·9999 992 | ·9999 987 | ·9999 981 | ·9999 973 |
| ·87 | ·9999 999 | ·9999 998 | ·9999 998 | ·9999 996 | ·9999 995− | ·9999 992 |
| ·88 | 1·0000 000 | 1·0000 000 | ·9999 999 | ·9999 999 | ·9999 999 | ·9999 998 |
| ·89 | | | 1·0000 000 | 1·0000 000 | 1·0000 000 | 1·0000 000 |

| $x$ | $p$ = 33 | $p$ = 34 | $p$ = 35 | $p$ = 36 | $p$ = 37 | $p$ = 38 |
|---|---|---|---|---|---|---|
| $B(p, q)$ = | ·1497 5182 $\times\frac{1}{10^{15}}$ | ·9151 5000 $\times\frac{1}{10^{16}}$ | ·5657 2909 $\times\frac{1}{10^{16}}$ | ·3535 8068 $\times\frac{1}{10^{16}}$ | ·2233 1411 $\times\frac{1}{10^{16}}$ | ·1424 5900 $\times\frac{1}{10^{16}}$ |
| ·27 | ·0000 001 | | | | | |
| ·28 | ·0000 002 | ·0000 001 | | | | |
| ·29 | ·0000 005+ | ·0000 002 | ·0000 001 | | | |
| ·30 | ·0000 012 | ·0000 006 | ·0000 003 | ·0000 001 | ·0000 001 | |
| ·31 | ·0000 027 | ·0000 013 | ·0000 006 | ·0000 003 | ·0000 001 | ·0000 001 |
| ·32 | ·0000 058 | ·0000 029 | ·0000 015− | ·0000 007 | ·0000 004 | ·0000 002 |
| ·33 | ·0000 121 | ·0000 063 | ·0000 032 | ·0000 016 | ·0000 008 | ·0000 004 |
| ·34 | ·0000 244 | ·0000 130 | ·0000 069 | ·0000 036 | ·0000 019 | ·0000 010 |
| ·35 | ·0000 476 | ·0000 261 | ·0000 142 | ·0000 077 | ·0000 041 | ·0000 022 |
| ·36 | ·0000 902 | ·0000 509 | ·0000 285− | ·0000 158 | ·0000 086 | ·0000 047 |
| ·37 | ·0001 658 | ·0000 961 | ·0000 552 | ·0000 314 | ·0000 177 | ·0000 099 |
| ·38 | ·0002 963 | ·0001 763 | ·0001 039 | ·0000 607 | ·0000 351 | ·0000 201 |
| ·39 | ·0005 156 | ·0003 146 | ·0001 901 | ·0001 138 | ·0000 676 | ·0000 398 |
| ·40 | ·0008 742 | ·0005 466 | ·0003 385+ | ·0002 077 | ·0001 263 | ·0000 762 |
| ·41 | ·0014 454 | ·0009 255+ | ·0005 870 | ·0003 689 | ·0002 298 | ·0001 420 |
| ·42 | ·0023 329 | ·0015 287 | ·0009 923 | ·0006 383 | ·0004 070 | ·0002 573 |
| ·43 | ·0036 786 | ·0024 654 | ·0016 368 | ·0010 769 | ·0007 024 | ·0004 544 |
| ·44 | ·0056 714 | ·0038 849 | ·0026 365− | ·0017 733 | ·0011 825− | ·0007 820 |
| ·45 | ·0085 549 | ·0059 861 | ·0041 501 | ·0028 518 | ·0019 429 | ·0013 129 |
| ·46 | ·0126 340 | ·0090 250+ | ·0063 883 | ·0044 823 | ·0031 185− | ·0021 520 |
| ·47 | ·0182 781 | ·0133 220 | ·0096 225− | ·0068 901 | ·0048 924 | ·0034 460 |
| ·48 | ·0259 194 | ·0192 645− | ·0141 910 | ·0103 642 | ·0075 069 | ·0053 940 |
| ·49 | ·0360 459 | ·0273 049 | ·0205 024 | ·0152 645− | ·0112 720 | ·0082 582 |
| ·50 | ·0491 853 | ·0379 524 | ·0290 321 | ·0220 233 | ·0165 720 | ·0123 730 |
| ·51 | ·0658 822 | ·0517 552 | ·0403 124 | ·0311 419 | ·0238 665+ | ·0181 505− |
| ·52 | ·0866 653 | ·0692 757 | ·0549 139 | ·0431 783 | ·0336 855− | ·0260 809 |
| ·53 | ·1120 090 | ·0910 547 | ·0734 162 | ·0587 257 | ·0466 143 | ·0367 255+ |
| ·54 | ·1422 885+ | ·1175 700 | ·0963 702 | ·0783 807 | ·0632 695+ | ·0506 986 |
| ·55 | ·1777 346 | ·1491 886 | ·1242 528 | ·1027 011 | ·0842 625+ | ·0686 395+ |
| ·56 | ·2183 906 | ·1861 187 | ·1574 156 | ·1321 570 | ·1101 539 | ·0911 717 |
| ·57 | ·2640 780 | ·2283 654 | ·1960 350− | ·1670 761 | ·1413 993 | ·1188 523 |
| ·58 | ·3143 743 | ·2756 955− | ·2400 655− | ·2075 913 | ·1782 924 | ·1521 134 |
| ·59 | ·3686 087 | ·3276 174 | ·2892 062 | ·2535 935− | ·2209 089 | ·1912 013 |
| ·60 | ·4258 780 | ·3833 803 | ·3428 826 | ·3046 990 | ·2690 609 | ·2361 198 |
| ·61 | ·4850 834 | ·4419 955− | ·4002 509 | ·3602 358 | ·3222 658 | ·2865 844 |
| ·62 | ·5449 864 | ·5022 798 | ·4602 263 | ·4192 543 | ·3797 390 | ·3419 965+ |
| ·63 | ·6042 812 | ·5629 205− | ·5215 354 | ·4805 642 | ·4404 124 | ·4014 435− |
| ·64 | ·6616 749 | ·6225 543 | ·5827 894 | ·5427 972 | ·5029 828 | ·4637 279 |
| ·65 | ·7159 701 | ·6798 555− | ·6425 724 | ·6044 904 | ·5659 852 | ·5274 281 |
| ·66 | ·7661 400 | ·7336 222 | ·6995 359 | ·6641 835+ | ·6278 884 | ·5909 852 |
| ·67 | ·8113 887 | ·7828 539 | ·7524 895− | ·7205 201 | ·6872 009 | ·6528 100 |
| ·68 | ·8511 909 | ·8268 096 | ·8004 777 | ·7723 403 | ·7425 771 | ·7113 981 |
| ·69 | ·8853 069 | ·8650 427 | ·8428 355+ | ·8187 566 | ·7929 119 | ·7654 397 |
| ·70 | ·9137 725− | ·8974 082 | ·8792 150+ | ·8592 026 | ·8374 112 | ·8139 118 |
| ·71 | ·9368 663 | ·9240 430 | ·9095 829 | ·8934 496 | ·8756 313 | ·8561 419 |
| ·72 | ·9550 602 | ·9453 237 | ·9341 897 | ·9215 923 | ·9074 833 | ·8918 337 |
| ·73 | ·9689 593 | ·9618 077 | ·9535 159 | ·9440 041 | ·9332 029 | ·9210 560 |
| ·74 | ·9792 385+ | ·9741 663 | ·9682 049 | ·9612 727 | ·9532 929 | ·9441 959 |
| ·75 | ·9865 844 | ·9831 181 | ·9789 890 | ·9741 226 | ·9684 451 | ·9618 854 |
| ·76 | ·9916 465+ | ·9893 693 | ·9866 206 | ·9833 379 | ·9794 570 | ·9749 132 |
| ·77 | ·9950 022 | ·9935 680 | ·9918 142 | ·9896 920 | ·9871 501 | ·9841 348 |
| ·78 | ·9971 363 | ·9962 730 | ·9952 037 | ·9938 930 | ·9923 026 | ·9903 916 |
| ·79 | ·9984 343 | ·9979 395− | ·9973 187 | ·9965 480 | ·9956 009 | ·9944 483 |
| ·80 | ·9991 866 | ·9989 176 | ·9985 759 | ·9981 462 | ·9976 116 | ·9969 526 |
| ·81 | ·9996 005− | ·9994 624 | ·9992 849 | ·9990 589 | ·9987 741 | ·9984 188 |
| ·82 | ·9998 155− | ·9997 490 | ·9996 624 | ·9995 509 | ·9994 086 | ·9992 288 |
| ·83 | ·9999 204 | ·9998 906 | ·9998 512 | ·9997 998 | ·9997 336 | ·9996 488 |
| ·84 | ·9999 682 | ·9999 558 | ·9999 392 | ·9999 174 | ·9998 888 | ·9998 518 |
| ·85 | ·9999 883 | ·9999 836 | ·9999 772 | ·9999 687 | ·9999 574 | ·9999 426 |
| ·86 | ·9999 961 | ·9999 945− | ·9999 922 | ·9999 892 | ·9999 852 | ·9999 799 |
| ·87 | ·9999 988 | ·9999 983 | ·9999 976 | ·9999 967 | ·9999 954 | ·9999 937 |
| ·88 | ·9999 997 | ·9999 996 | ·9999 994 | ·9999 991 | ·9999 987 | ·9999 982 |
| ·89 | ·9999 999 | ·9999 999 | ·9999 999 | ·9999 998 | ·9999 997 | ·9999 996 |
| ·90 | 1·0000 000 | 1·0000 000 | 1·0000 000 | 1·0000 000 | ·9999 999 | ·9999 999 |
| ·91 | | | | | 1·0000 000 | 1·0000 000 |

$x$ = ·32 to ·92 $q = 21$ $p$ = 39 to 44

| $x$ | $p = 39$ | $p = 40$ | $p = 41$ | $p = 42$ | $p = 43$ | $p = 44$ |
|---|---|---|---|---|---|---|
| $B(p, q)$ = | ·9175 3257 × $\frac{1}{10^{17}}$ | ·5963 9617 × $\frac{1}{10^{17}}$ | ·3910 7945 $\overset{+}{\times}$ $\frac{1}{10^{17}}$ | ·2586 1706 × $\frac{1}{10^{17}}$ | ·1724 1137 × $\frac{1}{10^{17}}$ | ·1158 3889 × $\frac{1}{10^{17}}$ |
| ·32 | ·0000 001 | | | | | |
| ·33 | ·0000 002 | ·0000 001 | ·0000 001 | | | |
| ·34 | ·0000 005⁻ | ·0000 002 | ·0000 001 | ·0000 001 | | |
| ·35 | ·0000 011 | ·0000 006 | ·0000 003 | ·0000 002 | ·0000 001 | |
| ·36 | ·0000 025⁺ | ·0000 014 | ·0000 007 | ·0000 004 | ·0000 002 | ·0000 001 |
| ·37 | ·0000 055⁻ | ·0000 030 | ·0000 016 | ·0000 009 | ·0000 005⁻ | ·0000 003 |
| ·38 | ·0000 115⁻ | ·0000 065⁻ | ·0000 036 | ·0000 020 | ·0000 011 | ·0000 006 |
| ·39 | ·0000 232 | ·0000 134 | ·0000 077 | ·0000 044 | ·0000 025⁻ | ·0000 014 |
| ·40 | ·0000 456 | ·0000 270 | ·0000 159 | ·0000 093 | ·0000 054 | ·0000 031 |
| ·41 | ·0000 870 | ·0000 529 | ·0000 319 | ·0000 191 | ·0000 114 | ·0000 067 |
| ·42 | ·0001 614 | ·0001 005⁻ | ·0000 621 | ·0000 381 | ·0000 232 | ·0000 140 |
| ·43 | ·0002 915⁺ | ·0001 856 | ·0001 173 | ·0000 736 | ·0000 459 | ·0000 284 |
| ·44 | ·0005 130 | ·0003 340 | ·0002 158 | ·0001 385⁻ | ·0000 883 | ·0000 559 |
| ·45 | ·0008 802 | ·0005 856 | ·0003 867 | ·0002 536 | ·0001 652 | ·0001 069 |
| ·46 | ·0014 734 | ·0010 012 | ·0006 754 | ·0004 524 | ·0003 010 | ·0001 989 |
| ·47 | ·0024 084 | ·0016 706 | ·0011 504 | ·0007 867 | ·0005 344 | ·0003 606 |
| ·48 | ·0038 460 | ·0027 219 | ·0019 126 | ·0013 346 | ·0009 251 | ·0006 371 |
| ·49 | ·0060 043 | ·0043 334 | ·0031 054 | ·0022 101 | ·0015 625⁻ | ·0010 976 |
| ·50 | ·0091 685⁺ | ·0067 446 | ·0049 267 | ·0035 744 | ·0025 762 | ·0018 450⁻ |
| ·51 | ·0137 010 | ·0102 680 | ·0076 418 | ·0056 491 | ·0041 488 | ·0030 279 |
| ·52 | ·0200 453 | ·0152 974 | ·0115 939 | ·0087 287 | ·0065 293 | ·0048 537 |
| ·53 | ·0287 260 | ·0223 120 | ·0172 129 | ·0131 920 | ·0100 462 | ·0076 034 |
| ·54 | ·0403 376 | ·0318 734 | ·0250 174 | ·0195 092 | ·0151 184 | ·0116 445⁺ |
| ·55 | ·0555 244 | ·0446 120 | ·0356 092 | ·0282 423 | ·0222 611 | ·0174 414 |
| ·56 | ·0749 470 | ·0612 017 | ·0496 557 | ·0400 359 | ·0320 835⁻ | ·0255 589 |
| ·57 | ·0992 364 | ·0823 214 | ·0678 591 | ·0555 944 | ·0452 745⁺ | ·0366 562 |
| ·58 | ·1289 381 | ·1086 032 | ·0909 119 | ·0756 456 | ·0625 748 | ·0514 676 |
| ·59 | ·1644 489 | ·1405 706 | ·1194 383 | ·1008 883 | ·0847 321 | ·0707 663 |
| ·60 | ·2059 527 | ·1785 702 | ·1539 260 | ·1319 266 | ·1124 413 | ·0953 124 |
| ·61 | ·2533 630 | ·2227 057 | ·1946 543 | ·1691 959 | ·1462 712 | ·1257 829 |
| ·62 | ·3062 802 | ·2727 802 | ·2416 251 | ·2128 865⁺ | ·1865 838 | ·1626 912 |
| ·63 | ·3639 717 | ·3282 583 | ·2945 091 | ·2628 754 | ·2334 561 | ·2063 014 |
| ·64 | ·4253 821 | ·3882 555⁻ | ·3526 139 | ·3186 762 | ·2866 131 | ·2565 484 |
| ·65 | ·4891 753 | ·4515 600 | ·4148 852 | ·3794 179 | ·3453 862 | ·3129 770 |
| ·66 | ·5538 100 | ·5166 920 | ·4799 451 | ·4438 615⁺ | ·4087 058 | ·3747 110 |
| ·67 | ·6176 409 | ·5819 938 | ·5461 685⁺ | ·5104 568 | ·4751 362 | ·4404 645⁺ |
| ·68 | ·6790 377 | ·6457 479 | ·6117 923 | ·5774 392 | ·5429 552 | ·5085 995⁺ |
| ·69 | ·7365 070 | ·7063 059 | ·6750 476 | ·6429 584 | ·6102 732 | ·5772 308 |
| ·70 | ·7888 046 | ·7622 167 | ·7342 994 | ·7052 246 | ·6751 806 | ·6443 681 |
| ·71 | ·8350 215⁻ | ·8123 360 | ·7881 760 | ·7626 548 | ·7359 061 | ·7080 813 |
| ·72 | ·8746 356 | ·8559 029 | ·8356 716 | ·8139 994 | ·7909 650⁺ | ·7666 664 |
| ·73 | ·9075 218 | ·8925 751 | ·8762 085⁻ | ·8584 327 | ·8392 773 | ·8187 898 |
| ·74 | ·9339 211 | ·9224 187 | ·9096 509 | ·8955 939 | ·8802 384 | ·8635 901 |
| ·75 | ·9543 761 | ·9458 561 | ·9362 710 | ·9255 754 | ·9137 339 | ·9007 220 |
| ·76 | ·9696 423 | ·9635 822 | ·9566 737 | ·9488 620 | ·9400 981 | ·9303 395⁺ |
| ·77 | ·9805 911 | ·9764 631 | ·9716 953 | ·9662 333 | ·9600 248 | ·9530 207 |
| ·78 | ·9881 166 | ·9854 320 | ·9822 910 | ·9786 459 | ·9744 489 | ·9696 525⁺ |
| ·79 | ·9930 584 | ·9913 974 | ·9894 291 | ·9871 156 | ·9844 176 | ·9812 949 |
| ·80 | ·9961 480 | ·9951 743 | ·9940 058 | ·9926 149 | ·9909 725⁻ | ·9890 473 |
| ·81 | ·9979 794 | ·9974 411 | ·9967 870 | ·9959 988 | ·9950 562 | ·9939 377 |
| ·82 | ·9990 038 | ·9987 247 | ·9983 814 | ·9979 625⁺ | ·9974 555⁺ | ·9968 465⁻ |
| ·83 | ·9995 414 | ·9994 066 | ·9992 387 | ·9990 314 | ·9987 775⁻ | ·9984 686 |
| ·84 | ·9998 045⁻ | ·9997 442 | ·9996 684 | ·9995 735⁺ | ·9994 560 | ·9993 112 |
| ·85 | ·9999 235⁻ | ·9998 988 | ·9998 674 | ·9998 277 | ·9997 779 | ·9997 158 |
| ·86 | ·9999 728 | ·9999 637 | ·9999 519 | ·9999 369 | ·9999 177 | ·9998 936 |
| ·87 | ·9999 914 | ·9999 883 | ·9999 844 | ·9999 793 | ·9999 727 | ·9999 644 |
| ·88 | ·9999 976 | ·9999 967 | ·9999 955⁺ | ·9999 940 | ·9999 921 | ·9999 895⁺ |
| ·89 | ·9999 994 | ·9999 992 | ·9999 989 | ·9999 985⁺ | ·9999 980 | ·9999 973 |
| ·90 | ·9999 999 | ·9999 998 | ·9999 998 | ·9999 997 | ·9999 996 | ·9999 994 |
| ·91 | 1·0000 000 | 1·0000 000 | 1·0000 000 | ·9999 999 | ·9999 999 | ·9999 999 |
| ·92 | | | | 1·0000 000 | 1·0000 000 | 1·0000 000 |

$x$ = ·36 to ·93 $q = 21$ $p$ = 45 to 50

| $x$ | $p = 45$ | $p = 46$ | $p = 47$ | $p = 48$ | $p = 49$ | $p = 50$ |
|---|---|---|---|---|---|---|
| $B(p, q) =$ | ·7841 4018×$\frac{1}{10^{18}}$ | ·5346 4103×$\frac{1}{10^{18}}$ | ·3670 6698×$\frac{1}{10^{18}}$ | ·2537 0806×$\frac{1}{10^{18}}$ | ·1764 9256×$\frac{1}{10^{18}}$ | ·1235 4479×$\frac{1}{10^{18}}$ |
| ·36 | ·0000 001 | | | | | |
| ·37 | ·0000 001 | ·0000 001 | | | | |
| ·38 | ·0000 003 | ·0000 002 | ·0000 001 | ·0000 001 | | |
| ·39 | ·0000 008 | ·0000 004 | ·0000 002 | ·0000 001 | ·0000 001 | |
| ·40 | ·0000 018 | ·0000 010 | ·0000 006 | ·0000 003 | ·0000 002 | ·0000 001 |
| ·41 | ·0000 039 | ·0000 023 | ·0000 013 | ·0000 008 | ·0000 004 | ·0000 003 |
| ·42 | ·0000 084 | ·0000 050+ | ·0000 030 | ·0000 018 | ·0000 010 | ·0000 006 |
| ·43 | ·0000 175− | ·0000 107 | ·0000 065− | ·0000 039 | ·0000 024 | ·0000 014 |
| ·44 | ·0000 351 | ·0000 220 | ·0000 136 | ·0000 084 | ·0000 052 | ·0000 032 |
| ·45 | ·0000 687 | ·0000 439 | ·0000 279 | ·0000 176 | ·0000 111 | ·0000 069 |
| ·46 | ·0001 306 | ·0000 853 | ·0000 553 | ·0000 357 | ·0000 229 | ·0000 146 |
| ·47 | ·0002 418 | ·0001 612 | ·0001 068 | ·0000 704 | ·0000 461 | ·0000 300 |
| ·48 | ·0004 360 | ·0002 966 | ·0002 006 | ·0001 349 | ·0000 902 | ·0000 600 |
| ·49 | ·0007 662 | ·0005 317 | ·0003 668 | ·0002 516 | ·0001 717 | ·0001 165+ |
| ·50 | ·0013 132 | ·0009 291 | ·0006 536 | ·0004 572 | ·0003 181 | ·0002 201 |
| ·51 | ·0021 963 | ·0015 837 | ·0011 354 | ·0008 096 | ·0005 741 | ·0004 050+ |
| ·52 | ·0035 863 | ·0026 344 | ·0019 242 | ·0013 977 | ·0010 099 | ·0007 259 |
| ·53 | ·0057 203 | ·0042 786 | ·0031 823 | ·0023 540 | ·0017 321 | ·0012 680 |
| ·54 | ·0089 160 | ·0067 877 | ·0051 387 | ·0038 694 | ·0028 984 | ·0021 600 |
| ·55 | ·0135 856 | ·0105 224 | ·0081 051 | ·0062 098 | ·0047 331 | ·0035 894 |
| ·56 | ·0202 442 | ·0159 453 | ·0124 912 | ·0097 338 | ·0075 462 | ·0058 212 |
| ·57 | ·0295 108 | ·0236 278 | ·0188 165+ | ·0149 071 | ·0117 503 | ·0092 164 |
| ·58 | ·0420 971 | ·0342 468 | ·0277 139 | ·0223 125+ | ·0178 743 | ·0142 495+ |
| ·59 | ·0587 812 | ·0485 672 | ·0399 209 | ·0326 487 | ·0265 704 | ·0215 205− |
| ·60 | ·0803 636 | ·0674 079 | ·0562 548 | ·0467 153 | ·0386 067 | ·0317 558 |
| ·61 | ·1076 042 | ·0915 868 | ·0775 680 | ·0653 777 | ·0548 432 | ·0457 943 |
| ·62 | ·1411 444 | ·1218 481 | ·1046 827 | ·0895 114 | ·0761 860 | ·0645 521 |
| ·63 | ·1814 178 | ·1587 737 | ·1383 054 | ·1199 232 | ·1035 170 | ·0889 624 |
| ·64 | ·2285 609 | ·2026 880 | ·1789 299 | ·1572 545− | ·1376 021 | ·1198 908 |
| ·65 | ·2823 355+ | ·2535 666 | ·2267 364 | ·2018 757 | ·1789 832 | ·1580 295− |
| ·66 | ·3420 761 | ·3109 640 | ·2815 023 | ·2537 832 | ·2278 659 | ·2037 786 |
| ·67 | ·4066 754 | ·3739 750+ | ·3425 400 | ·3125 162 | ·2840 187 | ·2571 321 |
| ·68 | ·4746 186 | ·4412 419 | ·4086 777 | ·3771 113 | ·3467 023 | ·3175 842 |
| ·69 | ·5440 682 | ·5110 161 | ·4782 946 | ·4461 091 | ·4146 479 | ·3840 790 |
| ·70 | ·6129 953 | ·5812 739 | ·5494 146 | ·5176 231 | ·4860 966 | ·4550 204 |
| ·71 | ·6793 458 | ·6498 760 | ·6198 552 | ·5894 704 | ·5589 083 | ·5283 524 |
| ·72 | ·7412 191 | ·7147 540 | ·6874 143 | ·6593 533 | ·6307 311 | ·6017 117 |
| ·73 | ·7970 359 | ·7740 976 | ·7500 722 | ·7250 705− | ·6992 145− | ·6726 356 |
| ·74 | ·8456 704 | ·8265 164 | ·8061 797 | ·7847 267 | ·7622 368 | ·7388 014 |
| ·75 | ·8865 273 | ·8711 496 | ·8546 019 | ·8369 099 | ·8181 122 | ·7982 595− |
| ·76 | ·9195 518 | ·9077 094 | ·8947 960 | ·8808 055+ | ·8657 425− | ·8496 219 |
| ·77 | ·9451 761 | ·9364 512 | ·9268 121 | ·9162 316 | ·9046 900 | ·8921 757 |
| ·78 | ·9642 109 | ·9580 799 | ·9512 186 | ·9435 894 | ·9351 591 | ·9258 996 |
| ·79 | ·9777 066 | ·9736 119 | ·9689 707 | ·9637 439 | ·9578 943 | ·9513 868 |
| ·80 | ·9868 071 | ·9842 185− | ·9812 472 | ·9778 586 | ·9740 182 | ·9696 919 |
| ·81 | ·9926 199 | ·9910 780 | ·9892 861 | ·9872 170 | ·9848 428 | ·9821 346 |
| ·82 | ·9961 200 | ·9952 595− | ·9942 471 | ·9930 637 | ·9916 890 | ·9901 016 |
| ·83 | ·9980 957 | ·9976 487 | ·9971 164 | ·9964 865− | ·9957 458 | ·9948 802 |
| ·84 | ·9991 344 | ·9989 199 | ·9986 614 | ·9983 518 | ·9979 833 | ·9975 475+ |
| ·85 | ·9996 391 | ·9995 448 | ·9994 299 | ·9992 907 | ·9991 230 | ·9989 223 |
| ·86 | ·9998 635− | ·9998 260 | ·9997 798 | ·9997 231 | ·9996 541 | ·9995 705+ |
| ·87 | ·9999 538 | ·9999 405+ | ·9999 239 | ·9999 033 | ·9998 780 | ·9998 469 |
| ·88 | ·9999 863 | ·9999 821 | ·9999 769 | ·9999 703 | ·9999 622 | ·9999 520 |
| ·89 | ·9999 965− | ·9999 954 | ·9999 940 | ·9999 922 | ·9999 899 | ·9999 871 |
| ·90 | ·9999 992 | ·9999 990 | ·9999 987 | ·9999 983 | ·9999 977 | ·9999 971 |
| ·91 | ·9999 999 | ·9999 998 | ·9999 998 | ·9999 997 | ·9999 996 | ·9999 995− |
| ·92 | 1·0000 000 | 1·0000 000 | 1·0000 000 | 1·0000 000 | ·9999 999 | ·9999 999 |
| ·93 | | | | | 1·0000 000 | 1·0000 000 |

$x = \cdot 16$ to $\cdot 87$ $q = 22$ $p = 22$ to $27$

| $x$ | $p = 22$ | $p = 23$ | $p = 24$ | $p = 25$ | $p = 26$ | $p = 27$ |
|---|---|---|---|---|---|---|
| $B(p, q) =$ | $\cdot 4320\ 5711 \times \frac{1}{10^{13}}$ | $\cdot 2160\ 2855^{+} \times \frac{1}{10^{13}}$ | $\cdot 1104\ 1459 \times \frac{1}{10^{13}}$ | $\cdot 5760\ 7614 \times \frac{1}{10^{14}}$ | $\cdot 3064\ 2348 \times \frac{1}{10^{14}}$ | $\cdot 1659\ 7938 \times \frac{1}{10^{14}}$ |
| ·16 | ·0000 001 | | | | | |
| ·17 | ·0000 003 | ·0000 001 | | | | |
| ·18 | ·0000 008 | ·0000 003 | ·0000 001 | | | |
| ·19 | ·0000 022 | ·0000 008 | ·0000 003 | ·0000 001 | | |
| ·20 | ·0000 052 | ·0000 020 | ·0000 007 | ·0000 003 | ·0000 001 | |
| ·21 | ·0000 120 | ·0000 048 | ·0000 018 | ·0000 007 | ·0000 003 | ·0000 001 |
| ·22 | ·0000 260 | ·0000 108 | ·0000 044 | ·0000 018 | ·0000 007 | ·0000 003 |
| ·23 | ·0000 537 | ·0000 233 | ·0000 099 | ·0000 041 | ·0000 017 | ·0000 007 |
| ·24 | ·0001 062 | ·0000 480 | ·0000 213 | ·0000 093 | ·0000 040 | ·0000 017 |
| ·25 | ·0002 015− | ·0000 948 | ·0000 438 | ·0000 199 | ·0000 089 | ·0000 039 |
| ·26 | ·0003 681 | ·0001 799 | ·0000 863 | ·0000 407 | ·0000 189 | ·0000 086 |
| ·27 | ·0006 491 | ·0003 291 | ·0001 638 | ·0000 801 | ·0000 386 | ·0000 183 |
| ·28 | ·0011 072 | ·0005 814 | ·0002 998 | ·0001 519 | ·0000 757− | ·0000 372 |
| ·29 | ·0018 311 | ·0009 945+ | ·0005 305− | ·0002 781 | ·0001 435− | ·0000 729 |
| ·30 | ·0029 409 | ·0016 501 | ·0009 093 | ·0004 926 | ·0002 626 | ·0001 379 |
| ·31 | ·0045 954 | ·0026 604 | ·0015 129 | ·0008 459 | ·0004 654 | ·0002 522 |
| ·32 | ·0069 962 | ·0041 743 | ·0024 469 | ·0014 104 | ·0008 001 | ·0004 471 |
| ·33 | ·0103 922 | ·0063 837 | ·0038 531 | ·0022 872 | ·0013 365− | ·0007 693 |
| ·34 | ·0150 801 | ·0095 270 | ·0059 150+ | ·0036 124 | ·0021 719 | ·0012 865+ |
| ·35 | ·0214 016 | ·0138 918 | ·0088 636 | ·0055 638 | ·0034 387 | ·0020 942 |
| ·36 | ·0297 364 | ·0198 132 | ·0129 792 | ·0083 662 | ·0053 106 | ·0033 220 |
| ·37 | ·0404 901 | ·0276 678 | ·0185 919 | ·0122 954 | ·0080 088 | ·0051 417 |
| ·38 | ·0540 775+ | ·0378 634 | ·0260 765− | ·0176 783 | ·0118 063 | ·0077 727 |
| ·39 | ·0709 007 | ·0508 230 | ·0358 433 | ·0248 893 | ·0170 288 | ·0114 872 |
| ·40 | ·0913 241 | ·0669 638 | ·0483 228 | ·0343 421 | ·0240 523 | ·0166 120 |
| ·41 | ·1156 477 | ·0866 726 | ·0639 460 | ·0464 749 | ·0332 948 | ·0235 263 |
| ·42 | ·1440 793 | ·1102 778 | ·0831 190 | ·0617 314 | ·0452 031 | ·0326 543 |
| ·43 | ·1767 100 | ·1380 209 | ·1061 949 | ·0805 351 | ·0602 332 | ·0444 523 |
| ·44 | ·2134 936 | ·1700 294 | ·1334 439 | ·1032 608 | ·0788 246 | ·0593 885− |
| ·45 | ·2542 330 | ·2062 938 | ·1650 244 | ·1302 034 | ·1013 716 | ·0779 180 |
| ·46 | ·2985 753 | ·2466 510 | ·2009 575+ | ·1615 469 | ·1281 898 | ·1004 521 |
| ·47 | ·3460 164 | ·2907 759 | ·2411 074 | ·1973 370 | ·1594 844 | ·1277 243 |
| ·48 | ·3959 159 | ·3381 839 | ·2851 708 | ·2374 590 | ·1953 200 | ·1587 563 |
| ·49 | ·4475 212 | ·3882 431 | ·3326 764 | ·2816 245− | ·2355 961 | ·1948 255+ |
| ·50 | ·5000 000$^{e}$ | ·4401 979 | ·3829 959 | ·3293 690 | ·2800 323 | ·2354 395+ |
| ·51 | ·5524 788 | ·4932 008 | ·4353 661 | ·3800 616 | ·3281 639 | ·2803 182 |
| ·52 | ·6040 841 | ·5463 520 | ·4889 212 | ·4329 261 | ·3793 501 | ·3289 886 |
| ·53 | ·6539 836 | ·5987 430 | ·5427 338 | ·4870 748 | ·4327 960 | ·3807 928 |
| ·54 | ·7014 247 | ·6495 003 | ·5958 602 | ·5415 496 | ·4875 865+ | ·4349 103 |
| ·55 | ·7457 670 | ·6978 278 | ·6473 875− | ·5953 708 | ·5427 300 | ·4903 929 |
| ·56 | ·7865 064 | ·7430 422 | ·6964 788 | ·6475 873 | ·5972 094 | ·5462 115+ |
| ·57 | ·8232 900 | ·7846 009 | ·7424 129 | ·6973 245− | ·6500 358 | ·6013 102 |
| ·58 | ·8559 207 | ·8221 192 | ·7846 141 | ·7438 275− | ·7002 999 | ·6546 629 |
| ·59 | ·8843 523 | ·8553 772 | ·8226 731 | ·7864 942 | ·7472 184 | ·7053 292 |
| ·60 | ·9086 759 | ·8843 155+ | ·8563 541 | ·8248 975− | ·7901 694 | ·7525 027 |
| ·61 | ·9290 993 | ·9090 216 | ·8855 918 | ·8587 940 | ·8287 161 | ·7955 495− |
| ·62 | ·9459 225− | ·9297 084 | ·9104 771 | ·8881 207 | ·8626 165− | ·8340 322 |
| ·63 | ·9595 099 | ·9466 875+ | ·9312 339 | ·9129 792 | ·8918 184 | ·8677 196 |
| ·64 | ·9702 636 | ·9603 404 | ·9481 910 | ·9336 117 | ·9164 431 | ·8965 804 |
| ·65 | ·9785 984 | ·9710 886 | ·9617 504 | ·9503 694 | ·9367 577 | ·9207 640 |
| ·66 | ·9849 199 | ·9793 667 | ·9723 553 | ·9636 786 | ·9531 417 | ·9405 703 |
| ·67 | ·9896 078 | ·9855 992 | ·9804 613 | ·9740 068 | ·9660 496 | ·9564 123 |
| ·68 | ·9930 038 | ·9901 820 | ·9865 111 | ·9818 308 | ·9759 748 | ·9687 764 |
| ·69 | ·9954 046 | ·9934 697 | ·9909 155+ | ·9876 111 | ·9834 158 | ·9781 829 |
| ·70 | ·9970 591 | ·9957 683 | ·9940 397 | ·9917 710 | ·9888 488 | ·9851 512 |
| ·71 | ·9981 689 | ·9973 324 | ·9961 962 | ·9946 836 | ·9927 076 | ·9901 714 |
| ·72 | ·9988 928 | ·9983 669 | ·9976 427 | ·9966 650− | ·9953 697 | ·9936 838 |
| ·73 | ·9993 509 | ·9990 309 | ·9985 841 | ·9979 724 | ·9971 508 | ·9960 666 |
| ·74 | ·9996 319 | ·9994 437 | ·9991 774 | ·9988 077 | ·9983 045− | ·9976 313 |
| ·75 | ·9997 985+ | ·9996 918 | ·9995 388 | ·9993 236 | ·9990 265+ | ·9986 238 |
| ·76 | ·9998 938 | ·9998 357 | ·9997 511 | ·9996 306 | ·9994 622 | ·9992 307 |
| ·77 | ·9999 463 | ·9999 159 | ·9998 711 | ·9998 065− | ·9997 149 | ·9995 874 |
| ·78 | ·9999 740 | ·9999 588 | ·9999 362 | ·9999 030 | ·9998 555− | ·9997 884 |
| ·79 | ·9999 880 | ·9999 808 | ·9999 699 | ·9999 537 | ·9999 302 | ·9998 967 |
| ·80 | ·9999 948 | ·9999 915+ | ·9999 865+ | ·9999 791 | ·9999 680 | ·9999 521 |
| ·81 | ·9999 978 | ·9999 965− | ·9999 943 | ·9999 911 | ·9999 862 | ·9999 791 |
| ·82 | ·9999 992 | ·9999 986 | ·9999 977 | ·9999 964 | ·9999 944 | ·9999 914 |
| ·83 | ·9999 997 | ·9999 995− | ·9999 992 | ·9999 987 | ·9999 979 | ·9999 967 |
| ·84 | ·9999 999 | ·9999 998 | ·9999 997 | ·9999 995+ | ·9999 993 | ·9999 988 |
| ·85 | 1·0000 000 | ·9999 999 | ·9999 999 | ·9999 999 | ·9999 998 | ·9999 996 |
| ·86 | | 1·0000 000 | 1·0000 000 | 1·0000 000 | ·9999 999 | ·9999 999 |
| ·87 | | | | | 1·0000 000 | 1·0000 000 |

$x$ = ·22 to ·89 $q = 22$ $p$ = 28 to 33

| $x$ | $p = 28$ | $p = 29$ | $p = 30$ | $p = 31$ | $p = 32$ | $p = 33$ |
|---|---|---|---|---|---|---|
| $B(p, q) =$ | ·9145 8028 $\times \frac{1}{10^{15}}$ | ·5121 6496 $\times \frac{1}{10^{15}}$ | ·2912 3105+ $\times \frac{1}{10^{15}}$ | ·1680 1792 $\times \frac{1}{10^{15}}$ | ·9827 4630 $\times \frac{1}{10^{16}}$ | ·5823 6818 $\times \frac{1}{10^{16}}$ |
| ·22 | ·0000 001 | | | | | |
| ·23 | ·0000 003 | ·0000 001 | | | | |
| ·24 | ·0000 007 | ·0000 003 | ·0000 001 | | | |
| ·25 | ·0000 017 | ·0000 007 | ·0000 003 | ·0000 001 | ·0000 001 | |
| ·26 | ·0000 039 | ·0000 017 | ·0000 008 | ·0000 003 | ·0000 001 | ·0000 001 |
| ·27 | ·0000 085+ | ·0000 039 | ·0000 018 | ·0000 008 | ·0000 004 | ·0000 002 |
| ·28 | ·0000 180 | ·0000 086 | ·0000 040 | ·0000 019 | ·0000 009 | ·0000 004 |
| ·29 | ·0000 365− | ·0000 180 | ·0000 088 | ·0000 042 | ·0000 020 | ·0000 009 |
| ·30 | ·0000 713 | ·0000 364 | ·0000 183 | ·0000 091 | ·0000 045− | ·0000 022 |
| ·31 | ·0001 347 | ·0000 710 | ·0000 369 | ·0000 190 | ·0000 096 | ·0000 048 |
| ·32 | ·0002 463 | ·0001 338 | ·0000 718 | ·0000 380 | ·0000 199 | ·0000 103 |
| ·33 | ·0004 366 | ·0002 444 | ·0001 351 | ·0000 737 | ·0000 398 | ·0000 212 |
| ·34 | ·0007 514 | ·0004 330 | ·0002 463 | ·0001 384 | ·0000 769 | ·0000 422 |
| ·35 | ·0012 576 | ·0007 451 | ·0004 359 | ·0002 520 | ·0001 439 | ·0000 813 |
| ·36 | ·0020 494 | ·0012 476 | ·0007 500− | ·0004 454 | ·0002 615− | ·0001 518 |
| ·37 | ·0032 558 | ·0020 347 | ·0012 557 | ·0007 657 | ·0004 616 | ·0002 752 |
| ·38 | ·0050 477 | ·0032 356 | ·0020 484 | ·0012 815− | ·0007 926 | ·0004 849 |
| ·39 | ·0076 449 | ·0050 226 | ·0032 593 | ·0020 903 | ·0013 255− | ·0008 315− |
| ·40 | ·0113 211 | ·0076 174 | ·0050 632 | ·0033 263 | ·0021 609 | ·0013 889 |
| ·41 | ·0164 061 | ·0112 974 | ·0076 860 | ·0051 689 | ·0034 378 | ·0022 623 |
| ·42 | ·0232 846 | ·0163 978 | ·0114 108 | ·0078 501 | ·0053 415+ | ·0035 965+ |
| ·43 | ·0323 887 | ·0233 108 | ·0165 807 | ·0116 609 | ·0081 124 | ·0055 851 |
| ·44 | ·0441 851 | ·0324 784 | ·0235 975+ | ·0169 546 | ·0120 517 | ·0084 788 |
| ·45 | ·0591 551 | ·0443 794 | ·0329 154 | ·0241 455− | ·0175 256 | ·0125 917 |
| ·46 | ·0777 688 | ·0595 087 | ·0450 266 | ·0337 016 | ·0249 631 | ·0183 054 |
| ·47 | ·1004 527 | ·0783 508 | ·0604 407 | ·0461 305− | ·0348 485− | ·0260 662 |
| ·48 | ·1275 552 | ·1013 463 | ·0796 562 | ·0619 571 | ·0477 065− | ·0363 772 |
| ·49 | ·1593 099 | ·1288 552 | ·1031 262 | ·0816 940 | ·0640 781 | ·0497 816 |
| ·50 | ·1958 015− | ·1611 182 | ·1312 188 | ·1058 043 | ·0844 889 | ·0668 371 |
| ·51 | ·2369 381 | ·1982 213 | ·1641 773 | ·1346 611 | ·1094 105− | ·0880 816 |
| ·52 | ·2824 321 | ·2400 658 | ·2020 822 | ·1685 047 | ·1392 164 | ·1139 919 |
| ·53 | ·3317 943 | ·2863 481 | ·2448 197 | ·2074 026 | ·1741 376 | ·1449 371 |
| ·54 | ·3843 411 | ·3365 533 | ·2920 611 | ·2512 173 | ·2142 207 | ·1811 319 |
| ·55 | ·4392 188 | ·3899 638 | ·3432 564 | ·2995 850+ | ·2592 946 | ·2225 926 |
| ·56 | ·4954 403 | ·4456 845+ | ·3976 444 | ·3519 103 | ·3089 496 | ·2691 037 |
| ·57 | ·5519 350− | ·5026 832 | ·4542 805+ | ·4073 784 | ·3625 339 | ·3201 979 |
| ·58 | ·6076 061 | ·5598 435− | ·5120 808 | ·4649 869 | ·4191 690 | ·3751 552 |
| ·59 | ·6613 921 | ·6160 271 | ·5698 799 | ·5235 942 | ·4777 864 | ·4330 235+ |
| ·60 | ·7123 250− | ·6701 383 | ·6264 970 | ·5819 828 | ·5371 814 | ·4926 601 |
| ·61 | ·7595 821 | ·7211 869 | ·6808 057 | ·6389 305− | ·5960 826 | ·5527 929 |
| ·62 | ·8025 259 | ·7683 416 | ·7317 998 | ·6932 847 | ·6532 290 | ·6120 968 |
| ·63 | ·8407 288 | ·8109 715+ | ·7786 490 | ·7440 315− | ·7074 486 | ·6692 767 |
| ·64 | ·8739 810 | ·8486 697 | ·8207 400 | ·7903 525− | ·7577 300 | ·7231 502 |
| ·65 | ·9022 824 | ·8812 596 | ·8576 995+ | ·8316 657 | ·8032 803 | ·7727 218 |
| ·66 | ·9258 199 | ·9087 831 | ·8893 965− | ·8676 447 | ·8435 633 | ·8172 394 |
| ·67 | ·9449 332 | ·9314 739 | ·9159 262 | ·8982 172 | ·8783 147 | ·8562 291 |
| ·68 | ·9600 743 | ·9497 189 | ·9375 780 | ·9235 431 | ·9075 344 | ·8895 045− |
| ·69 | ·9717 640 | ·9640 131 | ·9547 922 | ·9439 761 | ·9314 573 | ·9171 507 |
| ·70 | ·9805 498 | ·9749 130 | ·9681 099 | ·9600 143 | ·9505 085− | ·9394 877 |
| ·71 | ·9869 702 | ·9829 927 | ·9781 237 | ·9722 468 | ·9652 476 | ·9570 170 |
| ·72 | ·9915 259 | ·9888 069 | ·9854 316 | ·9813 002 | ·9763 106 | ·9703 605+ |
| ·73 | ·9946 596 | ·9928 621 | ·9905 998 | ·9877 922 | ·9843 543 | ·9801 977 |
| ·74 | ·9967 457 | ·9955 988 | ·9941 355+ | ·9922 947 | ·9900 097 | ·9872 092 |
| ·75 | ·9980 869 | ·9973 822 | ·9964 709 | ·9953 090 | ·9938 473 | ·9920 316 |
| ·76 | ·9989 179 | ·9985 020 | ·9979 569 | ·9972 527 | ·9963 550− | ·9952 250− |
| ·77 | ·9994 129 | ·9991 777 | ·9988 655+ | ·9984 568 | ·9979 290 | ·9972 558 |
| ·78 | ·9996 954 | ·9995 685+ | ·9993 978 | ·9991 715− | ·9988 753 | ·9984 927 |
| ·79 | ·9998 495+ | ·9997 844 | ·9996 956 | ·9995 764 | ·9994 184 | ·9992 118 |
| ·80 | ·9999 295+ | ·9998 978 | ·9998 541 | ·9997 947 | ·9997 150e | ·9996 093 |
| ·81 | ·9999 688 | ·9999 543 | ·9999 341 | ·9999 062 | ·9998 682 | ·9998 174 |
| ·82 | ·9999 871 | ·9999 809 | ·9999 721 | ·9999 598 | ·9999 429 | ·9999 200 |
| ·83 | ·9999 950+ | ·9999 925+ | ·9999 890 | ·9999 839 | ·9999 770 | ·9999 673 |
| ·84 | ·9999 982 | ·9999 973 | ·9999 960 | ·9999 941 | ·9999 914 | ·9999 877 |
| ·85 | ·9999 994 | ·9999 991 | ·9999 987 | ·9999 980 | ·9999 971 | ·9999 958 |
| ·86 | ·9999 998 | ·9999 997 | ·9999 996 | ·9999 994 | ·9999 991 | ·9999 987 |
| ·87 | 1·0000 000 | ·9999 999 | ·9999 999 | ·9999 998 | ·9999 998 | ·9999 996 |
| ·88 | | 1·0000 000 | 1·0000 000 | 1·0000 000 | ·9999 999 | ·9999 999 |
| ·89 | | | | | 1·0000 000 | 1·0000 000 |

$x$ = ·27 to ·90 $q$ = 22 $p$ = 34 to 39

| $x$ | $p$ = 34 | $p$ = 35 | $p$ = 36 | $p$ = 37 | $p$ = 38 | $p$ = 39 |
|---|---|---|---|---|---|---|
| $B(p, q)$ = | ·3494 2091 $\times \frac{1}{10^{16}}$ | ·2121 4841 $\times \frac{1}{10^{16}}$ | ·1302 6657 $\times \frac{1}{10^{16}}$ | ·8085 5110 $\times \frac{1}{10^{17}}$ | ·5070 5747 $\times \frac{1}{10^{17}}$ | ·3211 3640 $\times \frac{1}{10^{17}}$ |
| ·27 | ·0000 001 | | | | | |
| ·28 | ·0000 002 | ·0000 001 | | | | |
| ·29 | ·0000 004 | ·0000 002 | ·0000 001 | | | |
| ·30 | ·0000 010 | ·0000 005⁻ | ·0000 002 | ·0000 001 | ·0000 001 | |
| ·31 | ·0000 024 | ·0000 012 | ·0000 006 | ·0000 003 | ·0000 001 | ·0000 001 |
| ·32 | ·0000 053 | ·0000 027 | ·0000 013 | ·0000 007 | ·0000 003 | ·0000 002 |
| ·33 | ·0000 112 | ·0000 059 | ·0000 030 | ·0000 016 | ·0000 008 | ·0000 004 |
| ·34 | ·0000 229 | ·0000 123 | ·0000 066 | ·0000 035⁻ | ·0000 018 | ·0000 009 |
| ·35 | ·0000 455⁻ | ·0000 252 | ·0000 138 | ·0000 075⁻ | ·0000 040 | ·0000 021 |
| ·36 | ·0000 872 | ·0000 496 | ·0000 279 | ·0000 156 | ·0000 086 | ·0000 047 |
| ·37 | ·0001 624 | ·0000 949 | ·0000 549 | ·0000 315⁻ | ·0000 179 | ·0000 101 |
| ·38 | ·0002 936 | ·0001 760 | ·0001 045⁻ | ·0000 615⁻ | ·0000 358 | ·0000 207 |
| ·39 | ·0005 162 | ·0003 173 | ·0001 932 | ·0001 166 | ·0000 697 | ·0000 413 |
| ·40 | ·0008 835⁺ | ·0005 565⁺ | ·0003 472 | ·0002 147 | ·0001 316 | ·0000 800 |
| ·41 | ·0014 736 | ·0009 505⁺ | ·0006 074 | ·0003 846 | ·0002 414 | ·0001 503 |
| ·42 | ·0023 972 | ·0015 824 | ·0010 348 | ·0006 707 | ·0004 309 | ·0002 746 |
| ·43 | ·0038 069 | ·0025 699 | ·0017 189 | ·0011 395⁺ | ·0007 490 | ·0004 882 |
| ·44 | ·0059 062 | ·0040 752 | ·0027 861 | ·0018 881 | ·0012 687 | ·0008 455⁺ |
| ·45 | ·0089 586 | ·0063 139 | ·0044 097 | ·0030 530 | ·0020 960 | ·0014 273 |
| ·46 | ·0132 940 | ·0095 649 | ·0068 203 | ·0048 213 | ·0033 799 | ·0023 504 |
| ·47 | ·0193 118 | ·0141 764 | ·0103 147 | ·0074 409 | ·0053 236 | ·0037 785⁺ |
| ·48 | ·0274 786 | ·0205 691 | ·0152 626 | ·0112 296 | ·0081 951 | ·0059 336 |
| ·49 | ·0383 185⁺ | ·0292 323 | ·0221 087 | ·0165 820 | ·0123 369 | ·0091 072 |
| ·50 | ·0523 947 | ·0407 134 | ·0313 683 | ·0239 702 | ·0181 716 | ·0136 701 |
| ·51 | ·0702 817 | ·0555 968 | ·0436 139 | ·0339 377 | ·0262 019 | ·0200 765⁻ |
| ·52 | ·0925 281 | ·0744 733 | ·0594 517 | ·0470 839 | ·0370 025⁻ | ·0288 631 |
| ·53 | ·1196 124 | ·0979 002 | ·0794 882 | ·0640 375⁻ | ·0512 008 | ·0406 377 |
| ·54 | ·1518 934 | ·1263 527 | ·1042 856 | ·0854 182 | ·0694 472 | ·0560 568 |
| ·55 | ·1895 608 | ·1601 722 | ·1343 102 | ·1117 887 | ·0923 716 | ·0757 904 |
| ·56 | ·2325 903 | ·1995 134 | ·1698 766 | ·1435 986 | ·1205 308 | ·1004 739 |
| ·57 | ·2807 099 | ·2442 997 | ·2110 936 | ·1811 250⁺ | ·1543 477 | ·1306 499 |
| ·58 | ·3333 821 | ·2941 891 | ·2578 181 | ·2244 173 | ·1940 496 | ·1667 028 |
| ·59 | ·3898 070 | ·3485 607 | ·3096 241 | ·2732 509 | ·2396 106 | ·2087 943 |
| ·60 | ·4489 482 | ·4065 219 | ·3657 927 | ·3271 000 | ·2907 079 | ·2568 058 |
| ·61 | ·5095 819 | ·4669 428 | ·4253 271 | ·3851 332 | ·3466 991 | ·3102 981 |
| ·62 | ·5703 664 | ·5285 132 | ·4869 948 | ·4462 376 | ·4066 260 | ·3684 946 |
| ·63 | ·6299 249 | ·5898 208 | ·5493 959 | ·5090 720 | ·4692 494 | ·4302 967 |
| ·64 | ·6869 357 | ·6494 430 | ·6110 505⁺ | ·5721 461 | ·5331 156 | ·4943 315⁺ |
| ·65 | ·7402 186 | ·7060 424 | ·6704 992 | ·6339 193 | ·5966 474 | ·5590 321 |
| ·66 | ·7888 095⁺ | ·7584 565⁺ | ·7264 037 | ·6929 085⁺ | ·6582 546 | ·6227 434 |
| ·67 | ·8320 153 | ·8057 718 | ·7776 387 | ·7477 942 | ·7164 494 | ·6838 426 |
| ·68 | ·8694 421 | ·8473 735⁻ | ·8233 629 | ·7975 114 | ·7699 552 | ·7408 616 |
| ·69 | ·9009 972 | ·8829 671 | ·8630 618 | ·8413 154 | ·8177 939 | ·7925 950⁻ |
| ·70 | ·9268 638 | ·9125 692 | ·8965 592 | ·8788 148 | ·8593 439 | ·8381 821 |
| ·71 | ·9474 546 | ·9364 718 | ·9239 954 | ·9099 698 | ·8943 597 | ·8771 516 |
| ·72 | ·9633 502 | ·9551 852 | ·9457 792 | ·9350 564 | ·9229 540 | ·9094 248 |
| ·73 | ·9752 324 | ·9693 689 | ·9625 204 | ·9546 047 | ·9455 466 | ·9352 799 |
| ·74 | ·9838 181 | ·9797 586 | ·9749 523 | ·9693 208 | ·9627 884 | ·9552 829 |
| ·75 | ·9898 032 | ·9870 997 | ·9838 554 | ·9800 028 | ·9754 735⁻ | ·9701 991 |
| ·76 | ·9938 196 | ·9920 919 | ·9899 909 | ·9874 628 | ·9844 509 | ·9808 968 |
| ·77 | ·9964 076 | ·9953 511 | ·9940 495⁺ | ·9924 627 | ·9905 473 | ·9882 573 |
| ·78 | ·9980 044 | ·9973 882 | ·9966 193 | ·9956 696 | ·9945 085⁻ | ·9931 022 |
| ·79 | ·9989 446 | ·9986 031 | ·9981 716 | ·9976 317 | ·9969 632 | ·9961 432 |
| ·80 | ·9997 709 | ·9992 919 | ·9990 628 | ·9987 725⁺ | ·9984 086 | ·9979 565⁻ |
| ·81 | ·9997 499 | ·9996 616 | ·9995 471 | ·9994 002 | ·9992 137 | ·9989 792 |
| ·82 | ·9998 892 | ·9998 484 | ·9997 948 | ·9997 253 | ·9996 359 | ·9995 221 |
| ·83 | ·9999 543 | ·9999 368 | ·9999 135⁻ | ·9998 829 | ·9998 431 | ·9997 918 |
| ·84 | ·9999 826 | ·9999 757 | ·9999 663 | ·9999 539 | ·9999 376 | ·9999 163 |
| ·85 | ·9999 939 | ·9999 914 | ·9999 880 | ·9999 834 | ·9999 773 | ·9999 692 |
| ·86 | ·9999 981 | ·9999 973 | ·9999 961 | ·9999 946 | ·9999 925⁺ | ·9999 898 |
| ·87 | ·9999 995⁻ | ·9999 992 | ·9999 989 | ·9999 984 | ·9999 978 | ·9999 970 |
| ·88 | ·9999 999 | ·9999 998 | ·9999 997 | ·9999 996 | ·9999 994 | ·9999 992 |
| ·89 | 1·0000 000 | 1·0000 000 | ·9999 999 | ·9999 999 | ·9999 999 | ·9999 998 |
| ·90 | | | 1·0000 000 | 1·0000 000 | 1·0000 000 | 1·0000 000 |

$x$ = ·32 to ·91 $q = 22$ $p$ = 40 to 45

| $x$ | $p = 40$ | $p = 41$ | $p = 42$ | $p = 43$ | $p = 44$ | $p = 45$ |
|---|---|---|---|---|---|---|
| $B(p, q) =$ | ·2053 1671 $\times \frac{1}{10^{17}}$ | ·1324 6240 $\times \frac{1}{10^{17}}$ | ·8620 5686 $\times \frac{1}{10^{18}}$ | ·5657 2482 $\times \frac{1}{10^{18}}$ | ·3742 4872 $\times \frac{1}{10^{18}}$ | ·2494 9915 $\times \frac{1}{10^{18}}$ |
| ·32 | ·0000 001 | | | | | |
| ·33 | ·0000 002 | ·0000 001 | ·0000 001 | | | |
| ·34 | ·0000 005$^-$ | ·0000 002 | ·0000 001 | ·0000 001 | | |
| ·35 | ·0000 011 | ·0000 006 | ·0000 003 | ·0000 002 | ·0000 001 | |
| ·36 | ·0000 026 | ·0000 014 | ·0000 007 | ·0000 004 | ·0000 002 | ·0000 001 |
| ·37 | ·0000 056 | ·0000 031 | ·0000 017 | ·0000 009 | ·0000 005$^+$ | ·0000 003 |
| ·38 | ·0000 119 | ·0000 068 | ·0000 038 | ·0000 021 | ·0000 012 | ·0000 007 |
| ·39 | ·0000 243 | ·0000 142 | ·0000 082 | ·0000 047 | ·0000 027 | ·0000 015$^+$ |
| ·40 | ·0000 482 | ·0000 288 | ·0000 171 | ·0000 101 | ·0000 059 | ·0000 034 |
| ·41 | ·0000 928 | ·0000 569 | ·0000 346 | ·0000 209 | ·0000 125$^+$ | ·0000 075$^-$ |
| ·42 | ·0001 736 | ·0001 089 | ·0000 678 | ·0000 419 | ·0000 257 | ·0000 157 |
| ·43 | ·0003 158 | ·0002 026 | ·0001 291 | ·0000 817 | ·0000 513 | ·0000 320 |
| ·44 | ·0005 591 | ·0003 669 | ·0002 390 | ·0001 546 | ·0000 993 | ·0000 634 |
| ·45 | ·0009 644 | ·0006 467 | ·0004 305$^-$ | ·0002 846 | ·0001 868 | ·0001 219 |
| ·46 | ·0016 219 | ·0011 108 | ·0007 553 | ·0005 099 | ·0003 420 | ·0002 279 |
| ·47 | ·0026 613 | ·0018 606 | ·0012 914 | ·0008 902 | ·0006 095$^+$ | ·0004 146 |
| ·48 | ·0042 635$^+$ | ·0030 411 | ·0021 537 | ·0015 148 | ·0010 584 | ·0007 348 |
| ·49 | ·0066 726 | ·0048 533 | ·0035 052 | ·0025 144 | ·0017 918 | ·0012 688 |
| ·50 | ·0102 074 | ·0075 670 | ·0055 707 | ·0040 734 | ·0029 592 | ·0021 362 |
| ·51 | ·0152 703 | ·0115 323 | ·0086 495$^+$ | ·0064 442 | ·0047 702 | ·0035 090 |
| ·52 | ·0223 515$^+$ | ·0171 879 | ·0131 275$^-$ | ·0099 604 | ·0075 092 | ·0056 262 |
| ·53 | ·0320 246 | ·0250 631 | ·0194 837 | ·0150 481 | ·0115 491 | ·0088 095$^+$ |
| ·54 | ·0449 324 | ·0357 715$^+$ | ·0282 908 | ·0222 315$^+$ | ·0173 615$^-$ | ·0134 766 |
| ·55 | ·0617 601 | ·0499 922 | ·0402 048 | ·0321 301 | ·0255 202 | ·0201 496 |
| ·56 | ·0831 941 | ·0684 372 | ·0559 406 | ·0454 435$^-$ | ·0366 942 | ·0294 562 |
| ·57 | ·1098 686 | ·0918 045$^+$ | ·0762 342 | ·0629 215$^+$ | ·0516 274 | ·0421 173 |
| ·58 | ·1423 010 | ·1207 176 | ·1017 874 | ·0853 181 | ·0711 008 | ·0589 193 |
| ·59 | ·1808 226 | ·1556 550$^+$ | ·1332 006 | ·1133 285$^+$ | ·0958 780 | ·0806 683 |
| ·60 | ·2255 116 | ·1968 774 | ·1708 970 | ·1475 147 | ·1266 338 | ·1081 257 |
| ·61 | ·2761 370 | ·2443 588 | ·2150 452 | ·1882 234 | ·1638 716 | ·1419 273 |
| ·62 | ·3321 231 | ·2977 339 | ·2654 919 | ·2355 068 | ·2078 369 | ·1824 938 |
| ·63 | ·3925 425$^-$ | ·3562 701 | ·3217 141 | ·2890 586 | ·2584 385$^-$ | ·2299 408 |
| ·64 | ·4561 441 | ·4188 732 | ·3828 023 | ·3481 742 | ·3151 889 | ·2840 028 |
| ·65 | ·5214 169 | ·4841 308 | ·4474 813 | ·4117 480 | ·3771 782 | ·3439 832 |
| ·66 | ·5866 859 | ·5503 941 | ·5141 730 | ·4783 142 | ·4430 891 | ·4087 447 |
| ·67 | ·6502 325$^+$ | ·6158 914 | ·5810 980 | ·5461 306 | ·5112 608 | ·4767 476 |
| ·68 | ·7104 252 | ·6788 627 | ·6464 071 | ·6133 025$^+$ | ·5797 976 | ·5461 403 |
| ·69 | ·7658 453 | ·7376 981 | ·7083 288 | ·6779 315$^+$ | ·6467 143 | ·6148 940 |
| ·70 | ·8153 925$^-$ | ·7910 646 | ·7653 126 | ·7382 730 | ·7101 015$^+$ | ·6809 697 |
| ·71 | ·8583 551 | ·8380 032 | ·8161 521 | ·7928 808 | ·7682 889 | ·7424 954 |
| ·72 | ·8944 387 | ·8779 839 | ·8600 683 | ·8407 194 | ·8199 845$^+$ | ·7979 302 |
| ·73 | ·9237 496 | ·9109 135$^-$ | ·8967 437 | ·8812 278 | ·8643 695$^+$ | ·8461 894 |
| ·74 | ·9467 382 | ·9370 955$^-$ | ·9263 051 | ·9143 277 | ·9011 359 | ·8867 149 |
| ·75 | ·9641 134 | ·9571 528 | ·9492 584 | ·9403 773 | ·9304 635$^-$ | ·9194 794 |
| ·76 | ·9767 413 | ·9719 251 | ·9663 900 | ·9600 799 | ·9529 422 | ·9449 285$^+$ |
| ·77 | ·9855 447 | ·9823 593 | ·9786 503 | ·9743 664 | ·9694 569 | ·9638 723 |
| ·78 | ·9914 148 | ·9894 075$^+$ | ·9870 400 | ·9842 699 | ·9810 541 | ·9773 486 |
| ·79 | ·9951 466 | ·9939 459 | ·9925 115$^+$ | ·9908 118 | ·9888 133 | ·9864 809 |
| ·80 | ·9974 000 | ·9967 212 | ·9958 999 | ·9949 144 | ·9937 410 | ·9923 542 |
| ·81 | ·9986 870 | ·9983 260 | ·9978 838 | ·9973 466 | ·9966 989 | ·9959 239 |
| ·82 | ·9993 786 | ·9991 991 | ·9989 765$^+$ | ·9987 027 | ·9983 686 | ·9979 639 |
| ·83 | ·9997 263 | ·9996 434 | ·9995 394 | ·9994 099 | ·9992 499 | ·9990 536 |
| ·84 | ·9998 888 | ·9998 535$^-$ | ·9998 087 | ·9997 523 | ·9996 817 | ·9995 941 |
| ·85 | ·9999 587 | ·9999 450$^-$ | ·9999 274 | ·9999 050$^-$ | ·9998 766 | ·9998 410 |
| ·86 | ·9999 861 | ·9999 813 | ·9999 751 | ·9999 671 | ·9999 568 | ·9999 437 |
| ·87 | ·9999 959 | ·9999 944 | ·9999 924 | ·9999 899 | ·9999 865$^+$ | ·9999 823 |
| ·88 | ·9999 989 | ·9999 985$^+$ | ·9999 980 | ·9999 973 | ·9999 963 | ·9999 951 |
| ·89 | ·9999 998 | ·9999 997 | ·9999 995$^+$ | ·9999 994 | ·9999 991 | ·9999 989 |
| ·90 | 1·0000 000 | ·9999 999 | ·9999 999 | ·9999 999 | ·9999 998 | ·9999 998 |
| ·91 | | 1·0000 000 | 1·0000 000 | 1·0000 000 | 1·0000 000 | 1·0000 000 |

$x$ = ·36 to ·92 $q = 22$ $p$ = 46 to 50

| $x$ | $p = 46$ | $p = 47$ | $p = 48$ | $p = 49$ | $p = 50$ |
|---|---|---|---|---|---|
| $B(p, q) =$ | ·1675 7406 $\times \frac{1}{10^{18}}$ | ·1133 5892 $\times \frac{1}{10^{18}}$ | ·7721 5496 $\times \frac{1}{10^{19}}$ | ·5294 7769 $\times \frac{1}{10^{19}}$ | ·3654 1418 $\times \frac{1}{10^{19}}$ |
| ·36 | ·0000 001 | | | | |
| ·37 | ·0000 001 | ·0000 001 | | | |
| ·38 | ·0000 004 | ·0000 002 | ·0000 001 | ·0000 001 | |
| ·39 | ·0000 009 | ·0000 005− | ·0000 003 | ·0000 001 | ·0000 001 |
| ·40 | ·0000 020 | ·0000 011 | ·0000 006 | ·0000 004 | ·0000 002 |
| ·41 | ·0000 044 | ·0000 026 | ·0000 015+ | ·0000 009 | ·0000 005+ |
| ·42 | ·0000 095+ | ·0000 057 | ·0000 034 | ·0000 020 | ·0000 012 |
| ·43 | ·0000 198 | ·0000 122 | ·0000 075− | ·0000 046 | ·0000 028 |
| ·44 | ·0000 402 | ·0000 253 | ·0000 159 | ·0000 099 | ·0000 061 |
| ·45 | ·0000 790 | ·0000 509 | ·0000 326 | ·0000 207 | ·0000 131 |
| ·46 | ·0001 509 | ·0000 993 | ·0000 649 | ·0000 422 | ·0000 273 |
| ·47 | ·0002 803 | ·0001 884 | ·0001 258 | ·0000 836 | ·0000 552 |
| ·48 | ·0005 069 | ·0003 476 | ·0002 370 | ·0001 607 | ·0001 083 |
| ·49 | ·0008 928 | ·0006 246 | ·0004 344 | ·0003 004 | ·0002 066 |
| ·50 | ·0015 327 | ·0010 931 | ·0007 752 | ·0005 466 | ·0003 834 |
| ·51 | ·0025 656 | ·0018 648 | ·0013 477 | ·0009 687 | ·0006 925− |
| ·52 | ·0041 901 | ·0031 025− | ·0022 842 | ·0016 725+ | ·0012 181 |
| ·53 | ·0066 800 | ·0050 361 | ·0037 755+ | ·0028 151 | ·0020 880 |
| ·54 | ·0103 997 | ·0079 796 | ·0060 889 | ·0046 213 | ·0034 891 |
| ·55 | ·0158 173 | ·0123 468 | ·0095 852 | ·0074 018 | ·0056 862 |
| ·56 | ·0235 114 | ·0186 625+ | ·0147 339 | ·0115 713 | ·0090 413 |
| ·57 | ·0341 668 | ·0275 662 | ·0221 228 | ·0176 626 | ·0140 307 |
| ·58 | ·0485 568 | ·0398 028 | ·0324 569 | ·0263 322 | ·0212 575− |
| ·59 | ·0675 069 | ·0561 966 | ·0465 420 | ·0383 537 | ·0314 521 |
| ·60 | ·0918 386 | ·0776 051 | ·0652 491 | ·0545 922 | ·0454 576 |
| ·61 | ·1222 945− | ·1048 512 | ·0894 565+ | ·0759 573 | ·0641 938 |
| ·62 | ·1594 484 | ·1386 374 | ·1199 695+ | ·1033 318 | ·0885 955+ |
| ·63 | ·2036 090 | ·1794 467 | ·1574 230 | ·1374 778 | ·1195 271 |
| ·64 | ·2547 295− | ·2274 416 | ·2021 743 | ·1789 283 | ·1576 748 |
| ·65 | ·3123 374 | ·2823 771 | ·2542 016 | ·2278 752 | ·2034 292 |
| ·66 | ·3754 993 | ·3435 403 | ·3130 229 | ·2840 695+ | ·2567 706 |
| ·67 | ·4428 327 | ·4097 361 | ·3776 535+ | ·3467 540 | ·3171 788 |
| ·68 | ·5125 728 | ·4793 264 | ·4466 176 | ·4146 447 | ·3835 853 |
| ·69 | ·5826 919 | ·5503 287 | ·5180 206 | ·4859 751 | ·4543 873 |
| ·70 | ·6510 609 | ·6205 670 | ·5896 839 | ·5586 077 | ·5275 315− |
| ·71 | ·7156 358 | ·6878 594 | ·6593 266 | ·6302 053 | ·6006 680 |
| ·72 | ·7746 409 | ·7502 174 | ·7247 755− | ·6984 430 | ·6713 583 |
| ·73 | ·8267 246 | ·8060 285− | ·7841 698 | ·7612 319 | ·7373 109 |
| ·74 | ·8710 633 | ·8541 936 | ·8361 322 | ·8169 194 | ·7966 087 |
| ·75 | ·9073 970 | ·8941 982 | ·8798 761 | ·8644 351 | ·8478 912 |
| ·76 | ·9359 959 | ·9261 079 | ·9152 354 | ·9033 571 | ·8904 606 |
| ·77 | ·9575 655− | ·9504 922 | ·9426 122 | ·9338 901 | ·9242 958 |
| ·78 | ·9731 095− | ·9682 935− | ·9628 586 | ·9567 647 | ·9499 744 |
| ·79 | ·9837 784 | ·9806 688 | ·9771 146 | ·9730 783 | ·9685 231 |
| ·80 | ·9907 271 | ·9888 311 | ·9866 366 | ·9841 129 | ·9812 287 |
| ·81 | ·9950 032 | ·9939 169 | ·9926 439 | ·9911 617 | ·9894 466 |
| ·82 | ·9974 771 | ·9968 957 | ·9962 059 | ·9953 929 | ·9944 405− |
| ·83 | ·9988 148 | ·9985 261 | ·9981 793 | ·9977 656 | ·9972 751 |
| ·84 | ·9994 863 | ·9993 543 | ·9991 939 | ·9990 002 | ·9987 678 |
| ·85 | ·9997 965+ | ·9997 415+ | ·9996 739 | ·9995 913 | ·9994 909 |
| ·86 | ·9999 273 | ·9999 066 | ·9998 809 | ·9998 492 | ·9998 102 |
| ·87 | ·9999 769 | ·9999 700 | ·9999 613 | ·9999 505− | ·9999 370 |
| ·88 | ·9999 936 | ·9999 916 | ·9999 890 | ·9999 858 | ·9999 817 |
| ·89 | ·9999 985− | ·9999 980 | ·9999 973 | ·9999 965+ | ·9999 955− |
| ·90 | ·9999 997 | ·9999 996 | ·9999 995− | ·9999 993 | ·9999 991 |
| ·91 | 1·0000 000 | ·9999 999 | ·9999 999 | ·9999 999 | ·9999 998 |
| ·92 | | 1·0000 000 | 1·0000 000 | 1·0000 000 | 1·0000 000 |

$x$ = ·16 to ·87 $q = 23$ $p$ = 23 to 28

| $x$ | $p = 23$ | $p = 24$ | $p = 25$ | $p = 26$ | $p = 27$ | $p = 28$ |
|---|---|---|---|---|---|---|
| $B(p, q)$ = | ·1056 1396 × $\frac{1}{10^{13}}$ | ·5280 6980 × $\frac{1}{10^{14}}$ | ·2696 5266 × $\frac{1}{10^{14}}$ | ·1404 4409 × $\frac{1}{10^{14}}$ | ·7452 1356 × $\frac{1}{10^{15}}$ | ·4024 1532 × $\frac{1}{10^{15}}$ |
| ·16 | ·0000 001 | | | | | |
| ·17 | ·0000 002 | ·0000 001 | | | | |
| ·18 | ·0000 005− | ·0000 002 | ·0000 001 | | | |
| ·19 | ·0000 013 | ·0000 005− | ·0000 002 | ·0000 001 | | |
| ·20 | ·0000 033 | ·0000 012 | ·0000 005− | ·0000 002 | ·0000 001 | |
| ·21 | ·0000 078 | ·0000 031 | ·0000 012 | ·0000 005− | ·0000 002 | ·0000 001 |
| ·22 | ·0000 175− | ·0000 073 | ·0000 030 | ·0000 012 | ·0000 005− | ·0000 002 |
| ·23 | ·0000 373 | ·0000 162 | ·0000 069 | ·0000 029 | ·0000 012 | ·0000 005− |
| ·24 | ·0000 759 | ·0000 344 | ·0000 153 | ·0000 067 | ·0000 029 | ·0000 012 |
| ·25 | ·0001 481 | ·0000 699 | ·0000 324 | ·0000 147 | ·0000 066 | ·0000 029 |
| ·26 | ·0002 778 | ·0001 361 | ·0000 655+ | ·0000 310 | ·0000 144 | ·0000 066 |
| ·27 | ·0005 019 | ·0002 551 | ·0001 274 | ·0000 625+ | ·0000 302 | ·0000 144 |
| ·28 | ·0008 759 | ·0004 611 | ·0002 385− | ·0001 213 | ·0000 607 | ·0000 300 |
| ·29 | ·0014 797 | ·0008 057 | ·0004 311 | ·0002 269 | ·0001 175+ | ·0000 600 |
| ·30 | ·0024 246 | ·0013 639 | ·0007 540 | ·0004 100 | ·0002 195+ | ·0001 158 |
| ·31 | ·0038 601 | ·0022 405+ | ·0012 782 | ·0007 174 | ·0003 964 | ·0002 159 |
| ·32 | ·0059 803 | ·0035 776 | ·0021 039 | ·0012 174 | ·0006 936 | ·0003 894 |
| ·33 | ·0090 293 | ·0055 612 | ·0033 676 | ·0020 068 | ·0011 777 | ·0006 811 |
| ·34 | ·0133 031 | ·0084 270 | ·0052 494 | ·0032 182 | ·0019 433 | ·0011 566 |
| ·35 | ·0191 487 | ·0124 634 | ·0079 786 | ·0050 277 | ·0031 209 | ·0019 097 |
| ·36 | ·0269 579 | ·0180 115+ | ·0118 385+ | ·0076 606 | ·0048 839 | ·0030 698 |
| ·37 | ·0371 563 | ·0254 607 | ·0171 666 | ·0113 972 | ·0074 562 | ·0048 100 |
| ·38 | ·0501 861 | ·0352 382 | ·0243 511 | ·0165 733 | ·0111 169 | ·0073 540 |
| ·39 | ·0664 836 | ·0477 930 | ·0338 218 | ·0235 781 | ·0162 026 | ·0109 824 |
| ·40 | ·0864 520 | ·0635 745+ | ·0460 351 | ·0328 454 | ·0231 053 | ·0160 348 |
| ·41 | ·1104 322 | ·0830 053 | ·0614 524 | ·0448 394 | ·0322 646 | ·0229 081 |
| ·42 | ·1386 711 | ·1064 509 | ·0805 136 | ·0600 335+ | ·0441 536 | ·0320 496 |
| ·43 | ·1712 935+ | ·1341 873 | ·1036 056 | ·0788 833 | ·0592 576 | ·0439 423 |
| ·44 | ·2082 779 | ·1663 708 | ·1310 292 | ·1017 947 | ·0780 472 | ·0590 844 |
| ·45 | ·2494 391 | ·2030 110 | ·1629 668 | ·1290 894 | ·1009 451 | ·0779 606 |
| ·46 | ·2944 214 | ·2439 509 | ·1994 527 | ·1609 708 | ·1282 907 | ·1010 088 |
| ·47 | ·3427 020 | ·2888 569 | ·2403 514 | ·1974 919 | ·1603 031 | ·1285 824 |
| ·48 | ·3936 066 | ·3372 200 | ·2853 443 | ·2385 316 | ·1970 484 | ·1609 120 |
| ·49 | ·4463 356 | ·3883 694 | ·3339 295+ | ·2837 795− | ·2384 130 | ·1980 704 |
| ·50 | ·5000 000e | ·4414 980 | ·3854 335− | ·3327 329 | ·2840 862 | ·2399 438 |
| ·51 | ·5536 644 | ·4956 982 | ·4390 363 | ·3847 088 | ·3335 574 | ·2862 139 |
| ·52 | ·6063 934 | ·5500 067 | ·4938 081 | ·4388 682 | ·3861 260 | ·3363 529 |
| ·53 | ·6572 980 | ·6034 528 | ·5487 551 | ·4942 544 | ·4409 274 | ·3896 348 |
| ·54 | ·7055 786 | ·6551 082 | ·6028 712 | ·5498 403 | ·4969 725− | ·4451 621 |
| ·55 | ·7505 609 | ·7041 329 | ·6551 900 | ·6045 830 | ·5531 974 | ·5019 071 |
| ·56 | ·7917 221 | ·7498 151 | ·7048 348 | ·6574 797 | ·6085 217 | ·5587 659 |
| ·57 | ·8287 065− | ·7916 002 | ·7510 617 | ·7076 205+ | ·6619 071 | ·6146 191 |
| ·58 | ·8613 289 | ·8291 087 | ·7932 906 | ·7542 345− | ·7124 144 | ·6683 949 |
| ·59 | ·8895 678 | ·8621 410 | ·8311 258 | ·7967 237 | ·7592 520 | ·7191 294 |
| ·60 | ·9135 480 | ·8906 704 | ·8643 612 | ·8346 845− | ·8018 118 | ·7660 170 |
| ·61 | ·9335 164 | ·9148 258 | ·8929 734 | ·8679 131 | ·8396 913 | ·8084 487 |
| ·62 | ·9498 139 | ·9348 659 | ·9171 027 | ·8963 979 | ·8726 990 | ·8460 332 |
| ·63 | ·9628 437 | ·9511 480 | ·9370 256 | ·9202 989 | ·9008 446 | ·8786 017 |
| ·64 | ·9730 421 | ·9640 957 | ·9531 215− | ·9399 173 | ·9243 160 | ·9061 954 |
| ·65 | ·9808 513 | ·9741 660 | ·9658 372 | ·9556 594 | ·9434 460 | ·9290 387 |
| ·66 | ·9866 969 | ·9818 207 | ·9756 524 | ·9679 988 | ·9586 731 | ·9475 030 |
| ·67 | ·9909 707 | ·9875 026 | ·9830 490 | ·9774 392 | ·9705 003 | ·9620 632 |
| ·68 | ·9940 197 | ·9916 169 | ·9884 854 | ·9844 820 | ·9794 562 | ·9732 540 |
| ·69 | ·9961 399 | ·9945 204 | ·9923 785− | ·9896 000 | ·9860 607 | ·9816 287 |
| ·70 | ·9975 754 | ·9965 147 | ·9950 916 | ·9932 188 | ·9907 985− | ·9877 239 |
| ·71 | ·9985 203 | ·9978 463 | ·9969 291 | ·9957 049 | ·9941 002 | ·9920 325− |
| ·72 | ·9991 241 | ·9987 093 | ·9981 369 | ·9973 621 | ·9963 322 | ·9949 864 |
| ·73 | ·9994 981 | ·9992 513 | ·9989 060 | ·9984 321 | ·9977 934 | ·9969 473 |
| ·74 | ·9997 222 | ·9995 806 | ·9993 796 | ·9991 001 | ·9987 182 | ·9982 053 |
| ·75 | ·9998 519 | ·9997 736 | ·9996 611 | ·9995 024 | ·9992 828 | ·9989 838 |
| ·76 | ·9999 241 | ·9998 826 | ·9998 221 | ·9997 357 | ·9996 145+ | ·9994 473 |
| ·77 | ·9999 627 | ·9999 417 | ·9999 106 | ·9998 656 | ·9998 016 | ·9997 122 |
| ·78 | ·9999 825+ | ·9999 723 | ·9999 571 | ·9999 347 | ·9999 025+ | ·9998 570 |
| ·79 | ·9999 922 | ·9999 875+ | ·9999 804 | ·9999 699 | ·9999 545+ | ·9999 325− |
| ·80 | ·9999 967 | ·9999 947 | ·9999 916 | ·9999 869 | ·9999 799 | ·9999 698 |
| ·81 | ·9999 987 | ·9999 979 | ·9999 966 | ·9999 946 | ·9999 917 | ·9999 873 |
| ·82 | ·9999 995+ | ·9999 992 | ·9999 987 | ·9999 979 | ·9999 968 | ·9999 950+ |
| ·83 | ·9999 998 | ·9999 997 | ·9999 995+ | ·9999 993 | ·9999 988 | ·9999 982 |
| ·84 | ·9999 999 | ·9999 999 | ·9999 998 | ·9999 998 | ·9999 996 | ·9999 994 |
| ·85 | 1·0000 000 | 1·0000 000 | 1·0000 000 | ·9999 999 | ·9999 999 | ·9999 998 |
| ·86 | | | | 1·0000 000 | 1·0000 000 | ·9999 999 |
| ·87 | | | | | | 1·0000 000 |

$x$ = ·22 to ·88 $q = 23$ $p$ = 29 to 34

| $x$ | $p = 29$ | $p = 30$ | $p = 31$ | $p = 32$ | $p = 33$ | $p = 34$ |
|---|---|---|---|---|---|---|
| $B(p, q)$ = | ·2209 3390 $\times \frac{1}{10^{15}}$ | ·1232 1314 $\times \frac{1}{10^{15}}$ | ·6974 3286 $\times \frac{1}{10^{16}}$ | ·4003 7812 $\times \frac{1}{10^{16}}$ | ·2329 4727 $\times \frac{1}{10^{16}}$ | ·1372 7250 $\times \frac{1}{10^{16}}$ |
| ·22 | ·0000 001 | | | | | |
| ·23 | ·0000 002 | ·0000 001 | | | | |
| ·24 | ·0000 005+ | ·0000 002 | ·0000 001 | | | |
| ·25 | ·0000 013 | ·0000 005+ | ·0000 002 | ·0000 001 | | |
| ·26 | ·0000 030 | ·0000 013 | ·0000 006 | ·0000 003 | ·0000 001 | |
| ·27 | ·0000 067 | ·0000 031 | ·0000 014 | ·0000 006 | ·0000 003 | ·0000 001 |
| ·28 | ·0000 146 | ·0000 070 | ·0000 033 | ·0000 015+ | ·0000 007 | ·0000 003 |
| ·29 | ·0000 302 | ·0000 150− | ·0000 073 | ·0000 036 | ·0000 017 | ·0000 008 |
| ·30 | ·0000 602 | ·0000 309 | ·0000 157 | ·0000 078 | ·0000 039 | ·0000 019 |
| ·31 | ·0001 159 | ·0000 614 | ·0000 321 | ·0000 166 | ·0000 085− | ·0000 043 |
| ·32 | ·0002 156 | ·0001 178 | ·0000 636 | ·0000 339 | ·0000 178 | ·0000 093 |
| ·33 | ·0003 885+ | ·0002 187 | ·0001 216 | ·0000 668 | ·0000 363 | ·0000 195− |
| ·34 | ·0006 790 | ·0003 934 | ·0002 251 | ·0001 273 | ·0000 711 | ·0000 393 |
| ·35 | ·0011 527 | ·0006 868 | ·0004 042 | ·0002 350+ | ·0001 351 | ·0000 768 |
| ·36 | ·0019 036 | ·0011 653 | ·0007 046 | ·0004 210 | ·0002 487 | ·0001 454 |
| ·37 | ·0030 615+ | ·0019 238 | ·0011 942 | ·0007 327 | ·0004 445− | ·0002 668 |
| ·38 | ·0048 006 | ·0030 943 | ·0019 703 | ·0012 401 | ·0007 719 | ·0004 754 |
| ·39 | ·0073 469 | ·0048 535− | ·0031 679 | ·0020 440 | ·0013 044 | ·0008 236 |
| ·40 | ·0109 844 | ·0074 317 | ·0049 685− | ·0032 839 | ·0021 469 | ·0013 889 |
| ·41 | ·0160 578 | ·0111 184 | ·0076 082 | ·0051 477 | ·0034 453 | ·0022 820 |
| ·42 | ·0229 715+ | ·0162 663 | ·0113 849 | ·0078 797 | ·0053 955− | ·0036 565− |
| ·43 | ·0321 823 | ·0232 894 | ·0166 612 | ·0117 883 | ·0082 525+ | ·0057 185− |
| ·44 | ·0441 851 | ·0326 560 | ·0238 633 | ·0172 488 | ·0123 376 | ·0087 360 |
| ·45 | ·0594 909 | ·0448 744 | ·0334 735+ | ·0247 022 | ·0180 414 | ·0130 459 |
| ·46 | ·0785 988 | ·0604 698 | ·0460 150− | ·0346 469 | ·0258 225− | ·0190 571 |
| ·47 | ·1019 597 | ·0799 546 | ·0620 278 | ·0476 228 | ·0361 978 | ·0272 482 |
| ·48 | ·1299 378 | ·1037 914 | ·0820 376 | ·0641 854 | ·0497 251 | ·0381 569 |
| ·49 | ·1627 706 | ·1323 520 | ·1065 164 | ·0848 728 | ·0669 763 | ·0523 608 |
| ·50 | ·2005 310 | ·1658 749 | ·1358 396 | ·1101 642 | ·0885 007 | ·0704 477 |
| ·51 | ·2430 975+ | ·2044 266 | ·1702 415+ | ·1404 343 | ·1147 815− | ·0929 766 |
| ·52 | ·2901 351 | ·2478 697 | ·2097 745− | ·1759 066 | ·1461 875+ | ·1204 310 |
| ·53 | ·3410 901 | ·2958 430 | ·2542 760 | ·2166 109 | ·1829 243 | ·1531 677 |
| ·54 | ·3952 020 | ·3477 572 | ·3033 488 | ·2623 499 | ·2249 896 | ·1913 654 |
| ·55 | ·4515 326 | ·4028 083 | ·3563 578 | ·3126 794 | ·2721 403 | ·2349 795+ |
| ·56 | ·5090 101 | ·4600 092 | ·4124 457 | ·3669 074 | ·3238 738 | ·2837 090 |
| ·57 | ·5664 866 | ·5182 380 | ·4705 684 | ·4241 135+ | ·3794 298 | ·3369 802 |
| ·58 | ·6228 033 | ·5762 999 | ·5295 484 | ·4831 891 | ·4378 149 | ·3939 531 |
| ·59 | ·6768 575− | ·6329 967 | ·5881 417 | ·5428 960 | ·4978 483 | ·4535 514 |
| ·60 | ·7276 656 | ·6871 981 | ·6451 120 | ·6019 398 | ·5582 279 | ·5145 160 |
| ·61 | ·7744 166 | ·7379 083 | ·6993 070 | ·6590 495− | ·6176 094 | ·5754 787 |
| ·62 | ·8165 104 | ·7843 204 | ·7497 268 | ·7130 577 | ·6746 926 | ·6350 486 |
| ·63 | ·8535 786 | ·8258 546 | ·7955 800 | ·7629 714 | ·7283 044 | ·6919 040 |
| ·64 | ·8854 862 | ·8621 775+ | ·8363 205+ | ·8080 279 | ·7774 719 | ·7448 789 |
| ·65 | ·9123 160 | ·8932 003 | ·8716 631 | ·8477 292 | ·8214 766 | ·7930 363 |
| ·66 | ·9343 382 | ·9190 580 | ·9015 775− | ·8818 527 | ·8598 842 | ·8357 188 |
| ·67 | ·9519 687 | ·9400 747 | ·9262 617 | ·9104 392 | ·8925 499 | ·8725 735− |
| ·68 | ·9657 228 | ·9567 164 | ·9461 010 | ·9337 597 | ·9195 980 | ·9035 481 |
| ·69 | ·9761 679 | ·9695 414 | ·9616 161 | ·9522 669 | ·9413 809 | ·9288 619 |
| ·70 | ·9838 806 | ·9791 494 | ·9734 089 | ·9665 388 | ·9584 234 | ·9489 556 |
| ·71 | ·9894 110 | ·9861 377 | ·9821 093 | ·9772 194 | ·9713 607 | ·9644 280 |
| ·72 | ·9932 561 | ·9910 653 | ·9883 311 | ·9849 653 | ·9808 760 | ·9759 688 |
| ·73 | ·9958 443 | ·9944 283 | ·9926 365+ | ·9904 003 | ·9876 456 | ·9842 940 |
| ·74 | ·9975 276 | ·9966 457 | ·9955 144 | ·9940 832 | ·9922 960 | ·9900 917 |
| ·75 | ·9985 834 | ·9980 553 | ·9973 687 | ·9964 884 | ·9953 742 | ·9939 814 |
| ·76 | ·9992 205− | ·9989 172 | ·9985 177 | ·9979 987 | ·9973 330 | ·9964 898 |
| ·77 | ·9995 893 | ·9994 228 | ·9992 006 | ·9989 082 | ·9985 281 | ·9980 404 |
| ·78 | ·9997 935+ | ·9997 065− | ·9995 888 | ·9994 318 | ·9992 252 | ·9989 566 |
| ·79 | ·9999 014 | ·9998 582 | ·9997 990 | ·9997 191 | ·9996 125+ | ·9994 723 |
| ·80 | ·9999 554 | ·9999 352 | ·9999 071 | ·9998 686 | ·9998 168 | ·9997 476 |
| ·81 | ·9999 810 | ·9999 721 | ·9999 596 | ·9999 422 | ·9999 185+ | ·9998 865− |
| ·82 | ·9999 925− | ·9999 888 | ·9999 836 | ·9999 762 | ·9999 661 | ·9999 523 |
| ·83 | ·9999 972 | ·9999 958 | ·9999 938 | ·9999 909 | ·9999 869 | ·9999 814 |
| ·84 | ·9999 991 | ·9999 986 | ·9999 978 | ·9999 968 | ·9999 954 | ·9999 933 |
| ·85 | ·9999 997 | ·9999 995+ | ·9999 993 | ·9999 990 | ·9999 985− | ·9999 978 |
| ·86 | ·9999 999 | ·9999 999 | ·9999 998 | ·9999 997 | ·9999 996 | ·9999 994 |
| ·87 | 1·0000 000 | 1·0000 000 | 1·0000 000 | ·9999 999 | ·9999 999 | ·9999 998 |
| ·88 | | | | 1·0000 000 | 1·0000 000 | 1·0000 000 |

$x = \cdot 27$ to $\cdot 90$ $q = 23$ $p = 35$ to $40$

| $x$ | $p = 35$ | $p = 36$ | $p = 37$ | $p = 38$ | $p = 39$ | $p = 40$ |
|---|---|---|---|---|---|---|
| $B(p, q) =$ | $\cdot 8188\ 1842 \times \frac{1}{10^{17}}$ | $\cdot 4941\ 1456 \times \frac{1}{10^{17}}$ | $\cdot 3014\ 9363 \times \frac{1}{10^{17}}$ | $\cdot 1859\ 2107 \times \frac{1}{10^{17}}$ | $\cdot 1158\ 1968 \times \frac{1}{10^{17}}$ | $\cdot 7285\ 4318 \times \frac{1}{10^{18}}$ |
| ·27 | ·0000 001 | | | | | |
| ·28 | ·0000 001 | ·0000 001 | | | | |
| ·29 | ·0000 004 | ·0000 002 | ·0000 001 | | | |
| ·30 | ·0000 009 | ·0000 004 | ·0000 002 | ·0000 001 | | |
| ·31 | ·0000 021 | ·0000 011 | ·0000 005+ | ·0000 003 | ·0000 001 | ·0000 001 |
| ·32 | ·0000 048 | ·0000 024 | ·0000 012 | ·0000 006 | ·0000 003 | ·0000 002 |
| ·33 | ·0000 103 | ·0000 054 | ·0000 028 | ·0000 015− | ·0000 007 | ·0000 004 |
| ·34 | ·0000 215+ | ·0000 117 | ·0000 063 | ·0000 033 | ·0000 018 | ·0000 009 |
| ·35 | ·0000 432 | ·0000 241 | ·0000 133 | ·0000 073 | ·0000 039 | ·0000 021 |
| ·36 | ·0000 841 | ·0000 482 | ·0000 273 | ·0000 154 | ·0000 086 | ·0000 047 |
| ·37 | ·0001 585− | ·0000 932 | ·0000 543 | ·0000 313 | ·0000 179 | ·0000 102 |
| ·38 | ·0002 898 | ·0001 749 | ·0001 046 | ·0000 620 | ·0000 364 | ·0000 212 |
| ·39 | ·0005 148 | ·0003 186 | ·0001 954 | ·0001 187 | ·0000 715+ | ·0000 427 |
| ·40 | ·0008 895− | ·0005 641 | ·0003 545− | ·0002 208 | ·0001 363 | ·0000 834 |
| ·41 | ·0014 964 | ·0009 719 | ·0006 254 | ·0003 989 | ·0002 522 | ·0001 582 |
| ·42 | ·0024 535− | ·0016 307 | ·0010 739 | ·0007 010 | ·0004 537 | ·0002 912 |
| ·43 | ·0039 238 | ·0026 670 | ·0017 964 | ·0011 994 | ·0007 940 | ·0005 214 |
| ·44 | ·0061 259 | ·0042 556 | ·0029 298 | ·0019 996 | ·0013 533 | ·0009 085+ |
| ·45 | ·0093 433 | ·0066 298 | ·0046 626 | ·0032 509 | ·0022 479 | ·0015 420 |
| ·46 | ·0139 313 | ·0100 914 | ·0072 455+ | ·0051 581 | ·0036 419 | ·0025 511 |
| ·47 | ·0203 202 | ·0150 172 | ·0110 018 | ·0079 923 | ·0057 590 | ·0041 172 |
| ·48 | ·0290 112 | ·0218 619 | ·0163 331 | ·0121 013 | ·0088 941 | ·0064 862 |
| ·49 | ·0405 653 | ·0311 524 | ·0237 215+ | ·0179 154 | ·0134 232 | ·0099 804 |
| ·50 | ·0555 806 | ·0434 744 | ·0337 223 | ·0259 469 | ·0198 085+ | ·0150 079 |
| ·51 | ·0746 605− | ·0594 476 | ·0469 477 | ·0367 823 | ·0285 964 | ·0220 666 |
| ·52 | ·0983 713 | ·0796 899 | ·0640 390 | ·0510 615− | ·0404 062 | ·0317 400 |
| ·53 | ·1271 920 | ·1047 712 | ·0856 263 | ·0694 464 | ·0559 063 | ·0446 819 |
| ·54 | ·1614 596 | ·1351 595+ | ·1122 785+ | ·0925 761 | ·0757 772 | ·0615 886 |
| ·55 | ·2013 162 | ·1711 635+ | ·1444 449 | ·1210 119 | ·1006 622 | ·0831 563 |
| ·56 | ·2466 630 | ·2128 770 | ·1823 945− | ·1551 744 | ·1311 062 | ·1100 249 |
| ·57 | ·2971 276 | ·2601 330 | ·2261 595+ | ·1952 805− | ·1674 893 | ·1427 124 |
| ·58 | ·3520 522 | ·3124 739 | ·2754 901 | ·2412 851 | ·2099 605+ | ·1815 435+ |
| ·59 | ·4105 052 | ·3691 439 | ·3298 278 | ·2928 387 | ·2583 805+ | ·2265 818 |
| ·60 | ·4713 184 | ·4291 081 | ·3883 049 | ·3492 661 | ·3122 820 | ·2775 738 |
| ·61 | ·5331 497 | ·4910 989 | ·4497 723 | ·4095 737 | ·3708 562 | ·3339 157 |
| ·62 | ·5945 651 | ·5536 884 | ·5128 571 | ·4724 893 | ·4329 713 | ·3946 490 |
| ·63 | ·6541 332 | ·6153 804 | ·5760 463 | ·5365 314 | ·4972 246 | ·4584 922 |
| ·64 | ·7105 220 | ·6747 123 | ·6377 885− | ·6001 062 | ·5620 273 | ·5239 094 |
| ·65 | ·7625 884 | ·7303 572 | ·6966 039 | ·6616 191 | ·6257 137 | ·5892 098 |
| ·66 | ·8094 496 | ·7812 140 | ·7511 901 | ·7195 920 | ·6866 635+ | ·6526 711 |
| ·67 | ·8505 289 | ·8264 751 | ·8005 104 | ·7727 703 | ·7434 242 | ·7126 709 |
| ·68 | ·8855 722 | ·8656 652 | ·8438 560 | ·8202 077 | ·7948 170 | ·7678 116 |
| ·69 | ·9146 345+ | ·8986 470 | ·8808 742 | ·8613 193 | ·8400 148 | ·8170 222 |
| ·70 | ·9380 396 | ·9255 955+ | ·9115 613 | ·8958 961 | ·8785 819 | ·8596 251 |
| ·71 | ·9563 207 | ·9469 463 | ·9362 231 | ·9240 827 | ·9104 727 | ·8953 586 |
| ·72 | ·9701 494 | ·9633 257 | ·9554 103 | ·9463 225+ | ·9359 912 | ·9243 565− |
| ·73 | ·9802 642 | ·9754 734 | ·9698 389 | ·9632 799 | ·9557 199 | ·9470 879 |
| ·74 | ·9874 051 | ·9841 674 | ·9803 073 | ·9757 524 | ·9704 303 | ·9642 703 |
| ·75 | ·9922 610 | ·9901 596 | ·9876 204 | ·9845 837 | ·9809 875+ | ·9767 690 |
| ·76 | ·9954 343 | ·9941 279 | ·9925 282 | ·9905 897 | ·9882 634 | ·9854 981 |
| ·77 | ·9974 219 | ·9966 462 | ·9956 840 | ·9945 026 | ·9930 662 | ·9913 362 |
| ·78 | ·9986 116 | ·9981 733 | ·9976 225− | ·9969 374 | ·9960 937 | ·9950 643 |
| ·79 | ·9992 898 | ·9990 549 | ·9987 561 | ·9983 796 | ·9979 099 | ·9973 296 |
| ·80 | ·9996 565− | ·9995 377 | ·9993 847 | ·9991 895− | ·9989 429 | ·9986 343 |
| ·81 | ·9998 437 | ·9997 874 | ·9997 138 | ·9996 188 | ·9994 973 | ·9993 433 |
| ·82 | ·9999 336 | ·9999 086 | ·9998 756 | ·9998 325− | ·9997 766 | ·9997 050− |
| ·83 | ·9999 738 | ·9999 635+ | ·9999 498 | ·9999 317 | ·9999 079 | ·9998 770 |
| ·84 | ·9999 905− | ·9999 866 | ·9999 814 | ·9999 744 | ·9999 651 | ·9999 529 |
| ·85 | ·9999 969 | ·9999 955+ | ·9999 937 | ·9999 913 | ·9999 880 | ·9999 836 |
| ·86 | ·9999 991 | ·9999 987 | ·9999 981 | ·9999 973 | ·9999 963 | ·9999 948 |
| ·87 | ·9999 998 | ·9999 996 | ·9999 995− | ·9999 993 | ·9999 990 | ·9999 986 |
| ·88 | ·9999 999 | ·9999 999 | ·9999 999 | ·9999 998 | ·9999 998 | ·9999 997 |
| ·89 | 1·0000 000 | 1·0000 000 | 1·0000 000 | 1·0000 000 | 1·0000 000 | ·9999 999 |
| ·90 | | | | | | 1·0000 000 |

$x$ = ·32 to ·91 $q = 23$ $p$ = 41 to 45

| | $p = 41$ | $p = 42$ | $p = 43$ | $p = 44$ | $p = 45$ |
|---|---|---|---|---|---|
| $B(p, q) =$ | $\cdot 4625\ 6710 \times \frac{1}{10^{18}}$ | $\cdot 2963\ 3205 \times \frac{1}{10^{18}}$ | $\cdot 1914\ 7609 \times \frac{1}{10^{18}}$ | $\cdot 1247\ 4957 \times \frac{1}{10^{18}}$ | $\cdot 8192\ 5094 \times \frac{1}{10^{19}}$ |
| $x$ | | | | | |
| ·32 | ·0000 001 | | | | |
| ·33 | ·0000 002 | ·0000 001 | | | |
| ·34 | ·0000 005⁻ | ·0000 002 | ·0000 001 | ·0000 001 | |
| ·35 | ·0000 011 | ·0000 006 | ·0000 003 | ·0000 002 | ·0000 001 |
| ·36 | ·0000 026 | ·0000 014 | ·0000 008 | ·0000 004 | ·0000 002 |
| ·37 | ·0000 057 | ·0000 032 | ·0000 018 | ·0000 010 | ·0000 005⁺ |
| ·38 | ·0000 122 | ·0000 070 | ·0000 040 | ·0000 023 | ·0000 013 |
| ·39 | ·0000 253 | ·0000 149 | ·0000 087 | ·0000 050⁺ | ·0000 029 |
| ·40 | ·0000 507 | ·0000 306 | ·0000 183 | ·0000 109 | ·0000 064 |
| ·41 | ·0000 984 | ·0000 608 | ·0000 372 | ·0000 227 | ·0000 137 |
| ·42 | ·0001 855⁻ | ·0001 172 | ·0000 736 | ·0000 458 | ·0000 283 |
| ·43 | ·0003 397 | ·0002 197 | ·0001 410 | ·0000 899 | ·0000 569 |
| ·44 | ·0006 052 | ·0004 001 | ·0002 626 | ·0001 712 | ·0001 108 |
| ·45 | ·0010 495⁺ | ·0007 091 | ·0004 756 | ·0003 168 | ·0002 096 |
| ·46 | ·0017 733 | ·0012 236 | ·0008 382 | ·0005 703 | ·0003 854 |
| ·47 | ·0029 212 | ·0020 574 | ·0014 388 | ·0009 993 | ·0006 894 |
| ·48 | ·0046 947 | ·0033 734 | ·0024 069 | ·0017 057 | ·0012 008 |
| ·49 | ·0073 656 | ·0053 968 | ·0039 268 | ·0028 380 | ·0020 377 |
| ·50 | ·0112 875⁻ | ·0084 291 | ·0062 513 | ·0046 052 | ·0033 707 |
| ·51 | ·0169 048 | ·0128 597 | ·0097 161 | ·0072 926 | ·0054 386 |
| ·52 | ·0247 550⁻ | ·0191 738 | ·0147 513 | ·0112 751 | ·0085 636 |
| ·53 | ·0354 610 | ·0279 517 | ·0218 870 | ·0170 282 | ·0131 654 |
| ·54 | ·0497 127 | ·0398 587 | ·0317 502 | ·0251 314 | ·0197 702 |
| ·55 | ·0682 324 | ·0556 200 | ·0450 495⁺ | ·0362 613 | ·0290 111 |
| ·56 | ·0917 263 | ·0759 806 | ·0625 443 | ·0511 703 | ·0416 161 |
| ·57 | ·1208 220 | ·1016 493 | ·0849 963 | ·0706 477 | ·0583 796 |
| ·58 | ·1559 966 | ·1332 287 | ·1131 062 | ·0954 640 | ·0801 152 |
| ·59 | ·1975 018 | ·1711 383 | ·1474 363 | ·1262 974 | ·1075 895⁻ |
| ·60 | ·2452 953 | ·2155 360 | ·1883 275⁻ | ·1636 500⁺ | ·1414 403 |
| ·61 | ·2989 885⁺ | ·2662 506 | ·2358 200 | ·2077 601 | ·1820 853 |
| ·62 | ·3578 212 | ·3227 361 | ·2895 890 | ·2585 232 | ·2296 320 |
| ·63 | ·4206 700 | ·3840 563 | ·3489 072 | ·3154 338 | ·2838 014 |
| ·64 | ·4860 964 | ·4489 105⁺ | ·4126 454 | ·3775 611 | ·3438 801 |
| ·65 | ·5524 322 | ·5156 994 | ·4793 164 | ·4435 680 | ·4087 133 |
| ·66 | ·6178 969 | ·5826 308 | ·5471 631 | ·5117 780 | ·4767 466 |
| ·67 | ·6807 337 | ·6478 539 | ·6142 852 | ·5802 871 | ·5461 191 |
| ·68 | ·7393 479 | ·7096 069 | ·6787 895⁻ | ·6471 121 | ·6148 011 |
| ·69 | ·7924 317 | ·7663 598 | ·7389 470 | ·7103 549 | ·6807 620 |
| ·70 | ·8390 569 | ·8169 336 | ·7933 354 | ·7683 653 | ·7421 466 |
| ·71 | ·8787 255⁺ | ·8605 792 | ·8409 467 | ·8198 759 | ·7974 356 |
| ·72 | ·9113 722 | ·8970 071 | ·8812 465⁺ | ·8640 932 | ·8455 675⁺ |
| ·73 | ·9373 208 | ·9263 650⁻ | ·9141 779 | ·9007 297 | ·8860 038 |
| ·74 | ·9572 049 | ·9491 709 | ·9401 117 | ·9299 780 | ·9187 296 |
| ·75 | ·9718 649 | ·9662 133 | ·9597 543 | ·9524 316 | ·9441 936 |
| ·76 | ·9822 406 | ·9784 364 | ·9740 309 | ·9689 696 | ·9631 997 |
| ·77 | ·9892 715⁺ | ·9868 287 | ·9839 623 | ·9806 261 | ·9767 727 |
| ·78 | ·9938 198 | ·9923 282 | ·9905 554 | ·9884 651 | ·9860 195⁻ |
| ·79 | ·9966 190 | ·9957 565⁻ | ·9947 181 | ·9934 781 | ·9920 087 |
| ·80 | ·9982 517 | ·9977 813 | ·9972 079 | ·9965 146 | ·9956 825⁻ |
| ·81 | ·9991 500⁺ | ·9989 094 | ·9986 125⁻ | ·9982 489 | ·9978 071 |
| ·82 | ·9996 139 | ·9994 992 | ·9993 558 | ·9991 781 | ·9989 595⁺ |
| ·83 | ·9998 373 | ·9997 867 | ·9997 226 | ·9996 422 | ·9995 422 |
| ·84 | ·9999 370 | ·9999 165⁻ | ·9998 902 | ·9998 568 | ·9998 148 |
| ·85 | ·9999 778 | ·9999 702 | ·9999 604 | ·9999 479 | ·9999 318 |
| ·86 | ·9999 930 | ·9999 905⁻ | ·9999 872 | ·9999 829 | ·9999 774 |
| ·87 | ·9999 980 | ·9999 973 | ·9999 963 | ·9999 951 | ·9999 934 |
| ·88 | ·9999 995⁺ | ·9999 993 | ·9999 991 | ·9999 988 | ·9999 983 |
| ·89 | ·9999 999 | ·9999 999 | ·9999 998 | ·9999 997 | ·9999 996 |
| ·90 | 1·0000 000 | 1·0000 000 | 1·0000 000 | 1·0000 000 | ·9999 999 |
| ·91 | | | | | 1·0000 000 |

$x = \cdot 36$ to $\cdot 91$ $q = 23$ $p = 46$ to $50$

| | $p = 46$ | $p = 47$ | $p = 48$ | $p = 49$ | $p = 50$ |
|---|---|---|---|---|---|
| $B(p, q) =$ | ·5421 5136 × $\frac{1}{10^{19}}$ | ·3614 3424 × $\frac{1}{10^{19}}$ | ·2426 7727 × $\frac{1}{10^{19}}$ | ·1640 6351 × $\frac{1}{10^{19}}$ | ·1116 5433 × $\frac{1}{10^{19}}$ |
| $x$ | | | | | |
| ·36 | ·0000 001 | ·0000 001 | | | |
| ·37 | ·0000 003 | ·0000 002 | ·0000 001 | | |
| ·38 | ·0000 007 | ·0000 004 | ·0000 002 | ·0000 001 | ·0000 001 |
| ·39 | ·0000 017 | ·0000 009 | ·0000 005+ | ·0000 003 | ·0000 002 |
| ·40 | ·0000 038 | ·0000 022 | ·0000 013 | ·0000 007 | ·0000 004 |
| ·41 | ·0000 082 | ·0000 049 | ·0000 029 | ·0000 017 | ·0000 010 |
| ·42 | ·0000 174 | ·0000 106 | ·0000 065− | ·0000 039 | ·0000 023 |
| ·43 | ·0000 358 | ·0000 224 | ·0000 139 | ·0000 086 | ·0000 053 |
| ·44 | ·0000 712 | ·0000 455+ | ·0000 289 | ·0000 183 | ·0000 115− |
| ·45 | ·0001 377 | ·0000 900 | ·0000 584 | ·0000 377 | ·0000 242 |
| ·46 | ·0002 587 | ·0001 726 | ·0001 145− | ·0000 755− | ·0000 495− |
| ·47 | ·0004 726 | ·0003 219 | ·0002 180 | ·0001 467 | ·0000 982 |
| ·48 | ·0008 400 | ·0005 840 | ·0004 035+ | ·0002 772 | ·0001 894 |
| ·49 | ·0014 538 | ·0010 309 | ·0007 267 | ·0005 093 | ·0003 549 |
| ·50 | ·0024 517 | ·0017 724 | ·0012 738 | ·0009 102 | ·0006 468 |
| ·51 | ·0040 308 | ·0029 695− | ·0021 748 | ·0015 838 | ·0011 471 |
| ·52 | ·0064 644 | ·0048 507 | ·0036 188 | ·0026 846 | ·0019 807 |
| ·53 | ·0101 173 | ·0077 291 | ·0058 709 | ·0044 347 | ·0033 317 |
| ·54 | ·0154 597 | ·0120 189 | ·0092 911 | ·0071 430 | ·0054 622 |
| ·55 | ·0230 739 | ·0182 467 | ·0143 490 | ·0112 228 | ·0087 313 |
| ·56 | ·0336 500+ | ·0270 555+ | ·0216 340 | ·0172 064 | ·0136 138 |
| ·57 | ·0479 681 | ·0391 953 | ·0318 541 | ·0257 518 | ·0207 117 |
| ·58 | ·0668 607 | ·0554 964 | ·0458 198 | ·0376 350+ | ·0307 565− |
| ·59 | ·0911 556 | ·0768 224 | ·0644 075− | ·0537 254 | ·0445 933 |
| ·60 | ·1215 996 | ·1040 018 | ·0885 007 | ·0749 373 | ·0631 454 |
| ·61 | ·1587 669 | ·1377 398 | ·1189 093 | ·1021 580 | ·0873 520 |
| ·62 | ·2029 622 | ·1785 188 | ·1562 701 | ·1361 535+ | ·1180 815− |
| ·63 | ·2541 302 | ·2264 973 | ·2009 398 | ·1774 588 | ·1560 241 |
| ·64 | ·3117 859 | ·2814 220 | ·2528 928 | ·2262 656 | ·2015 729 |
| ·65 | ·3749 818 | ·3425 701 | ·3116 411 | ·2823 231 | ·2547 102 |
| ·66 | ·4423 225+ | ·4087 366 | ·3761 939 | ·3448 716 | ·3149 173 |
| ·67 | ·5120 346 | ·4782 761 | ·4450 706 | ·4126 262 | ·3811 285+ |
| ·68 | ·5820 881 | ·5492 043 | ·5163 766 | ·4838 224 | ·4517 465+ |
| ·69 | ·6503 603 | ·6193 505− | ·5879 382 | ·5563 297 | ·5247 275+ |
| ·70 | ·7148 209 | ·6865 447 | ·6574 865− | ·6278 228 | ·5977 354 |
| ·71 | ·7737 137 | ·7488 159 | ·7228 640 | ·6959 930 | ·6683 487 |
| ·72 | ·8257 081 | ·8045 707 | ·7822 280 | ·7587 682 | ·7342 934 |
| ·73 | ·8699 984 | ·8527 265+ | ·8342 162 | ·8145 105− | ·7936 666 |
| ·74 | ·9063 364 | ·8927 792 | ·8780 510 | ·8621 568 | ·8451 143 |
| ·75 | ·9349 944 | ·9247 954 | ·9135 655+ | ·9012 829 | ·8879 350+ |
| ·76 | ·9566 708 | ·9493 357 | ·9411 516 | ·9320 809 | ·9220 921 |
| ·77 | ·9723 551 | ·9673 266 | ·9616 423 | ·9552 593 | ·9481 377 |
| ·78 | ·9831 793 | ·9799 044 | ·9761 543 | ·9718 886 | ·9670 675− |
| ·79 | ·9902 803 | ·9882 619 | ·9859 210 | ·9832 240 | ·9801 368 |
| ·80 | ·9946 914 | ·9935 193 | ·9921 428 | ·9905 368 | ·9886 752 |
| ·81 | ·9972 744 | ·9966 365− | ·9958 770 | ·9949 818 | ·9939 301 |
| ·82 | ·9986 927 | ·9983 692 | ·9979 798 | ·9975 142 | ·9969 609 |
| ·83 | ·9994 185+ | ·9992 668 | ·9990 819 | ·9988 582 | ·9985 891 |
| ·84 | ·9997 622 | ·9996 970 | ·9996 165− | ·9995 179 | ·9993 978 |
| ·85 | ·9999 115+ | ·9998 860 | ·9998 542 | ·9998 148 | ·9997 662 |
| ·86 | ·9999 704 | ·9999 615− | ·9999 502 | ·9999 361 | ·9999 184 |
| ·87 | ·9999 912 | ·9999 885− | ·9999 849 | ·9999 805− | ·9999 748 |
| ·88 | ·9999 977 | ·9999 970 | ·9999 960 | ·9999 948 | ·9999 932 |
| ·89 | ·9999 995+ | ·9999 993 | ·9999 991 | ·9999 988 | ·9999 985− |
| ·90 | ·9999 999 | ·9999 999 | ·9999 998 | ·9999 998 | ·9999 997 |
| ·91 | 1·0000 000 | 1·0000 000 | 1·0000 000 | 1·0000 000 | 1·0000 000 |

$x$ = ·17 to ·86 $q$ = 24 $p$ = 24 to 29

| $x$ | $p$ = 24 | $p$ = 25 | $p$ = 26 | $p$ = 27 | $p$ = 28 | $p$ = 29 |
|---|---|---|---|---|---|---|
| $B(p, q)$ = | ·2584 1713 $\times \frac{1}{10^{14}}$ | ·1292 0857 $\times \frac{1}{10^{14}}$ | ·6592 2738 $\times \frac{1}{10^{15}}$ | ·3427 9824 $\times \frac{1}{10^{15}}$ | ·1814 8142 $\times \frac{1}{10^{15}}$ | ·9772 0765+ $\times \frac{1}{10^{16}}$ |
| ·17 | ·0000 001 | | | | | |
| ·18 | ·0000 003 | ·0000 001 | | | | |
| ·19 | ·0000 008 | ·0000 003 | ·0000 001 | | | |
| ·20 | ·0000 021 | ·0000 008 | ·0000 003 | ·0000 001 | | |
| ·21 | ·0000 051 | ·0000 020 | ·0000 008 | ·0000 003 | ·0000 001 | |
| ·22 | ·0000 118 | ·0000 049 | ·0000 020 | ·0000 008 | ·0000 003 | ·0000 001 |
| ·23 | ·0000 259 | ·0000 113 | ·0000 048 | ·0000 020 | ·0000 008 | ·0000 003 |
| ·24 | ·0000 543 | ·0000 247 | ·0000 110 | ·0000 048 | ·0000 021 | ·0000 009 |
| ·25 | ·0001 090 | ·0000 515+ | ·0000 239 | ·0000 109 | ·0000 049 | ·0000 022 |
| ·26 | ·0002 098 | ·0001 030 | ·0000 497 | ·0000 236 | ·0000 110 | ·0000 051 |
| ·27 | ·0003 884 | ·0001 978 | ·0000 991 | ·0000 488 | ·0000 237 | ·0000 113 |
| ·28 | ·0006 934 | ·0003 659 | ·0001 898 | ·0000 969 | ·0000 487 | ·0000 241 |
| ·29 | ·0011 967 | ·0006 531 | ·0003 505− | ·0001 851 | ·0000 963 | ·0000 493 |
| ·30 | ·0020 003 | ·0011 279 | ·0006 254 | ·0003 413 | ·0001 834 | ·0000 972 |
| ·31 | ·0032 446 | ·0018 878 | ·0010 802 | ·0006 084 | ·0003 375+ | ·0001 846 |
| ·32 | ·0051 153 | ·0030 676 | ·0018 094 | ·0010 507 | ·0006 010 | ·0003 390 |
| ·33 | ·0078 502 | ·0048 469 | ·0029 440 | ·0017 605+ | ·0010 373 | ·0006 026 |
| ·34 | ·0117 428 | ·0074 571 | ·0046 595− | ·0028 668 | ·0017 381 | ·0010 391 |
| ·35 | ·0171 431 | ·0111 862 | ·0071 831 | ·0045 427 | ·0028 313 | ·0017 402 |
| ·36 | ·0244 529 | ·0163 797 | ·0107 995− | ·0070 135+ | ·0044 896 | ·0028 346 |
| ·37 | ·0341 155− | ·0234 377 | ·0158 522 | ·0105 627 | ·0069 385− | ·0044 960 |
| ·38 | ·0465 986 | ·0328 051 | ·0227 414 | ·0155 342 | ·0104 625− | ·0069 521 |
| ·39 | ·0623 716 | ·0449 562 | ·0319 155+ | ·0223 307 | ·0154 082 | ·0104 908 |
| ·40 | ·0818 765+ | ·0603 716 | ·0438 559 | ·0314 056 | ·0221 831 | ·0154 638 |
| ·41 | ·1054 953 | ·0795 100 | ·0590 544 | ·0432 484 | ·0312 476 | ·0222 856 |
| ·42 | ·1335 158 | ·1027 745+ | ·0779 847 | ·0583 627 | ·0431 011 | ·0314 259 |
| ·43 | ·1660 987 | ·1304 776 | ·1010 688 | ·0772 364 | ·0582 588 | ·0433 952 |
| ·44 | ·2032 490 | ·1628 059 | ·1286 396 | ·1003 079 | ·0772 227 | ·0587 216 |
| ·45 | ·2447 963 | ·1997 901 | ·1609 047 | ·1279 269 | ·1004 455− | ·0779 205− |
| ·46 | ·2903 837 | ·2412 810 | ·1979 135− | ·1603 172 | ·1282 907 | ·1014 570 |
| ·47 | ·3394 713 | ·2869 377 | ·2395 314 | ·1975 404 | ·1609 926 | ·1297 052 |
| ·48 | ·3913 512 | ·3362 276 | ·2854 257 | ·2394 695+ | ·1986 196 | ·1629 051 |
| ·49 | ·4451 763 | ·3884 404 | ·3350 633 | ·2857 716 | ·2410 440 | ·2011 246 |
| ·50 | ·5000 000[e] | ·4427 167 | ·3877 248 | ·3359 055+ | ·2879 247 | ·2442 278 |
| ·51 | ·5548 237 | ·4980 879 | ·4425 321 | ·3891 345− | ·3387 034 | ·2918 565+ |
| ·52 | ·6086 488 | ·5535 253 | ·4984 899 | ·4445 552 | ·3926 181 | ·3434 263 |
| ·53 | ·6605 287 | ·6079 951 | ·5545 370 | ·5011 405− | ·4487 328 | ·3981 407 |
| ·54 | ·7096 163 | ·6605 135+ | ·6096 038 | ·5577 934 | ·5059 830 | ·4550 237 |
| ·55 | ·7552 037 | ·7101 975+ | ·6626 710 | ·6134 079 | ·5632 325+ | ·5129 675+ |
| ·56 | ·7967 510 | ·7563 079 | ·7128 235− | ·6669 307 | ·6193 382 | ·5707 938 |
| ·57 | ·8339 013 | ·7982 803 | ·7592 966 | ·7174 191 | ·6732 151 | ·6273 218 |
| ·58 | ·8664 842 | ·8357 429 | ·8015 094 | ·7640 895− | ·7238 978 | ·6814 381 |
| ·59 | ·8945 047 | ·8685 193 | ·8390 831 | ·8063 524 | ·7705 910 | ·7321 602 |
| ·60 | ·9181 235− | ·8966 186 | ·8718 449 | ·8438 317 | ·8127 058 | ·7786 897 |
| ·61 | ·9376 284 | ·9202 129 | ·8998 160 | ·8763 673 | ·8498 791 | ·8204 487 |
| ·62 | ·9534 014 | ·9396 079 | ·9231 881 | ·9040 022 | ·8819 740 | ·8570 978 |
| ·63 | ·9658 845+ | ·9552 067 | ·9422 908 | ·9269 557 | ·9090 647 | ·8885 349 |
| ·64 | ·9755 471 | ·9674 739 | ·9575 535− | ·9455 880 | ·9314 067 | ·9148 753 |
| ·65 | ·9828 569 | ·9769 000 | ·9694 658 | ·9603 589 | ·9493 968 | ·9364 185+ |
| ·66 | ·9882 572 | ·9839 716 | ·9785 408 | ·9717 858 | ·9635 296 | ·9536 046 |
| ·67 | ·9921 498 | ·9891 465+ | ·9852 831 | ·9804 048 | ·9743 520 | ·9669 655+ |
| ·68 | ·9948 847 | ·9928 369 | ·9901 633 | ·9867 370 | ·9824 225− | ·9770 786 |
| ·69 | ·9967 554 | ·9953 985+ | ·9936 010 | ·9912 635+ | ·9882 767 | ·9845 230 |
| ·70 | ·9979 997 | ·9971 272 | ·9959 547 | ·9944 078 | ·9924 026 | ·9898 460 |
| ·71 | ·9988 033 | ·9982 598 | ·9975 189 | ·9965 275− | ·9952 239 | ·9935 382 |
| ·72 | ·9993 066 | ·9989 791 | ·9985 263 | ·9979 120 | ·9970 928 | ·9960 185+ |
| ·73 | ·9996 116 | ·9994 211 | ·9991 541 | ·9987 867 | ·9982 901 | ·9976 297 |
| ·74 | ·9997 902 | ·9996 835− | ·9995 318 | ·9993 203 | ·9990 304 | ·9986 397 |
| ·75 | ·9998 910 | ·9998 335+ | ·9997 507 | ·9996 338 | ·9994 713 | ·9992 493 |
| ·76 | ·9999 457 | ·9999 160 | ·9998 727 | ·9998 108 | ·9997 235+ | ·9996 028 |
| ·77 | ·9999 741 | ·9999 595− | ·9999 379 | ·9999 065+ | ·9998 618 | ·9997 991 |
| ·78 | ·9999 882 | ·9999 814 | ·9999 711 | ·9999 560 | ·9999 342 | ·9999 033 |
| ·79 | ·9999 949 | ·9999 919 | ·9999 873 | ·9999 804 | ·9999 703 | ·9999 558 |
| ·80 | ·9999 979 | ·9999 967 | ·9999 947 | ·9999 917 | ·9999 874 | ·9999 810 |
| ·81 | ·9999 992 | ·9999 987 | ·9999 979 | ·9999 967 | ·9999 949 | ·9999 923 |
| ·82 | ·9999 997 | ·9999 995+ | ·9999 992 | ·9999 988 | ·9999 981 | ·9999 971 |
| ·83 | ·9999 999 | ·9999 998 | ·9999 997 | ·9999 996 | ·9999 993 | ·9999 990 |
| ·84 | 1·0000 000 | 1·0000 000 | ·9999 999 | ·9999 999 | ·9999 998 | ·9999 997 |
| ·85 | | | 1·0000 000 | 1·0000 000 | ·9999 999 | ·9999 999 |
| ·86 | | | | | 1·0000 000 | 1·0000 000 |

$x$ = ·23 to ·88 $q = 24$ $p$ = 30 to 35

| | $p = 30$ | $p = 31$ | $p = 32$ | $p = 33$ | $p = 34$ | $p = 35$ |
|---|---|---|---|---|---|---|
| $B(p, q) =$ | ·5346 9853 $\times \frac{1}{10^{16}}$ | ·2970 5474 $\times \frac{1}{10^{16}}$ | ·1674 3085$^{+}$ $\times \frac{1}{10^{16}}$ | ·9567 4772 $\times \frac{1}{10^{17}}$ | ·5539 0658 $\times \frac{1}{10^{17}}$ | ·3247 0386 $\times \frac{1}{10^{17}}$ |
| $x$ | | | | | | |
| ·23 | ·0000 001 | ·0000 001 | | | | |
| ·24 | ·0000 004 | ·0000 002 | ·0000 001 | | | |
| ·25 | ·0000 010 | ·0000 004 | ·0000 002 | ·0000 001 | | |
| ·26 | ·0000 023 | ·0000 010 | ·0000 005$^{-}$ | ·0000 002 | ·0000 001 | |
| ·27 | ·0000 053 | ·0000 025$^{-}$ | ·0000 011 | ·0000 005$^{+}$ | ·0000 002 | ·0000 001 |
| ·28 | ·0000 118 | ·0000 057 | ·0000 027 | ·0000 013 | ·0000 006 | ·0000 003 |
| ·29 | ·0000 249 | ·0000 124 | ·0000 061 | ·0000 030 | ·0000 014 | ·0000 007 |
| ·30 | ·0000 508 | ·0000 262 | ·0000 133 | ·0000 067 | ·0000 033 | ·0000 016 |
| ·31 | ·0000 996 | ·0000 530 | ·0000 279 | ·0000 145$^{-}$ | ·0000 074 | ·0000 038 |
| ·32 | ·0001 886 | ·0001 036 | ·0000 562 | ·0000 301 | ·0000 160 | ·0000 084 |
| ·33 | ·0003 454 | ·0001 954 | ·0001 092 | ·0000 603 | ·0000 330 | ·0000 178 |
| ·34 | ·0006 129 | ·0003 570 | ·0002 054 | ·0001 168 | ·0000 657 | ·0000 365$^{+}$ |
| ·35 | ·0010 555$^{+}$ | ·0006 321 | ·0003 740 | ·0002 187 | ·0001 265$^{-}$ | ·0000 724 |
| ·36 | ·0017 662 | ·0010 868 | ·0006 607 | ·0003 970 | ·0002 360 | ·0001 388 |
| ·37 | ·0028 755$^{+}$ | ·0018 163 | ·0011 336 | ·0006 995$^{-}$ | ·0004 269 | ·0002 578 |
| ·38 | ·0045 603 | ·0029 545$^{+}$ | ·0018 916 | ·0011 974 | ·0007 497 | ·0004 645$^{+}$ |
| ·39 | ·0070 521 | ·0046 827 | ·0030 731 | ·0019 942 | ·0012 801 | ·0008 132 |
| ·40 | ·0106 445$^{+}$ | ·0072 389 | ·0048 659 | ·0032 345$^{+}$ | ·0021 271 | ·0013 845$^{+}$ |
| ·41 | ·0156 970 | ·0109 246 | ·0075 162 | ·0051 144 | ·0034 433 | ·0022 946 |
| ·42 | ·0226 333 | ·0161 093 | ·0113 361 | ·0078 906 | ·0054 348 | ·0037 056 |
| ·43 | ·0319 349 | ·0232 289 | ·0167 078 | ·0118 883 | ·0083 715$^{+}$ | ·0058 363 |
| ·44 | ·0441 249 | ·0327 784 | ·0240 818 | ·0175 050$^{+}$ | ·0125 944 | ·0089 721 |
| ·45 | ·0597 451 | ·0452 957 | ·0339 693 | ·0252 090 | ·0185 193 | ·0134 725$^{-}$ |
| ·46 | ·0793 239 | ·0613 371 | ·0469 244 | ·0355 294 | ·0266 343 | ·0197 747 |
| ·47 | ·1033 374 | ·0814 433 | ·0635 184 | ·0490 385$^{-}$ | ·0374 896 | ·0283 898 |
| ·48 | ·1321 660 | ·1060 992 | ·0843 040 | ·0663 230 | ·0516 766 | ·0398 906 |
| ·49 | ·1660 505$^{+}$ | ·1356 881 | ·1097 723 | ·0879 464 | ·0697 977 | ·0548 892 |
| ·50 | ·2050 514 | ·1704 455$^{-}$ | ·1403 049 | ·1144 028 | ·0924 252 | ·0740 029 |
| ·51 | ·2490 159 | ·2104 164 | ·1761 252 | ·1460 668 | ·1200 526 | ·0978 105$^{-}$ |
| ·52 | ·2975 591 | ·2554 225$^{-}$ | ·2172 549 | ·1831 425$^{+}$ | ·1530 410 | ·1267 996 |
| ·53 | ·3500 608 | ·3050 419 | ·2634 794 | ·2256 185$^{-}$ | ·1915 666 | ·1613 105$^{+}$ |
| ·54 | ·4056 811 | ·3586 083 | ·3143 294 | ·2732 331 | ·2355 739 | ·2014 813 |
| ·55 | ·4633 959 | ·4152 288 | ·3690 815$^{+}$ | ·3254 580 | ·2847 427 | ·2472 008 |
| ·56 | ·5220 486 | ·4738 233 | ·4267 803 | ·3815 014 | ·3384 728 | ·2980 765$^{-}$ |
| ·57 | ·5804 158 | ·5331 814 | ·4862 822 | ·4403 357 | ·3958 928 | ·3534 238 |
| ·58 | ·6372 801 | ·5920 328 | ·5463 184 | ·5007 469 | ·4558 935$^{+}$ | ·4122 802 |
| ·59 | ·6915 032 | ·6491 249 | ·6055 711 | ·5614 047 | ·5171 848 | ·4734 462 |
| ·60 | ·7420 931 | ·7033 006 | ·6627 563 | ·6209 449 | ·5783 733 | ·5355 514 |
| ·61 | ·7882 580 | ·7535 671 | ·7167 052 | ·6780 579 | ·6380 520 | ·5971 401 |
| ·62 | ·8294 424 | ·7991 505$^{-}$ | ·7664 352 | ·7315 730 | ·6948 937 | ·6567 689 |
| ·63 | ·8653 432 | ·8395 308 | ·8112 038 | ·7805 310 | ·7477 390 | ·7131 049 |
| ·64 | ·8959 041 | ·8744 540 | ·8505 406 | ·8242 359 | ·7956 674 | ·7650 150$^{+}$ |
| ·65 | ·9212 921 | ·9039 220 | ·8842 545$^{+}$ | ·8622 822 | ·8380 461 | ·8116 359 |
| ·66 | ·9418 587 | ·9281 631 | ·9124 176 | ·8945 562 | ·8745 515$^{-}$ | ·8524 168 |
| ·67 | ·9580 916 | ·9475 877 | ·9353 287 | ·9212 117 | ·9051 611 | ·8871 325$^{-}$ |
| ·68 | ·9705 627 | ·9627 349 | ·9534 629 | ·9426 261 | ·9301 211 | ·9158 655$^{-}$ |
| ·69 | ·9798 787 | ·9742 173 | ·9674 127 | ·9593 428 | ·9498 937 | ·9389 634 |
| ·70 | ·9866 370 | ·9826 686 | ·9778 296 | ·9720 078 | ·9650 921 | ·9569 764 |
| ·71 | ·9913 920 | ·9887 000 | ·9853 707 | ·9813 078 | ·9764 126 | ·9705 860 |
| ·72 | ·9946 316 | ·9928 675$^{-}$ | ·9906 549 | ·9879 168 | ·9845 713 | ·9805 332 |
| ·73 | ·9967 653 | ·9956 505$^{+}$ | ·9942 330 | ·9924 544 | ·9902 511 | ·9875 546 |
| ·74 | ·9981 212 | ·9974 435$^{-}$ | ·9965 698 | ·9954 586 | ·9940 632 | ·9923 321 |
| ·75 | ·9989 508 | ·9985 553 | ·9980 386 | ·9973 725$^{-}$ | ·9965 247 | ·9954 588 |
| ·76 | ·9994 383 | ·9992 173 | ·9989 249 | ·9985 428 | ·9980 501 | ·9974 223 |
| ·77 | ·9997 126 | ·9995 948 | ·9994 369 | ·9992 279 | ·9989 548 | ·9986 022 |
| ·78 | ·9998 600 | ·9998 003 | ·9997 192 | ·9996 106 | ·9994 667 | ·9992 786 |
| ·79 | ·9999 353 | ·9999 067 | ·9998 673 | ·9998 138 | ·9997 421 | ·9996 471 |
| ·80 | ·9999 718 | ·9999 589 | ·9999 408 | ·9999 160 | ·9998 823 | ·9998 372 |
| ·81 | ·9999 885$^{-}$ | ·9999 830 | ·9999 752 | ·9999 645$^{-}$ | ·9999 496 | ·9999 295$^{+}$ |
| ·82 | ·9999 956 | ·9999 934 | ·9999 903 | ·9999 860 | ·9999 799 | ·9999 716 |
| ·83 | ·9999 984 | ·9999 977 | ·9999 965$^{+}$ | ·9999 949 | ·9999 926 | ·9999 894 |
| ·84 | ·9999 995$^{-}$ | ·9999 992 | ·9999 988 | ·9999 983 | ·9999 975$^{-}$ | ·9999 964 |
| ·85 | ·9999 999 | ·9999 998 | ·9999 997 | ·9999 995$^{-}$ | ·9999 992 | ·9999 989 |
| ·86 | 1·0000 000 | ·9999 999 | ·9999 999 | ·9999 999 | ·9999 998 | ·9999 997 |
| ·87 | | 1·0000 000 | 1·0000 000 | 1·0000 000 | ·9999 999 | ·9999 999 |
| ·88 | | | | | 1·0000 000 | 1·0000 000 |

$x$ = ·28 to ·89 $q$ = 24 $p$ = 36 to 40

| $x$ | $p$ = 36 | $p$ = 37 | $p$ = 38 | $p$ = 39 | $p$ = 40 |
|---|---|---|---|---|---|
| $B(p, q)$ = | ·1926 2093 $\times \frac{1}{10^{17}}$ | ·1155 7256 $\times \frac{1}{10^{17}}$ | ·7010 1388 $\times \frac{1}{10^{18}}$ | ·4296 5367 $\times \frac{1}{10^{18}}$ | ·2659 7608 $\times \frac{1}{10^{18}}$ |
| ·28 | ·0000 001 | ·0000 001 | | | |
| ·29 | ·0000 003 | ·0000 002 | ·0000 001 | | |
| ·30 | ·0000 008 | ·0000 004 | ·0000 002 | ·0000 001 | |
| ·31 | ·0000 019 | ·0000 009 | ·0000 005$^{-}$ | ·0000 002 | ·0000 001 |
| ·32 | ·0000 043 | ·0000 022 | ·0000 011 | ·0000 006 | ·0000 003 |
| ·33 | ·0000 095$^{+}$ | ·0000 050$^{+}$ | ·0000 026 | ·0000 014 | ·0000 007 |
| ·34 | ·0000 201 | ·0000 110 | ·0000 059 | ·0000 032 | ·0000 017 |
| ·35 | ·0000 410 | ·0000 230 | ·0000 128 | ·0000 070 | ·0000 038 |
| ·36 | ·0000 808 | ·0000 466 | ·0000 266 | ·0000 150$^{+}$ | ·0000 084 |
| ·37 | ·0001 541 | ·0000 912 | ·0000 535$^{+}$ | ·0000 311 | ·0000 179 |
| ·38 | ·0002 850$^{-}$ | ·0001 731 | ·0001 042 | ·0000 622 | ·0000 368 |
| ·39 | ·0005 115$^{+}$ | ·0003 187 | ·0001 967 | ·0001 203 | ·0000 730 |
| ·40 | ·0008 923 | ·0005 696 | ·0003 603 | ·0002 259 | ·0001 404 |
| ·41 | ·0015 142 | ·0009 898 | ·0006 411 | ·0004 117 | ·0002 621 |
| ·42 | ·0025 021 | ·0016 737 | ·0011 095$^{+}$ | ·0007 291 | ·0004 752 |
| ·43 | ·0040 298 | ·0027 567 | ·0018 690 | ·0012 563 | ·0008 374 |
| ·44 | ·0063 309 | ·0044 263 | ·0030 673 | ·0021 075$^{-}$ | ·0014 361 |
| ·45 | ·0097 090 | ·0069 335$^{-}$ | ·0049 081 | ·0034 450$^{-}$ | ·0023 983 |
| ·46 | ·0145 457 | ·0106 036 | ·0076 630 | ·0054 916 | ·0039 037 |
| ·47 | ·0213 023 | ·0158 430 | ·0116 822 | ·0085 429 | ·0061 973 |
| ·48 | ·0305 157 | ·0231 408 | ·0174 003 | ·0129 771 | ·0096 018 |
| ·49 | ·0427 834 | ·0330 619 | ·0253 371 | ·0192 610 | ·0145 279 |
| ·50 | ·0587 387 | ·0462 305$^{-}$ | ·0360 887 | ·0279 486 | ·0214 783 |
| ·51 | ·0790 127 | ·0633 009 | ·0503 068 | ·0396 687 | ·0310 437 |
| ·52 | ·1041 869 | ·0849 159 | ·0686 658 | ·0551 012 | ·0438 878 |
| ·53 | ·1347 370 | ·1116 550$^{+}$ | ·0918 169 | ·0749 389 | ·0607 181 |
| ·54 | ·1709 733 | ·1439 737 | ·1203 308 | ·0998 362 | ·0822 423 |
| ·55 | ·2129 840 | ·1821 414 | ·1546 331 | ·1303 462 | ·1091 107 |
| ·56 | ·2605 887 | ·2261 832 | ·1949 394 | ·1668 528 | ·1418 485$^{+}$ |
| ·57 | ·3133 087 | ·2758 346 | ·2411 963 | ·2095 023 | ·1807 826 |
| ·58 | ·3703 615$^{+}$ | ·3305 155$^{+}$ | ·2930 388 | ·2581 459 | ·2259 729 |
| ·59 | ·4306 823 | ·3893 319 | ·3497 697 | ·3123 001 | ·2771 556 |
| ·60 | ·4929 741 | ·4511 064 | ·4103 703 | ·3711 350$^{-}$ | ·3337 105$^{+}$ |
| ·61 | ·5557 840 | ·5144 394 | ·4735 418 | ·4334 944 | ·3946 587 |
| ·62 | ·6175 983 | ·5777 966 | ·5377 798 | ·4979 526 | ·4586 972 |
| ·63 | ·6769 468 | ·6396 136 | ·6014 732 | ·5629 012 | ·5242 699 |
| ·64 | ·7325 060 | ·6984 077 | ·6630 192 | ·6266 621 | ·5896 711 |
| ·65 | ·7831 884 | ·7528 838 | ·7209 412 | ·6876 116 | ·6531 709 |
| ·66 | ·8282 079 | ·8020 218 | ·7739 957 | ·7443 028 | ·7131 480 |
| ·67 | ·8671 155$^{+}$ | ·8451 359 | ·8212 552 | ·7955 710 | ·7682 140 |
| ·68 | ·8998 014 | ·8818 989 | ·8621 577 | ·8406 087 | ·8173 136 |
| ·69 | ·9264 653 | ·9123 321 | ·8965 181 | ·8790 021 | ·8597 883 |
| ·70 | ·9475 621 | ·9367 619 | ·9245 021 | ·9107 261 | ·8953 958 |
| ·71 | ·9637 305$^{-}$ | ·9557 533 | ·9465 689 | ·9361 010 | ·9242 857 |
| ·72 | ·9757 151 | ·9700 298 | ·9633 917 | ·9557 196 | ·9469 379 |
| ·73 | ·9842 927 | ·9803 902 | ·9757 704 | ·9703 568 | ·9640 742 |
| ·74 | ·9902 093 | ·9876 348 | ·9845 453 | ·9808 754 | ·9765 581 |
| ·75 | ·9941 340 | ·9925 056 | ·9905 251 | ·9881 407 | ·9852 978 |
| ·76 | ·9966 316 | ·9956 468 | ·9944 331 | ·9929 524 | ·9911 633 |
| ·77 | ·9981 523 | ·9975 846 | ·9968 757 | ·9959 995$^{+}$ | ·9949 270 |
| ·78 | ·9990 354 | ·9987 246 | ·9983 314 | ·9978 391 | ·9972 286 |
| ·79 | ·9995 227 | ·9993 617 | ·9991 555$^{-}$ | ·9988 939 | ·9985 654 |
| ·80 | ·9997 773 | ·9996 988 | ·9995 969 | ·9994 661 | ·9992 997 |
| ·81 | ·9999 025$^{+}$ | ·9998 667 | ·9998 196 | ·9997 583 | ·9996 795$^{-}$ |
| ·82 | ·9999 602 | ·9999 450$^{-}$ | ·9999 247 | ·9998 981 | ·9998 633 |
| ·83 | ·9999 850$^{-}$ | ·9999 790 | ·9999 710 | ·9999 603 | ·9999 461 |
| ·84 | ·9999 948 | ·9999 927 | ·9999 897 | ·9999 858 | ·9999 805$^{+}$ |
| ·85 | ·9999 984 | ·9999 977 | ·9999 967 | ·9999 954 | ·9999 936 |
| ·86 | ·9999 995$^{+}$ | ·9999 993 | ·9999 991 | ·9999 987 | ·9999 981 |
| ·87 | ·9999 999 | ·9999 998 | ·9999 998 | ·9999 997 | ·9999 995$^{+}$ |
| ·88 | 1·0000 000 | 1·0000 000 | ·9999 999 | ·9999 999 | ·9999 999 |
| ·89 | | | 1·0000 000 | 1·0000 000 | 1·0000 000 |

$x$ = ·31 to ·90 $q$ = 24 $p$ = 41 to 45

| $x$ | $p$ = 41 | $p$ = 42 | $p$ = 43 | $p$ = 44 | $p$ = 45 |
|---|---|---|---|---|---|
| $B(p, q)$ = | ·1662 3505$^{+}\times\frac{1}{10^{18}}$ | ·1048 5595$^{+}\times\frac{1}{10^{18}}$ | ·6672 6517$\times\frac{1}{10^{19}}$ | ·4282 4481$\times\frac{1}{10^{19}}$ | ·2770 9958$\times\frac{1}{10^{19}}$ |
| ·31 | ·0000 001 | | | | |
| ·32 | ·0000 001 | ·0000 001 | | | |
| ·33 | ·0000 004 | ·0000 002 | ·0000 001 | | |
| ·34 | ·0000 009 | ·0000 005$^-$ | ·0000 002 | ·0000 001 | ·0000 001 |
| ·35 | ·0000 021 | ·0000 011 | ·0000 006 | ·0000 003 | ·0000 002 |
| ·36 | ·0000 047 | ·0000 026 | ·0000 014 | ·0000 008 | ·0000 004 |
| ·37 | ·0000 102 | ·0000 058 | ·0000 033 | ·0000 018 | ·0000 010 |
| ·38 | ·0000 216 | ·0000 125$^+$ | ·0000 072 | ·0000 042 | ·0000 024 |
| ·39 | ·0000 439 | ·0000 262 | ·0000 155$^+$ | ·0000 091 | ·0000 053 |
| ·40 | ·0000 866 | ·0000 530 | ·0000 322 | ·0000 194 | ·0000 116 |
| ·41 | ·0001 655$^+$ | ·0001 037 | ·0000 645$^-$ | ·0000 398 | ·0000 244 |
| ·42 | ·0003 071 | ·0001 970 | ·0001 254 | ·0000 792 | ·0000 497 |
| ·43 | ·0005 537 | ·0003 633 | ·0002 366 | ·0001 530 | ·0000 982 |
| ·44 | ·0009 708 | ·0006 512 | ·0004 336 | ·0002 866 | ·0001 882 |
| ·45 | ·0016 565$^-$ | ·0011 354 | ·0007 725$^+$ | ·0005 219 | ·0003 501 |
| ·46 | ·0027 533 | ·0019 272 | ·0013 392 | ·0009 240 | ·0006 331 |
| ·47 | ·0044 610 | ·0031 871 | ·0022 605$^-$ | ·0015 920 | ·0011 137 |
| ·48 | ·0070 501 | ·0051 382 | ·0037 180 | ·0026 716 | ·0019 068 |
| ·49 | ·0108 751 | ·0080 812 | ·0059 624 | ·0043 690 | ·0031 800 |
| ·50 | ·0163 829 | ·0124 060 | ·0093 286 | ·0069 669 | ·0051 688 |
| ·51 | ·0241 157 | ·0186 003 | ·0142 470 | ·0108 394 | ·0081 930 |
| ·52 | ·0347 040 | ·0272 495$^+$ | ·0212 504 | ·0164 623 | ·0126 709 |
| ·53 | ·0488 473 | ·0390 264 | ·0309 709 | ·0244 178 | ·0191 292 |
| ·54 | ·0672 787 | ·0546 655$^-$ | ·0441 244 | ·0353 876 | ·0282 036 |
| ·55 | ·0907 155$^-$ | ·0749 225$^+$ | ·0614 797 | ·0501 314 | ·0406 273 |
| ·56 | ·1197 947 | ·1005 165$^+$ | ·0838 087 | ·0694 478 | ·0572 019 |
| ·57 | ·1549 996 | ·1320 589 | ·1118 220 | ·0941 171 | ·0787 500$^-$ |
| ·58 | ·1965 829 | ·1699 741 | ·1460 896 | ·1248 268 | ·1060 480 |
| ·59 | ·2444 975$^+$ | ·2144 203 | ·1869 568 | ·1620 865$^-$ | ·1397 427 |
| ·60 | ·2983 444 | ·2652 210 | ·2344 636 | ·2061 382 | ·1802 590 |
| ·61 | ·3573 474 | ·3218 196 | ·2882 798 | ·2568 771 | ·2277 083 |
| ·62 | ·4203 643 | ·3832 656 | ·3476 685$^-$ | ·3137 933 | ·2818 120 |
| ·63 | ·4859 379 | ·4482 418 | ·4114 880 | ·3759 479 | ·3418 537 |
| ·64 | ·5523 842 | ·5151 337 | ·4782 379 | ·4419 943 | ·4066 732 |
| ·65 | ·6179 124 | ·5821 378 | ·5461 503 | ·5102 465$^+$ | ·4747 099 |
| ·66 | ·6807 626 | ·6473 978 | ·6133 180 | ·5787 944 | ·5440 982 |
| ·67 | ·7393 455$^-$ | ·7091 533 | ·6778 468 | ·6456 521 | ·6128 062 |
| ·68 | ·7923 646 | ·7658 821 | ·7380 125$^-$ | ·7089 243 | ·6788 049 |
| ·69 | ·8389 078 | ·8164 179 | ·7924 019 | ·7669 674 | ·7402 437 |
| ·70 | ·8784 941 | ·8600 260 | ·8400 188 | ·8185 227 | ·7956 099 |
| ·71 | ·9110 732 | ·8964 300 | ·8803 398 | ·8628 053 | ·8438 481 |
| ·72 | ·9369 795$^+$ | ·9257 872 | ·9133 158 | ·8995 335$^-$ | ·8844 230 |
| ·73 | ·9568 508 | ·9486 196 | ·9393 204 | ·9289 009 | ·9173 187 |
| ·74 | ·9715 263 | ·9657 139 | ·9590 573 | ·9514 967 | ·9429 773 |
| ·75 | ·9819 396 | ·9780 080 | ·9734 446 | ·9681 913 | ·9621 919 |
| ·76 | ·9890 219 | ·9864 814 | ·9834 932 | ·9800 076 | ·9759 737 |
| ·77 | ·9936 262 | ·9920 628 | ·9901 997 | ·9879 978 | ·9854 160 |
| ·78 | ·9964 787 | ·9955 656 | ·9944 634 | ·9931 437 | ·9915 764 |
| ·79 | ·9981 567 | ·9976 526 | ·9970 364 | ·9962 891 | ·9953 902 |
| ·80 | ·9990 901 | ·9988 284 | ·9985 043 | ·9981 063 | ·9976 216 |
| ·81 | ·9995 789 | ·9994 517 | ·9992 922 | ·9990 940 | ·9988 495$^+$ |
| ·82 | ·9998 184 | ·9997 610 | ·9996 880 | ·9995 962 | ·9994 816 |
| ·83 | ·9999 276 | ·9999 037 | ·9998 729 | ·9998 337 | ·9997 841 |
| ·84 | ·9999 736 | ·9999 644 | ·9999 525$^+$ | ·9999 372 | ·9999 176 |
| ·85 | ·9999 912 | ·9999 881 | ·9999 839 | ·9999 785$^+$ | ·9999 715$^+$ |
| ·86 | ·9999 974 | ·9999 964 | ·9999 951 | ·9999 934 | ·9999 912 |
| ·87 | ·9999 993 | ·9999 991 | ·9999 987 | ·9999 982 | ·9999 976 |
| ·88 | ·9999 998 | ·9999 998 | ·9999 997 | ·9999 996 | ·9999 994 |
| ·89 | 1·0000 000 | 1·0000 000 | ·9999 999 | ·9999 999 | ·9999 999 |
| ·90 | | | 1·0000 000 | 1·0000 000 | 1·0000 000 |

$x$ = ·35 to ·91 $q = 24$ $p$ = 46 to 50

| | $p = 46$ | $p = 47$ | $p = 48$ | $p = 49$ | $p = 50$ |
|---|---|---|---|---|---|
| $B(p, q)$ = | ·1807 1712 $\times \frac{1}{10^{19}}$ | ·1187 5696 $\times \frac{1}{10^{19}}$ | ·7861 3765 $\times \frac{1}{10^{20}}$ | ·5240 9176 $\times \frac{1}{10^{20}}$ | ·3517 8762 $\times \frac{1}{10^{20}}$ |
| $x$ | | | | | |
| ·35 | ·0000 001 | | | | |
| ·36 | ·0000 002 | ·0000 001 | ·0000 001 | | |
| ·37 | ·0000 006 | ·0000 003 | ·0000 002 | ·0000 001 | |
| ·38 | ·0000 013 | ·0000 007 | ·0000 004 | ·0000 002 | ·0000 001 |
| ·39 | ·0000 031 | ·0000 018 | ·0000 010 | ·0000 006 | ·0000 003 |
| ·40 | ·0000 069 | ·0000 041 | ·0000 024 | ·0000 014 | ·0000 008 |
| ·41 | ·0000 149 | ·0000 090 | ·0000 054 | ·0000 032 | ·0000 019 |
| ·42 | ·0000 310 | ·0000 192 | ·0000 118 | ·0000 072 | ·0000 044 |
| ·43 | ·0000 626 | ·0000 397 | ·0000 250$^{-}$ | ·0000 156 | ·0000 097 |
| ·44 | ·0001 227 | ·0000 795$^{-}$ | ·0000 512 | ·0000 327 | ·0000 208 |
| ·45 | ·0002 333 | ·0001 545$^{-}$ | ·0001 016 | ·0000 665$^{-}$ | ·0000 432 |
| ·46 | ·0004 310 | ·0002 915$^{-}$ | ·0001 959 | ·0001 309 | ·0000 869 |
| ·47 | ·0007 739 | ·0005 344 | ·0003 667 | ·0002 501 | ·0001 696 |
| ·48 | ·0013 520 | ·0009 526 | ·0006 671 | ·0004 644 | ·0003 214 |
| ·49 | ·0022 996 | ·0016 526 | ·0011 804 | ·0008 381 | ·0005 917 |
| ·50 | ·0038 103 | ·0027 913 | ·0020 326 | ·0014 714 | ·0010 591 |
| ·51 | ·0061 536 | ·0045 934 | ·0034 083 | ·0025 143 | ·0018 443 |
| ·52 | ·0096 918 | ·0073 680 | ·0055 684 | ·0041 841 | ·0031 265$^{-}$ |
| ·53 | ·0148 936 | ·0115 263 | ·0088 683 | ·0067 845$^{-}$ | ·0051 617 |
| ·54 | ·0223 414 | ·0175 931 | ·0137 742 | ·0107 238 | ·0083 035$^{-}$ |
| ·55 | ·0327 282 | ·0262 116 | ·0208 734 | ·0165 306 | ·0130 209 |
| ·56 | ·0468 391 | ·0381 343 | ·0308 742 | ·0248 604 | ·0199 119 |
| ·57 | ·0655 138 | ·0541 968 | ·0445 895$^{-}$ | ·0364 893 | ·0297 049 |
| ·58 | ·0895 893 | ·0752 703 | ·0629 011 | ·0522 893 | ·0432 455$^{+}$ |
| ·59 | ·1198 220 | ·1021 922 | ·0867 004 | ·0731 807 | ·0614 599 |
| ·60 | ·1567 953 | ·1356 779 | ·1168 070 | ·1000 591 | ·0852 936 |
| ·61 | ·2008 212 | ·1762 194 | ·1538 685$^{-}$ | ·1337 014 | ·1156 251 |
| ·62 | ·2518 491 | ·2239 836 | ·1982 524 | ·1746 549 | ·1531 570 |
| ·63 | ·3093 960 | ·2787 235$^{-}$ | ·2499 435$^{+}$ | ·2231 242 | ·1982 971 |
| ·64 | ·3725 137 | ·3397 207 | ·3084 627 | ·2788 717 | ·2510 441 |
| ·65 | ·4398 051 | ·4057 728 | ·3728 267 | ·3411 505$^{-}$ | ·3108 964 |
| ·66 | ·5094 944 | ·4752 368 | ·4415 622 | ·4086 874 | ·3768 056 |
| ·67 | ·5795 516 | ·5461 307 | ·5127 809 | ·4797 298 | ·4471 914 |
| ·68 | ·6478 555$^{+}$ | ·6162 871 | ·5843 157 | ·5521 579 | ·5200 262 |
| ·69 | ·7123 798 | ·6835 407 | ·6539 040 | ·6236 559 | ·5929 881 |
| ·70 | ·7713 732 | ·7459 247 | ·7193 932 | ·6919 221 | ·6636 661 |
| ·71 | ·8235 091 | ·8018 481 | ·7789 427 | ·7548 873 | ·7297 911 |
| ·72 | ·8679 828 | ·8502 274 | ·8311 876 | ·8109 102 | ·7894 575$^{-}$ |
| ·73 | ·9045 422 | ·8905 520 | ·8753 413 | ·8589 170 | ·8412 994 |
| ·74 | ·9334 506 | ·9228 761 | ·9112 215$^{+}$ | ·8984 647 | ·8845 936 |
| ·75 | ·9553 925$^{+}$ | ·9477 432 | ·9391 988 | ·9297 198 | ·9192 736 |
| ·76 | ·9713 410 | ·9660 597 | ·9600 818 | ·9533 616 | ·9458 569 |
| ·77 | ·9824 120 | ·9789 423 | ·9749 633 | ·9704 314 | ·9653 039 |
| ·78 | ·9897 290 | ·9875 676 | ·9850 567 | ·9821 597 | ·9788 394 |
| ·79 | ·9943 172 | ·9930 455$^{+}$ | ·9915 494 | ·9898 011 | ·9877 716 |
| ·80 | ·9970 355$^{+}$ | ·9963 323 | ·9954 944 | ·9945 029 | ·9933 373 |
| ·81 | ·9985 502 | ·9981 866 | ·9977 480 | ·9972 224 | ·9965 969 |
| ·82 | ·9993 396 | ·9991 650$^{-}$ | ·9989 516 | ·9986 929 | ·9983 811 |
| ·83 | ·9997 220 | ·9996 446 | ·9995 489 | ·9994 315$^{+}$ | ·9992 883 |
| ·84 | ·9998 928 | ·9998 614 | ·9998 222 | ·9997 735$^{+}$ | ·9997 134 |
| ·85 | ·9999 625$^{+}$ | ·9999 510 | ·9999 365$^{+}$ | ·9999 183 | ·9998 955$^{-}$ |
| ·86 | ·9999 883 | ·9999 845$^{+}$ | ·9999 797 | ·9999 736 | ·9999 659 |
| ·87 | ·9999 968 | ·9999 957 | ·9999 943 | ·9999 925$^{-}$ | ·9999 902 |
| ·88 | ·9999 992 | ·9999 990 | ·9999 986 | ·9999 982 | ·9999 976 |
| ·89 | ·9999 998 | ·9999 998 | ·9999 997 | ·9999 996 | ·9999 995$^{-}$ |
| ·90 | 1·0000 000 | 1·0000 000 | 1·0000 000 | ·9999 999 | ·9999 999 |
| ·91 | | | | 1·0000 000 | 1·0000 000 |

$x = \cdot17$ to $\cdot85$ $q = 25$ $p = 25$ to $30$

| $x$ | $p = 25$ | $p = 26$ | $p = 27$ | $p = 28$ | $p = 29$ | $p = 30$ |
|---|---|---|---|---|---|---|
| $B(p, q) =$ | $\cdot6328\ 5829 \times \frac{1}{10^{15}}$ | $\cdot3164\ 2914 \times \frac{1}{10^{15}}$ | $\cdot1613\ 1682 \times \frac{1}{10^{15}}$ | $\cdot8376\ 0656 \times \frac{1}{10^{16}}$ | $\cdot4425\ 0912 \times \frac{1}{10^{16}}$ | $\cdot2376\ 4379 \times \frac{1}{10^{16}}$ |
| ·17 | ·0000 001 | | | | | |
| ·18 | ·0000 002 | ·0000 001 | | | | |
| ·19 | ·0000 005⁻ | ·0000 002 | ·0000 001 | | | |
| ·20 | ·0000 013 | ·0000 005⁻ | ·0000 002 | ·0000 001 | | |
| ·21 | ·0000 033 | ·0000 013 | ·0000 005⁺ | ·0000 002 | ·0000 001 | |
| ·22 | ·0000 079 | ·0000 033 | ·0000 014 | ·0000 006 | ·0000 002 | ·0000 001 |
| ·23 | ·0000 180 | ·0000 079 | ·0000 034 | ·0000 014 | ·0000 006 | ·0000 002 |
| ·24 | ·0000 389 | ·0000 177 | ·0000 079 | ·0000 035⁻ | ·0000 015⁺ | ·0000 006 |
| ·25 | ·0000 803 | ·0000 380 | ·0000 177 | ·0000 081 | ·0000 037 | ·0000 016 |
| ·26 | ·0001 585⁺ | ·0000 780 | ·0000 378 | ·0000 180 | ·0000 084 | ·0000 039 |
| ·27 | ·0003 007 | ·0001 535⁺ | ·0000 771 | ·0000 381 | ·0000 185⁺ | ·0000 089 |
| ·28 | ·0005 493 | ·0002 904 | ·0001 511 | ·0000 774 | ·0000 390 | ·0000 194 |
| ·29 | ·0009 684 | ·0005 297 | ·0002 850⁺ | ·0001 510 | ·0000 788 | ·0000 406 |
| ·30 | ·0016 514 | ·0009 332 | ·0005 188 | ·0002 841 | ·0001 532 | ·0000 815⁺ |
| ·31 | ·0027 290 | ·0015 914 | ·0009 131 | ·0005 160 | ·0002 873 | ·0001 578 |
| ·32 | ·0043 781 | ·0026 314 | ·0015 565⁺ | ·0009 068 | ·0005 207 | ·0002 948 |
| ·33 | ·0068 290 | ·0042 261 | ·0025 742 | ·0015 445⁻ | ·0009 134 | ·0005 328 |
| ·34 | ·0103 714 | ·0066 015⁺ | ·0041 366 | ·0025 536 | ·0015 540 | ·0009 329 |
| ·35 | ·0153 560 | ·0100 436 | ·0064 680 | ·0041 041 | ·0025 676 | ·0015 847 |
| ·36 | ·0221 924 | ·0149 009 | ·0098 530 | ·0064 204 | ·0041 255⁻ | ·0026 155⁺ |
| ·37 | ·0313 393 | ·0215 824 | ·0146 400 | ·0097 880 | ·0064 541 | ·0041 996 |
| ·38 | ·0432 882 | ·0305 492 | ·0212 399 | ·0145 579 | ·0098 423 | ·0065 674 |
| ·39 | ·0585 402 | ·0422 992 | ·0301 184 | ·0211 452 | ·0146 460 | ·0100 137 |
| ·40 | ·0775 756 | ·0573 438 | ·0417 808 | ·0300 222 | ·0212 872 | ·0149 016 |
| ·41 | ·1008 180 | ·0761 774 | ·0567 493 | ·0417 033 | ·0302 469 | ·0216 624 |
| ·42 | ·1285 972 | ·0992 420 | ·0755 320 | ·0567 220 | ·0420 503 | ·0307 885⁻ |
| ·43 | ·1611 117 | ·1268 873 | ·0985 863 | ·0755 996 | ·0572 431 | ·0428 174 |
| ·44 | ·1983 959 | ·1593 324 | ·1262 786 | ·0988 073 | ·0763 593 | ·0583 080 |
| ·45 | ·2402 957 | ·1966 306 | ·1588 436 | ·1267 246 | ·0998 823 | ·0778 069 |
| ·46 | ·2864 555⁺ | ·2386 428 | ·1963 470 | ·1595 966 | ·1282 012 | ·1018 075⁻ |
| ·47 | ·3363 193 | ·2850 217 | ·2386 566 | ·1974 947 | ·1615 663 | ·1307 049 |
| ·48 | ·3891 462 | ·3352 115⁺ | ·2854 257 | ·2402 865⁺ | ·2000 482 | ·1647 494 |
| ·49 | ·4440 416 | ·3884 627 | ·3360 902 | ·2876 167 | ·2435 057 | ·2040 036 |
| ·50 | ·5000 000ᵉ | ·4438 624 | ·3898 840 | ·3389 043 | ·2915 661 | ·2483 087 |
| ·51 | ·5559 584 | ·5003 795⁺ | ·4458 695⁻ | ·3933 581 | ·3436 223 | ·2972 652 |
| ·52 | ·6108 538 | ·5569 191 | ·5029 844 | ·4500 086 | ·3988 491 | ·3502 299 |
| ·53 | ·6636 807 | ·6123 832 | ·5600 991 | ·5077 570 | ·4562 373 | ·4063 344 |
| ·54 | ·7135 445⁻ | ·6657 318 | ·6160 801 | ·5654 354 | ·5146 461 | ·4645 222 |
| ·55 | ·7597 043 | ·7160 393 | ·6698 552 | ·6218 750⁻ | ·5728 666 | ·5236 048 |
| ·56 | ·8016 041 | ·7625 406 | ·7204 723 | ·6759 733 | ·6296 943 | ·5823 302 |
| ·57 | ·8388 883 | ·8046 639 | ·7671 486 | ·7267 572 | ·6840 000 | ·6394 588 |
| ·58 | ·8714 028 | ·8420 475⁺ | ·8093 052 | ·7734 341 | ·7347 958 | ·6938 392 |
| ·59 | ·8991 820 | ·8745 415⁻ | ·8465 839 | ·8154 268 | ·7812 875⁺ | ·7444 759 |
| ·60 | ·9224 244 | ·9021 926 | ·8788 482 | ·8523 913 | ·8229 107 | ·7905 836 |
| ·61 | ·9414 598 | ·9252 187 | ·9061 667 | ·8842 145⁺ | ·8593 458 | ·8316 216 |
| ·62 | ·9567 118 | ·9439 728 | ·9287 840 | ·9109 962 | ·8905 148 | ·8673 073 |
| ·63 | ·9686 607 | ·9589 039 | ·9470 831 | ·9330 163 | ·9165 582 | ·8976 086 |
| ·64 | ·9778 076 | ·9705 161 | ·9615 420 | ·9506 933 | ·9377 988 | ·9227 167 |
| ·65 | ·9846 440 | ·9793 316 | ·9726 911 | ·9645 381 | ·9546 963 | ·9430 048 |
| ·66 | ·9896 286 | ·9858 588 | ·9810 740 | ·9751 089 | ·9677 974 | ·9589 783 |
| ·67 | ·9931 710 | ·9905 680 | ·9872 141 | ·9829 696 | ·9776 883 | ·9712 214 |
| ·68 | ·9956 219 | ·9938 751 | ·9915 909 | ·9886 570 | ·9849 519 | ·9803 474 |
| ·69 | ·9972 710 | ·9961 333 | ·9946 236 | ·9926 561 | ·9901 348 | ·9869 554 |
| ·70 | ·9983 486 | ·9976 305⁻ | ·9966 637 | ·9953 854 | ·9937 236 | ·9915 976 |
| ·71 | ·9990 316 | ·9985 929 | ·9979 939 | ·9971 906 | ·9961 314 | ·9947 570 |
| ·72 | ·9994 507 | ·9991 919 | ·9988 335⁺ | ·9983 461 | ·9976 944 | ·9968 368 |
| ·73 | ·9996 993 | ·9995 521 | ·9993 454 | ·9990 605⁻ | ·9986 742 | ·9981 588 |
| ·74 | ·9998 415⁻ | ·9997 610 | ·9996 464 | ·9994 862 | ·9992 661 | ·9989 684 |
| ·75 | ·9999 197 | ·9998 775⁻ | ·9998 166 | ·9997 302 | ·9996 100 | ·9994 452 |
| ·76 | ·9999 611 | ·9999 399 | ·9999 089 | ·9998 644 | ·9998 016 | ·9997 144 |
| ·77 | ·9999 820 | ·9999 718 | ·9999 568 | ·9999 350⁻ | ·9999 037 | ·9998 598 |
| ·78 | ·9999 921 | ·9999 875⁻ | ·9999 806 | ·9999 704 | ·9999 556 | ·9999 346 |
| ·79 | ·9999 967 | ·9999 947 | ·9999 917 | ·9999 872 | ·9999 806 | ·9999 711 |
| ·80 | ·9999 987 | ·9999 979 | ·9999 967 | ·9999 948 | ·9999 920 | ·9999 880 |
| ·81 | ·9999 995⁺ | ·9999 992 | ·9999 988 | ·9999 980 | ·9999 969 | ·9999 953 |
| ·82 | ·9999 998 | ·9999 997 | ·9999 996 | ·9999 993 | ·9999 989 | ·9999 983 |
| ·83 | ·9999 999 | ·9999 999 | ·9999 999 | ·9999 998 | ·9999 996 | ·9999 994 |
| ·84 | 1·0000 000 | 1·0000 000 | 1·0000 000 | ·9999 999 | ·9999 999 | ·9999 998 |
| ·85 | | | | 1·0000 000 | 1·0000 000 | 1·0000 000 |

$x$ = ·23 to ·87 $q = 25$ $p$ = 31 to 36

| $x$ | $p = 31$ | $p = 32$ | $p = 33$ | $p = 34$ | $p = 35$ | $p = 36$ |
|---|---|---|---|---|---|---|
| $B(p, q) =$ | ·1296 2389 $\times \frac{1}{10^{16}}$ | ·7175 6079 $\times \frac{1}{10^{17}}$ | ·4028 4115 $\times \frac{1}{10^{17}}$ | ·2292 0272 $\times \frac{1}{10^{17}}$ | ·1320 8292 $\times \frac{1}{10^{17}}$ | ·7704 8372 $\times \frac{1}{10^{18}}$ |
| ·23 | ·0000 001 | | | | | |
| ·24 | ·0000 003 | ·0000 001 | | | | |
| ·25 | ·0000 007 | ·0000 003 | ·0000 001 | ·0000 001 | | |
| ·26 | ·0000 018 | ·0000 008 | ·0000 004 | ·0000 002 | ·0000 001 | |
| ·27 | ·0000 042 | ·0000 020 | ·0000 009 | ·0000 004 | ·0000 002 | ·0000 001 |
| ·28 | ·0000 095+ | ·0000 046 | ·0000 022 | ·0000 010 | ·0000 005− | ·0000 002 |
| ·29 | ·0000 206 | ·0000 103 | ·0000 051 | ·0000 025+ | ·0000 012 | ·0000 006 |
| ·30 | ·0000 428 | ·0000 222 | ·0000 114 | ·0000 057 | ·0000 029 | ·0000 014 |
| ·31 | ·0000 855+ | ·0000 458 | ·0000 242 | ·0000 126 | ·0000 065+ | ·0000 033 |
| ·32 | ·0001 648 | ·0000 909 | ·0000 496 | ·0000 267 | ·0000 142 | ·0000 075+ |
| ·33 | ·0003 068 | ·0001 744 | ·0000 980 | ·0000 544 | ·0000 299 | ·0000 162 |
| ·34 | ·0005 528 | ·0003 235− | ·0001 871 | ·0001 069 | ·0000 605− | ·0000 338 |
| ·35 | ·0009 655+ | ·0005 810 | ·0003 455+ | ·0002 032 | ·0001 182 | ·0000 680 |
| ·36 | ·0016 371 | ·0010 122 | ·0006 185− | ·0003 737 | ·0002 233 | ·0001 321 |
| ·37 | ·0026 981 | ·0017 125− | ·0010 743 | ·0006 664 | ·0004 090 | ·0002 484 |
| ·38 | ·0043 274 | ·0028 172 | ·0018 129 | ·0011 537 | ·0007 264 | ·0004 527 |
| ·39 | ·0067 618 | ·0045 117 | ·0029 760 | ·0019 415+ | ·0012 533 | ·0008 008 |
| ·40 | ·0103 040 | ·0070 411 | ·0047 572 | ·0031 791 | ·0021 024 | ·0013 763 |
| ·41 | ·0153 271 | ·0107 187 | ·0074 121 | ·0050 705− | ·0034 327 | ·0023 008 |
| ·42 | ·0222 745+ | ·0159 303 | ·0112 672 | ·0078 844 | ·0054 607 | ·0037 447 |
| ·43 | ·0316 519 | ·0231 338 | ·0167 238 | ·0119 630 | ·0084 708 | ·0059 394 |
| ·44 | ·0440 114 | ·0328 508 | ·0242 572 | ·0177 260 | ·0128 238 | ·0091 878 |
| ·45 | ·0599 258 | ·0456 497 | ·0344 073 | ·0256 689 | ·0189 609 | ·0138 724 |
| ·46 | ·0799 535− | ·0621 178 | ·0477 600 | ·0363 522 | ·0274 003 | ·0204 588 |
| ·47 | ·1045 963 | ·0828 250+ | ·0649 181 | ·0503 810 | ·0387 257 | ·0294 913 |
| ·48 | ·1342 513 | ·1082 787 | ·0864 618 | ·0683 735− | ·0535 624 | ·0415 781 |
| ·49 | ·1691 627 | ·1388 736 | ·1129 007 | ·0909 182 | ·0725 434 | ·0573 658 |
| ·50 | ·2093 771 | ·1748 410 | ·1446 219 | ·1185 236 | ·0962 632 | ·0775 010 |
| ·51 | ·2547 093 | ·2162 031 | ·1818 363 | ·1515 623 | ·1252 239 | ·1025 804 |
| ·52 | ·3047 223 | ·2627 380 | ·2245 322 | ·1902 164 | ·1597 763 | ·1330 934 |
| ·53 | ·3587 269 | ·3139 606 | ·2724 398 | ·2344 294 | ·2000 635+ | ·1693 601 |
| ·54 | ·4158 018 | ·3691 245− | ·3250 145− | ·2838 718 | ·2459 722 | ·2114 727 |
| ·55 | ·4748 356 | ·4272 463 | ·3814 415+ | ·3379 271 | ·2971 002 | ·2592 479 |
| ·56 | ·5345 871 | ·4871 521 | ·4406 658 | ·3957 009 | ·3527 461 | ·3121 969 |
| ·57 | ·5937 595+ | ·5475 443 | ·5014 446 | ·4560 573 | ·4119 249 | ·3695 199 |
| ·58 | ·6510 805− | ·6070 804 | ·5624 203 | ·5176 791 | ·4734 115+ | ·4301 305+ |
| ·59 | ·7053 820 | ·6644 595+ | ·6222 071 | ·5791 480 | ·5358 102 | ·4927 078 |
| ·60 | ·7556 704 | ·7185 048 | ·6794 808 | ·6390 378 | ·5976 432 | ·5557 756 |
| ·61 | ·8011 803 | ·7682 350+ | ·7330 659 | ·6960 105− | ·6574 510 | ·6178 009 |
| ·62 | ·8414 077 | ·8129 182 | ·7820 070 | ·7489 040 | ·7138 926 | ·6773 008 |
| ·63 | ·8761 198 | ·8521 009 | ·8256 201 | ·7968 041 | ·7658 354 | ·7329 466 |
| ·64 | ·9053 421 | ·8856 136 | ·8635 176 | ·8390 915+ | ·8124 240 | ·7836 535+ |
| ·65 | ·9293 258 | ·9135 509 | ·8956 068 | ·8754 606 | ·8531 220 | ·8286 452 |
| ·66 | ·9485 011 | ·9362 327 | ·9220 627 | ·9059 089 | ·8877 216 | ·8674 869 |
| ·67 | ·9634 223 | ·9541 514 | ·9432 813 | ·9307 016 | ·9163 238 | ·9000 851 |
| ·68 | ·9747 114 | ·9679 119 | ·9598 204 | ·9503 166 | ·9392 923 | ·9266 552 |
| ·69 | ·9830 066 | ·9781 725− | ·9723 353 | ·9653 784 | ·9571 898 | ·9476 652 |
| ·70 | ·9889 189 | ·9855 921 | ·9815 168 | ·9765 894 | ·9707 055− | ·9637 625− |
| ·71 | ·9930 005− | ·9907 878 | ·9880 386 | ·9846 671 | ·9805 836 | ·9756 962 |
| ·72 | ·9957 254 | ·9943 057 | ·9925 168 | ·9902 920 | ·9875 596 | ·9842 431 |
| ·73 | ·9974 816 | ·9966 044 | ·9954 839 | ·9940 711 | ·9923 116 | ·9901 465+ |
| ·74 | ·9985 719 | ·9980 514 | ·9973 773 | ·9965 156 | ·9954 279 | ·9940 711 |
| ·75 | ·9992 227 | ·9989 267 | ·9985 381 | ·9980 348 | ·9973 908 | ·9965 766 |
| ·76 | ·9995 951 | ·9994 343 | ·9992 203 | ·9989 395− | ·9985 753 | ·9981 088 |
| ·77 | ·9997 988 | ·9997 156 | ·9996 034 | ·9994 542 | ·9992 583 | ·9990 039 |
| ·78 | ·9999 050+ | ·9998 642 | ·9998 084 | ·9997 332 | ·9996 332 | ·9995 017 |
| ·79 | ·9999 576 | ·9999 386 | ·9999 124 | ·9998 766 | ·9998 284 | ·9997 642 |
| ·80 | ·9999 822 | ·9999 739 | ·9999 623 | ·9999 463 | ·9999 245− | ·9998 950+ |
| ·81 | ·9999 930 | ·9999 896 | ·9999 848 | ·9999 781 | ·9999 689 | ·9999 563 |
| ·82 | ·9999 974 | ·9999 962 | ·9999 943 | ·9999 917 | ·9999 881 | ·9999 831 |
| ·83 | ·9999 991 | ·9999 987 | ·9999 980 | ·9999 971 | ·9999 958 | ·9999 940 |
| ·84 | ·9999 997 | ·9999 996 | ·9999 994 | ·9999 991 | ·9999 986 | ·9999 980 |
| ·85 | ·9999 999 | ·9999 999 | ·9999 998 | ·9999 997 | ·9999 996 | ·9999 994 |
| ·86 | 1·0000 000 | 1·0000 000 | 1·0000 000 | ·9999 999 | ·9999 999 | ·9999 998 |
| ·87 | | | | 1·0000 000 | 1·0000 000 | 1·0000 000 |

$x$ = ·28 to ·89 $q = 25$ $p$ = 37 to 43

| $x$ | $p = 37$ | $p = 38$ | $p = 39$ | $p = 40$ | $p = 41$ | $p = 42$ | $p = 43$ |
|---|---|---|---|---|---|---|---|
| $B(p, q) =$ | $·4547\ 1171 \times \frac{1}{10^{18}}$ | $·2713\ 6021 \times \frac{1}{10^{18}}$ | $·1636\ 7759 \times \frac{1}{10^{18}}$ | $·9974\ 1030 \times \frac{1}{10^{19}}$ | $·6137\ 9095 \times \frac{1}{10^{19}}$ | $·3812\ 9438 \times \frac{1}{10^{19}}$ | $·2390\ 2036 \times \frac{1}{10^{19}}$ |
| ·28 | ·0000 001 | | | | | | |
| ·29 | ·0000 003 | ·0000 001 | ·0000 001 | | | | |
| ·30 | ·0000 007 | ·0000 003 | ·0000 002 | ·0000 001 | | | |
| ·31 | ·0000 017 | ·0000 008 | ·0000 004 | ·0000 002 | ·0000 001 | | |
| ·32 | ·0000 039 | ·0000 020 | ·0000 010 | ·0000 005+ | ·0000 003 | ·0000 001 | ·0000 001 |
| ·33 | ·0000 087 | ·0000 047 | ·0000 025− | ·0000 013 | ·0000 007 | ·0000 003 | ·0000 002 |
| ·34 | ·0000 187 | ·0000 103 | ·0000 056 | ·0000 030 | ·0000 016 | ·0000 008 | ·0000 004 |
| ·35 | ·0000 387 | ·0000 219 | ·0000 122 | ·0000 068 | ·0000 037 | ·0000 020 | ·0000 011 |
| ·36 | ·0000 774 | ·0000 449 | ·0000 258 | ·0000 147 | ·0000 083 | ·0000 046 | ·0000 026 |
| ·37 | ·0001 494 | ·0000 890 | ·0000 525+ | ·0000 307 | ·0000 178 | ·0000 102 | ·0000 058 |
| ·38 | ·0002 794 | ·0001 708 | ·0001 034 | ·0000 621 | ·0000 370 | ·0000 218 | ·0000 128 |
| ·39 | ·0005 067 | ·0003 176 | ·0001 973 | ·0001 215− | ·0000 742 | ·0000 449 | ·0000 270 |
| ·40 | ·0008 923 | ·0005 731 | ·0003 648 | ·0002 302 | ·0001 440 | ·0000 894 | ·0000 550+ |
| ·41 | ·0015 273 | ·0010 045− | ·0006 547 | ·0004 231 | ·0002 711 | ·0001 724 | ·0001 087 |
| ·42 | ·0025 436 | ·0017 118 | ·0011 419 | ·0007 552 | ·0004 953 | ·0003 223 | ·0002 081 |
| ·43 | ·0041 253 | ·0028 392 | ·0019 369 | ·0013 102 | ·0008 790 | ·0005 851 | ·0003 864 |
| ·44 | ·0065 213 | ·0045 871 | ·0031 985− | ·0022 115+ | ·0015 167 | ·0010 320 | ·0006 969 |
| ·45 | ·0100 560 | ·0072 246 | ·0051 458 | ·0036 347 | ·0025 467 | ·0017 705− | ·0012 216 |
| ·46 | ·0151 370 | ·0111 011 | ·0080 720 | ·0058 211 | ·0041 645+ | ·0029 564 | ·0020 831 |
| ·47 | ·0222 577 | ·0166 527 | ·0123 545− | ·0090 912 | ·0066 372 | ·0048 086 | ·0034 581 |
| ·48 | ·0319 907 | ·0244 037 | ·0184 618 | ·0138 546 | ·0103 163 | ·0076 237 | ·0055 927 |
| ·49 | ·0449 708 | ·0349 576 | ·0269 524 | ·0206 159 | ·0156 481 | ·0117 890 | ·0088 174 |
| ·50 | ·0618 657 | ·0489 772 | ·0384 629 | ·0299 706 | ·0231 767 | ·0177 914 | ·0135 600 |
| ·51 | ·0833 334 | ·0671 504 | ·0536 843 | ·0425 904 | ·0335 378 | ·0262 184 | ·0203 524 |
| ·52 | ·1099 682 | ·0901 430 | ·0733 230 | ·0591 941 | ·0474 389 | ·0377 480 | ·0298 291 |
| ·53 | ·1422 387 | ·1185 405− | ·0980 477 | ·0805 028 | ·0656 247 | ·0531 235− | ·0427 117 |
| ·54 | ·1804 232 | ·1527 807 | ·1284 262 | ·1071 816 | ·0888 263 | ·0731 123 | ·0597 779 |
| ·55 | ·2245 500− | ·1930 874 | ·1648 539 | ·1397 695− | ·1176 952 | ·0984 475− | ·0818 120 |
| ·56 | ·2743 509 | ·2394 098 | ·2074 847 | ·1786 048 | ·1527 284 | ·1297 552 | ·1095 387 |
| ·57 | ·3292 352 | ·2913 785− | ·2561 717 | ·2237 544 | ·1941 898 | ·1674 736 | ·1435 434 |
| ·58 | ·3882 922 | ·3482 858 | ·3104 270 | ·2749 563 | ·2420 395− | ·2117 720 | ·1841 854 |
| ·59 | ·4503 237 | ·4090 966 | ·3694 100 | ·3315 857 | ·2958 796 | ·2624 813 | ·2315 162 |
| ·60 | ·5139 079 | ·4724 928 | ·4319 497 | ·3926 540 | ·3549 302 | ·3190 465+ | ·2852 134 |
| ·61 | ·5774 899 | ·5369 502 | ·4966 024 | ·4568 444 | ·4180 405+ | ·3805 144 | ·3445 429 |
| ·62 | ·6394 892 | ·6008 396 | ·5617 426 | ·5225 853 | ·4837 414 | ·4455 606 | ·4083 616 |
| ·63 | ·6984 134 | ·6625 455+ | ·6256 771 | ·5881 564 | ·5503 356 | ·5125 609 | ·4751 639 |
| ·64 | ·7529 650+ | ·7205 845+ | ·6867 724 | ·6518 160 | ·6160 205+ | ·5797 013 | ·5431 745− |
| ·65 | ·8021 287 | ·7737 131 | ·7435 775+ | ·7119 352 | ·6790 272 | ·6451 159 | ·6104 780 |
| ·66 | ·8452 288 | ·8210 096 | ·7949 292 | ·7671 236 | ·7377 609 | ·7070 374 | ·6751 728 |
| ·67 | ·8819 518 | ·8619 219 | ·8400 261 | ·8163 281 | ·7909 238 | ·7639 395+ | ·7355 289 |
| ·68 | ·9123 332 | ·8962 770 | ·8784 632 | ·8588 953 | ·8376 055− | ·8146 540 | ·7901 287 |
| ·69 | ·9367 119 | ·9242 518 | ·9102 244 | ·8945 892 | ·8773 280 | ·8584 459 | ·8379 722 |
| ·70 | ·9556 623 | ·9463 142 | ·9356 378 | ·9235 652 | ·9100 439 | ·8950 385− | ·8785 326 |
| ·71 | ·9699 127 | ·9631 430 | ·9553 008 | ·9463 064 | ·9360 888 | ·9245 877 | ·9117 559 |
| ·72 | ·9802 634 | ·9755 393 | ·9699 898 | ·9635 353 | ·9560 997 | ·9476 122 | ·9380 092 |
| ·73 | ·9875 123 | ·9843 420 | ·9805 660 | ·9761 132 | ·9709 123 | ·9648 933 | ·9579 886 |
| ·74 | ·9923 976 | ·9903 560 | ·9878 911 | ·9849 445− | ·9814 558 | ·9773 629 | ·9726 034 |
| ·75 | ·9955 588 | ·9943 004 | ·9927 605− | ·9908 948 | ·9886 560 | ·9859 940 | ·9828 566 |
| ·76 | ·9975 180 | ·9967 776 | ·9958 595+ | ·9947 324 | ·9933 619 | ·9917 106 | ·9897 384 |
| ·77 | ·9986 775− | ·9982 631 | ·9977 425− | ·9970 949 | ·9962 971 | ·9953 232 | ·9941 448 |
| ·78 | ·9993 307 | ·9991 109 | ·9988 311 | ·9984 785+ | ·9980 386 | ·9974 945+ | ·9968 277 |
| ·79 | ·9996 797 | ·9995 696 | ·9994 277 | ·9992 466 | ·9990 177 | ·9987 311 | ·9983 752 |
| ·80 | ·9998 558 | ·9998 040 | ·9997 364 | ·9996 491 | ·9995 373 | ·9993 955+ | ·9992 173 |
| ·81 | ·9999 393 | ·9999 165+ | ·9998 865− | ·9998 471 | ·9997 962 | ·9997 307 | ·9996 474 |
| ·82 | ·9999 762 | ·9999 670 | ·9999 546 | ·9999 381 | ·9999 166 | ·9998 886 | ·9998 525− |
| ·83 | ·9999 914 | ·9999 879 | ·9999 832 | ·9999 769 | ·9999 685+ | ·9999 575+ | ·9999 431 |
| ·84 | ·9999 972 | ·9999 960 | ·9999 944 | ·9999 921 | ·9999 892 | ·9999 852 | ·9999 800 |
| ·85 | ·9999 992 | ·9999 988 | ·9999 983 | ·9999 976 | ·9999 966 | ·9999 954 | ·9999 936 |
| ·86 | ·9999 998 | ·9999 997 | ·9999 995+ | ·9999 993 | ·9999 991 | ·9999 987 | ·9999 982 |
| ·87 | ·9999 999 | ·9999 999 | ·9999 999 | ·9999 998 | ·9999 998 | ·9999 997 | ·9999 996 |
| ·88 | 1·0000 000 | 1·0000 000 | 1·0000 000 | 1·0000 000 | 1·0000 000 | ·9999 999 | ·9999 999 |
| ·89 | | | | | | 1·0000 000 | 1·0000 000 |

$x$ = ·33 to ·90 $q = 25$ $p$ = 44 to 50

| | $p = 44$ | $p = 45$ | $p = 46$ | $p = 47$ | $p = 48$ | $p = 49$ | $p = 50$ |
|---|---|---|---|---|---|---|---|
| $B(p, q)$ = | ·1511 4523 $\times \frac{1}{10^{19}}$ | ·9638 2463 $\times \frac{1}{10^{20}}$ | ·6196 0155 $\times \frac{1}{10^{20}}$ | ·4014 3199 $\times \frac{1}{10^{20}}$ | ·2620 4588 $\times \frac{1}{10^{20}}$ | ·1723 0414 $\times \frac{1}{10^{20}}$ | ·1140 9328 $\times \frac{1}{10^{20}}$ |
| $x$ | | | | | | | |
| ·33 | ·0000 001 | | | | | | |
| ·34 | ·0000 002 | ·0000 001 | ·0000 001 | | | | |
| ·35 | ·0000 006 | ·0000 003 | ·0000 002 | ·0000 001 | | | |
| ·36 | ·0000 014 | ·0000 008 | ·0000 004 | ·0000 002 | ·0000 001 | ·0000 001 | |
| ·37 | ·0000 033 | ·0000 019 | ·0000 010 | ·0000 006 | ·0000 003 | ·0000 002 | ·0000 001 |
| ·38 | ·0000 074 | ·0000 043 | ·0000 025− | ·0000 014 | ·0000 008 | ·0000 004 | ·0000 002 |
| ·39 | ·0000 161 | ·0000 095+ | ·0000 056 | ·0000 033 | ·0000 019 | ·0000 011 | ·0000 006 |
| ·40 | ·0000 336 | ·0000 204 | ·0000 123 | ·0000 074 | ·0000 044 | ·0000 026 | ·0000 015+ |
| ·41 | ·0000 681 | ·0000 423 | ·0000 261 | ·0000 160 | ·0000 097 | ·0000 059 | ·0000 035+ |
| ·42 | ·0001 334 | ·0000 849 | ·0000 536 | ·0000 336 | ·0000 210 | ·0000 130 | ·0000 080 |
| ·43 | ·0002 534 | ·0001 649 | ·0001 066 | ·0000 685− | ·0000 437 | ·0000 277 | ·0000 174 |
| ·44 | ·0004 671 | ·0003 109 | ·0002 055+ | ·0001 349 | ·0000 880 | ·0000 571 | ·0000 368 |
| ·45 | ·0008 367 | ·0005 691 | ·0003 844 | ·0002 579 | ·0001 720 | ·0001 139 | ·0000 750+ |
| ·46 | ·0014 572 | ·0010 122 | ·0006 983 | ·0004 786 | ·0003 259 | ·0002 206 | ·0001 484 |
| ·47 | ·0024 691 | ·0017 507 | ·0012 330 | ·0008 627 | ·0005 998 | ·0004 145− | ·0002 847 |
| ·48 | ·0040 737 | ·0029 469 | ·0021 176 | ·0015 118 | ·0010 726 | ·0007 563 | ·0005 301 |
| ·49 | ·0065 487 | ·0048 307 | ·0035 398 | ·0025 773 | ·0018 649 | ·0013 412 | ·0009 590 |
| ·50 | ·0102 635− | ·0077 161 | ·0057 632 | ·0042 773 | ·0031 549 | ·0023 131 | ·0016 861 |
| ·51 | ·0156 910 | ·0120 170 | ·0091 439 | ·0069 141 | ·0051 963 | ·0038 821 | ·0028 836 |
| ·52 | ·0234 130 | ·0182 568 | ·0141 456 | ·0108 924 | ·0083 369 | ·0063 436 | ·0047 994 |
| ·53 | ·0341 136 | ·0270 709 | ·0213 476 | ·0167 316 | ·0130 358 | ·0100 977 | ·0077 778 |
| ·54 | ·0485 584 | ·0391 952 | ·0314 424 | ·0250 717 | ·0198 749 | ·0156 654 | ·0122 789 |
| ·55 | ·0675 557 | ·0554 379 | ·0452 186 | ·0366 654 | ·0295 590 | ·0236 962 | ·0188 923 |
| ·56 | ·0918 987 | ·0766 321 | ·0635 232 | ·0523 521 | ·0429 018 | ·0349 636 | ·0283 408 |
| ·57 | ·1222 901 | ·1035 678 | ·0872 046 | ·0730 112 | ·0607 899 | ·0503 406 | ·0414 673 |
| ·58 | ·1592 548 | ·1369 079 | ·1170 341 | ·0994 933 | ·0841 246 | ·0707 538 | ·0592 003 |
| ·59 | ·2030 500+ | ·1770 940 | ·1536 125+ | ·1325 302 | ·1137 400 | ·0971 107 | ·0824 938 |
| ·60 | ·2535 833 | ·2242 536 | ·1972 702 | ·1726 333 | ·1503 028 | ·1302 053 | ·1122 407 |
| ·61 | ·3103 532 | ·2781 217 | ·2479 745− | ·2199 900 | ·1942 026 | ·1706 071 | ·1491 642 |
| ·62 | ·3724 256 | ·3379 924 | ·3052 580 | ·2743 737 | ·2454 476 | ·2185 464 | ·1936 984 |
| ·63 | ·4384 540 | ·4027 119 | ·3681 850− | ·3350 842 | ·3035 822 | ·2738 127 | ·2458 719 |
| ·64 | ·5067 496 | ·4707 221 | ·4353 671 | ·4009 344 | ·3676 446 | ·3356 863 | ·3052 151 |
| ·65 | ·5753 970 | ·5401 565− | ·5050 335− | ·4702 923 | ·4361 793 | ·4029 192 | ·3707 112 |
| ·66 | ·6424 042 | ·6089 801 | ·5751 550− | ·5411 828 | ·5073 118 | ·4737 795+ | ·4408 084 |
| ·67 | ·7058 696 | ·6751 587 | ·6436 083 | ·6114 407 | ·5788 829 | ·5461 624 | ·5135 020 |
| ·68 | ·7641 433 | ·7368 350+ | ·7083 616 | ·6788 978 | ·6486 315+ | ·6177 599 | ·5864 852 |
| ·69 | ·8159 607 | ·7924 884 | ·7676 548 | ·7415 794 | ·7144 000 | ·6862 694 | ·6573 522 |
| ·70 | ·8605 296 | ·8410 537 | ·8201 496 | ·7978 821 | ·7743 354 | ·7496 114 | ·7238 278 |
| ·71 | ·8975 602 | ·8819 837 | ·8650 261 | ·8467 044 | ·8270 535+ | ·8061 253 | ·7839 884 |
| ·72 | ·9272 360 | ·9152 484 | ·9020 140 | ·8875 138 | ·8717 424 | ·8547 094 | ·8364 389 |
| ·73 | ·9501 349 | ·9412 745+ | ·9313 568 | ·9203 395+ | ·9081 900 | ·8948 863 | ·8804 178 |
| ·74 | ·9671 157 | ·9608 397 | ·9537 185+ | ·9456 995− | ·9367 352 | ·9267 849 | ·9158 152 |
| ·75 | ·9791 903 | ·9749 407 | ·9700 537 | ·9644 761 | ·9581 567 | ·9510 475+ | ·9431 040 |
| ·76 | ·9874 030 | ·9846 600 | ·9814 634 | ·9777 665+ | ·9735 222 | ·9686 837 | ·9632 052 |
| ·77 | ·9927 310 | ·9910 485+ | ·9890 621 | ·9867 346 | ·9840 272 | ·9809 002 | ·9773 130 |
| ·78 | ·9960 172 | ·9950 402 | ·9938 718 | ·9924 848 | ·9908 507 | ·9889 386 | ·9867 168 |
| ·79 | ·9979 371 | ·9974 023 | ·9967 544 | ·9959 755+ | ·9950 460 | ·9939 446 | ·9926 483 |
| ·80 | ·9989 951 | ·9987 204 | ·9983 834 | ·9979 732 | ·9974 774 | ·9968 825+ | ·9961 735− |
| ·81 | ·9995 423 | ·9994 106 | ·9992 472 | ·9990 457 | ·9987 991 | ·9984 995+ | ·9981 380 |
| ·82 | ·9998 064 | ·9997 479 | ·9996 744 | ·9995 827 | ·9994 691 | ·9993 294 | ·9991 587 |
| ·83 | ·9999 245+ | ·9999 007 | ·9998 703 | ·9998 319 | ·9997 838 | ·9997 239 | ·9996 499 |
| ·84 | ·9999 731 | ·9999 643 | ·9999 528 | ·9999 382 | ·9999 196 | ·9998 963 | ·9998 670 |
| ·85 | ·9999 914 | ·9999 884 | ·9999 845+ | ·9999 795− | ·9999 730 | ·9999 648 | ·9999 544 |
| ·86 | ·9999 975+ | ·9999 966 | ·9999 955− | ·9999 939 | ·9999 920 | ·9999 894 | ·9999 861 |
| ·87 | ·9999 994 | ·9999 992 | ·9999 988 | ·9999 984 | ·9999 979 | ·9999 972 | ·9999 963 |
| ·88 | ·9999 999 | ·9999 998 | ·9999 997 | ·9999 997 | ·9999 995+ | ·9999 994 | ·9999 991 |
| ·89 | 1·0000 000 | 1·0000 000 | 1·0000 000 | ·9999 999 | ·9999 999 | ·9999 999 | ·9999 998 |
| ·90 | | | | 1·0000 000 | 1·0000 000 | 1·0000 000 | 1·0000 000 |

$x$ = ·18 to ·85 $q = 26$ $p$ = 26 to 31

| | $p = 26$ | $p = 27$ | $p = 28$ | $p = 29$ | $p = 30$ | $p = 31$ |
|---|---|---|---|---|---|---|
| $B(p, q)$ = | ·1551 1233$\times\frac{1}{10^{15}}$ | ·7755 6163$\times\frac{1}{10^{16}}$ | ·3950 9743$\times\frac{1}{10^{16}}$ | ·2048 6534$\times\frac{1}{10^{16}}$ | ·1080 1990$\times\frac{1}{10^{16}}$ | ·5786 7806$\times\frac{1}{10^{17}}$ |
| $x$ | | | | | | |
| ·18 | ·0000 001 | | | | | |
| ·19 | ·0000 003 | ·0000 001 | | | | |
| ·20 | ·0000 008 | ·0000 003 | ·0000 001 | | | |
| ·21 | ·0000 021 | ·0000 009 | ·0000 003 | ·0000 001 | | |
| ·22 | ·0000 053 | ·0000 022 | ·0000 009 | ·0000 004 | ·0000 002 | ·0000 001 |
| ·23 | ·0000 125$^{+}$ | ·0000 055$^{-}$ | ·0000 024 | ·0000 010 | ·0000 004 | ·0000 002 |
| ·24 | ·0000 279 | ·0000 127 | ·0000 057 | ·0000 025$^{+}$ | ·0000 011 | ·0000 005$^{-}$ |
| ·25 | ·0000 591 | ·0000 281 | ·0000 131 | ·0000 060 | ·0000 027 | ·0000 012 |
| ·26 | ·0001 199 | ·0000 591 | ·0000 287 | ·0000 137 | ·0000 065$^{-}$ | ·0000 030 |
| ·27 | ·0002 330 | ·0001 192 | ·0000 600 | ·0000 297 | ·0000 145$^{+}$ | ·0000 070 |
| ·28 | ·0004 354 | ·0002 307 | ·0001 203 | ·0000 618 | ·0000 313 | ·0000 156 |
| ·29 | ·0007 841 | ·0004 298 | ·0002 318 | ·0001 232 | ·0000 645$^{+}$ | ·0000 333 |
| ·30 | ·0013 641 | ·0007 724 | ·0004 306 | ·0002 364 | ·0001 280 | ·0000 683 |
| ·31 | ·0022 967 | ·0013 420 | ·0007 720 | ·0004 376 | ·0002 445$^{+}$ | ·0001 348 |
| ·32 | ·0037 493 | ·0022 582 | ·0013 392 | ·0007 826 | ·0004 509 | ·0002 563 |
| ·33 | ·0059 440 | ·0036 862 | ·0022 513 | ·0013 549 | ·0008 041 | ·0004 709 |
| ·34 | ·0091 650$^{+}$ | ·0058 463 | ·0036 731 | ·0022 745$^{+}$ | ·0013 890 | ·0008 371 |
| ·35 | ·0137 623 | ·0090 210 | ·0058 250$^{+}$ | ·0037 077 | ·0023 278 | ·0014 423 |
| ·36 | ·0201 508 | ·0135 602 | ·0089 907 | ·0058 770 | ·0037 897 | ·0024 120 |
| ·37 | ·0288 025$^{-}$ | ·0198 801 | ·0135 221 | ·0090 692 | ·0060 014 | ·0039 203 |
| ·38 | ·0402 308 | ·0284 565$^{-}$ | ·0198 394 | ·0136 412 | ·0092 555$^{-}$ | ·0062 001 |
| ·39 | ·0549 672 | ·0398 094 | ·0284 242 | ·0200 195$^{+}$ | ·0139 160 | ·0095 519 |
| ·40 | ·0735 292 | ·0544 802 | ·0398 054 | ·0286 945$^{-}$ | ·0204 187 | ·0143 499 |
| ·41 | ·0963 827 | ·0729 990 | ·0545 346 | ·0402 048 | ·0292 648 | ·0210 416 |
| ·42 | ·1239 004 | ·0958 467 | ·0731 544 | ·0551 140 | ·0410 052 | ·0301 414 |
| ·43 | ·1563 203 | ·1234 119 | ·0961 589 | ·0739 769 | ·0562 160 | ·0422 145$^{-}$ |
| ·44 | ·1937 082 | ·1559 477 | ·1239 491 | ·0972 988 | ·0754 640 | ·0578 506 |
| ·45 | ·2359 292 | ·1935 321 | ·1567 880 | ·1254 899 | ·0992 642 | ·0776 281 |
| ·46 | ·2826 305$^{-}$ | ·2360 374 | ·1947 593 | ·1588 180 | ·1280 323 | ·1020 697 |
| ·47 | ·3332 414 | ·2831 115$^{-}$ | ·2377 346 | ·1973 654 | ·1620 353 | ·1315 926 |
| ·48 | ·3869 888 | ·3341 760 | ·2853 535$^{-}$ | ·2409 947 | ·2013 472 | ·1664 573 |
| ·49 | ·4429 300 | ·3884 417 | ·3370 209 | ·2893 282 | ·2458 126 | ·2067 212 |
| ·50 | ·5000 000$^{e}$ | ·4449 420 | ·3919 232 | ·3417 446 | ·2950 267 | ·2522 019 |
| ·51 | ·5570 700 | ·5025 817 | ·4490 622 | ·3973 966 | ·3483 322 | ·3024 570 |
| ·52 | ·6130 112 | ·5601 983 | ·5073 073 | ·4552 473 | ·4048 390 | ·3567 830 |
| ·53 | ·6667 586 | ·6166 286 | ·5654 589 | ·5141 248 | ·4634 633 | ·4142 372 |
| ·54 | ·7173 695$^{+}$ | ·6707 764 | ·6223 196 | ·5727 897 | ·5229 867 | ·4736 816 |
| ·55 | ·7640 708 | ·7216 738 | ·6767 643 | ·6300 104 | ·5821 279 | ·5338 463 |
| ·56 | ·8062 918 | ·7685 312 | ·7278 057 | ·6846 367 | ·6396 218 | ·5934 066 |
| ·57 | ·8436 797 | ·8107 713 | ·7746 453 | ·7356 678 | ·6942 979 | ·6510 664 |
| ·58 | ·8760 996 | ·8480 459 | ·8167 090 | ·7823 054 | ·7451 496 | ·7056 406 |
| ·59 | ·9036 173 | ·8802 336 | ·8536 628 | ·8239 890 | ·7913 886 | ·7561 259 |
| ·60 | ·9264 708 | ·9074 218 | ·8854 096 | ·8604 100 | ·8324 795$^{-}$ | ·8017 558 |
| ·61 | ·9450 328 | ·9298 750$^{-}$ | ·9120 674 | ·8915 060 | ·8681 511 | ·8420 325$^{-}$ |
| ·62 | ·9597 692 | ·9479 948 | ·9339 353 | ·9174 355$^{+}$ | ·8983 868 | ·8767 347 |
| ·63 | ·9711 975$^{+}$ | ·9622 752 | ·9514 494 | ·9385 396 | ·9233 952 | ·9059 033 |
| ·64 | ·9798 492 | ·9732 586 | ·9651 351 | ·9552 940 | ·9435 662 | ·9298 055$^{+}$ |
| ·65 | ·9862 377 | ·9814 964 | ·9755 610 | ·9682 583 | ·9594 196 | ·9488 868 |
| ·66 | ·9908 350$^{-}$ | ·9875 162 | ·9832 977 | ·9780 276 | ·9715 508 | ·9637 139 |
| ·67 | ·9940 560 | ·9917 982 | ·9888 847 | ·9851 899 | ·9805 803 | ·9749 181 |
| ·68 | ·9962 507 | ·9947 596 | ·9928 068 | ·9902 932 | ·9871 105$^{+}$ | ·9831 428 |
| ·69 | ·9977 033 | ·9967 486 | ·9954 799 | ·9938 229 | ·9916 940 | ·9890 009 |
| ·70 | ·9986 359 | ·9980 442 | ·9972 466 | ·9961 897 | ·9948 121 | ·9930 441 |
| ·71 | ·9992 159 | ·9988 615$^{+}$ | ·9983 770 | ·9977 258 | ·9968 648 | ·9957 442 |
| ·72 | ·9995 646 | ·9993 599 | ·9990 760 | ·9986 892 | ·9981 705$^{+}$ | ·9974 859 |
| ·73 | ·9997 670 | ·9996 532 | ·9994 932 | ·9992 720 | ·9989 715$^{-}$ | ·9985 692 |
| ·74 | ·9998 801 | ·9998 193 | ·9997 327 | ·9996 114 | ·9994 442 | ·9992 174 |
| ·75 | ·9999 409 | ·9999 098 | ·9998 649 | ·9998 012 | ·9997 122 | ·9995 898 |
| ·76 | ·9999 721 | ·9999 569 | ·9999 347 | ·9999 028 | ·9998 576 | ·9997 946 |
| ·77 | ·9999 875$^{-}$ | ·9999 804 | ·9999 700 | ·9999 547 | ·9999 329 | ·9999 021 |
| ·78 | ·9999 947 | ·9999 916 | ·9999 869 | ·9999 800 | ·9999 700 | ·9999 557 |
| ·79 | ·9999 979 | ·9999 966 | ·9999 946 | ·9999 917 | ·9999 874 | ·9999 811 |
| ·80 | ·9999 992 | ·9999 987 | ·9999 979 | ·9999 967 | ·9999 950$^{-}$ | ·9999 924 |
| ·81 | ·9999 997 | ·9999 995$^{+}$ | ·9999 992 | ·9999 988 | ·9999 981 | ·9999 972 |
| ·82 | ·9999 999 | ·9999 998 | ·9999 997 | ·9999 996 | ·9999 994 | ·9999 990 |
| ·83 | 1·0000 000 | 1·0000 000 | ·9999 999 | ·9999 999 | ·9999 998 | ·9999 997 |
| ·84 | | | 1·0000 000 | 1·0000 000 | ·9999 999 | ·9999 999 |
| ·85 | | | | | 1·0000 000 | 1·0000 000 |

$x$ = ·23 to ·87 $q = 26$ $p$ = 32 to 37

| $x$ | $p = 32$ | $p = 33$ | $p = 34$ | $p = 35$ | $p = 36$ | $p = 37$ |
|---|---|---|---|---|---|---|
| $B(p, q) =$ | ·3147 1965 $\times \frac{1}{10^{17}}$ | ·1736 3843 $\times \frac{1}{10^{17}}$ | ·9711 9797 $\times \frac{1}{10^{18}}$ | ·5503 4552 $\times \frac{1}{10^{18}}$ | ·3157 7202 $\times \frac{1}{10^{18}}$ | ·1833 5149 $\times \frac{1}{10^{18}}$ |
| ·23 | ·0000 001 | | | | | |
| ·24 | ·0000 002 | ·0000 001 | | | | |
| ·25 | ·0000 005+ | ·0000 002 | ·0000 001 | | | |
| ·26 | ·0000 014 | ·0000 006 | ·0000 003 | ·0000 001 | ·0000 001 | |
| ·27 | ·0000 033 | ·0000 016 | ·0000 007 | ·0000 003 | ·0000 002 | ·0000 001 |
| ·28 | ·0000 077 | ·0000 037 | ·0000 018 | ·0000 009 | ·0000 004 | ·0000 002 |
| ·29 | ·0000 170 | ·0000 086 | ·0000 043 | ·0000 021 | ·0000 010 | ·0000 005− |
| ·30 | ·0000 360 | ·0000 188 | ·0000 096 | ·0000 049 | ·0000 025− | ·0000 012 |
| ·31 | ·0000 734 | ·0000 394 | ·0000 209 | ·0000 110 | ·0000 057 | ·0000 029 |
| ·32 | ·0001 439 | ·0000 797 | ·0000 437 | ·0000 237 | ·0000 127 | ·0000 067 |
| ·33 | ·0002 723 | ·0001 555− | ·0000 878 | ·0000 490 | ·0000 270 | ·0000 148 |
| ·34 | ·0004 981 | ·0002 928 | ·0001 701 | ·0000 978 | ·0000 556 | ·0000 313 |
| ·35 | ·0008 825− | ·0005 335− | ·0003 188 | ·0001 884 | ·0001 101 | ·0000 637 |
| ·36 | ·0015 161 | ·0009 416 | ·0005 781 | ·0003 511 | ·0002 109 | ·0001 254 |
| ·37 | ·0025 294 | ·0016 127 | ·0010 165+ | ·0006 338 | ·0003 910 | ·0002 388 |
| ·38 | ·0041 027 | ·0026 830 | ·0017 349 | ·0011 096 | ·0007 024 | ·0004 401 |
| ·39 | ·0064 774 | ·0043 416 | ·0028 775+ | ·0018 867 | ·0012 243 | ·0007 866 |
| ·40 | ·0099 646 | ·0068 402 | ·0046 435+ | ·0031 188 | ·0020 733 | ·0013 647 |
| ·41 | ·0149 511 | ·0105 031 | ·0072 979 | ·0050 174 | ·0034 146 | ·0023 011 |
| ·42 | ·0218 989 | ·0157 325+ | ·0111 806 | ·0078 631 | ·0054 744 | ·0037 745+ |
| ·43 | ·0313 385− | ·0230 081 | ·0167 124 | ·0120 147 | ·0085 518 | ·0060 287 |
| ·44 | ·0438 507 | ·0328 783 | ·0243 930 | ·0179 143 | ·0130 274 | ·0093 840 |
| ·45 | ·0600 400 | ·0459 420 | ·0347 918 | ·0260 848 | ·0193 679 | ·0142 464 |
| ·46 | ·0804 957 | ·0628 184 | ·0485 266 | ·0371 184 | ·0281 223 | ·0211 102 |
| ·47 | ·1057 458 | ·0841 071 | ·0662 323 | ·0516 538 | ·0399 077 | ·0305 532 |
| ·48 | ·1362 045− | ·1103 383 | ·0885 166 | ·0703 404 | ·0553 840 | ·0432 195− |
| ·49 | ·1721 189 | ·1419 176 | ·1159 079 | ·0937 920 | ·0752 147 | ·0597 903 |
| ·50 | ·2135 214 | ·1790 717 | ·1487 976 | ·1225 304 | ·1000 157 | ·0809 407 |
| ·51 | ·2601 926 | ·2217 980 | ·1873 825− | ·1569 248 | ·1302 960 | ·1072 843 |
| ·52 | ·3116 414 | ·2698 289 | ·2316 149 | ·1971 324 | ·1663 937 | ·1393 095− |
| ·53 | ·3671 072 | ·3226 135+ | ·2811 670 | ·2430 483 | ·2084 149 | ·1773 121 |
| ·54 | ·4255 853 | ·3793 227 | ·3354 153 | ·2942 715− | ·2561 840 | ·2213 340 |
| ·55 | ·4858 763 | ·4388 807 | ·3934 515+ | ·3500 935− | ·3092 130 | ·2711 146 |
| ·56 | ·5466 546 | ·5000 195+ | ·4541 193 | ·4095 151 | ·3666 951 | ·3260 636 |
| ·57 | ·6065 519 | ·5613 557 | ·5160 774 | ·4712 918 | ·4275 299 | ·3852 632 |
| ·58 | ·6642 453 | ·6214 788 | ·5778 829 | ·5340 049 | ·4903 777 | ·4475 018 |
| ·59 | ·7185 427 | ·6790 451 | ·6380 873 | ·5961 537 | ·5537 409 | ·5113 398 |
| ·60 | ·7684 554 | ·7328 656 | ·6953 345− | ·6562 580 | ·6160 650+ | ·5752 022 |
| ·61 | ·8132 515− | ·7819 791 | ·7484 513 | ·7129 612 | ·6758 487 | ·6374 888 |
| ·62 | ·8524 844 | ·8257 030 | ·7965 194 | ·7651 212 | ·7317 494 | ·6966 906 |
| ·63 | ·8859 964 | ·8636 572 | ·8389 215+ | ·8118 796 | ·7826 744 | ·7514 978 |
| ·64 | ·9138 964 | ·8957 601 | ·8753 594 | ·8527 026 | ·8278 450− | ·8008 882 |
| ·65 | ·9365 192 | ·9221 999 | ·9058 411 | ·8873 894 | ·8668 290 | ·8441 839 |
| ·66 | ·9543 703 | ·9433 857 | ·9306 436 | ·9160 501 | ·8995 386 | ·8810 733 |
| ·67 | ·9680 651 | ·9598 864 | ·9502 555− | ·9390 580 | ·9261 969 | ·9115 960 |
| ·68 | ·9782 689 | ·9723 654 | ·9653 098 | ·9569 842 | ·9472 789 | ·9360 963 |
| ·69 | ·9856 441 | ·9815 184 | ·9765 150− | ·9705 242 | ·9634 379 | ·9551 528 |
| ·70 | ·9908 085+ | ·9880 210 | ·9845 915+ | ·9804 257 | ·9754 267 | ·9694 974 |
| ·71 | ·9943 068 | ·9924 890 | ·9902 207 | ·9874 259 | ·9840 243 | ·9799 319 |
| ·72 | ·9965 954 | ·9954 534 | ·9940 082 | ·9922 026 | ·9899 739 | ·9872 550− |
| ·73 | ·9980 387 | ·9973 489 | ·9964 639 | ·9953 428 | ·9939 398 | ·9922 044 |
| ·74 | ·9989 143 | ·9985 146 | ·9979 949 | ·9973 275− | ·9964 808 | ·9954 192 |
| ·75 | ·9994 240 | ·9992 026 | ·9989 106 | ·9985 307 | ·9980 421 | ·9974 213 |
| ·76 | ·9997 081 | ·9995 910 | ·9994 347 | ·9992 284 | ·9989 597 | ·9986 137 |
| ·77 | ·9998 592 | ·9998 004 | ·9997 207 | ·9996 144 | ·9994 740 | ·9992 908 |
| ·78 | ·9999 356 | ·9999 076 | ·9998 692 | ·9998 173 | ·9997 479 | ·9996 561 |
| ·79 | ·9999 722 | ·9999 596 | ·9999 422 | ·9999 183 | ·9998 860 | ·9998 426 |
| ·80 | ·9999 887 | ·9999 834 | ·9999 760 | ·9999 657 | ·9999 516 | ·9999 324 |
| ·81 | ·9999 957 | ·9999 937 | ·9999 907 | ·9999 866 | ·9999 808 | ·9999 729 |
| ·82 | ·9999 985+ | ·9999 978 | ·9999 967 | ·9999 951 | ·9999 930 | ·9999 899 |
| ·83 | ·9999 995+ | ·9999 993 | ·9999 989 | ·9999 984 | ·9999 976 | ·9999 966 |
| ·84 | ·9999 999 | ·9999 998 | ·9999 997 | ·9999 995+ | ·9999 993 | ·9999 989 |
| ·85 | 1·0000 000 | ·9999 999 | ·9999 999 | ·9999 999 | ·9999 998 | ·9999 997 |
| ·86 | | 1·0000 000 | 1·0000 000 | 1·0000 000 | 1·0000 000 | ·9999 999 |
| ·87 | | | | | | 1·0000 000 |

$x$ = ·28 to ·88 $q = 26$ $p$ = 38 to 43

| $x$ | $p = 38$ | $p = 39$ | $p = 40$ | $p = 41$ | $p = 42$ | $p = 43$ |
|---|---|---|---|---|---|---|
| $B(p, q) =$ | ·1076 8262 $\times \frac{1}{10^{18}}$ | ·6393 6558 $\times \frac{1}{10^{19}}$ | ·3836 1935 $\times \frac{1}{10^{19}}$ | ·2324 9657 $\times \frac{1}{10^{19}}$ | ·1422 7402 $\times \frac{1}{10^{19}}$ | ·8787 5132 $\times \frac{1}{10^{20}}$ |
| ·28 | ·0000 001 | | | | | |
| ·29 | ·0000 002 | ·0000 001 | ·0000 001 | | | |
| ·30 | ·0000 006 | ·0000 003 | ·0000 001 | ·0000 001 | | |
| ·31 | ·0000 015− | ·0000 008 | ·0000 004 | ·0000 002 | ·0000 001 | |
| ·32 | ·0000 035+ | ·0000 018 | ·0000 009 | ·0000 005− | ·0000 002 | ·0000 001 |
| ·33 | ·0000 080 | ·0000 043 | ·0000 023 | ·0000 012 | ·0000 006 | ·0000 003 |
| ·34 | ·0000 174 | ·0000 096 | ·0000 053 | ·0000 028 | ·0000 015+ | ·0000 008 |
| ·35 | ·0000 365+ | ·0000 207 | ·0000 117 | ·0000 065$^{e}$ | ·0000 036 | ·0000 020 |
| ·36 | ·0000 739 | ·0000 431 | ·0000 249 | ·0000 143 | ·0000 081 | ·0000 046 |
| ·37 | ·0001 444 | ·0000 865+ | ·0000 514 | ·0000 302 | ·0000 176 | ·0000 102 |
| ·38 | ·0002 731 | ·0001 679 | ·0001 023 | ·0000 618 | ·0000 370 | ·0000 220 |
| ·39 | ·0005 005− | ·0003 155+ | ·0001 971 | ·0001 221 | ·0000 750+ | ·0000 457 |
| ·40 | ·0008 897 | ·0005 748 | ·0003 680 | ·0002 336 | ·0001 471 | ·0000 919 |
| ·41 | ·0015 361 | ·0010 161 | ·0006 662 | ·0004 331 | ·0002 793 | ·0001 786 |
| ·42 | ·0025 782 | ·0017 451 | ·0011 710 | ·0007 791 | ·0005 141 | ·0003 366 |
| ·43 | ·0042 107 | ·0029 147 | ·0020 001 | ·0013 611 | ·0009 188 | ·0006 153 |
| ·44 | ·0066 977 | ·0047 382 | ·0033 232 | ·0023 116 | ·0015 950+ | ·0010 921 |
| ·45 | ·0103 844 | ·0075 032 | ·0053 755+ | ·0038 197 | ·0026 926 | ·0018 836 |
| ·46 | ·0157 053 | ·0115 833 | ·0084 717 | ·0061 458 | ·0044 235+ | ·0031 597 |
| ·47 | ·0231 859 | ·0174 453 | ·0130 176 | ·0096 360 | ·0070 775− | ·0051 592 |
| ·48 | ·0334 353 | ·0256 491 | ·0195 160 | ·0147 321 | ·0110 357 | ·0082 054 |
| ·49 | ·0471 256 | ·0368 373 | ·0285 644 | ·0219 771 | ·0167 811 | ·0127 197 |
| ·50 | ·0649 590 | ·0517 109 | ·0408 408 | ·0320 088 | ·0249 001 | ·0192 300 |
| ·51 | ·0876 187 | ·0709 909 | ·0570 747 | ·0455 416 | ·0360 732 | ·0283 700 |
| ·52 | ·1157 096 | ·0953 640 | ·0780 024 | ·0633 319 | ·0510 516 | ·0408 648 |
| ·53 | ·1496 894 | ·1254 178 | ·1043 078 | ·0861 267 | ·0706 152 | ·0575 006 |
| ·54 | ·1897 995− | ·1615 678 | ·1365 501 | ·1145 972 | ·0955 141 | ·0790 755− |
| ·55 | ·2360 024 | ·2039 855+ | ·1750 883 | ·1492 614 | ·1263 951 | ·1063 327 |
| ·56 | ·2879 359 | ·2525 374 | ·2200 071 | ·1904 044 | ·1637 187 | ·1398 795+ |
| ·57 | ·3448 928 | ·3067 427 | ·2710 578 | ·2380 046 | ·2076 762 | ·1800 991 |
| ·58 | ·4058 311 | ·3657 614 | ·3276 232 | ·2916 780 | ·2581 175+ | ·2270 660 |
| ·59 | ·4694 201 | ·4284 160 | ·3887 156 | ·3506 528 | ·3145 025− | ·2804 781 |
| ·60 | ·5341 184 | ·4932 509 | ·4530 122 | ·4137 794 | ·3758 862 | ·3396 171 |
| ·61 | ·5982 787 | ·5586 250− | ·5189 305+ | ·4795 834 | ·4409 465+ | ·4033 491 |
| ·62 | ·6602 672 | ·6228 278 | ·5847 357 | ·5463 578 | ·5080 549 | ·4701 714 |
| ·63 | ·7185 855− | ·6842 094 | ·6486 698 | ·6122 861 | ·5753 878 | ·5383 050− |
| ·64 | ·7719 789 | ·7413 046 | ·7090 887 | ·6755 841 | ·6410 663 | ·6058 252 |
| ·65 | ·8195 191 | ·7929 396 | ·7645 880 | ·7346 418 | ·7033 077 | ·6708 173 |
| ·66 | ·8606 516 | ·8383 060 | ·8141 040 | ·7881 473 | ·7605 700 | ·7315 349 |
| ·67 | ·8952 036 | ·8769 950+ | ·8569 749 | ·8351 781 | ·8116 694 | ·7865 430 |
| ·68 | ·9233 541 | ·9089 890 | ·8929 590 | ·8752 459 | ·8558 565− | ·8348 236 |
| ·69 | ·9455 735+ | ·9346 153 | ·9222 072 | ·9082 946 | ·8928 415+ | ·8758 320 |
| ·70 | ·9625 425− | ·9544 711 | ·9451 993 | ·9346 527 | ·9227 684 | ·9094 977 |
| ·71 | ·9750 631 | ·9693 321 | ·9626 546 | ·9549 505+ | ·9461 453 | ·9361 724 |
| ·72 | ·9839 746 | ·9800 589 | ·9754 323 | ·9700 191 | ·9637 452 | ·9565 391 |
| ·73 | ·9900 815+ | ·9875 123 | ·9844 346 | ·9807 836 | ·9764 932 | ·9714 970 |
| ·74 | ·9941 028 | ·9924 877 | ·9905 265− | ·9881 681 | ·9853 587 | ·9820 424 |
| ·75 | ·9966 411 | ·9956 709 | ·9944 769 | ·9930 217 | ·9912 648 | ·9891 627 |
| ·76 | ·9981 730 | ·9976 178 | ·9969 253 | ·9960 701 | ·9950 238 | ·9937 553 |
| ·77 | ·9990 544 | ·9987 527 | ·9983 714 | ·9978 943 | ·9973 029 | ·9965 766 |
| ·78 | ·9995 361 | ·9993 810 | ·9991 825− | ·9989 308 | ·9986 148 | ·9982 216 |
| ·79 | ·9997 853 | ·9997 102 | ·9996 129 | ·9994 879 | ·9993 290 | ·9991 287 |
| ·80 | ·9999 067 | ·9998 727 | ·9998 280 | ·9997 698 | ·9996 950− | ·9995 994 |
| ·81 | ·9999 622 | ·9999 478 | ·9999 287 | ·9999 035+ | ·9998 707 | ·9998 283 |
| ·82 | ·9999 858 | ·9999 802 | ·9999 726 | ·9999 625+ | ·9999 492 | ·9999 318 |
| ·83 | ·9999 951 | ·9999 931 | ·9999 903 | ·9999 866 | ·9999 817 | ·9999 751 |
| ·84 | ·9999 985− | ·9999 978 | ·9999 969 | ·9999 957 | ·9999 940 | ·9999 917 |
| ·85 | ·9999 996 | ·9999 994 | ·9999 991 | ·9999 987 | ·9999 982 | ·9999 975+ |
| ·86 | ·9999 999 | ·9999 998 | ·9999 998 | ·9999 997 | ·9999 995+ | ·9999 994 |
| ·87 | 1·0000 000 | 1·0000 000 | ·9999 999 | ·9999 999 | ·9999 999 | ·9999 998 |
| ·88 | | | 1·0000 000 | 1·0000 000 | 1·0000 000 | 1·0000 000 |

$x$ = ·32 to ·90 $q = 26$ $p$ = 44 to 50

| | $p = 44$ | $p = 45$ | $p = 46$ | $p = 47$ | $p = 48$ | $p = 49$ | $p = 50$ |
|---|---|---|---|---|---|---|---|
| $B(p, q) =$ | $·5476\ 2763 \times \frac{1}{10^{20}}$ | $·3442\ 2308 \times \frac{1}{10^{20}}$ | $·2181\ 6956 \times \frac{1}{10^{20}}$ | $·1393\ 8611 \times \frac{1}{10^{20}}$ | $·8974\ 1741 \times \frac{1}{10^{21}}$ | $·5821\ 0859 \times \frac{1}{10^{21}}$ | $·3803\ 1094 \times \frac{1}{10^{21}}$ |
| $x$ | | | | | | | |
| ·32 | ·0000 001 | | | | | | |
| ·33 | ·0000 002 | ·0000 001 | | | | | |
| ·34 | ·0000 004 | ·0000 002 | ·0000 001 | ·0000 001 | | | |
| ·35 | ·0000 011 | ·0000 006 | ·0000 003 | ·0000 002 | ·0000 001 | | |
| ·36 | ·0000 026 | ·0000 014 | ·0000 008 | ·0000 004 | ·0000 002 | ·0000 001 | ·0000 001 |
| ·37 | ·0000 059 | ·0000 033 | ·0000 019 | ·0000 011 | ·0000 006 | ·0000 003 | ·0000 002 |
| ·38 | ·0000 130 | ·0000 076 | ·0000 044 | ·0000 025+ | ·0000 015− | ·0000 008 | ·0000 005− |
| ·39 | ·0000 276 | ·0000 166 | ·0000 099 | ·0000 058 | ·0000 034 | ·0000 020 | ·0000 012 |
| ·40 | ·0000 569 | ·0000 350+ | ·0000 214 | ·0000 130 | ·0000 078 | ·0000 047 | ·0000 028 |
| ·41 | ·0001 134 | ·0000 715− | ·0000 447 | ·0000 278 | ·0000 171 | ·0000 105+ | ·0000 064 |
| ·42 | ·0002 187 | ·0001 411 | ·0000 903 | ·0000 575− | ·0000 363 | ·0000 228 | ·0000 142 |
| ·43 | ·0004 090 | ·0002 699 | ·0001 768 | ·0001 150+ | ·0000 744 | ·0000 478 | ·0000 305− |
| ·44 | ·0007 421 | ·0005 006 | ·0003 354 | ·0002 231 | ·0001 475− | ·0000 968 | ·0000 632 |
| ·45 | ·0013 078 | ·0009 015+ | ·0006 171 | ·0004 196 | ·0002 834 | ·0001 902 | ·0001 269 |
| ·46 | ·0022 403 | ·0015 771 | ·0011 026 | ·0007 656 | ·0005 282 | ·0003 621 | ·0002 467 |
| ·47 | ·0037 334 | ·0026 826 | ·0019 143 | ·0013 570 | ·0009 557 | ·0006 688 | ·0004 652 |
| ·48 | ·0060 569 | ·0044 397 | ·0032 322 | ·0023 376 | ·0016 798 | ·0011 996 | ·0008 515− |
| ·49 | ·0095 725− | ·0071 542 | ·0053 109 | ·0039 168 | ·0028 703 | ·0020 905− | ·0015 134 |
| ·50 | ·0147 467 | ·0112 314 | ·0084 973 | ·0063 873 | ·0047 711 | ·0035 421 | ·0026 141 |
| ·51 | ·0221 573 | ·0171 886 | ·0132 467 | ·0101 437 | ·0077 195− | ·0058 392 | ·0043 909 |
| ·52 | ·0324 880 | ·0256 570 | ·0201 315+ | ·0156 967 | ·0121 640 | ·0093 702 | ·0071 762 |
| ·53 | ·0465 087 | ·0373 729 | ·0298 410 | ·0236 796 | ·0186 770 | ·0146 447 | ·0114 173 |
| ·54 | ·0650 376 | ·0531 501 | ·0431 646 | ·0348 419 | ·0279 570 | ·0223 029 | ·0176 918 |
| ·55 | ·0888 831 | ·0738 327 | ·0609 564 | ·0500 254 | ·0408 155+ | ·0331 119 | ·0267 131 |
| ·56 | ·1187 680 | ·1002 282 | ·0840 780 | ·0701 186 | ·0581 432 | ·0479 442 | ·0393 187 |
| ·57 | ·1552 412 | ·1330 217 | ·1133 203 | ·0959 874 | ·0808 525− | ·0677 324 | ·0564 384 |
| ·58 | ·1985 853 | ·1726 808 | ·1493 092 | ·1283 865+ | ·1097 965+ | ·0933 986 | ·0790 353 |
| ·59 | ·2487 326 | ·2193 608 | ·1924 040 | ·1678 557 | ·1456 683 | ·1257 597 | ·1080 207 |
| ·60 | ·3052 036 | ·2728 236 | ·2426 022 | ·2146 147 | ·1888 899 | ·1654 161 | ·1441 459 |
| ·61 | ·3670 807 | ·3323 867 | ·2994 659 | ·2684 703 | ·2395 059 | ·2126 354 | ·1878 816 |
| ·62 | ·4330 280 | ·3969 145− | ·3620 850+ | ·3287 547 | ·2970 980 | ·2672 484 | ·2392 994 |
| ·63 | ·5013 601 | ·4648 603 | ·4290 904 | ·3943 081 | ·3607 395+ | ·3285 766 | ·2979 759 |
| ·64 | ·5701 580 | ·5343 611 | ·4987 232 | ·4635 192 | ·4290 044 | ·3954 099 | ·3629 398 |
| ·65 | ·6374 202 | ·6033 779 | ·5689 574 | ·5344 246 | ·5000 387 | ·4660 469 | ·4326 794 |
| ·66 | ·7012 305− | ·6698 654 | ·6376 638 | ·6048 603 | ·5716 938 | ·5384 029 | ·5052 208 |
| ·67 | ·7599 208 | ·7319 493 | ·7027 968 | ·6726 493 | ·6417 064 | ·6101 769 | ·5782 742 |
| ·68 | ·8122 059 | ·7880 872 | ·7625 750+ | ·7357 983 | ·7079 049 | ·6790 585+ | ·6494 351 |
| ·69 | ·8572 719 | ·8371 891 | ·8156 334 | ·7926 767 | ·7684 109 | ·7429 470 | ·7164 126 |
| ·70 | ·8948 073 | ·8786 812 | ·8611 217 | ·8421 498 | ·8218 055+ | ·8001 473 | ·7772 514 |
| ·71 | ·9249 748 | ·9125 074 | ·8987 378 | ·8836 481 | ·8672 357 | ·8495 137 | ·8305 114 |
| ·72 | ·9483 342 | ·9390 702 | ·9286 945− | ·9171 639 | ·9044 459 | ·8905 197 | ·8753 770 |
| ·73 | ·9657 292 | ·9591 265− | ·9516 286 | ·9431 806 | ·9337 331 | ·9232 445− | ·9116 813 |
| ·74 | ·9781 614 | ·9736 578 | ·9684 736 | ·9625 523 | ·9558 399 | ·9482 856 | ·9398 433 |
| ·75 | ·9866 696 | ·9837 374 | ·9803 165− | ·9763 564 | ·9718 065− | ·9666 167 | ·9607 386 |
| ·76 | ·9922 308 | ·9904 138 | ·9882 657 | ·9857 459 | ·9828 122 | ·9794 214 | ·9755 295− |
| ·77 | ·9956 921 | ·9946 241 | ·9933 448 | ·9918 245− | ·9900 311 | ·9879 310 | ·9854 888 |
| ·78 | ·9977 367 | ·9971 434 | ·9964 237 | ·9955 571 | ·9945 217 | ·9932 934 | ·9918 466 |
| ·79 | ·9988 784 | ·9985 684 | ·9981 875− | ·9977 230 | ·9971 608 | ·9964 854 | ·9956 796 |
| ·80 | ·9994 786 | ·9993 269 | ·9991 382 | ·9989 052 | ·9986 197 | ·9982 722 | ·9978 525− |
| ·81 | ·9997 739 | ·9997 049 | ·9996 179 | ·9995 092 | ·9993 743 | ·9992 081 | ·9990 048 |
| ·82 | ·9999 092 | ·9998 802 | ·9998 431 | ·9997 963 | ·9997 374 | ·9996 639 | ·9995 730 |
| ·83 | ·9999 665+ | ·9999 553 | ·9999 409 | ·9999 223 | ·9998 988 | ·9998 691 | ·9998 318 |
| ·84 | ·9999 888 | ·9999 849 | ·9999 797 | ·9999 731 | ·9999 645+ | ·9999 536 | ·9999 398 |
| ·85 | ·9999 966 | ·9999 954 | ·9999 938 | ·9999 916 | ·9999 888 | ·9999 852 | ·9999 806 |
| ·86 | ·9999 991 | ·9999 988 | ·9999 983 | ·9999 977 | ·9999 969 | ·9999 958 | ·9999 945− |
| ·87 | ·9999 998 | ·9999 997 | ·9999 996 | ·9999 994 | ·9999 992 | ·9999 990 | ·9999 986 |
| ·88 | 1·0000 000 | ·9999 999 | ·9999 999 | ·9999 999 | ·9999 998 | ·9999 998 | ·9999 997 |
| ·89 | | 1·0000 000 | 1·0000 000 | 1·0000 000 | 1·0000 000 | 1·0000 000 | ·9999 999 |
| ·90 | | | | | | | 1·0000 000 |

$x = \cdot 18$ to $\cdot 84$ $q = 27$ $p = 27$ to $32$

| | $p = 27$ | $p = 28$ | $p = 29$ | $p = 30$ | $p = 31$ | $p = 32$ |
|---|---|---|---|---|---|---|
| $B(p, q) =$ | $\cdot 3804\ 6419 \times \frac{1}{10^{18}}$ | $\cdot 1902\ 3210 \times \frac{1}{10^{18}}$ | $\cdot 9684\ 5431 \times \frac{1}{10^{17}}$ | $\cdot 5015\ 2098 \times \frac{1}{10^{17}}$ | $\cdot 2639\ 5841 \times \frac{1}{10^{17}}$ | $\cdot 1410\ 8122 \times \frac{1}{10^{17}}$ |
| $x$ | | | | | | |
| ·18 | ·0000 001 | | | | | |
| ·19 | ·0000 002 | ·0000 001 | | | | |
| ·20 | ·0000 005+ | ·0000 002 | ·0000 001 | | | |
| ·21 | ·0000 014 | ·0000 006 | ·0000 002 | ·0000 001 | | |
| ·22 | ·0000 036 | ·0000 015+ | ·0000 006 | ·0000 003 | ·0000 001 | |
| ·23 | ·0000 087 | ·0000 038 | ·0000 016 | ·0000 007 | ·0000 003 | ·0000 001 |
| ·24 | ·0000 200 | ·0000 091 | ·0000 041 | ·0000 018 | ·0000 008 | ·0000 003 |
| ·25 | ·0000 436 | ·0000 207 | ·0000 097 | ·0000 045− | ·0000 020 | ·0000 009 |
| ·26 | ·0000 907 | ·0000 448 | ·0000 218 | ·0000 104 | ·0000 049 | ·0000 023 |
| ·27 | ·0001 806 | ·0000 926 | ·0000 467 | ·0000 232 | ·0000 114 | ·0000 055+ |
| ·28 | ·0003 453 | ·0001 833 | ·0000 958 | ·0000 493 | ·0000 251 | ·0000 126 |
| ·29 | ·0006 353 | ·0003 488 | ·0001 886 | ·0001 005+ | ·0000 528 | ·0000 274 |
| ·30 | ·0011 274 | ·0006 396 | ·0003 574 | ·0001 968 | ·0001 069 | ·0000 573 |
| ·31 | ·0019 339 | ·0011 322 | ·0006 529 | ·0003 711 | ·0002 081 | ·0001 151 |
| ·32 | ·0032 125+ | ·0019 387 | ·0011 525+ | ·0006 754 | ·0003 905− | ·0002 228 |
| ·33 | ·0051 764 | ·0032 165+ | ·0019 693 | ·0011 886 | ·0007 078 | ·0004 160 |
| ·34 | ·0081 030 | ·0051 793 | ·0032 621 | ·0020 259 | ·0012 413 | ·0007 508 |
| ·35 | ·0123 399 | ·0081 052 | ·0052 468 | ·0033 494 | ·0021 098 | ·0013 121 |
| ·36 | ·0183 054 | ·0123 440 | ·0082 051 | ·0053 792 | ·0034 802 | ·0022 232 |
| ·37 | ·0264 827 | ·0183 175+ | ·0124 911 | ·0084 026 | ·0055 788 | ·0036 576 |
| ·38 | ·0374 050− | ·0265 143 | ·0185 330 | ·0127 809 | ·0087 008 | ·0058 500− |
| ·39 | ·0516 325− | ·0374 755− | ·0268 273 | ·0189 514 | ·0132 177 | ·0091 061 |
| ·40 | ·0697 194 | ·0517 710 | ·0379 251 | ·0274 213 | ·0195 784 | ·0138 102 |
| ·41 | ·0921 736 | ·0699 666 | ·0524 071 | ·0387 532 | ·0283 033 | ·0204 255− |
| ·42 | ·1194 118 | ·0925 825+ | ·0708 508 | ·0535 403 | ·0399 690 | ·0294 884 |
| ·43 | ·1517 131 | ·1200 472 | ·0937 871 | ·0723 716 | ·0551 820 | ·0415 912 |
| ·44 | ·1891 770 | ·1526 494 | ·1216 531 | ·0957 872 | ·0745 427 | ·0573 552 |
| ·45 | ·2316 894 | ·1904 936 | ·1547 416 | ·1242 290 | ·0985 985+ | ·0773 913 |
| ·46 | ·2789 030 | ·2334 654 | ·1931 558 | ·1579 891 | ·1277 926 | ·1022 523 |
| ·47 | ·3302 336 | ·2812 091 | ·2367 719 | ·1971 615+ | ·1624 100 | ·1323 780 |
| ·48 | ·3848 763 | ·3331 244 | ·2852 170 | ·2416 047 | ·2025 281 | ·1680 398 |
| ·49 | ·4418 402 | ·3883 824 | ·3378 647 | ·2909 182 | ·2479 777 | ·2092 897 |
| ·50 | ·5000 000[e] | ·4459 616 | ·3938 531 | ·3444 399 | ·2983 209 | ·2559 212 |
| ·51 | ·5581 598 | ·5047 019 | ·4521 224 | ·4012 652 | ·3528 492 | ·3074 474 |
| ·52 | ·6151 237 | ·5633 718 | ·5114 721 | ·4602 882 | ·4106 057 | ·3631 028 |
| ·53 | ·6697 664 | ·6207 419 | ·5706 318 | ·5202 626 | ·4704 307 | ·4218 686 |
| ·54 | ·7210 970 | ·6756 594 | ·6283 393 | ·5798 771 | ·5310 272 | ·4825 239 |
| ·55 | ·7683 106 | ·7271 148 | ·6834 178 | ·6378 373 | ·5910 414 | ·5437 171 |
| ·56 | ·8108 230 | ·7742 954 | ·7348 456 | ·6929 471 | ·6491 493 | ·6040 516 |
| ·57 | ·8482 869 | ·8166 210 | ·7818 111 | ·7441 804 | ·7041 414 | ·6621 779 |
| ·58 | ·8805 882 | ·8537 589 | ·8237 484 | ·7907 369 | ·7549 965− | ·7168 810 |
| ·59 | ·9078 264 | ·8856 193 | ·8603 509 | ·8320 763 | ·8009 367 | ·7671 551 |
| ·60 | ·9302 806 | ·9123 322 | ·8915 633 | ·8679 298 | ·8414 602 | ·8122 583 |
| ·61 | ·9483 675+ | ·9342 105− | ·9175 557 | ·8982 879 | ·8763 483 | ·8517 405+ |
| ·62 | ·9625 950+ | ·9517 043 | ·9386 822 | ·9233 699 | ·9056 486 | ·8854 462 |
| ·63 | ·9735 173 | ·9653 522 | ·9554 315+ | ·9435 781 | ·9296 384 | ·9134 909 |
| ·64 | ·9816 946 | ·9757 331 | ·9683 751 | ·9594 439 | ·9487 741 | ·9362 181 |
| ·65 | ·9876 601 | ·9834 254 | ·9781 169 | ·9715 729 | ·9636 327 | ·9541 430 |
| ·66 | ·9918 970 | ·9889 733 | ·9852 517 | ·9805 934 | ·9748 544 | ·9678 898 |
| ·67 | ·9948 236 | ·9928 638 | ·9903 314 | ·9871 135+ | ·9830 891 | ·9781 312 |
| ·68 | ·9967 875− | ·9955 136 | ·9938 431 | ·9916 887 | ·9889 540 | ·9855 348 |
| ·69 | ·9980 661 | ·9972 644 | ·9961 975+ | ·9948 014 | ·9930 033 | ·9907 219 |
| ·70 | ·9988 726 | ·9983 848 | ·9977 262 | ·9968 520 | ·9957 096 | ·9942 393 |
| ·71 | ·9993 647 | ·9990 783 | ·9986 860 | ·9981 579 | ·9974 579 | ·9965 441 |
| ·72 | ·9996 547 | ·9994 927 | ·9992 677 | ·9989 605− | ·9985 476 | ·9980 010 |
| ·73 | ·9998 194 | ·9997 313 | ·9996 073 | ·9994 356 | ·9992 017 | ·9988 877 |
| ·74 | ·9999 093 | ·9998 634 | ·9997 979 | ·9997 059 | ·9995 789 | ·9994 061 |
| ·75 | ·9999 564 | ·9999 335+ | ·9999 004 | ·9998 534 | ·9997 875− | ·9996 966 |
| ·76 | ·9999 800 | ·9999 691 | ·9999 532 | ·9999 303 | ·9998 977 | ·9998 522 |
| ·77 | ·9999 913 | ·9999 864 | ·9999 791 | ·9999 685− | ·9999 532 | ·9999 316 |
| ·78 | ·9999 964 | ·9999 943 | ·9999 912 | ·9999 865+ | ·9999 797 | ·9999 700 |
| ·79 | ·9999 986 | ·9999 978 | ·9999 965− | ·9999 946 | ·9999 917 | ·9999 876 |
| ·80 | ·9999 995− | ·9999 992 | ·9999 987 | ·9999 979 | ·9999 968 | ·9999 952 |
| ·81 | ·9999 998 | ·9999 997 | ·9999 995+ | ·9999 993 | ·9999 989 | ·9999 983 |
| ·82 | ·9999 999 | ·9999 999 | ·9999 999 | ·9999 998 | ·9999 996 | ·9999 994 |
| ·83 | 1·0000 000 | 1·0000 000 | 1·0000 000 | ·9999 999 | ·9999 999 | ·9999 998 |
| ·84 | | | | 1·0000 000 | 1·0000 000 | 1·0000 000 |

$x$ = ·24 to ·86 $q = 27$ $p$ = 33 to 38

| $x$ | $p = 33$ | $p = 34$ | $p = 35$ | $p = 36$ | $p = 37$ | $p = 38$ |
|---|---|---|---|---|---|---|
| $B(p, q) =$ | ·7651 8628 $\times\frac{1}{10^{18}}$ | ·4208 5245 $\times\frac{1}{10^{18}}$ | ·2345 7350 $\times\frac{1}{10^{18}}$ | ·1324 2052 $\times\frac{1}{10^{18}}$ | ·7566 8871 $\times\frac{1}{10^{19}}$ | ·4374 6066 $\times\frac{1}{10^{19}}$ |
| ·24 | ·0000 001 | ·0000 001 | | | | |
| ·25 | ·0000 004 | ·0000 002 | ·0000 001 | | | |
| ·26 | ·0000 011 | ·0000 005− | ·0000 002 | ·0000 001 | | |
| ·27 | ·0000 026 | ·0000 012 | ·0000 006 | ·0000 003 | ·0000 001 | ·0000 001 |
| ·28 | ·0000 062 | ·0000 030 | ·0000 015− | ·0000 007 | ·0000 003 | ·0000 002 |
| ·29 | ·0000 140 | ·0000 071 | ·0000 035+ | ·0000 018 | ·0000 009 | ·0000 004 |
| ·30 | ·0000 303 | ·0000 158 | ·0000 082 | ·0000 042 | ·0000 021 | ·0000 011 |
| ·31 | ·0000 629 | ·0000 339 | ·0000 181 | ·0000 096 | ·0000 050− | ·0000 026 |
| ·32 | ·0001 255+ | ·0000 699 | ·0000 385− | ·0000 209 | ·0000 113 | ·0000 060 |
| ·33 | ·0002 415− | ·0001 385− | ·0000 785+ | ·0000 440 | ·0000 244 | ·0000 134 |
| ·34 | ·0004 485+ | ·0002 648 | ·0001 545+ | ·0000 892 | ·0000 510 | ·0000 288 |
| ·35 | ·0008 060 | ·0004 893 | ·0002 937 | ·0001 744 | ·0001 025− | ·0000 596 |
| ·36 | ·0014 030 | ·0008 751 | ·0005 397 | ·0003 293 | ·0001 988 | ·0001 189 |
| ·37 | ·0023 693 | ·0015 170 | ·0009 606 | ·0006 017 | ·0003 731 | ·0002 290 |
| ·38 | ·0038 865− | ·0025 525− | ·0016 579 | ·0010 655− | ·0006 777 | ·0004 269 |
| ·39 | ·0061 997 | ·0041 732 | ·0027 785− | ·0018 304 | ·0011 937 | ·0007 708 |
| ·40 | ·0096 282 | ·0066 374 | ·0045 263 | ·0030 545+ | ·0020 406 | ·0013 501 |
| ·41 | ·0145 713 | ·0102 800 | ·0071 751 | ·0049 564 | ·0033 898 | ·0022 961 |
| ·42 | ·0215 100 | ·0155 190 | ·0110 786 | ·0078 282 | ·0054 771 | ·0037 957 |
| ·43 | ·0309 989 | ·0228 556 | ·0166 763 | ·0120 453 | ·0086 158 | ·0061 049 |
| ·44 | ·0436 482 | ·0328 653 | ·0244 927 | ·0180 722 | ·0132 068 | ·0095 617 |
| ·45 | ·0600 942 | ·0461 779 | ·0351 267 | ·0264 594 | ·0197 422 | ·0145 954 |
| ·46 | ·0809 580 | ·0634 451 | ·0492 287 | ·0378 312 | ·0288 019 | ·0217 297 |
| ·47 | ·1067 944 | ·0852 965+ | ·0674 659 | ·0528 600 | ·0410 374 | ·0315 761 |
| ·48 | ·1380 350− | ·1122 854 | ·0904 740 | ·0722 272 | ·0571 432 | ·0448 151 |
| ·49 | ·1749 299 | ·1448 287 | ·1188 000 | ·0965 715− | ·0778 131 | ·0621 625− |
| ·50 | ·2174 964 | ·1831 470 | ·1528 387 | ·1264 272 | ·1036 839 | ·0843 215− |
| ·51 | ·2654 792 | ·2272 118 | ·1927 712 | ·1621 583 | ·1352 701 | ·1119 209 |
| ·52 | ·3183 314 | ·2767 075− | ·2385 114 | ·2038 949 | ·1728 939 | ·1454 454 |
| ·53 | ·3752 187 | ·3310 144 | ·2896 704 | ·2514 803 | ·2166 212 | ·1851 633 |
| ·54 | ·4350 514 | ·3892 188 | ·3455 430 | ·3044 379 | ·2662 101 | ·2310 612 |
| ·55 | ·4965 407 | ·4501 506 | ·4051 249 | ·3619 645+ | ·3210 821 | ·2827 962 |
| ·56 | ·5582 775+ | ·5124 479 | ·4671 575− | ·4229 540 | ·3803 223 | ·3396 723 |
| ·57 | ·6188 244 | ·5746 432 | ·5302 021 | ·4860 531 | ·4427 134 | ·4006 506 |
| ·58 | ·6768 121 | ·6352 618 | ·5927 339 | ·5497 443 | ·5068 025− | ·4643 945− |
| ·59 | ·7310 300 | ·6929 235− | ·6532 479 | ·6124 500− | ·5709 948 | ·5293 492 |
| ·60 | ·7805 012 | ·7464 345+ | ·7103 639 | ·6726 444 | ·6336 675− | ·5938 479 |
| ·61 | ·8245 336 | ·7948 615+ | ·7629 204 | ·7289 624 | ·6932 877 | ·6562 342 |
| ·62 | ·8627 438 | ·8375 785+ | ·8100 447 | ·7802 925+ | ·7485 238 | ·7149 863 |
| ·63 | ·8950 524 | ·8742 840 | ·8511 944 | ·8258 420 | ·7983 346 | ·7688 275− |
| ·64 | ·9216 532 | ·9049 874 | ·8861 649 | ·8651 697 | ·8420 284 | ·8168 106 |
| ·65 | ·9429 629 | ·9299 703 | ·9150 670 | ·8981 837 | ·8792 837 | ·8583 661 |
| ·66 | ·9595 584 | ·9497 274 | ·9382 771 | ·9251 060 | ·9101 349 | ·8933 106 |
| ·67 | ·9721 104 | ·9648 983 | ·9563 710 | ·9464 135+ | ·9349 238 | ·9218 161 |
| ·68 | ·9813 206 | ·9761 971 | ·9700 490 | ·9627 626 | ·9542 293 | ·9443 493 |
| ·69 | ·9878 688 | ·9843 491 | ·9800 634 | ·9749 095− | ·9687 849 | ·9615 894 |
| ·70 | ·9923 738 | ·9900 391 | ·9871 551 | ·9836 366 | ·9793 948 | ·9743 391 |
| ·71 | ·9953 681 | ·9938 754 | ·9920 050+ | ·9896 906 | ·9868 606 | ·9834 393 |
| ·72 | ·9972 877 | ·9963 694 | ·9952 027 | ·9937 387 | ·9919 232 | ·9896 976 |
| ·73 | ·9984 722 | ·9979 300 | ·9972 314 | ·9963 427 | ·9952 253 | ·9938 365+ |
| ·74 | ·9991 743 | ·9988 677 | ·9984 672 | ·9979 508 | ·9972 925+ | ·9964 632 |
| ·75 | ·9995 731 | ·9994 075− | ·9991 883 | ·9989 017 | ·9985 316 | ·9980 590 |
| ·76 | ·9997 895+ | ·9997 044 | ·9995 901 | ·9994 388 | ·9992 408 | ·9989 845+ |
| ·77 | ·9999 014 | ·9998 598 | ·9998 034 | ·9997 276 | ·9996 271 | ·9994 954 |
| ·78 | ·9999 563 | ·9999 371 | ·9999 108 | ·9998 749 | ·9998 268 | ·9997 628 |
| ·79 | ·9999 818 | ·9999 734 | ·9999 619 | ·9999 459 | ·9999 242 | ·9998 951 |
| ·80 | ·9999 929 | ·9999 895− | ·9999 847 | ·9999 781 | ·9999 690 | ·9999 565+ |
| ·81 | ·9999 974 | ·9999 961 | ·9999 943 | ·9999 918 | ·9999 882 | ·9999 832 |
| ·82 | ·9999 991 | ·9999 987 | ·9999 980 | ·9999 971 | ·9999 958 | ·9999 940 |
| ·83 | ·9999 997 | ·9999 996 | ·9999 994 | ·9999 991 | ·9999 987 | ·9999 981 |
| ·84 | ·9999 999 | ·9999 999 | ·9999 998 | ·9999 997 | ·9999 996 | ·9999 994 |
| ·85 | 1·0000 000 | 1·0000 000 | 1·0000 000 | ·9999 999 | ·9999 999 | ·9999 998 |
| ·86 | | | | 1·0000 000 | 1·0000 000 | 1·0000 000 |

$x$ = ·28 to ·88 $q = 27$ $p$ = 39 to 44

| $x$ | $p = 39$ | $p = 40$ | $p = 41$ | $p = 42$ | $p = 43$ | $p = 44$ |
|---|---|---|---|---|---|---|
| $B(p, q) =$ | $·2557\ 4623 \times \frac{1}{10^{19}}$ | $·1511\ 2277 \times \frac{1}{10^{19}}$ | $·9022\ 2551 \times \frac{1}{10^{20}}$ | $·5439\ 8891 \times \frac{1}{10^{20}}$ | $·3311\ 2368 \times \frac{1}{10^{20}}$ | $·2034\ 0455 \times \frac{1}{10^{20}}$ |
| ·28 | ·0000 001 | | | | | |
| ·29 | ·0000 002 | ·0000 001 | | | | |
| ·30 | ·0000 005+ | ·0000 003 | ·0000 001 | ·0000 001 | | |
| ·31 | ·0000 013 | ·0000 007 | ·0000 003 | ·0000 002 | ·0000 001 | |
| ·32 | ·0000 032 | ·0000 017 | ·0000 009 | ·0000 004 | ·0000 002 | ·0000 001 |
| ·33 | ·0000 073 | ·0000 039 | ·0000 021 | ·0000 011 | ·0000 006 | ·0000 003 |
| ·34 | ·0000 161 | ·0000 090 | ·0000 049 | ·0000 027 | ·0000 014 | ·0000 008 |
| ·35 | ·0000 343 | ·0000 196 | ·0000 111 | ·0000 062 | ·0000 035− | ·0000 019 |
| ·36 | ·0000 704 | ·0000 413 | ·0000 240 | ·0000 138 | ·0000 079 | ·0000 045− |
| ·37 | ·0001 392 | ·0000 839 | ·0000 501 | ·0000 296 | ·0000 174 | ·0000 101 |
| ·38 | ·0002 663 | ·0001 646 | ·0001 009 | ·0000 613 | ·0000 369 | ·0000 221 |
| ·39 | ·0004 931 | ·0003 126 | ·0001 964 | ·0001 224 | ·0000 756 | ·0000 463 |
| ·40 | ·0008 849 | ·0005 748 | ·0003 701 | ·0002 363 | ·0001 496 | ·0000 940 |
| ·41 | ·0015 409 | ·0010 248 | ·0006 757 | ·0004 418 | ·0002 865+ | ·0001 844 |
| ·42 | ·0026 064 | ·0017 738 | ·0011 969 | ·0008 009 | ·0005 316 | ·0003 501 |
| ·43 | ·0042 865− | ·0029 833 | ·0020 586 | ·0014 089 | ·0009 566 | ·0006 445− |
| ·44 | ·0068 605+ | ·0048 796 | ·0034 415+ | ·0024 075− | ·0016 709 | ·0011 508 |
| ·45 | ·0106 947 | ·0077 692 | ·0055 969 | ·0039 996 | ·0028 358 | ·0019 954 |
| ·46 | ·0162 507 | ·0120 500+ | ·0088 618 | ·0064 651 | ·0046 802 | ·0033 627 |
| ·47 | ·0240 868 | ·0182 201 | ·0136 705+ | ·0101 762 | ·0075 172 | ·0055 118 |
| ·48 | ·0348 488 | ·0268 757 | ·0205 610 | ·0156 079 | ·0117 586 | ·0087 937 |
| ·49 | ·0492 466 | ·0386 987 | ·0301 707 | ·0233 420 | ·0179 246 | ·0136 650+ |
| ·50 | ·0680 162 | ·0544 285− | ·0432 186 | ·0340 593 | ·0266 447 | ·0206 957 |
| ·51 | ·0918 652 | ·0748 178 | ·0604 725− | ·0485 168 | ·0386 449 | ·0305 660 |
| ·52 | ·1214 063 | ·1005 725− | ·0826 970 | ·0675 072 | ·0547 189 | ·0440 480 |
| ·53 | ·1570 829 | ·1322 786 | ·1105 872 | ·0918 004 | ·0756 795− | ·0619 692 |
| ·54 | ·1990 942 | ·1703 240 | ·1446 896 | ·1220 689 | ·1022 919 | ·0851 549 |
| ·55 | ·2473 314 | ·2148 220 | ·1853 197 | ·1588 037 | ·1351 917 | ·1143 528 |
| ·56 | ·3013 329 | ·2655 495+ | ·2324 857 | ·2022 282 | ·1747 948 | ·1501 430 |
| ·57 | ·3602 702 | ·3219 088 | ·2858 300 | ·2522 239 | ·2212 102 | ·1928 436 |
| ·58 | ·4229 686 | ·3829 235+ | ·3446 004 | ·3082 776 | ·2741 687 | ·2424 237 |
| ·59 | ·4879 666 | ·4472 736 | ·4076 591 | ·3694 649 | ·3329 803 | ·2984 387 |
| ·60 | ·5536 091 | ·5133 703 | ·4735 339 | ·4344 749 | ·3965 318 | ·3600 005− |
| ·61 | ·6181 666 | ·5794 645+ | ·5405 109 | ·5016 808 | ·4633 314 | ·4257 937 |
| ·62 | ·6799 661 | ·6437 785+ | ·6067 587 | ·5692 512 | ·5316 009 | ·4941 432 |
| ·63 | ·7375 188 | ·7046 446 | ·6704 720 | ·6352 908 | ·5994 061 | ·5631 291 |
| ·64 | ·7896 284 | ·7606 341 | ·7300 161 | ·6979 942 | ·6648 134 | ·6307 374 |
| ·65 | ·8354 668 | ·8106 592 | ·7840 531 | ·7557 922 | ·7260 510 | ·6950 302 |
| ·66 | ·8746 090 | ·8540 373 | ·8316 347 | ·8074 727 | ·7816 539 | ·7543 099 |
| ·67 | ·9070 252 | ·8905 086 | ·8722 495+ | ·8522 581 | ·8305 721 | ·8072 572 |
| ·68 | ·9330 340 | ·9202 100 | ·9058 215− | ·8898 327 | ·8722 298 | ·8530 221 |
| ·69 | ·9532 274 | ·9436 112 | ·9326 630 | ·9203 184 | ·9065 276 | ·8912 583 |
| ·70 | ·9683 787 | ·9614 249 | ·9533 932 | ·9442 058 | ·9337 933 | ·9220 975+ |
| ·71 | ·9793 482 | ·9745 071 | ·9688 357 | ·9622 555− | ·9546 914 | ·9460 736 |
| ·72 | ·9869 988 | ·9837 601 | ·9799 127 | ·9753 858 | ·9701 087 | ·9640 118 |
| ·73 | ·9921 290 | ·9900 515− | ·9875 492 | ·9845 640 | ·9810 359 | ·9769 031 |
| ·74 | ·9954 296 | ·9941 548 | ·9925 982 | ·9907 160 | ·9884 608 | ·9857 830 |
| ·75 | ·9974 620 | ·9967 157 | ·9957 922 | ·9946 603 | ·9932 859 | ·9916 319 |
| ·76 | ·9986 565+ | ·9982 410 | ·9977 200 | ·9970 729 | ·9962 767 | ·9953 057 |
| ·77 | ·9993 246 | ·9991 053 | ·9988 268 | ·9984 763 | ·9980 394 | ·9974 995− |
| ·78 | ·9996 788 | ·9995 696 | ·9994 291 | ·9992 500− | ·9990 237 | ·9987 406 |
| ·79 | ·9998 562 | ·9998 051 | ·9997 385+ | ·9996 525+ | ·9995 425− | ·9994 030 |
| ·80 | ·9999 398 | ·9999 174 | ·9998 879 | ·9998 493 | ·9997 993 | ·9997 352 |
| ·81 | ·9999 765+ | ·9999 674 | ·9999 553 | ·9999 392 | ·9999 181 | ·9998 907 |
| ·82 | ·9999 915+ | ·9999 881 | ·9999 835+ | ·9999 774 | ·9999 692 | ·9999 584 |
| ·83 | ·9999 972 | ·9999 960 | ·9999 944 | ·9999 923 | ·9999 894 | ·9999 855− |
| ·84 | ·9999 992 | ·9999 988 | ·9999 983 | ·9999 976 | ·9999 967 | ·9999 954 |
| ·85 | ·9999 998 | ·9999 997 | ·9999 995+ | ·9999 993 | ·9999 991 | ·9999 987 |
| ·86 | ·9999 999 | ·9999 999 | ·9999 999 | ·9999 998 | ·9999 998 | ·9999 997 |
| ·87 | 1·0000 000 | 1·0000 000 | 1·0000 000 | 1·0000 000 | ·9999 999 | ·9999 999 |
| ·88 | | | | | 1·0000 000 | 1·0000 000 |

$x$ = ·32 to ·89 $q = 27$ $p$ = 45 to 50

| $x$ | $p = 45$ | $p = 46$ | $p = 47$ | $p = 48$ | $p = 49$ | $p = 50$ |
|---|---|---|---|---|---|---|
| $B(p, q) =$ | $·1260\ 5352 \times \frac{1}{10^{20}}$ | $·7878\ 3452 \times \frac{1}{10^{21}}$ | $·4964\ 4367 \times \frac{1}{10^{21}}$ | $·3153\ 0882 \times \frac{1}{10^{21}}$ | $·2017\ 9764 \times \frac{1}{10^{21}}$ | $·1301\ 0638 \times \frac{1}{10^{21}}$ |
| ·32 | ·0000 001 | | | | | |
| ·33 | ·0000 002 | ·0000 001 | | | | |
| ·34 | ·0000 004 | ·0000 002 | ·0000 001 | ·0000 001 | | |
| ·35 | ·0000 010 | ·0000 006 | ·0000 003 | ·0000 002 | ·0000 001 | |
| ·36 | ·0000 025+ | ·0000 014 | ·0000 008 | ·0000 004 | ·0000 002 | ·0000 001 |
| ·37 | ·0000 059 | ·0000 034 | ·0000 019 | ·0000 011 | ·0000 006 | ·0000 003 |
| ·38 | ·0000 131 | ·0000 077 | ·0000 045+ | ·0000 026 | ·0000 015+ | ·0000 009 |
| ·39 | ·0000 282 | ·0000 170 | ·0000 102 | ·0000 061 | ·0000 036 | ·0000 021 |
| ·40 | ·0000 586 | ·0000 363 | ·0000 223 | ·0000 136 | ·0000 082 | ·0000 050− |
| ·41 | ·0001 178 | ·0000 747 | ·0000 470 | ·0000 294 | ·0000 182 | ·0000 113 |
| ·42 | ·0002 289 | ·0001 485+ | ·0000 957 | ·0000 613 | ·0000 389 | ·0000 246 |
| ·43 | ·0004 309 | ·0002 861 | ·0001 886 | ·0001 235− | ·0000 803 | ·0000 519 |
| ·44 | ·0007 867 | ·0005 340 | ·0003 599 | ·0002 409 | ·0001 602 | ·0001 059 |
| ·45 | ·0013 938 | ·0009 666 | ·0006 657 | ·0004 554 | ·0003 096 | ·0002 091 |
| ·46 | ·0023 985− | ·0016 987 | ·0011 948 | ·0008 349 | ·0005 796 | ·0003 998 |
| ·47 | ·0040 123 | ·0029 004 | ·0020 824 | ·0014 852 | ·0010 525+ | ·0007 413 |
| ·48 | ·0065 296 | ·0048 150− | ·0035 267 | ·0025 663 | ·0018 556 | ·0013 335− |
| ·49 | ·0103 445− | ·0077 773 | ·0058 084 | ·0043 100 | ·0031 780 | ·0023 291 |
| ·50 | ·0159 636 | ·0122 305− | ·0093 089 | ·0070 400 | ·0052 911 | ·0039 526 |
| ·51 | ·0240 111 | ·0187 365+ | ·0145 261 | ·0111 908 | ·0085 685+ | ·0065 215+ |
| ·52 | ·0352 203 | ·0279 777 | ·0220 828 | ·0173 218 | ·0135 050+ | ·0104 672 |
| ·53 | ·0504 090 | ·0407 420 | ·0327 227 | ·0261 212 | ·0207 273 | ·0163 516 |
| ·54 | ·0704 327 | ·0578 894 | ·0472 875+ | ·0383 955− | ·0309 929 | ·0248 744 |
| ·55 | ·0961 188 | ·0802 957 | ·0666 741 | ·0550 377 | ·0451 711 | ·0368 650− |
| ·56 | ·1281 805− | ·1087 754 | ·0917 664 | ·0769 722 | ·0641 998 | ·0532 521 |
| ·57 | ·1671 201 | ·1439 862 | ·1233 467 | ·1050 742 | ·0890 172 | ·0750 083 |
| ·58 | ·2131 317 | ·1863 262 | ·1619 916 | ·1400 696 | ·1204 678 | ·1030 661 |
| ·59 | ·2660 168 | ·2358 355+ | ·2079 638 | ·1824 226 | ·1591 908 | ·1382 111 |
| ·60 | ·3251 297 | ·2921 187 | ·2611 171 | ·2322 262 | ·2055 022 | ·1809 597 |
| ·61 | ·3893 649 | ·3543 043 | ·3208 291 | ·2891 130 | ·2592 868 | ·2314 388 |
| ·62 | ·4571 963 | ·4210 540 | ·3859 803 | ·3522 050+ | ·3199 215− | ·2892 851 |
| ·63 | ·5267 696 | ·4906 283 | ·4549 898 | ·4201 172 | ·3862 472 | ·3535 868 |
| ·64 | ·5960 419 | ·5610 072 | ·5259 115+ | ·4910 250− | ·4566 035+ | ·4228 846 |
| ·65 | ·6629 519 | ·6300 538 | ·5965 836 | ·5627 929 | ·5289 318 | ·4952 435− |
| ·66 | ·7255 988 | ·6957 009 | ·6648 151 | ·6331 538 | ·6009 385+ | ·5683 945+ |
| ·67 | ·7824 056 | ·7561 347 | ·7285 845− | ·6999 147 | ·6703 012 | ·6399 323 |
| ·68 | ·8322 430 | ·8099 492 | ·7862 209 | ·7611 598 | ·7348 874 | ·7075 426 |
| ·69 | ·8744 969 | ·8562 492 | ·8365 417 | ·8154 212 | ·7929 542 | ·7692 263 |
| ·70 | ·9090 726 | ·8946 873 | ·8789 261 | ·8617 899 | ·8432 971 | ·8234 834 |
| ·71 | ·9363 394 | ·9254 349 | ·9133 168 | ·8999 533 | ·8853 258 | ·8694 296 |
| ·72 | ·9570 282 | ·9490 947 | ·9401 541 | ·9301 558 | ·9190 577 | ·9068 271 |
| ·73 | ·9721 034 | ·9665 752 | ·9602 587 | ·9530 968 | ·9450 367 | ·9360 307 |
| ·74 | ·9826 304 | ·9789 496 | ·9746 863 | ·9697 863 | ·9641 961 | ·9578 643 |
| ·75 | ·9896 582 | ·9873 228 | ·9845 812 | ·9813 875+ | ·9776 948 | ·9734 557 |
| ·76 | ·9941 316 | ·9927 238 | ·9910 491 | ·9890 722 | ·9867 560 | ·9840 617 |
| ·77 | ·9968 381 | ·9960 347 | ·9950 663 | ·9939 082 | ·9925 335− | ·9909 132 |
| ·78 | ·9983 892 | ·9979 568 | ·9974 289 | ·9967 893 | ·9960 202 | ·9951 020 |
| ·79 | ·9992 278 | ·9990 093 | ·9987 392 | ·9984 077 | ·9980 040 | ·9975 159 |
| ·80 | ·9996 535+ | ·9995 505− | ·9994 214 | ·9992 611 | ·9990 633 | ·9988 211 |
| ·81 | ·9998 554 | ·9998 103 | ·9997 531 | ·9996 811 | ·9995 912 | ·9994 798 |
| ·82 | ·9999 443 | ·9999 261 | ·9999 027 | ·9998 730 | ·9998 353 | ·9997 881 |
| ·83 | ·9999 804 | ·9999 736 | ·9999 649 | ·9999 537 | ·9999 393 | ·9999 210 |
| ·84 | ·9999 937 | ·9999 915− | ·9999 885+ | ·9999 847 | ·9999 797 | ·9999 733 |
| ·85 | ·9999 982 | ·9999 975+ | ·9999 966 | ·9999 955− | ·9999 939 | ·9999 919 |
| ·86 | ·9999 995+ | ·9999 994 | ·9999 991 | ·9999 988 | ·9999 984 | ·9999 979 |
| ·87 | ·9999 999 | ·9999 999 | ·9999 998 | ·9999 997 | ·9999 996 | ·9999 995+ |
| ·88 | 1·0000 000 | 1·0000 000 | 1·0000 000 | 1·0000 000 | ·9999 999 | ·9999 999 |
| ·89 | | | | | 1·0000 000 | 1·0000 000 |

$x = \cdot 19$ to $\cdot 84$ $q = 28$ $p = 28$ to $33$

| $x$ | $p = 28$ | $p = 29$ | $p = 30$ | $p = 31$ | $p = 32$ | $p = 33$ |
|---|---|---|---|---|---|---|
| $B(p, q) =$ | $\cdot 9338\ 6666 \times \frac{1}{10^{17}}$ | $\cdot 4669\ 3333 \times \frac{1}{10^{17}}$ | $\cdot 2375\ 6257 \times \frac{1}{10^{17}}$ | $\cdot 1228\ 7719 \times \frac{1}{10^{17}}$ | $\cdot 6456\ 2592 \times \frac{1}{10^{18}}$ | $\cdot 3443\ 3383 \times \frac{1}{10^{18}}$ |
| ·19 | ·0000 001 | | | | | |
| ·20 | ·0000 003 | ·0000 001 | | | | |
| ·21 | ·0000 009 | ·0000 004 | ·0000 001 | ·0000 001 | | |
| ·22 | ·0000 024 | ·0000 010 | ·0000 004 | ·0000 002 | ·0000 001 | |
| ·23 | ·0000 061 | ·0000 027 | ·0000 012 | ·0000 005$^{-}$ | ·0000 002 | ·0000 001 |
| ·24 | ·0000 144 | ·0000 066 | ·0000 030 | ·0000 013 | ·0000 006 | ·0000 002 |
| ·25 | ·0000 322 | ·0000 153 | ·0000 072 | ·0000 033 | ·0000 015$^{+}$ | ·0000 007 |
| ·26 | ·0000 687 | ·0000 340 | ·0000 166 | ·0000 080 | ·0000 038 | ·0000 018 |
| ·27 | ·0001 401 | ·0000 719 | ·0000 364 | ·0000 181 | ·0000 089 | ·0000 043 |
| ·28 | ·0002 740 | ·0001 457 | ·0000 763 | ·0000 394 | ·0000 201 | ·0000 101 |
| ·29 | ·0005 150$^{-}$ | ·0002 833 | ·0001 535$^{+}$ | ·0000 820 | ·0000 432 | ·0000 225$^{-}$ |
| ·30 | ·0009 323 | ·0005 299 | ·0002 967 | ·0001 638 | ·0000 893 | ·0000 480 |
| ·31 | ·0016 293 | ·0009 556 | ·0005 523 | ·0003 148 | ·0001 770 | ·0000 983 |
| ·32 | ·0027 539 | ·0016 650$^{-}$ | ·0009 921 | ·0005 830 | ·0003 380 | ·0001 935$^{+}$ |
| ·33 | ·0045 100 | ·0028 077 | ·0017 229 | ·0010 428 | ·0006 228 | ·0003 673 |
| ·34 | ·0071 674 | ·0045 899 | ·0028 977 | ·0018 045$^{-}$ | ·0011 090 | ·0006 731 |
| ·35 | ·0110 695$^{+}$ | ·0072 848 | ·0047 268 | ·0030 258 | ·0019 119 | ·0011 930 |
| ·36 | ·0166 362 | ·0112 403 | ·0074 892 | ·0049 235$^{-}$ | ·0031 953 | ·0020 482 |
| ·37 | ·0243 597 | ·0168 825$^{-}$ | ·0115 401 | ·0077 845$^{-}$ | ·0051 846 | ·0034 110 |
| ·38 | ·0347 912 | ·0247 111 | ·0173 144 | ·0119 740 | ·0081 771 | ·0055 169 |
| ·39 | ·0485 180 | ·0352 867 | ·0253 222 | ·0179 385$^{-}$ | ·0125 507 | ·0086 766 |
| ·40 | ·0661 297 | ·0492 069 | ·0361 355$^{+}$ | ·0262 013 | ·0187 666 | ·0132 835$^{+}$ |
| ·41 | ·0881 764 | ·0670 725$^{+}$ | ·0503 641 | ·0373 482 | ·0273 638 | ·0198 162 |
| ·42 | ·1151 191 | ·0894 435$^{-}$ | ·0686 197 | ·0520 023 | ·0389 442 | ·0288 324 |
| ·43 | ·1472 799 | ·1167 890 | ·0914 711 | ·0707 863 | ·0541 451 | ·0409 517 |
| ·44 | ·1847 937 | ·1494 349 | ·1193 922 | ·0942 765$^{-}$ | ·0736 005$^{+}$ | ·0568 272 |
| ·45 | ·2275 699 | ·1875 143 | ·1527 074 | ·1229 475$^{+}$ | ·0978 916 | ·0771 030 |
| ·46 | ·2752 680 | ·2309 274 | ·1915 407 | ·1571 168 | ·1274 899 | ·1023 627 |
| ·47 | ·3272 922 | ·2793 165$^{-}$ | ·2357 744 | ·1968 913 | ·1626 992 | ·1330 697 |
| ·48 | ·3828 062 | ·3320 598 | ·2850 231 | ·2421 257 | ·2036 010 | ·1695 067 |
| ·49 | ·4407 711 | ·3882 888 | ·3386 298 | ·2923 972 | ·2500 124 | ·2117 203 |
| ·50 | ·5000 000$^{e}$ | ·4469 265$^{+}$ | ·3956 832 | ·3470 020 | ·3014 616 | ·2594 790 |
| ·51 | ·5592 289 | ·5067 467 | ·4550 608 | ·4049 771 | ·3571 876 | ·3122 505$^{-}$ |
| ·52 | ·6171 938 | ·5664 473 | ·5154 909 | ·4651 460 | ·4161 653 | ·3692 050$^{+}$ |
| ·53 | ·6727 078 | ·6247 321 | ·5756 315$^{-}$ | ·5261 871 | ·4771 574 | ·4292 462 |
| ·54 | ·7247 320 | ·6803 914 | ·6341 548 | ·5867 161 | ·5387 877 | ·4910 690 |
| ·55 | ·7724 301 | ·7323 746 | ·6898 328 | ·6453 767 | ·5996 299 | ·5532 397 |
| ·56 | ·8152 063 | ·7798 476 | ·7416 114 | ·7009 281 | ·6583 024 | ·6142 915$^{-}$ |
| ·57 | ·8527 201 | ·8222 292 | ·7886 683 | ·7523 217 | ·7135 599 | ·6728 236 |
| ·58 | ·8848 809 | ·8592 053 | ·8304 486 | ·7987 587 | ·7643 700 | ·7275 957 |
| ·59 | ·9118 236 | ·8907 198 | ·8666 760 | ·8397 229 | ·8099 701 | ·7776 047 |
| ·60 | ·9338 703 | ·9169 475$^{-}$ | ·8973 404 | ·8749 883 | ·8498 963 | ·8221 383 |
| ·61 | ·9514 820 | ·9382 508 | ·9226 653 | ·9046 016 | ·8839 858 | ·8607 994 |
| ·62 | ·9652 088 | ·9551 287 | ·9430 604 | ·9288 439 | ·9123 528 | ·8935 014 |
| ·63 | ·9756 403 | ·9681 630 | ·9590 666 | ·9481 782 | ·9353 439 | ·9204 360 |
| ·64 | ·9833 638 | ·9779 678 | ·9712 992 | ·9631 901 | ·9534 802 | ·9420 225$^{-}$ |
| ·65 | ·9889 305$^{-}$ | ·9851 457 | ·9803 952 | ·9745 284 | ·9673 935$^{-}$ | ·9588 428 |
| ·66 | ·9928 326 | ·9902 551 | ·9869 701 | ·9828 508 | ·9777 641 | ·9715 741 |
| ·67 | ·9954 900 | ·9937 876 | ·9915 852 | ·9887 815$^{-}$ | ·9852 669 | ·9809 252 |
| ·68 | ·9972 461 | ·9961 571 | ·9947 272 | ·9928 798 | ·9905 294 | ·9875 826 |
| ·69 | ·9983 707 | ·9976 970 | ·9967 994 | ·9956 226 | ·9941 034 | ·9921 707 |
| ·70 | ·9990 677 | ·9986 653 | ·9981 213 | ·9973 978 | ·9964 502 | ·9952 273 |
| ·71 | ·9994 850$^{+}$ | ·9992 533 | ·9989 356 | ·9985 071 | ·9979 378 | ·9971 926 |
| ·72 | ·9997 260 | ·9995 977 | ·9994 193 | ·9991 752 | ·9988 464 | ·9984 100 |
| ·73 | ·9998 599 | ·9997 917 | ·9996 955$^{+}$ | ·9995 622 | ·9993 801 | ·9991 349 |
| ·74 | ·9999 313 | ·9998 966 | ·9998 470 | ·9997 773 | ·9996 808 | ·9995 491 |
| ·75 | ·9999 678 | ·9999 510 | ·9999 266 | ·9998 918 | ·9998 430 | ·9997 755$^{+}$ |
| ·76 | ·9999 856 | ·9999 779 | ·9999 664 | ·9999 499 | ·9999 265$^{-}$ | ·9998 936 |
| ·77 | ·9999 939 | ·9999 905$^{+}$ | ·9999 854 | ·9999 780 | ·9999 673 | ·9999 522 |
| ·78 | ·9999 976 | ·9999 962 | ·9999 940 | ·9999 909 | ·9999 863 | ·9999 797 |
| ·79 | ·9999 991 | ·9999 985$^{+}$ | ·9999 977 | ·9999 965$^{-}$ | ·9999 946 | ·9999 919 |
| ·80 | ·9999 997 | ·9999 995$^{-}$ | ·9999 992 | ·9999 987 | ·9999 980 | ·9999 970 |
| ·81 | ·9999 999 | ·9999 998 | ·9999 997 | ·9999 996 | ·9999 993 | ·9999 990 |
| ·82 | 1·0000 000 | ·9999 999 | ·9999 999 | ·9999 999 | ·9999 998 | ·9999 997 |
| ·83 | | 1·0000 000 | 1·0000 000 | 1·0000 000 | ·9999 999 | ·9999 999 |
| ·84 | | | | | 1·0000 000 | 1·0000 000 |

$x = \cdot 24$ to $\cdot 86$ $q = 28$ $p = 34$ to $39$

| $x$ | $p = 34$ | $p = 35$ | $p = 36$ | $p = 37$ | $p = 38$ | $p = 39$ |
|---|---|---|---|---|---|---|
| $B(p, q) =$ | $\cdot 1862\ 7896 \times \frac{1}{10^{18}}$ | $\cdot 1021\ 5298 \times \frac{1}{10^{18}}$ | $\cdot 5675\ 1653 \times \frac{1}{10^{19}}$ | $\cdot 3192\ 2805 \times \frac{1}{10^{19}}$ | $\cdot 1817\ 1443 \times \frac{1}{10^{19}}$ | $\cdot 1046\ 2346 \times \frac{1}{10^{19}}$ |
| ·24 | ·0000 001 | | | | | |
| ·25 | ·0000 003 | ·0000 001 | ·0000 001 | | | |
| ·26 | ·0000 008 | ·0000 004 | ·0000 002 | ·0000 001 | | |
| ·27 | ·0000 021 | ·0000 010 | ·0000 005− | ·0000 002 | ·0000 001 | |
| ·28 | ·0000 050+ | ·0000 025− | ·0000 012 | ·0000 006 | ·0000 003 | ·0000 001 |
| ·29 | ·0000 116 | ·0000 059 | ·0000 029 | ·0000 015− | ·0000 007 | ·0000 003 |
| ·30 | ·0000 255− | ·0000 134 | ·0000 069 | ·0000 036 | ·0000 018 | ·0000 009 |
| ·31 | ·0000 539 | ·0000 292 | ·0000 156 | ·0000 083 | ·0000 043 | ·0000 023 |
| ·32 | ·0001 094 | ·0000 612 | ·0000 338 | ·0000 185− | ·0000 100 | ·0000 054 |
| ·33 | ·0002 140 | ·0001 232 | ·0000 702 | ·0000 395+ | ·0000 220 | ·0000 122 |
| ·34 | ·0004 036 | ·0002 392 | ·0001 402 | ·0000 813 | ·0000 467 | ·0000 265+ |
| ·35 | ·0007 356 | ·0004 484 | ·0002 703 | ·0001 612 | ·0000 952 | ·0000 556 |
| ·36 | ·0012 974 | ·0008 125− | ·0005 032 | ·0003 084 | ·0001 871 | ·0001 124 |
| ·37 | ·0022 178 | ·0014 257 | ·0009 066 | ·0005 705− | ·0003 554 | ·0002 192 |
| ·38 | ·0036 790 | ·0024 259 | ·0015 824 | ·0010 215+ | ·0006 528 | ·0004 132 |
| ·39 | ·0059 295+ | ·0040 074 | ·0026 794 | ·0017 731 | ·0011 617 | ·0007 539 |
| ·40 | ·0092 958 | ·0064 341 | ·0044 063 | ·0029 869 | ·0020 048 | ·0013 329 |
| ·41 | ·0141 898 | ·0100 511 | ·0070 452 | ·0048 885+ | ·0033 590 | ·0022 863 |
| ·42 | ·0211 106 | ·0152 920 | ·0109 630 | ·0077 812 | ·0054 696 | ·0038 089 |
| ·43 | ·0306 369 | ·0226 794 | ·0166 180 | ·0120 568 | ·0086 642 | ·0061 689 |
| ·44 | ·0434 085+ | ·0328 157 | ·0245 593 | ·0182 019 | ·0133 634 | ·0097 218 |
| ·45 | ·0600 942 | ·0463 620 | ·0354 156 | ·0267 952 | ·0200 853 | ·0149 205− |
| ·46 | ·0813 472 | ·0640 032 | ·0498 703 | ·0384 934 | ·0294 410 | ·0223 182 |
| ·47 | ·1077 499 | ·0863 994 | ·0686 235+ | ·0540 029 | ·0421 167 | ·0325 608 |
| ·48 | ·1397 516 | ·1141 273 | ·0923 392 | ·0740 373 | ·0588 417 | ·0463 654 |
| ·49 | ·1776 056 | ·1476 147 | ·1215 827 | ·0992 602 | ·0803 404 | ·0644 826 |
| ·50 | ·2213 130 | ·1870 759 | ·1567 515+ | ·1302 177 | ·1072 696 | ·0876 429 |
| ·51 | ·2705 815+ | ·2324 545− | ·1980 094 | ·1672 671 | ·1401 475− | ·1164 892 |
| ·52 | ·3248 062 | ·2833 847 | ·2452 296 | ·2105 085− | ·1792 782 | ·1514 997 |
| ·53 | ·3830 773 | ·3391 760 | ·2979 590 | ·2597 302 | ·2246 838 | ·1929 114 |
| ·54 | ·4442 179 | ·3988 275− | ·3554 083 | ·3143 771 | ·2760 518 | ·2406 513 |
| ·55 | ·5068 496 | ·4610 735− | ·4164 744 | ·3735 479 | ·3327 096 | ·2942 894 |
| ·56 | ·5694 803 | ·5244 583 | ·4797 964 | ·4360 278 | ·3936 314 | ·3530 201 |
| ·57 | ·6306 060 | ·5874 323 | ·5438 393 | ·5003 552 | ·4574 822 | ·4156 810 |
| ·58 | ·6888 155− | ·6484 612 | ·6070 001 | ·5649 171 | ·5226 976 | ·4808 114 |
| ·59 | ·7428 854 | ·7061 340 | ·6677 236 | ·6280 648 | ·5875 914 | ·5467 452 |
| ·60 | ·7918 568 | ·7592 596 | ·7246 135+ | ·6882 351 | ·6504 802 | ·6117 318 |
| ·61 | ·8350 836 | ·8069 400 | ·7765 287 | ·7440 647 | ·7098 108 | ·6740 696 |
| ·62 | ·8722 507 | ·8486 124 | ·8226 509 | ·7944 826 | ·7642 740 | ·7322 370 |
| ·63 | ·9033 598 | ·8840 586 | ·8625 184 | ·8387 704 | ·8128 915+ | ·7850 036 |
| ·64 | ·9286 899 | ·9133 809 | ·8960 249 | ·8765 861 | ·8550 669 | ·8315 091 |
| ·65 | ·9487 374 | ·9369 528 | ·9233 836 | ·9079 486 | ·8905 948 | ·8713 000 |
| ·66 | ·9641 462 | ·9553 507 | ·9450 675+ | ·9331 904 | ·9196 313 | ·9043 237 |
| ·67 | ·9756 363 | ·9692 788 | ·9617 332 | ·9528 861 | ·9426 330 | ·9308 824 |
| ·68 | ·9839 392 | ·9794 943 | ·9741 402 | ·9677 687 | ·9602 745− | ·9515 575+ |
| ·69 | ·9897 460 | ·9867 444 | ·9830 756 | ·9786 455− | ·9733 581 | ·9671 176 |
| ·70 | ·9936 709 | ·9917 161 | ·9892 923 | ·9863 230 | ·9827 279 | ·9784 231 |
| ·71 | ·9962 306 | ·9950 052 | ·9934 640 | ·9915 490 | ·9891 972 | ·9863 410 |
| ·72 | ·9978 386 | ·9971 006 | ·9961 592 | ·9949 731 | ·9934 960 | ·9916 768 |
| ·73 | ·9988 096 | ·9983 835− | ·9978 325− | ·9971 286 | ·9962 397 | ·9951 298 |
| ·74 | ·9993 720 | ·9991 367 | ·9988 284 | ·9984 291 | ·9979 179 | ·9972 710 |
| ·75 | ·9996 835+ | ·9995 597 | ·9993 952 | ·9991 793 | ·9988 992 | ·9985 399 |
| ·76 | ·9998 482 | ·9997 863 | ·9997 029 | ·9995 920 | ·9994 462 | ·9992 567 |
| ·77 | ·9999 309 | ·9999 016 | ·9998 616 | ·9998 077 | ·9997 358 | ·9996 413 |
| ·78 | ·9999 703 | ·9999 572 | ·9999 391 | ·9999 144 | ·9998 811 | ·9998 366 |
| ·79 | ·9999 880 | ·9999 825+ | ·9999 748 | ·9999 642 | ·9999 497 | ·9999 301 |
| ·80 | ·9999 955− | ·9999 933 | ·9999 903 | ·9999 860 | ·9999 801 | ·9999 721 |
| ·81 | ·9999 984 | ·9999 976 | ·9999 965+ | ·9999 949 | ·9999 927 | ·9999 896 |
| ·82 | ·9999 995− | ·9999 992 | ·9999 989 | ·9999 983 | ·9999 975+ | ·9999 965− |
| ·83 | ·9999 998 | ·9999 998 | ·9999 997 | ·9999 995− | ·9999 992 | ·9999 989 |
| ·84 | 1·0000 000 | ·9999 999 | ·9999 999 | ·9999 999 | ·9999 998 | ·9999 997 |
| ·85 | | 1·0000 000 | 1·0000 000 | 1·0000 000 | ·9999 999 | ·9999 999 |
| ·86 | | | | | 1·0000 000 | 1·0000 000 |

$x$ = ·28 to ·87 $q$ = 28 $p$ = 40 to 45

| $x$ | $p$ = 40 | $p$ = 41 | $p$ = 42 | $p$ = 43 | $p$ = 44 | $p$ = 45 |
|---|---|---|---|---|---|---|
| $B(p, q)$ = | ·6090 0222 $\times \frac{1}{10^{20}}$ | ·3582 3660 $\times \frac{1}{10^{20}}$ | ·2128 6523 $\times \frac{1}{10^{20}}$ | ·1277 1914 $\times \frac{1}{10^{20}}$ | ·7735 1026 $\times \frac{1}{10^{21}}$ | ·4727 0071 $\times \frac{1}{10^{21}}$ |
| ·28 | ·0000 001 | | | | | |
| ·29 | ·0000 002 | ·0000 001 | | | | |
| ·30 | ·0000 005⁻ | ·0000 002 | ·0000 001 | ·0000 001 | | |
| ·31 | ·0000 012 | ·0000 006 | ·0000 003 | ·0000 001 | ·0000 001 | |
| ·32 | ·0000 028 | ·0000 015⁻ | ·0000 008 | ·0000 004 | ·0000 002 | ·0000 001 |
| ·33 | ·0000 066 | ·0000 036 | ·0000 019 | ·0000 010 | ·0000 005⁺ | ·0000 003 |
| ·34 | ·0000 149 | ·0000 083 | ·0000 046 | ·0000 025⁺ | ·0000 014 | ·0000 007 |
| ·35 | ·0000 322 | ·0000 185⁻ | ·0000 105⁺ | ·0000 059 | ·0000 033 | ·0000 018 |
| ·36 | ·0000 669 | ·0000 394 | ·0000 230 | ·0000 134 | ·0000 077 | ·0000 044 |
| ·37 | ·0001 339 | ·0000 811 | ·0000 487 | ·0000 290 | ·0000 171 | ·0000 100 |
| ·38 | ·0002 591 | ·0001 610 | ·0000 992 | ·0000 606 | ·0000 367 | ·0000 221 |
| ·39 | ·0004 847 | ·0003 088 | ·0001 951 | ·0001 222 | ·0000 759 | ·0000 468 |
| ·40 | ·0008 780 | ·0005 732 | ·0003 711 | ·0002 382 | ·0001 517 | ·0000 958 |
| ·41 | ·0015 420 | ·0010 309 | ·0006 833 | ·0004 492 | ·0002 930 | ·0001 896 |
| ·42 | ·0026 286 | ·0017 982 | ·0012 198 | ·0008 206 | ·0005 478 | ·0003 628 |
| ·43 | ·0043 531 | ·0030 452 | ·0021 125⁺ | ·0014 536 | ·0009 924 | ·0006 724 |
| ·44 | ·0070 102 | ·0050 117 | ·0035 534 | ·0024 992 | ·0017 441 | ·0012 079 |
| ·45 | ·0109 873 | ·0080 226 | ·0058 099 | ·0041 741 | ·0029 758 | ·0021 057 |
| ·46 | ·0167 734 | ·0125 011 | ·0092 417 | ·0067 785⁻ | ·0049 339 | ·0035 648 |
| ·47 | ·0249 602 | ·0189 767 | ·0143 124 | ·0107 110 | ·0079 554 | ·0058 656 |
| ·48 | ·0362 308 | ·0280 825⁺ | ·0215 957 | ·0164 804 | ·0124 833 | ·0093 874 |
| ·49 | ·0513 328 | ·0405 401 | ·0317 691 | ·0247 084 | ·0190 763 | ·0146 231 |
| ·50 | ·0710 357 | ·0571 272 | ·0455 932 | ·0361 190 | ·0284 073 | ·0221 855⁻ |
| ·51 | ·0960 702 | ·0786 273 | ·0638 732 | ·0515 113 | ·0412 481 | ·0328 020 |
| ·52 | ·1270 546 | ·1057 630 | ·0874 002 | ·0717 132 | ·0584 339 | ·0472 914 |
| ·53 | ·1644 140 | ·1391 154 | ·1168 773 | ·0975 143 | ·0808 081 | ·0665 205⁺ |
| ·54 | ·2083 007 | ·1790 396 | ·1528 331 | ·1295 842 | ·1091 467 | ·0913 383 |
| ·55 | ·2585 291 | ·2255 849 | ·1955 333 | ·1683 796 | ·1440 676 | ·1224 906 |
| ·56 | ·3145 330 | ·2784 322 | ·2449 025⁻ | ·2140 551 | ·1859 338 | ·1605 223 |
| ·57 | ·3753 590 | ·3368 615⁺ | ·3004 673 | ·2663 868 | ·2347 632 | ·2056 767 |
| ·58 | ·4396 985⁺ | ·3997 573 | ·3613 358 | ·3247 257 | ·2901 588 | ·2578 074 |
| ·59 | ·5059 619 | ·4656 577 | ·4262 187 | ·3879 909 | ·3512 745⁺ | ·3163 189 |
| ·60 | ·5723 872 | ·5328 459 | ·4934 975⁻ | ·4547 112 | ·4168 269 | ·3801 480 |
| ·61 | ·6371 736 | ·5994 752 | ·5613 354 | ·5231 138 | ·4851 590 | ·4477 993 |
| ·62 | ·6986 228 | ·6637 144 | ·6278 184 | ·5912 557 | ·5543 530 | ·5174 334 |
| ·63 | ·7552 708 | ·7238 952 | ·6911 116 | ·6571 806 | ·6223 815⁺ | ·5870 051 |
| ·64 | ·8059 941 | ·7786 420 | ·7496 088 | ·7190 824 | ·6872 782 | ·6544 332 |
| ·65 | ·8500 757 | ·8269 678 | ·8020 564 | ·7754 545⁻ | ·7473 060 | ·7177 821 |
| ·66 | ·8872 263 | ·8683 252 | ·8476 354 | ·8252 016 | ·8010 985⁻ | ·7754 286 |
| ·67 | ·9175 591 | ·9026 069 | ·8859 918 | ·8677 033 | ·8477 561 | ·8261 904 |
| ·68 | ·9415 263 | ·9301 008 | ·9172 150⁻ | ·9028 197 | ·8868 845⁻ | ·8693 992 |
| ·69 | ·9598 306 | ·9514 086 | ·9417 707 | ·9308 453 | ·9185 734 | ·9049 096 |
| ·70 | ·9733 236 | ·9673 445⁺ | ·9604 029 | ·9524 200 | ·9433 233 | ·9330 481 |
| ·71 | ·9829 092 | ·9788 279 | ·9740 219 | ·9684 160 | ·9619 367 | ·9545 135⁻ |
| ·72 | ·9894 601 | ·9867 868 | ·9835 945⁺ | ·9798 185⁻ | ·9753 926 | ·9702 506 |
| ·73 | ·9937 587 | ·9920 821 | ·9900 522 | ·9876 178 | ·9847 249 | ·9813 171 |
| ·74 | ·9964 608 | ·9954 565⁻ | ·9942 240 | ·9927 255⁺ | ·9909 205⁻ | ·9887 651 |
| ·75 | ·9980 839 | ·9975 110 | ·9967 983 | ·9959 202 | ·9948 481 | ·9935 507 |
| ·76 | ·9990 129 | ·9987 026 | ·9983 115⁻ | ·9978 231 | ·9972 189 | ·9964 780 |
| ·77 | ·9995 180 | ·9993 590 | ·9991 560 | ·9988 992 | ·9985 772 | ·9981 773 |
| ·78 | ·9997 778 | ·9997 011 | ·9996 019 | ·9994 747 | ·9993 132 | ·9991 099 |
| ·79 | ·9999 038 | ·9998 691 | ·9998 236 | ·9997 646 | ·9996 887 | ·9995 919 |
| ·80 | ·9999 611 | ·9999 465⁻ | ·9999 270 | ·9999 015⁻ | ·9998 682 | ·9998 253 |
| ·81 | ·9999 854 | ·9999 797 | ·9999 720 | ·9999 618 | ·9999 483 | ·9999 306 |
| ·82 | ·9999 950⁻ | ·9999 929 | ·9999 901 | ·9999 863 | ·9999 813 | ·9999 746 |
| ·83 | ·9999 984 | ·9999 977 | ·9999 968 | ·9999 955⁺ | ·9999 938 | ·9999 915⁺ |
| ·84 | ·9999 995⁺ | ·9999 993 | ·9999 991 | ·9999 987 | ·9999 982 | ·9999 975⁻ |
| ·85 | ·9999 999 | ·9999 998 | ·9999 998 | ·9999 997 | ·9999 995⁺ | ·9999 993 |
| ·86 | 1·0000 000 | 1·0000 000 | ·9999 999 | ·9999 999 | ·9999 999 | ·9999 998 |
| ·87 | | | 1·0000 000 | 1·0000 000 | 1·0000 000 | 1·0000 000 |

$x$ = ·32 to ·88 $q = 28$ $p$ = 46 to 50

| | $p = 46$ | $p = 47$ | $p = 48$ | $p = 49$ | $p = 50$ |
|---|---|---|---|---|---|
| $B(p, q)$ = | ·2913 9085 $\times \frac{1}{10^{21}}$ | ·1811 3485 $\times \frac{1}{10^{21}}$ | ·1135 1117 $\times \frac{1}{10^{21}}$ | ·7169 1268 $\times \frac{1}{10^{22}}$ | ·4562 1716 $\times \frac{1}{10^{22}}$ |
| $x$ | | | | | |
| ·32 | ·0000 001 | | | | |
| ·33 | ·0000 001 | ·0000 001 | | | |
| ·34 | ·0000 004 | ·0000 002 | ·0000 001 | ·0000 001 | |
| ·35 | ·0000 010 | ·0000 006 | ·0000 003 | ·0000 002 | ·0000 001 |
| ·36 | ·0000 025⁻ | ·0000 014 | ·0000 008 | ·0000 004 | ·0000 002 |
| ·37 | ·0000 058 | ·0000 034 | ·0000 019 | ·0000 011 | ·0000 006 |
| ·38 | ·0000 132 | ·0000 078 | ·0000 046 | ·0000 027 | ·0000 016 |
| ·39 | ·0000 286 | ·0000 174 | ·0000 105⁻ | ·0000 063 | ·0000 037 |
| ·40 | ·0000 601 | ·0000 374 | ·0000 231 | ·0000 142 | ·0000 087 |
| ·41 | ·0001 218 | ·0000 777 | ·0000 492 | ·0000 309 | ·0000 193 |
| ·42 | ·0002 385⁺ | ·0001 557 | ·0001 009 | ·0000 650⁻ | ·0000 416 |
| ·43 | ·0004 522 | ·0003 019 | ·0002 002 | ·0001 319 | ·0000 863 |
| ·44 | ·0008 305⁺ | ·0005 670 | ·0003 844 | ·0002 589 | ·0001 732 |
| ·45 | ·0014 792 | ·0010 318 | ·0007 148 | ·0004 919 | ·0003 364 |
| ·46 | ·0025 571 | ·0018 215⁻ | ·0012 887 | ·0009 058 | ·0006 326 |
| ·47 | ·0042 940 | ·0031 218 | ·0022 544 | ·0016 174 | ·0011 531 |
| ·48 | ·0070 097 | ·0051 986 | ·0038 298 | ·0028 032 | ·0020 389 |
| ·49 | ·0111 317 | ·0084 169 | ·0063 224 | ·0047 188 | ·0035 000⁺ |
| ·50 | ·0172 080 | ·0132 584 | ·0101 492 | ·0077 201 | ·0058 364 |
| ·51 | ·0259 099 | ·0203 318 | ·0158 527 | ·0122 835⁻ | ·0094 601 |
| ·52 | ·0380 208 | ·0303 706 | ·0241 072 | ·0190 181 | ·0149 137 |
| ·53 | ·0544 046 | ·0442 141 | ·0357 104 | ·0286 683 | ·0228 795⁻ |
| ·54 | ·0759 518 | ·0627 662 | ·0515 557 | ·0420 968 | ·0341 745⁻ |
| ·55 | ·1035 029 | ·0869 299 | ·0725 784 | ·0602 451 | ·0497 241 |
| ·56 | ·1377 537 | ·1175 193 | ·0996 785⁺ | ·0840 679 | ·0705 090 |
| ·57 | ·1791 498 | ·1551 545⁻ | ·1336 200 | ·1144 408 | ·0974 848 |
| ·58 | ·2277 853 | ·2001 519 | ·1749 174 | ·1520 485⁺ | ·1314 759 |
| ·59 | ·2833 207 | ·2524 244 | ·2237 237 | ·1972 652 | ·1730 530 |
| ·60 | ·3449 363 | ·3114 086 | ·2797 357 | ·2500 423 | ·2224 092 |
| ·61 | ·4113 363 | ·3760 385⁻ | ·3421 375⁺ | ·3098 257 | ·2792 548 |
| ·62 | ·4808 093 | ·4447 742 | ·4095 979 | ·3755 209 | ·3427 513 |
| ·63 | ·5513 457 | ·5156 940 | ·4803 306 | ·4455 197 | ·4115 046 |
| ·64 | ·6207 998 | ·5866 400 | ·5522 186 | ·5177 972 | ·4836 286 |
| ·65 | ·6870 772 | ·6554 044 | ·6229 904 | ·5900 699 | ·5568 806 |
| ·66 | ·7483 212 | ·7199 291 | ·6904 255⁺ | ·6599 999 | ·6288 541 |
| ·67 | ·8030 720 | ·7784 911 | ·7525 609 | ·7254 152 | ·6972 059 |
| ·68 | ·8503 752 | ·8298 458 | ·8078 663 | ·7845 131 | ·7598 825⁺ |
| ·69 | ·8898 249 | ·8733 071 | ·8553 625⁻ | ·8360 159 | ·8153 111 |
| ·70 | ·9215 399 | ·9087 557 | ·8946 660 | ·8792 553 | ·8625 238 |
| ·71 | ·9460 807 | ·9365 792 | ·9259 577 | ·9141 744 | ·9011 984 |
| ·72 | ·9643 270 | ·9575 586 | ·9498 858 | ·9412 539 | ·9316 144 |
| ·73 | ·9773 368 | ·9727 257 | ·9674 259 | ·9613 808 | ·9545 363 |
| ·74 | ·9862 131 | ·9832 161 | ·9797 243 | ·9756 870 | ·9710 531 |
| ·75 | ·9919 937 | ·9901 406 | ·9879 523 | ·9853 879 | ·9824 049 |
| ·76 | ·9955 770 | ·9944 903 | ·9931 900 | ·9916 458 | ·9898 256 |
| ·77 | ·9976 845⁻ | ·9970 823 | ·9963 523 | ·9954 739 | ·9944 250⁻ |
| ·78 | ·9988 562 | ·9985 422 | ·9981 566 | ·9976 866 | ·9971 180 |
| ·79 | ·9994 695⁺ | ·9993 162 | ·9991 254 | ·9988 899 | ·9986 014 |
| ·80 | ·9997 703 | ·9997 005⁺ | ·9996 126 | ·9995 028 | ·9993 664 |
| ·81 | ·9999 078 | ·9998 784 | ·9998 409 | ·9997 935⁻ | ·9997 339 |
| ·82 | ·9999 659 | ·9999 545⁺ | ·9999 398 | ·9999 210 | ·9998 971 |
| ·83 | ·9999 885⁻ | ·9999 845⁻ | ·9999 792 | ·9999 725⁻ | ·9999 637 |
| ·84 | ·9999 965⁻ | ·9999 952 | ·9999 935⁺ | ·9999 913 | ·9999 884 |
| ·85 | ·9999 990 | ·9999 987 | ·9999 982 | ·9999 976 | ·9999 967 |
| ·86 | ·9999 998 | ·9999 997 | ·9999 996 | ·9999 994 | ·9999 992 |
| ·87 | 1·0000 000 | ·9999 999 | ·9999 999 | ·9999 999 | ·9999 998 |
| ·88 | | 1·0000 000 | 1·0000 000 | 1·0000 000 | 1·0000 000 |

$x = \cdot 19$ to $\cdot 83$ $q = 29$ $p = 29$ to $34$

| | $p = 29$ | $p = 30$ | $p = 31$ | $p = 32$ | $p = 33$ | $p = 34$ |
|---|---|---|---|---|---|---|
| $B(p, q) =$ | $\cdot 2293\ 7076 \times \frac{1}{10^{17}}$ | $\cdot 1146\ 8538 \times \frac{1}{10^{17}}$ | $\cdot 5831\ 4600 \times \frac{1}{10^{18}}$ | $\cdot 3012\ 9210 \times \frac{1}{10^{18}}$ | $\cdot 1580\ 5487 \times \frac{1}{10^{18}}$ | $\cdot 8412\ 5980 \times \frac{1}{10^{19}}$ |
| $x$ | | | | | | |
| ·19 | ·0000 001 | | | | | |
| ·20 | ·0000 002 | ·0000 001 | | | | |
| ·21 | ·0000 006 | ·0000 002 | ·0000 001 | | | |
| ·22 | ·0000 016 | ·0000 007 | ·0000 003 | ·0000 001 | | |
| ·23 | ·0000 042 | ·0000 019 | ·0000 008 | ·0000 003 | ·0000 001 | ·0000 001 |
| ·24 | ·0000 103 | ·0000 047 | ·0000 021 | ·0000 009 | ·0000 004 | ·0000 002 |
| ·25 | ·0000 237 | ·0000 113 | ·0000 053 | ·0000 025− | ·0000 011 | ·0000 005+ |
| ·26 | ·0000 520 | ·0000 258 | ·0000 126 | ·0000 061 | ·0000 029 | ·0000 014 |
| ·27 | ·0001 088 | ·0000 559 | ·0000 283 | ·0000 142 | ·0000 070 | ·0000 034 |
| ·28 | ·0002 176 | ·0001 159 | ·0000 608 | ·0000 315− | ·0000 161 | ·0000 081 |
| ·29 | ·0004 177 | ·0002 301 | ·0001 250− | ·0000 669 | ·0000 354 | ·0000 185− |
| ·30 | ·0007 713 | ·0004 391 | ·0002 464 | ·0001 364 | ·0000 745+ | ·0000 402 |
| ·31 | ·0013 733 | ·0008 068 | ·0004 673 | ·0002 670 | ·0001 506 | ·0000 839 |
| ·32 | ·0023 619 | ·0014 304 | ·0008 542 | ·0005 032 | ·0002 926 | ·0001 681 |
| ·33 | ·0039 312 | ·0024 517 | ·0015 077 | ·0009 148 | ·0005 480 | ·0003 242 |
| ·34 | ·0063 426 | ·0040 689 | ·0025 744 | ·0016 073 | ·0009 907 | ·0006 032 |
| ·35 | ·0099 341 | ·0065 494 | ·0042 590 | ·0027 334 | ·0017 321 | ·0010 844 |
| ·36 | ·0151 254 | ·0102 382 | ·0068 368 | ·0045 062 | ·0029 331 | ·0018 863 |
| ·37 | ·0224 156 | ·0155 641 | ·0106 629 | ·0072 116 | ·0048 172 | ·0031 796 |
| ·38 | ·0323 720 | ·0230 363 | ·0161 776 | ·0112 173 | ·0076 831 | ·0052 005+ |
| ·39 | ·0456 071 | ·0332 333 | ·0239 034 | ·0169 783 | ·0119 143 | ·0082 636 |
| ·40 | ·0627 452 | ·0467 794 | ·0344 325+ | ·0250 330 | ·0179 833 | ·0127 708 |
| ·41 | ·0843 777 | ·0643 097 | ·0484 024 | ·0359 896 | ·0264 473 | ·0192 154 |
| ·42 | ·1110 110 | ·0864 240 | ·0664 594 | ·0505 006 | ·0379 330 | ·0281 760 |
| ·43 | ·1430 112 | ·1136 333 | ·0892 105+ | ·0692 232 | ·0531 085− | ·0402 997 |
| ·44 | ·1805 506 | ·1463 019 | ·1171 677 | ·0927 701 | ·0726 421 | ·0562 713 |
| ·45 | ·2235 643 | ·1845 930 | ·1506 880 | ·1216 500− | ·0971 491 | ·0767 689 |
| ·46 | ·2717 208 | ·2284 236 | ·1899 179 | ·1562 068 | ·1271 310 | ·1024 077 |
| ·47 | ·3244 136 | ·2774 348 | ·2347 467 | ·1965 615+ | ·1629 109 | ·1336 756 |
| ·48 | ·3807 764 | ·3309 847 | ·2847 780 | ·2425 660 | ·2045 751 | ·1708 669 |
| ·49 | ·4397 214 | ·3881 647 | ·3393 233 | ·2937 747 | ·2519 270 | ·2140 231 |
| ·50 | ·5000 000[e] | ·4478 416 | ·3974 218 | ·3494 417 | ·3044 604 | ·2628 867 |
| ·51 | ·5602 786 | ·5087 219 | ·4578 869 | ·4085 442 | ·3613 603 | ·3168 787 |
| ·52 | ·6192 236 | ·5694 319 | ·5193 747 | ·4698 342 | ·4215 322 | ·3751 037 |
| ·53 | ·6755 864 | ·6286 076 | ·5804 699 | ·5319 130 | ·4836 596 | ·4363 859 |
| ·54 | ·7282 792 | ·6849 820 | ·6397 797 | ·5933 234 | ·5462 864 | ·4993 349 |
| ·55 | ·7764 357 | ·7374 644 | ·6960 249 | ·6526 471 | ·6079 138 | ·5624 349 |
| ·56 | ·8194 494 | ·7852 007 | ·7481 207 | ·7086 007 | ·6671 046 | ·6241 499 |
| ·57 | ·8569 888 | ·8276 110 | ·7952 366 | ·7601 156 | ·7225 801 | ·6830 312 |
| ·58 | ·8889 890 | ·8644 020 | ·8368 318 | ·8063 979 | ·7733 010 | ·7378 170 |
| ·59 | ·9156 223 | ·8955 543 | ·8726 634 | ·8469 592 | ·8185 238 | ·7875 121 |
| ·60 | ·9372 548 | ·9212 891 | ·9027 688 | ·8816 198 | ·8578 272 | ·8314 390 |
| ·61 | ·9543 929 | ·9420 191 | ·9274 263 | ·9104 845+ | ·8911 073 | ·8692 581 |
| ·62 | ·9676 280 | ·9582 923 | ·9471 019 | ·9338 972 | ·9185 468 | ·9009 543 |
| ·63 | ·9775 844 | ·9707 328 | ·9623 876 | ·9523 814 | ·9405 616 | ·9267 969 |
| ·64 | ·9848 746 | ·9799 875− | ·9739 404 | ·9665 748 | ·9577 359 | ·9472 794 |
| ·65 | ·9900 659 | ·9866 812 | ·9824 277 | ·9771 657 | ·9707 527 | ·9630 473 |
| ·66 | ·9936 574 | ·9913 837 | ·9884 825+ | ·9848 382 | ·9803 285− | ·9748 265+ |
| ·67 | ·9960 688 | ·9945 892 | ·9926 726 | ·9902 287 | ·9871 586 | ·9833 562 |
| ·68 | ·9976 381 | ·9967 066 | ·9954 820 | ·9938 972 | ·9918 765+ | ·9893 366 |
| ·69 | ·9986 267 | ·9980 602 | ·9973 046 | ·9963 122 | ·9950 283 | ·9933 908 |
| ·70 | ·9992 287 | ·9988 965− | ·9984 469 | ·9978 479 | ·9970 617 | ·9960 444 |
| ·71 | ·9995 823 | ·9993 948 | ·9991 374 | ·9987 895− | ·9983 264 | ·9977 186 |
| ·72 | ·9997 824 | ·9996 807 | ·9995 392 | ·9993 452 | ·9990 833 | ·9987 348 |
| ·73 | ·9998 912 | ·9998 384 | ·9997 638 | ·9996 602 | ·9995 184 | ·9993 270 |
| ·74 | ·9999 480 | ·9999 217 | ·9998 842 | ·9998 313 | ·9997 579 | ·9996 576 |
| ·75 | ·9999 763 | ·9999 638 | ·9999 458 | ·9999 201 | ·9998 840 | ·9998 339 |
| ·76 | ·9999 897 | ·9999 841 | ·9999 759 | ·9999 640 | ·9999 471 | ·9999 234 |
| ·77 | ·9999 958 | ·9999 934 | ·9999 899 | ·9999 847 | ·9999 772 | ·9999 666 |
| ·78 | ·9999 984 | ·9999 974 | ·9999 960 | ·9999 938 | ·9999 907 | ·9999 862 |
| ·79 | ·9999 994 | ·9999 990 | ·9999 985+ | ·9999 977 | ·9999 965− | ·9999 947 |
| ·80 | ·9999 998 | ·9999 997 | ·9999 995− | ·9999 992 | ·9999 987 | ·9999 981 |
| ·81 | ·9999 999 | ·9999 999 | ·9999 998 | ·9999 997 | ·9999 996 | ·9999 994 |
| ·82 | 1·0000 000 | 1·0000 000 | ·9999 999 | ·9999 999 | ·9999 999 | ·9999 998 |
| ·83 | | | 1·0000 000 | 1·0000 000 | 1·0000 000 | 1·0000 000 |

$x = \cdot 24$ to $\cdot 85$ $q = 29$ $p = 35$ to $40$

| | $p = 35$ | $p = 36$ | $p = 37$ | $p = 38$ | $p = 39$ | $p = 40$ |
|---|---|---|---|---|---|---|
| $B(p, q) =$ | ·4540 1322 $\times \frac{1}{10^{19}}$ | ·2482 8848 $\times \frac{1}{10^{19}}$ | ·1375 1362 $\times \frac{1}{10^{19}}$ | ·7709 0969 $\times \frac{1}{10^{20}}$ | ·4372 3236 $\times \frac{1}{10^{20}}$ | ·2507 6562 $\times \frac{1}{10^{20}}$ |
| $x$ | | | | | | |
| ·24 | ·0000 001 | | | | | |
| ·25 | ·0000 002 | ·0000 001 | | | | |
| ·26 | ·0000 006 | ·0000 003 | ·0000 001 | ·0000 001 | | |
| ·27 | ·0000 016 | ·0000 008 | ·0000 004 | ·0000 002 | ·0000 001 | |
| ·28 | ·0000 040 | ·0000 020 | ·0000 010 | ·0000 005− | ·0000 002 | ·0000 001 |
| ·29 | ·0000 095+ | ·0000 049 | ·0000 024 | ·0000 012 | ·0000 006 | ·0000 003 |
| ·30 | ·0000 214 | ·0000 113 | ·0000 059 | ·0000 030 | ·0000 015+ | ·0000 008 |
| ·31 | ·0000 461 | ·0000 251 | ·0000 135+ | ·0000 072 | ·0000 038 | ·0000 020 |
| ·32 | ·0000 954 | ·0000 535+ | ·0000 297 | ·0000 163 | ·0000 089 | ·0000 048 |
| ·33 | ·0001 896 | ·0001 096 | ·0000 626 | ·0000 354 | ·0000 198 | ·0000 110 |
| ·34 | ·0003 630 | ·0002 160 | ·0001 271 | ·0000 740 | ·0000 427 | ·0000 244 |
| ·35 | ·0006 710 | ·0004 105+ | ·0002 485− | ·0001 488 | ·0000 882 | ·0000 518 |
| ·36 | ·0011 991 | ·0007 537 | ·0004 687 | ·0002 885− | ·0001 758 | ·0001 061 |
| ·37 | ·0020 747 | ·0013 388 | ·0008 548 | ·0005 401 | ·0003 379 | ·0002 094 |
| ·38 | ·0034 803 | ·0023 036 | ·0015 087 | ·0009 781 | ·0006 279 | ·0003 992 |
| ·39 | ·0056 673 | ·0038 447 | ·0025 810 | ·0017 153 | ·0011 288 | ·0007 359 |
| ·40 | ·0089 688 | ·0062 313 | ·0042 847 | ·0029 168 | ·0019 664 | ·0013 134 |
| ·41 | ·0138 085− | ·0098 182 | ·0069 097 | ·0048 148 | ·0033 230 | ·0022 722 |
| ·42 | ·0207 033 | ·0150 539 | ·0108 357 | ·0077 234 | ·0054 530 | ·0038 148 |
| ·43 | ·0302 561 | ·0224 824 | ·0165 398 | ·0120 507 | ·0086 981 | ·0062 214 |
| ·44 | ·0431 361 | ·0327 331 | ·0245 956 | ·0183 056 | ·0134 987 | ·0098 651 |
| ·45 | ·0600 451 | ·0464 989 | ·0356 619 | ·0270 948 | ·0203 989 | ·0152 225− |
| ·46 | ·0816 693 | ·0644 978 | ·0504 554 | ·0391 076 | ·0300 413 | ·0228 767 |
| ·47 | ·1086 192 | ·0874 215− | ·0697 096 | ·0550 854 | ·0431 474 | ·0335 082 |
| ·48 | ·1413 623 | ·1158 703 | ·0941 171 | ·0757 739 | ·0604 815− | ·0478 711 |
| ·49 | ·1801 548 | ·1502 830 | ·1242 614 | ·1018 617 | ·0827 983 | ·0667 509 |
| ·50 | ·2249 813 | ·1908 664 | ·1605 421 | ·1339 058 | ·1107 744 | ·0909 050+ |
| ·51 | ·2755 108 | ·2375 351 | ·2031 038 | ·1722 552 | ·1449 298 | ·1209 886 |
| ·52 | ·3310 786 | ·2898 711 | ·2517 770 | ·2169 776 | ·1855 482 | ·1574 713 |
| ·53 | ·3906 973 | ·3471 103 | ·3060 417 | ·2678 035− | ·2326 042 | ·2005 548 |
| ·54 | ·4531 015− | ·4081 626 | ·3650 213 | ·3240 953 | ·2857 111 | ·2501 023 |
| ·55 | ·5168 223 | ·4716 657 | ·4275 127 | ·3848 513 | ·3440 985− | ·3055 922 |
| ·56 | ·5802 856 | ·5360 703 | ·4920 516 | ·4487 467 | ·4066 270 | ·3661 056 |
| ·57 | ·6419 237 | ·5997 474 | ·5570 087 | ·5142 123 | ·4718 439 | ·4303 554 |
| ·58 | ·7002 876 | ·6611 069 | ·6207 072 | ·5795 431 | ·5380 758 | ·4967 573 |
| ·59 | ·7541 471 | ·7187 134 | ·6815 475− | ·6430 255− | ·6035 506 | ·5635 392 |
| ·60 | ·8025 673 | ·7713 858 | ·7381 255− | ·7030 674 | ·6665 331 | ·6288 748 |
| ·61 | ·8449 540 | ·8182 682 | ·7893 288 | ·7583 168 | ·7254 604 | ·6910 285+ |
| ·62 | ·8810 644 | ·8588 672 | ·8344 011 | ·8077 528 | ·7790 568 | ·7484 918 |
| ·63 | ·9109 837 | ·8930 516 | ·8729 676 | ·8507 394 | ·8264 172 | ·8000 930 |
| ·64 | ·9350 759 | ·9210 175+ | ·9050 222 | ·8870 383 | ·8670 478 | ·8450 684 |
| ·65 | ·9539 142 | ·9432 285+ | ·9308 806 | ·9167 805+ | ·9008 623 | ·8830 870 |
| ·66 | ·9682 047 | ·9603 381 | ·9511 079 | ·9404 058 | ·9281 379 | ·9142 280 |
| ·67 | ·9787 107 | ·9731 081 | ·9664 349 | ·9585 803 | ·9494 400 | ·9389 193 |
| ·68 | ·9861 871 | ·9823 321 | ·9776 718 | ·9721 046 | ·9655 296 | ·9578 485+ |
| ·69 | ·9913 304 | ·9887 714 | ·9856 324 | ·9818 273 | ·9772 673 | ·9718 619 |
| ·70 | ·9947 459 | ·9931 098 | ·9910 738 | ·9885 700 | ·9855 259 | ·9818 653 |
| ·71 | ·9969 317 | ·9959 261 | ·9946 567 | ·9930 735− | ·9911 210 | ·9887 396 |
| ·72 | ·9982 772 | ·9976 842 | ·9969 251 | ·9959 649 | ·9947 643 | ·9932 791 |
| ·73 | ·9990 723 | ·9987 375+ | ·9983 031 | ·9977 460 | ·9970 396 | ·9961 538 |
| ·74 | ·9995 222 | ·9993 418 | ·9991 045− | ·9987 960 | ·9983 995− | ·9978 954 |
| ·75 | ·9997 653 | ·9996 728 | ·9995 494 | ·9993 869 | ·9991 751 | ·9989 023 |
| ·76 | ·9998 905− | ·9998 455− | ·9997 846 | ·9997 034 | ·9995 962 | ·9994 562 |
| ·77 | ·9999 516 | ·9999 309 | ·9999 026 | ·9998 642 | ·9998 129 | ·9997 451 |
| ·78 | ·9999 799 | ·9999 709 | ·9999 585− | ·9999 414 | ·9999 184 | ·9998 875− |
| ·79 | ·9999 921 | ·9999 885+ | ·9999 834 | ·9999 763 | ·9999 666 | ·9999 534 |
| ·80 | ·9999 971 | ·9999 958 | ·9999 938 | ·9999 911 | ·9999 873 | ·9999 820 |
| ·81 | ·9999 990 | ·9999 986 | ·9999 979 | ·9999 969 | ·9999 955+ | ·9999 936 |
| ·82 | ·9999 997 | ·9999 995+ | ·9999 993 | ·9999 990 | ·9999 985+ | ·9999 979 |
| ·83 | ·9999 999 | ·9999 999 | ·9999 998 | ·9999 997 | ·9999 996 | ·9999 994 |
| ·84 | 1·0000 000 | 1·0000 000 | ·9999 999 | ·9999 999 | ·9999 999 | ·9999 998 |
| ·85 | | | 1·0000 000 | 1·0000 000 | 1·0000 000 | 1·0000 000 |

$x = \cdot 29$ to $\cdot 87$ $q = 29$ $p = 41$ to $45$

| | $p=41$ | $p=42$ | $p=43$ | $p=44$ | $p=45$ |
|---|---|---|---|---|---|
| $B(p,q) =$ | $\cdot 1453\ 7137 \times \frac{1}{10^{20}}$ | $\cdot 8514\ 6090 \times \frac{1}{10^{21}}$ | $\cdot 5036\ 8110 \times \frac{1}{10^{21}}$ | $\cdot 3008\ 0954 \times \frac{1}{10^{21}}$ | $\cdot 1813\ 0986 \times \frac{1}{10^{21}}$ |
| $x$ | | | | | |
| ·29 | ·0000 001 | ·0000 001 | | | |
| ·30 | ·0000 004 | ·0000 002 | ·0000 001 | | |
| ·31 | ·0000 010 | ·0000 005+ | ·0000 003 | ·0000 001 | ·0000 001 |
| ·32 | ·0000 025+ | ·0000 013 | ·0000 007 | ·0000 004 | ·0000 002 |
| ·33 | ·0000 060 | ·0000 033 | ·0000 018 | ·0000 010 | ·0000 005+ |
| ·34 | ·0000 138 | ·0000 077 | ·0000 043 | ·0000 024 | ·0000 013 |
| ·35 | ·0000 301 | ·0000 174 | ·0000 099 | ·0000 056 | ·0000 032 |
| ·36 | ·0000 634 | ·0000 376 | ·0000 221 | ·0000 129 | ·0000 074 |
| ·37 | ·0001 286 | ·0000 782 | ·0000 472 | ·0000 282 | ·0000 168 |
| ·38 | ·0002 515+ | ·0001 571 | ·0000 972 | ·0000 597 | ·0000 364 |
| ·39 | ·0004 754 | ·0003 044 | ·0001 933 | ·0001 217 | ·0000 760 |
| ·40 | ·0008 693 | ·0005 704 | ·0003 711 | ·0002 394 | ·0001 533 |
| ·41 | ·0015 398 | ·0010 345+ | ·0006 892 | ·0004 554 | ·0002 986 |
| ·42 | ·0026 452 | ·0018 184 | ·0012 397 | ·0008 384 | ·0005 625+ |
| ·43 | ·0044 110 | ·0031 009 | ·0021 619 | ·0014 953 | ·0010 262 |
| ·44 | ·0071 472 | ·0051 347 | ·0036 588 | ·0025 865+ | ·0018 145+ |
| ·45 | ·0112 625+ | ·0082 636 | ·0060 144 | ·0043 432 | ·0031 126 |
| ·46 | ·0172 739 | ·0129 365− | ·0096 112 | ·0070 855− | ·0051 843 |
| ·47 | ·0258 065− | ·0197 146 | ·0149 427 | ·0112 394 | ·0083 913 |
| ·48 | ·0375 810 | ·0292 687 | ·0226 188 | ·0173 483 | ·0132 087 |
| ·49 | ·0533 834 | ·0423 601 | ·0333 577 | ·0260 742 | ·0202 341 |
| ·50 | ·0740 161 | ·0598 047 | ·0479 618 | ·0381 846 | ·0301 850+ |
| ·51 | ·1002 316 | ·0824 160 | ·0672 727 | ·0545 206 | ·0438 785− |
| ·52 | ·1326 513 | ·1109 308 | ·0921 063 | ·0759 436 | ·0621 905+ |
| ·53 | ·1716 783 | ·1459 218 | ·1231 703 | ·1032 601 | ·0859 925− |
| ·54 | ·2174 135− | ·1877 065+ | ·1609 702 | ·1371 314 | ·1160 666 |
| ·55 | ·2695 889 | ·2362 639 | ·2057 160 | ·1779 742 | ·1530 066 |
| ·56 | ·3275 293 | ·2911 735+ | ·2572 414 | ·2258 660 | ·1971 148 |
| ·57 | ·3901 530 | ·3515 882 | ·3149 516 | ·2804 706 | ·2483 092 |
| ·58 | ·4560 173 | ·4162 511 | ·3778 104 | ·3409 968 | ·3060 572 |
| ·59 | ·5234 078 | ·4835 603 | ·4443 768 | ·4062 049 | ·3693 516 |
| ·60 | ·5904 632 | ·5516 769 | ·5128 906 | ·4744 651 | ·4367 383 |
| ·61 | ·6553 227 | ·6186 676 | ·5814 017 | ·5438 670 | ·5064 006 |
| ·62 | ·7162 764 | ·6826 624 | ·6479 279 | ·6123 694 | ·5762 937 |
| ·63 | ·7718 998 | ·7420 082 | ·7106 220 | ·6779 730 | ·6443 149 |
| ·64 | ·8211 549 | ·7953 983 | ·7679 246 | ·7388 919 | ·7084 868 |
| ·65 | ·8634 453 | ·8419 592 | ·8186 825+ | ·7937 007 | ·7671 292 |
| ·66 | ·8986 210 | ·8812 859 | ·8622 173 | ·8414 369 | ·8189 940 |
| ·67 | ·9269 362 | ·9134 245+ | ·8983 365+ | ·8816 450− | ·8633 450− |
| ·68 | ·9489 692 | ·9388 079 | ·9272 917 | ·9143 614 | ·8999 735− |
| ·69 | ·9655 214 | ·9581 587 | ·9496 915+ | ·9400 449 | ·9291 530 |
| ·70 | ·9775 090 | ·9723 772 | ·9663 900 | ·9594 700 | ·9515 434 |
| ·71 | ·9858 652 | ·9824 306 | ·9783 664 | ·9736 018 | ·9680 662 |
| ·72 | ·9914 613 | ·9892 586 | ·9866 154 | ·9834 730 | ·9797 707 |
| ·73 | ·9950 544 | ·9937 038 | ·9920 606 | ·9900 799 | ·9877 140 |
| ·74 | ·9972 613 | ·9964 716 | ·9954 976 | ·9943 076 | ·9928 665+ |
| ·75 | ·9985 545− | ·9981 154 | ·9975 666 | ·9968 870 | ·9960 529 |
| ·76 | ·9992 753 | ·9990 440 | ·9987 510 | ·9983 833 | ·9979 260 |
| ·77 | ·9996 563 | ·9995 412 | ·9993 936 | ·9992 058 | ·9989 692 |
| ·78 | ·9998 465− | ·9997 926 | ·9997 227 | ·9996 326 | ·9995 176 |
| ·79 | ·9999 357 | ·9999 122 | ·9998 812 | ·9998 408 | ·9997 885− |
| ·80 | ·9999 749 | ·9999 654 | ·9999 526 | ·9999 357 | ·9999 136 |
| ·81 | ·9999 910 | ·9999 873 | ·9999 825− | ·9999 760 | ·9999 674 |
| ·82 | ·9999 970 | ·9999 958 | ·9999 941 | ·9999 918 | ·9999 887 |
| ·83 | ·9999 991 | ·9999 987 | ·9999 982 | ·9999 974 | ·9999 964 |
| ·84 | ·9999 998 | ·9999 996 | ·9999 995− | ·9999 993 | ·9999 990 |
| ·85 | ·9999 999 | ·9999 999 | ·9999 999 | ·9999 998 | ·9999 997 |
| ·86 | 1·0000 000 | 1·0000 000 | 1·0000 000 | 1·0000 000 | ·9999 999 |
| ·87 | | | | | 1·0000 000 |

$x$ = ·32 to ·88 $q = 29$ $p$ = 46 to 50

| | $p = 46$ | $p = 47$ | $p = 48$ | $p = 49$ | $p = 50$ |
|---|---|---|---|---|---|
| $B(p, q)$ = | ·1102 5600 $\times \frac{1}{10^{21}}$ | ·6762 3678 $\times \frac{1}{10^{22}}$ | ·4181 9906 $\times \frac{1}{10^{22}}$ | ·2606 9552 $\times \frac{1}{10^{22}}$ | ·1637 7026 $\times \frac{1}{10^{22}}$ |
| $x$ | | | | | |
| ·32 | ·0000 001 | | | | |
| ·33 | ·0000 003 | ·0000 001 | ·0000 001 | | |
| ·34 | ·0000 007 | ·0000 004 | ·0000 002 | ·0000 001 | ·0000 001 |
| ·35 | ·0000 018 | ·0000 010 | ·0000 005+ | ·0000 003 | ·0000 002 |
| ·36 | ·0000 043 | ·0000 024 | ·0000 014 | ·0000 008 | ·0000 004 |
| ·37 | ·0000 099 | ·0000 058 | ·0000 033 | ·0000 019 | ·0000 011 |
| ·38 | ·0000 220 | ·0000 132 | ·0000 079 | ·0000 046 | ·0000 027 |
| ·39 | ·0000 471 | ·0000 290 | ·0000 177 | ·0000 107 | ·0000 065− |
| ·40 | ·0000 974 | ·0000 614 | ·0000 384 | ·0000 239 | ·0000 148 |
| ·41 | ·0001 943 | ·0001 255− | ·0000 805− | ·0000 512 | ·0000 324 |
| ·42 | ·0003 746 | ·0002 476 | ·0001 625+ | ·0001 060 | ·0000 686 |
| ·43 | ·0006 990 | ·0004 727 | ·0003 174 | ·0002 116 | ·0001 402 |
| ·44 | ·0012 635− | ·0008 734 | ·0005 996 | ·0004 088 | ·0002 769 |
| ·45 | ·0022 142 | ·0015 639 | ·0010 969 | ·0007 641 | ·0005 289 |
| ·46 | ·0037 656 | ·0027 158 | ·0019 452 | ·0013 839 | ·0009 782 |
| ·47 | ·0062 197 | ·0045 779 | ·0033 464 | ·0024 301 | ·0017 533 |
| ·48 | ·0099 852 | ·0074 962 | ·0055 897 | ·0041 407 | ·0030 478 |
| ·49 | ·0155 919 | ·0119 326 | ·0090 714 | ·0068 515+ | ·0051 423 |
| ·50 | ·0236 965− | ·0184 775− | ·0143 133 | ·0110 167 | ·0084 265+ |
| ·51 | ·0350 739 | ·0278 503 | ·0219 715− | ·0172 243 | ·0134 199 |
| ·52 | ·0505 891 | ·0408 842 | ·0328 312 | ·0262 009 | ·0207 831 |
| ·53 | ·0711 462 | ·0584 881 | ·0477 826 | ·0387 989 | ·0313 168 |
| ·54 | ·0976 138 | ·0815 839 | ·0677 709 | ·0559 608 | ·0459 391 |
| ·55 | ·1307 299 | ·1110 199 | ·0937 213 | ·0786 570 | ·0656 372 |
| ·56 | ·1709 959 | ·1474 662 | ·1264 396 | ·1077 961 | ·0913 898 |
| ·57 | ·2185 706 | ·1913 017 | ·1664 985+ | ·1441 137 | ·1240 633 |
| ·58 | ·2731 830 | ·2425 100 | ·2141 211 | ·1880 506 | ·1642 893 |
| ·59 | ·3340 789 | ·3006 006 | ·2690 810 | ·2396 365+ | ·2123 373 |
| ·60 | ·4000 175− | ·3645 739 | ·3306 386 | ·2984 001 | ·2680 037 |
| ·61 | ·4693 255+ | ·4329 436 | ·3975 292 | ·3633 249 | ·3305 375+ |
| ·62 | ·5400 096 | ·5038 202 | ·4680 157 | ·4328 677 | ·3986 235− |
| ·63 | ·6099 163 | ·5750 540 | ·5400 064 | ·5050 463 | ·4704 359 |
| ·64 | ·6769 195− | ·6444 189 | ·6112 268 | ·5775 921 | ·5437 653 |
| ·65 | ·7391 110 | ·7098 137 | ·6794 255+ | ·6481 511 | ·6162 064 |
| ·66 | ·7949 653 | ·7694 530 | ·7425 836 | ·7145 052 | ·6853 838 |
| ·67 | ·8434 549 | ·8220 169 | ·7990 964 | ·7747 816 | ·7491 816 |
| ·68 | ·8841 020 | ·8667 401 | ·8479 005− | ·8276 165− | ·8059 416 |
| ·69 | ·9169 613 | ·9034 285− | ·8885 280 | ·8722 493 | ·8545 984 |
| ·70 | ·9425 424 | ·9324 064 | ·9210 843 | ·9085 356 | ·8947 320 |
| ·71 | ·9616 904 | ·9544 081 | ·9461 575+ | ·9368 824 | ·9265 340 |
| ·72 | ·9754 465− | ·9704 378 | ·9646 833 | ·9581 230 | ·9507 006 |
| ·73 | ·9849 121 | ·9816 218 | ·9777 889 | ·9733 587 | ·9682 766 |
| ·74 | ·9911 366 | ·9890 773 | ·9866 455+ | ·9837 963 | ·9804 831 |
| ·75 | ·9950 381 | ·9938 137 | ·9923 484 | ·9906 083 | ·9885 574 |
| ·76 | ·9973 623 | ·9966 730 | ·9958 371 | ·9948 312 | ·9936 298 |
| ·77 | ·9986 737 | ·9983 077 | ·9978 580 | ·9973 096 | ·9966 462 |
| ·78 | ·9993 721 | ·9991 895+ | ·9989 623 | ·9986 816 | ·9983 376 |
| ·79 | ·9997 215+ | ·9996 364 | ·9995 291 | ·9993 949 | ·9992 282 |
| ·80 | ·9998 850− | ·9998 481 | ·9998 010 | ·9997 413 | ·9996 664 |
| ·81 | ·9999 560 | ·9999 413 | ·9999 222 | ·9998 977 | ·9998 666 |
| ·82 | ·9999 846 | ·9999 792 | ·9999 721 | ·9999 629 | ·9999 511 |
| ·83 | ·9999 951 | ·9999 933 | ·9999 909 | ·9999 878 | ·9999 837 |
| ·84 | ·9999 986 | ·9999 980 | ·9999 973 | ·9999 964 | ·9999 951 |
| ·85 | ·9999 996 | ·9999 995− | ·9999 993 | ·9999 990 | ·9999 987 |
| ·86 | ·9999 999 | ·9999 999 | ·9999 998 | ·9999 998 | ·9999 997 |
| ·87 | 1·0000 000 | 1·0000 000 | 1·0000 000 | 1·0000 000 | ·9999 999 |
| ·88 | | | | | 1·0000 000 |

$x$ = ·20 to ·83 $q = 30$ $p$ = 30 to 35

| | $p = 30$ | $p = 31$ | $p = 32$ | $p = 33$ | $p = 34$ | $p = 35$ |
|---|---|---|---|---|---|---|
| $B(p, q) =$ | $\cdot 5637\,0780 \times \frac{1}{10^{18}}$ | $\cdot 2818\,5390 \times \frac{1}{10^{18}}$ | $\cdot 1432\,3723 \times \frac{1}{10^{18}}$ | $\cdot 7392\,8891 \times \frac{1}{10^{19}}$ | $\cdot 3872\,4657 \times \frac{1}{10^{19}}$ | $\cdot 2057\,2474 \times \frac{1}{10^{19}}$ |
| $x$ | | | | | | |
| ·20 | ·0000 001 | ·0000 001 | | | | |
| ·21 | ·0000 004 | ·0000 002 | ·0000 001 | | | |
| ·22 | ·0000 011 | ·0000 005− | ·0000 002 | ·0000 001 | | |
| ·23 | ·0000 030 | ·0000 013 | ·0000 006 | ·0000 002 | ·0000 001 | |
| ·24 | ·0000 074 | ·0000 034 | ·0000 015+ | ·0000 007 | ·0000 003 | ·0000 001 |
| ·25 | ·0000 175+ | ·0000 084 | ·0000 039 | ·0000 018 | ·0000 008 | ·0000 004 |
| ·26 | ·0000 394 | ·0000 196 | ·0000 096 | ·0000 046 | ·0000 022 | ·0000 010 |
| ·27 | ·0000 844 | ·0000 435− | ·0000 221 | ·0000 111 | ·0000 055− | ·0000 027 |
| ·28 | ·0001 728 | ·0000 922 | ·0000 485− | ·0000 252 | ·0000 129 | ·0000 065+ |
| ·29 | ·0003 389 | ·0001 870 | ·0001 018 | ·0000 546 | ·0000 290 | ·0000 152 |
| ·30 | ·0006 384 | ·0003 640 | ·0002 047 | ·0001 136 | ·0000 622 | ·0000 337 |
| ·31 | ·0011 580 | ·0006 814 | ·0003 955− | ·0002 265− | ·0001 281 | ·0000 715+ |
| ·32 | ·0020 266 | ·0012 293 | ·0007 356 | ·0004 344 | ·0002 533 | ·0001 459 |
| ·33 | ·0034 282 | ·0021 415− | ·0013 196 | ·0008 026 | ·0004 821 | ·0002 861 |
| ·34 | ·0056 150+ | ·0036 082 | ·0022 876 | ·0014 317 | ·0008 849 | ·0005 404 |
| ·35 | ·0089 187 | ·0058 899 | ·0038 381 | ·0024 692 | ·0015 691 | ·0009 853 |
| ·36 | ·0137 570 | ·0093 281 | ·0062 421 | ·0041 243 | ·0026 920 | ·0017 365+ |
| ·37 | ·0206 342 | ·0143 523 | ·0098 536 | ·0066 807 | ·0044 750− | ·0029 628 |
| ·38 | ·0301 314 | ·0214 801 | ·0151 172 | ·0105 080 | ·0072 174 | ·0049 004 |
| ·39 | ·0428 848 | ·0313 062 | ·0225 662 | ·0160 685+ | ·0113 075+ | ·0078 670 |
| ·40 | ·0595 520 | ·0444 803 | ·0328 119 | ·0239 147 | ·0172 284 | ·0122 726 |
| ·41 | ·0807 654 | ·0616 713 | ·0465 191 | ·0346 768 | ·0255 546 | ·0186 244 |
| ·42 | ·1070 771 | ·0835 188 | ·0643 682 | ·0490 358 | ·0369 371 | ·0275 215− |
| ·43 | ·1388 983 | ·1105 763 | ·0870 050+ | ·0676 840 | ·0520 750− | ·0396 382 |
| ·44 | ·1764 408 | ·1432 478 | ·1149 803 | ·0912 709 | ·0716 711 | ·0556 915+ |
| ·45 | ·2196 672 | ·1817 286 | ·1486 854 | ·1203 404 | ·0963 761 | ·0763 941 |
| ·46 | ·2682 570 | ·2259 539 | ·1882 905− | ·1552 643 | ·1267 218 | ·1023 934 |
| ·47 | ·3215 949 | ·2755 654 | ·2336 933 | ·1961 786 | ·1630 520 | ·1342 026 |
| ·48 | ·3787 847 | ·3299 012 | ·2844 869 | ·2429 328 | ·2054 585+ | ·1721 285− |
| ·49 | ·4386 903 | ·3880 131 | ·3399 515+ | ·2950 590 | ·2537 307 | ·2162 070 |
| ·50 | ·5000 000[e] | ·4487 109 | ·3990 763 | ·3517 683 | ·3073 275+ | ·2661 544 |
| ·51 | ·5613 097 | ·5106 325+ | ·4606 093 | ·4119 773 | ·3653 790 | ·3213 436 |
| ·52 | ·6212 153 | ·5723 318 | ·5231 329 | ·4743 646 | ·4267 194 | ·3808 118 |
| ·53 | ·6784 051 | ·6323 756 | ·5851 582 | ·5374 539 | ·4899 519 | ·4433 022 |
| ·54 | ·7317 430 | ·6894 399 | ·6452 263 | ·5997 139 | ·5535 396 | ·5073 380 |
| ·55 | ·7803 328 | ·7423 943 | ·7020 081 | ·6596 656 | ·6159 118 | ·5713 215+ |
| ·56 | ·8235 592 | ·7903 663 | ·7543 894 | ·7159 841 | ·6755 771 | ·6336 488 |
| ·57 | ·8611 017 | ·8327 797 | ·8015 342 | ·7675 839 | ·7312 262 | ·6928 262 |
| ·58 | ·8929 229 | ·8693 646 | ·8429 186 | ·8136 792 | ·7818 171 | ·7475 747 |
| ·59 | ·9192 346 | ·9001 404 | ·8783 361 | ·8538 131 | ·8266 297 | ·7969 118 |
| ·60 | ·9404 480 | ·9253 763 | ·9078 737 | ·8878 551 | ·8652 887 | ·8402 001 |
| ·61 | ·9571 152 | ·9455 365+ | ·9318 662 | ·9159 703 | ·8977 525+ | ·8771 611 |
| ·62 | ·9698 686 | ·9612 172 | ·9508 356 | ·9385 659 | ·9242 735− | ·9078 540 |
| ·63 | ·9793 658 | ·9730 838 | ·9654 239 | ·9562 249 | ·9453 365+ | ·9326 260 |
| ·64 | ·9862 430 | ·9818 141 | ·9763 279 | ·9696 348 | ·9615 868 | ·9520 429 |
| ·65 | ·9910 813 | ·9880 526 | ·9842 422 | ·9795 209 | ·9737 551 | ·9668 108 |
| ·66 | ·9943 850− | ·9923 781 | ·9898 146 | ·9865 893 | ·9825 899 | ·9776 990 |
| ·67 | ·9965 718 | ·9952 851 | ·9936 165− | ·9914 854 | ·9888 028 | ·9854 724 |
| ·68 | ·9979 734 | ·9971 762 | ·9961 269 | ·9947 668 | ·9930 291 | ·9908 397 |
| ·69 | ·9988 420 | ·9983 654 | ·9977 289 | ·9968 917 | ·9958 064 | ·9944 189 |
| ·70 | ·9993 616 | ·9990 872 | ·9987 154 | ·9982 193 | ·9975 668 | ·9967 206 |
| ·71 | ·9996 611 | ·9995 092 | ·9993 005− | ·9990 180 | ·9986 412 | ·9981 454 |
| ·72 | ·9998 272 | ·9997 465+ | ·9996 342 | ·9994 799 | ·9992 713 | ·9989 930 |
| ·73 | ·9999 156 | ·9998 746 | ·9998 167 | ·9997 362 | ·9996 257 | ·9994 763 |
| ·74 | ·9999 606 | ·9999 407 | ·9999 123 | ·9998 721 | ·9998 164 | ·9997 399 |
| ·75 | ·9999 825− | ·9999 733 | ·9999 600 | ·9999 410 | ·9999 142 | ·9998 770 |
| ·76 | ·9999 926 | ·9999 886 | ·9999 827 | ·9999 742 | ·9999 620 | ·9999 448 |
| ·77 | ·9999 970 | ·9999 954 | ·9999 929 | ·9999 893 | ·9999 841 | ·9999 766 |
| ·78 | ·9999 989 | ·9999 982 | ·9999 973 | ·9999 958 | ·9999 937 | ·9999 907 |
| ·79 | ·9999 996 | ·9999 994 | ·9999 990 | ·9999 985− | ·9999 977 | ·9999 965+ |
| ·80 | ·9999 999 | ·9999 998 | ·9999 997 | ·9999 995− | ·9999 992 | ·9999 988 |
| ·81 | 1·0000 000 | ·9999 999 | ·9999 999 | ·9999 998 | ·9999 998 | ·9999 996 |
| ·82 | | 1·0000 000 | 1·0000 000 | 1·0000 000 | ·9999 999 | ·9999 999 |
| ·83 | | | | | 1·0000 000 | 1·0000 000 |

$x = \cdot 24$ to $\cdot 85$ $\qquad q = 30 \qquad$ $p = 36$ to $40$

| $x$ | $p = 36$ | $p = 37$ | $p = 38$ | $p = 39$ | $p = 40$ |
|---|---|---|---|---|---|
| $B(p, q) =$ | $\cdot 1107\ 7486 \times \frac{1}{10^{19}}$ | $\cdot 6042\ 2652 \times \frac{1}{10^{20}}$ | $\cdot 3336\ 7733 \times \frac{1}{10^{20}}$ | $\cdot 1864\ 6674 \times \frac{1}{10^{20}}$ | $\cdot 1053\ 9425 \times \frac{1}{10^{20}}$ |
| ·24 | ·0000 001 | | | | |
| ·25 | ·0000 002 | ·0000 001 | | | |
| ·26 | ·0000 005− | ·0000 002 | ·0000 001 | | |
| ·27 | ·0000 013 | ·0000 006 | ·0000 003 | ·0000 001 | ·0000 001 |
| ·28 | ·0000 033 | ·0000 016 | ·0000 008 | ·0000 004 | ·0000 002 |
| ·29 | ·0000 078 | ·0000 040 | ·0000 020 | ·0000 010 | ·0000 005+ |
| ·30 | ·0000 180 | ·0000 095+ | ·0000 050− | ·0000 026 | ·0000 013 |
| ·31 | ·0000 395− | ·0000 216 | ·0000 116 | ·0000 062 | ·0000 033 |
| ·32 | ·0000 831 | ·0000 468 | ·0000 261 | ·0000 144 | ·0000 078 |
| ·33 | ·0001 678 | ·0000 973 | ·0000 559 | ·0000 317 | ·0000 178 |
| ·34 | ·0003 263 | ·0001 948 | ·0001 151 | ·0000 673 | ·0000 390 |
| ·35 | ·0006 117 | ·0003 756 | ·0002 282 | ·0001 372 | ·0000 817 |
| ·36 | ·0011 075+ | ·0006 987 | ·0004 362 | ·0002 695+ | ·0001 649 |
| ·37 | ·0019 397 | ·0012 562 | ·0008 051 | ·0005 108 | ·0003 209 |
| ·38 | ·0032 904 | ·0021 858 | ·0014 370 | ·0009 353 | ·0006 029 |
| ·39 | ·0054 134 | ·0036 857 | ·0024 837 | ·0016 572 | ·0010 952 |
| ·40 | ·0086 478 | ·0060 299 | ·0041 620 | ·0028 447 | ·0019 259 |
| ·41 | ·0134 287 | ·0095 825− | ·0067 695+ | ·0047 361 | ·0032 824 |
| ·42 | ·0202 903 | ·0148 066 | ·0106 983 | ·0076 560 | ·0054 281 |
| ·43 | ·0298 594 | ·0222 672 | ·0164 438 | ·0120 287 | ·0087 186 |
| ·44 | ·0428 348 | ·0326 209 | ·0246 043 | ·0183 851 | ·0136 139 |
| ·45 | ·0599 517 | ·0465 923 | ·0358 687 | ·0273 603 | ·0206 845+ |
| ·46 | ·0819 298 | ·0649 336 | ·0509 876 | ·0396 766 | ·0306 046 |
| ·47 | ·1094 086 | ·0883 681 | ·0707 283 | ·0561 104 | ·0441 312 |
| ·48 | ·1428 742 | ·1175 205+ | ·0958 123 | ·0774 403 | ·0620 643 |
| ·49 | ·1825 858 | ·1528 403 | ·1268 412 | ·1043 793 | ·0851 888 |
| ·50 | ·2285 104 | ·1945 262 | ·1642 160 | ·1374 952 | ·1142 001 |
| ·51 | ·2802 774 | ·2424 623 | ·2080 608 | ·1771 266 | ·1496 190 |
| ·52 | ·3371 602 | ·2961 763 | ·2581 609 | ·2233 068 | ·1917 058 |
| ·53 | ·3980 920 | ·3548 284 | ·3139 267 | ·2757 051 | ·2403 844 |
| ·54 | ·4617 173 | ·4172 371 | ·3743 919 | ·3335 987 | ·2951 904 |
| ·55 | ·5264 764 | ·4819 427 | ·4382 516 | ·3958 827 | ·3552 520 |
| ·56 | ·5907 143 | ·5473 027 | ·5039 381 | ·4611 212 | ·4193 143 |
| ·57 | ·6528 023 | ·6116 111 | ·5697 296 | ·5276 387 | ·4858 069 |
| ·58 | ·7112 582 | ·6732 268 | ·6338 797 | ·5936 420 | ·5529 505− |
| ·59 | ·7648 505− | ·7306 962 | ·6947 512 | ·6573 590 | ·6188 929 |
| ·60 | ·8126 744 | ·7828 548 | ·7509 398 | ·7171 772 | ·6818 562 |
| ·61 | ·8541 929 | ·8288 959 | ·8013 700 | ·7717 653 | ·7402 780 |
| ·62 | ·8892 390 | ·8684 006 | ·8453 544 | ·8201 613 | ·7929 269 |
| ·63 | ·9179 835− | ·9013 276 | ·8826 100 | ·8618 186 | ·8389 801 |
| ·64 | ·9408 738 | ·9279 672 | ·9132 328 | ·8966 062 | ·8780 526 |
| ·65 | ·9585 570 | ·9488 703 | ·9376 388 | ·9247 671 | ·9101 791 |
| ·66 | ·9717 963 | ·9647 622 | ·9564 811 | ·9468 444 | ·9357 548 |
| ·67 | ·9813 922 | ·9764 563 | ·9705 572 | ·9635 885− | ·9554 477 |
| ·68 | ·9881 172 | ·9847 747 | ·9807 203 | ·9758 592 | ·9700 958 |
| ·69 | ·9926 682 | ·9904 871 | ·9878 026 | ·9845 366 | ·9806 075− |
| ·70 | ·9956 373 | ·9942 683 | ·9925 588 | ·9904 489 | ·9878 738 |
| ·71 | ·9975 018 | ·9966 767 | ·9956 318 | ·9943 237 | ·9927 043 |
| ·72 | ·9986 265+ | ·9981 501 | ·9975 383 | ·9967 615+ | ·9957 865− |
| ·73 | ·9992 768 | ·9990 139 | ·9986 716 | ·9982 309 | ·9976 701 |
| ·74 | ·9996 364 | ·9994 981 | ·9993 155+ | ·9990 773 | ·9987 700 |
| ·75 | ·9998 259 | ·9997 568 | ·9996 643 | ·9995 420 | ·9993 821 |
| ·76 | ·9999 210 | ·9998 882 | ·9998 439 | ·9997 844 | ·9997 057 |
| ·77 | ·9999 661 | ·9999 515− | ·9999 314 | ·9999 042 | ·9998 676 |
| ·78 | ·9999 863 | ·9999 802 | ·9999 717 | ·9999 599 | ·9999 440 |
| ·79 | ·9999 948 | ·9999 924 | ·9999 891 | ·9999 843 | ·9999 779 |
| ·80 | ·9999 982 | ·9999 973 | ·9999 961 | ·9999 943 | ·9999 919 |
| ·81 | ·9999 994 | ·9999 991 | ·9999 987 | ·9999 981 | ·9999 972 |
| ·82 | ·9999 998 | ·9999 997 | ·9999 996 | ·9999 994 | ·9999 991 |
| ·83 | 1·0000 000 | ·9999 999 | ·9999 999 | ·9999 998 | ·9999 998 |
| ·84 | | 1·0000 000 | 1·0000 000 | 1·0000 000 | ·9999 999 |
| ·85 | | | | | 1·0000 000 |

$x = \cdot28$ to $\cdot86$ $q = 30$ $p = 41$ to $45$

| $x$ | $p = 41$ | $p = 42$ | $p = 43$ | $p = 44$ | $p = 45$ |
|---|---|---|---|---|---|
| $B(p, q) =$ | $\cdot6022\ 5283 \times \frac{1}{10^{21}}$ | $\cdot3477\ 7981 \times \frac{1}{10^{21}}$ | $\cdot2028\ 7155 \overset{+}{\times} \frac{1}{10^{21}}$ | $\cdot1194\ 9968 \times \frac{1}{10^{21}}$ | $\cdot7105\ 3865 \overset{+}{\times} \frac{1}{10^{22}}$ |
| ·28 | ·0000 001 | | | | |
| ·29 | ·0000 002 | ·0000 001 | ·0000 001 | | |
| ·30 | ·0000 007 | ·0000 003 | ·0000 002 | ·0000 001 | |
| ·31 | ·0000 017 | ·0000 009 | ·0000 005− | ·0000 002 | ·0000 001 |
| ·32 | ·0000 042 | ·0000 023 | ·0000 012 | ·0000 006 | ·0000 003 |
| ·33 | ·0000 099 | ·0000 055− | ·0000 030 | ·0000 016 | ·0000 009 |
| ·34 | ·0000 223 | ·0000 127 | ·0000 071 | ·0000 040 | ·0000 022 |
| ·35 | ·0000 482 | ·0000 282 | ·0000 163 | ·0000 094 | ·0000 053 |
| ·36 | ·0001 000 | ·0000 600 | ·0000 357 | ·0000 211 | ·0000 124 |
| ·37 | ·0001 998 | ·0001 232 | ·0000 753 | ·0000 457 | ·0000 275− |
| ·38 | ·0003 851 | ·0002 437 | ·0001 529 | ·0000 951 | ·0000 587 |
| ·39 | ·0007 171 | ·0004 654 | ·0002 994 | ·0001 910 | ·0001 209 |
| ·40 | ·0012 919 | ·0008 590 | ·0005 662 | ·0003 702 | ·0002 400 |
| ·41 | ·0022 543 | ·0015 346 | ·0010 358 | ·0006 934 | ·0004 605− |
| ·42 | ·0038 140 | ·0026 566 | ·0018 348 | ·0012 569 | ·0008 542 |
| ·43 | ·0062 633 | ·0044 607 | ·0031 504 | ·0022 070 | ·0015 340 |
| ·44 | ·0099 926 | ·0072 721 | ·0052 487 | ·0037 579 | ·0026 696 |
| ·45 | ·0155 024 | ·0115 211 | ·0084 924 | ·0062 103 | ·0045 066 |
| ·46 | ·0234 060 | ·0177 525− | ·0133 562 | ·0099 700 | ·0073 857 |
| ·47 | ·0344 191 | ·0266 257 | ·0204 337 | ·0155 607 | ·0117 609 |
| ·48 | ·0493 330 | ·0388 995+ | ·0304 335+ | ·0236 292 | ·0182 105+ |
| ·49 | ·0689 681 | ·0553 980 | ·0441 575− | ·0349 350− | ·0274 375+ |
| ·50 | ·0941 081 | ·0769 564 | ·0624 591 | ·0503 218 | ·0402 534 |
| ·51 | ·1254 192 | ·1043 476 | ·0861 809 | ·0706 674 | ·0575 408 |
| ·52 | ·1633 596 | ·1381 938 | ·1160 718 | ·0968 102 | ·0801 928 |
| ·53 | ·2080 925+ | ·1788 723 | ·1526 924 | ·1294 592 | ·1090 298 |
| ·54 | ·2594 130 | ·2264 280 | ·1963 174 | ·1690 919 | ·1447 002 |
| ·55 | ·3167 036 | ·2805 057 | ·2468 504 | ·2158 561 | ·1875 738 |
| ·56 | ·3789 289 | ·3403 165+ | ·3037 635− | ·2694 886 | ·2376 441 |
| ·57 | ·4446 757 | ·4046 481 | ·3660 786 | ·3292 671 | ·2944 552 |
| ·58 | ·5122 386 | ·4719 238 | ·4323 962 | ·3940 084 | ·3570 689 |
| ·59 | ·5797 440 | ·5403 087 | ·5009 766 | ·4621 202 | ·4240 851 |
| ·60 | ·6452 990 | ·6078 502 | ·5698 664 | ·5317 059 | ·4937 188 |
| ·61 | ·7071 454 | ·6726 391 | ·6370 565− | ·6007 126 | ·5639 309 |
| ·62 | ·7637 997 | ·7329 675+ | ·7006 525− | ·6671 049 | ·6325 967 |
| ·63 | ·8141 604 | ·7874 641 | ·7590 325+ | ·7290 405− | ·6976 920 |
| ·64 | ·8575 694 | ·8351 878 | ·8109 731 | ·7850 239 | ·7574 705− |
| ·65 | ·8938 222 | ·8756 702 | ·8557 245− | ·8340 162 | ·8106 057 |
| ·66 | ·9231 293 | ·9089 026 | ·8930 296 | ·8754 880 | ·8562 801 |
| ·67 | ·9460 389 | ·9352 761 | ·9230 861 | ·9094 105+ | ·8942 089 |
| ·68 | ·9633 353 | ·9554 865+ | ·9464 642 | ·9361 913 | ·9246 016 |
| ·69 | ·9759 308 | ·9704 214 | ·9639 952 | ·9565 706 | ·9480 711 |
| ·70 | ·9847 644 | ·9810 482 | ·9766 508 | ·9714 966 | ·9655 106 |
| ·71 | ·9907 209 | ·9883 168 | ·9854 311 | ·9820 006 | ·9779 596 |
| ·72 | ·9945 754 | ·9930 867 | ·9912 747 | ·9890 902 | ·9864 808 |
| ·73 | ·9969 639 | ·9960 837 | ·9949 974 | ·9936 697 | ·9920 617 |
| ·74 | ·9983 778 | ·9978 822 | ·9972 622 | ·9964 940 | ·9955 508 |
| ·75 | ·9991 752 | ·9989 103 | ·9985 744 | ·9981 525+ | ·9976 276 |
| ·76 | ·9996 024 | ·9994 684 | ·9992 962 | ·9990 771 | ·9988 008 |
| ·77 | ·9998 190 | ·9997 551 | ·9996 719 | ·9995 647 | ·9994 278 |
| ·78 | ·9999 225+ | ·9998 940 | ·9998 563 | ·9998 071 | ·9997 434 |
| ·79 | ·9999 690 | ·9999 571 | ·9999 411 | ·9999 201 | ·9998 924 |
| ·80 | ·9999 885− | ·9999 838 | ·9999 776 | ·9999 692 | ·9999 581 |
| ·81 | ·9999 960 | ·9999 944 | ·9999 921 | ·9999 891 | ·9999 849 |
| ·82 | ·9999 988 | ·9999 982 | ·9999 975− | ·9999 964 | ·9999 950+ |
| ·83 | ·9999 996 | ·9999 995− | ·9999 993 | ·9999 990 | ·9999 985+ |
| ·84 | ·9999 999 | ·9999 999 | ·9999 998 | ·9999 997 | ·9999 996 |
| ·85 | 1·0000 000 | 1·0000 000 | 1·0000 000 | ·9999 999 | ·9999 999 |
| ·86 | | | | 1·0000 000 | 1·0000 000 |

$x$ = ·31 to ·87 $q$ = 30 $p$ = 46 to 50

| | $p$ = 46 | $p$ = 47 | $p$ = 48 | $p$ = 49 | $p$ = 50 |
|---|---|---|---|---|---|
| $B(p, q)$ = | ·4263 2319 × $\frac{1}{10^{22}}$ | ·2580 3772 × $\frac{1}{10^{22}}$ | ·1575 0354 × $\frac{1}{10^{22}}$ | ·9692 5258 × $\frac{1}{10^{23}}$ | ·6011 8198 × $\frac{1}{10^{23}}$ |
| $x$ | | | | | |
| ·31 | ·0000 001 | | | | |
| ·32 | ·0000 002 | ·0000 001 | | | |
| ·33 | ·0000 005− | ·0000 002 | ·0000 001 | ·0000 001 | |
| ·34 | ·0000 012 | ·0000 007 | ·0000 004 | ·0000 002 | ·0000 001 |
| ·35 | ·0000 030 | ·0000 017 | ·0000 009 | ·0000 005+ | ·0000 003 |
| ·36 | ·0000 072 | ·0000 041 | ·0000 024 | ·0000 013 | ·0000 008 |
| ·37 | ·0000 164 | ·0000 097 | ·0000 057 | ·0000 033 | ·0000 019 |
| ·38 | ·0000 359 | ·0000 218 | ·0000 132 | ·0000 079 | ·0000 047 |
| ·39 | ·0000 759 | ·0000 473 | ·0000 292 | ·0000 180 | ·0000 109 |
| ·40 | ·0001 544 | ·0000 986 | ·0000 625+ | ·0000 393 | ·0000 246 |
| ·41 | ·0003 034 | ·0001 984 | ·0001 288 | ·0000 830 | ·0000 532 |
| ·42 | ·0005 760 | ·0003 856 | ·0002 562 | ·0001 691 | ·0001 108 |
| ·43 | ·0010 580 | ·0007 244 | ·0004 924 | ·0003 324 | ·0002 228 |
| ·44 | ·0018 822 | ·0013 173 | ·0009 154 | ·0006 317 | ·0004 330 |
| ·45 | ·0032 458 | ·0023 207 | ·0016 476 | ·0011 617 | ·0008 136 |
| ·46 | ·0054 309 | ·0039 647 | ·0028 741 | ·0020 694 | ·0014 801 |
| ·47 | ·0088 241 | ·0065 736 | ·0048 632 | ·0035 736 | ·0026 088 |
| ·48 | ·0139 334 | ·0105 860 | ·0079 879 | ·0059 874 | ·0044 588 |
| ·49 | ·0213 963 | ·0165 698 | ·0127 456 | ·0097 396 | ·0073 950− |
| ·50 | ·0319 750− | ·0252 262 | ·0197 698 | ·0153 932 | ·0119 099 |
| ·51 | ·0465 320 | ·0373 780 | ·0298 288 | ·0236 526 | ·0186 386 |
| ·52 | ·0659 830 | ·0539 356 | ·0438 055− | ·0353 553 | ·0283 606 |
| ·53 | ·0912 245+ | ·0758 384 | ·0626 522 | ·0514 411 | ·0419 827 |
| ·54 | ·1230 404 | ·1039 704 | ·0873 187 | ·0728 940 | ·0604 948 |
| ·55 | ·1619 941 | ·1390 557 | ·1186 552 | ·1006 560 | ·0848 975+ |
| ·56 | ·2083 189 | ·1815 437 | ·1572 979 | ·1355 171 | ·1161 011 |
| ·57 | ·2618 249 | ·2314 999 | ·2035 493 | ·1779 920 | ·1548 027 |
| ·58 | ·3218 369 | ·2885 196 | ·2572 722 | ·2281 991 | ·2013 570 |
| ·59 | ·3871 826 | ·3516 839 | ·3178 168 | ·2857 629 | ·2556 584 |
| ·60 | ·4562 383 | ·4195 726 | ·3839 990 | ·3497 594 | ·3170 572 |
| ·61 | ·5270 348 | ·4903 392 | ·4541 433 | ·4187 241 | ·3843 314 |
| ·62 | ·5974 136 | ·5618 481 | ·5261 918 | ·4907 286 | ·4557 287 |
| ·63 | ·6652 150− | ·6318 554 | ·5978 713 | ·5635 260 | ·5290 827 |
| ·64 | ·7284 721 | ·6982 130 | ·6668 979 | ·6347 478 | ·6019 941 |
| ·65 | ·7855 826 | ·7590 635− | ·7311 902 | ·7021 265− | ·6720 545− |
| ·66 | ·8354 330 | ·8129 998 | ·7890 583 | ·7637 102 | ·7370 793 |
| ·67 | ·8774 601 | ·8591 638 | ·8393 416 | ·8180 368 | ·7953 146 |
| ·68 | ·9116 417 | ·8972 732 | ·8814 739 | ·8642 395+ | ·8455 842 |
| ·69 | ·9384 271 | ·9275 775+ | ·9154 722 | ·9020 731 | ·8873 559 |
| ·70 | ·9586 201 | ·9507 560 | ·9418 545− | ·9318 588 | ·9207 208 |
| ·71 | ·9732 416 | ·9677 799 | ·9615 094 | ·9543 676 | ·9462 958 |
| ·72 | ·9833 912 | ·9797 642 | ·9755 416 | ·9706 644 | ·9650 745+ |
| ·73 | ·9901 313 | ·9878 337 | ·9851 216 | ·9819 456 | ·9782 550+ |
| ·74 | ·9944 031 | ·9930 184 | ·9913 615− | ·9893 945+ | ·9870 775+ |
| ·75 | ·9969 802 | ·9961 886 | ·9952 285+ | ·9940 735− | ·9926 945− |
| ·76 | ·9984 556 | ·9980 278 | ·9975 020 | ·9968 610 | ·9960 855+ |
| ·77 | ·9992 543 | ·9990 366 | ·9987 655+ | ·9984 307 | ·9980 202 |
| ·78 | ·9996 617 | ·9995 578 | ·9994 268 | ·9992 629 | ·9990 593 |
| ·79 | ·9998 565+ | ·9998 103 | ·9997 513 | ·9996 764 | ·9995 823 |
| ·80 | ·9999 435− | ·9999 244 | ·9998 997 | ·9998 680 | ·9998 277 |
| ·81 | ·9999 794 | ·9999 722 | ·9999 627 | ·9999 504 | ·9999 344 |
| ·82 | ·9999 932 | ·9999 906 | ·9999 873 | ·9999 829 | ·9999 772 |
| ·83 | ·9999 979 | ·9999 971 | ·9999 961 | ·9999 947 | ·9999 928 |
| ·84 | ·9999 994 | ·9999 992 | ·9999 989 | ·9999 985+ | ·9999 980 |
| ·85 | ·9999 999 | ·9999 998 | ·9999 997 | ·9999 996 | ·9999 995− |
| ·86 | 1·0000 000 | 1·0000 000 | ·9999 999 | ·9999 999 | ·9999 999 |
| ·87 | | | 1·0000 000 | 1·0000 000 | 1·0000 000 |

$x = \cdot 20$ to $\cdot 83$ $\qquad q = 31 \qquad p = 31$ to $36$

| | $p = 31$ | $p = 32$ | $p = 33$ | $p = 34$ | $p = 35$ | $p = 36$ |
|---|---|---|---|---|---|---|
| $B(p, q) =$ | $\cdot 1386\ 1667 \times \frac{1}{10^{18}}$ | $\cdot 6930\ 8336 \times \frac{1}{10^{19}}$ | $\cdot 3520\ 4234 \times \frac{1}{10^{19}}$ | $\cdot 1815\ 2183 \times \frac{1}{10^{19}}$ | $\cdot 9494\ 9881 \times \frac{1}{10^{20}}$ | $\cdot 5035\ 2210 \times \frac{1}{10^{20}}$ |
| $x$ | | | | | | |
| ·20 | ·0000 001 | | | | | |
| ·21 | ·0000 003 | ·0000 001 | | | | |
| ·22 | ·0000 008 | ·0000 003 | ·0000 001 | ·0000 001 | | |
| ·23 | ·0000 021 | ·0000 009 | ·0000 004 | ·0000 002 | ·0000 001 | |
| ·24 | ·0000 053 | ·0000 024 | ·0000 011 | ·0000 005$^{-}$ | ·0000 002 | ·0000 001 |
| ·25 | ·0000 130 | ·0000 062 | ·0000 029 | ·0000 014 | ·0000 006 | ·0000 003 |
| ·26 | ·0000 299 | ·0000 149 | ·0000 073 | ·0000 035$^{+}$ | ·0000 017 | ·0000 008 |
| ·27 | ·0000 656 | ·0000 338 | ·0000 172 | ·0000 086 | ·0000 043 | ·0000 021 |
| ·28 | ·0001 373 | ·0000 734 | ·0000 387 | ·0000 201 | ·0000 103 | ·0000 052 |
| ·29 | ·0002 751 | ·0001 520 | ·0000 829 | ·0000 446 | ·0000 237 | ·0000 124 |
| ·30 | ·0005 287 | ·0003 019 | ·0001 701 | ·0000 946 | ·0000 519 | ·0000 282 |
| ·31 | ·0009 769 | ·0005 757 | ·0003 348 | ·0001 921 | ·0001 089 | ·0000 610 |
| ·32 | ·0017 396 | ·0010 569 | ·0006 336 | ·0003 750$^{-}$ | ·0002 192 | ·0001 266 |
| ·33 | ·0029 907 | ·0018 711 | ·0011 552 | ·0007 042 | ·0004 241 | ·0002 524 |
| ·34 | ·0049 729 | ·0032 006 | ·0020 331 | ·0012 753 | ·0007 903 | ·0004 840 |
| ·35 | ·0080 100 | ·0052 983 | ·0034 594 | ·0022 307 | ·0014 212 | ·0008 950$^{+}$ |
| ·36 | ·0125 169 | ·0085 010 | ·0056 999 | ·0037 748 | ·0024 703 | ·0015 981 |
| ·37 | ·0190 009 | ·0132 381 | ·0091 069 | ·0061 888 | ·0041 564 | ·0027 599 |
| ·38 | ·0280 551 | ·0200 336 | ·0141 277 | ·0098 433 | ·0067 787 | ·0046 159 |
| ·39 | ·0403 375$^{+}$ | ·0294 970 | ·0213 056 | ·0152 068 | ·0107 295$^{+}$ | ·0074 867 |
| ·40 | ·0565 377 | ·0423 022 | ·0312 697 | ·0228 449 | ·0165 016 | ·0117 893 |
| ·41 | ·0773 285$^{-}$ | ·0591 509 | ·0447 112 | ·0334 088 | ·0246 860 | ·0180 442 |
| ·42 | ·1033 078 | ·0807 228 | ·0623 444 | ·0476 082 | ·0359 579 | ·0268 707 |
| ·43 | ·1349 332 | ·1076 141 | ·0848 540 | ·0661 699 | ·0510 468 | ·0389 700 |
| ·44 | ·1724 576 | ·1402 703 | ·1128 306 | ·0897 813 | ·0706 910 | ·0550 916 |
| ·45 | ·2158 733 | ·1789 199 | ·1467 012 | ·1190 224 | ·0955 768 | ·0759 830 |
| ·46 | ·2648 728 | ·2235 183 | ·1866 612 | ·1542 939 | ·1262 676 | ·1023 252 |
| ·47 | ·3188 331 | ·2737 090 | ·2326 179 | ·1957 480 | ·1631 289 | ·1346 572 |
| ·48 | ·3768 294 | ·3288 113 | ·2841 545$^{-}$ | ·2432 326 | ·2062 585$^{-}$ | ·1732 987 |
| ·49 | ·4376 767 | ·3878 369 | ·3405 202 | ·2962 575$^{+}$ | ·2554 318 | ·2182 803 |
| ·50 | ·5000 000$^{e}$ | ·4495 382 | ·4006 532 | ·3539 904 | ·3100 724 | ·2692 914 |
| ·51 | ·5623 233 | ·5124 834 | ·4632 354 | ·4152 857 | ·3692 541 | ·3256 555$^{+}$ |
| ·52 | ·6231 706 | ·5751 525$^{+}$ | ·5267 743 | ·4787 479 | ·4317 386 | ·3863 410 |
| ·53 | ·6811 669 | ·6360 428 | ·5897 060 | ·5428 216 | ·4960 475$^{-}$ | ·4500 084 |
| ·54 | ·7351 272 | ·6937 728 | ·6505 057 | ·6059 013 | ·5605 622 | ·5150 936 |
| ·55 | ·7841 267 | ·7471 733 | ·7077 949 | ·6664 475$^{-}$ | ·6236 408 | ·5799 168 |
| ·56 | ·8275 424 | ·7953 551 | ·7604 318 | ·7230 957 | ·6837 391 | ·6428 082 |
| ·57 | ·8650 668 | ·8377 478 | ·8075 773 | ·7747 463 | ·7395 207 | ·7022 318 |
| ·58 | ·8966 922 | ·8741 073 | ·8487 275$^{-}$ | ·8206 251 | ·7899 439 | ·7568 959 |
| ·59 | ·9226 715$^{+}$ | ·9044 940 | ·8837 148 | ·8603 099 | ·8343 167 | ·8058 355$^{+}$ |
| ·60 | ·9434 623 | ·9292 269 | ·9126 782 | ·8937 224 | ·8723 135$^{-}$ | ·8484 578 |
| ·61 | ·9596 625$^{-}$ | ·9488 219 | ·9360 098 | ·9210 894 | ·9039 574 | ·8845 492 |
| ·62 | ·9719 449 | ·9639 234 | ·9542 875$^{+}$ | ·9428 822 | ·9295 715$^{-}$ | ·9142 452 |
| ·63 | ·9809 991 | ·9752 363 | ·9682 021 | ·9597 418 | ·9497 090 | ·9379 705$^{+}$ |
| ·64 | ·9874 831 | ·9834 672 | ·9784 876 | ·9724 033 | ·9650 736 | ·9563 616 |
| ·65 | ·9919 900 | ·9892 782 | ·9858 631 | ·9816 253 | ·9764 403 | ·9701 811 |
| ·66 | ·9950 271 | ·9932 549 | ·9909 886 | ·9881 330 | ·9845 855$^{-}$ | ·9802 371 |
| ·67 | ·9970 093 | ·9958 897 | ·9944 363 | ·9925 772 | ·9902 326 | ·9873 153 |
| ·68 | ·9982 604 | ·9975 777 | ·9966 782 | ·9955 105$^{-}$ | ·9940 158 | ·9921 283 |
| ·69 | ·9990 231 | ·9986 219 | ·9980 855$^{+}$ | ·9973 790 | ·9964 614 | ·9952 855$^{-}$ |
| ·70 | ·9994 713 | ·9992 445$^{+}$ | ·9989 370 | ·9985 259 | ·9979 843 | ·9972 802 |
| ·71 | ·9997 249 | ·9996 018 | ·9994 325$^{+}$ | ·9992 030 | ·9988 963 | ·9984 919 |
| ·72 | ·9998 627 | ·9997 987 | ·9997 094 | ·9995 868 | ·9994 205$^{-}$ | ·9991 982 |
| ·73 | ·9999 344 | ·9999 026 | ·9998 577 | ·9997 950$^{+}$ | ·9997 090 | ·9995 923 |
| ·74 | ·9999 701 | ·9999 551 | ·9999 335$^{+}$ | ·9999 030 | ·9998 606 | ·9998 023 |
| ·75 | ·9999 870 | ·9999 803 | ·9999 705$^{-}$ | ·9999 564 | ·9999 366 | ·9999 089 |
| ·76 | ·9999 947 | ·9999 918 | ·9999 876 | ·9999 814 | ·9999 726 | ·9999 602 |
| ·77 | ·9999 979 | ·9999 968 | ·9999 951 | ·9999 925$^{+}$ | ·9999 889 | ·9999 836 |
| ·78 | ·9999 992 | ·9999 988 | ·9999 982 | ·9999 972 | ·9999 958 | ·9999 937 |
| ·79 | ·9999 997 | ·9999 996 | ·9999 994 | ·9999 990 | ·9999 985$^{-}$ | ·9999 977 |
| ·80 | ·9999 999 | ·9999 999 | ·9999 998 | ·9999 997 | ·9999 995$^{-}$ | ·9999 992 |
| ·81 | 1·0000 000 | 1·0000 000 | ·9999 999 | ·9999 999 | ·9999 998 | ·9999 998 |
| ·82 | | | 1·0000 000 | 1·0000 000 | 1·0000 000 | ·9999 999 |
| ·83 | | | | | | 1·0000 000 |

$x$ = ·25 to ·85 $q = 31$ $p$ = 37 to 43

| | $p = 37$ | $p = 38$ | $p = 39$ | $p = 40$ | $p = 41$ | $p = 42$ | $p = 43$ |
|---|---|---|---|---|---|---|---|
| $B(p, q)$ = | ·2705 4919 $\times \frac{1}{10^{20}}$ | ·1472 1059 $\times \frac{1}{10^{20}}$ | ·8107 2497 $\times \frac{1}{10^{21}}$ | ·4516 8963 $\times \frac{1}{10^{21}}$ | ·2544 7303 $\times \frac{1}{10^{21}}$ | ·1449 0825$^{+}$ $\times \frac{1}{10^{21}}$ | ·8337 1871 $\times \frac{1}{10^{22}}$ |
| $x$ | | | | | | | |
| ·25 | ·0000 001 | ·0000 001 | | | | | |
| ·26 | ·0000 004 | ·0000 002 | ·0000 001 | | | | |
| ·27 | ·0000 010 | ·0000 005⁻ | ·0000 002 | ·0000 001 | ·0000 001 | | |
| ·28 | ·0000 026 | ·0000 013 | ·0000 006 | ·0000 003 | ·0000 001 | ·0000 001 | |
| ·29 | ·0000 065⁻ | ·0000 033 | ·0000 017 | ·0000 008 | ·0000 004 | ·0000 002 | ·0000 001 |
| ·30 | ·0000 151 | ·0000 080 | ·0000 042 | ·0000 022 | ·0000 011 | ·0000 006 | ·0000 003 |
| ·31 | ·0000 338 | ·0000 185⁺ | ·0000 100 | ·0000 054 | ·0000 029 | ·0000 015⁺ | ·0000 008 |
| ·32 | ·0000 723 | ·0000 409 | ·0000 228 | ·0000 126 | ·0000 069 | ·0000 038 | ·0000 020 |
| ·33 | ·0001 485⁺ | ·0000 864 | ·0000 498 | ·0000 284 | ·0000 160 | ·0000 090 | ·0000 050⁻ |
| ·34 | ·0002 932 | ·0001 756 | ·0001 041 | ·0000 611 | ·0000 355⁺ | ·0000 205⁻ | ·0000 117 |
| ·35 | ·0005 574 | ·0003 434 | ·0002 094 | ·0001 264 | ·0000 756 | ·0000 448 | ·0000 263 |
| ·36 | ·0010 225⁻ | ·0006 472 | ·0004 055⁺ | ·0002 515⁺ | ·0001 545⁺ | ·0000 941 | ·0000 567 |
| ·37 | ·0018 125⁺ | ·0011 778 | ·0007 576 | ·0004 825⁻ | ·0003 044 | ·0001 902 | ·0001 178 |
| ·38 | ·0031 092 | ·0020 724 | ·0013 674 | ·0008 934 | ·0005 782 | ·0003 708 | ·0002 357 |
| ·39 | ·0051 681 | ·0035 306 | ·0023 878 | ·0015 993 | ·0010 612 | ·0006 977 | ·0004 547 |
| ·40 | ·0083 337 | ·0058 307 | ·0040 391 | ·0027 712 | ·0018 836 | ·0012 688 | ·0008 473 |
| ·41 | ·0130 518 | ·0093 453 | ·0066 258 | ·0046 532 | ·0032 379 | ·0022 329 | ·0015 266 |
| ·42 | ·0198 735⁺ | ·0145 519 | ·0105 523 | ·0075 803 | ·0053 958 | ·0038 071 | ·0026 631 |
| ·43 | ·0294 494 | ·0220 362 | ·0163 319 | ·0119 923 | ·0087 267 | ·0062 951 | ·0045 026 |
| ·44 | ·0425 080 | ·0324 819 | ·0245 877 | ·0184 424 | ·0137 105⁻ | ·0101 050⁺ | ·0073 855⁻ |
| ·45 | ·0598 181 | ·0466 459 | ·0360 388 | ·0275 940 | ·0209 436 | ·0157 612 | ·0117 634 |
| ·46 | ·0821 338 | ·0653 148 | ·0514 702 | ·0402 028 | ·0311 325⁺ | ·0239 073 | ·0182 097 |
| ·47 | ·1101 240 | ·0892 443 | ·0716 833 | ·0570 807 | ·0450 701 | ·0352 946 | ·0274 183 |
| ·48 | ·1442 940 | ·1190 835⁺ | ·0974 290 | ·0790 394 | ·0635 921 | ·0507 520 | ·0401 864 |
| ·49 | ·1849 059 | ·1552 929 | ·1293 270 | ·1068 165⁺ | ·0875 138 | ·0711 347 | ·0573 763 |
| ·50 | ·2319 088 | ·1980 624 | ·1677 788 | ·1409 895⁻ | ·1175 488 | ·0972 526 | ·0798 558 |
| ·51 | ·2848 909 | ·2472 442 | ·2128 866 | ·1818 855⁻ | ·1542 170 | ·1297 810 | ·1084 169 |
| ·52 | ·3430 619 | ·3023 094 | ·2643 882 | ·2295 006 | ·1977 529 | ·1691 645⁺ | ·1436 800 |
| ·53 | ·4052 738 | ·3623 406 | ·3216 219 | ·2834 403 | ·2480 269 | ·2155 242 | ·1859 932 |
| ·54 | ·4700 796 | ·4260 633 | ·3835 296 | ·3428 935⁺ | ·3044 925⁺ | ·2685 828 | ·2353 408 |
| ·55 | ·5358 285⁻ | ·4919 189 | ·4487 026 | ·4066 498 | ·3661 740 | ·3276 233 | ·2912 755⁻ |
| ·56 | ·6007 858 | ·5581 728 | ·5154 701 | ·4731 616 | ·4316 992 | ·3914 908 | ·3528 908 |
| ·57 | ·6632 649 | ·6230 447 | ·5820 201 | ·5406 484 | ·4993 802 | ·4586 454 | ·4188 416 |
| ·58 | ·7217 549 | ·6848 473 | ·6465 411 | ·6072 330 | ·5673 353 | ·5272 625⁺ | ·4874 187 |
| ·59 | ·7750 284 | ·7421 148 | ·7073 649 | ·6710 914 | ·6336 389 | ·5953 735⁺ | ·5566 708 |
| ·60 | ·8222 166 | ·7937 059 | ·7630 944 | ·7305 991 | ·6964 791 | ·6610 275⁺ | ·6245 630 |
| ·61 | ·8628 444 | ·8388 694 | ·8126 988 | ·7844 547 | ·7543 041 | ·7224 547 | ·6891 494 |
| ·62 | ·8968 242 | ·8772 657 | ·8555 660 | ·8317 632 | ·8059 371 | ·7782 086 | ·7487 373 |
| ·63 | ·9244 126 | ·9089 456 | ·8915 086 | ·8720 731 | ·8506 454 | ·8272 683 | ·8020 211 |
| ·64 | ·9461 396 | ·9342 932 | ·9207 259 | ·9053 635⁻ | ·8881 576 | ·8690 885⁺ | ·8481 670 |
| ·65 | ·9627 223 | ·9539 431 | ·9437 315⁻ | ·9319 881 | ·9186 301 | ·9035 941 | ·8868 397 |
| ·66 | ·9749 757 | ·9686 875⁻ | ·9612 608 | ·9525 888 | ·9425 726 | ·9311 248 | ·9181 724 |
| ·67 | ·9837 318 | ·9793 842 | ·9741 716 | ·9679 927 | ·9607 479 | ·9523 422 | ·9426 877 |
| ·68 | ·9897 751 | ·9868 776 | ·9833 517 | ·9791 098 | ·9740 620 | ·9681 178 | ·9611 887 |
| ·69 | ·9937 980 | ·9919 394 | ·9896 445⁺ | ·9868 430 | ·9834 602 | ·9794 182 | ·9746 371 |
| ·70 | ·9963 766 | ·9952 313 | ·9937 966 | ·9920 198 | ·9898 432 | ·9872 047 | ·9840 385⁺ |
| ·71 | ·9979 655⁺ | ·9972 887 | ·9964 289 | ·9953 487 | ·9940 067 | ·9923 566 | ·9903 482 |
| ·72 | ·9989 047 | ·9985 221 | ·9980 292 | ·9974 012 | ·9966 100 | ·9956 235⁻ | ·9944 058 |
| ·73 | ·9994 361 | ·9992 297 | ·9989 600 | ·9986 118 | ·9981 668 | ·9976 044 | ·9969 005⁻ |
| ·74 | ·9997 232 | ·9996 172 | ·9994 768 | ·9992 931 | ·9990 551 | ·9987 501 | ·9983 633 |
| ·75 | ·9998 709 | ·9998 192 | ·9997 499 | ·9996 580 | ·9995 373 | ·9993 805⁺ | ·9991 790 |
| ·76 | ·9999 430 | ·9999 192 | ·9998 868 | ·9998 434 | ·9997 855⁺ | ·9997 094 | ·9996 102 |
| ·77 | ·9999 762 | ·9999 659 | ·9999 517 | ·9999 324 | ·9999 063 | ·9998 715⁺ | ·9998 256 |
| ·78 | ·9999 907 | ·9999 865⁺ | ·9999 807 | ·9999 726 | ·9999 616 | ·9999 467 | ·9999 268 |
| ·79 | ·9999 966 | ·9999 950⁺ | ·9999 928 | ·9999 896 | ·9999 853 | ·9999 794 | ·9999 714 |
| ·80 | ·9999 988 | ·9999 983 | ·9999 975⁻ | ·9999 964 | ·9999 948 | ·9999 926 | ·9999 896 |
| ·81 | ·9999 996 | ·9999 995⁻ | ·9999 992 | ·9999 988 | ·9999 983 | ·9999 976 | ·9999 965⁺ |
| ·82 | ·9999 999 | ·9999 998 | ·9999 998 | ·9999 997 | ·9999 995⁻ | ·9999 993 | ·9999 989 |
| ·83 | 1·0000 000 | 1·0000 000 | ·9999 999 | ·9999 999 | ·9999 999 | ·9999 998 | ·9999 997 |
| ·84 | | | 1·0000 000 | 1·0000 000 | 1·0000 000 | 1·0000 000 | ·9999 999 |
| ·85 | | | | | | | 1·0000 000 |

$x$ = ·30 to ·86 $q$ = 31 $p$ = 44 to 50

| $x$ | $p$ = 44 | $p$ = 45 | $p$ = 46 | $p$ = 47 | $p$ = 48 | $p$ = 49 | $p$ = 50 |
|---|---|---|---|---|---|---|---|
| $B(p, q)$ = | ·4844 5817×$\frac{1}{10^{22}}$ | ·2842 1546×$\frac{1}{10^{22}}$ | ·1682 8547×$\frac{1}{10^{22}}$ | ·1005 3418×$\frac{1}{10^{22}}$ | ·6057 8286×$\frac{1}{10^{23}}$ | ·3680 7060×$\frac{1}{10^{23}}$ | ·2254 4324×$\frac{1}{10^{23}}$ |
| ·30 | ·0000 001 | ·0000 001 | | | | | |
| ·31 | ·0000 004 | ·0000 002 | ·0000 001 | ·0000 001 | | | |
| ·32 | ·0000 011 | ·0000 006 | ·0000 003 | ·0000 002 | ·0000 001 | | |
| ·33 | ·0000 027 | ·0000 015− | ·0000 008 | ·0000 004 | ·0000 002 | ·0000 001 | ·0000 001 |
| ·34 | ·0000 066 | ·0000 037 | ·0000 021 | ·0000 011 | ·0000 006 | ·0000 003 | ·0000 002 |
| ·35 | ·0000 153 | ·0000 088 | ·0000 050+ | ·0000 029 | ·0000 016 | ·0000 009 | ·0000 005− |
| ·36 | ·0000 339 | ·0000 201 | ·0000 118 | ·0000 069 | ·0000 040 | ·0000 023 | ·0000 013 |
| ·37 | ·0000 724 | ·0000 441 | ·0000 266 | ·0000 160 | ·0000 095− | ·0000 056 | ·0000 033 |
| ·38 | ·0001 485+ | ·0000 928 | ·0000 576 | ·0000 354 | ·0000 216 | ·0000 131 | ·0000 079 |
| ·39 | ·0002 938 | ·0001 883 | ·0001 197 | ·0000 755+ | ·0000 473 | ·0000 294 | ·0000 181 |
| ·40 | ·0005 610 | ·0003 684 | ·0002 400 | ·0001 552 | ·0000 996 | ·0000 634 | ·0000 401 |
| ·41 | ·0010 350+ | ·0006 960 | ·0004 644 | ·0003 075− | ·0002 021 | ·0001 318 | ·0000 854 |
| ·42 | ·0018 475+ | ·0012 714 | ·0008 681 | ·0005 882 | ·0003 957 | ·0002 642 | ·0001 752 |
| ·43 | ·0031 941 | ·0022 478 | ·0015 697 | ·0010 879 | ·0007 484 | ·0005 113 | ·0003 469 |
| ·44 | ·0053 540 | ·0038 507 | ·0027 483 | ·0019 469 | ·0013 692 | ·0009 562 | ·0006 632 |
| ·45 | ·0087 092 | ·0063 977 | ·0046 642 | ·0033 753 | ·0024 251 | ·0017 302 | ·0012 261 |
| ·46 | ·0137 602 | ·0103 180 | ·0076 789 | ·0056 733 | ·0041 617 | ·0030 319 | ·0021 939 |
| ·47 | ·0211 338 | ·0161 662 | ·0122 749 | ·0092 532 | ·0069 265− | ·0051 495− | ·0038 029 |
| ·48 | ·0315 767 | ·0246 263 | ·0190 660 | ·0146 564 | ·0111 888 | ·0084 840 | ·0063 909 |
| ·49 | ·0459 312 | ·0364 995− | ·0287 968 | ·0225 610 | ·0175 552 | ·0135 693 | ·0104 204 |
| ·50 | ·0650 888 | ·0526 711 | ·0423 230 | ·0337 746 | ·0267 722 | ·0210 827 | ·0164 963 |
| ·51 | ·0899 197 | ·0740 540 | ·0605 682 | ·0492 050− | ·0397 107 | ·0318 422 | ·0253 724 |
| ·52 | ·1211 825+ | ·1015 075− | ·0844 557 | ·0698 061 | ·0573 258 | ·0467 800 | ·0379 387 |
| ·53 | ·1594 822 | ·1357 378 | ·1148 166 | ·0964 968 | ·0805 898 | ·0668 900 | ·0551 835+ |
| ·54 | ·2048 663 | ·1771 899 | ·1522 811 | ·1300 582 | ·1103 980 | ·0931 462 | ·0781 265+ |
| ·55 | ·2573 367 | ·2259 434 | ·1971 662 | ·1710 165− | ·1474 539 | ·1263 949 | ·1077 211 |
| ·56 | ·3161 938 | ·2816 320 | ·2493 742 | ·2195 288 | ·1921 472 | ·1672 300 | ·1447 333 |
| ·57 | ·3803 246 | ·3434 008 | ·3083 231 | ·2752 891 | ·2444 410 | ·2158 679 | ·1896 099 |
| ·58 | ·4481 864 | ·4099 170 | ·3729 234 | ·3374 738 | ·3037 891 | ·2720 413 | ·2423 539 |
| ·59 | ·5179 050+ | ·4794 389 | ·4416 138 | ·4047 425+ | ·3691 030 | ·3349 335+ | ·3024 307 |
| ·60 | ·5874 202 | ·5499 396 | ·5124 591 | ·4753 045− | ·4387 823 | ·4031 731 | ·3687 267 |
| ·61 | ·6546 590 | ·6192 751 | ·5833 014 | ·5470 461 | ·5108 140 | ·4748 990 | ·4395 776 |
| ·62 | ·7177 170 | ·6853 713 | ·6519 473 | ·6177 096 | ·5829 329 | ·5478 952 | ·5128 719 |
| ·63 | ·7750 183 | ·7464 076 | ·7163 663 | ·6850 973 | ·6528 237 | ·6197 835+ | ·5862 242 |
| ·64 | ·8254 354 | ·8009 681 | ·7748 695+ | ·7472 732 | ·7183 381 | ·6882 456 | ·6571 951 |
| ·65 | ·8683 515− | ·8481 405− | ·8262 452 | ·8027 316 | ·7776 921 | ·7512 441 | ·7235 278 |
| ·66 | ·9036 597 | ·8875 506 | ·8698 307 | ·8505 082 | ·8296 152 | ·8072 075+ | ·7833 639 |
| ·67 | ·9317 062 | ·9193 321 | ·9055 143 | ·8902 187 | ·8734 292 | ·8551 497 | ·8354 041 |
| ·68 | ·9531 895+ | ·9440 414 | ·9336 735− | ·9220 254 | ·9090 489 | ·8947 099 | ·8789 897 |
| ·69 | ·9690 365− | ·9625 372 | ·9550 631 | ·9465 425+ | ·9369 107 | ·9261 111 | ·9140 970 |
| ·70 | ·9802 759 | ·9758 463 | ·9706 785− | ·9647 017 | ·9578 476 | ·9500 509 | ·9412 519 |
| ·71 | ·9879 274 | ·9850 368 | ·9816 162 | ·9776 036 | ·9729 363 | ·9675 514 | ·9613 873 |
| ·72 | ·9929 174 | ·9911 152 | ·9889 525− | ·9863 798 | ·9833 451 | ·9797 945− | ·9756 729 |
| ·73 | ·9960 282 | ·9949 572 | ·9936 542 | ·9920 827 | ·9902 032 | ·9879 737 | ·9853 496 |
| ·74 | ·9978 772 | ·9972 724 | ·9965 264 | ·9956 143 | ·9945 086 | ·9931 789 | ·9915 926 |
| ·75 | ·9989 224 | ·9985 987 | ·9981 941 | ·9976 927 | ·9970 767 | ·9963 259 | ·9954 180 |
| ·76 | ·9994 823 | ·9993 187 | ·9991 116 | ·9988 515− | ·9985 276 | ·9981 276 | ·9976 375+ |
| ·77 | ·9997 656 | ·9996 879 | ·9995 882 | ·9994 613 | ·9993 013 | ·9991 011 | ·9988 525− |
| ·78 | ·9999 005− | ·9998 659 | ·9998 210 | ·9997 631 | ·9996 891 | ·9995 953 | ·9994 774 |
| ·79 | ·9999 606 | ·9999 463 | ·9999 274 | ·9999 028 | ·9998 710 | ·9998 301 | ·9997 781 |
| ·80 | ·9999 855+ | ·9999 800 | ·9999 727 | ·9999 631 | ·9999 504 | ·9999 339 | ·9999 127 |
| ·81 | ·9999 951 | ·9999 932 | ·9999 906 | ·9999 871 | ·9999 824 | ·9999 763 | ·9999 684 |
| ·82 | ·9999 985− | ·9999 979 | ·9999 970 | ·9999 959 | ·9999 943 | ·9999 923 | ·9999 896 |
| ·83 | ·9999 996 | ·9999 994 | ·9999 992 | ·9999 988 | ·9999 983 | ·9999 977 | ·9999 969 |
| ·84 | ·9999 999 | ·9999 998 | ·9999 998 | ·9999 997 | ·9999 996 | ·9999 994 | ·9999 992 |
| ·85 | 1·0000 000 | 1·0000 000 | 1·0000 000 | ·9999 999 | ·9999 999 | ·9999 999 | ·9999 998 |
| ·86 | | | | 1·0000 000 | 1·0000 000 | 1·0000 000 | 1·0000 000 |

$x$ = ·20 to ·82 $q = 32$ $p$ = 32 to 37

| | $p = 32$ | $p = 33$ | $p = 34$ | $p = 35$ | $p = 36$ | $p = 37$ |
|---|---|---|---|---|---|---|
| $B(p, q)$ = | ·3410 4102$\times\frac{1}{10^{19}}$ | ·1705 2051$\times\frac{1}{10^{19}}$ | ·8657 1950$\overset{+}{\times}\frac{1}{10^{20}}$ | ·4459 7671$\times\frac{1}{10^{20}}$ | ·2329 7291$\times\frac{1}{10^{20}}$ | ·1233 3860$\times\frac{1}{10^{20}}$ |
| $x$ | | | | | | |
| ·20 | ·0000 001 | | | | | |
| ·21 | ·0000 002 | ·0000 001 | | | | |
| ·22 | ·0000 005+ | ·0000 002 | ·0000 001 | | | |
| ·23 | ·0000 014 | ·0000 006 | ·0000 003 | ·0000 001 | ·0000 001 | |
| ·24 | ·0000 038 | ·0000 018 | ·0000 008 | ·0000 004 | ·0000 002 | ·0000 001 |
| ·25 | ·0000 096 | ·0000 046 | ·0000 022 | ·0000 010 | ·0000 005− | ·0000 002 |
| ·26 | ·0000 227 | ·0000 113 | ·0000 055+ | ·0000 027 | ·0000 013 | ·0000 006 |
| ·27 | ·0000 510 | ·0000 263 | ·0000 134 | ·0000 067 | ·0000 033 | ·0000 016 |
| ·28 | 0001 092 | ·0000 584 | ·0000 308 | ·0000 161 | ·0000 083 | ·0000 042 |
| ·29 | ·0002 234 | ·0001 236 | ·0000 675+ | ·0000 364 | ·0000 194 | ·0000 102 |
| ·30 | ·0004 380 | ·0002 504 | ·0001 413 | ·0000 787 | ·0000 434 | ·0000 236 |
| ·31 | ·0008 245− | ·0004 866 | ·0002 834 | ·0001 630 | ·0000 926 | ·0000 520 |
| ·32 | ·0014 938 | ·0009 088 | ·0005 458 | ·0003 237 | ·0001 897 | ·0001 099 |
| ·33 | ·0026 100 | ·0016 353 | ·0010 115− | ·0006 179 | ·0003 730 | ·0002 226 |
| ·34 | ·0044 057 | ·0028 398 | ·0018 072 | ·0011 360 | ·0007 057 | ·0004 334 |
| ·35 | ·0071 965+ | ·0047 674 | ·0031 186 | ·0020 153 | ·0012 871 | ·0008 128 |
| ·36 | ·0113 924 | ·0077 492 | ·0052 056 | ·0034 550+ | ·0022 666 | ·0014 704 |
| ·37 | ·0175 026 | ·0122 133 | ·0084 179 | ·0057 331 | ·0038 600 | ·0025 701 |
| ·38 | ·0261 299 | ·0186 886 | ·0132 045+ | ·0092 205− | ·0063 657 | ·0043 467 |
| ·39 | ·0379 526 | ·0277 980 | ·0201 173 | ·0143 908 | ·0101 793 | ·0071 224 |
| ·40 | ·0536 906 | ·0402 381 | ·0298 022 | ·0218 218 | ·0158 023 | ·0113 211 |
| ·41 | ·0740 565+ | ·0567 428 | ·0429 757 | ·0321 848 | ·0238 418 | ·0174 757 |
| ·42 | ·0996 942 | ·0780 313 | ·0603 859 | ·0462 176 | ·0349 964 | ·0262 251 |
| ·43 | ·1311 085+ | ·1047 434 | ·0827 565+ | ·0646 820 | ·0500 262 | ·0382 974 |
| ·44 | ·1685 951 | ·1373 670 | ·1107 190 | ·0883 033 | ·0697 047 | ·0544 745+ |
| ·45 | ·2121 780 | ·1761 658 | ·1447 369 | ·1176 989 | ·0947 551 | ·0755 398 |
| ·46 | ·2615 644 | ·2211 167 | ·1850 324 | ·1532 994 | ·1257 733 | ·1022 080 |
| ·47 | ·3161 257 | ·2718 666 | ·2315 237 | ·1952 745− | ·1631 473 | ·1350 449 |
| ·48 | ·3749 087 | ·3277 165− | ·2837 849 | ·2434 711 | ·2069 814 | ·1743 840 |
| ·49 | ·4366 799 | ·3876 385− | ·3410 342 | ·2973 769 | ·2570 377 | ·2202 505− |
| ·50 | ·5000 000$^e$ | ·4503 266 | ·4021 585$^e$ | ·3561 154 | ·3127 034 | ·2723 061 |
| ·51 | ·5633 201 | ·5142 786 | ·4657 721 | ·4184 783 | ·3729 951 | ·3298 240 |
| ·52 | ·6250 913 | ·5778 992 | ·5303 066 | ·4829 939 | ·4366 005+ | ·3917 020 |
| ·53 | ·6838 743 | ·6396 152 | ·5941 222 | ·5480 271 | ·5019 583 | ·4565 166 |
| ·54 | ·7384 356 | ·6979 878 | ·6556 280 | ·6118 977 | ·5673 678 | ·5226 152 |
| ·55 | ·7878 220 | ·7518 098 | ·7133 968 | ·6730 066 | ·6311 162 | ·5882 367 |
| ·56 | ·8314 049 | ·8001 767 | ·7662 611 | ·7299 514 | ·6916 084 | ·6516 464 |
| ·57 | ·8688 915− | ·8425 264 | ·8133 810 | ·7816 210 | ·7474 836 | ·7112 696 |
| ·58 | ·9003 058 | ·8786 429 | ·8542 754 | ·8272 562 | ·7977 049 | ·7658 059 |
| ·59 | ·9259 435− | ·9086 297 | ·8888 186 | ·8664 728 | ·8416 115+ | ·8143 125− |
| ·60 | ·9463 094 | ·9328 569 | ·9172 031 | ·8992 472 | ·8789 315− | ·8562 455+ |
| ·61 | ·9620 474 | ·9518 927 | ·9398 794 | ·9258 698 | ·9097 548 | ·8914 598 |
| ·62 | ·9738 701 | ·9664 287 | ·9574 810 | ·9468 754 | ·9344 759 | ·9201 683 |
| ·63 | ·9824 974 | ·9772 081 | ·9707 456 | ·9629 620 | ·9537 152 | ·9428 732 |
| ·64 | ·9886 076 | ·9849 644 | ·9804 424 | ·9749 096 | ·9682 323 | ·9602 790 |
| ·65 | ·9928 035− | ·9903 744 | ·9873 122 | ·9835 070 | ·9788 430 | ·9732 007 |
| ·66 | ·9955 943 | ·9940 283 | ·9920 239 | ·9894 949 | ·9863 472 | ·9824 809 |
| ·67 | ·9973 900 | ·9964 152 | ·9951 487 | ·9935 264 | ·9914 767 | ·9889 209 |
| ·68 | ·9985 062 | ·9979 212 | ·9971 498 | ·9961 469 | ·9948 610 | ·9932 335− |
| ·69 | ·9991 755+ | ·9988 376 | ·9983 855− | ·9977 890 | ·9970 129 | ·9960 163 |
| ·70 | ·9995 620 | ·9993 745+ | ·9991 199 | ·9987 792 | ·9983 295+ | ·9977 437 |
| ·71 | ·9997 766 | ·9996 768 | ·9995 394 | ·9993 529 | ·9991 032 | ·9987 733 |
| ·72 | ·9998 908 | ·9998 400 | ·9997 691 | ·9996 715+ | ·9995 390 | ·9993 614 |
| ·73 | ·9999 490 | ·9999 244 | ·9998 894 | ·9998 407 | ·9997 736 | ·9996 825+ |
| ·74 | ·9999 773 | ·9999 659 | ·9999 496 | ·9999 265− | ·9998 942 | ·9998 497 |
| ·75 | ·9999 904 | ·9999 854 | ·9999 782 | ·9999 678 | ·9999 531 | ·9999 325+ |
| ·76 | ·9999 962 | ·9999 941 | ·9999 911 | ·9999 866 | ·9999 803 | ·9999 713 |
| ·77 | ·9999 986 | ·9999 978 | ·9999 966 | ·9999 948 | ·9999 922 | ·9999 886 |
| ·78 | ·9999 995− | ·9999 992 | ·9999 988 | ·9999 981 | ·9999 971 | ·9999 957 |
| ·79 | ·9999 998 | ·9999 997 | ·9999 996 | ·9999 994 | ·9999 990 | ·9999 985+ |
| ·80 | ·9999 999 | ·9999 999 | ·9999 999 | ·9999 998 | ·9999 997 | ·9999 995+ |
| ·81 | 1·0000 000 | 1·0000 000 | 1·0000 000 | ·9999 999 | ·9999 999 | ·9999 999 |
| ·82 | | | | 1·0000 000 | 1·0000 000 | 1·0000 000 |

$x$ = ·25 to ·84 $q = 32$ $p$ = 38 to 43

| | $p = 38$ | $p = 39$ | $p = 40$ | $p = 41$ | $p = 42$ | $p = 43$ |
|---|---|---|---|---|---|---|
| $B(p, q) =$ | $\cdot6613\ 8090\times\frac{1}{10^{21}}$ | $\cdot3590\ 3534\times\frac{1}{10^{21}}$ | $\cdot1972\ 1660\times\frac{1}{10^{21}}$ | $\cdot1095\ 6478\times\frac{1}{10^{21}}$ | $\cdot6153\ 6381\times\frac{1}{10^{22}}$ | $\cdot3492\ 6054\times\frac{1}{10^{22}}$ |
| $x$ | | | | | | |
| ·25 | ·0000 001 | | | | | |
| ·26 | ·0000 003 | ·0000 001 | ·0000 001 | | | |
| ·27 | ·0000 008 | ·0000 004 | ·0000 002 | ·0000 001 | | |
| ·28 | ·0000 021 | ·0000 011 | ·0000 005+ | ·0000 003 | ·0000 001 | ·0000 001 |
| ·29 | ·0000 053 | ·0000 027 | ·0000 014 | ·0000 007 | ·0000 004 | ·0000 002 |
| ·30 | ·0000 127 | ·0000 068 | ·0000 036 | ·0000 019 | ·0000 010 | ·0000 005− |
| ·31 | ·0000 289 | ·0000 159 | ·0000 086 | ·0000 046 | ·0000 025− | ·0000 013 |
| ·32 | ·0000 629 | ·0000 357 | ·0000 200 | ·0000 111 | ·0000 061 | ·0000 033 |
| ·33 | ·0001 314 | ·0000 767 | ·0000 443 | ·0000 254 | ·0000 144 | ·0000 081 |
| ·34 | ·0002 633 | ·0001 582 | ·0000 941 | ·0000 554 | ·0000 324 | ·0000 187 |
| ·35 | ·0005 077 | ·0003 138 | ·0001 920 | ·0001 163 | ·0000 698 | ·0000 415+ |
| ·36 | ·0009 436 | ·0005 992 | ·0003 767 | ·0002 345+ | ·0001 446 | ·0000 884 |
| ·37 | ·0016 930 | ·0011 037 | ·0007 123 | ·0004 553 | ·0002 883 | ·0001 809 |
| ·38 | ·0029 366 | ·0019 637 | ·0013 002 | ·0008 526 | ·0005 539 | ·0003 566 |
| ·39 | ·0049 314 | ·0033 798 | ·0022 937 | ·0015 419 | ·0010 269 | ·0006 779 |
| ·40 | ·0080 269 | ·0056 342 | ·0039 164 | ·0026 967 | ·0018 400 | ·0012 444 |
| ·41 | ·0126 787 | ·0091 075+ | ·0064 795− | ·0045 669 | ·0031 899 | ·0022 086 |
| ·42 | ·0194 547 | ·0142 913 | ·0103 989 | ·0074 971 | ·0053 569 | ·0037 945+ |
| ·43 | ·0290 285+ | ·0217 915− | ·0162 060 | ·0119 428 | ·0087 236 | ·0063 176 |
| ·44 | ·0421 587 | ·0323 189 | ·0245 481 | ·0184 790 | ·0137 896 | ·0102 033 |
| ·45 | ·0596 482 | ·0466 630 | ·0361 750+ | ·0277 978 | ·0211 777 | ·0159 998 |
| ·46 | ·0822 857 | ·0656 453 | ·0519 064 | ·0406 885− | ·0316 266 | ·0243 815− |
| ·47 | ·1107 706 | ·0900 543 | ·0725 783 | ·0579 989 | ·0459 656 | ·0361 356 |
| ·48 | ·1456 278 | ·1205 644 | ·0989 714 | ·0805 742 | ·0650 666 | ·0521 289 |
| ·49 | ·1871 221 | ·1576 466 | ·1317 233 | ·1091 764 | ·0897 752 | ·0732 518 |
| ·50 | ·2351 843 | ·2014 815+ | ·1712 355− | ·1443 921 | ·1208 224 | ·1003 391 |
| ·51 | ·2893 599 | ·2518 880 | ·2175 867 | ·1865 356 | ·1587 258 | ·1340 745− |
| ·52 | ·3487 936 | ·3082 790 | ·2704 654 | ·2355 634 | ·2036 920 | ·1748 862 |
| ·53 | ·4122 539 | ·3696 569 | ·3291 351 | ·2910 142 | ·2555 339 | ·2228 498 |
| ·54 | ·4782 013 | ·4346 523 | ·3924 433 | ·3519 859 | ·3136 205+ | ·2776 118 |
| ·55 | ·5448 937 | ·5016 077 | ·4588 767 | ·4171 605− | ·3768 687 | ·3383 518 |
| ·56 | ·6105 180 | ·5686 969 | ·5266 614 | ·4848 780 | ·4437 877 | ·4037 930 |
| ·57 | ·6733 329 | ·6340 684 | ·5938 978 | ·5532 552 | ·5125 730 | ·4722 685+ |
| ·58 | ·7318 033 | ·6959 932 | ·6587 139 | ·6203 349 | ·5812 445− | ·5418 376 |
| ·59 | ·7847 114 | ·7529 993 | ·7194 171 | ·6842 480 | ·6478 095− | ·6104 426 |
| ·60 | ·8312 297 | ·8039 756 | ·7746 250− | ·7433 666 | ·7104 310 | ·6760 838 |
| ·61 | ·8709 495+ | ·8482 317 | ·8233 587 | ·7964 274 | ·7675 780 | ·7369 909 |
| ·62 | ·9038 653 | ·8855 116 | ·8650 872 | ·8426 101 | ·8181 376 | ·7917 655− |
| ·63 | ·9303 200 | ·9159 598 | ·8997 217 | ·8815 635− | ·8614 743 | ·8394 766 |
| ·64 | ·9509 241 | ·9400 527 | ·9275 646 | ·9133 781 | ·8974 338 | ·8796 978 |
| ·65 | ·9664 606 | ·9585 054 | ·9492 244 | ·9385 164 | ·9262 936 | ·9124 847 |
| ·66 | ·9777 911 | ·9721 708 | ·9655 129 | ·9577 132 | ·9486 731 | ·9383 029 |
| ·67 | ·9857 738 | ·9819 451 | ·9773 408 | ·9718 651 | ·9654 226 | ·9579 201 |
| ·68 | ·9911 996 | ·9886 883 | ·9856 232 | ·9819 236 | ·9775 057 | ·9722 843 |
| ·69 | ·9947 524 | ·9931 690 | ·9912 079 | ·9888 062 | ·9858 959 | ·9824 057 |
| ·70 | ·9969 900 | ·9960 319 | ·9948 283 | ·9933 327 | ·9914 943 | ·9892 576 |
| ·71 | ·9983 428 | ·9977 877 | ·9970 804 | ·9961 890 | ·9950 776 | ·9937 061 |
| ·72 | ·9991 264 | ·9988 192 | ·9984 221 | ·9979 147 | ·9972 732 | ·9964 703 |
| ·73 | ·9995 602 | ·9993 982 | ·9991 859 | ·9989 107 | ·9985 580 | ·9981 105− |
| ·74 | ·9997 893 | ·9997 080 | ·9996 002 | ·9994 584 | ·9992 743 | ·9990 374 |
| ·75 | ·9999 042 | ·9998 656 | ·9998 137 | ·9997 446 | ·9996 536 | ·9995 349 |
| ·76 | ·9999 588 | ·9999 416 | ·9999 180 | ·9998 862 | ·9998 438 | ·9997 877 |
| ·77 | ·9999 834 | ·9999 761 | ·9999 660 | ·9999 523 | ·9999 337 | ·9999 089 |
| ·78 | ·9999 937 | ·9999 908 | ·9999 868 | ·9999 813 | ·9999 737 | ·9999 634 |
| ·79 | ·9999 978 | ·9999 967 | ·9999 952 | ·9999 932 | ·9999 903 | ·9999 863 |
| ·80 | ·9999 993 | ·9999 989 | ·9999 984 | ·9999 977 | ·9999 967 | ·9999 953 |
| ·81 | ·9999 998 | ·9999 997 | ·9999 995+ | ·9999 993 | ·9999 990 | ·9999 985− |
| ·82 | ·9999 999 | ·9999 999 | ·9999 999 | ·9999 998 | ·9999 997 | ·9999 996 |
| ·83 | 1·0000 000 | 1·0000 000 | 1·0000 000 | ·9999 999 | ·9999 999 | ·9999 999 |
| ·84 | | | | 1·0000 000 | 1·0000 000 | 1·0000 000 |

$x$ = ·29 to ·86 $q = 32$ $p$ = 44 to 50

| | $p = 44$ | $p = 45$ | $p = 46$ | $p = 47$ | $p = 48$ | $p = 49$ | $p = 50$ |
|---|---|---|---|---|---|---|---|
| $B(p, q)$ = | ·2002 4271 $\times \frac{1}{10^{22}}$ | ·1159 2999 $\times \frac{1}{10^{22}}$ | ·6775 1293 $\times \frac{1}{10^{23}}$ | ·3995 5891 $\times \frac{1}{10^{23}}$ | ·2377 1226 $\times \frac{1}{10^{23}}$ | ·1426 2736 $\times \frac{1}{10^{23}}$ | ·8628 0747 $\times \frac{1}{10^{24}}$ |
| $x$ | | | | | | | |
| ·29 | ·0000 001 | | | | | | |
| ·30 | ·0000 002 | ·0000 001 | ·0000 001 | | | | |
| ·31 | ·0000 007 | ·0000 004 | ·0000 002 | ·0000 001 | | | |
| ·32 | ·0000 018 | ·0000 010 | ·0000 005+ | ·0000 003 | ·0000 001 | ·0000 001 | |
| ·33 | ·0000 045− | ·0000 025− | ·0000 014 | ·0000 007 | ·0000 004 | ·0000 002 | ·0000 001 |
| ·34 | ·0000 107 | ·0000 061 | ·0000 034 | ·0000 019 | ·0000 011 | ·0000 006 | ·0000 003 |
| ·35 | ·0000 245− | ·0000 143 | ·0000 083 | ·0000 048 | ·0000 027 | ·0000 015+ | ·0000 009 |
| ·36 | ·0000 535+ | ·0000 321 | ·0000 191 | ·0000 113 | ·0000 066 | ·0000 039 | ·0000 022 |
| ·37 | ·0001 125+ | ·0000 694 | ·0000 425− | ·0000 258 | ·0000 155+ | ·0000 093 | ·0000 055− |
| ·38 | ·0002 276 | ·0001 441 | ·0000 904 | ·0000 563 | ·0000 348 | ·0000 213 | ·0000 130 |
| ·39 | ·0004 436 | ·0002 879 | ·0001 853 | ·0001 183 | ·0000 750+ | ·0000 472 | ·0000 295− |
| ·40 | ·0008 343 | ·0005 548 | ·0003 659 | ·0002 395− | ·0001 555+ | ·0001 003 | ·0000 642 |
| ·41 | ·0015 162 | ·0010 323 | ·0006 972 | ·0004 673 | ·0003 108 | ·0002 052 | ·0001 345+ |
| ·42 | ·0026 653 | ·0018 568 | ·0012 833 | ·0008 802 | ·0005 992 | ·0004 049 | ·0002 717 |
| ·43 | ·0045 372 | ·0032 323 | ·0022 846 | ·0016 024 | ·0011 157 | ·0007 712 | ·0005 293 |
| ·44 | ·0074 877 | ·0054 510 | ·0039 375+ | ·0028 228 | ·0020 088 | ·0014 193 | ·0009 959 |
| ·45 | ·0119 900 | ·0089 142 | ·0065 767 | ·0048 159 | ·0035 009 | ·0025 270 | ·0018 115+ |
| ·46 | ·0186 460 | ·0141 489 | ·0106 551 | ·0079 649 | ·0059 112 | ·0043 564 | ·0031 887 |
| ·47 | ·0281 846 | ·0218 149 | ·0167 587 | ·0127 808 | ·0096 780 | ·0072 779 | ·0054 362 |
| ·48 | ·0414 417 | ·0326 977 | ·0256 092 | ·0199 137 | ·0153 768 | ·0117 926 | ·0089 837 |
| ·49 | ·0593 183 | ·0476 807 | ·0380 499 | ·0301 506 | ·0237 269 | ·0185 465+ | ·0144 022 |
| ·50 | ·0827 140 | ·0676 926 | ·0550 078 | ·0443 912 | ·0355 817 | ·0283 322 | ·0224 143 |
| ·51 | ·1124 386 | ·0936 302 | ·0774 298 | ·0635 997 | ·0518 940 | ·0420 686 | ·0338 875− |
| ·52 | ·1491 085− | ·1262 600 | ·1061 939 | ·0887 277 | ·0736 548 | ·0607 549 | ·0498 031 |
| ·53 | ·1930 388 | ·1661 074 | ·1420 007 | ·1206 139 | ·1018 026 | ·0853 936 | ·0711 949 |
| ·54 | ·2441 489 | ·2133 477 | ·1852 571 | ·1598 656 | ·1371 105+ | ·1168 869 | ·0990 571 |
| ·55 | ·3018 950+ | ·2677 168 | ·2359 690 | ·2067 404 | ·1800 615− | ·1559 115− | ·1342 258 |
| ·56 | ·3652 494 | ·3284 577 | ·2936 610 | ·2610 428 | ·2307 288 | ·2027 893 | ·1772 446 |
| ·57 | ·4327 326 | ·3943 199 | ·3573 413 | ·3220 589 | ·2886 832 | ·2573 726 | ·2282 346 |
| ·58 | ·5025 041 | ·4636 175+ | ·4255 260 | ·3885 441 | ·3529 470 | ·3189 666 | ·2867 893 |
| ·59 | ·5725 022 | ·5343 462 | ·4963 259 | ·4587 767 | ·4220 105− | ·3863 089 | ·3519 189 |
| ·60 | ·6406 184 | ·6043 469 | ·5675 918 | ·5306 769 | ·4939 190 | ·4576 207 | ·4220 631 |
| ·61 | ·7048 815− | ·6714 950− | ·6370 995− | ·6019 787 | ·5664 245− | ·5307 295+ | ·4951 803 |
| ·62 | ·7636 270 | ·7338 898 | ·7027 517 | ·6704 357 | ·6371 846 | ·6032 547 | ·5689 092 |
| ·63 | ·8156 270 | ·7900 158 | ·7627 655− | ·7340 283 | ·7039 826 | ·6728 289 | ·6407 852 |
| ·64 | ·8601 633 | ·8388 530 | ·8158 190 | ·7911 425− | ·7649 329 | ·7373 255− | ·7084 785+ |
| ·65 | ·8970 381 | ·8799 239 | ·8611 364 | ·8406 947 | ·8186 438 | ·7950 539 | ·7700 198 |
| ·66 | ·9265 242 | ·9132 732 | ·8985 027 | ·8821 846 | ·8643 110 | ·8448 958 | ·8239 750− |
| ·67 | ·9492 695+ | ·9393 902 | ·9282 111 | ·9156 736 | ·9017 330 | ·8863 605− | ·8695 449 |
| ·68 | ·9661 740 | ·9590 915+ | ·9509 578 | ·9416 994 | ·9312 512 | ·9195 580 | ·9065 762 |
| ·69 | ·9782 612 | ·9733 868 | ·9677 064 | ·9611 456 | ·9536 328 | ·9451 011 | ·9354 898 |
| ·70 | ·9865 631 | ·9833 481 | ·9795 472 | ·9750 935+ | ·9699 198 | ·9639 591 | ·9571 469 |
| ·71 | ·9920 303 | ·9900 022 | ·9875 702 | ·9846 799 | ·9812 743 | ·9772 946 | ·9726 815− |
| ·72 | ·9954 755+ | ·9942 546 | ·9927 700 | ·9909 807 | ·9888 427 | ·9863 092 | ·9833 311 |
| ·73 | ·9975 483 | ·9968 487 | ·9959 862 | ·9949 322 | ·9936 554 | ·9921 213 | ·9902 930 |
| ·74 | ·9987 358 | ·9983 553 | ·9978 798 | ·9972 908 | ·9965 674 | ·9956 864 | ·9946 220 |
| ·75 | ·9993 818 | ·9991 860 | ·9989 380 | ·9986 267 | ·9982 392 | ·9977 609 | ·9971 751 |
| ·76 | ·9997 144 | ·9996 195− | ·9994 976 | ·9993 425+ | ·9991 469 | ·9989 023 | ·9985 987 |
| ·77 | ·9998 759 | ·9998 327 | ·9997 764 | ·9997 040 | ·9996 113 | ·9994 940 | ·9993 464 |
| ·78 | ·9999 495+ | ·9999 311 | ·9999 069 | ·9998 753 | ·9998 343 | ·9997 817 | ·9997 148 |
| ·79 | ·9999 809 | ·9999 736 | ·9999 639 | ·9999 511 | ·9999 343 | ·9999 124 | ·9998 842 |
| ·80 | ·9999 933 | ·9999 907 | ·9999 871 | ·9999 823 | ·9999 759 | ·9999 675− | ·9999 565+ |
| ·81 | ·9999 978 | ·9999 970 | ·9999 957 | ·9999 941 | ·9999 919 | ·9999 889 | ·9999 850+ |
| ·82 | ·9999 994 | ·9999 991 | ·9999 987 | ·9999 982 | ·9999 975+ | ·9999 966 | ·9999 953 |
| ·83 | ·9999 998 | ·9999 998 | ·9999 997 | ·9999 995+ | ·9999 993 | ·9999 990 | ·9999 987 |
| ·84 | 1·0000 000 | ·9999 999 | ·9999 999 | ·9999 999 | ·9999 998 | ·9999 998 | ·9999 997 |
| ·85 | | 1·0000 000 | 1·0000 000 | 1·0000 000 | 1·0000 000 | ·9999 999 | ·9999 999 |
| ·86 | | | | | | 1·0000 000 | 1·0000 000 |

$x$ = ·21 to ·82 $q = 33$ $p$ = 33 to 38

| $x$ | $p = 33$ | $p = 34$ | $p = 35$ | $p = 36$ | $p = 37$ | $p = 38$ |
|---|---|---|---|---|---|---|
| $B(p, q)$ = | ·8394 8558 $\times \frac{1}{10^{20}}$ | ·4197 4279 $\times \frac{1}{10^{20}}$ | ·2130 0380 $\times \frac{1}{10^{20}}$ | ·1096 3431 $\times \frac{1}{10^{20}}$ | ·5720 0510 $\times \frac{1}{10^{21}}$ | ·3023 4555 $\overset{+}{\times} \frac{1}{10^{21}}$ |
| ·21 | ·0000 001 | | | | | |
| ·22 | ·0000 003 | ·0000 001 | ·0000 001 | | | |
| ·23 | ·0000 010 | ·0000 004 | ·0000 002 | ·0000 001 | | |
| ·24 | ·0000 028 | ·0000 013 | ·0000 006 | ·0000 003 | ·0000 001 | |
| ·25 | ·0000 071 | ·0000 034 | ·0000 016 | ·0000 008 | ·0000 003 | ·0000 002 |
| ·26 | ·0000 172 | ·0000 086 | ·0000 042 | ·0000 020 | ·0000 010 | ·0000 005− |
| ·27 | ·0000 396 | ·0000 205− | ·0000 105− | ·0000 053 | ·0000 026 | ·0000 013 |
| ·28 | ·0000 868 | ·0000 465+ | ·0000 246 | ·0000 128 | ·0000 066 | ·0000 034 |
| ·29 | ·0001 815+ | ·0001 006 | ·0000 550+ | ·0000 297 | ·0000 159 | ·0000 084 |
| ·30 | ·0003 629 | ·0002 078 | ·0001 175− | ·0000 656 | ·0000 362 | ·0000 197 |
| ·31 | ·0006 961 | ·0004 113 | ·0002 400 | ·0001 383 | ·0000 788 | ·0000 444 |
| ·32 | ·0012 832 | ·0007 818 | ·0004 703 | ·0002 795− | ·0001 642 | ·0000 953 |
| ·33 | ·0022 786 | ·0014 296 | ·0008 858 | ·0005 422 | ·0003 281 | ·0001 903 |
| ·34 | ·0039 046 | ·0025 203 | ·0016 067 | ·0010 121 | ·0006 302 | ·0003 880 |
| ·35 | ·0064 678 | ·0042 908 | ·0028 117 | ·0018 207 | ·0011 656 | ·0007 379 |
| ·36 | ·0103 723 | ·0070 656 | ·0047 548 | ·0031 624 | ·0020 795+ | ·0013 525+ |
| ·37 | ·0161 274 | ·0112 704 | ·0077 819 | ·0053 111 | ·0035 843 | ·0023 927 |
| ·38 | ·0243 440 | ·0174 375+ | ·0123 430 | ·0086 370 | ·0059 770 | ·0040 920 |
| ·39 | ·0357 186 | ·0262 018 | ·0189 971 | ·0136 182 | ·0096 558 | ·0067 739 |
| ·40 | ·0510 001 | ·0382 813 | ·0284 056 | ·0208 436 | ·0151 301 | ·0108 682 |
| ·41 | ·0709 400 | ·0544 411 | ·0413 099 | ·0310 037 | ·0230 222 | ·0169 196 |
| ·42 | ·0962 281 | ·0754 396 | ·0584 909 | ·0448 641 | ·0340 535+ | ·0255 862 |
| ·43 | ·1274 174 | ·1019 607 | ·0807 118 | ·0632 210 | ·0490 146 | ·0376 225+ |
| ·44 | ·1648 478 | ·1345 357 | ·1086 456 | ·0868 387 | ·0687 148 | ·0538 434 |
| ·45 | ·2085 768 | ·1734 649 | ·1427 936 | ·1163 724 | ·0939 145− | ·0750 680 |
| ·46 | ·2583 286 | ·2187 486 | ·1834 061 | ·1522 844 | ·1252 431 | ·1020 461 |
| ·47 | ·3134 701 | ·2700 385+ | ·2304 136 | ·1947 625− | ·1631 122 | ·1353 711 |
| ·48 | ·3730 210 | ·3266 182 | ·2833 817 | ·2436 536 | ·2076 334 | ·1753 905− |
| ·49 | ·4356 991 | ·3874 200 | ·3414 980 | ·2984 232 | ·2585 551 | ·2221 243 |
| ·50 | ·5000 000$^{e}$ | ·4510 792 | ·4035 973 | ·3581 504 | ·3152 282 | ·2752 063 |
| ·51 | ·5643 009 | ·5160 218 | ·4682 255− | ·4215 626 | ·3766 107 | ·3338 578 |
| ·52 | ·6269 790 | ·5805 763 | ·5337 367 | ·4871 114 | ·4413 149 | ·3969 046 |
| ·53 | ·6865 299 | ·6430 983 | ·5984 148 | ·5530 803 | ·5076 953 | ·4628 379 |
| ·54 | ·7416 714 | ·7020 915− | ·6606 023 | ·6177 145− | ·5739 688 | ·5299 158 |
| ·55 | ·7914 232 | ·7563 113 | ·7188 242 | ·6793 556 | ·6383 521 | ·5962 958 |
| ·56 | ·8351 522 | ·8048 401 | ·7718 891 | ·7365 656 | ·6992 012 | ·6601 806 |
| ·57 | ·8725 826 | ·8471 259 | ·8189 588 | ·7882 245− | ·7551 339 | ·7199 594 |
| ·58 | ·9037 719 | ·8829 834 | ·8595 780 | ·8335 913 | ·8051 214 | ·7743 278 |
| ·59 | ·9290 600 | ·9125 610 | ·8936 648 | ·8723 230 | ·8485 387 | ·8223 695− |
| ·60 | ·9489 999 | ·9362 812 | ·9214 676 | ·9044 532 | ·8851 701 | ·8635 939 |
| ·61 | ·9642 814 | ·9547 646 | ·9434 957 | ·9303 367 | ·9151 747 | ·8979 269 |
| ·62 | ·9756 560 | ·9687 495+ | ·9604 374 | ·9505 720 | ·9390 186 | ·9256 603 |
| ·63 | ·9838 726 | ·9790 156 | ·9730 758 | ·9659 124 | ·9573 879 | ·9473 728 |
| ·64 | ·9896 277 | ·9863 210 | ·9822 129 | ·9771 799 | ·9710 956 | ·9638 338 |
| ·65 | ·9935 322 | ·9913 552 | ·9886 083 | ·9851 905− | ·9809 941 | ·9759 073 |
| ·66 | ·9960 954 | ·9947 111 | ·9929 376 | ·9906 969 | ·9879 034 | ·9844 653 |
| ·67 | ·9977 214 | ·9968 724 | ·9957 682 | ·9943 520 | ·9925 597 | ·9903 204 |
| ·68 | ·9987 168 | ·9982 154 | ·9975 535− | ·9966 919 | ·9955 852 | ·9941 818 |
| ·69 | ·9993 039 | ·9990 192 | ·9986 378 | ·9981 341 | ·9974 776 | ·9966 328 |
| ·70 | ·9996 371 | ·9994 819 | ·9992 711 | ·9989 886 | ·9986 152 | ·9981 276 |
| ·71 | ·9998 185− | ·9997 376 | ·9996 260 | ·9994 744 | ·9992 711 | ·9990 019 |
| ·72 | ·9999 132 | ·9998 728 | ·9998 165− | ·9997 388 | ·9996 331 | ·9994 912 |
| ·73 | ·9999 604 | ·9999 412 | ·9999 141 | ·9998 762 | ·9998 239 | ·9997 527 |
| ·74 | ·9999 828 | ·9999 742 | ·9999 618 | ·9999 442 | ·9999 196 | ·9998 857 |
| ·75 | ·9999 929 | ·9999 892 | ·9999 839 | ·9999 762 | ·9999 653 | ·9999 500− |
| ·76 | ·9999 972 | ·9999 958 | ·9999 936 | ·9999 904 | ·9999 858 | ·9999 793 |
| ·77 | ·9999 990 | ·9999 984 | ·9999 976 | ·9999 964 | ·9999 946 | ·9999 920 |
| ·78 | ·9999 997 | ·9999 995− | ·9999 992 | ·9999 987 | ·9999 981 | ·9999 971 |
| ·79 | ·9999 999 | ·9999 998 | ·9999 997 | ·9999 996 | ·9999 994 | ·9999 990 |
| ·80 | 1·0000 000 | ·9999 999 | ·9999 999 | ·9999 999 | ·9999 998 | ·9999 997 |
| ·81 | | 1·0000 000 | 1·0000 000 | 1·0000 000 | ·9999 999 | ·9999 999 |
| ·82 | | | | | 1·0000 000 | 1·0000 000 |

$x$ = ·25 to ·84 $q = 33$ $p$ = 39 to 44

| | $p = 39$ | $p = 40$ | $p = 41$ | $p = 42$ | $p = 43$ | $p = 44$ |
|---|---|---|---|---|---|---|
| $B(p, q)$ = | ·1618 1875 $^{-}\times\frac{1}{10^{21}}$ | ·8765 1821 $\times\frac{1}{10^{22}}$ | ·4802 8395 $^{+}\times\frac{1}{10^{22}}$ | ·2661 0327 $\times\frac{1}{10^{22}}$ | ·1490 1783 $\times\frac{1}{10^{22}}$ | ·8431 2720 $\times\frac{1}{10^{23}}$ |
| $x$ | | | | | | |
| ·25 | ·0000 001 | | | | | |
| ·26 | ·0000 002 | ·0000 001 | | | | |
| ·27 | ·0000 006 | ·0000 003 | ·0000 001 | ·0000 001 | | |
| ·28 | ·0000 017 | ·0000 008 | ·0000 004 | ·0000 002 | ·0000 001 | |
| ·29 | ·0000 044 | ·0000 023 | ·0000 012 | ·0000 006 | ·0000 003 | ·0000 001 |
| ·30 | ·0000 106 | ·0000 057 | ·0000 030 | ·0000 016 | ·0000 008 | ·0000 004 |
| ·31 | ·0000 247 | ·0000 136 | ·0000 074 | ·0000 040 | ·0000 021 | ·0000 011 |
| ·32 | ·0000 548 | ·0000 311 | ·0000 175+ | ·0000 098 | ·0000 054 | ·0000 029 |
| ·33 | ·0001 162 | ·0000 680 | ·0000 394 | ·0000 226 | ·0000 129 | ·0000 073 |
| ·34 | ·0002 364 | ·0001 425− | ·0000 850+ | ·0000 503 | ·0000 294 | ·0000 171 |
| ·35 | ·0004 622 | ·0002 866 | ·0001 759 | ·0001 069 | ·0000 644 | ·0000 384 |
| ·36 | ·0008 704 | ·0005 544 | ·0003 497 | ·0002 184 | ·0001 352 | ·0000 829 |
| ·37 | ·0015 806 | ·0010 336 | ·0006 693 | ·0004 293 | ·0002 728 | ·0001 718 |
| ·38 | ·0027 725− | ·0018 596 | ·0012 353 | ·0008 128 | ·0005 300 | ·0003 425+ |
| ·39 | ·0047 035+ | ·0032 336 | ·0022 016 | ·0014 851 | ·0009 927 | ·0006 578 |
| ·40 | ·0077 278 | ·0054 409 | ·0037 944 | ·0026 218 | ·0017 953 | ·0012 187 |
| ·41 | ·0123 105− | ·0088 702 | ·0063 313 | ·0044 778 | ·0031 390 | ·0021 815+ |
| ·42 | ·0190 352 | ·0140 261 | ·0102 393 | ·0074 075+ | ·0053 120 | ·0037 769 |
| ·43 | ·0285 988 | ·0215 349 | ·0160 674 | ·0118 814 | ·0087 101 | ·0063 316 |
| ·44 | ·0417 897 | ·0321 344 | ·0244 874 | ·0184 966 | ·0138 523 | ·0102 881 |
| ·45 | ·0594 453 | ·0466 466 | ·0362 798 | ·0279 736 | ·0213 880 | ·0162 191 |
| ·46 | ·0823 897 | ·0659 287 | ·0522 990 | ·0411 359 | ·0320 885+ | ·0248 296 |
| ·47 | ·1113 532 | ·0908 025+ | ·0734 166 | ·0588 676 | ·0468 196 | ·0369 431 |
| ·48 | ·1468 809 | ·1219 680 | ·1004 432 | ·0820 474 | ·0664 898 | ·0534 648 |
| ·49 | ·1892 407 | ·1599 068 | ·1340 343 | ·1114 622 | ·0919 749 | ·0753 200 |
| ·50 | ·2383 439 | ·2047 897 | ·1745 909 | ·1477 066 | ·1240 229 | ·1033 684 |
| ·51 | ·2936 926 | ·2564 007 | ·2221 668 | ·1910 807 | ·1631 477 | ·1383 002 |
| ·52 | ·3543 643 | ·3140 928 | ·2763 987 | ·2414 995− | ·2095 251 | ·1805 251 |
| ·53 | ·4190 428 | ·3767 862 | ·3364 734 | ·2984 318 | ·2629 083 | ·2300 696 |
| ·54 | ·4860 946 | ·4430 150− | ·4011 416 | ·3608 819 | ·3225 777 | ·2865 004 |
| ·55 | ·5536 862 | ·5110 219 | ·4687 843 | ·4274 223 | ·3873 406 | ·3488 901 |
| ·56 | ·6199 278 | ·5788 906 | ·5375 250+ | ·4962 806 | ·4555 861 | ·4158 379 |
| ·57 | ·6830 263 | ·6447 010 | ·6053 793 | ·5654 726 | ·5253 948 | ·4855 501 |
| ·58 | ·7414 273 | ·7066 876 | ·6704 195− | ·6329 660 | ·5946 921 | ·5559 731 |
| ·59 | ·7939 277 | ·7633 784 | ·7309 349 | ·6968 535− | ·6614 250+ | ·6249 667 |
| ·60 | ·8397 466 | ·8136 980 | ·7855 654 | ·7555 116 | ·7237 405+ | ·6904 916 |
| ·61 | ·8785 458 | ·8570 228 | ·8333 906 | ·8077 237 | ·7801 379 | ·7507 879 |
| ·62 | ·9104 038 | ·8931 835− | ·8739 656 | ·8527 510 | ·8295 765− | ·8045 157 |
| ·63 | ·9357 500− | ·9224 195+ | ·9073 028 | ·8903 462 | ·8715 245− | ·8508 424 |
| ·64 | ·9552 726 | ·9452 977 | ·9338 067 | ·9207 125− | ·9059 472 | ·8894 650− |
| ·65 | ·9698 167 | ·9626 093 | ·9541 768 | ·9444 177 | ·9332 411 | ·9205 701 |
| ·66 | ·9802 852 | ·9752 626 | ·9692 958 | ·9622 841 | ·9541 305− | ·9447 444 |
| ·67 | ·9875 565+ | ·9841 853 | ·9801 197 | ·9752 696 | ·9695 443 | ·9628 536 |
| ·68 | ·9924 239 | ·9902 476 | ·9875 839 | ·9843 589 | ·9804 950+ | ·9759 123 |
| ·69 | ·9955 590 | ·9942 102 | ·9925 349 | ·9904 768 | ·9879 748 | ·9849 636 |
| ·70 | ·9974 989 | ·9966 977 | ·9956 882 | ·9944 301 | ·9928 783 | ·9909 837 |
| ·71 | ·9986 498 | ·9981 947 | ·9976 130 | ·9968 778 | ·9959 580 | ·9948 190 |
| ·72 | ·9993 030 | ·9990 564 | ·9987 367 | ·9983 269 | ·9978 071 | ·9971 542 |
| ·73 | ·9996 570 | ·9995 298 | ·9993 626 | ·9991 454 | ·9988 660 | ·9985 102 |
| ·74 | ·9998 395+ | ·9997 773 | ·9996 944 | ·9995 852 | ·9994 427 | ·9992 589 |
| ·75 | ·9999 289 | ·9999 001 | ·9998 612 | ·9998 093 | ·9997 407 | ·9996 510 |
| ·76 | ·9999 703 | ·9999 577 | ·9999 406 | ·9999 173 | ·9998 862 | ·9998 450− |
| ·77 | ·9999 883 | ·9999 832 | ·9999 761 | ·9999 663 | ·9999 531 | ·9999 354 |
| ·78 | ·9999 957 | ·9999 938 | ·9999 910 | ·9999 872 | ·9999 820 | ·9999 748 |
| ·79 | ·9999 985+ | ·9999 978 | ·9999 969 | ·9999 955− | ·9999 935+ | ·9999 909 |
| ·80 | ·9999 995+ | ·9999 993 | ·9999 990 | ·9999 985+ | ·9999 979 | ·9999 970 |
| ·81 | ·9999 999 | ·9999 998 | ·9999 997 | ·9999 996 | ·9999 994 | ·9999 991 |
| ·82 | 1·0000 000 | ·9999 999 | ·9999 999 | ·9999 999 | ·9999 998 | ·9999 997 |
| ·83 | | 1·0000 000 | 1·0000 000 | 1·0000 000 | 1·0000 000 | ·9999 999 |
| ·84 | | | | | | 1·0000 000 |

$x = {\cdot}29$ to ${\cdot}85$ $q = 33$ $p = 45$ to $50$

| $x$ | $p = 45$ | $p = 46$ | $p = 47$ | $p = 48$ | $p = 49$ | $p = 50$ |
|---|---|---|---|---|---|---|
| $B(p, q) =$ | ·4817 8697 $\times \frac{1}{10^{23}}$ | ·2779 5402 $\times \frac{1}{10^{23}}$ | ·1618 4665 $\bar{\times} \frac{1}{10^{23}}$ | ·9508 4905 $\bar{\times} \frac{1}{10^{24}}$ | ·5634 6610 $\times \frac{1}{10^{24}}$ | ·3367 0535 $\overset{+}{\times} \frac{1}{10^{24}}$ |
| ·29 | ·0000 001 | | | | | |
| ·30 | ·0000 002 | ·0000 001 | ·0000 001 | | | |
| ·31 | ·0000 006 | ·0000 003 | ·0000 002 | ·0000 001 | | |
| ·32 | ·0000 016 | ·0000 009 | ·0000 005− | ·0000 002 | ·0000 001 | ·0000 001 |
| ·33 | ·0000 041 | ·0000 022 | ·0000 012 | ·0000 007 | ·0000 004 | ·0000 002 |
| ·34 | ·0000 098 | ·0000 056 | ·0000 032 | ·0000 018 | ·0000 010 | ·0000 005+ |
| ·35 | ·0000 227 | ·0000 133 | ·0000 078 | ·0000 045− | ·0000 026 | ·0000 015− |
| ·36 | ·0000 504 | ·0000 304 | ·0000 182 | ·0000 108 | ·0000 064 | ·0000 037 |
| ·37 | ·0001 073 | ·0000 664 | ·0000 408 | ·0000 249 | ·0000 150+ | ·0000 090 |
| ·38 | ·0002 195− | ·0001 395− | ·0000 879 | ·0000 550− | ·0000 341 | ·0000 210 |
| ·39 | ·0004 321 | ·0002 816 | ·0001 820 | ·0001 167 | ·0000 743 | ·0000 470 |
| ·40 | ·0008 204 | ·0005 477 | ·0003 628 | ·0002 384 | ·0001 555+ | ·0001 007 |
| ·41 | ·0015 035− | ·0010 278 | ·0006 971 | ·0004 692 | ·0003 134 | ·0002 079 |
| ·42 | ·0026 632 | ·0018 629 | ·0012 929 | ·0008 905+ | ·0006 089 | ·0004 133 |
| ·43 | ·0045 650− | ·0032 651 | ·0023 174 | ·0016 324 | ·0011 415+ | ·0007 926 |
| ·44 | ·0075 793 | ·0055 399 | ·0040 183 | ·0028 930 | ·0020 678 | ·0014 675+ |
| ·45 | ·0122 014 | ·0091 078 | ·0067 473 | ·0049 618 | ·0036 227 | ·0026 265+ |
| ·46 | ·0190 620 | ·0145 223 | ·0109 813 | ·0082 434 | ·0061 444 | ·0045 483 |
| ·47 | ·0289 251 | ·0224 769 | ·0173 379 | ·0132 782 | ·0100 980 | ·0076 272 |
| ·48 | ·0426 659 | ·0337 964 | ·0265 774 | ·0207 531 | ·0160 936 | ·0123 965− |
| ·49 | ·0612 240 | ·0494 052 | ·0395 853 | ·0314 976 | ·0248 925+ | ·0195 425− |
| ·50 | ·0855 305− | ·0702 691 | ·0573 302 | ·0464 559 | ·0373 941 | ·0299 042 |
| ·51 | ·1164 119 | ·0973 107 | ·0807 923 | ·0666 321 | ·0545 960 | ·0444 488 |
| ·52 | ·1544 778 | ·1313 016 | ·1108 661 | ·0930 047 | ·0775 248 | ·0642 184 |
| ·53 | ·2000 074 | ·1727 442 | ·1482 430 | ·1264 160 | ·1071 355− | ·0902 434 |
| ·54 | ·2528 502 | ·2217 574 | ·1932 872 | ·1674 459 | ·1441 888 | ·1234 282 |
| ·55 | ·3123 621 | ·2779 852 | ·2459 250+ | ·2162 864 | ·1891 177 | ·1644 163 |
| ·56 | ·3773 906 | ·3405 496 | ·3055 666 | ·2726 380 | ·2419 045+ | ·2134 542 |
| ·57 | ·4463 211 | ·4080 598 | ·3710 794 | ·3356 490 | ·3019 902 | ·2702 753 |
| ·58 | ·5171 838 | ·4786 875+ | ·4408 273 | ·4039 175+ | ·3682 381 | ·3340 296 |
| ·59 | ·5878 123 | ·5503 029 | ·5127 772 | ·4755 628 | ·4389 687 | ·4032 783 |
| ·60 | ·6560 337 | ·6206 569 | ·5846 649 | ·5483 666 | ·5120 682 | ·4760 662 |
| ·61 | ·7198 636 | ·6875 856 | ·6541 989 | ·6199 669 | ·5851 643 | ·5500 705+ |
| ·62 | ·7776 779 | ·7492 059 | ·7192 732 | ·6880 796 | ·6558 461 | ·6228 101 |
| ·63 | ·8283 366 | ·8040 753 | ·7781 579 | ·7507 130 | ·7218 959 | ·6918 849 |
| ·64 | ·8712 447 | ·8512 914 | ·8296 378 | ·8063 440 | ·7814 973 | ·7552 106 |
| ·65 | ·9063 439 | ·8905 213 | ·8730 820 | ·8540 286 | ·8333 875− | ·8112 088 |
| ·66 | ·9340 442 | ·9219 601 | ·9084 364 | ·8934 338 | ·8769 309 | ·8589 259 |
| ·67 | ·9551 107 | ·9462 338 | ·9361 490 | ·9247 917 | ·9121 094 | ·8980 631 |
| ·68 | ·9705 296 | ·9642 666 | ·9570 451 | ·9487 911 | ·9394 365+ | ·9289 212 |
| ·69 | ·9813 748 | ·9771 376 | ·9721 801 | ·9664 304 | ·9598 183 | ·9522 765− |
| ·70 | ·9886 930 | ·9859 493 | ·9826 926 | ·9788 607 | ·9743 902 | ·9692 172 |
| ·71 | ·9934 221 | ·9917 250+ | ·9896 820 | ·9872 437 | ·9843 585− | ·9809 722 |
| ·72 | ·9963 423 | ·9953 421 | ·9941 209 | ·9926 430 | ·9908 696 | ·9887 588 |
| ·73 | ·9980 616 | ·9975 012 | ·9968 076 | ·9959 565− | ·9949 210 | ·9936 714 |
| ·74 | ·9990 240 | ·9987 265− | ·9983 532 | ·9978 889 | ·9973 162 | ·9966 157 |
| ·75 | ·9995 347 | ·9993 856 | ·9991 959 | ·9989 567 | ·9986 577 | ·9982 871 |
| ·76 | ·9997 909 | ·9997 205− | ·9996 298 | ·9995 139 | ·9993 671 | ·9991 827 |
| ·77 | ·9999 117 | ·9998 806 | ·9998 400 | ·9997 874 | ·9997 199 | ·9996 340 |
| ·78 | ·9999 652 | ·9999 524 | ·9999 354 | ·9999 132 | ·9998 843 | ·9998 470 |
| ·79 | ·9999 873 | ·9999 824 | ·9999 758 | ·9999 671 | ·9999 556 | ·9999 406 |
| ·80 | ·9999 957 | ·9999 940 | ·9999 916 | ·9999 885− | ·9999 843 | ·9999 787 |
| ·81 | ·9999 987 | ·9999 981 | ·9999 974 | ·9999 963 | ·9999 949 | ·9999 930 |
| ·82 | ·9999 996 | ·9999 995− | ·9999 992 | ·9999 989 | ·9999 985+ | ·9999 979 |
| ·83 | ·9999 999 | ·9999 999 | ·9999 998 | ·9999 997 | ·9999 996 | ·9999 994 |
| ·84 | 1·0000 000 | 1·0000 000 | 1·0000 000 | ·9999 999 | ·9999 999 | ·9999 999 |
| ·85 | | | | 1·0000 000 | 1·0000 000 | 1·0000 000 |

$x$ = ·21 to ·82 $q$ = 34 $p$ = 34 to 39

| $x$ | $p$ = 34 | $p$ = 35 | $p$ = 36 | $p$ = 37 | $p$ = 38 | $p$ = 39 |
|---|---|---|---|---|---|---|
| $B(p, q)$ = | ·2067 3899 $\times\frac{1}{10^{20}}$ | ·1033 6949 $\times\frac{1}{10^{20}}$ | ·5243 3801 $\times\frac{1}{10^{21}}$ | ·2696 5955 $\times\frac{1}{10^{21}}$ | ·1405 2681 $\times\frac{1}{10^{21}}$ | ·7416 6925 $\times\frac{1}{10^{22}}$ |
| ·21 | ·0000 001 | | | | | |
| ·22 | ·0000 002 | ·0000 001 | | | | |
| ·23 | ·0000 007 | ·0000 003 | ·0000 001 | ·0000 001 | | |
| ·24 | ·0000 020 | ·0000 009 | ·0000 004 | ·0000 002 | ·0000 001 | |
| ·25 | ·0000 052 | ·0000 025+ | ·0000 012 | ·0000 006 | ·0000 003 | ·0000 001 |
| ·26 | ·0000 131 | ·0000 065+ | ·0000 032 | ·0000 016 | ·0000 008 | ·0000 004 |
| ·27 | ·0000 308 | ·0000 160 | ·0000 082 | ·0000 041 | ·0000 021 | ·0000 010 |
| ·28 | ·0000 691 | ·0000 370 | ·0000 196 | ·0000 103 | ·0000 053 | ·0000 027 |
| ·29 | ·0001 475+ | ·0000 818 | ·0000 448 | ·0000 243 | ·0000 130 | ·0000 069 |
| ·30 | ·0003 009 | ·0001 725− | ·0000 977 | ·0000 546 | ·0000 302 | ·0000 165+ |
| ·31 | ·0005 879 | ·0003 478 | ·0002 033 | ·0001 174 | ·0000 670 | ·0000 378 |
| ·32 | ·0011 027 | ·0006 727 | ·0004 053 | ·0002 413 | ·0001 421 | ·0000 827 |
| ·33 | ·0019 900 | ·0012 502 | ·0007 758 | ·0004 758 | ·0002 885+ | ·0001 730 |
| ·34 | ·0034 617 | ·0022 374 | ·0014 287 | ·0009 017 | ·0005 627 | ·0003 473 |
| ·35 | ·0058 147 | ·0038 627 | ·0025 354 | ·0016 450+ | ·0010 554 | ·0006 699 |
| ·36 | ·0094 465− | ·0064 438 | ·0043 437 | ·0028 946 | ·0019 077 | ·0012 438 |
| ·37 | ·0148 646 | ·0104 025+ | ·0071 949 | ·0049 202 | ·0033 279 | ·0022 271 |
| ·38 | ·0226 864 | ·0162 735− | ·0115 389 | ·0080 905+ | ·0056 114 | ·0038 513 |
| ·39 | ·0336 249 | ·0247 019 | ·0179 409 | ·0128 870 | ·0091 580 | ·0064 408 |
| ·40 | ·0484 563 | ·0364 259 | ·0270 765+ | ·0199 087 | ·0144 844 | ·0104 304 |
| ·41 | ·0679 702 | ·0522 406 | ·0397 108 | ·0298 645+ | ·0222 270 | ·0163 762 |
| ·42 | ·0929 020 | ·0729 435+ | ·0566 575− | ·0435 472 | ·0331 298 | ·0249 549 |
| ·43 | ·1238 535− | ·0992 627 | ·0787 190 | ·0617 874 | ·0480 134 | ·0369 471 |
| ·44 | ·1612 103 | ·1317 741 | ·1066 103 | ·0853 888 | ·0677 233 | ·0532 005− |
| ·45 | ·2050 656 | ·1708 160 | ·1408 720 | ·1150 454 | ·0930 578 | ·0745 709 |
| ·46 | ·2551 622 | ·2164 139 | ·1817 840 | ·1512 519 | ·1246 808 | ·1018 436 |
| ·47 | ·3108 642 | ·2682 254 | ·2292 900 | ·1942 158 | ·1630 281 | ·1356 404 |
| ·48 | ·3711 649 | ·3255 176 | ·2829 483 | ·2437 846 | ·2082 196 | ·1763 235+ |
| ·49 | ·4347 335+ | ·3871 834 | ·3419 157 | ·2994 018 | ·2599 903 | ·2239 080 |
| ·50 | ·5000 000[e] | ·4517 987 | ·4049 745+ | ·3601 014 | ·3176 538 | ·2779 989 |
| ·51 | ·5652 665− | ·5177 164 | ·4706 010 | ·4245 458 | ·3801 087 | ·3377 648 |
| ·52 | ·6288 351 | ·5831 879 | ·5370 712 | ·4911 081 | ·4458 905− | ·4019 579 |
| ·53 | ·6891 358 | ·6464 969 | ·6025 911 | ·5579 901 | ·5132 685+ | ·4689 824 |
| ·54 | ·7448 378 | ·7060 895− | ·6654 370 | ·6233 616 | ·5803 765+ | ·5370 068 |
| ·55 | ·7949 344 | ·7606 848 | ·7240 867 | ·6855 061 | ·6453 615− | ·6041 076 |
| ·56 | ·8387 897 | ·8093 534 | ·7773 268 | ·7429 515+ | ·7065 323 | ·6684 263 |
| ·57 | ·8761 465+ | ·8515 558 | ·8243 233 | ·7945 719 | ·7624 885+ | ·7283 198 |
| ·58 | ·9070 980 | ·8871 396 | ·8646 493 | ·8396 476 | ·8122 133 | ·7824 831 |
| ·59 | ·9320 298 | ·9163 002 | ·8982 695+ | ·8778 799 | ·8551 206 | ·8300 315+ |
| ·60 | ·9515 437 | ·9395 132 | ·9254 892 | ·9093 616 | ·8910 545+ | ·8705 313 |
| ·61 | ·9663 751 | ·9574 521 | ·9468 771 | ·9345 132 | ·9202 445+ | ·9039 820 |
| ·62 | ·9773 136 | ·9709 006 | ·9631 757 | ·9539 960 | ·9432 285− | ·9307 551 |
| ·63 | ·9851 354 | ·9806 734 | ·9752 118 | ·9686 169 | ·9607 566 | ·9515 042 |
| ·64 | ·9905 535+ | ·9875 509 | ·9838 173 | ·9792 375− | ·9736 922 | ·9670 611 |
| ·65 | ·9941 853 | ·9922 334 | ·9897 684 | ·9866 974 | ·9829 208 | ·9783 344 |
| ·66 | ·9965 383 | ·9953 141 | ·9937 442 | ·9917 584 | ·9892 787 | ·9862 209 |
| ·67 | ·9980 100 | ·9972 702 | ·9963 072 | ·9950 706 | ·9935 030 | ·9915 407 |
| ·68 | ·9988 973 | ·9984 673 | ·9978 991 | ·9971 587 | ·9962 061 | ·9949 958 |
| ·69 | ·9994 121 | ·9991 721 | ·9988 503 | ·9984 248 | ·9978 693 | ·9971 531 |
| ·70 | ·9996 991 | ·9995 707 | ·9993 961 | ·9991 618 | ·9988 515+ | ·9984 458 |
| ·71 | ·9998 525− | ·9997 868 | ·9996 962 | ·9995 729 | ·9994 073 | ·9991 876 |
| ·72 | ·9999 309 | ·9998 989 | ·9998 540 | ·9997 922 | ·9997 079 | ·9995 945+ |
| ·73 | ·9999 692 | ·9999 543 | ·9999 332 | ·9999 037 | ·9998 629 | ·9998 073 |
| ·74 | ·9999 869 | ·9999 804 | ·9999 710 | ·9999 576 | ·9999 389 | ·9999 131 |
| ·75 | ·9999 948 | ·9999 920 | ·9999 881 | ·9999 824 | ·9999 743 | ·9999 629 |
| ·76 | ·9999 980 | ·9999 970 | ·9999 954 | ·9999 931 | ·9999 898 | ·9999 851 |
| ·77 | ·9999 993 | ·9999 989 | ·9999 983 | ·9999 975− | ·9999 962 | ·9999 944 |
| ·78 | ·9999 998 | ·9999 996 | ·9999 994 | ·9999 991 | ·9999 987 | ·9999 980 |
| ·79 | ·9999 999 | ·9999 999 | ·9999 998 | ·9999 997 | ·9999 996 | ·9999 994 |
| ·80 | 1·0000 000 | 1·0000 000 | ·9999 999 | ·9999 999 | ·9999 999 | ·9999 998 |
| ·81 | | | 1·0000 000 | 1·0000 000 | 1·0000 000 | ·9999 999 |
| ·82 | | | | | | 1·0000 000 |

$x = \cdot 25$ to $\cdot 83$ $q = 34$ $p = 40$ to $45$

| | $p = 40$ | $p = 41$ | $p = 42$ | $p = 43$ | $p = 44$ | $p = 45$ |
|---|---|---|---|---|---|---|
| $B(p, q) =$ | ·3962 3426 $\times \frac{1}{10^{22}}$ | ·2141 8068 $\times \frac{1}{10^{22}}$ | ·1170 8544 $\times \frac{1}{10^{22}}$ | ·6470 5111 $\times \frac{1}{10^{23}}$ | ·3613 4023 $\times \frac{1}{10^{23}}$ | ·2038 3295 $\times \frac{1}{10^{23}}$ |
| $x$ | | | | | | |
| ·25 | ·0000 001 | | | | | |
| ·26 | ·0000 002 | ·0000 001 | | | | |
| ·27 | ·0000 005⁻ | ·0000 002 | ·0000 001 | ·0000 001 | | |
| ·28 | ·0000 014 | ·0000 007 | ·0000 003 | ·0000 002 | ·0000 001 | |
| ·29 | ·0000 036 | ·0000 019 | ·0000 010 | ·0000 005⁻ | ·0000 002 | ·0000 001 |
| ·30 | ·0000 089 | ·0000 048 | ·0000 025⁺ | ·0000 013 | ·0000 007 | ·0000 004 |
| ·31 | ·0000 211 | ·0000 117 | ·0000 064 | ·0000 035⁻ | ·0000 019 | ·0000 010 |
| ·32 | ·0000 476 | ·0000 272 | ·0000 153 | ·0000 086 | ·0000 047 | ·0000 026 |
| ·33 | ·0001 027 | ·0000 603 | ·0000 351 | ·0000 202 | ·0000 115⁺ | ·0000 065⁺ |
| ·34 | ·0002 121 | ·0001 283 | ·0000 768 | ·0000 455⁺ | ·0000 268 | ·0000 156 |
| ·35 | ·0004 207 | ·0002 616 | ·0001 611 | ·0000 982 | ·0000 594 | ·0000 356 |
| ·36 | ·0008 026 | ·0005 127 | ·0003 244 | ·0002 033 | ·0001 263 | ·0000 777 |
| ·37 | ·0014 752 | ·0009 675⁻ | ·0006 284 | ·0004 044 | ·0002 579 | ·0001 630 |
| ·38 | ·0026 165⁻ | ·0017 601 | ·0011 728 | ·0007 742 | ·0005 066 | ·0003 286 |
| ·39 | ·0044 844 | ·0030 919 | ·0021 117 | ·0014 291 | ·0009 586 | ·0006 375⁻ |
| ·40 | ·0074 367 | ·0052 513 | ·0036 736 | ·0025 466 | ·0017 499 | ·0011 922 |
| ·41 | ·0119 477 | ·0086 340 | ·0061 819 | ·0043 866 | ·0030 856 | ·0021 521 |
| ·42 | ·0186 162 | ·0137 576 | ·0100 746 | ·0073 123 | ·0052 617 | ·0037 546 |
| ·43 | ·0281 621 | ·0212 681 | ·0159 177 | ·0118 094 | ·0086 870 | ·0063 374 |
| ·44 | ·0414 035⁻ | ·0319 305⁻ | ·0244 075⁺ | ·0184 966 | ·0138 999 | ·0103 604 |
| ·45 | ·0592 126 | ·0465 995⁺ | ·0363 553 | ·0281 233 | ·0215 760 | ·0164 200 |
| ·46 | ·0824 495⁺ | ·0661 682 | ·0526 508 | ·0415 472 | ·0325 197 | ·0252 525⁺ |
| ·47 | ·1118 763 | ·0914 927 | ·0742 014 | ·0596 890 | ·0476 337 | ·0377 181 |
| ·48 | ·1480 586 | ·1232 986 | ·1018 480 | ·0834 617 | ·0678 633 | ·0547 607 |
| ·49 | ·1912 674 | ·1620 785⁺ | ·1362 642 | ·1136 766 | ·0941 147 | ·0773 404 |
| ·50 | ·2413 943 | ·2079 926 | ·1778 496 | ·1509 362 | ·1271 523 | ·1063 414 |
| ·51 | ·2978 964 | ·2607 889 | ·2266 319 | ·1955 246 | ·1674 847 | ·1424 590 |
| ·52 | ·3597 827 | ·3197 584 | ·2821 941 | ·2473 130 | ·2152 548 | ·1860 819 |
| ·53 | ·4256 502 | ·3837 371 | ·3436 436 | ·3056 980 | ·2701 527 | ·2371 844 |
| ·54 | ·4937 706 | ·4511 613 | ·4096 328 | ·3695 874 | ·3313 674 | ·2952 495⁻ |
| ·55 | ·5622 190 | ·5201 734 | ·4784 354 | ·4374 427 | ·3975 940 | ·3592 397 |
| ·56 | ·6290 306 | ·5887 681 | ·5480 736 | ·5073 791 | ·4671 010 | ·4276 284 |
| ·57 | ·6923 637 | ·6549 604 | ·6164 807 | ·5773 138 | ·5378 554 | ·4984 957 |
| ·58 | ·7506 490 | ·7169 526 | ·6816 782 | ·6451 441 | ·6076 923 | ·5696 787 |
| ·59 | ·8027 037 | ·7732 785⁺ | ·7419 443 | ·7089 314 | ·6745 058 | ·6389 615⁻ |
| ·60 | ·8477 980 | ·8229 050⁻ | ·7959 476 | ·7670 648 | ·7364 355⁺ | ·7042 748 |
| ·61 | ·8856 679 | ·8652 798 | ·8428 329 | ·8183 818 | ·7920 202 | ·7638 791 |
| ·62 | ·9164 779 | ·9003 232 | ·8822 458 | ·8622 314 | ·8402 994 | ·8165 032 |
| ·63 | ·9407 428 | ·9283 700 | ·9143 012 | ·8984 738 | ·8808 502 | ·8614 202 |
| ·64 | ·9592 263 | ·9500 752 | ·9395 046 | ·9274 239 | ·9137 587 | ·8984 537 |
| ·65 | ·9728 306 | ·9663 018 | ·9586 423 | ·9497 519 | ·9395 383 | ·9279 203 |
| ·66 | ·9824 951 | ·9780 073 | ·9726 614 | ·9663 609 | ·9590 113 | ·9505 225⁻ |
| ·67 | ·9891 134 | ·9861 455⁻ | ·9825 564 | ·9782 624 | ·9731 775⁺ | ·9672 155⁻ |
| ·68 | ·9934 764 | ·9915 908 | ·9892 766 | ·9864 665⁻ | ·9830 891 | ·9790 701 |
| ·69 | ·9962 408 | ·9950 920 | ·9936 613 | ·9918 985⁻ | ·9897 486 | ·9871 527 |
| ·70 | ·9979 213 | ·9972 514 | ·9964 050⁺ | ·9953 470 | ·9940 380 | ·9924 345⁺ |
| ·71 | ·9988 997 | ·9985 265⁺ | ·9980 484 | ·9974 422 | ·9966 814 | ·9957 362 |
| ·72 | ·9994 439 | ·9992 459 | ·9989 886 | ·9986 577 | ·9982 368 | ·9977 063 |
| ·73 | ·9997 324 | ·9996 325⁺ | ·9995 010 | ·9993 295⁺ | ·9991 083 | ·9988 257 |
| ·74 | ·9998 778 | ·9998 301 | ·9997 664 | ·9996 823 | ·9995 722 | ·9994 297 |
| ·75 | ·9999 472 | ·9999 257 | ·9998 966 | ·9998 576 | ·9998 060 | ·9997 382 |
| ·76 | ·9999 785⁺ | ·9999 694 | ·9999 569 | ·9999 399 | ·9999 172 | ·9998 868 |
| ·77 | ·9999 918 | ·9999 882 | ·9999 832 | ·9999 763 | ·9999 669 | ·9999 542 |
| ·78 | ·9999 971 | ·9999 958 | ·9999 939 | ·9999 912 | ·9999 876 | ·9999 827 |
| ·79 | ·9999 990 | ·9999 986 | ·9999 979 | ·9999 970 | ·9999 957 | ·9999 939 |
| ·80 | ·9999 997 | ·9999 996 | ·9999 994 | ·9999 991 | ·9999 986 | ·9999 980 |
| ·81 | ·9999 999 | ·9999 999 | ·9999 998 | ·9999 997 | ·9999 996 | ·9999 994 |
| ·82 | 1·0000 000 | 1·0000 000 | 1·0000 000 | ·9999 999 | ·9999 999 | ·9999 998 |
| ·83 | | | | 1·0000 000 | 1·0000 000 | 1·0000 000 |

$x$ = ·29 to ·85 $q$ = 34 $p$ = 46 to 50

| $x$ | $p$ = 46 | $p$ = 47 | $p$ = 48 | $p$ = 49 | $p$ = 50 |
|---|---|---|---|---|---|
| $B(p, q)$ = | $\cdot 1161\ 0738 \times \frac{1}{10^{23}}$ | $\cdot 6676\ 1742 \times \frac{1}{10^{24}}$ | $\cdot 3873\ 8294 \times \frac{1}{10^{24}}$ | $\cdot 2267\ 6075 \times \frac{1}{10^{24}}$ | $\cdot 1338\ 7080 \times \frac{1}{10^{24}}$ |
| ·29 | ·0000 001 | | | | |
| ·30 | ·0000 002 | ·0000 001 | | | |
| ·31 | ·0000 005+ | ·0000 003 | ·0000 001 | ·0000 001 | |
| ·32 | ·0000 014 | ·0000 008 | ·0000 004 | ·0000 002 | ·0000 001 |
| ·33 | ·0000 037 | ·0000 020 | ·0000 011 | ·0000 006 | ·0000 003 |
| ·34 | ·0000 090 | ·0000 051 | ·0000 029 | ·0000 016 | ·0000 009 |
| ·35 | ·0000 211 | ·0000 124 | ·0000 073 | ·0000 042 | ·0000 024 |
| ·36 | ·0000 474 | ·0000 287 | ·0000 172 | ·0000 103 | ·0000 061 |
| ·37 | ·0001 022 | ·0000 635+ | ·0000 392 | ·0000 240 | ·0000 146 |
| ·38 | ·0002 113 | ·0001 348 | ·0000 853 | ·0000 536 | ·0000 334 |
| ·39 | ·0004 204 | ·0002 750− | ·0001 784 | ·0001 149 | ·0000 735− |
| ·40 | ·0008 055− | ·0005 399 | ·0003 590 | ·0002 369 | ·0001 552 |
| ·41 | ·0014 888 | ·0010 217 | ·0006 957 | ·0004 702 | ·0003 154 |
| ·42 | ·0026 574 | ·0018 660 | ·0013 002 | ·0008 992 | ·0006 174 |
| ·43 | ·0045 862 | ·0032 930 | ·0023 465− | ·0016 596 | ·0011 654 |
| ·44 | ·0076 609 | ·0056 211 | ·0040 933 | ·0029 590 | ·0021 238 |
| ·45 | ·0123 983 | ·0092 902 | ·0069 096 | ·0051 018 | ·0037 404 |
| ·46 | ·0194 582 | ·0148 807 | ·0112 966 | ·0085 144 | ·0063 727 |
| ·47 | ·0296 403 | ·0231 200 | ·0179 039 | ·0137 668 | ·0105 128 |
| ·48 | ·0438 592 | ·0348 727 | ·0275 305− | ·0215 833 | ·0168 061 |
| ·49 | ·0630 935− | ·0511 043 | ·0411 049 | ·0328 366 | ·0260 566 |
| ·50 | ·0883 053 | ·0728 177 | ·0596 368 | ·0485 155− | ·0392 098 |
| ·51 | ·1203 363 | ·1009 597 | ·0841 392 | ·0696 631 | ·0573 081 |
| ·52 | ·1597 873 | ·1363 052 | ·1155 209 | ·0972 828 | ·0814 119 |
| ·53 | ·2068 975+ | ·1793 299 | ·1544 603 | ·1322 177 | ·1124 898 |
| ·54 | ·2614 431 | ·2300 914 | ·2012 745− | ·1750 150+ | ·1512 851 |
| ·55 | ·3226 752 | ·2881 376 | ·2558 046 | ·2257 955− | ·1981 749 |
| ·56 | ·3893 137 | ·3524 650+ | ·3173 411 | ·2841 490 | ·2530 433 |
| ·57 | ·4596 082 | ·4215 408 | ·3846 074 | ·3490 820 | ·3151 951 |
| ·58 | ·5314 624 | ·4933 956 | ·4558 148 | ·4190 326 | ·3833 314 |
| ·59 | ·6026 115− | ·5657 794 | ·5287 906 | ·4919 636 | ·4556 026 |
| ·60 | ·6708 277 | ·6363 626 | ·6011 642 | ·5655 258 | ·5297 419 |
| ·61 | ·7341 247 | ·7029 536 | ·6705 888 | ·6372 732 | ·6032 642 |
| ·62 | ·7909 303 | ·7637 006 | ·7349 646 | ·7048 996 | ·6737 056 |
| ·63 | ·8402 026 | ·8172 460 | ·7926 288 | ·7664 576 | ·7388 657 |
| ·64 | ·8814 753 | ·8628 138 | ·8424 847 | ·8205 292 | ·7970 145+ |
| ·65 | ·9148 306 | ·9002 186 | ·8840 521 | ·8663 195− | ·8470 307 |
| ·66 | ·9408 113 | ·9298 038 | ·9174 380 | ·9036 656 | ·8884 541 |
| ·67 | ·9602 915+ | ·9523 245− | ·9432 386 | ·9329 660 | ·9214 480 |
| ·68 | ·9743 330 | ·9688 009 | ·9623 977 | ·9550 501 | ·9466 889 |
| ·69 | ·9840 480 | ·9803 690 | ·9760 480 | ·9710 168 | ·9652 073 |
| ·70 | ·9904 890 | ·9881 500+ | ·9853 632 | ·9820 713 | ·9782 151 |
| ·71 | ·9945 730 | ·9931 546 | ·9914 404 | ·9893 867 | ·9869 465− |
| ·72 | ·9970 443 | ·9962 258 | ·9952 226 | ·9940 038 | ·9925 352 |
| ·73 | ·9984 681 | ·9980 198 | ·9974 627 | ·9967 764 | ·9959 381 |
| ·74 | ·9992 468 | ·9990 145− | ·9987 218 | ·9983 564 | ·9979 038 |
| ·75 | ·9996 500+ | ·9995 365− | ·9993 915+ | ·9992 081 | ·9989 778 |
| ·76 | ·9998 469 | ·9997 948 | ·9997 274 | ·9996 409 | ·9995 309 |
| ·77 | ·9999 373 | ·9999 149 | ·9998 856 | ·9998 475− | ·9997 984 |
| ·78 | ·9999 760 | ·9999 671 | ·9999 552 | ·9999 396 | ·9999 192 |
| ·79 | ·9999 915+ | ·9999 882 | ·9999 838 | ·9999 779 | ·9999 700 |
| ·80 | ·9999 972 | ·9999 961 | ·9999 946 | ·9999 925+ | ·9999 898 |
| ·81 | ·9999 992 | ·9999 988 | ·9999 984 | ·9999 977 | ·9999 968 |
| ·82 | ·9999 998 | ·9999 997 | ·9999 995+ | ·9999 994 | ·9999 991 |
| ·83 | ·9999 999 | ·9999 999 | ·9999 999 | ·9999 998 | ·9999 998 |
| ·84 | 1·0000 000 | 1·0000 000 | 1·0000 000 | 1·0000 000 | ·9999 999 |
| ·85 | | | | | 1·0000 000 |

$x$ = ·22 to ·81 $q = 35$ $p$ = 35 to 40

| | $p = 35$ | $p = 36$ | $p = 37$ | $p = 38$ | $p = 39$ | $p = 40$ |
|---|---|---|---|---|---|---|
| $B(p, q)$ = | ·5093 5692 $\times \frac{1}{10^{21}}$ | ·2546 7846 $\times \frac{1}{10^{21}}$ | ·1291 3274 $\times \frac{1}{10^{21}}$ | ·6635 9881 $\times \frac{1}{10^{22}}$ | ·3454 3500 $\times \frac{1}{10^{22}}$ | ·1820 5358 $\times \frac{1}{10^{22}}$ |
| $x$ | | | | | | |
| ·22 | ·0000 002 | ·0000 001 | | | | |
| ·23 | ·0000 005$^-$ | ·0000 002 | ·0000 001 | | | |
| ·24 | ·0000 014 | ·0000 007 | ·0000 003 | ·0000 001 | ·0000 001 | |
| ·25 | ·0000 039 | ·0000 019 | ·0000 009 | ·0000 004 | ·0000 002 | ·0000 001 |
| ·26 | ·0000 099 | ·0000 050$^-$ | ·0000 024 | ·0000 012 | ·0000 006 | ·0000 003 |
| ·27 | ·0000 240 | ·0000 124 | ·0000 064 | ·0000 032 | ·0000 016 | ·0000 008 |
| ·28 | ·0000 550$^-$ | ·0000 295$^+$ | ·0000 157 | ·0000 082 | ·0000 042 | ·0000 022 |
| ·29 | ·0001 199 | ·0000 666 | ·0000 365$^+$ | ·0000 198 | ·0000 106 | ·0000 056 |
| ·30 | ·0002 495$^+$ | ·0001 432 | ·0000 812 | ·0000 455$^+$ | ·0000 252 | ·0000 138 |
| ·31 | ·0004 967 | ·0002 942 | ·0001 722 | ·0000 996 | ·0000 570 | ·0000 322 |
| ·32 | ·0009 479 | ·0005 789 | ·0003 494 | ·0002 084 | ·0001 229 | ·0000 717 |
| ·33 | ·0017 384 | ·0010 935$^-$ | ·0006 797 | ·0004 176 | ·0002 538 | ·0001 525$^+$ |
| ·34 | ·0030 699 | ·0019 867 | ·0012 706 | ·0008 033 | ·0005 024 | ·0003 108 |
| ·35 | ·0052 291 | ·0034 782 | ·0022 866 | ·0014 863 | ·0009 556 | ·0006 079 |
| ·36 | ·0086 057 | ·0058 780 | ·0039 686 | ·0026 496 | ·0017 499 | ·0011 436 |
| ·37 | ·0137 044 | ·0096 034 | ·0066 530 | ·0045 582 | ·0030 896 | ·0020 725$^+$ |
| ·38 | ·0211 473 | ·0151 901 | ·0107 884 | ·0075 787 | ·0052 677 | ·0036 239 |
| ·39 | ·0316 618 | ·0232 920 | ·0169 449 | ·0121 949 | ·0086 849 | ·0061 226 |
| ·40 | ·0460 502 | ·0346 660 | ·0258 116 | ·0190 153 | ·0138 644 | ·0100 078 |
| ·41 | ·0651 389 | ·0501 363 | ·0381 760 | ·0287 661 | ·0214 561 | ·0158 461 |
| ·42 | ·0897 086 | ·0705 389 | ·0548 837 | ·0422 664 | ·0322 257 | ·0243 322 |
| ·43 | ·1204 108 | ·0966 464 | ·0767 768 | ·0603 817 | ·0470 240 | ·0362 727 |
| ·44 | ·1576 780 | ·1290 800 | ·1046 129 | ·0839 548 | ·0667 324 | ·0525 482 |
| ·45 | ·2016 406 | ·1682 179 | ·1389 730 | ·1137 197 | ·0921 878 | ·0740 514 |
| ·46 | ·2520 623 | ·2141 120 | ·1801 675$^+$ | ·1502 047 | ·1240 897 | ·1016 040 |
| ·47 | ·3083 059 | ·2664 275$^-$ | ·2281 553 | ·1936 379 | ·1628 992 | ·1358 571 |
| ·48 | ·3693 390 | ·3244 158 | ·2824 876 | ·2438 682 | ·2087 450$^-$ | ·1771 881 |
| ·49 | ·4337 825$^+$ | ·3869 305$^-$ | ·3422 909 | ·3003 176 | ·2613 487 | ·2256 072 |
| ·50 | ·5000 000$^e$ | ·4524 873 | ·4062 943 | ·3619 741 | ·3199 865$^-$ | ·2806 904 |
| ·51 | ·5662 175$^-$ | ·5193 654 | ·4729 038 | ·4274 342 | ·3834 962 | ·3415 523 |
| ·52 | ·6306 610 | ·5857 379 | ·5403 156 | ·4949 915$^+$ | ·4503 354 | ·4068 701 |
| ·53 | ·6916 941 | ·6498 157 | ·6066 577 | ·5627 648 | ·5186 871 | ·4749 597 |
| ·54 | ·7479 377 | ·7099 874 | ·6701 395$^+$ | ·6288 485$^+$ | ·5866 014 | ·5438 992 |
| ·55 | ·7983 594 | ·7649 367 | ·7291 929 | ·6914 688 | ·6521 562 | ·6116 845$^-$ |
| ·56 | ·8423 220 | ·8137 241 | ·7825 842 | ·7491 214 | ·7136 155$^+$ | ·6763 982 |
| ·57 | ·8795 892 | ·8558 249 | ·8294 861 | ·8006 771 | ·7695 635$^-$ | ·7363 676 |
| ·58 | ·9102 914 | ·8911 217 | ·8695 026 | ·8454 411 | ·8189 987 | ·7902 919 |
| ·59 | ·9348 611 | ·9198 586 | ·9026 473 | ·8831 614 | ·8613 781 | ·8373 216 |
| ·60 | ·9539 498 | ·9425 655$^+$ | ·9292 840 | ·9139 922 | ·8966 078 | ·8770 839 |
| ·61 | ·9683 382 | ·9599 683 | ·9500 408 | ·9384 203 | ·9249 894 | ·9096 540 |
| ·62 | ·9788 527 | ·9728 954 | ·9657 137 | ·9571 693 | ·9471 319 | ·9354 834 |
| ·63 | ·9862 956 | ·9821 946 | ·9771 708 | ·9710 976 | ·9638 480 | ·9552 991 |
| ·64 | ·9913 943 | ·9886 666 | ·9852 721 | ·9811 033 | ·9760 481 | ·9699 923 |
| ·65 | ·9947 709 | ·9930 200 | ·9908 071 | ·9880 469 | ·9846 475$^+$ | ·9805 116 |
| ·66 | ·9969 301 | ·9958 469 | ·9944 568 | ·9926 962 | ·9904 946 | ·9877 748 |
| ·67 | ·9982 616 | ·9976 166 | ·9967 764 | ·9956 962 | ·9943 249 | ·9926 051 |
| ·68 | ·9990 521 | ·9986 832 | ·9981 953 | ·9975 588 | ·9967 386 | ·9956 947 |
| ·69 | ·9995 033 | ·9993 009 | ·9990 293 | ·9986 697 | ·9981 995$^+$ | ·9975 923 |
| ·70 | ·9997 505$^-$ | ·9996 442 | ·9994 995$^-$ | ·9993 051 | ·9990 473 | ·9987 095$^-$ |
| ·71 | ·9998 801 | ·9998 267 | ·9997 531 | ·9996 529 | ·9995 179 | ·9993 386 |
| ·72 | ·9999 450$^+$ | ·9999 195$^+$ | ·9998 839 | ·9998 346 | ·9997 674 | ·9996 768 |
| ·73 | ·9999 760 | ·9999 645$^-$ | ·9999 480 | ·9999 251 | ·9998 933 | ·9998 498 |
| ·74 | ·9999 901 | ·9999 851 | ·9999 780 | ·9999 678 | ·9999 536 | ·9999 339 |
| ·75 | ·9999 961 | ·9999 941 | ·9999 912 | ·9999 869 | ·9999 809 | ·9999 725$^+$ |
| ·76 | ·9999 986 | ·9999 978 | ·9999 967 | ·9999 950$^+$ | ·9999 926 | ·9999 893 |
| ·77 | ·9999 995$^+$ | ·9999 992 | ·9999 988 | ·9999 982 | ·9999 973 | ·9999 961 |
| ·78 | ·9999 998 | ·9999 998 | ·9999 996 | ·9999 994 | ·9999 991 | ·9999 987 |
| ·79 | 1·0000 000 | ·9999 999 | ·9999 999 | ·9999 998 | ·9999 997 | ·9999 996 |
| ·80 | | 1·0000 000 | 1·0000 000 | ·9999 999 | ·9999 999 | ·9999 999 |
| ·81 | | | | 1·0000 000 | 1·0000 000 | 1·0000 000 |

$x$ = ·26 to ·83 $q = 35$ $p$ = 41 to 45

| | $p = 41$ | $p = 42$ | $p = 43$ | $p = 44$ | $p = 45$ |
|---|---|---|---|---|---|
| $B(p, q)$ = | ·9709 5242 $\times \frac{1}{10^{23}}$ | ·5238 0328 $\times \frac{1}{10^{23}}$ | ·2857 1088 $\times \frac{1}{10^{23}}$ | ·1575 0728 $\times \frac{1}{10^{23}}$ | ·8772 5573 $\times \frac{1}{10^{24}}$ |
| $x$ | | | | | |
| ·26 | ·0000 001 | ·0000 001 | | | |
| ·27 | ·0000 004 | ·0000 002 | ·0000 001 | | |
| ·28 | ·0000 011 | ·0000 006 | ·0000 003 | ·0000 001 | ·0000 001 |
| ·29 | ·0000 030 | ·0000 015+ | ·0000 008 | ·0000 004 | ·0000 002 |
| ·30 | ·0000 075− | ·0000 040 | ·0000 021 | ·0000 011 | ·0000 006 |
| ·31 | ·0000 180 | ·0000 100 | ·0000 055− | ·0000 030 | ·0000 016 |
| ·32 | ·0000 414 | ·0000 237 | ·0000 134 | ·0000 075+ | ·0000 042 |
| ·33 | ·0000 908 | ·0000 535− | ·0000 312 | ·0000 180 | ·0000 103 |
| ·34 | ·0001 903 | ·0001 154 | ·0000 693 | ·0000 412 | ·0000 243 |
| ·35 | ·0003 828 | ·0002 387 | ·0001 474 | ·0000 902 | ·0000 547 |
| ·36 | ·0007 399 | ·0004 740 | ·0003 007 | ·0001 891 | ·0001 178 |
| ·37 | ·0013 763 | ·0009 051 | ·0005 897 | ·0003 806 | ·0002 435+ |
| ·38 | ·0024 684 | ·0016 651 | ·0011 128 | ·0007 369 | ·0004 837 |
| ·39 | ·0042 739 | ·0029 550− | ·0020 242 | ·0013 742 | ·0009 248 |
| ·40 | ·0071 539 | ·0050 657 | ·0035 543 | ·0024 716 | ·0017 040 |
| ·41 | ·0115 910 | ·0083 996 | ·0060 319 | ·0042 936 | ·0030 301 |
| ·42 | ·0181 989 | ·0134 868 | ·0099 056 | ·0072 121 | ·0052 068 |
| ·43 | ·0277 201 | ·0209 927 | ·0157 582 | ·0117 276 | ·0086 552 |
| ·44 | ·0410 023 | ·0317 092 | ·0243 102 | ·0184 804 | ·0139 332 |
| ·45 | ·0589 529 | ·0465 242 | ·0364 037 | ·0282 484 | ·0217 428 |
| ·46 | ·0824 687 | ·0663 670 | ·0529 643 | ·0419 242 | ·0329 215− |
| ·47 | ·1123 440 | ·0921 284 | ·0749 355+ | ·0604 656 | ·0484 094 |
| ·48 | ·1491 655+ | ·1245 604 | ·1031 891 | ·0848 197 | ·0691 890 |
| ·49 | ·1932 076 | ·1641 664 | ·1384 166 | ·1158 227 | ·0961 967 |
| ·50 | ·2443 415− | ·2110 956 | ·1810 159 | ·1540 841 | ·1302 128 |
| ·51 | ·3019 782 | ·2650 585+ | ·2309 869 | ·1998 708 | ·1717 390 |
| ·52 | ·3650 565− | ·3252 825+ | ·2878 571 | ·2530 080 | ·2208 835− |
| ·53 | ·4320 851 | ·3905 176 | ·3506 524 | ·3128 175+ | ·2772 699 |
| ·54 | ·5012 398 | ·4591 005+ | ·4179 245− | ·3781 082 | ·3399 932 |
| ·55 | ·5705 045− | ·5290 734 | ·4878 396 | ·4472 291 | ·4076 338 |
| ·56 | ·6378 409 | ·5983 433 | ·5583 191 | ·5181 831 | ·4783 391 |
| ·57 | ·7013 625+ | ·6648 633 | ·6272 170 | ·5887 915+ | ·5499 643 |
| ·58 | ·7594 894 | ·7268 087 | ·6925 096 | ·6568 863 | ·6202 591 |
| ·59 | ·8110 639 | ·7827 249 | ·7524 695+ | ·7205 044 | ·6870 718 |
| ·60 | ·8554 123 | ·8316 265− | ·8058 018 | ·7780 553 | ·7485 431 |
| ·61 | ·8923 480 | ·8730 371 | ·8517 216 | ·8284 380 | ·8032 601 |
| ·62 | ·9221 225+ | ·9069 693 | ·8899 689 | ·8710 945+ | ·8503 498 |
| ·63 | ·9453 353 | ·9338 527 | ·9207 625+ | ·9059 950− | ·8895 023 |
| ·64 | ·9628 221 | ·9544 278 | ·9447 064 | ·9335 653 | ·9209 251 |
| ·65 | ·9755 382 | ·9696 246 | ·9626 692 | ·9545 734 | ·9452 448 |
| ·66 | ·9844 538 | ·9804 444 | ·9756 560 | ·9699 968 | ·9633 755+ |
| ·67 | ·9904 734 | ·9878 608 | ·9846 933 | ·9808 931 | ·9763 795− |
| ·68 | ·9943 814 | ·9927 479 | ·9907 378 | ·9882 903 | ·9853 398 |
| ·69 | ·9968 172 | ·9958 389 | ·9946 173 | ·9931 080 | ·9912 619 |
| ·70 | ·9982 721 | ·9977 119 | ·9970 025− | ·9961 131 | ·9950 095+ |
| ·71 | ·9991 031 | ·9987 973 | ·9984 043 | ·9979 047 | ·9972 758 |
| ·72 | ·9995 561 | ·9993 972 | ·9991 902 | ·9989 232 | ·9985 825− |
| ·73 | ·9997 912 | ·9997 128 | ·9996 093 | ·9994 741 | ·9992 990 |
| ·74 | ·9999 069 | ·9998 704 | ·9998 215− | ·9997 567 | ·9996 716 |
| ·75 | ·9999 608 | ·9999 448 | ·9999 230 | ·9998 937 | ·9998 548 |
| ·76 | ·9999 845− | ·9999 779 | ·9999 688 | ·9999 564 | ·9999 397 |
| ·77 | ·9999 943 | ·9999 917 | ·9999 882 | ·9999 833 | ·9999 766 |
| ·78 | ·9999 980 | ·9999 971 | ·9999 958 | ·9999 940 | ·9999 915+ |
| ·79 | ·9999 994 | ·9999 991 | ·9999 986 | ·9999 980 | ·9999 972 |
| ·80 | ·9999 998 | ·9999 997 | ·9999 996 | ·9999 994 | ·9999 991 |
| ·81 | 1·0000 000 | ·9999 999 | ·9999 999 | ·9999 998 | ·9999 998 |
| ·82 | | 1·0000 000 | 1·0000 000 | 1·0000 000 | ·9999 999 |
| ·83 | | | | | 1·0000 000 |

$x = \cdot 29$ to $\cdot 84$ $q = 35$ $p = 46$ to $50$

| | $p = 46$ | $p = 47$ | $p = 48$ | $p = 49$ | $p = 50$ |
|---|---|---|---|---|---|
| $B(p, q) =$ | $\cdot 4934\ 5635^{+} \times \frac{1}{10^{24}}$ | $\cdot 2802\ 3447 \times \frac{1}{10^{24}}$ | $\cdot 1606\ 2220 \times \frac{1}{10^{24}}$ | $\cdot 9288\ 9945^{+} \times \frac{1}{10^{25}}$ | $\cdot 5418\ 5801 \times \frac{1}{10^{25}}$ |
| $x$ | | | | | |
| ·29 | ·0000 001 | ·0000 001 | | | |
| ·30 | ·0000 003 | ·0000 002 | ·0000 001 | | |
| ·31 | ·0000 009 | ·0000 005⁻ | ·0000 002 | ·0000 001 | ·0000 001 |
| ·32 | ·0000 023 | ·0000 013 | ·0000 007 | ·0000 004 | ·0000 002 |
| ·33 | ·0000 059 | ·0000 033 | ·0000 018 | ·0000 010 | ·0000 006 |
| ·34 | ·0000 142 | ·0000 082 | ·0000 047 | ·0000 027 | ·0000 015⁺ |
| ·35 | ·0000 329 | ·0000 196 | ·0000 116 | ·0000 068 | ·0000 040 |
| ·36 | ·0000 728 | ·0000 446 | ·0000 271 | ·0000 163 | ·0000 098 |
| ·37 | ·0001 545⁻ | ·0000 972 | ·0000 606 | ·0000 375⁺ | ·0000 231 |
| ·38 | ·0003 148 | ·0002 032 | ·0001 301 | ·0000 827 | ·0000 521 |
| ·39 | ·0006 171 | ·0004 084 | ·0002 681 | ·0001 747 | ·0001 129 |
| ·40 | ·0011 649 | ·0007 899 | ·0005 313 | ·0003 547 | ·0002 350⁺ |
| ·41 | ·0021 207 | ·0014 723 | ·0010 141 | ·0006 932 | ·0004 703 |
| ·42 | ·0037 281 | ·0026 481 | ·0018 663 | ·0013 054 | ·0009 064 |
| ·43 | ·0063 359 | ·0046 014 | ·0033 161 | ·0023 719 | ·0016 842 |
| ·44 | ·0104 207 | ·0077 329 | ·0056 947 | ·0041 627 | ·0030 209 |
| ·45 | ·0166 033 | ·0125 811 | ·0094 618 | ·0070 638 | ·0052 359 |
| ·46 | ·0256 513 | ·0198 352 | ·0152 243 | ·0116 010 | ·0087 777 |
| ·47 | ·0384 618 | ·0303 307 | ·0237 445⁻ | ·0184 563 | ·0142 462 |
| ·48 | ·0560 175⁺ | ·0450 222 | ·0359 265⁺ | ·0284 681 | ·0224 039 |
| ·49 | ·0793 141 | ·0649 271 | ·0527 778 | ·0426 078 | ·0341 667 |
| ·50 | ·1092 590 | ·0910 384 | ·0753 376 | ·0619 265⁺ | ·0505 682 |
| ·51 | ·1465 517 | ·1242 116 | ·1045 762 | ·0874 687 | ·0726 900 |
| ·52 | ·1915 573 | ·1650 363 | ·1412 689 | ·1201 556 | ·1015 586 |
| ·53 | ·2441 949 | ·2137 083 | ·1858 618 | ·1606 490 | ·1380 142 |
| ·54 | ·3038 602 | ·2699 265⁺ | ·2383 466 | ·2092 141 | ·1825 667 |
| ·55 | ·3694 024 | ·3328 333 | ·2981 704 | ·2656 017 | ·2352 596 |
| ·56 | ·4391 679 | ·4010 186 | ·3642 005⁺ | ·3289 779 | ·2955 666 |
| ·57 | ·5111 112 | ·4725 959 | ·4347 608 | ·3979 189 | ·3623 477 |
| ·58 | ·5829 645⁻ | ·5453 456 | ·5077 427 | ·4704 845⁻ | ·4338 802 |
| ·59 | ·6524 431 | ·6169 110 | ·5807 816 | ·5443 662 | ·5079 732 |
| ·60 | ·7174 569 | ·6850 192 | ·6514 772 | ·6170 966 | ·5821 547 |
| ·61 | ·7762 973 | ·7476 932 | ·7176 225⁺ | ·6862 863 | ·6539 077 |
| ·62 | ·8277 704 | ·8034 239 | ·7774 094 | ·7498 556 | ·7209 186 |
| ·63 | ·8712 614 | ·8512 757 | ·8295 764 | ·8062 224 | ·7813 004 |
| ·64 | ·9067 231 | ·8909 158 | ·8734 806 | ·8544 181 | ·8337 528 |
| ·65 | ·9345 998 | ·9225 664 | ·9090 864 | ·8941 180 | ·8776 374 |
| ·66 | ·9557 037 | ·9468 977 | ·9368 814 | ·9255 880 | ·9129 625⁻ |
| ·67 | ·9710 705⁻ | ·9648 843 | ·9577 412 | ·9495 654 | ·9402 867 |
| ·68 | ·9818 176 | ·9776 523 | ·9727 708 | ·9671 002 | ·9605 686 |
| ·69 | ·9890 256 | ·9863 420 | ·9831 509 | ·9793 893 | ·9749 929 |
| ·70 | ·9936 534 | ·9920 024 | ·9900 106 | ·9876 288 | ·9848 047 |
| ·71 | ·9964 920 | ·9955 241 | ·9943 399 | ·9929 034 | ·9911 759 |
| ·72 | ·9981 518 | ·9976 125⁺ | ·9969 433 | ·9961 203 | ·9951 164 |
| ·73 | ·9990 747 | ·9987 898 | ·9984 315⁻ | ·9979 846 | ·9974 321 |
| ·74 | ·9995 612 | ·9994 190 | ·9992 378 | ·9990 086 | ·9987 214 |
| ·75 | ·9998 036 | ·9997 368 | ·9996 505⁺ | ·9995 399 | ·9993 994 |
| ·76 | ·9999 174 | ·9998 880 | ·9998 494 | ·9997 994 | ·9997 350⁻ |
| ·77 | ·9999 675⁺ | ·9999 554 | ·9999 394 | ·9999 182 | ·9998 907 |
| ·78 | ·9999 881 | ·9999 835⁻ | ·9999 773 | ·9999 690 | ·9999 580 |
| ·79 | ·9999 960 | ·9999 943 | ·9999 921 | ·9999 891 | ·9999 851 |
| ·80 | ·9999 987 | ·9999 982 | ·9999 975⁻ | ·9999 965⁺ | ·9999 952 |
| ·81 | ·9999 996 | ·9999 995⁻ | ·9999 993 | ·9999 990 | ·9999 986 |
| ·82 | ·9999 999 | ·9999 999 | ·9999 998 | ·9999 997 | ·9999 996 |
| ·83 | 1·0000 000 | 1·0000 000 | 1·0000 000 | ·9999 999 | ·9999 999 |
| ·84 | | | | 1·0000 000 | 1·0000 000 |

$x$ = ·22 to ·81 $q = 36$ $p$ = 36 to 40

| $x$ | $p = 36$ | $p = 37$ | $p = 38$ | $p = 39$ | $p = 40$ |
|---|---|---|---|---|---|
| $B(p, q) =$ | ·1255 4572 $\times \frac{1}{10^{21}}$ | ·6277 2860 $\times \frac{1}{10^{22}}$ | ·3181 6381 $\times \frac{1}{10^{22}}$ | ·1633 8142 $\times \frac{1}{10^{22}}$ | ·8495 8337 $\times \frac{1}{10^{23}}$ |
| ·22 | ·0000 001 | | | | |
| ·23 | ·0000 003 | ·0000 002 | ·0000 001 | | |
| ·24 | ·0000 010 | ·0000 005− | ·0000 002 | ·0000 001 | |
| ·25 | ·0000 029 | ·0000 014 | ·0000 007 | ·0000 003 | ·0000 001 |
| ·26 | ·0000 075+ | ·0000 038 | ·0000 019 | ·0000 009 | ·0000 004 |
| ·27 | ·0000 187 | ·0000 097 | ·0000 050− | ·0000 025+ | ·0000 013 |
| ·28 | ·0000 438 | ·0000 235+ | ·0000 125− | ·0000 066 | ·0000 034 |
| ·29 | ·0000 975+ | ·0000 542 | ·0000 298 | ·0000 162 | ·0000 087 |
| ·30 | ·0002 070 | ·0001 189 | ·0000 675+ | ·0000 379 | ·0000 210 |
| ·31 | ·0004 197 | ·0002 489 | ·0001 459 | ·0000 845+ | ·0000 485− |
| ·32 | ·0008 151 | ·0004 984 | ·0003 012 | ·0001 800 | ·0001 064 |
| ·33 | ·0015 192 | ·0009 567 | ·0005 955+ | ·0003 665+ | ·0002 232 |
| ·34 | ·0027 233 | ·0017 645− | ·0011 301 | ·0007 158 | ·0004 485+ |
| ·35 | ·0047 038 | ·0031 327 | ·0020 626 | ·0013 431 | ·0008 652 |
| ·36 | ·0078 420 | ·0053 630 | ·0036 265− | ·0024 255− | ·0016 051 |
| ·37 | ·0126 382 | ·0088 675+ | ·0061 526 | ·0042 229 | ·0028 682 |
| ·38 | ·0197 176 | ·0141 815− | ·0100 877 | ·0070 993 | ·0049 446 |
| ·39 | ·0298 205− | ·0219 664 | ·0160 058 | ·0115 400 | ·0082 354 |
| ·40 | ·0437 734 | ·0329 963 | ·0246 077 | ·0181 617 | ·0132 694 |
| ·41 | ·0624 384 | ·0481 236 | ·0367 027 | ·0277 072 | ·0207 091 |
| ·42 | ·0866 415− | ·0682 220 | ·0531 678 | ·0410 214 | ·0313 417 |
| ·43 | ·1170 838 | ·0941 088 | ·0748 843 | ·0590 039 | ·0460 471 |
| ·44 | ·1542 462 | ·1264 516 | ·1026 534 | ·0825 376 | ·0657 435+ |
| ·45 | ·1982 983 | ·1656 694 | ·1370 970 | ·1123 970 | ·0913 069 |
| ·46 | ·2490 263 | ·2118 426 | ·1785 581 | ·1491 452 | ·1234 729 |
| ·47 | ·3057 932 | ·2646 451 | ·2270 113 | ·1930 319 | ·1627 292 |
| ·48 | ·3675 420 | ·3233 137 | ·2820 021 | ·2439 084 | ·2092 138 |
| ·49 | ·4328 455− | ·3866 626 | ·3426 266 | ·3011 749 | ·2626 354 |
| ·50 | ·5000 000$^{e}$ | ·4531 472 | ·4075 606 | ·3637 735+ | ·3222 320 |
| ·51 | ·5671 545+ | ·5209 717 | ·4751 383 | ·4302 337 | ·3867 798 |
| ·52 | ·6324 580 | ·5882 296 | ·5434 753 | ·4987 682 | ·4546 571 |
| ·53 | ·6942 068 | ·6530 587 | ·6106 206 | ·5674 118 | ·5239 593 |
| ·54 | ·7509 737 | ·7137 900 | ·6747 169 | ·6341 838 | ·5926 529 |
| ·55 | ·8017 017 | ·7690 727 | ·7341 509 | ·6972 533 | ·6587 474 |
| ·56 | ·8457 538 | ·8179 592 | ·7876 705+ | ·7550 863 | ·7204 635+ |
| ·57 | ·8829 162 | ·8599 413 | ·8344 577 | ·8065 532 | ·7763 734 |
| ·58 | ·9133 585+ | ·8949 390 | ·8741 499 | ·8509 864 | ·8254 947 |
| ·59 | ·9375 616 | ·9232 467 | ·9068 117 | ·8881 839 | ·8673 304 |
| ·60 | ·9562 266 | ·9454 495+ | ·9328 666 | ·9183 631 | ·9018 514 |
| ·61 | ·9701 795+ | ·9623 254 | ·9530 024 | ·9420 773 | ·9294 322 |
| ·62 | ·9802 824 | ·9747 463 | ·9680 670 | ·9601 117 | ·9507 529 |
| ·63 | ·9873 618 | ·9835 912 | ·9789 685+ | ·9733 739 | ·9666 862 |
| ·64 | ·9921 580 | ·9896 791 | ·9865 918 | ·9827 961 | ·9781 867 |
| ·65 | ·9952 962 | ·9937 250− | ·9917 377 | ·9892 561 | ·9861 955+ |
| ·66 | ·9972 767 | ·9963 180 | ·9950 866 | ·9935 253 | ·9915 701 |
| ·67 | ·9984 808 | ·9979 184 | ·9971 851 | ·9962 412 | ·9950 413 |
| ·68 | ·9991 849 | ·9988 683 | ·9984 492 | ·9979 018 | ·9971 955+ |
| ·69 | ·9995 803 | ·9994 095− | ·9991 801 | ·9988 762 | ·9984 782 |
| ·70 | ·9997 930 | ·9997 049 | ·9995 850− | ·9994 237 | ·9992 094 |
| ·71 | ·9999 025− | ·9998 592 | ·9997 993 | ·9997 177 | ·9996 078 |
| ·72 | ·9999 562 | ·9999 360 | ·9999 076 | ·9998 683 | ·9998 147 |
| ·73 | ·9999 813 | ·9999 723 | ·9999 596 | ·9999 417 | ·9999 169 |
| ·74 | ·9999 925− | ·9999 887 | ·9999 833 | ·9999 755+ | ·9999 647 |
| ·75 | ·9999 971 | ·9999 956 | ·9999 935− | ·9999 903 | ·9999 859 |
| ·76 | ·9999 990 | ·9999 984 | ·9999 976 | ·9999 964 | ·9999 947 |
| ·77 | ·9999 997 | ·9999 995− | ·9999 992 | ·9999 988 | ·9999 981 |
| ·78 | ·9999 999 | ·9999 998 | ·9999 997 | ·9999 996 | ·9999 994 |
| ·79 | 1·0000 000 | 1·0000 000 | ·9999 999 | ·9999 999 | ·9999 998 |
| ·80 | | | 1·0000 000 | 1·0000 000 | ·9999 999 |
| ·81 | | | | | 1·0000 000 |

$x$ = ·25 to ·82 $q = 36$ $p$ = 41 to 45

| $x$ | $p = 41$ | $p = 42$ | $p = 43$ | $p = 44$ | $p = 45$ |
|---|---|---|---|---|---|
| $B(p, q)$ = | ·4471 4914 $\times\frac{1}{10^{23}}$ | ·2380 9240 $\times\frac{1}{10^{23}}$ | ·1282 0360 $\times\frac{1}{10^{23}}$ | ·6978 1706 $\times\frac{1}{10^{24}}$ | ·3837 9938 $\times\frac{1}{10^{24}}$ |
| ·25 | ·0000 001 | | | | |
| ·26 | ·0000 002 | ·0000 001 | | | |
| ·27 | ·0000 006 | ·0000 003 | ·0000 001 | ·0000 001 | |
| ·28 | ·0000 017 | ·0000 009 | ·0000 004 | ·0000 002 | ·0000 001 |
| ·29 | ·0000 046 | ·0000 024 | ·0000 013 | ·0000 007 | ·0000 003 |
| ·30 | ·0000 116 | ·0000 063 | ·0000 034 | ·0000 018 | ·0000 010 |
| ·31 | ·0000 275⁻ | ·0000 154 | ·0000 086 | ·0000 047 | ·0000 026 |
| ·32 | ·0000 622 | ·0000 360 | ·0000 206 | ·0000 117 | ·0000 066 |
| ·33 | ·0001 344 | ·0000 802 | ·0000 473 | ·0000 277 | ·0000 161 |
| ·34 | ·0002 781 | ·0001 707 | ·0001 038 | ·0000 625⁻ | ·0000 373 |
| ·35 | ·0005 517 | ·0003 482 | ·0002 177 | ·0001 348 | ·0000 827 |
| ·36 | ·0010 513 | ·0006 818 | ·0004 379 | ·0002 787 | ·0001 757 |
| ·37 | ·0019 283 | ·0012 837 | ·0008 465⁻ | ·0005 530 | ·0003 580 |
| ·38 | ·0034 093 | ·0023 279 | ·0015 745⁺ | ·0010 552 | ·0007 009 |
| ·39 | ·0058 189 | ·0040 719 | ·0028 228 | ·0019 391 | ·0013 204 |
| ·40 | ·0096 001 | ·0068 795⁻ | ·0048 844 | ·0034 367 | ·0023 971 |
| ·41 | ·0153 294 | ·0112 408 | ·0081 676 | ·0058 819 | ·0041 994 |
| ·42 | ·0237 189 | ·0177 843 | ·0132 146 | ·0097 332 | ·0071 079 |
| ·43 | ·0356 007 | ·0272 742 | ·0207 101 | ·0155 901 | ·0116 372 |
| ·44 | ·0518 884 | ·0405 881 | ·0314 724 | ·0241 969 | ·0184 492 |
| ·45 | ·0735 122 | ·0586 688 | ·0464 230 | ·0364 270 | ·0283 507 |
| ·46 | ·1013 306 | ·0824 503 | ·0665 279 | ·0532 419 | ·0422 689 |
| ·47 | ·1360 250⁺ | ·1127 598 | ·0927 129 | ·0756 218 | ·0611 993 |
| ·48 | ·1779 887 | ·1502 060 | ·1257 572 | ·1044 697 | ·0861 237 |
| ·49 | ·2272 272 | ·1950 662 | ·1661 749 | ·1404 953 | ·1179 030 |
| ·50 | ·2832 867 | ·2471 911 | ·2141 035⁺ | ·1840 938 | ·1571 533 |
| ·51 | ·3452 270 | ·3059 445⁻ | ·2692 153 | ·2352 365⁻ | ·2041 227 |
| ·52 | ·4116 488 | ·3701 930 | ·3306 718 | ·2933 932 | ·2585 885⁺ |
| ·53 | ·4807 784 | ·4383 558 | ·3971 352 | ·3575 059 | ·3197 950⁻ |
| ·54 | ·5506 028 | ·5085 118 | ·4668 416 | ·4260 243 | ·3864 500⁻ |
| ·55 | ·6190 381 | ·5785 540 | ·5377 325⁻ | ·4970 060 | ·4567 885⁺ |
| ·56 | ·6841 096 | ·6463 724 | ·6076 290 | ·5682 728 | ·5287 019 |
| ·57 | ·7441 187 | ·7100 389 | ·6744 255⁻ | ·6376 029 | ·5999 183 |
| ·58 | ·7977 725⁻ | ·7679 677 | ·7362 753 | ·7029 319 | ·6682 093 |
| ·59 | ·8442 611 | ·8190 313 | ·7917 410 | ·7625 340 | ·7315 945⁻ |
| ·60 | ·8832 758 | ·8626 160 | ·8398 903 | ·8151 563 | ·7885 110 |
| ·61 | ·9149 694 | ·8986 158 | ·8803 271 | ·8600 903 | ·8379 265⁺ |
| ·62 | ·9398 734 | ·9273 698 | ·9131 575⁻ | ·8971 736 | ·8793 806 |
| ·63 | ·9587 864 | ·9495 609 | ·9389 055⁺ | ·9267 286 | ·9129 549 |
| ·64 | ·9726 555⁻ | ·9660 935⁺ | ·9583 942 | ·9494 558 | ·9391 847 |
| ·65 | ·9824 655⁻ | ·9779 712 | ·9726 155⁻ | ·9663 007 | ·9589 311 |
| ·66 | ·9891 506 | ·9861 905⁻ | ·9826 088 | ·9783 207 | ·9732 394 |
| ·67 | ·9935 339 | ·9916 618 | ·9893 622 | ·9865 674 | ·9832 054 |
| ·68 | ·9962 950⁺ | ·9951 599 | ·9937 449 | ·9919 994 | ·9898 683 |
| ·69 | ·9979 633 | ·9973 047 | ·9964 716 | ·9954 289 | ·9941 371 |
| ·70 | ·9989 282 | ·9985 633 | ·9980 951 | ·9975 005⁻ | ·9967 532 |
| ·71 | ·9994 614 | ·9992 688 | ·9990 181 | ·9986 952 | ·9982 836 |
| ·72 | ·9997 423 | ·9996 457 | ·9995 181 | ·9993 516 | ·9991 362 |
| ·73 | ·9998 829 | ·9998 370 | ·9997 755⁺ | ·9996 941 | ·9995 874 |
| ·74 | ·9999 497 | ·9999 291 | ·9999 011 | ·9998 635⁺ | ·9998 136 |
| ·75 | ·9999 796 | ·9999 709 | ·9999 589 | ·9999 426 | ·9999 207 |
| ·76 | ·9999 922 | ·9999 888 | ·9999 840 | ·9999 774 | ·9999 683 |
| ·77 | ·9999 972 | ·9999 960 | ·9999 942 | ·9999 917 | ·9999 882 |
| ·78 | ·9999 991 | ·9999 987 | ·9999 980 | ·9999 971 | ·9999 959 |
| ·79 | ·9999 997 | ·9999 996 | ·9999 994 | ·9999 991 | ·9999 987 |
| ·80 | ·9999 999 | ·9999 999 | ·9999 998 | ·9999 997 | ·9999 996 |
| ·81 | 1·0000 000 | 1·0000 000 | 1·0000 000 | ·9999 999 | ·9999 999 |
| ·82 | | | | 1·0000 000 | 1·0000 000 |

$x$ = ·28 to ·83 $q = 36$ $p$ = 46 to 50

| $x$ | $p = 46$ | $p = 47$ | $p = 48$ | $p = 49$ | $p = 50$ |
|---|---|---|---|---|---|
| $B(p, q) =$ | ·2132 2188 $\times \frac{1}{10^{24}}$ | ·1196 1227 $\times \frac{1}{10^{24}}$ | ·6773 2252 $\times \frac{1}{10^{25}}$ | ·3870 4144 $\times \frac{1}{10^{25}}$ | ·2231 1801 $\times \frac{1}{10^{25}}$ |
| ·28 | ·0000 001 | | | | |
| ·29 | ·0000 002 | ·0000 001 | | | |
| ·30 | ·0000 005⁻ | ·0000 003 | ·0000 001 | ·0000 001 | |
| ·31 | ·0000 014 | ·0000 007 | ·0000 004 | ·0000 002 | ·0000 001 |
| ·32 | ·0000 037 | ·0000 020 | ·0000 011 | ·0000 006 | ·0000 003 |
| ·33 | ·0000 092 | ·0000 053 | ·0000 030 | ·0000 017 | ·0000 009 |
| ·34 | ·0000 220 | ·0000 129 | ·0000 075⁺ | ·0000 043 | ·0000 025⁻ |
| ·35 | ·0000 503 | ·0000 303 | ·0000 181 | ·0000 108 | ·0000 063 |
| ·36 | ·0001 098 | ·0000 681 | ·0000 418 | ·0000 255⁺ | ·0000 154 |
| ·37 | ·0002 298 | ·0001 462 | ·0000 923 | ·0000 578 | ·0000 359 |
| ·38 | ·0004 615⁺ | ·0003 014 | ·0001 952 | ·0001 254 | ·0000 800 |
| ·39 | ·0008 914 | ·0005 968 | ·0003 963 | ·0002 611 | ·0001 707 |
| ·40 | ·0016 578 | ·0011 370 | ·0007 736 | ·0005 223 | ·0003 499 |
| ·41 | ·0029 730 | ·0020 876 | ·0014 542 | ·0010 052 | ·0006 896 |
| ·42 | ·0051 476 | ·0036 979 | ·0026 356 | ·0018 641 | ·0013 086 |
| ·43 | ·0086 155⁻ | ·0063 275⁻ | ·0046 110 | ·0033 347 | ·0023 939 |
| ·44 | ·0139 532 | ·0104 699 | ·0077 958 | ·0057 613 | ·0042 267 |
| ·45 | ·0218 896 | ·0167 699 | ·0127 505⁻ | ·0096 228 | ·0072 100 |
| ·46 | ·0332 954 | ·0260 269 | ·0201 935⁻ | ·0155 535⁺ | ·0118 946 |
| ·47 | ·0491 484 | ·0391 750⁻ | ·0309 968 | ·0243 503 | ·0189 952 |
| ·48 | ·0704 685⁻ | ·0572 364 | ·0461 553 | ·0369 579 | ·0293 898 |
| ·49 | ·0982 227 | ·0812 419 | ·0667 252 | ·0544 253 | ·0440 934 |
| ·50 | ·1332 062 | ·1121 223 | ·0937 299 | ·0778 282 | ·0641 982 |
| ·51 | ·1759 129 | ·1505 793 | ·1280 378 | ·1081 589 | ·0907 792 |
| ·52 | ·2264 135⁺ | ·1969 525⁻ | ·1702 244 | ·1461 913 | ·1247 676 |
| ·53 | ·2842 630 | ·2511 023 | ·2204 392 | ·1923 379 | ·1668 067 |
| ·54 | ·3484 587 | ·3123 339 | ·2782 997 | ·2465 203 | ·2171 017 |
| ·55 | ·4174 648 | ·3793 806 | ·3428 360 | ·3080 805⁺ | ·2753 111 |
| ·56 | ·4893 070 | ·4504 601 | ·4125 059 | ·3757 536 | ·3404 713 |
| ·57 | ·5617 312 | ·5234 031 | ·4852 869 | ·4477 187 | ·4110 092 |
| ·58 | ·6324 065⁻ | ·5958 409 | ·5588 396 | ·5217 305⁻ | ·4848 333 |
| ·59 | ·6991 424 | ·6654 275⁺ | ·6307 227 | ·5953 166 | ·5595 058 |
| ·60 | ·7600 894 | ·7300 613 | ·6986 277 | ·6660 152 | ·6324 710 |
| ·61 | ·8138 911 | ·7880 739 | ·7605 979 | ·7316 164 | ·7013 100 |
| ·62 | ·8597 687 | ·8383 577 | ·8151 973 | ·7903 675⁻ | ·7639 769 |
| ·63 | ·8975 283 | ·8804 148 | ·8616 046 | ·8411 132 | ·8189 825⁻ |
| ·64 | ·9274 986 | ·9143 288 | ·8996 234 | ·8833 495⁻ | ·8654 947 |
| ·65 | ·9504 152 | ·9406 681 | ·9296 145⁺ | ·9171 907 | ·9033 471 |
| ·66 | ·9672 772 | ·9603 482 | ·9523 695⁻ | ·9432 638 | ·9329 613 |
| ·67 | ·9792 009 | ·9744 764 | ·9689 538 | ·9625 556 | ·9552 069 |
| ·68 | ·9872 921 | ·9842 074 | ·9805 477 | ·9762 445⁻ | ·9712 282 |
| ·69 | ·9925 526 | ·9906 273 | ·9883 096 | ·9855 443 | ·9822 734 |
| ·70 | ·9958 233 | ·9946 770 | ·9932 771 | ·9915 826 | ·9895 492 |
| ·71 | ·9977 640 | ·9971 145⁻ | ·9963 098 | ·9953 220 | ·9941 196 |
| ·72 | ·9988 606 | ·9985 111 | ·9980 721 | ·9975 256 | ·9968 510 |
| ·73 | ·9994 490 | ·9992 710 | ·9990 443 | ·9987 582 | ·9984 002 |
| ·74 | ·9997 480 | ·9996 625⁻ | ·9995 521 | ·9994 108 | ·9992 315⁺ |
| ·75 | ·9998 914 | ·9998 528 | ·9998 022 | ·9997 366 | ·9996 523 |
| ·76 | ·9999 561 | ·9999 398 | ·9999 181 | ·9998 896 | ·9998 525⁻ |
| ·77 | ·9999 834 | ·9999 770 | ·9999 683 | ·9999 568 | ·9999 416 |
| ·78 | ·9999 942 | ·9999 918 | ·9999 886 | ·9999 843 | ·9999 785⁺ |
| ·79 | ·9999 981 | ·9999 973 | ·9999 962 | ·9999 947 | ·9999 927 |
| ·80 | ·9999 994 | ·9999 992 | ·9999 989 | ·9999 984 | ·9999 977 |
| ·81 | ·9999 999 | ·9999 998 | ·9999 997 | ·9999 996 | ·9999 994 |
| ·82 | 1·0000 000 | ·9999 999 | ·9999 999 | ·9999 999 | ·9999 998 |
| ·83 | | 1·0000 000 | 1·0000 000 | 1·0000 000 | 1·0000 000 |

$x$ = ·22 to ·80 $q = 37$ $p$ = 37 to 41

| $x$ | $p = 37$ | $p = 38$ | $p = 39$ | $p = 40$ | $p = 41$ |
|---|---|---|---|---|---|
| $B(p, q)$ = | ·3095 6479$\times\frac{1}{10^{22}}$ | ·1547 8239$\times\frac{1}{10^{22}}$ | ·7842 3080$\times\frac{1}{10^{23}}$ | ·4024 3423$\times\frac{1}{10^{23}}$ | ·2090 5674$\times\frac{1}{10^{23}}$ |
| ·22 | ·0000 001 | | | | |
| ·23 | ·0000 002 | ·0000 001 | | | |
| ·24 | ·0000 007 | ·0000 003 | ·0000 002 | ·0000 001 | |
| ·25 | ·0000 021 | ·0000 010 | ·0000 005$^{-}$ | ·0000 002 | ·0000 001 |
| ·26 | ·0000 057 | ·0000 029 | ·0000 014 | ·0000 007 | ·0000 003 |
| ·27 | ·0000 145$^{+}$ | ·0000 075$^{+}$ | ·0000 039 | ·0000 020 | ·0000 010 |
| ·28 | ·0000 349 | ·0000 188 | ·0000 100 | ·0000 052 | ·0000 027 |
| ·29 | ·0000 794 | ·0000 442 | ·0000 243 | ·0000 132 | ·0000 071 |
| ·30 | ·0001 718 | ·0000 988 | ·0000 562 | ·0000 316 | ·0000 176 |
| ·31 | ·0003 548 | ·0002 107 | ·0001 236 | ·0000 718 | ·0000 412 |
| ·32 | ·0007 011 | ·0004 291 | ·0002 597 | ·0001 554 | ·0000 920 |
| ·33 | ·0013 279 | ·0008 372 | ·0005 219 | ·0003 217 | ·0001 962 |
| ·34 | ·0024 165$^{-}$ | ·0015 675$^{-}$ | ·0010 054 | ·0006 378 | ·0004 004 |
| ·35 | ·0042 325$^{-}$ | ·0028 220 | ·0018 607 | ·0012 136 | ·0007 834 |
| ·36 | ·0071 479 | ·0048 942 | ·0033 142 | ·0022 204 | ·0014 722 |
| ·37 | ·0116 578 | ·0081 896 | ·0056 906 | ·0039 125$^{-}$ | ·0026 625$^{-}$ |
| ·38 | ·0183 889 | ·0132 422 | ·0094 336 | ·0066 504 | ·0046 409 |
| ·39 | ·0280 926 | ·0207 196 | ·0151 201 | ·0109 204 | ·0078 085$^{-}$ |
| ·40 | ·0416 180 | ·0314 118 | ·0234 618 | ·0173 463 | ·0126 986 |
| ·41 | ·0598 618 | ·0461 979 | ·0352 884 | ·0266 866 | ·0199 859 |
| ·42 | ·0836 943 | ·0659 889 | ·0515 078 | ·0398 114 | ·0304 778 |
| ·43 | ·1138 673 | ·0916 470 | ·0730 405$^{-}$ | ·0576 543 | ·0450 838 |
| ·44 | ·1509 109 | ·1238 867 | ·1007 312 | ·0811 381 | ·0647 583 |
| ·45 | ·1950 354 | ·1631 693 | ·1352 445$^{+}$ | ·1110 789 | ·0904 172 |
| ·46 | ·2460 516 | ·2096 051 | ·1769 567 | ·1480 755$^{-}$ | ·1228 333 |
| ·47 | ·3033 243 | ·2628 784 | ·2258 597 | ·1924 006 | ·1625 215$^{+}$ |
| ·48 | ·3657 729 | ·3222 121 | ·2814 942 | ·2439 084 | ·2096 302 |
| ·49 | ·4319 218 | ·3863 813 | ·3429 260 | ·3019 778 | ·2638 550$^{-}$ |
| ·50 | ·5000 000$^{e}$ | ·4537 803 | ·4087 769 | ·3655 044 | ·3243 956 |
| ·51 | ·5680 782 | ·5225 376 | ·4773 087 | ·4329 495$^{+}$ | ·3899 655$^{-}$ |
| ·52 | ·6342 271 | ·5906 663 | ·5465 552 | ·5024 441 | ·4588 624 |
| ·53 | ·6966 757 | ·6562 298 | ·6144 853 | ·5719 381 | ·5290 931 |
| ·54 | ·7539 484 | ·7175 019 | ·6791 756 | ·6393 751 | ·5985 399 |
| ·55 | ·8049 646 | ·7730 984 | ·7389 681 | ·7028 688 | ·6651 450$^{-}$ |
| ·56 | ·8490 891 | ·8220 650$^{-}$ | ·7925 944 | ·7608 568 | ·7270 880 |
| ·57 | ·8861 327 | ·8639 125$^{-}$ | ·8392 480 | ·8122 119 | ·7829 318 |
| ·58 | ·9163 057 | ·8986 002 | ·8786 024 | ·8562 972 | ·8317 168 |
| ·59 | ·9401 382 | ·9264 744 | ·9107 753 | ·8929 629 | ·8729 952 |
| ·60 | ·9583 820 | ·9481 758 | ·9362 507 | ·9224 910 | ·9068 049 |
| ·61 | ·9719 074 | ·9645 345$^{-}$ | ·9557 762 | ·9455 020 | ·9335 943 |
| ·62 | ·9816 111 | ·9764 643 | ·9702 503 | ·9628 413 | ·9541 135$^{-}$ |
| ·63 | ·9883 422 | ·9848 739 | ·9806 189 | ·9754 638 | ·9692 932 |
| ·64 | ·9928 521 | ·9905 984 | ·9877 896 | ·9843 325$^{+}$ | ·9801 288 |
| ·65 | ·9957 675$^{+}$ | ·9943 571 | ·9925 717 | ·9903 401 | ·9875 839 |
| ·66 | ·9975 835$^{+}$ | ·9967 346 | ·9956 434 | ·9942 585$^{+}$ | ·9925 218 |
| ·67 | ·9986 721 | ·9981 814 | ·9975 411 | ·9967 162 | ·9956 660 |
| ·68 | ·9992 989 | ·9990 270 | ·9986 670 | ·9981 961 | ·9975 878 |
| ·69 | ·9996 452 | ·9995 010 | ·9993 073 | ·9990 503 | ·9987 133 |
| ·70 | ·9998 282 | ·9997 552 | ·9996 558 | ·9995 219 | ·9993 438 |
| ·71 | ·9999 206 | ·9998 855$^{-}$ | ·9998 368 | ·9997 704 | ·9996 808 |
| ·72 | ·9999 651 | ·9999 490 | ·9999 264 | ·9998 951 | ·9998 523 |
| ·73 | ·9999 855$^{-}$ | ·9999 785$^{-}$ | ·9999 685$^{+}$ | ·9999 546 | ·9999 352 |
| ·74 | ·9999 943 | ·9999 914 | ·9999 873 | ·9999 814 | ·9999 732 |
| ·75 | ·9999 979 | ·9999 968 | ·9999 952 | ·9999 928 | ·9999 895$^{+}$ |
| ·76 | ·9999 993 | ·9999 989 | ·9999 983 | ·9999 974 | ·9999 962 |
| ·77 | ·9999 998 | ·9999 996 | ·9999 994 | ·9999 991 | ·9999 987 |
| ·78 | ·9999 999 | ·9999 999 | ·9999 998 | ·9999 997 | ·9999 996 |
| ·79 | 1·0000 000 | 1·0000 000 | 1·0000 000 | ·9999 999 | ·9999 999 |
| ·80 | | | | 1·0000 000 | 1·0000 000 |

$x$ = ·26 to ·82 $q = 37$ $p$ = 42 to 46

| $x$ | $p = 42$ | $p = 43$ | $p = 44$ | $p = 45$ | $p = 46$ |
|---|---|---|---|---|---|
| $B(p, q)$ = | ·1098 8880 $\times \frac{1}{10^{23}}$ | ·5842 1894 $\times \frac{1}{10^{24}}$ | ·3140 1768 $\times \frac{1}{10^{24}}$ | ·1705 7750$^{+}$ $\times \frac{1}{10^{24}}$ | ·9360 9606 $\times \frac{1}{10^{25}}$ |
| ·26 | ·0000 002 | ·0000 001 | | | |
| ·27 | ·0000 005⁻ | ·0000 002 | ·0000 001 | ·0000 001 | |
| ·28 | ·0000 014 | ·0000 007 | ·0000 004 | ·0000 002 | ·0000 001 |
| ·29 | ·0000 038 | ·0000 020 | ·0000 010 | ·0000 005⁺ | ·0000 003 |
| ·30 | ·0000 097 | ·0000 053 | ·0000 028 | ·0000 015⁺ | ·0000 008 |
| ·31 | ·0000 234 | ·0000 132 | ·0000 073 | ·0000 040 | ·0000 022 |
| ·32 | ·0000 539 | ·0000 313 | ·0000 180 | ·0000 102 | ·0000 058 |
| ·33 | ·0001 185⁻ | ·0000 708 | ·0000 419 | ·0000 246 | ·0000 143 |
| ·34 | ·0002 488 | ·0001 531 | ·0000 933 | ·0000 563 | ·0000 337 |
| ·35 | ·0005 005⁺ | ·0003 167 | ·0001 985⁻ | ·0001 232 | ·0000 758 |
| ·36 | ·0009 664 | ·0006 282 | ·0004 045⁻ | ·0002 581 | ·0001 632 |
| ·37 | ·0017 939 | ·0011 970 | ·0007 913 | ·0005 183 | ·0003 366 |
| ·38 | ·0032 069 | ·0021 948 | ·0014 883 | ·0010 001 | ·0006 662 |
| ·39 | ·0055 291 | ·0038 783 | ·0026 954 | ·0018 566 | ·0012 678 |
| ·40 | ·0092 071 | ·0066 135⁻ | ·0047 074 | ·0033 212 | ·0023 231 |
| ·41 | ·0148 263 | ·0108 976 | ·0079 384 | ·0057 323 | ·0041 043 |
| ·42 | ·0231 155⁺ | ·0173 730 | ·0129 419 | ·0095 582 | ·0070 001 |
| ·43 | ·0349 323 | ·0268 256 | ·0204 214 | ·0154 144 | ·0115 390 |
| ·44 | ·0512 230 | ·0401 627 | ·0312 218 | ·0240 692 | ·0184 043 |
| ·45 | ·0729 556 | ·0583 627 | ·0462 980 | ·0364 270 | ·0284 314 |
| ·46 | ·1010 265⁻ | ·0823 972 | ·0666 533 | ·0534 857 | ·0425 829 |
| ·47 | ·1361 478 | ·1131 273 | ·0932 494 | ·0762 628 | ·0618 922 |
| ·48 | ·1787 296 | ·1511 839 | ·1268 925⁺ | ·1056 927 | ·0873 761 |
| ·49 | ·2287 727 | ·1968 478 | ·1681 080 | ·1425 035⁻ | ·1199 203 |
| ·50 | ·2857 934 | ·2499 484 | ·2170 211 | ·1870 872 | ·1601 467 |
| ·51 | ·3487 952 | ·3098 010 | ·2732 644 | ·2393 850⁻ | ·2082 837 |
| ·52 | ·4163 011 | ·3751 990 | ·3359 322 | ·2988 072 | ·2640 583 |
| ·53 | ·4864 466 | ·4444 702 | ·4035 970 | ·3642 101 | ·3266 349 |
| ·54 | ·5571 269 | ·5155 957 | ·4743 928 | ·4339 391 | ·3946 181 |
| ·55 | ·6261 790 | ·5863 781 | ·5461 606 | ·5059 432 | ·4661 279 |
| ·56 | ·6915 732 | ·6546 377 | ·6166 371 | ·5779 457 | ·5389 446 |
| ·57 | ·7515 879 | ·7184 080 | ·6836 618 | ·6476 521 | ·6107 061 |
| ·58 | ·8049 422 | ·7761 021 | ·7453 706 | ·7129 629 | ·6791 292 |
| ·59 | ·8508 700 | ·8266 271 | ·8003 489 | ·7721 596 | ·7422 226 |
| ·60 | ·8891 294 | ·8694 338 | ·8477 228 | ·8240 381 | ·7984 586 |
| ·61 | ·9199 527 | ·9044 987 | ·8871 794 | ·8679 708 | ·8468 797 |
| ·62 | ·9439 509 | ·9322 494 | ·9189 206 | ·9038 954 | ·8871 272 |
| ·63 | ·9619 923 | ·9534 502 | ·9435 632 | ·9322 381 | ·9193 955⁻ |
| ·64 | ·9750 761 | ·9690 706 | ·9620 093 | ·9537 924 | ·9443 266 |
| ·65 | ·9842 195⁻ | ·9801 581 | ·9753 080 | ·9695 761 | ·9628 698 |
| ·66 | ·9903 692 | ·9877 306 | ·9845 313 | ·9806 920 | ·9761 310 |
| ·67 | ·9943 446 | ·9927 004 | ·9906 765⁺ | ·9882 110 | ·9852 377 |
| ·68 | ·9968 109 | ·9958 297 | ·9946 040 | ·9930 886 | ·9912 337 |
| ·69 | ·9982 766 | ·9977 171 | ·9970 078 | ·9961 179 | ·9950 126 |
| ·70 | ·9991 096 | ·9988 053 | ·9984 138 | ·9979 156 | ·9972 879 |
| ·71 | ·9995 613 | ·9994 038 | ·9991 983 | ·9989 330 | ·9985 940 |
| ·72 | ·9997 945⁻ | ·9997 171 | ·9996 147 | ·9994 808 | ·9993 071 |
| ·73 | ·9999 087 | ·9998 728 | ·9998 245⁺ | ·9997 605⁺ | ·9996 764 |
| ·74 | ·9999 617 | ·9999 459 | ·9999 245⁺ | ·9998 957 | ·9998 573 |
| ·75 | ·9999 849 | ·9999 784 | ·9999 694 | ·9999 573 | ·9999 408 |
| ·76 | ·9999 944 | ·9999 919 | ·9999 884 | ·9999 836 | ·9999 770 |
| ·77 | ·9999 981 | ·9999 972 | ·9999 959 | ·9999 941 | ·9999 917 |
| ·78 | ·9999 994 | ·9999 991 | ·9999 987 | ·9999 981 | ·9999 972 |
| ·79 | ·9999 998 | ·9999 997 | ·9999 996 | ·9999 994 | ·9999 991 |
| ·80 | 1·0000 000 | ·9999 999 | ·9999 999 | ·9999 998 | ·9999 998 |
| ·81 | | 1·0000 000 | 1·0000 000 | 1·0000 000 | ·9999 999 |
| ·82 | | | | | 1·0000 000 |

$x$ = ·29 to ·83 $q = 37$ $p$ = 47 to 50

| $x$ | $p = 47$ | $p = 48$ | $p = 49$ | $p = 50$ |
|---|---|---|---|---|
| $B(p, q)$ = | ·5188 0022 × $\frac{1}{10^{25}}$ | ·2902 8108 × $\frac{1}{10^{25}}$ | ·1639 2343 × $\frac{1}{10^{25}}$ | ·9339 8235 × $\frac{1}{10^{26}}$ |
| ·29 | ·0000 001 | ·0000 001 | | |
| ·30 | ·0000 004 | ·0000 002 | ·0000 001 | ·0000 001 |
| ·31 | ·0000 012 | ·0000 006 | ·0000 003 | ·0000 002 |
| ·32 | ·0000 032 | ·0000 018 | ·0000 010 | ·0000 005+ |
| ·33 | ·0000 082 | ·0000 047 | ·0000 027 | ·0000 015+ |
| ·34 | ·0000 200 | ·0000 118 | ·0000 069 | ·0000 040 |
| ·35 | ·0000 463 | ·0000 280 | ·0000 168 | ·0000 100 |
| ·36 | ·0001 023 | ·0000 636 | ·0000 392 | ·0000 240 |
| ·37 | ·0002 167 | ·0001 383 | ·0000 876 | ·0000 550+ |
| ·38 | ·0004 400 | ·0002 882 | ·0001 873 | ·0001 208 |
| ·39 | ·0008 585− | ·0005 765+ | ·0003 841 | ·0002 539 |
| ·40 | ·0016 115− | ·0011 088 | ·0007 569 | ·0005 127 |
| ·41 | ·0029 144 | ·0020 529 | ·0014 348 | ·0009 951 |
| ·42 | ·0050 848 | ·0036 643 | ·0026 202 | ·0018 595− |
| ·43 | ·0085 684 | ·0063 127 | ·0046 152 | ·0033 491 |
| ·44 | ·0139 610 | ·0105 085− | ·0078 501 | ·0058 210 |
| ·45 | ·0220 176 | ·0169 207 | ·0129 068 | ·0097 736 |
| ·46 | ·0336 427 | ·0263 801 | ·0205 338 | ·0158 686 |
| ·47 | ·0498 521 | ·0398 588 | ·0316 393 | ·0249 379 |
| ·48 | ·0717 035− | ·0584 184 | ·0472 590 | ·0379 670 |
| ·49 | ·1001 943 | ·0831 251 | ·0684 882 | ·0560 468 |
| ·50 | ·1361 345− | ·1149 322 | ·0963 802 | ·0802 892 |
| ·51 | ·1800 085+ | ·1545 428 | ·1318 147 | ·1117 073 |
| ·52 | ·2318 475− | ·2022 684 | ·1753 514 | ·1510 712 |
| ·53 | ·2911 346 | ·2579 078 | ·2270 899 | ·1987 563 |
| ·54 | ·3567 674 | ·3206 722 | ·2865 624 | ·2546 104 |
| ·55 | ·4270 916 | ·3891 766 | ·3526 834 | ·3178 659 |
| ·56 | ·5000 114 | ·4615 090 | ·4237 766 | ·3871 223 |
| ·57 | ·5731 658 | ·5353 779 | ·4976 844 | ·4604 141 |
| ·58 | ·6441 481 | ·6083 186 | ·5719 516 | ·5353 619 |
| ·59 | ·7107 366 | ·6779 309 | ·6440 590 | ·6093 922 |
| ·60 | ·7710 997 | ·7421 109 | ·7116 726 | ·6799 920 |
| ·61 | ·8239 455− | ·7992 399 | ·7728 667 | ·7449 596 |
| ·62 | ·8685 948 | ·8483 038 | ·8262 880 | ·8026 098 |
| ·63 | ·9049 726 | ·8889 265− | ·8712 356 | ·8519 019 |
| ·64 | ·9335 274 | ·9213 220 | ·9076 519 | ·8924 753 |
| ·65 | ·9550 992 | ·9461 796 | ·9360 335− | ·9245 932 |
| ·66 | ·9707 648 | ·9645 104 | ·9572 866 | ·9490 160 |
| ·67 | ·9816 865− | ·9774 847 | ·9725 581 | ·9668 322 |
| ·68 | ·9889 853 | ·9862 852 | ·9830 722 | ·9792 821 |
| ·69 | ·9936 532 | ·9919 967 | ·9899 964 | ·9876 023 |
| ·70 | ·9965 046 | ·9955 364 | ·9943 502 | ·9929 099 |
| ·71 | ·9981 650− | ·9976 270 | ·9969 585+ | ·9961 352 |
| ·72 | ·9990 842 | ·9988 008 | ·9984 438 | ·9979 978 |
| ·73 | ·9995 670 | ·9994 258 | ·9992 456 | ·9990 173 |
| ·74 | ·9998 066 | ·9997 404 | ·9996 547 | ·9995 447 |
| ·75 | ·9999 188 | ·9998 896 | ·9998 514 | ·9998 016 |
| ·76 | ·9999 681 | ·9999 561 | ·9999 401 | ·9999 191 |
| ·77 | ·9999 883 | ·9999 837 | ·9999 775+ | ·9999 693 |
| ·78 | ·9999 960 | ·9999 944 | ·9999 922 | ·9999 892 |
| ·79 | ·9999 988 | ·9999 982 | ·9999 975− | ·9999 965− |
| ·80 | ·9999 996 | ·9999 995− | ·9999 993 | ·9999 990 |
| ·81 | ·9999 999 | ·9999 999 | ·9999 998 | ·9999 997 |
| ·82 | 1·0000 000 | 1·0000 000 | 1·0000 000 | ·9999 999 |
| ·83 | | | | 1·0000 000 |

$x$ = ·23 to ·80 $q = 38$ $p$ = 38 to 43

| $x$ | $p = 38$ | $p = 39$ | $p = 40$ | $p = 41$ | $p = 42$ | $p = 43$ |
|---|---|---|---|---|---|---|
| $B(p, q) =$ | ·7635 9315 $\times \frac{1}{10^{23}}$ | ·3817 9657 $\times \frac{1}{10^{23}}$ | ·1933 7749 $\times \frac{1}{10^{23}}$ | ·9916 7941 $\times \frac{1}{10^{24}}$ | ·5146 6906 $\times \frac{1}{10^{24}}$ | ·2702 0126 $\times \frac{1}{10^{24}}$ |
| ·23 | ·0000 002 | ·0000 001 | | | | |
| ·24 | ·0000 005$^{+}$ | ·0000 002 | ·0000 001 | | | |
| ·25 | ·0000 016 | ·0000 008 | ·0000 004 | ·0000 002 | ·0000 001 | |
| ·26 | ·0000 044 | ·0000 022 | ·0000 011 | ·0000 005$^{+}$ | ·0000 003 | ·0000 001 |
| ·27 | ·0000 113 | ·0000 059 | ·0000 030 | ·0000 015$^{+}$ | ·0000 008 | ·0000 004 |
| ·28 | ·0000 278 | ·0000 150$^{-}$ | ·0000 080 | ·0000 042 | ·0000 022 | ·0000 011 |
| ·29 | ·0000 646 | ·0000 360 | ·0000 198 | ·0000 108 | ·0000 058 | ·0000 031 |
| ·30 | ·0001 426 | ·0000 821 | ·0000 467 | ·0000 263 | ·0000 147 | ·0000 081 |
| ·31 | ·0003 000 | ·0001 783 | ·0001 048 | ·0000 609 | ·0000 350$^{+}$ | ·0000 200 |
| ·32 | ·0006 032 | ·0003 696 | ·0002 240 | ·0001 343 | ·0000 796 | ·0000 468 |
| ·33 | ·0011 611 | ·0007 328 | ·0004 574 | ·0002 824 | ·0001 726 | ·0001 044 |
| ·34 | ·0021 448 | ·0013 928 | ·0008 945$^{+}$ | ·0005 684 | ·0003 575$^{-}$ | ·0002 226 |
| ·35 | ·0038 093 | ·0025 427 | ·0016 788 | ·0010 968 | ·0007 092 | ·0004 541 |
| ·36 | ·0065 168 | ·0044 671 | ·0030 292 | ·0020 327 | ·0013 502 | ·0008 881 |
| ·37 | ·0107 561 | ·0075 648 | ·0052 638 | ·0036 250$^{-}$ | ·0024 714 | ·0016 685$^{+}$ |
| ·38 | ·0171 537 | ·0123 672 | ·0088 228 | ·0062 300 | ·0043 557 | ·0030 159 |
| ·39 | ·0264 705$^{+}$ | ·0195 468 | ·0142 847 | ·0103 342 | ·0074 031 | ·0052 530 |
| ·40 | ·0395 768 | ·0299 078 | ·0223 709 | ·0165 675$^{+}$ | ·0121 513 | ·0088 286 |
| ·41 | ·0574 023 | ·0443 551 | ·0339 307 | ·0257 032 | ·0192 858 | ·0143 368 |
| ·42 | ·0808 615$^{-}$ | ·0638 363 | ·0499 019 | ·0386 359 | ·0296 341 | ·0225 227 |
| ·43 | ·1107 564 | ·0892 583 | ·0712 440 | ·0563 327 | ·0441 345$^{-}$ | ·0342 684 |
| ·44 | ·1476 680 | ·1213 834 | ·0988 460 | ·0797 569 | ·0637 779 | ·0505 534 |
| ·45 | ·1918 488 | ·1607 165$^{-}$ | ·1334 158 | ·1097 666 | ·0895 205$^{+}$ | ·0723 837 |
| ·46 | ·2431 359 | ·2073 991 | ·1753 644 | ·1469 976 | ·1221 732 | ·1006 942 |
| ·47 | ·3008 976 | ·2611 275$^{+}$ | ·2247 022 | ·1917 465$^{-}$ | ·1622 792 | ·1362 287 |
| ·48 | ·3640 305$^{-}$ | ·3211 116 | ·2809 659 | ·2438 713 | ·2099 976 | ·1794 145$^{+}$ |
| ·49 | ·4310 110 | ·3860 877 | ·3431 916 | ·3027 299 | ·2650 118 | ·2302 482 |
| ·50 | ·5000 000$^{e}$ | ·4543 885$^{-}$ | ·4099 465$^{-}$ | ·3671 710 | ·3264 822 | ·2882 153 |
| ·51 | ·5689 890 | ·5240 656 | ·4794 187 | ·4355 866 | ·3930 588 | ·3522 625$^{+}$ |
| ·52 | ·6359 695$^{+}$ | ·5930 506 | ·5495 595$^{-}$ | ·5060 249 | ·4629 574 | ·4208 334 |
| ·53 | ·6991 024 | ·6593 324 | ·6182 571 | ·5763 500$^{-}$ | ·5340 954 | ·4919 716 |
| ·54 | ·7568 641 | ·7211 274 | ·6835 213 | ·6444 299 | ·6042 705$^{+}$ | ·5634 801 |
| ·55 | ·8081 512 | ·7770 188 | ·7436 513 | ·7083 234 | ·6713 584 | ·6331 173 |
| ·56 | ·8523 320 | ·8260 475$^{-}$ | ·7973 636 | ·7664 423 | ·7334 999 | ·6988 005$^{+}$ |
| ·57 | ·8892 436 | ·8677 455$^{-}$ | ·8438 660 | ·8176 643 | ·7892 514 | ·7587 888 |
| ·58 | ·9191 385$^{+}$ | ·9021 134 | ·8828 706 | ·8613 860 | ·8376 796 | ·8118 170 |
| ·59 | ·9425 977 | ·9295 505$^{+}$ | ·9145 496 | ·8975 123 | ·8783 889 | ·8571 666 |
| ·60 | ·9604 232 | ·9507 542 | ·9394 490 | ·9263 913 | ·9114 866 | ·8946 654 |
| ·61 | ·9735 295$^{-}$ | ·9666 057 | ·9583 753 | ·9487 107 | ·9374 951 | ·9246 265$^{-}$ |
| ·62 | ·9828 463 | ·9780 598 | ·9722 768 | ·9653 747 | ·9572 337 | ·9477 397 |
| ·63 | ·9892 439 | ·9860 527 | ·9821 348 | ·9773 834 | ·9716 887 | ·9649 404 |
| ·64 | ·9934 832 | ·9914 335$^{-}$ | ·9888 771 | ·9857 277 | ·9818 931 | ·9772 770 |
| ·65 | ·9961 907 | ·9949 240 | ·9933 196 | ·9913 121 | ·9888 297 | ·9857 946 |
| ·66 | ·9978 552 | ·9971 032 | ·9961 360 | ·9949 072 | ·9933 643 | ·9914 488 |
| ·67 | ·9988 389 | ·9984 106 | ·9978 515$^{-}$ | ·9971 303 | ·9962 110 | ·9950 525$^{+}$ |
| ·68 | ·9993 968 | ·9991 633 | ·9988 538 | ·9984 487 | ·9979 246 | ·9972 542 |
| ·69 | ·9997 000 | ·9995 782 | ·9994 146 | ·9991 972 | ·9989 118 | ·9985 414 |
| ·70 | ·9998 574 | ·9997 969 | ·9997 144 | ·9996 032 | ·9994 551 | ·9992 602 |
| ·71 | ·9999 354 | ·9999 068 | ·9998 673 | ·9998 132 | ·9997 402 | ·9996 426 |
| ·72 | ·9999 722 | ·9999 594 | ·9999 414 | ·9999 165$^{-}$ | ·9998 823 | ·9998 360 |
| ·73 | ·9999 887 | ·9999 832 | ·9999 755$^{+}$ | ·9999 646 | ·9999 495$^{+}$ | ·9999 288 |
| ·74 | ·9999 956 | ·9999 935$^{-}$ | ·9999 903 | ·9999 859 | ·9999 796 | ·9999 708 |
| ·75 | ·9999 984 | ·9999 976 | ·9999 964 | ·9999 947 | ·9999 922 | ·9999 888 |
| ·76 | ·9999 995$^{-}$ | ·9999 992 | ·9999 988 | ·9999 981 | ·9999 972 | ·9999 960 |
| ·77 | ·9999 998 | ·9999 997 | ·9999 996 | ·9999 994 | ·9999 991 | ·9999 986 |
| ·78 | 1·0000 000 | ·9999 999 | ·9999 999 | ·9999 998 | ·9999 997 | ·9999 996 |
| ·79 | | 1·0000 000 | 1·0000 000 | ·9999 999 | ·9999 999 | ·9999 999 |
| ·80 | | | | 1·0000 000 | 1·0000 000 | 1·0000 000 |

| | $p = 44$ | $p = 45$ | $p = 46$ | $p = 47$ | $p = 48$ | $p = 49$ | $p = 50$ |
|---|---|---|---|---|---|---|---|
| $B(p, q) =$ | $\cdot 1434\ 4017 \times \frac{1}{10^{24}}$ | $\cdot 7696\ 7898 \times \frac{1}{10^{25}}$ | $\cdot 4172\ 9583 \times \frac{1}{10^{25}}$ | $\cdot 2285\ 1915 \times \frac{1}{10^{25}}$ | $\cdot 1263\ 5765 \times \frac{1}{10^{25}}$ | $\cdot 7052\ 5198 \times \frac{1}{10^{26}}$ | $\cdot 3972\ 1088 \times \frac{1}{10^{26}}$ |
| $x$ | | | | | | | |
| ·26 | ·0000 001 | | | | | | |
| ·27 | ·0000 002 | ·0000 001 | | | | | |
| ·28 | ·0000 006 | ·0000 003 | ·0000 001 | ·0000 001 | | | |
| ·29 | ·0000 016 | ·0000 009 | ·0000 004 | ·0000 002 | ·0000 001 | ·0000 001 | |
| ·30 | ·0000 044 | ·0000 024 | ·0000 013 | ·0000 007 | ·0000 004 | ·0000 002 | ·0000 001 |
| ·31 | ·0000 112 | ·0000 063 | ·0000 035⁻ | ·0000 019 | ·0000 010 | ·0000 006 | ·0000 003 |
| ·32 | ·0000 272 | ·0000 157 | ·0000 089 | ·0000 051 | ·0000 028 | ·0000 016 | ·0000 009 |
| ·33 | ·0000 625⁺ | ·0000 371 | ·0000 218 | ·0000 127 | ·0000 074 | ·0000 042 | ·0000 024 |
| ·34 | ·0001 372 | ·0000 838 | ·0000 507 | ·0000 304 | ·0000 181 | ·0000 107 | ·0000 062 |
| ·35 | ·0002 879 | ·0001 809 | ·0001 126 | ·0000 695⁻ | ·0000 425⁺ | ·0000 258 | ·0000 155⁺ |
| ·36 | ·0005 786 | ·0003 735⁻ | ·0002 389 | ·0001 515⁻ | ·0000 952 | ·0000 594 | ·0000 367 |
| ·37 | ·0011 159 | ·0007 394 | ·0004 856 | ·0003 162 | ·0002 041 | ·0001 307 | ·0000 830 |
| ·38 | ·0020 688 | ·0014 062 | ·0009 474 | ·0006 328 | ·0004 192 | ·0002 754 | ·0001 795⁺ |
| ·39 | ·0036 928 | ·0025 728 | ·0017 767 | ·0012 166 | ·0008 262 | ·0005 565⁺ | ·0003 719 |
| ·40 | ·0063 559 | ·0045 351 | ·0032 079 | ·0022 500⁺ | ·0015 653 | ·0010 802 | ·0007 397 |
| ·41 | ·0105 617 | ·0077 124 | ·0055 836 | ·0040 088 | ·0028 548 | ·0020 170 | ·0014 141 |
| ·42 | ·0169 658 | ·0126 694 | ·0093 812 | ·0068 893 | ·0050 188 | ·0036 276 | ·0026 021 |
| ·43 | ·0263 756 | ·0201 277 | ·0152 322 | ·0114 338 | ·0085 148 | ·0062 920 | ·0046 146 |
| ·44 | ·0397 277 | ·0309 589 | ·0239 283 | ·0183 466 | ·0139 573 | ·0105 372 | ·0078 961 |
| ·45 | ·0580 366 | ·0461 513 | ·0364 054 | ·0284 921 | ·0221 278 | ·0170 563 | ·0130 508 |
| ·46 | ·0823 121 | ·0667 459 | ·0536 979 | ·0428 681 | ·0339 646 | ·0267 119 | ·0208 565⁺ |
| ·47 | ·1134 497 | ·0937 407 | ·0768 610 | ·0625 462 | ·0505 219 | ·0405 141 | ·0322 587 |
| ·48 | ·1521 031 | ·1279 697 | ·1068 611 | ·0885 791 | ·0728 956 | ·0595 645⁺ | ·0483 338 |
| ·49 | ·1985 567 | ·1699 696 | ·1444 444 | ·1218 769 | ·1021 134 | ·0849 646 | ·0702 165⁺ |
| ·50 | ·2526 182 | ·2198 527 | ·1899 997 | ·1630 671 | ·1389 996 | ·1176 899 | ·0989 896 |
| ·51 | ·3135 534 | ·2772 109 | ·2434 365⁺ | ·2123 568 | ·1840 280 | ·1584 435⁻ | ·1355 428 |
| ·52 | ·3800 808 | ·3410 695⁺ | ·3041 041 | ·2694 209 | ·2371 877 | ·2075 063 | ·1804 174 |
| ·53 | ·4504 355⁺ | ·4099 096 | ·3707 705⁻ | ·3333 416 | ·2978 877 | ·2646 127 | ·2336 602 |
| ·54 | ·5225 000⁻ | ·4817 620 | ·4416 758 | ·4026 179 | ·3649 229 | ·3288 771 | ·2947 144 |
| ·55 | ·5939 868 | ·5543 672 | ·5146 595⁻ | ·4752 539 | ·4365 191 | ·3987 930 | ·3623 758 |
| ·56 | ·6626 486 | ·6253 793 | ·5873 481 | ·5489 199 | ·5104 591 | ·4723 188 | ·4348 323 |
| ·57 | ·7264 842 | ·6925 864 | ·6573 779 | ·6211 667 | ·5842 775⁻ | ·5470 425⁻ | ·5097 923 |
| ·58 | ·7839 095⁺ | ·7541 119 | ·7226 192 | ·6896 613 | ·6554 974 | ·6204 081 | ·5846 887 |
| ·59 | ·8338 713 | ·8085 695⁺ | ·7813 673 | ·7524 087 | ·7218 728 | ·6899 691 | ·6569 326 |
| ·60 | ·8758 884 | ·8551 483 | ·8324 724 | ·8079 233 | ·7815 983 | ·7536 280 | ·7241 736 |
| ·61 | ·9100 221 | ·8936 221 | ·8753 926 | ·8553 282 | ·8334 538 | ·8098 248 | ·7845 274 |
| ·62 | ·9367 884 | ·9242 891 | ·9101 676 | ·8943 699 | ·8768 648 | ·8576 456 | ·8367 320 |
| ·63 | ·9570 309 | ·9478 575⁺ | ·9373 266 | ·9253 556 | ·9118 768 | ·8968 396 | ·8802 126 |
| ·64 | ·9717 806 | ·9653 049 | ·9577 527 | ·9490 316 | ·9390 561 | ·9277 506 | ·9150 515⁻ |
| ·65 | ·9821 243 | ·9777 324 | ·9725 305⁺ | ·9664 295⁺ | ·9593 420 | ·9511 840 | ·9418 773 |
| ·66 | ·9890 969 | ·9862 392 | ·9828 024 | ·9787 096 | ·9738 819 | ·9682 395⁻ | ·9617 035⁻ |
| ·67 | ·9936 084 | ·9918 273 | ·9896 527 | ·9870 239 | ·9838 759 | ·9801 410 | ·9757 491 |
| ·68 | ·9964 062 | ·9953 445⁺ | ·9940 291 | ·9924 151 | ·9904 535⁺ | ·9880 915⁺ | ·9852 725⁻ |
| ·69 | ·9980 660 | ·9974 621 | ·9967 027 | ·9957 574 | ·9945 916 | ·9931 671 | ·9914 420 |
| ·70 | ·9990 063 | ·9986 791 | ·9982 617 | ·9977 346 | ·9970 751 | ·9962 577 | ·9952 534 |
| ·71 | ·9995 138 | ·9993 454 | ·9991 275⁻ | ·9988 483 | ·9984 941 | ·9980 488 | ·9974 939 |
| ·72 | ·9997 741 | ·9996 920 | ·9995 842 | ·9994 442 | ·9992 641 | ·9990 344 | ·9987 441 |
| ·73 | ·9999 007 | ·9998 628 | ·9998 125⁻ | ·9997 462 | ·9996 597 | ·9995 479 | ·9994 046 |
| ·74 | ·9999 588 | ·9999 424 | ·9999 203 | ·9998 907 | ·9998 516 | ·9998 004 | ·9997 340 |
| ·75 | ·9999 839 | ·9999 773 | ·9999 682 | ·9999 558 | ·9999 393 | ·9999 173 | ·9998 884 |
| ·76 | ·9999 942 | ·9999 916 | ·9999 881 | ·9999 833 | ·9999 768 | ·9999 680 | ·9999 562 |
| ·77 | ·9999 980 | ·9999 971 | ·9999 959 | ·9999 941 | ·9999 917 | ·9999 885⁻ | ·9999 840 |
| ·78 | ·9999 994 | ·9999 991 | ·9999 987 | ·9999 981 | ·9999 973 | ·9999 962 | ·9999 946 |
| ·79 | ·9999 998 | ·9999 997 | ·9999 996 | ·9999 994 | ·9999 992 | ·9999 988 | ·9999 983 |
| ·80 | 1·0000 000 | ·9999 999 | ·9999 999 | ·9999 998 | ·9999 998 | ·9999 997 | ·9999 995⁺ |
| ·81 | | 1·0000 000 | 1·0000 000 | 1·0000 000 | ·9999 999 | ·9999 999 | ·9999 999 |
| ·82 | | | | | 1·0000 000 | 1·0000 000 | 1·0000 000 |

$x$ = ·23 to ·80 $q = 39$ $p$ = 39 to 44

| $x$ | $p = 39$ | $p = 40$ | $p = 41$ | $p = 42$ | $p = 43$ | $p = 44$ |
|---|---|---|---|---|---|---|
| $B(p, q)$ = | ·1884 1909 × $\frac{1}{10^{23}}$ | ·9420 9544 × $\frac{1}{10^{24}}$ | ·4770 1035+ × $\frac{1}{10^{24}}$ | ·2444 6780 × $\frac{1}{10^{24}}$ | ·1267 6108 × $\frac{1}{10^{24}}$ | ·6647 2276 × $\frac{1}{10^{25}}$ |
| ·23 | ·0000 001 | ·0000 001 | | | | |
| ·24 | ·0000 004 | ·0000 002 | ·0000 001 | | | |
| ·25 | ·0000 012 | ·0000 006 | ·0000 003 | ·0000 001 | ·0000 001 | |
| ·26 | ·0000 033 | ·0000 017 | ·0000 008 | ·0000 004 | ·0000 002 | ·0000 001 |
| ·27 | ·0000 088 | ·0000 046 | ·0000 024 | ·0000 012 | ·0000 006 | ·0000 003 |
| ·28 | ·0000 221 | ·0000 119 | ·0000 064 | ·0000 034 | ·0000 017 | ·0000 009 |
| ·29 | ·0000 526 | ·0000 293 | ·0000 162 | ·0000 088 | ·0000 048 | ·0000 025+ |
| ·30 | ·0001 184 | ·0000 682 | ·0000 389 | ·0000 219 | ·0000 122 | ·0000 068 |
| ·31 | ·0002 538 | ·0001 510 | ·0000 888 | ·0000 517 | ·0000 298 | ·0000 170 |
| ·32 | ·0005 191 | ·0003 184 | ·0001 932 | ·0001 160 | ·0000 689 | ·0000 405+ |
| ·33 | ·0010 155− | ·0006 415+ | ·0004 009 | ·0002 479 | ·0001 518 | ·0000 920 |
| ·34 | ·0019 041 | ·0012 378 | ·0007 960 | ·0005 066 | ·0003 191 | ·0001 991 |
| ·35 | ·0034 294 | ·0022 915+ | ·0015 149 | ·0009 912 | ·0006 421 | ·0004 119 |
| ·36 | ·0059 429 | ·0040 781 | ·0027 691 | ·0018 610 | ·0012 384 | ·0008 161 |
| ·37 | ·0099 263 | ·0069 890 | ·0048 696 | ·0033 587 | ·0022 939 | ·0015 517 |
| ·38 | ·0160 050− | ·0115 520 | ·0082 524 | ·0058 364 | ·0040 877 | ·0028 360 |
| ·39 | ·0249 473 | ·0184 431 | ·0134 967 | ·0097 796 | ·0070 184 | ·0049 898 |
| ·40 | ·0376 430 | ·0284 797 | ·0213 324 | ·0158 237 | ·0116 266 | ·0084 642 |
| ·41 | ·0550 538 | ·0425 912 | ·0326 273 | ·0247 558 | ·0186 086 | ·0138 609 |
| ·42 | ·0781 375− | ·0617 608 | ·0483 484 | ·0374 941 | ·0288 107 | ·0219 407 |
| ·43 | ·1077 467 | ·0869 402 | ·0694 939 | ·0550 390 | ·0431 998 | ·0336 100 |
| ·44 | ·1445 138 | ·1189 398 | ·0969 974 | ·0783 945− | ·0628 035− | ·0498 810 |
| ·45 | ·1887 356 | ·1583 097 | ·1316 110 | ·1084 612 | ·0886 186 | ·0717 985− |
| ·46 | ·2402 769 | ·2052 242 | ·1737 818 | ·1459 131 | ·1214 949 | ·1003 362 |
| ·47 | ·2985 114 | ·2593 925+ | ·2235 401 | ·1910 718 | ·1620 050− | ·1362 707 |
| ·48 | ·3623 137 | ·3200 129 | ·2804 193 | ·2438 000 | ·2103 196 | ·1800 470 |
| ·49 | ·4301 125+ | ·3857 829 | ·3434 259 | ·3034 347 | ·2661 096 | ·2316 576 |
| ·50 | ·5000 000[e] | ·4549 732 | ·4110 721 | ·3687 772 | ·3284 962 | ·2905 572 |
| ·51 | ·5698 875− | ·5255 578 | ·4814 719 | ·4381 495+ | ·3960 649 | ·3556 343 |
| ·52 | ·6376 863 | ·5953 854 | ·5524 924 | ·5095 156 | ·4669 481 | ·4252 518 |
| ·53 | ·7014 886 | ·6623 698 | ·6219 405− | ·5806 533 | ·5389 729 | ·4973 603 |
| ·54 | ·7597 231 | ·7246 703 | ·6877 597 | ·6493 547 | ·6098 524 | ·5696 703 |
| ·55 | ·8112 644 | ·7808 385− | ·7482 067 | ·7136 250− | ·6773 965+ | ·6398 621 |
| ·56 | ·8554 862 | ·8299 122 | ·8019 855− | ·7718 518 | ·7397 093 | ·7058 026 |
| ·57 | ·8922 533 | ·8714 468 | ·8483 203 | ·8229 207 | ·7953 440 | ·7657 343 |
| ·58 | ·9218 625+ | ·9054 859 | ·8869 639 | ·8662 645+ | ·8433 966 | ·8184 120 |
| ·59 | ·9449 462 | ·9324 836 | ·9181 454 | ·9018 452 | ·8835 270 | ·8631 682 |
| ·60 | ·9623 570 | ·9531 938 | ·9424 728 | ·9300 783 | ·9159 132 | ·8999 032 |
| ·61 | ·9750 527 | ·9685 485− | ·9608 117 | ·9517 182 | ·9411 525− | ·9290 116 |
| ·62 | ·9839 950+ | ·9795 421 | ·9741 585+ | ·9677 271 | ·9601 319 | ·9512 613 |
| ·63 | ·9900 737 | ·9871 363 | ·9835 277 | ·9791 473 | ·9738 908 | ·9676 526 |
| ·64 | ·9940 571 | ·9921 923 | ·9898 651 | ·9869 952 | ·9834 966 | ·9792 789 |
| ·65 | ·9965 706 | ·9954 328 | ·9939 906 | ·9921 843 | ·9899 479 | ·9872 096 |
| ·66 | ·9980 959 | ·9974 295+ | ·9965 719 | ·9954 813 | ·9941 103 | ·9924 057 |
| ·67 | ·9989 845+ | ·9986 106 | ·9981 221 | ·9974 914 | ·9966 866 | ·9956 708 |
| ·68 | ·9994 809 | ·9992 802 | ·9990 141 | ·9986 655− | ·9982 139 | ·9976 354 |
| ·69 | ·9997 462 | ·9996 434 | ·9995 051 | ·9993 212 | ·9990 794 | ·9987 653 |
| ·70 | ·9998 816 | ·9998 314 | ·9997 630 | ·9996 706 | ·9995 475− | ·9993 851 |
| ·71 | ·9999 474 | ·9999 242 | ·9998 920 | ·9998 480 | ·9997 884 | ·9997 088 |
| ·72 | ·9999 779 | ·9999 677 | ·9999 533 | ·9999 334 | ·9999 062 | ·9998 692 |
| ·73 | ·9999 912 | ·9999 869 | ·9999 809 | ·9999 724 | ·9999 607 | ·9999 445− |
| ·74 | ·9999 967 | ·9999 950+ | ·9999 927 | ·9999 893 | ·9999 845− | ·9999 778 |
| ·75 | ·9999 988 | ·9999 982 | ·9999 973 | ·9999 961 | ·9999 942 | ·9999 917 |
| ·76 | ·9999 996 | ·9999 994 | ·9999 991 | ·9999 987 | ·9999 980 | ·9999 971 |
| ·77 | ·9999 999 | ·9999 998 | ·9999 997 | ·9999 996 | ·9999 994 | ·9999 991 |
| ·78 | 1·0000 000 | ·9999 999 | ·9999 999 | ·9999 999 | ·9999 998 | ·9999 997 |
| ·79 | | 1·0000 000 | 1·0000 000 | 1·0000 000 | ·9999 999 | ·9999 999 |
| ·80 | | | | | 1·0000 000 | 1·0000 000 |

$x = \cdot 27$ to $\cdot 81$ $q = 39$ $p = 45$ to $50$

| | $p = 45$ | $p = 46$ | $p = 47$ | $p = 48$ | $p = 49$ | $p = 50$ |
|---|---|---|---|---|---|---|
| $B(p, q) =$ | $\cdot 3523\ 8315 \times \frac{1}{10^{25}}$ | $\cdot 1887\ 7669 \times \frac{1}{10^{25}}$ | $\cdot 1021\ 6150 \overset{+}{\times} \frac{1}{10^{25}}$ | $\cdot 5583\ 2448 \times \frac{1}{10^{26}}$ | $\cdot 3080\ 4109 \times \frac{1}{10^{26}}$ | $\cdot 1715\ 2288 \times \frac{1}{10^{26}}$ |
| $x$ | | | | | | |
| ·27 | ·0000 002 | ·0000 001 | | | | |
| ·28 | ·0000 005− | ·0000 002 | ·0000 001 | ·0000 001 | | |
| ·29 | ·0000 013 | ·0000 007 | ·0000 004 | ·0000 002 | ·0000 001 | ·0000 001 |
| ·30 | ·0000 037 | ·0000 020 | ·0000 011 | ·0000 006 | ·0000 003 | ·0000 002 |
| ·31 | ·0000 096 | ·0000 054 | ·0000 030 | ·0000 016 | ·0000 009 | ·0000 005− |
| ·32 | ·0000 236 | ·0000 136 | ·0000 078 | ·0000 044 | ·0000 025− | ·0000 014 |
| ·33 | ·0000 552 | ·0000 328 | ·0000 194 | ·0000 113 | ·0000 066 | ·0000 038 |
| ·34 | ·0001 230 | ·0000 753 | ·0000 457 | ·0000 275− | ·0000 164 | ·0000 097 |
| ·35 | ·0002 617 | ·0001 648 | ·0001 028 | ·0000 636 | ·0000 390 | ·0000 238 |
| ·36 | ·0005 328 | ·0003 447 | ·0002 210 | ·0001 405+ | ·0000 886 | ·0000 554 |
| ·37 | ·0010 400 | ·0006 907 | ·0004 548 | ·0002 969 | ·0001 922 | ·0001 234 |
| ·38 | ·0019 495+ | ·0013 282 | ·0008 971 | ·0006 008 | ·0003 990 | ·0002 629 |
| ·39 | ·0035 154 | ·0024 548 | ·0016 995+ | ·0011 668 | ·0007 945+ | ·0005 367 |
| ·40 | ·0061 067 | ·0043 674 | ·0030 970 | ·0021 780 | ·0015 193 | ·0010 516 |
| ·41 | ·0102 333 | ·0074 900 | ·0054 361 | ·0039 131 | ·0027 944 | ·0019 800 |
| ·42 | ·0165 633 | ·0123 977 | ·0092 028 | ·0067 761 | ·0049 499 | ·0035 882 |
| ·43 | ·0259 251 | ·0198 301 | ·0150 442 | ·0113 224 | ·0084 551 | ·0062 660 |
| ·44 | ·0392 847 | ·0306 851 | ·0237 755+ | ·0182 773 | ·0139 429 | ·0105 567 |
| ·45 | ·0576 925+ | ·0459 846 | ·0363 637 | ·0285 340 | ·0222 212 | ·0171 775− |
| ·46 | ·0821 974 | ·0668 077 | ·0538 803 | ·0431 258 | ·0342 623 | ·0270 232 |
| ·47 | ·1137 298 | ·0941 893 | ·0774 185− | ·0631 633 | ·0511 592 | ·0411 420 |
| ·48 | ·1529 668 | ·1289 918 | ·1079 772 | ·0897 347 | ·0740 462 | ·0606 758 |
| ·49 | ·2001 967 | ·1717 630 | ·1463 211 | ·1237 752 | ·1039 818 | ·0867 615− |
| ·50 | ·2552 050− | ·2226 023 | ·1928 347 | ·1659 172 | ·1418 036 | ·1203 966 |
| ·51 | ·3172 068 | ·2810 594 | ·2473 951 | ·2163 452 | ·1879 734 | ·1622 824 |
| ·52 | ·3848 443 | ·3460 890 | ·3092 883 | ·2746 800 | ·2424 366 | ·2126 674 |
| ·53 | ·4562 585− | ·4160 791 | ·3771 925− | ·3399 192 | ·3045 252 | ·2712 187 |
| ·54 | ·5292 325− | ·4889 564 | ·4492 407 | ·4104 545+ | ·3729 289 | ·3369 502 |
| ·55 | ·6013 894 | ·5623 609 | ·5231 628 | ·4841 731 | ·4457 521 | ·4082 328 |
| ·56 | ·6704 164 | ·6338 663 | ·5964 899 | ·5586 364 | ·5206 566 | ·4828 940 |
| ·57 | ·7342 807 | ·7012 125− | ·6667 928 | ·6313 112 | ·5950 757 | ·5584 038 |
| ·58 | ·7914 060 | ·7625 155+ | ·7319 168 | ·6998 206 | ·6664 674 | ·6321 203 |
| ·59 | ·8407 827 | ·8164 224 | ·7901 768 | ·7621 721 | ·7325 689 | ·7015 580 |
| ·60 | ·8820 013 | ·8621 897 | ·8404 832 | ·8169 292 | ·7916 088 | ·7646 347 |
| ·61 | ·9152 097 | ·8996 810 | ·8823 834 | ·8633 009 | ·8424 452 | ·8198 573 |
| ·62 | ·9410 119 | ·9292 910 | ·9160 210 | ·9011 416 | ·8846 131 | ·8664 183 |
| ·63 | ·9603 284 | ·9518 177 | ·9420 268 | ·9308 713 | ·9182 795+ | ·9041 948 |
| ·64 | ·9742 482 | ·9683 099 | ·9613 697 | ·9533 368 | ·9441 258 | ·9336 590 |
| ·65 | ·9838 926 | ·9799 159 | ·9751 957 | ·9696 469 | ·9631 849 | ·9557 272 |
| ·66 | ·9903 091 | ·9877 568 | ·9846 808 | ·9810 092 | ·9766 675− | ·9715 797 |
| ·67 | ·9944 024 | ·9928 350+ | ·9909 173 | ·9885 937 | ·9858 043 | ·9824 861 |
| ·68 | ·9969 023 | ·9959 829 | ·9948 412 | ·9934 371 | ·9917 265+ | ·9896 612 |
| ·69 | ·9983 613 | ·9978 471 | ·9971 993 | ·9963 909 | ·9953 915+ | ·9941 672 |
| ·70 | ·9991 733 | ·9988 998 | ·9985 503 | ·9981 077 | ·9975 527 | ·9968 629 |
| ·71 | ·9996 034 | ·9994 654 | ·9992 864 | ·9990 567 | ·9987 644 | ·9983 959 |
| ·72 | ·9998 196 | ·9997 537 | ·9996 670 | ·9995 542 | ·9994 086 | ·9992 226 |
| ·73 | ·9999 224 | ·9998 927 | ·9998 532 | ·9998 009 | ·9997 326 | ·9996 441 |
| ·74 | ·9999 686 | ·9999 560 | ·9999 390 | ·9999 163 | ·9998 862 | ·9998 466 |
| ·75 | ·9999 881 | ·9999 831 | ·9999 763 | ·9999 670 | ·9999 546 | ·9999 380 |
| ·76 | ·9999 958 | ·9999 939 | ·9999 914 | ·9999 879 | ·9999 831 | ·9999 766 |
| ·77 | ·9999 986 | ·9999 980 | ·9999 971 | ·9999 959 | ·9999 942 | ·9999 918 |
| ·78 | ·9999 996 | ·9999 994 | ·9999 991 | ·9999 987 | ·9999 981 | ·9999 974 |
| ·79 | ·9999 999 | ·9999 998 | ·9999 997 | ·9999 996 | ·9999 995− | ·9999 992 |
| ·80 | 1·0000 000 | 1·0000 000 | ·9999 999 | ·9999 999 | ·9999 999 | ·9999 998 |
| ·81 | | | 1·0000 000 | 1·0000 000 | 1·0000 000 | 1·0000 000 |

$x$ = ·23 to ·79 $q$ = 40 $p$ = 40 to 45

| | $p$ = 40 | $p$ = 41 | $p$ = 42 | $p$ = 43 | $p$ = 44 | $p$ = 45 |
|---|---|---|---|---|---|---|
| $B(p, q)$ = | ·4650 8509 $\times \frac{1}{10^{24}}$ | ·2325 4255 $\times \frac{1}{10^{24}}$ | ·1177 0672 $\times \frac{1}{10^{24}}$ | ·6028 8808 $\times \frac{1}{10^{25}}$ | ·3123 3961 $\times \frac{1}{10^{25}}$ | ·1636 0646 $\times \frac{1}{10^{25}}$ |
| $x$ | | | | | | |
| ·23 | ·0000 001 | | | | | |
| ·24 | ·0000 003 | ·0000 001 | ·0000 001 | | | |
| ·25 | ·0000 009 | ·0000 004 | ·0000 002 | ·0000 001 | | |
| ·26 | ·0000 025+ | ·0000 013 | ·0000 006 | ·0000 003 | ·0000 002 | ·0000 001 |
| ·27 | ·0000 069 | ·0000 036 | ·0000 018 | ·0000 009 | ·0000 005− | ·0000 002 |
| ·28 | ·0000 176 | ·0000 095+ | ·0000 051 | ·0000 027 | ·0000 014 | ·0000 007 |
| ·29 | ·0000 428 | ·0000 239 | ·0000 132 | ·0000 072 | ·0000 039 | ·0000 021 |
| ·30 | ·0000 983 | ·0000 567 | ·0000 324 | ·0000 183 | ·0000 102 | ·0000 057 |
| ·31 | ·0002 147 | ·0001 279 | ·0000 753 | ·0000 439 | ·0000 253 | ·0000 145− |
| ·32 | ·0004 468 | ·0002 744 | ·0001 667 | ·0001 002 | ·0000 596 | ·0000 351 |
| ·33 | ·0008 883 | ·0005 618 | ·0003 515+ | ·0002 177 | ·0001 335− | ·0000 810 |
| ·34 | ·0016 909 | ·0011 003 | ·0007 084 | ·0004 515− | ·0002 849 | ·0001 781 |
| ·35 | ·0030 880 | ·0020 655+ | ·0013 672 | ·0008 959 | ·0005 813 | ·0003 736 |
| ·36 | ·0054 208 | ·0037 237 | ·0025 316 | ·0017 039 | ·0011 357 | ·0007 499 |
| ·37 | ·0091 626 | ·0064 580 | ·0045 055− | ·0031 122 | ·0021 291 | ·0014 429 |
| ·38 | ·0149 363 | ·0107 923 | ·0077 196 | ·0054 678 | ·0038 361 | ·0026 664 |
| ·39 | ·0235 164 | ·0174 044 | ·0127 533 | ·0092 550− | ·0066 532 | ·0047 391 |
| ·40 | ·0358 103 | ·0271 236 | ·0203 437 | ·0151 134 | ·0111 239 | ·0081 136 |
| ·41 | ·0528 105+ | ·0409 024 | ·0313 759 | ·0238 432 | ·0179 537 | ·0133 986 |
| ·42 | ·0755 172 | ·0597 593 | ·0468 455− | ·0363 853 | ·0280 074 | ·0213 698 |
| ·43 | ·1048 338 | ·0846 900 | ·0677 890 | ·0537 731 | ·0422 801 | ·0329 578 |
| ·44 | ·1414 449 | ·1165 543 | ·0951 848 | ·0770 512 | ·0618 359 | ·0492 072 |
| ·45 | ·1856 930 | ·1559 479 | ·1298 302 | ·1071 638 | ·0877 129 | ·0712 017 |
| ·46 | ·2374 727 | ·2030 796 | ·1722 097 | ·1448 237 | ·1208 005− | ·0999 548 |
| ·47 | ·2961 642 | ·2576 734 | ·2223 746 | ·1903 787 | ·1617 014 | ·1362 765− |
| ·48 | ·3606 217 | ·3189 164 | ·2798 559 | ·2436 970 | ·2105 990 | ·1806 303 |
| ·49 | ·4292 260 | ·3854 679 | ·3436 310 | ·3040 950+ | ·2671 519 | ·2330 048 |
| ·50 | ·5000 000 $^e$ | ·4555 361 | ·4121 566 | ·3703 264 | ·3304 418 | ·2928 234 |
| ·51 | ·5707 740 | ·5270 160 | ·4834 714 | ·4406 422 | ·3989 883 | ·3589 154 |
| ·52 | ·6393 783 | ·5976 731 | ·5553 575− | ·5129 210 | ·4708 398 | ·4295 620 |
| ·53 | ·7038 358 | ·6653 450− | ·6255 399 | ·5848 534 | ·5437 317 | ·5026 193 |
| ·54 | ·7625 273 | ·7281 342 | ·6918 956 | ·6541 557 | ·6152 924 | ·5757 048 |
| ·55 | ·8143 070 | ·7845 619 | ·7526 403 | ·7187 806 | ·6832 673 | ·6464 222 |
| ·56 | ·8585 551 | ·8336 645− | ·8064 669 | ·7770 935+ | ·7457 255+ | ·7125 895− |
| ·57 | ·8951 662 | ·8750 225− | ·8526 187 | ·8279 906 | ·8012 204 | ·7724 363 |
| ·58 | ·9244 828 | ·9087 249 | ·8908 915+ | ·8709 437 | ·8488 804 | ·8247 411 |
| ·59 | ·9471 895− | ·9352 814 | ·9215 726 | ·9059 739 | ·8884 235+ | ·8688 908 |
| ·60 | ·9641 897 | ·9555 029 | ·9453 331 | ·9335 651 | ·9201 004 | ·9048 607 |
| ·61 | ·9764 836 | ·9703 716 | ·9630 968 | ·9545 385− | ·9445 830 | ·9331 274 |
| ·62 | ·9850 637 | ·9809 198 | ·9759 065+ | ·9699 122 | ·9628 248 | ·9545 359 |
| ·63 | ·9908 374 | ·9881 328 | ·9848 082 | ·9807 687 | ·9759 158 | ·9701 484 |
| ·64 | ·9945 792 | ·9928 821 | ·9907 628 | ·9881 470 | ·9849 545− | ·9811 002 |
| ·65 | ·9969 120 | ·9958 895− | ·9945 927 | ·9929 670 | ·9909 519 | ·9884 812 |
| ·66 | ·9983 091 | ·9977 185− | ·9969 578 | ·9959 897 | ·9947 711 | ·9932 540 |
| ·67 | ·9991 117 | ·9987 851 | ·9983 582 | ·9978 066 | ·9971 018 | ·9962 110 |
| ·68 | ·9995 532 | ·9993 807 | ·9991 518 | ·9988 517 | ·9984 625− | ·9979 632 |
| ·69 | ·9997 853 | ·9996 984 | ·9995 815− | ·9994 258 | ·9992 211 | ·9989 545+ |
| ·70 | ·9999 017 | ·9998 601 | ·9998 032 | ·9997 265+ | ·9996 241 | ·9994 889 |
| ·71 | ·9999 572 | ·9999 383 | ·9999 121 | ·9998 762 | ·9998 277 | ·9997 626 |
| ·72 | ·9999 824 | ·9999 742 | ·9999 628 | ·9999 469 | ·9999 252 | ·9998 956 |
| ·73 | ·9999 931 | ·9999 898 | ·9999 851 | ·9999 785+ | ·9999 693 | ·9999 567 |
| ·74 | ·9999 975− | ·9999 962 | ·9999 944 | ·9999 918 | ·9999 882 | ·9999 831 |
| ·75 | ·9999 991 | ·9999 987 | ·9999 980 | ·9999 971 | ·9999 957 | ·9999 938 |
| ·76 | ·9999 997 | ·9999 996 | ·9999 994 | ·9999 990 | ·9999 986 | ·9999 979 |
| ·77 | ·9999 999 | ·9999 999 | ·9999 998 | ·9999 997 | ·9999 996 | ·9999 993 |
| ·78 | 1·0000 000 | 1·0000 000 | ·9999 999 | ·9999 999 | ·9999 999 | ·9999 998 |
| ·79 | | | 1·0000 000 | 1·0000 000 | 1·0000 000 | 1·0000 000 |

$x$ = ·27 to ·81 $q$ = 40 $p$ = 46 to 50

| $x$ | $p$ = 46 | $p$ = 47 | $p$ = 48 | $p$ = 49 | $p$ = 50 |
|---|---|---|---|---|---|
| $B(p, q)$ = | ·8661 5186 $\times\frac{1}{10^{26}}$ | ·4632 9053 $\times\frac{1}{10^{26}}$ | ·2502 8339 $\times\frac{1}{10^{26}}$ | ·1365 1821 $\times\frac{1}{10^{26}}$ | ·7516 1712 $\times\frac{1}{10^{27}}$ |
| ·27 | ·0000 001 | ·0000 001 | | | |
| ·28 | ·0000 004 | ·0000 002 | ·0000 001 | | |
| ·29 | ·0000 011 | ·0000 006 | ·0000 003 | ·0000 002 | ·0000 001 |
| ·30 | ·0000 031 | ·0000 017 | ·0000 009 | ·0000 005− | ·0000 003 |
| ·31 | ·0000 082 | ·0000 046 | ·0000 026 | ·0000 014 | ·0000 008 |
| ·32 | ·0000 205+ | ·0000 119 | ·0000 068 | ·0000 039 | ·0000 022 |
| ·33 | ·0000 488 | ·0000 291 | ·0000 172 | ·0000 101 | ·0000 058 |
| ·34 | ·0001 103 | ·0000 676 | ·0000 411 | ·0000 248 | ·0000 148 |
| ·35 | ·0002 379 | ·0001 501 | ·0000 939 | ·0000 582 | ·0000 358 |
| ·36 | ·0004 906 | ·0003 181 | ·0002 044 | ·0001 303 | ·0000 824 |
| ·37 | ·0009 690 | ·0006 450+ | ·0004 257 | ·0002 786 | ·0001 808 |
| ·38 | ·0018 367 | ·0012 541 | ·0008 490 | ·0005 700 | ·0003 796 |
| ·39 | ·0033 457 | ·0023 415+ | ·0016 249 | ·0011 184 | ·0007 636 |
| ·40 | ·0058 659 | ·0042 046 | ·0029 886 | ·0021 070 | ·0014 737 |
| ·41 | ·0099 125+ | ·0072 714 | ·0052 900 | ·0038 176 | ·0027 334 |
| ·42 | ·0161 660 | ·0121 273 | ·0090 236 | ·0066 609 | ·0048 787 |
| ·43 | ·0254 750+ | ·0195 295− | ·0148 515− | ·0112 056 | ·0083 900 |
| ·44 | ·0388 348 | ·0304 016 | ·0236 120 | ·0181 973 | ·0139 185+ |
| ·45 | ·0573 323 | ·0457 996 | ·0363 035+ | ·0285 583 | ·0222 988 |
| ·46 | ·0820 554 | ·0668 408 | ·0540 347 | ·0433 576 | ·0345 370 |
| ·47 | ·1139 703 | ·0945 978 | ·0779 375+ | ·0637 450+ | ·0517 654 |
| ·48 | ·1537 783 | ·1299 617 | ·1090 437 | ·0908 450+ | ·0751 570 |
| ·49 | ·2017 715− | ·1734 918 | ·1481 363 | ·1256 175− | ·1058 009 |
| ·50 | ·2577 129 | ·2252 738 | ·1955 955− | ·1686 995+ | ·1445 480 |
| ·51 | ·3207 659 | ·2848 142 | ·2512 644 | ·2202 518 | ·1918 468 |
| ·52 | ·3894 949 | ·3509 958 | ·3143 642 | ·2798 390 | ·2475 966 |
| ·53 | ·4619 454 | ·4221 115+ | ·3834 811 | ·3463 718 | ·3110 498 |
| ·54 | ·5358 005+ | ·4959 830 | ·4566 399 | ·4181 328 | ·3807 888 |
| ·55 | ·6085 947 | ·5701 503 | ·5314 606 | ·4928 918 | ·4547 952 |
| ·56 | ·6779 513 | ·6421 083 | ·6053 806 | ·5681 021 | ·5306 105+ |
| ·57 | ·7418 101 | ·7095 526 | ·6759 088 | ·6411 506 | ·6055 694 |
| ·58 | ·7986 064 | ·7705 967 | ·7408 708 | ·7096 213 | ·6770 709 |
| ·59 | ·8473 787 | ·8239 259 | ·7986 069 | ·7715 313 | ·7428 423 |
| ·60 | ·8877 923 | ·8688 687 | ·8480 929 | ·8254 992 | ·8011 534 |
| ·61 | ·9200 833 | ·9053 804 | ·8889 694 | ·8708 249 | ·8509 476 |
| ·62 | ·9449 429 | ·9339 526 | ·9214 844 | ·9074 733 | ·8918 724 |
| ·63 | ·9633 661 | ·9554 705+ | ·9463 688 | ·9359 758 | ·9242 168 |
| ·64 | ·9764 957 | ·9710 503 | ·9646 735− | ·9572 763 | ·9487 741 |
| ·65 | ·9854 833 | ·9818 826 | ·9776 001 | ·9725 548 | ·9666 651 |
| ·66 | ·9913 850− | ·9891 056 | ·9863 528 | ·9830 598 | ·9791 566 |
| ·67 | ·9950 969 | ·9937 177 | ·9920 267 | ·9899 733 | ·9875 025+ |
| ·68 | ·9973 295+ | ·9965 332 | ·9955 425− | ·9943 214 | ·9928 301 |
| ·69 | ·9986 112 | ·9981 735+ | ·9976 209 | ·9969 298 | ·9960 734 |
| ·70 | ·9993 122 | ·9990 836 | ·9987 908 | ·9984 194 | ·9979 525− |
| ·71 | ·9996 764 | ·9995 633 | ·9994 164 | ·9992 273 | ·9989 862 |
| ·72 | ·9998 559 | ·9998 030 | ·9997 333 | ·9996 424 | ·9995 249 |
| ·73 | ·9999 394 | ·9999 161 | ·9998 850+ | ·9998 439 | ·9997 899 |
| ·74 | ·9999 760 | ·9999 664 | ·9999 534 | ·9999 359 | ·9999 127 |
| ·75 | ·9999 911 | ·9999 874 | ·9999 823 | ·9999 754 | ·9999 661 |
| ·76 | ·9999 969 | ·9999 956 | ·9999 938 | ·9999 912 | ·9999 877 |
| ·77 | ·9999 990 | ·9999 986 | ·9999 980 | ·9999 971 | ·9999 959 |
| ·78 | ·9999 997 | ·9999 996 | ·9999 994 | ·9999 991 | ·9999 987 |
| ·79 | ·9999 999 | ·9999 999 | ·9999 998 | ·9999 998 | ·9999 996 |
| ·80 | 1·0000 000 | 1·0000 000 | 1·0000 000 | ·9999 999 | ·9999 999 |
| ·81 | | | | 1·0000 000 | 1·0000 000 |

$x$ = ·23 to ·79 $q$ = 41 $p$ = 41 to 45

| | $p$ = 41 | $p$ = 42 | $p$ = 43 | $p$ = 44 | $p$ = 45 |
|---|---|---|---|---|---|
| $B(p, q)$ = | ·1148 3583 $\times\frac{1}{10^{24}}$ | ·5741 7913 $\times\frac{1}{10^{25}}$ | ·2905 4847 $\times\frac{1}{10^{25}}$ | ·1487 3315 $\times\frac{1}{10^{25}}$ | ·7699 1276 $\times\frac{1}{10^{26}}$ |
| $x$ | | | | | |
| ·23 | ·0000 001 | | | | |
| ·24 | ·0000 002 | ·0000 001 | | | |
| ·25 | ·0000 006 | ·0000 003 | ·0000 001 | ·0000 001 | |
| ·26 | ·0000 019 | ·0000 010 | ·0000 005− | ·0000 002 | ·0000 001 |
| ·27 | ·0000 054 | ·0000 028 | ·0000 014 | ·0000 007 | ·0000 004 |
| ·28 | ·0000 141 | ·0000 076 | ·0000 041 | ·0000 021 | ·0000 011 |
| ·29 | ·0000 349 | ·0000 195− | ·0000 108 | ·0000 059 | ·0000 032 |
| ·30 | ·0000 817 | ·0000 472 | ·0000 269 | ·0000 152 | ·0000 085+ |
| ·31 | ·0001 817 | ·0001 083 | ·0000 639 | ·0000 373 | ·0000 216 |
| ·32 | ·0003 847 | ·0002 364 | ·0001 438 | ·0000 866 | ·0000 516 |
| ·33 | ·0007 773 | ·0004 920 | ·0003 082 | ·0001 911 | ·0001 174 |
| ·34 | ·0015 019 | ·0009 782 | ·0006 306 | ·0004 024 | ·0002 543 |
| ·35 | ·0027 813 | ·0018 621 | ·0012 341 | ·0008 097 | ·0005 262 |
| ·36 | ·0049 456 | ·0034 006 | ·0023 147 | ·0015 602 | ·0010 416 |
| ·37 | ·0084 594 | ·0059 684 | ·0041 690 | ·0028 839 | ·0019 761 |
| ·38 | ·0139 417 | ·0100 840 | ·0072 220 | ·0051 227 | ·0035 998 |
| ·39 | ·0221 717 | ·0164 265− | ·0120 519 | ·0087 587 | ·0063 068 |
| ·40 | ·0340 730 | ·0258 354 | ·0194 022 | ·0144 352 | ·0106 422 |
| ·41 | ·0506 670 | ·0392 853 | ·0301 745− | ·0229 642 | ·0173 205+ |
| ·42 | ·0729 959 | ·0578 287 | ·0453 915− | ·0353 087 | ·0272 242 |
| ·43 | ·1020 137 | ·0825 056 | ·0661 281 | ·0525 348 | ·0413 759 |
| ·44 | ·1384 581 | ·1142 250+ | ·0934 077 | ·0757 275+ | ·0608 761 |
| ·45 | ·1827 185− | ·1536 299 | ·1280 736 | ·1058 752 | ·0868 048 |
| ·46 | ·2347 213 | ·2009 650+ | ·1706 487 | ·1437 307 | ·1200 917 |
| ·47 | ·2938 548 | ·2559 703 | ·2212 067 | ·1896 689 | ·1613 709 |
| ·48 | ·3589 535− | ·3178 227 | ·2792 773 | ·2435 646 | ·2108 387 |
| ·49 | ·4283 508 | ·3851 437 | ·3438 089 | ·3047 138 | ·2681 422 |
| ·50 | ·5000 000[e] | ·4560 783 | ·4132 024 | ·3718 221 | ·3323 227 |
| ·51 | ·5716 492 | ·5284 421 | ·4854 202 | ·4430 686 | ·4018 335− |
| ·52 | ·6410 465+ | ·5999 158 | ·5581 583 | ·5162 454 | ·4746 374 |
| ·53 | ·7061 452 | ·6682 607 | ·6290 593 | ·5889 553 | ·5483 774 |
| ·54 | ·7652 787 | ·7315 225− | ·6959 338 | ·6588 387 | ·6205 971 |
| ·55 | ·8172 815+ | ·7881 929 | ·7569 574 | ·7237 968 | ·6889 783 |
| ·56 | ·8615 419 | ·8373 089 | ·8108 142 | ·7821 752 | ·7515 575− |
| ·57 | ·8979 863 | ·8784 783 | ·8567 686 | ·8328 828 | ·8068 908 |
| ·58 | ·9270 041 | ·9118 368 | ·8946 617 | ·8754 335+ | ·8541 427 |
| ·59 | ·9493 330 | ·9379 512 | ·9248 405− | ·9099 095+ | ·8930 919 |
| ·60 | ·9659 270 | ·9576 895− | ·9480 397 | ·9368 640 | ·9240 627 |
| ·61 | ·9778 283 | ·9720 830 | ·9652 406 | ·9571 842 | ·9478 020 |
| ·62 | ·9860 583 | ·9822 006 | ·9775 310 | ·9719 427 | ·9653 281 |
| ·63 | ·9915 406 | ·9890 496 | ·9859 857 | ·9822 598 | ·9777 786 |
| ·64 | ·9950 544 | ·9935 094 | ·9915 790 | ·9891 941 | ·9862 803 |
| ·65 | ·9972 187 | ·9962 996 | ·9951 332 | ·9936 698 | ·9918 537 |
| ·66 | ·9984 981 | ·9979 744 | ·9972 996 | ·9964 399 | ·9953 567 |
| ·67 | ·9992 227 | ·9989 374 | ·9985 642 | ·9980 816 | ·9974 643 |
| ·68 | ·9996 153 | ·9994 669 | ·9992 701 | ·9990 116 | ·9986 761 |
| ·69 | ·9998 183 | ·9997 449 | ·9996 460 | ·9995 142 | ·9993 407 |
| ·70 | ·9999 183 | ·9998 838 | ·9998 366 | ·9997 728 | ·9996 877 |
| ·71 | ·9999 651 | ·9999 498 | ·9999 284 | ·9998 992 | ·9998 596 |
| ·72 | ·9999 859 | ·9999 795− | ·9999 704 | ·9999 577 | ·9999 403 |
| ·73 | ·9999 946 | ·9999 921 | ·9999 884 | ·9999 833 | ·9999 761 |
| ·74 | ·9999 981 | ·9999 971 | ·9999 958 | ·9999 938 | ·9999 910 |
| ·75 | ·9999 994 | ·9999 990 | ·9999 985+ | ·9999 978 | ·9999 968 |
| ·76 | ·9999 998 | ·9999 997 | ·9999 995+ | ·9999 993 | ·9999 990 |
| ·77 | ·9999 999 | ·9999 999 | ·9999 999 | ·9999 998 | ·9999 997 |
| ·78 | 1·0000 000 | 1·0000 000 | 1·0000 000 | ·9999 999 | ·9999 999 |
| ·79 | | | | 1·0000 000 | 1·0000 000 |

$x = \cdot 26$ to $\cdot 80$ $q = 41$ $p = 46$ to $50$

| | $p = 46$ | $p = 47$ | $p = 48$ | $p = 49$ | $p = 50$ |
|---|---|---|---|---|---|
| $B(p, q) =$ | $\cdot 4028\ 6133 \times \frac{1}{10^{26}}$ | $\cdot 2130\ 0714 \times \frac{1}{10^{26}}$ | $\cdot 1137\ 6518 \times \frac{1}{10^{26}}$ | $\cdot 6135\ 6500^{-} \times \frac{1}{10^{27}}$ | $\cdot 3340\ 5205^{+} \times \frac{1}{10^{27}}$ |
| $x$ | | | | | |
| ·26 | ·0000 001 | | | | |
| ·27 | ·0000 002 | ·0000 001 | | | |
| ·28 | ·0000 006 | ·0000 003 | ·0000 002 | ·0000 001 | |
| ·29 | ·0000 017 | ·0000 009 | ·0000 005⁻ | ·0000 003 | ·0000 001 |
| ·30 | ·0000 047 | ·0000 026 | ·0000 014 | ·0000 008 | ·0000 004 |
| ·31 | ·0000 123 | ·0000 070 | ·0000 039 | ·0000 022 | ·0000 012 |
| ·32 | ·0000 305⁻ | ·0000 178 | ·0000 103 | ·0000 059 | ·0000 034 |
| ·33 | ·0000 714 | ·0000 430 | ·0000 257 | ·0000 152 | ·0000 089 |
| ·34 | ·0001 592 | ·0000 988 | ·0000 607 | ·0000 370 | ·0000 224 |
| ·35 | ·0003 388 | ·0002 161 | ·0001 367 | ·0000 857 | ·0000 533 |
| ·36 | ·0006 889 | ·0004 516 | ·0002 934 | ·0001 890 | ·0001 208 |
| ·37 | ·0013 417 | ·0009 028 | ·0006 022 | ·0003 983 | ·0002 613 |
| ·38 | ·0025 067 | ·0017 301 | ·0011 838 | ·0008 033 | ·0005 406 |
| ·39 | ·0045 005⁺ | ·0031 835⁺ | ·0022 328 | ·0015 530 | ·0010 715⁻ |
| ·40 | ·0077 764 | ·0056 333 | ·0040 465⁻ | ·0028 828 | ·0020 374 |
| ·41 | ·0129 498 | ·0095 996 | ·0070 569 | ·0051 457 | ·0037 225⁻ |
| ·42 | ·0208 104 | ·0157 742 | ·0118 589 | ·0088 440 | ·0065 441 |
| ·43 | ·0323 124 | ·0250 261 | ·0192 266 | ·0146 546 | ·0110 838 |
| ·44 | ·0485 330 | ·0383 794 | ·0301 097 | ·0234 387 | ·0181 074 |
| ·45 | ·0705 949 | ·0569 575⁻ | ·0455 978 | ·0362 260 | ·0285 661 |
| ·46 | ·0995 521 | ·0818 880 | ·0668 472 | ·0541 628 | ·0435 649 |
| ·47 | ·1362 486 | ·1141 737 | ·0949 685⁺ | ·0784 201 | ·0642 931 |
| ·48 | ·1811 673 | ·1545 404 | ·1308 821 | ·1100 628 | ·0919 118 |
| ·49 | ·2342 931 | ·2032 844 | ·1751 589 | ·1498 928 | ·1274 059 |
| ·50 | ·2950 178 | ·2601 458 | ·2278 706 | ·1982 851 | ·1714 166 |
| ·51 | ·3621 104 | ·3242 353 | ·2884 796 | ·2550 480 | ·2240 794 |
| ·52 | ·4337 690 | ·3940 378 | ·3557 945⁺ | ·3193 358 | ·2849 010 |
| ·53 | ·5077 543 | ·4675 022 | ·4280 123 | ·3896 413 | ·3527 033 |
| ·54 | ·5815 907 | ·5422 112 | ·5028 484 | ·4638 792 | ·4256 576 |
| ·55 | ·6528 056 | ·6156 107 | ·5777 432 | ·5395 600 | ·5014 159 |
| ·56 | ·7191 707 | ·6852 633 | ·6501 149 | ·6140 293 | ·5773 250⁺ |
| ·57 | ·7789 061 | ·7490 841 | ·7176 187 | ·6847 374 | ·6506 952 |
| ·58 | ·8308 175⁻ | ·8055 248 | ·7783 701 | ·7494 956 | ·7190 772 |
| ·59 | ·8743 495⁻ | ·8536 758 | ·8310 976 | ·8066 754 | ·7805 038 |
| ·60 | ·9095 545⁺ | ·8932 802 | ·8752 053 | ·8553 228 | ·8336 551 |
| ·61 | ·9369 917 | ·9246 633 | ·9107 427 | ·8951 748 | ·8779 261 |
| ·62 | ·9575 817 | ·9486 026 | ·9382 977 | ·9265 844 | ·9133 939 |
| ·63 | ·9724 460 | ·9661 651 | ·9588 404 | ·9503 805⁺ | ·9407 000 |
| ·64 | ·9827 579 | ·9785 432 | ·9735 501 | ·9676 915⁺ | ·9608 812 |
| ·65 | ·9896 241 | ·9869 146 | ·9836 545⁺ | ·9797 696 | ·9751 831 |
| ·66 | ·9940 063 | ·9923 401 | ·9903 044 | ·9878 412 | ·9848 884 |
| ·67 | ·9966 831 | ·9957 045⁻ | ·9944 908 | ·9930 001 | ·9911 859 |
| ·68 | ·9982 452 | ·9976 974 | ·9970 078 | ·9961 482 | ·9950 864 |
| ·69 | ·9991 146 | ·9988 229 | ·9984 503 | ·9979 789 | ·9973 882 |
| ·70 | ·9995 750⁺ | ·9994 276 | ·9992 366 | ·9989 914 | ·9986 797 |
| ·71 | ·9998 065⁻ | ·9997 360 | ·9996 433 | ·9995 227 | ·9993 671 |
| ·72 | ·9999 167 | ·9998 848 | ·9998 424 | ·9997 864 | ·9997 132 |
| ·73 | ·9999 662 | ·9999 527 | ·9999 344 | ·9999 100 | ·9998 775⁺ |
| ·74 | ·9999 871 | ·9999 817 | ·9999 744 | ·9999 644 | ·9999 509 |
| ·75 | ·9999 954 | ·9999 934 | ·9999 906 | ·9999 868 | ·9999 816 |
| ·76 | ·9999 985⁻ | ·9999 978 | ·9999 968 | ·9999 955⁻ | ·9999 936 |
| ·77 | ·9999 995⁺ | ·9999 993 | ·9999 990 | ·9999 986 | ·9999 979 |
| ·78 | ·9999 999 | ·9999 998 | ·9999 997 | ·9999 996 | ·9999 994 |
| ·79 | 1·0000 000 | 1·0000 000 | ·9999 999 | ·9999 999 | ·9999 998 |
| ·80 | | | 1·0000 000 | 1·0000 000 | 1·0000 000 |

$x$ = ·24 to ·79 $q = 42$ $p$ = 42 to 46

| $x$ | $p = 42$ | $p = 43$ | $p = 44$ | $p = 45$ | $p = 46$ |
|---|---|---|---|---|---|
| $B(p, q) =$ | ·2836 3065 $\overset{+}{\times}\frac{1}{10^{25}}$ | ·1418 1533 $\times\frac{1}{10^{25}}$ | ·7174 1871 $\times\frac{1}{10^{26}}$ | ·3670 5143 $\times\frac{1}{10^{26}}$ | ·1898 5419 $\times\frac{1}{10^{26}}$ |
| ·24 | ·0000 001 | ·0000 001 | | | |
| ·25 | ·0000 005$^-$ | ·0000 002 | ·0000 001 | ·0000 001 | |
| ·26 | ·0000 015$^-$ | ·0000 007 | ·0000 004 | ·0000 002 | ·0000 001 |
| ·27 | ·0000 042 | ·0000 022 | ·0000 011 | ·0000 006 | ·0000 003 |
| ·28 | ·0000 112 | ·0000 061 | ·0000 032 | ·0000 017 | ·0000 009 |
| ·29 | ·0000 284 | ·0000 159 | ·0000 088 | ·0000 048 | ·0000 026 |
| ·30 | ·0000 679 | ·0000 392 | ·0000 224 | ·0000 127 | ·0000 071 |
| ·31 | ·0001 538 | ·0000 918 | ·0000 542 | ·0000 317 | ·0000 183 |
| ·32 | ·0003 314 | ·0002 038 | ·0001 241 | ·0000 748 | ·0000 446 |
| ·33 | ·0006 803 | ·0004 310 | ·0002 703 | ·0001 678 | ·0001 032 |
| ·34 | ·0013 343 | ·0008 699 | ·0005 614 | ·0003 587 | ·0002 271 |
| ·35 | ·0025 055$^+$ | ·0016 791 | ·0011 140 | ·0007 319 | ·0004 764 |
| ·36 | ·0045 130 | ·0031 061 | ·0021 167 | ·0014 286 | ·0009 552 |
| ·37 | ·0078 118 | ·0055 168 | ·0038 581 | ·0026 724 | ·0018 340 |
| ·38 | ·0130 159 | ·0094 237 | ·0067 571 | ·0047 996 | ·0033 780 |
| ·39 | ·0209 078 | ·0155 057 | ·0113 900 | ·0082 892 | ·0059 781 |
| ·40 | ·0324 254 | ·0246 115$^+$ | ·0185 057 | ·0137 876 | ·0101 809 |
| ·41 | ·0486 183 | ·0377 364 | ·0290 208 | ·0221 176 | ·0167 086 |
| ·42 | ·0705 692 | ·0559 661 | ·0439 848 | ·0342 636 | ·0264 608 |
| ·43 | ·0992 825$^+$ | ·0803 845$^-$ | ·0645 101 | ·0513 236 | ·0404 872 |
| ·44 | ·1355 501 | ·1119 504 | ·0916 656 | ·0744 235$^-$ | ·0599 248 |
| ·45 | ·1798 096 | ·1513 548 | ·1263 410 | ·1045 961 | ·0858 954 |
| ·46 | ·2320 208 | ·1988 799 | ·1690 993 | ·1426 352 | ·1193 703 |
| ·47 | ·2915 817 | ·2542 830 | ·2200 375$^+$ | ·1889 442 | ·1610 155$^+$ |
| ·48 | ·3573 082 | ·3167 322 | ·2786 850$^+$ | ·2434 050$^-$ | ·2110 414 |
| ·49 | ·4274 867 | ·3848 110 | ·3439 614 | ·3052 936 | ·2690 834 |
| ·50 | ·5000 000$^e$ | ·4566 012 | ·4142 116 | ·3732 672 | ·3341 425$^-$ |
| ·51 | ·5725 133 | ·5298 377 | ·4873 208 | ·4454 320 | ·4046 044 |
| ·52 | ·6426 918 | ·6021 157 | ·5608 980 | ·5194 929 | ·4783 454 |
| ·53 | ·7084 183 | ·6711 196 | ·6325 024 | ·5929 636 | ·5529 152 |
| ·54 | ·7679 792 | ·7348 383 | ·6998 785$^+$ | ·6634 091 | ·6257 726 |
| ·55 | ·8201 904 | ·7917 355$^+$ | ·7611 631 | ·7286 799 | ·6945 365$^+$ |
| ·56 | ·8644 499 | ·8408 502 | ·8150 332 | ·7871 039 | ·7572 133 |
| ·57 | ·9007 175$^-$ | ·8818 194 | ·8607 767 | ·8376 058 | ·8123 649 |
| ·58 | ·9294 308 | ·9148 278 | ·8982 822 | ·8797 436 | ·8591 946 |
| ·59 | ·9513 817 | ·9404 998 | ·9279 578 | ·9136 628 | ·8975 443 |
| ·60 | ·9675 746 | ·9597 606 | ·9506 020 | ·9399 863 | ·9278 136 |
| ·61 | ·9790 922 | ·9736 901 | ·9672 528 | ·9596 670 | ·9508 236 |
| ·62 | ·9869 841 | ·9833 920 | ·9790 412 | ·9738 302 | ·9676 558 |
| ·63 | ·9921 882 | ·9898 933 | ·9870 689 | ·9836 315$^-$ | ·9794 928 |
| ·64 | ·9954 870 | ·9940 801 | ·9923 212 | ·9901 465$^-$ | ·9874 866 |
| ·65 | ·9974 945$^-$ | ·9966 680 | ·9956 186 | ·9943 009 | ·9926 640 |
| ·66 | ·9986 657 | ·9982 012 | ·9976 024 | ·9968 388 | ·9958 758 |
| ·67 | ·9993 197 | ·9990 704 | ·9987 441 | ·9983 218 | ·9977 810 |
| ·68 | ·9996 686 | ·9995 411 | ·9993 717 | ·9991 491 | ·9988 599 |
| ·69 | ·9998 462 | ·9997 841 | ·9997 005$^-$ | ·9995 889 | ·9994 419 |
| ·70 | ·9999 321 | ·9999 035$^-$ | ·9998 643 | ·9998 113 | ·9997 404 |
| ·71 | ·9999 716 | ·9999 591 | ·9999 417 | ·9999 179 | ·9998 856 |
| ·72 | ·9999 888 | ·9999 836 | ·9999 764 | ·9999 663 | ·9999 524 |
| ·73 | ·9999 958 | ·9999 938 | ·9999 910 | ·9999 870 | ·9999 813 |
| ·74 | ·9999 985$^+$ | ·9999 978 | ·9999 968 | ·9999 953 | ·9999 931 |
| ·75 | ·9999 995$^+$ | ·9999 993 | ·9999 989 | ·9999 984 | ·9999 977 |
| ·76 | ·9999 999 | ·9999 998 | ·9999 997 | ·9999 995$^-$ | ·9999 993 |
| ·77 | 1·0000 000 | ·9999 999 | ·9999 999 | ·9999 999 | ·9999 998 |
| ·78 | | 1·0000 000 | 1·0000 000 | 1·0000 000 | ·9999 999 |
| ·79 | | | | | 1·0000 000 |

$x$ = ·27 to ·80 $q$ = 42 $p$ = 47 to 50

| | $p$ = 47 | $p$ = 48 | $p$ = 49 | $p$ = 50 |
|---|---|---|---|---|
| $B(p, q)$ = | ·9924 1962 × $\frac{1}{10^{27}}$ | ·5240 8677 × $\frac{1}{10^{27}}$ | ·2795 1294 × $\frac{1}{10^{27}}$ | ·1505 0697 × $\frac{1}{10^{27}}$ |
| $x$ | | | | |
| ·27 | ·0000 001 | ·0000 001 | | |
| ·28 | ·0000 005⁻ | ·0000 002 | ·0000 001 | ·0000 001 |
| ·29 | ·0000 014 | ·0000 007 | ·0000 004 | ·0000 002 |
| ·30 | ·0000 040 | ·000C 022 | ·0000 012 | ·0000 006 |
| ·31 | ·0000 105⁺ | ·0000 060 | ·0000 034 | ·0000 019 |
| ·32 | ·0000 264 | ·0000 155⁻ | ·0000 090 | ·0000 052 |
| ·33 | ·0000 629 | ·0000 380 | ·0000 227 | ·0000 135⁻ |
| ·34 | ·0001 424 | ·0000 885⁺ | ·0000 545⁺ | ·0000 333 |
| ·35 | ·0003 072 | ·0001 964 | ·0001 244 | ·0000 782 |
| ·36 | ·0006 329 | ·0004 156 | ·0002 706 | ·0001 747 |
| ·37 | ·0012 474 | ·0008 409 | ·0005 621 | ·0003 726 |
| ·38 | ·0023 563 | ·0016 294 | ·0011 172 | ·0007 597 |
| ·39 | ·0042 734 | ·0030 286 | ·0021 285⁻ | ·0014 837 |
| ·40 | ·0074 524 | ·0054 088 | ·0038 932 | ·0027 797 |
| ·41 | ·0125 143 | ·0092 945⁻ | ·0068 467 | ·0050 034 |
| ·42 | ·0202 626 | ·0153 884 | ·0115 927 | ·0086 645⁺ |
| ·43 | ·0316 744 | ·0245 791 | ·0189 221 | ·0144 543 |
| ·44 | ·0478 594 | ·0379 196 | ·0298 103 | ·0232 567 |
| ·45 | ·0699 796 | ·0565 696 | ·0453 806 | ·0361 326 |
| ·46 | ·0991 299 | ·0816 972 | ·0668 287 | ·0542 662 |
| ·47 | ·1361 893 | ·1143 423 | ·0953 035⁺ | ·0788 680 |
| ·48 | ·1816 609 | ·1552 559 | ·1317 555⁺ | ·1110 368 |
| ·49 | ·2355 259 | ·2047 387 | ·1767 673 | ·1515 930 |
| ·50 | ·2971 441 | ·2625 074 | ·2303 962 | ·2009 064 |
| ·51 | ·3652 235⁺ | ·3276 190 | ·2920 592 | ·2587 491 |
| ·52 | ·4378 778 | ·3984 778 | ·3604 897 | ·3242 071 |
| ·53 | ·5127 711 | ·4729 345⁻ | ·4337 867 | ·3956 775⁻ |
| ·54 | ·5873 344 | ·5484 708 | ·5095 587 | ·4709 642 |
| ·55 | ·6590 199 | ·6224 454 | ·5851 470 | ·5474 680 |
| ·56 | ·7255 553 | ·6923 615⁺ | ·6578 953 | ·6224 444 |
| ·57 | ·7851 542 | ·7561 139 | ·7254 220 | ·6932 895⁻ |
| ·58 | ·8366 533 | ·8121 743 | ·7858 493 | ·7578 050⁺ |
| ·59 | ·8795 582 | ·8596 894 | ·8379 536 | ·8143 992 |
| ·60 | ·9140 002 | ·8984 822 | ·8812 185⁻ | ·8621 931 |
| ·61 | ·9406 211 | ·9289 685⁺ | ·9157 890 | ·9010 225⁻ |
| ·62 | ·9604 156 | ·9520 108 | ·9423 488 | ·9313 459 |
| ·63 | ·9745 616 | ·9687 447 | ·9619 500⁻ | ·9540 875⁻ |
| ·64 | ·9842 669 | ·9804 089 | ·9758 306 | ·9704 488 |
| ·65 | ·9906 517 | ·9882 027 | ·9852 511 | ·9817 273 |
| ·66 | ·9946 736 | ·9931 881 | ·9913 702 | ·9891 664 |
| ·67 | ·9970 958 | ·9962 361 | ·9951 682 | ·9938 541 |
| ·68 | ·9984 879 | ·9980 142 | ·9974 171 | ·9966 713 |
| ·69 | ·9992 500⁻ | ·9990 021 | ·9986 849 | ·9982 829 |
| ·70 | ·9996 466 | ·9995 236 | ·9993 639 | ·9991 587 |
| ·71 | ·9998 422 | ·9997 845⁺ | ·9997 086 | ·9996 095⁺ |
| ·72 | ·9999 335⁻ | ·9999 080 | ·9998 739 | ·9998 289 |
| ·73 | ·9999 736 | ·9999 630 | ·9999 487 | ·9999 295⁻ |
| ·74 | ·9999 902 | ·9999 861 | ·9999 804 | ·9999 728 |
| ·75 | ·9999 966 | ·9999 951 | ·9999 930 | ·9999 902 |
| ·76 | ·9999 989 | ·9999 984 | ·9999 977 | ·9999 967 |
| ·77 | ·9999 997 | ·9999 995⁺ | ·9999 993 | ·9999 990 |
| ·78 | ·9999 999 | ·9999 999 | ·9999 998 | ·9999 997 |
| ·79 | 1·0000 000 | 1·0000 000 | 1·0000 000 | ·9999 999 |
| ·80 | | | | 1·0000 000 |

$x$ = ·24 to ·78 $q = 43$ $p$ = 43 to 47

| | $p = 43$ | $p = 44$ | $p = 45$ | $p = 46$ | $p = 47$ |
|---|---|---|---|---|---|
| $B(p, q)$ = | ·7007 3455$^{+}$ × $\frac{1}{10^{26}}$ | ·3503 6728 × $\frac{1}{10^{26}}$ | ·1771 9724 × $\frac{1}{10^{26}}$ | ·9061 2227 × $\frac{1}{10^{27}}$ | ·4683 3286 × $\frac{1}{10^{27}}$ |
| $x$ | | | | | |
| ·24 | ·0000 001 | | | | |
| ·25 | ·0000 004 | ·0000 002 | ·0000 001 | | |
| ·26 | ·0000 011 | ·0000 006 | ·0000 003 | ·0000 001 | ·0000 001 |
| ·27 | ·0000 033 | ·0000 017 | ·0000 009 | ·0000 004 | ·0000 002 |
| ·28 | ·0000 090 | ·0000 048 | ·0000 026 | ·0000 014 | ·0000 007 |
| ·29 | ·0000 231 | ·0000 129 | ·0000 072 | ·0000 039 | ·0000 021 |
| ·30 | ·0000 564 | ·0000 326 | ·0000 187 | ·0000 106 | ·0000 059 |
| ·31 | ·0001 302 | ·0000 778 | ·0000 460 | ·0000 269 | ·0000 156 |
| ·32 | ·0002 854 | ·0001 757 | ·0001 071 | ·0000 646 | ·0000 386 |
| ·33 | ·0005 955$^{+}$ | ·0003 776 | ·0002 371 | ·0001 474 | ·0000 908 |
| ·34 | ·0011 857 | ·0007 736 | ·0004 998 | ·0003 198 | ·0002 027 |
| ·35 | ·0022 576 | ·0015 143 | ·0010 058 | ·0006 617 | ·0004 313 |
| ·36 | ·0041 191 | ·0028 376 | ·0019 358 | ·0013 082 | ·0008 760 |
| ·37 | ·0072 151 | ·0051 002 | ·0035 707 | ·0024 766 | ·0017 022 |
| ·38 | ·0121 537 | ·0088 078 | ·0063 227 | ·0044 970 | ·0031 698 |
| ·39 | ·0197 193 | ·0146 384 | ·0107 654 | ·0078 452 | ·0056 664 |
| ·40 | ·0308 627 | ·0234 485$^{+}$ | ·0176 520 | ·0131 693 | ·0097 392 |
| ·41 | ·0466 596 | ·0362 527 | ·0279 130 | ·0213 024 | ·0161 174 |
| ·42 | ·0682 327 | ·0541 689 | ·0426 239 | ·0332 493 | ·0257 170 |
| ·43 | ·0966 368 | ·0783 246 | ·0629 340 | ·0501 393 | ·0396 143 |
| ·44 | ·1327 181 | ·1097 287 | ·0899 578 | ·0731 393 | ·0589 826 |
| ·45 | ·1769 642 | ·1491 214 | ·1246 325$^{-}$ | ·1033 271 | ·0849 860 |
| ·46 | ·2293 695$^{+}$ | ·1968 236 | ·1675 619 | ·1415 384 | ·1186 378 |
| ·47 | ·2893 438 | ·2526 115$^{-}$ | ·2188 678 | ·1882 061 | ·1606 372 |
| ·48 | ·3556 852 | ·3156 451 | ·2780 802 | ·2432 200 | ·2112 093 |
| ·49 | ·4266 331 | ·3844 706 | ·3440 903 | ·3058 368 | ·2699 782 |
| ·50 | ·5000 000$^{e}$ | ·4571 058 | ·4151 865$^{+}$ | ·3746 645$^{+}$ | ·3359 043 |
| ·51 | ·5733 669 | ·5312 043 | ·4891 759 | ·4477 359 | ·4073 048 |
| ·52 | ·6443 148 | ·6042 747 | ·5635 794 | ·5226 671 | ·4819 682 |
| ·53 | ·7106 562 | ·6739 239 | ·6358 726 | ·5968 826 | ·5573 502 |
| ·54 | ·7706 305$^{-}$ | ·7380 846 | ·7037 339 | ·6678 717 | ·6308 245$^{+}$ |
| ·55 | ·8230 358 | ·7951 931 | ·7652 622 | ·7334 356 | ·6999 486 |
| ·56 | ·8672 819 | ·8442 924 | ·8191 295$^{-}$ | ·7918 863 | ·7627 007 |
| ·57 | ·9033 632 | ·8850 510 | ·8646 496 | ·8421 672 | ·8176 516 |
| ·58 | ·9317 673 | ·9177 036 | ·9017 604 | ·8838 828 | ·8640 464 |
| ·59 | ·9533 404 | ·9429 335$^{+}$ | ·9309 325$^{+}$ | ·9172 434 | ·9017 925$^{-}$ |
| ·60 | ·9691 373 | ·9617 232 | ·9530 284 | ·9429 425$^{-}$ | ·9313 656 |
| ·61 | ·9802 807 | ·9751 998 | ·9691 420 | ·9619 978 | ·9536 609 |
| ·62 | ·9878 463 | ·9845 004 | ·9804 457 | ·9755 855$^{+}$ | ·9698 210 |
| ·63 | ·9927 849 | ·9906 700 | ·9880 657 | ·9848 938 | ·9810 708 |
| ·64 | ·9958 809 | ·9945 994 | ·9929 963 | ·9910 128 | ·9885 843 |
| ·65 | ·9977 424 | ·9969 991 | ·9960 547 | ·9948 680 | ·9933 923 |
| ·66 | ·9988 143 | ·9984 022 | ·9978 707 | ·9971 924 | ·9963 360 |
| ·67 | ·9994 045$^{-}$ | ·9991 865$^{+}$ | ·9989 012 | ·9985 315$^{+}$ | ·9980 577 |
| ·68 | ·9997 146 | ·9996 048 | ·9994 590 | ·9992 673 | ·9990 179 |
| ·69 | ·9998 698 | ·9998 173 | ·9997 465$^{-}$ | ·9996 521 | ·9995 274 |
| ·70 | ·9999 436 | ·9999 198 | ·9998 872 | ·9998 432 | ·9997 842 |
| ·71 | ·9999 769 | ·9999 667 | ·9999 525$^{+}$ | ·9999 331 | ·9999 068 |
| ·72 | ·9999 910 | ·9999 869 | ·9999 811 | ·9999 731 | ·9999 620 |
| ·73 | ·9999 967 | ·9999 952 | ·9999 930 | ·9999 898 | ·9999 854 |
| ·74 | ·9999 989 | ·9999 983 | ·9999 975$^{+}$ | ·9999 964 | ·9999 948 |
| ·75 | ·9999 996 | ·9999 995$^{-}$ | ·9999 992 | ·9999 988 | ·9999 983 |
| ·76 | ·9999 999 | ·9999 998 | ·9999 998 | ·9999 996 | ·9999 995$^{-}$ |
| ·77 | 1·0000 000 | 1·0000 000 | ·9999 999 | ·9999 999 | ·9999 998 |
| ·78 | | | 1·0000 000 | 1·0000 000 | 1·0000 000 |

$x$ = ·27 to ·79 $q$ = 43 $p$ = 48 to 50

| | $p = 48$ | $p = 49$ | $p = 50$ |
|---|---|---|---|
| $B(p, q)$ = | ·2445 7383 × $\frac{1}{10^{27}}$ | ·1290 0597 × $\frac{1}{10^{27}}$ | ·6870 9703 × $\frac{1}{10^{28}}$ |
| $x$ | | | |
| ·27 | ·0000 001 | ·0000 001 | |
| ·28 | ·0000 004 | ·0000 002 | ·0000 001 |
| ·29 | ·0000 011 | ·0000 006 | ·0000 003 |
| ·30 | ·0000 033 | ·0000 018 | ·0000 010 |
| ·31 | ·0000 090 | ·0000 051 | ·0000 029 |
| ·32 | ·0000 229 | ·0000 134 | ·0000 078 |
| ·33 | ·0000 554 | ·0000 335+ | ·0000 201 |
| ·34 | ·0001 273 | ·0000 793 | ·0000 489 |
| ·35 | ·0002 786 | ·0001 784 | ·0001 133 |
| ·36 | ·0005 814 | ·0003 825− | ·0002 495+ |
| ·37 | ·0011 596 | ·0007 832 | ·0005 245+ |
| ·38 | ·0022 147 | ·0015 343 | ·0010 540 |
| ·39 | ·0040 574 | ·0028 808 | ·0020 285+ |
| ·40 | ·0071 410 | ·0051 923 | ·0037 448 |
| ·41 | ·0120 919 | ·0089 972 | ·0066 409 |
| ·42 | ·0197 264 | ·0150 088 | ·0113 291 |
| ·43 | ·0310 443 | ·0241 347 | ·0186 168 |
| ·44 | ·0471 873 | ·0374 562 | ·0295 045− |
| ·45 | ·0693 570 | ·0561 700 | ·0451 494 |
| ·46 | ·0986 899 | ·0814 848 | ·0667 868 |
| ·47 | ·1361 009 | ·1144 783 | ·0956 048 |
| ·48 | ·1821 135+ | ·1559 274 | ·1325 843 |
| ·49 | ·2367 061 | ·2061 373 | ·1783 197 |
| ·50 | ·2992 059 | ·2648 010 | ·2328 537 |
| ·51 | ·3682 588 | ·3309 210 | ·2955 567 |
| ·52 | ·4418 928 | ·4028 193 | ·3650 855− |
| ·53 | ·5176 748 | ·4782 474 | ·4394 395+ |
| ·54 | ·5929 418 | ·5545 856 | ·5161 197 |
| ·55 | ·6650 721 | ·6291 058 | ·5923 688 |
| ·56 | ·7317 514 | ·6992 547 | ·6654 582 |
| ·57 | ·7911 904 | ·7629 100 | ·7329 732 |
| ·58 | ·8422 601 | ·8185 676 | ·7930 473 |
| ·59 | ·8845 302 | ·8654 338 | ·8445 096 |
| ·60 | ·9182 122 | ·9034 147 | ·8869 261 |
| ·61 | ·9440 309 | ·9330 165+ | ·9205 388 |
| ·62 | ·9630 531 | ·9551 855− | ·9461 264 |
| ·63 | ·9765 102 | ·9711 229 | ·9648 198 |
| ·64 | ·9856 411 | ·9821 094 | ·9779 116 |
| ·65 | ·9915 759 | ·9893 623 | ·9866 900 |
| ·66 | ·9952 657 | ·9939 412 | ·9923 178 |
| ·67 | ·9974 566 | ·9967 014 | ·9957 618 |
| ·68 | ·9986 967 | ·9982 872 | ·9977 701 |
| ·69 | ·9993 646 | ·9991 539 | ·9988 839 |
| ·70 | ·9997 060 | ·9996 034 | ·9994 700 |
| ·71 | ·9998 713 | ·9998 241 | ·9997 619 |
| ·72 | ·9999 469 | ·9999 264 | ·9998 991 |
| ·73 | ·9999 794 | ·9999 711 | ·9999 599 |
| ·74 | ·9999 925+ | ·9999 894 | ·9999 851 |
| ·75 | ·9999 975− | ·9999 964 | ·9999 948 |
| ·76 | ·9999 992 | ·9999 988 | ·9999 983 |
| ·77 | ·9999 998 | ·9999 997 | ·9999 995+ |
| ·78 | ·9999 999 | ·9999 999 | ·9999 999 |
| ·79 | 1·0000 000 | 1·0000 000 | 1·0000 000 |

$x$ = ·24 to ·79 $q = 44$ $p$ = 44 to 50

| $x$ | $p = 44$ | $p = 45$ | $p = 46$ | $p = 47$ | $p = 48$ | $p = 49$ | $p = 50$ |
|---|---|---|---|---|---|---|---|
| $B(p, q) =$ | ·1731 7003 $\times \frac{1}{10^{26}}$ | ·8658 5017 $\times \frac{1}{10^{27}}$ | ·4377 8941 $\times \frac{1}{10^{27}}$ | ·2237 5903 $\times \frac{1}{10^{27}}$ | ·1155 6785 $\times \frac{1}{10^{27}}$ | ·6029 6270 $\times \frac{1}{10^{28}}$ | ·3176 9003 $\times \frac{1}{10^{28}}$ |
| ·24 | ·0000 001 | | | | | | |
| ·25 | ·0000 003 | ·0000 001 | ·0000 001 | | | | |
| ·26 | ·0000 008 | ·0000 004 | ·0000 002 | ·0000 001 | ·0000 001 | | |
| ·27 | ·0000 025+ | ·0000 013 | ·0000 007 | ·0000 004 | ·0000 002 | ·0000 001 | |
| ·28 | ·0000 071 | ·0000 039 | ·0000 021 | ·0000 011 | ·0000 006 | ·0000 003 | ·0000 002 |
| ·29 | ·0000 189 | ·0000 106 | ·0000 059 | ·0000 032 | ·0000 017 | ·0000 009 | ·0000 005+ |
| ·30 | ·0000 469 | ·0000 271 | ·0000 156 | ·0000 088 | ·0000 050− | ·0000 028 | ·0000 015+ |
| ·31 | ·0001 103 | ·0000 659 | ·0000 390 | ·0000 228 | ·0000 133 | ·0000 076 | ·0000 043 |
| ·32 | ·0002 459 | ·0001 515+ | ·0000 924 | ·0000 559 | ·0000 334 | ·0000 198 | ·0000 117 |
| ·33 | ·0005 214 | ·0003 309 | ·0002 080 | ·0001 294 | ·0000 798 | ·0000 488 | ·0000 296 |
| ·34 | ·0010 538 | ·0006 882 | ·0004 450+ | ·0002 851 | ·0001 810 | ·0001 139 | ·0000 710 |
| ·35 | ·0020 346 | ·0013 658 | ·0009 081 | ·0005 982 | ·0003 904 | ·0002 526 | ·0001 620 |
| ·36 | ·0037 603 | ·0025 926 | ·0017 706 | ·0011 981 | ·0008 034 | ·0005 340 | ·0003 519 |
| ·37 | ·0066 652 | ·0047 157 | ·0033 050+ | ·0022 952 | ·0015 798 | ·0010 779 | ·0007 293 |
| ·38 | ·0113 507 | ·0082 333 | ·0059 168 | ·0042 136 | ·0029 743 | ·0020 815− | ·0014 445− |
| ·39 | ·0186 015+ | ·0138 215+ | ·0101 759 | ·0074 251 | ·0053 708 | ·0038 519 | ·0027 396 |
| ·40 | ·0293 798 | ·0223 431 | ·0168 389 | ·0125 790 | ·0093 162 | ·0068 419 | ·0049 836 |
| ·41 | ·0447 864 | ·0348 311 | ·0268 492 | ·0205 174 | ·0155 464 | ·0116 824 | ·0087 079 |
| ·42 | ·0659 825− | ·0524 345− | ·0413 071 | ·0322 648 | ·0249 925+ | ·0192 020 | ·0146 357 |
| ·43 | ·0940 731 | ·0763 238 | ·0613 987 | ·0489 816 | ·0387 573 | ·0304 224 | ·0236 932 |
| ·44 | ·1299 594 | ·1075 585− | ·0882 838 | ·0718 751 | ·0580 499 | ·0465 174 | ·0369 902 |
| ·45 | ·1741 799 | ·1469 288 | ·1229 479 | ·1020 688 | ·0840 773 | ·0687 283 | ·0557 599 |
| ·46 | ·2267 658 | ·1947 957 | ·1660 368 | ·1404 413 | ·1178 955+ | ·0982 338 | ·0812 524 |
| ·47 | ·2871 398 | ·2509 557 | ·2176 984 | ·1874 560 | ·1602 378 | ·1359 853 | ·1145 836 |
| ·48 | ·3540 836 | ·3145 619 | ·2774 641 | ·2430 116 | ·2113 446 | ·1825 276 | ·1565 571 |
| ·49 | ·4257 899 | ·3841 231 | ·3441 971 | ·3063 455− | ·2708 294 | ·2378 364 | ·2074 829 |
| ·50 | ·5000 000[e] | ·4575 933 | ·4161 289 | ·3760 166 | ·3376 112 | ·3012 061 | ·2670 299 |
| ·51 | ·5742 101 | ·5325 433 | ·4909 877 | ·4499 831 | ·4099 382 | ·3712 197 | ·3341 449 |
| ·52 | ·6459 164 | ·6063 947 | ·5662 054 | ·5257 716 | ·4855 098 | ·4458 183 | ·4070 666 |
| ·53 | ·7128 602 | ·6766 760 | ·6391 731 | ·6007 163 | ·5616 868 | ·5224 703 | ·4834 458 |
| ·54 | ·7732 342 | ·7412 640 | ·7075 036 | ·6722 312 | ·6357 581 | ·5984 187 | ·5605 612 |
| ·55 | ·8258 201 | ·7985 690 | ·7692 590 | ·7380 693 | ·7052 206 | ·6709 689 | ·6355 989 |
| ·56 | ·8700 406 | ·8476 397 | ·8231 082 | ·7965 289 | ·7680 268 | ·7377 671 | ·7059 512 |
| ·57 | ·9059 269 | ·8881 776 | ·8683 931 | ·8465 743 | ·8227 592 | ·7970 240 | ·7694 822 |
| ·58 | ·9340 175+ | ·9204 695+ | ·9051 031 | ·8878 593 | ·8687 076 | ·8476 488 | ·8247 162 |
| ·59 | ·9552 136 | ·9452 584 | ·9337 722 | ·9206 605+ | ·9058 471 | ·8892 776 | ·8709 227 |
| ·60 | ·9706 202 | ·9635 835− | ·9553 271 | ·9457 425− | ·9347 304 | ·9222 041 | ·9080 929 |
| ·61 | ·9813 985− | ·9766 185− | ·9709 164 | ·9641 868 | ·9563 260 | ·9472 353 | ·9368 237 |
| ·62 | ·9886 493 | ·9855 319 | ·9817 523 | ·9772 184 | ·9718 356 | ·9655 085+ | ·9581 433 |
| ·63 | ·9933 348 | ·9913 852 | ·9889 834 | ·9860 558 | ·9825 239 | ·9783 055+ | ·9733 158 |
| ·64 | ·9962 397 | ·9950 721 | ·9936 108 | ·9918 012 | ·9895 836 | ·9868 929 | ·9836 596 |
| ·65 | ·9979 654 | ·9972 967 | ·9964 466 | ·9953 776 | ·9940 470 | ·9924 074 | ·9904 063 |
| ·66 | ·9989 462 | ·9985 805+ | ·9981 086 | ·9975 059 | ·9967 442 | ·9957 912 | ·9946 103 |
| ·67 | ·9994 786 | ·9992 880 | ·9990 384 | ·9987 148 | ·9982 996 | ·9977 722 | ·9971 088 |
| ·68 | ·9997 541 | ·9996 596 | ·9995 341 | ·9993 689 | ·9991 538 | ·9988 765− | ·9985 225− |
| ·69 | ·9998 897 | ·9998 453 | ·9997 854 | ·9997 054 | ·9995 998 | ·9994 615+ | ·9992 825− |
| ·70 | ·9999 531 | ·9999 333 | ·9999 063 | ·9998 697 | ·9998 206 | ·9997 554 | ·9996 698 |
| ·71 | ·9999 811 | ·9999 728 | ·9999 613 | ·9999 455− | ·9999 240 | ·9998 950+ | ·9998 564 |
| ·72 | ·9999 929 | ·9999 896 | ·9999 850− | ·9999 785+ | ·9999 697 | ·9999 576 | ·9999 412 |
| ·73 | ·9999 975− | ·9999 962 | ·9999 945+ | ·9999 921 | ·9999 886 | ·9999 839 | ·9999 774 |
| ·74 | ·9999 992 | ·9999 987 | ·9999 981 | ·9999 973 | ·9999 960 | ·9999 943 | ·9999 919 |
| ·75 | ·9999 997 | ·9999 996 | ·9999 994 | ·9999 991 | ·9999 987 | ·9999 981 | ·9999 973 |
| ·76 | ·9999 999 | ·9999 999 | ·9999 998 | ·9999 997 | ·9999 996 | ·9999 994 | ·9999 992 |
| ·77 | 1·0000 000 | 1·0000 000 | 1·0000 000 | ·9999 999 | ·9999 999 | ·9999 998 | ·9999 998 |
| ·78 | | | | 1·0000 000 | 1·0000 000 | 1·0000 000 | ·9999 999 |
| ·79 | | | | | | | 1·0000 000 |

$x$ = ·24 to ·78 $q = 45$ $p$ = 45 to 50

| $x$ | $p = 45$ | $p = 46$ | $p = 47$ | $p = 48$ | $p = 49$ | $p = 50$ |
|---|---|---|---|---|---|---|
| $B(p, q)$ = | ·4280 6076 $\times \frac{1}{10^{27}}$ | ·2140 3038 $\times \frac{1}{10^{27}}$ | ·1081 9118 $\times \frac{1}{10^{27}}$ | ·5527 1581 $\times \frac{1}{10^{28}}$ | ·2852 7268 $\times \frac{1}{10^{28}}$ | ·1487 0597 $\times \frac{1}{10^{28}}$ |
| ·24 | ·0000 001 | | | | | |
| ·25 | ·0000 002 | ·0000 001 | | | | |
| ·26 | ·0000 006 | ·0000 003 | ·0000 002 | ·0000 001 | | |
| ·27 | ·0000 020 | ·0000 010 | ·0000 005$^{+}$ | ·0000 003 | ·0000 001 | ·0000 001 |
| ·28 | ·0000 057 | ·0000 031 | ·0000 017 | ·0000 009 | ·0000 005$^{-}$ | ·0000 002 |
| ·29 | ·0000 154 | ·0000 086 | ·0000 048 | ·0000 026 | ·0000 014 | ·0000 008 |
| ·30 | ·0000 390 | ·0000 226 | ·0000 130 | ·0000 074 | ·0000 041 | ·0000 023 |
| ·31 | ·0000 934 | ·0000 559 | ·0000 331 | ·0000 194 | ·0000 113 | ·0000 065$^{-}$ |
| ·32 | ·0002 119 | ·0001 307 | ·0000 798 | ·0000 483 | ·0000 289 | ·0000 172 |
| ·33 | ·0004 567 | ·0002 900 | ·0001 824 | ·0001 137 | ·0000 702 | ·0000 430 |
| ·34 | ·0009 368 | ·0006 122 | ·0003 963 | ·0002 542 | ·0001 616 | ·0001 018 |
| ·35 | ·0018 340 | ·0012 322 | ·0008 201 | ·0005 408 | ·0003 535$^{-}$ | ·0002 290 |
| ·36 | ·0034 333 | ·0023 692 | ·0016 197 | ·0010 972 | ·0007 368 | ·0004 905$^{-}$ |
| ·37 | ·0061 583 | ·0043 608 | ·0030 595$^{-}$ | ·0021 273 | ·0014 662 | ·0010 019 |
| ·38 | ·0106 025$^{+}$ | ·0076 974 | ·0055 375$^{-}$ | ·0039 483 | ·0027 909 | ·0019 561 |
| ·39 | ·0175 499 | ·0130 518 | ·0096 195$^{+}$ | ·0070 278 | ·0050 905$^{-}$ | ·0036 565$^{-}$ |
| ·40 | ·0279 725$^{-}$ | ·0212 923 | ·0160 643 | ·0120 155$^{-}$ | ·0089 113 | ·0065 547 |
| ·41 | ·0429 944 | ·0334 687 | ·0258 275$^{-}$ | ·0197 616 | ·0149 949 | ·0112 855$^{+}$ |
| ·42 | ·0638 148 | ·0507 603 | ·0400 329 | ·0313 095$^{+}$ | ·0242 872 | ·0186 893 |
| ·43 | ·0915 882 | ·0743 802 | ·0599 030 | ·0478 500$^{-}$ | ·0379 162 | ·0298 091 |
| ·44 | ·1272 713 | ·1054 383 | ·0866 429 | ·0706 308 | ·0571 273 | ·0458 505$^{+}$ |
| ·45 | ·1714 548 | ·1447 760 | ·1212 870 | ·1008 217 | ·0831 703 | ·0680 946 |
| ·46 | ·2242 082 | ·1927 956 | ·1645 243 | ·1393 448 | ·1171 448 | ·0977 629 |
| ·47 | ·2849 688 | ·2493 155$^{-}$ | ·2165 300 | ·1866 951 | ·1598 189 | ·1358 442 |
| ·48 | ·3525 027 | ·3134 826 | ·2768 377 | ·2427 813 | ·2114 494 | ·1829 054 |
| ·49 | ·4249 566 | ·3837 692 | ·3442 831 | ·3068 217 | ·2716 392 | ·2389 195$^{+}$ |
| ·50 | ·5000 000$^{e}$ | ·4580 644 | ·4170 405$^{+}$ | ·3773 259 | ·3392 660 | ·3031 480 |
| ·51 | ·5750 434 | ·5338 561 | ·4927 583 | ·4521 765$^{-}$ | ·4125 077 | ·3741 099 |
| ·52 | ·6474 973 | ·6084 772 | ·5687 785$^{-}$ | ·5288 095$^{+}$ | ·4889 737 | ·4496 583 |
| ·53 | ·7150 312 | ·6793 779 | ·6424 070 | ·6044 685$^{+}$ | ·5659 293 | ·5271 621 |
| ·54 | ·7757 918 | ·7443 792 | ·7111 911 | ·6764 919 | ·6405 783 | ·6037 704 |
| ·55 | ·8285 452 | ·8018 664 | ·7731 577 | ·7425 860 | ·7103 583 | ·6767 166 |
| ·56 | ·8727 287 | ·8508 957 | ·8269 743 | ·8010 374 | ·7731 985$^{-}$ | ·7436 097 |
| ·57 | ·9084 118 | ·8912 038 | ·8720 131 | ·8508 339 | ·8276 957 | ·8026 639 |
| ·58 | ·9361 852 | ·9231 307 | ·9083 167 | ·8916 809 | ·8731 874 | ·8528 295$^{-}$ |
| ·59 | ·9570 056 | ·9474 799 | ·9364 840 | ·9239 228 | ·9097 183 | ·8938 121 |
| ·60 | ·9720 275$^{+}$ | ·9653 473 | ·9575 054 | ·9483 954 | ·9379 189 | ·9259 885$^{-}$ |
| ·61 | ·9824 501 | ·9779 520 | ·9725 835$^{+}$ | ·9662 431 | ·9588 300 | ·9502 476 |
| ·62 | ·9893 975$^{-}$ | ·9864 923 | ·9829 682 | ·9787 378 | ·9737 107 | ·9677 951 |
| ·63 | ·9938 417 | ·9920 441 | ·9898 284 | ·9871 257 | ·9838 623 | ·9799 601 |
| ·64 | ·9965 667 | ·9955 026 | ·9941 701 | ·9925 190 | ·9904 936 | ·9880 333 |
| ·65 | ·9981 660 | ·9975 642 | ·9967 989 | ·9958 358 | ·9946 358 | ·9931 555$^{-}$ |
| ·66 | ·9990 632 | ·9987 386 | ·9983 195$^{-}$ | ·9977 839 | ·9971 064 | ·9962 577 |
| ·67 | ·9995 433 | ·9993 767 | ·9991 583 | ·9988 749 | ·9985 110 | ·9980 483 |
| ·68 | ·9997 881 | ·9997 068 | ·9995 987 | ·9994 563 | ·9992 708 | ·9990 313 |
| ·69 | ·9999 066 | ·9998 690 | ·9998 183 | ·9997 506 | ·9996 610 | ·9995 436 |
| ·70 | ·9999 610 | ·9999 446 | ·9999 221 | ·9998 916 | ·9998 508 | ·9997 965$^{-}$ |
| ·71 | ·9999 846 | ·9999 779 | ·9999 685$^{-}$ | ·9999 556 | ·9999 380 | ·9999 143 |
| ·72 | ·9999 943 | ·9999 917 | ·9999 880 | ·9999 829 | ·9999 758 | ·9999 661 |
| ·73 | ·9999 980 | ·9999 971 | ·9999 957 | ·9999 938 | ·9999 911 | ·9999 874 |
| ·74 | ·9999 994 | ·9999 990 | ·9999 986 | ·9999 979 | ·9999 970 | ·9999 956 |
| ·75 | ·9999 998 | ·9999 997 | ·9999 996 | ·9999 993 | ·9999 990 | ·9999 986 |
| ·76 | ·9999 999 | ·9999 999 | ·9999 999 | ·9999 998 | ·9999 997 | ·9999 996 |
| ·77 | 1·0000 000 | 1·0000 000 | 1·0000 000 | 1·0000 000 | ·9999 999 | ·9999 999 |
| ·78 | | | | | 1·0000 000 | 1·0000 000 |

$x$ = ·25 to ·78 $q = 46$ $p$ = 46 to 50

| $x$ | $p = 46$ | $p = 47$ | $p = 48$ | $p = 49$ | $p = 50$ |
|---|---|---|---|---|---|
| $B(p, q) =$ | ·1058 3920$\times\frac{1}{10^{27}}$ | ·5291 9599$\times\frac{1}{10^{28}}$ | ·2674 4313$\times\frac{1}{10^{28}}$ | ·1365 6671$\times\frac{1}{10^{28}}$ | ·7043 9670$\times\frac{1}{10^{29}}$ |
| ·25 | ·0000 001 | ·0000 001 | | | |
| ·26 | ·0000 005$^-$ | ·0000 002 | ·0000 001 | ·0000 001 | |
| ·27 | ·0000 015$^+$ | ·0000 008 | ·0000 004 | ·0000 002 | ·0000 001 |
| ·28 | ·0000 046 | ·0000 025$^-$ | ·0000 013 | ·0000 007 | ·0000 004 |
| ·29 | ·0000 125$^+$ | ·0000 070 | ·0000 039 | ·0000 021 | ·0000 012 |
| ·30 | ·0000 324 | ·0000 188 | ·0000 108 | ·0000 061 | ·0000 035$^-$ |
| ·31 | ·0000 791 | ·0000 474 | ·0000 281 | ·0000 165$^-$ | ·0000 096 |
| ·32 | ·0001 827 | ·0001 127 | ·0000 689 | ·0000 417 | ·0000 250$^+$ |
| ·33 | ·0004 000 | ·0002 542 | ·0001 601 | ·0000 999 | ·0000 618 |
| ·34 | ·0008 330 | ·0005 448 | ·0003 530 | ·0002 267 | ·0001 443 |
| ·35 | ·0016 534 | ·0011 117 | ·0007 406 | ·0004 890 | ·0003 200 |
| ·36 | ·0031 354 | ·0021 653 | ·0014 818 | ·0010 050$^-$ | ·0006 757 |
| ·37 | ·0056 910 | ·0040 331 | ·0028 324 | ·0019 717 | ·0013 607 |
| ·38 | ·0099 053 | ·0071 972 | ·0051 829 | ·0036 998 | ·0026 187 |
| ·39 | ·0165 603 | ·0123 264 | ·0090 943 | ·0066 520 | ·0048 247 |
| ·40 | ·0266 364 | ·0202 932 | ·0153 265$^+$ | ·0114 774 | ·0085 238 |
| ·41 | ·0412 798 | ·0321 629 | ·0248 461 | ·0190 339 | ·0144 624 |
| ·42 | ·0617 261 | ·0491 440 | ·0388 000 | ·0303 826 | ·0236 005$^-$ |
| ·43 | ·0891 791 | ·0724 917 | ·0584 459 | ·0467 440 | ·0370 911 |
| ·44 | ·1246 513 | ·1033 666 | ·0850 346 | ·0694 065$^+$ | ·0562 151 |
| ·45 | ·1687 869 | ·1426 620 | ·1196 498 | ·0995 861 | ·0822 658 |
| ·46 | ·2216 952 | ·1908 229 | ·1630 247 | ·1382 496 | ·1163 868 |
| ·47 | ·2828 296 | ·2476 908 | ·2153 631 | ·1859 247 | ·1593 820 |
| ·48 | ·3509 419 | ·3124 077 | ·2762 020 | ·2425 306 | ·2115 255$^-$ |
| ·49 | ·4241 328 | ·3834 095$^-$ | ·3443 497 | ·3072 674 | ·2724 100 |
| ·50 | ·5000 000$^e$ | ·4585 203 | ·4179 231 | ·3785 945$^+$ | ·3408 713 |
| ·51 | ·5758 672 | ·5351 439 | ·4944 898 | ·4543 186 | ·4150 163 |
| ·52 | ·6490 581 | ·6105 239 | ·5713 010 | ·5317 839 | ·4923 636 |
| ·53 | ·7171 704 | ·6820 316 | ·6455 769 | ·6081 426 | ·5700 817 |
| ·54 | ·7783 048 | ·7474 325$^+$ | ·7147 998 | ·6806 579 | ·6452 897 |
| ·55 | ·8312 131 | ·8050 882 | ·7769 622 | ·7469 905$^-$ | ·7153 672 |
| ·56 | ·8753 487 | ·8540 640 | ·8307 323 | ·8054 174 | ·7782 220 |
| ·57 | ·9108 209 | ·8941 336 | ·8755 147 | ·8549 525$^+$ | ·8324 684 |
| ·58 | ·9382 739 | ·9256 919 | ·9114 073 | ·8953 549 | ·8774 943 |
| ·59 | ·9587 202 | ·9496 034 | ·9390 743 | ·9270 384 | ·9134 156 |
| ·60 | ·9733 636 | ·9670 203 | ·9595 703 | ·9509 097 | ·9409 412 |
| ·61 | ·9834 397 | ·9792 058 | ·9741 503 | ·9681 754 | ·9611 836 |
| ·62 | ·9900 947 | ·9873 866 | ·9841 001 | ·9801 521 | ·9754 564 |
| ·63 | ·9943 090 | ·9926 512 | ·9906 068 | ·9881 113 | ·9850 954 |
| ·64 | ·9968 646 | ·9958 946 | ·9946 794 | ·9931 725$^-$ | ·9913 224 |
| ·65 | ·9983 466 | ·9978 049 | ·9971 157 | ·9962 477 | ·9951 655$^-$ |
| ·66 | ·9991 670 | ·9988 789 | ·9985 066 | ·9980 305$^+$ | ·9974 278 |
| ·67 | ·9996 000 | ·9994 542 | ·9992 630 | ·9990 149 | ·9986 959 |
| ·68 | ·9998 173 | ·9997 473 | ·9996 542 | ·9995 315$^+$ | ·9993 714 |
| ·69 | ·9999 209 | ·9998 891 | ·9998 461 | ·9997 887 | ·9997 127 |
| ·70 | ·9999 676 | ·9999 539 | ·9999 352 | ·9999 099 | ·9998 759 |
| ·71 | ·9999 875$^-$ | ·9999 820 | ·9999 743 | ·9999 638 | ·9999 494 |
| ·72 | ·9999 954 | ·9999 934 | ·9999 904 | ·9999 863 | ·9999 807 |
| ·73 | ·9999 985$^-$ | ·9999 977 | ·9999 967 | ·9999 952 | ·9999 931 |
| ·74 | ·9999 995$^+$ | ·9999 993 | ·9999 989 | ·9999 984 | ·9999 977 |
| ·75 | ·9999 999 | ·9999 998 | ·9999 997 | ·9999 995$^+$ | ·9999 993 |
| ·76 | 1·0000 000 | ·9999 999 | ·9999 999 | ·9999 999 | ·9999 998 |
| ·77 | | 1·0000 000 | 1·0000 000 | 1·0000 000 | ·9999 999 |
| ·78 | | | | | 1·0000 000 |

$x$ = ·25 to ·77 $q = 47$ $p$ = 47 to 50

| $x$ | $p = 47$ | $p = 48$ | $p = 49$ | $p = 50$ |
|---|---|---|---|---|
| $B(p, q) =$ | ·2617 5285$^{+}\times\frac{1}{10^{28}}$ | ·1308 7643$\times\frac{1}{10^{28}}$ | ·6612 7037$\times\frac{1}{10^{29}}$ | ·3375 2342$\times\frac{1}{10^{29}}$ |
| ·25 | ·0000 001 | ·0000 001 | | |
| ·26 | ·0000 004 | ·0000 002 | ·0000 001 | ·0000 001 |
| ·27 | ·0000 012 | ·0000 006 | ·0000 003 | ·0000 002 |
| ·28 | ·0000 036 | ·0000 020 | ·0000 011 | ·0000 006 |
| ·29 | ·0000 102 | ·0000 057 | ·0000 032 | ·0000 018 |
| ·30 | ·0000 270 | ·0000 156 | ·0000 090 | ·0000 051 |
| ·31 | ·0000 671 | ·0000 402 | ·0000 238 | ·0000 140 |
| ·32 | ·0001 575$^{+}$ | ·0000 972 | ·0000 595$^{-}$ | ·0000 361 |
| ·33 | ·0003 504 | ·0002 229 | ·0001 405$^{-}$ | ·0000 877 |
| ·34 | ·0007 407 | ·0004 848 | ·0003 144 | ·0002 021 |
| ·35 | ·0014 909 | ·0010 032 | ·0006 690 | ·0004 422 |
| ·36 | ·0028 637 | ·0019 793 | ·0013 557 | ·0009 205$^{-}$ |
| ·37 | ·0052 599 | ·0037 306 | ·0026 225$^{-}$ | ·0018 276 |
| ·38 | ·0092 554 | ·0067 304 | ·0048 515$^{-}$ | ·0034 672 |
| ·39 | ·0156 289 | ·0116 428 | ·0085 984 | ·0062 964 |
| ·40 | ·0253 678 | ·0193 430 | ·0146 237 | ·0109 637 |
| ·41 | ·0396 387 | ·0309 111 | ·0239 035$^{+}$ | ·0183 332 |
| ·42 | ·0597 130 | ·0475 834 | ·0376 069 | ·0294 832 |
| ·43 | ·0868 428 | ·0706 566 | ·0570 264 | ·0456 633 |
| ·44 | ·1220 972 | ·1013 421 | ·0834 582 | ·0682 021 |
| ·45 | ·1661 744 | ·1405 859 | ·1180 360 | ·0983 624 |
| ·46 | ·2192 254 | ·1888 770 | ·1615 382 | ·1371 564 |
| ·47 | ·2807 213 | ·2460 814 | ·2141 983 | ·1851 457 |
| ·48 | ·3494 006 | ·3113 373 | ·2755 578 | ·2422 610 |
| ·49 | ·4233 183 | ·3830 443 | ·3443 981 | ·3076 841 |
| ·50 | ·5000 000$^{e}$ | ·4589 615$^{+}$ | ·4187 780 | ·3798 246 |
| ·51 | ·5766 817 | ·5364 077 | ·4961 840 | ·4564 118 |
| ·52 | ·6505 994 | ·6125 362 | ·5737 751 | ·5346 976 |
| ·53 | ·7192 787 | ·6846 389 | ·6486 856 | ·6117 418 |
| ·54 | ·7807 746 | ·7504 262 | ·7183 328 | ·6847 329 |
| ·55 | ·8338 256 | ·8082 371 | ·7806 761 | ·7512 871 |
| ·56 | ·8779 028 | ·8571 478 | ·8343 864 | ·8096 741 |
| ·57 | ·9131 572 | ·8969 709 | ·8789 030 | ·8589 361 |
| ·58 | ·9402 870 | ·9281 575$^{+}$ | ·9143 804 | ·8988 882 |
| ·59 | ·9603 613 | ·9516 336 | ·9415 496 | ·9300 146 |
| ·60 | ·9746 322 | ·9686 075$^{-}$ | ·9615 284 | ·9532 935$^{+}$ |
| ·61 | ·9843 711 | ·9803 850$^{+}$ | ·9756 233 | ·9699 918 |
| ·62 | ·9907 446 | ·9882 197 | ·9851 540 | ·9814 689 |
| ·63 | ·9947 401 | ·9932 108 | ·9913 240 | ·9890 194 |
| ·64 | ·9971 363 | ·9962 518 | ·9951 432 | ·9937 677 |
| ·65 | ·9985 091 | ·9980 214 | ·9974 006 | ·9966 183 |
| ·66 | ·9992 593 | ·9990 033 | ·9986 726 | ·9982 493 |
| ·67 | ·9996 496 | ·9995 220 | ·9993 546 | ·9991 373 |
| ·68 | ·9998 425$^{-}$ | ·9997 822 | ·9997 020 | ·9995 962 |
| ·69 | ·9999 329 | ·9999 060 | ·9998 697 | ·9998 210 |
| ·70 | ·9999 730 | ·9999 617 | ·9999 462 | ·9999 251 |
| ·71 | ·9999 898 | ·9999 853 | ·9999 791 | ·9999 705$^{-}$ |
| ·72 | ·9999 964 | ·9999 947 | ·9999 924 | ·9999 891 |
| ·73 | ·9999 988 | ·9999 982 | ·9999 974 | ·9999 962 |
| ·74 | ·9999 996 | ·9999 994 | ·9999 992 | ·9999 988 |
| ·75 | ·9999 999 | ·9999 998 | ·9999 998 | ·9999 996 |
| ·76 | 1·0000 000 | 1·0000 000 | ·9999 999 | ·9999 999 |
| ·77 | | | 1·0000 000 | 1·0000 000 |

$x = \cdot 25$ to $\cdot 76$ $q = 48$ $p = 48$ to $50$

| $x$ | $p = 48$ | $p = 49$ | $p = 50$ |
|---|---|---|---|
| $B(p, q) =$ | $\cdot 6474\ 9390 \times \frac{1}{10^{29}}$ | $\cdot 3237\ 4695^{+} \times \frac{1}{10^{29}}$ | $\cdot 1635\ 4227 \times \frac{1}{10^{29}}$ |
| ·25 | ·0000 001 | | |
| ·26 | ·0000 003 | ·0000 001 | ·0000 001 |
| ·27 | ·0000 009 | ·0000 005⁻ | ·0000 003 |
| ·28 | ·0000 029 | ·0000 016 | ·0000 008 |
| ·29 | ·0000 083 | ·0000 047 | ·0000 026 |
| ·30 | ·0000 224 | ·0000 130 | ·0000 075⁻ |
| ·31 | ·0000 568 | ·0000 341 | ·0000 202 |
| ·32 | ·0001 358 | ·0000 839 | ·0000 514 |
| ·33 | ·0003 071 | ·0001 954 | ·0001 233 |
| ·34 | ·0006 589 | ·0004 315⁺ | ·0002 801 |
| ·35 | ·0013 446 | ·0009 055⁻ | ·0006 043 |
| ·36 | ·0026 161 | ·0018 095⁻ | ·0012 405⁺ |
| ·37 | ·0048 623 | ·0034 512 | ·0024 283 |
| ·38 | ·0086 494 | ·0062 947 | ·0045 416 |
| ·39 | ·0147 519 | ·0109 983 | ·0081 301 |
| ·40 | ·0241 628 | ·0184 393 | ·0139 540 |
| ·41 | ·0380 678 | ·0297 109 | ·0229 981 |
| ·42 | ·0577 722 | ·0460 764 | ·0364 523 |
| ·43 | ·0845 768 | ·0688 731 | ·0556 435⁻ |
| ·44 | ·1196 065⁺ | ·0993 635⁻ | ·0819 131 |
| ·45 | ·1636 156 | ·1385 468 | ·1164 454 |
| ·46 | ·2167 976 | ·1869 575⁺ | ·1600 649 |
| ·47 | ·2786 429 | ·2444 873 | ·2130 362 |
| ·48 | ·3478 780 | ·3102 715⁺ | ·2749 061 |
| ·49 | ·4225 129 | ·3826 743 | ·3444 293 |
| ·50 | ·5000 000[e] | ·4593 890 | ·4196 068 |
| ·51 | ·5774 871 | ·5376 486 | ·4978 426 |
| ·52 | ·6521 220 | ·6145 155⁻ | ·5762 029 |
| ·53 | ·7213 571 | ·6872 015⁺ | ·6517 354 |
| ·54 | ·7832 024 | ·7533 624 | ·7217 928 |
| ·55 | ·8363 844 | ·8113 157 | ·7843 028 |
| ·56 | ·8803 935⁻ | ·8601 504 | ·8379 408 |
| ·57 | ·9154 232 | ·8997 195⁺ | ·8821 827 |
| ·58 | ·9422 278 | ·9305 319 | ·9172 416 |
| ·59 | ·9619 322 | ·9535 753 | ·9439 155⁻ |
| ·60 | ·9758 372 | ·9701 136 | ·9633 856 |
| ·61 | ·9852 481 | ·9814 944 | ·9770 084 |
| ·62 | ·9913 506 | ·9889 959 | ·9861 357 |
| ·63 | ·9951 377 | ·9937 266 | ·9919 849 |
| ·64 | ·9973 839 | ·9965 773 | ·9955 658 |
| ·65 | ·9986 554 | ·9982 162 | ·9976 569 |
| ·66 | ·9993 411 | ·9991 138 | ·9988 199 |
| ·67 | ·9996 929 | ·9995 813 | ·9994 348 |
| ·68 | ·9998 642 | ·9998 123 | ·9997 432 |
| ·69 | ·9999 432 | ·9999 204 | ·9998 896 |
| ·70 | ·9999 776 | ·9999 681 | ·9999 552 |
| ·71 | ·9999 917 | ·9999 880 | ·9999 829 |
| ·72 | ·9999 971 | ·9999 958 | ·9999 939 |
| ·73 | ·9999 991 | ·9999 986 | ·9999 980 |
| ·74 | ·9999 997 | ·9999 996 | ·9999 994 |
| ·75 | ·9999 999 | ·9999 999 | ·9999 998 |
| ·76 | 1·0000 000 | 1·0000 000 | 1·0000 000 |

$x = \cdot 25$ to $\cdot 76$ $q = 49$ $p = 49$ to $50$

| $x$ | $p = 49$ | $p = 50$ |
|---|---|---|
| $B(p, q) =$ | $\cdot 1602\ 0468 \times \frac{1}{10^{28}}$ | $\cdot 8010\ 2339 \times \frac{1}{10^{30}}$ |
| ·25 | ·0000 001 | |
| ·26 | ·0000 002 | ·0000 001 |
| ·27 | ·0000 007 | ·0000 004 |
| ·28 | ·0000 023 | ·0000 013 |
| ·29 | ·0000 068 | ·0000 038 |
| ·30 | ·0000 187 | ·0000 108 |
| ·31 | ·0000 482 | ·0000 289 |
| ·32 | ·0001 171 | ·0000 724 |
| ·33 | ·0002 691 | ·0001 714 |
| ·34 | ·0005 861 | ·0003 842 |
| ·35 | ·0012 129 | ·0008 173 |
| ·36 | ·0023 902 | ·0016 544 |
| ·37 | ·0044 954 | ·0031 931 |
| ·38 | ·0080 843 | ·0058 878 |
| ·39 | ·0139 261 | ·0103 906 |
| ·40 | ·0230 181 | ·0175 796 |
| ·41 | ·0365 635$^+$ | ·0285 599 |
| ·42 | ·0559 009 | ·0446 207 |
| ·43 | ·0823 782 | ·0671 394 |
| ·44 | ·1171 774 | ·0974 294 |
| ·45 | ·1611 087 | ·1365 439 |
| ·46 | ·2144 104 | ·1850 638 |
| ·47 | ·2765 935$^+$ | ·2429 082 |
| ·48 | ·3463 738 | ·3092 106 |
| ·49 | ·4217 161 | ·3822 998 |
| ·50 | ·5000 000$^e$ | ·4598 034 |
| ·51 | ·5782 839 | ·5388 677 |
| ·52 | ·6536 262 | ·6164 630 |
| ·53 | ·7234 065$^-$ | ·6897 211 |
| ·54 | ·7855 896 | ·7562 431 |
| ·55 | ·8388 913 | ·8143 265$^-$ |
| ·56 | ·8828 226 | ·8630 746 |
| ·57 | ·9176 218 | ·9023 829 |
| ·58 | ·9440 991 | ·9328 189 |
| ·59 | ·9634 365$^-$ | ·9554 329 |
| ·60 | ·9769 891 | ·9715 434 |
| ·61 | ·9860 739 | ·9825 383 |
| ·62 | ·9919 157 | ·9897 193 |
| ·63 | ·9955 046 | ·9942 023 |
| ·64 | ·9976 098 | ·9968 739 |
| ·65 | ·9987 871 | ·9983 916 |
| ·66 | ·9994 139 | ·9992 119 |
| ·67 | ·9997 309 | ·9996 332 |
| ·68 | ·9998 829 | ·9998 382 |
| ·69 | ·9999 518 | ·9999 325$^+$ |
| ·70 | ·9999 813 | ·9999 735$^-$ |
| ·71 | ·9999 932 | ·9999 902 |
| ·72 | ·9999 977 | ·9999 966 |
| ·73 | ·9999 993 | ·9999 989 |
| ·74 | ·9999 998 | ·9999 997 |
| ·75 | ·9999 999 | ·9999 999 |
| ·76 | 1·0000 000 | 1·0000 000 |

$x = \cdot 26$ to $\cdot 75$ $q = 50$ $p = 50$

| $x$ | $p = 50$ |
|---|---|
| $B(p, q) =$ | $\cdot 3964\ 6612 \times \frac{1}{10^{30}}$ |
| ·26 | ·0000 002 |
| ·27 | ·0000 006 |
| ·28 | ·0000 019 |
| ·29 | ·0000 056 |
| ·30 | ·0000 155$^+$ |
| ·31 | ·0000 409 |
| ·32 | ·0001 010 |
| ·33 | ·0002 359 |
| ·34 | ·0005 215$^-$ |
| ·35 | ·0010 942 |
| ·36 | ·0021 842 |
| ·37 | ·0041 568 |
| ·38 | ·0075 571 |
| ·39 | ·0131 483 |
| ·40 | ·0219 304 |
| ·41 | ·0351 229 |
| ·42 | ·0540 961 |
| ·43 | ·0802 448 |
| ·44 | ·1148 076 |
| ·45 | ·1586 522 |
| ·46 | ·2120 626 |
| ·47 | ·2745 724 |
| ·48 | ·3448 872 |
| ·49 | ·4209 278 |
| ·50 | ·5000 000$^e$ |
| ·51 | ·5790 722 |
| ·52 | ·6551 128 |
| ·53 | ·7254 276 |
| ·54 | ·7879 374 |
| ·55 | ·8413 478 |
| ·56 | ·8851 924 |
| ·57 | ·9197 552 |
| ·58 | ·9459 039 |
| ·59 | ·9648 771 |
| ·60 | ·9780 696 |
| ·61 | ·9868 517 |
| ·62 | ·9924 429 |
| ·63 | ·9958 432 |
| ·64 | ·9978 158 |
| ·65 | ·9989 058 |
| ·66 | ·9994 785$^+$ |
| ·67 | ·9997 641 |
| ·68 | ·9998 990 |
| ·69 | ·9999 591 |
| ·70 | ·9999 845$^-$ |
| ·71 | ·9999 944 |
| ·72 | ·9999 981 |
| ·73 | ·9999 994 |
| ·74 | ·9999 998 |
| ·75 | 1·0000 000 |

# TABLE II

## CONSTANTS OF THE CURVES

Frequency constants of the distribution

$$f(x) = \{B(p, q)\}^{-1} x^{p-1}(1-x)^{q-1}, \quad 0 \leqslant x \leqslant 1$$

for various values of $p$ and $q$.

$$\bar{x} = \text{mean} = \frac{p}{p+q}, \quad \tilde{x} = \text{mode} = \frac{p-1}{p+q-2},$$

$$\sigma^2 = \mu_2 = \frac{pq}{(p+q)^2\,(p+q+1)},$$

$$\beta_1 = \frac{\mu_3^2}{\mu_2^3} = \frac{4(q-p)^2\,(p+q+1)}{pq(p+q+2)^2},$$

$$\beta_2 = \frac{\mu_4}{\mu_2^2} = \frac{3(p+q+1)\,\{2(p+q)^2+pq(p+q-6)\}}{pq(p+q+2)\,(p+q+3)}.$$

The constants are given to eight decimal places; an *e* is affixed when the *exact* value is provided by those eight decimal places.

$p = 0{\cdot}5$

| $q$ | $\bar{x}$ | $\tilde{x}$ | $\sigma$ | $\beta_1$ | $\beta_2$ |
|---|---|---|---|---|---|
| 0·5 | ·5000 0000$^{e}$ | *J*-curves with asymptote at $x=0$ | ·3535 5339 | ·0000 0000$^{e}$ | 1·5000 0000$^{e}$ |
| 1 | ·3333 3333 | | ·2981 4240 | ·4081 6327 | 2·1428 5714 |
| 1·5 | ·2500 0000$^{e}$ | | ·2500 0000$^{e}$ | 1·0000 0000$^{e}$ | 3·0000 0000$^{e}$ |
| 2 | ·2000 0000$^{e}$ | | ·2138 0899 | 1·5555 5556 | 3·8181 8182 |
| 2·5 | ·1666 6667 | | ·1863 3900 | 2·0480 0000$^{e}$ | 4·5600 0000$^{e}$ |
| 3 | ·1428 5714 | | ·1649 5722 | 2·4793 3884 | 5·2237 7622 |
| 3·5 | ·1250 0000$^{e}$ | | ·1479 0199 | 2·8571 4286 | 5·8163 2653 |
| 4 | ·1111 1111 | | ·1340 0504 | 3·1893 4911 | 6·3461 5385− |
| 4·5 | ·1000 0000$^{e}$ | | ·1224 7449 | 3·4829 9320 | 6·8214 2857 |
| 5 | ·0909 0909 | | ·1127 5885− | 3·7440 0000$^{e}$ | 7·2494 1176 |
| 5·5 | ·0833 3333 | | ·1044 6386 | 3·9772 7273 | 7·6363 6364 |
| 6 | ·0769 2308 | | ·0973 0085+ | 4·1868 5121 | 7·9876 1610 |
| 6·5 | ·0714 2857 | | ·0910 5392 | 4·3760 6838 | 8·3076 9231 |
| 7 | ·0666 6667 | | ·0855 5853 | 4·5476 8500+ | 8·6004 2965− |
| 7·5 | ·0625 0000$^{e}$ | | ·0806 8715+ | 4·7040 0000$^{e}$ | 8·8690 9091 |
| 8 | ·0588 2353 | | ·0763 3949 | 4·8469 3878 | 9·1164 5963 |
| 8·5 | ·0555 5556 | | ·0724 3558 | 4·9781 2348 | 9·3449 1979 |
| 9 | ·0526 3158 | | ·0689 1091 | 5·0989 2880 | 9·5565 2174 |
| 9·5 | ·0500 0000$^{e}$ | | ·0657 1287 | 5·2105 2632 | 9·7530 3644 |
| 10 | ·0476 1905− | | ·0627 9814 | 5·3139 2000$^{e}$ | 9·9360 0000$^{e}$ |
| 10·5 | ·0454 5455− | | ·0601 3071 | 5·4099 7464 | 10·1067 5039 |
| 11 | ·0434 7826 | | ·0576 8043 | 5·4994 3883 | 10·2664 5768 |
| 12 | ·0400 0000$^{e}$ | | ·0533 3333 | 5·6611 1772 | 10·5567 2970 |
| 13 | ·0370 3704 | | ·0495 9512 | 5·8032 4982 | 10·8136 7020 |
| 14 | ·0344 8276 | | ·0463 4631 | 5·9291 6175− | 11·0426 7161 |
| 15 | ·0322 5806 | | ·0434 9677 | 6·0414 6939 | 11·2480 3089 |
| 16 | ·0303 0303 | | ·0409 7718 | 6·1422 5712 | 11·4332 1206 |
| 17 | ·0285 7143 | | ·0387 3339 | 6·2332 0571 | 11·6010 3741 |
| 18 | ·0270 2703 | | ·0367 2250− | 6·3156 8511 | 11·7538 2870 |
| 19 | ·0256 4103 | | ·0349 1003 | 6·3908 2292 | 11·8935 1285+ |
| 20 | ·0243 9024 | | ·0322 6802 | 6·4595 5556 | 12·0217 0213 |
| 21 | ·0232 5581 | | ·0317 7349 | 6·5226 6701 | 12·1397 5560 |
| 22 | ·0222 2222 | | ·0304 0744 | 6·5808 1860 | 12·2488 2680 |
| 23 | ·0212 7660 | | ·0291 5399 | 6·6345 7199 | 12·3499 0108 |
| 24 | ·0204 0816 | | ·0279 9977 | 6·6844 0726 | 12·4438 2504 |
| 25 | ·0196 0784 | | ·0269 3344 | 6·7307 3719 | 12·5313 3014 |
| 26 | ·0188 6792 | | ·0259 4535− | 6·7739 1860 | 12·6130 5153 |
| 27 | ·0181 8182 | | ·0250 2717 | 6·8142 6155− | 12·6895 4339 |
| 28 | ·0175 4386 | | ·0241 7175+ | 6·8520 3670 | 12·7612 9140 |
| 29 | ·0169 4915+ | | ·0233 7287 | 6·8874 8143 | 12·8287 2300 |
| 30 | ·0163 9344 | | ·0226 2510 | 6·9208 0473 | 12·8922 1584 |
| 31 | ·0158 7302 | | ·0219 2369 | 6·9521 9138 | 12·9521 0483 |
| 32 | ·0153 8462 | | ·0212 6445+ | 6·9818 0529 | 13·0086 8800 |
| 33 | ·0149 2537 | | ·0206 4370 | 7·0097 9243 | 13·0622 3142 |
| 34 | ·0144 9275+ | | ·0200 5816 | 7·0362 8316 | 13·1129 7341 |
| 35 | ·0140 8451 | | ·0195 0492 | 7·0613 9429 | 13·1611 2801 |
| 36 | ·0136 9863 | | ·0189 8138 | 7·0852 3079 | 13·2068 8805− |
| 37 | ·0133 3333 | | ·0184 8520 | 7·1078 8725− | 13·2504 2764 |
| 38 | ·0129 8701 | | ·0180 1431 | 7·1294 4914 | 13·2919 0446 |
| 39 | ·0126 5823 | | ·0175 6680 | 7·1499 9386 | 13·3314 6159 |
| 40 | ·0123 4568 | | ·0171 4099 | 7·1695 9170 | 13·3692 2921 |
| 41 | ·0120 4819 | | ·0167 3533 | 7·1883 0660 | 13·4053 2597 |
| 42 | ·0117 6471 | | ·0163 4843 | 7·2061 9691 | 13·4398 6030 |
| 43 | ·0114 9425+ | | ·0159 7901 | 7·2233 1591 | 13·4729 3142 |
| 44 | ·0112 3596 | | ·0156 2592 | 7·2397 1242 | 13·5046 3034 |
| 45 | ·0109 8901 | | ·0152 8810 | 7·2554 3121 | 13·5350 4069 |
| 46 | ·0107 5269 | | ·0149 6457 | 7·2705 1343 | 13·5642 3944 |
| 47 | ·0105 2632 | | ·0146 5445− | 7·2849 9697 | 13·5922 9753 |
| 48 | ·0103 0928 | | ·0143 5692 | 7·2989 1677 | 13·6192 8050− |
| 49 | ·0101 0101 | | ·0140 7123 | 7·3123 0511 | 13·6452 4895− |
| 50 | ·0099 0099 | | ·0137 9669 | 7·3251 9184 | 13·6702 5901 |

$p = 1$

| $q$ | $\bar{x}$ | $\tilde{x}$ | $\sigma$ | $\beta_1$ | $\beta_2$ |
|---|---|---|---|---|---|
| 0·5 | ·6666 6667 | ·0000 0000 | ·2981 4240 | ·4081 6327 | 2·1428 5714 |
| 1 | ·5000 0000$^e$ | ·0000 0000 | ·2886 7513 | ·0000 0000$^e$ | 1·8000 0000$^e$ |
| 1·5 | ·4000 0000$^e$ | ·0000 0000 | ·2618 6147 | ·1152 2634 | 2·0505 0505$^+$ |
| 2 | ·3333 3333 | ·0000 0000 | ·2357 0226 | ·3200 0000$^e$ | 2·4000 0000$^e$ |
| 2·5 | ·2857 1429 | ·0000 0000 | ·2129 5885$^+$ | ·5355 3719 | 2·7566 4336 |
| 3 | ·2500 0000$^e$ | ·0000 0000 | ·1936 4917 | ·7407 4074 | 3·0952 3810 |
| 3·5 | ·2222 2222 | ·0000 0000 | ·1772 7201 | ·9298 3939 | 3·4087 9121 |
| 4 | ·2000 0000$^e$ | ·0000 0000 | ·1632 9932 | 1·1020 4082 | 3·6964 2857 |
| 4·5 | ·1818 1818 | ·0000 0000 | ·1512 8187 | 1·2582 7160 | 3·9594 7712 |
| 5 | ·1666 6667 | ·0000 0000 | ·1408 5904 | 1·4000 0000$^e$ | 4·2000 0000$^e$ |
| 5·5 | ·1538 4615$^+$ | ·0000 0000 | ·1317 4598 | 1·5287 8264 | 4·4202 0827 |
| 6 | ·1428 5714 | ·0000 0000 | ·1237 1791 | 1·6460 9053 | 4·6222 2222 |
| 6·5 | ·1333 3333 | ·0000 0000 | ·1165 9662 | 1·7532 4952 | 4·8079 8149 |
| 7 | ·1250 0000$^e$ | ·0000 0000 | ·1102 3964 | 1·8514 2857 | 4·9792 2078 |
| 7·5 | ·1176 4706 | ·0000 0000 | ·1045 3215$^+$ | 1·9416 4777 | 5·1374 7412 |
| 8 | ·1111 1111 | ·0000 0000 | ·0993 8080 | 2·0247 9339 | 5·2840 9091 |
| 8·5 | ·1052 6316 | ·0000 0000 | ·0947 0899 | 2·1016 3460 | 5·4202 5575$^+$ |
| 9 | ·1000 0000$^e$ | ·0000 0000 | ·0904 5340 | 2·1728 3951 | 5·5470 0855$^-$ |
| 9·5 | ·0952 3810 | ·0000 0000 | ·0865 6126 | 2·2389 8947 | 5·6652 6316 |
| 10 | ·0909 0909 | ·0000 0000 | ·0829 8827 | 2·3005 9172 | 5·7758 2418 |
| 10·5 | ·0869 5652 | ·0000 0000 | ·0796 9697 | 2·3580 9001 | 5·8794 0157 |
| 11 | ·0833 3333 | ·0000 0000 | ·0766 5552 | 2·4118 7384 | 5·9766 2338 |
| 12 | ·0769 2308 | ·0000 0000 | ·0712 1693 | 2·5096 2963 | 6·1541 6667 |
| 13 | ·0714 2857 | ·0000 0000 | ·0664 9638 | 2·5961 5385$^-$ | 6·3122 1719 |
| 14 | ·0666 6667 | ·0000 0000 | ·0623 6096 | 2·6732 5754 | 6·4537 8151 |
| 15 | ·0625 0000$^e$ | ·0000 0000 | ·0587 0853 | 2·7423 8683 | 6·5812 8655$^-$ |
| 16 | ·0588 2353 | ·0000 0000 | ·0554 5936 | 2·8047 0914 | 6·6967 1053 |
| 17 | ·0555 5556 | ·0000 0000 | ·0525 5029 | 2·8611 7647 | 6·8016 8067 |
| 18 | ·0526 3158 | ·0000 0000 | ·0499 3070 | 2·9125 7244 | 6·8975 4690 |
| 19 | ·0500 0000$^e$ | ·0000 0000 | ·0475 5949 | 2·9595 4763 | 6·9854 3790 |
| 20 | ·0476 1905$^-$ | ·0000 0000 | ·0454 0298 | 3·0026 4650$^+$ | 7·0663 0435$^-$ |
| 21 | ·0454 5455$^-$ | ·0000 0000 | ·0434 3332 | 3·0423 2804 | 7·1409 5238 |
| 22 | ·0434 7826 | ·0000 0000 | ·0416 2727 | 3·0789 8182 | 7·2100 6993 |
| 23 | ·0416 6667 | ·0000 0000 | ·0399 6526 | 3·1129 4057 | 7·2742 4749 |
| 24 | ·0400 0000$^e$ | ·0000 0000 | ·0384 3076 | 3·1444 9017 | 7·3339 9471 |
| 25 | ·0384 6154 | ·0000 0000 | ·0370 0963 | 3·1738 7755$^+$ | 7·3897 5369 |
| 26 | ·0370 3704 | ·0000 0000 | ·0356 8978 | 3·2013 1711 | 7·4419 0981 |
| 27 | ·0357 1429 | ·0000 0000 | ·0344 6076 | 3·2269 9588 | 7·4908 0048 |
| 28 | ·0344 8276 | ·0000 0000 | ·0333 1351 | 3·2510 7775$^-$ | 7·5367 2235$^+$ |
| 29 | ·0333 3333 | ·0000 0000 | ·0322 4014 | 3·2737 0690 | 7·5799 3730 |
| 30 | ·0322 5806 | ·0000 0000 | ·0312 3374 | 3·2950 1071 | 7·6206 7736 |
| 31 | ·0312 5000$^e$ | ·0000 0000 | ·0302 8823 | 3·3151 0213 | 7·6591 4882 |
| 32 | ·0303 0303 | ·0000 0000 | ·0293 9826 | 3·3340 8163 | 7·6955 3571 |
| 33 | ·0294 1176 | ·0000 0000 | ·0285 5907 | 3·3520 3891 | 7·7300 0273 |
| 34 | ·0285 7143 | ·0000 0000 | ·0277 6644 | 3·3690 5427 | 7·7626 9768 |
| 35 | ·0277 7778 | ·0000 0000 | ·0270 1660 | 3·3851 9984 | 7·7937 5361 |
| 36 | ·0270 2703 | ·0000 0000 | ·0263 0618 | 3·4005 4058 | 7·8232 9060 |
| 37 | ·0263 1579 | ·0000 0000 | ·0256 3215$^-$ | 3·4151 3514 | 7·8514 1727 |
| 38 | ·0256 4103 | ·0000 0000 | ·0249 9178 | 3·4290 3660 | 7·8782 3217 |
| 39 | ·0250 0000$^e$ | ·0000 0000 | ·0243 8262 | 3·4422 9316 | 7·9038 2486 |
| 40 | ·0243 9024 | ·0000 0000 | ·0238 0244 | 3·4549 4862 | 7·9282 7696 |
| 41 | ·0238 0952 | ·0000 0000 | ·0232 4922 | 3·4670 4293 | 7·9516 6297 |
| 42 | ·0232 5581 | ·0000 0000 | ·0227 2113 | 3·4786 1258 | 7·9740 5107 |
| 43 | ·0227 2727 | ·0000 0000 | ·0222 1648 | 3·4896 9095$^-$ | 7·9955 0373 |
| 44 | ·0222 2222 | ·0000 0000 | ·0217 3376 | 3·5003 0865$^+$ | 8·0160 7834 |
| 45 | ·0217 3913 | ·0000 0000 | ·0212 7157 | 3·5104 9383 | 8·0358 2766 |
| 46 | ·0212 7660 | ·0000 0000 | ·0208 2862 | 3·5202 7235$^+$ | 8·0548 0035$^+$ |
| 47 | ·0208 3333 | ·0000 0000 | ·0204 0373 | 3·5296 6809 | 8·0730 4130 |
| 48 | ·0204 0816 | ·0000 0000 | ·0199 9583 | 3·5387 0306 | 8·0905 9201 |
| 49 | ·0200 0000$^e$ | ·0000 0000 | ·0196 0392 | 3·5473 9766 | 8·1074 9089 |
| 50 | ·0196 0784 | ·0000 0000 | ·0192 2707 | 3·5557 7074 | 8·1237 7358 |

$p = 1·5$

| $q$ | $\bar{x}$ | $\tilde{x}$ | $\sigma$ | $\beta_1$ | $\beta_2$ |
|---|---|---|---|---|---|
| 0·5 | ·7500 0000$^e$ | *J*-curve | ·2500 0000$^e$ | 1·0000 0000$^e$ | 3·0000 0000$^e$ |
| 1 | ·6000 0000$^e$ | 1·0000 0000$^e$ | ·2618 6147 | ·1152 2634 | 2·0505 0505$^+$ |
| 1·5 | ·5000 0000$^e$ | ·5000 0000$^e$ | ·2500 0000$^e$ | ·0000 0000$^e$ | 2·0000 0000$^e$ |
| 2 | ·4285 7143 | ·3333 3333 | ·2332 8474 | ·0495 8678 | 2·1398 6014 |
| 2·5 | ·3750 0000$^e$ | ·2500 0000$^e$ | ·2165 0635$^+$ | ·1481 4815$^-$ | 2·3333 3333 |
| 3 | ·3333 3333 | ·2000 0000$^e$ | ·2010 0756 | ·2603 5503 | 2·5384 6154 |
| 3·5 | ·3000 0000$^e$ | ·1666 6667 | ·1870 8287 | ·3731 7784 | 2·7397 9592 |
| 4 | ·2727 2727 | ·1428 5714 | ·1746 8526 | ·4814 8148 | 2·9313 7255$^-$ |
| 4·5 | ·2500 0000$^e$ | ·1250 0000$^e$ | ·1636 6342 | ·5833 3333 | 3·1111 1111 |
| 5 | ·2307 6923 | ·1111 1111 | ·1538 4615$^+$ | ·6782 0069 | 3·2786 3777 |
| 5·5 | ·2142 8571 | ·1000 0000$^e$ | ·1450 7211 | ·7661 8032 | 3·4343 4343 |
| 6 | ·2000 0000$^e$ | ·0909 0909 | ·1371 9887 | ·8476 4543 | 3·5789 4737 |
| 6·5 | ·1875 0000$^e$ | ·0833 3333 | ·1301 0412 | ·9230 7692 | 3·7132 8671 |
| 7 | ·1764 7059 | ·0769 2308 | ·1236 8411 | ·9929 8132 | 3·8382 1355$^-$ |
| 7·5 | ·1666 6667 | ·0714 2857 | ·1178 5113 | 1·0578 5124 | 3·9545 4545$^+$ |
| 8 | ·1578 9474 | ·0666 6667 | ·1125 3105$^-$ | 1·1181 4745$^-$ | 4·0630 4348 |
| 8·5 | ·1500 0000$^e$ | ·0625 0000$^e$ | ·1076 6108 | 1·1742 9194 | 4·1644 0422 |
| 9 | ·1428 5714 | ·0588 2353 | ·1031 8787 | 1·2266 6667 | 4·2592 5926 |
| 9·5 | ·1363 6364 | ·0555 5556 | ·0990 6589 | 1·2756 1507 | 4·3481 7814 |
| 10 | ·1304 3478 | ·0526 3158 | ·0952 5610 | 1·3214 4490 | 4·4316 7305$^+$ |
| 10·5 | ·1250 0000$^e$ | ·0500 0000$^e$ | ·0917 2492 | 1·3644 3149 | 4·5102 0408 |
| 11 | ·1200 0000$^e$ | ·0476 1905$^-$ | ·0884 4333 | 1·4048 2110 | 4·5841 8445$^-$ |
| 12 | ·1111 1111 | ·0434 7826 | ·0825 3126 | 1·4786 6805$^+$ | 4·7199 4135$^-$ |
| 13 | ·1034 4828 | ·0400 0000$^e$ | ·0773 5412 | 1·5444 8918 | 4·8414 9184 |
| 14 | ·0967 7419 | ·0370 3704 | ·0727 8401 | 1·6034 9854 | 4·9509 1009 |
| 15 | ·0909 0909 | ·0344 8276 | ·0687 2081 | 1·6566 8371 | 5·0498 9605$^-$ |
| 16 | ·0857 1429 | ·0322 5806 | ·0650 8512 | 1·7048 5426 | 5·1398 5303 |
| 17 | ·0810 8108 | ·0303 0303 | ·0618 1318 | 1·7486 7901 | 5·2219 4788 |
| 18 | ·0769 2308 | ·0285 7143 | ·0588 5324 | 1·7887 1462 | 5·2971 5762 |
| 19 | ·0731 7073 | ·0270 2703 | ·0561 6288 | 1·8254 2777 | 5·3663 0584 |
| 20 | ·0697 6744 | ·0256 4103 | ·0537 0700 | 1·8592 1231 | 5·4300 9119 |
| 21 | ·0666 6667 | ·0243 9024 | ·0514 5633 | 1·8904 0281 | 5·4891 0993 |
| 22 | ·0638 2979 | ·0232 5581 | ·0493 8625$^+$ | 1·9192 8512 | 5·5438 7381 |
| 23 | ·0612 2449 | ·0222 2222 | ·0474 7592 | 1·9461 0491 | 5·5948 2437 |
| 24 | ·0588 2353 | ·0212 7660 | ·0457 0757 | 1·9710 7438 | 5·6423 4450$^-$ |
| 25 | ·0566 0377 | ·0204 0816 | ·0440 6598 | 1·9943 7776 | 5·6867 6777 |
| 26 | ·0545 4545$^+$ | ·0196 0784 | ·0425 3801 | 2·0161 7572 | 5·7283 8609 |
| 27 | ·0526 3158 | ·0188 6792 | ·0411 1229 | 2·0366 0904 | 5·7674 5598 |
| 28 | ·0508 4746 | ·0181 8182 | ·0397 7889 | 2·0558 0151 | 5·8042 0373 |
| 29 | ·0491 8033 | ·0175 4386 | ·0385 2916 | 2·0738 6248 | 5·8388 2972 |
| 30 | ·0476 1905$^-$ | ·0169 4915$^+$ | ·0373 5545$^+$ | 2·0908 8884 | 5·8715 1201 |
| 31 | ·0461 5385$^-$ | ·0163 9344 | ·0362 5106 | 2·1069 6678 | 5·9024 0931 |
| 32 | ·0447 7612 | ·0158 7302 | ·0352 1002 | 2·1221 7318 | 5·9316 6361 |
| 33 | ·0434 7826 | ·0153 8462 | ·0342 2704 | 2·1365 7688 | 5·9594 0224 |
| 34 | ·0422 5352 | ·0149 2537 | ·0332 9740 | 2·1502 3965$^+$ | 5·9857 3975$^+$ |
| 35 | ·0410 9589 | ·0144 9275$^+$ | ·0324 1688 | 2·1632 1712 | 6·0107 7946 |
| 36 | ·0400 0000$^e$ | ·0140 8451 | ·0315 8168 | 2·1755 5947 | 6·0346 1478 |
| 37 | ·0389 6104 | ·0136 9863 | ·0307 8841 | 2·1873 1214 | 6·0573 3042 |
| 38 | ·0379 7468 | ·0133 3333 | ·0300 3398 | 2·1985 1632 | 6·0790 0332 |
| 39 | ·0370 3704 | ·0129 8701 | ·0293 1561 | 2·2092 0948 | 6·0997 0354 |
| 40 | ·0361 4458 | ·0126 5823 | ·0286 3077 | 2·2194 2573 | 6·1194 9503 |
| 41 | ·0352 9412 | ·0123 4568 | ·0279 7718 | 2·2291 9624 | 6·1384 3624 |
| 42 | ·0344 8276 | ·0120 4819 | ·0273 5275$^+$ | 2·2385 4952 | 6·1565 8075$^-$ |
| 43 | ·0337 0787 | ·0117 6471 | ·0267 5557 | 2·2475 1170 | 6·1739 7770 |
| 44 | ·0329 6703 | ·0114 9425$^+$ | ·0261 8389 | 2·2561 0677 | 6·1906 7232 |
| 45 | ·0322 5806 | ·0112 3596 | ·0256 3611 | 2·2643 5682 | 6·2067 0624 |
| 46 | ·0315 7895$^-$ | ·0109 8901 | ·0251 1077 | 2·2722 8218 | 6·2221 1786 |
| 47 | ·0309 2784 | ·0107 5269 | ·0246 0652 | 2·2799 0164 | 6·2369 4270 |
| 48 | ·0303 0303 | ·0105 2632 | ·0241 2211 | 2·2872 3254 | 6·2512 1359 |
| 49 | ·0297 0297 | ·0103 0928 | ·0236 5640 | 2·2942 9096 | 6·2649 6099 |
| 50 | ·0291 2621 | ·0101 0101 | ·0232 0832 | 2·3010 9180 | 6·2782 1315$^+$ |

$p = 2$

| $q$ | $\bar{x}$ | $\tilde{x}$ | $\sigma$ | $\beta_1$ | $\beta_2$ |
|---|---|---|---|---|---|
| 0·5 | ·8000 0000$^{e}$ | *J*-curve | ·2138 0899 | 1·5555 5556 | 3·8181 8182 |
| 1 | ·6666 6667 | 1·0000 0000$^{e}$ | ·2357 0226 | ·3200 0000$^{e}$ | 2·4000 0000$^{e}$ |
| 1·5 | ·5714 2857 | ·6666 6667 | ·2332 8474 | ·0495 8678 | 2·1398 6014 |
| 2 | ·5000 0000$^{e}$ | ·5000 0000$^{e}$ | ·2236 0680 | ·0000 0000$^{e}$ | 2·1428 5714 |
| 2·5 | ·4444 4444 | ·4000 0000$^{e}$ | ·2118 8058 | ·0260 3550$^{+}$ | 2·2338 4615$^{+}$ |
| 3 | ·4000 0000$^{e}$ | ·3333 3333 | ·2000 0000$^{e}$ | ·0816 3265$^{+}$ | 2·3571 4286 |
| 3·5 | ·3636 3636 | ·2857 1429 | ·1886 8164 | ·1485 7143 | 2·4907 5630 |
| 4 | ·3333 3333 | ·2500 0000$^{e}$ | ·1781 7416 | ·2187 5000$^{e}$ | 2·6250 0000$^{e}$ |
| 4·5 | ·3076 9231 | ·2222 2222 | ·1685 3002 | ·2883 5063 | 2·7554 1796 |
| 5 | ·2857 1429 | ·2000 0000$^{e}$ | ·1597 1914 | ·3555 5556 | 2·8800 0000$^{e}$ |
| 5·5 | ·2666 6667 | ·1818 1818 | ·1516 7905$^{+}$ | ·4195 4168 | 2·9979 4942 |
| 6 | ·2500 0000$^{e}$ | ·1666 6667 | ·1443 3757 | ·4800 0000$^{e}$ | 3·1090 9091 |
| 6·5 | ·2352 9412 | ·1538 4615$^{+}$ | ·1376 2298 | ·5368 9168 | 3·2135 6904 |
| 7 | ·2222 2222 | ·1428 5714 | ·1314 6844 | ·5903 1877 | 3·3116 8831 |
| 7·5 | ·2105 2632 | ·1333 3333 | ·1258 1354 | ·6404 5369 | 3·4038 2609 |
| 8 | ·2000 0000$^{e}$ | ·1250 0000$^{e}$ | ·1206 0454 | ·6875 0000$^{e}$ | 3·4903 8462 |
| 8·5 | ·1904 7619 | ·1176 4706 | ·1157 9405$^{-}$ | ·7316 7059 | 3·5717 6471 |
| 9 | ·1818 1818 | ·1111 1111 | ·1113 4044 | ·7731 7554 | 3·6483 5165$^{-}$ |
| 9·5 | ·1739 1304 | ·1052 6316 | ·1072 0720 | ·8122 1572 | 3·7205 0817 |
| 10 | ·1666 6667 | ·1000 0000$^{e}$ | ·1033 6228 | ·8489 7959 | 3·7885 7143 |
| 10·5 | ·1600 0000$^{e}$ | ·0952 3810 | ·0997 7753 | ·8836 4192 | 3·8528 5238 |
| 11 | ·1538 4615$^{+}$ | ·0909 0909 | ·0964 2818 | ·9163 6364 | 3·9136 3636 |
| 12 | ·1428 5714 | ·0833 3333 | ·0903 5079 | ·9765 6250$^{e}$ | 4·0257 3529 |
| 13 | ·1333 3333 | ·0769 2308 | ·0849 8366 | 1·0306 0953 | 4·1266 9683 |
| 14 | ·1250 0000$^{e}$ | ·0714 2857 | ·0802 1112 | 1·0793 6508 | 4·2180 4511 |
| 15 | ·1176 4706 | ·0666 6667 | ·0759 4085$^{-}$ | 1·1235 4571 | 4·3010 5263 |
| 16 | ·1111 1111 | ·0625 0000$^{e}$ | ·0720 9841 | 1·1637 5000$^{e}$ | 4·3767 8571 |
| 17 | ·1052 6316 | ·0588 2353 | ·0686 2318 | 1·2004 8019 | 4·4461 4209 |
| 18 | ·1000 0000$^{e}$ | ·0555 5556 | ·0654 6537 | 1·2341 5978 | 4·5098 8142 |
| 19 | ·0952 3810 | ·0526 3158 | ·0625 8369 | 1·2651 4775$^{-}$ | 4·5686 4989 |
| 20 | ·0909 0909 | ·0500 0000$^{e}$ | ·0599 4368 | 1·2937 5000$^{e}$ | 4·6230 0000$^{e}$ |
| 21 | ·0869 5652 | ·0476 1905$^{-}$ | ·0575 1633 | 1·3202 2857 | 4·6734 0659 |
| 22 | ·0833 3333 | ·0454 5455$^{-}$ | ·0552 7708 | 1·3448 0904 | 4·7202 7972 |
| 23 | ·0800 0000$^{e}$ | ·0434 7826 | ·0532 0497 | 1·3676 8653 | 4·7639 7516 |
| 24 | ·0769 2308 | ·0416 6667 | ·0512 8205$^{+}$ | 1·3890 3061 | 4·8048 0296 |
| 25 | ·0740 7407 | ·0400 0000$^{e}$ | ·0494 9282 | 1·4089 8930 | 4·8430 3448 |
| 26 | ·0714 2857 | ·0384 6154 | ·0478 2386 | 1·4276 9231 | 4·8789 0819 |
| 27 | ·0689 6552 | ·0370 3704 | ·0462 6348 | 1·4452 5379 | 4·9126 3441 |
| 28 | ·0666 6667 | ·0357 1429 | ·0448 0143 | 1·4617 7455$^{+}$ | 4·9443 9935$^{+}$ |
| 29 | ·0645 1613 | ·0344 8276 | ·0434 2875$^{-}$ | 1·4773 4397 | 4·9743 6843 |
| 30 | ·0625 0000$^{e}$ | ·0333 3333 | ·0421 3749 | 1·4920 4152 | 5·0026 8908 |
| 31 | ·0606 0606 | ·0322 5806 | ·0409 2064 | 1·5059 3812 | 5·0294 9309 |
| 32 | ·0588 2353 | ·0312 5000$^{e}$ | ·0397 7196 | 1·5190 9722 | 5·0548 9865$^{-}$ |
| 33 | ·0571 4286 | ·0303 0303 | ·0386 8590 | 1·5315 7580 | 5·0790 1203 |
| 34 | ·0555 5556 | ·0294 1176 | ·0376 5747 | 1·5434 2513 | 5·1019 2903 |
| 35 | ·0540 5405$^{+}$ | ·0285 7143 | ·0366 8221 | 1·5546 9146 | 5·1237 3626 |
| 36 | ·0526 3158 | ·0277 7778 | ·0357 5612 | 1·5654 1667 | 5·1445 1220 |
| 37 | ·0512 8205$^{+}$ | ·0270 2703 | ·0348 7557 | 1·5756 3870 | 5·1643 2809 |
| 38 | ·0500 0000$^{e}$ | ·0263 1579 | ·0340 3728 | 1·5853 9205$^{+}$ | 5·1832 4882 |
| 39 | ·0487 8049 | ·0256 4103 | ·0332 3830 | 1·5947 0816 | 5·2013 3355$^{+}$ |
| 40 | ·0476 1905$^{-}$ | ·0250 0000$^{e}$ | ·0324 7592 | 1·6036 1570 | 5·2186 3636 |
| 41 | ·0465 1163 | ·0243 9024 | ·0317 4769 | 1·6121 4092 | 5·2352 0679 |
| 42 | ·0454 5455$^{-}$ | ·0238 0952 | ·0310 5137 | 1·6203 0786 | 5·2510 9026 |
| 43 | ·0444 4444 | ·0232 5581 | ·0303 8490 | 1·6281 3859 | 5·2663 2855$^{+}$ |
| 44 | ·0434 7826 | ·0227 2727 | ·0297 4641 | 1·6356 5341 | 5·2809 6011 |
| 45 | ·0425 5319 | ·0222 2222 | ·0291 3418 | 1·6428 7103 | 5·2950 2041 |
| 46 | ·0416 6667 | ·0217 3913 | ·0285 4662 | 1·6498 0870 | 5·3085 4220 |
| 47 | ·0408 1633 | ·0212 7660 | ·0279 8226 | 1·6564 8237 | 5·3215 5579 |
| 48 | ·0400 0000$^{e}$ | ·0208 3333 | ·0274 3974 | 1·6629 0680 | 5·3340 8926 |
| 49 | ·0392 1569 | ·0204 0816 | ·0269 1790 | 1·6690 9569 | 5·3461 6866 |
| 50 | ·0384 6154 | ·0200 0000$^{e}$ | ·0264 1549 | 1·6750 6173 | 5·3578 1818 |

$p = 2{\cdot}5$

| $q$ | $\bar{x}$ | $\tilde{x}$ | $\sigma$ | $\beta_1$ | $\beta_2$ |
|---|---|---|---|---|---|
| 0·5 | ·8333 3333 | *J*-curve | ·1863 3900 | 2·0480 0000$^e$ | 4·5600 0000$^e$ |
| 1 | ·7142 8571 | 1·0000 0000$^e$ | ·2129 5885+ | ·5355 3719 | 2·7566 4336 |
| 1·5 | ·6250 0000$^e$ | ·7500 0000$^e$ | ·2165 0635+ | ·1481 4815− | 2·3333 3333 |
| 2 | ·5555 5556 | ·6000 0000$^e$ | ·2118 8058 | ·0260 3550+ | 2·2338 4615+ |
| 2·5 | ·5000 0000$^e$ | ·5000 0000$^e$ | ·2041 2415− | ·0000 0000$^e$ | 2·2500 0000$^e$ |
| 3 | ·4545 4545+ | ·4285 7143 | ·1953 0406 | ·0154 0741 | 2·3145 0980 |
| 3·5 | ·4166 6667 | ·3750 0000$^e$ | ·1863 3900 | ·0500 0000$^e$ | 2·4000 0000$^e$ |
| 4 | ·3846 1538 | ·3333 3333 | ·1776 4624 | ·0934 2561 | 2·4938 0805− |
| 4·5 | ·3571 4286 | ·3000 0000$^e$ | ·1694 0773 | ·1404 6639 | 2·5896 2963 |
| 5 | ·3333 3333 | ·2727 2727 | ·1616 9042 | ·1883 6565+ | 2·6842 1053 |
| 5·5 | ·3125 0000$^e$ | ·2500 0000$^e$ | ·1545 0414 | ·2356 3636 | 2·7758 6777 |
| 6 | ·2941 1765− | ·2307 6923 | ·1478 3079 | ·2814 8148 | 2·8637 6812 |
| 6·5 | ·2777 7778 | ·2142 8571 | ·1416 3943 | ·3254 9269 | 2·9475 5245− |
| 7 | ·2631 5789 | ·2000 0000$^e$ | ·1358 9415+ | ·3674 8582 | 3·0271 3043 |
| 7·5 | ·2500 0000$^e$ | ·1875 0000$^e$ | ·1305 5824 | ·4074 0741 | 3·1025 6410 |
| 8 | ·2380 9524 | ·1764 7059 | ·1255 9628 | ·4452 8000$^e$ | 3·1740 0000$^e$ |
| 8·5 | ·2272 7273 | ·1666 6667 | ·1209 7515− | ·4811 6951 | 3·2416 2896 |
| 9 | ·2173 9130 | ·1578 9474 | ·1166 6442 | ·5151 6537 | 3·3056 6198 |
| 9·5 | ·2083 3333 | ·1500 0000$^e$ | ·1126 3643 | ·5473 6842 | 3·3663 1579 |
| 10 | ·2000 0000$^e$ | ·1428 5714 | ·1088 6621 | ·5778 8347 | 3·4238 0423 |
| 10·5 | ·1923 0769 | ·1363 6364 | ·1053 3126 | ·6068 1481 | 3·4783 3333 |
| 11 | ·1851 8519 | ·1304 3478 | ·1020 1137 | ·6342 6355+ | 3·5300 9864 |
| 12 | ·1724 1379 | ·1200 0000$^e$ | ·0959 4598 | ·6850 9336 | 3·6260 6061 |
| 13 | ·1612 9032 | ·1111 1111 | ·0905 4574 | ·7310 7692 | 3·7130 1455+ |
| 14 | ·1515 1515+ | ·1034 4828 | ·0857 0991 | ·7728 2688 | 3·7920 9979 |
| 15 | ·1428 5714 | ·0967 7419 | ·0813 5640 | ·8108 7004 | 3·8642 9018 |
| 16 | ·1351 3514 | ·0909 0909 | ·0774 1782 | ·8456 5735− | 3·9304 1690 |
| 17 | ·1282 0513 | ·0857 1429 | ·0738 3851 | ·8775 7452 | 3·9911 9015+ |
| 18 | ·1219 5122 | ·0810 8108 | ·0705 7212 | ·9069 5199 | 4·0472 1828 |
| 19 | ·1162 7907 | ·0769 2308 | ·0675 7983 | ·9340 7353 | 4·0990 2416 |
| 20 | ·1111 1111 | ·0731 7073 | ·0648 2889 | ·9591 8367 | 4·1470 5882 |
| 21 | ·1063 8298 | ·0697 6744 | ·0622 9149 | ·9824 9391 | 4·1917 1291 |
| 22 | ·1020 4082 | ·0666 6667 | ·0599 4392 | 1·0041 8784 | 4·2333 2606 |
| 23 | ·0980 3922 | ·0638 2979 | ·0577 6580 | 1·0244 2544 | 4·2721 9472 |
| 24 | ·0943 3962 | ·0612 2449 | ·0557 3954 | 1·0433 4667 | 4·3085 7865+ |
| 25 | ·0909 0909 | ·0588 2353 | ·0538 4990 | 1·0610 7440 | 4·3427 0631 |
| 26 | ·0877 1930 | ·0566 0377 | ·0520 8358 | 1·0777 1691 | 4·3747 7932 |
| 27 | ·0847 4576 | ·0545 4545+ | ·0504 2895+ | 1·0933 6991 | 4·4049 7626 |
| 28 | ·0819 6721 | ·0526 3158 | ·0488 7580 | 1·1081 1834 | 4·4334 5580 |
| 29 | ·0793 6508 | ·0508 4746 | ·0474 1511 | 1·1220 3778 | 4·4603 5937 |
| 30 | ·0769 2308 | ·0491 8033 | ·0460 3889 | 1·1351 9569 | 4·4858 1343 |
| 31 | ·0746 2687 | ·0476 1905− | ·0447 4005+ | 1·1476 5248 | 4·5099 3135+ |
| 32 | ·0724 6377 | ·0461 5385− | ·0435 1227 | 1·1594 6238 | 4·5328 1507 |
| 33 | ·0704 2254 | ·0447 7612 | ·0423 4988 | 1·1706 7421 | 4·5545 5647 |
| 34 | ·0684 9315+ | ·0434 7826 | ·0412 4782 | 1·1813 3204 | 4·5752 3861 |
| 35 | ·0666 6667 | ·0422 5352 | ·0402 0151 | 1·1914 7573 | 4·5949 3671 |
| 36 | ·0649 3506 | ·0410 9589 | ·0392 0685+ | 1·2011 4142 | 4·6137 1907 |
| 37 | ·0632 9114 | ·0400 0000$^e$ | ·0382 6011 | 1·2103 6200 | 4·6316 4786 |
| 38 | ·0617 2840 | ·0389 6104 | ·0373 5792 | 1·2191 6736 | 4·6487 7976 |
| 39 | ·0602 4096 | ·0379 7468 | ·0364 9722 | 1·2275 8485+ | 4·6651 6655+ |
| 40 | ·0588 2353 | ·0370 3704 | ·0356 7520 | 1·2356 3944 | 4·6808 5566 |
| 41 | ·0574 7126 | ·0361 4458 | ·0348 8933 | 1·2433 5402 | 4·6958 9058 |
| 42 | ·0561 7978 | ·0352 9412 | ·0341 3728 | 1·2507 4960 | 4·7103 1126 |
| 43 | ·0549 4505+ | ·0344 8276 | ·0334 1692 | 1·2578 4552 | 4·7241 5450+ |
| 44 | ·0537 6344 | ·0337 0787 | ·0327 2628 | 1·2646 5956 | 4·7374 5420 |
| 45 | ·0526 3158 | ·0329 6703 | ·0320 6356 | 1·2712 0815+ | 4·7502 4169 |
| 46 | ·0515 4639 | ·0322 5806 | ·0314 2712 | 1·2775 0647 | 4·7625 4592 |
| 47 | ·0505 0505+ | ·0315 7895− | ·0308 1542 | 1·2835 6855− | 4·7743 9372 |
| 48 | ·0495 0495+ | ·0309 2784 | ·0302 2704 | 1·2894 0741 | 4·7858 0997 |
| 49 | ·0485 4369 | ·0303 0303 | ·0296 6068 | 1·2950 3512 | 4·7968 1778 |
| 50 | ·0476 1905− | ·0297 0297 | ·0291 1514 | 1·3004 6292 | 4·8074 3863 |

$p = 3$

| $q$ | $\bar{x}$ | $\tilde{x}$ | $\sigma$ | $\beta_1$ | $\beta_2$ |
|---|---|---|---|---|---|
| 0·5 | ·8571 4286 | *J*-curve | ·1649 5722 | 2·4793 3884 | 5·2237 7622 |
| 1 | ·7500 0000$^{e}$ | 1·0000 0000$^{e}$ | ·1936 4917 | ·7407 4074 | 3·0952 3810 |
| 1·5 | ·6666 6667 | ·8000 0000$^{e}$ | ·2010 0756 | ·2603 5503 | 2·5384 6154 |
| 2 | ·6000 0000$^{e}$ | ·6666 6667 | ·2000 0000$^{e}$ | ·0816 3265$^{+}$ | 2·3571 4286 |
| 2·5 | ·5454 5455$^{-}$ | ·5714 2857 | ·1953 0406 | ·0154 0741 | 2·3145 0980 |
| 3 | ·5000 0000$^{e}$ | ·5000 0000$^{e}$ | ·1889 8224 | ·0000 0000$^{e}$ | 2·3333 3333 |
| 3·5 | ·4615 3846 | ·4444 4444 | ·1820 3322 | ·0098 8631 | 2·3816 8952 |
| 4 | ·4285 7143 | ·4000 0000$^{e}$ | ·1749 6355$^{+}$ | ·0329 2181 | 2·4444 4444 |
| 4·5 | ·4000 0000$^{e}$ | ·3636 3636 | ·1680 3361 | ·0627 8855$^{+}$ | 2·5137 8446 |
| 5 | ·3750 0000$^{e}$ | ·3333 3333 | ·1613 7431 | ·0960 0000$^{e}$ | 2·5854 5455$^{-}$ |
| 5·5 | ·3529 4118 | ·3076 9231 | ·1550 4624 | ·1305 5727 | 2·6570 6757 |
| 6 | ·3333 3333 | ·2857 1429 | ·1490 7120 | ·1652 8926 | 2·7272 7273 |
| 6·5 | ·3157 8947 | ·2666 6667 | ·1434 4950$^{+}$ | ·1995 0560 | 2·7953 1773 |
| 7 | ·3000 0000$^{e}$ | ·2500 0000$^{e}$ | ·1381 6986 | ·2328 0423 | 2·8608 0586 |
| 7·5 | ·2857 1429 | ·2352 9412 | ·1332 1497 | ·2649 6000$^{e}$ | 2·9235 5556 |
| 8 | ·2727 2727 | ·2222 2222 | ·1285 6487 | ·2958 5799 | 2·9835 1648 |
| 8·5 | ·2608 6957 | ·2105 2632 | ·1241 9876 | ·3254 5254 | 3·0407 1820 |
| 9 | ·2500 0000$^{e}$ | ·2000 0000$^{e}$ | ·1200 9612 | ·3537 4150$^{-}$ | 3·0952 3810 |
| 9·5 | ·2400 0000$^{e}$ | ·1904 7619 | ·1162 3731 | ·3807 4973 | 3·1471 8108 |
| 10 | ·2307 6923 | ·1818 1818 | ·1126 0385$^{+}$ | ·4065 1852 | 3·1966 6667 |
| 10·5 | ·2222 2222 | ·1739 1304 | ·1091 7860 | ·4310 9856 | 3·2438 2070 |
| 11 | ·2142 8571 | ·1666 6667 | ·1059 4569 | ·4545 4545$^{+}$ | 3·2887 7005$^{+}$ |
| 12 | ·2000 0000$^{e}$ | ·1538 4615$^{+}$ | ·1000 0000$^{e}$ | ·4982 6990 | 3·3725 4902 |
| 13 | ·1875 0000$^{e}$ | ·1428 5714 | ·0946 6466 | ·5381 4498 | 3·4489 4287 |
| 14 | ·1764 7059 | ·1333 3333 | ·0898 5443 | ·5745 9438 | 3·5187 9699 |
| 15 | ·1666 6667 | ·1250 0000$^{e}$ | ·0854 9820 | ·6080 0000$^{e}$ | 3·5828 5714 |
| 16 | ·1578 9474 | ·1176 4706 | ·0815 3649 | ·6386 9992 | 3·6417 7489 |
| 17 | ·1500 0000$^{e}$ | ·1111 1111 | ·0779 1937 | ·6669 9076 | 3·6961 1718 |
| 18 | ·1428 5714 | ·1052 6316 | ·0746 0471 | ·6931 3170 | 3·7463 7681 |
| 19 | ·1363 6364 | ·1000 0000$^{e}$ | ·0715 5679 | ·7173 4893 | 3·7929 8246 |
| 20 | ·1304 3478 | ·0952 3810 | ·0687 4517 | ·7398 4000$^{e}$ | 3·8363 0769 |
| 21 | ·1250 0000$^{e}$ | ·0909 0909 | ·0661 4378 | ·7607 7768 | 3·8766 7888 |
| 22 | ·1200 0000$^{e}$ | ·0869 5652 | ·0637 3020 | ·7803 1342 | 3·9143 8191 |
| 23 | ·1153 8462 | ·0833 3333 | ·0614 8502 | ·7985 8030 | 3·9496 6802 |
| 24 | ·1111 1111 | ·0800 0000$^{e}$ | ·0593 9139 | ·8156 9560 | 3·9827 5862 |
| 25 | ·1071 4286 | ·0769 2308 | ·0574 3460 | ·8317 6296 | 4·0138 4946 |
| 26 | ·1034 4828 | ·0740 7407 | ·0556 0178 | ·8468 7425$^{-}$ | 4·0431 1414 |
| 27 | ·1000 0000$^{e}$ | ·0714 2857 | ·0538 8159 | ·8611 1111 | 4·0707 0707 |
| 28 | ·0967 7419 | ·0689 6552 | ·0522 6404 | ·8745 4633 | 4·0967 6598 |
| 29 | ·0937 5000$^{e}$ | ·0666 6667 | ·0507 4026 | ·8872 4496 | 4·1214 1408 |
| 30 | ·0909 0909 | ·0645 1613 | ·0493 0238 | ·8992 6531 | 4·1447 6190 |
| 31 | ·0882 3529 | ·0625 0000$^{e}$ | ·0479 4337 | ·9106 5976 | 4·1669 0884 |
| 32 | ·0857 1429 | ·0606 0606 | ·0466 5695$^{-}$ | ·9214 7553 | 4·1879 4452 |
| 33 | ·0833 3333 | ·0588 2353 | ·0454 3748 | ·9317 5523 | 4·2079 4994 |
| 34 | ·0810 8108 | ·0571 4286 | ·0442 7989 | ·9415 3743 | 4·2269 9849 |
| 35 | ·0789 4737 | ·0555 5556 | ·0431 7961 | ·9508 5714 | 4·2451 5679 |
| 36 | ·0769 2308 | ·0540 5405$^{+}$ | ·0421 3250$^{+}$ | ·9597 4618 | 4·2624 8548 |
| 37 | ·0750 0000$^{e}$ | ·0526 3158 | ·0411 3482 | ·9682 3354 | 4·2790 3984 |
| 38 | ·0731 7073 | ·0512 8205$^{+}$ | ·0401 8314 | ·9763 4568 | 4·2948 7037 |
| 39 | ·0714 2857 | ·0500 0000$^{e}$ | ·0392 7439 | ·9841 0680 | 4·3100 2331 |
| 40 | ·0697 6744 | ·0487 8049 | ·0384 0571 | ·9915 3909 | 4·3245 4106 |
| 41 | ·0681 8182 | ·0476 1905$^{-}$ | ·0375 7454 | ·9986 6292 | 4·3384 6258 |
| 42 | ·0666 6667 | ·0465 1163 | ·0367 7849 | 1·0054 9699 | 4·3518 2371 |
| 43 | ·0652 1739 | ·0454 5455$^{-}$ | ·0360 1539 | 1·0120 5857 | 4·3646 5749 |
| 44 | ·0638 2979 | ·0444 4444 | ·0352 8324 | 1·0183 6356 | 4·3769 9443 |
| 45 | ·0625 0000$^{e}$ | ·0434 7826 | ·0345 8021 | 1·0244 2667 | 4·3888 6275$^{-}$ |
| 46 | ·0612 2449 | ·0425 5319 | ·0339 0459 | 1·0302 6149 | 4·4002 8854 |
| 47 | ·0600 0000$^{e}$ | ·0416 6667 | ·0332 5481 | 1·0358 8065$^{-}$ | 4·4112 9605$^{+}$ |
| 48 | ·0588 2353 | ·0408 1633 | ·0326 2942 | 1·0412 9583 | 4·4219 0776 |
| 49 | ·0576 9231 | ·0400 0000$^{e}$ | ·0320 2708 | 1·0465 1792 | 4·4321 4458 |
| 50 | ·0566 0377 | ·0392 1569 | ·0314 4654 | 1·0515 5702 | 4·4420 2597 |

$p = 3{\cdot}5$

| $q$ | $\bar{x}$ | $\tilde{x}$ | $\sigma$ | $\beta_1$ | $\beta_2$ |
|---|---|---|---|---|---|
| 0·5 | ·8750 0000$^e$ | $J$-curve | ·1479 0199 | 2·8571 4286 | 5·8163 2653 |
| 1 | ·7777 7778 | 1·0000 0000$^e$ | ·1772 7201 | ·9298 3939 | 3·4087 9121 |
| 1·5 | ·7000 0000$^e$ | ·8333 3333 | ·1870 8287 | ·3731 7784 | 2·7397 9592 |
| 2 | ·6363 6364 | ·7142 8571 | ·1886 8164 | ·1485 7143 | 2·4907 5630 |
| 2·5 | ·5833 3333 | ·6250 0000$^e$ | ·1863 3900 | ·0500 0000$^e$ | 2·4000 0000$^e$ |
| 3 | ·5384 6154 | ·5555 5556 | ·1820 3322 | ·0098 8631 | 2·3816 8952 |
| 3·5 | ·5000 0000$^e$ | ·5000 0000$^e$ | ·1767 7670 | ·0000 0000$^e$ | 2·4000 0000$^e$ |
| 4 | ·4666 6667 | ·4545 4545+ | ·1711 1705+ | ·0067 2734 | 2·4377 0140 |
| 4·5 | ·4375 0000$^e$ | ·4166 6667 | ·1653 5946 | ·0228 5714 | 2·4857 1429 |
| 5 | ·4117 6471 | ·3846 1538 | ·1596 7550+ | ·0443 1487 | 2·5389 5297 |
| 5·5 | ·3888 8889 | ·3571 4286 | ·1541 6041 | ·0686 9164 | 2·5944 5100 |
| 6 | ·3684 2105+ | ·3333 3333 | ·1488 6459 | ·0945 1796 | 2·6504 3478 |
| 6·5 | ·3500 0000$^e$ | ·3125 0000$^e$ | ·1438 1175− | ·1208 7912 | 2·7058 3263 |
| 7 | ·3333 3333 | ·2941 1765− | ·1390 0961 | ·1472 0000$^e$ | 2·7600 0000$^e$ |
| 7·5 | ·3181 8182 | ·2777 7778 | ·1344 5636 | ·1731 1919 | 2·8125 5887 |
| 8 | ·3043 4783 | ·2631 5789 | ·1301 4460 | ·1984 1270 | 2·8633 0049 |
| 8·5 | ·2916 6667 | ·2500 0000$^e$ | ·1260 6385− | ·2229 4632 | 2·9121 2485− |
| 9 | ·2800 0000$^e$ | ·2380 9524 | ·1222 0202 | ·2466 4515+ | 2·9590 0207 |
| 9·5 | ·2692 3077 | ·2272 7273 | ·1185 4642 | ·2694 7368 | 3·0039 4737 |
| 10 | ·2592 5926 | ·2173 9130 | ·1150 8435− | ·2914 2263 | 3·0470 0461 |
| 10·5 | ·2500 0000$^e$ | ·2083 3333 | ·1118 0340 | ·3125 0000$^e$ | 3·0882 3529 |
| 11 | ·2413 7931 | ·2000 0000$^e$ | ·1086 9173 | ·3327 2513 | 3·1277 1125− |
| 12 | ·2258 0645+ | ·1851 8519 | ·1029 3213 | ·3707 2886 | 3·2017 0987 |
| 13 | ·2121 2121 | ·1724 1379 | ·0977 2434 | ·4056 8635+ | 3·2696 3058 |
| 14 | ·2000 0000$^e$ | ·1612 9032 | ·0929 9811 | ·4378 6982 | 3·3320 8255+ |
| 15 | ·1891 8919 | ·1515 1515+ | ·0886 9326 | ·4675 4483 | 3·3896 2807 |
| 16 | ·1794 8718 | ·1428 5714 | ·0847 5835− | ·4949 5866 | 3·4427 7409 |
| 17 | ·1707 3171 | ·1351 3514 | ·0811 4941 | ·5203 3613 | 3·4919 7211 |
| 18 | ·1627 9070 | ·1282 0513 | ·0778 2884 | ·5438 7894 | 3·5376 2174 |
| 19 | ·1555 5556 | ·1219 5122 | ·0747 6432 | ·5657 6677 | 3·5800 7564 |
| 20 | ·1489 3617 | ·1162 7907 | ·0719 2802 | ·5861 5917 | 3·6196 4484 |
| 21 | ·1428 5714 | ·1111 1111 | ·0692 9589 | ·6051 9758 | 3·6566 0377 |
| 22 | ·1372 5490 | ·1063 8298 | ·0668 4705− | ·6230 0741 | 3·6911 9493 |
| 23 | ·1320 7547 | ·1020 4082 | ·0645 6330 | ·6396 9994 | 3·7236 3296 |
| 24 | ·1272 7273 | ·0980 3922 | ·0624 2873 | ·6553 7407 | 3·7541 0828 |
| 25 | ·1228 0702 | ·0943 3962 | ·0604 2938 | ·6701 1786 | 3·7827 9023 |
| 26 | ·1186 4407 | ·0909 0909 | ·0585 5295− | ·6840 0987 | 3·8098 2973 |
| 27 | ·1147 5410 | ·0877 1930 | ·0567 8855+ | ·6971 2032 | 3·8353 6165+ |
| 28 | ·1111 1111 | ·0847 4576 | ·0551 2655− | ·7095 1214 | 3·8595 0681 |
| 29 | ·1076 9231 | ·0819 6721 | ·0535 5836 | ·7212 4186 | 3·8823 7370 |
| 30 | ·1044 7761 | ·0793 6508 | ·0520 7632 | ·7323 6036 | 3·9040 5998 |
| 31 | ·1014 4928 | ·0769 2308 | ·0506 7358 | ·7429 1353 | 3·9246 5375− |
| 32 | ·0985 9155− | ·0746 2687 | ·0493 4399 | ·7529 4286 | 3·9442 3469 |
| 33 | ·0958 9041 | ·0724 6377 | ·0480 8200 | ·7624 8595+ | 3·9628 7503 |
| 34 | ·0933 3333 | ·0704 2254 | ·0468 8262 | ·7715 7695+ | 3·9806 4036 |
| 35 | ·0909 0909 | ·0684 9315+ | ·0457 4133 | ·7802 4691 | 3·9975 9036 |
| 36 | ·0886 0759 | ·0666 6667 | ·0446 5403 | ·7885 2415− | 4·0137 7949 |
| 37 | ·0864 1975+ | ·0649 3506 | ·0436 1700 | ·7964 3452 | 4·0292 5748 |
| 38 | ·0843 3735− | ·0632 9114 | ·0426 2685+ | ·8040 0168 | 4·0440 6988 |
| 39 | ·0823 5294 | ·0617 2840 | ·0416 8049 | ·8112 4733 | 4·0582 5845− |
| 40 | ·0804 5977 | ·0602 4096 | ·0407 7507 | ·8181 9138 | 4·0718 6155− |
| 41 | ·0786 5169 | ·0588 2353 | ·0399 0803 | ·8248 5216 | 4·0849 1449 |
| 42 | ·0769 2308 | ·0574 7126 | ·0390 7696 | ·8312 4654 | 4·0974 4981 |
| 43 | ·0752 6882 | ·0561 7978 | ·0382 7970 | ·8373 9009 | 4·1094 9755+ |
| 44 | ·0736 8421 | ·0549 4505+ | ·0375 1422 | ·8432 9720 | 4·1210 8549 |
| 45 | ·0721 6495− | ·0537 6344 | ·0367 7866 | ·8489 8119 | 4·1322 3933 |
| 46 | ·0707 0707 | ·0526 3158 | ·0360 7132 | ·8544 5441 | 4·1429 8292 |
| 47 | ·0693 0693 | ·0515 4639 | ·0353 9060 | ·8597 2830 | 4·1533 3839 |
| 48 | ·0679 6117 | ·0505 0505+ | ·0347 3504 | ·8648 1352 | 4·1633 2633 |
| 49 | ·0666 6667 | ·0495 0495+ | ·0341 0326 | ·8697 1996 | 4·1729 6589 |
| 50 | ·0654 2056 | ·0485 4369 | ·0334 9400 | ·8744 5685+ | 4·1822 7492 |

$p = 4$

| $q$ | $\bar{x}$ | $\tilde{x}$ | $\sigma$ | $\beta_1$ | $\beta_2$ |
|---|---|---|---|---|---|
| 0·5 | ·8888 8889 | *J*-curve | ·1340 0504 | 3·1893 4911 | 6·3461 5385$^{-}$ |
| 1 | ·8000 0000$^{e}$ | 1·0000 0000$^{e}$ | ·1632 9932 | 1·1020 4082 | 3·6964 2857 |
| 1·5 | ·7272 7273 | ·8571 4286 | ·1746 8526 | ·4814 8148 | 2·9313 7255$^{-}$ |
| 2 | ·6666 6667 | ·7500 0000$^{e}$ | ·1781 7416 | ·2187 5000$^{e}$ | 2·6250 0000$^{e}$ |
| 2·5 | ·6153 8462 | ·6666 6667 | ·1776 4624 | ·0934 2561 | 2·4938 0805$^{-}$ |
| 3 | ·5714 2857 | ·6000 0000$^{e}$ | ·1749 6355$^{+}$ | ·0329 2181 | 2·4444 4444 |
| 3·5 | ·5333 3333 | ·5454 5455$^{-}$ | ·1711 1705$^{+}$ | ·0067 2734 | 2·4377 0140 |
| 4 | ·5000 0000$^{e}$ | ·5000 0000$^{e}$ | ·1666 6667 | ·0000 0000$^{e}$ | 2·4545 4545$^{+}$ |
| 4·5 | ·4705 8824 | ·4615 3846 | ·1619 4052 | ·0047 8710 | 2·4848 1712 |
| 5 | ·4444 4444 | ·4285 7143 | ·1571 3484 | ·0165 2893 | 2·5227 2727 |
| 5·5 | ·4210 5263 | ·4000 0000$^{e}$ | ·1523 6776 | ·0324 7981 | 2·5648 2213 |
| 6 | ·4000 0000$^{e}$ | ·3750 0000$^{e}$ | ·1477 0979 | ·0509 2593 | 2·6089 7436 |
| 6·5 | ·3809 5238 | ·3529 4118 | ·1432 0179 | ·0707 6923 | 2·6538 4615$^{+}$ |
| 7 | ·3636 3636 | ·3333 3333 | ·1388 6593 | ·0912 9332 | 2·6985 8713 |
| 7·5 | ·3478 2609 | ·3157 8947 | ·1347 1246 | ·1120 2561 | 2·7426 5645$^{-}$ |
| 8 | ·3333 3333 | ·3000 0000$^{e}$ | ·1307 4409 | ·1326 5306 | 2·7857 1429 |
| 8·5 | ·3200 0000$^{e}$ | ·2857 1429 | ·1269 5873 | ·1529 6915$^{+}$ | 2·8275 5349 |
| 9 | ·3076 9231 | ·2727 2727 | ·1233 5134 | ·1728 3951 | 2·8680 5556 |
| 9·5 | ·2962 9630 | ·2608 6957 | ·1199 1514 | ·1921 7920 | 2·9071 6160 |
| 10 | ·2857 1429 | ·2500 0000$^{e}$ | ·1166 4237 | ·2109 3750$^{e}$ | 2·9448 5294 |
| 10·5 | ·2758 6207 | ·2400 0000$^{e}$ | ·1135 2481 | ·2290 8741 | 2·9811 3791 |
| 11 | ·2666 6667 | ·2307 6923 | ·1105 5416 | ·2466 1843 | 3·0160 4278 |
| 12 | ·2500 0000$^{e}$ | ·2142 8571 | ·1050 2101 | ·2798 3539 | 3·0818 7135$^{-}$ |
| 13 | ·2352 9412 | ·2000 0000$^{e}$ | ·0999 8077 | ·3106 7547 | 3·1427 1255$^{+}$ |
| 14 | ·2222 2222 | ·1875 0000$^{e}$ | ·0953 7723 | ·3392 8571 | 3·1989 7959 |
| 15 | ·2105 2632 | ·1764 7059 | ·0911 6057 | ·3658 3522 | 3·2510 8225$^{+}$ |
| 16 | ·2000 0000$^{e}$ | ·1666 6667 | ·0872 8716 | ·3904 9587 | 3·2994 0711 |
| 17 | ·1904 7619 | ·1578 9474 | ·0837 1896 | ·4134 3267 | 3·3443 0946 |
| 18 | ·1818 1818 | ·1500 0000$^{e}$ | ·0804 2289 | ·4347 9938 | 3·3861 1111 |
| 19 | ·1739 1304 | ·1428 5714 | ·0773 7013 | ·4547 3684 | 3·4251 0121 |
| 20 | ·1666 6667 | ·1363 6364 | ·0745 3560 | ·4733 7278 | 3·4615 3846 |
| 21 | ·1600 0000$^{e}$ | ·1304 3478 | ·0718 9736 | ·4908 2239 | 3·4956 5382 |
| 22 | ·1538 4615$^{+}$ | ·1250 0000$^{e}$ | ·0694 3623 | ·5071 8924 | 3·5276 5338 |
| 23 | ·1481 4815$^{-}$ | ·1200 0000$^{e}$ | ·0671 3533 | ·5225 6630 | 3·5577 2114 |
| 24 | ·1428 5714 | ·1153 8462 | ·0649 7983 | ·5370 3704 | 3·5860 2151 |
| 25 | ·1379 3103 | ·1111 1111 | ·0629 5662 | ·5506 7638 | 3·6127 0161 |
| 26 | ·1333 3333 | ·1071 4286 | ·0610 5406 | ·5635 5168 | 3·6378 9336 |
| 27 | ·1290 3226 | ·1034 4828 | ·0592 6185$^{-}$ | ·5757 2357 | 3·6617 1519 |
| 28 | ·1250 0000$^{e}$ | ·1000 0000$^{e}$ | ·0575 7077 | ·5872 4666 | 3·6842 7371 |
| 29 | ·1212 1212 | ·0967 7419 | ·0559 7261 | ·5981 7030 | 3·7056 6502 |
| 30 | ·1176 4706 | ·0937 5000$^{e}$ | ·0544 6001 | ·6085 3909 | 3·7259 7598 |
| 31 | ·1142 8571 | ·0909 0909 | ·0530 2633 | ·6183 9346 | 3·7452 8518 |
| 32 | ·1111 1111 | ·0882 3529 | ·0516 6562 | ·6277 7008 | 3·7636 6397 |
| 33 | ·1081 0811 | ·0857 1429 | ·0503 7249 | ·6367 0233 | 3·7811 7716 |
| 34 | ·1052 6316 | ·0833 3333 | ·0491 4208 | ·6452 2059 | 3·7978 8379 |
| 35 | ·1025 6410 | ·0810 8108 | ·0479 6997 | ·6533 5260 | 3·8138 3773 |
| 36 | ·1000 0000$^{e}$ | ·0789 4737 | ·0468 5213 | ·6611 2371 | 3·8290 8822 |
| 37 | ·0975 6098 | ·0769 2308 | ·0457 8491 | ·6685 5715$^{-}$ | 3·8436 8036 |
| 38 | ·0952 3810 | ·0750 0000$^{e}$ | ·0447 6497 | ·6756 7421 | 3·8576 5550$^{+}$ |
| 39 | ·0930 2326 | ·0731 7073 | ·0437 8925$^{-}$ | ·6824 9446 | 3·8710 5165$^{+}$ |
| 40 | ·0909 0909 | ·0714 2857 | ·0428 5496 | ·6890 3592 | 3·8839 0379 |
| 41 | ·0888 8889 | ·0697 6744 | ·0419 5952 | ·6953 1517 | 3·8962 4416 |
| 42 | ·0869 5652 | ·0681 8182 | ·0411 0057 | ·7013 4755$^{+}$ | 3·9081 0253 |
| 43 | ·0851 0638 | ·0666 6667 | ·0402 7595$^{-}$ | ·7071 4722 | 3·9195 0641 |
| 44 | ·0833 3333 | ·0652 1739 | ·0394 8363 | ·7127 2727 | 3·9304 8128 |
| 45 | ·0816 3265$^{+}$ | ·0638 2979 | ·0387 2177 | ·7180 9988 | 3·9410 5078 |
| 46 | ·0800 0000$^{e}$ | ·0625 0000$^{e}$ | ·0379 8865$^{-}$ | ·7232 7631 | 3·9512 3683 |
| 47 | ·0784 3137 | ·0612 2449 | ·0372 8268 | ·7282 6704 | 3·9610 5982 |
| 48 | ·0769 2308 | ·0600 0000$^{e}$ | ·0366 0238 | ·7330 8185$^{-}$ | 3·9705 3872 |
| 49 | ·0754 7179 | ·0588 2353 | ·0359 4639 | ·7377 2980 | 3·9796 9123 |
| 50 | ·0740 7407 | ·0576 9231 | ·0353 1343 | ·7422 1939 | 3·9885 3383 |

$p = 4{\cdot}5$

| $q$ | $\bar{x}$ | $\tilde{x}$ | $\sigma$ | $\beta_1$ | $\beta_2$ |
|---|---|---|---|---|---|
| 0·5 | ·9000 0000$^e$ | $J$-curve | ·1224 7449 | 3·4829 9320 | 6·8214 2857 |
| 1 | ·8181 8182 | 1·0000 0000$^e$ | ·1512 8187 | 1·2582 7160 | 3·9594 7712 |
| 1·5 | ·7500 0000$^e$ | ·8750 0000$^e$ | ·1636 6342 | ·5833 3333 | 3·1111 1111 |
| 2 | ·6923 0769 | ·7777 7778 | ·1685 3002 | ·2883 5063 | 2·7554 1796 |
| 2·5 | ·6428 5714 | ·7000 0000$^e$ | ·1694 0773 | ·1404 6639 | 2·5896 2963 |
| 3 | ·6000 0000$^e$ | ·6363 6364 | ·1680 3361 | ·0627 8855+ | 2·5137 8446 |
| 3·5 | ·5625 0000$^e$ | ·5833 3333 | ·1653 5946 | ·0228 5714 | 2·4857 1429 |
| 4 | ·5294 1176 | ·5384 6154 | ·1619 4052 | ·0047 8710 | 2·4848 1712 |
| 4·5 | ·5000 0000$^e$ | ·5000 0000$^e$ | ·1581 1388 | ·0000 0000$^e$ | 2·5000 0000$^e$ |
| 5 | ·4736 8421 | ·4666 6667 | ·1540 8949 | ·0035 2867 | 2·5248 6957 |
| 5·5 | ·4500 0000$^e$ | ·4375 0000$^e$ | ·1500 0000$^e$ | ·0123 4568 | 2·5555 5556 |
| 6 | ·4285 7143 | ·4117 6471 | ·1459 2969 | ·0245 3333 | 2·5896 2963 |
| 6·5 | ·4090 9091 | ·3888 8889 | ·1419 3177 | ·0388 4084 | 2·6255 2832 |
| 7 | ·3913 0435− | ·3684 2105+ | ·1380 3920 | ·0544 3421 | 2·6622 2709 |
| 7·5 | ·3750 0000$^e$ | ·3500 0000$^e$ | ·1342 7154 | ·0707 4830 | 2·6990 4762 |
| 8 | ·3600 0000$^e$ | ·3333 3333 | ·1306 3945+ | ·0873 9596 | 2·7355 3949 |
| 8·5 | ·3461 5385− | ·3181 8182 | ·1271 4765+ | ·1041 1038 | 2·7714 0523 |
| 9 | ·3333 3333 | ·3043 4783 | ·1237 9689 | ·1207 0760 | 2·8064 5161 |
| 9·5 | ·3214 2857 | ·2916 6667 | ·1205 8531 | ·1370 6140 | 2·8405 5728 |
| 10 | ·3103 4483 | ·2800 0000$^e$ | ·1175 0935− | ·1530 8642 | 2·8736 5079 |
| 10·5 | ·3000 0000$^e$ | ·2692 3077 | ·1145 6439 | ·1687 2631 | 2·9056 9561 |
| 11 | ·2903 2258 | ·2592 5926 | ·1117 4521 | ·1839 4558 | 2·9366 7954 |
| 12 | ·2727 2727 | ·2413 7931 | ·1064 6183 | ·2130 5089 | 2·9954 9550− |
| 13 | ·2571 4286 | ·2258 0645+ | ·1016 1411 | ·2403 5020 | 3·0502 5577 |
| 14 | ·2432 4324 | ·2121 2121 | ·0971 5860 | ·2658 8482 | 3·1012 0736 |
| 15 | ·2307 6923 | ·2000 0000$^e$ | ·0930 5514 | ·2897 4220 | 3·1486 3049 |
| 16 | ·2195 1220 | ·1891 8919 | ·0892 6746 | ·3120 3018 | 3·1928 0930 |
| 17 | ·2093 0233 | ·1794 8718 | ·0857 6320 | ·3328 6289 | 3·2340 1701 |
| 18 | ·2000 0000$^e$ | ·1707 3171 | ·0825 1370 | ·3523 5319 | 3·2725 0900 |
| 19 | ·1914 8936 | ·1627 9070 | ·0794 9360 | ·3706 0870 | 3·3085 2010 |
| 20 | ·1836 7347 | ·1555 5556 | ·0766 8053 | ·3877 2992 | 3·3422 6415+ |
| 21 | ·1764 7059 | ·1489 3617 | ·0740 5473 | ·4038 0952 | 3·3739 3484 |
| 22 | ·1698 1132 | ·1428 5714 | ·0715 9874 | ·4189 3232 | 3·4037 0701 |
| 23 | ·1636 3636 | ·1372 5490 | ·0692 9710 | ·4331 7554 | 3·4317 3828 |
| 24 | ·1578 9474 | ·1320 7547 | ·0671 3608 | ·4466 0933 | 3·4581 7070 |
| 25 | ·1525 4237 | ·1272 7273 | ·0651 0350− | ·4592 9733 | 3·4831 3228 |
| 26 | ·1475 4098 | ·1228 0702 | ·0631 8847 | ·4712 9722 | 3·5067 3850− |
| 27 | ·1428 5714 | ·1186 4407 | ·0613 8127 | ·4826 6132 | 3·5290 9366 |
| 28 | ·1384 6154 | ·1147 5410 | ·0596 7322 | ·4934 3709 | 3·5502 9209 |
| 29 | ·1343 2836 | ·1111 1111 | ·0580 5651 | ·5036 6763 | 3·5704 1921 |
| 30 | ·1304 3478 | ·1076 9231 | ·0565 2409 | ·5133 9213 | 3·5895 5251 |
| 31 | ·1267 6056 | ·1044 7761 | ·0550 6965+ | ·5226 4628 | 3·6077 6242 |
| 32 | ·1232 8767 | ·1014 4928 | ·0536 8745− | ·5314 6259 | 3·6251 1302 |
| 33 | ·1200 0000$^e$ | ·0985 9155− | ·0523 7229 | ·5398 7075− | 3·6416 6276 |
| 34 | ·1168 8312 | ·0958 9041 | ·0511 1948 | ·5478 9791 | 3·6574 6501 |
| 35 | ·1139 2405+ | ·0933 3333 | ·0499 2472 | ·5555 6892 | 3·6725 6859 |
| 36 | ·1111 1111 | ·0909 0909 | ·0487 8412 | ·5629 0657 | 3·6870 1826 |
| 37 | ·1084 3373 | ·0886 0759 | ·0476 9409 | ·5699 3180 | 3·7008 5506 |
| 38 | ·1058 8235+ | ·0864 1975+ | ·0466 5138 | ·5766 6385+ | 3·7141 1675+ |
| 39 | ·1034 4828 | ·0843 3735− | ·0456 5300 | ·5831 2046 | 3·7268 3809 |
| 40 | ·1011 2360 | ·0823 5294 | ·0446 9619 | ·5893 1797 | 3·7390 5112 |
| 41 | ·0989 0110 | ·0804 5977 | ·0437 7844 | ·5952 7149 | 3·7507 8544 |
| 42 | ·0967 7419 | ·0786 5169 | ·0428 9742 | ·6009 9499 | 3·7620 6840 |
| 43 | ·0947 3684 | ·0769 2308 | ·0420 5097 | ·6065 0142 | 3·7729 2535+ |
| 44 | ·0927 8351 | ·0752 6882 | ·0412 3711 | ·6118 0276 | 3·7833 7979 |
| 45 | ·0909 0909 | ·0736 8421 | ·0404 5401 | ·6169 1017 | 3·7934 5354 |
| 46 | ·0891 0891 | ·0721 6495− | ·0396 9997 | ·6218 3400 | 3·8031 6686 |
| 47 | ·0873 7864 | ·0707 0707 | ·0389 7340 | ·6265 8388 | 3·8125 3865+ |
| 48 | ·0857 1429 | ·0693 0693 | ·0382 7284 | ·6311 6881 | 3·8215 8649 |
| 49 | ·0841 1215− | ·0679 6117 | ·0375 9692 | ·6355 9717 | 3·8303 2681 |
| 50 | ·0825 6881 | ·0666 6667 | ·0369 4437 | ·6398 7678 | 3·8387 7491 |

$p = 5$

| $q$ | $\bar{x}$ | $\tilde{x}$ | $\sigma$ | $\beta_1$ | $\beta_2$ |
|---|---|---|---|---|---|
| 0·5 | ·9090 9091 | *J*-curve | ·1127 5885$^-$ | 3·7440 0000$^e$ | 7·2494 1176 |
| 1 | ·8333 3333 | 1·0000 0000$^e$ | ·1408 5904 | 1·4000 0000$^e$ | 4·2000 0000$^e$ |
| 1·5 | ·7692 3077 | ·8888 8889 | ·1538 4615$^+$ | ·6782 0069 | 3·2786 3777 |
| 2 | ·7142 8571 | ·8000 0000$^e$ | ·1597 1914 | ·3555 5556 | 2·8800 0000$^e$ |
| 2·5 | ·6666 6667 | ·7272 7273 | ·1616 9042 | ·1883 6565$^+$ | 2·6842 1053 |
| 3 | ·6250 0000$^e$ | ·6666 6667 | ·1613 7431 | ·0960 0000$^e$ | 2·5854 5455$^-$ |
| 3·5 | ·5882 3529 | ·6153 8462 | ·1596 7550$^+$ | ·0443 1487 | 2·5389 5297 |
| 4 | ·5555 5556 | ·5714 2857 | ·1571 3484 | ·0165 2893 | 2·5227 2727 |
| 4·5 | ·5263 1579 | ·5333 3333 | ·1540 8949 | ·0035 2867 | 2·5248 6957 |
| 5 | ·5000 0000$^e$ | ·5000 0000$^e$ | ·1507 5567 | ·0000 0000$^e$ | 2·5384 6154 |
| 5·5 | ·4761 9048 | ·4705 8824 | ·1472 7469 | ·0026 7636 | 2·5592 7273 |
| 6 | ·4545 4545$^+$ | ·4444 4444 | ·1437 3989 | ·0094 6746 | 2·5846 1538 |
| 6·5 | ·4347 8261 | ·4210 5263 | ·1402 1318 | ·0189 9335$^+$ | 2·6127 3210 |
| 7 | ·4166 6667 | ·4000 0000$^e$ | ·1367 3544 | ·0303 2070 | 2·6424 4898 |
| 7·5 | ·4000 0000$^e$ | ·3809 5238 | ·1333 3333 | ·0428 0618 | 2·6729 6997 |
| 8 | ·3846 1538 | ·3636 3636 | ·1300 2373 | ·0560 0000$^e$ | 2·7037 5000$^e$ |
| 8·5 | ·3703 7037 | ·3478 2609 | ·1268 1677 | ·0695 8438 | 2·7344 1435$^+$ |
| 9 | ·3571 4286 | ·3333 3333 | ·1237 1792 | ·0833 3333 | 2·7647 0588 |
| 9·5 | ·3448 2759 | ·3200 0000$^e$ | ·1207 2938 | ·0970 8569 | 2·7944 4976 |
| 10 | ·3333 3333 | ·3076 9231 | ·1178 5113 | ·1107 2664 | 2·8235 2941 |
| 10·5 | ·3225 8065$^-$ | ·2962 9630 | ·1150 8162 | ·1241 7493 | 2·8518 6983 |
| 11 | ·3125 0000$^e$ | ·2857 1429 | ·1124 1827 | ·1373 7374 | 2·8794 2584 |
| 12 | ·2941 1765$^-$ | ·2666 6667 | ·1073 9658 | ·1628 8089 | 2·9321 0526 |
| 13 | ·2777 7778 | ·2500 0000$^e$ | ·1027 5604 | ·1870 7692 | 2·9815 3846 |
| 14 | ·2631 5789 | ·2352 9412 | ·0984 6467 | ·2099 1254 | 3·0278 2931 |
| 15 | ·2500 0000$^e$ | ·2222 2222 | ·0944 9112 | ·2314 0496 | 3·0711 4625$^-$ |
| 16 | ·2380 9524 | ·2105 2632 | ·0908 0596 | ·2516 0681 | 3·1116 8478 |
| 17 | ·2272 7273 | ·2000 0000$^e$ | ·0873 8218 | ·2705 8824 | 3·1496 4706 |
| 18 | ·2173 9130 | ·1904 7619 | ·0841 9529 | ·2884 2667 | 3·1852 3077 |
| 19 | ·2083 3333 | ·1818 1818 | ·0812 2329 | ·3052 0087 | 3·2186 2348 |
| 20 | ·2000 0000$^e$ | ·1739 1304 | ·0784 4645$^+$ | ·3209 8765$^+$ | 3·2500 0000$^e$ |
| 21 | ·1923 0769 | ·1666 6667 | ·0758 4718 | ·3358 6006 | 3·2795 2146 |
| 22 | ·1851 8519 | ·1600 0000$^e$ | ·0734 0972 | ·3498 8650$^-$ | 3·3073 3542 |
| 23 | ·1785 7143 | ·1538 4615$^+$ | ·0711 2002 | ·3631 3043 | 3·3335 7644 |
| 24 | ·1724 1379 | ·1481 4815$^-$ | ·0689 6552 | ·3756 5036 | 3·3583 6694 |
| 25 | ·1666 6667 | ·1428 5714 | ·0669 3494 | ·3875 0000$^e$ | 3·3818 1818 |
| 26 | ·1612 9032 | ·1379 3103 | ·0650 1821 | ·3987 2854 | 3·4040 3126 |
| 27 | ·1562 5000$^e$ | ·1333 3333 | ·0632 0624 | ·4093 8101 | 3·4250 9804 |
| 28 | ·1515 1515$^+$ | ·1290 3226 | ·0614 9087 | ·4194 9854 | 3·4451 0204 |
| 29 | ·1470 5882 | ·1250 0000$^e$ | ·0598 6473 | ·4291 1877 | 3·4641 1929 |
| 30 | ·1428 5714 | ·1212 1212 | ·0583 2118 | ·4382 7611 | 3·4822 1906 |
| 31 | ·1388 8889 | ·1176 4706 | ·0568 5419 | ·4470 0206 | 3·4994 6454 |
| 32 | ·1351 3514 | ·1142 8571 | ·0554 5829 | ·4553 2544 | 3·5159 1346 |
| 33 | ·1315 7895$^-$ | ·1111 1111 | ·0541 2851 | ·4632 7273 | 3·5316 1863 |
| 34 | ·1282 0513 | ·1081 0811 | ·0528 6033 | ·4708 6818 | 3·5466 2841 |
| 35 | ·1250 0000$^e$ | ·1052 6316 | ·0516 4962 | ·4781 3411 | 3·5609 8719 |
| 36 | ·1219 5122 | ·1025 6410 | ·0504 9260 | ·4850 9104 | 3·5747 3573 |
| 37 | ·1190 4762 | ·1000 0000$^e$ | ·0493 8583 | ·4917 5787 | 3·5879 1155$^-$ |
| 38 | ·1162 7907 | ·0975 6098 | ·0483 2613 | ·4981 5205$^-$ | 3·6005 4920 |
| 39 | ·1136 3636 | ·0952 3810 | ·0473 1059 | ·5042 8966 | 3·6126 8057 |
| 40 | ·1111 1111 | ·0930 2326 | ·0463 3654 | ·5101 8560 | 3·6243 3511 |
| 41 | ·1086 9565$^+$ | ·0909 0909 | ·0454 0150$^-$ | ·5158 5366 | 3·6355 4007 |
| 42 | ·1063 8298 | ·0888 8889 | ·0445 0319 | ·5213 0660 | 3·6463 2070 |
| 43 | ·1041 6667 | ·0869 5652 | ·0436 3952 | ·5265 5628 | 3·6567 0041 |
| 44 | ·1020 4082 | ·0851 0638 | ·0428 0852 | ·5316 1372 | 3·6667 0095$^-$ |
| 45 | ·1000 0000$^e$ | ·0833 3333 | ·0420 0840 | ·5364 8915$^+$ | 3·6763 4253 |
| 46 | ·0980 3922 | ·0816 3265$^+$ | ·0412 3748 | ·5411 9213 | 3·6856 4397 |
| 47 | ·0961 5385$^-$ | ·0800 0000$^e$ | ·0404 9418 | ·5457 3155$^-$ | 3·6946 2282 |
| 48 | ·0943 3962 | ·0784 3137 | ·0397 7708 | ·5501 1570 | 3·7032 9545$^+$ |
| 49 | ·0925 9259 | ·0769 2308 | ·0390 8480 | ·5543 5235$^+$ | 3·7116 7715$^+$ |
| 50 | ·0909 0909 | ·0754 7179 | ·0384 1610 | ·5584 4875$^+$ | 3·7197 8221 |

$p = 5{\cdot}5$

| $q$ | $\bar{x}$ | $\tilde{x}$ | $\sigma$ | $\beta_1$ | $\beta_2$ |
|---|---|---|---|---|---|
| 0·5 | ·9166 6667 | *J*-curve | ·1044 6386 | 3·9772 7273 | 7·6363 6364 |
| 1 | ·8461 5385− | 1·0000 0000$^e$ | ·1317 4598 | 1·5287 8264 | 4·4202 0827 |
| 1·5 | ·7857 1429 | ·9000 0000$^e$ | ·1450 7211 | ·7661 8032 | 3·4343 4343 |
| 2 | ·7333 3333 | ·8181 8182 | ·1516 7905+ | ·4195 4168 | 2·9979 4942 |
| 2·5 | ·6875 0000$^e$ | ·7500 0000$^e$ | ·1545 0414 | ·2356 3636 | 2·7758 6777 |
| 3 | ·6470 5882 | ·6923 0769 | ·1550 4624 | ·1305 5727 | 2·6570 6757 |
| 3·5 | ·6111 1111 | ·6428 5714 | ·1541 6041 | ·0686 9164 | 2·5944 5100 |
| 4 | ·5789 4737 | ·6000 0000$^e$ | ·1523 6776 | ·0324 7981 | 2·5648 2213 |
| 4·5 | ·5500 0000$^e$ | ·5625 0000$^e$ | ·1500 0000$^e$ | ·0123 4568 | 2·5555 5556 |
| 5 | ·5238 0952 | ·5294 1176 | ·1472 7469 | ·0026 7636 | 2·5592 7273 |
| 5·5 | ·5000 0000$^e$ | ·5000 0000$^e$ | ·1443 3757 | ·0000 0000$^e$ | 2·5714 2857 |
| 6 | ·4782 6087 | ·4736 8421 | ·1412 8762 | ·0020 7840 | 2·5891 0949 |
| 6·5 | ·4583 3333 | ·4500 0000$^e$ | ·1381 9270 | ·0074 2115+ | 2·6103 8961 |
| 7 | ·4400 0000$^e$ | ·4285 7143 | ·1350 9941 | ·0150 0996 | 2·6339 6559 |
| 7·5 | ·4230 7692 | ·4090 9091 | ·1320 3972 | ·0241 3468 | 2·6589 3939 |
| 8 | ·4074 0741 | ·3913 0435− | ·1290 3531 | ·0342 9193 | 2·6846 8408 |
| 8·5 | ·3928 5714 | ·3750 0000$^e$ | ·1261 0060 | ·0451 2032 | 2·7107 5810 |
| 9 | ·3793 1034 | ·3600 0000$^e$ | ·1232 4484 | ·0563 5789 | 2·7368 4901 |
| 9·5 | ·3666 6667 | ·3461 5385− | ·1204 7360 | ·0678 1345− | 2·7627 3572 |
| 10 | ·3548 3871 | ·3333 3333 | ·1177 8980 | ·0793 4694 | 2·7882 6255− |
| 10·5 | ·3437 5000$^e$ | ·3214 2857 | ·1151 9445− | ·0908 5565− | 2·8133 2118 |
| 11 | ·3333 3333 | ·3103 4483 | ·1126 8723 | ·1022 6443 | 2·8378 3784 |
| 12 | ·3142 8571 | ·2903 2258 | ·1079 3146 | ·1245 7912 | 2·8850 7021 |
| 13 | ·2972 9730 | ·2727 2727 | ·1035 0563 | ·1460 1698 | 2·9297 6847 |
| 14 | ·2820 5128 | ·2571 4286 | ·0993 8797 | ·1664 5010 | 2·9719 1181 |
| 15 | ·2682 9268 | ·2432 4324 | ·0955 5510 | ·1858 3464 | 3·0115 7103 |
| 16 | ·2558 1395+ | ·2307 6923 | ·0919 8386 | ·2041 7507 | 3·0488 6413 |
| 17 | ·2444 4444 | ·2195 1220 | ·0886 5217 | ·2215 0307 | 3·0839 3090 |
| 18 | ·2340 4255+ | ·2093 0233 | ·0855 3950− | ·2378 6500+ | 3·1169 1835+ |
| 19 | ·2244 8980 | ·2000 0000$^e$ | ·0826 2700 | ·2533 1428 | 3·1479 7246 |
| 20 | ·2156 8627 | ·1914 8936 | ·0798 9751 | ·2679 0684 | 3·1772 3358 |
| 21 | ·2075 4717 | ·1836 7347 | ·0773 3548 | ·2816 9840 | 3·2048 3412 |
| 22 | ·2000 0000$^e$ | ·1764 7059 | ·0749 2686 | ·2947 4289 | 3·2308 9747 |
| 23 | ·1929 8246 | ·1698 1132 | ·0726 5900 | ·3070 9157 | 3·2555 3776 |
| 24 | ·1864 4068 | ·1636 3636 | ·0705 2044 | ·3187 9261 | 3·2788 6003 |
| 25 | ·1803 2787 | ·1578 9474 | ·0685 0089 | ·3298 9091 | 3·3009 6065+ |
| 26 | ·1746 0317 | ·1525 4237 | ·0665 9104 | ·3404 2812 | 3·3219 2791 |
| 27 | ·1692 3077 | ·1475 4098 | ·0647 8246 | ·3504 4275− | 3·3418 4259 |
| 28 | ·1641 7910 | ·1428 5714 | ·0630 6756 | ·3599 7032 | 3·3607 7862 |
| 29 | ·1594 2029 | ·1384 6154 | ·0614 3942 | ·3690 4358 | 3·3788 0362 |
| 30 | ·1549 2958 | ·1343 2836 | ·0598 9178 | ·3776 9266 | 3·3959 7954 |
| 31 | ·1506 8493 | ·1304 3478 | ·0584 1894 | ·3859 4532 | 3·4123 6309 |
| 32 | ·1466 6667 | ·1267 6056 | ·0570 1573 | ·3938 2711 | 3·4280 0633 |
| 33 | ·1428 5714 | ·1232 8767 | ·0556 7741 | ·4013 6158 | 3·4429 5701 |
| 34 | ·1392 4051 | ·1200 0000$^e$ | ·0543 9966 | ·4085 7043 | 3·4572 5905− |
| 35 | ·1358 0247 | ·1168 8312 | ·0531 7853 | ·4154 7369 | 3·4709 5282 |
| 36 | ·1325 3012 | ·1139 2405+ | ·0520 1038 | ·4220 8984 | 3·4840 7555− |
| 37 | ·1294 1176 | ·1111 1111 | ·0508 9190 | ·4284 3596 | 3·4966 6155+ |
| 38 | ·1264 3678 | ·1084 3373 | ·0498 2002 | ·4345 2788 | 3·5087 4253 |
| 39 | ·1235 9551 | ·1058 8235+ | ·0487 9192 | ·4403 8022 | 3·5203 4779 |
| 40 | ·1208 7912 | ·1034 4828 | ·0478 0502 | ·4460 0655− | 3·5315 0446 |
| 41 | ·1182 7957 | ·1011 2360 | ·0468 5690 | ·4514 1947 | 3·5422 3770 |
| 42 | ·1157 8947 | ·0989 0110 | ·0459 4534 | ·4566 3067 | 3·5525 7084 |
| 43 | ·1134 0206 | ·0967 7419 | ·0450 6831 | ·4616 5105− | 3·5625 2557 |
| 44 | ·1111 1111 | ·0947 3684 | ·0442 2388 | ·4664 9072 | 3·5721 2205+ |
| 45 | ·1089 1089 | ·0927 8351 | ·0434 1028 | ·4711 5912 | 3·5813 7908 |
| 46 | ·1067 9612 | ·0909 0909 | ·0426 2588 | ·4756 6506 | 3·5903 1416 |
| 47 | ·1047 6190 | ·0891 0891 | ·0418 6914 | ·4800 1677 | 3·5989 4363 |
| 48 | ·1028 0374 | ·0873 7864 | ·0411 3864 | ·4842 2194 | 3·6072 8277 |
| 49 | ·1009 1743 | ·0857 1429 | ·0404 3304 | ·4882 8774 | 3·6153 4585− |
| 50 | ·0990 9910 | ·0841 1215− | ·0397 5110 | ·4922 2093 | 3·6231 4625− |

$p = 6$

| $q$ | $\bar{x}$ | $\tilde{x}$ | $\sigma$ | $\beta_1$ | $\beta_2$ |
|---|---|---|---|---|---|
| 0·5 | ·9230 7692 | *J*-curve | ·0973 0085$^{+}$ | 4·1868 5121 | 7·9876 1610 |
| 1 | ·8571 4286 | 1·0000 0000$^{e}$ | ·1237 1791 | 1·6460 9053 | 4·6222 2222 |
| 1·5 | ·8000 0000$^{e}$ | ·9090 9091 | ·1371 9887 | ·8476 4543 | 3·5789 4737 |
| 2 | ·7500 0000$^{e}$ | ·8333 3333 | ·1443 3757 | ·4800 0000$^{e}$ | 3·1090 9091 |
| 2·5 | ·7058 8235$^{+}$ | ·7692 3077 | ·1478 3079 | ·2814 8148 | 2·8637 6812 |
| 3 | ·6666 6667 | ·7142 8571 | ·1490 7120 | ·1652 8926 | 2·7272 7273 |
| 3·5 | ·6315 7895$^{-}$ | ·6666 6667 | ·1488 6459 | ·0945 1796 | 2·6504 3478 |
| 4 | ·6000 0000$^{e}$ | ·6250 0000$^{e}$ | ·1477 0979 | ·0509 2593 | 2·6089 7436 |
| 4·5 | ·5714 2857 | ·5882 3529 | ·1459 2969 | ·0245 3333 | 2·5896 2963 |
| 5 | ·5454 5455$^{-}$ | ·5555 5556 | ·1437 3989 | ·0094 6746 | 2·5846 1538 |
| 5·5 | ·5217 3913 | ·5263 1579 | ·1412 8762 | ·0020 7840 | 2·5891 0949 |
| 6 | ·5000 0000$^{e}$ | ·5000 0000$^{e}$ | ·1386 7505$^{-}$ | ·0000 0000$^{e}$ | 2·6000 0000$^{e}$ |
| 6·5 | ·4800 0000$^{e}$ | ·4761 9048 | ·1359 7385$^{+}$ | ·0016 4639 | 2·6152 1349 |
| 7 | ·4615 3846 | ·4545 4545$^{+}$ | ·1332 3468 | ·0059 2593 | 2·6333 3333 |
| 7·5 | ·4444 4444 | ·4347 8261 | ·1304 9338 | ·0120 7076 | 2·6533 7243 |
| 8 | ·4285 7143 | ·4166 6667 | ·1277 7531 | ·0195 3125$^{e}$ | 2·6746 3235$^{+}$ |
| 8·5 | ·4137 9310 | ·4000 0000$^{e}$ | ·1250 9825$^{+}$ | ·0279 0832 | 2·6966 1319 |
| 9 | ·4000 0000$^{e}$ | ·3846 1538 | ·1224 7449 | ·0369 0888 | 2·7189 5425$^{-}$ |
| 9·5 | ·3870 9677 | ·3703 7037 | ·1199 1231 | ·0463 1579 | 2·7413 9403 |
| 10 | ·3750 0000$^{e}$ | ·3571 4286 | ·1174 1705$^{+}$ | ·0559 6708 | 2·7637 4269 |
| 10·5 | ·3636 3636 | ·3448 2759 | ·1149 9191 | ·0657 4142 | 2·7858 6279 |
| 11 | ·3529 4118 | ·3333 3333 | ·1126 3848 | ·0755 4772 | 2·8076 5550$^{+}$ |
| 12 | ·3333 3333 | ·3125 0000$^{e}$ | ·1081 4761 | ·0950 0000$^{e}$ | 2·8500 0000$^{e}$ |
| 13 | ·3157 8947 | ·2941 1765$^{-}$ | ·1039 3904 | ·1139 6011 | 2·8904 4289 |
| 14 | ·3000 0000$^{e}$ | ·2777 7778 | ·1000 0000$^{e}$ | ·1322 3140 | 2·9288 5375$^{+}$ |
| 15 | ·2857 1429 | ·2631 5789 | ·0963 1427 | ·1497 1645$^{-}$ | 2·9652 1739 |
| 16 | ·2727 2727 | ·2500 0000$^{e}$ | ·0928 6435$^{-}$ | ·1663 7731 | 2·9995 8333 |
| 17 | ·2608 6957 | ·2380 9524 | ·0896 3273 | ·1822 1176 | 3·0320 3620 |
| 18 | ·2500 0000$^{e}$ | ·2272 7273 | ·0866 0254 | ·1972 3866 | 3·0626 7806 |
| 19 | ·2400 0000$^{e}$ | ·2173 9130 | ·0837 5789 | ·2114 8894 | 3·0916 1793 |
| 20 | ·2307 6923 | ·2083 3333 | ·0810 8404 | ·2250 0000$^{e}$ | 3·1189 6552 |
| 21 | ·2222 2222 | ·2000 0000$^{e}$ | ·0785 6742 | ·2378 1213 | 3·1448 2759 |
| 22 | ·2142 8571 | ·1923 0769 | ·0761 9561 | ·2499 6633 | 3·1693 0596 |
| 23 | ·2068 9655$^{+}$ | ·1851 8519 | ·0739 5728 | ·2615 0296 | 3·1924 9649 |
| 24 | ·2000 0000$^{e}$ | ·1785 7143 | ·0718 4212 | ·2724 6094 | 3·2144 8864 |
| 25 | ·1935 4839 | ·1724 1379 | ·0698 4076 | ·2828 7726 | 3·2353 6542 |
| 26 | ·1875 0000$^{e}$ | ·1666 6667 | ·0679 4466 | ·2927 8680 | 3·2552 0362 |
| 27 | ·1818 1818 | ·1612 9032 | ·0661 4608 | ·3022 2222 | 3·2740 7407 |
| 28 | ·1764 7059 | ·1562 5000$^{e}$ | ·0644 3795$^{-}$ | ·3112 1399 | 3·2920 4204 |
| 29 | ·1714 2857 | ·1515 1515$^{+}$ | ·0628 1384 | ·3197 9043 | 3·3091 6761 |
| 30 | ·1666 6667 | ·1470 5882 | ·0612 6789 | ·3279 7784 | 3·3255 0607 |
| 31 | ·1621 6216 | ·1428 5714 | ·0597 9474 | ·3358 0058 | 3·3411 0835$^{+}$ |
| 32 | ·1578 9474 | ·1388 8889 | ·0583 8949 | ·3432 8125$^{e}$ | 3·3560 2134 |
| 33 | ·1538 4615$^{+}$ | ·1351 3514 | ·0570 4768 | ·3504 4075$^{+}$ | 3·3702 8825$^{-}$ |
| 34 | ·1500 0000$^{e}$ | ·1315 7895$^{-}$ | ·0557 6519 | ·3572 9847 | 3·3839 4893 |
| 35 | ·1463 4146 | ·1282 0513 | ·0545 3824 | ·3638 7236 | 3·3970 4017 |
| 36 | ·1428 5714 | ·1250 0000$^{e}$ | ·0533 6338 | ·3701 7906 | 3·4095 9596 |
| 37 | ·1395 3488 | ·1219 5122 | ·0522 3742 | ·3762 3401 | 3·4216 4773 |
| 38 | ·1363 6364 | ·1190 4762 | ·0511 5740 | ·3820 5154 | 3·4332 2460 |
| 39 | ·1333 3333 | ·1162 7907 | ·0501 2063 | ·3876 4495$^{-}$ | 3·4443 5352 |
| 40 | ·1304 3478 | ·1136 3636 | ·0491 2458 | ·3930 2662 | 3·4550 5952 |
| 41 | ·1276 5957 | ·1111 1111 | ·0481 6694 | ·3982 0806 | 3·4653 6585$^{+}$ |
| 42 | ·1250 0000$^{e}$ | ·1086 9565$^{+}$ | ·0472 4556 | ·4032 0000$^{e}$ | 3·4752 9412 |
| 43 | ·1224 4898 | ·1063 8298 | ·0463 5844 | ·4080 1242 | 3·4848 6443 |
| 44 | ·1200 0000$^{e}$ | ·1041 6667 | ·0455 0372 | ·4126 5465$^{+}$ | 3·4940 9553 |
| 45 | ·1176 4706 | ·1020 4082 | ·0446 7968 | ·4171 3540 | 3·5030 0489 |
| 46 | ·1153 8462 | ·1000 0000$^{e}$ | ·0438 8471 | ·4214 6279 | 3·5116 0884 |
| 47 | ·1132 0755$^{-}$ | ·0980 3922 | ·0431 1732 | ·4256 4445$^{+}$ | 3·5199 2263 |
| 48 | ·1111 1111 | ·0961 5385$^{-}$ | ·0423 7612 | ·4296 8750$^{e}$ | 3·5279 6053 |
| 49 | ·1090 9091 | ·0943 3962 | ·0416 5978 | ·4335 9862 | 3·5357 3589 |
| 50 | ·1071 4286 | ·0925 9259 | ·0409 6710 | ·4373 8407 | 3·5432 6125$^{+}$ |

$p = 6{\cdot}5$

| $q$ | $\bar{x}$ | $\tilde{x}$ | $\sigma$ | $\beta_1$ | $\beta_2$ |
|---|---|---|---|---|---|
| 0·5 | ·9285 7143 | *J*-curve | ·0910 5392 | 4·3760 6838 | 8·3076 9231 |
| 1 | ·8666 6667 | 1·0000 0000$^e$ | ·1165 9662 | 1·7532 4952 | 4·8079 8149 |
| 1·5 | ·8125 0000$^e$ | ·9166 6667 | ·1301 0412 | ·9230 7692 | 3·7132 8671 |
| 2 | ·7647 0588 | ·8461 5385− | ·1376 2298 | ·5368 9168 | 3·2135 6904 |
| 2·5 | ·7222 2222 | ·7857 1429 | ·1416 3943 | ·3254 9269 | 2·9475 5245− |
| 3 | ·6842 1053 | ·7333 3333 | ·1434 4950+ | ·1995 0560 | 2·7953 1773 |
| 3·5 | ·6500 0000$^e$ | ·6875 0000$^e$ | ·1438 1175− | ·1208 7912 | 2·7058 3263 |
| 4 | ·6190 4762 | ·6470 5882 | ·1432 0179 | ·0707 6923 | 2·6538 4615+ |
| 4·5 | ·5909 0909 | ·6111 1111 | ·1419 3177 | ·0388 4084 | 2·6255 2832 |
| 5 | ·5652 1739 | ·5789 4737 | ·1402 1318 | ·0189 9335+ | 2·6127 3210 |
| 5·5 | ·5416 6667 | ·5500 0000$^e$ | ·1381 9270 | ·0074 2115+ | 2·6103 8961 |
| 6 | ·5200 0000$^e$ | ·5238 0952 | ·1359 7385+ | ·0016 4639 | 2·6152 1349 |
| 6·5 | ·5000 0000$^e$ | ·5000 0000$^e$ | ·1336 3062 | ·0000 0000$^e$ | 2·6250 0000$^e$ |
| 7 | ·4814 8148 | ·4782 6087 | ·1312 1634 | ·0013 2646 | 2·6382 3274 |
| 7·5 | ·4642 8571 | ·4583 3333 | ·1287 6969 | ·0048 0769 | 2·6538 4615+ |
| 8 | ·4482 7586 | ·4400 0000$^e$ | ·1263 1875+ | ·0098 5378 | 2·6710 7892 |
| 8·5 | ·4333 3333 | ·4230 7692 | ·1238 8391 | ·0160 3282 | 2·6893 7982 |
| 9 | ·4193 5484 | ·4074 0741 | ·1214 7985+ | ·0230 2459 | 2·7083 4571 |
| 9·5 | ·4062 5000$^e$ | ·3928 5714 | ·1191 1708 | ·0305 8929 | 2·7276 7952 |
| 10 | ·3939 3939 | ·3793 1034 | ·1168 0292 | ·0385 4582 | 2·7471 6136 |
| 10·5 | ·3823 5294 | ·3666 6667 | ·1145 4237 | ·0467 5657 | 2·7666 2811 |
| 11 | ·3714 2857 | ·3548 3871 | ·1123 3863 | ·0551 1648 | 2·7859 5888 |
| 12 | ·3513 5135+ | ·3333 3333 | ·1081 0811 | ·0719 8096 | 2·8238 7975+ |
| 13 | ·3333 3333 | ·3142 8571 | ·1041 1584 | ·0886 9659 | 2·8604 6512 |
| 14 | ·3170 7317 | ·2972 9730 | ·1003 5698 | ·1050 0611 | 2·8954 8749 |
| 15 | ·3023 2558 | ·2820 5128 | ·0968 2167 | ·1207 6470 | 2·9288 5534 |
| 16 | ·2888 8889 | ·2682 9268 | ·0934 9755− | ·1358 9690 | 2·9605 5730 |
| 17 | ·2765 9574 | ·2558 1395+ | ·0903 7124 | ·1503 7029 | 2·9906 2881 |
| 18 | ·2653 0612 | ·2444 4444 | ·0874 2925+ | ·1641 7924 | 3·0191 3181 |
| 19 | ·2549 0196 | ·2340 4255+ | ·0846 5846 | ·1773 3463 | 3·0461 4222 |
| 20 | ·2452 8302 | ·2244 8980 | ·0820 4639 | ·1898 5723 | 3·0717 4226 |
| 21 | ·2363 6364 | ·2156 8627 | ·0795 8133 | ·2017 7352 | 3·0960 1568 |
| 22 | ·2280 7018 | ·2075 4717 | ·0772 5243 | ·2131 1287 | 3·1190 4489 |
| 23 | ·2203 3898 | ·2000 0000$^e$ | ·0750 4965− | ·2239 0584 | 3·1409 0926 |
| 24 | ·2131 1475+ | ·1929 8246 | ·0729 6376 | ·2341 8298 | 3·1616 8418 |
| 25 | ·2063 4921 | ·1864 4068 | ·0709 8629 | ·2439 7416 | 3·1814 4062 |
| 26 | ·2000 0000$^e$ | ·1803 2787 | ·0691 0947 | ·2533 0813 | 3·2002 4495− |
| 27 | ·1940 2985+ | ·1746 0317 | ·0673 2622 | ·2622 1225+ | 3·2181 5897 |
| 28 | ·1884 0580 | ·1692 3077 | ·0656 3000$^e$ | ·2707 1240 | 3·2352 4010 |
| 29 | ·1830 9859 | ·1641 7910 | ·0640 1487 | ·2788 3289 | 3·2515 4156 |
| 30 | ·1780 8219 | ·1594 2029 | ·0624 7537 | ·2865 9652 | 3·2671 1263 |
| 31 | ·1733 3333 | ·1549 2958 | ·0610 0647 | ·2940 2459 | 3·2819 9893 |
| 32 | ·1688 3117 | ·1506 8493 | ·0596 0359 | ·3011 3696 | 3·2962 4266 |
| 33 | ·1645 5696 | ·1466 6667 | ·0582 6249 | ·3079 5217 | 3·3098 8289 |
| 34 | ·1604 9383 | ·1428 5714 | ·0569 7929 | ·3144 8747 | 3·3229 5577 |
| 35 | ·1566 2651 | ·1392 4051 | ·0557 5042 | ·3207 5891 | 3·3354 9480 |
| 36 | ·1529 4118 | ·1358 0247 | ·0545 7257 | ·3267 8145+ | 3·3475 3103 |
| 37 | ·1494 2529 | ·1325 3012 | ·0534 4269 | ·3325 6903 | 3·3590 9326 |
| 38 | ·1460 6742 | ·1294 1176 | ·0523 5797 | ·3381 3462 | 3·3702 0820 |
| 39 | ·1428 5714 | ·1264 3678 | ·0513 1580 | ·3434 9030 | 3·3809 0071 |
| 40 | ·1397 8495− | ·1235 9551 | ·0503 1378 | ·3486 4737 | 3·3911 9386 |
| 41 | ·1368 4211 | ·1208 7912 | ·0493 4966 | ·3536 1633 | 3·4011 0915+ |
| 42 | ·1340 2062 | ·1182 7957 | ·0484 2136 | ·3584 0701 | 3·4106 6662 |
| 43 | ·1313 1313 | ·1157 8947 | ·0475 2696 | ·3630 2859 | 3·4198 8492 |
| 44 | ·1287 1287 | ·1134 0206 | ·0466 6465+ | ·3674 8965+ | 3·4287 8150− |
| 45 | ·1262 1359 | ·1111 1111 | ·0458 3276 | ·3717 9823 | 3·4373 7262 |
| 46 | ·1238 0952 | ·1089 1089 | ·0450 2974 | ·3759 6184 | 3·4456 7353 |
| 47 | ·1214 9533 | ·1067 9612 | ·0442 5411 | ·3799 8754 | 3·4536 9846 |
| 48 | ·1192 6606 | ·1047 6190 | ·0435 0451 | ·3838 8194 | 3·4614 6077 |
| 49 | ·1171 1712 | ·1028 0374 | ·0427 7968 | ·3876 5124 | 3·4689 7298 |
| 50 | ·1150 4425− | ·1009 1743 | ·0420 7841 | ·3913 0127 | 3·4762 4683 |

$p = 7$

| $q$ | $\bar{x}$ | $\tilde{x}$ | $\sigma$ | $\beta_1$ | $\beta_2$ |
|---|---|---|---|---|---|
| 0·5 | ·9333 3333 | *J*-curve | ·0855 5853 | 4·5476 8500$^+$ | 8·6004 2965$^-$ |
| 1 | ·8750 0000$^e$ | 1·0000 0000$^e$ | ·1102 3964 | 1·8514 2857 | 4·9792 2078 |
| 1·5 | ·8235 2941 | ·9230 7692 | ·1236 8411 | ·9929 8132 | 3·8382 1355$^-$ |
| 2 | ·7777 7778 | ·8571 4286 | ·1314 6844 | ·5903 1877 | 3·3116 8831 |
| 2·5 | ·7368 4211 | ·8000 0000$^e$ | ·1358 9415$^+$ | ·3674 8582 | 3·0271 3043 |
| 3 | ·7000 0000$^e$ | ·7500 0000$^e$ | ·1381 6986 | ·2328 0423 | 2·8608 0586 |
| 3·5 | ·6666 6667 | ·7058 8235$^+$ | ·1390 0961 | ·1472 0000$^e$ | 2·7600 0000$^e$ |
| 4 | ·6363 6364 | ·6666 6667 | ·1388 6593 | ·0912 9332 | 2·6985 8713 |
| 4·5 | ·6086 9565$^+$ | ·6315 7895$^-$ | ·1380 3920 | ·0544 3421 | 2·6622 2709 |
| 5 | ·5833 3333 | ·6000 0000$^e$ | ·1367 3544 | ·0303 2070 | 2·6424 4898 |
| 5·5 | ·5600 0000$^e$ | ·5714 2857 | ·1350 9941 | ·0150 0996 | 2·6339 6559 |
| 6 | ·5384 6154 | ·5454 5455$^-$ | ·1332 3468 | ·0059 2593 | 2·6333 3333 |
| 6·5 | ·5185 1852 | ·5217 3913 | ·1312 1634 | ·0013 2646 | 2·6382 3274 |
| 7 | ·5000 0000$^e$ | ·5000 0000$^e$ | ·1290 9944 | ·0000 0000$^e$ | 2·6470 5882 |
| 7·5 | ·4827 5862 | ·4800 0000$^e$ | ·1269 2460 | ·0010 8444 | 2·6586 7656 |
| 8 | ·4666 6667 | ·4615 3846 | ·1247 2191 | ·0039 5452 | 2·6722 6891 |
| 8·5 | ·4516 1290 | ·4444 4444 | ·1225 1374 | ·0081 4955$^-$ | 2·6872 3922 |
| 9 | ·4375 0000$^e$ | ·4285 7143 | ·1203 1668 | ·0133 2549 | 2·7031 4676 |
| 9·5 | ·4242 4242 | ·4137 9310 | ·1181 4296 | ·0192 2264 | 2·7196 6298 |
| 10 | ·4117 6471 | ·4000 0000$^e$ | ·1160 0156 | ·0256 4306 | 2·7365 4135$^+$ |
| 10·5 | ·4000 0000$^e$ | ·3870 9677 | ·1138 9896 | ·0324 3480 | 2·7535 9600 |
| 11 | ·3888 8889 | ·3750 0000$^e$ | ·1118 3972 | ·0394 8052 | 2·7706 8646 |
| 12 | ·3684 2105$^+$ | ·3529 4118 | ·1078 6264 | ·0539 8985$^-$ | 2·8045 7638 |
| 13 | ·3500 0000$^e$ | ·3333 3333 | ·1040 8330 | ·0686 5861 | 2·8376 4062 |
| 14 | ·3333 3333 | ·3157 8947 | ·1005 0378 | ·0831 7580 | 2·8695 6522 |
| 15 | ·3181 8182 | ·3000 0000$^e$ | ·0971 1986 | ·0973 5450$^-$ | 2·9001 9048 |
| 16 | ·3043 4783 | ·2857 1429 | ·0939 2378 | ·1110 8571 | 2·9294 5055$^-$ |
| 17 | ·2916 6667 | ·2727 2727 | ·0909 0593 | ·1243 1008 | 2·9573 3678 |
| 18 | ·2800 0000$^e$ | ·2608 6957 | ·0880 5593 | ·1370 0002 | 2·9838 7503 |
| 19 | ·2692 3077 | ·2500 0000$^e$ | ·0853 6318 | ·1491 4838 | 3·0091 1145$^-$ |
| 20 | ·2592 5926 | ·2400 0000$^e$ | ·0828 1733 | ·1607 6100 | 3·0331 0345$^-$ |
| 21 | ·2500 0000$^e$ | ·2307 6923 | ·0804 0844 | ·1718 5185$^+$ | 3·0559 1398 |
| 22 | ·2413 7931 | ·2222 2222 | ·0781 2711 | ·1824 3983 | 3·0776 0788 |
| 23 | ·2333 3333 | ·2142 8571 | ·0759 6453 | ·1925 4658 | 3·0982 4958 |
| 24 | ·2258 0645$^+$ | ·2068 9655$^+$ | ·0739 1251 | ·2021 9511 | 3·1179 0171 |
| 25 | ·2187 5000$^e$ | ·2000 0000$^e$ | ·0719 6347 | ·2114 0880 | 3·1366 2425$^-$ |
| 26 | ·2121 2121 | ·1935 4839 | ·0701 1038 | ·2202 1081 | 3·1544 7410 |
| 27 | ·2058 8235$^+$ | ·1875 0000$^e$ | ·0683 4676 | ·2286 2369 | 3·1715 0484 |
| 28 | ·2000 0000$^e$ | ·1818 1818 | ·0666 6667 | ·2366 6910 | 3·1877 6671 |
| 29 | ·1944 4444 | ·1764 7059 | ·0650 6458 | ·2443 6773 | 3·2033 0668 |
| 30 | ·1891 8919 | ·1714 2857 | ·0635 3545$^+$ | ·2517 3914 | 3·2181 6850$^-$ |
| 31 | ·1842 1053 | ·1666 6667 | ·0620 7461 | ·2588 0184 | 3·2323 9294 |
| 32 | ·1794 8718 | ·1621 6216 | ·0606 7774 | ·2655 7321 | 3·2460 1792 |
| 33 | ·1750 0000$^e$ | ·1578 9474 | ·0593 4089 | ·2720 6958 | 3·2590 7868 |
| 34 | ·1707 3171 | ·1538 4615$^+$ | ·0580 6039 | ·2783 0624 | 3·2716 0801 |
| 35 | ·1666 6667 | ·1500 0000$^e$ | ·0568 3286 | ·2842 9752 | 3·2836 3636 |
| 36 | ·1627 9070 | ·1463 4146 | ·0556 5517 | ·2900 5683 | 3·2951 9209 |
| 37 | ·1590 9091 | ·1428 5714 | ·0545 2441 | ·2955 9670 | 3·3063 0154 |
| 38 | ·1555 5556 | ·1395 3488 | ·0534 3790 | ·3009 2887 | 3·3169 8928 |
| 39 | ·1521 7391 | ·1363 6364 | ·0523 9316 | ·3060 6431 | 3·3272 7816 |
| 40 | ·1489 3617 | ·1333 3333 | ·0513 8786 | ·3110 1327 | 3·3371 8950$^+$ |
| 41 | ·1458 3333 | ·1304 3478 | ·0504 1986 | ·3157 8537 | 3·3467 4319 |
| 42 | ·1428 5714 | ·1276 5957 | ·0494 8717 | ·3203 8959 | 3·3559 5777 |
| 43 | ·1400 0000$^e$ | ·1250 0000$^e$ | ·0485 8790 | ·3248 3438 | 3·3648 5059 |
| 44 | ·1372 5490 | ·1224 4898 | ·0477 2034 | ·3291 2762 | 3·3734 3788 |
| 45 | ·1346 1538 | ·1200 0000$^e$ | ·0468 8285$^-$ | ·3332 7672 | 3·3817 3481 |
| 46 | ·1320 7547 | ·1176 4706 | ·0460 7392 | ·3372 8864 | 3·3897 5559 |
| 47 | ·1296 2963 | ·1153 8462 | ·0452 9214 | ·3411 6990 | 3·3975 1354 |
| 48 | ·1272 7273 | ·1132 0755$^-$ | ·0445 3618 | ·3449 2664 | 3·4050 2117 |
| 49 | ·1250 0000$^e$ | ·1111 1111 | ·0438 0479 | ·3485 6463 | 3·4122 9022 |
| 50 | ·1228 0702 | ·1090 9091 | ·0430 9681 | ·3520 8930 | 3·4193 3172 |

$p = 7{\cdot}5$

| $q$ | $\bar{x}$ | $\tilde{x}$ | $\sigma$ | $\beta_1$ | $\beta_2$ |
|---|---|---|---|---|---|
| 0·5 | ·9375 0000$^e$ | $J$-curve | ·0806 8715+ | 4·7040 0000$^e$ | 8·8690 9091 |
| 1 | ·8823 5294 | 1·0000 0000$^e$ | ·1045 3215+ | 1·9416 4777 | 5·1374 7412 |
| 1·5 | ·8333 3333 | ·9285 7143 | ·1178 5113 | 1·0578 5124 | 3·9545 4545+ |
| 2 | ·7894 7368 | ·8666 6667 | ·1258 1354 | ·6404 5369 | 3·4038 2609 |
| 2·5 | ·7500 0000$^e$ | ·8125 0000$^e$ | ·1305 5824 | ·4074 0741 | 3·1025 6410 |
| 3 | ·7142 8571 | ·7647 0588 | ·1332 1497 | ·2649 6000$^e$ | 2·9235 5556 |
| 3·5 | ·6818 1818 | ·7222 2222 | ·1344 5636 | ·1731 1919 | 2·8125 5887 |
| 4 | ·6521 7391 | ·6842 1053 | ·1347 1246 | ·1120 2561 | 2·7426 5645− |
| 4·5 | ·6250 0000$^e$ | ·6500 0000$^e$ | ·1342 7154 | ·0707 4830 | 2·6990 4762 |
| 5 | ·6000 0000$^e$ | ·6190 4762 | ·1333 3333 | ·0428 0618 | 2·6729 6997 |
| 5·5 | ·5769 2308 | ·5909 0909 | ·1320 3972 | ·0241 3468 | 2·6589 3939 |
| 6 | ·5555 5556 | ·5652 1739 | ·1304 9338 | ·0120 7076 | 2·6533 7243 |
| 6·5 | ·5357 1429 | ·5416 6667 | ·1287 6969 | ·0048 0769 | 2·6538 4615+ |
| 7 | ·5172 4138 | ·5200 0000$^e$ | ·1269 2460 | ·0010 8444 | 2·6586 7656 |
| 7·5 | ·5000 0000$^e$ | ·5000 0000$^e$ | ·1250 0000$^e$ | ·0000 0000$^e$ | 2·6666 6667 |
| 8 | ·4838 7097 | ·4814 8148 | ·1230 2743 | ·0008 9796 | 2·6769 4981 |
| 8·5 | ·4687 5000$^e$ | ·4642 8571 | ·1210 3073 | ·0032 9218 | 2·6888 8889 |
| 9 | ·4545 4545+ | ·4482 7586 | ·1190 2794 | ·0068 1763 | 2·7020 0970 |
| 9·5 | ·4411 7647 | ·4333 3333 | ·1170 3271 | ·0111 9697 | 2·7159 5568 |
| 10 | ·4285 7143 | ·4193 5484 | ·1150 5532 | ·0162 1740 | 2·7304 5654 |
| 10·5 | ·4166 6667 | ·4062 5000$^e$ | ·1131 0348 | ·0217 1429 | 2·7453 0612 |
| 11 | ·4054 0541 | ·3939 3939 | ·1111 8289 | ·0275 5935+ | 2·7603 4652 |
| 12 | ·3846 1538 | ·3714 2857 | ·1074 5081 | ·0399 1347 | 2·7905 4264 |
| 13 | ·3658 5366 | ·3513 5135+ | ·1038 7935− | ·0527 0529 | 2·8203 7461 |
| 14 | ·3488 3721 | ·3333 3333 | ·1004 7660 | ·0655 7589 | 2·8494 5103 |
| 15 | ·3333 3333 | ·3170 7317 | ·0972 4333 | ·0783 0071 | 2·8775 5102 |
| 16 | ·3191 4894 | ·3023 2558 | ·0941 7589 | ·0907 4074 | 2·9045 5975− |
| 17 | ·3061 2245− | ·2888 8889 | ·0912 6808 | ·1028 1239 | 2·9304 2882 |
| 18 | ·2941 1765− | ·2765 9574 | ·0885 1233 | ·1144 6832 | 2·9551 5152 |
| 19 | ·2830 1887 | ·2653 0612 | ·0859 0040 | ·1256 8510 | 2·9787 4705− |
| 20 | ·2727 2727 | ·2549 0196 | ·0834 2390 | ·1364 5504 | 3·0012 5035− |
| 21 | ·2631 5789 | ·2452 8302 | ·0810 7452 | ·1467 8082 | 3·0227 0548 |
| 22 | ·2542 3729 | ·2363 6364 | ·0788 4425+ | ·1566 7178 | 3·0431 6128 |
| 23 | ·2459 0164 | ·2280 7018 | ·0767 2545− | ·1661 4150− | 3·0626 6860 |
| 24 | ·2380 9524 | ·2203 3898 | ·0747 1091 | ·1752 0606 | 3·0812 7839 |
| 25 | ·2307 6923 | ·2131 1475+ | ·0727 9388 | ·1838 8294 | 3·0990 4062 |
| 26 | ·2238 8060 | ·2063 4921 | ·0709 6806 | ·1921 9019 | 3·1160 0350+ |
| 27 | ·2173 9130 | ·2000 0000$^e$ | ·0692 2759 | ·2001 4595+ | 3·1322 1309 |
| 28 | ·2112 6761 | ·1940 2985+ | ·0675 6704 | ·2077 6804 | 3·1477 1305− |
| 29 | ·2054 7945+ | ·1884 0580 | ·0659 8134 | ·2150 7378 | 3·1625 4457 |
| 30 | ·2000 0000$^e$ | ·1830 9859 | ·0644 6584 | ·2220 7979 | 3·1767 4637 |
| 31 | ·1948 0519 | ·1780 8219 | ·0630 1620 | ·2288 0200 | 3·1903 5473 |
| 32 | ·1898 7342 | ·1733 3333 | ·0616 2842 | ·2352 5548 | 3·2034 0361 |
| 33 | ·1851 8519 | ·1688 3117 | ·0602 9878 | ·2414 5455− | 3·2159 2476 |
| 34 | ·1807 2289 | ·1645 5696 | ·0590 2385− | ·2474 1269 | 3·2279 4782 |
| 35 | ·1764 7059 | ·1604 9383 | ·0578 0043 | ·2531 4264 | 3·2395 0047 |
| 36 | ·1724 1379 | ·1566 2651 | ·0566 2557 | ·2586 5636 | 3·2506 0853 |
| 37 | ·1685 3933 | ·1529 4118 | ·0554 9652 | ·2639 6511 | 3·2612 9613 |
| 38 | ·1648 3516 | ·1494 2529 | ·0544 1073 | ·2690 7946 | 3·2715 8580 |
| 39 | ·1612 9032 | ·1460 6742 | ·0533 6582 | ·2740 0934 | 3·2814 9857 |
| 40 | ·1578 9474 | ·1428 5714 | ·0523 5958 | ·2787 6407 | 3·2910 5411 |
| 41 | ·1546 3918 | ·1397 8495− | ·0513 8995+ | ·2833 5242 | 3·3002 7079 |
| 42 | ·1515 1515+ | ·1368 4211 | ·0504 5502 | ·2877 8261 | 3·3091 6584 |
| 43 | ·1485 1485+ | ·1340 2062 | ·0495 5299 | ·2920 6237 | 3·3177 5535+ |
| 44 | ·1456 3107 | ·1313 1313 | ·0486 8219 | ·2961 9895+ | 3·3260 5442 |
| 45 | ·1428 5714 | ·1287 1287 | ·0478 4105− | ·3001 9920 | 3·3340 7720 |
| 46 | ·1401 8692 | ·1262 1359 | ·0470 2811 | ·3040 6952 | 3·3418 3695+ |
| 47 | ·1376 1468 | ·1238 0952 | ·0462 4200 | ·3078 1597 | 3·3493 4615− |
| 48 | ·1351 3514 | ·1214 9533 | ·0454 8142 | ·3114 4423 | 3·3566 1650− |
| 49 | ·1327 4336 | ·1192 6606 | ·0447 4518 | ·3149 5968 | 3·3636 5902 |
| 50 | ·1304 3478 | ·1171 1712 | ·0440 3215− | ·3183 6735− | 3·3704 8406 |

$p = 8$

| $q$ | $\bar{x}$ | $\tilde{x}$ | $\sigma$ | $\beta_1$ | $\beta_2$ |
|---|---|---|---|---|---|
| 0·5 | ·9411 7647 | *J*-curve | ·0763 3949 | 4·8469 3878 | 9·1164 5963 |
| 1 | ·8888 8889 | 1·0000 0000$^e$ | ·0993 8080 | 2·0247 9339 | 5·2840 9091 |
| 1·5 | ·8421 0526 | ·9333 3333 | ·1125 3105− | 1·1181 4745− | 4·0630 4348 |
| 2 | ·8000 0000$^e$ | ·8750 0000$^e$ | ·1206 0454 | ·6875 0000$^e$ | 3·4903 8462 |
| 2·5 | ·7619 0476 | ·8235 2941 | ·1255 9628 | ·4452 8000$^e$ | 3·1740 0000$^e$ |
| 3 | ·7272 7273 | ·7777 7778 | ·1285 6487 | ·2958 5799 | 2·9835 1648 |
| 3·5 | ·6956 5217 | ·7368 4211 | ·1301 4460 | ·1984 1270 | 2·8633 0049 |
| 4 | ·6666 6667 | ·7000 0000$^e$ | ·1307 4409 | ·1326 5306 | 2·7857 1429 |
| 4·5 | ·6400 0000$^e$ | ·6666 6667 | ·1306 3945+ | ·0873 9596 | 2·7355 3949 |
| 5 | ·6153 8462 | ·6363 6364 | ·1300 2373 | ·0560 0000$^e$ | 2·7037 5000$^e$ |
| 5·5 | ·5925 9259 | ·6086 9565+ | ·1290 3531 | ·0342 9193 | 2·6846 8408 |
| 6 | ·5714 2857 | ·5833 3333 | ·1277 7531 | ·0195 3125$^e$ | 2·6746 3235+ |
| 6·5 | ·5517 2414 | ·5600 0000$^e$ | ·1263 1875+ | ·0098 5378 | 2·6710 7892 |
| 7 | ·5333 3333 | ·5384 6154 | ·1247 2191 | ·0039 5452 | 2·6722 6891 |
| 7·5 | ·5161 2903 | ·5185 1852 | ·1230 2743 | ·0008 9796 | 2·6769 4981 |
| 8 | ·5000 0000$^e$ | ·5000 0000$^e$ | ·1212 6781 | ·0000 0000$^e$ | 2·6842 1053 |
| 8·5 | ·4848 4848 | ·4827 5862 | ·1194 6797 | ·0007 5194 | 2·6933 7777 |
| 9 | ·4705 8824 | ·4666 6667 | ·1176 4706 | ·0027 7008 | 2·7039 4737 |
| 9·5 | ·4571 4286 | ·4516 1290 | ·1158 1982 | ·0057 6145− | 2·7155 3767 |
| 10 | ·4444 4444 | ·4375 0000$^e$ | ·1139 9759 | ·0095 0000$^e$ | 2·7278 5714 |
| 10·5 | ·4324 3243 | ·4242 4242 | ·1121 8909 | ·0138 0981 | 2·7406 8147 |
| 11 | ·4210 5263 | ·4117 6471 | ·1104 0093 | ·0185 5288 | 2·7538 3707 |
| 12 | ·4000 0000$^e$ | ·3888 8889 | ·1069 0450− | ·0289 2562 | 2·7806 3241 |
| 13 | ·3809 5238 | ·3684 2105+ | ·1035 3473 | ·0399 8837 | 2·8074 8328 |
| 14 | ·3636 3636 | ·3500 0000$^e$ | ·1003 0496 | ·0513 3929 | 2·8339 2857 |
| 15 | ·3478 2609 | ·3333 3333 | ·0972 2035− | ·0627 2000$^e$ | 2·8596 9231 |
| 16 | ·3333 3333 | ·3181 8182 | ·0942 8090 | ·0739 6450− | 2·8846 1538 |
| 17 | ·3200 0000$^e$ | ·3043 4783 | ·0914 8350− | ·0849 6732 | 2·9086 1345− |
| 18 | ·3076 9231 | ·2916 6667 | ·0888 2312 | ·0956 6327 | 2·9316 5025− |
| 19 | ·2962 9630 | ·2800 0000$^e$ | ·0862 9368 | ·1060 1414 | 2·9537 2051 |
| 20 | ·2857 1429 | ·2692 3077 | ·0838 8860 | ·1160 0000$^e$ | 2·9748 3871 |
| 21 | ·2758 6207 | ·2592 5926 | ·0816 0110 | ·1256 1320 | 2·9950 3168 |
| 22 | ·2666 6667 | ·2500 0000$^e$ | ·0794 2445+ | ·1348 5440 | 3·0143 3368 |
| 23 | ·2580 6452 | ·2413 7931 | ·0773 5212 | ·1437 2979 | 3·0327 8307 |
| 24 | ·2500 0000$^e$ | ·2333 3333 | ·0753 7784 | ·1522 4913 | 3·0504 2017 |
| 25 | ·2424 2424 | ·2258 0645+ | ·0734 9564 | ·1604 2449 | 3·0672 8571 |
| 26 | ·2352 9412 | ·2187 5000$^e$ | ·0716 9993 | ·1682 6923 | 3·0834 1996 |
| 27 | ·2285 7143 | ·2121 2121 | ·0699 8542 | ·1757 9742 | 3·0988 6202 |
| 28 | ·2222 2222 | ·2058 8235+ | ·0683 4719 | ·1830 2335− | 3·1136 4951 |
| 29 | ·2162 1622 | ·2000 0000$^e$ | ·0667 8062 | ·1899 6123 | 3·1278 1830 |
| 30 | ·2105 2632 | ·1944 4444 | ·0652 8144 | ·1966 2500$^e$ | 3·1414 0244 |
| 31 | ·2051 2821 | ·1891 8919 | ·0638 4564 | ·2030 2815+ | 3·1544 3408 |
| 32 | ·2000 0000$^e$ | ·1842 1053 | ·0624 6950+ | ·2091 8367 | 3·1669 4352 |
| 33 | ·1951 2195+ | ·1794 8718 | ·0611 4958 | ·2151 0399 | 3·1789 5925+ |
| 34 | ·1904 7619 | ·1750 0000$^e$ | ·0598 8264 | ·2208 0092 | 3·1905 0802 |
| 35 | ·1860 4651 | ·1707 3171 | ·0586 6570 | ·2262 8571 | 3·2016 1491 |
| 36 | ·1818 1818 | ·1666 6667 | ·0574 9596 | ·2315 6900 | 3·2123 0342 |
| 37 | ·1777 7778 | ·1627 9070 | ·0563 7083 | ·2366 6083 | 3·2225 9560 |
| 38 | ·1739 1304 | ·1590 9091 | ·0552 8789 | ·2415 7072 | 3·2325 1208 |
| 39 | ·1702 1277 | ·1555 5556 | ·0542 4489 | ·2463 0763 | 3·2420 7221 |
| 40 | ·1666 6667 | ·1521 7391 | ·0532 3971 | ·2508 8000$^e$ | 3·2512 9412 |
| 41 | ·1632 6531 | ·1489 3617 | ·0522 7040 | ·2552 9581 | 3·2601 9479 |
| 42 | ·1600 0000$^e$ | ·1458 3333 | ·0513 3512 | ·2595 6255+ | 3·2687 9017 |
| 43 | ·1568 6275− | ·1428 5714 | ·0504 3214 | ·2636 8732 | 3·2770 9522 |
| 44 | ·1538 4615+ | ·1400 0000$^e$ | ·0495 5986 | ·2676 7677 | 3·2851 2397 |
| 45 | ·1509 4340 | ·1372 5490 | ·0487 1677 | ·2715 3719 | 3·2928 8961 |
| 46 | ·1481 4815− | ·1346 1538 | ·0479 0147 | ·2752 7451 | 3·3004 0454 |
| 47 | ·1454 5455− | ·1320 7547 | ·0471 1262 | ·2788 9432 | 3·3076 8043 |
| 48 | ·1428 5714 | ·1296 2963 | ·0463 4898 | ·2824 0190 | 3·3147 2823 |
| 49 | ·1403 5088 | ·1272 7273 | ·0456 0938 | ·2858 0223 | 3·3215 5828 |
| 50 | ·1379 3103 | ·1250 0000$^e$ | ·0448 9273 | ·2891 0000$^e$ | 3·3281 8033 |

$p = 8{\cdot}5$

| $q$ | $\bar{x}$ | $\tilde{x}$ | $\sigma$ | $\beta_1$ | $\beta_2$ |
|---|---|---|---|---|---|
| 0·5 | ·9444 4444 | $J$-curve | ·0724 3558 | 4·9781 2348 | 9·3449 1979 |
| 1 | ·8947 3684 | 1·0000 0000$^e$ | ·0947 0899 | 2·1016 3460 | 5·4202 5575+ |
| 1·5 | ·8500 0000$^e$ | ·9375 0000$^e$ | ·1076 6108 | 1·1742 9194 | 4·1644 0422 |
| 2 | ·8095 2381 | ·8823 5294 | ·1157 9405− | ·7316 7059 | 3·5717 6471 |
| 2·5 | ·7727 2727 | ·8333 3333 | ·1209 7515− | ·4811 6951 | 3·2416 2896 |
| 3 | ·7391 3043 | ·7894 7368 | ·1241 9876 | ·3254 5254 | 3·0407 1820 |
| 3·5 | ·7083 3333 | ·7500 0000$^e$ | ·1260 6385− | ·2229 4632 | 2·9121 2485− |
| 4 | ·6800 0000$^e$ | ·7142 8571 | ·1269 5873 | ·1529 6915+ | 2·8275 5349 |
| 4·5 | ·6538 4615+ | ·6818 1818 | ·1271 4765+ | ·1041 1038 | 2·7714 0523 |
| 5 | ·6296 2963 | ·6521 7391 | ·1268 1677 | ·0695 8438 | 2·7344 1435+ |
| 5·5 | ·6071 4286 | ·6250 0000$^e$ | ·1261 0060 | ·0451 2032 | 2·7107 5810 |
| 6 | ·5862 0690 | ·6000 0000$^e$ | ·1250 9825+ | ·0279 0832 | 2·6966 1319 |
| 6·5 | ·5666 6667 | ·5769 2308 | ·1238 8391 | ·0160 3282 | 2·6893 7982 |
| 7 | ·5483 8710 | ·5555 5556 | ·1225 1374 | ·0081 4955− | 2·6872 3922 |
| 7·5 | ·5312 5000$^e$ | ·5357 1429 | ·1210 3073 | ·0032 9218 | 2·6888 8889 |
| 8 | ·5151 5152 | ·5172 4138 | ·1194 6797 | ·0007 5194 | 2·6933 7777 |
| 8·5 | ·5000 0000$^e$ | ·5000 0000$^e$ | ·1178 5113 | ·0000 0000$^e$ | 2·7000 0000$^e$ |
| 9 | ·4857 1429 | ·4838 7097 | ·1162 0018 | ·0006 3598 | 2·7082 2450+ |
| 9·5 | ·4722 2222 | ·4687 5000$^e$ | ·1145 3071 | ·0023 5294 | 2·7176 4706 |
| 10 | ·4594 5946 | ·4545 4545+ | ·1128 5490 | ·0049 1304 | 2·7279 5703 |
| 10·5 | ·4473 6842 | ·4411 7647 | ·1111 8226 | ·0081 3024 | 2·7389 1375− |
| 11 | ·4358 9744 | ·4285 7143 | ·1095 2021 | ·0118 5783 | 2·7503 2956 |
| 12 | ·4146 3415− | ·4054 0541 | ·1062 4952 | ·0204 0184 | 2·7739 8137 |
| 13 | ·3953 4884 | ·3846 1538 | ·1030 7454 | ·0298 6548 | 2·7980 7177 |
| 14 | ·3777 7778 | ·3658 5366 | ·1000 1313 | ·0398 0834 | 2·8220 7673 |
| 15 | ·3617 0213 | ·3488 3721 | ·0970 7428 | ·0499 4158 | 2·8456 7039 |
| 16 | ·3469 3878 | ·3333 3333 | ·0942 6127 | ·0600 7476 | 2·8686 5352 |
| 17 | ·3333 3333 | ·3191 4894 | ·0915 7371 | ·0700 8264 | 2·8909 0909 |
| 18 | ·3207 5472 | ·3061 2245− | ·0890 0891 | ·0798 8381 | 2·9123 7399 |
| 19 | ·3090 9091 | ·2941 1765− | ·0865 6277 | ·0894 2664 | 2·9330 2061 |
| 20 | ·2982 4561 | ·2830 1887 | ·0842 3035− | ·0986 7999 | 2·9528 4474 |
| 21 | ·2881 3559 | ·2727 2727 | ·0820 0629 | ·1076 2682 | 2·9718 5746 |
| 22 | ·2786 8852 | ·2631 5789 | ·0798 8508 | ·1162 5985− | 2·9900 7963 |
| 23 | ·2698 4127 | ·2542 3729 | ·0778 6121 | ·1245 7847 | 3·0075 3820 |
| 24 | ·2615 3846 | ·2459 0164 | ·0759 2929 | ·1325 8666 | 3·0242 6366 |
| 25 | ·2537 3134 | ·2380 9524 | ·0740 8412 | ·1402 9149 | 3·0402 8827 |
| 26 | ·2463 7681 | ·2307 6923 | ·0723 2074 | ·1477 0202 | 3·0556 4495+ |
| 27 | ·2394 3662 | ·2238 8060 | ·0706 3445− | ·1548 2856 | 3·0703 6641 |
| 28 | ·2328 7671 | ·2173 9130 | ·0690 2080 | ·1616 8215− | 3·0844 8466 |
| 29 | ·2266 6667 | ·2112 6761 | ·0674 7561 | ·1682 7412 | 3·0980 3066 |
| 30 | ·2207 7922 | ·2054 7945+ | ·0659 9497 | ·1746 1590 | 3·1110 3411 |
| 31 | ·2151 8987 | ·2000 0000$^e$ | ·0645 7519 | ·1807 1876 | 3·1235 2336 |
| 32 | ·2098 7654 | ·1948 0519 | ·0632 1285+ | ·1865 9373 | 3·1355 2530 |
| 33 | ·2048 1928 | ·1898 7342 | ·0619 0474 | ·1922 5149 | 3·1470 6538 |
| 34 | ·2000 0000$^e$ | ·1851 8519 | ·0606 4784 | ·1977 0231 | 3·1581 6768 |
| 35 | ·1954 0230 | ·1807 2289 | ·0594 3937 | ·2029 5604 | 3·1688 5484 |
| 36 | ·1910 1124 | ·1764 7059 | ·0582 7671 | ·2080 2208 | 3·1791 4822 |
| 37 | ·1868 1319 | ·1724 1379 | ·0571 5740 | ·2129 0938 | 3·1890 6790 |
| 38 | ·1827 9570 | ·1685 3933 | ·0560 7916 | ·2176 2643 | 3·1986 3278 |
| 39 | ·1789 4737 | ·1648 3516 | ·0550 3986 | ·2221 8129 | 3·2078 6061 |
| 40 | ·1752 5773 | ·1612 9032 | ·0540 3748 | ·2265 8159 | 3·2167 6807 |
| 41 | ·1717 1717 | ·1578 9474 | ·0530 7017 | ·2308 3457 | 3·2253 7087 |
| 42 | ·1683 1683 | ·1546 3918 | ·0521 3617 | ·2349 4706 | 3·2336 8375+ |
| 43 | ·1650 4854 | ·1515 1515+ | ·0512 3383 | ·2389 2552 | 3·2417 2060 |
| 44 | ·1619 0476 | ·1485 1485+ | ·0503 6161 | ·2427 7607 | 3·2494 9448 |
| 45 | ·1588 7850+ | ·1456 3107 | ·0495 1807 | ·2465 0448 | 3·2570 1766 |
| 46 | ·1559 6330 | ·1428 5714 | ·0487 0184 | ·2501 1622 | 3·2643 0173 |
| 47 | ·1531 5315+ | ·1401 8692 | ·0479 1166 | ·2536 1645+ | 3·2713 5759 |
| 48 | ·1504 4248 | ·1376 1468 | ·0471 4631 | ·2570 1007 | 3·2781 9552 |
| 49 | ·1478 2609 | ·1351 3514 | ·0464 0467 | ·2603 0168 | 3·2848 2520 |
| 50 | ·1452 9915− | ·1327 4336 | ·0456 8568 | ·2634 9566 | 3·2912 5580 |

$p = 9$

| $q$ | $\bar{x}$ | $\tilde{x}$ | $\sigma$ | $\beta_1$ | $\beta_2$ |
|---|---|---|---|---|---|
| 0·5 | ·9473 6842 | *J*-curve | ·0689 1091 | 5·0989 2880 | 9·5565 2174 |
| 1 | ·9000 0000$^e$ | 1·0000 0000$^e$ | ·0904 5340 | 2·1728 3951 | 5·5470 0855$^-$ |
| 1·5 | ·8571 4286 | ·9411 7647 | ·1031 8787 | 1·2266 6667 | 4·2592 5926 |
| 2 | ·8181 8182 | ·8888 8889 | ·1113 4044 | ·7731 7554 | 3·6483 5165$^-$ |
| 2·5 | ·7826 0870 | ·8421 0526 | ·1166 6442 | ·5151 6537 | 3·3056 6198 |
| 3 | ·7500 0000$^e$ | ·8000 0000$^e$ | ·1200 9612 | ·3537 4150$^-$ | 3·0952 3810 |
| 3·5 | ·7200 0000$^e$ | ·7619 0476 | ·1222 0202 | ·2466 4515$^+$ | 2·9590 0207 |
| 4 | ·6923 0769 | ·7272 7273 | ·1233 5134 | ·1728 3951 | 2·8680 5556 |
| 4·5 | ·6666 6667 | ·6956 5217 | ·1237 9689 | ·1207 0760 | 2·8064 5161 |
| 5 | ·6428 5714 | ·6666 6667 | ·1237 1792 | ·0833 3333 | 2·7647 0588 |
| 5·5 | ·6206 8966 | ·6400 0000$^e$ | ·1232 4484 | ·0563 5789 | 2·7368 4901 |
| 6 | ·6000 0000$^e$ | ·6153 8462 | ·1224 7449 | ·0369 0888 | 2·7189 5425$^-$ |
| 6·5 | ·5806 4516 | ·5925 9259 | ·1214 7985$^+$ | ·0230 2459 | 2·7083 4571 |
| 7 | ·5625 0000$^e$ | ·5714 2857 | ·1203 1668 | ·0133 2549 | 2·7031 4676 |
| 7·5 | ·5454 5455$^-$ | ·5517 2414 | ·1190 2794 | ·0068 1763 | 2·7020 0970 |
| 8 | ·5294 1176 | ·5333 3333 | ·1176 4706 | ·0027 7008 | 2·7039 4737 |
| 8·5 | ·5142 8571 | ·5161 2903 | ·1162 0018 | ·0006 3598 | 2·7082 2450$^+$ |
| 9 | ·5000 0000$^e$ | ·5000 0000$^e$ | ·1147 0787 | ·0000 0000$^e$ | 2·7142 8571 |
| 9·5 | ·4864 8649 | ·4848 4848 | ·1131 8634 | ·0005 4270 | 2·7217 0642 |
| 10 | ·4736 8421 | ·4705 8824 | ·1116 4844 | ·0020 1562 | 2·7301 5873 |
| 10·5 | ·4615 3846 | ·4571 4286 | ·1101 0432 | ·0042 2365$^-$ | 2·7393 8723 |
| 11 | ·4500 0000$^e$ | ·4444 4444 | ·1085 6203 | ·0070 1227 | 2·7491 9152 |
| 12 | ·4285 7143 | ·4210 5263 | ·1055 0699 | ·0138 6263 | 2·7699 2754 |
| 13 | ·4090 9091 | ·4000 0000$^e$ | ·1025 1947 | ·0218 4236 | 2·7914 5299 |
| 14 | ·3913 0435$^-$ | ·3809 5238 | ·0996 2121 | ·0304 7619 | 2·8131 8681 |
| 15 | ·3750 0000$^e$ | ·3636 3636 | ·0968 2458 | ·0394 4773 | 2·8347 5783 |
| 16 | ·3600 0000$^e$ | ·3478 2609 | ·0941 3574 | ·0485 4443 | 2·8559 3034 |
| 17 | ·3461 5385$^-$ | ·3333 3333 | ·0915 5677 | ·0576 2305$^-$ | 2·8765 5752 |
| 18 | ·3333 3333 | ·3200 0000$^e$ | ·0890 8708 | ·0665 8740 | 2·8965 5172 |
| 19 | ·3214 2857 | ·3076 9231 | ·0867 2434 | ·0753 7362 | 2·9158 6493 |
| 20 | ·3103 4483 | ·2962 9630 | ·0844 6516 | ·0839 4034 | 2·9344 7581 |
| 21 | ·3000 0000$^e$ | ·2857 1429 | ·0823 0549 | ·0922 6190 | 2·9523 8095$^+$ |
| 22 | ·2903 2258 | ·2758 6207 | ·0802 4092 | ·1003 2371 | 2·9695 8894 |
| 23 | ·2812 5000$^e$ | ·2666 6667 | ·0782 6692 | ·1081 1895$^+$ | 2·9861 1619 |
| 24 | ·2727 2727 | ·2580 6452 | ·0763 7891 | ·1156 4626 | 3·0019 8413 |
| 25 | ·2647 0588 | ·2500 0000$^e$ | ·0745 7243 | ·1229 0809 | 3·0172 1722 |
| 26 | ·2571 4286 | ·2424 2424 | ·0728 4314 | ·1299 0954 | 3·0318 4156 |
| 27 | ·2500 0000$^e$ | ·2352 9412 | ·0711 8685$^-$ | ·1366 5743 | 3·0458 8394 |
| 28 | ·2432 4324 | ·2285 7143 | ·0695 9960 | ·1431 5978 | 3·0593 7118 |
| 29 | ·2368 4211 | ·2222 2222 | ·0680 7762 | ·1494 2529 | 3·0723 2969 |
| 30 | ·2307 6923 | ·2162 1622 | ·0666 1734 | ·1554 6302 | 3·0847 8513 |
| 31 | ·2250 0000$^e$ | ·2105 2632 | ·0652 1540 | ·1612 8220 | 3·0967 6229 |
| 32 | ·2195 1220 | ·2051 2821 | ·0638 6865$^+$ | ·1668 9201 | 3·1082 8488 |
| 33 | ·2142 8571 | ·2000 0000$^e$ | ·0625 7411 | ·1723 0153 | 3·1193 7557 |
| 34 | ·2093 0233 | ·1951 2195$^+$ | ·0613 2901 | ·1775 1957 | 3·1300 5589 |
| 35 | ·2045 4545$^+$ | ·1904 7619 | ·0601 3071 | ·1825 5469 | 3·1403 4624 |
| 36 | ·2000 0000$^e$ | ·1860 4651 | ·0589 7678 | ·1874 1512 | 3·1502 6596 |
| 37 | ·1956 5217 | ·1818 1818 | ·0578 6492 | ·1921 0878 | 3·1598 3330 |
| 38 | ·1914 8936 | ·1777 7778 | ·0567 9297 | ·1966 4321 | 3·1690 6552 |
| 39 | ·1875 0000$^e$ | ·1739 1304 | ·0557 5891 | ·2010 2564 | 3·1779 7888 |
| 40 | ·1836 7347 | ·1702 1277 | ·0547 6085$^-$ | ·2052 6293 | 3·1865 8874 |
| 41 | ·1800 0000$^e$ | ·1666 6667 | ·0537 9700 | ·2093 6162 | 3·1949 0955$^+$ |
| 42 | ·1764 7059 | ·1632 6531 | ·0528 6571 | ·2133 2791 | 3·2029 5498 |
| 43 | ·1730 7692 | ·1600 0000$^e$ | ·0519 6539 | ·2171 6769 | 3·2107 3787 |
| 44 | ·1698 1132 | ·1568 6275$^-$ | ·0510 9458 | ·2208 8655$^+$ | 3·2182 7037 |
| 45 | ·1666 6667 | ·1538 4615$^+$ | ·0502 5189 | ·2244 8980 | 3·2255 6391 |
| 46 | ·1636 3636 | ·1509 4340 | ·0494 3602 | ·2279 8245$^-$ | 3·2326 2930 |
| 47 | ·1607 1429 | ·1481 4815$^-$ | ·0486 4573 | ·2313 6928 | 3·2394 7673 |
| 48 | ·1578 9474 | ·1454 5455$^-$ | ·0478 7988 | ·2346 5479 | 3·2461 1582 |
| 49 | ·1551 7241 | ·1428 5714 | ·0471 3736 | ·2378 4329 | 3·2525 5567 |
| 50 | ·1525 4237 | ·1403 5088 | ·0464 1717 | ·2409 3882 | 3·2588 0487 |

$p = 9{\cdot}5$

| $q$ | $\bar{x}$ | $\tilde{x}$ | $\sigma$ | $\beta_1$ | $\beta_2$ |
|---|---|---|---|---|---|
| 0·5 | ·9500 0000$^e$ | $J$-curve | ·0657 1287 | 5·2105 2632 | 9·7530 3644 |
| 1 | ·9047 6190 | 1·0000 0000$^e$ | ·0865 6126 | 2·2389 8947 | 5·6652 6316 |
| 1·5 | ·8636 3636 | ·9444 4444 | ·0990 6589 | 1·2756 1507 | 4·3481 7814 |
| 2 | ·8260 8696 | ·8947 3684 | ·1072 0720 | ·8122 1572 | 3·7205 0817 |
| 2·5 | ·7916 6667 | ·8500 0000$^e$ | ·1126 3643 | ·5473 6842 | 3·3663 1579 |
| 3 | ·7600 0000$^e$ | ·8095 2381 | ·1162 3731 | ·3807 4973 | 3·1471 8108 |
| 3·5 | ·7307 6923 | ·7727 2727 | ·1185 4642 | ·2694 7368 | 3·0039 4737 |
| 4 | ·7037 0370 | ·7391 3043 | ·1199 1514 | ·1921 7920 | 2·9071 6160 |
| 4·5 | ·6785 7143 | ·7083 3333 | ·1205 8531 | ·1370 6140 | 2·8405 5728 |
| 5 | ·6551 7241 | ·6800 0000$^e$ | ·1207 2938 | ·0970 8569 | 2·7944 4976 |
| 5·5 | ·6333 3333 | ·6538 4615$^+$ | ·1204 7360 | ·0678 1345$^-$ | 2·7627 3572 |
| 6 | ·6129 0323 | ·6296 2963 | ·1199 1231 | ·0463 1579 | 2·7413 9403 |
| 6·5 | ·5937 5000$^e$ | ·6071 4286 | ·1191 1708 | ·0305 8929 | 2·7276 7952 |
| 7 | ·5757 5758 | ·5862 0690 | ·1181 4296 | ·0192 2264 | 2·7196 6298 |
| 7·5 | ·5588 2353 | ·5666 6667 | ·1170 3271 | ·0111 9697 | 2·7159 5568 |
| 8 | ·5428 5714 | ·5483 8710 | ·1158 1982 | ·0057 6145$^-$ | 2·7155 3767 |
| 8·5 | ·5277 7778 | ·5312 5000$^e$ | ·1145 3071 | ·0023 5294 | 2·7176 4706 |
| 9 | ·5135 1351 | ·5151 5152 | ·1131 8634 | ·0005 4270 | 2·7217 0642 |
| 9·5 | ·5000 0000$^e$ | ·5000 0000$^e$ | ·1118 0340 | ·0000 0000$^e$ | 2·7272 7273 |
| 10 | ·4871 7949 | ·4857 1429 | ·1103 9522 | ·0004 6682 | 2·7340 0245$^-$ |
| 10·5 | ·4750 0000$^e$ | ·4722 2222 | ·1089 7247 | ·0017 3989 | 2·7416 2679 |
| 11 | ·4634 1463 | ·4594 5946 | ·1057 4372 | ·0036 5763 | 2·7499 3383 |
| 12 | ·4418 6047 | ·4358 9744 | ·1046 9422 | ·0089 3474 | 2·7679 5713 |
| 13 | ·4222 2222 | ·4146 3415 | ·1018 8659 | ·0155 3334 | 2·7870 9216 |
| 14 | ·4042 5532 | ·3953 4884 | ·0991 4594 | ·0229 4664 | 2·8067 0600 |
| 15 | ·3877 5510 | ·3777 7778 | ·0964 8755$^-$ | ·0308 3322 | 2·8263 8621 |
| 16 | ·3725 4902 | ·3617 0213 | ·0939 2013 | ·0389 6042 | 2·8458 6376 |
| 17 | ·3584 9057 | ·3469 3878 | ·0914 4796 | ·0471 6860 | 2·8649 6466 |
| 18 | ·3454 5455$^-$ | ·3333 3333 | ·0890 7235$^+$ | ·0553 4808 | 2·8835 7877 |
| 19 | ·3333 3333 | ·3207 5472 | ·0867 9261 | ·0634 2381 | 2·9016 3934 |
| 20 | ·3220 3390 | ·3090 9091 | ·0846 0672 | ·0713 4503 | 2·9191 0931 |
| 21 | ·3114 7541 | ·2982 4561 | ·0825 1185$^+$ | ·0790 7817 | 2·9359 7196 |
| 22 | ·3015 8730 | ·2881 3559 | ·0805 0464 | ·0866 0191 | 2·9522 2452 |
| 23 | ·2923 0769 | ·2786 8852 | ·0785 8144 | ·0939 0370 | 2·9678 7370 |
| 24 | ·2835 8209 | ·2698 4127 | ·0767 3846 | ·1009 7725$^-$ | 2·9829 3256 |
| 25 | ·2753 6232 | ·2615 3846 | ·0749 7190 | ·1078 2076 | 2·9974 1831 |
| 26 | ·2676 0563 | ·2537 3134 | ·0732 7801 | ·1144 3563 | 3·0113 5075$^+$ |
| 27 | ·2602 7397 | ·2463 7681 | ·0716 5313 | ·1208 2548 | 3·0247 5118 |
| 28 | ·2533 3333 | ·2394 3662 | ·0700 9373 | ·1269 9551 | 3·0376 4157 |
| 29 | ·2467 5325$^-$ | ·2328 7671 | ·0685 9643 | ·1329 5193 | 3·0500 4410 |
| 30 | ·2405 0633 | ·2266 6667 | ·0671 5801 | ·1387 0167 | 3·0619 8068 |
| 31 | ·2345 6790 | ·2207 7922 | ·0657 7541 | ·1442 5200 | 3·0734 7276 |
| 32 | ·2289 1566 | ·2151 8987 | ·0644 4574 | ·1496 1043 | 3·0845 4113 |
| 33 | ·2235 2941 | ·2098 7654 | ·0631 6627 | ·1547 8448 | 3·0952 0580 |
| 34 | ·2183 9080 | ·2048 1928 | ·0619 3441 | ·1597 8163 | 3·1054 8595$^-$ |
| 35 | ·2134 8315$^-$ | ·2000 0000$^e$ | ·0607 4774 | ·1646 0923 | 3·1153 9987 |
| 36 | ·2087 9121 | ·1954 0230 | ·0596 0397 | ·1692 7443 | 3·1249 6502 |
| 37 | ·2043 0108 | ·1910 1124 | ·0585 0095$^-$ | ·1737 8416 | 3·1341 9793 |
| 38 | ·2000 0000$^e$ | ·1868 1319 | ·0574 3665$^+$ | ·1781 4509 | 3·1431 1431 |
| 39 | ·1958 7629 | ·1827 9570 | ·0564 0918 | ·1823 6364 | 3·1517 2904 |
| 40 | ·1919 1919 | ·1789 4737 | ·0554 1673 | ·1864 4597 | 3·1600 5621 |
| 41 | ·1881 1881 | ·1752 5773 | ·0544 5763 | ·1903 9795$^-$ | 3·1681 0913 |
| 42 | ·1844 6602 | ·1717 1717 | ·0535 3029 | ·1942 2519 | 3·1759 0040 |
| 43 | ·1809 5238 | ·1683 1683 | ·0526 3321 | ·1979 3305$^+$ | 3·1834 4193 |
| 44 | ·1775 7009 | ·1650 4854 | ·0517 6499 | ·2015 2663 | 3·1907 4499 |
| 45 | ·1743 1193 | ·1619 0476 | ·0509 2430 | ·2050 1076 | 3·1978 2021 |
| 46 | ·1711 7117 | ·1588 7850$^+$ | ·0501 0990 | ·2083 9008 | 3·2046 7768 |
| 47 | ·1681 4159 | ·1559 6330 | ·0493 2060 | ·2116 6896 | 3·2113 2691 |
| 48 | ·1652 1739 | ·1531 5315$^+$ | ·0485 5530 | ·2148 5158 | 3·2177 7691 |
| 49 | ·1623 9316 | ·1504 4248 | ·0478 1294 | ·2179 4189 | 3·2240 3620 |
| 50 | ·1596 6387 | ·1478 2609 | ·0470 9254 | ·2209 4367 | 3·2301 1286 |

$p = 10$

| $q$ | $\bar{x}$ | $\tilde{x}$ | $\sigma$ | $\beta_1$ | $\beta_2$ |
|---|---|---|---|---|---|
| 0·5 | ·9523 8095$^+$ | *J*-curve | ·0627 9814 | 5·3139 2000$^e$ | 9·9360 0000$^e$ |
| 1 | ·9090 9091 | 1·0000 0000$^e$ | ·0829 8827 | 2·3005 9172 | 5·7758 2418 |
| 1·5 | ·8695 6522 | ·9473 6842 | ·0952 5610 | 1·3214 4490 | 4·4316 7305$^+$ |
| 2 | ·8333 3333 | ·9000 0000$^e$ | ·1033 6228 | ·8489 7959 | 3·7885 7143 |
| 2·5 | ·8000 0000$^e$ | ·8571 4286 | ·1088 6621 | ·5778 8347 | 3·4238 0423 |
| 3 | ·7692 3077 | ·8181 8182 | ·1126 0385$^+$ | ·4065 1852 | 3·1966 6667 |
| 3·5 | ·7407 4074 | ·7826 0870 | ·1150 8435$^-$ | ·2914 2263 | 3·0470 0461 |
| 4 | ·7142 8571 | ·7500 0000$^e$ | ·1166 4237 | ·2109 3750$^e$ | 2·9448 5294 |
| 4·5 | ·6896 5517 | ·7200 0000$^e$ | ·1175 0935$^-$ | ·1530 8642 | 2·8736 5079 |
| 5 | ·6666 6667 | ·6923 0769 | ·1178 5113 | ·1107 2664 | 2·8235 2941 |
| 5·5 | ·6451 6129 | ·6666 6667 | ·1177 8980 | ·0793 4694 | 2·7882 6255$^-$ |
| 6 | ·6250 0000$^e$ | ·6428 5714 | ·1174 1705$^+$ | ·0559 6708 | 2·7637 4269 |
| 6·5 | ·6060 6061 | ·6206 8966 | ·1168 0292 | ·0385 4582 | 2·7471 6136 |
| 7 | ·5882 3529 | ·6000 0000$^e$ | ·1160 0156 | ·0256 4306 | 2·7365 4135$^+$ |
| 7·5 | ·5714 2857 | ·5806 4516 | ·1150 5532 | ·0162 1740 | 2·7304 5654 |
| 8 | ·5555 5556 | ·5625 0000$^e$ | ·1139 9759 | ·0095 0000$^e$ | 2·7278 5714 |
| 8·5 | ·5405 4054 | ·5454 5455$^-$ | ·1128 5490 | ·0049 1304 | 2·7279 5703 |
| 9 | ·5263 1579 | ·5294 1176 | ·1116 4844 | ·0020 1562 | 2·7301 5873 |
| 9·5 | ·5128 2051 | ·5142 8571 | ·1103 9522 | ·0004 6682 | 2·7340 0245$^-$ |
| 10 | ·5000 0000$^e$ | ·5000 0000$^e$ | ·1091 0895$^-$ | ·0000 0000$^e$ | 2·7391 3043 |
| 10·5 | ·4878 0488 | ·4864 8649 | ·1078 0069 | ·0004 0447 | 2·7452 6174 |
| 11 | ·4761 9048 | ·4736 8421 | ·1064 7943 | ·0015 1229 | 2·7521 7391 |
| 12 | ·4545 4545$^+$ | ·4500 0000$^e$ | ·1038 2550$^-$ | ·0053 2407 | 2·7676 6667 |
| 13 | ·4347 8261 | ·4285 7143 | ·1011 9015$^-$ | ·0106 3385$^-$ | 2·7845 6805$^-$ |
| 14 | ·4166 6667 | ·4090 9091 | ·0986 0133 | ·0169 0617 | 2·8021 9780 |
| 15 | ·4000 0000$^e$ | ·3913 0435$^-$ | ·0960 7689 | ·0237 7686 | 2·8201 0582 |
| 16 | ·3846 1538 | ·3750 0000$^e$ | ·0936 2779 | ·0309 9490 | 2·8379 9261 |
| 17 | ·3703 7037 | ·3600 0000$^e$ | ·0912 6026 | ·0383 8568 | 2·8556 5923 |
| 18 | ·3571 4286 | ·3461 5385$^-$ | ·0889 7730 | ·0458 2716 | 2·8729 7491 |
| 19 | ·3448 2759 | ·3333 3333 | ·0867 7971 | ·0532 3402 | 2·8898 5569 |
| 20 | ·3333 3333 | ·3214 2857 | ·0846 6675$^+$ | ·0605 4688 | 2·9062 5000$^e$ |
| 21 | ·3225 8065$^-$ | ·3103 4483 | ·0826 3670 | ·0677 2487 | 2·9221 2885$^+$ |
| 22 | ·3125 0000$^e$ | ·3000 0000$^e$ | ·0806 8715$^+$ | ·0747 4048 | 2·9374 7899 |
| 23 | ·3030 3030 | ·2903 2258 | ·0788 1530 | ·0815 7587 | 2·9522 9814 |
| 24 | ·2941 1765$^-$ | ·2812 5000$^e$ | ·0770 1808 | ·0882 2016 | 2·9665 9159 |
| 25 | ·2857 1429 | ·2727 2727 | ·0752 9233 | ·0946 6764 | 2·9803 6984 |
| 26 | ·2777 7778 | ·2647 0588 | ·0736 3483 | ·1009 1626 | 2·9936 4684 |
| 27 | ·2702 7027 | ·2571 4286 | ·0720 4243 | ·1069 6666 | 3·0064 3875$^-$ |
| 28 | ·2631 5789 | ·2500 0000$^e$ | ·0705 1201 | ·1128 2143 | 3·0187 6307 |
| 29 | ·2564 1026 | ·2432 4324 | ·0690 4057 | ·1184 8448 | 3·0306 3799 |
| 30 | ·2500 0000$^e$ | ·2368 4211 | ·0676 2522 | ·1239 6070 | 3·0420 8195$^-$ |
| 31 | ·2439 0244 | ·2307 6923 | ·0662 6319 | ·1292 5557 | 3·0531 1328 |
| 32 | ·2380 9524 | ·2250 0000$^e$ | ·0649 5184 | ·1343 7500$^e$ | 3·0637 5000$^e$ |
| 33 | ·2325 5814 | ·2195 1220 | ·0636 8867 | ·1393 2510 | 3·0740 0966 |
| 34 | ·2272 7273 | ·2142 8571 | ·0624 7130 | ·1441 1209 | 3·0839 0923 |
| 35 | ·2222 2222 | ·2093 0233 | ·0612 9748 | ·1487 4216 | 3·0934 6505$^-$ |
| 36 | ·2173 9130 | ·2045 4545$^+$ | ·0601 6508 | ·1532 2145$^+$ | 3·1026 9274 |
| 37 | ·2127 6596 | ·2000 0000$^e$ | ·0590 7210 | ·1575 5597 | 3·1116 0728 |
| 38 | ·2083 3333 | ·1956 5217 | ·0580 1663 | ·1617 5158 | 3·1202 2291 |
| 39 | ·2040 8163 | ·1914 8936 | ·0569 9690 | ·1658 1394 | 3·1285 5320 |
| 40 | ·2000 0000$^e$ | ·1875 0000$^e$ | ·0560 1120 | ·1697 4852 | 3·1366 1103 |
| 41 | ·1960 7843 | ·1836 7347 | ·0550 5796 | ·1735 6059 | 3·1444 0865$^+$ |
| 42 | ·1923 0769 | ·1800 0000$^e$ | ·0541 3565$^+$ | ·1772 5521 | 3·1519 5767 |
| 43 | ·1886 7925$^-$ | ·1764 7059 | ·0532 4287 | ·1808 3721 | 3·1592 6910 |
| 44 | ·1851 8519 | ·1730 7692 | ·0523 7828 | ·1843 1122 | 3·1663 5338 |
| 45 | ·1818 1818 | ·1698 1132 | ·0515 4061 | ·1876 8168 | 3·1732 2041 |
| 46 | ·1785 7143 | ·1666 6667 | ·0507 2867 | ·1909 5280 | 3·1798 7955$^+$ |
| 47 | ·1754 3860 | ·1636 3636 | ·0499 4133 | ·1941 2861 | 3·1863 3970 |
| 48 | ·1724 1379 | ·1607 1429 | ·0491 7752 | ·1972 1296 | 3·1926 0929 |
| 49 | ·1694 9153 | ·1578 9474 | ·0484 3624 | ·2002 0951 | 3·1986 9630 |
| 50 | ·1666 6667 | ·1551 7241 | ·0477 1653 | ·2031 2175$^-$ | 3·2046 0829 |

$p = 10{\cdot}5$

| $q$ | $\bar{x}$ | $\tilde{x}$ | $\sigma$ | $\beta_1$ | $\beta_2$ |
|---|---|---|---|---|---|
| 0·5 | ·9545 4545$^+$ | *J*-curve | ·0601 3071 | 5·4099 7464 | 10·1067 5039 |
| 1 | ·9130 4348 | 1·0000 0000$^e$ | ·0796 9697 | 2·3580 9001 | 5·8794 0157 |
| 1·5 | ·8750 0000$^e$ | ·9500 0000$^e$ | ·0917 2492 | 1·3644 3149 | 4·5102 0408 |
| 2 | ·8400 0000$^e$ | ·9047 6190 | ·0997 7753 | ·8836 4192 | 3·8528 5238 |
| 2·5 | ·8076 9231 | ·8636 3636 | ·1053 3120 | ·6068 1481 | 3·4783 3333 |
| 3 | ·7777 7778 | ·8260 8696 | ·1091 7860 | ·4310 9856 | 3·2438 2070 |
| 3·5 | ·7500 0000$^e$ | ·7916 6667 | ·1118 0340 | ·3125 0000$^e$ | 3·0882 3529 |
| 4 | ·7241 3793 | ·7600 0000$^e$ | ·1135 2481 | ·2290 8741 | 2·9811 3791 |
| 4·5 | ·7000 0000$^e$ | ·7307 6923 | ·1145 6439 | ·1687 2631 | 2·9056 9561 |
| 5 | ·6774 1935$^+$ | ·7037 0370 | ·1150 8162 | ·1241 7493 | 2·8518 6983 |
| 5·5 | ·6562 5000$^e$ | ·6785 7143 | ·1151 9445$^-$ | ·0908 5565$^-$ | 2·8133 2118 |
| 6 | ·6363 6364 | ·6551 7241 | ·1149 9191 | ·0657 4142 | 2·7858 6279 |
| 6·5 | ·6176 4706 | ·6333 3333 | ·1145 4237 | ·0467 5657 | 2·7666 2811 |
| 7 | ·6000 0000$^e$ | ·6129 0323 | ·1138 9896 | ·0324 3480 | 2·7535 9600 |
| 7·5 | ·5833 3333 | ·5937 5000$^e$ | ·1131 0348 | ·0217 1429 | 2·7453 0612 |
| 8 | ·5675 6757 | ·5757 5758 | ·1121 8909 | ·0138 0981 | 2·7406 8147 |
| 8·5 | ·5526 3158 | ·5588 2353 | ·1111 8226 | ·0081 3024 | 2·7389 1375$^-$ |
| 9 | ·5384 6154 | ·5428 5714 | ·1101 0432 | ·0042 2365$^-$ | 2·7393 8723 |
| 9·5 | ·5250 0000$^e$ | ·5277 7778 | ·1089 7247 | ·0017 3989 | 2·7416 2679 |
| 10 | ·5121 9512 | ·5135 1351 | ·1078 0069 | ·0004 0447 | 2·7452 6174 |
| 10·5 | ·5000 0000$^e$ | ·5000 0000$^e$ | ·1066 0036 | ·0000 0000$^e$ | 2·7500 0000$^e$ |
| 11 | ·4883 7209 | ·4871 7949 | ·1053 8075$^-$ | ·0003 5275$^-$ | 2·7556 0957 |
| 12 | ·4666 6667 | ·4634 1463 | ·1029 1266 | ·0027 9645$^+$ | 2·7687 3607 |
| 13 | ·4468 0851 | ·4418 6047 | ·1004 4202 | ·0069 0070 | 2·7835 4535$^-$ |
| 14 | ·4285 7143 | ·4222 2222 | ·0979 9919 | ·0121 0395$^+$ | 2·7993 1389 |
| 15 | ·4117 6471 | ·4042 5532 | ·0956 0424 | ·0180 2125$^+$ | 2·8155 5707 |
| 16 | ·3962 2642 | ·3877 5510 | ·0932 7010 | ·0243 8479 | 2·8319 4745$^+$ |
| 17 | ·3818 1818 | ·3725 4902 | ·0910 0473 | ·0310 0625$^+$ | 2·8482 6317 |
| 18 | ·3684 2105$^+$ | ·3584 9057 | ·0888 1269 | ·0377 5227 | 2·8643 5448 |
| 19 | ·3559 3220 | ·3454 5455$^-$ | ·0866 9609 | ·0445 2805$^-$ | 2·8801 2155$^-$ |
| 20 | ·3442 6230 | ·3333 3333 | ·0846 5537 | ·0512 6627 | 2·8954 9943 |
| 21 | ·3333 3333 | ·3220 3390 | ·0826 8982 | ·0579 1936 | 2·9104 4776 |
| 22 | ·3230 7692 | ·3114 7541 | ·0807 9793 | ·0644 5406 | 2·9249 4360 |
| 23 | ·3134 3284 | ·3015 8730 | ·0789 7765$^+$ | ·0708 4762 | 2·9389 7632 |
| 24 | ·3043 4783 | ·2923 0769 | ·0772 2661 | ·0770 8495$^+$ | 2·9525 4403 |
| 25 | ·2957 7465$^-$ | ·2835 8209 | ·0755 4224 | ·0831 5666 | 2·9656 5096 |
| 26 | ·2876 7123 | ·2753 6232 | ·0739 2185$^-$ | ·0890 5751 | 2·9783 0560 |
| 27 | ·2800 0000$^e$ | ·2676 0563 | ·0723 6272 | ·0947 8538 | 2·9905 1935$^+$ |
| 28 | ·2727 2727 | ·2602 7397 | ·0708 6216 | ·1003 4040 | 3·0023 0552 |
| 29 | ·2658 2278 | ·2533 3333 | ·0694 1751 | ·1057 2434 | 3·0136 7859 |
| 30 | ·2592 5926 | ·2467 5325$^-$ | ·0680 2620 | ·1109 4019 | 3·0246 5372 |
| 31 | ·2530 1205$^-$ | ·2405 0633 | ·0666 8573 | ·1159 9176 | 3·0352 4634 |
| 32 | ·2470 5882 | ·2345 6790 | ·0653 9373 | ·1208 8346 | 3·0454 7188 |
| 33 | ·2413 7931 | ·2289 1566 | ·0641 4789 | ·1256 2006 | 3·0553 4558 |
| 34 | ·2359 5506 | ·2235 2941 | ·0629 4604 | ·1302 0660 | 3·0648 8232 |
| 35 | ·2307 6923 | ·2183 9080 | ·0617 8610 | ·1346 4820 | 3·0740 9658 |
| 36 | ·2258 0645$^+$ | ·2134 8315$^-$ | ·0606 6611 | ·1389 5004 | 3·0830 0234 |
| 37 | ·2210 5263 | ·2087 9121 | ·0595 8419 | ·1431 1727 | 3·0916 1302 |
| 38 | ·2164 9485$^-$ | ·2043 0108 | ·0585 3857 | ·1471 5497 | 3·0999 4153 |
| 39 | ·2121 2121 | ·2000 0000$^e$ | ·0575 2757 | ·1510 6809 | 3·1080 0018 |
| 40 | ·2079 2079 | ·1958 7629 | ·0565 4962 | ·1548 6146 | 3·1158 0075$^+$ |
| 41 | ·2038 8350$^-$ | ·1919 1919 | ·0556 0319 | ·1585 3978 | 3·1233 5445$^-$ |
| 42 | ·2000 0000$^e$ | ·1881 1881 | ·0546 8687 | ·1621 0757 | 3·1306 7196 |
| 43 | ·1962 6168 | ·1844 6602 | ·0537 9933 | ·1655 6918 | 3·1377 6344 |
| 44 | ·1926 6055$^+$ | ·1809 5238 | ·0529 3928 | ·1689 2881 | 3·1446 3855$^-$ |
| 45 | ·1891 8919 | ·1775 7009 | ·0521 0552 | ·1721 9048 | 3·1513 0647 |
| 46 | ·1858 4071 | ·1743 1193 | ·0512 9691 | ·1753 5804 | 3·1577 7593 |
| 47 | ·1826 0870 | ·1711 7117 | ·0505 1238 | ·1784 3519 | 3·1640 5521 |
| 48 | ·1794 8718 | ·1681 4159 | ·0497 5090 | ·1814 2545$^-$ | 3·1701 5219 |
| 49 | ·1764 7059 | ·1652 1739 | ·0490 1150$^+$ | ·1843 3220 | 3·1760 7433 |
| 50 | ·1735 5372 | ·1623 9316 | ·0482 9328 | ·1871 5867 | 3·1818 2875$^+$ |

$p = 11$

| $q$ | $\bar{x}$ | $\tilde{x}$ | $\sigma$ | $\beta_1$ | $\beta_2$ |
|---|---|---|---|---|---|
| 0·5 | ·9565 2174 | *J*-curve | ·0576 8043 | 5·4994 3883 | 10·2664 5768 |
| 1 | ·9166 6667 | 1·0000 0000$^e$ | ·0766 5552 | 2·4118 7384 | 5·9766 2338 |
| 1·5 | ·8800 0000$^e$ | ·9523 8095$^+$ | ·0884 4333 | 1·4048 2110 | 4·5841 8445$^-$ |
| 2 | ·8461 5385$^-$ | ·9090 9091 | ·0964 2818 | ·9163 6364 | 3·9136 3636 |
| 2·5 | ·8148 1481 | ·8695 6522 | ·1020 1137 | ·6342 6355$^+$ | 3·5300 9864 |
| 3 | ·7857 1429 | ·8333 3333 | ·1059 4569 | ·4545 4545$^+$ | 3·2887 7005$^+$ |
| 3·5 | ·7586 2069 | ·8000 0000$^e$ | ·1086 9173 | ·3327 2513 | 3·1277 1125$^-$ |
| 4 | ·7333 3333 | ·7692 3077 | ·1105 5416 | ·2466 1843 | 3·0160 4278 |
| 4·5 | ·7096 7742 | ·7407 4074 | ·1117 4521 | ·1839 4558 | 2·9366 7954 |
| 5 | ·6875 0000$^e$ | ·7142 8571 | ·1124 1827 | ·1373 7374 | 2·8794 2584 |
| 5·5 | ·6666 6667 | ·6896 5517 | ·1126 8723 | ·1022 6443 | 2·8378 3784 |
| 6 | ·6470 5882 | ·6666 6667 | ·1126 3848 | ·0755 4772 | 2·8076 5550$^+$ |
| 6·5 | ·6285 7143 | ·6451 6129 | ·1123 3863 | ·0551 1648 | 2·7859 5888 |
| 7 | ·6111 1111 | ·6250 0000$^e$ | ·1118 3972 | ·0394 8052 | 2·7706 8646 |
| 7·5 | ·5945 9459 | ·6060 6061 | ·1111 8289 | ·0275 5935$^+$ | 2·7603 4652 |
| 8 | ·5789 4737 | ·5882 3529 | ·1104 0093 | ·0185 5288 | 2·7538 3707 |
| 8·5 | ·5641 0256 | ·5714 2857 | ·1095 2021 | ·0118 5783 | 2·7503 2956 |
| 9 | ·5500 0000$^e$ | ·5555 5556 | ·1085 6203 | ·0070 1227 | 2·7491 9152 |
| 9·5 | ·5365 8537 | ·5405 4054 | ·1057 4372 | ·0036 5763 | 2·7499 3383 |
| 10 | ·5238 0952 | ·5263 1579 | ·1064 7943 | ·0015 1229 | 2·7521 7391 |
| 10·5 | ·5116 2791 | ·5128 2051 | ·1053 8075$^-$ | ·0003 5275$^-$ | 2·7556 0957 |
| 11 | ·5000 0000$^e$ | ·5000 0000$^e$ | ·1042 5721 | ·0000 0000$^e$ | 2·7600 0000$^e$ |
| 12 | ·4782 6087 | ·4761 9048 | ·1019 6556 | ·0011 6364 | 2·7709 0909 |
| 13 | ·4583 3333 | ·4545 4545$^+$ | ·0996 5217 | ·0041 3787 | 2·7837 5471 |
| 14 | ·4400 0000$^e$ | ·4347 8261 | ·0973 4949 | ·0083 3734 | 2·7977 7365$^+$ |
| 15 | ·4230 7692 | ·4166 6667 | ·0950 7947 | ·0133 5807 | 2·8124 4962 |
| 16 | ·4074 0741 | ·4000 0000$^e$ | ·0928 5677 | ·0189 1687 | 2·8274 2947 |
| 17 | ·3928 5714 | ·3846 1538 | ·0906 9091 | ·0248 1283 | 2·8424 7024 |
| 18 | ·3793 1034 | ·3703 7037 | ·0885 8781 | ·0309 0215$^+$ | 2·8574 0469 |
| 19 | ·3666 6667 | ·3571 4286 | ·0865 5079 | ·0370 8134 | 2·8721 1831 |
| 20 | ·3548 3871 | ·3448 2759 | ·0845 8136 | ·0432 7573 | 2·8865 3379 |
| 21 | ·3437 5000$^e$ | ·3333 3333 | ·0826 7973 | ·0494 3154 | 2·9006 0024 |
| 22 | ·3333 3333 | ·3225 8065$^-$ | ·0808 4521 | ·0555 1020 | 2·9142 8571 |
| 23 | ·3235 2941 | ·3125 0000$^e$ | ·0790 7648 | ·0614 8441 | 2·9275 7184 |
| 24 | ·3142 8571 | ·3030 3030 | ·0773 7179 | ·0673 3515$^-$ | 2·9404 5002 |
| 25 | ·3055 5556 | ·2941 1765$^-$ | ·0757 2913 | ·0730 4961 | 2·9529 1866 |
| 26 | ·2972 9730 | ·2857 1429 | ·0741 4630 | ·0786 1961 | 2·9649 8117 |
| 27 | ·2894 7368 | ·2777 7778 | ·0726 2102 | ·0840 4040 | 2·9766 4449 |
| 28 | ·2820 5128 | ·2702 7027 | ·0711 5096 | ·0893 0986 | 2·9879 1801 |
| 29 | ·2750 0000$^e$ | ·2631 5789 | ·0697 3381 | ·0944 2774 | 2·9988 1273 |
| 30 | ·2682 9268 | ·2564 1026 | ·0683 6731 | ·0993 9525$^+$ | 3·0093 4076 |
| 31 | ·2619 0476 | ·2500 0000$^e$ | ·0670 4921 | ·1042 1463 | 3·0195 1480 |
| 32 | ·2558 1395$^+$ | ·2439 0244 | ·0657 7737 | ·1088 8889 | 3·0293 4783 |
| 33 | ·2500 0000$^e$ | ·2380 9524 | ·0645 4972 | ·1134 2155$^+$ | 3·0388 5291 |
| 34 | ·2444 4444 | ·2325 5814 | ·0633 6426 | ·1178 1652 | 3·0480 4301 |
| 35 | ·2391 3043 | ·2272 7273 | ·0622 1908 | ·1220 7792 | 3·0569 3082 |
| 36 | ·2340 4255$^+$ | ·2222 2222 | ·0611 1237 | ·1262 1004 | 3·0655 2876 |
| 37 | ·2291 6667 | ·2173 9130 | ·0600 4238 | ·1302 1720 | 3·0738 4882 |
| 38 | ·2244 8980 | ·2127 6596 | ·0590 0748 | ·1341 0374 | 3·0819 0262 |
| 39 | ·2200 0000$^e$ | ·2083 3333 | ·0580 0608 | ·1378 7396 | 3·0897 0130 |
| 40 | ·2156 8627 | ·2040 8163 | ·0570 3671 | ·1415 3209 | 3·0972 5557 |
| 41 | ·2115 3846 | ·2000 0000$^e$ | ·0560 9795$^+$ | ·1450 8226 | 3·1045 7569 |
| 42 | ·2075 4717 | ·1960 7843 | ·0551 8846 | ·1485 2850$^-$ | 3·1116 7145$^-$ |
| 43 | ·2037 0370 | ·1923 0769 | ·0543 0696 | ·1518 7470 | 3·1185 5219 |
| 44 | ·2000 0000$^e$ | ·1886 7925$^-$ | ·0534 5225$^-$ | ·1551 2465$^+$ | 3·1252 2686 |
| 45 | ·1964 2857 | ·1851 8519 | ·0526 2319 | ·1582 8199 | 3·1317 0395$^-$ |
| 46 | ·1929 8246 | ·1818 1818 | ·0518 1869 | ·1613 5021 | 3·1379 9156 |
| 47 | ·1896 5517 | ·1785 7143 | ·0510 3772 | ·1643 3269 | 3·1440 9741 |
| 48 | ·1864 4068 | ·1754 3860 | ·0502 7931 | ·1672 3266 | 3·1500 2884 |
| 49 | ·1833 3333 | ·1724 1379 | ·0495 4253 | ·1700 5323 | 3·1557 9286 |
| 50 | ·1803 2787 | ·1694 9153 | ·0488 2651 | ·1727 9736 | 3·1613 9610 |

$p = 12$

| $q$ | $\bar{x}$ | $\tilde{x}$ | $\sigma$ | $\beta_1$ | $\beta_2$ |
|---|---|---|---|---|---|
| 0·5 | ·9600 0000$^e$ | $J$-curve | ·0533 3333 | 5·6611 1772 | 10·5567 2970 |
| 1 | ·9230 7692 | 1·0000 0000$^e$ | ·0712 1693 | 2·5096 2963 | 6·1541 6667 |
| 1·5 | ·8888 8889 | ·9565 2174 | ·0825 3126 | 1·4786 6805$^+$ | 4·7199 4135$^-$ |
| 2 | ·8571 4286 | ·9166 6667 | ·0903 5079 | ·9765 6250$^e$ | 4·0257 3529 |
| 2·5 | ·8275 8621 | ·8800 0000$^e$ | ·0959 4598 | ·6850 9336 | 3·6260 6061 |
| 3 | ·8000 0000$^e$ | ·8461 5385$^-$ | ·1000 0000$^e$ | ·4982 6990 | 3·3725 4902 |
| 3·5 | ·7741 9355$^-$ | ·8148 1481 | ·1029 3213 | ·3707 2886 | 3·2017 0987 |
| 4 | ·7500 0000$^e$ | ·7857 1429 | ·1050 2101 | ·2798 3539 | 3·0818 7134 |
| 4·5 | ·7272 7273 | ·7586 2069 | ·1064 6183 | ·2130 5089 | 2·9954 9550$^-$ |
| 5 | ·7058 8235$^+$ | ·7333 3333 | ·1073 9658 | ·1628 8089 | 2·9321 0526 |
| 5·5 | ·6857 1429 | ·7096 7742 | ·1079 3146 | ·1245 7912 | 2·8850 7021 |
| 6 | ·6666 6667 | ·6875 0000$^e$ | ·1081 4761 | ·0950 0000$^e$ | 2·8500 0000$^e$ |
| 6·5 | ·6486 4865$^-$ | ·6666 6667 | ·1081 0811 | ·0719 8096 | 2·8238 7975$^+$ |
| 7 | ·6315 7895$^-$ | ·6470 5882 | ·1078 6264 | ·0539 8985$^-$ | 2·8045 7638 |
| 7·5 | ·6153 8462 | ·6285 7143 | ·1074 5081 | ·0399 1347 | 2·7905 4264 |
| 8 | ·6000 0000$^e$ | ·6111 1111 | ·1069 0450$^-$ | ·0289 2562 | 2·7806 3241 |
| 8·5 | ·5853 6585$^+$ | ·5945 9459 | ·1062 4952 | ·0204 0184 | 2·7739 8137 |
| 9 | ·5714 2857 | ·5789 4737 | ·1055 0699 | ·0138 6263 | 2·7699 2754 |
| 9·5 | ·5581 3953 | ·5641 0256 | ·1046 9422 | ·0089 3474 | 2·7679 5713 |
| 10 | ·5454 5455$^-$ | ·5500 0000$^e$ | ·1038 2550$^-$ | ·0053 2407 | 2·7676 6667 |
| 10·5 | ·5333 3333 | ·5365 8537 | ·1029 1266 | ·0027 9645$^+$ | 2·7687 3607 |
| 11 | ·5217 3913 | ·5238 0952 | ·1019 6556 | ·0011 6364 | 2·7709 0909 |
| 12 | ·5000 0000$^e$ | ·5000 0000$^e$ | ·1000 0000$^e$ | ·0000 0000$^e$ | 2·7777 7778 |
| 13 | ·4800 0000$^e$ | ·4782 6087 | ·0979 7959 | ·0009 1449 | 2·7870 3704 |
| 14 | ·4615 3846 | ·4583 3333 | ·0959 3993 | ·0032 7988 | 2·7978 5362 |
| 15 | ·4444 4444 | ·4400 0000$^e$ | ·0939 0603 | ·0066 5874 | 2·8096 5517 |
| 16 | ·4285 7143 | ·4230 7692 | ·0918 9536 | ·0107 4074 | 2·8220 4301 |
| 17 | ·4137 9310 | ·4074 0741 | ·0899 2003 | ·0153 0269 | 2·8347 3672 |
| 18 | ·4000 0000$^e$ | ·3928 5714 | ·0879 8827 | ·0201 8229 | 2·8475 3788 |
| 19 | ·3870 9677 | ·3793 1034 | ·0861 0547 | ·0252 6058 | 2·8603 0584 |
| 20 | ·3750 0000$^e$ | ·3666 6667 | ·0842 7498 | ·0304 4983 | 2·8729 4118 |
| 21 | ·3636 3636 | ·3548 3871 | ·0824 9866 | ·0356 8513 | 2·8853 7415$^-$ |
| 22 | ·3529 4118 | ·3437 5000$^e$ | ·0807 7724 | ·0409 1844 | 2·8975 5665$^-$ |
| 23 | ·3428 5714 | ·3333 3333 | ·0791 1070 | ·0461 1427 | 2·9094 5637 |
| 24 | ·3333 3333 | ·3235 2941 | ·0774 9843 | ·0512 4654 | 2·9210 5263 |
| 25 | ·3243 2432 | ·3142 8571 | ·0759 3939 | ·0562 9630 | 2·9323 3333 |
| 26 | ·3157 8947 | ·3055 5556 | ·0744 3229 | ·0612 5000$^e$ | 2·9432 9268 |
| 27 | ·3076 9231 | ·2972 9730 | ·0729 7564 | ·0660 9822 | 2·9539 2954 |
| 28 | ·3000 0000$^e$ | ·2894 7368 | ·0715 6781 | ·0708 3468 | 2·9642 4616 |
| 29 | ·2926 8293 | ·2820 5128 | ·0702 0711 | ·0754 5551 | 2·9742 4728 |
| 30 | ·2857 1429 | ·2750 0000$^e$ | ·0688 9183 | ·0799 5868 | 2·9839 3939 |
| 31 | ·2790 6977 | ·2682 9268 | ·0676 2024 | ·0843 4356 | 2·9933 3022 |
| 32 | ·2727 2727 | ·2619 0476 | ·0663 9061 | ·0886 1059 | 3·0024 2831 |
| 33 | ·2666 6667 | ·2558 1395$^+$ | ·0652 0129 | ·0927 6102 | 3·0112 4275$^-$ |
| 34 | ·2608 6957 | ·2500 0000$^e$ | ·0640 5062 | ·0967 9670 | 3·0197 8291 |
| 35 | ·2553 1915$^-$ | ·2444 4444 | ·0629 3702 | ·1007 1994 | 3·0280 5831 |
| 36 | ·2500 0000$^e$ | ·2391 3043 | ·0618 5896 | ·1045 3333 | 3·0360 7843 |
| 37 | ·2448 9796 | ·2340 4255$^+$ | ·0608 1496 | ·1082 3973 | 3·0438 5268 |
| 38 | ·2400 0000$^e$ | ·2291 6667 | ·0598 0360 | ·1118 4211 | 3·0513 9027 |
| 39 | ·2352 9412 | ·2244 8980 | ·0588 2353 | ·1153 4354 | 3·0587 0021 |
| 40 | ·2307 6923 | ·2200 0000$^e$ | ·0578 7345$^-$ | ·1187 4714 | 3·0657 9125$^-$ |
| 41 | ·2264 1509 | ·2156 8627 | ·0569 5211 | ·1220 5604 | 3·0726 7184 |
| 42 | ·2222 2222 | ·2115 3846 | ·0560 5833 | ·1252 7332 | 3·0793 5016 |
| 43 | ·2181 8182 | ·2075 4717 | ·0551 9099 | ·1284 0206 | 3·0858 3407 |
| 44 | ·2142 8571 | ·2037 0370 | ·0543 4899 | ·1314 4525$^-$ | 3·0921 3113 |
| 45 | ·2105 2632 | ·2000 0000$^e$ | ·0535 3133 | ·1344 0582 | 3·0982 4859 |
| 46 | ·2068 9655$^+$ | ·1964 2857 | ·0527 3701 | ·1372 8663 | 3·1041 9340 |
| 47 | ·2033 8983 | ·1929 8246 | ·0519 6512 | ·1400 9046 | 3·1115 9560 |
| 48 | ·2000 0000$^e$ | ·1896 5517 | ·0512 1475$^+$ | ·1428 1998 | 3·1155 9140 |
| 49 | ·1967 2131 | ·1864 4068 | ·0504 8507 | ·1454 7779 | 3·1210 5705$^+$ |
| 50 | ·1935 4839 | ·1833 3333 | ·0497 7527 | ·1480 6641 | 3·1263 7500$^-$ |

$p = 13$

| $q$ | $\bar{x}$ | $\tilde{x}$ | $\sigma$ | $\beta_1$ | $\beta_2$ |
|---|---|---|---|---|---|
| 0·5 | ·9629 6296 | *J*-curve | ·0495 9512 | 5·8032 4982 | 10·8136 7020 |
| 1 | ·9285 7143 | 1·0000 0000$^{e}$ | ·0664 9638 | 2·5961 5385$^{-}$ | 6·3122 1719 |
| 1·5 | ·8965 5172 | ·9600 0000$^{e}$ | ·0773 5412 | 1·5444 8918 | 4·8414 9184 |
| 2 | ·8666 6667 | ·9230 7692 | ·0849 8366 | 1·0306 0953 | 4·1266 9683 |
| 2·5 | ·8387 0968 | ·8888 8889 | ·0905 4574 | ·7310 7692 | 3·7130 1455$^{+}$ |
| 3 | ·8125 0000$^{e}$ | ·8571 4286 | ·0946 6466 | ·5381 4498 | 3·4489 4287 |
| 3·5 | ·7878 7879 | ·8275 8621 | ·0977 2434 | ·4056 8635$^{+}$ | 3·2696 3058 |
| 4 | ·7647 0588 | ·8000 0000$^{e}$ | ·0999 8077 | ·3106 7547 | 3·1427 1255$^{+}$ |
| 4·5 | ·7428 5714 | ·7741 9355$^{-}$ | ·1016 1411 | ·2403 5020 | 3·0502 5577 |
| 5 | ·7222 2222 | ·7500 0000$^{e}$ | ·1027 5604 | ·1870 7692 | 2·9815 3846 |
| 5·5 | ·7027 0270 | ·7272 7273 | ·1035 0563 | ·1460 1698 | 2·9297 6847 |
| 6 | ·6842 1053 | ·7058 8235$^{+}$ | ·1039 3904 | ·1139 6011 | 2·8904 4289 |
| 6·5 | ·6666 6667 | ·6857 1429 | ·1041 1584 | ·0886 9659 | 2·8604 6512 |
| 7 | ·6500 0000$^{e}$ | ·6666 6667 | ·1040 8330 | ·0686 5861 | 2·8376 4062 |
| 7·5 | ·6341 4634 | ·6486 4865$^{-}$ | ·1038 7935$^{-}$ | ·0527 0529 | 2·8203 7461 |
| 8 | ·6190 4762 | ·6315 7895$^{-}$ | ·1035 3473 | ·0399 8837 | 2·8074 8328 |
| 8·5 | ·6046 5116 | ·6153 8462 | ·1030 7454 | ·0298 6548 | 2·7980 7177 |
| 9 | ·5909 0909 | ·6000 0000$^{e}$ | ·1025 1947 | ·0218 4236 | 2·7914 5299 |
| 9·5 | ·5777 7778 | ·5853 6585$^{+}$ | ·1018 8659 | ·0155 3334 | 2·7870 9216 |
| 10 | ·5652 1739 | ·5714 2857 | ·1011 9015$^{-}$ | ·0106 3385$^{-}$ | 2·7845 6805$^{-}$ |
| 10·5 | ·5531 9149 | ·5581 3953 | ·1004 4202 | ·0069 0070 | 2·7835 4535$^{-}$ |
| 11 | ·5416 6667 | ·5454 5455$^{-}$ | ·0996 5217 | ·0041 3787 | 2·7837 5471 |
| 12 | ·5200 0000$^{e}$ | ·5217 3913 | ·0979 7959 | ·0009 1449 | 2·7870 3704 |
| 13 | ·5000 0000$^{e}$ | ·5000 0000$^{e}$ | ·0962 2504 | ·0000 0000$^{e}$ | 2·7931 0345$^{-}$ |
| 14 | ·4814 8148 | ·4800 0000$^{e}$ | ·0944 2629 | ·0007 3173 | 2·8010 6101 |
| 15 | ·4642 8571 | ·4615 3846 | ·0926 1051 | ·0026 4387 | 2·8102 8950$^{-}$ |
| 16 | ·4482 7586 | ·4444 4444 | ·0907 9732 | ·0054 0303 | 2·8203 5127 |
| 17 | ·4333 3333 | ·4285 7143 | ·0890 0083 | ·0087 6697 | 2·8309 3377 |
| 18 | ·4193 5484 | ·4137 9310 | ·0872 3108 | ·0125 5759 | 2·8418 1178 |
| 19 | ·4062 5000$^{e}$ | ·4000 0000$^{e}$ | ·0854 9516 | ·0166 4262 | 2·8528 2210 |
| 20 | ·3939 3939 | ·3870 9677 | ·0837 9793 | ·0209 2308 | 2·8638 4615$^{+}$ |
| 21 | ·3823 5294 | ·3750 0000$^{e}$ | ·0821 4259 | ·0253 2447 | 2·8747 9787 |
| 22 | ·3714 2857 | ·3636 3636 | ·0805 3112 | ·0297 9052 | 2·8856 1510 |
| 23 | ·3611 1111 | ·3529 4118 | ·0789 6456 | ·0342 7862 | 2·8962 5337 |
| 24 | ·3513 5135$^{+}$ | ·3428 5714 | ·0774 4329 | ·0387 5655$^{+}$ | 2·9066 8146 |
| 25 | ·3421 0526 | ·3333 3333 | ·0759 6714 | ·0432 0000$^{e}$ | 2·9168 7805$^{-}$ |
| 26 | ·3333 3333 | ·3243 2432 | ·0745 3560 | ·0475 9072 | 2·9268 2927 |
| 27 | ·3250 0000$^{e}$ | ·3157 8947 | ·0731 4786 | ·0519 1516 | 2·9365 2687 |
| 28 | ·3170 7317 | ·3076 9231 | ·0718 0293 | ·0561 6341 | 2·9459 6682 |
| 29 | ·3095 2381 | ·3000 0000$^{e}$ | ·0704 9968 | ·0603 2839 | 2·9551 4830 |
| 30 | ·3023 2558 | ·2926 8293 | ·0692 3688 | ·0644 0519 | 2·9640 7284 |
| 31 | ·2954 5455$^{-}$ | ·2857 1429 | ·0680 1326 | ·0683 9066 | 2·9727 4374 |
| 32 | ·2888 8889 | ·2790 6977 | ·0668 2750$^{+}$ | ·0722 8297 | 2·9811 6561 |
| 33 | ·2826 0870 | ·2727 2727 | ·0656 7830 | ·0760 8133 | 2·9893 4399 |
| 34 | ·2765 9574 | ·2666 6667 | ·0645 6433 | ·0797 8576 | 2·9972 8507 |
| 35 | ·2708 3333 | ·2608 6957 | ·0634 8431 | ·0833 9692 | 3·0049 9548 |
| 36 | ·2653 0612 | ·2553 1915$^{-}$ | ·0624 3697 | ·0869 1595$^{-}$ | 3·0124 8211 |
| 37 | ·2600 0000$^{e}$ | ·2500 0000$^{e}$ | ·0614 2108 | ·0903 4433 | 3·0197 5203 |
| 38 | ·2549 0196 | ·2448 9796 | ·0604 3543 | ·0936 8384 | 3·0268 1231 |
| 39 | ·2500 0000$^{e}$ | ·2400 0000$^{e}$ | ·0594 7887 | ·0969 3644 | 3·0336 7003 |
| 40 | ·2452 8302 | ·2352 9412 | ·0585 5027 | ·1001 0426 | 3·0403 3217 |
| 41 | ·2407 4074 | ·2307 6923 | ·0576 4856 | ·1031 8949 | 3·0468 0557 |
| 42 | ·2363 6364 | ·2264 1509 | ·0567 7271 | ·1061 9441 | 3·0530 9693 |
| 43 | ·2321 4286 | ·2222 2222 | ·0559 2171 | ·1091 2131 | 3·0592 1278 |
| 44 | ·2280 7018 | ·2181 8182 | ·0550 9462 | ·1119 7249 | 3·0651 5942 |
| 45 | ·2241 3793 | ·2142 8571 | ·0542 9051 | ·1147 5024 | 3·0709 4297 |
| 46 | ·2203 3898 | ·2105 2632 | ·0535 0852 | ·1174 5683 | 3·0765 6935$^{+}$ |
| 47 | ·2166 6667 | ·2068 9655$^{+}$ | ·0527 4781 | ·1200 9449 | 3·0820 4424 |
| 48 | ·2131 1475$^{+}$ | ·2033 8983 | ·0520 0758 | ·1226 6540 | 3·0873 7313 |
| 49 | ·2096 7742 | ·2000 0000$^{e}$ | ·0512 8705$^{+}$ | ·1251 7170 | 3·0925 6128 |
| 50 | ·2063 4921 | ·1967 2131 | ·0505 8551 | ·1276 1548 | 3·0976 1377 |

$p = 14$

| $q$ | $\bar{x}$ | $\tilde{x}$ | $\sigma$ | $\beta_1$ | $\beta_2$ |
|---|---|---|---|---|---|
| 0·5 | ·9655 1724 | $J$-curve | ·0463 4631 | 5·9291 6175$^{-}$ | 11·0426 7161 |
| 1 | ·9333 3333 | 1·0000 0000$^{e}$ | ·0623 6096 | 2·6732 5754 | 6·4537 8151 |
| 1·5 | ·9032 2581 | ·9629 6296 | ·0727 8401 | 1·6034 9854 | 4·9509 1009 |
| 2 | ·8750 0000$^{e}$ | ·9285 7143 | ·0802 1112 | 1·0793 6508 | 4·2180 4511 |
| 2·5 | ·8484 8485$^{-}$ | ·8965 5172 | ·0857 0991 | ·7728 2688 | 3·7920 9979 |
| 3 | ·8235 2941 | ·8666 6667 | ·0898 5443 | ·5745 9438 | 3·5187 9699 |
| 3·5 | ·8000 0000$^{e}$ | ·8387 0968 | ·0929 9811 | ·4378 6982 | 3·3320 8255$^{+}$ |
| 4 | ·7777 7778 | ·8125 0000$^{e}$ | ·0953 7723 | ·3392 8571 | 3·1989 7959 |
| 4·5 | ·7567 5676 | ·7878 7879 | ·0971 5860 | ·2658 8482 | 3·1012 0736 |
| 5 | ·7368 4211 | ·7647 0588 | ·0984 6467 | ·2099 1254 | 3·0278 2931 |
| 5·5 | ·7179 4872 | ·7428 5714 | ·0993 8797 | ·1664 5010 | 2·9719 1181 |
| 6 | ·7000 0000$^{e}$ | ·7222 2222 | ·1000 0000$^{e}$ | ·1322 3140 | 2·9288 5375$^{+}$ |
| 6·5 | ·6829 2683 | ·7027 0270 | ·1003 5698 | ·1050 0611 | 2·8954 8749 |
| 7 | ·6666 6667 | ·6842 1053 | ·1005 0378 | ·0831 7580 | 2·8695 6522 |
| 7·5 | ·6511 6279 | ·6666 6667 | ·1004 7660 | ·0655 7589 | 2·8494 5103 |
| 8 | ·6363 6364 | ·6500 0000$^{e}$ | ·1003 0496 | ·0513 3929 | 2·8339 2857 |
| 8·5 | ·6222 2222 | ·6341 4634 | ·1000 1313 | ·0398 0834 | 2·8220 7673 |
| 9 | ·6086 9565$^{+}$ | ·6190 4762 | ·0996 2121 | ·0304 7619 | 2·8131 8681 |
| 9·5 | ·5957 4468 | ·6046 5116 | ·0991 4594 | ·0229 4664 | 2·8067 0600 |
| 10 | ·5833 3333 | ·5909 0909 | ·0986 0133 | ·0169 0617 | 2·8021 9780 |
| 10·5 | ·5714 2857 | ·5777 7778 | ·0979 9919 | ·0121 0395$^{+}$ | 2·7993 1389 |
| 11 | ·5600 0000$^{e}$ | ·5652 1739 | ·0973 4949 | ·0083 3734 | 2·7977 7365$^{+}$ |
| 12 | ·5384 6154 | ·5416 6667 | ·0959 3993 | ·0032 7988 | 2·7978 5362 |
| 13 | ·5185 1852 | ·5200 0000$^{e}$ | ·0944 2629 | ·0007 3173 | 2·8010 6101 |
| 14 | ·5000 0000$^{e}$ | ·5000 0000$^{e}$ | ·0928 4767 | ·0000 0000$^{e}$ | 2·8064 5161 |
| 15 | ·4827 5862 | ·4814 8148 | ·0912 3280 | ·0005 9462 | 2·8133 6406 |
| 16 | ·4666 6667 | ·4642 8571 | ·0896 0287 | ·0021 6239 | 2·8213 2711 |
| 17 | ·4516 1290 | ·4482 7586 | ·0879 7348 | ·0044 4475$^{+}$ | 2·8300 0045$^{-}$ |
| 18 | ·4375 0000$^{e}$ | ·4333 3333 | ·0863 5616 | ·0072 4996 | 2·8391 3565$^{+}$ |
| 19 | ·4242 4242 | ·4193 5484 | ·0847 5931 | ·0104 3425$^{-}$ | 2·8485 4995$^{-}$ |
| 20 | ·4117 6471 | ·4062 5000$^{e}$ | ·0831 8903 | ·0138 8889 | 2·8581 0811 |
| 21 | ·4000 0000$^{e}$ | ·3939 3939 | ·0816 4966 | ·0175 3104 | 2·8677 0982 |
| 22 | ·3888 8889 | ·3823 5294 | ·0801 4418 | ·0212 9726 | 2·8772 8061 |
| 23 | ·3783 7838 | ·3714 2857 | ·0786 7458 | ·0251 3874 | 2·8867 6541 |
| 24 | ·3684 2105$^{+}$ | ·3611 1111 | ·0772 4204 | ·0290 1786 | 2·8961 2369 |
| 25 | ·3589 7436 | ·3513 5135$^{+}$ | ·0758 4718 | ·0329 0558 | 2·9053 2603 |
| 26 | ·3500 0000$^{e}$ | ·3421 0526 | ·0744 9014 | ·0367 7955$^{-}$ | 2·9143 5143 |
| 27 | ·3414 6341 | ·3333 3333 | ·0731 7073 | ·0406 2256 | 2·9231 8534 |
| 28 | ·3333 3333 | ·3250 0000$^{e}$ | ·0718 8852 | ·0444 2149 | 2·9318 1818 |
| 29 | ·3255 8140 | ·3170 7317 | ·0706 4285$^{+}$ | ·0481 6639 | 2·9402 4416 |
| 30 | ·3181 8182 | ·3095 2381 | ·0694 3296 | ·0518 4985$^{+}$ | 2·9484 6042 |
| 31 | ·3111 1111 | ·3023 2558 | ·0682 5798 | ·0554 6643 | 2·9564 6632 |
| 32 | ·3043 4783 | ·2954 5455$^{-}$ | ·0671 1696 | ·0590 1228 | 2·9642 6294 |
| 33 | ·2978 7234 | ·2888 8889 | ·0660 0890 | ·0624 8479 | 2·9718 5264 |
| 34 | ·2916 6667 | ·2826 0870 | ·0649 3281 | ·0658 8235$^{+}$ | 2·9792 3875$^{+}$ |
| 35 | ·2857 1429 | ·2765 9574 | ·0638 8766 | ·0692 0415$^{+}$ | 2·9864 2534 |
| 36 | ·2800 0000$^{e}$ | ·2708 3333 | ·0628 7242 | ·0724 4999 | 2·9934 1696 |
| 37 | ·2745 0980 | ·2653 0612 | ·0618 8609 | ·0756 2015$^{-}$ | 3·0002 1855$^{-}$ |
| 38 | ·2692 3077 | ·2600 0000$^{e}$ | ·0609 2767 | ·0787 1531 | 3·0068 3527 |
| 39 | ·2641 5094 | ·2549 0196 | ·0599 9618 | ·0817 3645$^{-}$ | 3·0132 7244 |
| 40 | ·2592 5926 | ·2500 0000$^{e}$ | ·0590 9067 | ·0846 8477 | 3·0195 3545$^{-}$ |
| 41 | ·2545 4545$^{+}$ | ·2452 8302 | ·0582 1022 | ·0875 6165$^{+}$ | 3·0256 2968 |
| 42 | ·2500 0000$^{e}$ | ·2407 4074 | ·0573 5393 | ·0903 6861 | 3·0315 6049 |
| 43 | ·2456 1404 | ·2363 6364 | ·0565 2094 | ·0931 0724 | 3·0373 3318 |
| 44 | ·2413 7931 | ·2321 4286 | ·0557 1041 | ·0957 7922 | 3·0429 5295$^{-}$ |
| 45 | ·2372 8814 | ·2280 7018 | ·0549 2153 | ·0983 8625$^{+}$ | 3·0484 2487 |
| 46 | ·2333 3333 | ·2241 3793 | ·0541 5353 | ·1009 3006 | 3·0537 5390 |
| 47 | ·2295 0820 | ·2203 3898 | ·0534 0566 | ·1034 1239 | 3·0589 4485$^{+}$ |
| 48 | ·2258 0645$^{+}$ | ·2166 6667 | ·0526 7720 | ·1058 3496 | 3·0640 0240 |
| 49 | ·2222 2222 | ·2131 1475$^{+}$ | ·0519 6746 | ·1081 9949 | 3·0689 3107 |
| 50 | ·2187 5000$^{e}$ | ·2096 7742 | ·0512 7579 | ·1105 0767 | 3·0737 3522 |

$p = 15$

| $q$ | $\bar{x}$ | $\tilde{x}$ | $\sigma$ | $\beta_1$ | $\beta_2$ |
|---|---|---|---|---|---|
| 0·5 | ·9677 4194 | *J*-curve | ·0434 9677 | 6·0414 6939 | 11·2480 3089 |
| 1 | ·9375 0000$^{e}$ | 1·0000 0000$^{e}$ | ·0587 0853 | 2·7423 8683 | 6·5812 8655$^{-}$ |
| 1·5 | ·9090 9091 | ·9655 1724 | ·0687 2081 | 1·6566 8371 | 5·0498 9605$^{-}$ |
| 2 | ·8823 5294 | ·9333 3333 | ·0759 4085$^{-}$ | 1·1235 4571 | 4·3010 5263 |
| 2·5 | ·8571 4286 | ·9032 2581 | ·0813 5640 | ·8108 7004 | 3·8642 9018 |
| 3 | ·8333 3333 | ·8750 0000$^{e}$ | ·0854 9820 | ·6080 0000$^{e}$ | 3·5828 5714 |
| 3·5 | ·8108 1081 | ·8484 8485$^{-}$ | ·0886 9326 | ·4675 4483 | 3·3896 2807 |
| 4 | ·7894 7368 | ·8235 2941 | ·0911 6057 | ·3658 3522 | 3·2510 8225$^{+}$ |
| 4·5 | ·7692 3077 | ·8000 0000$^{e}$ | ·0930 5514 | ·2897 4220 | 3·1486 3049 |
| 5 | ·7500 0000$^{e}$ | ·7777 7778 | ·0944 9112 | ·2314 0496 | 3·0711 4625$^{-}$ |
| 5·5 | ·7317 0732 | ·7567 5676 | ·0955 5510 | ·1858 3464 | 3·0115 7103 |
| 6 | ·7142 8571 | ·7368 4211 | ·0963 1427 | ·1497 1645$^{-}$ | 2·9652 1739 |
| 6·5 | ·6976 7442 | ·7179 4872 | ·0968 2167 | ·1207 6470 | 2·9288 5534 |
| 7 | ·6818 1818 | ·7000 0000$^{e}$ | ·0971 1986 | ·0973 5450$^{-}$ | 2·9001 9048 |
| 7·5 | ·6666 6667 | ·6829 2683 | ·0972 4333 | ·0783 0071 | 2·8775 5102 |
| 8 | ·6521 7391 | ·6666 6667 | ·0972 2035$^{-}$ | ·0627 2000$^{e}$ | 2·8596 9231 |
| 8·5 | ·6382 9787 | ·6511 6279 | ·0970 7428 | ·0499 4158 | 2·8456 7039 |
| 9 | ·6250 0000$^{e}$ | ·6363 6364 | ·0968 2458 | ·0394 4773 | 2·8347 5783 |
| 9·5 | ·6122 4490 | ·6222 2222 | ·0964 8755$^{-}$ | ·0308 3322 | 2·8263 8621 |
| 10 | ·6000 0000$^{e}$ | ·6086 9565$^{+}$ | ·0960 7689 | ·0237 7686 | 2·8201 0582 |
| 10·5 | ·5882 3529 | ·5957 4468 | ·0956 0424 | ·0180 2125$^{+}$ | 2·8155 5707 |
| 11 | ·5769 2308 | ·5833 3333 | ·0950 7947 | ·0133 5807 | 2·8124 4962 |
| 12 | ·5555 5556 | ·5600 0000$^{e}$ | ·0939 0603 | ·0066 5874 | 2·8096 5517 |
| 13 | ·5357 1429 | ·5384 6154 | ·0926 1051 | ·0026 4387 | 2·8102 8950$^{-}$ |
| 14 | ·5172 4138 | ·5185 1852 | ·0912 3280 | ·0005 9462 | 2·8133 6406 |
| 15 | ·5000 0000$^{e}$ | ·5000 0000$^{e}$ | ·0898 0265$^{+}$ | ·0000 0000$^{e}$ | 2·8181 8182 |
| 16 | ·4838 7097 | ·4827 5862 | ·0883 4235$^{-}$ | ·0004 8975$^{-}$ | 2·8242 4242 |
| 17 | ·4687 5000$^{e}$ | ·4666 6667 | ·0868 6866 | ·0017 9117 | 2·8311 8141 |
| 18 | ·4545 4545$^{+}$ | ·4516 1290 | ·0853 9422 | ·0037 0068 | 2·8387 3016 |
| 19 | ·4411 7647 | ·4375 0000$^{e}$ | ·0839 2851 | ·0060 6454 | 2·8466 8879 |
| 20 | ·4285 7143 | ·4242 4242 | ·0824 7861 | ·0087 6552 | 2·8549 0754 |
| 21 | ·4166 6667 | ·4117 6471 | ·0810 4979 | ·0117 1349 | 2·8632 7357 |
| 22 | ·4054 0541 | ·4000 0000$^{e}$ | ·0796 4590 | ·0148 3872 | 2·8717 0163 |
| 23 | ·3947 3684 | ·3888 8889 | ·0782 6970 | ·0180 8696 | 2·8801 2725$^{+}$ |
| 24 | ·3846 1538 | ·3783 7838 | ·0769 2308 | ·0214 1582 | 2·8885 0174 |
| 25 | ·3750 0000$^{e}$ | ·3684 2105$^{+}$ | ·0756 0730 | ·0247 9214 | 2·8967 8848 |
| 26 | ·3658 5366 | ·3589 7436 | ·0743 2310 | ·0281 8987 | 2·9049 6016 |
| 27 | ·3571 4286 | ·3500 0000$^{e}$ | ·0730 7082 | ·0315 8861 | 2·9129 9663 |
| 28 | ·3488 3721 | ·3414 6341 | ·0718 5051 | ·0349 7237 | 2·9208 8337 |
| 29 | ·3409 0909 | ·3333 3333 | ·0706 6196 | ·0383 2866 | 2·9286 1016 |
| 30 | ·3333 3333 | ·3255 8140 | ·0695 0480 | ·0416 4780 | 2·9361 7021 |
| 31 | ·3260 8696 | ·3181 8182 | ·0683 7852 | ·0449 2234 | 2·9435 5936 |
| 32 | ·3191 4894 | ·3111 1111 | ·0672 8250$^{+}$ | ·0481 4661 | 2·9507 7551 |
| 33 | ·3125 0000$^{e}$ | ·3043 4783 | ·0662 1606 | ·0513 1636 | 2·9578 1818 |
| 34 | ·3061 2245$^{-}$ | ·2978 7234 | ·0651 7845$^{-}$ | ·0544 2854 | 2·9646 8814 |
| 35 | ·3000 0000$^{e}$ | ·2916 6667 | ·0641 6889 | ·0574 8098 | 2·9713 8710 |
| 36 | ·2941 1765$^{-}$ | ·2857 1429 | ·0631 8661 | ·0604 7229 | 2·9779 1754 |
| 37 | ·2884 6154 | ·2800 0000$^{e}$ | ·0622 3078 | ·0634 0167 | 2·9842 8246 |
| 38 | ·2830 1887 | ·2745 0980 | ·0613 0059 | ·0662 6881 | 2·9904 8530 |
| 39 | ·2777 7778 | ·2692 3077 | ·0603 9526 | ·0690 7378 | 2·9965 2979 |
| 40 | ·2727 2727 | ·2641 5094 | ·0595 1397 | ·0718 1697 | 3·0024 1984 |
| 41 | ·2678 5714 | ·2592 5926 | ·0586 5595$^{+}$ | ·0744 9900 | 3·0081 5954 |
| 42 | ·2631 5789 | ·2545 4545$^{+}$ | ·0578 2044 | ·0771 2070 | 3·0137 5303 |
| 43 | ·2586 2069 | ·2500 0000$^{e}$ | ·0570 0669 | ·0796 8303 | 3·0192 0447 |
| 44 | ·2542 3729 | ·2456 1404 | ·0562 1398 | ·0821 8710 | 3·0245 1805$^{+}$ |
| 45 | ·2500 0000$^{e}$ | ·2413 7931 | ·0554 4160 | ·0846 3406 | 3·0296 9790 |
| 46 | ·2459 0164 | ·2372 8814 | ·0546 8886 | ·0870 2517 | 3·0347 4810 |
| 47 | ·2419 3548 | ·2333 3333 | ·0539 5511 | ·0893 6170 | 3·0396 7267 |
| 48 | ·2380 9524 | ·2295 0820 | ·0532 3971 | ·0916 4497 | 3·0444 7552 |
| 49 | ·2343 7500$^{e}$ | ·2258 0645$^{+}$ | ·0525 4205$^{+}$ | ·0938 7630 | 3·0491 6051 |
| 50 | ·2307 6923 | ·2222 2222 | ·0518 6153 | ·0960 5703 | 3·0537 3134 |

$p = 16$

| $q$ | $\bar{x}$ | $\tilde{x}$ | $\sigma$ | $\beta_1$ | $\beta_2$ |
|---|---|---|---|---|---|
| 0·5 | ·9696 9697 | $J$-curve | ·0409 7718 | 6·1422 5712 | 11·4332 1206 |
| 1 | ·9411 7647 | 1·0000 0000$^e$ | ·0554 5936 | 2·8047 0914 | 6·6967 1053 |
| 1·5 | ·9142 8571 | ·9677 4194 | ·0650 8512 | 1·7048 5426 | 5·1398 5303 |
| 2 | ·8888 8889 | ·9375 0000$^e$ | ·0720 9841 | 1·1637 5000$^e$ | 4·3767 8571 |
| 2·5 | ·8648 6486 | ·9090 9091 | ·0774 1782 | ·8456 5735− | 3·9304 1690 |
| 3 | ·8421 0526 | ·8823 5294 | ·0815 3649 | ·6386 9992 | 3·6417 7489 |
| 3·5 | ·8205 1282 | ·8571 4286 | ·0847 5835− | ·4949 5866 | 3·4427 7409 |
| 4 | ·8000 0000$^e$ | ·8333 3333 | ·0872 8716 | ·3904 9587 | 3·2994 0711 |
| 4·5 | ·7804 8780 | ·8108 1081 | ·0892 6746 | ·3120 3018 | 3·1928 0930 |
| 5 | ·7619 0476 | ·7894 7368 | ·0908 0596 | ·2516 0681 | 3·1116 8478 |
| 5·5 | ·7441 8605− | ·7692 3077 | ·0919 8386 | ·2041 7507 | 3·0488 6413 |
| 6 | ·7272 7273 | ·7500 0000$^e$ | ·0928 6435− | ·1663 7731 | 2·9995 8333 |
| 6·5 | ·7111 1111 | ·7317 0732 | ·0934 9755− | ·1358 9690 | 2·9605 5730 |
| 7 | ·6956 5217 | ·7142 8571 | ·0939 2378 | ·1110 8571 | 2·9294 5055− |
| 7·5 | ·6808 5106 | ·6976 7442 | ·0941 7589 | ·0907 4074 | 2·9045 5975− |
| 8 | ·6666 6667 | ·6818 1818 | ·0942 8090 | ·0739 6450− | 2·8846 1538 |
| 8·5 | ·6530 6122 | ·6666 6667 | ·0942 6127 | ·0600 7476 | 2·8686 5352 |
| 9 | ·6400 0000$^e$ | ·6521 7391 | ·0941 3574 | ·0485 4443 | 2·8559 3034 |
| 9·5 | ·6274 5098 | ·6382 9787 | ·0939 2013 | ·0389 6042 | 2·8458 6376 |
| 10 | ·6153 8462 | ·6250 0000$^e$ | ·0936 2779 | ·0309 9490 | 2·8379 9261 |
| 10·5 | ·6037 7358 | ·6122 4490 | ·0932 7010 | ·0243 8479 | 2·8319 4745+ |
| 11 | ·5925 9259 | ·6000 0000$^e$ | ·0928 5677 | ·0189 1687 | 2·8274 2947 |
| 12 | ·5714 2857 | ·5769 2308 | ·0918 9536 | ·0107 4074 | 2·8220 4301 |
| 13 | ·5517 2414 | ·5555 5556 | ·0907 9732 | ·0054 0303 | 2·8203 5127 |
| 14 | ·5333 3333 | ·5357 1429 | ·0896 0287 | ·0021 6239 | 2·8213 2711 |
| 15 | ·5161 2903 | ·5172 4138 | ·0883 4235− | ·0004 8975− | 2·8242 4242 |
| 16 | ·5000 0000$^e$ | ·5000 0000$^e$ | ·0870 3883 | ·0000 0000$^e$ | 2·8285 7143 |
| 17 | ·4848 4848 | ·4838 7097 | ·0857 0991 | ·0004 0816 | 2·8339 2857 |
| 18 | ·4705 8824 | ·4687 5000$^e$ | ·0843 6908 | ·0015 0034 | 2·8400 2753 |
| 19 | ·4571 4286 | ·4545 4545+ | ·0830 2665− | ·0031 1407 | 2·8466 5344 |
| 20 | ·4444 4444 | ·4411 7647 | ·0816 9051 | ·0051 2465+ | 2·8536 4372 |
| 21 | ·4324 3243 | ·4285 7143 | ·0803 6670 | ·0074 3558 | 2·8608 7454 |
| 22 | ·4210 5263 | ·4166 6667 | ·0790 5975− | ·0099 7159 | 2·8682 5111 |
| 23 | ·4102 5641 | ·4054 0541 | ·0777 7308 | ·0126 7362 | 2·8757 0065+ |
| 24 | ·4000 0000$^e$ | ·3947 3684 | ·0765 0921 | ·0154 9509 | 2·8831 6722 |
| 25 | ·3902 4390 | ·3846 1538 | ·0752 6993 | ·0183 9913 | 2·8906 0782 |
| 26 | ·3809 5238 | ·3750 0000$^e$ | ·0740 5649 | ·0213 5648 | 2·8979 8951 |
| 27 | ·3720 9302 | ·3658 5366 | ·0728 6972 | ·0243 4385+ | 2·9052 8717 |
| 28 | ·3636 3636 | ·3571 4286 | ·0717 1006 | ·0273 4270 | 2·9124 8183 |
| 29 | ·3555 5556 | ·3488 3721 | ·0705 7771 | ·0303 3827 | 2·9195 5934 |
| 30 | ·3478 2609 | ·3409 0909 | ·0694 7265− | ·0333 1887 | 2·9265 0935+ |
| 31 | ·3404 2553 | ·3333 3333 | ·0683 9469 | ·0362 7521 | 2·9333 2456 |
| 32 | ·3333 3333 | ·3260 8696 | ·0673 4350+ | ·0392 0000$^e$ | 2·9400 0000$^e$ |
| 33 | ·3265 3061 | ·3191 4894 | ·0663 1868 | ·0420 8754 | 2·9465 3263 |
| 34 | ·3200 0000$^e$ | ·3125 0000$^e$ | ·0653 1973 | ·0449 3343 | 2·9529 2090 |
| 35 | ·3137 2549 | ·3061 2245− | ·0643 4609 | ·0477 3432 | 2·9591 6442 |
| 36 | ·3076 9231 | ·3000 0000$^e$ | ·0633 9718 | ·0504 8773 | 2·9652 6375− |
| 37 | ·3018 8679 | ·2941 1765− | ·0624 7239 | ·0531 9187 | 2·9712 2017 |
| 38 | ·2962 9630 | ·2884 6154 | ·0615 7107 | ·0558 4553 | 2·9770 3552 |
| 39 | ·2909 0909 | ·2830 1887 | ·0606 9258 | ·0584 4796 | 2·9827 1209 |
| 40 | ·2857 1429 | ·2777 7778 | ·0598 3627 | ·0609 9881 | 2·9882 5248 |
| 41 | ·2807 0175+ | ·2727 2727 | ·0590 0150+ | ·0634 9801 | 2·9936 5957 |
| 42 | ·2758 6207 | ·2678 5714 | ·0581 8763 | ·0659 4577 | 2·9989 3638 |
| 43 | ·2711 8644 | ·2631 5789 | ·0573 9401 | ·0683 4247 | 3·0040 8606 |
| 44 | ·2666 6667 | ·2586 2069 | ·0566 2004 | ·0706 8868 | 3·0091 1186 |
| 45 | ·2622 9508 | ·2542 3729 | ·0558 6510 | ·0729 8508 | 3·0140 1703 |
| 46 | ·2580 6452 | ·2500 0000$^e$ | ·0551 2860 | ·0752 3246 | 3·0188 0487 |
| 47 | ·2539 6825+ | ·2459 0164 | ·0544 0996 | ·0774 3170 | 3·0234 7865− |
| 48 | ·2500 0000$^e$ | ·2419 3548 | ·0537 0862 | ·0795 8372 | 3·0280 4161 |
| 49 | ·2461 5385− | ·2380 9524 | ·0530 2403 | ·0816 8948 | 3·0324 9695+ |
| 50 | ·2424 2424 | ·2343 7500$^e$ | ·0523 5566 | ·0837 5000$^e$ | 3·0368 4783 |

$p = 17$

| $q$ | $\bar{x}$ | $\tilde{x}$ | $\sigma$ | $\beta_1$ | $\beta_2$ |
|---|---|---|---|---|---|
| 0·5 | ·9714 2857 | *J*-curve | ·0387 3339 | 6·2332 0571 | 11·6010 3741 |
| 1 | ·9444 4444 | 1·0000 0000$^e$ | ·0525 5029 | 2·8611 7647 | 6·8016 8067 |
| 1·5 | ·9189 1892 | ·9696 9697 | ·0618 1318 | 1·7486 7901 | 5·2219 4788 |
| 2 | ·8947 3684 | ·9411 7647 | ·0686 2318 | 1·2004 8019 | 4·4461 4209 |
| 2·5 | ·8717 9487 | ·9142 8571 | ·0738 3851 | ·8775 7452 | 3·9911 9015$^+$ |
| 3 | ·8500 0000$^e$ | ·8888 8889 | ·0779 1937 | ·6669 9076 | 3·6961 1718 |
| 3·5 | ·8292 6829 | ·8648 6486 | ·0811 4941 | ·5203 3613 | 3·4919 7211 |
| 4 | ·8095 2381 | ·8421 0526 | ·0837 1896 | ·4134 3267 | 3·3443 0946 |
| 4·5 | ·7906 9767 | ·8205 1282 | ·0857 6320 | ·3328 6289 | 3·2340 1701 |
| 5 | ·7727 2727 | ·8000 0000$^e$ | ·0873 8218 | ·2705 8824 | 3·1496 4706 |
| 5·5 | ·7555 5556 | ·7804 8780 | ·0886 5217 | ·2215 0307 | 3·0839 3090 |
| 6 | ·7391 3043 | ·7619 0476 | ·0896 3273 | ·1822 1176 | 3·0320 3620 |
| 6·5 | ·7234 0426 | ·7441 8605$^-$ | ·0903 7124 | ·1503 7029 | 2·9906 2881 |
| 7 | ·7083 3333 | ·7272 7273 | ·0909 0593 | ·1243 1008 | 2·9573 3678 |
| 7·5 | ·6938 7755$^+$ | ·7111 1111 | ·0912 6808 | ·1028 1239 | 2·9304 2882 |
| 8 | ·6800 0000$^e$ | ·6956 5217 | ·0914 8350$^-$ | ·0849 6732 | 2·9086 1345$^-$ |
| 8·5 | ·6666 6667 | ·6808 5106 | ·0915 7371 | ·0700 8264 | 2·8909 0909 |
| 9 | ·6538 4615$^+$ | ·6666 6667 | ·0915 5677 | ·0576 2305$^-$ | 2·8765 5752 |
| 9·5 | ·6415 0943 | ·6530 6122 | ·0914 4796 | ·0471 6860 | 2·8649 6466 |
| 10 | ·6296 2963 | ·6400 0000$^e$ | ·0912 6026 | ·0383 8568 | 2·8556 5923 |
| 10·5 | ·6181 8182 | ·6274 5098 | ·0910 0473 | ·0310 0625$^+$ | 2·8482 6317 |
| 11 | ·6071 4286 | ·6153 8462 | ·0906 9091 | ·0248 1283 | 2·8424 7024 |
| 12 | ·5862 0690 | ·5925 9259 | ·0899 2003 | ·0153 0269 | 2·8347 3672 |
| 13 | ·5666 6667 | ·5714 2857 | ·0890 0083 | ·0087 6697 | 2·8309 3377 |
| 14 | ·5483 8710 | ·5517 2414 | ·0879 7348 | ·0044 4475$^+$ | 2·8300 0045$^-$ |
| 15 | ·5312 5000$^e$ | ·5333 3333 | ·0868 6866 | ·0017 9117 | 2·8311 8141 |
| 16 | ·5151 5152 | ·5161 2903 | ·0857 0991 | ·0004 0816 | 2·8339 2857 |
| 17 | ·5000 0000$^e$ | ·5000 0000$^e$ | ·0845 1543 | ·0000 0000$^e$ | 2·8378 3784 |
| 18 | ·4857 1429 | ·4848 4848 | ·0832 9931 | ·0003 4375$^-$ | 2·8426 0731 |
| 19 | ·4722 2222 | ·4705 8824 | ·0820 7254 | ·0012 6926 | 2·8480 0892 |
| 20 | ·4594 5946 | ·4571 4286 | ·0808 4365$^+$ | ·0026 4532 | 2·8538 6878 |
| 21 | ·4473 6842 | ·4444 4444 | ·0796 1927 | ·0043 6975$^-$ | 2·8600 5329 |
| 22 | ·4358 9744 | ·4324 3243 | ·0784 0454 | ·0063 6240 | 2·8664 5922 |
| 23 | ·4250 0000$^e$ | ·4210 5263 | ·0772 0341 | ·0085 5995$^-$ | 2·8730 0643 |
| 24 | ·4146 3415$^-$ | ·4102 5641 | ·0760 1890 | ·0109 1210 | 2·8796 3251 |
| 25 | ·4047 6190 | ·4000 0000$^e$ | ·0748 5330 | ·0133 7871 | 2·8862 8877 |
| 26 | ·3953 4884 | ·3902 4390 | ·0737 0829 | ·0159 2760 | 2·8929 3724 |
| 27 | ·3863 6364 | ·3809 5238 | ·0725 8509 | ·0185 3293 | 2·8995 4835$^-$ |
| 28 | ·3777 7778 | ·3720 9302 | ·0714 8453 | ·0211 7388 | 2·9060 9914 |
| 29 | ·3695 6522 | ·3636 3636 | ·0704 0714 | ·0238 3367 | 2·9125 7193 |
| 30 | ·3617 0213 | ·3555 5556 | ·0693 5322 | ·0264 9876 | 2·9189 5318 |
| 31 | ·3541 6667 | ·3478 2609 | ·0683 2286 | ·0291 5825$^+$ | 2·9252 3273 |
| 32 | ·3469 3878 | ·3404 2553 | ·0673 1601 | ·0318 0338 | 2·9314 0305$^-$ |
| 33 | ·3400 0000$^e$ | ·3333 3333 | ·0663 3250$^-$ | ·0344 2711 | 2·9374 5877 |
| 34 | ·3333 3333 | ·3265 3061 | ·0653 7205$^-$ | ·0370 2385$^+$ | 2·9433 9623 |
| 35 | ·3269 2308 | ·3200 0000$^e$ | ·0644 3431 | ·0395 8917 | 2·9492 1314 |
| 36 | ·3207 5472 | ·3137 2549 | ·0635 1890 | ·0421 1959 | 2·9549 0833 |
| 37 | ·3148 1481 | ·3076 9231 | ·0626 2536 | ·0446 1244 | 2·9604 8149 |
| 38 | ·3090 9091 | ·3018 8679 | ·0617 5321 | ·0470 6568 | 2·9659 3303 |
| 39 | ·3035 7143 | ·2962 9630 | ·0609 0197 | ·0494 7784 | 2·9712 6393 |
| 40 | ·2982 4561 | ·2909 0909 | ·0600 7110 | ·0518 4785$^-$ | 2·9764 7557 |
| 41 | ·2931 0345$^-$ | ·2857 1429 | ·0592 6010 | ·0541 7504 | 2·9815 6973 |
| 42 | ·2881 3559 | ·2807 0175$^+$ | ·0584 6843 | ·0564 5903 | 2·9865 4840 |
| 43 | ·2833 3333 | ·2758 6207 | ·0576 9558 | ·0586 9968 | 2·9914 1382 |
| 44 | ·2786 8852 | ·2711 8644 | ·0569 4100 | ·0608 9709 | 2·9961 6835$^+$ |
| 45 | ·2741 9355$^-$ | ·2666 6667 | ·0562 0420 | ·0630 5147 | 3·0008 1448 |
| 46 | ·2698 4127 | ·2622 9508 | ·0554 8465$^-$ | ·0651 6321 | 3·0053 5475$^-$ |
| 47 | ·2656 2500$^e$ | ·2580 6452 | ·0547 8186 | ·0672 3280 | 3·0097 9175$^+$ |
| 48 | ·2615 3846 | ·2539 6825$^+$ | ·0540 9533 | ·0692 6081 | 3·0141 2811 |
| 49 | ·2575 7576 | ·2500 0000$^e$ | ·0534 2459 | ·0712 4788 | 3·0183 6643 |
| 50 | ·2537 3134 | ·2461 5385$^-$ | ·0527 6916 | ·0731 9471 | 3·0225 0932 |

$p = 18$

| $q$ | $\bar{x}$ | $\tilde{x}$ | $\sigma$ | $\beta_1$ | $\beta_2$ |
|---|---|---|---|---|---|
| 0·5 | ·9729 7297 | $J$-curve | ·0367 2250$^-$ | 6·3156 8511 | 11·7538 2870 |
| 1 | ·9473 6842 | 1·0000 0000$^e$ | ·0499 3070 | 2·9125 7244 | 6·8975 4690 |
| 1·5 | ·9230 7692 | ·9714 2857 | ·0588 5324 | 1·7887 1462 | 5·2971 5762 |
| 2 | ·9000 0000$^e$ | ·9444 4444 | ·0654 6537 | 1·2341 5978 | 4·5098 8142 |
| 2·5 | ·8780 4878 | ·9189 1892 | ·0705 7212 | ·9069 5199 | 4·0472 1828 |
| 3 | ·8571 4286 | ·8947 3684 | ·0746 0471 | ·6931 3170 | 3·7463 7681 |
| 3·5 | ·8372 0930 | ·8717 9487 | ·0778 2884 | ·5438 7894 | 3·5376 2174 |
| 4 | ·8181 8182 | ·8500 0000$^e$ | ·0804 2289 | ·4347 9938 | 3·3861 1111 |
| 4·5 | ·8000 0000$^e$ | ·8292 6829 | ·0825 1370 | ·3523 5319 | 3·2725 0900 |
| 5 | ·7826 0870 | ·8095 2381 | ·0841 9529 | ·2884 2667 | 3·1852 3077 |
| 5·5 | ·7659 5745$^-$ | ·7906 9767 | ·0855 3950$^-$ | ·2378 6500$^+$ | 3·1169 1835$^+$ |
| 6 | ·7500 0000$^e$ | ·7727 2727 | ·0866 0254 | ·1972 3866 | 3·0626 7806 |
| 6·5 | ·7346 9388 | ·7555 5556 | ·0874 2925$^+$ | ·1641 7924 | 3·0191 3181 |
| 7 | ·7200 0000$^e$ | ·7391 3043 | ·0880 5593 | ·1370 0002 | 2·9838 7503 |
| 7·5 | ·7058 8235$^+$ | ·7234 0426 | ·0885 1233 | ·1144 6832 | 2·9551 5152 |
| 8 | ·6923 0769 | ·7083 3333 | ·0888 2312 | ·0956 6327 | 2·9316 5025$^-$ |
| 8·5 | ·6792 4528 | ·6938 7755$^+$ | ·0890 0891 | ·0798 8381 | 2·9123 7399 |
| 9 | ·6666 6667 | ·6800 0000$^e$ | ·0890 8708 | ·0665 8740 | 2·8965 5172 |
| 9·5 | ·6545 4545$^+$ | ·6666 6667 | ·0890 7235$^+$ | ·0553 4808 | 2·8835 7877 |
| 10 | ·6428 5714 | ·6538 4615$^+$ | ·0889 7730 | ·0458 2716 | 2·8729 7491 |
| 10·5 | ·6315 7895$^-$ | ·6415 0943 | ·0888 1269 | ·0377 5227 | 2·8643 5448 |
| 11 | ·6206 8966 | ·6296 2963 | ·0885 8781 | ·0309 0215$^+$ | 2·8574 0469 |
| 12 | ·6000 0000$^e$ | ·6071 4286 | ·0879 8827 | ·0201 8229 | 2·8475 3788 |
| 13 | ·5806 4516 | ·5862 0690 | ·0872 3108 | ·0125 5759 | 2·8418 1178 |
| 14 | ·5625 0000$^e$ | ·5666 6667 | ·0863 5616 | ·0072 4996 | 2·8391 3565$^+$ |
| 15 | ·5454 5455$^-$ | ·5483 8710 | ·0853 9422 | ·0037 0068 | 2·8387 3016 |
| 16 | ·5294 1176 | ·5312 5000$^e$ | ·0843 6908 | ·0015 0034 | 2·8400 2753 |
| 17 | ·5142 8571 | ·5151 5152 | ·0832 9931 | ·0003 4375$^-$ | 2·8426 0731 |
| 18 | ·5000 0000$^e$ | ·5000 0000$^e$ | ·0821 9949 | ·0000 0000$^e$ | 2·8461 5385$^-$ |
| 19 | ·4864 8649 | ·4857 1429 | ·0810 8108 | ·0002 9221 | 2·8504 2735$^+$ |
| 20 | ·4736 8421 | ·4722 2222 | ·0799 5311 | ·0010 8333 | 2·8552 4390 |
| 21 | ·4615 3846 | ·4594 5946 | ·0788 2270 | ·0022 6622 | 2·8604 6126 |
| 22 | ·4500 0000$^e$ | ·4473 6842 | ·0776 9547 | ·0037 5638 | 2·8659 6866 |
| 23 | ·4390 2439 | ·4358 9744 | ·0765 7582 | ·0054 8671 | 2·8716 7938 |
| 24 | ·4285 7143 | ·4250 0000$^e$ | ·0754 6722 | ·0074 0358 | 2·8775 2525$^+$ |
| 25 | ·4186 0465$^+$ | ·4146 3415$^-$ | ·0743 7234 | ·0094 6392 | 2·8834 5250$^-$ |
| 26 | ·4090 9091 | ·4047 6190 | ·0732 9325$^+$ | ·0116 3298 | 2·8894 1863 |
| 27 | ·4000 0000$^e$ | ·3953 4884 | ·0722 3151 | ·0138 8260 | 2·8953 9007 |
| 28 | ·3913 0435$^-$ | ·3863 6364 | ·0711 8828 | ·0161 8993 | 2·9013 4030 |
| 29 | ·3829 7872 | ·3777 7778 | ·0701 6439 | ·0185 3634 | 2·9072 4842 |
| 30 | ·3750 0000$^e$ | ·3695 6522 | ·0691 6042 | ·0209 0667 | 2·9130 9804 |
| 31 | ·3673 4694 | ·3617 0213 | ·0681 7671 | ·0232 8853 | 2·9188 7640 |
| 32 | ·3600 0000$^e$ | ·3541 6667 | ·0672 1344 | ·0256 7184 | 2·9245 7366 |
| 33 | ·3529 4118 | ·3469 3878 | ·0662 7067 | ·0280 4837 | 2·9301 8233 |
| 34 | ·3461 5385$^-$ | ·3400 0000$^e$ | ·0653 4832 | ·0304 1143 | 2·9356 9684 |
| 35 | ·3396 2264 | ·3333 3333 | ·0644 4623 | ·0327 5561 | 2·9411 1317 |
| 36 | ·3333 3333 | ·3269 2308 | ·0635 6417 | ·0350 7653 | 2·9464 2857 |
| 37 | ·3272 7273 | ·3207 5472 | ·0627 0186 | ·0373 7070 | 2·9516 4130 |
| 38 | ·3214 2857 | ·3148 1481 | ·0618 5896 | ·0396 3535$^+$ | 2·9567 5044 |
| 39 | ·3157 8947 | ·3090 9091 | ·0610 3511 | ·0418 6831 | 2·9617 5576 |
| 40 | ·3103 4483 | ·3035 7143 | ·0602 2991 | ·0440 6790 | 2·9666 5756 |
| 41 | ·3050 8475$^-$ | ·2982 4561 | ·0594 4298 | ·0462 3287 | 2·9714 5658 |
| 42 | ·3000 0000$^e$ | ·2931 0345$^-$ | ·0586 7387 | ·0483 6232 | 2·9761 5390 |
| 43 | ·2950 8197 | ·2881 3559 | ·0579 2217 | ·0504 5563 | 2·9807 5089 |
| 44 | ·2903 2258 | ·2833 3333 | ·0571 8744 | ·0525 1243 | 2·9852 4913 |
| 45 | ·2857 1429 | ·2786 8852 | ·0564 6924 | ·0545 3254 | 2·9896 5035$^-$ |
| 46 | ·2812 5000$^e$ | ·2741 9355$^-$ | ·0557 6716 | ·0565 1597 | 2·9939 5644 |
| 47 | ·2769 2308 | ·2698 4127 | ·0550 8075$^+$ | ·0584 6285$^-$ | 2·9981 6935$^+$ |
| 48 | ·2727 2727 | ·2656 2500$^e$ | ·0544 0960 | ·0603 7341 | 3·0022 9113 |
| 49 | ·2686 5672 | ·2615 3846 | ·0537 5329 | ·0622 4802 | 3·0063 2385$^+$ |
| 50 | ·2647 0588 | ·2575 7576 | ·0531 1141 | ·0640 8707 | 3·0102 6962 |

$p = 19$

| $q$ | $\bar{x}$ | $\tilde{x}$ | $\sigma$ | $\beta_1$ | $\beta_2$ |
|---|---|---|---|---|---|
| 0·5 | ·9743 5897 | *J*-curve | ·0349 1003 | 6·3908 2292 | 11·8935 1285$^+$ |
| 1 | ·9500 0000$^e$ | 1·0000 0000$^e$ | ·0475 5949 | 2·9595 4763 | 6·9854 3790 |
| 1·5 | ·9268 2927 | ·9729 7297 | ·0561 6288 | 1·8254 2777 | 5·3663 0584 |
| 2 | ·9047 6190 | ·9473 6842 | ·0625 8369 | 1·2651 4775$^-$ | 4·5686 4989 |
| 2·5 | ·8837 2093 | ·9230 7692 | ·0675 7983 | ·9340 7353 | 4·0990 2416 |
| 3 | ·8636 3636 | ·9000 0000$^e$ | ·0715 5679 | ·7173 4893 | 3·7929 8246 |
| 3·5 | ·8444 4444 | ·8780 4878 | ·0747 6432 | ·5657 6677 | 3·5800 7564 |
| 4 | ·8260 8696 | ·8571 4286 | ·0773 7013 | ·4547 3684 | 3·4251 0121 |
| 4·5 | ·8085 1064 | ·8372 0930 | ·0794 9360 | ·3706 0870 | 3·3085 2010 |
| 5 | ·7916 6667 | ·8181 8182 | ·0812 2329 | ·3052 0087 | 3·2186 2348 |
| 5·5 | ·7755 1020 | ·8000 0000$^e$ | ·0826 2700 | ·2533 1428 | 3·1479 7246 |
| 6 | ·7600 0000$^e$ | ·7826 0870 | ·0837 5789 | ·2114 8894 | 3·0916 1793 |
| 6·5 | ·7450 9804 | ·7659 5745$^-$ | ·0846 5846 | ·1773 3463 | 3·0461 4222 |
| 7 | ·7307 6923 | ·7500 0000$^e$ | ·0853 6318 | ·1491 4838 | 3·0091 1145$^-$ |
| 7·5 | ·7169 8113 | ·7346 9388 | ·0859 0040 | ·1256 8510 | 2·9787 4705$^-$ |
| 8 | ·7037 0370 | ·7200 0000$^e$ | ·0862 9368 | ·1060 1414 | 2·9537 2051 |
| 8·5 | ·6909 0909 | ·7058 8235$^+$ | ·0865 6277 | ·0894 2664 | 2·9330 2061 |
| 9 | ·6785 7143 | ·6923 0769 | ·0867 2434 | ·0753 7362 | 2·9158 6493 |
| 9·5 | ·6666 6667 | ·6792 4528 | ·0867 9261 | ·0634 2381 | 2·9016 3934 |
| 10 | ·6551 7241 | ·6666 6667 | ·0867 7971 | ·0532 3402 | 2·8898 5569 |
| 10·5 | ·6440 6780 | ·6545 4545$^+$ | ·0866 9609 | ·0445 2805$^-$ | 2·8801 2155$^-$ |
| 11 | ·6333 3333 | ·6428 5714 | ·0865 5079 | ·0370 8134 | 2·8721 1831 |
| 12 | ·6129 0323 | ·6206 8966 | ·0861 0547 | ·0252 6058 | 2·8603 0584 |
| 13 | ·5937 5000$^e$ | ·6000 0000$^e$ | ·0854 9516 | ·0166 4262 | 2·8528 2210 |
| 14 | ·5757 5758 | ·5806 4516 | ·0847 5931 | ·0104 3425$^-$ | 2·8485 4995$^-$ |
| 15 | ·5588 2353 | ·5625 0000$^e$ | ·0839 2851 | ·0060 6454 | 2·8466 8879 |
| 16 | ·5428 5714 | ·5454 5455$^-$ | ·0830 2665$^-$ | ·0031 1407 | 2·8466 5344 |
| 17 | ·5277 7778 | ·5294 1176 | ·0820 7254 | ·0012 6926 | 2·8480 0892 |
| 18 | ·5135 1351 | ·5142 8571 | ·0810 8108 | ·0002 9221 | 2·8504 2735$^+$ |
| 19 | ·5000 0000$^e$ | ·5000 0000$^e$ | ·0800 6408 | ·0000 0000$^e$ | 2·8536 5854 |
| 20 | ·4871 7949 | ·4864 8649 | ·0790 3095$^-$ | ·0002 5048 | 2·8575 0963 |
| 21 | ·4750 0000$^e$ | ·4736 8421 | ·0779 8921 | ·0009 3204 | 2·8618 3066 |
| 22 | ·4634 1463 | ·4615 3846 | ·0769 4486 | ·0019 5631 | 2·8665 0414 |
| 23 | ·4523 8095$^+$ | ·4500 0000$^e$ | ·0759 0270 | ·0032 5283 | 2·8714 3749 |
| 24 | ·4418 6047 | ·4390 2439 | ·0748 6652 | ·0047 6500$^-$ | 2·8765 5734 |
| 25 | ·4318 1818 | ·4285 7143 | ·0738 3935$^+$ | ·0064 4712 | 2·8818 0535$^-$ |
| 26 | ·4222 2222 | ·4186 0465$^+$ | ·0728 2358 | ·0082 6212 | 2·8871 3498 |
| 27 | ·4130 4348 | ·4090 9091 | ·0718 2108 | ·0101 7977 | 2·8925 0905$^+$ |
| 28 | ·4042 5532 | ·4000 0000$^e$ | ·0708 3328 | ·0121 7538 | 2·8978 9781 |
| 29 | ·3958 3333 | ·3913 0435$^-$ | ·0698 6128 | ·0142 2868 | 2·9032 7746 |
| 30 | ·3877 5510 | ·3829 7872 | ·0689 0589 | ·0163 2301 | 2·9086 2904 |
| 31 | ·3800 0000$^e$ | ·3750 0000$^e$ | ·0679 6770 | ·0184 4466 | 2·9139 3743 |
| 32 | ·3725 4902 | ·3673 4694 | ·0670 4709 | ·0205 8234 | 2·9191 9067 |
| 33 | ·3653 8462 | ·3600 0000$^e$ | ·0661 4431 | ·0227 2673 | 2·9243 7936 |
| 34 | ·3584 9057 | ·3529 4118 | ·0652 5947 | ·0248 7015$^-$ | 2·9294 9620 |
| 35 | ·3518 5185$^+$ | ·3461 5385$^-$ | ·0643 9256 | ·0270 0629 | 2·9345 3559 |
| 36 | ·3454 5455$^-$ | ·3396 2264 | ·0635 4353 | ·0291 2997 | 2·9394 9332 |
| 37 | ·3392 8571 | ·3333 3333 | ·0627 1222 | ·0312 3694 | 2·9443 6634 |
| 38 | ·3333 3333 | ·3272 7273 | ·0618 9845$^-$ | ·0333 2376 | 2·9491 5254 |
| 39 | ·3275 8621 | ·3214 2857 | ·0611 0195$^+$ | ·0353 8761 | 2·9538 5058 |
| 40 | ·3220 3390 | ·3157 8947 | ·0603 2248 | ·0374 2627 | 2·9584 5974 |
| 41 | ·3166 6667 | ·3103 4483 | ·0595 5971 | ·0394 3795$^+$ | 2·9629 7983 |
| 42 | ·3114 7541 | ·3050 8475$^-$ | ·0588 1333 | ·0414 2127 | 2·9674 1109 |
| 43 | ·3064 5161 | ·3000 0000$^e$ | ·0580 8300 | ·0433 7515$^+$ | 2·9717 5407 |
| 44 | ·3015 8730 | ·2950 8197 | ·0573 6838 | ·0452 9883 | 2·9760 0964 |
| 45 | ·2968 7500$^e$ | ·2903 2258 | ·0566 6911 | ·0471 9175$^-$ | 2·9801 7885$^+$ |
| 46 | ·2923 0769 | ·2857 1429 | ·0559 8483 | ·0490 5355$^-$ | 2·9842 6296 |
| 47 | ·2878 7879 | ·2812 5000$^e$ | ·0553 1519 | ·0508 8404 | 2·9882 6336 |
| 48 | ·2835 8209 | ·2769 2308 | ·0546 5982 | ·0526 8317 | 2·9921 8154 |
| 49 | ·2794 1176 | ·2727 2727 | ·0540 1836 | ·0544 5100 | 2·9960 1908 |
| 50 | ·2753 6232 | ·2686 5672 | ·0533 9046 | ·0561 8768 | 2·9997 7761 |

$p = 20$

| $q$ | $\bar{x}$ | $\tilde{x}$ | $\sigma$ | $\beta_1$ | $\beta_2$ |
|---|---|---|---|---|---|
| 0·5 | ·9756 0976 | *J*-curve | ·0332 6802 | 6·4595 5556 | 12·0217 0213 |
| 1 | ·9523 8095+ | 1·0000 0000$^e$ | ·0454 0298 | 3·0026 4650+ | 7·0663 0435− |
| 1·5 | ·9302 3256 | ·9743 5897 | ·0537 0700 | 1·8592 1231 | 5·4300 9119 |
| 2 | ·9090 9091 | ·9500 0000$^e$ | ·0599 4368 | 1·2937 5000$^e$ | 4·6230 0000$^e$ |
| 2·5 | ·8888 8889 | ·9268 2927 | ·0648 2889 | ·9591 8367 | 4·1470 5882 |
| 3 | ·8695 6522 | ·9047 6190 | ·0687 4517 | ·7398 4000$^e$ | 3·8363 0769 |
| 3·5 | ·8510 6383 | ·8837 2093 | ·0719 2802 | ·5861 5917 | 3·6196 4484 |
| 4 | ·8333 3333 | ·8636 3636 | ·0745 3560 | ·4733 7278 | 3·4615 3846 |
| 4·5 | ·8163 2653 | ·8444 4444 | ·0766 8053 | ·3877 2992 | 3·3422 6415+ |
| 5 | ·8000 0000$^e$ | ·8260 8696 | ·0784 4645+ | ·3209 8765+ | 3·2500 0000$^e$ |
| 5·5 | ·7843 1373 | ·8085 1064 | ·0798 9751 | ·2679 0684 | 3·1772 3358 |
| 6 | ·7692 3077 | ·7916 6667 | ·0810 8404 | ·2250 0000$^e$ | 3·1189 6552 |
| 6·5 | ·7547 1698 | ·7755 1020 | ·0820 4639 | ·1898 5723 | 3·0717 4226 |
| 7 | ·7407 4074 | ·7600 0000$^e$ | ·0828 1733 | ·1607 6100 | 3·0331 0345− |
| 7·5 | ·7272 7273 | ·7450 9804 | ·0834 2390 | ·1364 5504 | 3·0012 5035− |
| 8 | ·7142 8571 | ·7307 6923 | ·0838 8860 | ·1160 0000$^e$ | 2·9748 3871 |
| 8·5 | ·7017 5439 | ·7169 8113 | ·0842 3035− | ·0986 7999 | 2·9528 4474 |
| 9 | ·6896 5517 | ·7037 0370 | ·0844 6516 | ·0839 4034 | 2·9344 7581 |
| 9·5 | ·6779 6610 | ·6909 0909 | ·0846 0672 | ·0713 4503 | 2·9191 0931 |
| 10 | ·6666 6667 | ·6785 7143 | ·0846 6675+ | ·0605 4688 | 2·9062 5000$^e$ |
| 10·5 | ·6557 3770 | ·6666 6667 | ·0846 5537 | ·0512 6627 | 2·8954 9943 |
| 11 | ·6451 6129 | ·6551 7241 | ·0845 8136 | ·0432 7573 | 2·8865 3379 |
| 12 | ·6250 0000$^e$ | ·6333 3333 | ·0842 7498 | ·0304 4983 | 2·8729 4118 |
| 13 | ·6060 6061 | ·6129 0323 | ·0837 9793 | ·0209 2308 | 2·8638 4615+ |
| 14 | ·5882 3529 | ·5937 5000$^e$ | ·0831 8903 | ·0138 8889 | 2·8581 0811 |
| 15 | ·5714 2857 | ·5757 5758 | ·0824 7861 | ·0087 6552 | 2·8549 0754 |
| 16 | ·5555 5556 | ·5588 2353 | ·0816 9051 | ·0051 2465+ | 2·8536 4372 |
| 17 | ·5405 4054 | ·5428 5714 | ·0808 4365+ | ·0026 4532 | 2·8538 6878 |
| 18 | ·5263 1579 | ·5277 7778 | ·0799 5311 | ·0010 8333 | 2·8552 4390 |
| 19 | ·5128 2051 | ·5135 1351 | ·0790 3095− | ·0002 5048 | 2·8575 0963 |
| 20 | ·5000 0000$^e$ | ·5000 0000$^e$ | ·0780 8688 | ·0000 0000$^e$ | 2·8604 6512 |
| 21 | ·4878 0488 | ·4871 7949 | ·0771 2872 | ·0002 1633 | 2·8639 5349 |
| 22 | ·4761 9048 | ·4750 0000$^e$ | ·0761 6279 | ·0008 0766 | 2·8678 5124 |
| 23 | ·4651 1628 | ·4634 1463 | ·0751 9416 | ·0017 0048 | 2·8720 6049 |
| 24 | ·4545 4545+ | ·4523 8095+ | ·0742 2696 | ·0028 3554 | 2·8765 0324 |
| 25 | ·4444 4444 | ·4418 6047 | ·0732 6450− | ·0041 6478 | 2·8811 1702 |
| 26 | ·4347 8261 | ·4318 1818 | ·0723 0943 | ·0056 4904 | 2·8858 5165− |
| 27 | ·4255 3191 | ·4222 2222 | ·0713 6387 | ·0072 5624 | 2·8906 6667 |
| 28 | ·4166 6667 | ·4130 4348 | ·0704 2952 | ·0089 6000$^e$ | 2·8955 2941 |
| 29 | ·4081 6327 | ·4042 5532 | ·0695 0770 | ·0107 3858 | 2·9004 1348 |
| 30 | ·4000 0000$^e$ | ·3958 3333 | ·0685 9943 | ·0125 7396 | 2·9052 9753 |
| 31 | ·3921 5686 | ·3877 5510 | ·0677 0551 | ·0144 5125− | 2·9101 6433 |
| 32 | ·3846 1538 | ·3800 0000$^e$ | ·0668 2650− | ·0163 5802 | 2·9150 0000$^e$ |
| 33 | ·3773 5849 | ·3725 4902 | ·0659 6282 | ·0182 8400 | 2·9197 9339 |
| 34 | ·3703 7037 | ·3653 8462 | ·0651 1475− | ·0202 2059 | 2·9245 3560 |
| 35 | ·3636 3636 | ·3584 9057 | ·0642 8243 | ·0221 6066 | 2·9292 1960 |
| 36 | ·3571 4286 | ·3518 5185+ | ·0634 6595+ | ·0240 9830 | 2·9338 3986 |
| 37 | ·3508 7719 | ·3454 5455− | ·0626 6528 | ·0260 2856 | 2·9383 9212 |
| 38 | ·3448 2759 | ·3392 8571 | ·0618 8036 | ·0279 4737 | 2·9428 7317 |
| 39 | ·3389 8305+ | ·3333 3333 | ·0611 1104 | ·0298 5136 | 2·9472 8064 |
| 40 | ·3333 3333 | ·3275 8621 | ·0603 5716 | ·0317 3777 | 2·9516 1290 |
| 41 | ·3278 6885+ | ·3220 3390 | ·0596 1853 | ·0336 0434 | 2·9558 6890 |
| 42 | ·3225 8065− | ·3166 6667 | ·0588 9490 | ·0354 4922 | 2·9600 4808 |
| 43 | ·3174 6032 | ·3114 7541 | ·0581 8602 | ·0372 7095+ | 2·9641 5027 |
| 44 | ·3125 0000$^e$ | ·3064 5161 | ·0574 9164 | ·0390 6837 | 2·9681 7565+ |
| 45 | ·3076 9231 | ·3015 8730 | ·0568 1146 | ·0408 4057 | 2·9721 2467 |
| 46 | ·3030 3030 | ·2968 7500$^e$ | ·0561 4519 | ·0425 8688 | 2·9759 9800 |
| 47 | ·2985 0746 | ·2923 0769 | ·0554 9255− | ·0443 0680 | 2·9797 9648 |
| 48 | ·2941 1765− | ·2878 7879 | ·0548 5322 | ·0460 0000$^e$ | 2·9835 2113 |
| 49 | ·2898 5507 | ·2835 8209 | ·0542 2692 | ·0476 6628 | 2·9871 7304 |
| 50 | ·2857 1429 | ·2794 1176 | ·0536 1333 | ·0493 0556 | 2·9907 5342 |

$p = 21$

| $q$ | $\bar{x}$ | $\tilde{x}$ | $\sigma$ | $\beta_1$ | $\beta_2$ |
|---|---|---|---|---|---|
| 0·5 | ·9767 4419 | *J*-curve | ·0317 7349 | 6·5226 6701 | 12·1397 5560 |
| 1 | ·9545 4545$^+$ | 1·0000 0000$^e$ | ·0434 3332 | 3·0423 2804 | 7·1409 5238 |
| 1·5 | ·9333 3333 | ·9756 0976 | ·0514 5633 | 1·8904 0281 | 5·4891 0993 |
| 2 | ·9130 4348 | ·9523 8095$^+$ | ·0575 1633 | 1·3202 2857 | 4·6734 0659 |
| 2·5 | ·8936 1702 | ·9302 3256 | ·0622 9149 | ·9824 9391 | 4·1917 1291 |
| 3 | ·8750 0000$^e$ | ·9090 9091 | ·0661 4378 | ·7607 7768 | 3·8766 7888 |
| 3·5 | ·8571 4286 | ·8888 8889 | ·0692 9589 | ·6051 9758 | 3·6566 0377 |
| 4 | ·8400 0000$^e$ | ·8695 6522 | ·0718 9736 | ·4908 2239 | 3·4956 5382 |
| 4·5 | ·8235 2941 | ·8510 6383 | ·0740 5473 | ·4038 0952 | 3·3739 3484 |
| 5 | ·8076 9231 | ·8333 3333 | ·0758 4718 | ·3358 6006 | 3·2795 2146 |
| 5·5 | ·7924 5283 | ·8163 2653 | ·0773 3548 | ·2816 9840 | 3·2048 3412 |
| 6 | ·7777 7778 | ·8000 0000$^e$ | ·0785 6742 | ·2378 1213 | 3·1448 2759 |
| 6·5 | ·7636 3636 | ·7843 1373 | ·0795 8133 | ·2017 7352 | 3·0960 1568 |
| 7 | ·7500 0000$^e$ | ·7692 3077 | ·0804 0844 | ·1718 5185$^+$ | 3·0559 1398 |
| 7·5 | ·7368 4211 | ·7547 1698 | ·0810 7452 | ·1467 8082 | 3·0227 0548 |
| 8 | ·7241 3793 | ·7407 4074 | ·0816 0110 | ·1256 1320 | 2·9950 3168 |
| 8·5 | ·7118 6441 | ·7272 7273 | ·0820 0629 | ·1076 2682 | 2·9718 5746 |
| 9 | ·7000 0000$^e$ | ·7142 8571 | ·0823 0549 | ·0922 6190 | 2·9523 8095$^+$ |
| 9·5 | ·6885 2459 | ·7017 5439 | ·0825 1185$^+$ | ·0790 7817 | 2·9359 7196 |
| 10 | ·6774 1935$^+$ | ·6896 5517 | ·0826 3670 | ·0677 2487 | 2·9221 2885$^+$ |
| 10·5 | ·6666 6667 | ·6779 6610 | ·0826 8982 | ·0579 1936 | 2·9104 4776 |
| 11 | ·6562 5000$^e$ | ·6666 6667 | ·0826 7973 | ·0494 3154 | 2·9006 0024 |
| 12 | ·6363 6364 | ·6451 6129 | ·0824 9866 | ·0356 8513 | 2·8853 7415$^-$ |
| 13 | ·6176 4706 | ·6250 0000$^e$ | ·0821 4259 | ·0253 2447 | 2·8747 9787 |
| 14 | ·6000 0000$^e$ | ·6060 6061 | ·0816 4966 | ·0175 3104 | 2·8677 0982 |
| 15 | ·5833 3333 | ·5882 3529 | ·0810 4979 | ·0117 1349 | 2·8632 7357 |
| 16 | ·5675 6757 | ·5714 2857 | ·0803 6670 | ·0074 3558 | 2·8608 7454 |
| 17 | ·5526 3158 | ·5555 5556 | ·0796 1927 | ·0043 6975$^-$ | 2·8600 5329 |
| 18 | ·5384 6154 | ·5405 4054 | ·0788 2270 | ·0022 6622 | 2·8604 6126 |
| 19 | ·5250 0000$^e$ | ·5263 1579 | ·0779 8921 | ·0009 3204 | 2·8618 3066 |
| 20 | ·5121 9512 | ·5128 2051 | ·0771 2872 | ·0002 1633 | 2·8639 5349 |
| 21 | ·5000 0000$^e$ | ·5000 0000$^e$ | ·0762 4929 | ·0000 0000$^e$ | 2·8666 6667 |
| 22 | ·4883 7209 | ·4878 0488 | ·0753 5745$^-$ | ·0001 8812 | 2·8698 4127 |
| 23 | ·4772 7273 | ·4761 9048 | ·0744 5856 | ·0007 0448 | 2·8733 7466 |
| 24 | ·4666 6667 | ·4651 1628 | ·0735 5697 | ·0014 8742 | 2·8771 8465$^+$ |
| 25 | ·4565 2174 | ·4545 4545$^+$ | ·0726 5624 | ·0024 8677 | 2·8812 0505$^+$ |
| 26 | ·4468 0851 | ·4444 4444 | ·0717 5924 | ·0036 6148 | 2·8853 8237 |
| 27 | ·4375 0000$^e$ | ·4347 8261 | ·0708 6834 | ·0049 7778 | 2·8896 7320 |
| 28 | ·4285 7143 | ·4255 3191 | ·0699 8542 | ·0064 0779 | 2·8940 4223 |
| 29 | ·4200 0000$^e$ | ·4166 6667 | ·0691 1201 | ·0079 2841 | 2·8984 6068 |
| 30 | ·4117 6471 | ·4081 6327 | ·0682 4934 | ·0095 2042 | 2·9029 0506 |
| 31 | ·4038 4615$^+$ | ·4000 0000$^e$ | ·0673 9834 | ·0111 6779 | 2·9073 5620 |
| 32 | ·3962 2642 | ·3921 5686 | ·0665 5978 | ·0128 5714 | 2·9117 9847 |
| 33 | ·3888 8889 | ·3846 1538 | ·0657 3422 | ·0145 7726 | 2·9162 1912 |
| 34 | ·3818 1818 | ·3773 5849 | ·0649 2208 | ·0163 1875$^-$ | 2·9206 0781 |
| 35 | ·3750 0000$^e$ | ·3703 7037 | ·0641 2365$^-$ | ·0180 7372 | 2·9249 5617 |
| 36 | ·3684 2105$^+$ | ·3636 3636 | ·0633 3912 | ·0198 3557 | 2·9292 5747 |
| 37 | ·3620 6897 | ·3571 4286 | ·0625 6861 | ·0215 9874 | 2·9335 0634 |
| 38 | ·3559 3220 | ·3508 7719 | ·0618 1214 | ·0233 5858 | 2·9376 9856 |
| 39 | ·3500 0000$^e$ | ·3448 2759 | ·0610 6970 | ·0251 1121 | 2·9418 3083 |
| 40 | ·3442 6230 | ·3389 8305$^+$ | ·0603 4120 | ·0268 5335$^+$ | 2·9459 0065$^+$ |
| 41 | ·3387 0968 | ·3333 3333 | ·0596 2654 | ·0285 8232 | 2·9499 0619 |
| 42 | ·3333 3333 | ·3278 6885$^+$ | ·0589 2557 | ·0302 9586 | 2·9538 4615$^+$ |
| 43 | ·3281 2500$^e$ | ·3225 8065$^-$ | ·0582 3810 | ·0319 9213 | 2·9577 1971 |
| 44 | ·3230 7692 | ·3174 6032 | ·0575 6396 | ·0336 6961 | 2·9615 2640 |
| 45 | ·3181 8182 | ·3125 0000$^e$ | ·0569 0292 | ·0353 2707 | 2·9652 6611 |
| 46 | ·3134 3284 | ·3076 9231 | ·0562 5476 | ·0369 6354 | 2·9689 3896 |
| 47 | ·3088 2353 | ·3030 3030 | ·0556 1925$^-$ | ·0385 7825$^+$ | 2·9725 4530 |
| 48 | ·3043 4783 | ·2985 0746 | ·0549 9613 | ·0401 7060 | 2·9760 8568 |
| 49 | ·3000 0000$^e$ | ·2941 1765$^-$ | ·0543 8517 | ·0417 4015$^+$ | 2·9795 6077 |
| 50 | ·2957 7465$^-$ | ·2898 5507 | ·0537 8610 | ·0432 8660 | 2·9829 7139 |

$p = 22$

| $q$ | $\bar{x}$ | $\tilde{x}$ | $\sigma$ | $\beta_1$ | $\beta_2$ |
|---|---|---|---|---|---|
| 0·5 | ·9777 7778 | *J*-curve | ·0304 0744 | 6·5808 1860 | 12·2488 2680 |
| 1 | ·9565 2174 | 1·0000 0000$^{e}$ | ·0416 2727 | 3·0789 8182 | 7·2100 6993 |
| 1·5 | ·9361 7021 | ·9767 4419 | ·0493 8625$^{+}$ | 1·9192 8512 | 5·5438 7381 |
| 2 | ·9166 6667 | ·9545 4545$^{+}$ | ·0552 7708 | 1·3448 0904 | 4·7202 7972 |
| 2·5 | ·8979 5918 | ·9333 3333 | ·0599 4392 | 1·0041 8784 | 4·2333 2606 |
| 3 | ·8800 0000$^{e}$ | ·9130 4348 | ·0637 3020 | ·7803 1342 | 3·9143 8191 |
| 3·5 | ·8627 4510 | ·8936 1702 | ·0668 4705$^{-}$ | ·6230 0741 | 3·6911 9493 |
| 4 | ·8461 5385$^{-}$ | ·8750 0000$^{e}$ | ·0694 3623 | ·5071 8924 | 3·5276 5338 |
| 4·5 | ·8301 8868 | ·8571 4286 | ·0715 9874 | ·4189 3232 | 3·4037 0701 |
| 5 | ·8148 1481 | ·8400 0000$^{e}$ | ·0734 0972 | ·3498 8650$^{-}$ | 3·3073 3542 |
| 5·5 | ·8000 0000$^{e}$ | ·8235 2941 | ·0749 2686 | ·2947 4289 | 3·2308 9747 |
| 6 | ·7857 1429 | ·8076 9231 | ·0761 9561 | ·2499 6633 | 3·1693 0596 |
| 6·5 | ·7719 2982 | ·7924 5283 | ·0772 5243 | ·2131 1287 | 3·1190 4489 |
| 7 | ·7586 2069 | ·7777 7778 | ·0781 2711 | ·1824 3983 | 3·0776 0788 |
| 7·5 | ·7457 6271 | ·7636 3636 | ·0788 4425$^{+}$ | ·1566 7178 | 3·0431 6128 |
| 8 | ·7333 3333 | ·7500 0000$^{e}$ | ·0794 2445$^{+}$ | ·1348 5440 | 3·0143 3368 |
| 8·5 | ·7213 1148 | ·7368 4211 | ·0798 8508 | ·1162 5985$^{-}$ | 2·9900 7963 |
| 9 | ·7096 7742 | ·7241 3793 | ·0802 4092 | ·1003 2371 | 2·9695 8894 |
| 9·5 | ·6984 1270 | ·7118 6441 | ·0805 0464 | ·0866 0191 | 2·9522 2452 |
| 10 | ·6875 0000$^{e}$ | ·7000 0000$^{e}$ | ·0806 8715$^{+}$ | ·0747 4048 | 2·9374 7899 |
| 10·5 | ·6769 2308 | ·6885 2459 | ·0807 9793 | ·0644 5406 | 2·9249 4360 |
| 11 | ·6666 6667 | ·6774 1935$^{+}$ | ·0808 4521 | ·0555 1020 | 2·9142 8571 |
| 12 | ·6470 5882 | ·6562 5000$^{e}$ | ·0807 7724 | ·0409 1844 | 2·8975 5665$^{-}$ |
| 13 | ·6285 7143 | ·6363 6364 | ·0805 3112 | ·0297 9052 | 2·8856 1510 |
| 14 | ·6111 1111 | ·6176 4706 | ·0801 4418 | ·0212 9726 | 2·8772 8061 |
| 15 | ·5945 9459 | ·6000 0000$^{e}$ | ·0796 4590 | ·0148 3872 | 2·8717 0163 |
| 16 | ·5789 4737 | ·5833 3333 | ·0790 5975$^{-}$ | ·0099 7159 | 2·8682 5111 |
| 17 | ·5641 0256 | ·5675 6757 | ·0784 0454 | ·0063 6240 | 2·8664 5922 |
| 18 | ·5500 0000$^{e}$ | ·5526 3158 | ·0776 9547 | ·0037 5638 | 2·8659 6866 |
| 19 | ·5365 8537 | ·5384 6154 | ·0769 4486 | ·0019 5631 | 2·8665 0414 |
| 20 | ·5238 0952 | ·5250 0000$^{e}$ | ·0761 6279 | ·0008 0766 | 2·8678 5124 |
| 21 | ·5116 2791 | ·5121 9512 | ·0753 5745$^{-}$ | ·0001 8812 | 2·8698 4127 |
| 22 | ·5000 0000$^{e}$ | ·5000 0000$^{e}$ | ·0745 3560 | ·0000 0000$^{e}$ | 2·8723 4043 |
| 23 | ·4888 8889 | ·4883 7209 | ·0737 0277 | ·0001 6462 | 2·8752 4178 |
| 24 | ·4782 6087 | ·4772 7273 | ·0728 6353 | ·0006 1816 | 2·8784 5934 |
| 25 | ·4680 8511 | ·4666 6667 | ·0720 2162 | ·0013 0855$^{-}$ | 2·8819 2356 |
| 26 | ·4583 3333 | ·4565 2174 | ·0711 8012 | ·0021 9301 | 2·8855 7795$^{+}$ |
| 27 | ·4489 7959 | ·4468 0851 | ·0703 4158 | ·0032 3626 | 2·8893 7642 |
| 28 | ·4400 0000$^{e}$ | ·4375 0000$^{e}$ | ·0695 0808 | ·0044 0905$^{+}$ | 2·8932 8125$^{-}$ |
| 29 | ·4313 7255$^{-}$ | ·4285 7143 | ·0686 8130 | ·0056 8705$^{-}$ | 2·8972 6149 |
| 30 | ·4230 7692 | ·4200 0000$^{e}$ | ·0678 6263 | ·0070 4992 | 2·9012 9170 |
| 31 | ·4150 9434 | ·4117 6471 | ·0670 5319 | ·0084 8065$^{-}$ | 2·9053 5095$^{+}$ |
| 32 | ·4074 0741 | ·4038 4615$^{+}$ | ·0662 5387 | ·0099 6492 | 2·9094 2199 |
| 33 | ·4000 0000$^{e}$ | ·3962 2642 | ·0654 6537 | ·0114 9072 | 2·9134 9062 |
| 34 | ·3928 5714 | ·3888 8889 | ·0646 8824 | ·0130 4787 | 2·9175 4517 |
| 35 | ·3859 6491 | ·3818 1818 | ·0639 2291 | ·0146 2783 | 2·9215 7605$^{+}$ |
| 36 | ·3793 1034 | ·3750 0000$^{e}$ | ·0631 6967 | ·0162 2334 | 2·9255 7543 |
| 37 | ·3728 8136 | ·3684 2105$^{+}$ | ·0624 2874 | ·0178 2829 | 2·9295 3691 |
| 38 | ·3666 6667 | ·3620 6897 | ·0617 0026 | ·0194 3749 | 2·9334 5534 |
| 39 | ·3606 5574 | ·3559 3220 | ·0609 8429 | ·0210 4656 | 2·9373 2656 |
| 40 | ·3548 3871 | ·3500 0000$^{e}$ | ·0602 8085$^{+}$ | ·0226 5181 | 2·9411 4729 |
| 41 | ·3492 0635$^{-}$ | ·3442 6230 | ·0595 8991 | ·0242 5012 | 2·9449 1495$^{+}$ |
| 42 | ·3437 5000$^{e}$ | ·3387 0968 | ·0589 1140 | ·0258 3887 | 2·9486 2758 |
| 43 | ·3384 6154 | ·3333 3333 | ·0582 4521 | ·0274 1585$^{+}$ | 2·9522 8372 |
| 44 | ·3333 3333 | ·3281 2500$^{e}$ | ·0575 9123 | ·0289 7924 | 2·9558 8235$^{+}$ |
| 45 | ·3283 5821 | ·3230 7692 | ·0569 4929 | ·0305 2750$^{-}$ | 2·9594 2280 |
| 46 | ·3235 2941 | ·3181 8182 | ·0563 1924 | ·0320 5937 | 2·9629 0470 |
| 47 | ·3188 4058 | ·3134 3284 | ·0557 0089 | ·0335 7382 | 2·9663 2795$^{-}$ |
| 48 | ·3142 8571 | ·3088 2353 | ·0550 9406 | ·0350 7003 | 2·9696 9265$^{-}$ |
| 49 | ·3098 5915$^{+}$ | ·3043 4783 | ·0544 9853 | ·0365 4732 | 2·9729 9907 |
| 50 | ·3055 5556 | ·3000 0000$^{e}$ | ·0539 1411 | ·0380 0518 | 2·9762 4767 |

$p = 23$

| $q$ | $\bar{x}$ | $\tilde{x}$ | $\sigma$ | $\beta_1$ | $\beta_2$ |
|---|---|---|---|---|---|
| 0·5 | ·9787 2340 | *J*-curve | ·0291 5399 | 6·6345 7199 | 12·3499 0108 |
| 1 | ·9583 3333 | 1·0000 0000$^e$ | ·0399 6526 | 3·1129 4057 | 7·2742 4749 |
| 1·5 | ·9387 7551 | ·9777 7778 | ·0474 7592 | 1·9461 0491 | 5·5948 2437 |
| 2 | ·9200 0000$^e$ | ·9565 2174 | ·0532 0497 | 1·3676 8653 | 4·7639 7516 |
| 2·5 | ·9019 6078 | ·9361 7021 | ·0577 6580 | 1·0244 2544 | 4·2721 9472 |
| 3 | ·8846 1538 | ·9166 6667 | ·0614 8502 | ·7985 8030 | 3·9496 6802 |
| 3·5 | ·8679 2453 | ·8979 5918 | ·0645 6330 | ·6396 9994 | 3·7236 3296 |
| 4 | ·8518 5185+ | ·8800 0000$^e$ | ·0671 3533 | ·5225 6630 | 3·5577 2114 |
| 4·5 | ·8363 6364 | ·8627 4510 | ·0692 9710 | ·4331 7554 | 3·4317 3828 |
| 5 | ·8214 2857 | ·8461 5385− | ·0711 2002 | ·3631 3043 | 3·3335 7644 |
| 5·5 | ·8070 1754 | ·8301 8868 | ·0726 5900 | ·3070 9157 | 3·2555 3776 |
| 6 | ·7931 0345− | ·8148 1481 | ·0739 5728 | ·2615 0296 | 3·1924 9649 |
| 6·5 | ·7796 6102 | ·8000 0000$^e$ | ·0750 4965− | ·2239 0584 | 3·1409 0926 |
| 7 | ·7666 6667 | ·7857 1429 | ·0759 6453 | ·1925 4658 | 3·0982 4958 |
| 7·5 | ·7540 9836 | ·7719 2982 | ·0767 2545− | ·1661 4150− | 3·0626 6860 |
| 8 | ·7419 3548 | ·7586 2069 | ·0773 5212 | ·1437 2979 | 3·0327 8307 |
| 8·5 | ·7301 5873 | ·7457 6271 | ·0778 6121 | ·1245 7847 | 3·0075 3820 |
| 9 | ·7187 5000$^e$ | ·7333 3333 | ·0782 6692 | ·1081 1895+ | 2·9861 1619 |
| 9·5 | ·7076 9231 | ·7213 1148 | ·0785 8144 | ·0939 0370 | 2·9678 7370 |
| 10 | ·6969 6970 | ·7096 7742 | ·0788 1530 | ·0815 7587 | 2·9522 9814 |
| 10·5 | ·6865 6716 | ·6984 1270 | ·0789 7765+ | ·0708 4762 | 2·9389 7632 |
| 11 | ·6764 7059 | ·6875 0000$^e$ | ·0790 7648 | ·0614 8441 | 2·9275 7184 |
| 12 | ·6571 4286 | ·6666 6667 | ·0791 1070 | ·0461 1427 | 2·9094 5637 |
| 13 | ·6388 8889 | ·6470 5882 | ·0789 6456 | ·0342 7862 | 2·8962 5337 |
| 14 | ·6216 2162 | ·6285 7143 | ·0786 7458 | ·0251 3874 | 2·8867 6541 |
| 15 | ·6052 6316 | ·6111 1111 | ·0782 6970 | ·0180 8696 | 2·8801 2725+ |
| 16 | ·5897 4359 | ·5945 9459 | ·0777 7308 | ·0126 7362 | 2·8757 0065+ |
| 17 | ·5750 0000$^e$ | ·5789 4737 | ·0772 0341 | ·0085 5995− | 2·8730 0643 |
| 18 | ·5609 7561 | ·5641 0256 | ·0765 7582 | ·0054 8671 | 2·8716 7938 |
| 19 | ·5476 1905− | ·5500 0000$^e$ | ·0759 0270 | ·0032 5283 | 2·8714 3749 |
| 20 | ·5348 8372 | ·5365 8537 | ·0751 9416 | ·0017 0048 | 2·8720 6049 |
| 21 | ·5227 2727 | ·5238 0952 | ·0744 5856 | ·0007 0448 | 2·8733 7466 |
| 22 | ·5111 1111 | ·5116 2791 | ·0737 0277 | ·0001 6462 | 2·8752 4178 |
| 23 | ·5000 0000$^e$ | ·5000 0000$^e$ | ·0729 3250− | ·0000 0000$^e$ | 2·8775 5102 |
| 24 | ·4893 6170 | ·4888 8889 | ·0721 5245− | ·0001 4487 | 2·8802 1295+ |
| 25 | ·4791 6667 | ·4782 6087 | ·0713 6654 | ·0005 4539 | 2·8831 5499 |
| 26 | ·4693 8776 | ·4680 8511 | ·0705 7803 | ·0011 5726 | 2·8863 1789 |
| 27 | ·4600 0000$^e$ | ·4583 3333 | ·0697 8960 | ·0019 4380 | 2·8896 5314 |
| 28 | ·4509 8039 | ·4489 7959 | ·0690 0349 | ·0028 7452 | 2·8931 2083 |
| 29 | ·4423 0769 | ·4400 0000$^e$ | ·0682 2156 | ·0039 2396 | 2·8966 8802 |
| 30 | ·4339 6226 | ·4313 7255− | ·0674 4532 | ·0050 7079 | 2·9003 2750− |
| 31 | ·4259 2593 | ·4230 7692 | ·0666 7602 | ·0062 9705− | 2·9040 1670 |
| 32 | ·4181 8182 | ·4150 9434 | ·0659 1469 | ·0075 8762 | 2·9077 3692 |
| 33 | ·4107 1429 | ·4074 0741 | ·0651 6215+ | ·0089 2970 | 2·9114 7261 |
| 34 | ·4035 0877 | ·4000 0000$^e$ | ·0644 1908 | ·0103 1247 | 2·9152 1089 |
| 35 | ·3965 5172 | ·3928 5714 | ·0636 8600 | ·0117 2671 | 2·9189 4104 |
| 36 | ·3898 3051 | ·3859 6491 | ·0629 6331 | ·0131 6461 | 2·9226 5422 |
| 37 | ·3833 3333 | ·3793 1034 | ·0622 5133 | ·0146 1951 | 2·9263 4309 |
| 38 | ·3770 4918 | ·3728 8136 | ·0615 5028 | ·0160 8576 | 2·9300 0163 |
| 39 | ·3709 6774 | ·3666 6667 | ·0608 6032 | ·0175 5853 | 2·9336 2490 |
| 40 | ·3650 7937 | ·3606 5574 | ·0601 8155+ | ·0190 3370 | 2·9372 0888 |
| 41 | ·3593 7500$^e$ | ·3548 3871 | ·0595 1401 | ·0205 0779 | 2·9407 5031 |
| 42 | ·3538 4615+ | ·3492 0635− | ·0588 5769 | ·0219 7781 | 2·9442 4661 |
| 43 | ·3484 8485− | ·3437 5000$^e$ | ·0582 1257 | ·0234 4124 | 2·9476 9575+ |
| 44 | ·3432 8358 | ·3384 6154 | ·0575 7859 | ·0248 9596 | 2·9510 9616 |
| 45 | ·3382 3529 | ·3333 3333 | ·0569 5564 | ·0263 4014 | 2·9544 4668 |
| 46 | ·3333 3333 | ·3283 5821 | ·0563 4362 | ·0277 7227 | 2·9577 4648 |
| 47 | ·3285 7143 | ·3235 2941 | ·0557 4240 | ·0291 9108 | 2·9609 9502 |
| 48 | ·3239 4366 | ·3188 4058 | ·0551 5183 | ·0305 9551 | 2·9641 9201 |
| 49 | ·3194 4444 | ·3142 8571 | ·0545 7177 | ·0319 8469 | 2·9673 3735− |
| 50 | ·3150 6849 | ·3098 5915+ | ·0540 0204 | ·0333 5791 | 2·9704 3112 |

$p = 24$

| $q$ | $\bar{x}$ | $\tilde{x}$ | $\sigma$ | $\beta_1$ | $\beta_2$ |
|---|---|---|---|---|---|
| 0·5 | ·9795 9184 | *J*-curve | ·0279 9977 | 6·6844 0726 | 12·4438 2504 |
| 1 | ·9600 0000$^e$ | 1·0000 0000$^e$ | ·0384 3076 | 3·1444 9017 | 7·3339 9471 |
| 1·5 | ·9411 7647 | ·9787 2340 | ·0457 0757 | 1·9710 7438 | 5·7423 4450− |
| 2 | ·9230 7692 | ·9583 3333 | ·0512 8205+ | 1·3890 3061 | 4·8048 0296 |
| 2·5 | ·9056 6038 | ·9387 7551 | ·0557 3954 | 1·0433 4667 | 4·3085 7865+ |
| 3 | ·8888 8889 | ·9200 0000$^e$ | ·0593 9139 | ·8156 9560 | 3·9827 5862 |
| 3·5 | ·8727 2727 | ·9019 6078 | ·0624 2873 | ·6553 7407 | 3·7541 0828 |
| 4 | ·8571 4286 | ·8846 1538 | ·0649 7983 | ·5370 3704 | 3·5860 2151 |
| 4·5 | ·8421 0526 | ·8679 2453 | ·0671 3608 | ·4466 0933 | 3·4581 7070 |
| 5 | ·8275 8621 | ·8518 5185+ | ·0689 6552 | ·3756 5036 | 3·3583 6694 |
| 5·5 | ·8135 5932 | ·8363 6364 | ·0705 2044 | ·3187 9261 | 3·2788 6003 |
| 6 | ·8000 0000$^e$ | ·8214 2857 | ·0718 4212 | ·2724 6094 | 3·2144 8864 |
| 6·5 | ·7868 8525− | ·8070 1754 | ·0729 6376 | ·2341 8298 | 3·1616 8418 |
| 7 | ·7741 9355− | ·7931 0345− | ·0739 1251 | ·2021 9511 | 3·1179 0171 |
| 7·5 | ·7619 0476 | ·7796 6102 | ·0747 1091 | ·1752 0606 | 3·0812 7839 |
| 8 | ·7500 0000$^e$ | ·7666 6667 | ·0753 7784 | ·1522 4913 | 3·0504 2017 |
| 8·5 | ·7384 6154 | ·7540 9836 | ·0759 2929 | ·1325 8666 | 3·0242 6366 |
| 9 | ·7272 7273 | ·7419 3548 | ·0763 7891 | ·1156 4626 | 3·0019 8413 |
| 9·5 | ·7164 1791 | ·7301 5873 | ·0767 3846 | ·1009 7725− | 2·9829 3256 |
| 10 | ·7058 8235+ | ·7187 5000$^e$ | ·0770 1808 | ·0882 2016 | 2·9665 9159 |
| 10·5 | ·6956 5217 | ·7076 9231 | ·0772 2661 | ·0770 8495+ | 2·9525 4403 |
| 11 | ·6857 1429 | ·6969 6970 | ·0773 7179 | ·0673 3515− | 2·9404 5002 |
| 12 | ·6666 6667 | ·6764 7059 | ·0774 9843 | ·0512 4654 | 2·9210 5263 |
| 13 | ·6486 4865− | ·6571 4286 | ·0774 4329 | ·0387 5655+ | 2·9066 8146 |
| 14 | ·6315 7895− | ·6388 8889 | ·0772 4204 | ·0290 1786 | 2·8961 2369 |
| 15 | ·6153 8462 | ·6216 2162 | ·0769 2308 | ·0214 1582 | 2·8885 0174 |
| 16 | ·6000 0000$^e$ | ·6052 6316 | ·0765 0921 | ·0154 9509 | 2·8831 6722 |
| 17 | ·5853 6585+ | ·5897 4359 | ·0760 1890 | ·0109 1210 | 2·8796 3251 |
| 18 | ·5714 2857 | ·5750 0000$^e$ | ·0754 6722 | ·0074 0358 | 2·8775 2525+ |
| 19 | ·5581 3953 | ·5609 7561 | ·0748 6652 | ·0047 6500− | 2·8765 5734 |
| 20 | ·5454 5455− | ·5476 1905− | ·0742 2696 | ·0028 3554 | 2·8765 0324 |
| 21 | ·5333 3333 | ·5348 8372 | ·0735 5697 | ·0014 8742 | 2·8771 8465+ |
| 22 | ·5217 3913 | ·5227 2727 | ·0728 6353 | ·0006 1816 | 2·8784 5934 |
| 23 | ·5106 3830 | ·5111 1111 | ·0721 5245− | ·0001 4487 | 2·8802 1295+ |
| 24 | ·5000 0000$^e$ | ·5000 0000$^e$ | ·0714 2857 | ·0000 0000$^e$ | 2·8823 5294 |
| 25 | ·4897 9592 | ·4893 6170 | ·0706 9595+ | ·0001 2816 | 2·8848 0392 |
| 26 | ·4800 0000$^e$ | ·4791 6667 | ·0699 5797 | ·0004 8361 | 2·8875 0419 |
| 27 | ·4705 8824 | ·4693 8776 | ·0692 1746 | ·0010 2844 | 2·8904 0298 |
| 28 | ·4615 3846 | ·4600 0000$^e$ | ·0684 7678 | ·0017 3101 | 2·8934 5839 |
| 29 | ·4528 3019 | ·4509 8039 | ·0677 3792 | ·0025 6483 | 2·8966 3569 |
| 30 | ·4444 4444 | ·4423 0769 | ·0670 0252 | ·0035 0765+ | 2·8999 0602 |
| 31 | ·4363 6364 | ·4339 6226 | ·0662 7195+ | ·0045 4069 | 2·9032 4532 |
| 32 | ·4285 7143 | ·4259 2593 | ·0655 4735+ | ·0056 4804 | 2·9066 3355− |
| 33 | ·4210 5263 | ·4181 8182 | ·0648 2966 | ·0068 1622 | 2·9100 5393 |
| 34 | ·4137 9310 | ·4107 1429 | ·0641 1964 | ·0080 3377 | 2·9134 9245− |
| 35 | ·4067 7966 | ·4035 0877 | ·0634 1792 | ·0092 9090 | 2·9169 3737 |
| 36 | ·4000 0000$^e$ | ·3965 5172 | ·0627 2500+ | ·0105 7926 | 2·9203 7890 |
| 37 | ·3934 4262 | ·3898 3051 | ·0620 4129 | ·0118 9171 | 2·9238 0885+ |
| 38 | ·3870 9677 | ·3833 3333 | ·0613 6708 | ·0132 2214 | 2·9272 2039 |
| 39 | ·3809 5238 | ·3770 4918 | ·0607 0261 | ·0145 6532 | 2·9306 0785+ |
| 40 | ·3750 0000$^e$ | ·3709 6774 | ·0600 4806 | ·0159 1674 | 2·9339 6653 |
| 41 | ·3692 3077 | ·3650 7937 | ·0594 0353 | ·0172 7257 | 2·9372 9255+ |
| 42 | ·3636 3636 | ·3593 7500$^e$ | ·0587 6909 | ·0186 2951 | 2·9405 8275+ |
| 43 | ·3582 0896 | ·3538 4615+ | ·0581 4476 | ·0199 8473 | 2·9438 3456 |
| 44 | ·3529 4118 | ·3484 8485− | ·0575 3055− | ·0213 3581 | 2·9470 4591 |
| 45 | ·3478 2609 | ·3432 8358 | ·0569 2641 | ·0226 8069 | 2·9502 1518 |
| 46 | ·3428 5714 | ·3382 3529 | ·0563 3228 | ·0240 1760 | 2·9533 4111 |
| 47 | ·3380 2817 | ·3333 3333 | ·0557 4808 | ·0253 4506 | 2·9564 2276 |
| 48 | ·3333 3333 | ·3285 7143 | ·0551 7373 | ·0266 6180 | 2·9594 5946 |
| 49 | ·3287 6712 | ·3239 4366 | ·0546 0909 | ·0279 6674 | 2·9624 5077 |
| 50 | ·3243 2432 | ·3194 4444 | ·0540 5405+ | ·0292 5900 | 2·9653 9645− |

$p = 25$

| $q$ | $\bar{x}$ | $\tilde{x}$ | $\sigma$ | $\beta_1$ | $\beta_2$ |
|---|---|---|---|---|---|
| 0·5 | ·9803 9216 | *J*-curve | ·0269 3344 | 6·7307 3719 | 12·5313 3014 |
| 1 | ·9615 3846 | 1·0000 0000$^e$ | ·0370 0963 | 3·1738 7755$^+$ | 7·3897 5369 |
| 1·5 | ·9433 9623 | ·9795 9184 | ·0440 6598 | 1·9943 7776 | 5·6867 6777 |
| 2 | ·9259 2593 | ·9600 0000$^e$ | ·0494 9282 | 1·4089 8930 | 4·8430 3448 |
| 2·5 | ·9090 9091 | ·9411 7647 | ·0538 4990 | 1·0610 7440 | 4·3427 0631 |
| 3 | ·8928 5714 | ·9230 7692 | ·0574 3460 | ·8317 6296 | 4·0138 4946 |
| 3·5 | ·8771 9298 | ·9056 6038 | ·0604 2938 | ·6701 1786 | 3·7827 9023 |
| 4 | ·8620 6897 | ·8888 8889 | ·0629 5662 | ·5506 7638 | 3·6127 0161 |
| 4·5 | ·8474 5763 | ·8727 2727 | ·0651 0350$^-$ | ·4592 9733 | 3·4831 3228 |
| 5 | ·8333 3333 | ·8571 4286 | ·0669 3494 | ·3875 0000$^e$ | 3·3818 1818 |
| 5·5 | ·8196 7213 | ·8421 0526 | ·0685 0089 | ·3298 9091 | 3·3009 6065$^+$ |
| 6 | ·8064 5161 | ·8275 8621 | ·0698 4076 | ·2828 7726 | 3·2353 6542 |
| 6·5 | ·7936 5079 | ·8135 5932 | ·0709 8629 | ·2439 7416 | 3·1814 4062 |
| 7 | ·7812 5000$^e$ | ·8000 0000$^e$ | ·0719 6347 | ·2114 0880 | 3·1366 2425$^-$ |
| 7·5 | ·7692 3077 | ·7868 8525$^-$ | ·0727 9388 | ·1838 8294 | 3·0990 4062 |
| 8 | ·7575 7576 | ·7741 9355$^-$ | ·0734 9564 | ·1604 2449 | 3·0672 8571 |
| 8·5 | ·7462 6866 | ·7619 0476 | ·0740 8412 | ·1402 9149 | 3·0402 8827 |
| 9 | ·7352 9412 | ·7500 0000$^e$ | ·0745 7243 | ·1229 0809 | 3·0172 1722 |
| 9·5 | ·7246 3768 | ·7384 6154 | ·0749 7190 | ·1078 2076 | 2·9974 1831 |
| 10 | ·7142 8571 | ·7272 7273 | ·0752 9233 | ·0946 6764 | 2·9803 6984 |
| 10·5 | ·7042 2535$^+$ | ·7164 1791 | ·0755 4224 | ·0831 5666 | 2·9656 5096 |
| 11 | ·6944 4444 | ·7058 8235$^+$ | ·0757 2913 | ·0730 4961 | 2·9529 1866 |
| 12 | ·6756 7568 | ·6857 1429 | ·0759 3929 | ·0562 9630 | 2·9323 3333 |
| 13 | ·6578 9474 | ·6666 6667 | ·0759 6714 | ·0432 0000$^e$ | 2·9168 7805$^-$ |
| 14 | ·6410 2564 | ·6486 4865$^-$ | ·0758 4718 | ·0329 0558 | 2·9053 2603 |
| 15 | ·6250 0000$^e$ | ·6315 7895$^-$ | ·0756 0730 | ·0247 9214 | 2·8967 8848 |
| 16 | ·6097 5610 | ·6153 8462 | ·0752 6993 | ·0183 9913 | 2·8906 0782 |
| 17 | ·5952 3810 | ·6000 0000$^e$ | ·0748 5330 | ·0133 7871 | 2·8862 8877 |
| 18 | ·5813 9535$^-$ | ·5853 6585$^+$ | ·0743 7234 | ·0094 6392 | 2·8834 5250$^-$ |
| 19 | ·5681 8182 | ·5714 2857 | ·0738 3935$^+$ | ·0064 4712 | 2·8818 0535$^-$ |
| 20 | ·5555 5556 | ·5581 3953 | ·0732 6450$^-$ | ·0041 6478 | 2·8811 1702 |
| 21 | ·5434 7826 | ·5454 5455$^-$ | ·0726 5624 | ·0024 8677 | 2·8812 0505$^+$ |
| 22 | ·5319 1489 | ·5333 3333 | ·0720 2162 | ·0013 0855$^-$ | 2·8819 2356 |
| 23 | ·5208 3333 | ·5217 3913 | ·0713 6654 | ·0005 4539 | 2·8831 5499 |
| 24 | ·5102 0408 | ·5106 3830 | ·0706 9595$^+$ | ·0001 2816 | 2·8848 0392 |
| 25 | ·5000 0000$^e$ | ·5000 0000$^e$ | ·0700 1400 | ·0000 0000$^e$ | 2·8867 9245$^+$ |
| 26 | ·4901 9608 | ·4897 9592 | ·0693 2419 | ·0001 1392 | 2·8890 5660 |
| 27 | ·4807 6923 | ·4800 0000$^e$ | ·0686 2946 | ·0004 3083 | 2·8915 4358 |
| 28 | ·4716 9811 | ·4705 8824 | ·0679 3229 | ·0009 1806 | 2·8942 0965$^-$ |
| 29 | ·4629 6296 | ·4615 3846 | ·0672 3477 | ·0015 4821 | 2·8970 1841 |
| 30 | ·4545 4545$^+$ | ·4528 3019 | ·0665 3864 | ·0022 9814 | 2·8999 3950$^+$ |
| 31 | ·4464 2857 | ·4444 4444 | ·0658 4539 | ·0031 4833 | 2·9029 4753 |
| 32 | ·4385 9649 | ·4363 6364 | ·0651 5626 | ·0040 8216 | 2·9060 2119 |
| 33 | ·4310 3448 | ·4285 7143 | ·0644 7227 | ·0050 8552 | 2·9091 4257 |
| 34 | ·4237 2881 | ·4210 5263 | ·0637 9429 | ·0061 4636 | 2·9122 9664 |
| 35 | ·4166 6667 | ·4137 9310 | ·0631 2303 | ·0072 5435$^-$ | 2·9154 7070 |
| 36 | ·4098 3607 | ·4067 7966 | ·0624 5908 | ·0084 0066 | 2·9186 5410 |
| 37 | ·4032 2581 | ·4000 0000$^e$ | ·0618 0291 | ·0095 7770 | 2·9218 3784 |
| 38 | ·3968 2540 | ·3934 4262 | ·0611 5490 | ·0107 7895$^-$ | 2·9250 1435$^+$ |
| 39 | ·3906 2500$^e$ | ·3870 9677 | ·0605 1536 | ·0119 9878 | 2·9281 7730 |
| 40 | ·3846 1538 | ·3809 5238 | ·0598 8453 | ·0132 3235$^-$ | 2·9313 2133 |
| 41 | ·3787 8788 | ·3750 0000$^e$ | ·0592 6259 | ·0144 7548 | 2·9344 4202 |
| 42 | ·3731 3433 | ·3692 3077 | ·0586 4965$^-$ | ·0157 2459 | 2·9375 3564 |
| 43 | ·3676 4706 | ·3636 3636 | ·0580 4580 | ·0169 7655$^+$ | 2·9405 9913 |
| 44 | ·3623 1884 | ·3582 0896 | ·0574 5108 | ·0182 2871 | 2·9436 2996 |
| 45 | ·3571 4286 | ·3529 4118 | ·0568 6552 | ·0194 7874 | 2·9466 2608 |
| 46 | ·3521 1268 | ·3478 2609 | ·0562 8910 | ·0207 2466 | 2·9495 8582 |
| 47 | ·3472 2222 | ·3428 5714 | ·0557 2178 | ·0219 6478 | 2·9525 0788 |
| 48 | ·3424 6575$^+$ | ·3380 2817 | ·0551 6351 | ·0231 9763 | 2·9553 9123 |
| 49 | ·3378 3784 | ·3333 3333 | ·0546 1421 | ·0244 2196 | 2·9582 3511 |
| 50 | ·3333 3333 | ·3287 6712 | ·0540 7381 | ·0256 3670 | 2·9610 3896 |

$p = 26$

| $q$ | $\bar{x}$ | $\tilde{x}$ | $\sigma$ | $\beta_1$ | $\beta_2$ |
|---|---|---|---|---|---|
| 0·5 | ·9811 3208 | *J*-curve | ·0259 4535$^{-}$ | 6·7739 1860 | 12·6130 5153 |
| 1 | ·9629 6296 | 1·0000 0000$^{e}$ | ·0356 8978 | 3·2013 1711 | 7·4419 0981 |
| 1·5 | ·9454 5455$^{-}$ | ·9803 9216 | ·0425 3801 | 2·0161 7572 | 5·7283 8609 |
| 2 | ·9285 7143 | ·9615 3846 | ·0478 2386 | 1·4276 9231 | 4·8789 0819 |
| 2·5 | ·9122 8070 | ·9433 9623 | ·0520 8358 | 1·0777 1691 | 4·3747 7932 |
| 3 | ·8965 5172 | ·9259 2593 | ·0556 0178 | ·8468 7425$^{-}$ | 4·0431 1414 |
| 3·5 | ·8813 5593 | ·9090 9091 | ·0585 5295$^{-}$ | ·6840 0987 | 3·8098 2973 |
| 4 | ·8666 6667 | ·8928 5714 | ·0610 5406 | ·5635 5168 | 3·6378 9336 |
| 4·5 | ·8524 5902 | ·8771 9298 | ·0631 8847 | ·4712 9722 | 3·5067 3850$^{-}$ |
| 5 | ·8387 0968 | ·8620 6897 | ·0650 1821 | ·3987 2854 | 3·4040 3126 |
| 5·5 | ·8253 9683 | ·8474 5763 | ·0665 9104 | ·3404 2812 | 3·3219 2791 |
| 6 | ·8125 0000$^{e}$ | ·8333 3333 | ·0679 4466 | ·2927 8680 | 3·2552 0362 |
| 6·5 | ·8000 0000$^{e}$ | ·8196 7213 | ·0691 0947 | ·2533 0813 | 3·2002 4495$^{-}$ |
| 7 | ·7878 7879 | ·8064 5161 | ·0701 1038 | ·2202 1081 | 3·1544 7410 |
| 7·5 | ·7761 1940 | ·7936 5079 | ·0709 6806 | ·1921 9019 | 3·1160 0350$^{+}$ |
| 8 | ·7647 0588 | ·7812 5000$^{e}$ | ·0716 9993 | ·1682 6923 | 3·0834 1996 |
| 8·5 | ·7536 2319 | ·7692 3077 | ·0723 2074 | ·1477 0202 | 3·0556 4495$^{+}$ |
| 9 | ·7428 5714 | ·7575 7576 | ·0728 4314 | ·1299 0954 | 3·0318 4156 |
| 9·5 | ·7323 9437 | ·7462 6866 | ·0732 7801 | ·1144 3563 | 3·0113 5075$^{+}$ |
| 10 | ·7222 2222 | ·7352 9412 | ·0736 3483 | ·1009 1626 | 2·9936 4684 |
| 10·5 | ·7123 2877 | ·7246 3768 | ·0739 2185$^{-}$ | ·0890 5751 | 2·9783 0560 |
| 11 | ·7027 0270 | ·7142 8571 | ·0741 4630 | ·0786 1961 | 2·9649 8117 |
| 12 | ·6842 1053 | ·6944 4444 | ·0744 3229 | ·0612 5000$^{e}$ | 2·9432 9268 |
| 13 | ·6666 6667 | ·6756 7568 | ·0745 3560 | ·0475 9072 | 2·9268 2927 |
| 14 | ·6500 0000$^{e}$ | ·6578 9474 | ·0744 9014 | ·0367 7955$^{-}$ | 2·9143 5143 |
| 15 | ·6341 4634 | ·6410 2564 | ·0743 2310 | ·0281 8987 | 2·9049 6016 |
| 16 | ·6190 4762 | ·6250 0000$^{e}$ | ·0740 5649 | ·0213 5648 | 2·8979 8951 |
| 17 | ·6046 5116 | ·6097 5610 | ·0737 0829 | ·0159 2760 | 2·8929 3724 |
| 18 | ·5909 0909 | ·5952 3810 | ·0732 9325$^{+}$ | ·0116 3298 | 2·8894 1863 |
| 19 | ·5777 7778 | ·5813 9535$^{-}$ | ·0728 2358 | ·0082 6212 | 2·8871 3498 |
| 20 | ·5652 1739 | ·5681 8182 | ·0723 0943 | ·0056 4904 | 2·8858 5165$^{-}$ |
| 21 | ·5531 9149 | ·5555 5556 | ·0717 5924 | ·0036 6148 | 2·8853 8237 |
| 22 | ·5416 6667 | ·5434 7826 | ·0711 8012 | ·0021 9301 | 2·8855 7795$^{+}$ |
| 23 | ·5306 1224 | ·5319 1489 | ·0705 7803 | ·0011 5726 | 2·8863 1789 |
| 24 | ·5200 0000$^{e}$ | ·5208 3333 | ·0699 5797 | ·0004 8361 | 2·8875 0419 |
| 25 | ·5098 0392 | ·5102 0408 | ·0693 2419 | ·0001 1392 | 2·8890 5660 |
| 26 | ·5000 0000$^{e}$ | ·5000 0000$^{e}$ | ·0686 8028 | ·0000 0000$^{e}$ | 2·8909 0909 |
| 27 | ·4905 6604 | ·4901 9608 | ·0680 2927 | ·0001 0172 | 2·8930 0699 |
| 28 | ·4814 8148 | ·4807 6923 | ·0673 7373 | ·0003 8546 | 2·8953 0488 |
| 29 | ·4727 2727 | ·4716 9811 | ·0667 1584 | ·0008 2294 | 2·8977 6485$^{+}$ |
| 30 | ·4642 8571 | ·4629 6296 | ·0660 5746 | ·0013 9029 | 2·9003 5517 |
| 31 | ·4561 4035$^{+}$ | ·4545 4545$^{+}$ | ·0654 0014 | ·0020 6723 | 2·9030 4917 |
| 32 | ·4482 7586 | ·4464 2857 | ·0647 4521 | ·0028 3654 | 2·9058 2440 |
| 33 | ·4406 7797 | ·4385 9649 | ·0640 9380 | ·0036 8350$^{-}$ | 2·9086 6194 |
| 34 | ·4333 3333 | ·4310 3448 | ·0634 4684 | ·0045 9551 | 2·9115 4576 |
| 35 | ·4262 2951 | ·4237 2881 | ·0628 0512 | ·0055 6179 | 2·9144 6232 |
| 36 | ·4193 5484 | ·4166 6667 | ·0621 6930 | ·0065 7302 | 2·9174 0015$^{-}$ |
| 37 | ·4126 9841 | ·4098 3607 | ·0615 3993 | ·0076 2120 | 2·9203 4951 |
| 38 | ·4062 5000$^{e}$ | ·4032 2581 | ·0609 1746 | ·0086 9943 | 2·9233 0215$^{-}$ |
| 39 | ·4000 0000$^{e}$ | ·3968 2540 | ·0603 0227 | ·0098 0174 | 2·9262 5110 |
| 40 | ·3939 3939 | ·3906 2500$^{e}$ | ·0596 9464 | ·0109 2294 | 2·9291 9044 |
| 41 | ·3880 5970 | ·3846 1538 | ·0590 9481 | ·0120 5858 | 2·9321 1518 |
| 42 | ·3823 5294 | ·3787 8788 | ·0585 0296 | ·0132 0475$^{+}$ | 2·9350 2112 |
| 43 | ·3768 1159 | ·3731 3433 | ·0579 1923 | ·0143 5811 | 2·9379 0471 |
| 44 | ·3714 2857 | ·3676 4706 | ·0573 4371 | ·0155 1573 | 2·9407 6300 |
| 45 | ·3661 9718 | ·3623 1884 | ·0567 7646 | ·0166 7509 | 2·9435 9354 |
| 46 | ·3611 1111 | ·3571 4286 | ·0562 1752 | ·0178 3398 | 2·9463 9429 |
| 47 | ·3561 6438 | ·3521 1268 | ·0556 6688 | ·0189 9051 | 2·9491 6358 |
| 48 | ·3513 5135$^{+}$ | ·3472 2222 | ·0551 2454 | ·0201 4303 | 2·9519 0007 |
| 49 | ·3466 6667 | ·3424 6575$^{+}$ | ·0545 9045$^{-}$ | ·0212 9013 | 2·9546 0270 |
| 50 | ·3421 0526 | ·3378 3784 | ·0540 6457 | ·0224 3059 | 2·9572 7062 |

$p = 27$

| $q$ | $\bar{x}$ | $\tilde{x}$ | $\sigma$ | $\beta_1$ | $\beta_2$ |
|---|---|---|---|---|---|
| 0·5 | ·9818 1818 | *J*-curve | ·0250 2717 | 6·8142 6155$^-$ | 12·6895 4339 |
| 1 | ·9642 8571 | 1·0000 0000$^e$ | ·0344 6076 | 3·2269 9588 | 7·4908 0048 |
| 1·5 | ·9473 6842 | ·9811 3208 | ·0411 1229 | 2·0366 0904 | 5·7674 5598 |
| 2 | ·9310 3448 | ·9629 6296 | ·0462 6348 | 1·4452 5379 | 4·9126 3441 |
| 2·5 | ·9152 5424 | ·9454 5455$^-$ | ·0504 2895$^+$ | 1·0933 6991 | 4·4049 7626 |
| 3 | ·9000 0000$^e$ | ·9285 7143 | ·0538 8159 | ·8611 1111 | 4·0707 0707 |
| 3·5 | ·8852 4590 | ·9122 8070 | ·0567 8855$^+$ | ·6971 2032 | 3·8353 6165$^+$ |
| 4 | ·8709 6774 | ·8965 5172 | ·0592 6185$^-$ | ·5757 2357 | 3·6617 1519 |
| 4·5 | ·8571 4286 | ·8813 5593 | ·0613 8127 | ·4826 6132 | 3·5290 9366 |
| 5 | ·8437 5000$^e$ | ·8666 6667 | ·0632 0624 | ·4093 8101 | 3·4250 9804 |
| 5·5 | ·8307 6923 | ·8524 5902 | ·0647 8246 | ·3504 4275$^-$ | 3·3418 4259 |
| 6 | ·8181 8181 | ·8387 0968 | ·0661 4608 | ·3022 2222 | 3·2740 7407 |
| 6·5 | ·8059 7015$^-$ | ·8253 9683 | ·0673 2622 | ·2622 1225$^+$ | 3·2181 5897 |
| 7 | ·7941 1765$^-$ | ·8125 0000$^e$ | ·0683 4676 | ·2286 2369 | 3·1715 0484 |
| 7·5 | ·7826 0870 | ·8000 0000$^e$ | ·0692 2759 | ·2001 4595$^+$ | 3·1322 1309 |
| 8 | ·7714 2857 | ·7878 7879 | ·0699 8542 | ·1757 9742 | 3·0988 6202 |
| 8·5 | ·7605 6338 | ·7761 1940 | ·0706 3445$^-$ | ·1548 2856 | 3·0703 6641 |
| 9 | ·7500 0000$^e$ | ·7647 0588 | ·0711 8685$^-$ | ·1366 5743 | 3·0458 8394 |
| 9·5 | ·7397 2603 | ·7536 2319 | ·0716 5313 | ·1208 2548 | 3·0247 5118 |
| 10 | ·7297 2973 | ·7428 5714 | ·0720 4243 | ·1069 6666 | 3·0064 3875$^-$ |
| 10·5 | ·7200 0000$^e$ | ·7323 9437 | ·0723 6272 | ·0947 8538 | 2·9905 1935$^+$ |
| 11 | ·7105 2632 | ·7222 2222 | ·0726 2102 | ·0840 4040 | 2·9766 4449 |
| 12 | ·6923 0769 | ·7027 0270 | ·0729 7564 | ·0660 9822 | 2·9539 2954 |
| 13 | ·6750 0000$^e$ | ·6842 1053 | ·0731 4786 | ·0519 1516 | 2·9365 2687 |
| 14 | ·6585 3659 | ·6666 6667 | ·0731 7073 | ·0406 2256 | 2·9231 8534 |
| 15 | ·6428 5714 | ·6500 0000$^e$ | ·0730 7082 | ·0315 8861 | 2·9129 9663 |
| 16 | ·6279 0698 | ·6341 4634 | ·0728 6972 | ·0243 4385$^+$ | 2·9052 8717 |
| 17 | ·6136 3636 | ·6190 4762 | ·0725 8509 | ·0185 3293 | 2·8995 4835$^-$ |
| 18 | ·6000 0000$^e$ | ·6046 5116 | ·0722 3151 | ·0138 8260 | 2·8953 9007 |
| 19 | ·5869 5652 | ·5909 0909 | ·0718 2108 | ·0101 7977 | 2·8925 0905$^+$ |
| 20 | ·5744 6809 | ·5777 7778 | ·0713 6387 | ·0072 5624 | 2·8906 6667 |
| 21 | ·5625 0000$^e$ | ·5652 1739 | ·0708 6834 | ·0049 7778 | 2·8896 7320 |
| 22 | ·5510 2041 | ·5531 9149 | ·0703 4158 | ·0032 3626 | 2·8893 7642 |
| 23 | ·5400 0000$^e$ | ·5416 6667 | ·0697 8960 | ·0019 4380 | 2·8896 5314 |
| 24 | ·5294 1176 | ·5306 1224 | ·0692 1746 | ·0010 2844 | 2·8904 0298 |
| 25 | ·5192 3077 | ·5200 0000$^e$ | ·0686 2946 | ·0004 3083 | 2·8915 4358 |
| 26 | ·5094 3396 | ·5098 0392 | ·0680 2927 | ·0001 0172 | 2·8930 0699 |
| 27 | ·5000 0000$^e$ | ·5000 0000$^e$ | ·0674 1999 | ·0000 0000$^e$ | 2·8947 3684 |
| 28 | ·4909 0909 | ·4905 6604 | ·0668 0427 | ·0000 9120 | 2·8966 8616 |
| 29 | ·4821 4286 | ·4814 8148 | ·0661 8437 | ·0003 4624 | 2·8988 1564 |
| 30 | ·4736 8421 | ·4727 2727 | ·0655 6222 | ·0007 4053 | 2·9010 9228 |
| 31 | ·4655 1724 | ·4642 8571 | ·0649 3947 | ·0012 5315$^+$ | 2·9034 8826 |
| 32 | ·4576 2712 | ·4561 4035$^+$ | ·0643 1751 | ·0018 6628 | 2·9059 8008 |
| 33 | ·4500 0000$^e$ | ·4482 7586 | ·0636 9754 | ·0025 6467 | 2·9085 4784 |
| 34 | ·4426 2295$^+$ | ·4406 7797 | ·0630 8058 | ·0033 3522 | 2·9111 7466 |
| 35 | ·4354 8387 | ·4333 3333 | ·0624 6747 | ·0041 6667 | 2·9138 4615$^+$ |
| 36 | ·4285 7143 | ·4262 2951 | ·0618 5896 | ·0050 4931 | 2·9165 5012 |
| 37 | ·4218 7500$^e$ | ·4193 5484 | ·0612 5564 | ·0059 7475$^+$ | 2·9192 7613 |
| 38 | ·4153 8462 | ·4126 9841 | ·0606 5804 | ·0069 3573 | 2·9220 1531 |
| 39 | ·4090 9091 | ·4062 5000$^e$ | ·0600 6657 | ·0079 2595$^-$ | 2·9247 6009 |
| 40 | ·4029 8507 | ·4000 0000$^e$ | ·0594 8160 | ·0089 3992 | 2·9275 0403 |
| 41 | ·3970 5882 | ·3939 3939 | ·0589 0340 | ·0099 7290 | 2·9302 4161 |
| 42 | ·3913 0435$^-$ | ·3880 5970 | ·0583 3221 | ·0110 2074 | 2·9329 6818 |
| 43 | ·3857 1429 | ·3823 5294 | ·0577 6821 | ·0120 7984 | 2·9356 7976 |
| 44 | ·3802 8169 | ·3768 1159 | ·0572 1154 | ·0131 4705$^-$ | 2·9383 7299 |
| 45 | ·3750 0000$^e$ | ·3714 2857 | ·0566 6230 | ·0142 1963 | 2·9410 4505$^-$ |
| 46 | ·3698 6301 | ·3661 9718 | ·0561 2057 | ·0152 9519 | 2·9436 9353 |
| 47 | ·3648 6486 | ·3611 1111 | ·0555 8639 | ·0163 7165$^-$ | 2·9463 1647 |
| 48 | ·3600 0000$^e$ | ·3561 6438 | ·0550 5978 | ·0174 4720 | 2·9489 1220 |
| 49 | ·3552 6316 | ·3513 5135$^+$ | ·0545 4073 | ·0185 2026 | 2·9514 7937 |
| 50 | ·3506 4935$^+$ | ·3466 6667 | ·0540 2924 | ·0195 8945$^+$ | 2·9540 1688 |

$p = 28$

| $q$ | $\bar{x}$ | $\tilde{x}$ | $\sigma$ | $\beta_1$ | $\beta_2$ |
|---|---|---|---|---|---|
| 0·5 | ·9824 5614 | $J$-curve | ·0241 7175$^{+}$ | 6·8520 3670 | 12·7612 9140 |
| 1 | ·9655 1724 | 1·0000 0000$^{e}$ | ·0333 1351 | 3·2510 7775$^{-}$ | 7·5367 2235$^{+}$ |
| 1·5 | ·9491 5254 | ·9818 1818 | ·0397 7889 | 2·0558 0151 | 5·8042 0373 |
| 2 | ·9333 3333 | ·9642 8571 | ·0448 0143 | 1·4617 7455$^{+}$ | 4·9443 9935$^{+}$ |
| 2·5 | ·9180 3279 | ·9473 6842 | ·0488 7580 | 1·1081 1834 | 4·4334 5580 |
| 3 | ·9032 2581 | ·9310 3448 | ·0522 6404 | ·8745 4633 | 4·0967 6598 |
| 3·5 | ·8888 8889 | ·9152 5424 | ·0551 2655$^{-}$ | ·7095 1214 | 3·8595 0681 |
| 4 | ·8750 0000$^{e}$ | ·9000 0000$^{e}$ | ·0575 7077 | ·5872 4666 | 3·6842 7371 |
| 4·5 | ·8615 3846 | ·8852 4590 | ·0596 7322 | ·4934 3709 | 3·5502 9209 |
| 5 | ·8484 8485$^{-}$ | ·8709 6774 | ·0614 9087 | ·4194 9854 | 3·4451 0204 |
| 5·5 | ·8358 2090 | ·8571 4286 | ·0630 6756 | ·3599 7032 | 3·3607 7862 |
| 6 | ·8235 2941 | ·8437 5000$^{e}$ | ·0644 3795$^{-}$ | ·3112 1399 | 3·2920 4204 |
| 6·5 | ·8115 9420 | ·8307 6923 | ·0656 3000 | ·2707 1240 | 3·2352 4010 |
| 7 | ·8000 0000$^{e}$ | ·8181 8181 | ·0666 6667 | ·2366 6910 | 3·1877 6671 |
| 7·5 | ·7887 3239 | ·8059 7015$^{-}$ | ·0675 6704 | ·2077 6804 | 3·1477 1305$^{-}$ |
| 8 | ·7777 7778 | ·7941 1765$^{-}$ | ·0683 4719 | ·1830 2335$^{-}$ | 3·1136 4951 |
| 8·5 | ·7671 2329 | ·7826 0870 | ·0690 2080 | ·1616 8215$^{-}$ | 3·0844 8466 |
| 9 | ·7567 5676 | ·7714 2857 | ·0695 9960 | ·1431 5978 | 3·0593 7118 |
| 9·5 | ·7466 6667 | ·7605 6338 | ·0700 9373 | ·1269 9551 | 3·0376 4157 |
| 10 | ·7368 4211 | ·7500 0000$^{e}$ | ·0705 1201 | ·1128 2143 | 3·0187 6307 |
| 10·5 | ·7272 7273 | ·7397 2603 | ·0708 6216 | ·1003 4040 | 3·0023 0552 |
| 11 | ·7179 4872 | ·7297 2973 | ·0711 5096 | ·0893 0986 | 2·9879 1801 |
| 12 | ·7000 0000$^{e}$ | ·7105 2632 | ·0715 6781 | ·0708 3468 | 2·9642 4616 |
| 13 | ·6829 2683 | ·6923 0769 | ·0718 0293 | ·0561 6341 | 2·9459 6682 |
| 14 | ·6666 6667 | ·6750 0000$^{e}$ | ·0718 8852 | ·0444 2149 | 2·9318 1818 |
| 15 | ·6511 6279 | ·6585 3659 | ·0718 5051 | ·0349 7237 | 2·9208 8337 |
| 16 | ·6363 6364 | ·6428 5714 | ·0717 1006 | ·0273 4270 | 2·9124 8183 |
| 17 | ·6222 2222 | ·6279 0698 | ·0714 8453 | ·0211 7388 | 2·9060 9914 |
| 18 | ·6086 9565$^{+}$ | ·6136 3636 | ·0711 8828 | ·0161 8993 | 2·9013 4030 |
| 19 | ·5957 4468 | ·6000 0000$^{e}$ | ·0708 3328 | ·0121 7538 | 2·8978 9781 |
| 20 | ·5833 3333 | ·5869 5652 | ·0704 2952 | ·0089 6000$^{e}$ | 2·8955 2941 |
| 21 | ·5714 2857 | ·5744 6809 | ·0699 8542 | ·0064 0779 | 2·8940 4223 |
| 22 | ·5600 0000$^{e}$ | ·5625 0000$^{e}$ | ·0695 0808 | ·0044 0905$^{+}$ | 2·8932 8125$^{-}$ |
| 23 | ·5490 1961 | ·5510 2041 | ·0690 0349 | ·0028 7452 | 2·8931 2083 |
| 24 | ·5384 6154 | ·5400 0000$^{e}$ | ·0684 7678 | ·0017 3101 | 2·8934 5839 |
| 25 | ·5283 0189 | ·5294 1176 | ·0679 3229 | ·0009 1806 | 2·8942 0965$^{-}$ |
| 26 | ·5185 1852 | ·5192 3077 | ·0673 7373 | ·0003 8546 | 2·8953 0488 |
| 27 | ·5090 9091 | ·5094 3396 | ·0668 0427 | ·0000 9120 | 2·8966 8616 |
| 28 | ·5000 0000$^{e}$ | ·5000 0000$^{e}$ | ·0662 2662 | ·0000 0000$^{e}$ | 2·8983 0508 |
| 29 | ·4912 2807 | ·4909 0909 | ·0656 4311 | ·0000 8208 | 2·9001 2107 |
| 30 | ·4827 5862 | ·4821 4286 | ·0650 5574 | ·0003 1217 | 2·9020 9992 |
| 31 | ·4745 7627 | ·4736 8421 | ·0644 6622 | ·0006 6877 | 2·9042 1278 |
| 32 | ·4666 6667 | ·4655 1724 | ·0638 7602 | ·0011 3349 | 2·9064 3515$^{+}$ |
| 33 | ·4590 1639 | ·4576 2712 | ·0632 8639 | ·0016 9059 | 2·9087 4626 |
| 34 | ·4516 1290 | ·4500 0000$^{e}$ | ·0626 9841 | ·0023 2652 | 2·9111 2839 |
| 35 | ·4444 4444 | ·4426 2295$^{+}$ | ·0621 1300 | ·0030 2959 | 2·9135 6643 |
| 36 | ·4375 0000$^{e}$ | ·4354 8387 | ·0615 3095$^{-}$ | ·0037 8970 | 2·9160 4747 |
| 37 | ·4307 6923 | ·4285 7143 | ·0609 5293 | ·0045 9811 | 2·9185 6045$^{-}$ |
| 38 | ·4242 4242 | ·4218 7500$^{e}$ | ·0603 7950$^{-}$ | ·0054 4723 | 2·9210 9590 |
| 39 | ·4179 1045$^{-}$ | ·4153 8462 | ·0598 1114 | ·0063 3043 | 2·9236 4571 |
| 40 | ·4117 6471 | ·4090 9091 | ·0592 4825$^{+}$ | ·0072 4198 | 2·9262 0293 |
| 41 | ·4057 9710 | ·4029 8507 | ·0586 9118 | ·0081 7685$^{+}$ | 2·9287 6159 |
| 42 | ·4000 0000$^{e}$ | ·3970 5882 | ·0581 4019 | ·0091 3066 | 2·9313 1659 |
| 43 | ·3943 6620 | ·3913 0435$^{-}$ | ·0575 9551 | ·0100 9957 | 2·9338 6355$^{-}$ |
| 44 | ·3888 8889 | ·3857 1429 | ·0570 5733 | ·0110 8023 | 2·9363 9874 |
| 45 | ·3835 6164 | ·3802 8169 | ·0565 2578 | ·0120 6970 | 2·9389 1896 |
| 46 | ·3783 7838 | ·3750 0000$^{e}$ | ·0560 0097 | ·0130 6542 | 2·9414 2152 |
| 47 | ·3733 3333 | ·3698 6301 | ·0554 8299 | ·0140 6512 | 2·9439 0412 |
| 48 | ·3684 2105$^{+}$ | ·3648 6486 | ·0549 7189 | ·0150 6684 | 2·9463 6482 |
| 49 | ·3636 3636 | ·3600 0000$^{e}$ | ·0544 6770 | ·0160 6885$^{+}$ | 2·9488 0199 |
| 50 | ·3589 7436 | ·3552 6316 | ·0539 7043 | ·0170 6964 | 2·9512 1429 |

$p = 29$

| $q$ | $\bar{x}$ | $\tilde{x}$ | $\sigma$ | $\beta_1$ | $\beta_2$ |
|---|---|---|---|---|---|
| 0·5 | ·9830 5085$^-$ | *J*-curve | ·0233 7287 | 6·8874 8143 | 12·8287 2300 |
| 1 | ·9666 6667 | 1·0000 0000$^e$ | ·0322 4014 | 3·2737 0690 | 7·5799 3730 |
| 1·5 | ·9508 1967 | ·9824 5614 | ·0385 2916 | 2·0738 6248 | 5·8388 2972 |
| 2 | ·9354 8387 | ·9655 1724 | ·0434 2875$^-$ | 1·4773 4397 | 4·9743 6843 |
| 2·5 | ·9206 3492 | ·9491 5254 | ·0474 1511 | 1·1220 3778 | 4·4603 5937 |
| 3 | ·9062 5000$^e$ | ·9333 3333 | ·0507 4026 | ·8872 4496 | 4·1214 1408 |
| 3·5 | ·8923 0769 | ·9180 3279 | ·0535 5836 | ·7212 4186 | 3·8823 7370 |
| 4 | ·8787 8788 | ·9032 2581 | ·0559 7261 | ·5981 7030 | 3·7056 6502 |
| 4·5 | ·8656 7164 | ·8888 8889 | ·0580 5651 | ·5036 6763 | 3·5704 1921 |
| 5 | ·8529 4118 | ·8750 0000$^e$ | ·0598 6473 | ·4291 1877 | 3·4641 1929 |
| 5·5 | ·8405 7971 | ·8615 3846 | ·0614 3942 | ·3690 4358 | 3·3788 0362 |
| 6 | ·8285 7143 | ·8484 8485$^-$ | ·0628 1384 | ·3197 9043 | 3·3091 6761 |
| 6·5 | ·8169 0141 | ·8358 2090 | ·0640 1487 | ·2788 3289 | 3·2515 4156 |
| 7 | ·8055 5556 | ·8235 2941 | ·0650 6458 | ·2443 6773 | 3·2033 0668 |
| 7·5 | ·7945 2055$^-$ | ·8115 9420 | ·0659 8134 | ·2150 7378 | 3·1625 4457 |
| 8 | ·7837 8378 | ·8000 0000$^e$ | ·0667 8062 | ·1899 6123 | 3·1278 1830 |
| 8·5 | ·7733 3333 | ·7887 3239 | ·0674 7561 | ·1682 7412 | 3·0980 3066 |
| 9 | ·7631 5789 | ·7777 7778 | ·0680 7762 | ·1494 2529 | 3·0723 2969 |
| 9·5 | ·7532 4675$^+$ | ·7671 2329 | ·0685 9643 | ·1329 5193 | 3·0500 4410 |
| 10 | ·7435 8974 | ·7567 5676 | ·0690 4057 | ·1184 8448 | 3·0306 3799 |
| 10·5 | ·7341 7722 | ·7466 6667 | ·0694 1751 | ·1057 2434 | 3·0136 7859 |
| 11 | ·7250 0000$^e$ | ·7368 4211 | ·0697 3381 | ·0944 2774 | 2·9988 1273 |
| 12 | ·7073 1707 | ·7179 4872 | ·0702 0711 | ·0754 5551 | 2·9742 4728 |
| 13 | ·6904 7619 | ·7000 0000$^e$ | ·0704 9968 | ·0603 2839 | 2·9551 4830 |
| 14 | ·6744 1860 | ·6829 2683 | ·0706 4285$^+$ | ·0481 6639 | 2·9402 4416 |
| 15 | ·6590 9091 | ·6666 6667 | ·0706 6196 | ·0383 2866 | 2·9286 1016 |
| 16 | ·6444 4444 | ·6511 6279 | ·0705 7771 | ·0303 3827 | 2·9195 5934 |
| 17 | ·6304 3478 | ·6363 6364 | ·0704 0714 | ·0238 3367 | 2·9125 7193 |
| 18 | ·6170 2128 | ·6222 2222 | ·0701 6439 | ·0185 3634 | 2·9072 4842 |
| 19 | ·6041 6667 | ·6086 9565$^+$ | ·0698 6128 | ·0142 2868 | 2·9032 7746 |
| 20 | ·5918 3673 | ·5957 4468 | ·0695 0770 | ·0107 3858 | 2·9004 1348 |
| 21 | ·5800 0000$^e$ | ·5833 3333 | ·0691 1201 | ·0079 2841 | 2·8984 6068 |
| 22 | ·5686 2745$^+$ | ·5714 2857 | ·0686 8130 | ·0056 8705$^-$ | 2·8972 6149 |
| 23 | ·5576 9231 | ·5600 0000$^e$ | ·0682 2156 | ·0039 2396 | 2·8966 8802 |
| 24 | ·5471 6981 | ·5490 1961 | ·0677 3792 | ·0025 6483 | 2·8966 3569 |
| 25 | ·5370 3704 | ·5384 6154 | ·0672 3477 | ·0015 4821 | 2·8970 1841 |
| 26 | ·5272 7273 | ·5283 0189 | ·0667 1584 | ·0008 2294 | 2·8977 6485$^+$ |
| 27 | ·5178 5714 | ·5185 1852 | ·0661 8437 | ·0003 4624 | 2·8988 1564 |
| 28 | ·5087 7193 | ·5090 9091 | ·0656 4311 | ·0000 8208 | 2·9001 2107 |
| 29 | ·5000 0000$^e$ | ·5000 0000$^e$ | ·0650 9446 | ·0000 0000$^e$ | 2·9016 3934 |
| 30 | ·4915 2542 | ·4912 2807 | ·0645 4045$^+$ | ·0000 7414 | 2·9033 3522 |
| 31 | ·4833 3333 | ·4827 5862 | ·0639 8286 | ·0002 8243 | 2·9051 7882 |
| 32 | ·4754 0984 | ·4745 7627 | ·0634 2322 | ·0006 0599 | 2·9071 4478 |
| 33 | ·4677 4194 | ·4666 6667 | ·0628 6284 | ·0010 2861 | 2·9092 1148 |
| 34 | ·4603 1746 | ·4590 1639 | ·0623 0285$^+$ | ·0015 3630 | 2·9113 6044 |
| 35 | ·4531 2500$^e$ | ·4516 1290 | ·0617 4423 | ·0021 1701 | 2·9135 7587 |
| 36 | ·4461 5385$^-$ | ·4444 4444 | ·0611 8781 | ·0027 6026 | 2·9158 4421 |
| 37 | ·4393 9394 | ·4375 0000$^e$ | ·0606 3432 | ·0034 5698 | 2·9181 5380 |
| 38 | ·4328 3582 | ·4307 6923 | ·0600 8437 | ·0041 9927 | 2·9204 9464 |
| 39 | ·4264 7059 | ·4242 4242 | ·0595 3849 | ·0049 8024 | 2·9228 5810 |
| 40 | ·4202 8986 | ·4179 1045$^-$ | ·0589 9713 | ·0057 9387 | 2·9252 3677 |
| 41 | ·4142 8571 | ·4117 6471 | ·0584 6066 | ·0066 3489 | 2·9276 2423 |
| 42 | ·4084 5070 | ·4057 9710 | ·0579 2940 | ·0074 9870 | 2·9300 1497 |
| 43 | ·4027 7778 | ·4000 0000$^e$ | ·0574 0362 | ·0083 8125$^+$ | 2·9324 0426 |
| 44 | ·3972 6027 | ·3943 6620 | ·0568 8354 | ·0092 7900 | 2·9347 8799 |
| 45 | ·3918 9189 | ·3888 8889 | ·0563 6933 | ·0101 8881 | 2·9371 6266 |
| 46 | ·3866 6667 | ·3835 6164 | ·0558 6115$^-$ | ·0111 0796 | 2·9395 2524 |
| 47 | ·3815 7895$^-$ | ·3783 7838 | ·0553 5909 | ·0120 3402 | 2·9418 7317 |
| 48 | ·3766 2338 | ·3733 3333 | ·0548 6326 | ·0129 6488 | 2·9442 0422 |
| 49 | ·3717 9487 | ·3684 2105$^+$ | ·0543 7370 | ·0138 9866 | 2·9465 1654 |
| 50 | ·3670 8861 | ·3636 3636 | ·0538 9046 | ·0148 3374 | 2·9488 0852 |

$p = 30$

| $q$ | $\bar{x}$ | $\tilde{x}$ | $\sigma$ | $\beta_1$ | $\beta_2$ |
|---|---|---|---|---|---|
| 0·5 | ·9836 0656 | $J$-curve | ·0226 2510 | 6·9208 0473 | 12·8922 1584 |
| 1 | ·9677 4194 | 1·0000 0000$^e$ | ·0312 3374 | 3·2950 1071 | 7·6206 7736 |
| 1·5 | ·9523 8095$^+$ | ·9830 5085$^-$ | ·0373 5545$^+$ | 2·0908 8884 | 5·8715 1201 |
| 2 | ·9375 0000$^e$ | ·9666 6667 | ·0421 3749 | 1·4920 4152 | 5·0026 8908 |
| 2·5 | ·9230 7692 | ·9508 1967 | ·0460 3889 | 1·1351 9569 | 4·4858 1343 |
| 3 | ·9090 9091 | ·9354 8387 | ·0493 0238 | ·8992 6531 | 4·1447 6190 |
| 3·5 | ·8955 2239 | ·9206 3492 | ·0520 7632 | ·7323 6036 | 3·9040 5998 |
| 4 | ·8823 5294 | ·9062 5000$^e$ | ·0544 6001 | ·6085 3909 | 3·7259 7598 |
| 4·5 | ·8695 6522 | ·8923 0769 | ·0565 2409 | ·5133 9213 | 3·5895 5251 |
| 5 | ·8571 4286 | ·8787 8788 | ·0583 2118 | ·4382 7611 | 3·4822 1906 |
| 5·5 | ·8450 7042 | ·8656 7164 | ·0598 9178 | ·3776 9266 | 3·3959 7954 |
| 6 | ·8333 3333 | ·8529 4118 | ·0612 6789 | ·3279 7784 | 3·3255 0607 |
| 6·5 | ·8219 1781 | ·8405 7971 | ·0624 7537 | ·2865 9652 | 3·2671 1263 |
| 7 | ·8108 1081 | ·8285 7143 | ·0635 3545$^+$ | ·2517 3914 | 3·2181 6850$^-$ |
| 7·5 | ·8000 0000$^e$ | ·8169 0141 | ·0644 6584 | ·2220 7979 | 3·1767 4637 |
| 8 | ·7894 7368 | ·8055 5556 | ·0652 8144 | ·1966 2500$^e$ | 3·1414 0244 |
| 8·5 | ·7792 2078 | ·7945 2055$^-$ | ·0659 9497 | ·1746 1590 | 3·1110 3411 |
| 9 | ·7692 3077 | ·7837 8378 | ·0666 1734 | ·1554 6302 | 3·0847 8513 |
| 9·5 | ·7594 9367 | ·7733 3333 | ·0671 5801 | ·1387 0167 | 3·0619 8068 |
| 10 | ·7500 0000$^e$ | ·7631 5789 | ·0676 2522 | ·1239 6070 | 3·0420 8195$^-$ |
| 10·5 | ·7407 4074 | ·7532 4675$^+$ | ·0680 2620 | ·1109 4019 | 3·0246 5372 |
| 11 | ·7317 0732 | ·7435 8974 | ·0683 6731 | ·0993 9525$^+$ | 3·0093 4076 |
| 12 | ·7142 8571 | ·7250 0000$^e$ | ·0688 9183 | ·0799 5868 | 2·9839 3939 |
| 13 | ·6976 7442 | ·7073 1707 | ·0692 3688 | ·0644 0519 | 2·9640 7284 |
| 14 | ·6818 1818 | ·6904 7619 | ·0694 3296 | ·0518 4985$^+$ | 2·9484 6042 |
| 15 | ·6666 6667 | ·6744 1860 | ·0695 0480 | ·0416 4780 | 2·9361 7021 |
| 16 | ·6521 7391 | ·6590 9091 | ·0694 7265$^-$ | ·0333 1887 | 2·9265 0935$^+$ |
| 17 | ·6382 9787 | ·6444 4444 | ·0693 5322 | ·0264 9876 | 2·9189 5318 |
| 18 | ·6250 0000$^e$ | ·6304 3478 | ·0691 6042 | ·0209 0667 | 2·9130 9804 |
| 19 | ·6122 4490 | ·6170 2128 | ·0689 0589 | ·0163 2301 | 2·9086 2904 |
| 20 | ·6000 0000$^e$ | ·6041 6667 | ·0685 9943 | ·0125 7396 | 2·9052 9753 |
| 21 | ·5882 3529 | ·5918 3673 | ·0682 4934 | ·0095 2042 | 2·9029 0506 |
| 22 | ·5769 2308 | ·5800 0000$^e$ | ·0678 6263 | ·0070 4992 | 2·9012 9170 |
| 23 | ·5660 3774 | ·5686 2745$^+$ | ·0674 4532 | ·0050 7079 | 2·9003 2750$^-$ |
| 24 | ·5555 5556 | ·5576 9231 | ·0670 0252 | ·0035 0765$^+$ | 2·8999 0602 |
| 25 | ·5454 5455$^-$ | ·5471 6981 | ·0665 3864 | ·0022 9814 | 2·8999 3950$^+$ |
| 26 | ·5357 1429 | ·5370 3704 | ·0660 5746 | ·0013 9029 | 2·9003 5517 |
| 27 | ·5263 1579 | ·5272 7273 | ·0655 6222 | ·0007 4053 | 2·9010 9228 |
| 28 | ·5172 4138 | ·5178 5714 | ·0650 5574 | ·0003 1217 | 2·9020 9992 |
| 29 | ·5084 7458 | ·5087 7193 | ·0645 4045$^+$ | ·0000 7414 | 2·9033 3522 |
| 30 | ·5000 0000$^e$ | ·5000 0000$^e$ | ·0640 1844 | ·0000 0000$^e$ | 2·9047 6190 |
| 31 | ·4918 0328 | ·4915 2542 | ·0634 9153 | ·0000 6719 | 2·9063 4921 |
| 32 | ·4838 7097 | ·4833 3333 | ·0629 6130 | ·0002 5635$^-$ | 2·9080 7091 |
| 33 | ·4761 9048 | ·4754 0984 | ·0624 2910 | ·0005 5083 | 2·9099 0464 |
| 34 | ·4687 5000$^e$ | ·4677 4194 | ·0618 9612 | ·0009 3628 | 2·9118 3122 |
| 35 | ·4615 3846 | ·4603 1746 | ·0613 6339 | ·0014 0025$^-$ | 2·9138 3419 |
| 36 | ·4545 4545$^+$ | ·4531 2500$^e$ | ·0608 3178 | ·0019 3195$^-$ | 2·9158 9940 |
| 37 | ·4477 6119 | ·4461 5385$^-$ | ·0603 0207 | ·0025 2199 | 2·9180 1466 |
| 38 | ·4411 7647 | ·4393 9394 | ·0597 7492 | ·0031 6219 | 2·9201 6944 |
| 39 | ·4347 8261 | ·4328 3582 | ·0592 5088 | ·0038 4539 | 2·9223 5464 |
| 40 | ·4285 7143 | ·4264 7059 | ·0587 3046 | ·0045 6533 | 2·9245 6240 |
| 41 | ·4225 3521 | ·4202 8986 | ·0582 1407 | ·0053 1651 | 2·9267 8592 |
| 42 | ·4166 6667 | ·4142 8571 | ·0577 0206 | ·0060 9413 | 2·9290 1931 |
| 43 | ·4109 5890 | ·4084 5070 | ·0571 9475$^-$ | ·0068 9392 | 2·9312 5745$^-$ |
| 44 | ·4054 0541 | ·4027 7778 | ·0566 9237 | ·0077 1216 | 2·9334 9593 |
| 45 | ·4000 0000$^e$ | ·3972 6027 | ·0561 9515$^-$ | ·0085 4557 | 2·9357 3094 |
| 46 | ·3947 3684 | ·3918 9189 | ·0557 0326 | ·0093 9123 | 2·9379 5916 |
| 47 | ·3896 1039 | ·3866 6667 | ·0552 1684 | ·0102 4658 | 2·9401 7775$^+$ |
| 48 | ·3846 1538 | ·3815 7895$^-$ | ·0547 3601 | ·0111 0938 | 2·9423 8426 |
| 49 | ·3797 4684 | ·3766 2338 | ·0542 6085$^-$ | ·0119 7760 | 2·9445 7657 |
| 50 | ·3750 0000$^e$ | ·3717 9487 | ·0537 9144 | ·0128 4949 | 2·9467 5287 |

$p = 31$

| $q$ | $\bar{x}$ | $\tilde{x}$ | $\sigma$ | $\beta_1$ | $\beta_2$ |
|---|---|---|---|---|---|
| 0·5 | ·9841 2698 | *J*-curve | ·0219 2369 | 6·9521 9138 | 12·9521 0483 |
| 1 | ·9687 5000$^e$ | 1·0000 0000$^e$ | ·0302 8823 | 3·3151 0213 | 7·6591 4882 |
| 1·5 | ·9538 4615+ | ·9836 0656 | ·0362 5106 | 2·1069 6678 | 5·9024 0931 |
| 2 | ·9393 9394 | ·9677 4194 | ·0409 2064 | 1·5059 3812 | 5·0294 9309 |
| 2·5 | ·9253 7313 | ·9523 8095+ | ·0447 4005+ | 1·1476 5248 | 4·5099 3135+ |
| 3 | ·9117 6471 | ·9375 0000$^e$ | ·0479 4337 | ·9106 5976 | 4·1669 0884 |
| 3·5 | ·8985 5072 | ·9230 7692 | ·0506 7358 | ·7429 1353 | 3·9246 5375− |
| 4 | ·8857 1429 | ·9090 9091 | ·0530 2633 | ·6183 9346 | 3·7452 8518 |
| 4·5 | ·8732 3944 | ·8955 2239 | ·0550 6965+ | ·5226 4628 | 3·6077 6242 |
| 5 | ·8611 1111 | ·8823 5294 | ·0568 5419 | ·4470 0206 | 3·4994 6454 |
| 5·5 | ·8493 1507 | ·8695 6522 | ·0584 1894 | ·3859 4532 | 3·4123 6309 |
| 6 | ·8378 3784 | ·8571 4286 | ·0597 9474 | ·3358 0058 | 3·3411 0835+ |
| 6·5 | ·8266 6667 | ·8450 7042 | ·0610 0647 | ·2940 2459 | 3·2819 9893 |
| 7 | ·8157 8947 | ·8333 3333 | ·0620 7461 | ·2588 0184 | 3·2323 9294 |
| 7·5 | ·8051 9481 | ·8219 1781 | ·0630 1620 | ·2288 0200 | 3·1903 5473 |
| 8 | ·7948 7179 | ·8108 1081 | ·0638 4564 | ·2030 2815+ | 3·1544 3408 |
| 8·5 | ·7848 1013 | ·8000 0000$^e$ | ·0645 7519 | ·1807 1876 | 3·1235 2336 |
| 9 | ·7750 0000$^e$ | ·7894 7368 | ·0652 1540 | ·1612 8220 | 3·0967 6229 |
| 9·5 | ·7654 3210 | ·7792 2078 | ·0657 7541 | ·1442 5200 | 3·0734 7276 |
| 10 | ·7560 9756 | ·7692 3077 | ·0662 6319 | ·1292 5557 | 3·0531 1328 |
| 10·5 | ·7469 8795+ | ·7594 9367 | ·0666 8573 | ·1159 9176 | 3·0352 4634 |
| 11 | ·7380 9524 | ·7500 0000$^e$ | ·0670 4921 | ·1042 1463 | 3·0195 1480 |
| 12 | ·7209 3023 | ·7317 0732 | ·0676 2024 | ·0843 4356 | 2·9933 3022 |
| 13 | ·7045 4545+ | ·7142 8571 | ·0680 1326 | ·0683 9066 | 2·9727 4374 |
| 14 | ·6888 8889 | ·6976 7442 | ·0682 5798 | ·0554 6643 | 2·9564 6632 |
| 15 | ·6739 1304 | ·6818 1818 | ·0683 7852 | ·0449 2234 | 2·9435 5936 |
| 16 | ·6595 7447 | ·6666 6667 | ·0683 9469 | ·0362 7521 | 2·9333 2456 |
| 17 | ·6458 3333 | ·6521 7391 | ·0683 2286 | ·0291 5825+ | 2·9252 3273 |
| 18 | ·6326 5306 | ·6382 9787 | ·0681 7671 | ·0232 8853 | 2·9188 7640 |
| 19 | ·6200 0000$^e$ | ·6250 0000$^e$ | ·0679 6770 | ·0184 4466 | 2·9139 3743 |
| 20 | ·6078 4314 | ·6122 4490 | ·0677 0551 | ·0144 5125− | 2·9101 6433 |
| 21 | ·5961 5385− | ·6000 0000$^e$ | ·0673 9834 | ·0111 6779 | 2·9073 5620 |
| 22 | ·5849 0566 | ·5882 3529 | ·0670 5319 | ·0084 8065− | 2·9053 5095+ |
| 23 | ·5740 7407 | ·5769 2308 | ·0666 7602 | ·0062 9705− | 2·9040 1670 |
| 24 | ·5636 3636 | ·5660 3774 | ·0662 7195+ | ·0045 4069 | 2·9032 4532 |
| 25 | ·5535 7143 | ·5555 5556 | ·0658 4539 | ·0031 4833 | 2·9029 4753 |
| 26 | ·5438 5965− | ·5454 5455− | ·0654 0014 | ·0020 6723 | 2·9030 4917 |
| 27 | ·5344 8276 | ·5357 1429 | ·0649 3947 | ·0012 5315+ | 2·9034 8826 |
| 28 | ·5254 2373 | ·5263 1579 | ·0644 6622 | ·0006 6877 | 2·9042 1278 |
| 29 | ·5166 6667 | ·5172 4138 | ·0639 8286 | ·0002 8243 | 2·9051 7882 |
| 30 | ·5081 9672 | ·5084 7458 | ·0634 9153 | ·0000 6719 | 2·9063 4921 |
| 31 | ·5000 0000$^e$ | ·5000 0000$^e$ | ·0629 9408 | ·0000 0000$^e$ | 2·9076 9231 |
| 32 | ·4920 6349 | ·4918 0328 | ·0624 9213 | ·0000 6108 | 2·9091 8114 |
| 33 | ·4843 7500$^e$ | ·4838 7097 | ·0619 8708 | ·0002 3338 | 2·9107 9261 |
| 34 | ·4769 2308 | ·4761 9048 | ·0614 8016 | ·0005 0218 | 2·9125 0689 |
| 35 | ·4696 9697 | ·4687 5000$^e$ | ·0609 7243 | ·0008 5469 | 2·9143 0693 |
| 36 | ·4626 8657 | ·4615 3846 | ·0604 6483 | ·0012 7981 | 2·9161 7801 |
| 37 | ·4558 8235+ | ·4545 4545+ | ·0599 5815+ | ·0017 6788 | 2·9181 0742 |
| 38 | ·4492 7536 | ·4477 6119 | ·0594 5310 | ·0023 1043 | 2·9200 8417 |
| 39 | ·4428 5714 | ·4411 7647 | ·0589 5029 | ·0029 0006 | 2·9220 9872 |
| 40 | ·4366 1972 | ·4347 8261 | ·0584 5023 | ·0035 3029 | 2·9241 4279 |
| 41 | ·4305 5556 | ·4285 7143 | ·0579 5339 | ·0041 9540 | 2·9262 0920 |
| 42 | ·4246 5753 | ·4225 3521 | ·0574 6015+ | ·0048 9039 | 2·9282 9170 |
| 43 | ·4189 1892 | ·4166 6667 | ·0569 7085+ | ·0056 1082 | 2·9303 8485+ |
| 44 | ·4133 3333 | ·4109 5890 | ·0564 8578 | ·0063 5279 | 2·9324 8394 |
| 45 | ·4078 9474 | ·4054 0541 | ·0560 0517 | ·0071 1285− | 2·9345 8485− |
| 46 | ·4025 9740 | ·4000 0000$^e$ | ·0555 2924 | ·0078 8794 | 2·9366 8401 |
| 47 | ·3974 3590 | ·3947 3684 | ·0550 5815+ | ·0086 7536 | 2·9387 7831 |
| 48 | ·3924 0506 | ·3896 1039 | ·0545 9205− | ·0094 7272 | 2·9408 6507 |
| 49 | ·3875 0000$^e$ | ·3846 1538 | ·0541 3104 | ·0102 7790 | 2·9429 4195+ |
| 50 | ·3827 1605− | ·3797 4684 | ·0536 7523 | ·0110 8902 | 2·9450 0694 |

$p = 32$

| $q$ | $\bar{x}$ | $\tilde{x}$ | $\sigma$ | $\beta_1$ | $\beta_2$ |
|---|---|---|---|---|---|
| 0·5 | ·9846 1538 | $J$-curve | ·0212 6445$^+$ | 6·9818 0529 | 13·0086 8800 |
| 1 | ·9696 9697 | 1·0000 0000$^e$ | ·0293 9826 | 3·3340 8163 | 7·6955 3571 |
| 1·5 | ·9552 2388 | ·9841 2698 | ·0352 1002 | 2·1221 7318 | 5·9316 6361 |
| 2 | ·9411 7647 | ·9687 5000$^e$ | ·0397 7196 | 1·5190 9722 | 5·0548 9865$^-$ |
| 2·5 | ·9275 3623 | ·9538 4615$^+$ | ·0435 1227 | 1·1594 6238 | 4·5328 1507 |
| 3 | ·9142 8571 | ·9393 9394 | ·0466 5695$^-$ | ·9214 7553 | 4·1879 4452 |
| 3·5 | ·9014 0845$^+$ | ·9253 7313 | ·0493 4399 | ·7529 4286 | 3·9442 3469 |
| 4 | ·8888 8889 | ·9117 6471 | ·0516 6562 | ·6277 7008 | 3·7636 6397 |
| 4·5 | ·8767 1233 | ·8985 5072 | ·0536 8745$^-$ | ·5314 6259 | 3·6251 1302 |
| 5 | ·8648 6486 | ·8857 1429 | ·0554 5829 | ·4553 2544 | 3·5159 1346 |
| 5·5 | ·8533 3333 | ·8732 3944 | ·0570 1573 | ·3938 2711 | 3·4280 0633 |
| 6 | ·8421 0526 | ·8611 1111 | ·0583 8949 | ·3432 8125$^e$ | 3·3560 2134 |
| 6·5 | ·8311 6883 | ·8493 1507 | ·0596 0359 | ·3011 3696 | 3·2962 4266 |
| 7 | ·8205 1282 | ·8378 3784 | ·0606 7774 | ·2655 7321 | 3·2460 1792 |
| 7·5 | ·8101 2658 | ·8266 6667 | ·0616 2842 | ·2352 5548 | 3·2034 0361 |
| 8 | ·8000 0000$^e$ | ·8157 8947 | ·0624 6950$^+$ | ·2091 8367 | 3·1669 4352 |
| 8·5 | ·7901 2346 | ·8051 9481 | ·0632 1285$^+$ | ·1865 9373 | 3·1355 2530 |
| 9 | ·7804 8780 | ·7948 7179 | ·0638 6865$^+$ | ·1668 9201 | 3·1082 8488 |
| 9·5 | ·7710 8434 | ·7848 1013 | ·0644 4574 | ·1496 1043 | 3·0845 4113 |
| 10 | ·7619 0476 | ·7750 0000$^e$ | ·0649 5184 | ·1343 7500$^e$ | 3·0637 5000$^e$ |
| 10·5 | ·7529 4118 | ·7654 3210 | ·0653 9373 | ·1208 8346 | 3·0454 7188 |
| 11 | ·7441 8605$^-$ | ·7560 9756 | ·0657 7737 | ·1088 8889 | 3·0293 4783 |
| 12 | ·7272 7273 | ·7380 9524 | ·0663 9061 | ·0886 1059 | 3·0024 2831 |
| 13 | ·7111 1111 | ·7209 3023 | ·0668 2750$^+$ | ·0722 8297 | 2·9811 6561 |
| 14 | ·6956 5217 | ·7045 4545$^+$ | ·0671 1696 | ·0590 1228 | 2·9642 6294 |
| 15 | ·6808 5106 | ·6888 8889 | ·0672 8250$^+$ | ·0481 4661 | 2·9507 7551 |
| 16 | ·6666 6667 | ·6739 1304 | ·0673 4350$^+$ | ·0392 0000$^e$ | 2·9400 0000$^e$ |
| 17 | ·6530 6122 | ·6595 7447 | ·0673 1601 | ·0318 0338 | 2·9314 0305$^-$ |
| 18 | ·6400 0000$^e$ | ·6458 3333 | ·0672 1344 | ·0256 7184 | 2·9245 7366 |
| 19 | ·6274 5098 | ·6326 5306 | ·0670 4709 | ·0205 8234 | 2·9191 9067 |
| 20 | ·6153 8462 | ·6200 0000$^e$ | ·0668 2650$^-$ | ·0163 5802 | 2·9150 0000$^e$ |
| 21 | ·6037 7358 | ·6078 4314 | ·0665 5978 | ·0128 5714 | 2·9117 9847 |
| 22 | ·5925 9259 | ·5961 5385$^-$ | ·0662 5387 | ·0099 6492 | 2·9094 2198 |
| 23 | ·5818 1818 | ·5849 0566 | ·0659 1469 | ·0075 8762 | 2·9077 3692 |
| 24 | ·5714 2857 | ·5740 7407 | ·0655 4735$^+$ | ·0056 4804 | 2·9066 3355$^-$ |
| 25 | ·5614 0351 | ·5636 3636 | ·0651 5626 | ·0040 8216 | 2·9060 2119 |
| 26 | ·5517 2414 | ·5535 7143 | ·0647 4521 | ·0028 3654 | 2·9058 2440 |
| 27 | ·5423 7288 | ·5438 5965$^-$ | ·0643 1751 | ·0018 6628 | 2·9059 8008 |
| 28 | ·5333 3333 | ·5344 8276 | ·0638 7602 | ·0011 3349 | 2·9064 3515$^+$ |
| 29 | ·5245 9016 | ·5254 2373 | ·0634 2322 | ·0006 0599 | 2·9071 4478 |
| 30 | ·5161 2903 | ·5166 6667 | ·0629 6130 | ·0002 5635$^-$ | 2·9080 7091 |
| 31 | ·5079 3651 | ·5081 9672 | ·0624 9213 | ·0000 6108 | 2·9091 8114 |
| 32 | ·5000 0000$^e$ | ·5000 0000$^e$ | ·0620 1737 | ·0000 0000$^e$ | 2·9104 4776 |
| 33 | ·4923 0769 | ·4920 6349 | ·0615 3846 | ·0000 5569 | 2·9118 4702 |
| 34 | ·4848 4848 | ·4843 7500$^e$ | ·0610 5667 | ·0002 1308 | 2·9133 5847 |
| 35 | ·4776 1194 | ·4769 2308 | ·0605 7309 | ·0004 5909 | 2·9149 6451 |
| 36 | ·4705 8824 | ·4696 9697 | ·0600 8870 | ·0007 8231 | 2·9166 4990 |
| 37 | ·4637 6812 | ·4626 8657 | ·0596 0432 | ·0011 7282 | 2·9184 0146 |
| 38 | ·4571 4286 | ·4558 8235$^+$ | ·0591 2070 | ·0016 2189 | 2·9202 0773 |
| 39 | ·4507 0423 | ·4492 7536 | ·0586 3848 | ·0021 2192 | 2·9220 5878 |
| 40 | ·4444 4444 | ·4428 5714 | ·0581 5821 | ·0026 6618 | 2·9239 4595$^-$ |
| 41 | ·4383 5616 | ·4366 1972 | ·0576 8039 | ·0032 4878 | 2·9258 6168 |
| 42 | ·4324 3243 | ·4305 5556 | ·0572 0543 | ·0038 6451 | 2·9277 9941 |
| 43 | ·4266 6667 | ·4246 5753 | ·0567 3371 | ·0045 0878 | 2·9297 5339 |
| 44 | ·4210 5263 | ·4189 1892 | ·0562 6552 | ·0051 7751 | 2·9317 1860 |
| 45 | ·4155 8442 | ·4133 3333 | ·0558 0116 | ·0058 6712 | 2·9336 9066 |
| 46 | ·4102 5641 | ·4078 9474 | ·0553 4084 | ·0065 7439 | 2·9356 6576 |
| 47 | ·4050 6329 | ·4025 9740 | ·0548 8477 | ·0072 9650$^+$ | 2·9376 4055$^-$ |
| 48 | ·4000 0000$^e$ | ·3974 3590 | ·0544 3311 | ·0080 3093 | 2·9396 1211 |
| 49 | ·3950 6173 | ·3924 0506 | ·0539 8599 | ·0087 7545$^+$ | 2·9415 7790 |
| 50 | ·3902 4390 | ·3875 0000$^e$ | ·0535 4354 | ·0095 2806 | 2·9435 3571 |

$p = 33$

| $q$ | $\bar{x}$ | $\tilde{x}$ | $\sigma$ | $\beta_1$ | $\beta_2$ |
|---|---|---|---|---|---|
| 0·5 | ·9850 7463 | *J*-curve | ·0206 4370 | 7·0097 9243 | 13·0622 3142 |
| 1 | ·9705 8824 | 1·0000 0000$^{e}$ | ·0285 5907 | 3·3520 3891 | 7·7300 0273 |
| 1·5 | ·9565 2174 | ·9846 1538 | ·0342 2704 | 2·1365 7688 | 5·9594 0224 |
| 2 | ·9428 5714 | ·9696 9697 | ·0386 8590 | 1·5315 7580 | 5·0790 1203 |
| 2·5 | ·9295 7746 | ·9552 2388 | ·0423 4988 | 1·1706 7421 | 4·5545 5647 |
| 3 | ·9166 6667 | ·9411 7647 | ·0454 3748 | ·9317 5523 | 4·2079 4994 |
| 3·5 | ·9041 0959 | ·9275 3623 | ·0480 8200 | ·7624 8595$^{+}$ | 3·9628 7503 |
| 4 | ·8918 9189 | ·9142 8571 | ·0503 7249 | ·6367 0233 | 3·7811 7716 |
| 4·5 | ·8800 0000$^{e}$ | ·9014 0845$^{+}$ | ·0523 7229 | ·5398 7075$^{-}$ | 3·6416 6276 |
| 5 | ·8684 2105$^{+}$ | ·8888 8889 | ·0541 2851 | ·4632 7273 | 3·5316 1863 |
| 5·5 | ·8571 4286 | ·8767 1233 | ·0556 7741 | ·4013 6158 | 3·4429 5701 |
| 6 | ·8461 5385$^{-}$ | ·8648 6486 | ·0570 4768 | ·3504 4075$^{+}$ | 3·3702 8825$^{-}$ |
| 6·5 | ·8354 4304 | ·8533 3333 | ·0582 6249 | ·3079 5217 | 3·3098 8289 |
| 7 | ·8250 0000$^{e}$ | ·8421 0526 | ·0593 4089 | ·2720 6958 | 3·2590 7868 |
| 7·5 | ·8148 1481 | ·8311 6883 | ·0602 9878 | ·2414 5455$^{-}$ | 3·2159 2476 |
| 8 | ·8048 7805$^{-}$ | ·8205 1282 | ·0611 4958 | ·2151 0399 | 3·1789 5925$^{+}$ |
| 8·5 | ·7951 8072 | ·8101 2658 | ·0619 0474 | ·1922 5149 | 3·1470 6538 |
| 9 | ·7857 1429 | ·8000 0000$^{e}$ | ·0625 7411 | ·1723 0153 | 3·1193 7557 |
| 9·5 | ·7764 7059 | ·7901 2346 | ·0631 6627 | ·1547 8448 | 3·0952 0580 |
| 10 | ·7674 4186 | ·7804 8780 | ·0636 8867 | ·1393 2510 | 3·0740 0966 |
| 10·5 | ·7586 2069 | ·7710 8434 | ·0641 4789 | ·1256 2006 | 3·0553 4558 |
| 11 | ·7500 0000$^{e}$ | ·7619 0476 | ·0645 4972 | ·1134 2155$^{+}$ | 3·0388 5291 |
| 12 | ·7333 3333 | ·7441 8605$^{-}$ | ·0652 0129 | ·0927 6102 | 3·0112 4275$^{-}$ |
| 13 | ·7173 9130 | ·7272 7273 | ·0656 7830 | ·0760 8133 | 2·9893 4399 |
| 14 | ·7021 2766 | ·7111 1111 | ·0660 0890 | ·0624 8479 | 2·9718 5264 |
| 15 | ·6875 0000$^{e}$ | ·6956 5217 | ·0662 1606 | ·0513 1636 | 2·9578 1818 |
| 16 | ·6734 6939 | ·6808 5106 | ·0663 1868 | ·0420 8754 | 2·9465 3263 |
| 17 | ·6600 0000$^{e}$ | ·6666 6667 | ·0663 3250$^{-}$ | ·0344 2711 | 2·9374 5877 |
| 18 | ·6470 5882 | ·6530 6122 | ·0662 7067 | ·0280 4837 | 2·9301 8233 |
| 19 | ·6346 1538 | ·6400 0000$^{e}$ | ·0661 4431 | ·0227 2673 | 2·9243 7936 |
| 20 | ·6226 4151 | ·6274 5098 | ·0659 6282 | ·0182 8400 | 2·9197 9339 |
| 21 | ·6111 1111 | ·6153 8462 | ·0657 3422 | ·0145 7726 | 2·9162 1912 |
| 22 | ·6000 0000$^{e}$ | ·6037 7358 | ·0654 6537 | ·0114 9072 | 2·9134 9062 |
| 23 | ·5892 8571 | ·5925 9259 | ·0651 6215$^{+}$ | ·0089 2970 | 2·9114 7261 |
| 24 | ·5789 4737 | ·5818 1818 | ·0648 2966 | ·0068 1622 | 2·9100 5393 |
| 25 | ·5689 6552 | ·5714 2857 | ·0644 7227 | ·0050 8552 | 2·9091 4257 |
| 26 | ·5593 2203 | ·5614 0351 | ·0640 9380 | ·0036 8350$^{-}$ | 2·9086 6194 |
| 27 | ·5500 0000$^{e}$ | ·5517 2414 | ·0636 9754 | ·0025 6467 | 2·9085 4784 |
| 28 | ·5409 8361 | ·5423 7288 | ·0632 8639 | ·0016 9059 | 2·9087 4626 |
| 29 | ·5322 5806 | ·5333 3333 | ·0628 6284 | ·0010 2861 | 2·9092 1148 |
| 30 | ·5238 0952 | ·5245 9016 | ·0624 2910 | ·0005 5083 | 2·9099 0464 |
| 31 | ·5156 2500$^{e}$ | ·5161 2903 | ·0619 8708 | ·0002 3338 | 2·9107 9261 |
| 32 | ·5076 9231 | ·5079 3651 | ·0615 3846 | ·0000 5569 | 2·9118 4702 |
| 33 | ·5000 0000$^{e}$ | ·5000 0000$^{e}$ | ·0610 8472 | ·0000 0000$^{e}$ | 2·9130 4348 |
| 34 | ·4925 3731 | ·4923 0769 | ·0606 2715$^{+}$ | ·0000 5092 | 2·9143 6100 |
| 35 | ·4852 9412 | ·4848 4848 | ·0601 6689 | ·0001 9507 | 2·9157 8144 |
| 36 | ·4782 6087 | ·4776 1194 | ·0597 0492 | ·0004 2079 | 2·9172 8909 |
| 37 | ·4714 2857 | ·4705 8824 | ·0592 4212 | ·0007 1789 | 2·9188 7030 |
| 38 | ·4647 8873 | ·4637 6812 | ·0587 7927 | ·0010 7743 | 2·9205 1322 |
| 39 | ·4583 3333 | ·4571 4286 | ·0583 1702 | ·0014 9157 | 2·9222 0752 |
| 40 | ·4520 5479 | ·4507 0423 | ·0578 5598 | ·0019 5340 | 2·9239 4418 |
| 41 | ·4459 4595$^{-}$ | ·4444 4444 | ·0573 9665$^{+}$ | ·0024 5684 | 2·9257 1532 |
| 42 | ·4400 0000$^{e}$ | ·4383 5616 | ·0569 3949 | ·0029 9650$^{-}$ | 2·9275 1404 |
| 43 | ·4342 1053 | ·4324 3243 | ·0564 8488 | ·0035 6762 | 2·9293 3433 |
| 44 | ·4285 7143 | ·4266 6667 | ·0560 3318 | ·0041 6600 | 2·9311 7089 |
| 45 | ·4230 7692 | ·4210 5263 | ·0555 8468 | ·0047 8788 | 2·9330 1908 |
| 46 | ·4177 2152 | ·4155 8442 | ·0551 3962 | ·0054 2994 | 2·9348 7485$^{-}$ |
| 47 | ·4125 0000$^{e}$ | ·4102 5641 | ·0546 9825$^{-}$ | ·0060 8922 | 2·9367 3463 |
| 48 | ·4074 0741 | ·4050 6329 | ·0542 6073 | ·0067 6309 | 2·9385 9529 |
| 49 | ·4024 3902 | ·4000 0000$^{e}$ | ·0538 2724 | ·0074 4920 | 2·9404 5410 |
| 50 | ·3975 9036 | ·3950 6173 | ·0533 9791 | ·0081 4545$^{+}$ | 2·9423 0867 |

## $p = 34$

| $q$ | $\bar{x}$ | $\tilde{x}$ | $\sigma$ | $\beta_1$ | $\beta_2$ |
|---|---|---|---|---|---|
| 0·5 | ·9855 0725$^{-}$ | *J*-curve | ·0200 5816 | 7·0362 8316 | 13·1129 7341 |
| 1 | ·9714 2857 | 1·0000 0000$^{e}$ | ·0277 6644 | 3·3690 5427 | 7·7626 9768 |
| 1·5 | ·9577 4648 | ·9850 7463 | ·0332 9740 | 2·1502 3965$^{+}$ | 5·9857 3975$^{+}$ |
| 2 | ·9444 4444 | ·9705 8824 | ·0376 5747 | 1·5434 2513 | 5·1019 2903 |
| 2·5 | ·9315 0685$^{-}$ | ·9565 2174 | ·0412 4782 | 1·1813 3204 | 4·5752 3861 |
| 3 | ·9189 1892 | ·9428 5714 | ·0442 7989 | ·9415 3743 | 4·2269 9849 |
| 3·5 | ·9066 6667 | ·9295 7746 | ·0468 8262 | ·7715 7695$^{+}$ | 3·9806 4036 |
| 4 | ·8947 3684 | ·9166 6667 | ·0491 4208 | ·6452 2059 | 3·7978 8379 |
| 4·5 | ·8831 1688 | ·9041 0959 | ·0511 1948 | ·5478 9791 | 3·6574 6501 |
| 5 | ·8717 9487 | ·8918 9189 | ·0528 6033 | ·4708 6818 | 3·5466 2841 |
| 5·5 | ·8607 5949 | ·8800 0000$^{e}$ | ·0543 9966 | ·4085 7043 | 3·4572 5905$^{-}$ |
| 6 | ·8500 0000$^{e}$ | ·8684 2105$^{+}$ | ·0557 6519 | ·3572 9847 | 3·3839 4893 |
| 6·5 | ·8395 0617 | ·8571 4286 | ·0569 7929 | ·3144 8747 | 3·3229 5577 |
| 7 | ·8292 6829 | ·8461 5385$^{-}$ | ·0580 6039 | ·2783 0624 | 3·2716 0801 |
| 7·5 | ·8192 7711 | ·8354 4304 | ·0590 2385$^{-}$ | ·2474 1269 | 3·2279 4782 |
| 8 | ·8095 2381 | ·8250 0000$^{e}$ | ·0598 8264 | ·2208 0092 | 3·1905 0802 |
| 8·5 | ·8000 0000$^{e}$ | ·8148 1481 | ·0606 4784 | ·1977 0231 | 3·1581 6768 |
| 9 | ·7906 9767 | ·8048 7805$^{-}$ | ·0613 2901 | ·1775 1957 | 3·1300 5589 |
| 9·5 | ·7816 0920 | ·7951 8072 | ·0619 3441 | ·1597 8163 | 3·1054 8595$^{-}$ |
| 10 | ·7727 2727 | ·7857 1429 | ·0624 7130 | ·1441 1209 | 3·0839 0923 |
| 10·5 | ·7640 4494 | ·7764 7059 | ·0629 4604 | ·1302 0660 | 3·0648 8232 |
| 11 | ·7555 5556 | ·7674 4186 | ·0633 6426 | ·1178 1652 | 3·0480 4301 |
| 12 | ·7391 3043 | ·7500 0000$^{e}$ | ·0640 5062 | ·0967 9670 | 3·0197 8291 |
| 13 | ·7234 0426 | ·7333 3333 | ·0645 6433 | ·0797 8576 | 2·9972 8507 |
| 14 | ·7083 3333 | ·7173 9130 | ·0649 3281 | ·0658 8235$^{+}$ | 2·9792 3875$^{+}$ |
| 15 | ·6938 7755$^{+}$ | ·7021 2766 | ·0651 7845$^{-}$ | ·0544 2854 | 2·9646 8814 |
| 16 | ·6800 0000$^{e}$ | ·6875 0000$^{e}$ | ·0653 1973 | ·0449 3343 | 2·9529 2090 |
| 17 | ·6666 6667 | ·6734 6939 | ·0653 7205$^{-}$ | ·0370 2385$^{+}$ | 2·9433 9623 |
| 18 | ·6538 4615$^{+}$ | ·6600 0000$^{e}$ | ·0653 4832 | ·0304 1143 | 2·9356 9684 |
| 19 | ·6415 0943 | ·6470 5882 | ·0652 5947 | ·0248 7015$^{-}$ | 2·9294 9620 |
| 20 | ·6296 2963 | ·6346 1538 | ·0651 1475$^{-}$ | ·0202 2059 | 2·9245 3560 |
| 21 | ·6181 8182 | ·6226 4151 | ·0649 2208 | ·0163 1875$^{-}$ | 2·9206 0781 |
| 22 | ·6071 4286 | ·6111 1111 | ·0646 8824 | ·0130 4787 | 2·9175 4517 |
| 23 | ·5964 9123 | ·6000 0000$^{e}$ | ·0644 1908 | ·0103 1247 | 2·9152 1089 |
| 24 | ·5862 0690 | ·5892 8571 | ·0641 1964 | ·0080 3377 | 2·9134 9245$^{-}$ |
| 25 | ·5762 7119 | ·5789 4737 | ·0637 9429 | ·0061 4636 | 2·9122 9664 |
| 26 | ·5666 6667 | ·5689 6552 | ·0634 4684 | ·0045 9551 | 2·9115 4576 |
| 27 | ·5573 7705$^{-}$ | ·5593 2203 | ·0630 8058 | ·0033 3522 | 2·9111 7466 |
| 28 | ·5483 8710 | ·5500 0000$^{e}$ | ·0626 9841 | ·0023 2652 | 2·9111 2839 |
| 29 | ·5396 8254 | ·5409 8361 | ·0623 0285$^{+}$ | ·0015 3630 | 2·9113 6044 |
| 30 | ·5312 5000$^{e}$ | ·5322 5806 | ·0618 9612 | ·0009 3628 | 2·9118 3122 |
| 31 | ·5230 7692 | ·5238 0952 | ·0614 8016 | ·0005 0218 | 2·9125 0689 |
| 32 | ·5151 5152 | ·5156 2500$^{e}$ | ·0610 5667 | ·0002 1308 | 2·9133 5847 |
| 33 | ·5074 6269 | ·5076 9231 | ·0606 2715$^{+}$ | ·0000 5092 | 2·9143 6100 |
| 34 | ·5000 0000$^{e}$ | ·5000 0000$^{e}$ | ·0601 9293 | ·0000 0000$^{e}$ | 2·9154 9296 |
| 35 | ·4927 5362 | ·4925 3731 | ·0597 5515$^{+}$ | ·0000 4668 | 2·9167 3571 |
| 36 | ·4857 1429 | ·4852 9412 | ·0593 1486 | ·0001 7903 | 2·9180 7309 |
| 37 | ·4788 7324 | ·4782 6087 | ·0588 7294 | ·0003 8664 | 2·9194 9104 |
| 38 | ·4722 2222 | ·4714 2857 | ·0584 3019 | ·0006 6035$^{+}$ | 2·9209 7732 |
| 39 | ·4657 5342 | ·4647 8873 | ·0579 8732 | ·0009 9212 | 2·9225 2124 |
| 40 | ·4594 5946 | ·4583 3333 | ·0575 4493 | ·0013 7486 | 2·9241 1343 |
| 41 | ·4533 3333 | ·4520 5479 | ·0571 0358 | ·0018 0229 | 2·9257 4570 |
| 42 | ·4473 6842 | ·4459 4595$^{-}$ | ·0566 6373 | ·0022 6889 | 2·9274 1089 |
| 43 | ·4415 5844 | ·4400 0000$^{e}$ | ·0562 2580 | ·0027 6973 | 2·9291 0267 |
| 44 | ·4358 9744 | ·4342 1053 | ·0557 9017 | ·0033 0047 | 2·9308 1551 |
| 45 | ·4303 7975$^{-}$ | ·4285 7143 | ·0553 5714 | ·0038 5722 | 2·9325 4453 |
| 46 | ·4250 0000$^{e}$ | ·4230 7692 | ·0549 2700 | ·0044 3653 | 2·9342 8545$^{+}$ |
| 47 | ·4197 5309 | ·4177 2152 | ·0544 9999 | ·0050 3533 | 2·9360 3450$^{+}$ |
| 48 | ·4146 3415$^{-}$ | ·4125 0000$^{e}$ | ·0540 7633 | ·0056 5087 | 2·9377 8835$^{+}$ |
| 49 | ·4096 3855$^{+}$ | ·4074 0741 | ·0536 5618 | ·0062 8071 | 2·9395 4408 |
| 50 | ·4047 6190 | ·4024 3902 | ·0532 3971 | ·0069 2266 | 2·9412 9912 |

$p = 35$

| $q$ | $\bar{x}$ | $\tilde{x}$ | $\sigma$ | $\beta_1$ | $\beta_2$ |
|---|---|---|---|---|---|
| 0·5 | ·9859 1549 | *J*-curve | ·0195 0492 | 7·0613 9429 | 13·1611 2801 |
| 1 | ·9722 2222 | 1·0000 0000$^{e}$ | ·0270 1660 | 3·3851 9984 | 7·7937 5361 |
| 1·5 | ·9589 0411 | ·9855 0725$^{-}$ | ·0324 1688 | 2·1632 1712 | 6·0107 7946 |
| 2 | ·9459 4595$^{-}$ | ·9714 2857 | ·0366 8221 | 1·5546 9146 | 5·1237 3626 |
| 2·5 | ·9333 3333 | ·9577 4648 | ·0402 0151 | 1·1914 7573 | 4·5949 3671 |
| 3 | ·9210 5263 | ·9444 4444 | ·0431 7961 | ·9508 5714 | 4·2451 5679 |
| 3·5 | ·9090 9091 | ·9315 0685$^{-}$ | ·0457 4133 | ·7802 4691 | 3·9975 9036 |
| 4 | ·8974 3590 | ·9189 1892 | ·0479 6997 | ·6533 5260 | 3·8138 3773 |
| 4·5 | ·8860 7595$^{-}$ | ·9066 6667 | ·0499 2472 | ·5555 6892 | 3·6725 6859 |
| 5 | ·8750 0000$^{e}$ | ·8947 3684 | ·0516 4962 | ·4781 3411 | 3·5609 8719 |
| 5·5 | ·8641 9753 | ·8831 1688 | ·0531 7853 | ·4154 7369 | 3·4709 5282 |
| 6 | ·8536 5854 | ·8717 9487 | ·0545 3824 | ·3638 7236 | 3·3970 4017 |
| 6·5 | ·8433 7349 | ·8607 5949 | ·0557 5042 | ·3207 5891 | 3·3354 9480 |
| 7 | ·8333 3333 | ·8500 0000$^{e}$ | ·0568 3286 | ·2842 9752 | 3·2836 3636 |
| 7·5 | ·8235 2941 | ·8395 0617 | ·0578 0043 | ·2531 4264 | 3·2395 0047 |
| 8 | ·8139 5349 | ·8292 6829 | ·0586 6570 | ·2262 8571 | 3·2016 1491 |
| 8·5 | ·8045 9770 | ·8192 7711 | ·0594 3937 | ·2029 5604 | 3·1688 5484 |
| 9 | ·7954 5455$^{-}$ | ·8095 2381 | ·0601 3071 | ·1825 5469 | 3·1403 4624 |
| 9·5 | ·7865 1685$^{+}$ | ·8000 0000$^{e}$ | ·0607 4774 | ·1646 0923 | 3·1153 9987 |
| 10 | ·7777 7778 | ·7906 9767 | ·0612 9748 | ·1487 4216 | 3·0934 6505$^{-}$ |
| 10·5 | ·7692 3077 | ·7816 0920 | ·0617 8610 | ·1346 4820 | 3·0740 9658 |
| 11 | ·7608 6957 | ·7727 2727 | ·0622 1908 | ·1220 7792 | 3·0569 3082 |
| 12 | ·7446 8085$^{+}$ | ·7555 5556 | ·0629 3702 | ·1007 1994 | 3·0280 5831 |
| 13 | ·7291 6667 | ·7391 3043 | ·0634 8431 | ·0833 9692 | 3·0049 9548 |
| 14 | ·7142 8571 | ·7234 0426 | ·0638 8766 | ·0692 0415$^{+}$ | 2·9864 2534 |
| 15 | ·7000 0000$^{e}$ | ·7083 3333 | ·0641 6889 | ·0574 8098 | 2·9713 8710 |
| 16 | ·6862 7451 | ·6938 7755$^{+}$ | ·0643 4609 | ·0477 3432 | 2·9591 6442 |
| 17 | ·6730 7692 | ·6800 0000$^{e}$ | ·0644 3431 | ·0395 8917 | 2·9492 1314 |
| 18 | ·6603 7736 | ·6666 6667 | ·0644 4623 | ·0327 5561 | 2·9411 1317 |
| 19 | ·6481 4815$^{-}$ | ·6538 4615$^{+}$ | ·0643 9256 | ·0270 0629 | 2·9345 3559 |
| 20 | ·6363 6364 | ·6415 0943 | ·0642 8243 | ·0221 6066 | 2·9292 1960 |
| 21 | ·6250 0000$^{e}$ | ·6296 2963 | ·0641 2365$^{-}$ | ·0180 7372 | 2·9249 5617 |
| 22 | ·6140 3509 | ·6181 8182 | ·0639 2291 | ·0146 2783 | 2·9215 7605$^{+}$ |
| 23 | ·6034 4828 | ·6071 4286 | ·0636 8600 | ·0117 2671 | 2·9189 4104 |
| 24 | ·5932 2034 | ·5964 9123 | ·0634 1792 | ·0092 9090 | 2·9169 3737 |
| 25 | ·5833 3333 | ·5862 0690 | ·0631 2303 | ·0072 5435$^{-}$ | 2·9154 7070 |
| 26 | ·5737 7049 | ·5762 7119 | ·0628 0512 | ·0055 6179 | 2·9144 6232 |
| 27 | ·5645 1613 | ·5666 6667 | ·0624 6747 | ·0041 6667 | 2·9138 4615$^{+}$ |
| 28 | ·5555 5556 | ·5573 7705$^{-}$ | ·0621 1300 | ·0030 2959 | 2·9135 6643 |
| 29 | ·5468 7500$^{e}$ | ·5483 8710 | ·0617 4423 | ·0021 1701 | 2·9135 7587 |
| 30 | ·5384 6154 | ·5396 8254 | ·0613 6339 | ·0014 0025$^{-}$ | 2·9138 3419 |
| 31 | ·5303 0303 | ·5312 5000$^{e}$ | ·0609 7243 | ·0008 5469 | 2·9143 0693 |
| 32 | ·5223 8806 | ·5230 7692 | ·0605 7309 | ·0004 5909 | 2·9149 6451 |
| 33 | ·5147 0588 | ·5151 5152 | ·0601 6689 | ·0001 9507 | 2·9157 8144 |
| 34 | ·5072 4638 | ·5074 6269 | ·0597 5515$^{+}$ | ·0000 4668 | 2·9167 3571 |
| 35 | ·5000 0000$^{e}$ | ·5000 0000$^{e}$ | ·0593 3908 | ·0000 0000$^{e}$ | 2·9178 0822 |
| 36 | ·4929 5775$^{-}$ | ·4927 5362 | ·0589 1972 | ·0000 4289 | 2·9189 8239 |
| 37 | ·4861 1111 | ·4857 1429 | ·0584 9799 | ·0001 6471 | 2·9202 4377 |
| 38 | ·4794 5205$^{+}$ | ·4788 7324 | ·0580 7472 | ·0003 5609 | 2·9215 7974 |
| 39 | ·4729 7297 | ·4722 2222 | ·0576 5062 | ·0006 0881 | 2·9229 7928 |
| 40 | ·4666 6667 | ·4657 5342 | ·0572 2634 | ·0009 1560 | 2·9244 3271 |
| 41 | ·4605 2632 | ·4594 5946 | ·0568 0244 | ·0012 7002 | 2·9259 3156 |
| 42 | ·4545 4545$^{+}$ | ·4533 3333 | ·0563 7942 | ·0016 6640 | 2·9274 6835$^{+}$ |
| 43 | ·4487 1795$^{-}$ | ·4473 6842 | ·0559 5773 | ·0020 9967 | 2·9290 3654 |
| 44 | ·4430 3797 | ·4415 5844 | ·0555 3775$^{-}$ | ·0025 6534 | 2·9306 3035$^{-}$ |
| 45 | ·4375 0000$^{e}$ | ·4358 9744 | ·0551 1982 | ·0030 5940 | 2·9322 4466 |
| 46 | ·4320 9877 | ·4303 7975$^{-}$ | ·0547 0424 | ·0035 7830 | 2·9338 7499 |
| 47 | ·4268 2927 | ·4250 0000$^{e}$ | ·0542 9128 | ·0041 1885$^{+}$ | 2·9355 1736 |
| 48 | ·4216 8675$^{-}$ | ·4197 5309 | ·0538 8116 | ·0046 7820 | 2·9371 6826 |
| 49 | ·4166 6667 | ·4146 3415$^{-}$ | ·0534 7408 | ·0052 5381 | 2·9388 2461 |
| 50 | ·4117 6471 | ·4096 3855$^{+}$ | ·0530 7022 | ·0058 4338 | 2·9404 8365$^{+}$ |

$p = 36$

| $q$ | $\bar{x}$ | $\tilde{x}$ | $\sigma$ | $\beta_1$ | $\beta_2$ |
|---|---|---|---|---|---|
| 0·5 | ·9863 0137 | *J*-curve | ·0189 8138 | 7·0852 3079 | 13·2068 8805− |
| 1 | ·9729 7297 | 1·0000 0000$^e$ | ·0263 0618 | 3·4005 4058 | 7·8232 9060 |
| 1·5 | ·9600 0000$^e$ | ·9859 1549 | ·0315 8168 | 2·1755 5947 | 6·0346 1478 |
| 2 | ·9473 6842 | ·9722 2222 | ·0357 5612 | 1·5654 1667 | 5·1445 1220 |
| 2·5 | ·9350 6494 | ·9589 0411 | ·0392 0685+ | 1·2011 4142 | 4·6137 1907 |
| 3 | ·9230 7692 | ·9459 4595− | ·0421 3250+ | ·9597 4618 | 4·2624 8548 |
| 3·5 | ·9113 9241 | ·9333 3333 | ·0446 5403 | ·7885 2415− | 4·0137 7949 |
| 4 | ·9000 0000$^e$ | ·9210 5263 | ·0468 5213 | ·6611 2371 | 3·8290 8822 |
| 4·5 | ·8888 8889 | ·9090 9091 | ·0487 8412 | ·5629 0657 | 3·6870 1826 |
| 5 | ·8780 4878 | ·8974 3590 | ·0504 9260 | ·4850 9104 | 3·5747 3573 |
| 5·5 | ·8674 6988 | ·8860 7595− | ·0520 1038 | ·4220 8984 | 3·4840 7555− |
| 6 | ·8571 4286 | ·8750 0000$^e$ | ·0533 6338 | ·3701 7906 | 3·4095 9596 |
| 6·5 | ·8470 5882 | ·8641 9753 | ·0545 7257 | ·3267 8145+ | 3·3475 3103 |
| 7 | ·8372 0930 | ·8536 5854 | ·0556 5517 | ·2900 5683 | 3·2951 9209 |
| 7·5 | ·8275 8621 | ·8433 7349 | ·0566 2557 | ·2586 5636 | 3·2506 0853 |
| 8 | ·8181 8182 | ·8333 3333 | ·0574 9596 | ·2315 6900 | 3·2123 0342 |
| 8·5 | ·8089 8876 | ·8235 2941 | ·0582 7671 | ·2080 2208 | 3·1791 4822 |
| 9 | ·8000 0000$^e$ | ·8139 5349 | ·0589 7678 | ·1874 1512 | 3·1502 6596 |
| 9·5 | ·7912 0879 | ·8045 9770 | ·0596 0397 | ·1692 7443 | 3·1249 6502 |
| 10 | ·7826 0870 | ·7954 5455− | ·0601 6508 | ·1532 2145+ | 3·1026 9274 |
| 10·5 | ·7741 9355− | ·7865 1685+ | ·0606 6611 | ·1389 5004 | 3·0830 0234 |
| 11 | ·7659 5745− | ·7777 7778 | ·0611 1237 | ·1262 1004 | 3·0655 2876 |
| 12 | ·7500 0000$^e$ | ·7608 6957 | ·0618 5896 | ·1045 3333 | 3·0360 7843 |
| 13 | ·7346 9388 | ·7446 8085+ | ·0624 3697 | ·0869 1595− | 3·0124 8211 |
| 14 | ·7200 0000$^e$ | ·7291 6667 | ·0628 7242 | ·0724 4999 | 2·9934 1696 |
| 15 | ·7058 8235+ | ·7142 8571 | ·0631 8661 | ·0604 7229 | 2·9779 1754 |
| 16 | ·6923 0769 | ·7000 0000$^e$ | ·0633 9718 | ·0504 8773 | 2·9652 6375− |
| 17 | ·6792 4528 | ·6862 7451 | ·0635 1890 | ·0421 1959 | 2·9549 0833 |
| 18 | ·6666 6667 | ·6730 7692 | ·0635 6417 | ·0350 7653 | 2·9464 2857 |
| 19 | ·6545 4545+ | ·6603 7736 | ·0635 4353 | ·0291 2997 | 2·9394 9332 |
| 20 | ·6428 5714 | ·6481 4815− | ·0634 6595+ | ·0240 9830 | 2·9338 3986 |
| 21 | ·6315 7895− | ·6363 6364 | ·0633 3912 | ·0198 3557 | 2·9292 5747 |
| 22 | ·6206 8966 | ·6250 0000$^e$ | ·0631 6967 | ·0162 2334 | 2·9255 7543 |
| 23 | ·6101 6949 | ·6140 3509 | ·0629 6331 | ·0131 6461 | 2·9226 5422 |
| 24 | ·6000 0000$^e$ | ·6034 4828 | ·0627 2500+ | ·0105 7926 | 2·9203 7890 |
| 25 | ·5901 6393 | ·5932 2034 | ·0624 5908 | ·0084 0066 | 2·9186 5410 |
| 26 | ·5806 4516 | ·5833 3333 | ·0621 6930 | ·0065 7302 | 2·9174 0015− |
| 27 | ·5714 2857 | ·5737 7049 | ·0618 5896 | ·0050 4931 | 2·9165 5012 |
| 28 | ·5625 0000$^e$ | ·5645 1613 | ·0615 3095− | ·0037 8970 | 2·9160 4747 |
| 29 | ·5538 4615+ | ·5555 5556 | ·0611 8781 | ·0027 6026 | 2·9158 4421 |
| 30 | ·5454 5455− | ·5468 7500$^e$ | ·0608 3178 | ·0019 3195− | 2·9158 9940 |
| 31 | ·5373 1343 | ·5384 6154 | ·0604 6483 | ·0012 7981 | 2·9161 7801 |
| 32 | ·5294 1176 | ·5303 0303 | ·0600 8870 | ·0007 8231 | 2·9166 4990 |
| 33 | ·5217 3913 | ·5223 8806 | ·0597 0492 | ·0004 2079 | 2·9172 8909 |
| 34 | ·5142 8571 | ·5147 0588 | ·0593 1486 | ·0001 7903 | 2·9180 7309 |
| 35 | ·5070 4225+ | ·5072 4638 | ·0589 1972 | ·0000 4289 | 2·9189 8239 |
| 36 | ·5000 0000$^e$ | ·5000 0000$^e$ | ·0585 2057 | ·0000 0000$^e$ | 2·9200 0000$^e$ |
| 37 | ·4931 5068 | ·4929 5775− | ·0581 1837 | ·0000 3951 | 2·9211 1111 |
| 38 | ·4864 8649 | ·4861 1111 | ·0577 1394 | ·0001 5187 | 2·9223 0277 |
| 39 | ·4800 0000$^e$ | ·4794 5205+ | ·0573 0803 | ·0003 2868 | 2·9235 6362 |
| 40 | ·4736 8421 | ·4729 7297 | ·0569 0131 | ·0005 6250− | 2·9248 8370 |
| 41 | ·4675 3247 | ·4666 6667 | ·0564 9437 | ·0008 4675− | 2·9262 5425− |
| 42 | ·4615 3846 | ·4605 2632 | ·0560 8771 | ·0011 7560 | 2·9276 6755− |
| 43 | ·4556 9620 | ·4545 4545+ | ·0556 8182 | ·0015 4385− | 2·9291 1680 |
| 44 | ·4500 0000$^e$ | ·4487 1795− | ·0552 7708 | ·0019 4689 | 2·9305 9600 |
| 45 | ·4444 4444 | ·4430 3797 | ·0548 7387 | ·0023 8061 | 2·9320 9983 |
| 46 | ·4390 2439 | ·4375 0000$^e$ | ·0544 7249 | ·0028 4131 | 2·9336 2359 |
| 47 | ·4337 3494 | ·4320 9877 | ·0540 7325− | ·0033 2573 | 2·9351 6314 |
| 48 | ·4285 7143 | ·4268 2927 | ·0536 7637 | ·0038 3090 | 2·9367 1478 |
| 49 | ·4235 2941 | ·4216 8675− | ·0532 8208 | ·0043 5420 | 2·9382 7526 |
| 50 | ·4186 0465+ | ·4166 6667 | ·0528 9056 | ·0048 9325+ | 2·9398 4168 |

$p = 37$

| $q$ | $\bar{x}$ | $\tilde{x}$ | $\sigma$ | $\beta_1$ | $\beta_2$ |
|---|---|---|---|---|---|
| 0·5 | ·9866 6667 | *J*-curve | ·0184 8520 | 7·1078 8725$^{-}$ | 13·2504 2764 |
| 1 | ·9736 8421 | 1·0000 0000$^{e}$ | ·0256 3215$^{-}$ | 3·4151 3514 | 7·8514 1727 |
| 1·5 | ·9610 3896 | ·9863 0137 | ·0307 8841 | 2·1873 1214 | 6·0573 3042 |
| 2 | ·9487 1795$^{-}$ | ·9729 7297 | ·0348 7557 | 1·5756 3870 | 5·1643 2809 |
| 2·5 | ·9367 0886 | ·9600 0000$^{e}$ | ·0382 6011 | 1·2103 6200 | 4·6316 4786 |
| 3 | ·9250 0000$^{e}$ | ·9473 6842 | ·0411 3482 | ·9682 3354 | 4·2790 3984 |
| 3·5 | ·9135 8025$^{-}$ | ·9350 6494 | ·0436 1700 | ·7964 3452 | 4·0292 5748 |
| 4 | ·9024 3902 | ·9230 7692 | ·0457 8491 | ·6685 5715$^{-}$ | 3·8436 8036 |
| 4·5 | ·8915 6627 | ·9113 9241 | ·0476 9409 | ·5699 3180 | 3·7008 5506 |
| 5 | ·8809 5238 | ·9000 0000$^{e}$ | ·0493 8583 | ·4917 5787 | 3·5879 1155$^{-}$ |
| 5·5 | ·8705 8824 | ·8888 8889 | ·0508 9190 | ·4284 3596 | 3·4966 6155$^{+}$ |
| 6 | ·8604 6512 | ·8780 4878 | ·0522 3742 | ·3762 3401 | 3·4216 4773 |
| 6·5 | ·8505 7471 | ·8674 6988 | ·0534 4269 | ·3325 6903 | 3·3590 9326 |
| 7 | ·8409 0909 | ·8571 4286 | ·0545 2441 | ·2955 9670 | 3·3063 0154 |
| 7·5 | ·8314 6067 | ·8470 5882 | ·0554 9652 | ·2639 6511 | 3·2612 9613 |
| 8 | ·8222 2222 | ·8372 0930 | ·0563 7083 | ·2366 6083 | 3·2225 9560 |
| 8·5 | ·8131 8681 | ·8275 8621 | ·0571 5740 | ·2129 0938 | 3·1890 6790 |
| 9 | ·8043 4783 | ·8181 8182 | ·0578 6492 | ·1921 0878 | 3·1598 3330 |
| 9·5 | ·7956 9892 | ·8089 8876 | ·0585 0095$^{-}$ | ·1737 8416 | 3·1341 9793 |
| 10 | ·7872 3404 | ·8000 0000$^{e}$ | ·0590 7210 | ·1575 5597 | 3·1116 0728 |
| 10·5 | ·7789 4737 | ·7912 0879 | ·0595 8419 | ·1431 1727 | 3·0916 1302 |
| 11 | ·7708 3333 | ·7826 0870 | ·0600 4238 | ·1302 1720 | 3·0738 4882 |
| 12 | ·7551 0204 | ·7659 5745$^{-}$ | ·0608 1496 | ·1082 3973 | 3·0438 5268 |
| 13 | ·7400 0000$^{e}$ | ·7500 0000$^{e}$ | ·0614 2108 | ·0903 4433 | 3·0197 5203 |
| 14 | ·7254 9020 | ·7346 9388 | ·0618 8609 | ·0756 2015$^{-}$ | 3·0002 1855$^{-}$ |
| 15 | ·7115 3846 | ·7200 0000$^{e}$ | ·0622 3078 | ·0634 0167 | 2·9842 8246 |
| 16 | ·6981 1321 | ·7058 8235$^{+}$ | ·0624 7239 | ·0531 9187 | 2·9712 2017 |
| 17 | ·6851 8519 | ·6923 0769 | ·0626 2536 | ·0446 1244 | 2·9604 8149 |
| 18 | ·6727 2727 | ·6792 4528 | ·0627 0186 | ·0373 7070 | 2·9516 4130 |
| 19 | ·6607 1429 | ·6666 6667 | ·0627 1222 | ·0312 3694 | 2·9443 6634 |
| 20 | ·6491 2281 | ·6545 4545$^{+}$ | ·0626 6528 | ·0260 2856 | 2·9383 9212 |
| 21 | ·6379 3103 | ·6428 5714 | ·0625 6861 | ·0215 9874 | 2·9335 0634 |
| 22 | ·6271 1864 | ·6315 7895$^{-}$ | ·0624 2874 | ·0178 2829 | 2·9295 3691 |
| 23 | ·6166 6667 | ·6206 8966 | ·0622 5133 | ·0146 1951 | 2·9263 4309 |
| 24 | ·6065 5738 | ·6101 6949 | ·0620 4129 | ·0118 9171 | 2·9238 0885$^{+}$ |
| 25 | ·5967 7419 | ·6000 0000$^{e}$ | ·0618 0291 | ·0095 7770 | 2·9218 3784 |
| 26 | ·5873 0159 | ·5901 6393 | ·0615 3993 | ·0076 2120 | 2·9203 4951 |
| 27 | ·5781 2500$^{e}$ | ·5806 4516 | ·0612 5564 | ·0059 7475$^{+}$ | 2·9192 7613 |
| 28 | ·5692 3077 | ·5714 2857 | ·0609 5293 | ·0045 9811 | 2·9185 6045$^{-}$ |
| 29 | ·5606 0606 | ·5625 0000$^{e}$ | ·0606 3432 | ·0034 5698 | 2·9181 5380 |
| 30 | ·5522 3881 | ·5538 4615$^{+}$ | ·0603 0207 | ·0025 2199 | 2·9180 1466 |
| 31 | ·5441 1765$^{-}$ | ·5454 5455$^{-}$ | ·0599 5815$^{+}$ | ·0017 6788 | 2·9181 0742 |
| 32 | ·5362 3188 | ·5373 1343 | ·0596 0432 | ·0011 7282 | 2·9184 0146 |
| 33 | ·5285 7143 | ·5294 1176 | ·0592 4212 | ·0007 1789 | 2·9188 7030 |
| 34 | ·5211 2676 | ·5217 3913 | ·0588 7294 | ·0003 8664 | 2·9194 9104 |
| 35 | ·5138 8889 | ·5142 8571 | ·0584 9799 | ·0001 6471 | 2·9202 4377 |
| 36 | ·5068 4932 | ·5070 4225$^{+}$ | ·0581 1837 | ·0000 3951 | 2·9211 1111 |
| 37 | ·5000 0000$^{e}$ | ·5000 0000$^{e}$ | ·0577 3503 | ·0000 0000$^{e}$ | 2·9220 7792 |
| 38 | ·4933 3333 | ·4931 5068 | ·0573 4884 | ·0000 3647 | 2·9231 3092 |
| 39 | ·4868 4211 | ·4864 8649 | ·0569 6055$^{+}$ | ·0001 4033 | 2·9242 5847 |
| 40 | ·4805 1948 | ·4800 0000$^{e}$ | ·0565 7087 | ·0003 0401 | 2·9254 5031 |
| 41 | ·4743 5897 | ·4736 8421 | ·0561 8038 | ·0005 2076 | 2·9266 9743 |
| 42 | ·4683 5443 | ·4675 3247 | ·0557 8962 | ·0007 8464 | 2·9279 9187 |
| 43 | ·4625 0000$^{e}$ | ·4615 3846 | ·0553 9909 | ·0010 9031 | 2·9293 2660 |
| 44 | ·4567 9012 | ·4556 9620 | ·0550 0919 | ·0014 3304 | 2·9306 9540 |
| 45 | ·4512 1951 | ·4500 0000$^{e}$ | ·0546 2032 | ·0018 0861 | 2·9320 9277 |
| 46 | ·4457 8313 | ·4444 4444 | ·0542 3280 | ·0022 1323 | 2·9335 1380 |
| 47 | ·4404 7619 | ·4390 2439 | ·0538 4694 | ·0026 4352 | 2·9349 5418 |
| 48 | ·4352 9412 | ·4337 3494 | ·0534 6300 | ·0030 9644 | 2·9364 1006 |
| 49 | ·4302 3256 | ·4285 7143 | ·0530 8121 | ·0035 6926 | 2·9378 7801 |
| 50 | ·4252 8736 | ·4235 2941 | ·0527 0178 | ·0040 5955$^{-}$ | 2·9393 5500$^{-}$ |

$p = 38$

| $q$ | $\bar{x}$ | $\tilde{x}$ | $\sigma$ | $\beta_1$ | $\beta_2$ |
|---|---|---|---|---|---|
| 0·5 | ·9870 1299 | *J*-curve | ·0180 1431 | 7·1294 4914 | 13·2919 0446 |
| 1 | ·9743 5897 | 1·0000 0000$^e$ | ·0249 9178 | 3·4290 3660 | 7·8782 3217 |
| 1·5 | ·9620 2532 | ·9866 6667 | ·0300 3398 | 2·1985 1632 | 6·0790 0332 |
| 2 | ·9500 0000$^e$ | ·9736 8421 | ·0340 3728 | 1·5853 9205$^+$ | 5·1832 4882 |
| 2·5 | ·9382 7160 | ·9610 3896 | ·0373 5792 | 1·2191 6736 | 4·6487 7976 |
| 3 | ·9268 2927 | ·9487 1795$^-$ | ·0401 8314 | ·9763 4568 | 4·2948 7037 |
| 3·5 | ·9156 6265$^+$ | ·9367 0886 | ·0426 2685$^+$ | ·8040 0168 | 4·0440 6988 |
| 4 | ·9047 6190 | ·9250 0000$^e$ | ·0447 6497 | ·6756 7421 | 3·8576 5550$^+$ |
| 4·5 | ·8941 1765$^-$ | ·9135 8025$^-$ | ·0466 5138 | ·5766 6385$^+$ | 3·7141 1675$^+$ |
| 5 | ·8837 2093 | ·9024 3902 | ·0483 2613 | ·4981 5205$^-$ | 3·6005 4920 |
| 5·5 | ·8735 6322 | ·8915 6627 | ·0498 2002 | ·4345 2788 | 3·5087 4253 |
| 6 | ·8636 3636 | ·8809 5238 | ·0511 5740 | ·3820 5154 | 3·4332 2460 |
| 6·5 | ·8539 3258 | ·8705 8824 | ·0523 5797 | ·3381 3462 | 3·3702 0820 |
| 7 | ·8444 4444 | ·8604 6512 | ·0534 3790 | ·3009 2887 | 3·3169 8928 |
| 7·5 | ·8351 6484 | ·8505 7471 | ·0544 1073 | ·2690 7946 | 3·2715 8580 |
| 8 | ·8260 8696 | ·8409 0909 | ·0552 8789 | ·2415 7072 | 3·2325 1208 |
| 8·5 | ·8172 0430 | ·8314 6067 | ·0560 7916 | ·2176 2643 | 3·1986 3278 |
| 9 | ·8085 1064 | ·8222 2222 | ·0567 9297 | ·1966 4321 | 3·1690 6552 |
| 9·5 | ·8000 0000$^e$ | ·8131 8681 | ·0574 3665$^+$ | ·1781 4509 | 3·1431 1431 |
| 10 | ·7916 6667 | ·8043 4783 | ·0580 1663 | ·1617 5158 | 3·1202 2291 |
| 10·5 | ·7835 0515$^+$ | ·7956 9892 | ·0585 3857 | ·1471 5497 | 3·0999 4153 |
| 11 | ·7755 1020 | ·7872 3404 | ·0590 0748 | ·1341 0374 | 3·0819 0262 |
| 12 | ·7600 0000$^e$ | ·7708 3333 | ·0598 0360 | ·1118 4211 | 3·0513 9027 |
| 13 | ·7450 9804 | ·7551 0204 | ·0604 3543 | ·0936 8384 | 3·0268 1231 |
| 14 | ·7307 6923 | ·7400 0000$^e$ | ·0609 2767 | ·0787 1531 | 3·0068 3527 |
| 15 | ·7169 8113 | ·7254 9020 | ·0613 0059 | ·0662 6881 | 2·9904 8530 |
| 16 | ·7037 0370 | ·7115 3846 | ·0615 7107 | ·0558 4553 | 2·9770 3552 |
| 17 | ·6909 0909 | ·6981 1321 | ·0617 5321 | ·0470 6568 | 2·9659 3303 |
| 18 | ·6785 7143 | ·6851 8519 | ·0618 5896 | ·0396 3535$^+$ | 2·9567 5044 |
| 19 | ·6666 6667 | ·6727 2727 | ·0618 9845$^-$ | ·0333 2376 | 2·9491 5254 |
| 20 | ·6551 7241 | ·6607 1429 | ·0618 8036 | ·0279 4737 | 2·9428 7317 |
| 21 | ·6440 6780 | ·6491 2281 | ·0618 1214 | ·0233 5858 | 2·9376 9856 |
| 22 | ·6333 3333 | ·6379 3103 | ·0617 0026 | ·0194 3749 | 2·9334 5534 |
| 23 | ·6229 5082 | ·6271 1864 | ·0615 5028 | ·0160 8576 | 2·9300 0163 |
| 24 | ·6129 0323 | ·6166 6667 | ·0613 6708 | ·0132 2214 | 2·9272 2039 |
| 25 | ·6031 7460 | ·6065 5738 | ·0611 5490 | ·0107 7895$^-$ | 2·9250 1435$^+$ |
| 26 | ·5937 5000$^e$ | ·5967 7419 | ·0609 1746 | ·0086 9943 | 2·9233 0215$^-$ |
| 27 | ·5846 1538 | ·5873 0159 | ·0606 5804 | ·0069 3573 | 2·9220 1531 |
| 28 | ·5757 5758 | ·5781 2500$^e$ | ·0603 7950$^-$ | ·0054 4723 | 2·9210 9590 |
| 29 | ·5671 6418 | ·5692 3077 | ·0600 8437 | ·0041 9927 | 2·9204 9464 |
| 30 | ·5588 2353 | ·5606 0606 | ·0597 7492 | ·0031 6219 | 2·9201 6944 |
| 31 | ·5507 2464 | ·5522 3881 | ·0594 5310 | ·0023 1043 | 2·9200 8417 |
| 32 | ·5428 5714 | ·5441 1765$^-$ | ·0591 2070 | ·0016 2189 | 2·9202 0773 |
| 33 | ·5352 1127 | ·5362 3188 | ·0587 7927 | ·0010 7743 | 2·9205 1322 |
| 34 | ·5277 7778 | ·5285 7143 | ·0584 3019 | ·0006 6035$^+$ | 2·9209 7732 |
| 35 | ·5205 4795$^-$ | ·5211 2676 | ·0580 7472 | ·0003 5609 | 2·9215 7974 |
| 36 | ·5135 1351 | ·5138 8889 | ·0577 1394 | ·0001 5187 | 2·9223 0277 |
| 37 | ·5066 6667 | ·5068 4932 | ·0573 4884 | ·0000 3647 | 2·9231 3092 |
| 38 | ·5000 0000$^e$ | ·5000 0000$^e$ | ·0569 8029 | ·0000 0000$^e$ | 2·9240 5063 |
| 39 | ·4935 0649 | ·4933 3333 | ·0566 0908 | ·0000 3373 | 2·9250 4997 |
| 40 | ·4871 7949 | ·4868 4211 | ·0562 3590 | ·0001 2993 | 2·9261 1842 |
| 41 | ·4810 1266 | ·4805 1948 | ·0558 6138 | ·0002 8174 | 2·9272 4673 |
| 42 | ·4750 0000$^e$ | ·4743 5897 | ·0554 8607 | ·0004 8306 | 2·9284 2671 |
| 43 | ·4691 3580 | ·4683 5443 | ·0551 1047 | ·0007 2846 | 2·9296 5111 |
| 44 | ·4634 1463 | ·4625 0000$^e$ | ·0547 3501 | ·0010 1308 | 2·9309 1351 |
| 45 | ·4578 3133 | ·4567 9012 | ·0543 6011 | ·0013 3260 | 2·9322 0822 |
| 46 | ·4523 8095$^-$ | ·4512 1951 | ·0539 8610 | ·0016 8314 | 2·9335 3017 |
| 47 | ·4470 5882 | ·4457 8313 | ·0536 1330 | ·0020 6122 | 2·9348 7487 |
| 48 | ·4418 6047 | ·4404 7619 | ·0532 4200 | ·0024 6371 | 2·9362 3831 |
| 49 | ·4367 8161 | ·4352 9412 | ·0528 7243 | ·0028 8781 | 2·9376 1692 |
| 50 | ·4318 1818 | ·4302 3256 | ·0525 0481 | ·0033 3099 | 2·9390 0752 |

$p = 39$

| $q$ | $\bar{x}$ | $\tilde{x}$ | $\sigma$ | $\beta_1$ | $\beta_2$ |
|---|---|---|---|---|---|
| 0·5 | ·9873 4177 | *J*-curve | ·0175 6680 | 7·1499 9386 | 13·3314 6159 |
| 1 | ·9750 0000$^e$ | 1·0000 0000$^e$ | ·0243 8262 | 3·4422 9316 | 7·9038 2486 |
| 1·5 | ·9629 6296 | ·9870 1299 | ·0293 1561 | 2·2092 0948 | 6·0997 0354 |
| 2 | ·9512 1951 | ·9743 5897 | ·0332 3830 | 1·5947 0816 | 5·2013 3355$^+$ |
| 2·5 | ·9397 5904 | ·9620 2532 | ·0364 9722 | 1·2275 8485$^+$ | 4·6651 6655$^+$ |
| 3 | ·9285 7143 | ·9500 0000$^e$ | ·0392 7439 | ·9841 0680 | 4·3100 2331 |
| 3·5 | ·9176 4706 | ·9382 7160 | ·0416 8049 | ·8112 4733 | 4·0582 5845$^-$ |
| 4 | ·9069 7674 | ·9268 2927 | ·0437 8925$^-$ | ·6824 9446 | 3·8710 5165$^+$ |
| 4·5 | ·8965 5172 | ·9156 6265$^+$ | ·0456 5300 | ·5831 2046 | 3·7268 3809 |
| 5 | ·8863 6364 | ·9047 6190 | ·0473 1059 | ·5042 8966 | 3·6126 8057 |
| 5·5 | ·8764 0449 | ·8941 1765$^-$ | ·0487 9192 | ·4403 8022 | 3·5203 4779 |
| 6 | ·8666 6667 | ·8837 2093 | ·0501 2063 | ·3876 4495$^-$ | 3·4443 5352 |
| 6·5 | ·8571 4286 | ·8735 6322 | ·0513 1580 | ·3434 9030 | 3·3809 0071 |
| 7 | ·8478 2609 | ·8636 3636 | ·0523 9316 | ·3060 6431 | 3·3272 7816 |
| 7·5 | ·8387 0968 | ·8539 3258 | ·0533 6582 | ·2740 0934 | 3·2814 9857 |
| 8 | ·8297 8723 | ·8444 4444 | ·0542 4489 | ·2463 0763 | 3·2420 7221 |
| 8·5 | ·8210 5263 | ·8351 6484 | ·0550 3986 | ·2221 8129 | 3·2078 6061 |
| 9 | ·8125 0000$^e$ | ·8260 8696 | ·0557 5891 | ·2010 2564 | 3·1779 7888 |
| 9·5 | ·8041 2371 | ·8172 0430 | ·0564 0918 | ·1823 6364 | 3·1517 2904 |
| 10 | ·7959 1837 | ·8085 1064 | ·0569 9690 | ·1658 1394 | 3·1285 5320 |
| 10·5 | ·7878 7879 | ·8000 0000$^e$ | ·0575 2757 | ·1510 6809 | 3·1080 0018 |
| 11 | ·7800 0000$^e$ | ·7916 6667 | ·0580 0608 | ·1378 7396 | 3·0897 0130 |
| 12 | ·7647 0588 | ·7755 1020 | ·0588 2353 | ·1153 4354 | 3·0587 0021 |
| 13 | ·7500 0000$^e$ | ·7600 0000$^e$ | ·0594 7887 | ·0969 3644 | 3·0336 7003 |
| 14 | ·7358 4906 | ·7450 9804 | ·0599 9618 | ·0817 3645$^-$ | 3·0132 7244 |
| 15 | ·7222 2222 | ·7307 6923 | ·0603 9526 | ·0690 7378 | 2·9965 2979 |
| 16 | ·7090 9091 | ·7169 8113 | ·0606 9258 | ·0584 4796 | 2·9827 1209 |
| 17 | ·6964 2857 | ·7037 0370 | ·0609 0197 | ·0494 7784 | 2·9712 6393 |
| 18 | ·6842 1053 | ·6909 0909 | ·0610 3511 | ·0418 6831 | 2·9617 5576 |
| 19 | ·6724 1379 | ·6785 7143 | ·0611 0195$^+$ | ·0353 8761 | 2·9538 5058 |
| 20 | ·6610 1695$^-$ | ·6666 6667 | ·0611 1104 | ·0298 5136 | 2·9472 8064 |
| 21 | ·6500 0000$^e$ | ·6551 7241 | ·0610 6970 | ·0251 1121 | 2·9418 3083 |
| 22 | ·6393 4426 | ·6440 6780 | ·0609 8429 | ·0210 4656 | 2·9373 2656 |
| 23 | ·6290 3226 | ·6333 3333 | ·0608 6032 | ·0175 5853 | 2·9336 2490 |
| 24 | ·6190 4762 | ·6229 5082 | ·0607 0261 | ·0145 6532 | 2·9306 0785$^+$ |
| 25 | ·6093 7500$^e$ | ·6129 0323 | ·0605 1536 | ·0119 9878 | 2·9281 7730 |
| 26 | ·6000 0000$^e$ | ·6031 7460 | ·0603 0227 | ·0098 0174 | 2·9262 5110 |
| 27 | ·5909 0909 | ·5937 5000$^e$ | ·0600 6657 | ·0079 2595$^-$ | 2·9247 6009 |
| 28 | ·5820 8955$^+$ | ·5846 1538 | ·0598 1114 | ·0063 3043 | 2·9236 4571 |
| 29 | ·5735 2941 | ·5757 5758 | ·0595 3849 | ·0049 8024 | 2·9228 5810 |
| 30 | ·5652 1739 | ·5671 6418 | ·0592 5088 | ·0038 4539 | 2·9223 5464 |
| 31 | ·5571 4286 | ·5588 2353 | ·0589 5029 | ·0029 0006 | 2·9220 9872 |
| 32 | ·5492 9577 | ·5507 2464 | ·0586 3848 | ·0021 2192 | 2·9220 5878 |
| 33 | ·5416 6667 | ·5428 5714 | ·0583 1702 | ·0014 9157 | 2·9222 0752 |
| 34 | ·5342 4658 | ·5352 1127 | ·0579 8732 | ·0009 9212 | 2·9225 2124 |
| 35 | ·5270 2703 | ·5277 7778 | ·0576 5062 | ·0006 0881 | 2·9229 7928 |
| 36 | ·5200 0000$^e$ | ·5205 4795$^-$ | ·0573 0803 | ·0003 2868 | 2·9235 6362 |
| 37 | ·5131 5789 | ·5135 1351 | ·0569 6055$^+$ | ·0001 4033 | 2·9242 5847 |
| 38 | ·5064 9351 | ·5066 6667 | ·0566 0908 | ·0000 3373 | 2·9250 4997 |
| 39 | ·5000 0000$^e$ | ·5000 0000$^e$ | ·0562 5440 | ·0000 0000$^e$ | 2·9259 2593 |
| 40 | ·4936 7089 | ·4935 0649 | ·0558 9722 | ·0000 3126 | 2·9268 7559 |
| 41 | ·4875 0000$^e$ | ·4871 7949 | ·0555 3819 | ·0001 2054 | 2·9278 8947 |
| 42 | ·4814 8148 | ·4810 1266 | ·0551 7788 | ·0002 6161 | 2·9289 5916 |
| 43 | ·4756 0976 | ·4750 0000$^e$ | ·0548 1679 | ·0004 4892 | 2·9300 7722 |
| 44 | ·4698 7952 | ·4691 3580 | ·0544 5539 | ·0006 7752 | 2·9312 3703 |
| 45 | ·4642 8571 | ·4634 1463 | ·0540 9409 | ·0009 4299 | 2·9324 3271 |
| 46 | ·4588 2353 | ·4578 3133 | ·0537 3324 | ·0012 4135$^-$ | 2·9336 5904 |
| 47 | ·4534 8837 | ·4523 8095$^-$ | ·0533 7319 | ·0015 6903 | 2·9349 1137 |
| 48 | ·4482 7586 | ·4470 5882 | ·0530 1421 | ·0019 2283 | 2·9361 8554 |
| 49 | ·4431 8182 | ·4418 6047 | ·0526 5658 | ·0022 9988 | 2·9374 7784 |
| 50 | ·4382 0225$^-$ | ·4367 8161 | ·0523 0052 | ·0026 9756 | 2·9387 8496 |

$p = 40$

| $q$ | $\bar{x}$ | $\tilde{x}$ | $\sigma$ | $\beta_1$ | $\beta_2$ |
|---|---|---|---|---|---|
| 0·5 | ·9876 5432 | $J$-curve | ·0171 4099 | 7·1695 9170 | 13·3692 2921 |
| 1 | ·9756 0976 | 1·0000 0000$^{e}$ | ·0238 0244 | 3·4549 4862 | 7·9282 7696 |
| 1·5 | ·9638 5542 | ·9873 4177 | ·0286 3077 | 2·2194 2573 | 6·1194 9503 |
| 2 | ·9523 8095+ | ·9750 0000$^{e}$ | ·0324 7592 | 1·6036 1570 | 5·2186 3636 |
| 2·5 | ·9411 7647 | ·9629 6296 | ·0356 7520 | 1·2356 3944 | 4·6808 5566 |
| 3 | ·9302 3256 | ·9512 1951 | ·0384 0571 | ·9915 3909 | 4·3245 4106 |
| 3·5 | ·9195 4023 | ·9397 5904 | ·0407 7507 | ·8181 9138 | 4·0718 6155− |
| 4 | ·9090 9091 | ·9285 7143 | ·0428 5496 | ·6890 3592 | 3·8839 0379 |
| 4·5 | ·8988 7640 | ·9176 4706 | ·0446 9619 | ·5893 1797 | 3·7390 5112 |
| 5 | ·8888 8889 | ·9069 7674 | ·0463 3654 | ·5101 8560 | 3·6243 3511 |
| 5·5 | ·8791 2088 | ·8965 5172 | ·0478 0502 | ·4460 0655− | 3·5315 0446 |
| 6 | ·8695 6522 | ·8863 6364 | ·0491 2458 | ·3930 2662 | 3·4550 5952 |
| 6·5 | ·8602 1505+ | ·8764 0449 | ·0503 1378 | ·3486 4737 | 3·3911 9386 |
| 7 | ·8510 6383 | ·8666 6667 | ·0513 8786 | ·3110 1327 | 3·3371 8950+ |
| 7·5 | ·8421 0526 | ·8571 4286 | ·0523 5958 | ·2787 6407 | 3·2910 5411 |
| 8 | ·8333 3333 | ·8478 2609 | ·0532 3971 | ·2508 8000$^{e}$ | 3·2512 9412 |
| 8·5 | ·8247 4227 | ·8387 0968 | ·0540 3748 | ·2265 8159 | 3·2167 6807 |
| 9 | ·8163 2653 | ·8297 8723 | ·0547 6085− | ·2052 6293 | 3·1865 8874 |
| 9·5 | ·8080 8081 | ·8210 5263 | ·0554 1673 | ·1864 4597 | 3·1600 5621 |
| 10 | ·8000 0000$^{e}$ | ·8125 0000$^{e}$ | ·0560 1120 | ·1697 4852 | 3·1366 1103 |
| 10·5 | ·7920 7921 | ·8041 2371 | ·0565 4962 | ·1548 6146 | 3·1158 0075+ |
| 11 | ·7843 1373 | ·7959 1837 | ·0570 3671 | ·1415 3209 | 3·0972 5557 |
| 12 | ·7692 3077 | ·7800 0000$^{e}$ | ·0578 7345− | ·1187 4714 | 3·0657 9125− |
| 13 | ·7547 1698 | ·7647 0588 | ·0585 5027 | ·1001 0426 | 3·0403 3217 |
| 14 | ·7407 4074 | ·7500 0000$^{e}$ | ·0590 9067 | ·0846 8477 | 3·0195 3545− |
| 15 | ·7272 7273 | ·7358 4906 | ·0595 1397 | ·0718 1697 | 3·0024 1984 |
| 16 | ·7142 8571 | ·7222 2222 | ·0598 3627 | ·0609 9881 | 2·9882 5248 |
| 17 | ·7017 5439 | ·7090 9091 | ·0600 7110 | ·0518 4785− | 2·9764 7557 |
| 18 | ·6896 5517 | ·6964 2857 | ·0602 2991 | ·0440 6790 | 2·9666 5756 |
| 19 | ·6779 6610 | ·6842 1053 | ·0603 2248 | ·0374 2627 | 2·9584 5974 |
| 20 | ·6666 6667 | ·6724 1379 | ·0603 5716 | ·0317 3777 | 2·9516 1290 |
| 21 | ·6557 3770 | ·6610 1695− | ·0603 4120 | ·0268 5335+ | 2·9459 0065+ |
| 22 | ·6451 6129 | ·6500 0000$^{e}$ | ·0602 8085+ | ·0226 5181 | 2·9411 4729 |
| 23 | ·6349 2063 | ·6393 4426 | ·0601 8155+ | ·0190 3370 | 2·9372 0888 |
| 24 | ·6250 0000$^{e}$ | ·6290 3226 | ·0600 4806 | ·0159 1674 | 2·9339 6653 |
| 25 | ·6153 8462 | ·6190 4762 | ·0598 8453 | ·0132 3235− | 2·9313 2133 |
| 26 | ·6060 6061 | ·6093 7500$^{e}$ | ·0596 9464 | ·0109 2294 | 2·9291 9044 |
| 27 | ·5970 1493 | ·6000 0000$^{e}$ | ·0594 8160 | ·0089 3992 | 2·9275 0403 |
| 28 | ·5882 3529 | ·5909 0909 | ·0592 4825+ | ·0072 4198 | 2·9262 0293 |
| 29 | ·5797 1014 | ·5820 8955+ | ·0589 9713 | ·0057 9387 | 2·9252 3677 |
| 30 | ·5714 2857 | ·5735 2941 | ·0587 3046 | ·0045 6533 | 2·9245 6240 |
| 31 | ·5633 8028 | ·5652 1739 | ·0584 5023 | ·0035 3029 | 2·9241 4279 |
| 32 | ·5555 5556 | ·5571 4286 | ·0581 5821 | ·0026 6618 | 2·9239 4595− |
| 33 | ·5479 4521 | ·5492 9577 | ·0578 5598 | ·0019 5340 | 2·9239 4418 |
| 34 | ·5405 4054 | ·5416 6667 | ·0575 4493 | ·0013 7486 | 2·9241 1343 |
| 35 | ·5333 3333 | ·5342 4658 | ·0572 2634 | ·0009 1560 | 2·9244 3271 |
| 36 | ·5263 1579 | ·5270 2703 | ·0569 0131 | ·0005 6250− | 2·9248 8370 |
| 37 | ·5194 8052 | ·5200 0000$^{e}$ | ·0565 7087 | ·0003 0401 | 2·9254 5031 |
| 38 | ·5128 2051 | ·5131 5789 | ·0562 3590 | ·0001 2993 | 2·9261 1842 |
| 39 | ·5063 2911 | ·5064 9351 | ·0558 9722 | ·0000 3126 | 2·9268 7559 |
| 40 | ·5000 0000$^{e}$ | ·5000 0000$^{e}$ | ·0555 5556 | ·0000 0000$^{e}$ | 2·9277 1084 |
| 41 | ·4938 2716 | ·4936 7089 | ·0552 1156 | ·0000 2903 | 2·9286 1446 |
| 42 | ·4878 0488 | ·4875 0000$^{e}$ | ·0548 6580 | ·0001 1203 | 2·9295 7783 |
| 43 | ·4819 2771 | ·4814 8148 | ·0545 1883 | ·0002 4334 | 2·9305 9333 |
| 44 | ·4761 9048 | ·4756 0976 | ·0541 7109 | ·0004 1792 | 2·9316 5415+ |
| 45 | ·4705 8824 | ·4698 7952 | ·0538 2302 | ·0006 3123 | 2·9327 5427 |
| 46 | ·4651 1628 | ·4642 8571 | ·0534 7501 | ·0008 7922 | 2·9338 8828 |
| 47 | ·4597 7011 | ·4588 2353 | ·0531 2737 | ·0011 5825− | 2·9350 5140 |
| 48 | ·4545 4545+ | ·4534 8837 | ·0527 8043 | ·0014 6502 | 2·9362 3932 |
| 49 | ·4494 3820 | ·4482 7586 | ·0524 3446 | ·0017 9659 | 2·9374 4820 |
| 50 | ·4444 4444 | ·4431 8182 | ·0520 8969 | ·0021 5028 | 2·9386 7461 |

$p = 41$

| $q$ | $\bar{x}$ | $\tilde{x}$ | $\sigma$ | $\beta_1$ | $\beta_2$ |
|---|---|---|---|---|---|
| 0·5 | ·9879 5181 | *J*-curve | ·0167 3533 | 7·1883 0660 | 13·4053 2597 |
| 1 | ·9761 9048 | 1·0000 0000$^e$ | ·0232 4922 | 3·4670 4293 | 7·9516 6297 |
| 1·5 | ·9647 0588 | ·9876 5432 | ·0279 7718 | 2·2291 9624 | 6·1384 3624 |
| 2 | ·9534 8837 | ·9756 0976 | ·0317 4769 | 1·6121 4092 | 5·2352 0679 |
| 2·5 | ·9425 2874 | ·9638 5542 | ·0348 8933 | 1·2433 5402 | 4·6958 9058 |
| 3 | ·9318 1818 | ·9523 8095$^+$ | ·0375 7454 | ·9986 6292 | 4·3384 6258 |
| 3·5 | ·9213 4831 | ·9411 7647 | ·0399 0803 | ·8248 5216 | 4·0849 1449 |
| 4 | ·9111 1111 | ·9302 3256 | ·0419 5952 | ·6953 1517 | 3·8962 4416 |
| 4·5 | ·9010 9890 | ·9195 4023 | ·0437 7844 | ·5952 7149 | 3·7507 8544 |
| 5 | ·8913 0435$^-$ | ·9090 9091 | ·0454 0150$^-$ | ·5158 5366 | 3·6355 4007 |
| 5·5 | ·8817 2043 | ·8988 7640 | ·0468 5690 | ·4514 1947 | 3·5422 3770 |
| 6 | ·8723 4043 | ·8888 8889 | ·0481 6694 | ·3982 0806 | 3·4653 6585$^+$ |
| 6·5 | ·8631 5789 | ·8791 2088 | ·0493 4966 | ·3536 1633 | 3·4011 0915$^+$ |
| 7 | ·8541 6667 | ·8695 6522 | ·0504 1986 | ·3157 8537 | 3·3467 4319 |
| 7·5 | ·8453 6082 | ·8602 1505$^+$ | ·0513 8995$^+$ | ·2833 5242 | 3·3002 7079 |
| 8 | ·8367 3469 | ·8510 6383 | ·0522 7040 | ·2552 9581 | 3·2601 9479 |
| 8·5 | ·8282 8283 | ·8421 0526 | ·0530 7017 | ·2308 3457 | 3·2253 7087 |
| 9 | ·8200 0000$^e$ | ·8333 3333 | ·0537 9700 | ·2093 6162 | 3·1949 0955$^+$ |
| 9·5 | ·8118 8119 | ·8247 4227 | ·0544 5763 | ·1903 9795$^-$ | 3·1681 0913 |
| 10 | ·8039 2157 | ·8163 2653 | ·0550 5796 | ·1735 6059 | 3·1444 0865$^+$ |
| 10·5 | ·7961 1650$^+$ | ·8080 8081 | ·0556 0319 | ·1585 3978 | 3·1233 5445$^-$ |
| 11 | ·7884 6154 | ·8000 0000$^e$ | ·0560 9795$^+$ | ·1450 8226 | 3·1045 7569 |
| 12 | ·7735 8491 | ·7843 1373 | ·0569 5211 | ·1220 5604 | 3·0726 7184 |
| 13 | ·7592 5926 | ·7692 3077 | ·0576 4856 | ·1031 8949 | 3·0468 0557 |
| 14 | ·7454 5455$^-$ | ·7547 1698 | ·0582 1022 | ·0875 6165$^+$ | 3·0256 2968 |
| 15 | ·7321 4286 | ·7407 4074 | ·0586 5595$^+$ | ·0744 9900 | 3·0081 5954 |
| 16 | ·7192 9825$^-$ | ·7272 7273 | ·0590 0150$^+$ | ·0634 9801 | 2·9936 5957 |
| 17 | ·7068 9655$^+$ | ·7142 8571 | ·0592 6010 | ·0541 7504 | 2·9815 6973 |
| 18 | ·6949 1525$^+$ | ·7017 5439 | ·0594 4298 | ·0462 3287 | 2·9714 5658 |
| 19 | ·6833 3333 | ·6896 5517 | ·0595 5971 | ·0394 3795$^+$ | 2·9629 7983 |
| 20 | ·6721 3115$^-$ | ·6779 6610 | ·0596 1853 | ·0336 0434 | 2·9558 6890 |
| 21 | ·6612 9032 | ·6666 6667 | ·0596 2654 | ·0285 8232 | 2·9499 0619 |
| 22 | ·6507 9365$^+$ | ·6557 3770 | ·0595 8991 | ·0242 5012 | 2·9449 1495$^+$ |
| 23 | ·6406 2500$^e$ | ·6451 6129 | ·0595 1401 | ·0205 0779 | 2·9407 5031 |
| 24 | ·6307 6923 | ·6349 2063 | ·0594 0353 | ·0172 7257 | 2·9372 9255$^+$ |
| 25 | ·6212 1212 | ·6250 0000$^e$ | ·0592 6259 | ·0144 7548 | 2·9344 4202 |
| 26 | ·6119 4030 | ·6153 8462 | ·0590 9481 | ·0120 5858 | 2·9321 1518 |
| 27 | ·6029 4118 | ·6060 6061 | ·0589 0340 | ·0099 7290 | 2·9302 4161 |
| 28 | ·5942 0290 | ·5970 1493 | ·0586 9118 | ·0081 7685$^+$ | 2·9287 6159 |
| 29 | ·5857 1429 | ·5882 3529 | ·0584 6066 | ·0066 3489 | 2·9276 2423 |
| 30 | ·5774 6479 | ·5797 1014 | ·0582 1407 | ·0053 1651 | 2·9267 8592 |
| 31 | ·5694 4444 | ·5714 2857 | ·0579 5339 | ·0041 9540 | 2·9262 0920 |
| 32 | ·5616 4384 | ·5633 8028 | ·0576 8039 | ·0032 4878 | 2·9258 6168 |
| 33 | ·5540 5405$^+$ | ·5555 5556 | ·0573 9665$^+$ | ·0024 5684 | 2·9257 1532 |
| 34 | ·5466 6667 | ·5479 4521 | ·0571 0358 | ·0018 0229 | 2·9257 4570 |
| 35 | ·5394 7368 | ·5405 4054 | ·0568 0244 | ·0012 7002 | 2·9259 3156 |
| 36 | ·5324 6753 | ·5333 3333 | ·0564 9437 | ·0008 4675$^-$ | 2·9262 5425$^-$ |
| 37 | ·5256 4103 | ·5263 1579 | ·0561 8038 | ·0005 2076 | 2·9266 9743 |
| 38 | ·5189 8734 | ·5194 8052 | ·0558 6138 | ·0002 8174 | 2·9272 4673 |
| 39 | ·5125 0000$^e$ | ·5128 2051 | ·0555 3819 | ·0001 2054 | 2·9278 8947 |
| 40 | ·5061 7284 | ·5063 2911 | ·0552 1156 | ·0000 2903 | 2·9286 1446 |
| 41 | ·5000 0000$^e$ | ·5000 0000$^e$ | ·0548 8213 | ·0000 0000$^e$ | 2·9294 1176 |
| 42 | ·4939 7590 | ·4938 2716 | ·0545 5051 | ·0000 2701 | 2·9302 7260 |
| 43 | ·4880 9524 | ·4878 0488 | ·0542 1724 | ·0001 0430 | 2·9311 8914 |
| 44 | ·4823 5294 | ·4819 2771 | ·0538 8279 | ·0002 2674 | 2·9321 5443 |
| 45 | ·4767 4419 | ·4761 9048 | ·0535 4761 | ·0003 8971 | 2·9331 6226 |
| 46 | ·4712 6437 | ·4705 8824 | ·0532 1208 | ·0005 8906 | 2·9342 0711 |
| 47 | ·4659 0909 | ·4651 1628 | ·0528 7656 | ·0008 2108 | 2·9352 8402 |
| 48 | ·4606 7416 | ·4597 7011 | ·0525 4136 | ·0010 8241 | 2·9363 8857 |
| 49 | ·4555 5556 | ·4545 4545$^+$ | ·0522 0676 | ·0013 7002 | 2·9375 1680 |
| 50 | ·4505 4945$^-$ | ·4494 3820 | ·0518 7303 | ·0016 8118 | 2·9386 6515$^-$ |

$p = 42$

| $q$ | $\bar{x}$ | $\tilde{x}$ | $\sigma$ | $\beta_1$ | $\beta_2$ |
|---|---|---|---|---|---|
| 0·5 | ·9882 3529 | *J*-curve | ·0163 4843 | 7·2061 9691 | 13·4398 6030 |
| 1 | ·9767 4419 | 1·0000 0000$^e$ | ·0227 2113 | 3·4786 1258 | 7·9740 5107 |
| 1·5 | ·9655 1724 | ·9879 5181 | ·0273 5275+ | 2·2385 4952 | 6·1565 8075− |
| 2 | ·9545 4545+ | ·9761 9048 | ·0310 5137 | 1·6203 0786 | 5·2510 9026 |
| 2·5 | ·9438 2022 | ·9647 0588 | ·0341 3728 | 1·2507 4960 | 4·7103 1126 |
| 3 | ·9333 3333 | ·9534 8837 | ·0367 7849 | 1·0054 9699 | 4·3518 2371 |
| 3·5 | ·9230 7692 | ·9425 2874 | ·0390 7696 | ·8312 4654 | 4·0974 4981 |
| 4 | ·9130 4348 | ·9318 1818 | ·0411 0057 | ·7013 4755+ | 3·9081 0253 |
| 4·5 | ·9032 2581 | ·9213 4831 | ·0428 9742 | ·6009 9499 | 3·7620 6840 |
| 5 | ·8936 1702 | ·9111 1111 | ·0445 0319 | ·5213 0660 | 3·6463 2070 |
| 5·5 | ·8842 1053 | ·9010 9890 | ·0459 4534 | ·4566 3067 | 3·5525 7084 |
| 6 | ·8750 0000$^e$ | ·8913 0435− | ·0472 4556 | ·4032 0000$^e$ | 3·4752 9412 |
| 6·5 | ·8659 7938 | ·8817 2043 | ·0484 2136 | ·3584 0701 | 3·4106 6662 |
| 7 | ·8571 4286 | ·8723 4043 | ·0494 8717 | ·3203 8959 | 3·3559 5777 |
| 7·5 | ·8484 8485− | ·8631 5789 | ·0504 5502 | ·2877 8261 | 3·3091 6584 |
| 8 | ·8400 0000$^e$ | ·8541 6667 | ·0513 3512 | ·2595 6255+ | 3·2687 9017 |
| 8·5 | ·8316 8317 | ·8453 6082 | ·0521 3617 | ·2349 4706 | 3·2336 8375+ |
| 9 | ·8235 2941 | ·8367 3469 | ·0528 6571 | ·2133 2791 | 3·2029 5498 |
| 9·5 | ·8155 3398 | ·8282 8283 | ·0535 3029 | ·1942 2519 | 3·1759 0040 |
| 10 | ·8076 9231 | ·8200 0000$^e$ | ·0541 3565+ | ·1772 5521 | 3·1519 5767 |
| 10·5 | ·8000 0000$^e$ | ·8118 8119 | ·0546 8687 | ·1621 0757 | 3·1306 7196 |
| 11 | ·7924 5283 | ·8039 2157 | ·0551 8846 | ·1485 2850− | 3·1116 7145− |
| 12 | ·7777 7778 | ·7884 6154 | ·0560 5833 | ·1252 7332 | 3·0793 5016 |
| 13 | ·7636 3636 | ·7735 8491 | ·0567 7271 | ·1061 9441 | 3·0530 9693 |
| 14 | ·7500 0000$^e$ | ·7592 5926 | ·0573 5393 | ·0903 6861 | 3·0315 6049 |
| 15 | ·7368 4211 | ·7454 5455− | ·0578 2044 | ·0771 2070 | 3·0137 5303 |
| 16 | ·7241 3793 | ·7321 4286 | ·0581 8763 | ·0659 4577 | 2·9989 3638 |
| 17 | ·7118 6441 | ·7192 9825− | ·0584 6844 | ·0564 5903 | 2·9865 4840 |
| 18 | ·7000 0000$^e$ | ·7068 9655+ | ·0586 7387 | ·0483 6232 | 2·9761 5390 |
| 19 | ·6885 2459 | ·6949 1525+ | ·0588 1333 | ·0414 2127 | 2·9674 1109 |
| 20 | ·6774 1935+ | ·6833 3333 | ·0588 9490 | ·0354 4922 | 2·9600 4808 |
| 21 | ·6666 6667 | ·6721 3115− | ·0589 2557 | ·0302 9586 | 2·9538 4615+ |
| 22 | ·6562 5000$^e$ | ·6612 9032 | ·0589 1140 | ·0258 3887 | 2·9486 2758 |
| 23 | ·6461 5385− | ·6507 9365+ | ·0588 5769 | ·0219 7781 | 2·9442 4661 |
| 24 | ·6363 6364 | ·6406 2500$^e$ | ·0587 6909 | ·0186 2951 | 2·9405 8275+ |
| 25 | ·6268 6567 | ·6307 6923 | ·0586 4965− | ·0157 2459 | 2·9375 3564 |
| 26 | ·6176 4706 | ·6212 1212 | ·0585 0296 | ·0132 0475+ | 2·9350 2112 |
| 27 | ·6086 9565+ | ·6119 4030 | ·0583 3221 | ·0110 2074 | 2·9329 6818 |
| 28 | ·6000 0000$^e$ | ·6029 4118 | ·0581 4019 | ·0091 3066 | 2·9313 1659 |
| 29 | ·5915 4930 | ·5942 0290 | ·0579 2940 | ·0074 9870 | 2·9300 1497 |
| 30 | ·5833 3333 | ·5857 1429 | ·0577 0206 | ·0060 9413 | 2·9290 1931 |
| 31 | ·5753 4247 | ·5774 6479 | ·0574 6015+ | ·0048 9039 | 2·9282 9170 |
| 32 | ·5675 6757 | ·5694 4444 | ·0572 0543 | ·0038 6451 | 2·9277 9941 |
| 33 | ·5600 0000$^e$ | ·5616 4384 | ·0569 3949 | ·0029 9650− | 2·9275 1404 |
| 34 | ·5526 3158 | ·5540 5405+ | ·0566 6373 | ·0022 6889 | 2·9274 1089 |
| 35 | ·5454 5455− | ·5466 6667 | ·0563 7942 | ·0016 6640 | 2·9274 6835+ |
| 36 | ·5384 6154 | ·5394 7368 | ·0560 8771 | ·0011 7560 | 2·9276 6755− |
| 37 | ·5316 4557 | ·5324 6753 | ·0557 8962 | ·0007 8464 | 2·9279 9187 |
| 38 | ·5250 0000$^e$ | ·5256 4103 | ·0554 8607 | ·0004 8306 | 2·9284 2671 |
| 39 | ·5185 1852 | ·5189 8734 | ·0551 7788 | ·0002 6161 | 2·9289 5916 |
| 40 | ·5121 9512 | ·5125 0000$^e$ | ·0548 6580 | ·0001 1203 | 2·9295 7783 |
| 41 | ·5060 2410 | ·5061 7284 | ·0545 5051 | ·0000 2701 | 2·9302 7260 |
| 42 | ·5000 0000$^e$ | ·5000 0000$^e$ | ·0542 3261 | ·0000 0000$^e$ | 2·9310 3448 |
| 43 | ·4941 1765− | ·4939 7590 | ·0539 1266 | ·0000 2517 | 2·9318 5550+ |
| 44 | ·4883 7209 | ·4880 9524 | ·0535 9113 | ·0000 9727 | 2·9327 2853 |
| 45 | ·4827 5862 | ·4823 5294 | ·0532 6848 | ·0002 1161 | 2·9336 4723 |
| 46 | ·4772 7273 | ·4767 4419 | ·0529 4511 | ·0003 6398 | 2·9346 0590 |
| 47 | ·4719 1011 | ·4712 6437 | ·0526 2139 | ·0005 5057 | 2·9355 9949 |
| 48 | ·4666 6667 | ·4659 0909 | ·0522 9764 | ·0007 6796 | 2·9366 2342 |
| 49 | ·4615 3846 | ·4606 7416 | ·0519 7415− | ·0010 1305+ | 2·9376 7363 |
| 50 | ·4565 2174 | ·4555 5556 | ·0516 5119 | ·0012 8306 | 2·9387 4644 |

$p = 43$

| $q$ | $\bar{x}$ | $\tilde{x}$ | $\sigma$ | $\beta_1$ | $\beta_2$ |
|---|---|---|---|---|---|
| 0·5 | ·9885 0575$^{-}$ | *J*-curve | ·0159 7901 | 7·2233 1591 | 13·4729 3142 |
| 1 | ·9772 7273 | 1·0000 0000$^{e}$ | ·0222 1648 | 3·4896 9095$^{-}$ | 7·9955 0373 |
| 1·5 | ·9662 9213 | ·9882 3529 | ·0267 5557 | 2·2475 1170 | 6·1739 7770 |
| 2 | ·9555 5556 | ·9767 4419 | ·0303 8490 | 1·6281 3859 | 5·2663 2855$^{+}$ |
| 2·5 | ·9450 5495$^{-}$ | ·9655 1724 | ·0334 1692 | 1·2578 4552 | 4·7241 5450$^{+}$ |
| 3 | ·9347 8261 | ·9545 4545$^{+}$ | ·0360 1539 | 1·0120 5857 | 4·3646 5749 |
| 3·5 | ·9247 3118 | ·9438 2022 | ·0382 7970 | ·8373 9009 | 4·1094 9755$^{+}$ |
| 4 | ·9148 9362 | ·9333 3333 | ·0402 7595$^{-}$ | ·7071 4722 | 3·9195 0641 |
| 4·5 | ·9052 6316 | ·9230 7692 | ·0420 5097 | ·6065 0142 | 3·7729 2535$^{+}$ |
| 5 | ·8958 3333 | ·9130 4348 | ·0436 3952 | ·5265 5628 | 3·6567 0041 |
| 5·5 | ·8865 9794 | ·9032 2581 | ·0450 6831 | ·4616 5105$^{-}$ | 3·5625 2557 |
| 6 | ·8775 5102 | ·8936 1702 | ·0463 5844 | ·4080 1242 | 3·4848 6443 |
| 6·5 | ·8686 8687 | ·8842 1053 | ·0475 2696 | ·3630 2859 | 3·4198 8492 |
| 7 | ·8600 0000$^{e}$ | ·8750 0000$^{e}$ | ·0485 8790 | ·3248 3438 | 3·3648 5059 |
| 7·5 | ·8514 8515$^{-}$ | ·8659 7938 | ·0495 5299 | ·2920 6237 | 3·3177 5535$^{+}$ |
| 8 | ·8431 3725$^{+}$ | ·8571 4286 | ·0504 3214 | ·2636 8732 | 3·2770 9522 |
| 8·5 | ·8349 5146 | ·8484 8485$^{-}$ | ·0512 3383 | ·2389 2552 | 3·2417 2060 |
| 9 | ·8269 2308 | ·8400 0000$^{e}$ | ·0519 6539 | ·2171 6769 | 3·2107 3787 |
| 9·5 | ·8190 4762 | ·8316 8317 | ·0526 3321 | ·1979 3305$^{+}$ | 3·1834 4193 |
| 10 | ·8113 2075$^{+}$ | ·8235 2941 | ·0532 4287 | ·1808 3721 | 3·1592 6910 |
| 10·5 | ·8037 3832 | ·8155 3398 | ·0537 9933 | ·1655 6918 | 3·1377 6344 |
| 11 | ·7962 9630 | ·8076 9231 | ·0543 0696 | ·1518 7470 | 3·1185 5219 |
| 12 | ·7818 1818 | ·7924 5283 | ·0551 9099 | ·1284 0206 | 3·0858 3407 |
| 13 | ·7678 5714 | ·7777 7778 | ·0559 2171 | ·1091 2131 | 3·0592 1278 |
| 14 | ·7543 8596 | ·7636 3636 | ·0565 2094 | ·0931 0724 | 3·0373 3318 |
| 15 | ·7413 7931 | ·7500 0000$^{e}$ | ·0570 0669 | ·0796 8303 | 3·0192 0447 |
| 16 | ·7288 1356 | ·7368 4211 | ·0573 9401 | ·0683 4247 | 3·0040 8606 |
| 17 | ·7166 6667 | ·7241 3793 | ·0576 9558 | ·0586 9968 | 2·9914 1382 |
| 18 | ·7049 1803 | ·7118 6441 | ·0579 2217 | ·0504 5563 | 2·9807 5089 |
| 19 | ·6935 4839 | ·7000 0000$^{e}$ | ·0580 8300 | ·0433 7515$^{+}$ | 2·9717 5407 |
| 20 | ·6825 3968 | ·6885 2459 | ·0581 8602 | ·0372 7095$^{+}$ | 2·9641 5027 |
| 21 | ·6718 7500$^{e}$ | ·6774 1935$^{+}$ | ·0582 3810 | ·0319 9213 | 2·9577 1971 |
| 22 | ·6615 3846 | ·6666 6667 | ·0582 4521 | ·0274 1585$^{+}$ | 2·9522 8372 |
| 23 | ·6515 1515$^{+}$ | ·6562 5000$^{e}$ | ·0582 1257 | ·0234 4124 | 2·9476 9575$^{+}$ |
| 24 | ·6417 9104 | ·6461 5385$^{-}$ | ·0581 4476 | ·0199 8473 | 2·9438 3456 |
| 25 | ·6323 5294 | ·6363 6364 | ·0580 4580 | ·0169 7655$^{+}$ | 2·9405 9913 |
| 26 | ·6231 8841 | ·6268 6567 | ·0579 1923 | ·0143 5811 | 2·9379 0471 |
| 27 | ·6142 8571 | ·6176 4706 | ·0577 6821 | ·0120 7984 | 2·9356 7976 |
| 28 | ·6056 3380 | ·6086 9565$^{+}$ | ·0575 9551 | ·0100 9957 | 2·9338 6355$^{-}$ |
| 29 | ·5972 2222 | ·6000 0000$^{e}$ | ·0574 0362 | ·0083 8125$^{+}$ | 2·9324 0426 |
| 30 | ·5890 4110 | ·5915 4930 | ·0571 9475$^{-}$ | ·0068 9392 | 2·9312 5745$^{-}$ |
| 31 | ·5810 8108 | ·5833 3333 | ·0569 7085$^{+}$ | ·0056 1082 | 2·9303 8485$^{+}$ |
| 32 | ·5733 3333 | ·5753 4247 | ·0567 3371 | ·0045 0878 | 2·9297 5339 |
| 33 | ·5657 8947 | ·5675 6757 | ·0564 8488 | ·0035 6762 | 2·9293 3433 |
| 34 | ·5584 4156 | ·5600 0000$^{e}$ | ·0562 2580 | ·0027 6973 | 2·9291 0267 |
| 35 | ·5512 8205$^{+}$ | ·5526 3158 | ·0559 5773 | ·0020 9967 | 2·9290 3654 |
| 36 | ·5443 0380 | ·5454 5455$^{-}$ | ·0556 8182 | ·0015 4385$^{-}$ | 2·9291 1680 |
| 37 | ·5375 0000$^{e}$ | ·5384 6154 | ·0553 9909 | ·0010 9031 | 2·9293 2660 |
| 38 | ·5308 6420 | ·5316 4557 | ·0551 1047 | ·0007 2846 | 2·9296 5111 |
| 39 | ·5243 9024 | ·5250 0000$^{e}$ | ·0548 1679 | ·0004 4892 | 2·9300 7722 |
| 40 | ·5180 7229 | ·5185 1852 | ·0545 1883 | ·0002 4334 | 2·9305 9333 |
| 41 | ·5119 0476 | ·5121 9512 | ·0542 1724 | ·0001 0430 | 2·9311 8914 |
| 42 | ·5058 8235$^{+}$ | ·5060 2410 | ·0539 1266 | ·0000 2517 | 2·9318 5550$^{+}$ |
| 43 | ·5000 0000$^{e}$ | ·5000 0000$^{e}$ | ·0536 0563 | ·0000 0000$^{e}$ | 2·9325 8427 |
| 44 | ·4942 5287 | ·4941 1765$^{-}$ | ·0532 9666 | ·0000 2349 | 2·9333 6817 |
| 45 | ·4886 3636 | ·4883 7209 | ·0529 8620 | ·0000 9085$^{+}$ | 2·9342 0072 |
| 46 | ·4831 4607 | ·4827 5862 | ·0526 7468 | ·0001 9780 | 2·9350 7609 |
| 47 | ·4777 7778 | ·4772 7273 | ·0523 6245$^{-}$ | ·0003 4047 | 2·9359 8909 |
| 48 | ·4725 2747 | ·4719 1011 | ·0520 4986 | ·0005 1536 | 2·9369 3503 |
| 49 | ·4673 9130 | ·4666 6667 | ·0517 3721 | ·0007 1933 | 2·9379 0974 |
| 50 | ·4623 6559 | ·4615 3846 | ·0514 2477 | ·0009 4951 | 2·9389 0942 |

$p = 44$

| $q$ | $\bar{x}$ | $\tilde{x}$ | $\sigma$ | $\beta_1$ | $\beta_2$ |
|---|---|---|---|---|---|
| 0·5 | ·9887 6404 | $J$-curve | ·0156 2592 | 7·2397 1242 | 13·5046 3034 |
| 1 | ·9777 7778 | 1·0000 0000$^{e}$ | ·0217 3376 | 3·5003 0865$^{+}$ | 8·0160 7834 |
| 1·5 | ·9670 3297 | ·9885 0575$^{-}$ | ·0261 8389 | 2·2561 0677 | 6·1906 7232 |
| 2 | ·9565 2174 | ·9772 7273 | ·0297 4641 | 1·6356 5341 | 5·2809 6011 |
| 2·5 | ·9462 3656 | ·9662 9213 | ·0327 2628 | 1·2646 5956 | 4·7374 5420 |
| 3 | ·9361 7021 | ·9555 5556 | ·0352 8324 | 1·0183 6356 | 4·3769 9443 |
| 3·5 | ·9263 1579 | ·9450 5495$^{-}$ | ·0375 1422 | ·8432 9720 | 4·1210 8549 |
| 4 | ·9166 6667 | ·9347 8261 | ·0394 8363 | ·7127 2727 | 3·9304 8128 |
| 4·5 | ·9072 1649 | ·9247 3118 | ·0412 3711 | ·6118 0276 | 3·7833 7979 |
| 5 | ·8979 5918 | ·9148 9362 | ·0428 0852 | ·5316 1372 | 3·6667 0095$^{-}$ |
| 5·5 | ·8888 8889 | ·9052 6316 | ·0442 2388 | ·4664 9072 | 3·5721 2205$^{+}$ |
| 6 | ·8800 0000$^{e}$ | ·8958 3333 | ·0455 0372 | ·4126 5465$^{+}$ | 3·4940 9553 |
| 6·5 | ·8712 8713 | ·8865 9794 | ·0466 6465$^{+}$ | ·3674 8965$^{+}$ | 3·4287 8150$^{-}$ |
| 7 | ·8627 4510 | ·8775 5102 | ·0477 2034 | ·3291 2762 | 3·3734 3788 |
| 7·5 | ·8543 6893 | ·8686 8687 | ·0486 8219 | ·2961 9895$^{+}$ | 3·3260 5442 |
| 8 | ·8461 5385$^{-}$ | ·8600 0000$^{e}$ | ·0495 5986 | ·2676 7677 | 3·2851 2397 |
| 8·5 | ·8380 9524 | ·8514 8515$^{-}$ | ·0503 6161 | ·2427 7607 | 3·2494 9448 |
| 9 | ·8301 8868 | ·8431 3725$^{+}$ | ·0510 9458 | ·2208 8655$^{+}$ | 3·2182 7037 |
| 9·5 | ·8224 2991 | ·8349 5146 | ·0517 6499 | ·2015 2663 | 3·1907 4499 |
| 10 | ·8148 1481 | ·8269 2308 | ·0523 7828 | ·1843 1122 | 3·1663 5338 |
| 10·5 | ·8073 3945$^{-}$ | ·8190 4762 | ·0529 3928 | ·1689 2881 | 3·1446 3855$^{-}$ |
| 11 | ·8000 0000$^{e}$ | ·8113 2075$^{+}$ | ·0534 5225$^{-}$ | ·1551 2465$^{+}$ | 3·1252 2686 |
| 12 | ·7857 1429 | ·7962 9630 | ·0543 4899 | ·1314 4525$^{-}$ | 3·0921 3113 |
| 13 | ·7719 2982 | ·7818 1818 | ·0550 9462 | ·1119 7249 | 3·0651 5942 |
| 14 | ·7586 2069 | ·7678 5714 | ·0557 1041 | ·0957 7922 | 3·0429 5295$^{-}$ |
| 15 | ·7457 6271 | ·7543 8596 | ·0562 1398 | ·0821 8710 | 3·0245 1805$^{+}$ |
| 16 | ·7333 3333 | ·7413 7931 | ·0566 2004 | ·0706 8868 | 3·0091 1186 |
| 17 | ·7213 1148 | ·7288 1356 | ·0569 4100 | ·0608 9709 | 2·9961 6835$^{+}$ |
| 18 | ·7096 7742 | ·7166 6667 | ·0571 8744 | ·0525 1243 | 2·9852 4913 |
| 19 | ·6984 1270 | ·7049 1803 | ·0573 6838 | ·0452 9883 | 2·9760 0964 |
| 20 | ·6875 0000$^{e}$ | ·6935 4839 | ·0574 9164 | ·0390 6837 | 2·9681 7565$^{+}$ |
| 21 | ·6769 2308 | ·6825 3968 | ·0575 6396 | ·0336 6961 | 2·9615 2640 |
| 22 | ·6666 6667 | ·6718 7500$^{e}$ | ·0575 9123 | ·0289 7924 | 2·9558 8235$^{+}$ |
| 23 | ·6567 1642 | ·6615 3846 | ·0575 7859 | ·0248 9596 | 2·9510 9616 |
| 24 | ·6470 5882 | ·6515 1515$^{+}$ | ·0575 3055$^{-}$ | ·0213 3581 | 2·9470 4591 |
| 25 | ·6376 8116 | ·6417 9104 | ·0574 5108 | ·0182 2871 | 2·9436 2996 |
| 26 | ·6285 7143 | ·6323 5294 | ·0573 4371 | ·0155 1573 | 2·9407 6300 |
| 27 | ·6197 1831 | ·6231 8841 | ·0572 1154 | ·0131 4705$^{-}$ | 2·9383 7299 |
| 28 | ·6111 1111 | ·6142 8571 | ·0570 5733 | ·0110 8023 | 2·9363 9874 |
| 29 | ·6027 3973 | ·6056 3380 | ·0568 8354 | ·0092 7900 | 2·9347 8799 |
| 30 | ·5945 9459 | ·5972 2222 | ·0566 9237 | ·0077 1216 | 2·9334 9593 |
| 31 | ·5866 6667 | ·5890 4110 | ·0564 8578 | ·0063 5279 | 2·9324 8394 |
| 32 | ·5789 4737 | ·5810 8108 | ·0562 6552 | ·0051 7751 | 2·9317 1860 |
| 33 | ·5714 2857 | ·5733 3333 | ·0560 3318 | ·0041 6600 | 2·9311 7089 |
| 34 | ·5641 0256 | ·5657 8947 | ·0557 9017 | ·0033 0047 | 2·9308 1551 |
| 35 | ·5569 6203 | ·5584 4156 | ·0555 3775$^{-}$ | ·0025 6534 | 2·9306 3035$^{-}$ |
| 36 | ·5500 0000$^{e}$ | ·5512 8205$^{+}$ | ·0552 7708 | ·0019 4689 | 2·9305 9600 |
| 37 | ·5432 0988 | ·5443 0380 | ·0550 0919 | ·0014 3304 | 2·9306 9540 |
| 38 | ·5365 8537 | ·5375 0000$^{e}$ | ·0547 3501 | ·0010 1308 | 2·9309 1351 |
| 39 | ·5301 2048 | ·5308 6420 | ·0544 5539 | ·0006 7752 | 2·9312 3703 |
| 40 | ·5238 0952 | ·5243 9024 | ·0541 7109 | ·0004 1792 | 2·9316 5415$^{+}$ |
| 41 | ·5176 4706 | ·5180 7229 | ·0538 8279 | ·0002 2674 | 2·9321 5443 |
| 42 | ·5116 2791 | ·5119 0476 | ·0535 9113 | ·0000 9727 | 2·9327 2853 |
| 43 | ·5057 4713 | ·5058 8235$^{+}$ | ·0532 9666 | ·0000 2349 | 2·9333 6817 |
| 44 | ·5000 0000$^{e}$ | ·5000 0000$^{e}$ | ·0529 9989 | ·0000 0000$^{e}$ | 2·9340 6593 |
| 45 | ·4943 8202 | ·4942 5287 | ·0527 0130 | ·0000 2196 | 2·9348 1518 |
| 46 | ·4888 8889 | ·4886 3636 | ·0524 0130 | ·0000 8499 | 2·9356 0999 |
| 47 | ·4835 1648 | ·4831 4607 | ·0521 0027 | ·0001 8517 | 2·9364 4501 |
| 48 | ·4782 6087 | ·4777 7778 | ·0517 9856 | ·0003 1894 | 2·9373 1548 |
| 49 | ·4731 1828 | ·4725 2747 | ·0514 9648 | ·0004 8309 | 2·9382 1709 |
| 50 | ·4680 8511 | ·4673 9130 | ·0511 9431 | ·0006 7472 | 2·9391 4597 |

$p = 45$

| $q$ | $\bar{x}$ | $\tilde{x}$ | $\sigma$ | $\beta_1$ | $\beta_2$ |
|---|---|---|---|---|---|
| 0·5 | ·9890 1099 | *J*-curve | ·0152 8810 | 7·2554 3121 | 13·5350 4069 |
| 1 | ·9782 6087 | 1·0000 0000$^e$ | ·0212 7157 | 3·5104 9383 | 8·0358 2766 |
| 1·5 | ·9677 4194 | ·9887 6404 | ·0256 3611 | 2·2643 5682 | 6·2067 0624 |
| 2 | ·9574 4681 | ·9777 7778 | ·0291 3418 | 1·6428 7103 | 5·2950 2041 |
| 2·5 | ·9473 6842 | ·9670 3297 | ·0320 6356 | 1·2712 0815$^+$ | 4·7502 4169 |
| 3 | ·9375 0000$^e$ | ·9565 2174 | ·0345 8021 | 1·0244 2667 | 4·3888 6275$^-$ |
| 3·5 | ·9278 3505$^+$ | ·9462 3656 | ·0367 7866 | ·8489 8119 | 4·1322 3933 |
| 4 | ·9183 6735$^-$ | ·9361 7021 | ·0387 2177 | ·7180 9988 | 3·9410 5078 |
| 4·5 | ·9090 9091 | ·9263 1579 | ·0404 5401 | ·6169 1017 | 3·7934 5354 |
| 5 | ·9000 0000$^e$ | ·9166 6667 | ·0420 0840 | ·5364 8915$^+$ | 3·6763 4253 |
| 5·5 | ·8910 8911 | ·9072 1649 | ·0434 1028 | ·4711 5912 | 3·5813 7908 |
| 6 | ·8823 5294 | ·8979 5918 | ·0446 7968 | ·4171 3540 | 3·5030 0489 |
| 6·5 | ·8737 8641 | ·8888 8889 | ·0458 3276 | ·3717 9823 | 3·4373 7262 |
| 7 | ·8653 8462 | ·8800 0000$^e$ | ·0468 8285$^-$ | ·3332 7672 | 3·3817 3481 |
| 7·5 | ·8571 4286 | ·8712 8713 | ·0478 4105$^-$ | ·3001 9920 | 3·3340 7720 |
| 8 | ·8490 5660 | ·8627 4510 | ·0487 1677 | ·2715 3719 | 3·2928 8961 |
| 8·5 | ·8411 2150$^-$ | ·8543 6893 | ·0495 1807 | ·2465 0448 | 3·2570 1766 |
| 9 | ·8333 3333 | ·8461 5385$^-$ | ·0502 5189 | ·2244 8980 | 3·2255 6391 |
| 9·5 | ·8256 8807 | ·8380 9524 | ·0509 2430 | ·2050 1076 | 3·1978 2021 |
| 10 | ·8181 8182 | ·8301 8868 | ·0515 4061 | ·1876 8168 | 3·1732 2041 |
| 10·5 | ·8108 1081 | ·8224 2991 | ·0521 0552 | ·1721 9048 | 3·1513 0647 |
| 11 | ·8035 7143 | ·8148 1481 | ·0526 2319 | ·1582 8199 | 3·1317 0395$^-$ |
| 12 | ·7894 7368 | ·8000 0000$^e$ | ·0535 3133 | ·1344 0582 | 3·0982 4859 |
| 13 | ·7758 6207 | ·7857 1429 | ·0542 9051 | ·1147 5024 | 3·0709 4297 |
| 14 | ·7627 1186 | ·7719 2982 | ·0549 2153 | ·0983 8625$^+$ | 3·0484 2487 |
| 15 | ·7500 0000$^e$ | ·7586 2069 | ·0554 4160 | ·0846 3406 | 3·0296 9790 |
| 16 | ·7377 0492 | ·7457 6271 | ·0558 6510 | ·0729 8508 | 3·0140 1703 |
| 17 | ·7258 0645$^+$ | ·7333 3333 | ·0562 0420 | ·0630 5147 | 3·0008 1448 |
| 18 | ·7142 8571 | ·7213 1148 | ·0564 6924 | ·0545 3254 | 2·9896 5035$^-$ |
| 19 | ·7031 2500$^e$ | ·7096 7742 | ·0566 6911 | ·0471 9175$^-$ | 2·9801 7885$^+$ |
| 20 | ·6923 0769 | ·6984 1270 | ·0568 1146 | ·0408 4057 | 2·9721 2467 |
| 21 | ·6818 1818 | ·6875 0000$^e$ | ·0569 0292 | ·0353 2707 | 2·9652 6611 |
| 22 | ·6716 4179 | ·6769 2308 | ·0569 4929 | ·0305 2750$^-$ | 2·9594 2280 |
| 23 | ·6617 6471 | ·6666 6667 | ·0569 5564 | ·0263 4014 | 2·9544 4668 |
| 24 | ·6521 7391 | ·6567 1642 | ·0569 2641 | ·0226 8069 | 2·9502 1518 |
| 25 | ·6428 5714 | ·6470 5882 | ·0568 6552 | ·0194 7874 | 2·9466 2608 |
| 26 | ·6338 0282 | ·6376 8116 | ·0567 7646 | ·0166 7509 | 2·9435 9354 |
| 27 | ·6250 0000$^e$ | ·6285 7143 | ·0566 6230 | ·0142 1963 | 2·9410 4505$^-$ |
| 28 | ·6164 3836 | ·6197 1831 | ·0565 2578 | ·0120 6970 | 2·9389 1896 |
| 29 | ·6081 0811 | ·6111 1111 | ·0563 6933 | ·0101 8881 | 2·9371 6266 |
| 30 | ·6000 0000$^e$ | ·6027 3973 | ·0561 9515$^-$ | ·0085 4557 | 2·9357 3094 |
| 31 | ·5921 0526 | ·5945 9459 | ·0560 0517 | ·0071 1285$^-$ | 2·9345 8485$^-$ |
| 32 | ·5844 1558 | ·5866 6667 | ·0558 0116 | ·0058 6712 | 2·9336 9066 |
| 33 | ·5769 2308 | ·5789 4737 | ·0555 8468 | ·0047 8788 | 2·9330 1908 |
| 34 | ·5696 2025$^+$ | ·5714 2857 | ·0553 5714 | ·0038 5722 | 2·9325 4453 |
| 35 | ·5625 0000$^e$ | ·5641 0256 | ·0551 1982 | ·0030 5940 | 2·9322 4466 |
| 36 | ·5555 5556 | ·5569 6203 | ·0548 7387 | ·0023 8061 | 2·9320 9983 |
| 37 | ·5487 8049 | ·5500 0000$^e$ | ·0546 2032 | ·0018 0861 | 2·9320 9277 |
| 38 | ·5421 6867 | ·5432 0988 | ·0543 6011 | ·0013 3260 | 2·9322 0822 |
| 39 | ·5357 1429 | ·5365 8537 | ·0540 9409 | ·0009 4299 | 2·9324 3271 |
| 40 | ·5294 1176 | ·5301 2048 | ·0538 2302 | ·0006 3123 | 2·9327 5427 |
| 41 | ·5232 5581 | ·5238 0952 | ·0535 4761 | ·0003 8971 | 2·9331 6226 |
| 42 | ·5172 4138 | ·5176 4706 | ·0532 6848 | ·0002 1161 | 2·9336 4723 |
| 43 | ·5113 6364 | ·5116 2791 | ·0529 8620 | ·0000 9085$^+$ | 2·9342 0072 |
| 44 | ·5056 1798 | ·5057 4713 | ·0527 0130 | ·0000 2196 | 2·9348 1518 |
| 45 | ·5000 0000$^e$ | ·5000 0000$^e$ | ·0524 1424 | ·0000 0000$^e$ | 2·9354 8387 |
| 46 | ·4945 0549 | ·4943 8202 | ·0521 2546 | ·0000 2055$^+$ | 2·9362 0072 |
| 47 | ·4891 3043 | ·4888 8889 | ·0518 3533 | ·0000 7962 | 2·9369 6028 |
| 48 | ·4838 7097 | ·4835 1648 | ·0515 4422 | ·0001 7359 | 2·9377 5768 |
| 49 | ·4787 2340 | ·4782 6087 | ·0512 5245$^+$ | ·0002 9919 | 2·9385 8849 |
| 50 | ·4736 8421 | ·4731 1828 | ·0509 6031 | ·0004 5347 | 2·9394 4877 |

$p = 46$

| $q$ | $\bar{x}$ | $\tilde{x}$ | $\sigma$ | $\beta_1$ | $\beta_2$ |
|---|---|---|---|---|---|
| 0·5 | ·9892 4731 | *J*-curve | ·0149 6457 | 7·2705 1343 | 13·5642 3944 |
| 1 | ·9787 2340 | 1·0000 0000$^e$ | ·0208 2862 | 3·5202 7235+ | 8·0548 0035+ |
| 1·5 | ·9684 2105+ | ·9890 1099 | ·0251 1077 | 2·2722 8218 | 6·2221 1786 |
| 2 | ·9583 3333 | ·9782 6087 | ·0285 4662 | 1·6498 0870 | 5·3085 4220 |
| 2·5 | ·9484 5361 | ·9677 4194 | ·0314 2712 | 1·2775 0647 | 4·7625 4592 |
| 3 | ·9387 7551 | ·9574 4681 | ·0339 0459 | 1·0302 6149 | 4·4002 8854 |
| 3·5 | ·9292 9293 | ·9473 6842 | ·0360 7132 | ·8544 5441 | 4·1429 8292 |
| 4 | ·9200 0000$^e$ | ·9375 0000$^e$ | ·0379 8865− | ·7232 7631 | 3·9512 3683 |
| 4·5 | ·9108 9109 | ·9278 3505+ | ·0396 9997 | ·6218 3400 | 3·8031 6686 |
| 5 | ·9019 6078 | ·9183 6735− | ·0412 3748 | ·5411 9213 | 3·6856 4397 |
| 5·5 | ·8932 0388 | ·9090 9091 | ·0426 2588 | ·4756 6506 | 3·5903 1416 |
| 6 | ·8846 1538 | ·9000 0000$^e$ | ·0438 8471 | ·4214 6279 | 3·5116 0884 |
| 6·5 | ·8761 9048 | ·8910 8911 | ·0450 2974 | ·3759 6184 | 3·4456 7353 |
| 7 | ·8679 2453 | ·8823 5294 | ·0460 7392 | ·3372 8864 | 3·3897 5559 |
| 7·5 | ·8598 1308 | ·8737 8641 | ·0470 2811 | ·3040 6952 | 3·3418 3695+ |
| 8 | ·8518 5185+ | ·8653 8462 | ·0479 0147 | ·2752 7451 | 3·3004 0454 |
| 8·5 | ·8440 3670 | ·8571 4286 | ·0487 0184 | ·2501 1622 | 3·2643 0173 |
| 9 | ·8363 6364 | ·8490 5660 | ·0494 3602 | ·2279 8245− | 3·2326 2930 |
| 9·5 | ·8288 2883 | ·8411 2150− | ·0501 0990 | ·2083 9008 | 3·2046 7768 |
| 10 | ·8214 2857 | ·8333 3333 | ·0507 2867 | ·1909 5280 | 3·1798 7955+ |
| 10·5 | ·8141 5929 | ·8256 8807 | ·0512 9691 | ·1753 5804 | 3·1577 7593 |
| 11 | ·8070 1754 | ·8181 8182 | ·0518 1869 | ·1613 5021 | 3·1379 9156 |
| 12 | ·7931 0345− | ·8035 7143 | ·0527 3701 | ·1372 8663 | 3·1041 9340 |
| 13 | ·7796 6102 | ·7894 7368 | ·0535 0852 | ·1174 5683 | 3·0765 6935+ |
| 14 | ·7666 6667 | ·7758 6207 | ·0541 5353 | ·1009 3006 | 3·0537 5390 |
| 15 | ·7540 9836 | ·7627 1186 | ·0546 8886 | ·0870 2517 | 3·0347 4810 |
| 16 | ·7419 3548 | ·7500 0000$^e$ | ·0551 2860 | ·0752 3246 | 3·0188 0487 |
| 17 | ·7301 5873 | ·7377 0492 | ·0554 8465− | ·0651 6321 | 3·0053 5475− |
| 18 | ·7187 5000$^e$ | ·7258 0645+ | ·0557 6716 | ·0565 1597 | 2·9939 5644 |
| 19 | ·7076 9231 | ·7142 8571 | ·0559 8483 | ·0490 5355− | 2·9842 6296 |
| 20 | ·6969 6970 | ·7031 2500$^e$ | ·0561 4519 | ·0425 8688 | 2·9759 9800 |
| 21 | ·6865 6716 | ·6923 0769 | ·0562 5476 | ·0369 6354 | 2·9689 3896 |
| 22 | ·6764 7059 | ·6818 1818 | ·0563 1924 | ·0320 5937 | 2·9629 0470 |
| 23 | ·6666 6667 | ·6716 4179 | ·0563 4362 | ·0277 7227 | 2·9577 4648 |
| 24 | ·6571 4286 | ·6617 6471 | ·0563 3228 | ·0240 1760 | 2·9533 4111 |
| 25 | ·6478 8732 | ·6521 7391 | ·0562 8910 | ·0207 2466 | 2·9495 8582 |
| 26 | ·6388 8889 | ·6428 5714 | ·0562 1752 | ·0178 3398 | 2·9463 9429 |
| 27 | ·6301 3699 | ·6338 0282 | ·0561 2057 | ·0152 9519 | 2·9436 9353 |
| 28 | ·6216 2162 | ·6250 0000$^e$ | ·0560 0097 | ·0130 6542 | 2·9414 2152 |
| 29 | ·6133 3333 | ·6164 3836 | ·0558 6115− | ·0111 0796 | 2·9395 2524 |
| 30 | ·6052 6316 | ·6081 0811 | ·0557 0326 | ·0093 9123 | 2·9379 5916 |
| 31 | ·5974 0260 | ·6000 0000$^e$ | ·0555 2924 | ·0078 8794 | 2·9366 8401 |
| 32 | ·5897 4359 | ·5921 0526 | ·0553 4084 | ·0065 7439 | 2·9356 6576 |
| 33 | ·5822 7848 | ·5844 1558 | ·0551 3962 | ·0054 2994 | 2·9348 7485− |
| 34 | ·5750 0000$^e$ | ·5769 2308 | ·0549 2700 | ·0044 3653 | 2·9342 8545+ |
| 35 | ·5679 0123 | ·5696 2025+ | ·0547 0424 | ·0035 7830 | 2·9338 7499 |
| 36 | ·5609 7561 | ·5625 0000$^e$ | ·0544 7249 | ·0028 4131 | 2·9336 2359 |
| 37 | ·5542 1687 | ·5555 5556 | ·0542 3280 | ·0022 1323 | 2·9335 1380 |
| 38 | ·5476 1905− | ·5487 8049 | ·0539 8610 | ·0016 8314 | 2·9335 3017 |
| 39 | ·5411 7647 | ·5421 6867 | ·0537 3324 | ·0012 4135− | 2·9336 5904 |
| 40 | ·5348 8372 | ·5357 1429 | ·0534 7501 | ·0008 7922 | 2·9338 8828 |
| 41 | ·5287 3563 | ·5294 1176 | ·0532 1208 | ·0005 8906 | 2·9342 0711 |
| 42 | ·5227 2727 | ·5232 5581 | ·0529 4511 | ·0003 6398 | 2·9346 0590 |
| 43 | ·5168 5393 | ·5172 4138 | ·0526 7468 | ·0001 9780 | 2·9350 7609 |
| 44 | ·5111 1111 | ·5113 6364 | ·0524 0130 | ·0000 8499 | 2·9356 0999 |
| 45 | ·5054 9451 | ·5056 1798 | ·0521 2546 | ·0000 2055+ | 2·9362 0072 |
| 46 | ·5000 0000$^e$ | ·5000 0000$^e$ | ·0518 4758 | ·0000 0000$^e$ | 2·9368 4211 |
| 47 | ·4946 2366 | ·4945 0549 | ·0515 6808 | ·0000 1927 | 2·9375 2860 |
| 48 | ·4893 6170 | ·4891 3043 | ·0512 8730 | ·0000 7470 | 2·9382 5522 |
| 49 | ·4842 1053 | ·4838 7097 | ·0510 0559 | ·0001 6296 | 2·9390 1745+ |
| 50 | ·4791 6667 | ·4787 2340 | ·0507 2322 | ·0002 8104 | 2·9398 1124 |

$p = 47$

| $q$ | $\bar{x}$ | $\tilde{x}$ | $\sigma$ | $\beta_1$ | $\beta_2$ |
|---|---|---|---|---|---|
| 0·5 | ·9894 7368 | *J*-curve | ·0146 5445$^{-}$ | 7·2849 9697 | 13·5922 9753 |
| 1 | ·9791 6667 | 1·0000 0000$^{e}$ | ·0204 0373 | 3·5296 6809 | 8·0730 4130 |
| 1·5 | ·9690 7216 | ·9892 4731 | ·0246 0652 | 2·2799 0164 | 6·2369 4270 |
| 2 | ·9591 8367 | ·9787 2340 | ·0279 8226 | 1·6564 8237 | 5·3215 5579 |
| 2·5 | ·9494 9495$^{-}$ | ·9684 2105$^{+}$ | ·0308 1542 | 1·2835 6855$^{-}$ | 4·7743 9372 |
| 3 | ·9400 0000$^{e}$ | ·9583 3333 | ·0332 5481 | 1·0358 8065$^{-}$ | 4·4112 9605$^{+}$ |
| 3·5 | ·9306 9307 | ·9484 5361 | ·0353 9060 | ·8597 2830 | 4·1533 3839 |
| 4 | ·9215 6863 | ·9387 7551 | ·0372 8268 | ·7282 6704 | 3·9610 5982 |
| 4·5 | ·9126 2136 | ·9292 9293 | ·0389 7340 | ·6265 8388 | 3·8125 3865$^{+}$ |
| 5 | ·9038 4615$^{+}$ | ·9200 0000$^{e}$ | ·0404 9418 | ·5457 3155$^{-}$ | 3·6946 2282 |
| 5·5 | ·8952 3810 | ·9108 9109 | ·0418 6914 | ·4800 1677 | 3·5989 4363 |
| 6 | ·8867 9245$^{+}$ | ·9019 6078 | ·0431 1732 | ·4256 4445$^{+}$ | 3·5199 2263 |
| 6·5 | ·8785 0467 | ·8932 0388 | ·0442 5411 | ·3799 8754 | 3·4536 9846 |
| 7 | ·8703 7037 | ·8846 1538 | ·0452 9214 | ·3411 6990 | 3·3975 1354 |
| 7·5 | ·8623 8532 | ·8761 9048 | ·0462 4200 | ·3078 1597 | 3·3493 4615$^{-}$ |
| 8 | ·8545 4545$^{+}$ | ·8679 2453 | ·0471 1262 | ·2788 9432 | 3·3076 8043 |
| 8·5 | ·8468 4685$^{-}$ | ·8598 1308 | ·0479 1166 | ·2536 1645$^{+}$ | 3·2713 5759 |
| 9 | ·8392 8571 | ·8518 5185$^{+}$ | ·0486 4573 | ·2313 6928 | 3·2394 7673 |
| 9·5 | ·8318 5841 | ·8440 3670 | ·0493 2060 | ·2116 6896 | 3·2113 2691 |
| 10 | ·8245 6140 | ·8363 6364 | ·0499 4133 | ·1941 2861 | 3·1863 3970 |
| 10·5 | ·8173 9130 | ·8288 2883 | ·0505 1238 | ·1784 3519 | 3·1640 5521 |
| 11 | ·8103 4483 | ·8214 2857 | ·0510 3772 | ·1643 3269 | 3·1440 9741 |
| 12 | ·7966 1017 | ·8070 1754 | ·0519 6512 | ·1400 9046 | 3·1099 7221 |
| 13 | ·7833 3333 | ·7931 0345$^{-}$ | ·0527 4781 | ·1200 9449 | 3·0820 4424 |
| 14 | ·7704 9180 | ·7796 6102 | ·0534 0566 | ·1034 1239 | 3·0589 4485$^{+}$ |
| 15 | ·7580 6452 | ·7666 6667 | ·0539 5511 | ·0893 6170 | 3·0396 7267 |
| 16 | ·7460 3175$^{-}$ | ·7540 9836 | ·0544 0996 | ·0774 3170 | 3·0234 7865$^{-}$ |
| 17 | ·7343 7500$^{e}$ | ·7419 3548 | ·0547 8186 | ·0672 3280 | 3·0097 9175$^{+}$ |
| 18 | ·7230 7692 | ·7301 5873 | ·0550 8075$^{+}$ | ·0584 6285$^{-}$ | 2·9981 6935$^{+}$ |
| 19 | ·7121 2121 | ·7187 5000$^{e}$ | ·0553 1519 | ·0508 8404 | 2·9882 6336 |
| 20 | ·7014 9254 | ·7076 9231 | ·0554 9255$^{-}$ | ·0443 0680 | 2·9797 9648 |
| 21 | ·6911 7647 | ·6969 6970 | ·0556 1925$^{-}$ | ·0385 7825$^{+}$ | 2·9725 4530 |
| 22 | ·6811 5942 | ·6865 6716 | ·0557 0089 | ·0335 7382 | 2·9663 2795$^{-}$ |
| 23 | ·6714 2857 | ·6764 7059 | ·0557 4240 | ·0291 9108 | 2·9609 9502 |
| 24 | ·6619 7183 | ·6666 6667 | ·0557 4808 | ·0253 4506 | 2·9564 2276 |
| 25 | ·6527 7778 | ·6571 4286 | ·0557 2178 | ·0219 6478 | 2·9525 0788 |
| 26 | ·6438 3562 | ·6478 8732 | ·0556 6688 | ·0189 9051 | 2·9491 6358 |
| 27 | ·6351 3514 | ·6388 8889 | ·0555 8639 | ·0163 7165$^{-}$ | 2·9463 1647 |
| 28 | ·6266 6667 | ·6301 3699 | ·0554 8299 | ·0140 6512 | 2·9439 0412 |
| 29 | ·6184 2105$^{+}$ | ·6216 2162 | ·0553 5909 | ·0120 3402 | 2·9418 7317 |
| 30 | ·6103 8961 | ·6133 3333 | ·0552 1684 | ·0102 4658 | 2·9401 7775$^{+}$ |
| 31 | ·6025 6410 | ·6052 6316 | ·0550 5815$^{+}$ | ·0086 7536 | 2·9387 7831 |
| 32 | ·5949 3671 | ·5974 0260 | ·0548 8477 | ·0072 9650$^{+}$ | 2·9376 4055$^{-}$ |
| 33 | ·5875 0000$^{e}$ | ·5897 4359 | ·0546 9825$^{-}$ | ·0060 8922 | 2·9367 3463 |
| 34 | ·5802 4691 | ·5822 7848 | ·0544 9999 | ·0050 3533 | 2·9360 3450$^{+}$ |
| 35 | ·5731 7073 | ·5750 0000$^{e}$ | ·0542 9128 | ·0041 1885$^{+}$ | 2·9355 1736 |
| 36 | ·5662 6506 | ·5679 0123 | ·0540 7325$^{-}$ | ·0033 2573 | 2·9351 6314 |
| 37 | ·5595 2381 | ·5609 7561 | ·0538 4694 | ·0026 4352 | 2·9349 5418 |
| 38 | ·5529 4118 | ·5542 1687 | ·0536 1330 | ·0020 6122 | 2·9348 7487 |
| 39 | ·5465 1163 | ·5476 1905$^{-}$ | ·0533 7319 | ·0015 6903 | 2·9349 1137 |
| 40 | ·5402 2989 | ·5411 7647 | ·0531 2737 | ·0011 5825$^{-}$ | 2·9350 5140 |
| 41 | ·5340 9091 | ·5348 8372 | ·0528 7656 | ·0008 2108 | 2·9352 8402 |
| 42 | ·5280 8989 | ·5287 3563 | ·0526 2139 | ·0005 5057 | 2·9355 9949 |
| 43 | ·5222 2222 | ·5227 2727 | ·0523 6245$^{-}$ | ·0003 4047 | 2·9359 8909 |
| 44 | ·5164 8352 | ·5168 5393 | ·0521 0027 | ·0001 8517 | 2·9364 4501 |
| 45 | ·5108 6957 | ·5111 1111 | ·0518 3533 | ·0000 7962 | 2·9369 6028 |
| 46 | ·5053 7634 | ·5054 9451 | ·0515 6808 | ·0000 1927 | 2·9375 2860 |
| 47 | ·5000 0000$^{e}$ | ·5000 0000$^{e}$ | ·0512 9892 | ·0000 0000$^{e}$ | 2·9381 4433 |
| 48 | ·4947 3684 | ·4946 2366 | ·0510 2821 | ·0000 1809 | 2·9388 0237 |
| 49 | ·4895 8333 | ·4893 6170 | ·0507 5629 | ·0000 7017 | 2·9394 9813 |
| 50 | ·4845 3608 | ·4842 1053 | ·0504 8347 | ·0001 5318 | 2·9402 2747 |

$p = 48$

| $q$ | $\bar{x}$ | $\tilde{x}$ | $\sigma$ | $\beta_1$ | $\beta_2$ |
|---|---|---|---|---|---|
| 0·5 | ·9896 9072 | *J*-curve | ·0143 5692 | 7·2989 1677 | 13·6192 8050− |
| 1 | ·9795 9184 | 1·0000 0000$^e$ | ·0199 9583 | 3·5387 0306 | 8·0905 9201 |
| 1·5 | ·9696 9697 | ·9894 7368 | ·0241 2211 | 2·2872 3254 | 6·2512 1359 |
| 2 | ·9600 0000$^e$ | ·9791 6667 | ·0274 3977 | 1·6629 0680 | 5·3340 8926 |
| 2·5 | ·9504 9505− | ·9690 7216 | ·0302 2704 | 1·2894 0741 | 4·7858 0997 |
| 3 | ·9411 7647 | ·9591 8367 | ·0326 2942 | 1·0412 9583 | 4·4219 0776 |
| 3·5 | ·9320 3883 | ·9494 9495− | ·0347 3504 | ·8648 1352 | 4·1633 2633 |
| 4 | ·9230 7692 | ·9400 0000$^e$ | ·0366 0238 | ·7330 8185− | 3·9705 3872 |
| 4·5 | ·9142 8571 | ·9306 9307 | ·0382 7284 | ·6311 6881 | 3·8215 8649 |
| 5 | ·9056 6038 | ·9215 6863 | ·0397 7708 | ·5501 1570 | 3·7032 9545+ |
| 5·5 | ·8971 9626 | ·9126 2136 | ·0411 3864 | ·4842 2194 | 3·6072 8277 |
| 6 | ·8888 8889 | ·9038 4615+ | ·0423 7612 | ·4296 8750$^e$ | 3·5279 6053 |
| 6·5 | ·8807 3394 | ·8952 3810 | ·0435 0451 | ·3838 8194 | 3·4614 6077 |
| 7 | ·8727 2727 | ·8867 9245+ | ·0445 3618 | ·3449 2664 | 3·4050 2117 |
| 7·5 | ·8648 6486 | ·8785 0467 | ·0454 8142 | ·3114 4423 | 3·3566 1650− |
| 8 | ·8571 4286 | ·8703 7037 | ·0463 4898 | ·2824 0190 | 3·3147 2823 |
| 8·5 | ·8495 5752 | ·8623 8532 | ·0471 4631 | ·2570 1007 | 3·2781 9552 |
| 9 | ·8421 0526 | ·8545 4545+ | ·0478 7988 | ·2346 5479 | 3·2461 1582 |
| 9·5 | ·8347 8261 | ·8468 4685− | ·0485 5530 | ·2148 5158 | 3·2177 7691 |
| 10 | ·8275 8621 | ·8392 8571 | ·0491 7752 | ·1972 1296 | 3·1926 0929 |
| 10·5 | ·8205 1282 | ·8318 5841 | ·0497 5090 | ·1814 2545− | 3·1701 5219 |
| 11 | ·8135 5932 | ·8245 6140 | ·0502 7931 | ·1672 3266 | 3·1500 2884 |
| 12 | ·8000 0000$^e$ | ·8103 4483 | ·0512 1475+ | ·1428 1998 | 3·1155 9140 |
| 13 | ·7868 8525− | ·7966 1017 | ·0520 0758 | ·1226 6540 | 3·0873 7313 |
| 14 | ·7741 9355− | ·7833 3333 | ·0526 7720 | ·1058 3496 | 3·0640 0240 |
| 15 | 7619 0476 | ·7704 9180 | ·0532 3971 | ·0916 4497 | 3·0444 7552 |
| 16 | ·7500 0000$^e$ | ·7580 6452 | ·0537 0862 | ·0795 8372 | 3·0280 4161 |
| 17 | ·7384 6154 | ·7460 3175− | ·0540 9533 | ·0692 6081 | 3·0141 2811 |
| 18 | ·7272 7273 | ·7343 7500$^e$ | ·0544 0960 | ·0603 7341 | 3·0022 9113 |
| 19 | ·7164 1791 | ·7230 7692 | ·0546 5982 | ·0526 8317 | 2·9921 8154 |
| 20 | ·7058 8235+ | ·7121 2121 | ·0548 5322 | ·0460 0000$^e$ | 2·9835 2113 |
| 21 | ·6956 5217 | ·7014 9254 | ·0549 9613 | ·0401 7060 | 2·9760 8568 |
| 22 | ·6857 1429 | ·6911 7647 | ·0550 9406 | ·0350 7003 | 2·9696 9265− |
| 23 | ·6760 5634 | ·6811 5942 | ·0551 5183 | ·0305 9551 | 2·9641 9201 |
| 24 | ·6666 6667 | ·6714 2857 | ·0551 7373 | ·0266 6180 | 2·9594 5946 |
| 25 | ·6575 3425− | ·6619 7183 | ·0551 6351 | ·0231 9763 | 2·9553 9123 |
| 26 | ·6486 4865− | ·6527 7778 | ·0551 2454 | ·0201 4303 | 2·9519 0007 |
| 27 | ·6400 0000$^e$ | ·6438 3562 | ·0550 5978 | ·0174 4720 | 2·9489 1220 |
| 28 | ·6315 7895− | ·6351 3514 | ·0549 7189 | ·0150 6684 | 2·9463 6482 |
| 29 | ·6233 7662 | ·6266 6667 | ·0548 6326 | ·0129 6488 | 2·9442 0422 |
| 30 | ·6153 8462 | ·6184 2105+ | ·0547 3601 | ·0111 0938 | 2·9423 8426 |
| 31 | ·6075 9494 | ·6103 8961 | ·0545 9205− | ·0094 7272 | 2·9408 6507 |
| 32 | ·6000 0000$^e$ | ·6025 6410 | ·0544 3311 | ·0080 3093 | 2·9396 1211 |
| 33 | ·5925 9259 | ·5949 3671 | ·0542 6073 | ·0067 6309 | 2·9385 9529 |
| 34 | ·5853 6585+ | ·5875 0000$^e$ | ·0540 7633 | ·0056 5087 | 2·9377 8835+ |
| 35 | ·5783 1325+ | ·5802 4691 | ·0538 8116 | ·0046 7820 | 2·9371 6826 |
| 36 | ·5714 2857 | ·5731 7073 | ·0536 7637 | ·0038 3090 | 2·9367 1478 |
| 37 | ·5647 0588 | ·5662 6506 | ·0534 6300 | ·0030 9644 | 2·9364 1006 |
| 38 | ·5581 3953 | ·5595 2381 | ·0532 4200 | ·0024 6371 | 2·9362 3831 |
| 39 | ·5517 2414 | ·5529 4118 | ·0530 1421 | ·0019 2283 | 2·9361 8554 |
| 40 | ·5454 5455− | ·5465 1163 | ·0527 8043 | ·0014 6502 | 2·9362 3932 |
| 41 | ·5393 2584 | ·5402 2989 | ·0525 4136 | ·0010 8241 | 2·9363 8857 |
| 42 | ·5333 3333 | ·5340 9091 | ·0522 9764 | ·0007 6796 | 2·9366 2342 |
| 43 | ·5274 7253 | ·5280 8989 | ·0520 4986 | ·0005 1536 | 2·9369 3503 |
| 44 | ·5217 3913 | ·5222 2222 | ·0517 9856 | ·0003 1894 | 2·9373 1548 |
| 45 | ·5161 2903 | ·5164 8352 | ·0515 4422 | ·0001 7359 | 2·9377 5768 |
| 46 | ·5106 3830 | ·5108 6957 | ·0512 8730 | ·0000 7470 | 2·9382 5522 |
| 47 | ·5052 6316 | ·5053 7634 | ·0510 2821 | ·0000 1809 | 2·9388 0237 |
| 48 | ·5000 0000$^e$ | ·5000 0000$^e$ | ·0507 6731 | ·0000 0000$^e$ | 2·9393 9394 |
| 49 | ·4948 4536 | ·4947 3684 | ·0505 0494 | ·0000 1701 | 2·9400 2525+ |
| 50 | ·4897 9592 | ·4895 8333 | ·0502 4142 | ·0000 6600$^e$ | 2·9406 9208 |

$p = 49$

| $q$ | $\bar{x}$ | $\tilde{x}$ | $\sigma$ | $\beta_1$ | $\beta_2$ |
|---|---|---|---|---|---|
| 0·5 | ·9898 9899 | *J*-curve | ·0140 7123 | 7·3123 0511 | 13·6452 4895$^-$ |
| 1 | ·9800 0000$^e$ | 1·0000 0000$^e$ | ·0196 0392 | 3·5473 9766 | 8·1074 9089 |
| 1·5 | ·9702 9703 | ·9896 9072 | ·0236 5640 | 2·2942 9096 | 6·2649 6099 |
| 2 | ·9607 8431 | ·9795 9184 | ·0269 1790 | 1·6690 9569 | 5·3461 6866 |
| 2·5 | ·9514 5631 | ·9696 9697 | ·0296 6068 | 1·2950 3512 | 4·7968 1778 |
| 3 | ·9423 0769 | ·9600 0000$^e$ | ·0320 2708 | 1·0465 1792 | 4·4321 4458 |
| 3·5 | ·9333 3333 | ·9504 9505$^-$ | ·0341 0326 | ·8697 1996 | 4·1729 6589 |
| 4 | ·9245 2830 | ·9411 7647 | ·0359 4639 | ·7377 2980 | 3·9796 9123 |
| 4·5 | ·9158 8785$^+$ | ·9320 3883 | ·0375 9692 | ·6355 9717 | 3·8303 2681 |
| 5 | ·9074 0741 | ·9230 7692 | ·0390 8480 | ·5543 5235$^+$ | 3·7116 7715$^+$ |
| 5·5 | ·8990 8257 | ·9142 8571 | ·0404 3304 | ·4882 8774 | 3·6153 4585$^-$ |
| 6 | ·8909 0909 | ·9056 6038 | ·0416 5978 | ·4335 9862 | 3·5357 3589 |
| 6·5 | ·8828 8288 | ·8971 9626 | ·0427 7968 | ·3876 5124 | 3·4689 7298 |
| 7 | ·8750 0000$^e$ | ·8888 8889 | ·0438 0479 | ·3485 6463 | 3·4122 9022 |
| 7·5 | ·8672 5664 | ·8807 3394 | ·0447 4518 | ·3149 5968 | 3·3636 5902 |
| 8 | ·8596 4912 | ·8727 2727 | ·0456 0938 | ·2858 0223 | 3·3215 5828 |
| 8·5 | ·8521 7391 | ·8648 6486 | ·0464 0467 | ·2603 0168 | 3·2848 2520 |
| 9 | ·8448 2759 | ·8571 4286 | ·0471 3736 | ·2378 4329 | 3·2525 5567 |
| 9·5 | ·8376 0684 | ·8495 5752 | ·0478 1294 | ·2179 4189 | 3·2240 3620 |
| 10 | ·8305 0847 | ·8421 0526 | ·0484 3624 | ·2002 0951 | 3·1986 9630 |
| 10·5 | ·8235 2941 | ·8347 8261 | ·0490 1150$^+$ | ·1843 3220 | 3·1760 7433 |
| 11 | ·8166 6667 | ·8275 8621 | ·0495 4253 | ·1700 5323 | 3·1557 9286 |
| 12 | ·8032 7869 | ·8135 5932 | ·0504 8507 | ·1454 7779 | 3·1210 5705$^+$ |
| 13 | ·7903 2258 | ·8000 0000$^e$ | ·0512 8705$^+$ | ·1251 7170 | 3·0925 6128 |
| 14 | ·7777 7778 | ·7868 8525$^-$ | ·0519 6746 | ·1081 9949 | 3·0689 3107 |
| 15 | ·7656 2500$^e$ | ·7741 9355$^-$ | ·0525 4205$^+$ | ·0938 7630 | 3·0491 6051 |
| 16 | ·7538 4615$^+$ | ·7619 0476 | ·0530 2403 | ·0816 8948 | 3·0324 9695$^+$ |
| 17 | ·7424 2424 | ·7500 0000$^e$ | ·0534 2459 | ·0712 4788 | 3·0183 6643 |
| 18 | ·7313 4328 | ·7384 6154 | ·0537 5329 | ·0622 4802 | 3·0063 2385$^+$ |
| 19 | ·7205 8824 | ·7272 7273 | ·0540 1836 | ·0544 5100 | 2·9960 1908 |
| 20 | ·7101 4493 | ·7164 1791 | ·0542 2692 | ·0476 6628 | 2·9871 7304 |
| 21 | ·7000 0000$^e$ | ·7058 8235$^+$ | ·0543 8517 | ·0417 4015$^+$ | 2·9795 6077 |
| 22 | ·6901 4085$^-$ | ·6956 5217 | ·0544 9853 | ·0365 4732 | 2·9729 9907 |
| 23 | ·6805 5556 | ·6857 1429 | ·0545 7177 | ·0319 8469 | 2·9673 3735$^-$ |
| 24 | ·6712 3288 | ·6760 5634 | ·0546 0909 | ·0279 6674 | 2·9624 5077 |
| 25 | ·6621 6216 | ·6666 6667 | ·0546 1421 | ·0244 2196 | 2·9582 3511 |
| 26 | ·6533 3333 | ·6575 3425$^-$ | ·0545 9045$^-$ | ·0212 9013 | 2·9546 0270 |
| 27 | ·6447 3684 | ·6486 4865$^-$ | ·0545 4073 | ·0185 2026 | 2·9514 7937 |
| 28 | ·6363 6364 | ·6400 0000$^e$ | ·0544 6770 | ·0160 6885$^+$ | 2·9488 0199 |
| 29 | ·6282 0513 | ·6315 7895$^-$ | ·0543 7370 | ·0138 9866 | 2·9465 1654 |
| 30 | ·6202 5316 | ·6233 7662 | ·0542 6085$^-$ | ·0119 7760 | 2·9445 7657 |
| 31 | ·6125 0000$^e$ | ·6153 8462 | ·0541 3104 | ·0102 7790 | 2·9429 4195$^+$ |
| 32 | ·6049 3827 | ·6075 9494 | ·0539 8599 | ·0087 7545$^+$ | 2·9415 7790 |
| 33 | ·5975 6098 | ·6000 0000$^e$ | ·0538 2724 | ·0074 4920 | 2·9404 5410 |
| 34 | ·5903 6145$^-$ | ·5925 9259 | ·0536 5618 | ·0062 8071 | 2·9395 4408 |
| 35 | ·5833 3333 | ·5853 6585$^+$ | ·0534 7408 | ·0052 5381 | 2·9388 2461 |
| 36 | ·5764 7059 | ·5783 1325$^+$ | ·0532 8208 | ·0043 5420 | 2·9382 7526 |
| 37 | ·5697 6744 | ·5714 2857 | ·0530 8121 | ·0035 6926 | 2·9378 7801 |
| 38 | ·5632 1839 | ·5647 0588 | ·0528 7243 | ·0028 8781 | 2·9376 1692 |
| 39 | ·5568 1818 | ·5581 3953 | ·0526 5658 | ·0022 9988 | 2·9374 7784 |
| 40 | ·5505 6180 | ·5517 2414 | ·0524 3446 | ·0017 9659 | 2·9374 4820 |
| 41 | ·5444 4444 | ·5454 5455$^-$ | ·0522 0676 | ·0013 7002 | 2·9375 1680 |
| 42 | ·5384 6154 | ·5393 2584 | ·0519 7415$^-$ | ·0010 1305$^+$ | 2·9376 7363 |
| 43 | ·5326 0870 | ·5333 3333 | ·0517 3721 | ·0007 1933 | 2·9379 0974 |
| 44 | ·5268 8172 | ·5274 7253 | ·0514 9648 | ·0004 8309 | 2·9382 1709 |
| 45 | ·5212 7660 | ·5217 3913 | ·0512 5245$^+$ | ·0002 9919 | 2·9385 8849 |
| 46 | ·5157 8947 | ·5161 2903 | ·0510 0559 | ·0001 6296 | 2·9390 1745$^+$ |
| 47 | ·5104 1667 | ·5106 3830 | ·0507 5629 | ·0000 7017 | 2·9394 9813 |
| 48 | ·5051 5464 | ·5052 6316 | ·0505 0494 | ·0000 1701 | 2·9400 2525$^+$ |
| 49 | ·5000 0000$^e$ | ·5000 0000$^e$ | ·0502 5189 | ·0000 0000$^e$ | 2·9405 9406 |
| 50 | ·4949 4949 | ·4948 4536 | ·0499 9745$^-$ | ·0000 1600 | 2·9412 0024 |

$p = 50$

| $q$ | $\bar{x}$ | $\tilde{x}$ | $\sigma$ | $\beta_1$ | $\beta_2$ |
|---|---|---|---|---|---|
| 0·5 | ·9900 9901 | $J$-curve | ·0137 9669 | 7·3251 9184 | 13·6702 5901 |
| 1 | ·9803 9216 | 1·0000 0000$^e$ | ·0192 2707 | 3·5557 7074 | 8·1237 7358 |
| 1·5 | ·9708 7379 | ·9898 9899 | ·0232 0832 | 2·3010 9180 | 6·2782 1315$^+$ |
| 2 | ·9615 3846 | ·9800 0000$^e$ | ·0264 1549 | 1·6750 6173 | 5·3578 1818 |
| 2·5 | ·9523 8095$^+$ | ·9702 9703 | ·0291 1514 | 1·3004 6292 | 4·8074 3863 |
| 3 | ·9433 9623 | ·9607 8431 | ·0314 4654 | 1·0515 5702 | 4·4420 2597 |
| 3·5 | ·9345 7944 | ·9514 5631 | ·0334 9400 | ·8744 5685$^+$ | 4·1822 7492 |
| 4 | ·9259 2593 | ·9423 0769 | ·0353 1343 | ·7422 1939 | 3·9885 3383 |
| 4·5 | ·9174 3119 | ·9333 3333 | ·0369 4437 | ·6398 7678 | 3·8387 7491 |
| 5 | ·9090 9091 | ·9245 2830 | ·0384 1610 | ·5584 4875$^+$ | 3·7197 8221 |
| 5·5 | ·9009 0090 | ·9158 8785$^+$ | ·0397 5110 | ·4922 2093 | 3·6231 4625$^-$ |
| 6 | ·8928 5714 | ·9074 0741 | ·0409 6710 | ·4373 8407 | 3·5432 6125$^+$ |
| 6·5 | ·8849 5575$^+$ | ·8990 8257 | ·0420 7841 | ·3913 0127 | 3·4762 4683 |
| 7 | ·8771 9298 | ·8909 0909 | ·0430 9681 | ·3520 8930 | 3·4193 3172 |
| 7·5 | ·8695 6522 | ·8828 8288 | ·0440 3215$^-$ | ·3183 6735$^-$ | 3·3704 8406 |
| 8 | ·8620 6897 | ·8750 0000$^e$ | ·0448 9273 | ·2891 0000$^e$ | 3·3281 8033 |
| 8·5 | ·8547 0085$^+$ | ·8672 5664 | ·0456 8568 | ·2634 9566 | 3·2912 5580 |
| 9 | ·8474 5763 | ·8596 4912 | ·0464 1717 | ·2409 3882 | 3·2588 0487 |
| 9·5 | ·8403 3613 | ·8521 7391 | ·0470 9254 | ·2209 4367 | 3·2301 1286 |
| 10 | ·8333 3333 | ·8448 2759 | ·0477 1653 | ·2031 2175$^-$ | 3·2046 0829 |
| 10·5 | ·8264 4628 | ·8376 0684 | ·0482 9328 | ·1871 5867 | 3·1818 2875$^+$ |
| 11 | ·8196 7213 | ·8305 0847 | ·0488 2651 | ·1727 9736 | 3·1613 9610 |
| 12 | ·8064 5161 | ·8166 6667 | ·0497 7527 | ·1480 6641 | 3·1263 7500$^e$ |
| 13 | ·7936 5079 | ·8032 7869 | ·0505 8551 | ·1276 1548 | 3·0976 1377 |
| 14 | ·7812 5000$^e$ | ·7903 2258 | ·0512 7579 | ·1105 0767 | 3·0737 3522 |
| 15 | ·7692 3077 | ·7777 7778 | ·0518 6153 | ·0960 5703 | 3·0537 3134 |
| 16 | ·7575 7576 | ·7656 2500$^e$ | ·0523 5566 | ·0837 5000$^e$ | 3·0368 4783 |
| 17 | ·7462 6866 | ·7538 4615$^+$ | ·0527 6916 | ·0731 9471 | 3·0225 0932 |
| 18 | ·7352 9412 | ·7424 2424 | ·0531 1141 | ·0640 8707 | 3·0102 6962 |
| 19 | ·7246 3768 | ·7313 4328 | ·0533 9046 | ·0561 8768 | 2·9997 7761 |
| 20 | ·7142 8571 | ·7205 8824 | ·0536 1333 | ·0493 0556 | 2·9907 5342 |
| 21 | ·7042 2535$^+$ | ·7101 4493 | ·0537 8610 | ·0432 8660 | 2·9829 7139 |
| 22 | ·6944 4444 | ·7000 0000$^e$ | ·0539 1411 | ·0380 0518 | 2·9762 4767 |
| 23 | ·6849 3151 | ·6901 4085$^-$ | ·0540 0204 | ·0333 5791 | 2·9704 3112 |
| 24 | ·6756 7568 | ·6805 5556 | ·0540 5405$^+$ | ·0292 5900 | 2·9653 9645$^-$ |
| 25 | ·6666 6667 | ·6712 3288 | ·0540 7381 | ·0256 3670 | 2·9610 3896 |
| 26 | ·6578 9474 | ·6621 6216 | ·0540 6457 | ·0224 3059 | 2·9572 7062 |
| 27 | ·6493 5065$^-$ | ·6533 3333 | ·0540 2924 | ·0195 8945$^+$ | 2·9540 1688 |
| 28 | ·6410 2564 | ·6447 3684 | ·0539 7043 | ·0170 6964 | 2·9512 1429 |
| 29 | ·6329 1139 | ·6363 6364 | ·0538 9046 | ·0148 3374 | 2·9488 0852 |
| 30 | ·6250 0000$^e$ | ·6282 0513 | ·0537 9144 | ·0128 4949 | 2·9467 5287 |
| 31 | ·6172 8395$^+$ | ·6202 5316 | ·0536 7523 | ·0110 8902 | 2·9450 0694 |
| 32 | ·6097 5610 | ·6125 0000$^e$ | ·0535 4354 | ·0095 2806 | 2·9435 3571 |
| 33 | ·6024 0964 | ·6049 3827 | ·0533 9791 | ·0081 4545$^+$ | 2·9423 0867 |
| 34 | ·5952 3810 | ·5975 6098 | ·0532 3971 | ·0069 2266 | 2·9412 9912 |
| 35 | ·5882 3529 | ·5903 6145$^-$ | ·0530 7022 | ·0058 4338 | 2·9404 8365$^+$ |
| 36 | ·5813 9535$^-$ | ·5833 3333 | ·0528 9056 | ·0048 9325$^+$ | 2·9398 4168 |
| 37 | ·5747 1264 | ·5764 7059 | ·0527 0178 | ·0040 5955$^-$ | 2·9393 5500$^-$ |
| 38 | ·5681 8182 | ·5697 6744 | ·0525 0481 | ·0033 3099 | 2·9390 0752 |
| 39 | ·5617 9775$^+$ | ·5632 1839 | ·0523 0052 | ·0026 9756 | 2·9387 8496 |
| 40 | ·5555 5556 | ·5568 1818 | ·0520 8969 | ·0021 5028 | 2·9386 7461 |
| 41 | ·5494 5055$^-$ | ·5505 6180 | ·0518 7303 | ·0016 8118 | 2·9386 6515$^-$ |
| 42 | ·5434 7826 | ·5444 4444 | ·0516 5119 | ·0012 8306 | 2·9387 4644 |
| 43 | ·5376 3441 | ·5384 6154 | ·0514 2477 | ·0009 4951 | 2·9389 0942 |
| 44 | ·5319 1489 | ·5326 0870 | ·0511 9431 | ·0006 7472 | 2·9391 4597 |
| 45 | ·5263 1579 | ·5268 8172 | ·0509 6031 | ·0004 5347 | 2·9394 4877 |
| 46 | ·5208 3333 | ·5212 7660 | ·0507 2322 | ·0002 8104 | 2·9398 1124 |
| 47 | ·5154 6392 | ·5157 8947 | ·0504 8347 | ·0001 5318 | 2·9402 2747 |
| 48 | ·5102 0408 | ·5104 1667 | ·0502 4142 | ·0000 6600$^e$ | 2·9406 9208 |
| 49 | ·5050 5051 | ·5051 5464 | ·0499 9745$^-$ | ·0000 1600 | 2·9412 0024 |
| 50 | ·5000 0000$^e$ | ·5000 0000$^e$ | ·0497 5186 | ·0000 0000$^e$ | 2·9417 4757 |

# TABLE III

## VALUES OF $I_x(p, 0\cdot5)$ FOR $p = 11\cdot5, 12\cdot5, 13\cdot5, 14\cdot5$

$q = 0{\cdot}5$

| | $p = 11{\cdot}5$ | $p = 12{\cdot}5$ | $p = 13{\cdot}5$ | $p = 14{\cdot}5$ |
|---|---|---|---|---|
| $B(p, q)$ | ·5283 7848 | ·5063 6271 | ·4868 8722 | ·4694 9839 |
| $x$ | | | | |
| ·27 | ·0000 001 | | | |
| ·28 | ·0000 001 | | | |
| ·29 | ·0000 001 | | | |
| ·30 | ·0000 002 | ·0000 001 | | |
| ·31 | ·0000 003 | ·0000 001 | | |
| ·32 | ·0000 004 | ·0000 001 | | |
| ·33 | ·0000 006 | ·0000 002 | ·0000 001 | |
| ·34 | ·0000 008 | ·0000 003 | ·0000 001 | |
| ·35 | ·0000 011 | ·0000 004 | ·0000 001 | |
| ·36 | ·0000 016 | ·0000 006 | ·0000 002 | ·0000 001 |
| ·37 | ·0000 022 | ·0000 008 | ·0000 003 | ·0000 001 |
| ·38 | ·0000 030 | ·0000 011 | ·0000 004 | ·0000 001 |
| ·39 | ·0000 041 | ·0000 015$^{+}$ | ·0000 006 | ·0000 002 |
| ·40 | ·0000 055$^{-}$ | ·0000 021 | ·0000 008 | ·0000 003 |
| ·41 | ·0000 074 | ·0000 029 | ·0000 011 | ·0000 005$^{-}$ |
| ·42 | ·0000 098 | ·0000 039 | ·0000 016 | ·0000 006 |
| ·43 | ·0000 129 | ·0000 053 | ·0000 022 | ·0000 009 |
| ·44 | ·0000 170 | ·0000 072 | ·0000 030 | ·0000 013 |
| ·45 | ·0000 221 | ·0000 096 | ·0000 042 | ·0000 018 |
| ·46 | ·0000 287 | ·0000 127 | ·0000 056 | ·0000 025$^{+}$ |
| ·47 | ·0000 371 | ·0000 168 | ·0000 076 | ·0000 035$^{-}$ |
| ·48 | ·0000 476 | ·0000 220 | ·0000 102 | ·0000 047 |
| ·49 | ·0000 609 | ·0000 287 | ·0000 136 | ·0000 064 |
| ·50 | ·0000 774 | ·0000 373 | ·0000 180 | ·0000 087 |

| | $p = 11{\cdot}5$ | $p = 12{\cdot}5$ | $p = 13{\cdot}5$ | $p = 14{\cdot}5$ |
|---|---|---|---|---|
| $B(p, q)$ | ·5283 7848 | ·5063 6271 | ·4868 8722 | ·4694 9839 |
| $x$ | | | | |
| ·51 | ·0000 981 | ·0000 482 | ·0000 237 | ·0000 117 |
| ·52 | ·0001 237 | ·0000 619 | ·0000 311 | ·0000 156 |
| ·53 | ·0001 555$^{-}$ | ·0000 793 | ·0000 406 | ·0000 208 |
| ·54 | ·0001 945$^{+}$ | ·0001 011 | ·0000 527 | ·0000 276 |
| ·55 | ·0002 425$^{-}$ | ·0001 284 | ·0000 682 | ·0000 363 |
| ·56 | ·0003 012 | ·0001 624 | ·0000 878 | ·0000 476 |
| ·57 | ·0003 728 | ·0002 047 | ·0001 126 | ·0000 622 |
| ·58 | ·0004 599 | ·0002 569 | ·0001 439 | ·0000 808 |
| ·59 | ·0005 655$^{+}$ | ·0003 214 | ·0001 832 | ·0001 046 |
| ·60 | ·0006 933 | ·0004 008 | ·0002 323 | ·0001 349 |
| ·61 | ·0008 473 | ·0004 981 | ·0002 935$^{+}$ | ·0001 734 |
| ·62 | ·0010 327 | ·0006 171 | ·0003 696 | ·0002 219 |
| ·63 | ·0012 552 | ·0007 622 | ·0004 640 | ·0002 831 |
| ·64 | ·0015 217 | ·0009 388 | ·0005 807 | ·0003 599 |
| ·65 | ·0018 400 | ·0011 531 | ·0007 244 | ·0004 561 |
| ·66 | ·0022 194 | ·0014 125$^{-}$ | ·0009 012 | ·0005 762 |
| ·67 | ·0026 710 | ·0017 259 | ·0011 180 | ·0007 258 |
| ·68 | ·0032 071 | ·0021 036 | ·0013 832 | ·0009 115$^{-}$ |
| ·69 | ·0038 427 | ·0025 580 | ·0017 070 | ·0011 415$^{+}$ |
| ·70 | ·0045 947 | ·0031 034 | ·0021 013 | ·0014 258 |
| ·71 | ·0054 830 | ·0037 571 | ·0025 807 | ·0017 764 |
| ·72 | ·0065 309 | ·0045 391 | ·0031 623 | ·0022 077 |
| ·73 | ·0077 651 | ·0054 729 | ·0038 666 | ·0027 374 |
| ·74 | ·0092 169 | ·0065 866 | ·0047 180 | ·0033 865$^{-}$ |
| ·75 | ·0109 225$^{-}$ | ·0079 127 | ·0057 457 | ·0041 806 |
| ·76 | ·0129 240 | ·0094 899 | ·0069 843 | ·0051 506 |
| ·77 | ·0152 704 | ·0113 632 | ·0084 750$^{+}$ | ·0063 335$^{-}$ |
| ·78 | ·0180 187 | ·0135 862 | ·0102 670 | ·0077 740 |
| ·79 | ·0212 354 | ·0162 214 | ·0124 189 | ·0095 261 |
| ·80 | ·0249 978 | ·0193 431 | ·0150 003 | ·0116 548 |
| ·81 | ·0293 965$^{+}$ | ·0230 387 | ·0180 948 | ·0142 386 |
| ·82 | ·0345 377 | ·0274 116 | ·0218 019 | ·0173 724 |
| ·83 | ·0405 459 | ·0325 847 | ·0262 412 | ·0211 711 |
| ·84 | ·0475 680 | ·0387 040 | ·0315 561 | ·0257 742 |
| ·85 | ·0557 779 | ·0459 441 | ·0379 198 | ·0313 517 |
| ·86 | ·0653 825$^{+}$ | ·0545 144 | ·0455 417 | ·0381 110 |
| ·87 | ·0766 295$^{+}$ | ·0646 676 | ·0546 770 | ·0463 071 |
| ·88 | ·0898 176 | ·0767 107 | ·0656 380 | ·0562 550$^{-}$ |
| ·89 | ·1053 104 | ·0910 202 | ·0788 106 | ·0683 466 |
| ·90 | ·1235 557 | ·1080 624 | ·0946 761 | ·0830 747 |
| ·91 | ·1451 131 | ·1284 232 | ·1138 428 | ·1010 662 |
| ·92 | ·1706 939 | ·1528 512 | ·1370 924 | ·1231 314 |
| ·93 | ·2012 248 | ·1823 249 | ·1654 510 | ·1503 396 |
| ·94 | ·2379 498 | ·2181 617 | ·2003 050$^{-}$ | ·1841 413 |
| ·95 | ·2826 105$^{+}$ | ·2622 090 | ·2436 027 | ·2265 815$^{-}$ |
| ·96 | ·3377 927 | ·3172 098 | ·2982 406 | ·2807 046 |
| ·97 | ·4076 856 | ·3876 043 | ·3689 045$^{+}$ | ·3514 376 |
| ·98 | ·5001 161 | ·4816 672 | ·4643 104 | ·4479 307 |
| ·99 | ·6343 615$^{-}$ | ·6197 006 | ·6057 668 | ·5924 833 |
| 1·00 | 1·0000 000 | 1·0000 000 | 1·0000 000 | 1·0000 000 |

*Note.* For $x > 0{\cdot}97$, when interpolation becomes difficult or impossible, pp. 498–500 of Table IV may be used with one of the mid-panel Lagrangian interpolation formulae, (9)–(11), given on p. xvii. For many purposes the simple four-point equation will suffice.

# TABLE IV

ADDITIONAL VALUES OF $I_x(p, q)$ FOR $x$ NEAR UNITY AND

$q = 0{\cdot}5(0{\cdot}5)\,3{\cdot}0,\quad p = 0{\cdot}5(0{\cdot}5)\,11{\cdot}0(1{\cdot}0)16{\cdot}0$

$q = 0{\cdot}5$

| p \ x | 0·9750 | 0·9800 | 0·9850 | 0·9880 | 0·9885 | 0·9890 | 0·9895 |
|---|---|---|---|---|---|---|---|
| 0·5 | ·8989 174 | ·9096 655+ | ·9218 341 | ·9301 2156 | ·9315 9860 | ·9331 0773 | ·9346 5118 |
| 1·0 | ·8418 861 | ·8585 786 | ·8775 255+ | ·8904 5549 | ·8927 6195− | ·8951 1912 | ·8975 3049 |
| 1·5 | ·7995 251 | ·8205 388 | ·8444 514 | ·8608 0305+ | ·8637 2242 | ·8667 0674 | ·8697 6045+ |
| 2·0 | ·7648 056 | ·7892 822 | ·8172 068 | ·8363 4050− | ·8397 5954 | ·8432 5552 | ·8468 3370 |
| 2·5 | ·7349 202 | ·7623 093 | ·7936 367 | ·8151 4526 | ·8189 9202 | ·8229 2634 | ·8269 5420 |
| 3·0 | ·7084 405− | ·7383 493 | ·7726 464 | ·7962 4129 | ·8004 6488 | ·8047 8569 | ·8092 1035− |
| 3·5 | ·6845 283 | ·7166 574 | ·7535 948 | ·7790 5734 | ·7836 1922 | ·7882 8730 | ·7930 6877 |
| 4·0 | ·6626 438 | ·6967 541 | ·7360 697 | ·7632 2628 | ·7680 9590 | ·7730 8015− | ·7781 8676 |
| 4·5 | ·6424 151 | ·6783 097 | ·7197 880 | ·7484 9603 | ·7536 4835+ | ·7589 2328 | ·7643 2908 |
| 5·0 | ·6235 735+ | ·6610 862 | ·7045 452 | ·7346 8480 | ·7400 9876 | ·7456 4296 | ·7513 2615− |
| 5·5 | ·6059 170 | ·6449 047 | ·6901 882 | ·7216 5641 | ·7273 1395− | ·7331 0905+ | ·7390 5093 |
| 6·0 | ·5892 893 | ·6296 271 | ·6765 987 | ·7093 0572 | ·7151 9111 | ·7212 2112 | ·7274 0543 |
| 6·5 | ·5735 664 | ·6151 439 | ·6636 830 | ·6975 4955− | ·7036 4890 | ·7098 9971 | ·7163 1209 |
| 7·0 | ·5586 479 | ·6013 664 | ·6513 653 | ·6863 2073 | ·6926 2165+ | ·6990 8069 | ·7057 0834 |
| 7·5 | ·5444 509 | ·5882 217 | ·6395 836 | ·6755 6409 | ·6820 5545− | ·6887 1137 | ·6955 4278 |
| 8·0 | ·5309 064 | ·5756 491 | ·6282 858 | ·6652 3364 | ·6719 0532 | ·6787 4786 | ·6857 7259 |
| 8·5 | ·5179 558 | ·5635 970 | ·6174 282 | ·6552 9057 | ·6621 3333 | ·6691 5312 | ·6763 6163 |
| 9·0 | ·5055 489 | ·5520 214 | ·6069 733 | ·6457 0173 | ·6527 0710 | ·6598 9552 | ·6672 7906 |
| 9·5 | ·4936 426 | ·5408 843 | ·5968 889 | ·6364 3857 | ·6435 9873 | ·6509 4784 | ·6584 9833 |
| 10·0 | ·4821 990 | ·5301 526 | ·5871 468 | ·6274 7628 | ·6347 8396 | ·6422 8639 | ·6499 9635+ |
| 10·5 | ·4711 849 | ·5197 973 | ·5777 225− | ·6187 9311 | ·6262 4156 | ·6338 9044 | ·6417 5291 |
| 11·0 | ·4605 710 | ·5097 927 | ·5685 941 | ·6103 6988 | ·6179 5279 | ·6257 4173 | ·6337 5017 |
| 12·0 | ·4404 423 | ·4907 470 | ·5511 504 | ·5942 3698 | ·6020 7144 | ·6101 2281 | ·6184 0529 |
| 13·0 | ·4216 346 | ·4728 599 | ·5346 842 | ·5789 6182 | ·5870 2683 | ·5953 1934 | ·6038 5418 |
| 14·0 | ·4040 023 | ·4560 048 | ·5190 889 | ·5644 5042 | ·5727 2722 | ·5812 4180 | ·5900 0964 |
| 15·0 | ·3874 248 | ·4400 767 | ·5042 761 | ·5506 2520 | ·5590 9688 | ·5678 1636 | ·5767 9972 |
| 16·0 | ·3718 005− | ·4249 874 | ·4901 718 | ·5374 2119 | ·5460 7241 | ·5549 8118 | ·5641 6422 |

| p \ x | 0·9900 | 0·9905 | 0·9910 | 0·9915 | 0·9920 | 0·9925 |
|---|---|---|---|---|---|---|
| 0·5 | ·9362 3144 | ·9378 5131 | ·9395 1398 | ·9412 2309 | ·9429 8280 | ·9447 9796 |
| 1·0 | ·9000 0000 | ·9025 3206 | ·9051 3167 | ·9078 0456 | ·9105 5728 | ·9133 9746 |
| 1·5 | ·8728 8857 | ·8760 9673 | ·8793 9132 | ·8827 7962 | ·8862 7002 | ·8898 7221 |
| 2·0 | ·8505 0000 | ·8542 6106 | ·8581 2441 | ·8620 9866 | ·8661 9369 | ·8704 2095− |
| 2·5 | ·8310 8228 | ·8353 1813 | ·8396 7028 | ·8441 4850− | ·8487 6397 | ·8535 2967 |
| 3·0 | ·8137 4625 | ·8184 0174 | ·8231 8627 | ·8281 1062 | ·8331 8718 | ·8384 3031 |
| 3·5 | ·7979 7170 | ·8030 0516 | ·8081 7944 | ·8135 0628 | ·8189 9916 | ·8246 7369 |
| 4·0 | ·7834 2441 | ·7888 0286 | ·7943 3318 | ·8000 2800 | ·8059 0180 | ·8119 7138 |
| 4·5 | ·7698 7500 | ·7755 7145− | ·7814 3022 | ·7874 6478 | ·7936 9057 | ·8001 2550+ |
| 5·0 | ·7571 5811 | ·7631 4988 | ·7693 1395+ | ·7756 6457 | ·7822 1809 | ·7889 9346 |
| 5·5 | ·7451 4991 | ·7514 1759 | ·7578 6713 | ·7645 1354 | ·7713 7402 | ·7784 6854 |
| 6·0 | ·7337 5484 | ·7402 8153 | ·7469 9930 | ·7539 2386 | ·7610 7327 | ·7684 6843 |
| 6·5 | ·7228 9733 | ·7296 6814 | ·7366 3893 | ·7438 2612 | ·7512 4855+ | ·7589 2806 |
| 7·0 | ·7125 1637 | ·7195 1803 | ·7267 2830 | ·7341 6428 | ·7418 4558 | ·7497 9493 |
| 7·5 | ·7025 6189 | ·7097 8245− | ·7172 2003 | ·7248 9236 | ·7328 1981 | ·7410 2597 |
| 8·0 | ·6929 9215− | ·7004 2079 | ·7080 7464 | ·7159 7205+ | ·7241 3414 | ·7325 8530 |
| 8·5 | ·6837 7194 | ·6913 9879 | ·6992 5883 | ·7073 7106 | ·7157 5726 | ·7244 4267 |
| 9·0 | ·6748 7123 | ·6826 8722 | ·6907 4422 | ·6990 6181 | ·7076 6249 | ·7165 7227 |
| 9·5 | ·6662 6414 | ·6742 6090 | ·6825 0632 | ·6910 2059 | ·6998 2686 | ·7089 5191 |
| 10·0 | ·6579 2818 | ·6660 9795+ | ·6745 2390 | ·6832 2678 | ·6922 3039 | ·7015 6229 |
| 10·5 | ·6498 4366 | ·6581 7924 | ·6667 7836 | ·6756 6234 | ·6848 5563 | ·6943 8652 |
| 11·0 | ·6419 9323 | ·6504 8787 | ·6592 5329 | ·6683 1137 | ·6776 8718 | ·6874 0975− |
| 12·0 | ·6269 3471 | ·6357 2889 | ·6448 0799 | ·6541 9494 | ·6639 1608 | ·6740 0182 |
| 13·0 | ·6126 4793 | ·6217 1924 | ·6310 8916 | ·6407 8170 | ·6508 2435+ | ·6612 4893 |
| 14·0 | ·5990 4802 | ·6083 7639 | ·6180 1670 | ·6279 9397 | ·6383 3686 | ·6490 7850+ |
| 15·0 | ·5860 6496 | ·5956 3230 | ·6055 2457 | ·6157 6776 | ·6263 9168 | ·6374 3075+ |
| 16·0 | ·5736 4018 | ·5834 3005− | ·5935 5752 | ·6040 4956 | ·6149 3705+ | ·6262 5570 |

$q = 0{\cdot}5$ (*continued*)

| $p$ \ $x$ | 0·9930 | 0·9935 | 0·9940 | 0·9945 | 0·9950 | 0·9955 |
|---|---|---|---|---|---|---|
| 0·5 | ·9466 7423 | ·9486 1831 | ·9506 3820 | ·9527 4363 | ·9549 4659 | ·9572 6215+ |
| 1·0 | ·9163 3400 | ·9193 7742 | ·9225 4033 | ·9258 3802 | ·9292 8932 | ·9329 1796 |
| 1·5 | ·8935 9755− | ·8974 5946 | ·9014 7400 | ·9056 6066 | ·9100 4345+ | ·9146 5260 |
| 2·0 | ·8747 9383 | ·8793 2816 | ·8840 4288 | ·8889 6097 | ·8941 1076 | ·8995 2788 |
| 2·5 | ·8584 6079 | ·8635 7525+ | ·8688 9453 | ·8744 4465− | ·8802 5770 | ·8863 7405+ |
| 3·0 | ·8438 5679 | ·8494 8645− | ·8553 4303 | ·8614 5530 | ·8678 5876 | ·8745 9800 |
| 3·5 | ·8305 4814 | ·8366 4408 | ·8429 8733 | ·8496 0919 | ·8565 4825+ | ·8638 5302 |
| 4·0 | ·8182 5638 | ·8247 8000 | ·8315 6998 | ·8386 5998 | ·8460 9147 | ·8539 1660 |
| 4·5 | ·8067 9049 | ·8137 1027 | ·8209 1440 | ·8284 3874 | ·8363 2747 | ·8446 3615− |
| 5·0 | ·7960 1283 | ·8033 0238 | ·8108 9338 | ·8188 2377 | ·8271 4033 | ·8359 0180 |
| 5·5 | ·7858 2041 | ·7934 5716 | ·8014 1174 | ·8097 2405+ | ·8184 4332 | ·8276 3135+ |
| 6·0 | ·7761 3377 | ·7840 9816 | ·7923 9609 | ·8010 6938 | ·8101 6958 | ·8197 6144 |
| 6·5 | ·7668 9014 | ·7751 6493 | ·7837 8842 | ·7928 0428 | ·8022 6629 | ·8122 4201 |
| 7·0 | ·7580 3886 | ·7666 0872 | ·7755 4197 | ·7848 8403 | ·7946 9084 | ·8050 3269 |
| 7·5 | ·7495 3837 | ·7583 8954 | ·7676 1835+ | ·7772 7193 | ·7874 0831 | ·7981 0039 |
| 8·0 | ·7413 5406 | ·7504 7408 | ·7599 8562 | ·7699 3744 | ·7803 8960 | ·7914 1754 |
| 8·5 | ·7334 5674 | ·7428 3428 | ·7526 1684 | ·7628 5480 | ·7736 1020 | ·7849 6093 |
| 9·0 | ·7258 2155− | ·7354 4618 | ·7454 8905+ | ·7560 0208 | ·7670 4922 | ·7787 1078 |
| 9·5 | ·7184 2705− | ·7282 8920 | ·7385 8249 | ·7493 6037 | ·7606 8867 | ·7726 5004 |
| 10·0 | ·7112 5464 | ·7213 4542 | ·7318 7999 | ·7429 1329 | ·7545 1296 | ·7667 6395− |
| 10·5 | ·7042 8806 | ·7145 9921 | ·7253 6655+ | ·7366 4648 | ·7485 0844 | ·7610 3957 |
| 11·0 | ·6975 1295+ | ·7080 3677 | ·7190 2896 | ·7305 4733 | ·7426 6307 | ·7554 6553 |
| 12·0 | ·6844 8770 | ·6954 1563 | ·7068 3566 | ·7188 0838 | ·7314 0836 | ·7447 2921 |
| 13·0 | ·6720 9255+ | ·6833 9900 | ·6952 2053 | ·7076 2042 | ·7206 7653 | ·7344 8654 |
| 14·0 | ·6602 5757 | ·6719 1964 | ·6841 1915− | ·6969 2194 | ·7104 0906 | ·7246 8214 |
| 15·0 | ·6489 2514 | ·6609 2221 | ·6734 7847 | ·6866 6229 | ·7005 5779 | ·7152 7043 |
| 16·0 | ·6380 4715+ | ·6503 6047 | ·6632 5420 | ·6767 9917 | ·6910 8250+ | ·7062 1340 |

| $p$ \ $x$ | 0·9960 | 0·9965 | 0·9970 | 0·9975 | 0·9980 | 0·9985 |
|---|---|---|---|---|---|---|
| 0·5 | ·9597 0974 | ·9623 1506 | ·9651 1344 | ·9681 5573 | ·9715 2000 | ·9753 3765+ |
| 1·0 | ·9367 5445− | ·9408 3920 | ·9452 2774 | ·9500 0000 | ·9552 7864 | ·9612 7017 |
| 1·5 | ·9195 2698 | ·9247 1810 | ·9302 9668 | ·9363 6456 | ·9430 7798 | ·9506 9998 |
| 2·0 | ·9052 5816 | ·9113 6233 | ·9179 2377 | ·9250 6250 | ·9329 6268 | ·9419 3430 |
| 2·5 | ·8928 4562 | ·8997 4118 | ·9071 5514 | ·9152 2343 | ·9241 5456 | ·9342 9949 |
| 3·0 | ·8817 3044 | ·8893 3206 | ·8975 0723 | ·9064 0613 | ·9162 5919 | ·9274 5415− |
| 3·5 | ·8715 8592 | ·8798 2958 | ·8886 9745+ | ·8983 5280 | ·9090 4610 | ·9211 9879 |
| 4·0 | ·8622 0242 | ·8710 3775+ | ·8805 4449 | ·8908 9803 | ·9023 6745− | ·9154 0546 |
| 4·5 | ·8534 3621 | ·8628 2223 | ·8729 2404 | ·8839 2842 | ·8961 2189 | ·9099 8646 |
| 5·0 | ·8451 8376 | ·8550 8626 | ·8657 4661 | ·8773 6236 | ·8902 3649 | ·9048 7867 |
| 5·5 | ·8373 6766 | ·8477 5750− | ·8589 4529 | ·8711 3880 | ·8846 5668 | ·9000 3489 |
| 6·0 | ·8299 2823 | ·8407 8017 | ·8524 6848 | ·8652 1071 | ·8793 4046 | ·8954 1877 |
| 6·5 | ·8228 1833 | ·8341 1022 | ·8462 7545+ | ·8595 4094 | ·8742 5461 | ·8910 0158 |
| 7·0 | ·8159 9993 | ·8277 1214 | ·8403 3337 | ·8540 9955+ | ·8693 7240 | ·8867 6020 |
| 7·5 | ·8094 4189 | ·8215 5682 | ·8346 1531 | ·8488 6199 | ·8646 7190 | ·8826 7565+ |
| 8·0 | ·8031 1825− | ·8156 2002 | ·8290 9886 | ·8438 0784 | ·8601 3487 | ·8787 3216 |
| 8·5 | ·7970 0715+ | ·8098 8133 | ·8237 6516 | ·8389 1989 | ·8557 4593 | ·8749 1644 |
| 9·0 | ·7910 8997 | ·8043 2334 | ·8185 9811 | ·8341 8348 | ·8514 9200 | ·8712 1716 |
| 9·5 | ·7853 5069 | ·7989 3109 | ·8135 8389 | ·8295 8601 | ·8473 6182 | ·8676 2460 |
| 10·0 | ·7797 7538 | ·7936 9159 | ·8087 1048 | ·8251 1653 | ·8433 4561 | ·8641 3031 |
| 10·5 | ·7743 5189 | ·7885 9348 | ·8039 6741 | ·8207 6549 | ·8394 3486 | ·8607 2691 |
| 11·0 | ·7690 6951 | ·7836 2677 | ·7993 4542 | ·8165 2446 | ·8356 2203 | ·8574 0790 |
| 12·0 | ·7588 9114 | ·7740 5308 | ·7904 3285+ | ·8083 4345− | ·8282 6426 | ·8510 0068 |
| 13·0 | ·7491 7589 | ·7649 1040 | ·7819 1727 | ·8005 2291 | ·8212 2716 | ·8448 6964 |
| 14·0 | ·7398 7168 | ·7561 5013 | ·7737 5377 | ·7930 2196 | ·8144 7426 | ·8389 8325+ |
| 15·0 | ·7309 3564 | ·7477 3229 | ·7659 0545− | ·7858 0698 | ·8079 7556 | ·8333 1560 |
| 16·0 | ·7223 3202 | ·7396 2353 | ·7583 4149 | ·7788 4994 | ·8017 0604 | ·8278 4510 |

q = 0·5 (*continued*)

| p \ x | 0·9988 | 0·9989 | 0·9990 | 0·9991 | 0·9992 | 0·9993 |
|---|---|---|---|---|---|---|
| 0·5 | ·9779 4243 | ·9788 8184 | ·9798 6496 | ·9808 9854 | ·9819 9127 | ·9831 5466 |
| 1·0 | ·9653 5898 | ·9668 3375+ | ·9683 7722 | ·9700 0000 | ·9717 1573 | ·9735 4249 |
| 1·5 | ·9559 0251 | ·9577 7916 | ·9597 4334 | ·9618 0854 | ·9639 9215− | ·9663 1718 |
| 2·0 | ·9480 5926 | ·9502 6887 | ·9525 8165− | ·9550 1350 | ·9575 8491 | ·9603 2299 |
| 2·5 | ·9412 2686 | ·9437 2619 | ·9463 4235− | ·9490 9333 | ·9520 0233 | ·9551 0005− |
| 3·0 | ·9351 0004 | ·9378 5887 | ·9407 4681 | ·9437 8374 | ·9469 9527 | ·9504 1531 |
| 3·5 | ·9295 0043 | ·9324 9618 | ·9356 3227 | ·9389 3032 | ·9424 1816 | ·9461 3263 |
| 4·0 | ·9243 1364 | ·9275 2859 | ·9308 9431 | ·9344 3403 | ·9381 7763 | ·9421 6469 |
| 4·5 | ·9194 6127 | ·9228 8104 | ·9264 6138 | ·9302 2701 | ·9342 0972 | ·9384 5165− |
| 5·0 | ·9148 8688 | ·9184 9953 | ·9222 8199 | ·9262 6040 | ·9304 6837 | ·9349 5045− |
| 5·5 | ·9105 4828 | ·9143 4366 | ·9183 1764 | ·9224 9769 | ·9269 1916 | ·9316 2889 |
| 6·0 | ·9064 1297 | ·9103 8231 | ·9145 3866 | ·9189 1075− | ·9235 3558 | ·9284 6217 |
| 6·5 | ·9024 5528 | ·9065 9093 | ·9109 2163 | ·9154 7736 | ·9202 9669 | ·9254 3072 |
| 7·0 | ·8986 5454 | ·9029 4972 | ·9074 4770 | ·9121 7963 | ·9171 8562 | ·9225 1875+ |
| 7·5 | ·8949 9378 | ·8994 4245− | ·9041 0138 | ·9090 0289 | ·9141 8853 | ·9197 1334 |
| 8·0 | ·8914 5892 | ·8960 5562 | ·9008 6982 | ·9059 3494 | ·9112 9393 | ·9170 0373 |
| 8·5 | ·8880 3808 | ·8927 7786 | ·8977 4218 | ·9029 6549 | ·9084 9214 | ·9143 8086 |
| 9·0 | ·8847 2113 | ·8895 9950+ | ·8947 0923 | ·9000 8580 | ·9057 7490 | ·9118 3701 |
| 9·5 | ·8814 9939 | ·8865 1221 | ·8917 6303 | ·8972 8835− | ·9031 3512 | ·9093 6556 |
| 10·0 | ·8783 6530 | ·8835 0877 | ·8888 9671 | ·8945 6659 | ·9005 6665+ | ·9069 6075− |
| 10·5 | ·8753 1227 | ·8805 8287 | ·8861 0424 | ·8919 1484 | ·8980 6411 | ·9046 1755+ |
| 11·0 | ·8723 3451 | ·8777 2894 | ·8833 8034 | ·8893 2806 | ·8956 2277 | ·9023 3154 |
| 12·0 | ·8665 8475− | ·8722 1790 | ·8781 1998 | ·8843 3215− | ·8909 0739 | ·8979 1585− |
| 13·0 | ·8610 8117 | ·8669 4229 | ·8730 8384 | ·8795 4870 | ·8863 9209 | ·8936 8710 |
| 14·0 | ·8557 9562 | ·8618 7518 | ·8682 4624 | ·8749 5338 | ·8820 5394 | ·8896 2385− |
| 15·0 | ·8507 0496 | ·8569 9440 | ·8635 8607 | ·8705 2616 | ·8778 7407 | ·8857 0845+ |
| 16·0 | ·8457 8989 | ·8522 8151 | ·8590 8576 | ·8662 5037 | ·8738 3675+ | ·8819 2622 |

| p \ x | 0·9994 | 0·9995 | 0·9996 | 0·9997 | 0·9998 | 0·9999 |
|---|---|---|---|---|---|---|
| 0·5 | ·9844 0450+ | ·9857 6356 | ·9872 6676 | ·9889 7287 | ·9909 9654 | ·9936 3370 |
| 1·0 | ·9755 0510 | ·9776 3932 | ·9800 0000 | ·9826 7949 | ·9858 5786 | ·9900 0000 |
| 1·5 | ·9688 1525− | ·9715 3187 | ·9745 3691 | ·9779 4795− | ·9819 9427 | ·9872 6782 |
| 2·0 | ·9632 6500+ | ·9664 6457 | ·9700 0400 | ·9740 2184 | ·9787 8821 | ·9850 0050 |
| 2·5 | ·9584 2864 | ·9620 4882 | ·9660 5374 | ·9706 0020 | ·9759 9397 | ·9830 2432 |
| 3·0 | ·9540 9044 | ·9580 8770 | ·9625 1000 | ·9675 3054 | ·9734 8703 | ·9812 5125 |
| 3·5 | ·9501 2435− | ·9544 6617 | ·9592 6991 | ·9647 2377 | ·9711 9468 | ·9796 2986 |
| 4·0 | ·9464 4955− | ·9511 1046 | ·9562 6750− | ·9621 2275+ | ·9690 7026 | ·9781 2719 |
| 4·5 | ·9430 1065+ | ·9479 7001 | ·9534 5753 | ·9596 8834 | ·9670 8183 | ·9767 2062 |
| 5·0 | ·9397 6779 | ·9450 0843 | ·9508 0749 | ·9573 9236 | ·9652 0637 | ·9753 9391 |
| 5·5 | ·9366 9116 | ·9421 9853 | ·9482 9304 | ·9552 1374 | ·9634 2669 | ·9741 3489 |
| 6·0 | ·9337 5781 | ·9395 1935+ | ·9458 9545+ | ·9531 3628 | ·9617 2955+ | ·9729 3420 |
| 6·5 | ·9309 4961 | ·9369 5436 | ·9435 9993 | ·9511 4715+ | ·9601 0450+ | ·9717 8445+ |
| 7·0 | ·9282 5197 | ·9344 9021 | ·9413 9455 | ·9492 3604 | ·9585 4311 | ·9706 7969 |
| 7·5 | ·9256 5290 | ·9321 1600 | ·9392 6955+ | ·9473 9450− | ·9570 3848 | ·9696 1504 |
| 8·0 | ·9231 4247 | ·9298 2263 | ·9372 1681 | ·9456 1548 | ·9555 8487 | ·9685 8643 |
| 8·5 | ·9207 1227 | ·9276 0246 | ·9352 2949 | ·9438 9307 | ·9541 7744 | ·9675 9045+ |
| 9·0 | ·9183 5518 | ·9254 4896 | ·9333 0175+ | ·9422 2223 | ·9528 1207 | ·9666 2420 |
| 9·5 | ·9160 6506 | ·9233 5654 | ·9314 2859 | ·9405 9860 | ·9514 8522 | ·9656 8515+ |
| 10·0 | ·9138 3658 | ·9213 2034 | ·9296 0567 | ·9390 1844 | ·9501 9384 | ·9647 7116 |
| 10·5 | ·9116 6508 | ·9193 3611 | ·9278 2919 | ·9374 7846 | ·9489 3522 | ·9638 8031 |
| 11·0 | ·9095 4647 | ·9174 0011 | ·9260 9580 | ·9359 7576 | ·9477 0701 | ·9630 1095− |
| 12·0 | ·9054 5383 | ·9136 5994 | ·9227 4681 | ·9330 7226 | ·9453 3370 | ·9613 3092 |
| 13·0 | ·9015 3407 | ·9100 7740 | ·9195 3865− | ·9302 9057 | ·9430 5973 | ·9597 2105− |
| 14·0 | ·8977 6733 | ·9066 3438 | ·9164 5510 | ·9276 1667 | ·9408 7366 | ·9581 7325+ |
| 15·0 | ·8941 3730 | ·9033 1598 | ·9134 8288 | ·9250 3904 | ·9387 6609 | ·9566 8088 |
| 16·0 | ·8906 3037 | ·9001 0980 | ·9106 1088 | ·9225 4808 | ·9367 2917 | ·9552 3840 |

$q = 1{\cdot}0$

| $p$ \ $x$ | 0·975 | 0·980 | 0·985 | 0·988 | 0·989 | 0·990 | 0·991 | 0·992 |
|---|---|---|---|---|---|---|---|---|
| 0·5 | ·9874 209 | ·9899 495$^{-}$ | ·9924 717 | ·9939 8189 | ·9944 8479 | ·9949 8744 | ·9954 8983 | ·9959 9197 |
| 1·0 | ·9750 000 | ·9800 000 | ·9850 000 | ·9880 0000 | ·9890 0000 | ·9900 0000 | ·9910 0000 | ·9920 0000 |
| 1·5 | ·9627 354 | ·9701 505$^{+}$ | ·9775 846 | ·9820 5411 | ·9835 4546 | ·9850 3756 | ·9865 3042 | ·9880 2403 |
| 2·0 | ·9506 250 | ·9604 000 | ·9702 250 | ·9761 4400 | ·9781 2100 | ·9801 0000 | ·9820 8100 | ·9840 6400 |
| 2·5 | ·9386 670 | ·9507 475$^{-}$ | ·9629 208 | ·9702 6946 | ·9727 2646 | ·9751 8719 | ·9776 5165$^{-}$ | ·9801 1984 |
| 3·0 | ·9268 594 | ·9411 920 | ·9556 716 | ·9644 3027 | ·9673 6167 | ·9702 9900 | ·9732 4227 | ·9761 9149 |
| 3·5 | ·9152 003 | ·9317 325$^{+}$ | ·9484 770 | ·9586 2623 | ·9620 2647 | ·9654 3532 | ·9688 5278 | ·9722 7888 |
| 4·0 | ·9036 879 | ·9223 682 | ·9413 366 | ·9528 5711 | ·9567 2069 | ·9605 9601 | ·9644 8309 | ·9683 8196 |
| 4·5 | ·8923 203 | ·9130 979 | ·9342 499 | ·9471 2271 | ·9514 4418 | ·9557 8096 | ·9601 3311 | ·9645 0065$^{+}$ |
| 5·0 | ·8810 957 | ·9039 208 | ·9272 165$^{+}$ | ·9414 2282 | ·9461 9676 | ·9509 9005$^{-}$ | ·9558 0274 | ·9606 3490 |
| 5·5 | ·8700 123 | ·8948 359 | ·9202 361 | ·9357 5724 | ·9409 7829 | ·9462 2315$^{+}$ | ·9514 9191 | ·9567 8464 |
| 6·0 | ·8590 683 | ·8858 424 | ·9133 083 | ·9301 2575$^{-}$ | ·9357 8860 | ·9414 8015$^{-}$ | ·9472 0052 | ·9529 4982 |
| 6·5 | ·8482 620 | ·8769 392 | ·9064 326 | ·9245 2815$^{+}$ | ·9306 2753 | ·9367 6092 | ·9429 2848 | ·9491 3037 |
| 7·0 | ·8375 916 | ·8681 255$^{+}$ | ·8996 086 | ·9189 6424 | ·9254 9492 | ·9320 6535$^{-}$ | ·9386 7571 | ·9453 2622 |
| 7·5 | ·8270 554 | ·8594 004 | ·8928 361 | ·9134 3381 | ·9203 9063 | ·9273 9331 | ·9344 4213 | ·9415 3732 |
| 8·0 | ·8166 518 | ·8507 630 | ·8861 145$^{+}$ | ·9079 3667 | ·9153 1448 | ·9227 4469 | ·9302 2763 | ·9377 6361 |
| 8·5 | ·8063 790 | ·8422 124 | ·8794 435$^{+}$ | ·9024 7261 | ·9102 6633 | ·9181 1938 | ·9260 3215$^{-}$ | ·9340 0503 |
| 9·0 | ·7962 355 | ·8337 478 | ·8728 228 | ·8970 4143 | ·9052 4602 | ·9135 1725$^{-}$ | ·9218 5558 | ·9302 6150$^{+}$ |
| 9·5 | ·7862 196 | ·8253 682 | ·8662 519 | ·8916 4294 | ·9002 5340 | ·9089 3818 | ·9176 9786 | ·9265 3299 |
| 10·0 | ·7763 296 | ·8170 728 | ·8597 304 | ·8862 7693 | ·8952 8831 | ·9043 8208 | ·9135 5888 | ·9228 1941 |
| 10·5 | ·7665 641 | ·8088 608 | ·8532 581 | ·8809 4322 | ·8903 5061 | ·8998 4880 | ·9094 3858 | ·9191 2072 |
| 11·0 | ·7569 214 | ·8007 314 | ·8468 345$^{-}$ | ·8756 4161 | ·8854 4014 | ·8953 3825$^{+}$ | ·9053 3685$^{+}$ | ·9154 3686 |
| 12·0 | ·7379 983 | ·7847 167 | ·8341 320 | ·8651 3391 | ·8757 0030 | ·8863 8487 | ·8971 8882 | ·9081 1336 |
| 13·0 | ·7195 484 | ·7690 224 | ·8216 200 | ·8547 5230 | ·8660 6760 | ·8775 2102 | ·8891 1412 | ·9008 4845$^{+}$ |
| 14·0 | ·7015 597 | ·7536 419 | ·8092 957 | ·8444 9528 | ·8565 4085$^{+}$ | ·8687 4581 | ·8811 1209 | ·8936 4167 |
| 15·0 | ·6840 207 | ·7385 691 | ·7971 563 | ·8343 6133 | ·8471 1891 | ·8600 5835$^{+}$ | ·8731 8209 | ·8864 9253 |
| 16·0 | ·6669 202 | ·7237 977 | ·7851 989 | ·8243 4900 | ·8378 0060 | ·8514 5777 | ·8653 2345$^{-}$ | ·8794 0059 |

| $p$ \ $x$ | 0·993 | 0·994 | 0·995 | 0·996 | 0·997 | 0·998 | 0·999 |
|---|---|---|---|---|---|---|---|
| 0·5 | ·9964 9385$^{+}$ | ·9969 9549 | ·9974 9687 | ·9979 9800 | ·9984 9887 | ·9989 9950$^{-}$ | ·9994 9987 |
| 1·0 | ·9930 0000 | ·9940 0000 | ·9950 0000 | ·9960 0000 | ·9970 0000 | ·9980 0000 | ·9990 0000 |
| 1·5 | ·9895 1840 | ·9910 1351 | ·9925 0938 | ·9940 0600 | ·9955 0338 | ·9970 0150$^{+}$ | ·9985 0038 |
| 2·0 | ·9860 4900 | ·9880 3600 | ·9900 2500 | ·9920 1600 | ·9940 0900 | ·9960 0400 | ·9980 0100 |
| 2·5 | ·9825 9177 | ·9850 6743 | ·9875 4684 | ·9900 2998 | ·9925 1687 | ·9950 0750$^{-}$ | ·9975 0187 |
| 3·0 | ·9791 4666 | ·9821 0778 | ·9850 7488 | ·9880 4794 | ·9910 2697 | ·9940 1199 | ·9970 0300 |
| 3·5 | ·9757 1363 | ·9791 5703 | ·9826 0910 | ·9860 6986 | ·9895 3932 | ·9930 1748 | ·9965 0437 |
| 4·0 | ·9722 9263 | ·9762 1514 | ·9801 4950$^{+}$ | ·9840 9574 | ·9880 5389 | ·9920 2397 | ·9960 0600 |
| 4·5 | ·9688 8363 | ·9732 8209 | ·9776 9606 | ·9821 2558 | ·9865 7070 | ·9910 3145$^{-}$ | ·9955 0787 |
| 5·0 | ·9654 8658 | ·9703 5785$^{-}$ | ·9752 4875$^{+}$ | ·9801 5936 | ·9850 8973 | ·9900 3992 | ·9950 0999 |
| 5·5 | ·9621 0144 | ·9674 4239 | ·9728 0758 | ·9781 9708 | ·9836 1099 | ·9890 4938 | ·9945 1236 |
| 6·0 | ·9587 2818 | ·9645 3570 | ·9703 7251 | ·9762 3872 | ·9821 3446 | ·9880 5984 | ·9940 1498 |
| 6·5 | ·9553 6673 | ·9616 3774 | ·9679 4354 | ·9742 8429 | ·9806 6015$^{+}$ | ·9870 7129 | ·9935 1785$^{-}$ |
| 7·0 | ·9520 1708 | ·9587 4849 | ·9655 2065$^{-}$ | ·9723 3377 | ·9791 8806 | ·9860 8372 | ·9930 2096 |
| 7·5 | ·9486 7917 | ·9558 6791 | ·9631 0382 | ·9703 8715$^{+}$ | ·9777 1817 | ·9850 9714 | ·9925 2433 |
| 8·0 | ·9453 5296 | ·9529 9599 | ·9606 9304 | ·9684 4443 | ·9762 5049 | ·9841 1155$^{+}$ | ·9920 2794 |
| 8·5 | ·9420 3841 | ·9501 3270 | ·9582 8830 | ·9665 0560 | ·9747 8502 | ·9831 2695$^{-}$ | ·9915 3181 |
| 9·0 | ·9387 3549 | ·9472 7802 | ·9558 8958 | ·9645 7066 | ·9733 2174 | ·9821 4333 | ·9910 3592 |
| 9·5 | ·9354 4414 | ·9444 3191 | ·9534 9686 | ·9626 3958 | ·9718 6066 | ·9811 6070 | ·9905 4027 |
| 10·0 | ·9321 6434 | ·9415 9435$^{+}$ | ·9511 1013 | ·9607 1237 | ·9704 0178 | ·9801 7904 | ·9900 4488 |
| 10·5 | ·9288 9604 | ·9387 6532 | ·9487 2938 | ·9587 8902 | ·9689 4508 | ·9791 9837 | ·9895 4973 |
| 11·0 | ·9256 3919 | ·9359 4478 | ·9463 5458 | ·9568 6952 | ·9674 9057 | ·9782 1869 | ·9890 5484 |
| 12·0 | ·9191 5972 | ·9303 2912 | ·9416 2281 | ·9530 4205$^{-}$ | ·9645 8810 | ·9762 6225$^{-}$ | ·9880 6578 |
| 13·0 | ·9127 2560 | ·9247 4714 | ·9369 1469 | ·9492 2988 | ·9616 9434 | ·9743 0972 | ·9870 7771 |
| 14·0 | ·9063 3652 | ·9191 9866 | ·9322 3012 | ·9454 3296 | ·9588 0925$^{+}$ | ·9723 6110 | ·9860 9064 |
| 15·0 | ·8999 9216 | ·9136 8347 | ·9275 6897 | ·9416 5123 | ·9559 3282 | ·9704 1638 | ·9851 0455$^{-}$ |
| 16·0 | ·8936 9222 | ·9082 0137 | ·9229 3112 | ·9378 8462 | ·9530 6503 | ·9684 7555$^{-}$ | ·9841 1944 |

$q = 1{\cdot}5$

| $p$ \ $x$ | 0·975 | 0·980 | 0·985 | 0·988 | 0·989 | 0·990 | 0·991 | 0·992 |
|---|---|---|---|---|---|---|---|---|
| 0·5 | ·9983 096 | ·9987 923 | ·9992 168 | ·9994 4007 | ·9995 0873 | ·9995 7431 | ·9996 3665− | ·9996 9558 |
| 1·0 | ·9960 472 | ·9971 716 | ·9981 629 | ·9986 8547 | ·9988 4631 | ·9990 0000 | ·9991 4618 | ·9992 8446 |
| 1·5 | ·9933 400 | ·9952 272 | ·9968 953 | ·9977 7643 | ·9980 4791 | ·9983 0745− | ·9985 5444 | ·9987 8818 |
| 2·0 | ·9902 661 | ·9930 138 | ·9954 485+ | ·9967 3733 | ·9971 3481 | ·9975 1500 | ·9978 7699 | ·9982 1973 |
| 2·5 | ·9868 795− | ·9905 689 | ·9938 464 | ·9955 8485+ | ·9961 2157 | ·9966 3520 | ·9971 2448 | ·9975 8798 |
| 3·0 | ·9832 205− | ·9879 205− | ·9921 065+ | ·9943 3137 | ·9950 1897 | ·9956 7731 | ·9963 0477 | ·9968 9947 |
| 3·5 | ·9793 207 | ·9850 906 | ·9902 426 | ·9929 8652 | ·9938 3540 | ·9946 4856 | ·9954 2398 | ·9961 5927 |
| 4·0 | ·9752 061 | ·9820 972 | ·9882 660 | ·9915 5811 | ·9925 7764 | ·9935 5478 | ·9944 8703 | ·9953 7149 |
| 4·5 | ·9708 981 | ·9789 550+ | ·9861 858 | ·9900 5264 | ·9912 5136 | ·9924 0083 | ·9934 9803 | ·9945 3952 |
| 5·0 | ·9664 152 | ·9756 769 | ·9840 101 | ·9884 7563 | ·9898 6136 | ·9911 9082 | ·9924 6047 | ·9936 6626 |
| 5·5 | ·9617 736 | ·9722 740 | ·9817 458 | ·9868 3188 | ·9884 1180 | ·9899 2832 | ·9913 7735+ | ·9927 5420 |
| 6·0 | ·9569 871 | ·9687 560 | ·9793 990 | ·9851 2559 | ·9869 0632 | ·9886 1646 | ·9902 5132 | ·9918 0552 |
| 6·5 | ·9520 684 | ·9651 314 | ·9769 749 | ·9833 6050− | ·9853 4817 | ·9872 5801 | ·9890 8471 | ·9908 2216 |
| 7·0 | ·9470 286 | ·9614 082 | ·9744 785− | ·9815 3994 | ·9837 4024 | ·9858 5546 | ·9878 7961 | ·9898 0584 |
| 7·5 | ·9418 779 | ·9575 932 | ·9719 140 | ·9796 6694 | ·9820 8516 | ·9844 1105− | ·9866 3793 | ·9887 5814 |
| 8·0 | ·9366 256 | ·9536 930 | ·9692 856 | ·9777 4426 | ·9803 8532 | ·9829 2682 | ·9853 6137 | ·9876 8047 |
| 8·5 | ·9312 799 | ·9497 133 | ·9665 967 | ·9757 7442 | ·9786 4291 | ·9814 0466 | ·9840 5151 | ·9865 7413 |
| 9·0 | ·9258 486 | ·9456 595+ | ·9638 509 | ·9737 5975− | ·9768 5993 | ·9798 4627 | ·9827 0981 | ·9854 4032 |
| 9·5 | ·9203 390 | ·9415 367 | ·9610 511 | ·9717 0239 | ·9750 3826 | ·9782 5325+ | ·9813 3761 | ·9842 8015+ |
| 10·0 | ·9147 574 | ·9373 494 | ·9582 003 | ·9696 0435− | ·9731 7963 | ·9766 2709 | ·9799 3614 | ·9830 9464 |
| 10·5 | ·9091 101 | ·9331 019 | ·9553 012 | ·9674 6748 | ·9712 8564 | ·9749 6915+ | ·9785 0657 | ·9818 8476 |
| 11·0 | ·9034 027 | ·9287 983 | ·9523 562 | ·9652 9353 | ·9693 5781 | ·9732 8075− | ·9770 4999 | ·9806 5138 |
| 12·0 | ·8918 287 | ·9200 372 | ·9463 382 | ·9608 4086 | ·9654 0622 | ·9698 1729 | ·9740 5982 | ·9781 1750+ |
| 13·0 | ·8800 739 | ·9110 937 | ·9401 633 | ·9562 5831 | ·9613 3527 | ·9662 4559 | ·9709 7308 | ·9754 9916 |
| 14·0 | ·8681 721 | ·9019 919 | ·9338 472 | ·9515 5661 | ·9571 5424 | ·9625 7362 | ·9677 9647 | ·9728 0186 |
| 15·0 | ·8561 534 | ·8927 536 | ·9274 037 | ·9467 4544 | ·9528 7152 | ·9588 0853 | ·9645 3603 | ·9700 3058 |
| 16·0 | ·8440 446 | ·8833 983 | ·9208 452 | ·9418 3355− | ·9484 9473 | ·9549 5685− | ·9611 9722 | ·9671 8983 |

| $p$ \ $x$ | 0·993 | 0·994 | 0·995 | 0·996 | 0·997 | 0·998 | 0·999 |
|---|---|---|---|---|---|---|---|
| 0·5 | ·9997 5091 | ·9998 0239 | ·9998 4972 | ·9998 9250+ | ·9999 3020 | ·9999 6202 | ·9999 8657 |
| 1·0 | ·9994 1434 | ·9995 3524 | ·9996 4645− | ·9997 4702 | ·9998 3568 | ·9999 1056 | ·9999 6838 |
| 1·5 | ·9990 0784 | ·9992 1242 | ·9994 0069 | ·9995 7104 | ·9997 2130 | ·9998 4825− | ·9999 4633 |
| 2·0 | ·9985 4199 | ·9988 4229 | ·9991 1877 | ·9993 6906 | ·9995 8995− | ·9997 7666 | ·9999 2099 |
| 2·5 | ·9980 2401 | ·9984 3052 | ·9988 0498 | ·9991 4414 | ·9994 4360 | ·9996 9686 | ·9998 9273 |
| 3·0 | ·9974 5920 | ·9979 8129 | ·9984 6247 | ·9988 9851 | ·9992 8370 | ·9996 0963 | ·9998 6182 |
| 3·5 | ·9968 5168 | ·9974 9786 | ·9980 9369 | ·9986 3391 | ·9991 1136 | ·9995 1556 | ·9998 2847 |
| 4·0 | ·9962 0478 | ·9969 8282 | ·9977 0061 | ·9983 5172 | ·9989 2748 | ·9994 1514 | ·9997 9285− |
| 4·5 | ·9955 2125+ | ·9964 3836 | ·9972 8486 | ·9980 5311 | ·9987 3280 | ·9993 0877 | ·9997 5510 |
| 5·0 | ·9948 0343 | ·9958 6629 | ·9968 4781 | ·9977 3905+ | ·9985 2794 | ·9991 9678 | ·9997 1534 |
| 5·5 | ·9940 5334 | ·9952 6820 | ·9963 9065 | ·9974 1037 | ·9983 1344 | ·9990 7947 | ·9996 7366 |
| 6·0 | ·9932 7275− | ·9946 4547 | ·9959 1442 | ·9970 6781 | ·9980 8976 | ·9989 5707 | ·9996 3016 |
| 6·5 | ·9924 6320 | ·9939 9932 | ·9954 2003 | ·9967 1200 | ·9978 5732 | ·9988 2982 | ·9995 8491 |
| 7·0 | ·9916 2611 | ·9933 3085− | ·9949 0830 | ·9963 4354 | ·9976 1649 | ·9986 9790 | ·9995 3798 |
| 7·5 | ·9907 6273 | ·9926 4103 | ·9943 7997 | ·9959 6292 | ·9973 6760 | ·9985 6150+ | ·9994 8943 |
| 8·0 | ·9898 7421 | ·9919 3077 | ·9938 3571 | ·9955 7064 | ·9971 1094 | ·9984 2078 | ·9994 3931 |
| 8·5 | ·9889 6159 | ·9912 0089 | ·9932 7612 | ·9951 6710 | ·9968 4679 | ·9982 7587 | ·9993 8768 |
| 9·0 | ·9880 2584 | ·9904 5212 | ·9927 0178 | ·9947 5271 | ·9965 7540 | ·9981 2692 | ·9993 3458 |
| 9·5 | ·9870 6784 | ·9896 8518 | ·9921 1319 | ·9943 2784 | ·9962 9700 | ·9979 7404 | ·9992 8006 |
| 10·0 | ·9860 8844 | ·9889 0069 | ·9915 1083 | ·9938 9280 | ·9960 1181 | ·9978 1736 | ·9992 2414 |
| 10·5 | ·9850 8839 | ·9880 9927 | ·9908 9516 | ·9934 4793 | ·9957 2001 | ·9976 5696 | ·9991 6688 |
| 11·0 | ·9840 6841 | ·9872 8146 | ·9902 6659 | ·9929 9351 | ·9954 2181 | ·9974 9297 | ·9991 0830 |
| 12·0 | ·9819 7135− | ·9855 9879 | ·9889 7230 | ·9920 5710 | ·9948 0684 | ·9971 5451 | ·9989 8731 |
| 13·0 | ·9798 0219 | ·9838 5652 | ·9876 3082 | ·9910 8558 | ·9941 6817 | ·9968 0265 | ·9988 6141 |
| 14·0 | ·9775 6538 | ·9820 5809 | ·9862 4471 | ·9900 8072 | ·9935 0693 | ·9964 3800 | ·9987 3079 |
| 15·0 | ·9752 6490 | ·9802 0661 | ·9848 1628 | ·9890 4414 | ·9928 2412 | ·9960 6107 | ·9985 9565− |
| 16·0 | ·9729 0438 | ·9783 0490 | ·9833 4761 | ·9879 7729 | ·9921 2068 | ·9956 7236 | ·9984 5614 |

$q = 2·0$

| $p$ \ $x$ | 0·975 | 0·980 | 0·985 | 0·988 | 0·989 | 0·990 | 0·991 | 0·992 |
|---|---|---|---|---|---|---|---|---|
| 0·5 | ·9997 636 | ·9998 490 | ·9999 152 | ·9999 4578 | ·9999 5446 | ·9999 6237 | ·9999 6953 | ·9999 7594 |
| 1·0 | ·9993 750 | ·9996 000 | ·9997 750 | ·9998 5600 | ·9998 7900 | ·9999 0000 | ·9999 1900 | ·9999 3600 |
| 1·5 | ·9988 379 | ·9992 550$^{+}$ | ·9995 802 | ·9997 3108 | ·9997 7396 | ·9998 1313 | ·9998 4858 | ·9998 8032 |
| 2·0 | ·9981 562 | ·9988 160 | ·9993 318 | ·9995 7146 | ·9996 3966 | ·9997 0200 | ·9997 5846 | ·9998 0902 |
| 2·5 | ·9973 337 | ·9982 849 | ·9990 303 | ·9993 7754 | ·9994 7644 | ·9995 6687 | ·9996 4881 | ·9997 2224 |
| 3·0 | ·9963 738 | ·9976 635$^{+}$ | ·9986 768 | ·9991 4976 | ·9992 8460 | ·9994 0797 | ·9995 1981 | ·9996 2008 |
| 3·5 | ·9952 803 | ·9969 538 | ·9982 720 | ·9988 8853 | ·9990 6449 | ·9992 2555$^{+}$ | ·9993 7164 | ·9995 0269 |
| 4·0 | ·9940 567 | ·9961 576 | ·9978 167 | ·9985 9425 | ·9988 1640 | ·9990 1985$^{+}$ | ·9992 0448 | ·9993 7018 |
| 4·5 | ·9927 063 | ·9952 767 | ·9973 117 | ·9982 6734 | ·9985 4066 | ·9987 9111 | ·9990 1850$^{-}$ | ·9992 2267 |
| 5·0 | ·9912 327 | ·9943 129 | ·9967 577 | ·9979 0819 | ·9982 3758 | ·9985 3955$^{+}$ | ·9988 1387 | ·9990 6030 |
| 5·5 | ·9896 390 | ·9932 679 | ·9961 556 | ·9975 1722 | ·9979 0748 | ·9982 6543 | ·9985 9076 | ·9988 8317 |
| 6·0 | ·9879 285$^{+}$ | ·9921 435$^{-}$ | ·9955 060 | ·9970 9480 | ·9975 5065$^{-}$ | ·9979 6896 | ·9983 4935$^{-}$ | ·9986 9141 |
| 6·5 | ·9861 046 | ·9909 413 | ·9948 097 | ·9966 4135$^{-}$ | ·9971 6740 | ·9976 5038 | ·9980 8980 | ·9984 8515$^{-}$ |
| 7·0 | ·9841 701 | ·9896 631 | ·9940 675$^{+}$ | ·9961 5724 | ·9967 5803 | ·9973 0992 | ·9978 1228 | ·9982 6449 |
| 7·5 | ·9821 283 | ·9883 105$^{-}$ | ·9932 801 | ·9956 4286 | ·9963 2285$^{+}$ | ·9969 4781 | ·9975 1697 | ·9980 2956 |
| 8·0 | ·9799 822 | ·9868 851 | ·9924 482 | ·9950 9859 | ·9958 6215$^{+}$ | ·9965 6427 | ·9972 0402 | ·9977 8048 |
| 8·5 | ·9777 346 | ·9853 885$^{+}$ | ·9915 726 | ·9945 2481 | ·9953 7623 | ·9961 5953 | ·9968 7361 | ·9975 1737 |
| 9·0 | ·9753 885$^{-}$ | ·9838 224 | ·9906 539 | ·9939 2190 | ·9948 6538 | ·9957 3380 | ·9965 2589 | ·9972 4033 |
| 9·5 | ·9729 467 | ·9821 881 | ·9896 928 | ·9932 9023 | ·9943 2988 | ·9952 8731 | ·9961 6102 | ·9969 4949 |
| 10·0 | ·9704 120 | ·9804 874 | ·9886 900 | ·9926 3016 | ·9937 7003 | ·9948 2028 | ·9957 7918 | ·9966 4496 |
| 10·5 | ·9677 872 | ·9787 216 | ·9876 463 | ·9919 4207 | ·9931 8611 | ·9943 3293 | ·9953 8052 | ·9963 2686 |
| 11·0 | ·9650 748 | ·9768 922 | ·9865 622 | ·9912 2630 | ·9925 7840 | ·9938 2546 | ·9949 6520 | ·9959 9530 |
| 12·0 | ·9593 978 | ·9730 487 | ·9842 757 | ·9897 1319 | ·9912 9274 | ·9927 5106 | ·9940 8521 | ·9952 9224 |
| 13·0 | ·9534 016 | ·9689 682 | ·9818 359 | ·9880 9366 | ·9899 1526 | ·9915 9876 | ·9931 4047 | ·9945 3669 |
| 14·0 | ·9471 056 | ·9646 617 | ·9792 478 | ·9863 7048 | ·9884 4815$^{-}$ | ·9903 7023 | ·9921 3222 | ·9937 2953 |
| 15·0 | ·9405 284 | ·9601 398 | ·9765 164 | ·9845 4637 | ·9868 9352 | ·9890 6711 | ·9910 6167 | ·9928 7164 |
| 16·0 | ·9336 882 | ·9554 130 | ·9736 466 | ·9826 2400 | ·9852 5350$^{+}$ | ·9876 9101 | ·9899 3002 | ·9919 6387 |

| $p$ \ $x$ | 0·993 | 0·994 | 0·995 | 0·996 | 0·997 | 0·998 | 0·999 |
|---|---|---|---|---|---|---|---|
| 0·5 | ·9999 8158 | ·9999 8647 | ·9999 9061 | ·9999 9399 | ·9999 9662 | ·9999 9850$^{-}$ | ·9999 9962 |
| 1·0 | ·9999 5100 | ·9999 6400 | ·9999 7500 | ·9999 8400 | ·9999 9100 | ·9999 9600 | ·9999 9900 |
| 1·5 | ·9999 0834 | ·9999 3264 | ·9999 5320 | ·9999 7004 | ·9999 8314 | ·9999 9250$^{+}$ | ·9999 9813 |
| 2·0 | ·9998 5369 | ·9998 9243 | ·9999 2525 | ·9999 5213 | ·9999 7305$^{+}$ | ·9999 8802 | ·9999 9700 |
| 2·5 | ·9997 8712 | ·9998 4344 | ·9998 9117 | ·9999 3028 | ·9999 6074 | ·9999 8254 | ·9999 9563 |
| 3·0 | ·9997 0874 | ·9997 8572 | ·9998 5100 | ·9999 0451 | ·9999 4622 | ·9999 7606 | ·9999 9401 |
| 3·5 | ·9996 1861 | ·9997 1933 | ·9998 0476 | ·9998 7484 | ·9999 2948 | ·9999 6860 | ·9999 9214 |
| 4·0 | ·9995 1682 | ·9996 4430 | ·9997 5249 | ·9998 4128 | ·9999 1054 | ·9999 6016 | ·9999 9002 |
| 4·5 | ·9994 0346 | ·9995 6070 | ·9996 9422 | ·9998 0384 | ·9998 8940 | ·9999 5073 | ·9999 8765$^{+}$ |
| 5·0 | ·9992 7861 | ·9994 6858 | ·9996 2997 | ·9997 6255$^{-}$ | ·9998 6608 | ·9999 4032 | ·9999 8504 |
| 5·5 | ·9991 4235$^{+}$ | ·9993 6799 | ·9995 5978 | ·9997 1741 | ·9998 4057 | ·9999 2893 | ·9999 8218 |
| 6·0 | ·9989 9476 | ·9992 5898 | ·9994 8368 | ·9996 6845$^{+}$ | ·9998 1288 | ·9999 1656 | ·9999 7907 |
| 6·5 | ·9988 3592 | ·9991 4161 | ·9994 0170 | ·9996 1568 | ·9997 8303 | ·9999 0321 | ·9999 7571 |
| 7·0 | ·9986 6592 | ·9990 1592 | ·9993 1387 | ·9995 5911 | ·9997 5101 | ·9998 8889 | ·9999 7211 |
| 7·5 | ·9984 8482 | ·9988 8197 | ·9992 2021 | ·9994 9877 | ·9997 1683 | ·9998 7360 | ·9999 6826 |
| 8·0 | ·9982 9272 | ·9987 3980 | ·9991 2077 | ·9994 3466 | ·9996 8051 | ·9998 5734 | ·9999 6417 |
| 8·5 | ·9980 8970 | ·9985 8947 | ·9990 1555$^{+}$ | ·9993 6679 | ·9996 4204 | ·9998 4011 | ·9999 5983 |
| 9·0 | ·9978 7582 | ·9984 3103 | ·9989 0461 | ·9992 9520 | ·9996 0143 | ·9998 2191 | ·9999 5524 |
| 9·5 | ·9976 5118 | ·9982 6453 | ·9987 8796 | ·9992 1989 | ·9995 5869 | ·9998 0275$^{-}$ | ·9999 5041 |
| 10·0 | ·9974 1584 | ·9980 9001 | ·9986 6564 | ·9991 4087 | ·9995 1383 | ·9997 8262 | ·9999 4533 |
| 10·5 | ·9971 6989 | ·9979 0753 | ·9985 3767 | ·9990 5816 | ·9994 6685$^{+}$ | ·9997 6154 | ·9999 4001 |
| 11·0 | ·9969 1341 | ·9977 1714 | ·9984 0408 | ·9989 7178 | ·9994 1776 | ·9997 3950$^{-}$ | ·9999 3444 |
| 12·0 | ·9963 6913 | ·9973 1281 | ·9981 2018 | ·9987 8806 | ·9993 1327 | ·9996 9254 | ·9999 2257 |
| 13·0 | ·9957 8363 | ·9968 7742 | ·9978 1415$^{-}$ | ·9985 8983 | ·9992 0041 | ·9996 4178 | ·9999 0972 |
| 14·0 | ·9951 5750$^{-}$ | ·9964 1135$^{-}$ | ·9974 8623 | ·9983 7720 | ·9990 7924 | ·9995 8721 | ·9998 9591 |
| 15·0 | ·9944 9134 | ·9959 1498 | ·9971 3664 | ·9981 5030 | ·9989 4980 | ·9995 2887 | ·9998 8111 |
| 16·0 | ·9937 8575$^{-}$ | ·9953 8870 | ·9967 6561 | ·9979 0924 | ·9988 1215$^{-}$ | ·9994 6677 | ·9998 6535 |

$q = 2\cdot5$

| $p$ \ $x$ | 0·975 | 0·980 | 0·985 | 0·988 | 0·989 | 0·990 | 0·991 | 0·992 |
|---|---|---|---|---|---|---|---|---|
| 0·5 | ·9999 661 | ·9999 807 | ·9999 906 | ·9999 9462 | ·9999 9567 | ·9999 9659 | ·9999 9738 | ·9999 9805$^{+}$ |
| 1·0 | ·9999 012 | ·9999 434 | ·9999 724 | ·9999 8423 | ·9999 8731 | ·9999 9000 | ·9999 9232 | ·9999 9428 |
| 1·5 | ·9998 005$^{-}$ | ·9998 856 | ·9999 442 | ·9999 6800 | ·9999 7425$^{-}$ | ·9999 7970 | ·9999 8440 | ·9999 8837 |
| 2·0 | ·9996 603 | ·9998 048 | ·9999 046 | ·9999 4526 | ·9999 5593 | ·9999 6525 | ·9999 7328 | ·9999 8008 |
| 2·5 | ·9994 775$^{-}$ | ·9996 993 | ·9998 527 | ·9999 1540 | ·9999 3187 | ·9999 4626 | ·9999 5866 | ·9999 6917 |
| 3·0 | ·9992 493 | ·9995 672 | ·9997 876 | ·9998 7790 | ·9999 0163 | ·9999 2237 | ·9999 4026 | ·9999 5543 |
| 3·5 | ·9989 735$^{+}$ | ·9994 071 | ·9997 085$^{+}$ | ·9998 3226 | ·9998 6481 | ·9998 9328 | ·9999 1784 | ·9999 3869 |
| 4·0 | ·9986 482 | ·9992 178 | ·9996 148 | ·9997 7806 | ·9998 2106 | ·9998 5869 | ·9998 9118 | ·9999 1876 |
| 4·5 | ·9982 717 | ·9989 980 | ·9995 057 | ·9997 1490 | ·9997 7006 | ·9998 1835$^{+}$ | ·9998 6007 | ·9998 9550$^{-}$ |
| 5·0 | ·9978 424 | ·9987 469 | ·9993 807 | ·9996 4243 | ·9997 1150$^{+}$ | ·9997 7202 | ·9998 2431 | ·9998 6874 |
| 5·5 | ·9973 592 | ·9984 636 | ·9992 393 | ·9995 6031 | ·9996 4512 | ·9997 1945$^{+}$ | ·9997 8372 | ·9998 3837 |
| 6·0 | ·9968 210 | ·9981 471 | ·9990 810 | ·9994 6823 | ·9995 7065$^{-}$ | ·9996 6046 | ·9997 3815$^{-}$ | ·9998 0424 |
| 6·5 | ·9962 269 | ·9977 969 | ·9989 053 | ·9993 6591 | ·9994 8785$^{+}$ | ·9995 9484 | ·9996 8743 | ·9997 6624 |
| 7·0 | ·9955 762 | ·9974 123 | ·9987 120 | ·9992 5309 | ·9993 9651 | ·9995 2241 | ·9996 3142 | ·9997 2425$^{-}$ |
| 7·5 | ·9948 682 | ·9969 928 | ·9985 005$^{-}$ | ·9991 2952 | ·9992 9642 | ·9994 4300 | ·9995 6999 | ·9996 7817 |
| 8·0 | ·9941 024 | ·9965 379 | ·9982 706 | ·9989 9498 | ·9991 8739 | ·9993 5645$^{+}$ | ·9995 0299 | ·9996 2790 |
| 8·5 | ·9932 785$^{+}$ | ·9960 473 | ·9980 219 | ·9988 4926 | ·9990 6923 | ·9992 6262 | ·9994 3032 | ·9995 7334 |
| 9·0 | ·9923 961 | ·9955 204 | ·9977 543 | ·9986 9216 | ·9989 4179 | ·9991 6135$^{+}$ | ·9993 5185$^{+}$ | ·9995 1440 |
| 9·5 | ·9914 550$^{+}$ | ·9949 570 | ·9974 674 | ·9985 2350$^{-}$ | ·9988 0489 | ·9990 5252 | ·9992 6748 | ·9994 5099 |
| 10·0 | ·9904 551 | ·9943 570 | ·9971 610 | ·9983 4311 | ·9986 5840 | ·9989 3601 | ·9991 7711 | ·9993 8304 |
| 10·5 | ·9893 964 | ·9937 199 | ·9968 349 | ·9981 5083 | ·9985 0218 | ·9988 1169 | ·9990 8064 | ·9993 1046 |
| 11·0 | ·9882 788 | ·9930 458 | ·9964 889 | ·9979 4651 | ·9983 3610 | ·9986 7946 | ·9989 7797 | ·9992 3319 |
| 12·0 | ·9858 676 | ·9915 856 | ·9957 366 | ·9975 0124 | ·9979 7387 | ·9983 9084 | ·9987 5370 | ·9990 6426 |
| 13·0 | ·9832 227 | ·9899 758 | ·9949 030 | ·9970 0633 | ·9975 7084 | ·9980 6938 | ·9985 0368 | ·9988 7574 |
| 14·0 | ·9803 465$^{-}$ | ·9882 161 | ·9939 872 | ·9964 6093 | ·9971 2626 | ·9977 1443 | ·9982 2731 | ·9986 6715$^{+}$ |
| 15·0 | ·9772 416 | ·9863 069 | ·9929 884 | ·9958 6435$^{-}$ | ·9966 3946 | ·9973 2537 | ·9979 2409 | ·9984 3806 |
| 16·0 | ·9739 117 | ·9842 487 | ·9919 063 | ·9952 1598 | ·9961 0987 | ·9969 0168 | ·9975 9355$^{+}$ | ·9981 8807 |

| $p$ \ $x$ | 0·993 | 0·994 | 0·995 | 0·996 | 0·997 | 0·998 | 0·999 |
|---|---|---|---|---|---|---|---|
| 0·5 | ·9999 9860 | ·9999 9905$^{+}$ | ·9999 9940 | ·9999 9966 | ·9999 9983 | ·9999 9994 | ·9999 9999 |
| 1·0 | ·9999 9590 | ·9999 9721 | ·9999 9823 | ·9999 9899 | ·9999 9951 | ·9999 9982 | ·9999 9997 |
| 1·5 | ·9999 9167 | ·9999 9433 | ·9999 9641 | ·9999 9794 | ·9999 9900 | ·9999 9964 | ·9999 9994 |
| 2·0 | ·9999 8572 | ·9999 9028 | ·9999 9383 | ·9999 9647 | ·9999 9828 | ·9999 9937 | ·9999 9989 |
| 2·5 | ·9999 7790 | ·9999 8495$^{-}$ | ·9999 9045$^{-}$ | ·9999 9453 | ·9999 9733 | ·9999 9903 | ·9999 9983 |
| 3·0 | ·9999 6804 | ·9999 7823 | ·9999 8618 | ·9999 9208 | ·9999 9613 | ·9999 9860 | ·9999 9975$^{+}$ |
| 3·5 | ·9999 5601 | ·9999 7003 | ·9999 8096 | ·9999 8908 | ·9999 9467 | ·9999 9806 | ·9999 9966 |
| 4·0 | ·9999 4169 | ·9999 6026 | ·9999 7475$^{+}$ | ·9999 8552 | ·9999 9293 | ·9999 9743 | ·9999 9954 |
| 4·5 | ·9999 2497 | ·9999 4884 | ·9999 6748 | ·9999 8134 | ·9999 9089 | ·9999 9668 | ·9999 9941 |
| 5·0 | ·9999 0573 | ·9999 3569 | ·9999 5912 | ·9999 7653 | ·9999 8853 | ·9999 9583 | ·9999 9926 |
| 5·5 | ·9998 8387 | ·9999 2075$^{+}$ | ·9999 4960 | ·9999 7106 | ·9999 8585$^{+}$ | ·9999 9485$^{+}$ | ·9999 9909 |
| 6·0 | ·9998 5930 | ·9999 0395$^{+}$ | ·9999 3889 | ·9999 6489 | ·9999 8284 | ·9999 9375$^{-}$ | ·9999 9889 |
| 6·5 | ·9998 3192 | ·9998 8522 | ·9999 2695$^{+}$ | ·9999 5802 | ·9999 7947 | ·9999 9252 | ·9999 9867 |
| 7·0 | ·9998 0166 | ·9998 6451 | ·9999 1374 | ·9999 5041 | ·9999 7574 | ·9999 9116 | ·9999 9843 |
| 7·5 | ·9997 6844 | ·9998 4176 | ·9998 9922 | ·9999 4204 | ·9999 7163 | ·9999 8966 | ·9999 9816 |
| 8·0 | ·9997 3217 | ·9998 1691 | ·9998 8335$^{-}$ | ·9999 3289 | ·9999 6714 | ·9999 8802 | ·9999 9787 |
| 8·5 | ·9996 9279 | ·9997 8991 | ·9998 6610 | ·9999 2294 | ·9999 6226 | ·9999 8623 | ·9999 9755$^{+}$ |
| 9·0 | ·9996 5023 | ·9997 6072 | ·9998 4744 | ·9999 1217 | ·9999 5697 | ·9999 8429 | ·9999 9721 |
| 9·5 | ·9996 0442 | ·9997 2928 | ·9998 2734 | ·9999 0056 | ·9999 5126 | ·9999 8221 | ·9999 9684 |
| 10·0 | ·9995 5529 | ·9996 9556 | ·9998 0576 | ·9998 8809 | ·9999 4513 | ·9999 7996 | ·9999 9643 |
| 10·5 | ·9995 0280 | ·9996 5950$^{+}$ | ·9997 8268 | ·9998 7475$^{-}$ | ·9999 3857 | ·9999 7756 | ·9999 9601 |
| 11·0 | ·9994 4689 | ·9996 2108 | ·9997 5806 | ·9998 6051 | ·9999 3156 | ·9999 7499 | ·9999 9555$^{-}$ |
| 12·0 | ·9993 2456 | ·9995 3694 | ·9997 0413 | ·9998 2930 | ·9999 1619 | ·9999 6935$^{-}$ | ·9999 9454 |
| 13·0 | ·9991 8790 | ·9994 4286 | ·9996 4377 | ·9997 9432 | ·9998 9894 | ·9999 6301 | ·9999 9341 |
| 14·0 | ·9990 3654 | ·9993 3855$^{+}$ | ·9995 7677 | ·9997 5547 | ·9998 7977 | ·9999 5596 | ·9999 9214 |
| 15·0 | ·9988 7014 | ·9992 2376 | ·9995 0297 | ·9997 1263 | ·9998 5860 | ·9999 4817 | ·9999 9075$^{-}$ |
| 16·0 | ·9986 8838 | ·9990 9825$^{-}$ | ·9994 2219 | ·9996 6568 | ·9998 3539 | ·9999 3962 | ·9999 8921 |

$q = 3{\cdot}0$

| $p$ \ $x$ | 0·975 | 0·980 | 0·985 | 0·988 | 0·989 | 0·990 | 0·991 | 0·992 |
|---|---|---|---|---|---|---|---|---|
| 0·5 | ·9999 951 | ·9999 975$^-$ | ·9999 989 | ·9999 9946 | ·9999 9958 | ·9999 9969 | ·9999 9977 | ·9999 9984 |
| 1·0 | ·9999 844 | ·9999 920 | ·9999 966 | ·9999 9827 | ·9999 9867 | ·9999 9900 | ·9999 9927 | ·9999 9949 |
| 1·5 | ·9999 661 | ·9999 826 | ·9999 927 | ·9999 9624 | ·9999 9710 | ·9999 9782 | ·9999 9841 | ·9999 9888 |
| 2·0 | ·9999 387 | ·9999 685$^-$ | ·9999 867 | ·9999 9315$^+$ | ·9999 9472 | ·9999 9603 | ·9999 9710 | ·9999 9796 |
| 2·5 | ·9999 003 | ·9999 487 | ·9999 782 | ·9999 8881 | ·9999 9137 | ·9999 9351 | ·9999 9526 | ·9999 9667 |
| 3·0 | ·9998 496 | ·9999 224 | ·9999 670 | ·9999 8303 | ·9999 8691 | ·9999 9015$^-$ | ·9999 9281 | ·9999 9494 |
| 3·5 | ·9997 848 | ·9998 888 | ·9999 526 | ·9999 7561 | ·9999 8118 | ·9999 8583 | ·9999 8965$^+$ | ·9999 9272 |
| 4·0 | ·9997 047 | ·9998 471 | ·9990 348 | ·9999 6636 | ·9999 7403 | ·9999 8045$^-$ | ·9999 8571 | ·9999 8994 |
| 4·5 | ·9996 079 | ·9997 965$^+$ | ·9999 130 | ·9999 5511 | ·9999 6533 | ·9999 7388 | ·9999 8091 | ·9999 8656 |
| 5·0 | ·9994 929 | ·9997 364 | ·9998 871 | ·9999 4167 | ·9999 5493 | ·9999 6604 | ·9999 7517 | ·9999 8251 |
| 5·5 | ·9993 586 | ·9996 660 | ·9998 567 | ·9999 2586 | ·9999 4270 | ·9999 5680 | ·9999 6840 | ·9999 7773 |
| 6·0 | ·9992 038 | ·9995 845$^+$ | ·9998 214 | ·9999 0750$^+$ | ·9999 2848 | ·9999 4607 | ·9999 6053 | ·9999 7218 |
| 6·5 | ·9990 273 | ·9994 915$^-$ | ·9997 810 | ·9998 8644 | ·9999 1217 | ·9999 3374 | ·9999 5149 | ·9999 6579 |
| 7·0 | ·9988 280 | ·9993 861 | ·9997 351 | ·9998 6250$^+$ | ·9998 9361 | ·9999 1971 | ·9999 4120 | ·9999 5852 |
| 7·5 | ·9986 048 | ·9992 679 | ·9996 834 | ·9998 3552 | ·9998 7268 | ·9999 0388 | ·9999 2958 | ·9999 5030 |
| 8·0 | ·9983 568 | ·9991 361 | ·9996 258 | ·9998 0533 | ·9998 4926 | ·9998 8615$^+$ | ·9999 1657 | ·9999 4109 |
| 8·5 | ·9980 831 | ·9989 903 | ·9995 618 | ·9997 7179 | ·9998 2322 | ·9998 6643 | ·9999 0208 | ·9999 3084 |
| 9·0 | ·9977 826 | ·9988 298 | ·9994 912 | ·9997 3473 | ·9997 9444 | ·9998 4463 | ·9998 8605$^-$ | ·9999 1949 |
| 9·0 | ·9974 547 | ·9986 542 | ·9994 137 | ·9996 9401 | ·9997 6280 | ·9998 2064 | ·9998 6841 | ·9999 0699 |
| 10·5 | ·9970 984 | ·9984 630 | ·9993 292 | ·9996 4948 | ·9997 2817 | ·9997 9438 | ·9998 4909 | ·9998 9329 |
| 10·5 | ·9967 130 | ·9982 556 | ·9992 372 | ·9996 0099 | ·9996 9045$^+$ | ·9997 6576 | ·9998 2802 | ·9998 7835$^-$ |
| 11·0 | ·9962 978 | ·9980 316 | ·9991 377 | ·9995 4840 | ·9996 4953 | ·9997 3469 | ·9998 0513 | ·9998 6211 |
| 12·0 | ·9953 753 | ·9975 319 | ·9989 147 | ·9994 3038 | ·9995 5760 | ·9996 6486 | ·9997 5365$^+$ | ·9998 2555$^-$ |
| 13·0 | ·9943 259 | ·9969 606 | ·9986 586 | ·9992 9434 | ·9994 5153 | ·9995 8420 | ·9996 9413 | ·9997 8324 |
| 14·0 | ·9931 454 | ·9963 146 | ·9983 674 | ·9991 3925$^+$ | ·9993 3050$^-$ | ·9994 9206 | ·9996 2608 | ·9997 3481 |
| 15·0 | ·9918 300 | ·9955 912 | ·9980 396 | ·9989 6414 | ·9991 9369 | ·9993 8781 | ·9995 4900 | ·9996 7990 |
| 16·0 | ·9903 765$^-$ | ·9947 876 | ·9976 737 | ·9987 6805$^+$ | ·9990 4035$^-$ | ·9992 7084 | ·9994 6243 | ·9996 1817 |

| $p$ \ $x$ | 0·993 | 0·994 | 0·995 | 0·996 | 0·997 | 0·998 | 0·999 |
|---|---|---|---|---|---|---|---|
| 0·5 | ·9999 9989 | ·9999 9993 | ·9999 9996 | ·9999 9998 | ·9999 9999 | 1·0000 0000 | 1·0000 0000 |
| 1·0 | ·9999 9966 | ·9999 9978 | ·9999 9988 | ·9999 9994 | ·9999 9997 | 0·9999 9999 | 1·0000 0000 |
| 1·5 | ·9999 9925$^+$ | ·9999 9953 | ·9999 9973 | ·9999 9986 | ·9999 9994 | ·9999 9998 | 1·0000 0000 |
| 2·0 | ·9999 9864 | ·9999 9914 | ·9999 9950$^+$ | ·9999 9974 | ·9999 9989 | ·9999 9997 | 1·0000 0000 |
| 2·5 | ·9999 9777 | ·9999 9859 | ·9999 9918 | ·9999 9958 | ·9999 9982 | ·9999 9995$^-$ | 0·9999 9999 |
| 3·0 | ·9999 9661 | ·9999 9786 | ·9999 9876 | ·9999 9936 | ·9999 9973 | ·9999 9992 | ·9999 9999 |
| 3·5 | ·9999 9511 | ·9999 9692 | ·9999 9821 | ·9999 9908 | ·9999 9961 | ·9999 9988 | ·9999 9999 |
| 4·0 | ·9999 9325$^-$ | ·9999 9574 | ·9999 9753 | ·9999 9873 | ·9999 9946 | ·9999 9984 | ·9999 9998 |
| 4·5 | ·9999 9097 | ·9999 9430 | ·9999 9669 | ·9999 9830 | ·9999 9928 | ·9999 9979 | ·9999 9997 |
| 5·0 | ·9999 8824 | ·9999 9258 | ·9999 9569 | ·9999 9779 | ·9999 9906 | ·9999 9972 | ·9999 9997 |
| 5·5 | ·9999 8503 | ·9999 9054 | ·9999 9451 | ·9999 9718 | ·9999 9881 | ·9999 9964 | ·9999 9996 |
| 6·0 | ·9999 8129 | ·9999 8817 | ·9999 9313 | ·9999 9647 | ·9999 9850$^+$ | ·9999 9956 | ·9999 9994 |
| 6·5 | ·9999 7699 | ·9999 8545$^-$ | ·9999 9154 | ·9999 9565$^+$ | ·9999 9816 | ·9999 9945$^+$ | ·9999 9993 |
| 7·0 | ·9999 7208 | ·9999 8234 | ·9999 8973 | ·9999 9472 | ·9999 9776 | ·9999 9933 | ·9999 9992 |
| 7·5 | ·9999 6654 | ·9999 7883 | ·9999 8769 | ·9999 9366 | ·9999 9731 | ·9999 9920 | ·9999 9990 |
| 8·0 | ·9999 6033 | ·9999 7488 | ·9999 8539 | ·9999 9248 | ·9999 9681 | ·9999 9905 | ·9999 9988 |
| 8·5 | ·9999 5340 | ·9999 7049 | ·9999 8283 | ·9999 9116 | ·9999 9625$^-$ | ·9999 9888 | ·9999 9986 |
| 9·0 | ·9999 4574 | ·9999 6562 | ·9999 7999 | ·9999 8969 | ·9999 9562 | ·9999 9870 | ·9999 9984 |
| 9·5 | ·9999 3729 | ·9999 6026 | ·9999 7685$^+$ | ·9999 8807 | ·9999 9494 | ·9999 9849 | ·9999 9981 |
| 10·0 | ·9999 2803 | ·9999 5437 | ·9999 7341 | ·9999 8630 | ·9999 9418 | ·9999 9826 | ·9999 9978 |
| 10·5 | ·9999 1792 | ·9999 4794 | ·9999 6966 | ·9999 8435$^+$ | ·9999 9335$^+$ | ·9999 9802 | ·9999 9975$^+$ |
| 11·0 | ·9999 0692 | ·9999 4094 | ·9999 6557 | ·9999 8224 | ·9999 9245$^-$ | ·9999 9775$^-$ | ·9999 9972 |
| 12·0 | ·9998 8216 | ·9999 2518 | ·9999 5634 | ·9999 7746 | ·9999 9041 | ·9999 9714 | ·9999 9964 |
| 13·0 | ·9998 5347 | ·9999 0689 | ·9999 4563 | ·9999 7191 | ·9999 8804 | ·9999 9642 | ·9999 9955$^-$ |
| 14·0 | ·9998 2060 | ·9998 8592 | ·9999 3333 | ·9999 6553 | ·9999 8532 | ·9999 9561 | ·9999 9945$^-$ |
| 15·0 | ·9997 8329 | ·9998 6209 | ·9999 1935$^-$ | ·9999 5827 | ·9999 8221 | ·9999 9467 | ·9999 9933 |
| 16·0 | ·9997 4131 | ·9998 3525$^+$ | ·9999 0358 | ·9999 5007 | ·9999 7870 | ·9999 9362 | ·9999 9919 |